McGraw-Hill Machining and Metalworking Handbook

Ronald A. Walsh
(deceased)

Denis R. Cormier
Associate Professor
Department of Industrial Engineering
North Carolina State University

Third Edition

McGraw-Hill

New York   Chicago   San Francisco   Lisbon   London   Madrid
Mexico City   Milan   New Delhi   San Juan   Seoul
Singapore   Sydney   Toronto

The McGraw·Hill Companies

Library of Congress Cataloging-in-Publication Data

Walsh, Ronald A.
   McGraw-Hill machining and metalworking handbook / Ronald A. Walsh.
and Denis R. Cormier—3rd ed.
     p. cm.
   ISBN 0-07-145787-9
   1. Machining—Handbooks, manuals, etc. 2. Metal-work—Handbooks,
manuals, etc. I. Cormier, Denis R. II. Title.
   TJ1185.W35 2005

                                   2005051055

1 2 3 4 5 6 7 8 9 0 DOC/DOC 0 1 0 9 8 7 6 5

ISBN 0-07-145787-9

*The sponsoring editor for this book was Kenneth McCombs, the editing
supervisor was Caroline Levine, and the production supervisor was
Pamela A. Pelton. The art director for the cover was Handel Low. It
was set in Century Schoolbook by Wayne A. Palmer of McGraw-Hill
Professional's Hightstown, N.J., composition unit.*

Printed and bound by R. R. Donnelley & Sons Company.

McGraw-Hill books are available at special quantity discounts to use as
premiums and sales promotions, or for use in corporate training programs. For more information, please write to the Director of Special
Sales, McGraw-Hill, Professional Publishing, 2 Penn Plaza, New York,
NY 10121-2298. Or contact your local bookstore.

Contents

Chapter 6   Computer-Aided Design, Manufacturing, and Engineering Systems   297

Chapter 7   Machining, Machine Tools, and Practices   325

Preface

Expanding on the already enormous amount of information contained in the Second Edition of the *McGraw-Hill Machining and Metalworking Handbook*, this Third Edition provides new material of prime importance to the machining and metalworking industries. Like the original material, this new data are presented in an easy-to-read and easy-to-understand format that the authors hope will prove valuable to readers. The new information touches on many chapters of the *Handbook*. Additions include expanded coverage of modern machining equipment, computer-aided design (CAD) systems, computer-aided manufacturing (CAM) systems, computer-aided engineering (CAE) systems, product data management (PDM), statistical process control, powder coating technology, and an entirely new chapter on solid freeform fabrications (SFF) processes with an emphasis on direct-metal SFF processes. Many new photographs illustrate recent advances in technology. Likewise, information on various standards has been updated, as has contact information for a wide variety of standards organizations and professional trade groups.

Credits and acknowledgments are due the following organizations, associations, societies, and institutes who shared data vital to producing this Third Edition: ASTM, SAE, SMI, AGMA, ANSI, AFBMA, CDA, ASM, AWS, AISI, IFI, and NEMA. Manufacturers providing valuable support included Valenite, Kennametal, Boston Gear, Waldes Truarc, PEM, Reid Tool Supply, Ruland Manufacturing, The Timken Company, Tinnerman, T. B. Woods Sons, The Torrington Company, and Industrial Information Headquarters.

The entire machining and metalworking industry, together with the American standards, has advanced and changed greatly in the past 50 years. Many new machines and new and improved processes and materials evolved during those years. The mountain of data contained in this book thus represents the contributions of thousands of talented people—ranging from machine operators to industrial leaders—throughout the world over a period of many decades. We wish to thank all those largely anonymous individuals, whose talents and efforts made this *Handbook* possible.

Acknowledgments

The author and publisher wish to express their gratitude to the following individuals and organizations whose data, assistance, technical expertise, and copyright permissions made production of this *Handbook* possible: R. Aman, research assistant, North Carolina State University; R. Siegel, president, Powercon Corporation, Severn, MD; J. Beylis, manager, manufacturing and tool engineering; A. Feygelman, design engineer; B. Sabintsev, product and tool design engineer; A. Korsunsky, design engineer; L. Becker, design engineer; H. Crawford, editor-in-chief, McGraw-Hill, Inc.; and M. E. Walsh and K. T. Walsh.

The author gratefully acknowledges the permission granted by Valenite, Inc., to reproduce technical information appearing in Valenite's copyrighted catalogues. Technical assistance for machining applications using Valenite insert cutting tools may be obtained by contacting Valenite's Technical Support Center at (800) 488-9073. The Valenite main offices are located at 31700 Research Park Drive, PO. Box 9636, Madison Heights, MI 48071-9636.

The author also acknowledges the assistance of the following organizations: American Institute of Steel Construction (AISC), American Society for Testing and Materials (ASTM), Society of Automotive Engineers (SAE), American National Standards Institute (ANSI), American Iron and Steel Institute (AISI), Industrial Fasteners Institute (IFI), Spring Manufacturers Institute (SMI), American Gear Manufacturers Association (AGMA), Anti-Friction Bearing Manufacturers Association (AFBMA), IMO Industries, Boston Gear Division, The Timken Company, The Torrington Company, and the Industrial Information Headquarters.

Introduction

Machines, machining, and metalworking practices are among the most important elements of modern technology. Almost every modern product manufactured worldwide relies on the vast array of disciplines practiced in these three major elements.

The topics and data presented in this *Handbook* offer the modern product design engineer, machinist, tool designer, tool maker, and general metalworker a broad base of study and basic information for the practice of their profession or trade.

The author has attempted to assemble all those basic disciplines, data, and practices that are of prime importance to those associated with the manufacturing and metalworking industries. To be able to understand and practice the basics of the vast amount of data and principles required in these industries is beneficial to all those associated with these industries. On-the-job training and experience, together with knowing the basics, complete the individual's skills. Specialization in the metalworking industries is, of course, required, but a firm understanding of all the important basic aspects can only be beneficial to any individual and his company.

This *Handbook* provides a broad overview of modern manufacturing and metalworking industrial practices for students in technical teaching facilities and also for the inexperienced product designer and metalworker.

Computers and computer numerical control (CNC) machine tools and other metalworking machines such as five-axis machining centers, double-spindle turning and machining centers, fast-wire EDMs with independent axes, and random access memory (RAM) EDM machines have made modern metalworking faster and more accurate. Much labor has been reduced in the diemaking processes and other classifications. The conventional manual engine lathes and milling machines have been made more productive with digital control and read-out panels. The present-day 16-bit microprocessors and controllers are being replaced gradually with the new

32-bit CNC units that will allow machining times to be reduced by a factor of 2 to 5.

New cutting tool materials and tool geometries are advancing at a steady pace to improve production rates. The new cryogenic treatment of cutting tools now allows the tools to remain sharp two to four times longer than was possible in the past.

The earlier technologies of water-jet cutting, laser cutting, explosive forming, plasma cutting, lost-foam casting, and EDM cutting all have improved to their present, highly efficient state of the art.

This all translates to better productivity and improved manufacturing costs and possible improvement in the quality of the product. Improving productivity and lowering manufacturing costs do not necessarily indicate an improvement in quality. The term *quality* is being constantly debated in the industry today. Many of the authors of engineering and manufacturing papers and articles today seem to have a problem defining the word *quality* and what it means to different people. For purposes of this discussion, it would be appropriate to define *quality* in the broad term *customer satisfaction and performance*. The skills and knowledge of the design engineer, tool designer, diemaker, machinist, and metalworker all dictate, to a great extent, productivity and quality of a product; the satisfied or dissatisfied customer eventually determines the extent of the success of a product manufacturer. Few experienced people have difficulty in recognizing a "quality" product.

With these thoughts in mind, it is the author's and publisher's wish that this *Handbook* will prove useful and valuable to those users who are or will be associated with the metalworking industries in the United States and in developing countries worldwide.

Modern Metalworking Machinery, Tools, and Measuring Devices

Metalworking machinery, tools, and measuring instruments have advanced considerably over the past 50 years. This chapter will show some of the new machines, tools, and instruments used throughout industry today that allow us to produce parts faster and more accurately than was possible in the past. The widespread use and implementation of microprocessors to control the actions of metalworking machinery is evident in many of the photographs of modern equipment shown in this chapter. Photographs of other modern metalworking machinery appear throughout this *Handbook*.

1.1 Metalworking Process Overview

When a metal part is fabricated, the part blank either can come from a near-net-shape manufacturing process or it can come in the form of bars, rods, plates, etc. Metal casting processes such as die casting, sand casting, and investment casting are the most common methods of producing a part blank that is close to its final shape (i.e., near net shape). Recent years also have seen a flood of new solid freeform fabrication (SFF) processes that are capable of directly producing near-net-shape functional metal parts without the need for molds, dies, etc. (see Chap. 10). In the case of near-net-shape processes, rough

machining of large amounts of stock is not necessary. Instead, it is only necessary to finish machine those features that are critical to the function of a part.

1.1.1   Primary processes

Die casting.   Small or medium-sized parts in nonferrous alloys such as magnesium, aluminum, and zinc are injected under pressure into a steel die. A machining allowance of 0.25 to 0.5 mm (0.010 to 0.020 in) for critical features is typical.

Sand casting.   Molten metal is cast into a packed-sand mold. Parts weighing from just a few ounces to several tons can be sand cast. The most commonly sand-cast metals include irons, stainless steels, aluminum, and nickel alloys. Since the surface of the cast part is textured, a machining allowance typically is provided for critical features. Recommended machining allowances for a variety of metals are provided in Table 1.1.

Investment casting.   Both ferrous and nonferrous metals may be investment cast into a single-use refractory ceramic mold. High-temperature-reactive metals such as titanium typically are vacuum investment cast.

Forging.   Metals such as nonferrous alloys (e.g., aluminum, magnesium, and brass), steels, and nickel alloys are relatively easy to forge. The slugs are essentially hammered by a die such that the metal deforms to the shape of the die. Recommended machining allowances for a variety of metals are provided in Table 1.2.

Powder metallurgy.   Metal powder is compacted by a die and then sintered to hold its shape. The resulting parts are porous and optionally are infiltrated to 100 percent density.

Extrusion.   A heated billet is forced through a die opening such that the length of the billet takes on the cross-sectional shape of the die opening.

1.1.2   Metal-cutting processes

CNC machining.   The two most versatile machines in the modern machining industry are the computer numerical control (CNC)

TABLE 1.1   Sand Casting Allowances for Each Side

| | Casting size, mm (in)[*] | Allowance, mm (in) | |
		Drag and sides	Cope surface
Gray iron	Up to 150 (up to 6)	2.3 ($^3/_{32}$)	3 ($^1/_8$)
	150–300 (6–12)	3 ($^1/_8$)	4 ($^5/_{32}$)
	300–600 (12–24)	5 ($^3/_{16}$)	6 ($^1/_4$)
	600–900 (24–36)	6 ($^1/_4$)	8 ($^5/_{16}$)
	900–1500 (36–60)	8 ($^5/_{16}$)	10 ($^3/_8$)
	1500–2100 (60–84)	10 ($^3/_8$)	13 ($^1/_2$)
	2100–3000 (84–120)	11 ($^7/_{16}$)	16 ($^5/_8$)
Cast steel	Up to 150 (up to 6)	3 ($^1/_8$)	6 ($^1/_4$)
	150–300 (6–12)	5 ($^3/_{16}$)	6 ($^1/_4$)
	300–600 (12–24)	6 ($^1/_4$)	8 ($^5/_{16}$)
	600–900 (24–36)	8 ($^5/_{16}$)	10 ($^3/_8$)
	900–1500 (36–60)	10 ($^3/_8$)	13 ($^1/_2$)
	1500–2100 (60–84)	11 ($^7/_{16}$)	14 ($^9/_{16}$)
	2100–3000 (84–120)	13 ($^1/_2$)	19 ($^3/_4$)
Malleable iron	Up to 75 (up to 3)	1.5 ($^1/_{16}$)	2.3 ($^3/_{32}$)
	75–300 (3–12)	2.3 ($^3/_{32}$)	3 ($^1/_8$)
	300–450 (12–18)	3 ($^1/_8$)	4 ($^5/_{32}$)
	450–600 (18–24)	4 ($^5/_{32}$)	5 ($^3/_{16}$)
Ductile iron	Up to 150 (up to 6)	2.3 ($^3/_{32}$)	6 ($^1/_4$)
	150–300 (6–12)	3 ($^1/_8$)	10 ($^3/_8$)
	300–600 (12–24)	5 ($^3/_{16}$)	19 ($^3/_4$)
	600–900 (24–36)	6 ($^1/_4$)	19 ($^3/_4$)
	900–1500 (36–60)	8 ($^5/_{16}$)	25 (1)
	1500–2100 (60–84)	10 ($^3/_8$)	28 (1$^1/_8$)
	2100–3000 (84–120)	11 ($^7/_{16}$)	32 (1$^1/_4$)
Nonferrous metals	Up to 150 (up to 6)	1.6 ($^1/_{16}$)	2.3 ($^3/_{32}$)
	150–300 (6–12)	2.3 ($^3/_{32}$)	3 ($^1/_8$)
	300–600 (12–24)	3 ($^1/_8$)	4 ($^5/_{32}$)
	600–900 (24–36)	4 ($^5/_{32}$)	5 ($^3/_{16}$)

[*]*Casting size* refers to the overall length of the casting and not to the length of a particular measurement.

SOURCE: Bralla, J., *Design for Manufacturability Handbook.* New York: McGraw-Hill, 1999.

milling machine (Fig. 1.1) and the CNC lathe (Fig. 1.2). A key to the versatility of these machines is the automatic tool changer. Vertical machining centers (VMCs) such as the one shown in Fig. 1.1 include a carousel that holds many different cutting tools such as milling cutters, drills, reamers, and taps. The automatic tool changer changes cutting tools between machining operations without any

TABLE 1.2   Typical Machining Allowances for Forgings

Alloy family	Forging size: Projected area at parting line, mm (in)		
	To 640 cm² (100 in²)	To 2600 cm² (400 in²)	Over 2600 cm² (400 in²)
Aluminum	0.5–1.5 (0.020–0.060)	1.0–2.0 (0.040–0.080)	1.5–3.0 (0.060–0.120)
Magnesium	0.5–1.5 (0.020–0.060)	1.0–2.0 (0.040–0.080)	1.5–3.0 (0.060–0.120)
Brass	0.5–1.5 (0.020–0.060)	1.0–2.0 (0.040–0.080)	1.5–3.0 (0.060–0.120)
Steel	0.5–1.5 (0.020–0.060)	1.5–3.0 (0.060–0.120)	3.0–6.0 (0.120–0.240)
Stainless steel	0.5–1.5 (0.020–0.060)	1.5–2.5 (0.060–0.100)	1.5–5.0 (0.060–0.200)
Titanium	0.8–1.5 (0.030–0.060)	—	2.0–6.0 (0.080–0.240)
Niobium	0.8–2.5 (0.030–0.100)	—	—
Tantalum	0.8–2.5 (0.030–0.100)	—	—
Molybdenum	0.8–2.0 (0.030–0.080)	2.0–3.0 (0.080–0.120)	—

SOURCE: Bralla, J., *Design for Manufacturability Handbook.* New York: McGraw-Hill, 1999.

user intervention, thus allowing several machining operations to be executed in a single workpiece setup. Likewise, the CNC lathe in Fig. 1.2 incorporates an automatic tool changer that can switch between tools that perform facing, knurling, grooving, boring, and many other turning operations.

Electric discharge machining (EDM).   EDM comes in two forms—sinker EDM and wire EDM. Sinker EDM uses spark erosion to machine a workpiece with a graphite or copper electrode whose shape is the negative of the cavity being machined. Wire EDM uses spark erosion with a wire to cut two-dimensional (2D) profiles.

Laser machining.   A powerful laser beam coupled with a CNC motion-control system is used to cut 2D profiles in sheet or plate material.

Figure 1.1 Vertical machining center.

Figure 1.2 CNC lathe.

Complex, thin parts whose quantity does not warrant a hard die are produced using this method.

Chemical milling.   Large masses of metal may be removed effectively in producing a part using the etching action of chemicals. Very thin and delicate parts also may be produced with chemical milling or etching. A tough photoresistive substance covers the parts of the metal that are not to be removed. Printed circuit board production is actually a chemical milling operation.

Waterjet machining.   A very high pressure jet of water, loaded with microfine abrasives, is used to cut the sheet or plate material of metal, plastic, glass, or other composition. As is the case with laser machining, waterjet machining is useful when the production volumes do not warrant a hard die. The absence of a heat-affected zone is advantageous as well. Figure 1.3 shows a nested pattern of sheet metal parts being waterjet machined. Figure 1.4 shows a complex geometric shape cut from plate.

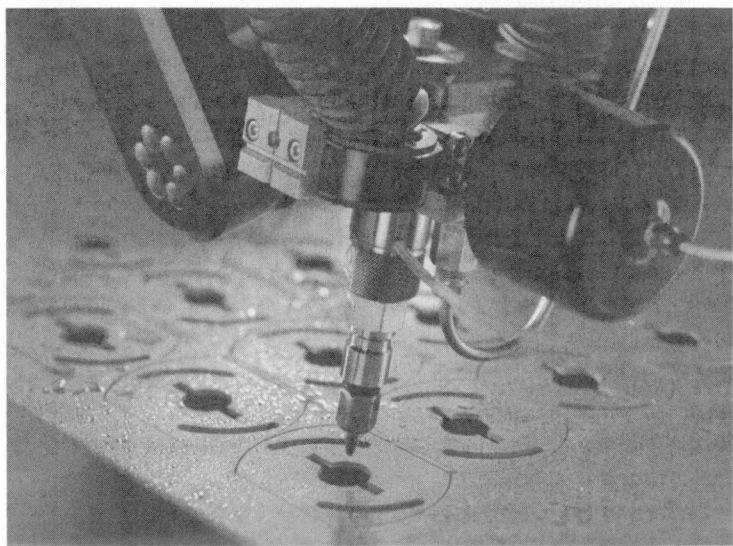

Figure 1.3   Waterjet machining operation. *(Image courtesy of OMAX Corporation, www.omax.com.)*

Figure 1.4 Complex waterjet-machined plate. *(Image courtesy of OMAX Corporation, www.omax.com.)*

1.1.3 Sheet metal parts fabrication methods

Hard dies. A die set is used to stamp out the part in flat pattern. Progressive dies also bend the part into the required shape after it is stamped in flat pattern. This is the most common, economical method devised to mass produce large quantities of parts to great accuracy.

Punch press. Large sheet metal parts may be made to accurate standards using modern computer-controlled automatic multistation punch presses. Programmers write the direct numerical control (DNC) programs for these machines, which are then loaded into the machine's computer or controller. The machine operator starts the program and stands back to watch the machine go through the sequence of operations required to produce the finished part in flat pattern.

Roll forming.   Flat strips of sheet metal are fed into the roll-forming machine, where they progress through a set of sequenced rollers to produce a long sheet metal part of constant cross-sectional shape.

Hydropressing.   A sheet metal flat-pattern part is placed on a set of forming dies, being located correctly with locator pins, and is then pressed into shape by the action of the hydropress. Many aircraft sheet metal parts are produced in this manner. Lightening holes and shrink flutes are produced simultaneously with the part to control the metal along curved surfaces.

Hydraulic brakes.   In this machine, a flat-pattern sheet metal part is given flanges or webs to produce the finished part. The modern brakes have automatic back gauges and material-handling devices to assist the operator in making the various bends and flanges required on the part.

Hydraulic shears.   The standard hydraulic shear cuts sheet metal according to the back gauge set by the machine operator and his or her accuracy in placing the sheet into the machine.

1.2   Measurement and Gauging

The preceding section provided an overview of many types of metalworking and machining processes. In a production environment, parts typically are fabricated according to specifications on the computer-aided design (CAD) drawing using one or more of the aforementioned processes. At certain points during the fabrication process, parts are inspected to verify that they satisfy the required geometric and dimensional tolerances. In some cases, 100 percent of the parts are inspected. In many instances, however, it is sufficient to inspect a subset of parts using a statistical sampling scheme. This section describes some of the instruments used to perform component inspection.

1.2.1   Coordinate measuring machines (CMMs)

CMMs are highly versatile inspection machines. Although CMMs are available in numerous configurations, the typical CMM consists of a probe that is positioned beneath a gantry. Depending on the type

of CMM, the probe can be moved manually by the operator's hand, or it can be moved automatically via a motion-control system. The workpiece being inspected is rigidly clamped to the CMM's granite table.

In manual mode, the operator tells the computer which feature(s) he or she is going to inspect, and the control computer will then instruct the user as to what points need to be probed for a given feature. To measure the distance between two faces, for instance, the operator must touch the probe to at least three points on the first face (i.e., three points define the plane) and one point on the second face (i.e., the perpendicular distance from a point on the second surface to the plane defined by the three points on the first surface). To measure the diameter of a hole, the user is prompted to touch the probe to three or more points around the perimeter of the hole. For each feature, the CMM control computer prompts the user to touch the probe to the appropriate number of points for the feature being inspected.

Fully automated CMMs are also available. With automated CMMs, the inspection planner starts with the geometric and/or dimensional tolerances (GD&Ts) specified in the CAD model by the mechanical designer. CMM software packages such as *PC-DMIS* are now available that are capable of extracting GD&T specifications from a CAD model. Using this software, the inspection planner identifies each feature in the CAD model to be inspected in a given setup on the bed of the CMM. The software then automatically generates an inspection plan for that setup on the CMM. The process is very much like generating toolpaths for a computer numerical control (CNC) milling machine. In this case, the touch probe rather than a rotating cutting tool automatically follows the prescribed path. After the workpiece has been inspected, the CMM software generates an inspection report. In many instances, companies will store these inspection results in a central database for purposes of traceability.

Both rigid and touch probes are available on CMMs. With a rigid probe, the operator must press a button manually so that the CMM can capture the x,y,z coordinates of the probe at that instant. With a touch probe, the probe automatically senses when it has touched the part, and the x,y,z coordinates are sent to the control computer immediately. Motorized touch probes are also available that can tilt and swivel in order to inspect features that otherwise would not be accessible in a given setup orientation.

1.2.2　Handheld measurement and gauging devices

Definitions

Precision. For any measuring device, *precision* is an indication of how much variation one will observe when one measures the same dimension on the same part using the same measuring device. The terms *precision* and *repeatability* are often used interchangeably. The sample standard deviation of multiple measurements taken on the same feature with the same device by the same operator is an indicator of precision. The smaller the standard deviation, the higher is the precision.

Accuracy. *Accuracy* is an indication of how close the measured dimension is to the true value for that dimension. Note that accuracy and precision are not the same thing. A device can be highly precise but very inaccurate. In other words, it can consistently give the same wrong measurement.

Resolution. This is the smallest unit of measure that can be displayed by the measuring device. If a digital caliper displays measurements to four decimal places, then the *resolution* is 0.0001 in (or millimeters).

Devices

Micrometers. Micrometers are used to measure thickness or diameter (Fig. 1.5).

Calipers. Calipers are highly versatile devices used to measure length (Fig. 1.6). They are designed to allow the operator to measure outside dimensions, inside dimensions, and depths. Digital readouts

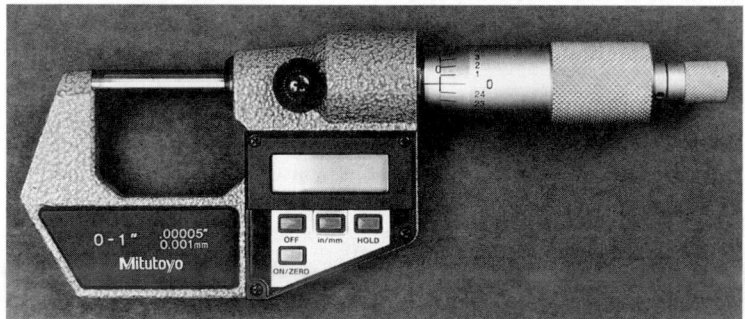

Figure 1.5 Digital micrometer.

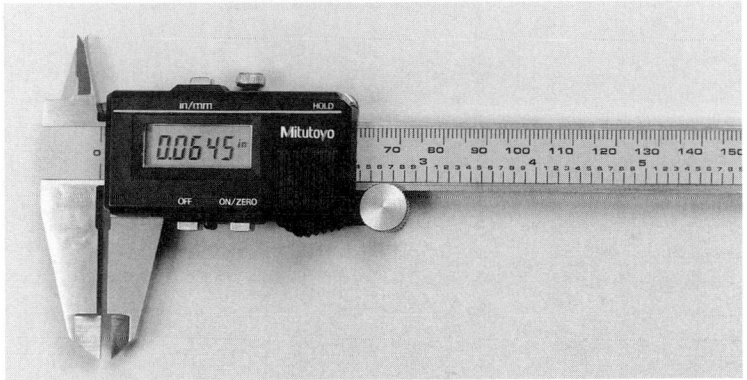

Figure 1.6 Digital caliper.

permit the operator to toggle between millimeters and inches as needed.

Dial indicators. Dial indicators show linear displacement of a stylus as it is moved across the surface of a part or vice versa (Fig. 1.7). They can be used to measure features such as the roundness of a rotational part, the flatness of a surface, or the depth of a hole.

Height gages. Height gauges measure the height of a feature, as the name implies (Fig. 1.8).

1.3   Statistical Process Control

Modern handheld digital measurement devices can be interfaced with computers on the shop floor for use with statistical process control (SPC) programs. Measurements collected from these devices do much more than indicate whether any individual part is within specifications. When the measurements for a succession of parts are plotted graphically, the machine operator can detect any nonrandom trends in machine performance visually and then take corrective action if necessary.

1.3.1   Process capability

When a process planner selects machines to perform a given operation on a part, he or she must know whether or not the machine

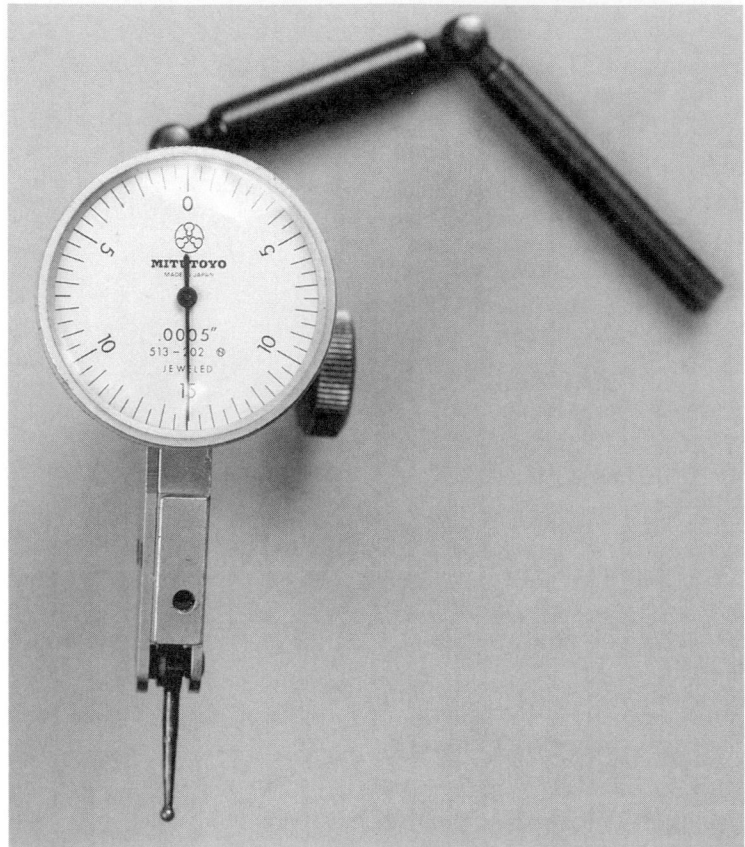

Figure 1.7 Dial indicator.

is capable of satisfying the tolerances specified for that part. The process capability study is used to determine whether or not this is the case. For a given feature, a target dimension is specified along with upper and lower tolerance values. For instance, the specification

$$2.5 \pm \frac{0.003}{0.003}$$

has a target value of 2.500 in, an upper specification limit (USL) of 2.503 in, and a lower specification limit (LSL) of 2.497 in.

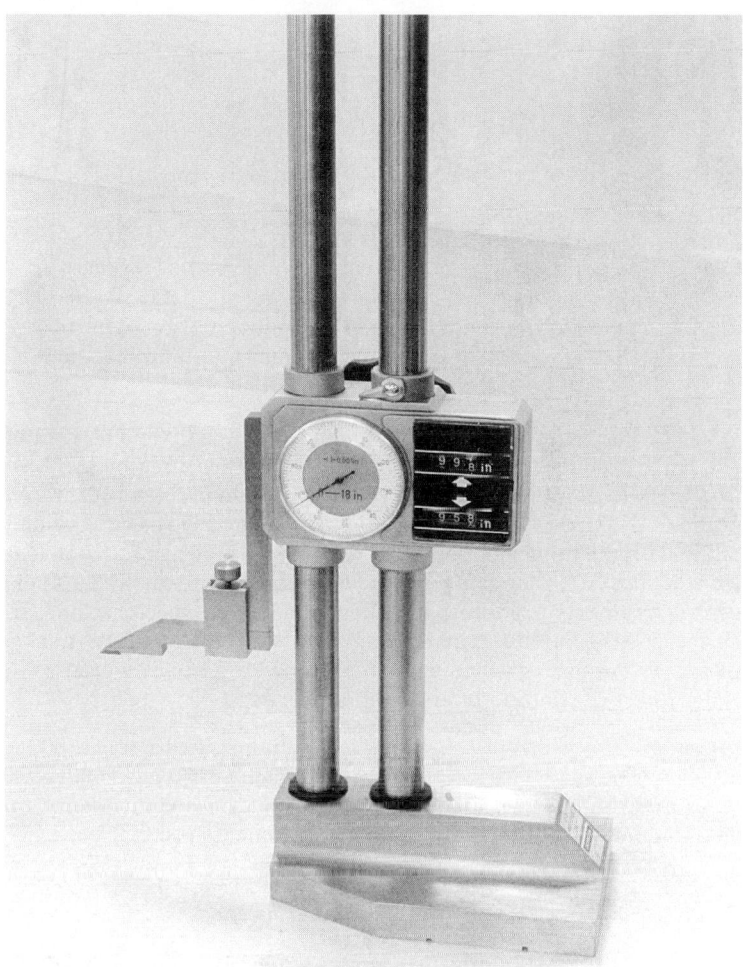

Figure 1.8   Height gauge.

CpK is one measure of process capability that provides an indication of both accuracy and precision:

$$CpK = \min\left\{\frac{USL - \bar{X}}{3\sigma}, \frac{\bar{X} - LSL}{3\sigma}\right\}$$

where σ is usually estimated by S:

$$S = \sqrt{\frac{\sum_{i=1}^{n}\left(x_i - \bar{X}\right)^2}{n - 1}}$$

$$\bar{X} = \frac{\sum_{i=1}^{n} x_i}{n}$$

where USL = upper limit on the tolerance
  LSL = lower limit on the tolerance
  $\bar{X}$ = process mean, or average value of a set of measurements
  σ = standard deviation of entire population of parts
  S = standard deviation of measurements from a sampling of n parts

When $CpK \geq 1$, then one can conclude that at least 99.73 percent of the parts produced will fall within the range specified by the LSL and USL. In plain English, this means that the process is centered sufficiently close to the target dimension value and that the spread of measurements is smaller than the tolerance range for that feature. If $CpK < 1$, then one can conclude that fewer than 99.73 percent of the parts produced will meet the design specifications. In this case, the manufacturing engineer can consider alternative processes, or he or she can work to improve the existing process in order to get the defects to an acceptable rate.

Example The width of a slot has a design specification of 2.500 ± 0.003 in. The slot width for each part in a batch of 30 parts has been measured, and the average value of the 30 measurements is 2.501 in. The standard deviation of these 30 measurements is 0.0008.

$$CpK = \min\left\{\frac{2.503 - 2.501}{3(0.0008)}, \frac{2.501 - 2.497}{3(0.0008)}\right\}$$

$$= \min\{0.833, 1.667\}$$

$$= 0.833$$

The CpK value of 0.833 indicates that the defect rate for this process will be unacceptably high if this company is striving for 99.73 percent acceptance rate (i.e., 6σ manufacturing).

1.3.2 Control charts

Variations in the dimensions produced by a process can either be *random* or *assignable*. All metalworking processes inherently produce

a certain amount of *random variation* in feature sizes that are produced. The magnitude of this random variation is what determines whether or not a machine is capable of meeting required tolerances for a given part. There also may be *assignable variation* present. Assignable variability refers to variations that can be attributed specifically to a particular cause. For example, a metal chip may be lodged beneath parallels supporting a part in a vise. The chip will raise the height of the part, thus increasing the depth of cut beyond what was intended. This is an assignable cause of variation that can be identified and eliminated. Control charts are an extremely valuable tool. They allow the machine operator to see graphically both sudden and gradual shifts in the process.

Many different types of control charts are available, and interested readers are encouraged to consult books dedicated to statistical process control. In its simplest form, a process control chart graphically plots the measured dimensions for the last 20 parts (typically) to be measured. The target value, USL, and LSL are also indicated on the chart. Much more sophisticated SPC tools are available, but even this simple control chart allows a machine operator to detect either sudden or gradual shifts in process performance. For example, Fig. 1.9 shows measurements of a feature for 20 parts. The nominal (target) measurement is 2.500 in, with an allowable tolerance of ±0.003 in. This chart clearly shows an upward nonrandom trend in the size of this feature. On seeing a nonrandom (i.e.,

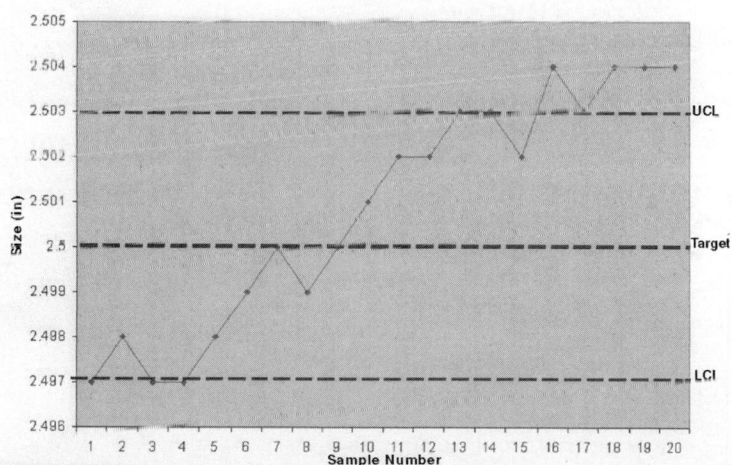

Figure 1.9   Control chart indicating upward trend.

assignable) cause of variation, the machine operator would know that he or she should stop the machine and investigate the root cause of this variation before parts are produced outside specifications. The root cause could be a cutting tool that is shifting in its collett or any number of other problems.

Mathematics for Machinists and Metalworkers

This chapter covers all the basic and special mathematical procedures of value to the modern machinist and metalworker. Geometry and plane trigonometry are of prime importance, as are the basic algebraic manipulations. Solutions to many basic and complex machining and metalworking operations would be difficult or impossible without the use of these branches of mathematics. In this chapter and other subsections of this *Handbook*, all the basic and important aspects of these branches of mathematics will be covered in detail.

2.1 General Mathematics, Algebra, and Trigonometry

2.1.1 General mathematics and algebraic procedures

If $A/B = C/D$, then

$$A = \frac{BC}{D} \quad B = \frac{AD}{C} \quad C = \frac{AD}{B} \quad \text{and} \quad D = \frac{BC}{A}$$

Transposing an equation. We may solve for any one unknown if all other variables are known. The given equation is

$$R = \frac{Gd^4}{8ND^3}$$

an equation with five variables, shown in terms of R. Solving for G gives

$$Gd^4 = R8ND^3 \text{ (cross-multiplied)}$$
$$G = \frac{8RND^3}{d^4}$$

Solving for d gives

$$d = \sqrt[4]{\frac{8RND^3}{G}}$$

Solving for D gives

$$D = \sqrt[3]{\frac{GD^4}{8RN}}$$

Solve for N using the same transposition procedures just shown.

Solving a typical algebraic equation.   An algebraic equation can be solved by substituting the numerical values assigned to the variables, which are denoted by letters, and then finding the unknown value.

Example:

$$L = 2C + 1.57(D + d) + \frac{(D - d)^2}{4C} \qquad \text{(belt-length equation)}$$

If $C = 16$, $D = 5.56$, and $d = 3.12$ (the variables), solve for L (by substituting the values of the variables into the equation):

$$L = 2(16) + 1.57(5.56 + 3.12) + \frac{(5.56 - 3.12)^2}{4(16)} = 45.721$$

Most of the equations shown in this *Handbook* are solved in a similar manner, i.e., by substituting known values for the variables in the equations and solving for the unknown quantity using standard algebraic and trigonometric rules and procedures.

Ratios and proportions.   If $a/b = c/d$, then

$$\frac{a+b}{b} = \frac{c+d}{d} \quad \text{and} \quad \frac{a-b}{b} = \frac{c-d}{d} \quad \text{and} \quad \frac{a-b}{a+b} = \frac{c-d}{c+d}$$

Quadratic equations. Any quadratic equation may be reduced to the form

$$ax^2 + bx + c = 0$$

The two roots, x_1 and x_2, equal

$$\frac{-b \pm \sqrt{b^2 - 4ac}}{2a}$$

When a, b, and c are real, if $b^2 - 4ac$ is positive, the roots are real and unequal. If $b^2 - 4ac$ is zero, the roots are real and equal. If $b^2 - 4ac$ is negative, the roots are imaginary and unequal.

2.1.2 Plane trigonometry

There are six trigonometric functions: sine, cosine, tangent, cotangent, secant, and cosecant. The relationship of the trigonometric functions is shown in Fig. 2.1. Trigonometric functions shown for angle A (right-angled triangle) include

$$\sin A = a/c \quad \text{(sine)}$$

$$\cos A = b/c \quad \text{(cosine)}$$

$$\tan A = a/b \quad \text{(tangent)}$$

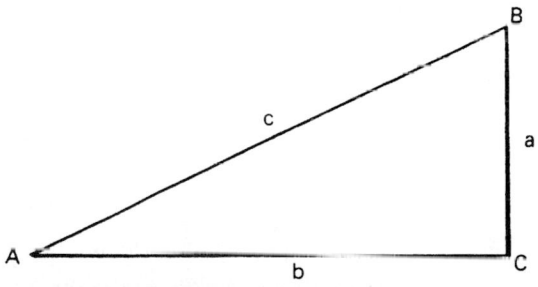

Figure 2.1 Right-angled triangle.

$$\cot A = b/a \ \text{(cotangent)}$$

$$\sec A = c/b \ \text{(secant)}$$

$$\csc A = c/a \ \text{(cosecant)}$$

For angle B, the functions would become

$$\sin B = b/c \ \text{(sine)}$$

$$\cos B = a/c \ \text{(cosine)}$$

$$\tan B = b/a \ \text{(tangent)}$$

$$\cot B = a/b \ \text{(cotangent)}$$

$$\sec B = c/a \ \text{(secant)}$$

$$\csc B = c/b \ \text{(cosecant)}$$

As can be seen from the preceding, the sine of a given angle is always the side opposite the given angle divided by the hypotenuse of the triangle, the cosine is always the side adjacent to the given angle divided by the hypotenuse, and the tangent is always the side opposite the given angle divided by the side adjacent to the angle. These relationships must be remembered at all times when performing trigonometric operations. Also,

$$\sin A = 1/\csc A$$

$$\cos A = 1/\sec A$$

$$\tan A = 1/\cot A$$

This reflects the important fact that the cosecant, secant, and cotangent are the reciprocals of the sine, cosine, and tangent, respectively. This fact also must be remembered when performing trigonometric operations.

Also, in any right-angled triangle,

$$\sin x = \cos (90° - x)$$

$$\cos x = \sin (90° - x) \qquad (x \text{ is the given angle other than } 90°)$$

$$\tan x = \cot (90° - x)$$

Equivalent expressions. The following trigonometric expressions are mathematically equivalent and may be used to advantage in solving many trigonometric problems. It is wise to try to remember as many of these expressions as possible, although they may be referred to in this chapter of the *Handbook* as required.

$$\tan x = \frac{\sin x}{\cos x}$$

$$\cot x = \frac{\cos x}{\sin x}$$

$$\sin^2 x + \cos^2 x = 1$$

$$\sin x = \pm\sqrt{1 - \cos^2 x}$$

$$\cos x = \pm\sqrt{1 - \sin^2 x}$$

$$\tan x = \pm\sqrt{\sec^2 x - 1}$$

$$\cot x = \pm\sqrt{\csc^2 x - 1}$$

$$\sec x = \pm\sqrt{\tan^2 x + 1}$$

$$\csc x = \pm\sqrt{\cot^2 x + 1}$$

Note: The choice of the ± sign is determined by which quadrant the angle x is situated in (see "Signs and Limits of Trigonometric Functions" below).

Signs and limits of the trigonometric functions. The following coordinate chart shows the sign of the function in each quadrant and its numerical limits. As an example, the sine of any angle between 0 and 90° will always be positive, and its numerical value will range between 0 and 1, whereas the cosine of any angle between 90 and 180° will always be negative, and its numerical value will range between 0 and 1. Each quadrant contains 90°; thus the fourth quadrant ranges between 270 and 360°.

Quadrant II	y	Quadrant I
(1–0) + sin		sin + (0–1)
(0–1) − cos		cos + (1–0)
(∞–0) − tan		tan + (0–∞)
(0–∞) − cot		cot + (∞–0)
(∞–1) − sec		sec + (1–∞)
(1–∞) + csc		csc + (∞–1)
x'————————		————————x
Quadrant III	0	Quadrant IV
(0–1) − sin		sin − (1–0)
(1–0) − cos		cos + (0–1)
(0–∞) + tan		tan − (∞–0)
(∞–0) + cot		cot − (0–?)
(1–∞) − sec		sec + (∞–1)
(∞–1) − csc	y'	csc − (1–∞)

Trigonometric laws.   The trigonometric "laws" show the relationships between the sides and angles of non-right-angle triangles or acute and obtuse triangles and allow us to calculate the unknown parts of the triangle when certain values are known. Refer to Fig. 2.2 for illustrations of the trigonometric laws that follow.

The law of sines (see Fig. 2.2)

$$\frac{a}{\sin A} = \frac{b}{\sin B} = \frac{c}{\sin C}$$

$$\frac{a}{b} = \frac{\sin A}{\sin B} \qquad \frac{b}{c} = \frac{\sin B}{\sin C} \qquad \frac{a}{c} = \frac{\sin A}{\sin C}$$

The law of cosines (see Fig. 2.2)

$$a^2 = b^2 + c^2 - 2bc\,\cos A$$

$$b^2 = a^2 + c^2 - 2ac\,\cos B$$

$$c^2 = a^2 + b^2 - 2ab\,\cos C$$

The law of tangents (see Fig. 2.2)

$$\frac{a+b}{a-b} = \frac{\tan\dfrac{A+B}{2}}{\tan\dfrac{A-B}{2}}$$

With the preceding laws, the trigonometric functions for right-angled triangles, the Pythagorean theorem, and the following triangle

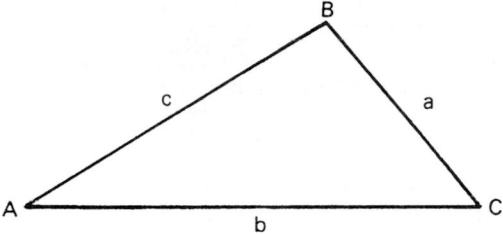

Figure 2.2 Oblique triangle.

solution chart, it will be possible to find the solution to any plane triangle problem, provided the correct parts are specified.

The Solution of Triangles

In right-angled triangles	To solve
Known: Any two sides	Use the Pythagorean theorem to solve unknown side; then use the trigonometric functions to solve the two unknown angles. The third angle is 90°.
Known: Any one side and either angle that is not 90°	Use trigonometric functions to solve the two unknown sides. The third angle is 180°—the sum of two known angles.
Known: Three angles and no sides (*all* triangles)	Cannot be solved because there are an infinite number of triangles that satisfy three known internal angles.
Known: Three sides	Use trigonometric functions to solve the two unknown angles.

In oblique triangles	To solve
Known: Two sides and any one of two nonincluded angles	Use the law of sines to solve the second unknown angle. The third angle is 180°—the sum of two known angles. Then find the other sides using the law of sines or the law of tangents.
Known: Two sides and the included angle	Use the law of cosines for one side and the law of sines for the two angles.
Known: Two angles and any one side	Use the law of sines to solve the other sides or the law of tangents. The third angle is 180°—the sum of two known angles.
Known: Three sides	Use the law of cosines to solve two of the unknown angles. The third angle is 180°—the sum of two known angles.

The Solution of Triangles (*Continued*)

In oblique triangles	To solve
Known: One angle and one side (non-right triangle)	Cannot be solved except under certain conditions. If the triangle is equilateral or isosceles, it may be solved if the known angle is opposite the known side.

Finding heights of non-right-angled triangles. The height x shown in Fig. 2.3 and Fig. 2.4 is found from

$$x = b\frac{\sin A \sin C}{\sin(A+C)} = \frac{b}{\cot A + \cot C} \qquad \text{(for Fig. 2.3)}$$

$$x = b\frac{\sin A \sin C}{\sin(C'-A)} = \frac{b}{\cot A - \cot C'} \qquad \text{(for Fig. 2.4)}$$

Areas of triangles (see Fig. 2.5)

$$A = \tfrac{1}{2}bh$$

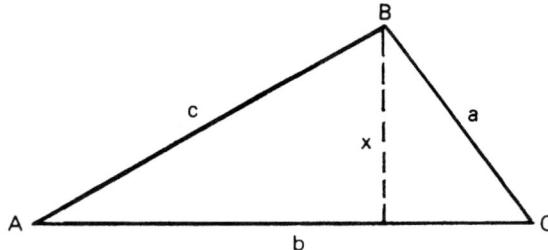

Figure 2.3   Height of triangle x.

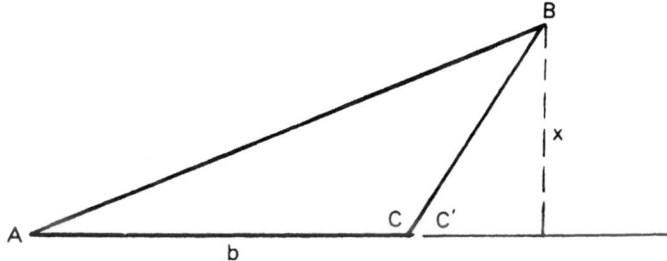

Figure 2.4   Height of triangle x.

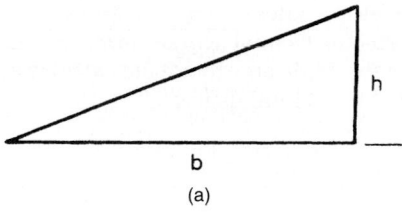

(a)

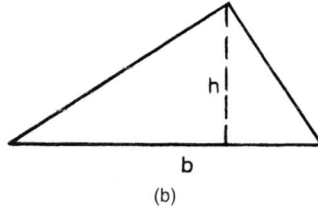

(b)

Figure 2.5 Triangles: (*a*) right triangle, (*b*) oblique triangle.

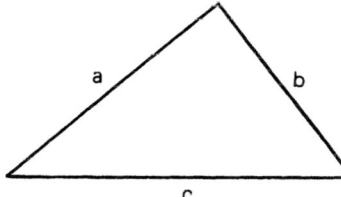

Figure 2.6 Triangle.

The area when the three sides are known is (see Fig. 2.6) (this holds true for any triangle)

$$A = \sqrt{s(s-a)(s-b)(s-c)}$$

where $s = \dfrac{a+b+c}{2}$

The Pythagorean theorem (for right-angled triangles)

$$c^2 = a^2 + b^2$$

$$b^2 = c^2 - a^2$$

$$a^2 = c^2 - b^2$$

Note: Side *c* is the hypotenuse.

Converting angles to decimal degrees. Angles given in degrees, min-utes, and seconds must be converted to decimal degrees prior to find-ing the trigonometric functions of the angle on a handheld calculator.
  Procedure: Convert 26°41′26″ to decimal degrees.

Degrees =   26.000000   in decimal degrees

Minutes = 41/60 = 0.683333   in decimal parts of a degree

Seconds = 26/3600 = 0.007222   in decimal parts of a degree

The angle in decimal degrees is then

$$26.000000 + 0.683333 + 0.007222 = 26.690555°$$

Converting decimal degrees to degrees, minutes, and seconds
  Procedure: Convert 56.5675 decimal degrees to degrees, minutes, and seconds.

$$\text{Degrees} = 56 \text{ degrees}$$

$$\text{Minutes} = 0.5675 \times 60 = 34.05 = 34 \text{ minutes}$$

$$\text{Seconds} = 0.05 \text{ (minutes)} \times 60 = 3 \text{ seconds}$$

The answer, therefore, is 56°34′3″.

Samples of solutions to triangles
  Solving right-angled triangles by trigonometry. *Required*: Any one side and angle A or angle B (see Fig. 2.7). Solve for side a:

$$\sin A = \frac{a}{c}$$

$$\sin 33.162° = \frac{a}{3.625}$$

$$a = 1.9829$$

Solve for side b:

$$\cos A = \frac{b}{c}$$

$$\cos 33.162° = \frac{b}{3.625}$$

$$b = 3.0345$$

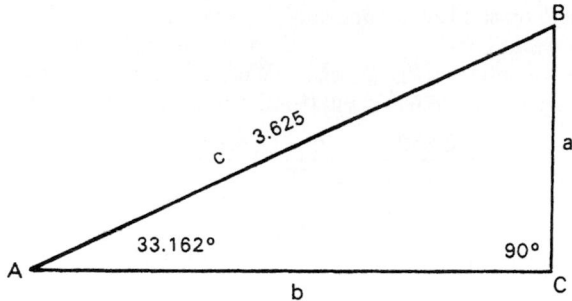

Figure 2.7 Solve the triangle.

Then angle $B = 180° - ($angle $A + 90°) = 180° - 123.162° = 56.838°$. We now know sides a, b, and c and angles A, B, and C.

Solving non-right-angled triangles using the trigonometric laws. Solve the triangle in Fig. 2.8 given two angles and one side:

$$A = 45°$$

$$B = 109°$$

$$a = 3.250$$

First, find angle C:

$$\text{Angle } C = 180° - (\text{angle } A + \text{angle } B)$$

$$= 180° - (45° + 109°)$$

$$= 180° - 154°$$

$$= 26°$$

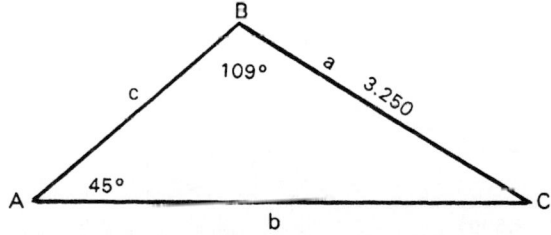

Figure 2.8 Solve the triangle.

Second, find side b by the law of sines:

$$\frac{a}{\sin A} = \frac{b}{\sin B}$$

$$\frac{3.250}{0.07071} = \frac{b}{0.9455}$$

$$b = 4.3457$$

Third, find side c by the law of sines:

$$\frac{a}{\sin A} = \frac{c}{\sin C}$$

$$\frac{3.250}{0.07071} = \frac{c}{0.4384}$$

$$c = 2.0150$$

Solve the triangle in Fig. 2.9 given two sides and one angle:

$$\text{Angle } A = 16°$$

$$a = 1.562$$

$$b = 2.509$$

First, find angle B from the law of sines:

$$\frac{a}{\sin A} = \frac{b}{\sin B}$$

$$\frac{1.562}{\sin 16} = \frac{2.509}{B}$$

$$B = 26.276°$$

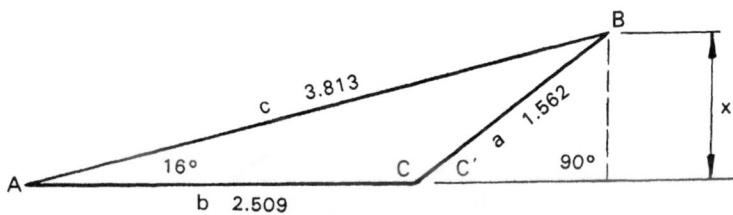

Figure 2.9   Solve the triangle.

Second, find angle C:

$$\text{Angle } C = 180° - (\text{angle } A + \text{angle } B)$$

$$= 180° - 42.276°$$

$$= 137.724°$$

Third, find side c from the law of sines:

$$\frac{a}{\sin A} = \frac{c}{\sin C}$$

$$\frac{1.562}{0.2756} = \frac{c}{0.6727}$$

$$c = 3.813$$

We may now find the altitude or height x of this triangle (see Fig. 2.9). Refer to Fig. 2.4.

$$x = b\frac{\sin A \sin C}{\sin(C' - A)} \qquad \text{(where angle } C' = 180° - 137.724° = 42.276°)$$

$$= 2.509 \times \frac{0.2756 \times 0.6727}{\sin(42.276 - 16)}$$

$$= 1.051$$

This height x also can be found from the sine function of angle C' when side a is known, as shown below:

$$\sin C' = \frac{x}{1.562}$$

$$x = 1.562 \sin C' = 1.562 \times 0.6727 = 1.051$$

Both methods yield the same numerical solution of 1.051.

Solve the triangle in Fig. 2.9a given three sides and no angles. According to the preceding triangle solution chart, solving this triangle requires use of the law of cosines. Proceed as follows: First, solve for any angle (we will take angle C first):

$$c^2 = a^2 + b^2 - 2ab \cos C$$

$$(1.75)^2 = (1.1875)^2 + (2.4375)^2 - 2(1.1875 \times 2.4375) \cos C$$

$$\cos C = 0.7409$$

$$\text{arc cos } C = 42.192°$$

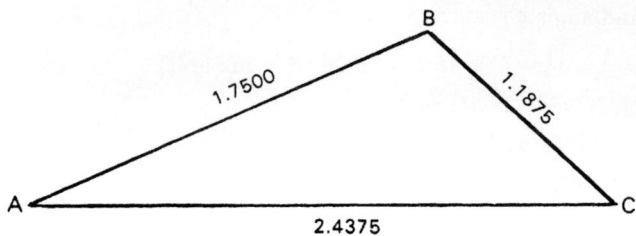

Figure 2.9a Solve the triangle.

Second, by the law of cosines, find angle *B:*

$$b^2 = a^2 + c^2 - 2ac \cos B$$

$$(2.4375)^2 = (1.1875)^2 + (1.7500)^2 - 2(1.1875 \times 1.7500) \cos B$$

$$\cos B = -0.3534$$

$$\text{arc} \cos B = 110.695°$$

Then angle *A* is found from

$$\text{Angle } A = 180 - (42.192 + 110.695)$$

$$= 27.113°$$

Converting degrees to radians.   To convert from degrees to radians, you must first find the degrees as decimal degrees. If *R* represents radians, then

$$2\pi R = 360° \qquad \text{or} \qquad \pi R = 180°$$

From this,

$$1 \text{ radian} = 57.2957795°$$

and

$$1° = 0.0174533 \text{ radian}$$

Example: Convert 56.785° to radians.

$$56.785° \times 0.0174533 = 0.9911 \text{ radian}$$

Example: Convert 2.0978*R* to decimal degrees.

$$57.2957795 \times 2.0978 = 120.0591°$$

2.1.3   Important mathematical constants

$$\pi = 3.1415926535898$$

$$1 \text{ radian} = 57.295779513082°$$

$$1° = 0.0174532925199 \text{ radian}$$

$$2\pi R = 360°$$

$$\pi R = 180°$$

$$e = 2.718281828 \text{ (base of natural logarithms)}$$

2.1.4   Summary of trigonometric procedures for triangles

There are four possible cases in the solution of oblique triangles:

Case 1: Given one side and two angles: *a, A, B*

Case 2: Given two sides and the angle opposite them: *a, b, A* or *B*

Case 3: Given two sides and their included angle: *a, b, C*

Case 4: Given the three sides: *a, b, c*

All oblique (non-right-angle) triangles can be solved by use of natural trigonometric functions: the law of sines, the law of cosines, and the angle formula: angle *A* + angle *B* + angle *C* = 180°. This may be done in the following manner:

Case 1: Given *a, A,* and *B,* angle *C* may be found from the angle formula, and then sides *b* and *c* may be found by using the law of sines twice.

Case 2: Given *a, b,* and *A,* angle *B* may be found by the law of sines, angle *C* from the angle formula, and side *c* by the law of sines again.

Case 3: Given *a, b,* and *C,* side *c* may be found by the law of cosines, and angles *A* and *B* may be found by the law of sines used twice or angle *A* from the law of sines and angle *B* from the angle formula.

Case 4: Given *a, b,* and *c,* the angles all may be found by the law of cosines, or angle *A* may be found from the law of cosines, and angles *B* and *C* from the law of sines, or angle *A* from the law of cosines, angle *B* from the law of sines, and angle *C* from the angle formula.

Note: Case 2 is called the *ambiguous case,* in which there may be one solution, two solutions, or *no* solution, given *a, b,* and *A.*

- If angle $A < 90°$ and $a < b \sin A$, there is *no* solution.

- If angle $A < 90°$ and $a = b \sin A$, there is one solution—a right triangle.

- If angle $A < 90°$ and $b > a > b \sin A$, there are two solutions—oblique triangles.

- If angle $A < 90°$ and $a \geq b$, there is one solution—an oblique triangle.

- If angle $A < 90°$ and $a \leq b$, there is *no* solution.

- If angle $A > 90°$ and $a > b$, there is one solution—an oblique triangle.

Special half-angle formulas. In case 4 triangles where only the three sides a, b, and c are known, the sets of half-angle formulas shown below may be used to find the angles:

$$\sin\frac{A}{2} = \sqrt{\frac{(s-b)(s-c)}{bc}}$$

$$\sin\frac{B}{2} = \sqrt{\frac{(s-c)(s-a)}{ca}}$$

$$\sin\frac{C}{2} = \sqrt{\frac{(s-a)(s-b)}{ab}}$$

$$\cos\frac{A}{2} = \sqrt{\frac{s(s-a)}{bc}}$$

$$\cos\frac{B}{2} = \sqrt{\frac{s(s-b)}{ac}}$$

$$\cos\frac{C}{2} = \sqrt{\frac{s(s-c)}{ab}}$$

$$\tan\frac{A}{2} = \sqrt{\frac{(s-b)(s-c)}{s(s-a)}}$$

$$\tan\frac{B}{2} = \sqrt{\frac{(s-c)(s-a)}{s(s-b)}}$$

$$\tan\frac{C}{2} = \sqrt{\frac{(s-a)(s-b)}{s(s-c)}}$$

where $s = \sqrt{\dfrac{a+b+c}{2}}$

2.1.5   Powers-of-10 notation

Numbers written in the form 1.875×10^5 or 3.452×10^{-6} are so stated in powers-of-10 notation. Arithmetic operations on numbers that are either very large or very small are processed easily and conveniently using the powers-of-10 notation and procedures. If you will note, on the handheld scientific calculator this process is carried out automatically by the calculator. If the calculated answer is larger or smaller than the digital display can handle, the answer will be given in powers-of-10 notation.

This method of handling numbers is always used in scientific and engineering calculations when the values of the numbers so dictate. Engineering notation usually is given in multiples of 3, such as 1.246×10^3, 6.983×10^{-6}, etc.

How to calculate with powers-of-10 notation.   Numbers with many digits may be expressed more conveniently in powers-of-10 notation, as shown below:

$$0.000001389 = 1.389 \times 10^{-6}$$

$$3,768,145 = 3.768145 \times 10^6$$

You are actually counting the number of places that the decimal point is shifted, either to the right or to the left. Shifting to the right produces a negative exponent, and shifting to the left produces a positive exponent.

Multiplication, division, exponents, and radicals in powers-of-10 notation are handled easily, as shown below.

$$(1.246 \times 10^4) \times (2.573 \times 10^{-4}) = 3.206 \times 10^0 = 3.206 \quad \textbf{(Note: } 10^0 = 1)$$

$$(1.785 \times 10^7) \div (1.039 \times 10^{-4}) = (1.785/1.039) \times 10^{7-(-4)} = 1.718 \times 10^{11}$$

$$(1.447 \times 10^5)^2 = (1.447)^2 \times 10^{10} = 2.094 \times 10^{10}$$

$$\sqrt{1.391 \times 10^8} = 1.391^{1/2} \times 10^{8/2} = 1.179 \times 10^4$$

In the preceding examples, you must use the standard algebraic rules for addition, subtraction, multiplication, and division of exponents or powers of numbers. Thus

- Exponents are algebraically added for multiplication.

- Exponents are algebraically subtracted for division.

- Exponents are algebraically multiplied for power raising.

- Exponents are algebraically divided for taking roots.

2.2 Geometric Principles

In any triangle, angle A + angle B + angle $C = 180°$, and angle $A = 180° -$ (angle A + angle B), and so on (see Fig. 2.10). If three sides of one triangle are proportional to the corresponding sides of another triangle, the triangles are similar. Also, if $a:b:c = a':b':c'$, then angle A = angle A', angle B = angle B', angle C = angle C', and $a/a' = b/b' = c/c'$. Conversely, if the angles of one triangle are equal to the respective angles of another triangle, the triangles are similar and their sides proportional; thus, if angle A = angle A', angle B = angle B', and angle C = angle C', then $a:b:c = a':b':c'$ and $a/a' = b/b' = c/c'$ (see Fig. 2.11).

- *Isosceles triangle* (see Fig. 2.12). If side c = side b, then angle A = angle B.

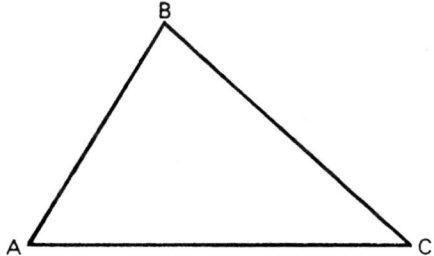

Figure 2.10 Triangle.

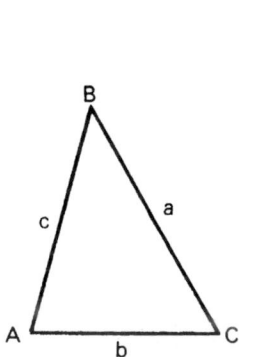
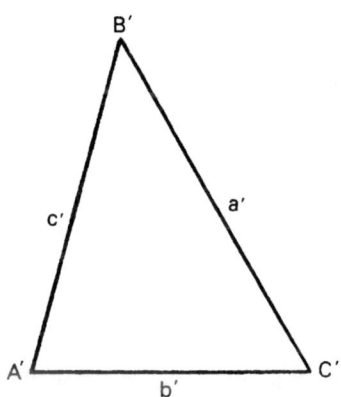

Figure 2.11 Similar triangles.

- *Equilateral triangle* (see Fig. 2.13). If side a = side b = side c, angles A, B, and C are equal (60°)
- *Right triangle* (see Fig. 2.14). $c^2 = a^2 + b^2$ and $c = (a^2 + b^2)^{1/2}$ when angle $C = 90°$. Therefore, $a = (c^2 - b^2)^{1/2}$ and b = $(c^2 - a^2)^{1/2}$. This relationship in all right angle triangles is called the Pythagorean theorem.
- *Exterior angle of a triangle* (see Fig. 2.15). Angle C = angle A + angle B.
- *Intersecting straight lines* (see Fig. 2.16). Angle A = angle A', and angle B = angle B'.

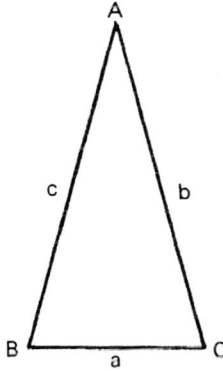

Figure 2.12   Isosceles triangle.

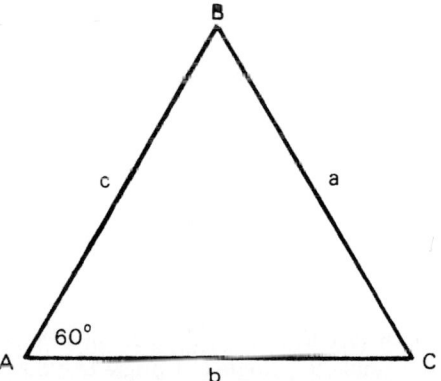

Figure 2.13   Equilateral triangle.

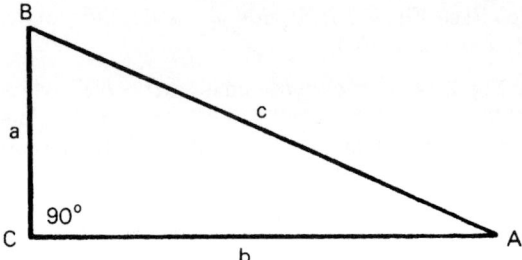

Figure 2.14 Right-angled triangle.

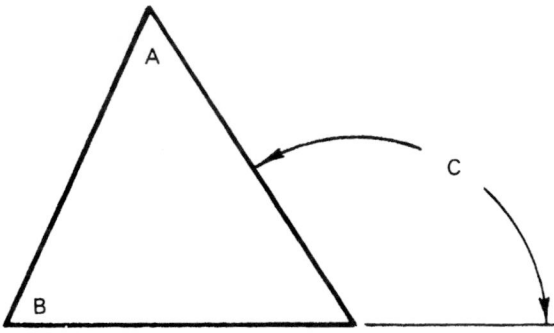

Figure 2.15 Exterior angle of a triangle.

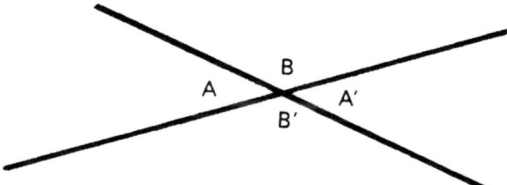

Figure 2.16 Intersecting straight lines.

■ *Two parallel lines intersected by a straight line* (see Fig. 2.17). Alternate interior and exterior angles are equal: Angle A = angle A', and angle B = angle B'.

■ *Any four-sided geometric figure* (see Fig. 2.18). The sum of all interior angles = 360°; angle A + angle B + angle C + angle D = 360°.

■ *A line tangent to a point on a circle is at 90°, or normal, to a radial line drawn to the tangent point* (see Fig. 2.19).

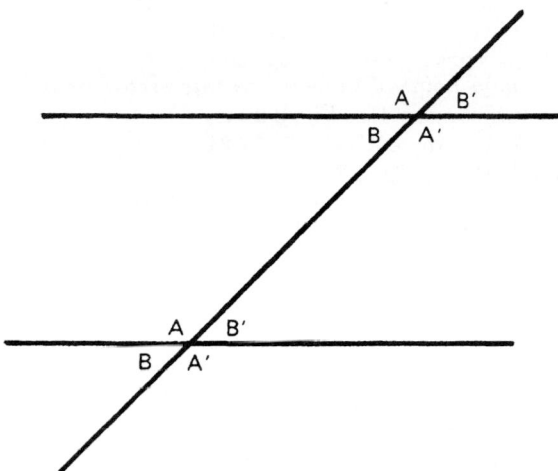

Figure 2.17 Straight line intersecting two parallel lines.

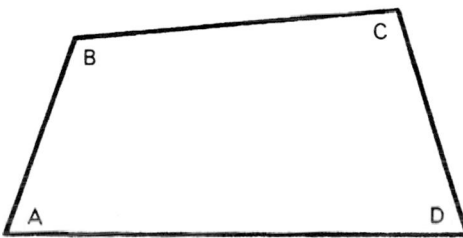

Figure 2.18 Quadrilateral (four-sided figure).

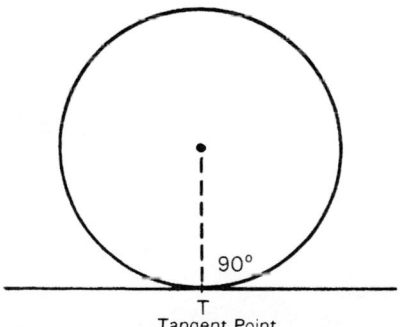

Figure 2.19 Tangent at a point on a circle.

■ *Two circles' common point of tangency is intersected by a line drawn between their centers* (see Fig. 2.20).
Side $a = a'$; angle A = angle A' (see Fig. 2.21).
Angle $A = \frac{1}{2}$ angle B (see Fig. 2.22).
Angle A = angle B = angle C. All perimeter angles of a chord are equal (see Fig. 2.23).
Angle $B = \frac{1}{2}$ angle A (see Fig. 2.24).
$a^2 = bc$ (see Fig. 2.25).

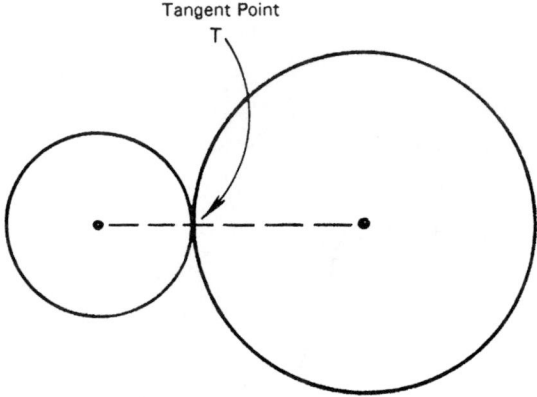

Figure 2.20   Common point of tangency.

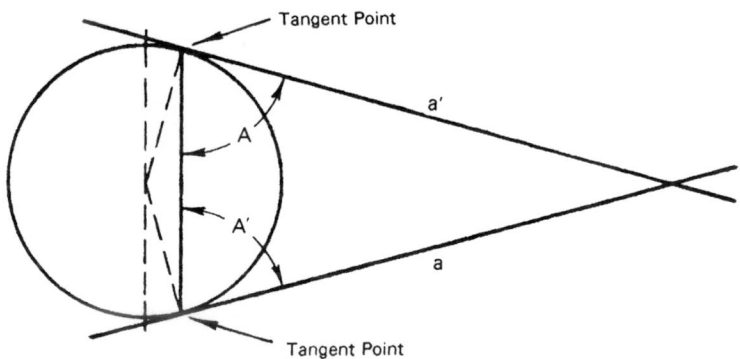

Figure 2.21   Tangents and angles.

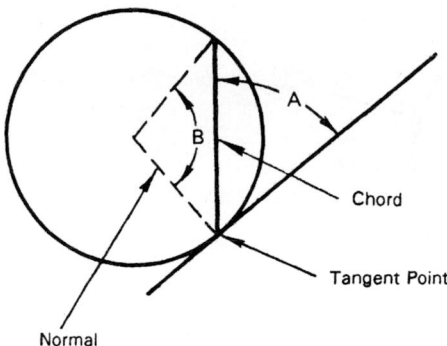

Figure 2.22 Half-angle (A)

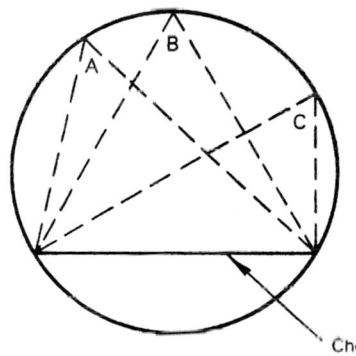

Figure 2.23 Perimeter angles of a chord.

- *All perimeter angles in a circle, drawn from the diameter, are 90°* (see Fig. 2.26).

- *Arcs are proportional to internal angles* (see Fig. 2.27). Angle A : angle B = a:b. Thus, if angle A = 89°, angle B = 30°, and arc a = 2.15 minutes, arc b would be calculated as

$$\frac{\text{Angle } A}{\text{Angle } B} = \frac{a}{b}$$

$$\frac{89}{30} = \frac{2.15}{b}$$

$$b = 64.5 = 0.7247 \, \text{min}$$

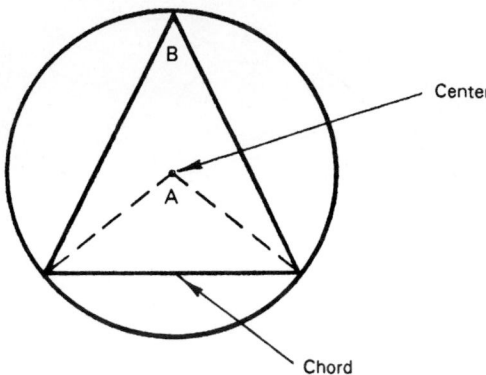

Figure 2.24 Half-angle (*B*).

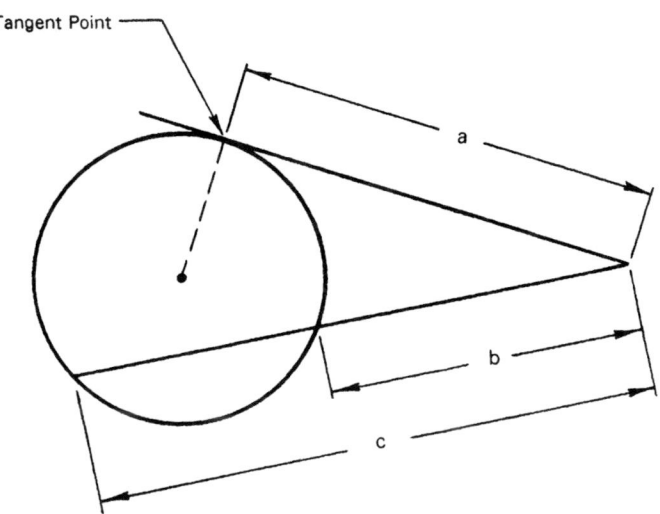

Figure 2.25 Line and circle relationship ($a^2 = bc$).

Note: The angles may be given in decimal degrees or radians, consistently.

- *Circumferences are proportional to their respective radii* (see Fig. 2.28). *C : C′ = r : R,* and areas are proportional to the squares of the respective radii.

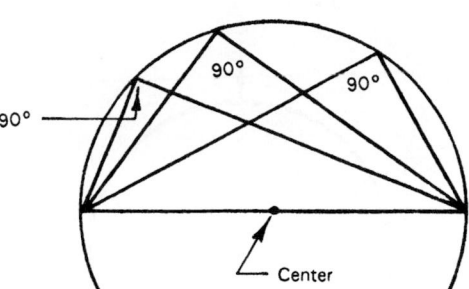

Figure 2.26   Ninety-degree perimeter angles.

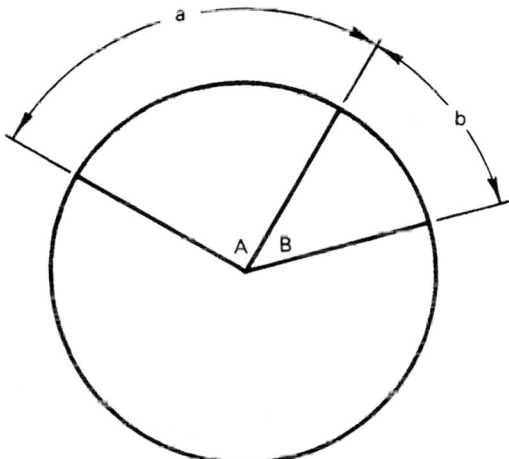

Figure 2.27   Proportional arcs and angles.

2.3   Geometric Construction

The following figures show the methods used to perform most of the basic geometric constructions used in standard drawing practices. Many of those constructions have widespread use in the machine shop and sheet metal shop.

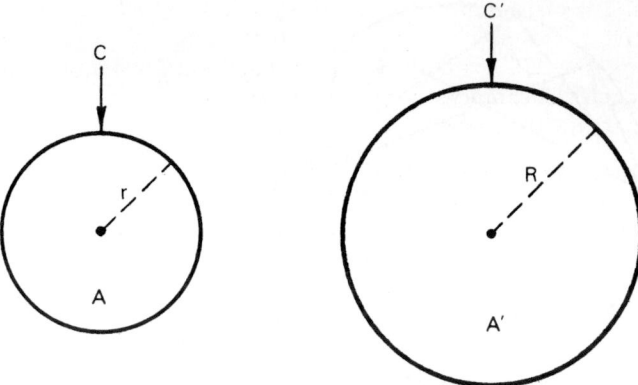

Figure 2.28 Circumference and radii proportionality.

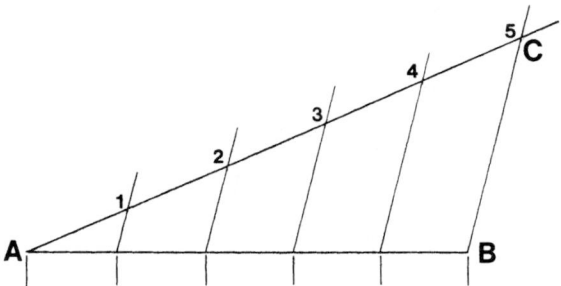

Figure 2.29 Dividing a line equally.

- *To divide any straight line into any number of equal spaces* (Fig. 2.29). To divide line *AB* into five equal spaces, draw line *AC* at any convenient angle such as angle *BAC*. With a divider or compass, mark off five equal spaces along line *AC*. Now connect point 5 on line *AC* with the endpoint of line *AB*. Draw line *CB*, and parallel transfer the other points along line *AC* to intersect line *AB*, thus dividing it into five equal spaces.

- *To bisect any angle BAC* (Fig. 2.30), swing an arc from point *A* through points *d* and *e*. Swing an arc from point *d* and another equal arc from point *e*. The intersection of these two arcs will be at point *f*. Draw a line from point *A* to point *f*, forming the bisector line *AD*.

- *To divide any line into two equal parts and erect a perpendicular* (Fig. 2.31), draw an arc from point *A* that is more than half the length of line *AB*. Using the same arc length, draw another arc from point *B*. The intersection points of the two arcs meet at points *c* and *d*. Draw the perpendicular bisector line *cd*.

- *To erect a perpendicular line through any point along a line* (Fig. 2.32), from point *c* along line *AB,* mark points 1 and 2 equidistant from point *c*. Select an arc length on the compass greater than the distance from points 1 to *c* or points 2 to *c*. Swing this arc from point 1 and point 2. The intersection of the arcs is at point *f*. Draw a line from point *f* to point *c*, which is perpendicular to line *AB*.

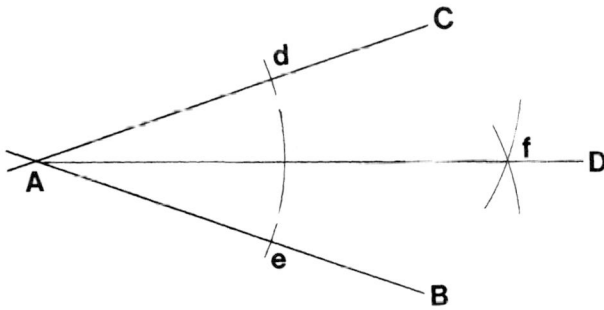

Figure 2.30   Bisecting an angle.

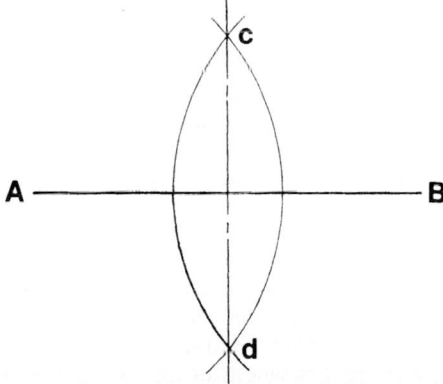

Figure 2.31   Erecting a perpendicular.

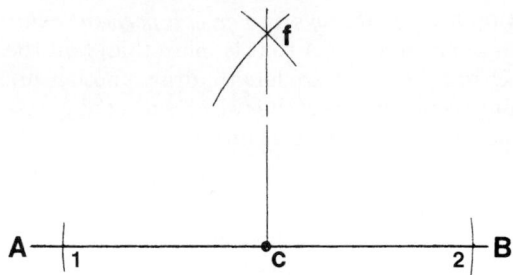

Figure 2.32 Perpendicular to a point.

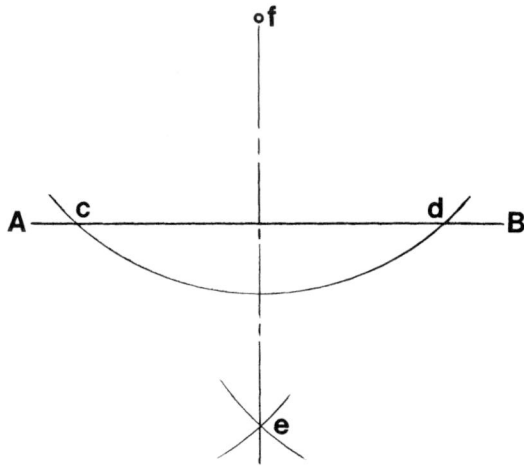

Figure 2.33 Drawing a perpendicular to a line.

- *To draw a perpendicular to a line AB from a point f, a distance from it* (Fig. 2.33), with point *f* as a center, draw a circular arc intersecting line *AB* at points *c* and *d*. With points *c* and *d* as centers, draw circular arcs with radii longer than half the distance between points *c* and *d*. These arcs intersect at point *e*, and line *fe* is the required perpendicular.

- *To draw a circular arc with a given radius through two given points* (Fig. 2.34), with points *A* and *B* as centers and the set given radius, draw circular arcs intersecting at point *f*. With point *f* as a center, draw the circular arc that will intersect both points *A* and *B*.

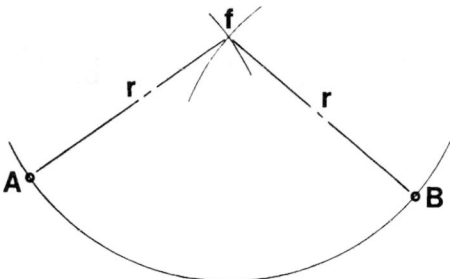

Figure 2.34   Drawing a circular arc through given points.

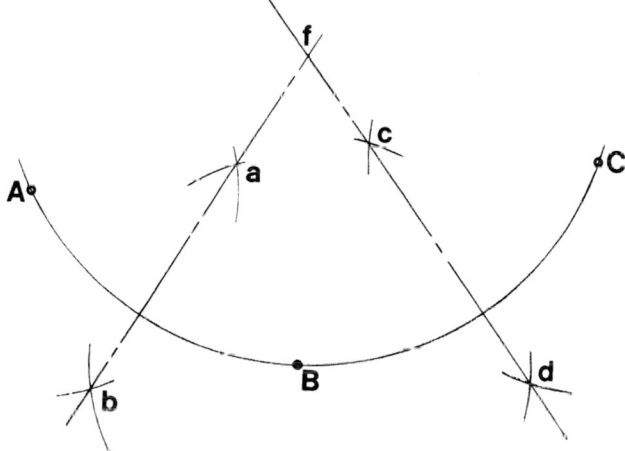

Figure 2.35   Finding the center of a circle.

- *To find the center of a circle or the arc of a circle* (Fig. 2.35), select three points on the perimeter of the given circle such as *A, B,* and *C.* With each of these points as a center and the same radius, describe arcs that intersect each other. Through the points of intersection, draw lines *fb* and *fd.* The intersection point of these two lines is the center of the circle or circular arc.

- *To draw a tangent to a circle from any given point on the circumference* (Fig. 2.36), through the tangent point *f,* draw a radial line *OA.* At point *f,* draw a line *CD* at right angles to *OA.* Line *CD* is the required tangent to point *f* on the circle.

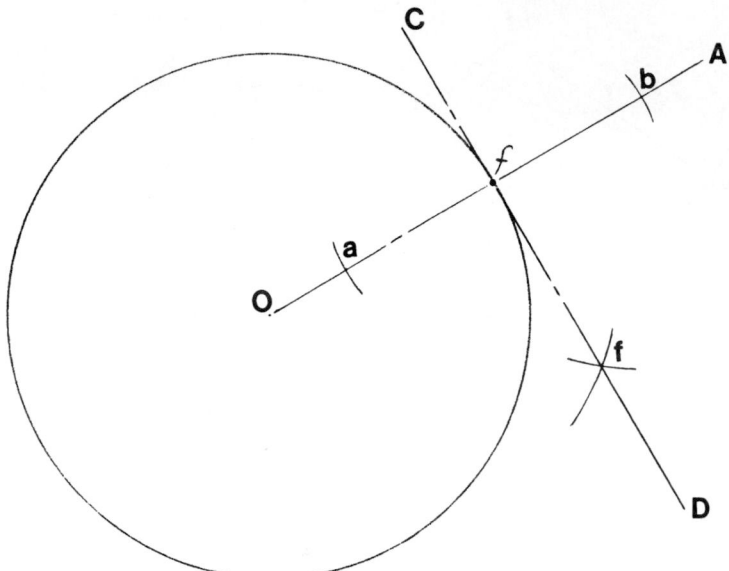

Figure 2.36 Drawing a tangent to a given point on a circle.

- *To draw an ellipse given the major and minor axes using the concentric circle method* (Fig. 2.37), on the two principal diameters *ef* and *cd* that intersect at point *O*, draw circles. From a number of points on the outer circle, such as *g* and *h*, draw radii *Og* and *Oh* intersecting the inner circle at points *g′* and *h′*. From *g* and *h*, draw lines parallel to *Oa*, and from *g′* and *h′*, draw lines parallel to *Od*. The intersection of the lines through *g* and *g′* and *h* and *h′* describe points on the ellipse. Each quadrant of the concentric circles may be divided into as many equal angles as required or as dictated by the size and accuracy required.

- *To draw an ellipse using the parallelogram method* (Fig. 2.38), on the axes *ab* and *cd*, construct a parallelogram. Divide *aO* into any number of equal parts, and divide *ae* into the same number of equal parts. Draw lines through points 1 through 4 from points *c* and *d*. The intersection of these lines will be points on the ellipse.

- *To draw a parabola using the parallelogram method* (Fig. 2.39), divide *Oa* and *ba* into the same number of equal parts. From the divisions on *ab*, draw lines converging at *O*. Lines drawn parallel

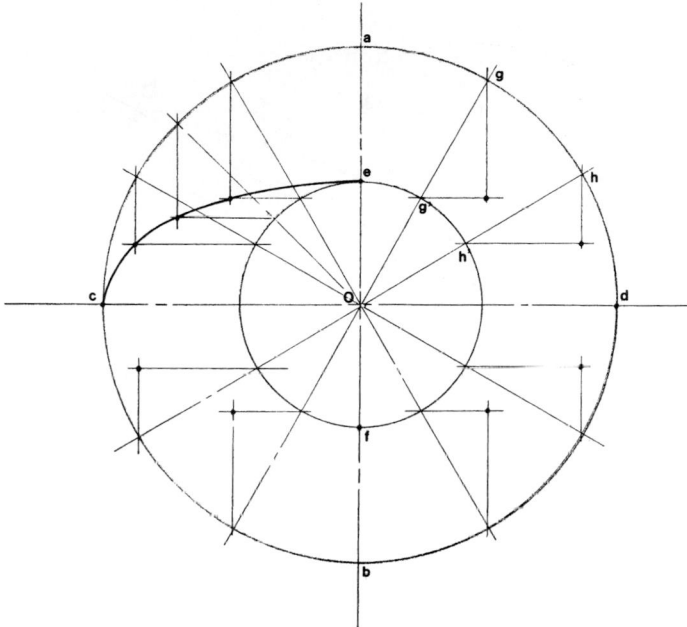

Figure 2.37 Drawing an ellipse.

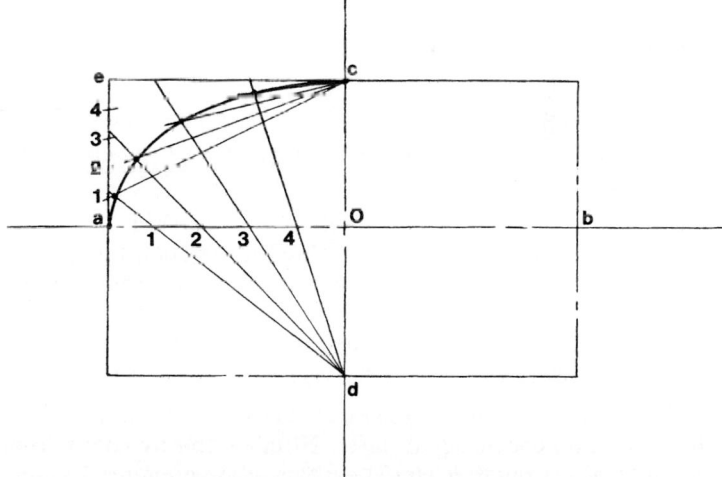

Figure 2.38 An ellipse by the parallelogram method.

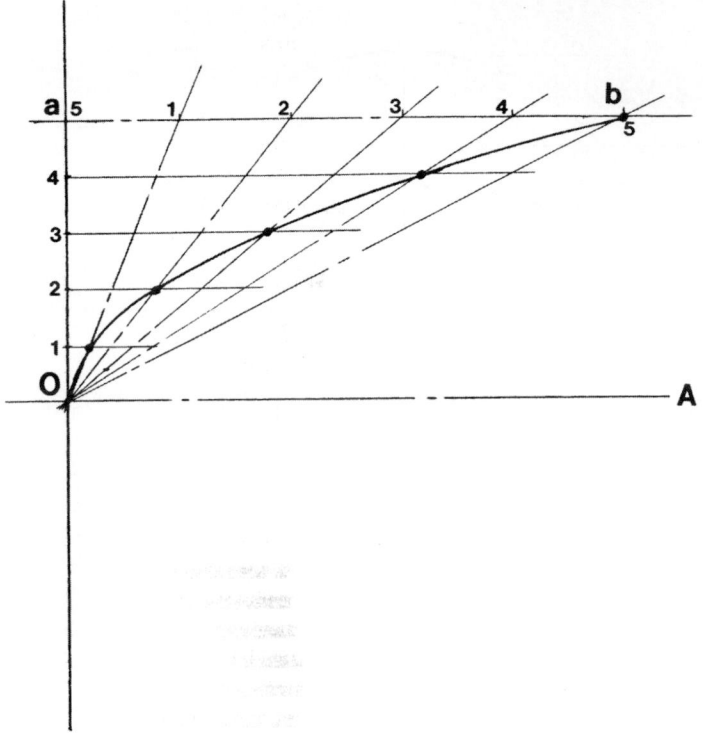

Figure 2.39 A parabola by the parallelogram method.

to line *OA* and intersecting the divisions on *Oa* will intersect the lines drawn from point *O*. These intersections are points on the parabola.

- *To draw a parabola using the offset method* (Fig. 2.40), the parabola may be plotted by computing the offsets from line *O*5. These offsets vary as the square of their distance from point *O*. If *O*5 is divided into five equal parts, distance 1*e* will be ½₅ distance 5*a*. Offset 2*d* will be ½₅ distance 5*a;* offset 3*c* will be ⅗₅ distance 5*a;* etc.

- *To draw a parabolic envelope* (Fig. 2.41), divide *Oa* and *Ob* into the same number of equal parts. Number the divisions from *Oa* and *Ob*, 1 through 6, etc. The intersection of points 1 and 6, 2 and 5, 3 and 4, 4 and 3, 5 and 2, and 6 and 1 will be points

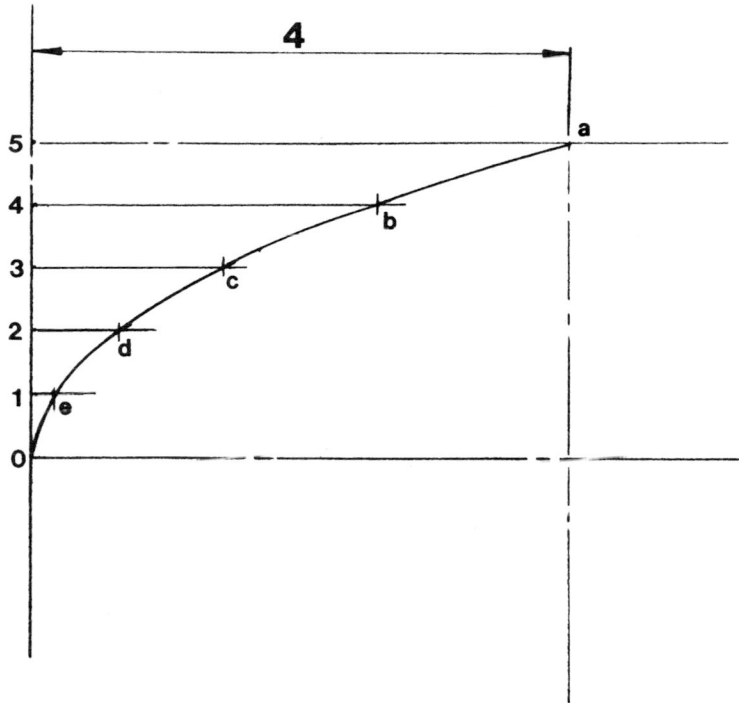

Figure 2.40   A parabola by the offset method.

on the parabola. This parabola's axis is not parallel to either ordinate.

- *To draw a parabola when the focus and directorix are given* (Fig. 2.42), draw axis *Op* through point *f* and perpendicular to directorix *AB*. Through any point *k* on the axis *Op*, draw lines parallel to *AB*. With distance *kO* as a radius and *f* as a center, draw an arc intersecting the line through *k*, thus locating a point on the parabola. Repeat for *Oj*, *Oi*, etc.

2.4   Mensuration

Mensuration is the mathematical name for calculating the areas, volumes, lengths of sides, and other geometric parts of standard geometric shapes such as circles, spheres, polygons, prisms, cylinders,

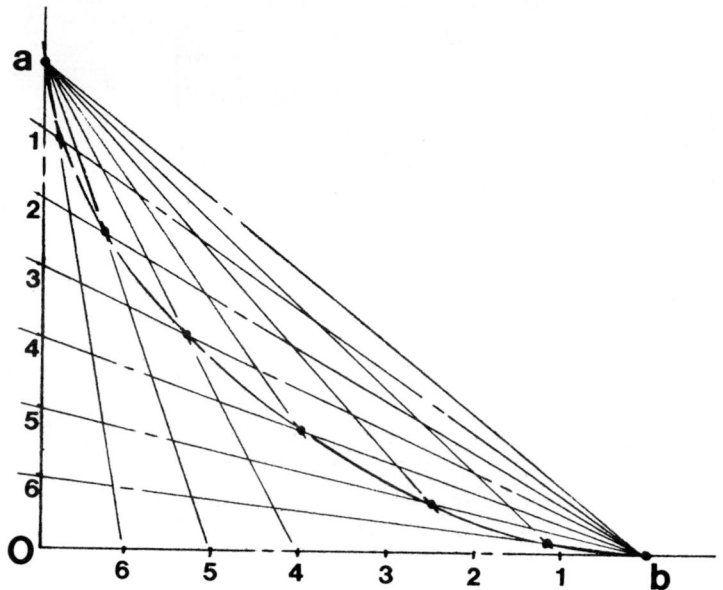

Figure 2.41 A parabolic envelope.

cones, etc. through the use of mathematical equations or formulas. Included here are the most frequently used and important mensuration formulas for the common geometric figures, both plane and solid.

$$A = \tfrac{1}{2}bh \qquad \text{(Fig. 2.43)}$$

$$A = \tfrac{1}{2}ab\sin C \qquad \text{(Fig. 2.44)}$$

$$A = \sqrt{s(s-a)(s-b)(s-c)} \qquad \text{where } s = \tfrac{1}{2}(a+b+c)$$

$$A = ab \qquad \text{(Fig. 2.45)}$$

$$A = bh \qquad \text{(Fig. 2.46)}$$

$$A = \tfrac{1}{2}cd \qquad \text{(Fig. 2.47)}$$

$$A = \tfrac{1}{2}(a+b)h \qquad \text{(Fig. 2.48)}$$

$$A = \frac{(H+h)a + bh + cH}{2} \qquad \text{(Fig. 2.49)}$$

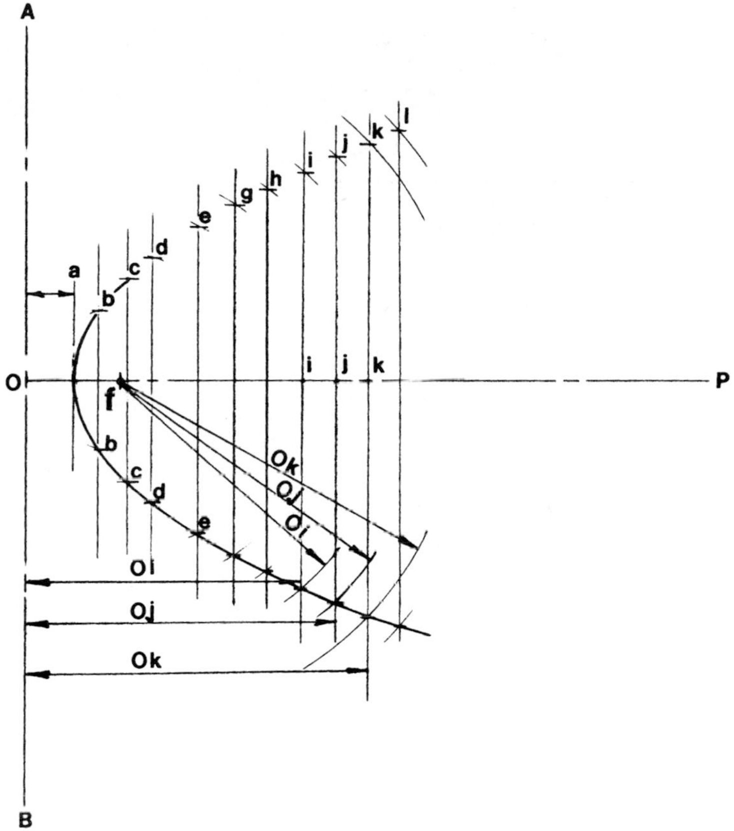

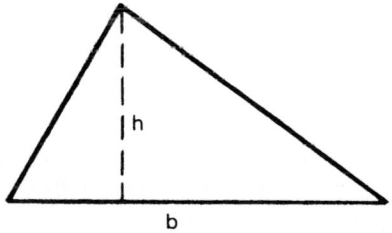

Figure 2.42 A parabolic curve.

Figure 2.43 Triangle.

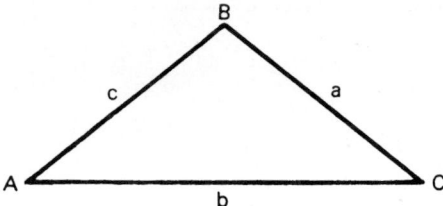

Figure 2.44   Triangle.

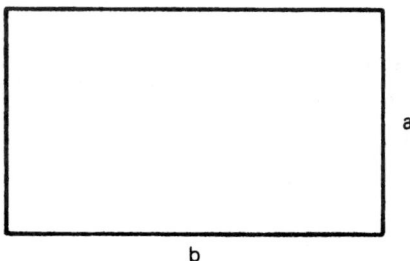

Figure 2.45   Rectangle.

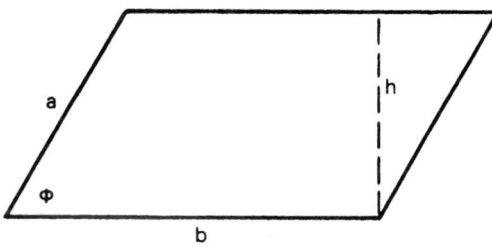

Figure 2.46   Parallelogram.

$$A = \tfrac{1}{4} n L^2 \cot \frac{180}{n} \qquad \text{(Fig. 2.50)}$$

where n = number of sides, and L = length of a side. In a polygon of n sides, each of which is L, the radius of the inscribed circle is

$$r = \frac{L}{2} \cot \frac{180}{n}$$

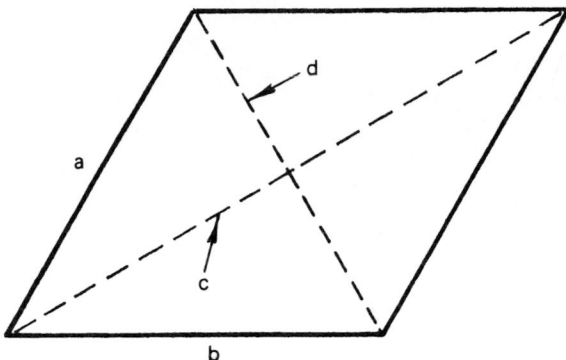

Figure 2.47 Rhombus.

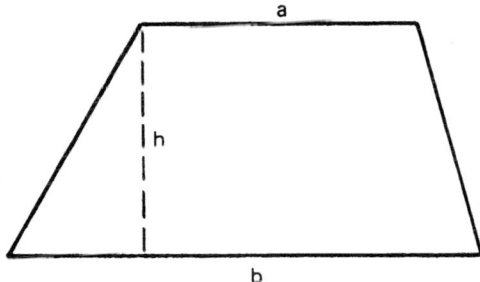

Figure 2.48 Trapezoid.

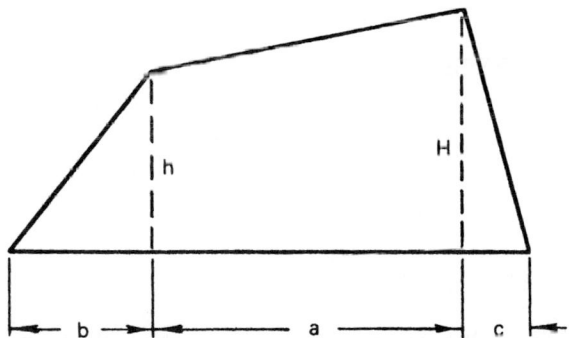

Figure 2.49 Trapezium.

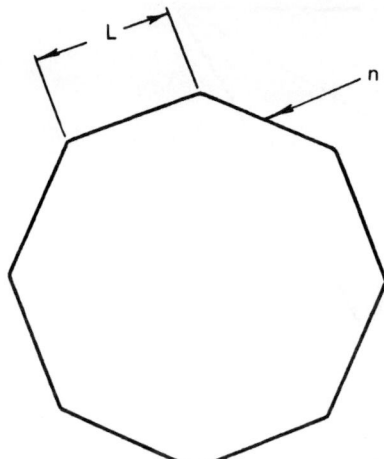

Figure 2.50 Polygon.

and the radius of the circumscribed circle is

$$r_1 = \frac{L}{2}\operatorname{cosec}\frac{180}{n}$$

The radius of a circle inscribed in any triangle whose sides are a, b, and c is

$$r = \frac{\sqrt{s(s-a)(s-b)(s-c)}}{s} \qquad \text{(Fig. 2.51)}$$

where $s = \frac{1}{2}(a + b + c)$.

The radius of the circumscribed circle is

$$R = \frac{abc}{4s(s-a)(s-b)(s-c)} \qquad \text{(Fig. 2.52)}$$

Area of an inscribed polygon is

$$A = \frac{1}{2}nr^2\sin\frac{2\pi}{n} \qquad \text{(Fig. 2.53)}$$

where r is the radius of the circumscribed circle, and n is the number of sides.

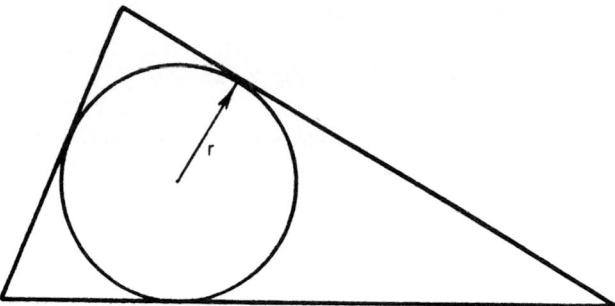

Figure 2.51 Circle inscribed in a triangle.

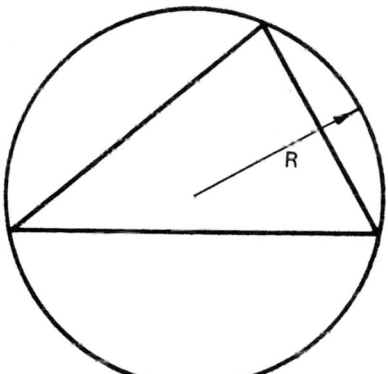

Figure 2.52 Circle circumscribed around a triangle

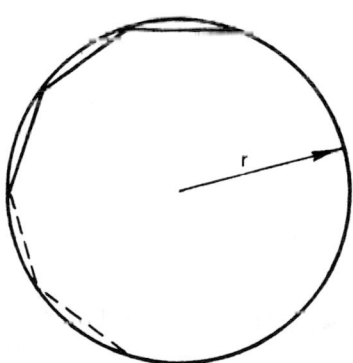

Figure 2.53 Inscribed polygon.

Area of a circumscribed polygon is

$$A = nR^2 \tan \frac{\pi}{n} \qquad \text{(Fig. 2.54)}$$

where R is radius of the inscribed circle, and n is the number of sides.

Circumference of a circle is

$$C = 2\pi r = \pi d \qquad \text{(Fig. 2.55)}$$

Area of a circle is

$$A = \pi r^2 = \tfrac{1}{4}\pi d^2 \qquad \text{(Fig. 2.56)}$$

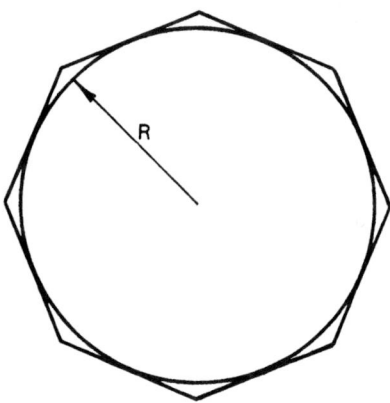

Figure 2.54 Circumscribed polygon.

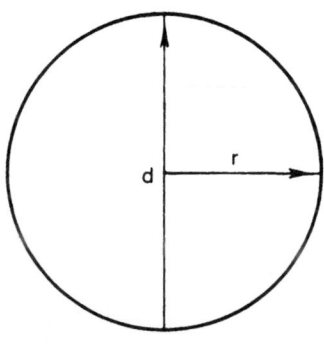

Figure 2.55 Circle.

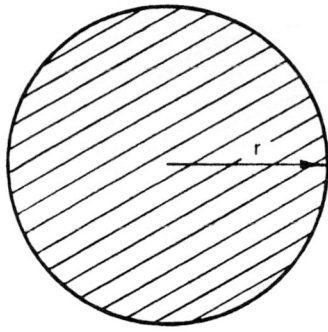

Figure 2.56 Area of a circle.

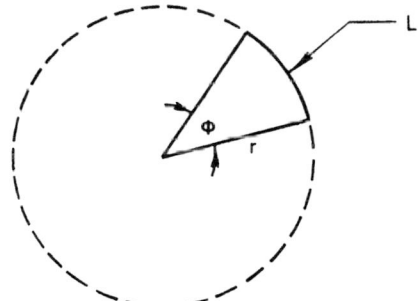

Figure 2.57 Arc length.

Length of arc L is

$$L = \frac{\pi r \phi}{180} \qquad \text{(Fig. 2.57)}$$

where ϕ is in radians, $L = r\phi$.

Length of chord AB is

$$AB = 2r \sin(\tfrac{1}{2}\phi) \qquad \text{(Fig. 2.58)}$$

Area of the sector is

$$A = \frac{r^2 \phi}{360} \qquad \text{or} \qquad A - \frac{rL}{2}$$

where L is the length of the arc.

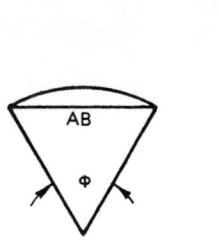

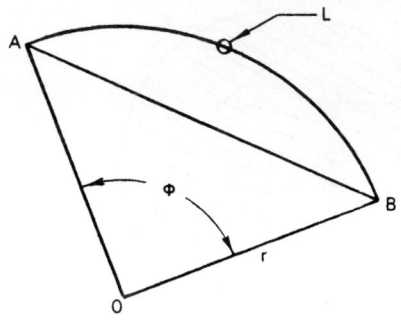

Figure 2.58 Chord.

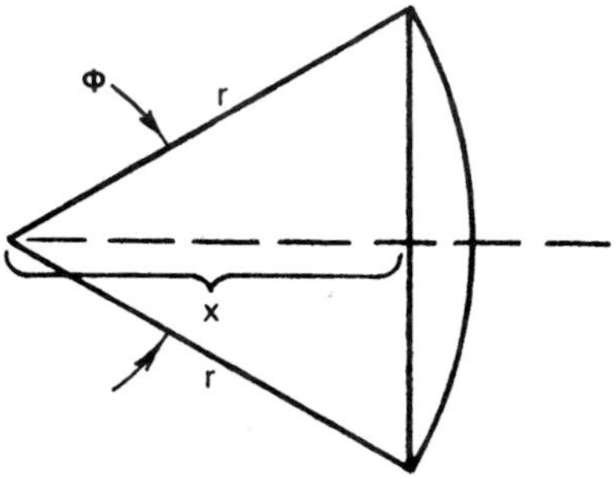

Figure 2.59 Segment.

Area of a segment of a circle is

$$A = \frac{\pi r^2 \phi}{360} - \frac{r^2 \sin \phi}{2} \qquad \text{(Fig. 2.59)}$$

where $\phi = 180° - 2 \arcsin x/r$, and $x =$ perpendicular distance, center to chord. If ϕ is in radians, $A = \frac{1}{2}r^2(\phi - \sin\phi)$.

Area of the ring between circles is

$$A = \pi(R + r)(R - r) \qquad \text{(Fig. 2.60)}$$

Note: The circles need not be concentric.

Circumference of an ellipse is

$$C = 2\pi \sqrt{\frac{a^2 + b^2}{2}} \text{ (approximate)} \qquad \text{(Fig. 2.61)}$$

$$A = \pi ab$$

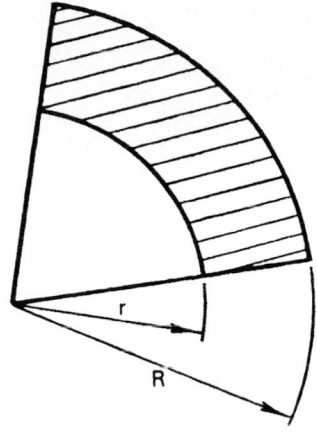

Figure 2.60 Ring between circles.

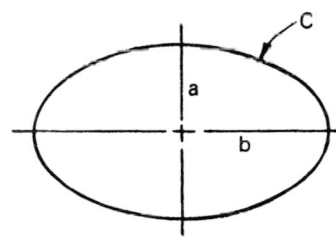

Figure 2.61 Ellipse.

Surface and volume of a sphere are

$$S = 4\pi r^2 = \pi d^2 \quad \text{(Fig. 2.62)}$$

$$V = \frac{4}{3}\pi r^3 = \frac{1}{6}\pi d^3$$

Surface and volume of a cylinder are

$$S = 2\pi rh \quad \text{(Fig. 2.63)}$$

$$V = \pi r^2 h$$

Surface and volume of a cone are

$$S = \pi r \sqrt{r^2 + h^2} \quad \text{(Fig. 2.64)}$$

$$V = (\pi/3)r^2 h$$

Area and volume of curved surface of spherical segment are

$$A = 2\pi rh \quad \text{(Fig. 2.65)}$$

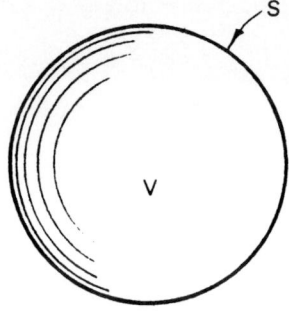

Figure 2.62 Sphere.

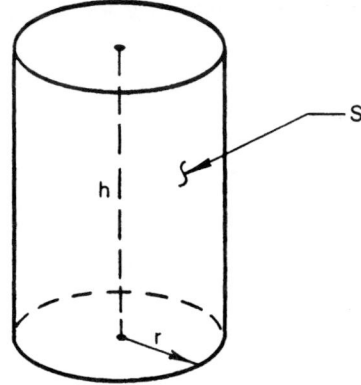

Figure 2.63 Cylinder.

$$V = (\tfrac{1}{3}\pi h^2)(3r - h)$$

When a is the radius of the base of the segment,

$$V = (\tfrac{1}{6}\pi h)(h^2 + 3a^2)$$

Surface area and volume of the frustum of a cone are

$$S = \pi(r_1 + r_2)\sqrt{h^2 + (r_1 - r_2)^2} \qquad \text{(Fig. 2.66)}$$

$$V = \frac{h}{3}(r_1^2 + r_1 r_2 + r_2^2)\pi$$

Area and volume of a truncated cylinder are

$$A = \pi r(h_1 + h_2) \qquad \text{(Fig. 2.67)}$$

$$V = 1.5708 r^2(h_1 + h_2)$$

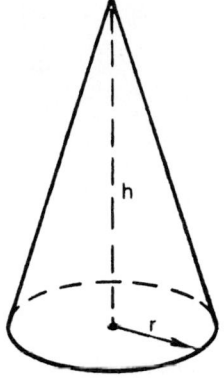

Figure 2.64 Cone.

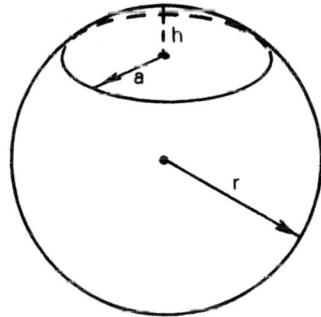

Figure 2.65 Spherical segment.

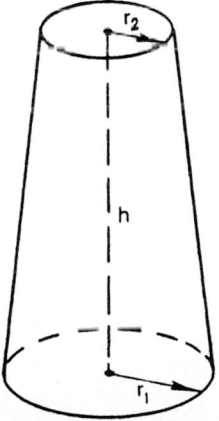

Figure 2.66 Frustum of a cone.

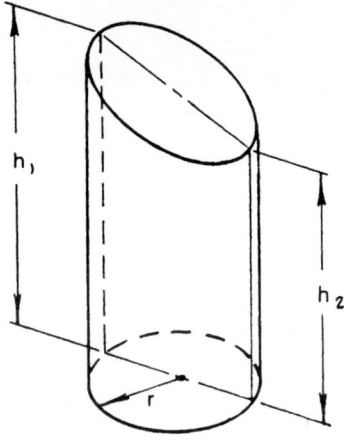

Figure 2.67 Truncated cylinder.

Area and volume of a portion of a cylinder (base edge = diameter) are

$$A = 2rh \qquad \text{(Fig. 2.68)}$$

$$V = \tfrac{2}{3}r^2h$$

Area and volume of a portion of a cylinder (special cases) are

$$A = \frac{h(ad \pm c \times \text{perimeter of base})}{r \pm c} \qquad \text{(Fig. 2.69)}$$

$$V = \frac{h(\tfrac{2}{3}a^3 \pm cA)}{r \pm c}$$

where d = diameter of base circle. Use $+c$ when the base area is larger than half the base circle; use $-c$ when the base area is smaller than half the base circle.

Volume of a wedge is

$$V = \frac{(2b+c)ah}{6} \qquad \text{(Fig. 2.70)}$$

Area and volume of a spherical zone are

$$A = 2\pi rh \qquad \text{(Fig. 2.71)}$$

$$V = 0.5236h\left(\frac{3c_1^2}{4} + \frac{3c_2^2}{4} + h^2\right)$$

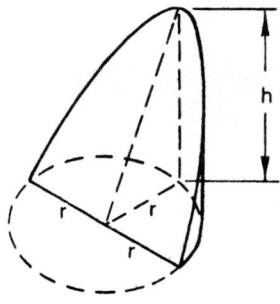

Figure 2.68   Portion of a cylinder.

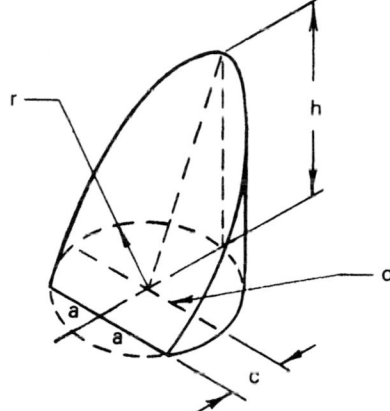

Figure 2.69   Portion of a cylinder.

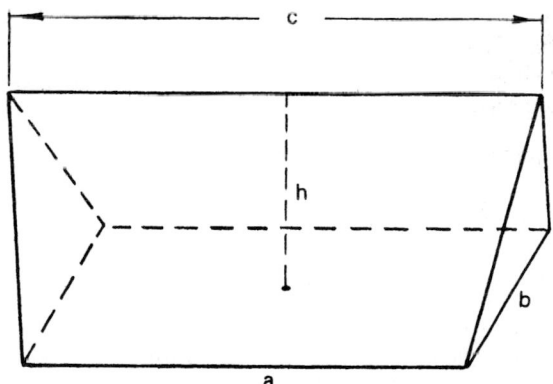

Figure 2.70   Wedge.

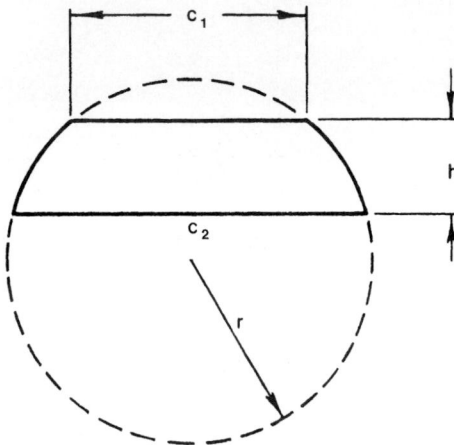

Figure 2.71 Spherical zone.

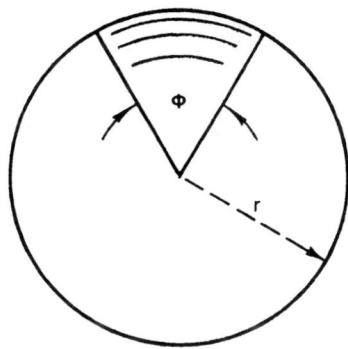

Figure 2.72   Spherical wedge.

Area and volume of a spherical wedge are

$$A = \frac{\phi}{360} 4\pi r^2 \qquad \text{(Fig. 2.72)}$$

$$V = \frac{\phi}{360} \frac{4\pi r^3}{3}$$

The volume of a paraboloid is

$$V = \frac{\pi r^2 h}{2} \qquad \text{(Fig. 2.73)}$$

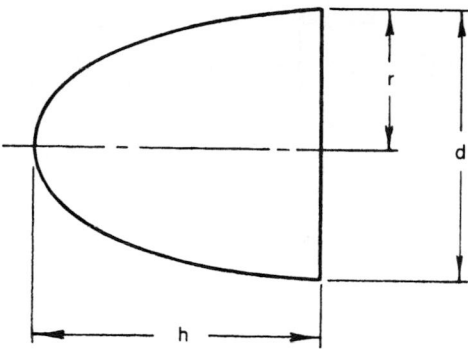

Figure 2.73 Paraboloid.

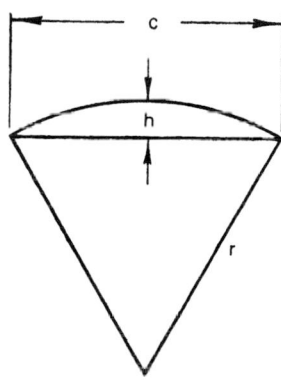

Figure 2.74 Spherical sector.

Area and volume of a spherical sector are

$$A = r\left(2h + \frac{c}{2}\right) \quad \text{(total area)} \quad \text{(Fig. 2.74)}$$

$$V = \frac{2\pi r^2 h}{3}$$

where $c = 2\sqrt{h(2r - h)}$.

Area and volume of a spherical segment are

$$A = 2\pi rh \quad \text{(spherical surface)} \quad \text{(Fig. 2.75)}$$

$$A = \pi\left(\frac{c^2}{4} + h^2\right) \qquad c = 2\sqrt{h(2r - h)} \qquad r = \frac{c^2 + 4h^2}{8h}$$

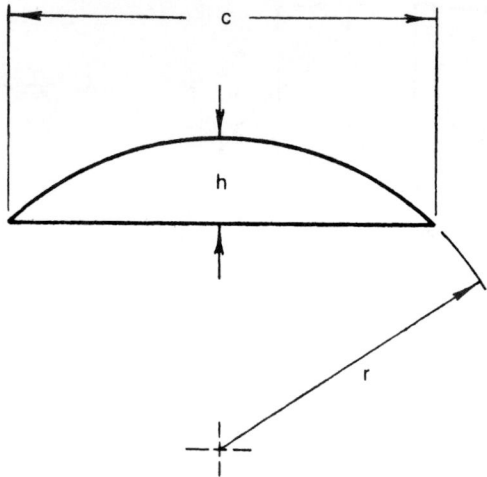

Figure 2.75 Spherical segment.

$$V = \pi h^2 \left(r - \frac{h}{3} \right)$$

Area and volume of a torus are

$$A = 4\pi^2 cr \quad \text{(total surface)} \quad \text{(Figs. 2.76 and 2.77)}$$

$$V = 2\pi^2 cr^2 \quad \text{(total volume)}$$

For properties of the circle, see Fig. 2.78.

2.5   Percentage Calculations

Percentage calculation procedures have many applications in machining, design, and metalworking problems. Although the procedures are relatively simple, it is easy to make mistakes in the manipulations of the numbers involved.

Ordinarily, 100 percent of any quantity is represented by the number 1.00, meaning the total quantity. Thus, if we take 50 percent of any quantity, or any multiple of 100 percent, it *must* be expressed as a decimal:

$$1\% = 0.01$$

$$10\% = 0.10$$

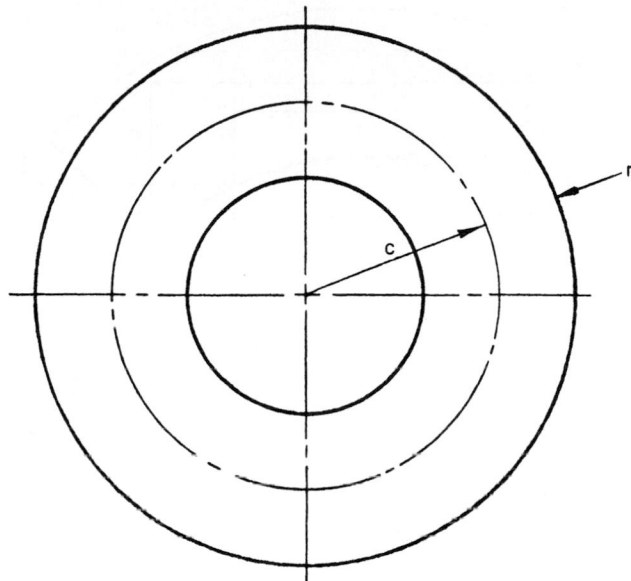

Figure 2.76 Torus.

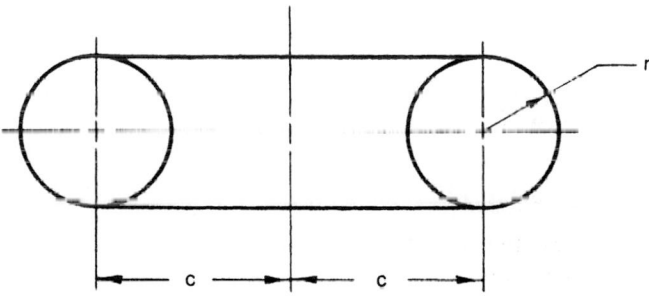

Figure 2.77 Torus.

$$65.5\% = 0.655$$

$$145\% = 1.45$$

In effect, we are dividing the percentage figure, such as 65.5 percent, by 100 to arrive at the decimal equivalent required for calculations.

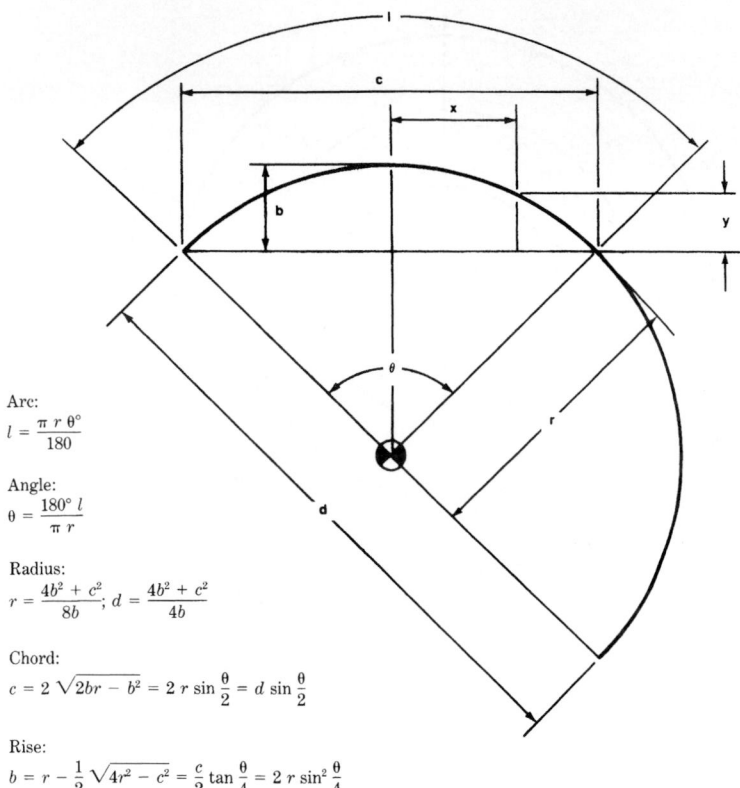

Arc:
$$l = \frac{\pi \, r \, \theta°}{180}$$

Angle:
$$\theta = \frac{180° \, l}{\pi \, r}$$

Radius:
$$r = \frac{4b^2 + c^2}{8b}; \; d = \frac{4b^2 + c^2}{4b}$$

Chord:
$$c = 2 \sqrt{2br - b^2} = 2 \, r \sin \frac{\theta}{2} = d \sin \frac{\theta}{2}$$

Rise:
$$b = r - \frac{1}{2} \sqrt{4r^2 - c^2} = \frac{c}{2} \tan \frac{\theta}{4} = 2 \, r \sin^2 \frac{\theta}{4}$$

Rise:
$$b = r + y - \sqrt{r^2 - x^2}$$
where $y = b - r + \sqrt{r^2 - x^2}$ and $x = \sqrt{r^2 - (r + y - b)^2}$.

Figure 2.78 Properties of the circle.

Let us take a percentage of a given number:

$$45\% \text{ of } 136.5 = 0.45 \times 136.5 = 61.425$$

$$33.5\% \text{ of } 235.7 = 0.335 \times 235.7 = 78.9595$$

Let us now compare two arbitrary numbers, 33 and 52, as an illustration:

$$\frac{52 - 33}{33} = 0.5758$$

Thus the number 52 is 57.58 percent larger than the number 33. We also can say that 33 increased by 57.58 percent is equal to 52; that is, $0.5758 \times 33 + 33 = 52$. Now

$$\frac{52 - 33}{52} = 0.3654$$

Thus the number 52 minus 36.54 percent of itself is 33. We also can say that 33 is 36.54 percent less than 52; that is, $0.3654 \times 52 = 19$ and $52 - 19 = 33$. The number 33 is what percent of 52? That is, $33/52 = 0.6346$. Therefore, 33 is 63.46 percent of 52.

2.6 Decimal Equivalents and Millimeter Chart (see Fig. 2.79)

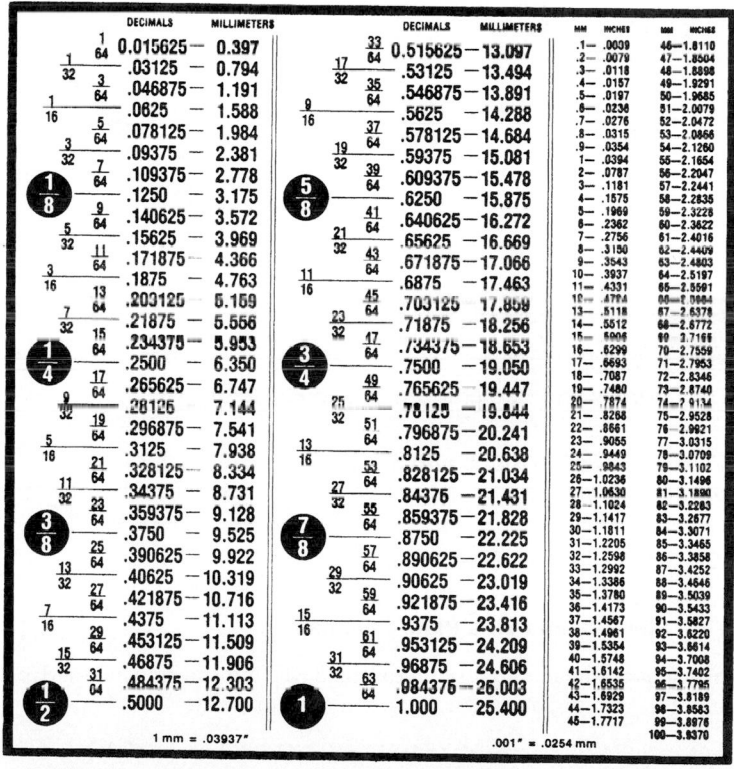

Figure 2.79 Decimal equivalents and millimeters.

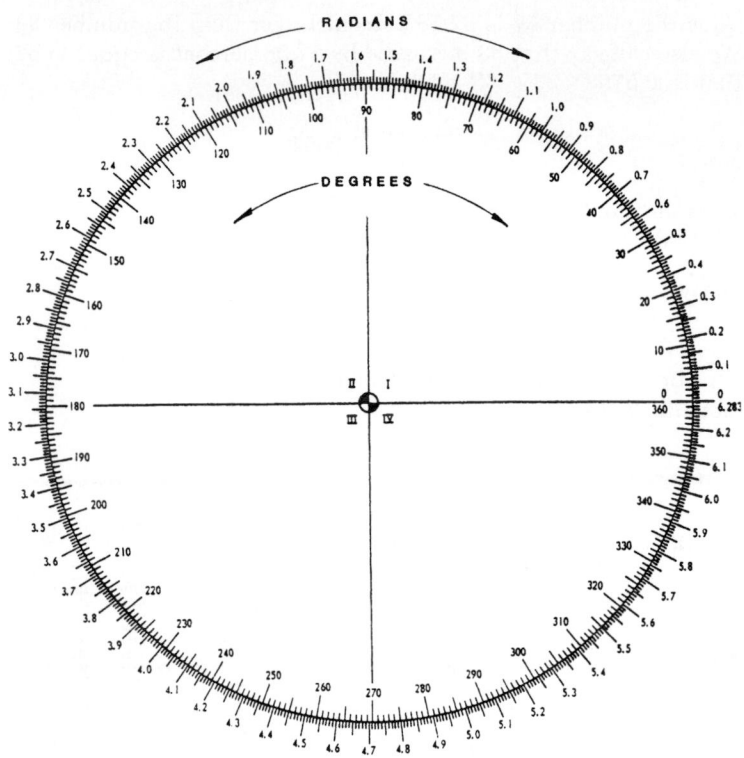

Figure 2.80 Degrees to radians conversion chart.

2.7 Degrees and Radians Chart (see Fig. 2.80)

2.8 Mathematical Signs and Symbols
(see Table 2.1)

2.9 Greek Alphabet (see Table 2.2)

2.10 Sine Bar and Sine Plate Calculations

Sine bar procedures.   Referring to Fig. 2.81, find the sine bar–setting height for an angle of 34°25′ using a 5-in sine bar.

$$\sin 34°25′ = x/5 \qquad (34°25′ = 34.416667 \text{ decimal degrees})$$

TABLE 2.1　Mathematical Signs and Symbols

$+$	Plus, positive
$-$	Minus, negative
$\times$ or $\cdot$	Times, multiplied by
$\div$ or $/$	Divided by
$=$	Is equal to
$\equiv$	Is identical to
$\cong$	Is congruent to or approximately equal to
$\sim$	Is approximately equal to or is similar to
$<$ and $\not<$	Is less than, is not less than
$>$ and $\not>$	Is greater than, is not greater than
$\neq$	Is not equal to
$\pm$	Plus or minus, respectively
$\mp$	Minus or plus, respectively
α	Is proportional to
$\rightarrow$	Approaches, e.g., as $x \rightarrow 0$
$\leq, \leqq$	Less than or equal to
$\geq, \geqq$	More than or equal to
$\therefore$	Therefore
$:$	Is to, is proportional to
Q.E.D.	Which was to be proved, end of proof
%	Percent
#	Number
@	At
$\angle$ or $\sphericalangle$	Angle
$\circ\ '\ ''$	Degrees, minutes, seconds
$\|, /\!/$	Parallel to
$\perp$	Perpendicular to
e	Base of natural logs, 2.71828 . . .
π	Pi, 3.14159 . . .
( )	Parentheses
[ ]	Brackets
{ }	Braces
$'$	Prime, $f'(x)$
$''$	Double prime, $f''(x)$
$\sqrt{\ }, \sqrt[n]{\ }$	Square root, nth root
$1/x$ or x^{-1}	Reciprocal of x
!	Factorial
∞	Infinity
Δ	Delta, increment of
∂	Curly "d," partial differentiation
Σ	Sigma, summation of terms
Π	The product of terms, product
arc	As in arcsine (the angle whose sinc is)
f	Function, as $f(x)$
rms	Root mean square
$|x|$	Absolute value of x
i	For -1
j	Operator, equal to -1

TABLE 2.2　The Greek Alphabet

α	A	alpha	ι	I	iota	ρ	P	rho
β	B	beta	κ	κ	kappa	σ	Σ	sigma
γ	Γ	gamma	λ	Λ	lambda	τ	T	tau
δ	Δ	delta	μ	M	mu	υ	Y	upsilon
ε	E	epsilon	ν	N	nu	ϕ	Φ	phi
ζ	Z	zeta	φ	Ξ	xi	χ	X	chi
η	H	eta	o	O	omicron	ψ	Ψ	psi
θ	Θ	theta	π	Π	pi	ω	Ω	omega

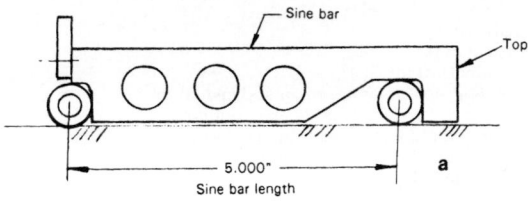

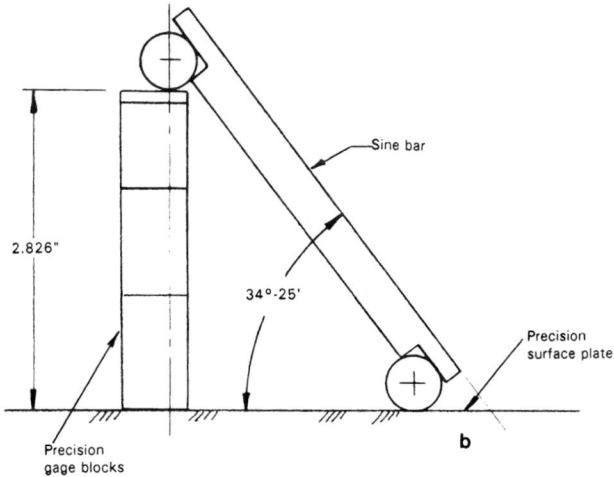

Figure 2.81 (*a*) Sine bar. (*b*) Sine bar.

$$\sin 34.416667° = x/5$$

$$x = 5 \times 0.565207$$

$$x = 2.826 \text{ in}$$

Set the sine bar height with Jo-blocks or precision blocks to 2.826 in.

From this example it is apparent that the setting height can be found for any sine bar length simply by multiplying the length of the sine bar times the natural sine value of the required angle. The simplicity, speed, and accuracy possible for setting sine bars with the aid of the pocket calculator render sine bar tables obsolete. No sine bar table will give you the required setting height for such an angle as 42°17′26″, but by using the calculator proce-

dure, this becomes a routine, simple process with less chance for errors.

Method

1. Convert the required angle to decimal degrees.

2. Find the natural sine of the required angle.

3. Multiply the natural sine of the angle by the length of the sine bar to find the bar-setting height (see Fig. 2.81).

Formulas for finding angles.   Refer to Fig. 2.82 when angles α and ϕ are known to find angles $X, A, B,$ and C.

$$\tan X = \tan \alpha \, \cos \phi$$

$$\sin C = \cos \alpha / \cos X$$

$$\text{Angle } B = 180° - (\text{angle } A + \text{angle } C)$$

$$\tan A = \frac{\sin \alpha \sin C}{\sin \phi - (\sin \alpha \cos C)}$$

$$D = \text{true angle}$$

$$\tan D = \tan \phi \, \sin T$$

$$\tan C = \frac{\sin D}{\tan \theta}$$

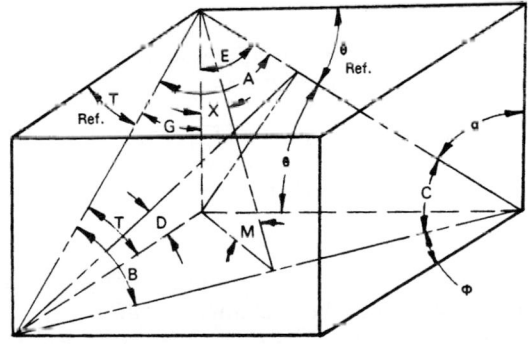

Figure 2.82   Finding the unknown angles.

$$\tan M = \sqrt{(\tan \theta)^2 + (\tan T)^2}$$

$$\cos A = \cos E \cos G$$

$$\cos A = \sin \theta \sin T$$

Formulas for finding true and apparent angles.   See Fig. 2.83a, where α = apparent angle, θ = true angle, and ϕ = angle of rotation.

Note: Apparent angle α is OA triangle projected onto plane OB. See also Fig. 2.83b.

$$\tan \theta = K/L$$

$$\tan \alpha = K/(L \cos \phi)$$

$$\tan \alpha \cos \phi = K/L$$

$$K/L = \tan \theta = \cos \phi \tan \alpha$$

or

$$\tan \theta = \cos \phi \tan \alpha$$

and

$$\tan \alpha = \tan \theta /\cos \phi$$

The three-dimensional relationships shown for the angles and triangles in the preceding figures and formulas are of importance and should be understood. This will help in the setting of compound sine plates when it is required to set a compound angle.

Setting compound sine plates.   For setting two known angles at 90° to each other, proceed as shown in Fig. 2.84.

Example: First angle = 22.45°. Second angle = 38.58° (see Fig. 2.84). To find the amount the intermediate plate must be raised from the base plate (X dimension in Fig. 2.84b) to obtain the desired first angle,

1. Find the natural cosine of the second angle (38.58°), and multiply this times the natural tangent of the first angle (22.45°).

2. Find the arctangent of this product, and then find the natural sine of this angle.

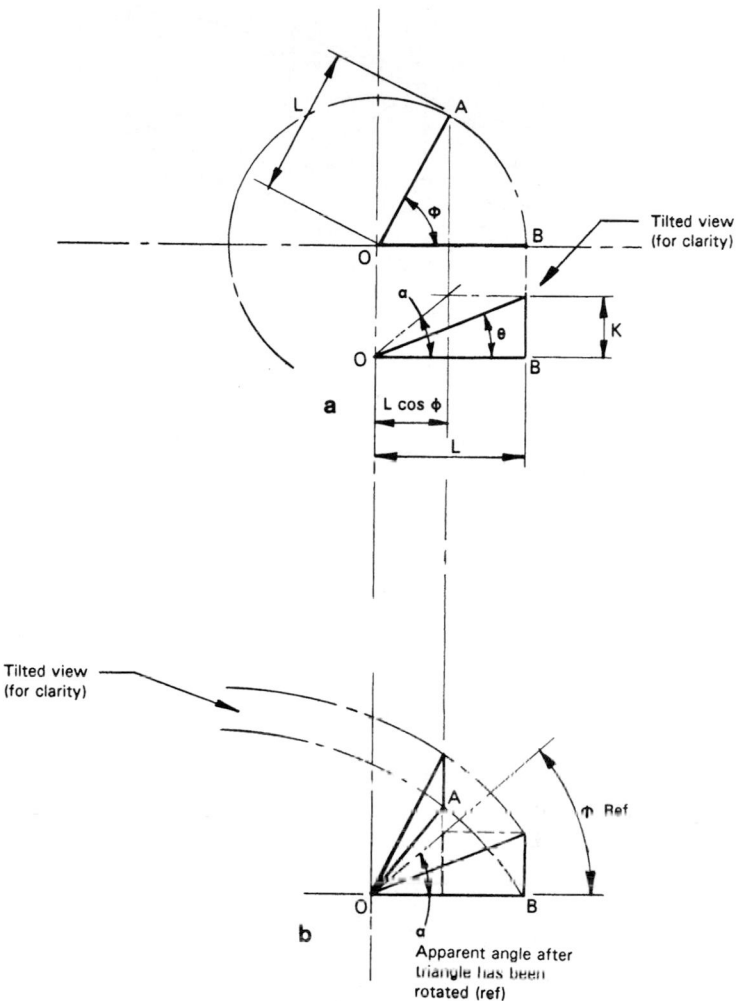

Figure 2.83 True and apparent angles.

3. This natural sine is now multiplied by the length of the sine plate to find the X dimension in Fig. 2.84b to which the intermediate plate must be set.

4. Set up the Jo-blocks to equal the X dimension, and set them in position between the base plate and the intermediate plate.

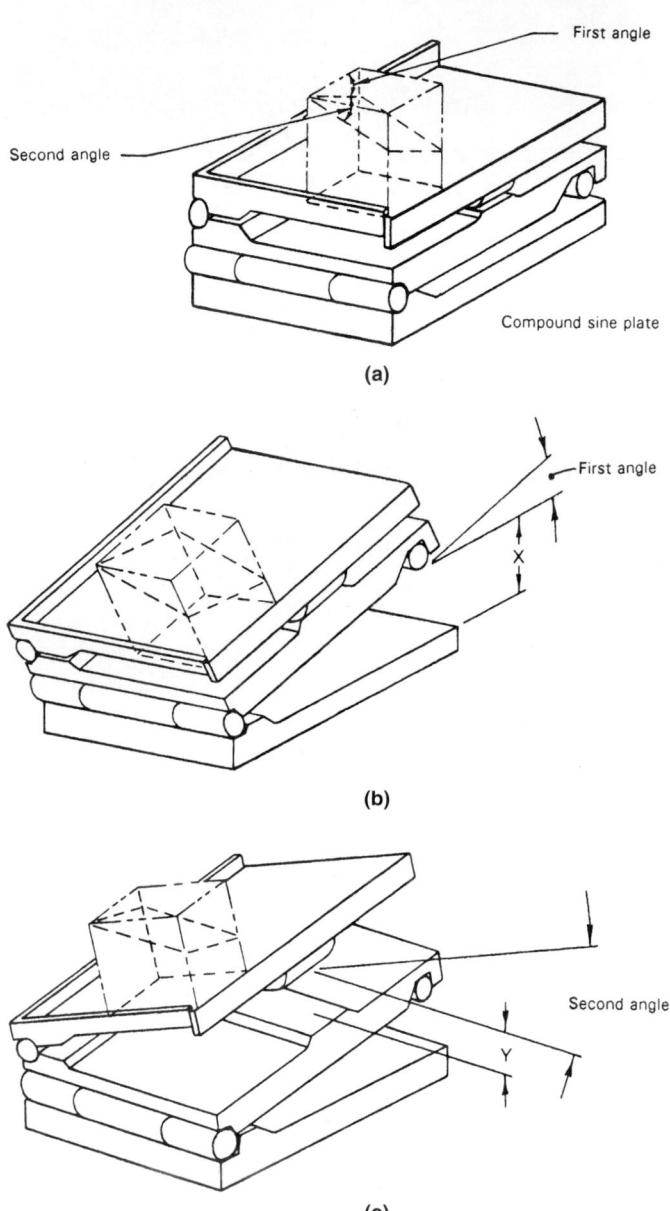

First angle

Second angle

Compound sine plate

(a)

First angle

X

(b)

Second angle

Y

(c)

Figure 2.84   Setting angles on a sine plate.

$$\cos 38.58° = 0.781738$$

$$\tan 22.45° = 0.413192$$

$$0.781738 \times 0.413192 = 0.323008$$

$$\text{arctan } 0.323008 = 17.900872°$$

$$\sin 17.900872° = 0.307371$$

$$0.307371 \times 10 \text{ in (for 10-in sine plate)} = 3.0737 \text{ in}$$

Therefore, set X dimension to 3.074 in (to three decimal places).

To find the amount the top plate must be raised (the Y dimension in Fig. 2.84c) above the intermediate plate to obtain the desired second angle,

1. Find the natural sine of the second angle, and multiply this by the length of the sine plate.

2. Set up the Jo-blocks to equal the Y dimension, and set them in position between the top plate and the intermediate plate.

$$\sin 38.58° = 0.632607$$

$$0.632607 \times 10 \text{ in (for 10-in sine plate)} = 6.32607$$

Therefore, set the Y dimension to 6.326 in (to three decimal places).

2.11 Solutions to Problems In Machining and Metalworking

The following sample problems will show in detail the importance of trigonometry and basic algebraic operations as they apply to machining and metalworking. By using the methods and procedures shown in this chapter of the *Handbook*, you will be able to solve many basic and complex machining and metalworking problems.

Taper (Fig. 2.85).   Solve for x if y is given; solve for y if x is given; solve for d. Use the tangent function:

$$\tan A = y/x$$

$$d = D - 2y$$

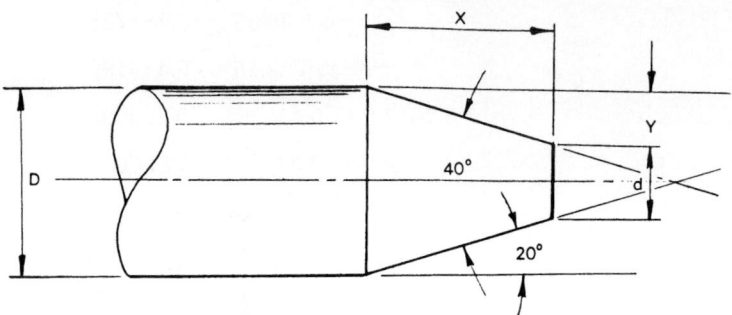

Figure 2.85 Taper.

where A = taper angle
  D = outside diameter of rod
  d = diameter at end of taper
  x = length of taper
  y = drop of taper

Example: If the rod diameter = 0.9375 diameter, taper length = 0.875 = x, and taper angle = 20° = angle A, find y and d from

$$\tan 20° = y/x$$

$$y = x \tan 20°$$

$$= 0.875(0.36397) = 0.318$$

$$d = D - 2y$$

$$= 0.9375 - 2(0.318)$$

$$= 0.9375 - 0.636$$

$$= 0.3015$$

Countersink depths (three methods for calculating)
Method 1: To find the tool, travel y from the top surface of the part for a given countersink finished diameter at the part surface:

$$y = \frac{D/2}{\tan A/2} \qquad \text{(Fig. 2.86)}$$

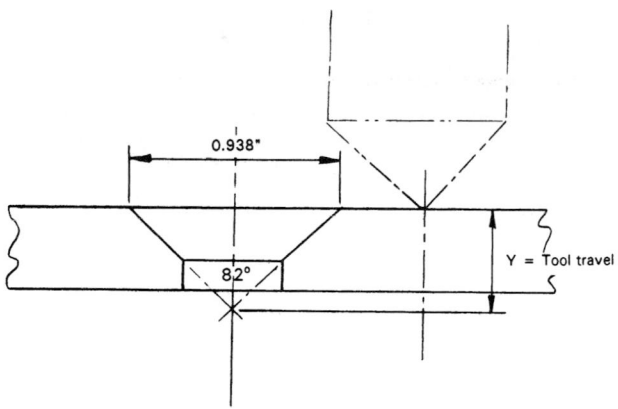

Figure 2.86 Countersink depth.

where D = finished countersink diameter
A = countersink angle
y = tool advance from surface of part

$$y = \frac{0.938/2}{\tan 41°} = \frac{0.469}{0.869} = 0.5397, \text{ or } 0.540$$

Method 2: To find the tool travel from the edge of the hole (Fig. 2.87), where D = finished countersink diameter, H = hole diameter, and A = ½ countersink angle, 41°,

$$\tan A = x/y$$

$$y = x/\tan A \qquad \text{or} \qquad x/(\text{½ countersink angle})$$

First, find x from

$$D = H + 2x$$

If D = 0.875 and H = 0.500,

$$0.875 = 0.500 + 2x$$

$$2x = 0.375$$

$$x = 0.1875$$

Now solve for y, the tool advance:

$$y = x/\tan A$$

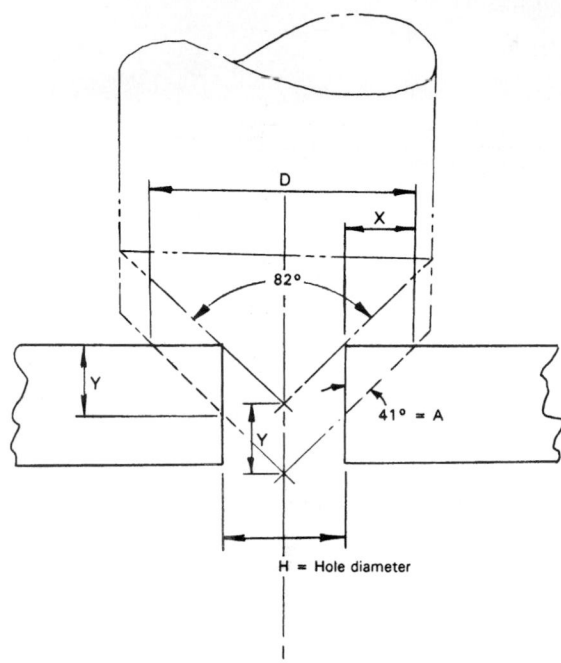

Figure 2.87 Tool travel in countersinking.

$$= 0.1875/\tan 41°$$

$$= 0.1875/0.8693$$

$$= 0.2157, \text{ or } 0.216 \text{ (tool advance from edge of hole)}$$

Method 3: To find tool travel from the edge of the hole (Fig. 2.88), where D = finished countersink diameter, d = hole diameter, ϕ = ½ countersink angle, and H = countersink tool advance from edge of hole,

$$H = ½(D - d) \cot\!an \phi \qquad \text{or} \qquad H = \frac{D - d}{2\tan\theta}$$

(Remember that cotan ϕ = 1/tan ϕ or tan ϕ = 1/cotan ϕ.)

Finding taper angle α. Given dimensions shown in Fig. 2.89, find angle α and length x. First, find angle α from

$$y = \frac{1.875 - 0.500}{2} = \frac{1.375}{2} = 0.6875$$

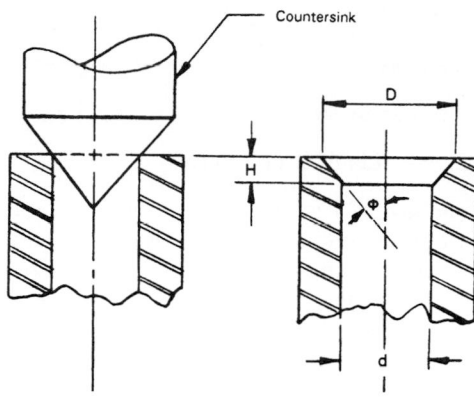

Figure 2.88 Tool travel from the edge of the hole.

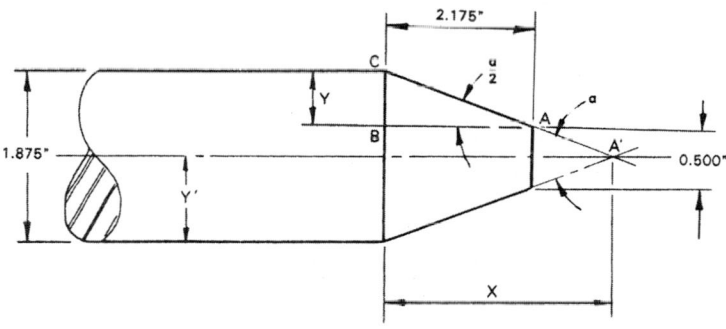

Figure 2.89 Finding taper angle α.

Then solve triangle ABC for ½ angle α:

$$\tan \frac{1}{2}\alpha = \frac{0.6875}{2.175} = 0.316092$$

$$\text{Arctan } \tfrac{1}{2}\alpha = 0.316092$$

$$\tfrac{1}{2}\alpha = 17.541326°$$

$$\text{Angle } \alpha = 35.082652°$$

Then solve triangle $A'B'C'$, where $y' = 0.9375$ or ½ diameter of rod:

$$\text{Angle } C = 90° - 17.541326°$$

$$= 72.458674°$$

Now the x dimension is found from

$$\tan \tfrac{1}{2}\alpha = 0.9375/x$$

$$x = 0.9375/\tan \tfrac{1}{2}\alpha$$

$$= 2.966 \ (\text{side } A'B' \text{ or length } x)$$

Compound angles. A solid geometric figure consisting of several plane right triangles contains a compound angle. Each of these right triangles contains one of the angles of the compound angle, and one of these angles is called the *true angle*. The true angle of inclination of a plane to a reference plane is called the *dihedral angle*. Figure 2.90 shows a typical triangular pyramid in which all four faces are right triangles. Most of the solids encountered in actual practice may be reduced to this type of pyramid by drawing a plane of symmetry. The unknown angle then may be calculated from any two known angles that lie in adjacent faces of the pyramid.

The trigonometric solutions for all the compound angles of Fig. 2.90 may be calculated from the following trigonometric relationships:

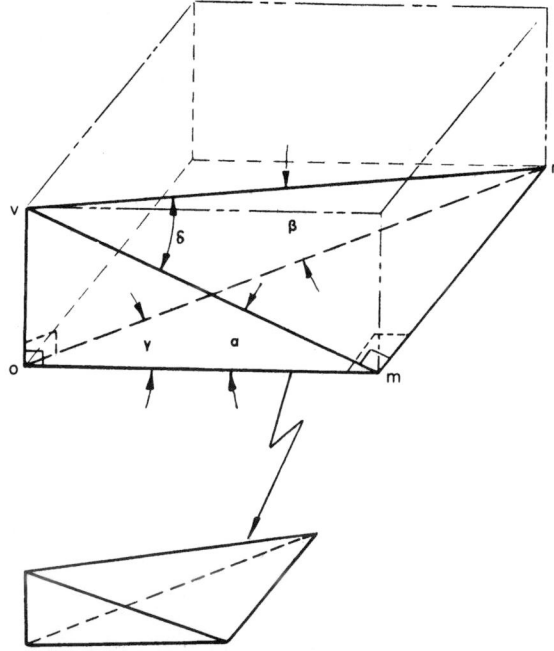

Figure 2.90   Compound angles.

Solutions To Compound Angles

Given	To find	Equation
α and β	γ	$\cos\gamma = \dfrac{\tan\beta}{\tan\alpha}$
α and β	δ	$\cos\delta = \dfrac{\sin\beta}{\sin\alpha}$
α and γ	β	$\tan\beta = \cos\gamma\,\tan\alpha$
α and γ	δ	$\tan\delta = \cos\alpha\,\tan\gamma$
α and δ	β	$\sin\beta = \sin\alpha\,\cos\delta$
α and δ	γ	$\tan\gamma = \dfrac{\tan\delta}{\cos\alpha}$
β and γ	α	$\tan\alpha = \dfrac{\tan\beta}{\cos\gamma}$
β and γ	δ	$\sin\delta = \cos\beta\,\sin\gamma$
β and δ	α	$\sin\alpha = \dfrac{\sin\beta}{\cos\delta}$
β and δ	γ	$\sin\gamma = \dfrac{\sin\delta}{\cos\beta}$
γ and δ	α	$\cos\alpha = \dfrac{\tan\delta}{\tan\gamma}$
γ and δ	β	$\cos\beta = \dfrac{\sin\delta}{\sin\gamma}$

U.S. Customary and Metric (SI) Measures and Conversions

3.1 Conversions for Length, Pressure, Velocity, Volume, and Weight

To convert from:	To:	Multiply by:
	Length	
Centimeters	Inches	0.3937
Centimeters	Yards	0.01094
Feet	Inches	12.0
Feet	Meters	0.30481
Feet	Yards	0.333
Inches	Centimeters	2.540
Inches	Feet	0.08333
Inches	Meters	0.02540
Inches	Microns	25,400.
Inches	Millimeters	25.400
Inches	Yards	0.02778
Meters	Feet	3.2809
Meters	Yards	1.0936
Microns	Inches	0.0000394
Microns	Meters	0.000001
Millimeters	Inches	0.03937

To convert from:	to:	Multiply by:
Rods	Meters	5.0292
Yards	Centimeters	91.44
Yards	Feet	3.0
Yards	Inches	36.0
Yards	Meters	0.9144
	Pressure	
Dynes per cm^2	Pascals	0.1000
Grams per cm^3	Ounces per in^3	0.5780
Kilograms per cm^2	Pounds per in^2	14.223
Kilograms per cm^2	Pascals	98,066.5
Kilograms per m^2	Pascals	9.8066
Kilograms per m^2	Pounds per ft^2	0.2048
Kilograms per m^2	Pounds per yd^2	1.8433
Kilograms per m^3	Pounds per ft^3	0.06243
Ounces per in^3	Grams per cm^3	1.7300
Pounds per ft^3	Kilograms per cm^3	16.019
Pounds per ft^2	Kilograms per m^2	4.8824
Pounds per ft^2	Pascals	47.880
Pounds per in^2	Kilograms per cm^2	0.0703
Pounds per in^2	Pascals	6,894.76
Pounds per yd^2	Kilograms per m^2	0.5425
	Velocity	
Feet per minute	Meters per second	0.00508
Feet per second	Meters per second	0.3048
Inches per second	Meters per second	0.0254
Miles per hour	Meters per second	0.4470
	Volume	
Cubic centimeters	Cubic inches	0.06102
Cubic feet	Cubic inches	1,728.0
Cubic feet	Cubic meters	0.0283
Cubic feet	Cubic yards	0.0370
Cubic feet	Gallons	7.481
Cubic feet	Liters	28.32
Cubic feet	Quarts	29.9222
Cubic inches	Cubic centimeters	16.39
Cubic inches	Cubic feet	0.0005787
Cubic inches	Cubic meters	0.00001639

Cubic inches	Liters	0.0164
Cubic inches	Gallons	0.004329
Cubic inches	Quarts	0.01732
Cubic meters	Cubic feet	35.31
Cubic meters	Cubic inches	61,023.
Cubic meters	Cubic yards	1.3087
Cubic yards	Cubic feet	27.0
Cubic yards	Cubic meters	0.7641
Gallons	Cubic feet	0.1337
Gallons	Cubic inches	231.0
Gallons	Cubic meters	0.003785
Gallons	Liters	3.785
Gallons	Quarts	4.0
Liters	Cubic feet	0.0351
Liters	Cubic inches	61.017
Liters	Gallons	0.2642
Liters	Pints	2.1133
Liters	Quarts	1.057
Liters	Cubic meters	0.0010
Pints	Cubic meters	0.004732
Pints	Liters	0.4732
Pints	Quarts	0.50
Quarts	Cubic feet	0.03342
Quarts	Cubic inches	57.75
Quarts	Cubic meters	0.0009464
Quarts	Gallons	0.25
Quarts	Liters	0.9464
Quarts	Pints	2.0
	Weight	
Grams	Kilograms	0.001
Grams	Ounces	0.03527
Grams	Pounds	0.002205
Kilograms	Ounces	35.274
Kilograms	Pounds	2.2046
Ounces	Grams	28.35
Ounces	Kilograms	0.02835
Ounces	Pounds	0.0625
Pounds	Grams	453.6
Pounds	Kilograms	0.4536
Pounds	Ounces	16.0

3.2 Standard Conversion Table: Measures Are Found from the Table

Multiply:	By:	To obtain:
Atmospheres	76.0	Centimeters of mercury
Atmospheres	29.92	Inches of mercury
Atmospheres	33.90	Feet of water
Atmospheres	10,333	Kilograms per square meter
Atmospheres	14.70	Pounds per square inch
Atmospheres	1.058	Tons per square foot
Barrels—oil	42	Gallons—oil
British thermal units (Btu)	0.2520	Kilogram-calories
British thermal units	777.5	Foot pounds
British thermal units	3.927×10^{-4}	Horsepower-hours
British thermal units	107.5	Kilogram-meters
British thermal units	2.928×10^{-4}	Kilowatt-hours
Btu/min	12.96	Foot pounds per second
Btu/min	0.02356	Horsepower
Btu/min	0.01757	Kilowatts
Btu/min	17.57	Watts
Centares (centiares)	1	Square meters
Centigrams	0.01	Grams
Centiliters	0.01	Liters
Centimeters	0.3937	Inches
Centimeters	0.01	Meters
Centimeters	10	Millimeters
Centimeters of mercury	0.01316	Atmospheres
Centimeters of mercury	0.4461	Feet of water
Centimeters of mercury	136.0	Kilograms per square meter
Centimeters of mercury	27.85	Pounds per square foot
Centimeters of mercury	0.1934	Pounds per square inch
Centimeters per second	1.969	Feet per minute
Centimeters per second	0.03281	Feet per second
Centimeters per second	0.036	Kilometers per hour
Centimeters per second	0.6	Meters per minute
Centimeters per second	0.02237	Miles per hour
Centimeters per second	3.728×10^{-4}	Miles per minute
Centimeters per second per second	0.03281	Feet per second per second
Cubic centimeters	3.531×10^{-5}	Cubic feet
Cubic centimeters	6.102×10^{-2}	Cubic inches
Cubic centimeters	10^{-6}	Cubic meters
Cubic centimeters	1.308×10^{-6}	Cubic yards
Cubic centimeters	2.642×10^{-4}	Gallons
Cubic centimeters	10^{-3}	Liters

Cubic centimeters	2.113×10^{-3}	Pints (liq.)
Cubic centimeters	1.057×10^{-3}	Quarts (liq.)
Cubic feet	2.832×10^{4}	Cubic centimeters
Cubic feet	1728	Cubic inches
Cubic feet	0.02832	Cubic meters
Cubic feet	0.03704	Cubic yards
Cubic feet	7.48052	Gallons
Cubic feet	28.32	Liters
Cubic feet	59.84	Pints (liq.)
Cubic feet	29.92	Quarts (liq.)
Cubic feet per minute	472.0	Cubic centimeters per second
Cubic feet per minute	0.1247	Gallons per second
Cubic feet per minute	0.4720	Liters per second
Cubic feet per minute	62.43	Pounds of water per minute
Cubic feet per second	0.646317	Million gallons per day
Cubic feet per second	448.831	Gallons per minute
Cubic inches	16.39	Cubic centimeters
Cubic inches	5.787×10^{-4}	Cubic feet
Cubic inches	4.329×10^{-3}	Gallons
Cubic inches	1.639×10^{-2}	Liters
Cubic inches	0.03463	Pints (liq.)
Cubic inches	0.01732	Quarts (liq.)
Cubic meters	10^{6}	Cubic centimeters
Cubic meters	35.31	Cubic feet
Cubic meters	61.023×10^{3}	Cubic inches
Cubic meters	1.308	Cubic yards
Cubic meters	264.2	Gallons
Cubic meters	10^{3}	Liters
Cubic meters	2113	Pints (liq.)
Cubic meters	1057	Quarts (liq.)
Cubic yards	7.646×10^{5}	Cubic centimeters
Cubic yards	27	Cubic feet
Cubic yards	46,656	Cubic inches
Cubic yards	0.7646	Cubic meters
Cubic yards	202.0	Gallons
Cubic yards	764.6	Liters
Cubic yards	1616	Pints (liq.)
Cubic yards	807.9	Quarts (liq.)
Cubic yards per minute	0.45	Cubic feet per second
Cubic yards per minute	3.367	Gallons per second
Cubic yards per minute	12.74	Liters per second
Decigrams	0.1	Grams
Deciliters	0.1	Liters
Decimeters	0.1	Meters
Degrees (angle)	60	Minutes
Degrees (angle)	0.01745	Radians

Multiply:	By:	To obtain:
Degrees (angle)	3600	Seconds
Degrees per second	0.01745	Radians per second
Degrees per second	0.1667	Revolutions per minute
Degrees per second	0.002778	Revolutions per second
Drams	0.0625	Ounces
Drams	1.771845	Grams
Feet	30.48	Centimeters
Feet	12	Inches
Feet	0.3048	Meters
Feet	⅓	Yards
Feet of water	0.02950	Atmospheres
Feet per minute	0.5080	Centimeters per second
Feet per minute	0.01667	Feet per second
Feet per minute	0.01829	Kilometers per hour
Feet per minute	0.3048	Meters per minute
Feet per second	18.29	Meters per minute
Gallons	3785	Cubic centimeters
Gallons	0.1337	Cubic feet
Gallons	231	Cubic inches
Gallons	3.785×10^{-3}	Cubic meters
Gallons	4.95×10^{-3}	Cubic yards
Gallons	3.785	Liters
Gallons	8	Pints (liq.)
Gallons	4	Quarts (liq.)
Gallons—Imperial	1.20095	U.S. gallons
Gallons—U.S.	0.83267	Imperial gallons
Gallons water	8.3453	Pounds of water
Gallons per minute	2.228×10^{-3}	Cubic feet per second
Gallons per minute	0.06308	Liters per second
Gallons per minute	8.0208	Cubic feet per hour
Grams	980.7	Dynes
Grams	15.43	Grains
Grams	10^{-3}	Kilograms
Grams	10^3	Milligrams
Grams	0.03527	Ounces
Grams	0.03215	Ounces (troy)
Grams	2.205×10^{-3}	Pounds
Grams per centimeter	5.600×10^{-3}	Pounds per inch
Grams per cubic centimeter	62.43	Pounds per cubic foot
Grams per cubic centimeter	0.03613	Pounds per cubic inch
Grams per liter	0.062427	Pounds per cubic foot
Horsepower	42.44	Btu per minute
Horsepower	33,000	Foot pounds per minute
Horsepower	550	Foot pounds per second

Horsepower	1.014	Horsepower (metric)
Horsepower	10.70	Kilogram-calories per minute
Horsepower	0.7457	Kilowatts
Horsepower	745.7	Watts
Horsepower (boiler)	33,479	Btu per hour
Horsepower (boiler)	9803	Kilowatts
Horsepower-hours	2547	British thermal units
Horsepower-hours	1.98×10^6	Foot pounds
Horsepower-hours	641.7	Kilogram-calories
Horsepower-hours	2.737×10^5	Kilogram-meters
Horsepower-hours	0.7457	Kilowatthours
Inches	2.540	Centimeters
Inches of mercury	0.03342	Atmospheres
Inches of mercury	1.133	Feet of water
Inches of mercury	345.3	Kilograms per square meter
Inches of mercury	70.73	Pounds per square foot
Inches of mercury	0.4912	Pounds per square inch
Inches of water	0.002458	Atmospheres
Inches of water	0.07355	Inches of mercury
Inches of water	25.40	Kilograms per square meter
Inches of water	0.5781	Ounces per square inch
Inches of water	5.202	Pounds per square foot
Inches of water	0.03613	Pounds per square inch
Kilograms	980,665	Dynes
Kilograms	2.205	Pounds
Kilograms	1.102×10^{-3}	Tons (short)
Kilograms	10^3	Grams
Kilogram-calories	3.968	British thermal units
Kilogram-calories	3086	Foot pounds
Kilogram-calories	1.550×10^{-3}	Horsepower-hours
Kilogram-calories	1.162×10^{-3}	Kilowatthours
Kilogram-calories per minute	51.43	Foot pounds per second
Kilogram-calories per minute	0.09351	Horsepower
Kilogram-calories per minute	0.06972	Kilowatts
Kilograms per meter	0.6720	Pounds per foot
Kilograms per square meter	9.678×10^{-5}	Atmospheres
Kilograms per square meter	3.281×10^{-3}	Feet of water
Kilograms per square meter	2.896×10^{-3}	Inches of mercury
Kilograms per square meter	0.2048	Pounds per square foot

Multiply:	By:	To obtain:
Kilograms per square meter	1.422×10^{-3}	Pounds per square inch
Kilograms per square millimeter	10^6	Kilograms per square meter
Kiloliters	10^3	Liters
Kilometers	10^5	Centimeters
Kilometers	3281	Feet
Kilometers	10^3	Meters
Kilometers	0.6214	Miles
Kilometers	1094	Yards
Kilometers per hour	27.78	Centimeters per second
Kilometers per hour	54.68	Feet per minute
Kilometers per hour	0.9113	Feet per second
Kilometers per hour	0.5396	Knots
Kilometers per hour	16.67	Meters per minute
Kilometers per hour	0.6214	Miles per hour
Kilograms per hour per second	27.78	Centimeters per second per second
Kilograms per hour per second	0.9113	Feet per second per second
Kilograms per hour per second	0.2778	Meters per second per second
Kilowatts	56.92	Btu per minute
Kilowatts	4.425×10^4	Foot pounds per minute
Kilowatts	737.6	Foot pounds per second
Kilowatts	1.341	Horsepower
Kilowatts	14.34	Kilogram-calories per minute
Kilowatts	10^3	Watts
Kilowatthours	3415	British thermal units
Kilowatthours	2.655×10^6	Foot pounds
Kilowatthours	1.341	Horsepower-hours
Kilowatthours	860.5	Kilogram-calories
Kilowatthours	3.671×10^5	Kilogram-meters
Liters	10^3	Cubic centimeters
Liters	0.03531	Cubic feet
Liters	61.02	Cubic inches
Liters	10^3	Cubic meters
Liters	1.308×10^{-3}	Cubic yards
Liters	0.2642	Gallons
Liters	2.113	Pints (liq.)
Liters	1.057	Quarts (liq.)
Liters per minute	5.886×10^{-4}	Cubic feet per second
Liters per minute	4.403×10^{-3}	Gallons per second
Meters	100	Centimeters
Meters	3.281	Feet
Meters	39.37	Inches
Meters	10^3	Millimeters

Meters	1.094	Yards
Meters per minute	1.667	Centimeters per second
Meters per minute	3.281	Feet per minute
Meters per minute	0.05468	Feet per second
Meters per second	196.8	Feet per minute
Meters per second	3.281	Feet per second
Microns	10^{-6}	Meters
Milligrams	10^{-3}	Grams
Milliliters	10^{-3}	Liters
Millimeters	0.1	Centimeters
Millimeters	0.03937	Inches
Million gallons per day	1.54723	Cubic feet per second
Minutes (angle)	2.909×10^{-4}	Radians
Ounces	16	Drams
Ounces	437.5	Grains
Ounces	0.0625	Pounds
Ounces	28.349527	Grams
Ounces	0.9115	Ounces (troy)
Ounces	2.790×10^{-5}	Tons (long)
Ounces	2.835×10^{-5}	Tons (metric)
Ounces (troy)	480	Grains
Ounces (troy)	20	Pennyweights (troy)
Ounces (troy)	0.08333	Pounds (troy)
Ounces (troy)	31.103481	Grams
Ounces (troy)	1.09714	Ounces (avoir.)
Ounces (fluid)	1.805	Cubic inches
Ounces (fluid)	0.02957	Liters
Ounces per square inch	0.0625	Pounds per square inch
Pennyweights (troy)	24	Grains
Pennyweights (troy)	1.55517	Grams
Pennyweights (troy)	0.05	Ounces (troy)
Pennyweights (troy)	4.1667×10^{-3}	Pounds (troy)
Pounds	16	Ounces
Pounds	256	Drams
Pounds	7000	Grains
Pounds	453.5924	Grams
Pounds	1.21528	Pounds (troy)
Pounds	14.5833	Ounces (troy)
Pounds (troy)	5760	Grains
Pounds (troy)	240	Pennyweights (troy)
Pounds (troy)	12	Ounces (troy)
Pounds (troy)	373.24177	Grams
Pounds (troy)	0.822857	Pounds (avoir.)
Pounds (troy)	13.1657	Ounces (avoir.)
Pounds of water	0.01602	Cubic feet
Pounds of water	27.68	Cubic inches
Pounds of water	0.1198	Gallons
Pounds of water per minute	2.670×10^{-4}	Cubic feet per second

Multiply:	By:	To obtain:
Pounds per cubic foot	0.01602	Grams per cubic centimeter
Pounds per cubic foot	16.02	Kilograms per cubic meter
Pounds per cubic foot	5.787×10^{-4}	Pounds per cubic inch
Pounds per cubic inch	27.68	Grams per cubic centimeter
Pounds per cubic inch	2.768×10^4	Kilograms per cubic meter
Pounds per cubic inch	1728	Pounds per cubic foot
Pounds per foot	1.488	Kilograms per meter
Pounds per inch	178.6	Grams per centimeter
Pounds per square foot	0.01602	Feet of water
Pounds per square foot	4.883	Kilograms per square meter
Pounds per square foot	6.945×10^{-3}	Pounds per square inch
Pounds per square inch	0.06804	Atmospheres
Pounds per square inch	2.307	Feet of water
Pounds per square inch	2.036	Inches of mercury
Pounds per square inch	703.1	Kilograms per square meter
Quarts (dry)	67.20	Cubic inches
Quarts (liq.)	57.75	Cubic inches
Radians	57.30	Degrees
Radians	3438	Minutes
Radians per second	57.30	Degrees per second
Radians	0.1592	Revolutions per second
Radians per second	9.549	Revolutions per minute
Radians per second	573.0	Revolutions per minute
Radians per second	0.1592	Revolutions per second
Revolutions	360	Degrees
Revolutions	6.283	Radians
Revolutions per minute	6	Degrees per second
Revolutions per minute	0.1047	Radians per second
Revolutions per minute	0.01667	Revolutions per second
Revolutions per minute	1.745×10^{-3}	Radians per second
Revolutions per minute	2.778×10^{-4}	Revolutions per second
Revolutions per second	360	Degrees per second
Revolutions per second	6.283	Radians per second
Revolutions per second	60	Revolutions per minute
Revolutions per second	6.283	Radians per second
Revolutions per second	3600	Revolutions per minute
Seconds (angle)	4.848×10^{-6}	Radians
Temperature (°C) + 273	1	Abs. temp. (°C)
Temperature (°C) + 17.78	1.8	Temp. (°F)
Temperature (°F) + 460	1	Abs. temp. (°F)
Temperature (°F) − 32	5⁄9	Temp. (°C)
Watts	0.05692	Btu per minute

Watts	44.26	Foot pounds per minute
Watts	0.7376	Foot pounds per second
Watts	1.341×10^{-3}	Horsepower
Watts	0.01434	Kilogram-calories per minute
Watts	10^{-3}	Kilowatts
Watthours	3.415	British thermal units
Watthours	2655	Foot pounds
Watthours	1.341×10^{-3}	Horsepower-hours
Watthours	0.8605	Kilogram-calories
Watthours	367.1	Kilogram-meters
Watthours	10^{-3}	Kilowatthours
Yards	91.44	Centimeters
Yards	3	Feet
Yards	36	Inches
Yards	0.9144	Meters

3.3   Temperature Systems and Conversions

There are four common temperature systems used in engineering and design calculations: (°F) Fahrenheit, (°C) Celsius (formerly centigrade), (°K) Kelvin, and (°R) Rankine. The conversion equation for Celsius to Fahrenheit or Fahrenheit to Celsius is

$$\frac{5}{9} = \frac{°C}{°F - 32}$$

This exact relational equation is all that you need to convert from either system. Enter the known temperature, and solve the equation for the unknown value.

Example: You wish to convert 66°C to Fahrenheit

$$\frac{5}{9} = \frac{66}{°F - 32}$$

$$5°F - 160 = 594$$

$$°F = 150.8$$

The other two systems, Kelvin and Rankine, are converted as described here. The Kelvin and Celsius scales are related by the equation

$$K = 273.18 + °C$$

Thus 0°C = 273.18 K. Absolute zero is equal to −273.18°C.

Example: A temperature of $-75°C = 273.18 + (-75°C) = 198.18$ K.

The Rankine and Fahrenheit scales are related by the equation

$$°R = 459.69 + °F$$

Thus $0°F = 459.69°R$. Absolute zero is equal to $-459.69°F$.

Example: A temperature of $75°F = 459.69 + (+75°F) = 534.69°R$.

3.4   Small Weight Equivalents: U.S. Customary (Grains and Ounces) versus Metric (Grams)

1 gram	= 15.43 grains
1 gram	= 15,430 milligrains
1 gram	= 15,430,000 micrograins
1 pound	= 7,000 grains
1 ounce	= 437.5 grains
1 ounce	= 28.35 grams
1 grain	= 0.0648 grams
1 grain	= 64.8 milligrams
0.1 grain	= 0.00648 grams
0.1 grain	= 6.48 milligrams
1 micrograin	= 0.0000000648 grams
1 micrograin	= 0.0000648 milligrams
1000 micrograins	= 0.0000648 grams
1000 micrograins	= 0.0648 milligrams
1 grain	= 0.002286 ounces
1 ounce	= 437.5 grains
10 grains	= 0.02286 ounces or 0.648 grams
100 grains	= 0.2286 ounces or 6.48 grams

Example: To obtain the weight in grams, multiply the weight in grains by 0.0648, or divide the weight in grains by 15.43.

Example: To obtain the weight in grams, multiply the weight in grams by 15.43, or divide the weight in grams by 0.0648.

Materials: Physical Properties, Characteristics, and Uses

Materials, including metals, alloys, plastics, and other composite compounds, are of prime importance to the mechanical designer, tool designer, machinist, and metalworker. The most important characteristics of materials to those who design and manufacture parts are the physical and chemical properties and the various uses to which the materials may be applied. This chapter discusses a great number of metals, alloys, plastics, and compounds, including elastomers. Included are composition, physical properties, hardness, heat-treatment temperatures, and other characteristics useful for design and metalworking practices.

In design and metalworking practices, sometimes a material defect or failure occurs for unknown reasons. With the information provided in this chapter and in other appropriate American Society for Testing and Materials (ASTM), Society of Automotive Engineers (SAE), American Iron and Steel Institute (AISI), and American Gear Manufacturers Association (AGMA) standards, you may decide to have the material analyzed at a metallurgical and chemical laboratory to check the properties and compositions, or you may have the material checked mechanically at your company to determine if it meets the requirements of the material standards listed herein.

When producing the engineering drawings or specifications for a particular part, the appropriate American standard material designation or military/federal specification always must be indicated

on the drawings or specifications. For example, if you are preparing a drawing for a mechanical spring, and you wish to use music wire in its construction, you must indicate on the drawing that the material is to meet ASTM A-228 or appropriate SAE specifications. You also may request a certified material analysis data sheet from the supplier of the material, whether it comes from a mill, a foundry, a forge, or a materials processing plant. It is the responsibility of material suppliers to provide design engineers and purchasing departments with these analysis sheets when so directed, which is the usual procedure.

Chapter 7 of this *Handbook* lists the machining characteristics and machine tool cutting and drilling speeds for many common and popular steels, alloys, other metals, plastics, and composites. In this chapter you also will find master cross-reference tables of all current hardness scales or systems such as Rockwell, Brinell, and Vickers. With these tables, you will be able to convert the hardness numbers for all the common and currently used hardness measurement systems relative to each other.

Materials specifications and characteristics or properties tables throughout this and other chapters of this *Handbook* are extracted from the latest standards of the ASTM, SAE, and AGMA.

There are a tremendous number of different engineering materials for which the ASTM and SAE have listings: metallic, plastic, and composite. Steels alone account for hundreds of different alloys for wrought products, castings, and forgings. Although these alloys are listed and have specifications for composition and physical properties, they may not be readily available, except as special order "mill run" quantities. A typical example would be austenitic stainless steel sheet in light gauges, cold-rolled to three-quarters hard, spring temper. This material is listed in the various 300 series stainless steels but may only be available from stock in 301 grade, three-quarters hard, spring temper. If you require 304 grade in your design application, it may be available only in mill-run quantities of a minimum of 500 to 2000 lb per single order.

As a general rule, during the early design stages of a particular part, alternate materials must be investigated. Owing to limited availability of your chosen material, you may be forced to use another material that is stocked by the material vendors. When your anticipated material quantities are large, you then will have control of the material type as well as its physical size and special characteristics, such as finish, temper, and gauge. For example, some of the larger companies that order hundreds of tons of hot-rolled sheet steel per year may specify to the mill that they want to run light on

the minus side of a particular gauge. The tolerance on hot-rolled sheet steel is such that when it is run on the minus side of the gauge limits, many thousands of pounds of steel can be saved. The same is true for wrought-copper products such as bus bars and copper sheet, where the savings may be even more advantageous.

The list of steel alloys is so large that some authorities and regulating organizations have contemplated restricting the material standards to controlled listings, wherein redundant materials are eliminated. The large listings of alloys, plastics, and composites create a stocking problem for distributors of mill products, plastics, and composite materials.

For designers, machinists, tool makers, and metalworkers employed at small to medium-sized companies, a good approach to the aforementioned materials problems is to obtain the stock materials catalogs available from the large distributors and manufacturers such as Ryerson, Vincent, Atlantic, Alcoa, Reynolds, Bethlehem Steel, U.S. Steel, Anaconda, General Electric, Dupont, Monsanto, and others. The design and fabrication data for a great number of materials are listed in this and other chapters of this *Handbook*, together with typical uses and applications for these materials.

4.1 Steels

This section lists carbon and alloy steels, as well as the stainless steels, in their wrought form, that is to say, in the hot-rolled, cold rolled, or cold-drawn forms. The usual shapes are sheets, plates, bars or strips, rounds, hexagons, tube, pipe, and structural configurations (beams, angles, channels, tees, square and rectangular tubes, and zees). Cast irons and steels and other casting materials are listed in Chap. 12.

When carbon is added to iron in small quantities, carbon steel is produced. Besides carbon, a number of metallic elements can be added to iron to give the characteristics inherent in the various types of steels. The usual alloying elements are

- Aluminum, which controls grain size in the steel

- Boron, which improves hardenability

- Chromium, which increases response to heat treatment as well as toughness (Chromium is used in stainless steels alone or with nickel.)

- Columbium, which is used in 18-8 stainless steels and welding electrodes

- Copper, which controls atmospheric corrosion and increases yield strengths

- Lead, which greatly improves machinability

- Manganese, which imparts strength and response to heat treatment

- Molybdenum, which increases depth of hardness and toughness

- Nickel, which increases strength and toughness but is not effective in improving hardenability

- Phosphorus, which is present in all steels and increases yield strength

- Silicon, which improves tensile strength and can improve hardenability

- Sulfur, which improves machinability but is detrimental to hot-forming properties

- Tellurium, which improves machinability in leaded steels

- Titanium, which is added to 18-8 stainless steels to prevent carbide precipitation

- Tungsten, which is used in good tool steels, making a fine-grain structure when used in small amounts (When used in amounts from 17 to 20 percent, it produces a high-speed steel that retains hardness at high temperatures.)

- Vanadium, which is used to improve the shock strength of steels and retards grain growth even after hardening from high temperatures

4.1.1 Glossary of steel terms

Age hardening   A process of aging that increases hardness and strength. Age hardening usually follows rapid cooling or cold working.

Annealing   A process involving heat and cooling, usually applied to induce softening.

Austenite   A term designating a metallurgical phase of steels, i.e., austenitic stainless steel.

Carburizing   To introduce carbon while the steel is molten or while it is in the solid state by heating it in contact with carbonaceous material below its melting point.

Case hardening   A process of hardening a ferrous alloy so that the case, or surface layer, is much harder than the interior of the part. The typical case-hardening processes are carburizing and quenching, cyaniding, carbonitriding, nitriding, induction hardening, and flame hardening. Cases of Rockwell C 55 to 60 are readily obtained in medium- to high-carbon steels.

Ductility   The property of a material that allows it to be drawn out of shape before fracturing by stress.

Elastic limit   The maximum stress a material is capable of sustaining without a permanent set or deformation.

Fatigue   The tendency for a metal to break after repeated or cyclic loadings that are below the ultimate tensile strength. Also known as *fatigue-endurance limit.*

Flame hardening   A process of hardening a ferrous alloy by heating it above the transformation range by means of a flame and then cooling as required.

Hardenability   The property of a ferrous alloy that determines the depth and distribution of hardness induced by quenching.

Induction hardening   A process of hardening a ferrous alloy by heating it above the transformation range by means of electrical induction and then cooling as required.

Killed steel   Steel that is deoxidized with silicon or aluminum in order to reduce the oxygen content so that no reaction occurs between carbon and oxygen during solidification.

Martensite   An unstable constituent in quenched steel formed without diffusion only during cooling below a certain temperature. Martensite is the hardest of the transformation products of austenite.

Nitriding   A process of case hardening in which a ferrous alloy, usually of special composition, is heated in an atmosphere of ammonia or in direct contact with a nitrogenous material to produce surface hardening by absorbing nitrogen without quenching.

Normalizing   Heating a steel part of heavy section to a temperature 100°F above the critical range and then cooling in still air.

Pickling   Chemical or electrochemical removal of surface scale and oxides.

Quench hardening   Heating a steel within or above the transformation range and cooling at a rate faster than the critical rate

to increase the hardness substantially. Usually involves the formation of martensite.

Solution heat treatment   A process in which an alloy is heated to a suitable temperature, held at this temperature long enough for certain constituents to enter into solid solution, and then cooled rapidly to hold the constituent in solution. The metal is left in a supersaturated state that is unstable and subsequently may exhibit age hardening.

Spheroidizing   Any process of heating and cooling that produces a round or globular form of carbide in steels.

Strain hardening   An increase in hardness and strength caused by plastic deformation at temperatures lower than the recrystallization range.

Stress relieving   A process of reducing residual stresses in a metal part by heating the part to a suitable temperature and holding this temperature for a sufficient time. This process is applied to relieve stresses induced by casting, quenching, normalizing, machining, cold working (i.e., springs), or welding.

Temper   A condition produced in a metal or alloy by mechanical or thermal treatment and having characteristic structure and mechanical properties. In addition to the annealed temper, conditions produced by thermal treatment are the solution heat-treated temper and the heat-treated and artificially aged temper.

Yield strength   The stress at which a material exhibits a specified limited deviation from proportionality of stress to strain. In most steels, there is a proportionality between the amount of stress that produces a certain amount of strain. This phenomenon is known as *Hooke's law.* When a material passes its yield point, a lesser amount of stress produces a greater amount of strain until the ultimate strength point is reached, where the material breaks.

4.1.2   Carbon, alloy, stainless steel, and tool and die steels: Physical properties, compositions, heat treatment, and uses

Tables 4.1 through 4.20 contain the numbering system for identification of the various types of steels given in the SAE and Unified Numbering System (UNS) designations. Figure 4.1 shows properties and heat treatments for carbon, alloy, and stainless steel. Table 4.21 is an approximate equivalent hardness number table for Brinell hardness numbers for steel, with cross-references to other hardness

designation systems. Table 4.22 is an approximate equivalent hardness number table for Rockwell C hardness numbers for steel, with cross-references to other hardness designation systems.

Typical uses for the various steels.   Following is a listing of the popular and readily available steels that are usually stocked by suppliers and which have vast applications in industry. Tables 4.1 through 4.20 show the physical properties and heat-treatment processes for a great number of American standard steels. The following list of applications will prove useful to many designers and mechanical engineers, as well as to machinists, tool engineers, tool and die makers, and other metalworkers throughout industry. The following list is not all-inclusive but rather is indicative of the much-used and readily available types of carbon, alloy, and stainless steels.

(Text continued on page 150.)

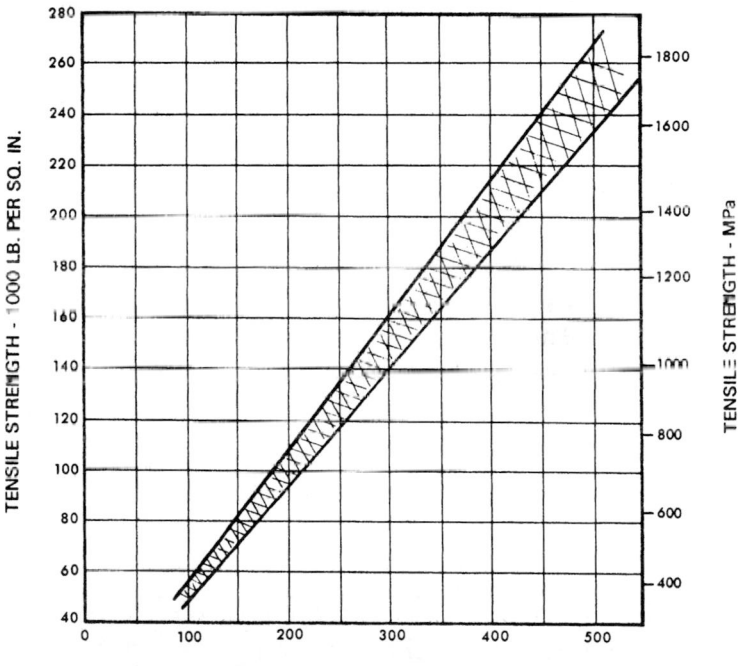

BRINELL HARDNESS - (HB)

Figure 4.1 Brinell hardness numbers indicating approximate tensile strengths of steels.

TABLE 4.1 Designation System for Steels (SAE and UNS)

Numerals and digits		Type of identifying elements	Refer to SAE Standard or Information Report—JXXX for composition limits
UNS	SAE		
		Carbon Steels	
G10XX0	10XX*	Nonresulfurized, manganese 1.00% maximum	403 and 1249
G11XX0	11XX	Resulfurized	403 and 1249
G12XX0	12XX	Rephosphorized and resulfurized	403
		Alloy Steels	
G13XX0	13XX	Manganese steels	404 and 1249
G23XX0	23XX	Nickel steels	1249
G25XX0	25XX	Nickel steels	1249
G31XX0	31XX	Nickel-chromium steels	1249
G32XX0	32XX	Nickel-chromium steels	1249
G33XX0	33XX	Nickel-chromium steels	1249
G34XX0	34XX	Nickel-chromium steels	1249
G40XX0	40XX	Molybdenum steels	404 and 1249
G41XX0	41XX	Chromium-molybdenum steels	404 and 1249
G43XX0	43XX	Nickel-chromium-molybdenum steels	404 and 1249
G44XX0	44XX	Molybdenum steels	404 and 1249
G46XX0	46XX	Nickel-molybdenum steels	404 and 1249
G47XX0	47XX	Nickel-chromium-molybdenum steels	404
G48XX0	48XX	Nickel-molybdenum steels	404 and 1249
G50XX0	50XX	Chromium steels	404 and 1249
G51XX0	51XX	Chromium steels	404 and 1249
G50XX6	50XXX	Chromium steels	404
G51XX6	51XXX	Chromium steels	404

G52XXX6	52XXX6	Chromium steels	404
G61XX0	61XX	Chromium-vanadium steels	404 and 1249
G71XX0	71XXX	Tungsten-chromium steels	1249
G72XX0	72XX	Tungsten-chromium steels	1249
G81XX0	81XX	Nickel-chromium-molybdenum steels	404
G86XX0	86XX	Nickel-chromium-molybdenum steels	404 and 1249
G87XX0	87XX	Nickel-chromium-molybdenum steels	404 and 1249
G88XX0	88XX	Nickel-chromium-molybdenum steels	404
G92XX0	92XX	Silicon-manganese steels	404 and 1249
G93XX0	93XX	Nickel-chromium-molybdenum steels	404 and 1249
G94XX0	94XX	Nickel-chromium-molybdenum steels	404 and 1249
G97XX0	97XX	Nickel-chromium-molybdenum steels	1249
G98XX0	98XX	Nickel-chromium-molybdenum steels	1249

Carbon and Alloy Steels

GXXXX1	XXBXX	B denotes boron steels	403 and 404
GXXXX4	XXLXX	L denotes leaded steels	403 and 404

Stainless Steels

S2XXXX	302XX	Chromium-nickel steels	405
S3XXXX	303XX	Chromium-nickel steels	405
S4XXXX	514XX	Chromium steels	405
S5XXXX	515XX	Chromium steels	405

Experimental Steels

None	EX—	SAE experimental steels	1081

* SAE J403 describes UNS G 15XX0 and SAE 15XX as applicable only to semifinished products for forging, for hot-rolled and cold-finished bars, for wire rods, and for seamless tubing.
SOURCE: Reprinted with permission copyright 1992, Society of Automotive Engineers.

TABLE 4.2 Mechanical Requirements-Stainless Steels

Type	Condition	Finish	Diameter or thickness, in (mm)	Tensile strength, min ksi	Tensile strength, min MPa	Yield strength,[a] min ksi	Yield strength,[a] min MPa	Elongation in 2 in or 50 mm,[b] min, %	Reduction of area,[c] min, %	Brinell hardness, max
Austenitic Grades										
201, 202	A	Hot finished or cold finished	All	75	515	40	275	40	45	—
S20161	A	Hot finished or cold finished	All	125	860	50	345	40	40	255
205	A	Hot finished or cold finished	All	100	690	60	414	40	50	—
XM-19	A	Hot finished or cold finished	All	100	690	55	380	35	55	—
XM-19	As hot rolled	Hot finished or cold finished	Up to 2 (50.8), incl	135	930	105	725	20	50	—
			Over 2 to 3 (50.8 to 76.2), incl	115	795	75	515	25	50	—
			Over 3 to 8 (76.2 to 203.2), incl	100	690	60	415	30	55	—
S21800	A	Hot finished or cold finished	All	95	655	50	345	35	55	241
XM-10, XM-11	A	Hot finished or cold finished	All	90	620	50	345	45	60	—
XM-29	A	Hot finished or cold finished	All	100	690	55	380	30	50	—
XM-28	A	Hot finished or cold finished	All	100	690	55	380	30	50	—
S 28200	A	Hot finished or cold finished	All	110	760	60	410	35	55	—
302, 302B, 304, 304LN, 305, 308, 309, 309S, 309CB, 310, 310S, 310CB, 314, 316, 316LN, 316CB, 316Ti, 317, 321, 347, 348	A	Hot finished	All	75[e]	515	30[e]	205	40[g]	50	—
		Cold finished	Up to ½ (12.70) incl	90	620	45	310	30	40	—
			Over ½ (12.70)	75[e]	515	30[e]	205	30	40	—
304L, 316L	A	Hot finished	All	70	480	25	170	40[g]	50	—
		Cold finished	Up to ½ (12.70) incl	90	620	45	310	30	40	—

Grade		Condition	Size							
304N, 316N	A	Hot finished or cold finished	Over ½ (12.70)	70	480	25	170	30	40	—
			All	80	550	35	240	30	—	—
202, 302, 304, 304N, 316, 316N	B	Cold finished	Up to ¾ (19.05) incl	125	860	100	690	12	35	—
			Over ¾ (19.05) to 1 (25.40)	115	795	80	550	15	35	—
			Over 1 (25.40) to 1¼ (31.75)	105	725	65	450	20	35	—
			Over 1¼ (31.75) to 1½ (38.10)	100	690	50	345	24	45	—
			Over 1½ (38.10) to 1¾ (44.45)	95	655	45	310	28	45	—
304, 304N, 316, 316N	S	Cold finished	Up to 2 (50.8) incl	95	650	75	515	25	40	—
			Over 2 to 2½ (50.8 to 63.5) incl	90	620	65	450	30	40	—
			Over 2½ to 3 (63.5 to 76.2) incl	80	550	55	380	30	40	—
XM-21, S 30454, S 31654	A	Hot finished or cold finished	All	90	620	50	345	30	50	—
XM-21, S 30454 S 31654	B	Cold finished	Up to 1 (25.40) incl	145	1000	125	860	15	45	—
			Over 1 (25.40) to 1¼ (31.75)	135	930	115	795	16	45	—
			Over 1¼ (31.75) to 1½ (38.10)	135	895	105	725	17	45	—
			Over 1½ (38.10) to 1¾ (44.45)	125	860	100	690	18	45	—
XM-7	A	Hot finished or cold finished	All	70	480	25	170	40	50	—
S 30815	A	Hot finished or cold finished	All	87	600	45	310	40	50	—
S 31254	A	Hot finished or cold finished	All	95	650	44	300	35	50	—
S 31725	A	Hot finished or cold finished	All	75	515	30	205	40	—	—
S 31726	A	Hot finished or cold finished	All	80	550	35	240	40	—	—
Austenitic-Ferritic Grades										
XM-26	A	Hot finished or cold finished	All	90	620	65	450	20	55	—
S 31803	A	Hot finished	All	90	620	65	448	25	—	290
		Cold finished	All	90	620	65	448	25	—	290

TABLE 4.2 Mechanical Requirements-Stainless Steels (Continued)

Type	Condition	Finish	Diameter or thickness, in (mm)	Tensile strength, min — ksi	MPa	Yield strength,[a] min — ksi	MPa	Elongation in 2 in or 50 mm,[b] min, %	Reduction of area,[c] min, %	Brinell hardness, max
			Ferritic Grades							
405[f]	A	Hot finished	All	—	—	—	—	—	—	207
		Cold finished	All	—	—	—	—	—	—	217
429	A	Hot finished	All	70	480	40	275	20	45	—
		Cold finished	All	70	480	40	275	16	45	—
430	A	Hot finished	All	70	480	40	275	20	45	—
		Hot finished or cold finished	All	60	415	30	207	20	45	—
S 44400	A	Hot finished	All	60	415	45	310	20	45	217
		Cold finished	All	60	415	45	310	16	45	217
446, XM-27	A	Hot finished	All	65	450	40	275	20	45	219
		Cold finished	All	65	450	40	275	16	45	219
S 44700	A	Hot finished	All	70	480	55	380	20	40	—
		Cold finished	All	75	520	60	415	15	30	—
S 44800	A	Hot finished	All	70	480	55	380	20	40	—
		Cold finished	All	75	520	60	415	15	30	—
			Martensitic Grades							
403, 410	A	Hot finished	All	70	480	40	275	20	45	—
		Cold finished	All	70	480	40	275	16	45	—
403, 410	T	Hot finished	All	100	690	80	550	15	45	—
		Cold finished	All	100	690	80	550	12	40	—
XM-30	T	Hot finished	All	125	860	100	690	13	45	302
		Cold finished	All	125	860	100	690	12	35	—
414	T	Hot finished or cold finished	All	115	790	90	620	15	45	—
403, 410	H	Hot finished	All	120	830	90	620	12	40	—
		Cold finished	All (rounds only)	120	830	90	620	12	40	—
XM-30	A	Hot finished	All	70	480	40	275	13	45	235
		Cold finished	All	70	480	40	275	12	35	—

414	A	Hot finished or cold finished	All	—	—	—	—	—	—	298
S 41500	T	Hot finished or cold finished	All	115	795	90	620	15	45	295
420	A	Hot finished	All	—	—	—	—	—	—	241
		Cold finished	All	—	—	—	—	—	—	255
S 42010	A	Hot finished or cold finished	All	—	—	—	—	—	—	235
		Cold finished	All	—	—	—	—	—	—	255
431	A	Hot finished or cold finished	All	—	—	—	—	—	—	285
440A, 440B, and 440C	A	Hot finished	All	—	—	—	—	—	—	269
		Cold finished	All	—	—	—	—	—	—	285
9 (S 50400)	A	Hot finished or cold finished	All	60	415	30	207	30	45	179
	T	Hot finished or cold finished	All	100	690	80	550	14	35	241

[a] Yield strength shall be determined by the 0.2% offset method in accordance with Test Methods and Definitions A 370. An alternative method of determining yield strength may be used based on a total extension under load of 0.5%.

[b] For some specific products, it may not be practicable to use a 2-in or 50-mm gage length. The use of sub-size test specimens, when necessary, is permissible in accordance with Test Methods and Definitions A 370.

[c] Reduction of area does not apply on flat bars $\frac{3}{16}$ in (4.76 mm) and under in thickness as this determination is not generally made in this product size.

[d] Or equivalent Rockwell hardness.

[e] For extruded shapes of all Cr-Ni grades of condition A, the yield strength shall be 25 ksi (170 MPa) min and tensile strength shall be 70 ksi (480 MPa) min.

[f] Material shall be capable of being heat treated to a maximum Brinell hardness of 250 when oil quenched from 1750°F (953°C).

[g] For shapes having section thickness of $\frac{1}{4}$ in (6.5 mm) or less, 30% min. elongation is acceptable.

SOURCE: Reprinted with permission, from the Annual Book of ASTM Standards, copyright 1992, American Society for Testing and Materials.

TABLE 4.3   Response to Heat Treatment

Type[a]	Heat treatment temperature °F (°C), min	Quenchant	Hardness HRC, min
403	1750 (955)	Air	35
410	1750 (955)	Air	35
414	1750 (955)	Oil	42
420	1825 (995)	Air	50
S42010	1850 (1010)	Oil	48
431	1875 (1020)	Oil	40
440A	1875 (1020)	Air	55
440B	1875 (1020)	Oil	56
440C	1875 (1020)	Air	58

[a] Samples for testing shall be in the form of a section not exceeding ⅜ in (9.50 mm) in thickness.

SOURCE: Reprinted with permission, from the Annual Book of ASTM Standards, copyright 1992, American Society for Testing and Materials.

TABLE 4.4   Tensile Property Requirements[a]

		\multicolumn{6}{c}{Annealed}				
		\multicolumn{2}{c}{Tensile strength, min}	\multicolumn{2}{c}{Yield strength, min}	Elongation in 2 in or 50 mm,	\multicolumn{2}{c}{Hardness, max}
Type	UNS designation	psi	MPa	psi	MPa	min, %	Brinell	Rockwell B
201-1[b]	S20100 Class 1	95 000	655	38 000	260	40	217	95
201-2	S20100 Class 2	95 000	655	45 000	310	40	255	100
	S20200	90 000	620	38 000	260	40	—	—
205	S20500	115 000	790	65 000	450	40	255	100
301	S30100	90 000	620	30 000	205	40	217	95
302	S30200	75 000	515	30 000	205	40	201	92
304	S30400	75 000	515	30 000	205	40	201	92
304L	S30403	70 000	485	25 000	170	40	183	88
304N	S30451	80 000	550	35 000	240	40	217	95
304LN	—	80 000	550	35 000	240	40	217	95
316	S31600	75 000	515	30 000	205	40	217	95
316L	S31603	70 000	485	25 000	170	40	217	95
316N	S31651	80 000	550	35 000	240	40	217	95
	S21904							
	Sheet, strip	100 000	690	60 000	415	40	—	—
	Plate	90 000	620	50 000	345	45	—	—
	S21460	105 000	725	55 000	380	40	—	—

		\multicolumn{5}{c}{$\frac{1}{16}$ Hard[c]}		
		\multicolumn{2}{c}{Tensile strength, min}	\multicolumn{2}{c}{Yield strength, min}	\multicolumn{3}{c}{Elongation in 2 in or 50 mm, min, %}
Type	UNS designation	psi	MPa	psi	MPa	<0.015 in	≥0.015 to ≤0.030 in	>0.030 in
201	S20100 PSS[d]	95 000	655	45 000	310	40	40	40
	FB[e]	75 000	515	40 000	275	—	—	40
205	S20500	115 000	790	65 000	450	40	10	10
301	S30100	90 000	620	45 000	310	40	40	40
302	S30200 PSS	85 000	585	45 000	310	40	40	40
	FB	90 000	620	45 000	310		—	40
304	S30400 PSS	80 000	550	45 000	310	35	35	35
	FB	90 000	620	45 000	310	—	—	40
304L	S30403	80 000	550	45 000	310	40	40	40
304N	S30451	90 000	620	45 000	310	40	40	40
304LN	—	90 000	620	45 000	310	40	40	40
316	S31600 PSS	85 000	585	45 000	310	35	35	35
	FB	90 000	620	45 000	310	—	—	40
316L	S31603	85 000	585	45 000	310	35	35	35
316N	S31651	90 000	620	45 000	310	35	35	35

TABLE 4.4   Tensile Property Requirements[a] (*Continued*)

⅛ Hard[c]

Type	UNS designation	Tensile strength, min psi	MPa	Yield strength, min psi	MPa	Elongation in 2 in or 50 mm, min, % <0.015 in	≥0.015 to ≤0.030 in	>0.030 in
201	S20100	100 000	690	55 000	380	45	45	45
205	S20500	115 000	790	65 000	450	40	40	40
301	S30100	100 000	690	55 000	380	40	40	40
302	S30200	100 000	690	55 000	380	35	35	35
304	S30400	100 000	690	55 000	380	35	35	35
304L	S30403	100 000	690	55 000	380	30	30	30
304N	S30451	100 000	690	55 000	380	37	37	37
304LN	—	100 000	690	55 000	380	33	33	33
316	S31600	100 000	690	55 000	380	30	30	30
316L	S31603	100 000	690	55 000	380	25	25	25
316N	S31651	100 000	690	55 000	380	32	32	32

¼ Hard

Type	UNS designation	Tensile strength, min psi	MPa	Yield strength, min psi	MPa	Elongation in 2 in or 50 mm, min, % <0.015 in	≥0.015 to ≤0.030 in	>0.030 in
201	S20100	125 000	860	75 000	515	25	25	25
202	S20200	125 000	860	75 000	515	12	12	—
205	S20500	125 000	860	75 000	515	45	45	45
301	S30100	125 000	860	75 000	515	25	25	25
302	S30200	125 000	860	75 000	515	10	10	12
304	S30400	125 000	860	75 000	515	10	10	12
304L	S30403	125 000	860	75 000	515	8	8	10
304N	S30451	125 000	860	75 000	515	12	12	12
304LN	—	125 000	860	75 000	515	10	10	12
316	S31600	125 000	860	75 000	515	10	10	10
316L	S31603	125 000	860	75 000	515	8	8	8
316N	S31651	125 000	860	75 000	515	12	12	12
XM-11	S21904	130 000	895	115 000	795	15	15	—

½ Hard

Type	UNS designation	Tensile strength, min psi	MPa	Yield strength, min psi	MPa	Elongation in 2 in or 50 mm, min, % <0.015 in	≥0.015 to ≤0.030 in	>0.030 in
201	S20100	150 000	1035	110 000	760	15	18	18
205	S20500	150 000	1035	110 000	760	15	18	18
301	S30100	150 000	1035	110 000	760	15	18	18

TABLE 4.4   Tensile Property Requirements[a] (*Continued*)

½ Hard

Type	UNS designation	Tensile strength, min		Yield strength, min		Elongation in 2 in or 50 mm, min, %		
		psi	MPa	psi	MPa	<0.015 in	≥0.015 to ≤0.030 in	>0.030 in
302	S30200	150 000	1035	110 000	760	9	10	10
304	S30400	150 000	1035	110 000	760	6	7	7
304L	S30403	150 000	1035	110 000	760	5	6	6
304N	S30451	150 000	1035	110 000	760	6	8	8
304LN	—	150 000	1035	110 000	760	6	7	7
316	S31600	150 000	1035	110 000	760	6	7	7
316L	S31603	150 000	1035	110 000	760	5	6	6
316N	S31651	150 000	1035	110 000	760	6	8	8

¾ Hard

Type	UNS designation	Tensile strength, min		Yield strength, min		Elongation in 2 in or 50 mm, min, %		
		psi	MPa	psi	MPa	<0.015 in	≥0.015 to ≤0.030 in	>0.030 in
201	S20100	175 000	1205	135 000	930	10	12	12
205	S20500	175 000	1205	135 000	930	15	15	15
301	S30100	175 000	1205	135 000	930	10	12	12
302	S30200	175 000	1205	135 000	930	5	6	6

Full Hard

Type	UNS designation	Tensile strength, min		Yield strength, min		Elongation in 2 in or 50 mm, min, %		
		psi	MPa	psi	MPa	<0.015 in	≥0.015 to ≤0.030 in	>0.030 in
201	S20100	185 000	1275	140 000	965	8	9	9
205	S20500	185 000	1275	140 000	965	10	10	10
301	S30100	185 000	1275	140 000	965	8	9	9
302	S30200	185 000	1275	140 000	965	3	4	4

[a] This specification defines minimum properties only and does not imply a range. Depending on the work hardening characteristics of the particular grade, either the yield or the tensile strength can be the controlling factor in meeting the properties. The noncontrolling factor normally will exceed considerably the specified minimum.

[b] Type 201 is generally produced with a chemical composition balanced for rich side (type 201-1) or lean side (type 201-2), austenite stability depending on the properties required for specific applications.

[c] Annealed material that naturally meets mechanical properties may be applied.

[d] PSS means plate, strip, sheet.

[e] FB means flat bar.

SOURCE: Reprinted with permission, from the Annual Book of ASTM Standards, copyright 1992, American Society for Testing and Materials.

TABLE 4.5 Estimated Mechanical Properties and Machinability Ratings of Nonresulfurized Carbon Steel Bars, Manganese 1.00% Maximum

UNS no.	SAE and/or AISI no.	Type of processing	Tensile strength		Yield strength		Elongation in 2 in, %	Reduction in area, %	Brinell hardness	Average machinability rating (cold drawn 1212 = 100%)
			psi	MPa	psi	MPa				
G10060	1006	Hot rolled	43 000	300	24 000	170	30	55	86	
		Cold drawn	48 000	330	41 000	280	20	45	95	50
G10080	1008	Hot rolled	44 000	303	24 500	170	30	55	86	
		Cold drawn	49 000	340	41 500	290	20	45	95	55
G10100	1010	Hot rolled	47 000	320	26 000	180	28	50	95	
		Cold drawn	53 000	370	44 000	300	20	40	105	55
G10120	1012	Hot rolled	48 000	330	26 500	180	28	50	95	
		Cold drawn	54 000	370	45 000	310	19	40	105	55
G10150	1015	Hot rolled	50 000	340	27 500	190	28	50	101	
		Cold drawn	56 000	390	47 000	320	18	40	111	60
G10160	1016	Hot rolled	55 000	380	30 000	210	25	50	111	
		Cold drawn	61 000	420	51 000	350	18	40	121	70
G10170	1017	Hot rolled	53 000	370	29 000	200	26	50	105	
		Cold drawn	59 000	410	49 000	340	18	40	116	65
G10180	1018	Hot rolled	58 000	400	32 000	220	25	50	116	
		Cold drawn	64 000	440	54 000	370	15	40	126	70
G10190	1019	Hot rolled	59 000	410	32 500	220	25	50	116	
		Cold drawn	66 000	460	55 000	380	15	40	131	70
G10200	1020	Hot rolled	55 000	380	30 000	210	25	50	111	
		Cold drawn	61 000	420	51 000	350	15	40	121	65
G10210	1021	Hot rolled	61 000	420	33 000	230	24	48	116	
		Cold drawn	68 000	470	57 000	390	15	40	131	70
G10220	1022	Hot rolled	62 000	430	34 000	230	23	47	151	
		Cold drawn	69 000	480	58 000	400	15	40	137	70
G10230	1023	Hot rolled	56 000	370	31 000	210	25	50	111	
		Cold drawn	62 000	430	52 500	360	15	40	121	65

G-number	SAE	Condition								
G10250	1025	Hot rolled	58 000	400	32 000	220	25	50	116	
		Cold drawn	64 000	440	54 000	370	15	40	126	65
G10260	1026	Hot rolled	64 000	440	35 000	240	24	49	126	
		Cold drawn	71 000	490	60 000	410	15	40	143	75
G10300	1030	Hot rolled	68 000	470	37 500	260	20	42	137	
		Cold drawn	76 000	520	64 000	440	12	35	149	70
G10350	1035	Hot rolled	72 000	500	39 500	270	18	40	143	
		Cold drawn	80 000	550	67 000	460	12	35	163	65
G10370	1037	Hot rolled	74 000	510	40 500	280	18	40	143	
		Cold drawn	82 000	570	69 000	480	12	35	167	65
G10380	1038	Hot rolled	75 000	520	41 000	280	13	40	149	
		Cold drawn	83 000	570	70 000	480	12	35	163	65
G10390	1039	Hot rolled	79 000	540	43 500	300	16	40	156	
		Cold drawn	88 000	610	74 000	510	12	35	179	60
G10400	1040	Hot rolled	76 000	520	42 000	290	18	40	149	
		Cold drawn	85 000	590	71 000	490	12	35	170	60
G10420	1042	Hot rolled	80 000	550	44 000	300	16	40	163	
		Cold drawn	89 000	610	75 000	520	12	35	179	60
		NCD†	85 000	590	73 000	500	12	45	179	70
G10430	1043	Hot rolled	82 000	570	45 000	310	16	40	163	
		Cold drawn	91 000	630	77 000	530	12	35	179	60
		NCD†	87 000	600	75 000	520	12	45	179	70
G10440	1044	Hot rolled	80 000	550	44 000	300	16	40	163	
G10450	1045	Hot rolled	82 000	570	45 000	310	16	40	163	
		Cold drawn	91 000	630	77 000	530	12	35	179	55
		ACD≈	85 000	590	73 000	500	12	45	170	65
G10460	1046	Hot rolled	85 000	590	47 000	320	15	40	170	
		Cold drawn	94 000	650	79 000	540	12	35	187	55
		ACD*	90 000	620	75 000	520	12	45	179	65
G10490	1049	Hot rolled	87 000	600	48 000	330	15	35	179	
		Cold drawn	97 000	670	81 500	560	10	30	197	45
		ACD*	92 000	630	77 000	530	10	40	187	55

TABLE 4.5 Estimated Mechanical Properties and Machinability Ratings of Nonresulfurized Carbon Steel Bars, Manganese 1.00% Maximum (Continued)

UNS no.	SAE and/or AISI no.	Type of processing	Tensile strength		Yield strength		Elongation in 2 in, %	Reduction in area, %	Brinell hardness	Average machinability rating (cold drawn 1212 = 100%)
			psi	MPa	psi	MPa				
G10500	1050	Hot rolled	90 000	620	49 500	340	15	35	179	45
		Cold drawn	100 000	690	84 000	580	10	30	197	55
		ACD*	95 000	660	80 000	550	10	40	189	
G10550	1055	Hot rolled	94 000	650	51 500	360	12	30	192	55
		ACD*	96 000	660	81 000	560	10	40	197	
G10600	1060	Hot rolled	98 000	680	54 000	370	12	30	201	60
		SACD‡	90 000	620	70 000	480	10	45	183	
G10640	1064	Hot rolled	97 000	670	53 500	370	12	30	201	60
		SACD‡	89 000	610	69 000	480	10	45	183	
G10650	1065	Hot rolled	100 000	690	55 000	380	12	30	207	60
		SACD‡	92 000	630	71 000	490	10	45	187	
G10700	1070	Hot rolled	102 000	700	56 000	390	12	30	212	55
		SACD‡	93 000	640	72 000	500	10	45	192	
G10740	1074	Hot rolled	105 000	720	58 000	400	12	30	217	55
		SACD‡	94 500	650	73 000	500	10	40	192	
G10780	1078	Hot rolled	100 000	690	55 000	380	12	30	207	55
		SACD‡	94 000	650	72 500	500	10	40	192	
G10800	1080	Hot rolled	112 000	770	61 500	420	10	25	229	45
		SACD‡	98 000	680	75 000	520	10	40	192	
G10840	1084	Hot rolled	119 000	820	65 500	450	10	25	241	45
		SACD‡	100 000	690	77 000	530	10	40	192	
G10850	1085	Hot rolled	121 000	830	66 500	460	10	25	248	45
		SACD‡	100 500	690	78 000	540	10	40	192	
G10860	1086	Hot rolled	112 000	770	61 500	420	10	25	229	45
		SACD‡	97 000	670	74 000	510	10	40	192	

G1090C	1090	Hot rolled	122 000	840	67 000	460	10	25	248	45
		SACD‡	101 000	700	78 000	540	10	40	197	
G1095C	1095	Hot rolled	120 000	830	66 000	460	10	25	248	45
		SACD‡	99 000	680	76 000	520	10	40	197	

* ACD represents annealed cold drawn.
† NCD represents normalized cold drawn.
‡ SACD represents spheroidized annealed cold drawn.
SOURCE: Reprinted with permission, copyright 1992, Society of Automotive Engineers.

TABLE 4.6 Estimated Mechanical Properties and Machinability Ratings of Resulfurized Carbon Steel Bars*

UNS no.	SAE and/or AISI no.	Type of processing	Tensile strength		Yield strength		Elongation in 2 in, %	Reduction in area, %	Brinell hardness	Average machinability rating (cold drawn 1212 = 100%)
			psi	MPa	psi	MPa				
G11080	1108	Hot rolled	50 000	340	27 500	190	30	50	101	
		Cold drawn	56 000	390	47 000	320	20	40	121	80
G11170	1117	Hot rolled	62 000	430	34 000	230	23	47	121	
		Cold drawn	69 000	480	58 000	400	15	40	137	90
G11320	1132	Hot rolled	83 000	570	45 500	310	16	40	167	
		Cold drawn	92 000	630	77 000	530	12	35	183	75
G11370	1137	Hot rolled	88 000	610	48 000	330	15	35	179	
		Cold drawn	98 000	680	82 000	570	10	30	197	70
G11400	1140	Hot rolled	79 000	540	43 500	300	16	40	156	
		Cold drawn	88 000	610	74 000	510	12	35	170	70
G11410	1141	Hot rolled	94 000	650	51 500	360	15	35	187	
		Cold drawn	105 100	720	88 000	610	10	30	212	70
G11440	1144	Hot rolled	97 000	670	53 000	370	15	35	197	
		Cold drawn	108 000	740	90 000	620	10	30	217	80
G11460	1146	Hot rolled	85 000	590	47 000	320	15	40	170	
		Cold drawn	94 000	650	80 000	550	12	35	187	70
G11510	1151	Hot rolled	92 000	630	50 500	340	15	35	187	
		Cold drawn	102 000	700	86 000	590	10	30	207	65
G12110	1211	Hot rolled	55 000	380	33 000	230	25	45	121	
		Cold drawn	75 000	520	58 000	400	10	35	163	95
G12120	1212	Hot rolled	56 000	390	33 500	230	25	45	121	
		Cold drawn	78 000	540	60 000	410	10	35	167	100
G12130	1213	Hot rolled	56 000	390	33 500	230	25	45	121	
		Cold drawn	78 000	540	60 000	410	10	35	167	135
G12144	12L14	Hot rolled	57 000	390	34 000	230	22	45	121	
		Cold drawn	78 000	540	60 000	410	10	35	163	160

* All 1100 and 1200 series steels are rated on the basis of 0.10% max. silicon or coarse grain melting practice.
SOURCE: Reprinted with permission, copyright 1992, Society of Automotive Engineers.

TABLE 4.7 Estimated Mechanical Properties and Machinability Ratings of Nonresulfurized Carbon Steel Bars, Manganese Maximum over 1.00%

UNS no.	SAE and/or AISI no.	Type of processing	Tensile strength psi	Tensile strength MPa	Yield strength psi	Yield strength MPa	Elongation in 2 in, %	Reduction in area, %	Brinell hardness	Average machinability rating (cold drawn 1212 = 100%)
							Estimated minimum values			
G15240	1524	Hot rolled	74 000	510	41 000	280	20	42	149	
		Cold drawn	82 000	570	69 000	480	12	35	163	60
G15270	1527	Hot rolled	75 000	520	41 000	280	13	40	149	
		Cold drawn	83 000	570	70 000	480	12	35	163	65
G15360	1536	Hot rolled	83 000	570	45 500	310	16	40	163	
		Cold drawn	92 000	630	77 500	530	12	35	187	55
G15410	1541	Hot rolled	92 000	630	51 000	350	15	40	187	
		Cold drawn	102 500	710	87 000	600	10	30	207	
		ACD*	94 000	650	80 000	550	10	45	184	45
G15480	1548	Hot rolled	96 000	650	53 000	370	14	33	197	60
		Cold drawn	106 500	730	89 500	620	10	28	217	45
		ACD*	93 500	640	78 500	540	10	35	192	50
G15520	1552	Hot rolled	108 000	740	59 500	410	12	30	217	50
		ACD*	98 000	680	83 000	570	10	40	193	

* ACD represents annealed cold drawn.

SOURCE: Reprinted with permission, copyright 1992, Society of Automotive Engineers.

TABLE 4.8 Machinability of Alloy Steels

UNS no.	AISI and/or SAE no.	Machinability rating	Condition	Range of typical hardness HB	Microstructure type
G13300	1330	55	Annealed and cold drawn	179/235	A
G13350	1335	55	Annealed and cold drawn	179/235	A
G13400	1340	50	Annealed and cold drawn	183/241	A
G13450	1345	45	Annealed and cold drawn	183/241	A
G40230	4023	70	Cold drawn	156/207	C
G40240	4024	75	Cold drawn	156/207	C
G40270	4027	70	Annealed and cold drawn	167/212	A
G40280	4028	75	Annealed and cold drawn	167/212	A
G40320	4032	70	Annealed and cold drawn	174/217	A
G40370	4037	70	Annealed and cold drawn	174/217	A
G40420	4042	65	Annealed and cold drawn	179/229	A
G40470	4047	65	Annealed and cold drawn	179/229	A
G41180	4118	60	Cold drawn	170/207	C
G41300	4130	70	Annealed and cold drawn	187/229	A
G41350	4135	70	Annealed and cold drawn	187/229	A
G41370	4137	70	Annealed and cold drawn	187/229	A
G41400	4140	65	Annealed and cold drawn	187/229	A
G41420	4142	65	Annealed and cold drawn	187/229	A
G41450	4145	60	Annealed and cold drawn	187/229	A
G41470	4147	60	Annealed and cold drawn	187/235	A
G41500	4150	55	Annealed and cold drawn	187/241	A, B
G41610	4161	50	Spheroidized and cold drawn	187/241	B, A
G43200	4320	60	Annealed and cold drawn	187/229	D, B, A
G43400	4340	50	Annealed and cold drawn	187/241	B, A
G43406	E4340	50	Annealed and cold drawn	187/241	B, A
G44220	4422	65	Cold drawn	170/212	C
G44270	4427	65	Annealed and cold drawn	170/212	A

G4615C	4615	65	Cold drawn	174/223	C
G4617C	4617	65	Cold drawn	174/223	C
G46200	4620	65	Cold drawn	183/229	C
G46260	4626	70	Cold drawn	170/212	C
G47180	4718	60	Cold drawn	187/229	C
G47200	4720	65	Cold drawn	187/229	C
G48150	4815	50	Annealed and cold drawn	187/229	D, B
G48170	4817	50	Annealed and cold drawn	187/229	D, B
G48200	4820	50	Annealed and cold drawn	187/229	D, B
G50401	50B40	65	Annealed and cold drawn	174/223	A
G50441	50B44	65	Annealed and cold drawn	174/223	A
G50460	5046	60	Annealed and cold drawn	174/223	A
G50461	50B46	60	Annealed and cold drawn	174/223	A
G50501	50B50	55	Annealed and cold drawn	183/235	B
G50600	5060	55	Spheroidized annealed and cold drawn	170/212	B
G50601	50B60	55	Spheroidized annealed and cold drawn	170/212	B
G51150	5115	65	Cold drawn	163/201	C
G51200	5120	70	Cold drawn	163/201	C
G51300	5130	70	Annealed and cold drawn	174/212	A
G51320	5132	70	Annealed and cold drawn	174/212	A
G51350	5135	70	Annealed and cold drawn	179/217	A
G51400	5140	65	Annealed and cold drawn	179/217	A
G51470	5147	55	Annealed and cold drawn	179/229	A
G51500	5150	50	Annealed and cold drawn	183/235	A, B
G51550	5155	55	Annealed and cold drawn	183/235	A, B
G51600	5160	55	Spheroidized annealed and cold drawn	179/217	B
G51601	51B60	55	Spheroidized annealed and cold drawn	179/217	B
G50986	50100	40	Spheroidized annealed and cold drawn	183/241	B
G51986	51100	40	Spheroidized annealed and cold drawn	183/241	B
G52986	52100	40	Spheroidized annealed and cold drawn	183/241	B
G61180	6118	60	Cold drawn	179/217	C
G61500	6150	55	Annealed and cold drawn	183/241	B, A
G81150	8115	65	Cold drawn	163/202	C

121

TABLE 4.8 Machinability of Alloy Steels (*Continued*)

UNS no.	AISI and/or SAE no.	Machinability rating	Condition	Range of typical hardness HB	Microstructure type
G81451	81B45	65	Annealed and cold drawn	179/223	A
G86150	8615	70	Cold drawn	179/235	C
G86170	8617	70	Cold drawn	179/235	C
G86200	8620	65	Cold drawn	179/235	C
G86220	8622	65	Cold drawn	179/235	C
G86250	8625	60	Annealed and cold drawn	179/223	A
G86270	8627	60	Annealed and cold drawn	179/223	A
G86300	8630	70	Annealed and cold drawn	179/229	A
G86370	8637	65	Annealed and cold drawn	179/229	A
G86400	8640	65	Annealed and cold drawn	184/229	A
G86420	8642	65	Annealed and cold drawn	184/229	A
G86450	8645	65	Annealed and cold drawn	184/235	A
G86451	86B45	65	Annealed and cold drawn	184/235	A
G86500	8650	60	Annealed and cold drawn	187/248	A, B
G86550	8655	55	Annealed and cold drawn	187/248	A, B
G86600	8660	55	Spheroidized annealed and cold drawn	179/217	B
G87200	8720	65	Cold drawn	179/235	C
G87400	8740	65	Annealed and cold drawn	184/235	A
G88220	8822	55	Cold drawn	179/223	B
G92540	9254	45	Spheroidized annealed and cold drawn	187/241	B
G92600	9260	40	Spheroidized annealed and cold drawn	184/235	B
G93106	9310	50	Annealed and cold drawn	184/229	D
G94151	94B15	70	Cold drawn	163/202	C
G94171	94B17	70	Cold drawn	163/202	C
G94301	94B30	70	Annealed and cold drawn	170/223	A

SOURCE: Reprinted with permission, copyright 1992, Society of Automotive Engineers.

TABLE 4.9 Typical Treatments for Case Hardening Grades of Carbon Steels

UNS no.	SAE steels*	Carburizing temperature, °F	°C	Cooling medium	Reheat temperature, °F	°C	Cooling medium	Carbonitriding temperature, °F	°C	Cooling medium	Temper °F	°C‡
G10100	1010	—	—	—	—	—	—	1450–1650	790–900	Oil	250–400	120–205
G10150	1015	—	—	—	—	—	—	1450–1650	790–900	Oil	250–400	120–205
G10160	1016	1650–1700	900–925	Water or caustic	—	—	—	1450–1650	790–900	Oil	250–400	120–205
G10180	1018	1650–1700	900–925	Water or caustic	1450	790	Water or caustic§	1450–1650	790–900	Oil	250–400	120–205
G10200	1020	1650–1700	900–925	Water or caustic	1450	790	Water or caustic§	1450–1650	790–900	Oil	250–400	120–205
G10220	1022	1650–1700	900–925	Water or caustic	1450	790	Water or caustic§	1450–1650	790–900	Oil	250–400	120–205
G10260	1026	1650–1700	900–925	Water or caustic	1450	790	Water or caustic§	1450–1650	790–900	Oil	250–400	120–205
G10300	1030	1650–1700	900–925	Water or caustic	1450	790	Water or caustic§	1450–1650	790–900	Oil	250–400	120–205
G11170	1117	1650–1700	900–925	Water or oil	1450–1600	790–870	Water or caustic§	1450–1650	790–900	Oil	250–400	120–205
G11180	1118	1650–1700	900–925	Oil	1450–1600	790–870	Oil	1450–1650	790–900	—	250–400	120–205
G15130	1513	1650–1700	900–925	Oil	1450	790	Oil	—	—	—	250–400	120–205
G15220	1522	1650–1700	900–925	Oil	1450	790	Oil	—	—	—	250–400	120–205
G15240	1524	1650–1700	900–925	Oil	1450	790	Oil	—	—	—	250–400	120–205
G15260	1526	1650–1700	900–925	Oil	1450	790	Oil	—	—	—	250–400	120–205
G15270	1527	1650–1700	900–925	Oil	1450	790	Oil	—	—	—	250–400	120–205

* Generally, it is not necessary to normalize the carbon grades for fulfilling either dimensional or machinability requirements of parts made from the steel grades listed in the table although, where dimension is of vital importance, normalizing temperatures of at least 50°F above the carburizing temperatures are sometimes required.

† The higher manganese steels such as 1118 and the 1500 series are not usually carbonitrided. If carbonitriding is performed, care must be taken to limit the nitrogen content because high nitrogen will increase their tendency to retain austenite

‡ Even where recommended tempering temperatures are shown, the temper is not mandatory on many applications. Tempering is generally employed for a partial stress relief and improves resistance to cracking from grinding operations. Higher temperatures than those shown may be employed where the hardness specification on the finished parts permits.

§ 3% sodium hydroxide.

SOURCE: Reprinted with permission, copyright 1992, Society of Automotive Engineers.

TABLE 4.10 Typical Treatments for Heat Treating Grades of Carbon Steels

UNS no.	SAE steels	Normalizing temperature, °F	°C	Annealing temperature, °F	°C	Hardening temperature, °F	°C	Quenching medium*
G10300	1030	—	—	—	—	1575–1600	855–870	Water or caustic
G10350	1035	—	—	—	—	1550–1600	840–870	Water or caustic
G10380†	1038†	—	—	—	—	1525–1575	830–855	Water or caustic
G10390†	1039†	—	—	—	—	1525–1575	830–855	Water or caustic
G10400†	1040†	—	—	—	—	1525–1575	830–855	Water or caustic
G10420	1042	—	—	—	—	1500–1550	815–845	Water or caustic
G10430†	1043†	—	—	—	—	1500–1550	815–845	Water or caustic
G10450†	1045†	—	—	—	—	1500–1550	815–845	Water or caustic
G10460†	1046†	—	—	—	—	1500–1550	815–845	Water or caustic
G10500†	1050†	1600–1700	870–925	—	—	1500–1550	815–845	Water or caustic
G10530	1053	1600–1700	870–925	—	—	1500–1550	815–845	Water or caustic
G10600	1060	1600–1700	870–925	1400–1500	760–815	1575–1625	855–885	Oil
G10740	1074	1550–1650	870–900	1400–1500	760–815	1575–1625	855–885	Oil
G10800	1080	1550–1650	845–900	1400–1500‡	760–815‡	1575–1625	855–885	Oil§
G10900	1090	1550–1650	845–900	1400–1500‡	760–815‡	1575–1625	855–885	Oil§
G10950	1095	1550–1650	845–900	1400–1500‡	760–815‡	1575–1625	855–885	Water and oil
G11370	1137	—	—	—	—	1550–1600	845–870	Oil
G11410	1141	—	—	1400–1500	760–815	1500–1550	815–845	Oil
G11440	1144	1600–1700	870–925	1400–1500	760–815	1500–1550	815–845	Oil
G11450	1145	—	—	—	—	1475–1500	800–815	Water or oil
G11460	1146	—	—	—	—	1475–1500	800–815	Water or oil
G15410	1541	1600–1700	870–925	1400–1500	760–815	1500–1550	815–845	Water or oil
G15480	1548	1600–1700	870–925	—	—	1500–1550	815–845	Oil
G15520	1552	1600–1700	870–925	—	—	1500–1550	815–845	Oil
G15660	1566	1600–1700	870–925	—	—	1575–1625	855–885	Oil

* All steels are tempered to desired hardness; however, tempering is not mandatory on many applications. Tempering is generally employed for a partial stress relief and improves resistance to cracking from grinding operations. Higher temperatures than those shown may be employed where the hardness specification on the finished parts permits.

† Commonly used on parts where induction hardening is employed. However, all steels from SAE 1030 up may have induction hardening applications.

‡ Spheroidal structures are often required for machining purposes and should be cooled very slowly or be isothermally transformed to produce the desired structure.

§ May be water or brine quenched by special techniques such as partial immersion or time quenched; otherwise they are subject to quench cracking.

SOURCE: Reprinted with permission, copyright 1992, Society of Automotive Engineers.

TABLE 4.11 Typical Heat Treatments for Carburizing Grades of Alloy Steels

UNS no.	SAE steels[a]	Pretreatments			Carburizing[e] temperature, °F	°C	Cooling method	Reheat[i] temperature, °F	°C	Quenching medium	Tempering[f] temperature, °F	°C
		Normalize[b]	Normalize and temper[c]	Cycle anneal[d]								
G40120	4012											
G40230	4023											
G40270	4027	Yes	—	—	1650–1700	900–925	Quench in oil[g]	—	—	—	250–350	120–175
G40280	4028											
G40320	4032											
G41180	4118	Yes	—	—	1650–1700	900–925	Quench in oil[g]	—	—	—	250–350	120–175
G43200	4320	Yes	—	Yes	1650–1700	900–925	Quench in oil[g]	—	—	—	250–350	120–175
					1650–1700	900–925	Cool slowly	1525–1550[h]	830–845	Oil	250–350	120–175
G44220	4422	Yes	—	Yes	1650–1700	900–925	Quench in oil[g]	—	—	—	250–350	120–175
G44270	4427											
G46200	4620	Yes	—	Yes	1650–1700	900–925	Cool slowly	1500–1550[h]	815–845	Oil	250–350	120–175
G47200	4720	Yes	—	Yes	1650–1700	900–925	Quench in oil	1500–1550[h]	815–845	Oil	250–350	120–175
G48150	4815	—	Yes	Yes	1650–1700	900–925	Quench in oil[g]	—	—	—	250–325	120–175
G48200	4820				1650–1700	900–925	Quench in oil	1475–1525[h]	800–830	Oil	250–325	120–175
G51200	5120	Yes	—	—	1650–1700	900–925	Quench in oil[g]	—	—	—	250–350	120–175
G86150	8615	Yes	—	—								
G86170	8617	Yes	—	—								
G86200	8620				1650–1700	900–925	Quench in oil[g]	1550–1600[h]	845–870	—	250–350	120–175
G86220	8622				1650–1700	900–925	Cool slowly	1550–1600[h]	845–870	Oil	250–350	120–175
G86250	8625				1650–1700	900–925	Quench in oil	1550–1600[h]	845–870	Oil	250–350	120–175
G86270	8627	Yes	—	Yes								
G87200	8720											
G88220	8822											
G93100	9310	—	Yes	—	1600–1700	900–925	Quench in oil	1450–1525[h]	790–830	Oil	250–325	120–175
					1600–1700	900–925	Cool slowly	1450–1525[h]	790–830	Oil	250–325	120–175
G94151	94B15	Yes	—	—	1650–1700	900–925	Quench in oils	—	—	—	250–350	120–175
G94171	94B17											

a These steels are fine grain. Heat treatments are not necessarily correct for coarse grain.

b Normalizing temperature should be at least as high as the carburizing temperature followed by air cooling.

c After normalizing, reheat to temperature of 1100–1200°F and hold at temperature approximately 1 h per in of maximum section or 4 h minimum time.

d Where cycle annealing is desired, heat to at least as high as the carburizing temperature, hold for uniformity, cool rapidly to 1000–1250°F, hold 1 to 3 h, then air cool or furnace cool to obtain a structure suitable for machining and finish.

e It is general practice to reduce carburizing temperatures to approximately 1550°F before quenching to minimize distortion and retained austenite. For 4800 series steels, the carburizing temperature is reduced to approximately 1500°F before quenching.

f Tempering treatment is optional. Tempering is generally employed for partial stress relief and improved resistance to cracking from grinding operations. Temperatures higher than those shown are used in some instances where application requires.

g This treatment is most commonly used and generally produces a minimum of distortion.

h This treatment is used where the maximum grain refinement is required and/or where parts are subsequently ground on critical dimensions. A combination of good case and core properties is secured with somewhat greater distortion than is obtained by a single quench from the carburizing treatment.

i In this treatment the parts are slowly cooled, preferably under a protective atmosphere. They are then reheated and oil quenched. A tempering operation follows as required. This treatment is used when machining must be done between carburizing and hardening or when facilities for quenching from the carburizing cycle are not available. Distortion is at least equal to that obtained by a single quench from the carburizing cycle, as described in note e.

SOURCE: Reprinted with permission, copyright 1992, Society of Automotive Engineers.

TABLE 4.12 Typical Heat Treatments for Directly Hardenable Grades of Alloy Steels

UNS no.	SAE steels[a]	Normalizing temperature, °F	°C	Annealing[d] temperature, °F	°C	Hardening[e] temperature, °F	°C	Quenching medium[f]
G13300	1330	1600–1700[b]	870–925[b]	1550–1650	845–900	1525–1575	830–855	Water or oil
G13350	1335	1600–1700[b]	870–925[b]	1550–1650	845–900	1500–1550	815–845	Oil
G13400	1340							
G40370	4037	—		1500–1575	815–855	1525–1575	830–855	Oil
G40470	4047	—		1450–1550	790–845	1500–1575	815–855	Oil
G41300	4130	1600–1700[b]	870–925[b]	1450–1550	790–845	1500–1600	815–870	Water or oil
G41370	4137							
G41400	4140	—		1450–1550	790–845	1550–1600	845–900	Oil
G41420	4142							
G41450	4145							
G41470	4147	—		1450–1550	790–845	1500–1550	815–845	Oil
G41450	4140							
G43400	4340	1600–1700[b,c]	870–925[b,c]	1450–1550	790–845	1500–1550	815–845	Oil
G50461	5046	1600–1700[b]	870–925[b]	1500–1600	815–870	1500–1550	815–845	Oil
G51300	5130	1600–1700[b]	870–925[b]	1450–1550	790–845	1525–1575	830–855	Water, caustic solution, or oil
G51320	5132							
G51400	5140	1600–1700[b]	870–925[b]	1500–1600	815–870	1500–1550	815–845	Oil
G51500	5150							
G51600	5160	1600–1700[b]	870–925[b]	1500–1600	815–870	1475–1550	800–845	Oil
G51601	5160							
G61500	6150	—		1550–1650	845–900	1550–1625	845–885	Oil

G86300	8630	1600–1700[b]	870–925[b]	1450–1550	790–845	1525–1600	830–870	Water or oil
G86370	8637	—		1500–1600	815–870	1525–1575	830–855	Oil
G86400	8640							
G86450	8645	—		1500–1600	815–870	1500–1575	815–855	Oil
G92600	9260	—		—	—	1500–1650	815–900	Oil

[a] These steels are fine grain unless otherwise specified.
[b] These steels should be either normalized or annealed for optimum machinability.
[c] Temper at 1100–1225°F (595–665°C).
[d] The specific annealing cycle is dependent on the alloy content of the steel, the type of subsequent machining operations, and the desired surface finish.
[e] Frequently, these steels, with the exception of 4340, 50100, 51100, and 52100, are hardened and tempered to a final machinable hardness without preliminary heat treatment.
[f] All steels are tempered to desired hardness.

SOURCE: Reprinted with permission, copyright 1992, Society of Automotive Engineers.

TABLE 4.13   Typical Heat Treatments for Grades of Chromium-Nickel Austenitic Steels Not Hardenable by Thermal Treatment

SAE steels	AISI no.	Annealing temperature, °F*	°C
20201	201	1850–2050	1010–1120
20202	202	1850–2050	1010–1120
30301	301	1850–2050	1010–1120
30302	302	1850–2050	1010–1120
30303	303	1850–2050	1010–1120
30304	304	1850–2050	1010–1120
30305	305	1850–2050	1010–1120
30309	309	1900–2050	1040–1120
30310	310	1900–2100	1040–1150
30316	316	1850–2050	1010–1120
30317	317	1850–2050	1010–1120
30321	321	1750–2050	955–1120
30325	325	1800–2100	980–1150
30330	—	1950–2150	1065–1175
30347	347	1850–2050	1010–1120

* Quench to produce full austenitic structure using water or air in accordance with thickness of section. Annealing temperatures given cover process and full annealing as already established and used by industry, the lower end of the range being used for process annealing. All steels are quenched in air.

SOURCE: Reprinted with permission, copyright 1992, Society of Automotive Engineers.

TABLE 4.14 Typical Heat Treatments for Stainless Chromium Steels

SAE steels	AISI no.	Subcritical annealing temperature, °F	°C	Full annealing temperature, °F*	°C	Hardening temperature, °F	°C	Quenching medium¶
				1625	885			Air
51409	—	—	—			—	—	
51410	410	1300–1350†	705–730†	1500–1650	815–900	1700–1850	925–1010	Oil or air
51414	414	1200–1250†	650–675†	—	—	1800–1900	980–1040	Oil or air
51416	416	1300–1350§	705–730†	1500–1650	815–900	1700–1850	925–1010	Oil or air
51420	420	1350–1450†	730–790†	1550–1650	845–900	1800–1900	980–1040	Oil or air
51420F*	—	1350–1450†	730–790†	1550–1650	845–900	1800–1900	980–1040	Oil or air
51430	430	1400–1500§	760–815§	—	—	—	—	—
51430F‡	—	1250–1400§	675–760§	—	—	—	—	—
51431	431	1150–1225†	620–665†	—	—	1800–1950	980–1065	Oil or air
51434	—	1400–1600§	760–870§	—	—	—	—	—
51436	—	1400–1600§	760–870§	—	—	—	—	—
51440A‡	440A	1350–1440†	730–780†	1550–1650	845–900	1850–1950	1010–1065	Oil or air
51440B‡	440B	1350–1440†	730–780†	1550–1650	845–900	1850–1950	1010–1065	Oil or air
51440C‡	440C	1350–1440†	730–780†	1550–1650	845–900	1850–1950	1010–1065	Oil or air
51440F‡	—	1350–1440†	730–780†	1550–1650	845–900	1850–1950	1010–1065	Oil or air
51442	—	1350–1500§	730–815§	—	—	—	—	—
51446	—	1450–1600§	790–870†	—	—	—	—	—
51501	501	1325–1375§	720–745§	1525–1600	830–870	1600–1700	870–925	Oil or air

* Cool slowly in furnace.
† Usually air cooled but may be furnace cooled.
‡ Suffixes A, B, and C denote three types of steel differing only in carbon content. Suffix F denotes a free machining steel.
§ Cool rapidly in air.
¶ All steels are tempered to desired hardness.
SOURCE: Reprinted with permission, copyright 1992, Society of Automotive Engineers.

TABLE 4.15   Typical Heat Treatments for Wrought Stainless Steels of Special Machinability

Proprietary designation	Subcritical annealing temperature, °F	°C	Full annealing temperature, °F	°C	Quenching medium
203-EZ	—	—	1850–2050*	1010–1120*	Water or air
303 Ma	—	—	1850–2050*	1010–1120*	Water or air
303 Cu	—	—	1850–2050*	1010–1120*	Water or air
303 Plus X	—	—	1850–2050*	1010–1120*	Water or air
416 Plus X	1300–1350†	705–730†	1550–1650‡	845–900*	—

* Quench to produce full austenitic structure using water or air in accordance with thickness of section. Annealing temperatures given cover process and full annealing as already established and used by industry, the lower end of the range being used for process annealing.

† Usually air cooled but may be furnace cooled.

‡ Cool slowly in the furnace.

SOURCE: Reprinted with permission, copyright 1992, Society of Automotive Engineers.

TABLE 4.16 Monotonic Stress-Strain Properties of Selected Metals (Sort: Steel, A1, SAE Spec., Increasing True Fracture Strength)

SAE spec	BHn	Grain dir	Process description	Ultimate strength, ksi (MPa)	Yield strength, ksi (MPa)	True fracture strength, ksi (MPa)	%RA	True fracture ductility	Strain hard'G exponent	Strength coefficient, ksi (MPa)
A-538-A	405	L	Sol tr & aged	220 (1517)	215 (1482)	275 (1896)	67	1.10	0.030	
A-538-B	460	L	Sol tr & aged	270 (1862)	260 (1793)	310 (2137)	56	0.82	0.020	
A-538-C	480	L	Sol tr & aged	290 (1999)	280 (1931)	325 (2241)	55	0.81	0.015	
AM-350†		L	HR & annealed	191 (1317)	64 (441)	298 (2055)	52	0.74		
AM-350¶	496		CD	276 (1903)	270 (1862)	316 (2179)	20	0.23		
Gainex¶		LT	HR sheet	77 (531)	58 (400)	117 (807)	58	0.86	0.20	
Gainex‡		L	HR sheet	74 (510)	57 (393)	118 (814)	64	1.02	0.20	
H-11	660	LT	Ausformed	375 (2586)	295 (2034)	460 (3172)	33	0.40	0.120	
R-100§	236	LT	As rec plate	177 (1220)	117 (807)	214 (1475)				
R-100§	236	L	As rec plate	169 (1165)	112 (772)	236 (1627)				
RQC-100*	290	LT	HR plate	136 (938)	130 (896)	155 (1069)	43	0.56	0.06	170 (1172)
RQC-100*	290	L	HR plate	135 (931)	128 (883)	193 (1331)	67	1.02	0.06	170 (1172)
10B62	430	L	Q&T	238 (1641)	219 (1510)	253 (1779)	38	0.89	0.042	260 (1793)
1005–1009	90	LT	HR sheet	52 (359)	39 (269)	104 (717)	73	1.3	0.12	73 (503)
1005–1009	125	LT	CD sheet	68 (469)	65 (448)	108 (745)	66	1.09	0.029	78 (538)
1005–1009	125		CD sheet	60 (414)	58 (400)	122 (841)	64	1.02	0.049	76 (524)
1005–1009	90	L	HR sheet	50 (345)	38 (262)	123 (848)	80	1.6	0.16	77 (531)
1015	80	L	Normalized	60 (414)	33 (228)	105 (724)	68	1.14	0.26	
1020	108	L	HR plate	64 (441)	38 (262)	103 (710)	62	0.96	0.19	
1040	225	L	As forged	90 (621)	50 (345)	152 (1048)	60	0.93	0.22	107 (738)
1045	225	L	Q&T	105 (724)	92 (634)	178 (1227)	65	1.04	0.13	166 (1145)
1045	410	L	Q&T	210 (1448)	198 (1365)	270 (1862)	51	0.72	0.076	302 (2082)
1045	390	L	Q&T	195 (1344)	185 (1276)	270 (1862)	59	0.89	0.044	
1045	450	L	Q&T	230 (1586)	220 (1517)	305 (2103)	55	0.81	0.041	
1045	500	L	Q&T	265 (1827)	245 (1689)	330 (2275)	51	0.71	0.047	
1045	595	L	Q&T	325 (2241)	270 (1862)	395 (2723)	41	0.52	0.071	
1080 + Mn	326	L	HR plate	162 (1117)	92 (634)	181 (1248)	17	0.17		
1080 + Mn	375	L	Q&T	189 (1303)	166 (1145)	235 (1620)	31	0.37		

TABLE 4.16 Monotonic Stress-Strain Properties of Selected Metals (Sort: Steel, A1, SAE Spec., Increasing True Fracture Strength) (Continued)

SAE spec	BHn	Grain dir	Process description	Ultimate strength, ksi (MPa)	Yield strength, ksi (MPa)	True fracture strength, ksi (MPa)	%RA	True fracture ductility	Strain hard'G exponent	Strength coefficient, ksi (MPa)
1080 + Mn	415	L	Q&T	206 (1420)	180 (1241)	243 (1675)	31	0.36		
1080 + Mn	505	L	Q&T	265 (1827)	235 (1620)	295 (2034)	30	0.36		
1080 + Mn	555	L	Q&T	309 (2130)	273 (1882)	339 (2337)	17	0.18		
1144	265	L	CD strain rel	135 (931)	104 (717)	220 (1517)	33	0.51		
1144	305	L	Drawn at temp	150 (1034)	148 (1020)	168 (1158)	25	0.29		
1541F	290	L	Q&T forging	138 (951)	129 (889)	185 (1276)	49	0.68	0.12	
1541F	260	L	Q&T forging	129 (889)	114 (786)	185 (1276)	60	0.93	0.13	
30304	160	L	HR & annealed	108 (745)	37 (255)	228 (1572)	74	1.37		
30304	327	L	CD	138 (951)	108 (745)	246 (1696)	69	1.16		
30310	145	L	HR & annealed	93 (641)	32 (221)	168 (1158)	64	1.01		
4130	258	L	Q&T	130 (896)	113 (779)	206 (1420)	67	1.12		
4130	365	L	Q&T	207 (1427)	197 (1358)	264 (1820)	55	0.79		
4140	310	L	Q&T drawn at temp	156 (1076)	140 (965)	221 (1524)	60	0.69		
4142	310	L	Drawn at temp	154 (1062)	152 (1048)	162 (1117)	29	0.35	0.051	
4142	335	L	Drawn at temp	181 (1248)	179 (1234)	246 (1696)	28	0.34	0.032	
4142	380	L	Q&T	205 (1413)	200 (1379)	265 (1827)	48	0.66	0.043	
4142	400	L	Q&T and deformed	225 (1551)	210 (1448)	275 (1896)	47	0.63	0.01	
4142	450	L	Q&T	255 (1758)	230 (1586)	290 (1999)	42	0.54	0.016	
4142	475	L	Q&T and deformed	295 (2034)	275 (1896)	300 (2068)	20	0.22	0.048	
4142	450	L	Q&T and deformed	280 (1931)	270 (1862)	305 (2103)	37	0.46		
4142	475	L	Q&T and deformed	280 (1931)	250 (1724)	315 (2172)	35	0.43		
4142	670	L	As quenched	355 (2448)	235 (1620)	375 (2586)	6	0.06	0.136	
4142	560	L	Q&T	325 (2241)	245 (1689)	385 (2654)	27	0.31	0.091	
4340	243	L	HR & annealed	120 (827)	92 (634)	158 (1089)	43	0.57		
4340	409	L	Q&T	213 (1469)	199 (1372)	226 (1558)	38	0.48		
4340	350	L	Q&T	180 (1241)	170 (1172)	240 (1655)	57	0.84	0.066	229 (1579)
5160	430	L	Q&T	242 (1669)	222 (1531)	280 (1931)	42	0.87	0.055	308 (2124)
52100	518	L	Sol heat Q&T	292 (2013)	279 (1924)	318 (2193)	11	0.12		

Grade	Condition	Dir.								
9262	Annealed	L	260	134 (924)	66 (455)	151 (1041)	14	0.16	0.22	253 (1744)
9262	Q&T	L	280	145 (1000)	114 (786)	177 (1220)	33	0.41	0.14	283 (1951)
9262	Q&T	L	410	227 (1565)	200 (1379)	269 (1855)	32	0.38	0.06	134 (924)
950C	HR plate	LT	159	82 (565)	46 (317)	135 (931)	64	1.03	0.19	
950C	HR bar	L	150	82 (565)	47 (324)	145 (1000)	69	1.19	0.21	
950X	Plate channel	L	150	64 (441)	50 (345)	139 (752)	65	1.06	0.16	98 (676)
950X	HR plate	L	156	77 (531)	48 (331)	145 (1000)	72	1.24	0.19	131 (903)
980X	Plate channel	L	225	101 (696)	82 (565)	177 (1220)	68	1.15	0.13	181 (1248)
1100 AI	As received	L	26	16 (110)	14 (97)		88	2.09		
2014-T6	Sol tr & artif age	L	155	74 (510)	67 (462)	37 (600)	25	0.29	0.032	66 (455)
2024-T351	Sol tr strn harden	L		68 (469)	55 (379)	31 (558)	25	0.28		
2024-T4	Sol tr & RT age	L		69 (476)	44 (303)	92 (634)	35	0.43	0.20	117 (807)
5456-H311	Strain hardened	L	95	58 (400)	34 (234)	76 (524)	35	0.42		
7075-T6	Sol tr & artif age	L		84 (579)	68 (469)	108 (745)	33	0.41	0.113	120 (827)

* Trade name—Bethlehem Steel Corp.
† ASTM specification.
‡ Trade name—Armco Steel Corp.
§ Trade name—Republic Steel Corp.
¶ Grade number—Allegheny Ludlum Steel Corp.
SOURCE: Reprinted with permission, copyright 1992, Society of Automotive Engineers.

TABLE 4.17 Fundamental Quality Description* of Carbon and Alloy Steels†

Carbon steels	Alloy steels

Semifinished for forging
  Forging quality
  Special hardenability
  Special internal soundness
  Nonmetallic inclusion require-
    ment
  Special surface
Carbon steel structural sections
  Structural quality
Carbon steel plates
  Regular quality
  Structural quality
  Cold-drawing quality
  Cold-pressing quality
  Cold-flanging quality
  Forging quality
  Pressure vessel quality
Hot-rolled carbon steel bars
  Merchant quality
  Special quality
  Special hardenability
  Special internal soundness
  Nonmetallic inclusion require-
    ment
  Special surface
  Scrapless nut quality
  Axle shaft quality
  Cold-extrusion quality
  Cold-heading and cold-forging
    quality
	Alloy steel plates
	  Drawing quality
	  Pressure vessel quality
	  Structural quality
	  Aircraft physical quality
	Hot-rolled alloy steel bars
	  Regular quality
	  Aircraft quality or steel subject to
	    magnetic particle inspection
	  Axle shaft quality
	  Bearing quality
	  Cold-heading quality
	  Special cold-heading quality
	  Rifle barrel quality, gun quality,
	    shell or A.P. shot quality
	Alloy steel wire
	  Aircraft quality
	  Bearing quality
	  Special surface quality
	Cold-finished alloy steel bars
	  Regular quality
	  Aircraft quality or steel subject to
	    magnetic particle inspection
	  Axle shaft quality
	  Bearing shaft quality
	  Cold-heading quality
	  Special cold-heading quality
	  Rifle barrel quality, gun quality,
	    shell or A.P. shot quality

Hot-rolled sheets
  Commercial quality
  Drawing quality
  Drawing quality special killed
  Structural quality
Cold-rolled sheets
  Commercial quality
  Drawing quality
  Drawing quality special killed
  Structural quality
Porcelain enameling sheets
  Commercial quality
  Drawing quality
  Drawing quality special killed
Long-terne sheets
  Commercial quality
  Drawing quality
  Drawing quality special killed
  Structural quality
Galvanized sheets
  Commercial quality
  Drawing quality
  Drawing quality special killed
  Lock forming quality
Electrolytic zinc-coated sheets
  Commercial quality
  Drawing quality
  Drawing quality special killed
  Structural quality
	Tin mill products
	  Specific quality descriptions are not
	    applicable to tin mill products.
	Carbon steel wire
	  Industrial quality wire
	  Cold-extrusion wires
	  Heading, forging, and roll threading
	    wires
	  Mechanical spring wires
	  Upholstery spring construction wires
	  Welding wire
	Carbon steel flat wire
	  Stitching wire
	  Stapling wire
	Carbon steel pipe
	  Structural tubing
	  Line pipe
	  Oil country tubular goods
	  Steel specialty tubular products
	  Pressure tubing
	  Mechanical tubing
	  Aircraft tubing
	Hot-rolled carbon steel wire rods
	  Industrial quality
	  Rods for manufacture of wire
	    intended for electric welded chain
	  Rods for heading, forging, and roll-
	    threading wire

Cold-finished carbon steel bars	Hot-rolled strip	Rods for upholstery spring wire	Line pipe
Standard quality	Commercial quality	Rods for welding wire	Oil country tubular goods
Special hardenability	Drawing quality		Steel specialty tubular goods
Special internal soundness	Drawing quality special killed		Pressure tubing
Nonmetallic inclusion require-	Structural quality		Mechanical tubing
ment	Cold-rolled strip		Stainless and heat-resisting pipe,
Special surface	Specific quality descriptions are not		pressure tubing, and mechanical
Cold-heading and cold-forging	provided in cold-rolled strip, since		tubing
quality	this product is largely produced for		Aircraft tubing
Cold-extrusion quality	specific end use.		Pipe
	Rods for lock washer wire		
	Rods for scrapless nut wire		

* In the case of certain qualities, phosphorus and sulfur are ordinarily furnished to lower limits than the specified maximum. For details, refer to the appropriate AISI Manual.

† Detailed description of many of the categories listed in this table appear in an appropriate section of the AISI manual.

SOURCE: Reprinted with permission, copyright 1992, Society of Automotive Engineers.

TABLE 4.18 Comparison of Tool Steels on the Basis of Properties Affecting Selection

SAE steel designation	Nondeforming properties	Safety in hardening	Depth of hardening*	Toughness	Resistance to softening effect of heat	Wear resistance	Machinability
Water-hardening tool steels							
W108	Poor	Fair	Shallow	Good†	Poor	Fair	Best
W109	Poor	Fair	Shallow	Good†	Poor	Fair	Best
W110	Poor	Fair	Shallow	Good†	Poor	Good	Best
W112	Poor	Fair	Shallow	Good†	Poor	Good	Best
W209	Poor	Fair	Shallow	Good†	Poor	Fair	Best
W210	Poor	Fair	Shallow	Good	Poor	Good	Best
W310	Poor	Fair	Shallow	Good	Poor	Good	Best
Shock-resisting tool steels							
S1—Chromium-tungsten	Fair	Good	Medium	Good	Fair	Fair	Fair
S2—Silicon-molybdenum	W Poor‡ O Fair‡	W Poor‡ O Good‡	Medium	Best	Fair	Fair	Good
S5—Silicon-manganese	W Poor‡ O Fair‡	W Poor‡ O Good‡	Medium	Best	Fair	Fair	Fair
Cold-work tool steels							
Oil-hardening types							
O1—Low manganese	Good	Good	Medium	Fair	Poor	Good	Good
O2—High manganese	Good	Good	Medium	Fair	Poor	Good	Good
O6—Molybdenum graphitic	Fair	Good	Medium	Fair	Poor	Good	Best
Medium-alloy air-hardening types							
A2—5% Chromium air hard	Best	Best	Deep	Fair	Fair	Good	Fair
High-carbon–high-chromium types							
D2—High-carbon–high-chromium (air)	Best	Best	Deep	Fair	Fair	Best	Poor
D3—High-carbon–high-chromium (oil)	Good	Good	Deep	Poor	Fair	Best	Poor
D5—High-carbon–high-chromium–cobalt	Best	Best	Deep	Fair	Fair	Best	Poor
D7—High-carbon–high-chromium–high-vanadium	Best	Best	Deep	Poor	Fair	Best	Poor

Hot-work tool steels						
Chromium base types						
H11—Chromium-molybdenum-V	Good	Good	Deep	Good	Good	Fair	Fair
H12—Chromium-molybdenum-tungsten	Good	Good	Deep	Good	Good	Fair	Fair
H13—Chromium-molybdenum-VV	Good	Good	Deep	Good	Good	Fair	Fair
Tungsten base types						
H21—Tungsten	Good	Good	Deep	Good	Good	Fair	Fair
High-speed tool steels						
Tungsten base types						
T1—Tungsten 18-4-1	Good	Good	Deep	Poor	Good	Good	Fair
T2—Tungsten 18-4-2	Good	Good	Deep	Poor	Good	Good	Fair
T4—Cobalt-tungsten 18-4-1-5	Good	Fair	Deep	Poor	Best	Good	Fair
T5—Cobalt-tungsten 18-4-2-8	Good	Fair	Deep	Poor	Best	Good	Fair
T8—Cobalt-tungsten 14-4-2-5	Good	Fair	Deep	Poor	Best	Good	Fair
Molybdenum base types						
M1—Molybdenum 8-2-1	Good	Fair	Deep	Poor	Good	Good	Fair
M2—Molybdenum-tungsten 6-6-2	Good	Fair	Deep	Poor	Good	Good	Fair
M3—Molybdenum-tungsten 6-6-3	Good	Fair	Deep	Poor	Good	Best	Fair
M4—Molybdenum-tungsten 6-6-4	Good	Fair	Deep	Poor	Good	Best	Fair
Special-purpose tool steels						
Low-alloy types						
L6—Nickel-chromium	Fair	Good	Medium	Fair	Poor	Fair	Fair
L7—Chromium	Fair	Good	Medium	Fair	Poor	Good	Fair

* These are intended to emphasize major differences between the groups of steels and do not account for the minor differences in depths of hardening that exist between steels of the same group. This is particularly true of the water-hardening W steels which are frequently furnished with varying degrees of hardenability as listed in Table 4.19.

† Toughness decreases somewhat with increasing depth of hardening.

‡ W as shown here indicates water quench O as shown here indicates oil quench.

SOURCE Reprinted with permission, copyright 1992, Society of Automotive Engineers.

TABLE 4.19 Approximate Comparison of Tool and Die Steels on the Basis of Some Heat-Treating Characteristics

SAE steel designation	Quench medium	Preheat temperature, °F	Hardening temperature range, °F*	Hardness after quenching, Rockwell C	Tempering temperature range, °F*	Hardness after tempering, Rockwell C	Decarburization (prevention of during heat treatment)
Water-hardening tool steel							
W108	Water	—†	1420–1450	65–67	350–525	65–56	—‡
W109	Water	—†	1420–1450	65–67	350–525	65–56	—‡
W110	Water	—†	1420–1450	65–67	350–525	65–56	—‡
W112	Water	—†	1420–1500	65–67	350–525	65–56	—‡
W209	Water	—†	1420–1500	65–67	350–525	65–56	—‡
W210	Water	—†	1420–1500	65–67	350–525	65–56	—‡
W310	Water	—†	1420–1500	65–67	350–525	65–56	—‡
Shock-resisting tool steels							
S1—Chromium-tungsten	Oil	1200–1300	1650–1800	57–59	300–1000	57–45	—§
S2—Silicon-molybdenum	Water	—†	1550–1575	60–62	300–500	60–54	—‡
	Oil	—†	1600–1625	58–60	300–500	58–54	—‡
S5—Silicon-manganese	Water	—†	1550–1600	60–62	300–650	60–54	—‡
	Oil	—†	1600–1675	58–60	300–650	58–54	—‡
Cold-work tool steels							
Oil-hardening types							
O1—Low manganese	Oil	—†	1450–1500	63–65	300–800	62–50	—‡
O2—High manganese	Oil	—†	1420–1450	63–65	375–500	62–57	—‡
O6—Molybdenum graphitic	Oil	—†	1450–1500	63–65	300–800	63–50	—‡
Medium-alloy air-hardening types							
A2—5% Chromium air hard	Air	1200–1300	1725–1775	61–63	400–700	60–57	—§
High-carbon–high-chromium types							
D2—High-carbon–high-chromium	Air	1200–1300	1800–1875	61–63	400–700	60–58	—§
D3—High-carbon–high-chromium	Oil	1200–1300	1750–1800	62–64	400–700	62–58	—§
D5—High-carbon–high-chromium–cobalt	Air	1200–1300	1800–1875	60–62	400–700	59–57	—§
D7—High-carbon–high-chromium–high-vanadium	Air	1200–1300	1850–1950	63–65	300–500 / 850–1000	65–63 / 62–58	—§

Hot-work tool steels
Chromium base types

H11—Chromium-molybdenum-V	Air	1450–1500	1825–1875	53–55	1000–1100	51–43	—§
H12—Chromium-molybdenum-tungsten	Oil, Air	1450–1500	1800–1900	53–55	1000–1100	51–43	—§
H13—Chromium-molybdenum-VV	Air	1400–1450	1825–1875	53–55	1000–1100	51–43	—§

Tungsten base types

H21—Tungsten	Oil, Air	1500–1550	2100–2150	50–52	950–1150	50–47	—§

High-speed tool steels
Tungsten base types

T1—Tungsten 18-4-1	Oil, air, salt	1500–1550	2300–2375	63–65	1025–1100	65–63	—§
T2—Tungsten 18-4-2	Oil, air, salt	1500–1550	2300–2375	63–65	1025–1100	65–63	—§
T4—Cobalt-tungsten 18-4-1-5	Oil, air, salt	1500–1550	2300–2375	63–65	1025–1100	65–63	—§
T5—Cobalt-tungsten 18-4-2-8	Oil, air, salt	1500–1550	2300–2400	63–65	1050–1100	65–63	—§
T8—Cobalt-tungsten 14-4-2-5	Oil, air, salt	1500–1550	2300–2375	63–65	1025–1100	65–63	—§

Molybdenum base types

M1—Molybdenum 8-2-1	Oil, air, salt	1400–1500	2150–2250	63–65	1025–1050	65–63	—§
M2—Molybdenum-tungsten 6-6-2	Oil, air, salt	1450–1500	2175–2250	63–65	1025–1075	65–63	—§
M3—Molybdenum-tungsten 6-6-3	Oil, air, salt	1450–1500	2150–2225	63–65	1025–1075	65–63	—§
M4—Molybdenum-tungsten 6-6-4	Oil, air, salt	1450–1500	2150–2225	63–65	1025–1075	65–63	—§

Special-purpose tool steels
Low-alloy types

L6—Nickel-chromium	Oil	—†	1500–1600	62–64	400–800	62–48	—‡
L7—Chromium	Oil	—†	1525–1550	63–65	350–500	62–60	—‡

* The purpose of these columns is to show the usual ranges of temperature employed in hardening and tempering and is not to be used as a specification.

† For large tools and tools having intricate sections, preheating at 1050 to 1200°F is recommended.

‡ Use moderately oxidizing atmosphere in furnace or a suitable neutral salt bath.

§ Use protective pack from which volatile matter has been removed, carefully balanced neutral salt bath, or atmosphere controlled furnaces. In the latter case, the furnace atmosphere should be in equilibrium with the carbon content of the steel being treated. Furnace atmosphere dew point is considered a reliable method for measuring and controlling this equilibrium.

SOURCE: Reprinted with permission, copyright 1922, Society of Automotive Engineers.

TABLE 4.20 Forging, Normalizing, and Annealing Treatments of Tool and Die Steels

SAE steel designation*	Forging†			Normalizing‡			Annealing§		
	Heat slowly to	Start forging at	Do not forge below	Heat slowly to	Hold at	Temperature	Maximum rate of cooling, °F/h	Approximate Brinell hardness	Approximate Rockwell B
Water-hardening tool steels									
W108	1450	1800–1950	1500	1450	1500	1400–1450	75	159–202	84–94
W109	1450	1800–1950	1500	1450	1500	1375–1425	75	159–202	84–94
W110	1450	1800–1900	1500	1450	1550	1400–1450	75	159–202	84–94
W112	1450	1800–1900	1500	1450	1625	1400–1450	75	159–202	84–94
W209	1450	1800–1950	1500	1450	1500	1375–1425	75	159–202	84–94
W210	1450	1800–1900	1500	1450	1550	1400–1450	75	159–202	84–94
W310	1450	1800–1900	1500	1450	1550	1400–1450	75	159–202	84–94
Shock-resisting tool steels									
S1—Chromium-tungsten	1500	1800–2000	1600	Do not normalize		1450–1500	50	192–235	92–99
S2—Silicon-molybdenum	1500	1900–2100	1600	1500	1650	1400–1450	50	192–229	92–98
S5—Silicon-manganese	1500	1900–2050	1600	1500	1600	1400–1450	50	192–229	92–98
Cold-work tool steels									
Oil-hardening types									
O1—Low manganese	1500	1750–1900	1550	1500	1600	1425–1475	50	183–212	90–96
O2—High manganese	1500	1750–1900	1550	1500	1550	1375–1425	50	183–212	90–96
O6—Molybdenum graphitic	1500	1750–1900	1500	1500	1625	1425–1475	20	183–217	90–96
Medium-alloy air-hardening types									
A2—5% Chromium air hard	1600	1850–2000	1650	Do not normalize		1550–1600	40	202–229	94–98
High-carbon–high-chromium types									
D2—High-carbon–high-chromium (air)	1650	1850–2000	1650	Do not normalize		1600–1650	40	207–255	95–102
D3—High-carbon–high-chromium (oil)	1650	1850–2000	1650	Do not normalize		1600–1650	50	212–255	96–102
D5—High-carbon–high-chromium-cobalt	1600	1850–2000	1650	Do not normalize		1600–1650	40	207–255	95–102
D7—High-carbon–high-chromium-high-vanadium	1650	2050–2125	1800	Do not normalize		1600–1650	50	235–262	99–103

Hot-work tool steels

Chromium base types							
H11—Chromium-molybdenum-V	1650	1950–2100	1650	Do not normalize	1550–1600	50	192–229	92–98
H12—Chromium-molybdenum-tungsten	1650	1950–2100	1650	Do not normalize	1600–1650	50	192–229	92–98
H13—Chromium-molybdenum-V	1650	1950–2100	1650	Do not normalize	1550–1600	50	192–229	92–98
Tungsten base types							
H21—Tungsten	1600	2000–2150	1650	Do not normalize	1600–1650	50	202–235	94–99

High-speed tool steels

Tungsten base types							
T1—Tungsten 18-4-1	1500	1950–2100	1750	Do not normalize	1600–1650	50	217–255	96–102
T2—Tungsten 18-4-2	1500	2000–2150	1750	Do not normalize	1600–1650	50	223–255	97–102
T4—Cobalt-tungsten 18-4-1-5	1600	2000–2150	1750	Do not normalize	1600–1650	50	229–255	98–102
T5—Cobalt-tungsten 18-4-2-8	1600	2000–2150	1800	Do not normalize	1600–1650	50	248–293	102–106
T8—Cobalt-tungsten 14-4-2-5	1600	2000–2150	1750	Do not normalize	1600–1650	50	229–255	98–102
Molybdenum base types							
M1—Molybdenum 8-2-1	1500	1900–2050	1700	Do not normalize	1525–1600	50	207–248	95–102
M2—Molybdenum-tungsten 6-6-2	1500	1950–2100	1700	Do not normalize	1550–1625	50	217–248	96–102
M3—Molybdenum-tungsten 6-6-3	1500	2000–2150	1700	Do not normalize	1550–1625	50	223–255	97–102
M4—Molybdenum-tungsten 6-6-4	1500	2000–2150	1700	Do not normalize	1550–1625	50	229–255	98–102

Special-purpose tool steels

Low-alloy types								
L6—Nickel-chromium	1500	1800–2000	1600	1550	1650	1400–1450	50	183–212	90–96
L7—Chromium	1500	1800–2000	1550	1550	1650	1450–1500	50	174–212	88–96

* These tool and die steels are the same as those listed in Table 4.18.

† The temperature at which to start forging is given as a range, the higher side of which should be used for large sections and heavy or rapid reductions and the lower side for smaller sections and lighter reductions. As the alloy content of the steel increases, the time of soaking at forging temperature increases proportionately. Likewise, as the alloy content increases, it becomes necessary to cool slowly from the forging temperature. With very high alloy steels, such as high speed or air hardening steels, this slow cooling is imperative in order to prevent cracking and to leave the steel in a semisoft condition. Either furnace cooling or burying in an insulating medium, such as lime, mica, or silocel, is satisfactory.

‡ The length of time the steel is held after being uniformly heated through at the normalizing temperature varies from about 15 min for a small section to about 1 h for large sizes. Cooling from the normalizing temperature is done in still air. The purpose of normalizing after forging is to refine the grain structure and to produce a uniform structure throughout the forging. Normalizing should not be confused with low temperature (about 1200°F) annealing used for the relief of residual stresses resulting from heavy machining, bending, and forming.

§ The annealing temperature is given as a range, the upper limit of which should be used for large sections and the lower limit for smaller sections. The length of time the steel is held after being uniformly heated through at the annealing temperature varies from about 1 h for light sections and small furnace charges of carbon or low alloy steel to about 4 hr for heavy sections and large furnace charges of high alloy steel.

For information on the forging and heat treating of tool steels, see ASM Handbook, 1948 edition, pp. 653–655.

SOURCE: Reprinted with permission, copyright 1992, Society of Automotive Engineers.

TABLE 4.21 Approximate Equivalent Hardness Numbers* for Brinell Hardness Numbers† for Steel

| Brinell indentation dia, mm | Brinell hardness no.,† 10-mm ball, 3000-kg load | | Vickers hardness no. | Rockwell hardness no.† | | | | Rockwell superficial hardness no., superficial Brale penetrator | | | Shore scleroscope hardness no. | Tensile strength (approximate) in MPa (1000 psi) | Brinell indentation dia, mm |
| | Standard ball | Tungsten-carbide ball | | A-scale, 60-kg load, Brale penetrator | B-scale, 100-kg load, (1⁄16-in) dia ball | C-scale, 150-kg load, Brale penetrator | D-scale, 100-kg load, Brale penetrator | 15-N scale, 15-kg load | 30-N scale, 30-kg load | 45-N scale, 45-kg load | | | |
Col. 1	Col. 2	Col. 3	Col. 4	Col. 5	Col. 6	Col. 7	Col. 8	Col. 9	Col. 10	Col. 11	Col. 12	Col. 13	Col. 14
—	—	—	940	85.6	—	68.0	76.9	93.2	84.4	75.4	97	—	—
—	—	—	920	85.3	—	67.5	76.5	93.0	84.0	74.8	96	—	—
—	—	(767)	900	85.0	—	67.0	76.1	92.9	83.6	74.2	95	—	—
—	—	(757)	880	84.7	—	66.4	75.7	92.7	83.1	73.6	93	—	—
—	—	—	860	84.4	—	65.9	75.3	92.5	82.7	73.1	92	—	—
2.25	—	(745)	840	84.1	—	65.3	74.8	92.3	82.2	72.2	91	—	2.25
—	—	(733)	820	83.8	—	64.7	74.3	92.1	81.7	71.8	90	—	—
—	—	(722)	800	83.4	—	64.0	73.8	91.8	81.1	71.0	88	—	—
2.30	—	(712)	—	—	—	—	—	—	—	—	—	—	2.30
—	—	(710)	780	83.0	—	63.3	73.3	91.5	80.4	70.2	87	—	—
—	—	(698)	760	82.6	—	62.5	72.6	91.2	79.7	69.4	86	—	—
—	—	(684)	740	82.2	—	61.8	72.1	91.0	79.1	68.6	—	—	—
2.35	—	**(682)**	**737**	**82.2**	—	**61.7**	**72.0**	**91.0**	**79.0**	**68.5**	84	—	**2.35**
—	—	(670)	720	81.8	—	61.0	71.5	90.7	78.4	67.7	83	—	—
—	—	(656)	700	81.3	—	60.1	70.8	90.3	77.6	66.7	—	—	—
2.40	—	**(653)**	**697**	**81.2**	—	**60.0**	**70.7**	**90.2**	**77.5**	**66.5**	81	—	**2.40**
—	—	(647)	690	81.1	—	59.7	70.5	90.1	77.2	66.2	—	—	—
—	—	(638)	680	80.8	—	59.2	70.1	89.8	76.8	65.7	80	—	—
—	—	630	670	80.6	—	58.8	69.8	89.7	76.4	65.3	—	—	—
2.45	—	**627**	**667**	**80.5**	—	**58.7**	**69.7**	**89.6**	**76.3**	**65.1**	79	—	**2.45**
—	—	—	**677**	**80.7**	—	**59.1**	**70.0**	**89.8**	**76.8**	**65.7**	—	—	—
2.50	—	**601**	**640**	**79.8**	—	**57.3**	**68.7**	**89.0**	**75.1**	**63.5**	77	—	**2.50**

2.55–3.55													2.55–3.55
2.55			63.5	75.1	89.0	68.7	57.3		79.8	640			2.55
		75	62.1	73.9	88.4	67.7	56.0		79.1	615	578		
2.60	2055 (298)		61.6	73.5	88.1	67.4	55.6		78.8	607			2.60
	2015 (292)	73	60.6	72.7	87.8	66.7	54.7		78.4	591	555		
2.65	1985 (288)		59.8	72.0	87.5	66.1	54.0		78.0	579			2.65
	1915 (278)	71	59.2	71.6	87.2	65.8	53.5		77.8	569	534		
2.70	1890 (274)		58.0	70.7	86.7	65.0	52.5		77.1	553			2.70
	1855 (269)	70	57.6	70.3	86.5	64.7	52.1		76.9	547	514		
2.75	1825 (265)		56.9	69.9	86.3	64.3	51.6		76.7	539			2.75
	1820 (264)		56.2	69.5	86.0	63.9	51.1		76.4	530			
	1780 (258)	68	56.1	69.4	85.9	63.8	51.0		76.3	528	495	(495)	
2.80	1740 (252)		55.2	68.7	85.6	63.2	50.3		75.9	516			2.80
	1740 (252)		54.5	68.2	85.3	62.7	49.6		75.6	508			
	1680 (244)	66	54.5	68.2	85.3	62.7	49.6		75.6	508	477	(477)	
2.85	1670 (242)		53.5	67.4	84.9	61.9	48.8		75.1	495			2.85
	1670 (242)		53.2	67.2	84.7	61.7	48.5		74.9	491			
	1595 (231)	65	53.2	67.2	84.7	61.7	48.5		74.9	491	461	(461)	
2.90	1585 (230)		51.7	66.0	84.1	61.0	47.2		74.3	474			2.90
	1585 (230)		51.5	65.8	84.0	60.8	47.1		74.2	472			
	1510 (219)	63	51.5	65.8	84.0	60.8	47.1		74.2	472	444	444	
2.95	1460 (212)	61	49.9	64.6	83.4	59.7	45.7		73.4	455	429	429	2.95
3.00	1390 (202)	59	48.4	63.5	82.8	58.8	44.5		72.8	440	415	415	3.00
3.05	1330 (193)	58	46.9	62.3	82.0	57.8	43.1		72.0	425	401	401	3.05
3.10	1270 (184)	56	45.3	61.1	81.4	56.8	41.8		71.4	410	388	388	3.10
3.15	1220 (177)	54	43.6	59.9	80.6	55.7	40.4		70.6	396	375	375	3.15
3.20	1180 (171)	52	42.0	58.7	80.0	54.6	39.1		70.0	383	363	363	3.20
3.25	1130 (164)	51	40.5	57.6	79.3	53.8	37.9	(110.0)	69.3	372	352	352	3.25
3.30	1095 (159)	50	39.1	56.4	78.6	52.8	36.6	(109.0)	68.7	360	341	341	3.30
3.35	1060 (154)	48	37.8	55.4	78.0	51.9	35.5	(108.5)	68.1	350	331	331	3.35
3.40	1025 (149)	47	36.4	54.3	77.3	51.0	34.3	(108.0)	67.5	338	321	321	3.40
3.45	1005 (146)	46	34.4	53.3	76.7	50.0	33.1	(107.5)	66.9	328	311	311	3.45
3.50	970 (141)	45	33.8	52.2	76.1	49.3	32.1	(107.0)	66.3	318	302	302	3.50
3.55		43	32.4	51.2	75.5	48.3	30.9	(106.0)	65.7	309	293	293	3.55

TABLE 4.21 Approximate Equivalent Hardness Numbers* for Brinell Hardness Numbers† for Steel (Continued)

| Brinell indentation dia, mm | Brinell hardness no.,† 10-mm ball, 3000-kg load | | Vickers hardness no. | Rockwell hardness no.† | | | | Rockwell superficial hardness no., superficial Brale penetrator | | | Shore scleroscope hardness no. | Tensile strength (approximate) in MPa (1000 psi) | Brinell indentation dia, mm |
| | Standard ball | Tungsten-carbide ball | | A-scale, 60-kg load, Brale penetrator | B-scale, 100-kg load, 1.6-mm (1/16-in) dia ball | C-scale, 150-kg load, Brale penetrator | D-scale, 100-kg load, Brale penetrator | 15-N scale, 15-kg load | 30-N scale, 30-kg load | 45-N scale, 45-kg load | | | |
Col. 1	Col. 2	Col. 3	Col. 4	Col. 5	Col. 6	Col. 7	Col. 8	Col. 9	Col. 10	Col. 11	Col. 12	Col. 13	Col. 14
3.60	**285**	**285**	**301**	**65.3**	**(105.5)**	**29.9**	**47.6**	**75.0**	**50.3**	**31.2**	—	950 (138)	**3.60**
3.65	**277**	**277**	**292**	**64.6**	**(104.5)**	**28.8**	**46.7**	**74.4**	**49.3**	**29.9**	41	925 (134)	**3.65**
3.70	**269**	**269**	**284**	**64.1**	**(104.0)**	**27.6**	**45.9**	**73.7**	**48.3**	**28.5**	40	895 (130)	**3.70**
3.75	**262**	**262**	**276**	**63.6**	**(103.0)**	**26.6**	**45.0**	**73.1**	**47.3**	**27.3**	39	875 (127)	**3.75**
3.80	**255**	**255**	**269**	**63.0**	**(102.0)**	**25.4**	**44.2**	**72.5**	**46.2**	**26.0**	38	850 (123)	**3.80**
3.85	**248**	**248**	**261**	**62.5**	**(101.0)**	**24.2**	**43.2**	**71.7**	**45.1**	**24.5**	37	825 (120)	**3.85**
3.90	**241**	**241**	**253**	**61.8**	**100.0**	**22.8**	**42.0**	**70.9**	**43.9**	**22.8**	36	800 (116)	**3.90**
3.95	**235**	**235**	**247**	**61.4**	**99.0**	**21.7**	**41.4**	**70.3**	**42.9**	**21.5**	35	785 (114)	**3.95**
4.00	**229**	**229**	**241**	**60.8**	**98.2**	**20.5**	**40.5**	**69.7**	**41.9**	**20.1**	34	765 (111)	**4.00**
4.05	223	223	234	—	97.3	(18.8)	—	—	—	—	—	—	4.05
4.10	217	217	228	—	96.4	(17.5)	—	—	—	—	33	725 (105)	4.10
4.15	212	212	222	—	95.5	(16.0)	—	—	—	—	—	705 (102)	4.15
4.20	207	207	218	—	94.6	(15.2)	—	—	—	—	32	690 (100)	4.20
4.25	201	201	212	—	93.8	(13.8)	—	—	—	—	31	675 (98)	4.25
4.30	197	197	207	—	92.8	(12.7)	—	—	—	—	30	655 (95)	4.30
4.35	192	192	202	—	91.9	(11.5)	—	—	—	—	29	640 (93)	4.35
4.40	187	187	196	—	90.7	(10.0)	—	—	—	—	—	620 (90)	4.40
4.45	183	183	192	—	90.0	(9.0)	—	—	—	—	28	615 (89)	4.45
4.50	179	179	188	—	89.0	(8.0)	—	—	—	—	27	600 (87)	4.50
4.55	174	174	182	—	87.8	(6.4)	—	—	—	—	—	585 (85)	4.55
4.60	170	170	178	—	86.8	(5.4)	—	—	—	—	26	570 (83)	4.60
4.65	167	167	175	—	86.0	(4.4)	—	—	—	—	—	560 (81)	4.65
4.70	163	163	171	—	85.0	(3.3)	—	—	—	—	25	545 (79)	4.70

4.80	156	156	163	—	82.9	(0.9)	—	—	—	—	—	525 (76)	4.80
4.90	149	149	156	—	80.8	—	—	—	—	—	23	505 (73)	4.90
5.00	143	143	150	—	78.7	—	—	—	—	—	22	490 (71)	5.00
5.10	137	137	143	—	76.4	—	—	—	—	—	21	460 (67)	5.10
5.20	131	131	137	—	74.0	—	—	—	—	—	—	450 (65)	5.20
5.30	126	126	132	—	72.0	—	—	—	—	—	20	435 (63)	5.30
5.40	121	121	127	—	69.8	—	—	—	—	—	19	415 (60)	5.40
5.50	116	116	122	—	67.6	—	—	—	—	—	18	400 (58)	5.50
5.60	111	111	117	—	65.7	—	—	—	—	—	15	385 (56)	5.60

* This table corresponds to the table in ASM Metals Handbook, 8th Edition, Vol. 1, page 1235. It has been modified to add metric equivalents for approximate tensile strength values and to indicate Brinell hardness values that are beyond the recommended range for this test.

† Values in ( ) are beyond normal range and are given for information only.

SOURCE: Reprinted with permission, copyright 1992, Society of Automotive Engineers.

TABLE 4.22 Approximate Equivalent Hardness Numbers* for Rockwell C Hardness Numbers for Steel

Rockwell C-scale hardness no.[†]	Vickers hardness no.	Brinell hardness no., 10-mm ball, 3000-kg load — Standard ball	Brinell hardness no., 10-mm ball, 3000-kg load — Tungsten-carbide ball	Rockwell hardness no.[†] — A-scale, 60-kg load, Brale penetrator	Rockwell hardness no.[†] — B-scale, 100-kg load, 1.6-mm ($\frac{1}{16}$-in) dia ball	Rockwell hardness no.[†] — D-scale, 100-kg load, Brale penetrator	Rockwell superficial hardness no., superficial Brale penetrator — 15-N scale, 15-kg load	Rockwell superficial hardness no., superficial Brale penetrator — 30-N scale, 30-kg load	Rockwell superficial hardness no., superficial Brale penetrator — 45-N scale, 45-kg load	Shore scleroscope hardness no.	Tensile strength (approximate) in MPa (1000 psi)	Rockwell C-scale hardness no.[†]
Col. 1	Col. 2	Col. 3	Col. 4	Col. 5	Col. 6	Col. 7	Col. 8	Col. 9	Col. 10	Col. 11	Col. 12	Col. 13
68	940	—	—	85.6	—	76.9	93.2	84.4	75.4	97	—	68
67	900	—	—	85.0	—	76.1	92.9	83.6	74.2	95	—	67
66	865	—	—	84.5	—	75.4	92.5	82.8	73.3	92	—	66
65	832	—	(739)	83.9	—	74.5	92.2	81.9	72.0	91	—	65
64	800	—	(722)	83.4	—	73.8	91.8	81.1	71.0	88	—	64
63	772	—	(705)	82.8	—	73.0	91.4	80.1	69.9	87	—	63
62	746	—	(688)	82.3	—	72.2	91.1	79.3	68.8	85	—	62
61	720	—	(670)	81.8	—	71.5	90.7	78.4	67.7	83	—	61
60	697	—	(654)	81.2	—	70.7	90.2	77.5	66.6	81	—	60
59	674	—	(634)	80.7	—	69.9	89.8	76.6	65.5	80	—	59
58	653	—	615	80.1	—	69.2	89.3	75.7	64.3	78	—	58
57	633	—	595	79.6	—	68.5	88.9	74.8	63.2	76	—	57
56	613	—	577	79.0	—	67.7	88.3	73.9	62.0	75	—	56
55	595	—	560	78.5	—	66.9	87.9	73.0	60.9	74	2075 (301)	55
54	577	—	543	78.0	—	66.1	87.4	72.0	59.8	72	2015 (292)	54
53	560	(500)	525	77.4	—	65.4	86.9	71.2	58.6	71	1950 (283)	53
52	544	(487)	512	76.8	—	64.6	86.4	70.2	57.4	69	1880 (273)	52
51	528	(475)	496	76.3	—	63.8	85.9	69.4	56.1	68	1820 (264)	51
50	513	(464)	481	75.9	—	63.1	85.5	68.5	55.0	67	1760 (255)	50
49	498	451	469	75.2	—	62.1	85.0	67.6	53.8	66	1695 (246)	49
48	484	442	455	74.7	—	61.4	84.5	66.7	52.5	64	1635 (237)	48
47	471	432	443	74.1	—	60.8	83.9	65.8	51.4	63	1580 (229)	47
46	458	—	432	73.6	—	60.0	83.5	64.8	50.3	62	1530 (222)	46
45	446	421	421	73.1	—	59.2	83.0	64.0	49.0	60	1480 (215)	45
44	434	409	409	72.5	—	58.5	82.5	63.1	47.8	58	1435 (208)	44

148

43	423	400	400	72.0	—	57.7	82.0	62.2	46.7	57	1385 (201)	43
42	412	390	390	71.5	—	56.9	81.5	61.3	45.5	56	1340 (194)	42
41	402	381	381	70.9	—	56.2	80.9	60.4	44.3	55	1295 (188)	41
40	392	371	371	70.4	—	55.4	80.4	59.5	43.1	54	1250 (181)	40
39	382	362	362	69.9	—	54.6	79.9	58.6	41.9	52	1215 (176)	39
38	372	353	355	69.4	—	53.8	79.4	57.7	40.8	51	1180 (171)	38
37	363	344	344	68.9	—	53.1	78.8	56.8	39.6	50	1160 (168)	37
36	354	336	336	68.4	(109.0)	52.3	78.3	55.9	38.4	49	1115 (162)	36
35	345	327	327	67.9	(108.5)	51.5	77.7	55.0	37.2	48	1080 (157)	35
34	336	319	319	67.4	(108.0)	50.8	77.2	54.2	36.1	47	1055 (153)	34
33	327	311	311	66.8	(107.5)	50.0	76.6	53.3	34.9	46	1025 (149)	33
32	318	301	301	66.3	(107.0)	49.2	76.1	52.1	33.7	44	1000 (145)	32
31	310	294	294	65.8	(106.0)	48.4	75.6	51.3	32.5	43	980 (142)	31
30	302	286	286	65.3	(105.5)	47.7	75.0	50.4	31.3	42	950 (138)	30
29	294	279	279	64.7	(104.5)	47.0	74.5	49.5	30.1	41	930 (135)	29
28	286	271	271	64.3	(104.0)	46.1	73.9	48.6	28.9	41	910 (132)	28
27	279	264	264	63.8	(103.0)	45.2	73.3	47.7	27.8	40	880 (128)	27
26	272	258	258	63.3	(102.5)	44.6	72.8	46.8	26.7	38	860 (125)	26
25	266	253	253	62.8	(101.5)	43.8	72.2	45.9	25.5	38	840 (122)	25
24	260	247	247	62.4	(101.0)	43.1	71.6	45.0	24.3	37	825 (120)	24
23	254	243	243	62.0	100.0	42.1	71.0	44.0	23.1	36	805 (117)	23
22	248	237	237	61.5	99.0	41.6	70.5	43.2	22.0	35	785 (114)	22
21	243	231	231	61.0	98.5	40.9	69.9	42.3	20.7	35	770 (112)	21
20	238	226	226	60.5	97.8	40.1	69.4	41.5	19.6	34	760 (110)	20
(18)	230	219	219	—	96.7	—	—	—	—	33	730 (106)	(18)
(16)	222	212	212	—	95.5	—	—	—	—	32	705 (102)	(16)
(14)	213	203	203	—	93.9	—	—	—	—	31	675 (98)	(14)
(12)	204	194	194	—	92.3	—	—	—	—	29	650 (94)	(12)
(10)	196	187	187	—	90.7	—	—	—	—	28	620 (90)	(10)
(8)	188	179	179	—	89.5	—	—	—	—	27	600 (87)	(8)
(6)	180	171	171	—	87.1	—	—	—	—	26	580 (84)	(6)
(4)	173	165	165	—	85.5	—	—	—	—	25	550 (80)	(4)
(2)	166	158	158	—	83.5	—	—	—	—	24	530 (77)	(2)
(0)	160	152	152	—	81.7	—	—	—	—	24	515 (75)	(0)

* The values in this table shown in **boldface** type correspond to the values shown in the corresponding joint SAE-ASM-ASTM Committee on Hardness Conversions as printed in ASTM E 140, Table 1.

† Values in ( ) are beyond normal range and are given for information only.

SOURCE: Reprinted with permission, copyright 1992, Society of Automotive Engineers.

Carbon and alloy steels: Uses by SAE/AISI number

SAE/AISI 1006 through 1015.   Low-carbon, high-formability, weldable. Used for sheet metal, strip, wire, and rod. Excellent drawing qualities. Low-strength applications. Sheet metal structures, body and fender work, deep-drawing parts of sheet steel.

SAE/AISI 1016 through 1030.   Increased strength over the low-carbon group. These are known as the *case-hardening* or *carburizing grades*. The higher-manganese grades machine well. The higher-carbon types are used for thicker sections where a stronger core is desired. Type 1018 is used for a great many applications, may be easily case-hardened, and is readily available. Grades 1020 and 1025 are used for low-strength bolts. All these steels are readily welded.

SAE/AISI 1030 through 1052.   These are the medium-carbon types of steels used where higher strength than the lower-carbon grades is required. All these steels are used for forgings. Axles and shafts are made from the 1038 to 1045 group. Widely used for machined parts, both heat-treated and non-heat-treated. Welding is possible with precautions taken during the cooling process.

SAE/AISI 1055 through 1095.   These are the high-carbon grades of carbon steel. Used for flat stampings, spring wire, cutting tools, flat springs, and many other high-strength applications. These steels are usually heat-treated for their particular application and provide excellent wear resistance. Not recommended for welding applications.

Specific applications

SAE/AISI 1018.   Low-carbon, medium-manganese steel. Quality bar for carburized parts: gears, shafts, bolts, pins, etc.

SAE/AISI 1035.   Medium-carbon, special-quality steel used for bolts, nuts, shafts, pins, etc. Can be heat-treated.

SAE/AISI 1045.   Medium-carbon, special-quality steel used for shafts, axles, gears, splines, etc. Can be heat-treated.

SAE/AISI 12L14.   Low-carbon, resulfurized, rephosphorized, and leaded steel used for screw-machine parts such as studs, nuts, and various fasteners. Can be case-hardened.

SAE/AISI 1215.   Similar to 12L14 and low-carbon steel used for studs, nuts, and fasteners. Can be case-hardened.

SAE/AISI 12L15.   Leaded version of 1215 used for screw-machine parts. Can be case-hardened.

SAE / AISI 1117.   Low-carbon, resulfurized, free-cutting steel used for shafts, gears, pins, nuts, etc. Can be carburized.

SAE / AISI 11L17.   Leaded version of 1117 used for screw-machine parts, gears, shafts, pins, etc. Can be carburized.

SAE / AISI 1141.   Medium-carbon, resulfurized, free-cutting steel used for shafts, nuts, bolts, etc. Can be hardened by heat treatment.

SAE / AISI 4140.   Medium-carbon chromium-molybdenum alloy steel used for studs, nuts, bolts, gears, wrenches, shafts, etc. Can be heat-treated.

SAE / AISI 41L40.   Leaded version of 4140, free-machining type. Can be heat-treated.

SAE / AISI 4145.   Medium-carbon chromium-molybdenum steel used for studs, nuts, shafts, wrenches, gears, bolts, etc. Can be heat-treated.

SAE / AISI 41L45.   Leaded version of 4145, free-machining type. Can be heat-treated.

SAE / AISI 4620.   Low-carbon nickel-molybdenum steel used for gears, cams, pinions, and shafts. Excellent carburizing grade.

SAE / AISI 46L20.   Leaded version of 4620, free-machining type. Can be carburized.

SAE / AISI 8620.   Low-carbon nickel-chromium-molybdenum alloy steel used for gears, cams, shafts, and pinions. Excellent carburizing grade.

SAE / AISI 86L20.   Leaded version of 8620, free-machining type. Can be carburized.

SAE / AISI 4340.   Medium-carbon nickel-chromium-molybdenum alloy steel used for gears and shafting. Has high hardenability.

SAE / AISI 8642.   Similar to 4340. Has high hardenability.

EF 4130.   Aircraft-quality alloy steel.

EF 4140.   Aircraft-quality alloy steel.

E 4340, EF 4620, EF 8740, and E 9310.   All aircraft-quality alloy steels.

Rather than going into a lengthy description of all the characteristics and heat-treating properties of all the various grades of carbon and alloy steels, the preceding tables can be used by the engineer or designer to determine strength, ductility, hardness, and heat-treatment temperatures for each SAE/AISI and ASTM steel listed. A metallurgist should be consulted prior to making a final design choice about the various steels used in important or critical applications.

Stainless steels: Uses by AISI number

Note: The stainless steels listed in the preceding tables typically are known by their three-digit SAE numbers, such as 201, 302, 304, 440, etc. The last three digits of the listed SAE numbers are the standard industry identification numbers and are used as such in the following usage summary.

Chromium-nickel stainless steels (austenitic)

201. Low nickel, good corrosion resistance. High work-hardening rate. Excellent weldability.

202. General-purpose type equivalent to 302.

203 EZ. Superior machinability. Good corrosion resistance.

216. Most corrosion resistant of all chromium-nickel-manganese stainless steels.

301. High work-hardening rate. Used in structural applications where high strength and resistance to atmospheric corrosion are required.

302. General-purpose stainless steel with good strength properties. Resistant to many corrosive conditions.

303. Free-machining type used in corrosive atmospheres, strong chemical solutions, many organic chemicals, most dyes, nitric acid, and foods.

303 Pb. Leaded version of 303 used for high-volume automatic machining operations.

304. Low-carbon variation of 302. Weldable with caution. Excellent resistance to a high number of corrosive conditions and chemicals.

304 L. Extra-low-carbon version of 304. Low carbon content prevents carbide precipitation during welding, which can produce cracks at the weld joints. Excellent weldability.

305. Good fabrication stainless for spinning, deep-drawing, and cold-heading operations.

309. Used in high-temperature applications. Resistant to most acids.

310. Higher alloy content improves the basic characteristics. Improved over 309 and 304 for corrosion resistance.

316. Best corrosion resistance of the standard stainless steels. Resists pitting and most chemicals. Used for paper-mill machinery parts and photographic industry parts and containers. High-temperature strength.

316 L. Low-carbon version of 316 that is welded more easily without carbide precipitation.

317.   Higher alloy content than 316, providing more corrosion resistance.

321.   This alloy is stabilized with titanium for weldments subject to severe corrosion. Excellent corrosion resistance to a wide variety of organic and inorganic substances.

347.   Stabilized with Cb and Ta for use in the carbide precipitation range of 800 to 1500°F, with no impairment to corrosion resistance.

Chromium stainless steels (ferritic)

409.   Developed for automotive muffler service and used in noncritical exterior parts. Economical and easily fabricated.

430.   Most widely used of the nonhardenable types of chromium stainless steel. Good mechanical properties and heat resistance. Resistant to nitric acid, sulfur gases, and many organic chemicals, including foods.

430 F.   Free-machining version of 430. Similar in properties to 430.

446.   High resistance to corrosion and scaling at high temperatures. Excellent in sulfuric atmospheres.

Chromium stainless steels (martensitic)

403.   Excellent for highly stressed parts such as turbine blades. Good resistance to water and atmospheric corrosion.

405.   Variation of 410 for improved weldability. Same corrosion resistance as 403.

410.   Low-cost general-purpose stainless steel that is heat-treatable. Used where corrosion is not severe.

410 S.   Same as 410 except lower carbon range for improved weldability.

414.   Modification of 410 with more nickel to improve corrosion resistance. Can be heat treated to Rockwell C 25 to C 43.

416.   Free-machining version of 410 with corrosion resistance to food acids, basic salts, water, and most atmospheric corrosion products.

440 A, B, C, and F.   Series of high-carbon stainless steels. All are the same basic composition except carbon content. Can be heat-treated for high strength and high hardness. These steels are corrosion resistant only in the hardened conditions. 440 F is a free-machining type used in many applications.

Low-chromium stainless steels

PH13-8Mo

15-5PH

PH15-7Mo

17-4PH

17-7PH

All these grades provide excellent corrosion resistance similar to the austenitic stainless steels and are capable of being heat-treated to various high-strength conditions with minimum distortion. They are generally furnished in the annealed condition for ease of machining. They develop their high-strength properties by aging at selected temperatures. They may be provided as vacuum arc-remelted steels for more demanding applications. They are used in many high-strength, anticorrosion applications. Among other uses, 17-7PH (ASTM A313) round wire is used for helical spring applications, and 17-7PH (ASTM A693) strip is used for flat spring applications.

Tool steels: SAE designations. Refer to Tables 4.17 through 4.19 for physical properties and heat-treating procedures. Also see Chap. 8.

4.1.3 High-strength, low-alloy (HSLA) steels

Typical HSLA steels, with data indicating their mechanical properties, are shown in Table 4.23. Although these are low-alloy grades of steel, they have excellent tensile strength and other properties, making them valuable in many applications. HSLA steels were developed more than 75 years ago by Krupp in Germany. Some of the older German ordnance steels could be classified as modern HSLA steels, and they also were copper-bearing to improve their resistance to atmospheric corrosion. The famous Mauser M1898 rifles produced for various South American countries had bare polished-metal receivers and bolts.

Many of the HSLA steels rely on carbon, manganese, and silicon contents to achieve the strengths associated with these special materials. Other alloying elements such as nickel, chromium, and copper are also present in some of these steels.

4.1.4 Ultra-high-strength steels

Structural steels with very high tensile strengths are referred to as ultra-high-strength steels. An arbitrary minimum yield strength level of 200 ksi has been established for these grades of steels. (Some ultra-high-strength steels fall below this arbitrary minimum.) Tables 4.24 through 4.30 give the mechanical properties of

(Text continued on page 164.)

TABLE 4.23 Mechanical Properties of HSLA Steel Grades Described in ASTM Specifications

ASTM specification	Type, grade, or condition	UNS designation	Min tensile strength MPa	Min tensile strength ksi	Min yield strength MPa	Min yield strength ksi	Min elongation, % In 200 mm or 8 in	Min elongation, % In 50 mm or 2 in
A242	Type 1	K11510	435 to 480	63 to 70	290 to 345	42 to 50	18	21
	Type 2	K12010	435 to 480	63 to 70	290 to 345	42 to 50	18	21
A440	—	K12810	435 to 485	63 to 70	290 to 345	42 to 50	18	21
A441	—	K12211	415 to 485	60 to 70	275 to 345	40 to 50	18	21
A572	Grade 42	—	415	60	290	42	20	24
	Grade 45	—	415	60	310	45	19	22
	Grade 50	—	450	65	345	50	18	21
	Grade 55	—	485	70	380	55	17	20
	Grade 60	—	520	75	415	60	16	18
	Grade 65	—	550	80	450	65	15	17
A588	Grades A to J	—	435 to 485	63 to 70	290 to 345	42 to 50	18	21
A606	Hot rolled	—	480	70	345	50	—	22
	Hot rolled and annealed or normalized	—	450	65	310	45	—	22
	Cold rolled	—	450	65	310	45	—	22
A607	Grade 45	—	410	60	310	45	—	22 to 25
	Grade 50	—	450	65	345	50	—	20 to 22
	Grade 55	—	480	70	380	55	—	18 to 20
	Grade 60	—	520	75	415	60	—	16 to 18
	Grade 65	—	550	80	450	65	—	15 to 16
	Grade 70	—	590	85	485	70	—	14
A613	Grade I	K02601	483	70	345	50	19	22
	Grade II	K12609	483	70	345	50	18	22
	Grade III	K12700	448	65	345	50	18	20

155

TABLE 4.23 Mechanical Properties of HSLA Steel Grades Described in ASTM Specifications (*Continued*)

ASTM specification	Type, grade, or condition	UNS designation	Min tensile strength MPa	Min tensile strength ksi	Min yield strength MPa	Min yield strength ksi	Min elongation, % In 200 mm or 8 in	Min elongation, % In 50 mm or 2 in
A633	Grades A and B	—	430 to 570	63 to 83	290	42	18	23
	Grades C and D	—	450 to 620	65 to 90	315 to 345	46 to 50	18	23
	Grade E	K12202	520 to 690	75 to 100	380 to 415	55 to 60	18	23
A656	Grades 1 and 2	(e)	655 to 793	95 to 115	552	80	12	—
A690	—	K12249	485	70	345	50	18	—
A715	Grade 50	—	415	60	345	50	—	22 to 24
	Grade 60	—	485	70	415	60	—	20 to 22
	Grade 70	—	550	80	485	70	—	18 to 20
	Grade 80	—	620	90	550	80	—	16 to 18

TABLE 4.24 Typical Mechanical Properties of Heat-Treated 4130 Steel

Tempering temperature		Tensile strength		Yield strength		Elongation in 50 mm or 2 in, %	Reduction in area, %	Hardness, HB	Izod impact energy	
°C	°F	MPa	ksi	MPa	ksi				J	ft · lb
Water-quenched and tempered*										
205	400	1765	256	1520	220	10.0	33.0	475	18	13
260	500	1670	242	1430	208	11.5	37.0	455	14	10
315	600	1570	228	1340	195	13.0	41.0	425	14	10
370	700	1475	214	1250	182	15.0	45.0	400	20	15
425	800	1380	200	1170	170	16.5	49.0	375	34	25
540	1000	1170	170	1000	145	20.0	56.0	325	81	60
650	1200	965	140	830	120	22.0	63.0	270	135	100
Oil-quenched and tempered†										
205	400	1550	225	1340	195	11.0	38.0	450	—	—
260	500	1500	218	1275	185	11.5	40.0	440	—	—
315	600	1420	206	1210	175	12.5	43.0	418	—	—
370	700	1320	192	1120	162	14.5	48.0	385	—	—
425	800	1230	178	1030	150	16.5	54.0	360	—	—
540	1000	1030	150	840	122	20.0	60.0	305	—	—
650	1200	830	120	670	97	24.0	67.0	250	—	—

* 25-mm (1-in) dia round bars quenched from 845 to 870°C (1550 to 1600°F).
† 25-mm (1-in) dia round bars quenched from 860°C (1575°F).

TABLE 4.25 Typical Mechanical Properties of Heat-Treated 4140 Steel*

Tempering temperature		Tensile strength		Yield strength		Elongation in 50 mm or 2 in, %	Reduction in area, %	Hardness, HB	Izod impact energy	
°C	°F	MPa	ksi	MPa	ksi				J	ft · lb
205	400	1965	285	1740	252	11.0	42	578	15	11
260	500	1860	270	1650	240	11.0	44	534	11	8
315	600	1720	250	1570	228	11.5	46	495	9	7
370	700	1590	231	1460	212	12.5	48	461	15	11
425	800	1450	210	1340	195	15.0	50	429	28	21
480	900	1300	188	1210	175	16.0	52	388	46	34
540	1000	1150	167	1050	152	17.5	55	341	65	48
595	1100	1020	148	910	132	19.0	58	311	93	69
650	1200	900	130	790	114	21.0	61	277	112	83
705	1300	810	117	690	100	23.0	65	235	136	100

* 12.7-mm (0.5-in) dia round bars, oil-quenched from 845°C (1550°F).

TABLE 4.26 Typical Mechanical Properties of 4340 Steel*

Tempering temperature		Tensile strength		Yield strength		Elongation in 50 mm or 2 in, %	Reduction in area, %	Hardness		Izod impact energy	
°C	°F	MPa	ksi	MPa	ksi			HB	HRC	J	ft · lb
205	400	1980	287	1860	270	11	39	520	53	20	15
315	600	1760	255	1620	235	12	44	490	49.5	14	10
425	800	1500	217	1365	198	14	48	440	46	16	12
540	1000	1240	180	1160	168	17	53	360	39	47	35
650	1200	1020	148	860	125	20	60	290	31	100	74
705	1300	860	125	740	108	23	63	250	24	102	75

* Oil-quenched from 845°C (1550°F) and tempered at various temperatures.

TABLE 4.27 Room-Temperature Mechanical Properties of Heat-Treated 6150 Steel

Tempering temperature		Tensile strength		Yield strength		Elongation in 50 mm or 2 in, %	Reduction in area, %	Hardness, HB	Izod impact energy	
°C	°F	MPa	ksi	MPa	ksi				J	ft · lb
Round bars, 14 mm (0.55 in) in dia*										
205	400	2050	298	1810	263	1	5	610	—	—
260	500	2070	300	1810	263	4	12	570	—	—
315	600	1950	283	1720	250	7	27	540	—	—
370	700	1770	257	1620	235	10	37	505	9	7
425	800	1585	230	1490	216	11	42	470	14	10
480	900	1410	204	1340	195	12	44	420	16	12
540	1000	1250	182	1210	175	13	46	380	20	15
595	1100	1150	167	1080	157	16	47	350	28	21
Round bars, 25 mm (1 in) in dia†										
425	800	1570	228	1450	210	10	37	461	—	—
480	900	1360	197	1210	175	11	41	401	—	—
540	1000	1180	171	1030	150	12	45	341	—	—
595	1100	1030	150	875	127	15	50	302	—	—
650	1200	920	133	760	110	19	55	262	—	—
705	1300	810	118	660	96	23	61	235	—	—

* Normalized at 870°C (1600°F). Oil-quenched from 860°C (1575°F) and tempered at various temperatures.
† Oil-quenched from 860°C and tempered at various temperatures.

TABLE 4.28 Typical Room-Temperature Mechanical Properties of 8640 Steel

Tempering temperature		Tensile strength		Yield strength		Elongation in 50 mm or 2 in, %	Reduction in area, %	Impact energy		Hardness	
°C	°F	MPa	ksi	MPa	ksi			J	ft · lb	HB	HRC
Round bars, 13.5 mm (0.53 in) in dia*											
205	400	1810	263	1670	242	8.0	25.8	11.5†	8.5†	555	55
315	600	1585	230	1430	208	9.0	37.3	15.6†	11.5†	461	48
425	800	1380	200	1230	179	10.5	46.3	27.8†	20.5†	415	44
540	1000	1170	170	1050	152	14.0	53.3	56.3†	41.5†	341	37
650	1200	870	126	760	110	20.5	61.0	96.9†	71.5†	269	28
Round bars, 25 mm (1 in.) in dia*											
425	800	1382	200.5	1230	179	10	46	27‡	20‡	415	44
480	900	1250	181	1120	162	13	51	41‡	30‡	388	42
540	1000	1070	155	940	137	17	56	54‡	40⁼	331	36
595	1100	1020	148	910	132	16	57	73‡	54‡	302	32
650	1200	865	125.5	760	110.5	20	61	83‡	61‡	269	28

* Oil-quenched from 830°C (1525°F) and tempered at indicated temperature.
† Izod.
‡ Charpy V-notch.

TABLE 4.29 Typical Longitudinal Mechanical Properties of H11 Mod Steel*

Tempering temperature		Tensile strength		Yield strength		Elongation in 50 mm or 2 in, %	Reduction in area, %	Charpy V-notch impact energy		Hardness, HRC
°C	°F	MPa	ksi	MPa	ksi			J	ft·lb	
510	950	2120	308	1710	248	5.9	29.5	13.6	10.0	56.5
540	1000	2010	291	1675	243	9.6	30.6	21.0	15.5	56.0
565	1050	1850	269	1565	227	11.0	34.5	26.4	19.5	52.0
595	1100	1540	223	1320	192	13.1	39.3	31.2	23.0	45.0
650	1200	1060	154	850	124	14.1	41.2	40.0	29.5	33.0
705	1300	940	136	700	101	16.4	42.2	90.6	66.8	29.0

* Air-cooled from 1010°C (1850°F) and double-tempered, 2 + 2 h at indicated temperature.

TABLE 4.30 Typical Longitudinal Room-Temperature Mechanical Properties of H13 Steel*

Tempering temperature		Tensile strength		Yield strength		Elongation in 4 D, %	Reduction in area, %	Charpy V-notch impact energy		Hardness, HRC
°C	°F	MPa	ksi	MPa	ksi			J	ft · lb	
527	980	1960	284	1570	228	13.0	46.2	16	12	52
555	1030	1835	266	1530	222	13.1	50.1	24	18	50
575	1065	1730	251	1470	213	13.5	52.4	27	20	48
593	1100	1580	229	1365	198	14.4	53.7	28.5	21	46
605	1120	1495	217	1290	187	15.4	54.0	30	22	44

* Round bars, oil-quenched from 1010°C (1850°F) and double tempered, 2 – 2 h at indicated temperature.

some ultra-high-strength steels. Many of these are considered high-quality specialty steels, with 4340 being used as the reference by which all other types of ultra-high-strength steels are measured. AISI type 4340 is used in critical applications, such as key components for aerospace vehicles and commercial and military aircraft.

4.2   Aluminum and Aluminum Alloys

Aluminum and its alloys are among the most used metallic materials, with countless applications and an extremely broad range of physical and chemical properties. Pure aluminum is a silvery white metal, light in weight, nontoxic, and easily cast, forged, and machined. The pure metal was first isolated in the laboratory in 1827. The method of extracting the metal by electrolysis of alumina dissolved in cryolite was discovered by Hall in 1886 in the United States. This method is still in use today and is known as the *Hall process.* Aluminum in its natural forms is the third most abundant material in the earth's crust, exceeded only by oxygen and silicon.

Aluminum alloys are mandatory in many design applications of modern technologies. Many of the modern aluminum alloys are stronger than some steels on a volume basis and weigh only 34 percent (or one-third) as much as steel. The average density of aluminum alloys is 0.098 lb/in^3 and that of steel is 0.282 lb/in^3. The electrical conductivity of aluminum is 60 percent that of an equal cross-sectional area of copper, which is the second best conductor of electric current. In order of electrical conductivity, the best four elements are silver, copper, gold, and aluminum, respectively.

Tables 4.31 through 4.38 show the physical properties of all the present aluminum alloys. Also included are the typical uses for all the wrought and cast types of aluminum alloys in use today. Figure 4.2 shows part of the SAE standard delineating alloy and temper designation systems for aluminum.

4.3   Copper and Copper Alloys

Copper and copper alloys are also among the most important and most used metallic materials. Almost all electrical components and electrical products contain parts made of copper or one or more of its

important alloys. All the electrical industries worldwide depend on copper and copper alloys. Many new copper alloys have been developed over the past 30 or 40 years, with beryllium-copper alloys as one of the preferred materials in many electrical applications as a replacement for the phosphor bronzes. The phosphor, silicon, and manganese bronzes have many applications where strength and current-carrying ability, combined with corrosion resistance and nonmagnetic properties, are desired. Springs with a high fatigue-endurance limit and good electrical properties are made from the beryllium-copper alloys.

One of the most important uses for copper is in the electrical power distribution industries, where ETP no. 110 copper bus conductors are used to carry the electrical power for all electrical applications. The power distribution industries use such equipment as transformers, power stations, power transmission lines, and electrical switch gear and control equipment. The electric motor industries are another large user of copper products.

Copper is one of the most important elements, with a specific gravity of 8.96 g/cm^3 and weight of 0.324 lb/in^3. It is the second-best conductor of electric current, exceeded only by silver. The copper metal is smelted from oxide, sulfide, and carbonate compounds that are found in their natural states as cuprite, malachite, azurite, and bornite. The most important compounds are the oxides and the sulfates (blue vitriol), the latter being used for agricultural poisons and water purification.

Tables 4.39 through 4.42 list the properties and uses for the many alloys of copper. Table 4.43 lists the standard specifications for brass sheet, strip, plate, and rolled bar (ASTM B-36). Table 4.44 lists the standard specifications for phosphor bronze plate, sheet, strip, and rolled bar (ASTM B-103). Tables 4.45 through 4.47 list the standard specifications for beryllium-copper sheet, plate, strip, and rolled bar (ASTM B-194). Table 4.48 presents the ASTM classification of coppers.

4.4   Magnesium Alloys

Magnesium and magnesium alloys have many applications where moderate strength and light weight are required. These alloys find

(Text continued on page 203.)

TABLE 4.31 Typical Heat Treatments for Aluminum Alloy Mill Products[a]

Alloy	Product	Solution heat treatment[b] Metal temperature[c] °C	Solution heat treatment[b] Temper designation	Precipitation heat treatment Metal temperature[c] °C	Precipitation heat treatment Approx. time of temperature,[d] h	Precipitation heat treatment Temper designation
2036	Sheet	e	T4			
2038	Sheet	e	T4			
6009	Sheet	e	T4	200–210	1	T6[f]
6010	Sheet	e	T4	200–210	1	T6[f]
6061	Sheet	515–550	T4	155–165	18	T6
	Plate	515–550	T451	155–165	18	T651
	Extrusions, tube, rod, bar forgings	515–550[g]	T4	170–180	8	T6
6063	Extrusions, tube	g	T1	175–185[h]	3	T5
		515–525[g]	T4	170–180[i]	8	T6
6111	Sheet	e	T4	200–210	1	T6
6463	Extrusions	g	T1	175–185[h]	3	T5
		515–525[g]	T4	170–180[i]	8	T6
7021	Sheet	395–405[j]	W	i	i	T61
7029	Extrusions	480–520[k]	W	m	m	T5, T6
7116	Extrusions	425–540	W	n	n	T5
7129	Extrusions	480–520	S	m	m	T5, T6

[a] The times and temperatures shown are typical for various forms, sizes, and methods of manufacture and may not exactly describe the optimum treatment for a specific item.

[b] Material should be quenched in water or by high-velocity fans from the solution heat-treating temperature as rapidly as possible and with minimum delay after removal from the furnace. Unless otherwise indicated, when material is quenched by total immersion in water, the water should be at room temperature and suitably cooled to remain below 35°C during the quench cycle. The use of high-velocity, high-volume jets of cold water is also effective for some material.

166

[c] The metal temperature should be attained as rapidly as possible. Where a temperature range exceeding 10°C is shown, a temperature range of 10° within the listed range should be selected and maintained during the time at temperature.

[d] The time at temperature will depend on the time required for load to reach temperature. The times shown are based on rapid heating with soaking time measured from the time the load reaches the 10°C range listed or selected.

[e] These alloys are supplied in the solution heat-treated condition. For optimum properties, subsequent reheat treatment is not recommended.

[f] Mechanical properties of material will meet tensile property limits of T6 temper.

[g] By suitable control of extrusion temperature, product may be quenched directly from extrusion press to provide specified properties for this temper. Some products may be adequately quenched in air blast at room temperature.

[h] An alternate treatment comprised of 1–2 h at 200–210°C may be used.

[i] An alternate treatment comprised of 6 h at 175–185°C may be used.

[j] Quenched at a minimum average cooling rate of 35°C/s as measured over the range 385–205°C.

[k] 10-min soak at temperature followed by cold water quench.

[l] A minimum of 8 h at room temperature followed by 2 h at 95–105°C plus 4 h at 155–165°C.

[m] 5 h at 95–105°C plus 5 h at 155–165°C.

[n] 5 h at 95–105°C plus 5 h at 160–170°C.

SOURCE: Reprinted with permission, copyright 1992, Society of Automotive Engineers.

167

TABLE 4.32 Typical Mechanical Properties and Comparative Characteristics of Aluminum Alloys

Alloy and temper[b]	Typical mechanical properties[a,c]						Comparative characteristics							
	Tension		Tension	Shear	Fatigue	Modulus	Resistance to corrosion		Tough-ness[n]	Work-ability (cold)[p]	Machin-ability[p]	Weldability[q]		
	Strength		Elongation											
	Ultimate, MPa	Yield 0.2% offset, MPa	Percent in 50 mm (1.60 mm thick specimen)	Ultimate shearing strength, MPa	Endur-ance limit, MPa	Modulus of elasticity, MPa × 10^3 [g]	General[h]	Stress-corrosion cracking[i]				Gas	Arc	Resistance spot and seam
1100-0	90	35	35	60	35[d]	69	A	A		A	E	A	A	B
2017-T4	425	275	20[k]	260	125[d]	75	D	C		C	B	N/A	N/A	N/A
2024-T4	470	325	20	285	140[d]	73	D	C		C	B	N/A	N/A	N/A
2036-T4	340	195	24	205	125[e]	71	C	A2[l]		B	C	—	B	B
2038-T4	325	170	25	205	125[d]	71	C	A2[l]	A	B	C	—	B	B
2117-T4	295	165	24[k]	195	95[d]	71	C	A		B	C	N/A	N/A	N/A
3002-0	95	40	33	70		69	A	A		A	E	A	A	A
3003-H14, H24	150	145	8	95	60[d]	69	A	A		B	D	A	A	B
3004-0	180	70	20	110	95[d]	69	A	A	B	B	D	B	A	A
-H32	215	170	10	115	105[d]	69	A	A		A	D	B	A	B
5005-0	125	40	25	75		69	A	A		A	E	A	A	B
5052-0	195	90	25	125	110[d]	70	A	A	A	B	D	A	A	B
-H32	230	195	12	140	115[d]	70	A	A		B	D	A	A	A
-H34	260	215	10	145	125[d]	70	A	A	B	C	C	A	A	A
5083-H321, H116	315	230	14[k]		160[d]	71	A	A		B	D	C	A	A
5086-H32	290	205	12			71	A	B		B	D	C	A	B
-H34	325	255	10		185[d]	71	A	A		A	D	C	A	B
-H112	270	130	14			71	A	B		A	D	C	A	B
5182-0	275	130	21	165		71	A-B	A		B	D	C	A	B
-02	270	125	23		140[e]	71	A-B	A2[m]		A	D	C	A	B
5252-H25	235	170	11	145		69	A	A2[m]		B	D	A	A	A
5454-0	250	115	22	160		70	A	A	A	A	D	C	A	B
-H32	275	205	10	165		70	A	A	B	B	D	C	A	A
-H34	305	240	10	180		70	A	A		A	D	C	A	A
5457-H25	180	160	12	110		69	A	A		B	E	A	A	B
5657-H25	160	140	12	95		69	A	A		A	D	A	A	A
6009-T4	220	125	25	150	115[e]	69	B	A2	A	A	C	A	A	A
6010-T4	290	165	24	195	125[e]	69	B	A2	A	B	C	A	A	A

Alloy and temper	Tensile strength, MPa	Yield strength, MPa	Elongation, %	Shearing strength, MPa	Fatigue endurance limit, MPa	Modulus of elasticity, GPa	Workability (p)	Machinability (p)	Stress-corrosion cracking (i)	Weldability (q)
6053-T61	—	—	—	—	—	69	A	C	A	N/A
6061-T4	240	145	22	165	95[d]	69	A	C	A	A
6061-T6, T651	310	275	12	205	95[c]	69	B	C	A	A
6063-T1	150	90	20	95	60[d]	69	A	D	A	A
6063-T5	185	145	12	115	70[d]	69	A	C	A	A
6063-T6	240	215	12	150	69	69	B	C	A	A
6111-T1	290	130	26	—	—	69	A	C	A	A
6463-T521	185	145	12	—	—	69	A	C	A	A
7021-T61	430	380	13	270	140[d]	71	B	B	A2[j]	C
7029-T5, T6	430	380	15	—	—	70	B	B	A2[j]	C
7116-T5	360	315	16	—	—	70	B	B	A1[j]	C
7129-T6	430	380	15	270	145[f]	70	B	B	A2[j]	C

[a] Typical properties are not guaranteed since in most cases they are averages for various sizes, product forms and methods of manufacture and may not be exactly representative of any particular product or size. These data are intended only as a basis for comparing alloys and tempers and should not be specified as engineering requirements or used for design purposes.

[b] Only the commonly used tempers are listed for the alloys shown. Other tempers of these alloys are available.

[c] The indicated typical mechanical properties for all except the O temper material are higher than the specified minimum properties. For O temper products, typical ultimate and yield values are slightly lower than specified minimum values.

[d] Based on 500,000,000 cycles of completely reversed stress using the RR Moore type of machine and specimen.

[e] Based on a single series of tests, 10,000,000 cycles sheet flexural specimens.

[f] Based on 50,000,000 cycles in a single series of tests using the RR Moore type of machine and specimen.

[g] Average of tension and compression moduli. Compression modulus is about 2% greater than tension modulus.

[h] General corrosion ratings are based on exposures to sodium chloride solution by intermittent spraying or immersion. Ratings A through D are relative ratings in decreasing order of merit. The ratings do not necessarily imply acceptable performance in the intended application.

[i] Stress-corrosion cracking ratings are based on service experience and on laboratory tests of specimens exposed to the 35% sodium chloride alternate immersion test for 2XXX, 6XXX, and copper containing 1XXX series alloys and total immersion in boiling in sodium chloride solution for 96 h for copper free 7XXX series alloys.

A   No known instance of failure in service or in laboratory tests.

A2   Insufficient service experience; no known instance of failure in laboratory tests.

B   No known instance of failure in service; limited failures in laboratory tests of short transverse specimens.

B2   Insufficient service experience; limited failures in laboratory service.

C   Service failures with sustained tension stress acting in short transverse direction relative to grain structure; limited failures in laboratory tests of long transverse specimens.

D   Limited service failures with sustained longitudinal or long transverse stress.

[j] Improved resistance to stress corrosion cracking can be realized by using controlled quenching and artificial aging practices in heat-treatable 7XXX aluminum alloys.

[k] Elongation in 50 mm apply for thicknesses up through 12.50 mm and in $5D$ ($5.65\sqrt{A}$) for thicknesses over 12.50 mm, where D and A are the diameter and cross-sectional area of the specimen. Values for elongations in $5D$ ($5.65\sqrt{A}$) are shown in brackets.

[l] This rating would be B2 for material exposed to elevated temperatures.

[m] This rating may be different for material held at elevated temperatures for long periods.

[n] Toughness ratings are based upon Kahn tear test of 1.60-mm-thick sheet specimens in both longitudinal and transverse directions. These data are based on a limited number of tests and should be used for general comparisons only.

Ratings: A   over 175 000 N·m/m²
            B   over 140 000 thru 175 000 N·m/m²
            C   over 105 000 thru 140 000 N·m/m²
            D   0 through 105 000 N·m/m²

[p] Ratings A through D for workability (cold) and A through E for machinability are relative ratings in decreasing order of merit.

[q] Ratings A through D for weldability are relative ratings as follows:

A   Generally weldable by all commercial procedures and methods.

B   Weldable with special techniques.

C   Limited weldability due to crack sensitivity, loss in corrosion resistances, loss in mechanical properties.

D   No commonly used welding methods have been developed.

N/A   Rating not applicable for end use application requirements, that is, rivets.

SOURCE: Reprinted with permission, copyright 1992, Society of Automotive Engineers.

TABLE 4.33 Typical Physical Properties of SAE Casting Alloys

Alloy UNS	Alloy ANSI	Temper	Density lb/in³	Density kg/m³	Approximate melting range§ °F	Approximate melting range§ °C	Elec. cond., % IACS	Therm. cond., W/(m·K)	Coeff. of thermal expan., ×10⁻⁶ 68–212°F per°F	20–100°C per°C	68–572°F per°F	20–300°C per°C
A02010	201.0	T6	0.101	2800	995–1200	535–650	30	121	10.7	19.3	13.7	24.7
		T7	0.101	2800	995–1200	535–650	30	121	10.7	19.3	13.7	24.7
A02060	206.0	T4	0.101	2800	1010–1200	542–650	—	121	10.7	19.3	—	—
A02080	208.0	F	0.101	2800	970–1160	521–627	31	125	12.4	22.3	13.4	24.1
		T4	0.101	2800	970–1160	521–627	—	—	12.4	22.3	13.4	24.1
		T55	0.101	2800	970–1160	521–627	—	—	12.4	22.3	13.4	24.1
		T6	0.101	2800	970–1160	521–627	—	—	12.4	22.3	13.4	24.1
		T7	0.101	2800	970–1160	521–627	—	—	12.4	22.3	13.4	24.1
A02220	222.0	0	0.107	2960	965–1155	518–624	—	—	12.3	22.1	13.1	23.6
		T551	0.107	2960	965–1155	518–624	—	—	12.3	22.1	13.1	23.6
		T61	0.107	2960	965–1155	518–624	33	130	12.3	22.1	13.1	23.6
		T65	0.107	2960	965–1155	518–624	—	—	12.3	22.1	13.1	23.6
A02420	242.0	0	0.102	2820	990–1175	532–635	—	—	12.6	22.7	13.6	24.5
		T571*	0.102	2820	990–1175	532–635	34	134	12.6	22.7	13.6	24.5
		T61	0.102	2820	990–1175	532–635	—	—	12.6	22.7	13.6	24.5
		T77	0.102	2820	990–1175	532–635	38	151	12.6	22.7	13.6	24.5
A02950	295.0	T4	0.102	2820	970–1190	521–643	—	138	12.7	22.9	13.8	24.8
		T6	0.102	2820	970–1190	521–643	35	138	12.7	22.9	13.8	24.8
		T62	0.102	2820	970–1190	521–643	—	138	12.7	22.9	13.8	24.8
		T7	0.102	2820	970–1190	521–643	—	—	12.7	22.9	13.8	24.8
A02960	296.0	T4	0.101	2800	970–1170	521–632	—	130	12.2	22.0	13.3	23.9
		T6*	0.101	2800	970–1170	521–632	33	130	12.2	22.0	13.3	23.9
		T7	0.101	2800	970–1170	521–632	—	—	12.2	22.0	13.3	23.9
A03190	319.0	F	0.101	2800	960–1120	516–604	27	109	11.9	21.4	12.7	22.9
		T5	0.101	2800	960–1120	516–604	—	—	11.9	21.4	12.7	22.9
		T6	0.101	2800	960–1120	516–604	—	—	11.9	21.4	12.7	22.9
		T61	0.101	2800	960–1120	516–604	—	—	11.9	21.4	12.7	22.9
A23190	B319.0	T5	—	—	—	—	—	—	—	—	—	—
		T6	—	—	—	—	—	—	—	—	—	—

A03280	328.0	F	0.098	2720	1025–1105	552–596	30	121	11.9	21.4	12.9	23.2
A03320	332.0	T6	0.098	2720	1025–1105	552–596	—	—	11.9	21.4	12.9	23.2
A03330	333.0	T5*	0.100	2770	970–1080	521–582	26	104	11.5	20.7	12.4	22.3
		F*	0.100	2770	960–1085	516–585	26	104	11.4	20.5	12.4	22.3
		T5*	0.100	2770	960–1085	516–585	29	117	11.4	20.5	12.4	22.3
		T6*	0.100	2770	960–1085	516–585	35	117	11.4	20.5	12.4	22.3
		T7*	0.100	2770	960–1085	516–585	29	138	11.4	20.5	12.4	22.3
A03360	336.0	T551*	0.098	2720	1000–1050	538–566	—	117	11.0	19.8	12.0	21.6
		T65	0.098	2720	1000–1050	538–566	32	—	11.0	19.8	12.0	21.6
A03390	339.0	T551*	0.098	2720	—	—	43	117	—	—	—	—
A03540	354.0	T61	0.098	2720	1000–1105	538–596	36	125	11.6	20.9	12.7	22.9
A03550	355.0	T51	0.098	2720	1015–1150	546–621	36	167	12.4	22.3	13.7	24.7
		T6	0.098	2720	1015–1150	546–621	42	142	12.4	22.3	13.7	24.7
		T62*	0.098	2720	1015–1150	546–621	39	142	12.4	22.3	13.7	24.7
		T7	0.098	2720	1015–1150	546–621	36	163	12.4	22.3	13.7	24.7
		T71	0.098	2720	1015–1150	546–621	37	151	12.4	22.3	13.7	24.7
A33550	C355.0	T6	0.098	2720	1015–1150	546–621	—	142	12.4	22.3	13.7	24.7
		T61	0.098	2720	1015–1150	546–621	43	146	12.4	22.3	13.7	24.7
A03560	356.0	F	0.097	2685	1035–1135	557–613	39	—	11.9	21.4	12.9	23.2
		T51	0.097	2685	1035–1135	557–613	40	167	11.9	21.4	12.9	23.2
		T6	0.097	2685	1035–1135	557–613	—	151	11.9	21.4	12.9	23.2
		T7	0.097	2685	1035–1135	557–613	39	155	11.9	21.4	12.9	23.2
		T71	0.097	2685	1035–1135	557–613	—	—	11.9	21.4	12.9	23.2
A13560	A356.0	T6	0.097	2685	1035–1135	557–613	—	151	11.9	21.4	12.9	23.2
		T61	0.097	2685	1035–1135	557–613	39	—	11.9	21.4	12.9	23.2
		T7	0.097	2685	1035–1135	557–613	39	—	11.9	21.4	12.9	23.2
		T71	0.097	2685	1035–1135	557–613	35	151	11.9	21.4	12.9	23.2
A03570	357.0	T6	0.097	2685	1035–1135	557–613	29	151	11.6	20.9	12.7	22.9
		T61	0.097	2685	1035–1135	557–613	23	138	—	—	—	—
A13570	A357.0	T61	0.097	2685	1035–1135	557–613	—	113	—	—	—	—
A03590	359.0	T61	0.097	2630	1045–1115	563–602	23	96	—	—	—	—
A03600	360.0	F	0.095	2630	1035–1105	557–596	23	100	12.2†	22.0†	—	—
A13600	A360.0	F	0.095	2630	1035–1105	557–596	—	96	12.2†	22.0†	—	—
A03800	380.0	F	0.098	2720	1000–1100	538–593	—	96	12.1†	21.8†	—	—
A13830	A380.0	F	0.098	2720	1000–1100	538–593	—	—	—	—	—	—
A03830	383.0	F	0.098	2720	960–1080	516–582	—	—	11.7†	21.1†	—	—
A03840	384.0	F	0.098	2720	960–1080	516–582	—	—	11.7†	21.1†	—	—

TABLE 4.33 Typical Physical Properties of SAE Casting Alloys (Continued)

Alloy UNS	Alloy ANSI	Temper	Density lb/in³	Density kg/m³	Approximate melting range§ °F	Approximate melting range§ °C	Elec. cond., % IACS	Therm. cond., W/(m·K)	Coeff. of thermal expan., ×10⁻⁶ 68–212°F per °F	20–100°C per °C	68–572°F per °F	20–300°C per °C
A03900	390.0	F	—	—	—	—	—	—	—	—	—	—
A13900	A390.0	T5	0.099	2740	945–1200	507–649	25	134	10.0	18.0	—	—
		T6	0.099	2740	945–1200	507–649	—	—	10.0	18.0	—	—
		T7	0.099	2740	945–1200	507–649	—	—	10.0	18.0	—	—
A23900	B390.0	F	—	—	—	—	—	—	—	—	—	—
A04130	413.0	F	0.096	2660	1065–1080	574–582	31	121	11.9†	21.4†	—	—
A14130	A413.0	F	0.096	2660	1065–1080	574–582	37	146	11.9†	21.4†	—	—
A24430	B443.0	F	0.097	2685	1065–1170	574–632	37	142	12.3	22.1	13.4	24.1
A34430	C443.0	F	0.097	2685	1065–1170	574–632	41	159	12.9†	23.2†	—	—
A14440	A444.0	F	0.095	2635	1065–1145	574–618	35	138	12.1	21.8	13.2	23.8
A05140	514.0	F	0.096	2660	1085–1185	585–640	21	88	13.4	24.1	14.5	26.1
A05200	520.0	T4	0.093	2570	840–1120	449–604	23	96	13.7	24.7	14.8	26.6
A05350	535.0	F	0.095	2635	1020–1165	548–629	25	104	13.1	23.6	14.8	26.6
A07050	705.0	T5	0.100	2770	1105–1180	596–638	25	104	13.1	23.6	14.3	25.7
A07070	707.0	T5	0.100	2770	1085–1165	585–629	—	—	13.2	23.8	14.4	25.9
		T7	0.100	2770	1085–1165	585–629	—	—	13.2	23.8	14.4	25.9
A07100	710.0	T5	0.102	2820	1105–1195	596–646	35	138	13.4	24.1	14.6	26.3
A07120	712.0	T5	0.101	2800	1135–1200	613–649	35	138	13.7	24.7	14.8‡	26.6ᶜ
A07130	713.0	T5	0.102	2810	1100–1180	593–638	30	121	13.4‡	24.1‡	14.6‡	26.3ᶜ

* Chill cast samples; all other samples cast in green sand molds.
† For die cast alloys, data valid for temperature range of 68–392°F (20–200°C).
‡ Estimated value.
§ The approximate melting range data shown are a practical parameter of the alloy—not concise values. Normal and common composition and process variations can cause deviations from the values given.
SOURCE: Reprinted with permission, copyright 1992, Society of Automotive Engineers.

TABLE 4.34 Mechanical Property Limits of SAE Sand Casting Alloys*

UNS	ANSI	Temper	Min. tensile strength ksi	Min. tensile strength MPa	Min. yield strength (0.2% offset) ksi	Min. yield strength (0.2% offset) MPa	Elongation % Min. in 4D	Brinell hardness† (500 kg)
A02010	201.0	T6	60.0	415	50.0	345	5.0	115–145
		T7	60.0	415	50.0	345	3.0	115–145
A02060	206.0	T4	40.0	275	24.0	165	8.0	—
A02080	208.0	F	19.0	130	12.0	85	1.5	40–70
		T55	21.0	145	—	—	—	—
A02220	222.0	O	23.0	160	—	—	—	—
		T61	30.0	205	—	—	—	100–130
A02420	242.0	O	23.0	160	—	—	—	—
		T571	29.0	200	—	—	—	—
		T61	32.0	220	20.0	140	1.0	90–120
		T77	24.0	165	13.0	90	6.0	—
A02950	295.0	T4	29.0	200	13.0	90	6.0	45–75
		T6	32.0	220	20.0	140	3.0	60–90
		T62	36.0	250	28.0	195	—	80–110
		T7	29.0	200	16.0	110	3.0	55–85
A03190	319.0	F	23.0	160	13.0	90	1.5	55–85
		T5	25.0	170	—	—	—	—
		T6	31.0	215	20.0	140	1.5	65–95
A23190	B319.0	T5	26.0‡	180‡	—	—	—	—
		T6	32.0‡	220‡	21.0‡	145‡	1.0‡	70–100‡
A03280	328.0	F	25.0	170	14.0	95	1.0	45–75
		T6	34.0	235	21.0	145	1.0	65–95

TABLE 4.34 Mechanical Property Limits of SAE Sand Casting Alloys* (Continued)

| Alloy | | Temper | Min. tensile strength | | Min. yield strength (0.2% offset) | | Elongation % Min. in 4D | Brinell hardness[†] (500 kg) |
UNS	ANSI		ksi	MPa	ksi	MPa		
A03550	355.0	T51	25.0	170	18.0	125	—	50–80
		T6	32.0	220	20.0	140	2.0	65–95
		T7	35.0	240	—	—	—	—
		T71	30.0	205	22.0	150	—	60–90
A33550	C355.0	T6	36.0	250	25.0	170	2.5	—
		T61	36.0[‡]	250[‡]	30.0[‡]	205[‡]	1.0[‡]	70–100[‡]
A03560	356.0	F	19.0	130	—	—	2.0	40–70
		T51	23.0	160	16.0	110	—	45–75
		T6	30.0	205	20.0	140	3.0	55–85
		T7	31.0	215	29.0	200	—	60–90
		T71	25.0	170	18.0	125	3.0	45–75
A13560	A356.0	T6	34.0	235	24.0	165	3.5	55–85
		T7	32.0[‡]	220[‡]	30.0[‡]	205[‡]	—	—
		T71	26.0[‡]	180[‡]	19.0[‡]	130[‡]	4.0[‡]	—
A03570	357.0	T6[§]	—	—	—	—	—	—
A13570	A357.0	T61[§]	—	—	—	—	—	—
A03590	359.0	T61[§]	—	—	—	—	—	—
A13900	A390.0	F	26.0[‡]	180[‡]	26.0[‡]	180[‡]	—	85–115[‡]
		T5	26.0[‡]	180[‡]	26.0[‡]	180[‡]	—	85–115[‡]
		T6	40.0[‡]	275[‡]	40.0[‡]	275[‡]	—	125–155[‡]
		T7	36.0[‡]	250[‡]	36.0[‡]	250[‡]	—	100–130[‡]
B24430	B443.0	F	17.0	115	6.0	40	3.0	25–55
A14440	A444.0	F	18.0[‡]	125[‡]	7.0[‡]	50[‡]	8.0[‡]	35–65[‡]
A05140	514.0	F	22.0	150	9.0	60	6.0	35–65

A05200	520.0	T4	42.0	290	22.0	150	12.0	60–90
A05350	535.0	F	35.0	240	18.0	125	9.0	60–90
A07050	705.0	T5	30.0	205	17.0	115	5.0	50–80
A07070	707.0	T7	33.0	230	22.0	150	2.0	60–90
		T7	37.0	255	30.0	205	1.0	65–95
A07100	710.0	T5	32.0	220	20.0	140	2.0	60–90
A07120	712.0	T5	34.0	235	25.0	170	4.0	60–90
A07130	713.0	T5	32.0	220	22.0	150	3.0	60–90

* Values represent properties obtained from 0.500 in diameter separately cast test bars as depicted in Fig. 8 of ASTM B 557, cast in green sand molds, and tested in accordance with the procedures of ASTM B 557.

† Hardness values are given for information only; not required for acceptance.

‡ Preliminary value.

§ Mechanical properties for these alloys are dependent on casting process and heat treat procedures set for individual casting requirements. These alloys have generally been used in premium quality applications, and process techniques have not been standardized. Consult individual foundry for applicable property limits.

SOURCE: Reprinted with permission, copyright 1992, Society of Automotive Engineers.

TABLE 4.35   Typical Uses of SAE Aluminum Casting Alloys and Similar Specifications

| Alloy designations | | | Type of casting* | Similar specifications | | | Typical uses and general data |
UNS	ANSI	Former SAE		ASTM	Federal	AMS	
A02010	201.0	382	S PM	B26 —	— —	— 4229	Very high strength at room and elevated temperature; good impact strength and ductility; high-cost premium casting alloy.
A02060	206.0	—	S PM	— —	— —	— 4237	High tensile and yield strength with moderate ductility; good fracture toughness in T4 temper, structural parts for automotive and aerospace applications.
A02080	208.0	380	S PM	B26 B108	QQ-A-601 —	— —	Manifolds, valve bodies, and similar castings requiring pressure tightness.
A02220	222.0	34	S PM	B26 B108	QQ-A-601 QQ-A-596	— —	Primarily a piston alloy, but also used for aircooled cylinder heads and valve tappet guides.
A02420	242.0	39	S PM	B26 B108	QQ-A-601 QQ-A-596	4222 —	Used primarily for aircooled cylinder heads, but also for pistons in high-performance gasoline engines.
A02950	295.0	38	S	B26	QQ-A-601	4231	General structural castings requiring high strength and shock resistance.
A02960	296.0	—	PM	B108	QQ-A-596	4282	Modification of alloy 295.0 for use in permanent molds.
A03190	319.0	326	S PM	B26 B108	QQ-A-601 QQ-A-596	— —	General-purpose low-cost alloy; good foundry characteristics.
A23190	B319.0	329	S PM	— —	— —	— —	General-purpose alloy similar to 319.0, but with lower ductility and improved machinability.
A03280	328.0	327	S	B26	QQ-A-601	—	Similar to alloys 355.0 and 356.0, but lower ductility.
A03320	332.0	332	PM	B108	QQ-A-596	—	Primarily used for automotive and compressor pistons.

UNS	Alloy		Form	ASTM	Federal	AMS	Application
A03330	333.0	331	PM	B108	QQ-A-596	—	General-purpose low-cost permanent mold alloy used for engine parts, motor housings, flywheel housings, and regulator parts.
A03360	336.0	321	PM	B108	QQ-A-596	—	Piston alloy having low expansion.
A03390	339.0	334	PM	—	—	—	Piston alloy.
A03540	354.0	—	PM	B108 B686	—	—	High-strength premium quality casting alloy.
A03550	355.0	322	S PM	B26 B108	QQ-A-601 QQ-A-596	4210 4212 4214 4280 4281	General use where high strength, medium ductility, and pressure tightness are required, such as pump bodies and liquid-cooled cylinder heads.
A33550	C355.0	335	S PM	B26 B108 B686	QQ-A-601 QQ-A-596	4215	Similar to alloy 355.0, but has greater ductility.
A03560	356.0	323	S PM	B26 B108	QQ-A-601 QQ-A-596	4217 4284 4286	For intricate castings requiring good strength and ductility.
A13560	A356.0	336	S PM	B26 B108 B686	QQ-A-601 QQ-A-596	4218	Similar to alloy 356.0, but has greater ductility.
A03570	357.0	—	S PM	B108	QQ-A-596	—	Similar to alloy A357.0, but has greater ductility.
A13570	A357.0	—	S PM	B108 B686	—	4219	High-strength structural alloy with good ductility.
A03590	359.0	—	S PM	B108	—	—	High-strength structural alloy with good ductility.
A03600	360.0	—	D	B85	—	—	Very good casting characteristics; good corrosion resistance; used in place of alloy 413 where higher mechanical properties are required.

TABLE 4.35 Typical Uses of SAE Aluminum Casting Alloys and Similar Specifications (Continued)

Alloy designations			Type of casting*	Similar specifications				Typical uses and general data
UNS	ANSI	Former SAE		ASTM	Federal	AMS		
A13600	A360.0	309	D	B85	QQ-A-591	4290		Excellent casting characteristics; suited for use in thin-walled or intricate castings produced in cold-chamber casting machine; high corrosion resistance; slightly higher mechanical properties than alloy 360.0.
A03800	380.0	308	D	B85	QQ-A-591	—		Similar to alloy A380.0, but suitable for use in either cold-chamber or gooseneck machines.
A13800	A380.0	306	D	B85	QQ-A-591	4291		Good casting characteristics and fair resistance to corrosion; not especially suited for thin sections; limited to cold-chamber machines.
A03830	383.0	383	D	B85	QQ-A-591	—		Similar to alloy 380.0, but with improved castability.
A03840	384.0	303	D	B85	QQ-A-591	—		General-purpose alloy with high fluidity; used for thin-walled castings or castings with large areas.
A03900	390.0	—	D	—	—	—		High wear resistance; used for cylinder blocks, transmission pump and air compressor housings, small engine crankcases, and air conditioner pistons.
A13900	A390.0	—	S PM	—	—	—		Similar to 390.0, but formulated for sand and permanent mold casting.
A23900	B390.0	—	D	—	—	—		Similar to alloy 390.0.
A04130	413.0	—	D	B85	—	—		Good for large thin-wall die castings, difficult to machine and finish.
A14130	A413.0	305	D	B85	QQ-A-591	—		High corrosion resistance; excellent castability; used for complicated castings with thin sections, also difficult to machine and finish.
A24430	B443.0	35	S PM	B26 B108	QQ-A-601 QQ-A-596	— —		Used for intricate castings having thin sections; good corrosion resistance; fair strength and good ductility.

UNS	Alloy	No.	Form	ASTM	QQ	SAE	Characteristics
A34430	C443.0	304	D	B85	QQ-A-591	—	Good casting characteristics and resistance to corrosion.
A14440	A444.0	—	S	—	—	—	Good castability; excellent ductility for impact absorption; used for bridge railing posts and turbocharger compressor housings.
A05140	514.0	320	S	B26	QQ-A-601	—	Moderate strength; very high corrosion resistance.
A05200	520.0	324	S	B26	QQ-A-601	4240	High-strength structural alloy; requires special foundry and heat treat practice; susceptible to stress corrosion failure.
A05350	535.0	—	S	B26	QQ-A-601	—	Excellent shock and corrosion resistance, dimensional stability, and machinability; used in computer components, frame sections, optical equipment, and applications where stress rupture is a factor.
A07050	705.0	311	S / PM	B26 / B108	QQ-A-601 / QQ-A-596	— / —	High-strength general-purpose alloy; excellent machinability and dimensional stability; high corrosion resistance; can be anodized.
A07070	707.0	312	S / PM	B26 / B108	QQ-A-601 / QQ-A-596	— / —	Similar to alloy 705.0, but higher strength and lower ductility.
A07100	710.0	313	S	B26	QQ-A-601	—	High-strength general-purpose alloy similar to alloys 705.0 and 707.0; easily polished.
A07120	712.0	310	S	B26	QQ-A-601	—	General-purpose structural castings developing strengths equivalent to alloy 295.0 without requiring heat treatment, but casting characteristics slightly poorer than alloy 295.0.
A07130	713.0	315	S / PM	B26 / B108	QQ-A-601 / QQ-A-596	— / —	Similar to alloy 710.0.

* S—sand cast; PM—permanent mold; D—die cast.

SOURCE: Reprinted with permission, copyright 1992, Society of Automotive Engineers.

TABLE 4.36 SAE Aluminum Alloy Characteristics

Alloy designations				Pattern shrinkage allowance[b]		Foundry characteristics[a]			
UNS	ANSI	Former SAE	Type of casting	in/ft	%	Resistance to hot cracking[c]	Pressure tightness	Fluidity[d]	Solidification shrinkage tendency[e]
A02010	201.0	382	S	5/32	1.30	4	3	3	4
			PM	b	b	4	3	3	4
A02060	206.0	—	S	5/32	1.30	4	3	3	4
			PM	b	b	4	3	3	4
A02080	208.0	380	S	5/32	1.30	4	3	3	3
			PM	b	b	4	3	3	3
A02220	222.0	34	S	5/32	1.30	4	3	3	3
			PM	b	b	3	4	3	4
A02420	242.0	39	S	5/32	1.30	4	3	3	4
			PM	b	b	4	4	3	4
A02950	295.0	38	S	5/32	1.30	4	4	3	3
			PM	b	b	4	3	3	3
A02960	296.0	—	S	5/32	1.30	4	3	2	2
			PM	b	b	2	2	2	3
A03190	319.0	326	S	5/32	1.30	2	2	2	2
			PM	b	b	2	2	2	2
A23190	B319.0	329	S	5/32	1.30	1	1	1	1
			PM	b	b	2	2	1	2
A03280	328.0	327	S	5/32	1.30	1	2	1	3
			PM	b	b				3
A03320	332.0	332	PM	b	b	2	2	1	2
A03330	333.0	331	PM	b	b	1	2	1	3
A03360	336.0	321	PM	b	b	2	1	1	1
A03390	339.0	334	PM	b	b		1	2	2
A03540	354.0	—	PM	b	b		1		1
A03550	355.0	322	S	5/32	1.30	1	1	1	2
			PM	b	b	1	1	2	1
A33550	C355.0	335	S	5/32	1.30	1	1	1	1
			PM	b	b	1	1	2	2
A03560	356.0	323	S	5/32	1.30	1	1	1	1
			PM	b	b	1	1	2	1

Catalog No.	Alloy	No.	Process						
A13560	A356.0	336	S, PM	5/32 b	1.30 b	1	1	1	1
A03570	357.0	—	S, PM	b	b	1	2	1	1
A13570	A357.0	—	S, PM	5/32 b	1.30 b	1	1	1	1
A03590	359.0	—	S, PM	b	b	1	2	1	1
A03600	360.0	—	S, PM	5/32 b	1.30 b	1	1	1	1
A13600	A360.0	309	D	b	b	2	2	2	2
A03800	380.0	308	D	b	b	2	1	2	2
A13800	A380.0	306	D	b	b	—	2	1	1
A03830	383.0	383	D	b	b	—	1	1	1
A03840	384.0	—	D	b	b	—	1	1	1
A03900	390.0	—	D	b	b	—	1	1	1
A13900	A390.0	—	S, PM	5/32 b	1.30 b	—	1	1	1
A23900	B390.0	—	D	b	b	3	1	3	3
A04130	413.0	—	D	b	b	3	1	3	3
A14130	A413.0	305	D	b	b	—	1	3	3
A24430	B443.0	35	S, PM	5/32 b	1.30 b	3	1	3	3
A34430	C443.0	304	D	b	b	—	1	3	1
A14440	A444.0	—	S	b	b	—	1	2	1
A05140	514.0	320	S	5/32 b	1.30 1.30	1	1	2	1
A05200	520.0	324	S	5/32 1/10	0.83 0.83	2	1	1	2
A05350	535.0	—	S	1/10 3/16	1.56 b	4	1	1	4
A07050	705.0	311	S	3/16 b	1.56 b	4	3	3	4
A07050	705.0	311	S	3/16 b	1.56 1.56	4	5	4	4
A07070	707.0	312	PM	3/16 3/16	1.56 b	3	5	5	3
A07100	710.0	313	S			4	4	5	5
A07120	712.0	310	S			5	3	5	5
A07130	713.0	315	PM			4	4	3	5

TABLE 4.36 SAE Aluminum Alloy Characteristics (Continued)

Alloy							Other characteristics				
UNS	ANSI	Normally heat treated	Resistance to corrosion[f]	Machining[g]	Polishing[h]	Electroplating[i]	Anodized appearance[j]	Chemical oxide coating[k] (protection)	Strength at elevated temperature[l]	Suitability for welding[m]	Suitability for brazing[n]
A02010	201.0	Yes	4	1	1	1	2	2	1	4	No
A02060	206.0	Yes	4	1	1	1	2	2	1	4	No
A02080	208.0	Yes	4	3	2	1	3	2	2	4	No
A02220	222.0	Yes	4	1	2	1	3	4	1	4	No
A02420	242.0	Yes	4	2	2	1	3	4	3	4	No
A02950	295.0	Yes	3	2	2	1	2	3	3	3	No
A02960	296.0	Yes	4	3	4	2	4	3	3	2	No
A03190	319.0	Yes	3	3	4	2	4	3	3	2	No
A23190	B319.0	Yes	3	3	4	2	4	3	3	2	No
A03280	328.0	Yes	3	4	5	2	5	2	2	2	No
A03320	332.0	Aged only	3	3	4	3	4	3	3	2	No
A03330	333.0	Yes	3	2	3	2	5	3	2	3	No
A03360	336.0	Yes	3	4	5	4	5	2	2	2	No
A03390	339.0	Aged only	3	3	4	3	5	3	3	2	No
A03540	354.0	Yes	3	4	4	2	4	3	2	3	No
A03550	355.0	Yes	3	3	3	1	4	2	2	3	No
A33550	C355.0	Yes	3	3	3	2	4	2	2	2	No
A03560	356.0	Yes	2	4	3	2	4	2	3	2	No
A13560	A356.0	Yes	2	3	3	1	4	2	3	1	No
A03570	357.0	Yes	2	3	3	1	4	2	3	1	No
A13570	A357.0	Yes	2	3	3	2	4	2	2	1	No
A03590	359.0	Yes	2	4	4	2	4	2	2	1	No
A03600	360.0	No	2	4	4	1	4	3	2	3	No
A13600	A360.0	No	3	3	3	1	4	3	2	3	No
A03800	380.0	No	4	3	3	1	4	5	2	4	No
A13800	A380.0	No	4	3	3	1	4	5	2	4	No
A03830	383.0	No	4	3	3	1	4	5	2	4	No
A03840	384.0	No	4	3	3	1	4	5	3	4	No

UNS No.	Alloy	(aged)	c	d	e	f	g	h	i	j	k	l	m
A03900	390.0	No	3	4	3	—	5	—	1	4	No
A13900	A390.0	Yes	3	4	3	—	5	—	1	4	No
A23900	B390.0	No	3	4	3	—	5	—	1	4	No
A04130	413.0	No	2	4	5	3	5	3	3	3	No
A14130	A413.0	No	3	5	5	3	5	3	3	3	No
A24430	443.0	No	3	5	5	2	5	2	4	1	Ltd.
A34430	B443.0	No	2	5	5	3	4	3	5	1	No
—	C443.0	No	2	4	4	2	4	2	3	1	No
A14440	A444.0	No	1	1	1	—	4	—	2	1	No
A05140	514.0	Yes	1	1	1	5	1	5	1	1	No
A05200	520.0	Opt	1	1	1	4	1	4	1	1	No
A05350	535.0	Aged only	2	1	1	—	1	—	1	2[p]	Yes
A07050	705.0	Aged only	2	1	1	3	2	3	2	3	Yes
A07070	707.0	Yes	2	1	1	3	2	3	2	5	Yes
A07100	710.0	Aged only	2	1	1	2	2	2	3	5	Yes
A07120	712.0	Aged only	2	1	1	2	2	2	3	5	Yes
A07130	713.0	Aged only	2	1	1	2	2	2	3	5	Yes

NOTE: Type of casting: S—sand cast; PM—permanent mold; D—die cast.

[a] 1 indicates best of group; 5 indicates poorest of group.

[b] Not applicable to permanent mold and die castings. Allowances are for average sand castings. Shrinkage requirements will vary with intricacy of design and dimensions.

[c] Ability of alloy to withstand contraction stresses while cooling through hot-short or brittle temperature range.

[d] Ability of liquid alloy to flow readily in mold and fill thin sections.

[e] Decrease in volume accompanying freezing of alloy and measure of amount of compensating feed metal required in form of risers.

[f] Based on alloy resistance in 5% salt spray test (ASTM B117).

[g] Composite rating based on ease of cutting, chip characteristics, quality of finishing, and tool life. Ratings, in the case of heat treatable alloys, based on T6 temper. Other tempers, particularly the annealed temper, may have lower rating.

[h] Composite rating based on ease and speed of polishing and quality of finish provided by typical polishing procedure.

[i] Ability of casting to take and hold on electroplate applied by present standard methods.

[j] Rated on lightness of color, brightness, and uniformity of clear anodized coating applied in sulfuric acid electrolyte.

[k] Rated on combined resistance of coating and base alloy to corrosion.

[l] Rating based on tensile and yield strengths of temperature up to 500°F (260°C), after prolonged heating at testing temperatures.

[m] Based on ability of material to be fusion welded with filler rod of same alloy.

[o] Refers to suitability of alloy to withstand brazing temperatures without excessive distortion or melting.

[p] Not recommended for service at temperatures exceeding 20°F (93°C).

SOURCE: Reprinted with permission, copyright 1992, Society of Automotive Engineers.

183

TABLE 4.37   Mechanical Properties of Aluminum: Typical Properties

Alloy and temper	Tension Strength, psi Ultimate	Tension Strength, psi Yield	% Elong. in 2 in $\frac{1}{16}$ in (thick)	% Elong. in 2 in $\frac{1}{2}$ in (thick)	Brinell hardness[a]	Ultimate shear strength, psi	Endurance limit, psi	Mod. of elast., psi
EC-O	12,000	4,000	—	—	—	8,000	—	10.0×10^6
EC-H14	16,000	14,000	—	—	—	10,000	—	10.0×10^6
EC-H19	27,000	24,000	—	—	—	15,000	7,000	10.0×10^6
1060-O	10,000	4,000	43	—	19	7,000	3,000	10.0×10^6
1060-H14	14,000	13,000	12	—	26	9,000	5,000	10.0×10^6
1100-O	13,000	5,000	35	45	23	9,000	5,000	10.0×10^6
1100-H12	16,000	15,000	12	25	28	10,000	6,000	10.0×10^6
1100-H14	18,000	17,000	9	20	32	11,000	7,000	10.0×10^6
1100-H16	21,000	20,000	6	17	38	12,000	9,000	10.0×10^6
1100-H18	24,000	22,000	5	15	44	13,000	9,000	10.0×10^6
2011-T3[b]	55,000	43,000	—	15	95	32,000	18,000	10.2×10^6
2011-T8	59,000	45,000	—	12	100	35,000	18,000	10.2×10^6
2014-O	27,000	14,000	—	18	45	18,000	13,000	10.6×10^6
2014-T4, T451	62,000	42,000	—	20	105	38,000	20,000	10.6×10^6
2014-T6, T651[c]	70,000	60,000	—	13	135	42,000	18,000	10.6×10^6
2017-T4, T451	62,000	40,000	—	22	105	38,000	18,000	10.5×10^6
2024-O	27,000	11,000	20	22	47	18,000	13,000	10.6×10^6
2024-T3	70,000	50,000	18	—	120	41,000	20,000	10.6×10^6
2024-T4, T351[c]	68,000	47,000	20	19	120	41,000	20,000	10.6×10^6
2024-T36	72,000	57,000	13	—	130	42,000	18,000	10.6×10^6
2024-T81, T851	70,000	65,000	6	—	128	43,000	18,000	10.6×10^6
2024-O[d]	26,000	11,000	20	—	—	18,000	—	10.6×10^6
2024-T3[d,e]	65,000	45,000	18	—	—	40,000	—	10.6×10^6
2024-T4, T351[d,e]	64,000	42,000	19	—	—	40,000	—	10.6×10^6
2024-T36[d,e]	67,000	53,000	11	—	—	41,000	—	10.6×10^6
2024-T81, T851[d,e]	65,000	60,000	6	—	—	40,000	—	10.6×10^6
2024-T86[d,e]	70,000	66,000	6	—	—	42,000	—	10.6×10^6
2219-O[f]	25,000	11,000	18	—	—	—	—	10.6×10^6
2219-T31, T351[f]	52,000	36,000	17	—	—	—	—	10.6×10^6
2219-T81, T851	66,000	51,000	10	—	—	—	15,000	10.6×10^6
3003-O	16,000	6,000	30	40	28	11,000	7,000	10.0×10^6
3003-H12	19,000	18,000	10	20	35	12,000	8,000	10.0×10^6
3003-H14	22,000	21,000	8	16	40	14,000	9,000	10.0×10^6
3003-H16	26,000	25,000	5	14	47	15,000	10,000	10.0×10^6
3003-H18	29,000	27,000	4	10	55	16,000	10,000	10.0×10^6
3004-O	26,000	10,000	20	25	45	16,000	14,000	10.0×10^6
3004-H32	31,000	25,000	10	17	52	17,000	15,000	10.0×10^6
3004-H34	35,000	29,000	9	12	63	18,000	15,000	10.0×10^6
3004-H36	38,000	33,000	5	9	70	20,000	16,000	10.0×10^6
3004-H38	41,000	36,000	5	6	77	21,000	16,000	10.0×10^6
3004-O[d]	26,000	10,000	20	25	—	16,000	—	10.0×10^6
3004-H32[d]	31,000	25,000	10	17	—	17,000	—	10.0×10^6
3004-H34[d]	35,000	29,000	9	12	—	18,000	—	10.0×10^6
3004-H36[d]	38,000	33,000	5	9	—	20,000	—	10.0×10^6
3004-H38[d]	41,000	36,000	5	6	—	21,000	—	10.0×10^6

TABLE 4.37   Mechanical Properties of Aluminum: Typical Properties (*Continued*)

4032-T6	55,000	46,000	—	9	120	38,000	16,000	11.4×10^6
5005-O	18,000	6,000	25	—	28	11,000	—	10.0×10^6
5005-H14	23,000	22,000	6	—	—	19,000	—	10.0×10^6
5005-H32	20,000	17,000	11	—	36	14,000	—	10.0×10^6
5005-H34	23,000	20,000	8	—	41	14,000	—	10.0×10^6
5005-H36	26,000	24,000	6	—	46	15,000	—	10.0×10^6
5005-H38	29,000	27,000	5	—	51	16,000	—	10.0×10^6
5050-O	21,000	8,000	24	—	36	15,000	12,000	10.0×10^6
5050-H32	25,000	21,000	9	—	46	17,000	13,000	10.0×10^6
5050-H34	28,000	24,000	8	—	53	18,000	13,000	10.0×10^6
5050-H36	30,000	26,000	7	—	58	19,000	14,000	10.0×10^6
5050-H38	32,000	29,000	6	—	63	20,000	14,000	10.0×10^6
5052-O	28,000	13,000	25	30	47	18,000	16,000	10.2×10^6
5052-H32	33,000	28,000	12	18	60	20,000	17,000	10.2×10^6
5052-H34	38,000	31,000	10	14	68	21,000	18,000	10.2×10^6
5052-H36	40,000	35,000	8	10	73	23,000	19,000	10.2×10^6
5052-H38	42,000	37,000	7	8	77	24,000	20,000	10.2×10^6
5083-O	42,000	21,000	22	25	67	25,000	22,000	10.3×10^6
5083-H112	44,000	28,000	16	—	80	26,000	22,000	10.3×10^6
5083-H113	46,000	33,000	16	—	82	—	23,000	10.3×10^6
5083-H321	46,000	33,000	—	16	—	—	23,000	10.3×10^6
5083-H323	47,000	36,000	10	—	84	27,000	—	10.3×10^6
5086-O	38,000	17,000	22	30	60	23,000	21,000	10.3×10^6
5086-H32	42,000	30,000	12	16	72	25,000	22,000	10.3×10^6
5086-H34	47,000	37,000	10	14	82	27,000	23,000[h]	10.3×10^6
5086-H112	39,000	19,000	14	—	64	23,000	21,000[h]	10.3×10^6
5154-O	35,000	17,000	27	30	58	22,000	17,000	10.2×10^6
5154-H32	39,000	30,000	15	18	67	22,000	18,000	10.2×10^6
5154-H34	42,000	33,000	13	16	73	24,000	19,000	10.2×10^6
5154-H36	45,000	36,000	12	14	78	26,000	20,000	10.2×10^6
5154-H38	48,000	39,000	10	—	80	28,000	21,000	10.2×10^6
5154-H112	35,000	17,000	25	—	63	—	17,000	10.2×10^6
5454-O	36,000	17,000	22	25	62	23,000	—	10.2×10^6
5454-H32	40,000	30,000	10	18	73	24,000	—	10.2×10^6
5454-H34	44,000	35,000	10	16	81	26,000	—	10.2×10^6
5454-H112	36,000	18,000	18	—	62	23,000	—	10.2×10^6
5454-H311	38,000	26,000	14	—	70	23,000	—	10.2×10^6
5456-O	45,000	23,000	24	20	70	27,000	22,000	10.3×10^6
5456-H112	45,000	24,000	22	—	70	27,000	—	10.3×10^6
5456-H311	47,000	33,000	18	—	75	27,000	24,000	10.3×10^6
5456-H321	51,000	37,000	16	—	90	30,000	23,000	10.3×10^6
5456-H343	56,000	43,000	8	—	94	33,000	—	10.3×10^6
5457-O	19,000	7,000	22	—	32	12,000	—	10.0×10^6
5457-H32	22,000	19,000	9	—	40	14,000	—	10.0×10^6
5457-H34	25,000	22,000	8	—	45	15,000	—	10.0×10^6
5457-H36	28,000	26,000	7	—	51	17,000	—	10.0×10^6
5457-H38	30,000	27,000	6	—	55	18,000	—	10.0×10^6
6061-O	18,000	8,000	25	30	30	12,000	9,000	10.0×10^6
6061-T4, T451	35,000	21,000	22	25	65	24,000	14,000	10.0×10^6
6061-T6, T651	45,000	40,000	12	17	95	30,000	14,000	10.0×10^6

TABLE 4.37   Mechanical Properties of Aluminum: Typical Properties (*Continued*)

Alloy and temper	Tension Strength, psi Ultimate	Yield	% Elong. in 2 in $\frac{1}{16}$ in	$\frac{1}{2}$ in (thick)	Brinell hardness[a]	Ultimate shear strength, psi	Endurance limit, psi	Mod. of elast., psi
6061-O[d]	17,000	7,000	25	—	—	11,000	—	10.0×10^6
6061-T4, T451[d]	33,000	19,000	22	—	—	22,000	—	10.0×10^6
6061-T6, T651[d]	42,000	37,000	12	—	—	27,000	—	10.0×10^6
6063-O	13,000	7,000	—	—	25	10,000	8,000	10.0×10^6
6063-T4	25,000	13,000	22	—	—	16,000	—	10.0×10^6
6063-T5	27,000	21,000	12	22	60	17,000	10,000	10.0×10^6
6063-T6	35,000	31,000	12	18	73	22,000	10,000	10.0×10^6
6063-T832	42,000	39,000	12	—	95	27,000	—	10.0×10^6
6262-T9	58,000	55,000	—	10	120	35,000	13,000	10.0×10^6
7075-O	33,000	15,000	17	16	60	22,000	17,000	10.4×10^6
7075-T6, T651[g]	83,000	73,000	11	11	150	48,000	23,000	10.4×10^6
7075-O[d]	32,000	14,000	17	—	—	22,000	—	10.4×10^6
7075-T6, T651[d]	76,000	67,000	11	—	—	46,000	—	10.4×10^6
7079-T6, T651	78,000	68,000	—	14	145	45,000	23,000	10.4×10^6
7178-O	33,000	15,000	15	16	60	22,000	—	10.4×10^6
7178-T6, T651[c]	88,000	78,000	10	11	160	52,000	22,000	10.4×10^6

[a] 500-kg load; 10-mm ball.

[b] Sizes larger than $1\frac{1}{2}$ in will have strengths slightly less than shown.

[c] Extruded shapes over $\frac{3}{4}$ in thick have strengths 15–20% higher than shown.

[d] These alloys are Alclad.

[e] Sheets thicker than 0.062 in will have strengths slightly higher than shown.

[f] Properties for sheets and plates only.

[g] Extruded products will have strengths approx. 10% higher.

[h] Applies to sheets and plates only.

SOURCE: Ryerson, Inc.

TABLE 4.38 Typical Characteristics and Applications of Wrought Aluminum

Alloy and temper	Resistance to corrosion General[a]	Resistance to corrosion Stress-corrosion cracking[b]	Workability (cold)[e]	Machinability[e]	Brazability[f]	Weldability[f] Gas	Weldability[f] Arc	Weldability[f] Resistance spot and seam	Some applications of alloy
1060–0	A	A	A	E	A	A	A	B	Chemical equipment, railroad tank cars
H12	A	A	A	E	A	A	A	A	
H14	A	A	A	D	A	A	A	A	
H16	A	A	B	D	A	A	A	A	
H18	A	A	B	D	A	A	A	A	
1100–0	A	A	A	E	A	A	A	B	Sheet metal work, spun holloware, fin stock
H12	A	A	A	E	A	A	A	A	
H14	A	A	A	D	A	A	A	A	
H16	A	A	B	D	A	A	A	A	
H18	A	A	C	D	A	A	A	A	
1350–0	A	A	A	E	A	A	A	B	Electrical conductors
H12, H111	A	A	A	E	A	A	A	A	
H14, H24	A	A	A	D	A	A	A	A	
H16, H26	A	A	B	D	A	A	A	A	
H18	A	A	B	D	A	A	A	A	
2011–T3	D[c]	D	C	A	D	D	D	D	Screw-machine products
T4, T451	D[c]	D	B	A	D	D	D	D	
T8	D	B	D	A	D	D	D	D	
2014–0	—	—	—	D	D	D	D	B	Truck frames, aircraft structures
T3, T4, T451	D[c]	C	C	B	D	D	B	B	
T6, T651, T6510, T6511	D	C	D	B	D	D	B	B	
2017–T4, T451	D[c]	C	C	B	D	D	B	B	Screw-machine products, fittings

TABLE 4.38 Typical Characteristics and Applications of Wrought Aluminum (*Continued*)

Alloy and temper	Resistance to corrosion		Workability (cold)[e]	Machinability[e]	Brazability[f]	Weldability[f]			Some applications of alloy
	General[a]	Stress-corrosion cracking[b]				Gas	Arc	Resistance spot and seam	
2018–T61	—	—	—	B	—	—	—	—	Aircraft engine cylinders, heads, and pistons
2024–0	—	—	—	D	D	D	D	D	Truck wheels, screw-machine products, aircraft structures
T4, T3, T351, T3510, T3511	D[c]	C	C	B	D	C	B	B	
T361	D[c]	C	D	B	D	D	C	B	
T6	D	B	C	B	D	D	C	B	
T861, T81, T851, T8510, T8511	D	B	D	B	D	D	C	B	
T72	—	—	—	B	—	—	—	—	
2025–T6	D	C	—	B	D	D	B	B	Forgings, aircraft propellers
2117–T4	C	A	B	C	D	D	B	B	
2218–T61	D	C	—	—	—	—	—	C	Jet engine impellers and rings
T72	D	C	—	B	D	D	C	B	
2618–T61	D	C	—	B	D	D	C	B	Aircraft engines
3003–0	A	A	A	E	A	A	A	A	Cooking utensils, chemical equipment, pressure vessels, sheet metal work, builder's hardware, storage tanks
H12	A	A	A	E	A	A	A	A	
H14	A	A	B	D	A	A	A	A	
H16	A	A	C	D	A	A	A	A	
H18	A	A	C	D	A	A	A	A	
H25	A	A	B	D	A	A	A	A	

Alloy/Temper									Applications
3004-0	B	A	B	B	D	A	A	A	Sheet metal work, storage tanks
H32	A	A	B	B	D	B	A	A	
H34	A	A	B	B	C	B	A	A	
H36	A	A	B	B	C	C	A	A	
H38	A	A	B	B	C	C	A	A	
3105-0	B	A	B	B	E	A	A	A	Residential siding, mobile homes, rain-carrying goods, sheet metal work
H12	A	A	B	B	E	B	A	A	
H14	A	A	B	B	D	B	A	A	
H16	A	A	B	B	D	C	A	A	
H18	A	A	B	B	D	C	A	A	
H25	A	A	B	B	D	B	A	A	
4032-T6	C	B	D	D	B	—	B	C	Pistons
5005-0	B	A	A	B	E	A	A	A	Appliances, utensils, architectural, electrical conductor
H12	A	A	A	B	E	A	A	A	
H14	A	A	A	B	D	B	A	A	
H16	A	A	A	B	D	C	A	A	
H18	A	A	A	B	D	C	A	A	
H32	A	A	A	B	E	A	A	A	
H34	A	A	A	B	D	B	A	A	
H36	A	A	A	B	D	C	A	A	
H38	A	A	A	B	D	C	A	A	
5050-0	B	A	A	B	E	A	A	A	Builder's hardware, refrigerator trim, coiled tubes
H32	A	A	A	B	D	A	A	A	
H34	A	A	A	B	D	B	A	A	
H36	A	A	A	B	C	C	A	A	
H38	A	A	A	B	C	C	A	A	
5052-0	B	A	A	C	D	A	A	A	Sheet metal work, hydraulic tube, appliances
H32	A	A	A	C	D	B	A	A	
H34	A	A	A	C	C	B	A	A	
H36	A	A	A	C	C	C	A	A	
H38	A	A	A	C	C	C	A	A	

TABLE 4.38 Typical Characteristics and Applications of Wrought Aluminum (Continued)

Alloy and temper	Resistance to corrosion General[a]	Resistance to corrosion Stress-corrosion cracking[b]	Work-ability (cold)[e]	Machin-ability[e]	Brazability[f]	Weldability[f] Gas	Weldability[f] Arc	Weldability[f] Resistance spot and seam	Some applications of alloy
5056–0	A[d]	B[d]	A	D	D	C	A	B	Cable sheathing, rivets for magnesium, screen wire, zippers
H111	A[d]	B[d]	A	D	D	C	A	A	
H12, H32	A[d]	B[d]	B	D	D	C	A	A	
H14, H34	A[d]	B[d]	B	C	D	C	A	A	
H18, H38	A[d]	C[d]	C	C	D	C	A	A	
H192	B[d]	D[d]	D	B	D	C	A	A	
H392	B[d]	D[d]	D	B	D	C	A	A	
5083–0	A[d]	B[d]	B	D	D	C	A	B	Unfired, welded pressure vessels, marine, auto aircraft cryogenics, TV towers, drilling rigs, transportation equipment, missile components
H321	A[d]	B[d]	C	D	D	C	A	A	
H111	A[d]	B[d]	C	D	D	C	A	A	
5086–0	A[d]	A[d]	A	D	D	C	A	B	
H32, H116, H117	A[d]	A[d]	B	D	D	C	A	A	
H34	A[d]	B[d]	B	C	D	C	A	A	
H36	A[d]	B[d]	C	C	D	C	A	A	
H38	A[d]	B[d]	C	C	D	C	A	A	
H111	A[d]	A[d]	B	D	D	C	A	A	
5154–0	A[d]	A[d]	A	D	D	C	A	B	Welded structures, storage tanks, pressure vessels, saltwater service
H32	A[d]	A[d]	B	D	D	C	A	A	
H34	A[d]	A[d]	B	C	D	C	A	A	
H36	A[d]	A[d]	C	C	D	C	A	A	
H38	A[d]	A[d]	C	C	D	C	A	A	
5252-H24	A	A	B	D	C	A	A	A	Automotive and appliance trim
H25	A	A	B	C	C	A	A	A	
H28	A	A	C	C	C	A	A	A	

Alloy and temper	1	2	3	4	5	6	7	8	Applications
5454–0	B	A	C	D	D	A	A	A	Welded structures, pressure vessels, marine service
H32	A	A	C	D	D	B	A	A	
H34	A	A	C	C	D	B	A	A	
H111	A	A	C	D	D	B	A	A	
5456–0	A[d]	B[d]	B	D	D	C	A	A	High-strength welded structures, storage tanks, pressure vessels, marine applications
H111	A[d]	B[d]	C	D	D	C	A	A	
H321[g], H1116	A[d]	B[d]	C	D	D	C	A	A	
5457–0	A	A	A	B	E	A	A	B	Anodized automotive and appliance trim
5657–H241	A	A	A	B	D	A	A	A	
H25	A	A	A	B	D	B	A	A	
H26	A	A	A	B	D	B	A	A	
H28	A	A	A	B	D	C	A	A	
6005–T5	A	A	C	C	A	A	A	B	Heavy-duty structures requiring good corrosion resistance—truck and marine, railroad cars, furniture, pipelines
6053–0	A	A	A	E	—	—	—	A	Wire and rod for rivets
T6, T61	A	A	A	C	C	—	A	A	
6061–0	B	A	A	A	D	A	A	B	Heavy-duty structures requiring good corrosion resistance—truck and marine, railroad cars, furniture, pipelines
T4, T451, T4510, T4511	A	A	A	A	C	B	B	B	
T6, T651, T652, T6510, T6511	A	A	A	A	C	C	A	B	
6063–T1	A	A	A	A	D	B	A	A	Pipe railing, furniture architectural extrusions
T4	A	A	A	A	D	B	A	A	
T5, T52	A	A	A	A	C	B	A	A	
T6	A	A	A	A	C	C	A	A	
T83, T831, T832	A	A	A	A	C	C	A	A	

TABLE 4.38 Typical Characteristics and Applications of Wrought Aluminum (*Continued*)

Alloy and temper	Resistance to corrosion — General[a]	Resistance to corrosion — Stress-corrosion cracking[b]	Work-ability (cold)[e]	Machin-ability[e]	Brazability[f]	Weldability[f] — Gas	Weldability[f] — Arc	Weldability[f] — Resistance spot and seam	Some applications of alloy
6151–T6, T652	—	—	—	—	—	—	—	—	Moderate strength intricate forgings for machine and automotive parts
6262–T6, T651, T6510, T6511	B	A	C	B	A	A	A	A	Screw-machine products
T9	B	A	D	B	A	A	A	A	
6463–T1	A	A	B	D	A	A	A	A	Extruded architectural and trim sections
T5	A	A	B	C	A	A	A	A	
T6	A	A	C	C	A	A	A	A	
7075–0	—	—	—	D	D	D	C	B	Aircraft and other structures
T6, T651, T652, T6510, T6511	C[c]	C	D	B	D	D	C	B	
T73, T7351	C	B	D	B	D	D	C	B	

[a] Ratings A through E are relative ratings in decreasing order of merit, based on exposures to sodium chloride solution by intermittent spraying or immersion. Alloys with A and B ratings can be used in industrial and seacoast atmospheres without protection. Alloys with C, D, and E ratings generally should be protected at least on faying surfaces.

[b] Stress-corrosion cracking ratings are based on service experience and on laboratory tests of specimens exposed to the 3.5% sodium chloride alternate immersion test.

A = No known instance of failure in service or in laboratory tests.

B = No known instance of failure in service; limited failures in laboratory tests of short transverse specimens.

C = Service failures with sustained tension stress acting in short transverse direction relative to grain structure; limited failures in laboratory tests of long transverse specimens.

D = Limited service failures with sustained longitudinal or long transverse stress.

^{c} In relatively thick sections the rating would be E.

^{d} This rating may be different for material held at elevated temperature for long periods.

^{e} Ratings A through E for workability (cold) and A through E for machinability are relative ratings in decreasing order of merit.

^{f} Ratings A through D for weldability and brazability are relative ratings defined as follows:

    A = Generally weldable by all commercial procedures and methods.

    B = Weldable with special techniques or for specific applications which justify preliminary trials or testing to develop welding procedure and weld performance.

    C = Limited weldability because of crack sensitivity or loss in resistance to corrosion and mechanical properties.

    D = No commonly used welding methods have been developed.

^{g} Material in this temper is not recommended for, and should not be used in, applications requiring exposure to sea water.

SOURCE: Reprinted with permission, copyright 1992 Society of Automotive Engineers.

193

1. Scope—This standard provides systems for designating wrought aluminum and wrought aluminum alloys, aluminum and aluminum alloys in the form of castings and foundry ingot, and the tempers in which aluminum and aluminum alloy wrought products and aluminum alloy castings are produced.

2. Wrought Aluminum and Aluminum Alloy Designation System (see Note 5.1)—A system of four-digit numerical designations is used to identify wrought aluminum and wrought aluminum alloys. The first digit indicates the alloy group as shown in Table 1. The last two digits identify the aluminum alloy or indicate the aluminum purity. The second digit indicates modifications of the original alloy or impurity limits.

2.1 Aluminum—In the 1xxx group for minimum aluminum purities of 99.00% and greater, the last two of the four digits in the designation indicate the minimum aluminum percentage (Note 5.2). These digits are the same as the two digits to the right of the decimal point in the minimum aluminum percentage when it is expressed to the nearest 0.01%. The second digit in the designation indicates modifications in impurity limits. If the second digit in the designation is zero, it indicates that there is no special control on individual impurities; integers 1 through 9, which are assigned consecutively as needed, indicate special control of one or more individual impurities or alloying elements.

2.2 Aluminum Alloys—In the 2xxx through 8xxx alloy groups, the last two of the four digits in the designation have no special significance but serve only to identify the different aluminum alloys in the group. The second digit in the alloy designation indicates alloy modifications (Note 5.3). If the second digit in the designation is zero, it indicates the original alloy; integers 1 through 9, which are assigned consecutively, indicate alloy modifications.

2.3 Experimental Alloys—Experimental alloys are also designated in accordance with this system, but they are indicated by the prefix X. The prefix is dropped when the alloy is no longer experimental. During development and before they are designated as experimental, new alloys are identified by serial numbers assigned by their originators. Use of the serial number is discontinued when the X number is assigned.

2.4 National Variations—National variations (Note 5.4) of wrought aluminum and wrought aluminum alloys registered by another country in accordance with this system are identified by a serial letter (Note 5) before the numerical designation.

3. Cast Aluminum and Aluminum Alloy Designation System[1] (see Note 5.1)—A system of four-digit numerical designations is used to identify aluminum and aluminum alloys in the form of castings and foundry ingot. The first digit indicates the alloy group, as shown in Table 2. The second two digits identify the aluminum alloy or indicate the

[1] The castings and ingot alloy designation system described herein is not currently in use for some SAE cast aluminum alloys. It is applicable to Aluminum Association (AA) and American National Standard Institute (ANSI), and other, specification systems. Although the chemical composition limits shown in most SAE reports conform to the limits shown for comparable castings and ingots covered in AA and ANSI publications, the designation system described herein is not currently used in SAE Standards and Information Reports.

Figure 4.2 Alloy and temper designation systems for aluminum (SAE J993). (Report of Nonferrous Metals Committee, approved July 1967 and last revised September 1973. Conforms to American National Standard H35, 1-1972. Reaffirmed January 1989; reprinted with permission; copyright © 1992, Society of Automotive Engineers.)

TABLE 1—DESIGNATION SYSTEM FOR WROUGHT ALUMINUM AND ALUMINUM ALLOY

Composition	Alloy No.
Aluminum, 99.0% min and greater	1xxx
Aluminum alloys grouped by major alloying element[a,b,c]	
Copper	2xxx
Manganese	3xxx
Silicon	4xxx
Magnesium	5xxx
Magnesium and silicon	6xxx
Zinc	7xxx
Other element	8xxx
Unused series	9xxx

[a] For codification purposes, an alloying element is any element which is intentionally added for any purpose other than grain refinement and for which minimum and maximum limits are specified.

[b] Standard limits for alloying elements and impurities are expressed to the following places:

Less than 1/1000%	0.000X
1/1000 up to 1/100%	0.00X
1/100 up to 1/10%	
Unalloyed aluminum made by a refining process	0.0XX
Alloys and unalloyed aluminum not made by a	
refining process	0.0X
1/10 through ½%	0.XX
Over ½%	0.X, X.X, etc.

[c] Standard limits for alloying elements and impurities are expressed in the following sequence: silicon; iron; copper; manganese; magnesium; chromium; nickel; zinc (Note 1); titanium; other elements (each); other elements (Total); aluminum (Note 2).

Note 1—Additional specified elements having limits are inserted in alphabetical order of their chemical symbols between zinc and titanium, or are specified in footnotes.

Note 2—Aluminum is specified as minimum for unalloyed aluminum, and as a remainder for aluminum alloys.

aluminum purity. The last digit, which is separated from the others by a decimal point, indicates the product form, that is, castings or ingot. A modification of the original alloy or impurity limits is indicated by a serial letter (Note 5.6) before the numerical designation.

3.1 Aluminum Castings and Ingot In the 1xx.x group for minimum aluminum purities of 99.00% and greater, the second two of the four digits in the designation indicate the minimum aluminum percentage (Note 5.2).

These digits are the same as the two digits to the right of the decimal point in the minimum aluminum percentage when it is expressed to the nearest 0.01%. The last digit, which is to the right of the decimal point, indicates the product form: 1xx.0 indicates castings, and 1xx.1 indicates ingot. Special control of one or more individual elements other than aluminum is indicated by a serial letter (Note 5.6) before the numerical designation.

3.2 Aluminum Alloy Castings and Ingot—In the 2xx.x through 9xx.x alloy groups, the second two of the four digits in the designation have no special significance but serve only to identify the different aluminum alloys in the group. The last digit, which is to the right of the decimal point, indicates the product form: xxx.0 indicates castings, xxx.1 indicates ingot which has chemical composition limits conforming to paragraph 3.2.1, and xxx.2 indicates ingot which has chemical com-

Figure 4.2 (*Continued*)

TABLE 2—DESIGNATION SYSTEM FOR CAST ALUMINUM AND ALUMINUM ALLOY

Composition	Alloy No.
Aluminum, 99.00% min and greater	1xx.x
Aluminum alloy group by major alloying element[a,b,c]	
Copper	2xx.x
Silicon, with added copper and/or magnesium	3xx.x
Silicon	4xx.x
Magnesium	5xx.x
Zinc	7xx.x
Tin	8xx.x
Other element	9xx.x
Unused series	6xx.x

[a] For codification purposes, an alloying element is any element which is intentionally added for any purpose other than grain refinement and for which minimum and maximum limits are specified.

[b] Standard limits for alloying elements and impurities are expressed to the following places:

Less than 1/1000%	0.000X
1/1000 up to 1/100%	0.00X
1/100 up to 1/10%	
Unalloyed aluminum made by a refining process	0.0XX
Alloys and unalloyed aluminum not made by a refining process	0.0X
1/10 through ½%	0.XX
Over ½%	0.X, X.X, etc.

[c] Standard limits for alloying elements and impurities are expressed in the following sequence: silicon; iron; copper; manganese; magnesium; chromium; nickel; zinc (Note 1); titanium; other elements (each); other elements (Total); aluminum (Note 2).

Note 1—Additional specified elements having limits are inserted in alphabetical order of their chemical symbols between zinc and titanium, or are specified in footnotes.

Note 2—Aluminum is specified as minimum for unalloyed aluminum, and as a remainder for aluminum alloys.

position limits that differ but fall within the limits for xxx.1 ingot. Alloy modifications (Note 5.3) are indicated by a serial letter (Note 5.9) before the numerical designation.

3.2.1 Limits for alloying elements and impurities for xxx.1 ingot are the same as for the alloy in the form of castings, except for the limits noted in Table 3.

3.3 Experimental Alloys—Experimental alloys are also designated in accordance with this system, but they are indicated by the prefix X. The prefix is dropped when the alloy is no longer experimental. During development and before they are designated as experimental, new alloys are identified by serial numbers assigned by their originators. Use of the serial number is discontinued when the X number is assigned.

4. Temper Designation System—The temper designation system is used for all forms of wrought and cast aluminum and aluminum alloys except ingot. It is based on the sequences of basic treatments used to produce the various tempers. The temper designation follows the alloy designation, the two being separated by a hyphen. Basic temper designations consist of letters. Subdivisions of the basic tempers, where required, are indicated by one or more digits following the letter. These designate specific sequences of basic treatments; but only operations recognized as significantly influencing the characteristics of the product are indicated. Should some other variation of the same sequence of ba-

Figure 4.2 (*Continued*)

TABLE 3		
Element, %	For Castings	For Ingot

Iron, max	Sand and permanent mold:
	  Up thru 0.15
	  Over 0.15 thru 0.25
	  Over 0.25 thru 0.6
	  Over 0.6 thru 1.0
	  Over 1.0
		0.03 less than castings
		0.05 less than castings
		0.10 less than castings
		0.2 less than castings
		0.3 less than castings

	Die
	  Up thru 1.3
	  Over 1.3
		0.3 less than castings
		1.1 maximum

Magnesium, min	All
	  Less than 0.50
	  0.5 and greater
		0.05 more than castings°
		0.1 more than castings°

Zinc, max	Die
	  Over 0.25 thru 0.6
	  Over 0.6
		0.10 less than castings
		0.1 less than castings

°Applicable only when the specified magnesium range for castings is greater than 0.15%.

sic operations be applied to the same alloy, resulting in different characteristics, then additional digits are added to the designation.

4.1 Basic Temper Designations

F As Fabricated—Applies to the products of shaping processes in which no special control over thermal conditions or strain-hardening is employed. For wrought products, there are no mechanical property limits.

O Annealed (Wrought Products Only)—Applies to wrought products which are fully annealed to obtain the lowest strength condition.

H Strain Hardened (Wrought Products Only)—Applies to products which have their strength increased by strain-hardening, with or without supplementary thermal treatments to produce some reduction in strength. The H is always followed by two or more digits.

W Solution Heat-Treated—An unstable temper applicable only to alloys which spontaneously age at room temperature after solution heat-treatment. This designation is specific only when the period of natural aging is indicated; for example, W ½ h.

T Thermally Treated To Produce Stable Tempers Other Than **F, O,** or **H**—Applies to products which are thermally treated, with or without supplementary strain-hardening, to produce stable tempers. The T is always followed by one or more digits.

4.2 Subdivisions of Basic Tempers

4.2.1 Subdivisions of H Temper: Strain Hardened

4.2.1.1 The first digit following the H indicates the specific combination of basic operations, as follows:

H1 Strain Hardened Only—Applies to products which are strain hardened to obtain the desired strength without supplementary thermal treatment. The number following this designation indicates the degree of strain hardening.

H2 Strain Hardened and Partially Annealed—Applies to products which are strain hardened more than the desired final

Figure 4.2 *(Continued)*

amount and then reduced in strength to the desired level by partial annealing. For alloys that age soften at room temperature, the H2 tempers have the same minimum ultimate tensile strength as the corresponding H3 tempers. For other alloys, the H2 tempers have the same minimum ultimate tensile strength as the corresponding H1 tempers and slightly higher elongation. The number following this designation indicates the degree of strain hardening remaining after the product has been partially annealed.

H3 STRAIN HARDENED AND STABILIZED—Applies to products which are strain hardened and whose mechanical properties are stabilized by a low-temperature thermal treatment which results in slightly lowered tensile strength and improved ductility. This designation is applicable only to those alloys which, unless stabilized, gradually age soften at room temperature. The number following this designation indicates the degree of strain hardening before the stabilization treatment.

4.2.1.2 The digit following the designations H1, H2, and H3 indicates the degree of strain hardening. Numeral 8 has been assigned to indicate tempers having an ultimate tensile strength equivalent to that achieved by a cold reduction (temperature during reduction not to exceed 120°F (49°C) of approximately 75% following a full anneal. Tempers between 0 (annealed) and 8 are designated by numerals 1 through 7. Material having an ultimate tensile strength about midway between that of the 0 temper and that of the 8 temper is designated by the numeral 4; about midway between the 0 and 4 tempers by the numeral 2; and about midway between the 4 and 8 tempers by the numeral 6. Numeral 9 designates tempers whose minimum ultimate tensile strength exceeds that of the 8 temper by 2.0 ksi (14 MPa) or more. For two-digit H tempers whose second digit is odd, the standard limits for ultimate tensile strength are exactly midway between those of the adjacent two-digit H tempers whose second digits are even.

NOTE: For alloys which cannot be cold reduced, an amount sufficient to establish an ultimate tensile strength applicable to the 8 temper (75% cold reduction after full anneal), the 6 temper tensile strength may be established by a cold reduction of approximately 55% following a full anneal, or the 4 temper tensile strength may be established by a cold reduction of approximately 35% after a full anneal.

4.2.1.3 The third digit (Note 10), when used, indicates a variation of a two-digit temper. It is used when the degree of control of temper or the mechanical properties are different from, but close to, those for the two-digit H temper designation to which it is added, or when some other characteristic is significantly affected. (See Appendix for three-digit H tempers.)

NOTE: The minimum ultimate tensile strength of a three-digit H temper is at least as close to that of the corresponding two-digit H temper as it is to the adjacent two-digit H tempers.

4.2.2 SUBDIVISIONS OF T TEMPER: THERMALLY TREATED

4.2.2.1 Numerals 1 through 10 following the T indicate specific sequences of basic treatments, as follows (Note 5.8):

T1 COOLED FROM AN ELEVATED TEMPERATURE SHAPING PROCESS AND NATURALLY AGED TO A SUBSTANTIALLY STABLE CONDITION—Applies to products for which the rate of cooling from an elevated temperature shaping process, such as casting or ex-

Figure 4.2 (*Continued*)

trusion, is such that their strength is increased by room temperature aging.

T2 ANNEALED (CAST PRODUCTS ONLY)—Applies to cast products which are annealed to improve ductility and dimensional stability.

T3 SOLUTION HEAT TREATED AND THEN COLD WORKED—Applies to products which are cold worked to improve strength, or in which the effect of cold work in flattening or straightening is recognized in mechanical property limits.

T4 SOLUTION HEAT TREATED AND NATURALLY AGED TO A SUBSTANTIALLY STABLE CONDITION—Applies to products which are not cold worked after solution heat treatment, or in which the effect or cold work in flattening or straightening may not be recognized in mechanical property limits.

T5 COOLED FROM AN ELEVATED TEMPERATURE SHAPING PROCESS AND THEN ARTIFICIALLY AGED—Applies to products which are cooled from an elevated temperature shaping process, such as casting or extrusion, and then artificially aged to improve mechanical properties or dimensional stability or both.

T6 SOLUTION HEAT TREATED AND THEN ARTIFICIALLY AGED—Applies to products which are not cold worked after solution heat treatment, or in which the effect of cold work in flattening or straightening may not be recognized in mechanical property limits.

T7 SOLUTION HEAT TREATED AND THEN STABILIZED—Applies to products which are stabilized to carry them beyond the point of maximum strength to provide control of some special characteristics.

T8 SOLUTION HEAT TREATED, COLD WORKED, AND THEN ARTIFICIALLY AGED—Applies to products which are cold worked to improve strength, or in which the effect of cold work in flattening or straightening is recognized in mechanical property limits.

T9 SOLUTION HEAT TREATED, ARTIFICIALLY AGED, AND THEN COLD WORKED—Applies to products which are cold worked to improve strength.

T10 COOLED FROM AN ELEVATED TEMPERATURE SHAPING PROCESS, ARTIFICIALLY AGED, AND THEN COLD WORKED—Applies to products which are artificially aged after cooling from an elevated temperature shaping process, such as casting or extrusion, and then cold worked to improve strength further.

4.2.2.2 Additional digits (Note 5.9), the first of which shall not be zero, may be added to designations T1 through T10 to indicate a variation in treatment which significantly alters the characteristics of the product. (See Appendix for specific additional digits for T tempers.)

5. Notes

5.1 Producers of wrought aluminum and wrought aluminum alloys, and aluminum and aluminum alloy castings and foundry ingot, may register chemical composition limits and designations conforming to this standard with the Aluminum Association (AA) provided the aluminum or aluminum alloy is offered for sale, the complete chemical composition limits are registered, and the composition is significantly different from that of any aluminum or aluminum alloy for which a numerical designation already has been assigned. A numerical designation

Figure 4.2 (*Continued*)

assigned in conformance with this standard should be used only to indicate an aluminum or aluminum alloy having chemical composition limits identical to those registered with AA for that aluminum or aluminum alloy.

5.2 The aluminum content for unalloyed aluminum made by a refining process is the difference between 100.000% and the sum of all other metallic elements present in amounts of 0.0010% or more each, expressed to the third decimal; for unalloyed aluminum not made by a refining process, it is the difference between 100.00% and the sum of all other metallic elements present in amounts of 0.010% or more each, expressed to the second decimal.

5.3 A modification of the original alloy is limited to any one or a combination of the following:

(a) Change of not more than the following amounts in the arithmetic mean of the limits for an alloying element:

Arithmetic Mean of Limits for Alloying Elements in Original Alloy, %	Maximum Change, %
Up thru 1.0	0.15
Over 1.0 thru 2.0	0.20
Over 2.0 thru 3.0	0.25
Over 3.0 thru 4.0	0.30
Over 4.0 thru 5.0	0.35
Over 5.0 thru 6.0	0.40
Over 6.0	0.50

To determine compliance when limits are specified for a combination of two or more elements in one alloy composition, the mean of such a combination should be compared to the sum of the mean values of the same individual elements, or any combination thereof, in another alloy composition.

(b) Addition or deletion of not more than one alloying element with limits having an arithmetic mean of not more than 0.30%.

(c) Substitution of one alloying element for another element serving the same purpose.

(d) Change in limits for impurities.

(e) Change in limits for grain refining elements.

(f) Distinctive iron or silicon limits, or both, reflecting high purity base metal.

An alloy shall not be registered as a modification if it meets the requirements for a national variation.

5.4 A national variation has composition limits which are similar but not identical to those registered by another country, with differences such as:

(a) Differences in the arithmetic mean of limits for alloying elements not exceeding the following amounts:

Arithmetic Mean of Limits for Alloying Elements in Original Alloy or Modification, %	Maximum Difference, %
Up thru 1.0	0.15
Over 1.0 thru 2.0	0.20
Over 2.0 thru 3.0	0.25
Over 3.0 thru 4.0	0.30
Over 4.0 thru 5.0	0.35
Over 5.0 thru 6.0	0.40
Over 6.0	0.50

Figure 4.2 *(Continued)*

To determine compliance when limits are specified for a combination of two or more elements in one alloy composition, the mean of such a combination should be compared to the sum of the mean values of the same individual elements, or any combination thereof, in another alloy composition.

(b) Substitution of one alloying element for another element serving the same purpose.

(c) Different limits on impurities except for low iron. Low iron, reflecting high purity base metal, should be considered an alloy modification. See paragraph 5.3 (f).

(d) Different limits on grain refining elements.

(e) Inclusion of a minimum limit for iron or silicon, or both.

Wrought aluminum and wrought aluminum alloys meeting these requirements shall not be registered as a new alloy or alloy modification.

5.5 The serial letters are assigned internationally in alphabetical sequence starting with A but omitting I, O, and Q.

5.6 The serial letters are assigned in alphabetical sequence starting with A but omitting I, O, Q, and X, the X being reserved for experimental alloys.

5.7 Numerals 1 through 9 may be arbitrarily assigned as the third digit and registered with AA for an alloy and product to indicate a variation of a two-digit H temper provided the temper is used or is available for use by more than one user, mechanical property limits are registered, the characteristics of the temper are significantly different from those of all other tempers which have the same sequence of basic treatments and for which designations already have been assigned for the same alloy and product, and the following are also registered if characteristics other than mechanical properties are considered significant: (a) test methods and limits for the characteristics, or (b) the specific practices used to produce the temper. Zero has been assigned to indicate variations negotiated between the manufacturer and purchaser which are not used widely enough to justify registration.

5.8 A period of natural aging at room temperature may occur between or after the operations listed for tempers T3 through T10. Control of this period is exercised when it is metallurgically important.

5.9 Additional digits may be arbitrarily assigned and registered with AA for an alloy and product to indicate a variation of tempers T1 through T10 provided the temper is used or is available for use by more than one user; mechanical property limits are registered, the characteristics of the temper are significantly different from those of all other tempers which have the same sequence of basic treatments and for which designations already have been assigned for the same alloy and product, and the following are also registered if characteristics other than mechanical properties are considered significant: a. test methods and limits for the characteristics, or b. the specific practices used to provide the temper. Variations in treatment which do not alter the characteristics of the product are considered alternate treatments for which additional digits are not assigned.

APPENDIX

A1. Three-Digit H Tempers

A1.1 The following three-digit H temper designations have been assigned for wrought products in all alloys:

Figure 4.2 (*Continued*)

H111 Applies to products which are strain hardened less than the amount required for a controlled H11 temper.

H112 Applies to products which acquire some temper from shaping processes not having special control over the amount of strain hardening or thermal treatment, but for which there are mechanical property limits.

A1.2 The following three-digit H temper designations have been assigned for wrought products in alloys containing over a nominal 4% magnesium.

H311 Applies to products which are strain hardened less than the amount required for a controlled H31 temper.

H321 Applies to products which are strain hardened less than the amount required for a controlled H32 temper.

H323 Applies to products which are specially fabricated to have
H343 acceptable resistance to stress corrosion cracking.

A1.3 The following three-digit H temper designations have been assigned for:

Patterned or Embossed Sheet	Fabricated from
H114	0 temper
H124, H224, H324	H11, H21, H31 temper, respectively
H134, H234, H334	H12, H22, H32 temper, respectively
H144, H244, H344	H13, H23, H33 temper, respectively
H154, H254, H354	H14, H24, H34 temper, respectively
H164, H264, H364	H15, H25, H35 temper, respectively
H174, H274, H374	H16, H26, H36 temper, respectively
H184, H284, H384	H17, H27, H37 temper, respectively
H194, H294, H394	H18, H28, H38 temper, respectively
H195, H295, H395	H19, H29, H39 temper, respectively

A2. Additional Digits for T Tempers

A2.1 The following specific additional digits have been assigned for stress-relieved tempers of wrought products:

T51 Stress Relieved by Stretching—Applies to the following products when stretched the indicated amounts after solution heat treatment or cooling from an elevated temperature shaping process.

Product	Stretch, Permanent Set, %
Plate	1.5-3
Rod, bar, shapes	
extruded tube	1-3
Drawn tube	0.5-3

Applies directly to plate and rolled or cold-finished rod and bar. These products receive no further straightening after stretching.

Applies to extruded rod, bar, shapes, and tube and to drawn tube when designated as follows:

T510 Products that receive no further straightening after stretching.

T511 Products that may receive minor straightening after stretching to comply with standard tolerances.

T52 STRESS RELIEVED BY COMPRESSING—Applies to products which are stress relieved by compressing after solution heat treat-

Figure 4.2 (*Continued*)

ment, or cooling from an elevated temperature shaping process to produce a permanent set of 1-5%.

T54 STRESS RELIEVED BY COMBINED STRETCHING AND COMPRESSING—Applies to die forgings which are stress relieved by restriking cold in the finish die.

A2.2 The following temper designations have been assigned for wrought products heat treated from O or F temper to demonstrate response to heat-treatment.

T42 SOLUTION HEAT TREATED FROM THE O OR F TEMPER—To demonstrate response to heat treatment, and naturally aged to a substantially stable condition.

T62 SOLUTION HEAT TREATED FROM THE O OR F TEMPER—To demonstrate response to heat treatment, and artificially aged.

Temper designations T42 and T62 may also be applied to wrought products heat treated from any temper by the user when such heat treatment results in the mechanical properties applicable to these tempers.

Figure 4.2 (*Continued*)

many applications in military and commercial aircraft, as well as in rockets and missiles (aerospace vehicles).

Magnesium is the eighth most abundant element in the earth's crust. The metal is obtained by electrolysis of fused magnesium chloride derived from brines, wells, and seawater. Magnesium is a silvery white, lightweight metal that is also fairly tough. It is one-third lighter than aluminum. The metal tarnishes in air and burns easily when in the finely divided state with a dazzling white flame. The specific gravity of magnesium is 1.738 g/cm^3, and its density is 0.063 lb/in^3. Caution must be used when handling magnesium because of its flammability in air. Water should not be used to extinguish a magnesium fire.

Table 4.49 lists the specifications of wrought magnesium alloys. Table 4.50 lists the physical properties and characteristics of magnesium cast alloys. Table 4.51 lists the strength requirements of extruded magnesium bar, rod, and shapes (ASTM B-107).

4.5 Titanium Alloys

Titanium alloys have a high strength-to-weight ratio. They are used as armor plate in modern war planes and propeller shafts, rigging, and other parts of ships where high strength and resistance to salt water are required. Titanium metal is produced by reducing titanium

(*Text continued on page 242.*)

TABLE 4.39 Typical Physical Properties of Wrought Copper Alloys

Customary units

Copper or copper alloy UNS no.	Melting point, °F		Density[a]	Coefficient of thermal expansion[b]			Thermal conductivity[c]	Electrical resistivity[d]	Electrical conductivity[e]	Thermal capacity[f]	Modulus	
	Liquidus	Solidus		68–212°F	68–392°F	68–572°F					Elastic[g]	Rigid[h]
C10200	1981	—	.323	9.4	9.6	9.8	226	10.3	101	.092	17	6.4
C11000	1981	1949	.322	9.4	9.6	9.8	226	10.3	101	.092	17	6.4
C11100	1981	—	.322	9.4	9.6	9.8	224	10.3	101	.092	17	6.4
C11300	1981	—	.322	9.4	9.6	9.8	224	10.3	100	.092	17	6.4
C11400	1981	—	.322	9.4	9.6	9.8	224	10.4	100	.092	17	6.4
C11500	1981	—	.322	9.4	9.6	9.8	224	10.4	100	.092	17	6.4
C11600	1981	—	.322	9.4	9.6	9.8	224	10.4	99	.092	17	6.4
C12000	1981	—	.323	9.4	9.6	9.8	223	10.7	97	.092	17	6.4
C12200	1981	—	.323	9.4	9.5	9.8	196	12.2	85	.092	17	6.4
C14500	1960	1931[i]	.323	9.4	9.6	9.8	205	10.9	95	.092	17	6.4
C14700	1970	1953	.323	9.4	9.6	9.8	216	10.9	95	.092	17	6.4
C15000	1979	—	.323	9.4	9.6	9.8	212[k]	11.2[k]	93[k]	.092	17	6.4
C16200	1969	—	.321	9.4	9.6	9.8	208	11.9	87	.092	17	6.4
C17000	1800	1600	.298	9.3	9.4	9.9	—	47.2	22	—	19	7.3
C17200	1800	1600	.298	9.3	9.4	9.9	—	47.2	22	.100	19	7.3
C17500	1955	1885	.316	—	9.8	—	—	23.1	45	—	18	6.8
C17600	1930	1850	.316	—	9.8	—	—	19.0	50	—	18	6.8
C18400	1967	—	.321	9.4	9.6	9.8	187[k]	13.0[k]	80[k]	.092	19	7.2
C18700	1976	1947[j]	.323	9.4	9.6	9.8	218	10.6	98	.092	17	6.4
C19200	1983	—	.320	9.0	—	—	125	20.8	50	.092	17	6.4
C21000	1950	1920	.320	—	—	10.0	135	18.5	00	.090	17	6.4
C22000	1910	1870	.318	—	—	10.2	109	23.6	44	.090	17	6.4
C23000	1880	1810	.316	—	—	10.4	92	28.0	37	.090	17	6.4

UNS No.												
C24000	1830	1770	.313	—	—	10.6	81	32.4	32	.090	16	6.0
C26000	1750	1680	.308	—	—	11.1	70	37.0	28	.090	16	6.0
C26800	1710	1660	.306	—	—	11.3	67	38.4	27	.090	15	5.6
C27000	1710	1660	.306	—	—	11.3	67	38.4	27	.090	15	5.6
C33000	1720	1660	.307	—	—	11.2	67	39.9	26	.090	15	5.6
C33100	1720	1660	.307	—	—	11.2	67	39.9	26	.090	15	5.6
C34200	1670	1630	.306	—	—	11.3	67	39.9	26	.090	15	5.6
C34500	1650	1625	.305	—	—	11.4	69	39.9	26	—	10	—
C35000	1650	1630	.305	—	—	11.4	67	39.9	26	.090	14	5.3
C36000	1650	1630	.307	—	—	11.4	67	39.9	26	.090	14	5.3
C37700	1640	1620	.305	—	—	11.5	69	38.4	27	.090	15	5.6
C46400	1650	1630	.304	—	—	11.8	57	39.9	26	.090	15	5.6
C46500	1650	1630	.304	—	—	11.8	57	39.9	26	.090	15	5.6
C46600	1650	1630	.304	—	—	11.8	67	39.9	26	.090	15	5.6
C46700	1650	1630	.304	—	—	11.8	67	39.9	26	.090	15	5.6
C51000	1920	1750	.320	—	—	9.9	40	69.1	15	.090	16	6.0
C51100	1945	1785	.320	—	—	9.9	48	52.0	20	.090	16	6.0
C52100	1880	1620	.318	—	—	10.1	36	79.8	13	.090	16	6.0
C52400	1830	1550	.317	—	—	10.2	29	94.3	11	.090	16	6.0
C54400	1830	1700	.321	—	—	9.6	50	54.6	19	.090	15	5.6
C60800	1945	1920	.295	—	—	10.0	46	60.0	17	.090	17.5	6.6
C61300	1915	1905	.285	—	9.0	9.0	39	74.1	14	.090	17	6.4
C61400	1915	1905	.285	—	9.0	9.0	39	74.1	14	.090	17	6.4
C61800	1910	1900	.274	—	9.0	9.0	37	79.8	13	—	17	—
C62300	1910	1890	.274	—	—	9.4	31	79.3	13	—	16	—
C62400	1910	1895	.274	—	—	9.2	34	79.3	13	—	16	—
C63000	1930	1890	.274	—	—	9.4	22	138.0	8	.090	17	6.4
C64200	1840	1800	.278	—	—	10.0	26	113.0	8	.090	16	6.0
C65500	1880	1780	.308	—	—	10.0	21	148.0	7	.090	15	5.6
C67000	1710	1665	.282	—	—	11.0	14	86.4	12	—	15	—
C67300	1620	1555	.299	—	—	11.0	—	86.4	12	—	15	—
C67400	1625	1550	.292	—	—	11.0	58	86.4	12	—	14	—

TABLE 4.39 Typical Physical Properties of Wrought Copper Alloys (*Continued*)

Copper or copper alloy UNS no.	Melting point, °F Liquidus	Melting point, °F Solidus	Density[a]	Coefficient of thermal expansion[b] 68–212°F	Coefficient of thermal expansion[b] 68–392°F	Coefficient of thermal expansion[b] 68–572°F	Thermal conductivity[c]	Electrical resistivity[d]	Electrical conductivity[e]	Thermal capacity[f]	Modulus Elastic[g]	Modulus Rigid[h]
C67500	1630	1590	.302	—	—	11.8	61	43.2	24	.090	15	5.6
C70600	2100	2010	.323	—	—	9.5	26	115.0	9	.090	18	6.8
C71000	2192	2066	.323	—	—	9.1	21	160.0	6	.090	20	7.5
C71500	2260	2140	.323	—	—	9.0	17	225.0	5	.090	22	8.3
C75200	2030	1960	.316	—	—	9.0	19	173.0	6	.090	18	6.8
C77000	1930	—	.314	—	—	9.3	17	189.0	6	.090	18	6.8

Customary units

[a] lb/in³ at 68°F.
[b] Per °F at temperature range indicated (multiply factor given by 10⁻⁶).
[c] Btu/ft²/ft·h·°F at 68°F.
[d] (Annealed) ohms (circular mil/ft) at 68°F.
[e] (Annealed) percent IACS at 68°F (volume basis).
[f] (Specific heat) Btu/lb/°F at 68°F.
[g] (Tension) psi (multiply factor given by 10⁶).
[h] Psi (multiply factor given by 10⁶).
[i] Small amount of tellurium-rich constituent remains liquid down to 1575°F.
[j] Small amount of lead-rich constituent remains liquid down to 619°F.
[k] After precipitation-hardening heat treatment.
SOURCE: Reprinted with permission, copyright 1992, Society of Automotive Engineers.

TABLE 4.40 Fabrication Properties, Other Characteristics, and Typical Uses of Copper

Copper or copper alloy UNS no.	Approximate relative suitability for being worked*		Best temperature for hot working, °C	Approximate relative suitability for being joined by									Machinability†	Type of chip†	Typical uses	Characteristics
	Cold	Hot		Soldering	Brazing	Oxyacetylene welding	Carbon arc welding	Gas-shielded arc welding	Coated-metal arc welding	Resistance welding Spot	Seam	Butt				
C10200	E	E	760–870	E	E	F	F	G	NR	NR	NR	G	20	L	Thermal and electrical conductors, electronic parts, glass-to-metal seals.	Oxygen-free 100% minimum electrical conductivity, excellent ductility, high purity, no out gassing. Not subject to hydrogen embrittlement. Designated for use where processing involves heating in a reducing atmosphere.
C11000	E	E	760–870	E	G	NR	F	F	NR	NR	NR	G	20	L	Electrical wiring and components, radiator fins, gaskets, washers, cold-heading wire, water deflectors, heat plugs, clock cases, plating anodes, screen wire.	Minimum electrical conductivity 100%, highest electrical conductivity of any metal except silver, has very high ductility. Will embrittle when heated to redness in a reducing atmosphere.
C11100	E	E	760–870	E	G	NR	F	F	NR	NR	NR	G	20	L	Radiator fins.	Has a softening temperature higher than the silver bearing coppers and electrolytic tough pitch copper.

TABLE 4.40 Fabrication Properties, Other Characteristics, and Typical Uses of Copper (Continued)

Copper or copper alloy UNS no.	Approximate relative suitability for being worked*		Best temperature for hot working, °C	Approximate relative suitability* for being joined by						Resistance welding			Machinability†	Type of chip†	Typical uses	Characteristics
	Cold	Hot		Soldering	Brazing	Oxyacetylene welding	Carbon arc welding	Gas-shielded arc welding	Coated-metal arc welding	Spot	Seam	Butt				
C11300	E	E	760–780	E	G	NR	F	F	NR	NR	NR	G	20	L	Commutator bars, segments, collector rings and contacts, core and fin stock for radiators.	Minimum electrical conductivity 98%. Resistance to softening increased by presence of silver. Effect increases with increased silver added. Higher-silver-content copper used for continued exposure to somewhat higher temperature.
C11400	E	E	760–780	E	G	NR	F	F	NR	NR	NR	G	20	L		
C11500	E	E	760–870	E	G	NR	F	F	NR	NR	NR	G	20	L		
C11600	E	E	760–870	E	G	NR	F	F	NR	NR	NR	G	20	L		
C12000	E	E	760–870	E	E	G	G	E	NR	NR	NR	G	20	L	Electrical conductors, applications involving welding or brazing.	Regarded as an alternate to Copper UNS no. C10200. More resistant to embrittlement than Copper UNS no. C11000 high electrical conductivity.
C12200	E	E	760–870	E	E	G	G	E	NR	NR	NR	G	20	L	Tube, all types of hydraulic systems, fuel lines, vacuum lines, air conditioning, heat exchangers, anodes, air, gasoline, hydraulic, and oil lines, oil coolers, gauge lines.	Slightly improved mechanical properties. Electrical conductivity about 85%. Not subject to hydrogen embrittlement.

TABLE 4.40 Fabrication Properties, Other Characteristics, and Typical Uses of Copper (Continued)

Copper or copper alloy UNS no.	Approximate relative suitability for being worked*		Best temperature for hot working, °C	Approximate relative suitability for being joined by*						Resistance welding			Machinability†	Type of chip‡	Typical uses	Characteristics
	Cold	Hot		Soldering	Brazing	Oxyacetylene welding	Carbon arc welding	Gas-shielded arc welding	Coated-metal arc welding	Spot	Seam	Butt				
C17600	G	G	760–925	G	G	NR	F	G	NR	NR	NR	G	20	L	Clips, welder tips, and wheels.	Precipitation hardened. Fairly high electrical conductivity. Resistance to softening at elevated temperatures.
C18400	E	E	900–925	G	G	NR	F	G	NR	NR	NR	G	20	L	Spot welding electrodes and wheels, flash welding dies, and commutator segments.	
C18700	G	NR	—	E	G	NR	NR	NR	NR	NR	NR	F	80	S	Screw machine parts requiring high electrical and thermal conductivity.	Free-machining copper, high electrical conductivity. Unsuited for hot working.
C19200	E	E	815–950	E	E	G		E	NR	NR	NR	G	20		Flexible hose, electrical terminals, fuse clips, gaskets, air-conditioning and heat-exchanger tubing.	Resistance to softening and also stress corrosion.
C21000	E	G	760–870	E	E	G	F	G	NR	NR	NR	G	20	L	Emblems, vitreous enamel base, ornamental trim, and jewelry.	Copper Alloy UNS nos. C21000, C22000, and C23000 are generally

UNS No.			Annealing temp, °C												Applications	Characteristics
C14500	E	E	760–845	E	E	F	F	G	NR	NR	NR	G	80	S	Forgings and screw machine parts requiring high electrical and thermal conductivity. Electrical connectors, motor and switch parts, soldering coppers, and welding torch tips.	Free-machining copper, combined with high electrical conductivity (90–96%).
C14700	E	E	760–870	E	E	NR	NR	NR	NR	NR	NR	G	80	S	Transformer and circuit-breaker terminals, studs, bolts, nuts, and current-carrying parts requiring fine machining.	Free-machining copper, combined with high electrical conductivity (90–96%).
C15000	E	E	760–870	E	G	F	F	F	NR	NR	NR	G	20	L	Resistance welding electrodes. Miscellaneous current-carrying components at elevated temperatures.	Precipitation hardened. Combined high strength and conductivity and resistance to softening at elevated temperatures.
C16200	G	E	760–870	E	E	F	F	G	NR	NR	NR	G	20	L	Electrical contacts and terminals, signal relays. Hard temper used for spring contact in small apparatus and resistance welding electrodes.	Moderate high strength and high electrical conductivity.
C17000	G	G	705–775	G	G	NR	F	G	NR	NR	NR	G	20	L	Leaf springs, electrical contacts, coil springs and bellows requiring severe forming distributor breaker arm, welding tips, and welding wheels.	Copper Alloy UNS nos. C17000, C17200, C17500, and C17600 can develop the highest mechanical properties by heat treatment Complete range of properties.
C17200	G	G	705–775	G	G	NR	F	G	NR	NR	NR	G	20	L		
C17500	G	G	760–925	G	G	NR	F	G	NR	NR	NR	G	20	L		

Alloy														Typical uses	Remarks
C22000	E	G	760–870	E	E	G	F	G	NR	NR	G	20	L	Emblems, vitreous enamel base, ornamental trim, jewelry, expansion plugs, valve parts, escutcheon fasteners, and spring clips.	reddish in color, soft and malleable, higher annealing point than copper and slightly stronger and similar in corrosion resistance. Good for drawing and forming. Resistance to dezincification and season cracking is excellent.
C23000	E	G	790–900	E	E	G	F	G	F	NR	G	30	L	Radiator parts, heat-exchanger tubes, tube bends.	
C24000	E	F	815–900	E	E	G	F	G	F	NR	G	30	L	Bellows and water temperature switch housing, flexible hose, pump lines.	Color is light golden, strength and ductility continue to increase.
C26000	E	F	730–845	E	E	G	F	F	G	NR	G	30	L	Radiator tanks and lock-seam tubes, header plates, reflectors, lamp bases, terminals, ground straps, baffles, ammeter shells and speedometer counter-weights, washers, wheel covers, trim, carburetor parts.	Color is brass yellow. Greatest ductility of the copper-zinc series. Strength is higher than any of the preceding copper-zinc alloys.
C26800	E	NR	—	E	E	G	F	F	G	NR	G	30	L	Radiator cores and tanks, lamp fixtures, socket shells, eyelets, fasteners and grommets, hinges, locks, pins, rivets, screws, and springs.	Strength increases and ductility decreases, but is still very good.
C27000	E	NR	—	E	E	G	F	F	G	NR	G	30	L		
C33000	E	NR	—	G	G	F	F	F	F	NR	F	60	M	Tube carburetor parts, oil cooler tube, radiator and ornamental work, pump and power cylinders and liners.	Provides some degree of machinability, together with moderate cold-working properties.

TABLE 4.40 Fabrication Properties, Other Characteristics, and Typical Uses of Copper (*Continued*)

Copper or copper alloy UNS no.	Approximate relative suitability for being worked*		Best temperature for hot working, °C	Approximate relative suitability* for being joined by						Resistance welding			Machinability[‡]	Type of chip[†]	Typical uses	Characteristics
	Cold	Hot		Soldering	Brazing	Oxyacetylene welding	Carbon arc welding	Gas-shielded arc welding	Coated-metal arc welding	Spot	Seam	Butt				
C33100	E	NR	—	E	G	NR	NR	NR	NR	NR	NR	F	70	M	Keys.	Intended for blanking, piercing, and machining.
C34200	E	NR	—	E	G	NR	NR	NR	NR	NR	NR	F	90	S	Clock plates and nuts, clock and watch backs, keys, gears, and wheels.	Provides increased machinability with moderate cold-working properties.
C34500	F	F	705–790	E	G	NR	NR	NR	NR	NR	NR	F	90	S	Screw-machine parts requiring roll threads, knurls, or staking operations.	Best combination of machinability and cold-working properties.
C35000	F	NR	—	E	G	NR	NR	NR	NR	NR	NR	F	70	M	Keys.	Intended for blanking, piercing, and machining.
C36000	NR	F	709–790	E	G	NR	NR	NR	NR	NR	NR	F	100	S	Automatic screw-machine parts and carburetor, magneto parts, radiator drums and other fittings, plugs, inserts, gears, pinions, locks.	The standard free-cutting brass and its machinability has become the standard by which other alloys are rated.
C37700	NR	E	650–815	E	G	NR	NR	NR	NR	NR	NR	F	80	S	Forgings and pressings of all kinds. Headings, air-conditioning tube fittings, convertible top hardware (latches, hinges, etc.), forged valve bodies.	Excellent hot-working properties and widely used as forging rod. At ordinary temperatures it is strong, hard, and free-cutting.

Alloy	1	2	3	4	5	6	7	8	9	10	11	12	13	14	Typical uses	Remarks
C16400	F	E	**E**	650–815	G	E	F	F	NR	G	F	G	30	L	Aircraft turnbuckle barrels and balls, cold headed parts, forgings, screw-machine parts, marine hardware, condenser plates, welding rod, nozzles, and fittings.	Excellent hot and fair cold-working properties of somewhat higher strength, good saltwater corrosion resistance.
C46500	F	E	**E**	650–815	G	E	F	F	NR	G	F	G	30	L		
C46600	F	E	**E**	650–815	G	E	F	F	NR	G	F	G	30	L		
C46700	F	E	**E**	650–815	G	E	F	F	NR	G	F	G	30	L		
C51000	E	NR	E	—	F	E	G	G	F	G	F	E	20	L	Springs, bearings, clips, contacts, switch parts, diaphragms, welding rod, thermostats, bellows, clutch disks, lock washers, fasteners.	C51000 and C52100 have a remarkable combination of strength, ductility and resilience, and fatigue resistance.
C51100	E	NR	E	—	F	E	G	G	F	G	F	E	20	L		
C52100	G	NR	E	—	F	E	G	G	F	G	F	E	20	L	Springs, clips, contacts, terminal wire and bushings, diaphragms, and bellows.	
C52400	G	NR	E	—	F	E	G	G	F	G	F	E	20	L		
C54400	G	NR	E	—	F	G	NR	NR	NR	NR	NR	F	80	S	Bearings, bushings, gears, pinions, shafts, thrust washers, valve parts.	Free-cutting, good cold-working properties, also suitable for blanking, forming, and bending.
C60800	G	F	F	790–870	F	NR	—	G	G	G	—	G	20	—	Condenser, evaporator, and heat exchanger tubes, ferrules.	
C61300	G	G	F	785–925	F	NR	G	G	G	G	G	G	20	L	Gibs, wear strips, gears, bushings, nuts, bolts, and threaded members.	Good cold-working properties and corrosion resistance. High strength and ductility.
C61400	G	G	F	785–925	F	NR	G	G	G	G	G	G	20	L	Gibs, wear strips, gears, bushings, nuts, bolts, and threaded members.	Good cold-working properties and corrosion resistance. High strength and ductility.

TABLE 4.40 Fabrication Properties, Other Characteristics, and Typical Uses of Copper (Continued)

Copper or copper alloy UNS no.	Approximate relative suitability for being worked*		Best temperature for hot working, °C	Approximate relative suitability* for being joined by						Resistance welding			Machinability‡	Type of chip†	Typical uses	Characteristics
	Cold	Hot		Soldering	Brazing	Oxyacetylene welding	Carbon arc welding	Gas-shielded arc welding	Coated-metal arc welding	Spot	Seam	Butt				
C61800	F	G	760–885	F	G	NR	—	G	G	G	G	G	40	—	Bushings, bearings, corrosion applications, welding rod.	
C62300	F	G	730–815	F	F	NR	G	§	G	G	G	G	30	L	Valve guides, spark plug inserts, gears, valve seat inserts, oil plugs, and shifter forks.	Good hot-working properties; high strength retained well at elevated temperatures; acid and oxidation resistant.
C62400	NR	E	720–775	F	F	NR	G	§	G	G	G	G	30	L	Valve guides, gears, spark plug inserts, gears, valve seat inserts, oil plugs, shifter forks, wear strips, ball bearings, and hydraulic valve components.	Excellent hot-working, poor cold-working properties; heat treated for high mechanical properties.
C63000	NR	G	705–760	F	F	NR	G	§	G	G	G	G	20	L	Retractable landing gear, propeller gears, large valve seat inserts, spacer bearings, high-pressure pump components.	Very high mechanical properties in the heat-treated condition; difficult to cold work; good hot-working properties, excellent corrosion resistance.
C64200	NR	E	705–760	F	F	NR	—	F	F	F	F	F	60	—	Valve stems, gears, bolts, nuts, valve bodies, and components.	Free machining, high strength, high corrosion resistance.

Alloy																Applications	Characteristics
C65500	NR	E	705–760	G	E	G	G	E	F	E	E	E	E	30	L	Hydraulic pressure lines, bolts, clamps, piston rings, rivets, and shafting.	Relatively high strength, marked ductility and capability for being both hot- and cold-worked and joined by all procedures. Excellent corrosion resistance
C67300	NR	E	565–745	NR	F	NR	NR	S	G	G	F	G	G	30	S	Diesel injector nozzles; high pressure hydraulic applications, cams, pistons, and other components involving high mechanical loads and sliding contact.	High strength and good wear-resistant properties.
C67300	F	E	625–745	NR	G	NR	NR	NR	NR	NR	F	NR	F	70	S	Forged water pump impellers; gears, axial piston pump components, bushings, and bearings.	Hot-forgeable, free-cutting alloy having fairly high strength and good corrosion-resistant properties.
C67400	F	E	565–745	NR	F	NR	NR	S	G	F	G	G	F	30	L	Connecting rods, transmission synchronizing stop ring, door striker plates, shifter shoes, differential idler pins, forged water pump impellers, axial piston pump parts, bushings, and bearings.	Hot-forgeable, high-strength alloy with good wear-resistant properties and good corrosion resistance.
C67500	NR	E	325–790	E	E	G	F	F	NR	G	G	G	F	30	L	Clutch disks, pump rods, shafting, balls, valve stems, and bodies.	Strong, rigid, and abrasion resistant; adopted to hot forging and pressing, hot heading, and upsetting.

TABLE 4.40 Fabrication Properties, Other Characteristics, and Typical Uses of Copper (*Continued*)

Copper or copper alloy UNS no.	Approximate relative suitability for being worked*		Best temperature for hot working, °C	Approximate relative suitability* for being joined by						Resistance welding			Machinability‡	Type of chip†	Typical uses	Characteristics
	Cold	Hot		Soldering	Brazing	Oxyacetylene welding	Carbon arc welding	Gas-shielded arc welding	Coated-metal arc welding	Spot	Seam	Butt				
C70600	G	G	760–980	E	E	F	NR	E	G	G	G	E	20	L	Condenser and heat-exchanger tubes.	Used where requirements are severe. Strong, tough, and very resistant to general corrosion as well as stress-corrosion cracking; also serviceable at higher temperatures than copper and brasses. Well suited for condenser and heat exchanger tube.
C71000	G	G	760–980	E	E	G	NR	E	E	E	E	E	20	L	Condenser and heat-exchanger tubes, ferrules.	Copper Alloy UNS nos. C71000 and C71500 are used where requirements are severe. Strong, tough, and very resistant to general corrosion as well as stress-corrosion cracking; also serviceable at higher temperatures than copper and brasses. Well suited for condenser and heat-exchanger tube.
C71500	G	G	925–1035	E	E	G	NR	E	E	E	E	E	20	L	Automatic oil coolers, heat-exchanger tubes.	

C75200	E	NR	—	E	E	G	NE	F	NR	G	F	G	20	L	Rivets, screws, name plates, radio dials, etching stock, trim.	Copper Alloy UNS nos. C75200 and C77000 are manufactured in a wide range of nickel contents. Higher the nickel, the more silver white the alloy. 65% copper alloys have good cold-working properties and are used for cold drawing, spinning, forming, and stamping. The lower-copper-content
C77000	G	NR	—	E	E	G	NR	F	NR	G	F	G	30	L	Springs, resistance wire.	

* E = Excellent; G = Good; F = Fair; NR = Not recommended.

† S = Short; M = Medium; L = Long.

‡ Approximate relative machinability rating (free-cutting brass = 100).

§ Consumable electrode excellent. Tungsten are good, with ac preferred.

SOURCE: Reprinted with permission, copyright 1992, Society of Automotive Engineers.

TABLE 4.41 Mechanical Properties of Cast Copper Alloys

Copper alloy UNS no.*	SAE suffix†§	ASTM standard no.	Casting method‡§ and condition	Tensile strength, min MPa	Tensile strength, min ksi	Yield strength, min MPa (0.5% ext. under load)	Yield strength, min ksi (0.5% ext. under load)	Elongation, min, % in 50 mm (2 in)
C83600	A	B271, B584	Sand, centrifugal	205	30	95	14	20
C83600	B	B505	Continuous	250	36	130	19	15
C83600	C		Continuous	345	50	170	15	12
C83800	A	B271, B584	Sand, centrifugal	205	30	90	13	20
C83800	B	B505	Continuous	205	30	95	15	16
C85200		B271, B584	Sand, centrifugal	240	35	85	12	25
C85400		B271, B584	Sand, centrifugal	205	30	75	11	20
						0.2% Offset		
C85800		B176	Die¶	380	55	205	30	15
C86200		B271, B505, B584	Sand, centrifugal, cont.	620	90	310	45	18
C86300	A	B271, B584	Sand, centrifugal	760	110	415	60	12
C86300	B	B505	Continuous	760	110	425	62	14
C86500	A	B271, B584	Sand, centrifugal	450	65	170	25	20
C86500	B	B505	Continuous	485	70	170	25	25
						0.5% Ext. Under Load		
C87200		B271, B584	Sand, centrifugal	310	45	125	18	20
C87400		B271, B584	Sand, centrifugal	345	50	145	21	18
C87500		B271, B584	Sand, centrifugal	415	60	165	24	16
						0.2% Offset		
C87800		B176	Die¶	585	85	345	50	25
C87900		B176	Die¶	485	70	240	35	25

Alloy		Spec	Casting			0.5% Ext. Under Load		
C90300	A	B271, B584	Sand, centrifugal	275	40	125	18	20
C90300	B	B505	Continuous	305	44	150	22	18
C90500	A	B271, B584	Sand, centrifugal	275	40	125	18	20
C90500	B	B505	Continuous	305	44	170	25	10
C90700	A	B505	Sand	240	35	125	18	10
C90700	B		Continuous	275	40	170	25	10
C92200	A	B271, B584	Sand, centrifugal	235	34	110	16	24
C92200	B	B505	Continuous	260	38	130	19	18
C92300	A	B271, B584	Sand, centrifugal	250	36	110	16	18
C92300	B	B505	Continuous	275	40	130	19	16
C92500	A	B505	Sand	240	35	125	18	10
C92500	B		Continuous	275	40	165	24	10
C92700	A	B505	Sand	240	35	125	18	10
C92700	B		Continuous	260	38	140	20	8
C92900		B247, B505	Sand, continuous	310	45	170	25	8
C93200	A	B271, B584	Sand, centrifugal	205	30	95	14	15
C93200	B	B505	Continuous	240	35	140	20	10
C93500	A	B271, B584	Sand, centrifugal	195	28	85	12	15
C93500	B	B505	Continuous	205	30	110	16	12
C93700	A	B271, B584	Sand, centrifugal	205	30	85	12	15
C93700	B	B505	Continuous	240	35	140	20	6
C93700	C	B505	Continuous	275	40	170	25	6
C93800	A	B271, B584	Sand, centrifugal	180	26	95	14	12
C93800	B	B505	Continuous	170	25	110	16	5
C94300	A	B271, B584	Sand, centrifugal	145	21	—	—	10
C94300	B	B505	Continuous	145	21	95	15	7
C94700	A	B505, B584	Sand, continuous	310	45	140	20	25
C94700	B	B505, B584	Sand, continuous (HT)	515	75	345	50	5
C94800		B505, B584	Sand, continuous	275	40	140	20	20

219

TABLE 4.41 Mechanical Properties of Cast Copper Alloys (Continued)

Copper alloy UNS no.*	SAE suffix†§	ASTM standard no.	Casting method‡§ and condition	Tensile strength, min		Yield strength, min 0.5% ext. under load		Elongation, min, % in 50 mm (2 in)
				MPa	ksi	MPa	ksi	
C95200	A	B148, B271	Sand, centrifugal	450	65	170	25	20
C95200	B	B505	Continuous	470	68	180	26	20
C95300	A	B148, B271	Sand, centrifugal	450	65	170	25	20
C95300	B	B505	Continuous	485	70	180	26	25
C95300	C	B148, B271, B505	Sand, centrifugal, cont. (HT)	550	80	275	40	12
C95400	A	B148, B271	Sand, centrifugal	515	75	205	30	12
C95400	B	B505	Continuous	585	85	220	32	12
C95400	C	B148, B271	Sand, centrifugal (HT)	620	90	310	45	6
C95400	D	B505	Continuous (HT)	655	95	310	45	10
C95500	A	B148, B271	Sand, centrifugal	620	90	275	40	6
C95500	B	B505	Continuous	655	95	290	42	10
C95500	C	B148, B271	Sand, centrifugal (HT)	760	110	415	60	5
C95500	D	B505	Continuous (HT)	760	110	425	62	8
C95800	A	B148, B271	Sand, centrifugal	585	85	240	35	15
C95800	B	B505	Continuous (3)	620	90	260	38	18
C96200		B369	Sand	310	45	170	25	20

* Unified Numbering System. For cross-reference to SAE, former SAE, former ASTM, and former trade names, see SAE Information Report for Wrought and Cast Copper Alloys, SAE J461.

† Suffix symbols may be specified to distinguish between two or more sets of mechanical properties, heat treatment, conditions, etc. as applicable.

‡ All alloys listed are in the "as cast" condition except those designated as heat treated (HT) and copper alloy UNS No. C95800 which is temper annealed.

§ Most commonly used method of casting is shown for each alloy. However, unless the purchaser specifies the method of casting or the mechanical properties by supplement to the UNS number, the supplier may use any method which will develop the properties indicated.

¶ Mechanical properties shown for die castings are typical, not minimum.

SOURCE: Reprinted with permission, copyright 1992, Society of Automotive Engineers.

TABLE 4.42 Typical Physical Properties of Cast Copper Alloys (Customary Units)

| Copper alloy UNS no. | Melting point, °F | | Density, lb/in³ | Specific gravity | Coefficient of thermal expansion, 10^6 in/in/°F (20–400)°F | Thermal conductivity, % of Cu* | Electrical conductivity, % IACS[†] | Modulus of elasticity, 10^6 psi |
	Liquidus	Solidus						
C83600	1840	1570	0.318	8.83	10.0	18	15	14
C83800	1840	1550	0.312	8.60	10.0	18	15	13
C85200	1725	1700	0.307	8.50	11.5	21	18	11
C85400	1725	1700	0.305	8.45	11.2	23	20	12
C85800	1650	1500	0.305	8.40	12.0	—	20	15
C86200	1725	1650	0.288	7.85	12.0	9	8	15
C86300	1690	1625	0.283	7.84	12.0	9	8	14
C86500	1620	1585	0.301	8.30	11.3	22	22	15
C87200	1780	1580	0.302	8.40	9.2	7	6	15
C87400	1680	1510	0.300	8.27	10.9	7	7	15
C87500	1680	1510	0.300	8.27	10.9	7	7	15
C87800	1680	1510	0.300	8.27	10.9	7	7	20
C87900	1700	1350	0.308	8.50	12.0	—	15	15
C90300	1830	1570	0.318	8.70	13.0	19	12	15
C90500	1830	1570	0.315	8.72	11.0	19	11	15
C90700	1830	1528	0.317	8.78	10.2	19	10	14
C92200	1810	1520	0.312	8.65	10.0	18	14	14
C92300	1830	1570	0.317	8.80	10.0	19	12	14
C92500	1830	1570	0.317	8.85	10.0	—	11	13
C92700	1800	1550	0.317	8.80	10.1	12	11	13

TABLE 4.42 Typical Physical Properties of Cast Copper Alloys (Customary Units) (Continued)

Copper alloy UNS no.	Melting point, °F		Density, lb/in³	Specific gravity	Coefficient of thermal expansion, 10⁶ in/in/°F (20–400°F)	Thermal conductivity, % of Cu*	Electrical conductivity, % IACS†	Modulus of elasticity, 10⁶ psi
	Liquidus	Solidus						
C92900	1880	1575	0.320	8.79	9.5	15	9	14
C93200	1800	1570	0.322	8.93	10.0	15	12	14
C93500	1830	1570	0.320	8.87	10.0	18	15	14.5
C93700	1705	1400	0.320	8.95	10.3	12	10	11
C93800	1730	1570	0.334	9.25	10.3	13	12	10
C94300	1750	1650	0.336	9.29	10.0	16	9	11
C94700	1880	1660	0.320	8.80	11.0	14	12	15
C94700 (HT)	1880	1660	0.320	8.80	11.0	15	15	15
C94800	1880	1660	0.320	8.80	11.0	10	12	15
C95200	1913	1907	0.276	7.64	9.0	13	11	15
C95300	1913	1904	0.272	7.53	9.0	16	15	16
C95300 (HT)	1913	1904	0.272	7.53	9.0	16	13	15
C95400	1900	1880	0.269	7.45	9.0	15	13	15.5
C95400 (HT)	1900	1880	0.269	7.45	9.2	15	12	16
C95500	1930	1900	0.272	7.53	9.0	11	8.5	16
C95500 (HT)	1930	1900	0.272	7.53	9.0	11	8	17
C95800	1940	1910	0.276	7.64	9.0	9	7	16.5
C96200	2100	2010	0.323	8.94	9.5	11	11	18

* Cu = 226 Btu/ft²/ft/h/°F at 68°F.
† International annealed copper standard.
SOURCE: Reprinted with permission, copyright 1992, Society of Automotive Engineers.

TABLE 4.43 Tensile Strength Requirements and Approximate Rockwell Hardness Values for Rolled Tempers

Rolled temper Temper designation		Tensile strength, ksi*		Tensile strength, MPa†		Approximate Rockwell hardness‡							
						B scale				Superficial 30-T			
						0.020 (0.508) to 0.036 in (0.914 mm) incl		Over 0.036 in (0.914 mm)		0.012 (0.305) to 0.028 in (0.711 mm) incl		Over 0.028 in (0.711 mm)	
Standard	Former	Min	Max	Min	Max	Min	Max	Min	Max	Min	Max	Min	Max
						Copper Alloy UNS No. C21000							
M20	As hot rolled	32	42	220	290	—	—	—	—	—	—	—	—
H01	Quarter hard	37	47	255	325	20	48	24	52	34	51	37	54
H02	Half hard	42	52	290	355	40	56	44	60	46	57	48	59
E03	Three-quarter hard	46	56	315	385	50	61	53	64	52	60	54	62
H04	Hard	50	59	345	405	57	64	60	67	57	62	59	64
H06	Extra hard	56	64	385	440	64	70	66	72	62	66	63	67
H08	Spring	60	68	415	470	68	73	70	75	64	68	65	69
H10	Extra spring	61	69	420	475	69	74	71	76	65	69	66	70
						Copper Alloy UNS No. C22000							
M20	As hot rolled	33	43	230	295	—	—	—	—	—	—	—	—
H01	Quarter hard	40	50	275	345	27	52	31	56	34	51	37	54
H02	Half hard	47	57	325	395	50	63	53	66	50	59	52	61
H03	Three-quarter hard	52	62	355	425	59	68	62	71	55	62	58	64
H04	Hard	57	66	395	455	65	72	68	75	60	65	62	67
H06	Extra hard	64	72	440	495	72	77	74	79	64	68	66	69
H08	Spring	69	77	475	530	76	79	78	81	67	69	68	70
H10	Extra spring	73	80	495	550	78	81	80	83	68	70	69	71

TABLE 4.43 Tensile Strength Requirements and Approximate Rockwell Hardness Values for Rolled Tempers (*Continued*)

Rolled temper		Tensile strength, ksi*		Tensile strength, MPa†		Approximate Rockwell hardness‡							
Standard	Former (Temper designation)					B scale				Superficial 30-T			
						0.020 (0.508) to 0.036 in (0.914 mm) incl		Over 0.036 in (0.914 mm)		0.012 (0.305) to 0.028 in (0.711 mm) incl		Over 0.028 in (0.711 mm)	
		Min	Max	Min	Max	Min	Max	Min	Max	Min	Max	Min	Max
				Copper Alloy UNS No. C22600									
H01	Quarter hard	42	52	290	355	29	58	29	58	39	58	39	58
H02	Half hard	48	58	330	400	52	68	52	68	54	64	54	64
H03	Three-quarter hard	53	63	365	435	61	73	61	73	59	68	59	68
H04	Hard	58	67	400	460	67	77	67	77	64	70	64	70
H06	Extra hard	65	73	450	505	74	81	74	81	68	73	68	73
H08	Spring	70	78	485	540	78	83	78	83	71	74	71	74
H10	Extra spring	74	82	510	565	81	86	81	86	73	76	73	76
				Copper Alloy UNS No. C23000									
M20	As hot rolled	37	47	255	325	—	—	—	—	—	—	—	—
H01	Quarter hard	44	54	305	370	33	58	37	62	42	57	45	60
H02	Half hard	51	61	350	420	56	68	59	71	56	64	58	66
H03	Three-quarter hard	57	67	395	460	66	73	69	76	63	68	65	70
H04	Hard	63	72	435	495	72	78	74	80	67	71	68	72
H06	Extra hard	72	80	495	550	78	83	80	85	70	74	71	75
H08	Spring	78	86	540	595	82	85	84	87	74	76	75	77
H10	Extra spring	82	90	565	620	84	87	86	89	75	77	76	78

Copper Alloy UNS No. C24000

M20	As hot rolled	41	51	285	350	—	—	—	—	—	—	—	—
H01	Quarter hard	48	58	330	400	38	61	42	65	42	57	45	60
E02	Half hard	55	65	380	450	59	70	62	73	56	64	58	66
E03	Three-quarter hard	61	71	420	490	69	76	72	79	63	68	65	70
H04	Hard	68	77	470	530	76	82	78	84	68	72	69	73
H06	Extra hard	78	87	540	600	83	87	85	89	72	75	73	76
H08	Spring	85	93	585	640	87	90	89	92	75	77	76	78
H10	Extra spring	89	97	615	670	88	91	90	93	76	78	77	79

Copper Alloy UNS No. C26000

M20	As hot rolled	41	51	285	350	—	—	—	—	—	—	—	—
H01	Quarter hard	49	59	340	405	40	61	44	65	43	57	46	60
H02	Half hard	57	67	395	460	60	74	63	77	56	66	58	68
H03	Three-quarter hard	64	74	440	510	72	79	75	82	65	70	67	72
H04	Hard	71	81	490	560	79	84	81	86	70	73	71	74
H06	Extra hard	83	92	570	635	85	89	87	91	74	76	75	77
H08	Spring	91	100	625	690	89	92	90	93	76	78	76	78
H10	Extra spring	95	104	655	715	91	94	92	95	77	79	77	79

Copper Alloy UNS No. C26800

M20	As hot rolled	40	50	275	345	—	—	—	—	—	—	—	—
H01	Quarter hard	49	59	340	405	40	61	44	65	43	57	46	60
H02	Half hard	55	65	380	450	57	71	60	74	54	64	56	66
H03	Three-quarter hard	65	72	425	495	70	77	73	80	65	69	67	71
H04	Hard	68	78	470	540	76	82	78	84	68	72	69	73
H06	Extra hard	79	89	545	615	83	87	85	89	73	75	74	76
H08	Spring	86	95	595	655	87	90	89	92	75	77	76	78
H10	Extra spring	90	99	620	685	88	91	90	93	76	78	77	79

TABLE 4.43 Tensile Strength Requirements and Approximate Rockwell Hardness Values for Rolled Tempers (Continued)

Rolled temper		Tensile strength, ksi*		Tensile strength, MPa†		Approximate Rockwell hardness‡							
						B scale				Superficial 30-T			
						0.020 (0.508) to 0.036 in (0.914 mm) incl		Over 0.036 in (0.914 mm)		0.012 (0.305) to 0.028 in (0.711 mm) incl		Over 0.028 in (0.711 mm)	
Standard	Former	Min	Max	Min	Max	Min	Max	Min	Max	Min	Max	Min	Max
					Copper Alloy UNS No. C27200								
M20	As hot rolled	41	51	285	350	—	—	—	—	—	—	—	—
H01	Quarter hard	49	59	340	405	40	61	44	65	43	57	46	60
H02	Half hard	56	66	385	455	57	74	60	76	54	67	56	68
H03	Three-quarter hard	63	73	435	505	71	78	74	81	64	70	66	71
H04	Hard	70	80	485	550	76	82	78	84	67	72	68	73
H06	Extra hard	81	91	560	625	82	87	85	89	71	75	72	76
					Copper Alloy UNS No. C28000								
M20	As hot rolled	40	55	275	380	—	—	—	—	—	—	—	—
H01	Quarter hard	50	62	345	425	40	65	45	70	45	65	45	70
H02	Half hard	58	70	400	485	50	75	52	80	50	70	50	75
H03	Three-quarter hard	60	75	415	515	55	80	55	82	52	78	55	80
H04	Hard	70	85	485	585	60	85	60	87	55	80	55	82
H06	Extra hard	82	95	565	655	65	92	65	90	60	85	60	85

NOTE: Plate is generally available in only the as hot-rolled (M20) temper. Required properties for other tempers shall be agreed upon between the manufacturer and the purchaser at the time of placing the order.

* ksi = 1000 psi.

† MPa (megapascals).

‡ Rockwell hardness values apply as follows: the B scale values apply to metal 0.020 in (0.508 mm) and over in thickness, and the 30-T scale values apply to metal 0.012 in (0.305 mm) and over in thickness.

SOURCE: Reprinted with permission from the *Annual Book of ASTM Standards*, copyright 1992, American Society for Testing and Materials.

TABLE 4.44 Tensile Strength Requirements and Approximate Rockwell Hardness Values

Temper designation[†]			Tensile strength, ksi* (MPa)		Approximate Rockwell hardness	
Standard	Former	Thickness, in (mm)	Min	Max	B scale	Superficial 30-T
		Copper Alloy UNS No. C51000				
M20	As hot rolled	Over 0.188 (4.775)	40 (275)	60 (415)	—	—
O60	Soft	Over 0.039 (0.991)	43 (295)	58 (400)	16–64	—
		Over 0.029 (0.737)			—	32–59
		Over 0.020 (0.508) to 0.039 (0.991) incl			12–60	—
		Over 0.010 (0.254) to 0.029 (0.737) incl			—	24–53
		0.003 (0.076) to 0.010 (0.254) incl				
H02	Half hard	Over 0.039 (0.991)	58 (400)	73 (505)	64–85	—
		Over 0.029 (0.737)			—	59–73
		Over 0.020 (0.508) to 0.039 (0.991) incl			60–82	—
		Over 0.010 (0.254) to 0.029 (0.737) incl			—	53–69
		0.003 (0.076) to 0.010 (0.254) incl				
H04	Hard	Over 0.039 (0.991)	76 (525)	91 (625)	86–93	—
		Over 0.029 (0.737)			—	73–78
		Over 0.020 (0.508) to 0.039 (0.991) incl			84–91	—
		Over 0.010 (0.254) to 0.029 (0.737) incl			—	71–75
		0.003 (0.076) to 0.010 (0.254) incl				
H06	Extra hard	Over 0.039 (0.991)	88 (605)	103 (710)	92–96	—
		Over 0.029 (0.737)			—	77–81
		Over 0.020 (0.508) to 0.039 (0.991) incl			89–95	—
		Over 0.010 (0.254) to 0.029 (0.737) incl			—	74–78
		0.003 (0.076) to 0.010 (0.254) incl				
H08	Spring	Over 0.039 (0.991)	95 (655)	110 (760)	94–98	—
		Over 0.029 (0.737)			—	79–82
		Over 0.020 (0.508) to 0.039 (0.991) incl			92–97	—

227

TABLE 4.44 Tensile Strength Requirements and Approximate Rockwell Hardness Values (*Continued*)

Temper designation†		Thickness, in (mm)	Tensile strength, ksi* (MPa)		Approximate Rockwell hardness	
Standard	Former		Min	Max	B scale	Superficial 30-T
		Copper Alloy UNS No. C51000 (*Continued*)				
H10	Extra spring	Over 0.010 (0.254) to 0.029 (0.737) incl			—	76–80
		0.003 (0.076) to 0.010 (0.254) incl				
		Over 0.039 (0.991)	100 (690)	114 (790)	95–99	—
		Over 0.029 (0.737)			—	80–83
		Over 0.020 (0.508) to 0.039 (0.991) incl			94–98	—
		Over 0.010 (0.254) to 0.029 (0.737) incl			—	77–81
		0.003 (0.076) to 0.010 (0.254) incl				
		Copper Alloy UNS Nos. C51100, C53200, C53400, and C54400				
M20	As hot rolled	Over 0.188 (4.775)	40 (275)	58 (415)	—	—
O60	Soft	Over 0.039 (0.991)	40 (275)	55 (380)	7–50	24–50
		Over 0.029 (0.737)			—	—
		Over 0.020 (0.508) to 0.039 (0.991) incl			0–45	16–46
		Over 0.010 (0.254) to 0.029 (0.737) incl			—	—
H02	Half hard	Over 0.039 (0.991)	55 (380)	70 (485)	60–81	—
		Over 0.029 (0.737)			—	57–73
		Over 0.020 (0.508) to 0.039 (0.991) incl			53–78	—
		Over 0.010 (0.254) to 0.029 (0.737) incl			—	52–71
H04	Hard	Over 0.039 (0.991)	72 (495)	87 (600)	82–90	—
		Over 0.029 (0.737)			—	71–77
		Over 0.020 (0.508) to 0.039 (0.991) incl			80–88	—
		Over 0.010 (0.254) to 0.029 (0.737) incl			—	69–75
H06	Extra hard	Over 0.039 (0.991)	84 (580)	99 (685)	88–94	—
		Over 0.029 (0.737)			—	75–80

Temper		Thickness, in. (mm)	Tensile strength, ksi (MPa)		Rockwell hardness	
		Over 0.020 (0.508) to 0.039 (0.991) incl			86–92	—
		Over 0.010 (0.254) to 0.029 (0.737) incl			—	73–78
H08	Spring	Over 0.039 (0.991)	91 (625)	105 (720)	90–96	—
		Over 0.029 (0.737)			—	77–81
		Over 0.020 (0.508) to 0.039 (0.991) incl			88–94	—
		Over 0.010 (0.254) to 0.029 (0.737) incl			—	75–79
H10	Extra spring	Over 0.039 (0.991)	96 (660)	109 (750)	92–97	—
		Over 0.029 (0.737)			—	78–82
		Over 0.020 (0.508) to 0.039 (0.991) incl			89–94	—
		Over 0.010 (0.254) to 0.029 (0.737) incl			—	76–80
Copper Alloy UNS No. C51900						
O50	Soft	Over 0.039 (0.991)	48 (330)	63 (435)	22–66	—
		Over 0.029 (0.737)			—	35–64
		Over 0.020 (0.508) to 0.039 (0.991) incl			18–63	—
		Over 0.010 (0.254) to 0.029 (0.737) incl			—	25–57
H02	Half hard	Over 0.039 (0.991)	64 (440)	79 (545)	70–88	—
		Over 0.029 (0.737)			—	63–76
		Over 0.020 (0.508) to 0.039 (0.991) incl			65–85	—
		Over 0.010 (0.254) to 0.029 (0.737) incl			—	58–72
H04	Hard	Over 0.039 (0.991)	80 (550)	96 (660)	89–95	—
		Over 0.029 (0.737)			—	74–80
		Over 0.020 (0.508) to 0.039 (0.991) incl			86–93	—
		Over 0.010 (0.254) to 0.029 (0.737) incl			—	72–78
Copper Alloy UNS No. C52100						
M20	As hot rolled	Over 0.188 (4.775)	50 (345)‡	78 (485)	—	—
O60	Soft	Over 0.039 (0.991)	53 (365)	67 (460)	29–70	—
		Over 0.029 (0.737)			—	38–68
		Over 0.020 (0.508) to 0.039 (0.991) incl			20–66	—
		Over 0.010 (0.254) to 0.029 (0.737) incl			—	27–62

TABLE 4.44 Tensile Strength Requirements and Approximate Rockwell Hardness Values (Continued)

| Temper designation[†] | | Thickness, in (mm) | Tensile strength, ksi* (MPa) | | Approximate Rockwell hardness | |
Standard	Former		Min	Max	B scale	Superficial 30-T
		Copper Alloy UNS No. C52100 (Continued)				
H02	Half hard	Over 0.039 (0.991)	69 (475)	84 (580)	76–91	—
		Over 0.029 (0.737)			—	67–78
		Over 0.020 (0.508) to 0.039 (0.991) incl			69–88	—
		Over 0.010 (0.254) to 0.029 (0.737) incl			—	63–75
H04	Hard	Over 0.039 (0.991)	85 (585)	100 (690)	91–97	—
		Over 0.029 (0.737)			—	76–81
		Over 0.020 (0.508) to 0.039 (0.991) incl			89–95	—
		Over 0.010 (0.254) to 0.029 (0.737) incl			—	73–80
H06	Extra hard	Over 0.039 (0.991)	97 (670)	112 (770)	95–100	—
		Over 0.029 (0.737)			—	78–83
		Over 0.020 (0.508) to 0.039 (0.991) incl			93–98	—
		Over 0.010 (0.254) to 0.029 (0.737) incl			—	77–82
H08	Spring	Over 0.039 (0.991)	105 (720)	119 (820)	97–102	—
		Over 0.029 (0.737)			—	79–84
		Over 0.020 (0.508) to 0.039 (0.991) incl			95–100	—
		Over 0.010 (0.254) to 0.029 (0.737) incl			—	78–83
H10	Extra spring	Over 0.039 (0.991)	110 (760)	122 (830)	98–103	—
		Over 0.029 (0.737)			—	80–84
		Over 0.020 (0.508) to 0.039 (0.991) incl			96–101	—
		Over 0.010 (0.254) to 0.029 (0.737) incl			—	79–83
		Copper Alloy UNS No. C52400				
M20	As hot rolled	Over 0.188 (4.775)	55 (380)	75 (515)	—	—
O60	Soft	Over 0.039 (0.991)	58 (400)	73 (505)	35–75	—
		Over 0.029 (0.737)			—	40–78

Designation	Temper	Thickness, in. (mm)	Tensile strength, ksi (MPa)		Rockwell hardness	
	(preceding temper, continued)	Over 0.029 (0.737)				40–78
		Over 0.020 (0.508) to 0.039 (0.991) incl			25–71	
		Over 0.010 (0.254) to 0.029 (0.737) incl				29–84
H02	Half hard	Over 0.039 (0.991)	76 (525)	91 (625)	78–95	67–80
		Over 0.029 (0.737)			74–93	63–77
H04	Hard	Over 0.039 (0.991)	94 (650)	109 (750)	94–101	78–82
		Over 0.029 (0.737)			92–100	75–81
H06	Extra hard	Over 0.039 (0.991)	107 (740)	122 (830)	98–103	80–84
		Over 0.029 (0.737)			97–102	79–83
H08	Spring	Over 0.039 (0.991)	115 (790)	129 (890)	99–104	81–85
		Over 0.029 (0.737)			98–103	80–84
H10	Extra spring	Over 0.039 (0.991)	120 (830)	133 (920)	100–105	82–86
		Over 0.029 (0.737)			99–104	81–85

NOTE: Plate is generally available in only the "As hot-rolled" (M20) temper. Required properties for other tempers shall be agreed upon between the manufacturer and the purchaser at the time of placing the order.

≈ ksi = 1000 psi.

 Standard designations defined in Practice B 601.

 Editorially corrected.

SOURCE: Reprinted with permission from the *Annual Book of ASTM Standards*, copyright 1992, American Society for Testing and Materials.

TABLE 4.45 Mechanical Property Requirements for Material in the Solution Heat-Treated or Solution Heat-Treated and Cold-Worked Condition

Temper designation§		Tensile strength, ksi* (MPa)	Elongation† in 2 in or 50 mm, min, %	Rockwell hardness‡		
Standard	Former			B scale	30T scale	15T scale
TB00	A	60–78 (410–540)	35	45–78	46–67	75–85
TD01	¼ H	75–88 (520–610)	10	68–90	62–75	83–89
TD02	½ H	85–100 (590–690)	5	88–96	74–79	88–91
TD04	H	100–120 (690–830)	2	96–102	79–83	91–94

* ksi = 1000 psi.
† Elongation requirement applies only to strip 0.004 in (0.102 mm) and thicker.
‡ The thickness of material that may be tested by use of the Rockwell hardness scales is as follows:

    B scale   0.040 in (1.016 mm) and over
    30T scale   0.020 to 0.040 in (0.508 to 1.016 mm), excl.
    15T scale   0.015 to 0.020 in (0.381 to 0.508 mm), excl.

Hardness values shown apply only to direct determinations, not converted values.
§ Standard designations defined in Practice B 601.
SOURCE: Reprinted with permission from the *Annual Book of ASTM Standards*, copyright 1992, American Society for Testing and Materials.

TABLE 4.46 Mechanical Property Requirements After Precipitation Heat Treatment*

Temper designation		Tensile strength, ksi[†] (MPa)	Yield strength, ksi (MPa), min, 0.2% offset	Elongation in 2 in (50 mm), min, %[‡]	Rockwell hardness[§], min		
Standard	Former				C scale	30N scale	15N scale
Copper Alloy UNS No. C17000							
TF00	AT	150–180¶ (1030–1240)	130 (890)	3	33	53	76.5
TF01	¼ HT	160–190¶ (1100–1310)	135 (930)	2.5	35	55	77
TH02	½ HT	170–200¶ (1170–1380)	145 (1000)	1	37	57	78.5
TH04	HT	180–210¶ (1240–1450)	155 (1070)	1	38	58	79.5
Copper Alloy UNS No. C17200							
TF00	AT	165–195¶ (1140–1340)	140 (960)	3	36	56	78
TH01	¼ HT	175–205¶ (1210–1410)	150 (1030)	2.5	36	56	79
TH02	½ HT	185–215¶ (1280–1480)	160 (1130)	1	38	58	79.5
TH04	HT	190–220¶ (1310–1520)	165 (1140)	1	38	58	80

* These values apply to mill products (Section 11). See 11.3 for exceptions in end products.
† ksi = 1000 psi.
‡ Applicable to material 0.004 in (0.102 mm) and thicker.
§ The thickness of material that may be tested by use of the Rockwell hardness scales is as follows:

C scale   0.040 in (1.016 mm) and over
30N scale   0.020 to 0.040 in (0.508 to 1.016 mm), excl.
15N scale   0.015 to 0.02 in (0.381 to 0.508 mm), excl.

Hardness values shown apply only to direct determinations, not converted values.
¶ The upper limits in the tensile strength column are for design guidance only.

SOURCE: Reprinted with permission from the *Annual Book of ASTM Standards*, copyright 1992, American Society for Testing and Materials.

TABLE 4.47 Mechanical Property Requirements—Mill-Hardened Condition*

Temper designation		Tensile strength, ksi[†] (MPa)	Yield strength, ksi (MPa), 0.2% offset	Elongation in 2 in (50 mm), min, %[‡]	Rockwell hardness,[§] min		
Standard	Former[†]				C scale	30N scale	15N scale
Copper Alloy UNS No. C17000							
TM00	AM	100–110¶ (690–760)	70–95 (480–660)	18	18	37	67.5
TM01	¼ HM	110–120¶ (760–830)	80–110 (550–760)	15	20	42	70
TM02	½ HM	120–135¶ (830–930)	95–125 (660–860)	12	24	45	72
TM04	HM	135–150¶ (930–1040)	110–135 (760–930)	9	28	48	75
TM05	SHM	150–160¶ (1030–1100)	125–140 (860–970)	9	31	52	75.5
TM06	XHM	155–175¶ (1070–1210)	135–165 (930–1140)	3	32	52	76
Copper Alloy UNS No. C17200							
TM00	AM	100–110¶ (690–760)	70–95 (480–660)	16	R_B95	37	67.5
TM01	¼ HM	110–120¶ (760–830)	80–110 (550–760)	15	20	42	70
TM02	½ HM	120–135¶ (830–930)	95–125 (660–860)	12	23	44	72
TM04	HM	135–150¶ (930–1030)	110–135 (760–930)	9	28	48	75

TM05	SHM	150–160[1] (1030–1100)	125–140 (860–970)	9	31	52	75.5
TM06	XHM	155–175[1] (1070–1210)	135–170 (930–1170)	4	32	52	76
TM08	XHMS	175–190[¶] (1210–1310)	150–180 (1030–1240)	3	33	53	76.5

* These values apply to mill products (Section 11). See 11.3 for exceptions in end products.

[†] ksi = 1000 psi.

[‡] Applicable to material 0.004 in (0.102 mm) and thicker.

[§] The thickness of material that may be tested by use of the Rockwell hardness scales is as follows:

  C scale   0.040 in (1.016 mm) and over

  30N scale 0.020 to 0.040 in (0.508 to 1.016 mm), excl.

  15N scale 0.015 to 0.020 in (0.381 to 0.508 mm), excl.

Hardness values shown apply only to direct determinations, not converted values.

[¶] The upper limits in the tensile strength column are for design guidance only.

SOURCE: Reprinted with permission from the *Annual Book of ASTM Standards*, copyright 1992, American Society for Testing and Materials.

TABLE 4.48 Classification of Coppers

| | | | Form in which copper is available[b] | | | | | | | |
| | | | From refiners | | | | From fabricators | | | |
Designations	Type of copper	UNS nos.[a]	Wire bars	Billets	Cakes	Ingots and ingot bars	Flat products	Pipe and tube	Rod and wire	Shapes
CATH	Electrolytic cathode		Cathodes only							
Tough-Pitch Coppers										
ETP	Electrolytic tough pitch	C11000	x	x	x	x	x	x	x	x
RHC	Remelted, high-conductivity tough pitch	C11010	x	x	x	x	x	x	x	x
ETP	Electrolytic tough pitch (anneal resist)	C11100	x	x	x	x	x	x	x	x
CRTP	Chemically refined tough pitch	C11030	x	x	x	x	x	x	x	x
FRHC	Fire-refined, high-conductivity tough pitch	C11020	x	x	x	x	x	x	x	x
ETP[c]	Silver-bearing, tough pitch	C11300, C11400, C11500, C11600	x	x	x	x	x	x	x	x
FRTP	Fire-refined, tough pitch	C12500		x	x	x	x	x		x
FRSTP	Fire-refined tough pitch with silver	C12700, C12800, C12900, C1300		x	x	x	x	x	x	x
Oxygen-Free Coppers (Without Use of Deoxidants)										
OFE	Oxygen-free, electronic	C10100	x	x	x	x	x	x	x	x
OF	Oxygen-free	C10200	x	x	x	x	x	x	x	x
OFS	Oxygen-free, silver-bearing	C10400, C10500, C10700	x	x	x	x	x	x	x	x
OFXLP	Oxygen-free, extra low phosphorus	C10300	x	x	x	x	x	x	x	x
OFLP	Oxygen-free, low phosphorus	C10800	x	x	x	x	x	x	x	x

			Deoxidized Coppers					
DLP	Phosphorized, low-residual phosphorus	C12000	x	x	x	x	x	x
DLPS[d]	Phosphorized, low-residual phosphorus silver-bearing	C12100	x	x	x	x	x	x
DHP[e]	Phosphorized, high-residual phosphorus	C12200	x	x	x	x	x	x
DHPS[d]	Phosphorized, high-residual phosphorus silver-bearing	C12300	x	x	x	x	x	x
DPA	Phosphorized, arsenic-bearing	C14200	x	x	x	x		
DPTE[f]	Phosphorized, tellurium-bearing	C14500	x				x	
			Other Coppers					
	Sulfur-bearing	C14700	x	x		x		
	Zirconium-bearing	C15000	x	x	x			x
PTE	Tellurium-bearing	C14500	x					x

[a] The chemical compositions associated with these numbers are listed in the product specifications and in the Standard Designations for Copper and Copper Alloys.

[b] The x in the table indicates commercial availability.

[c] This includes types ETP, CRTP, and FRHC coppers to which silver has been added in amounts agreed upon.

[d] This includes oxygen-free copper to which phosphorus and silver have been added in amounts agreed upon.

[e] This includes oxygen-free copper to which phosphorus has been added.

[f] This includes oxygen-free tellurium-bearing copper to which phosphorus has been added in amounts agreed upon.

SOURCE: Reprinted with permission from the Annual Book of ASTM Standards, copyright 1992, American Society for Testing and Materials.

TABLE 4.49   **Specifications of Magnesium Wrought Alloys**

Alloy designation						
UNS	ASTM and SAE	Old SAE	Form	ASTM	AMS	Military or federal
M11311	AZ31B	510	Sheet and plate	B90	4375, 4376, 4377	QQ-M-44
			Bar, rod, shapes	B107	—	QQ-M-31
			Tube	B107	—	WW-T-825
			Forgings	B91	—	QQ-M-40
M11610	AZ61A	520	Bar, rod, shapes	B107	4350	QQ-M-31
			Tube	B107	4350	WW-T-825
			Wire (welding rod)	—	—	Mil-R-6944
			Forgings	B91	4358*	QQ-M-40
M11800	AZ80A	523	Bar, rod, shapes	B107	—	QQ-M-31
			Forgings	B91	4360*	QQ-M-40
M14141	LA141A		Sheet and plate	B90	—	—
M13310	HK31A	507	Sheet and plate	B90	4384, 4385	Mil-M-26075
M13210	HM21A		Sheet and plate	B90	4383, 4390	Mil-M-8917
			Forgings	—	4363	QQ-M-40
M13312	HM31A		Bar, rod, shapes	—	4388, 4389	Mil-M-8916
M15100	M1A	522	Bar, rod, shapes	B107	—	QQ-M-31
			Forgings	—	—	QQ-M-40
M16100	ZE10A	534	Sheet and plate	B90	—	Mil-M-46037
M16400	ZK40A		Bar, rod, shapes	B107	—	—
M16600	ZK60A	524	Bar, rod, shapes	B107	4352	QQ-M-31
			Tube	B107	4352	WW-T-825
			Forgings	B91	4362	QQ-M-40

* Noncurrent specifications.

SOURCE: Reprinted with permission from the *Annual Book of ASTM Standards,* copyright 1992, American Society for Testing and Materials.

TABLE 4.50 Physical Properties and Characteristics of Magnesium Sand-Casting Alloys

Alloy designation			Approximate melting range, °F (°C)			Pattern shrinkage allowance, in./ft (mm/m)[b]	Foundry characteristics[c]					Other characteristics				
UNS	ASTM and SAE	Old SAE	Nonequilibrium solidus[a]	Solidus	Liquidus		Pressure tightness	Fluidity[d]	Microporosity tendency[e]	Normally heat treated	Castability	Machining[f]	Electroplating[g]	Surface treatment[h]	Suitability to brazing[i]	Suitability to welding[j]
M10100[m]	AM100A	502	810 (432)	867 (464)	1100 (593)	5/32 (13.0)	2	1	2	Yes	2	1	2	2	No	1
M11630	AZ63A	50	655 (363)	850 (454)	1130 (610)	5/32 (13.0)	3	1	3	Yes	3	1	1	1	No	3
M11810[m]	AZ81A	505	730 (421)	882 (472)	1115 (602)	5/32 (13.0)	2	1	2	Yes	1	1	2	2	No	1
M11914[m]	AZ91C	504	755 (418)	875 (468)	1105 (596)	5/32 (13.0)	2	1	2	Yes	1	1	2	2	No	2
M11920[m]	AZ92A	500	770 (410)	830 (443)	1100 (593)	5/32 (13.0)	2	1	2	Yes	2	1	2	2	No	2
M12330[n]	EZ33A	506	—	1010 (543)	1189 (643)	5/16 (15.5)	1	2	1	Yes	1	1	1	1	No	1
M13310[m]	HK31A	507	—	1092 (589)	1204 (651)	7/32 (18.0)	1	2	1	Yes	1	1	1	1	—[k]	1
M13320[n]	HZ32A	—	—	1026 (552)	1198 (648)	3/16 (15.5)	1	2	1	Yes	1	1	1[k]	2	—[k]	2
M18010[p]	K1A	—	—	—	1205 (652)	3/16 (15.5)	1[k]	2	—	No	2	1	3	2	—[k]	1
M18210	QH21A	—	—	1004 (539)	1184 (640)	3/16 (15.5)	2	2	2	Yes	1	1	3	1	No[k]	2
M18220[m]	QE22A	—	—	1020 (549)	1190 (643)	5/32 (13.0)	2	2	2	Yes	1	1	2	1	—[k]	1
M16410[p]	ZE41A	—	—	950 (510)	1184 (640)	3/16 (15.5)	—[k]	2	—[k]	Yes	1	1	1[k]	1	—[k]	2
M16630[p]	ZE63A	—	—	510 (266)	950 (510)	3/16 (15.5)	1	2	1	Yes	1	1	1	1	No	1[k]
M16620	ZH62A	508	—	—	1169 (632)	5/32 (13.0)	2	2	2	Yes	2	1	1	2	No	—
M16510	ZK51A	509	—	1020 (549)	1185 (641)	5/32 (13.0)	3	2	3	Yes	3	1	2	2	No	3
M16610	ZK61A	513	—	985 (529)	1175 (635)	5/32 (13.0)	3	2	3	Yes	3	1	2	1	No	3

[a] As measured on metal solidified under normal casting conditions.

[b] Allowance for average castings. Shrinkage requirements will vary with intricacy of design and dimensions. (1 in/ft × 8.333 = % shrinkage.)

[c] Rating of 1 indicates best of group; 3 indicates poorest of group.

[d] Ability of liquid alloy to flow readily in mold and fill thin sections.

[e] Based on radiographic evidence.

[f] Composite rating based on ease of cutting, chip characteristics, quality of finish, and tool life. Ratings, in the case of heat-treatable alloys, based on –T6 type temper. Other tempers, particularly the annealed temper, may have lower ratings.

[g] Ability of casting to take and hold an electroplate applied by present standard methods

[h] Ability of castings to be cleaned in standard pickle solutions and to be conditioned for best paint adhesion.

[i] Refers to suitability of alloy to withstand brazing temperature without excessive distortion or melting.

[j] Based on ability of material to be fusion welded with filler rod of same alloy.

[k] Inexperience with these alloys under wide production conditions makes it undesirable to supply ratings at this time.

[m] Properties applicable for permanent mold and investment castings.

[n] Properties applicable for permanent mold castings also.

[p] Properties applicable for investment castings also.

SOURCE: Reprinted with permission from the *Annual Book of ASTM Standards,* copyright 1992, American Society for Testing and Materials.

239

TABLE 4.51 Tensile Requirements

UNS no.	ASTM no.	Temper	Form	Specified diameter or thickness, in[a,c]	Specified cross-sectional area, in[2] or OD of tube, in	Tensile strength, min ksi	Tensile strength, min (MPa)[d]	Yield strength (0.2% offset), min ksi	Yield strength (0.2% offset), min (MPa)[d]	Elongation in 2 in or 4 × dia., min, %
M11311	AZ31B	F	Bars, rods, shapes, and wire	0.249 and under	All	35.0	(241)	21.0	(145)	7
				0.250–1.499	All	35.0	(241)	22.0	(152)	7
				1.500–2.499	All	34.0	(234)	22.0	(152)	7
				2.500–4.999	All	32.0	(221)	20.0	(138)	7
			Hollow shapes	All	All	32.0	(221)	16.0	(110)	8
			Tubes	0.028–0.500	6.000 and under	32.0	(221)	16.0	(110)	8
				0.251–0.750		32.0	(221)	16.0	(110)	4
M11610	AZ61A	F	Bars, rods, shapes, and wire	0.249 and under	All	38.0	(262)	21.0	(145)	8
				0.250–2.499	All	39.0	(269)	24.0	(165)	9
				2.500–4.999	All	40.0	(276)	22.0	(152)	7
			Hollow shapes	All	All	36.0	(248)	16.0	(110)	7
			Tubes	0.028–0.750	6.000 and under	36.0	(248)	16.0	(110)	7
M11800	AZ80A	F	Bars, rods, shapes, and wire	0.249 and under	All	43.0	(296)	28.0	(193)	9
				0.250–1.499	All	43.0	(296)	28.0	(193)	8
				1.500–2.499	All	43.0	(296)	28.0	(193)	6
				2.500–4.999	All	42.0	(290)	27.0	(186)	4
M11800	AZ80A	T5	Bars, rods, shapes, and wire	0.249 and under	All	47.0	(324)	30.0	(207)	4
				0.250–2.499	All	48.0	(331)	33.0	(228)	4
				2.500–4.999	All	45.0	(310)	30.0	(207)	2
M15100	M1A	F	Bars, rods, shapes, and wire	0.249 and under	All	30.0	(207)	b	—	2
				0.250–1.499	All	32.0	(221)	b	—	3
				1.500–2.499	All	32.0	(221)	b	—	2

UNS No.	Alloy	Temper	Product form			ksi	(MPa)	ksi	(MPa)	%
				2.500–4.999	All	29.0	(200)	b	—	2
			Hollow shapes	All	All	28.0	(193)	b	—	2
			Tubes	0.028–0.750	6.000 and under	28.0	(193)	b	—	2
M16400	ZK40A	T5	Bars, rods, shapes, and wire	All	4.999 and under	40.0	(276)	37.0	(255)	4.0
			Hollow shapes	All	All	40.0	(276)	37.0	(255)	4.0
			Tubes	0.062–0.500	3.000 and under	40.0	(276)	36.0	(248)	4.0
M16600	ZK60A	F	Bars, rods, shapes, and wire	All	4.999 and under	43.0	(296)	31.0	(214)	5
					5.000–39.999	43.0	(296)	31.0	(214)	4
			Hollow shapes	All	All	40.0	(276)	28.0	(193)	5
			Tubes	0.028–0.750	3.000 and under	40.0	(276)	28.0	(193)	5
M16600	ZK60A	T5	Bars, rods, shapes, and wire	All	4.999 and under	45.0	(310)	36.0	(248)	4
			Hollow shapes	All	All	46.0	(317)	38.0	(262)	4
			Tubes	0.028–0.250	3.000 and under	46.0	(317)	38.0	(262)	4
				0.094–1.188	3.031–8.500	44.0	(303)	33.0	(228)	4

NOTE: For purposes of determining conformance with this specification, each value for tensile strength and yield strength shall be rounded to the nearest 100 psi and each value for elongation shall be rounded to the nearest 0.5%, both in accordance with the rounding method of Recommended Practice E 29.

a Intermediate dimensions shall be rounded off to the third decimal place in accordance with Recommended Practice E 29.

b Not required.

c Wall thickness of tubes.

d The values in megapascals are included for information only.

e Elongation of full-section and machined sheet-type specimens is measured in 2 in; of machined round specimens, in 4 × specimen dia.

f For material of such dimensions that a standard test specimen cannot be obtained, for wire less than 0.125 in diameter, or for material thinner than 0.062 in the test for elongation is not required.

SOURCE: Reprinted with permission from the Annual Book of ASTM Standards, copyright 1992, American Society for Testing and Materials.

tetrachloride with magnesium. Titanium is a lustrous, white metal with low density and good strength. The specific gravity of titanium is 4.54 g/cm^3, with a weight of 0.164 lb/in^3. The pure metal burns in air and is the only element that burns in nitrogen gas.

Titanium is used extensively in modern aircraft and aerospace vehicles, where a lightweight alloy with high strength finds many applications. Titanium is the ninth most abundant element in the earth's crust.

Table 4.52 lists the tensile and bend requirements of titanium alloys for annealed titanium plate, sheet, and strip, which are classified as follows:

Grades 1 through 4: Unalloyed titanium

Grade 5: Titanium alloy (6 percent aluminum, 4 percent vanadium)

Grade 6: Titanium alloy (5 percent aluminum, 2.5 percent tin)

Grade 7: Unalloyed titanium plus palladium

Grade 9: Titanium alloy (3 percent aluminum, 2.5 percent vanadium)

Grade 11: Unalloyed titanium plus palladium

Grade 12: Titanium alloy (0.3 percent molybdenum, 0.8 percent nickel)

Table 4.53 lists the mechanical properties of wrought titanium alloys.

4.6   The Unified Numbering System (UNS) for Metals and Alloys

The Unified Numbering System (UNS) provides a means of coordinating many nationally used numbering systems currently administered by societies, trade associations, and the producers of metals and alloys, thereby avoiding confusion caused by use of more than one identification number for the same material or by having the same number assigned to two or more entirely different materials. Table 4.54 shows the primary series of numbers, and Table 4.55 lists the secondary division of some series of numbers. When you know the UNS number for a metal or alloy, you may use these

(Text continued on page 248.)

TABLE 4.52 Tensile and Bend Requirements of Titanium Alloys

| Grade | Tensile strength,* min | | Yield strength,* 0.2% offset | | | | Elongation in 2 in or 50 mm, min, % | Bend test† | |
| | | | min | | max | | | Under 0.070 in (1.8 mm) in thickness | 0.070 to 0.187 in (1.8 to 4.75 mm) in thickness |
	ksi	MPa	ksi	MPa	ksi	MPa			
1	35	240	25	170	45	310	24	3T	4T
2	50	345	40	275	65	450	20	4T	5T
3	65	450	55	380	80	550	18	4T	5T
4	80	550	70	485	95	655	15	5T	6T
5	130	895	120	830	—	—	10‡,§	9T	10T
6	120	830	115	795	—	—	10‡	8T	9T
7	50	345	40	275	65	450	20	4T	5T
9	90	620	70	485	—	—	15¶	5T	6T
11	35	240	25	170	45	310	24	3T	4T
12	70	483	50	345	—	—	18	4T	5T

* Minimum and maximum limits apply to tests taken both longitudinal and transverse to the direction of rolling. Mechanical properties for conditions other than annealed or plate thickness over 1 in (25 mm) may be established by agreement between the manufacturer and the purchaser.

† T equals the thickness of the bend test specimen. Bend tests are not applicable to material over 0.187 in (4.75 mm) in thickness.

‡ For grades 5 and 6 the elongation on materials under 0.025 in (0.635 mm) in thickness may be obtained only by negotiation.

§ For grade 5, the elongation will be 8% minimum for thicknesses between 0.025 and 0.063 in.

¶ Elongation for continuous rolled and annealed (strip product from coil) for grade 9 shall be 12% minimum in the longitudinal direction and 8% minimum in the transverse direction.

SOURCE: Reprinted with permission from the *Annual Book of ASTM Standards*, copyright 1992, American Society for Testing and Materials.

TABLE 4.53 Mechanical Properties of Wrought Titanium Alloys

Nominal composition (%)	Condition	Tensile strength (ksi)	Room temperature Yield strength (ksi)	Elongation (%)	Reduction in area (%)
Commercially Pure					
99.5 Ti	Annealed	48	35	30	55
99.2 Ti	Annealed	63	50	28	50
99.1 Ti	Annealed	75	65	25	45
99.0 Ti	Annealed	96	85	20	40
99.2 Ti[a]	Annealed	63	50	28	50
98.9[b]	Annealed	75	65	25	42
Alpha Alloys					
5 Al, 2.5 Sn	Annealed	125	117	16	40
5 Al, 2.5 Sn (low O_2)	Annealed	117	108	16	—
Near Alpha Alloys					
8 Al, 1 Mo, 1 V	Duplex annealed	145	138	15	28
11 Sn, 1 Mo, 2.25 Al, 5.0 Zr, 1 Mo, 0.2 Si	Duplex annealed	160	144	15	35
6 Al, 2 Sn, 4 Zr, 2 Mo	Duplex annealed	142	130	15	35
5 Al, 5 Sn, 2 Zr, 2 Mo, 0.25 Si	975°C (1785°F) (½ h), AC + 595°C (1100°F) (2 h), AC	152	140	13	—
6 Al, 2 Nb, 1 Ta, 1 Mo	As-rolled 2.5-cm (1-in) plate	124	110	13	34
6 Al, 2 Sn, 1.5 Zr, 1 Mo, 0.35 Bi, 0.1 Si	Beta forge + duplex anneal	147	137	11	—
Alpha-Beta Alloys					
8 Mn	Annealed	137	125	15	32
3 Al, 2.5 V	Annealed	100	85	20	—
6 Al, 4 V	Annealed	144	134	14	30
	Solution + age	170	160	10	25
6 Al, 4 V (low O_2)	Annealed	130	120	15	35
6 Al, 6 V, 2 Sn	Annealed	155	145	14	30
	Solution + age	185	170	10	20
7 Al, 4 Mo	Solution + age	160	150	16	22
6 Al, 2 Sn, 4 Zr, 6 Mo	Solution + age	184	170	10	23
6 Al, 2 Sn, 2 Zr, 2 Mo, 2 Cr, 0.25 Si	Solution + age	185	165	11	33
10 V, 2 Fe, 3 Al	Solution + age	185	174	10	19
Beta Alloys					
13 V, 11 Cr, 3 Al	Solution + age	177	170	8	—
	Solution + age	185	175	8	—
8 Mo, 8 V, 2 Fe, 3 Al	Solution + age	190	180	8	—
3 Al, 8 V, 6 Cr, 4 Mo, 4 Zr	Solution + age	210	200	7	—
	Annealed	128	121	15	—
11.5 Mo, 6 Zr, 4.5 Sn	Solution + age	201	191	11	—

[a] Also contains 0.2 Pd.
[b] Also contains 0.8 Ni and 0.3 Mo.
SOURCE: Titanium Metals Corp. of America and RMI Co.

TABLE 4.54   **Primary Series of Numbers**

Nonferrous Metals and Alloys	
A00001–A99999	Aluminum and aluminum alloys
C00001–C99999	Copper and copper alloys
E00001–E99999	Rare earth and rare earth-like metals and alloys
L00001–L99999	Low-melting metals and alloys
M00001–M99999	Miscellaneous nonferrous metals and alloys
N00001–N99999	Nickel and nickel alloys
P00001–P99999	Precious metals and alloys
R00001–R99999	Reactive and refractory metals and alloys
Z00001–Z99999	Zinc and zinc alloys
Ferrous Metals and Alloys	
D00001–D99999	Specified mechanical properties steels
F00001–F99999	Cast irons and cast steels
G00001–G99999	AISI and SAE carbon and alloy steels
H00001 H99999	AISI H-steels
J00001–J99999	Cast steels (except tool steels)
K00001–K99999	Miscellaneous steels and ferrous alloys
S00001–S99999	Heat and corrosion resistant (stainless) steels
T00001–T99999	Tool steels
Specialized Metals and Alloys	
W00001–W99999	Welding filler metals, covered and tubular electrodes, classified by weld deposit composition

SOURCE: Reprinted with permission from the *Annual Book of ASTM Standards,* copyright 1992, American Society for Testing and Materials.

TABLE 4.55 Secondary Division of Some Series of Numbers

E00001–E99999 Rare Earth and Rare Earth–Like Metals and Alloys	
E00000–E00999	Actinium
E01000–E20999	Cerium
E21000–E45999	Mixed rare earths*
E46000–E47999	Dysprosium
E48000–E49999	Erbium
E50000–E51999	Europium
E52000–E55999	Gadolinium
E56000–E57999	Nolmium
E58000–E67999	Lanthanum
E68000–E68999	Lutetium
E69000–E73999	Neodymium
E74000–E77999	Praseodymium
E78000–E78999	Promethium
E79000–E82999	Samarium
E83000–E84999	Scandium
E85000–E86999	Terbium
E87000–E87999	Thulium
E88000–E89999	Ytterbium
E90000–E99999	Yttrium

F00001–F9999 Cast Irons
K00001–K99999 Miscellanenous Steels and Ferrous Alloys
L00001–L99999 Low-Melting Metals and Alloys

L00001–L00999	Bismuth
L01001–L01999	Cadmium
L02001–L02999	Cesium
L03001–L03999	Gallium
L04001–L04999	Indium
L05001–L05999	Lead
L06001–L06999	Lithium
L07001–L07999	Mercury
L08001–L08999	Potassium
L09001–L09999	Tubidium
L10001–L10999	Selenium
L11001–L11999	Sodium
L12001–L12999	Thallium
L13001–L13999	Tin

M00001–M99999 Miscellaneous Nonferrous Metals and Alloys

M00001–M00999	Antimony
M01001–M01999	Arsenic
M02001–M02999	Barium
M03001–M03999	Calcium
M04001–M04999	Germanium
M05001–M05999	Plutonium
M06001–M06999	Strontium

M07001–M07999	Tellurium
M08001–M08999	Uranium
M10001–M19999	Magnesium
M20001–M29999	Manganese
M30001–M39999	Silicon

P00001–P99999 Precious Metals and Alloys

P00001–P00999	Gold
P01001–P01999	Iridium
P02001–P02999	Osmium
P03001–P03999	Palladium
P04001–P04999	Platinum
P05001–P05999	Rhodium
P06001–P06999	Ruthenium
P07001–P07999	Silver

R00001–R99999 Reactive and Refractory Metals and Alloys

R01001–R01999	Boron
R02001–R02999	Hafnium
R03001–R03999	Molybdenum
R04001–R04999	Niubium (columbium)
R05001–R05999	Tantalum
R06001–R06999	Thorium
R07001–R07999	Tungsten
R08001–R08999	Vanadium
R10001–R19999	Beryllium
R20001–R29999	Chromium
R30001–R39999	Cobalt
R40001–R49999	Rhenium
R50001–R59999	Titanium
R60001–R69999	Zirconium

W00001–W99999 Welding Filler Metals Classified
by Weld Deposit Composition

W00001–W09999	Carbon steel with no significant alloying elements
W10000–W19999	Manganese-molybdenum low-alloy steels
W20000–W29999	Nickel low-alloy steels
W30000–W39999	Austenitic stainless steels
W40000–W49999	Ferritic stainless steels
W50000–W59999	Chromium low-alloy steels
W60000–W69999	Copper-base alloys
W70000–W79999	Surfacing alloys
W80000–W89999	Nickel-base alloys

Z00001–Z99999 Zinc and Zinc Alloys

* Alloys in which the rare earths are used in the ratio of their natural occurrence (that is, unseparated rare earths). In this mixture, cerium is the most abundant of the rare earth elements.

source: Reprinted with permission from the *Annual Book of ASTM Standards,* copyright 1992, American Society for Testing and Materials.

tables to determine the prime material (metal) or classification of alloy represented by the UNS number.

4.7   Hardness Tests and Hardness Number Conversions

4.7.1   Brinell hardness numbers (HB)

The Brinell hardness system is one of the most widely used systems for indicating the hardness of metals and alloys. The Brinell hardness number of a material may be calculated from the following equation:

$$\text{HB} = \frac{P}{\frac{\pi D}{2}\left(D - \sqrt{D^2 - d^2}\right)}$$

where HB = Brinell hardness number
  P = load applied to the test ball, kg
  D = diameter of ball, mm
  d = measured diameter at the rim of the impression in the material, mm

The standard ball is 10 mm in diameter, and standard loads are 500, 1500, and 3000 kg. The test is not valid when the hardness of the test material is above the anticipated Brinell hardness number of 630.

4.7.2   Vickers hardness numbers (HV)

The Vickers hardness is determined by forcing a square-base diamond pyramid having an apex angle of 136° into the test specimen under a load ranging from 3 to 50 kg and then measuring the diagonals of the indentation created. The Vickers hardness is defined as the load per unit area of surface contact in kilograms per square millimeter and may be calculated from the average diagonal using the following equation:

$$\text{HV} = \frac{2L \sin \dfrac{\alpha}{2}}{d^2}$$

where HV = Vickers hardness number
  d = length of average diagonal, mm
  α = apex angle, 136°
  L = load, kg

4.8 Plastics (Thermoplastics and Thermoset Plastics)

Plastic materials are derived mainly from petroleum products. The types, trade names, and compositions of the various modern plastics form a long list, with more being developed as required to meet specific design and application needs in industry.

A *thermoplastic* is a plastic in which the finished molded part may be remelted for remolding. A *thermoset plastic* is a plastic in which the chemical reaction cannot be reversed, thus allowing the part to be cast only once. Thermoplastics are extruded, injection-molded, and cast in dies. Thermoset plastics usually are compression-molded. Some of the thermoplastics are also formulated for thermoset applications, such as the urethanes. Table 4.56 lists the common trade names, suppliers, SAE symbols, and plastic "family" names for most plastics.

Common plastics and compositions.   Listed here are some of the more prevalent plastics and compositions.

ABS (acrylonitrile-butadiene-styrene)

Acetal (Delrin, Celcon)

Acetate (cellulose)

Acrylic (Lucite, Plexiglas)

Benelex

Epoxy, epoxy glass

Diallyl phthalate, Melamine

Mylar (polyester film)

Nylon

Phenol formaldehyde

Phenolic laminates

Polycarbonate (Lexan)

Polyester glass

Polyethylene

Polypropylene

(*Text continued on page 263.*)

TABLE 4.56 Common Plastics

Common/trade name	Supplier	SAE symbol	Plastic "family" name
ABS	Generic Name	ABS	Acrylonitrile/butadiene/styrene
Absafil	Wilson Fiberfil International	ABS	Acrylonitrile/butadiene/styrene
Acetal	Generic Name	POM	Polyoxymethylene: polyformaldehyde
Aclar	Ausimont	PCTFE	Polychlorotrifluoroethylene
Acrylic	Generic Name	PMMA	Poly (methyl methacrylate)
Acrylite	Cyro Industries	PMMA	Poly (methyl methacrylate)—"acrylic"
Adiprane	Uniroyal Inc.	PUR	Polyurethane, thermoset (unsaturated)
Alathon	E.I. Dupont de Nemours & Co.	PE	Polyethylene
Alton	International Polymers	PPS + PTFE	Polyphenylene sulfide + polytetrafluoroethylene
Ampol	American Polymers Inc.	CA	Cellulose acetate
Andrez	Unknown	SB	Styrene-butadiene
Apex	Teknor Apex Co.	PVC	Poly (vinyl chloride)
Araldrite	Ciba-Geigy Corp.	EP	Epoxide: epoxy
Aramid	Generic Name	PARA	Polyarylamid (polyaramide)
Ardel	Union Carbide Corp.	PAT	Polyester, thermoplastic—"polyarylate"
Arloy	Arco Chemical Co.	PC + SMA	Polycarbonate + styrene maleic anhydride
Arylon	E.I. Dupont de Nemours & Co.	PAT	Polyester, thermoplastic—"polyarylate"
Astrel	Amoco Chemicals Corp.	PASU	Polyarylsulfone
Azdel	Azdel Inc.	PP	Polypropylene
Bakelite	Union Carbide Corp.	PF	Phenol-formaldehyde
Bayblend	Mobay Chemical Corp.	ABS + PC	Acrylonitrile/butadiene/styrene + polycarbonate
Bayflex	Mobay Chemical Corp.	PUR	Polyurethane, thermoset
Bayflex	Mobay Chemical Corp.	TPU	Thermoplastic elastomer
Bexloy C	E.I. Dupont de Nemours & Co.	PA+	Polyamide (amorphous) blend
Bexloy K & J	E.I. Dupont de Nemours & Co.	PBT + PET	Polybutylene terephthalate + polyethylene terephthalate

BMC	Generic Name	Polyester, thermoset (unsaturated)
Butvar	Monsanto Co.	PVB	Poly (vinyl butyral)
Cadon	Monsanto Co.	ABS	Acrylonitrile/butadiene/styrene
Cadon	Monsanto Co.	ABS + SMA	Acrylonitrile/butadiene/styrene + styrene maleic anhydride
Cadon	Monsanto Co.	SMA	Styrene maleic anhydride
Calibre	Dow Chemical Co.	PC	Polycarbonate
Capron	Allied Engineered Plastics	PA	Polyamide—"nylon"
Castethane	Dow Chemical Co.	PUR	Polyurethane, thermoset
Celanese	Celanese Engineering Resins	PA	Polyamide—"nylon"
Celanese N	Celanese Engineering Resins	PA + RUBBER	Polyamide + rubber
Celanex	Celanese Engineering Resins	PBT	Polyester, thermoplastic—"polybutylene terephthalate"
Celanex	Celanese Engineering Resins	PBT + PET	Polybuthylene terephthalate + polyethylene terephthalate
Celanex	Celanese Engineering Resins	PBT + RUBBER	Polybutylene terephthalate + rubber
Celcon	Celanese Engineering Resins	POM	Polyoxymethylene: polyformaldehyde—"acetal"
Celcon C	Celanese Engineering Resins	POM + RUBBER	Polyoxymethylene + rubber
Cellidor	Bayer AG	CA	Cellulose acetate
Cellidor	Bayer AG	CAP	Cellulose acetate propionate
Cellidor	Bayer AG	CP	Cellulose propionate
Centrex	Monsanto Co.	ASA	Acrylonitrile/styrene/acrylate
Corvel	Polymer Corp.	EP	Epoxide
Cyanaprene	American Cyanamid Co.	PUR	Polyurethane, thermoset (unsaturated)
Cycolac	Borg-Warner Chemicals Inc.	ABS	Acrylonitrile/butadiene/styrene
Cycoloy	Borg-Warner Chemicals Inc.	ABS + TPU	Acrylonitrile/butadiene/styrene + thermoplastic polyurethane
Cycoloy EHA	Borg-Warner Chemicals Inc.	ABS + PC	Acrylonitrile/butadiene/styrene + polycarbonate
Cycovin K	Borg-Warner Chemicals Inc.	ABS + PVC	Acrylonitrile/butadiene/styrene + polyvinyl chloride
Cymel	American Cyanamid Co.	MF	Melamine-formaldehyde
Cytor	Unknown	TPU	Polyurethane, thermoplastic
Daplen	Chemie Linz AG	PP	Polypropylene
Dapon	Chemie Linz AG	DAP	Poly (diallyl phthalate)

TABLE 4.56 Common Plastics (*Continued*)

Common/trade name	Supplier	SAE symbol	Plastic "family" name
Delrin	E.I. Dupont de Nemours & Co.	POM	Polyoxymethylene: polyformaldehyde—"acetal"
Delrin AF	E.I. Dupont de Nemours & Co.	POM + PTFE	Polyoxymethylene + polytetra-fluoroethylene
Delrin ST	E.I. Dupont de Nemours & Co.	POM + RUBBER	Polyoxymethylene + rubber
Delrin T	E.I. Dupont de Nemours & Co.	POM + RUBBER	Polyoxymethylene + rubber
Derakane	Dow Chemical Co.	UP	Polyester, thermoset (unsaturated)
Desmopan	Bayer AG	PUR	Polyurethane, thermoset (unsaturated)
Diakon	ICI Hyde Group	PMMA	Poly (methyl methacrylate)—"acrylic"
Diaron	Reichold Chemicals Inc.	MF	Melamine-formaldehyde
DKE + 450	Sumitoma Corp. of America	PVC + PMMA	Polyvinyl chloride = polymethyl methacrylate
Duraflex	Shell Chemical Co.	PB	Polybutene-1
Durathon	Bayer AG	PS	Polystyrene
Durel	Celanese Engineering Resins	PAT	Polyester, thermoplastic—"polyarylate"
Durethan	Bayer AG	PA	Polyamide—"nylon"
Durez	Occidental Chemical Corp.	DAP	Poly (diallyl phthalate)
Durez	Occidental Chemical Corp.	PF	Phenol-formaldehyde
Durilite	Unknown	EC	Ethyl cellulose
Dylan	Arco Chemical Co.	PE	Polyethylene
Dylark	Arco Chemical Co.	SMA	Styrene/maleic anhydride
Dylark	Arco Chemical Co.	SMA + PS	Styrene/maleic anhydride + high impact polystrene
Dylen	Arco Chemical Co.	PS	Polystyrene
Dynyl	Phone Poulenc Inc.	PEBA	Thermoplastic polyester—"polyether block amide"
Econol	Sohio Chemical Co.	PAT	Polyester, thermoplastic—"polyarylate"
Elkcel	Carborundum Co.	POB	Poly-p-oxybenzoate
Elemid	Borg-Warner Chemicals Inc.	ABS + PA	Acrylonitrile/butadiene/styrene + polyamide
Elexar	Shell Chemical Co.	SB	Styrene-butadiene

Elexar	Shell Chemical Co.	TES	Thermoplastic elastomer—"styrene block copolymer"
Elvax	E.I. Dupont de Nemours & Co.	EVAC	Ethylene/vinyl acetate
Envex	Rogers Corp.	PI	Polyamide
Epon	Shell Chemical Co.	EP	Epoxide: epoxy
Epotuf	Reichold Chemicals Inc.	EP	Epoxide: epoxy
Escorene	Exxon Chemical Americas	PE	Polyethylene
Escorene	Exxon Chemical Americas	PP	Polypropylene
Estamid (ester)	Dow Chemical Co.	PEBA	Polyester, thermoplastic—"polyether block amide"
Estane	B.G. Goodrich Chemical Group	ABS + TPU	Acrylonitrile/butadiene/styrene + thermoplastic polyurethane
Estane	B.G. Goodrich Chemical Group	TPU	Thermoplastic elastomer—"polyurethane"
Ethocel	Dow Chemical Co.	CA	Cellulose acetate
Ethocel	Dow Chemical Co.	CAB	Cellulose acetate butyrate
Ethocel	Dow Chemical Co.	CAP	Cellulose acetate propionate
Ethocel	Dow Chemical Co.	CP	Cellulose propionate
Ethocel	Dow Chemical Co.	EC	Ethyl cellulose
Evanol	Unknown	PVAL	Poly (vinyl alcohol)
Ferroflex	Ferro Corp.	TPO	Thermoplastic elastomer—"polyolefinic"
Fluon	ICI Americas Inc.	PTFE	Polytetrafluoroethylene
Fluorocomp	LNP Corp./ICI Americas	FEP	Perfluoro (ethylene/propylene)
Formion	A. Schulman Inc.	EMA	Ethylene methacrylate acid
Formvar	Monsanto Co.	PVFM	Poly (vinyl formal)
Forsacryl	Unknown	SAN	Styrene-acrylonitrile
Fortiflex	Soltex Polymer Corp.	PE	Polyethylene
Fortron	Celanese Engineering Resins	PPS	Poly (phenylene sulfide)
Geloy 1200	General Electric Co.	ASA + PVC	Acrylonitrile/styrene/acrylate + polyvinyl chloride
Geloy 1320	General Electric Co.	ASA + PMMA	Acrylonitrile/styrene/acrylate + polymethyl methacrylate
Gelva	Monsanto Co.	PVAC	Poly (vinyl acetate)
Gelvatol	Monsanto Co.	PVAL	Poly (vinyl alcohol)
Gemax	General Electric Co.	PBT + PPE	Polybutylene terephthalate + polyphenylene ether

TABLE 4.56 Common Plastics (*Continued*)

Common/trade name	Supplier	SAE symbol	Plastic "family" name
Gemon	General Electric Co.	PI	Polyimide
Geon	B.G. Goodrich Chemical Group	PVC	Poly (vinyl chloride)
Geon	B.G. Goodrich Chemical Group	VCVDC	Vinyl chloride/vinylidene chloride
Gracon	W.R. Grace Co.	PVC	Poly (vinyl chloride)
Grilon	Emser Industries Inc.	PA + RUBBER	Polyamide + rubber
Halon	Allied Engineered Plastics	PCTFE	Polychlorotrifluoroethylene
Halon	Allied Engineered Plastics	PTFE	Polytetrafluoroethylene
Hi-Fax	Himont U.S.A. Inc.	PE	Polyethylene
Hostadur	American Hoechst Corp.	PET	Polyester, thermoplastic—"polyethylene terephthalate"
Hostaflon	American Hoechst Corp.	PTFE	Polytetrafluoroethylene
Hostaform	American Hoechst Corp.	POM	Polyoxymethylene: polyformaldehyde—"acetal"
Hostalen	American Hoechst Corp.	PE	Polyethylene
Hostalen	American Hoechst Corp.	PEOX	Poly (ethylene oxide)
Hostalen	American Hoechst Corp.	PP	Polypropylene
Hostalite Z	American Hoechst Corp.	PVC + CPE	Polyvinyl chloride + chlorinated polyethylene
Hycar	B.G. Goodrich Chemical Group	PVC + NBR	Polyvinyl chloride + nitrile butadiene rubber
Hytrel	E.I. Dupont de Nemours & Co.	TEEE	Thermoplastic elastomer—"ether ester block copolymer"
Impet	Celanese Engineering Resins	PET	Polyester thermoplastic—"poly (ethylene terephthalate)"
Ionomer	Generic name	EMA	Ethylene/methacrylic acid
Isomin	Unknown	MF	Melamine-formaldehyde
Isoplast	Dow Chemical Co.	RTPU	Thermoplastic elastomer—"rigid polyurethane"
Kadel	Amoco Chemicals Corp.	PEEK	Polyether-etherketone
Kamax	Rohm & Haas Co.	PI	Polyamide
Kel-F	3M Company	PCTFE	Polychlorotrifluoroethylene
Kinel	Rhone Poulenc Inc.	PI	Polyamide

Kralastic FVM	Uniroyal Inc.	ABS + PVC	Acrylonitrile/butadiene/styrene + polyvinyl chloride
Kraton	Shell Chemical Co.	TES	Thermoplastic elastomer—"styrene block copolymer"
Kydex	Rohm & Haas Co.	PVC + PMMA	Polyvinyl chloride + polymethyl methacrylate
Kynar	Pennwalt Corp.	PVDF	Poly (vinylidene fluoride)
K-Resins	Phillips Chemical Co.	SB	Styrene-butadiene
Lexan	General Electric Co.	PC	Polycarbonate plastics
Lexan	General Electric Co.	PC + PE	Polycarbonate + polyethylene
Lomod	General Electric Co.	TEEE	Thermoplastic elastomer—"ether ester block copolymer"
Lupolen	BASF Wyandotte Corp.	PE	Polyethylene
Luran	BASF Wyandotte Corp.	ASA	Acrylonitrile/styrene/acrylate
Luran	BASF Wyandotte Corp.	SAN	Styrene/acrylonitrile
Lustran	Monsanto Co.	ABS	Acrylonitrile/butadiene/styrene
Lustran	Monsanto Co.	ABS + PVC	Acrylonitrile/butadiene/styrene + polyvinyl chloride
Lustran	Monsanto Co.	SAN	Styrene/acrylonitrile
Lustrex	Polysar Inc.	PS	Polystyrene
Luvican	BASF Wyandotte Corp.	PVK	Polyvinylcarbazole
Macroblend	Mobay Chemical Corp.	PC + PBT	Polycarbonate + polybutylene terephthalate
Macroblend	Mobay Chemical Corp.	PC + PET	Polycarbonate + polyethylene terephthalate
Macrolon	Mobay Chemical Corp.	PC	Polycarbonate plastics
Marlex	Phillips Chemical Co.	PE	Polyethylene
Marlex	Phillips Chemical Co.	PP	Polypropylene
Melmac	Phillips Chemical Co.	MF	Melamine-formaldehyde
Merlon	Mobay Chemical Corp.	PC	Polycarbonate plastics
Merlon T	Mobay Chemical Corp.	PC + PE	Polycarbonate + polyethylene
Microthane	Unknown	EVAC	Ethylene/vinyl acetate
Mindel	Union Carbide Corp.	PARA	Polyarylamid (polyaramide)
Mindel	Union Carbide Corp.	PPSU	Poly (phenylene sulfone)
Mindel A	Union Carbide Corp.	ABS + PPSU	Acrylonitrile/butadiene/styrene + polyphenylene sulfone
Mindel B	Union Carbide Corp.	PET + PPSU	Polyethylene terephthalate + polyphenylene sulfone

TABLE 4.56 Common Plastics (Continued)

Common/trade name	Supplier	SAE symbol	Plastic "family" name
Moplen	Himont-Italia	PP	Polypropylene
N5	Thermofil Inc.	PA + SAN	Polyamide + styrene/acrylonitrile
Nitrocellulose	Generic name	CN	Cellulose nitrate
None	—	AB	Acrylonitrile/butadiene
None	—	ABA	Acrylonitrile/butadiene/acrylate
None	—	AMMA	Acrylonitrile/methyl methacrylate
None	—	CF	Cresol formaldehyde
None	—	CMC	Carboxymethyl cellulose
None	—	CPVC	Chlorinated poly (vinyl chloride)
None	—	CS	Casein
None	—	EEA	Ethylene/ethyl acrylate
None	—	ETFE	Ethylene/tetrafluoroethylene
None	—	FF	Fluran formaldehyde
None	—	PA + EMA	Polyamide + ethylene methacrylic acid (ionomer)
None	—	PAE	Polyarylether
None	—	PB	Polybutene-1
None	—	PFF	Phenol furfurol
None	—	PPOX	Poly (propylene oxide)
None	—	PVCA	Poly (vinyl chloride acetate)
None	—	PVP	Polyvinylpyrrolidone
None	—	SI	Silicone
None	—	SMS	Styrene/A-methylstyrene
None	—	VCE	Vinyl chloride/ethylene
None	—	VCEMA	Vinyl chloride/ethylene/methyl acrylate
None	—	VCMA	Vinyl chloride/methyl acrylate

Noryl	PPE	Polyphenylene ether plastics
Noryl	PPE + PS	Polyphenylene ether + high impact polystyrene
Noryl GTX	PA + PPE	Polyamide + polyphenylene ether
Novodur	ABS	Acrylonitrile/butadiene/styrene
Novolen	PP	Polypropylene
Nycoa 1900	PA + EMA	Polyamide + ethylene methacrylic acid—"ionomer"
Nydur BC	PA + RUBBER	Polyamide + rubber
Nydur KL	PA + RUBBER	Polyamide + rubber
Nyion	PA	Polyamide—"nylon"
Oleflo	PP	Polypropylene
Oppanol B	PIB	Polyisobutylene
Oroglas	PMMA	Poly (methyl methacrylate)—"acrylic"
Orthane	TPU	Polyurethane, thermoplastic
Ozo	PVC + NBR	Polyvinyl chloride + nitrile butadiene rubber
Peracril	PVC + NBR	Polyvinyl chloride + nitrile butadiene rubber
Paxon	PE	Polyethylene
Pebax	PEBA	Polyester, thermoplastic—"polyether block amide"
Pellethane	ABS + TPU	Acrylonitrile/butadiene/styrene + thermoplastic polyurethane
Pellethane	TPU	Thermoplastic elastomer—"polyurethane"
Penton	CPE	Chlorinated polyethylene
Petlon	PBT	Polyester, thermoplastic—"polybutylene terephthalate"
Petlon	PET	Polyester, thermoplastic—"polyethylene terephthalate"
Fetra	PBT	Polyester, thermoplastic—"polybuthylene terephthalate"
Petra	PET	Polyester, thermoplastic—"polyethylene terephthalate"
Phenolic	PF	Phenol-formaldehyde
Piso	PISU	Polymidesulfone
Plaskon	UF	Urea-formaldehyde
Plenco	PF	Phenol-formaldehyde
Plexiglas	PMMA	Poly (methyl methacrylate)—"acrylic"

Additional column (manufacturer) between trade name and abbreviation:

Noryl	General Electric Co.
Noryl	General Electric Co.
Noryl GTX	General Electric Co.
Novodur	Mobay Chemical Corp.
Novolen	BASF Wyandotte Corp.
Nycoa 1900	Nylon Corp. of America
Nydur BC	Unknown
Nydur KL	Unknown
Nyion	Generic name
Oleflo	Avisun
Oppanol B	BASF Wyandotte Corp.
Oroglas	Rohm & Haas Co.
Orthane	Eagle Pichor Plastics Div.
Ozo	Unknown
Peracril	Uniroyal Inc.
Paxon	Allied Engineered Plastics
Pebax	Atochem U.S.A. Inc.
Pellethane	Dow Chemical Co.
Pellethane	Dow Chemical Co
Penton	Hercules Inc.
Petlon	Mobay Chemical Corp.
Petlon	Mobay Chemical Corp.
Fetra	Allied Engineered Plastics
Petra	Allied Engineered Plastics
Phenolic	Generic name
Piso	Celanese Engineering Resins
Plaskon	Plaskon Electronic Mtls. Inc.
Plenco	Plastics Engineering Co.
Plexiglas	Rohm & Haas Co.

TABLE 4.56 Common Plastics (Continued)

Common/trade name	Supplier	SAE symbol	Plastic "family" name
Pliolite	Goodyear Tire & Rubber Co.	SB	Styrene-butadiene
Pliovic	Goodyear Tire & Rubber Co.	PVC	Poly (vinyl chloride)
Pocan	Bayer AG/Mobay	PBT + RUBBER	"Polybutylene terephthalate" + rubber
Pocan B	Bayer AG/Mobay	PBT	Polyester, thermoplastic—"polybutylene terephthalate"
Polycomp	LNP Corp./ICI Americas	PPS + PTFE	Polyphenylene sulfide + fluoroethylene
Polyfort	A. Schulman Inc.	PE	Polyethylene
Polyfort	A. Schulman Inc.	PP	Polypropylene
Polypur	A. Schulman Inc.	TPU	Thermoplastic elastomer—"polyurethane"
Polystyrol	BASF Wyandotte Corp.	PS	Polystyrene
Polystyrol	BASF Wyandotte Corp.	SB	Styrene/butadiene
Polystyrol SB	BASF Wyandotte Corp.	PS	Polystyrene, high impact
Polytrope	A. Schulman Inc.	TPO	Thermoplastic elastomer—"polyolefinic"
Polyvin	A. Schulman Inc.	PVC	Poly (vinyl chloride)
Poly-Dap	DAP	DAP	Poly (diallyl phthalate)
Premi-Glas	Premix Inc.	UP	Polyester, unsaturated thermoset
Prevex	Borg-Warner Chemicals Inc.	PPE	Polyphenylene ether plastics
Prevex	Borg-Warner Chemicals Inc.	PPE + PS	Polyphenylene ether + high impact polystyrene
Profax	Himont U.S.A. Inc.	PP	Polypropylene
Proloy	Borg-Warner Chemicals Inc.	ABS + PC	Acrylonitrile/butadiene/styrene + polycarbonate
Pyralin	E.I. Dupont de Nemours & Co.	PI	Polyimide
Radel	Union Carbide Corp.	PPSU	Polyphenylene sulfone
Renflex	Research Polymers Inc.	TPO	Thermoplastic elastomer—"polyolefinic"
Rilsan	Atochem U.S.A. Inc.	PA	Polyamide—"nylon"
Riteflex BP	Celanese Engineering Resins	TEEE	Thermoplastic elastomer—"ether ester block copolymer"
Ropet	Rohm & Haas Co.	PET + PMMA	Polyethylene terephthalate + polymethyl methacrylate

Rozel	Dow Chemical Co.	AES	Acrylonitrile/ethylene/styrene
Rynite	E.I. Dupont de Nemours & Co.	PBT	Polyester, thermoplastic—"polybutylene terephthalate"
Rynite	E.I. Dupont de Nemours & Co.	PET	Polyester, thermoplastic—"polybutylene terephthalate"
Rynite SST	E.I. Dupont de Nemours & Co.	PET + RUBBER	Polyethylene terephthalate + rubber
Ryton	Phillips Chemical Co.	PPS	Poly (phenylene sulfide)
Santoprene	Monsanto Co.	TPO	Thermoplastic elastomer—"polyolefinic"
Saran	Dow Chemical Co.	PVDC	Poly (vinylidene chloride)
Selar	E.I. Dupont de Nemours & Co.	PA + PE	Polyamide + polyethylene
Selectron	PFG Industries Inc.	UP	Polyester unsaturated thermoset
Skanopal	Penstorp Inc.	UF	Urea-formaldehyde
Skybond	Monsanto Co.	PI	Polyimide
SMC	Generic name	UP	Polyester, thermoset (unsaturated)
Solair	Soltex Polymer Corp.	PE	Polyethylene
Spectrim	Dow Chemical Co.	PUR	Polyurethane, thermoset
Styrolux	Westlake Plastics Co.	SB	Styrene-butadiene
Styron	Dow Chemical Co	PS	Polystyrene
Surlyn	E.I. Dupont de Nemours & Co.	EMA	Ethylene/methacrylic acid—"ionomer"
Tedlar	E.I. Dupont de Nemours & Co.	PVF	Poly (vinyl fluoride)
Teflon	E.I. Dupont de Nemours & Co.	FEP	Perfluoro (ethylene/propylene)
Teflon	E.I. Dupont de Nemours & Co.	PTFE	Polytetrafluoroethylene
Teflonz	E.I. Dupont de Nemours & Co.	FEP	Tetrafluoroethylene/hexafluoro propylene
Teflonz	E.I. Dupont de Nemours & Co.	FEP	Tetrafluoroethylene/hexafluoro propylene
Telcar	Teknor Apex Co.	TPO	Thermoplastic elastomer—"polyolefinic"
Tenite	Eastman Chemical Products Inc.	CA	Cellulose acetate
Tenite	Eastman Chemical Products Inc.	CAB	Cellulose acetate butyrate
Tenite	Eastman Chemical Products Inc.	CAP	Cellulose acetate propionate
Tenite	Eastman Chemical Products Inc.	CP	Cellulose propionate
Tenite	Eastman Chemical Products Inc.	PP	Polypropylene
Terblend SKR	BASF Wyandotte Corp.	ASA + PC	Acrylonitrile/styrene/acrylate + polycarbonate

TABLE 4.56 Common Plastics (Continued)

Common/trade name	Supplier	SAE symbol	Plastic "family" name
Terlukan	Bayer AG	ABS	Acrylonitrile/butadiene/styrene
Tetran	Pennwalt Corp.	PTFE	Polytetrafluoroethylene
Texin	Mobay Chemical Corp.	PC + TPU	Polycarbonate + thermoplastic polyurethane
Texin	Mobay Chemical Corp.	TPU	Polyurethane, thermoplastic
Thermo	Unknown	CPE	Chlorinated polyethylene
Thermocomp AL	LNP Corp./ICI Americas	AS + PTFE	Acrylonitrile/butadiene/styrene + polytetra fluoroethylene
Thermocomp KL	LNP Corp./ICI Americas	POM + PTFE	Polyoxymethylene + polytetra fluoroethylene
Torlon	Amoco Chemicals Corp.	PAI	Polyamide-imide
TPE	Generic name	—	Thermoplastic elastomer
TPX	Mitsui & Co.	PMP	Poly (4-methylpentene-1)
Triax 1000	Monsanto Co.	ABS + PA	Acrylonitrile/butadiene/styrene + polyamide
Trosiplast	Kay-Fries Inc.	PVC	Poly (vinyl chloride)
Tuf-Flex	American Hoechst Corp.	PS	Polystyrene, high impact
Tyril	Dow Chemical Co.	SAN	Styrene/acrylonitrile
Tyrin	Dow Chemical Co.	CPE	Chlorinated polyethylene
Udel	Union Carbide Corp.	PPSU	Poly (phenylene sulfone)
Ultradur	BASF Wyandotte Corp.	PET	Poly (ethylene terephthalate)
Ultaform	BASF Wyandotte Corp.	POM	Polyoxymethylene: polyformaldehyde—"acetal"
Ultamid KR	BASF Wyandotte Corp.	PA + RUBBER	Polyamide + rubber
Ultem	General Electric Co.	PEI	Polyetherimide
Ultradur	BASF Wyandotte Corp.	PBT	Polyester, thermoplastic—"polybutylene terephthalate"
Ultradur KR	BASF Wyandotte Corp.	PBT + RUBBER	Polybutylene terephthalate + rubber
Ultramid	BASF Wyandotte Corp.	PA	Polyamide—"nylon"

Ultrason	BASF Wyandotte Corp.	PESU	Polyether sulfone
Ultrason	BASF Wyandotte Corp.	PPSU	Poly (phenylene sulfone)
Ultrex	Spiratex Co.	PE	Polyethylene
Unichem	Colorite Plastics Co.	PVC	Poly (vinyl chloride)
Uyex	Eastman Chemical Products Inc.	CAB	Cellulose acetate butyrate
Valox	General Electric Co.	PBT	Polyester, thermoplastic—"polybutylene terephthalate"
Valox	General Electric Co.	PBT + PET	Polybutylene terephthalate + "polyethylene terephthalate"
Valox	General Electric Co.	PBT + RUBBER	Polybutylene terephthalate + rubber
Valox	General Electric Co.	PC + PBT	Polycarbonate + polybutylene terephthalate
Valox	General Electric Co.	PC + PET	Polycarbonate + terephthalate
Valox	General Electric Co.	PET	Polyester, thermoplastic—"polyethylene terephthalate"
Vancor PB	Celanese Engineering Resins	PBT + RUBBER	Polybutylene terephthalate + rubber
Vectra	Celanese Engineering Resins	ARP	Polyester, thermoplastic—"liquid crystal polymer"
Vedril	Vedril Spa (Spanish)	PMMA	Poly (methyl methacrylate)—"acrylic"
Vespel	E.I. Dupont de Nemours & Co.	PARA	Polyarylamid (polyaramide)—"aramid"
Vespel	E.I. Dupont de Nemours & Co.	PI	Polyimide
Vibrin-Mat	U.S Rubber Co.	UP	Polyester, thermoset (unsaturated)
Vicrex	ICI Americas Inc.	PESU	Polyethersulfone
Victrex Peek	ICI Americas Inc.	PEEK	Polyetherketone
Victrex Pek	ICI Americas Inc.	PEK	Polyetherketone
Vinoflex	BASF Wyandotte Corp.	PVC	Poly (vinyl chloride)
Vinylite	Canadian Resins & Chemical Ltd	PVAC	Poly (vinyl acetate)
Vinylite	Canadian Resins & Chemical Ltd	PVB	Poly (vinyl butyral)
Vinylite	Canadian Resins & Chemical Ltd	PVC	Poly (vinyl chloride)
Vinylite	Canadian Resins & Chemical Ltd	VCVAC	Vinyl chloride/vinyl acetate
Vistaflex	Esso Chemicals (Europe)	TPO	Thermoplastic elastomer—"polyolefinic"

TABLE 4.56 Common Plastics (*Continued*)

Common/trade name	Supplier	SAE symbol	Plastic "family" name
Vydyne	Monsanto Co.	PA	Polyamide "nylon"
Vynite	Allied Engineered Plastics	PVC + NBR	Polyvinyl chloride + nitrile butadiene rubber
Vythene	Alpha Chemicals & Plastics CRP	PVC + PU	Polyvinyl chloride + polyurethane
Wellamid	Wellman Inc.	PA	Polyamide—"nylon"
Wellite	Wellman Inc.	PBT	Polyester, thermoplastic—"polybutylene terephthalate"
Wellpet	Wellman Inc.	PET	Polyester, thermoplastic—"polyethylene terephthalate"
Xenoy	General Electric Co.	PC + PBT	Polycarbonate + polybutylene terephthalate
XMC	Generic name	UP	Polyester, thermoset (unsaturated)—50% glass
Xydar	Dartco	ARP	Polyester, thermoplastic—"liquid crystal polymer"
Zytel	E.I. Dupont de Nemours & Co.	PA	Polyamide—"nylon"
Zytel ST	E.I. Dupont de Nemours & Co.	PA + RUBBER	Polyamide + rubber

SOURCE: Reprinted with permission, copyright 1992, Society of Automotive Engineers.

Polyimide

Polystyrene

Polysulfone

Polyurethane

Polyvinyl chloride (PVC)

RTV (room-temperature vulcanizing) silicones, Styrofoam (polystyrene)

Teflon (PTFE, polytetrafluoroethylene), urea-formaldehyde

Common plastics and typical uses

Acetal (Delrin, Celcon). *Properties:* High modulus of elasticity, low coefficient of friction, excellent abrasion and impact resistance, low moisture absorption, excellent machinability, ablative. *Typical uses*: Bearings, gears, antifriction parts, electrical components, washers, seals, insulators, and cams.

Acetate (Cellulose). *Properties*: Odorless, tasteless, nontoxic, grease resistant, high impact strength. *Typical uses*: Badges, blister packaging, displays, optical covers, and book covers.

Acrylic (Plexiglas, Lucite). *Properties*: Unusual optical clarity, high tensile strength, weatherability, good electrical properties, ablative. *Typical uses*: Displays, signs, models, lenses, and electrical and electronic parts.

Benelox (Laminate). *Properties*: High compressive strength, machinable, resists corrosion (alkalis or acids), good electrical insulation, high flexural, shear, and tensile strength. *Typical uses*: Work surfaces, electrical panels and switch gear, bus braces (low voltage only), and neutron shielding.

Diallyl phthalate, Melamine. *Properties*: High strength, chemical resistant, low water absorption, medium-high temperature use. *Typical uses*: Terminal blocks and strips, dishware, automotive applications, and aerospace applications.

Epoxy glass. *Properties*: High strength, high-temperature applications, flame retardant, low coefficient of thermal expansion, low water absorption. *Typical uses*: High-quality printed circuit boards, microwave stripline applications, VHF and UHF applications, electrical insulation, and service in temperature range −400 to 500°F.

Mylar (polyester film, polyethylene terephthalate). *Properties*: High dielectric strength, chemical resistance, high mechanical strength, moisture resistant, temperature range 70 to 105°C, does not embrittle with age. *Typical uses*: Electrical and industrial applications and graphic arts applications.

Nylon. *Properties*: Wear resistant, low friction, high tensile strength, excellent impact resistance, high fatigue resistance, easy machining, corrosion resistant, lightweight. *Typical uses*: Bearings, bushings, valve seats, washers, seals, cams, gears, guides, wheels, insulators, and wear parts.

Phenol formaldehyde (Bakelite). *Properties*: Wear resistant, rigid, moldable to precise dimensions, strong, excellent electrical properties, economical, will not support combustion. *Typical uses*: Electrical and electronic parts, handles, housings, insulator parts, mechanism parts, and parts that are to resist temperatures to 250°C.

Phenolic laminates. *Properties*: Immune to common solvents, lightweight, strong, easily machined. *Typical uses*: Bearings, machined parts, insulation, gears, cams, sleeves, and electrical and electronic parts.

Polycarbonate (Lexan). *Properties*: Virtually unbreakable, weather resistant, optically clear, lightweight, self-extinguishing, thermoformable, machinable, solvent cementable. *Typical uses*: High-voltage insulation, impact resistant injection moldings, glazing, bulletproof, glazing, and plumbing fittings. The strongest thermoplastic.

Polyester glass. *Properties*: Extremely tough, high dielectric strength, heat resistant, low water absorption, antitracking electrically, self-extinguishing, machinable. *Typical uses*: Insulators and bus braces, switch phase barriers, general electrical insulation, mechanical insulated push rods for switches and breakers, contact blocks, and terminal blocks.

Polyethylene. *Properties*: Transparent in thin sheets, water resistant. *Typical uses*: Bags for food storage, vapor barriers in construction, trays, rollers, gaskets, seals, and radiation shielding.

Polypropylene. *Properties*: Good tensile strength, low water absorption, excellent chemical resistance, stress-crack resistant, electrical properties. *Typical uses*: Tanks, ducts, exhaust systems, gaskets, laboratory and hospital ware, wire coating, and sporting goods.

Polystyrene. *Properties*: Outstanding electrical properties, excellent machinability, ease of fabrication, excellent chemical resistance, oil resistant, clarity, rigidity, hardness, dimensional stability. *Typical uses*: Lighting panels, tote boxes, electronic components, door panels (refrigerators), drip pans, displays, and furniture components.

Polysulfone. *Properties*: Tough, rigid, high-strength, high-temperature thermoplastic, temperature range −150 to +300°F, excellent electrical characteristics, good chemical resistance, low creep and cold-flow properties, capable of being autoclaved repeatedly. *Typical uses*: Food-processing and medical industries, electrical and electronics, appliance, automotive, aircraft, and aerospace uses.

Polyurethane. *Properties*: Elastomeric to rock-hard forms available, high physical characteristics, toughness, durability, broad hardness range, withstands severe use, abrasion resistant, weather resistant, radiation resistant, temperature range −80 to 250°F, resistant to common solvents, available also in foam types. *Typical uses*: Replaces a host of materials that are not performing well; extremely broad range of usage; replaces rubber parts, plastic parts, and some metallic parts.

Polyvinyl chloride (PVC). *Properties*: Corrosion resistant, formable, lightweight, excellent electrical properties, impact resistant, low water absorption, cementable, machinable, weldable. *Typical uses*: Machined parts, nuts, bolts, PVC pipe and fittings, valves, and strainers.

RTV silicone rubber. *Properties*: Resistant to temperature extremes (−75 to 400°F), excellent electrical characteristics, weather resistant, good chemical resistance; FDA, USDA, and UL approved. *Typical uses*: General-purpose high-quality sealant, gasket cement, food contact surfaces, electrical insulation, bonding agent, glass-tank construction, and countless other applications.

Styrofoam. *Properties*: Low water absorption, floats, thermal insulator, extremely lightweight. *Typical uses*: Insulation board for homes and buildings, cups, containers, thermos containers, shock-absorbing packaging, plates (food), and flotation logs.

Teflon (PTFE). *Properties*: Unexcelled chemical resistance, cryogenic service, electrical insulation, very low friction, high dielectric strength, very low dissipation factor, very high resistivity, machinability. *Typical uses*: Valve components, gaskets (with caution, due

to cold flow), pump parts, seal rings, insulators (electrical), terminals, bearings, rollers, bushings, electrical tapes, plumbing tapes, and machined parts; bondable with special etchant preparations.

Urea-formaldehyde. *Properties*: Hard, strong, molds accurately, low water absorption, excellent electrical properties, ablative, economical, will not support combustion. *Typical uses*: Electrical and electronic parts, insulators, small parts, and housings.

4.9   Properties of Materials: General and Specific

Table 4.57 lists the properties of various metals and alloys. Table 4.58 lists the properties of other metals and alloys. Table 4.59 lists the coefficients of linear expansion for common materials. Table 4.60 lists the chemical symbols for metals.

4.10   Material Specification Sheets and Analysis Reports

All manufacturers of plastic materials provide material specification sheets when so requested. Material analysis sheets likewise are supplied by metal providers, forges, and foundries when so requested. The material test report is generated by one of the various materials test laboratories nationwide, but it must be purchased by the company requesting the analysis. Owing to the rise in imported materials that may or may not meet the requirements of the SAE, ASTM, AISI, ASM, and ANSI standards, material confirmation by way of laboratory analysis is necessary in many instances. Materials that do not conform to the standards applicable for a given material, as specified on the engineering drawings, pose a serious safety problem for manufacturers of industrial and consumer products and their users. When in doubt about a material's performance, have the material analyzed at a testing laboratory. The material supplier is responsible for the specifications of the material that it sells. If you order a type 304L stainless steel per SAE or ASTM specifications, the material must conform to these specifications both chemically and mechanically.

TABLE 4.57   Properties of Various Metals and Alloys

Material	Relative resistivity* at 20°C	Density, g/cm³	Thermal conductivity at 20°C, W/cm·°C	Thermal expansion × 10⁻⁶/°C, in	Melting, °C
Aluminum	1.54	2.70	2.22	23.6	660
Beryllium	2.3	1.85	1.46	11.6	1277
Bismuth	67.0	9.80	0.084	13.3	271
Brass, yellow	3.7	8.47	1.17	20.3	930
Cadmium	4.3	8.65	0.92	29.8	321
Carbon, graphite	790	2.25	0.24	0.6–4.3	Sublimes
Chromium	7.4	7.19	0.67	6.2	1875
Cobalt	3.6	8.85	0.69	13.8	1495
Columbium (see Niobium)					
Constantan	29.0	8.9	0.21	14.9	1290
Copper, hard drawn	1.03	8.94	3.91	16.8	1083
Gallium	4.7	5.91	0.29	18.0	30
Germanium	2.7×10^6	5.33	0.59	5.75	937
Gold	1.36	19.32	2.96	14.2	1063
Inconel, 17–16 8	56.9	8.51	0.15	11.5	1425
Indium	4.9	7.31	0.24	33.0	156
Invar, 64–36	46.0	8.00	0.11	0–2	1425
Iron	5.6	7.87	0.75	11.8	1536
Lead	12.0	11.34	0.35	29.3	327
Magnesium	2.58	1.74	1.53	27.1	650
Mercury	55.6	13.55	0.082	—	-38.9
Molybdenum	3.3	10.22	1.42	4.9	2610
Monel, 67–30	27.9	8.84	0.26	14.0	1325
Nichrome, 80–20	62.5	8.4	0.134	13.0	1400
Nickel	5.5	8.89	0.61	13.3	1440
Niobium	7.2	8.57	0.52	7.31	2468
Palladium	6.3	12.02	0.70	11.8	1552
Phosphor bronze 95–5	6.4	8.86	0 71	17.8	1000
Platinum, 99.9%	6.16	21.45	0.69	8.9	1769
Silicon	10^{11}	2.33	1.25	2.5	1420
Silver	0.922	10.49	4.18	19.7	961
Steel, 0.4–0.5 C	7–1?	7.8	0.5	11.0	1480
Steel, stainless 304	42	7.9	0.16	17.0	1430
Steel, stainless 410	33	7.7	0.24	11	1500
Tantalum	7.4	16.6	0.54	6.6	3000
Thorium	8.1	11.6	0.37	12.5	1750
Tin	7.0	7.30	0.63	23	232
Titanium	24.2	4.51	0.41	8.4	1670
Tungsten	3.2	19.3	1.67	4.6	3410
Uranium	17.5	18.7	0.3	6.8–14.1	1132
Zinc	3.5	7.14	1.10	27	420
Zirconium	23	6.5	0.21	5.8	1852

* Standard resistivity of 100% IACS copper at 20°C = 1.7241×10^{-6} Ω·cm.

TABLE 4.58   Other Common Metals and Alloys

Metal	Density, g/cm^3	Weight, lb/in^3	Young's modulus (E), tension, lb/in^2	Torsional modulus (G), lb/in^2	Poisson's ratio	Electrical resistivity, $\Omega \cdot$cm
Aluminum (pure)	2.70	0.098	9×10^6	—	0.33	2.6×10^{-6}
Aluminum alloy (high strength)	2.78	0.101	$10\text{–}11 \times 10^6$	—	0.33	2.8×10^{-6}
Antimony	6.69	0.242	11.3×10^6	2.9×10^6	—	3.1×10^{-6}
Beryllium copper (C170) alloy	8.4	0.303	19×10^6	—	0.29	3.1×10^{-6}
Bismuth	9.75	0.352	—	—	—	119×10^{-6}
Brass (80Cu/20Zn)	8.6	0.311	16×10^6	6×10^6	0.34	7×10^{-6}
Cadmium	8.65	0.312	10.1×10^6	3.5×10^6	—	7.5×-6
Cast iron (gray)	7.2	0.260	14×10^6	5.6×10^6	0.21	—
Chromium	7.19	0.260	—	—	—	2.6×10^{-6}
Copper	8.96	0.324	17×10^6	6.4×10^6	0.34	1.72×10^{-6}
Gold	19.32	0.698	11.4×10^6	—	—	2.44×10^{-6}
Iron	7.87	0.284	28×10^6	11.2×10^6	0.30	10×10^{-6}
Iron (malleable)	7.85	0.284	25×10^6	11.5×10^6	0.17	—
Lead	11.35	0.410	2.4×10^6	—	0.43	22×10^{-6}
Lithium	0.53	0.019	—	—	—	9×10^{-6}
Magnesium (cast)	1.74	0.063	6.5×10^6	2.4×10^6	0.35	44×10^{-6}
Mercury	13.55	0.490	—	—	—	96×10^{-6}
Molybdenum	10.22	0.369	—	—	—	5.7×10^{-6}
Nickel	8.90	0.322	30×10^6	10.6×10^6	0.32	7.8×10^{-6}
Palladium	12.02	0.434	—	—	—	11×10^{-6}
Phosphor bronze	8.90	0.322	16×10^6	6×10^6	0.35	10×10^{-6}
Platinum	21.45	0.775	24.2×10^6	9.3×10^6	—	10×10^{-6}
Silver	10.50	0.379	11.2×10^6	3.8×10^6	—	1.6×10^{-6}
Steel (medium) carbon)	7.80	0.282	30×10^6	11.4×10^6	0.30	15×10^{-6}
Steel (stainless 300 type)	8.03	0.290	28×10^6	11.4×10^6	0.28	30×10^{-6}
Steel (stainless 400 type)	7.75	0.280	29×10^6	12.6×10^6	0.28	30×10^{-6}
Tin	7.31	0.264	7×10^6	2.4×10^6	—	11.5×10^{-6}
Titanium	4.54	0.164	16×10^6	—	—	—
Tungsten	19.30	0.697	51.5×10^6	21.5×10^6	—	5.51×10^{-6}
Zinc	7.13	0.258	14.5×10^6	5×10^6	0.11	6×10^{-6}

TABLE 4.59   Coefficients of Linear
Expansion for Common Materials

Metal, alloy, or other material	Linear expansion	
	in per 1°F	in per 1°C
Aluminum, wrought	0.0000128	0.0000231
Brass	0.0000104	0.0000188
Bronze	0.0000101	0.0000181
Copper	0.0000093	0.0000168
Cast iron, gray	0.0000059	0.0000106
Wrought iron	0.0000067	0.0000120
Lead	0.0000159	0.0000286
Magnesium alloy	0.0000160	0.0000290
Nickel	0.0000070	0.0000126
Cast steel	0.0000061	0.0000110
Hard steel	0.0000073	0.0000132
Medium steel	0.0000067	0.0000120
Soft steel	0.0000061	0.0000110
Stainless steel	0.0000099	0.0000178
Zinc, rolled	0.0000173	0.0000263
Concrete	0.0000079	0.0000143
Granite	0.0000047	0.0000084
Marble	0.0000056	0.0000100
Plaster	0.0000092	0.0000166
Slate	0.0000058	0.0000104
Fir	0.0000021	0.0000037
Maple	0.0000036	0.0000064
Oak	0.0000027	0.0000049
Pine	0.0000030	0.0000054
Plate glass	0.0000050	0.0000089
Hard rubber	0.0000044	0.0000080
Porcelain	0.0000009	0.0000016
Silver	0.0000104	0.0000188
Tin	0.0000148	0.0000269
Tungsten	0.0000024	0.0000043

TABLE 4.60   Chemical Symbols for Metals

Metal	Symbol	Metal	Symbol
Aluminum	Al	Manganese	Mn
Antimony	Sb	Mercury	Hg
Beryllium	Be	Molybdenum	Mo
Bismuth	Bi	Nickel	Ni
Boron	B	Platinum	Pt
Cadmium	Cd	Selenium	Se
Carbon	C	Silicon	Si
Chromium	Cr	Silver	Ag
Cobalt	Co	Tellurium	Te
Copper	Cu	Tin	Sn
Gold	Au	Titanium	Ti
Iridium	Ir	Tungsten	W
Iron	Fe	Vanadium	V
Lead	Pb	Zinc	Zn
Lithium	Li	Zirconium	Zr
Magnesium	Mg		

Modern Engineering Drawing Practices

Modern engineering drawing practices have been greatly influenced by the widespread acceptance of computer-aided design (CAD) packages. Although these systems have greatly simplified the process of documenting dimensions, tolerances, and other important information, having a basic understanding of drawing practices is still essential. This chapter also covers procedures for determining limits and fits on hole/shaft systems because this is an important activity related to dimensioning and tolerancing.

5.1 General Dimensioning and Tolerancing Practices

Our definition of *tolerance* is the amount of dimensional variation allowed on a part or assembly of parts. Tolerance is equal to the difference between maximum and minimum limits of the specified dimension. For example:

$$5.000 \pm 0.010 \text{ has a tolerance of } 0.020$$

$$\frac{5.000}{4.995} \text{ has a tolerance of } 0.005$$

$$\frac{6.129}{6.121} \text{ has a tolerance of } 0.008$$

Unilateral and bilateral tolerances. Unilateral tolerance is used to relate the total tolerance to a basic dimension in one direction only. For example:

$$5.000 {+0.002 \atop -0.000} \quad \text{(positive unilateral)}$$

$$5.000 {+0.000 \atop -0.002} \quad \text{(negative unilateral)}$$

Bilateral tolerance is used to relate the total tolerance to a basic dimension in both plus and minus directions. For example:

$$5.000 \pm 0.005 \quad \text{(tolerance} = 0.010)$$

$$5.000 {+0.002 \atop -0.006} \quad \text{(tolerance} = 0.008)$$

Showing dimensional tolerances in an over-and-under form is as illustrated:

For shafts: $\dfrac{5.005}{4.995}$ (larger dimension "over") $T = 0.010$

For holes: $\dfrac{5.002}{5.007}$ (smaller dimension "over") $T = 0.005$

Tolerancing practices. Tolerances are applied to show the permissible variation in the direction that is the least critical. If variation in either direction is equally critical, the bilateral tolerance is used. If a variation in one direction is more critical than a variation in another direction, the unilateral tolerance should be given in the least critical direction.

For example, if you want the diameter of a round rod not to exceed 3.000 in but will allow it to go under to 2.995 in, this is the same as stating

$$3.000 {+0.000 \atop -0.005}$$

Tolerances on the centerline distances between holes in a part are usually bilateral.

For nonmating surfaces, unilateral or bilateral tolerances can be used, but on mating surfaces, the tolerances should be unilateral in all but a few cases.

Locating toleranced dimensions.   A common baseline or point in each plane of a part should be used to establish dimensions.

5.1.1   Tolerance studies

At many companies, the design and checking groups perform on assembly drawings what is commonly referred to as a *tolerance study*. This is an in-depth analysis of all dimensions and tolerances on the detail parts to confirm the interchangeability of mass-produced parts going into the final product or subassembly.

The method generally followed to do the study involves comparing in combination all matching surfaces and hole-pattern dimensional tolerances at maximum and minimum conditions. If interference or mismatch occurs, the tolerances must be balanced to afford interchangeability of all parts and subassemblies.

This study is essential and critical when producing great quantities of parts to eliminate special fitting, which would destroy the ability to interchange parts.

A thorough designer will perform a tolerance study on the mechanism or assembly of parts and balance his or her tolerances on all dimensions to create interchangeability. For standard tolerances and fits, both English and metric, refer to ANSI B4.1-1967 (R 1999) and ANSI B4.2-1999.

Parts also must be toleranced to allow for electrodeposited finishes such as zinc plating, hard-coat anodizing, and Teflon coating. (See Chap. 13 for plating thicknesses.)

General tolerancing procedures, force fits, sliding fits, and so on are described in detail in the following standards references:

Dimensioning and tolerancing: ANSI/American Society of Mechanical Engineers (ASME) Y14.5M-1994

Standard limits and fits: ANSI B4.1-1967 (R 1999)

Basic metric fits: ANSI B4.2-1978 (R1999)

Geometric tolerancing: ANSI/ASME Y14.5M-1994

Surface texture: ANSI/ASME B46.1-2002

Drawing practices (surface texture): ANSI Y14.36-1978

5.1.2   Tolerance accumulation

The tolerance values resulting from the three methods of dimensioning parts or assemblies as shown in Fig. 5.1a–c are as follows. We will then have the following conditions:

1. *Chain dimensioning.* The maximum variation between two features is equal to the sum of the tolerances on the intermediate distances. This results in the greatest tolerance accumulation (see Fig. 5.1a).

2. *Baseline dimensioning.* The maximum variation between two features is equal to the sum of the tolerances on the two dimensions from their origin to the features. This results in a reduction of the tolerance accumulation (see Fig. 5.1b).

3. *Direct dimensioning.* The maximum variation between two features is controlled by the tolerance on the dimension between the features. This results in the least tolerance (see Fig. 5.1c).

5.1.3   Dimensioning per ANSI/ASME Y14.5M-1994

ANSI Y14.5M-1994 covers dimensioning, tolerancing, and related practices for use on engineering drawings and related documents. The International System of Units (SI) is featured in this standard because SI units are expected to supersede U.S. customary units specified on engineering drawings. Nevertheless, it should be understood that U.S. customary units are equally applicable to this standard. Many U.S. companies and manufacturing facilities of all types still employ the U.S. customary or inch system.

Decimal dimensioning will be used on engineering drawings except where certain commercial commodities, such as pipe size, lumber size, and wire sizes, are identified.

Millimeter dimensioning

- When the dimension is less than 1 mm, a zero precedes the decimal point.

- Whole-number dimensions use neither a decimal point nor a zero following.

- On dimensions that exceed a whole number, the last digit to the right of the decimal point is not followed by a zero.

Note: The exception to the last rule is for bilateral tolerances or limits, where a zero is used by itself as part of the tolerance or limit.

Decimal-inch dimensioning.   The decimal-inch system is explained in ANSI B87.1, and the following will be observed when specifying decimal-inch dimensions on engineering drawings:

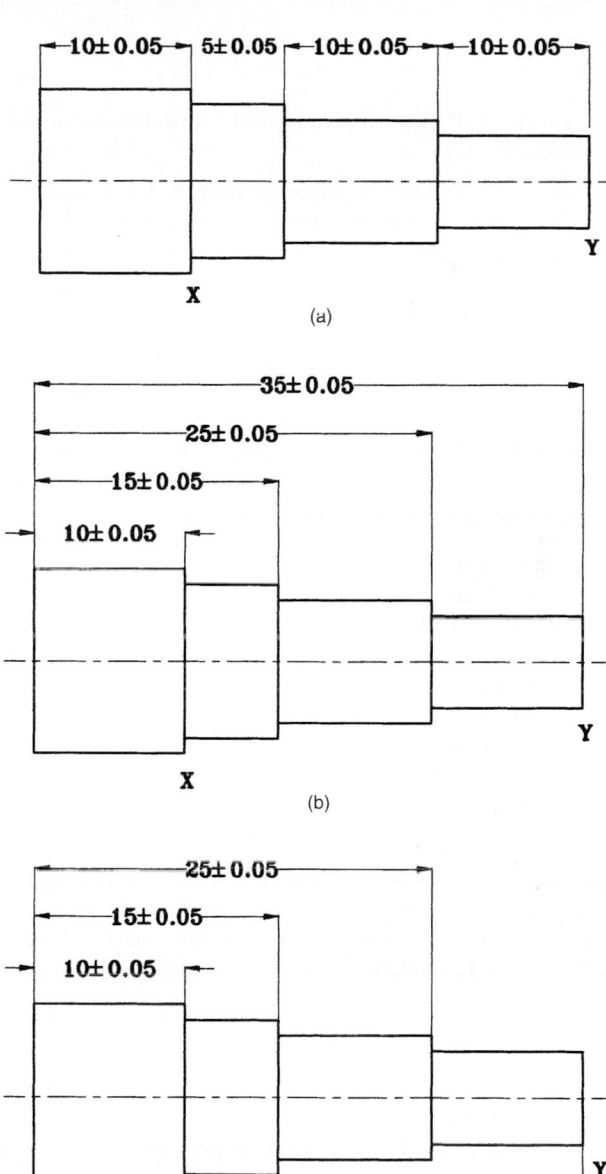

Figure 5.1 (*a*) Chain dimensioning—greatest tolerance accumulation between *X* and *Y*. (*b*) Baseline dimensioning—lesser tolerance accumulation between *X* and *Y*. (*c*) Direct dimensioning—least tolerance accumulation between *X* and *Y*.

- A zero is not used *before* the decimal point for dimensional values of less than 1 in.

- A dimension is expressed to the same number of decimal places as its tolerance. Zeros are added to the right of the decimal point where necessary.

- Decimal points must be uniform and large enough to be clearly visible and meet the requirements of ANSI Y14.2M.

- The conversion and rounding of U.S. customary units are covered in ANSI Z210.1.

Glossary of terms—dimensioning and tolerancing.   The following glossary of terms will be useful to the design engineer in that the use or recognition of these terms related to dimensioning and tolerancing will prevent misunderstanding between individuals and companies in relation to the meaning of these terms.

Actual size   The measured size.

Basic dimension   A numerical value used to describe the theoretically exact size, profile, orientation, or location of a feature or datum target. It is the basis from which permissible variations are established by tolerances on other dimensions, in notes, or in feature control frames.

Bilateral tolerance   A tolerance in which variation is permitted in both directions from the specified dimension.

Datum   A theoretically exact point, axis, or plane derived from the true geometric counterpart of a specified datum feature. A datum is the origin from which the location or geometric characteristics of features of a part are established.

Datum feature   An actual feature of a part that is used to establish a datum.

Datum target   A specified point, line, or area on a part used to establish a datum.

Dimension   A numerical value expressed in appropriate units of measure and indicated on a drawing and in other documents along with lines, symbols, and notes to define the size or geometric characteristic, or both, of a part or part feature.

Feature   The general term applied to a physical portion of a part, such as a surface, hole, or slot.

Feature of size   One cylindrical or spherical surface or a set of two plane parallel surfaces, each of which is associated with a size dimension.

Full indicator movement (FIM)   The total movement of an indicator when applied appropriately to a surface to measure its variations.

Geometric tolerance   The general term applied to the category of tolerances used to control form, profile, orientation, location, and runout.

Least-material condition (LMC)   The condition in which a feature of size contains the least amount of material within the stated limits of size—for example, maximum hole diameter, minimum shaft diameter.

Limits of size   The specified maximum and minimum sizes.

Maximum material condition (MMC)   The condition in which a feature of size contains the maximum amount of material within the stated limits of size—for example, minimum hole diameter, maximum shaft diameter.

Reference dimension   A dimension, usually without tolerance, used for information purposes only. It is considered auxiliary information and does not govern production or inspection operations. A reference dimension is a repeat of a dimension or is derived from other values shown on the drawing or on related drawings.

Regardless of feature size (RFS)   The term used to indicate that a geometric tolerance or datum reference applies at any increment of size of the feature within its size tolerance.

Tolerance   The total amount by which a specific dimension is permitted to vary. The tolerance is the difference between the maximum and minimum limits.

True position   The theoretically exact location of a feature established by basic dimensions.

Unilateral tolerance   A tolerance in which variation is permitted in one direction from the specified dimension.

Virtual condition   The boundary generated by the collective effects of the specified MMC limit of size of a feature and any applicable geometric tolerances.

Applications of dimensions.   Figures 5.2 through 5.24 show the applications of dimensions and dimension lines.

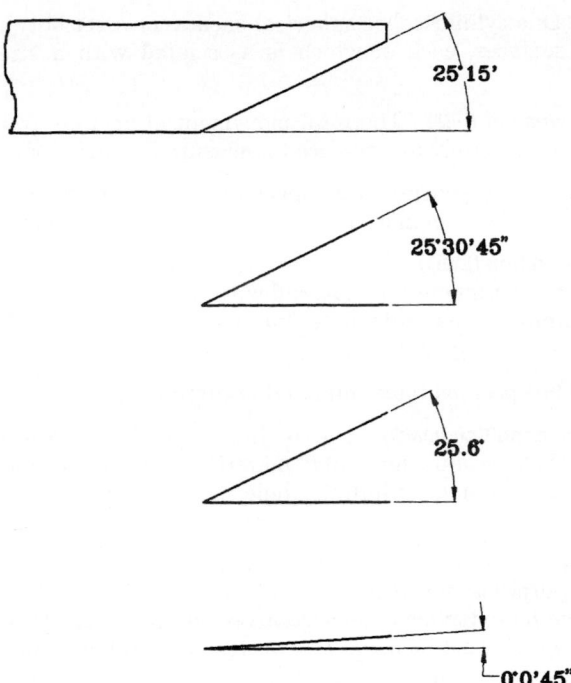

Figure 5.2 Angular units.

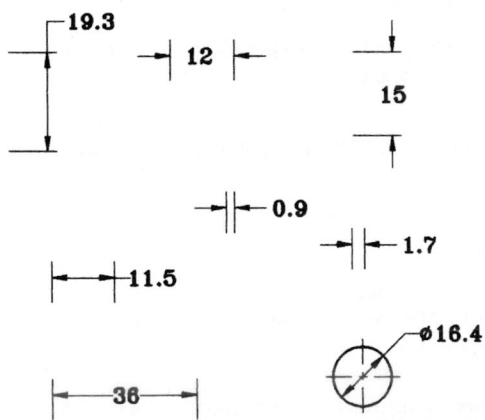

Figure 5.3 Millimeter dimensions.

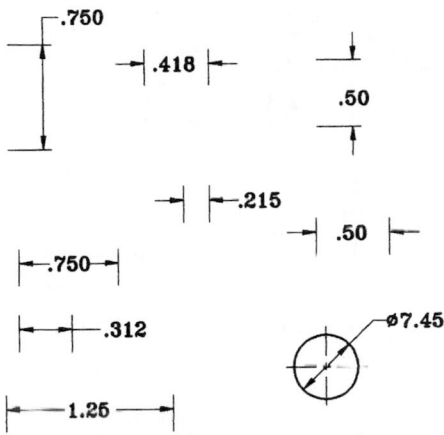

Figure 5.4 Decimal-inch dimensions.

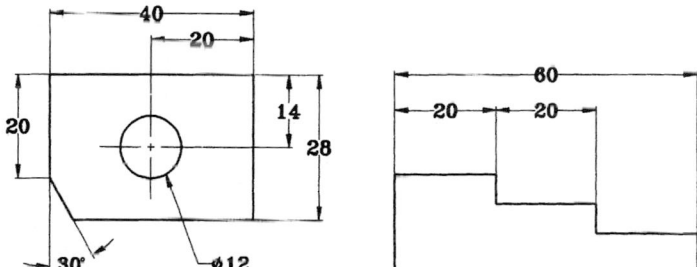

Figure 5.5 Application of dimensions.

Figure 5.6 Grouping of dimensions.

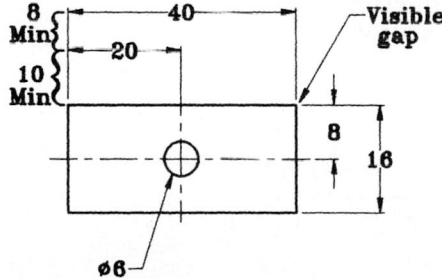

Figure 5.7 Spacing of dimensions.

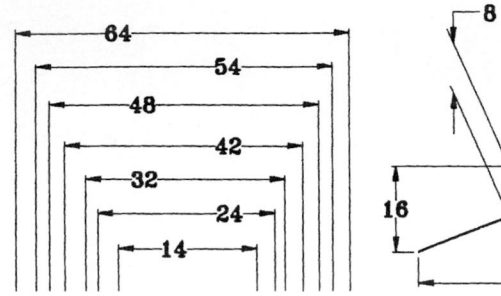

Figure 5.8 Staggered dimensions.

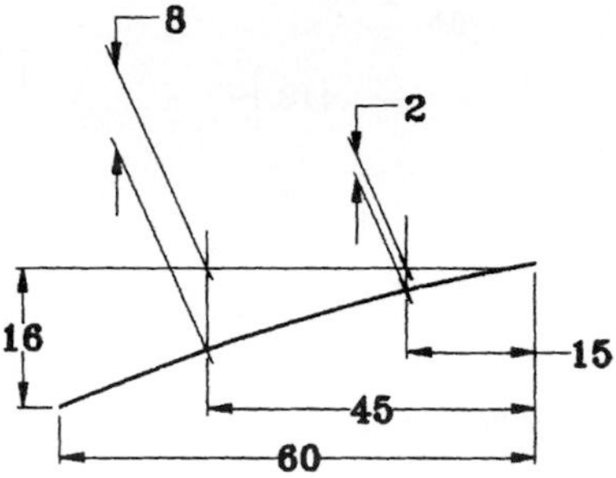

Figure 5.9 Oblique extension lines.

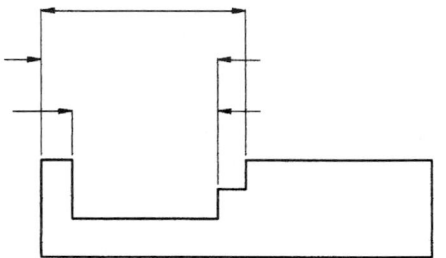

Figure 5.10 Breaks in dimension lines.

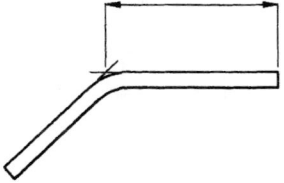

Figure 5.11 Point locations.

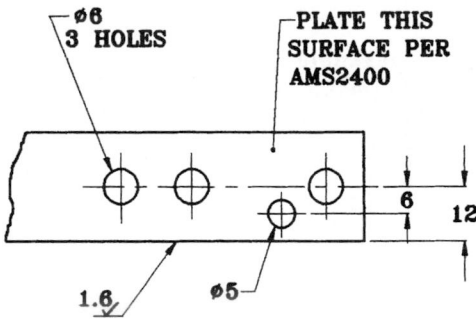

Figure 5.12 Leaders.

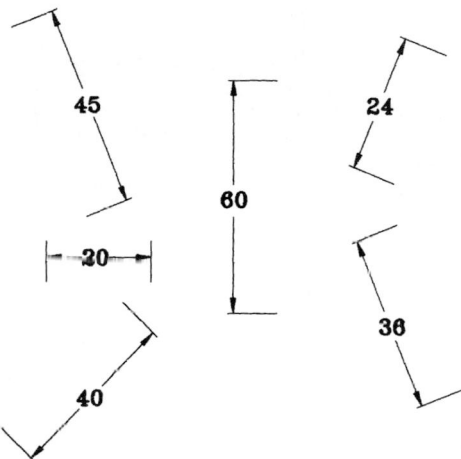

Figure 5.13 Reading directions of dimensions.

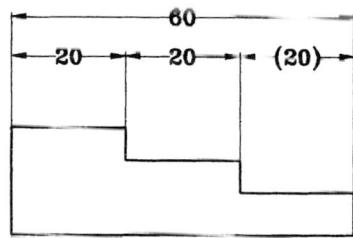

Figure 5.14 The intermediate reference dimension.

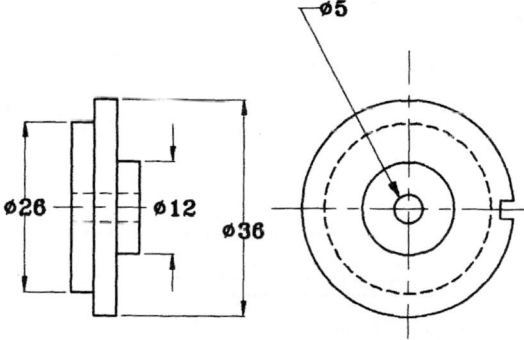

Figure 5.15 Dimensioning diameters.

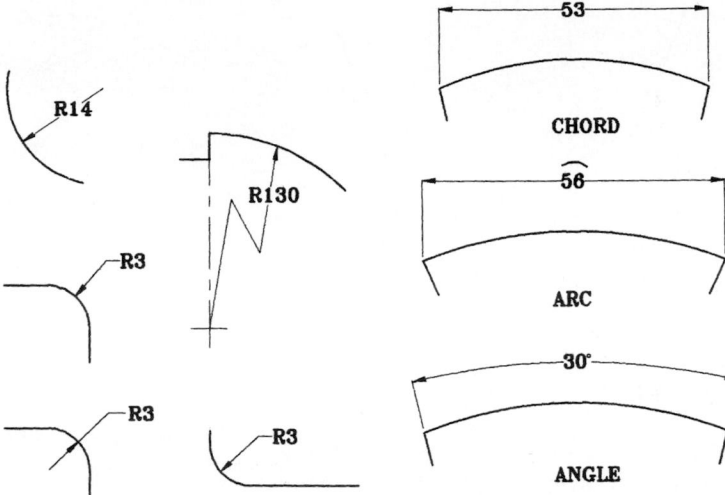

Figure 5.16 Dimensioning radii.

Figure 5.17 Dimensioning chords, arcs, and angles.

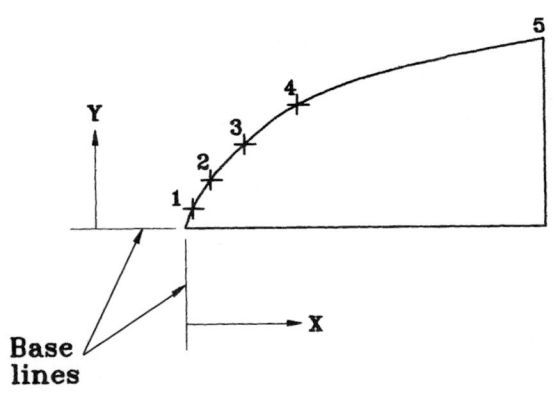

STATION	1	2	3	4	5
X	2	5	15	28	57
Y	4.3	8.5	14	19	24.5

Figure 5.18 Tabulated outline dimensions.

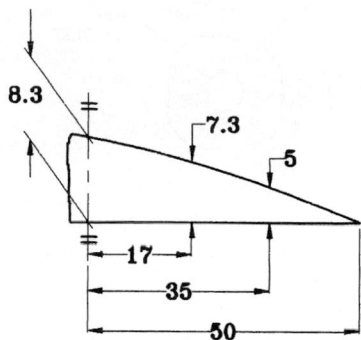

Figure 5.19 Symmetric outlines.

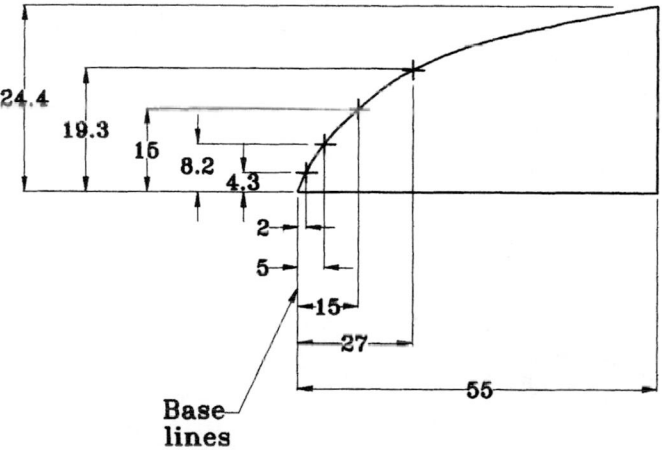

Figure 5.20 Coordinate or offset outline dimensions.

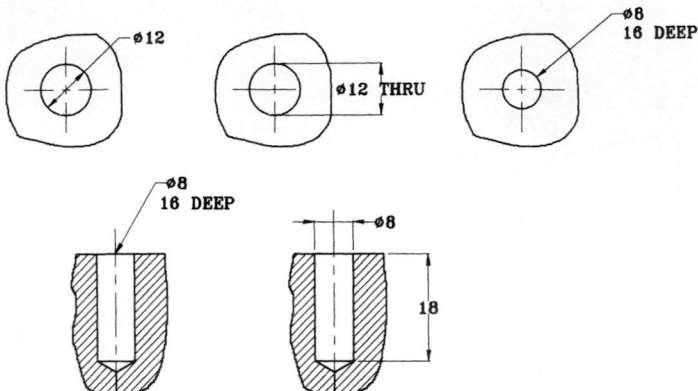

Figure 5.21   Dimensioning round holes.

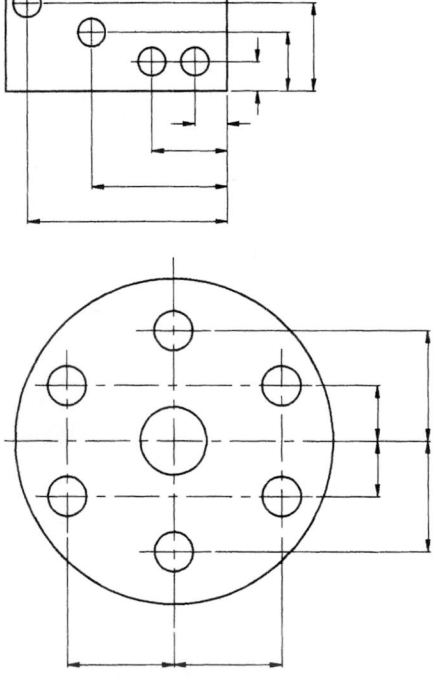

Figure 5.22   Rectangular coordinate dimensioning.

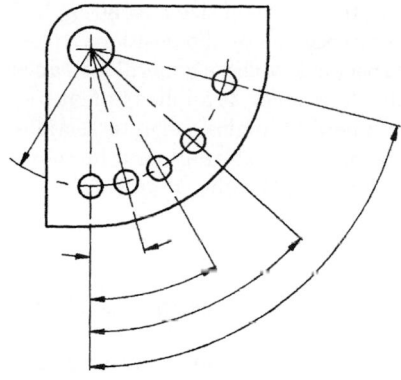

Figure 5.23 Polar coordinate dimensioning.

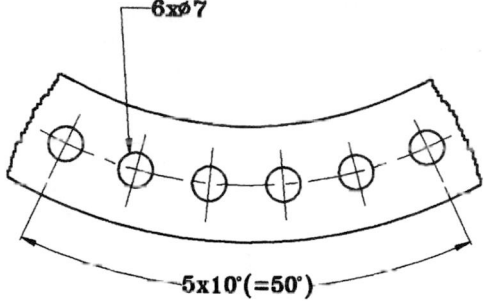

Figure 5.24 Dimensioning repetitive features.

5.1.4 ANSI Y14.5M-1994 (R1999) tolerancing practices

Per ANSI Y14.5M-1994 (R1999), tolerances may be expressed as follows:

- As direct limits or as tolerance values applied directly to a dimension
- As a geometric tolerance
- In a note referring to specific dimensions
- As specified in other documents referenced on the engineering drawing for specific features or processes
- In a general tolerance block referring to all dimensions on the engineering drawing, unless specified otherwise

Tolerances on dimensions that locate features of size may be applied directly to the locating dimensions or specified by the positional tolerancing method. Unless otherwise specified, where a general tolerance note on a drawing includes angular tolerances, it applies to features shown at specified angles and at implied 90° angles, i.e., intersections of centerlines, corners of parts (internal and external), or other obvious areas not specifically shown to have angles other than 90°.

5.1.5 Direct tolerancing methods

Limits and directly applied tolerance values are specified as follows:

- *Limit dimensions.* The high limit or maximum value is placed above the low limit or minimum value. As a single-line callout, the low limit precedes the high limit, with a dash separating the values.

- *Plus and minus tolerancing.* The basic dimension is given first, followed by a plus-and-minus expression of tolerance.

- *Tolerance limits.* All tolerance limits are absolute.

- *Dimensional limits before or after plating.* For plated or coated parts, the engineering drawing or referenced document will state whether the dimensions are before or after plating; e.g., "Dimensional limits apply before plating" or "Dimensional limits apply after plating."

5.1.6 Positional tolerancing

Positional or *location tolerancing* defines a zone within which the center, axis, or center plane of a feature of size is permitted to vary from the true or exact position.

Geometric tolerancing is the general term applied to the category of tolerances used to control form, profile, orientation, location, and runout.

5.1.7 Examples of ANSI Y14.5M-1994 (1999) dimensioning and tolerancing practices

Figure 5.25 shows a typical engineering drawing using the ANSI form for dimensioning, tolerancing, and positioning. For a complete description and operational instructions for the use of ANSI standard dimensioning and tolerancing practices, see ANSI Y14.5M-1994 (1999), which may be obtained directly from ANSI, Inc. See Chap. 16

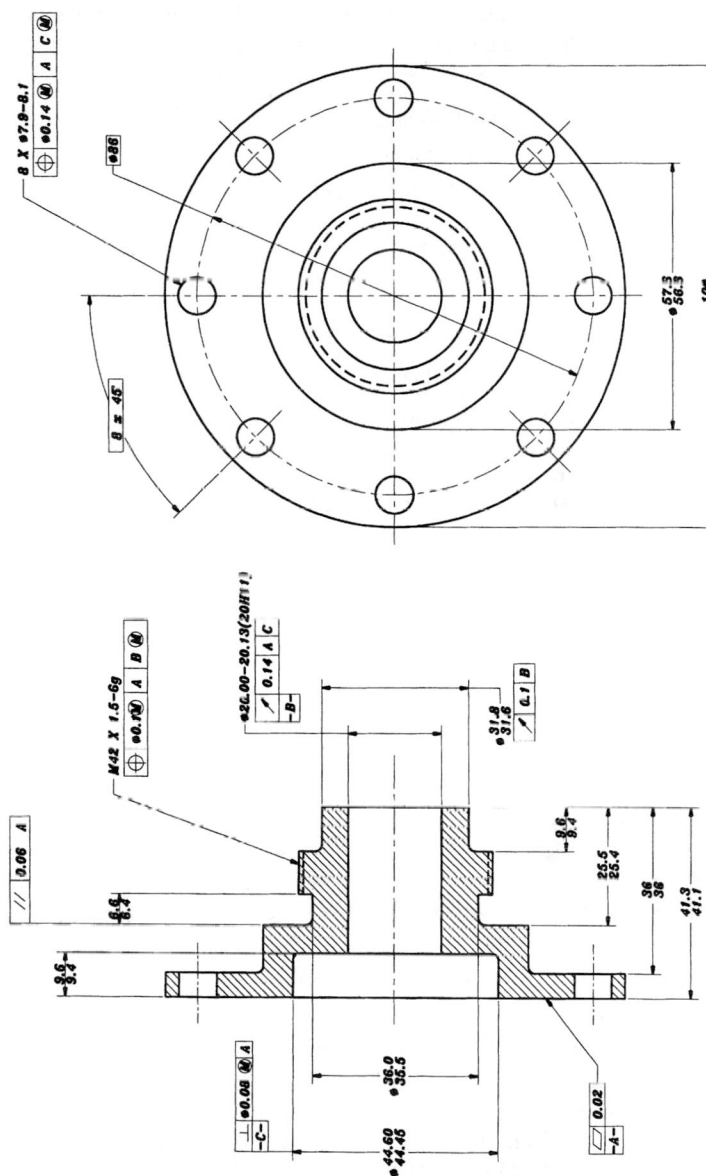

Figure 5.25 Feature control frame placement dimensions per ANSI Y14.5M-1994 (R1999).

287

for addresses and acronyms of American standards organizations, specification authorities, societies, and institutes.

5.1.8 Design notes on dimensioning and tolerancing

From an electromechanical design standpoint, the dimensioning and tolerancing practices used on engineering drawings for the design and manufacturing of parts, subassemblies, and assemblies should take the following points into consideration:

- Close tolerances add cost to a finished product.

- The tolerance should be balanced to the function of the part.

- Arbitrary selection of a general tolerance can cause design and fit problems on the finished product or create unnecessary work.

- Use care in the selection of bilateral and unilateral tolerances.

- Remember that modern computer numerical controlled (CNC) turning centers, machining centers, electric discharge machining (EDM) machines, CNC punch presses, and other CNC equipment are capable of producing parts with closer tolerances than were possible in the past. Spindle accuracies are higher and CNC movement controls are very accurate on modern machine tools and equipment.

- Use tables of preferred limits and fits only when applicable.

- Select plating thickness limits (range) carefully so as to prevent dimensional interference between mating parts.

- To control a large tolerance spread owing to many parts in a dynamic assembly, design an adjustment means in the mechanism at one or more critical positions.

- Tooling fixtures and tooled parts help control tolerance ranges to a great extent on assemblies and complex mechanisms.

- Machined finishes are surface textures and therefore can be considered to have a tolerance [root-mean-square (rms)] value. Therefore, specify only a machined or tooled surface finish that is functional to the part.

The experienced electromechanical designer or product design engineer must be proficient in dimensioning and tolerancing practices to be most effective and successful at the design function. Many

design and manufacturing problems occur when the engineering drawings are not dimensioned and toleranced properly and effectively. I therefore recommend that a thorough study be made of dimensioning and tolerancing practices through the use of ANSI YI4.5M-1982 (reaffirmed 1988) or a later revision.

5.1.9   Symbols used in ANSI Y14.5M-1994 (R1999) and ISO dimensioning and tolerancing

See Figure 5.26 for the ANSI and International Standards Organization (ISO) symbols currently used for dimensioning and tolerancing.

5.2   Typical Industrial Design Engineering Drawings

There are certain exceptions to the recommended dimensioning and tolerancing practices described in this chapter.

■ Engineering drawings that are used exclusively within the confines of any particular company or organization may use drawing styles and dimensioning and tolerancing practices that differ from those shown in the preceding sections of this chapter. Some companies are still using the English fractional system of dimensioning, even though the fractional system was discontinued from general industrial use in the early 1950s.

■ Different symbolisms for the different manufacturing processes such as welding, brazing, assembly, and so forth are also used by different companies.

■ Tooling drawings such as that shown in Fig. 5.27 also may differ from the ANSI and ISO standards. Figure 5.27 shows a part that is to be die-stamped, and there are no tolerances on the drawing, with the exception of the hole location, which is produced at a different stage of manufacturing as a separate operation. The EDM machine that is used to cut the dies for this part will allow the part to be stamped exactly to the dimensions shown, i.e., to the precise three-place decimals indicated on the drawing.

Although there are exceptions to the modern dimensioning and tolerancing systems, as indicated previously, the advent of ANSI Y14.5M, SI, ISO 9000, and other national and international standards will make it difficult for American companies to compete in the

SYMBOL FOR:	ANSI Y14.5	ISO
STRAIGHTNESS	—	—
FLATNESS	▱	▱
CIRCULARITY	○	○
CYLINDRICITY	⌭	⌭
PROFILE OF A LINE	⌒	⌒
PROFILE OF SURFACE	⌓	⌓
ALL AROUND PROFILE	⌒○	NONE
ANGULARITY	∠	∠
PERPENDICULARITY	⊥	⊥
PARALLELISM	//	//
POSITION	⌖	⌖
CONCENTRICITY/COAXIALITY	◎	◎
SYMMETRY	NONE	⌯
CIRCULAR RUNOUT	↗	↗
TOTAL RUNOUT	⌰	⌰
AT MAXIMUM MATERIAL CONDITION	Ⓜ	Ⓜ
AT LEAST MATERIAL CONDITION	Ⓛ	NONE
REGARDLESS OF FEATURE SIZE	Ⓢ	NONE
PROJECTED TOLERANCE ZONE	Ⓟ	Ⓟ
DIAMETER	Ø	Ø
BASIC DIMENSION	[50]	[50]
REFERENCE DIMENSION	(50)	(50)
DATUM FEATURE	[-A-]	⊿ OR ⊿[A]
DATUM TARGET	Ø2/A1	Ø2/A1
TARGET POINT	X	X
DIMENSION ORIGIN	⌾→	NONE
FEATURE CONTROL PLANE	[⌖|Ø0.5Ⓜ|A|B|C]	[⌖|Ø0.5Ⓜ|A|B|C]
CONICAL TAPER	⌲	⌲
SLOPE	⌳	⌳
COUTERBORE/SPOTFACE	⌴	NONE
COUNTERSINC	⌵	NONE
DEPTH/DEEP	↧	NONE
SQUARE (SHAPE)	□	□
DIMENSION NOT TO SCALE	15 (underlined)	15 (underlined)
NUMBER OF TIMES/PLACES	8X	8X
ARC LENGTH	⌒105	NONE
RADIUS	R	R
SPHERICAL RADIUS	SR	NONE
SPHERICAL DIAMETER	SØ	NONE

Figure 5.26 ANSI and ISO dimensioning and tolerancing symbols.

world marketplace unless they adhere to these worldwide standards. One requirement of these systems is drawing, dimensioning, and tolerancing practices and engineering documentation control. Also, it is to be noted that when any particular company's engineering drawings are sent to outside vendors or subcontractors, these drawings

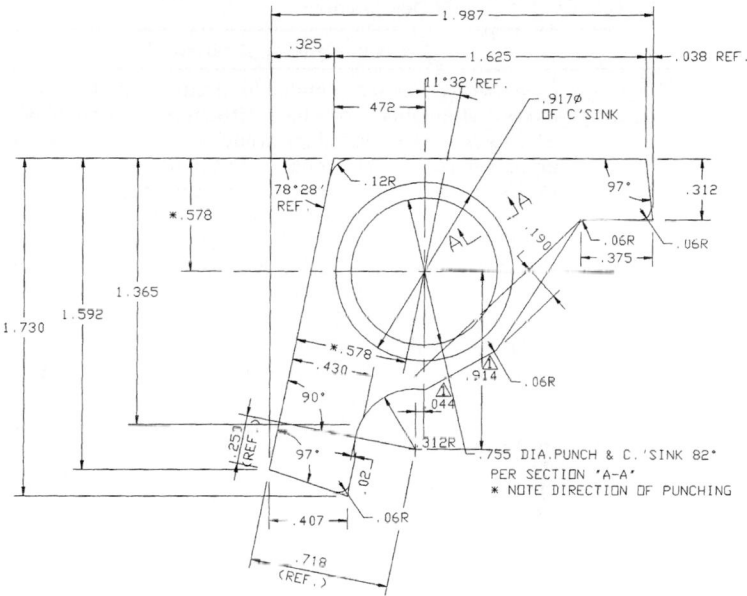

Figure 5.27 AutoCAD drawing showing typical industrial practice on a drawing for a tooled part.

always should be prepared to the national and international standards; otherwise, misinterpretations of the drawings may occur, causing losses, rejections, and possible lawsuits.

5.2.1   Limits and Fits, U.S. Customary and Metric Standards

Limits and fits of shafts and holes are important design and manufacturing considerations. Fits should be selected carefully according to function. The fits outlined in this section are all on a unilateral hole basis. Table 5.1 describes the various U.S. Customary fit designations. Classes RC9, LC10, and LC11 are described in the ANSI standards but are not included here. Table 5.1 is valid for sizes up to approximately 20 in in diameter and is in accordance with American, British, and Canadian recommendations.

The coefficients C listed in Table 5.2 are to be used with the equation $L = CD^{1/3}$, where L is the limit in thousandths of an inch corresponding to the coefficients C and the basic size D in inches.

TABLE 5.1   U.S. Customary Fit Designations

Designation	Name and application
RC1	*Close sliding fits* are intended for accurate location of parts that must be assembled without perceptible play.
RC2	*Sliding fits* are intended for accurate location, but with greater maximum clearance than the RC1 fit.
RC3	*Precision running fits* are the loosest fits that can be expected to run freely. They are intended for precision work at slow speeds and light pressures, but are not suited for temperature differences.
RC4	*Close running fits* are intended for running fits on accurate machinery with moderate speeds and pressures. They exhibit minimum play.
RC5	*Medium running fits* are intended for higher running speeds or heavy journal pressures or both.
RC6	*Medium running fits* are for use where more play than RC5 is required.
RC7	*Free running fits* are for use where accuracy is not essential or where large temperature variations may occur or both.
RC8	*Loose running fits* are intended where wide commercial tolerances may be necessary, together with an allowance on the hole.
LC1 to LC9	*Locational clearance fits* are required for parts that are normally stationary but can be freely assembled and disassembled. Snug fits are for accuracy of location. Medium fits are for parts such as ball, race, and housing. The looser fastener fits are used where freedom of assembly is important.
LT1 to LT6	*Locational transitional fits* are a compromise between clearance and interference fits where accuracy of location is important, but either a small amount of clearance or interference is permitted.
LN1 to LN3	*Locational interference fits* are for accuracy of location and for parts requiring rigidity and alignment, with no special requirement for bore pressure. Not intended for parts that must transmit frictional loads to one another.
FN1	*Light drive fits* require light assembly pressures and produce permanent assemblies. Suitable for thin sections or long fits or in cast-iron external members.
FN2	*Medium drive fits* are for ordinary steel parts or shrink fits on light sections. They are the tightest fits that can be used with high-grade cast-iron external members.
FN3	*Heavy drive fits* are for heavier steel parts or for shrink fits in medium sections.
FN4 and FN5	*Force fits* are suitable for parts that can be highly stressed or for shrink fits where the heavy pressing forces required are not practical.

The resulting calculated values of L are then summed algebraically to the basic shaft size to obtain the four limiting dimensions for the shaft and hole. The limits obtained by the preceding equation and Table 5.2 are very close approximations to the standards and are applicable in all cases except where exact conformance to the standards is required by specifications.

Example A "precision running fit" is required for a nominal 1.5000-in-diameter shaft (designated as an RC3 fit per Table 5.1).

Lower limit for the hole:

$L_1 = CD^{1/3}/1000$

$L_1 = 0\ (1.5)^{1/3}/1000$

$L_1 = 0$

$d_L = 0 + 1.5000$

$d_L = 1.50000$

Upper limit for the hole:

$L_2 = CD^{1/3}/1000$

$L_2 = +0.907\ (1.5)^{1/3}/1000$

$L_2 = 0.001038$

$d_U = 0.001038 + 1.5000$

$d_U = 1.50104$

Lower limit for the shaft:

$L_3 = CD^{1/3}/1000$

$L_3 = -1.542\ (1.5)^{1/3}/1000$

$L_3 = -0.00176513$

$D_L = 1.500 + (-0.00176513)$

$D_L = 1.49823$

Upper limit for the shaft:

$L_4 = CD^{1/3}/1000$

$L_4 = -0.971\ (1.5)^{1/3}/1000$

$L_4 = -0.0011115$

$D_U = 1.500 + (-0.0011115)$

$D_U = 1.49889$

Therefore, the hole and shaft limits are as follows:

Hole size = 1.50000/1.50104 diameter

Shaft size = 1.49889/1.49823 diameter

Table 5.3 shows the metric preferred fits for the cylindrical parts in holes. The procedures for calculating the limits of fit for the metric standards are shown in the ANSI standards. An alternative to this procedure would be to correlate the type of fit between the metric standard fits shown in Table 5.3 and the U.S. Customary fits shown in Table 5.1 and proceed to convert the metric measurements in millimeters to inches and then calculate the limits of fit according to the method shown in this section for the U.S. Customary system. The calculated answers then would be converted back to millimeters.

There should be no technical problem with this procedure except conflict with mandatory specifications, in which case you will need to concur with ANSI B4.2-1978(R1999) for the metric standard. The U.S. Customary standard for preferred limits and fits is ANSI B4.1-1967(R1999).

TABLE 5.2   Coefficients C

Class of fit	Hole limits		Shaft Limits	
	Lower	Upper	Lower	Upper
RCI	0	+0.392	−0.588	−0.308
RC2	0	+0.571	−0.700	−0.308
RC3	0	+0.907	−1.542	−0.971
RC4	0	+1.413	−1.879	−0.971
RC5	0	+1.413	−2.840	−1.932
RC6	0	+2.278	−3.345	−1.932
RC7	0	+2.278	−4.631	−3.218
RC8	0	+3.570	−7.531	−5.253
LCI	0	+0.571	−0.392	0
LC2	0	+0.907	−0.571	0
LC3	0	+1.413	−0.907	0
LC4	0	+3.570	−2.278	0
LC5	0	+0.907	−0.879	−0.308
LC6	0	+2.278	−2.384	−0.971
LC7	0	+3.570	−4.211	−1.933
LC8	0	+3.570	−5.496	−3.218
LC9	0	+5.697	−8.823	−5.253
LT1	0	+0.907	−0.281	+0.290
LT2	0	+1.413	−0.442	+0.465
LT3*	0	+0.907	+0.083	+0.654
LT4*	0	+1.413	+0.083	+0.990
LT5	0	+0.907	+0.656	+1.227
LT6	0	+0.907	+0.656	+1.563
LN1	0	+0.571	+0.656	+1.048
LN2	0	+0.907	+0.994	+1.565
LN3	0	+0.907	+1.582	+2.153
FN1	0	+0.571	+1.660	+2.052
FN2	0	+0.907	+2.717	+3.288
FN3[†]	0	+0.907	+3.739	+4.310
FN4	0	+0.907	+5.440	+6.011
FN5	0	+1.413	+7.701	+8.608

Note: Coefficients are for use with the equation $L = CD^{1/3}$.
*Not for sizes under 0.24 in.
[†]Not for sizes under 0.95 in.
Source: J. Shigley and C. Mischke (eds.), *Standard Handbook of Machine Design.*
New York: McGraw-Hill, 1996, p. 19.11.

The preceding procedures for limits and fits are mandatory practice for design engineers and tool design engineers in order for parts to function according to their intended design requirements. Assigning arbitrary or "rule of thumb" procedures for the fitting of cylindrical parts in holes is not good design practice and can create many problems in the finished product.

TABLE 5.3   **Metric Preferred Fits for Cylindrical Parts in Holes**

Type	Hole basis	Shaft basis	Name and application
Clearance	H11/c11	C11/h11	*Loose running fit* is for wide commercial tolerances or allowances on external parts.
	H9/d9	D9/h9	*Free running fit* is not for use where accuracy is essential, but good for large temperature variations, high running speeds, or heavy journal pressures.
	H8/f7	F8/h7	*Close running fit* is for running on accurate machines and accurate location at moderate speeds and journal pressures.
	H7/g6	G7/h6	*Sliding fit* is not intended for running freely, but to move and turn freely and locate accurately.
	H7/h6	H7/h6	*Locational clearance fit* provides snug fit for locating stationary parts, but can be freely assembled and disassembled.
Transition	H7/k6	K7/h6	*Locational transition fit* is for accurate location, a compromise between clearance and interference.
	H7/n6	N7/h6	*Locational transition fit* is for more accurate location where greater interference is permitted.
Interference	H7/p6	P7/h6	*Locational interference fit* is for parts requiring rigidity and alignment with prime accuracy of location but with special bore pressures required.
	H7/s6	S7/h6	*Medium drive fit* is for ordinary steel parts or shrink fits on light sections, the tightest fit usable with cast iron.
	H7/u6	U7/h6	*Force fit* is suitable for parts that can be highly stressed or for shrink fits where the heavy pressing forces required are not practical.

6

Computer-Aided Design, Manufacturing, and Engineering Systems

Regardless of what is being produced, the fabrication process nearly always begins with a computer-aided design drawing. While paper prints are still a necessity in some instances, the machining and metalworking industries have been transformed by the evolution of computer-aided design (CAD) and computer-aided manufacturing (CAM) systems. Two-dimensional drafting packages are largely giving way to three-dimensional solid-modeling packages. Although mechanical design typically is not the responsibility of the machinist or metalworker, machinists and metalworkers increasingly are required to be fluent in the use of a solid-modeling system for purposes of designing tooling, generating toolpaths, and conducting engineering analysis studies.

6.1 Computer-Aided Design (CAD)

6.1.1 File formats

Given the widely distributed nature of design and manufacturing, it is often the case that the machinist will not be working with the same CAD package as the mechanical designer. In these cases, either the machinist's CAD system must be capable of importing the CAD format used by the mechanical designer, or the two must agree on a neutral interchange format.

Some of the more commonly encountered CAD file formats can be found in Table 6.1. Of the formats shown on this list, IGES and STEP are examples of neutral data-exchange file formats developed by standards committees. The Parasolid and ACIS formats are commercial formats, but they are also of interest in the sense that most of the CAD systems represented in Table 6.1 are built on either the Parasolid or the ACIS kernel. If the mechanical designer and machinist both happen to use CAD systems built on the same kernel, then they typically have the option of swapping files using either the Parasolid or the ACIS file formats. Having said this, many CAD vendors have built import-export support for competitors' file formats into their systems as a matter of corporate strategy. Having the ability to import files from a competing system makes it easier for a company to switch CAD systems without having to redesign all of its CAD models. Support for multiple file formats also facilitates file transfers between design firms and the array of external suppliers manufacturing the components or subsystems.

It is worth noting that when a CAD file is imported from a different system, the workpiece geometry typically is the only information that gets transferred. Embedded notes, tolerances, and other types

TABLE 6.1   Commonly Used CAD File Formats

CAD system or standard	Associated file format(s)
ACIS	*.SAT
AutoCAD	*.DWG; *.DXF
Autodesk Inventor	*.IPT
CADKEY	*.PRT
CATIA	*.CGR
IGES	*.IGS; *.IGES
Parasolid	*.X_B; *.XMT_BIN (binary files)
	*.X_T; *.XMT_TXT (text files)
ProEngineer	*.PRT; *.ASM
SolidEdge	*.PAR; *.ASM
SolidWorks	*.SLDPRT; *.SLDASM; *.SLDDRW
STEP	*.STP; *.STEP
STL	*.STL
Unigraphics	*.PRT
VRML	*.WRL

of nongeometric information often are lost when a CAD model is ported into another file format. Models consisting of complex surfaces that have been stitched or otherwise blended together do not always transfer cleanly between systems either.

6.1.2 CAD vendors

There are numerous suppliers of CAD software throughout the United States and other countries. A list of these software suppliers is shown in Table 6.2.

6.1.3 CAD terminology

As mentioned previously, the machinist or metalworker often is called on to design molds, dies, and/or machining fixtures that are needed to fabricate a given part. Although it is well beyond the scope of this *Handbook* to describe how tools, dies, etc. are designed in a given CAD system, it is worthwhile to outline the basic terminology associated with these systems, as well as the general approaches that may be employed by machinists and metalworkers.

Feature-based design. This refers to the act of designing parts by successively adding one solid feature after another. The features may either add volume to the model (i.e., an additive feature) or remove volume from the model (i.e., a subtractive feature). Examples of features include holes, slots, chamfers, fillets, shells, etc.

Parametric CAD. The size and location of each feature in the CAD system is defined by parameters (e.g., length, width, etc.). The solid

TABLE 6.2 Selected CAD Vendors

Company	Web site	Product(s)
Autodesk, Inc.	*www.autodesk.com*	AutoCAD Inventor
Bentley Systems, Inc.	*www.bentley.com*	Microstation
Dassault Systems	*www.3ds.com*	Catia
Parametric Technology Corp.	*www.ptc.com*	Pro/Engineer
Solidworks Corp.	*www.solidworks.com*	SolidWorks
TurboCAD	*www.turbocad.com*	TurboCAD
UGS	*www.ugs.com*	NX
	www.solidedge.com	Solid Edge

features will change size, shape, and/or location depending on the values that you assign to the parameters. If you increase the value of length parameter for a cube, the distance between the two end faces will increase accordingly. More important, if you have specified the distance of a hole from one end of the cube, then it will remain at that distance regardless of how long or short the cube is. The CAD modeler need not worry about repositioning every feature if the size of the cube changes.

Parametric associativity.   This is among the most important aspects of modern CAD systems. It means that you can link, or associate, the size and/or position of a feature in one CAD model with the size and/or location of a feature in another CAD model. Suppose a machining fixture has been modeled that consists of a steel plate with a number of drill bushings. The steel plate will have holes machined in it that correspond to the locations of the drill bushings. Likewise, the size of the holes must be related to the outer diameter (OD) of the drill bushings. It is possible to link the hole diameter in the steel plate's CAD model with the OD of the drill bushing. If a different-diameter drill bushing is selected, then the associated hole diameter in the steel plate will change automatically when the CAD model is rebuilt. Associativity can go far beyond this simple example. The dimensions on a part drawing will update automatically when the solid model for that part changes. Toolpaths generated in CAM packages can be associated with CAD model geometry. If the curvature on a mold surface is modified slightly by the designer, then a parametrically associated CAM system will automatically regenerate its toolpaths accordingly.

Assembly layout sketch.   This is a sketch that shows the overall size and location of parts within an assembly.

Top-down design.   With top-down modeling, the designer will generate assembly layout sketches in an assembly file that shows key dimensions, shapes, and locations of components in the assembly. Then the individual components are modeled with their vertices, edges, and faces referring back to the sketches in the top-level assembly file. Any time the modeler changes the assembly layout drawing, the components that are linked to the layout sketch update automatically. In other words, the design starts with the assembly and then proceeds to the individual parts.

Bottom-up design.   With bottom-up modeling, each individual component is modeled first. The components are then brought together in

an assembly file. Of course, it is possible to mix the two approaches as well.

6.1.4 Solid modeling techniques

Creating base features. The process of modeling an individual component typically begins with a two-dimensional sketch. This sketch may be extruded linearly, revolved, or swept along a guide path to create a three-dimensional *base feature*. Figure 6.1 shows how the same rectangular sketch can be extruded, revolved, or swept to produce completely different base features.

To put these three approaches in the context of activities that might be performed by a machinist, the linear extrude would be used to model the basic shape of a rectangular mold base. The revolve would be used to model cylindrical components such as ejector pins. The sweep could be used to model a U-shaped path followed by a cooling channel.

Once a base feature is created, additional features can be created by making new sketches that are extruded, revolved, or swept. These subsequent features can be used to add volume to the existing part, or they can be used to cut material away from an existing part. Figure 6.2a and b show how an extruded circle can be used to create a hole or a boss depending on whether the CAD modeler specifies an additive or subtractive feature. In both cases, a circle was sketched on the front face of the plate. The circle was extruded away from the plate in Fig. 6.2a to add a boss. The circle was cut-extruded into the plate in Fig. 6.2b to create a hole.

When creating a feature from a sketch, the following procedure is generally followed:

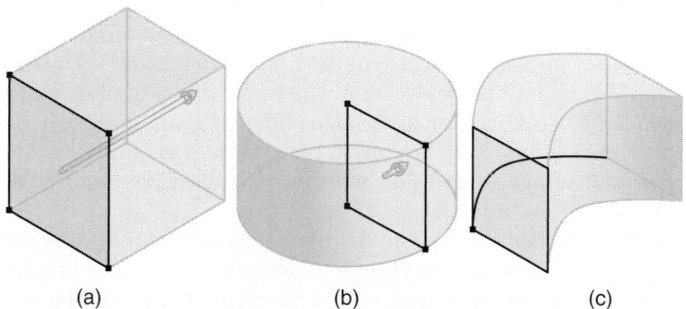

(a)	(b)	(c)

Figure 6.1 (*a*) Extruded rectangle. (*b*) Revolved rectangle. (*c*) Swept rectangle.

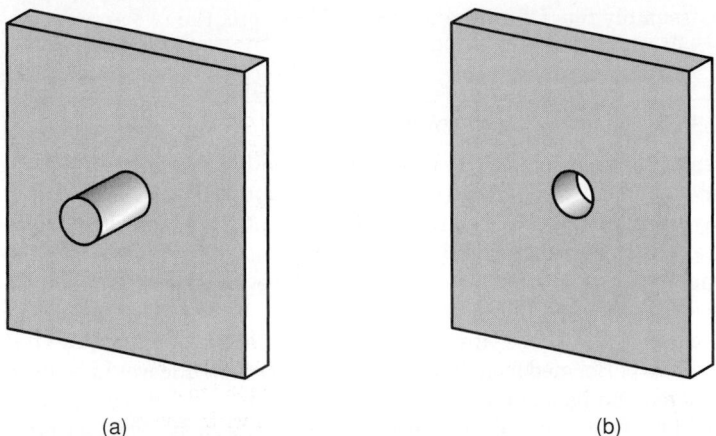

(a)   (b)

Figure 6.2   (*a*) Additive feature. (*b*) Subtractive feature.

- Quickly sketch the desired shape using lines, arcs, circles, splines, etc. Just sketch the approximate shape, and do not be concerned with whether the sketch entities are the correct size.

- Apply geometric relationships that constrain the size and/or location of the sketched entities.

- Apply dimensions.

- Extrude, revolve, or sweep the sketch to add or subtract volume from the model.

Capturing the design intent.   Many CAD modelers fail to appreciate how important it is to use good practices when assigning geometric relationships and dimensions to a sketch. The term *design intent* is used frequently in connection with these steps. When an individual models the "design intent," he or she is creating the model such that the model will update properly when the inevitable design changes are made. In a sense, the modeler also should make the models as error-proof as possible. This is important because it is quite possible that other individuals not familiar with the model may be required to modify it at a future time.

One of the simplest ways to model the design intent and to make a model as error-proof as possible is to specify geometric relations wherever possible instead of explicitly defining dimensions. Table 6.3 shows different types of sketch relationships that are commonly available in most solid-modeling CAD systems.

TABLE 6.3   Geometric Relationships

Type of relationship	Description
Coincident	The selected sketch point lies on a line or arc.
Collinear	A pair of selected lines lies on the same infinite line.
Concentric	A pair of selected arcs has the same center point.
Coradial	A pair of selected arcs has the same center point and radius value.
Equal	The selected entities have the same length or radius value.
Horizontal	The selected entities (points, lines) are aligned horizontally with respect to the coordinate system.
Perpendicular	The selected lines are oriented 90° to one another.
Parallel	The selected lines are parallel with each other.
Tangent	The selected line is tangent to the selected circle or arc.
Vertical	The selected entities (points, lines) are aligned vertically with respect to the coordinate system.

Example: Consider the mold insert plate shown in Fig. 6.3. The plate has a hole on each of the four corners. Suppose that the intent is for all four of these holes to be 0.250 in in diameter, 0.500 in deep, and 0.375 in from each of the two edges that bound the corner they are closest to. Because these four holes lie on the same plane, they can be created with a single sketch containing four circles. One approach to constraining this sketch would be to apply a 0.250-in-diameter dimension to each of the four holes. Then the vertical and horizontal distance of 0.375 in from each hole to the edges it is closest to could be specified (i.e., eight distance dimensions for the four holes). Now suppose that the size of the hole changes from 0.250 to 0.375 in. The modeler will have to go back and explicitly change the diameter dimension for every one of the four holes. Suppose that the 0.375-in side offset changes to 0.4063 in. The modeler will have to go back and change all eight distance dimensions.

A much more efficient way of specifying the design intent for this plate would be to apply the following relations:

- Apply an *equal diameter* relation to holes 1, 2, 3, and 4.

- Apply a *horizontal* relation to holes 1 and 2.

- Apply a *horizontal* relation to holes 3 and 4.

- Apply a *vertical* relation to holes 1 and 3.

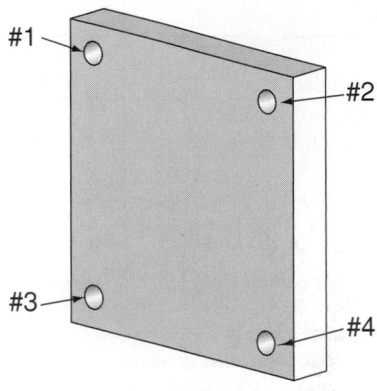

#1

#2

#3

#4

Figure 6.3 CAD model of plate with four holes.

- Apply a *vertical* relation to holes 2 and 4.

- Apply an *equal* relation that says the distance of hole 1 from the left edge is equal to the distance of hole 1 to the top edge.

- Apply an *equal* relation that says the distance of hole 1 from the top edge is equal to the distance of hole 4 to the bottom edge.

- Apply an *equal* relation that says the distance of hole 1 from the left edge is equal to the distance of hole 4 to the right edge.

If the size of hole 1 is now dimensioned to 0.250 in, then the diameter of the other three holes will change automatically to match that of hole 1. If the 0.375-in distance of hole 1 from the left edge then is dimensioned, all four holes will be located automatically the appropriate distance from their edges by virtue of the relations. If the hole diameters need to be changed for any reason, the modeler simply changes a single diameter (hole 1), and the rest of the holes will update automatically. Likewise, if the offset distance from the holes to the edge of the plate changes, the modeler simply has to change a single dimension rather than the eight dimensions using the previously described modeling approach. This is a very simple example that dramatically illustrates how important proper modeling technique can be. Without these parametric relationships, it is very easy to make mistakes when design changes are made. If these mistakes go unnoticed all the way to the shop floor (and they often do), then the consequences can be quite costly.

Using predefined features. In addition to creating new geometry using extruded/revolved/swept features, CAD systems also make available libraries of the most commonly used features. Predefined features made available typically include

ountersunk, counterbored, etc.)

with a defined wall thickness)

gular)

sence or
ould be
reusable
etal fab-
vill vary
e louver
oth, and
it when
ers.

the CAD
ll create
D1, D2,
critical to
somcone
h, depth,
ch better
arameter

ng as far
rgence of
D models
rge from
oad. For
found by
eb site at
mold and
andards,
ous time
facturing

nese pro-
an addi-
han just
parts to
that help

ks or ser-
s such as
nese com-

Start any model with a base feature that
shape of the component. Then add those
w this component will be assembled with
features, such as chamfers, fillets, and
portant role in the mechanical function-
erally should be added last in a model.

hen a component has symmetric features
n, or front/back directions, then model a
about a midplane to the other side. The
t saves a tremendous amount of modeling
del more error-proof because the geometry
ked to the geometry of the originating fea-
the CAD model will be reduced because
twice.

tiple instances of a feature are needed in
create a linear or circular array from one
creating the feature multiple times. This
result in smaller file sizes.

ies.   Tool and die design is a repetitive
ric shapes often are modeled repeatedly,

perhaps with slight variations in overall size and the pre
absence of certain features. These geometric shapes s▌
developed parametrically and then filed away as custom
features. For instance, louvers are used widely in sheet m
rication, yet the width, depth, and bottom fillet radius ▼
considerably from one part to the next. In this case, a sing
should be modeled with three basic parameters—*width, de*
bottom_fillet_radius. File the feature away, and then reuse
needed simply by changing any or all three of the parame

Use meaningful parameter names. When you create a feature,
system has no notion of what the feature is. It therefore w
a default name for each newly created parameter such as
etc. It is good practice to rename all parameters that are c
the basic function and understanding of the model. When
else examines your model and sees parameter names of *widt*
and *bottom_fillet_radius,* then he or she likely will have mu
luck figuring out how to edit/update the model than if the p.
names were *D1, D2,* and *D3,* respectively.

6.1.5 Standard part/assembly libraries

One of the most significant advances in CAD solid modeli
as the machinist or metalworker is concerned is the eme
standard part and assembly libraries. In many instances, CA
for standard tooling components are available free of cha
the manufacturer or its distributor via Internet downl
example, CAD models from over 200 manufacturers can be
searching the available catalogs of the Part Solutions W
www.part-solutions.com. The available catalogs encompass
die systems, workholding systems, national/international s
and more. These vendor-supplied CAD models are tremen
savers, and they often can reduce the occurrence of manu
errors.

In some cases, add-in programs are available for a fee. T
grams embed themselves within a given CAD system as
tional menu option. They often have more functionality
providing the machinist or metalworker with a library o
choose from. Some of them offer so-called design wizards
streamline the mechanical design process.

Table 6.4 lists selected companies that provide produc
vices related to mechanical component libraries for item
fasteners, gears, bearings, sprockets, etc. In some cases, t▌

TABLE 6.4   Selected CAD Component Library Providers

Company	URL	Related product(s)
Alamar Systems	*www.edgeparts.com*	ASAP
BCT Technology AG	*www.bct-technology.com*	3D Pool
Bogner and Herac GmbH	*www.solidedge.at*	ISL
Cadalog, Inc.	*www.cadalog-inc.com*	SE-PartsXL
CADBAS, GmbH	*www.cadbas.de*	CADBAS-PartExplorer
CADENAS	*www.cadenas.de*	PARTsolutions
CAD-Partner	*www.cadpartner.de*	CAD-Partner
CadParts, Inc	*www.cadparts.com*	Fastener Library
Camnetics, Inc.	*www.camnetics.com*	GearTrax and CAMTrax
Engineering S.P.A.	*www.engineering.it*	Standard Parts
Ideal Industrial	*www.ideal-parts.com*	IDEAL-PARTS
ISAP AG	*www.isap.de*	Part Manager
Jacobs Associates, LLC	*www.jacobsassociates.com*	Jacobs Professional Productivity Suite
Logopress	*www.logopress3.com*	123Go
Modularis Software AG	*www.starvars.de*	CAD-Symbols
PARTsolutions LLC	*www.part-solutions.com*	PARTSolutions
Scibit	*www.scibit.com*	Scibit
SolidLib.com Corp.	*www.solidlib.com*	SolidLib.com
SolidPartners, Inc.	*www.solidpartners.com*	SolidParts
Solidworks Corp.	*www.solidworks.com*	Solidworks Toolbox
Stream Design ApS	*www.stream-design.com*	Systems/Parts
TDS	*www.tds-technik.cz/en/*	TDS-Technical
Teamworks GmbH	*www.pwnorm.de*	PowerWorks Norm
TRACE PARTS	*www.traceparts.com*	Trace Parts
Web2CAD AG	*www.web2cad.com*	PowerParts

panies provide libraries that are based on an applicable standard (ANSI, DIN, etc.). Other vendors provide libraries of parts based on actual components supplied by manufacturers.

6.1.6   Designing sheet metal flat patterns

Much of the preceding discussion of CAD modeling techniques was generic in nature. However, modern CAD systems have advanced considerably the ease with which sheet metal components and their associated tooling are designed. Chapter 9 discusses analytic methods for determining factors such as setbacks and bend radii in relation to the composition and thickness of the sheet being used.

Modern CAD systems now have much of this expertise built in. In some cases, sheet metal capabilities are built into the CAD system. In other cases, optional sheet metal modules include databases for a wide variety of materials and thicknesses.

The sheet metal modules present in CAD systems work by providing the designer with built-in features that are specific to sheet metal (flanges, louvers, relief cuts, etc.). Because the CAD system recognizes each bend and cut as a sheet metal feature, it is able to "flatten" the sheet metal part into the flat pattern that must be cut from a sheet to produce the part blank. Flat patterns for one or more parts are then imported into special nesting programs that nest parts as tightly as possible on a given sheet in order to maximize material utilization. This nested pattern then can be fed to a laser cutting machine.

An example can be seen in Fig. 6.4. This figure shows a boxlike sheet metal part with bends, mitered flanges, a cut through a flange, and louvers. The part was modeled in the SolidWorks CAD system using built-in sheet metal features. For example, the miter flange feature creates all three flanges on the rear side of the part along with the respective miters. The user merely clicks on the edge(s) to receive a flange, specifies the flange length, and accepts or edits the default bend radius. SolidWorks then automatically creates all three flanges along with their miters. Likewise, the louvers are built-in parametric features. The user drags and drops a louver on the desired face, positions and orients it, and then specifies the dimensions. SolidWorks then automatically constructs the solid louver features. The sheet metal module includes default tables for bend factors, and those tables can be customized by the user. Figure 6.5 shows the box in its flat-pattern state. This is accomplished with a single click of the mouse button. Note that the bend lines are shown as well.

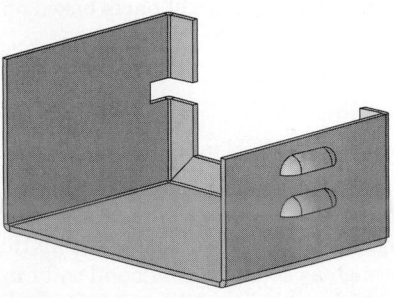

Figure 6.4 Folded sheet metal example.

Figure 6.5   Flattened sheet metal example.

6.2   Computer-Aided Manufacturing (CAM)

In most cases, the term *computer-aided manufacturing* refers to the process of semiautomatically generating computer numerical control (CNC) machine toolpaths from a CAD model. In reality, CAM packages encompass a much broader range of activities. For example, the increasing use of laser and waterjet systems to cut sheet and plate materials has led to new developments in nesting software. These systems nest patterns on a sheet of specified size such that the largest number of parts can be cut from the smallest amount of material.

When selecting a CAM package for CNC machining, it is necessary to consider many different factors:

- *Stand-alone versus integrated CAD.* Most CAM systems are available in a stand-alone version. Some are also available as add-ons to a particular CAD system (i.e., they appear as an extra menu item in the CAD system). If toolpaths are to be generated from a mix of different CAD systems, then there may not be much advantage to using an integrated CAM system. If, however, virtually all toolpaths will be generated from the same CAD system, then it is definitely an option to consider. A valuable advantage of integrated CAD/CAM is toolpath associativity. As mentioned previously, this is where toolpaths are updated automatically when changes to a part's geometry are made. It is also the case that CAD models often are built using predefined features such as holes, slots, etc. Modern CAM systems have intelligent machining wizards that attempt to semiautomatically recognize machining features in a part. For instance, they might recognize a pocket and then attempt to semiautomatically generate roughing and finishing operations using appropriate end mills. For some types of parts, this can result in significant time savings.

- *Network licensing.* For large companies that purchase many sets of a CAM package, it is usually much simpler for the information technology (IT) department to install the software once along with a floating network license. For smaller installations, support for network licensing is less important.

- *Number of simultaneous axes.* CNC machines are configured most often with three, four, or five axes of motion. Not all CAM packages support fourth- or fifth-axis commands. Some CAM packages do support commands for fourth and possibly fifth axes, but they do not allow *simultaneous* four- or five-axis motion. If true simultaneous four- or five-axis motion is needed for a machine, then the CAM vendor should be asked to clarify if this is supported.

- *Types of processes supported.* Some CAM packages can generate paths for other types of processes such as wire or sinker electric discharge machining (EDM) machines, waterjet cutters, etc.

- *Availability of postprocessors.* G code is not as standard as it might appear, particularly when four- or five-axis motion is involved. Different machine controllers expect part programs to be formatted in different ways, and not all G or M codes are supported by every controller. Consequently, CAM systems must "post" (postprocess) their toolpaths to match the specific machine the code will be run on. The reseller should be asked whether or not the necessary postprocessors are available and whether or not there is an extra charge for them. If they are not available, the reseller should be asked how much it would cost to write the new post.

- *Technical support and training.* Learning a new CAM system is not always easy, particularly when four- or five-axis motion is involved. Although the market share of a CAM system should not be a primary factor in selecting a system, it can provide an indicator of how easy or difficult it will be to hire trained CAM operators. The larger the user community is for any system, the more resources generally are available, such as third-party training manuals and short courses, online user groups, etc.

- *File import-export capability.* For obvious reasons, it is important to verify that the CAM package supports the CAD file format(s) of interest.

- *Graphic toolpath verification.* Graphic toolpath verification refers to software that simulates the part program on the computer display. The raw stock for the workpiece is modeled along with (optionally) any workholding devices. The program then

graphically displays material being removed from the part blank as the cutting tool progresses through the part program. Toolpath verification is a very inexpensive way to discover programming errors. The alternative of discovering a toolpath error on the machine can be catastrophic. Not all CAM systems provide graphic toolpath verification, and some only provide it as an extra option. It is important to clarify whether or not this feature is provided when a CAM system is purchased.

- *Tool and material databases.* Many CAM programs come equipped with user-customizable cutting tool and material databases. For any given machining operation, the user specifies the specific alloy being machined, as well as the type of cutting tool used. The CAM program will then apply default cutting parameters from its database. The material and machining parameter databases typically can be edited by the user. This allows users to add new materials or to enter machining parameters recommended by the manufacturing of the cutting tools in use.

Table 6.5 provides a listing of selected CAM software vendors, their Web sites, and related product offerings.

6.2.1   CNC part programming fundamentals

A *part program* is the text file that contains machining instructions for the CNC machine to produce a particular part. Whether a part program is generated via a CAM package or is written manually using a text editor, it must adhere to the syntax and format for the machine on which it will be run. ISO 6983-1:1982 and ANSI/EIA RS-274-D are standards that define the syntax for numerical control of machines. While the vast majority of machine tool makers have adopted these standards, a small number of machine tool makers use their own propriety standards.

In order to write or read a part program, it is necessary to understand some fundamental concepts and terminology.

G codes   These are codes that instruct the cutting tool to move to the specified location in the prescribed manner (linear, circular, etc.). Selected G codes for CNC milling are shown in Table 6.6. Selected G codes for CNC turning are shown in Table 6.7.

M codes   These are miscellaneous codes that instruct the machine to perform nonmovement actions such as turning on/off the coolant. Selected M codes for CNC milling are shown in Table 6.6. Selected M codes for CNC turning are shown in Table 6.7.

TABLE 6.5 Selected CAM Vendors

Company	URL	Related product(s)
AccuMetria	*www.agile.com*	CMMWorks
Advanced Tubular Technologies	*www.advancedtubular.com*	Benderlink
Anderson CAD/CAM, Inc	*www.acadcam.com*	GeoWORKS
Antech Micro Systems	*www.antechmicro.com*	CADlink
Applied Production	*www.appliedproduction.com*	ProFab Series
AUTON srl	*www.auton.it*	AUTON Cam Processor
CamSoft Corp.	*www.camsoftcorp.com*	Advanced System 3000
Camtek, Ltd.	*www.peps.com*	PEPS SolidCut
CN Industries	*www.cn-industries.com*	Goélan V4
CNC Software, Inc.	*www.mastercam.com*	MasterCAM
COSCOM Computer	*www.millit.com*	Millit
Delcam International	*www.delcam.com*	PowerMILL
DP Technology Corp.	*www.dptechnology.com*	ESPRIT
EFICAD	*www.eficad.fr*	EFICN SW
EMBLEM Corp.	*www.emblem.co.jp*	PowerEdge/Mill
Engineering Geometry Systems	*www.featurecam.com*	FeatureCAM
FeatureCAM/Engineering Geometry Systems	*www.featurecam.com*	FeatureCAM
Gibbs and Associates	*www.GibbsCAM.com*	GibbsCAM
IMAO Corp.	*www.imao.co.jp*	IMAO CAD/CAF
IMCS, Inc./PartMaker	*www.partmaker.com*	PartMaker
Jetcam International Holding, Ltd.	*www.jetcam.com*	JETCAM Expert Systems
Licom Systems	*www.licom.com*	AlphaCAM
MecSoft Corp.	*www.mecsoft.com*	VisualMill
NCCS	*www.nccs.com*	NCL

Opus	*www.opus-cam.de*	OPUS CAM
Pathtrace Engineering Systems	*www.edgecamforsolidworks.com*	EdgeCAM
QARM Pty, Ltd.	*www.onecnc.com*	OneCNC
SolidCAM, Ltd.	*www.solidcam.com*	SolidCAM
SPRUT Technology	*www.sprutcam.com*	SprutCAM
Surfware, Inc.	*www.surfwcre.com*	SURFCAM
TekSoft CAD/CAM System, Inc.	*www.teksoft.com*	CAMWorks, ProCAM
UGS	*www.ugs.con*	NX Machining
Vero International Software UL	*www.vero-software.com*	VISI-Series
Vision Numeric	*www.type3.com*	Type3
Wittlock Engineering	*www.wittlockeng.com*	WE-CIM

TABLE 6.6 Selected CNC Milling G Codes and M Codes

G Code	Description	M Code	Description
G00	Rapid positioning	M00	Program stop
G01	Linear interpolation	M02	Program end
G02	Clockwise circular interpolation	M03	Spindle on clockwise
G03	Counter clockwise circular interpolation	M04	Spindle on counter clockwise
G04	Dwell	M05	Spindle stop
G40	Cancel cutter compensation	M06	Tool change
G41	Left cutter compensation	M08	Coolant on
G42	Right cutter compensation	M09	Coolant off
G43	Tool length compensation	M30	End program and reset to start
G44	Cancel tool length compensation		
G45	Increase tool offset		
G46	Decrease tool offset		
G70	Inch units		
G71	Metric units		
G81	Drilling cycle		
G90	Absolute positioning		
G91	Incremental positioning		

TABLE 6.7 Selected CNC Turning G Codes and M Codes

G Code	Description	M Code	Description
G00	Rapid positioning	M00	Program stop
G01	Linear interpolation	M02	Program end
G02	Clockwise circular interpolation	M03	Spindle on clockwise
G03	Counter clockwise circular interpolation	M04	Spindle on counter clockwise
G04	Dwell	M05	Spindle stop
G20	Inch units	M06	Tool change
G21	Metric units	M08	Coolant on
G40	Cancel tool nose radius compensation	M09	Coolant off
G41	Left tool nose radius compensation	M30	End program and reset to start
G42	Right tool nose radius compensation		
G70	Finishing cycle		
G71	Turning cycle		
G72	Facing cycle		
G76	Threading cycle		
G90	Absolute positioning		
G91	Incremental positioning		

Absolute versus incremental positioning   G codes that govern motion of the cutting tool determine the (x, y, z) coordinates of the cutting tool at the end of the move. If a G90 code is specified at the start of the part program, then any (x, y, z) coordinates specified in the program are absolute locations relative to the machine's $(0, 0, 0)$ position. If a G91 code is specified at the start of the part program, then any (x, y, z) coordinates specified in the program are distances relative to where the cutting tool was at the start of the command.

x,y,z   These are the values that specify the desired location for the cutting tool at the end of the move.

i, j, k   These are (x, y, z) coordinates for the center of an arc during a move.

S   This is the spindle speed in revolutions per minute.

F   This is the feed rate in inches per minute or millimeters per minute.

T   This is the tool number to load during a tool change command.

N   Line number.

Sample part programming commands for milling

%

:0526	(Program 0526)
N10 G90 G70	(Absolute coordinates, units in inches)
N20 M06 T6	(Load tool 6 from the tool changer into the spindle)
N30 M03 S1000	(Turn spindle on CW at 1000 rev/min)
N40 G00 X2 Y0	(Rapid move to absolute $x = 2, y = 0$)
N50 G01 Z1 F30	(Linear move to $z = 1$ at 30 in/min)
N60 G01 Z-0.125 F10	(Linear feed to $Z = -0.125$ at 10 in/min)
N70 G02 X-2 Y0 I0 J0	(Cut CW arc from starting point of $x = 2$, $y = 0$. To $x = -2, y = 0$, with the center at $x = 0, y = 0$)
N80 G00 Z1	(Rapid move up to $Z = 1$)
N90 M05	(Turn off the spindle)
N100 M30	(Program stop)

6.2.2   CAM toolpath generation procedure

Although the details and terminology vary with every CAM system, the basic steps are fairly similar in most cases. The first step is to

import the workpiece geometry. If the part is to be held in place with toggle clamps or other fixturing devices in the general vicinity of the cutting tool, then it's a good idea to import geometry for any fixturing devices that could conceivably interfere with the cutting tool.

After the geometry has been imported, the workpiece is oriented and positioned such that it is aligned with the coordinate system of the machine tool on which it will be machined. At this point, the machining operations can be specified. Some of the more common machining operations and their definitions include

Pocketing. This refers to the act of removing *all* material inside a closed boundary to a uniform depth.

Contouring. This refers to removing material exactly along a contour. The stock inside and outside the contour is not machined.

Face milling. This refers to flattening the face of a workpiece. This is frequently the first operation performed when working with a plate because it establishes a reference height for subsequent features.

Drilling. One or more holes are drilled according to the parameter specified.

For each type of machining operation specified, the user must define an appropriate set of parameters. Default values based on the workpiece material and specified cutting tool streamline the process. However, the user still must specify certain pieces of information, such as how much stock to leave in a roughing operation, the type of cutting tool to use (flat end mill, ball end mill, twist drill, etc.), the type of lead-in and lead-out to use, the depth per pass, etc.

When all of the machining operations and their parameters have been specified, it is customary to conduct a graphic toolpath verification for each separate machining operation to ensure that the cutting tools cut geometry as intended and that they don't collide with anything. However, almost all CNC machine operators can say from painful experience that graphic toolpath verification is not foolproof. Most CNC machines have a wide variety of tool offsets, fixture offsets, machine offsets, etc. For example, a toolpath is generated assuming that the (0, 0, 0) reference point is the top, left, rear corner of a block that is clamped in a vise. When the part is clamped in the vise on the actual machine, the operator must remember to measure the height of the top surface of the block relative to the machine's coordinate system. Note that the top of the part is almost always

higher than the machine's $z = 0$ position. This z-offset value must be entered. If the operator forgets to do this, the machine will run the program at a lower z-height than was intended. In the simplest case, this leads to a broken tool. In the extreme, it can lead to a collision that damages the spindle of the machine.

Figure 6.6 shows a Solidworks CAD model of a connecting rod that has been mounted on a modular fixturing plate. The fixturing plate, locating pins, and toggle clamps all can be downloaded from online databases. For this particular machining setup, the top face of the boss on either end of the connecting rod is to be faced, and the holes are to be reamed. This CAD assembly was imported into the Surf-CAM software package, where the CNC toolpath was generated. For illustrative purposes, the clearance plane for the cutting tool was intentionally set too low so that the cutting tool would collide with the vertical toggle clamp that holds the part down. Figure 6.7 shows a snapshot from the graphic toolpath verification with the cutting tool above the left boss. Note the unintended slot through the vertical toggle clamp that resulted from the improper z-clearance plane. Although this example was contrived, it shows the value of including both the part to be machined and any workholding devices in the toolpath simulation procedure.

6.3   Computer-Aided Engineering

Computer-aided engineering (CAE) refers to a host of different software packages that help engineers and machinists to design

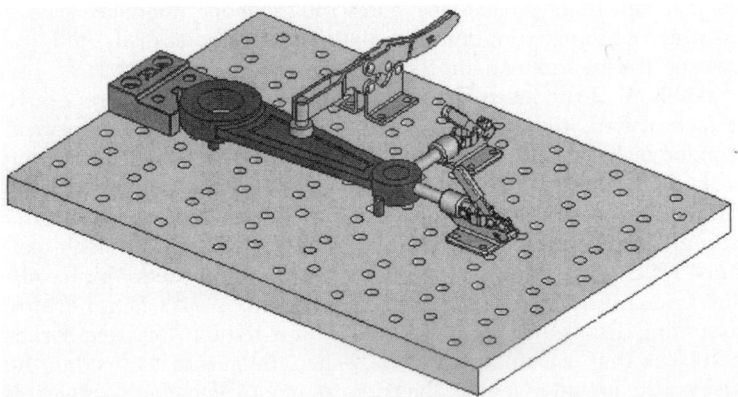

Figure 6.6   Connecting rod and machining fixture.

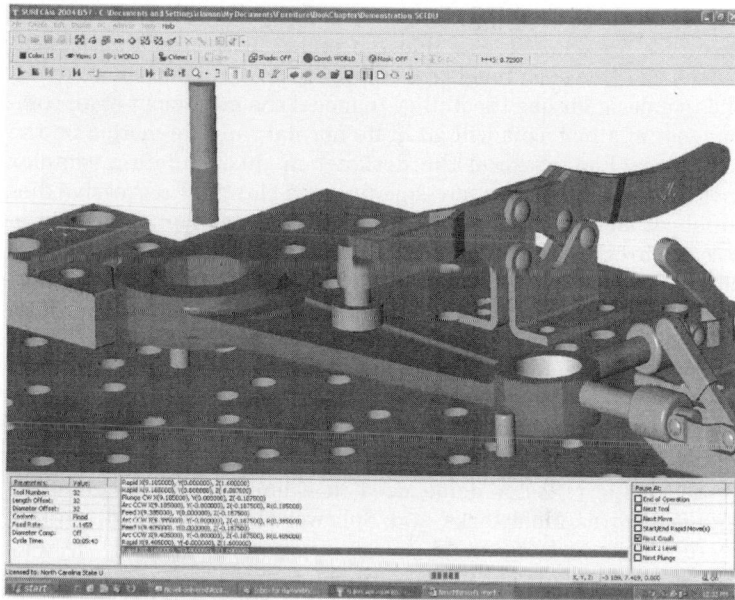

Figure 6.7 Graphic toolpath verification.

TABLE 6.8 CAE Design Automation Software

Company	URL	Related product(s)
CAPVIDIA	www.capvidia.com	FaceWorks
EasyMold	www.easymold.de	EasyMold
Manusoft Technologies Pte, Ltd.	www.imold.com	IMOLD
Moldex3D	www.moldex3d.com	Moldex3DWorks
Moldflow Corp.	www.moldflow.com	Moldflow Plastics Advisors
R&B, Ltd., MoldWorks	www.RnBUSA.com	MoldWorks and SplitWorks

and analyze systems. Of particular interest to machinists and metalworkers are software packages that help to automate the tool and die design process. Table 6.8 provides a listing of several companies that provide CAE software for a variety of tasks of interest to machinists and metalworkers.

Even without using a third-party CAE package, most CAD solid modelers provide the tool and die maker with numerous useful CAE tools. One such tool is motion analysis. With motion analysis, the tool designer has the ability to model the movement of all components in a tool as it will go in its normal range of motion on the shop floor. For instance, the designer might simulate a complex mold with slides, lifters, etc. opening and closing. Progressive dies for sheet metal are also complex tools whose motion can be simulated. Throughout the motion simulation, the CAD system can check for interferences between components and can warn the designer of possible collisions. This sort of virtual testing can be done before a single metal chip has been cut. More advanced finite-element simulation models can even simulate expansion of components as they heat up in operation.

Many third-party software packages are aimed at streamlining the tool design process. Figure 6.8 shows the CAD model for a small plastic reflector dome used in a handheld flashlight. Programs such as Moldworks and Splitworks are designed to automate portions of the mold design process in the CAD system. Moldflow Mold Advisor is an example of a program that will analyze plastic parts and their molds to predict problems such as trapped air, short shots, etc. For a given part, the software will recommend the best gate location(s). When the user has specified one or more gate locations, the software then simulates the cavity filling with plastic, as shown in Fig. 6.9. Based on the simulation of how the mold fills, the software then will warn the designer of potential problems with warping, sink marks, or trapped air. If a potential problem exists, the designer can use the analysis to redesign the

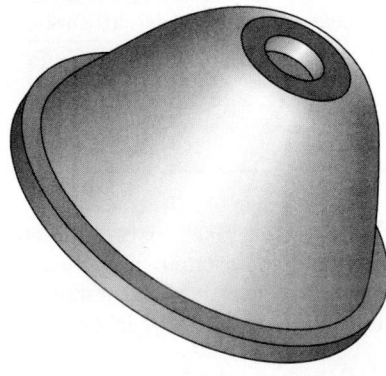

Figure 6.8 CAD model of plastic reflector dome.

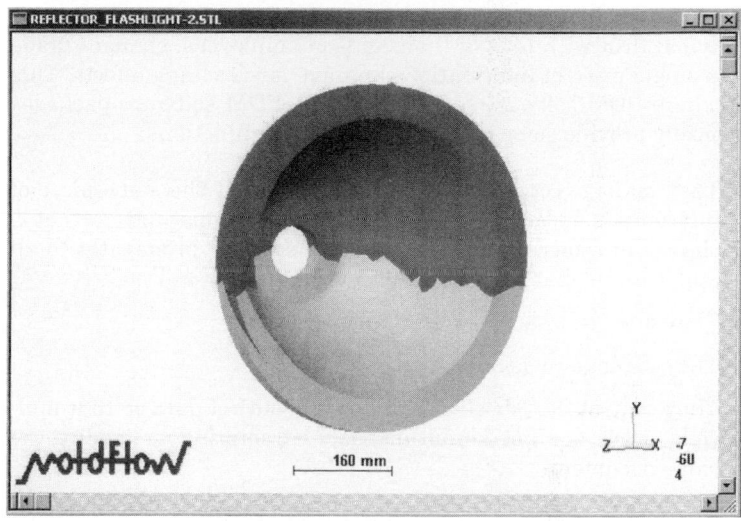

Figure 6.9 Moldflow filling simulation analysis.

mold. For instance, he or she may choose to relocate a gate, add vents, etc.

6.4 Product Data Management

Product data management (PDM) means different things to different people. A great deal of information is generated in any product-development effort. The CAD model represents just a small fraction of the information associated with a new product. Each major department within a company uses information in a different way, yet the departments often share the same information. As a simple example, the mechanical designer may generate a CAD model. He or she is concerned primarily with the mechanical function of the product. The manufacturing department will use those CAD models to design tooling. The purchasing department will use the CAD models to determine which components must be purchased and what the specifications are for those "buy" items. Sales and marketing might derive costing data from CAD models generated by both the design and manufacturing departments.

What is clear from this discussion is that the same information is used by a wide variety of departments within a company. Whether

it is a small firm with fewer than 20 employees or a large multi-
national firm with tens of thousands of employees, changes made
to a single piece of information can have far-reaching effects. This
is where PDM software comes into play. PDM software packages
typically provide some or all of the following functions:

- They manage consistency of the data so that the same piece of
  information is not stored in multiple locations and so that a
  change in a piece of information immediately propagates to all
  applications that use that information.

- They prevent unauthorized changes to data.

- They manage revision histories.

- They control the "check-in" and "check-out" of data so that multiple users are not simultaneously attempting to modify the
  same document.

Some PDM packages are designed for workgroup-level imple-
mentations that are appropriate for use in small companies that are
confined to a single building. These packages can cost as little as a
few thousand dollars. Other PDM packages provide enterprise-wide
solutions for global companies. These systems can cost hundreds of
thousands to even millions of dollars for turnkey installations. A
listing of selected PDM vendors is provided in Table 6.9.

TABLE 6.9 Selected PDM Vendors

Company	URL	Related product(s)
Agile Software Corp.	*www.agile.com*	Agile PLM, Agile Configurator
ARC Solutions, Ltd.	*www.arcsolutions.de*	REMARC
Arena Solutions	*www.arenasolutions.com*	Arena SolidWorks Integration Adapter
Assetium	*www.assetium.com*	@UDROS
Auto-Trol	*www.auto-trol.com*	Centra 2000
Baan PLM	*www.baan.com*	iBaan PLM Integration for SolidWorks, Baan PDM
CAD-Partner	*www.cadpartner.de*	SMAP 3D PartFinder and Organizer
CadWorks Software Oy, Ltd.	*www.cadworkssoftware.com*	CustomWorks
Cegid	*www.cegid.fr*	ISOFLEX
Centare Group, Ltd.	*www.centare.com*	SimplePDM
CONTACT Software GmbH	*www.contact.de*	CIM DATABASE
CYCO Software BV	*www.cyco.com*	AutoManager Meridian
Delta Group, Ltd.	*www.deltagl.com*	TreeWorks
EIGNER PARTNER AG	*www.ep-ag.com*	Axalant, CADIM/EDB
Elton Solutions, Ltd.	*www.eltonsolutions.com*	RevZone
EMCAD	*www.emcad.nu*	EasyWorks
Engineering PLM Solutions srl	*www.engineeringplmsolutions.com*	EDM 2004
Engineering S.P.A.	*www.engineering.it*	Engineering DataCenter Manager
ERP Competec Oy	*www.competec.com*	PD Spiner for SolidWorks
GCS Scandinavia AB	*www.conisio.com*	Conisio
Gedas Deutschland GmbH	*www.gedas.com*	SAP R/3 Integration for SolidWorks, SAP R/3 Integration for Solid Edge
ICON Informationssyteme GmbH	*www.swgain.de*	SwGain
ICoolSoft	*www.cool-soft.com*	CoolWorks
Integrated Dynamic Solutions	*www.integrateddynamicsolutions.com*	DataValve
Itworks, Inc.	*www.itworksaok.com*	DataBridge

TABLE 6.9 Selected PDM Vendors (Continued)

Company	URL	Related product(s)
Keytech software GmbH	www.keytech.de	ProfilDB
Lascom	www.lascom	Advitium
Logic Design Corp.	www.ldcglobal.com	Global Edge
Logotec Engineering S.A.	www.logotec.com.pl	PDM9000
MatrixOne, Inc.	www.matrixone.com	eMatrix SolidWorks, Matrix One
Maxxsoft GmbH	www.maxxsoft.com	MaxxDatabase
MechWorks	www.mechworks.com	DBWorks
Modultek, Inc.	www.solidpdm.com	SolidPDM
OLE Technology Co., Ltd.	www.ole-tech.com	SmarGroup
PROCAD GmbH & Co. KG	www.procad.de	PRO.FILE
Product Sight Corporation	www.productsight.com	Find View
Proteus Systems, Inc.	www.proteus-global.com	CDD Systems
Questica	www.questica.com	Questica 2000
ReadySolutions	readysolutions.it	Ready2Works
SDH Development	www.tooworks.info	ToolWorks ERP Link
SIGHT Informationssysteme GmbH	www.sight.de	Sight/EDM
SMARTEAM/Dassault Systemes	www.smarteam.com	SMARTEAM
SofTech, Inc.	www.softech.com	ProductCenter
SOLID Applications, Ltd.	www.solidapps.co.uk	Drawing ManEdger
SolidWorks	www.solidworks.com	PDMWorks

Chapter

7

Machining, Machine Tools, and Practices

Machine tools and cutting tools have advanced in great strides in the past 50 years. What at one time was a difficult operation on a machine tool such as a lathe or milling machine has become commonplace and greatly simplified with the advent of microprocessor controls, advanced positioning and control techniques, and modern cutting tool materials. Chapter 1 described some of the modern equipment that is standard in many companies throughout the nation and the world. This chapter discusses other types of equipment, controls, and cutting tools, together with their applications to machining and metalworking practices.

Over the past 50 years, cutting tools have progressed from high-carbon steel, high-speed steel, cobalt matrix, and solid carbides to cemented or sintered carbides, ceramics, silicon nitride, cubic boron nitride, and special cutting tool coatings such as titanium nitride, titanium carbide, aluminum oxide, and others. The advantages of these new technologies in cutting tools have been increased production rates coupled with cost savings. Cutting tool technology is changing constantly, with new types of cutters, mills, drills, reamers, etc. being introduced to the market at a steady pace.

This chapter will show the latest types of cutting tools, together with the important speeds and feeds recommended for them. Also

shown are the typical speeds and feeds for the popular and widely used high-speed-steel (HSS) and cobalt-matrix tools. The speeds and feeds tables for the various materials and machining methods are provided by the manufacturers who produce the cutting tools. Their research and development efforts, coupled with their experience and expertise, will be appreciated by those who use these tables.

7.1   Turning and Boring

Turning is the machining process used to generate external cylindrical forms by removing material, usually with a single-point cutting tool. *Boring* is essentially internal turning to generate internal shapes. The common turning machines include engine lathes, single-spindle automatic lathes, horizontal-turret lathes, automatic screw machines, Swiss-type automatic screw machines, multiple-spindle automatic bar and chucking machines, and computer-controlled automatic turning centers. The single-point tool is moved parallel to the machine spindle for straight or contour turning of the outside diameter and turning or boring of an internal surface. Form tools, both flat and circular, were at one time fed into the workpiece to produce the desired contour on the part. Knurling produces a controlled rough surface pattern on the periphery of the part, either diamond-shaped or straight, in coarse, medium, or fine texture.

7.1.1   Turning and boring tool materials

High-speed steel (HSS), cast nonferrous alloys (Stellite and Tantung), and cemented carbides (sintered carbides) are among the most widely used turning, milling, and boring tool bit materials. The more advanced materials used for turning, boring, and milling operations include the following families:

- Tungsten carbides, coated (aluminum oxide and titanium nitride) and uncoated

- Cermets (ceramics with metallic binders)—titanium carbide and titanium nitride with a metallic binder

- Ceramics—alumina-base (aluminum oxide) and silicon nitride–base ceramics

- Polycrystallines—polycrystalline diamond and cubic boron nitride

Popular grades of HSS and Tantung (cast-alloy) tool bits

HSS and cobalt ground tool bits

M-2 HSS. General-purpose tool bit material. Works extremely well on mild steel, alloys, and tool steels and is an excellent finishing tool. Hardness: Rockwell C 63 to C 65.

5% cobalt. For heavy cuts on castings and forgings. A T-4 type of tungsten HHS that withstands higher heat than M-2 types. Hardness: Rockwell C 64 to C 66.

10% cobalt. A superior type of HHS used for heavy cuts in hard materials. Hardness: Rockwell C 64.5 to C 67.5.

Cobalt M-34. For heavy cuts on castings and forgings. A T-4 type of tungsten HSS that withstands higher heat than M-2 types.

Cobalt M-43. A superior type of HSS that is excellent for high-heat applications and for heavy cuts in hard materials.

Tantung tool bits

Tantung G. A cast-alloy cutting tool material composed of chromium, tungsten, columbium, and carbon in a cobalt matrix. These materials have the ability to retain cutting hardness at temperatures up to 1500°F. As a cutting tool, Tantung G is excellent for all turning, boring, facing, milling, and cutoff applications on most metal and nonmetallic materials. Performs best at cutting speeds of 100 to 250 surface feet per minute (sfpm). Hardness: Rockwell C 60 to C 63 and transverse rupture strength of 300,000 psi, minimum.

Note: Typical modern high-speed turning and boring tool materials and insert forms are shown following the speeds and feeds tables in this section.

7.1.2   Turning tool types, terms and definitions, and grinding

Single-point tools.   A typical set of turning tools of HSS is shown in Fig. 7.1, together with a straight-shank lathe tool holder. The unground and ground tool bits are shown under the tool holder, followed by the different configurations of basic cutting bits. The cutting bit types are as follows:

A   Left-hand turning tool

B   Round-nose turning tool

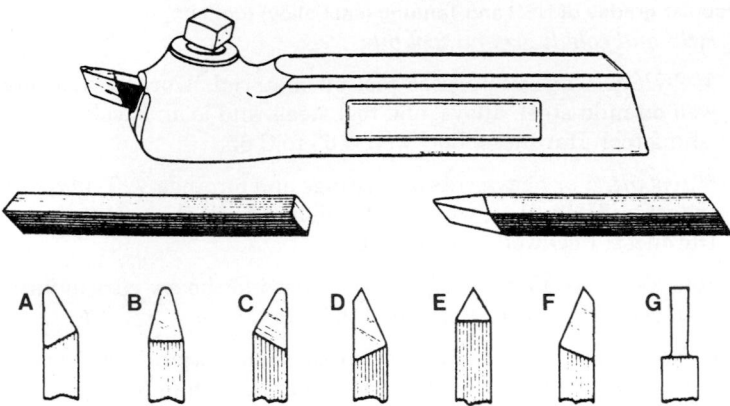

Figure 7.1 Standard lathe cutting tools.

C   Right-hand turning tool

D   Left-hand facing tool

E   Threading tool

F   Right-hand facing tool

G   Cutoff tool

Applications for the cutting tools are shown in Fig. 7.2 and include turning, facing, threading, cutoff, boring, and inside threading. The important geometric form angles relating to single-point cutting tools that are ground from HSS and cast-alloy bit-stock materials (Tantung, Tantung G, Stellite, etc.) are shown in Fig. 7.3.

7.1.3 Tool nomenclature

A single-point tool contains several geometric elements that are classified and defined by the American National Standards Institute (ANSI) B94.50-1975 (R1986), *Single-Point Cutting Tools, Basic Nomenclature and Definitions for*. The basic definitions may be summarized as follows:

The *size* of a tool of square or rectangular section is expressed by giving, in the order named, the width of the shank w, the height of shank h, and the total tool length l in inches, such as 0.75 × 1.50 × 8 in.

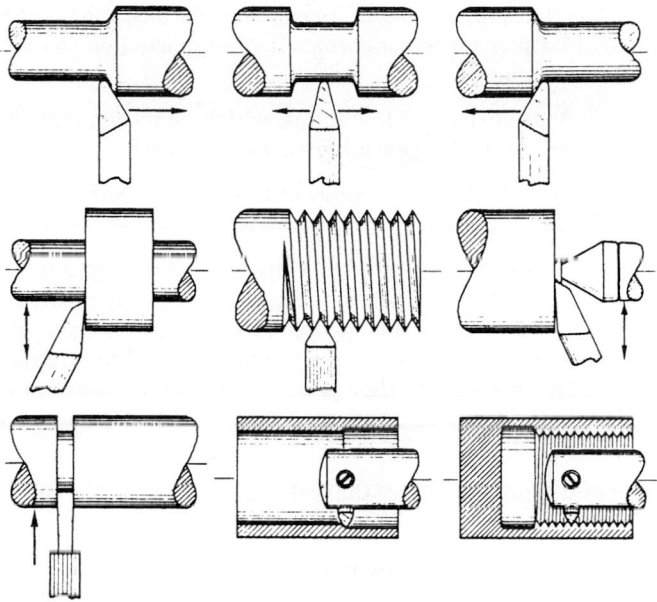

Figure 7.2   Turning operations.

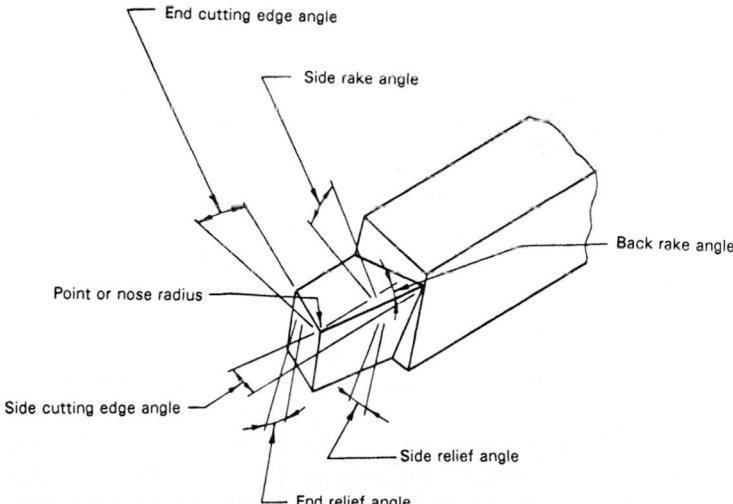

Figure 7.3   Form angles of a single-point cutting tool.

The *shank* is that part of the tool on one end of which the point is formed or the tip or bit is supported. It is supported on the tool post of the machine.

The *base* is that surface of the shank which bears against the support and takes the tangential pressure of the cut.

The *heel* consists of the areas adjacent to the intersection of the base and flank.

The *face* is the surface on which the chip apparently impinges as it is separated from the workpiece. It may be provided with a narrow land ground along the cutting edge to support the built-up edge. This land is usually of less rake than that of the balance of the face. The face rake is then often greater than normal. For sintered-carbide tools, the land may be ground to a negative rake.

The *tool point* is all that part of the tool which is shaped to produce the cutting edges and face.

The *cutting edge* is the portion of the face edge along which the chip is separated from the workpiece. It consists usually of the side-cutting edge, the nose, and the end-cutting edge.

The *nose* is the corner, arc, or chamfer joining the side-cutting and the end-cutting edges.

The *flank* of the tool is the surface or surfaces below and adjacent to the cutting edge.

The *neck* is an extension of the shank of reduced cross-sectional area. A small point, as required in boring operations, is sometimes attached to the shank by a neck.

The *flat* or *drag* is the straight portion of the end-cutting edge at 0° intended to eliminate feed marks and produce a smooth machined surface.

A typical single-point carbide-tipped turning tool is shown in Fig. 7.3, and the angles shown are normal, that is, taken with reference to the cutting edges, since these are the ones specified in grinding a single-point tool. If a land is used, the rake of the land is given, followed by the rake of the face, such as −3 (12). This indicates a land rake of negative 3° with a positive side rake of 12°.

7.1.4   Tool angles

The tool angles shown in Fig. 7.4 are defined as follows:

Back-rake angle. The angle between the face of the tool and the base of the shank or holder. This is usually measured in a plane through the side-cutting edge and at right angles to the base. It is positive if the face slopes downward from the point toward the shank, tending to reduce the included angle of the tool point. It is negative if the face slopes upward toward the shank.

Side-rake angle. The angle between the face of the tool and the base of the shank or holder. It is usually measured in a plane perpendicular to the base and to the side-cutting edge and hence is normal side rake.

Side-relief angle. The angle between the portion of the side flank immediately below the side-cutting edge and a line drawn through this cutting edge perpendicular to the base. It is usually measured in a plane at right angles to the side flank and hence is normal side relief.

End-relief angle. The angle between the portion of the end flank immediately below the end-cutting edge and a line drawn

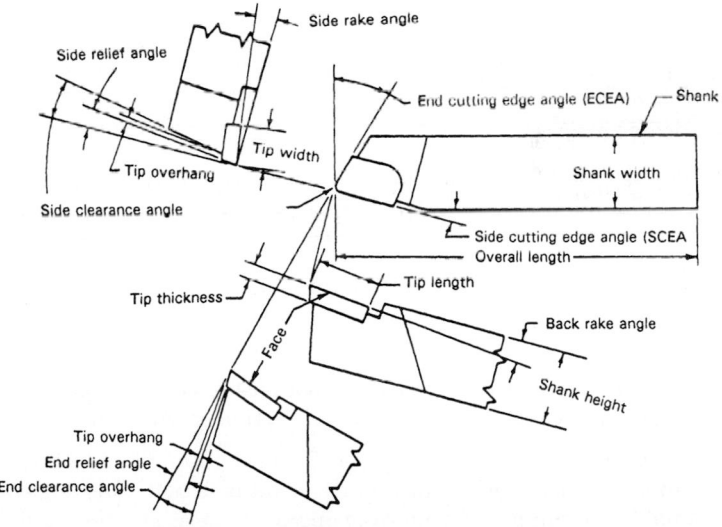

Figure 7.4   Tool angles and definitions.

through that cutting edge perpendicular to the base. It is usually measured in a plane at right angles to the end flank and hence is normal end relief.

Clearance angle. The angle between a plane perpendicular to the base of the tool or holder and that portion of the flank immediately below the relieved flank. Side-clearance angle is measured in the plane of the back-rake angle. The clearance angle is greater than its corresponding relief angle, except when only one plane exists on the flank, in which case the clearance and relief angles coincide.

Side-cutting-edge angle. The angle between the straight side-cutting edge and the side of the tool shank or holder.

End-cutting-edge angle. The angle between the cutting edge on the end of the tool and a line at right angles to the side of the tool shank.

Note: As the setting of the tool or holder is changed in relation to the workpiece, the *effective working angles* will no longer agree with the tool angles.

Tool character.   An abbreviated, convenient designation system for specifying the angles of single-point cutting tools is shown in Fig. 7.5. These angles are considered normal as used in most tool specifications and in grinding operations.

Working angles.   Working angles are the angles between the tool and the workpiece. These angles depend on the shape of the tool, as well as on its position relative to the workpiece. These angles are defined as follows:

Setting angle. The angle made by the straight portion of the shank of a tool or holder with the machined portion of the workpiece, commonly 90°.

Entering angle. The angle that the side-cutting edge of a tool makes with the machined surface of the workpiece, which is 90° in the case of a tool with 0° side-cutting-edge angle effective.

True rake angle. The slope of the tool face toward the tool base, from the active cutting edge in the direction of chip flow.

Cutting angle. The angle between the face of the tool and a tangent to the machined surface at the point of action. This angle is equal to 90° minus the true rake angle.

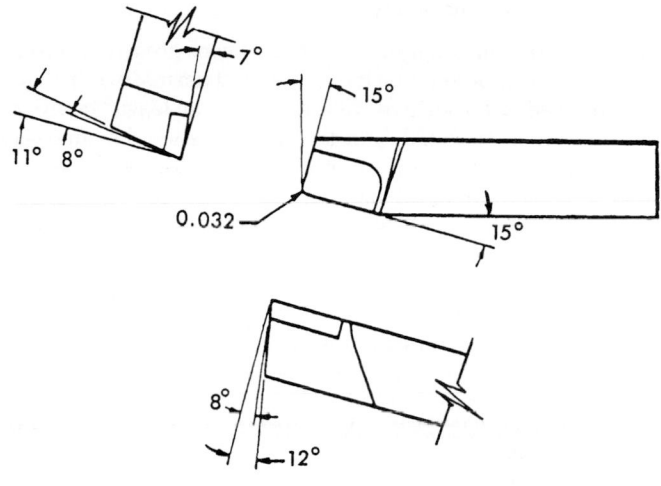

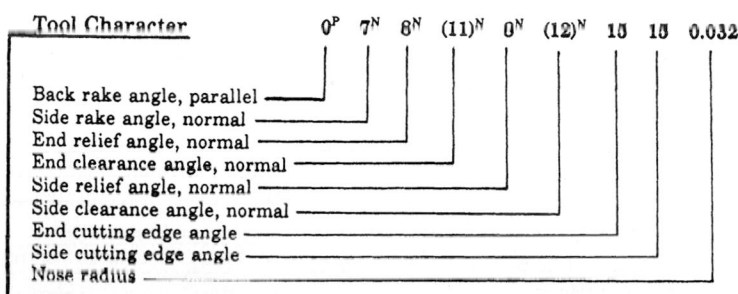

Tool Character	0^P	7^N	8^N	$(11)^N$	8^N	$(12)^N$	15	15	0.032

Back rake angle, parallel
Side rake angle, normal
End relief angle, normal
End clearance angle, normal
Side relief angle, normal
Side clearance angle, normal
End cutting edge angle
Side cutting edge angle
Nose radius

Figure 7.5   Tool character system.

Lip angle. The included angle of the tool material between the face and the relieved flank. According to the direction of measurement, it may represent the end lip angle, side lip angle, or true lip angle.

Working-relief angle. The angle between the relieved ground flank of the tool and a line tangent to the machined surface passing through the active cutting edge.

Working end-cutting-edge angle. The angle between the end-cutting edge and a plane tangent to the machined surface at the point of cutting.

7.1.5   Selection of tool geometry

Rake angles.   The rake-angle combination varies greatly according to tool material, workpiece material, and cost. Thermally efficient cutting is improved with positive rake angles or at least positive rake angles with respect to chip flow or true rake. It should be understood that a high positive rake results in a more fragile cutting edge and often must be compromised for tool durability. As cutting speeds increase, the rake angle has less effect on tool pressures. Therefore, in using tool materials that can be operated at higher cutting speeds, it becomes possible to use less positive rake angles or even negative rakes to increase tool strength and economize on tool maintenance. Prismatic-insert and throwaway-insert tools are greatly simplified in use and maintenance by using negative rake angles, although more cutting power and increased cutting forces are required.

Combining negative back rake with positive side rake allows safer cutting through slots or keyways, thus placing the initial impact loads on a portion of the cutting edge removed from the nose of the tool bit.

Relief angles.   Side and end relief angles vary between 5° and 15° for cutting metals and may run higher for some of the nonmetallic materials. Increased relief angles reduce cutting forces and result in a cleaner cut on low-tensile-strength materials. Reduced relief angles give more support to the cutting edge of the tool and are indicated for cutting high-strength metals. A wear land of excessive width may be the result of too small a relief angle, and tool breakage downward from the cutting edge may result from too large a relief angle.

Cutting-edge angles.   The shape of the workpiece often determines the side-cutting-edge angle. Side-cutting-edge angles of approximately 15° will reduce the cutting power requirement, increase tool life, and aid in controlling the chip. A greater angle increases the chip thickness or makes an increase in feed possible. A decreased side-cutting-edge angle reduces the force at right angles to the workpiece and is required on thin-walled or long, slender workpieces.

End-cutting-edge angles of 6° to 15° are common. An angle of less than 6° may cause excessive forces normal to the work surface and lead to rapid dulling of the cutting tool. Angles greater than 15° may weaken the tool point but are used for tracing tools that must "in-feed" while operating. In all cases, the end-cutting-edge angle should be a minimum of 4° in relation to the surface of the workpiece.

A modification of the end-cutting-edge angle to 0° for a distance a little greater than the feed will produce a drag, or flattening, effect on the tool feed mark. If this modification is used, it must have a relief angle under it.

Nose radius.   A small nose radius reduces cutting forces and is indicated on long, slender parts or thin-walled sections. Large nose radii make the tool stronger and are indicated for roughing operations. Large radii also generally are used for machining cast iron and similar materials that produce a crumbling chip.

Chipbreaker.   A chipbreaker is a small step or groove in the face of a tool or a separate piece attached to the tool or toolholder that causes the chip to break into small sections or curl. The three common types of chipbreakers are those which are ground into the tool point, those which are attached to the tool point, and those which are preformed into an insert. Figure 7.6 shows a typical removable insert toolholder and support for an engine lathe. The removable cutting insert shown mounted on the end of the toolholder has a chip breaker groove formed into its perimeter.

Figure 7.7 shows a typical modern engine lathe with its ACU-RITE III digital readout panel at the upper left. Sensors on the lathe send digital electronic signals to the readout panel so that the

Figure 7.6   Close-up view of a turning insert and toolholder.

Figure 7.7 Typical engine lathe with digital control panel.

operator can set all the controls and movements to a precision of ±0.0005 in without reading the vernier dials on the machine. This allows higher productivity, with less chance of machine operator errors owing to incorrect reading of the mechanical verniers on the machine's controls. The toolholder clamp knob for clamping the toolholder shown in Fig. 7.7 can be seen directly to the right of the three-jaw chuck.

Cutting angles for single-point cutting tools (HSS, cast-alloy, and carbide)

Cutting angles for HHS cutting tools. Figure 7.8 shows the recommended cutting angles for HSS lathe tool bits. Figure 7.9 shows the recommended cutting angles for cast-alloy (Stellite, Tantung/Tantung G, etc.) lathe tool bits. Figure 7.10 shows the recommended cutting angles for carbide lathe bit tools.

7.1.6 Grinding/sharpening of HSS, cast-alloy, and carbide tool bits

Proper grinding and sharpening of standard tool-bit materials should produce a surface condition that promotes a substantial

Recommended Angles for HSS (High-speed-steel) Single-point Tools - (Tabular data in degrees)

Material	Side-relief angle	Front-relief angle	Back-rake angle	Side-rake angle
High-speed, alloy and high-carbon tool steels and stainless steels	7 - 9 (8)	6 - 8 (8)	0 - 7 (0)	8 - 10 (8)
SAE steels:				
1020, 1035, 1040	8 - 10 (8)	8 - 10 (8)	0 - 12 (0)	8 - 12 (8)
1045, 1095	7 - 9 (8)	8 - 10 (8)	0 - 12 (0)	8 - 12 (8)
1112, 1120	7 - 9 (8)	7 - 9 (8)	0 - 14 (0)	10 - 14 (10)
1314, 1315	7 - 9 (8)	7 - 9 (8)	0 - 14 (0)	10 - 16 (10)
1335	7 - 9 (8)	7 - 9 (8)	0 - 14 (0)	10 - 16 (10)
3115, 3120, 3130	7 - 9 (8)	7 - 9 (8)	0 - 10 (0)	8 - 12 (8)
3135, 3140	7 - 9 (8)	7 - 9 (8)	0 - 10 (0)	8 - 10 (8)
3250, 4140, 4340	7 - 9 (8)	7 - 9 (8)	0 - 8 (0)	8 - 10 (8)
6140, 6145	7 - 9 (8)	7 - 9 (8)	0 - 8 (0)	8 - 10 (8)
Aluminum & alloys	12 - 14 (14)	(14)	(0)	(15)
Phenol formaldahyde (Bakelite)	(14)	(14)	(0)	(10)
Brass, free-cutting	10 - 12 (10)	8 - 10 (10)	(0)	1 - 8 (8)
Bronzes-cast, red, yellow	8 - 10 (10)	8 - 10 (10)	(0)	-4 to +6 (+6)
Bronze, free-cutting	8 - 10 (10)	8 - 10 (10)	(0)	2 - 6 (6)
Phosphor-bronze, hard	8 - 10 (10)	6 - 10 (10)	(0)	0 - 6 (6)
Cast iron, gray	8 - 10 (8)	6 - 8 (8)	0 - 5 (0)	8 - 12 (8)
Copper	12 - 14 (12)	12 - 14 (12)	0 - 16 (0)	12 - 20 (12)
Copper alloys:				
Hard	3 - 10 (10)	6 - 10 (10)	(0)	0 - 8 (8)
Soft	10 - 12 (12)	8 - 12 (12)	0 - 2 (0)	0 - 10 (10)
Fiber	14 - 16 (14)	12 - 14 (14)	0 - 2 (0)	0 - 10 (10)
Formica	14 - 16 (14)	10 - 14 (14)	0 - 16 (0)	10 - 12 (12)
Nickel iron	14 - 16 (14)	10 - 14 (14)	0 - 8 (0)	12 - 15 (15)
Micarta	14 - 16 (14)	10 - 14 (14)	0 - 16 (0)	10 - 15 (15)
Monel and nickel	14 - 16 (14)	12 - 14 (14)	0 - 10 (0)	12 - 15 (15)
Nickel silvers	10 - 14 (14)	10 - 14 (14)	0 - 10 (0)	0 - 10 (10)
Rubber, hard and plastics	18 - 20 (20)	14 - 20 (20)	(0)	0 - 20 (20)

NOTE: Angles in parentheses are recommended as a starting point.

Figure 7.8 Angles for HSS single-point tools.

Recommended Cutting Angles for Cast-Alloy Tools * - (Tabular data in degrees)

Material	Back-rake angle	Side-rake angle	Side-relief angle	Front-relief angle	Side-cutting-edge angle	End-cutting-edge angle
Steel	8 - 20 **	8 - 20 **	6	6	10	15
Cast steel	8	8	6	6	10	15
Cast iron	0	4	6	6	10	15
Bronze	4	4	6	6	10	10
Stainless steel	8 - 20 **	8 - 20 **	6	6	10	15

* Stellite 98M-2 turning tools & Tantung/Tantung "G"
** Angles depend on grade and type of steel. Soft materials require more positive rake than hard materials.
Boring tools use the same rake but greater relief to clear the workpiece.

Figure 7.9 Angles for cast-alloy tools.

Recommended Angles for Carbide Single-point Tools - (Tabular data in degrees)

Material	Normal end relief	Normal side relief	Normal back rake	Normal side rake
Aluminum & magnesium alloys	6 - 10 (10)	6 - 10 (10)	0 to 10 (10)	10 to 20 (15)
Copper	6 - 8 (8)	6 - 8 (8)	0 to 4 (0)	15 to 20 (15)
Brass and bronze	6 - 8 (8)	6 - 8 (8)	0 to -5 (0)	+8 to -5 (+8)
Cast iron	5 - 8 (6)	5 - 8 (6)	0 to -7 (0)	+6 to -7 (-6)
Low carbon steels to SAE 1020	5 - 10 (6)	5 - 10 (6)	0 to -7 (0)	+6 to -7 (+6)
Carbon steels, SAE 1025 and above	5 - 8 (6)	5 - 8 (6)	0 to -7 (0)	+6 to -7 (+6)
Alloy steels	5 - 8 (6)	5 - 8 (6)	0 to -7 (0)	+6 to -7 (+6)
Free machining steels SAE 1100 and 1300	5 - 10 (6)	5 - 10 (6)	0 to -7 (0)	+6 to -7 (+6)
Stainless steels, austenitic	5 - 10 (6)	5 - 10 (6)	0 to -7 (0)	+6 to -7 (+6)
Stainless steels, hardenable grades	5 - 8 (6)	5 - 8 (6)	0 to -7 (0)	+6 to -7 (+6)
High nickel alloys (Monel, Inconel, etc.).	5 - 10 (8)	5 - 10 (8)	0 to -3 (0)	+6 to +10 (+10)
Titanium alloys	5 - 8 (6)	5 - 8 (6)	0 to -5 (0)	+6 to -5 (+6)

NOTE: Angles in parantheses are recommended as a starting point.

Figure 7.10 Angles for carbide tools.

increase in the number of workpieces produced per grind. First, grind the tool bit to the proper angles using an aluminum oxide wheel of 40 to 60 grit, followed by a 320-grit aluminum oxide wheel. The second step should remove only a minimal amount of material, on the order of 0.0005 to 0.001 in. If hand honing of the tool is attempted, a hard Arkansas or medium India stone should be used. An experienced operator is required for hand honing tool bits.

7.1.7   Turning operation calculations

Cutting speed.   Cutting speed is given in surface feet per minute (sfpm) and is the speed of the workpiece in relation to the stationary tool bit at the cutting point surface. The cutting speed is given by the simple relation

$$S = \frac{\pi d_f (\text{rpm})}{12} \qquad \text{for inch units}$$

and

$$S = \frac{\pi d_f (\text{rpm})}{1000} \qquad \text{for metric units}$$

where S = cutting speed, sfpm or m/min
d_f = diameter of work, in or mm
rpm = revolutions per minute of the workpiece

When the cutting speed (sfpm) is given for the material, the revolutions per minute (rpm) of the workpiece or lathe spindle can be found from

$$\text{rpm} = \frac{12S}{\pi d_f} \qquad \text{for inch units}$$

and

$$\text{rpm} = \frac{1000S}{\pi d_f} \qquad \text{for metric units}$$

Example: A 2-in-diameter metal rod has an allowable cutting speed of 300 sfpm for a given depth of cut and feed. At what revolutions per minute (rpm) should the machine be set to rotate the work?

$$\text{rpm} = \frac{12S}{\pi d_f} = \frac{12(300)}{3.14 \times 2} = 573 \text{ rpm}$$

Set the machine speed to the next lowest even speed that the machine is capable of attaining.

Lathe cutting time.   The time required to make any particular cut on a lathe may be found using two methods. When the cutting speed is given, the following simple relation may be used:

$$T = \frac{\pi d_f L}{12FS} \qquad \text{for inch units}$$

and

$$T = \frac{\pi d_f L}{1000FS} \qquad \text{for metric units}$$

where T = time for the cut, min
 d_f = diameter of work, in or mm
 L = length of cut, in or mm
 F = feed, ipr (inches per revolution) or mmpr (millimeters per revolution)
 S = cutting speed, sfpm (surface feet per minute) or m/min

Example: What is the cutting time in minutes for one pass over a 10-in length of 2.25-in-diameter rod when the cutting speed allowable is 250 sfpm with a feed of 0.03 ipr?

$$T = \frac{\pi d_f L}{12FS} = \frac{3.14(2.25)(10)}{12(0.03)(250)} = 0.785 \text{ min or } 47 \text{ s}$$

When the revolutions per minute of the machine are known, the cutting time may be found from

$$T = \frac{L}{F(\text{rpm})}$$

where L = length of work, in
 T = cutting time, min
 F = feed, ipr
 rpm = spindle speed or workpiece speed, rpm

Volume of metal removed.   The volume of metal removed during a lathe cutting operation can be calculated as follows:

$$V_r = 12C_d FS \qquad \text{for inch units}$$

and

$$V_r = C_d FS \qquad \text{for metric units}$$

where V_r = volume of metal removed, in³ or cm³
$\quad C_d$ = depth of cut, in or mm
$\quad F$ = feed, ipr or mmpr
$\quad S$ = cutting speed, sfpm or m/min

Note: 1 in³ = 16.387 cm³.

Example: With a depth of cut of 0.25 in and a feed of 0.125 in, what volume of material is removed in 1 min when the cutting speed is 120 sfpm?

$$V_r = 12C_dFS = 12 \times 0.25 \times 0.125 \times 120 = 45 \text{ in}^3$$

For convenience, the chart in Fig. 7.11 may be used for quick calculations of volume of material removed for various depths of cut, feeds, and speeds.

Machine power requirements (horsepower or kilowatts). It is often necessary to know the machine power requirements for an anticipated feed, speed, and depth of cut for a particular material or class of materials to see if the machine is capable of sustaining the desired production rate. The following simple formulas for calculating required horsepower are approximate only because of the complex nature and many variables involved in cutting any material.

The following formula is for approximating machine power requirements for making a particular cut:

$$\text{hp} = dfSC$$

where hp = required machine horsepower
$\quad d$ = depth of cut, in
$\quad f$ = feed, ipr
$\quad S$ = cutting speed, sfpm
$\quad C$ = power constant for the particular material (see Fig. 7.12)

Example: With a depth of cut of 0.06 in and a feed of 0.025 in, what is the power requirement for turning aluminum alloy bar stock at a speed of 350 sfpm?

$$\text{hp} = dfSC = 0.06 \times 0.025 \times 350 \times 4 \quad \text{(see Fig. 7.12)}$$

$$= 2.1$$

For the metric system, the kilowatt requirement is 2.1 × 0.746 = 1.76 kW.

Note: 0.746 kW = 1 hp or 746 W = 1 hp.

The national manufacturers of cutting tools provide the users of their materials with various devices for quickly approximating the

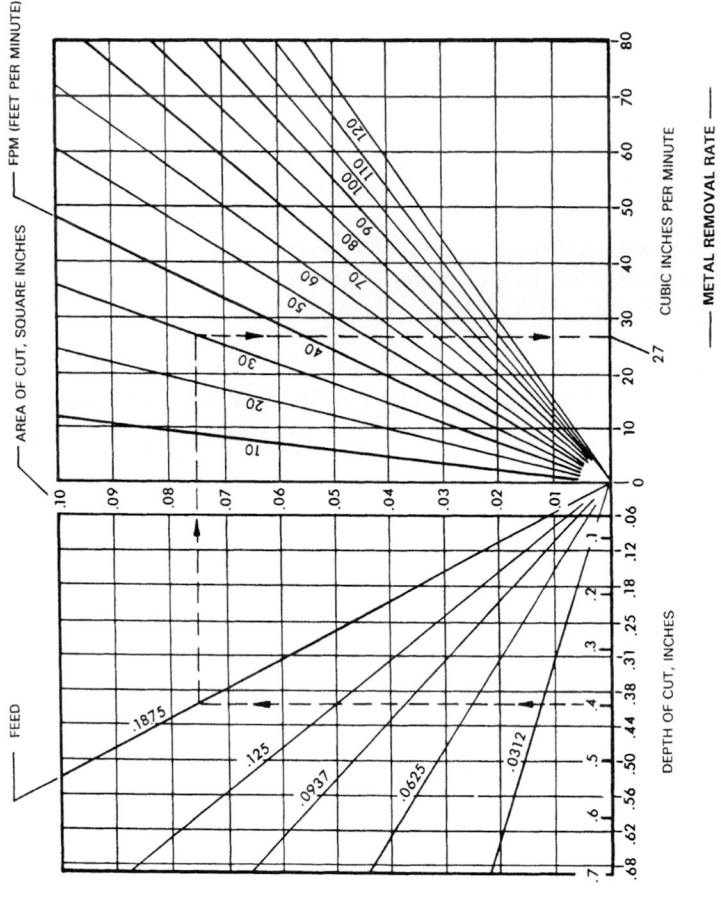

ESTIMATING METAL REMOVAL RATES AT VARIOUS CUTS, FEEDS AND SPEEDS

Figure 7.11 Metal-removal rate (mrr) chart.

Power Constants for Various Metals and Alloys

Material	Constant	Material	Constant
SAE Steels:		**Titanium & alloys:**	
1005 - 1029	6	Pure	4
1030 - 1050	7	Alpha alloys	7
1053 - 1095	8	Beta alloys	8
1211 - 1215	6		
1314 - 1345	6	Copper	4
1330 - 1350	9		
1524 - 1552	9	Zinc alloys	3
4130 - 4820	9		
5120 - 52100	10	Monel	12
Cast steels	9		
		Brass & bronze:	
SAE Stainless steels:		Hard	10
30303, 51403, 51410, 51416		Soft	4
51431, 51430F, 51440F	10		
30302, 30304, 30309,30316		**Aluminum alloys:**	
30321, 51431, 51501	11	Cast	3
51420, 51420F, 51440A, B, C	12	Bar stock	4
Cast irons:		Magnesium alloys	3
Hard	4		
Medium	3		
Soft	3		
Semi-steel	3		
Malleable irons:			
Hard	5		
Medium	4		
Soft	3		

Figure 7.12 Power constant table

various machining calculations shown in the preceding formulas. Although formulas and calculators are available for doing the various machining calculations, it is to be cautioned that these calculations are approximations and that the following factors must be taken into consideration when metals and other materials are cut at high powers and speeds using modern cutting tools:

1. Available machine power

2. Condition of the machine

3. Size, strength, and rigidity of the workpiece

4. Size, strength, and rigidity of the cutting tool

Prior to beginning a large production run of turned parts, sample pieces are run in order to determine the exact feeds and speeds required for a particular material and cutting tool combination.

Power constants.   Figure 7.12 shows a table of constants for various materials that may be used when calculating the approximate power requirements of the cutting machines.

7.1.8   Speeds, cuts, and feeds

HSS, cast-alloy, and carbide tools.   The surface speed (sfpm), depth of cut (in), and feed (ipr) for various materials using HSS, cast-alloy, and carbide cutting tools are shown in Fig. 7.13. In all cases, especially where combinations of values are selected that have not been used previously on a given machine, the selected values should have their required horsepower or kilowatts calculated. Use the approximate calculations shown previously, or use one of the machining calculators available from the cutting tool manufacturers. The method indicated earlier for calculating the required horsepower gives a conservative value that is higher than the actual power required. In any event, on a manually controlled machine, the machinist or machine operator will know if the selected speed, depth of cut, and feed are more than the given machine can tolerate and can make corrections accordingly. On CNC/DNC automatic turning centers and other automatic machines, the cutting parameters must be selected carefully, with the machine operator carefully watching the first trial program run so that he or she may intervene if problems of overloading or tool damage occur.

Recommended Cuts, Feeds and Speeds for Metals and Nonmetallic Turning - (For HSS, Cast-Alloys and Carbide Tools)

Class	Material or SAE No.	Cutting Tool Material	Depth of Cut: 0.005-0.015 Feed: 0.002-0.005	Depth of Cut: 0.015-0.094 Feed: 0.005-0.015	Depth of Cut: 0.094-0.187 Feed: 0.015-0.030	Depth of Cut: 0.187-0.375 Feed: 0.030-0.050	Depth of Cut: 0.375-0.750 Feed: 0.050-0.090
Free-cutting steels	1112, 1L17	HSS	......	250 - 350	175 - 250	80 - 150	55 - 75
	1120	Cast-alloys	......	425 - 550	315 - 400	215 - 300	100 - 210
	1315, etc.	Sintered carbides	750 - 1,500	600 - 750	450 - 600	353 - 450	175 - 350
Carbon & low-alloy steels	1010, 1020	HSS	......	225 - 300	150 - 200	75 - 125	45 - 65
	1025, etc.	Cast-alloys	......	375 - 500	275 - 350	180 - 250	100 - 175
		Sintered carbides	700 - 1,200	550 - 700	400 - 550	300 - 400	150 - 300
Medium alloy steels	1030	HSS	------	200 - 275	125 - 175	70 - 120	40 - 60
	1050	Cast-alloys	......	325 - 400	230 - 300	160 - 225	80 - 150
		Sintered carbides	600 - 1,000	450 - 600	350 - 450	250 - 350	125 - 250
High-alloy steels	1060	HSS	......	175 - 250	125 - 175	65 - 100	35 - 55
	1095	Cast-alloys	......	250 - 350	200 - 250	150 - 200	65 - 150
	1350	Sintered carbides	500 - 750	400 - 500	300 - 400	200 - 300	100 - 200
Chromium-nickel steels	3120, 3450	HSS	......	150 - 200	100 - 125	50 - 75	30 - 50
	5140, 52100	Cast-alloy	......	230 - 315	165 - 225	110 - 160	55 - 110
		Sintered carbides	425 - 550	325 - 425	250 - 325	175 - 250	75 - 175
Molybdenum steels	4130	HSS	......	160 - 210	110 - 140	60 - 80	35 - 55
	4615	Cast-alloys	......	250 - 325	160 - 225	120 - 160	65 - 100
		Sintered carbides	475 - 650	350 - 475	275 - 350	200 - 275	100 - 200
Chrome, vanadium and stainless steels	6120	HSS	......	100 - 150	80 - 100	50 - 75	30 - 50
	6150, 18Cr-5Ni	Cast-alloys	......	210 - 250	170 - 200	110 - 165	55 - 100
	6195	Sintered carbides	375 - 500	300 - 375	250 - 300	175 - 250	75 - 175
Tungsten steels	7260	HSS	......	120 - 150	75 - 120	40 - 75	25 - 40
	18-4-1 annealed	Cast-alloys	......	130 - 175	110 - 130	80 - 100	35 - 80
		Sintered carbides	325 - 400	250 - 325	200 - 250	150 - 200	50 - 160
Special steels	12-14% Mn.	HSS	......	......	......	......	......
		Cast-alloys	......	......	......	......	......
		Sintered carbides	200 - 250	25 - 200	75 - 125	50 - 75	......

Figure 7.13   Cuts, feeds, and speeds table.

Material	Description	Tool					
Special steels	Si. elect., sheet ingot iron, etc.	HSS	400 - 500	300 - 400	200 - 300	150 - 200	....
		Cast-alloys	....	500 - 600	350 - 450	250 - 300	....
		Sintered carbide	1,000 - 1,200	800 - 1,000	600 - 800	500 - 600	
Cast iron	Soft gray	HSS	....	120 - 150	90 - 120	75 - 90	35 - 75
		Cast-alloys	....	225 - 300	160 - 220	125 - 160	70 - 125
		Sintered carbides	450 - 600	350 - 450	250 - 350	200 - 250	100 - 200
	Medium and malleable	HSS	....	120 - 150	90 - 120	60 - 90	30 - 60
		Cast-alloys	....	100 - 225	150 - 190	120 - 150	60 - 120
		Sintered carbides	350 - 450	250 - 350	200 - 250	150 - 200	75 - 150
	Hard alloys	HSS	....	90 - 125	60 - 90	40 - 60	20 - 40
		Cast-alloys	....	120 - 170	80 - 120	55 - 80	35 - 55
		Sintered carbides	250 - 300	150 - 250	100 - 150	75 - 100	50 - 75
	Chilled	HSS	10 - 15	....	....	....	....
		Cast-alloys	....	....	....	....	....
		Sintered carbides	30 - 50	10 - 30	....	....	
Copper base alloys	Leaded, free - cutting, soft brass and bronze	HSS	....	300 - 400	225 - 300	150 - 255	100 - 150
		Cast-alloys	....	500 - 600	400 - 500	325 - 400	200 - 325
		Sintered carbides	1,000 - 1,250	800 - 1,000	650 - 800	500 - 650	300 - 500
	Normal brass, bronze low alloy	HSS	....	275 - 350	225 - 275	150 - 225	100 - 150
		Cast-alloys	....	375 - 425	325 - 375	250 - 325	175 - 250
		Sintered carbides	700 - 800	600 - 700	500 - 600	400 - 500	200 - 400
	Tough copper, high tin & alum. bronzes, gilding.	HSS	....	100 - 150	75 - 100	50 - 75	35 - 50
		Cast-alloys	....	225 - 300	180 - 225	125 - 180	75 - 125
		Sintered carbides	500 - 600	400 - 500	300 - 400	200 - 300	100 - 200
Light alloys	Magnesium	HSS	500 - 750	350 - 500	275 - 350	200 - 275	125 - 200
		Cast-alloys	700 - 1,000	500 - 700	400 - 500	300 - 400	200 - 300
		Sintered carbides	1,250 - 2,000	800 - 1,250	600 - 800	500 - 600	300 - 500
	Aluminum	HSS	350 - 500	225 - 350	150 - 225	100 - 150	50 - 100
		Cast-alloys	450 - 650	300 - 450	225 - 300	150 - 225	75 - 150
		Sintered carbides	700 - 1,000	450 - 700	300 - 450	200 - 300	100 - 200
Titanium	Pure & low alloys	HSS	....	100 - 160	70 - 110	50 - 75	....
		Cast-alloys	....	165 - 375	110 - 250	75 - 165	....
		Sintered carbides	550 - 900	375 - 600	250 - 400	165 - 265	

Material	Tool material	Col 1	Col 2	Col 3	Col 4
Alpha alloys	HSS	....	....	30 - 75	....
	Cast-alloys	....	....	75 - 110	....
	Sintered carbides	165 - 450	110 - 300	75 - 200	50 - 135
Beta alloys	HSS	....	....	30 - 40	20 - 25
	Cast-alloys	....	....	40 - 90	25 - 60
	Sintered carbides	125 - 225	90 - 150	60 - 100	40 - 70
Plastics — Thermoplastic, thermosetting	HSS	....	....	....	....
	Cast-alloys	....	....	....	....
	Sintered carbides	650 - 1,000	400 - 650	250 - 400	150 - 250
Abrasives — Glass, hard rubber, green ceramics, marble	HSS	....	....	....	....
	Cast-alloys	....	....	....	....
	Sintered carbides	150 - 250	75 - 150	....	....

NOTE: It is possible that a combination of speeds, feeds and cuts may be so selected for a given application, that a higher horsepower may be required than is available at the lathe spindle. In all cases, especially where combinations of values are selected that have not been used previously on a given machine, the selected values should have their required horsepower calculated. See the subsection (Horsepower requirements) for calculating required horsepower when speed, feed and cut are given. Values of depth of cut are in inches; feeds are given in ipr (inches per revolution) and speed is given in sfpm (surface feet per minute).

(Tabular values are in sfpm).

Figure 7.13 (*Continued*)

Charts of speeds, feeds, and depths of cut for turning, boring, milling, threading, and grooving or cut off using modern machining inserts

---
WARNING

Cutting tools may chip or fragment in use. Always use machine guards, protective clothing, and safety glasses to prevent burns or other injury to body or eyes from flying particles or chips. Wet or dry grinding of cutting tools produces potentially hazardous dusts or mists; to avoid adverse health affects, use adequate ventilation and read material safety data sheet for applicable tool material or grade before grinding.

©1997 Valenite, Inc. AEROTEC, Centre-Dex, Econo-Mizer, MasterMill, Mini-Mills, Ripper, SpectraMill, SpectraSystem, Turn-Tech, V-CutPlus, V-Thread, VAL-U-ADD, VAL-U-CLEAN, VAL-U-DEX, VALCOOL, VALTAP, Valenite, and VNT are Trademarks of Valenite Inc.

Modern high-speed insert-type cutting tools. Figures 7.14 through 7.44 show the recommended ranges of starting speeds, depths of cut, and feeds for modern types of insert cutting tools. The figures shown are reprinted with permission from Valenite, Inc. Included in these figures or charts are Valenite-designated insert numbers for the various machining operations of turning, boring, milling, threading, and grooving or cut off. Figures 7.45 and 7.46 are insert cutting tool comparison charts for various manufacturers' cutting inserts, the equivalents of which are cross-referenced to the Valenite insert numbers.

(Text continued on page 383.)

CARBON/ALLOY STEELS >35RC

1095, 4340, 52100 Types

Starting Recommendations:

MACHINING OPERATIONS	GRADE ▼	CHIPBREAKER ▼	Speed - sfm (M/min.)	Feed - ipr (mm/rev.)	Feed - ipt (mm/tooth)
			- RECOMMENDATIONS -		
T U R N I N G Roughing	CBN	---	250-600 (75-180)	.003-.010 (0,07-0,25)	---
General Purpose	C32	---	300-600 (90-180)	.003-.008 (0,07-0,20)	---
Finishing	C32	---	400-700 (120-210)	.003-.006 (0,07-0,15)	---
M I L L I N G Roughing	●	---	---	---	---
General Purpose	●	---	---	---	---
Finishing	●	---	---	---	---
Grooving & Cut Off ......⟳ ALL CASES	●	---	---	---	---
Threading ......⟳	●	---	---	---	---

NOTE: sfm = Surface feet/minute (Meters/Minute); ipr = Inches/revolution (mm/revolution); ipt = Inches/tooth (mm/tooth).
● Recommendations for these operations need to consider all the variables in the application (machine, tool age, coolant type, rigidity of the set up, overhangs, etc.) For product and application assistance, please contact Valenite's Technical Support Center. Phone number:(800) 488-9073.
© Valenite Inc.

Figure 7.14

CARBON/ALLOY STEELS <35 RC   1010, 1020, 1050, 1095, 4140, 4150, 4340, 6150, 8620 Types

Starting recommendations:

- RECOMMENDATIONS -

MACHINING OPERATIONS		GRADE ▼	CHIPBREAKER ▼	Speed - sfm (M/min.)	Feed - ipr (mm/rev.)	Feed - ipt (mm/tooth)
T & U B R O N R I I N N G G	Roughing	SV325	GM/GR	300-500 (90-150)	.015-.030 (0,37-0,75)	----
	General Purpose	SV315	LM/GM	400-700 (120-210)	.012-.025 (0,30-0,62)	----
	Finishing	SV310	LF/GF	500-800 (150-240)	.004-.010 (0,10-0,25)	----
M I L L I N G	Roughing	V1N	----	<400 (<120)	----	.008-.016 (0,20-0,40)
	General Purpose	VC935M	----	300-600 (90-180)	----	.006-.012 (0,15-0,30)
	Finishing	VC630	----	400-700 (120-210)	----	.004-.007 (0,10-0,17)
Grooving & Cut Off ......⟶ ALL CASES		VC121	----	300-700 (90-210)	.004-.008 (0,10-0,20)	----
Threading ......⟶		VN8	----	300-700 (90-210)	----	----

Note: sfm = Surface feet/minute (Meters/Minute); ipr = Inches/revolution (mm/revolution); ipt = Inches/tooth (mm/tooth).
© Valenite Inc.

Figure 7.15

CAST IRONS - ALLOYED

CLASS 40 THROUGH CLASS 60
SAE 60-40-18, SAE 80-60-3

Starting Recommendations:

- RECOMMENDATIONS -

MACHINING OPERATIONS		GRADE ▼	CHIPBREAKER ▼	Speed - sfm (M/min.)	Feed - ipr (mm/rev.)	Feed - ipt (mm/tooth)
T & U B R O I N N G G	Roughing	V1N	---	250-400 (75-120)	.016-.025 (0,40-0,62)	---
	General Purpose	SV315	GM	350-500 (105-150)	.012-.018 (0,30-0,45)	---
	Finishing	SV310/VC929	---	400-600 (120-180)	.004-.008 (0,10-0,20)	---
M I L L I N G	Roughing	V1N	---	200-400 (60-120)	---	.008-.016 (0,20-0,40)
	General Purpose	VC928	---	300-600 (90-180)	---	.006-.012 (0,15-0,30)
	Finishing	VC630/VC929	---	400-700 (120-210)	---	.004-.010 (0,10-0,25)
Grooving & Cut Off ......⟳ ALL CASES		SV221	---	300-600 (90-180)	.004-.010 (0,10-0,25)	---
Threading ......⟳		VN8	---	300-600 (90-180)	---	---

Note: sfm = Surface feet/minute (Meters/Minute); ipr = Inches/revolution (mm/revolution); ipt = Inches/tooth (mm/tooth).

© Valenite Inc.

Figure 7.16

CAST IRONS - GREY

CLASS 20 THROUGH CLASS 35

Starting Recommendations:

- RECOMMENDATIONS -

MACHINING OPERATIONS		GRADE ▼	CHIPBREAKER ▼	Speed - sfm (M/min.)	Feed - ipr (mm/rev.)	Feed - ipt (mm/tooth)
T & U B R O N R I I N N G G	Roughing	V1N	----	300-450 (90-135)	.016-.025 (0,40-0,62)	----
	General Purpose	SV315	GM	400-650 (120-195)	.012-.018 (0,30-0,45)	----
	Finishing	SV310/VC929	----	500-700 (150-210)	.004-.008 (0,10-0,20)	----
M I L L I N G	Roughing	V1N	----	300-500 (90-150)	----	.008-.016 (0,20-0,40)
	General Purpose	VC928	----	400-650 (120-195)	----	.006-.012 (0,15-0,30)
	Finishing	VC929/VC630	----	500-800 (150-240)	----	.004-.010 (0,10-0,25)
Grooving & Cut Off ......⟳	ALL CASES	SV221	----	300-700 (90-210)	.004-.010 (0,10-0,25)	----
Threading ......⟳		VN8	----	300-700 (90-210)	----	----

Note: sfm = Surface feet/minute (Meters/Minute); ipr = Inches/revolution (mm/revolution); ipt = Inches/tooth (mm/tooth).
© Valenite Inc.

Figure 7.17

STAINLESS STEELS - SERIES 200/300

202, 204, 304, 310, 316, 321, 347

Starting Recommendations:

	MACHINING OPERATIONS	GRADE ▼	CHIPBREAKER ▼	Speed - sfm (M/min.)	Feed - ipr (mm/rev.)	Feed - ipt (mm/tooth)
				- RECOMMENDATIONS -		
T & B U R R O N R I I N N G G	Roughing	VC90?	LM/GM	200-400 (60-120)	.016-.025 (0,40-0,62)	----
	General Purpose	VC929	GM	300-550 (90-165)	.012-.018 (0,30-0,45)	----
	Finishing	SV310	LF/GF	400-700 (120-210)	.004-.008 (0,10-0,20)	----
M I L L I N G	Roughing	V1N	----	200-350 (60-105)	----	.008-.016 (0,20-0,40)
	General Purpose	VC901	----	250-500 (75-150)	----	.006-.012 (0,15-0,30)
	Finishing	VC929	----	400-600 (120-180)	----	.004-.010 (0,10-0,25)
	Grooving & Cut Off .....⇔ ALL CASES	VC231	----	150-300 (45-90)	.004-.010 (0,10-0,25)	----
	Threading .....⇔	VN8	----	150-300 (45-90)	----	----

Note: sfm = Surface feet/minute (Meters/Minute); ipr = Inches/revolution (mm/revolution); ipt = Inches/tooth (mm/tooth).

© Valenite Inc.

Figure 7.18

STAINLESS STEELS - SERIES 400/500/PH

403, 410, 416, 431, 440
501, 502, 17-4PH, 155MD, 138

Starting Recommendations:

MACHINING OPERATIONS		GRADE ▼	CHIPBREAKER ▼	- RECOMMENDATIONS -		
				Speed - sfm (M/min.)	Feed - ipr (mm/rev.)	Feed - ipt (mm/tooth)
T & U B R O N R I I N N G G	Roughing	VC901	LM/GM	250-450 (75-135)	.015-.020 (0,37-0,50)	----
	General Purpose	VC929	GM	350-600 (105-180)	.008-.016 (0,20-0,40)	----
	Finishing	SV310	LF/GF	400-700 (120-210)	.004-.008 (0,10-0,20)	----
M I L L I N G	Roughing	V1N	----	200-350 (60-105)	----	.008-.012 (0,20-0,30)
	General Purpose	VC901	----	250-450 (75-135)	----	.006-.012 (0,15-0,30)
	Finishing	VC929	----	300-600 (90-180)	----	.004-.008 (0,10-0,20)
Grooving & Cut Off ......↩		VC231	----	250-500 (75-150)	.004-.010 (0,10-0,25)	----
ALL CASES Threading ......↩		VN8	----	250-400 (75-120)	----	----

Note: sfm = Surface feet/minute (Meters/Minute); ipr = Inches/revolution (mm/revolution); ipt = Inches/tooth (mm/tooth).
© Valenite Inc.

Figure 7.19

354

NON-FERROUS MATERIALS

ALUMINUM, BRASS, COPPER, COMPOSITES, PLASTICS

Starting Recommendations:

- RECOMMENDATIONS -

MACHINING OPERATIONS		GRADE ▼	CHIPBREAKER ▼	Speed - sfm (M/min.)	Feed - ipr (mm/rev.)	Feed - ipt (mm/tooth)
T & **U B** **R O** **N R** **I -** **N N** **G G**	Roughing	VC2	---	200-1200 (60-360)	.008-.016 (0,20-0,40)	---
	General Purpose	VC2	---	400-1400 (120-420)	.008-.016 (0,20-0,40)	---
	Finishing	VC3	---	600 UP (180 UP)	.004-.008 (0,10-0,20)	---
M **I** **L** **L** **I** **N** **G**	Roughing	VC2	---	200-1200 (60-360)	---	.008-.012 (0,20-0,30)
	General Purpose	VC2	---	400-1400 (120-420)	---	.006-.012 (0,15-0,30)
	Finishing	VC29	---	600 UP (180 UP)	---	.004-.008 (0,10-0,20)
Grooving & Cut Off ......↺ **ALL CASES**		VC121	---	400-1200 (120-360)	.006-.014 (0,15-0,35)	---
Threading ......↺		VC2	---	400-1200 (120-360)	---	---

Note: sfm = Surface feet/minute (Meters/Minute); ipr = Inches/revolution (mm/revolution); ipt = Inches/tooth (mm/tooth).
© Valenite Inc.

Figure 7.20

EXOTIC MATERIALS

INCONEL 625, INCONEL 718, MONELS,
RENE 95, WASPALLOY, HASTALLOY

Starting Recommendations:

- RECOMMENDATIONS -

MACHINING OPERATIONS	GRADE ▼	CHIPBREAKER ▼	Speed - sfm (M/min.)	Feed - ipr (mm/rev.)	Feed - ipt (mm/tooth)
T & U B R O N R I I N N G G Roughing	V1N	GM/LM	75-150 (22-45)	.008-.012 (0,20-0,30)	-----
General Purpose	VC901	LM/GF	100-200 (30-60)	.006-.010 (0,15-0,25)	-----
Finishing	VC929	LF/GF	150-300 (45-90)	.004-.008 (0,10-0,20)	-----
M I L L I N G Roughing	●	-----	-----	-----	-----
General Purpose	VC901	-----	75-150 (22-45)	-----	.006-.010 (0,15-0,25)
Finishing	VC929	-----	100-200 (30-60)	-----	.004-.008 (0,10-0,20)
Grooving & Cut Off ....↩ ALL	●	-----	-----	-----	-----
CASES Threading ....↩	●	-----	-----	-----	-----

Note: sfm = Surface feet/minute (Meters/Minute); ipr = Inches/revolution (mm/revolution); ipt = Inches/tooth (mm/tooth).
● Recommendations for these operations need to consider all the variables in the application (machine, tool age, coolant type, rigidity of set up, overhangs, etc.). For product and application assistance, please contact Valenite's Technical Support Center. Phone number (800) 488-9073.
© **Valenite Inc.**

Figure 7.21

Uncoated Carbide

Valenite Grade	ISO Class	Industry Class	Application	Materials	Working Methods & Conditions
VC2	M10-20 K10-20	C2	Turning, Boring & Milling	Cast-iron, copper, brass, non-ferrous Alloys, high-temp. Exotics, stone & plastics	General purpose grade of high toughness and resistance against flank wear at low to medium cutting speeds.
VC3	K10-05	C3, C4	Precision Turning, Boring, Milling	Cast-iron, aluminum high-temperature exotics and non-ferrous materials.	Wear-resistant grade for finishing cuts, low to medium feed rates under rigid conditions.
VC5	P20-30 M20-40	C5	Turning, Boring and Milling	Steel, cast steel, malleable cast-iron, 400/500 series stainless steels.	General purpose grade with many applications, low to medium cutting spec, high feeds and depths of cut. Has good deformation resistance.
VC7	P05-15	C7	Turning, Boring, Threading and Grooving.	Steel, cast-steel, malleable cast-iron, 400/500 series stainless steels.	Light roughing to finishing at low to moderate feeds. Good crater and deformation resistance.
VC27	P15-30 M15-30 K20-30	C2	Turning and Milling	Steel, cast-steel, alloyed cast-irons, cast alloys and exotics.	General purpose fine grain grade with improved toughness and wear resistance for turning and milling.
VC28	M20-30 K15-30	C2	Milling	Cast and alloy irons.	General purpose grade for roughing to finishing in cast-irons.
VC29	M10-20 K10-20	C2, C3	Turning, Boring and Milling	Stainless steels, irons, exotics and non-ferrous materials.	Fine grain grade for finishing of exotic irons and non-ferrous metals.
VC35	P20-35	C5	Milling	Carbon, alloy steel and stainless steel.	General purpose steel milling grade for moderate roughing and finishing.
VC101	M30-40 K30-40	C1	Turning, Boring and Milling	Iron 200/300 stainless steel and exotics.	Fine grain heavy-duty grade for roughing at low to moderate speeds.
VC121	K05-30	C2	V-Cut plus Cut Off & Grooving	Cast-iron, aluminum, non-ferrous alloys and exotics	Medium cutting speeds and feeds.
VC135	P25-50	C5	V-Cut Plus Cut Off and Grooving.	Steel, stainless steel.	Medium speeds, heavy feeds in unfavorable conditions.

© Valenite Inc.

Figure 7.22

Physical Vapor Deposition (PVD) Coated Carbides

Valenite Grade	ISO Class	Industry Class	Application	Materials	Working Methods & Conditions
VC901	M20-30 K20-30	C1	Turning, Boring, Milling, Threading, Grooving & Cut-Off	200 and 300 stainless steels, titanium alloys, nickel-based alloys, Inconel 600 and 718, Hastelloy C, Waspalloy, cobalt-based alloys.	A micro-grained material for severe cutting conditions at low speeds. VC901 exhibits outstanding wear resistance on difficult to machine materials.
VC902	M10-20 K10-20	C2	Turning, Boring, Milling and Threading	Gray cast-iron, nickel based alloys, titanium alloys, wrought and cast aluminum.	A very tough general purpose grade with medium feed finishing cuts. Requires a very rigid set-up for optimum performance.
VC903	K01	C3, C4	Precision Turning, Boring and Milling	Gray cast-iron, nodular and malleable iron-nickel bases alloys, wrought and cast aluminum.	A highly wear resistant grade for low to medium feed finishing cuts. Requires a very rigid set-up for optimum performance.
VC905	P20-30 M20-40	C5	Turning, Boring, Threading and Grooving	Low to medium carbon steels, alloy steels, stainless steels.	A general purpose grade with a wide application range. Suitable for low to high speeds at high feeds and depths of cut.
VC907	P05-10 M20-40	C7	Turning, Boring, Threading and Grooving	Low to medium carbon steels, alloy steels, stainless steels.	Excellent wear and crater resistance. Recommended for medium speeds and high feeds.
VC927*	P10-20 M10-20 K10-20	C2	Turning, Boring, Threading and Grooving	Alloyed Cast Irons, stainless steels, aluminum.	Provides higher edge wear resistance than VC905 at medium cutting feeds and speeds. Especially good for alloyed cast-irons.
VC928*	M20-30 K15-30	C2	Milling	Alloyed cast-irons, 300 series stainless steels, aluminum	Recommended for medium to heavy machining operations at medium speeds and high feeds.
VC929*	M10-15 K05-15	C2, C3	Turning, Boring, Milling, Threading, and Grooving	Nickel and cobalt based alloys, titanium alloys, aluminum, copper, brass, and zinc alloys.	A fine grained material for light cutting conditions. Recommended for high speed applications where flank wear resistance is required.
VC935	P15-35	C5, C6	Milling	Low carbon to medium carbon steels, heat treated alloy steels, and stainless steels.	A standard milling grade for medium to high cutting speeds, high feeds and high depths of cut under unfavorable conditions.

*New products.   © Valenite Inc.

Figure 7.23

358

Chemical Vapor Deposition (CVD). Coated Carbide

Valenite Grade	ISO Class	Industry Class	Application	Materials	Working Methods & Conditions
VN5	P10-25 M15-20	C5-C7	Turning and Boring	Steel, cast steel, malleable cast-iron, stainless steel.	A TiN coated grade for roughing and finishing. Has excellent crater and deformation resistance.
VN8	P10-30 M20-30 K10-30	C5-C7 C2-C3	Turning, Boring, Milling, Threading and Grooving	Cast-iron, steels, 300 & 400 series stainless steels, PH stainless steels.	A very well balanced TiN coated grade suitable for a broad range of applications. Has outstanding crater and impact resistance at low to high speeds.
VO1	P01-30 M10-30 K01-30	C8-C5 C4-C2	Turning, Boring and Milling	Cast-iron, stainless steels, alloyed steels, carbon steels.	A composite ceramic coated grade providing maximum resistance to built-up edge. Suitable for operations ranging from roughing to finishing at medium to high speeds.
V1N	P30-45 M30-40 K25-45	C5 C1, C2	Milling, Threading and Grooving	Cast-iron, steels, high temperature exotics, 300 & 400 series stainless steels, and PH stainless steels.	A TiN coated heavy-duty grade used in severe roughing and interrupted cuts at slow speeds.
V88	P05-10 M10-30 K05-30	C6-C7 C3-C1	Turning, Boring and Milling	Cast-iron, steel, and alloyed steels.	A TiN coated grade with excellent flank wear resistance for use in applications where abrasive wear is the primary failure mode.
VX8	P15-30 M15-30 K15-30	C6-C7 C2	Turning, Boring, Threading and Grooving	Cast-iron, steel, high temperature exotics, and stainless steels.	A TiC and TiN coated grade for moderate to heavy cuts with medium to heavy feeds. Optimized for flank wear resistance.

© Valenite Inc.

Figure 7.24

SpectraSystem® Coated Inserts

Valenite Grade	ISO Class	Industry Class	Application	Materials	Working Methods & Conditions
SV221	P10-25	C5, C6	Cut-off, Grooving, Turning & Slotting V-Cut Plus	Carbon steel, alloy steel, steel castings, malleable cast-iron, stainless steel and free-cutting steel.	Roughing and semi-finishing. Used in V-Cut Plus blades and toolholders.
SV231	P20-35	C5	Cut-Off, Grooving, Turning and Slotting V-Cut Plus	Stainless steel, carbon steel, and steel castings.	Roughing at medium and low speeds and high feeds. Used in V-Cut Plus blades and toolholders.
SV235	P20-45 M15-35	C5	Turning, Boring	Carbon steel, alloy steel, and stainless steel.	SV235 is the toughest of the SpectraSystem grades, engineered for maximum shock and impact resistance. It provides excellent wear resistance on all types of steel.
SV310	P01-15 M01-15 K01-15	C7,C8 C3, C4	Turning, Boring	Steels, stainless steels, iron, and some exotics.	SV310 is a high-speed finishing grade with high wear and crater resistance, high surface lubricity, and excellent wear resistance to built-up edge.
SV315	P10-30 M05-25 K01-15	C6, C7 C2, C3	Turning, Boring	Low to medium carbon steels, alloy steels, and stainless steels.	SV315 covers applications ranging from light roughing to high-speed finishing. It combines excellent wear resistance with high deformation resistance for high-speed operations.
SV325	P15-35 M10-30 K15-30	C5, C6 C1, C2	Turning, Boring	Carbon steel, alloy steel, and stainless steel.	SV325 is a general purpose grade that delivers outstanding performance in moderate roughing to semi-finishing applications, especially in interrupted cuts.

© Valenite Inc.

Figure 7.25

600 Series Cermets

Valenite Grade	ISO Class	Industry Class	Application	Materials	Working Methods & Conditions
VC605	P01-05 K01-05	C8 C3-C4	Turning, Boring	Cast irons, powder metal and steels.	A high-speed finishing grade for use at low feed rates.
VC610	P01-10 K05-10	C8, C7 C3-C4	Turning, Boring	Steels, stainless steels, carbon and alloy steels, and cast-iron.	A general purpose grade recommended for high and moderate speed finishing operations. Has better wear resistance than VC671.
VC630	P10-30	C8-C6	Milling	Gray cast-iron, nodular and malleable iron-nickel based alloys, both wrought and cast aluminum.	A relatively tough grade suitable for light roughing in turning operations with a maximum depth of cut of 0.025 in. Can be used for finish milling on interrupted surfaces at high speeds.
VC671	P10-15	C7	Turning, Boring, Milling, Threading	Low to medium carbon steels, alloy steels, stainless steels.	A general purpose grade suitable for high-speed finishing at low feed rates. Somewhat tougher than VC610, but less wear resistant.

© Valenite Inc.

Figure 7.26

Ceramics

Valenite Grade	ISO Class	Industry Class	Application	Materials	Working Methods & Conditions
Q6	K01-K20	C4-C2	Turning, Boring, and Milling	Cast -iron	Specifically recommended for operations ranging from heavy roughing to finishing of cast-iron. Suitable for high-speeds, heavy feeds, and interrupted cuts.
Q32	P05-15 N05-15	C7, C8	Turning, Boring, and Milling	Gray cast-iron, nickel based alloys, titanium alloys, both wrought and cast.	Cold-pressed ceramic used for light finishing operations on continuous cuts under rigid conditions.
V44	P01-10 K01-10	C7, C8 C3, C4	Turning, Boring	Cast-irons, steels, stainless steels with hardness below 35RC.	Recommended for continuous light roughing and finishing cuts on very rigid machines.

© Valenite Inc.

Figure 7.27

PCBN

Valenite Grade	ISO Class	Industry Class	Application	Materials	Working Methods & Conditions
VC722	-----	-----	Turning, Boring and Milling	All cast-irons, super alloys & powder metals.	PCBN tipped inserts for general purpose roughing and finishing at high-speeds and low-speeds.
VC734	-----	-----	Turning, Boring and Milling	Martensitic cast irons (N-HARD), high chrome irons, chilled cast-irons, cold work tool steels, nickel and iron based alloys, fully pearlitic cast-irons, case hardened steel, and martensitic stainless steels.	A solid PCBN insert intended for heavy machining and high stock removal rates. Provides excellent wear and shock resistance at high-speeds. Used without coolant.
VC721*	-----	-----	Hard Turning and Boring	Hardened steels above 60RC.	A tipped PCBN insert for a wide range of applications. Superior performance at high speeds in both continuous and interrupted cuts.

* New products
© Valenite Inc.

Figure 7.28

PCD

Valenite Grade	ISO Class	Industry Class	Application	Materials	Working Methods & Conditions
VC727	-----	-----	Turning, Boring and Milling	Aluminum alloys with less than 16% silicon. Copper, fiberglass, carbon epoxies, chipboard and fiberboard.	A medium-grained, general purpose material for semi-finishing to finishing.
VC728	-----	-----	Turning, Boring and Milling	Aluminum alloys with more than 16% silicon. Non-ferrous alloys and nonmetallics.	A coarse-grained material suitable for high-speed roughing and semi-finishing operations.
VC746	-----	-----	Turning, Boring, Milling & Drilling	Aluminum, plastics, precious metal, lead alloys, other moderately abrasive materials and wood composites.	A fine-grained material for high-speed, low-feed finishing where excellent surface finishes are required.

© Valenite Inc,

Figure 7.29

364

Recommended Starting Grades for Types of Workpiece Materials

VALENITE STARTING GRADES		STEELS		CAST IRONS		STAINLESS STL.		NON-FERROUS MATERIALS	EXOTIC MATERIALS
		<35RC	>35RC	GREY	ALLOY	200/300	400/500		
Turning & Boring	Roughing	VC5● VC27♦	--- ---	VC2● VC111♦	VC5● ---	VC2● VC101♦	VC5● ---	VC2● ---	VC2● VC101♦
	Finishing	VC7● VC5♦	VC3● VC3♦	VC3● VC29♦	VC7● VC5♦	VC29● VC2♦	VC7● VC27♦	VC3● VC29♦	VC29● VC2♦
Milling		VC7● VC35♦	--- ---	VC2● VC28♦	VC7● VC35♦	VC2● VC2♦	VC7● VC27♦	VC2● VC2♦	VC2● VC2♦
Grooving & Cut Off		VC135●	---	VC121●	VC135●	VC121●	VC135●	VC121●	VC121●
Threading		VC7●	---	VC2●	VC7●	VC2●	VC27●	VC2●	VC2●

Note: ● Primary choice has most wear resistance with medium impact strength or best general purpose grade. ♦ Alternate choice for more severe applications that require higher edge strength and toughness.
© Valenite Inc.

Figure 7.30

365

Recommended Starting Grades for Types of Workpiece Materials

VALENTE STARTING GRADES		STEELS		CAST IRONS		STAINLESS STL.		NON-FERROUS MATERIALS	EXOTIC MATERIALS
		<35RC	>35RC	GREY	ALLOY	200/300	400/500		
Turning & Boring	Roughing	VC905● VC927♦	----- -----	VC902● VC927♦	VC905● VC927♦	VC901● VC928♦	VC927● VC905♦	VC902● VC927♦	VC901● VC927♦
	Finishing	VC907● VC927♦	VC929● -----	VC903● VC929♦	VC927● VC905♦	VC929● VC902♦	VC907● VC927♦	VC903● VC929♦	VC929● VC927♦
Milling		VC935● VC901♦	----- -----	VC928● VC902♦	VC935● VC905♦	VC928● -----	VC935● VC905♦	VC902● VC929♦	VC902● VC901♦
Grooving & Cut Off		VC901●	-----	VC901●	VC901●	VC901●	VC901●	VC901●	VC901●
Threading		VC905●	-----	VC902●	VC907●	VC907●	VC902●	VC90●	VC902●

Note: ● Primary choice has most wear resistance with medium impact strength or best general purpose grade. ♦ Alternate choice for more severe applications that require higher edge strength and toughness.

© Valenite Inc.

Figure 7.31

Recommended Starting Grades for Types of Workpiece Materials

VALENITE STARTING GRADES		STEELS		CAST IRONS		STAINLESS STL.		NON-FERROUS MATERIALS	EXOTIC MATERIALS
		<35RC	>35RC	GREY	ALLOY	200/300	400/500		
Turning & Boring	Roughing	VN8● VN5◆	-----	VO1● -----	VO1● -----	VN8● VN5◆	VN8● VN5◆	SEE UNCOATED PVD & CERMET	VN8● -----
	Finishing	VO1● VN8◆	-----	VO1● -----	VO1● -----	VN8● VN5◆	VNO1● VN8◆	"	VN8● -----
Milling		V1N● VO1◆	-----	VN5● V1N◆	VN5● V1N◆	V1N● VN8◆	V1N● VN8◆	"	V1N● V1N◆
Grooving & Cut Off		VN8●	-----	VN8●	VN8●	VN8●	VN8●	"	VN8●
Threading		VN8	-----	VN8●	VN8●	VN8●	VN25●	SEE UNCOATED PVD & CERMET	VN8●

Note: ● Primary choice has more wear resistance with medium impact strength or best general purpose grade. ◆ Alternate choice for more severe applications that require higher edge strength and toughness.
© Valenite Inc,

Figure 7.32

Recommended Starting Grades for Types of Workpiece Materials

VALENITE STARTING GRADES		STEELS		CAST IRONS		STAINLESS STL.			NON-FERROUS MATERIALS	EXOTIC MATERIALS
		<35RC	>35RC	GREY	ALLOY	200/300	400/500			
Turning & Boring	Roughing	SV325● SV325◆	-----	SV315● SV325◆	SV315● SV325◆	SV315● SV315◆	SV325● SV325◆	SEE PVD & CERMET GRADES	SEE PVD & CERMET GRADES	
	Finishing	SV310● SV315◆	-----	SV310● SV315◆	SV310● SV315◆	SV310● SV315◆	SV310● SV315◆			
Milling				SEE PVD/CVD/CERMET GRADES						
Grooving & Cut Off		SV231●	-----	SV221●	SV221●	SV231●	SV231●	-----	SV221●	
Threading				SEE PVD/CVD/CERMET GRADES						

Note: ● Primary choice has most wear resistance with medium impact strength or best general purpose grade. ◆ Alternate choice for more severe applications that require higher edge strength and toughness.
© Valenite Inc.

Figure 7.33

Recommended Starting Grades for Types of Workpiece Materials

VALENITE STARTING GRADES		STEELS		CAST IRONS		STAINLESS STL.		NON-FERROUS MATERIALS	EXOTIC MATERIALS
		<35RC	>35RC	GREY	ALLOY	200/300	400/500		
Turning & Boring	Roughing	VC610● VC671♦	---- ----	VC610● VC671♦	VC610● VC671♦	VC610● VC671♦	VC610● VC671♦	VC610● VC671♦	VC610● VC671♦
	Finishing	VC605● VC610♦	VC505● VC610♦	VC605● VC610♦	VC605● VC610♦	VC610● ----	VC610● ----	VC605● VC610♦	VC610● ----
Milling		VC630● VC610♦	---- ----	VC630● VC610♦	VC630● VC610♦	VC630● VC610♦	VC630● VC610♦	VC610● VC630♦	VC610● VC610♦
Grooving & Cut Off		SEE SPECTRA, CVD & PVD GRADES							
Threading									

Note: ● Primary choice has most wear resistance with medium impact strength or best general purpose grade. ♦ Alternate choice for more severe applications that require higher edge strength and toughness.
© Valenite Inc.

Figure 7.34

Recommended Starting Grades for Types of Workpiece Materials

VALENITE STARTING GRADES		STEELS		CAST IRONS		STAINLESS STL.		NON-FERROUS MATERIALS	EXOTIC MATERIALS
		<35RC	>35RC	GREY	ALLOY	200/300	400/500		
Turning & Boring	Roughing	VC610● VC671◆	------ ------	VC610● VC671◆	VC610● VC671◆	VC610● VC671◆	VC610● VC671◆	VC610● VC671◆	VC610● VC671◆
	Finishing	VC605● VC610◆	VC605● VC610◆	VC605● VC610◆	VC605● VC610◆	VC610● ------	VC610● ------	VC605● VC610◆	VC610● ------
Milling		VC630● VC610◆	------ ------	VC630● VC610◆	VC630● VC610◆	VC630● VC610◆	VC630● VC610◆	VC610● VC630◆	VC610● VC610◆
Grooving & Cut Off		SEE SPECTRA, CVD & PVD GRADES							
Threading									

Note: ● Primary choice has most wear resistance with medium impact strength or best general purpose grade. ◆ Alternate choice for more severe applications that require higher edge strength and toughness.
© Valenite Inc.

Figure 7.35

Recommended Starting Grades for Types of Workpiece Materials

VALENITE STARTING GRADES		STEELS		CAST IRONS		STAINLESS STL.		NON-FERROUS MATERIALS	EXOTIC MATERIALS
		<35RC	>35RC	GREY	ALLOY	200/300	400/500		
Turning & Boring	Roughing	VC722● VC721♦	----	VC734● VC722♦	VC734● VC722♦	VC772● VC734♦	VC772● VC734♦	VC773● VC727♦	VC772● VC101♦
	Finishing	---- VC722♦	----	VC734● VC722♦	VC734● VC722♦	VC722● ----	VC722● ----	VC743● VC743♦	VC722● VC722♦
Milling		VC722● ----	----	VC734● VC722♦	VC734● VC722♦	VC722● ----	VC722● ----	VC723●	VC722●
Grooving & Cut Off				SEE SPECTRA/PVD/CERMET GRADES					
Threading		VC722●	----	VC722●	VC722●	VC722●	VC722●	VC722●	VC722●

Note: ● Primary choice has most wear resistance with medium impact strength or best general purpose grade. ♦ Alternate choice for more severe applications that require higher edge strength and toughness.
© Valenite Inc.

Figure 7.36

ROUGHING CHIPBREAKERS

TYPE	STYLE	D.O.C. In. (mm)	FEED RANGE ipr (mm/rev)	DESCRIPTION
3C		.050 to .300 (1,25-7,50)	.012 - .036 (0,30 - 0,90)	3C chipbreakers provide a neutral rake angle with land for high edge strength in medium duty applications on a wide range of materials. Recommended for interrupted cuts and heavy operations.
DC		.050 to .400 (1,25 - 10,00)	.010 - .043 (0,25 - 1,10)	DC chipbreakers provide for extremely heavy duty applications using a neutral rake with land for greater edge strength. Recommended for machining of steels and other materials
VI		.050 to .400 (1,25 - 10,00)	.010 - .043 (0,25 - 1,10)	VI chipbreakers provide for extremely heavy duty applications using a neutral rake with land for greater strength. Recommended for machining of steels and other materials.
GR		.080 to .400 (2,00 - 10,00)	.014 - .038 (0,35 - 0,95)	GR chipbreakers have a variable land and double step primary/secondary chipbreaker island to provide strength in heavy cuts. Recommended for interrupted cuts and heavy operations.
HS		.125 to .500 (3,12 - 12,50)	.020 - .055 (0,50 - 1,40)	HS chipbreaker provides positive cutting action for reduced cutting force and significantly less chatter in this application range. Recommended for high-feed machining on all types of metals.

© Valenite Inc.

Figure 7.37

GENERAL PURPOSE CHIPBREAKERS

TYPE	STYLE	D.O.C. In. (mm)	FEED RANGE ipr (mm/rev)	DESCFIPTION
2B		.005 to .025 (0,12 - 0,62)	.005 - .025 (0,12 - 0,62)	2B chipbreakers provide a positive shear angle geometry for light to medium duty machining conditions in all types of steels. Recommended for general purpose use.
E		.050 to .200 (1,25 - 5,00)	.006 - .022 (0,15 - 0,55)	E chipbreakers provide a low, positive rake angle with land for high edge strength in medium duty applications on a wide range of materials. Recommended for general purpose use on all types of steel.
ER/EL		.050 to .200 (1,25 - 5,00)	.007 - .022 (0,17 - 0,55)	The POS/NEG chipbreakers provide excellent chip control in the medium duty application ranges using a high positive shear angle. Recommended for machining of steels and other materials.
LM		.025 to .20C (0,62 - 5,00)	.007 - .022 (0,17 - 0,55)	LM chipbreakers provide excellent chip control with low cutting forces and very freecutting action over a broad range of light duty applications. Recommended for light-duty use on carbon, alloy and stainless steels.
GM		.050 to .250 (1,25 - 6,25)	.010 - .024 (0,25 - 0.60)	GM chipbreakers feature a neutral land with shear secondary angle for a near optimum combination of roughness and edge strength. Recommended for general purpose use on all types of metal.
2M		.005 to .075 (0,12 - 1,87)	.003 - .025 (0,07 - 0,62)	2M chipbreakers are specifically designed to control chips using a combination of high positive radial and axial rake angles. Recommended for low carbon steels, and high ductility materials.
2N		.076 to .635 (1,9 - 5,87)	.010 - .025 (0,25 - C,62)	2N is designed to provide superior chip control on a round geometry for bi-directional contouring operations. The 2N design allows for 360 degree indexing with no change in performance on medium to high feed applications.
1W		.010 to .025 (0,25 - 0,62)	.006 - .022 (0,15 - 0,55)	1W is an excellent choice for thin-walled fabricated parts and shaft turning operations where high cutting pressures can cause chatter. For work-hardened materials.

© Valenite Inc.

Figure 7.38

FINISHING CHIPBREAKERS

TYPE	STYLE	D.O.C. In. (mm)	FEED RANGE ipr (mm/rev)	DESCRIPTION
LF		.005 to .080 (0,12 - 2,00)	.002 - .012 (0,05 - 0,30)	LF chipbreakers are engineered for light finishing operations at high speeds in the 0.002 to 0.012 ipr (0,05mm to 0,3mm) feed range at depths of cut between 0.005 & 0.080" (0,12mm to 2,03mm)
GF		.010 to .125 (0,25 - 3,12)	.006 - .016 (0,15 - 0,40)	GF chipbreakers provide low cutting forces with good chip control and excellent surface finishes at low feed rates and depths of cut. Recommended for light feed finishing and profiling applications.
FM		.010 to .090 (0,25 - 2,25)	.006 - .013 (0,15 - 0,32)	Light duty Pos/neg inserts provide excellent chip control in light feed ranges using high positive shear angles. Recommended for machining of steels and other materials.
6G		.002 to .004 (0,05 - 1,00)	.002 - .012 (0,05 - 0,30)	6G chipbreakers provide excellent chip control in light finishing operations. Recommended for finishing steels and cast-iron.
6K		.012 to .118 (0,30 - 2,95)	.002 - .012 (0,05 - 0,30)	6K chipbreakers combined with the performance of cermets provide for efficient chip control in finishing and light machining operations. Recommended for finishing steels & cast-iron.
7K		.010 to .080 (0,25 - 2,00)	.002 - .012 (0,05 - 0,30)	7K chipbreakers provide positive cutting action in light finishing operations at moderate to high speeds. Recommended for finishing steels and cast iron.
2T		.010 to .150 (0,25 - 3,57)	.003 - .017 (0,07 - 0,42)	2T chipbreakers provide a combination of medium and high positive rake angles for optimum chip control in light duty applications. Recommended for Inconel, stainless steel, copper and aluminum.

© Valenite Inc.

Figure 7.39

POSITIVE CHIPBREAKERS

TYPE	STYLE	D.O.C. In. (mm)	FEED RANGE ipr (mm/rev)	DESCRIPTION
1A	 ◉	.010 to .150 (0,25 - 3,75)	.005 - .013 (0,12 - 0,32)	1A chipbreakers provide positive shear angle cutting action for reduced cutting forces in light duty operations. Recommended for low carbon steels, non-ferrous and high-temperature alloys.
2A	 ◉	.005 to .060 (0,12 - 1,50)	.0025 - .011 (0,06 - 0,27)	2A chipbreakers provide positive shear angle cutting action for reduced cutting forces in light duty operations. Recommended for low carbon steels, non-ferrous and high-temperature alloys.

© Valenite Inc.

Figure 7.40

Primary Selection by Workpiece Material

AREA		STEELS	IRONS	ALUMINUM	ALUMINUM	NON-FERROUS	STAINLESS STEELS	HIGH-TEMP. ALLOYS
2	INSERT MATERIAL	CVD COATED INSERTS ————		UNCOATED INSERTS ———— CVD COATED INSERTS ———— UNCOATED INSERTS	PVD COATED INSERTS	DIAMOND INSERTS		
		CERMET INSERTS ————					CERAMIC INS.	CERAMIC INS.
		CERAMIC INSERTS ————					CERMET INS.	CERMET INS.
1	WORK-PIECE MATERIAL	400 series s/s Carbon Steels Alloy Steels Tool Steels	Grey Ductile/ Nodular	High Silicon (Automotive)	Low Silicon (Aircraft)	Copper Brass Other	300 Series PH Series	Nickel Base Iron Base Cobalt Base Titanium- Alloys
3	VALCOOL CUTTING FLUID	VNT-800 ———— VNT-900 ———— VNT-910 ———— VNT-920 ————		TURN TECH &VNT-800 ————		AEROTECH ————	AEROTECH ———— VNT-900 ———— VNT-910 ————	

© Valenite Inc.

Figure 7.41

Secondary Cutting Fluid Selection by Other Criteria

AREA	CRITERIA		PRODUCT TYPE					CONDITIONS	
			SYNTHETIC	SEMI - SYNTHETIC	HYBRID	SOLUBLE - OIL	STRAIGHT OIL	HARD WATER	CHLORINE FREE
4	VALCOOL CUTTING FLUID		VNT-900	AEROTECH	TURN TECH		(VNT-LT)	VNT-700	AEROTECH
			VNT-910		VNT-800			VNT-800	VNT-700
			VNT-920					VNT-900	VNT-900
			(VNT-LT)					VNT-910	VNT-910
								VNT-920	VNT-920

© Valenite Inc.

Figure 7.42

Valenite Standard Grade Designations

Material Group	Material Type		Available Grades
Carbide	Uncoated		VC2, VC3, VC5, VC6, VC7, VC8, VC27, VC28, VC29, VC35M, VC101, VC111, VC121, VC135
	CVD Coated		V1N, V88, VN5, VN8, VO1, VX8
	SpectraSystem PVD Coated (900 Series)		VC901, VC902, VC903, VC905, VC907, VC927, VC928, VC929, VC935
	SpectraSystem Coated (200 & 300 Series)		SV210, SV221, SV230, SV231, SV235, SV310, SV315, SV325
	Cermet (600 Series)		VC605, VC610, VC630, VC671, VC673
High Performance Materials	Ceramic	Silicon Nitride	Q6
		Aluminum Oxide	Q32
	Polycrystaline (700 Series)	PCBN	VC721, VC722, VC734
		PCD	VC706, VC725, VC726, VC727, VC728, VC746

© Valenite Inc.

Figure 7.43

VALCOOL® EASY SELECTION CHART

By Material Being Cut

By Operation

STEELS	IRONS	ALUMINUM	ALUMINUM	NON-FERROUS	STAINLESS STEEL	HIGH-TEMP ALLOY	GRINDING	MACHINING
400 SERIES STAINLESS	GREY	HIGH SILICON (AUTOMOTIVE)	LOW SILICON (AIRCRAFT)	COPPER	300 SERIES	NICKEL BASE	VNT-920 VNT-800 TURN TECH VNT-700 VNT-910	VNT-800 TURN TECH VNT-700 AEROTECH VNT-910 VNT-900 VNT-920
CARBON STEELS	DUCTILE NODULAR			BRASS	PH SERIES	IRON BASE		
ALLOY STEELS				OTHER		COBALT BASE		
TOOL STEELS						TITANIUM BASE		
VNT-800 TURN TECH VNT-910 VNT-700	VNT-800 TURN TECH VNT-920 VNT-700	VNT-800 TURN TECH AEROTECH VNT-700	VNT-800 TURN TECH AEROTECH VNT-700	VNT-800 TURN TECH AEROTECH VNT-700	VNT-800 TURN TECH VNT-910 VNT-900	VNT-800 TURN TECH VNT-910 VNT-900		

© Valenite Inc.

Figure 7.44

TURNING GRADE COMPARISON CHART

VALCOOL (■)	ISO/ANSI CLASS	VALENITE COATED	VALENITE UNCTD.	GREENLEAF COATED	GREENLEAF UNCTD.	ISCAR COATED	ISCAR UNCTD.	KENNAMETAL COATED	KENNAMETAL UNCTD.	MITSUBISHI COATED	MITSUBISHI UNCTD.	NEWCOMER COATED	NEWCOMER UNCTD.
TURN TECH VNT-800	P50 P-40 C5	SV326 SV235 V1N	VC111 VC56	T15 T16 GA56	G01 G50	IC635 IC656	IC54	KC850 KC9045 KC9040 KC935	K2S				N52 N55
VNT-900	P40 P30 C5-C6	SV325 SV315 SV225	VC27 VC5	GA6 GA60	G53 G54 G52	IC825	IC50 IC70	KC935 KC9025	K2S	U6025 U625	Uti20T	NP1000	N50 N60
	P30 P20 C6-C7	SV315 SV310	VC7	G1 +	G60 G74 G70	IC805 IC815	IC78 IC60T	KC9010	K45	U6010 U610		NN55 NA02 NN6	N70 N72
VNT-910	P20 P10 C7-C8	SV310 V01	VC7			IC848		KC9010	K45	U505 U66	ST110T		N80 N93 N95
	P10 P01 C8	V01 SV310	VC8				IC80T	KC9010		U505 U66			N10 N52 N55
TURN TECH VNT-800	M40 M30	V1N VC901 SV235	VC111 VC101	GA56 GA60	G50	IC635		KC9045 KC9040 KC250	K2S	U625			N10 N52 N55
VNT-900	M30 M20	SV325 SV225 VC901 SV915	VC101 VC29	T15 GA60 GA56	G10 G52 G60	IC825		KC850 KC935 KC730 KC9025	K21 K313 K68	U6025 US735 U625 UP35N	UTI20T	NP1000	N22 N25 N30
VNT-910	M15 M10	SV315 VC929 SV310	VC29 VC2	G1 + T16 GA6	G02 G74	IC825		KC9010 KC910 KC730				NT25	N20 N60
	K40 K30 C1	V1N SV325	VC111 VC101		G01M	IC635	IC28	KC9040 KC250 KC850	K1				N10
VNT-800	K30 K20 C1-C2	SV325 SV315 VC928	VC28 VC2 VC29	GA56	G10 G02M	IC848 IC805		KC250 KC9025 KC720	K6 K40 K313	U6025 UP20M U625			N22 N20
	K20 K10 C2-C3	SV315 SV310 VC929	VC29 VC3	G1 + GA2 TI2 TI6	G02 G30 G23	IC848 IC805	IC20 IC2 IC28	KC9010 KC990 KC730	K68 K6	U6010 U610	Uti20T HTi10T	NT25 NT2	N30 N20
VNT-920	K05 K01 C4	SV310 V01	VC3 VC29	G2	G40 G25		IC4	KC9010 KC910		U66	HTi05T		N40 N20

K = Grey cast-iron; M = Stainless steel, Nodular iron; P = Steels. (■) = Use VNT-800 in harder water. © Valenite Inc.

TURNING GRADE COMPARISON CHART (Continued)

ISO/ANSI CLASS	VALENITE COATED	VALENITE UNCTD.	RTW COATED	RTW UNCTD.	SANDVIK COATED	SANDVIK UNCTD.	SUMITOMO COATED	SUMITOMO UNCTD.	TELEDYNE COATED	TELEDYNE UNCTD.	TOSHIBA COATED	TOSHIBA UNCTD.	V. R. WESSON COATED	V. R. WESSON UNCTD.
P60-P40 C5	SV325 SV235 V1N	VC111 VC55	955 755	CY55	GC2035 GC235 GC435	S6		ET40E	TC41	T04	T370 T823	TX4■	643 623 650 653	VR79 VR77
P40-P30 C5-C6	SV325 SV315 SV225	VC27 VC5	925 716	CY16 CY17	GC4025 GC1025 GC425	S4	AC25 AC2030	ET30E	MP51	T14 T12	T822 T813 T803	UX30 TX25	630 663	VR75 VR73
P30-P20 C6-C7	SV315 SV310	VC7	918 715 714		GC4015 GC415	S4	AC2000 AC10 AC15	ET20E	MP37 MP26	T25 SD5 T24	T801 T802 T823	TX20 TX10	680	
P20-P10 C7-C8	SV310 V01	VC7		CY31 CY14	GC3015 GC4015	S1P	AC5 AC10	ET10P	MP15	T25	T822		690	
P10-P01 C8	V01 SV310	VC8			GC301E		AC05 AC10	ET10P	MP15	SD3	T5010		690	
M40-M30	V1N VC901 SV235	VC111 VC101			GC235 GC435 GC4035		AC720	G10E					550 653	RAMET1 VR82 VR73
M30-M20	SV325 SV225 VC901 SV315	VC101 VC29	925	CO12	GC4025 G2425	S6 H13A	AC720 AC25				T802 T370 T812 T803 T823		630 653 650 663	RAMET1 VR82 VR73
M15-M10	SV315 VC929 SV310	VC29 VC2	918 718		GC4015 GC4025 GC425 GC415	H10A H13A	AC10 AC2000 AC15	G10E	MP37		T802TB 03 T812 T813		630 680 633	VR52 VR65
K40-K30 C1	V1N SV325	VC111 VC101		CO12	GC4035 GC435	H10A H13A					T370	C3		RAMET1
K30-K20 C1-C2	SV325 SV315 VC928	VC28 VC2 VC29	925	CO2	GC4035 GC4025 GC435	H10A H13A	AC10G AC25 AC2000		MP21	H17 H91	T813	C2 C2F	633 660	
K20-K10 C2-C3	SV315 SV310 VC929	VC29 VC3	918	CO4 CO3	GC4015 GC3015 GC415	H10A H1P H13A H20	AC105 AC10G	G10E H10E	HN+ C46	HTA H21 HA H56		T-10 C1F	630 680 690	2A5 VR82 2A7
K05-K01 C4	SV310 V01	VC3 VC29			GC3015	H1P	AC05 AC105		HN+	SD3 HF	T5010	T■03		VR52 VR65

© Valenite Inc.

Figure 7.45 Insert cutting tool comparison chart.

TURNING CHIPBREAKER COMPARISON CHART

C'BRKR. TYPE	VALENITE	GREENLEAF	ISCAR	KENNAMT'L	MITSUBISHI	NEWCOMER	RTW	SANDVIK	SUMITOMO	TELEDYNE	TOSHIBA	VR. WESSON
LIGHT FINISHING	LF		PP	MG-UF	F, PK		FF	QF	EFP, ENK EFK		O1, 11	
GENERAL FINISHING	GF	FF	NF	MP-K MG-K	SA, SH, C		FF	PF, MF -23	EUP, ENS EFM, EUU	2B, 2M, 3F	27, SS	
LIGHT MACHINING	LM	GP	NP	MS GG-P MG-P	MA	S	GP	PM, -61	ENG, ENJ EFJ, ENZ	3G	32, 32X 28, SA	MF, MS
GENERAL MACHINING	GM	MR	TNM, NG	MG-P MG-MG	MS, D	MN	GA	PM, OM	EMU, EMX	3S	33, 37 38, 21	W
GENERAL ROUGHING	GR	HR		MG	GH	PG	GR	MR		4T, 4U 4W, SW, SX		MH
HIGH-FEED ROUGHING	HS	MM	NM	MM, MM-M MM-P MM-MR, MH	HA		HR	PR, QR, HR	EHP, ENP		57, 65 81, 82	
POSITIVE INSERT C'BRKRS.	1A, 2A FL FH		14, 19, 16 MR, GR	MR, MT, GM GT, GT-HP	MR, SQ, MQ	MR, MT	CF1	UF, UM, UR	EFK, ENK EFM, EMF	1K, 2C 2K, 3K	23, 24	MP, MM, GP
POS-NEG C'BRKRS.	FM, MM MP, 2B, 1W		MM, MS-12 MP-14	MP, GP MS, GS	MP, KMS 1G, PK	MZ, MX F, Y	MP, MX MZ, MS	MP	Q, T	MP-3D	P, S	

© Valenite Inc.

Figure 7.46 Insert cutting tool comparison chart.

Machining insert material types. The following describes currently available machining insert materials and their applications. This information, together with the machining charts and remedies for cutting or machining problems appearing in this section of the *Handbook,* will enable you to select the proper type of insert and insert material for specific machining applications and requirements. (Information courtesy of Valenite, Inc.)

Uncoated carbide. This tungsten carbide–based material with a cobalt binder, frequently alloyed with combinations of titanium and tantalum carbides, is available in all ANSI and International Standards Organization (ISO) grade groups for metal-cutting applications. These inserts cut all materials, including irons, steels, and superalloys; aluminum and brass alloys; and nonmetallics, including plastics, phenolics, and composites. Applications range from heavy roughing to finishing on almost all machining operations.

CVD-coated carbide. This insert material is available in a complete range of grades, with substrates engineered for maximum compatibility with coatings. Coatings include

- Titanium carbide (TiC)
- Titanium nitride (TiN)
- Composite ceramic (Al_2O_3)
- Multiple coatings

These grades are excellent for machining most materials, including cast irons, steels, stainless steels, and some superalloys—in turning, boring, and milling operations.

PVD-coated carbide. Inserts coated with a thin film (2 to 4 microns) of titanium nitride (TiN) by physical vapor deposition (PVD) are characterized by good wear resistance, low coefficient of friction, and resistance to edge buildup. PVD (TiN) coatings are applied at temperatures substantially below those used for chemical vapor deposition (CVD), the process used most commonly for coating carbide cutting tools. The result is less reaction with the substrate and thus better edge-strength retention. Low coating temperatures allow greater control over deposition rate and result in closer tolerances and more grain-size control.

Silicon nitride. This hot-pressed monolithic silicon nitride–based ceramic is designed to do one thing exceptionally well—cut cast

iron materials. It offers outstanding toughness and thermal resistance and handles turning, boring, and milling operations at speeds from 350 to 5000 sfm. Because Quantum 6 retains its hardness, toughness, and abrasion resistance at elevated temperatures, it can provide excellent tool life at these speeds.

PCBN. Polycrystalline cubic boron nitride (PCBN) inserts, available in both tipped and solid forms, are designed to machine ferrous workpiece materials in the 40- to 70-RC range. PCBN is significantly harder than cemented carbide or ceramic materials and is the only material that combines a high degree of toughness with exceptional hot-hardness. PCBN excels in turning of hardened steels, chilled and gray cast irons, and Ni-Hard materials.

PCD. Polycrystalline diamond (PCD) inserts provide premium performance when cutting highly abrasive, low-tensile-strength, nonferrous, nonmetallic, and aluminum composite work materials. The exceptionally hard, wear-resistant diamond cutting tip retains its sharpness even in demanding applications, running up to 7500 sfm and beyond. As a result, PCD inserts provide up to 100 times the tool life of carbide. Excellent for roughing and high-speed finishing on turning, boring, and milling operations.

SpectraSystem coated carbide. Spectra-coated carbide grades represent the latest development in advanced coating technology and offer unique substrates for turning and milling. This combination allows Spectra-coated inserts to provide increased productivity at higher cutting speeds with greater wear resistance over a broad range of applications and work materials. Applications range from heavy roughing to precision finishing on work materials that include all steels, stainless steels, many superalloys, and alloyed cast irons.

Cermet. This type includes composite titanium carbide– and titanium nitride–based materials designed for light roughing to high-speed finishing in turning, boring, and milling operations. These grades show excellent results on steels, stainless steels, cast irons, and aluminum. They are capable of running at moderate to high speeds and are best used with light feeds and depths of cut. They feature excellent abrasion resistance and thermal properties, as well as lubricity for crater resistance.

Ceramic. Quantum 32 (Q32) is a metal oxide composite with edge strength well suited for materials with hardness from 35 to 65 RC,

including some hardened steels, irons, and some superalloys. Quantum 32 is available in a wide range of shapes, sizes, and geometries.

Material characteristics and insert tool application considerations. The following listings provide extensive information relating to insert tool machining of various materials. Each item describes machining characteristics for the specific material, notes various conditions that may be encountered while machining it, and provides remedies for correcting the conditions or problems listed. See Fig. 7.75 for milling recommendations on different milling operations.

Low-carbon and free-machining steels.   1100 series, 1008, 1010, 1018, and 1200 series, leaded steels, and steels with free-machining additives.

Material characteristics: Low-carbon, soft and gummy; difficult chip control; torn finish is common; high-speed capability. Free machining, easy to machine; high-speed capability; high depths of cut achievable; easy chip control.

Tool-application considerations

Condition: Torn finish (built-up edge)

Remedy: Increase speed; use cermets; use PVD-coated grade with sharp edge; use positive-rake inserts; increase coolant concentration.

Condition: Difficult chip control

Remedy: Increase speed; change lead angle to redirect chip away from workpiece.

Condition: Burrs and sharp edges

Remedy: Change cutting path; use PVD-coated grade with sharp edge; use positive-rake or high-positive-rake inserts.

Medium/high-carbon steels and alloy steels.   1045, 1085, 1541, 1561, 1572, and 4000 series, 4300 series, 5100 series, 8600 series, 52100, and 300M.

Material characteristics: Higher chrome, nickel, and molybdenum content, work hardening; higher carbon content, abrasive but achieves surface finish easier; higher nickel content, more carbon and alloy, more difficult to machine.

Tool-application considerations

Condition: Crater wear

Remedy: Use aluminum oxide–coated grade; reduce speed; use more free-cutting chip control; use cermets; increase coolant pressure; reduce feed.

Condition: Flank wear

Remedy: Use TiC/alumina ceramics; use aluminum oxide–coated grade; increase feed; increase coolant concentration; increase depth of cut.

Condition: Thermal deformation

Remedy: Use more wear-resistant grade; reduce speed; reduce feed.

Tool steels

Material characteristics: Abrasive, work hardening, and tough; hard-to-break chips.

Tool-application considerations

Condition: Flank wear

Remedy: Use aluminum oxide–coated grade; reduce speed; increase feed.

Condition: Crater wear

Remedy: Use aluminum oxide–coated grade; use cermets; reduce speed; reduce feed.

Condition: Dull surface finish

Remedy: Increase speed; use positive-rake inserts (PVD-coated); use cermets; increase coolant concentration.

Gray cast iron.   Gray, classes 20, 25, 30, 45, and 60, high tensile.

Material characteristics: Abrasive, potential scale and inclusions, tendency to break out during exit from cut, potential for chatter on thin-wall sections.

Tool-application considerations

Condition: Excessive edge wear

Remedy: Use harder grade or aluminum oxide–coated grades; use ceramics; increase feed to reduce in-cut time.

Condition: Chipping

Remedy: Increase lead angle; use stronger grade; use prehoned inserts or inserts with edge preparation.

Condition: Workpiece breakout

Remedy: Use sharp inserts or PVD-coated grades; use lower feed rate during exit; increase speed; increase lead angle; use positive geometries for finish cuts.

Condition: Workpiece chatter

Remedy: Use sharp inserts or PVD-coated grades; use positive rake inserts; increase lead angle; increase feed to stabilize workpiece.

Ductile cast iron. Nodular/ductile, malleable.

Material characteristics: Abrasive, difficult to machine, good surface finishes difficult to achieve.

Tool-application considerations

Condition: Edge wear

Remedy: Use harder grades; apply aluminum oxide–coated grades; consider ceramics and cermets for finishing; increase feed to reduce in-cut time.

Condition: Crater wear

Remedy: Apply aluminum oxide–coated grades; apply geometries that are free-cutting; lower speed (sfpm).

Condition: Torn or dull surface finish

Remedy: Apply ceramics, cermets, or PVDs at higher speeds and lower feeds; increase lead angle; increase coolant concentration.

Hardened irons

Material characteristics: Highly abrasive, high cutting forces.

Tool-application considerations

Condition: Excessive edge wear

Remedy: Apply alumina-based ceramics or cubic boron nitride (CBN) grades; feed above 0.004 ipr; increase lead angles or use round inserts; check coolant application.

Condition: Catastrophic breakage

Remedy: Increase lead angles or use round inserts; use large nose radii; repair toolholder; reduce feeds.

Austenitic stainless steels. 200 and 300 series.

Material characteristics: Work hardens rapidly; small depths of cut are difficult; usually abrasive rather than hard; tough and stringy chips.

Tool-application considerations

Condition: Depth-of-cut notch

Remedy: Increase lead angle; use tougher grade; feed over 0.005 ipr.

Condition: Built-up edge

Remedy: Increase speed; use cermets; use positive-rake inserts (PVD-coated).

Condition: Workpiece glazing

Remedy: Increase depth of cut; reduce nose radius; use sharp PVD-coated grades.

Condition: Dull surface finish

Remedy: Increase speed; reduce feed; increase positive rake.

Ferritic/martensitic stainless steels. 400 series and precipitation-hardening PH grades.

Material characteristics: Brittle, stringy chips; high cutting forces; high work hardening (especially PH stainless grades).

Tool-application considerations

Condition: Built-up edge

Remedy: Increase speed; use cermets; use positive-rake inserts (PVD-coated); increase coolant concentration.

Condition: Workpiece glazing

Remedy: Increase depth of cut; reduce nose radius; use positive-rake inserts (PVD-coated); increase coolant concentration.

Condition: Depth-of-cut notch

Remedy: Increase lead angle; use tougher grade; feed over 0.005 ipr.

Condition: Torn or dull surface finish

Remedy: Increase speed; reduce feed; use positive-rake inserts (PVD-coated); increase coolant concentration.

High-temperature alloys.   Iron, nickel, and cobalt base.

Material characteristics: Relatively poor tool life; small depths of cut are difficult; work hardens rapidly; usually abrasive rather than hard; tough and stringy chips.

Tool-application considerations

Condition: Depth-of-cut notch

Remedy: Increase lead angle; use tougher grade; use 0.015-in or greater depth of cut; feed over 0.005 ipr; increase coolant concentration.

Condition: Built-up edge

Remedy: Increase speed; change cutting tool material; use positive-rake inserts (PVD-coated); do not overrun cutting edge; increase coolant concentration.

Condition: Workpiece glazing

Remedy: Increase depth of cut; increase feed rate; reduce nose radius; use positive-rake inserts (PVD-coated).

Condition: Torn or dull surface finish

Remedy: Increase speed; reduce feed; use positive-rake inserts (PVD-coated); increase coolant concentration.

Titanium alloys.   Pure, alpha, alpha-beta, and beta alloys.

Material characteristics: Relatively poor tool life; chips tend to gall and weld to cutting edge; requires lower cutting speeds than steels of equal hardness; usually abrasive; tough and stringy; low thermal conductivity.

Tool-application considerations

Condition: Depth-of-cut notch

Remedy: Increase lead angle; use 0.015-in or greater depth of cut; remove embrittled surface from mill stock; feed over 0.005 ipr; increase coolant concentration.

Condition: Built-up edge

Remedy: Keep a sharp edge; use positive-rake inserts (PVD-coated); increase coolant concentration.

Condition: Workpiece glazing

Remedy: Increase depth of cut; reduce nose radius; use positive-rake inserts (PVD-coated); index cutting edge.

Condition: Torn or dull surface finish

Remedy: Index cutting edge; reduce feed; use positive-rake inserts (PVD-coated); increase speed; increase coolant concentration.

High-silicon and free-machining aluminum

Free-machining aluminum: 1000, 1100, 1200, and 1300 series; 2011 through 2024 series; 3000 series. 1050 to 1350 are almost pure. 2011 to 2024: 2024 is popular (screw-machine products, aircraft parts); contains copper, manganese, and magnesium.

Material characteristics: Soft and gummy

Tool-application considerations

Condition: Built-up edge

Remedy: Increase speed; use positive-rake inserts (PVD-coated); use coolants designed for machining aluminum; J polish top of insert.

Condition: Torn or dull surface finish

Remedy: Increase speed; reduce feed; use positive-rake inserts (PVD-coated); use coolant designed for aluminum; use polycrystalline diamond tools; J polish top of insert.

High-silicon aluminum: 4000, 5000, 6000, and 7000 series (7000 series uses zinc, zirconium, and titanium as alloying additives). 4000 series: Contains silicon for heat resistance and wear. 5000 series: Contains chromium. 6000 series: 6061 typical, contains silicon, copper, magnesium, and chromium. 7000 series: Contains copper, magnesium, chromium, and zinc (aircraft parts).

Material characteristics: Abrasive and tough.

Tool-application considerations

Condition: Edge wear

Remedy: Use coated grade; use polycrystalline diamond tools; increase coolant concentration.

Condition: Built-up edge/surface finish

Remedy: Increase speed; increase coolant concentration.

Nonferrous alloys and nonmetallics

Nonferrous metals: Free-machining and high-strength copper alloys, brass, zinc, and magnesium

Material characteristics: Easily machined; high machining speeds; positive chip control with magnesium to reduce fire hazards.

Caution: Fire hazard is present when machining magnesium at high speeds.

Nonmetallics: Nylons, acrylics, phenolics, and resin materials.

Material characteristics: Easily machined; high machining speeds; machine with polycrystalline diamonds whenever possible; center height of cutting is important.

Tool-application considerations

Condition: Built-up edge

Remedy: Increase speed; use positive-rake inserts; J polish top of inserts.

Condition: Poor surface finish

Remedy: Increase speed; use polycrystalline diamond inserts; use positive-rake inserts; J polish top of inserts; use positive chip control.

7.1.9   Grades of cutting inserts (for use with Figs. 7.14 through 7.44)

The insert grades described in the charts are those of Valenite, Inc., with the figures also listing the ANSI and ISO equivalents as reference.

Selection of speed, feed, and depth of cut.   Use the preceding speed, feed, and depth-of-cut figures as a basis for these choices. Useful tool life is influenced most by cutting speed. The feed rate is the

next most influential factor in tool life, followed by the depth of cut (doc).

When the depth of cut exceeds approximately 10 times the feed rate, a further increase in depth of cut has little effect on tool life. In selecting the cutting conditions for a turning or boring operation, the first step is to select the depth of cut, followed by selection of the feed rate and then the cutting speed. Use the preceding horsepower/kilowatt equations to determine the approximate power requirements for a particular depth of cut, feed rate, and cutting speed to see if the machine can handle the power required. You also may use one of the cutting tool manufacturers' machining calculators to determine appropriate machining parameters. Many production and tooling engineers prefer to use the calculators provided by the manufacturers who supply their cutting tools, together with production trial runs on any particular part, in determining the final machining parameters.

Select the heaviest depth of cut and feed rate that the machine can sustain, considering its horsepower or kilowatt rating, in conjunction with the required surface finish desired on the workpiece.

Relation of speed to feed.   The following general rules apply to most turning and boring operations:

- If the tool shows a built-up edge, increase feed or increase speed.

- If the tool shows excessive cratering, reduce feed or reduce speed.

- If the tool shows excessive edge wear, increase feed or reduce speed.

Caution: The productivity settings from the machining calculators and any handbook speed and feed tables are suggestions and guides only. A safety hazard may exist if the user calculates or uses a table-selected machine setting without also considering the machine power and the condition, size, strength, and rigidity of the workpiece, machine, and cutting tools.

Turning, boring, and threading inserts.   Figure 7.47*a* shows a sample of some of the typical cutting inserts available today. The versatile 80° diamond inserts are shown in the center area of the figure (two inserts) to the left and right of the lower center triangle. Various types and styles of tool holders for the inserts are produced by the insert manufacturers to hold the disposable inserts rigidly in position during the cutting operation. Figure 7.47*b* shows samples of typical turning and boring toolholders, and Fig. 7.47*c* shows a modern

(a)

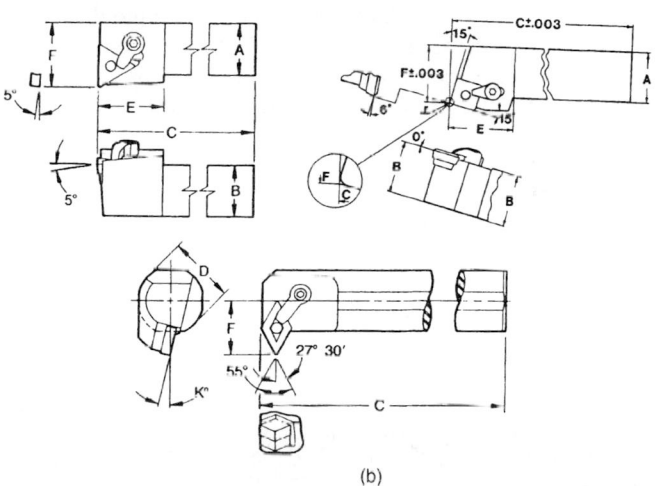

(b)

(c)

Figure 7.47 Cutting inserts (*a*); insert tool holders (*b*); a convertible insert and tool holder (*c*).

tool holder with a multipurpose, double-ended cutting insert. With the one insert, finishing, facing, and cutoff operations may be performed according to the orientation angle of the tool holder in relation to the workpiece.

7.1.10   ANSI and ISO identification systems for turning, boring, and milling inserts

Removable turning, boring, and milling inserts are identified and specified according to the American National Standards Institute (ANSI) and the International Standards Organization (ISO) systems shown in Figs. 7.48 and 7.49, respectively. Sample drawings of various types of inserts are shown in Fig. 7.50. The identification code is listed at the top right of each insert and may be referenced back to the ANSI identification system (see Fig. 7.48).

7.1.11   Thread turning

Thread-turning inserts are available in different styles or types for turning external and internal thread systems such as UN series, 60° metric, Whitworth (BSW), Acme, ISO, American buttress, etc. Figure 7.51 shows some of the typical thread-cutting inserts.

The defining dimensions and forms for various thread systems are shown in Fig. 7.52 with indications of their normal industrial uses. The dimensions in the figure are in U.S. Customary and metric systems as indicated. In all parts of the figure, P = pitch, reciprocal of threads per inch (for U.S. Customary) or millimeters (for metric).

Figure 7.52a defines the ISO thread system: M (metric) and UN (unified national). Typical uses: all branches of the mechanical industries. Figure 7.52b defines the UNJ thread system (controlled-root radii). Typical uses: aerospace industries. Figure 7.52c defines the Whitworth system (BSW). Typical uses: fittings and pipe couplings for water, sewer, and gas lines; presently replaced by ISO system. Figure 7.52d defines the American buttress system, 7° face. Typical uses: machine design. Figure 7.52e defines the American national pipe thread (NPT) system. Typical uses: pipe threads, fittings, and couplings. Figure 7.52f defines the British standard pipe thread (BSPT) system. Typical uses: pipe thread for water, gas, and steam lines. Figure 7.52g defines the Acme thread system, 29°. Typical uses: mechanical industries for motion-transmission screws.

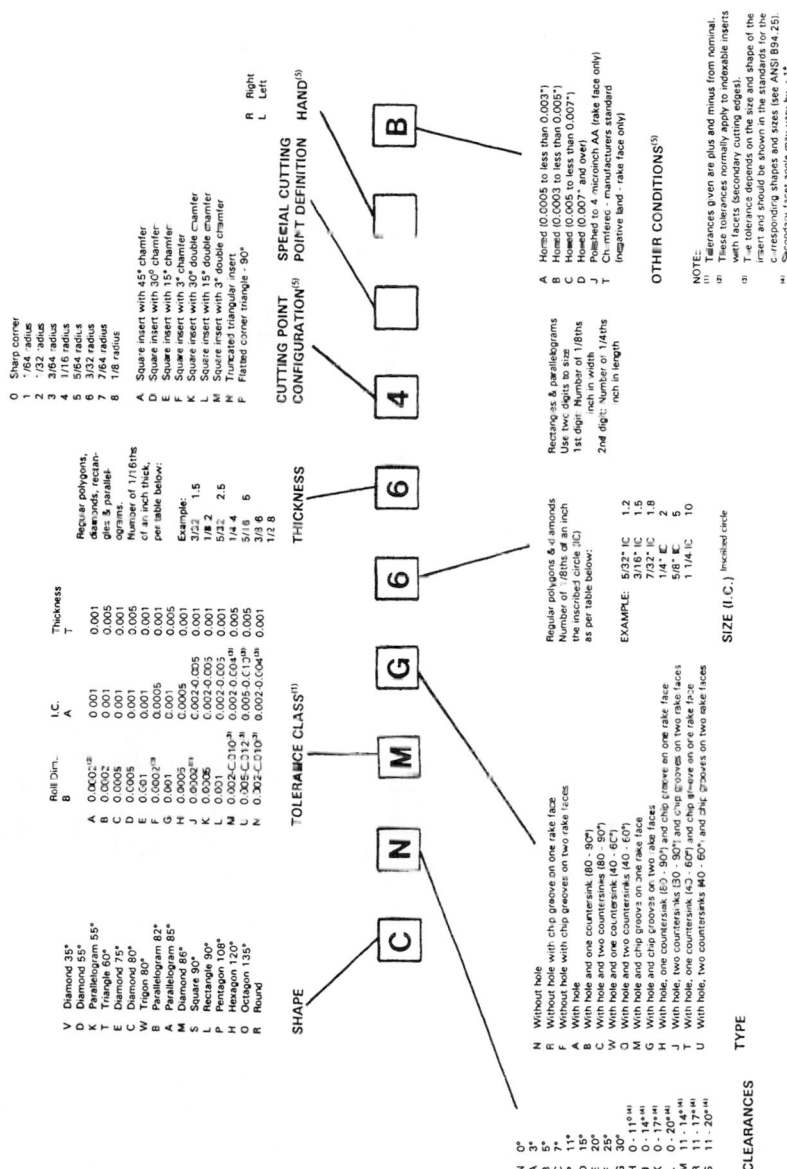

Figure 7.48 ANSI cutting insert identification system.

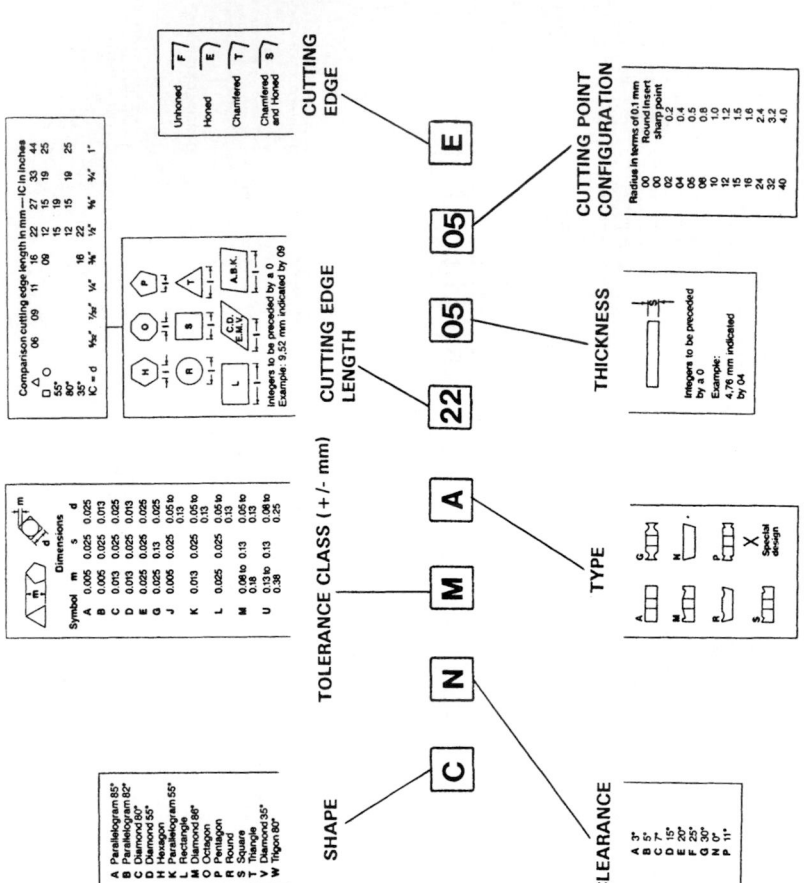

Figure 7.49 ISO cutting insert identification system.

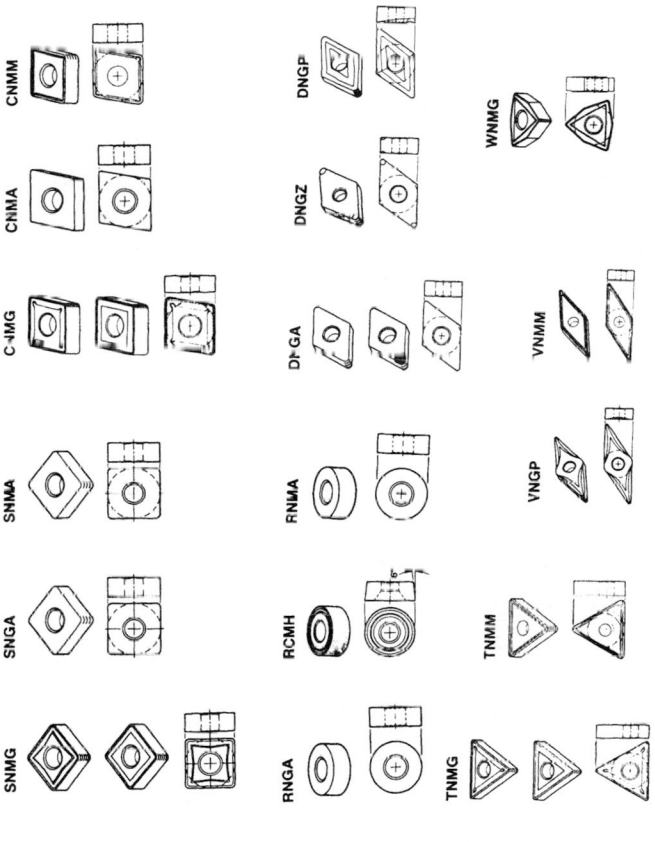

Figure 7.50 Sample inserts with identification codes.

397

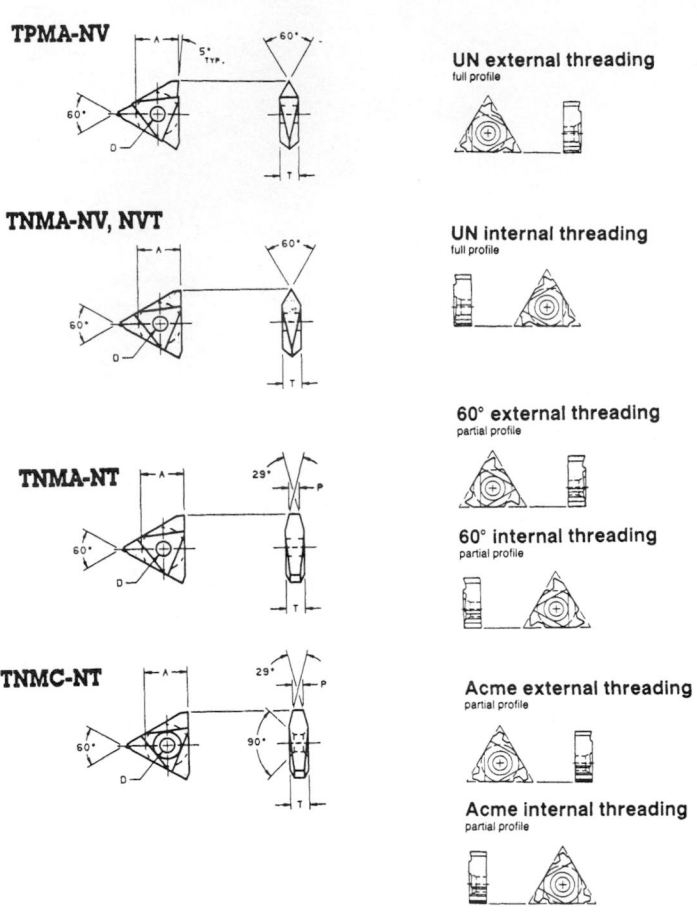

Figure 7.51   Sample inserts and threat-cutting inserts.

Figure 7.52*h* defines the stub Acme thread system, 29°. Typical uses: same as Acme, but used where normal Acme thread is too deep. Figure 7.52*i* defines the API 1:6 tapered-thread system. Typical uses: petroleum industries. Figure 7.52*j* defines the TR DIN 103 thread system. Typical uses: mechanical industries for motion-transmission screws. Figure 7.52*k* defines the RD DIN 405 (round) thread system. Typical uses: pipe couplings and fittings in the fire-protection and food industries.

(Text continued on page 409.)

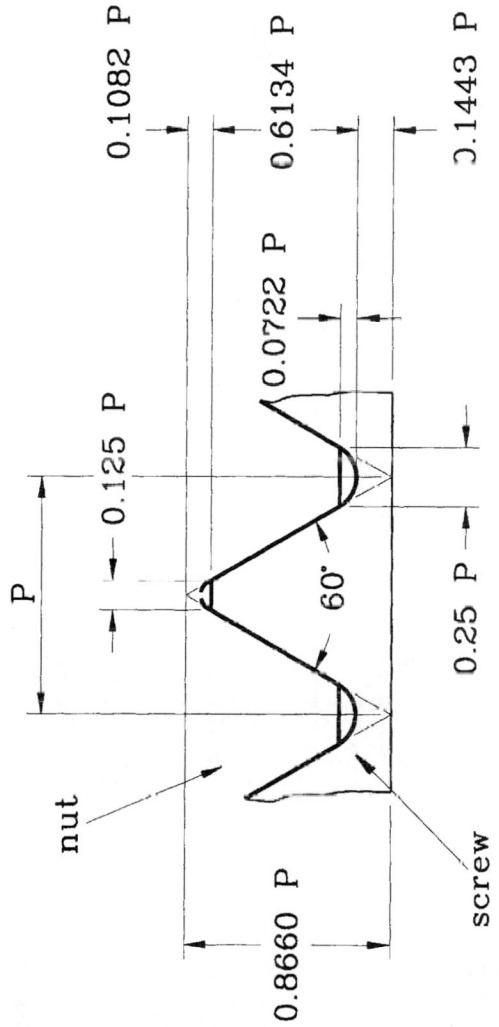

(a) ISO - M (Metric)
(UN) (Unified National)

Figure 7.52 Thread systems and dimensional geometry.

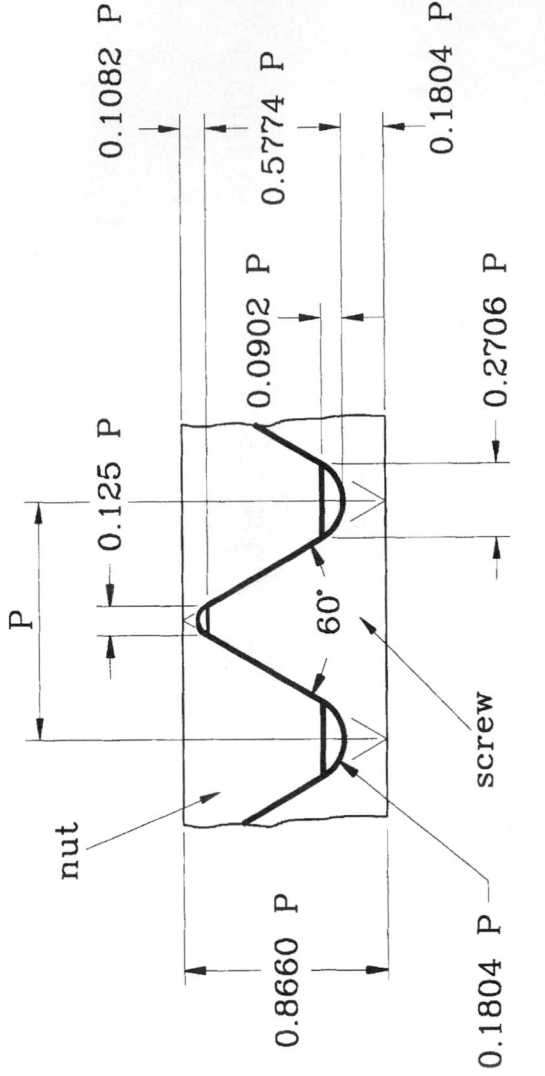

(b) UNJ
(Controlled root radii)

Figure 7.52 (Continued)

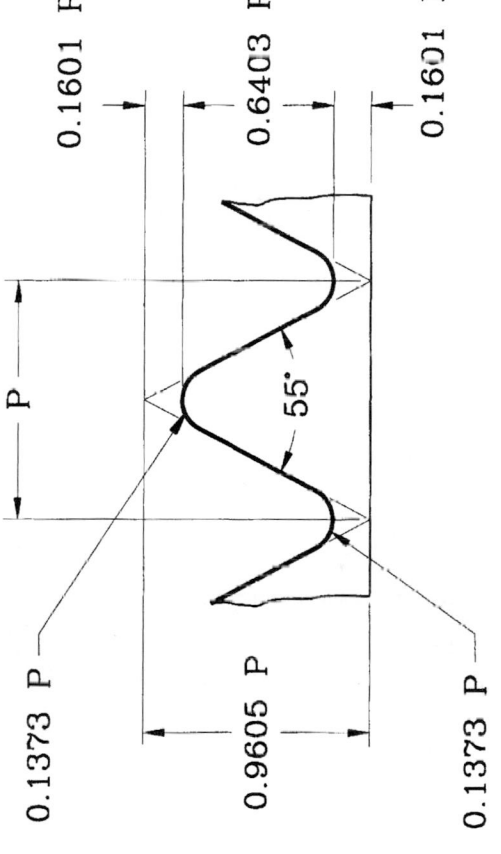

(c) Whitworth (BSW)

Figure 7.52 (*Continued*)

401

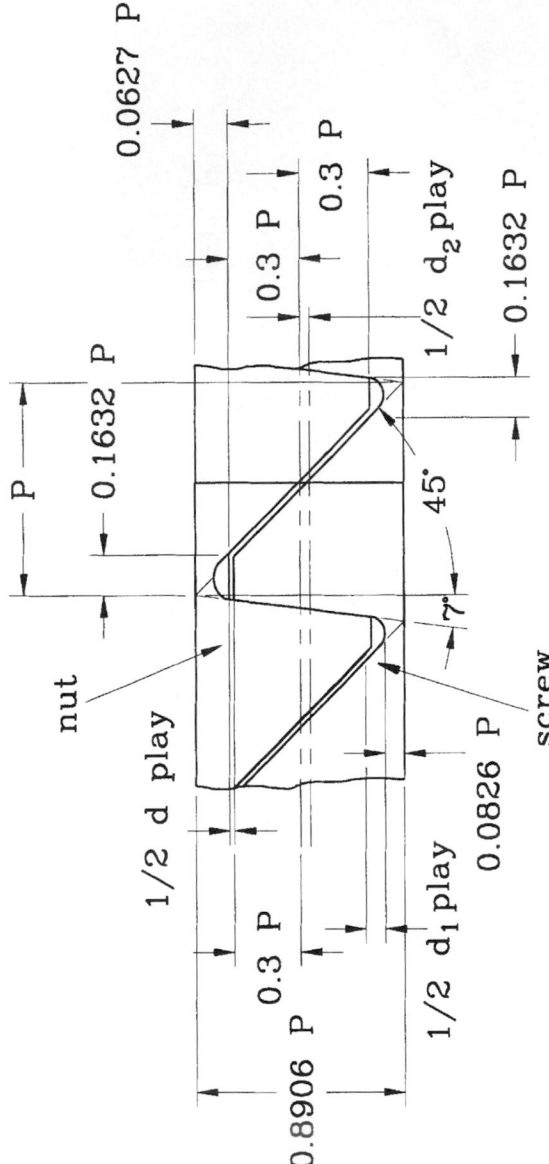

(d) American Buttress (7° face)

Figure 7.52 (*Continued*)

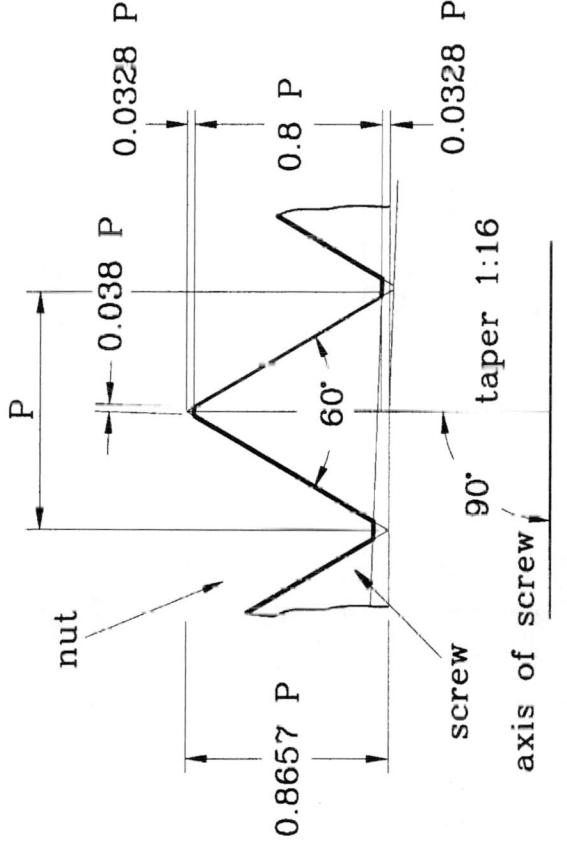

**(e) NPT
(American National Pipe Thread)**

Figure 7.52 *(Continued)*

403

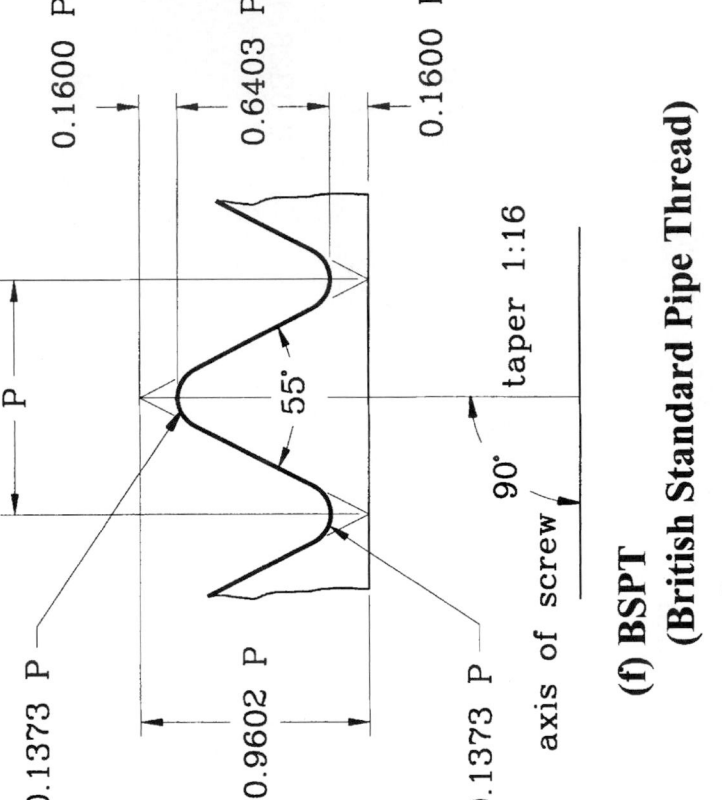

**(f) BSPT
(British Standard Pipe Thread)**

Figure 7.52 *(Continued)*

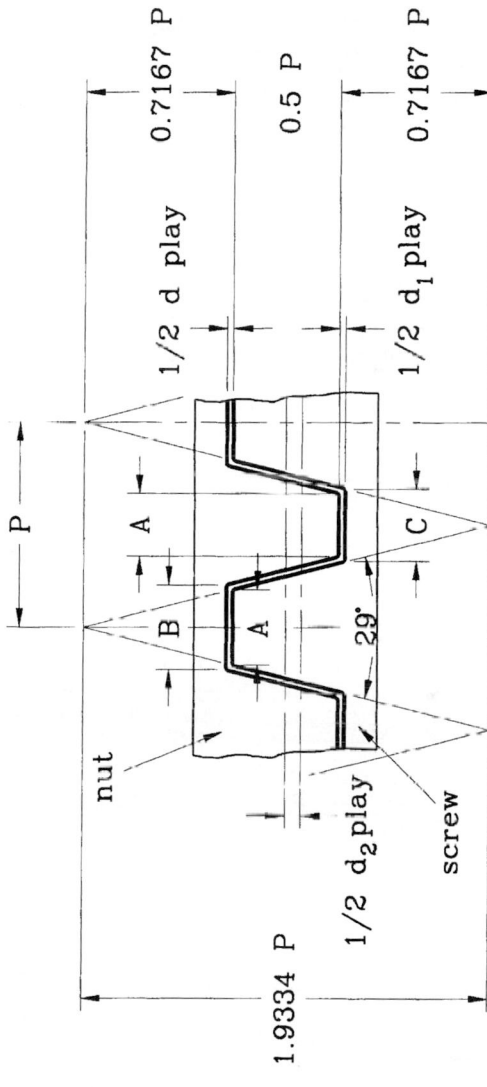

$A = 0.3707\ P$

$B = 0.3707\ P - 0.259 \times d\ \text{play}$

$C = 0.3707\ P - 0.259\ (d_1\ \text{play} - d_2\ \text{play})$

(g) Acme (29°)

Figure 7.52 (*Continued*)

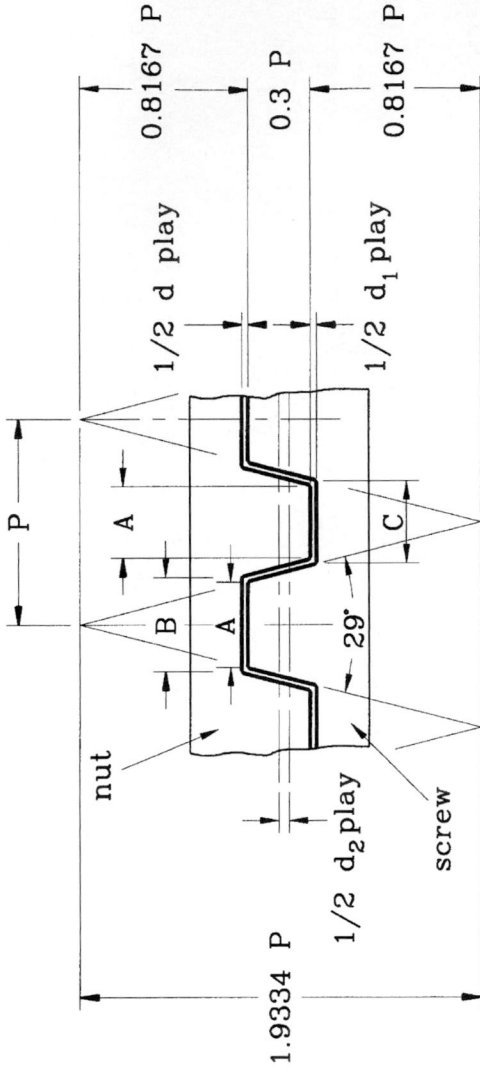

$A = 0.4224 \ P$

$B = 0.4224 \ P - 0.259 \times d$ play

$C = 0.4224 \ P - 0.259 \ (d_1 \text{ play} - d_2 \text{ play})$

(h) Acme Stub (29°)

Figure 7.52 *(Continued)*

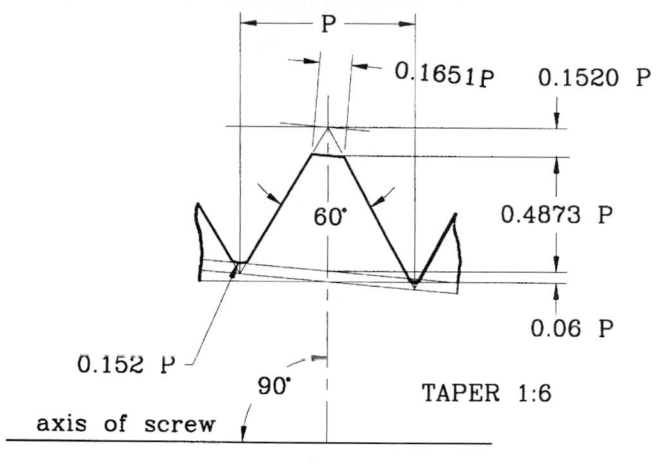

(i) API
Taper: 1:6 (V-0.38"R)

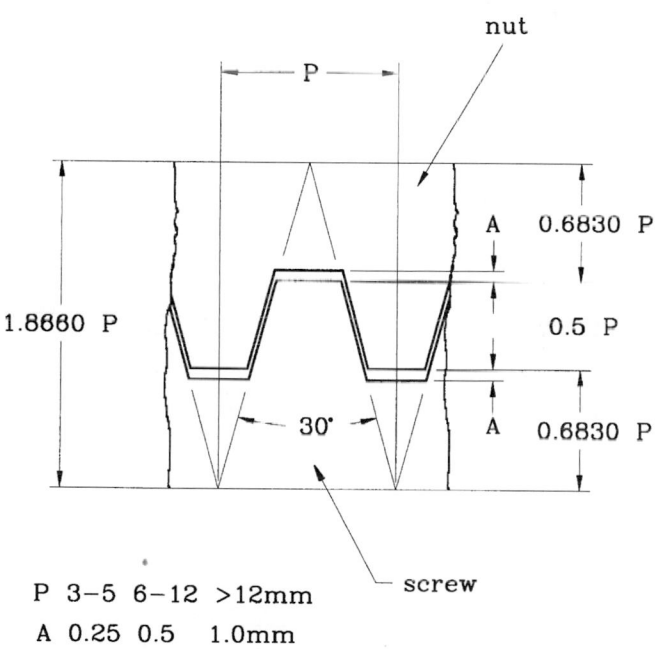

P 3–5 6–12 >12mm
A 0.25 0.5   1.0mm

(j) TR DIN 103

Figure 7.52 *(Continued)*

407

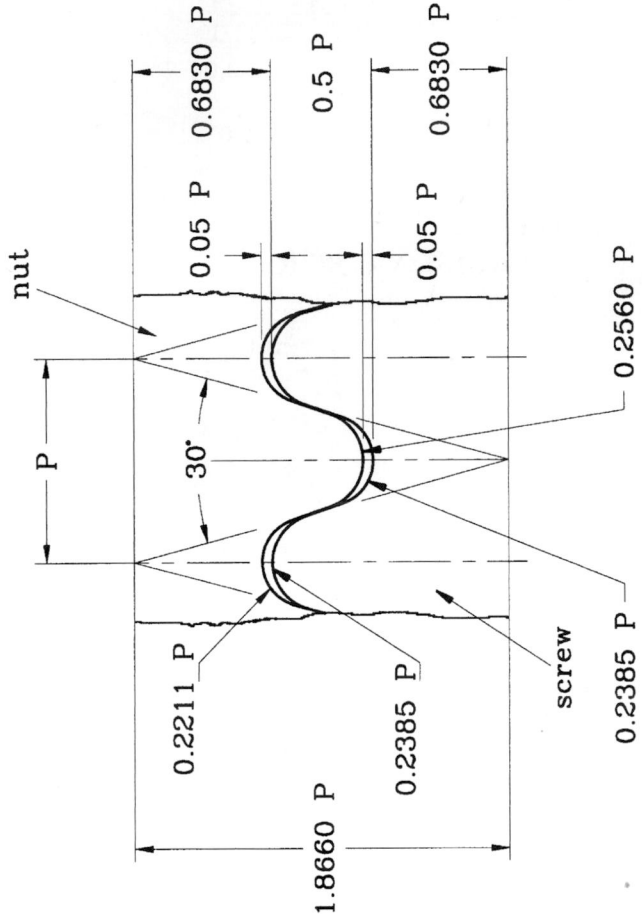

(k) RD DIN 405 (Round)

Figure 7.52 (*Continued*)

Threading operations. Prior to cutting (turning) any particular thread, the following should be determined:

- Machining toward the spindle (standard helix)
- Machining away from the spindle (reverse helix)
- Helix angle (see following equation)
- Insert and toolholder
- Insert grade
- Speed (sfpm)
- Number of thread passes
- Method of in-feed

Calculating the thread helix angle. To calculate the helix angle of a given thread system, use the following simple equation (see Fig. 7.53):

$$\tan \alpha = \frac{p}{\pi D_e}$$

where tan α = natural tangent of the helix angle, degrees
D_e = effective diameter of thread, in or mm
π = 3.1416
p = pitch of thread, in or mm

Example: Find the helix angle of a unified national coarse 0.375-16 thread.

$$p = 1/16 = 0.0625$$

(The pitch is the reciprocal of the number of threads per inch in the U.S. Customary system.)

$$D_e = 0.375 \text{ in}$$

Therefore,

$$\tan \alpha = \frac{0.0625}{3.1416 \times 0.375} = 0.05305$$

$$\arctan 0.05305 = 3.037°$$

The helix angle of any helical thread system can be found by using the preceding procedure.

Cutting procedures for external and internal threads. Figure 7.54 illustrates the methods for turning the external thread systems

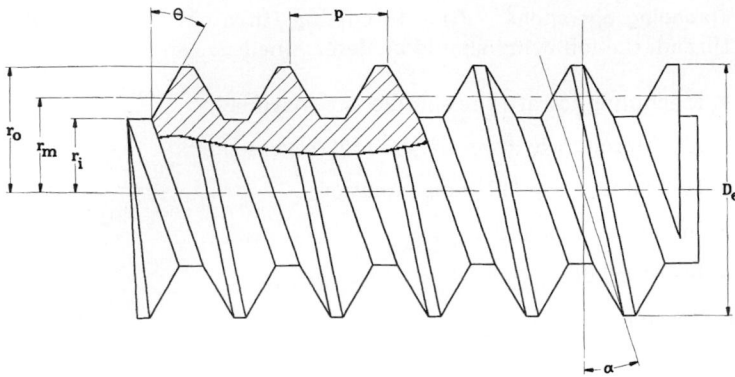

Figure 7.53 Calculating the helix angle.

(standard and reverse helix). Figure 7.55 illustrates the methods for turning the internal thread systems (standard and reverse helix).

Problems in thread cutting

Problem	Possible remedy
Burr on crest of thread	Increase surface feet per minute (rpm).
	Use positive rake.
	Use full-profile insert (NTC type).
Poor tool life	Increase surface feet per minute (rpm).
	Increase chip load.
	Use more wear-resistant tool.
	Increase surface feet per minute (rpm).
Built-up edge	Increase chip load.
	Use positive rake, sharp tool.
	Use coolant or increase concentration.
Torn threads on workpiece	Use neutral rake.
	Alter in-feed angle.
	Decrease chip load.
	Increase coolant concentration.
	Increase surface feet per minute (rpm).

In-feed methods for thread cutting

0° in-feed angle. When the in-feed angle is 0°, cutting occurs on both sides of the thread form, which places all the cutting edge in

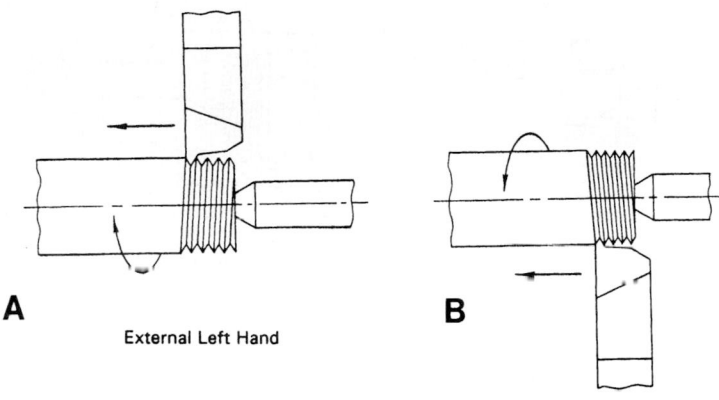

A

External Left Hand

B

External Right Hand

Feed Direction Towards Spindle (Standard Helix)

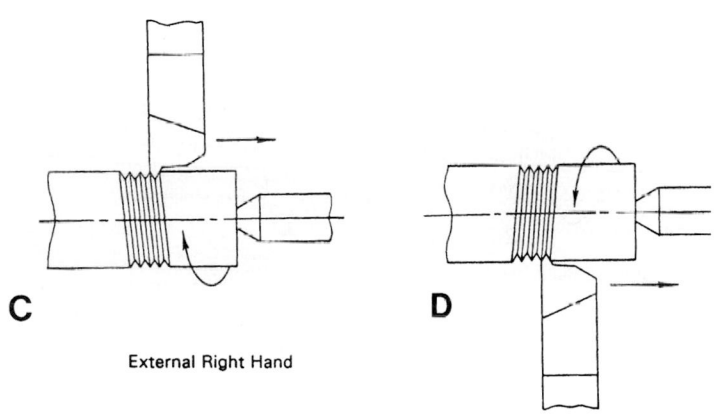

C

External Right Hand

D

External Left Hand

Feed Direction Away from Spindle (Reverse Helix)

Figure 7.54 Thread-cutting procedures, external.

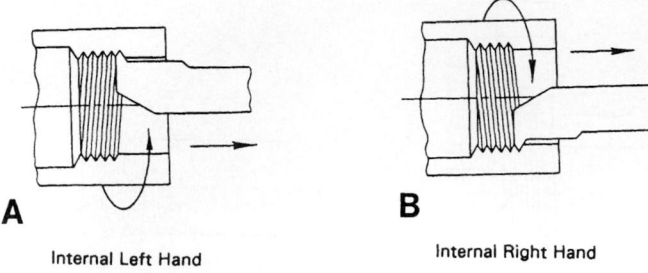

A

Internal Left Hand

B

Internal Right Hand

Feed Direction Away from Spindle (Reverse Helix)

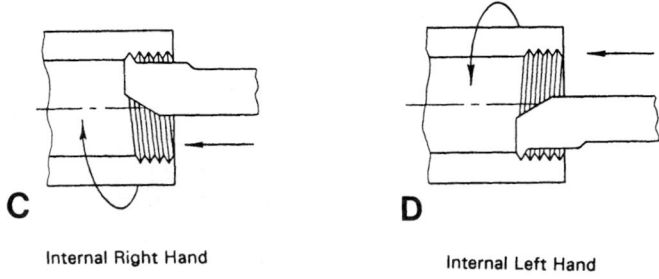

C

Internal Right Hand

D

Internal Left Hand

Feed Direction Towards Spindle (Standard Helix)

Figure 7.55 Thread-cutting procedures, internal.

the cut and helps protect the edge from chipping. *Disadvantages:* May produce difficult-to-handle chip; the tip of the tool may chip when cutting high-tensile-strength materials; burring is increased; and chatter problems may develop.

Large in-feed angle (15° to 30°). Cutting with the leading edge of the tool produces an easy-to-handle chip, and reduced burr problems occur on the trailing edge of the tool. *Disadvantages:* Trailing edge of tool may drag and tend to chip; torn or poor surface occurs when cutting soft, gummy materials such as low-carbon steels (1020, etc.) and low-alloy aluminums or pure aluminum.

Small in-feed angle (5° to 10°). Tool cuts from both sides and is protected from chipping; chip is easy to handle. *Disadvantages:* Similar to 0° in-feed but less in magnitude because cutting forces are equalized, and chip is easy to control.

Standard thread systems and standard hardware. See Chap. 17 for ANSI standards applicable to machining and metalworking. Also see Chap. 14 on fastening and joining techniques and hardware.

7.1.12 Modern turning centers and standard lathe machine tools

The modern direct numerical control (DNC) turning center has replaced many types of older machine tools such as automatic screw machines, small turret lathes, and profiling equipment. Figure 7.56 shows a typical automatic CNC turning center.

A sample of typical parts produced on this type of machine tool is shown in Fig. 7.57. The controllers of these machines can be seen in the photographs. The spindle and coolant line can be seen near the center of the machine shown in Fig. 7.56. The bar-stock material is fed through the spindle and clamped in position for cutting by a collet-type mechanism at the spindle.

Figure 7.58 shows a typical small turret lathe used to produce small parts when the production rates do not warrant the use of a DNC turning center. Figure 7.59 shows a typical geared-head engine lathe equipped with digital readout panel used for setting the controls and movements of the machine. The digital readout panel is shown in the upper left of the photograph and reads to four decimal places (±0.0005 in). Engine lathes of this type have been the major basic machines used in the metalworking industries for

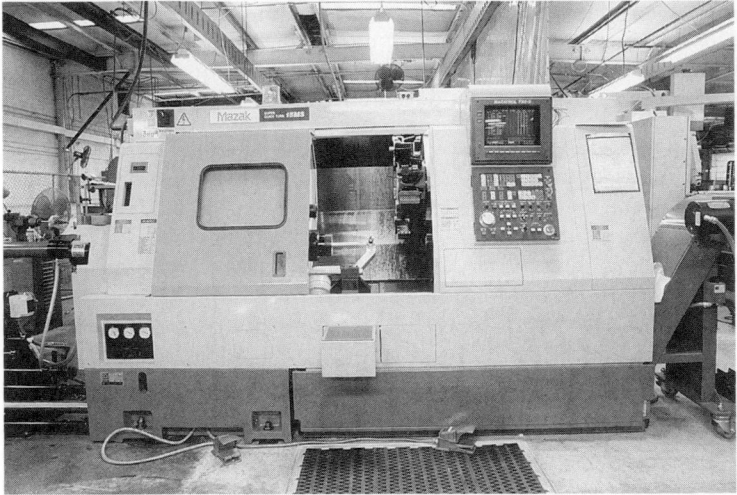

Figure 7.56   A modern CNC turning center.

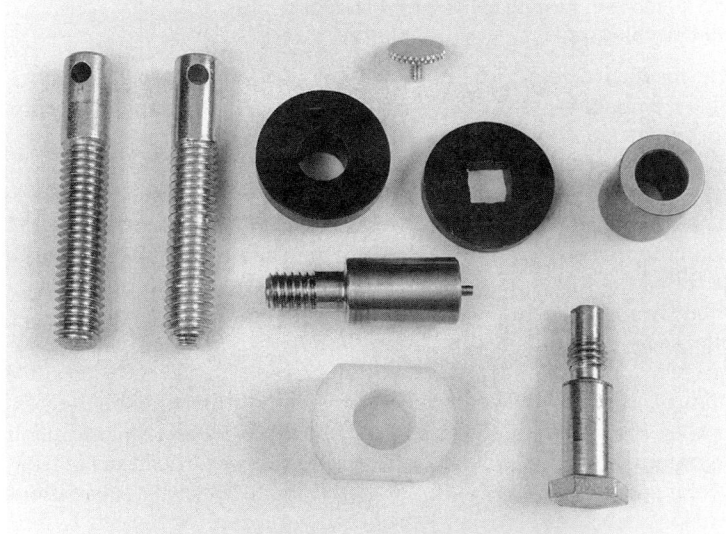

Figure 7.57   Typical CNC machined parts (turned, milled, etc.).

Figure 7.58 A small production manual turret lathe.

Figure 7.59 Large geared-head engine lathe with digital panel.

many years, together with the standard milling machines. It has been said often that the engine lathe can produce almost any type of detail part requiring machining when equipped with the proper accessories and auxiliary devices.

7.2 Milling

Milling is a machining process for generating machined surfaces by removing a predetermined amount of material progressively from the workpiece. The milling process employs relative motion between the workpiece and the rotating cutting tool to generate the required surfaces. In some applications the workpiece is stationary, and the cutting tool moves, whereas in others the cutting tool and the workpiece are moved in relation to each other and to the machine. A characteristic feature of the milling process is that each tooth of the cutting tool takes a portion of the stock in the form of small, individual chips.

Typical cutting tool types for milling machine operations are shown in Figs. 7.60 and 7.61.

Milling methods

- Peripheral milling (slab milling)
- Face milling and straddle milling
- End milling
- Single-piece milling
- String or "gang" milling
- Slot milling
- Profile milling
- Thread milling
- Worm milling
- Gear milling

Modern milling machines have many forms, but the most common types are shown in Fig. 7.62. The well-known and highly popular Bridgeport-type milling machine is shown in Fig. 7.63. The Bridgeport machine is often used in tool and die making operations

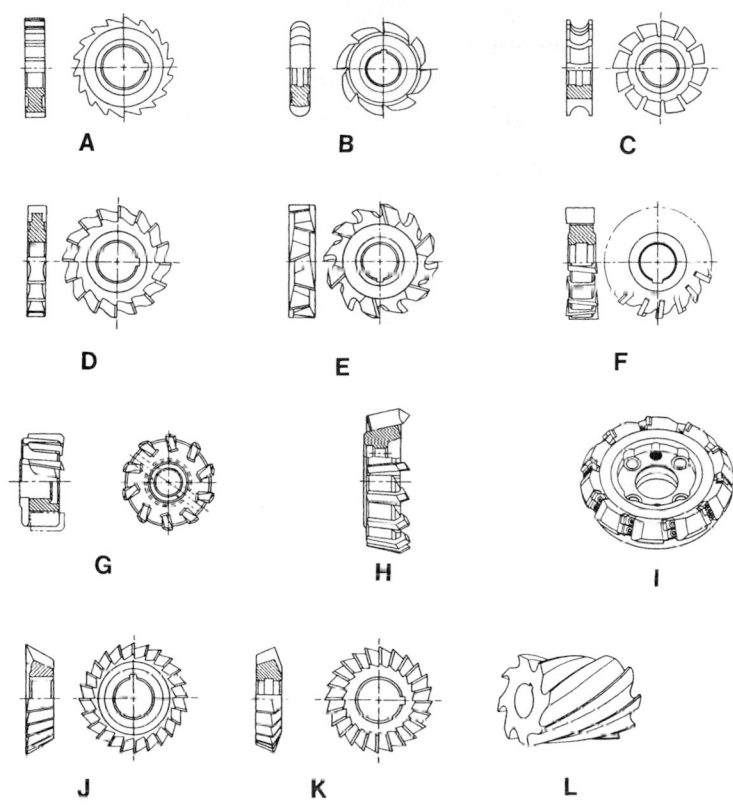

Milling Cutter Styles - High-Speed Steel and Carbide Insert

A   Disk type milling cutter
B   Convex half-round milling cutter
C   Concave half-round milling cutter
D   Three-side milling cutter
E   Staggered-tooth milling cutter
F   Inserted blade milling cutter

G   Face milling cutter
H   Face milling head
I   Double-angle carbide insert milling cutter
J   Single-angle milling cutter
K   Double-angle milling cutter
L   Left hand slab milling cutter

Figure 7.60   Typical milling cutters.

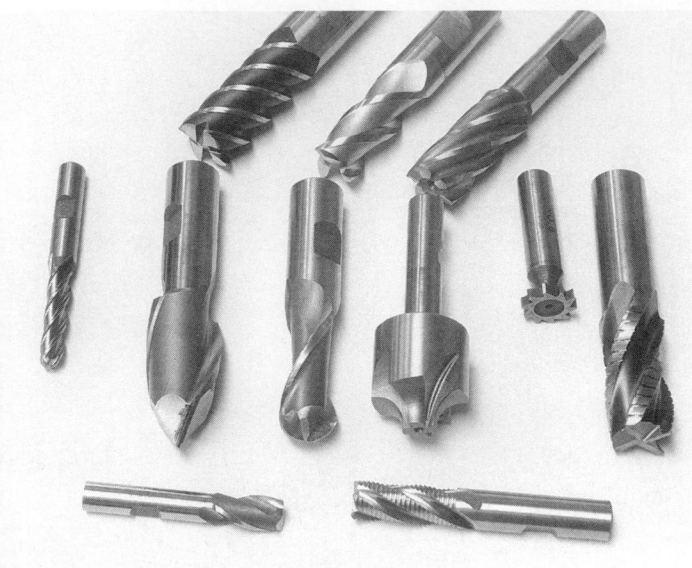

Figure 7.61 Samples of end-milling cutters.

and in model shops, where prototype work is done. The great stability and accuracy of the Bridgeport makes this machine popular with many experienced machinists and die makers. The Bridgeport shown in Fig. 7.63 is equipped with digital sensing controls and readout panel, reading to ±0.0005 in.

A ball-end milling operation is shown in Fig. 7.64, where an aluminum alloy part is being cut on a Bridgeport-type machine. This aluminum part is being manufactured without the use of a cutting solution or coolant. Many of the better grades of aluminum alloys, such as 2024, 6061, and 7075, can be considered free-machining and often are cut without coolants and at high surface speeds (see speeds and feeds section). When cutting solutions and coolants are used for turning or milling aluminum alloys, special formulations are available and may be used (see coolants and cutting solutions section).

A large vertical milling machine is shown in Fig. 7.65. This machine is set up to straddle-mill copper alloy castings using face mills in a parallel, stacked array in order to make two cuts on two surfaces with one pass of the part through the cutters. In this application, the cast

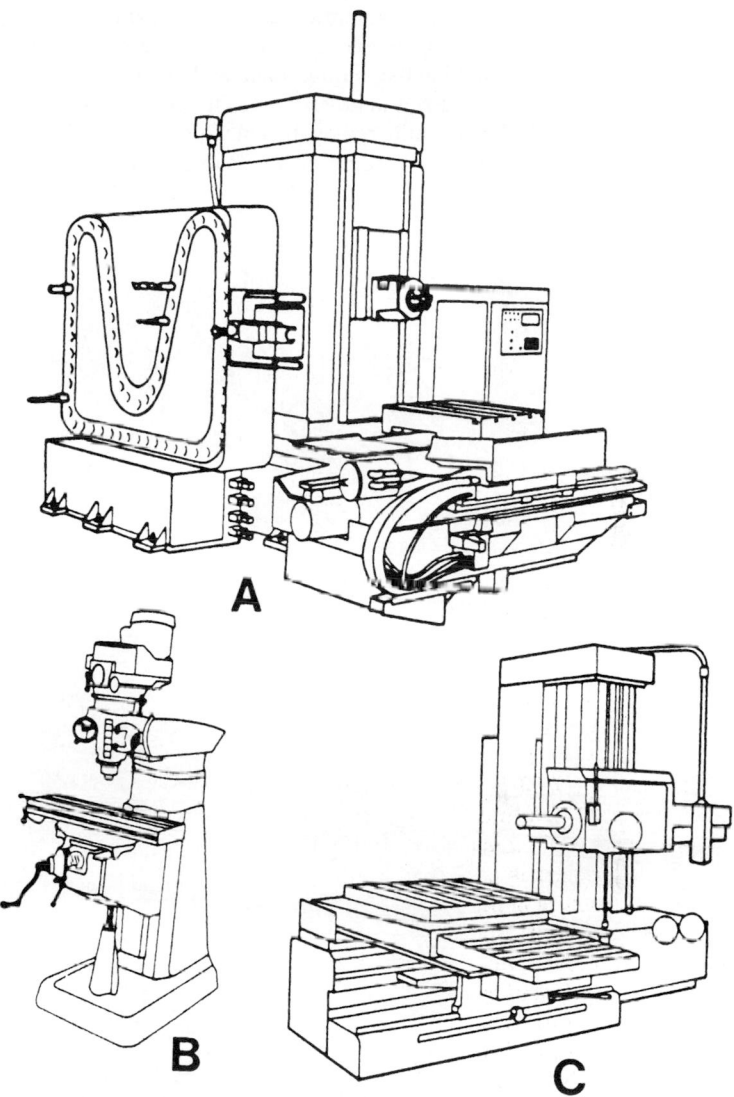

Figure 7.62 Types of milling machines.

copper parts are positioned in pneumatic clamps in a "gang" or string arrangement. Many parts thus are cut in a single traverse of the horizontal table. Mass-production techniques such as these allow parts to be manufactured more quickly and at less cost. The milling cutters on this machine are of the facemill removable-carbide insert type, the inserts being made of tungsten carbide with a titanium nitride or titanium carbide coating.

(a)

Figure 7.63 (a) The popular Bridgeport universal milling machine. (b) Close-up of milling operation showing digital panel.

(b)

Figure 7.63 (*Continued*)

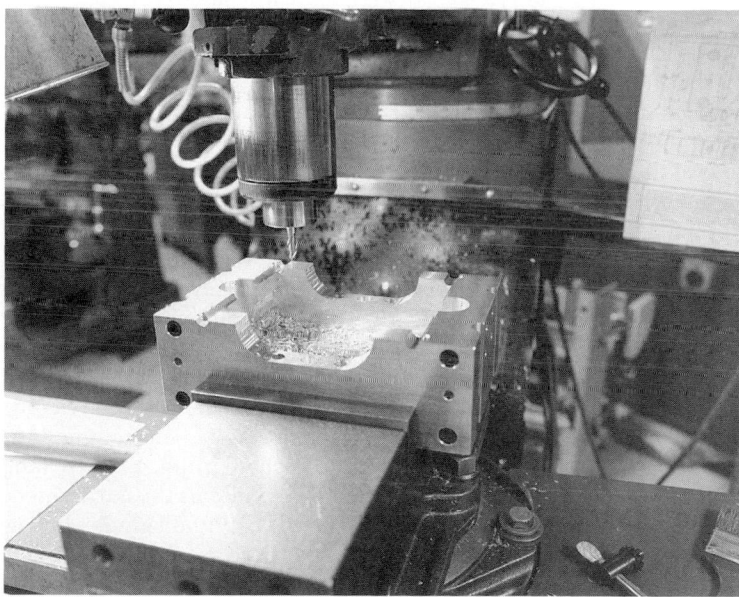

Figure 7.64 Close-up of ball-milling operation on an aluminum alloy part.

Figure 7.65   Large milling machine shown straddle-milling production parts.

The modern machining center is being used to replace the conventional milling machine in many industrial applications. Figure 7.66 shows a machining center with its control panel at the right side of the machine. Machines such as these generally cost $250,000 or more, depending on the accessories and auxiliary equipment obtained with the machine. These machines are the modern "workhorses" of industry and cannot remain idle for long periods owing to their cost. As described in Chap. 1, these machines are computer controlled and make their own tool changes automatically during ongoing machining operations. Figure 7.67 shows a typical "gang-milling" operation of aluminum cast parts that are finish bored, drilled, and then tapped while being held in a pneumatically actuated clamping fixture. The pneumatic line can be seen coming into the fixture at the lower right side of the photograph. Four coolant lines are shown directed at the machine spindle in the cutting tool location. These coolant lines move with the tool and spindle during the cutting operation. One needs to see these machines in actual operation to appreciate the great speed and accuracy with which they perform their programmed (CNC) machining functions.

Figure 7.66 Vertical machining center in operation.

Figure 7.67 Close-up of production milling, pallet-mounted parts.

The modern machining center may be equipped for three-, four-, or five-axis operation. The normal or common operations usually call for three-axis machining, whereas more involved machining procedures require four- or even five-axis operation. Three-axis operation consists of x and y table movements and z-axis vertical spindle movements. The four-axis operation includes the addition of spindle rotation with three-axis operation. Five-axis operation includes a horizontal fixture for rotating the workpiece on a horizontal axis at a predetermined speed (rpms), together with the functions of the four-axis machine. This allows all types of screw threads to be machined on the part and other operations such as producing a worm for worm-gear applications, segment cuts, arcs, etc. Very complex parts may be mass produced economically on a three-, four-, or five-axis machining center, all automatically, using CNC.

The control panels on these machining centers contain a microprocessor that is, in turn, controlled by a host computer, generally located in the tool or manufacturing engineering office; the host computer controls one or more machines with direct numerical control (DNC) or distributed numerical control. Various machining programs are available for writing the operational instructions sent to the controller on the machining center.

7.2.1 Milling calculations

The following calculation methods and procedures for milling operations are intended to be guidelines and not absolute because of the many variables encountered in actual practice.

Metal-removal rates. The metal-removal rate R (sometimes mrr) for all types of milling is equal to the volume of metal removed by the cutting process in a given time, usually expressed as cubic inches per minute (in^3/min). Thus

$$R = WHf$$

where R = metal-removal rate, in^3/min
W = width of cut, in
H = depth of cut, in
f = feed rate, ipm (in/min)

In peripheral or slab milling, W is measured parallel to the cutter axis and H perpendicular to the axis. In face milling, W is measured perpendicular to the axis and H parallel to the axis.

Feed rate. The speed or rate at which the workpiece moves past the cutter is the feed rate f, which is measured in inches per minute (ipm). Thus

$$f = F_t N C_{rpm}$$

where f = feed rate, ipm

F_t = feed per tooth (chip thickness), in (also called cpt)

N = number of cutter teeth

C_{rpm} = rotation of the cutter, rpm

Feed per tooth. Production rates of milled parts are directly related to the feed rate that can be used. The feed rate should be as high as possible, considering machine rigidity and power available at the cutter. To prevent overloading the machine drive motor, the allowable feed per tooth F_t may be calculated from

$$F_t = \frac{K hp_c}{N C_{rpm} WH}$$

where hp_c = horsepower available at the cutter (80 to 90 percent of motor rating); i.e., if motor nameplate states 15 hp, then the horsepower available at the cutter is 0.8 to 0.9×15 (80 to 90 percent represents motor efficiency)

K = machinability factor (see Fig. 7.68)

Other symbols are as in the preceding equation.

Figure 7.69 gives the suggested feed per tooth for milling using high-speed-steel (HSS) cutters for the various cutter types. For carbide, cermets, and ceramic tools, see the figures in the feeds and speeds section.

Cutting speed. The cutting speed of a milling cutter is the peripheral linear speed resulting from rotation of the cutter. The cutting speed is expressed in feet per minute (fpm, or ft/min) or surface feet per minute (sfpm or sfm) and is determined from

$$S = \frac{\pi D (rpm)}{12}$$

where S = cutting speed, fpm or sfpm (sfpm is also termed spm)

D = outside diameter of the cutter, in

rpm = rotational speed of cutter, rpm

The required rotational speed of the cutter may be found from the following simple equation:

(K) Factor for Various Materials

Material	K(in³/min/(hp_c))
Cold drawn steel, SAE 1112, 1120, 1315	1.0
Forged and alloy steel, SAE 3120, 1020, 2320, 2345, 150-300 BHN	0.63 - 0.87
Alloy steel, 300 - 400 BHN	0.5
Malleable iron and cold drawn steel, SAE 6140	0.9
Cast irons:	
Soft	1.5
Medium	0.8 - 1.0
Hard	0.6 - 0.8
Stainless steel, AISI 416, free-machining	1.1
Stainless steel, austenitic, AISI 303, free-machining	0.83
Stainless steel, austenitic, AISI 304	0.72
Tool steel	0.51
Bronze and brass:	
Soft	1.7- 2.5
Medium	1.0 - 1.4
Hard	0.6 - 1.0
Aluminum and magnesium	2.5 - 4.0
Monel metal	0.55
Copper, annealed	0.84
Nickel	0.53
Titanium & alloys	0.75 - 1.1

NOTE: "K" values are in cubic inches per minute per cutter horsepower (in³/min/hp_c)

Figure 7.68 K-factor table.

Suggested Feed per Tooth for Milling - High-Speed Steel Cutters (Tabulated Data in Inches)

Material	Face mills	Helical mills	Slot/side mills	End mills	Form-relieved cutters
Magnesium & alloys	0.022	0.018	0.013	0.011	0.007
Aluminum & Alloys	0.022	0.018	0.013	0.011	0.007
Free-cutting brasses & bronzes	0.022	0.018	0.013	0.011	0.007
Medium brasses & bronzes	0.014	0.011	0.008	0.007	0.004
Hard brasses & bronzes	0.009	0.007	0.006	0.005	0.003
Copper	0.012	0.010	0.007	0.006	0.004
Cast iron, soft (150-180 Bhn)	0.016	0.013	0.009	0.008	0.005
Cast iron, medium (180-220 Bhn)	0.013	0.010	0.007	0.007	0.004
Cast iron, hard (220-300 Bhn)	0.011	0.008	0.006	0.006	0.003
Malleable iron	0.012	0.010	0.007	0.006	0.004
Cast steel	0.012	0.010	0.007	0.006	0.004
Low-carbon steel, free-machining	0.012	0.010	0.007	0.006	0.004
Low-carbon steels	0.010	0.008	0.006	0.005	0.003
Medium-carbon steels	0.010	0.008	0.006	0.005	0.003
Alloy steel, ann'ld (180-220 Bhn)	0.008	0.007	0.005	0.004	0.003
Alloy steel, tough (220-300 Bhn)	0.006	0.005	0.004	0.003	0.002
Alloy steel, hard (300-400 Bhn)	0.004	0.003	0.003	0.002	0.002
Stainless steels, free-machining	0.010	0.008	0.006	0.005	0.003
Stainless steels	0.006	0.005	0.004	0.003	0.002
Monel metal	0.008	0.007	0.005	0.004	0.003
Titanium & alloys	0.008	0.007	0.005	0.004	0.003
Machinable plastics	0.013	0.010	0.008	0.007	0.004

NOTE: Tabular data in inches. For feed per tooth in millimeters, multiply tabular data by 25.4. For carbon-steel cutter multiply tabular data by 0.50 or divide by 2. Source: Cincinnati Milicron, Inc.

Figure 7.69 Milling feed table, HSS.

$$\text{rpm} = \frac{S}{(D/12)\pi} \quad \text{or} \quad \frac{S}{0.26D}$$

When it is necessary to increase the production rate, it is better to change the cutter material rather than to increase the cutting speed. Increasing the cutting speed alone may shorten the life of the cutter because the cutter is usually being operated at its maximum speed for optimal productivity.

General rules for selection of the cutting speed

- Use lower cutting speeds for longer tool life.

- Take into account the Brinell hardness of the material.

- Use the lower range of recommended cutting speeds when starting a job.

- For a fine finish, use a lower feed rate in preference to a higher cutting speed.

Number of cutter teeth. The number of cutter teeth N required for a particular application may be found from the simple expression (not applicable to carbide or other high-speed cutters)

$$N = \frac{f}{F_t C_{\text{rpm}}}$$

where f = feed rate, ipm
F_t = feed per tooth (chip thickness), in
C_{rpm} = rotational speed of cutter, rpms
N = number of cutter teeth

An industry-recommended equation for calculating the number of cutter teeth required for a particular operation is

$$N = 19.5\sqrt{R} - 5.8$$

where N = number of cutter teeth
R = radius of cutter, in

This simple equation is suitable for HSS cutters only and is not valid for carbide, cobalt cast alloy, or other high-speed cutting tool materials.

Figure 7.70 gives recommended cutting speed ranges (sfpm) for HSS cutters. See the figures in feeds and speeds section for carbide, cermet, ceramic, and other high-speed advanced cutting materials.

Milling horsepower. Ratios for metal removal per horsepower (cubic inches per minute per horsepower at the milling cutter) have been given for various materials (see Fig. 7.68). The general equation is

$$K = \frac{\text{in}^3/\text{min}}{\text{hp}_c} = \frac{WHf}{\text{hp}_c}$$

where K = metal removal factor, $\text{in}^3/\text{min}/\text{hp}_c$ (see Fig. 7.68)
hp_c = horsepower at the cutter

Milling Cutting Speeds for Various Materials
(sfpm) Surface feet per minute (High-speed steel tools only)

Material	High-speed steel tools	
	Rough	Finish
Cast iron................................	50 - 60	80 - 110
Semisteel..............................	40 - 50	65 - 90
Malleable iron......................	80 - 100	110 - 130
Cast steel.............................	45 - 60	70 - 90
Copper.................................	100 - 150	150 - 200
Brass...................................	200 - 300	200 - 300
Bronze.................................	100 - 150	150 - 180
Aluminum.............................	400 - 450	700 - 750
* Magnesium........................	600 - 800	1,000 - 1,500
SAE steels:		
1020 (coarse feed), low-carbon.......	60 - 80	60 - 80
1020 (fine feed), low-carbon..........	100 - 120	100 - 120
1035, medium-carbon..............	75 - 90	90 - 120
1330, alloy steel...................	90 - 110	90 - 110
1050, Med-high-carbon..........	60 - 80	100 - 125
2315, nickel steel.................	90 - 110	90 - 110
3150, nickel-chromium...........	50 - 60	70 - 90
4150, chrome-molybdenum.....	40 - 50	70 - 90
4340, nickel-chrome-molybdenum..	40 - 50	60 - 70
Stainless steel.....................	60 - 80	100 - 120
Titanium, hard alloy..............	80 - 100	110 - 130

NOTE: Tabular data ranges are in sfpm (surface feet per minute
for HSS cutters only).
* A fire hazard is present when machining magnesium at high-speeds.

Figure 7.70 Milling cutting speeds, HSS.

W = width of cut, in
H = depth of cut, in
f = feed rate, ipm

The total horsepower required at the cutter may then be expressed as

$$\text{hp}_c = \frac{\text{in}^3/\text{min}}{K} \quad \text{or} \quad \frac{WHf}{K}$$

The K factor varies with type and hardness of material and for the same material varies with the feed per tooth, increasing as the chip thickness increases. The K factor represents a particular rate of metal removal and not a general or average rate. For a quick approximation of total power requirements at the machine motor, see Fig. 7.71, which gives the maximum metal-removal rates for different horsepower-rated milling machines cutting different materials.

7.2.2 Feeds and speeds for milling with advanced cutting tool materials

Figures 7.14 through 7.44 present the feeds and speeds with which materials may be milled using the carbide, cermet, ceramic, and advanced cutting tool materials such as cubic boron nitride (CBN)

Milling-Machine Horsepower Ratings - for maximum metal removal rates (in³/min) for HSS (high-speed steel cutters)

Workpiece Material	Rated hp of Machine									
	3	5	7.5	10	15	20	25	30	40	50
	Max. Metal Removal (in³/min)									
Aluminum	2.7	5.5	8.7	12	18	27	37	48	69	91
Brass, soft	2.4	4.7	7.5	10	16	24	32	41	60	79
Bronze, hard	1.7	3.3	5.3	7.3	11	17	23	30	43	56
Bronze, very hard	0.78	1.6	2.5	3.4	5.3	7.8	11	15	20	26
Cast iron, soft	1.6	3.2	5.2	7.1	11	16	22	28	41	54
Cast iron, hard	1	2	3.3	4.6	7	10	14	18	26	35
Cast iron, chilled	0.78	1.6	2.5	3.4	5.3	7.8	10	13	19	26
Malleable iron	1	2.1	3.4	4.7	7.3	11	14	10	20	00
Steel, soft	1	2	3.3	4.6	7	10	14	18	26	35
Steel, medium	0.78	1.6	2.5	3.4	5.3	7.8	10	13	19	26
Steel, hard	0.56	1.1	1.8	2.5	3.9	5.7	7.7	10	14	19

NOTE: Data source - Kearney & Trecker Corp.

Figure 7.71 Milling machine horsepower ratings.

that are used widely in industry today. Cutting tool technology has advanced rapidly, and new tools and materials are being made available at a rapid pace. Nevertheless, the data presented here are the latest available at the date of publication of this *Handbook*.

Modern theory of milling. The key characteristics of the milling process are

- Simultaneous motion of cutter rotation and feed movement of the workpiece
- Interrupted cut
- Production of tapered chips

It was common practice for many years in the industry to mill against the direction of feed. This was due to the type of tool materials then available (HSS) and the absence of antibacklash devices on the machines. This method became known as "conventional" or "up milling" and is illustrated in Fig. 7.72*b*. "Climb milling" or "down milling" is now the preferred method of milling with advanced cutting tool materials such as carbides, cermets, CBN, etc. Climb milling is illustrated in Fig. 7.72*a*. Here, the insert enters the cut with some chip load and proceeds to produce a chip that thins as it progresses toward the end of the cut. This allows the heat generated in the cutting process to dissipate into the chip. Climb-milling forces push the workpiece toward the clamping fixture, in the direction of the feed. Conventional-milling (up-milling)

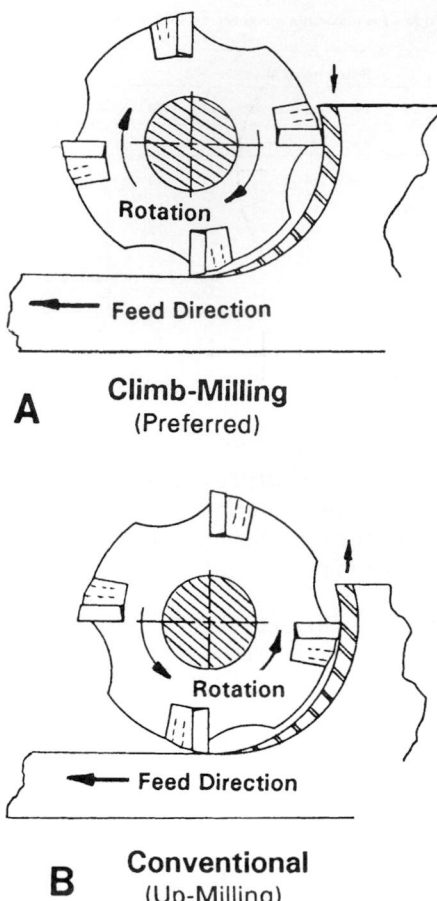

A   **Climb-Milling**
(Preferred)

B   **Conventional**
(Up-Milling)

Figure 7.72   Climb and up milling.

forces are against the direction of feed and produce a lifting force on the workpiece and clamping fixture.

The angle of entry is determined by the position of the cutter centerline in relation to the edge of the workpiece. A negative angle of entry β is preferred and is illustrated in Fig. 7.73b, where the centerline of the cutter is below the edge of the workpiece. A negative angle is preferred because it ensures contact with the workpiece at the strongest point of the insert cutter. A positive angle of entry

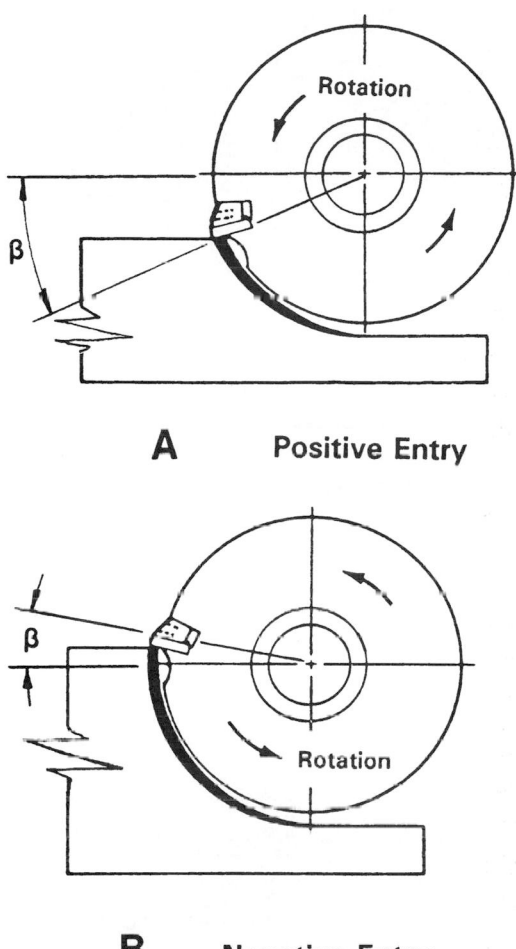

A   **Positive Entry**

B   **Negative Entry**

Figure 7.73   Milling entry angles.

will increase insert chipping. If a positive angle of entry must be employed, use an insert with a honed or negative land.

Figure 7.74a shows an eight-tooth cutter climb milling a workpiece using a negative angle of entry, and the feed, or advance, per revolution is 0.048 in with a chip load per tooth of 0.006 in. The following milling formulas will allow you to calculate the various milling parameters.

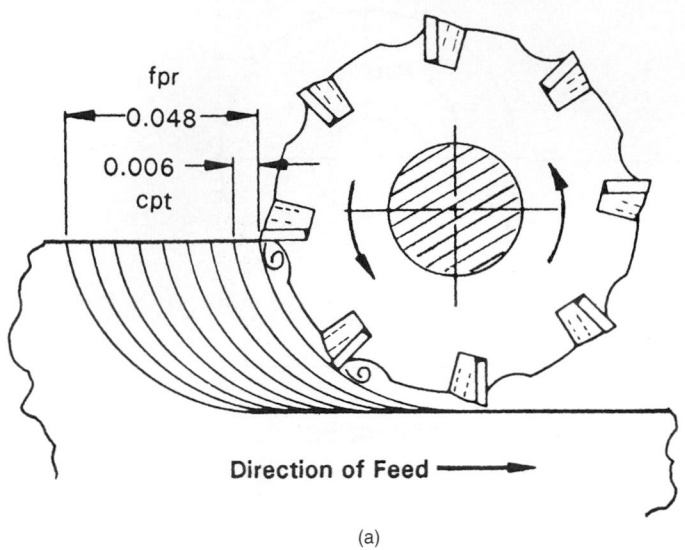

(a)

Power Constant Factor (P$_f$) for Milling Various Materials

P$_f$ = cubic inches of metal removal per single horsepower

Material	P$_f$ Factor
Free-machining aluminum alloys	0.20
Gray cast irons	0.25
Non-ferrous free-machining alloys	0.45
Alloy cast irons and ductile irons	0.50
Martensitic stainless steels	0.81
Titanium	0.87
Free-machining carbon steel	0.89
Standard carbon steel	0.92
Alloy steels	0.95
Austenitic stainless steel	0.96
High-temperature alloys	1.00
Tool steels	1.05
Cobalt based alloys	1.20

Note: The P$_f$ factors will vary per feed rate (ipm) and Brinell hardness (Bhn).
The P$_f$ factors in the table are for normal feed rates and material hardness ranges to 285 Bhn.
A 15 hp spindle-rated mill should be able to machine 15 x 0.92 = 13.8 cubic inches of standard plain carbon steel per minute.

(b)

Figure 7.74 (a) Milling principle. (b) Power constants for milling.

In the following formulas,

n_t = number of teeth or inserts in the cutter

cpt = chip load per tooth or insert, in

ipm = feed, inches per minute

fpr = feed (advance) per revolution, in

D = cutter effective cutting diameter, in

rpm = revolutions per minute

sfpm = surface feet per minute (also termed sfm)

$$\text{sfpm} = \frac{\pi D(\text{rpm})}{12} \qquad \text{rpm} = \frac{12(\text{sfpm})}{\pi D} \qquad \text{fpr} = \frac{\text{ipm}}{\text{rpm}}$$

$$\text{ipm} = \text{cpt} \times n_t \times \text{rpm} \qquad \text{cpt} = \frac{\text{ipm}}{n_t(\text{rpm})} \qquad \text{or} \qquad \frac{\text{fpr}}{n_t}$$

Example: Given a cutter of 5 in diameter, eight teeth, 500 sfpm, and 0.007 cpt,

$$\text{rpm} = \frac{12 \times 500}{3.1415 \times 5} = 382$$

$$\text{ipm} = 0.007 \times 8 \times 382 = 21.4$$

$$\text{fpr} = \frac{21.4}{382} = 0.056$$

Slotting. Special consideration is given for slot milling, and the following equations may be used effectively to calculate chip load per tooth (cpt) and inches per minute (ipm):

$$\text{cpt} = \frac{\left(\dfrac{\sqrt{(D-x)x}}{r} \right)\left(\dfrac{\text{ipm}}{\text{rpm}} \right)}{\text{number of effective teeth}}$$

$$\text{ipm} = \text{rpm} \times \text{number of effective teeth}\left[\frac{\text{cpt}}{\dfrac{\sqrt{(D-x)x}}{r}} \right]$$

where D = diameter of slot cutter, in
r = radius of cutter, in
x = depth of slot, in
cpt = chip load per tooth, in
ipm = feed, inches per minute
rpm = rotational speed of cutter, rpm

Milling horsepower for advanced cutting tool materials

Horsepower consumption. It is advantageous to calculate the milling operational horsepower requirements before starting a job. Lower-horsepower machining centers take advantage of the ability of the modern cutting tools to cut at extremely high surface speeds (sfpm). Knowing your machine's speed and feed limits could be critical to your obtaining the desired productivity goals. The condition of your milling machine is also critical to obtaining these productivity goals. Older machines with low-spindle-speed capability should use the uncoated grades of carbide cutters and inserts.

Horsepower calculation. A popular equation used in industry for calculating horsepower at the spindle is

$$hp = \frac{M_{rr}P_f}{E_s}$$

where M_{rr} = metal removal rate, in^3/min
P_f = power constant factor (see Fig. 7.74b)
E_s = spindle efficiency, 0.80 to 0.90 (80 to 90 percent)

Note: The spindle efficiency is a reflection of losses from the machine's motor to actual power delivered at the cutter and must be taken into account, as the equation shows.

A table of P_f factors is shown in Fig. 7.74b and represents the number of cubic inches of material that may be removed per each (1) horsepower input at the spindle or cutter for different types of materials.

Note: The metal removal rate M_{rr} = depth of cut $\times$ width of cut $\times$ ipm = in^3/min.

Axial cutting forces at various lead angles. Axial cutting forces vary as you change the lead angle of the cutting insert. The 0° lead angle produces the minimum axial force into the part. This is advantageous for weak fixtures and thin web sections. The 45° lead angle loads the spindle with the maximum axial force, which is advantageous when using the older machines.

Tangential cutting forces. The use of a tangential-force equation is appropriate for finding the approximate forces that fixtures, part walls, or webs and the spindle bearings are subjected to during the milling operation. The tangential force is calculated easily when you have determined the horsepower being used at the spindle or cutter. It is important to remember that the tangential forces decrease as the spindle speed (rpms) increases, i.e., at higher surface feet per minute. The ability of the newer advanced cutting tools to operate at higher speeds thus produces fewer fixture- and web-deflecting forces with a decrease in horsepower requirements for any particular machine. Some of the new high-speed cutter inserts can operate efficiently at speeds of 10,000 sfpm or higher when machining such materials as free-machining aluminum and magnesium alloys.

The tangential force developed during the milling operations may be calculated from

$$t_f = \frac{126,000\text{hp}}{\pi D(\text{rpm})}$$

where t_f = tangential force, pounds force
  hp = horsepower at the spindle or cutter
  D = effective diameter of cutter, in
  rpm = rotational speed in revolutions per minute

The preceding calculation procedure for finding the tangential forces developed on the workpiece being cut may be used in conjunction with the clamping fixture types and clamping calculations shown in Chap. 8, "Tooling Practices."

7.2.3 Feeds and speeds tables: Advanced cutting tools and Inserts

See Sec. 7.1.8 and Figs. 7.14 through 7.44.

How to apply the range of conditions for the preceding feeds and speeds tables is shown in Fig. 7.75. A chart of carbide insert-grade comparisons for different manufacturers is shown in Fig. 7.76. Insert-grade comparisons may be made using this chart, although the chart is a guide only and indicates those grades having similar properties under most conditions. It is not intended to imply that all cross-referenced grades are exact duplicates in physical and metallurgical characteristics or that they perform equally in the same applications. Figure 7.76 is used for older inserts. See Figs. 7.45 and 7.46 for newer inserts.

-GENERAL APPLICATIONS FOR CUTTING CONDITIONS-

CONDITION	-SPEED-	-FEED-
Roughing	⇩	⇧
Finishing	⇧	⇩
End Milling	⇧	⇩
Slotting	⇧	⇩
Hard Material	⇩	⇨
Soft Material	⇧	⇧
Scale	⇩	⇧
Tool Life	⇩	⇨
Heavy d.o.c.	⇩	⇩

Higher- ⇧

Lower- ⇩

Same- ⇨

Figure 7.75 Applying range of conditions—milling tables.

Cutter speed (rpm) from surface speed (sfpm). A time-saving table of surface speed versus cutter speed is shown in Fig. 7.77 for cutter diameters from 0.25 through 5 in. For cutter speed (rpm) values when the surface speed is greater than 200 sfpm, use the simple equation

$$\text{rpm} = \frac{12(\text{sfpm})}{\pi D}$$

where D is the effective diameter of cutter in inches.

7.2.4 Milling accessories

Many types of accessories are available for various milling capabilities, and a few basic ones are shown here. Figure 7.78 shows the

(Obsolete Carbide Grade Comparison)

Kenna-metal	IMCO	Adamas	Carboloy	Carmet	New-comber	Sandvic	SECO	Firth Sterling	TRW	Valenite	VR/Wesson	Walmet
Uncoated Grades												
K1	IC-10	B	44A	CA-3	N-10			H-6	CQ-12	VC-1	VR-54	WA-1
K68	MG-10	PWX	820	CA-310		H10	G-27	H-17	CQ-22	VC-27	RAM-1	WA-10
K68	IC-20	A	883	CA-4	N-22	H20	H-13	H-21	CQ-2	VC-2	2A5	WA-2
K313	IC-40	AAA	999	CA-8	N-40	H05		HF	CQ-4	VC-4		WA-4
K420	IC-50	434	390	CA-740	N-50		S6	T-04	CY-12		2A7	WA-54
K420	IC-55	499	370	CA-720	N-55	S6	S6	T-12	CY-16	VC-5	VR-75	WA-5
K2884	IC-67	548		CA-717	N-60	SIP	S2	T-25	CY-14	VC-6	VR-73	WA-47
K313	IC-70	490	350		N-70	SIP	SIG	T-25	CY-31	VC-8		WA-73
Coated Grades												
KC730	IC-202	ACT-2	523	CA9443	NT-2	GC315	TP-15	TC	027	VN2	630	P2
KC810	IC-552	ACT-5	550	CA9720	NT-5	GC1025	TP-35	HN	715	VN5	660	P5
KC850	IC-672	ACT-7	518		NT-8		TP-25	TC1	714		670	P47
KC910	IC-554	ROXIDE	570		NA02	GC015		CC46	918	V01		A60
KC950	IC-0014		585			GC415		CC44		V03	680	A62

NOTE: This grade selection chart is intended to be used as a guide to grade selection of various manufacturers. It indicates those grades having similar characteristics under most general conditions. It is not intended to imply that all comparable grades are exact duplicates in physical and metallurgical properties, or that they perform exactly the same on all applications.

Figure 7.76   Carbide insert-grade comparison for various manufacturers.

CUTTER SPEEDS IN REVOLUTIONS PER MINUTE

Diameter of cutter (in.)	Surface speed (ft. per min.)																
	25	30	35	40	50	55	60	70	75	80	90	100	120	140	160	180	200
	Cutter revolutions per minute																
1/4	382	458	535	611	764	851	917	1,070	1,147	1,222	1,376	1,528	1,834	2,139	2,445	2,750	3,056
5/16	306	367	428	489	611	672	733	856	917	978	1,100	1,222	1,466	1,711	1,955	2,200	2,444
3/8	255	306	357	408	509	560	611	713	764	815	916	1,018	1,222	1,425	1,629	1,832	2,036
7/16	218	262	306	349	437	481	524	611	656	699	786	874	1,049	1,224	1,398	1,573	1,748
1/2	191	229	268	306	382	420	459	535	573	611	688	764	917	1,070	1,222	1,375	1,528
5/8	153	184	214	245	306	337	367	428	459	489	552	612	736	857	979	1,102	1,224
3/4	127	153	178	203	254	279	306	357	381	408	458	508	610	711	813	914	1,016
7/8	109	131	153	175	219	241	262	306	329	349	392	438	526	613	701	788	876
1	95.5	115	134	153	191	210	229	267	287	306	344	382	458	535	611	688	764
1-1/4	76.3	91.8	107	123	153	168	183	214	230	245	274	306	367	428	490	551	612
1-1/2	63.7	76.3	89.2	102	127	140	153	178	191	204	230	254	305	356	406	457	508
1-3/4	54.5	65.5	76.4	87.3	109	120	131	153	164	175	196	218	262	305	349	392	436
2	47.8	57.3	66.9	76.4	95.5	105	115	134	143	153	172	191	229	267	306	344	382
2-1/2	38.2	45.8	53.5	61.2	76.3	84.2	91.7	107	114	122	138	153	184	213	245	275	306
3	31.8	38.2	44.6	51	63.7	69.9	76.4	89.1	95.3	102	114	127	152	178	208	228	254
3-1/2	27.3	32.7	38.2	44.6	54.5	60	65.5	76.4	81.8	87.4	98.1	109	131	153	174	196	218
4	23.9	28.7	33.4	38.2	47.8	52.6	57.3	66.9	71.7	76.4	86	95.6	115	134	153	172	191
5	19.1	22.9	26.7	30.6	38.2	42	45.9	53.5	57.3	61.1	68.8	76.4	91.7	107	122	138	153

Figure 7.77   Cutter revolutions per minute from surface speed.

widely used horizontal dividing and angle-setting tool. With this accessory, you may divide the circle into an equal number of divisions or set any horizontal angle from a particular baseline or starting point. The vernier on this device will set an angle within ±15 minutes of arc. The clamping table or surface of this device can be rotated a full 360°. Figure 7.79 shows a vertical dividing head for milling operations.

Figure 7.78 Horizontal dividing and angle-setting head.

Figure 7.79 Vertical dividing head for milling operations.

Horizontal angle control

Clamping Table

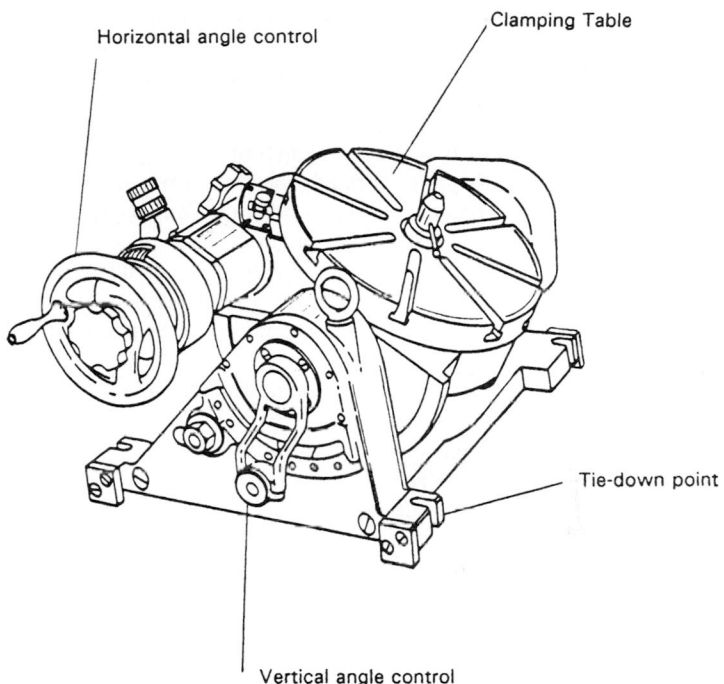

Tie-down point

Vertical angle control

Figure 7.80 Vertical and horizontal angle-setting device.

The device shown in Fig. 7.80 will allow the setting of a compound angle because the clamping table on this device can be rotated both horizontally and vertically, and both directions are controlled by a vernier setting. This device is similar to a compound sine plate but is perhaps more convenient and easier to use, although the compound sine plate is more accurate owing to the fact that it is set using Jo-blocks. Previous chapters of this *Handbook* explain the use and setting practices of compound sine plates and simple sine plates.

Quick answers to milling calculations can be obtained by using the various cutting tool manufacturers' calculators. With these devices, all the basic milling calculations can be done, including the required horsepower needed for a particular milling operation. When a device such as this is not available to the tool or manufacturing engineer or machinist, the calculations may be performed by using the equations and charts presented in this section. This device is similar to the turning and boring calculator shown in the turning

and boring section (Sec. 7.1), except that it has been designed for use in milling operations.

Standard dividing and indexing head procedures are shown in Sec. 7.12.8.

Machining calculations. Although the modern machining centers can set angles and compound angles through the programmed controller on these machines, the basic accessories are nonetheless important when used on manually set machines, such as those used for small production runs and in prototype and model shops. Many of the machining procedures and calculations would indeed be complex without the use of the modern machining center equipped for four- or five-axis operations.

It should be apparent that the basic milling machine and machining center also can be used for drilling and jig-boring operations, although on a limited scale in relation to the available horsepower and physical size of the machine. Figure 7.81 illustrates some of the basic cutting tools used on mills and machining centers. In the figure, parts a, b, and c are drill bits that have been coated with titanium nitride, part d is a typical ball-end mill, part e is a high-speed close-spiral tap, whereas parts f, g, and h are a

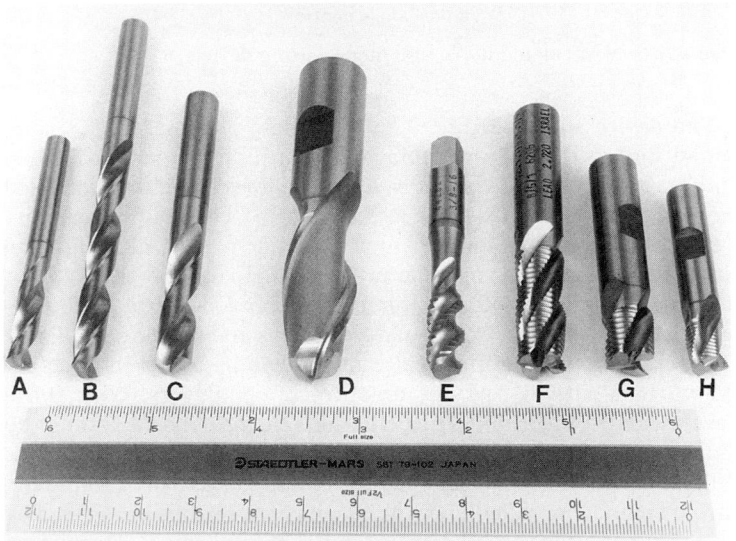

Figure 7.81   Titanium nitride–coated drills and end mills.

newer type end-mill design used for roughing operations at high speed, where large volumes of material are removed very rapidly. In parts *f*, *g*, and *h*, notice the threadlike grooves in the spiral flutes, which allow a rapid roughing operation to be performed. A close examination of the photograph will reveal the line on the tools that shows where the titanium nitride coating ends. The titanium nitride coating imparts a gold-colored finish on these cutting tools that is not apparent in the black and white photograph.

7.3 Drills and Drilling

Drilling is a machining operation for producing round holes in metallic and nonmetallic materials. A drill is a rotary-end cutting tool with one or more cutting edges or lips and one or more straight or helical grooves or flutes for the passage of chips and cutting fluids and coolants. When the depth of a drilled hole reaches three or four times the drill diameter, a reduction must be made in the drilling feed and speed. A coolant-hole drill can produce drilled depths to eight or more times the diameter of the drill. The gun drill can produce an accurate hole to depths of more than 100 times the diameter of the drill with great precision.

Enlarging a drilled hole for a portion of its depth is called *countersinking*, whereas a counterbore for cleaning the surface a small amount around the hole is called *spotfacing*. Cutting an angular bevel at the perimeter of a drilled hole is also termed *countersinking*. Countersinking tools are available to produce 82°, 90°, and 100° countersinks and other special angles.

Drills are classified by material, length, shape, number, type of helix or flute, shank, point characteristics, and size series. Most drills are made for right-hand rotation. Right-hand drills, as viewed from their point, with the shank facing away from your view, are rotated in a counterclockwise direction in order to cut. Left-hand drills cut when rotated clockwise in a similar manner.

7.3.1 Drill terminology

Figure 7.81*a* to *c* shows common twist drills made of HSS steel and coated with titanium nitride. The line just above the flutes shows the limit of the titanium nitride coating. Figure 7.82 shows the standard twist-drill form with the appropriate terminology describing its characteristic features.

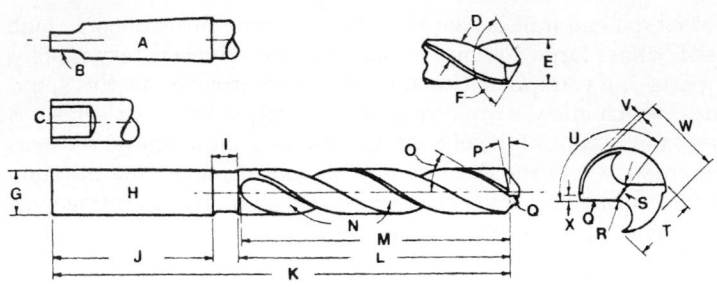

Twist Drill Features

A	Taper shank	M	Flute length
B	Tang	N	Flutes
C	Tang drive	O	Helix angle
D	Land width	P	Lip relief angle
E	Drill diameter	Q	Lip
F	Point angle	R	Web
G	Shank diameter	S	Chisel edge
H	Straight shank	T	Land
I	Neck	U	Chisel edge angle
J	Shank length	V	Body diameter clearance
K	Overall length	W	Clearance diameter
L	Body	X	Margin

Figure 7.82 Twist-drill features.

Drill types or styles

- HSS jobber drills

- Solid-carbide jobber drills

- Carbide-tipped screw-machine drills

- HSS screw-machine drills

- Carbide-tipped glass drills

- HSS extralong straight-shank drills (24 in)

- Taper-shank drills (0 through number 7 ANSI taper)

- Core drills

- Coolant-hole drills

- HSS taper-shank extralong drills (24 in)

- Aircraft extension drills (6 and 12 in)

- Gun drills

- HSS half-round jobber drills
- Spotting and centering drills
- Parabolic drills
- S-point drills
- Square solid-carbide die drills
- Spade drills
- Miniature drills
- Microdrills and microtools

American national standard tapers. Figure 7.83 shows the American national standard taper geometry and dimensions for ANSI tapers 1 through 7. Taper number 0 is not listed in the national standards. Table 7.1 accompanies Fig. 7.83 and lists the detail dimensions.

7.3.2   Drill point styles and angles

Over a period of many years, the metalworking industry has developed many different drill point styles for a wide variety of applications from drilling soft plastics to drilling the hardest types of metal alloys. All the standard point styles and special points are shown in Fig. 7.84, including the important point angles that differentiate these different points. New drill styles are being introduced

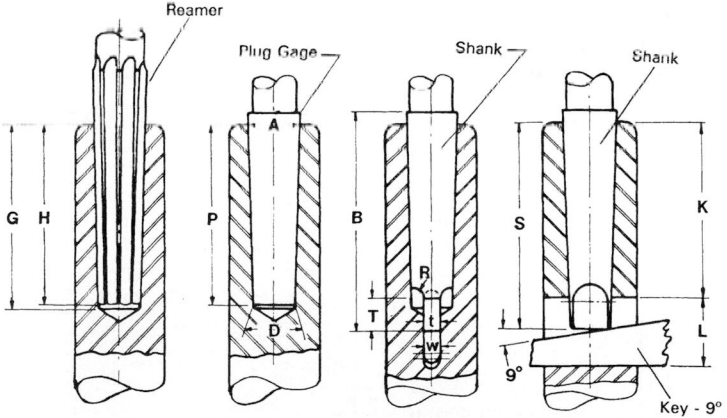

Figure 7.83 American national standard tapers.

TABLE 7.1 Detail Dimensions* for American National Standard Tapers

American national standard taper number	Shank								Tang				Tang Slot			Taper per inch	Taper per foot	American national standard taper number
	Diam. of plug at small end D	Diam. at end of socket A	Whole length B	Depth S	Depth of drilled hole G	Depth of reamed hole H	Standard plug depth P	Thickness t	Length T	Radius R	Radius a	Width W	Length L	End of socket to tang slot K			
0†	0.25200	0.35610	2 11/32	2 7/32	2 1/16	2 3/32	2	5/32	1/4	5/32	3/64	11/64	9/16	1 15/16	0.052050	0.62460	0†
1	0.36900	0.47500	2 9/16	2 7/16	2 3/16	2 5/32	2 1/8	13/64	3/8	3/16	3/64	7/32	3/4	2 1/16	0.049882	0.59858	1
2	0.57200	0.70000	3 1/8	2 15/16	2 21/32	2 39/64	2 9/16	1/4	7/16	1/4	1/16	17/64	7/8	2 1/2	0.049951	0.59941	2
3	0.77800	0.93800	3 7/8	3 11/16	3 5/16	3 1/4	3 3/16	5/16	9/16	9/32	5/64	21/64	1 3/16	3 1/16	0.050196	0.60235	3
4	1.02000	1.23100	4 7/8	4 5/8	4 3/16	4 1/8	4 1/16	15/32	5/8	5/16	3/32	31/64	1 1/4	3 7/8	0.051938	0.62326	4
4½	1.26600	1.50000	5 3/8	5 1/8	4 5/8	4 9/16	4 1/2	9/16	11/16	3/8	1/8	37/64	1 3/8	4 5/16	0.052000	0.62400	4½
5	1.47500	1.74800	6 1/8	5 7/8	5 5/16	5 1/4	5 3/16	5/8	3/4	3/8	1/8	21/32	1 1/2	4 15/16	0.052626	0.63151	5
6	2.11600	2.49400	8 9/16	8 1/4	7 13/32	7 21/64	7 1/4	3/4	1 1/8	1/2	5/32	25/32	1 3/4	7	0.052138	0.62565	6
7	2.75000	3.27000	11 5/8	11 1/4	10 5/32	10 5/64	10	1 1/8	1 3/8	3/4	3/16	1 5/32	2 5/8	9 1/2	0.052000	0.62400	7

* Table agrees with American national standards for taper shanks except for angle and undercut of tang.
† Size 0 taper shank not listed in American national standards.

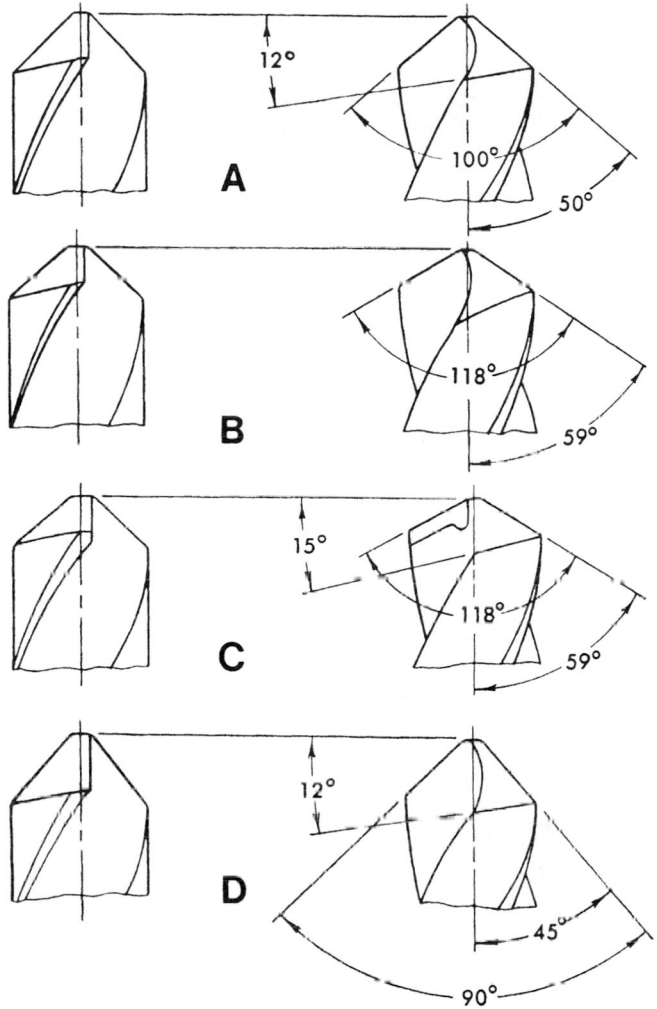

Figure 7.84 Drill-point styles and angles.

periodically, but the styles shown in Fig. 7.84 include some of the newer types, as well as the commonly used older configurations.

The old practice of grinding drill points by hand and eye is, at the least, ineffective with today's modern drills and materials. For a drill to perform accurately and efficiently, modern drill-grinding

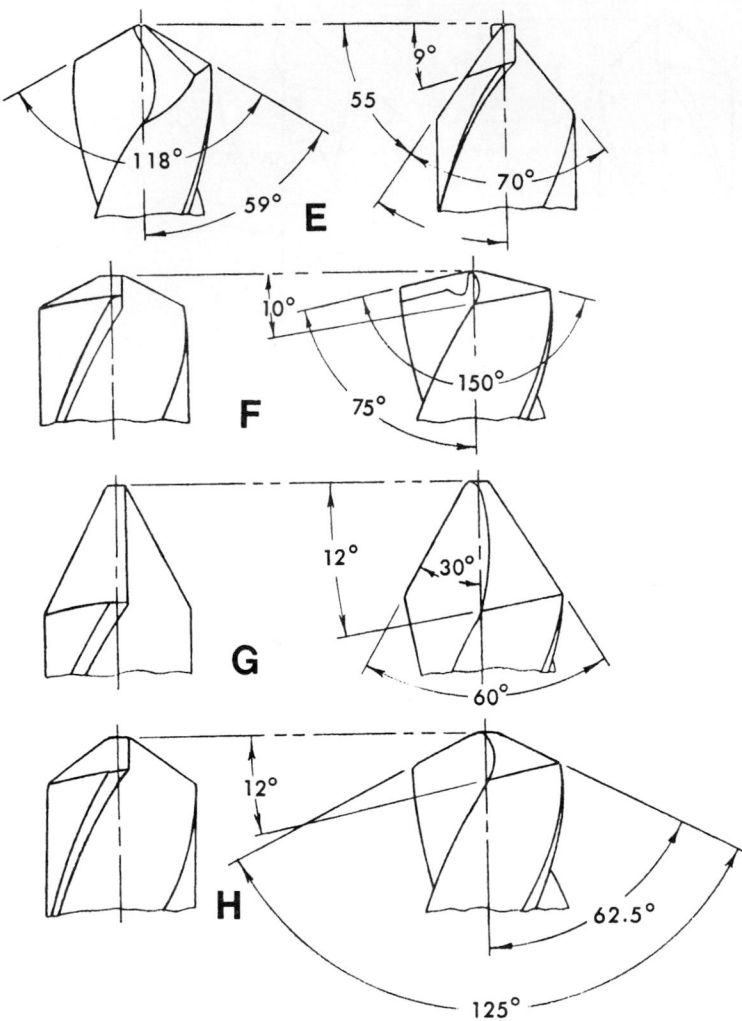

Figure 7.84   (*Continued*)

machines such as the models produced by the Darex Corporation are required. Models are also produced that are capable of sharpening taps, reamers, end mills, and countersinks. Metalcutting tool sharpening and grinding practices are not detailed in this *Handbook*. The tool suppliers and tool-sharpening vendors are equipped

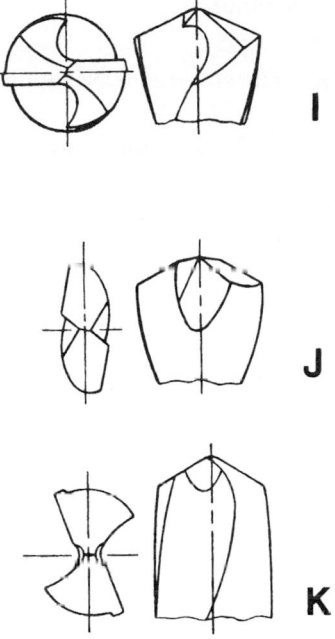

I

J

K

Figure 7.84 (*Continued*)

to sharpen all the cutting tools they distribute in a more accurate and efficient manner than can be done by hand in modern machine shops. HSS- and cobalt-based turning bits are an exception to this modern grinding and sharpening practice, in that the machinist or machine operator occasionally dresses or hones such a cutting tool.

Recommended general uses for drill point angles shown in Fig. 7.84 are shown below. Figure 7.84*k* illustrates web thinning of a standard twist drill.

Typical uses

A	Copper and medium to soft copper alloys
B	Molded plastics, Bakelite, etc.
C	Brasses and soft bronzes
D	Alternate for G, cast irons, die castings, and aluminum
E	Crankshafts and deep holes
F	Manganese steel and hard alloys (point angle 125° to 135°)

G   Wood, fiber, hard rubber, and aluminum

H   Heat-treated steels and drop forgings

I   Split point, 118° or 135° point, self-centering (CNC applications)

J   Parabolic flute for accurate, deep holes and rapid cutting

K   Web thinning (thin the web as the drill wears from resharpening; this restores the chisel point to its proper length)

Other drill styles that are used today include the helical or S-point, which is self-centering and permits higher feed rates, and the chamfered point, which is effective in reducing burr generation in many materials.

Drills are produced from HSS and solid carbide or with carbide-brazed inserts. Drill systems are made by many of the leading tool manufacturers that allow the use of removable inserts of carbide, cermet, ceramics, and cubic boron nitride (CBN). Many of the HSS twist drills used today have coatings such as titanium nitride, titanium carbide, aluminum oxide, and other tremendously hard and wear-resistant coatings. These coatings can increase drill life by as much as three to five times over premium HHS and plain carbide drills.

7.3.3 Classification of high-speed steels

Figure 7.85 shows the chemical composition and characteristics of the M and T series of HSSs. The classification shown applies to drills, turning tools, mills, and other tools made of HSS and is of great value in selecting the proper type of HSS tool for the intended application in machining.

7.3.4 Conversion of surface speed to revolutions per minute for drills

Fractional drill sizes. Figure 7.86 shows the standard fractional drill sizes and the revolutions per minute of each fractional drill size for various surface speeds. The drilling-speed tables that follow give the allowable drilling speed (sfpm) of the various materials. From these values, the correct rpm setting for drilling can be ascertained using the speed/rpm tables given here.

Wire drill sizes (1 through 80). See Fig. 7.87a and b.

Letter drill sizes. See Fig. 7.88.

CLASSIFICATION of HIGH-SPEED STEELS

	Chemical Composition					Comparative Physical Properties *			
AISI - SAE Grade	Tungsten	Chromium	Vanadium	Molybdenum	Cobalt	Red Hardness	Abrasion Resistance	Edge Strength	Grinding Ability
M - 1	1.5	4.0	1.0	8.5	-	10	10	18	18
M - 2	6.0	4.0	2.0	5.0	-	10	11	17	17
M - 3-1	6.0	4.0	2.4	6.0	-	10	14	12	8
M - 3-2	6.0	4.0	3.0	6.0	-	10	17	11	6
M - 4	5.5	4.5	4.0	4.5	-	10	18	8	5
M - 7	1.7	4.0	2.0	8.75	-	10	12	17	14
M - 10	-	4.0	2.0	8.0	-	10	10	17	17
M - 15	6.5	4.5	5.0	3.0	5.0	13	19	9	2
M - 30	2.0	4.0	1.0	8.0	5.0	13	10	14	14
M - 34	2.0	4.0	2.0	8.5	8.0	15	11	14	14
M - 36	6.0	4.0	2.0	6.0	9.0	16	11	14	14
M - 42	1.5	3.75	1.15	9.5	8.0	17	19	15	14
T - 1	18.0	4.0	1.0	-	-	10	10	16	18
T - 2	18.0	4.0	2.0	-	-	10	11	15	17
T - 3	18.0	4.0	3.0	-	-	10	14	8	7
T - 4	18.0	4.0	1.0	-	5.0	13	10	11	14
T - 5	18.0	4.0	2.0	-	8.0	15	11	11	14
T - 6	22.0	4.5	1.5	-	12.0	20	10	11	14
T - 15	13.0	4.5	5.0	-	5.0	17	20	9	2

* Comparative Range of Physical Properties:
0 - 4 = Poor; 5 - 8 = Fair; 9 - 12 = Good; 13 - 16 = Very Good; 17 - 20 = Excellent

Figure 7.85 Classification of HSSs.

FRACTIONAL SIZE DRILLS
Surface Feet per Minute

Diam. Inches	10'	12'	15'	20'	25'	30'	35'	40'	45'	50'	60'	70'	80'	90'	100'
							Revolutions per Minute								
1/64	2445	2934	3667	4889	6112	7334	8556	9778	11001	12223	14668	17112	19557	22001	24446
1/32	1222	1467	1833	2445	3056	3667	4278	4889	5500	6112	7334	8556	9778	11001	12223
3/64	815	978	1222	1630	2037	2445	2852	3259	3667	4074	4889	5704	6519	7334	8149
1/16	611	733	917	1222	1528	1833	2139	2445	2750	3056	3667	4278	4889	5500	6112
5/64	489	587	733	978	1222	1467	1711	1956	2200	2445	2934	3422	3911	4400	4889
3/32	407	489	611	815	1019	1222	1426	1630	1833	2037	2445	2852	3259	3667	4074
7/64	349	419	524	698	873	1048	1222	1397	1572	1746	2095	2445	2794	3143	3492
1/8	306	367	458	611	764	917	1070	1222	1375	1528	1833	2139	2445	2750	3056
9/64	272	326	407	543	679	815	951	1086	1222	1358	1630	1901	2173	2445	2716
5/32	244	293	367	489	611	733	856	978	1100	1222	1467	1711	1956	2200	2445
11/64	222	267	333	444	556	667	778	889	1000	1111	1333	1556	1778	2000	2222
3/16	204	244	306	407	509	611	713	815	917	1019	1222	1426	1630	1833	2037
13/64	188	226	282	376	470	564	658	752	846	940	1128	1316	1504	1692	1880
7/32	175	210	262	349	437	524	611	698	786	873	1048	1222	1397	1572	1746
15/64	163	196	244	326	407	489	570	652	733	815	978	1141	1304	1467	1630
1/4	153	183	229	306	382	458	535	611	688	764	917	1070	1222	1375	1528
9/32	136	163	204	272	340	407	475	543	611	679	815	951	1086	1222	1358
5/16	122	147	183	244	306	367	428	489	550	611	733	856	978	1100	1222
11/32	111	133	167	222	278	333	389	444	500	556	667	778	889	1000	1111
3/8	102	122	153	204	255	306	357	407	458	509	611	713	815	917	1019
13/32	94	113	141	188	235	282	329	376	423	470	564	658	752	846	940
7/16	87	105	131	175	218	262	306	349	393	437	524	611	698	786	873
15/32	81	98	122	163	204	244	285	326	367	407	489	570	652	733	815
1/2	76	92	115	153	191	229	267	306	344	382	458	535	611	688	764
9/16	68	81	102	136	170	204	238	272	306	340	407	475	543	611	679
5/8	61	73	92	122	153	183	214	244	275	306	367	428	489	550	611
11/16	56	67	83	111	139	167	194	222	250	278	333	389	444	500	556
3/4	51	61	76	102	127	153	178	204	229	255	306	357	407	458	509
13/16	47	56	71	94	118	141	165	188	212	235	282	329	376	423	470
7/8	44	52	65	87	109	131	153	175	196	218	262	306	349	393	437
15/16	41	49	61	81	102	122	143	163	183	204	244	285	326	367	407
1	38	46	57	76	95	115	134	153	172	191	229	267	306	344	382
1-1/8	34	41	51	68	85	102	119	136	153	170	204	238	272	306	340
1-1/4	31	37	46	61	76	92	107	122	138	153	183	214	244	275	306
1-3/8	28	33	42	56	69	83	97	111	125	139	167	194	222	250	278
1-1/2	25	31	38	51	64	76	89	102	115	127	153	178	204	229	255
1-5/8	24	28	35	47	59	71	82	94	106	118	141	165	188	212	235
1-3/4	22	26	33	44	55	65	76	87	98	109	131	153	175	196	218
1-7/8	20	24	31	41	51	61	71	81	92	102	122	143	163	183	204
2	19	23	29	38	48	57	67	76	86	95	115	134	153	172	191
2-1/4	17	20	25	34	42	51	59	68	76	85	102	119	136	153	170
2-1/2	15	18	23	31	38	46	53	61	69	76	92	107	122	138	153
2-3/4	14	17	21	28	35	42	49	56	63	69	83	97	111	125	139
3	13	15	19	25	32	38	45	51	57	64	76	89	102	115	127
3-1/2	11	13	16	22	27	33	38	44	49	55	65	76	87	98	109

For speeds higher than tabulated, multiply all values by 10 or 100 For speeds lower than tabulated, divide all values by 10

Figure 7.86   Drill rpm/surface speed, fractional drills.

WIRE SIZE DRILLS

Surface Feet per Minute

	10'	12'	15'	20'	25'	30'	35'	40'	45'	50'	60'	70'	80'	90'	100'
Diam. No.							Revolutions per Minute								
1	168	201	251	335	419	503	586	670	754	838	1005	1173	1340	1508	1675
2	173	207	259	346	432	519	605	691	778	864	1037	1210	1382	1555	1728
3	179	215	269	359	448	538	628	717	807	897	1076	1255	1434	1614	1793
4	183	219	274	366	457	548	640	731	822	914	1097	1280	1462	1645	1828
5	186	223	279	372	465	558	651	743	836	930	1115	1301	1487	1673	1859
6	187	225	281	374	468	562	655	749	843	936	1123	1310	1498	1685	1872
7	190	228	285	380	475	570	665	760	855	950	1140	1330	1520	1710	1900
8	192	230	288	384	480	576	672	768	864	960	1151	1343	1535	1727	1919
9	195	234	000	000	487	585	682	780	877	975	1169	1364	1559	1754	1949
10	197	237	296	395	494	592	691	790	888	987	1184	1382	1579	1777	1974
11	200	240	300	400	500	600	700	800	900	1000	1200	1400	1600	1800	2000
12	202	243	303	404	505	606	707	808	909	1010	1213	1415	1617	1819	2021
13	206	248	310	413	516	619	723	826	929	1032	1239	1450	1652	1859	2065
14	210	252	315	420	525	630	735	839	944	1050	1259	1469	1679	1889	2099
15	212	255	318	424	531	637	743	849	955	1064	1276	1489	1702	1914	2127
16	216	259	324	432	540	647	755	863	971	1079	1295	1511	1726	1942	2158
17	221	265	331	442	552	662	773	883	994	1104	1325	1546	1766	1987	2208
18	225	270	338	451	563	676	789	901	1014	1130	1356	1582	1808	2034	2260
19	230	276	345	460	575	690	805	920	1035	1151	1381	1611	1841	2071	2301
20	237	285	356	474	593	712	830	949	1068	1186	1423	1660	1898	2135	2372
21	240	288	360	480	601	721	841	961	1081	1201	1441	1681	1922	2162	2402
22	243	292	365	487	608	730	852	973	1095	1217	1460	1703	1946	2190	2433
23	248	298	372	496	620	744	868	992	1116	1240	1488	1736	1984	2232	2480
24	251	302	377	503	628	754	880	1005	1131	1257	1508	1759	2010	2262	2513
25	255	307	383	511	639	766	894	1022	1150	1276	1533	1789	2044	2300	2555
26	260	312	390	520	650	780	909	1039	1169	1299	1559	1819	2078	2338	2598
27	265	318	398	531	663	790	920	1061	1104	1327	1592	1857	2122	2388	2653
28	272	326	408	544	680	816	952	1087	1223	1360	1631	1903	2175	2447	2719
29	281	337	421	562	702	843	983	1123	1264	1405	1685	1966	2247	2528	2809
30	297	357	446	595	743	892	1040	1189	1338	1487	1784	2081	2378	2676	2973
31	318	382	477	637	796	955	1114	1273	1432	1592	1910	2228	2546	2865	3183
32	329	395	494	659	823	988	1152	1317	1482	1647	1976	2305	2634	2964	3293
33	338	406	507	676	845	1014	1183	1352	1521	1690	2028	2366	2704	3042	3380
34	344	413	516	688	860	1032	1204	1376	1640	1721	2065	2409	2753	3097	3442
35	347	417	521	694	868	1042	1215	1389	1563	1736	2083	2430	2778	3125	3472
36	359	430	538	717	897	1076	1255	1435	1614	1794	2152	2511	2870	3228	3587
37	367	441	551	735	918	1102	1285	1469	1653	1837	2204	2571	2938	3306	3673
38	376	452	564	753	941	1129	1317	1505	1693	1882	2258	2634	3010	3387	3763
39	384	461	576	768	960	1152	1344	1536	1728	1920	2303	2687	3071	3455	3839
40	390	468	585	780	974	1169	1364	1559	1754	1949	2339	2729	3118	3508	3898

For speeds higher than tabulated, multiply all values by 10 or 100. For speeds lower than tabulated, divide all values by 10.

Figure 7.87   Drill rpm/surface speed, wire-size drills.

WIRE SIZE DRILLS

Surface Feet per Minute

Diam. No.	10'	12'	15'	20'	25'	30'	35'	40'	45'	50'	60'	70'	80'	90'	100'
							Revolutions per Minute								
41	398	477	597	796	995	1194	1393	1592	1790	1990	2387	2785	3183	3581	3979
42	409	490	613	817	1021	1226	1430	1634	1838	2043	2451	2860	3268	3677	4085
43	429	515	644	858	1073	1288	1502	1717	1931	2146	2575	3004	3434	3863	4292
44	444	533	666	888	1110	1332	1555	1777	1999	2221	2665	3109	3554	3999	4442
45	466	559	699	932	1165	1397	1630	1863	2096	2329	2795	3261	3726	4192	4658
46	472	566	707	943	1179	1415	1650	1886	2122	2358	2830	3301	3773	4244	4716
47	487	584	730	973	1216	1460	1703	1946	2190	2433	2920	3406	3893	4379	4866
48	503	603	754	1005	1256	1508	1759	2010	2262	2513	3016	3518	4021	4523	5026
49	523	628	785	1046	1308	1570	1831	2093	2355	2617	3140	3663	4186	4710	5233
50	546	655	819	1091	1364	1637	1910	2183	2456	2729	3274	3820	4366	4911	5457
51	570	684	855	1140	1425	1710	1995	2280	2565	2851	3421	3991	4561	5131	5701
52	602	722	902	1203	1504	1805	2105	2406	2707	3008	3609	4211	4812	5414	6015
53	642	770	963	1284	1605	1926	2247	2568	2889	3207	3848	4490	5131	5773	6414
54	694	833	1042	1389	1736	2083	2431	2778	3125	3473	4167	4862	5556	6251	6945
55	735	881	1102	1469	1836	2204	2571	2938	3306	3673	4408	5142	5877	6611	7346
56	821	986	1232	1643	2054	2464	2875	3286	3696	4108	4929	5751	6572	7394	8215
57	888	1066	1332	1777	2221	2665	3109	3553	3997	4452	5342	6232	7122	8013	8903
58	909	1091	1364	1819	2274	2728	3183	3638	4093	4547	5456	6367	7275	8186	9095
59	932	1118	1397	1863	2329	2795	3261	3726	4192	4658	5590	6521	7453	8388	9316
60	955	1146	1432	1910	2387	2865	3342	3820	4297	4775	5729	6684	7639	8594	9549
61	979	1175	1469	1959	2449	2938	3428	3918	4407	4897	5876	6856	7835	8815	9794
62	1005	1206	1508	2010	2513	3016	3518	4021	4523	5025	6030	7035	8040	9045	10050
63	1032	1239	1549	2065	2581	3097	3613	4129	4646	5160	6192	7224	8256	9288	10320
64	1061	1273	1592	2122	2653	3183	3714	4244	4775	5305	6366	7427	8488	9549	10610
65	1091	1310	1637	2183	2728	3274	3820	4365	4911	5455	6546	7637	8728	9819	10910
66	1157	1389	1736	2315	2894	3472	4051	4630	5207	5790	6948	8106	9264	10422	11580
67	1194	1432	1790	2387	2984	3581	4178	4775	5371	5970	7164	8358	9552	10746	11940
68	1232	1479	1848	2464	3080	3696	4313	4929	5545	6160	7392	8624	9856	11088	12320
69	1308	1570	1962	2616	3270	3924	4578	5232	5887	6530	7836	9142	10488	11754	13060
70	1364	1637	2046	2728	3410	4093	4775	5457	6139	6820	8184	9548	10912	12276	13640
71	1469	1763	2204	2938	3673	4407	5142	5876	6611	7365	8838	10311	11784	13257	14730
72	1528	1833	2272	3056	3820	4584	5348	6112	6875	7640	9168	10696	12224	13752	15280
73	1592	1910	2387	3183	3979	4775	5570	6366	7162	7960	9552	11144	12736	14328	15920
74	1698	2037	2546	3395	4244	5093	5942	6791	7639	8510	10212	11914	13616	15318	17020
75	1819	2183	2728	3638	4547	5457	6366	7276	8185	9095	10914	12733	14552	16371	18190
76	1910	2292	2865	3820	4775	5730	6684	7639	8594	9550	11460	13370	15280	17190	19100
77	2122	2546	3183	4244	5305	6366	7427	8488	9549	10610	12732	14854	16976	19098	21220
78	2387	2865	3581	4775	5968	7162	8356	9549	10743	11935	14322	16709	19096	21483	23870
79	2634	3161	3951	5269	6586	7903	9220	10537	11854	13170	15804	18438	21072	23706	26340
80	2829	3395	4244	5659	7074	8488	9903	11318	12732	14150	16980	19810	22640	25470	28300

For speeds higher than tabulated, multiply all values by 10 or 100. For speeds lower than tabulated, divide all values by 10

Figure 7.87   (*Continued*)

LETTER SIZE DRILLS
Surface Feet per Minute

Letter Size	10'	12'	15'	20'	25'	30'	35'	40'	45'	50'	60'	70'	80'	90'	100'
							Revolutions per Minute								
A	163	196	245	326	408	490	571	653	735	818	982	1145	1309	1472	1636
B	160	193	241	321	401	481	562	642	722	803	963	1124	1284	1445	1605
C	158	189	237	316	395	473	552	631	710	789	947	1105	1262	1420	1578
D	155	186	233	311	388	466	543	621	699	778	934	1089	1245	1400	1556
E	153	183	229	306	382	458	535	611	687	764	917	1070	1222	1375	1528
F	149	178	223	297	372	446	520	595	669	743	892	1040	1189	1337	1486
G	146	176	220	293	366	439	512	585	659	732	878	1024	1170	1317	1463
H	144	172	215	287	359	431	503	574	646	718	862	1005	1149	1294	1436
I	140	169	211	281	351	421	492	562	632	702	842	983	1123	1264	1404
J	138	165	207	276	345	414	483	552	621	690	827	965	1103	1241	1379
K	136	163	204	272	340	408	476	544	612	680	815	951	1087	1223	1359
L	132	158	198	263	329	395	461	527	593	659	790	922	1054	1185	1317
M	129	155	194	259	324	388	453	518	583	648	777	907	1036	1166	1295
N	126	152	190	253	316	379	442	506	569	633	759	886	1012	1139	1265
O	121	145	181	242	302	363	423	484	544	605	725	846	967	1088	1209
P	118	142	177	237	296	355	414	473	532	592	710	828	946	1065	1183
Q	115	138	173	230	288	345	403	460	518	575	690	805	920	1035	1150
R	113	135	169	225	282	338	394	451	507	564	676	789	902	1014	1127
S	110	132	165	220	274	329	384	439	494	549	659	769	878	988	1098
T	107	128	160	213	267	320	373	427	480	533	640	746	853	959	1066
U	104	125	156	208	259	311	363	415	467	519	623	727	830	934	1038
V	101	122	152	203	253	304	355	405	456	507	608	709	810	912	1013
W	99	119	148	198	247	297	346	396	445	495	594	693	792	891	989
X	96	115	144	192	240	289	337	385	433	481	576	672	769	865	962
Y	95	113	142	189	236	284	331	378	425	473	567	662	756	851	945
Z	92	111	139	185	231	277	324	370	416	462	555	647	740	832	925

For speeds higher than tabulated, multiply all values by 10 or 100. For speeds lower than tabulated divide all values by 10.

Figure 7.88 Drill rpm/surface speed, letter-size drills.

7.3.5 Tap drill sizes for producing unified inch screw threads, metric and pipe threads

Tap drills for unified inch screw threads. See Fig. 7.89.

Tap-drill sizes for producing metric screw threads. See Fig. 7.90.

Tap-drill sizes for pipe threads (taper and straight pipe). See Fig. 7.91.

Equation for obtaining tap-drill sizes for cutting taps

$$D_h = D_{bm} - 0.0130\left(\frac{\%\ \text{of full size thread desired}}{n_i}\right)$$ for unified inch-size threads

$$D_{h1} = D_{bm1} - \left(\frac{\%\ \text{of full size thread desired}}{76.98}\right)$$ for metric series threads

where D_h = drilled hole size, in

D_{h1} = drilled hole size, mm
D_{bm} = basic major diameter of thread, in
D_{bm1} = basic major diameter of thread, mm
n_i = number of threads per inch

Note: In the preceding equations, use the percentage whole number; i.e., for 84 percent, use 84.

Example: What is the drilled hole size in inches for a ⅜-16 tapped thread with 84 percent of full thread?

$$D_h = 0.375 - 0.0130 \times \left(\frac{84}{16} \right) = 0.30675 \text{ in}$$

0.30675 in is then the decimal equivalent of the required tap drill for 84 percent of full thread. Use the next-closest drill size, which would be letter size N (0.302 in). The diameters of the American standard wire and letter-size drills are shown in Fig. 7.92. For metric tapped holes, see Fig. 7.93.

When producing a tapped hole, be sure that the correct class of fit is satisfied, i.e., class 2B, 3B, interference fit, etc. The different classes of fits for the thread systems are shown in the section on ANSI and American Society of Mechanical Engineers (ASME) standards; see Chap. 17 on ANSI standards applicable to machining and metalworking. Certain fit classes allow for electroplating and other characteristics that are determined by the design engineer, who selects the class of fit for a particular application.

7.3.6   Speeds and feeds, drill geometry, and cutting recommendations for drills

The composite drilling table shown in Fig. 7.94 has been derived from data originated by the Society of Manufacturing Engineers (SME) and various major drill manufacturers. Figure 7.95 illustrates the standard drill point and split point together with the chisel-edge angle range in common use.

7.3.7   Spade drills

Spade drills are used to produce holes ranging from 1 to over 6 in in diameter. Very deep holes can be produced with spade drills,

(Text continued on page 465.)

Tap size	Tap drill size	Decimal equiv. of tap drill, in	Theoretical percent of thread, %	Probable mean over-size, in	Probable hole size, in	Probable percent of thread,* %
0–80	56	0.0465	83	0.0015	0.0480	74
	3⁄64	0.0469	81	0.0015	0.0484	71
	1.20 mm	0.0472	79	0.0015	0.0487	69
	1.25 mm	0.0492	67	0.0015	0.0507	57
1–64	54	0.0550	89	0.0015	0.0565	81
	1.45 mm	0.0571	78	0.0015	0.0586	71
	53	0.0595	67	0.0015	0.0610	59
1–72	1.5 mm	0.0591	77	0.0015	0.0606	68
	53	0.0595	75	0.0015	0.0610	67
	1.55 mm	0610	67	0.0015	0.0606	68
2–56	51	0.0670	82	0.0017	0.0687	74
	1.75 mm	0.0689	73	0.0017	0.0706	66
	50	0.0700	69	0.0017	0.0717	62
	1.80 mm	0.0709	65	0.0017	0.0726	58
2–64	50	0.0700	79	0.0017	0.0717	70
	1.80 mm	0.0709	74	0.0017	0.0726	66
	49	0.0730	64	0.0017	0.0747	56
3–48	48	0.0760	85	0.0019	0.0779	78
	5⁄64	0.0781	77	0.0019	0.0800	70
	47	0.0785	76	0.0019	0.0804	69
	2.00 mm	0.0787	75	0.0019	0.0806	68
	46	0.0810	67	0.0019	0.0829	60
	45	0.0820	63	0.0019	0.0839	56
3–56	46	0.0810	78	0.0019	0.0829	69
	45	0.0820	73	0.0019	0.0839	65
	2.10 mm	0.0827	70	0.0019	0.0846	62
	2.15 mm	0.0846	62	0.0019	0.0865	54
4–40	44	0.0860	80	0.0020	0.0880	74
	2.20 mm	0.0866	78	0.0020	0.0886	72
	43	0.0890	71	0.0020	0.0910	65
	2.30 mm	0.0906	66	0.0020	0.0926	60
4–48	2.35 mm	0.0925	72	0.0020	0.0926	72
	42	0.0935	68	0.0020	0.0955	61
	3⁄32	0.0938	68	0.0020	0.0958	60
	2.40 mm	0.0945	65	0.0020	0.0965	57
5–40	40	0.0980	83	0.0023	0.1003	76
	39	0.0995	79	0.0023	0.1018	71
	38	0.1015	72	0.0023	0.1038	65
	2.60 mm	0.1024	70	0.0023	0.1047	63

Figure 7.89 Tap-drill sizes, unified inch screw threads.

Tap size	Tap drill size	Decimal equiv. of tap drill, in	Theoretical percent of thread, %	Probable mean over-size, in	Probable hole size, in	Probable percent of thread,* %
5–44	38	0.1015	79	0.0023	0.1038	72
	2.60 mm	0.1024	77	0.0023	0.1047	69
	37	0.1040	71	0.0023	0.1063	63
6–32	37	0.1040	84	0.0023	0.1063	78
	36	0.1065	78	0.0023	0.1088	72
	$\frac{7}{64}$	0.1095	70	0.0026	0.1120	64
	35	0.1100	69	0.0026	0.1126	63
	34	0.1100	67	0.0026	0.1136	60
6–40	34	0.1110	83	0.0026	0.1136	75
	33	0.1130	77	0.0026	0.1156	69
	2.90 mm	0.1142	73	0.0026	0.1168	65
	32	0.1160	68	0.0026	0.1186	60
8–32	3.40 mm	0.1339	74	0.0029	0.1368	67
	29	0.1360	69	0.0029	0.1389	62
8–36	29	0.1360	78	0.0029	0.1389	70
	3.5 mm	0.1378	72	0.0029	0.1407	65
10–24	27	0.1440	85	0.0032	0.1472	79
	3.70 mm	0.1457	82	0.0032	0.1489	76
	26	0.1470	79	0.0032	0.1502	74
	25	0.1495	75	0.0032	0.1527	69
	24	0.1520	70	0.0032	0.1552	64
10–32	$\frac{5}{32}$	0.1563	83	0.0032	0.1595	75
	22	0.1570	81	0.0032	0.1602	73
	21	0.1590	76	0.0032	0.1622	68
12–24	$\frac{11}{64}$	0.1719	82	0.0035	0.1754	75
	17	0.1730	79	0.0035	0.1765	73
	16	0.1770	72	0.0035	0.1805	66
12–28	16	0.1770	84	0.0035	0.1805	77
	15	0.1800	78	0.0035	0.1835	70
	4.60 mm	0.1811	75	0.0035	0.1846	67
	14	0.1820	73	0.0035	0.1855	66
¼–20	9	0.1960	83	0.0038	0.1998	77
	8	0.1990	79	0.0038	0.2028	73
	7	0.2010	75	0.0038	0.2048	70
	$\frac{13}{64}$	0.2031	72	0.0038	0.2069	66
¼–28	5.40 mm	0.2126	81	0.0038	0.2164	72
	3	0.2130	80	0.0038	0.2168	72
$\frac{5}{16}$–18	F	0.2570	77	0.0038	0.2608	72
	6.60 mm	0.2598	73	0.0038	0.2636	68
	G	0.2610	71	0.0041	0.2651	66
$\frac{5}{16}$–24	H	0.2660	86	0.0041	0.2701	78
	6.80 mm	0.2677	83	0.0041	0.2718	75

Figure 7.89 *(Continued)*

Tap size	Tap drill size	Decimal equiv. of tap drill, in	Theoretical percent of thread, %	Probable mean over-size, in	Probable hole size, in	Probable percent of thread,* %
	I	0.2720	75	0.0041	0.2761	67
⅜–16	7.80 mm	0.3071	84	0.0044	0.3115	78
	7.90 mm	0.3110	79	0.0044	0.3154	73
	⁵⁄₁₆	0.3125	77	0.0044	0.3169	72
	O	0.3160	73	0.0044	0.3204	68
8–24	²¹⁄₆₄	0.3281	87	0.0044	0.3325	79
	8.40 mm	0.3307	82	0.0044	0.3351	74
	Q	0.3320	79	0.0044	0.3364	71
	8.50 mm	0.3346	75	0.0044	0.3390	67
⁷⁄₁₆–14	T	0.3580	86	0.0046	0.3626	81
	²³⁄₆₄	0.3594	84	0.0046	0.3640	79
	9.20 mm	0.3622	81	0.0046	0.3668	76
	9.30 mm	0.3661	77	0.0046	0.3707	72
	U	0.3680	75	0.0046	0.3726	70
	9.40 mm	0.3701	73	0.0046	0.3747	68
⁷⁄₁₆–20	W	0.3860	79	0.0046	0.3906	72
	²⁵⁄₆₄	0.3906	72	0.0046	0.3952	65
½–13	10.50 mm	0.4134	87	0.0047	0.4181	82
	²⁷⁄₆₄	0.4219	78	0.0047	0.4266	73
½–20	²⁹⁄₆₄	0.4531	72	0.0047	0.4578	65
⁹⁄₁₆–12	¹⁵⁄₃₂	0.4688	87	0.0048	0.4736	82
	³¹⁄₆₄	0.4844	72	0.0048	0.4892	68
⁹⁄₁₆–18	½	0.5000	87	0.0048	0.5048	80
⅝–11	¹⁷⁄₃₂	0.5313	79	0.0049	0.5362	75
⅝–18	⁹⁄₁₆	0.5625	87	0.0049	0.5674	80
¾–10	⁴¹⁄₆₄	0.6406	84	0.0050	0.6456	80
	²¹⁄₃₂	0.6563	72	0.0050	0.6613	68
¾–16	¹¹⁄₁₆	0.6875	77	0.0050	0.6925	71
	17.50 mm	0.6890	75	0.0050	0.6940	69
⅞–9	⁴⁹⁄₆₄	0.7656	76	0.0052	0.7708	72
⅞–14	⁵¹⁄₆₄	0.7969	84	0.0052	0.8021	79
1–8	⁵⁵⁄₆₄	0.8594	87	0.0059	0.8653	83
	⅞	0.8750	77	0.0059	0.8809	73
1–12	²⁹⁄₃₂	0.9063	87	0.0059	0.9122	81
	⁵⁹⁄₆₄	0.9219	72	0.0060	0.9279	67
1–14	⁵⁹⁄₆₄	0.9219	84	0.0060	0.9279	78
1⅛–7	³¹⁄₃₂	0.9688	84	0.0062	0.9750	81
	⁶³⁄₆₄	0.9844	76	0.0067	0.9911	72
1⅛–12	1¹⁄₃₂	1.0313	87	0.0071	1.0384	80

Figure 7.89 *(Continued)*

Metric Tap size	Tap drill size	Decimal equiv. of tap drill, in	Theoretical percent of thread, %	Probable mean over- size, in	Probable hole size, in	Probable percent of thread,* %
M1.6 × 0.35	1.20 mm	0.0472	88	0.0014	0.0486	80
	1.25 mm	0.0492	77	0.0014	0.0506	69
M2 × 0.4	1⁄16	0.0625	79	0.0015	0.0640	72
	1.60 mm	0.0630	77	0.0017	0.0647	69
	52	0.0635	74	0.0017	0.0652	66
M2.5 × 0.45	2.05 mm	0.0807	77	0.0019	0.0826	69
	46	0.0810	76	0.0019	0.0829	67
	45	0.0820	71	0.0019	0.0839	63
M3 × 0.5	40	0.0980	79	0.0023	0.1003	70
	2.5 mm	0.0984	77	0.0023	0.1007	68
	39	0.0995	73	0.0023	0.1018	64
M3.5 × 0.6	33	0.1130	81	0.0026	0.1156	72
	2.9 mm	0.1142	77	0.0026	0.1163	68
	32	0.1160	71	0.0026	0.1186	63
M4 × 0.7	3.2 mm	0.1260	88	0.0029	0.1289	80
	30	0.1285	81	0.0029	0.1314	73
	3.3 mm	0.1299	77	0.0029	0.1328	69
M4.5 × 0.75	3.7 mm	0.1457	82	0.0032	0.1489	74
	26	0.1470	79	0.0032	0.1502	70
	25	0.1495	72	0.0032	0.1527	64
M5 × 0.8	4.2 mm	0.1654	77	0.0032	0.1686	69
	19	0.1660	75	0.0032	0.1692	68
M6 × 1	10	0.1935	84	0.0038	0.1973	76
	9	0.1960	79	0.0038	0.1998	71
	5 mm	0.1968	77	0.0038	0.2006	70
	8	0.1990	73	0.0038	0.2028	65
M7 × 1	A	0.2340	81	0.0038	0.2378	74
	6 mm	0.2362	77	0.0038	0.2400	70
	B	0.2380	74	0.0038	0.2418	66
M8 × 1.25	6.7 mm	0.2638	80	0.0041	0.2679	74
	17⁄64	0.2656	77	0.0041	0.2697	71
	H	0.2660	77	0.0041	0.2701	70
	6.8 mm	0.2677	74	0.0041	0.2718	68
M10 × 1.5	8.4 mm	0.3307	82	0.0044	0.3351	76
	Q	0.3320	80	0.0044	0.3364	75
	8.5 mm	0.3346	77	0.0044	0.3390	71
M12 × 1.75	10.25 mm	0.4035	77	0.0047	0.4082	72
	Y	0.4040	76	0.0047	0.4087	71
	13⁄32	0.4062	74	0.0047	0.4109	69
M14 × 2	15⁄32	0.4688	81	0.0048	0.4736	76
	12 mm	0.4724	77	0.0048	0.4772	72

Figure 7.90 Tap-drill sizes, metric screw threads.

Metric Tap size	Tap drill size	Decimal equiv. of tap drill, in	Theoretical percent of thread, %	Probable mean over-size, in	Probable hole size, in	Probable percent of thread,* %
M16 × 2	³⁵⁄₆₄	0.5469	81	0.0049	0.5518	76
	14 mm	0.5512	77	0.0049	0.5561	72
M20 × 2.5	¹¹⁄₁₆	0.6875	78	0.0050	0.6925	74
	17.5 mm	0.6890	77	0.0052	0.6942	73
M24 × 3	¹³⁄₁₆	0.8125	86	0.0052	0.8177	82
	21 mm	0.8268	76	0.0054	0.8322	73
	⁵³⁄₆₄	0.8281	76	0.0054	0.8335	73
M30 × 3.5	1¹⁄₃₂	1.0312	83	0.0071	1.0383	80
	25.1 mm	1.0394	79	0.0071	1.0465	75
	1³⁄₆₄	1.0469	75	0.0072	1.0541	70
M36 × 4	1¹⁷⁄₆₄	1.2656	74	Reaming recommended		

Figure 7.90 (*Continued*)

Taper pipe		Straight pipe	
Thread	Drill	Thread	Drill
⅛–27	R	⅛–27	S
¼–18	⁷⁄₁₆	¼–18	²⁹⁄₆₄
⅜–18	³⁷⁄₆₄	⅜–18	¹⁹⁄₃₂
½–14	²³⁄₃₂	½–14	⁴⁷⁄₆₄
¾–14	⁵⁹⁄₆₄	¾–14	¹⁵⁄₁₆
1–11½	1⁵⁄₃₂	1–11½	1³⁄₁₆
1¼–11½	1½	1¼–11½	1³³⁄₆₄
1½–1½	1⁴⁷⁄₆₄	1½–11½	1¾
2–11½	2⁷⁄₃₂	2–11½	2⁷⁄₃₂
2½–8	2⅝	2½–8	2²¹⁄₃₂
3–8	3¼	3–8	3⁹⁄₃₂
3½–8	3¾	3½–8	3²⁵⁄₃₂
4–8	4¼	4–8	4⁹⁄₃₂

Figure 7.91 Pipe taps.

DRILL NO.	DECIMAL	DRILL NO.	DECIMAL	DRILL NO.	DECIMAL
97	0.0059	56	0.0465	15	0.180
96	0.0063	55	0.052	14	0.182
95	0.0067	54	0.055	13	0.185
94	0.0071	53	0.0595	12	0.189
93	0.0075	52	0.0635	11	0.191
92	0.0079	51	0.067	10	0.1935
91	0.0083	50	0.070	9	0.196
90	0.0087	49	0.073	8	0.199
89	0.0091	48	0.076	7	0.201
88	0.0095	47	0.0785	6	0.204
87	0.010	46	0.076	5	0.2055
86	0.0105	45	0.082	4	0.209
85	0.011	44	0.086	3	0.213
84	0.0115	43	0.089	2	0.221
83	0.012	42	0.0935	1	0.228
82	0.0125	41	0.096	A	0.234
81	0.013	40	0.098	B	0.238
80	0.0135	39	0.0995	C	0.242
79	0.0145	38	0.1015	D	0.246
78	0.016	37	0.104	E	0.250
77	0.018	36	0.1065	F	0.257
76	0.020	35	0.110	G	0.261
75	0.021	34	0.111	H	0.266
74	0.0225	33	0.113	I	0.272
73	0.024	32	0.116	J	0.277
72	0.025	31	0.120	K	0.281
71	0.026	30	0.1285	L	0.290
70	0.028	29	0.136	M	0.295
69	0.0292	28	0.1405	N	0.302
68	0.031	27	0.144	O	0.316
67	0.032	26	0.147	P	0.323
66	0.033	25	0.1495	Q	0.332
65	0.035	24	0.152	R	0.339
64	0.036	23	0.154	S	0.348
63	0.037	22	0.157	T	0.358
62	0.038	21	0.159	U	0.368
61	0.039	20	0.161	V	0.377
60	0.040	19	0.166	W	0.386
59	0.041	18	0.1695	X	0.397
58	0.042	17	0.173	Y	0.404
57	0.043	16	0.177	Z	0.413

Figure 7.92 Drill sizes (American national standard).

Drill	Decimal	Drill	Decimal	Drill	Decimal	Drill	Decimal	Drill	Decimal	Drill	Decimal
.35mm	.0138	1.75mm	.0689	3.70mm	.1457	6.40mm	.2520	9.00mm	.3543	17.00mm	.6693
.38mm	.0150	1.80mm	.0709	3.75mm	.1477	6.50mm	.2559	9.10mm	.3583	17.50mm	.6890
.40mm	.0157	1.85mm	.0728	3.80mm	.1496	6.60mm	.2598	9.20mm	.3622	18.00mm	.7087
.42mm	.0165	1.90mm	.0748	3.90mm	.1535	6.70mm	.2638	9.25mm	.3642	18.50mm	.7283
.45mm	.0177	1.95mm	.0768	4.00mm	.1575	6.75mm	.2658	9.30mm	.3661	19.00mm	.7480
.48mm	.0189	2.00mm	.0787	4.10mm	.1614	6.80mm	.2677	9.40mm	.3701	19.50mm	.7677
.50mm	.0197	2.05mm	.0807	4.20mm	.1654	6.90mm	.2716	9.50mm	.3740	20.00mm	.7874
.55mm	.0217	2.10mm	.0827	4.25mm	.1674	7.00mm	.2756	9.60mm	.3780	20.50mm	.8071
.60mm	.0236	2.15mm	.0846	4.50mm	.1771	7.10mm	.2795	9.70mm	.3819	21.00mm	.8268
.65mm	.0256	2.20mm	.0866	4.60mm	.1811	7.20mm	.2835	9.75mm	.3839	21.50mm	.8465
.70mm	.0276	2.25mm	.0886	4.70mm	.1850	7.25mm	.2855	9.80mm	.3858	22.00mm	.8661
.75mm	.0295	2.30mm	.0905	4.75mm	.1870	7.30mm	.2874	9.90mm	.3898	22.50mm	.8858
.80mm	.0315	2.35mm	.0925	4.80mm	.1890	7.40mm	.2913	10.00mm	.3937	23.00mm	.9055
.85mm	.0335	2.40mm	.0945	4.90mm	.1929	7.50mm	.2953	10.20mm	.4016	23.50mm	.9252
.90mm	.0354	2.45mm	.0965	5.00mm	.1968	7.60mm	.2992	10.50mm	.4134	24.00mm	.9449
.95mm	.0374	2.50mm	.0984	5.10mm	.2008	7.70mm	.3031	10.80mm	.4252	24.50mm	.9646
1.00mm	.0394	2.55mm	.1004	5.20mm	.2047	7.75mm	.3051	11.00mm	.4330	25.00mm	.9843
1.05mm	.0413	2.60mm	.1024	5.25mm	.2067	7.80mm	.3071	11.20mm	.4409	+1.00mm increments	
1.10mm	.0433	2.65mm	.1043	5.30mm	.2087	7.90mm	.3110	11.50mm	.4528	up to 48mm.	
1.15mm	.0453	2.70mm	.1063	5.40mm	.2126	8.00mm	.3150	11.80mm	.4646		
1.20mm	.0472	2.75mm	.1083	5.50mm	.2165	8.10mm	.3189	12.00mm	.4724	+5.00mm increments	
1.25mm	.0492	2.80mm	.1102	5.60mm	.2205	8.20mm	.3228	12.20mm	.4803	from 50mm up to	
1.30mm	.0512	2.90mm	.1142	5.70mm	.2244	8.25mm	.3248	12.50mm	.4921	105mm.	
1.35mm	.0531	3.00mm	.1181	5.75mm	.2264	8.30mm	.3268	13.00mm	.5118		
1.40mm	.0551	3.10mm	.1220	5.80mm	.2283	8.40mm	.3307	13.50mm	.5315		
1.45mm	.0571	3.20mm	.1260	5.90mm	.2323	8.50mm	.3346	14.00mm	.5512		
1.50mm	.0591	3.25mm	.1280	6.00mm	.2362	8.60mm	.3386	14.50mm	.5709		
1.55mm	.0610	3.30mm	.1299	6.10mm	.2401	8.70mm	.3425	15.00mm	.5906		
1.60mm	.0629	3.40mm	.1339	6.20mm	.2441	8.75mm	.3445	15.50mm	.6102		
1.65mm	.0650	3.50mm	.1378	6.25mm	.2461	8.80mm	.3465	16.00mm	.6299		
1.70mm	.0669	3.60mm	.1417	6.30mm	.2480	8.90mm	.3504	16.50mm	.6496		

Figure 7.93 Drill sizes (metric).

Speeds, Feeds, Drill Geometry, and Cutting Recommendations for Standard Drill Types *

Material Type	Hardness Bhn	Tool grade	Drill type	PA-deg.	LRf deg.	HA-deg.	Point type	Speed, sfpm	Feed, ipr x 10³
Low-alloy steels 4130, 4340, 4140	to 300	M-1, M-2	A	118-135	7-10	25-30	Split	50-60	3-7
	300-400	M-1, M-2	A	118-135	7-10	25-30	split	40-50	2-6
	400-500	Cobalt	B	118-135	7-10	25-30	split	25-40	1-4
	over 500	C-2	C	118	7-10	0	notched	75-100	0.5-2
Die steels Hot-work	to 300	M-1, M-2	A	118-135	7-10	25-30	split	45-55	3-7
	300-400	M-1, M-2	A	118-135	7-10	25-30	split	35-50	2-6
	400-500	Cobalt	B	118-135	7-10	25-30	split	25-35	1-4
	over 500	C-2	C	118	7-10	0	Notched	70-90	0.5-2
Stainless steels (Austenitic) 300	135-185	M-1, M-2	A	118-135	7-10	25-30	split	70-90	2-6
Stainless steels (Martensitic) 400	150-250	M-1, M-2	A	118-135	7-10	25-30	split	50-70	3-7
	250-450	M-1, M-2	A	118-135	7-10	25-30	split	30-40	2-6
	over 450	Cobalt	B	118-135	7-10	25-30	split	20-30	1-4
Stainless steels Precipitation hardening 17-7PH, etc.	to 200	M-1, M-2	A	118-135	7-10	25-30	split	50-60	3-7
	200-350	M-1, M-2	A	118-135	7-10	25-30	split	35-45	2-6
	over 350	Cobalt	B	118-135	7-10	25-30	split	20-30	1-4
Nickel-cobalt steels High-strength	to 400	M-1, M-2	A	118-135	7-10	25-30	split	55-65	2-6
	400-500	Cobalt	B	118-135	7-10	25-30	split	30-40	1-4
	over 500	C-2	C	118	7-10	0	notched	70-90	0.5-2
High-temperature Cobalt-base alloys	to 300	Cobalt	B	118-135	7-10	25-30	split	15-25	1-4
High-temperature Iron-base alloys	to 250	Cobalt	B	118-135	7-10	25-30	split	20-30	2-6
	over 250	Cobalt	B	118-135	7-10	25-30	split	15-25	2-6
High-temperature Nickel-base alloys	to 265	Cobalt	B	118-135	7-10	25-30	split	20-30	2-6
	265-330	Cobalt	B	118-135	7-10	25-30	split	20-25	2-5
	over 330	Cobalt	B	118-135	7-10	25-30	split	15-20	1-4

Material	Condition	Tool	Type				Point		
Magnesium & alloys	All	M-1, M-2	A	118-135	7-10	25-30	split	150-350	2-7
Aluminum & alloys 2024, 6061, 7075, etc.	All	M-1, M-2	A	118-135	7-10	25-30	split	175-400	2-7
Titanium	to 250	M-34, M-42	B	135	7-12	30-38	split	25-30	5-7
Titanium Alpha alloys	250-300	M-34, M-42	B	135	7-12	30-38	split	20-25	5-7
Titanium Alpha-Beta alloys	to 350	M-34, M-42	B	135	7-12	30-38	split	20-25	5-7
	over 350	M-34, M-42	B	135	7-12	30-38	split	15-25	5-7
Titanium Beta alloys	to 350	M-34, M-42	B	135	7-12	30-38	split	15-20	1-4
	over 350	M-34, M-42	B	135	7-12	30-38	split	15-17	0.5-2
Beryllium copper	250	C-2	D	90-118	10-15	25-30	split	30-45	2-8
Tungsten & alloys	to 350	C-2, C-3	D	90-118	7-10	25-30	notched	200-250	1-4
Brass, free-machining	All	M-1, M-2	A	118	7-10	25-30	standard	100-250	4-10
Bronzes, common	All	M-1, M-2	A	118	7-10	25-30	standard	200-250	3-15
Bronze, phosphur	Hard	M-1, M-2	A	118	7-10	25-30	notched	75-150	2-6
Copper	All	M-1, M-2	A	90-118	7-10	25-30	standard	100-250	1-5
Cast iron Soft to medium	soft-med.	M-1, M-2	A	118	7-10	25-30	std or split	75-150	2-8
Cast iron Hard	Hard	C-2	D	118	7-10	25-30	std or split	40-75	1-5
Zinc	All	M-1, M-2	A	118	7-10	25-30	standard	200-250	3-10
Low-carbon steels	to 300	M-1, M-2	A	118	7-10	25-30	standard	80-100	3-10
Thermoplastics	Medium	M-1, M-2	E	60-90	12-15	17	standard	100-150	2-15

Figure 7.94 Drilling recommendation table.

463

| Thermosetting plastics | Soft | M-1, M-2 | E | 60-90 | 12-16 | 17 | standard | 150 | 3-8 |
| | Hard | M-1, M-2 | E | 60-90 | 12-16 | 17 | standard | 100 | 2-6 |

NOTE:
Drill Types: A = AIAA type B or C; B = Heavy duty cobalt; C = Carbide tipped; D = Solid carbide; E = Standard with wide, polished flutes.

Coolants and cutting fluids are recommended in the Coolant & Cutting Fluid Section of the handbook.

* Tabular data in the table is for drills of 0.125 through 0.500" diameter and hole depths of 1 to 3 drill diameters. Adjustments must be made for other conditions, by interpolation or trial drilling. (Smaller drills have a lower ipr feed rate; larger drills have a higher ipr feed rate).

Drill geometry: PA = Point angle, degrees; LRf = Lip relief angle, degrees; HA = helix angle, degrees

Tabular data for ipr - Feed is given in powers of ten notation, i.e. 2 = .002", 6 = .006"; 0.5 = 0.0005", etc.

Figure 7.94 (*Continued*)

Standard and Split Drill Points

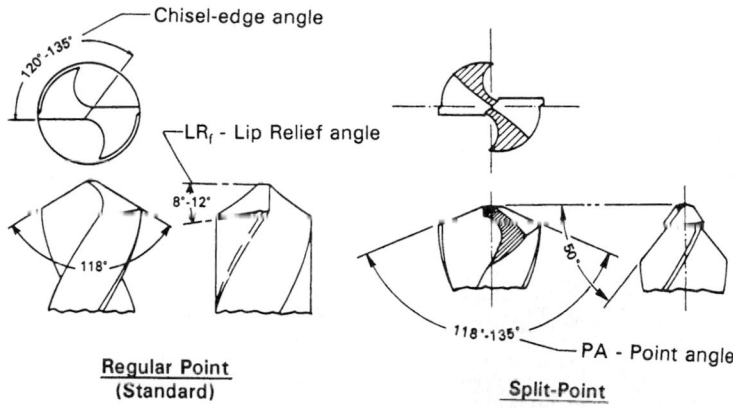

Figure 7.95 Standard and split points.

including core drilling, counterboring, and bottoming to a flat or other shape. The spade drill consists of the spade drill bit and holder. The holder may contain coolant holes through which coolant can be delivered to the cutting edges, under pressure, which cools the spade and flushes the chips from the drilled hole. Typical spade drills, holders, and terminology and spade geometry are shown in Fig. 7.96.

The standard point angle on a spade drill is 130°. The rake angle ranges from 10° to 12° for average-hardness materials. The rake angle should be 5° to 7° for hard steels and 15° to 20° for soft, ductile materials. The back-taper angle should be 0.001 to 0.002 in per inch of blade depth. The outside diameter clearance angle is generally between 7° and 10°.

The cutting speeds for spade drills are normally 10 to 15 percent lower than those for standard twist drills. See the tables of drill speeds and feeds in the preceding section for approximate starting speeds. Heavy feed rates should be used with spade drilling. The table shown in Fig. 7.97 gives recommended feed rates for spade drilling various materials.

Horsepower and thrust forces for spade drilling. The following simplified equations will allow you to calculate the approximate horse-

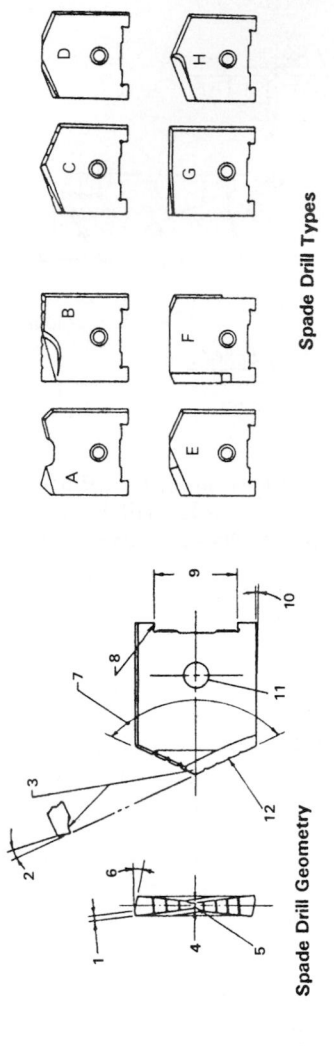

Spade Drill Types

A- Spur-core blade
B- Facing blade (flat bottom)
C- Regular blade
D- Standard core blade
E- Standard carbide core blade
F- Core blade with carbide edges
G- Counterbore
H- Regular carbide blade

Spade Drill Holder

Spade Drill Geometry

1- Circular OD land, radial land or margin
2- Front clearance/lip relief
3- Chip curler/gullet
4- Web
5- Chisel edge
6- OD clearance/radial relief
7- Point angle
8- Locating ears
9- Locating slot
10- Back taper
11- Screw hole
12- Chipbreaker grooves
13- Flute length
14- Locating flats
15- Coolant inlet (pipe thread)
16- Coolant holes
17- Shank
18- Coolant inductor
19- Blade screw
20- Body diameter

Figure 7.96 Spade drills.

power requirements and thrust needed to spade drill various materials with different diameter spade drills. In order to do this, you must find the feed rate for your particular spade drill diameter, as shown in Fig. 7.97, and then select the P factor for your material, as tabulated in Fig. 7.98.

The following equations then may be used to estimate the required horsepower at the machine's motor and the thrust required in pounds force for the drilling process.

$$C_{\text{hp}} = P\left(\frac{\pi D^2}{4}\right) FN$$

$$T_p = 148,500 PFD$$

$$M_{\text{hp}} = \frac{C_{\text{hp}}}{e}$$

$$F = \frac{f_m}{N} \quad \text{and} \quad f_m = FN$$

where C_{hp} = horsepower at the cutter
M_{hp} = required motor horsepower
T_p = thrust for spade drilling, pounds force
D = drill diameter, in
F = feed, ipr (see Fig. 7.97 for ipr/diameter/material)
P = power factor constant (see Fig. 7.98)
f_m = feed, ipm
N = spindle speed, rpm
e = drive motor efficiency factor (0.90 for direct belt drive to the spindle; 0.80 for geared head drive to the spindle)

Note: The P factors must be increased by 40 to 50 percent for dull tools, although dull cutters should not be used if productivity is to remain high.

7.3.8 Microdrills

Drills below 0.020 in in diameter are considered to be microdrills. Microdrills as small as 0.0016 in in diameter are available as stock items from the drill supply vendors. Microdrills are manufactured from HSS, cobalt-base carbides, and solid-tungsten carbides, and a limited selection is available in diamond and CBN surfaces. Some

Recommended Feed Rates for Spade Drilling- (ipr, inches per revolution)

Material	Hardness Bhn	Feed - ipr Spade Drill Diameter, inches					
		1-1.25	1.25-2	2-3	3-4	4-5	5-8
Plain carbon steels	100-225	.012	.015	.018	.022	.025	.030
	225-275	.010	.013	.015	.018	.020	.025
	275-325	.008	.010	.013	.015	.018	.020
Free-machining steels	100-240	.014	.016	.018	.022	.025	.030
	240-325	.010	.014	.016	.020	.022	.025
Free-machining alloy steels	150-250	.014	.016	.018	.022	.025	.030
	250-325	.012	.014	.016	.018	.020	.025
	325-375	.010	.012	.014	.016	.018	.020
Alloy steels	125-180	.012	.015	.018	.022	.025	.030
	180-225	.010	.012	.016	.018	.022	.025
	225-325	.009	.010	.013	.015	.018	.020
	325-400	.006	.006	.010	.012	.014	.016
Grey cast iron	110-160	.020	.022	.026	.028	.030	.034
	160-240	.012	.014	.016	.018	.020	.022
	240-325	.010	.012	.016	.018	.018	.018
Ductile & nodular iron	140-190	.014	.016	.018	.020	.022	.024
	190-250	.012	.014	.016	.018	.018	.020
	250-325	.010	.012	.016	.018	.018	.018
Malleable iron- ferritic	110-160	.014	.016	.018	.020	.022	.024
Malleable iron- pearlitic	160-280	.011	.013	.015	.018	.018	.018
Stainless steels- free-machining							
Ferritic & austenitic (screw stock)	------	.016	.018	.020	.022	.024	.026
Martensitic (440F, etc.)	------	.012	.014	.016	.016	.018	.020

	Hardness						
Stainless steels							
Ferritic & austenitic (200 & 300 series)	------	.012	.014	.016	.018	.020	.020
Martensitic (400 series)	------	.010	.012	.012	.014	.016	.018
Copper alloys							
Soft (ETP-110, etc)	------	.016	.018	.020	.026	.028	.030
Hard (bronzes)	------	.010	.012	.014	.016	.018	.018
Aluminum alloys (free-machining)	------	.020	.022	.024	.028	.030	.040
Magnesium alloys (general) **	------	.024	.026	.030	.034	.040	.050
High-temperature alloys (general)	------	.008	.010	.012	.012	.014	.014
Titanium alloys (general)	------	.008	.010	.012	.014	.014	.016
Tool steels							
Water-hardening & shock resisting	150-250	.012	.14	.016	.018	.020	.022
Cold work	200-250	.007	.008	.009	.010	.011	.012
Hot work	150-250	.012	.013	.015	.016	.018	.020
High-speed steels	200-250	.010	.012	.013	.015	.017	.018

NOTE: Hardness ranges are Brinell hardness numbers. Tabular data is in inches of feed per revolution, ipr.
** A fire hazard exists when machining or drilling magnesium & alloys.

Figure 7.97 Feed rates for spade drills.

"P" Factors for Spade Drilling Various Materials

Material	Hardness Bhn	"P" Factor
Plain carbon & alloy steels	90-200	0.75
	200-275	0.92
	300-375	1.02
	375-450	1.18
	45-52R_C	1.45
Gray cast irons	-------	0.25
Alloy cast irons & ductile irons	-------	0.50
Stainless steel (austenitic)	-------	0.96
Stainless steels (martensitic)	-------	0.81
Titanium alloys	-------	0.87
Aluminum alloys	-------	0.20
Magnesium alloys	-------	0.15
Copper alloys	Soft - R_B 20-80	0.42
	Hard - R_B 80-100	0.75
Tool steels	-------	1.10
Cobalt based alloys	-------	1.25
High-temperature alloys	-------	1.45
Non-ferrous free-machining alloys	-------	0.45

Note: Where no hardness range is given, the maximum hardness is 300 Bhn. For harder materials, use a higher "P" factor.

Figure 7.98 *P*-factor table.

of the standard commercially available microdrill sizes are listed in Fig. 7.99.

In order to use microdrills, special machines have been developed for controlling these extremely small and fragile tools. Holes of great precision may be made in many materials using the appropriate machine tools and drills designed for microdrilling.

Needless to say, a microdrilling operation is conducted under a high-magnification viewing device, with the hand motions of the

Standard Size Microdrills - High-Speed Steel and Cobalt

High-Speed Steel Microdrills

Diameter inches	mm	Flute lth.
0.0059	0.15	0.0469
0.0063	0.16	0.0469
0.0067	0.17	0.0625
0.0071	0.18	0.0625
0.0075	0.19	0.0625
0.0079	0.20	0.0625
0.0083	0.21	0.0625
0.0087	0.22	0.0625
0.0091	0.23	0.0625
0.0094	0.24	0.0781
0.0098	0.25	0.0781
0.0102	0.26	0.0781
0.0106	0.27	0.0781
0.0110	0.28	0.0938
0.0114	0.29	0.0938
0.0118	0.30	0.0938
0.0126	0.32	0.0938
0.0134	0.34	0.0938
0.0142	0.36	0.1875
0.0150	0.38	0.1875
0.0157	0.40	0.1875
0.0165	0.42	0.1875
0.0173	0.44	0.1875
0.0181	0.46	0.1875
0.0189	0.48	0.1875
0.0197	0.50	0.1875

Note: Drill points are 118°. Shank size is same as drill diameter. Drills 0.38mm and less are 0.75" long, all others 0.875".

Cobalt Microdrills

Diameter inches	mm	Flute lth.
0.0019	0.05	0.0158
0.0024	0.06	0.0158
0.0028	0.07	0.0158
0.0032	0.08	0.0197
0.0035	0.09	0.0197
0.0039	0.10	0.0197
0.0043	0.11	0.0276
0.0047	0.12	0.0276
0.0051	0.13	0.0276
0.0055	0.14	0.0394
0.0059	0.15	0.0394
0.0063	0.16	0.0394
0.0067	0.17	0.0552
0.0071	0.18	0.0552
0.0075	0.19	0.0552
0.0079	0.20	0.0552
0.0083	0.21	0.0709
0.0087	0.22	0.0709
0.0091	0.23	0.0709
0.0094	0.24	0.0709
0.0098	0.25	0.0867
0.0102	0.26	0.0867
0.0106	0.27	0.0867
0.0110	0.28	0.0867
0.0114	0.29	0.0867
0.0118	0.30	0.0867
0.0122	0.31	0.1103
0.0126	0.32	0.1103
0.0130	0.33	0.1103
0.0134	0.34	0.1103
0.0138	0.35	0.1103

Diameter inches	mm	Flute lth.
0.0142	0.36	0.1103
0.0146	0.37	0.1103
0.0150	0.38	0.1103
0.0154	0.39	0.1142
0.0157	0.40	0.1142
0.0161	0.41	0.1142
0.0165	0.42	0.1142
0.0169	0.43	0.1142
0.0173	0.44	0.1142
0.0177	0.45	0.1142
0.0181	0.46	0.1142
0.0185	0.47	0.1142
0.0189	0.48	0.1142
0.0193	0.49	0.1576
0.0197	0.50	0.1576

Note: Cobalt drills are for extremely accurate drilling jobs. Right hand spiral, 118° points and 25° helix angle. All shank diameters are 1mm and total length 25mm.

Figure 7.99 Microdrills.

operator greatly reduced in magnitude by linkage translators on the microdrilling machine.

Specialty flat-style microdrill geometry and terminology are shown in Fig. 7.100, together with the various types of microdrilling tools. The microdrilled hole usually requires additional operations such as countersinking, reaming, counterboring, etc. Figure 7.101 gives the preferred dimensions for the special flat-type microdrills that usually are custom made for various applications. Materials for these special microdrills may be HSS, cobalt-base carbides, or tungsten carbides.

Solid-carbide square die drills. For cold drilling extremely accurate holes in hard steels from Rockwell C 40 to C 70, the carbide square die drill is standardized in sizes from 0.09375 through 0.5000 in in diameter. These are available from national drill distributors and are ideally suited for mold and die repair and maintenance.

7.3.9 Drilling problems and solutions

Problem	Causes	Remedy
Outer corners breakdown	Revolutions per minute too high Poor lip relief	Reduce feed and speed Check lip relief
Cutting lips chip	Feed too high High lip relief	Reduce feed Check lip relief
Cracks in cutting lips	Running too hot	Repoint drill Check feed and speed
Drill breaks	Improper point Flutes clogging	Use proper point Check feed
Drill splits up center	Feed too high Lip relief wrong	Reduce feed Correct relief
Drill will not penetrate	Reverse chisel	Check chisel angle
Rough hole	Dull point No lubricant	Repoint drill Use lubricant
Oversize hole	Unequal length on cutting lips	Repoint drill
Loose spindle		Inspect spindle
Chips change form while drilling	Improper point	Repoint drill
Margins chip	Oversize bushing	Change bushing

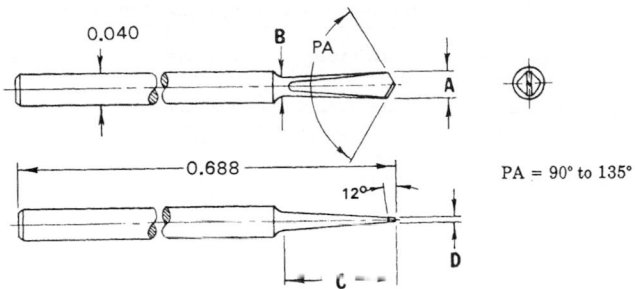

Flat Microdrill Geometry

(See Figure 7.101 for dimensions)

PA = 90° to 135°

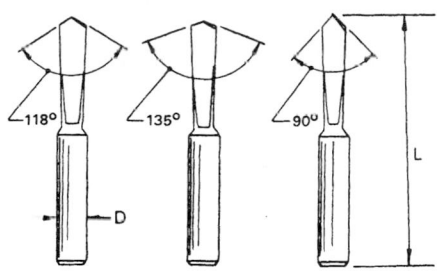

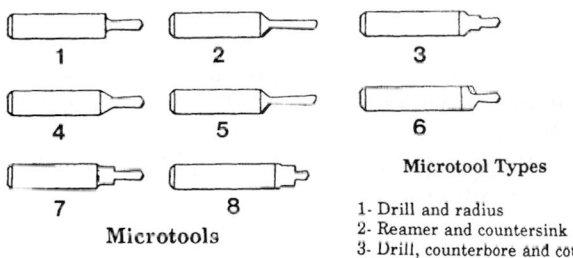

Microtools

Microtool Types

1- Drill and radius
2- Reamer and countersink
3- Drill, counterbore and countersink
4- Drill and countersink
5- Reamer and radius
6- Drill, radius and countersink
7- Drill and counterbore
8- Drill, counterbore and radius

Microdrills and Microtools

Figure 7.100 Geometry of microdrills and microtools.

Flat Microdrill Dimensions - Refer to Figure 7.100 for drill geometry.

Diameter A	Back Taper B	Length C	Web Thickness D
0.0010	0.0001	0.0070	0.0005-0.00075
0.0020	0.0001	0.0140	0.0011-0.0015
0.0030	0.0002	0.0210	0.0016-0.0020
0.0040	0.0002	0.0280	0.0018-0.0022
0.0050	0.0003	0.0350	0.0020-0.0024
0.0060	0.0003	0.0420	0.0022-0.0026
0.0070	0.0004	0.0490	0.0024-0.0028
0.0080	0.0004	0.0560	0.0026-0.0030
0.0090	0.0005	0.0630	0.0028-0.0032
0.0100	0.0005	0.0700	0.0030-0.0034
0.0110	0.0006	0.0770	0.0032-0.0036
0.0120	0.0006	0.0840	0.0038-0.0042
0.0130	0.0007	0.0910	0.0042-0.0046
0.0140	0.0007	0.0980	0.0044-0.0048
0.0150	0.0008	0.1050	0.0046-0.0050
0.0160	0.0008	0.1120	0.0048-0.0052
0.0170	0.0009	0.1190	0.0050-0.0054
0.0180	0.0009	0.1260	0.0053-0.0058
0.0190	0.0010	0.1330	0.0056-0.0060
0.0200	0.0010	0.1400	0.0060-0.0064

Note: These microdrill sizes and dimensions are for custom made flat microdrills.
See Figure 7.99 for modern standard size microdrills of HSS and cobalt alloy.

Figure 7.101   Flat microdrill dimensions.

7.4   Reaming

A reamer is a rotary cutting tool, either cylindrical or conical in shape, used for enlarging drilled holes to accurate dimensions, normally on the order of ±0.0001 in and closer. Reamers usually have two or more flutes that may be straight or spiral in either left-hand or right-hand spiral. Reamers are made for manual or machine operation. Figure 7.102 shows reamer geometry and terminology.

Reamers are made in various forms, including

- Hand reamers
- Machine reamers
- Left-hand flute
- Right-hand flute
- Expansion reamers
- Chucking reamers

Reamer Features

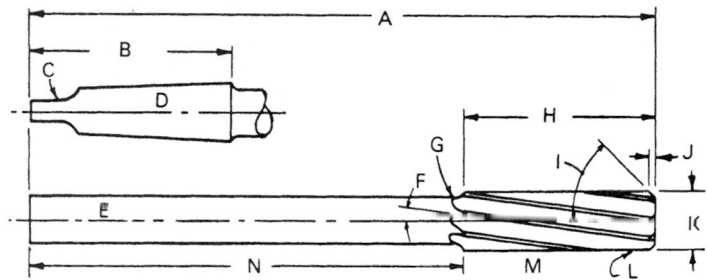

Chucking Reamer - Straight & Taper Shank

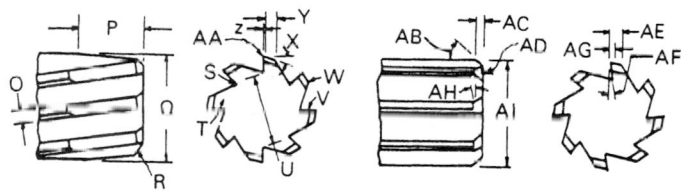

Hand Reamer   Machine Reamer

A	Overall length		S	Flute
B	Shank length		T	Land
C	Tang		U	Core diameter
D	Taper shank		V	Cutting face
E	Straight shank		W	Heel
F	Helix angle		X	Relief angle
G	Cutter sweep		Y	Relieved land
H	Flute length		Z	Margin
I	Chamfer angle		AA	Cutting edge
J	Chamfer length		AB	Chamfer angle
K	Actual size		AC	Chamfer length
L	Body		AD	Chamfer relief
M	Helix flutes, Right hand shown		AE	Land width
N	Shank length		AF	Radial rake angle
O	Land width		AG	Margin
P	Starting taper		AH	Chamfer relief angle
Q	Actual size diameter		AI	Actual size diameter
R	Bevel			

Figure 7.102 Reamer features.

- Stub screw-machine reamers

- End-cutting reamers

- Jobber reamers

- Shell reamers

- Combined drill and reamer

Most reamers are produced from premium-grade HSS. Reamers are also produced in cobalt alloys, and these may be run at speeds 25 percent faster than HSS reamers. Reamer feeds depend on the type of reamer, the material and amount to be removed, and the final finish required. Material-removal rates depend on the size of the reamer and material, but general figures may be used on a trial basis and are summarized below:

Hole diameter	Material to be removed
Up to 0.500 in diameter	0.005 in for finishing
More than 0.500 in diameter	0.015 in for finishing
Up to 0.500 in diameter	0.015 in for semifinished holes
More than 0.500 in diameter	0.030 in for semifinished holes

This is an important consideration when using the expansion reamer owing to the maximum amount of expansion allowed by the adjustment on the expansion reamer.

7.4.1 Machine speeds and feeds for HSS reamers

Note: Cobalt alloy and carbide reamers may be run at speeds 25 percent faster than those shown in Fig. 7.103.

Carbide-tipped and solid-carbide chucking reamers are also available and afford greater effective life than HHS and cobalt reamers without losing their nominal size dimensions. Speeds and feeds for carbide reamers generally are similar to those for the cobalt alloy types.

7.4.2 Sharpening reamers

It would be difficult, if not impossible, to hand sharpen any type of reamer. Reamer sharpening machines are produced by various man-

Machine Speeds and Feeds for HSS Reamers- (sfpm and ipr)

Material	Speed (sfpm)	Feed Code (ipr)
Steel - 150 Bhn	80	1
Steel - 200 Bhn	55	2
Steel - 250 Bhn	35	3
Steel - 300 Bhn	30	3
Steel - 350 Bhn	17	4
Steel - 400 Bhn	10	4
Steel, cast	25	3
Steel, forged alloys	30	3
Steel, low carbon	75	2
Steel, high carbon	45	4
Steel, stainless	15	3
Steel, tool	35	4
Titanium	40	1
Zinc alloy	150	1
Aluminum & alloys	150	1
Brass, leaded	175	1
Brass, red & yellow	150	1
Bronzes	160	1
Copper	45	3
Cast iron, chilled	10	4
Cast iron, hard	50	3
Cast iron, pearlitic	60	1
Cast iron, soft	95	1
Malleable iron	65	2
Monels	30	3
Nickels	40	3
Plastic, hard	50	1
Plastic, soft	65	3

Feed Code, ipr (inches per revolution)

Reamer Diameter	Code 1	Code 2	Code 3	Code 4
0.125"	0.006	0.005	0.004	0.003
0.500"	0.012	0.010	0.007	0.005
1.00"	0.020	0.015	0.012	0.008
2.00"	0.032	0.025	0.020	0.012
2.25 - 2.50"	0.043	0.035	0.028	0.018
2.75 - 3.00"	0.055	0.045	0.035	0.024

Note: Reamer feeds may be interpolated for intermediate sizes than those shown in the table. Cobalt reamers may be run at speeds 25% faster than those shown in the table for HSS.

Figure 7.103 Speeds and feeds for HSS reamers.

ufacturers (Darex), and sharpening facilities are available nationwide for this purpose.

Standard reamer sizes are produced and may be purchased separately or in sets for various applications. Some of the more common types of reamers are shown in Fig. 7.104.

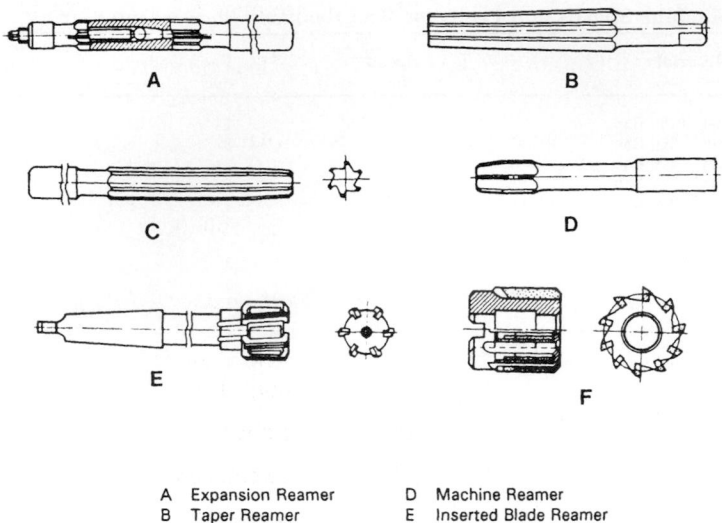

A   Expansion Reamer   D   Machine Reamer
B   Taper Reamer   E   Inserted Blade Reamer
C   Hand Reamer   F   Shell Reamer

Figure 7.104   Reamer types.

7.4.3   Forms of reamers

Other forms of reamers include the following:

Morse taper reamers. These reamers are used to produce and maintain holes for American standard Morse taper shanks. They usually come in a set of two, one for roughing and the other for finishing the tapered hole.

Taper-pin reamers. Taper-pin reamers are produced in HSS with straight, spiral, and helical flutes. They range in size from pin size 7/0 through 14 and include 21 different sizes to accommodate all standard taper pins.

Dowel-pin reamers. Dowel-pin reamers are produced in HSS for standard length and jobber's lengths in 14 different sizes from 0.125 through 0.500 in. The nominal reamer size is slightly smaller than the pin diameter to afford a force fit.

Helical-flute die-maker's reamers. These reamers are used as milling cutters to join closely drilled holes. They are produced from HSS and are available in 16 sizes ranging from size AAA through O.

Reamer blanks. Reamer blanks are available for use as gauges, guide pins, or punches. They are made of HSS in jobber's lengths from 0.015 through 0.500 in in diameter. Fractional sizes through 1.00 in in diameter and wire-gauge sizes also are available.

Shell reamers (see Fig. 7.104*f*). These reamers are designed for mounting on arbors and are best suited for sizing and finishing operations. Most shell reamers are produced from HSS. The inside hole in the shell reamer is tapered ⅛ in per foot and fits the taper on the reamer arbor.

Expansion reamers (see Fig. 7.104*a*). The hand expansion reamer has an adjusting screw at the cutting end that allows the reamer flutes to expand within certain limits. The recommended expansion limits are listed below for sizes through 1.00 in in diameter:

Reamer size: 0.25 to 0.625 in diameter Expansion limit = 0.010 in

Reamer size: 0.75 to 1.000 in diameter Expansion limit = 0.013 in

Note: Expansion reamer stock sizes up to 3.00 in in diameter are available.

7.5 Broaching

Broaching is a precision machining operation wherein a broach tool is either pulled or pushed through a hole in a workpiece or over the surface of a workpiece to produce a very accurate shape such as round, square, hexagonal, spline, keyway, and so on. Keyways in gear and sprocket hubs are broached to an exact dimension so that the key will fit with very little clearance between the hub of the gear or sprocket and the shaft. The cutting teeth on broaches are increased in size along the axis of the broach so that as the broach is pushed or pulled through the workpiece, a progressive series of cuts is made to the finished size in a single pass.

Broaches are driven or pulled by manual arbor presses and horizontal or vertical broaching machines. A single stroke of the broaching tool completes the machining operation. Broaches are commonly made from premium-quality HSS and are supplied either in single tools or as sets in graduated sizes and different shapes. Figure 7.105 shows a number of different types of broaches such as keyway, square, and hexagonal.

Broaches may be used to cut internal or external shapes on workpieces. Blind holes also can be broached with specially designed

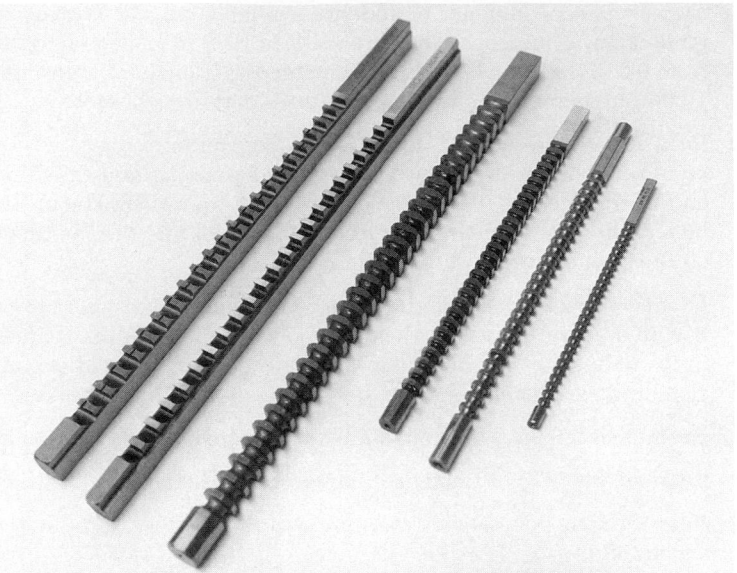

Figure 7.105   Keyway, square, and hexagonal broaches.

broaching tools. The broaching tool teeth along the length of the broach normally are divided into three separate sections. The teeth of a broach include roughing teeth, semifinishing teeth, and finishing teeth. All finishing teeth of a broach are the same size, whereas the semifinishing and roughing teeth are progressive in size up to the finishing teeth.

Figure 7.106 shows the terminology and geometry of broaching teeth. A broaching tool must have sufficient strength and stock-removal and chip-carrying capacity for its intended operation. An interval-pull broach must have sufficient tensile strength to withstand the maximum pulling forces that occur during the pulling operation. An internal-push broach must have sufficient compressive strength, as well as the ability to withstand buckling or breaking, under the pushing forces that occur during the pushing operation.

Broaches are produced in sizes ranging from 0.050 in to as large as 20 in or more. The term *button broach* was used for broaching tools that produced the spiral lands that form the "rifling" in gun barrels from small to large caliber. Broaches may be rotated to produce a predetermined spiral angle during the pull or push operations.

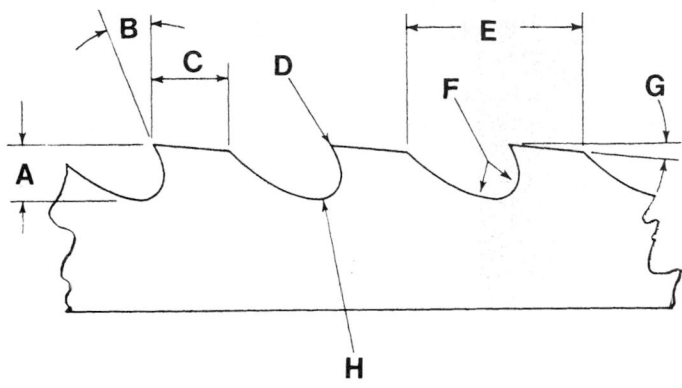

Broach Terminology and Geometry

A   Gullet Depth   E   Pitch
B   Face Angle   F   Tooth Gullet
C   Land   G   Rake Angle
D   Cutting Edge   H   Root Radius

Figure 7.106   Broach terminology.

7.5.1   Calculation of pull forces during broaching

The allowable pulling force P is determined by first calculating the cross-sectional area at the minimum root of the broach. The allowable pull in pounds force is determined from

$$P = \frac{A_r F_y}{f_s}$$

where A_r = minimum tool cross section, in²
F_y = tensile yield strength or yield point of tool steel, psi
f_s = factor of safety (generally 3 for pull broaching)

The minimum root cross section for a round broach is

$$A_r = \frac{\pi D_r^2}{4} \qquad \text{or} \qquad 0.7854 D_r^2$$

where D_r is the minimum root diameter in inches. The minimum pull-end cross section A_p is

$$A_p = \frac{\pi}{4} D_p^2 - WD_p \qquad \text{or} \qquad 0.7854 D_p^2 - WD_p$$

where D_p = pull-end diameter, in
W = pull-slot width, in

7.5.2 Calculation of push forces during broaching

Knowing the length L and the compressive yield point of the tool steel used in a broach, the following relations may be used in designing or determining the maximum push forces allowed in push broaching. If the length of the broach is L and the minimum tool diameter is D_r, the ratio L/D_r should be less than 25 so that the tool will not bend under maximum load. Most push broaches are short enough that the maximum compressive strength of the broach material will allow much greater forces than the forces applied during the broaching operation.

If the L/D_r ratio is greater than 25, compressive broaching forces may bend or break the broach tool if they exceed the maximum allowable force for the tool. The maximum allowable compressive force (pounds force) for a long push broach is determined from the following equation:

$$P = \frac{5.6 \times 10^7 D_r^4}{(f_s) L^2}$$

where L is measured from the push end to the first tooth in inches.

7.5.3 Minimum forces required for broaching different materials

For flat-surface broaches,

$$F = WnR\psi$$

For round-hole internal broaches,

$$F = \frac{\pi DnR}{2}\psi$$

For spline-hole broaches,

$$F = \frac{nSWR}{2}\psi$$

where F = minimum pulling or pushing force required, lbf (pounds force)

  W = width of cut per tooth or spline, in

  D = hole diameter before broaching, in

  R = rise per tooth, in

  n = maximum number of broach teeth engaged in the workpiece

  S = number of splines (for splined holes only)

  ψ = broaching constant (see Fig. 7.107 for values)

Referring to Fig. 7.106, typical rake and relief angles are specified in Figure 7.108.

7.6 Vertical Boring and Jig Boring

The increased demand for accuracy in producing large parts initiated the refined development of modern vertical and jig boring machines. Although the modern CNC machining centers can handle small to medium-sized jig boring operations, very large and heavy work of high precision is done on modern CNC jig boring machines or vertical boring machines. Also, any size work that requires extreme accuracy is usually jig-bored. The modern jig boring machines are equipped with high-precision spindles and x,y-coordinate table movements of high precision and may be CNC

Broaching Constants (ψ) for Various Materials

Material	Value of ψ
Aluminum	200,000 - 300,000
Babbitt	25,000 - 35,000
Brass	200,000 - 300,000
Bronze	300,000 - 350,000
Cast irons	200,000 - 350,000
High-temperature alloys	350,000 - 600,000
Mild steels	350,000 - 450,000
Steel castings	350,000 - 400,000
Titanium	325,000 - 375,000
Zinc alloys	200,000 - 250,000

Note: The tabular values given in the table have a limited value due to the many variables involved in broaching, such as chipbreakers, lubricating and cutting fluid effects and other factors which tend to increase or reduce the required cutting force as calculated using the preceding equations.

Figure 7.107 Broaching constants.

Typical HSS Broach Rake Angles and Relief Angles - Degrees)

Material	Rake Angle (Degrees)	Relief Angle (Degrees)
Aluminum	6 - 10	------
Babbitt	8 - 10	------
Brass	-5 to 5	2 - 3
Bronze	0	1 - 2
Cast iron	6 - 10	2 - 5
Copper	12 - 15	2 - 3
Aluminum bronze	12 - 15	2 - 3
Steels: SAE		
1035	15	1 - 2
1112	15	2 - 3
4140	8 - 15	1 - 3
5140	18 - 20 (finishing)	1 - 2
Stainless steels: AISI		
303	15	0.5 - 2
304, 304L	15	0.5 - 2
316	12 - 15	0.5 - 2
410	12 - 15	0.5 - 2
430	15 - 20	0.5 - 2
440	12- 15	0.5 - 2
Titanium:		
Soft	5 - 15	2 - 4
Hard	12 - 15	2.5 - 3
Zinc	6	------

Figure 7.108 Broach rake angles.

controlled with digital readout panels. For a modern jig boring operation that is CNC/DNC controlled, the circle diameter and number of equally spaced holes or other geometric pattern are entered into the DNC program, and the computer calculates all the coordinates and orientation of the holes from a reference point. This information is either sent to the CNC jig boring machine's controller, or the machine operator can load this information into the controller, which controls the machine movements to complete the machining operation. A typical jig boring machine is illustrated in Fig. 7.109.

Extensive tables of jig boring coordinates are not necessary with the modern CNC jig boring or vertical boring machines. Figures 7.110 and 7.111 are for manually controlled machines, where the machine operator makes the movements and coordinate settings manually.

Vertical boring machines with tables up to 192 in in diameter are produced for machining very large and heavy workpieces. For manually controlled machines with vernier or digital readouts, a table

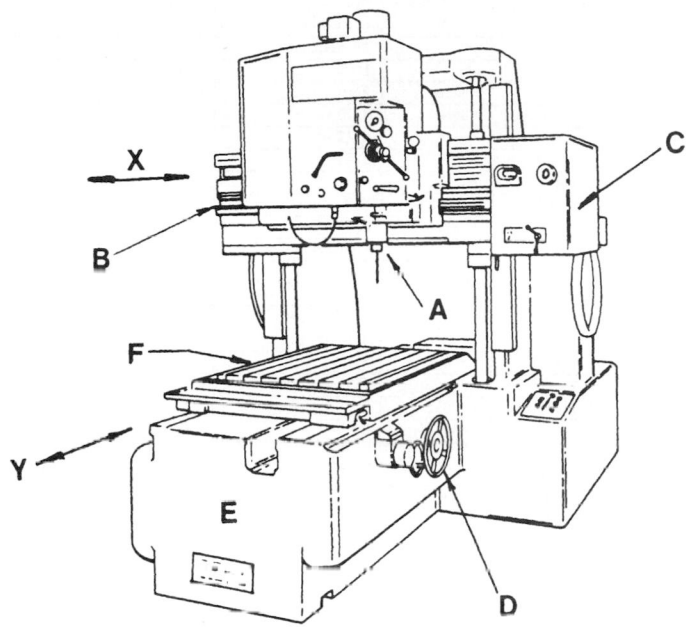

Typical Jig Boring Machine- NC or CNC

A   High precision spindle
B   X axis ways
C   Controller
D   Manual table positioning control
E   Front
F   Horizontal clamping table
X   X axis movement
Y   Y axis movement

Figure 7.109   Jig boring machine.

of jig boring dimensional coordinates is shown in Fig. 7.110 for dividing a 1-in circle into a number of equal divisions. Since the dimensions or coordinates given in the table are for x,y table movements, the machine operator may use these directly to make the appropriate machine settings after converting the coordinates for the required circle diameter to be divided.

Jig Boring Coordinates for Dividing the Circle

Hole No.	Horizontal X	Vertical Y	Hole No.	Horizontal X	Vertical Y
Three holes:			Thirteen holes:		
1	0.50000		1	0.50000	
2	0.75000	0.43301	2	0.05727	0.23236
3	----------	0.86602	3, 13	0.15870	0.17913
Five holes:			4, 12	0.22376	0.08486
1	0.50000		5, 11	0.23757	0.02885
2	0.34549	0.47553	6, 10	0.19695	0.13594
3, 5	0.55902	0.18164	7, 9	0.11121	0.21190
4	----------	0.58778	8	----------	0.23932
Six holes:			Fourteen holes:		
1, 3, 6	0.50000		1	0.50000	
2, 4, 5	0.25000	0.43301	2, 8, 9	0.04951	0.21694
Seven holes:			3, 7, 10, 14	0.13875	0.17397
1	0.50000		4, 6, 11, 13	0.20048	0.09655
2	0.18826	0.39091	5, 12	0.22252	
3, 7	0.42300	0.09655	Fifteen holes:		
4, 6	0.33923	0.27052	1	0.50000	
5	----------	0.43388	2	0.04323	0.20337
Eight holes:			3, 15	0.12221	0.16820
1	0.50000		4, 14	0.18005	0.10396
2, 5, 6	0.14645	0.35355	5, 13	0.20677	0.02173
3, 4, 7, 8	0.35355	0.14645	6, 12	0.19774	0.06425
Nine holes:			7, 11	0.15451	0.13912
1	0.50000		8, 10	0.08456	0.18994
2	0.11698	0.32139	9	----------	0.20790
3, 9	0.29620	0.17101	Sixteen holes:		
4, 8	0.33682	0.05939	1	0.50000	
5, 7	0.21084	0.26200	2, 9, 10	0.03806	0.19134
6	----------	0.34202	3, 8, 11, 16	0.10839	0.16221
Ten holes:			4, 7, 12, 15	0.16221	0.10839
1	0.50000		5, 6, 13, 14	0.19134	0.03806
2, 6, 7	0.09549	0.29389	Seventeen holes:		
3, 5, 8, 10	0.25000	0.18164	1	0.50000	
4, 9	0.30902		2	0.03377	0.18062
Eleven holes:			3, 17	0.09672	0.15623
1	0.50000		4, 16	0.14664	0.11073
2	0.07937	0.27032	5, 15	0.17674	0.05028
3, 11	0.21292	0.18450	6, 14	0.18296	0.01695
4, 10	0.27887	0.04009	7, 13	0.16449	0.08190
5, 9	0.25626	0.11704	8, 12	0.12379	0.13580
6, 8	0.15233	0.23701	9, 11	0.06637	0.17134
7	----------	0.28172	10	----------	0.18374
Twelve holes:			Eighteen holes:		
1	0.50000		1	0.50000	
2, 7, 8	0.06699	0.25000	2, 10, 11	0.03016	0.17101
3, 6, 9, 12	0.18301	0.18301	3, 9, 12, 18	0.08682	0.15038
4, 5, 10, 11	0.25000	0.06699	4, 8, 13, 17	0.13302	0.11162
			5, 7, 14, 16	0.16318	0.05939
			6, 15	0.17364	

Figure 7.110 Jig boring coordinates.

Jig Boring Coordinates for Dividing the Circle

Hole No.	Horizontal X	Vertical Y	Hole No.	Horizontal X	Vertical Y
Nineteen holes:			11, 15	0.07076	0.11634
1	0.50000		12, 14	0.03673	0.13112
2	0.02709	0.16235	13	----------	0.13616
3, 10	0.07834	0.14475	Twenty-four holes:		
4, 18	0.12110	0.11148	1	0.50000	
5, 17	0.15073	0.06612	2, 13, 14	0.01704	0.12941
6, 16	0.16403	0.01358	3, 12, 15, 24	0.04995	0.12059
7, 15	0.15956	0.04039	4, 11, 16, 23	0.07946	0.10355
8, 14	0.13779	0.09003	5, 10, 17, 22	0.10355	0.07946
9, 13	0.10110	0.12989	6, 9, 18, 21	0.12059	0.04995
10, 12	0.05344	0.15567	7, 8, 19, 20	0.12941	0.01704
11	----------	0.16460	Twenty-five holes:		
Twenty holes:			1	0.50000	
1	0.50000		2	0.01508	0.12434
2, 11, 12	0.02447	0.15451	3, 25	0.04677	0.11653
3, 10, 13, 20	0.07102	0.13938	4, 24	0.07367	0.10140
4, 9, 14, 19	0.11062	0.11062	5, 23	0.09657	0.07989
5, 8, 15, 18	0.13938	0.07102	6, 22	0.11340	0.05337
6, 7, 16, 17	0.15451	0.02447	7, 21	0.12312	0.02348
Twenty-one holes:			8, 20	0.12508	0.00787
1	0.50000		9, 19	0.11920	0.03873
2	0.02221	0.14738	10, 15	0.10582	0.06716
3, 21	0.06467	0.13428	11, 17	0.08580	0.09136
4, 20	0.10138	0.10923	12, 16	0.06038	0.10983
5, 19	0.12908	0.07452	13, 15	0.03116	0.12140
6, 18	0.14530	0.03317	14	----------	0.12532
7, 17	0.14862	0.01114	Twenty-six holes:		
8, 16	0.13874	0.05445	1	0.50000	
9, 15	0.11652	0.09293	2, 14, 15	0.01454	0.11966
10, 14	0.08397	0.12314	3, 13, 16, 26	0.04273	0.11270
11, 13	0.04393	0.14242	4, 12, 17, 25	0.06848	0.09920
12	----------	0.14904	5, 11, 18, 24	0.09022	0.07993
Twenty-two holes:			6, 10, 19, 23	0.10673	0.05601
1	0.50000		7, 9, 20, 22	0.11703	0.02885
2, 12, 13	0.02025	0.14086	8, 21	0.12054	
3, 11, 14, 22	0.05912	0.12946	Twenty-seven holes		
4, 10, 15, 21	0.09321	0.10755	1	0.50000	
5, 9, 16, 20	0.11971	0.07695	2	0.01348	0.11530
6, 8, 17, 19	0.13655	0.04009	3, 27	0.03971	0.10910
7, 18	0.14232		4, 26	0.06379	0.09699
Twenty-three holes:			5, 25	0.08444	0.07967
1	0.50000		6, 24	0.10054	0.05805
2	0.01854	0.13490	7, 23	0.11121	0.03329
3, 23	0.05425	0.12480	8, 22	0.11580	0.00675
4, 22	0.08593	0.10562	9, 21	0.11433	0.02016
5, 21	0.11125	0.07853	10, 20	0.10660	0.04598
6, 20	0.12830	0.04560	11, 19	0.09312	0.06933
7, 19	0.13585	0.00930	12, 18	0.07462	0.08893
8, 18	0.13331	0.02771	13, 17	0.05210	0.10374
9, 17	0.12091	0.06264	14, 16	0.02678	0.11297
10, 16	0.09951	0.09295	15	----------	0.11608

Figure 7.110 (*Continued*)

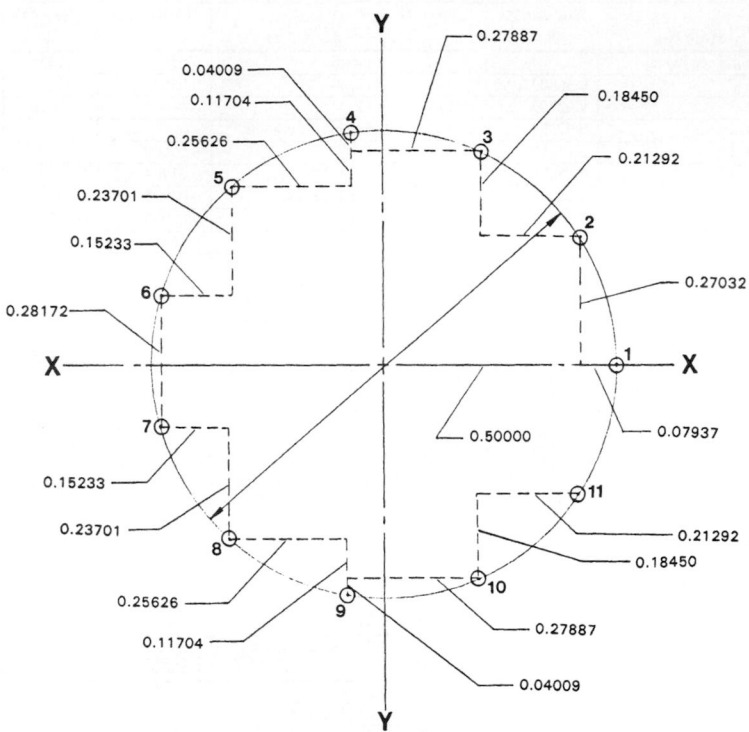

Figure 7.111 Coordinate diagram.

Figure 7.111 is a coordinate diagram of a jig bore layout for 11 equally spaced holes on a 1-in-diameter circle. The coordinates are taken from the table in Fig. 7.110. If a different-diameter circle is to be divided, simply multiply the coordinate values in the table by the diameter of the required circle; i.e., for an 11-hole circle of 5-in diameter, multiply the coordinates for the 11-hole circle by 5. Thus the first hole x dimension would be $5 \times 0.50000 = 2.50000$ in, and so on. Figure 7.112 shows a typical boring head for removable inserts.

7.7 Grinding, Lapping, Honing, and Superfinishing (Surface Finishes)

7.7.1 Grinding

The grinding process is an abrasive machining operation where material is removed from a workpiece in small chips or particles by

Figure 7.112 A modern removable-insert boring head.

the mechanical action of abrasive particles of irregular shape, size, and hardness. Grinding can be a rough or a precision operation for producing smooth surfaces, either flat, cylindrical, or irregularly shaped. The medium of the grinding operation is the grinding wheel, which is used for both external and internal grinding procedures. Grinding wheel shapes and other specifications are defined by the following ANSI standards:

ANSI B74.2-1982, Shapes and Sizes of Grinding Wheels

ANSI B74.3-1986, Shapes and Sizes of Diamond and Cubic Boron Nitride Abrasive Products

ANSI B74.13-1990, Markings for Identifying Grinding Wheels and Other Bonded Abrasives

Other ANSI standards define chemical analysis, bulk density, size of abrasive grains, and other specifications for grinding products and testing procedures.

An ideal grinding abrasive has the ability to fracture before serious dulling occurs and offers maximum resistance to point wear. Each abrasive has a special crystal structure and fracture characteristics, making it suitable for grinding operations on specific materials.

Grinding wheels are composed of abrasive grains of preselected size bonded together with different bonding media. Five important

considerations must be given to the selection of a grinding wheel to suit a particular application:

1. Abrasive type

2. Grain size

3. Bonding media

4. Grade or hardness of the wheel

5. Structure

Wheel abrasives

Natural abrasives. Corundum, emery, and diamond (corundum is natural aluminum oxide containing varying amounts of impurities).

Manufactured abrasives. Synthesized diamond, silicon carbide, aluminum oxide, and cubic boron nitride (CBN). These abrasives all have well-defined physical and chemical characteristics.

Grain size. Abrasive grains vary from 6 to 8 coarse grit to 1000 to 2000 grit for polishing and lapping.

Designation of grinding wheels.   A standard marking system defined by ANSI 74.13-1990 for the identification of grinding wheels (excluding diamond and CBN) is shown in Fig. 7.113. From this marking system, you may determine the characteristics of the grinding wheel from its markings.

A standard marking system defined by ANSI B74.3-1986 for the identification of diamond and CBN is shown in Fig. 7.114. From this marking system, you may determine the characteristics of diamond and CBN grinding wheels.

Grinding wheel speeds.   The most efficient operating speeds in surface feet per minute (sfpm) for general use are summarized in Fig. 7.115. The manufacturer of your particular wheel may recommend a different surface speed based on its experience with its product. Too low a speed will result in wasted abrasive and lower efficiency, and too high a speed may result in too hard grinding action and breakage of the grinding wheel. Do not exceed the maximum speed in revolutions per minute that is marked on each wheel. Severe injury can result from the flying fragments of a broken grinding wheel.

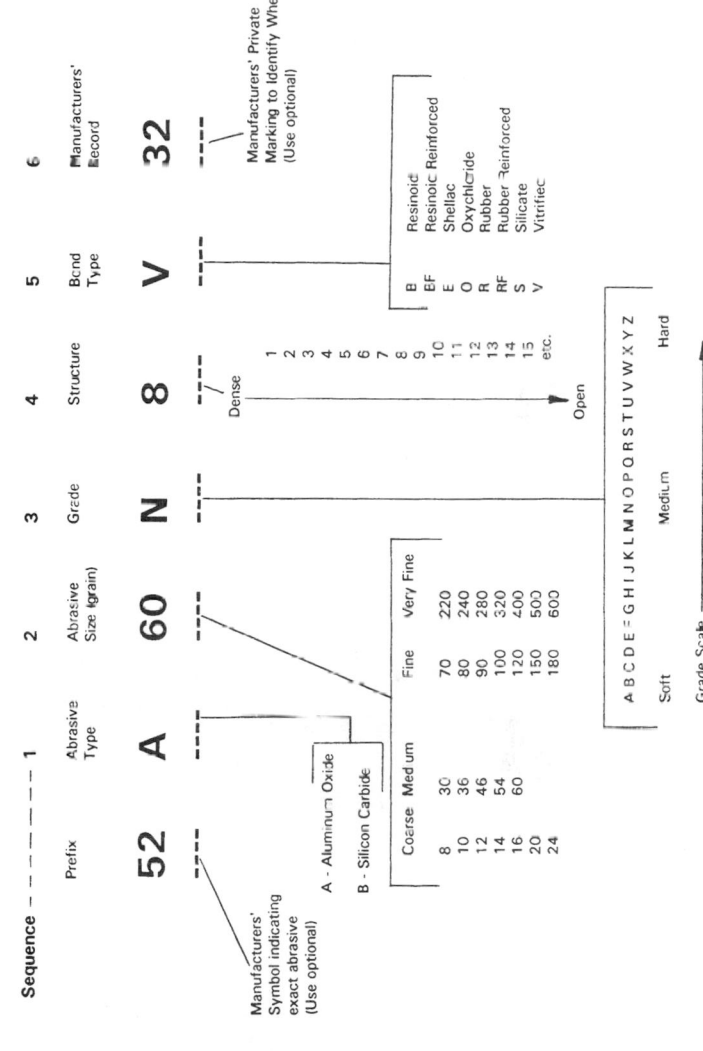

Figure 7.113 Grinding wheel marking system.

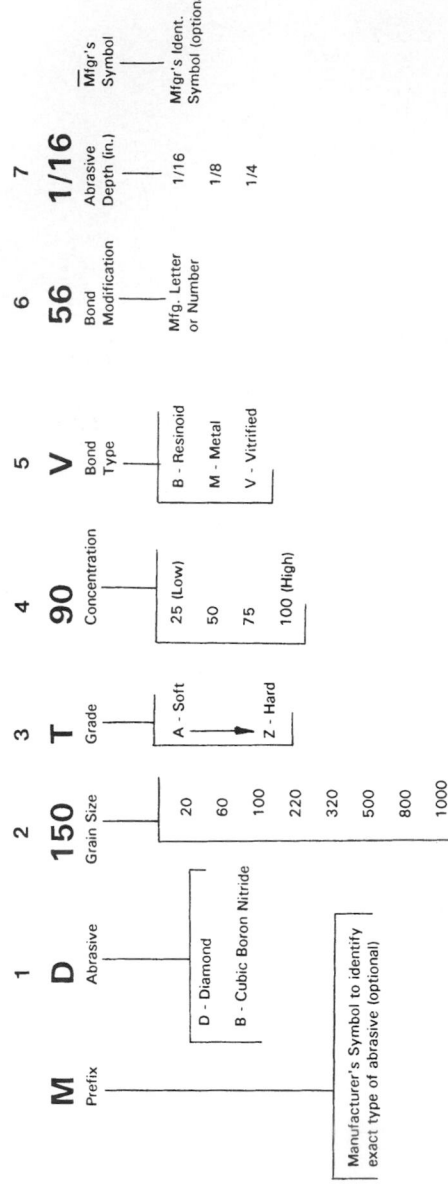

Standard Marking System for Diamond and Cubic Boron Nitride Grinding Wheels.
(ANSI B74.3-1986)

1	2	3	4	5	6	7
M **D**	**150**	**T**	**90**	**V**	**56**	**1/16** $\overline{\text{Mfg'r's}}$ **Symbol**
Prefix Abrasive	Grain Size	Grade	Concentration	Bond Type	Bond Modification	Abrasive Depth (in.) / Mfgr's Ident. Symbol (optional)

D - Diamond
B - Cubic Boron Nitride

Grain Size:
20
60
100
220
320
500
800
1000

Grade:
A - Soft
Z - Hard

Concentration:
25 (Low)
50
75
100 (High)

Bond Type:
B - Resinoid
M - Metal
V - Vitrified

Bond Modification:
Mfg. Letter or Number

Abrasive Depth (in.):
1/16
1/8
1/4

Mfgr's Ident. Symbol (optional)

Manufacturer's Symbol to identify exact type of abrasive (optional)

Figure 7.114 Marking system for diamond and CBN.

Efficient Grinding Wheel Operating Speeds in Surface Feet Per Minute

Type of Operation	sfpm
Cutlery (large-offhand)	4,000-5,000
Cut-off (Rubber, shellac, resinoid)	9,000-16,000
Cylinders (Including hemming)	2,500-5,000
Cylindrical grinding	5,000-12,000
Disc grinding	4,000-5,500
Internal grinding	4,000-12,000
Knife grinding (machine knives)	3,500-4,500
Portable grinding	6,500-12,500
Snagging (vitrified small hole)	5,000-6,000
Snagging (resinoid and rubber)	7,000-9,500 - 12,500
Surface grinding	4,000-6,500
Tool grinding	5,000-6,000
Weld grinding	9,500-14,200

Note: The higher speeds are permitted only where bearings, protection devices
and machine rigidity are adequate for the intended operation.
To convert sfpm to rpm use the following equation: rpm = (12 x sfpm)/(π x D)
Where: sfpm = surface feet per minute; π = 3.1416 and D = wheel diameter, inches.
CAUTION: Do not operate the grinding wheel above its marked maximum rpm limit.

Figure 7.115 Grinding wheel speeds.

Some of the standard shapes of grinding wheels are shown in
Fig. 7.116. ANSI B74.2-1982 defines shapes and sizes, although
different forms and shapes may be produced by some wheel manufacturers. Standard types and shapes of diamond or CBN grinding
wheels are shown in Fig. 7.117. As can be seen from these figures,
many shapes and sizes are available for a multitude of grinding
operations, from roughing to tool and die finishes.

It is difficult in a modern machining and metalworking handbook to give the exact type or number of grinding wheel to use for
any specific application because of the many variables that arise in
actual production situations. The data given in this section are for
reference and approximate applicational uses. The manufacturer
of the grinding wheels and abrasives employed should be contacted
for precise applications on any grinding operation.

Modern machine grinding equipment has been developed to a high
degree over the past 40 years. Figure 7.118 shows a modern surface
grinder used in tool and die making and other precision applications.
In this machine, the grinding wheel is stationary, and the table traverses from left to right and reverses itself in a continuous travel
until the final finish is produced on the workpiece. The grinding
wheel moves vertically in exact increments to produce the required
cut and surface finish. This type of machine is used for flat-surface
grinding only.

Standard Types of Grinding Wheels -

Key to Letter Dimensions:

A   Radial width of flat at periphery
B   Depth of blind hole bushing
D   Diameter - overall
E   Thickness of hole
F   Depth of recess, one side
G   Depth of recess, other side
H   Hole diameter
J   Diameter of outside flat
K   Diameter of inside flat

N   Depth of relief one side
O   Depth of relief other side
P   Diameter of recess
R   Radius
S   Length of cylindrical section
T   Thickness, overall
U   Width of edge
V   Face angle
W   Wall thickness at grinding face

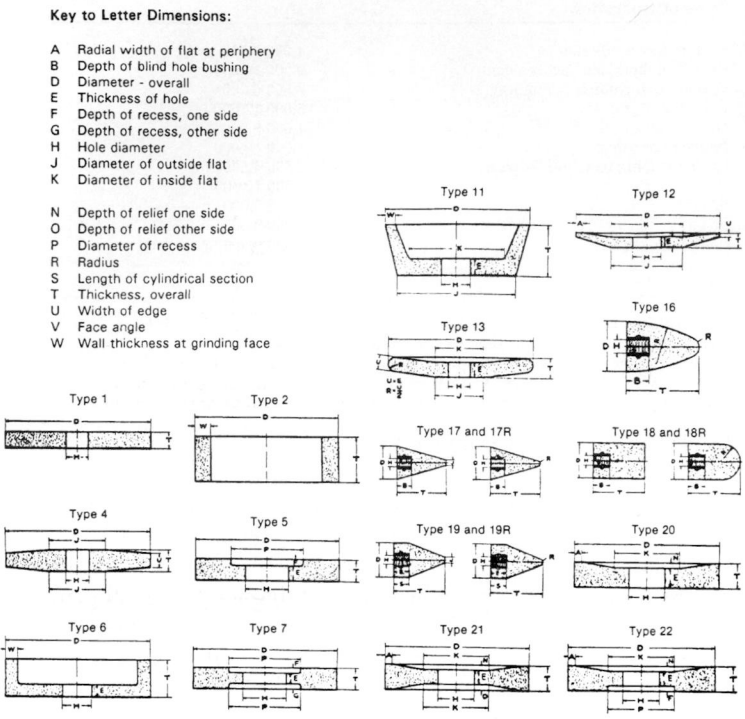

Figure 7.116   Standard grinding wheels.

Figure 7.119 shows a Brown and Sharpe cylindrical grinding machine in which the workpiece is rotated at the same time the grinding wheel is making the required surface finish on the cylindrical part.

The final quality or surface texture of machined and ground parts is usually given on the engineering drawing in root mean square (rms) numbers. An illustration of the finish texture according to the rms system is shown in Fig. 7.120. These finishes range from 500 to 2 rms, with the numbers representing microinches (μin) average. 500 rms would represent a roughing operation on a milling machine or shaper to a superfinishing operation of 2 μm, average or rms (root mean square). Finishes finer than 2 rms can

Standard Types of Diamond or CBN Wheels - Norton Company (Abrasives Marketing Group) Worcester, Massachusetts

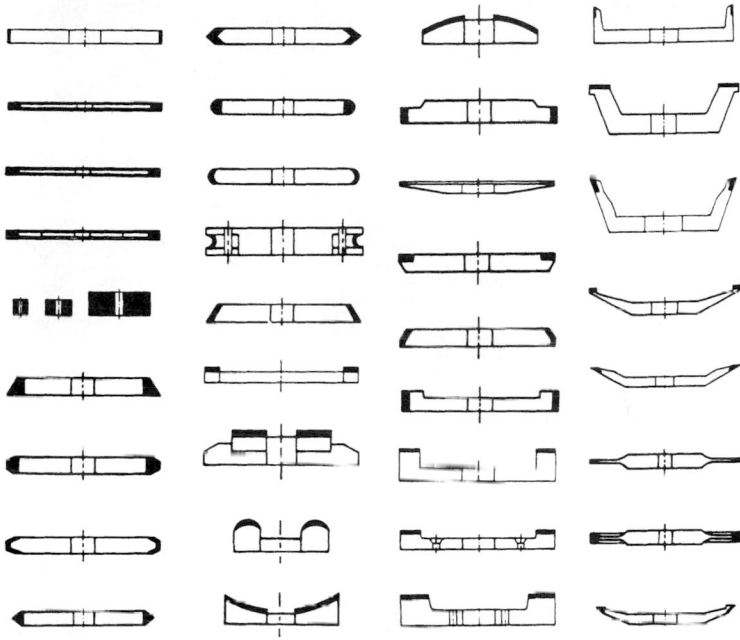

Figure 7.117 Diamond and CBN wheels.

be made using polishing compounds such as cerium oxide or rouge. Cerium oxide and rouge are used extensively in the optical industries for polishing lenses and reflecting mirrors. For finishes of higher rms values (ranging from 5 to 3 μin), compounds such as aluminum oxide powder may be used.

A convenient table of surface feet per minute converted to revolutions per minute of the grinding wheel is shown in Fig. 7.121. The figures in the table may be calculated using the following equation for converting surface feet per minute (sfpm) to revolutions per minute (rpm):

$$\text{sfpm} = \frac{\pi D(\text{rpm})}{12} \qquad \text{and} \qquad \text{rpm} = \frac{12(\text{sfpm})}{\pi D}$$

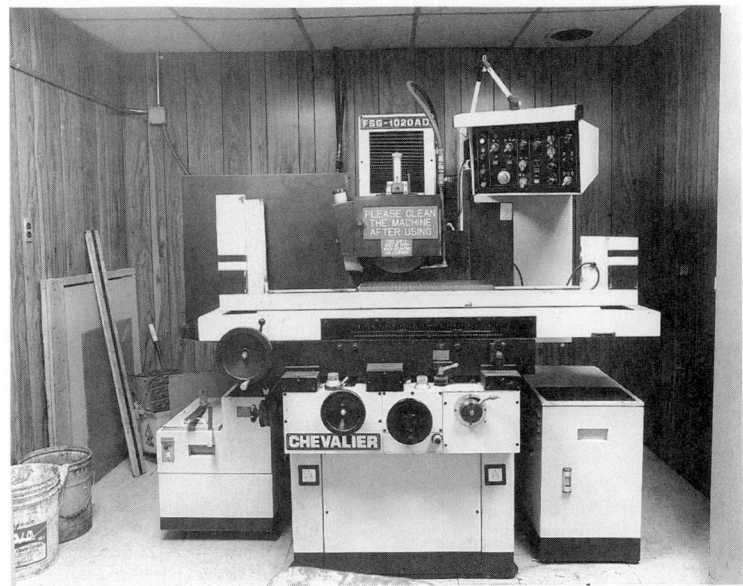

Figure 7.118 Modern automatic surface grinder.

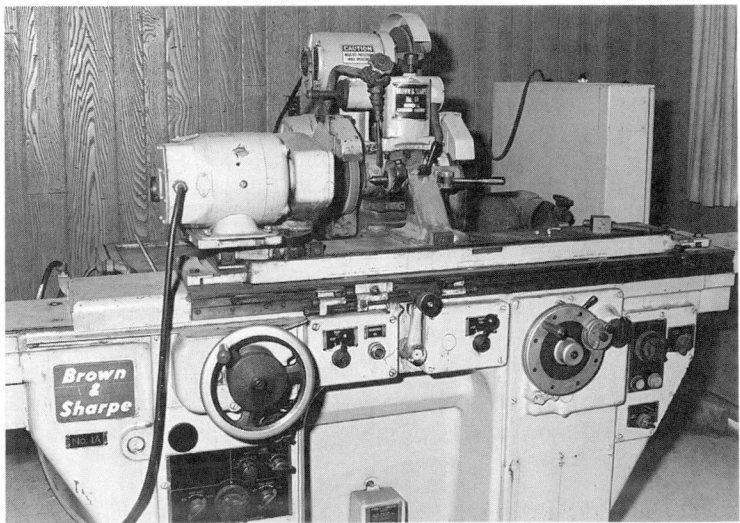

Figure 7.119 Cylindrical grinding machine.

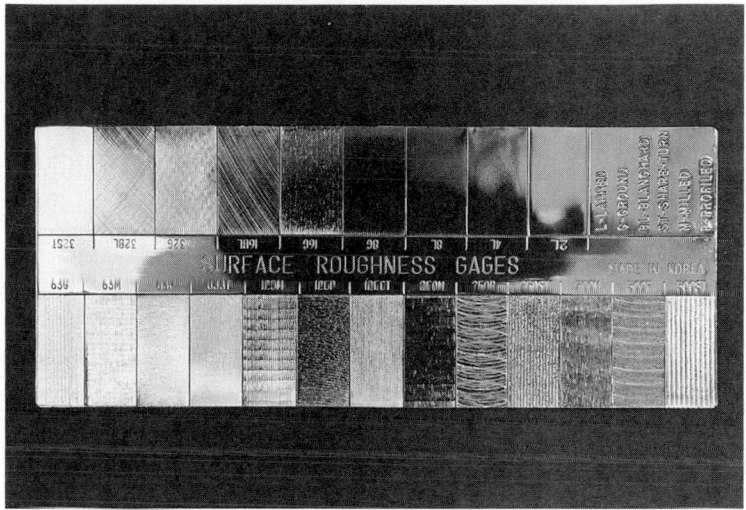

Figure 7.120 Surface roughness gauges.

Characteristics of grinding

Wheel speed. As wheel speed is increased, less work is required of each individual abrasive grain, and this promotes slower wheel wear.

Work speed. This is the speed at which the workpiece traverses the wheel or rotates about a center.

In-feed rate. The rate at which the wheel enters the workpiece during the grinding action. High in-feed rates increase wheel wear and produce a rougher finish than low in-feed rates.

Traverse or cross-feed. This is the rate at which the workpiece is moved across the face of the wheel. It is not the same as work speed.

Material to be ground.   Materials to be ground are either metallic or nonmetallic, and the metallics are divided into low- or high-tensile types. Aluminum oxide wheels generally are used for grinding metallic materials, whereas diamond and CBN are used to grind the extremely hard metallics, as well as ceramics and other hard nonmetallics. Silicon carbide wheels generally are used to grind the

Revolutions per Minute for Various Grinding Speeds in sfm - (Surface feet per minute)

Wheel Dia. Inches	4000	4500	5000	5500	6000	Peripheral 6,500 Rev	surface 7,000 per	speed 7,500 Minute	"sfm" 8,000 "rpm"	8500	9000	9500	10000	12000	14000	16000	Wheel Dia. Inches
1	15,279	17,189	19,099	21,008	22,918	24,828	26,738	28,648	30,558	32,468	34,377	36,287	38,197	45,837	53,476	61,115	1
2	7,639	8,594	9,549	10,504	11,459	12,414	13,369	14,324	15,279	16,234	17,189	18,144	19,099	22,918	26,738	30,558	2
3	5,093	5,730	6,366	7,003	7,639	8,276	8,913	9,549	10,186	10,823	11,459	12,096	12,732	15,279	17,825	20,372	3
4	3,820	4,297	4,775	5,252	5,730	6,207	6,685	7,162	7,639	8,117	8,594	9,072	9,549	11,459	13,369	15,279	4
5	3,056	3,438	3,820	4,202	4,584	4,966	5,348	5,730	6,112	6,494	6,875	7,257	7,639	9,167	10,695	12,223	5
6	2,546	2,865	3,183	3,501	3,820	4,138	4,456	4,775	5,093	5,411	5,730	6,048	6,366	7,639	8,913	10,186	6
7	2,183	2,456	2,728	3,001	3,274	3,547	3,820	4,093	4,365	4,638	4,911	5,184	5,457	6,548	7,639	8,731	7
8	1,910	2,149	2,387	2,626	2,865	3,104	3,342	3,581	3,820	4,058	4,297	4,536	4,775	5,730	6,685	7,639	8
9	1,698	1,910	2,122	2,334	2,546	2,759	2,971	3,183	3,395	3,608	3,820	4,032	4,244	5,093	5,942	6,791	9
10	1,528	1,719	1,910	2,101	2,292	2,483	2,674	2,865	3,056	3,247	3,438	3,629	3,820	4,584	5,348	6,112	10
12	1,273	1,432	1,592	1,751	1,910	2,069	2,228	2,387	2,546	2,706	2,865	3,024	3,183	3,820	4,456	5,093	12
14	1,091	1,228	1,364	1,501	1,637	1,773	1,910	2,046	2,183	2,319	2,456	2,592	2,728	3,274	3,820	4,365	14
16	955	1,074	1,194	1,313	1,432	1,552	1,671	1,790	1,910	2,029	2,149	2,268	2,387	2,865	3,342	3,820	16
18	849	955	1,061	1,167	1,273	1,379	1,485	1,592	1,698	1,804	1,910	2,016	2,122	2,546	2,971	3,395	18
20	764	859	955	1,050	1,146	1,241	1,337	1,432	1,528	1,6623	1,719	1,814	1,910	2,292	2,674	3,056	20
22	694	781	868	955	1,042	1,129	1,215	1,302	1,389	1,476	1,563	1,649	1,736	2,083	2,431	2,778	22
24	637	716	796	875	955	1,035	1,114	1,194	1,273	1,353	1,432	1,512	1,592	1,910	2,228	2,546	24
26	588	661	735	808	881	955	1,028	1,102	1,175	1,249	1,322	1,396	1,469	1,763	2,057	2,351	26
28	546	614	682	750	819	887	955	1,023	1,091	1,160	1,228	1,296	1,364	1,637	1,910	2,183	28
30	509	573	637	700	764	828	891	955	1,019	1,082	1,146	1,210	1,273	1,528	1,783	2,037	30
32	477	537	597	657	716	776	836	895	955	1,015	1,074	1,134	1,194	1,432	1,671	1,910	32
34	449	506	562	618	674	730	786	843	899	955	1,011	1,067	1,123	1,348	1,573	1,798	34
36	424	477	531	584	637	690	743	796	849	902	955	1,008	1,061	1,273	1,485	1,698	36
38	402	452	503	553	603	653	704	754	804	854	905	955	1,005	1,206	1,407	1,608	38

Figure 7.121 Rpm/surface-speed table.

softer-grade nonmetallics. Specific and specialized grinding applications should be referred to the grinding wheel manufacturer. Arbitrarily selecting a grinding wheel or relying on handbook listings of specific grinding wheels for exact operations may not be the best solution to your grinding requirements, especially as far as productivity and wheel wear are concerned.

Grinding wheel dressing. There are various forms of grinding wheel dressing tools available. Some of the many types of grinding wheel dressers include

- Helical hooded dressers
- Abrasive wheel dressers
- Single-stone diamond dressers
- Multiple-point diamond dressers
- T-type diamond hand dressers
- Diamond-stick dressers (silicon carbide)
- Dressing sticks for diamond and CBN grinding wheels
- Ball bearing dressers (for extremely accurate dressing)
- Diamond surface grinder wheel dressers (diamond nib)

7.7.2 Lapping

Lapping is a final finishing operation that results in four major improvements to a workpiece:

1. Extreme accuracy of dimension
2. Correction of minor imperfections
3. Better surface finish
4. Close fit between mating or faying surfaces

In normal lapping operations, less heat is generated than in most other finishing operations. In manual and semiautomatic machine lapping, the end results depend on the following major factors:

1. Type of lap material
2. Type of lapping medium or compound

3. Speed of the lapping motion

4. The material to be lapped

Lap materials.   Cast iron is the most efficient machine lapping material. Other materials used for hand lapping include soft steel, bronze, brass, lead, leather, and various cloths. Leather and cloth are used for polishing. The material of the lap should be softer than the material that is being lapped.

Lapping media

■ Silicon carbide—for rapid stock removal

■ Fused aluminum oxide—for soft steel and nonferrous alloys

■ Unfused aluminum oxide—for excellent polishing action

■ Diamond—for precious stones and tungsten carbides

The manufacturers who produce the lapping media also provide the proper mixtures and viscosities of the lapping solutions.

Lapping speeds.   Efficient lapping speeds range from 300 to 8000 sfpm. Higher speeds will improve the surface finish. Lapping pressures range between 1 and 3 psi for soft materials and 10 psi for hard materials. In manual lapping, the final surface finish depends on the skill of the operator and the lapping medium. New materials that can be used for lapping include cerium oxide and microfine aluminum oxides of optical grade. There are no known materials other than optical rouge and cerium oxide for producing ultrafine finishes on glass and metallic materials.

7.7.3   Honing

Honing is a low-speed abrading process using bonded abrasive sticks for removing stock from metallic and nonmetallic materials. Honing corrects surface errors produced by other machining or grinding operations. Honing has its most important function in the final finishing of internal cylindrical surfaces.

Honing speeds.   Figure 7.122 gives the approximate honing speeds for cast irons and steels, which are the most commonly honed materials. The combined rotation and reciprocation of a hone produces a cross-hatched surface finish on an internal cylindrical part.

Honing Speeds - sfpm (surface feet per minute)

Material	Hardness R_C	Bore Character	sfpm
Steel	15-35	Interrupted	150
	15-35	Plain	80
	35-50	Interrupted	80
	35-50	Plain	60
	50-65	Interrupted	80
	50-65	Plain	50
Cast Irons	15-50	Interrupted	200
	15-50	Plain	110
	50-65	Interrupted	110
	50-65	Plain	60

Note:Tabular values are approximate and should be used as a guide.
Observe the honing action and take appropriate steps to refine the operation.

Figure 7.122   Honing speeds.

Honing abrasives.   Honing sticks are produced with the following abrasives: silicon carbide, aluminum oxide, CBN, and diamond. Silicon carbide is used to hone cast irons and nonferrous materials, aluminum oxide is used to hone steels, and diamond and CBN are used on surfaces that have been chromium plated and other extremely hard materials.

Surfaces as fine as 3 to 4 μin are obtainable using 500 grit silicon carbide on steel parts. Grain sizes for manual and power stroke honing range from 150 to 1200 grit, according to the honing medium and the application.

7.7.4 Superfinishing

Using a bonded stick for cylindrical parts or a cup wheel for flat and spherical workpieces with an abrasive action, superfinishing may be performed. Superfinishing produces a highly wear-resistant finish on parts that are applicable for the superfinishing process. The objective in superfinishing is the removal of fragmentation or smear metal irregularities to restore surface geometry and the surface of the workpiece by eliminating surface stresses and burns. Stock removal may range from 0.0002 to 0.001 in. Scratch patterns of 30 μin rms or more to a mirror finish may be produced.

General-purpose and high-production superfinishing machines are available to produce a superfinish on almost any symmetric type of workpiece. The superfinishing stone is ground to the contour of the part to be superfinished, such as a cylindrical outside surface.

Superfinishing is possible on the hardest of steels and other metallic alloys. Very fine mesh aluminum oxide is employed for many superfinishing applications.

Superfinishing speeds. The recommended speed for most superfinishing operations is from 50 to 60 sfpm at a pressure on the stone contact area of 10 to 35 psi. Superfinishing times are suprisingly fast, with steel of hardness Rockwell C 35 being finished from 20 to 1 μin in approximately 2 minutes under average pressure and with 500 mesh aluminum oxide as the abrasive medium. The use of optical-grade cerium oxide also may improve the finish after the aluminum oxide operation is completed.

7.8   Files and Sharpening Stones

7.8.1   Files

Common hand files generally are divided into three categories:

- American-pattern machinist's files
- Swiss-pattern files
- Special-purpose files

The correct selection and proper use of hand files require extensive experience to produce first-class results. The cutting efficiency of hand files is a function of tooth design, construction, material, and pattern of the teeth. Most files are made of high-carbon steel, heat treated to extreme hardness.

The standard files are either single-cut or double-cut, with some patterns being wavy or curved. Among the many types and designs of hand files, a new class of files has been introduced within recent years. These include the sintered-diamond Swiss-pattern files that contain diamond crystals on their surfaces of varying grades of fineness. These relatively new files will cut any material efficiently and are excellent for tool and die hand-working procedures on extremely hard steels.

Figure 7.123 shows the American-pattern standard machinist's files and their common uses and characteristics. Figure 7.124 shows the standard Swiss-pattern files with their characteristics and uses. Files are characterized by coarseness grade and type of cut.

American Pattern Machinists' Files - Description and Uses

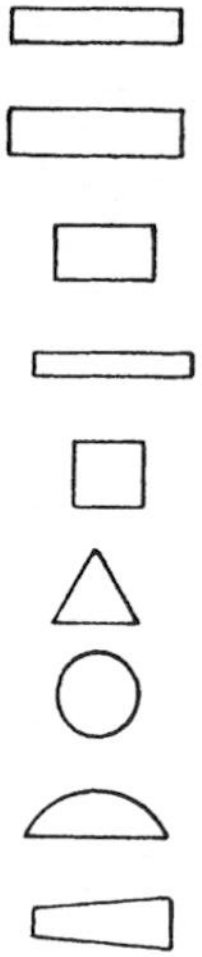

File Cross-Section	Name	Tooth Characteristic	General Uses
	Flat	Usually bastard cut. Also second-cut and smooth.	General purpose and lathe filing
	Hand	One edge smooth. Bastard, second-cut and smooth.	Finishing flat surfaces
	Pillar	One edge smooth. Bastard, second-cut and smooth	Keyways, slots and narrow work
	Warding	Usually bastard. also second-cut and smooth	Filing ward notches in keys and narrow work.
	Square	Bastard, second-cut and smooth.	Enlarging holes or recesses, mortises, keyways and splines.
	Three-Square	Sharp edges. Bastard, second-cut and smooth.	Filing acute angles, corners, grooves and notches.
	Round	Usually bastard. Also second-cut and smooth	Enlarging holes and shaping curved surfaces.
	Half-Round	Usually bastard. Also second-cut and smooth	Concave corners, crevices and round holes.
	Knife	Usually bastard. Also second-cut and smooth	Cleaning out acute angles, corners and slots.

Note: Many variations are available, including long angle lathe files, aluminum flat files, aluminum half-round files, regular taper saw files, chain saw files and cantsaw files.

Figure 7.123 Machinist's files.

Swiss Pattern Files - Description and Uses

File Cross-Section	Name	Tooth Characteristic	General Uses
	Hand	Double-cut on two flat faces and one edge. Other edge smooth.	Flat surfaces
	Pillar	Double-cut on two flat faces. Both edges smooth.	Flat surfaces and slots
	Warding	Double-cut on two flat faces. Single-cut on two edges.	Slots
	Square	Double-cut	Corners and holes.
	Three-Square	Double-cut three faces. Single-cut on edges	Corners and holes
	Round	Double-cut	Corners and holes
	Half-Round	Double-cut	Corners and holes
	Knife	Double-cut on flat faces. Single-cut on edges	Slots
	Crossing	Double-cut	Corners and holes
	Barrette	Cut on wide flat face. Other	Corners, flat surfaces, burring gear teeth.
	Crochet	Double-cut	Slots, flat surfaces and rounded corners.
	Cant	Double-cut three faces. Single cut on two sharp edges	Corners
	Pippin	Double-cut	Rounded corners and holes.

Figure 7.124 Swiss-pattern files.

Coarseness grades

- Bastard cut—for heavy material removal with coarse finish
- Second cut—for light removal with fair finish
- Smooth cut—for fine finishing work

Cut types

- Single cut—used with light pressure for smooth finishes or to sharpen cutting surfaces
- Double cut—for use with heavier pressure to produce a rougher finish

Die sinker's files. These specialized files are used for dressing and finishing dies and are supplied in cut 0 and cut 2 with the same general types as Swiss-pattern files.

Die sinker's rifler files. These specialized files are used in die and mold work, instrument work, and other fine filing jobs. Figure 7.125 shows die sinker's rifler files with their associated trade number designations. These files are available in cut numbers 0, 2, and 4. Most Swiss-pattern and die sinker's files are available in various coarseness grades ranging from number 00 through 6. Number 00 is the coarsest and number 6 is the finest grade. Presently, these specialized files are produced with sintered-diamond and CBN surfaces for working the hardest steel dies and molds.

Figure 7.126 shows an assortment of Swiss-pattern files in 4-in length. The file at the top of the figure is a sintered-diamond equaling needle file. The diamond needle files come in the following shapes:

- Equaling, flat
- Half-round
- Three-square
- Round
- Knife

The diamond needle files have the following grades of coarseness:

- Fine—200/300 grit

Die Sinkers' Rifler Files - Shape and Trade Number are Shown at Each File

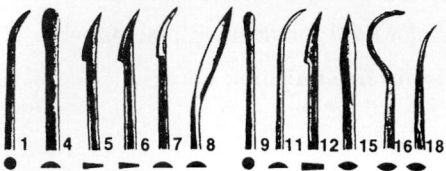

Note: Files are double-ended and approximately 6.5 inches length
Available in Cut Numbers 0, 2 or 4.

Figure 7.125   Die sinker's rifler files.

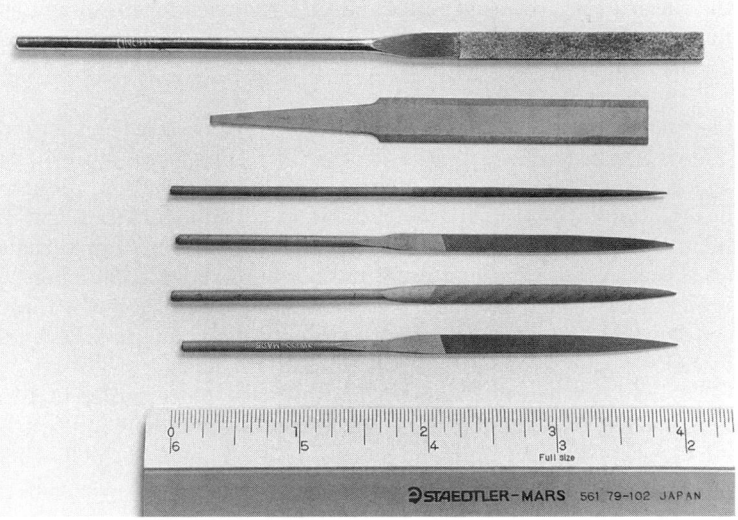

Figure 7.126   Swiss-pattern file samples, 4-in size (top is a diamond grit file).

- Medium—140/170 grit
- Coarse—100/120 grit

Rotary files and burrs.   Rotary files and burrs are made from high-grade carbon steel as well as solid carbide. They are available in various shank diameters, including ³⁄₃₂, ⅛, and ¼ in. Rotary files and burrs are made in the following shapes: ball, cone, 60° cone, 90° cone,

inverted cone, cylindrical flat end, cylindrical radius end, flame, oval, 14° taper radius end, tree-shape pointed end, and treeshape radius end. Both rotary files and burrs are made in single-cut and double-cut forms, with various degrees of coarseness.

7.8.2 Sharpening stones

Sharpening stones are made from the following abrasive types:

India stone (aluminum oxide). Used where close tolerances and smooth cutting edges are required.

Silicon carbide. Fastest cutting sharpening stone; used where speed is essential and moderate tolerances are permitted.

Arkansas. A natural stone recommended for final finishing that produces the highest precision edge possible.

Boron carbide. Next to diamond in hardness and will cut any material except diamond; used to dress carbide cutting tools.

Sharpening stones are made in file form also, with many different shapes available, such as those shown for the Swiss-pattern files in Fig. 7.124.

7.9 Knurling

Knurls are usually hob-cut to obtain sharp, perfectly formed teeth. Most knurls, either diamond or straight, are made from quality HSS (type M-4) or cobalt alloy (M 18). The knurls fit either revolving- or stationary-head knurling tools for use on lathe machines. The knurling tools are made in different sizes for use on lathes with swings of 7 through 36 in.

Standard-face diamond knurls are available in sizes ranging from 12, 16, 20, 24, 25, 30, 40, 50, to 80 teeth per inch. Standard straight knurls are available in sizes ranging from 12, 16, 20, 24, 25, 30, 40, 50, to 80 teeth per inch.

Straight knurling is a form of serrating. Both diamond and straight knurls are machined into a part either for gripping purposes, ornamental purposes, or both. Straight or diamond knurls are a necessity for parts such as thumb screws and the like, where the knurl provides a firm gripping surface. Knurl patterns are available in the following forms:

- Straight

- 30° diagonal right-hand and left-hand (in sets to produce a diamond pattern)

- 30° diamond male

- 30° diamond female (indented knurl)

- Diametral pitch knurls

The knurl is machined into the part on the lathe as either a narrow band, where the knurling tool is in-fed into the workpiece, or as a continuous knurl, using the lathe carriage to automatically traverse the workpiece as in turning a screw thread.

7.10   JIC and ISO Carbide Codes

The Joint Industrial Council (JIC) carbide classification code system is shown in Fig. 7.127. Since this system evolved around the early cemented and sintered carbide grades, no provision was made for the newer, more advanced cutting tool materials. Also, the wide variety of cemented carbide compositions prevented the universal acceptance of a single classification system. Nevertheless, two grade-classification systems are presently accepted: the JIC system and the International Standards Organization (ISO) system. The ISO carbide grade-classification system is shown in Fig. 7.128.

Reference may be made to all the modern advanced cutting tool materials in Secs. 7.1, "Turning and Boring," and 7.2, "Milling." The Kennametal material tables also reference the JIC and ISO classifications for the cutting grades shown in the tables.

7.11   Cutting Fluids and Coolants for Machining Operations

Cutting fluids and machining coolants have been developed over the past 20 to 25 years that are very different from those used previously in industry. Many of the modern cutting fluids and coolants are designed and formulated to be environmentally safe and biodegradable. Some cutting fluids have been developed for use on specific types of materials, whereas others are suitable for a very wide range of different materials and cutting conditions. Some of

JIC Carbide-Classification Code

Code	Application	Carbide Characteristics
C-1	Roughing	Medium-high shock resistance Medium-low wear resistance
C-2	General purpose	Medium shock resistance Medium wear resistance
C-3	Finishing	Medium-low shock resistance Medium-high wear resistance
C-4	Precision finishing	Low shock resistance High wear resistance
C-5	Roughing	High resistance to cutting temperatures Shock and cutting load Medium wear resistance
C-6	General purpose	Medium-high shock resistance Medium wear resistance Medium shock resistance
C-7	Finishing	Medium shock resistance Medium wear resistance
C-8	Precision finishing	Very high wear resistance Low shock resistance

Note: Hardness increases from top to bottom, toughness increases from bottom to top

Figure 7.127 JIC carbide code.

the modern cutting fluids are supplied in aerosol cans and are sprayed on the workpiece while the machining operation is being performed. These spray-application fluids adhere strongly to the workpiece and allow easier machinability while preventing corrosion on the workpiece, such as rusting on ferrous materials.

Combination cutting fluids and coolants are being produced that may be used for most of the high-speed machining operations afforded by the advanced cutting tool materials. With the advent of universal-type cutting-coolant combination fluids, the inventory of cutting fluids and coolants can be kept to a minimum within a manufacturing facility or machine shop.

Modern coolants/cutting fluids. Cutting oils are divided into three basic types: petroleum oils, fixed (animal or vegetable) oils, and synthetic oils. Chemical additives give oils additional or enhanced properties such as resistance to oxidation and foaming and the ability to perform under extreme pressures and temperatures.

ISO Carbide Classification System - International Standards Organization

Main Machining Group	Color Marking	Application Group	Operations and Working conditions
P Group: steel cast steel, long chip malleables	Blue	P01	High precision turning and boring. High cutting speeds. Good surface finish and vibration-free machining
		P10	Turning, thread cutting and milling. High cutting speeds. Small/medium chip cross-section
		P20	Turning, milling, medium cutting speeds and medium chip cross-section. Planing - small chip.
		P30	Turning, milling, planing. Medium to low cutting speeds. Medium/large chip cross-section. Unfavorable conditions
		P40	Planing, turning, milling, shaping. Low cutting speeds. Large chip cross-section. High rake angles. Unfavorable conditions
		P50	Where highest demands are made on carbide toughness. Turning, planing and shaping. Low cutting speeds. Large chips and high rakes. Unfavorable conditions.
M Group: steel, cast steel austenitic manganese steels, cast iron alloys, austenitic steels, malleable cast iron, free-cutting steels.	Yellow	M10	Turning. Medium-high cutting speeds. Small to medium chip cross-section.
		M20	Turning, milling. Medium cutting speeds and medium chip cross-section.
		M30	Turning, milling, planing. Medium cutting speeds. Medium/large chip cross-section.
		M40	Turning, form turning, parting off and recessing - for automatics.
K Group: cast iron, chilled cast irons, short chip malleables hardened steels, nonferrous metals and nonmetallic materials.	Red	K01	Turning, precision turning/boring. Finish milling and scraping.
		K10	Turning, milling, boring, countersinking. reaming, scraping and broaching.
		K20	Turning, milling, planing, countersinking. scraping, reaming and broaching under tougher conditions than K10.
		K30	Turning, milling, planing, shaping under unfavorable conditions. High rakes.
		K40	Turning, milling, planing, shaping under unfavorable conditions. High rakes.

Note: Cutting speed and wear resistance increase from bottom to top; feed and carbide toughness increase from top to bottom.

Figure 7.128   ISO carbide classification system.

Some of the modern coolants/cutting fluids and compounds include

Nonhazardous cutting compounds. Biodegradable, contain no 1,1,1-trichloroethane.

Plumber's lard oil. Used for mixing with mineral oils.

Wax cut cutting oil. Chlorinated waxes suspended in clear oil.

Medium-duty soluble oil. Used where tool life and surface finish are critical.

Thred Kut. Heavy-duty brown sulfochlorinated cutting oil, anti-weld, antiwear solution.

Sulfur-base cutting oils. Used for all metals and high-alloy steels.

Kleen Kut soluble oil. Water-soluble emulsion; economical; mixed with water at 5% to 10% concentration.

Thred Kut 99. A dark, heavy sulfochlorinated fatty oil for machining and grinding soft, tough, and stringy metals such as stainless steels, low-carbon steels, jet-engine alloys, and monel metals.

Trampol-X cutting fluid. Economical, water-soluble, and efficient cutting fluid/coolant for all types of machining operations; designed to withstand recycling in central coolant systems.

Synthetic coolant. Heavy-duty synthetic cutting and grinding concentrate for machining ferrous alloys, aluminum, and brasses.

Blasocut. A high-efficiency cutting-coolant fluid for all types of machining operations; used in central coolant systems and fed at high pressures into the tool and workpiece cutting area; water-soluble; made in Switzerland; frequently used on turning center and machining center CNC machines.

7.12   Calculations for Common Machining Problems

7.12.1   Drill point advance

When drilling a hole, it is often useful to know the distance from the cylindrical end of the drilled hole to the point of the drill for any angle point and any diameter drill. Refer to Fig. 7.129, where the advance *t* is calculated from

$$\tan\left(\frac{180-\alpha}{2}\right) = \frac{t}{D/2}$$

Then

$$t = \frac{D}{2}\tan\left(\frac{180-\alpha}{2}\right)$$

where D = diameter of drill, in
α = drill point angle

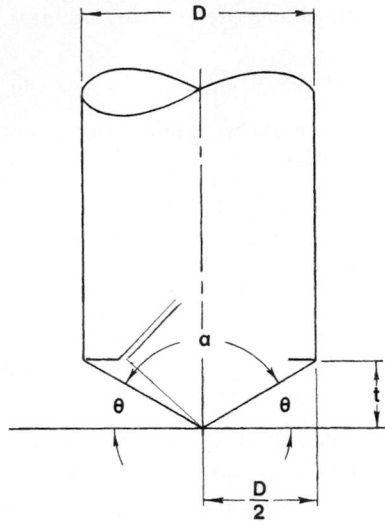

Figure 7.129 Drill advance.

Example: What is the advance t for a 0.875-in-diameter drill with a 118° point angle?

$$t = \frac{0.875}{2} \tan\left(\frac{180 - 18}{2}\right) = 0.2629 \text{ in}$$

Note : $\left(\dfrac{180 - \alpha}{2}\right)$ = angle θ (reference)

7.12.2 Tapers

Finding taper angles under a variety of given conditions is an essential part of machining mathematics. Following are a variety of taper problems with their associated equations and solutions.

For taper in inches per foot, see Fig. 7.130a. If the taper in inches per foot is denoted by T, then

$$T = \frac{12(D_1 - D_2)}{L}$$

where D_1 = diameter of larger end, in
$\quad D_2$ = diameter of smaller end, in
$\quad L$ = length of tapered part along axis, in
$\quad T$ = taper, in/ft

Also, to find the angle θ, use the relationship

$$\tan \theta = \frac{12 - (D_1 - D_2)}{L}$$

and then find arctan θ for angle θ.

Example: $D_1 = 1.255$ in, $D_2 = 0.875$ in, and $L = 3.5$ in. Find angle θ.

$$\tan \theta = \frac{1.255 - 0.875}{3.5} = 0.43429$$

$$\arctan 0.43429 = 23.475^\circ$$

Figure 7.130*b* shows a taper angle of 27.5° in 1 in, and the taper per inch is therefore 0.4894. This is found simply by solving the triangle formed by the axis line, which is 1 in long, and half the taper angle,

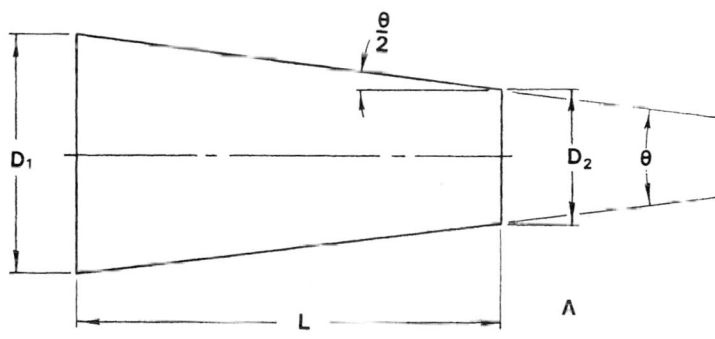

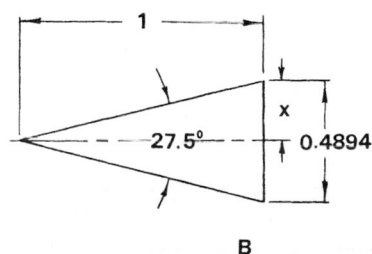

Figure 7.130 Taper angles.

which is 13.75°. Solve one of the right-angled triangles formed by the tangent function:

$$\tan 13.75° = x/1$$

and

$$x = \tan 13.75° = 0.2447$$

and

$$2 \times 0.2447 = 0.4894$$

as shown in Figure 7.130*b*. The taper in inches per foot is equal to 12 times the taper in inches per inch. Thus, in Fig. 7.130*b*, the taper per foot is 12 × 0.4894 = 5.8728 in.

7.12.3 Typical taper problems

1. Set two disks of known diameter and a required taper angle at the correct center distance *L* (see Fig. 7.131).
   Given: Two disks of known diameter *d* and *D* and the required angle θ. Solve for *L*.

$$L = \frac{D-d}{2\left(\sin\dfrac{\theta}{2}\right)}$$

2. Find the angle of the taper when given the taper per foot (see Fig. 7.132).
   Given: Taper per foot *T*. Solve for angle θ.

$$\theta = 2\left(\arctan\frac{T}{24}\right)$$

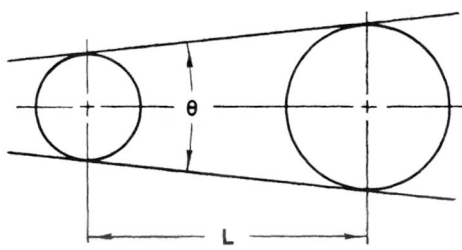

Figure 7.131   Taper.

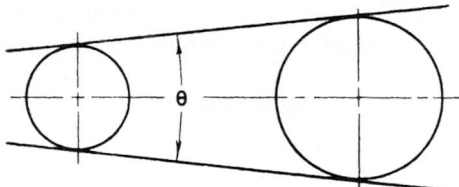

Figure 7.132   Angle of taper.

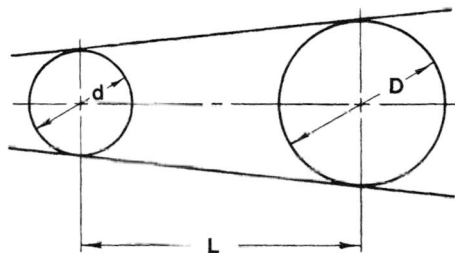

Figure 7.133   Taper per foot.

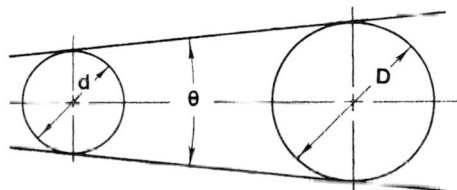

Figure 7.134   Angle of taper.

3. Find the taper per foot when the diameters of the disks and the length between them are known (see Fig. 7.133).
   Given: d, D, and L. Solve for T.

$$T = \tan\left(\arcsin \frac{D-d}{L} \right) \times 24$$

4. Find the angle of the taper when the disk dimensions and their center distance are known (see Fig. 7.134).
   Given: d, D, and L. Solve for angle θ.

$$\theta = 2\left(\arcsin \frac{D-d}{2L} \right)$$

5. Find the taper in inches per foot measured at right angles to one side when the disk diameters and their center distance are known (see Fig. 7.135).
   Given: d, D, and L. Solve for T, in inches per foot.

$$T = \tan\left[2\left(\arcsin\frac{D-d}{2\left(\sin\frac{\theta}{2}\right)}\right)\right] \times 12$$

6. Set a given angle with two disks in contact when the diameter of the smaller disk is known (see Fig. 7.136).
   Given: d and θ. Solve for D, diameter of the larger disk.

$$D = \left(\frac{2d\sin\frac{\theta}{2}}{1-\sin\frac{\theta}{2}}\right) + d$$

Figure 7.137 shows an angle-setting template that may be constructed easily in any machine shop. Angles of extreme precision are possible to set using this type of tool. The diameters of the disks may be machined precisely, and the center distances between the disks may be set with a gauge or Jo-blocks. Also, any angle may be repeated when a record is kept of the disk diameters and the precise

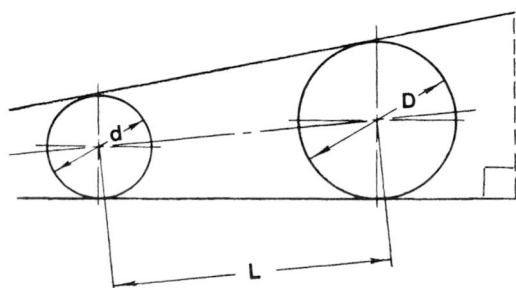

Figure 7.135 Taper in inches per foot.

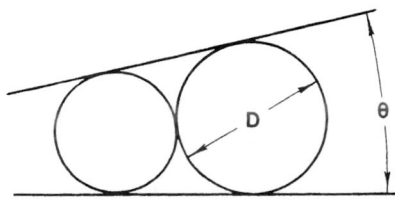

Figure 7.136 Setting a given angle.

center distance. The angle θ, taper per inch or taper per foot, may be calculated using some of the preceding equations.

7.12.4 Checking angles and notches with plugs

A machined plug may be used to check the correct width of an angular opening or machined notch or to check templates or parts that have corners cut off or in which the body is notched with a right angle. This is done using the following techniques and simple equations.

In Figs. 7.138 through 7.140, $D = a + b - c$ (right-angle notches). To check the width of a notched opening, see Fig. 7.141 and the following equation:

$$D = W \tan\left(45^\circ - \frac{\theta}{2}\right)$$

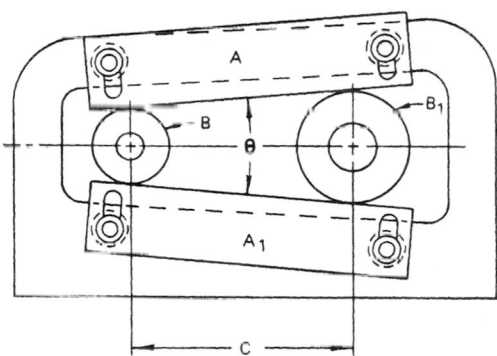

Figure 7.137 Angle-setting template.

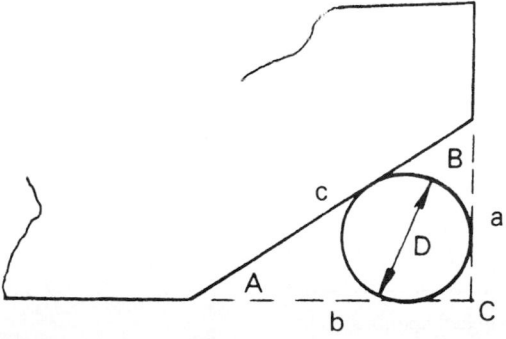

Figure 7.138 Right-angle notch.

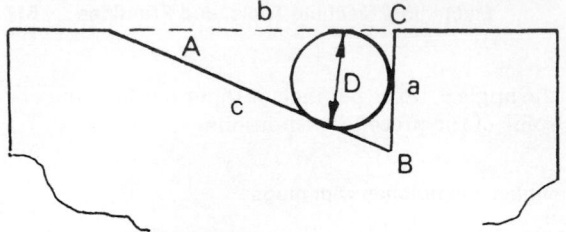

Figure 7.139 Right-angle notch.

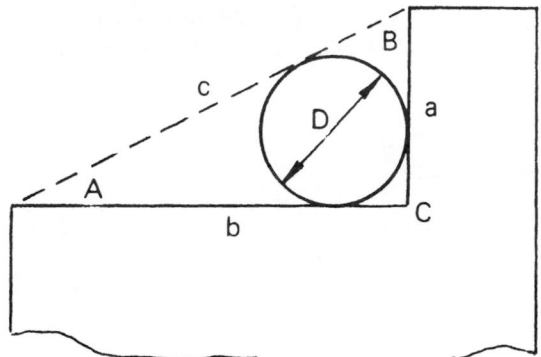

Figure 7.140 Right-angle notch.

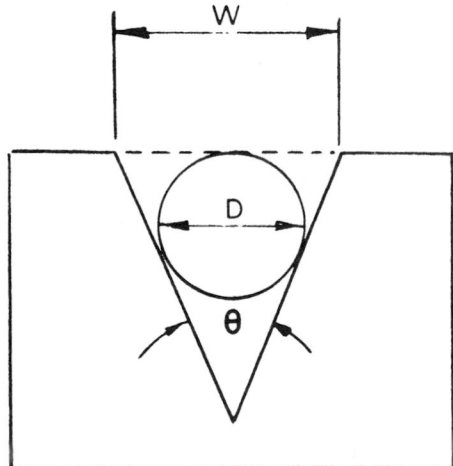

Figure 7.141 Width of notched opening.

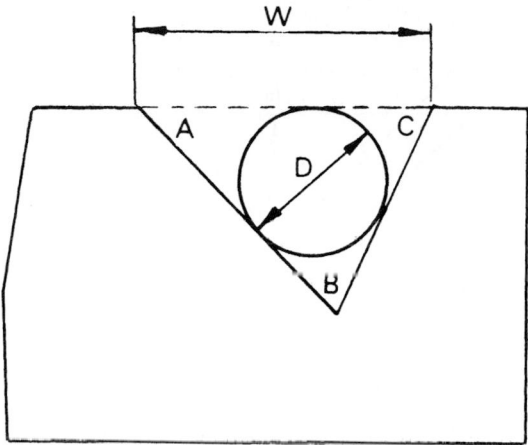

Figure 7.142   Finding plug diameter.

When the correct size plug is inserted into the notch, it should be tangent to the opening indicated by the dashed line.

Also, the equation for finding the correct plug diameter that will contact all sides of an oblique or non-right-angle triangular notch is as follows (see Fig. 7.142):

$$D = \frac{2W}{\left(\cot \dfrac{A}{2} \right) + \left(\cot \dfrac{B}{2} \right)}$$

where W = width of notch, in
A = angle A
B = angle B

7.12.5   Finding diameters

When the diameter of a part is too large to measure accurately with a micrometer or vernier caliper, you may use a 90° or any convenient included angle on the tool (which determines angle A) and measure the height H as shown in Fig. 7.143. The simple equation for calculating the diameter D for any angle A is as follows:

$$D = H \frac{2}{\csc A - 1}$$

Note: Cosecant 45° = 1.4142.

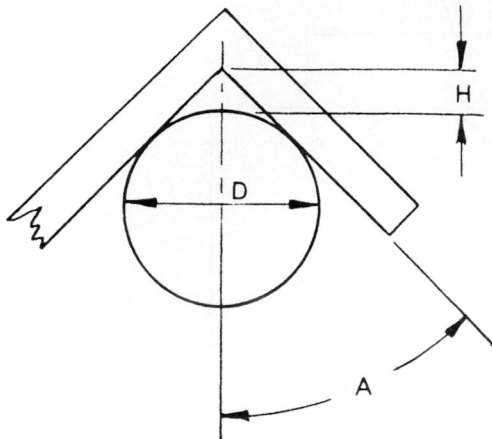

Figure 7.143 Finding the diameter.

Thus the equation for measuring the diameter D with a 90° square reduces to

$$D = 4.828H$$

Then, if the height H measured was 2.655 in, the diameter of the part would be

$$D = 4.828 \times 2.655 = 12.818 \text{ in}$$

When measuring large gears, a more convenient angle for the measuring tool would be 60°, as shown in Fig. 7.144. In this case, the calculation becomes simple. When the measuring angle of the tool is 60° (angle $A = 30°$), the diameter D of the part is $2H$.

For measuring either inside or outside radii on any type of part such as a casting or a broken segment of a wheel, the calculation for the radius of the part is as follows (see Figs. 7.145 and 7.146):

$$r = \frac{4b^2 + c^2}{8b}$$

where r = radius of part, in
b = chordal height, in
c = chord length, in

The chord should be made from a precisely measured piece of tool steel flat, and the chordal height b may be measured with an inside telescoping gauge or micrometer.

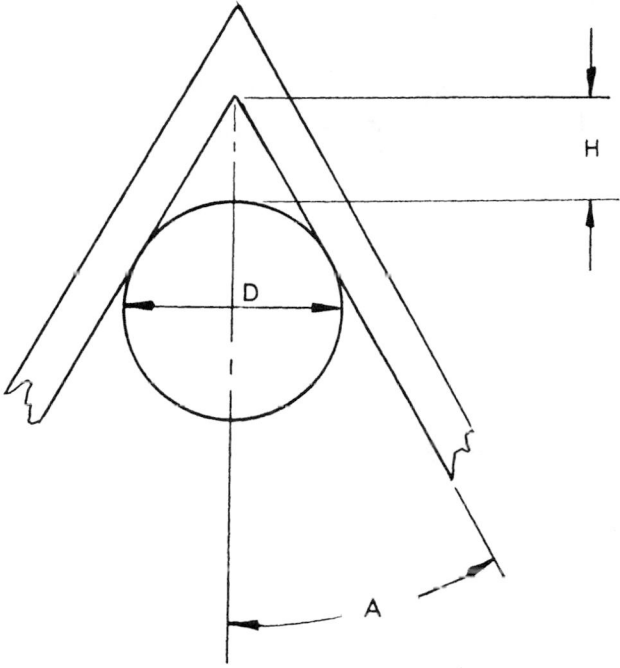

Figure 7.144 Finding the diameter.

7.12.6 Measuring radius of arc by measuring over rolls or plugs

Another accurate method of finding or checking the radius on a part is illustrated in Figs. 7.147 and 7.148. In this method, we may calculate either an inside or an outside radius by the following equations:

$$r = \frac{(L+D)^2}{8D} \qquad \text{for convex radii (Fig. 7.147)}$$

$$r = \frac{(L+D)^2}{8(h-D)} + \frac{h}{2} \qquad \text{for concave radii (Fig. 7.148)}$$

where L = length over rolls or plugs, in
D = diameter of rolls or plugs, in
h = height of concave high point above the rolls or plugs, in

For accuracy, the rolls or plugs must be placed on a tool plate or plane table and the distance L across the rolls measured accurately.

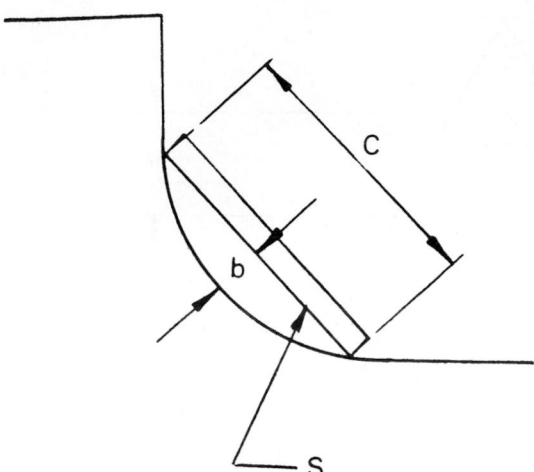

Figure 7.145   Finding the radius.

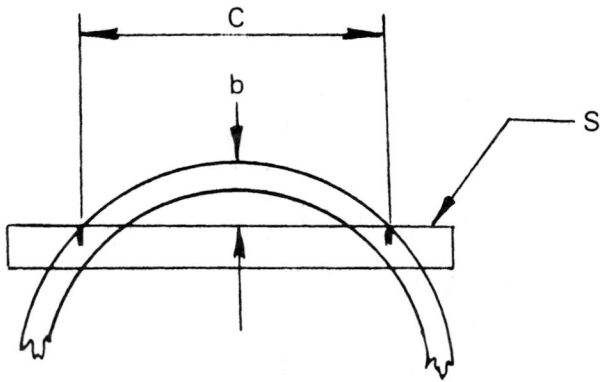

Figure 7.146   Finding the radius.

The diameter D of the rolls or plugs also must be measured precisely and the height h measured with a telescoping gauge or inside micrometers.

7.12.7   Measuring dovetail slides

The accuracy of machining of dovetail slides and their given widths may be checked using cylindrical rolls (such as a drill rod) or wires and the following equations (see Fig. 7.149):

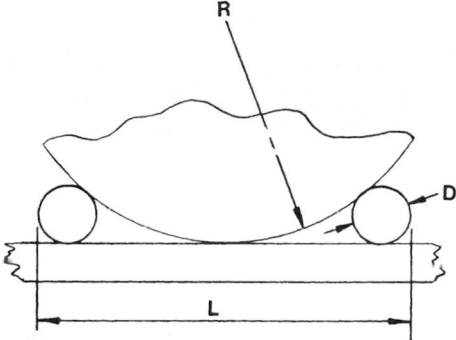

Figure 7.147 Finding the radius.

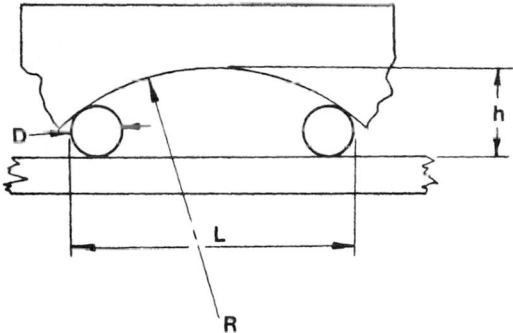

Figure 7.148 Finding the radius

$$x = D\left(\cot\frac{\theta}{2}\right) + \alpha \qquad \text{for male dovetails (Fig. 7.149}a)$$

$$y = b - D\left(1 + \cot\frac{\theta}{2}\right) \qquad \text{for female dovetails (Fig. 7.149}b)$$

Note: $c = h \cot \theta$. Also, the diameter of the rolls or wire should be sized so that the point of contact θ is below the corner or edge of the dovetail.

7.12.8 Universal dividing heads

The precision universal dividing head is a precision milling attachment used to divide the circumference of circular work and to equally

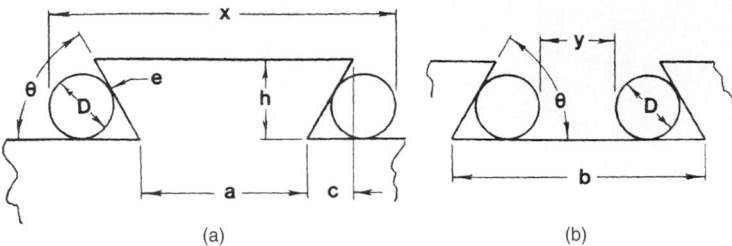

Figure 7.149 Measuring dovetail slides.

space the divisions. This milling tool may be used for milling spur gear teeth (using the appropriate cutter), pinion gears, splines, and spacing holes on a given circle. Other types of milled parts that may be produced using the universal dividing head include hexagonal nuts, octagons, pentagons, and other polygon shapes.

Refer to Fig. 7.150, in which A denotes the dividing head, B the indexing plates, and C the tail stock center. Work may be mounted between the head and tail stock centers or held in a chuck or collet, either of which is placed on the head spindle. Rotating the head crank causes the head stock spindle to rotate at a ratio of 40:1. This ratio and an index plate with a series of equally spaced holes (to measure and stop the crank rotation) make it possible to divide the circumference of the workpiece into the required number of equal divisions or sections.

The index plates (B) have different hole series, such as 15, 16, 17, 18, 19, 20, 21, 23, 27, 29, 31, 33, 37, 39, 41, 43, 47, and 49. An index chart of all possible divisions is supplied with the universal dividing head. These dividing heads usually are supplied in either left-hand or right-hand models, according to your particular setup and requirements. The precision dividing head usually can hold the angular divisions to within 10 seconds of arc.

Some dividing-head models are equipped with index plates and a series of change gears to alter the ratio of divisions. Instructions for the use of each particular dividing head must be obtained from the manufacturer. A precision dividing head is a necessary machine tool accessory in any model shop, machine shop, or tool and die operation that requires precise division of a circle.

7.13   Taps and Dies for Threading Operations

The tapping of threaded holes and the die cutting of external threads with the use of taps and dies are performed most efficiently and eco-

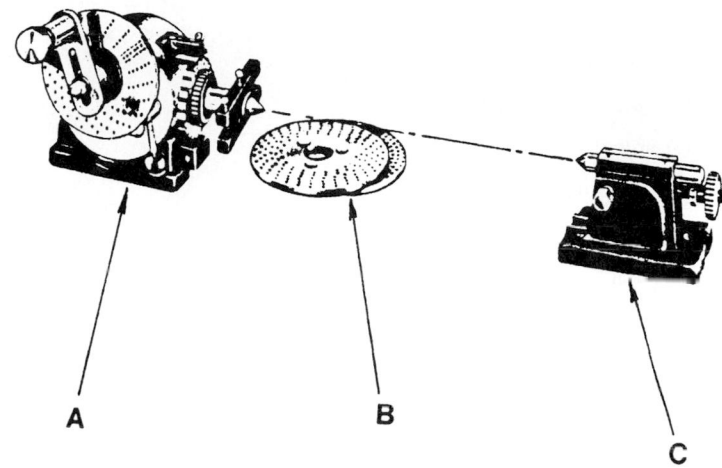

Figure 7.150 Universal dividing head.

nomically when the correct type of tap or die is used. The number of flutes on the tap and the style of flutes are important in the efficient removal of material that is cut in the tapping process. Four flute taps dispose of the cut material in shallow or large-diameter holes or when the material breaks into small chips but are not efficient in cutting stringy materials for deep or small-diameter holes.

Taps are available in spiral-point form, which allows the tap to push the cut material ahead of the tap on through holes. Spiral-flute and high-spiral-flute designs are made to pull the cut material up and out of the tapped hole during the tapping procedure. Thread-forming taps are also available that do not produce chips because the forming of the thread is a metal deforming or displacement process.

Some of the typical standard types of taps and dies are shown in Fig. 7.151. To produce a full thread to the bottom of a drilled hole, a standard set of taps is used, which includes the taper tap, plug tap, and bottoming tap used in the order listed to produce a fully threaded blind hole.

7.13.1 Available types of taps and dies

Taps

- Two-flute taps
- Three-flute taps
- Four-flute taps

- Cut-thread taps
- Ground taps
- 30° spiral-flute taps
- 52° spiral-flute taps
- Spiral-point taps
- Straight and tapered pipe taps
- Interrupted-thread taps
- Pulley taps
- Nut taps
- Combined tap and drill

Dies

- Round adjustable dies
- Thread-cone dies (acorn type)
- Hexagon rethreading dies (standard thread and pipe thread)
- Square bolt dies
- Square pipe dies
- Locknut-thread round adjustable dies

Taps and dies are made from carbon steel, HSS, and solid carbide (plain or titanium nitride coated). Taps and dies are made for American standard unified threads, ISO metric threads, Acme threads, BA (British Association) threads, and other special thread forms used in industry. Most taps and dies are available in either right-hand or left-hand thread.

Although taps are produced in both cut-thread and ground-thread forms, the ground-thread tap is preferred for the following reasons:

1. The ground-thread tap will produce more than five times the number of holes than will the cut tap.

2. Although the ground-thread tap is more costly, it is more economical owing to its long wear ability.

3. A ground tap will produce a more accurate tapped hole.

4. It requires less machine power to use the ground tap on a size-to-size basis.

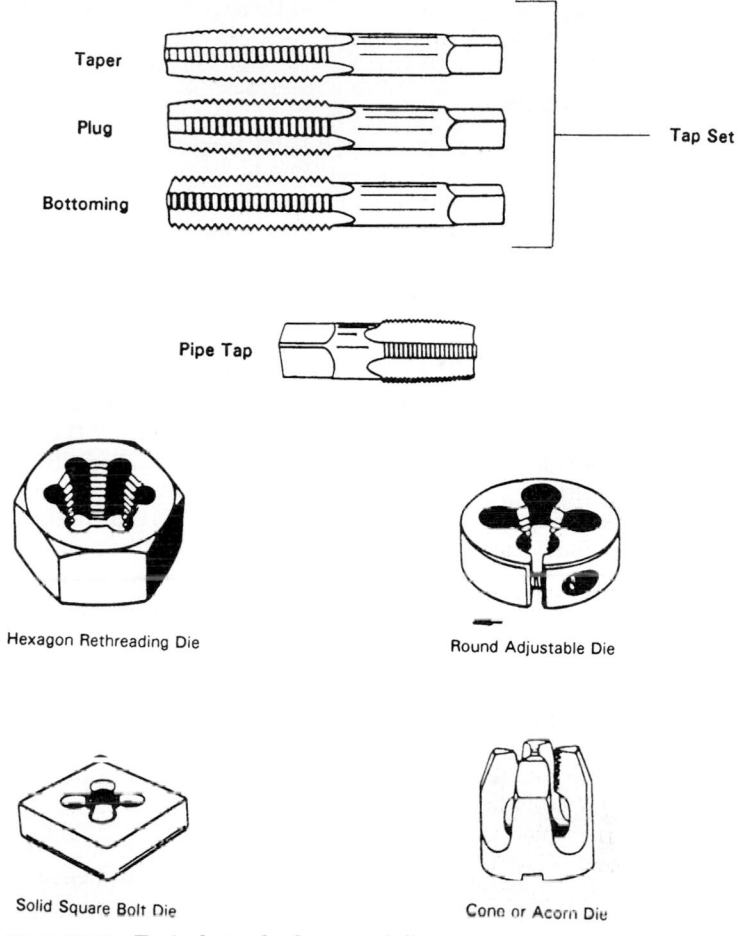

Figure 7.151 Typical standard taps and dies.

5. The cut-thread tap is usually not suitable for producing class 3B tapped holes.

The application of the proper cutting fluid is critical to economical and successful tapping operations. Various modern types of cutting fluids are produced for tapping and die-cutting threads on different materials. It is essential to use the proper type of cutting fluid for each specific type of tapping and die-cutting operation. The proper cutting fluids and cutting fluid direction and pressure are important

on high-speed tapping operations, such as those performed on CNC turning centers and machining centers.

7.13.2 Speeds for tapping

The normal tapping speed for any particular material when using an HSS ground-thread or carbide tap should be the same as the drilling speed for the equivalent drill diameter. See Sec. 7.3 for the cutting speeds and revolutions per minute required for different materials and drill diameters to arrive at the required tapping speed or revolutions per minute.

7.13.3 Thread lead tolerances

Ground-thread taps always should be used when a threaded assembly must carry a heavy load. The thread lead tolerance on cut-thread taps is much broader than the close-tolerance thread lead produced by the ground-thread tap and will cause the threads to not maintain a full thread-bearing condition, which is detrimental to the strength of the tapped and threaded assembly. In other words, if the thread leads of the male thread do not coincide closely with the thread leads of the nut or tapped hole, a weak threaded assembly results or may not assemble easily.

The thread lead tolerances of cut-thread and ground-thread taps is as follows:

- Cut-thread tap lead tolerance = ±0.0030 in/in.

- Ground-thread tap lead tolerance = ±0.0005 in/in.

Pitch-diameter tolerance ranges of cut-thread and ground thread taps are as follows (actual tolerance depends on the thread size):

- Cut-thread pitch-diameter tolerance range = 0.001 to 0.0055 in/in.

- Ground-thread pitch-diameter tolerance range = 0.0005 to 0.0025 in/in.

7.13.4 Limit numbers for ground-thread taps (American UN thread system, ANSI)

All American standard listed ground-thread taps are marked with the capital letter G to designate ground thread. The letter G is followed by either the letter H or the letter L. The letter H designates

above the basic pitch-diameter limits, and the letter L designates below the basic pitch-diameter limits. The number following the letter H or L designates the pitch-diameter limits. The number is a code number that references the actual dimensional limits of the basic pitch diameter. Figure 7.152 shows the H or L code number with the associated pitch-diameter limit dimensions.

Example: A tap marked GH2 indicates a ground-thread tap with pitch diameter limits of 0.0005 to 0.0010 in over the basic pitch diameter on sizes 1 in and under.

Note: Do not confuse these limit codes with the class of fit for a particular thread size. These tap-limit numbers are not related to the class of fit.

7.13.5   Limit numbers for ground-thread ISO metric taps

When the tap basic pitch diameter is over the basic thread pitch diameter by even multiples of 0.0005 in, the tap will be marked with the letter D. When the tap basic pitch diameter is under the basic thread pitch diameter by even multiples of 0.0005 in, the tap will be marked with the letters DU. The letters D and DU are followed by the limit number, which is determined as follows:

The D limit number. Amount the tap pitch diameter high limit is over basic pitch diameter, divided by 0.0005 in.

Limit Number Codes for Ground-Thread Taps - Pitch Diameter Limits

Limit Number Code	Pitch Diameter Limits	Tap Size
L1	Basic to basic minus 0.0005"	All to 1" diameter
H1	Basic to basic plus 0.0005"	"
H2	Basic plus 0.0005" to 0.001"	"
H3	Basic plus 0.001" to 0.0015"	"
H4	Basic plus 0.0015" to 0.002"	"
H5	Basic plus 0.002" to 0.0025"	"
H6	Basic plus 0.0025" to 0.003"	"
H7	Basic plus 0.003" to 0.0035"	"
H4	Basic plus 0.001" to 0.002"	Over 1" to 1.500" diameter

Note: The limit code number divided by 2 equals in thousandths of an inch, the amount that the maximum tap pitch diameter is above basic in the H series and the amount that the minimum tap pitch diameter is under basic in the L series.

Figure 7.152   Limit number codes for ground-thread taps.

The DU limit number. Amount the tap pitch diameter low limit is under basic pitch diameter, divided by 0.0005 in.

Examples: M12 × 1.75, marked D6 limit: The maximum tap pitch diameter equals basic pitch diameter plus 0.0030 in (6 × 0.0005 = 0.0030 in). Tap pitch-diameter tolerance is minus 0.0012 in.

M6 × 1, marked DU4 limit: The minimum tap pitch diameter equals basic pitch diameter minus 0.0020 in (4 × 0.0005 = 0.0020 in). Tap pitch-diameter tolerance is plus 0.0010 in.

Note: See ANSI/ASME Standard B94.9-1987, *Taps, Cut and Ground Threads.*

Tap markings. All taps are marked with the nominal thread size, number of threads per inch, and a symbol to identify the thread form. On multiple-thread lead taps, the lead is marked in fractions and type of lead, such as double, triple, etc. Also, the hand of the thread is marked on the tap, i.e., LH for left hand and RH for right hand.

Example: A ¼-20 unified national coarse double-lead left-hand tap would be marked

¼-20UNC Double LH 1/10 in lead

See Chap. 14, "Fastening and Joining Techniques and Hardware."

7.13.6   Thread gauges (external and internal)

As shown in the preceding section on taps and dies, taps are made with ground thread forms that have pitch-diameter limits. Note also that the round adjustable dies for cutting external threads may be set with their set screw to cut the thread larger by a few thousandths of an inch. With these capabilities on the taps and dies, any particular class of fit for a particular thread size and pitch may be produced (i.e., 1, 2, 3, 1A, 2A, 3A, 1B, 2B, and 3B). Classes 1, 1A, and 1B are loose-tolerance threads; 2, 2A, and 2B are moderate- or common-tolerance threads; and 3, 3A, and 3B are close-tolerance thread fits.

In all production applications of thread cutting such as tapped holes, threaded rods, or hardware such as bolts, screws, and nuts, a method for controlling the class of fit must be employed. To accomplish this, inspection procedures using thread gauges are employed.

Some of the gauges used for the quality-control inspection of threaded parts include

- Work-plug gauges—thread-setting plugs [American (inch) and metric]

- Thread-ring gauges [American (inch) and metric]

- Working thread-plug gauges [American (inch) and metric]

Note: Thread-setting plugs are used to check the ring gauges for wear.

7.14   Drill-Rod Types and Sizes for Machining and Tooling Applications

7.14.1   U.S. Customary or inch-size drill rods

U.S. Customary or inch-system drill rods generally are supplied in 36-in lengths in the following grades of material:

0-1. Oil-hardening for general toolroom use; good nondeforming characteristics; moderate resistance to abrasion; machinability 90 percent.

S-7. Air-hardening molybdenum, chromium; shock resisting; high impact strength with moderate wear resistance; machinability 95 percent.

A-2. Air-hardening; maximum abrasion resistance; machinability 65 percent.

D-2. Air-hardening; high carbon, high chrome; high wear resistance; machinability 50 percent.

W-1. Water-hardening offers a combination of high abrasion resistance and toughness; machinability 100 percent.

The five grades listed may be heat-treated and drawn or tempered to various hardness classes by using the heat-treating data shown in Chap. 11.

Available diameters.   U.S. Customary drill rods are available in the following diameters (inches): 0.0625, 0.1250, 0.1875, 0.2500, 0.3125, 0.3750, 0.4375, 0.5000, 0.5625, 0.6250, 0.7500, 0.8750, 1.000, 1.1250, 1.2500, and 1.5000. Other diameters are also available. W-1 is the most economical, and D-2 is the most expensive, with the other grades falling between these two limits.

Tolerances on drill rod

Up to 0.125 in in diameter: ±0.0003 in

0.1875 to 0.4375 in in diameter: ±0.0005 in

0.5000 to 1.5000 in in diameter: ±0.001 in

7.14.2   Metric drill rods

Metric drill rods are made of chrome vanadium alloy steel in 1-m lengths (approximately 39 in). Accuracy is metric h9. Heating for hardening: 1490–1580°F quench in oil; 1470–1510°F quench in water. Drawing temperatures: HRC-62, 400°F; HRC-60, 485°F; HRC-56, 570°F.

Available diameters.   Metric drill rods are available in diameters ranging from 1 through 50 mm, in various steps or increments. Accuracy is metric h9.

8

Tooling, Die Making, Molds, Jigs, and Fixtures

8.1 Definitions

Tooling. *Tooling* is a general term used to encompass many different processes involving the design and manufacture of special tools, dies, molds, jigs, and fixtures. The most common type of tooling consists of dies that are used to stamp or blank sheet metal parts for mass production. The use of a stamping or blanking die makes it possible to produce thousands of parts with consistent dimensional accuracy at a rapid pace. This ensures that the die-produced parts will fit correctly into their next assembly stage, such as in a complex mechanical device or mechanism, and have interchangeability.

The production of detail parts from a tooling device such as a stamping die reduces the overall cost of producing industrial and consumer products in relation to the cost of the die versus the number of parts to be made on the die. Complex tooling devices are relatively expensive to produce and usually must be justified by the number of parts they produce during a set time interval. As a general rule, thousands of parts should be produced yearly to justify tooling costs and maintenance of the tools.

Die making. One of the most common tooling procedures is the design and production of dies, i.e., die making. Industry uses a great variety of dies, including

- Stamping, blanking, and punching dies
- Bending and forming dies
- Combination or progressive dies
- Forging dies
- Dies for die casting metals and metal alloys
- Beading dies (a form of bending die)
- Extrusion dies
- Drawing dies (for wire, rod, and bar)
- Wire-form dies

Molds.   There are many different types of molds that fall under the general definition of tooling. Molds are used to produce metal and metal-alloy parts and various types of plastic parts. In die casting, the mold is referred to as a "die." The various types of molds made from metals, usually tool steels, include

- Die-casting molds
- Chill-casting molds
- Permanent casting molds
- Slush-casting molds
- Thermoplastic injection molds
- Thermoset plastic compression molds

Molds are also made from special tooling epoxies (see Sec. 8.6.4).

Jigs.   *Jigs* are defined as tooling devices used as patterns (templates) for producing parts, match plates for drilling holes, and other devices used to guide or control a machining process. The process of pin routing uses a template or jig to guide the action of the high-speed routing cutter while it is cutting a stack of thin-gauge sheet metal parts. A drilling or reaming jig may be designed to clamp previously die-stamped parts that are to receive a secondary machining operation such as reaming or additional drilling.

Specially hardened drill bushings are incorporated as part of the drill jig to guide the drill accurately as the part is jig drilled. Jigs allow parts to be produced with dimensional accuracy and consis-

tency. Clamping and clamping devices are an important part of the jigging process. A subsection detailing clamping devices and clamping calculations will be presented in a later part of this chapter (see Sec. 8.8).

Fixtures.   One of the most important aspects of tooling is the design and application of fixtures. A tooling *fixture* may be defined as a manufacturing aid or assembly device that is necessary in the production of mass-produced parts and assemblies. Various common types of fixtures are used in industrial applications, including

- Subassembly fixtures

- Assembly fixtures

- Welding fixtures

- Machining fixtures

- Wiring fixtures (for wire harnesses)

The design of tooling fixtures is limited only by the imagination and ingenuity of the tool designer. Fixtures for subassembly of small parts may weigh only a few ounces, whereas the assembly fixtures used to produce large aircraft and aerospace vehicles may weigh more than 100 tons. Tooling fixtures are an absolute necessity for the interchangeability of complex mass-produced products. Vast sums of money are spent each year by the automotive industry alone to produce the tooling fixtures required to assemble automotive equipment. The application of clamps and holding devices on tooling fixtures is an important part of tool design. The clamps and holding devices may be mechanical, electromechanical, pneumatic, or hydraulic and must be an integral part of the complete fixture. Safety and versatility must be kept in mind when designing the clamping and holding devices used on tooling fixtures.

8.2   Tool-Steel Characteristics, Heat Treatment, and Selection

To ensure that the proper tool steel is selected for each different type of tooling device and that its capabilities are developed to full potential, the producers of the various tool steels are prepared to advise tool designers or engineers on the mechanical properties and conditions of service under which the finished tools operate

best. The word *tool* throughout this chapter denotes any of the tooling devices previously described.

The majority of tool-steel applications may be categorized under a small number of groups or types of operations: cutting, shearing, forming, drawing, bending, extruding, rolling, and ramming or battering. For each of these groups, certain metallurgic characteristics are of utmost importance. Cutting tools must possess high hardness and extreme strength, high resistance to the effects of elevated temperatures, and high wear resistance. Shearing tools require high wear resistance combined with toughness, and these characteristics must be balanced properly, depending on the tool design, thickness of the stock being sheared, and temperature of the shearing operation. Forming tools must possess high toughness and high strength, and many require high resistance to the softening effect of elevated temperatures. In battering or ramming tools, high toughness is the most important characteristic.

Heat treatment plays an important role in attaining these desired characteristics. It should be kept in mind that tools actually are being processed, not steels. To ensure that proper heat-treatment procedures are employed, this section will present tables that list guidelines for forging, annealing, hardening, tempering, and normalizing the commonly used tool steels to which American Iron and Steel Institute (AISI) designations have been assigned. These guidelines, which include temperature ranges, heating rates, cooling rates, and time at temperature, should be adapted to fit the specific application.

8.2.1   Identification and classification of tool steels

Figure 8.1 provides a list of the main tool-steel groups. Table 8.1 shows the identification and type classification of tool steels.

8.2.2   Heat treatment of tool steels

Tables 8.2 and 8.3 list the typical heat treatment of tool steels along with supplement notes and explanations (also see Sec. 4.1.1, "Tool Steels").

Tool steels are either carbon or alloy steels, which are capable of being hardened and tempered, and they are usually melted in electric furnaces and produced under tool-steel practices to meet their special requirements. Special care should be taken when using the

Type of Tool-Steel	Designation
High-speed tool steels	M - molybdenum types T - tungsten types
Hot-work tool-steels	H H1 to H19 - chromium types H20 to H39 - tungsten types H40 to H59 - molybdenum types
Cold-work tool-steels	D - high-carbon high-chromium types A - medium-alloy air-hardening types O - oil-hardening types
Shock-resisting tool-steels	S
Mold tool-steels	P
Special-purpose tool-steels	L - low-alloy types F - carbon-tungsten types
Water-hardening tool-steels	W

Figure 8.1 Main tool-steel groups.

heat-treatment data shown in Tables 8.2 and 8.3, and this is outlined as follows.

Forging. The information shown in the tables is for those occasions when a consumer is unable to purchase the desired size and must forge from a larger size. The temperature at which to start forging is given as a range, the higher side of which should be used for larger sections and heavy or rapid reductions and the lower side for smaller sections and lighter reductions. As the alloy content of the steel increases, the time of soaking at forging temperatures increases proportionately. Likewise, as the alloy content increases, it becomes more necessary to cool slowly from the forging temperature. With very high alloy steels, such as high-speed steels and air-hardening steels, this slow cooling is imperative in order to prevent cracking. Either furnace cooling or burying in an insulating material, such as lime, mica, or silocel, is satisfactory.

Annealing. The information shown in the tables on annealing is intended for those instances where a finished tool must be softened for additional machining or where the user has performed a forging operation. The annealing temperature is given as a range, the

(Text continued on page 544.)

TABLE 8.1 Identification and Type Classification of Tool Steels

Type	Identifying elements, %					
	C	W	Mo	Cr	V	Co
High Speed: Tungsten Types (1)						
T1	0.75 (2)	18.00	—	4.00	1.00	—
T2	0.80	18.00	—	4.00	2.00	—
T4	0.75	18.00	—	4.00	1.00	5.00
T5	0.80	18.00	—	4.00	2.00	8.00
T6	0.80	20.00	—	4.50	1.50	12.00
T8	0.75	14.00	—	4.00	2.00	5.00
T15	1.50	12.00	—	4.00	5.00	5.00
High Speed: Molybdenum Types (1)						
M1	0.85 (2)	1.50	8.50	4.00	1.00	—
M2	0.85: 1.00 (2)	6.00	5.00	4.00	2.00	—
M3 class 1	1.05	6.00	5.00	4.00	2.40	—
M3 class 2	1.20	6.00	5.00	4.00	3.00	—
M4	1.30	5.50	4.50	4.00	4.00	—
M6	.80	4.00	5.00	4.00	1.50	12.00
M7	1.00	1.75	8.75	4.00	2.00	—
M10	0.85: 1.00 (2)	—	8.00	4.00	2.00	—
M30	0.80	2.00	8.00	4.00	1.25	5.00
M33	0.90	1.50	9.50	4.00	1.15	8.00
M34	0.90	2.00	8.00	4.00	2.00	8.00
M36	0.80	6.00	5.00	4.00	2.00	8.00
M41	1.10	6.75	3.75	4.25	2.00	5.00
M42	1.10	1.50	9.50	3.75	1.15	8.00
M43	1.20	2.75	8.00	3.75	1.60	8.25
M44	1.15	5.25	6.25	4.25	2.00	12.00
M46	1.25	2.00	8.25	4.00	3.20	8.25
M47	1.10	1.50	9.50	3.75	1.25	5.00

Type	Identifying elements, %						
	C	Mn or Si	W	Mo	Cr	V	Co	Other
Cold Work: Medium-Alloy Air-Hardening Types (1)								
A2	1.00	—	—	1.00	5.00	—	—	—
A3	1.25	—	—	1.00	5.00	1.00	—	—
A4	1.00	2.00 Mn	—	1.00	1.00	—	—	—
A6	0.70	2.00 Mn	—	1.25	1.00	—	—	—
A7	2.25	—	1.00 (3)	1.00	5.25	4.75	—	—
A8	0.55	—	1.25	1.25	5.00	—	—	—
A9	0.50	—	—	1.40	5.00	1.00	—	1.50 Ni
A10 (4)	1.35	1.80 Mn: 1.25 Si	—	1.50	—	—	—	1.80 Ni
Cold Work: Oil-Hardening Types								
O1	0.90	1.00	.50	—	.50	—	—	—
O2	0.90	1.60	—	—	—	—	—	—
O6 (4)	1.45	0.80 Mn: 1.00 Si	—	25	—	—	—	—
O7	1.20	—	1.75	—	.75	—	—	—
—								
W5	1.10	—	—	—	0.50	—	—	—
Shock-Resisting Types								
S1	0.50	—	2.50	—	1.50	—	—	—
S2	0.50	1.00 Si	—	0.50	—	—	—	—
S5	0.55	0.80 Mn: 2.00 Si	—	0.40	—	—	—	—
S6	0.45	1.40 Mn: 2.25 Si	—	0.40	1.50	—	—	—
S7	0.50	—	—	1.40	3.25	—	—	—

Hot Work: Chromium Types (1)						
H10	0.40	—	2.50	3.25	0.40	—
H11	0.35	—	1.50	5.00	0.40	—
H12	0.35	1.50	1.50	5.00	0.40	—
H13	0.35	—	1.50	5.00	1.00	—
H14	0.40	5.00	—	5.00	—	—
H19	0.40	4.25	—	4.25	2.00	4.25

Hot Work: Tungsten Types						
H21	0.35	9.00	—	3.50	—	—
H22	0.35	11.00	—	2.00	—	—
H23	0.30	12.00	—	12.00	—	—
H24	0.45	15.00	—	3.00	—	—
H25	0.25	15.00	—	4.00	—	—
H26	0.50	18.00	—	4.00	1.00	—

Hot Work: Molybdenum Types						
H42	0.60	6.00	5.00	4.00	2.00	—

Cold Work: High-Carbon High-Chromium Types (1)						
D2	1.50	—	1.00	12.00	1.00	—
D3	2.25	—	—	12.00	—	—
D4	2.25	—	1.00	12.00	—	—
D5	1.50	—	1.00	12.00	—	3.00
D7	2.35	—	1.00	12.00	4.00	—

Special-Purpose Low-Alloy Types						
L2	0.50/1.10 (5)	—	—	1.00	0.20	—
L6	0.70	—	0.25 (3)	0.75	—	1.50 Ni

Water-Hardening Types						
W1	0.60/1.40 (5)	—	—	—	—	—
W2	0.60/1.40 (5)	—	—	—	0.25	—
W5	1.10	—	—	0.50	—	—

Mold Types						
P2	0.07	—	0.20	2.00	—	0.50 Ni
P3	0.10	—	—	0.60	—	1.25 Ni
P4	0.07	—	0.75	5.00	—	—
P5	0.10	—	—	2.25	—	—
P6	0.10	—	—	1.50	—	3.50 Ni
P20	0.35	—	0.40	1.70	—	—
P21	0.20	—	—	—	—	4.00 Ni: 1.20 Al

(1) Some of the types can be produced with a sulfur addition to improve machinability.
(2) Other carbon contents may be available.
(3) Optional.
(4) Contains free graphite in the microstructure to improve machinability.
(5) Various carbon contents are available.

TABLE 8.2   Typical Heat Treatments for Tool Steels

Tool steel type	Forging		Annealing			Hardening		
	Start, °F	Do not forge below °F	Temp., °F	Cooling rate, °F Max/h	Brinell hardness	Heating rate	Preheat temp., °F	Hardening temp., °F
High Speed: Tungsten Types								
T1	1950–2150	1750	1600–1650	40	217–255	Rapid from preheat	1500–1600	2300–2375(a)
T2	1950–2150	1750	1600–1650	40	223–255	Rapid from preheat	1500–1600	2300–2375(a)
T4	1950–2150	1750	1600–1650	40	229–269	Rapid from preheat	1500–1600	2300–2375(a)
T5	1950–2150	1800	1600–1650	40	235–285	Rapid from preheat	1500–1600	2325–2375(a)
T6	1950–2150	1800	1600–1650	40	248–302	Rapid from preheat	1500–1600	2325–2375(a)
T8	1950–2150	1800	1600–1650	40	229–255	Rapid from preheat	1500–1600	2300–2375(a)
T15	1950–2150	1800	1600–1650	40	241–277	Rapid from preheat	1500–1600	2200–2300(a)
High Speed: Molybdenum Types								
M1	1900–2100	1700	1500–1600	40	207–235	Rapid from preheat	1350–1550	2150–2225(a)
M2	1900–2100	1700	1600–1650	40	212–241	Rapid from preheat	1350–1550	2175–2250(a)
M3 (1; 2)	1900–2100	1700	1600–1650	40	223–255	Rapid from preheat	1350–1550	2200–2250(a)
M4	1900–2100	1700	1600–1650	40	223–255	Rapid from preheat	1350–1550	2200–2250(a)
M6	1900–2100	1700	1600	40	248–277	Rapid from preheat	1450	2150–2200(a)
M7	1900–2100	1700	1500–1600	40	217–255	Rapid from preheat	1350–1550	2150–2225(a)
M10	1900–2100	1700	1500–1600	40	207–255	Rapid from preheat	1350–1550	2150–2225(a)
M30	1900–2100	1700	1600–1650	40	235–269	Rapid from preheat	1350–1550	2200–2250(a)
M33	1900–2100	1700	1600–1650	40	235–269	Rapid from preheat	1350–1550	2200–2250(a)
M34	1900–2100	1700	1600–1650	40	235–269	Rapid from preheat	1350–1500	2200–2250(a)
M36	1900–2100	1700	1600–1650	40	235–269	Rapid from preheat	1350–1550	2225–2275(a)
M41	1900–2100	1700	1600–1650	40	235–269	Rapid from preheat	1350–1550	2175–2220(a)
M42	1900–2100	1700	1600–1650	40	235–269	Rapid from preheat	1350–1550	2125–2175(a)
M43	1900–2100	1700	1600–1650	40	248–269	Rapid from preheat	1350–1550	2100–2150(a)
M44	1900–2100	1700	1600–1650	40	248–285	Rapid from preheat	1350–1550	2190–2240(a)
M46	1900–2100	1700	1600–1650	40	235–269	Rapid from preheat	1350–1550	2175–2225(a)
M47	1900–2100	1700	1600–1650	40	235–269	Rapid from preheat	1350–1550	2150–2200(a)
Hot Work: Chromium Types								
H10	1950–2100	1650	1550–1650	40	192–229	Moderate from preheat	1500	1850–1900
H11	1950–2100	1650	1550–1650	40	192–235	Moderate from preheat	1500	1825–1875

TABLE 8.2 Typical Heat Treatments for Tool Steels (*Continued*)

Hardening		Tempering					
Minutes at temp.	Quench medium (b)	Temp., °F	Tempered Rc hardness	Depth of hardening	Heat-treat distortion	Hardening safety	Resist. to decarb.
High Speed: Tungsten Types							
2–5	O, A, or S	1000–1100(c)	65–60	Deep	A or S = low; O = medium	High	High
2–5	O, A, or S	1000–1100(c)	66–61	Deep	A or S = low; O = medium	High	High
2–5	O, A, or S	1000–1100(c)	66–62	Deep	A or S = low; O = medium	Medium	Medium to high
2–5	O, A, or S	1000–1100(c)	65–60	Deep	A or S = low; O = medium	Medium	Low
2–5	O, A, or S	1000–1100(c)	65–60	Deep	A or S = low; O = medium	Medium	Low
2–5	O, A, or S	1000–1100 (c)	65–60	Deep	A or S = low; O = medium	Medium	Medium
2–5	O, A, or S	1000–1200(c)	68–63	Deep	A or S = low; O = medium	Medium	Medium
High Speed: Molybdenum Types							
2–5	O, A, or S	1000–1100(c)	65–60	Deep	A or S = low; O = medium	Medium	Low
2–5	O, A, or S	1000–1100(c)	65–60	Deep	A or S = low; O = medium	Medium	Medium
2–5	O, A, or S	1000–1100(c)	66–61	Deep	A or S = low; O = medium	Medium	Medium
2–5	O, A, or S	1000–1100(c)	66–61	Deep	A or S = low; O = medium	Medium	Medium
2–5	O, A, or S	1000–1100(c)	66–61	Deep	A or S = low; O = medium	Medium	Low
2–5	O, A, or S	1000–1100(c)	66–61	Deep	A or S = low; O = medium	Medium	Low
2–5	O, A, or S	1000–1100(c)	65–60	Deep	A or S = low; O = medium	Medium	Low
2–5	O, A, or S	1000–1100(c)	65–60	Deep	A or S = low; O = medium	Medium	Low
2–5	O, A, or S	1000–1100(c)	65–60	Deep	A or S = low; O = medium	Medium	Low
2–5	O, A, or S	1000–1100(c)	65–60	Deep	A or S = low; O = medium	Medium	Low
2–5	O, A, or S	1000–1100(c)	65–60	Deep	A or S = low; O = medium	Medium	Low
2–5	O, A, or S	1000–1100(d)	70–65	Deep	A or S = low; O = medium	Medium	Low
2–5	O, A, or S	950–1100(d)	70–65	Deep	A or S = low; O = medium	Medium	Low
2–5	O, A, or S	950–1100(d)	70–65	Deep	A or S = low; O = medium	Medium	Low
2–5	O, A, or S	1000–1160(d)	70–62	Deep	A or S = low; O = medium	Medium	Low
2–5	O, A, or S	975–1050(d)	69–67	Deep	A or S = low; O = medium	Medium	Low
2–5	O, A, or S	975–1100(d)	70–65	Deep	A or S = low; O = medium	Medium	Low
Hot Work: Chromium Types							
15–40(e)	A	1000–1200(c)	56–39	Deep	Very low	Highest	Medium
15–40(e)	A	1000–1200(c)	54–38	Deep	Very low	Highest	Medium

TABLE 8.2   Typical Heat Treatments for Tool Steels (*Continued*)

	Forging		Annealing			Hardening		
Tool steel type	Start, °F	Do not forge below °F	Temp., °F	Cooling rate, °F Max/h	Brinell hardness	Heating rate	Preheat temp., °F	Hardening temp., °F
H12	1950–2100	1650	1550–1650	40	192–235	Moderate from preheat	1500	1825–1875
H13	1950–2100	1650	1550–1650	40	192–229	Moderate from preheat	1500	1825–1900
H14	1950–2150	1700	1600–1650	40	207–235	Moderate from preheat	1500	1850–1950
H19	1900–2100	1650	1600–1650	40	207–241	Rapid from preheat	1500	2000–2200(a)
Hot Work: Tungsten Types								
H21	1950–2150	1650	1600–1650	40	207–235	Rapid from preheat	1500	2000–2200(a)
H22	1950–2150	1650	1600–1650	40	207–235	Rapid from preheat	1500	2000–2200(a)
H23	1950–2150	1800	1600–1650	40	212–255	Rapid from preheat	1550	2200–2300(a)
H24	1950–2150	1750	1600–1650	40	217–241	Rapid from preheat	1500	2000–2250(a)
H25	1950–2150	1700	1600–1650	40	207–235	Rapid from preheat	1500	2100–2300(a)
H26	1950–2150	1750	1600–1650	40	217–241	Rapid from preheat	1600	2150–2300(a)
Hot Work: Molybdenum Type								
H42	1900–2050	1700	1550–1650	40	207–235	Rapid from preheat	1350–1550	2050–2225(a)
Cold Work: High-Carbon High-Chromium Types								
D2	1850–2000	1700	1600–1650	40	217–255	Very slowly	1500	1800–1875
D3	1850–2000	1700	1600–1650	40	217–255	Very slowly	1500	1700–1800
D4	1850–2000	1700	1600–1650	40	217–255	Very slowly	1500	1775–1850
D5	1850–2000	1700	1600–1650	40	223–255	Very slowly	1500	1800–1875
D7	2050–2125	1800	1600–1650	40	235–262	Very slowly	1500	1850–1950
Cold Work: Medium-Alloy Air-Hardening Types								
A2	1850–2000	1650	1550–1600	40	201–235	Slowly	1450	1700–1800
A3	1850–1950	1650	1550–1600	40	207–229	Slowly	1450	1750–1850
A4	1850–2000	1650	1360–1400	25	200–241	Slowly	1250	1500–1600
A6	1900–2050	1600	1350–1375	25	217–248	Slowly	1200	1525–1600
A7	1925–2100	1800	1600–1650	25	235–269	Very slowly	1500	1750–1800
A8	1950–2100	1700	1550–1600	40	192–241	Slowly	1450	1800–1850
A9	1950–2100	1700	1550–1600	25	212–248	Slowly	1450	1800–1875
A10	1800–1925	1600	1410–1460	15	235–269	Slowly	1200	1450–1500
Cold Work: Oil-Hardening Types								
O1	1800–1950	1550	1400–1450	40	183–212	Slowly	1200	1450–1500
O2	1800–1925	1550	1375–1425	40	183–217	Slowly	1200	1400–1475
O6	1800–1950	1500	1410–1450	20	183–217	Slowly	—	1450–1500
O7	1800–2000	1600	1450–1500	40	192–217	Slowly	1200	W=1450–1525(b) O=1500–1625(b)
Shock-Resisting Types								
S1	1850–2050	1600	1450–1500	40	183–229	Slowly	1200	1650–1750
S2	1850–2050	1600	1400–1450	40	192–217	Slowly	1200	1550–1650
S5	1850–2050	1600	1425–1475	25	192–229	Slowly	1400	1600–1700
S6	1850–2050	1600	1475–1525	25	192–229	Slowly	1400	1650–1750
S7	1950–2050	1700	1500–1550	25	187–223	Slowly	1200–1300	1700–1750

TABLE 8.2   Typical Heat Treatments for Tool Steels (*Continued*)

Hardening		Tempering					
Minutes at temp.	Quench medium (b)	Temp., °F	Tempered Rc hardness	Depth of hardening	Heat-treat distortion	Hardening safety	Resist. to decarb.
15–40(e)	A	1000–1200(c)	55–38	Deep	Very low	Highest	Medium
15–40(e)	A	1000–1200(c)	53–38	Deep	Very low	Highest	Medium
15–40(e)	A	1100–1200(c)	47–40	Deep	Low	Highest	Medium
2–5	A or O	1000–1300(c)	57–40	Deep	A = low; O = medium	High	Medium
Hot Work: Tungsten Types							
2–5	A or O	1100–1250	54–36	Deep	A = low; O = medium	High	Medium
2–5	A or O	1100–1250	52–39	Deep	A = low; O = medium	High	Medium
2–5	O	1200–1350(c)	47–34	Medium	Medium	High	Medium
2–5	O	1050–1200(c)	55–45	Deep	A = low; O = medium	High	Medium
2–5	A or O	1050–1250(c)	44–35	Deep	A = low; O = medium	High	Medium
2–5	O, A, or S	1050–1250(c)	58–43	Deep	S, A = low; O = medium	High	Medium
Hot Work: Molybdenum Type							
15–45(e)	O, A, or S	1050–1200(c)	60–50	Deep	S, A = low; O = medium	Medium	Medium
Cold Work: High-Carbon High-Chromium Types							
15–45(e)	A	400–1000	61–54	Deep	Lowest	Highest	Medium
15–45(e)	O	400–1000	61–54	Deep	Very low	High	Medium
15–45(e)	A	400–1000	61–54	Deep	Lowest	Highest	Medium
15–45(e)	A	400–1000	61–54	Deep	Lowest	Highest	Medium
30–60(e)	A	300–1000	65–58	Deep	Lowest	Highest	Medium
Cold Work: Medium-Alloy Air-Hardening Types							
20–45(e)	A	350–1000	62–57	Deep	Lowest	Highest	Medium
25–60(c)	A	350–1000	65–57	Deep	Lowest	Highest	Medium
15–90(e)	A	350–800	62–54	Deep	Lowest	Highest	Medium to high
20–45(c)	A	300–800	60–54	Deep	Lowest	Highest	Medium to high
30–60(e)	A	300–1000	67–57	Deep	Lowest	Highest	Medium
20–45(e)	A	350–1100	60–50	Deep	Lowest	Highest	Medium
20–45(e)	A	950–1150	56–35	Deep	Lowest	Highest	Medium
30–60(e)	A	350–80	62–55	Deep	Lowest	Highest	Medium to high
Cold Work: Oil-Hardening Types							
10–30	O	350–500	62–57	Medium	Very low	Very high	High
5–20	O	350–500	62–57	Medium	Very low	Very high	High
10–30	O	350–600	63–58	Medium	Very low	Very high	High
10–30	O or W	350–550	64–58	Medium	W = high O = very low	W = low O = very high	High
Shock-Resisting Types							
15–45	O	400–1200	58–40	Medium	Medium	High	Medium
5–20	B or W	350–800	60–50	Medium	High	Low	Low
5–20	O	350–800	60–50	Medium	Medium	High	Low
10–30	O	400–600	56–54	Medium	Medium	High	Low
15–45(e)	A or O	400–1150	57–45	Deep	A = lowest; O = low	A = highest; O = high	Medium

TABLE 8.2   Typical Heat Treatments for Tool Steels (*Continued*)

Tool steel type	Forging		Annealing			Hardening		
	Start, °F	Do not forge below °F	Temp., °F	Cooling rate, °F Max/h	Brinell hardness	Heating rate	Preheat temp., °F	Hardening temp., °F
Special Purpose: Low-Alloy Types								
L2	1800–2000	1550	1400–1450	40	163–197	Slowly	—	W=1450–1550 O=1550–1700
L6	1800–2000	1550	1400–1450	40	183–255	Slowly	—	1450–1550
Water-Hardening Types								
W1	1800–1950(f)	1500	1360–1450(f)	40	156–201	Slowly	(g)	1400–1550(h)
W2	1800–1950(f)	1500	1360–1450(f)	40	156–201	Slowly	(g)	1400–1550(h)
W5	1800–1950(f)	1500	1360–1450(f)	40	163–201	Slowly	(g)	1400–1525(h)
						Hardening		
						Carburizing temp., °F		Hardening temp., °F
Mold Types								
P2	1850–2050	1550	1350–1500	40	103–123	1650–1700		1525–1550(i)
P3	1850–2050		1350–1500	40	109–137	1650–1700		1475–1525(i)
P4	1850–2050	1600	1600–1650	25	116–128	1775–1825		1775–1825(i)
P5	1850–2050	1550	1550–1600	40	105–131	1650–1700		1550–1600(i)
P6	1950–2150	1700	1550	15	183–217	1650–1700		1450–1500(i)
P20	1850–2050	1600	1400–1450	40	149–212	1600–1650		1500–1600
						Solution Treating		
						Heating rate	Preheat temp.	Solution temp., °F
P21	2000–2100	1750		Do not anneal		Slowly	Do not preheat	1300–1350

upper limit of which should be used for large sections and the lower limit for smaller sections. The length of time the steel is held, after being uniformly heated through at the annealing temperature, varies from about 1 hour for light sections and small furnace charges of carbon or low-alloy tool steel to about 4 hours for heavy sections and large furnace charges of high-alloy steel.

Normalizing.   The purpose of normalizing after forging is to refine the grain structure and to produce a uniform structure throughout the forging. Normalizing should not be confused with low-temperature (about 1200°F) annealing, which is used for relief of residual stresses resulting from heavy machining, bending, and forming.

Steels that can be normalized are O1 (1600°F), O2 (1550°F), O6 (1600°F), and O7 (1650°F); L2 (1600-1650°F) and L6 (1600°F); WI, W2, and W5 (1450-1700°F); and P20 and P21 (1650°F).

TABLE 8.2   Typical Heat Treatments for Tool Steels (*Continued*)

Hardening		Tempering					
Minutes at temp.	Quench medium (b)	Temp., °F	Tempered Rc hardness	Depth of hardening	Heat-treat distortion	Hardening safety	Resist. to decarb.
Special Purpose: Low-Alloy Types							
10–30	O or W	350–1000	63–45	O = medium	O = medium W = high	O = medium W = low	High
10–30	O	350–1000	62–45	Medium	Low	High	High
Water-Hardening Types							
10–30	B or W	350–650	64–50	Shallow	High	Low	Highest
10–30	B or W	350–650	64–50	Shallow	High	Low	Highest
10–30	B or W	350–650	64–50	Shallow	High	Low	Highest
Mold Types							
15	O	350–500	64–58(k)	Medium	Low	High	High
15	O	350–500	64–58(k)	Medium	Low	High	High
15	A	350–900	64–58(k)	Medium	Very low	High	High
15	O or W	350–500	64–58(k)		W = high O = low	High	High
15	A or O	350–450	61–58(k)		A = very low, O = low	High	High
15	O	900–1100	37–28	Medium	Low	High	High

		Aging					
		Temp., °F	Hardness Rc				
60–180	A or O	950–1025	40–30	Deep	Lowest	Highest	High

The length of time the steel is held after being uniformly heated through at the normalizing temperature varies from about 15 minutes for a small section to about 1 hour for large sizes. Cooling from the normalizing temperature is done in still air.

Other factors of concern to heat treaters.   Other properties of the tool steels that are of concern to the heat treater are depth of hardening, distortion in heat treating, safety in hardening, and resistance to decarburization. A qualitative evaluation of these properties is also presented in Tables 8.2 and 8.3, comparing all the tool steels relative to one another rather than within any particular class. The following general observations may be helpful.

Depth of hardening.   This relates to the depth of hardness penetration of the individual tool steels. The hardenability ratings are based

TABLE 8.3   Notes and Explanations for Table 8.2

(a) When the high temperature heating is carried out in a salt bath, the range of temperatures should be about 25°F lower than that shown.

(b) A = air quench
    B = brine quench
    O = oil quench
    S = salt bath quench
    W = water quench

(c) Double tempering recommended for not less than 1 h at temperature each temper.

(d) Triple tempering recommended for not less than 1 h at temperature each temper.

(e) Times shown apply to open furnace heat treatment. For pack hardening, a common rule is to heat for ½ h per inch of cross section of the pack.

(f) Forging, normalizing, and annealing temperatures of carbon tool steels are given as ranges because they vary with carbon content. The following temperatures are recommended:

Forging
    0.50 to 1.25% C: the range shown
    1.25 to 1.40% C: low side of range

Normalizing
    0.60 to 0.75% C: 1500°F
    0.75 to 0.90% C: 1450°F
    0.90 to 1.10% C: 1600°F
    1.10 to 1.40% C: 1600 to 1700°F

Annealing
    0.60 to 0.90% C: 1360 to 1400°F
    0.90 to 1.40% C: 1400 to 1450°F

(g) For large tools and tools having intricate sections, preheating at 1050–1200°F is recommended.

(h) Varies with carbon content as follows:
    0.60–0.80% C: 1450–1550°F
    0.85–1.05% C: 1425–1550°F
    1.10–1.40% C: 1400–1525°F

(i) After carburizing

(k) Carburized case hardness

SOURCE: "AISI Steel Products Manual—Tool Steels," April 1976.

on the use of the particular quenching medium recommended. Carbon tool steels in group W are very shallow hardening steels and generally are quenched in water. The hardenability increases in tool steels as the alloy content increases (except for cobalt and tungsten). The hardenability increases with carbon content until excess carbide appears, and then it begins to decrease with further increases in carbon. For large tool or die sections, it is imperative that a high-alloy steel be selected if high strength is to be developed throughout the section in the finished part. For the P steels that are used in the carburized and hardened condition, the core hardenability is rated as low, medium, or high.

Distortion in heat treating. This rates the tool steels on the basis of the distortion normally obtained in hardening the respective grades

from the normally recommended hardening temperatures. Distortion in heat treating is important in intricately designed tools that must maintain their shape after hardening. The steels rated lowest or very low usually can be machined very close to finish size prior to heat treatment so that little grinding will be required after the hardening operation.

In general, water-hardening steels exhibit the most distortion, whereas air-hardening steels exhibit the least. It is important to note, however, that carbon tool steels and other shallow-hardening types may distort very little in heat treatment when the hardened case is small in comparison with the unhardened core for any particular tool or die design.

Safety in hardening. This deals with the overall freedom from heat-treating difficulties that may be experienced in handling the tool-steel grades in question. The ratings apply particularly to freedom from cracking when complicated or intricate sections are being hardened. The air- and oil-hardening steels prove superior to the water-hardening steels in this respect.

When designing machine parts and tools that require heat treatment, the design engineer should allow generous filleting at changes in section and eliminate the use of sharp angles as far as possible. The greater the symmetry of design, the less hazardous is the heat treatment.

Resistance to carburization. This influences the type of heat-treating equipment selected, as well as the amount of material that should be removed from the surface after hardening. Both these factors concern the economics of tool-steel use. The ratings are intended to apply at the hardening temperature normally recommended. Those steels which are rated low must be protected in some way from decarburization during the heating cycle.

Additional heat-treatment data. See Sec. 4.1.1.

8.2.3   Selection of tool steels

The tool steels listed in the preceding tables are generally available. The relative availability within a group, however, will vary depending on whether the steel is designed for specific applications or finds use in general applications. Experience indicates that in most cases the choice is not limited to a single type of tool steel or even to a particular family of tool steels for a workable solution to an individual

tooling problem. It is desirable to select a tool steel that will give the most economical overall performance when weighed against expected productivity, ease of fabrication, and cost.

The majority of tool-steel applications can be divided into a small number of groups or types of operations, such as cutting, shearing, forming, drawing, extruding, rolling, and battering. The following characteristic descriptions of the tool steels are intended as a guide, and it must be understood that the proper selection of a tool steel for a specific application cannot always be made with 100 percent assurance. Because of this, consultation with the steel producer is recommended. The following list describes the characteristics of the major tool-steel types.

A. Cold-work tool steels (medium alloy, air hardening).   These steels show a minimum of distortion during heat treatment and have greater safety in hardening than the oil-hardening grades. Dies that could crack during heat treatment when made from oil-hardening grades generally can be handled safely when made from the air-hardening group. The low-carbon types A8 and A9 offer greater shock resistance than the other steels in this group but are lower in wear resistance. Type A7 exhibits maximum abrasion resistance but should be restricted to lower-toughness applications.

D. Cold-work tool steels (high carbon, high chromium).   These steels are more wear resistant than the medium-alloy air-hardening steels used where long-run dies are required. Of the group, Dl, D2, and D5 have the greater toughness but the least abrasion resistance. These steels, being very wear resistant, are difficult to machine and grind, and a minimum of stock should be allowed for grinding after hardening. These grades are all air hardening and show little distortion or movement during heat treatment. Wear resistance increases as the carbon and vanadium contents increase.

E. Special-purpose tool steels (carbon-tungsten types).   These are often called *finishing steels* and provide relatively high wear resistance in water-hardening steels. These steels typically are used for wire, bar, or tube cold-drawing dies, finishing tools, and fine-cutting-edge tools such as broaches, taps, and reamers. F2 is recommended where high wear resistance and sharp-cutting-edge capability are required. These steels are somewhat difficult to grind after hardening because of their wear and abrasion resistance.

H. Hot-work tool steels.   These steels are chromium-, tungsten-, and molybdenum-based. The chromium-based types are air hardening, showing very little distortion during heat treatment. The tungsten- and molybdenum-based steels are either air or oil hardening and show more distortion during oil quenching. These are high-heat-resisting steels with moderate wear resistance. The tungsten types (H2O to H39) are for those hot-work applications where resistance to the softening effect of elevated temperature is of greatest importance and a lesser degree of toughness is allowed. The molybdenum types (H40 to H59) offer excellent high-heat resistance but should be restricted to those applications where less ductility is acceptable. A wide range of hot-work tools are made from these steels for die-casting dies, extrusion-press parts, forging dies, hot punches, and shears.

I. Special-purpose tool steels (low alloy).   These steels have deeper hardening characteristics than the W group and have similar properties to O1 and O2 as general-purpose tool steels but exhibit a greater tendency to distort during heat treatment. Of the group, L6 has greater toughness, at the sacrifice of wear resistance. High carbon type L3 may be used for short-run and special-purpose tools and dies.

M. High-speed tool steels (molybdenum base).   These steels are used commonly for high-speed cutting tool applications. They decarburize better than the tungsten type and are somewhat more critical in their heat treatment. Vanadium and cobalt are added, as is the case with the tungsten types, with corresponding advantages and disadvantages. The M2 type is considered the molybdenum counterpart of T1 as a general-purpose cutting tool material. The series beginning with M41 has the characteristic of exceptionally high hardness in heat treatment. These steels are also used for cold header die inserts, punches, and thread-rolling and blanking dies. The steels in this group are under-hardened for these applications in order to increase toughness.

O. Cold-work tool steels (oil hardening).   All these steels listed under the symbol O are low-alloy types that must be oil quenched in heat treatment. Types O1 and O2 are widely used tool steels for many types of die applications, having greater hardenability and generally less distortion during heat treatment than the water-hardening steels. These steels will harden throughout in sections up to 2.50 in thick with relative freedom from distortion and cracking during heat treatment. Type O7 attains the greatest wear or abrasion resistance but is usually used in special applications.

P. Special-purpose tool steels (mold steels). These steels are generally considered as plastic-mold steels but have found other applications outside the mold-making field. These steels are usually supplied at very low hardness to facilitate machining and other processing prior to carburizing to develop the required surface properties for injection and compression molds for plastics. Types P20 and P21 are usually supplied in the hardened condition so that they may be machined and put into immediate service. These types are suitable for plastic molds, zinc die-casting dies, and holder blocks.

S. Shock-resisting tool steels. These steels contain less carbon but higher alloy content and therefore are tougher than water-hardening grades. They have less wear resistance than the water-hardening steels but are used where punching, shearing, or trimming is being performed. Type Sl has the greatest wear resistance in the group and may be carburized for even greater hardness and wear resistance, but with a loss in toughness. Because of the moderately high hardening temperatures and rapid quenching required, these steels are subject to distortion during heat treatment.

T. High-speed tool steels (tungsten base). These steels generally are used for metal-cutting tools, although type T1 is sometimes used as a die material because of its high strength and wear resistance. Type T1 has been an established general-purpose high-speed steel (HSS) for many years.

W. Water-hardening tool steels. Being low in alloy content, these steels are shallow hardening, developing a tough case with a hard core. They are quenched in water or brine, are subject to distortion, and do not retain their properties above 350°F. Water-hardening tool steels are used for short production runs in all types of tools and dies. Their shallow hardenability adapts them well for cold-heading dies. Although used for many applications, they are limited in their properties and capabilities.

8.3 Dies, Molds, and Die-Making Procedures

The stamping, punching, or blanking die is the most common type of tooling employed by industry. These types of dies consist of a male punch and a female die block. The material to be punched, stamped, or blanked is placed between the die sections, and a force

is applied to the punch or male portion of the die set. As the material is being punched, compressive and tensile stresses are developed within the stock material until a cutting and then shearing action takes place, forcing the punched portion of the material out of the female die block. Figure 8.2 illustrates a typical punching operation.

A portion of the stock is cut cleanly, whereas the remainder is broken by the shear forces developed within the material. Figure 8.3 shows a typical punched edge with the cut and sheared sections clearly defined. The shiny section is the cut portion, whereas the remainder is shear-fractured. The proportion of the cleanly cut section in relation to the irregular shear-fractured section is a function of the material hardness and the die clearance. Soft materials will have approximately 30 percent or more of the stock cut and the remainder shear-fractured, whereas hard materials will have only

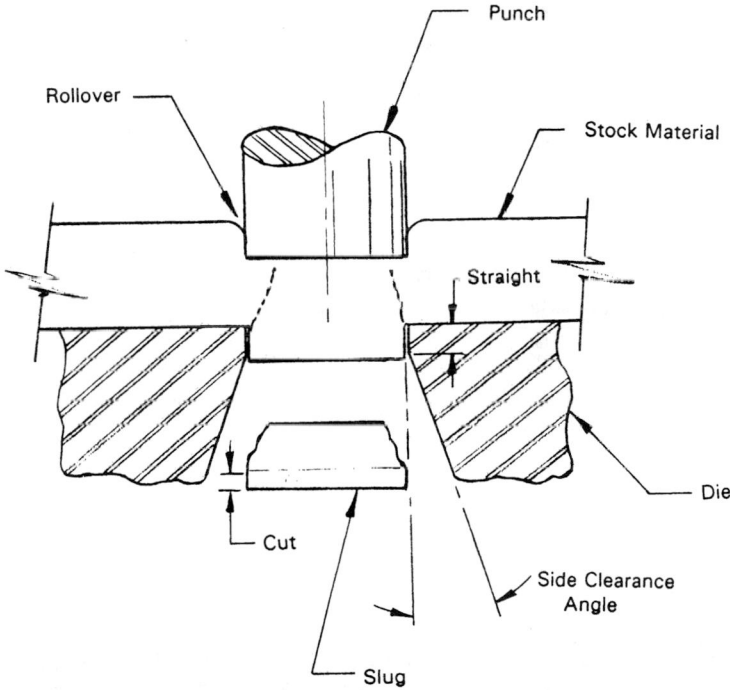

Figure 8.2 Punch action.

Figure 8.3   Punched edge.

10 to 20 percent of the stock thickness cleanly cut and the remainder shear-fractured.

The female die block must have an angular side clearance below the die-cutting straight in order for the punched part to fall out or clear the female die after the punching operation is completed. Figure 8.4 shows a typical blanking die set, separated to show the parts. In the figure, the parts of the die set are labeled and defined as follows:

A: Male portion of die

B: Top plate of the die set

C: Guidepost bushing

D: Bottom plate of the die set

E: Gate (for stock material)

F: Stripper or shoulder bolt

G: Opening in stripper plate

H: Guidepost

I: Die spring

J: Female die

K: Stripper plate

The die set shown in Fig. 8.4 was designed and built for blanking polycarbonate (Lexan) plastic material, 0.05 in thick. The plastic

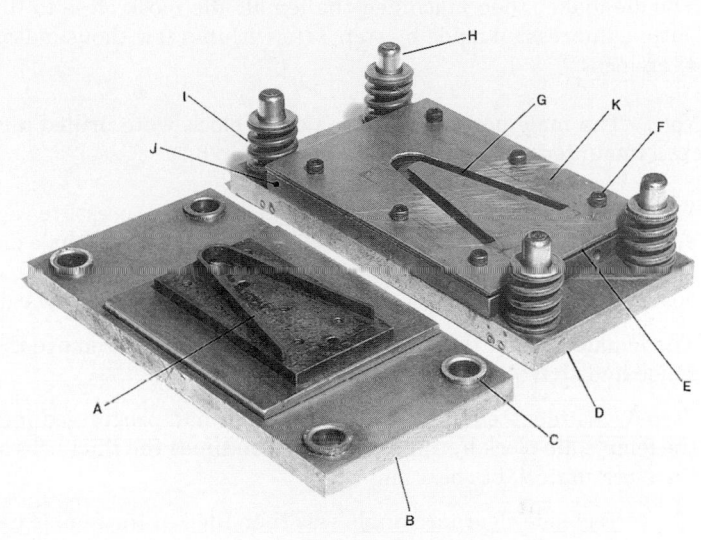

Figure 8.4 Blanking die, separated.

stock was cut in uniform strips to fit the die gate so that it could be fed progressively into the die during each blanking operation with the production of a minimum amount of scrap.

8.3.1 Manual die making (blanking dies)

Prior to the modern electric discharge machining (EDM) methods of punching and blanking, the die maker used the manual method, which consisted of the following process steps:

1. The male portion or punch section of the die was machined to the proper shape and dimensions from the selected tool steel.

2. The finished male portion or punch was then heat-treated to the required hardness and temper.

3. The heat-treated male portion was then set against the face or surface of the soft or annealed female die block and a high pressure applied. This step produced a sharp outline of the male portion into the surface of the female die block.

4. The die maker then machined the female die block close to the outline impression made by step 3 (to within a few thousandths of an inch).

Note: The male portion and female die block were drilled and pinned (mounted) into the die set prior to step 3.

5. With the male and female portions aligned and set, pressure was applied to the die set, forcing the male portion into the female die block a small distance. This procedure shaved the close machined edges of the female die block and rolled the die material inward.

6. The female portion then was hand filed carefully to remove the rolled and shaved burrs produced by step 5.

7. Step 6 was repeated until the male portion had progressed into the female die block by approximately two times the thickness of the stock material to be stamped.

8. The correct side clearance angle was then filed in the female die block, below the cutting straight. Angular side clearances range between 0.5 and 2.00 per side, according to stock thickness. The height of the cutting straight is 0.125 in for all stock 0.125 in thick and less and usually equal to the stock thickness for all stock above 0.125 in.

9. The proper die clearance was then applied to the male punch or female die block by hand filing the appropriate die member.

Note: As a general rule in die stamping or punching,

- For punching holes in parts, the punch (male) is made to the exact size of the punched hole or area, and the correct die clearance is filed or cut in the female die block (Fig. 8.5a).

- For stamping or blanking a part outline, the female die block is made to the exact part size, and the punch (male) is filed smaller according to the required die clearance (Fig. 8.5b).

8.3.2   EDM (electric-discharge machining) die-making procedures

Electric-discharge machining (EDM) has replaced much of the work of the die maker in many companies where EDM machines are employed. In the EDM process, which includes both wire and ram

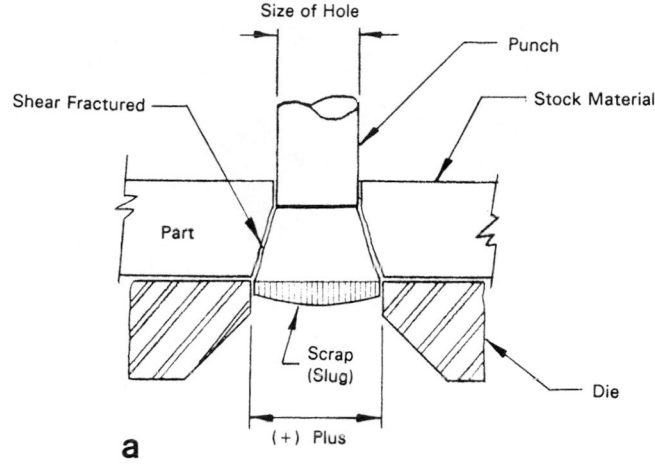

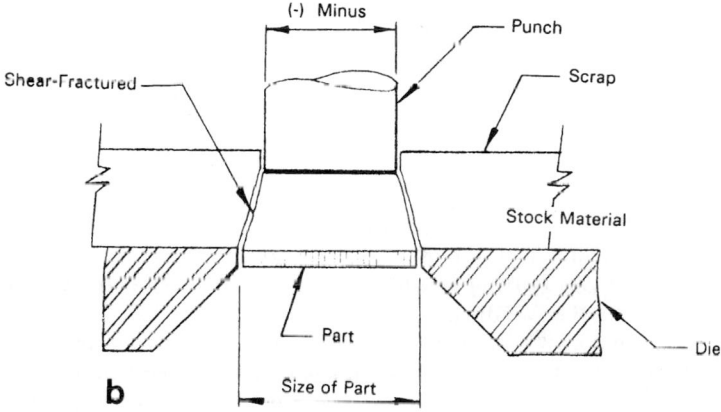

Figure 8.5 (*a*) Punching holes. (*b*) Punching a part

methods, an electrically charged electrode (wire or solid form) does the actual cutting of the metal (usually tool steel). The hand production of dies is still practiced where the cost of an EDM machine cannot be justified on a production basis.

The wire EDM machine is usually employed when making punching and blanking dies with complex outlines, where extreme accuracy is required. As an example of the accuracy of the EDM wire process,

Figure 8.6 A sample die set, wire EDM cut from hardened tool steel.

see Fig. 8.6. The die set in the figure was manufactured (wire EDM cut) of prehardened tool steel and was produced in less than 8 hours. The hand production of a small die set with a complex shape such as that shown in the figure would have taken an expert die maker as long as 5 to 8 days of tedious work. This small sample die set was cut with a total die clearance of 0.0005 in and is capable of punching a thin sheet of paper. The only additional work required after wire EDM cutting of the male and female portions was to surface grind the two die parts. The two portions of the die are separated and shown with a thin piece of paper, which was cut using hand pressure. A die set such as this would be difficult to produce by manual methods of die making owing to its small size and intricate pattern.

Another example of EDM is given in Fig. 8.7, which shows the teeth of a ratchet that was wire cut from 0.188-in-thick cold-rolled steel sheet stock. The sharp edges of the ratchet teeth are typical of an EDM-cut material. The movements of the EDM machine are determined by a computer numerical control (CNC) controller on the machine that has been programmed by the tool engineering department. The program to cut this ratchet was loaded into the EDM machine's controller memory and the program run to produce the part to great accuracy. The holes shown in the ratchet were die-punched after the teeth were EDM cut.

Figure 8.7 A ratchet plate, CNC EDM wire cut from mild steel. Note the sharp, accurate cut edges.

The wire EDM machines produced in the 1980s have been improved to the point where the cutting speeds have increased two to four times their original rates. The cutting section of a typical wire EDM machine is shown in Fig. 8.8. The wire can be seen running vertically between the wire guides and feed mechanism. During the cutting operation, cutting/cooling fluid is allowed to flow over the part in the vicinity of the wire, where it contacts the material being cut. When tool steels are being cut on the wire EDM machine, they may be cut in the annealed condition and later heat-treated, or they may be cut in the fully hardened condition to prevent the distortion that may occur during heat treatment. This characteristic of wire EDM cutting allows the tool engineer to use grades or types of tool steels for die making that otherwise would distort in heat treatment by cutting the dies in the fully hardened condition.

The newer types of wire EDM machines are designed so that the wire guides are controlled independently and are able to move so that tapered sides may be cut into the die. Thus a conical slug or other taper-sided outline can be cut from a die block.

EDM die-making procedures are similar to the manual method of making dies, except that tool-steel die members (punch and die block) are produced on the wire EDM machine, with the tool steel

Figure 8.8 Close-up of wire EDM. Note the wire guides.

either being heat treated or in the soft condition. It is usually found that the wire EDM machine will cut the tool-steel die members cleaner when they are in the fully heat-treated condition.

After the die members have been cut with the proper die clearance, the die parts are mounted in the die set with shims placed around the mated male and female die members to ensure that the die clearance is the same all around. The dies are then drilled and pinned, with additional shoulder bolts or socket-head cap screws employed besides the alignment pins to hold all the members securely in alignment.

The basic die set consisting of top and bottom plates (die shoe), guide-posts, and guide bushings plus an optional top shank may be ordered in many different styles and sizes from suppliers such as Danly or Dieco and others. Photographs and illustrations of typical standard and special die sets are shown in other subsections.

8.3.3 EDM mold making (ram EDM)

Molds that are machined from tool steels are presently being made using ram EDM machines. In this process, the cavity of the mold is produced in a positive shape from a carbon-type material. This form, or "positive," is then used in the ram EDM process to machine the cavity of the mold by the electric-discharge process. The ram, or electrode (positive form), is placed under a cutting/cooling fluid with the tool-steel mold block and is moved downward into the tool steel block as the electric-discharge cutting operation progresses. A hollow or cavity with the exact shape of the ram or electrode thus is produced to great accuracy with a fine surface finish on the tool-steel mold block.

Molds or dies are produced with this method for manufacturing plastic parts (injection or compression molding process), as well as for die casting and powder-metal processing. Large amounts of material are machined accurately at a relatively rapid rate in the newer ram EDM process. When the mold cavities have been cut, the die or mold maker must polish the mold and add gates, ejectors, and other requirements to complete the final mold assembly. The P-type special-purpose tool steels usually are selected for production of molds for plastic parts, whereas other types of tool steels are selected for die-casting dies, permanent steel molds, and powder-metal dies.

8.3.4 Samples of dies, drawings, and stamped or blanked parts

A typical high-quality die set is shown in Fig. 8.9. This is a heavy-duty, long-production-run type of stamping die for producing heavy-gauge cold-rolled steel parts. The male and female die element tool-steel materials were selected and heat treated with great care and assembled accurately by the die maker. The tool-steel die elements were cut on the wire EDM machine using a sophisticated CNC program designed by the tool engineering department. The CNC program was then loaded into the EDM machine's controller and the dies accurately cut, with the tool-steel die elements having been heat treated previously.

Figure 8.9 A typical high-quality blanking die set.

Figure 8.10 shows another finished die set with its required protective shield surrounding the die. Industry regulations call for protective shields on dies where an operator may accidentally come into hand contact with the moving parts of the die during the blanking, punching, forming, or drawing operations. The design and addition of the protective die shields are an important part of die design today and are necessary for worker safety.

Figure 8.11 shows two typical blanked parts. The upper part in the figure is made of a low-strength aluminum alloy (3003-H14), and the 50 percent cut or burnished section can be seen over the shear-fractured section. The part below is made of C-1018 cold-rolled steel, and a double cut or burnish is evident. The double cut alternating between the shear-fractured areas is characteristic of a close die clearance per side of 1 to 2 percent of the total stock thickness. This type of blanked edge may not be suitable for certain types of applications, and a secondary operation such as shaving may be required to finish the part edges.

In Sec. 8.5.1 figures will be presented for calculating the tonnage required on the press for stamping parts. To calculate the tonnage requirements, we must know the perimeter of the part in linear

Figure 8.10 Finished die set with required hand protection shield (clear Lexan plastic).

Figure 8.11 Typical blanked parts, edge characteristics.

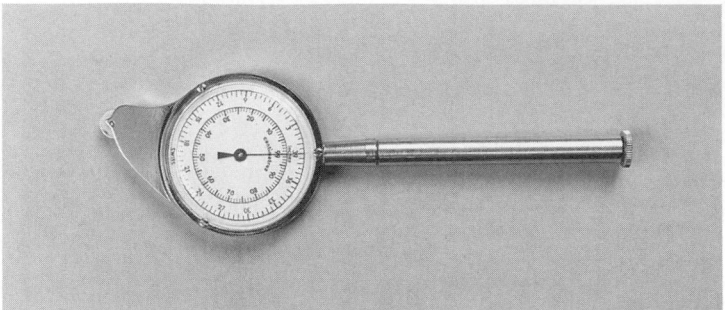

Figure 8.12   Linear measuring device.

inches or millimeters. Figure 8.12 shows a measuring tool used to determine the perimeter of irregularly shaped parts. This tool is sometimes called a *map measuring tool*. To use the tool, the scale is zeroed, and the small wheel at the end of the tool is placed at a starting point on the perimeter of the part to be measured. Tracing the tool around the perimeter of the part by rolling the wheel on the part outline will give a direct scale measurement in inches or millimeters on the dial of the tool. From this measurement, the tonnage requirements may be calculated using procedures shown in a later section.

The modern die-making and design processes use CNC and computer programs not only to control the EDM machines but also to produce the die drawings. A computer-generated drawing is possible using the various computer-aided design (CAD) systems available today. The CAD program sends the output information for the drawing to a plotter or laser printer, where a clean, neat drawing is produced. Figure 8.13 presents an example of a CAD drawing produced by a modern tool and die design and engineering department.

8.3.5   Steel-rule dies

Steel-rule dies make use of low-cost materials and are employed in many industrial applications. These are single-element tools that consist of a steel-rule cutting section only. Printers rule or similar steel strip is employed and bent to the shape of the part to be punched or cut. The sharp edges of the steel-rule members are operated against a flat metal platen or a hard wood or plastic block and produce the blanked part by a cutting or cleaving action.

Steel-rule dies are used to blank paper, cardboard, fiber, rubber, felt, leather, and similar materials (gaskets are a good example).

SECTION -A-A

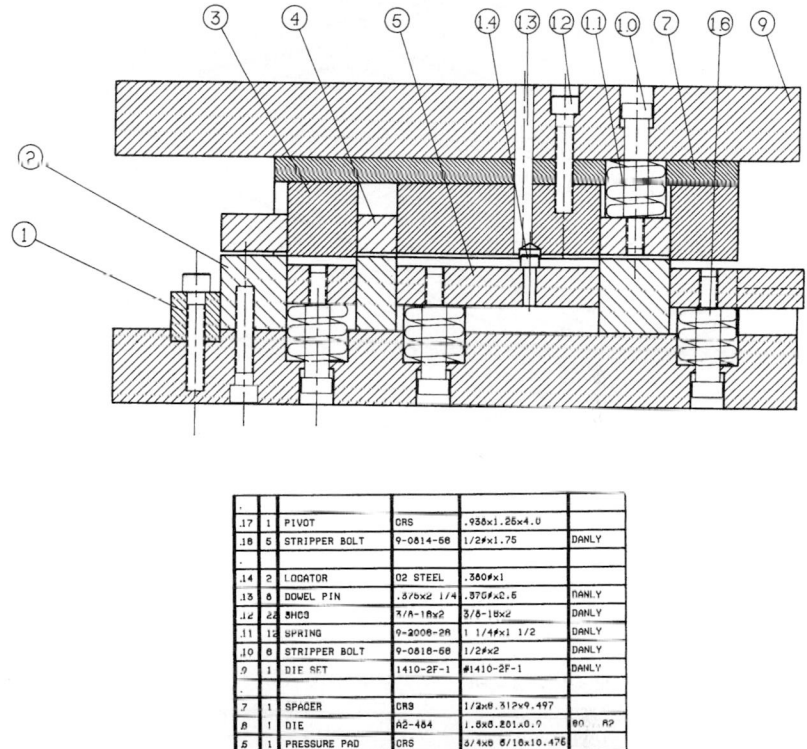

POS	QTY	PART NAME	MATERIAL	DIMENSIONS	HARDNESS
17	1	PIVOT	CRS	.938x1.25x4.0	
16	5	STRIPPER BOLT	9-0814-66	1/2#x1.75	DANLY
14	2	LOCATOR	O2 STEEL	.360#x1	
13	6	DOWEL PIN	.375x2 1/4	.375#x2.5	DANLY
12	22	SHCS	3/8-18x2	3/8-18x2	DANLY
11	12	SPRING	9-2006-26	1 1/4#x1 1/2	DANLY
10	6	STRIPPER BOLT	9-0818-66	1/2#x2	DANLY
9	1	DIE SET	1410-2F-1	#1410-2F-1	DANLY
7	1	SPACER	CRS	1/2x6.312x9.497	
6	1	DIE	A2-464	1.5x6.251x0.?	60...A2
5	1	PRESSURE PAD	CRS	3/4x6 6/16x10.476	
4	1	STRIPPER	A2-464	3/4x7 5/16x9.187	
3	1	PUNCH	A2-464	1.5x6.376x9.497	58...60
2	1	DIE	A2-464	1.5x3.29x9.187	60...62
1	4	KEY	CRS	1x1x6	

Figure 8.13 Typical CAD drawing of a die set.

Figure 8.14 shows an exploded view of steel-rule die construction. The steel rules that do the actual cutting should be hardened to Rockwell C 52 to C 55, although this high a hardness may not be required for some materials. The stripper portion of the die is usually 0.312 to 0.375 in thick, extending slightly above the height of the steel-rule cutters, and is made of neoprene, cork sheet, or rubber.

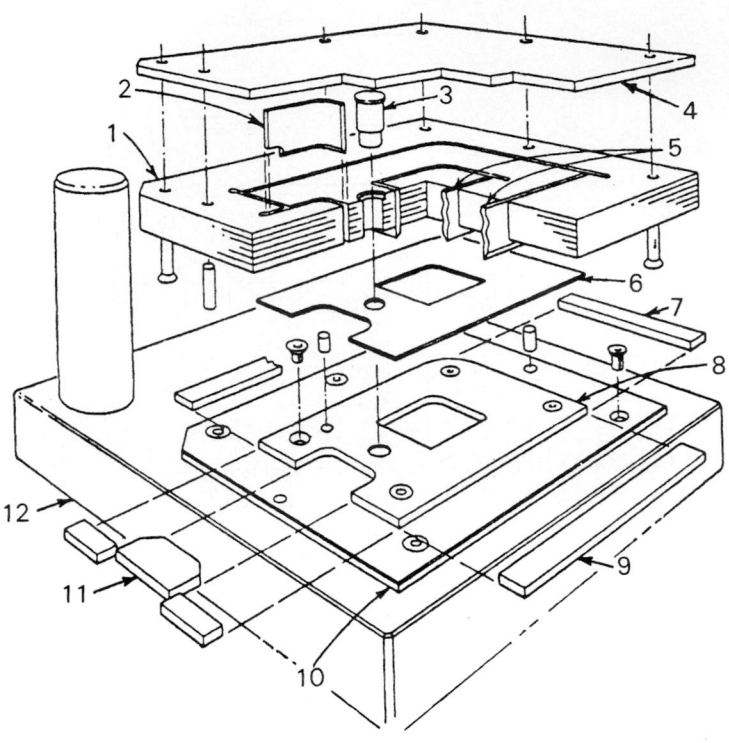

1	Plywood	7	Neoprene
2	Steel rule	8	Die plate
3	Punch	9	Neoprene
4	Steel subplate	10	Steel subplate
5	Steel rule	11	Neoprene
6	Blank	12	Die shoe

Note: Neoprene inserts are cut and glued in those areas
where stripping action is required.

Figure 8.14 Typical steel-rule die.

8.4   Die Clearances and Stamping Data

Figure 8.15 shows the different types of edges produced on punched
and blanked parts with different die clearances. This figure is typical of the low-carbon steels, which are used frequently for many
types of die-stamped, pierced, or blanked parts.

Punching of Low-Carbon Steel

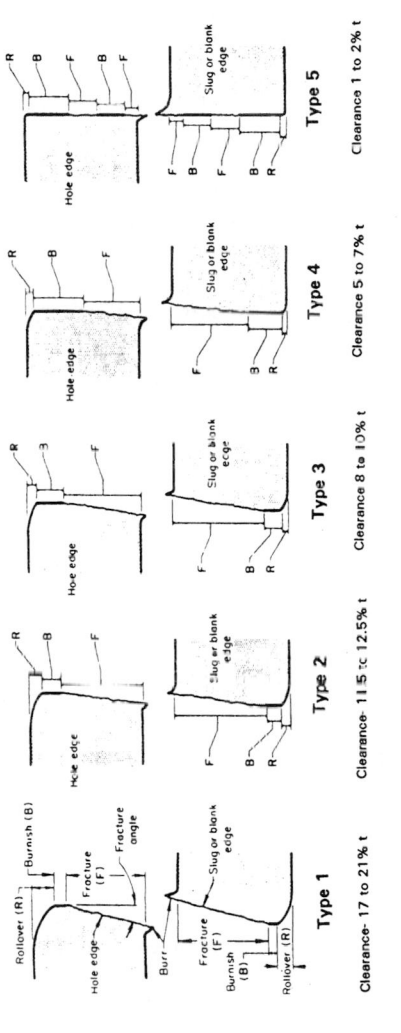

Edge Characteristic	Type 1	Type 2	Type 3	Type 4	Type 5
	Clearance- 17 to 21% t	Clearance- 11.5 to 12.5% t	Clearance 8 to 10% t	Clearance 5 to 7% t	Clearance 1 to 2% t
Fracture angle	14 to 16°	8 to 11°	7 to 11°	6 to 11°	......
Rollover (1)	10 to 20% t	8 to 10% t	6 to 8% t	4 to 7% t	2 to 5% t
Burnish (1)	10 to 20% t (2)	15 to 25% t	25 to 40% t	35 to 55% t (3)	50 to 70% t (4)
Fracture	70 to 80% t	60 to 75% t	50 to 60% t	35 to 50% t (5)	25 to 45% t (6)
Burr	Large, tensile & part distortion	Normal, tensile only	Normal, tensile only	Medium, tensile + compressive (7)	Large, tensile + compressive (7)

Note: (1) Rollover plus burnish approx. equals punch penetration before fracture, (2) Burnish on edge of slug or blank may be small and irregular or absent, (3) With spotty secondary shear, (4) In two separate portions, alternating with fracture, (5) With rough surface, (6) In two separate portions, alternating with burnish, (7) Amount of compressive burr depends on die sharpness.

Figure 8.15 Edge characteristics.

Figure 8.16 shows samples of typical slugs produced in the piercing or punching of hard copper (ETP 110, hard drawn). The figure clearly shows the effect of die clearance and the characteristic edges produced by the cutting (burnishing) and shear-fracture actions. (The cut or burnished section is indicated by a and the shear-fracture section by b in the figure.) These slugs are typical of type 1 and type 2 edges, as shown in Fig. 8.15.

Figure 8.17 gives punch-to-die clearances for piercing or blanking various metals and alloys to produce the five types of edge characteristics shown in Fig. 8.15. Clearances shown in Fig. 8.17 are based on data published on piercing by Danly Machine Corporation and on blanking by the American Society of Tool and Manufacturing Engineers (ASTME).

The tabular die-clearance data shown in Fig. 8.17 are given per side as a percentage of the stock thickness. For example, if we were punching stainless steel whose stock thickness was 0.050 in and we desired a type 3 edge, the total die clearance would be $0.05 \times (0.09$ to $0.11) \times 2 = 0.009$ to 0.011 in in total die clearance, or 0.0045 to 0.0055 in in clearance per side.

Another method that has been used for many years in industry for die clearances is outlined below.

Die Clearances per Side as a Percentage of Stock Thickness (Group Method)

Material group	Clearance per side, percent of stock thickness t
Group 1: 1100 and 5052 aluminum alloys (all tempers); 3003-0; etc.	4.5% t
Group 2: 2024 and 6061 aluminum alloys (all tempers); brass (all tempers); cold-rolled steel, soft; stainless steel, soft; copper, soft; BeCu, soft	6.0% t
Group 3: Cold-rolled steel, half hard; stainless steel, half hard and full hard; copper, hard	7.5% t
Group 4: Nonmetallic materials (general)	1.25% t

Note: Angular side clearance (one side) = 0.5 to 2 percent. Angular side clearance is cut below the die straight (see previous data for die-straight heights or thicknesses).

8.4.1   Calculation of punch dimensions

Punch dimensions may be estimated from the following: When the diameter of a pierced round hole is equal to the stock thickness, the

Figure 8.16 Typical die-punched slugs.

unit compressive stress on the punch is four times the unit shear stress on the cut area of the stock from the equation

$$1 = \frac{4S_s t}{S_c d}$$

where S_c = unit compressive stress on the punch, psi
  S_s = unit shear stress on the stock, psi
  t = stock thickness, in
  d = diameter of punched hole, in

The diameters of most punched holes are greater than stock thickness, and a value for the ratio d/t of 1.1 is recommended. The maximum allowable length of a punch can be calculated from the following equation:

$$L = \frac{\pi d}{8} \sqrt{\frac{Ed}{S_s t}}$$

where d/t = 1.1 or higher
  E = modulus of elasticity of the punch material, psi

Other terms were defined previously.

  The punching of holes with diameters less than stock thickness is generally achieved by using high-compressive-strength tool steels, guided punches, greater than average die clearances, shear added

Work Piece Material	Edge Type 1	Edge Type 2	Clearance per side, % of stock thickness * Edge Type 3	Edge Type 4	Edge Type 5
Low-carbon steel	21 max	11.5-12.5	8-10	5-7	1-2
High-carbon steel	25 max	17-19	14-16	11-13	2.5-5
Stainless steel	23 max	12.5-13.5	9-11	3-5	1-2
Aluminum alloys: Up to 33,000 psi TS	17 max	8-10	6-8	2-4	0.5-1
Over 33,000 psi TS	20 max	12.5-14	9-10	5-6	0.5-1
Brass, annealed	21 max	8-10	6-8	2-3	0.5-1
Brass, half-hard	24 max	9-11	6-8	3-5	0.5-1.5
Phosphor bronze	25 max	12.5-13.5	10-12	3.5-5	1.5-2.5
Copper, annealed	25 max	8-9	5-7	2-4	0.5-1
Copper, half-hard	25 max	9-11	6-8	3-5	1-2
Lead	22 max	8-10	6.5-7.5	4-6	1.5-2.5
Magnesium alloys	16 max	5-7	3.5-4.5	1.5-2.5	0.5-1

NOTE: * Tabular data is clearance per side as % of stock thickness. Also, for clearances that produce edges of types 1, 2 and 3, it is usually necessary to use ejector punches or other devices to prevent the slug or blanked part from adhering to the punch.

Figure 8.17 Die clearances for piercing of various metals and alloys (for five types of edge characteristics as shown in Fig. 8.15).

to the punch or dies, and prevention of stock slippage or movement during punching.

8.4.2   Standard die sets

Die sets are available from the major die supply manufacturers such as Danly and Dieco. Figures 8.18 and 8.19 show some of the most commonly used die sets that are available as stock items. In Fig. 8.18, the die sets are identified as follows:

   a: Round series, back-post AS-31

   b: Round series, center-post AS-32

   c: Four-post series AS-37

   d: Diagonal-post AS-35

   e: Long, narrow series, two-post AS-41 and three-post AS-42

In Fig. 8.19, the die sets are identified as follows:

   a: Four-post series AS-67

   b: Two-post AS-63

   c: Demountable two-post AS-39 or four-post AS-40

The top and bottom plates of a die set are called the *top shoe* and *bottom shoe*, respectively.

If a special size or configuration for a die set is not available commercially, the die set may be custom made by the die maker. In this case, the material and size of the top and bottom shoes and the optional shank are determined by the tool engineer. To complete the basic die set, guideposts and guide bushings must be selected. The guide bushings are available in plain or ball-bearing types.

8.4.3   Guide pins and guide bushings for die sets

When a die maker is building a custom die set, the guide pins and guide bushings are selected from commercially available types, which may be either plain or ball-bearing types. Guide pins and guide bushings are produced in various diameters and lengths, which must be determined by the design of the die set and the press used for the die operation.

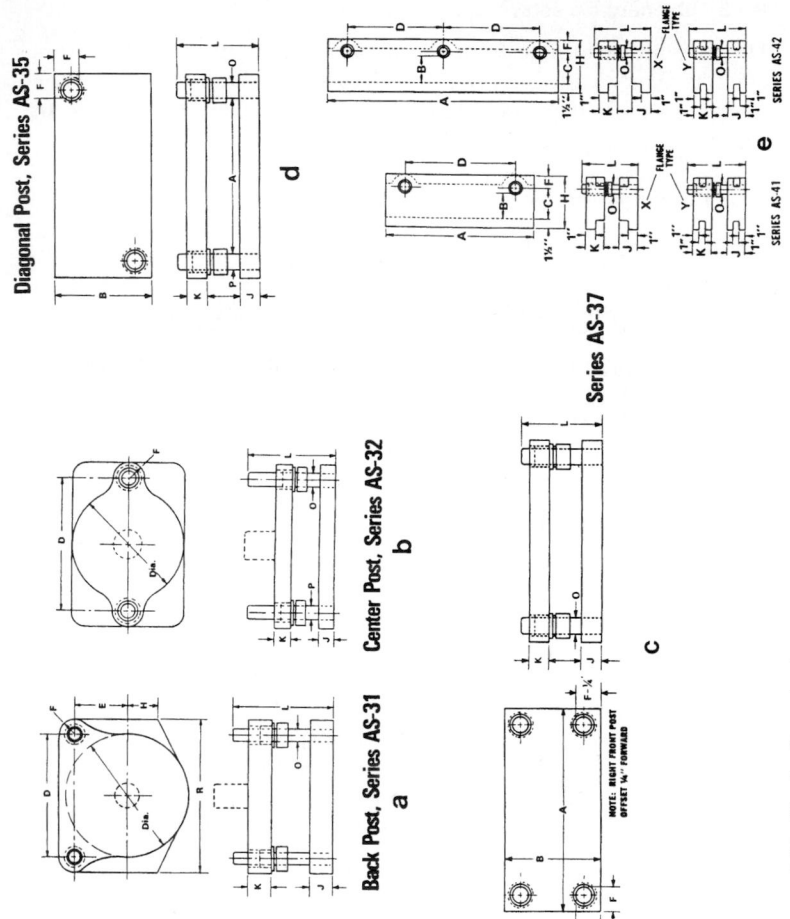

Figure 8.18 Danly die sets, types.

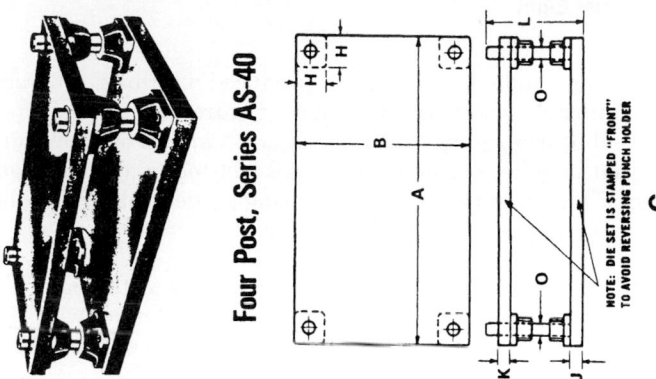

Four Post, Series AS-40

NOTE: DIE SET IS STAMPED "FRONT" TO AVOID REVERSING PUNCH HOLDER

C

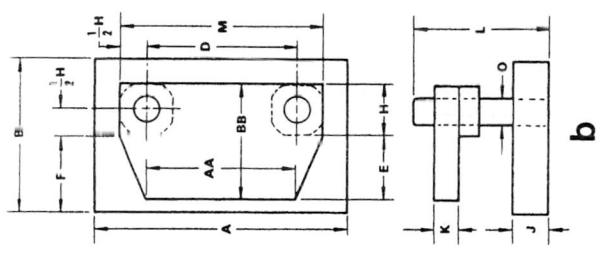

b

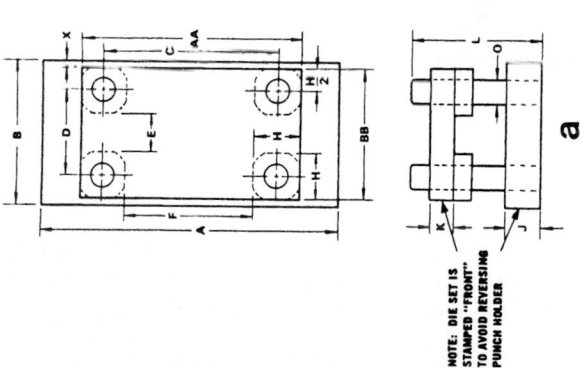

a

NOTE: DIE SET IS STAMPED "FRONT" TO AVOID REVERSING PUNCH HOLDER

Figure 8.19 Danly die sets, types.

571

Ball-bearing guide bushings are produced in three different types for different operating conditions. Figure 8.20 shows three types of ball-bearing guide bushings (*b*) and samples of typical die sets (*a*) that may use either plain or ball-bearing guide bushings. Selection of the correct type of ball-bearing guide bushing may be determined from the following descriptions:

Type I: Preloaded type. Recommended for use with high-production, long-life dies. All the balls remain in contact with the post and bushing in preloaded conditions throughout the press stroke.

Type II: Relieving type. Recommended when it is desirable that the ball cage does not leave the bushing at any time. Provides for safe operation and prevents foreign objects from falling into the bushing.

Type III: Disengaging type. Used where the ball cage can be permitted to leave the bushing with each stroke. This is the most economical type because of the short length, and it is used for general applications that require long press strokes.

These types of bushings should be ordered as complete sets (guidepost, ball cage, and bushing) to ensure proper fit of the components.

Important factors in designing and building any die set are

- Available stroke of the press
- Open height of the die set
- Minimum shut height (at depleted punch and die life)
- Maximum shut height (new punch and die sections)
- Dimension of the punch holder
- Dimension of the die holder
- Guidepost length
- Bushing length

Figure 8.21 shows the relations of these factors. The tool design engineer and die maker must determine the dimensions of all the factors listed and shown in this figure to produce the die correctly.

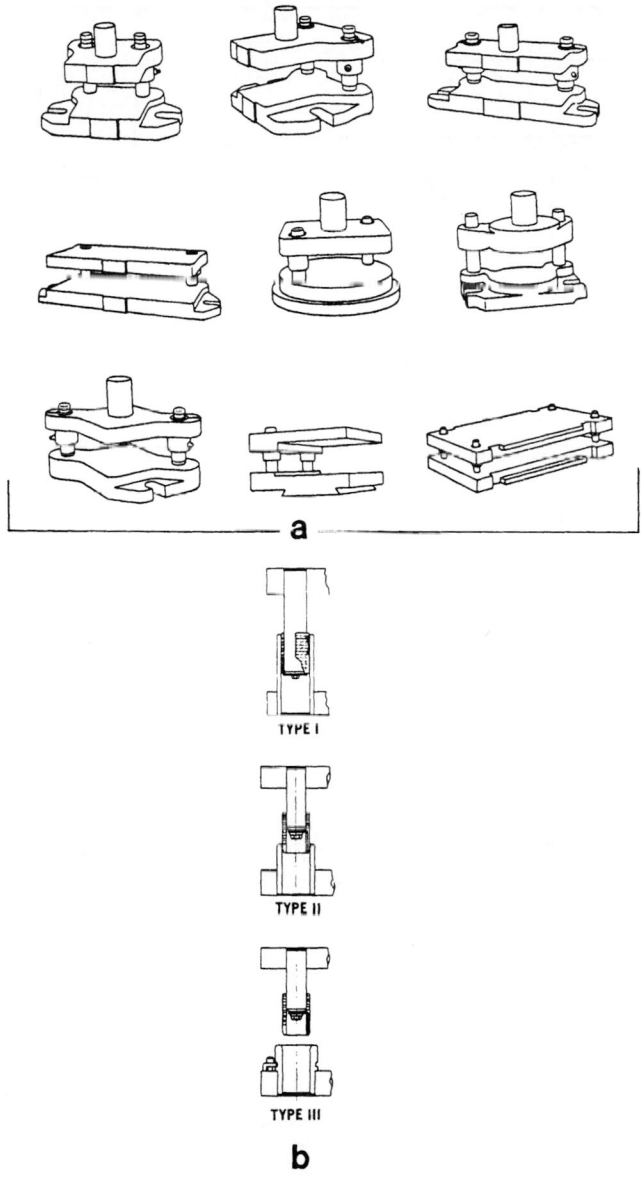

Figure 8.20 (*a*) Die sets. (*b*) Ball-bearing guide bushings.

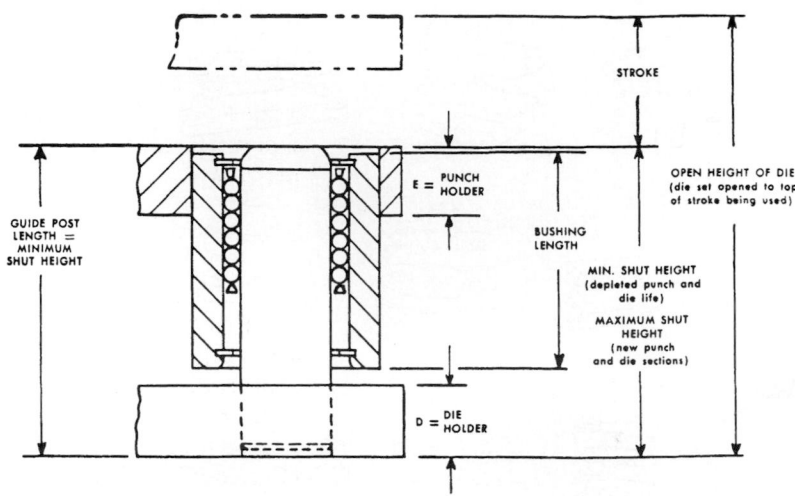

Figure 8.21   Die set variable characteristics.

8.5   Punching and Blanking Forces

8.5.1   Force required for punching or blanking

The simple equation for calculating the punching or blanking force P in pounds for a given material and thickness is given as

$$P = SLt \qquad \text{for any shape or aperture}$$

$$P = S\pi Dt \quad \text{for round holes}$$

where P = force required to punch or blank, pounds force (lbf)
S = shear strength of material, psi (see Fig. 8.22)
L = sheared length, in
D = diameter of hole, in
t = thickness of material, in

8.5.2   Stripping forces

Stripping forces vary from 2.5 to 20 percent of the punching or blanking forces. A frequently used equation for determining the stripping forces is

$$F_s = 3500Lt$$

where F_s = stripping force, lbf
L = perimeter of cut (sheared length), in
t = thickness of material, in

Note: This equation is approximate and may not be suitable for all conditions of punching and blanking owing to the many variables encountered in this type of metalworking.

8.5.3 Shear strengths of various materials

The shear strength (in pounds per square inch) of the material to be punched or blanked is required in order to calculate the force required to punch or blank any particular part. Figure 8.22 lists the average shear strengths of various materials, both metallic and nonmetallic. If you require the shear strength of a material that is not listed in Fig. 8.22 or elsewhere in this *Handbook*, then there are several Web sites such as *www.matweb.com* where shear strengths for a wide variety of materials may be obtained.

Manufacturers' standard gauges for steel sheets. The decimal equivalents of the American standard manufacturers' gauges for steel sheets is shown in Fig. 8.23. Sheet steels in the United States are purchased to these gauge equivalents, and tools and dies are designed for this standard gauging system.

8.6 Bending, Forming, and Progressive Die Operations and Data

Bending and forming dies are used to shape sheet metal parts by the action of bending or drawing the metal. Of concern to the die designer are allowable inside bend radii for various materials and gauges and permissible limits in the drawing or progressive forming of shapes with regular or irregular cross sections.

The die cavities used for forming metals may be ram EDM cut, as explained in an earlier section, or they may be cut using the CNC machining centers in profiling operations. The male portion of the die may be machined by various methods such as turning and milling on the modern CNC turning and machining centers or on engine lathes and manually controlled milling machines such as the Bridgeport machine.

Material	Shear Strength, psi
Carbon Steels:	
SAE 1010 HR	21,500
SAE 1020 HR	32,000
SAE 1045 QT	55,000
SAE 1045 A	44,000
SAE 1095 QT	90,000
SAE 1095 A	63,000
SAE 1117 HR	32,000
Alloy Steels:	
SAE 4130 N	43,500
SAE 4130 T (150,000)	90,000
SAE 4140 N	66,500
SAE 3120 HT-D (800°F)	95,000
SAE 3140 HT-D (800°F)	130,000
SAE 3250 HT-D (800°F)	165,000
Stainless Steels:	
AISI 201	52,000
AISI 301	50,000
AISI 302	41,000
AISI 304	38,500
AISI 310	42,750
AISI 316	38,250
AISI 321	38,250
Cold rolled S/S strip (full hard)	
AISI 300 Series	112,000
Stainless Steels: Annealed	
AISI 410	33,750
AISI 416	33,750
AISI 440C	49,500
AISI 430	33,750
Monel Metal:	
70,000 UTS	42,900
110,000 UTS	65,500
K Monel:	
155,500 UTS	98,500
Nickel:	
68,000 UTS	52,300
121,000 UTS	75,300
Inconel Alloys:	
80,000 UTS	59,000
100,000 UTS	66,000
150,000 UTS	80,000
175,000 UTS	87,000
Copper and Alloys:	
CA 110 (ETP 110)	22,000-28,000
CA 210 (Guilding)	26,000-37,000
CA 220 (Bronze)	28,000-38,000
CA 230 (Red brass)	31,000-42,000
CA 260 (Cartridge brass)	33,000-44,000
CA 268 (Yellow brass)	33,000-43,000

Figure 8.22 Shear strength of metallic and nonmetallic materials, psi.

Beryllium copper: Strip & sheet
C 17200 (25)	34,200-54,000
C 17000 (165)	34,200-94,500
C 17510 (3)	24,750-67,500
C 17500 (10)	24,750-67,500
C 17410 (174) HT	58,500

Beryllium Nickel:
UNS-N033 HT	123,750

Aluminum and Alloys:
1100-O	9,000
1100-H18	13,000
2014-O	18,000
2014 T4, T451	38,000
2014-T6, T651	42,000
2024-O	18,000
2024-T3, T4, T351	41,000
3003-O	11,000
3003-H14	14,000
3003-H18	16,000
5052-O	18,000
5052-H32, H38	60,000-77,000
6061-O	12,000
6061-T4, T451	24,000
6061-T6, T651	30,000
7075-O	22,000
7075-T6, T651	48,000
7178-O	23,000
7178 T6, T651	53,300

Magnesium Alloys:
Soft (annealed)	10,000
Hardened	28,500 max.

Titanium & Alloys:
Pure	27,000-49,500
Typical alloys	45,000-77,000

Nonmetallics:

Polyester-glass (GPO-1, 2 & 3)	12,000-17,000
Polycarbonate (Lexan)	6,000-10,000
Cycolac	4,400-7,400
ABS (Acrylonitrile Butadene Styrene)	1,500-4,000
Acetal (Delrin)	3,000-6,000
Acetate (Cellulose)	2,000-4,000
Epoxy-glass	4,000-10,000
Nylon	3,000-12,000
Phenolic resins (cloth)	26,000 (Hot-blanked)
Paper	3,500-6,400
Mica	10,000
Teflon, rigid (TFE)	1,500-3,000
Hard rubber	20,000
Polystyrene	10,000 max.
Asbestos board	5,000

Notes: For metallic materials- when the tabulated shear values are given in ranges, the shear values run from the annealed or soft condition to the hardest condition. Interpolate intermediate values between ranges. For nonmetallic materials- the shear value ranges are given from soft to hard grades or glass-filled grades.

A = annealed; HR = hot-rolled; N = normalized; T = tempered; HT-D = heat-treated and drawn (tempered); QT = quenched and tempered; UTS = ultimate tensile strength; HT = heat-treated.

Figure 8.22   (*Continued*)

Progressive forming, bending, and blanking die design and die making are complex arts as well as technologies that require considerable skill, knowledge, and practical experience. This section will present some of the basic information and data important to die design and die making.

8.6.1 Typical forming and progressive dies and parts

Forming and bending dies may be made for very simple operations or for complex, multistaged operations used to blank, form, or bend complex sheet metal parts. Figure 8.24 shows a very simple forming die used to produce a raised indent on an electrical contact finger made of beryllium-copper alloy. In the figure, the small forming die is shown at the left, with samples of two finger contacts. Part usage per year did not justify the cost of producing a complex blanking and bending progressive die to produce these parts. Production of the parts shown required the following sequence of operations:

- Beryllium-copper strip was purchased in correct slit widths (soft condition).

- The two holes were strippit punched from the correct cut length of the part.

- The raised indent was then formed on the end of the part using the small forming die shown in the figure.

- The punched and indented part was then bent to shape on a small press brake.

- The part was then heat-treated to correct hardness (for spring action).

- The part was deburred by tumbling.

- Silver plating was then applied to the finished part.

Typical examples of punched, blanked, and formed sheet metal parts are shown in Figs. 8.25 and 8.26. Figure 8.25 shows a stainless steel sheet metal part that required spring action. The material selected for this application was AISI 301 spring-temper stainless steel. This part was preslit to proper width, shear-cut to length, punched, beaded for stiffness, and finally bent on the press brake.

Standard Gage Number	Ounces/Ft²	Lb/Ft²	Thickness (Inches)
3	160	10.0000	0.2391
4	150	9.3750	0.2242
5	140	8.7500	0.2092
6	130	8.1250	0.1943
7	120	7.5000	0.1793
8	110	6.8750	0.1644
9	100	6.2500	0.1495
10	90	5.6250	0.1345
11	80	5.0000	0.1196
12	70	4.3750	0.1046
13	60	3.7500	0.0897
14	50	3.1250	0.0747
15	45	2.8125	0.0673
16	40	2.5000	0.0598
17	36	2.2500	0.0538
18	32	2.0000	0.0478
19	28	1.7500	0.0418
20	24	1.5000	0.0359
21	22	1.3750	0.0329
22	20	1.2500	0.0299
23	18	1.1250	0.0269
24	16	1.0000	0.0239
25	14	0.87500	0.0209
26	12	0.75000	0.0179
27	11	0.68750	0.0164
28	10	0.62500	0.0149
29	9	0.56250	0.0135
30	8	0.50000	0.0120
31	7	0.43750	0.0105
32	6.5	0.40625	0.0097
33	6	0.37500	0.0090
34	5.5	0.34375	0.0082
35	5	0.31250	0.0075
36	4.5	0.28125	0.0067
37	4.25	0.26562	0.0064
38	4	0.25000	0.0060

Note: Thickness equivalents are based on 0.0014945 in. per ounce per sq. ft.; 0.023912 in. per pound per sq. ft. (Reciprocal of 41.820 lb. per sq. ft. per inch thick); 3.443329 in. per lb. per sq. in.

Figure 8.23 Manufacturers' standard gauges for steel sheets.

Figure 8.26 shows a part made from no. 11 gauge low-carbon steel in a multistaged progressive die. This part required punching, blanking, bossing, and bending operations. The progressive die used for this type of part is complex and costly. The shear and forming marks are clearly evident on the surfaces of the part.

Figure 8.24 A small forming die and parts.

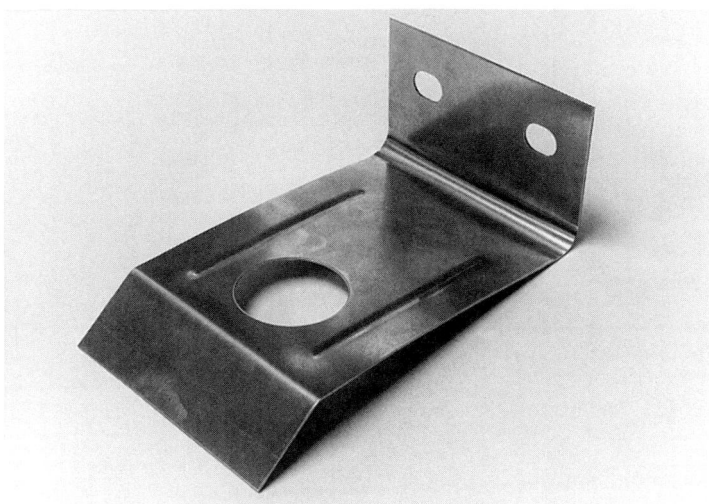

Figure 8.25 Die-punched stainless steel part.

Figure 8.26   Die-punched mild-carbon-steel part.

A simple progressive punching and blanking die is shown in Fig. 8.27. The part strip and piece part are shown at the bottom of the figure; the dimensions are in millimeters. The punching, blanking, and part dropoff slot are shown in section A–A, progressing from right to left. A heavy-duty die set with an upper shank was used to make the die set shown in the figure.

A drawing die is shown in Fig. 8.28a, with the parts identified as follows:

1. Drawing punch

2. Drawing die block

3. Clamp ring (or holder)

A compound die for piercing and drawing is shown in Fig. 8.28b, with the parts identified as follows:

1. Shear punch and drawing-die block

2. Piercing die

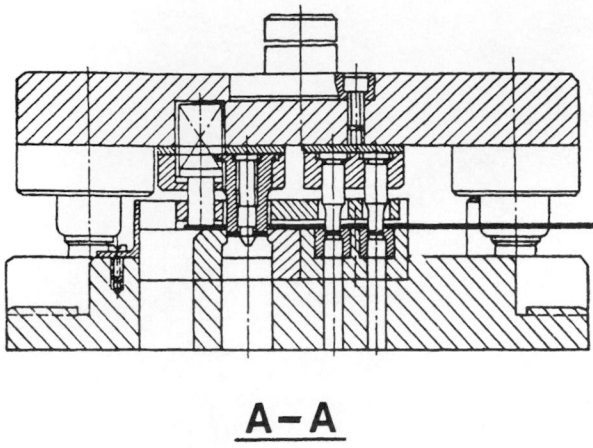

A–A

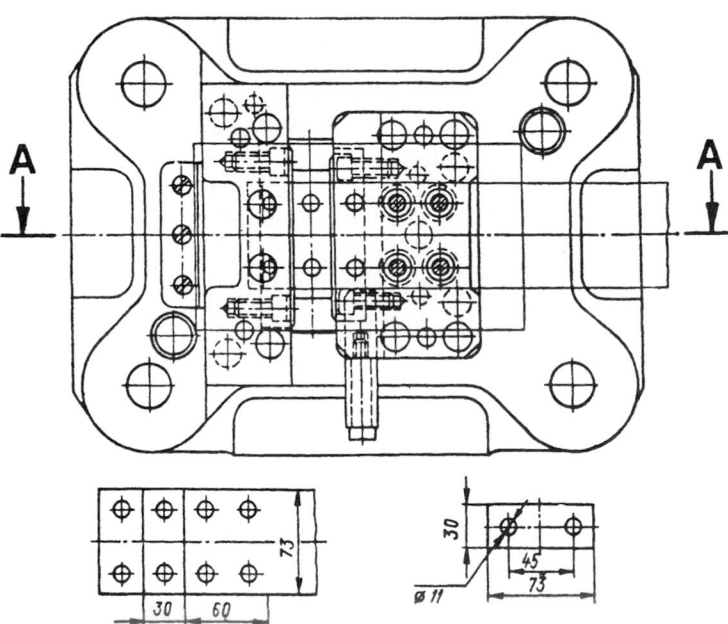

Figure 8.27 Progressive die with part samples.

3. Shearing-die block

4. Clamping ring

5. Draw punch

6. Dimensioned finished part (millimeters)

See Sec. 8.6.5 for drawing and forming shapes and appropriate equations for drawing and forming calculations.

Methods for applying springs to die sets are shown in Fig. 8.29. Here, the springs are shown with methods of applying preload, with adjustment. Parts *c* and *e* of the figure illustrate the use of Belleville disk springs in place of the standard helical die springs. Helical die springs may be made with either round or rectangular spring wire, although the rectangular wire types are usually employed in American die-making practice because of their higher power-to-weight ratios.

8.6.2   Bend radii for various metals and alloys

The minimum bend radii allowed for forming metals and alloys are important factors in die design and die making. A bend radius that is too small for a particular metal and gauge may cause cracking or rupture of the part being formed. A radius that is too large may not meet the requirements of the part design. This is especially true when a hole is close to the edge or flange of a part and would interfere with the inside bend radius by running past the tangent point of the radius. This would cause "pulling" of the hole when the part is bent, with subsequent hole deformation.

Minimum bend radii for various metals and alloys are shown in Fig. 8.30. These bend radii are for cold bending the metals at room temperature (70°F) and at a 90° angle. These minimum bend radii are for an axis of the bend that is at 90° to the direction of rolling of the metal (commonly called *grain direction*). Larger bend radii are required when the bend axis is at 45° or parallel to the direction of rolling of the metal. The tables in Fig. 8.30 list the minimum bend radii as a multiple of the metal thickness. Thus a tabulated bend radius of 3.5 for 0.063-in-thick material indicates that the bend radius is actually equal to $3.5 \times 0.063 = 0.22$ in. In other words, we are multiplying the tabulated values times the material thickness to arrive at the minimum bend radius in inches.

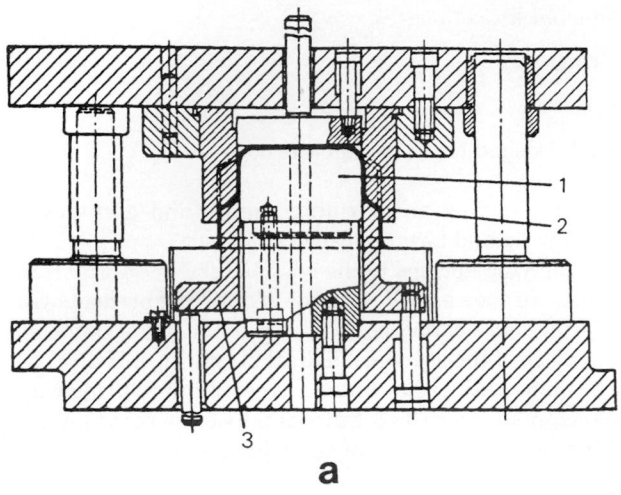

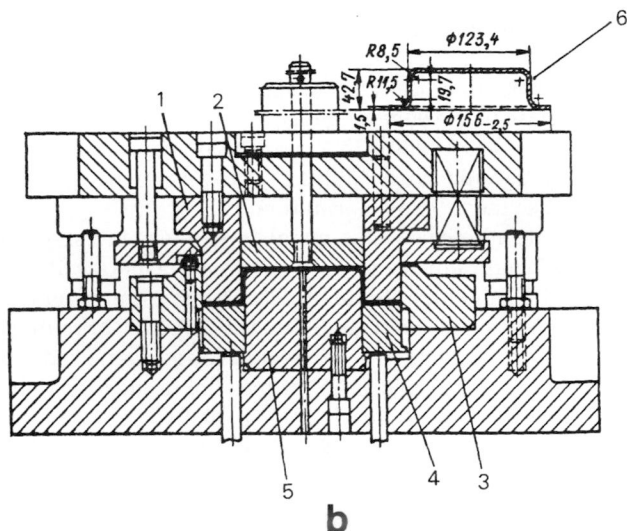

Figure 8.28 A complex drawing die.

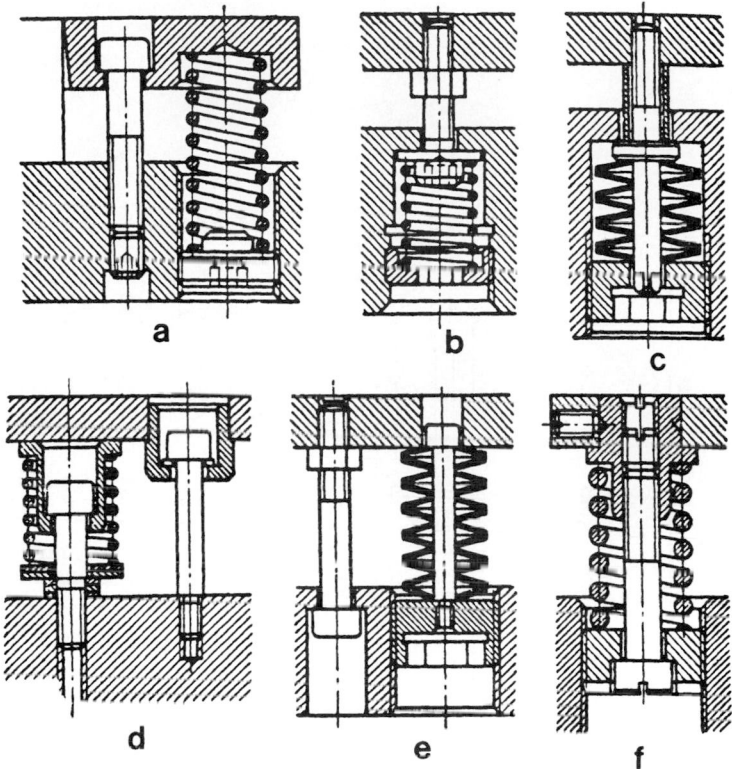

Figure 8.29 Die spring applications.

8.6.3 Springback

When a sheet of metal is bent, either in a die or on a press brake, the bent flange will tend to open to a larger angle when the bending pressure is removed (springback). In order to compensate for this characteristic, the flange usually is overbent by a certain number of degrees so that the metal will assume the correct angle of bend when the bending pressure is removed.

The harder the material, the greater is the amount of springback on the workpiece. Quarter-hard cold-rolled steel will exhibit approximately 1° to 2° of springback, half-hard has approximately 3° to 4°, hard steels have more than 5°, and annealed spring steels

Minimum Bend Radii for Metals and Alloys- Given in Multiples of Material Thickness (inches)

Material	Thickness, inches						
	0.015	0.031	0.063	0.093	0.125	0.188	0.250
Carbon steels:							
SAE 1010	S	S	S	S	S	0.5	0.5
SAE 1020-1025	0.5	0.5	1.0	1.0	1.0	1.1	1.25
SAE 1070 & 1095	3.75	3.0	2.6	2.7	2.5	2.7	2.8
Alloy steels:							
SAE 4130 & 8630	0.5	2.0	1.5	1.7	1.5	1.7	1.9
Stainless steels:							
AISI 301, 302, 304 (A)	0.5 ————					0.5	0.75
AISI 316 (A)	0.5 ————					0.5	0.75
AISI 410, 430 (A)	1.0 ————					1.0	1.25
AISI 301, 302, 304 (CR) ¼ H	0.5 ————		0.5	1.0 ————		1.0	1.25
AISI 316 ¼ H	1.0 ————					1.0	1.25
AISI 301, 302, 304 ½ H	1.0 ————					1.0	1.25
AISI 316 ½ H	2.0	2.0	3.0	2.0	2.0	2.0	2.5
AISI 301, 302, 304 H	2.0	2.0	1.5	1.5	1.5	1.5	1.5
Aluminum alloys:							
1100 O	0	0	0	0	0	0	0
H12	0	0	0	0	0	3.0	6.0
H14	0	0	0	0	0	3.0	6.0
H1	0	0	2.0	3.0	4.0	8.0	16.0
H18	1.0	2.0	4.0	6.0	8.0	16.0	24.0
2014 & Alclad O	0	0	0	0	0	3.0	6.0
T6	2.0	4.0	8.0	15.0	20.0	36.0	64.0
2024 & Alclad O	0	0	0	0	0	3.0	6.0
T3	2.0	4.0	8.0	15.0	20.0	30.0	48.0
3003, 5005, 5357, 5457 O	0	0	0	0	0	0	0
H12/H32	0	0	0	0	0	3.0	6.0
H14/H34	0	0	0	1.0	2.0	4.0	8.0

Table of minimum bend radii

Alloy	Temper							
(continued)	H16/H36	0	1.0	3.0	5.0	6.0	12.0	≤4.0
	H18/H38	1.0	2.0	5.0	9.0	12.0	24.0	≈0.0
5050, 5052, 5652	O	0	0	0	0	2.0	3.0	1.0
	H32	0	0	2.0	3.0	4.0	6.0	12.0
	H34	0	0	2.0	4.0	5.0	9.0	16.0
	H36	1.0	1.0	4.0	5.0	8.0	18.0	24.0
	H38	1.0	2.0	6.0	9.0	12.0	24.0	40.0
6061	O	0	0	0	0	2.0	3.0	4.0
	T6	1.0	2.0	4.0	6.0	9.0	18.0	28.0
7075 & Alclad	O	0	0	2.0	3.0	5.0	9.0	18.0
	T6	2.0	4.0	12.0	18.0	24.0	36.0	64.0
7178	O	0	0	2.0	3.0	5.0	9.0	18.0
	T6	2.0	4.0	12.0	21.0	28.0	42.0	80.0
Copper & alloys: ETP #110	soft	S	S	S	S	0.5	0.5	1.0
	hard	S	1.0	1.5	2.0	2.0	2.0	2.0
Alloy 210	¼ H	S	S	S	S	S	0.5	1.0
	½ H	S	S	S	S	S	1.0	1.5
	H	S	S	S	S	S	…	…
	EH	S	0.5	0.5	0.5	0.5	…	…
Alloy 260	¼ H	S	S	S	S	S	0.5	1.0
	½ H	S	S	S	0.3	0.3	…	…
	H	S	0.5	0.5	0.5	1.0	…	…
	EH	2.0	2.0	1.5	2.0	2.0	…	…
Alloy 353	¼ H	S	S	S	S	S	0.5	1.0
	½ H	0.5	0.5	0.5	0.7	0.3	…	…
	H	2.0	1.5	1.5	2.0	2.0	…	…
	EH	6.0	4.0	4.0	4.0	4.0	…	…
Magnesium sheet @ 70°F:	AZ31B-O (S.B.)	3.0 ———						
	AZ31B-O	5.5 ———						

Figure 8.30 Table of minimum bend radii.

Material						
AZ31B-H24						8.0
HK31A-O						6.0
HK31A-H24						13.0
HM21A-T8						9.0
HM21A-T81						10.0
LA141A-O						3.0
ZE10A-O						5.5
ZE10A-H24						8.0
Titanium & alloys @ 70°F:						
Pure (A)	3.0	3.0	3.5	3.5	3.5	3.5
Ti-8Mn (A)	4.0	4.0	4.0	5.0	5.0	5.0
Ti-5Al-2.5Sn (A)	5.5	5.5	5.5	6.0	6.0	6.0
Ti-6Al-4V (A)	4.5	4.5	5.0	5.0	5.0	5.0
Ti-6Al-4V (ST)	7.0					7.0
Ti-6Al-6V-2Sn (A)	4.0					4.0
Ti-13V-11Cr-3Al (A)	3.0	3.0	3.5	3.5	3.5	3.5
Ti-4Al-3Mo-1V (A)	3.5	3.5	4.0	4.0	4.0	4.0
Ti-4Al-3Mo-1V (ST)	5.5	5.5	6.0	6.0	6.0	6.0

Note: S = sharp bend; 0 = sharp bend; S.B. = special bending quality; A = annealed; ST = solution treated; H = hard; EH = extra hard; Magnesium sheet may be bent at temperatures to 800°F; titanium may be bent at temperatures to 1,000°F; on copper and alloys, direction of bending is at 90° to direction of rolling (bend radii must be increased 10-20% @ 45° and 25-35% parallel to direction of rolling). The tabulated values of the minimum bend radii are given in multiples of the material thickness. The values of the bend radii should be tested on a test specimen prior to die-design or production bending finished parts.

Figure 8.30 (*Continued*)

may have as much as 15° to 20° of springback. The size of the bend radius on the part also influences the amount of springback on all materials.

The amount of springback will depend on the following factors:

- Type of material
- Hardness of the material and temper
- Radius of bend
- Gauge of the material

For practical die-making procedures, which must make allowances for springback of bent materials, trial pieces of material usually are bent to the desired radius in a press brake at 90°, and the actual springback is measured with a protractor. The 90° bend angle will open to a larger angle when the bending pressure is removed from the part.

Springback is not a problem when bending sheet metal parts or bar stock on a press brake because the brake operator can adjust the stroke of the press brake to form the correct angle. Once the stroke of the brake is determined and set, all parts made of the same material will bend to the correct angle, provided that the gauge or hardness of the material does not change during the operation.

To eliminate springback in forming-die designs, the punch or die may be shaped and angled to provide overbend of the part, thus eliminating springback.

8.6.4 Nonmetallic dies and materials

Plastic materials are used for some die and mold applications. Their use is not recommended unless the specifications for the plastic materials are within the physical limits of the application requirements. Figure 8.31 shows the range of properties available in cast unfilled plastics that may be used for tooling. Shrinkage during hardening, weathering, permanence, and dimensional stability must be considered when using plastic materials for tooling applications.

Phenolics are used for die models, stretch-form dies, duplicate masters, and semipermanent molds. The phenolics are limited to the casting applications and are brittle and react with various other materials.

Properties of Major Tooling Plastics ♦

Property	Phenolic	Epoxy	Ethyl-cellulose
Specific gravity	1.30-1.32	1.11-1.23	1.09-1.17
Tensile strength, 10^3 psi	6-9	9-12	2-8
Modulus of elasticity, 10^6 psi	4.5	4.5	1.0-3.0
Compressive strength, 10^3 psi	12-15	15-18	10-35
Flexural strength, 10^3 psi	11-17	14-19	4-12
Impact (Izod) ft-lbs/in notch	0.25-0.40	0.45-1.7	2.0-8.0
Hardness, Rockwell	M93-120	M80-100	R50-115
Dimensional stability	Fair	Excellent	Good
Machinability	Excellent	Good	Good
Acid resistance	Good	Good	Poor
Alkali resistance	Poor	Excellent	Excellent
Solvent resistance	Fair	Excellent	Poor

Note: ♦ Per ASTM test methods.

Figure 8.31   Tooling plastics.

Epoxies are used for capped forming tools, metal-bonding tools, vacuum-forming tools, molding dies, and all plastic impact tools. Drawing dies having a metallic core and capped with a working face of epoxy are used in the appliance, automotive, and aerospace industries. Epoxy molding dies are being made to mold cycloaliphatic resins, which are two-part epoxy systems. Cycloaliphatic resins (epoxies) are now used widely for electrical insulators and support details and structures for live current-carrying parts in medium- and high-voltage applications in the switchgear industry. In effect, epoxy materials are molded in specially formulated epoxy molds. Special release agents are required to separate the cast epoxy part from the mold cavity. For long production runs, metal molds are made from tool steels. The plastic molding processes are discussed in Chap. 12, "Castings, Moldings, Extrusions, and Powder-Metal Technology."

Ethyl-cellulose compounds, which are thermoplastic materials, have been used for drop-hammer dies and other applications.

8.6.5   Drawing and forming shapes and equations

The diameter of the circular blank for producing simple drawn cylindrical parts may be approximated from

$$D = 1.13\sqrt{F} = 1.13\sqrt{\Sigma f}$$

where D = diameter of blank, in or mm
   F = total area of part to be drawn, in^2 or mm^2
   Σf = sum of the individual areas of the drawn part, in^2 or mm^2

Equations for determining the blank diameters for many different drawn-part configurations are compiled in Fig. 8.32. These configurations are intended for material thickness and blank diameter ranges determined by the ratio $S/D \times 100 = 0.06$ to 2.0 using ductile metals or alloys in the soft or annealed condition (where S = material thickness, in or mm, and D = blank diameter, in or mm). The ratio of material thickness and blank diameter times 100 must fall within the range of 0.06 to 2.0 in order for the blank diameter D equations shown in Fig. 8.32 to be valid.

Example: A blank that is 5.00 in in diameter and 0.125 in thick would fall within the range of $0.125/5.00 \times 100 = 2.5$, which is *outside* the range of 0.06 to 2.0, as explained previously. A blank that is 4.00 in in diameter and 0.062 in thick would fall within the range of $0.062/4.00 \times 100 = 1.55$, which is *inside* the range of 0.06 to 2.0, making the equations valid for this combination of material thickness and blank diameter.

Figure 8.33 shows the step equations used to determine the drawn height at each step in the drawing process when we know the blank diameter, the drawing coefficients $(m_1, \ldots, m_n)$, the workpiece diameter $(d_1, \ldots, d_n)$, and the other variables indicated in the equations. Figures 8.34 through 8.39 give the drawing coefficients when we know the $S/D \times 100$ range, as well as approximations for the first drawing step height (h_1). The designations for the symbols shown in the equations in Fig. 8.33 are as follows:

$$D = \text{diameter of the blank, in or mm}$$
$$d_1, d_2, \ldots, d_n = \text{workpiece diameter for each operation, in or}$$
$$\text{mm, proceeding from larger to smaller diameter}$$
$$\text{as drawing steps increase}$$
$$a_1, a_2, \ldots, a_n = \text{bevel dimensions for each operation (profile 3 only)}$$
$$m_1, m_2, \ldots, m_n = \text{drawing coefficients for each operation (see Figs.}$$
$$8.34 \text{ through } 8.39)$$
$$S, S_1, \ldots, S_n = \text{blank thickness and wall thickness for each}$$
$$\text{operation, in or mm}$$
$$d\phi = \text{flange diameter, in or mm (for profile 6 only)}$$
$$r_1, r_2, \ldots, r_n = \text{inside radius or corner radius for each operation,}$$
$$\text{in or mm}$$

The equations shown in Fig. 8.33 allow the tool designer to calculate the dimensions for the dies required in each step of the drawing operation. Some profile heights will require only one or two steps,

(Text continued on page 598.)

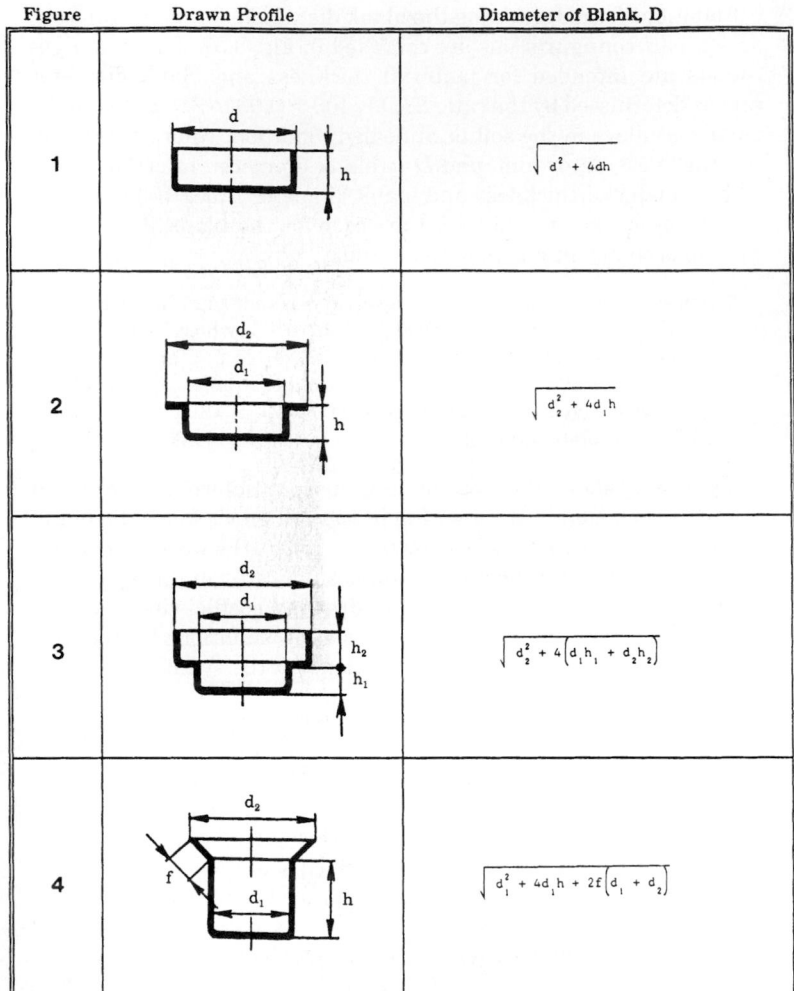

Figure	Drawn Profile	Diameter of Blank, D
1		$\sqrt{d^2 + 4dh}$
2		$\sqrt{d_2^2 + 4d_1 h}$
3		$\sqrt{d_2^2 + 4\left(d_1 h_1 + d_2 h_2\right)}$
4		$\sqrt{d_1^2 + 4d_1 h + 2f\left(d_1 + d_2\right)}$

Figure 8.32 Equations for determining blank diameters, drawing operations.

Figure	Drawn Profile	Diameter of Blank, D
5		$\sqrt{d_1^2 + 2\pi r d_1 + 8r^2}$
6		$\sqrt{d_1^2 + 2\pi r d_1 + 8r^2 + d_3^2 - d_2^2}$
7		$\sqrt{d_1^2 + 2\pi r d_1 + 8r^2 + 4d_2 h + d_3^2 - d_2^2}$
8		$\sqrt{d_1^2 + 2\pi r d_1 + 8r^2 + 4d_2 h + 2f\left(d_2 + d_3\right)}$

Figure 8.32 (*Continued*)

Figure	Drawn Profile	Diameter of Blank, D
9		$\sqrt{d_1^2 + 2\pi r d_1 + 8r^2 + 4d_2 h}$ or $\sqrt{d_2^2 + 4d_2 H - 1.72 r d_2 - 0.56 r^2}$ *
10		$\sqrt{d_1^2 + 2\pi r \left(d_1 + d_2 \right) + 4\pi r^2}$
11		$\sqrt{d_1^2 + 2\pi r \left(d_1 + d_2 \right) + 4d_2 h + 4\pi r^2}$
12		$\sqrt{d_1^2 + 4d_2 h + 2\pi r \left(d_1 + d_2 \right) + 4\pi r^2 + d_1^2 - d_3^2}$ or $\sqrt{d_4^2 + 4d_2 H - 3.44 r d_2}$ *

Figure 8.32 (*Continued*)

Figure	Drawn Profile	Diameter of Blank, D
13	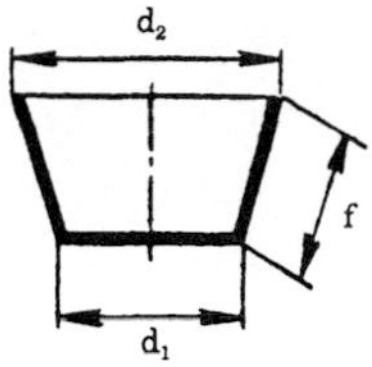	$$\sqrt{d_1^2 + 2f\left(d_1 + d_2\right)}$$
14	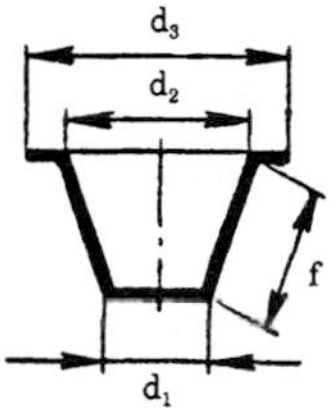	$$\sqrt{d_1^2 + 2f\left(d_1 + d_2\right) + d_3^2 - d_2^2}$$
15	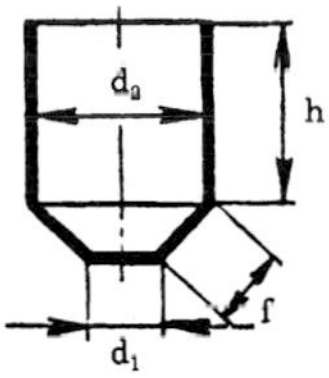	$$\sqrt{d_1^2 + 2f\left(d_1 + d_2\right) + 4d_2 h}$$
16	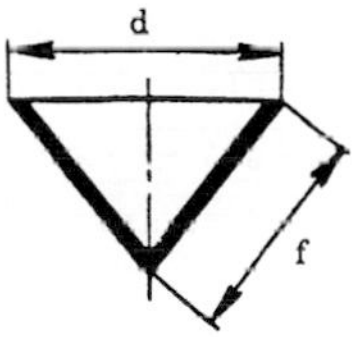	$$\sqrt{2df}$$

Figure 8.32 (*Continued*)

Figure	Drawn Profile	Diameter of Blank, D
17		$d\sqrt{2} = 1.414d$
18		$\sqrt{d^2 + d_1^2}$
19		$1.414\sqrt{d^2 + f(d + d_1)}$
20		$1.414\sqrt{d^2 + 2dh}$ or $2\sqrt{dH}$ *

Figure 8.32 (*Continued*)

Figure	Drawn Profile	Diameter of Blank, D
21		$\sqrt{d_1^2 + d_2^2 + 4d_1 h}$
22		$\sqrt{d^2 + 4h^2}$
23		$\sqrt{d_2^2 + 4h^2}$
24		$\sqrt{d^2 + 4\left(h_1^2 + dh_2\right)}$

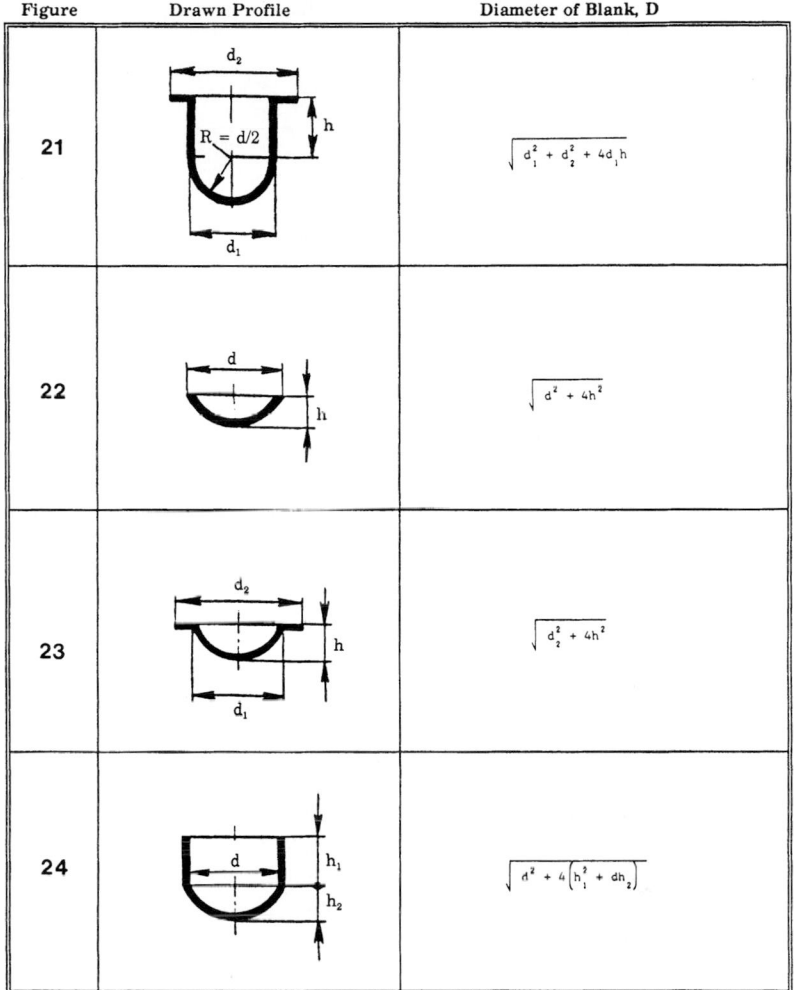

Figure 8.32 *(Continued)*

Figure	Drawn Profile	Diameter of Blank, D
25	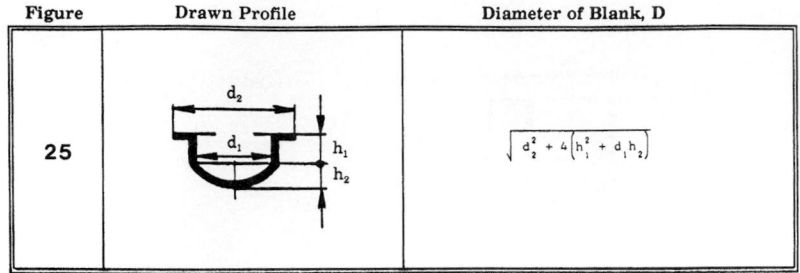	$\sqrt{d_2^2 + 4\left(h_1^2 + d_1 h_2\right)}$

Note: Equations marked (*) were derived for using the full height of the part, H.

Figure 8.32 *(Continued)*

whereas others will require more than two successive drawing steps. The most critical variables used in the drawing operations are the drawing coefficients $(m_1, m_2, \ldots, m_n)$ and the first-step relative value of h/d for flanged parts. These values must be determined accurately using the tables in Figs. 8.34 through 8.39.

The drawing-process calculations for cylindrical parts consist of

1. Determination of the deformation limit (coefficients $m_1, m_2, \ldots, m_n$).

2. Determination of the required number of steps.

3. Calculation of the drawn part dimensions for every step.

The deformation limits are determined by practical drawing coefficients m that were derived from experiments and proved by actual practice. The drawing coefficient for the first step is

$$m_1 = \frac{d_1}{D}$$

For the second step,

$$m_2 = \frac{d_2}{d_1}$$

Dimensions for the first step are

$$d_1 = m_1 D$$

(Text continued on page 603.)

Figure	Workpiece Profile	Operation	Equation for Profile Height
1	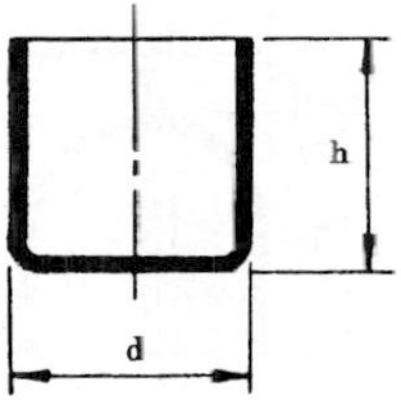	Step 1	$h_1 = 0.25 \left(\dfrac{D}{m_1} - d_1 \right)$
		Step 2	$h_2 = 0.25 \left(\dfrac{D}{m_1 m_2} - d_2 \right)$
		Step n	$h_n = 0.25 \left(\dfrac{D}{m_1 m_2 \cdots m_n} - d_n \right)$
2	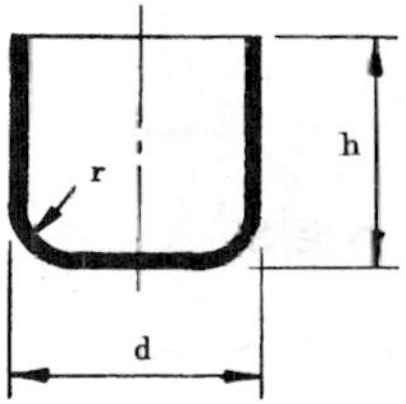	Step 1	$h_1 = 0.25 \left(\dfrac{D}{m_1} - d_1 \right) + 0.43 \dfrac{r_1}{d_1} \left(d_1 - 0.32 \, r_1 \right)$
		Step 2	$h_2 = 0.25 \left(\dfrac{D}{m_1 m_2} - d_2 \right) + 0.43 \dfrac{r_2}{d_2} \left(d_2 + 0.32 \, r_2 \right)$
		Step n	$h_n = 0.25 \left(\dfrac{D}{m_1 m_2 \cdots m_n} - d_n \right) + 0.43 \dfrac{r_n}{d_n} \left(d_n + 0.32 \, r_n \right)$

Figure 8.33 Equations for calculating the profile height h at each drawing step.

Figure	Workpiece Profile	Operation	Equation for Profile Height
3		Step 1	$h_1 - 0.25\left(\dfrac{D}{m_1} - d_1\right) + 0.57\dfrac{a_1}{d_1}\left(d_1 + 0.86\,a_1\right)$
		Step 2	$h_2 - 0.25\left(\dfrac{D}{m_1 m_2} - d_2\right) + 0.57\dfrac{a_2}{d_2}\left(d_2 + 0.86\,a_2\right)$
		Step n	$h_n - 0.25\left(\dfrac{D}{m_1 m_2 \cdots m_n} - d_n\right) + 0.57\dfrac{a_n}{d_n}\left(d_n + 0.86\,a_n\right)$
4		Step 1	$h_1 - 0.25\dfrac{D}{m_1}$
		Step 2	$h_2 - 0.25\dfrac{D}{m_1 m_2}$
		Step n	$h_n - 0.25\dfrac{D}{m_1 m_2 \cdots m_n}$

Figure 8.33 *(Continued)*

Figure	Workpiece Profile	Operation	Equation for Profile Height
5		Step 1	$h_1 - 0.25\left(\dfrac{D}{m_1} - d_1\right)\dfrac{s}{s_1} + s$
		Step 2	$h_2 - 0.25\left(\dfrac{D}{m_1 m_2} - d_2\right)\dfrac{s}{s_2} + s$
		Step n	$h_n - 0.25\left(\dfrac{D}{m_1 m_2 \ldots m_n} - d_n\right)\dfrac{s}{s_n} + s$
6		Step 1	$h_1 - 0.25\left(\dfrac{D}{m_1} - \dfrac{d\phi^2}{d_1} + 3.44\, r_1\right)$
		Step 2	$h_2 - 0.25\left(\dfrac{D}{m_1 m_2} - \dfrac{d\phi^2}{d_2} + 3.44\, r_2\right)$
		Step n	$h_n - 0.25\left(\dfrac{D}{m_1 m_2 \ldots m_n} - \dfrac{d\phi^2}{d_n} + 3.44\, r_n\right)$

Figure 8.33 (*Continued*)

601

| Coefficient | Coefficient Values for Relative Blank Thickness S/D x 100 | | | | |
Step	2.0 - 1.5	1.5 - 1.0	1.0 - 0.5	0.5 - 0.2	0.2 - 0.06
m_1	0.46-0.50	0.50-0.53	0.53-0.56	0.56-0.58	0.58-0.60
m_2	0.70-0.72	0.72-0.74	0.74-0.76	0.76-0.78	0.78-0.80
m_3	0.72-0.74	0.74-0.76	0.76-0.78	0.78-0.80	0.80-0.82
m_4	0.74-0.76	0.76-0.78	0.78-0.80	0.80-0.82	0.82-0.84

Note: Tabulated values are the drawing coefficients. S = material thickness and D = blank diameter.

Figure 8.34 Drawing coefficients for cylindrical parts without flanges (profiles 1–5 of Fig. 8.33).

| Relative Dia. of | Coefficient Values for Relative Blank Thickness S/D x 100 | | | | |
Flange, dφ/d	2.0 - 1.5	1.5 - 1.0	1.0 - 0.5	0.5 - 0.2	0.2 - 0.06
1.1	0.50	0.53	0.55	0.57	0.59
1.3	0.49	0.51	0.53	0.54	0.55
1.5	0.47	0.49	0.50	0.51	0.52
1.8	0.45	0.46	0.47	0.48	0.48
2.0	0.42	0.43	0.44	0.45	0.45
2.2	0.40	0.41	0.42	0.42	0.42
2.5	0.37	0.38	0.38	0.38	0.38
2.8	0.33	0.34	0.34	0.35	0.35

Note: Tabulated values are the drawing coefficients. S = material thickness and D = blank diameter.

Figure 8.35 Drawing coefficients for cylindrical parts with flanges (profile 6 of Fig. 8.33).

| Relative Dia. of | Value of (h/d) for Relative Blank Thickness S/D x 100 | | | | |
Flange dφ/d	2.0 - 1.5	1.5 - 1.0	1.0 - 0.5	0.5 - 0.2	0.2 - 0.06
1.1	0.90-0.75	0.82-0.60	0.70-0.57	0.62-0.50	0.52-0.45
1.3	0.80-0.65	0.72-0.56	0.60-0.50	0.53-0.45	0.47-0.40
1.5	0.70-0.58	0.63-0.50	0.53-0.45	0.48-0.40	0.42-0.35
1.8	0.58-0.48	0.53-0.42	0.44-0.37	0.39-0.34	0.35-0.29
2.0	0.51-0.42	0.46-0.36	0.38-0.32	0.34-0.29	0.30-0.25
2.2	0.45-0.35	0.40-0.31	0.38-0.27	0.29-0.25	0.26-0.22
2.5	0.35-0.28	0.32-0.25	0.27-0.22	0.23-0.20	0.21-0.17
2.8	0.27-0.22	0.24-0.19	0.21-0.17	0.18-0.15	0.16-0.13

Note: Tabulated values are the values of h/d. (see profile 6 of Figure 9-29).
The larger tabulated values correspond to the larger radii r = (10-12)S for S/D x 100 = 2.0-1.5; r = (20-25)S for S/D x 100 = 0.2-0.06. The smaller tabulated values correspond to the smaller radii for bottom and flange where r = approx. (4-8)S. S is the material thickness and D is the blank diameter.

Figure 8.36 Approximate value of the first step relative depth (h/d) for cylindrical parts with flanges (profile 6 of Fig. 8.33)

Blank Diameter	Height of Part (h_1) in 1st Step using Coefficients m_1						
(millimeters)	0.45	0.48	0.50	0.53	0.55	0.58	0.60 - (coefficients)
30	13	12	11	10	9.5	9	....
40	18	16	15	14	13	12	....
50	22	20	19	17	16	15	....
60	26	24	22	20	19	18	....
70	31	28	26	24	22	20	19
80	35	32	30	27	26	23	22
90	40	36	34	30	29	26	24
100	44	40	37	34	32	29	27
120	....	48	45	40	38	35	32
150	....	60	55	50	48	44	40
180	....	72	67	60	56	52	50
200	....	80	75	68	64	58	55

Note: Tabulated values are the height of the part in 1st drawing step, millimeters. (for profile 1, Fig.9-29)

Figure 8.37   Height of cylindrical part (h_1) for the listed drawing coefficients (profile 1 of Fig. 8.33).

Blank Diameter	Height of Part (h_1) in 1st Step using Coefficients m_1						
(millimeters)	0.45	0.48	0.50	0.53	0.55	0.59	0.60 - (coefficients)
30	14	13	12	11	10	10	....
40	19	17	16	15	14	13	....
50	23	21	20	17	16	15	....
60	28	26	24	22	21	20	....
70	33	30	28	26	24	22	21
80	37	34	32	29	28	25	24
90	42	38	36	32	32	29	27
100	47	43	40	37	35	32	30
120	....	52	49	44	42	39	36
150		65	60	55	53	49	45
180	....	77	72	65	63	57	55
200	....	88	81	74	70	64	61

Note:Tabulated values are the height of the part in 1st drawing step, millimeters. (for profile 2, Fig. 9-29)

Figure 8.38   Height of cylindrical part (h_1) for the listed drawing coefficients (profile 2 of Fig. 8.33).

For the second step,

$$D_2 = m_2 d_2$$

where D = blank diameter, in or mm
     d = part diameter, in or mm
     m = drawing coefficient

Blank Diameter	Height of Part (h_1) in 1st Step using Coefficients m_1						
(millimeters)	0.48	0.50	0.53	0.55	0.58	0.60	0.62 - (coefficients)
120	53	50	45	43	40	37	....
150	66	61	56	52	50	46	....
180	79	74	67	65	59	57	....
200	88	83	76	72	66	63	58
250	110	103	95	90	83	78	73
300	132	122	114	108	98	92	86
350	154	144	134	126	116	108	102
400	176	166	152	144	132	123	115
450	198	184	172	162	148	140	130
500	220	205	190	180	165	155	145
550	242	226	210	198	182	172	160
600	264	244	228	216	200	185	173

Note: Tabulated values are the height of the part in 1st drawing step, millimeters. (for profile 3, Fig. 9-29)

Figure 8.39   Height of cylindrical part (h_1) for the listed drawing coefficients (profile 3 of Fig. 8.33)

In this way, knowing the coefficient m, we can determine the dimensions and number of steps.

Note: The dimensions used in Figs. 8.34 through 8.39 are given in millimeters. The drawing coefficients are dimensionless values (multipliers).

8.6.6   General rules for die stamping and forming operations

Some of the basic rules and dimensions for stamping and forming operations are shown in Figs. 8.40 through 8.44. The edge-distance recommendations shown in Fig. 8.40 for die-stamping metals are easy to apply during punching or stamping operations, but these distances (A and B) often are reduced when the gauge of the metal is below 0.020 in.

Figure 8.41 shows the common methods of relieving the bending stresses on intermediately flanged parts and parts with angle-notched ends. Figure 8.42 shows methods for minimum flanges and stiffening ribs on the heel of die-formed parts. Light-gauge sheet metal parts can be made quite strong with the addition of heel-applied stiffening webs. The load-carrying ability of thin-gauge sheet metal parts such as angles is greatly increased with the addition of these die-formed stiffening webs.

Figure 8.43 shows a 90° formed bead, which is common in thin sheet metal parts for obtaining stiffness without adding another

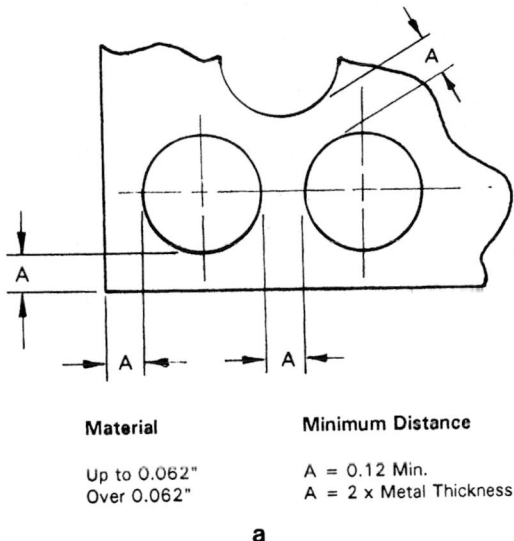

Material

Up to 0.062"
Over 0.062"

Minimum Distance

A = 0.12 Min.
A = 2 x Metal Thickness

a

Material

Up to 0.090"
Over 0.090"

Minimum Distance

B = 0.18"
B = 2 x Metal Thickness

b

Figure 8.40 (*a*) Minimum edge. (*b*) Minimum edge.

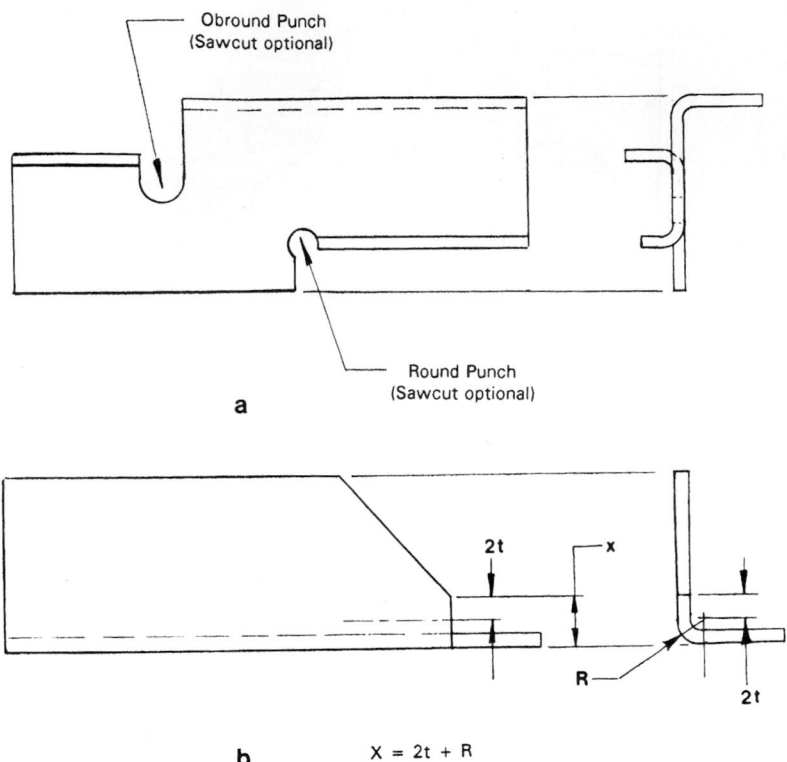

Figure 8.41   (*a*) Bend relief notches. (*b*) Notch limits.

member to the part. Automotive and aerospace sheet metal parts frequently contain stiffening beads and "lightening holes." See Chap. 9, "Sheet Metal Practices and Layout," for additional information on sheet metal bending and forming.

Figure 8.44 shows a 60° flat-bottomed bead. Such beads are normally larger than the 90° types shown in Fig. 8.43 and are used on large sheet metal parts.

Forces required for die bending.   The force for die forming a flange (edge bending) may be calculated from

$$P = \frac{lt^2}{2W}(S_t) \qquad \text{(see Fig. 8.42)}$$

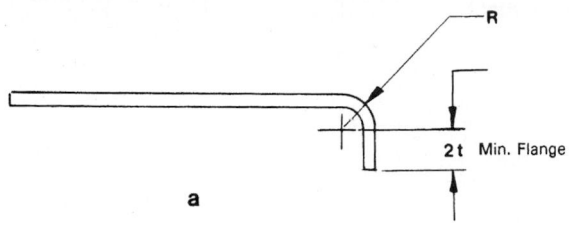

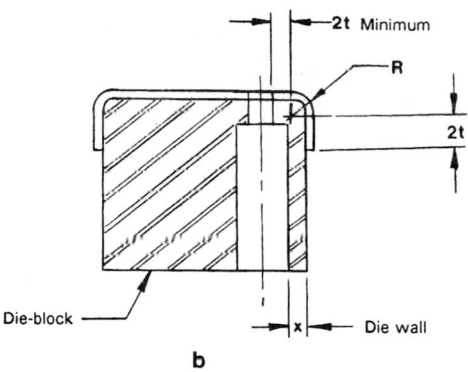

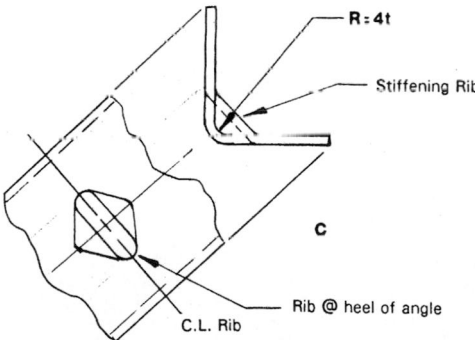

Figure 8.42 (*a*) Minimum flange. (*b*) Edge distance. (*c*) Rib.

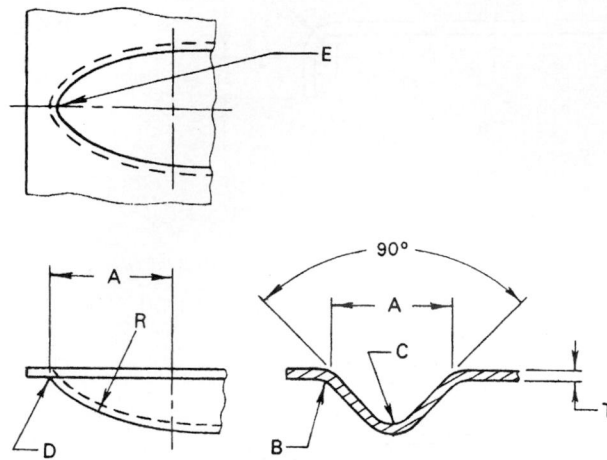

A	B (Radius)	C (Radius)	D (Radius)	E (Radius)
0.25	2T	2T	4T	T
0.38	2T	2T	4T	T
0.50	2T	2T	4T	2T
0.62	4T	4T	4T	2T
0.75	5T	5T	4T	3T
1.00	5T	5T	4T	3T

Figure 8.43 Stiffening beads.

The force required for V bending with a centrally located load may be calculated from

$$P = \frac{lt^2}{W}(S_t)$$

The force for bending channel and U bends may be calculated from

$$P = \frac{2lt^2}{W}(S_t)$$

where P = bending force, lbf
l = length of bend, in
t = material thickness, in
W = width of unsupported material, in
S_t = ultimate tensile strength of the material, psi

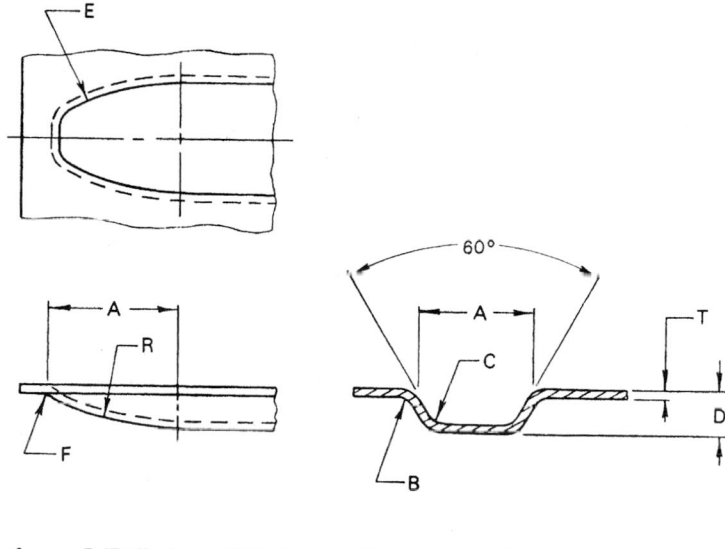

A	B (Radius)	C (Radius)	D	E (Radius)	F (Radius)
1.00	3T	2T	0.25	4T	4T
1.50	3T	2T	0.31	4T	4T

Figure 8.44 Stiffening beads.

8.7 Jigs and Fixtures

A *jig* is a tool that may be used as a guide to form (cut, grind, or file) the outline of a part. Jigs are also used with drill bushings to accurately drill and/or ream holes in a part. The jig is positioned accurately on the part to be drilled prior to drilling the holes. Stamped parts thus may be mass produced with additional drilled or reamed holes, making the parts accurate and interchangeable at the next assembly stage of manufacture. A drilling-guide tool is sometimes called a *drill fixture*. The terms *jig* and *fixture* sometimes are used interchangeably in industry when describing a drilling or reaming operation.

Different size drill bushings used in jigs and fixtures are shown in Fig. 8.45. The drill bushing is one of the most common parts used in tooling work. The drill bushing is available in different types and sizes to accommodate the different size drills and reamers. The bushing type may be plain cylindrical, flanged, press fit, or lock-in,

according to the intended application. The correct definitions of drill bushing types are as follows:

- Headless press fit, type P
- Standard head, type H
- Slip/fixed, type SF

Figure 8.45 shows the plain bushing (P), flanged bushing (H), and lock-in bushing (SF). The lock-in or slip/fixed type of drill bushing is used when the production rates are high so that the bushing may be replaced easily as it wears during operation. Drill bushings are made of quality-grade tool steels with hardness ranges of Rockwell C 50 to C 60.

8.7.1 Tooling fixtures

A *tooling fixture* is a device used to accurately and efficiently aid the manufacturing processes. Fixtures are made in countless varieties and designs, including those for

- Assembly and subassembly
- Welding

Figure 8.45 Drill bushings.

- Machining
- Machining pallets (quick-interchangeable fixtures)
- Heat treating
- Plating
- Wiring and cable harnesses

Tooling fixtures may weigh less than 1 oz or more than 100 tons. Large assembly fixtures, such as those used in the final assembly of aerospace vehicles or construction machinery, require the use of transits, optical/laser levels, and theodolites in order to position the critical parts accurately during construction of the fixture.

The properly designed tooling fixture has many requirements, such as

- Dimensional stability
- Dimensional accuracy
- Ease of manufacture
- Efficiency in use
- Practicality
- Cost-effectiveness
- Safety in operation

A typical small assembly fixture is shown in Fig. 8.46. The pneumatically actuated clamps used on this fixture are evident in the photograph. Holding clamps are important elements in the design of many fixtures. Section 8.8 will be useful to the tooling fixture designer and shows various clamp designs and calculation techniques for their application. Safety to manufacturing personnel is of prime importance when clamping devices are used on any tooling fixture, especially clamping devices that are pneumatically or hydraulically actuated.

Figure 8.47 shows an assembly fixture for pouring babbitt and epoxy potting compound into porcelain bushings used on a piece of electrical equipment. A brazed solid copper contact is first babbitted into place inside the porcelain bushing. While that is cooling, the epoxy potting compound is added to hold the porcelains to a polyester/glass mounting plate. This assembly fixture allows the assembly of two complete sets of bushings, comprised

Figure 8.46 Pneumatic assembly fixture.

Figure 8.47 Compound assembly fixture for babbitting and epoxy-bonding a porcelain bushing assembly.

of six contacts, six porcelain bushings, and two mounting plates. This completes the primary connection set for one breaker cell used on high-voltage switchgear. This newly designed assembly fixture allows the work to proceed at four times the original pace, thus saving time and money. Most of this assembly fixture is made from precision aluminum tooling plates. The sealing gaskets are held in place automatically by the fixture during the pouring processes. After the epoxy has set, the entire assembly is oven-heated at 350°F for 1 hour. After cooling, the completed primary electrical contact assemblies are easily removed from the fixture and assembled into the electrical equipment.

Figure 8.48 shows a drawing of a typical machining fixture that is actuated by a manual screw mechanism. The fixture drawing was generated by a CAD system and drawn on an automatic ink-pen plotter.

The design and application of tooling fixtures are as important to manufacturing processes as the design of the product that requires these fixtures for production, and in most cases, the fixtures are more difficult to design and have closer tolerances than the product itself.

8.7.2 Pallets

A *pallet* is a quickly removable indexed machining fixture that is used when maximum productivity of machined parts is required. In order for a palletizing system to be cost-effective, the cost of multiple pallets is weighed against the number of machined parts required for any particular job or production run.

The use of multiple pallets reduces machine "down time" drastically and makes the machined parts cost effective when quantities are large enough to warrant a palletizing system.

In a palletized system, the machine operator loads multiple parts into different pallets while the machine tool is machining parts on a previously loaded pallet. All the pallets are pin indexed to fit the bed of the machine tool at a precise location relative to the zero position of the cutting tool or spindle centerline. In effect, the machine operator is using the normal idle time (while the CNC machine tool is operating) to load pallets (machining fixtures that hold multiple parts and which are quick changing).

A palletizing system is usually expensive to implement but is totally justified if it produces parts that are cost-effective when the cost to produce the parts meets or exceeds expected price goals and specifications.

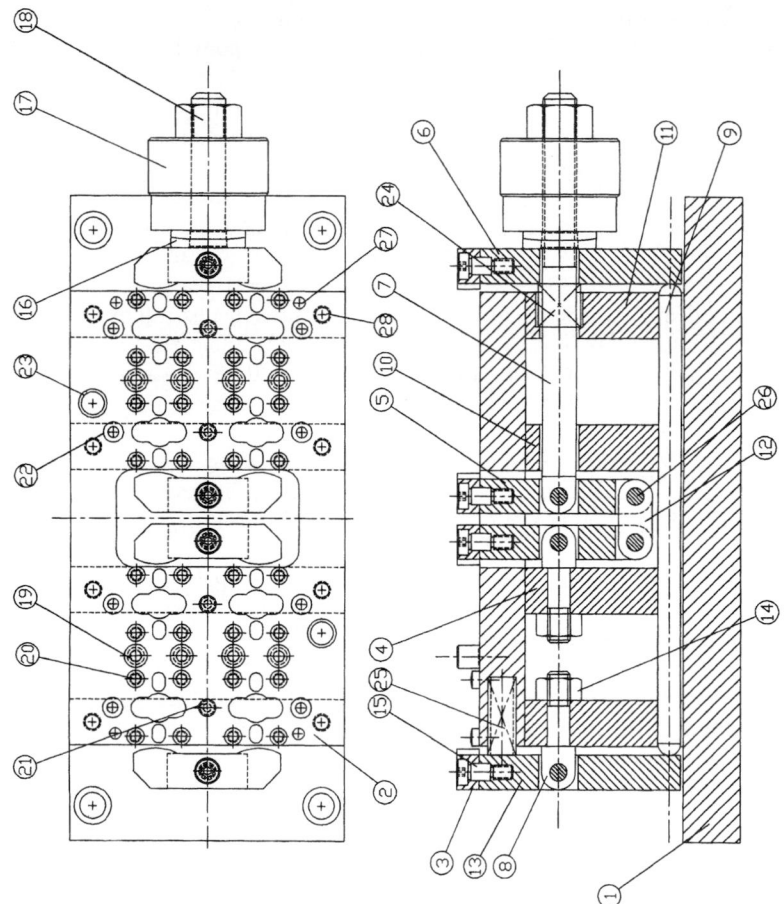

Figure 8.48 CAD drawing of a typical machining fixture.

POS	QTY	PART NAME	MATERIAL	DIMENSIONS	HARDNESS
28	8	SHCS	9-1236-41	3/8x4 1/2	DANLY
27	2	DOWEL PIN	7-0832-1	Ø1/4x2	DANLY
26	5	DOWEL PIN	7-1228-1	Ø3/8x1 3/4	DANLY
25	1	SPRING	LC-072H-6		LEE
24	1	SPRING	LC-082M-3		LEE
23	2	BUSHING	10063	.5x.625	STEVENS
22	8	LOCATING PIN	CL-7A-SLP-3		CARR LANE
21	4	REST BUTTON	CL-14-RB	3/8x1/2	CARR LANE
20	32	LOCATOR BUTTON	CL-2-SLB	3/8x3/16	CARR LANE
19	8	REST BUTTON	CL-1-RB	1/2x1/2	CARR LANE
18	1	LOCK NUT	LN-750	3/4-10	VLIER
17	1	HOLLOW CYLINDER	CY2129-25	7415 LB	VLIER
16	1	SPHERICAL WASHER	CL-5-SW	3/4 DIA	CARR LANE
15	4	SHOULDER SCREW	CL-14-SS	1/2x.375	CARR LANE
14	2	NUT		3/8-13	
13	1	CLAMP LEVER	1018 STEEL	.75x1.75x4 7/16	
12	1	LINK	1045 STEEL	.75x.75x1.75	RC 38-40
11	1	SUPPORT BAR	1018 STEEL	1x3.5x6	
10	1	SUPPORT BAR	1018 STEEL	1x3.5x6	
9	1	PUSH ROD	O1 STEEL	Ø1/2x10.25	RC 38-40
8	2	SWING BOLT	O1 STEEL	Ø3/4x2.625	
7	1	PULL ROD	O1 STEEL	Ø3/4x9 1/4	
6	1	CLAMP LEVER	1018 STEEL	.75x1.75x4 7/16	
5	2	CLAMP LEVER	1018 STEEL	.75x1.75x3 13/16	
4	2	SUPPORT BAR	1018 STEEL	1x3.5x6	
3	4	ADJUSTIBLE CLAMP	1045 STEEL	.5x.875x3.218	RC 40-45
2	1	TOP PLATE	1018 STEEL	1x6x9.875	
1	1	BASE PLATE	1018 STEEL	1 1/4x6x14	

Figure 8.48   (*Continued*)

8.8   Clamping Mechanisms and Calculation Procedures

Clamping mechanisms are an integral part of nearly all tooling fixtures. Countless numbers of clamping designs may be used by the tooling fixture designer and tool maker, but only the basic types are described in this section. With these basic clamp types, it is possible

to design a vast number of different tools. Both manual and pneu-matic/hydraulic clamping mechanisms are shown, together with the equations used to calculate each basic type.

The basic clamping mechanisms used by many tooling fixture designers are outlined in Fig. 8.49 (types 1 through 12). These basic clamping mechanisms also may be used for other mechanical design applications. Figure 8.50 shows a typical standard clamp set used on a milling machine.

Eccentric clamp, round.   The eccentric clamp, such as that shown in Fig. 8.49 (type 12) is a fast-action clamp compared with threaded clamps, but threaded clamps have higher clamping forces. The eccentric clamp usually develops clamping forces that are 10 to 15 times higher than the force applied to the handle.

The ratio of the handle length to the eccentric radius normally does not exceed 5 to 6, whereas for a swinging clamp or strap clamp (threaded clamps), the ratio of the handle length to the thread pitch diameter is 12 to 15. The round eccentrics are relatively cheap and have a wide range of applications in tooling.

The angle α in Fig. 8.49 (type 12) is the rising angle of the round eccentric clamp. Because this angle changes with rotation of the eccentric, the clamping force is not proportional at all handle rota-tion angles. The clamping stroke of the round eccentric at 90° of its handle rotation equals the roller eccentricity e. The machining allowance for the clamped part or blank x must be less than the eccentricity e. To provide secure clamping, eccentricity $e \geq x$ to $1.5x$ is suggested.

The round eccentric clamp is supposed to have a self-holding char-acteristic to prevent loosening in operation. This property is gained by choosing the correct ratio of the roller diameter D to the eccentric-ity e. The holding ability depends on the coefficient of static friction. In design practice, the coefficient of friction f normally would be 0.1 to 0.15, and the self-holding quality is maintained when f exceeds $\tan \alpha$.

The equation for determining the clamping force P is

$$P = Ql \frac{l}{[\tan(\alpha + \phi_1)\tan \phi_2]r}$$

Then the necessary handle torque $(M = Pl)$ is

$$M = P[\tan(\alpha + \phi_1) + \tan \phi_2]r$$

(Text continued on page 623.)

Type	Geometry	Equation
1	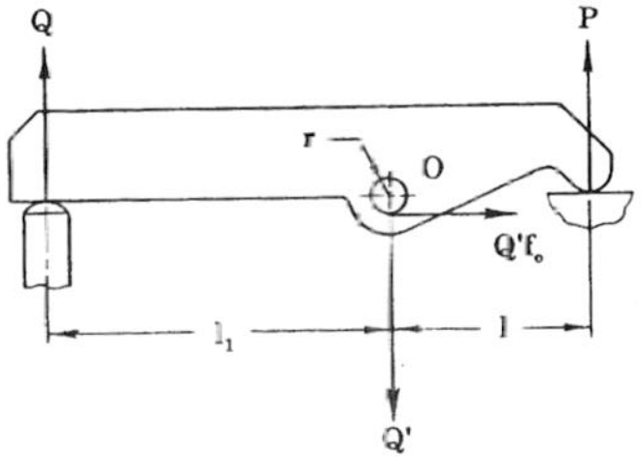	$$Q = P \frac{1 + rf_o}{l_1 - rf_o}$$ $$P = \left[\left(1 + rf_o \right) \div \left(1 - rf_o \right) \right] Q$$
2	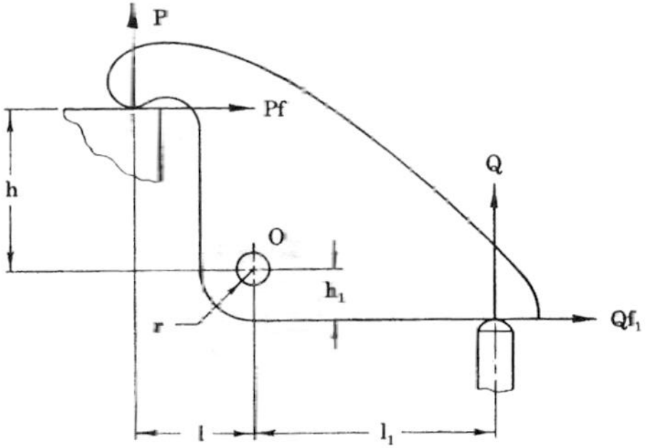	$$Q = P \frac{1 + hf + rf_o}{l_1 - h_1 f_1 - rf_o}$$ when: $l_1 \geq 1$ and $P \geq Q$

Figure 8.49 Basic clamping mechanisms (lever, cam, screw, pneumatic, and hydraulic).

617

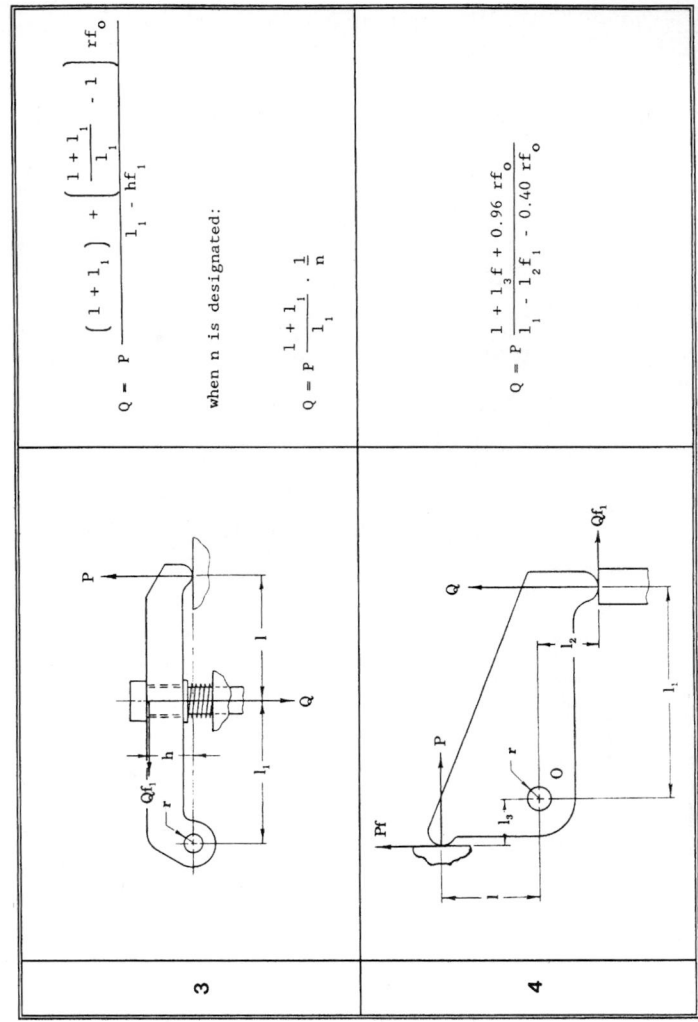

3

$$Q = P \frac{\left(1 + l_1\right) + \left(\frac{1 + l_1}{l_1} - 1\right) r f_o}{l_1 - h f_1}$$

when n is designated:

$$Q = P \frac{1 + l_1}{l_1} \cdot \frac{1}{n}$$

4

$$Q = P \frac{1 + l f + 0.96 \, r f_o}{l_1 - l_2 f_1 - 0.40 \, r f_o}$$

Figure 8.49 *(Continued)*

618

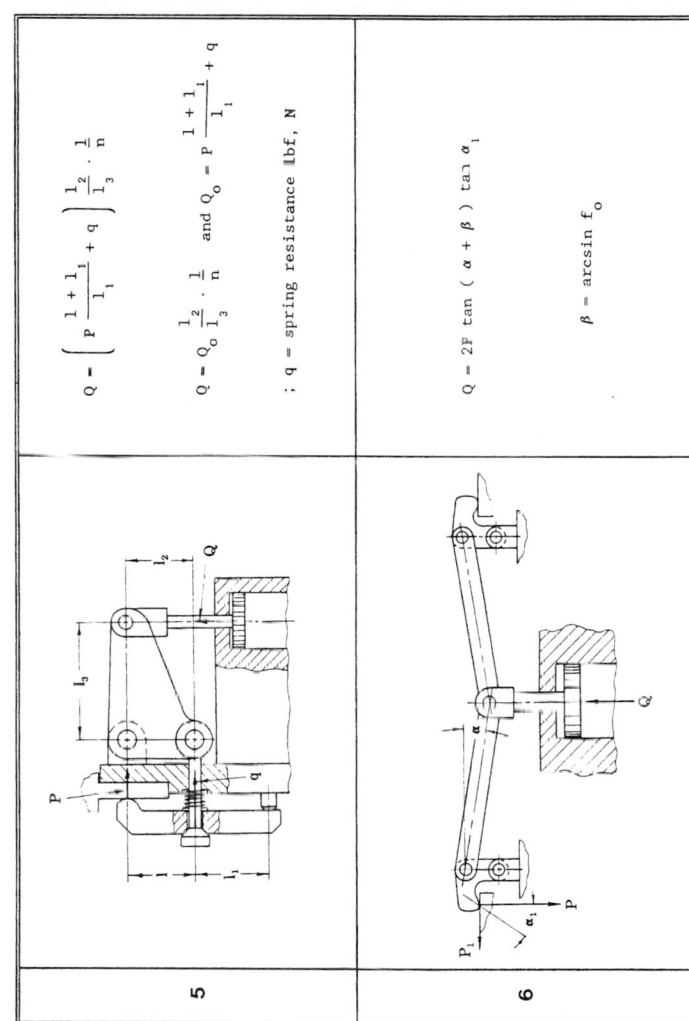

5	$Q = \left(P \dfrac{1+l_1}{l_1} + q \right) \dfrac{l_2}{l_3} \cdot \dfrac{1}{n}$ $Q = Q_o \dfrac{l_2}{l_3} \cdot \dfrac{1}{n}$ and $Q_o = P \dfrac{1+l_1}{l_1} + q$; q = spring resistance lbf, N
6	$Q = 2F \tan(\alpha + \beta) \tan \alpha_1$ $\beta = \arcsin f_o$

Figure 8.49 (*Continued*)

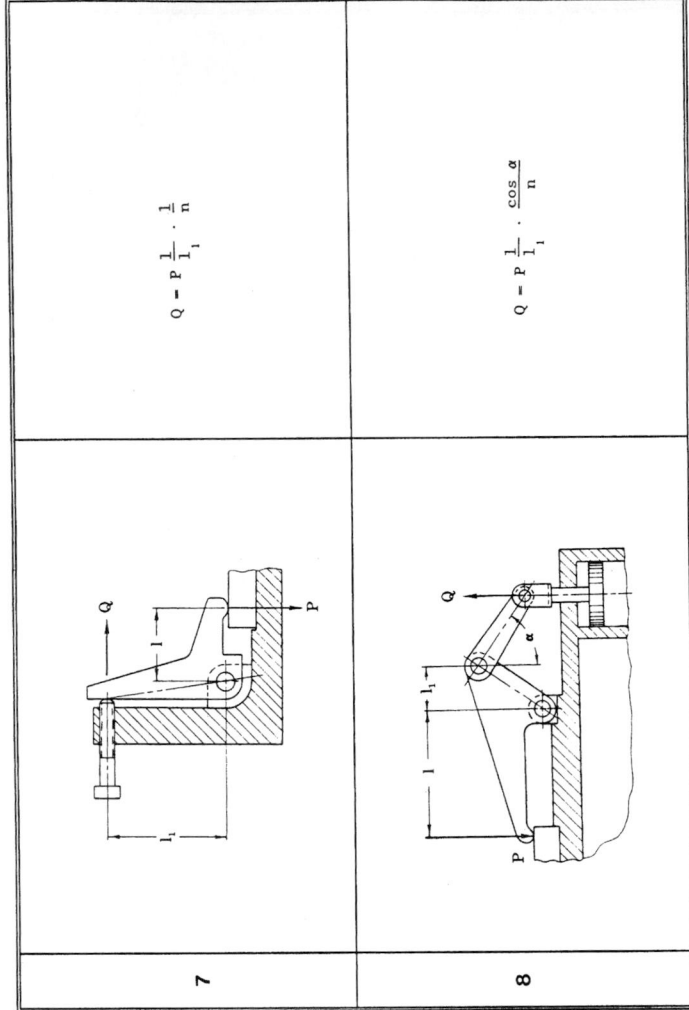

7	$Q = P \dfrac{l}{l_1} \cdot \dfrac{1}{n}$
8	$Q = P \dfrac{l}{l_1} \cdot \dfrac{\cos \alpha}{n}$

Figure 8.49 (*Continued*)

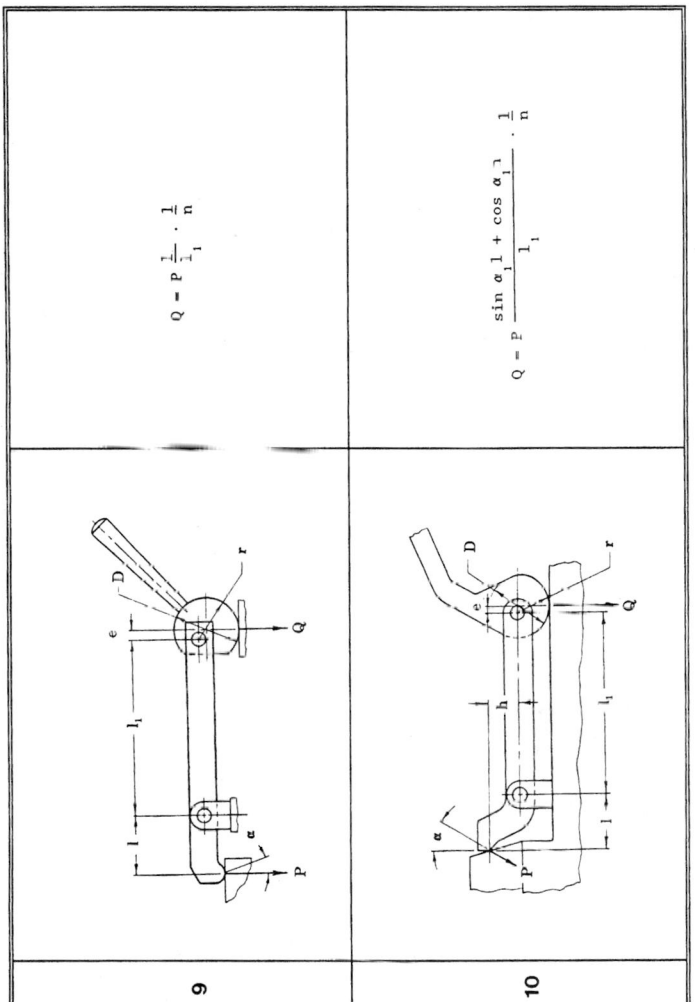

| 9 | $Q = P \dfrac{l}{l_1} \cdot \dfrac{1}{n}$ |
| 10 | $Q = P \dfrac{\sin \alpha_1 \, l + \cos \alpha_1 \, r}{l_1} \cdot \dfrac{1}{n}$ |

Figure 8.49 (*Continued*)

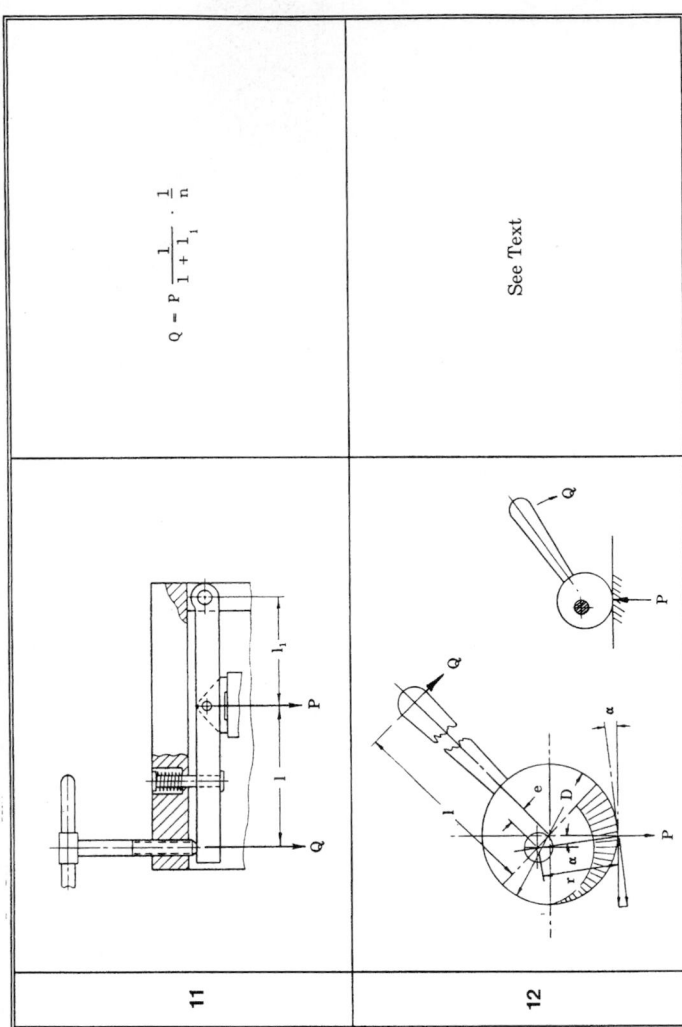

| 11 | | $Q = P \dfrac{l}{l + l_1} \cdot \dfrac{1}{n}$ |
| 12 | | See Text |

NOTE: f_o = coefficient of friction (axels and pivot pins) = 0.1 to 0.15 ; f = coefficient of friction of clamped surface = tan ϕ ; ϕ_1 = arctan f ; n = efficiency coefficient, 0.98 to 0.84, (determined by frictional losses in pivots and bearings, 0.98 for the best bearings through 0.84 for no bearings, (in order to avoid the use of complex, lengthy equations, the value of "n" can be taken as a mean between the limits shown) ; q = spring resistance or force, pounds or Newtons, (lbf or N).

Figure 8.49 *(Continued)*

622

Figure 8.50   Standard basic machining clamps for milling operations.

where r = distance from pivot point to contact point of the eccentric
and the machined part surface, in or mm

α = rotation angle of the eccentric at clamping (reference only)

$\tan \phi_1$ = friction coefficient at the clamping point

$\tan \phi_2$ = friction coefficient in the pivot axle

l = handle length, in or mm

Q = force applied to handle, lbf or newtons

D = diameter of eccentric blank or disk, in or mm

P = clamping force, lbf or newtons

Note: $\tan(\alpha + \phi_1) \approx 0.2$ and $\tan \phi_2 \approx 0.05$ in actual practice. See
Fig. 8.51 for listed clamping forces for the eccentric clamp shown in
Fig. 8.49 (type 12).

The cam lock.   Another clamping device that may be used instead
of the eccentric clamp is the standard cam lock. In this type of
clamping device, the clamping action is more uniform than in the
round eccentric, although it is more difficult to manufacture. A true
camming action is produced with this type of clamping device.
The method for producing the cam geometry is shown in Fig. 8.52.
The layout shown is for a cam surface generated in 90° of rotation

D mm (inches)	Clamping Force, P (Newtons)						
	490	735	980	1225	1470	1715	1960
40 (1.58")	2.65	3.97	5.40	6.67	8.00	9.37	10.64
50 (1.98")	3.34	5.00	6.67	8.39	10.01	11.77	13.68
60 (2.36")	4.02	6.03	8.00	10.01	11.97	14.03	16.48
70 (2.76")	4.71	7.06	9.42	11.77	14.08	16.48	18.79

NOTE: Tabulated values are torques (Newton-meters).
To convert clamping forces in Newtons to pounds-force, multiply table values by 0.2248, (i.e. 1960 N = 1960 x 0.2248 = 441 lbf).
To convert tabulated torques in Newton-meters to pounds-feet, multiply values by 0.7376, (i.e. 18.79 N-m = 18.79 x 0.7376 = 13.9 lbs-ft).

Figure 8.51 Torque values for listed clamping forces —eccentric clamps (type 12, Fig. 8.49).

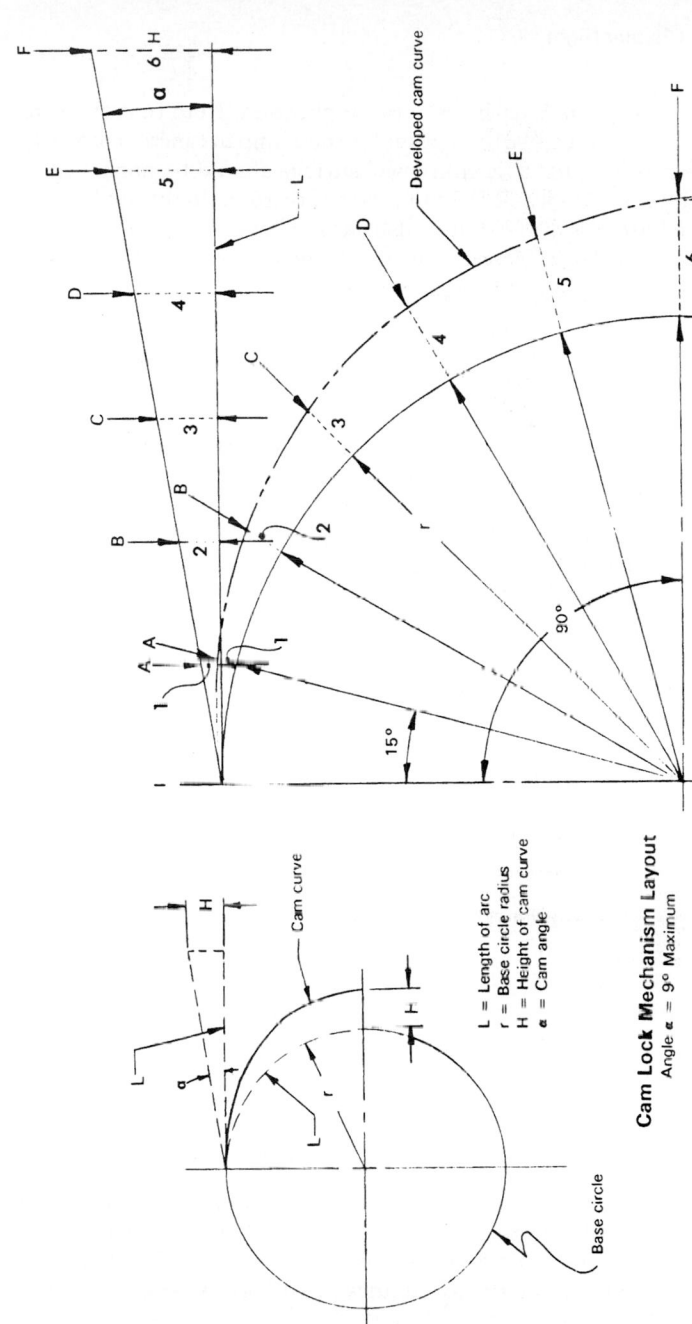

L = Length of arc
r = Base circle radius
H = Height of cam curve
α = Cam angle

Cam Lock Mechanism Layout
Angle α = 9° Maximum

Figure 8.52 Determining cam geometry.

of the device, which is the general application. Note that the cam angle should not exceed 9° in order for the clamp to function properly and be self-holding. The cam wear surface should be hardened to approximately Rockwell C 30 to C 50 or according to the application and the hardness of the materials that are being clamped. The cam geometry may be developed using CAD, and the program for machining the cam lock may be loaded into the CNC of a wire EDM machine.

8.9   Molds

Tooling practice also includes the design and manufacture of various types of molds. Molds are made for casting metals and alloys and for injection and compression molding plastics. Injection molding is used for thermoplastics, whereas compression molding is used for the thermoset plastics. See Chap. 4, "Materials and Their Uses," and Chap. 12, "Castings, Moldings, Extrusions, and Powder-Metal Technology," for more information on the various casting alloys and plastics used in molding processes.

Figure 8.53 shows a typical machined form used to make the tooling epoxy mold to mold other types of epoxy plastics. Forms or master patterns such as that shown in this figure are normally made of 2024 or 7075 aluminum alloys, although other materials may be used. The high-quality aluminum alloys were used on this master pattern owing to their excellent strength and easy machinability. This particular master pattern was machined on a CNC machining center and assembled by a master die maker.

Figure 8.54 shows another master pattern used for making tooling epoxy molds. In this example, the pattern was produced on a CNC turning center using a program written by the tool engineering department. It should be noted that no additional finishing was required on this master pattern. The extremely smooth and polished surfaces were produced with carbide cutting tools in a multiple-pass operation.

8.10   Force Gauge for Tooling Applications

It is often necessary to know the pressures or forces actually required to operate specific types of tooling devices. This information is of benefit to the tool design engineer and master die maker. Strain gauges and "load cells" normally are required to determine accurately the

Figure 8.53 A mold form.

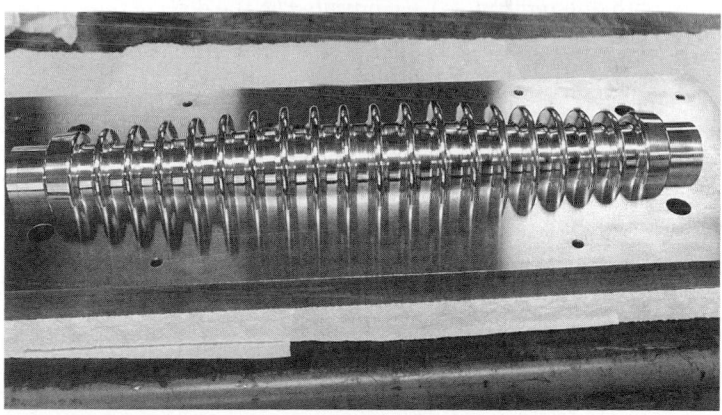

Figure 8.54 Master metal pattern for producing plastic molds.

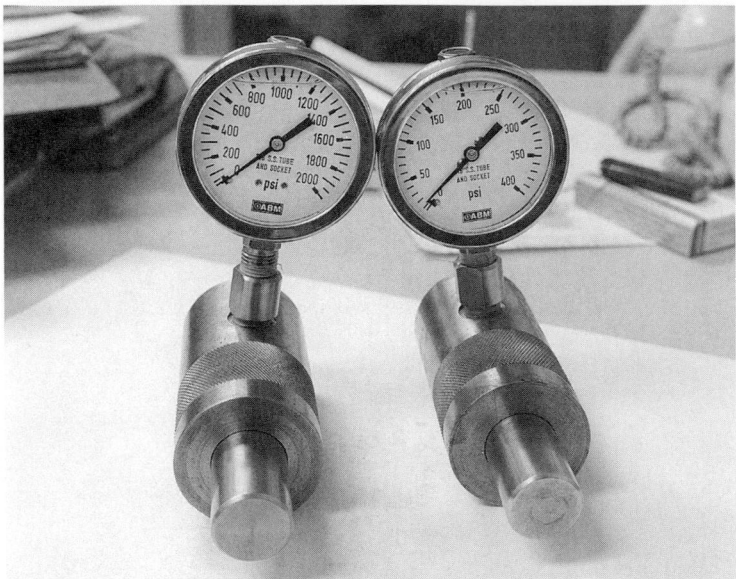

Figure 8.55   Two hydraulic force gauges.

forces applied to various components in operation. A simple, inexpensive type of hydraulic force gauge may be constructed by the die maker for this application. Figure 8.55 shows two hydraulic force gauges with different ranges, as indicated on the dials of the gauges.

8.11   Complex Molds for Plastic Parts

The molds for intricate plastic parts are difficult to design and produce, but the processes are greatly aided by modern machining centers and turning centers that are CNC controlled. A large amount of tooling expertise and experience is required to produce molds such as that shown in Fig. 8.56. The figure illustrates a compound mold for producing the parts shown in Fig. 8.57, which are electrical equipment bus-bar bushings made from cycloaliphatic epoxy resin. Figure 8.58 pictures another complex, compound mold, used for producing the parts shown in Figure 8.59. These parts are made of clear acrylic plastic and are shown assembled into an electrical power interruption unit known as an "arc-chute." The arc-

Figure 8.56   A complex, compound mold for the parts shown in Fig. 8.57.

chute assemblies are parts of a three-phase high-voltage inter-rupter switch used in the electrical power distribution industry. Refer to the tool-steel selection section of this chapter (Sec. 8.2) to determine the types of tool steels generally used to produce such molds.

8.12   The Five Major Rules of Die Making

Tooling practices and die making are exacting arts and sciences. The tool design engineer and tool and die maker require great skill, knowledge, and experience. Several designs may be possible for any particular tooling and die-making problem, but there is usually only one solution that is best for the given conditions and tooling objectives. With this in mind, the following five major tooling and die-making rules will help the new tool designer and die maker to achieve his or her goals:

1 Define the stamping or die-making problem correctly.

2. Select the design that is best for all conditions.

Figure 8.57 Plastic parts (cycloaliphatic epoxy) made from the mold shown in Fig. 8.56.

Figure 8.58 A complex, compound mold for the acrylic plastic parts shown in Fig. 8.59.

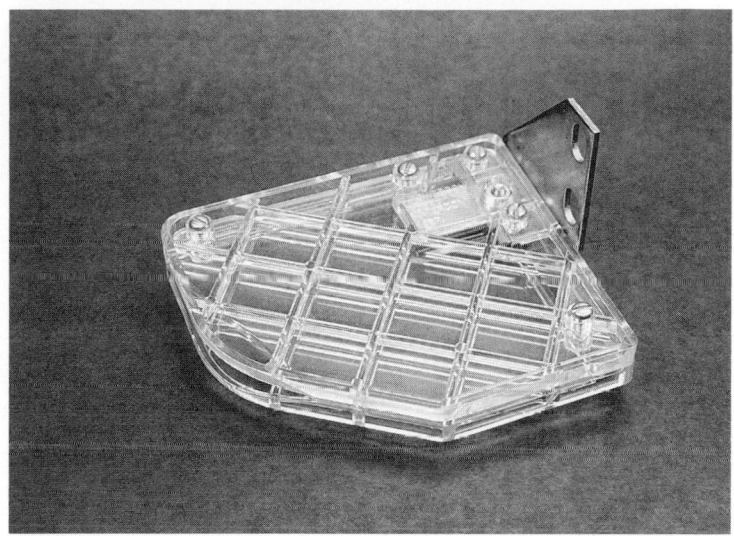

Figure 8.59   Acrylic plastic parts made from the mold shown in Figure 8.58.

3. Permit the fabrication of the die parts without difficulty.

4. The finished parts of the die must be assembled easily by the die maker.

5. The tool or die must function well and solve the stamping, tool, or die problem exactly as intended.

Clean and precise working drawings of the die and all its parts must be produced by the tool engineering department, and the die maker must transform these drawings into an accurate working die using the information provided by the tooling engineer and the die maker's own experience and skill.

The production of functional and accurate parts, through tooling, is the beginning of and the most important part of a sometimes long and complex process that determines the success of the finished product.

Sheet Metal Practices and Layout

The branch of metalworking known as sheet metal comprises a large and important element. Sheet metal parts are used in countless commercial and military products. Sheet metal parts are found on almost every product produced by the metalworking industries throughout the world.

Sheet metal gauges run from under 0.001 to 0.500 in. Hot-rolled steel products can run from ½ in thick to no. 18 gauge (0.0478 in) and still be considered as "sheet." Cold-rolled steel sheets generally are available from stock in sizes from no. 10 gauge (0.1345 in) down to no. 28 gauge (0.0148 in). Other sheet thicknesses are available as special-order "mill run" products when the order is large enough. Large manufacturers who use vast tonnages of steel products such as the automobile makers, switch-gear producers, and other sheet metal fabricators may order their steel to their own specifications (composition, gauges, and physical properties).

The steel sheets are supplied in flat form or rolled into coils. Flat-form sheets are made to specific standard sizes unless ordered to special nonstandard dimensions.

9.1 Carbon and Low-Alloy Steel Sheets

Carbon steel sheets and coils are produced in the following grades or classes:

Hot rolled

- Low carbon (commercial quality)
- Pickled and oiled
- 0.40/0.50 carbon
- Abrasion resistant
- Hi-Form (A715), high strength/low alloy (grades 50 and 80)
- A607 specification, high strength/low alloy (INX 45, INX 50, ExTen 50)
- A606 specification, high strength/low alloy (Cor-Ten)

Note: The code numbers indicate the yield strength of the high strength/low-alloy steels; i.e., INX 45 = 45,000 psi yield; grade 80 = 80,000 psi yield.

Cold rolled

- Low carbon (commercial quality)
- Special killed (drawing quality)
- Auto prototype (special killed drawing quality)
- Vitreous enameling
- Plating quality
- Stretcher leveled

The applications for the previously listed sheet steels are as follows:

Hot-rolled applications

Low carbon (commercial quality). Conforms to American Society for Testing and Materials (ASTM) A569 and is used for tanks, barrels, farm implements, and other applications where surface quality is not critical or important.

Pickled and oiled. Conforms to ASTM A569 and is used for automotive parts, switch gear, appliances, toys, and other applications where a better surface quality is required and paint and enamel adhere well. Carbon content is 10 percent maximum, and this material may be formed easily and welded the same as low-carbon sheet (commercial quality).

0.40/0.50 carbon. Has 50 percent more yield strength and abrasion resistance than low-carbon sheets. May be heat-treated for more strength and hardness. Used for scrapers, blades, tools, and other applications requiring a strong, moderate-cost steel sheet.

Abrasion resistant. Medium carbon content and higher manganese greatly improve resistance to abrasion. Brinell hardness = 210 minimum. Uses include scrapers, liners, chutes, conveyors, and other applications requiring a strong, abrasion-resistant steel sheet. Formability is moderate.

A607 specification. Lowest-cost low-alloy steel sheet. Low carbon content ensures good formability. Excellent weldability. Typical uses include utility poles, transmission towers, automotive parts, truck trailers, and other applications requiring a low-cost, high strength alloy steel sheet.

A606 specification. Five times more resistant to atmospheric corrosion than low-carbon steel. Excellent weldability and formability.

A715 specification. Fine-grained columbium-bearing series of high-strength steel. Enhanced bending and forming properties. Tough and fatigue resistant, with excellent weldability using all welding processes. Yield-point levels range from 40,000 to 80,000 psi.

Cold-rolled applications

Low carbon (commercial quality). Produced with a high degree of gauge accuracy and uniform physical characteristics. Excellent surface for painting (enamel or lacquer). Good for stamping and moderate drawing applications. Improved welding and forming characteristics, with uses such as household appliances, truck bodies, signs, panels, and many other applications.

Special killed (drawing quality). Used for severe forming and drawing applications. Freedom from age hardening and fluting. Conforms to ASTM A365 specification.

Auto prototype (special killed drawing quality). Used for prototype work and other deep drawing applications. Closely controlled gauge thickness with better tolerances compared with commercial-quality grades.

Vitreous enameling. Cold rolled from commercially pure iron ingots for porcelain-enameled products. Textured surface and suitable for

forming and moderate drawing applications and flatwork. Conforms to ASTM A424, grade A.

Plating quality. Two finishes are provided that are suitable for most plating applications: commercial bright and extralight matte.

Stretcher leveled. Uniform, high-quality matte sheets, further processed by stretching to provide superior flatness. Furnished resquared or not resquared. Resquared sheets have the stretching gripper marks removed. Used in the manufacture of table tops, cabinets, truck body panels, partitions, templates, and many other applications. Conforms to ASTM A336 specification.

Galvanized sheet and coil. The galvanic coating is zinc, and it is applied to standard steel sheets or coils in two basic methods: hot-dipped galvanized and electrogalvanized. The hot-dipped galvanized processes are known as *Ti-Co galvanized, galvanized bonderized, galvannealed, galvannealed A,* and *hot-dipped galvanized.* Galvanizing specifications are found in ASTM A526 and A527. Some of the hot-dipped sheets may have 1.25 oz zinc per square foot of surface area and others a lighter deposit.

Electrogalvanized sheets are cold-rolled steel sheets coated with zinc by electrolytic deposition and conform to ASTM A591. These sheets should be painted if they are to be exposed to outdoor conditions. These sheets can be formed, rolled, or stamped without flaking, peeling, or cracking of the zinc coating. These galvanized sheets have the same gauge thickness as cold-rolled sheets. Applications include cabinets, signs, light fixtures, and others where an excellent finish is required. Coating weight is typically 0.1 oz/ft^2, or each side is 0.00008 in thick. Trade names include Paint-Lok, Bethzin, Gripcoat, Lifecote 1, Weirzin Bonderized, and others.

Aluminized and long-terne sheets. Sheet steel is also aluminized and produced in long-terne sheets. Aluminized steel sheet is hot-dip coated on both sides with aluminum-silicon alloy by the continuous method. Strong and corrosion resistant, aluminized sheet is also inexpensive. The aluminum coating is typically 0.001 in thick on both sides, or 0.40 oz/ft^2. Aluminized sheet conforms to the ASTM A463 specification. Applications include dry kiln fan walls, dryers, incinerators, mufflers, and oven and space-heater components. Long-terne sheet is a soft steel coated with an 85 percent lead and 15 percent tin alloy for maximum ease in soldering. These long-terne sheets conform to ASTM A308 and are used for soldered

tanks, automotive accessories, hood and radiator work, and many other stamped and formed products.

As can be seen from the preceding descriptions of sheet steels that are available commercially, the selection of a particular steel for a particular sheet metal application is relatively easy. Not only are there a great number of different sheet metal stocks available, but special sheet steels may be ordered to your specifications when quantities are large enough to justify their production by American steel makers.

9.2 Nonferrous Sheet Metal

The nonferrous sheet metals include aluminum and aluminum alloys, copper and copper alloys, magnesium alloys, titanium alloys, and other special alloys. See Chap. 4, "Materials and Their Uses," for data and specifications on the nonferrous as well as the ferrous materials that are specified by the American Society for Testing and Materials (ASTM) and the Society of Automotive Engineers (SAE). Supplier catalogs are also available from companies such as Ryerson, Vincent, Atlantic, Alcoa, Reynolds, Anaconda, Chase, and others, and they may be selected from an industrial supplier master index such as the *Thomas Register of American Manufacturers.*

9.3 Machinery for Sheet Metal Fabrication

Some of the typical machinery found in a large manufacturing plant for processing and producing sheet metal parts includes

- Shears, hydraulic and squaring

- Press brakes

- Leaf brakes

- Roll-forming machines

- Automatic [computer numerical controlled (CNC)] multistation punch presses

- Single-die punch presses, strippit and unipunch setups

- Slitting machines

- Stretcher-bending machines

- Hydropresses (Marforming presses, Martin-Marietta Corp.)

- Pin routers

- Yoder hammers

- Spin forming machines

- Tumbling and deburring machines

- Sand-blasting equipment

- Explosive forming facilities

- Ironworkers (for structural shapes)

The designer and tool engineer should be familiar with all machinery used to manufacture parts in a factory. These specialists must know the limitations of the machinery that will produce the parts as designed and tooled. Coordination of design with the tooling and manufacturing departments within a company is essential to the quality and economics of the products that are manufactured. Our modern machinery has been designed and is being improved constantly to allow us to manufacture quality products at affordable prices to consumers. Medium- to large-sized companies can no longer afford to manufacture products whose quality standards do not meet the demands and requirements of the end user.

9.3.1   Modern sheet metal manufacturing machinery

The processing of sheet metal begins with the hydraulic shear, where the material is squared and cut to size for the next operation. Figure 9.1 shows a typical hydraulic shear with the capacity to do a 120-in cut in up to no. 7 gauge steel sheet. These types of machines are the "workhorses" of the typical sheet metal department because all operations on sheet metal parts start at the shear.

Figure 9.2 shows a Wiedemann Optishear, which shears and squares sheet metal to a high degree of accuracy. Blanks used in blanking, punching, and forming dies are produced on this machine, as are other flat and accurate pieces that proceed to the next stage of manufacture.

The flat, sheared sheet metal parts then may be routed to punch presses, where holes of various sizes and patterns are produced. Figure 9.3 shows a medium-sized CNC multistation turret punch press, which is both highly accurate and very fast. (See Chap. 1, "Modern Metalworking Machinery and Measuring Devices," for other types of machinery used by the metalworking industry.)

Figure 9.1 The hydraulic shear used to cut sheet metal.

Figure 9.2 A highly accurate squaring shear that is computer numerically controlled (CNC).

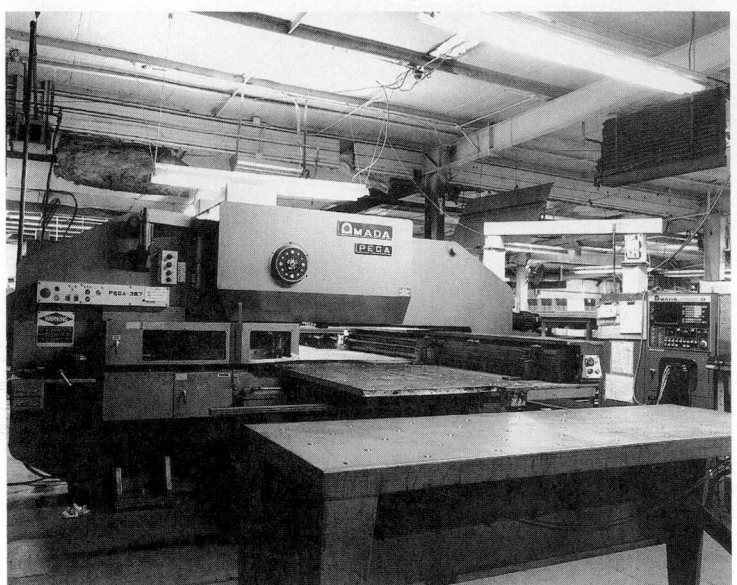

Figure 9.3 The Amada Pega high-speed CNC multistation punch press.

After a sheet metal part is sheared and punched, it may require a press-brake operation to form flanges or produce hemmed edges. Figure 9.4 shows a small press brake that has a digital readout for the "back-gauge" dimension. A small press brake such as this may be used for bending small sheet metal parts, flat bar stock, and copper and aluminum bus bars.

To quickly check a sheet metal gauge, either ferrous or nonferrous, a tool similar to that shown in Fig. 9.5 is used frequently in the sheet metal department. The tool gauge shown in the figure is for nonferrous metals such as stainless steel sheets and nonferrous wires.

9.4 Gauging Systems

To specify the thickness of different metal products such as steel sheets, wire, strip and tubing, music wire, and others, a host of gauging systems were developed over the course of many years. Shown in Fig. 9.6 are the common gauging systems used for commercial steel sheets, strip and tubing, and brass and steel wire. The steel sheets column in the figure lists the gauges and equivalent thicknesses

Figure 9.4 A small press brake with digital back gauge.

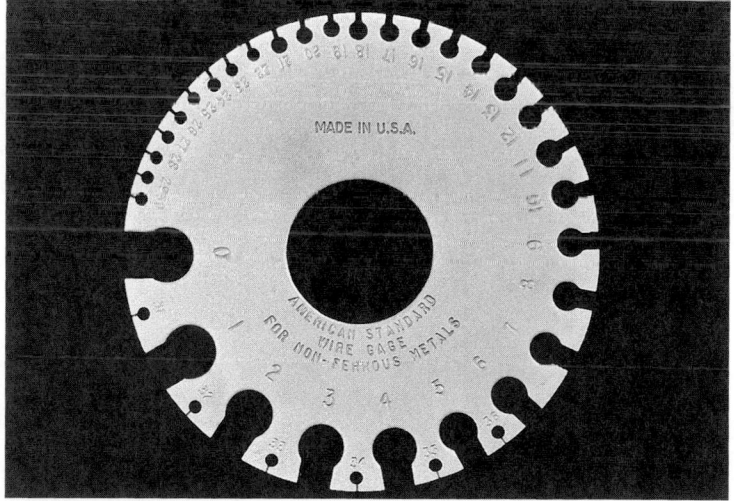

Figure 9.5 American standard wire-gauge-measuring tool.

Gauge No.	Brass (Brown & Sharpe)	Steel Sheets *	Strip & Tubing	Steel Wire Ga. ♦
6-0	0.5800	--------	--------	0.4615
5-0	0.5165	--------	0.500	0.4305
4-0	0.4600	--------	0.454	0.3938
3-0	0.4096	--------	0.425	0.3625
2-0	0.3648	--------	0.380	0.3310
0	0.3249	--------	0.340	0.3065
1	0.2893	--------	0.300	0.2830
2	0.2576	--------	0.284	0.2625
3	0.2294	0.2391	0.259	0.2437
4	0.2043	0.2242	0.238	0.2253
5	0.1819	0.2092	0.220	0.2070
6	0.1620	0.1943	0.203	0.1920
7	0.1443	0.1793	0.180	0.1770
8	0.1285	0.1644	0.165	0.1620
9	0.1144	0.1495	0.148	0.1483
10	0.1019	0.1345	0.134	0.1350
11	0.0907	0.1196	0.120	0.1205
12	0.0808	0.1046	0.109	0.1055
13	0.0720	0.0897	0.095	0.0915
14	0.0641	0.0747	0.083	0.0800
15	0.0571	0.0673	0.072	0.0720
16	0.0508	0.0598	0.065	0.0625
17	0.0453	0.0538	0.058	0.0540
18	0.0403	0.0478	0.049	0.0475
19	0.0359	0.0418	0.042	0.0410
20	0.0320	0.0359	0.035	0.0348
21	0.0285	0.0329	0.032	0.0317
22	0.0253	0.0299	0.028	0.0286
23	0.0226	0.0269	0.025	0.0258
24	0.0201	0.0239	0.022	0.0230
25	0.0179	0.0209	0.020	0.0204
26	0.0159	0.0179	0.018	0.0181
27	0.0142	0.0164	0.016	0.0173
28	0.0126	0.0149	0.014	0.0162
29	0.0113	0.0135	0.013	0.0150
30	0.0100	0.0120	0.012	0.0140
31	0.0089	0.0105	0.010	0.0132
32	0.0080	0.0097	0.009	0.0128
33	0.0071	0.0090	0.008	0.0118
34	0.0063	0.0082	0.007	0.0104
35	0.0056	0.0075	0.005	0.0095
36	0.0050	0.0067	0.004	0.0090
37	0.0045	0.0064	--------	0.0085
38	0.0040	0.0060	--------	0.0080

*= Common Commercial Standard; ♦ = Reference only

Figure 9.6   Gauging decimal systems table.

used by American steel sheet manufacturers and steel makers. This gauging system can be recognized immediately by its no. 11 gauge equivalent of 0.1196 in, which is standard today for this very common and high-use gauge of sheet steel.

Figure 9.7 is a table of gauging systems that were used widely in the past, although some are still in use today, including the American

or Brown and Sharpe system. The Brown and Sharpe system is also shown in Fig. 9.6, but there it is indicated in only four-place decimal equivalents.

9.4.1 Aluminum sheet metal standard thicknesses

Aluminum is used widely in the aerospace industry, and over the years, the gauge thicknesses of aluminum sheets have developed on their own. Aluminum sheet is now generally available in the thicknesses shown in Fig. 9.8. The fact that the final weight of an aerospace vehicle is very critical to its performance has played an important role in the development of the standard aluminum sheet gauges, wherein the strength-to-weight ratio is critical and a few thousandths of an inch extra on an aluminum sheet can mean more final weight of an aerospace vehicle. The gauges or thicknesses of other metals and alloys, together with their stock tolerances, are given in Chap. 4, "Materials and Their Uses."

9.5 Methods of Sheet Metal Fabrication

Many methods have been developed for working or fabricating sheet metal parts, including those employed to cut, punch, and form sheet metal parts.

9.5.1 Sheet metal cutting methods

Shearing. The sheet metal stock sheet is cut on a hydraulic-powered shear to its appropriate blank size or minimum size for punching or profiling the part. See Figs. 9.1 and 9.2 for hydraulic shears.

Slitting. The sheet metal stock sheet is run through a slitting machine, where accurate widths are slit with slitting knives to various lengths. The slit edges are sharp and accurate, with excellent straightness along both edges. Slit sheet metal is usually fed directly into the feed gates of punching or bending dies or roll-forming equipment. The roll-forming machine produces a specific type of cross-sectional shape to the finished part. Samples of roll-formed parts can be seen in Fig. 9.15.

Welding-process cutting and water-jet cutting. The sheet metal stock sheet may be oxyacetylene torch cut with an automatic torch cutting machine, or it may be plasma cut, arc cut, or waterjet cut.

Number of wire gauge	American or Brown & Sharpe	Birmingham or Stubs' Iron wire	Washburn & Moen, Worcester, Mass.	W. & M. steel music wire	American S. & W. Co's. music wire gauge	Stubs' steel wire	U.S. Standard gauge for sheet and plate iron and steel	Number of wire gauge
00000000				0.0083				00000000
0000000				0.0087				0000000
000000				0.0095	0.004		0.46875	000000
00000				0.010	0.005		0.4375	00000
0000	0.460	0.454	0.3938	0.011	0.006		0.40625	0000
000	0.40964	0.425	0.3625	0.012	0.007		0.375	000
00	0.3648	0.380	0.3310	0.0133	0.008		0.34375	00
0	0.32486	0.340	0.3065	0.0144	0.009		0.3125	0
1	0.2893	0.300	0.2830	0.0156	0.010	0.227	0.28125	1
2	0.025763	0.284	0.2625	0.0166	0.011	0.219	0.265625	2
3	0.22942	0.259	0.2437	0.0178	0.012	0.212	0.250	3
4	0.20431	0.238	0.2253	0.0188	0.013	0.207	0.234375	4
5	0.18194	0.220	0.2070	0.0202	0.014	0.204	0.21875	5
6	0.16202	0.203	0.1920	0.0215	0.016	0.201	0.203125	6
7	0.14428	0.180	0.1770	0.023	0.018	0.199	0.1875	7
8	0.12849	0.165	0.1620	0.0243	0.020	0.197	0.171875	8
9	0.11443	0.148	0.1483	0.0256	0.022	0.194	0.15625	9
10	0.10189	0.134	0.1350	0.027	0.024	0.191	0.140625	10
11	0.090742	0.120	0.1205	0.0284	0.026	0.188	0.125	11
12	0.080808	0.109	0.1055	0.0296	0.029	0.185	0.109375	12
13	0.071961	0.095	0.0915	0.0314	0.031	0.182	0.09375	13
14	0.064084	0.083	0.0800	0.0326	0.033	0.180	0.078125	14

15	0.057068	0.072	0.0720	0.0345	0.035	0.178	0.0703125	15
16	0.05082	0.065	0.0625	0.036	0.037	0.175	0.0625	16
17	0.045257	0.058	0.0540	0.0377	0.039	0.172	0.05625	17
18	0.040303	0.049	0.0475	0.0395	0.041	0.168	0.050	18
19	0.03589	0.042	0.0410	0.0414	0.043	0.164	0.04375	19
20	0.031961	0.035	0.0348	0.0434	0.045	0.161	0.0375	20
21	0.028462	0.032	0.03175	0.046	0.047	0.157	0.034375	21
22	0.025347	0.028	0.0286	0.0483	0.049	0.155	0.03125	22
23	0.022571	0.025	0.0258	0.051	0.051	0.153	0.028125	23
24	0.0201	0.022	0.0230	0.055	0.055	0.151	0.025	24
25	0.0179	0.020	0.0204	0.0586	0.059	0.148	0.021875	25
26	0.01594	0.018	0.0181	0.0626	0.063	0.146	0.01875	26
27	0.014195	0.016	0.0173	0.0658	0.067	0.143	0.0171875	27
28	0.012641	0.014	0.0162	0.072	0.071	0.139	0.015625	28
29	0.011257	0.013	0.0150	0.76	0.075	0.134	0.0140625	29
30	0.010025	0.012	0.0140	0.80	0.080	0.127	0.0125	30
31	0.008928	0.010	0.0132		0.085	0.120	0.0109375	31
32	0.00795	0.009	0.0128		0.090	0.115	0.01015625	32
33	0.00708	0.008	0.0118		0.095	0.112	0.009375	33
34	0.006304	0.007	0.0104			0.110	0.00859375	34
35	0.005614	0.005	0.0095			0.108	0.0078125	35
36	0.005	0.004	0.0090			0.106	0.00703125	36
37	0.004453					0.103	0.006640625	37
38	0.003965					0.101	0.00625	38
39	0.003531					0.099		39
40	0.003144					0.097		40

Figure 9.7 Other gauging systems.

Standard Thickness, in.	Weight, lbs/sq. ft.
0.010	0.141
0.016	0.226
0.020	0.282
0.025	0.353
0.032	0.452
0.040	0.564
0.050	0.706
0.063	0.889
0.071	1.002
0.080	1.129
0.090	1.270
0.100	1.411
0.125	1.764
0.160	2.258
0.190	2.681
0.250	3.528

Weight based on an average aluminum weight of 0.098 lb/in^3.

Figure 9.8   Standard aluminum sheet metal thicknesses and weights.

All these methods are used today to cut both metallic and nonmetallic flat sheet parts.

Laser-beam cutting. The technology of laser-beam cutting of sheet metal flat stock has progressed dramatically in the past 5 years. The older laser-cutting machines produced a ragged edge, which made parts processing difficult. The newer laser-cutting machines can cut a variety of sheet metal stocks, including steel, aluminum, stainless steel, and nonmetallic parts, while achieving excellent accuracy and edge qualities. Laser cutting allows sheet metal parts to be cut to a level of accuracy that would not be possible using conventional punch presses or other equipment.

Figure 9.9 shows a laser-cutting machine. Note the CNC control panel on the right side of the illustration of the machine. Figure 9.10 shows a close-up view of the laser-cutting machine actually cutting a steel sheet metal part. A drawing of this part is shown in Fig. 9.11. The cleanly cut edges of this part are shown in Fig. 9.12. Accuracy of the laser-cut parts is excellent, as are the edge qualities.

Laser-cut parts may be processed on the laser machine by using programs that will allow coordinates of AutoCad drawings of flat sheet metal or bar stock parts to be placed into the CNC controller of the laser machine, wherein the machine will exactly duplicate

Figure 9.9   A modern laser-cutting machine with CNC controller.

Figure 9.10   Close-up view of laser-cutting machine cutting the part shown in Figs. 9.11 and 9.12.

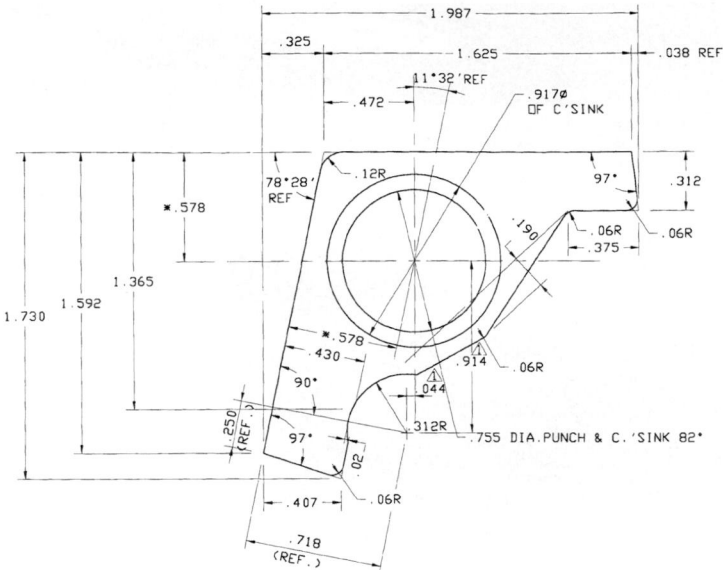

Figure 9.11   A scaled AutoCad drawing for a laser-cutting operation.

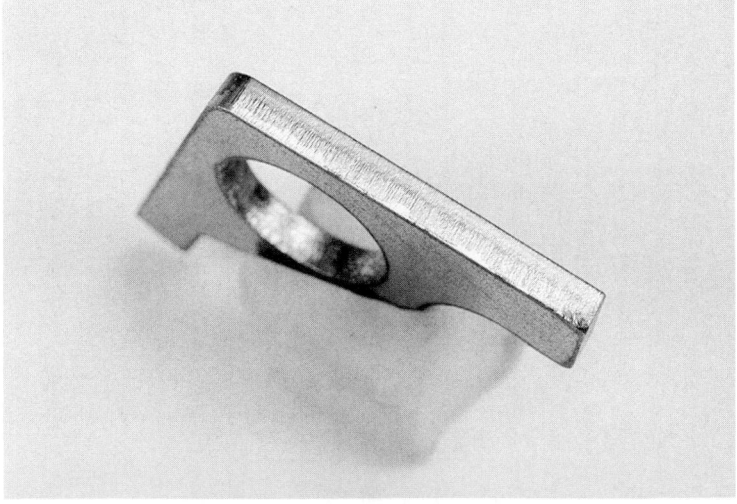

Figure 9.12   A laser-cut part based on the drawing in Fig. 9.11, showing the smooth, accurate cut.

the parts per the AutoCad drawing. The AutoCad drawings must be drawn to scale to allow this procedure to take place. Using this process, intricate sheet metal parts can be produced economically in quantities too small to justify use of a hard die. The laser-cutting machine shown in Fig. 9.9 will cut different steel sheet metal, plate, or bar stock up to 0.375 in thick. Larger laser-cutting machines can cut even thicker stock material.

Laser-cutting machines allow the design engineer and tool engineer to design and produce parts that would not have been economically possible using older methods of cutting and punching. Figure 9.13 is a drawing of a very complex flat pattern of a sheet metal part that normally would have to be produced from a hard die. The laser-cutting machine can produce this part in no.7 gauge sheet steel in about 1 minute.

Caution: You should wear protective plastic glasses when watching a laser-cutting machine in operation. Glass lenses will not block the radiation leaking from the machine. Normal plastic lenses are suitable. Laser-cutting machines display a caution warning sign in the area of the cutting head.

Other methods of cutting sheet metal. Additional processes include hard-die punching, strippit and turret-press punching, and pin

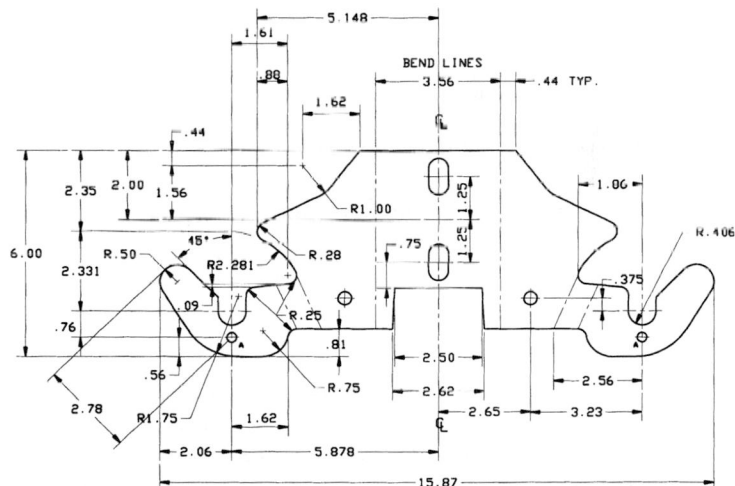

Figure 9.13 An AutoCad scale drawing for a complex laser-cut sheet metal part.

routing. Pin routing is used to cut a stack of thin aluminum alloy sheets from a pattern made from a loft. The pin-routing procedure is actually a tracer-milling operation. This practice is used in the aircraft and aerospace industries.

9.5.2 Sheet metal punching methods

Turret punch. The sheet metal stock sheet or blank is punched with various holes of different shapes by the punching dies contained on the punch press. Punch presses usually have a revolving turret that contains punches of different sizes and shapes, and these are interchangeable. The modern multistation punch presses are often CNC controlled and high speed.

Die punch. The sheet metal blank is placed in a punching die, where a pattern of holes is punched simultaneously with one stroke of the punching press. The high-tonnage brake press is often used to provide the power stroke required on the punching die block.

Strippit punch. The sheet metal blank is punched, one hole at a time, on a strippit punch press. The punching dies may be changed quickly for punching different sizes or shapes of holes, such as round, square, rectangular, oval, or obround, or other special shapes for which the dies are designed.

9.5.3 Sheet metal forming methods

Press brake. This type of machine tool is found in every sheet metal department and is used to bend flanges, hems, and other special shapes. Figure 9.14 shows the various bending abilities of the press brake when equipped with the proper tooling.

Die bending, forming, and molding. Hard dies are produced for making bends and molds and for forming and drawing sheet metal parts. The forming or drawing dies may be all metal or a combination of metal and neoprene pads, which force the metal against the die block or male form. This process is used widely in the aerospace industry, where aluminum sheet metal parts are formed on hydropresses. (Marforming is a hydropressing process originated at the Martin Company, Middle River, MD.) Large lead-alloy form blocks are also used in the aircraft industry to form large, compound curved surfaces in sheet metal, generally aluminum alloys.

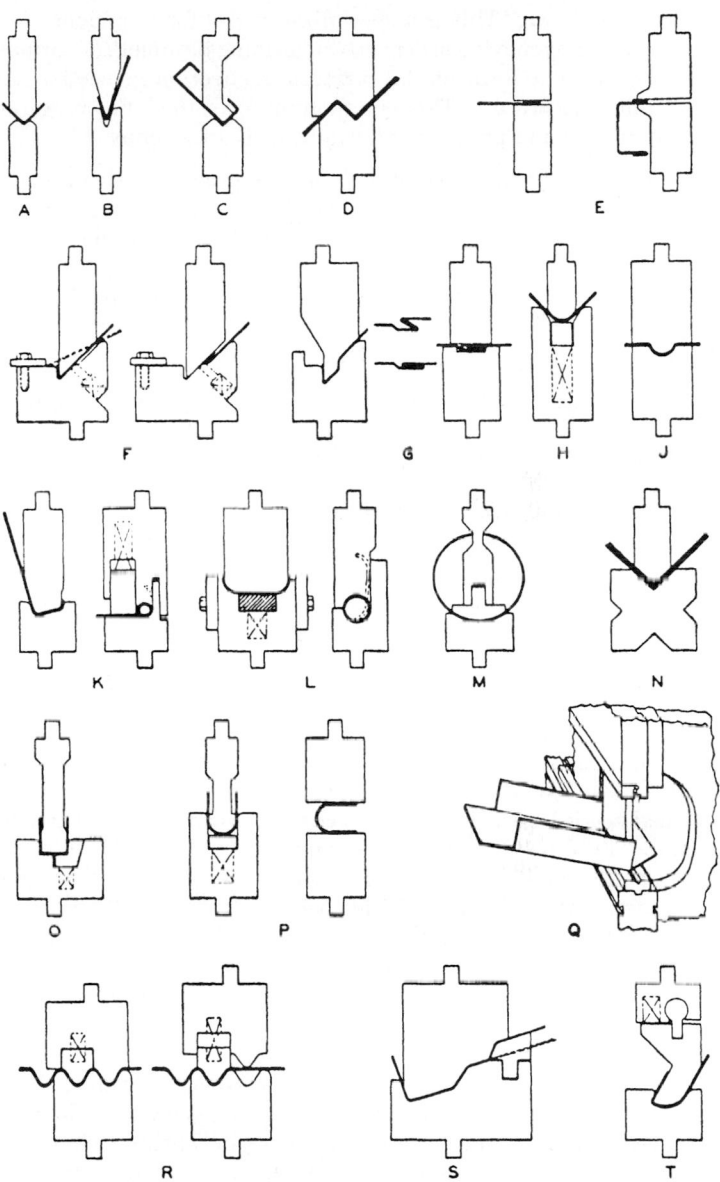

Figure 9.14 Press brake bending dies showing bending capabilities of the press brake.

Yoder hammering. This is a specialized metal forming operation where a rapidly moving set of vertical forming hammers of various shapes is used to form special surfaces on sheet metal sections in a hammering process. This is a manual operation that requires operator skill and practice and is relatively rare today.

Spin forming. In this sheet metal forming process, a sheet metal disk is rotated in a special type of lathe tool while a special forming tool is pressed against the rotating disk of metal in a rotary swaging operation, thus forming or spin-drawing the part to the required shape. The process is limited to metal sections that are of a symmetric rotated section of revolution, such as bell shapes, cones, parabolic sections, cylinders, etc.

Explosive forming. In this process of metal forming, a shaped charge of explosive is suspended above a sheet metal flat pattern, both of which are submerged under water. When the explosive charge is detonated, the shock waves from the explosion exert a hydrostatic pressure against the sheet metal, forcing it against a forming die block almost instantaneously. The sheet metal part conforms to the shape of the die block. Very complex sheet metal parts may be formed with this process. Some companies in the United States are devoted entirely to this specialized form of sheet metal fabrication. This process dates back to the 1950s, when it was used by the Martin Company, Middle River, MD, in its aircraft manufacturing facilities.

Stretcher bending. This process for sheet metal forming uses a machine known as a *stretcher bender,* wherein a sheet metal formed section such as an angle, channel, or Z section is pulled against a radiused die block to accurately form structural frame sections to a specified radius of curvature. The frame structures of rockets and aerospace vehicles, which have a single radius of curvature, are formed on the stretcher-bending machine. This is the only cost-effective method known for producing such sheet metal frame parts quickly and accurately on a relatively simple machine. In this operation, the straight frame section that was press-brake bent is held in a set of grippers at each end of the part. The part is stretched and pulled against the radiused die block simultaneously by the gripper arms, thus forming the part to the specified radius. Allowance is made for "springback" of the formed part by overbending and then allowing the metal to spring back or return to the correct form.

Roll forming. In the roll-forming process, a flat strip of sheet metal is fed into the roll-forming machine, which has a series of rolling dies whose shape gradually changes as the metal is being fed past each stage of rolls until the final roll-formed section is completed. The number of different cross sections of roll-formed sheet metal parts is limitless. The roll-formed part is usually made to a specific length or stock length or may be produced to any special length required. Figure 9.15 is a sample page of roll-formed sections taken from the Dahlstrom catalog of molded and rolled sections (Dahlstrom Manufacturing Corporation, Jamestown, NY).

A sample of sheet metal parts that have been press-brake bent and die formed may be seen in photographs shown in Chap. 8, "Tooling, Die Making, Molds, Jigs, and Fixtures."

9.6   Sheet Metal Flat Patterns

The correct determination of the flat-pattern dimensions of a sheet metal part that is formed or bent is of prime importance to sheet metal workers, designers, and design drafters. There are three methods for performing the calculations to determine flat patterns that are considered normal practice. The method chosen also can determine the accuracy of the results. The three common methods employed for doing the work include

- By bend deduction (BD) or setback
- By bend allowance (BA)
- By inside dimensions or inside mold line (IML) for sharply bent parts only

Other methods are also used for calculating the flat-pattern length of sheet metal parts. Some take into consideration the ductility of the material, and others are based on extensive experimental data for determining the bend allowances. The methods included in this section are accurate when the bend radius has been selected properly for each particular gauge and condition of the material. When the proper bend radius is selected, there is no stretching of the *neutral axis* within the part (the neutral axis is generally accepted as being located $0.445 \times$ material thickness inside the inside mold line (IML) (see Fig. 9.17).

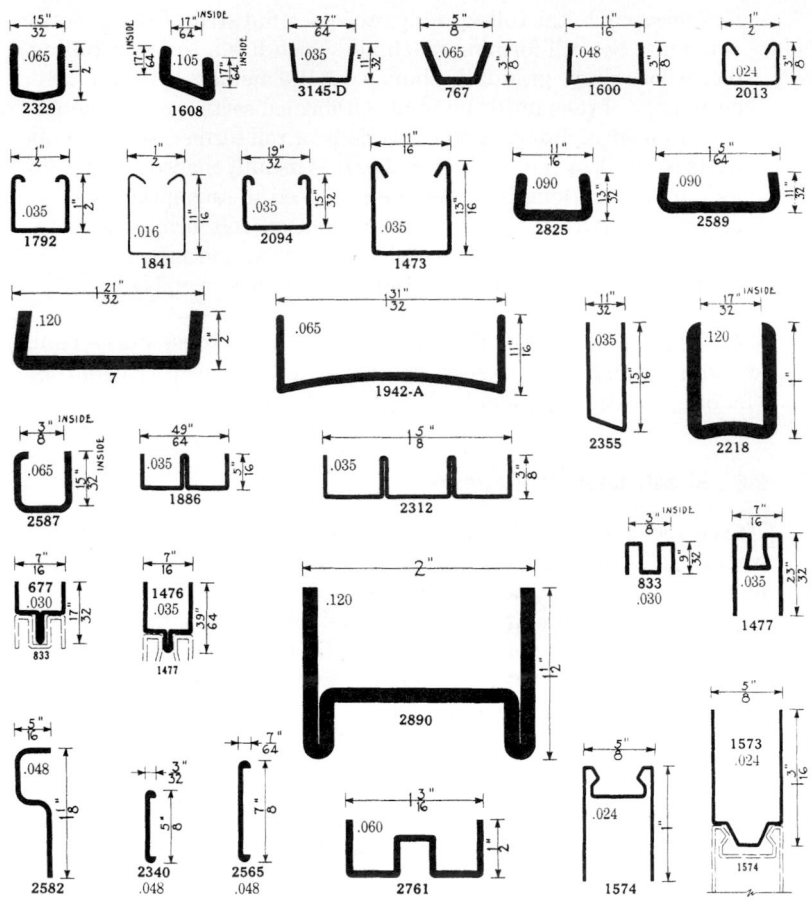

Figure 9.15 Samples of roll-formed sections of sheet metal.

Methods of determining flat patterns.   Refer to Fig. 9.16.

Method 1: By bend deduction (BD) or setback

$$L = a + b - \text{setback}$$

Method 2: By bend allowance (BA)

$$L = a' + b' + c$$

where c is the bend allowance or length along the neutral axis.

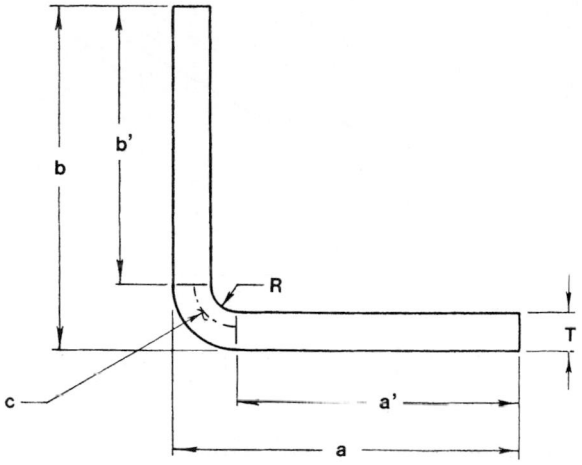

Figure 9.16 Sample of a bent sheet metal angle.

Method 3: By inside dimensions or inside mold line (IML)

$$L = (a - T) + (b - T)$$

Calculation of bend allowance and bend deduction (setback) is keyed to Fig. 9.17 and is as follows:

$$\text{Bend allowance (BA)} = A(0.01745R + 0.00778T)$$

$$\text{Bend deduction (BD)} = (2 \tan \tfrac{1}{2}A)(R + T) - (\text{BA})$$

$$X = (\tan \tfrac{1}{2}A)(R + T)$$

$$Z = T(\tan \tfrac{1}{2}A)$$

$$Y = X - Z \text{ or } R(\tan \tfrac{1}{2}A)$$

On "open" angles that are bent less than 90° (Fig. 9.18),

$$X = (\tan \tfrac{1}{2}A)(R + T)$$

The method used to calculate the sheet metal flat pattern may be determined by designer option, company standards, order of accuracy, and method of manufacture. For soft steel (1010, etc.), the inside dimension method (method 3) is used whenever the material may be bent with a sharp or minimal inside bend radius (0.062 in or less). The inside bend method is accurate enough for all gauges

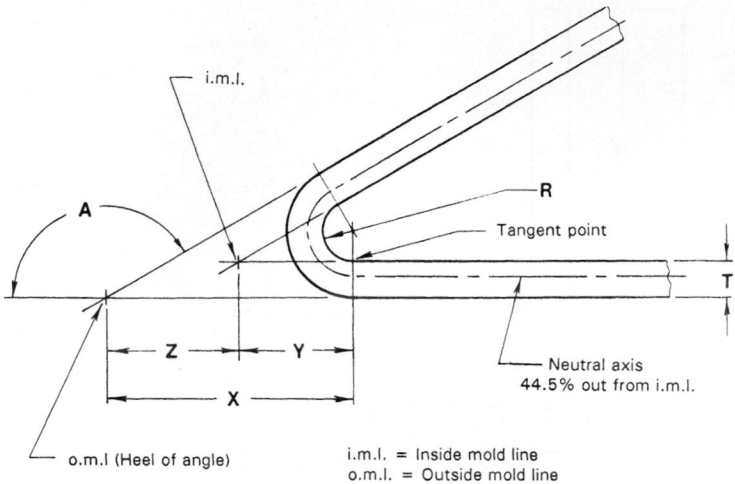

Figure 9.17 Geometry of a bend angle.

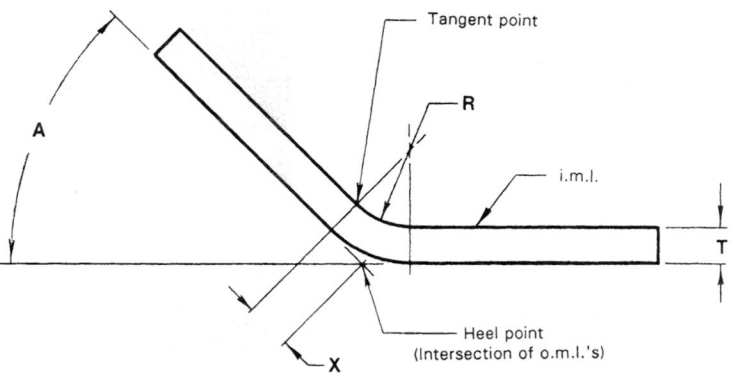

Figure 9.18 An open angle.

up to and including 0.375-in-thick stock where the tolerance of bent parts is ±0.032 in. On stock thicknesses from no. 7 gauge to 0.375 in, 0.062 in is added to the sum of the inside bend dimensions and divided across the bend, 0.032 in going into each leg or flange inside dimension. The 0.062-in allowance is added for each 90° bend in the part. This method is popular in such industries as the electrical switch-gear industry, the appliance industry, and others where great accuracy is not required.

For very accurate flat-pattern dimensions intended for aerospace vehicles, automotive work, appliances, and other consumer products, the bend deduction or setback method is used in *lofting* procedures and standard sheet metal tooling drawings. The tooling department is usually responsible for generating the flat-pattern drawings of parts produced with stamping dies, punching dies, and drawing dies. The engineering department is usually responsible for generating the lofting drawings for flat patterns of regular and irregular shapes. In lofting, the part is drawn very accurately in flat pattern on flat metal sheets with specially prepared surfaces or directly onto heavy Mylar drafting film. The loft, or drawing, is then photographed, and the pattern is transferred to another metal sheet in full scale.

The part is then accurately cut out and becomes a master pattern. A stack of sheets then can be pin-routed or tracer-milled using the master template as a guide or jig. The cutout parts then are sent to the forming dies or the brakes for the final bending or forming operations. With modern CNC equipment, the part outline may be programmed and then cut automatically on the appropriate machine prior to the bending or forming operations.

When the sheet metal part is to be press-brake bent only on radiused bending dies, the bend allowance method can be used to calculate the flat pattern whenever accuracy and a specific inside bend radius are required. Bend deduction or setback also may be used in this case.

9.6.1 Setback or J chart for determining bend deductions

Figure 9.19 shows a form of bend deduction (BD) or setback chart known as a *J chart*. You may use this chart to determine bend deduction or setback when the angle of bend, material thickness, and inside bend radius are known. The chart in the figure shows a sample line running from the top to the bottom and drawn through the ³⁄₁₆-in radius and the material thickness of 0.075 in. For a 90° bend, read across from the right to where the line intersects the closest curved line in the body of the chart. In this case, it can be seen that the line intersects the curve whose value is 0.18. This value is then the required setback or bend deduction for a bend of 90° in a part whose thickness is 0.075 in with an inside bend radius of ³⁄₁₆-in. If we compute this setback or bend deduction value using the appropriate equations shown previously, we can check the value given by the J chart.

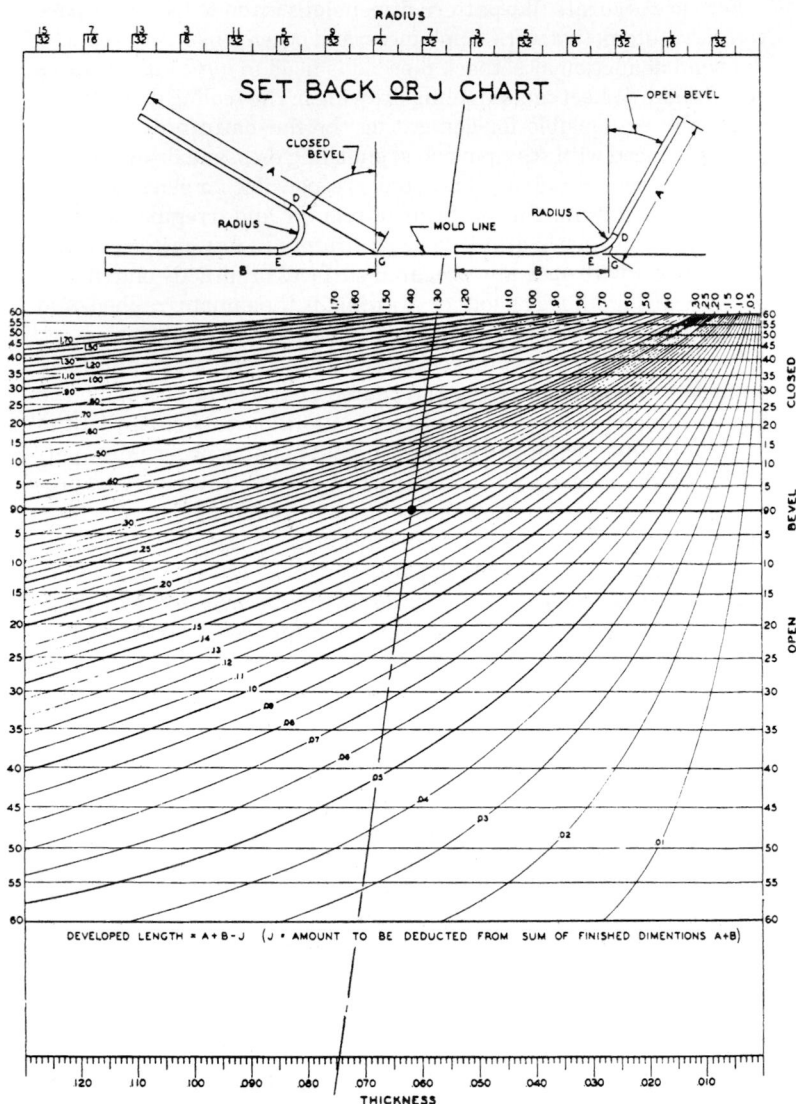

Figure 9.19 Setback or J chart.

Checking: Bend deduction (BD) or setback is given as

Bend deduction or setback = $(2 \tan \frac{1}{2}A)(R + T) - (BA)$

We first must find the bend allowance (BA) from

Bend allowance = $A(0.01745R + 0.00778T)$

$$= 90(0.01745 \times 0.1875 + 0.00778 \times 0.075)$$

$$= 90(0.003855)$$

$$= 0.34695$$

Now, substituting the bend allowance of 0.34695 in into the bend deduction equation yields

Bend deduction or setback = $[2 \tan \frac{1}{2}(90)](0.1875 + 0.075) - 0.34695$

$$= 0.178 \text{ or } 0.18, \text{ as shown in the chart}$$
(Fig. 9.19)

The J chart in Figure 9.19 is thus an important tool for determining the bend deduction or setback of sheet metal flat patterns without recourse to tedious calculations. The accuracy of this chart has been shown to be of a high order. This chart as well as the equations were developed after extensive experimentation and practical working experience in the aerospace industry.

9.6.2   Bend radii for aluminum-alloy and steel sheet (average)

Figures 9.20 and 9.21 show average bend radii for various aluminum alloys and steel sheets.

9.7   Sheet Metal Developments and Transitions

The layout of sheet metal as required in "development and transition" parts is an important phase of sheet metal design and practice. The methods included here will prove useful in many design and working applications. These methods have application in duct work, aerospace vehicles, automotive equipment, and other areas of product design and development requiring the use of transitions and developments.

When sheet metal is to be formed into a curved section, it may be laid out, or *developed,* with reasonable accuracy by triangulation if it

Material Gauge	Steel Designation	
	AISI 1020	302-303-304S/S
0.010	0.031	0.031
0.020	0.031	0.031
0.030	0.031	0.031
0.040	0.031	0.031
0.050	0.031	0.031
0.060	0.031	0.062
0.070	0.031	0.062
0.080	0.031	0.062
0.090	0.062	0.062
0.120	0.062	0.125
0.190	0.125	0.250
0.250	0.125	0.250

Figure 9.20 Bend radius table for sheet metal.

Material Gauge	Aluminum Alloy Designation		
	6061-T6 5052-H36 1100-H18	5052-H22 3003-H14	2024-T3
0.010	0.062	0.031	0.062
0.020	0.062	0.031	0.062
0.030	0.062	0.031	0.125
0.040	0.125	0.031	0.250
0.050	0.125	0.031	0.250
0.070	0.250	0.062	0.250
0.080	0.250	0.062	0.375
0.090	0.375	0.125	0.375
0.120	0.375	0.125	0.500
0.190	0.750	0.250	0.750
0.250	1.000	0.500	1.000

Figure 9.21 Bend radius table for aluminum sheet.

forms a simple curved surface without compound curves or curves in multiple directions. Sheet metal curved sections are found on many products, and if a straight edge can be placed flat against elements of the curved section, accurate layout or development is possible using the methods shown in this section.

On double-curved surfaces such as are found on automobile and truck bodies and aircraft, forming dies are created from a full-scale model in order to duplicate these compound curved surfaces in sheet metal. The full-scale models used in aerospace vehicle manufacturing facilities are commonly called *mock-ups,* and the models used to transfer the compound curved surfaces are made by tool makers in the tooling department.

9.8   Developing Flat Patterns

Developing flat patterns can be done by bend deduction or setback. Figure 9.22 shows a type of sheet metal part that may be bent on a press brake. The flat-pattern part is bent on the brake, with the center-of-bend line (CBL) held on the bending die centerline. The machine's back gauge is set by the operator in order to form the part. If you study the figure closely, you can see how the dimensions progress: The bend deduction is drawn in, and the next dimension is taken from the end of the first bend deduction. The next dimension is then measured, the bend deduction is drawn in for that bend, and then the next dimension is taken from the end of the second bend deduction, etc. Notice that the second bend deduction is larger because of the larger radius of the second bend (0.16R).

9.9   Stiffening Sheet Metal Parts

On many sheet metal parts that have large areas, stiffening can be achieved by creasing the metal in an X configuration by means of brake bending. On certain parts where great stiffness and rigidity are required, a method called *beading* is employed. The beading is carried out at the same time as the part is being hydropressed, Marformed, or hard-die formed. See Fig. 9.23b and Chap. 8, "Tooling, Die Making, Molds, Jigs, and Fixtures," for more data on beading sheet metal parts.

Another method for stiffening the edge of a long sheet metal part is to hem or "Dutch bend" the edge, as shown in Fig. 9.23a. In aerospace and automotive sheet metal parts, flanged lightening holes are used as shown in Fig. 9.23c. The lightening hole makes the part not only lighter in weight but also more rigid. This method is used commonly in wing ribs, airframes, and gussets or brackets. The lightening hole need not be circular but can take any convenient shape as required by the application.

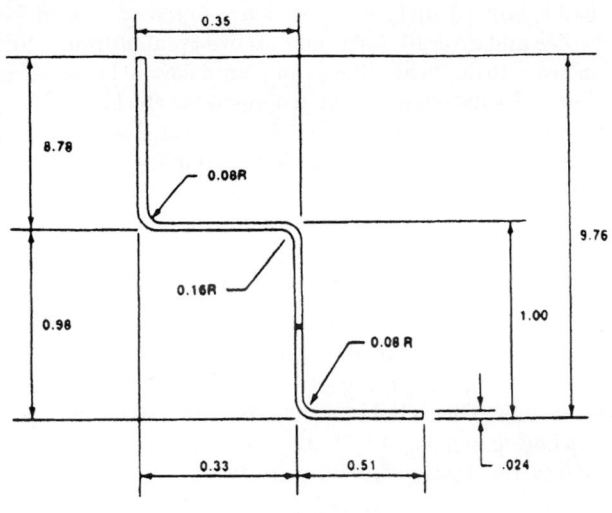

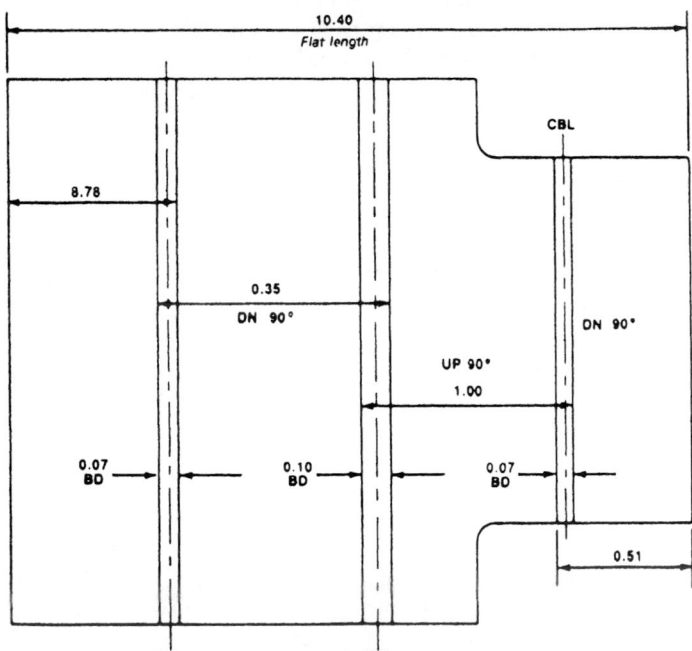

Figure 9.22 Layout of bent sheet metal by the bend deduction method.

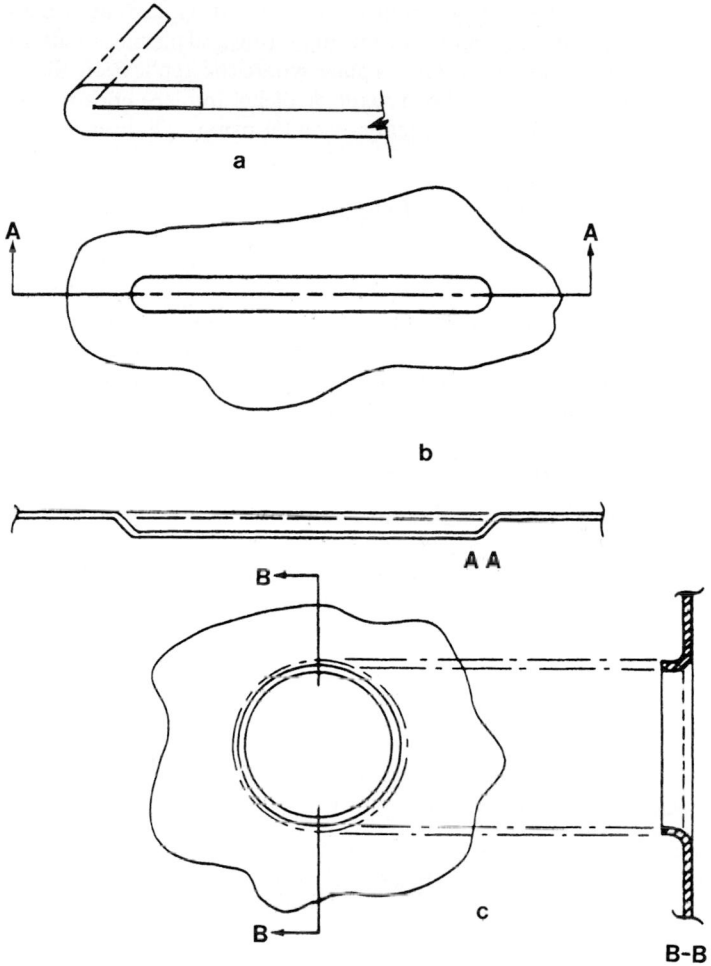

Figure 9.23 (*a*) A hemmed edge. (*b*) A stiffening bead. (*c*) A lightening hole.

9.10   Sheet Metal Faying Surfaces

Faying surfaces are those where two sheet metal parts come into contact with each other, such as the joining of a flange with a web section or the joining of two flanges. All faying surfaces should be primer painted or finished in some manner to prevent corrosion. A standard finish for many applications is zinc chromate primer.

Other finishes include zinc plating, nickel plating, cadmium plating, chrome plating, etc. Since cadmium plating and plating solutions are toxic, zinc plating is used in its place where the application allows this. New designs should specify zinc or nickel in place of cadmium unless there is a specific technical reason for using cadmium.

9.11   Design Points for Sheet Metal Parts

■ Do not specify flanges that are too narrow for the type of bending operation. That is, do not design a part with a 0.375-in flange width unless you have the equipment or dies to produce such a short flange on the press brake or other type of machine. Bottoming dies can enable the bending of such a flange in very light gauges up to no. 11 gauge.

■ Do not specify gauges that are heavier than necessary for the function of the part. An exception to this rule is code specifications, such as for electrical switch-gear equipment, where a minimum of no. 11 gauge steel is specified for certain sections of the equipment whether it is structurally required or not. The heavy gauge limit in this application is for the prevention of the spread of fires between adjacent units if there is an electrical fault.

■ Use proper edge distances for hardware components, keeping the flanges as narrow as practical with respect to their required rigidity or strength.

■ Do not design a part that is impractical to bend on the type of machinery with which your operation is equipped.

■ Keep brake-formed parts in a size range where the parts can be handled manually by the brake operators, unless your operation is equipped with automatic machinery or other special-handling equipment.

■ When using hot-rolled steel sheets, use tolerances on your parts that are functionally related to the equipment used in your operation. That is, do not expect to hold dimensions to $\pm\frac{1}{64}$ in when you are producing no. 11 gauge hot-rolled sheet metal parts on a press brake. Normal shop tolerance on general sheet metal parts is usually $\pm\frac{1}{32}$ in for parts under 3 ft in length in both directions. One of the reasons that hot-rolled steel sheet metal parts between nos. 16 and 7 gauges need a generous tolerance is that the steel sheet metal thickness varies from sheet to sheet. This

variation can reach $\pm\frac{1}{2}$ a gauge step on a batch of steel of the same gauge. The variation of the gauge causes the flanges to be bent to different outside dimensions even though the back-gauge dimension is the same and is very apparent when more than two bends are made on the same part.

In high-speed, mass-production operations there is no remedy for this condition of variable thickness within the same gauge of hot-rolled steel sheet material except a favorable tolerance spread on the dimensions of the parts. Some of the larger companies specify the gauge-variation limits when they order steel sheet metal, but the mill order must be large in order to do this. These problems do not occur when using gauge-accurate cold-rolled steel sheets.

9.12   Typical Transitions and Developments

The following transitions and developments are the most common types, and learning or using them for reference will prove helpful in many industrial applications. Using the principles shown will enable you to apply these to many different variations or geometric forms. The principles shown and described in Sec. 9.7 will prove helpful in trying to understand and put into practice the transitions and developments shown in this section. Construction of the true-length diagrams is of particular importance.

9.12.1   To develop a truncated right cylinder

Refer to Fig. 9.24. The development of a cylinder is similar to the development of a prism. Draw two projections of the cylinder:

1. A normal view of a right section

2. A normal view of the elements

In rolling the cylinder out on a tangent plane, the base or right section, being perpendicular to the axis, will develop into a straight line. For convenience in drawing, divide the normal view of the base, shown here in the bottom view, into a number of equal parts by points that represent elements. These divisions should be spaced so that the chordal distances approximate the arc closely enough to make the stretch-out practically equal to the periphery of the base or right section.

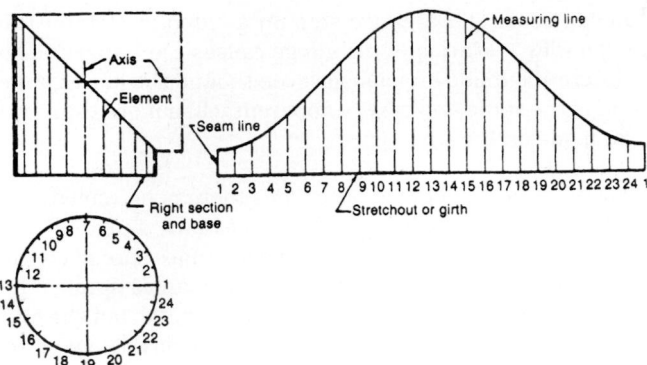

Figure 9.24   A truncated right cylinder.

Project these elements to the front view. Draw the stretch-out and measuring lines, the cylinder now being treated as a many-sided prism. Transfer the lengths of the elements in order, either by projection or by using dividers, and join the points thus found by a smooth curve. Sketch the curve in very lightly, freehand, before fitting the French curve or ship's curve to it. This development might be the pattern for one-half of a two-piece elbow.

Three-piece, four-piece, and five-piece elbows may be drawn similarly, as illustrated in Fig. 9.25. Since the base is symmetric, only one-half of it need be drawn. In these cases, the intermediate pieces such as B, C, and D are developed on a stretch-out line formed by laying off the perimeter of a right section. If the right section is taken through the middle of the piece, the stretch-out line becomes the center of the development. Evidently, any elbow could be cut from a single sheet without waste if the seams were made alternately on the long and short sides.

9.12.2   Conical connection between two cylindrical pipes

Refer to Fig. 9.26. The method used in drawing the pattern is the application of the development of an oblique cone. One-half the elliptical base is shown in true size in an auxiliary view (here attached to the front view). Find the true size of the base from its major and minor axes; divide it into a number of equal parts so that the sum of these chordal distances closely approximates the periphery of

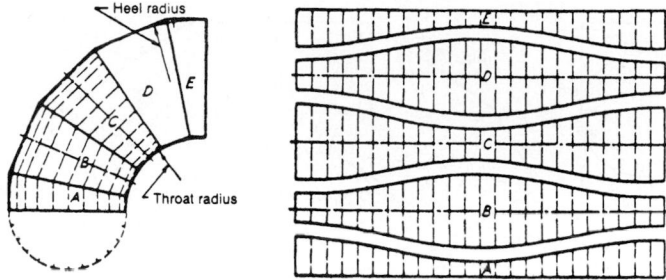

Figure 9.25 A five-piece elbow.

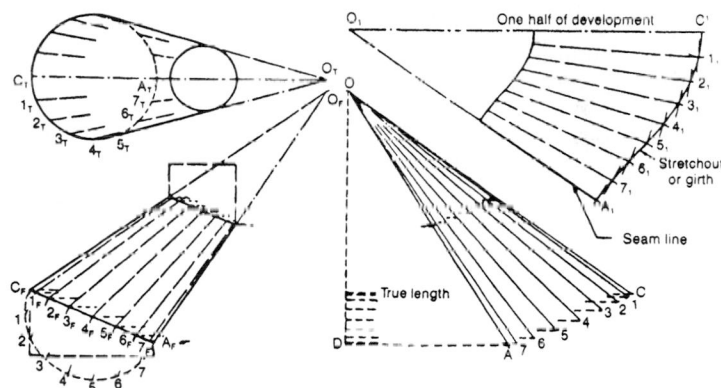

Figure 9.26 A conical connection between two cylinders.

the curve. Project these points to the front and top views. Draw the elements in each view through these points, and find the vertex O by extending the contour elements until they intersect.

The true length of each element is found by using the vertical distance between its ends as the vertical leg of the diagram and its horizontal projection as the other leg. As each true length from vertex to base is found, project the upper end of the intercept horizontally across from the front view to the true length of the corresponding element to find the true length of the intercept. The development is drawn by laying out each triangle in turn, from vertex to base, starting on the centerline O_1C_1, and then measuring on each element its intercept length. Draw smooth curves through these points to complete the pattern.

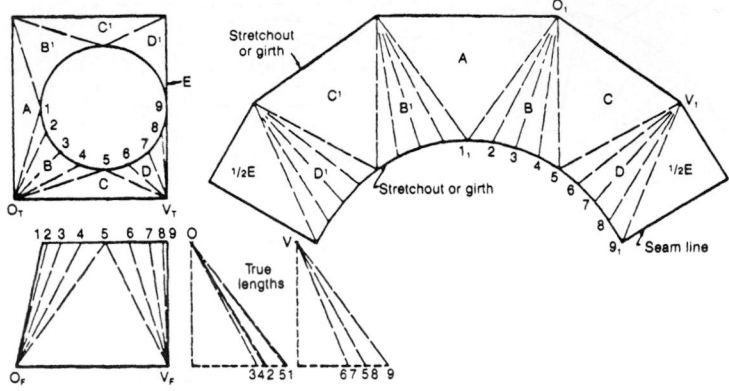

Figure 9.27   A transition piece.

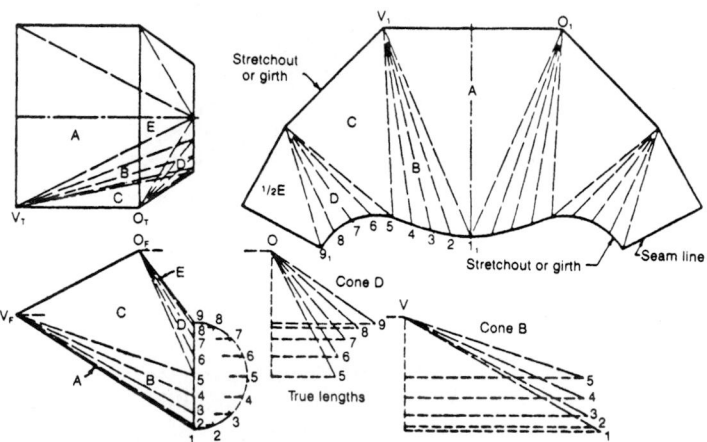

Figure 9.28   A transition piece.

9.12.3   To develop transition pieces

Refer to Figs. 9.27 and 9.28. Transitions are used to connect pipes or openings of different shapes or cross sections. Figure 9.27, showing a transition piece for connecting a round pipe and a rectangular pipe, is typical. These pieces are always developed by triangulation. The piece shown in the figure is evidently made up of four triangu-

lar planes whose bases are the sides of the rectangle and four parts of oblique cones whose common bases are arcs of the circle and whose vertices are at the corners of the rectangle. To develop the piece, make a true-length diagram as shown in Sec. 9.12.2. The true length of O_1 being found, all the sides of triangle A will be known. Attach the developments of cones B and B^1 and then those of triangle C and C^1 and so on.

Figure 9.28 is another transition piece joining a rectangle to a circular pipe whose axes are not parallel. By using a partial right-side view of the round opening, the divisions of the bases of the oblique cones can be found. (Since the object is symmetric, only one-half the opening need be divided.) The true lengths of the elements are obtained as shown in Fig. 9.27.

9.12.4   Triangulation of warped surfaces

The approximate development of a warped surface is made by dividing it into a number of narrow quadrilaterals and then splitting each of these into two triangles by a diagonal line, which is assumed to be a straight line, although it is really a curve. Figure 9.29 shows a warped transition piece that connects on ovular (upper) pipe with a right-circular cylindrical pipe (lower). Find the true size of one-half the elliptical base by rotating it until horizontal about an axis through 1, when its true shape will be seen. The major axis is $1 - 7_R$, and the minor axis through 4_R will be equal to the diameter of the lower pipe.

Divide the semiellipse into a sufficient number of equal parts, and project these to the top and front views. Divide the top semicircle into the same number of equal parts, and connect similar points on each end, thus dividing the surface into approximate quadrilaterals. Cut each into two triangles by a diagonal. On true-length diagrams, find the lengths of the elements and the diagonals, and draw the development by constructing the true sizes of the triangles in regular order.

9.12.5   Angled, flanged corner notching procedures and calculations

Sheet metal parts sometimes have angled flanges that must be bent up for an exact angular fit. Figure 9.30a shows a typical sheet metal part with 45° bent-up flanges. In order to lay out the corner

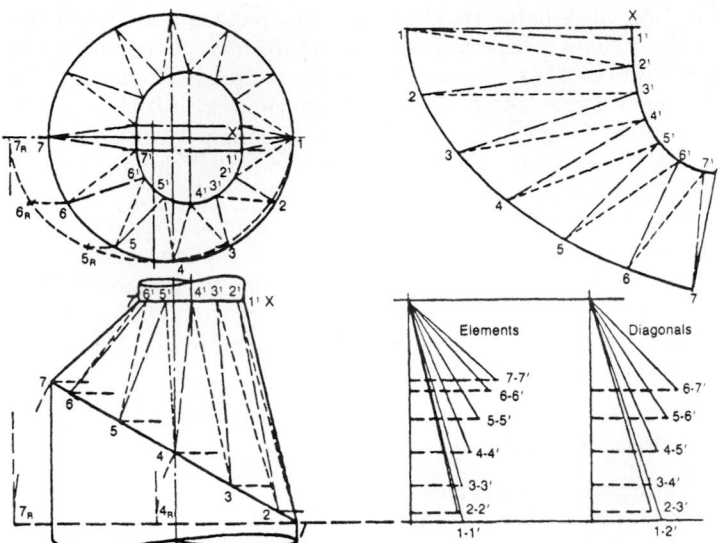

Figure 9.29 A warped transition piece.

notch angle for this type of part, you may use computer programs such as AutoCad to find the correct dimensions and angular cut at the corners, or you may calculate the corner angular cut by using trigonometry. To calculate the corner angular notch trigonometrically, proceed as follows: From Fig. 9.30a, sketch the flat-pattern edges and true lengths as shown in Fig. 9.30b, forming a triangle ABC, which may now be solved using the law of cosines first to find side b and then the law of sines to determine the corner half-notch angle, angle C.

The triangle ABC begins with known dimensions: $a = 4$, angle B = 45°, and $c = 0.828$. This is a triangle where you know two sides and the included angle B. You will need to first find side b using the law of cosines as follows:

$$b^2 = a^2 + c^2 - 2ac \cos B$$

$$b^2 = (4)^2 + (0.828)^2 - 2 \times 4 \times 0.828 \times 0.707 \qquad \text{by the law of cosines}$$

$$b^2 = 12.00242$$

$$b = 3.464$$

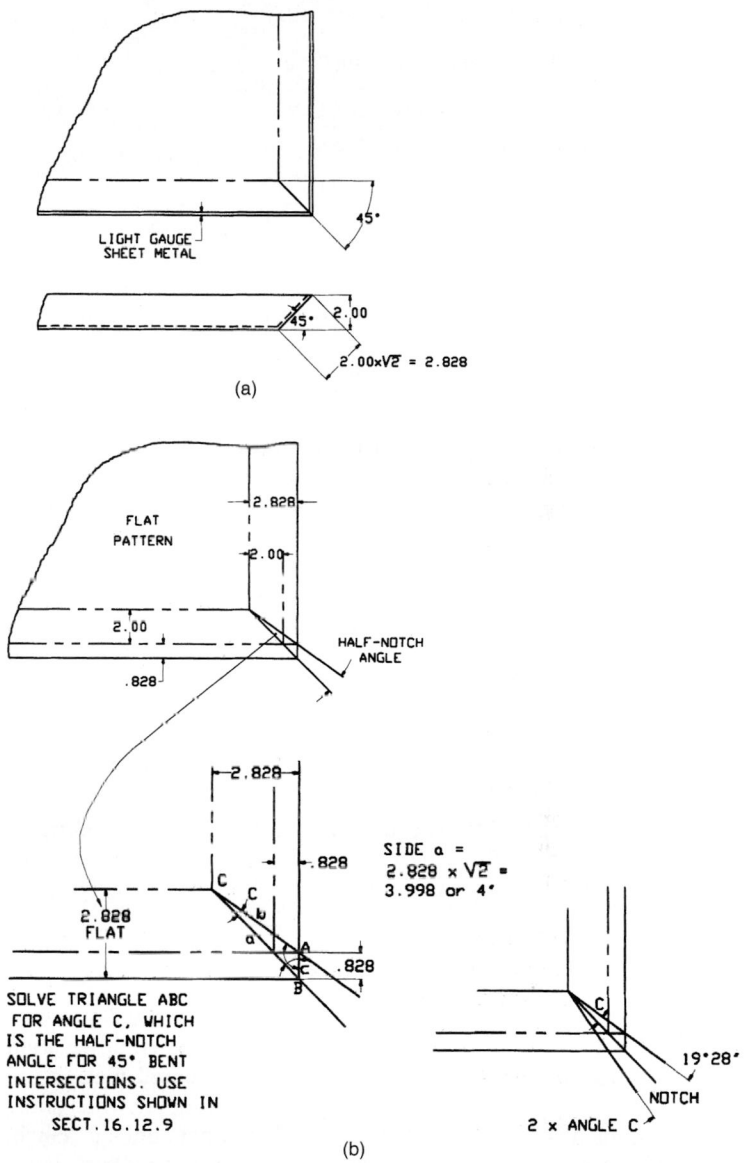

Figure 9.30 (*a*) Developing angular flange notching. (*b*) Flat-pattern development of angular bent-flange corner notches.

Then find angle C using the law of sines:

$$\sin B / b = \sin C / c$$

$$\frac{\sin B}{3.464} = \frac{\sin C}{0.828}$$

$$\frac{\sin 45}{3.464} = \frac{\sin C}{0.828}$$

Solving for sin C, we have

$$\sin C = 0.16899$$

$$\text{arcsin } 0.16899 = 9.729°$$

Therefore, the notch angle $= 2 \times 9.729° = 19.458°$.

This procedure may be used for determining the notch angle for flanges bent on any angle. The angle just given as 19.458° is valid for any flange length as long as the bent-up angle is 45°. This notch angle will increase as bent-up flanges approach 90° until the angle of notch becomes 90° for a bent-up angle of 90°.

Note: The triangle *ABC* shown in this example is actually the overlap angle of the metal flanges as they become bent-up 45°, which must be removed as the corner notch. On thicker sheet metal such as no. 16 through no. 7 gauge, you should do measurements and calculations from the inside mold line (IML) of the flat-pattern sheet metal. Also, the flange height, shown as 2 in in Fig. 9.30*a*, could have been 1 in or any other dimension in order to do the calculations. Thus the corner notch angle is a constant angle for every given bent-up angle; i.e., the angular notch for all 45° bent-up flanges is always 19.458°. It will always be a different constant angle for every different bent-up flange angle.

9.13   Sheet Metal Fabrication Practices

Figures 9.31 through 9.38 illustrate some of the methods used in sheet metal design and fabrication practices. A detailed description of the figures follows: Figure 9.31 shows the accepted methods of relieving the corner stresses and deformation that occur when sheet metal flanges meet at a 90° corner. Circular punch, oblong punch, and saw cuts are illustrated. Figure 9.32 shows that a sheet metal angle may be offset (joggled) to fit over another sheet metal part

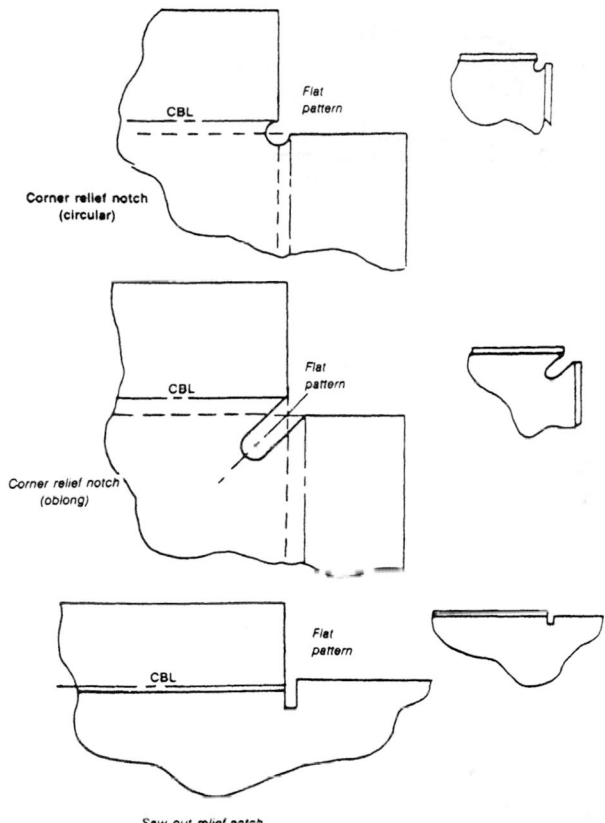

Figure 9.31 Sheet metal corners.

(common practice in the aerospace industry). Corners are formed and fastened by welding directly or with gusset plates.

In Fig. 9.33, a partially closed sheet metal part that normally would be roll-formed is produced on a press brake using this sequence. The outer flanges are bent; then the center is bent to form a W section; finally, the section is completed with a flattening die. Figure 9.34 shows that a return flange on a sheet metal part is possible only when a "gooseneck" die is used in the operation. This forming operation is normally done in a press brake.

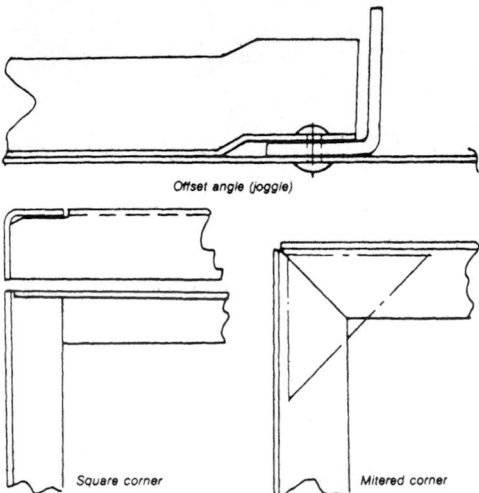

Figure 9.32 Sheet metal intersections.

A sheet metal stake is shown at left in Fig. 9.35. Stakes are used as stops as well as for spring anchors. An integral flange, at right in Fig. 9.35, is produced in the web of a sheet metal part. The sheet metal must first be die-punched prior to the bending operation.

The punch tap is an economical method for producing tapped holes in sheet metal parts. The hole is first punched and extruded by the same die and then tapped with the appropriate thread (Fig. 9.36). The maximum practical size of a punch-tapped hole is generally 0.375-16 in no. 11 gauge sheet steel. Figure 9.37a shows the common corner break or stiffening notch that is formed in the heel section of a sheet metal angle. These simple metal deformation methods add a great amount of strength and stiffness to light-gauge sheet metal parts. Figure 9.38 illustrates some of the common methods of applying gaskets to sheet metal doors and the sheet metal configurations required in these gasketing methods.

9.14   Light-Gauge Sheet Metal Structural Forms: Dimensions and Strengths

Tables 9.1 through 9.11 show the complete dimensions and properties of sheet metal structural shapes that have a wide range of uses in industrial applications. Light-gauge cold-formed steel sheet

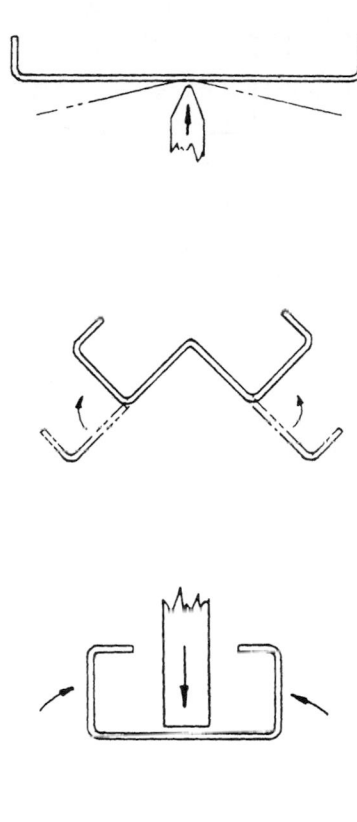

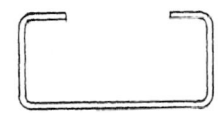

Figure 9.33 Press brake forming a partially closed section.

metal shapes have a high strength-to-weight ratio and are used in countless applications. The tables were extracted from the *Light Gauge Cold-Formed Steel Design Manual*, which was produced by the American Iron and Steel Institute (AISI). The latest edition of this engineering design manual may be obtained from the AISI (see Chap. 16, "Societies, Associations, Institutes, and Specification Authorities," for the address).

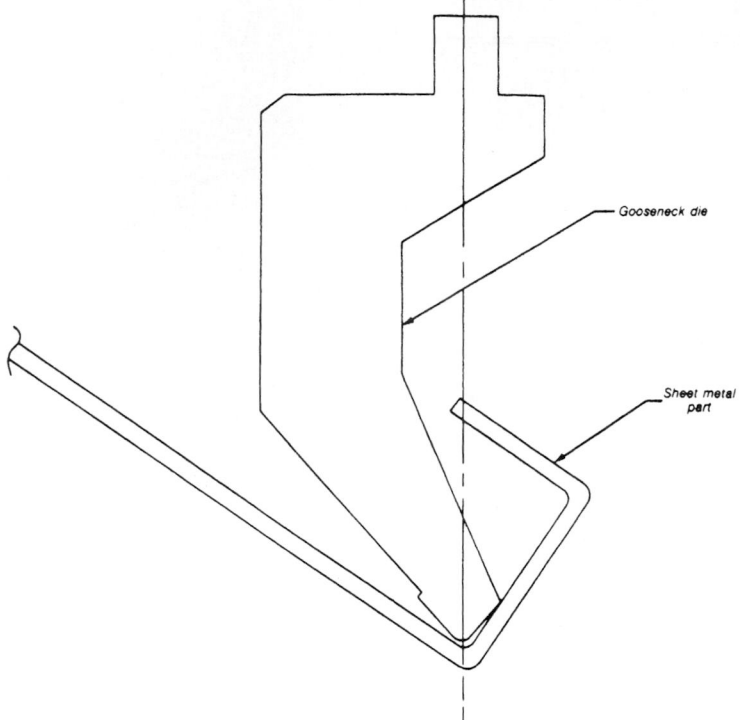

Figure 9.34   Return flange bending.

In the tables, the properties about the *xx* and *yy* axes are most important from a design standpoint for calculating the simple strength capabilities of each particular shape. The following symbolism is defined in the tables:

$I_{x,y}$ = moment of inertia, in^4 (about the *xx* or *yy* axes)

$S_{x,y}$ = section modulus, in^3 (about the *xx* or *yy* axes)

$r_{x,y}$ = radius of gyration, in (about the *xx* or *yy* axes)

x = location of centroid, in (center of gravity of the section)

The equations for calculating the deflection under a given load and the maximum stress imposed on the member for various conditions of loading may be found in other chapters of this *Handbook*. Also, column-bending calculations may be found herein (see Index).

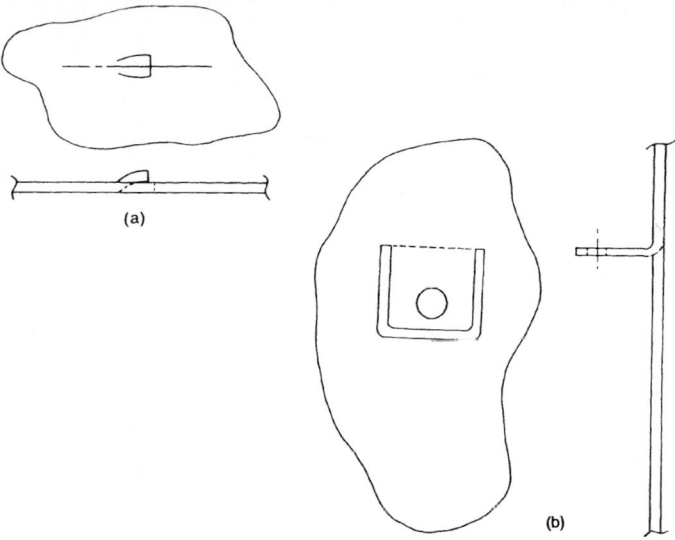

Figure 9.35 Stakes and integral flanges.

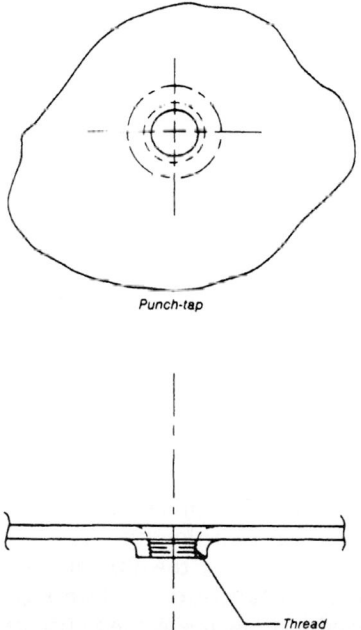

Punch-tap

Thread

Figure 9.36 A punch-tapped hole.

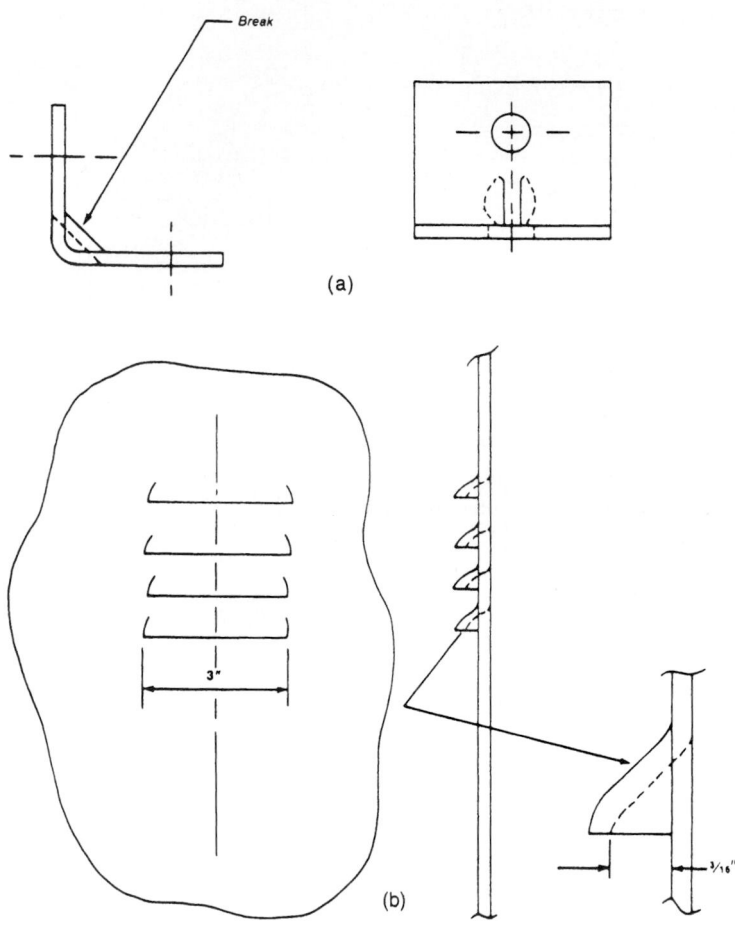

Figure 9.37 (*a*) Corner break. (*b*) Louvers (air vents).

9.15 The Effects of Cold-Working Steel Sheet Metal

It has long been known that any cold working, such as cold stretch-ing, bending, twisting, etc., affects the mechanical properties of steel. Generally, such operations produce strain hardening; i.e., they increase the yield strength and, to a lesser degree, the ultimate ten-sile strength of steels while decreasing the ductility. Cold working of one sort or another occurs in all cold-forming operations, such as roll forming or forming in press brakes. The properties of the cold-

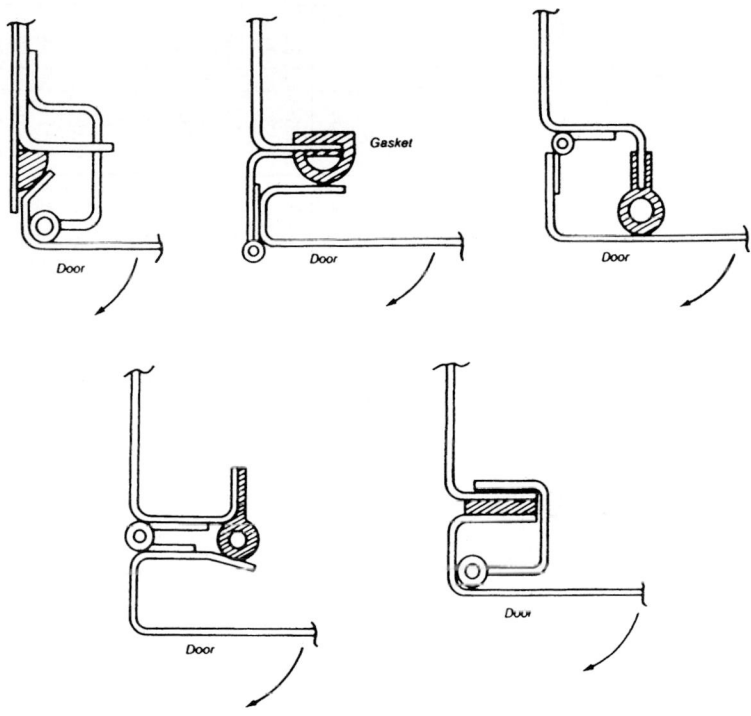

Figure 9.38 Methods of applying door gaskets.

worked parts thus are different from the metal prior to the cold-forming operations. The effects of cold forming depend strongly on the details of the particular cold-forming process. The effects of cold forming are also much more evident in the bent corners than in the flat sections. Metallurgically different kinds of structural carbon steels react differently to the same cold-forming process. The actual effects of any cold-forming operation on sheet steels are, of course, of an extremely complex nature. Unusual or excessive cold working of structural sheet steel may render the formed section unsuitable for a particular application. In other words, the cold-worked section may be weakened by excessive cold-forming operations.

Bending and buckling of sheet metal sections used as columns and bracing or support beams may be calculated by referring to the chapters of this *Handbook* that cover engineering subjects (also see the Index).

TABLE 9.1 Channel or Zee with Stiffened Flanges

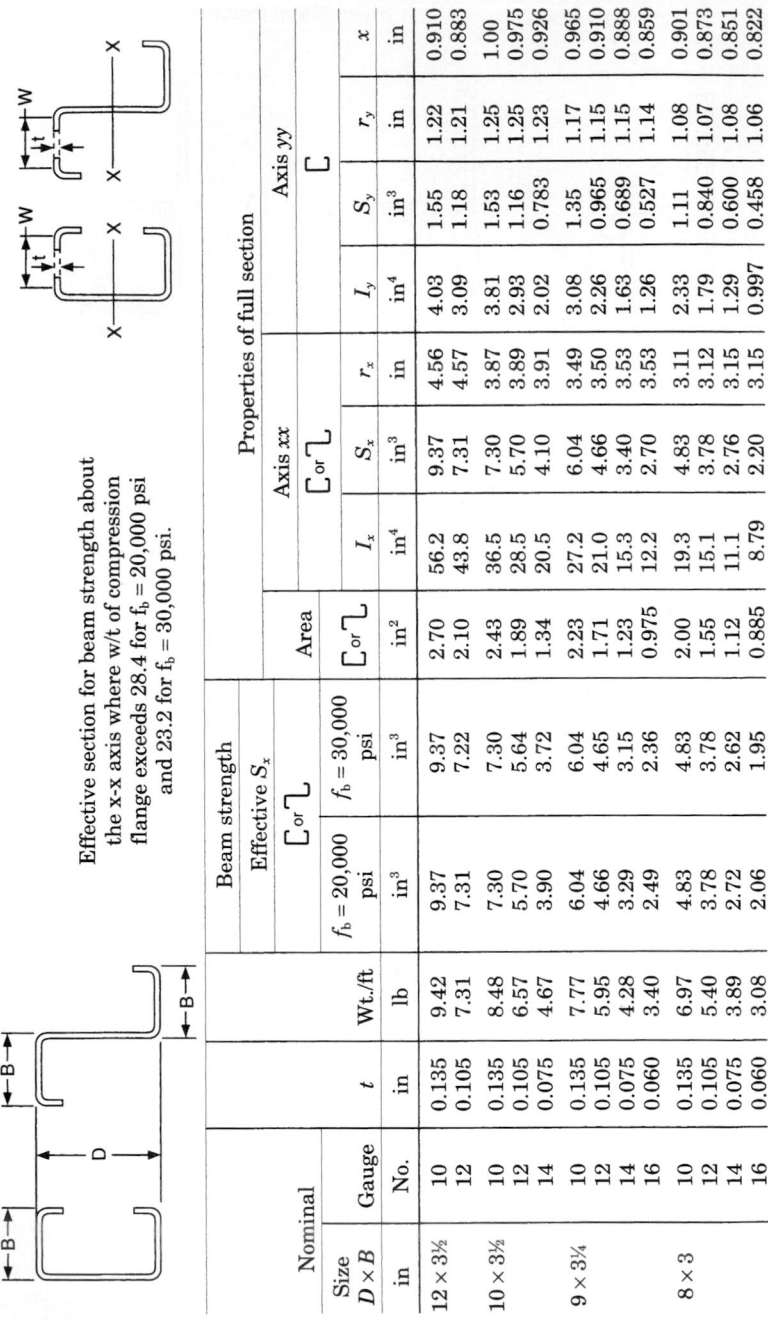

Effective section for beam strength about the x-x axis where w/t of compression flange exceeds 28.4 for f_b = 20,000 psi and 23.2 for f_b = 30,000 psi.

Nominal Size $D \times B$	Gauge No.	t	Wt./ft	Beam strength Effective S_x [C or ⌐] f_b = 20,000 psi	f_b = 30,000 psi	Area [C or ⌐]	Axis xx [C or ⌐] I_x	S_x	r_x	Axis yy [C] I_y	S_y	r_y	x
in		in	lb	in³	in³	in²	in⁴	in³	in	in⁴	in³	in	in
12 × 3½	10	0.135	9.42	9.37	9.37	2.70	56.2	9.37	4.56	4.03	1.55	1.22	0.910
	12	0.105	7.31	7.31	7.22	2.10	43.8	7.31	4.57	3.09	1.18	1.21	0.883
10 × 3½	10	0.135	8.48	7.30	7.30	2.43	36.5	7.30	3.87	3.81	1.53	1.25	1.00
	12	0.105	6.57	5.70	5.64	1.89	28.5	5.70	3.89	2.93	1.16	1.25	0.975
	14	0.075	4.67	3.90	3.72	1.34	20.5	4.10	3.91	2.02	0.783	1.23	0.926
9 × 3¼	10	0.135	7.77	6.04	6.04	2.23	27.2	6.04	3.49	3.08	1.35	1.17	0.965
	12	0.105	5.95	4.66	4.65	1.71	21.0	4.66	3.50	2.26	0.965	1.15	0.910
	14	0.075	4.28	3.29	3.15	1.23	15.3	3.40	3.53	1.63	0.689	1.15	0.888
	16	0.060	3.40	2.49	2.36	0.975	12.2	2.70	3.53	1.26	0.527	1.14	0.859
8 × 3	10	0.135	6.97	4.83	4.83	2.00	19.3	4.83	3.11	2.33	1.11	1.08	0.901
	12	0.105	5.40	3.78	3.78	1.55	15.1	3.78	3.12	1.79	0.840	1.07	0.873
	14	0.075	3.89	2.72	2.62	1.12	11.1	2.76	3.15	1.29	0.600	1.08	0.851
	16	0.060	3.08	2.06	1.95	0.885	8.79	2.20	3.15	0.997	0.458	1.06	0.822

Size	Gage	t												
7 × 2¾	10	0.135	6.17	3.75	3.75	3.75	1.77	13.1	3.75	2.74	1.71	0.893	0.982	0.837
	12	0.105	4.86	2.98	2.98	2.98	1.39	10.4	2.98	2.74	1.38	0.723	0.996	0.837
	14	0.075	3.50	2.18	2.11	2.11	1.00	7.66	2.19	2.76	1.00	0.517	0.999	0.815
	16	0.060	2.77	1.67	1.59	1.59	0.795	6.10	1.74	2.75	0.773	0.393	0.986	0.786
6 × 2½	10	0.135	5.37	2.81	2.81	2.81	1.54	8.42	2.81	2.34	1.21	0.700	0.885	0.774
	12	0.105	4.23	2.24	2.24	2.24	1.21	6.72	2.24	2.35	0.983	0.570	0.900	0.774
	14	0.075	3.10	1.68	1.65	1.68	0.891	5.04	1.68	2.38	0.756	0.440	0.921	0.780
	16	0.060	2.46	1.31	1.26	1.34	0.705	4.01	1.34	2.35	0.583	0.333	0.910	0.751
5 × 2	10	0.135	4.43	1.88	1.88	1.88	1.27	4.69	1.88	1.92	0.651	0.480	0.715	0.644
	12	0.105	3.50	1.51	1.51	1.51	1.00	3.76	1.51	1.94	0.534	0.394	0.729	0.643
	14	0.075	2.53	1.12	1.12	1.12	0.726	2.80	1.12	1.96	0.390	0.283	0.733	0.622
	16	0.060	2.00	0.890	0.881	0.881	0.573	2.23	0.991	1.97	0.298	0.212	0.721	0.594
	18	0.048	1.61	0.706	0.681	0.681	0.461	1.80	0.722	1.98	0.244	0.173	0.727	0.594
4 × 2	10	0.135	3.96	1.38	1.38	1.38	1.14	2.76	1.38	1.56	0.601	0.466	0.727	0.712
	12	0.105	3.14	1.11	1.11	1.11	0.900	2.22	1.11	1.57	0.493	0.383	0.740	0.712
	14	0.075	2.27	0.832	0.832	0.832	0.651	1.67	0.832	1.60	0.361	0.276	0.745	0.689
	16	0.060	1.79	0.565	0.655	0.655	0.513	1.33	0.665	1.61	0.277	0.206	0.735	0.660
	18	0.048	1.44	0.529	0.508	0.508	0.413	1.08	0.540	1.62	0.226	0.169	0.740	0.660
3½ × 2	10	0.135	3.73	1.15	1.15	1.15	1.07	2.01	1.15	1.37	0.571	0.458	0.730	0.753
	12	0.105	2.95	0.927	0.927	0.927	0.847	1.62	0.927	1.38	0.469	0.376	0.744	0.753
	14	0.075	2.14	0.699	0.699	0.699	0.613	1.22	0.699	1.41	0.344	0.271	0.750	0.729
	16	0.060	1.68	0.559	0.551	0.551	0.483	0.979	0.560	1.42	0.264	0.203	0.740	0.699
	18	0.048	1.36	0.444	0.426	0.426	0.389	0.795	0.455	1.43	0.216	0.166	0.745	0.699
3 × 1¾	12	0.105	2.59	0.679	0.679	0.679	0.742	1.02	0.679	1.17	0.319	0.300	0.655	0.689
	14	0.075	1.82	0.509	0.509	0.509	0.523	0.764	0.509	1.21	0.219	0.196	0.647	0.635
	16	0.060	1.47	0.416	0.416	0.416	0.423	0.624	0.416	1.22	0.181	0.162	0.654	0.635
	18	0.048	1.16	0.331	0.322	0.322	0.331	0.498	0.332	1.23	0.137	0.119	0.642	0.604

NOTE: The effective section moduli in bending about the yy axis have not been tabulated. When one of the webs acts as a compression flange, the section modulus should be calculated on the basis of its effective width as provided in Section 2.3 of the Design Specification. When the web acts as a tension flange, the section modulus of the full section is effective. For all of the sections listed in this table the moment of inertia I_x of the full section may be used in deflection calculations without appreciable error.

681

TABLE 9.1 Channel or Zee with Stiffened Flanges (Continued)

Properties of full section (cont'd)							Dimensions of sections ⌐ or ⌙							
Axis yy			Axis zz	Product of inertia ⌙	Column factor Q ⌐ or ⌙									
I_y	S_y	r_y	r min.	I_{xy}	$f_b = 20,000$ psi	$f_b = 30,000$ psi	D	B	d	t	R	m	Gauge No.	Size $D \times B$
in⁴	in³	in	in	in⁴	psi	psi	in	in	in	in	in	in		in
5.94	1.73	1.48	1.00	13.1	0.751	0.703	12.0	3.50	1.0	0.135	³⁄₁₆	1.41	10	12 × 3½
4.54	1.32	1.47	1.00	10.1	0.689	0.640	12.0	3.50	0.9	0.105	³⁄₁₆	1.41	12	
5.94	1.73	1.56	1.01	10.8	0.819	0.770	10.0	3.50	1.0	0.135	³⁄₁₆	1.50	10	10 × 3½
4.54	1.32	1.55	1.01	8.36	0.756	0.705	10.0	3.50	0.9	0.105	³⁄₁₆	1.49	12	
3.07	0.888	1.51	0.992	5.80	0.634	0.567	10.0	3.50	0.7	0.075	³⁄₃₂	1.43	14	
4.87	1.53	1.48	0.947	8.51	0.852	0.803	9.0	3.25	1.0	0.135	³⁄₁₆	1.43	10	9 × 3¼
3.51	1.10	1.43	0.925	6.33	0.785	0.737	9.0	3.25	0.8	0.105	³⁄₁₆	1.38	12	

Size														
8 × 3	14	1.35	³⁄₃₂	0.075	0.7	3.25	9.0	0.607	0.674	4.56	0.928	1.43	0.783	2.52
	16	1.32	³⁄₃₂	0.060	0.6	3.25	9.0	0.522	0.590	3.55	0.916	1.41	0.599	1.93
7 × 2¾	10	1.32	⁹⁄₆₄	0.135	0.9	3.00	8.0	0.837	0.887	6.29	0.864	1.36	1.27	3.72
	12	1.31	⁹⁄₆₄	0.105	0.8	3.00	8.0	0.775	0.821	4.85	0.860	1.35	0.960	2.83
	14	1.27	³⁄₃₂	0.075	0.7	3.00	8.0	0.655	0.719	3.50	0.863	1.35	0.685	2.03
	16	1.25	³⁄₃₂	0.060	0.6	3.00	8.0	0.563	0.635	2.73	0.851	1.32	0.523	1.55
6 × 2½	10	1.22	⁹⁄₆₄	0.135	0.8	2.75	7.0	0.876	0.924	4.49	0.782	1.25	1.03	2.76
	12	1.24	⁹⁄₆₄	0.105	0.8	2.75	7.0	0.814	0.861	3.61	0.793	1.27	0.830	2.24
	14	1.20	³⁄₃₂	0.075	0.7	2.75	7.0	0.701	0.764	2.61	0.796	1.27	0.593	1.61
	16	1.18	³⁄₃₂	0.060	0.6	2.75	7.0	0.615	0.686	2.03	0.785	1.24	0.451	1.23
6 × 2½	10	1.12	⁹⁄₆₄	0.135	0.7	2.50	6.0	0.915	0.962	3.06	0.699	1.13	0.813	1.98
	12	1.14	⁹⁄₆₄	0.105	0.7	2.50	6.0	0.858	0.905	2.48	0.710	1.15	0.660	1.62
	14	1.13	³⁄₃₂	0.075	0.7	2.50	6.0	0.759	0.815	1.88	0.729	1.18	0.507	1.25
	16	1.11	³⁄₃₂	0.060	0.6	2.50	6.0	0.674	0.744	1.46	0.718	1.16	0.385	0.950
5 × 2	10	0.913	⁹⁄₆₄	0.135	0.7	2.00	5.0	0.964	0.994	1.68	0.570	0.919	0.555	1.07
	12	0.936	⁹⁄₆₄	0.105	0.7	2.00	5.0	0.907	0.951	1.37	0.580	0.939	0.454	0.885
	14	0.900	³⁄₃₂	0.075	0.6	2.00	5.0	0.811	0.858	0.998	0.583	0.938	0.325	0.638
	16	0.878	³⁄₃₂	0.060	0.5	2.00	5.0	0.745	0.801	0.772	0.573	0.915	0.244	0.480
	18	0.890	³⁄₃₂	0.048	0.5	2.00	5.0	0.670	0.737	0.629	0.577	0.924	0.199	0.393
4 × 2	10	0.976	⁹⁄₆₄	0.135	0.7	2.00	4.0	0.998	1.000	1.31	0.558	0.971	0.555	1.07
	12	0.998	⁹⁄₆₄	0.105	0.7	2.00	4.0	0.968	0.994	1.07	0.568	0.992	0.454	0.885
	14	0.954	³⁄₃₂	0.075	0.6	2.00	4.0	0.885	0.926	0.786	0.571	0.990	0.325	0.638
	16	0.934	³⁄₃₂	0.060	0.5	2.00	4.0	0.821	0.875	0.610	0.561	0.967	0.244	0.480
	18	0.942	³⁄₃₂	0.048	0.5	2.00	4.0	0.738	0.810	0.498	0.565	0.976	0.199	0.393
3½ × 2	10	1.01	⁹⁄₆₄	0.135	0.7	2.00	3.5	1.000	1.000	1.13	0.544	1.00	0.555	1.07
	12	1.03	⁹⁄₆₄	0.105	0.7	2.00	3.5	0.991	1.000	0.923	0.553	1.02	0.454	0.885
	14	0.992	³⁄₃₂	0.075	0.6	2.00	3.5	0.922	0.959	0.681	0.556	1.02	0.325	0.638

TABLE 9.1 Channel or Zee with Stiffened Flanges (Continued)

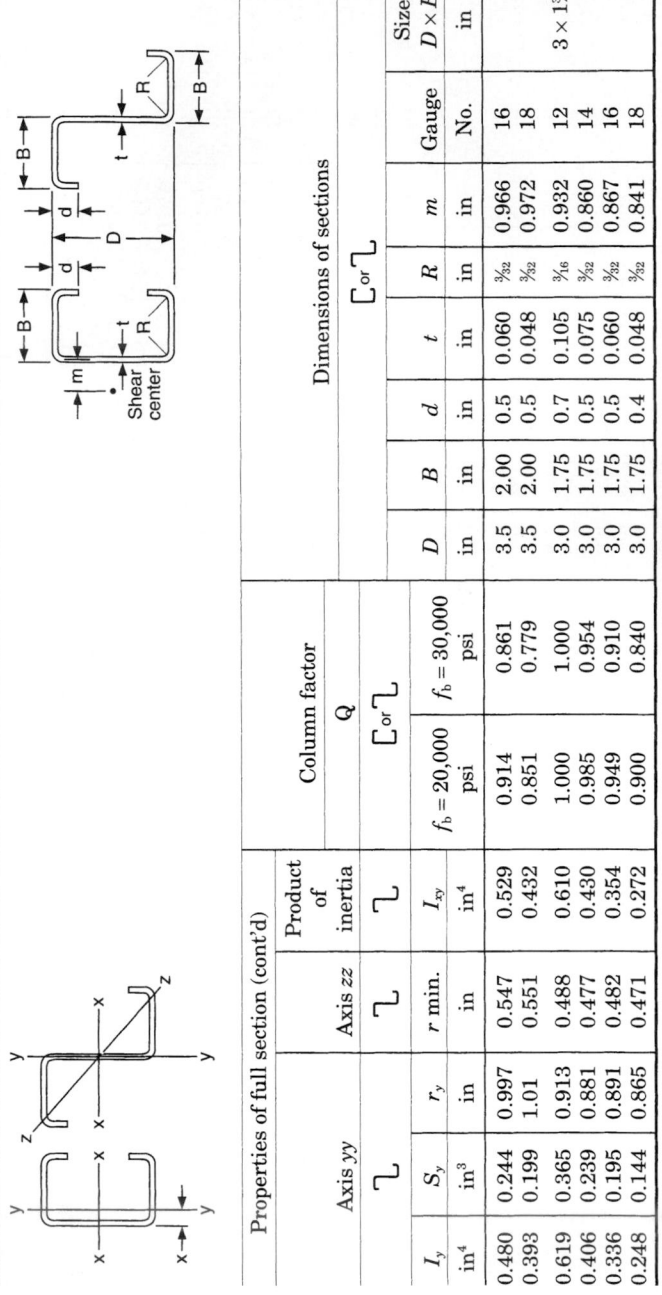

Properties of full section (cont'd)						**Column factor Q, [or L**			**Dimensions of sections, [or L**							
Axis yy			Axis zz		Product of inertia											Size $D \times B$
I_y	S_y	r_y	r min.	I_{xy}		$f_b = 20{,}000$ psi	$f_b = 30{,}000$ psi	D	B	d	t	R	m	Gauge		
in⁴	in³	in	in	in⁴		psi	psi	in	in	in	in	in	in	No.		in
0.480	0.244	0.997	0.547	0.529		0.914	0.861	3.5	2.00	0.5	0.060	9/32	0.966	16		3 × 1¾
0.393	0.199	1.01	0.551	0.432		0.851	0.779	3.5	2.00	0.5	0.048	9/32	0.972	18		
0.619	0.365	0.913	0.488	0.610		1.000	1.000	3.0	1.75	0.7	0.105	3/16	0.932	12		
0.406	0.239	0.881	0.477	0.430		0.985	0.954	3.0	1.75	0.5	0.075	9/32	0.860	14		
0.336	0.195	0.891	0.482	0.354		0.949	0.910	3.0	1.75	0.5	0.060	9/32	0.867	16		
0.248	0.144	0.865	0.471	0.272		0.900	0.840	3.0	1.75	0.4	0.048	9/32	0.841	18		

DIMENSIONS: Equipment and forming practices vary with different manufacturers, resulting in minor variations in some of these dimensions. These minor variations do not affect the published properties. Consult the manufacturer for actual weight per foot and actual dimensions.

TABLE 9.2 Channel or Zee with Unstiffened Flanges

Nominal				Area	Axis xx			Axis yy						
Size $D \times B$	Gauge	Wt./ft	t	$[$ or $]$	$[$ or $]$			$[$				$]$		
	No.				I_x	S_x	r_x	I_y	S_y	r_y	x	I_y	S_y	r_y
in		lb	in	in²	in⁴	in³	ir	in⁴	in³	in	in	in⁴	in³	in
8 × 2	10	5.38	0.135	1.55	12.9	3.24	2.89	0.465	0.293	0.548	0.383	0.621	0.327	0.634
	12	4.23	0.105	1.21	10.3	2.57	2.91	0.379	0.236	0.559	0.376	0.510	0.263	0.648
	14	3.08	0.075	0.884	7.66	1.91	2.94	0.294	0.178	0.577	0.373	0.396	0.199	0.669
	16	2.44	0.060	0.699	6.02	1.50	2.93	0.211	0.132	0.549	0.343	0.279	0.146	0.632
7 × 1½	10	4.39	0.135	1.26	7.54	2.15	2.45	0.170	0.148	0.368	0.260	0.214	0.161	0.413
	12	3.50	0.105	1.00	6.18	1.77	2.48	0.160	0.131	0.399	0.268	0.209	0.145	0.456
	14	2.53	0.075	0.725	4.54	1.30	2.50	0.113	0.0930	0.395	0.248	0.144	0.101	0.446
	16	1.99	0.060	0.572	3.54	1.01	2.49	0.077	0.0669	0.368	0.221	0.985	0.0730	0.415
6 × 1½	10	3.91	0.135	1.12	5.12	1.70	2.13	0.164	0.146	0.382	0.283	0.214	0.161	0.437
	12	3.13	0.105	0.898	4.22	1.40	2.17	0.155	0.130	0.415	0.293	0.209	0.145	0.482
	14	2.26	0.075	0.650	3.10	1.03	2.19	0.109	0.0918	0.410	0.272	0.144	0.101	0.471
	16	1.78	0.060	0.512	2.42	0.806	2.17	0.075	0.0660	0.383	0.243	0.0985	0.0730	0.439
	18	1.43	0.048	0.409	1.94	0.646	2.13	0.059	0.0519	0.378	0.234	0.0764	0.0572	0.432
5 × 1¼	12	2.58	0.105	0.741	2.38	0.953	1.73	0.087	0.0884	0.343	0.252	0.118	0.0991	0.398
	14	1.84	0.075	0.528	1.71	0.683	1.80	0.053	0.0563	0.316	0.214	0.0690	0.0620	0.362

Properties of full section

TABLE 9.2 Channel or Zee with Unstiffened Flanges (Continued)

Nominal					Axis xx [or ⌐			Properties of full section — Axis yy [				Axis yy ⌐		
Size D×B (in)	Gauge No.	t (in)	Wt./ft (lb)	Area [or ⌐ (in²)	I_x (in⁴)	S_x (in³)	r_x (in)	I_y (in⁴)	S_y (in³)	r_y (in)	x (in)	I_y (in⁴)	S_y (in³)	r_y (in)
	16	0.060	1.47	0.422	1.37	0.547	1.80	0.041	0.0439	0.311	0.202	0.0533	0.0485	0.355
	18	0.048	1.15	0.331	1.06	0.423	1.79	0.027	0.0307	0.284	0.176	0.0346	0.0337	0.323
4 × 1⅝	12	0.105	2.17	0.623	1.33	0.663	1.46	0.071	0.0779	0.337	0.265	0.0981	0.0878	0.397
	14	0.075	1.61	0.462	1.02	0.512	1.49	0.058	0.0611	0.355	0.262	0.0807	0.0689	0.418
	16	0.060	1.26	0.362	0.792	0.396	1.48	0.039	0.0430	0.327	0.230	0.0533	0.0485	0.384
	18	0.048	1.01	0.289	0.635	0.318	1.48	0.030	0.0337	0.322	0.220	0.0410	0.0378	0.377
3 × 1⅝	12	0.105	1.80	0.518	0.658	0.439	1.13	0.065	0.0750	0.354	0.308	0.0980	0.0877	0.435
	14	0.075	1.35	0.387	0.515	0.344	1.15	0.054	0.0590	0.372	0.306	0.0807	0.0688	0.457
	16	0.060	1.05	0.302	0.398	0.265	1.15	0.036	0.0416	0.344	0.270	0.0533	0.0485	0.420
	18	0.048	0.841	0.241	0.319	0.213	1.15	0.028	0.0326	0.339	0.259	0.0410	0.0378	0.412
2 × 1⅝	12	0.105	1.44	0.413	0.250	0.250	0.779	0.056	0.0702	0.369	0.374	0.0979	0.0876	0.487
	14	0.075	1.09	0.312	0.200	0.200	0.800	0.047	0.0556	0.387	0.370	0.0807	0.0688	0.509
	16	0.060	0.843	0.242	0.154	0.154	0.799	0.031	0.0392	0.360	0.330	0.0533	0.0484	0.469
	18	0.048	0.673	0.193	0.124	0.124	0.802	0.024	0.0308	0.356	0.318	0.0410	0.0378	0.461

NOTE: The effective section moduli in bending about the yy axis have not been tabulated. When one of the webs acts as a compression flange, the section modulus should be calculated on the basis of its effective width as provided in Section 2.3 of the Design Specification. When the web acts as a tension flange, the section modulus of the full section is effective.

Properties of full section (cont'd)		Allowable beam stress f_c [or]		Column factor Q [or]		Dimensions of sections [or]					Gauge	Size $D \times B$
Axis zz [	Product of inertia]											
r min.	I_{xy}	$f_b = 20,000$ psi	$f_b = 30,000$ psi	$f_b = 20,000$ psi	$f_b = 50,000$ psi	D	B	t	R	m	No.	
in	in	psi	psi			in	in	in	in	in		in
0.461	1.92	18,960	27,490	0.817	0.736	8.0	1.97	0.135	³⁄₁₆	0.531	10	8 × 2
0.470	1.56	17,100	22,980			8.0	1.99	0.105	³⁄₁₆	0.552	12	
0.486	1.18	13,040	13,150			8.0	2.03	0.075	³⁄₃₂	0.581	14	
0.463	0.869	11,610	11,610			8.0	1.94	0.060	³⁄₃₂	0.550	16	
0.312	0.825	20,000	33,000	0.893	0.825	7.0	1.40	0.135	³⁄₁₆	0.329	10	7 × 1½
0.341	0.750	19,340	28,400			7.0	1.49	0.105	³⁄₁₆	0.376	12	
0.336	0.526	16,610	21,790			7.0	1.46	0.075	³⁄₃₂	0.376	14	
0.316	0.380	15,090	18,130			7.0	1.38	0.060	³⁄₃₂	0.350	16	
0.320	0.705	20,000	30,000	0.948	0.889	6.0	1.40	0.135	³⁄₁₆	0.353	10	6 × 1½
0.348	0.641	19,340	28,400	0.849	0.774	6.0	1.49	0.105	³⁄₁₆	0.402	12	
0.345	0.450	16,610	21,790			6.0	1.46	0.075	³⁄₃₂	0.402	14	
0.324	0.325	15,090	18,130			6.0	1.38	0.060	³⁄₃₂	0.375	16	
0.320	0.255	12,840	12,840			6.0	1.36	0.048	³⁄₃₂	0.372	18	

TABLE 9.2 Channel or Zee with Unstiffened Flanges (Continued)

Properties of full section (cont'd)		Allowable beam stress f_c [or]		Column factor Q [or]		Dimensions of sections					Gauge	Size $D \times B$
Axis zz — r min.	Product of inertia — I_{xy}	$f_b = 20{,}000$ psi	$f_b = 30{,}000$ psi	$f_b = 20{,}000$ psi	$f_b = 30{,}000$ psi	D	B	t	R	m	No.	
in	in	psi	psi			in	in	in	in	in		in
0.287	0.363	20,000	30,000	0.934	0.874	5.0	1.24	0.105	$\frac{3}{16}$	0.327	12	$5 \times 1\frac{1}{4}$
0.267	0.229	18,550	26,490			5.0	1.15	0.075	$\frac{3}{32}$	0.303	14	
0.263	0.180	17,050	22,870			5.0	1.13	0.060	$\frac{3}{32}$	0.301	16	
0.243	0.125	15,810	19,850			5.0	1.05	0.048	$\frac{3}{32}$	0.273	18	
0.274	0.256	20,000	30,000	0.991	0.953	4.0	1.17	0.105	$\frac{3}{16}$	0.327	12	$4 \times 1\frac{1}{8}$
0.291	0.203	18,180	25,580	0.827	0.734	4.0	1.21	0.075	$\frac{3}{32}$	0.357	14	
0.271	0.143	17,050	22,870			4.0	1.13	0.060	$\frac{3}{32}$	0.330	16	
0.267	0.112	15,220	18,430			4.0	1.11	0.048	$\frac{3}{32}$	0.326	18	
0.276	0.191	20,000	30,000	1.000	1.000	3.0	1.17	0.105	$\frac{3}{16}$	0.362	12	$3 \times 1\frac{1}{8}$
0.294	0.151	18,180	25,580	0.897	0.815	3.0	1.21	0.075	$\frac{3}{32}$	0.394	14	

0.275	0.107	17,050	22,870	0.808	0.694	7.0	1.13	0.060	3/32	0.365	16	
0.272	0.0836	15,220	18,430	0.687	0.538	3.0	1.11	0.048	3/32	0.361	18	
0.262	0.124	20,906	30,000	1.000	1.000	2.0	1.17	0.105	3/16	0.405	12	2 × 1%
0.280	0.0993	18,180	25,580	0.909	0.853	2.0	1.21	0.075	3/32	0.439	14	
0.266	0.0705	17,050	22,870	0.853	0.762	2.0	1.13	0.060	3/32	0.408	16	
0.264	0.0553	15,220	18,430	0.759	0.608	2.0	1.11	0.048	3/32	0.404	18	

DIMENSIONS: Equipment and forming practices vary with different manufacturers, resulting in minor variations in some of these dimensions. These minor variations do not affect the published properties. Consult the manufacturer for actual weight per foot and actual dimensions. Column form factors Q for members having webs with w/t ratios in excess of 60 are not shown. See limitations of Section 2.3.3(a) of the Specification applicable to element stiffered by simple lip.

TABLE 9.3 Equal Leg Angle with Unstiffened Legs

Nominal					Properties of full section						Beam strength $f_b = 20{,}000$ psi		
Size	Gauge	t	Wt./ft	Area	Axis xx and Axis yy				Axis zz			M max. Comp. tension (x⌐x)	M max. Comp. tension (x⌐x)
	No.				I	S	r	$x=y$	I	r	f_c		
in		in	lb	in²	in⁴	in³	in	in	in⁴	in	psi	in·lb	in·lb
4 × 4	10	0.135	3.66	1.05	1.715	0.582	1.28	1.07	0.662	0.794	12,300	7,160	11,640
3 × 3	10	0.135	2.72	0.781	0.712	0.324	0.955	0.819	0.271	0.589	15,340	4,970	6,480
	12	0.105	2.16	0.620	0.586	0.262	0.972	0.817	0.224	0.601	12,600	3,300	5,240
2½ × 2½	10	0.135	2.25	0.646	0.407	0.223	0.793	0.694	0.153	0.487	17,080	3,810	4,460
	12	0.105	1.79	0.515	0.338	0.182	0.811	0.692	0.128	0.499	14,600	2,660	3,640
2 × 2	10	0.135	1.78	0.511	0.204	0.141	0.632	0.569	0.0756	0.385	18,830	2,650	2,820
	12	0.105	1.43	0.410	0.173	0.116	0.649	0.567	0.0643	0.396	16,830	1,950	2,320
	14	0.075	1.08	0.311	0.144	0.092	0.680	0.570	0.0555	0.423	12,590	1,160	1,840
	16	0.060	0.840	0.241	0.104	0.069	0.658	0.545	0.0404	0.409	11,040	760	1,380

NOTE: The allowable bending moments shown in this table apply only when the sections are adequately braced laterally. Where the vertical legs of the angles are in compression, M max is based on the values of f_c (see 3.2 of Design Specification) indicated: where the vertical legs of the angles are in tension, M max is based on f_b (tension) since the compression stress is always less than f_c for the sections listed. Because it is virtually impossible to load single angle struts concentrically, the design of any such strut should take the eccentricity into account.

Beam strength (cont'd)

$f_b = 30,000$ psi

f_c psi	M max. Comp. tension (in·lb)	M max. Comp. tension (in·lb)	Column factor Q $f_b = 20,000$ psi	Column factor Q $f_b = 30,000$ psi	B in	t in	R in	Wt./ft lb	Gauge No.	Size in
12300	7160	17460	0.542	0.361	4.01	0.135	³⁄₁₆	3.66	10	4 × 4
18730	6070	9720	0.767	0.624	3.01	0.135	³⁄₁₆	2.72	10	3 × 3
12600	3300	7860	0.587	0.391	3.05	0.105	³⁄₁₆	2.16	12	
22950	5120	6690	0.854	0.764	2.51	0.135	³⁄₁₆	2.25	10	2½ × 2½
16920	3080	5460	0.730	0.563	2.55	0.105	³⁄₁₆	1.79	12	
27150	3830	4230	0.941	0.904	2.01	0.135	³⁄₁₆	1.78	10	2 × 2
22330	2590	3480	0.842	0.745	2.05	0.105	³⁄₁₆	1.43	12	
12590	1160	2760	0.586	0.390	2.14	0.075	³⁄₃₂	1.08	14	
11040	760	2070	0.401	0.267	2.06	0.060	³⁄₃₂	0.840	16	

Dimensions of sections

DIMENSIONS: Equipment and forming practices vary with different manufacturers, resulting in minor variations in some of these dimensions. These minor variations do not affect the published properties. Consult the manufacturer for actual weight per foot and actual dimensions.

TABLE 9.4 Equal Leg Angle with Stiffened Legs

Nominal				Area	Properties of full section					
Size	Gauge No.	t	Wt./ft		Axis xx and Axis yy				Axis zz	
					I	S	r	$x = y$	I	r
in		in	lb	in^2	in^4	in^3	in	in	in^4	in
4 × 4	10	0.135	4.46	1.28	2.62	0.962	1.43	1.29	1.25	0.988
3 × 3	10	0.135	3.34	0.957	1.08	0.536	1.06	0.993	0.531	0.745
	12	0.105	2.59	0.743	0.864	0.416	1.08	0.973	0.407	0.740
2½ × 2½	10	0.135	2.68	0.768	0.579	0.342	0.868	0.818	0.277	0.600
	12	0.105	2.15	0.617	0.494	0.286	0.895	0.824	0.236	0.618
2 × 2	10	0.135	2.21	0.633	0.306	0.233	0.695	0.695	0.159	0.501
	12	0.105	1.78	0.512	0.267	0.198	0.722	0.701	0.137	0.517
	14	0.075	1.33	0.381	0.222	0.154	0.763	0.693	0.107	0.530
	16	0.060	1.00	0.287	0.153	0.108	0.731	0.646	0.071	0.497

NOTE: The properties listed in this table apply only when the sections are adequately braced laterally. Unless lipped angle compression struts are checked for torsional buckling, these Q values apply only to situations where such torsional buckling is prevented, as for instance when two angles are connected back to back. Because it is virtually impossible to load single angle struts concentrically, the design of any such strut should take the eccentricity into account.

| Column form factor | | Dimensions of sections | | | | | Gauge | Size |
| Q | | B | d | t | R | Wt./ft | No. | in |
f_b = 20,000 psi	f_b = 30,000 psi	in	in	in	in	lb		
1.000	0.997	4.01	1.1	0.135	3/16	4.46	10	4 × 4
1.000	1.000	3.01	0.5	0.135	3/16	3.34	10	3 × 3
1.000	1.000	3.05	0.8	0.105	3/16	2.59	12	
1.000	1.000	2.51	0.7	0.135	3/16	2.68	10	2½ × 2½
1.000	1.000	2.35	0.7	0.105	3/16	2.15	12	
1.000	1.000	2.01	0.7	0.135	3/16	2.21	10	2 × 2
1.000	1.000	2.05	0.7	0.105	3/16	1.78	12	
1.000	1.000	2.14	0.6	0.075	3/32	1.33	14	
1.000	0.985	2.06	0.5	0.060	3/32	1.00	16	

DIMENSIONS: Equipment and forming practices vary with different manufacturers, resulting in minor variations in some of these dimensions. These minor variations do not affect the published properties. Consult the manufacturer for actual weight per foot and actual dimensions.

TABLE 9.5 Two Channels with Stiffened Flanges Back to Back

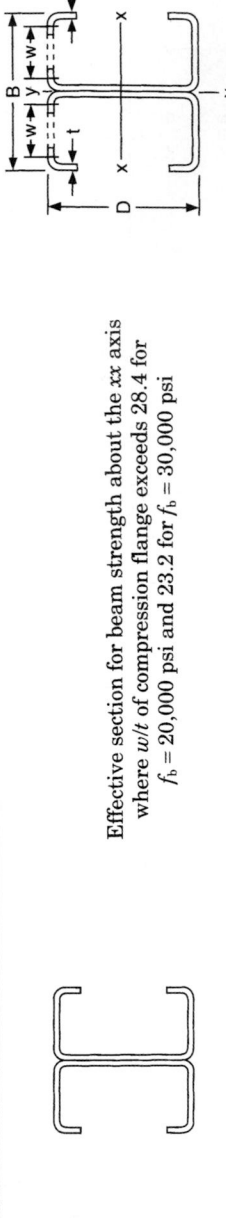

Effective section for beam strength about the xx axis where w/t of compression flange exceeds 28.4 for $f_b = 20,000$ psi and 23.2 for $f_b = 30,000$ psi

Nominal				Beam strength			Properties of full section			
				Effective S_x		S_y		Axis xx		
Size $D \times B$	Gauge	t	Wt./ft	$f_b = 20,000$ psi	$f_b = 30,000$ psi	$f_b = 20,000$ $f_b = 30,000$	Area	I_x	S_x	r_x
in	No.	in	lb	in^3	in^3	in^3	in^2	in^4	in^3	in
12 × 7	10	0.135	18.8	18.7	18.7	3.58	5.40	112.0	18.7	4.56
	12	0.105	14.6	14.6	14.5	2.70	4.20	87.7	14.6	4.57
10 × 7	10	0.135	17.0	14.6	14.6	3.58	4.86	73.0	14.6	3.87
	12	0.105	13.1	11.4	11.3	2.70	3.78	57.0	11.4	3.89
	14	0.075	9.34	7.80	7.44	1.81	2.68	41.0	8.20	3.91
9 × 6½	10	0.135	15.5	12.1	12.1	3.17	4.46	54.3	12.1	3.49
	12	0.105	11.9	9.32	9.30	2.26	3.42	41.9	9.32	3.50
	14	0.075	8.56	6.56	6.30	1.60	2.46	30.6	6.80	3.53
	16	0.060	6.80	4.95	4.72	1.22	1.95	24.3	5.40	3.53
8 × 6	10	0.135	13.9	9.66	9.66	2.63	4.00	38.6	9.66	3.11
	12	0.105	10.8	7.56	7.56	1.98	3.10	30.2	7.56	3.12

Section	Gauge	Thickness								
7 × 5½	14	0.075	7.78	5.42	5.24	1.40	2.24	22.1	5.52	3.15
	16	0.060	6.16	4.11	3.91	1.06	1.77	17.6	4.40	3.15
6 × 5	10	0.135	12.3	7.50	7.50	2.14	3.54	26.2	7.50	2.72
	12	0.105	9.72	5.96	5.96	1.71	2.78	20.9	5.96	2.74
	14	0.075	7.00	4.35	4.22	1.21	2.00	15.3	4.38	2.76
	16	0.060	5.54	3.33	3.18	0.919	1.59	12.2	3.48	2.77
6 × 5	10	0.135	10.7	5.62	5.62	1.71	3.08	16.8	5.62	2.34
	12	0.105	8.46	4.48	4.48	1.37	2.42	13.4	4.48	2.35
	14	0.075	6.29	3.35	3.30	1.04	1.78	10.1	3.36	2.38
	16	0.060	4.92	2.61	2.51	0.785	1.41	8.02	2.68	2.39
5 × 4	10	0.135	8.86	3.76	3.76	1.18	2.54	9.38	3.76	1.92
	12	0.105	7.00	3.02	3.02	0.950	2.00	7.53	3.02	1.94
	14	0.075	5.06	2.24	2.24	0.671	1.45	5.60	2.24	1.96
	16	0.060	4.00	1.78	1.76	0.500	1.15	4.45	1.78	1.97
	18	0.048	3.22	1.41	1.36	0.406	0.922	3.61	1.44	1.98
4 × 4	10	0.135	7.92	2.76	2.76	1.18	2.28	5.51	2.76	1.56
	12	0.105	6.28	2.22	2.22	0.950	1.80	4.44	2.22	1.57
	14	0.075	4.54	1.66	1.66	0.670	1.30	3.33	1.66	1.60
	16	0.060	3.58	1.33	1.31	0.500	1.03	2.66	1.33	1.61
	18	0.048	2.88	1.06	1.02	0.406	0.826	2.16	1.08	1.62
3½ × 4	10	0.135	7.46	2.30	2.30	1.18	2.14	4.01	2.30	1.37
	12	0.105	5.90	1.85	1.85	0.950	1.69	3.24	1.85	1.38
	14	0.075	4.28	1.40	1.40	0.670	1.23	2.45	1.40	1.41
	16	0.060	3.36	1.12	1.10	0.500	0.966	1.96	1.12	1.42
	18	0.048	2.72	0.888	0.852	0.406	0.778	1.59	0.910	1.43
3 × 3½	12	0.105	5.18	1.36	1.36	0.767	1.48	2.04	1.36	1.17
	14	0.075	3.64	1.02	1.02	0.491	1.05	1.53	1.02	1.21
	16	0.060	2.94	0.832	0.832	0.402	0.846	1.25	0.832	1.22
	18	0.048	2.32	0.661	0.644	0.294	0.662	0.995	0.664	1.23

NOTE: The properties of this table apply only when the channels are adequately joined together. See Section 4 of Design Specification.

TABLE 9.5 Two Channels with Stiffened Flanges Back to Back (Continued)

Properties of full section (cont'd) Axis yy			Column factor Q		Dimensions of sections							Nominal	
I_y	S_y	r_y	$f_b = 20,000$	$f_b = 30,000$	D	B	d	t	R	I_y	r_y	Gauge	Size $D \times B$
in⁴	in³	in	psi	psi	in	in	in	in	in	in⁴	in	No.	in
12.52	3.58	1.52	0.751	0.703	12.0	7.0	1.0	0.135	³⁄₁₆	44.28	2.86	10	12 × 7
9.45	2.70	1.50	0.689	0.640	12.0	7.0	0.9	0.105	³⁄₁₆	34.95	2.88	12	
12.52	3.58	1.60	0.819	0.770	10.0	7.0	1.0	0.135	³⁄₁₆	38.00	2.80	10	10 × 7
9.45	2.70	1.58	0.756	0.705	10.0	7.0	0.9	0.105	³⁄₁₆	29.96	2.81	12	
6.33	1.81	1.54	0.634	0.567	10.0	7.0	0.7	0.075	³⁄₃₂	21.80	2.85	14	
10.31	3.17	1.52	0.852	0.803	9.0	6.5	1.0	0.135	³⁄₁₆	29.45	2.57	10	9 × 6½
7.34	2.26	1.47	0.785	0.737	9.0	6.5	0.8	0.105	³⁄₁₆	23.25	2.60	12	
5.19	1.60	1.45	0.674	0.607	9.0	6.5	0.7	0.075	³⁄₃₂	16.98	2.62	14	
3.96	1.22	1.42	0.590	0.522	9.0	6.5	0.6	0.060	³⁄₃₂	13.67	2.65	16	
7.90	2.63	1.41	0.887	0.837	8.0	6.0	0.9	0.135	³⁄₁₆	22.28	2.36	10	8 × 6
5.94	1.98	1.38	0.821	0.774	8.0	6.0	0.8	0.105	³⁄₁₆	17.60	2.38	12	
4.20	1.40	1.37	0.719	0.653	8.0	6.0	0.7	0.075	³⁄₃₂	12.92	2.40	14	
3.19	1.06	1.34	0.635	0.565	8.0	6.0	0.6	0.060	³⁄₃₂	10.39	2.42	16	

Size													
7 × 5½	10	2.15	16.38	³⁄₁₆	0.135	0.8	5.5	7.0	0.876	0.924	1.29	2.14	5.90
	12	2.16	12.93	³⁄₁₆	0.105	0.8	5.5	7.0	0.814	0.861	1.30	1.71	4.72
	14	2.18	9.49	³⁄₃₂	0.075	0.7	5.5	7.0	0.701	0.764	1.29	1.21	3.33
	16	2.20	7.68	³⁄₃₂	0.060	0.6	5.5	7.0	0.615	0.686	1.26	0.919	2.53
6 × 5	10	1.94	11.60	³⁄₁₆	0.135	0.7	5.0	6.0	0.919	0.962	1.18	1.71	4.26
	12	1.95	9.18	³⁄₁₆	0.105	0.7	5.0	3.0	0.858	0.905	1.19	1.37	3.42
	14	1.95	6.78	³⁄₃₂	0.075	0.7	5.0	6.0	0.759	0.815	1.21	1.04	2.60
	16	1.97	5.48	³⁄₃₂	0.060	0.6	5.0	6.0	0.674	0.744	1.18	0.785	1.96
5 × 4	10	1.53	5.97	³⁄₁₆	0.135	0.7	4.0	5.0	0.964	0.994	0.962	1.18	2.36
	12	1.54	4.75	³⁄₁₆	0.105	0.7	4.0	5.0	0.907	0.951	0.973	0.950	1.90
	14	1.56	3.54	³⁄₃₂	0.075	0.6	4.0	5.0	0.811	0.858	0.961	0.671	1.34
	16	1.58	2.86	³⁄₃₂	0.060	0.5	4.0	5.0	0.745	0.801	0.934	0.500	1.00
	18	1.58	2.31	³⁄₃₂	0.048	0.5	4.0	5.0	0.670	0.737	0.939	0.406	0.813
4 × 4	10	1.48	4.98	³⁄₁₆	0.135	0.7	4.0	4.0	0.998	1.000	1.02	1.18	2.35
	12	1.49	3.97	³⁄₁₆	0.105	0.7	4.0	4.0	0.968	0.954	1.03	0.950	1.90
	14	1.51	2.96	³⁄₃₂	0.075	0.6	4.0	4.0	0.885	0.926	1.02	0.670	1.34
	16	1.53	2.40	³⁄₃₂	0.060	0.5	4.0	4.0	0.821	0.875	0.987	0.500	1.00
	18	1.53	1.94	³⁄₃₂	0.048	0.5	4.0	4.0	0.738	0.810	0.992	0.406	0.812
3½ × 4	10	1.45	4.47	³⁄₁₆	0.135	0.7	4.0	3.5	1.000	1.000	1.05	1.18	2.35
	12	1.45	3.57	³⁄₁₆	0.105	0.7	4.0	3.5	0.991	1.000	1.06	0.950	1.90
	14	1.48	2.67	³⁄₃₂	0.075	0.6	4.0	3.5	0.922	0.959	1.05	0.670	1.34
	16	1.50	2.16	³⁄₃₂	0.060	0.5	4.0	3.5	0.861	0.914	1.02	0.500	1.00
	18	1.50	1.75	³⁄₃₂	0.048	0.5	4.0	3.5	0.779	0.851	1.02	0.406	0.812
3 × 3½	12	1.25	2.31	³⁄₁₆	0.105	0.7	3.5	3.0	1.000	1.000	0.951	0.767	1.34
	14	1.29	1.74	³⁄₃₂	0.075	0.5	3.5	3.0	0.954	0.985	0.906	0.491	0.860
	16	1.29	1.41	³⁄₃₂	0.060	0.5	3.5	3.0	0.910	0.949	0.912	0.402	0.703
	18	1.31	1.14	³⁄₃₂	0.048	0.4	2.5	3.0	0.840	0.900	0.881	0.294	0.515

DIMENSIONS: Equipment and forming practices vary with different manufacturers, resulting in minor variations in some of these dimensions. These minor variations do not affect the published properties. Consult the manufacturer for actual weight per foot and actual dimensions.

TABLE 9.6 Two Channels with Unstiffened Flanges Back to Back

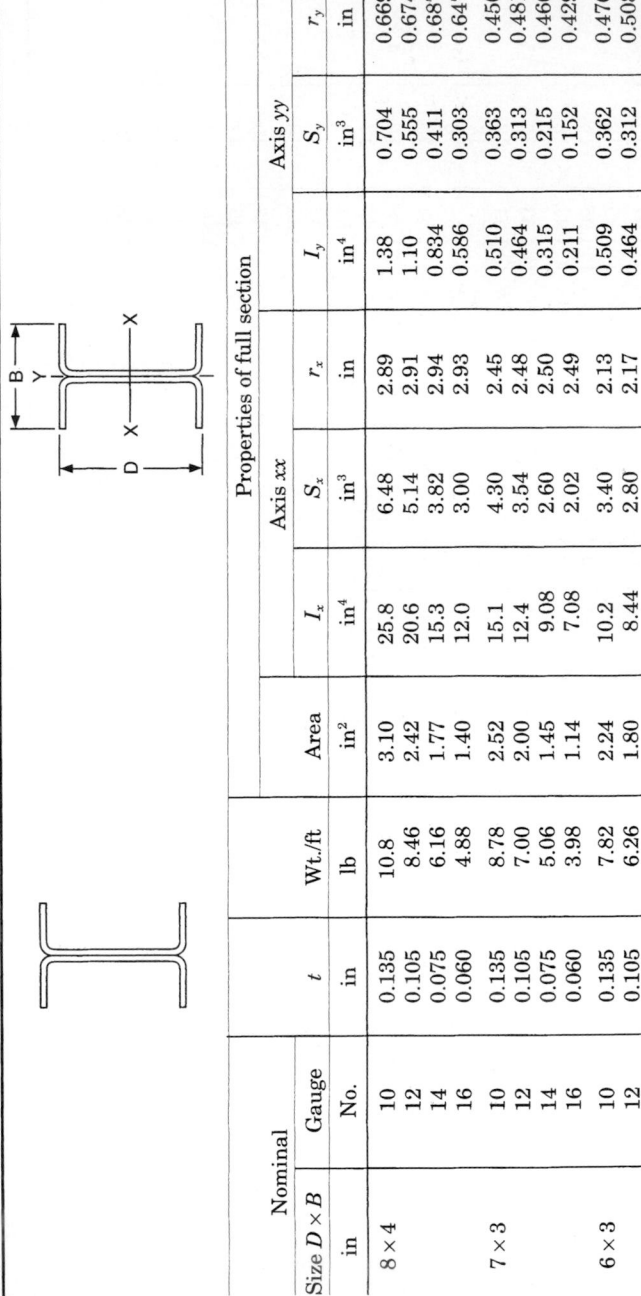

Nominal					Properties of full section					
Size $D \times B$	Gauge	Wt./ft	t	Area	Axis xx			Axis yy		
	No.				I_x	S_x	r_x	I_y	S_y	r_y
in		lb	in	in²	in⁴	in³	in	in⁴	in³	in
8 × 4	10	10.8	0.135	3.10	25.8	6.48	2.89	1.38	0.704	0.669
	12	8.46	0.105	2.42	20.6	5.14	2.91	1.10	0.555	0.674
	14	6.16	0.075	1.77	15.3	3.82	2.94	0.834	0.411	0.687
	16	4.88	0.060	1.40	12.0	3.00	2.93	0.586	0.303	0.647
7 × 3	10	8.78	0.135	2.52	15.1	4.30	2.45	0.510	0.363	0.450
	12	7.00	0.105	2.00	12.4	3.54	2.48	0.464	0.313	0.481
	14	5.06	0.075	1.45	9.08	2.60	2.50	0.315	0.215	0.466
	16	3.98	0.060	1.14	7.08	2.02	2.49	0.211	0.152	0.429
6 × 3	10	7.82	0.135	2.24	10.2	3.40	2.13	0.509	0.362	0.476
	12	6.26	0.105	1.80	8.44	2.80	2.17	0.464	0.312	0.508
	14	4.52	0.075	1.30	6.20	2.06	2.19	0.315	0.215	0.492
	16	3.56	0.060	1.02	4.84	1.61	2.17	0.211	0.152	0.454
	18	2.86	0.048	0.818	3.88	1.29	2.18	0.162	0.119	0.445

Size	Gauge	t								
5 × 2½	12	0.105	5.16	1.48	4.76	1.91	1.79	0.268	0.217	0.425
	14	0.075	3.68	1.06	3.42	1.37	1.80	0.154	0.134	0.382
	16	0.060	2.94	0.844	2.74	1.09	1.80	0.116	0.103	0.370
	18	0.048	2.30	0.664	2.12	0.846	1.79	0.074	0.071	0.335
4 × 2½	12	0.105	4.34	1.25	2.66	1.33	1.46	0.229	0.195	0.429
	14	0.075	3.22	0.924	2.04	1.02	1.49	0.180	0.148	0.441
	16	0.060	2.52	0.724	1.58	0.792	1.48	0.116	0.103	0.400
	18	0.048	2.02	0.578	1.27	0.636	1.48	0.088	0.079	0.390
3 × 2½	12	0.105	3.60	1.04	1.32	0.878	1.13	0.228	0.195	0.470
	14	0.075	2.70	0.774	1.03	0.688	1.15	0.180	0.148	0.481
	16	0.060	2.10	0.604	0.796	0.530	1.15	0.115	0.102	0.437
	18	0.048	1.68	0.482	0.638	0.426	1.15	0.088	0.079	0.427
2 × 2½	12	0.105	2.88	0.826	0.500	0.500	0.779	0.227	0.194	0.525
	14	0.075	2.18	0.624	0.400	0.400	0.800	0.179	0.148	0.536
	16	0.060	1.69	0.484	0.308	0.308	0.799	0.115	0.102	0.488
	18	0.048	1.35	0.386	0.248	0.248	0.802	0.088	0.079	0.477

NOTE: The properties of this table apply only when the channels are adequately joined together. See Section 4 of Design Specification.

TABLE 9.6 Two Channels with Unstiffened Flanges Back to Back (Continued)

| Allowable beam stress f_c, psi | | Column factor Q | | Dimensions of sections | | | | | | |
$f_b = 20,000$ psi	$f_b = 30,000$ psi	$f_b = 20,000$ psi	$f_b = 30,000$ psi	D in	B in	t in	R in	Wt./ft lb	Gauge No.	Size $D \times B$ in
18960	27490	0.817	0.736	8.0	3.934	0.135	3/16	10.8	10	8 × 4
17100	22980			8.0	3.972	0.105	3/16	8.46	12	
13040	13150			8.0	4.052	0.075	3/32	6.16	14	
11610	11610			8.0	3.882	0.060	3/32	4.88	16	
20000	30000	0.893	0.825	7.0	2.810	0.135	3/16	8.78	10	7 × 3
19340	28400			7.0	2.972	0.105	3/16	7.00	12	
16610	21790			7.0	2.928	0.075	3/32	5.06	14	
15090	18130			7.0	2.758	0.060	3/32	3.98	16	
20000	30000	0.948	0.889	6.0	2.810	0.135	3/16	7.82	10	6 × 3
19340	28400	0.849	0.774	6.0	2.972	0.105	3/16	6.26	12	
16610	21790			6.0	2.928	0.075	3/32	4.52	14	
15090	18130			6.0	2.758	0.060	3/32	3.56	16	
12840	12840			6.0	2.722	0.048	3/32	2.86	18	

Size	Gauge									
5 × 2½	12	5.16	3/16	0.105	2.472	5.0	0.934	0.874	30000	20000
	14	3.68	5/32	0.075	2.302	5.0			26490	18550
	16	2.94	3/32	0.060	2.258	5.0			22870	17050
	18	2.30	3/32	0.048	2.098	5.0			19850	15810
4 × 2¼	12	4.34	3/16	0.105	2.346	4.0	0.991	0.953	30000	20000
	14	3.22	5/32	0.075	2.428	4.0	0.827	0.734	25580	18180
	16	2.52	3/32	0.060	2.258	4.0			22870	17050
	18	2.02	3/32	0.048	2.222	4.0			18430	15220
3 × 2¼	12	3.60	3/16	0.105	2.346	3.0	1.000	1.000	30000	20000
	14	2.70	5/32	0.075	2.428	3.0	0.897	0.815	25580	18180
	16	2.10	3/32	0.060	2.258	3.0	0.808	0.694	22870	17050
	18	1.68	3/32	0.048	2.222	3.0	0.687	0.538	18430	15220
2 × 2¼	12	2.88	3/16	0.105	2.346	2.0	1.000	1.000	30000	20000
	14	2.18	5/32	0.075	2.428	2.0	0.909	0.853	25580	18180
	16	1.69	3/32	0.060	2.258	2.0	0.853	0.762	22870	17050
	18	1.35	3/32	0.048	2.222	2.0	0.759	0.608	18430	15220

DIMENSIONS: Equipment and forming practices vary with different manufacturers, resulting in minor variations in some of these dimensions. These minor variations do not affect the published properties. Consult the manufacturer for actual weight per foot and actual dimensions. Column form factors Q for members having webs with w/t-ratios in excess of 60 are not shown. See limitations of Section 2.3.3(a) of the Specification applicable to element stiffered by simple lip.

TABLE 9.7 Two Equal-Leg Angles, Back to Back, Unstiffened Legs

Full theoretical outline			Column properties		Dimensions				
			Q						
Area	r_x	r_y	$f_b = 20,000$	$f_b = 30,000$	B	Thickness t	Radius R	Weight per ft	Nominal size
in²	in	in	psi	psi	in	in	in	lb	in
2.10	1.28	1.67	0.542	0.361	8.030	.135	3/16	7.32	4 × 4
1.56	0.955	1.26	0.767	0.624	6.030	.135	3/16	5.44	3 × 3
1.24	0.972	1.27	0.587	0.391	6.110	.105	3/16	4.32	
1.29	0.793	1.05	0.854	0.764	5.030	.135	3/16	4.50	2½ × 2½
1.03	0.811	1.07	0.730	0.563	5.110	.105	3/16	3.58	
1.02	0.632	0.850	0.941	0.904	4.030	.135	3/16	3.56	2 × 2
0.820	0.649	0.862	0.842	0.745	4.110	.105	3/16	2.86	
0.622	0.680	0.887	0.586	0.390	4.276	.075	3/32	2.16	
0.482	0.658	0.855	0.401	0.267	4.128	.060	3/32	1.68	

DIMENSIONS: Equipment and forming practices vary with different manufacturers, resulting in minor variations in some of these dimensions. These minor variations do not affect the published properties. Consult the manufacturer for actual weight per foot and actual dimensions.

Nominal (one angle)			Section modulus based on full theoretical outline S_z	Beam strength						Deflection	
Size	Gauge No.	Thickness		f_b = 20,000 psi			f_b = 30,000 psi			Any grade of steel	
				f_c	M max. Comp. $x\!-\!\!\!\perp\!-\!x$ tension	M max. Comp. $\top\!-\!\!\!\mid\!-\!x$ tension	f_c	M max. Comp. $x\!-\!\!\!\perp\!-\!x$ tension	M max. Comp. $\top\!-\!\!\!\mid\!-\!x$ tension	y	I_x
in	No.	in	in³	psi	in·lb	in·lb	psi	in·lb	in·lb	in	in⁴
4 × 4	10	.135	1.164	12300	14320	23280	12300	14320	34920	1.069	3.430
3 × 3	10	.135	0.648	15340	9940	12960	18730	12140	19440	0.819	1.424
	12	.105	0.524	12600	6600	10480	12600	6600	15720	0.817	1.172
2½ × 2½	10	.135	0.446	17090	7620	8920	22950	10230	13380	0.694	0.814
	12	.105	0.364	14600	5310	7280	16920	6160	10920	0.692	0.676
2 × 2	10	.135	0.282	18830	5310	5640	27150	7660	8460	0.569	0.408
	12	.105	0.232	16830	3910	4640	22330	5180	6960	0.567	0.346
	14	.075	0.183	12590	2320	3680	12590	2320	5520	0.570	0.288
	16	.060	0.137	11040	1520	2760	11040	1520	4140	0.546	0.208

NOTE: The properties of this table may be used only when the angles are adequately joined and adequately braced laterally. Q is the column factor (Sec. 3.6.1, Design Specification). Where the vertical legs of the angles are in compression M_{max} is based on the values of f_c (Sec. 3.2 of Design Specification) indicated; where the vertical legs of the angles are in tension M_{max} is based on f_b (tension) since the compression stress is always less than f_c for the sections listed.

TABLE 9.8 Two Equal-Leg Angles, Back to Back, Stiffened Legs

Column form factor Q		Dimensions of sections					Gauge	Size
f_b = 20,000 psi	f_b = 30,000 psi	B	d	t	R	Wt./ft	No.	
		in	in	in	in	lb		in
1.000	0.997	8.030	1.1	0.135	3/16	8.92	10	4 × 4
1.000	1.000	6.030	0.9	0.135	3/16	6.68	10	3 × 3
1.000	1.000	6.110	0.8	0.105	3/16	5.18	12	
1.000	1.000	5.030	0.7	0.135	3/16	5.36	10	2½ × 2½
1.000	1.000	5.110	0.7	0.105	3/16	4.30	12	
1.000	1.000	4.030	0.7	0.135	3/16	4.42	10	2 × 2
1.000	1.000	4.110	0.7	0.105	3/16	3.56	12	
1.000	1.000	4.276	0.6	0.075	3/32	2.66	14	
1.000	0.985	4.128	0.5	0.060	3/32	2.00	16	

DIMENSIONS: Equipment and forming practices vary with different manufacturers, resulting in minor variations in some of these dimensions. These minor variations do not affect the published properties. Consult the manufacturer for actual weight per foot and actual dimensions.

704

Nominal (one angle)					Properties of full section					
Size	Gauge	Thickness	Wt./ft	Area	Axis xx				Axis yy	
	No.				I	S	r	y	I	r
in		in	lb	in^2	in^4	in^3	in	in	in^4	in
4 × 4	10	0.135	8.92	2.56	5.23	1.92	1.43	1.29	9.51	1.93
3 × 3	10	0.135	6.68	1.91	2.16	1.07	1.05	0.993	4.05	1.46
	12	0.105	5.18	1.49	1.73	0.832	1.08	0.973	3.13	1.45
2½ × 2½	10	0.135	5.36	1.54	1.16	0.684	0.868	0.818	2.19	1.19
	12	0.105	4.30	1.23	0.989	0.572	0.855	0.824	1.83	1.22
2 × 2	10	0.135	4.42	1.27	0.612	0.466	0.695	0.695	1.22	0.981
	12	0.105	3.56	1.02	0.534	0.396	0.722	0.701	1.04	1.008
	14	0.075	2.66	0.762	0.444	0.308	0.763	0.693	0.810	1.031
	16	0.060	2.00	0.574	0.306	0.216	0.731	0.646	0.546	0.975

NOTE: The properties listed in this table may be used only when the angles are adequately joined and adequately braced laterally.

TABLE 9.9 Hat Sections

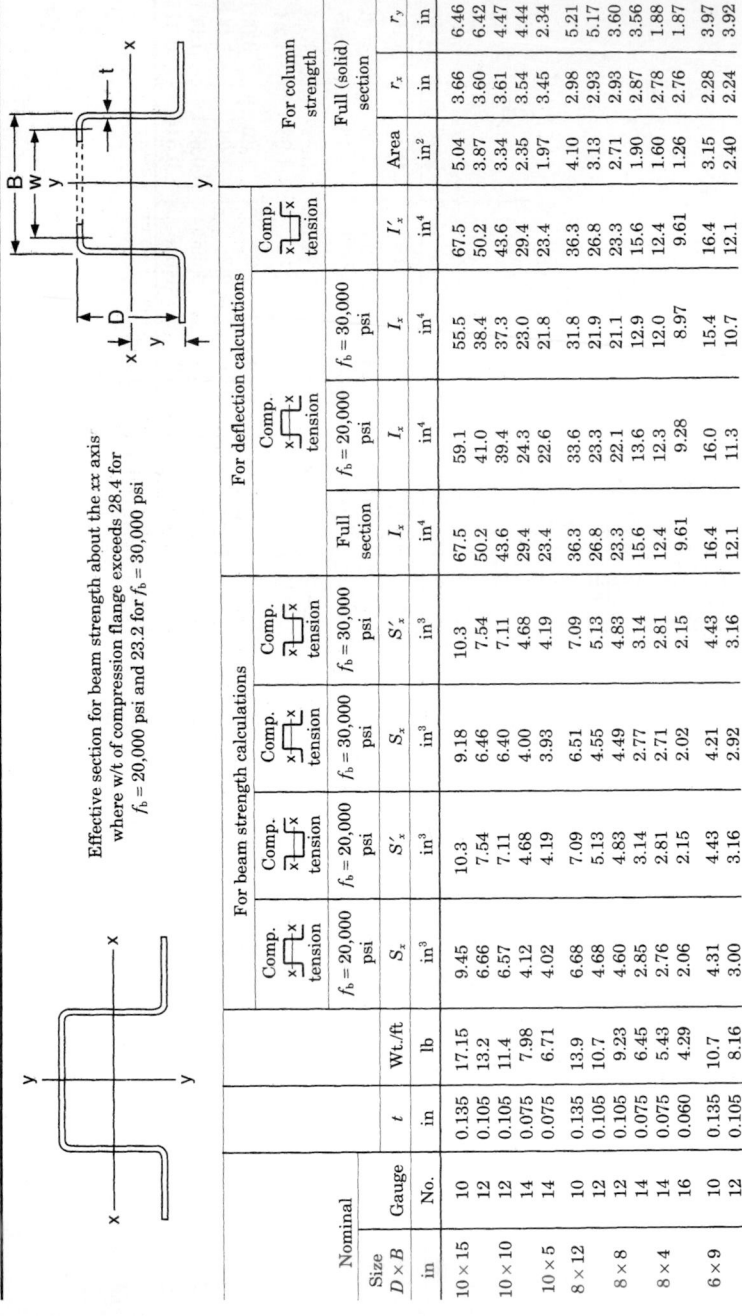

Effective section for beam strength about the xx axis where w/t of compression flange exceeds 28.4 for $f_b = 20,000$ psi and 23.2 for $f_b = 30,000$ psi

Nominal Size $D \times B$	Gauge No.	t	Wt./ft	For beam strength calculations				For deflection calculations				For column strength Full (solid) section		
				Comp. x⊓x tension f_b=20,000 psi S_x	Comp. x⊔x tension f_b=20,000 psi S'_x	Comp. x⊓x tension f_b=30,000 psi S_x	Comp. x⊔x tension f_b=30,000 psi S'_x	Full section I_x	Comp. x⊓x tension f_b=20,000 psi I_x	f_b=30,000 psi I_x	Comp. x⊓x tension I'_x	Area	r_x	r_y
in	No.	in	lb	in³	in³	in³	in³	in⁴	in⁴	in⁴	in⁴	in²	in	in
10 × 15	10	0.135	17.15	9.45	10.3	9.18	10.3	67.5	59.1	55.5	67.5	5.04	3.66	6.46
	12	0.105	13.2	6.66	7.54	6.46	7.54	50.2	41.0	38.4	50.2	3.87	3.60	6.42
10 × 10	12	0.105	11.4	6.57	7.11	6.40	7.11	43.6	39.4	37.3	43.6	3.34	3.61	4.47
	14	0.075	7.98	4.12	4.68	4.00	4.68	29.4	24.3	23.0	29.4	2.35	3.54	4.44
10 × 5	14	0.075	6.71	4.02	4.19	3.93	4.19	23.4	22.6	21.8	23.4	1.97	3.45	2.34
8 × 12	10	0.135	13.9	6.68	7.09	6.51	7.09	36.3	33.6	31.8	36.3	4.10	2.98	5.21
	12	0.105	10.7	4.68	5.13	4.55	5.13	26.8	23.3	21.9	26.8	3.13	2.93	5.17
8 × 8	12	0.105	9.23	4.60	4.83	4.49	4.83	23.3	22.1	21.1	23.3	2.71	2.93	3.60
	14	0.075	6.45	2.85	3.14	2.77	3.14	15.6	13.6	12.9	15.6	1.90	2.87	3.56
8 × 4	14	0.075	5.43	2.76	2.81	2.71	2.81	12.4	12.3	12.0	12.4	1.60	2.78	1.88
	16	0.060	4.29	2.06	2.15	2.02	2.15	9.61	9.28	8.97	9.61	1.26	2.76	1.87
6 × 9	10	0.135	10.7	4.31	4.43	4.21	4.43	16.4	16.0	15.4	16.4	3.15	2.28	3.97
	12	0.105	8.16	3.00	3.16	2.92	3.16	12.1	11.3	10.7	12.1	2.40	2.24	3.92

Size	Ga	t													
6×6	12	0.105	7.09	2.92	2.98	2.87	2.98	2.98	10.4	10.0	10.3	10.4	2.08	2.24	2.74
	14	0.075	4.92	1.75	1.90	1.75	1.90	1.90	6.91	6.16	6.47	6.91	1.45	2.18	2.69
6×3	14	0.075	4.16	1.70	1.70	1.63	1.70	1.70	5.48	5.46	5.44	5.48	1.22	2.12	1.43
	16	0.060	3.27	1.27	1.29	1.25	1.29	1.29	4.23	4.13	4.21	4.23	0.963	2.10	1.42
	18	0.048	2.59	0.954	0.984	0.93	0.986	0.986	3.28	3.10	3.20	3.28	0.761	2.08	1.41
4×6	10	0.135	7.51	2.34	2.35	2.31	2.35	2.35	5.43	5.40	5.43	5.43	2.21	1.57	2.76
	12	0.105	5.66	1.62	1.65	1.59	1.65	1.65	3.96	3.83	3.94	3.96	1.66	1.54	2.69
4×4	12	0.105	4.94	1.55	1.55	1.54	1.55	1.55	3.39	3.39	3.39	3.39	1.45	1.53	1.91
	14	0.075	3.39	0.945	0.961	0.928	0.961	0.961	2.23	2.16	2.22	2.23	0.998	1.49	1.83
4×2	14	0.075	2.88	0.863	0.863	0.863	0.863	0.863	1.75	1.75	1.75	1.75	0.848	1.44	0.996
	16	0.060	2.25	0.639	0.643	0.641	0.645	0.645	1.34	1.34	1.34	1.34	0.663	1.42	0.972
	18	0.048	1.77	0.482	0.483	0.477	0.483	0.483	1.03	1.03	1.03	1.03	0.520	1.41	0.956
3×4½	10	0.135	5.90	1.51	1.52	1.52	1.51	1.52	2.47	2.46	2.47	2.47	1.736	1.19	2.19
	12	0.105	4.41	1.05	1.05	1.04	1.05	1.05	1.81	1.80	1.81	1.81	1.297	1.18	2.09
3×3	12	0.105	3.87	0.993	0.993	0.993	0.993	0.993	1.53	1.53	1.53	1.53	1.139	1.16	1.52
	14	0.075	2.63	0.503	0.604	0.597	0.604	0.604	1.01	1.00	1.01	1.01	0.773	1.14	1.41
3×1½	14	0.075	2.25	0.504	0.504	0.504	0.504	0.504	0.784	0.784	0.784	0.784	0.660	1.09	0.793
	16	0.060	1.74	0.400	0.400	0.400	0.400	0.400	0.599	0.599	0.599	0.599	0.513	1.08	0.760
	18	0.048	1.36	0.297	0.297	0.297	0.297	0.297	0.459	0.459	0.459	0.459	0.400	1.07	0.738
2×4	12	0.105	3.52	0.596	0.596	0.593	0.596	0.596	0.672	0.671	0.672	0.672	1.03	0.806	1.90
	14	0.075	2.37	0.350	0.353	0.345	0.353	0.353	0.433	0.423	0.432	0.433	0.698	0.787	1.77
2×2	14	0.075	1.86	0.321	0.321	0.321	0.321	0.321	0.329	0.329	0.329	0.329	0.548	0.774	1.01
	16	0.060	1.44	0.233	0.233	0.233	0.233	0.233	0.250	0.248	0.248	0.250	0.423	0.768	0.973
2×1	16	0.060	1.23	0.178	0.178	0.178	0.178	0.178	0.193	0.193	0.193	0.193	0.363	0.729	0.566
	18	0.048	0.953	0.142	0.142	0.142	0.142	0.142	0.147	0.147	0.147	0.147	0.280	0.725	0.532
1½×3	12	0.105	2.80	0.390	0.390	0.390	0.390	0.390	0.304	0.304	0.304	0.304	0.824	0.607	1.54
	14	0.075	1.86	0.250	0.230	0.228	0.230	0.230	0.198	0.198	0.198	0.198	0.548	0.602	1.39
1½×1½	14	0.075	1.48	0.187	0.187	0.187	0.187	0.187	0.148	0.148	0.148	0.148	0.435	0.583	0.832
	16	0.060	1.13	0.150	0.150	0.150	0.150	0.150	0.113	0.113	0.113	0.113	0.333	0.582	0.779
1½×¾	16	0.060	0.979	0.0993	0.0993	0.0993	0.0993	0.0993	0.0856	0.086	0.0856	0.0856	0.288	0.545	0.484
	18	0.048	0.749	0.0795	0.0795	0.0795	0.0795	0.0795	0.0657	0.066	0.0657	0.0657	0.220	0.546	0.442

NOTE: $S'x$ and $I'x$ in this table are properties of full (unreduced) section and are applicable as indicated. The effective section moduli in bending about the yy axis have not been tabulated. When one of the webs acts as a compression flange the section modulus should be calculated on the basis of its effective width as provided in Section 2.3 of the Design Specification. When the web acts as a tension flange the section modulus of the full section is effective.

TABLE 9.9 Hat Sections (Continued)

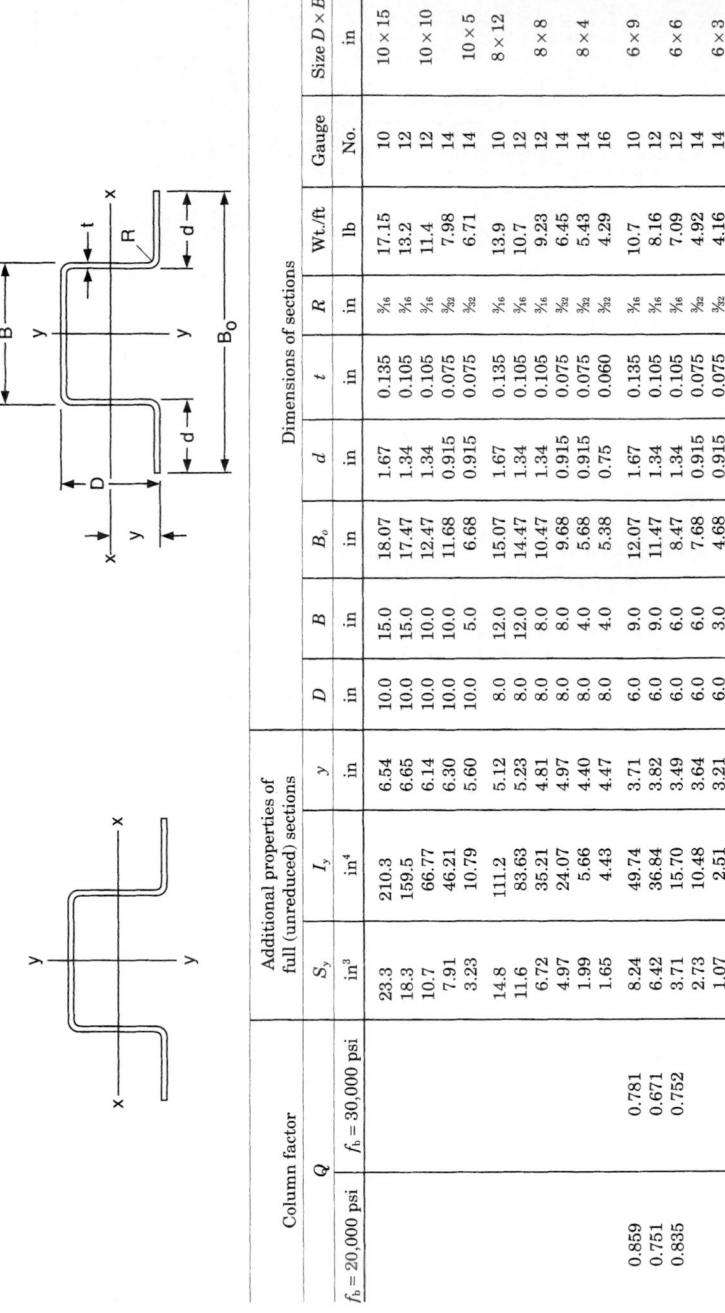

Column factor		Additional properties of full (unreduced) sections			Dimensions of sections								
$f_b = 20{,}000$ psi	$f_b = 30{,}000$ psi												
Q		S_y	I_y	y	D	B	B_o	d	t	R	Wt./ft	Gauge	Size $D \times B$
		in³	in⁴	in	in	in	in	in	in	in	lb	No.	in
		23.3	210.3	6.54	10.0	15.0	18.07	1.67	0.135	9/16	17.15	10	10 × 15
		18.3	159.5	6.65	10.0	15.0	17.47	1.34	0.105	9/16	13.2	12	10 × 15
		10.7	66.77	6.14	10.0	10.0	12.47	1.34	0.105	9/16	11.4	12	10 × 10
		7.91	46.21	6.30	10.0	10.0	11.68	0.915	0.075	9/32	7.98	14	10 × 10
		3.23	10.79	5.60	10.0	5.0	6.68	0.915	0.075	9/32	6.71	14	10 × 5
		14.8	111.2	5.12	8.0	12.0	15.07	1.67	0.135	9/16	13.9	10	8 × 12
		11.6	83.63	5.23	8.0	12.0	14.47	1.34	0.105	9/16	10.7	12	8 × 12
		6.72	35.21	4.81	8.0	8.0	10.47	1.34	0.105	9/16	9.23	12	8 × 8
		4.97	24.07	4.97	8.0	8.0	9.68	0.915	0.075	9/32	6.45	14	8 × 8
		1.99	5.66	4.40	8.0	4.0	5.68	0.915	0.075	9/32	5.43	14	8 × 4
		1.65	4.43	4.47	8.0	4.0	5.38	0.75	0.060	9/32	4.29	16	8 × 4
0.859	0.781	8.24	49.74	3.71	6.0	9.0	12.07	1.67	0.135	9/16	10.7	10	6 × 9
0.751	0.671	6.42	36.84	3.82	6.0	9.0	11.47	1.34	0.105	9/16	8.16	12	6 × 9
0.835	0.752	3.71	15.70	3.49	6.0	6.0	8.47	1.34	0.105	9/16	7.09	12	6 × 6
		2.73	10.48	3.64	6.0	6.0	7.68	0.915	0.075	9/32	4.92	14	6 × 6
		1.07	2.51	3.21	6.0	3.0	4.68	0.915	0.075	9/32	4.16	14	6 × 3

(1)	(2)	(3)	(4)	(5)	(6)	(7)	(8)	(9)	(10)	(11)	(12)	Ga.	Size
		0.883	1.93	3.28	6.0	3.0	4.38	0.75	0.060	3/32	3.27	16	4 × 6
		0.728	1.51	3.33	6.0	3.0	4.14	0.618	0.048	3/32	2.59	18	
		3.72	16.86	2.31	4.0	6.0	9.07	1.67	0.135	3/16	7.51	10	4 × 4
		2.84	12.05	2.41	4.0	6.0	8.47	1.34	0.135	3/16	5.66	12	
0.974	0.942	1.64	5.31	2.19	4.0	4.0	6.47	1.34	0.135	3/16	4.94	12	4 × 2
0.924	0.861	1.18	3.35	2.32	4.0	4.0	5.68	0.915	0.075	3/32	3.39	14	
0.988	0.935	0.457	0.842	2.03	4.0	2.0	3.68	0.915	0.075	3/32	2.88	14	
0.856	0.773	0.369	0.626	2.09	4.0	2.0	3.38	0.75	0.060	3/32	2.25	16	
0.887	0.822	0.303	0.476	2.14	4.0	2.0	3.14	0.618	0.048	3/32	1.77	18	
1.000	0.990	2.19	3.28	1.63	3.0	4.5	7.57	1.67	0.135	3/16	5.90	10	3 × 4½
0.982	0.955	1.63	5.69	1.71	3.0	4.5	6.97	1.34	0.105	3/16	4.41	12	
1.000	1.000	0.956	2.62	1.54	3.0	3.0	5.47	1.34	0.105	3/16	3.87	12	3 × 3
0.969	0.907	0.659	1.54	1.67	3.0	3.0	4.68	0.915	0.075	3/32	2.63	14	
0.976	0.928	0.261	0.415	1.45	3.0	1.5	3.18	0.915	0.075	3/32	2.25	14	3 × 1½
0.915	0.853	0.205	0.296	1.50	3.0	1.5	2.88	0.75	0.060	3/32	1.74	16	
0.839	0.772	0.165	0.218	1.55	3.0	1.5	2.64	0.618	0.045	3/32	1.36	18	
0.994	0.971	1.15	3.72	1.13	2.0	4.0	6.47	1.34	0.105	3/16	3.52	12	2 × 4
0.931	0.892	0.772	2.19	1.22	2.0	4.0	5.68	0.915	0.075	3/32	2.37	14	
1.000	1.000	0.307	0.564	1.02	2.0	2.0	3.68	0.915	0.075	3/32	1.86	14	2 × 2
1.000	0.978	0.237	0.400	1.07	2.0	2.0	3.33	0.75	0.060	3/32	1.44	16	
1.000	0.982	0.0976	0.116	0.920	2.0	1.0	2.38	0.75	0.060	3/32	1.23	16	2 × 1
0.975	0.928	0.0743	0.0795	0.961	2.0	1.0	2.14	0.618	0.048	3/32	0.953	18	
1.000	1.000	0.715	1.96	0.778	1.5	3.0	5.47	1.34	0.105	3/16	2.80	12	1½ × 3
0.986	0.956	0.454	1.06	0.864	1.5	3.0	4.68	0.915	0.075	3/32	1.86	14	
1.000	1.000	0.189	0.501	0.709	1.5	1.5	3.18	0.915	0.075	3/32	1.48	14	1½ × 1½
1.000	1.000	0.141	0.202	0.750	1.5	1.5	2.88	0.75	0.060	3/32	1.13	16	
1.000	1.000	0.0633	0.0674	0.637	1.5	0.75	2.13	0.75	0.060	3/32	0.979	16	1½ × ¾
1.000	0.996	0.0455	0.0430	0.673	1.5	0.75	1.89	0.618	0.048	3/32	0.749	18	

DIMENSIONS: Equipment and forming practices vary with different manufacturers, resulting in minor variations in some of these dimensions. These minor variations do not affect the published properties. Consult the manufacturer for actual weight per foot and actual dimensions. Column form factors Q for members having webs with w/t-ratios in excess of 60 are not shown. See limitations of Section 2.3.3(a) of the Specification applicable to element stiffened by a simple lip.

TABLE 9.10 One-Flange Stiffener (Which Includes One 90° Corner): Properties and Dimensions

Stock Width of Blank Taken at $t/3$ Distance From Inner Surface.

Nominal Gauge No.	Thickness t (in)	Depth d (in)	xx Axis I_x (in⁴)	xx Axis y (in)	yy Axis I_y (in⁴)	yy Axis x (in)	Area (in²)	Max. flange B (in)	Blank width (in)	Thickness t (in)	Radius R (in)	Depth d (in)	Nominal Gauge No.
10	0.135	1.0	0.01255	0.4737	0.000796	0.1005	0.14554	3.87	1.043	0.135	³⁄₁₆	1.0	10
		0.9	0.00916	0.4250	0.000759	0.1039	0.13204	3.25	0.943	0.135	³⁄₁₆	0.9	
		0.8	0.00643	0.3766	0.000719	0.1080	0.11854	2.81	0.843	0.135	³⁄₁₆	0.8	
		0.7	0.00430	0.3286	0.000674	0.1133	0.10504	2.53	0.743	0.135	³⁄₁₆	0.7	
		0.6	0.00269	0.2811	0.000621	0.1200	0.09154	2.37	0.643	0.135	³⁄₁₆	0.6	
		0.5	0.00153	0.2346	0.000556	0.1291	0.07804	2.30	0.543	0.135	³⁄₁₆	0.5	
		0.4	0.00076	0.1896	0.000474	0.1420	0.06454	2.27	0.443	0.135	³⁄₁₆	0.4	
12	0.105	0.9	0.00735	0.4205	0.000475	0.0850	0.10337	4.28	0.957	0.105	³⁄₁₆	0.9	12
		0.8	0.00518	0.3719	0.000453	0.0886	0.09287	3.34	0.857	0.105	³⁄₁₆	0.8	
		0.7	0.00348	0.3237	0.000428	0.0932	0.08237	2.66	0.757	0.105	³⁄₁₆	0.7	
		0.6	0.00219	0.2761	0.000398	0.0992	0.07187	2.22	0.657	0.105	³⁄₁₆	0.6	
		0.5	0.00126	0.2292	0.000362	0.1072	0.06137	1.98	0.557	0.105	³⁄₁₆	0.5	
		0.4	0.00064	0.1836	0.000314	0.1185	0.05087	1.88	0.457	0.105	³⁄₁₆	0.4	

14	0.9	3/32	0.075	0.918	6.79	0.07031	0.0475	0.000081	0.4351	0.00493	0.9	0.075	14
	0.8	3/32	0.075	0.818	4.94	0.06281	0.0487	0.000076	0.3355	0.00348	0.8		
	0.7	3/32	0.075	0.718	3.50	0.05531	0.0502	0.000072	0.3361	0.00234	0.7		
	0.6	3/32	0.075	0.618	2.46	0.04781	0.0522	0.000067	0.2869	0.00148	0.6		
	0.5	3/32	0.075	0.518	1.77	0.04031	0.0549	0.000062	0.2379	0.00086	0.5		
	0.4	3/32	0.075	0.418	1.40	0.03281	0.0589	0.000055	0.1894	0.00044	0.4		
16	0.8	3/32	0.060	0.825	7.50	0.05044	0.0400	0.000048	0.3836	0.00283	0.8	0.060	16
	0.7	3/32	0.060	0.725	5.19	0.04444	0.0414	0.000045	0.3341	0.00191	0.7		
	0.6	3/32	0.060	0.625	3.46	0.03844	0.0435	0.000043	0.2848	0.00121	0.6		
	0.5	3/32	0.060	0.525	2.23	0.03244	0.0456	0.000040	0.2357	0.00071	0.5		
	0.4	3/32	0.060	0.425	1.48	0.02644	0.0492	0.000036	0.1871	0.00036	0.4		
18	0.7	3/32	0.048	0.731	7.96	0.03567	0.0344	0.000030	0.3325	0.00155	0.7	0.048	18
	0.6	3/32	0.048	0.631	5.19	0.03087	0.0360	0.000028	0.2831	0.00099	0.6		
	0.5	3/32	0.048	0.531	3.20	0.02607	0.0382	0.000026	0.2340	0.00058	0.5		
	0.4	3/32	0.048	0.431	1.87	0.02127	0.0414	0.000024	0.1853	0.00030	0.4		

TABLE 9.11   One 90° Corner: Properties and Dimensions

Stock Width of Blank Taken at $t/3$ Distance From Inner Surface.

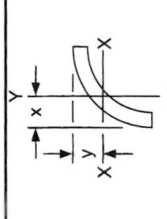

Nominal Gauge	Thickness	Moment of inertia	Centroid coordinates			Dimensions		Nominal
						Thickness	Radius inside	Gauge
No.	t	$I_x = I_y$	$x = y$	Area	Blank width	t	R	No.
	in	in^4	in	in^2	in	in	in	
10	0.135	0.0003889	0.1564	0.05407	0.3652	0.135	0.1875	10
12	0.105	0.0002408	0.1373	0.03958	0.3495	0.105	0.1875	12
14	0.075	0.0000301	0.0829	0.01546	0.1865	0.075	0.0938	14
16	0.060	0.0000193	0.0734	0.01166	0.1787	0.060	0.0938	16
18	0.048	0.0000128	0.0658	0.00888	0.1724	0.048	0.0938	18
20	0.036	0.00000313	0.0464	0.00452	0.1170	0.036	0.0625	20

10

Solid Freeform Fabrication

Solid freeform *fabrication* (SFF) refers to a category of manufacturing processes in which parts are built by depositing one cross-sectional layer of material on top of the next. It is very much like "printing" a succession of slice images one on top of the next so that thickness is gradually built up (Fig. 10.1). Owing to the fact that material is deposited in thin cross-sectional layers, there are typically no concerns about tool collisions, parting lines, undercuts, etc. It is a toolless and fixtureless approach to making parts. Fabricating highly complex geometric shapes via layered manufacturing processes therefore is no more difficult than fabricating simple geometric shapes such as cubes or cylinders. This is evident in Fig. 10.2, which shows a very impressive geometric sculpture designed by artist Bathsheba Grossman (*www.bathsheba.com*). The sculpture was fabricated using ProMetal's Direct Metal Printing process.

Rapid prototyping (RP) and *rapid manufacturing* (RM) are terms that typically are associated with SFF. In both cases, the parts being fabricated are built layer by layer. The distinction lies simply with the end use of the part being produced. RP produces parts that will be used as part of the iterative design process. RM produces a functional end item to be delivered and used by the customer. Note that the end item can be a steel mold or die if the customer is a molding, casting, or stamping shop.

The terms *rapid prototyping* and *rapid manufacturing* can be somewhat misleading owing to the fact that layered processes

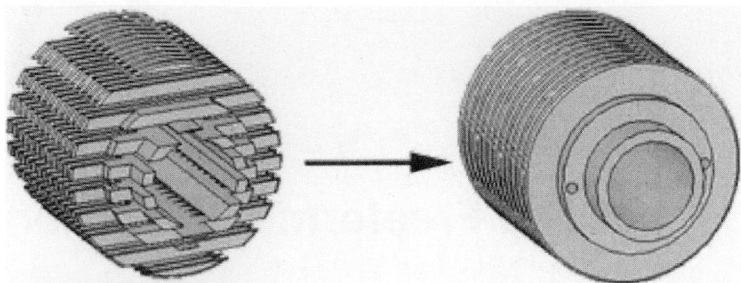

Figure 10.1 Principle of layer-additive manufacturing.

Figure 10.2 Complex freeform structure. *(Photo courtesy of Bathsheba Grossman.)*

generally are much slower than mass-production processes such as die casting or injection molding. However, layered manufacturing processes are capable of building parts with complex geometries much more quickly than generally would be possible when using conventional fabrication techniques. When production quantities are not sufficiently high to justify the expense of tooling and fixturing, then one of the SFF processes can be cost-effective.

The vast majority of early SFF processes were developed to produce nonmetallic parts. Some of the early processes included stereolithography (3D Systems), fused deposition modeling (Stratasys), and selective laser sintering (3D Systems). The SLS process also has been used to make sintered parts from both metal and ceramic powders.

More recently, literally dozens of relatively inexpensive layered manufacturing processes have been commercialized. In order to obtain fully dense metal parts from one of these processes, it is generally necessary to use a two-step process in which the prototype part is used as the pattern for a casting or forming process. For example, polymer prototype parts often are used as patterns for investment casting or sand casting.

10.1 Direct-Metal Rapid Manufacturing Processes

Recent years have witnessed a flurry of research and development aimed at "direct-metal" RM processes. These processes are of particular interest to machinists and metalworkers. Rather than using an indirect approach such as investment casting to produce a metal part, these processes directly produce near-net-shape functional metal components.

Owing to the considerable expense usually associated with direct-metal RM processes, they typically are used for one of several reasons.

Material cost. RM processes are selected most often to produce parts from very expensive alloys with special material properties. Because RM processes produce components to near-net shape, very little material is wasted in finish machining. In many cases these materials are quite difficult to machine; hence producing them to near-net shape dramatically speeds up fabrication of the final part and reduces material costs. Although titanium and its alloys (most notably Ti-6Al-4V) seem to have attracted the most interest, there also has been considerable development work involving a wide variety of metal-matrix composites (MMCs) involving TiB, TiC, SiC, and other reinforcements. The majority of RM processes are powder-based, so particulate-reinforced MMCs are employed rather than continuously reinforced MMCs.

Ease of joining. Some alloys are notoriously difficult to join via welding, brazing, or similar thermal methods. With traditional manufacturing methods, it is often the case that components in an assembly are designed as separate pieces merely to accommodate the production method. Because geometric complexity is not a major concern when RM processes are used, it is often possible to merge several parts into one part. In some cases the elimination of assembly joints also can have implications for improved component reliability.

Design optimization. RM processes can produce complex geometric shapes that would be extremely difficult and/or costly to fabricate via more conventional means. Owing to the fact that the designer need not be concerned with demolding a part or (in some cases) cutting-tool access, it is possible to optimize components for special form or function with little regard for the manufacturing process. Some examples include

Internal wiring. Electrical wiring can be routed through channels that are fabricated in the walls of a part so that the wiring is protected from external moving components.

Conformal cooling channels. Conventional mold/die cooling channels typically lie in a flat plane. With RM, it is possible to fabricate cooling channels that conform to or flow with the actual shape of the core/cavity surfaces in each of the three dimensions. By routing coolant close to the mold/die surface, more uniform cooling can be achieved, thus lowering stresses brought about by uneven cooling. Furthermore, cycle times can be reduced. In addition to having cooling channels that conform to the shape of the tool surface, it is also possible to vary the cross section of the channel itself. It need not be round. For instance, it can fan out to a larger surface area as it passes near the mold surface. Likewise, baffles can be built in to increase turbulence if so desired.

Weight optimization. Through finite element analysis (FEA) codes, mechanical designers are able to determine exactly where material can be removed from a component without adversely affecting the function and safety of a part. However, designing a weight-optimized part and fabricating it are two separate issues. In many instances, an RM method can fabricate optimized components that would be very difficult to fabricate via conventional methods. For this reason, the aerospace, automotive, and military industries have been among the early adopters of these direct-metal technologies. Figure 10.3 shows a geometrically complex component for a head-mounted display. In order to reduce weight, the three "tentacles" coming off the device have been hollowed out.

Compositional optimization. Some RM processes have the ability to independently deposit multiple materials in varying amounts. Much like a printer is able to print thousands of colors with just three or four print heads, these systems can locally vary the material composition of a component to achieve material properties that cannot be had any other way.

Figure 10.3 Ti-6Al-4V head-mounted display component built via electron-beam melting (EBM). *(CAD design courtesy of Dr. Jannick Rolland, University of Central Florida.)*

10.1.1 Electron-beam melting

Electron-beam melting (EBM) is a process developed by Arcam AB of Sweden (*www.arcam.com*). Alloys such as H13 steel, Ti-6Al-4V, TiAl, and Inconel can be fabricated with this process. A thin layer of metal powder, typically 100 μm thick, is first spread over a plate. An electron-beam gun is selectively scanned over the powder bed with enough power to melt the metal powder that it comes in contact with. The platform is lowered, a new layer of powder is spread, and the process repeats itself. The entire process takes place in vacuum, which helps to prevent unwanted oxidation of metal powders at elevated temperatures. On completion of the final layer, the part is removed from the bed of loose unmelted powder that surrounds it. The unmelted powder surrounding the part also serves to support any downward-facing surfaces.

Parts fabricated via most powder-based RM processes, including EBM, possess textured surfaces somewhat resembling those of sand

castings. For example, Fig. 10.3 shows an as-processed Ti-6Al-4V head-mounted display component built via EBM prior to finish machining. As is the case with sand castings, critical features are subjected to any necessary finish machining operations. Approximately 0.5 to 1.5 mm of material typically is recommended as a machining allowance. Finish machining of Ti-6Al-4V parts built with this process has been achieved using TiAlN-coated end mills with a cutting speed of 50 ft/min and a feed rate of 0.007 in/tooth. Figure 10.4 shows a finish-machined nozzle component built via the EBM process in the Ti-6Al-4V alloy.

Despite the fact that metal is melted in layers, components fabricated via EBM possess excellent interlayer bonding. The fact that each layer is so thin also means that melt solidification takes place very rapidly, thus leading to a refined microstructure. High-performance materials that are difficult to process, such as titanium aluminide, can be processed via the EBM process owing primarily to the fact that the entire part is kept at an elevated temperature during processing. This significantly reduces thermal stresses that would be caused by depositing molten material on top of (relatively) cold previously solidified material. Thermal stresses generally lead to warping and/or cracking in the workpiece; hence proper thermal management is critical for any of the direct-metal RM processes.

One particularly intriguing aspect of EBM and other direct-metal RM processes is that locating features easily can be built into each component to facilitate finish machining.

Figure 10.4   Finish-machined titanium component. *(Photo courtesy of Arcam AB.)*

10.1.2 Selective laser sintering (SLS)

3D System's SLS process (*www.3dsystems.com*) is capable of processing a wide variety of polymer, metal, and ceramic powders. Of particular interest to the machinist or metalworker is the Laser-Form A6 steel material. The LaserForm material consists of A6 steel particles that have been coated with a polymeric binder. A thin layer of the powder is spread over a platform, and a 100-W (typical) laser scans the cross-sectional area of the first layer to be produced. Heat from the laser causes the polymer binder to melt, thus agglomerating together the metal particles for that layer. The process is repeated one layer after the next until the part is complete. The green polymer-bound part is removed from the powder bed and placed in a furnace. The polymer binder is vaporized under careful control, and the remaining steel particles are sintered. The resulting component is porous sintered A6 steel. If full density is desired, the part can be infiltrated with a secondary metal such as bronze. Figure 10.5 shows a photograph of a well-known rook chess piece that was built in steel and infiltrated with bronze using the SLS process.

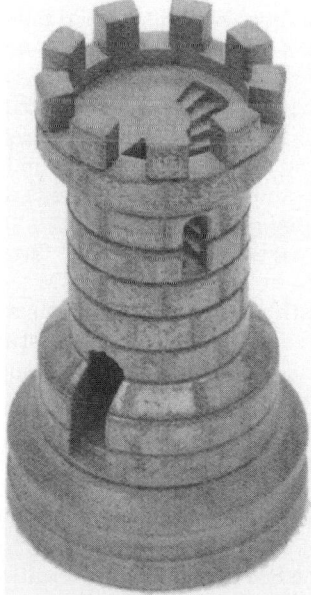

Figure 10.5 SLS rook chess piece in bronze-infiltrated tool steel.

10.1.3 Laser additive manufacturing (LAM)

Rapid prototyping processes based on laser forming generally involve using a high-power laser to start a melt pool. One or more powder feeders then are used to direct metal powder into the melt pool at a predetermined rate. The melt pool typically is shielded with inert gas such as argon to inhibit oxidation. The laser and powder are moved simultaneously above the surface of the substrate, thus producing a moving melt pool into which new powder material is constantly added. Slightly varying versions of laser additive manufacturing processes have been commercialized by Optomec (*www.optomec.com*), Aeromet (*www.aerometcorp.com*), and POM (*www.pomgroup.com*).

A particularly attractive advantage of LAM processes is that multiple powder feeders can be used. By using multiple powder feeders and varying the rate with which each powder is added to the melt pool, it is possible to locally vary the material composition of a component to match its mechanical requirements in service. For instance, the composition of a component might be tailored to provide a tough, fracture-resistant core that is surrounded by a hard, wear-resistant outer skin. Figure 10.6 shows a laser deposition nozzle with multiple powder feeders.

Because there typically is no powder bed surrounding the part(s) being fabricated, production of components with downward-facing surfaces can be a challenge. Systems equipped with three-axis motion controllers can deposit material only on gradually tapered downward-facing surfaces. Surfaces with steeper downward-facing slopes generally must be fabricated on systems equipped with five-axis positioning systems and specialized software to generate material deposition toolpaths that will not result in collisions between the powder feeders and previously deposited materials.

Although fabricating complex geometries with downward-facing surfaces generally is more difficult to accomplish with LAM systems that do not surround the part with powder, LAM systems are very well suited to repairing damaged or worn surfaces on existing parts. Because of their ability to control chemical composition, they are also used to deposit special coatings on existing parts.

10.2 Rapid Tooling

Rapid tooling can be defined in many ways. In the simplest sense, *rapid tooling* refers to the practice of either directly fabricating a tool (mold, die, etc.) from any one of the RM processes or fabricating a tool using an RP/RM component as a pattern.

Figure 10.6 Laser-engineered net shaping with multiple powder feed nozzles. *(Photo used with permission of Sandia National Laboratories.)*

10.2.1 RTV silicone molds

One of the most widely used approaches to rapid tooling involves making room temperature vulcanizing (RTV) silicone molds from a pattern made on any one of the RP/RM processes. The RTV mold may be fabricated in two pieces, or it may be a one-piece cut mold. Regardless of how the RTV mold is made, polyurethane or other two-part resins are then cast into the mold to produce a plastic part. The urethane may include dyes and/or fillers to modify its appearance and material properties. Common fillers include glass beads to reduce part weight and to reduce the volume (and cost) of the resin being used. Chopped fiberglass strands sometimes are added to increase part strength. Metal powders such as bronze are also added at times to make the part look and feel like a metal part. Provided the powder loading is sufficiently high, it is possible to buff the parts in order to achieve a somewhat shiny metallic look. Depending on the resin being cast, the workpiece geometry, and any fillers/additives, the typical RTV mold usually is good for approximately 30 to 50 castings.

Fabricating One-Piece Cut RTV Molds. The process for fabricating a one-piece cut mold is relatively straightforward.

1. Any of the RP processes is used to fabricate a pattern of the part to be molded. Shrinkage allowance for the resin used must be factored into the pattern dimensions.

2. Visual inspection of the part determines the parting line for the two halves of the mold. Note that the RTV is flexible; hence small to moderate undercut features generally may be molded without difficulty. To aid in cutting the mold apart in a later step, the parting line is often drawn onto the part using a permanent marker.

3. The RP part is suspended by wires and/or pins inside of a box with approximately 25 mm (1 in) of clearance on all sides. Care must be taken to ensure that the part is held firmly in place. When the RTV silicone is poured into the box, a loosely attached part can break free and either float or sink depending on its buoyancy in the RTV silicone. At this point, it is helpful to visually project the parting line from the part out to the walls of the box. A permanent marker is then used to sketch the location of the projected parting line around the perimeter of the box.

4. The RTV silicone rubber is mixed and degassed according to the manufacturer's recommendations and then is poured slowly into the box until the liquid level is approximately 25 mm (1 in) above the top of the part. The box is set aside and allowed to vulcanize (cure) for approximately 24 hours. Note that there are a wide variety of RTV rubbers that vary according to their translucency, hardness, tear strength, etc. It is generally much easier to make one-piece cut molds using clear or somewhat translucent RTV materials than opaque ones.

5. Once the rubber has cured, the walls of the box are removed. A number of knives then are used to cut the rubber down to the parting line of the part. Once the parting line has been located, the knives are used to follow the parting line around the part. Note that a slight zigzag pattern should be cut as the rubber halves are separated. Doing so will provide self-interlocking features so that the reassembled mold halves will align properly during casting.

6. Once the mold is cut apart, the RP pattern is removed, thus leaving a cavity in the rubber mold.

7. The knife is used to cut a pouring basin into the rubber, along with any vent channels that may be needed. The urethane generally is poured into the highest point on the part. It is helpful to visualize the urethane filling up the mold. Air vents should be cut into the mold at any locations where air is likely to become trapped during the filling process.

Fabricating Two-Piece RTV Molds.   The process for fabricating a two-piece RTV mold is only slightly more complicated than that used to make one-piece cut molds.

1. Any of the RP processes is used to fabricate a pattern of the part to be molded. Shrinkage allowance for the resin used must be factored into the pattern dimensions.

2. Visual inspection of the part determines the parting line for the two halves of the mold. The part is then embedded in modeling clay or a similar material. Various hand tools are used to build the clay up to the parting line for the part. At this point, the upper half of the part protrudes above the clay along the parting line.

3. The clay-supported pattern is then placed inside a box. A suitable release agent is sprayed on the exposed part, clay, and inner box walls.

4. The RTV silicone rubber is mixed and degassed according to the manufacturer's recommendations and then is poured slowly onto the exposed part and clay surface until the liquid level is approximately 25 mm (1 in) above the top of the part. The box is set aside and is allowed to vulcanize (cure) for approximately 24 hours.

5. The following day, the box is flipped over, and the clay that that part was embedded in from beneath is removed. At this point, the upper half of the part is now embedded in the cured RTV silicone.

6. A knife is used to cut pyramidal (or other) chunks of cured RTV silicone away around the perimeter of the part. When the second batch of RTV is poured, this will lead to the formation of interlocking features that keep the two mold halves aligned.

7. The exposed part and RTV silicone surfaces are sprayed with a suitable release agent.

8. A second batch of RTV silicone rubber is mixed and poured over the exposed part and RTV surfaces. This is allowed to cure for approximately 24 hours.

9. Once the rubber has cured, the walls of the box are removed. Provided an appropriate release agent has been used, the two cured halves of RTV silicone should just peel apart along the parting line without any need to cut them with a knife.

10. The RP pattern is removed, thus leaving a cavity in the rubber mold.

11. The knife is used to cut a pouring basin into the rubber along with any vent channels that may be needed.

Casting the Polyurethane

1. Regardless of which type of mold has been produced, all mold surfaces are sprayed with a release agent that has been matched to the urethane being cast. The mold halves then are clamped together between a pair of flat boards. Hand clamps or even rubber bands can be used to hold the assembled mold together. Since the RTV rubber is compressible, care must be taken to avoid distorting the cavity when the mold halves are clamped.

2. The urethane and any colorants/fillers/etc. are mixed together and poured slowly into the cavity. Once the urethane has hardened (generally 1 to 15 minutes depending on the urethane formulation), the mold is opened up, and the part is removed.

3. The process can be repeated until the polyurethane begins to stick to the mold despite the use of a mold releasing agent.

Caution: Be sure to follow the polyurethane manufacturer's recommended safety precautions regarding proper ventilation as well as skin and eye protection. Many of the uncured polyurethane compounds are considered carcinogenic. Also take note of the fact that the reaction taking place during curing is exothermic. The amount of heat released can be sufficient to burn skin in some cases.

10.2.2   Direct-inject stereolithography tooling

Direct-inject SLA tooling refers to the practice of using photosensitive resins with the stereolithography process (*www.3dsystems.com*) to

produce an injection-mold insert. The insert goes into a mold base in an injection-molding press. SLA tools are not intended for production use, but they are a fast and relatively inexpensive way to produce mold inserts that can be used to injection-mold small batches of functional plastic parts in the desired resin. Depending on the melting temperature of the resin being injected and the presence or absence of fillers (e.g., glass-filled nylon), direct-inject tooling may last from just a few shots to several hundred shots.

When the molded part has complex geometry, it is customary to produce a "hand mold." Rather than assembling a complex tool with hydraulic slides and other moving components, the direct-inject inserts with removable cores are assembled by hand and placed in the mold base. The inserts are then removed from the mold base following each shot, and the cores are removed from the part by hand. The tooling then is reassembled, and the next shot is made. This is obviously not a suitable approach for mass production, but it is quite feasible to produce tens or even hundreds of sample plastic parts by this method.

Figure 10.7 shows a close up photograph of a direct-inject six-cavity SLA insert used to mold miniature worm gears. The mating insert is not shown in the photograph because it is nearly identical to the one shown in Fig. 10.7. Each insert is 35 mm (1⅜ in) square and 9.5 mm (⅜ in) deep. It took approximately 1 hour of machine time to fabricate this pair of inserts on a high-resolution stereolithography machine without any need to program computer numerical

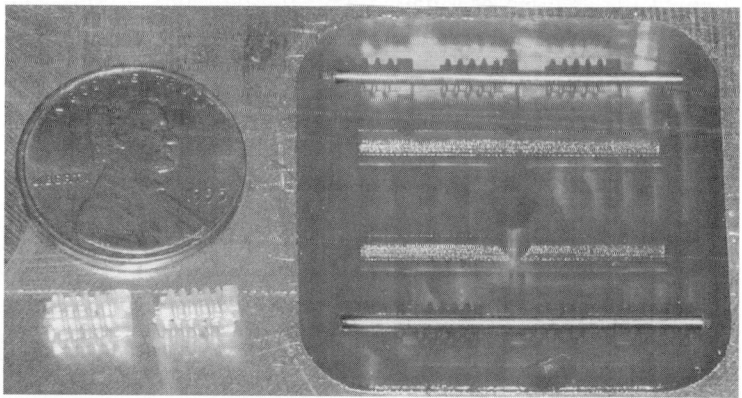

Figure 10.7   Direct-inject SLA inserts for molding worm gears.

control (CNC) toolpaths or to design an electrode. A steel pin insert is placed in the mold prior to injection, and the worms are slid off the pin following injection. The worms are approximately 6 mm long, with a diameter of just under 3 mm. A pair of molded worms is shown in the lower left of the photograph. The penny is included simply to provide the reader with an idea of the scale involved.

10.2.3 Electroplated prototypes

For instances where it is helpful to have metal-like parts that do not necessarily need the same material properties as a final production metal part, it is possible to electroplate prototype components. Chapter 13 provides details on various plating processes, although plating nonconducting polymer components can be tricky.

A simple though not terribly reliable approach to obtaining an electrically conducting surface on the plastic prior to electroplating is to manually paint it with conductive paints. Paints containing silver or copper powders are readily available for this purpose. This approach will work in simple cases, but the adhesion of the paint to the plastic surface during subsequent electroplating is often inadequate. The result is blistering or flaking of the plated surface.

A more reliable approach to preparing plastics for plating is as follows: First, the components to be plated must be completely cleaned of any oils, solvents, etc. In order to improve adhesion of the plating to the plastic, the surface of the plastic often is etched in a chromic acid bath. Next, the surfaces of the parts are sensitized through the addition of a palladium chloride solution. This step allows palladium metal to deposit on the etched surface of the plastic. Either electroless nickel or copper coatings may then be applied, with the palladium acting as a catalyst for nickel or copper deposits. A very thin (≤ 1 μm) electroless nickel coating is common. At this point, the surface of the plastic has been made to be electrically conducting with reasonably good adhesion properties. The surface then can be electroplated using techniques described in Chap. 13. When the part is to undergo mechanical stress, a copper plating often is recommended prior to nickel plating. Nickel platings can be followed by a chrome plate if desired for aesthetic purposes.

Many electroplating compounds are highly toxic and are highly regulated. Consequently, prototype plastic parts often are sent out to specialty shops for plating. Figure 10.8 shows a batch of complex

Figure 10.8 SLArmor air pump components. *(Photo courtesy of Fineline Prototyping, Inc.)*

handheld air compressor components that were fabricated in a DSM Somos Prototool material on a high-resolution stereolithography machine by Fineline Prototyping, Inc. The parts then were given a proprietary SLArmor coating that consists primarily of a copper strike coat followed by an electroless nickel plating. With minimal hand finishing, these components were assembled into a fully functional handheld air compressor.

10.2.4 Investment casting via 3D printing

Three-dimensional (3D) printing is a process that was developed at the Massachusetts Institute of Technology (MIT) and subsequently brought to market by several companies, including Z Corporation, ExtrudeHone, Soligen, and Therics. Z Corporation has introduced a special powder and binder formulation for fabricating investment casting shells. The company refers to this process as *ZCast*. At the present time, it is used for casting nonferrous alloys such as zinc or aluminum. As is the case with many RP processes, Z Corporation's version of 3D printing starts by spreading a thin layer of powder. In this case, the powder being spread is a specially formulated casting investment. An inkjet print system is then scanned across the powder bed. The printer prints an image of the current cross-sectional slice of the investment casting shell being produced. The moisture in the ink is the binder that causes the casting investment to

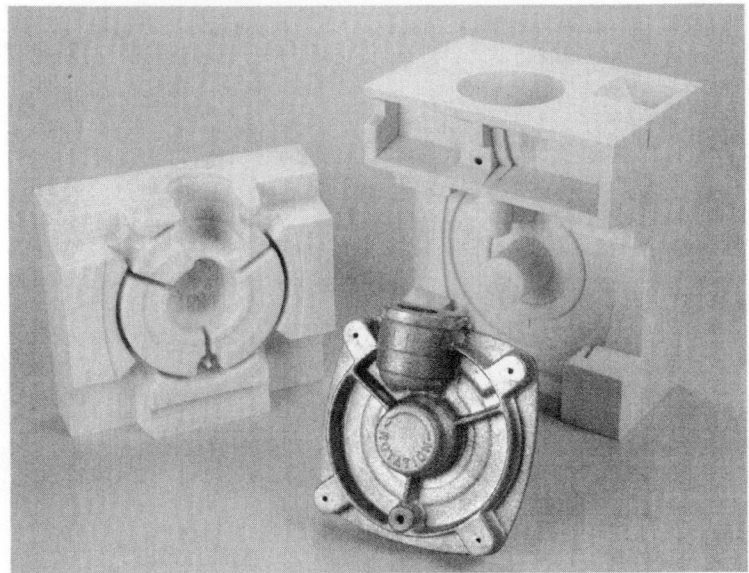

Figure 10.9 ZCast cover with mold. *(Photo courtesy of ZCorp.)*

harden. This is akin to mixing water with plaster, except on a much smaller scale. This process is used to fabricate the shell in two pieces, as well as any cores that are needed. Figure 10.9 shows a completed cover housing along with the ZCast tooling needed to produce such a casting.

Hardening and Tempering Steels and Nonferrous Alloys

The hardening and tempering of steels and nonferrous alloys are important aspects of metalworking. Carbon and alloy steels are relied on to perform a great number of services in the metalworking industries. Some of the nonferrous metals and alloys are also capable of being hardened above their normal condition either through heat treatment or cold working and find countless uses in product design and manufacturing.

11.1 Standard Steels and Steel-Making Practices

Steel is the generic name for a large group of iron-carbon alloys. The basic materials used in steel making are iron ore, coke, and limestone. A blast furnace converts these materials into a product known as *pig iron,* which contains considerable amounts of carbon, manganese, sulfur, phosphorus, and silicon. Basic oxygen furnaces are also employed in steel making, as are other methods. Steel making's basic constituent, pig iron, is hard and brittle and unsuitable for processing into usable wrought iron or steel products. Steel making is the process of refining pig iron and iron and steel scrap by removing unwanted elements and adding the desirable elements in predetermined and controlled amounts. Most steel-making processes cause a combination of carbon and oxygen to form a gas. When a steel is

deoxidized strongly with a deoxidizing agent, no gas forms, and the steel is called *killed steel.* The degree of deoxidation affects some of the properties of the steel, and the degree of gas evolution characterizes steels that are known as *semikilled, capped,* and *rimmed.* In addition to oxygen, fused steel contains small amounts of hydrogen and nitrogen. Special deoxidation practices, including vacuum treatment, may be used to control the amount of dissolved gases in the steel.

The carbon content of the common steel grades ranges from a few hundredths of 1 percent to 0.95 percent. All common steels contain manganese, sulfur, phosphorus, and silicon to some degree.

Wrought steels are the most common and widely used engineering materials. There is no other single material that offers such a broad range of practical applications as the various types and grades of steels.

The unified numbering system for steels is shown in Chap. 4, "Materials and Their Uses." This system classifies the various types of steels and provides identification numbers.

11.2   Constituents of Steel: Phases

When carbon steels are heated to various temperatures, changes take place in the structure that are known as *phases.* Figure 11.1 is a diagram showing the relationship between temperature and the amount of carbon in the steel that affects the basic structure. The diagram is presented strictly for academic reasons and serves the purpose of describing the different states that the steel assumes when the temperature and carbon content are varied. The actions and reactions that occur during the heating and cooling of carbon and alloy steels are complex and form the basis for the controlled heat treatments that are performed on the various types of steels.

The diagram in Fig. 11.1 indicates the phase transformations that occur in steels with up to 6.67 percent carbon content, where the main form of the steel above 2066°F is called *cementite.* The other forms or states/phases are indicated by the various letters and combinations of letters shown below the figure. Some of the other forms that occur are shown below the letter designations, the most important of which is *martensite.* Martensite is the phase or form of carbon steel that is produced in the hardening process, which will be described in a later section.

Carbon steels with less than 0.85 percent carbon are called *hypoeutectoid,* and those with carbon contents greater than 0.85

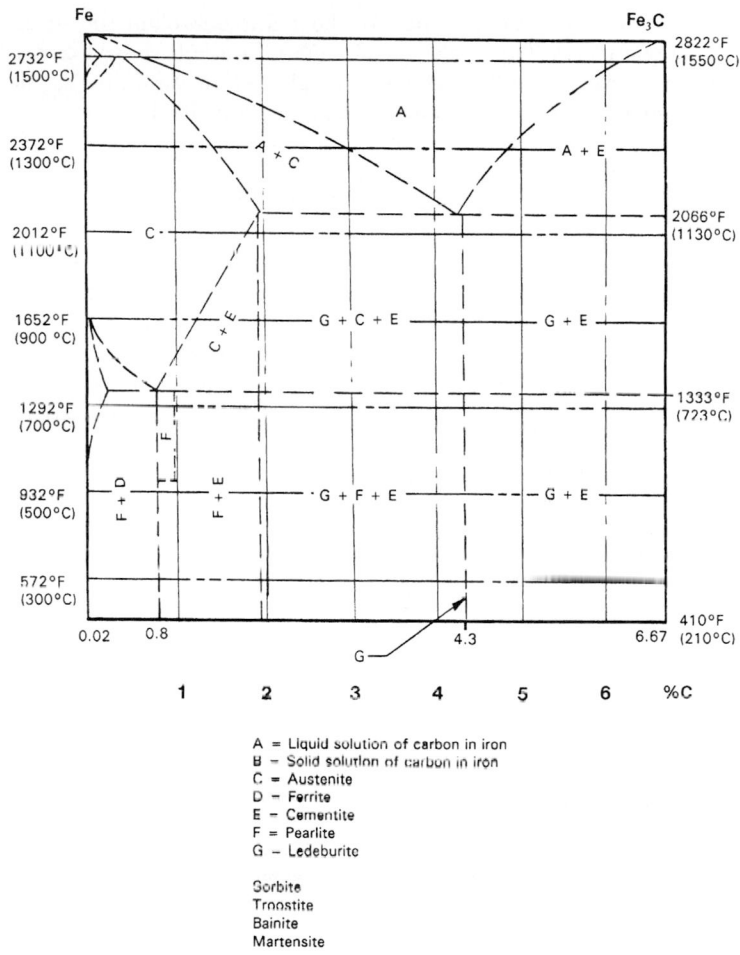

A = Liquid solution of carbon in iron
B = Solid solution of carbon in iron
C = Austenite
D = Ferrite
E = Cementite
F = Pearlite
G = Ledeburite

Sorbite
Troostite
Bainite
Martensite

Iron melting temperature = 2797°F, (1536°C)

Figure 11.1 Iron, iron-carbide diagram.

percent are called *hypereutectoid*. In binary-alloy systems, a *eutectoid alloy* is a mechanical mixture of two phases that form simultaneously from a solid solution when it cools through the eutectoid temperature. Alloys leaner or richer in one of the constituents or metals undergo transformation from the solid-solution phase over

a range of temperatures beginning above and ending at the eutectoid temperature. The structure of such alloys will consist of primary particles of one of the stable phases in addition to the eutectoid, e.g., ferrite and pearlite in low-carbon steel.

11.3   Standard Definitions of Terms Pertaining to Heat Treatment of Metals

Many terms are associated with the heat treatment of metals, and the American Society for Testing and Materials (ASTM) has classified these terms in ASTM Standard E44-84. Figure 11.2 outlines the terms and definitions as described in ASTM Standard E44-84 and is reproduced with permission from the ASTM.

By studying the diagram shown in Fig. 11.1 and the definitions of terms in Fig. 11.2, you will gain an excellent insight into the practice of heat treatment of metals.

11.4   Heat Treatment of Steels

In fully annealed carbon steels, the percentage of carbon determines the structural constitution of the steel. Figure 11.1 shows the constitution or phases for varying carbon content versus temperature.

Effect of heating fully annealed carbon steels.   When fully annealed carbon steels are heated above the lower critical point, which ranges between 1335 and 1355°F, depending on the carbon content, austenite is formed. If the temperature continues to rise, the steel structure will change completely to austenite. The temperature at which excess ferrite and cementite are completely dissolved in the austenite is called the *upper critical point.* This critical temperature varies with the carbon content of the steel. If the steel is cooled slowly to ambient temperature, the steel returns to its original condition.

Effect of rapid cooling or quenching carbon steels.   Simply stated, when carbon steels are heated to a certain temperature, suddenly cooling the steel at the proper cooling rate causes the austenite to transform to martensite, which has very high hardness. Thus the operation of hardening steels consists of two steps. The first step is to heat the steel at least 100°F higher than its transformation point, and the second step is to cool the steel at some rate that is

faster than the critical rate. (The *critical rate* is determined by or depends on the carbon content and alloying elements present in the steel.) The hardness of a martensitic steel depends on its carbon content and ranges from 460 Brinell at 0.20 percent carbon to 710 Brinell at 0.50 percent carbon. Ferrite has a hardness of approximately 90 Brinell, pearlite approximately 240 Brinell, and cementite approximately 550 Brinell.

Standard Definitions of Terms Relating to Heat Treatment of Metals
ASTM E44-84

(Temperatures have been omitted from these definitions, which are not intended as specifications)

Ac_{cm}, Ac_1, Ac_3, Ac_4—See **transformation temperature.**

age hardening—hardening by aging, usually after rapid cooling or cold working. See **aging.**

aging—a change in the properties of certain metals and alloys that occurs at ambient or moderately elevated temperatures after hot working or a heat treatment (quench aging in ferrous alloys, natural or artificial aging in ferrous and nonferrous alloys) or after a cold-working operation (strain aging). The change in properties is often, but not always, due to a phase change (precipitation) but never involves a change in chemical composition of the metals or alloys. See also **age hardening, artificial aging, natural aging, overaging, precipitation hardening, precipitation heat treatment, progressive aging, quench aging,** and **strain aging.**

annealing—heating to and holding at a suitable temperature and then cooling at a suitable rate, for such purposes as reducing hardness, improving machinability, facilitating cold working, producing a desired microstructure, or obtaining desired mechanical, physical, or other properties. When applicable, the following more specific terms should be used:

black annealing
blue annealing
box annealing
bright annealing
flame annealing
full annealing
graphitizing
intermediate annealing
isothermal annealing
malleablcizing
process annealing
quench annealing
recrystallization annealing
spheroidizing

Definitions of the above terms are given below in their alphabetical positions.

When applied to ferrous alloys, the term "annealing,"

without qualification, implies full annealing.

When applied to nonferrous alloys, the term "annealing" implies a heat treatment designed to soften a cold-worked structure by recrystallization or subsequent grain growth or to soften an age-hardened alloy by causing a nearly complete precipitation of the second phase in relatively coarse form.

Any process of annealing will usually reduce stresses but if the treatment is applied for the sole purpose of such relief it should be designated **stress relieving.**

Ar_{cm}, Ar_1, Ar_3, Ar_4—See **transformation temperature.**

artificial aging—aging above room temperature. See **aging** and **precipitation heat treatment.** Compare with **natural aging.**

austempering—quenching a ferrous alloy from a temperature above the transformation range in a medium having a rate of heat abstraction high enough to prevent the formation of high-temperature transformation products, and then holding the alloy, until transformation is complete, at a temperature below that of pearlite formation and above that of martensite formation.

austenitizing—forming austenite by heating a ferrous alloy into the transformation range (partial austenitizing) or above the transformation range (complete austenitizing).

baking—heating to a low temperature in order to remove gases.

black annealing—box annealing or pot annealing ferrous alloy sheet, strip, or wire. See **box annealing.**

blank carburizing—simulating the carburizing operation without introducing carbon. This is usually accomplished by using an inert material in place of the carburizing agent, or by applying a suitable protective coating to the ferrous alloy.

blank nitriding—simulating the nitriding operation without introducing nitrogen. This is usually accomplished by using an inert material in place of the nitriding agent, or by applying a suitable protective coating to the ferrous alloy.

blue annealing—heating hot-rolled ferrous sheet in an open furnace to a temperature within the transformation range and then cooling in air, in order to soften the metal. The formation of a bluish oxide on the surface is incidental.

bluing—subjecting the scale-free surface of a ferrous alloy to the action of air, steam or other agents at a suitable temperature, thus forming a thin blue film of oxide and improving the appearance and resistance to corrosion.

[1] These definitions are under the jurisdiction of ASTM Committee A-1 on Steel, Stainless Steel, and Related Alloys and are the direct responsibility of Subcommittee A1.92 on Terminology.

Current edition approved Oct. 26, 1984. Published December 1984. Originally published as E 44 – 42 T. Last previous edition E 44 – 83.

Figure 11.2 Standard terms relating to the heat treatment of metals. *(Reprinted with permission from the Annual Book of ASTM Standards, copyright 1992, American Society for Testing and Materials.)*

NOTE—This term is ordinarily applied to sheet, strip, or finished parts. It is used also to denote the heating of springs after fabrication, in order to improve their properties.

box annealing—annealing a metal or alloy in a sealed container under conditions that minimize oxidation. In box annealing a ferrous alloy, the charge is usually heated slowly to a temperature below the transformation range, but sometimes above or within it, and is then cooled slowly; this process is also called "close annealing" or "pot annealing." See **black annealing.**

bright annealing—annealing in a protective medium to prevent discoloration of the bright surface.

burning (burnt, burned)—a term applied to metal which has been permanently damaged by having been heated to a temperature close to or within the melting range. This results in a structure exhibiting incipient melting or intergranular oxidation.

carbonitriding—a case hardening process in which a suitable ferrous material is heated above the lower transformation temperature in a gaseous atmosphere of such composition as to cause simultaneous absorption of carbon and nitrogen by the surface and, by diffusion, create a concentration gradient. The process is completed by cooling at a rate that produces the desired properties in the workpiece.

carbon potential—a measure of the ability of an environment containing active carbon to alter or maintain, under prescribed conditions, the carbon content of the steel.

NOTE—In any particular environment, the carbon level attained will depend on such factors as temperature, time, and steel composition.

carbon restoration—replacing the carbon lost in the surface layer from previous processing by carburizing this layer to substantially the original carbon level.

carburizing—a process in which an austenitized ferrous material is brought into contact with a carbonaceous atmosphere of sufficient carbon potential to cause absorption of carbon at the surface and, by diffusion, create a concentration gradient.

case—in a ferrous alloy, the outer portion that has been made harder than the inner portion (see **core**) as a result of altered composition, or structure, or both, from treatments such as carburizing, nitriding, and induction hardening.

case hardening—a generic term covering several processes applicable to steel that change the chemical composition of the surface layer by absorption of carbon, nitrogen, or a mixture of the two and, by diffusion, create a concentration gradient. The processes commonly used are: **carburizing** and **quench hardening; cyaniding; nitriding;** and **carbonitriding.** The use of the applicable specific process name is preferred.

cementation—the introduction of one or more elements into the outer portion of a metal object by means of diffusion at high temperature.

close annealing—See **box annealing.**

cold treatment—exposing to subzero temperatures for the purpose of obtaining desired conditions or properties, such as dimensional or structural stability. When the treatment involves transformation of retained austenite, it is usually followed by a tempering treatment.

conditioning heat treatment—a preliminary heat treatment used to prepare a material for a desired reaction to a subsequent heat treatment. For the term to be meaningful, the exact treatment must be specified.

controlled cooling—cooling from an elevated temperature in a predetermined manner to avoid hardening, cracking, or internal damage or to produce a desired microstructure or mechanical properties.

core—(1) case hardening—interior portion of unaltered composition, or microstructure, or both, of a case-hardened steel article.

(2) clad products—the central portion of a multilayer composite metallic material.

critical cooling rate—the minimum rate of continuous cooling to prevent undesirable transformations. For steel, unless otherwise specified, it is the slowest rate at which austenite can be cooled from above critical temperature to prevent its transformation above the M_s temperature.

critical temperature range—synonymous with **transformation range.** The term is of historic significance only, and its use is discouraged.

cyaniding—introducing carbon and nitrogen into a solid ferrous alloy by holding above Ac_1 in contact with molten cyanide of suitable composition. The cyanided alloy is usually quench hardened.

cycle annealing—an annealing process employing a predetermined and closely controlled time-temperature cycle to produce specific properties or microstructure.

decarburization—the loss of carbon from the surface of a ferrous alloy as a result of heating in a medium that reacts with the carbon.

differential heating—heating that intentionally produces a temperature gradient within an object such that, after cooling, a desired stress distribution or variation in properties is present within the object.

diffusion coating—any process whereby a basis metal or alloy is either: (1) coated with another metal or alloy and heated to a sufficient temperature in a suitable environment, or (2) exposed to a gaseous or liquid medium containing the other metal or alloy, thus causing diffusion of the coating or of the other metal or alloy into the basis metal with resultant change in the composition and properties of its surface.

direct quenching—quenching carburized parts directly from the carburizing operation.

double aging—employment of two different aging treatments to control the type of precipitate formed from a supersaturated alloy matrix in order to obtain the desired properties. The first aging treatment, sometimes referred to as intermediate or stabilizing, is usually carried out at a higher temperature than the second.

double tempering—a treatment in which quench-hardened steel is given two complete tempering cycles at substantially the same temperature for the purpose of assuring completion of the tempering reaction and promoting stability of the resulting microstructure.

drawing—a misnomer for **tempering.**

ferritizing anneal—the process of producing a predominantly ferritic matrix in a ferrous alloy through an appropriate heat treatment.

flame annealing—annealing in which the heat is applied directly by a flame.

Figure 11.2 (*Continued*)

flame hardening—a surface hardening process in which only the surface layer of a suitable workpiece is heated by a suitably intense flame to above the upper transformation temperature and immediately quenched.

fog quenching—quenching in a mist.

full annealing—annealing a ferrous alloy by austenitizing and then cooling slowly through the transformation range. The austenitizing temperature for hypoeutectoid steel is usually above Ac_3 and for hypereutectoid steel usually between Ac_1 and Ac_{cm}.

gas cyaniding—a misnomer for carbonitriding.

grain growth—an increase in the grain size of a metal, usually as a result of heating at an elevated temperature.

grain size—the dimensions of the grains or crystals in a polycrystalline metal exclusive of twinned regions and subgrains when present. Grain size is usually estimated or measured on the cross section of an aggregate of grains. Common units are: (*1*) average diameter, (*2*) average area, (*3*) number of grains per linear unit, (*4*) number of grains per unit area, and (*5*) number of grains per unit volume. See Methods E 112, for Determining The Average Grain Size.[2]

(*1*) ASTM grain size number—a grain size designation bearing a relationship to average intercept distance at 100 diameters magnification according to the equation: $G =$ ASTM Grain Size Number $= 10.00 - 2 \log_2 \bar{L}$, where $\bar{L}$ is the average intercept distance in millimetres at 100 diameters magnification.

(*2*) average grain diameter—the mean diameter of an equiaxed grain section whose size is representative of all the grain sections in the aggregate being measured.

graphitizing—annealing a ferrous alloy in such a way that some or all of the carbon is precipitated as graphite.

hardenability—in a ferrous alloy, the property that determines the depth and distribution of hardness induced by quenching.

hardening—increasing the hardness by suitable treatment, usually involving heating and cooling. When applicable, the following more specific terms should be used: **age hardening, case hardening, flame hardening, induction hardening, precipitation hardening,** and **quench hardening.**

heat treatment—heating and cooling a solid metal or alloy in such a way as to obtain desired conditions or properties. Heating for the sole purpose of hot working is excluded from the meaning of this definition.

homogeneous carburizing—a process that converts a low-carbon ferrous alloy to one of substantially uniform and higher carbon content throughout the section, so that a specific response to hardening may be obtained.

homogenizing—holding at high temperature to eliminate or decrease chemical segregation by diffusion.

hot-cold working—mechanical deformation of austenitic and precipitation hardening alloys at a temperature just below the recrystallization range to increase the yield strength and hardness by either plastic deformation or precipitation hardening effects induced by plastic deformation, or both.

hot quenching—an imprecise term used to cover a variety of quenching procedures in which a quenching medium is maintained at a prescribed temperature above 160°F (71°C).

induction hardening—a surface hardening process in which only the surface layer of a suitable ferrous workpiece is heated by electrical induction to above the upper transformation temperature and immediately quenched.

induction heating—heating by electrical induction.

intermediate annealing—annealing wrought metals at one or more stages during manufacture and before final thermal treatment.

interrupted aging—aging at two or more temperatures, by steps, and cooling to room temperature after each step. See aging and compare with progressive aging.

interrupted quenching—quenching in which the metal object being quenched is removed from the quenching medium while the object is at a temperature substantially higher than that of the quenching medium. See also **time quenching.**

isothermal annealing—austenitizing a ferrous alloy and then cooling to and holding at a temperature at which austenite transforms to a relatively soft ferrite-carbide aggregate.

isothermal transformation—a change in phase at any constant temperature.

malleableizing—a process in which the ascast malleable-type (white) iron is thermally treated for the purpose of converting most or all of the carbon in Fe_3C to graphite (temper carbon) to produce a family of products with improved ductility.

maraging—a precipitation hardening treatment applied to a special group of iron-base alloys to precipitate one or more intermetallic compounds in a matrix of essentially carbon-free martensite.

NOTE—The first developed series of maraging steels contained, in addition to iron, more than 10 % nickel and one or more supplemental hardening elements. In this series, the aging is done at approximately 900°F (482°C).

martempering—quenching an austenitized ferrous alloy in a medium at a temperature in the upper part of the martensite range, or slightly above that range, and holding in the medium until the temperature throughout the alloy is substantially uniform. The alloy is then allowed to cool in air through the martensite range.

martensite range—the temperature interval between M_s and M_f.

M_f—See **transformation temperature.**

M_s—See **transformation temperature.**

natural aging—spontaneous aging of a super-saturated solid solution at room temperature. See aging and compare with **artificial aging.**

nitriding—introducing nitrogen into a solid ferrous alloy by holding at a suitable temperature (below Ac_1 for ferritic steels) in contact with a nitrogenous material, usually ammonia or molten cyanide of appropriate composition. Quenching is not required to produce a hard case.

normalizing—heating a ferrous alloy to a suitable temperature above the transformation range and then cooling in air to a temperature substantially below the transformation range.

Figure 11.2 (*Continued*)

overaging—aging under conditions of time and temperature greater than those required to obtain maximum change in a certain property, so that the property is altered in the direction of the initial value. See **aging.**

overheating—*(1)in ferrous alloys,* heating to an excessively high temperature such that the properties/structure undergo modification. The resulting structure is very coarsegrained. Unlike burning, it may be possible to restore the original properties/structure by further heat treatment or mechanical working, or a combination thereof.

(2) in aluminum alloys, overheating produces structures that show areas of resolidified eutectic or other evidence that indicates the metal has been heated within the melting range.

patenting—in wire making, a heat treatment applied to medium-carbon or high-carbon steel before the drawing of wire or between drafts. This process consists in heating to a temperature above the transformation range and then cooling to a temperature below Ae_1 in air or in a bath of molten lead or salt.

postheating—heating weldments immediately after welding, for tempering, for stress relieving, or for providing a controlled rate of cooling to prevent formation of a hard or brittle structure.

pot annealing—See **box annealing.**

precipitation hardening—hardening caused by the precipitation of a constituent from a supersaturated solid solution. See also **age hardening** and **aging.**

precipitation heat treatment—artificial aging in which a constituent precipitates from a supersaturated solid solution. See **artificial aging, interrupted aging,** and **progressive aging.**

preheating—heating before some further thermal or mechanical treatment. For tool steel, heating to an intermediate temperature immediately before final austenitizing. For some nonferrous alloys, heating to a high temperature for a long time in order to homogenize the structure before working.

process annealing—in the sheet and wire industries, heating a ferrous alloy to a temperature close to, but below, the lower limit of the transformation range and then cooling, in order to soften the alloy for further cold working.

progressive aging—aging by increasing the temperature in steps or continuously during the aging cycle. See **aging** and compare with **interrupted aging.**

pseudocarburizing—See **blank carburizing.**

pseudonitriding—See **blank nitriding.**

quench aging—aging induced by rapid cooling after **solution heat treatment.**

quench annealing—annealing an austenitic ferrous alloy by **solution heat treatment.**

quench hardening—hardening a ferrous alloy by austenitizing and then cooling rapidly enough so that some or all of the austenite transforms to martensite. The austenitizing temperature for hypoeutectoid steels is usually above Ac_3 and for hypereutectoid steels usually between Ac_1 and Ac_{cm}.

quenching—rapid cooling. When applicable, the following more specific terms should be used: **direct quenching, fog quenching, hot quenching, interrupted quenching, selective quenching, spray quenching,** and **time quenching.**

recrystallization—the formation of a new grain structure through nucleation and growth commonly produced by subjecting a metal, which may be strained, to suitable conditions of time and temperature.

recrystallization annealing—annealing cold-worked metal to produce a new grain structure without phase change.

recrystallization temperature—the approximate minimum temperature at which recrystallization of a cold worked metal occurs within a specified time.

secondary hardening—the hardening phenomenon that occurs during high-temperature tempering of certain steels containing one or more carbide-forming alloying elements. Up to an optimum combination of tempering time and temperature, the reaction results either in the retention of hardness or an actual increase in hardness.

selective heating—intentionally heating only certain portions of an object.

selective quenching—quenching only certain portions of an object.

shell hardening—a surface hardening process in which a suitable steel workpiece, when heated through and quench hardened, develops a martensitic layer or shell that closely follows the contour of the piece and surrounds a core of essentially pearlitic transformation product. This result is accomplished by a proper balance between section size, steel hardenability, and severity of quench.

slack quenching—the incomplete hardening of steel due to quenching from the austenitizing temperature at a rate slower than the *critical cooling rate* for the particular steel, resulting in the formation of one or more transformation products in addition to martensite.

snap temper—a precautionary interim stress-relieving treatment applied to high-hardenability steels immediately after quenching to prevent cracking because of delay in tempering them at the prescribed higher temperature.

soaking—prolonged holding at a selected temperature.

solution heat treatment—heating an alloy to a suitable temperature, holding at that temperature long enough to cause one or more constituents to enter into solid solution and then cooling rapidly enough to hold these constituents in solution.

spheroidizing—heating and cooling to produce a spheroidal or globular form of carbide in steel. Spheroidizing methods frequently used are:

(1) Prolonged holding at a temperature just below Ae_1.

(2) Heating and cooling alternately between temperatures that are just above and just below Ae_1.

(3) Heating to a temperature above Ae_1 or Ae_3 and then cooling very slowly in the furnace or holding at a temperature just below Ae_1.

(4) Cooling at a suitable rate from the minimum temperature at which all carbide is dissolved, to prevent the re-formation of a carbide network, and then reheating in accordance with method *(1)* or *(2)* above. (Applicable to hypereutectoid steel containing a carbide network.)

spinodal decomposition—mechanism of a phase separation from a solid solution into two homogeneous phases of different chemical composition, each having the same crystal structure as the parent metal.

spray quenching—quenching in a spray of liquid.

Figure 11.2 *(Continued)*

stabilizing treatment—any treatment intended to stabilize the structure of an alloy or the dimensions of a part.

(*1*) heating austenitic stainless steels that contain titanium, columbium, or tantalum to a suitable temperature below that of a full anneal in order to inactivate the maximum amount of carbon by precipitation as a carbide of titanium, columbium, or tantalum.

(*2*) transforming retained austenite in parts made from tool steel.

(*3*) precipitating a constituent from a non-ferrous solid solution to improve the workability, to decrease the tendency of certain alloys to age harden at room temperature, or to obtain dimensional stability.

strain aging—aging induced by cold working. See **aging**.

stress relieving—heating to a suitable temperature, holding long enough to reduce residual stresses and then cooling slowly enough to minimize the development of new residual stresses.

surface hardening—a generic term covering several processes applicable to a suitable ferrous alloy that produces by quench hardening only, a surface layer that is harder or more wear resistant than the core. There is no significant alteration of the chemical composition of the surface layer. The processes commonly used are induction hardening, flame hardening, and shell hardening. Use of the applicable specific process name is preferred.

temper brittleness—brittleness that results when certain steels are held within, or are cooled slowly through, a certain range of temperature below the transformation range. The brittleness is revealed by notched-bar impact tests at or below room temperature.

tempering—(*1*) reheating a quench hardened or normalized ferrous alloy to a temperature below the transformation range (Ac_1), and then cooling at any desired rate. (*2*) a term used in conjunction with a qualifying adjective to designate the relative properties of a particular metal or alloy induced by cold work or heat treatment, or both.

time quenching—interrupted quenching in which the duration of holding in the quenching medium is controlled.

transformation ranges or **transformation temperature ranges**—those ranges of temperature within which austenite forms during heating and transforms during cooling. The two ranges are distinct, sometimes overlapping but never coinciding. The limiting temperatures of the ranges depend on the composition of the alloy and on the rate of change of temperature, particularly during cooling. See **transformation temperature**.

transformation temperature—the temperature at which a change in phase occurs. The term is sometimes used to denote the limiting temperature of a transformation range. The following symbols are used for iron and steels:

Ac_{cm}—in hypereutectoid steel, the temperature at which the solution of cementite in austenite is completed during heating.

Ac_1—the temperature at which austenite begins to form during heating.

Ac_3—the temperature at which transformation of ferrite to austenite is completed during heating.

Ac_4—the temperature at which austenite transforms to delta ferrite during heating.

Ae_1, Ae_3, Ae_{cm}, Ae_4—the temperatures of phase changes at equilibrium.

Ar_{cm}—in hypereutectoid steel, the temperature at which precipitation of cementite starts during cooling.

Ar_1—the temperature at which transformation of austenite to ferrite or to ferrite plus cementite is completed during cooling.

Ar_3—the temperature at which austenite begins to transform to ferrite during cooling.

Ar_4—the temperature at which delta ferrite transforms to austenite during cooling.

M_s—the temperature at which transformation of austenite to martensite starts during cooling.

M_f—the temperature, during cooling, at which transformation of austenite to martensite is substantially completed.

NOTE—All these changes except the formation of martensite occur at lower temperatures during cooling than during heating, and depend on the rate of change of temperature.

Figure 11.2   (*Continued*)

The critical temperature points for hardening steel are also called the *decalescence* and *recalescence points*. These critical temperature points have a direct relation to the hardening of steel. Unless the hardening temperature passes the decalescence point, no hardening can take place, and unless the steel is cooled suddenly before its temperature reaches the recalescence point, no hardening can take place. These critical temperature points (decalescence and recalescence) vary for different types of steels and must be determined by tests. This variation in critical temperature points makes it necessary to heat different steels to different hardening temperatures together with the proper quench, or cooling rate.

The maximum temperature to which a steel is heated before quenching to harden is called the *hardening temperature*. The hardening temperatures for steels of various carbon contents may be summarized generally by the following:

Carbon content, %	Hardening temperature, °F
0.65–0.80	1450–1550
0.80–0.95	1410–1460
0.95–1.10	1390–1430
1.10 and over	1380–1420

Average hardening and tempering temperatures of steels.   Generally, the average hardening temperature range for carbon and alloy steels is from 1375 to 1575°F. The average tempering range for steels is from 300 to 700°F.

11.4.1   Treatments for heat-treating grades of carbon steels

Figure 11.3 lists data for heat treating the common-usage heat-treating grades of carbon steels. The tempering temperatures are not shown, and tempering is not mandatory on many applications. When tempering is required, the tempering range varies from 250 to 450°F.

11.4.2   Heat treatments for directly hardenable grades of alloy steels

Figure 11.4 lists data for heat treating the common-usage directly hardenable grades of alloy steels. Tempering temperatures range from 250 to 350°F.

11.4.3   Heat treatment of tool steels

Tool steels are an important family of steels that have many applications throughout industry. The heat treatment and application of the different grades or classes of tool steels are covered in Sec. 8.2.2, "Heat Treatment of Tool Steels."

11.4.4   Quenching baths and temperatures

The purpose of a quenching bath is to cool the heated steel at a rate that is faster than the critical cooling rate. To obtain different cooling rates, different quenching baths are used, including

- Water bath (70 to 100°F bath temperature)
- Water spray (ambient temperature)

SAE Steel	Normalizing Temperature (°F)	Annealing Temperature (°F)	Hardening Temperature (°F)	Quenching Medium
1025 1030	.....	.....	1575-1650	Water or brine
1035	.....	.....	1525-1575	oil or water
1036	1600-1700	.....	1525-1575	oil or water
1038 1039 1040	1600-1700	.....	1525-1575	oil or water
1041	1600-1700 and/or	1400-1500	1474-1550	oil
1042 1043* 1045* 1046* 1050*	1600-1700	.....	1475-1550	oil or water,
1052 1055 1060 1065 1070 1074	1550-1650 and/or	1400-1500	1475-1550	oil
1078	.....	1400-1500	1450-1500	water or brine
1080 1090	1550-1650 and/or	1400-1500■	1450-1500	oil ♦
1095	....	1400-1500■ 1400-1500■	1450-1500 1500-1600	oil, water or brine oil
1137	1600-1700 and/or	1400-1500	1525-1575	oil or water
1140	1600-1700	.....	1500-1550	oil or water
1141 1144	1600-1700	1400-1500	1475-1550	oil
1146	1600-1700	.....	1475-1550	oil or water

NOTES: * = Commonly used on parts where induction hardening is employed. However, all steels from SAE 1030 up may have induction hardening applications. ♦ = May be water or brine quenched by special techniques such as partial immersion, or time-quenched; otherwise, they are subject to quench cracking. ■ = Spheroidal structures are often required on these high-carbon steels for machining purposes and should be cooled very slowly or be isothermally transformed to produce the desired structure.

Figure 11.3 Table of treatments for heat-treating grades of carbon steel.

- Brine solution (9% by weight or 0.75 lb sodium chloride per gallon of water)

- Caustic soda solution [5% solution by weight potassium hydroxide (lye)]

- Oil (ambient or 90 to 140°F)

SAE Steel	Normalizing Temperature (°F)	Annealing Temperature (°F)	Hardening Temperature (°F)	Quenching Medium
1330	1600-1700 and/or	1500-1600	1525-1575	water or oil
1335 1340	1600-1700 and/or	1500-1600	1500-1550	oil
4037	.....	1525-1575	1500-1575	oil
4047	.....	1450-1550	1500-1575	oil
4130	1600-1700 and/or	1450-1550	1600-1650	water or oil
4137 4140 4145 4150	1600-1700 and/or	1450-1550	1550-1600	oil
4340	1600-1700 & temper	1100-1225	1475-1525	oil
5130 5132 5140 5150	1650-1750 and/or	1450-1550	1500-1550 oil oil	water or caustic
50100 51100 52100	....	1350-1450	1425-1475 1500-1600	water oil
6150	1650-1750 and/or	1550-1650	1600-1650	oil
9260	.....	.....	1500-1650	oil
8630	1600-1700	1450-1550	1550-1650	water or oil
8637 8640	1600-1700 and/or	1450-1550	1525-1575	oil
8645 8650	1600-1700 and/or	1450-1550	1500-1550	oil

NOTE: Except as noted, the steel is to be tempered to the required hardness. The exceptions are gears made from steels 4037 and 5150; temper at 350 to 400°F.

Figure 11.4   Table of heat treatments for directly hardenable grades of alloy steels.

- Oil over water
- Molten salt bath (nitrate salts) to 700°F
- Forced air

11.5 Tempering and Case Hardening Steels

The purpose of tempering or drawing is to reduce the brittleness of hardened steel and to remove the internal stresses caused by the

sudden cooling or quenching done in the hardening operation. In the tempering operation, the hardened steel is heated to a certain temperature and then quench cooled. Steel in the fully hardened condition consists mainly of martensite. When reheated to a temperature of approximately 300 to 750°F, a softer and tougher structure called *troostite* is formed. When the hardened steel is reheated to a temperature from 750 to 1290°F, a structure called *sorbite* is formed, which has less strength than troostite but is much more ductile.

11.5.1 Tempering temperatures for steels

The color of the oxide coating on the reheated steel was used in the past as a means of determining the tempering temperature. Since this color may be affected by the composition of the steel, it is not considered to be the most reliable method of determining the tempering temperature, although it served satisfactorily for many years. Today, the optical (infrared) pyrometer is used to determine the temperature to very accurate limits. Figure 11.5 shows the tempering temperatures of plain carbon steels as indicated by their surface colors.

Figure 11.6 shows the temperatures of metals by the color of the radiation or glow of the metal. Metals below 900°F do not produce a radiated color but only the color of the oxide coating, if present.

This color scale may be used in an emergency situation, where a fair indication of hardening temperature may be made for hardening

Degrees Fahrenheit	Degrees Celsius	Color of Steel
430	221.1	Very pale yellow
440	226.7	Light yellow
450	232.2	Pale straw-yellow
460	237.8	Straw-yellow
470	243.3	Deep straw-yellow
480	248.9	Dark-yellow
490	254.4	Yellow-brown
500	260.0	Brown-yellow
510	265.6	Spotted red-brown
520	271.1	Brown-purple
530	276.7	Light purple
540	282.2	Full purple
550	287.8	Dark purple
560	293.3	Full blue
570	298.9	Dark blue
640	337.8	Light blue

Figure 11.5 Tempering temperatures by color indication for plain carbon steels.

Temperature (°F)	Temperature (°C)	Color
932 - 1,022	500 - 550	Dull red
1,202 - 1,382	650 - 750	Dark red
1,562 - 1,742	850 - 950	Bright red
1,922 - 2,102	1,050 - 1,150	Yellowish red
2,282 - 2,462	1,250 - 1,350	Dull white
2,642 - 2,822	1,450 - 1,550	Bright white

Figure 11.6   Color scale for metal temperatures based on radiated color (glow).

plain carbon steels, and Figure 11.5 surface color equivalent temperatures may be used for tempering.

Additional data on hardening and tempering steels may be obtained from Chap. 4, "Materials and Their Uses." A good understanding of the hardening and tempering operations used today may be obtained by reviewing the heat-treatment terms and definitions in Sec. 11.3.

11.5.2   Case hardening carbon and alloy steels

The case-hardening operation is performed on steels when a tough outer layer is required on the part and through-hardening of the steel is not required. This process has applications on parts that must be hardened to prevent wear on the outer surfaces only, whereas the core remains at its normal condition. Case hardening of low-carbon steels is a two-step operation. First, the outer layer of the part is carburized by introducing carbon into the surface, and second, the outer carburized layer is heat-treated to form a hard "case." Surface- or case-hardening processes include

- Carburization

- Cyanide hardening

- Nitriding

- Carbonitriding

- Induction hardening

- Flame hardening

Low-carbon steels containing 0.10 to 0.20 percent carbon are suitable for carburized case-hardening operations. Additional heat-treating data may be found in Chap. 4, "Materials and Their Uses."

Case-hardening depths

Light case: 0.003 to 0.015 in

Medium case: 0.015 to 0.040 in

Heavy case: 0.040 to 0.250 in

Case depth measurements. A production method for determining the depth of the case-hardened layer consists of air cooling a test pin from the carburizing temperature, reheating to 1475°F, and quenching in oil. This treatment refines the effective case and leaves the core material coarse-grained. Etching the fractured surface of the test pin in an aqueous solution of 7% concentrated nitric acid produces good contrast between the case-hardened layer and the core material. The test pin is made of the same material as the case-hardened parts being checked and consists of a cylindrical pin 0.500 in in diameter, 3.5 in long, with a 0.468-in-diameter groove (0.060 in wide), approximately 0.30 in in from the end of the pin. Test rings usually are 1.500 in outside diameter with a 0.75-in hole through the center of the ring and a thickness of 0.375 in.

Other methods of determining the case depth include

- Reading of case depth with a Brinell microscope
- Step grinding and reading the Rockwell hardness at different depths
- Bluing of the fractured surface
- Carbon-cut analysis
- The martensite-start method

Cyanide case-hardening precautions. Cyanide compounds are deadly poisons. Care should be taken so that even minute portions are not consumed accidentally. In storage, keep cyanide-containing salts separate from acids because mixing these chemicals will produce deadly hydrocyanic gas. Do not add nitrate or nitrite salts to those containing cyanide because of the danger of explosion.

Surface hardness versus case depth. Figure 11.7 shows several different common steels and the case depth to be expected from the Brinell hardness reading taken on the surface of the case-hardened part.

Case Depth as Determined by Brinell Hardness - Tabulated values are in inches and indicate case depths.

Steel	SAE 1020		SAE 6118		SAE 1020		SAE 4617		SAE 8620		SAE 4820	
Quench Temp (°F)	1675	1500	1675	1500	1675	1500	1675	1500	1675	1500	1675	1500
Quench Medium	Oil	Oil	Oil	Oil	Water	Water	Oil	Oil	Oil	Oil	Oil	Oil
Brinell												
156	....	0.000	....	....	....	....	....	....	....	....	....	....
167	....	0.010	....	....	....	....	....	....	....	....	....	....
170	....	0.014	....	....	....	....	....	....	....	....	....	....
179	0.000	0.019	....	....	....	....	....	....	....	....	....	....
187	0.007	0.024	....	....	....	0.000	....	....	....	....	....	....
192	0.009	0.026	....	....	....	0.003	....	....	....	....	....	....
197	0.011	0.029	....	....	....	0.005	....	....	....	....	....	....
207	0.016	0.035	....	....	0.000	0.009	....	....	....	....	....	....
217	0.023	0.041	....	....	0.005	0.013	....	0.000	....	....	....	....
229	0.030	0.049	....	....	0.008	0.018	....	0.004	....	....	....	....
241	0.035	0.069	....	0.000	0.015	0.023	....	0.009	....	....	....	....
255	0.040	0.077	....	0.003	0.022	0.027	....	0.015	....	....	....	....
269	0.046	....	....	0.007	0.027	0.034	0.000	0.021	....	....	....	....
285	0.053	....	....	0.013	0.034	0.039	0.007	0.024	....	0.000	....	....
302	0.060	....	....	0.018	0.041	0.046	0.010	0.028	....	0.005	....	....
321	0.074	....	0.000	0.023	0.048	0.054	0.013	0.032	0.000	0.009	....	....
341	....	....	0.004	0.029	0.056	0.062	0.017	0.035	0.003	0.012	....	0.000
363	....	....	0.009	0.036	0.062	....	0.021	0.039	0.005	0.016	0.000	0.004
387	....	....	0.014	0.043	....	....	0.024	0.041	0.010	0.022	0.008	0.010
412	....	....	0.019	0.048	....	....	0.030	0.045	0.016	0.027	0.012	0.015
444	....	....	0.026	0.060	....	....	0.033	0.048	0.022	0.035	0.024	0.022
460	....	....	0.030	....	....	....	0.035	0.051	0.026	0.039	0.028	0.026
477	....	....	0.034	....	....	....	0.037	0.054	0.030	0.043	0.033	0.030
495	....	....	0.038	....	....	....	0.039	0.056	0.034	0.047	0.038	0.035
512	....	....	0.042	....	....	....	0.051	0.058	0.037	0.050	0.043	0.039
532	....	....	0.046	....	....	....	0.064	0.060	0.042	0.055	0.049	0.043
555	....	....	0.051	....	....	....	....	....	0.047	0.061	0.055	0.049
578	....	....	0.057	....	....	....	....	....	0.051	0.066	0.063	0.054
600	....	....	0.062	....	....	....	....	....	0.055	....	....	0.060
627	....	....	....	....	....	....	....	....	0.061	....	....	....

Figure 11.7 Case-depth table.

11.5.3   Heat treatments for carburizing grades of carbon steels

Figure 11.8 shows the heat-treating process for carburizing grades of some of the common carbon steels.

11.5.4   Heat treatments for carburizing grades of alloy steels

Figure 11.9 shows the heat-treating process for some of the common carburizing grades of alloy steels.

Note: Extensive data on hardening all types of steels are contained in Chap. 4, "Materials and Their Uses." Also, in Sec. 8.2.2, "Heat Treatment of Tool Steels," data are shown for hardening these classes of steels.

11.6   Heat Treatment of Aluminum Alloys

The heat treatment of aluminum alloys is a precision operation. The temperature/time cycles are critical and must be carried out by using the proper equipment and controls. The general types of heat treatments applied to aluminum and its alloys are

1. Preheating or homogenizing to reduce chemical segregation of cast structures and to improve their workability

2. Annealing to soften strain-hardened (work-hardened) and heat-treated alloy structures to relieve stresses and to stabilize properties and dimensions

3. Solution heat treatments to effect solid solution of alloying constituents and improve mechanical properties

4. Precipitation heat treatments to provide hardening by precipitation of constituents from solid solution

11.6.1   Solution heat treatment of aluminum alloys

The solution heat treatment of aluminum alloys improves mechanical properties by developing the maximum practical concentration of the hardening constituents in solid solution. This requires heating the aluminum alloy part to a temperature close to the eutectic temperature, holding it there long enough to effect the desired solution, and then quenching fast enough to retain the desired solid solution.

SAE Steel	Normalizing Temperature (°F)	Carburizing Temperature (°F)	Cooling Method	Reheat Temp. (°F)	Cooling Medium	Second Reheat (°F)	Cooling medium	Temper (°F)
1010	.....	1650-1700	water/brine	.....	.....	.....	.....	250-400
1015	.....	1650-1700	oil or water	1400-1450	water/brine	.....	.....	250-400
1018	.....	1650-1700	cool slowly	1650-1700	oil/water	1400-1450	water/brine	250-400
1020	.....	1500-1650	air/oil	.....	.....	.....	.....	optional
1024	1650-1750	1650-1700	oil	.....	.....	.....	.....	250-400
1025	.....	1650-1700	water/brine	.....	.....	.....	.....	250-400
1026	.....	1500-1650	oil/water	.....	.....	.....	.....	optional
1027	.....	1350-1575	air/oil	.....	.....	.....	.....	optional
1030	.....	1500-1650	oil/water	.....	.....	.....	.....	optional
1117	.....	1650-1700	cool slowly	1650-1700	oil/water	1400-1450	water/brine	250-400
1118	.....	1500-1650	oil/water	.....	.....	.....	.....	optional

NOTE: Even when the tempering temperatures are shown, the tempering operation is not mandatory on many applications. Tempering is generally employed for a partial stress relief and improves resistance to cracking from grinding operations. Higher temperatures than those shown may be used where the hardness specification on the finished parts permits.

Figure 11.8 Heat-treatment table for carburizing grades of common carbon steels.

SAE Steel	Alternate Pretreatments Normalize■	+ Temper♦	Cycle Anneal	Carburizing Temperature (°F)	Cooling Method	Reheat Temp (°F)	Cooling Medium	Tempering Temp (°F)♦
4023 4027 4028	Yes	…..	Yes	1650-1700	oil quench	…..	…..	250-350
4118	Yes	…..	…..	1650-1700	oil quench	…..	…..	250-350
4320 4620	Yes	…..	Yes	1650-1700 1650-1700	oil quench cool slowly	1425-1475 1475-1525	oil oil	250-350 250-350
4820	…..	…..	…..	1650-1700	oil quench	…..	…..	250-350
5120	Yes	…..	…..	1650-1700	cool slowly	1500-1550	oil	250-350
8615 8617 8620 8622 8720	Yes	…..	Yes	1650-1700 1500-1650	oil quench cool slowly cool slowly oil quench	1475-1525 1525-1575 1475-1525 1525-1575 …..	oil …..	250-350

NOTE: ■ Normalizing temperatures should be not less than 50°F higher than the carburizing temperature, followed by air cooling. ♦ After normalizing, reheat to temperature of 1000 to 1250°F and hold for approximately 4 hours ● The tempering treatment is optional.

Figure 11.9 Heat-treatment table for carburizing grades of common alloy steels.

The usual quenching medium for aluminum alloys is water. In quenching some products, water below 100°F provides the required quench rate for optimal properties of the alloy being heat treated. In other alloys or products, the water may be heated to temperatures above 100°F or even to the boiling point to control distortion and residual internal stresses. Hot oil is also used for some applications because it provides a quenching rate that improves resistance to stress corrosion. The water quench is either by total immersion or by spray.

The recommended conditions of temperature and time for heat treating some of the common aluminum alloys, produced by various methods, are given in Figs. 11.10 through 11.15.

Alloy Designation	Temper	Solution Heat Treatment Metal Temperature (°F)	Precipitation Heat Treatment Metal Temp. (°F)	Time @ Temp. Hours
2014	T4, T6, T651	935	320	18
2024	T3, T4, T6	920	375	16
6061	T4, T6	985	320	18
7075	T6, T651	900	250	24
7178	T6, T651	875	250	24

Figure 11.10   Heat-treatment table for aluminum alloy sheet and plate.

Alloy Designation	Temper	Solution Heat Treatment Metal Temperature (°F)	Precipitation Heat Treatment Metal Temp. (°F)	Time @ Temp. Hours
2014	T4, T6, T651	935	320	18
2024	T351	920	.....	.....
6061	T4, T6, T651	980	350	8
7075	T6, T651	870	250	24
7178	T6, T651	870	250	24

Figure 11.11   Heat-treatment table for extruded aluminum alloy rod, bar, shapes, and tube.

Alloy Designation	Temper	Solution Heat Treatment Metal Temperature (°F)	Precipitation Heat Treatment Metal Temp. (°F)	Time @ Temp. Hours
2011	T3	975	.....	.....
2014	T4, T6, T651	935	320	18
2024	T3, T4, T6	920 920	 375	 16
6061	T4, T6, T651	985 985	 320	 18
7075	T6, T651	915	250	24
7178	T6, T651	900	250	24

Figure 11.12 Heat-treatment table for rolled or drawn aluminum alloy wire, rod, bar, shapes, and tube.

Alloy Designation	Temper	Solution Heat Treatment Metal Temperature (°F) (Quench water temp.)		Precipitation Heat Treatment Metal Temp. (°F)	Time @ Temp. Hours
2014	T4 T6	935 935	(140-160) (140-160)	 340	 10
2024	T4 T6	920 920	(Room) (Room)	 375	 18
6061	T6	985	(Room)	350	8
7075	T6	880	(140-160)	250	24
7178	T6	870	(Room)	250	24

Figure 11.13 Heat-treatment table for aluminum alloy forgings.

11.6.2 Precipitation heat treatment of aluminum alloys

The rate of precipitation from the supersaturated solid solution existing immediately after quenching increases as the temperature of the metal is raised above room temperature. Precipitation occurs naturally at room temperature, providing useful degrees of precipitation hardening for some alloys. Precipitation heat treatment generally denotes treatment at an elevated temperature and often

Alloy Designation	Temper	Solution Heat Treatment		Precipitation Heat Treatment	
		Metal Temp. (°F)	Time Hours	Metal Temp. (°F)	Time Hours
142	T61	960	6	450	2
220	T4	810	16	.....	.....
319	T6	940	10	310	4
355	T6	980	10	310	4
356	T6	1000	10	310	4
357	T6	1000	10	350	6
363	T6	940	10	310	4
A750	T5	.....	.....	430	8

NOTE: Many of these alloys are covered by ASTM B26. Solution heat treatment is followed by quenching in water at 150 to 212°F. A boiling water quench is recommended when minimum quenching stresses and distortion are required. Maintain steady furnace temperatures for the solution heat treatment process.

Figure 11.14   Heat-treatment table for sand- and plaster-cast aluminum alloys.

Alloy Designation	Temper	Solution Heat Treatment		Precipitation Heat Treatment	
		Metal Temp. (°F)	Time Hours	Metal Temp. (°F)	Time Hours
142	T61	960	6	400	4
319	T6	940	6	310	4
354	T61	980	8	310	10
355	T6	980	6	310	4
356	T6	1000	8	310	4
357	T6	1000	8	350	6
359	T61	1000	10	310	10
750/A750	T5	.....	.....	430	8

NOTE: Many of these alloys are covered by ASTM B108. The solution heat treatment is followed by quenching in water at 150 to 212°F. Boiling water is recommended when minimum quenching stresses and distortion are required. Maintain steady furnace temperatures for the solution heat treatment process.

Figure 11.15   Heat-treatment table for permanent-mold cast aluminum alloys.

is called *artificial aging.* See Figs. 11.10 through Figure 11.15 for the precipitation-hardening temperature/times for those alloys which may be precipitation hardened (artificially aged).

11.7   Heat Treating Beryllium-Copper Alloys

Beryllium-copper alloys are important metallic products that over recent years have found many applications in various industries, including the electrical, electronics, and mechanical industries. Beryllium-copper alloys have many desirable characteristics, including corrosion resistance, good electrical conductivity, good spring properties, good heat conductivity, and excellent stress fatigue resistance. Beryllium is also alloyed with nickel, called *beryllium-nickel alloys,* and their tensile strengths are quite high when the alloys are cold worked. Many products that formerly used spring-tempered phosphor-bronzes are now using the beryllium-copper alloys.

The element beryllium is toxic, as are its alloys. The Occupational Safety and Health Administration (OSHA) safety regulations proscribe the conditions under which these alloys should be processed. The percentage of the population to which beryllium causes toxicity is small, but safeguards must be taken. The toxic effects of beryllium are seen as chronic pulmonary (lung) problems.

In the heat treatment of beryllium-copper alloys, the product is furnace heated to a prescribed temperature and held at this temperature for a number of hours, followed by still-air cooling to room temperature. Figure 11.16 shows the popular and common beryllium-copper alloys and their heat-treatment procedures. Figure 11.17 defines the ASTM temper designations used in Fig. 11.16.

11.8   Cold Work Hardening of Metals and Alloys

Many metals and alloys obtain their physical properties through the process of work hardening. The work-hardening processes include drawing, rolling, stretching, and hammering. Materials such as phosphor-bronze acquire their spring characteristics through the action of cold rolling or drawing. The percentage of reduction of the cross-sectional area of the material determines the degree of "temper" in the material. Figure 11.17 lists the percentage of reduction that determines the temper designation for beryllium-copper

Alloy Designation (UNS system)	Temper (*)	Heat Treatment Time/Temperature	Tensile Strength Kpsi
25	(TF00)	3 hrs @ 600 °F	140-175
(C17200)	(TH01)	2 hrs @ 600 °F	150-185
	(TH02)	2 hrs @ 600 °F	185-215
	(TH04)	2 hrs @ 600 °F	190-220
190	(TM00)	Mill	100-110
(C17200)	(TM01)	Mill	110-120
	(TM02)	Mill	120-135
	(TM04)	Mill	135-150
	(TM05)	Mill	150-160
	(TM06)	Mill	155-175
	(TM08)	Mill	175-190
290	(TM00)	Mill	100 minimum
(C17200)	(TM02)	Mill	120 minimum
	(TM04)	Mill	140 minimum
	(TM06)	Mill	155 minimum
	(TM08)	Mill	175 minimum
165	(TF00)	3 hrs @ 600 °F	150-180
(C17000)	(TH01)	2 hrs @ 600 °F	160-190
	(TH02)	2 hrs @ 600 °F	170-200
	(Th04)	2 hrs @ 600 °F	180-210
	(TM00)	Mill	100-110
	(TM01)	Mill	110-120
	(TM02)	Mill	120-135
	(TM04)	Mill	135-150
	(TM05)	Mill	150-160
	(TM06)	Mill	155-175
3 (C17510)	(TF00)	2-3 hrs @ 900 °F	100-130
and	(TH04)	2-3 hrs @ 900 °F	110-135
10 (C17500)			
174	(TH04)	Mill	110-130
(C17410)			

NOTE: (*) = ASTM alphanumeric code for product tempers. Mill = product normally heat treated at the mill. For heat treatment, the part must be fully annealed, then accurately held at the indicated temperatures for the times shown. Uniform heating of the parts within the furnace MUST be maintained for a successful heat treatment. Annealed strip which is "dead-soft" offers maximum deep drawing and die-forming capabilities.

Figure 11.16   Heat-treatment table for beryllium-copper alloys.

alloys. The phosphor-bronze alloys also receive their temper designations from the percentage of reduction that takes place in the cold-rolling operation (quarter hard, half hard, full hard, spring, and extra spring).

Stainless steels such as AISI types 303, 304, and others may acquire very high tensile strengths and spring properties through the cold-rolling process. The 300 series stainless steels are termed

ASTM Designation	Description	Cold Rolled Thickness Reduction in Percent
TB00	Solution annealed	0
TD01	Quarter hard	11
TD02	Half-hard	21
TD03	Three-quarters hard	29
TD04	Hard	37
TF00	The suffix "T" added to temper designations indicates that the	
TH01	material has been age-hardened by the standard heat treatment.	
TH02		
TH03		
TH04		
TM00	Mill hardened to specific property ranges,	
TM01	no further heat treatment required.	
TM02		
TM04		
TM05		
TM06		
TM08		

NOTE: Temper designations are defined in the specification ASTM B601, "Standard practice for temper designations for copper and copper alloys".

Figure 11.17 ASTM temper designations (for Fig. 11.16).

austenitic and cannot be heat treated by using conventional heat-treating processes. The spring properties of these and many other metals and alloys are obtainable only through cold-working processes. The martensitic stainless steels (400 series) may be heat treated using conventional heat-treating methods. The 440 series stainless steels are popular martensitic stainless steels that may be hardened to a high degree by heating and quenching at specified temperatures and cooling rates (see Chap. 4, "Materials and Their Uses," for hardening procedures for all classes of steels).

Castings, Moldings, Extrusions, and Powder-Metal Technology

12.1 Castings

The casting of metals and alloys is an important branch of the metal-working industry. Castings allow parts to be made at a rapid pace with controlled accuracy. Castings replace parts that otherwise would be difficult or impossible to machine and very costly to manufacture. However, castings cannot replace many types of machined parts because of material, configuration, and other physical considerations. Many types of processes are used in the casting and foundry industries to produce cast parts in different materials and for various dimensional accuracy requirements. The casting processes or methods that are in use today include

- Sand casting

- Shell casting

- Carbon dioxide casting

- Fluid sand casting

- Composite-mold casting

- Plaster-mold casting

- Slush casting

- Evaporative pattern casting (EPC)
- Die casting
- Permanent-mold casting
- Ceramic-mold casting
- Investment casting (lost-wax process)

The casting method or process chosen by the design engineer and the foundry is determined by the following factors:

- Type of metal to be cast
- Size of part to be cast
- Required cast accuracy of the part
- Economics
- Required secondary operations such as machining, hardening, welding, and plating

12.1.1   Sand casting

In the sand-casting process, a wooden, plastic, or metal pattern is packed in a special sand, which is dampened with water and then removed, leaving a hollow space having the part's shape. The pattern is purposely made larger than the size of the cast part to allow for shrinkage of the casting as it cools.

The mold consists of two steel frames, which are called the *cope* (top half) and the *drag* (bottom half). Figure 12.1 is an illustration of a typical sand-casting setup.

Sand cores may be placed in the cavity to produce holes in the part where required. Once the cope and drag are clamped together, the molten metal is poured into the *gate* of the mold. Vent holes placed appropriately in the mold allow hot gases to escape from the mold cavity during pouring. The pouring temperature of the metal is always made a few hundred degrees higher than the melting point so that the metal has good fluidity during the pour and does not cool prematurely, causing voids in the part.

Sand casting is the least expensive of all the casting processes on a part-to-part basis, but a need for secondary machining operations may indicate the use of one of the other casting processes. Figure 12.2 shows a sand-cast part made of a copper alloy. This part would be difficult to machine from solid stock.

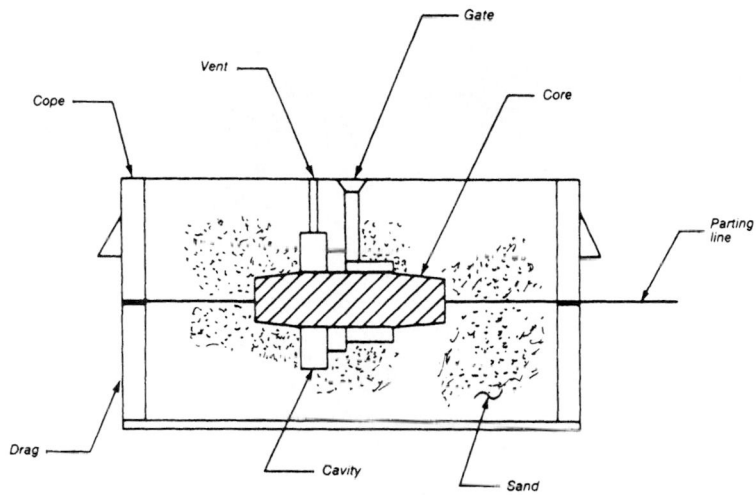

Figure 12.1   A typical sand-casting setup.

Figure 12.2   A typical sand-cast copper alloy part prior to machining.

Figure 12.3 An intricate sand-cast aluminum alloy part.

Figure 12.3 shows another sand-cast part of intricate design. This part is made of 356-T6 aluminum alloy (heat treated). Note that at the center section of the part there is a sprocket for number 40 American National Standards Institute (ANSI) roller chain. The sprocket in this application does not need a high degree of accuracy because the sprocket and chain action is only intermittent. A part of such design would be extremely difficult and costly to manufacture using the machining processes or methods.

Figure 12.4 shows some of the gating and venting methods employed in sand casting and similar casting processes. Figure 12.5 shows a typical combination casting-machining engineering drawing. This drawing provides enough information for the pattern maker, foundry, and machine shop to be able to produce the part.

12.1.2 Shell casting

This process entails forming a mold from a mixture of sand and a thermosetting-resin binder. The sand and thermoset mix are placed against a heated pattern, which causes the resin to bind the sand particles and form a strong shell. After the shell is cured and stripped

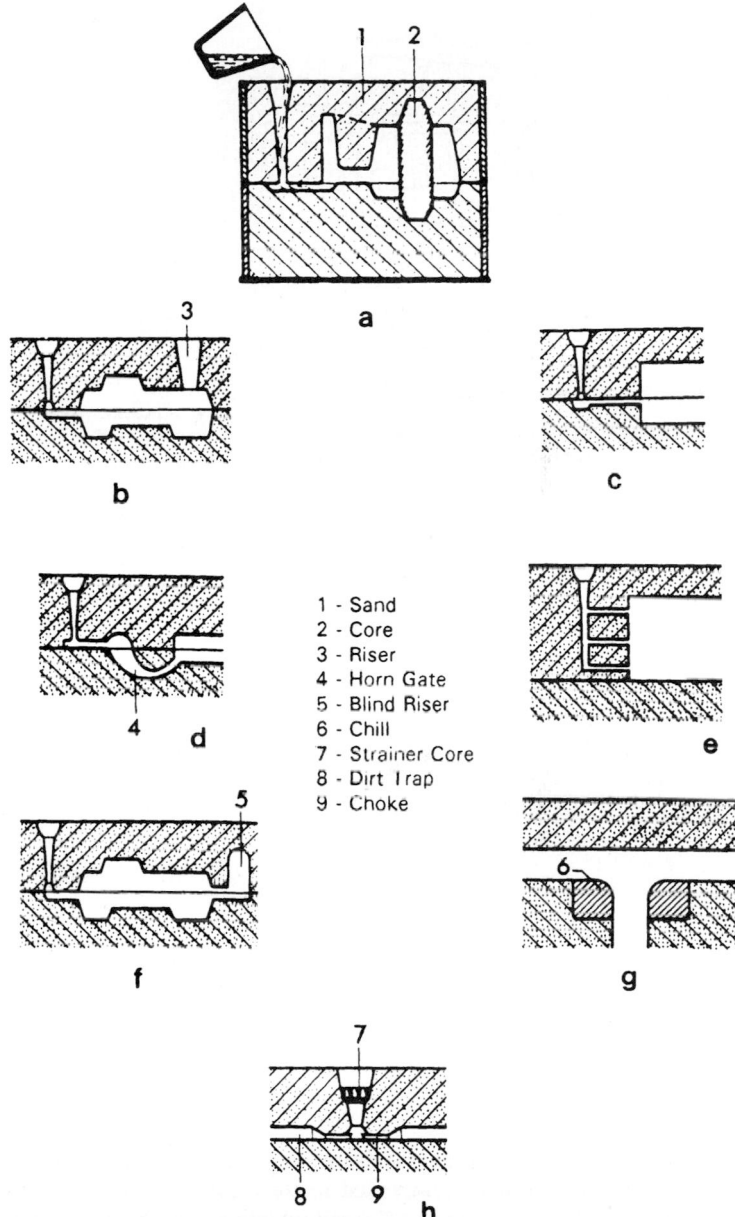

1 - Sand
2 - Core
3 - Riser
4 - Horn Gate
5 - Blind Riser
6 - Chill
7 - Strainer Core
8 - Dirt Trap
9 - Choke

Figure 12.4 Gating, venting, and chills.

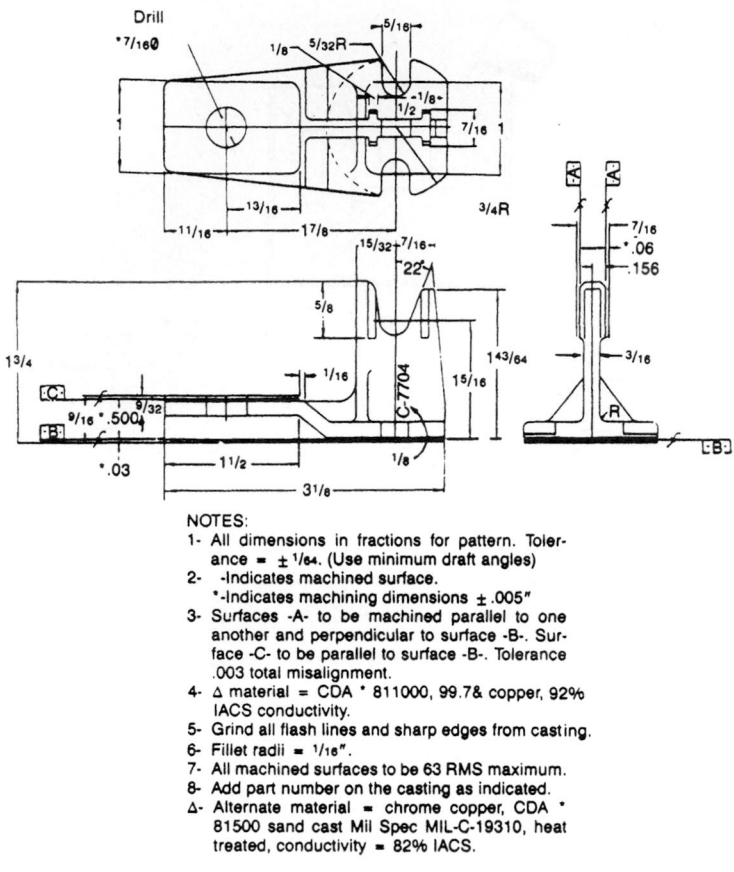

NOTES:

1- All dimensions in fractions for pattern. Tolerance = ± 1/64. (Use minimum draft angles)

2-   -Indicates machined surface.
   * -Indicates machining dimensions ± .005″

3- Surfaces -A- to be machined parallel to one another and perpendicular to surface -B-. Surface -C- to be parallel to surface -B-. Tolerance .003 total misalignment.

4- Δ material = CDA * 811000, 99.7& copper, 92% IACS conductivity.

5- Grind all flash lines and sharp edges from casting.

6- Fillet radii = 1/16″.

7- All machined surfaces to be 63 RMS maximum.

8- Add part number on the casting as indicated.

Δ- Alternate material = chrome copper, CDA * 81500 sand cast Mil Spec MIL-C-19310, heat treated, conductivity = 82% IACS.

Casting & Machining Drawing Sample

Figure 12.5   Typical casting-machining drawing.

from the pattern, cores may be set. The cope and drag are secured together and placed in a flask with added backup material. The mold is then ready for pouring the molten metal.

12.1.3   Carbon dioxide casting

The carbon dioxide process uses sodium silicate (water glass) binders instead of the clay binders employed in conventional sand casting. Treated with carbon dioxide gas, the sodium silicate–sand mixture is dried and strengthened.

Ready-for-use cores or molds can be made using this process in a few minutes with no baking required. Excellent dimensional accuracy is obtained, even while making cores rapidly. This process is used more often to make cores than to make molds.

12.1.4 Plaster-mold casting

This process is used when nonferrous metals must be cast more accurately than is possible with conventional sand casting. The four recognized plaster-mold processes are

- Conventional mold
- Match-plate pattern
- The Antioch method
- The foamed plaster method

Castings produced by the plaster-mold process have smoother surfaces, better accuracy, and finer detail than those made by conventional sand casting, but they are also more expensive.

12.1.5 Composite-mold casting

Composite-mold casting uses different sections of the mold and cores made by different methods so that the greatest advantage is obtained from each process in the appropriate section. This process is usually chosen for aluminum parts and is sometimes called *premium-quality casting* or *engineered casting*. The use of plaster mold sections affords accuracy and stability wherever they are essential on the cast part. Composite molds are used for the following reasons:

- Decreased cost of mold material
- Increased casting accuracy
- Decreased amount of gassing
- Improved finish or surface
- Quicker processing time

12.1.6 Investment casting

In this casting process, an expendable pattern is coated with a refractory slurry that sets at ambient temperature. The expendable

pattern (wax or plastic) is then melted out of the refractory shell. Ceramic cores are used as required.

Two distinct processes or methods are followed in investment casting: shell investment and solid investment. Investment casting is also known as the *lost-wax process* and *precision casting.* Cast parts can be produced in almost any pourable metal or alloy. The finished parts are dimensionally accurate and generally are used "as cast." The process is used for high-accuracy, mass-produced parts and for jewelry making.

The economics of this process must be weighed against the complexity of the part. Simple parts generally are not economical to produce with this process. The process is advantageous for applicable parts not in excess of 10 lb weight, although heavier parts frequently are investment cast.

12.1.7   Ceramic-mold casting

Ceramic-mold techniques are proprietary processes. They use permanent patterns and fine-grain zircon and calcined, high-alumina mullite slurries for molding.

As in other processes, the molds are constructed as a cope and a drag. Fine detail may be produced with high dimensional accuracy. The refractory mold allows the pouring of high-melting-point metals and alloys. The ferrous alloys and metals are cast more commonly with this process. Aluminum, beryllium-copper, titanium, ductile iron, carbon and low-alloy steels, and tool steels are cast using ceramic-mold processes.

Two of the proprietary processes for ceramic-mold casting are

- The Shaw method
- The unicast method

12.1.8   Permanent-mold casting

In this process, a metal mold consisting of two or more parts produces the cast parts. Metal, sand, or plaster cores are also used, in which case the process is known as *semi-permanent-mold casting.*

Intricate castings can be produced, but mold cost is high. Not all metals and alloys can be cast, and some shapes cannot be made because of the parting line or the difficulty of removing the part from the mold. Suitable casting metals include

- Aluminum alloys, up to 30 lb

- Magnesium alloys, up to 15 lb

- Copper alloys, up to 20 lb

- Zinc alloys, up to 20 lb

- Gray iron (hypereutectic), up to 30 lb

A variation of this process chills the cast metal rapidly, producing enhanced properties with regard to grain configuration and size. Surface qualities of permanent-mold castings are better than conventional sand castings, but the mold cost always must be evaluated with respect to quantity of parts produced and secondary operations required (machining) to produce a finished part. Many parts cast with this process can be used "as cast" when close tolerances are not required. Figure 12.6 shows a relatively complex part that was cast by the permanent-mold chill-cast process in an aluminum-bronze high-strength alloy.

The types of metals that can be permanent-mold cast are limited to those whose melting points are not above the copper-base alloys. Pouring higher-temperature alloys causes permanent damage to the steel molds.

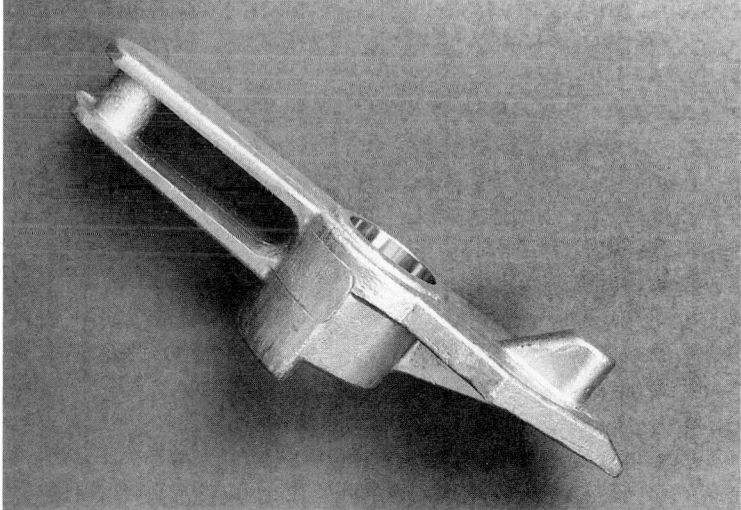

Figure 12.6 A permanent-mold chill-cast copper alloy part.

12.1.9   Die casting

Die castings are made by forcing molten metal under high pressure into permanent molds called *dies*. The advantages of die casting include

- Complex shapes possible
- Thin-walled sections possible
- High production rates
- High dimensional accuracy
- High volume of parts with little change in the dies
- Minimum surface preparation required for plating

The disadvantages of die casting include

- Casting size limited: 50 lb seldom exceeded
- Air entrapment and porosity difficulties in complex shapes
- Expensive machinery and dies
- Limited to metals having melting points no higher than copper-base alloys

Figure 12.7 shows some typical die-cast parts made of zinc alloy. The intricate designs shown are produced easily with the die-casting process.

12.1.10   Evaporative pattern casting (EPC)

Although this process was known and patents were issued as early as 1958, it has not developed until very recently. In this process, a plastic expendable pattern (usually styrofoam) is coated with a refractory slurry and cured. The composition of the coating is critical to the casting process because it must allow outgassing of the vaporized styrofoam during pouring of the molten metal. When the molten metal pours into the mold, the styrofoam vaporizes and gases out of the mold and through the sand, leaving the molten metal in the void left by the styrofoam. The process allows the coated styrofoam pattern to be packed in sand as in conventional sand casting, except that a cope and drag are not needed because the pattern is not removed mechanically but vaporizes.

Figure 12.7 Typical die-cast zinc alloy parts.

This process permits the casting of any pourable metal or alloy. Complexity of parts is generally not a problem, and cores may be used as in the investment process. Complex shapes are devised by gluing sections of the styrofoam patterns together. Risers and gates form part of the pattern. The entire refractory-coated assembly is packed in sand and then cast. This process is highly applicable to some of the automated systems and robotization.

12.1.11 Slush casting

This process is limited to hollow castings. Zinc- or lead-base alloys generally are used. Products made by this process include lamp bases and parts and consumer novelty items.

In this process, the molten metal is poured into a split bronze mold and allowed to set a specified time, after which the mold is inverted, and the remaining liquid metal is poured out. Remaining is a thin shell that has hardened at the mold surface, thus producing a hollow cast shape.

12.2 Ferrous Metal Alloys Used in Casting

The general ferrous metals and alloys used to produce castings include white iron, gray iron, malleable iron (ferritic and pearlitic), carbon steels, alloy steels, and stainless steels.

12.2.1   Gray iron castings

Carbon content in this iron ranges from 2.8 to 4 percent. Gray iron is poured at the lowest casting temperature and has the best castability and least shrinkage of all the ferrous metal cast alloys.

Gray iron may be heat treated in softening for better machinability or hardening for wear resistance. Pouring temperatures range from 2500 to 2700°F. Gray iron castings may be repaired by welding (shielded metal arc or oxyacetylene gas).

12.2.2   Ductile iron castings

The composition and handling of ductile irons are very similar to those of gray irons. The difference between the two irons is that in ductile iron, solidified graphite is spherical, whereas it is in flake form in gray iron. The metallurgy of the two irons is similar. Pouring temperatures range from 2500 to 2700°F. Ductile iron, as the name implies, has high ductility with resulting high impact strength for shock-load applications.

12.2.3   Malleable iron castings

This iron is produced from base metal having the following general composition:

Carbon: 2 to 3 percent

Silicon: 1 to 1.8 percent

Manganese: 0.2 to 0.5 percent

There are traces of sulfur, phosphorus, boron, and aluminum, with the remainder being iron.

The pouring range is 2700 to 2900°F. Ferritic malleable iron and pearlitic malleable iron receive different heat treatments to induce their metallurgic differences. Ferritic malleable iron must have a carbon-free matrix, and pearlitic malleable iron must have a matrix containing a controlled amount of carbon in the combined form.

Liquid-quenched and tempered malleable iron is made by two processes, both of which produce high-quality, high-strength cast irons.

Note that white cast iron is frequently formed on purpose at high wear points of a casting by incorporating *chills* in the mold. A chill is an insert that causes rapid cooling of the cast iron, with the consequent formation of white iron at the points selected. White cast iron is brittle but extremely hard and wear resistant. Figure 12.4g shows the position of chills in a sand casting mold (6).

12.2.4 Steel and alloy-steel castings

Green sand, dry sand, shell, investment, and ceramic molds are all used to produce steel and alloy-steel castings. Plaster molds cannot be used because the high pouring temperature of steels destroys plaster molds. Green sand is the most widely used method for producing steel castings. Pouring temperature of steels for casting commonly reaches as high as 3200°F.

Large steel castings are commonly produced in dry sand molds and may range from 1 ton to over 100 tons in weight. Steel castings are welded for repairs and for joining two or more castings into a structural assembly.

12.3 Representative Casting Metals and Alloys

Shown in the following figures is a selection of data on commonly used engineering alloys for casting processes. These represent only a small sample of the materials that are available but include many favorite engineering alloys that have been proven by many design applications. Chapter 4, "Materials and Their Uses," also should be consulted for many other metallic alloys used throughout industry for a broad range of applications.

Representative casting metals and alloys.   Figure 12.8 shows the physical properties of the common cast irons. Figure 12.9 shows the mechanical properties of cast structural steels. Figure 12.10 shows the designations and properties of cast stainless steels. Figure 12.11 shows the designations, types of processes, and physical properties of common cast aluminum alloys. Figure 12.12 shows the designations and mechanical properties of cast magnesium alloys. Figure 12.13 shows the Copper Development Association (CDA) designations and properties and application data for the common cast copper alloys.

	Tensile Strength psi	Yield Strength psi	Elongation Elongation % in 2 in.	Impact Impact ft-lbs	Young's Modulus psi
Ductile iron	90,000	60,000	15	2	22 × 10⁶
ASTM A536-77△	150,000	125,000	40	10	25 × 10⁶
Gray iron*	20,000	12,000	*	0.5	12 × 10⁶
ASTM A48-76	65,000	40,000		1.0	20 × 10⁶
Malleable iron	50,000	30,000	70	10	25 × 10⁶
(ferritic)	55,000	35,000	90	18	
Malleable iron	60,000	40,000	20	1	26 × 10⁶
(pearlitic)	100,000	90,000	35	10	28 × 10⁶
White iron	20,000	—	3	—	—
	50,000		10		

*Gray iron is not generally used for impact applications.
△Ductile iron is also Austempered (ADI), which approximately doubles its strength.

Figure 12.8 Physical properties of cast irons.

Grade	Tensile Strength, psi	Yield Strength, psi	Elongation % in 2 in	Impact ft-lbs	Endurance Limit, psi
60,000	63,000	35,000	30	12	30,000
65,000	68,000	38,000	28	35	30,000
70,000	75,000	42,000	27	30	35,000

Modulus of elasticity, typical: 29 × 10⁶ to 30 × 10⁶.

Figure 12.9 Mechanical properties of cast steels.

12.4   ASTM Listed Cast Irons and Steels

The American Society for Testing and Materials (ASTM) cast irons and steels shown in the following subsection have physical and chemical characteristics that make them suitable for the types of applications and services listed below:

- High-temperature applications
- Low-temperature applications
- Impact resistance
- Structural applications
- Pressure-vessel service

Designation †	Tensile Strength, psi	Yield Strength, psi	Elongation % in 2 in	Impact Charpy	Young's Modulus, psi
CA-15	115,000	100,000	30	35	29×10^6
CC-50 *(HC)	70,000 110,000	65,000	18	45	29×10^6
CE-30 *(HE)	87,000 92,000	63,000	18	10	25×10^6
CH-20 *(HH)	80,000 88,000	50,000	38	15	28×10^6

Typical physical properties of cast alloy steels

Melting temperature	2,700° to 2,800°F
Thermal conductivity (B.t.u-ft/hr-ft²-°F)	18 to 27
Density, lb/in³	0.283 to 0.284
Electrical resistivity (micro-ohm cm)	227

ASTM and SAE specifications—cast alloy steels

General mechanical	A27, A148
Low temperature applications	A352, A757
Weldability	A216
Automotive applications	SAE J435

*Heat-resistant grades. †ACI (Alloy Casting Institute) designations.

Figure 12.10 Properties of cast stainless steels.

- High ductility
- Corrosion resistance

12.4.1 ASTM specifications for selected cast irons and cast steels

ASTM A159-83 (R1988), *Standard Specification for Automotive Gray Iron Castings.* See Fig. 12.14 for mechanical properties. The grades of gray cast iron consist of the following:

G1800: ferritic-pearlitic

G2500: pearlitic-ferritic

G3000: pearlitic

G3500: pearlitic

Alloy Designation	Type	Uses and Typical Strengths
UNS A02010 (ANSI 201.0)	S PM	Very high strength, high impact, High ductility, high cost US = 60 YS = 50 El = 5.0
UNS A02060 (ANSI 206.0)	S PM	High tensile and yield, structural parts, automotive and aerospace US = 40 YS = 24 El = 8.0
UNS A02080 (ANSI 208.0)	S PM	Manifolds, valve bodies, pressure-tightness applciations US = 20 YS = 12 El = 1.5
UNS A02220 (ANSI 222.0)	S PM	Pistons and air-cooled cylinder heads US = 30/40
UNS A03190 (ANSI 319.0)	S	Low cost, general purpose alloy US = 31 YS = 20 El = 1.5
UNS A03540 (ANSI 354.0)	PM	High strength premium alloy US = 48 YS = 37 El = 3.0
UNS A03560 (ANSI 356.0)	S PM	For intricate work, good strength and ductility US = 33 YS = 22 El = 3.0
UNS A03600 (ANSI 360.0)	D	Very good casting and strength US = 44 YS = 25 El = 2.5
UNS A13600 (ANSI A360)	D	Excellent casting, corrosion resistance, thin walls, intricate parts US = 46 YS = 24 El = 3.5
UNS A03840 (ANSI 384.0)	D	General purpose alloy, thin sect. US = 48 YS = 24 El = 2.5
UNS A03900 (ANSI 390.0)	D	High wear resistance, cylinder heads, pistons, engine crankcases US = 41 YS = 35 El = 1.0

In the preceding table: US = ultimate strength, YS = yield strength in kpsi, El = elongation, % in 2 in, S = sand casting alloy, PM = permanent mold casting alloy, D = die casting alloy.

Figure 12.11   Aluminum casting alloys.

G4000: pearlitic

Figure 12.15 lists the typical applications for the automotive gray cast irons.

ASTM A 148/A 148M-89a, *Standard Specification for Steel Castings, High Strength, for Structural Applications*. Figure 12.16 lists the tensile strength properties.

ASTM A 297/A 297M-89, *Standard Specification for Steel Castings, Iron-Chromium, Iron-Chromium-Nickel, Heat Resistant, for General Applications*. Figure 12.17 lists the physical properties.

Designation	Temper	US	YS	Elongation % in 2 in
ΔM10100	F	20,000	—	—
(ASTM-AM 100A)	T4	34,000	—	6
SAE 502	T6	35,000	17,000	—
ΔM11630	F	26,000	11,000	4
(ASTM-AZ 63A)	T4	34,000	11,000	7
SAE 50	T6	34,000	16,000	3
*M11910	—	34,000	23,000	3
(ASTM AZ 91A)				
SAE 501				
*M11912	—	34,000	23,000	3
(ASTM-AZ 91B)				
SAE 501A				
ΔM11920	F	23,000	11,000	—
(ASTM-AZ 92A)	T4	34,000	11,000	6
	T6	34,000	18,000	1
ΔM16630	T6	40,000	27,000	5
(ASTM-ZE 63A)				

*Automotive die-casting alloys
ΔSand castings alloys

Figure 12.12 Magnesium casting alloys.

ASTM A 352/A 352M-89, *Standard Specification for Steel Castings, Ferritic and Martensitic, for Pressure Containing Parts, Suitable for Low-Temperature Service.* Table 12.1 lists the chemical, tensile, and impact properties.

ASTM A 436-84, *Standard Specification for Austenitic Gray Iron Castings.* Figure 12.18 lists the chemical and mechanical requirements.

ASTM A 439-83, *Standard Specification for Austenitic Ductile Iron Castings.* Figure 12.19 lists the chemical and mechanical properties.

ASTM A 487/A 487M-89a, *Standard Specification for Steel Castings Suitable for Pressure Service.* Tables 12.2 and 12.3 list the heat-treatment and mechanical requirements.

ASTM A 743/743M-89, *Standard Specification for Castings, Iron-Chromium, Iron-Chromium-Nickel, Corrosion Resistant, for General Applications.* Tables 12.4 and 12.5 list the heat-treatment and mechanical requirements.

Alloy No. C80100
Composition: 99.95% copper, 0.05 trace elements
Conductivity: 100% IACS
Tensile strength: 19,000 psi min. to 25,000 psi typical
Yield strength: 6,500 psi min. to 9,000 psi typical (0.2% offset)
Typical uses: Electrical and thermal conductors, corrosion resistance applications.
Similar alloys: C80300, C80500, C80700, C80900
All are difficult to cast, with low casting yields.

Alloy No. C81500 (chrome-copper)
Composition: 1% chromium, balance copper
Conductivity: 82% IACS
Tensile strength: 45,000 psi min. to 51,000 psi typical (heat treated)
Yield strength: 35,000 psi. to 40,000 psi typical (0.5% exten. under load)
Typical uses: Electrical and thermal conductors where high strength and hardness are required. A premium quality alloy.

Alloy No. C81700
Composition: 0.4% beryllium, 0.9 cobalt, 0.9 nickel, 1.0 silver
Conductivity: 48% IACS
Tensile strength: 85,000 psi min. to 92,000 psi typical
Yield strength: 62,000 psi min. to 68,000 psi typical (0.2% offset)
Typical uses: Electrical and thermal conductors where high strength and hardness are required.

Alloy No. C82400 (former name: 165C)
Composition: 1.7% beryllium, 0.25 cobalt, remainder copper
Conductivity: 25% IACS
Tensile strength: 145,000 psi min. to 150,000 psi typical (heat treated)
Yield strength: 135,000 psi min. to 140,000 psi typical (0.2% offset)
Typical uses: Molds, cams, bushings, gears, bearings
Similar alloys: C82800, C82700, C82600

Alloy No. C83600 (115, leaded red brass, composition bronze)
Composition: 85% copper, 5.0 lead, 5.0 tin, 5.0 zinc
Conductivity: 15% IACS
Tensile strength: 30,000 psi min. to 37,000 psi typical (as cast)
Yield strength: 14,000 psi min. to 17,000 psi typical (0.5% exten. under load)
Typical uses: Pipie fittings, pump castings, impellers, small gears
Similar alloys: C83800, C84200, C84400, C84500

Alloy No. C85200 (400, leaded yellow brass)
Composition: 72% copper, 3.0 lead, 1.0 tin, 24.0 zinc
Conductivity: 18% IACS
Tensile strength: 35,000 psi min. to 38,000 psi typical (as cast)
Yield strength: 12,000 psi min. to 13,000 psi typical (0.5% exten. under load)
Typical uses: Ferrules, valves, hardware, plumbing fittings
Similar alloys: C85300, C85400, C85500

Figure 12.13 CDA designations and properties of cast copper alloys.

Brinell hardness measurements for castings and other applications.
Calculation of the Brinell hardness number (BHN) can be performed using the equation shown in Sec. 4.8 of this *Handbook*. Table 12.6 is a table of Brinell hardness numbers as determined by the diameter of the indentation of a 10-mm ball at applied loads of 500, 1500, and

Alloy No. C86100 (high-strength yellow brass, 90 kpsi manganese bronze)
Composition: 5.0 aluminum, 67.0 copper, 3.0 iron, 4.0 manganese, 21.0 zinc
Conductivity: 7.5% IACS
Tensile strength: 90,000 psi min. to 95,000 psi typical
Yield strength: 45,000 psi min. to 50,000 psi typical (0.2% offset)
Typical uses: Marine castings, gears, bushings, bearings
Similar alloys: C86200, C86300, C86400, C86500

Alloy No. C87800 (die-cast silicon brass)
Composition: 82.0 copper, 4.0 silicon, 14.0 zinc
Conductivity: 6.7% IACS
Tensile strength: 85,000 psi typical
Yield strength: 50,000 psi typical (0.2% offset)
Typical uses: High-strength die castings, lever arms, brackets, hex nuts, clamps
Similar alloys: C87900

Alloy No. C94400 (phosphor bronze, 312)
Composition: 81.0 copper, 11.0 lead, .35 phosphorus, 8.0 tin
Conductivity: 10% IACS
Tensile strength: 32,000 psi typical
Yield strength: 16,000 psi typical (0.5% exten. under load)
Typical uses: Bushings, bearings, electrical items

Alloy No. C95400 (aluminum-bronze 9C, 415)
Composition: 11.0 aluminum, 85.0 copper, 4.0 iron
Conductivity: 13% IACS
Tensile strength: 75,000 psi min. to 85,000 psi typical
Yield strength: 30,000 psi min. to 35,000 psi typical (0.5% exten. under load)
Typical uses: Bearings, worms, gears, bushings, valve guides
Similar alloys: C95200, C95300, C95500

Alloy No. C95700 (manganese aluminum bronze)
Composition: 8.0 aluminum, 75.0 copper, 3.0 iron, 12.0 manganese, 2.0 nickel
Conductivity: 3.1% IACS
Tensile strength: 90,000 psi min. to 95,000 psi typical
Yield strength: 40,000 psi min. to 45,000 psi typical (0.5% exten. under load)
Typical uses: Impellers, propellers, safety tools, valves, pump castings

Note: The preceding alloy numbers refer to the UNS numbering system for copper cast alloys. For a complete listing of copper cast alloys, refer to the CDA handbook of cast copper alloys available from the Copper Development Association.

Figure 12.13 (*Continued*)

3000 kgf (kilogram-force). The table may be used for determining the hardness of most metals and alloys.

12.5 Plastic Moldings

There are two classifications of plastics and their moldings: thermoplastics and thermoset plastics. Thermoplastics are basically the same chemically after molding as they were in the raw form. This means that once molded, they may be reused, in most cases, by chopping the parts into small pieces and remelting. Thermoset

Mechanical Properties for Design Purposes

Grade	Hardness Range[A]	Tensile Strength, min, psi (kgf/mm^2)	Transverse Strength, min, lb (kg)[B]	Deflection, min, in. (mm)[B]
G1800	HB 143-187 5.0-4.4 BID	18 000 (14)	1720 (780)	0.14 (3.6)
G2500	HB 170-229 4.6-4.0 BID	25 000 (17.5)	2000 (910)	0.17 (4.3)
G3000	HB 187-241 4.4-3.9 BID	30 000 (21)	2200 (1000)	0.20 (5.1)
G3500	HB 207-255 4.2-3.8 BID	35 000 (24.5)	2450 (1090)	0.24 (6.1)
G4000	HB 217-269 4.1-3.7 BID	40 000 (28)	2600 (1180)	0.27 (6.9)

[A] Brinell impression diameter (BID) is the diameter in millimetres of the impression of a 10-mm ball at 3000-kg load.
[B] See Method A 438 for information concerning the B transverse test bar and the transverse test.

Figure 12.14 Mechanical properties of gray-iron castings. (*Reprinted with permission from the Annual Book of ASTM Standards, copyright 1992, American Society for Testing and Materials.*)

Grade	General Data
G1800	Miscellaneous soft iron castings (as cast or annealed) in which strength is not of primary consideration. Exhaust manifolds may be made of this grade of iron, alloyed or unalloyed. These may be annealed castings for exhaust manifolds in order to avoid growth and cracking due to heat.
G2500	Small cylinder blocks, cylinder heads, air cooled cylinders, pistons, clutch plates, oil pump bodies, transmission cases, gear boxes, clutch housings, and light-duty brake drums.
G3000	Automobile and diesel cylinder blocks, cylinder heads, flywheels, differential carries castings, pistons, medium-duty brake drums, and clutch plates.
G3500	Diesel engine blocks, truck and tractor cylinder blocks and heads, heavy flywheels, tractor transmission cases, and heavy gear boxes.
G4000	Diesel engine castings, liners, cylinders, and pistons.

Figure 12.15 Typical applications of gray iron for automotive castings. *(Reprinted with permission from the Annual Book of ASTM Standards, copyright 1992, American Society for Testing and Materials.)*

plastics, once molded, cannot be remolded or reprocessed because they have a one-way chemistry that alters their "as molded" characteristics from their raw constituents.

Types of thermoplastics include

- ABS (acrylonitrile-butadiene-styrene)
- Acetal (Delrin)
- Acrylic (Lucite, Plexiglas)
- Cellulosics (acetates)
- Fluoroplastics (PTFE, FEP, PFA, CTFE, ETFE, PVDF)
- Nylon
- Phenylene oxide
- Polycarbonate (Lexan)
- Polyester (Mylar)
- Polyethylene
- Polyimide
- Polyphenylene sulfide
- Polypropylene
- Polystyrene

Tensile Requirements

Grade	Tensile strength min. ksi [MPa]	Yield point min. ksi [MPa]	Elongation in 2 in. or 50 mm, min. %[A]	Reduction of Area, min. %
80-40 [550-275]	80 [550]	40 [275]	18	30
80-50 [550-345]	80 [550]	50 [345]	22	35
90-60 [620-415]	90 [620]	60 [415]	20	40
105-85 [725-585]	105 [725]	85 [585]	17	35
115-95 [795-655]	115 [795]	95 [655]	14	30
130-115 [895-795]	130 [895]	115 [795]	11	25
135-125 [930-860]	135 [930]	125 [860]	9	22
150-135 [1035-930]	150 [1035]	135 [930]	7	18
160-145 [1105-1000]	160 [1105]	145 [1000]	6	12
165-150 [1140-1035]	165 [1140]	150 [1035]	5	20
165-150L [1140-1035L][B]	165 [1140]	150 [1035]	5	20
210-180 [1450-1240]	210 [1450]	180 [1240]	4	15
210-180L [1450-1240L][B]	210 [1450]	180 [1240]	4	15
260-210 [1795-1450]	260 [1795]	210 [1450]	3	6
260-210L [1795-1450L][B]	260 [1795]	210 [1450]	3	6

[A] When ICI test bars are used in tensile testing as provided for in this specification, the gage length to reduced section diameter ratio shall be 4 to 1.
[B] These grades must be charpy impact tested.

Figure 12.16 Tensile-strength properties of steel castings. *(Reprinted with permission from the Annual Book of ASTM Standards, copyright 1992, American Society for Testing and Materials.)*

Tensile Requirements

Grade	Type	Tensile Strength, min		Yield Point, min		Elongation in 2 in. [50 mm], min, %[A]
		ksi	[MPa]	ksi	[MPa]	
HF	19 Chromium, 9 Nickel	70	485	35	240	25
HH	25 Chromium, 12 Nickel	75	515	35	240	10
HI	28 Chromium, 15 Nickel	70	485	35	240	10
HK	25 Chromium, 20 Nickel	65	450	35	240	10
HE	29 Chromium, 9 Nickel	85	585	40	275	9
HT	15 Chromium, 35 Nickel	65	450	. . .	. . .	4
HU	19 Chromium, 39 Nickel	65	450	. . .	. . .	4
HW	12 Chromium, 60 Nickel	60	415	. . .	. . .	. . .
HX	17 Chromium, 66 Nickel	60	415	. . .	. . .	. . .
HC	28 Chromium	55	380	. . .	. . .	. . .
HD	28 Chromium, 5 Nickel	75	515	35	240	8
HL	29 Chromium, 20 Nickel	65	450	35	240	10
HN	20 Chromium, 25 Nickel	63	435	. . .	. . .	8
HP	26 Chromium, 35 Nickel	62.5	430	34	235	4.5

[A] When ICI test bars are used in tensile testing as provided for in this specification, the gage length to reduced section diameter ratio shall be 4 to 1.

Figure 12.17 Tensile strengths for steel, iron-chromium-nickel, and heat-resisting castings. *(Reprinted with permission from the Annual Book of ASTM Standards, copyright 1992, American Society for Testing and Materials.)*

- Polysulfone
- Polyurethane
- Polyvinyl chloride

Types of thermoset plastics include

- Alkyd
- Allyl (diallyl phthalate)
- Amino (urea, melamine)
- Epoxy (including cycloaliphatic)
- Phenolic
- Polyester
- Polyurethane
- Silicone

Polyesters and polyurethanes include thermosets that are also thermoplastics. Thermoset grades usually are filled with reinforcing

(Text continued on page 793.)

TABLE 12.1 Chemical, Tensile, and Impact Requirements

Grade	Carbon steel	Carbon steel	Carbon-manganese steel	Carbon-molybdenum steel	2½% nickel steel	Nickel-chromium-molybdenum steel	3½% nickel steel	4½% nickel steel	9% nickel steel	12½% chromium, nickel-molybdenum steel
Element, % (max, except where range is given)	LCA	LCB[a]	LCC	LC1	LC2	LC2-1	LC3	LC4	LC9	CA6NM
Carbon	0.25[a]	0.30	0.25[a]	0.25	0.25	0.22	0.15	0.15	0.13	0.06
Silicon	0.60	0.60	0.60	0.60	0.60	0.50	0.60	0.60	0.45	1.00
Manganese	0.70[a]	1.00	1.20[a]	0.50–0.80	0.50–0.80	0.55–0.75	0.50–0.80	0.50–0.80	0.90	1.00
Phosphorus	0.04	0.04	0.04	0.04	0.04	0.04	0.04	0.04	0.04	0.04
Sulfur	0.045	0.045	0.045	0.045	0.045	0.045	0.045	0.045	0.045	0.03
Nickel	0.50[b]	0.50[b]	0.50[b]	—	2.00–3.00	2.50–3.50	3.00–4.00	4.00–5.00	8.50–10.0	3.5–4.5
Chromium	0.50[b]	0.50[b]	0.50[b]	—	—	1.35–1.85	—	—	0.50	11.5–14.0
Molybdenum	0.20[b]	0.20[b]	0.20[b]	0.45–0.65	—	0.30–0.60	—	—	0.20	0.4–1.0
Copper	0.30[b]	0.30[b]	0.30[b]	—	—	—	—	—	0.30	—
Vanadium	0.03[b]	0.03[b]	0.03[b]	—	—	—	—	—	0.03	—
Tensile Requirements:[c]										
Tensile strength, ksi (MPa)	60.0–85.0 (415–585)	65.0–90.0 (450–620)	70.0–95.0 (485–655)	65.0–90.0 (450–620)	70.0–95.0 (485–655)	105.0–130.0 (725–895)	70.0–95.0 (485–655)	70.0–95.0 (485–655)	85.0 (585)	110.0–135.0 (760–930)
Yield strength,[d] min, ksi (MPa)	30.0 (205)	35.0 (240)	40.0 (275)	35.0 (240)	40.0 (275)	80.0 (550)	40.0 (275)	40.0 (275)	75.0 (515)	80.0 (550)
Elongation in 2 in or 50 mm, min, %[e]	24	24	22	24	24	18	24	24	20	15
Reduction of area, min, %	35	35	35	35	35	30	35	35	30	35

Impact requirements Charpy V-notch[e,f]										
Energy value, ft·lbf (J), min value for two specimens and min avg of three specimens	13 (18)	13 (18)	15 (20)	13 (18)	15 (20)	30 (41)	15 (20)	15 (20)	15 (20)	20 (27)	20 (27)
Energy value, ft·lbf (J), min for single specimen	10 (14)	10 (14)	12 (16)	10 (14)	12 (16)	25 (34)	12 (16)	12 (16)	12 (16)	15 (20)	15 (20)
Testing temperature, °F (°C)	−25 (−32)	−50 (−46)	−50 (−46)	−75 (−59)	−100 (−73)	−100 (−73)	−150 (−101)	−175 (−115)	−320 (−196)	−100 (−73)

[a] For each reduction of 0.01% below the specified maximum carbon content, an increase of 0.04% manganese above the specified maximum will be permitted up to a maximum of 1.10% for LCA, 1.28% for LCB, and 1.40% for LCC.

[b] Specified Residual Elements—The total content of these elements is 1.00% maximum.

[c] See 1.2.

[d] Determine by either 0.2% offset method or 0.5% extension-under-load method.

[e] When ICI test bars are used in tensile testing as provided for in Specification A 703/A 703M, the gauge length to reduced section diameter ratio shall be 4 to 1.

[f] See Appendix X1.

SOURCE: Reprinted with permission from the *Annual Book of ASTM Standards*, copyright 1992, American Society of Testing and Materials.

Chemical Requirements

Element	Composition, %							
	Type 1	Type 1b	Type 2	Type 2b	Type 3	Type 4	Type 5	Type 6
Carbon, total, max	3.00	3.00	3.00	3.00	2.60	2.60	2.40	3.00
Silicon	1.00–2.80	1.00–2.80	1.00–2.80	1.00–2.80	1.00–2.00	5.00–6.00	1.00–2.00	1.50–2.50
Manganese	0.5–1.5	0.5–1.5	0.5–1.5	0.5–1.5	0.5–1.5	0.5–1.5	0.5–1.5	0.5–1.5
Nickel	13.50–17.50	13.50–17.50	18.00–22.00	18.00–22.00	28.00–32.00	29.00–32.00	34.00–36.00	18.00–22.00
Copper	5.50–7.50	5.50–7.50	0.50 max	0.50 max	0.50 max	0.50 max	0.50 max	3.50–5.50
Chromium	1.5–2.5	2.50–3.50	1.5–2.5	3.00–6.00[A]	2.50–3.50	4.50–5.50	0.10 max	1.00–2.00
Sulfur, max	0.12	0.12	0.12	0.12	0.12	0.12	0.12	0.12
Molybdenum, max	...	...	...	...	...	...	...	1.00

[A] Where some machining is required, the 3.00–4.00 % chromium range is recommended.

Mechanical Requirements

	Type 1	Type 1b	Type 2	Type 2b	Type 3	Type 4	Type 5	Type 6
Tensile strength, min, ksi (MPa)	25 (172)	30 (207)	25 (172)	30 (207)	25 (172)	25 (172)	20 (138)	25 (172)
Brinell hardness (3000 kg)	131 183	149 212	118 174	171 248	118 159	149 212	99 124	124 174

Figure 12.18 Chemical and mechanical requirements for austenitic gray-iron castings. (*Reprinted with permission from the Annual Book of ASTM Standards, copyright 1992, American Society for Testing and Materials.*)

Chemical Requirements

Element	Type								
	D-2[A]	D-2B	D-2C	D-3[A]	D-3A	D-4	D-5	D-5B	D-5S
	Composition, %								
Total carbon, max	3.00	3.00	2.90	2.60	2.60	2.60	2.40	2.40	2.30
Silicon	1.50-3.00	1.50-3.00	1.00-3.00	1.00-2.80	1.00-2.80	5.00-6.00	1.00-2.80	1.00-2.80	4.90-5.50
Manganese	0.70-1.25	0.70-1.25	1.80-2.40	1.00 max[B]	1.00 max[B]	1.00 max[B]	1.00 max[B]	1.00 max[B]	1.00 max
Phosphorus, max	0.08	0.08	0.03	0.08	0.08	0.08	0.08	0.08	0.08
Nickel	18.00-22.00	18.00-22.00	21.00-24.00	28.00-32.00	28.00-32.00	28.00-32.00	34.00-36.00	34.00-36.00	34.00-37.00
Chromium	1.75-2.75	2.75-4.00	0.50 max[B]	2.50-3.50	1.00-1.50	4.50-5.50	0.10 max	2.00-3.00	1.75-2.25

[A] Additions of 0.7 to 1.0 % of molybdenum will increase the mechanical properties above 800°F (425°C).
[B] Not intentionally added.

Mechanical Requirements

Element	Type								
	D-2	D-2B	D-2C	D-3	D-3A	D-4	D-5	D-5B	D-5S
	Properties								
Tensile strength, min. ksi (MPa)	58 (400)	53 (400)	58 (400)	55 (379)	55 (379)	60 (414)	55 (379)	55 (379)	65 (449)
Yield strength (0.2 percent offset), min. ksi (MPa)	30 (207)	30 (207)	28 (193)	30 (207)	30 (207)	...	30 (207)	30 (207)	30 (207)
Elongation in 2 in. or 50 mm, min. %	8.0	7.0	20.0	6.0	10.0	...	20.0	6.0	10
Brinell hardness (3000 kg)	139-202	148-211	121-171	139-202	131-193	202-273	131-185	139-193	131-193

Figure 12.19 Specifications for austenitic ductile iron castings. *(Reprinted with permission from the Annual Book of ASTM Standards, copyright 1992, American Society for Testing and Materials.)*

TABLE 12.2   Heat-Treatment Requirements

Grade	Class	Austenitizing temperature, min, °F (°C)	Media*	Quenching cool below, °F (°C)	Tempering temperature, °F (°C)[†]
1	A	1600 (870)	A	450 (230)	1100 (595)
1	B	1600 (870)	L	500 (260)	1100 (595)
1	C	1600 (870)	A or L	500 (260)	1150 (620)
2	A	1600 (870)	A	450 (230)	1100 (595)
2	B	1600 (870)	L	500 (260)	1100 (595)
2	C	1600 (870)	A or L	500 (260)	1150 (620)
4	A	1600 (870)	A or L	500 (260)	1100 (595)
4	B	1600 (870)	L	500 (260)	1100 (595)
4	C	1600 (870)	A or L	500 (260)	1150 (620)
4	D	1600 (870)	L	500 (260)	1150 (620)
4	E	1600 (870)	L	500 (260)	1100 (595)
6	A	1550 (845)	A	500 (260)	1100 (595)
6	B	1550 (845)	L	500 (260)	1100 (595)
7	A	1650 (900)	L	600 (315)	1100 (595)
8	A	1750 (955)	A	500 (260)	1250 (675)
8	B	1750 (955)	L	500 (260)	1250 (675)
8	C	1750 (955)	L	500 (260)	1250 (675)
9	A	1600 (870)	A or L	500 (260)	1100 (595)
9	B	1600 (870)	L	500 (260)	1100 (595)
9	C	1600 (870)	A or L	500 (260)	1150 (620)
9	D	1600 (870)	L	500 (260)	1150 (620)
9	E	1600 (870)	L	500 (260)	1100 (595)
10	A	1550 (845)	A	500 (260)	1100 (595)
10	B	1550 (845)	L	500 (260)	1100 (595)
11	A	1650 (900)	A	600 (315)	1100 (595)
11	B	1650 (900)	L	600 (315)	1100 (595)
12	A	1750 (955)	A	600 (315)	1100 (595)
12	B	1750 (955)	L	400 (205)	1100 (595)
13	A	1550 (845)	A	500 (260)	1100 (595)
13	B	1550 (845)	L	500 (260)	1100 (595)
14	A	1550 (845)	L	500 (260)	1100 (595)
16	A	1600 (870)[‡]	A	600 (315)	1100 (595)
CA15	A	1750 (955)	A or L	400 (205)	900 (480)
CA15	B	1750 (955)	A or L	400 (205)	1100 (595)
CA15	C	1750 (955)	A or L	400 (205)	1150 (620)[§]
CA15	D	1750 (955)	A or L	400 (205)	1150 (260)[§]
CA15M	A	1750 (955)	A or L	400 (205)	1100 (595)
CA6NM	A	1850 (1010)	A or L	200 (95)	1050–1150 (565–620)
CA6NM	B	1850 (1010)	A or L	200 (95)	1225–1275 (665–690)[§,¶] 1050–1150 (565–620)

* A = air, L = liquid.
[†] Minimum temperature unless range is specified.
[‡] Double austenitize.
[§] Double temper.
[¶] Air cool to below 200°F (95°C) after first temper.
SOURCE: Reprinted with permission from the *Annual Book of ASTM Standards,* copyright 1992, American Society of Testing and Materials.

TABLE 12.3 Required Mechanical Properties

Previous designation	Grade	Class	Tensile strength,* min, ksi (MPa)	Yield strength, min, ksi (MPa), at 0.2% offset	Elongation, 2 in (50 mm) or 4d, min, %	Reduction of area, min %	Hardness max, HRC	Max thickness, in (mm)
1N	1	A	85 (585)–110 (760)	55 (380)	22	40		
1Q	1	B	90 (620)–115 (795)	65 (450)	22	45		
	1	C	90 (620)	65 (450)	22	45	22 (235)	
2N	2	A	85 (585)–110 (760)	53 (365)	22	35		
2Q	2	B	90 (620)–115 (795)	65 (450)	22	40	22 (235)	
	2	C	90 (620)	65 (450)	22	40		
4N	4	A	90 (620)–115 (795)	60 (415)	18	40		
4Q	4	B	105 (725)–130 (895)	85 (585)	17	35		
	4	C	90 (620)	60 (415)	18	35	22 (235)	
	4	D	100 (690)	75 (515)	17	35	22 (235)	
4QA	4	E	115 (795)	95 (655)	15	35		
6N	6	A	115 (795)	80 (550)	18	30		
6Q	6	B	120 (825)	95 (655)	12	25		
7Q	7	A	115 (795)	100 (690)	15	30		2.5 (63.5)
8N	8	A	85 (585)–110 (760)	55 (380)	20	35		
8Q	8	B	105 (725)	85 (585)	17	30		
	8	C	100 (690)	75 (515)	17	35	22 (235)	
9N	9	A	90 (620)	60 (415)	18	35	22 (235)	
9Q	9	B	105 (725)	85 (585)	16	35	22 (235)	
	9	C	90 (620)	60 (415)	18	35		
	9	D	100 (690)	75 (515)	17	35		
	9	E	115 (795)	95 (655)	15	35		
10N	10	A	100 (690)	70 (485)	18	35		
10Q	10	B	125 (860)	100 (690)	15	35		

TABLE 12.3 Required Mechanical Properties (*Continued*)

Previous designation	Grade	Class	Tensile strength,* min, ksi (MPa)	Yield strength, min, ksi (MPa), at 0.2% offset	Elongation, 2 in (50 mm) or 4*d*, min, %	Reduction of area, min %	Hardness max, HRC	Max thickness, in (mm)
11N	11	A	70 (484)–95 (655)	40 (275)	20	35		
11Q	11	B	105 (725)–130 (895)	85 (585)	17	35		
12N	12	A	70 (485)–95 (655)	40 (275)	20	35		
12Q	12	B	105 (725)–130 (895)	85 (585)	17	35		
13N	13	A	90 (620)–115 (795)	60 (415)	18	35		
13Q	13	B	105 (725)–130 (895)	85 (585)	17	35		
14Q	14	A	120 (825)–145 (1000)	95 (655)	14	30		
16N	16	A	70 (485)–95 (655)	40 (275)	22	35		
CA15A	CA15	A	140 (965)–170 (1170)	110 (760)–130 (895)	10	25		
CA15	CA15	B	90 (620)–115 (795)	65 (450)	18	30		
	CA15	C	90 (620)	60 (415)	18	35	22 (235)	
	CA15	D	100 (690)	75 (515)	17	35	22 (235)	
CA15M	CA15M	A	90 (620)–115 (795)	65 (450)	18	30		
CA6NM	CA6NM	A	110 (760)–135 (930)	80 (515)	15	35		
CA6NM	CA6NM	B	100 (690)	75 (520)	17	35	23 (255)[†]	

* Minimum ksi, unless range is given.

[†] Test Methods and Definitions A 370, Table 3a does not apply to CA6NM. The conversion given is based on CA6NM test coupons. (For example, see ASTM STP 756.)

SOURCE: Reprinted with permission from the *Annual Book of ASTM Standards*, copyright 1992, American Society of Testing and Materials.

TABLE 12.4 Heat-Treatment Requirements

Grade	Heat treatment
CF-8, CG-8M, CG-12, CF-20, CF-8M CF-8C, CF-16F, CF-16Fa CH-20, CE-30, CK-20	Heat to 1900°F (1040°C) minimum, hold for sufficient time to heat casting to temperature, quench in water or rapid cool by other means so as to develop acceptable corrosion resistance.
	Heat to 2000°F (1093°C) minimum, hold for sufficient time to heat casting to temperature, quench in water or rapid cool by other means so as to develop acceptable corrosion resistance.
CA-15, CA-15M, CA-40, CA-40F	(1) Heat to 1750°F (955°C) minimum, air cool and temper at 1100°F (595°C) minimum, or
	(2) Anneal at 1450°F (790°C) minimum.
CB-30, CC-50	(1) Heat to 1450°F (790°C) minimum, and air cool, or
	(2) Heat to 1450°F (790°C) minimum, and furnace cool.
CF-3, CF-3M, CF-3MN	(1) Heat to 1900°F (1040°C) minimum, hold for sufficient time to heat casting to temperature, and cool rapidly so as to develop acceptable corrosion resistance, or
	(2) As cast if corrosion resistance is acceptable.
CN-3M	Heat to 2150°F (1175°C) minimum, hold for sufficient time to heat casting to temperature, quench in water or rapid cool by other means so as to develop acceptable corrosion resistance.
CN-7M, CG-6MMN	Heat to 2050°F (1120°C) minimum, hold for sufficient time to heat casting to temperature, quench in water or rapid cool by other means so as to develop acceptable corrosion resistance.
CN-7MS	Heat to 2100°F (1150°C) minimum, 2150°F (1180°C) maximum, hold for sufficient time (2 h minimum) to heat casting to temperature and quench in water to develop acceptable corrosion resistance.
CA-6NM	Heat to 1750°F (955°C) minimum, air cool to 200°F (95°C) or lower prior to any optional intermediate temper and prior to the final temper. The final temper shall be between 1050°F (565°C) and 1150°F (620°C).
CD-4MCu	Heat to 1900°F (1040°C) minimum, hold for sufficient time to heat casting uniformly to temperature, quench in water or rapid cool by other means to develop acceptable corrosion resistance.
CA-6N	Heat to 1900°F (1040°C), air cool, reheat to 1500°F (815°C), air cool, and age at 800°F (425°C), holding at each temperature sufficient time to heat casting uniformly to temperature.
CF10SMnN	Heat to 1950°F (1065°C) minimum, hold for sufficient time to heat casting to temperature, quench in water or rapid cool by other means so as to develop acceptable corrosion resistance.
CA-28MWV	(1) Heat to 1875–1925°F (1025–1050°C), quench in air or oil, and temper at 1150°F (620°C) minimum, or
	(2) Anneal at 1400°F (760°C) minimum.
CK-3MCuN	Heat to 2100°F (1150°C) minimum, hold for sufficient time to heat casting to temperature, quench in water or rapid cool by other means so as to develop acceptable corrosion resistance.

SOURCE: Reprinted with permission from the *Annual Book of ASTM Standards*, copyright 1992, American Society of Testing and Materials.

TABLE 12.5 Preheat and Tensile Requirements

	Minimum preheat temperatures	
Grade	°F	°C
CA-15, CA-15M, CA-40, CA-28MWV	400	[205]
Others	50	[10]

Tensile requirements

Grade	Type	Tensile strength, min		Yield strength, min		Elongation in 2 in (50 mm), min, %*	Reduction of area, min, %
		ksi	MPa	ksi	MPa		
CF-8	19 Chromium, 9 Nickel	70†	485†	30†	205†	35	—
CG-12	22 Chromium, 12 Nickel	70	485	28	195	35	—
CF-20	19 Chromium, 9 Nickel	70	485	30	205	30	—
CF-8M	19 Chromium, 10 Nickel, with molybdenum	70	485	30	205	30	—
CF-8C	19 Chromium, 10 Nickel with columbium	70	485	30	205	30	—
CF-16 and CF-16Fa	19 Chromium, 9 Nickel, free machining	70	485	30	205	30	—
CH-20 and CH-10	25 Chromium, 12 Nickel	70	485	30	205	30	—
CK-20	25 Chromium, 20 Nickel	65	450	28	195	30	—
CE-30	29 Chromium, 9 Nickel	80	550	40	275	10	—
CA-15 and CA-15M	12 Chromium	90	620	65	450	18	30

Designation	Composition	Tensile strength, ksi	Tensile strength, MPa	Yield strength, ksi	Yield strength, MPa	Elongation, %	Reduction, %
CB-30	20 Chromium	65	450	30	205	—	—
CC-50	28 Chromium	55	380	—	—	—	—
CA-40	12 Chromium	100	690	70	485	15	25
CA-40F	12 Chromium, free machining	100	690	70	485	12	—
CF-3	19 Chromium, 9 nickel	70	485	30	205	35	—
CF10SMnN	17 Chromium, 8.5 nickel with nitrogen 9 Nickel	85	585	42	290	30	—
CF-3M	19 Chromium, 10 Nickel, with molybdenum	70	485	30	205	30	—
CF-3MN	19 Chromium, 10 Nickel, with molybdenum, and nitrogen	75	515	37	255	35	—
CG6MMN	Chromium-nickel-manganese-molybdenum	85	585	42	290	30	—
CG-8M	19 Chromium, 11 Nickel, with molybdenum	75	520	35	240	25	—
CN-3M	20 Chromium,	63	435	25	170	30	—
CN-7M	29 Nickel, with copper and molybdenum	62	425	25	170	35	—
CN-7MS	19 Chromium, 24 Nickel, with copper and molybdenum	70	485	30	205	35	—
CA-6NM	12 Chromium, 4 nickel	110	755	80	550	15	35
CD-4MCu		100	690	70	485	16	—
CA-6N	11 Chromium, 7 nickel	140	965	135	930	15	50
CA-28MWV‡	12 Chromium, with molybdenum, tungsten, and vanadium	140	965	110	760	10	24
CK-3MCuN	20 Chromium 18 Nickel, with copper and molybdenum	80	550	38	260	35	—

* When ICI test bars are used in tensile testing as provided for in this specification, the gage length to reduced section diameter ratio shall be 4:1.

† For low ferrite or nonmagnetic castings of this grade, the following values shall apply; tensile strength, min, 65 ksi (450 MPa); yield point, min, 28 ksi (195 MPa).

‡ These mechanical properties apply only when heat treatment (1) has been used.

SOURCE: Reprinted with permission from the Annual Book of ASTM Standards, copyright 1992, American Society of Testing and Materials.

TABLE 12.6 Brinell Hardness Numbers*

(Ball 10 mm in diameter, applied loads of 500, 1500, and 3000 kgf)

Diameter of indentation, mm	500-kgf load	1500-kgf load	3000-kgf load	Diameter of indentation, mm	500-kgf load	1500-kgf load	3000-kgf load	Diameter of indentation, mm	500-kgf load	1500-kgf load	3000-kgf load	Diameter of indentation, mm	500-kgf load	1500-kgf load	3000-kgf load
2.00	158	473	945	2.60	92.6	278	555	3.20	60.5	182	363	3.80	42.4	127	255
2.01	156	468	936	2.61	91.8	276	551	3.21	60.1	180	361	3.81	42.2	127	253
2.02	154	463	926	2.62	91.1	273	547	3.22	59.8	179	359	3.82	42.0	126	252
2.03	153	459	917	2.63	90.4	271	543	3.23	59.4	178	356	3.83	41.7	125	250
2.04	151	454	908	2.64	89.7	269	538	3.24	59.0	177	354	3.84	41.5	125	249
2.05	150	450	899	2.65	89.0	267	534	3.25	58.6	176	352	3.85	41.3	124	248
2.06	148	445	890	2.66	88.4	265	530	3.26	58.3	175	350	3.86	41.1	123	246
2.07	147	441	882	2.67	87.7	263	526	3.27	57.9	174	347	3.87	40.9	123	245
2.08	146	437	873	2.68	87.0	261	522	3.28	57.5	173	345	3.88	40.6	122	244
2.09	144	432	865	2.69	86.4	259	518	3.29	57.2	172	343	3.89	40.4	121	242
2.10	143	428	856	2.70	85.7	257	514	3.30	56.8	170	341	3.90	40.2	121	241
2.11	141	424	848	2.71	85.1	255	510	3.31	56.5	169	339	3.91	40.0	120	240
2.12	140	420	840	2.72	84.4	253	507	3.32	56.1	168	337	3.92	39.8	119	239
2.13	139	416	832	2.73	83.8	251	503	3.33	55.8	167	335	3.93	39.6	119	237
2.14	137	412	824	2.74	83.2	250	499	3.34	55.4	166	333	3.94	39.4	118	236
2.15	136	408	817	2.75	82.6	248	495	3.35	55.1	165	331	3.95	39.1	117	235
2.16	135	404	809	2.76	81.9	246	492	3.36	54.8	164	329	3.96	38.9	117	234
2.17	134	401	802	2.77	81.3	244	488	3.37	54.4	163	326	3.97	38.7	116	232
2.18	132	397	794	2.78	80.8	242	485	3.38	54.1	162	325	3.98	38.5	116	231
2.19	131	393	787	2.79	80.2	240	481	3.39	53.8	161	323	3.99	38.3	115	230
2.20	130	390	780	2.80	79.6	239	477	3.40	53.4	160	321	4.00	38.1	114	229
2.21	129	386	772	2.81	79.0	237	474	3.41	53.1	159	319	4.01	37.9	114	228

2.22	128	383	765	2.82	78.4	235	470	3.42	52.8	158	317	4.02	37.7	113	226
2.23	126	379	758	2.83	77.9	234	467	3.43	52.5	157	315	4.03	37.5	113	225
2.24	125	376	752	2.84	77.3	232	464	3.44	52.2	156	313	4.04	37.3	112	224
2.25	124	372	745	2.85	76.8	230	461	3.45	51.8	156	311	4.05	37.1	111	223
2.26	123	369	738	2.86	76.2	229	457	3.46	51.5	155	309	4.06	37.0	111	222
2.27	122	366	732	2.87	75.7	227	454	3.47	51.2	154	307	4.07	36.8	110	221
2.28	121	363	725	2.88	75.1	225	451	3.48	50.9	153	306	4.08	36.6	110	219
2.29	120	359	719	2.89	74.6	224	448	3.49	50.6	152	304	4.09	36.4	109	218
2.30	119	356	712	2.90	74.1	222	444	3.50	50.3	151	302	4.10	36.2	109	217
2.31	118	353	706	2.91	73.6	221	441	3.51	50.0	150	300	4.11	36.0	108	216
2.32	117	350	700	2.92	73.0	219	438	3.52	49.7	149	298	4.12	35.8	108	215
2.33	116	347	694	2.93	72.5	218	435	3.53	49.4	149	297	4.13	35.7	107	214
2.34	115	344	688	2.94	72.0	216	432	3.54	49.2	148	295	4.14	35.5	106	213
2.35	114	341	682	2.95	71.5	215	429	3.55	48.9	147	293	4.15	35.3	106	212
2.36	113	338	676	2.96	71.0	213	426	3.56	48.6	147	292	4.16	35.1	105	211
2.37	112	335	670	2.97	70.5	212	423	3.57	48.3	146	290	4.17	34.9	105	210
2.38	111	332	665	2.98	70.1	210	420	3.58	48.0	145	288	4.18	34.8	104	209
2.39	110	330	659	2.99	69.6	209	417	3.59	47.7	144	286	4.19	34.6	104	208
2.40	109	327	653	3.00	69.1	207	415	3.60	47.5	143	285	4.20	34.4	103	207
2.41	108	324	648	3.01	68.6	206	412	3.61	47.2	142	283	4.21	34.2	103	205
2.42	107	322	643	3.02	68.2	205	409	3.62	46.9	142	282	4.22	34.1	102	204
2.43	106	319	637	3.03	67.7	203	406	3.63	46.7	141	280	4.23	33.9	102	203
2.44	105	316	632	3.04	67.3	202	404	3.64	46.4	140	278	4.24	33.7	101	202
2.45	104	313	627	3.05	66.8	200	401	3.65	46.1	139	277	4.25	33.6	101	201
2.46	104	311	621	3.06	66.4	199	398	3.66	45.9	138	275	4.26	33.4	100	200
2.47	103	308	616	3.07	65.9	198	395	3.67	45.6	138	274	4.27	33.2	99.7	199
2.48	102	306	611	3.08	65.5	196	393	3.68	45.4	137	272	4.28	33.1	99.2	198
2.49	101	303	606	3.09	65.0	195	390	3.69	45.1	136	271	4.29	32.9	98.8	198
2.50	100	301	601	3.10	64.6	194	388	3.70	44.9	135	269	4.30	32.8	98.3	197
2.51	99.4	298	597	3.11	64.2	193	385	3.71	44.6	135	268	4.31	32.6	97.8	196
2.52	98.6	296	592	3.12	63.8	191	383	3.72	44.4	134	266	4.32	32.4	97.3	195
2.53	97.8	294	587	3.13	63.3	190	380	3.73	44.1	133	265	4.33	32.3	96.8	194
2.54	97.1	291	582	3.14	62.9	189	378	3.74	43.9	132	263	4.34	32.1	96.4	193
2.55	96.3	289	578	3.15	62.5	188	375	3.75	43.6	132	262	4.35	32.0	95.9	192

TABLE 12.6 Brinell Hardness Numbers* (Continued)

Diameter of indentation, mm	Brinell hardness number 500-kgf load	Brinell hardness number 1500-kgf load	Brinell hardness number 3000-kgf load	Diameter of indentation, mm	Brinell hardness number 500-kgf load	Brinell hardness number 1500-kgf load	Brinell hardness number 3000-kgf load	Diameter of indentation, mm	Brinell hardness number 500-kgf load	Brinell hardness number 1500-kgf load	Brinell hardness number 3000-kgf load	Diameter of indentation, mm	Brinell hardness number 500-kgf load	Brinell hardness number 1500-kgf load	Brinell hardness number 3000-kgf load
2.56	95.5	287	573	3.16	62.1	186	373	3.76	43.4	130	260	4.36	31.8	95.5	191
2.57	94.8	284	569	3.17	61.7	185	370	3.77	43.1	129	259	4.37	31.7	95.0	190
2.58	94.0	282	564	3.18	61.3	184	368	3.78	42.9	129	257	4.38	31.5	94.5	189
2.59	93.3	280	560	3.19	60.9	183	366	3.79	42.7	128	256	4.39	31.4	94.1	188
4.40	31.2	93.6	187	5.05	23.3	69.8	140	5.70	17.8	53.5	107	6.35	14.0	42.0	84.0
4.41	31.1	93.2	186	5.06	23.2	69.5	139	5.71	17.8	53.3	107	6.36	13.9	41.8	83.7
4.42	30.9	92.7	185	5.07	23.1	69.2	138	5.72	17.7	53.1	106	6.37	13.9	41.7	83.4
4.43	30.8	92.3	185	5.08	23.0	68.9	138	5.73	17.6	52.9	106	6.38	13.8	41.5	83.1
4.44	30.6	91.8	184	5.09	22.9	68.6	137	5.74	17.6	52.7	105	6.39	13.8	41.4	82.8
4.45	30.5	91.4	183	5.10	22.8	68.3	137	5.75	17.5	52.5	105	6.40	13.7	41.2	82.5
4.46	30.3	91.0	182	5.11	22.7	68.0	136	5.76	17.4	52.3	105	6.41	13.7	41.1	82.2
4.47	30.2	90.5	181	5.12	22.6	67.7	135	5.77	17.4	52.1	104	6.42	13.6	40.9	81.9
4.48	30.0	90.1	180	5.13	22.5	67.4	135	5.78	17.3	51.9	104	6.43	13.6	40.8	81.6
4.49	29.9	89.7	179	5.14	22.4	67.1	134	5.79	17.2	51.7	103	6.44	13.5	40.6	81.3
4.50	29.8	89.3	179	5.15	22.3	66.9	134	5.80	17.2	51.5	103	6.45	13.5	40.5	81.0
4.51	29.6	88.8	178	5.16	22.2	66.6	133	5.81	17.1	51.3	103	6.46	13.4	40.4	80.7
4.52	29.5	88.4	177	5.17	22.1	66.3	133	5.82	17.0	51.1	102	6.47	13.4	40.2	80.4
4.53	29.3	88.0	176	5.18	22.0	66.0	132	5.83	17.0	50.9	102	6.48	13.4	40.1	80.1
4.54	29.2	87.6	175	5.19	21.9	65.8	132	5.84	16.9	50.7	101	6.49	13.3	39.9	79.8
4.55	29.1	87.2	174	5.20	21.8	65.5	131	5.85	16.8	50.5	101	6.50	13.3	39.8	79.6
4.56	28.9	86.8	174	5.21	21.7	65.2	130	5.86	16.8	50.3	101	6.51	13.2	39.6	79.3
4.57	28.8	86.4	173	5.22	21.6	64.9	130	5.87	16.7	50.2	100	6.52	13.2	39.5	79.0

4.58	28.7	86.0	172	5.23	21.6	64.7	129	5.88	16.7	50.0	99.9	6.53	13.1	39.4	78.7
4.59	28.5	85.6	171	5.24	21.5	64.4	129	5.89	16.6	49.8	99.5	6.54	13.1	39.2	78.4
4.60	28.4	85.4	170	5.25	21.4	64.1	128	5.90	16.5	49.6	99.2	6.55	13.0	39.1	78.2
4.61	28.3	84.8	169	5.26	21.3	63.9	128	5.91	16.5	49.4	98.8	6.56	13.0	38.9	78.0
4.62	28.1	84.4	168	5.27	21.2	63.6	127	5.92	16.4	49.2	98.4	6.57	12.9	38.8	77.6
4.63	28.0	84.0	167	5.28	21.1	63.3	127	5.93	16.3	49.0	98.0	6.58	12.9	38.7	77.3
4.64	27.9	83.6	167	5.29	21.0	63.1	126	5.94	16.3	48.8	97.7	6.59	12.8	38.5	77.1
4.65	27.8	83.3	166	5.30	20.9	62.8	126	5.95	16.2	48.7	97.3	6.60	12.8	38.4	76.8
4.66	27.6	82.9	165	5.31	20.9	62.6	125	5.96	16.2	48.5	96.9	6.61	12.8	38.3	76.5
4.67	27.5	82.5	164	5.32	20.8	62.3	125	5.97	16.1	48.3	96.6	6.62	12.7	38.1	76.2
4.68	27.4	82.1	164	5.33	20.7	62.1	124	5.98	16.0	48.1	96.2	6.63	12.7	38.0	76.0
4.69	27.3	81.8	163	5.34	20.6	61.8	124	5.99	16.0	47.9	95.9	6.64	12.6	37.9	75.7
4.70	27.1	81.4	162	5.35	20.5	61.5	123	6.00	15.9	47.7	95.5	6.65	12.6	37.7	75.4
4.71	27.0	81.0	161	5.36	20.4	61.3	123	6.01	15.9	47.6	95.1	6.66	12.5	37.6	75.2
4.72	26.9	80.7	161	5.37	20.3	61.0	122	6.02	15.8	47.4	94.8	6.67	12.5	37.5	74.9
4.73	26.8	80.3	160	5.38	20.3	60.8	122	6.03	15.7	47.2	94.4	6.68	12.4	37.3	74.7
4.74	26.6	79.9	159	5.39	20.2	60.6	121	6.04	15.7	47.0	94.1	6.69	12.4	37.2	74.4
4.75	26.5	79.6	158	5.40	20.1	60.3	121	6.05	15.6	46.8	93.7	6.70	12.4	37.1	74.1
4.76	26.4	79.2	158	5.41	20.0	60.1	120	6.06	15.6	46.7	93.4	6.71	12.3	36.9	73.9
4.77	26.3	78.9	157	5.42	19.9	59.8	120	6.07	15.5	46.5	93.0	6.72	12.3	36.8	73.6
4.78	26.1	78.5	156	5.43	19.8	59.6	119	6.08	15.4	46.3	92.7	6.73	12.2	36.7	73.4
4.79	25.9	78.2	156	5.44	19.8	59.3	119	6.09	15.4	46.2	92.3	6.74	12.2	36.6	73.1
4.80	25.8	77.8	155	5.45	19.7	59.1	118	6.10	15.3	46.0	92.0	6.75	12.1	36.4	72.8
4.81	25.7	77.5	154	5.46	19.6	58.9	118	6.11	15.3	45.8	91.7	6.76	12.1	36.3	72.6
4.82	25.6	77.1	154	5.47	19.5	58.6	117	6.12	15.2	45.7	91.3	6.77	12.1	36.2	72.3
4.83	25.5	76.8	153	5.48	19.5	58.4	117	6.13	15.2	45.5	91.0	6.78	12.0	36.0	72.1
4.84	25.4	76.4	152	5.49	19.4	58.2	116	6.14	15.1	45.3	90.6	6.79	12.0	35.9	71.8
4.85	25.3	76.1	152	5.50	19.3	57.9	116	6.15	15.1	45.2	90.3	6.80	11.9	35.8	71.6
4.86	25.1	75.8	151	5.51	19.2	57.7	115	6.16	15.0	45.0	90.0	6.81	11.9	35.7	71.3
4.87	25.0	75.4	150	5.52	19.2	57.5	115	6.17	14.9	44.8	89.6	6.82	11.8	35.5	71.1
4.88	25.0	75.1	150	5.53	19.1	57.2	114	6.18	14.9	44.7	89.3	6.83	11.8	35.4	70.8
4.89	24.9	74.8	149	5.54	19.0	57.0	114	6.19	14.8	44.5	89.0	6.84	11.8	35.3	70.6
4.90	24.8	74.4	149	5.55	18.9	56.8	114	6.20	14.7	44.3	88.7	6.86	11.7	35.2	70.4
4.91	24.7	74.1	148	5.56	18.9	56.6	113	6.21	14.7	44.2	88.3	6.86	11.7	35.1	70.1

TABLE 12.6 Brinell Hardness Numbers* (Continued)

Brinell hardness number				Brinell hardness number				Brinell hardness number				Brinell hardness number			
Diameter of indentation, mm	500-kgf load	1500-kgf load	3000-kgf load	Diameter of indentation, mm	500-kgf load	1500-kgf load	3000-kgf load	Diameter of indentation, mm	500-kgf load	1500-kgf load	3000-kgf load	Diameter of indentation, mm	500-kgf load	1500-kgf load	3000-kgf load
4.92	24.6	73.8	148	5.57	18.8	56.3	113	6.22	14.7	44.0	88.0	6.87	11.6	34.9	69.9
4.93	24.5	73.5	147	5.58	18.7	56.1	112	6.23	14.6	43.8	87.7	6.88	11.6	34.8	69.6
4.94	24.4	73.2	146	5.59	18.6	55.9	112	6.24	14.6	43.7	87.4	6.89	11.6	34.7	69.4
4.95	24.3	72.8	146	5.60	18.6	55.7	111	6.25	14.5	43.5	87.1	6.90	11.5	34.6	69.2
4.96	24.2	72.5	145	5.61	18.5	55.5	111	6.26	14.5	43.4	86.7	6.91	11.5	34.5	68.9
4.97	24.1	72.2	144	5.62	18.4	55.2	110	6.27	14.4	43.2	86.4	6.92	11.4	34.3	68.7
4.98	24.0	71.9	144	5.63	18.3	55.0	110	6.28	14.4	43.1	86.1	6.93	11.4	34.2	68.4
4.99	23.9	71.6	143	5.64	18.3	54.8	110	6.29	14.3	42.9	85.8	6.94	11.4	34.1	68.2
5.00	23.8	71.3	143	5.65	18.2	54.6	109	6.30	14.2	42.7	85.5	6.95	11.3	34.0	68.0
5.01	23.7	71.0	142	5.66	18.1	54.4	109	6.31	14.2	42.6	85.2	6.96	11.3	33.9	67.7
5.02	23.6	70.7	141	5.67	18.1	54.2	108	6.32	14.1	42.4	84.9	6.97	11.3	33.8	67.5
5.03	23.5	70.4	141	5.68	18.0	54.0	108	6.33	14.1	42.3	84.6	6.98	11.2	33.6	67.3
5.04	23.4	70.1	140	5.69	17.9	53.7	107	6.34	14.0	42.1	84.3	6.99	11.2	33.5	67.0

* Prepared by the Engineering Mechanics Section, Institute for Standards Technology.
SOURCE: Reprinted with permission, from the Annual Book of ASTM Standards, copyright 1992, American Society of Testing and Materials.

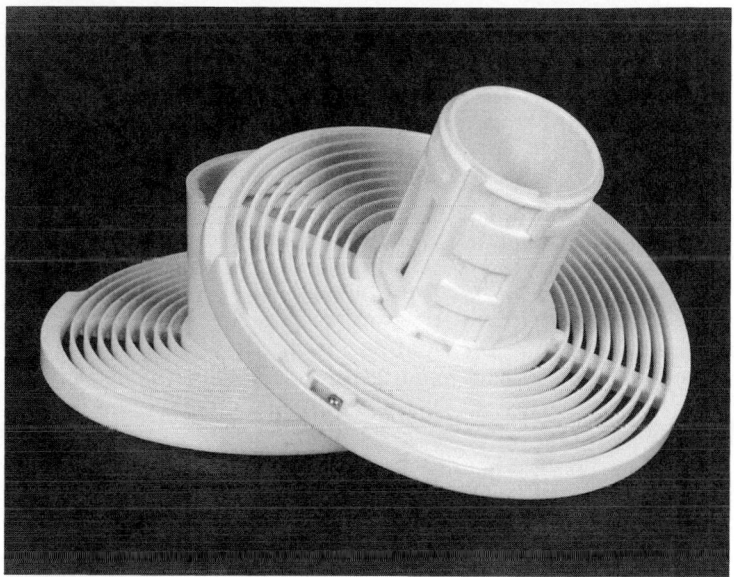

Figure 12.20 Typical intricate plastic-cast parts.

materials such as glass, carbon, and mineral fibers for added strength. Thermoset plastics usually are more dimensionally stable and heat resistant and have better electrical properties than thermoplastics.

Complex molded shapes of the plastics are analyzed today using the advanced finite-element analysis (FEA) techniques available for the personal computer and engineering design stations. Figure 12.20 shows a typical intricate plastic part that must be dimensionally accurate as well as chemical resistant.

12.5.1 Prototypes of the plastics

Building a prototype plastic part is a compromise because the part is usually machined from plastic blocks and slabs and will not duplicate the exact performance of the finished part, which is made in a mold. The closest duplicate to a plastic production part is made by molding a prototype in mild-steel molds made specifically for the prototype. This method is expensive but is the closest method to use if you wish to avoid expensive rework or changes to a finished production mold. This approach will give the most accurate test results prior to building the actual production mold.

12.5.2 Properties and characteristics of modern plastics

Widely used plastics are discussed in Chap. 4, "Materials and Their Uses." The applications of the modern plastics are given there, as well as the trade names and families of all plastics manufactured today.

12.5.3 Design of molded plastic parts

Final selection of plastic type and part configuration or design should be reviewed and coordinated with the mold maker and plastic part manufacturer before the final design drawings are made. Mold makers and molded-part manufacturers can alert the designer to the many problem areas that are prevalent on preliminary designs for plastic parts.

Plastic part design handbooks are available from all the leading producers of plastic materials, such as DuPont, General Electric, Monsanto, etc. These manuals or handbooks cover detail design, appropriate calculation techniques, and complete chemical, physical, and electrical properties of the materials. The design handbooks may be secured by writing directly to the plastics sections of the large suppliers or their distributors.

12.5.4 Plastics molding machinery and molds

A great many machines are made for molding plastic parts, most of which are expensive and require specialized techniques for operation. Figures 12.21 and 12.22 show a typical group of machines required for producing parts made of cycloaliphatic epoxy. This class of thermoset plastic has gained wide recognition in the electrical power distribution industry for parts that support or brace high-voltage current-carrying busses and parts of switching devices such as breakers and switches of all classes up to 34.5 kV. Figure 12.21 shows the mixing and dispensing machine, and Fig. 12.22 shows the complete group of equipment, with the mold-clamping machine at the front of the photograph.

Because of the high injection pressures developed in this plastic molding process, the clamping machine must exert a high load on the mold to prevent it from opening during the injection process. In this process, the hot, molten plastic is injected under pressure into the mold, where it is held for the "curing" or setting time prior to being released from the mold. In the molding of cycloaliphatic epoxy after the proper cure, the plastic part is removed from the mold,

Figure 12.21 Mixing and dispensing sections of epoxy casting equipment.

Figure 12.22 Complete assembly of epoxy casting machinery: (*far left*) oven, (*center*) die-clamping machine, (*far right*) dispensing section, (*center*) background process control panel (CNC).

placed in an oven set to a preselected temperature, and baked for a predetermined time interval. Cycloaliphatic epoxy is being used in the electrical industries as a substitute for wet-process porcelain, which for many years was one of the few materials available for this type of service. Glass and polyester-glass thermosets are also used in electrical applications, as are other electrical-grade epoxies.

The tonnage and size of the clamping machine for injection molding plastics are determined by the volume of plastic to be injected at one time (a single *shot*). A machine that handles parts up to 20 in^3 would be considered of moderate size, whereas a machine required to injection mold parts with a volume of 60 in^3 would be considered large. Clamping machines are made in sizes up to hundreds of tons of clamping capacity, and these machines usually are found at the larger plastic molding manufacturers.

12.6 Extrusions

An extruded shape is one in which the material being extruded is pushed, under high pressure and high temperature, through a set of extrusion dies that have the shape of a cross section of the part. Extrusion is similar to drawing, except that in drawing the drawn part is pulled out of or through the dies or series of dies of progressive sizes. The material is cold worked more effectively when it is drawn rather than extruded.

Aluminum alloys are particularly well suited for the extrusion processes. Very large, high-tonnage extrusion presses are required for the extrusion of aluminum and its alloys. Generally, two classes or grades of aluminum extrusions are widely available: structural aluminum and architectural aluminum shapes. These two classes of extrusions are available in the following alloys and tempers:

Class	Shape	Alloy, temper, and specification
Structural	Angles	6061-T6, ASTM B308
	Channels	6061-T6, ASTM B308
	Tees	6061-T6, ASTM B308
	Zees	6061-T6, ASTM B308
	H beams and I beams	6061-T6, ASTM B308
	Wide-flange beams	6061-T6, ASTM B308
Architectural	Angles	6063-T52, ASTM B221
	Channels	6063-T52, ASTM B221
	Tees	6063-T52, ASTM B221
	Zees	6063-T52, ASTM B221

Aluminum alloy extrusions are also available in the following standard shapes:

- Round rod
- Square bar
- Rectangular bar
- Hexagonal bar
- Round tube
- Square tube
- Rectangular tube

The alloy designations of the above-listed shapes may be one of the following:

- 2011
- 2017
- 2024
- 6061
- 6063

An aluminum alloy extrusion may be designed to any practical shape that the extrusion dies may accommodate, and shapes usually are limited to the designer's imagination and ingenuity and the size of the part. Some of the standard extruded shapes that are available from various sources nationwide are shown in Fig. 12.23. The temper designations and lengths available are also shown in the figure.

A table of distances across flats for squares, hexagons, and octagons is given in Fig. 12.24. This will prove useful and time saving when these dimensions need to be known and you do not wish to calculate them by using the figures shown in Chap. 2 of this *Handbook*.

Plastic extrusions. A great many plastic-extruded shapes are available from the plastic manufacturers, whose catalogs may be obtained through such trade magazines as *The American Machinist, Machine Design, Product Design and Development, Modern Machine Shop,* and others. If you work with or need plastic-extruded shapes, keep a

STANDARD ANGLES ①

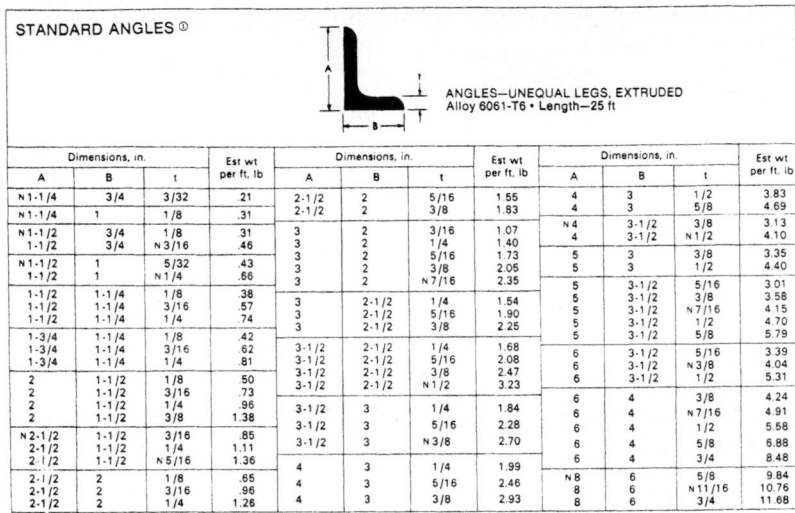

ANGLES—UNEQUAL LEGS, EXTRUDED
Alloy 6061-T6 • Length—25 ft

Dimensions, in.			Est wt per ft, lb	Dimensions, in.			Est wt per ft, lb	Dimensions, in.			Est wt per ft, lb
A	B	t		A	B	t		A	B	t	
N 1-1/4	3/4	3/32	.21	2-1/2	2	5/16	1.55	4	3	1/2	3.83
N 1-1/4	1	1/8	.31	2-1/2	2	3/8	1.83	4	3	5/8	4.69
N 1-1/2	3/4	1/8	.31	3	2	3/16	1.07	N 4	3-1/2	3/8	3.13
1-1/2	3/4	N 3/16	.46	3	2	1/4	1.40	4	3-1/2	N 1/2	4.10
N 1-1/2	1	5/32	.43	3	2	5/16	1.73	5	3	3/8	3.35
1-1/2	1	N 1/4	.66	3	2	3/8	2.05	5	3	1/2	4.40
1-1/2	1-1/4	1/8	.38	3	2	N 7/16	2.35	5	3-1/2	5/16	3.01
1-1/2	1-1/4	3/16	.57					5	3-1/2	3/8	3.58
1-1/2	1-1/4	1/4	.74	3	2-1/2	1/4	1.54	5	3-1/2	N 7/16	4.15
1-3/4	1-1/4	1/8	.42	3	2-1/2	5/16	1.90	5	3-1/2	1/2	4.70
1-3/4	1-1/4	3/16	.62	3	2-1/2	3/8	2.25	5	3-1/2	5/8	5.79
1-3/4	1-1/4	1/4	.81	3-1/2	2-1/2	1/4	1.68	6	3-1/2	3/8	3.39
2	1-1/2	1/8	.50	3-1/2	2-1/2	5/16	2.08	6	3-1/2	N 3/8	4.04
2	1-1/2	3/16	.73	3-1/2	2-1/2	3/8	2.47	6	3-1/2	1/2	5.31
2	1-1/2	1/4	.96	3-1/2	2-1/2	N 1/2	3.23				
2	1-1/2	3/8	1.38	3-1/2	3	1/4	1.84	6	4	3/8	4.24
N 2-1/2	1-1/2	3/16	.85	3-1/2	3	5/16	2.28	6	4	N 7/16	4.91
2-1/2	1-1/2	1/4	1.11	3-1/2	3	N 3/8	2.70	6	4	1/2	5.58
2-1/2	1-1/2	N 5/16	1.36					6	4	5/8	6.88
2-1/2	2	1/8	.65	4	3	1/4	1.99	6	4	3/4	8.48
2-1/2	2	3/16	.96	4	3	5/16	2.46	N 8	6	5/8	9.84
2-1/2	2	1/4	1.26	4	3	3/8	2.93	8	6	N 11/16	10.76
								8	6	3/4	11.68

① Angle sections listed as "standard" are approximations of American Standard sections. For elements of sections and detailed dimensions, consult *Alcoa Structural Handbook*.
N Not stocked at plant.

ZEES, EXTRUDED

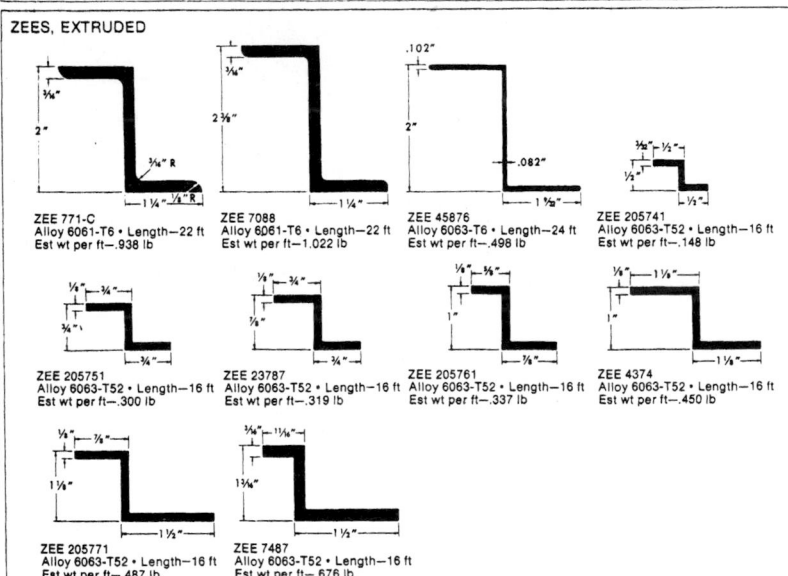

Figure 12.23 Samples of standard aluminum-extruded shapes.

DISTANCES ACROSS CORNERS OF
SQUARES, HEXAGONS & OCTAGONS

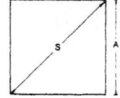      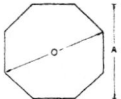

S = 1.414A   H = 1.155A   O = 1.082A

A Size in Inches	Distance Across Corners in inches			A Size in Inches	Distance Across Corners in inches		
	S Square	H Hexagon	O Octagon		S Square	H Hexagon	O Octagon
⅛	.177	.144	.135	2⅛	3.005	2.454	2.300
³⁄₁₆	.265	.217	.203	2³⁄₁₆	3.094	2.526	2.368
¼	.354	.289	.271	2¼	3.182	2.598	2.435
⁵⁄₁₆	.442	.361	.338	2⁵⁄₁₆	3.270	2.670	2.503
⅜	.530	.433	.406	2⅜	3.359	2.742	2.571
⁷⁄₁₆	.619	.505	.474	2⁷⁄₁₆	3.447	2.815	2.638
½	.707	.577	.541	2½	3.536	2.887	2.706
⁹⁄₁₆	.795	.650	.609	2⁹⁄₁₆	3.624	2.959	2.774
⅝	.884	.722	.677	2⅝	3.712	3.031	2.841
¹¹⁄₁₆	.972	.794	.744	2¹¹⁄₁₆	3.801	3.103	2.909
¾	1.061	.866	.812	2¾	3.889	3.175	2.977
¹³⁄₁₆	1.149	.938	.879	2¹³⁄₁₆	3.977	3.248	3.044
⅞	1.237	1.010	.947	2⅞	4.066	3.320	3.112
¹⁵⁄₁₆	1.326	1.083	1.015	2¹⁵⁄₁₆	4.154	3.392	3.180
1	1.414	1.155	1.082	3	4.243	3.464	3.247
1¹⁄₁₆	1.503	1.227	1.150				
1⅛	1.591	1.299	1.218	3⅛	4.419	3.608	3.383
1³⁄₁₆	1.679	1.371	1.285	3¼	4.596	3.753	3.518
1¼	1.768	1.443	1.353	3⅜	4.773	3.897	3.653
1⁵⁄₁₆	1.856	1.516	1.421	3½	4.950	4.041	3.788
1⅜	1.945	1.588	1.488	3⅝	5.126	4.186	3.924
1⁷⁄₁₆	2.033	1.660	1.556				
1½	2.121	1.732	1.624	3¾	5.303	4.330	4.059
1⁹⁄₁₆	2.210	1.804	1.691	3⅞	5.480	4.474	4.194
1⅝	2.298	1.876	1.759	4	5.657	4.619	4.330
1¹¹⁄₁₆	2.386	1.949	1.827	4¼	6.010	4.907	4.600
1¾	2.475	2.021	1.894	4½	6.364	5.196	4.871
1¹³⁄₁₆	2.563	2.093	1.962	4¾	6.717	5.485	5.141
1⅞	2.652	2.165	2.031	5	7.071	5.774	5.412
1¹⁵⁄₁₆	2.740	2.237	2.097	5¼	7.425	6.062	5.683
				5½	7.778	6.351	5.953
2	2.828	2.309	2.165	5¾	8.132	6.640	6.224
2¹⁄₁₆	2.917	2.382	2.232	6	8.485	6.928	6.494

Figure 12.24   Dimensions of squares, hexagons, and octagons.

series of these catalogs in your reference files. The plastic-extruded shapes not only make the design job easier, but they also enhance the appearance of the final product and facilitate assembly in many cases.

12.7   Powder-Metal Technology

Powder-metal parts are made by compressing a highly purified metallic powder in a set of dies under extremely high pressure and then fusing the particles in an oven under controlled high temperature. This compression process compacts the powder metal until the part is approximately 90 percent the density of the solid metal or alloy. The density of the part can be controlled to an extent that the open pores in the powder-metal part, after fusion, may be impregnated with lubricants or filler resins and binders. The well-known sintered-bronze journal bearings with impregnated lubrication are prime examples of powder-metal technology.

Powder-metal parts may be made of aluminum alloys, copper alloys, steels, and stainless steels. Other metals and alloys also can be used in powder-metal processing.

The modern powder-metal part may be impregnated with resins and binders, which make possible the application of electroplated finishes such as copper, zinc, nickel, and chromium. If the part is made of one of the corrosion-resistant stainless steels, the plating process may be eliminated.

Parts may be produced using powder metal that normally would be difficult to machine. An example of powder-metal parts is shown in Fig. 12.25. Here are shown a group of metal parts, most of which would be difficult to make using other processes. The parts shown could be made using the investment casting process or machining techniques but would be costly and time-consuming to produce. Figure 12.26 is a closer view of the three small three-pointed parts that shows a minute hole in the center of the parts. Each division on the scale shown in Fig. 12.26 is equal to 0.020 in (1/50 inch). One of the parts shown in Fig. 12.26 had secondary machining operations performed on it (threading). All the parts shown in both figures are easy and economical to produce using powder-metal technology.

Design of powder-metal parts.   The design and manufacturing of powder-metal parts are not difficult if the basic rules of powder-metal part design are followed and if the powder-metal part manufacturer is consulted during the design stages. The part manufacturer will advise you if the part can be produced as designed and what remedial

Figure 12.25 Samples of powder-metal parts.

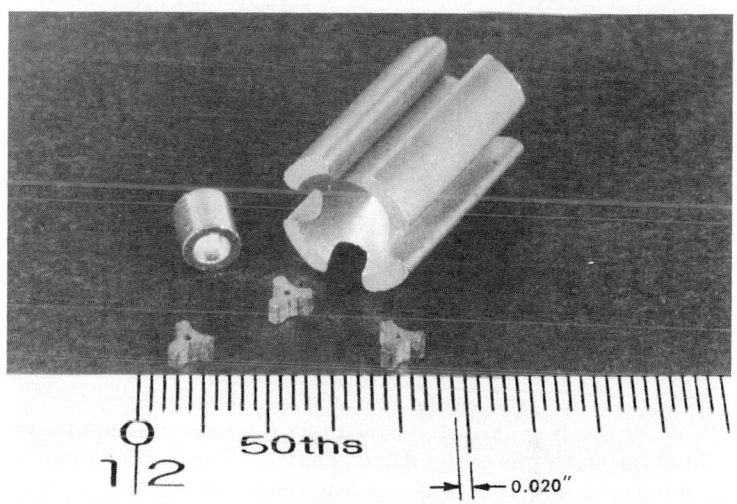

Figure 12.26 Close-up view of powder-metal parts. Note scale.

Typical Properties of Powder-Metal Parts

MATERIAL	CONDITION	DENSITY	ULTIMATE	YIELD	ELONGATION	IMPACT. FT-LBS.	HARDNESS
Low Alloy Steel	S/HT	6.8	110	110	<0.5	4.0	32HRC
Low Alloy Steel	S/HT	6.8	130	130	<0.5	6.0	34HRC
Stainless Steel	S	6.4	39	32	0.5	3.5	62HRB
Stainless Steel	S	6.5	60	39	10.0	28.0	62HRB
Stainless Steel	HT	6.5	105	105	<0.5	2.5	23HRC
Iron-Nickel	S	6.9	50	30	2.5	12	59HRB
Iron-Nickel	S/HT	7.0	135	135	<0.5	6.5	31HRC
Iron-Copper	S	6.7	50	45	<1.0	5.0	60HRB
Iron-Copper	S/HT	6.8	90	90	<0.5	4.5	35HRC

S = Sintered; HT = Heat-Treated; Density, g/cc; Ultimate strength = x1000 psi,; Yield = x 1000 psi; Elongation = %;
Impact Unnotched Charpy, ft-lbs.; Hardness = Rockwell B or C scale, as indicated.

Figure 12.27 Typical properties of powder-metal parts.

actions to take or design changes are needed to produce the part. The manufacturer also will advise you of the availability of the different metals and alloys for producing the part.

Basic rules for powder-metal part design

- Keep the outline elements of the part on lines parallel to the part axis or in the direction of compression of the male die.

- Do not design parts with reverse angles along the axis (the part cannot be extracted from the dies if this occurs).

- Keep wall sections or webs as thick as possible.

- Single tapered holes in the part are possible when the direction of taper allows extraction from the dies after compression of the powder metal.

- Limit the size or volume of the part to that which may be produced with available machinery.

- Do not specify electroplated finishes unless necessary (the part may be produced in a corrosion-resistant alloy if necessary).

- Holes in the part may be controlled to close tolerances.

- Check with the part manufacturer to ascertain if the part can sustain the imposed stress loads anticipated.

- The outline accuracy of parts can be closely controlled.

Some of the powder-metal part producers can provide design manuals or brochures to the design engineer that outline in more detail the design procedures for powder-metal parts.

Small-part design always should be reviewed to see if the design requirements can be met using powder-metal technology. Powder-metal technology is much further advanced today than in the past; then it was used mainly to produce journal bearings with impregnated lubrication.

12.7.1 Typical powder-metal parts and properties

Powder-metal technology has advanced greatly over the past 10 years. Although powder-metal parts have been in use for many years in the form of sintered bronze bearings, the technology now is able to produce parts with a drastic increase in compositions as well as strengths. Figure 12.27 (see previous page) shows some of the types

of materials now available through the powder-metal processes and their relative strengths for use in various types of parts.

Standards for powder metal parts are available from the Metal Powder Industries Federation (MPIF). MPIF Standard 35 lists the compositions and physical properties of available powder metals.

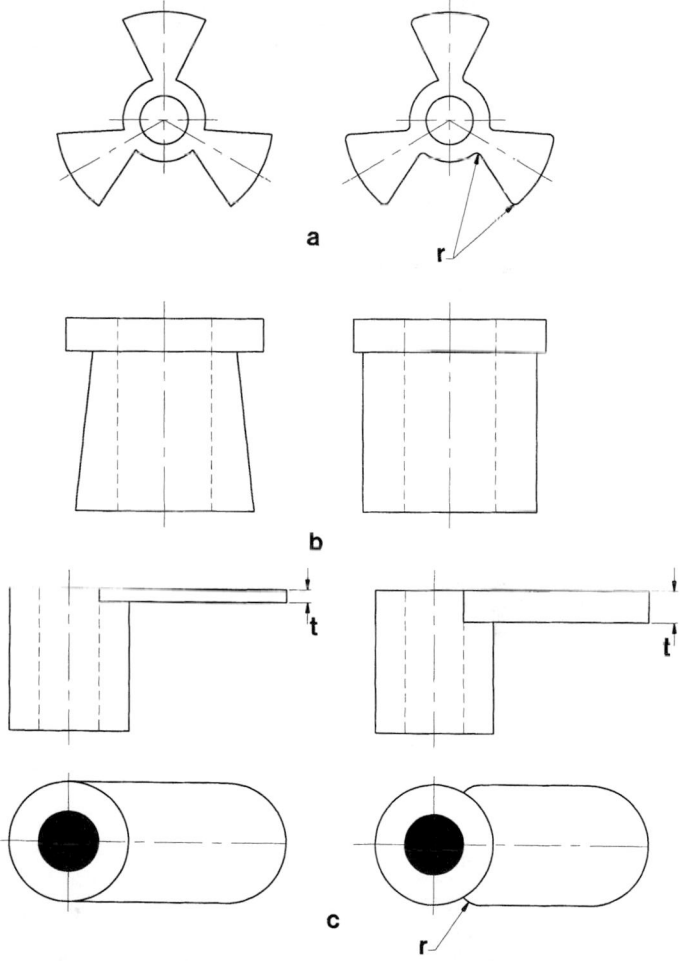

Figure 12.28 Typical powder-metal part design features. Parts shown on the left side are not acceptable; parts shown on the right side embody good design and are acceptable.

For those interested in obtaining the full standards listings, the address of the MPIF is shown in Chap. 16.

The design of powder-metal parts is limited by the process used for compression of the different powder metals. Some examples of design considerations are shown in Fig. 12.28. The figures on the left are not recommended; the figures on the right are recommended. Notice that generous radiuses are required for powder-metal parts design. For more detailed information on the design of powder-metal parts, consult powder-metal manufacturers or the MPIF.

13

Plating Practices and Finishes for Metal

Commercial products and equipment in all classifications require finishes of one type or other. These finishes range from basic oxide coatings to the various paints and plastics to electrodeposited metals such as copper, chromium, nickel, etc.

The finish or plating used on any particular part should contribute to the engineering qualities of the final product and not merely to its cosmetic appeal. The desired finish could include weather protection, resistance to corrosive chemicals, heat resistance, electrical conductivity, wear resistance, and improved lubrication qualities.

It is the design engineer's responsibility to specify the finish characteristics and specifications on a part or assembly. Designers should be aware of the types of finishes and plating processes that are available commercially and how to specify them on the design and detail part drawings. Arbitrary selection of a finish or plating and its thickness range can lead to many design problems relating to corrosion, cost, and dimensional interference.

This chapter will familiarize the designer, engineer, and other personnel in the metalworking industries with the common finishing processes, procedures for specifying thicknesses of platings, and the appropriate industrial standard specifications that control these finishes.

13.1 Finishes

The finishing processes and methods in common use for engineering applications are as follows:

Mechanical finishing

- Sanding or grinding
- Brushing (scratch and satin)
- Sandblasting
- Shot peening and tumbling (metal or ceramic ball)
- Burnishing
- Mechanical powder plating
- Polishing

Chemical finishing

- Etching
- Bonderite
- Alodine
- Iridite
- Phosphatizing
- Passivating (stainless steels)
- Black oxide
- Blueing
- Teflon coating
- Other plastic coatings

Electrolytic oxides

- Sulfuric acid anodize (Alumilite, Alcoa process)
- Chromic acid anodize (aerospace and other applications), Martin hard-coat anodize (Martin-Marietta process)
- Electropolishing

Note: Anodizing may be performed on aluminum, magnesium, titanium, and zinc.

Electroplating

- Copper plate
- Cadmium plate
- Chromium plate (bright, hard, and black)
- Nickel plate (bright, black, and chromium-nickel)
- Gold plate
- Silver plate
- Tin plate (tin-cadmium, tin-lead, and tin-nickel)
- Indium plate
- Rhodium plate
- Palladium plate
- Zinc plate (nickel-zinc, zinc with chromate)
- Lead plate

The base metals and their alloys that are commonly electroplated are iron, steel, stainless steel (more commonly passivated), aluminum, copper, brass, bronze, titanium, and magnesium.

Hot-dip plating

- Hot-dip tin
- Hot-dip zinc
- Hot-dip aluminum
- Hot-dip lead

13.2   Corrosion of Metals: Principles

The design engineer has four choices when planning for corrosion resistance of a metal part:

1. Use a corrosion-resistant alloy steel.
2. Use a nonferrous, noncorroding metal alloy.
3. Use plated steel or anodized alloys.

4. Use a chemically coated, painted, or plastic-coated metal part (ferrous or nonferrous).

The third choice is usually the most economical, although it may not be the best choice. Plated-steel parts are the most common and suffice for most applications, although the choice of a more expensive alloy may be mandatory as a result of required antimagnetic properties, low electrical resistivity, restrictions on the weight of the part (such as in aerospace vehicles), and other considerations.

Basic principles of corrosion of metals and metallic alloys. All metals and alloys have a specific relative electrical potential. When metals of different electrical potentials, such as steel and copper, are in contact in the presence of moisture (electrolyte), a low-energy electric current flows from the metal having the higher potential to the one having the lower potential. This is called *galvanic action*. One result is that corrosion of the metal having the higher potential (steel in our example) is accelerated.

The mechanism is an anode reaction, a cathode reaction, the conduction of electrons through the metal from anode to cathode, and the conduction of ions through the electrolyte solution. Corrosion occurs in the anode area, whereas the cathode area is protected.

It is important to know from which of two metals current will flow. Figure 13.1 shows the galvanic or electromotive series for common engineering materials that gives potential differences (in volts) with respect to the hydrogen electrode. This chart will immediately show the difference in potential for these materials. The greater the potential difference, the greater is the galvanic action or the faster will the corrosion progress. By convention, the anode is considered positive and the cathode negative. Gold is the most noble metal, and magnesium one of the least noble metals.

If you study Fig. 13.1 closely, you will see why the different electrochemical batteries were derived. The potential differences between nickel, cadmium, zinc, iron, lead, and silver are such that electrochemical battery systems were able to evolve around these elements when used with the proper electrolytic solutions.

Corrosion has been estimated as causing over $125 billion in damages per year in the United States alone. Corrosion is at work 24 hours a day, 365 days a year and is a never-ending problem. Authorities on corrosion problems believe that much of this damage and loss can be prevented if design engineers pay closer attention to

Galvanic/Electromotive Series - (for common engineering materials)

Engineering Material	Potential
Magnesium & magnesium alloys	-2.37v
Aluminum (1100 series)	-1.66
Zinc	-0.76
Chromium	-0.74
Type 304 stainless steel (active)	------
Type 316 stainless steel (active)	------
Steel, iron, cast iron	-0.41
Aluminum (2024)	------
Cadmium	-0.40
Nickel	-0.23
Tin	-0.14
Lead	-0.13
Copper	+0.34
Silver solder	------
Nickel (passive)	------
Chromium-iron (passive)	------
Type 304 & 316 Stainless steel (passive)	------
Silver	+0.80
Titanium	------
Platinum	+1.20
Gold	+1.50

Figure 13.1   Galvanic/electromotive series.

potential corrosion problems during the design process. Thinking about corrosion problems while selecting the design materials is the first step a product designer should undertake.

For example, the selection of high-strength steels to go into an environment containing sulfides or raw hydrogen is a grave design mistake. The high-strength steels will fail in these environments, although the authorities are not quite sure why this takes place, except that hydrogen embrittlement will occur (*Source:* J. Kruger, professor of materials science and engineering, Johns Hopkins University).

The use of dissimilar metals is another often-cited cause of corrosion problems. The welding of stainless steels that are not low in carbon content is another cause of stress-induced corrosion. Austenitic stainless steels with high nickel content and low carbon are recommended for welding applications (AISI 304, AISI 304L, AISI 316, and others).

One of the best books for design engineers to use as a reference for corrosion problems is *Design and Corrosion Control,* by V. R. Pludek (Macmillan, 1977).

Electrochemical equivalents for various metals. The electrochemical equivalents and other related data for the various plating metals and alloys are shown in Fig. 13.2. If you study the data for chromium, you will see why this metal is a difficult metal to plate because it takes much electrical energy for thin coatings. Cadmium, on the other hand, is an easy metal to plate, but it is a toxic element, as are its salts.

Electroplating practice. The table in Fig. 13.3 lists the common plating metals and alloys and their possible baths, uses, and plating requirements. Procedures for actually performing different plating operations will be shown in a later section. Many of the plating processes and procedures are proprietary, their procedures being closely guarded trade secrets. Platings such as bright chromium, black chromium, gold, nickel-chromium, and others use proprietary procedures that are not published in any book. Plating has been called a "black art," and you will understand why when you try to obtain information on the exact processes and procedures used in some of the industrial plating practices. Plating procedures are difficult and must be controlled closely to exacting standards if the finished article is to be of high quality.

13.3   Electroplating Data and Specifications

13.3.1   Electroplating and oxide layer thickness ranges

Many plating thickness specifications call for plating thicknesses in micrometers (μm), but others may be specified in mils (0.001 in = 1 mil).

$$1 \ \mu m = 0.00004 \text{ in or } 0.04 \text{ mils}$$

$$25.4 \ \mu m = 0.001 \text{ in} = 1 \text{ mil}$$

$$5 \ \mu m = 0.0002 \text{ in} = 0.2 \text{ mil}$$

Both measurement units are given in the following listings, which summarize in outline form thickness specifications for various platings and coatings on a number of base-metal types.

Electrochemical Equivalents and Related Data - Based on 100% Current Efficiency

Metal	mg/coulomb	gm/amp hr	oz/amp hr	Specific Gravity	oz/sq ft for 0.001 in	amp hr to deposit 0.001"/sq ft ◆	Symbol
Cadmium	0.58	2.10	0.074	8.64	0.72	9.73	Cd
Chromium	0.09	0.32	0.011	7.10	0.59	51.8	Cr
Copper	0.33	1.19	0.042	8.92	0.74	17.7	Cu
Gold	0.68	2.45	0.087	19.3	1.61	18.6	Au
Iron	0.29	1.04	0.037	7.90	0.66	17.9	Fe
Lead	1.07	3.87	0.136	11.3	0.94	6.91	Pb
Nickel	0.30	1.10	0.039	8.90	0.74	19.0	Ni
Palladium	0.28	0.10	0.035	12.0	0.10	28.6	Pd
Platinum	0.51	1.82	0.065	21.4	1.78	27.6	Pt
Rhodium	0.27	0.10	0.034	12.5	1.04	30.8	Rh
Silver	1.12	4.02	0.142	10.5	0.88	6.20	Ag
Tin	0.31	1.11	0.039	7.30	0.61	15.6	Sn
Zinc	0.34	1.22	0.43	7.10	0.59	13.7	Zn

NOTE: ◆ Equals specific gravity x 0.08323

Figure 13.2 Electrochemical equivalents and related data.

Electroplating Practice - Average operating conditions

Plating Metal	Typical Uses	Plating Solution Type	Plating Bath Temp. °F	Current Density Amps/ft²	Volts DC	Throwing Power	Time to Deposit 0.001"
Cadmium	Protection	Cyanide	70-95	15-45	1-4	Good	20 min.
Chromium	Decorative, Engineering (Hard), Cylinder liners	Chromic acid	120	200-250	6-8	Poor	2 hours
Copper	Electroforming, Undercoat, Stop-off for casehardening	Acid Cyanide Rochelle	75-120 75-100 140-160	15-40 5-15 20-60	1-2 1.5-3 2-3	Fair Good Good	35 min 90 min 45 min
Gold	Decorative, Electronics, PC boards	Cyanide Proprietary	120-160	5-15	2-6	Good	....
Indium	Bearing surfaces	Cyanide Sulfate Fluoborate	70-75 70-75 70-90	10-150 20 50-100	....	Good Poor Good	....
Iron	Electroforming, Repair	Chloride Sulfate	190 70-75	60 20	....	....	20 min 1 hour
Lead	Protection, Bearing surfaces	Fluoborate	70-75	10-80	0.5	Good	40 min
Nickel	Protection, Decorative, Base coat for chromium	Sulfate-chloride Sulfamate Fluoborate	75-100	Varies	0.5-3	Fair	30 min
Rhodium	Decorative, Optical	Sulfate Phosphate	110-120	10-80	2.5-5	.....	.....
Silver	Protective, Decorative, Electrical contacts	Cyanide	80	5-15	1	Good	.....
Tin	Protective,	Sulfate	70-75	40	1-3	Fair	15 min

	Food & dairy, Bearings,	Fluoborate Stannate	75-10● 150-1●0	50 40	 4-8	Good Excellent	10 min 30 min
Zinc	Protective	Sulfate Cyanide	75-1●● 100	15-400 10-50	....	Fair Good	10 min 40 min

Alloys

Metal	Application	Electrolyte					
Brass	Rubber bonding, Decorative	Cyanide	75-1●●	3-10	2-3	Good	.....
Bronze	Decorative, Base for chromium, Stop-off for steel	Cyanide-stannate	155	20-100	3-6	Excellent	30 min
Lead-tin	Bearings Solderability, Electrotyping	Fluoborate	70-7≡	60	1-2	Good	.....
Tin-zinc	Solderability	Cyanide-stannate	150	10-75	4-5	Excellent	30 min
Tin-nickel	Printed circuits	Chloride-fluoride	150	25	1-2	Excellent	30 min

Figure 13.3 Electroplating practices table.

13.3.2   Anodic coating thickness (anodized parts)

Aluminum

ASTM B 580-79

Type A hard coat	2.0 mil (50 μm)
B architectural	0.7 mil (18 μm)
D auto exterior	0.3 mil (8 μm)
E interior, a	0.2 mil (5 μm)
F interior, b	0.1 mil (3 μm)
G chromic acid	0.04 mil (1 μm)

Mil-A-8625C

Chromic acid	0.05–0.3 mil (1.3–8 μm)
Sulfuric acid	0.1–1.0 mil (3–30 μm)
Hard coat	0.5–4.5 mil (13–114 μm)

Magnesium

AMS 2478B

Acid, full coat	0.9–1.6 mil (23–41 μm)

Mil-M-45202C

Light HAE	0.1–0.3 mil (3–8 μm)
Heavy HAE	1.3–1.7 mil (33–43 μm)

13.3.3   Electroplating thicknesses and specifications

Cadmium plate

Range: 0.2–1.0 mil (5–25 μm)

Specification reference: ASTM A165-80

Note: Cadmium metal is toxic, as are its salts. Use may be restricted.

Chromium plate

Range: 0.01–6.0 mil (0.3–150 μm)

Decorative: 0.01 mil (0.3 μm)

Engineering: 2 mil (50 μm)

Specify hard chrome on the drawing for corrosion resistance.

Heavy chrome plating (4–6 mil) range will have porosity.

Specification reference: Fed QQ-C-320B

Copper plate

Range: 0.1–5 mil (3–125 μm)

Specification reference: Mil-C-14550A

Used as a base for other platings

Gold plate

Range: 0.02–1.5 mil (0.5–38 μm) in eight thickness classes

Specification reference: Mil-G-45204B

Nickel plate

Range: 0.2–7.9 mil (5–200 μm) engineering coatings

Specification reference: ASTM B689-81 (may be soft or hard deposited within other specifications)

Nickel with chromium plate on steel

Range: 0.4–1.6 mil nickel; 0.01 mil chromium

Kinds of coatings include bright, dull, and layered.

Specification reference: ASTM B456-79

Also covers nickel with chromium on copper and copper alloys

Rhodium plate

Range: 0.008–0.25 mil (0.2–6.4 μm) engineering coatings

Specification reference: ASTM B634-78

Palladium plate

Range: 0.05–0.2 mil (1.3–5 μm)

Specification reference: ASTM B679-80

Silver plate

Range: 0.04–1.6 mil (1–40 μm)

Types include mat finish, bright finish, tarnish resistant (chromate-treated)

Specification reference: ASTM B700-81

Note: Silver plating of electric current–carrying parts such as switch blades, breaker contacts, and other sliding parts is generally given as 0.2 mil (5 μm), except where industry specifications may call for 3 mil (76 μm). The 3-mil coating is more effective on surfaces subject to heavy friction, such as sliding electrical contacts.

Tin plate

Range: 0.2–1.2 mil (5–30 μm)

Specification reference: ASTM B545-72

Note: On electrical switch gear and other industrial equipment, 3-mil thickness is usually specified for service at paper mills and other environments where hydrogen sulfide in the atmosphere corrodes silver and other platings.

Zinc plating on steel

Range: 0.2–1 mil (5–25 μm) service conditions 1 through 4

The standard general thickness on most parts is 0.2 mil (0.0002 in). For severe service, the plating thickness should be 13 to 15 μm or 0.00057 in or ⁹⁄₁₀ mil (0.0006 in). Hot-dip coatings of 1 mil (0.001 in) are given to parts that are subject to extremely severe service.

Specification reference: ASTM B633-78

Note: Bright zinc plating on steel parts (usually sheet metal and machined parts) is often given a yellow chromate conversion coating after the zinc plate. This process imparts a gold, iridescent appearance. Many commercial parts are given this treatment for corrosion resistance in lieu of painting. Conversion coatings are also available for the following metals: silver, copper and its alloys, tin, aluminum, magnesium, zinc-base die-casting alloys, and electroplated chromium.

During the conversion process, a complex chromium metal gel forms on the part surface that contains hexavalent and trivalent chromium. When the soft gel coating dries, it becomes hard and

somewhat abrasion-resistant. This then forms what is the preferred plating and coating for many engineering applications.

13.3.4 Plating metals: Characteristics and properties

Copper

Melting point: 1083°C (1981°F)

Specific gravity: 8.96 at 20°C

Reddish colored, malleable, and ductile metal. Excellent conductor of heat and electric current. Good corrosion resistance. Used as a plating base on irons and steels and for other plating metals such as chromium, tin, and nickel.

Cadmium

Melting point: 320.9°C (609.6°F)

Specific gravity: 8.65 at 20°C

Soft, bluish-white metal, similar to zinc in appearance. Cadmium is toxic, as are its compounds. Being replaced by zinc in many plating applications owing to its toxic nature. Used to plate mechanical fasteners, etc., but should not be specified unless required by special application.

Chromium

Melting point: 1890°C (3434°F)

Specific gravity: 7.19 at 20°C

Steel-gray, lustrous, and very hard metal. On a scale of 1 to 10, it would be 9, with only diamond at 10. Hexavalent chromium compounds are toxic. Excellent corrosion resistance for many applications. Attacked by hydrochloric acid, in which it dissolves. Heavily plated deposits become porous on the outer layers. Used to plate hand tools, surgical instruments, decorative trim, cutting tools and drills, and wear-resistant surfaces.

Nickel

Melting point: 1453°C (2647°F)

Specific gravity: 8.90 at 20°C

Nickel is hard, malleable, ductile, slightly ferromagnetic, and a fair conductor of heat and electric current. Silvery-white metal that will take a high polish. Excellent corrosion resistance. Used to plate handguns, hand tools, decorative trim, and corrosion-resistant containers.

Gold

Melting point: 1063°C (1945°F)

Specific gravity: 19.32 at 20°C

Yellow, soft metal that is extremely malleable and ductile. Excellent conductor of heat and electric current. Excellent corrosion resistance. The metal alloys are measured in karats, 24 karat being pure gold. 18 karat = $\frac{18}{24}$ = 0.75, or 75 percent pure gold; 14 karat = $\frac{14}{24}$ = 0.583, or 58.3 percent pure gold, etc. Used to plate electrical contacts, printed circuits, watches, and other jewelry.

Silver

Melting point: 960.8°C (1761.4°F)

Specific gravity: 10.50 at 20°C

Brilliant white, lustrous metal. Harder than gold but still very malleable and ductile. Highest thermal and electrical conductivity of all metals. Stable in normal air but tarnishes when exposed to ozone, hydrogen sulfide, and air containing sulfur. Used to plate electrical contacts, conducting surfaces such as bus bars, eating utensils, and jewelry.

Silver was at one time used to plate printed circuit boards, but it exhibited a detrimental property known as *silver migration.* Thin whiskers of silver would form on printed circuit board conductor foil in the presence of an electric current and short out the circuits on the printed circuit board by bridging the spaces between the conductor patterns. This property of silver was not discovered until some time after the printed circuit boards were put into service on many military and commercial products. The result was many millions of dollars in damage, which could only be corrected by replacing all the defective printed circuit boards using gold- or tin-plated conductor foil.

Tin

Melting point: 231.9°C (449.4°F)

Specific gravity: (white tin) 7.31, (gray tin) 5.75 at 20°C

Silvery-white metal that is malleable and slightly ductile. Resists seawater and tap water but is attacked by strong acids, alkalies, and acid salts. Used to plate electrical contacts, bus bars, and corrosion-resistant containers and tubing.

Indium

Melting point: 156.6°C (313.9°F)

Specific gravity: 7.31 at 20°C

Very soft, silvery-white metal. May be alloyed to produce very low melting point alloys. The metal is toxic.

Rhodium

Melting point: 1966°C (3570°F)

Specific gravity: 12.41 at 20°C

Silvery-white metal with low electrical resistance. Highly resistant to corrosion. Plated rhodium is extremely hard. The base metal is also ductile. Used to plate electrical contacts.

Palladium

Melting point: 1552°C (2825.6°F)

Specific gravity: 12.02 at 20°C

Steel-white metal that is soft and ductile when annealed. Very hard when cold worked. Does not tarnish in air. Uses include engineering plating and plating of electrical contacts.

Zinc

Melting point: 419.4°C (786.9°F)

Specific gravity: 7.13 at 20°C

Bluish-white lustrous, brittle metal. Fair conductor of electricity. Will burn in air at a red heat, producing clouds of white zinc oxide. Good corrosion resistance and used widely to plate irons and steels.

13.3.5   Summary of plating and finishing

Figure 13.4 lists most of the presently available and widely used plating processes and protective finishes, excluding paints and plastic coatings. These finishes have a wide range of uses in aerospace,

Summary of Plating and Finishing Types and Applications

Finish	Application	Appearance
Brass	Used on high strength fasteners where a decorative finish is required	Gold iridescent
Cadmium	General purpose plating. High degree of corrosion resistance especially in salt atmospheres. Soft finish not for threaded applications requiring frequent tightening cycles. Cadmium is toxic. Zinc recommended as replacement.	Silver-gray
Cadmium and clear chromate	Suited for applications for hand contact. Corrosion resistant	Clear
Cadmium and black chromate	Recommended where black color is desired. Note: chromate can be applied in red, green, blue or other color as suits the application.	Black or color
Cadmium and bronze chromate	Normally used in military applications for identification of plating type.	Gold iridescent
Cadmium and yellow iridescent chromate	Similar to bronze	Yellow iridescent
Cadmium and olive drab chromate	Good for paint bonding applications.	Dull olive
Cadmium and phosphate	Similar to cadmium and olive drab chromate	Black
Copper	Used on Allen threaded products. Also for high heat applications.	Copper-matte
Chromium, decorative	Decorative and corrosion applications. Plating is usually very thin.	Bright or satin
Chromium, hard	A most durable plating which is extremely hard. Expensive. Engineering applications. Cutting tools and drills.	Satin
Lead	For soldering applications.	Matte
Nickel	Decorative appearance, excellent wear and corrosion resistance.	Matte or bright
Silver	Electrical properties, utensils, appearance.	Matte or bright
Tin	Anti-seize properties, corrosion resistance, non-toxic.	Grayish-white
Zinc	Popular plating for many applications; hardware, steel sheets, machined parts.	Bluish-white
Zinc and clear chromate	Similar to cadmium and clear chromate. Corrosion resistance, economical.	Clear
Zinc and black chromate	Similar to cadmium and black chromate. Where black color is desired.	Black

Finish	Description	Color
Zinc and bronze chromate	Similar to cadmium and bronze chromate. Color applications.	Gold iridescent
Zinc and yellow iridescent chromate	Similar to cadmium and yellow iridescent chromate. Color applications. Popular engineering finish. Economical and attractive for industrial applications.	
Zinc and olive drab chromate	Similar to cadmium and olive drab chromate. Good paint adhesion properties.	Dull olive
Zinc and phosphate	Similar to cadmium and phosphate. Black for color applications.	Black
Anodic	Finishes for aluminum, magnesium, and titanium. Many types and thicknesses available.	Many colors
Black oxide	For threaded fasteners such as set-screws. Only mildly corrosion and rust resistant. Normally applied to irons and steels. Often oiled.	Black
Iron phosphate	For hardware which is severely handled. Decorative gray finish.	Oily gray-black (oil finish)
Manganese phosphate	For thread lubrication applications. For frictional contact applications. Supplied oiled or waxed for additional protection	Gray-black (oil finish)
Black nitrate	For stainless steel fasteners which require uniform black finish.	Dull or luster black
Passivation	For stainless steel fasteners and parts to prevent corrosion. Acid treatment.	Bright or matte
Zinc phosphate	For exposed fasteners requiring good paint adhesion properties. Dry or lubricated.	Dull gray (oil coating)
Teflon coating	Corrosion resistant and high lubricity. For non-galling anti-seizing applications. Used for bearings. New processes incorporate anodizing with teflon impregnation.	Gray to black

Figure 13.4 Summary of plating and finishing.

commercial aircraft, automobiles, consumer products, and other industrial applications. The name of the finish, its applications, and its final appearance are all listed in the figure. For further information concerning the latest military specifications and American standard specifications for these finishes, consult the plating companies or finishers.

The effective corrosion resistance of selective platings and finishes is shown in Fig. 13.5. The thicknesses or ranges are shown in inches. As can be seen from the data in the figure, hard chromium is one of the best engineering finishes for metal products, although it is expensive compared with the other common platings.

13.4   Plating Baths and Procedures

This section discusses the chemical compositions of various plating baths and the plating voltages, current densities, and other requirements for plating the following metals:

- Cadmium
- Chromium (bright, decorative, and hard, for engineering applications)
- Copper
- Gold
- Nickel
- Rhodium
- Silver
- Tin
- Zinc
- Woods nickel strike
- Silver strike

13.4.1   Cadmium plating

Figure 13.6 shows the composition of the bath and operating conditions for cadmium plating. Cadmium metal has a toxic effect that is similar to that of mercury. Mercury is highly toxic and has contributed to a large extent to pollution of the environment, especially

Effective Corrosion Resistance of Selective Platings - Minimum thicknesses are given in inches

Plating or Surface Treatment	Specification	Type	Class	Minimum Thickness	Salt Spray Test (minimum hours)	
					White Corrosion	Rust
Anodize	Mil-C-5541A	...	...	...	168	...
	Mil-A-8625B	I, II	...	...	240	...
Cadmium and Chromate	AMS-2400L1	...	...	0.0002-0.0003	100	...
	AMS-2400L2	...	...	0.0002-0.0004	100	...
	AMS-2400L3	...	...	0.0003-0.0005	150	...
Cadmium and Chromate	QQ-P-416a	II	1	0.0005	96	...
	QQ-P-416a	II	2	0.0003	96	...
	QQ-P-416a	II	3	0.0002	96	...
	Mil-C-8837	II	1	0.0006	96	...
	Mil-C-8837	II	2	0.0003	96	...
	Mil-C-8837	II	3	0.0002	96	...
Chromium, hard	Mil-C-11436	...	...	0.0012	...	100
Lead	Mil-L-13808	I, II	3	0.00025	...	24
	Mil-L-13808	I, II	2	0.0005	...	48
Manganese Phosphate, dry	Mil-P-16232C	M	3	0.002	...	1.5
Manganese Phosphate, oil	Mil-P-16232C	M	2	0.002	...	24
Tin	Mil-T-10727A	I	...	0.0002	...	24
Zinc	Mil-Z-325a	I	2	0.0005	...	96
	Mil-Z-325a	I	3	0.0002	...	36
Zinc Phosphate dry	Mil-P-16232C	Z	3	0.0002	...	2
	Mil-P-16232C	Z	2	0.0002	...	48
Zinc and Chromate	QQ-Z-325a	II	2	0.0005	96	...
	AMS-2402E	II	...	0.0002-0.0003	100	...
	QQ-Z-325a	II	3	0.0002	96	...
	AMS-2402E	...	...	0.0003-0.0005	150	...
Zinc and Phosphate	QQ-Z-325a	III	2	0.0005	...	96
	QQ-Z-325a	III	3	0.0002	...	36

Figure 13.5 Corrosion resistance of platings.

Plating Bath Composition and Operating Conditions for Cadmium

Composition	g/L	oz/gal
Cadmium fluoborate	240.0	32.2
Fluoboric acid		to pH, (see pH below)
Boric acid	22.5	3.0
Ammonium fluoborate	60.0	8.0
Licorice	1.0	0.134
Cadmium present	95.0	12.6
pH, calorimetric		3.0-3.5

Conditions:
Temperature   70-100°F (20-38°C)
Current density   30-60 amp/ft² (3-6 amp/dm²)

Agitation, cathode   Preferred
Cathode efficiency, %   100
Ratio of anode to cathode area   2:1
Anode material   Cadmium

Volts   4-6
(barrel)   6-12

Filtration: as required

Uses: Rust preventive coating
Application: Over all ferrous base metals. Freedom from hydrogen embrittlement.
Note: Excess free acid can reduce efficiency and cause hydrogen embrittlement. Contamination with copper, lead or zinc can effect color and corrosion resistance of the deposited plating.

Figure 13.6 Cadmium plating.

in the water systems worldwide, where mercurous and cadmium solutions are disposed of on a regular basis. Zinc plating should be used to replace cadmium-plated parts, except where the specifications limit the choice to cadmium. The fumes that result when cadmium is used are toxic, as are those of mercury. Grave physical damage is caused by the ingestion of either metal (elements).

Note: The key numbers for the materials of construction for cadmium plating include 3, 8, 12, 13, and 14. See Fig. 13.17, which shows the key numbers and the respective materials.

13.4.2 Copper plating

Figure 13.7 shows the composition of the bath and operating conditions for one of the popular and easy-to-use methods of applying

Plating Bath Composition and Operating Conditions for Copper

Composition	g/L	oz/gal
Copper sulfate	200-250	28-34
Sulfuric acid	30-75	4-10
Copper content	40-50	5.2-6.6
Sulfuric acid	30-75	4-10

Conditions:	
Temperature:	70-120°F (21-49°C)
Current density:	20-100 amp/ft² (2-10 amp/dm²)
Agitation:	Preferred
Cathode efficiency, %	95-100
Ratio of anode to cathode area	1:1
Anode material	Copper
Volts	> 6 or higher for some applications

Filtration: continuous, especially for heavy coatings.

Uses: Heavy copper deposits to any required thickness.
Application: To all ferrous metals over copper strike. Under nickel and chromium deposits for protective coatings. For electroforming-electrotypes, printing rolls for textiles and rotogravure and other applications.
Note: check the acid content for most problems.

Figure 13.7 Copper plating

copper plating to base metals. Both copper and nickel are used for *strikes,* which are required before a plating metal is applied to a base metal such as iron or steel and other metals. Copper plating is applied most frequently over irons and steels prior to the final plating, which may be cadmium, nickel, chromium, tin, or another metal. Copper sulfate is poisonous and also is used to pressure-treat lumber and to provide a fungus-resistant coating to many articles, as well as a great variety of other commercial applications.

Note: The key numbers for the materials of construction for copper plating include 5, 8, 13, 14, and 15. See Fig. 13.17, which shows the key numbers and the respective materials.

13.4.3 Chromium plating

Figure 13.8 shows the composition of the bath and operating conditions for chromium plating. Chromium plating may be applied directly over steel tools when proper procedures are followed. This application of chromium plating had been used to coat cutting tools and drills. The present technology coats cutting tools and drills with titanium nitride and titanium carbide, which are harder and

Plating Bath Composition and Operating Conditions for Chromium

Composition	g/L	oz/gal
Chromic acid	398	53.0
Sulfuric acid (concentrated)	4.0	0.53
Ratio CrO_3/H_2SO_4	100:1	

Conditions:
Temperature: 110-120°F (43-49°C)
Current density   0.7-1.5 amp/in^2 (10-22
amp/dm^2)Cathode efficiency, %   13-18
Anode material   Pure lead
Ratio of anode to cathode area   2:1 or more (variable)
Volts   6-12

Agitation: Not normally used.
Filtration: Uncommon.
Note: Trivalent chromium is usually required for normal operation of the bath (about 1% of the chromic acid content)

Uses: Bright decorative deposits 0.01 to 0.03 mil. Hard deposits are 0.1 mil and heavier.
Applications: Decorative coating over copper and/or nickel protective coatings on all base metals. Industrial or hard coating over ferrous base metals for tools and dies, cutting tools and drills. Piston rings and cylinder liners.
Note: Problems ocurr with improper trivalent chromium levels, contamination with iron, copper or nickel. Avoid contamination or problems will ocurr. Current densities must be maintained because of the poor throwing power of chromium plating baths.

Figure 13.8 Chromium plating.

more wear resistant than chromium. Chromium is a very hard metal with the hardness of sapphire, but the newer titanium coatings are found more often on modern cutting tools and drills. Chromium plating enhances the appearance and corrosion resistance of many articles, but by modern standards it is considered an expensive electroplating process. Many applications still require chromium plate when the engineering specifications so direct. Chromium plating is difficult owing to the high current densities required and the poor "throwing" power of chromium. The shape and area of the anode in chrome plating are important, as is its position relative to the cathode or article being plated.

Note: The key numbers for the materials of construction for chromium plating include 1, 5, 9, 10, 14, and 15. See Fig. 13.17, which shows the key numbers and the respective materials.

13.4.4 Gold plating

Figure 13.9 shows the composition of the baths and operating conditions for two types of gold plating. Gold plating has many industrial

Plating Bath Composition and Operating Conditions for Gold

Composition	Matte		Bright	
	g/L	oz/gal	g/L	oz/gal
Gold as cyanide	2-12	0.25-1.5	4-16	0.5-2
Potassium cyanide	15-45	2-6	15-90	2-12
Potassium carbonate	0-45	0-6	0-30	0-4
Potassium phosphate	0-45	0-6	0-45	0-6
Potassium hydroxide	10-30	1.3-4	10-30	1.3-4
Brighteners	0	0	0.1-10	0.01-1

Conditions:
Temperature:	120-160°F (50-70°C)	60-75°F (15-25°C)
pH:	11-13	9-13
Current density	1-5 amp/ft^2 (0.1-0.5 amp/dm^2)	3-15 amp/ft^2 (0.3-1.5 amp/dm^2)
Agitation	Moderate	Rapid
Anodes	Platinum, stainless steel, gold	

Uses: Plating for corrosion prevention and electrical conductivity. Decorative.
Applications: Jewelry, printed circuit boards, integrated circuits, electrical contacts, relays.
Note: Problems with contamination. Gold solution baths are extremely sensitive to contamination, either organic or metallic. Nickel is used as the barrier or base layer (see nickel strike baths, Figure 13-10).

Figure 13.9 Gold plating.

applications, including printed circuits, high-quality relay points, integrated-circuit applications, and general electrical contact plating, where a high-conductivity, corrosion-resistant coating is required. The applications in jewelry are numerous but are not detailed in this section.

Note: The key numbers for the materials of construction for gold plating include 3, 9, and 14. See Fig. 13.17, which shows the key numbers and the respective materials.

13.4.5 Nickel plating

Nickel is a lustrous, highly corrosion-resistant metal with countless applications in industry. The metal is used in alloying steels and beryllium-copper (beryllium-nickel) alloys, as well as in the construction of nickel-cadmium electrochemical cells for batteries. Figure 13.10 shows the composition of the baths and operating conditions for three types of nickel plating. Nickel plating is used in *nickel strikes* as a preplating deposit applied subsequent to the plating of other metals. A popular nickel strike called *Woods nickel strike* is shown in Fig. 13.15.

Plating Bath Composition and Operating Conditions for Nickel - Also nickel strikes.

Composition	Watts Type g/L	oz/gal	High Chloride g/L	oz/gal	All Chloride g/L	oz/gal
Nickel sulfate	300	40	240	32	.....	.....
Nickel chloride	45	6	90	12	240	32
Boric acid	30-38	4-5	30-38	4-5	30	4
Nickel content	77.0	10.3	75	10	75	10
pH range (electrometric)	2.0-2.5		2.0-2.5		0.9-1.1	
Hydrogen peroxide to give free oxygen, ppm (Engineering applications)	5-10		5-10		5-10	

Conditions:			
Temperature:	130°F (55°C)	130°F (55°C)	130°F (55°C)
Current density:	10-60 amp/ft^2 (1-6 amp/dm^2)	10-60 amp/ft^2 (1-6 amp/dm^2)	50-100 amp/ft^2
Agitation:	Cathode		
Ratio of anode to cathode	1:1	1:1	1:1
Filtration:	Continuous		
Volts:	6-12	6-12	6-12
Anode material	Nickel, bagged, cast or rolled, depolarized or carbon type.		

Uses: Bath 1 and 2 (dull nickel 0.1-2.0 mil). For bright nickel proprietary brighteners are required. Bath 3 for hard, dull nickel, any thickness.
Applications: For decorative and protective coatings, copper and/or nickel-chromium, on most base metals use baths 1 or 2. For a nickel strike, especially over steel, copper and copper alloys prior to silver palting, use baths 1 or 3 (sometimes less concentrated).

Figure 13.10   Nickel plating.

Note: The key numbers for the materials of construction for nickel plating include 3, 4, 5, 8, 13, 14, and 15. See Fig. 13.17, which shows the key numbers and the respective materials.

13.4.6   Rhodium plating

Figure 13.11 shows the composition of the baths and operating conditions for three types of rhodium plating. Rhodium is a highly corrosion-resistant metal with a high luster used in special corrosion-resistant applications. See preceding sections for a description of this metal and its various applications.

Note: The key numbers for the materials of construction for rhodium plating include 8, 9,12,13, and 14. See Fig. 13.17, which shows the key numbers and the respective materials.

13.4.7   Silver plating

Figure 13.12 shows the composition of the baths and operating conditions for two types of silver plating. Silver plating should receive

Plating Bath Composition and Operating Conditions for Rhodium

Composition	Bath 1 g/L	Bath 2 g/L	Bath 3 g/L
Rhodium metal ♦	2	2	10-20
Sulfuric acid, pure, % vol	2.0	.....	2.5
Phosphoric acid, 85%, % vol	.....	2.0	.....
Conditions:			
Temperature:	104-113°F	104-113°F	122°F
Current density:	10-100 amp/ft^2	10-100 amp/ft^2	10-20 amp/ft^2
Agitation:	As required for all baths		
Cathode efficiency:	80%	80%	85%
Ratio of area anode			
to cathode:	1:1	1:1	1:1
Anode material:	Platinum for all baths		
Volts:	6	6	6

Note: nickel undercoat required, except for gold and platinum group.
Uses: Thin, bright decorative and corrosion resistant coating applications.
Decorative thickness: 0.001-0.006 mil; Scratch and corrosion resistance: 0.06 to over 1.0 mils.

Note: ♦ Metal is added in form of concentrate, as required.

Figure 13.11 Rhodium plating.

Plating Bath Composition and Operating Conditions for Silver

Composition	Bath 1		Bath 2	
	g/L	oz/gal	g/L	oz/gal
Silver cyanide	36	4.8	105	14.0
Cyanide potassium	60	8.0	113	15.0
Carbonate, potassium	45	6.0	15 115	2-15
Potassium hydroxide	.....	.....	30	4.0
Carbon disulfide	0.00075	0.0001		
Silver content	28.7	3.5 (Troy)	84	10.2 (Troy)
Free potassium cyanide	43	5.7	62	8.2

Conditions:		
Temperature	75-90°F (24-32°C)	110-130°F (43-55°C)
Current density	5-15 amp/ft^2 (0.5-1.5 amp/dm^2)	60-150 amp/ft^2 (6-15 amp/dm^2)
Agitation	Cathode	Cathode and solution
Cathode efficiency	100%	100%
Ratio of anode to cathode area	1:1	2:1-4:1
Anode material	Silver	Silver
Volts	< 6	< 6
Filtration	As required	Continuous through paper

Uses: Bath 1 for bright silver deposits following silver strike, 0.1-1.3 mil. Bath 2 for heavy silver deposits after silver strike, up to 60 mil.
Applications: Bath 1 decorative and protective. Bath 2 for engineering applications
Problems: Avoid problems by closely controlling current densities, cleanliness of solution and prestrike quality. Avoid organic contamination. Proprietary solutions for full bright processing are available from the plating suppliers.

Figure 13.12 Silver plating.

a *silver strike* prior to the finished plating being applied. Silver strike plating solutions are shown in Fig. 13.16.

Note: The key numbers for the materials of construction for silver plating include 1, 2, 3, 9, 10, 13, and 14. See Fig. 13.17, which shows the key numbers and the respective materials.

13.4.8 Tin plating

Figure 13.13 shows the composition of the baths and operating conditions for three types of tin plating. Tin plating has many applications in industry, including corrosion-resistant finishes, electrical conductivity, and solderability.

Note: The key numbers for the materials of construction for tin plating include 1, 2, 3, 13, and 14. See Fig. 13.17, which shows the key numbers and the respective materials.

13.4.9 Zinc plating

Figure 13.14 shows the composition of the baths and operating conditions for two types of zinc plating for rack and barrel applications.

Plating Bath Composition and Operating Conditions for Tin

Composition	Bath 1		Bath 2		Bath 3	
	g/L	oz/gal	g/L	oz/gal	g/L	oz/gal
Sodium stannate	90	12	80	10.6	140	19
Sodium hydroxide	7.5	1	.....	.....	15	2
Potassium hydroxide	.....	.....	30	4.0	.....	.....
Hydrogen peroxide, 100 vol	As required (0.07)					
Tin content	38	5.0	29	3.8	60	8.0

Conditions:			
Temperature	140°F (60°C)	185°F (85°C)	200°F (95°C)
Current density	10-25 amp/ft^2 (1-2.5 a/dm^2)	40 amp/ft^2 (4 amp/dm^2)	55 amp/ft^2 (5.5 amp/dm^2)
Agitation, cathode & solution	Optional	Prefered	Not required
Cathode efficiency	60-90%	80-100%	90%
Ratio of anode to cathode area	1:1	1:1	1:1
Anode material	Tin	Tin	Tin
Filtration:	As required..		
Volts	6	6	6

Uses: Average tin coating thickness 0.03-0.3 mil. For protective coatings 1.0-3.0 mil.
Applications: Baths 2 and 3 are ffor electro tin plate. Bath 1 is used for plating all base metals for ease of soldering, protective coatings or good throwing power.

Figure 13.13 Tin plating.

Plating Bath Composition and Operating Conditions for Zinc

Composition	Rack		Barrel	
	g/L	oz/gal	g/L	oz/gal
Ammoniated Solutions:				
Zinc chloride	75-95	10-12.5	35-65	4.5-8.5
Ammonium chloride	90-120	12-16	112-225	15-30
Sodium chloride	45-68	6-9	.....	.....
Boric acid or ammonium acetate	19-26	2.5-3.5	19-26	2.5-3.5
pH	4.4-5.6		4.4-5.6	
Temperature	60-110°F (15-45°C)		60-110°F (15-45°C)	

Conditions: Anode material Cathode current density	High purity zinc. Anode current density: 5-30 amp/ft² 2-60 amp/ft²			
Potassium solutions:				
Zinc chloride	75-85	10-11.3	50-85	6.7-11.3
Potassium chloride	200-270	27-36	180-270	24-36
Boric acid or Potassium acetate	7.5-12	1.0-1.6	7.5-12	1.0-1.6
pH	4.4-5.6		4.4-5.6	
Temperature	60-110°F (15-45°C)		60-110°F (15-45°C)	
Conditions:	Same as above baths			

Uses and applications. General corrosion resistant coating 0.2 to 3.0 mils for irons and steels. Bright zinc produced with these solutions. A yellow chromate coating is often applied over the deposited zinc plate.

Figure 13.14   Zinc plating.

Zinc is the most widely used of all plating metals because of its good corrosion resistance, quick and easy application, and low cost.

Note: The key number for the materials of construction for zinc plating is 14 [linings of polyvinyl chloride (PVC) or polypropylene].

13.4.10   Woods nickel strike solution

Figure 13.15 shows the composition of the striking bath and operating conditions for Woods nickel strike. This nickel strike can be applied on many base metals prior to other plating operations.

Plating Bath Composition and Operating Conditions - Woods nickel strike

Composition	g/L	oz/gal
Nickel chloride	180	24
Hydrochloric acid (commercial)	10 to 12% by volume.............	
Conditions:		
Temperature	100°F (35°C)	
Current density	20-100 amp/ft^2 (2-10 amp/dm^2)	
Anodes	Nickel	

Uses and applications: This is a widely used strike which is effective on stainless steels, nickel, high-alloy steels and carburized and hardened steels. Time of strike: 30 seconds to 2 minutes. Nickel content will climb in this bath and should be maintained at 45-60 g/L (6-8 oz/gal), by decanting part of the solution and replacing it with hydrochloric acid.

Figure 13.15   Woods nickel strike.

Note: The key numbers for the materials of construction for Woods nickel strike include 3, 4, 8, 12, 13, 14, and 15. See Fig. 13.17, which shows the key numbers and the respective materials.

13.4.11   Silver strike solutions

Figure 13.16 shows the composition of the baths and operating conditions for three silver strike solutions. These strikes should be applied prior to the finished silver plating processes shown in Fig. 13.12.

Note: The key numbers for the materials of construction for silver strikes include the following: For baths 1 and 2, use 1, 2, 3, 9, 10, 13, and 14; for bath 3, use 8, 9, 10, 12, 13, 14, and 15. See Fig. 13.17, which shows the key numbers and the respective materials.

13.4.12   Materials of construction for electroplating processes

Figure 13.17 presents the materials of construction for the previously listed plating processes. The key numbers are referenced to the respective applicable materials.

Precautionary note: The materials and chemicals used in the electroplating processes are almost all toxic to some extent, with some of the compounds being extremely poisonous (such as the cyanides). Extreme caution should be followed when handling these materials and chemicals. Also, some of the fumes created in the plating processes are also toxic to a high degree. The proper equipment and

Plating Bath Composition and Operating Conditions for Silver Striking

Composition	Bath 1		Bath 2		Bath 3	
	g/L	oz/gal	g/L	oz/gal	g/L	oz/gal
Silver cyanide	1.7	0.23	7	0.9	.....	.....
Potassium cyanide	75	10	75	10		
Copper cyanide	15	2	.....	.....	.....	.....
Nickel chloride	.....	.....	.....	.....	240	32
Hydrochloric acid Specific gravity 1.18					10.0%/volume	
Silver content	1.6	0.2	6	0.7		
Copper	10.5	1.4	.....	.....	.....	.....
Nickel	.....	.....	.....	.....	60	8
Free cyanide, KCN	53	7	71	9.5	.....	.....
Conditions:						
Temperature	75°F (24°C)		75°F (24°C)		75°F (24°C)	
Current density	30 amp/ft^2 (3 amp/dm^2)		30 amp/ft^2 (3 amp/dm^2)		150 amp/ft^2 (15 amp/dm^2)	
Time	0.33 min.		1 min		1-2 min	
Anode material	Steel		Steel		Nickel, bagged	
Volts	< 6		< 6		6	
Filtration	As required		As required		Constant	

Uses: Bath 1, first strike on steel; bath 2, second strike on steel or silver strike over nickel or other base metal; bath 3, first strike on steel or other base metal.
Applications: Strikes 1, 2 and 3 for regular silver plating. Strike 3 for engineering applications.

Figure 13.16   Silver strike.

ventilation must be used with the electroplating processes. As an example, the fumes from the chromium plating processes that use chromic acid are not only toxic but also corrosive. Any plating process that uses the cyanide compounds should be watched closely and handled with extreme caution, paying particular attention to proper venting of the fumes.

Because of the very high current densities used with some of the plating processes, the electrodes (anode and cathode) must be guarded against short-circuits. The operating voltages are low, but the power involved can be very high owing to the large electric currents required in certain processes (chromium plating).

13.5   Coloring Processes for Metals, Alloys, and Phosphating Steels

This section lists some of the coloring processes used to color metals and alloys. Coloring of metals may be for appearance as well as for additional protection against corrosion and rusting. These processes have evolved over a span of many years and are used widely in the

Materials of Construction for Plating Applications

Key	Material	Uses
1	Steel, low carbon	Tanks, filters, pumps, pipe, fittings and heating coils
2	Cast iron	Pumps, filters, valves and fittings
3	Stainless steel (316)	Tanks, pumps and filters
4	High silicon cast iron	Pumps, pipe, fittings and heat exchangers
5	Lead (6% antimony alloy)	Tank linings, pipe
6	Copper	Heating coils
7	Nickel	Heating coils
8	Carbon (Karbate)	Heaters and heat exchangers, pumps and air diffusers
9	Glass (Pyrex or tempered)	Tanks, heat exchangers and pumps
10	Chemical stoneware	Tanks, tower concentrators and tower packing
11	"Haveg"	Tanks and pipes
12	Hard rubber	Pipe, fittings and pumps
13	Rubber (Approved types)	Tank linings and hose
14	Plastics (Approved types)	Tank linings, hose, pipe, fittings, barrels, heating coils
15	Acid resistant brick	Tank linings
16	Wood	Tanks

Note: Plastic tanks or liners can influence the current distribution patterns within the plating tanks. Tests must be run to determine the proper plastic for each particular application. Polyurethane, polyethylene, polystyrene, polypropylene, ABS (Acrylonitrile-Butadene-Styrene), acetals, acrylics, nylons and polyvinyl chlorides may be usable for many applications.

Figure 13.17 Materials of construction for plating applications.

metal finishing industries. Most of the processes involve the use of chemical action together with the application of heat.

It should be noted that the finished appearance of the metal being colored is affected to a great extent by how well the bare metal itself is finished and polished. For example, in the blueing process for handguns and rifles, the base metal is finished very carefully and is highly polished prior to blueing. If the metal parts are not finished and highly polished, the blueing process will not hide the defects in the base metal.

13.5.1 Coloring metals (ferrous and nonferrous)

Black oxide. Irons and steels may be given a black finish or black oxide finish by either of two processes:

1. Heat the ferrous metal part to approximately 500 to 700°F, and plunge it into a container of good-quality machine oil. Reheat and plunge the part into the oil a number of times until the desired depth of black is obtained. Use precautions on hardened and tempered parts.

2. A thin black oxide coating also may be applied to iron and steel parts by immersing the parts in a boiling solution of sodium hydroxide and mixtures of nitrates and nitrites (sodium nitrate, potassium nitrate, etc.).

Blueing irons and steels. Many ferrous metal parts receive a blueing operation to enhance their appearance and also as a rust-preventive coating. Handguns, rifles, and shotguns normally receive a blueing operation unless chromium or satin nickel is specified. The following blueing process imparts a fine blue to blue-black finish: Clean the metal with a potassium bichromate–sulfuric acid solution, and then wash with ammonium hydroxide solution and wipe dry with a clean, lint-free cloth. Apply ammonium polysulfide until the desired depth of blueing is obtained. The finish may be made nearly black by repeating the process of applying the ammonium polysulfide. A light machine oil or silicone cloth wipe then will provide the surface with excellent corrosion and rust resistance.

Phosphating irons and steels. Three types of phosphate coatings (conversion coatings) are given to ferrous metals: zinc phosphate, iron phosphate, and manganese phosphate. Zinc phosphate coatings vary in color from light to dark gray; iron phosphate, dark gray; and manganese phosphate, dark gray, becoming black with service. The phosphate coatings are used for paint bases, for an aid in cold working the metal, and for rust prevention.

The phosphatizing solution contains approximately 3% to 5% phosphoric acid by volume. The part to be phosphatized is first thoroughly chemically cleaned and dipped into a solution of 2% by volume hydrochloric acid for approximately 15 seconds and then water rinsed and dried. The part is then immersed in the phosphatizing bath for the time required to impart the desired coating. Zinc plates and zinc parts are also phosphatized in the same manner. Stainless steels and certain alloy steels cannot be phosphatized.

Passivation of copper and copper alloys. A passivation coating may be applied to copper and its alloys that imparts a blue-green color or patina. This coating is corrosion resistant. The passivation solution may be made using the following solution composition:

1. 6 lb ammonium sulfate

2. 3 oz copper sulfate

3. 1.4 fluid oz technical ammonia (specific gravity = 0.09)

4. 6.5 gal water

Apply the solution with a spray to the chemically cleaned copper or copper alloy parts in six applications, drying between applications. The patina appears after approximately 6 hours and continues as the part weathers.

Copper and copper alloy blackening (alloys with more than 85 percent copper). To color copper and its alloys black, use the following procedure: Make a solution consisting of 4 oz arsenious oxide, 8 fluid oz hydrochloric acid (specific gravity = 1.16), and 1 gal water. Then

1. Heat the solution to 175 to 200°F.

2. Immerse the copper or copper alloy parts until a uniform black color is obtained.

3. Brush the parts while wet, dry the parts, and apply a protective clear finish such as polyurethane varnish.

Coloring brasses. Brass may be given a green color with the following procedure: Mix a solution containing 1 oz ferric nitrate, 6 oz sodium thiosulfate, and 1 gal water. Then

1. Heat the solution to 160 to 180°F.

2. Immerse the brass parts in the solution until the desired color is obtained.

3. Dry and coat with a clear protective finish such as clear lacquer or polyurethane varnish.

13.6   Metal Finishing, Phosphating, and Cleaning Solutions and Formulas

The following solutions and formulas for cleaning and finishing metals are for use by metal fabricating facilities for mechanisms and small, moving detail parts generated mainly for use in mechanisms and other ferrous metal products. These solutions, formulas, and

finishing processes should be used on all ferrous metal parts where a zinc, zinc and yellow chromate, or other electroplated finish is neither desired nor required. Zinc-plated and chromated parts and other parts that are electroplated usually increase in size owing to the thickness of the electroplated deposit. A study should be made, and the recommended parts that do not require a zinc, zinc and yellow chromate, or other electroplated finish should be phosphated in lieu of electroplating.

Phosphated metals (most ferrous metals and ferrous alloys) do not gain a significant plated thickness to interfere with mating parts on close assembly work. The simple phosphate coatings (which use only phosphoric acid and water as a 2% to 4% solution) are normally only 1.5 to 2.0 μin thick. Many industrial parts and products may receive a metallic phosphate finish in lieu of a zinc and yellow chromate or other electroplated finish, which may be expensive and require the services of outside vendors (platers). Since the phosphated finishes are not electrodeposited metals and use nontoxic chemicals, most metal fabricators possibly could apply these coatings and finishes in-house. The resulting cost savings could amount to thousands of dollars a year and at the same time give the fabricator complete control over the scheduling and manufacturing of the small parts.

The chemicals, metals, residues, and solutions used in phosphating processes are relatively nontoxic and may be neutralized easily for disposal into sewage systems or carried to disposal stations. Because they involve coatings and not electrodeposited processes (some of which are highly toxic and can cause environmental problems), the phosphating processes may not require a plating license.

13.6.1 Pickling formulas

Pickling solutions are used to chemically remove surface scale and oxides from metals and alloys. Typical formulas include

- One part sulfuric acid to 5 or 10 parts water by volume. Use for cleaning steel and other ferrous metals.

- Sulfuric acid and nitric acid; 1 part of each acid with 5 to 10 parts water by volume. Use for bright-cleaning brass alloys.

- Sulfuric acid and anhydrous ferric sulfate; 1 part acid to 5 or 10 parts water by volume, plus a small amount of ferric sulfate ($\frac{1}{10}$ part or less).

- Chromium phosphate and triethanolamine. Use for preparing aluminum for painting.

- Phosphoric acid (75 percent); 1 part acid to 25 to 35 parts water by volume. Use for cleaning and phosphating steels in preparation for painting.

Note: Pickling baths must be maintained at a temperature of 145 to 185°F. Keep the pickling times short because hydrogen embrittlement can occur if the metal remains in the pickling bath for an excessive amount of time. Pickling-bath times vary from 3 to 15 minutes, with the shorter times preferred. As a precaution to avoid hydrogen embrittlement, the pickled parts may be oven heated at 300°F for 30 minutes.

13.6.2   Phosphate coatings (ferrous metals and alloys)

The four major types of phosphate coatings for ferrous metals and alloys include

1. Zinc phosphate

2. Iron phosphate

3. Manganese phosphate

4. Lead phosphate (a relatively new process)

Formulas for phosphating solutions and directions for their proper use are given in the following expositions and procedures.

Zinc phosphating.   Zinc phosphate coatings are used on many military and commercial articles, including small arms, knives, heavy weapons and field artillery, tools, general mechanism parts, and a host of other fabricated metal applications. Zinc phosphating is also called *granodizing.*

To prepare 16 fluid oz (480 ml) of zinc phosphate concentrate, obtain the following ingredients:

- 33 g powdered zinc

- 436 ml phosphoric acid (75% to 85% solution)

- 110 ml distilled water

- 10 to 12 g sodium nitrite (accelerator). (The amount of sodium nitrite may need to be adjusted to suit the application; potassium nitrite may be used in lieu of sodium nitrite.)

For a smaller base quantity of concentrate, use the following measures:

- 16.5 g powdered zinc
- 218 ml phosphoric acid (75% to 85% solution)
- 55 ml distilled water
- 5 to 6 g sodium nitrite (accelerator). (The amount of sodium nitrite may need to be adjusted to suit the application; potassium nitrite may be used in lieu of sodium nitrite.)

Procedure.   Dissolve the powdered zinc in the phosphoric acid, and then add the distilled water and sodium nitrite. Mix thoroughly and filter through a stainless steel filter screen or paper filter. (Heat and outgassing occur during the acid/zinc mixing and dissolution.)

The finished zinc phosphate concentrate then may be mixed in the following phosphating solutions:

Solution I (normal): 16 fluid oz concentrate to 6.75 gal of water or 1 fluid oz of concentrate to 54 oz of water

Solution II (heavy): 16 fluid oz concentrate to 5 gal of water or 1 fluid oz of concentrate to 40 oz of water

Processing temperature is 170 to 180°F (recommended temperature is 175°F). Distilled water will yield the best results, but "soft" tap water is sufficient.

Processing time is 3 to 10 minutes. Longer processing times (10 minutes or more) deposit darker-colored and denser coatings and may be necessary for hardened steel parts such as firearm receivers and bolts. The phosphating process is usually self-terminating; chemical action will stop when the part is sufficiently coated.

Immediately rinse the phosphated parts in cold running water for 30 seconds. Do not allow the phosphating solution to dry on the phosphated part; otherwise, difficult-to-remove white deposits of zinc salts will form on the parts.

The parts then may be dipped into a mild solution of chromic acid and water (1 oz chromic acid to 1 gal of water). Then rinse the parts again in cold running water.

Thoroughly dry the parts with compressed or blown hot air. Bake critical parts as outlined herein.

Coat the parts with a light oil, an oil-kerosene mix, or WD40 lubricant. Then remove or drain excess oil from the parts. The parts now may be assembled or stored for future use.

Note: As a safety precaution, hardened and highly stressed parts that have been phosphate treated for 10 minutes or longer should be baked at 300°F for 30 minutes to prevent hydrogen embrittlement.

Iron phosphating (reference only). Iron phosphating, also known as *coslettizing,* uses powdered iron and phosphoric acid in the solution. Accelerator chemicals such as sodium nitrite, potassium nitrate, and potassium nitrite are not normally required for iron phosphating processes.

Manganese phosphating (reference only). Manganese phosphating, also called *parkerizing*, uses powdered iron, manganese phosphate, and phosphoric acid in the solution.

Bonderizing. Bonderizing is performed on ferrous metals and alloys in a solution consisting of phosphoric acid, a catalyst, and water. This process is used to impart a rough, tough, frosted surface providing excellent paint adhesion. The phosphated coating is rust resistant.

Note: Phosphate coatings are on the general order of 0.0002 in (0.2 mil) thick. All phosphating solutions are used or processed at temperatures of 145 to 200°F. Immediate washing in cold water is recommended after the phosphate coating is applied to prevent white streaks or blemishes from appearing on the surface of the metal or alloy.

Phosphate coatings are most applicable for low-, medium-, and high-carbon steels in both the annealed and hardened conditions. Ferrous metal parts that have been highly hardened and some of the higher-alloy steels are more difficult to phosphate coat. On these steels, the phosphate coating will be light gray in color unless the part is immersion-coated for longer periods (10 minutes or longer). Stainless steels and other high-nickel-chromium steels cannot be phosphated.

The more iron a ferrous material contains, the thicker and darker will be the phosphate coating. Heavy zinc phosphate coatings are the easiest to apply, with some of the processes requiring only 3 to 5 minutes to complete. Ferrous materials that are case hardened normally cannot be phosphatized. If high-alloy ferrous parts need a

heavier or darker coating, they may be processed in phosphating solutions up to double the time normally required, i.e., up to 10 to 15 minutes for the solutions used in the zinc phosphating process detailed in this disclosure.

13.7   Etching Metals

The solutions used to etch various metals are often called *mordants,* and some of the solutions used in the etching processes are as follows.

13.7.1   Etching irons, steels, and zinc plate

Nitric acid solution.   A popular solution for etching irons, steels, and zinc plates consists of a solution of 2 oz 50% nitric acid mixed into 15 oz water. This is a 1:16 solution. Always pour the acid into the water; never pour water into the acid. Pouring water into acid may cause a mixing reaction that could result in acid splashing from the mixing container. The solution is relatively slow acting unless used with a splash-type etching machine. For hand etching, a stronger solution can be used, but the etching cut becomes rough at the edges of the etching action. Solutions as strong as 1:8 can be used: 1 part concentrated nitric acid to 7 parts water by volume.

13.7.2   Etching copper and copper alloys

Ferric chloride is often used to etch copper and its alloys. A 40°Bé solution of ferric chloride used at 75 to 80°F etches copper cleanly and not too rapidly so as to produce a rough etched edge. A 40°Bé solution is made by mixing 20 oz ferric chloride (anhydrous) with water to make a final volume of 1 liter (1000 ml). The specific gravity of this solution can be from 1.37 to 1.38. The designation 40°Bé (Baumé) is pronounced "40 degrees 'bow-may'" and is defined with respect to specific gravity by the following equation:

$$\text{Specific gravity} = \frac{145}{145 - {}^\circ\text{Bé}}$$

With this equation, you may determine the degrees Baumé if you know what specific gravity you want or have, or you may determine the specific gravity required of the solution if you know the degrees Baumé you wish to produce.

Note: The reaction of ferric chloride produces a great amount of heat when it is dissolved in water. The water temperature should be between 50 and 75°F prior to mixing the ferric chloride solution.

Solutions of ferric chloride as low as 300°Bé are used for fast etching of copper and its alloys, while the 40 to 42°Bé solutions are used for etching intaglio printing plates or photogravure work. These solutions also have been used to etch stainless steels. The ferric chloride bath should be contained in a glass, plastic, or wooden tank because the solution is highly corrosive. Do not dispose of these solutions directly into standard drainage systems. The solutions should be diluted in water and neutralized with a base chemical such as sodium bicarbonate. Do not allow the solutions to come into contact with the skin; rubber gloves should be worn at all times during use of these solutions.

Printed circuit boards for electronic applications have been produced using ferric chloride solutions for many years. Using the lower °Bé solutions produces a fast and accurate etching action or cut, and the ferric chloride is economical and long lasting. Perchlorate chemicals are also used for etching printed circuit boards.

Frank Short's etching solution. The Frank Short etching solution has been used by printing plate makers and other metalworkers for years and is a slow-acting but very accurate etchant (mordant). The solution may be made as follows:

- 88 parts by volume distilled water
- 2 parts by volume potassium chloride
- 10 parts by volume hydrochloric acid (concentrated)

This is a two-part mixture that must be made by mixing two solutions and then pouring the two solutions together.

Solution 1. Mix 1 oz potassium chloride into 10 oz water at 190 to 200°F

Solution 2. Mix 5 oz hydrochloric acid into 34 oz distilled water

Then pour solution 1 into solution 2 in a well-ventilated area.

Caution: Chlorine gas is evolved when the two solutions are mixed. Allow the solution to sit for 30 minutes to 1 hour before using. Store in a glass bottle with a rubber or glass stopper.

The part to be etched is lowered into a shallow glass pan or tank so that the solution covers the part by approximately ½ in. The

etching action may be observed directly because the etching solution is clear and transparent.

Copper and its alloys, irons and steels, and chromium-plated parts may be etched with this solution. Use adequate ventilation because chlorine gas in small amounts is evolved during the etching action. This solution is often used to check the etching mask for holes prior to fast etching in ferric chloride solutions. A copper part will begin to show a frosting effect a few seconds after immersion in the solution. Holes in the etching mask then may be seen and repaired prior to the final etching.

As a final note, titanium heaters are used in etching baths for ferric chloride solutions.

13.7.3   Etching aluminum and aluminum alloys

Most aluminum alloys can be etched using sodium hydroxide solutions of varying strengths. Sodium hydroxide is commonly called *caustic soda* or *lye* and is a low-cost chemical. Sodium hydroxide is very corrosive or caustic and is poisonous. When aluminums are etched using sodium hydroxide solutions, one of the reaction products is hydrogen gas, which is highly explosive when mixed with normal air. Proper venting must be employed when using this process for etching aluminum alloys.

The etching action is relatively rapid, especially if the sodium hydroxide solution is above 1:16 (1 part sodium hydroxide to 15 parts water). Strong solutions produce a violent foaming reaction and substantial amounts of hydrogen gas.

Conversion formulas for chemical solutions.   Figure 13.18 shows the conversion formulas for solutions having concentrations expressed in various ways.

13.8   Anodizing

Anodizing is a process of oxidation produced in an electrolytic bath. Aluminum and its alloys are most commonly anodized, although the anodizing process may be performed on magnesium, zinc, and titanium. This section will cover the anodize coatings and processes for aluminum and its alloys only. The anodic coating, being aluminum oxide, is extremely hard and abrasive, especially aluminum hardcoat anodized surfaces. The classifications for aluminum and aluminum alloy anodic coatings are as follows.

CONVERSION FORMULAE FOR SOLUTIONS HAVING CONCENTRATIONS EXPRESSED IN VARIOUS WAYS

A = Weight per cent of solute
B = Molecular weight of solvent
E = Molecular weight of solute
F = Grams of solute per liter of solution

G = Molality
M = Molarity
N = Mole fraction
R = Density of solution grams per cc

Concentration of solute—SOUGHT	Concentration of solute—GIVEN				
	A	B	G	M	F
A	—	$\dfrac{100N \times E}{N \times E + (1-N)B}$	$\dfrac{100G \times E}{1000 + G \times E}$	$\dfrac{M \times E}{10R}$	$\dfrac{F}{10R}$
B	$\dfrac{\dfrac{A}{E}}{\dfrac{A}{E} + \dfrac{100-A}{B}}$	—	$\dfrac{B \times G}{B \times G + 1000}$	$\dfrac{B \times M}{M(B-E) + 1000R}$	$\dfrac{B \times F}{F(B-E) + 1000R \times E}$
G	$\dfrac{1000A}{E(100-A)}$	$\dfrac{100N}{B - N \times B}$	—	$\dfrac{1000M}{1000R - (M \times E)}$	$\dfrac{1000F}{E(1000R - F)}$
M	$\dfrac{10R \times A}{E}$	$\dfrac{10R \times N}{N \times E + (1-N)B}$	$\dfrac{1000R \times G}{1000 + E \times G}$	—	$\dfrac{F}{E}$
F	$10AR$	$\dfrac{10R \times N \times E}{N \times E + (1-N)B}$	$\dfrac{1000R \times G \times E}{1000 + G \times E}$	$M \times E$	—

Figure 13.18 Solution conversion equations. *(Source: Handbook of Chemistry and Physics, 50th ed., Chemical Rubber Publishing Company, Cleveland, Ohio.)*

Classifications

Type I. Chromic acid anodizing, conventional coatings produced from chromic acid baths. Shall not be applied to alloys containing over 5% copper or over 7% silicon or when alloying elements exceed 7.5%.

Type IB. Chromic acid anodizing, low-voltage process, 20 V. Heat-treatable alloys, such as T4, T6, etc., should be tempered prior to anodizing.

Type II. Sulfuric acid anodizing, conventional coatings produced from sulfuric acid baths. Heat-treatable alloys, such as T4, T6, etc., should be tempered prior to anodizing.

Type III. Hard anodic coatings (hardcoat). Shall not be applied to alloys containing over 5% copper or over 8% silicon unless agreed on by the supplier. Heat-treatable alloys, such as T4, T6, etc., should be tempered.

Classes

Class 1. Nondyed, natural, including dichromate sealing

Class 2. Dyed

Standard specifications for anodized aluminum and aluminum alloys

ASTM B244, *Thickness of Anodic Coatings, Measurement of*

ANSI/ASTM B137, *Weight of Coatings on Anodized Aluminum, Measurement of*

ASTM B117, *Method of Salt Spray (Fog) Testing*

Sealing anodized aluminum and aluminum alloys. All types of anodizing must be sealed using any of the following methods after the electrolytic anodized coating is applied:

- Immersion in an aqueous solution of 5% sodium dichromate (15 min at 90 to 100°C)

- Immersion in deionized water (15 min at 100°C)

- Immersion in an aqueous solution of nickel or cobalt acetate (100°C for 15 min)

- Teflon impregnation processes (for sealing and lubricity)

Anodic coating design data

Radii of curvature on anodized parts

Nominal coating thickness, in	Radius of curvature (outside/inside), in
0.001	approx. 0.032
0.002	approx. 0.062
0.003	approx. 0.093
0.004	approx. 0.125

Thickness ranges of anodic coatings on aluminum and aluminum alloys

Coating type	Thickness range, in
I and IB	0.00002 to 0.0003
II	0.00007 to 0.0010
III	0.0005 to 0.0045

Minimum thickness (typical) of anodic coatings on aluminum and alloys.
See Fig. 13.19 for minimum typical anodic coating thicknesses on
various aluminum alloys per type.

Maximum thicknesses of anodic coatings on various aluminum alloys.
Figure 13.20 shows the maximum attainable thicknesses of anodic
coatings on selected aluminum alloys.

Design notes for anodized parts

1. A 2-mil hardcoat anodic coating (0.002 in) will penetrate the part
   0.001 in and protrude from the part 0.001 in. A 1.000-in-diameter
   part that is anodized 2 mils (0.002 in) will have a finished diameter of 1.002 in. Half the coating is inside the part, and half is on
   the outside.

2. Avoid blind holes in parts.

3. Avoid hollow weldments (drill 0.250-in-diameter weep holes in
   the part).

4. Avoid steel inserts.

5. Avoid sharp corners (see radii chart above).

6. Avoid heavy to thin cross sections on the part.

7. Allow for the anodic coating in your design tolerances on the part.

Minimum Typical Anodic Coating Thicknesses for Various Aluminum Alloys

Alloy Designation	Thickness of Coating, inches	
	Type I and IB	Type II
1100	0.000029	0.000093
2024-T4	......	0.000125
2024-T6	0.000044	......
3003	0.000035	0.000103
5052	0.000033	0.000098
5056	0.000021	......
6061-T6	0.000034	0.000099
7075-T6	0.000040	......
Alclad 2014-T6	0.000045	......
Alclad 7075-T6	0.000041	......
295-T6	......	0.000107
356 T6	......	0.000102
514	......	0.000086

Note: Anodic coating types I, IB and II are normally applied as thin coats, while type III (Hardcoat) is normally applied as the thicker coatings. See table of anodic thicknesses in Figure 13-20.

Figure 13.19   Anodic coating thickness.

Anodic coating specifications

Name	Hardcoat anodize	Chromic anodize	Sulfuric anodize
Army/Navy	Mil-A-8625, Ty-3	Mil-A-8625, Ty-1	Mil-A-8625, Ty-2
G.E.	AMS-2468D	AMS-2470H	AMS-2471D
Boeing	Code-302	Code-300	Code-301
IBM	41-207	41-204	41-203
Grumman	G-9031	9030B	G-9032

Hardcoat anodize processes

- Martin
- Alumilite
- Alpha
- Mae

Maximum Anodic Coating Thicknesses on Various Aluminum Alloys

Alloy Designation	Anodic Coating Thickness - inches	Color
380	0.0006	Gray
360	0.001	Gray
319	0.0014	Gray
1100	0.0017	Bronze
2011	0.0021	Light gray
2014	0.0025	Gray
2019	0.0028	Bronze
6262	0.0031	Black
6061	0.0035	Black
2024	0.0040	Bronze
2017	0.0045	Bronze
6063	0.0050	Black
5052	0.0053	Black
2618	0.0055	Gray
2219	0.0058	Gray
218	0.0062	Gray
7079	0.0065	Dark gray
355	0.0075	Gray
7075	0.0082	Bronze
Almag 35	0.0100	Gray
356	0.0120	Light gray

Note: All above coatings may be dyed black. Coatings over 0.003 inches thick tend to chip and become milky in color and should be used only in the salvage of parts.
Source: Anodic, Inc., Stevenson, CT 06491.

Figure 13.20 Maximum anodic coating thickness.

- Sanford
- Boeing
- Scionic
- Hardas
- Imperv-X

Hardcoat/Teflon processes

- Amphodize

- Hardtef

- Analon

- Tufram

- Polylube

- Lukon

- Nituf

- Kalon

- PTFE

- Smoothcoat

- Sanfordize

- Hardlube

Hardcoat anodize physical data

Hardness	65 to 70 Rockwell C Scale (Harder than Hard Chrome)
Color	Dark gray to black
Dielectric strength	800 V/mil
Machining	Grinding, lapping, polishing, and honing
Sealing	Dichromate, nickel, or cobalt acetate, hot water or Teflon impregnation
Resistivity	10^8 to 10^{12} $\Omega \cdot$cm

Note: Parts that have been anodized cannot be reanodized, nor can the anodic coating be made thicker.

13.9   Paint Finishing

Paints are the most used of all finishes for metal products. The types and different varieties and colors available are limitless. Paint technology improves and changes constantly. Attesting to the quality and long-lasting ability of modern paints for metals are automotive vehicles and aircraft. Not only must modern industrial paints withstand the weather and corrosive elements, but many also are protected chemically from the harmful effects of ultraviolet radiation from the sun.

Common industrial paints include

- Enamels (baked and air drying)
- Lacquers
- Epoxies (one- and two-part systems)
- Varnishes (glyptols, polyurethanes, and special chemical resistant)
- Latex/water-based

Pigments such as zinc and titanium oxide, carbon black, zinc chromate, Prussian blue, cobalt blue, and many others can be added to the paint for coloring and to improve durability.

Products that are built to customer or military specifications frequently have the type of finish listed in the specifications, together with the method of base-metal preparation, such as phosphatizing, zinc chromate primer, zinc primer, etc. The color and the required minimum dry-film paint thickness also frequently are set out in the equipment specifications.

Some of the high-quality equipment manufacturers in the United States apply multiple coatings to their products, producing a finish that is not only attractive but also long-lasting and durable. Large equipment, primarily made of sheet steel, as emphasized in the switch-gear manufacturing industry, often will be multiply coated. Some of the manufacturers use a phosphatizing process on the bare metal, followed by a zinc-based primer, which is then epoxy coated and finally given a high-quality polyurethane paint finish. Equipment of this caliber can be expected to withstand outdoor weathering for many years without finish problems.

When the base metal of equipment is improperly prepared and then given a few coats of low-quality finish, corrosion usually begins soon after the equipment is exposed to the elements. Finish problems on large, fabricated metal equipment and structures are often a serious defect, and it becomes difficult to prevent further corrosion, even though repainting is undertaken. If the base-metal preparation and primers are not selected or applied correctly, corrosion problems always will be prevalent.

Specific design questions and requirements for the paint finish must be directed to the proper paint manufacturer if high-quality results are expected. High-quality paints and correct base-metal preparation techniques are expensive and require the proper equipment and facilities. There are many government regulations in force

today that stipulate the procedures to be followed for environmental protection relative to the use of paint finishes. These regulations are formulated and enforced by the Environmental Protection Agency (EPA) and the Occupational Safety and Health Administration (OSHA). Chapter 15, "Safety Practices in Industry," lists the OSHA standards that help to protect workers from industrial hazards caused by many substances and improper manufacturing practices, including paint ingredients and by-products.

13.9.1 Estimating paint-film thickness and coverage

Estimating the quantity of materials for a painting project is not as simple as one might expect. The surface-area coverage of any particular paint is usually listed on the paint container. But this coverage is for average wet-film area and does not take into consideration the dry-film thickness required on the product.

To determine the coverage for a 1-mil (0.001-in) dry-film thickness for a paint containing less than 100 percent solids, multiply the percentage of nonvolatile solids by 1604 and divide by 100. The percentage of nonvolatile solids is found on the paint container label or from the paint manufacturer's literature. The wet-film thickness required for a specific dry-film thickness is found by dividing the desired dry-film thickness by the percentage of solids by volume of the coating to be used. (Wet-film thickness is measured during application by a wet-film thickness gauge.)

A fast and accurate estimate of wet-film thickness required to obtain a specified dry-film thickness and theoretical coverage in square feet of area per gallon can be made using the nomograph shown in Fig. 13.21. To use the nomograph, connect the dry-film thickness on scale D with the percentage of solids on scale S, and read the wet-film thickness required on scale T together with the square feet of coverage per gallon.

13.10   Powder Coating

Powder coating is one of the newer finishing technologies; however, its use has increased in recent years to the point where it now accounts for approximately 10 percent of all industrial finishing applications. Powder coating generally involves spraying electrostatically charged powder consisting of pigment and resin onto the surface of a grounded metal part. The powder sticks to the surface

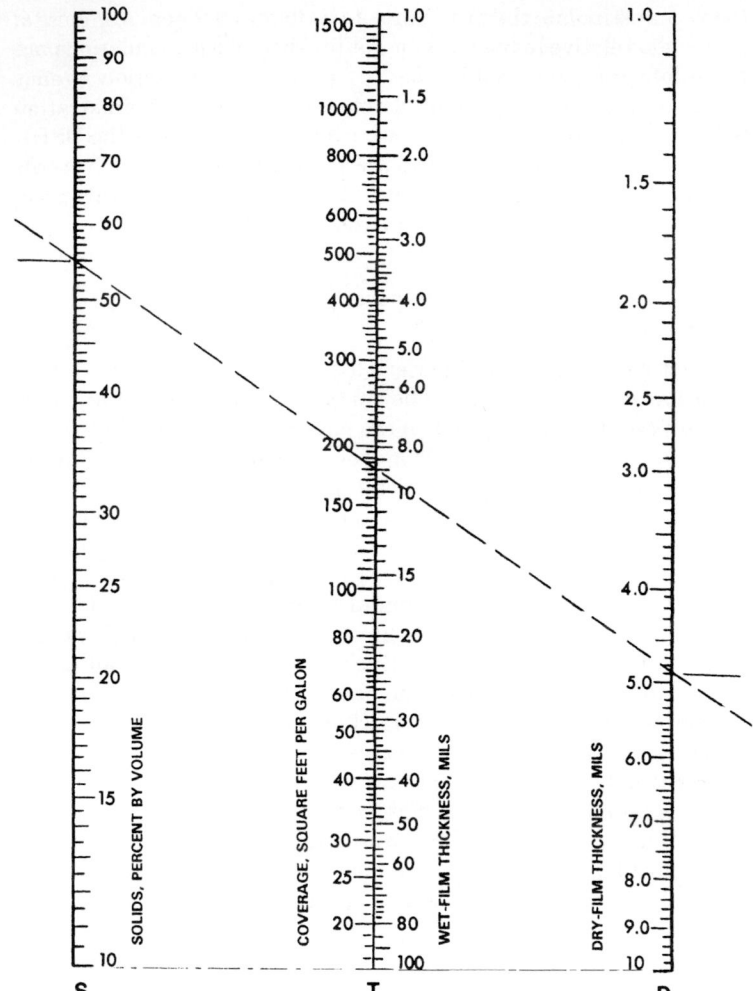

Figure 13.21 Wet-film to dry-film paint-finish nomograph.

of the part owing to the electrostatic attraction. Two types of powders are used: thermoplastics and cross-linked thermosets. Following the spraying operation, the coated part is passed through a furnace to melt (thermoplastic) or cure (thermoset) the powder. The parts being coated often are hung from a closed-loop overhead conveyor chain that moves the parts from one finishing step to the next.

There are two very significant advantages to powder coatings compared with more conventional paint-spray techniques. Because the powders are not suspended in solvents as is the case with most paints, the powder-coating process produces few, if any, volatile organic compounds. It is therefore a relatively environmentally friendly coating process. Another advantage of the process is its high material utilization. Any powder that does not adhere to the surface of the part is filtered and reused. Material utilization therefore approaches 95 percent. This is much higher than is the case with spray painting, where the overspray is generally lost.

Powder coating of steels.   As is the case with finishing any metal, the first step is to clean any greases, oils, or other contaminants from the surface. Parts intended strictly for interior use are then given an iron phosphate conversion coating as described in Sec. 13.6.2. The conversion coating and a subsequent acid rinse serve to enhance adhesion of the powder coating. Parts intended for outside use also must be galvanized to improve corrosion resistance. This is done by first applying a grain refiner that is followed by a zinc phosphate dip (see Sec. 13.6.2) and an acid rinse. The grain refiner produces a smoother surface. The part is then ready to accept its powder coating.

Powder coating of aluminum alloys.   The parts to be powder coated are first cleaned thoroughly with a detergent solution. The surface is then etched as described elsewhere in this chapter to remove the oxide skin. A chromate or phosphate conversion coating is then applied to enhance adhesion of the powder coating. The aluminum part is then ready for the application of powder coating.

Fastening and Joining Techniques and Hardware

14.1 Bolts, Screws, Nuts, and Washers

Bolts and screws are the most commonly used types of fastening devices. The thread on a bolt or screw may be compared with an inclined plane wrapped around a cylinder, thus assuming the form of one of the basic machines. Many different thread-form standards are used throughout the world, but this chapter addresses only the forms in common use, such as the 60° V thread form for the U.S. Customary (inch) system and the metric [International Standards Organization (ISO)] standard. The other basic thread forms and their geometry are shown in Chap. 7, "Machining, Machine Tools, and Practices."

14.1.1 Dimensions of bolts, screws, nuts, and washers

Basic dimensions for the most commonly used American standard bolts are shown in Fig. 14.1. Basic dimensions for the most commonly used American standard machine screws are shown in Fig. 14.2. Basic dimensions for American standard miniature screws are shown in Fig. 14.3.

The dimensions shown in these figures are the most important ones needed for machining clearances and tool/design engineering purposes. The complete dimensional specifications and applicable

Figure 14.1 Dimensions for American standard bolts.

Nominal size	Hex Bolts				Hex Cap Screw (Finished Hex Bolts)				Round Head Square Neck Bolts				Countersunk Bolts			
Basic Dia.	d	F	P	h	d	F	P	h	D	F	h	k	d	D	h	W ◆
0.2500 - 1/4	0.260	0.438	0.505	0.188	0.2500	0.438	0.505	0.163	0.594	0.260	0.145	0.156	0.260	0.493	0.150	0.064
0.3125 - 5/16	0.324	0.500	0.577	0.235	0.3125	0.500	0.577	0.211	0.719	0.324	0.176	0.187	0.324	0.618	0.189	0.072
0.3750 - 3/8	0.388	0.562	0.650	0.268	0.3750	0.562	0.650	0.243	0.844	0.388	0.208	0.219	0.388	0.740	0.225	0.081
0.4375 - 7/16	0.452	0.625	0.722	0.316	0.4375	0.625	0.722	0.291	0.969	0.452	0.239	0.250	0.452	0.803	0.226	0.081
0.5000 - 1/2	0.515	0.750	0.866	0.364	0.5000	0.750	0.866	0.323	1.094	0.515	0.270	0.281	0.515	0.935	0.269	0.091
0.5625 - 9/16	----	----	----	----	0.5625	0.812	0.938	0.371	----	----	----	----	----	----	----	-----
0.6250 - 5/8	0.642	0.938	1.083	0.444	0.6250	0.938	1.083	0.403	1.344	0.642	0.344	0.344	0.642	1.169	0.336	0.116
0.7500 - 3/4	0.768	1.125	1.299	0.524	0.7500	1.125	1.299	0.483	1.594	0.768	0.406	0.406	0.768	1.402	0.403	0.131
0.8750 - 7/8	0.895	1.312	1.516	0.604	0.8750	1.312	1.516	0.563	1.844	0.895	0.469	0.469	0.895	1.637	0.470	0.147
1.0000 - 1	1.022	1.500	1.732	0.700	1.0000	1.500	1.732	0.627	2.094	1.022	0.531	0.531	1.022	1.869	0.537	0.166
1.1250 - 1-1/8	1.149	1.688	1.949	0.780	1.1250	1.688	1.949	0.718	----	----	----	----	1.149	2.104	0.604	0.178
1.2500 - 1-1/4	1.277	1.875	2.165	0.876	1.2500	1.875	2.165	0.813	----	----	----	----	1.277	2.337	0.671	0.193
1.3750 - 1-3/8	1.404	2.062	2.382	0.940	1.3750	2.062	2.382	0.878	----	----	----	----	1.404	2.571	0.738	0.208
1.5000 - 1-1/2	1.531	2.250	2.598	1.036	1.5000	2.250	2.598	0.974	----	----	----	----	1.531	2.804	0.805	0.240
1.7500 - 1-3/4	1.785	2.625	3.031	1.196	1.7500	2.625	3.031	1.134	----	----	----	----	----	----	----	----
2.0000 - 2	2.039	3.000	3.464	1.388	2.0000	3.000	3.464	1.263	----	----	----	----	----	----	----	----

Note: ◆ Minimum dimensions of slot widths. All other tabulated values are maximum dimensions. ◻ diameters same as hex bolts (d) of same basic size.

Nominal Size		Fillister Head			Binding Head			Pan Head			Countersunk			Undercut C'sunk		
	Basic Dia.	D	h	k♦	D	h	k♦	D	h	k♦	D	h	k♦	D	h	k♦
0 -	0.0600	0.096	0.055	0.016	0.126	0.032	0.016	0.116	0.039	0.016	0.119	0.035	0.016	0.119	0.025	0.016
1 -	0.0730	0.118	0.066	0.019	0.153	0.041	0.019	0.142	0.046	0.019	0.146	0.043	0.019	0.146	0.031	0.019
2 -	0.0860	0.140	0.083	0.023	0.181	0.050	0.023	0.167	0.053	0.023	0.172	0.051	0.023	0.172	0.036	0.023
3 -	0.0990	0.161	0.095	0.027	0.208	0.059	0.027	0.193	0.060	0.027	0.199	0.059	0.027	0.199	0.042	0.027
4 -	0.1120	0.183	0.107	0.031	0.235	0.068	0.031	0.219	0.068	0.031	0.225	0.067	0.031	0.225	0.047	0.031
5 -	0.1250	0.205	0.120	0.035	0.263	0.078	0.035	0.245	0.075	0.035	0.252	0.075	0.035	0.252	0.053	0.035
6 -	0.1390	0.226	0.132	0.039	0.290	0.087	0.039	0.270	0.082	0.039	0.279	0.083	0.039	0.279	0.059	0.039
8 -	0.1640	0.270	0.156	0.045	0.344	0.105	0.045	0.322	0.096	0.045	0.332	0.100	0.045	0.332	0.070	0.045
10 -	0.1900	0.313	0.180	0.050	0.399	0.123	0.050	0.373	0.110	0.050	0.385	0.116	0.050	0.385	0.081	0.050
12 -	0.2160	0.357	0.205	0.056	0.454	0.141	0.056	0.425	0.125	0.056	0.438	0.132	0.056	0.438	0.092	0.056
1/4-	0.2500	0.414	0.237	0.064	0.525	0.165	0.064	0.492	0.144	0.064	0.507	0.153	0.064	0.507	0.107	0.064
5/16-	0.3125	0.518	0.295	0.072	0.656	0.209	0.072	0.615	0.178	0.072	0.635	0.191	0.072	0.635	0.134	0.072
3/8-	0.3750	0.622	0.335	0.081	0.788	0.253	0.081	0.740	0.212	0.081	0.762	0.230	0.081	0.762	0.161	0.081
7/16-	0.4375	0.625	0.368	0.081	---	---	---	0.863	0.247	0.081	0.812	0.223	0.081	0.812	0.156	0.081
1/2-	0.5000	0.750	0.412	0.091	---	---	---	0.937	0.281	0.091	0.875	0.223	0.091	0.875	0.156	0.091

Note: ♦ Minimum dimensions of slot widths. All other tabulated values are maximum dimensions.

Figure 14.2 Dimensions for standard machine screws.

Size	Thds/in	Basic dia.	Fillister			Binding			Pan			100° Flat Head		
			D	h	k♦	D	h	k♦	D	h	k♦	D	h	k♦
30 UNM	318	0.0118	0.021	0.012	0.003	—	—	—	0.025	0.010	0.003	0.023	0.007	0.003
40 UNM	254	0.0157	0.025	0.016	0.003	0.041	0.010	0.004	0.033	0.012	0.004	0.029	0.008	0.003
50 UNM	203	0.0197	0.033	0.020	0.004	0.051	0.012	0.005	0.041	0.016	0.005	0.037	0.011	0.004
60 UNM	169	0.0236	0.041	0.025	0.005	0.062	0.016	0.007	0.051	0.020	0.007	0.045	0.013	0.005
80 UNM	127	0.0315	0.051	0.032	0.007	0.082	0.020	0.008	0.062	0.025	0.008	0.056	0.016	0.007
100 UNM	102	0.0394	0.062	0.040	0.008	0.103	0.025	0.012	0.082	0.032	0.012	0.072	0.019	0.008
120 UNM	102	0.0472	0.082	0.050	0.012	0.124	0.032	0.015	0.103	0.040	0.015	0.092	0.025	0.012

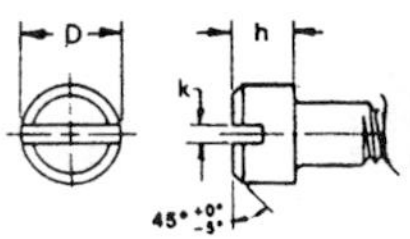

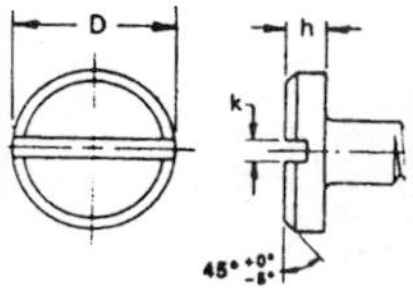

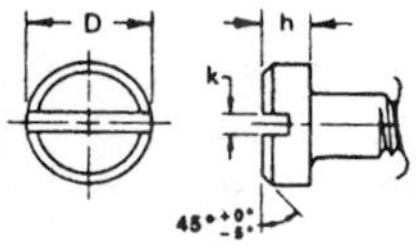

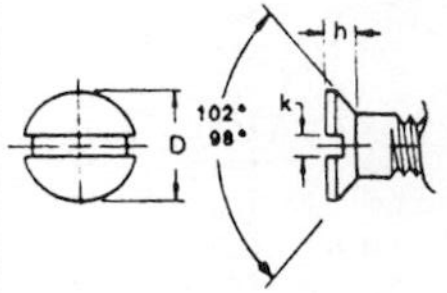

Note: ♦ Minimum dimensions of slot widths. All other tabulated values are maximum dimensions.

Figure 14.3 Dimensions for standard miniature screws.

tolerances for all standard hardware or fasteners should be obtained from the fastener handbooks distributed by the Industrial Fasteners Institute (IFI) (see Sec. 16.1).

The design specifications for strength of all types of threaded fasteners are shown in a subsection to follow. In addition, the normal tightening torques required for bolts and screws will be shown in another subsection to follow. Calculation procedures for bolts and screws likewise will be presented in a later subsection.

Figure 14.4 shows the basic dimensions for metric hex-cap screws. Figure 14.5 shows the dimensions of American standard hex nuts and jamb nuts. Figure 14.6 shows the dimensions of metric standard nuts, type 1. Figure 14.7 shows the dimensions of American standard flat washers. Figure 14.8 shows dimensions of metric standard flat washers.

Figure 14.9 shows dimensions of American standard shoulder bolts and socket-head cap screws, which are used widely in tooling applications, as well as in other mechanical design applications.

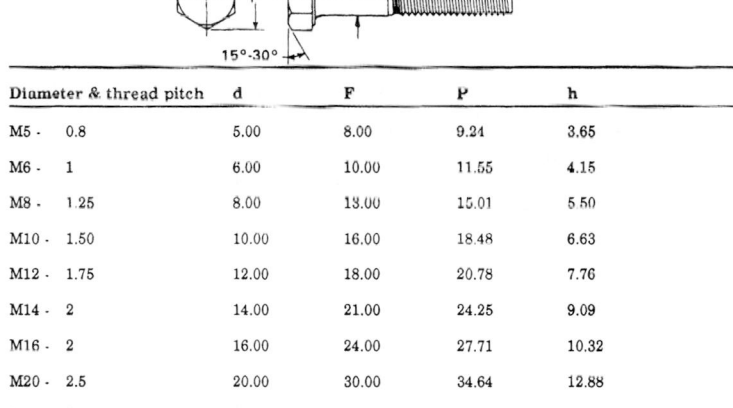

Diameter & thread pitch	d	F	P	h
M5 · 0.8	5.00	8.00	9.24	3.65
M6 · 1	6.00	10.00	11.55	4.15
M8 · 1.25	8.00	13.00	15.01	5.50
M10 · 1.50	10.00	16.00	18.48	6.63
M12 · 1.75	12.00	18.00	20.78	7.76
M14 · 2	14.00	21.00	24.25	9.09
M16 · 2	16.00	24.00	27.71	10.32
M20 · 2.5	20.00	30.00	34.64	12.88
M24 · 3	24.00	36.00	41.57	15.44
M30 · 3.5	30.00	46.00	53.12	19.48
M36 · 4	36.00	55.00	63.51	23.38

Note: Tabulated dimensions are in millimeters and are maximum values.

Figure 14.4   Dimensions for metric hex-cap screws.

Size	Flats	Points	Thickness
#0	5/32	0.180	0.050
#1	5/32	0.180	0.050
#2	3/16	0.217	0.066
#3	3/16	0.217	0.066
#4	1/4	0.289	0.098
#5	5/16	0.361	0.114
#6	5/16	0.361	0.114
#8	11/32	0.397	0.130
#10	3/8	0.433	0.130
1/4	7/16	0.505	0.226
			0.163, jamb nut
5/16	1/2	0.577	0.273
			0.195, jamb nut
3/8	9/16	0.650	0.337
			0.227, jamb nut
7/16	11/16	0.794	0.385
			0.260, jamb nut
1/2	3/4	0.866	0.448
			0.323, jamb nut
5/8	15/16	1.083	0.559
			0.387, jamb nut
3/4	1-1/8	1.299	0.665
			0.446, jamb nut
7/8	1-5/16	1.516	0.776
			0.510, jamb nut
1	1-1/2	1.732	0.887
			0.575, jamb nut

Figure 14.5 Dimensions for American standard hex and jamb nuts.

Standard shoulder screw mechanical data

Thread class	UNC-3A (ANSI/ASME B1.3) (No plating allowance is provided.)
Material	Alloys of chrome, nickel, molybdenum, or vanadium
Hardness	32 to 43 Rockwell C at the surface
Ultimate tensile strength	140,000 lb/in^2 based on minimum thread neck area
Shear strength	84,000 lb/in^2 in thread neck and shoulder areas

14.1.2 Set, self-tapping, thread-forming, and wood screws

Figures 14.10 and 14.11 show the American standard socket set screws with dimensions and applications data. The size of the hex wrench required for each set-screw size is also shown. Figure 14.12

Size	Flats	Points	Thickness
M1.6x0.35	3.20	3.70	1.30
M2x0.4	4.00	4.62	1.60
M2.5x0.45	5.00	5.77	2.00
M3x0.5	5.50	6.35	2.40
M3.5x0.6	6.00	6.93	2.80
M4x0.7	7.00	8.08	3.20
M5x0.8	8.00	9.24	4.70
M6x1	10.00	11.55	5.20
M8x1.25	13.00	15.01	6.80
M10x1.5	16.00	18.48	8.40
M12x1.75	18.00	20.78	10.80
M14x2	21.00	24.25	12.80
M16x2	24.00	27.71	14.80
M20x2.5	30.00	34.64	18.00
M24x3	36.00	41.57	21.50
M30x3.5	46.00	53.12	25.60
M36x4	55.00	63.51	31.00

NOTE: Tabulated dimensions are in millimeters. 1mm = 0.03937 inches.

Figure 14.6 Dimensions for metric standard nuts, hexagonal.

Size	Outside Diameter	Thickness
#0	3/16	0.028
#1	7/32	0.028
#2	1/4	0.036
#3	5/16	0.036
#4	3/8	0.045
#5	13/32	0.045
#6	7/16	0.045
#8	1/2	0.045
#10	9/16	0.045
¼	47/64	0.071
5/16	7/8	0.071
3/8	1	0.071
7/16	1-1/8	0.071
½	1-1/4	0.112
5/8	1-3/4	0.112
¾	2	0.112
7/8	2-1/4	0.174
1	2-1/2	0.174

Figure 14.7 Dimensions for American standard flat washers.

shows a chart of the standard self-tapping and thread-forming screws. Types AB, A, B, BF, and C are thread-forming, whereas types D, F, G, T, BF, and BT are thread-cutting or self-tapping. Type U is a spiral screw type that is driven or press-fit into the appropriately sized hole. All thread forms shown are 60° V thread.

Size	Series	Outside Diameter	Thickness
1.6	Narrow	4.00	0.70
	Regular	5.00	0.70
	Wide	6.00	0.90
2	Narrow	5.00	0.90
	Regular	6.00	0.90
	Wide	8.00	0.90
2.5	Narrow	6.00	0.90
	Regular	8.00	0.90
	Wide	10.00	1.20
3	Narrow	7.00	0.90
	Regular	10.00	1.20
	Wide	12.00	1.40
3.5	Narrow	9.00	1.20
	Regular	10.00	1.40
	Wide	15.00	1.75
4	Narrow	10.00	1.20
	Regular	12.00	1.40
	Wide	16.00	2.30
5	Narrow	11.00	1.40
	Regular	15.00	1.75
	Wide	20.00	2.30
6	Narrow	13.00	1.75
	Regular	18.80	1.75
	Wide	25.40	2.30
8	Narrow	18.80	2.30
	Regular	25.40	2.30
	Wide	32.00	2.80
10	Narrow	20.00	2.30
	Regular	28.00	2.80
	Wide	39.00	3.50
12	Narrow	25.40	2.80
	Regular	34.00	3.50
	Wide	44.00	3.50
14	Narrow	28.00	2.80
	Regular	39.00	3.50
	Wide	50.00	4.00
16	Narrow	32.00	3.50
	Regular	44.00	4.00
	Wide	56.00	4.60
20	Narrow	39.00	4.00
	Regular	50.00	4.60
	Wide	66.00	5.10
24	Narrow	44.00	4.60
	Regular	56.00	5.10
	Wide	72.00	5.60
30	Narrow	56.00	5.10
	Regular	72.00	5.60
	Wide	90.00	6.40
36	Narrow	66.00	5.60
	Regular	90.00	6.40
	Wide	110.00	8.50

Note: Tabulated dimensions are in millimeters.

Figure 14.8 Dimensions for metric standard flat washers.

Figure 14.9 American standard socket-head shoulder screws and socket-head cap screws.

Nominal size		Socket-head shoulder screw d	D	h	s	T	L	Socket-head cap screw d	D	h	s
4	0.1120	—	—	—	—	—	—	0.1120	0.183	0.112	0.094
5	0.1250	—	—	—	—	—	—	0.1250	0.205	0.125	0.094
6	0.1380	—	—	—	—	—	—	0.1380	0.226	0.138	0.109
8	0.1640	—	—	—	—	—	—	0.1640	0.270	0.164	0.141
10	0.1900	—	—	—	—	—	—	0.1900	0.312	0.190	0.156
1/4	0.2500	0.2480	0.375	0.188	0.125	0.190-24	0.375	0.2500	0.375	0.250	0.188
5/16	0.3125	0.3105	0.438	0.219	0.156	0.250-20	0.438	0.3125	0.469	0.312	0.250
3/8	0.3750	0.3730	0.562	0.250	0.188	0.312-18	0.500	0.3750	0.562	0.375	0.312
7/16	0.4375	—	—	—	—	—	—	0.4375	0.656	0.438	0.375
1/2	0.5000	0.4980	0.750	0.312	0.250	0.375-16	0.625	0.5000	0.750	0.500	0.375
5/8	0.6250	0.6230	0.875	0.375	0.312	0.500-13	0.750	0.6250	0.938	0.625	0.500
3/4	0.7500	0.7480	1.000	0.500	0.375	0.625-11	0.875	0.7500	1.125	0.750	0.625
7/8	0.8750	—	—	—	—	—	—	0.8750	1.312	0.875	0.750
1	1.0000	0.9980	1.312	0.625	0.500	0.750-10	1.00C	1.0000	1.500	1.000	0.750
1-1/8	1.1250	—	—	—	—	—	—	1.1250	1.688	1.125	0.875
1-1/4	1.2500	1.2480	1.750	0.750	0.625	0.875-9	1.12#	1.2500	1.875	1.250	0.875
1-3/8	1.3750	—	—	—	—	—	—	1.3750	2.062	1.375	1.000
1-1/2	1.5000	1.4980	2.125	1.000	0.875	1.125-7	1.50#	1.5000	2.250	1.500	1.000
1-3/4	1.7500	1.7480	2.375	1.125	1.000	1.250-7	1.75#	1.7500	2.625	1.750	1.250
2	2.0000	1.9980	2.750	1.250	1.250	1.500-6	2.0C#	2.0000	3.000	2.000	1.500

Note: All tabulated dimensions are maximum values. See text for materials, hardness, etc.

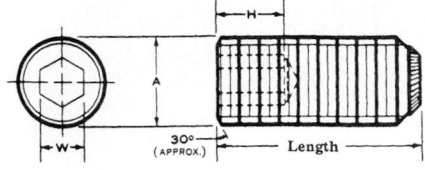

Dimensions

size	threads per inch UNC	UNF	A max.	A min. UNC	A min. UNF	C max.	C min.	D max.	D min.	F max.	F min.	H min.	W nom.
#0	....	80	.0600	....	.0568	.033	.027	.040	.037	.017	.013	.022	.028
#1	64	72	.0730	.0692	.0695	.040	.033	.049	.045	.021	.017	.028	.035
#2	56	64	.0860	.0819	.0822	.047	.039	.057	.053	.024	.020	.028	.035
#3	48	56	.0990	.0945	.0949	.054	.045	.066	.062	.027	.023	.040	.050
#4	40	48	.1120	.1069	.1075	.061	.051	.075	.070	.030	.026	.040	.050
#5	40	44	.1250	.1199	.1202	.067	.057	.083	.078	.033	.027	.050	.0625
#6	32	40	.1380	.1320	.1329	.074	.064	.092	.087	.038	.032	.050	.0625
#8	32	36	.1640	.1580	.1585	.087	.076	.109	.103	.043	.037	.062	.0781
#10	24	32	.1900	.1828	.1840	.102	.088	.127	.120	.049	.041	.075	.0937
¼	20	28	.2500	.2419	.2435	.132	.118	.156	.149	.0565	.0585	.100	.125
⁵⁄₁₆	18	24	.3125	.3038	.3053	.172	.156	.203	.195	.082	.074	.125	.1562
⅜	16	24	.3750	.3656	.3678	.212	.194	.250	.241	.0987	.0887	.150	.1875
⁷⁄₁₆	14	20	.4375	.4272	.4294	.252	.232	.296	.287	.114	.104	.175	.2187
½	13	20	.5000	.4891	.4919	.291	.270	.343	.334	.130	.120	.200	.250
⁹⁄₁₆	12	18	.5625	.5511	.5538	.332	.309	.390	.379	.1456	.1356	.200	.250
⅝	11	18	.6250	.6129	.6163	.371	.347	.468	.456	.164	.148	.250	.3125
¾	10	16	.7500	.7371	.7406	.450	.425	.562	.549	.1955	.1795	.300	.375
⅞	9	14	.8750	.8611	.8647	.530	.502	.656	.642	.2267	.2107	.400	.500
1	8	12	1.0000	.9850	.9886	.609	.579	.750	.734	.260	.240	.450	.5625
1⅛	7	12	1.1250	1.1086	1.1136	.689	.655	.843	.826	.291	.271	.450	.5625
1¼	7	12	1.2500	1.2336	1.2386	.767	.733	.937	.920	.3225	.3025	.500	.625
1⅜	6	12	1.3750	1.3568	1.3636	.848	.808	1.031	1.011	.3537	.3337	.500	.625
1½	6	12	1.5000	1.4818	1.4886	.926	.886	1.125	1.105	.385	.365	.600	.750

Figure 14.10   Socket set-screw dimensions.

Holes for self-tapping and thread-forming screws.   Extensive tables of recommended hole sizes for self-tapping and thread-forming screws are available from screw manufacturers. Because of the great variety of materials and thickness ranges possible for applying these types of screws, no tables are given here. In actual practice, you may measure the outside diameter of the screw thread and the root diameter and then take the mean difference to find an approximate hole diameter for the particular screw and material combination. A few trial combinations will give you the exact drill size to use for your application.

Figure 14.13 shows American standard wood screws and their dimensions. Wood screws are used by pattern makers and in wooden form blocks for vacuum-forming equipment applications (vacuum-

Point Types

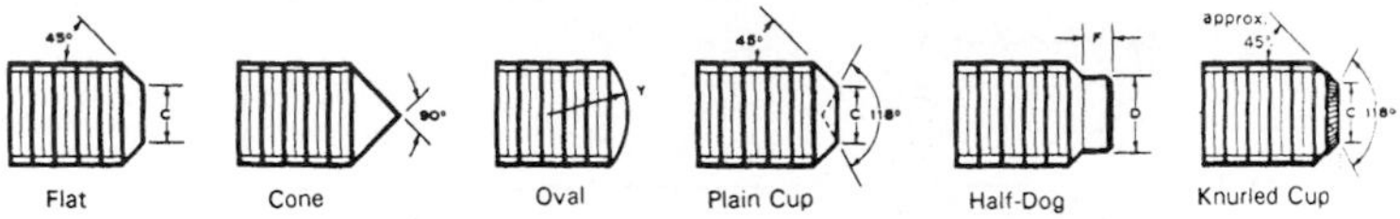
45°
90°
Y
45°
C 118°
P
D
approx. 45°
C 118°
Flat
Cone
Oval
Plain Cup
Half-Dog
Knurled Cup

Application Data

Y	tap drill size			recommended seating torques – Pound-inches								
				regular and LOC-WEL screws			self-locking with NYLOK					
rad.	size	UNC	UNF	min. screw length	alloy steel	stain-less	min. screw length	alloy steel	stain-less	min. screw length	alloy steel	stain-less
.047	#0	....	1.25mm	3/32	.5	.4	3/32	.4	.3	1/8	.5	.4
.055	#1	1.5mm	1.5mm	3/32	1.5	1.2	1/8	1.2	1.0	5/32	1.5	1.2
.062	#2	#50	1.85mm	3/32	1.5	1.2	1/8	1.2	1.0	5/32	1.5	1.2
.078	#3	#46	2.1mm	3/32	5	4	5/32-3/16	4	3	7/32	5	4
.084	#4	2.3mm	#42	3/32	5	4	5/32-3/16	4	3	7/32	5	4
.093	#5	#37	#37	5/32	9	7	3/16	7	6	1/4	9	7
.109	#6	#33	#32	1/8	9	7	3/16	7	6	1/4	9	7
.125	#8	#29	3.5mm	1/8	20	16	1/4	16	13	5/16	20	16
.141	#10	#24	#20	5/32	33	26	1/4-5/16	28	22	3/8	33	26
.188	1/4	#6	5.5mm	3/16	87	70	1/4-7/16	52	42	1/2	87	70
.234	5/16	G	I	1/4	165	130	5/16-7/16	90	72	5/8	165	130
.281	3/8	O	8.6mm	5/16	290	230	3/8-5/8	200	160	3/4	290	230
.328	7/16	9.1mm	25/64	3/8	430	340	7/16-3/4	300	240	7/8	430	340
.375	1/2	27/64	11.5mm	1/2	620	[illegible]	[illegible]	[illegible]	400	7/8	620	500
.422	9/16	31/64	1/2	1/2	620	500	1/2-3/4	500	400	[illegible]	[illegible]	[illegible]
.468	5/8	17/32	14.5mm	9/16	1,225	980	5/8-7/8	980	780	1	1,225	980
.562	3/4	21/32	17.5mm	5/8	2,125	1,700	3/4-1 1/8	1,700	1,360	1 1/4	2,125	1,700
.656	7/8	49/64	20.5mm	3/4	5,000	4,000	3/4-1 1/4	3,650	2,920	1 5/16	5,000	4,000
.750	1	7/8	23.5	7/8	7,000	5,600	7/8-1 7/16	5,200	4,160	1 1/2	7,000	5,600
.844	1 1/8	25mm	1 3/64	1	7,000	5,600	....	....	....	....	....	....
.938	1 1/4	1 7/64	1 11/64	1 1/8	9,600	7,700	....	....	....	....	....	....
1.032	1 3/8	1 7/32	1 19/64	1 1/4	9,600	7,700	....	....	....	....	....	....
1.125	1 1/2	34mm	36mm	1 1/4	11,320	9,100	....	....	....	....	....	....

Note: Materials: High grade alloy steel, austenitic stainless steel.
Hardness: Rc 45-53 for alloy steel grades. Thread class: 3A.

Figure 14.11 Set-screw point types.

forming plastics). Head styles of the various machine screws are shown in Fig. 14.14.

14.2 Thread Systems: American Standard and Metric (60° V)

The international standard screw threads consist of the unified inch series and the metric series. The metric series is standardized into the M and MJ profiles. The unified series is designated as UN (unified national). Another unified profile is designated as UNR, which has a rounded root on the external thread. The metric profile MJ also has a rounded root on the external thread and a larger

Type	ANSI Standard
	AB
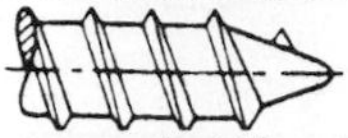 Not Recommended-Use type AB	A
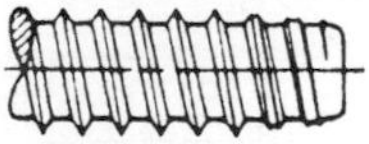	B
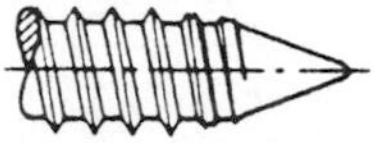	BP
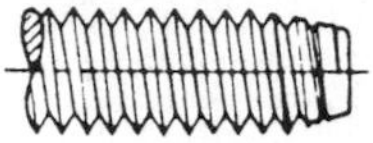	C
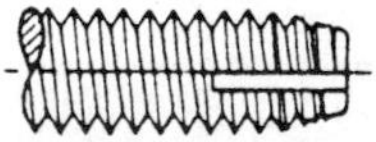	D
	F
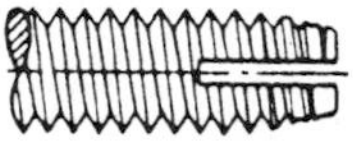	G
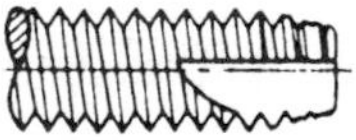	T
	BF
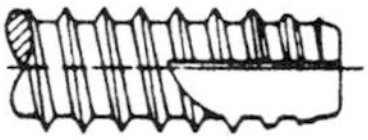	BT
	U

Figure 14.12 Self-tapping and thread-forming screw types.

Flat Head

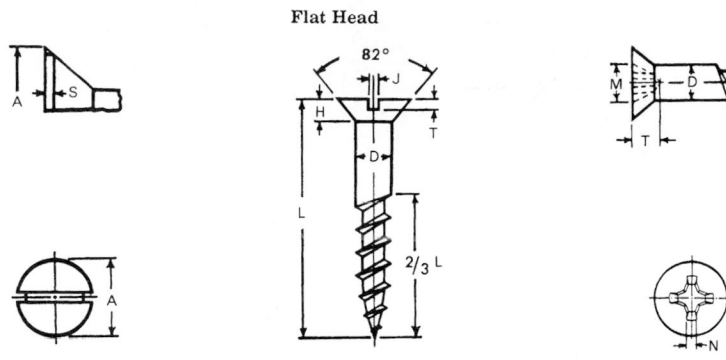

SLOTTED   PHILLIPS

Nominal Size	Basic Diameter D	Head Diameter A Max Edge	Head Diameter A Min Edge	Head Diameter A Absolute Min Edge Max E	Flat on Min B	Height of Head H Max	Height of Head H Min	Width of Slot J Max	Width of Slot J Min	Depth of Slot T Max	Depth of Slot T Min	Number Threads per inch	Basic Diameter of Screw D	Diameter of Recess M Max	Diameter of Recess M Min	Depth of Recess T Max	Width of Recess N Min	Driver Size
0	0.060	0.119	0.105	0.099	0.002	0.035	0.026	0.023	0.016	0.015	0.010	32	0.060	0.069	0.056	0.043	0.014	0
1	0.073	0.146	0.130	0.123	0.003	0.043	0.033	0.026	0.019	0.019	0.012	28	0.073	0.077	0.064	0.051	0.015	0
2	0.086	0.172	0.156	0.147	0.003	0.051	0.040	0.031	0.023	0.023	0.015	26	0.086	0.102	0.089	0.063	0.017	1
3	0.099	0.199	0.181	0.171	0.004	0.059	0.048	0.035	0.027	0.027	0.017	24	0.099	0.107	0.094	0.068	0.018	1
4	0.112	0.225	0.207	0.195	0.004	0.067	0.055	0.039	0.031	0.030	0.020	22	0.112	0.128	0.115	0.089	0.018	1
5	0.125	0.252	0.232	0.220	0.005	0.075	0.062	0.043	0.035	0.034	0.022	20	0.125	0.154	0.141	0.086	0.027	2
6	0.138	0.279	0.257	0.244	0.005	0.083	0.069	0.048	0.039	0.038	0.024	18	0.138	0.174	0.161	0.106	0.029	2
7	0.151	0.305	0.283	0.268	0.005	0.091	0.076	0.048	0.039	0.041	0.027	16	0.151	0.189	0.176	0.121	0.031	2
8	0.164	0.332	0.308	0.292	0.006	0.100	0.084	0.054	0.045	0.045	0.029	15	0.164	0.204	0.191	0.136	0.032	2
9	0.177	0.358	0.334	0.316	0.006	0.108	0.091	0.054	0.045	0.049	0.032	14	0.177	0.214	0.201	0.146	0.033	2
10	0.190	0.385	0.359	0.340	0.007	0.116	0.098	0.060	0.050	0.053	0.034	13	0.190	0.258	0.245	0.146	0.034	3
12	0.216	0.438	0.410	0.389	0.008	0.132	0.112	0.067	0.056	0.060	0.039	11	0.216	0.283	0.270	0.171	0.036	3
14	0.242	0.491	0.461	0.437	0.009	0.148	0.127	0.075	0.064	0.068	0.044	10	0.242	0.303	0.290	0.191	0.039	3
16	0.268	0.544	0.512	0.485	0.010	0.164	0.141	0.075	0.064	0.075	0.049	9	0.268	0.327	0.314	0.216	0.045	3
18	0.294	0.597	0.563	0.534	0.011	0.180	0.155	0.084	0.072	0.083	0.054	8	0.294	0.378	0.365	0.230	0.062	4
20	0.320	0.650	0.614	0.582	0.012	0.196	0.170	0.084	0.072	0.090	0.059	8	0.320	0.393	0.380	0.245	0.065	4
24	0.372	0.756	0.716	0.679	0.013	0.228	0.198	0.094	0.081	0.105	0.069	7	0.372	0.424	0.411	0.276	0.069	4

Figure 14.13 American standard wood screw dimensions.

minor diameter of both the internal and external threads. Both the UNR and MJ profiles are used for applications requiring high fatigue strength; they are also employed in aerospace applications.

A constant-pitch unified series is also standardized and consists of 4, 6, 8, 12, 16, 20, 28, and 32 threads per inch. These are used for sizes over 1 in in diameter, and 8UN, 12UN, and 16UN are the

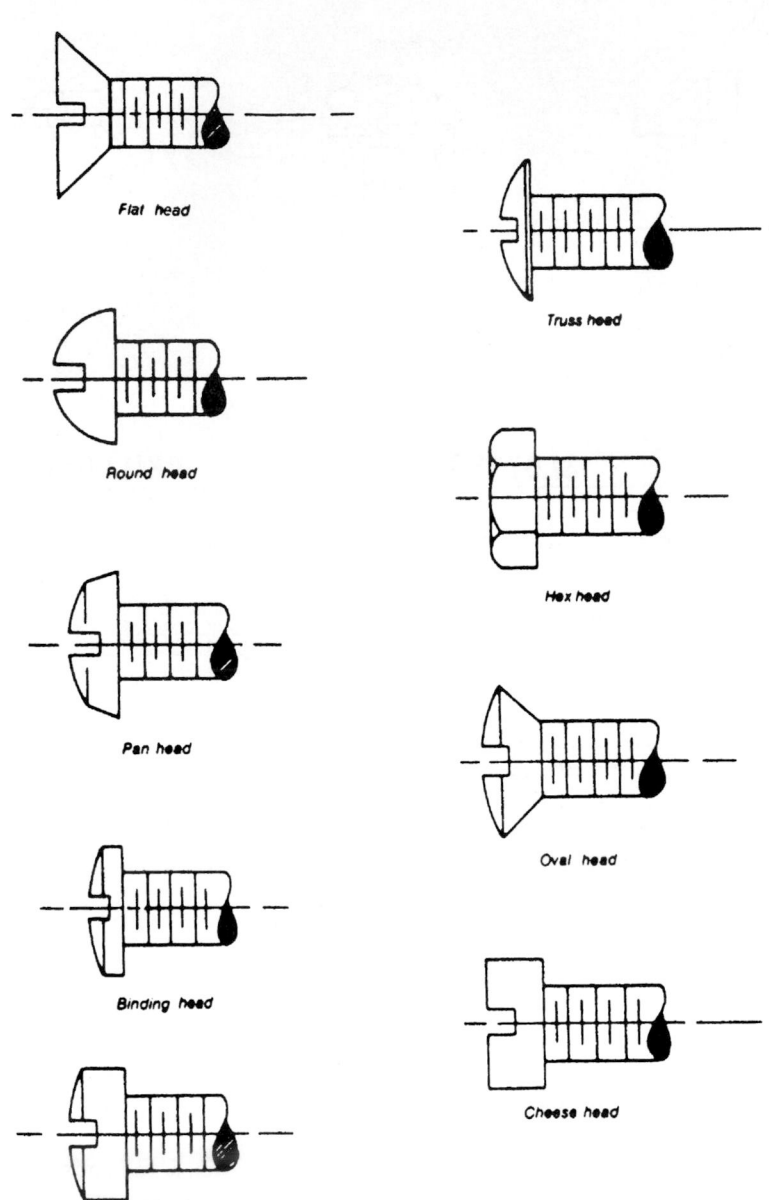

Figure 14.14 Head styles of standard machine screws.

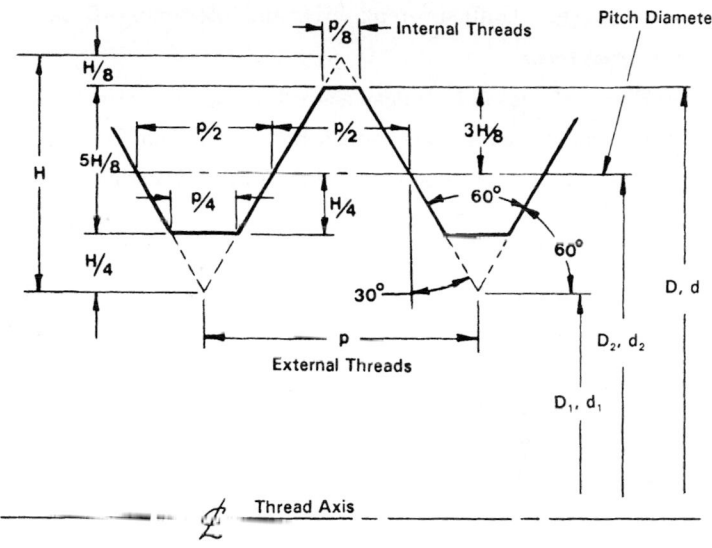

Basic Thread Profile for Unified (UN) and Metric (M) Threads (ISO 68)

D,(d) = basic major diameter of internal (external) thread
$D_1,(d_1)$ = basic minor diameter of internal (external) thread
$D_2,(d_2)$ = basic pitch diameter of internal (external) thread
p = pitch
$H = 0.5\sqrt{3}\,p$

Figure 14.15   Basic thread profiles (unified and metric).

preferred pitches. Figure 14.15 shows that both the unified and metric M series use the same profile geometry. The other common screw thread systems include the Acme, stub Acme, Whitworth, buttress, and others.

The unified threads are further classified as UN A for external threads and UN B for internal threads. The UN series contains three fit classes, 1A, 2A, and 3A for external threads and 1B, 2B, and 3B for internal threads. A typical engineering drawing callout for a coarse external class 2 thread in ¼-20 size would be

0.250-20UNC-2A

A ¼-20 size class 2 tapped hole would be shown as

0.250-20UNC-2B

Thread fit classes.   Unified thread series and interference-fit threads.

Unified thread series

1A: External, loose fit for easy assembly and noncritical uses

2A: External, general applications where plating may be applied

3A: External, tight fit used for great accuracy; no plating allowance provided

1B: Internal, loose fit for easy assembly and noncritical uses

2B: Internal, general applications where plating may be applied

3B: Internal, tight fit used for great accuracy; no plating allowance provided

Class 5 interference-fit threads

Class 5 external

NC5 HF: For driving in hard ferrous materials over 160 Brinell hardness number (BHN)

NC5 CSF: For driving in copper alloys and soft ferrous materials under 160 BHN

NC5 ONF: For other nonferrous materials, any hardness

Class 5 internal

NC5 IF: Entire ferrous material range

NC5 INF: Entire nonferrous material range

Interference-fit threads are used commonly on threaded studs to ensure a tight, vibration-resistant fit. The internal thread of a class 5 interference-fit application should be lubricated for best results when torquing the stud or externally threaded part into the class 5 fit tapped hole.

14.2.1   Unified and metric thread data

Unified national tap drill sizes.   You may calculate the tap drill diameter for 75 and 100 percent thread for the unified and metric M profile series using the following equations. For unified national threads,

$$D_{td} = D_m - \frac{0.947}{n}$$ (Tap drill diameter for 75 percent thread)

$$D_{td} = D_m - \frac{1.299}{n}$$ (Tap drill diameter for 100 percent thread)

For metric threads, M profile,

$$D_d = D_M - \frac{(\%T)(p)}{76.98}$$

$$\%T = \frac{76.98}{p}(D_M - D_d)$$

In the preceding equations, symbolism is as follows: For unified threads,

D_{td} = diameter of tap drill, in
D_m = major diameter of external thread or tap diameter, in
 n = number of threads per inch

For metric threads,

 D_d = drilled hole diameter, mm
D_M = basic major diameter, mm
$\%T$ = percent of thread (i.e., 70 percent = 0.70, etc.)
 p = pitch of thread, mm

Note: Recommended percent of thread is usually taken as 70 to 75 percent for a general class 2 type fit, which allows for electroplatings such as zinc, cadmium, chrome, nickel, etc.

Unified national coarse (UNC) screw thread data are presented in Fig. 14.16. Unified national fine (UNF) screw thread data are

Screw Thread Data, Unified National Coarse, (UNC)

Thread	Tap Drill	Decimal, in.	Stress Area Sq. inches	Basic Pitch Diameter, inches
#1-64	#53	0.0595	0.0026	0.0629
#2-56	#50	0.0700	0.0037	0.0744
#3-48	#47	0.0785	0.0048	0.0855
#4-40	#43	0.0890	0.0060	0.0958
#5-40	#38	0.1015	0.0080	0.1088
#6-32	#36	0.1065	0.0090	0.1177
#8-32	#29	0.1360	0.0140	0.1437
#10-24	#25	0.1495	0.0175	0.1629
1/4-20	#7	0.2010	0.0318	0.2175
5/16-18	F	0.2570	0.0524	0.2764
3/8-16	5/16	0.3125	0.0775	0.3344
7/16-14	T	0.3580	0.1063	0.3911
1/2-13	27/64	0.4219	0.1419	0.4500
9/16-12	31/64	0.4844	0.1820	0.5084
5/8-11	17/32	0.5312	0.2260	0.5660
3/4-10	41/64	0.6406	0.3340	0.6850
7/8-9	49/64	0.7656	0.4620	0.8028
1-8	7/8	0.8750	0.6060	0.9188

Figure 14.16 Screw thread data (UNC).

Screw Thread Data, Unified National Fine, (UNF)

Thread	Tap Drill	Decimal, in.	Stress Area, Sq. inches	Basic Pitch Diameter, inches
#0-80	3/64	0.0469	0.0018	0.0519
#1-72	#53	0.0595	0.0027	0.0640
#2-64	#50	0.0700	0.0039	0.0759
#3-56	#45	0.0820	0.0052	0.0874
#4-48	#42	0.0935	0.0066	0.0985
#5-44	#37	0.1040	0.0083	0.1102
#6-40	#33	0.1130	0.0102	0.1218
#8-36	#29	0.1360	0.0147	0.1460
#10-32	#21	0.1590	0.0200	0.1697
¼-28	#3	0.2130	0.0364	0.22268
5/16-24	I	0.2720	0.0580	0.2854
3/8-24	Q	0.3320	0.0878	0.3479
7/16-20	25/64	0.3906	0.1187	0.4050
½-20	29/64	0.4531	0.1599	0.4675
9/16-18	33/64	0.5156	0.2030	0.5264
5/8-18	9/16	0.5625	0.2560	0.5889
¾-16	11/16	0.6875	0.3730	0.7094
7/8-14	13/16	0.8125	0.5090	0.8286
1-12	29/32	0.9063	0.6630	0.9459

Figure 14.17   Screw thread data (UNF).

presented in Fig. 14.17. Metric M profile thread data are presented
in Fig. 14.18. American national drill sizes are shown in Fig. 14.19
and metric sizes in Fig. 14.20. Both number and letter size drills
are indicated with their decimal equivalents.

The standard limits of size for both American and unified national
series screw threads are fully covered in Table 14.1. The four sections of the table are extracted from National Bureau of Standards
Handbook H-28 and are in general agreement with current American
National Standards Institute (ANSI) standards. The table covers
or includes sizes 0-80 through 3-16UN.

Engagement of threads.   The length of engagement of a stud end or
bolt end E can be stated in terms of the major diameter D of the
thread. In general,

■ For a steel stud in cast iron or steel, $E = 1.50D$.

■ For a steel stud in hardened steel or high-strength bronze, $E = D$.

■ For a steel stud in aluminum or magnesium alloys subjected to
  shock loads, $E = 2.00D + 0.062$.

■ For a steel stud as above subjected to normal loads, $E = 1.50D
  + 0.062$.

Metric Thread Data, M Profile, Internal and External

Thread Designation Dia. X Pitch (Millimeters)	Tap Drill, mm	Pitch Dia. 6H, Internal, mm	Pitch Dia. 6G, External, mm
M1.6 x 0.35	1.25	1.373	1.291
M2 x 0.4	1.60	1.740	1.654
M2.5 x 0.45	2.05	2.208	2.117
M3 x 0.5	2.50	2.675	2.580
M3.5 x 0.6	2.90	3.110	3.004
M4 x 0.7	3.30	3.545	3.433
M5 x 0.8	4.20	4.480	4.361
M6 x 1	5.00	5.350	5.212
M8 x 1.25	6.70	7.188	7.042
M8 x 1	7.00	7.350	7.212
M10 x 1.5	8.50	9.026	8.862
M10 x 1.25	8.70	9.188	9.042
M10 x 0.75	------	9.513	9.391
M12 x 1.75	10.20	10.863	10.679
M12 x 1.5	------	11.026	10.854
M12 x 1.25	10.80	11.188	11.028
M12 x 1	------	11.350	11.206
M14 x 2	12.00	12.701	12.503
M14 x 1.5	12.50	13.026	12.854
M15 x 1		14.350	14.206
M16 x 2	14.00	14.701	14.503
M16 x 1.5	14.50	15.026	14.854
M17 x 1	------	16.350	16.206
M18 x 1.5	16.50	17.026	16.854
M20 x 2.5	17.50	18.376	18.164
M20 x 1.5	18.50	19.026	18.854
M20 x 1	------	19.350	19.206
M22 x 2.5	19.50	20.376	20.164
M22 x 1.5	20.50	21.026	20.854
M24 x 3	21.00	22.051	21.803
M24 x 2	22.00	22.701	22.493
M25 x 1.5	------	24.026	23.854

Figure 14.18 Metric thread data.

Load to break a threaded section. For screws or bolts,

$$P_b = SA_{ts}$$

where P_b = load to break the screw or bolt, pounds-force (lbf)
S = ultimate tensile strength of screw or bolt material, lb/in^2
A_{ts} = tensile stress area of screw or bolt thread, in^2

Tensile stress area calculation. The tensile stress area A_{ts} of screws and bolts is derived from

$$A_{ts} = \frac{\pi}{4}\left(D - \frac{0.9743}{n}\right)^2 \qquad \text{for inch-series threads}$$

where A_{ts} = tensile stress area of screw or bolt thread, in^2
$\quad\quad D$ = basic major diameter of thread, in
$\quad\quad n$ = number of threads per inch

Note: You may select the stress areas for unified bolts or screws by using Figs. 14.16 and 14.17, whereas the metric stress areas may be derived by converting millimeters to inches for each metric fastener and using the preceding equation.

(Text continued on page 888.)

DRILL NO.	DECIMAL	DRILL NO.	DECIMAL	DRILL NO.	DECIMAL
97	0.0059	56	0.0465	15	0.180
96	0.0063	55	0.052	14	0.182
95	0.0067	54	0.055	13	0.185
94	0.0071	53	0.0595	12	0.189
93	0.0075	52	0.0635	11	0.191
92	0.0079	51	0.067	10	0.1935
91	0.0083	50	0.070	9	0.196
90	0.0087	49	0.073	8	0.199
89	0.0091	48	0.076	7	0.201
88	0.0095	47	0.0785	6	0.204
87	0.010	46	0.076	5	0.2055
86	0.0105	45	0.082	4	0.209
85	0.011	44	0.086	3	0.213
84	0.0115	43	0.089	2	0.221
83	0.012	42	0.0935	1	0.228
82	0.0125	41	0.096	A	0.234
81	0.013	40	0.098	B	0.238
80	0.0135	39	0.0995	C	0.242
79	0.0145	38	0.1015	D	0.246
78	0.016	37	0.104	E	0.250
77	0.018	36	0.1065	F	0.257
76	0.020	35	0.110	G	0.261
75	0.021	34	0.111	H	0.266
74	0.0225	33	0.113	I	0.272
73	0.024	32	0.116	J	0.277
72	0.025	31	0.120	K	0.281
71	0.026	30	0.1285	L	0.290
70	0.028	29	0.136	M	0.295
69	0.0292	28	0.1405	N	0.302
68	0.031	27	0.144	O	0.316
67	0.032	26	0.147	P	0.323
66	0.033	25	0.1495	Q	0.332
65	0.035	24	0.152	R	0.339
64	0.036	23	0.154	S	0.348
63	0.037	22	0.157	T	0.358
62	0.038	21	0.159	U	0.368
61	0.039	20	0.161	V	0.377
60	0.040	19	0.166	W	0.386
59	0.041	18	0.1695	X	0.397
58	0.042	17	0.173	Y	0.404
57	0.043	16	0.177	Z	0.413

Figure 14.19 American national standard drill sizes.

Metric Drill Sizes

Drill	Decimal	Drill	Decimal	Drill	Decimal	Drill	Decimal	Drill	Decimal	Drill	Decimal
.35mm	.0138	1.75mm	.0650	3.70mm	.1457	6.40mm	.2520	9.00mm	.3543	17.00mm	.6693
.38mm	.0150	1.80mm	.0709	3.75mm	.1477	6.50mm	.2559	9.10mm	.3583	17.50mm	.6890
.40mm	.0157	1.85mm	.0728	3.80mm	.1496	6.60mm	.2598	9.20mm	.3622	18.00mm	.7087
.42mm	.0165	1.90mm	.0748	3.90mm	.1535	6.70mm	.2638	9.25mm	.3642	18.50mm	.7283
.45mm	.0177	1.95mm	.0768	4.00mm	.1575	6.75mm	.2658	9.30mm	.3661	19.00mm	.7480
.48mm	.0189	2.00mm	.0787	4.10mm	.1614	6.80mm	.2677	9.40mm	.3701	19.50mm	.7677
.50mm	.0197	2.05mm	.0807	4.20mm	.1654	6.90mm	.2716	9.50mm	.3740	20.00mm	.7874
.55mm	.0217	2.10mm	.0827	4.25mm	.1674	7.00mm	.2756	9.60mm	.3780	20.50mm	.8071
.60mm	.0236	2.15mm	.0846	4.50mm	.1771	7.10mm	.2795	9.70mm	.3819	21.00mm	.8268
.65mm	.0256	2.20mm	.0856	4.60mm	.1811	7.20mm	.2835	9.75mm	.3839	21.50mm	.8465
.70mm	.0276	2.25mm	.0886	4.70mm	.1850	7.25mm	.2855	9.80mm	.3858	22.00mm	.8661
.75mm	.0295	2.30mm	.0905	4.75mm	.1870	7.30mm	.2874	9.90mm	.3898	22.50mm	.8858
.80mm	.0315	2.35mm	.0925	4.80mm	.1890	7.40mm	.2913	10.00mm	.3937	23.00mm	.9055
.85mm	.0335	2.40mm	.0945	4.90mm	.1929	7.50mm	.2953	10.20mm	.4016	23.50mm	.9252
.90mm	.0354	2.45mm	.0965	5.00mm	.1968	7.60mm	.2992	10.50mm	.4134	24.00mm	.9449
.95mm	.0374	2.50mm	.0984	5.10mm	.2008	7.70mm	.3031	10.80mm	.4252	24.50mm	.9646
1.00mm	.0394	2.55mm	.1004	5.20mm	.2047	7.75mm	.3051	11.00mm	.4330	25.00mm	.9843
1.05mm	.0413	2.60mm	.1024	5.25mm	.2067	7.80mm	.3071	11.20mm	.4409		
1.10mm	.0433	2.65mm	.1043	5.30mm	.2087	7.90mm	.3110	11.50mm	.4528		
1.15mm	.0453	2.70mm	.1063	5.40mm	.2125	8.00mm	.3150	11.80mm	.4646	+1.00mm increments	
1.20mm	.0472	2.75mm	.1083	5.50mm	.2165	8.10mm	.3189	12.00mm	.4724	up to 48mm.	
1.25mm	.0492	2.80mm	.1102	5.60mm	.2205	8.20mm	.3228	12.20mm	.4803		
1.30mm	.0512	2.90mm	.1142	5.70mm	.2244	8.25mm	.3248	12.50mm	.4921	+5.00mm increments	
1.35mm	.0531	3.00mm	.1181	5.75mm	.2264	8.30mm	.3268	13.00mm	.5118	from 50mm up to	
1.40mm	.0551	3.10mm	.1220	5.80mm	.2283	8.40mm	.3307	13.50mm	.5315	105mm.	
1.45mm	.0571	3.20mm	.1260	5.90mm	.2323	8.50mm	.3346	14.00mm	.5512		
1.50mm	.0591	3.25mm	.1280	6.00mm	.2362	8.60mm	.3386	14.50mm	.5709		
1.55mm	.0610	3.30mm	.1299	6.10mm	.2401	8.70mm	.3425	15.00mm	.5906		
1.60mm	.0629	3.40mm	.1339	6.20mm	.2441	8.75mm	.3445	15.50mm	.6102		
1.65mm	.0650	3.50mm	.1378	6.25mm	.2451	8.83mm	.3465	16.00mm	.6299		
1.70mm	.0669	3.60mm	.1417	6.30mm	.2480	8.90mm	.3504	16.50mm	.6496		

Figure 14.20 Metric drill sizes.

TABLE 14.1 Standard Limits of Size: Unified and American Screw Threads

Nominal size and threads per inch	Series desig-nation	Class	Allow-ance, in	Major Max*	Major Min	Major Min†	Pitch Max*	Pitch Min	Pitch Toler-ance	Minor diam-eter, in	Class	Minor Min	Minor Max	Pitch Min	Pitch Max	Pitch Toler-ance	Major diameter Min
							No. 0–80 to ½–13										
0–80	NF	2A	0.0005	0.0595	0.0563		0.0514	0.0496	0.0018	0.0442	2B	0.0465	0.0514	0.0519	0.0542	0.0023	0.0600
		3A	0.0000	0.0600	0.0568		0.0519	0.0506	0.0013	0.0447	3B	0.0465	0.0514	0.0519	0.0536	0.0017	0.0600
1–64	NC	2A	0.0006	0.0724	0.0686		0.0623	0.0603	0.0020	0.0532	2B	0.0561	0.0623	0.0629	0.0655	0.0026	0.0730
		3A	0.0000	0.0730	0.0692		0.0629	0.0614	0.0015	0.0538	3B	0.0561	0.0623	0.0629	0.0648	0.0019	0.0730
1–72	NF	2A	0.0006	0.0724	0.0689		0.0634	0.0615	0.0019	0.0554	2B	0.0580	0.0635	0.0640	0.0665	0.0025	0.0730
		3A	0.0000	0.0730	0.0695		0.0640	0.0626	0.0014	0.0560	3B	0.0580	0.0635	0.0640	0.0659	0.0019	0.0730
2–56	NC	2A	0.0006	0.0854	0.0813		0.0738	0.0717	0.0021	0.0635	2B	0.0667	0.0737	0.0744	0.0772	0.0028	0.0860
		3A	0.0000	0.0860	0.0819		0.0744	0.0728	0.0016	0.0641	3B	0.0667	0.0737	0.0744	0.0765	0.0021	0.0860
2–64	NF	2A	0.0006	0.0854	0.0816		0.0753	0.0733	0.0020	0.0662	2B	0.0691	0.0753	0.0759	0.0786	0.0027	0.0860
		3A	0.0000	0.0860	0.0822		0.0759	0.0744	0.0015	0.0668	3B	0.0691	0.0753	0.0759	0.0779	0.0020	0.0860
3–48	NC	2A	0.0007	0.0983	0.0938		0.0848	0.0825	0.0023	0.0727	2B	0.0764	0.0845	0.0855	0.0885	0.0030	0.0990
		3A	0.0000	0.0990	0.0945		0.0855	0.0838	0.0017	0.0734	3B	0.0764	0.0845	0.0855	0.0877	0.0022	0.0990
3–56	NF	2A	0.0007	0.0983	0.0942		0.0867	0.0845	0.0022	0.0764	2B	0.0797	0.0865	0.0874	0.0902	0.0028	0.0990
		3A	0.0000	0.0990	0.0949		0.0874	0.0858	0.0016	0.0771	3B	0.0797	0.0865	0.0874	0.0895	0.0021	0.0990
4–40	NC	2A	0.0008	0.1112	0.1061		0.0950	0.0925	0.0025	0.0805	2B	0.0849	0.0939	0.0958	0.0991	0.0033	0.1120
		3A	0.0000	0.1120	0.1069		0.0958	0.0939	0.0019	0.0813	3B	0.0849	0.0939	0.0958	0.0982	0.0024	0.1120
4–48	NF	2A	0.0007	0.1113	0.1068		0.0985	0.0967	0.0024	0.0857	2B	0.0894	0.0968	0.0985	0.1016	0.0031	0.1120
		3A	0.0000	0.1120	0.1075		0.0985	0.0967	0.0018	0.0864	3B	0.0894	0.0968	0.0985	0.1008	0.0023	0.1120
5–40	NC	2A	0.0008	0.1242	0.1191		0.1080	0.1054	0.0026	0.0935	2B	0.0979	0.1062	0.1088	0.1121	0.0033	0.1250
		3A	0.0000	0.1250	0.1199		0.1088	0.1069	0.0019	0.0943	3B	0.0979	0.1062	0.1088	0.1113	0.0025	0.1250
5–44	NF	2A	0.0007	0.1243	0.1195		0.1095	0.1070	0.0025	0.0964	2B	0.1004	0.1079	0.1102	0.1134	0.0032	0.1250
		3A	0.0000	0.1250	0.1202		0.1102	0.1083	0.0019	0.0971	3B	0.1004	0.1079	0.1102	0.1126	0.0024	0.1250
6–32	NC	2A	0.0008	0.1372	0.1312		0.1169	0.1141	0.0028	0.0989	2B	0.104	0.114	0.1177	0.1214	0.0037	0.1380
		3A	0.0000	0.1380	0.1320		0.1177	0.1156	0.0021	0.0997	3B	0.1040	0.1140	0.1177	0.1204	0.0027	0.1380
6–40	NF	2A	0.0008	0.1372	0.1321		0.1210	0.1184	0.0026	0.1065	2B	0.111	0.119	0.1218	0.1252	0.0034	0.1380
		3A	0.0000	0.1380	0.1329		0.1218	0.1198	0.0020	0.1073	3B	0.1110	0.1186	0.1218	0.1243	0.0025	0.1380

External column group: Major diameter limits (Max, Min, Min†), Pitch diameter limits (Max*, Min, Tolerance), Minor diameter. Internal column group: Minor diameter limits (Min, Max), Pitch diameter limits (Min, Max, Tolerance), Major diameter (Min).*

8-32	NC	2A	0.0009	0.1631	0.1571		0.1428	0.1399	0.0029	0.1248	2B	0.130	0.139	0.1437	0.1475	0.0038	0.1640
		3A	0.0000	0.1640	0.1580		0.1437	0.1415	0.0022	0.1257	3B	0.1300	0.1389	0.1437	0.1465	0.0028	0.1640
8-36	NF	2A	0.0008	0.1632	0.1577		0.1452	0.1424	0.0023	0.1291	2B	0.134	0.142	0.1460	0.1496	0.0036	0.1640
		3A	0.0000	0.1640	0.1585		0.1460	0.1439	0.0021	0.1299	3B	0.1340	0.1416	0.1460	0.1487	0.0027	0.1640
10-24	NC	2A	0.0010	0.1890	0.1818		0.1619	0.1586	0.0033	0.1379	2B	0.145	0.156	0.1629	0.1672	0.0043	0.1900
		3A	0.0000	0.1900	0.1828		0.1629	0.1604	0.0025	0.1389	3B	0.1450	0.1555	0.1629	0.1661	0.0032	0.1900
10-32	NF	2A	0.0009	0.1891	0.1831		0.1688	0.1658	0.0030	0.1508	2B	0.156	0.164	0.1697	0.1736	0.0039	0.1900
		3A	0.0000	0.1900	0.1840		0.1697	0.1674	0.0023	0.1517	3B	0.1560	0.1641	0.1697	0.1726	0.0029	0.1900
12-24	NC	2A	0.0010	0.2150	0.2078		0.1879	0.1845	0.0034	0.1639	2B	0.171	0.181	0.1889	0.1933	0.0044	0.2160
		3A	0.0000	0.2160	0.2088		0.1889	0.1863	0.0026	0.1649	3B	0.1710	0.1807	0.1889	0.1922	0.0033	0.2160
12-28	NF	2A	0.0010	0.2150	0.2085		0.1918	0.1886	0.0032	0.1712	2B	0.177	0.186	0.1928	0.1970	0.0042	0.2160
		3A	0.0000	0.2160	0.2095		0.1928	0.1904	0.0024	0.1722	3B	0.1770	0.1857	0.1928	0.1959	0.0031	0.2160
12-32	NEF	2A	0.0009	0.2151	0.2091		0.1948	0.1917	0.0031	0.1758	2B	0.182	0.190	0.1957	0.1998	0.0041	0.2160
		3A	0.0000	0.2160	0.2100		0.1957	0.1933	0.0024	0.1777	3B	0.1820	0.1895	0.1957	0.1988	0.0031	0.2160
1/4-20	UNC	1A	0.0011	0.2489	0.2367	0.2367	0.2164	0.2108	0.0056	0.1876	1B	0.196	0.207	0.2175	0.2248	0.0073	0.2500
		2A	0.0011	0.2489	0.2408		0.2164	0.2127	0.0037	0.1876	2B	0.196	0.207	0.2175	0.2223	0.0048	0.2500
		3A	0.0000	0.2500	0.2419		0.2175	0.2147	0.0028	0.1887	3B	0.1960	0.2067	0.2175	0.2211	0.0036	0.2500
1/4-28	UNF	1A	0.0010	0.2490	0.2392	0.2382	0.2258	0.2208	0.0050	0.2052	1B	0.211	0.220	0.2268	0.2333	0.0065	0.2500
		2A	0.0010	0.2490	0.2425		0.2258	0.2225	0.0033	0.2052	2B	0.211	0.220	0.2268	0.2311	0.0043	0.2500
		3A	0.0000	0.2500	0.2435		0.2268	0.2243	0.0025	0.2062	3B	0.2110	0.2190	0.2268	0.2300	0.0032	0.2500
1/4-32	NEF	2A	0.0010	0.2490	0.2430		0.2287	0.2255	0.0032	0.2107	2B	0.216	0.224	0.2297	0.2339	0.0042	0.2500
		3A	0.0000	0.2500	0.2440		0.2297	0.2273	0.0024	0.2117	3B	0.2160	0.2229	0.2297	0.2328	0.0031	0.2500
5/16-18	UNC	1A	0.0012	0.3113	0.2982		0.2752	0.2691	0.0061	0.2431	1B	0.252	0.265	0.2764	0.2843	0.0079	0.3125
		2A	0.0012	0.3113	0.3026		0.2752	0.2712	0.0040	0.2431	2B	0.252	0.265	0.2764	0.2817	0.0053	0.3125
		3A	0.0000	0.3125	0.3038		0.2764	0.2734	0.0030	0.2443	3B	0.2520	0.2630	0.2764	0.2803	0.0039	0.3125
5/16-24	UNF	1A	0.0011	0.3114	0.3006		0.2843	0.2788	0.0055	0.2603	1B	0.267	0.277	0.2854	0.2925	0.0071	0.3125
		2A	0.0011	0.3114	0.3042		0.2843	0.2806	0.0037	0.2603	2B	0.267	0.277	0.2854	0.2902	0.0048	0.3125
		3A	0.0000	0.3125	0.3053		0.2854	0.2827	0.0027	0.2614	3B	0.2670	0.2754	0.2854	0.2890	0.0036	0.3125

TABLE 14.1 Standard Limits of Size: Unified and American Screw Threads (Continued)

No. 0–80 to ½–13

Nominal size and threads per inch	Series desig-nation	Ext Class	Ext Allow-ance, in	Ext Major diam. Max*	Ext Major diam. Min	Ext Min†	Ext Pitch diam. Max*	Ext Pitch diam. Min	Ext Pitch Toler-ance	Ext Minor diam-eter, in	Int Class	Int Minor diam. Min	Int Minor diam. Max	Int Pitch diam. Min	Int Pitch diam. Max	Int Pitch Toler-ance	Int Major diam. Min
5/16–32	NEF	2A	0.0010	0.3115	0.3055		0.2912	0.2880	0.0032	0.2732	2B	0.279	0.286	0.2922	0.2964	0.0042	0.3125
		3A	0.0000	0.3125	0.3065		0.2922	0.2898	0.0024	0.2742	3B	0.2790	0.2847	0.2922	0.2953	0.0031	0.3125
3/8–16	UNC	1A	0.0013	0.3737	0.3595	0.3595	0.3331	0.3266	0.0065	0.2970	1B	0.307	0.321	0.3344	0.3429	0.0085	0.3750
		2A	0.0013	0.3737	0.3643		0.3331	0.3287	0.0044	0.2970	2B	0.307	0.321	0.3344	0.3401	0.0057	0.3750
		3A	0.0000	0.3750	0.3656		0.3344	0.3311	0.0033	0.2983	3B	0.3070	0.3182	0.3344	0.3387	0.0043	0.3750
3/8–24	UNF	1A	0.0011	0.3739	0.3631		0.3468	0.3411	0.0057	0.3228	1B	0.330	0.340	0.3479	0.3553	0.0074	0.3750
		2A	0.0011	0.3739	0.3667		0.3468	0.3430	0.0038	0.3228	2B	0.330	0.340	0.3479	0.3528	0.0049	0.3750
		3A	0.0000	0.3750	0.3678		0.3479	0.3450	0.0029	0.3239	3B	0.3300	0.3372	0.3479	0.3516	0.0037	0.3750
3/8–32	NEF	2A	0.0010	0.3740	0.3680		0.3537	0.3503	0.0034	0.3357	2B	0.341	0.349	0.3547	0.3591	0.0044	0.3750
		3A	0.0000	0.3750	0.3690		0.3547	0.3522	0.0025	0.3367	3B	0.3410	0.3469	0.3547	0.3580	0.0033	0.3750
7/16–14	UNC	1A	0.0014	0.4361	0.4206	0.4206	0.3897	0.3826	0.0071	0.3485	1B	0.360	0.376	0.3911	0.4003	0.0092	0.4375
		2A	0.0014	0.4361	0.4258		0.3897	0.3850	0.0047	0.3485	2B	0.360	0.376	0.3911	0.3972	0.0061	0.4375
		3A	0.0000	0.4375	0.4272		0.3911	0.3876	0.0035	0.3499	3B	0.3600	0.3717	0.3911	0.3957	0.0046	0.4375
7/16–20	UNF	1A	0.0013	0.4362	0.4240		0.4037	0.3975	0.0062	0.3749	1B	0.383	0.395	0.4050	0.4131	0.0081	0.4375
		2A	0.0013	0.4362	0.4281		0.4037	0.3995	0.0042	0.3749	2B	0.383	0.395	0.4050	0.4104	0.0054	0.4375
		3A	0.0000	0.4375	0.4294		0.4050	0.4019	0.0031	0.3762	3B	0.3830	0.3916	0.4050	0.4091	0.0041	0.4375
7/16–28	UNEF	2A	0.0011	0.4364	0.4299		0.4132	0.4096	0.0036	0.3926	2B	0.399	0.407	0.4143	0.4189	0.0046	0.4375
		3A	0.0000	0.4375	0.4310		0.4143	0.4116	0.0027	0.3937	3B	0.3990	0.4051	0.4143	0.4178	0.0035	0.4375
1/2–12	N	2A	0.0016	0.4984	0.4870		0.4443	0.4389	0.0054	0.3962	2B	0.410	0.428	0.4459	0.4529	0.0070	0.5000
		3A	0.0000	0.5000	0.4886		0.4459	0.4419	0.0040	0.3978	3B	0.4100	0.4223	0.4459	0.4511	0.0052	0.5000
1/2–13	UNC	1A	0.0015	0.4985	0.4822	0.4822	0.4485	0.4411	0.0074	0.4041	1B	0.417	0.434	0.4500	0.4597	0.0097	0.5000
		2A	0.0015	0.4985	0.4876		0.4485	0.4435	0.0050	0.4041	2B	0.417	0.434	0.4500	0.4565	0.0065	0.5000
		3A	0.0000	0.5000	0.4891		0.4500	0.4463	0.0037	0.4056	3B	0.4170	0.4284	0.4500	0.4548	0.0048	0.5000

Size	Series	Class									Class						
½–20	UNF	1A	0.0013	0.4987	0.4565		0.4662	0.4598	0.0064	0.4374	1B	0.446	0.457	0.4675	0.4759	0.0084	0.5000
		2A	0.0013	0.4987	0.4906		0.4662	0.4619	0.0043	0.4374	2B	0.446	0.457	0.4675	0.4731	0.0056	0.5000
		3A	0.0000	0.5000	0.4919		0.4675	0.4643	0.0032	0.4357	3B	0.460	0.470	0.4675	0.4717	0.0042	0.5000
½–28	UNEF	2A	0.0011	0.4989	0.4924		0.4757	0.4720	0.0037	0.4551	2B	0.461	0.4537	0.4768	0.4816	0.0048	0.5000
		3A	0.0000	0.5000	0.4935	0.5427	0.4768	0.4740	0.0028	0.4562	3B	0.4610	0.4676	0.4768	0.4804	0.0036	0.5000
9/16–12	UNC	1A	0.0016	0.5609	0.5437		0.5068	0.4990	0.0078	0.4587	1B	0.472	0.490	0.5084	0.5186	0.0102	0.5625
		2A	0.0016	0.5609	0.5495		0.5068	0.5016	0.0052	0.4587	2B	0.472	0.490	0.5084	0.5152	0.0068	0.5625
		3A	0.0000	0.5625	0.5511		0.5084	0.5045	0.0039	0.4603	3B	0.4720	0.4843	0.5084	0.5135	0.0051	0.5625
9/16–18	UNF	1A	0.0014	0.5611	0.5480		0.5250	0.5182	0.0068	0.4929	1B	0.502	0.515	0.5264	0.5353	0.0089	0.5625
		2A	0.0014	0.5611	0.5524		0.5250	0.5205	0.0045	0.4929	2B	0.502	0.515	0.5264	0.5323	0.0059	0.5625
		3A	0.0000	0.5625	0.5538		0.5254	0.5230	0.0034	0.4943	3B	0.5020	0.5106	0.5264	0.5308	0.0044	0.5625
9/16–24	NEF	2A	0.0012	0.5613	0.5541		0.5342	0.5303	0.0039	0.5102	2B	0.517	0.527	0.5354	0.5405	0.0051	0.5625
		3A	0.0000	0.5625	0.5553	0.6052	0.5354	0.5325	0.0029	0.5114	3B	0.5170	0.5244	0.5354	0.5392	0.0038	0.5625
⅝–11	UNC	1A	0.0016	0.6234	0.6052		0.5644	0.5561	0.0083	0.5119	1B	0.527	0.546	0.5660	0.5767	0.0107	0.6250
		2A	0.0016	0.6234	0.6113		0.5644	0.5589	0.0055	0.5119	2B	0.527	0.546	0.5660	0.5732	0.0072	0.6250
		3A	0.0000	0.6250	0.6129		0.5663	0.5619	0.0041	0.5135	3B	0.5270	0.5391	0.5660	0.5714	0.0054	0.6250
⅝–12	N	2A	0.0016	0.6234	0.6120		0.5693	0.5639	0.0054	0.5212	2B	0.535	0.553	0.5709	0.5780	0.0071	0.6250
		3A	0.0000	0.6250	0.6136		0.5709	0.5668	0.0041	0.5228	3B	0.5350	0.5463	0.5709	0.5762	0.0053	0.6250
⅝–18	UNF	1A	0.0014	0.6236	0.6105		0.5875	0.5805	0.0070	0.5554	1B	0.565	0.578	0.5889	0.5980	0.0091	0.6250
		2A	0.0014	0.6236	0.6149		0.5875	0.5828	0.0047	0.5554	2B	0.565	0.578	0.5889	0.5949	0.0060	0.6250
		3A	0.0000	0.6250	0.6163		0.5889	0.5854	0.0035	0.5558	3B	0.5650	0.5730	0.5889	0.5934	0.0045	0.6250
⅝–24	NEF	2A	0.0012	0.6238	0.6166		0.5967	0.5927	0.0040	0.5727	2B	0.580	0.590	0.5979	0.6031	0.0052	0.6250
		3A	0.0000	0.6250	0.6178		0.5979	0.5949	0.0030	0.5739	3B	0.5800	0.5809	0.5979	0.6018	0.0039	0.6250
11/16–12	N	2A	0.0016	0.6859	0.6745		0.6318	0.6264	0.0054	0.5837	2B	0.597	0.615	0.6334	0.6405	0.0071	0.6875
		3A	0.0000	0.6875	0.6761	0.7288	0.6334	0.6293	0.0041	0.5853	3B	0.5970	0.6085	0.6334	0.6387	0.0053	0.6875
11/16–24	NEF	2A	0.0012	0.6863	0.6791		0.6592	0.6552	0.0040	0.6352	2B	0.642	0.652	0.6604	0.6656	0.0052	0.6875
		3A	0.0000	0.6875	0.6803		0.6604	0.6574	0.0030	0.6364	3B	0.6420	0.6494	0.6604	0.6643	0.0039	0.6875
¾–10	UNC	1A	0.0018	0.7482	0.7288		0.6832	0.6744	0.0088	0.6255	1B	0.642	0.663	0.6850	0.6965	0.0115	0.7500
		2A	0.0018	0.7482	0.7353		0.6832	0.6773	0.0059	0.6255	2B	0.642	0.663	0.6850	0.6927	0.0077	0.7500
		3A	0.0000	0.7500	0.7371		0.6850	0.6806	0.0044	0.6273	3B	0.6420	0.6545	0.6850	0.6907	0.0057	0.7500
¾–12	N	2A	0.0017	0.7483	0.7369		0.6942	0.6887	0.0055	0.6461	2B	0.660	0.678	0.6959	0.7031	0.0072	0.7500
		3A	0.0000	0.7500	0.7386		0.6959	0.6918	0.0041	0.6478	3B	0.6600	0.6707	0.6959	0.7013	0.0054	0.7500

TABLE 14.1 Standard Limits of Size: Unified and American Screw Threads (*Continued*)

Nominal size and threads per inch	Series designation	External										Internal						
		Class	Allowance, in	Major diameter limits, in			Pitch diameter limits, in			Minor diameter, in	Class	Minor diameter limits, in		Pitch diameter limits, in			Major diameter, in
				Max*	Min	Min†	Max*	Min	Tolerance			Min	Max	Min	Max	Tolerance	Min
½–20 to 1⅛–12																	
¾–16	UNF	1A	0.0015	0.7485	0.7343		0.7079	0.7004	0.0075	0.6718	1B	0.682	0.696	0.7094	0.7192	0.0098	0.7500
		2A	0.0015	0.7485	0.7391		0.7079	0.7029	0.0050	0.6718	2B	0.682	0.696	0.7094	0.7159	0.0065	0.7500
		3A	0.0000	0.7500	0.7406		0.7094	0.7056	0.0038	0.6733	3B	0.6820	0.6908	0.7094	0.7143	0.0049	0.7500
¾–20	UNEF	2A	0.0013	0.7487	0.7406		0.7162	0.7118	0.0044	0.6874	2B	0.696	0.707	0.7175	0.7232	0.0057	0.7500
		3A	0.0000	0.7500	0.7419		0.7175	0.7142	0.0033	0.6887	3B	0.6960	0.7037	0.7175	0.7218	0.0043	0.7500
13/16–12	N	2A	0.0017	0.8108	0.7994		0.7567	0.7512	0.0055	0.7086	2B	0.722	0.740	0.7584	0.7656	0.0072	0.8125
		3A	0.0000	0.8125	0.8011		0.7584	0.7543	0.0041	0.7103	3B	0.7220	0.7329	0.7584	0.7638	0.0054	0.8125
13/16–16	UN	2A	0.0015	0.8110	0.8016		0.7704	0.7655	0.0049	0.7343	2B	0.745	0.759	0.7719	0.7782	0.0063	0.8125
		3A	0.0000	0.8125	0.8031		0.7719	0.7683	0.0036	0.7358	3B	0.7450	0.7533	0.7719	0.7766	0.0047	0.8125
13/16–20	UNEF	2A	0.0013	0.8112	0.8031		0.7787	0.7743	0.0044	0.7498	2B	0.758	0.770	0.7800	0.7857	0.0057	0.8125
		3A	0.0000	0.8125	0.8044		0.7800	0.7767	0.0033	0.7512	3B	0.7580	0.7662	0.7800	0.7843	0.0043	0.8125
⅞–9	UNC	1A	0.0019	0.8731	0.8523		0.8009	0.7914	0.0095	0.7368	1B	0.755	0.778	0.8028	0.8151	0.0123	0.8750
		2A	0.0019	0.8731	0.8592	0.8523	0.8009	0.7946	0.0063	0.7368	2B	0.755	0.778	0.8028	0.8110	0.0082	0.8750
		3A	0.0000	0.8750	0.8611		0.8028	0.7981	0.0047	0.7387	3B	0.7550	0.7681	0.8028	0.8089	0.0061	0.8750
⅞–12	N	2A	0.0017	0.8733	0.8619		0.8192	0.8137	0.0055	0.7711	2B	0.785	0.803	0.8209	0.8281	0.0072	0.8750
		3A	0.0000	0.8750	0.8636		0.8209	0.8168	0.0041	0.7728	3B	0.7850	0.7952	0.8209	0.8263	0.0054	0.8750
⅞–14	UNF	1A	0.0016	0.8734	0.8579		0.8270	0.8189	0.0081	0.7858	1B	0.798	0.814	0.8286	0.8392	0.0106	0.8750
		2A	0.0016	0.8734	0.8631		0.8270	0.8216	0.0054	0.7858	2B	0.798	0.814	0.8286	0.8356	0.0070	0.8750
		3A	0.0000	0.8750	0.8647		0.8286	0.8245	0.0041	0.7874	3B	0.7980	0.8068	0.8286	0.8339	0.0053	0.8750
⅞–16	UN	2A	0.0015	0.8735	0.8641		0.8329	0.8280	0.0049	0.7968	2B	0.807	0.821	0.8344	0.8407	0.0063	0.8750
		3A	0.0000	0.8750	0.8656		0.8344	0.8308	0.0036	0.7983	3B	0.8070	0.8158	0.8344	0.8391	0.0047	0.8750
⅞–20	UNEF	2A	0.0013	0.8737	0.8656		0.8412	0.8368	0.0044	0.8124	2B	0.821	0.832	0.8425	0.8482	0.0057	0.8750
		3A	0.0000	0.8750	0.8669		0.8425	0.8392	0.0033	0.8137	3B	0.8210	0.8287	0.8425	0.8468	0.0043	0.8750
15/16–12	UN	2A	0.0017	0.9358	0.9244		0.8817	0.8760	0.0057	0.8336	2B	0.847	0.865	0.8834	0.8908	0.0074	0.9375
		3A	0.0000	0.9375	0.9261		0.8834	0.8793	0.0011	0.8353	3B	0.8470	0.8575	0.8834	0.8889	0.0055	0.9375
5/16–16	UN	2A	0.0015	0.9360	0.9266		0.8954	0.8904	0.0050	0.8593	2B	0.870	0.884	0.8969	0.9034	0.0065	0.9375
		3A	0.0000	0.9375	0.9281		0.8969	0.8932	0.0037	0.8608	3B	0.8700	0.8783	0.8969	0.9013	0.0049	0.9375

Size	Series	Class										Class						
13/16-20	UNEF	2A	0.0014	0.9361	0.9280		0.9036	0.8991	0.0045	0.8748		2B	0.833	0.895	0.9050	0.9109	0.0059	0.9375
		3A	0.0000	0.9375	0.9294		0.9050	0.9016	0.0034	0.8762		3B	0.8330	0.8912	0.9050	0.9094	0.0044	0.9375
1-8	UNC	1A	0.0020	0.9980	0.9735		0.9168	0.9067	0.0101	0.8446		1B	0.865	0.890	0.9188	0.9320	0.0132	1.0000
		2A	0.0020	0.9980	0.9830		0.9188	0.9100	0.0088	0.8446		2B	0.865	0.890	0.9188	0.9276	0.0088	1.0000
		3A	0.0000	1.0000	0.9850	0.3755	0.9188	0.9137	0.0051	0.8465		3B	0.8650	0.8797	0.9188	0.9254	0.0066	1.0000
1-12	UNF	1A	0.0018	0.9982	0.9810		0.9441	0.9353	0.0088	0.8960		1B	0.910	0.928	0.9459	0.9573	0.0114	1.0000
		2A	0.0018	0.9982	0.9868		0.9441	0.9382	0.0059	0.8960		2B	0.910	0.928	0.9459	0.9535	0.0076	1.0000
		3A	0.0000	1.0000	0.9886		0.9459	0.9415	0.0044	0.8978		3B	0.9100	0.9198	0.9459	0.9516	0.0057	1.0000
1-16	UN	2A	0.0015	0.9985	0.9891		0.9579	0.9529	0.0050	0.9218		2B	0.932	0.946	0.9594	0.9659	0.0065	1.0000
		3A	0.0000	1.0000	0.9906		0.9594	0.9557	0.0037	0.9233		3B	0.9320	0.9408	0.9594	0.9643	0.0049	1.0000
1-20	UNEF	2A	0.0014	0.9986	0.9905		0.9661	0.9616	0.0045	0.9373		2B	0.946	0.957	0.9675	0.9734	0.0059	1.0000
		3A	0.0000	1.0000	0.9919		0.9675	0.9641	0.0034	0.9387		3B	0.9460	0.9537	0.9675	0.9719	0.0044	1.0000
1 1/16-12	UN	2A	0.0017	1.0608	1.0494		1.0067	1.0010	0.0057	0.9586		2B	0.972	0.990	1.0084	1.0158	0.0074	1.0625
		3A	0.0000	1.0625	1.0511		1.0084	1.0042	0.0042	0.9603		3B	0.9720	0.9823	1.0084	1.0139	0.0055	1.0625
1 1/16-16	UN	2A	0.0015	1.0610	1.0516		1.0204	1.0154	0.0050	0.9843		2B	0.995	1.009	1.0219	1.0284	0.0065	1.0625
		3A	0.0000	1.0625	1.0531		1.0219	1.0182	0.0037	0.9853		3B	0.9950	1.0033	1.0219	1.0268	0.0049	1.0625
1 1/16-18	NEF	2A	0.0014	1.0611	1.0524		1.0250	1.0203	0.0047	0.9929		2B	1.002	1.015	1.0264	1.0326	0.0062	1.0625
		3A	0.0000	1.0625	1.0538		1.0264	1.0228	0.0036	0.9943		3B	1.0020	1.0105	1.0264	1.0310	0.0046	1.0625
1 1/8-7	UNC	1A	0.0022	1.1228	1.0982	1.0932	1.0300	1.0191	0.0109	0.9475		1B	0.970	0.998	1.0322	1.0416	0.0141	1.1250
		2A	0.0022	1.1228	1.1064		1.0300	1.0228	0.0072	0.9475		2B	0.970	0.998	1.0322	1.0393	0.0094	1.1250
		3A	0.0000	1.1250	1.1086	1.1004	1.0322	1.0268	0.0054	0.9497		3B	0.9700	0.9875	1.0322	1.0438	0.0071	1.1250
1 1/8-8	N	2A	0.0021	1.1229	1.1079		1.0417	1.0348	0.0069	0.9695		2B	0.900	1.015	1.0438	1.0528	0.0090	1.1250
		3A	0.0000	1.1250	1.1100		1.0438	1.0386	0.0052	0.9716		3B	0.9900	1.0047	1.0438	1.0505	0.0067	1.1250
1 1/8-12	UNF	1A	0.0018	1.1232	1.1118		1.0691	1.0601	0.0090	1.0210		1B	1.035	1.053	1.0709	1.0826	0.0117	1.1250
		2A	0.0018	1.1232	1.1118		1.0691	1.0631	0.0060	1.0210		2B	1.035	1.053	1.0709	1.0787	0.0078	1.1250
		3A	0.0000	1.1250	1.1136		1.0709	1.0664	0.0045	1.0228		3B	1.0350	1.0448	1.0709	1.0768	0.0059	1.1250

1⅛-16 to 1³⁄₁₆-12

Size	Series	Class										Class						
1⅛-16	UN	2A	0.0015	1.1235	1.1141		1.0829	1.0779	0.0050	1.0468		2B	1.057	1.071	1.0844	1.0909	0.0065	1.1250
		3A	0.0000	1.1250	1.1156		1.0844	1.0807	0.0037	1.0483		3B	1.0570	1.0658	1.0844	1.0893	0.0049	1.1250
1⅛-18	NEF	2A	0.0014	1.1236	1.1149		1.0875	1.0828	0.0047	1.0554		2B	1.065	1.078	1.0889	1.0951	0.0062	1.1250
		3A	0.0000	1.1250	1.1163		1.0889	1.0853	0.0036	1.0558		3B	1.0650	1.0730	1.0889	1.0935	0.0046	1.1250
1³⁄₁₆-12	UN	2A	0.0017	1.1858	1.1744		1.1317	1.1259	0.0058	1.0836		2B	1.097	1.115	1.1334	1.1409	0.0075	1.1875
		3A	0.0000	1.1875	1.1761		1.1334	1.1291	0.0043	1.0853		3B	1.0970	1.1073	1.1334	1.1390	0.0056	1.1875
1³⁄₁₆-16	UN	2A	0.0015	1.1860	1.1766		1.1454	1.1403	0.0051	1.1093		2B	1.120	1.134	1.1469	1.1535	0.0066	1.1875
		3A	0.0000	1.1875	1.1781		1.1469	1.1431	0.0038	1.1108		3B	1.1200	1.1283	1.1469	1.1519	0.0050	1.1875
1³⁄₁₆-18	NEF	2A	0.0015	1.1860	1.1773		1.1499	1.1450	0.0049	1.1178		2B	1.127	1.140	1.1514	1.1577	0.0063	1.1875
		3A	0.0000	1.1875	1.1788		1.1514	1.1478	0.0036	1.1193		3B	1.1270	1.1355	1.1514	1.1561	0.0047	1.1875

TABLE 14.1 Standard Limits of Size: Unified and American Screw Threads (Continued)

Range: 1¼–16 to 1⅜–12

Nominal size and threads per inch	Series designation	Class	Allowance, in	External — Major diameter limits, in · Max*	· Min	External — Pitch diameter limits, in · Max*	· Min†	· Min	· Tolerance	Minor diameter, in	Class	Internal — Minor diameter limits, in · Min	· Max	Internal — Pitch diameter limits, in · Min	· Max	· Tolerance	Major diameter, in · Min
1¼–7	UNC	1A	0.0022	1.2478	1.2232	1.1550		1.1439	0.0111	1.0725	1B	1.095	1.123	1.1572	1.1716	0.0144	1.2500
		2A	0.0022	1.2478	1.2314	1.1550		1.1476	0.0074	1.0725	2B	1.095	1.123	1.1572	1.1668	0.0096	1.2500
		3A	0.0000	1.2500	1.2336	1.1572		1.1517	0.0055	1.0747	3B	1.0950	1.1125	1.1572	1.1644	0.0072	1.2500
1¼–8	N	2A	0.0021	1.2479	1.2329	1.1667	1.2232	1.1597	0.0070	1.0945	2B	1.115	1.140	1.1688	1.1780	0.0092	1.2500
		3A	0.0000	1.2500	1.2350	1.1688	1.2254	1.1635	0.0053	1.0966	3B	1.1150	1.1297	1.1688	1.1757	0.0069	1.2500
1¼–12	UNF	1A	0.0018	1.2482	1.2310	1.1941		1.1849	0.0092	1.1460	1B	1.160	1.178	1.1959	1.2079	0.0120	1.2500
		2A	0.0018	1.2482	1.2368	1.1941		1.1879	0.0062	1.1460	2B	1.160	1.178	1.1959	1.2039	0.0080	1.2500
		3A	0.0000	1.2500	1.2386	1.1959		1.1913	0.0046	1.1478	3B	1.1600	1.1698	1.1959	1.2019	0.0060	1.2500
1¼–16	UN	2A	0.0015	1.2485	1.2391	1.2079		1.2028	0.0051	1.1718	2B	1.182	1.196	1.2094	1.2160	0.0066	1.2500
		3A	0.0000	1.2500	1.2406	1.2094		1.2056	0.0038	1.1733	3B	1.1820	1.1908	1.2094	1.2144	0.0050	1.2500
1¼–18	NEF	2A	0.0015	1.2485	1.2398	1.2124		1.2075	0.0049	1.1803	2B	1.190	1.203	1.2139	1.2202	0.0063	1.2500
		3A	0.0000	1.2500	1.2413	1.2139		1.2103	0.0036	1.1818	3B	1.1900	1.1980	1.2139	1.2186	0.0047	1.2500
1⁵⁄₁₆–12	UN	2A	0.0017	1.3108	1.2994	1.2567		1.2509	0.0058	1.2086	2B	1.222	1.240	1.2584	1.2659	0.0075	1.3125
		3A	0.0000	1.3125	1.3011	1.2584		1.2541	0.0043	1.2103	3B	1.2220	1.2323	1.2584	1.2640	0.0056	1.3125
1⁵⁄₁₆–16	UN	2A	0.0015	1.3110	1.3016	1.2704		1.2653	0.0051	1.2343	2B	1.245	1.259	1.2719	1.2785	0.0066	1.3125
		5A	0.0000	1.3125	1.3031	1.2719		1.2681	0.0038	1.2358	5B	1.2450	1.2533	1.2719	1.2769	0.0050	1.3125
1⁵⁄₁₆–18	NEF	2A	0.0015	1.3110	1.3023	1.2749		1.2700	0.0049	1.2428	2B	1.252	1.265	1.2764	1.2827	0.0063	1.3125
		3A	0.0000	1.3125	1.3038	1.2764		1.2728	0.0036	1.2443	3B	1.2520	1.2605	1.2764	1.2811	0.0047	1.3125
1⅜–6	UNC	1A	0.0024	1.3726	1.3453	1.2643		1.2523	0.0120	1.1681	1B	1.195	1.225	1.2667	†1.2822	*0.0155	1.3750
		2A	0.0024	1.3726	1.3544	1.2643		1.2563	0.0080	1.1681	2B	1.195	1.225	1.2667	1.2771	0.0104	1.3750
		3A	0.0000	1.3750	1.3568	1.2667		1.2607	0.0060	1.1705	3B	1.1950	1.2146	1.2667	1.2745	0.0078	1.3750
1⅜–8	N	2A	0.0022	1.3728	1.3578	1.2916	1.3453	1.2844	0.0072	1.2194	2B	1.240	1.265	1.2938	1.3031	0.0093	1.3750
		3A	0.0000	1.3750	1.3600	1.2938	1.3503	1.2884	0.0054	1.2216	3B	1.2400	1.2547	1.2938	1.3008	0.0070	1.3750

1⅜–12	UNF	1A	0.0019	1.3731	1.3559		1.3190	1.3096	0.00[illegible]4	1.2709	1B	1.285	1.303	1.3209	1.3332	0.0123	1.3750
		2A	0.0019	1.3731	1.3617		1.3190	1.3127	0.00[illegible]3	1.2709	2B	1.285	1.303	1.3209	1.3291	0.0082	1.3750
		3A	0.0000	1.3750	1.3636		1.3209	1.3162	0.00[illegible]7	1.2728	3B	1.2850	1.2948	1.3209	1.3270	0.0061	1.3750
1⅜–16	UN	2A	0.0015	1.3735	1.3641		1.3329	1.3278	0.00[illegible]1	1.2968	2B	1.307	1.321	1.3344	1.3410	0.0066	1.3750
		3A	0.0000	1.3750	1.3656		1.3344	1.3306	0.00[illegible]8	1.2983	3B	1.3070	1.3158	1.3344	1.3394	0.0050	1.3750
1⅜–18	NEF	2A	0.0015	1.3735	1.3648		1.3374	1.3325	0.00[illegible]9	1.3053	2B	1.315	1.328	1.3389	1.3452	0.0063	1.3750
		3A	0.0000	1.3750	1.3663		1.3389	1.3353	0.00[illegible]6	1.3068	3B	1.3150	1.3230	1.3389	1.3436	0.0047	1.3750
1$\frac{7}{16}$–12	UN	2A	0.0018	1.4357	1.4243		1.3816	1.3757	0.00[illegible]9	1.3335	2B	1.347	1.365	1.3834	1.3910	0.0076	1.4375
		3A	0.0000	1.4375	1.4261		1.3834	1.3790	0.00[illegible]4	1.3353	3B	1.3470	1.3573	1.3834	1.3891	.0057	1.4375
1$\frac{7}{16}$–16	UN	2A	0.0016	1.4359	1.4265		1.3953	1.3901	0.00[illegible]2	1.3592	2B	1.370	1.384	1.3969	1.4037	0.0068	1.4375
		3A	0.0000	1.4375	1.4281		1.3969	1.3930	0.00[illegible]9	1.3608	3B	1.3700	1.3783	1.3969	1.4020	0.0051	1.4375
1$\frac{7}{16}$–18	NEF	2A	0.0015	1.4360	1.4273		1.3999	1.3949	0.00[illegible]0	1.3678	2B	1.377	1.390	1.4014	1.4079	0.0065	1.4375
		3A	0.0000	1.4375	1.4288		1.4014	1.3977	0.00[illegible]7	1.3693	3B	1.3770	1.3855	1.4014	1.4062	0.0048	1.4375
1½–6	UNC	1A	0.0024	1.4976	1.4703		1.3893	1.3772	0.01[illegible]1	1.2931	1B	1.320	1.350	1.3917	1.4075	0.0158	1.5000
		2A	0.0024	1.4976	1.4794	1.4703	1.3893	1.3812	0.00[illegible]1	1.2931	2B	1.320	1.350	1.3917	1.4022	0.0105	1.5000
		3A	0.0000	1.5000	1.4818		1.3917	1.3856	0.00[illegible]1	1.2955	3B	1.3200	1.3396	1.3917	1.3996	0.0079	1.5000
1½–8	N	2A	0.0022	1.4978	1.4828	1.4753	1.4166	1.4093	0.00[illegible]3	1.3444	2B	1.365	1.390	1.4188	1.4283	0.0095	1.5000
		3A	0.0000	1.5000	1.4850		1.4188	1.4133	0.00[illegible]5	1.3466	3B	1.3650	1.3797	1.4188	1.4259	0.0071	1.5000
1½–12	UNF	1A	0.0019	1.4981	1.4809		1.4440	1.4344	0.00[illegible]6	1.3959	1B	1.410	1.428	1.4459	1.4584	0.0125	1.5000
		2A	0.0019	1.4981	1.4867		1.4440	1.4376	0.00[illegible]4	1.3959	2B	1.410	1.428	1.4459	1.4542	0.0083	1.5000
		3A	0.0000	1.5000	1.4886		1.4459	1.4411	0.00[illegible]8	1.3978	3B	1.4100	1.4198	1.4459	1.4522	0.0063	1.5000
1½–16	UN	2A	0.0016	1.4984	1.4890		1.4578	1.4526	0.00[illegible]2	1.4217	2B	1.432	1.446	1.4594	1.4662	0.0068	1.5000
		3A	0.0000	1.5000	1.4906		1.4594	1.4555	0.00[illegible]9	1.4233	3B	1.4320	1.4408	1.4594	1.4645	0.0051	1.5000
1½–18	NEF	2A	0.0015	1.4985	1.4898		1.4624	1.4574	0.00[illegible]0	1.4303	2B	1.440	1.452	1.4639	1.4704	0.0065	1.5000
		3A	0.0000	1.5000	1.4913		1.4639	1.4602	0.00[illegible]7	1.4318	3B	1.4400	1.4480	1.4639	1.4687	0.0048	1.5000
1$\frac{9}{16}$–16	N	2A	0.0016	1.5609	1.5515		1.5203	1.5151	0.00[illegible]2	1.4842	2B	1.495	1.509	1.5219	1.5287	0.0068	1.5625
		3A	0.0000	1.5625	1.5531		1.5219	1.5180	0.00[illegible]9	1.4858	3B	1.4950	1.5033	1.5219	1.5270	0.0051	1.5625
1$\frac{9}{16}$–18	NEF	2A	0.0015	1.5610	1.5523		1.5249	1.5199	0.00[illegible]0	1.4928	2B	1.502	1.515	1.5264	1.5329	0.0065	1.5625
		3A	0.0000	1.5625	1.5538		1.5264	1.5227	0.00[illegible]7	1.4943	3B	1.5020	1.5105	1.5264	1.5312	0.0048	1.5625

TABLE 14.1 Standard Limits of Size: Unified and American Screw Threads (*Continued*)

Nominal size and threads per inch	Series designation	External									Internal						
		Class	Allowance, in	Major diameter limits, in			Pitch diameter limits, in			Minor diameter, in	Class	Minor diameter limits, in		Pitch diameter limits, in			Major diameter, in
				Max*	Min	Min†	Max*	Min	Tolerance			Min	Max	Min	Max	Tolerance	Min
1⅝–16 to 1¾–12																	
1⅝–8	N	2A	0.0022	1.6228	1.6078	1.6003	1.5416	1.5342	0.0074	1.4694	2B	1.490	1.515	1.5438	1.5535	0.0097	1.6250
		3A	0.0000	1.6250	1.6100		1.5438	1.5382	0.0056	1.4716	3B	1.4900	1.5047	1.5438	1.5510	0.0072	1.6250
1⅝–12	UN	2A	0.0018	1.6232	1.6118		1.5691	1.5632	0.0059	1.5210	2B	1.535	1.553	1.5709	1.5785	0.0076	1.6250
		3A	0.0000	1.6250	1.6136		1.5709	1.5665	0.0044	1.5228	3B	1.5350	1.5448	1.5709	1.5766	0.0057	1.6250
1⅝–16	UN	2A	0.0016	1.6234	1.6140		1.5828	1.5776	0.0052	1.5467	2B	1.557	1.571	1.5844	1.5912	0.0068	1.6250
		3A	0.0000	1.6250	1.6156		1.5814	1.5805	0.0039	1.5483	3B	1.5570	1.5658	1.5844	1.5895	0.0051	1.6250
1⅝–18	NEF	2A	0.0015	1.6235	1.6148		1.5874	1.5824	0.0050	1.5553	2B	1.565	1.578	1.5889	1.5954	0.0065	1.6250
		3A	0.0000	1.6250	1.6163		1.5889	1.5852	0.0037	1.5568	3B	1.5650	1.5730	1.5889	1.5937	0.0048	1.6250
1 11/16–16	N	2A	0.0016	1.6859	1.6765		1.6453	1.6400	0.0053	1.6092	2B	1.620	1.634	1.6469	1.6538	0.0069	1.6875
		3A	0.0000	1.6875	1.6781		1.6469	1.6429	0.0040	1.6108	3B	1.6200	1.6283	1.6469	1.6521	0.0052	1.6875
1 11/16–18	NEF	2A	0.0015	1.6860	1.6773		1.6499	1.6448	0.0051	1.6178	2B	1.627	1.640	1.6514	1.6580	0.0066	1.6875
		3A	0.0000	1.6875	1.6788		1.6514	1.6476	0.0038	1.6193	3B	1.6270	1.6355	1.6514	1.6563	0.0049	1.6875
1¾–5	UNC	1A	0.0027	1.7473	1.7165		1.6174	1.6040	0.0134	1.5019	1B	1.534	1.568	1.6201	1.6375	0.0174	1.7500
		2A	0.0027	1.7473	1.7268	1.7165	1.6174	1.6085	0.0089	1.5019	2B	1.534	1.568	1.6201	1.6317	0.0116	1.7500
		3A	0.0000	1.7500	1.7295		1.6201	1.6134	0.0067	1.5046	3B	1.5340	1.5575	1.6201	1.6288	0.0087	1.7500
1¾–8	N	2A	0.0023	1.7477	1.7327	1.7252	1.6665	1.6590	0.0075	1.5943	2B	1.615	1.640	1.6688	1.6786	0.0098	1.7500
		3A	0.0000	1.7300	1.7350		1.6688	1.6632	0.0056	1.5966	3B	1.6150	1.6297	1.6688	1.6762	0.0074	1.7500
1¾–12	UN	2A	0.0018	1.7482	1.7368		1.6941	1.6881	0.0060	1.6460	2B	1.660	1.678	1.6959	1.7037	0.0078	1.7500
		3A	0.0000	1.7500	1.7386		1.6959	1.6914	0.0045	1.6478	3B	1.6600	1.6698	1.6959	1.7017	0.0058	1.7500
1¾–16 to 3–16																	
1¾–16	UNEF	2A	0.0016	1.7484	1.7390		1.7078	1.7025	0.0053	1.6717	2B	1.682	1.696	1.7094	1.7163	0.0069	1.7500
		3A	0.0000	1.7500	1.7406		1.7094	1.7054	0.0040	1.6733	3B	1.6820	1.6906	1.7094	1.7146	0.0052	1.7500
1 13/16–16	N	2A	0.0016	1.8109	1.8015		1.7703	1.7650	0.0053	1.7342	2B	1.745	1.759	1.7719	1.7788	0.0069	1.8125
		3A	0.0000	1.8125	1.8031		1.7719	1.7679	0.0040	1.7358	3B	1.7450	1.7533	1.7719	1.7771	0.0052	1.8125
1⅞–8	N	2A	0.0023	1.8727	1.8577	1.8502	1.7915	1.7838	0.0077	1.7193	2B	1.740	1.765	1.7938	1.8038	0.0100	1.8750
		3A	0.0000	1.8750	1.8600		1.7938	1.7881	0.0057	1.7216	3B	1.7400	1.7547	1.7938	1.8013	0.0075	1.8750

1⅞–12	UN	2A	0.0018	1.8732	1.8618		1.8191	1.8131	0.0060	1.7710	2B	1.785	1.803	1.8209	1.8287	0.0078	1.8750
		3A	0.0000	1.8750	1.8636		1.8209	1.8164	0.0045	1.7728	3B	1.7850	1.7948	1.8209	1.8267	0.0058	1.8750
1⅞–16	UN	2A	0.0016	1.8734	1.8640		1.8328	1.8275	0.0053	1.7967	2B	1.807	1.821	1.8344	1.8413	0.0069	1.8750
		3A	0.0000	1.8750	1.8656		1.8344	1.8304	0.0040	1.7983	3B	1.8070	1.8158	1.8344	1.8396	0.0052	1.8750
1$\frac{15}{16}$–16	N	2A	0.0016	1.9359	1.9265		1.8953	1.8899	0.0054	1.8592	2B	1.870	1.884	1.8969	1.9039	0.0070	1.9375
		3A	0.0000	1.9375	1.9281		1.8969	1.8929	0.0040	1.8608	3B	1.8700	1.8783	1.8969	1.9021	.0052	1.9375
		1A	0.0029	1.9971	1.9641		1.8528	1.8385	0.0143	1.7245	1B	1.759	1.795	1.8557	1.8743	0.0186	2.0000
2–4½	UNC	2A	0.0029	1.9971	1.9751	1.9641	1.8528	1.8433	0.0095	1.7245	2B	1.759	1.795	1.8557	1.8681	0.0124	2.0000
		3A	0.0000	2.0000	1.9780		1.8557	1.8486	0.0071	1.7274	3B	1.7590	1.7861	1.8557	1.8650	0.0093	2.0000
2–8	N	2A	0.0023	1.9977	1.9827	1.9752	1.9165	1.9087	0.0078	1.8443	2B	1.865	1.890	1.9188	1.9289	0.0101	2.0000
		3A	0.0000	2.0000	1.9850		1.9188	1.9130	0.0058	1.8466	3B	1.8650	1.8797	1.9188	1.9264	0.0076	2.0000
2–12	UN	2A	0.0018	1.9982	1.9868		1.9441	1.9380	0.0061	1.8960	2B	1.910	1.928	1.9459	1.9538	0.0079	2.0000
		3A	0.0000	2.0000	1.9886		1.9459	1.9414	0.0045	1.8978	3B	1.9100	1.9198	1.9459	1.9518	0.0059	2.0000
2–16	UNEF	2A	0.0016	1.9984	1.9890		1.9578	1.9524	0.0054	1.9217	2B	1.932	1.946	1.9594	1.0664	0.0070	2.0000
		3A	0.0000	2.0000	1.9906		1.9594	1.9554	0.0040	1.9233	3B	1.9320	1.9408	1.9594	1.9646	0.0052	2.0000
2$\frac{1}{16}$–16	N	2A	0.0016	2.0609	2.0515		2.0203	2.0149	0.0054	1.9842	2B	1.995	2.009	2.0219	2.0289	0.0070	2.0625
		3A	0.0000	2.0625	2.0531		2.0219	2.0179	0.0040	1.9858	3B	1.9950	2.0033	2.0219	2.0271	0.0052	2.0625
2⅛–8	N	2A	0.0024	2.1226	2.1076	2.1001	2.0414	2.0335	0.0079	1.9692	2B	1.990	2.015	2.0438	2.0540	0.0102	2.1250
		3A	0.0000	2.1250	2.1100		2.0438	2.0379	0.0059	1.9716	3B	1.9900	2.0047	2.0438	2.0515	0.0077	2.1250
2⅛–12	UN	2A	0.0018	2.1232	2.1118		2.0691	2.0630	0.0061	2.0210	2B	2.035	2.053	2.0709	2.0788	0.0079	2.1250
		3A	0.0000	2.1250	2.1136		2.0709	2.0664	0.0045	2.0228	3B	2.0350	2.0448	2.0709	2.0768	0.0059	2.1250
2⅛–16	UN	2A	0.0016	2.1234	2.1140		2.0828	2.0774	0.0054	2.0467	2B	2.057	2.071	2.0844	2.0914	0.0070	2.1250
		3A	0.0000	2.1250	2.1156		2.0844	2.0803	0.0041	2.0483	3B	2.0570	2.0658	2.0844	2.0896	0.0052	2.1250
2$\frac{3}{16}$–16	N	2A	0.0016	2.1859	2.1765		2.1453	2.1399	0.0054	2.1092	2B	2.120	2.134	2.1469	2.1539	0.0070	2.1875
		3A	0.0000	2.1875	2.1781		2.1469	2.1428	0.0041	2.1108	3B	2.1200	2.1283	2.1469	2.1521	0.0052	2.1875
		1A	0.0029	2.2471	2.2141		2.1028	2.0882	0.0146	1.9745	1B	2.009	2.045	2.1057	2.1247	0.0190	2.2500
2¼–4½	UNC	2A	0.0029	2.2471	2.2251	2.2141	2.1028	2.0931	0.0097	1.9745	2B	2.009	2.045	2.1057	2.1183	0.0126	2.2500
		3A	0.0000	2.2500	2.2280		2.1057	2.0984	0.0073	1.9774	3B	2.0090	2.0361	2.1057	2.1152	0.0095	2.2500
2¼–8	N	2A	0.0024	2.2476	2.2326	2.2251	2.1664	2.1584	0.0080	2.0942	2B	2.115	2.140	2.1688	2.1792	0.0104	2.2500
		3A	0.0000	2.2500	2.2350		2.1688	2.1628	0.0060	2.0966	3B	2.1150	2.1297	2.1688	2.1766	0.0078	2.2500
2¼–12	UN	2A	0.0018	2.2482	2.2368		2.1941	2.1880	0.0061	2.1460	2B	2.160	2.178	2.1959	2.2038	0.0079	2.2500
		3A	0.0000	2.2500	2.2386		2.1959	2.1914	0.0045	2.1478	3B	2.1600	2.1698	2.1959	2.2018	0.0059	2.2500

TABLE 14.1 Standard Limits of Size: Unified and American Screw Threads (*Continued*)

Nominal size and threads per inch	Series designation	External									Internal						
		Class	Allowance, in	Major diameter limits, in			Pitch diameter limits, in			Minor diameter, in	Class	Minor diameter limits, in		Pitch diameter limits, in			Major diameter, in
				Max*	Min	Min†	Max*	Min	Tolerance			Min	Max	Min	Max	Tolerance	Min
								$1\frac{3}{4}$–16 to 3–16									
$2\frac{1}{4}$–16	UN	2A	0.0016	2.2484	2.2390		2.2078	2.2024	0.0054	2.1717	2B	2.182	2.196	2.2094	2.2164	0.0070	2.2500
		3A	0.0000	2.2500	2.2406		2.2094	2.2053	0.0041	2.1733	3B	2.1820	2.1908	2.2094	2.2146	0.0052	2.2500
$2\frac{5}{16}$–16	N	2A	0.0017	2.3108	2.3014		2.2702	2.2647	0.0055	2.2341	2B	2.245	2.259	2.2719	2.2791	0.0072	2.3125
		3A	0.0000	2.3125	2.3031		2.2719	2.2678	0.0041	2.2358	3B	2.2450	2.2533	2.2719	2.2773	0.0054	2.3125
$2\frac{3}{8}$–12	UN	2A	0.0019	2.3731	2.3617		2.3190	2.3128	0.0062	2.2709	2B	2.285	2.303	2.3209	2.3290	0.0081	2.3750
		3A	0.0000	2.3750	2.3636		2.3209	2.3163	0.0046	2.2728	3B	2.2850	2.2948	2.3209	2.3269	0.0060	2.3750
$2\frac{3}{8}$–16	UN	2A	0.0017	2.3733	2.3639		2.3327	2.3272	0.0055	2.2966	2B	2.307	2.321	2.3344	2.3416	0.0072	2.3750
		3A	0.0000	2.3750	2.3656		2.3344	2.3303	0.0041	2.2983	3B	2.3070	2.3158	2.3344	2.3398	0.0054	2.3750
$2\frac{7}{16}$–16	N	2A	0.0017	2.4358	2.4264		2.3953	2.3897	0.0055	2.3591	2B	2.370	2.384	2.3969	2.4041	0.0072	2.4375
		3A	0.0000	2.4375	2.4281		2.3969	2.3928	0.0041	2.3608	3B	2.3700	2.3783	2.3969	2.4023	0.0054	2.4375
$2\frac{1}{2}$–4	UNC	1A	0.0031	2.4969	2.4612		2.3345	2.3190	0.0155	2.1902	1B	2.229	2.267	2.3376	2.3578	0.0202	2.5000
		2A	0.0031	2.4969	2.4731	2.4612	2.3345	2.3241	0.0104	2.1902	2B	2.229	2.267	2.3376	2.3511	0.0135	2.5000
		3A	0.0000	2.5000	2.4762		2.3376	2.3298	0.0078	2.1933	3B	2.2290	2.2594	2.3376	2.3477	0.0101	2.5000
$2\frac{1}{2}$–8	N	2A	0.0024	2.4976	2.4826	2.4751	2.4164	2.4082	0.0082	2.3442	2B	2.365	2.390	2.4188	2.4294	0.0106	2.5000
		3A	0.0000	2.5000	2.4850		2.4188	2.4127	0.0061	2.3466	3B	2.3650	2.3797	2.4188	2.4268	0.0080	2.5000
$2\frac{1}{2}$–12	UN	2A	0.0019	2.4981	2.4867		2.4440	2.4378	0.0062	2.3959	2B	2.410	2.428	2.4459	2.4540	0.0081	2.5000
		3A	0.0000	2.5000	2.4886		2.4459	2.4413	0.0046	2.3978	3B	2.4100	2.4198	2.4459	2.4519	0.0060	2.5000
$2\frac{1}{2}$–16	UN	2A	0.0017	2.4983	2.4889		2.4577	2.4522	0.0055	2.4216	2B	2.432	2.446	2.4594	2.4666	0.0072	2.5000
		3A	0.0000	2.5000	2.4906		2.4594	2.4553	0.0041	2.4233	3B	2.4320	2.4408	2.4594	2.4648	0.0054	2.5000
$2\frac{5}{8}$–12	UN	2A	0.0019	2.6231	2.6117		2.5690	2.5628	0.0062	2.5209	2B	2.535	2.553	2.5709	2.5790	0.0081	2.6250
		3A	0.0000	2.6250	2.6136		2.5709	2.5663	0.0046	2.5228	3B	2.5350	2.5448	2.5709	2.5769	0.0060	2.6250
$2\frac{5}{8}$–16	UN	2A	0.0017	2.6233	2.6139		2.5827	2.5772	0.0055	2.5466	2B	2.557	2.571	2.5844	2.5916	0.0072	2.6250
		3A	0.0000	2.6250	2.6156		2.5844	2.5803	0.0041	2.5483	3B	2.5570	2.5658	2.5844	2.5898	0.0054	2.6250
$2\frac{3}{4}$–4	UNC	1A	0.0032	2.7468	2.7111		2.5844	2.5686	0.0158	2.4401	1B	2.479	2.517	2.5876	2.6082	0.0206	2.7500
		2A	0.0032	2.7468	2.7230	2.7111	2.5844	2.5739	0.0105	2.4401	2B	2.479	2.517	2.5876	2.6013	0.0137	2.7500
		3A	0.0000	2.7500	2.7262		2.5876	2.5797	0.0079	2.4433	3B	2.4790	2.5094	2.5876	2.5979	0.0103	2.7500

Nominal Size	Series	Class	Allowance	Major Dia Max	Major Dia Min	Max*	Pitch Dia Max	Pitch Dia Min	Tol	Minor Dia Max	Class	Minor Dia Min	Minor Dia Max	Pitch Dia Min	Pitch Dia Max	Tol	Major Dia Min
2¾-8	N	2A	0.0025	2.7475	2.7325	2.7250	2.6663	2.6580	0.0083	2.5841	2B	2.615	2.640	2.6688	2.6796	0.0108	2.7500
		3A	0.0000	2.7500	2.7350		2.6688	2.6625	0.0063	2.5866	3B	2.6150	2.6297	2.6688	2.6769	0.0081	2.7500
2¾-12	UN	2A	0.0019	2.7481	2.7367		2.6940	2.6878	0.0062	2.6459	2B	2.660	2.678	2.6959	2.7040	0.0081	2.7500
		3A	0.0000	2.7500	2.7336		2.6959	2.6913	0.0046	2.6478	3B	2.6600	2.6698	2.6959	2.7019	0.0060	2.7500
2¾-16	UN	2A	0.0017	2.7483	2.7339		2.7077	2.7022	0.0055	2.6716	2B	2.682	2.696	2.7094	2.7166	0.0072	2.7500
		3A	0.0000	2.7500	2.7436		2.7094	2.7053	0.0041	2.6733	3B	2.6820	2.6908	2.7094	2.7148	0.0054	2.7500
2⅞-12	UN	2A	0.0019	2.8731	2.8617		2.8190	2.8127	0.0063	2.7709	2B	2.785	2.803	2.8209	2.8291	0.0082	2.8750
		3A	0.0000	2.8750	2.8636		2.8209	2.8162	0.0047	2.7728	3B	2.7850	2.7948	2.8209	2.8271	0.0062	2.8750
2⅞-16	UN	2A	0.0017	2.8733	2.8589		2.8327	2.8271	0.0056	2.7966	2B	2.807	2.821	2.8344	2.8417	0.0073	2.8750
		3A	0.0000	2.8750	2.8556		2.8344	2.8302	0.0042	2.7983	3B	2.8070	2.8158	2.8344	2.8399	0.0055	2.8750
3-4	UNC	1A	0.0032	2.9968	2.9611	2.9611	2.8344	2.8183	0.0161	2.6901	1B	2.729	2.767	2.8376	2.8585	0.0209	3.0000
		2A	0.0032	2.9968	2.9730		2.8344	2.8237	0.0107	2.6901	2B	2.729	2.767	2.8376	2.8515	0.0139	3.0000
		3A	0.0000	3.0000	2.9762		2.8375	2.8296	0.0080	2.6933	3B	2.7290	2.7594	2.8376	2.8480	0.0104	3.0000
3-8	N	2A	0.0026	2.9974	2.9624	2.9749	2.9162	2.9077	0.0085	2.8440	2B	2.865	2.890	2.9188	2.9299	0.0111	3.0000
		3A	0.0000	3.0000	2.9850		2.9188	2.9124	0.0064	2.8466	3B	2.8530	2.8797	2.9188	2.9271	0.0083	3.0000
3-12	UN	2A	0.0019	2.9981	2.9867		2.9440	2.9377	0.0063	2.8953	2B	2.910	2.928	2.9459	2.9541	0.0082	3.0000
		3A	0.0000	3.0000	2.9886		2.9459	2.9412	0.0047	2.8978	3B	2.9100	2.9198	2.9459	2.9521	0.0062	3.0000
3-16	UN	2A	0.0017	2.9983	2.9889		2.9577	2.9521	0.0056	2.9216	2B	2.932	2.946	2.9594	2.9667	0.0073	3.0000
		3A	0.0000	3.0000	2.9906		2.9594	2.9552	0.0042	2.9233	3B	2.9320	2.9408	2.9594	2.9649	0.0055	3.0000

* For class 2A threads having an additive finish, the maximum is increased to the basic size, the value being the same as for class 3A shown in this column.

† For unfinished hot-rolled material.

SOURCE: Extracted from National Bureau of Standards Handbook H28 (1967), Part I (Screw-Thread Standards for Federal Services), which is in general agreement with *American Standard Unified Screw Threads* (ASA B1.1-1930).

Thread engagement to prevent stripping. The calculation approach depends on the materials selected.

1. *Same materials* chosen for both external threaded part and internal threaded part:

$$E_L = \frac{2A_{ts}}{\pi D_m \{ \frac{1}{2} + [n(p_d - D_m)/\sqrt{3}] \}}$$

where E_L = length of engagement of the thread, in
  D_m = maximum minor diameter of internal thread, in
  n = number of threads per in
  A_{ts} = tensile stress area of screw thread as given in previous equation
  P_d = minimum pitch diameter of external thread, in

2. *Different materials,* i.e., internal threaded part of lower strength than external threaded part: Determine relative strength of external thread and internal thread from

$$R = \frac{A_{se}(S_e)}{A_{si}(S_i)}$$

where R = relative strength factor
  A_{se} = shear area of external thread, in^2
  A_{si} = shear area of internal thread, in^2
  S_e = tensile strength of external thread material, psi
  S_i = tensile strength of internal thread material, psi

If $R \leq 1$, the length of engagement as determined by the equation in item 1 (above) is adequate to prevent stripping of the internal thread. If $R > 1$, the length of engagement G to prevent internal thread strip is

$$G = E_L R$$

In the immediately preceding equation, A_{se} and A_{si} are the shear areas and are calculated as follows:

$$A_{se} = \pi n E_L D_m \left(\frac{1}{2n} + \frac{p_d - D_m}{\sqrt{3}} \right)$$

$$A_{si} = \pi n E_L D_m \left(\frac{1}{2n} + \frac{D_M - D_p}{\sqrt{3}} \right)$$

where D_p = maximum pitch diameter of internal thread, in
$\quad\quad D_M$ = minimum major diameter of external thread, in

Other symbols have been defined previously.

14.2.2   Calculating the pitch diameter of unified and metric threads

It is often necessary to find the pitch diameter of various unified (UN) and metric (M) thread sizes. This would be necessary for threads that are not listed in the preceding tables of thread sizes and where the thread is larger than that normally listed in handbooks. These would include threads on large bolts and threads on jack screws and lead screws used on various machinery or machine tools. To calculate the pitch diameters, refer to Fig. 14.15.

$$H = 0.5\sqrt{3} \times p = 0.866025p$$

where p is the pitch of the thread. In the unified system this is equal to the reciprocal of the number of threads per in (i.e., for a ⅜-16 thread, the pitch would be $1/16 = 0.0625$ in). For the metric system, the pitch is given in millimeters on the thread listing (i.e., on an M12 × 1.5 metric thread, the pitch would be 1.5 mm or 1.5 × 0.03937 in = 0.059055 in). In this equation, in addition, d is the basic diameter of the external thread (i.e., ⅜-16 would be 0.375 in; a no. 8-32 would be 0.164 in, etc.).

Example: Find the pitch diameter of a 0.375-16 UNC-3A thread using Fig. 14.15, where d is the basic outside diameter of the thread (= 0.375 in) and $H = 0.866025 \times p = 0.866025 \times 0.0625 = 0.054127$ in (for this case only).

We would next perform the following:

Pitch diameter = $[d/2 - (5H/8) + (H/4)] \times 2$

Pitch diameter = $[0.375/2 - (5 \times 0.054127/8) + (0.054127/4)] \times 2$

$\quad\quad\quad\quad = 0.3344$

If you check the basic pitch diameter for this thread in a table of pitch diameters, you will find that this is the correct answer when the thread is a class 3A and the pitch diameter is maximum. This may be verified by checking Table 14.1. Thus you may calculate any pitch diameter for the different classes of fits on any UN or M profile thread because the thread geometry is shown in Fig. 14.15. Pitch diameters for other classes or types of thread systems may be calculated when you know the basic thread geometry, as in this case for the UN and M thread systems.

The various thread systems used worldwide are shown in Fig. 7.52 and include ISO-M and UN, UNJ (controlled root radii), Whitworth (BSW), American buttress (7° face), NPT (American National Pipe Thread), BSPT (British Standard Pipe Thread), Acme (29°), Acme (stub 29°), API (taper 1:6), TR DIN 103, and RD DIN 405 (round).

14.3   Rivets

Rivets form a large class of fastening devices and are manufactured in many types and varieties and various materials. Rivets are made from carbon steels, aluminum alloys, brass, copper, bronze, and other materials agreed on by the manufacturer and purchaser.

Head forms for standard rivets include button, truss, brazier, coopers, oval, and flush. Flush-head rivets are provided in the following countersunk head angles: 60°, 78°, 90°, 100°, 120°, 144°, and 150°. The most common countersunk forms are the 90° and 100° types. To employ these types of rivets, a countersinking tool is used to cut the recess for the flush head. When the material is thin, processes known as *dimpling* and *double-dimpling* are employed.

Figure 14.21 shows the dimpling methods, in which a dimpling tool is used to produce the countersunk recess for the rivet. This method is used on aerospace vehicles to attach the "skins," or outer metal layers, on the craft. Modern methods of adhesive bonding are being used extensively in conjunction with spot welding to manufacture large sections of aerospace vehicles in lieu of riveting. But riveting is still used for many fastening applications throughout industry and the construction trades.

Rivets are made for blind-hole applications in the form of pull-stem pop rivets, drive-stem rivets, and explosive rivets. Other forms include solid-stem, tubular, and semitubular rivets.

Rivet edge distance.   The position or location of rivets from the edge of a part is normally $2d$, where d is the rivet shank diameter, with an absolute minimum of $1.5d$. The lateral spacing between rivets is known as *pitch* and is determined by the load requirements of the riveted joint. See Fig. 14.22 for an illustration of edge distance.

Rivet symbols on engineering drawings take the forms shown in Fig. 14.23. These symbols are encountered frequently on aerospace vehicle assembly drawings.

Other manufacturing techniques such as welding, spot welding, and structural adhesive bonding have replaced riveting in many applications. Nevertheless, riveting is still employed in a great number of industrial applications. The types and varieties of rivets are

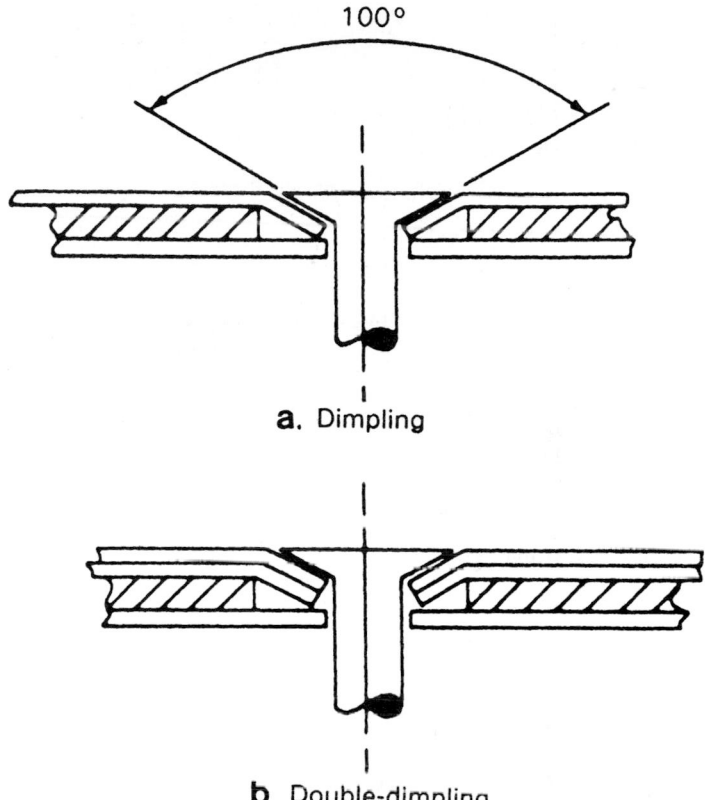

Figure 14.21 (*a*) Dimpling. (*b*) Double dimpling.

numerous, and a complete description and dimensional data, as well as materials specifications, are contained in the IFI handbooks on fasteners in both U.S. Customary and metric units.

14.3.1 General sizing of rivets

For rivet sizing (general, noncritical applications), the following approximations may be used:

$$d = 1.25\sqrt{t} \qquad \text{to} \qquad 1.45\sqrt{t}$$

where d is the diameter of the rivet shank in inches or millimeters (use next larger size), and t is the material thickness in inches.

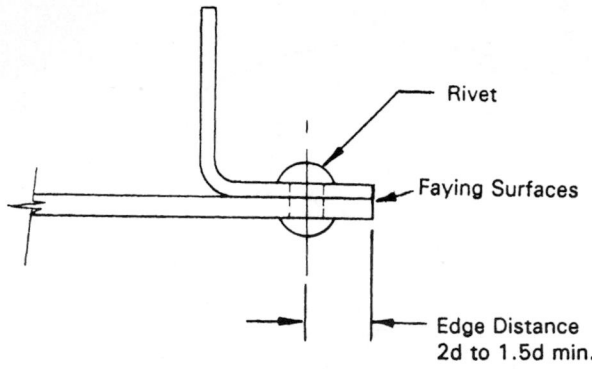

d = Rivet shank diameter

Figure 14.22   Rivet edge distance.

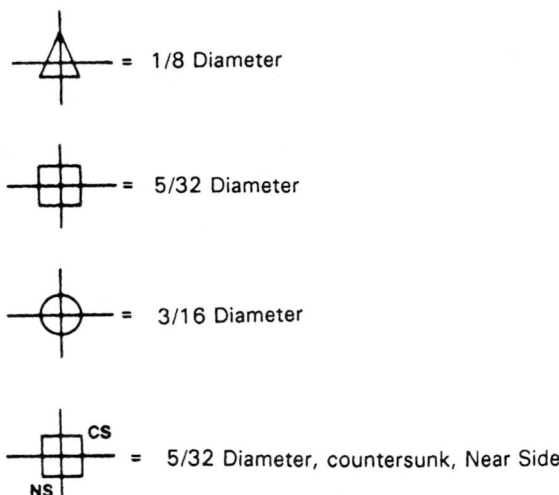

Figure 14.23   Standard rivet symbols.

14.4   Pins

Pins of various types are used throughout industry. Pins are available in different sizes, styles, and materials. This section details the most popular pins used in all types of design and assembly applications. The common-usage pins include

- Clevis pins

- Cotter pins

- Spring pins (roll pins)

- Spiral spring pins

- Taper pins

- Dowel pins

- Grooved pins

- Quick-release pins

14.4.1   Clevis pins

Clevis pins are used where a quickly detachable pin is of benefit from a design and manufacturing standpoint. Figure 14.24 shows the sizes and dimensions of standard clevis pins. The clevis pin may be made of various materials, such as carbon steel, stainless steel, aluminum alloy, and other materials.

14.4.2   Cotter pins

Cotter pins are very common and economical fastening devices that see countless uses and applications throughout industry. The two most common types of cotter pins and their sizes and dimensions are shown in Fig. 14.25. The cotter pin is normally used with the standard clevis pin.

14.4.3   Spring pins (roll pins)

The slotted type of spring pin (sometimes called a *roll pin*) is used in applications where economy is important. This type of pin has been used to replace the tapered pin, dowel pin, and grooved pin in many applications because of its low cost and ease of preparation and assembly into holes with loose tolerances. Spring-temper carbon steel is usually used for these pins, but they are also available in other materials. Figure 14.26 shows the sizes, dimensions, and recommended hole diameters for spring-pin applications, along with the double-shear values for design reference. Placement of the slot in relation to shock loads for this type of pin can affect its performance and shock-absorbing qualities.

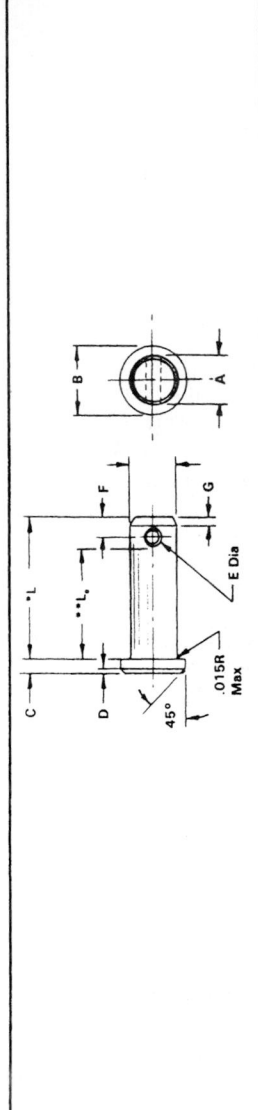

| Basic Pin Diameter | A Shank Diameter | | B Head Diameter | C Head Height | D Head Chamfer | E Hole Diameter | | F | G | Cotter Pin |
	Max	Min	Max	Max	± 0.01	Max	Min			Size
0.188	0.186	0.181	0.32	0.07	0.02	0.088	0.073	0.09	0.055	1/16
0.250	0.248	0.243	0.38	0.10	0.03	0.088	0.073	0.09	0.055	1/16
0.312	0.311	0.306	0.44	0.10	0.03	0.119	0.104	0.12	0.071	3/32
0.375	0.373	0.368	0.51	0.13	0.03	0.119	0.104	0.12	0.071	3/32
0.438	0.436	0.431	0.57	0.16	0.04	0.119	0.104	0.12	0.071	3/32
0.500	0.496	0.491	0.63	0.16	0.04	0.151	0.136	0.15	0.089	1/8
0.625	0.621	0.616	0.82	0.21	0.06	0.151	0.136	0.15	0.089	1/8
0.750	0.746	0.741	0.94	0.26	0.07	0.182	0.167	0.18	0.110	5/32
0.875	0.871	0.866	1.04	0.32	0.09	0.182	0.167	0.18	0.110	5/32
1.000	0.996	0.991	1.19	0.35	0.10	0.182	0.167	0.18	0.110	5/32

*L = Total length under head
**L_e = Length of effective grip
Effective grip lengths must be selected from manufacturer's catalogs.

Figure 14.24 Standard clevis-pin dimensions.

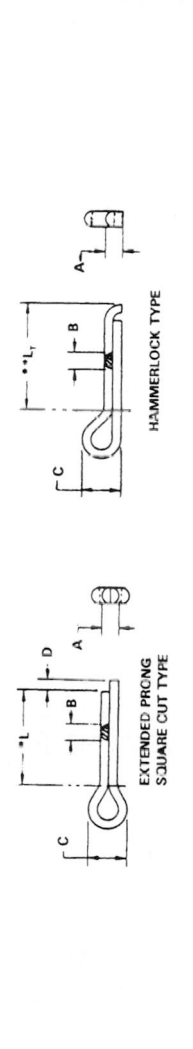

EXTENDED PRONG
SQUARE CUT TYPE

HAMMERLOCK TYPE

Basic Pin Diameter	Total Shank Diameter A	Wire Width Maximum B	Head Diameter Minimum C	Extended Prong Length, Minimum D	Recommended Hole Size
0.031	0.032	0.032	0.06	0.01	0.047
0.047	0.048	0.043	0.09	0.02	0.062
0.062	0.060	0.060	0.12	0.03	0.078
0.078	0.076	0.076	0.16	0.04	0.094
0.094	0.090	0.090	0.19	0.04	0.109
0.109	0.104	0.104	0.22	0.05	0.125
0.125	0.120	0.120	0.25	0.06	0.141
0.141	0.134	0.134	0.28	0.06	0.156
0.156	0.150	0.150	0.31	0.06	0.172
0.188	0.176	0.176	0.38	0.07	0.203
0.219	0.207	0.207	0.44	0.09	0.234
0.250	0.225	0.225	0.50	0.10	0.266
0.312	0.280	0.280	0.62	0.11	0.312
0.375	0.335	0.335	0.75	0.14	0.375
0.438	0.406	0.406	0.88	0.16	0.438
0.500	0.473	0.473	1.00	0.20	0.500
0.625	0.598	0.598	1.25	0.23	0.625
0.750	0.723	0.723	1.50	0.30	0.750
				0.36	

* L = Length
** L_T = Total length
Allow extra length for spreading and securing
Available lengths to be selected from manufacturer's catalogs.

Figure 14.25 Cotter-pin dimensions.

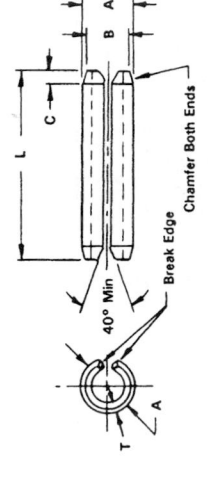

Basic Pin Diameter	Pin Diameter Max	Min	Chamfer Dia.	Chamfer Lth.	Stock Thickness	Hole Diameter Recommended Max	Min	Double Shear Load, Lb. ♦ AISI 1070 1095 & 420	AISI 302	Beryllium Copper
	A		B	C	T					
0.062	0.069	0.066	0.059	0.028	0.012	0.065	0.062	425	350	270
0.078	0.086	0.083	0.075	0.032	0.018	0.081	0.078	650	550	400
0.094	0.103	0.099	0.091	0.038	0.022	0.097	0.094	1,000	800	660
0.125	0.135	0.131	0.122	0.044	0.028	0.129	0.125	2,100	1,500	1,200
0.141	0.149	0.145	0.137	0.044	0.028	0.144	0.140	2,200	1,600	1,400
1.156	0.167	0.162	0.151	0.048	0.032	0.160	0.156	3,000	2,000	1,800
0.188	0.199	0.194	0.182	0.055	0.040	0.192	0.187	4,400	2,800	2,600
0.219	0.232	0.226	0.214	0.065	0.048	0.224	0.219	5,700	3,550	3,700
0.250	0.264	0.258	0.245	0.065	0.048	0.256	0.250	7,700	4,600	4,500
0.312	0.328	0.321	0.306	0.080	0.062	0.318	0.312	11,500	7,100	6,800
0.375	0.392	0.385	0.368	0.095	0.077	0.382	0.375	17,600	10,000	10,100
0.438	0.456	0.448	0.430	0.095	0.077	0.445	0.437	20,000	12,000	12,200
0.500	0.521	0.513	0.485	0.110	0.094	0.510	0.500	25,800	15,500	16,800
0.625	0.650	0.640	0.608	0.125	0.125	0.636	0.625	46,000	18,800	.....
0.750	0.780	0.769	0.730	0.150	0.150	0.764	0.750	66,000	23,200	.....

Length L, is selected from manufacturer's catalogs. ♦ Other materials may be available.

Figure 14.26 Spring-pin dimensions.

14.4.4   Spiral spring pins (coiled spring pins)

The spiral spring pin was developed after the standard slotted spring pin in order to provide more shock resistance and a tighter fit in drilled holes. These pins are made for standard, light-duty, and heavy-duty applications and are produced in various materials to suit the application. Studies of slotted spring pins and spiral spring pins have found that the heavy-duty slotted spring pin will withstand more shock-loading cycles than the standard spiral spring pin. Figure 14.27 shows the sizes, dimensions, recommended hole sizes, and double-shear values for these pins.

14.4.5   Taper pins

The standard taper pin was used widely before the advent of spring pins and grooved pins. This pin type is still used in some industrial applications. Figure 14.28 shows the standard sizes and dimensions of taper pins.

14.4.6   Dowel pins (hardened and ground machine type)

The hardened and ground dowel pin is used widely in tooling applications and a wide range of indexing applications where great accuracy is required. Figure 14.29 shows the available standard sizes and dimensions of this important class of pin.

14.4.7   Grooved pins

The solid groove pin is one of the most popular fastening devices in use today. Its applications are limitless, and its use is economical, practical, and easy to implement. The hole diameter required for its application is not as critical as that for other solid pins. Seven standard styles or types are recognized as American national standards. Figure 14.30 shows the different types, sizes, and dimensions. Figure 14.31 shows the standard lengths. Figure 14.32 shows the recommended hole sizes.

14.4.8   Quick-release pins

The quick-release pin is available in many types and finds use in design applications where a quick release action of a fastened joint

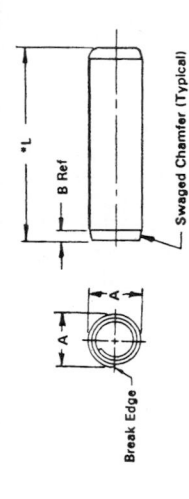

Figure 14.27 Spiral-spring-pin dimensions.

Basic Pin Diameter	Pin Diameter Std Duty Max	Pin Diameter Heavy Duty Max	Pin Diameter Light Duty Max	Chamfer Length Ref.	Hole Size Max	Hole Size Min	Double Shear Load, Min, Lb ♦ Std Duty 1070, 1095 and 420	Std Duty 302	Heavy Duty 1070, 1095 and 420	Heavy Duty 302	Light Duty 1070, 1095 and 420	Light Duty 302
0.031	0.035	.....	.....	0.024	0.032	0.031	75	60	.....	.....	.....	.....
0.047	0.052	.....	.....	0.024	0.048	0.046	170	140	.....	.....	.....	.....
0.062	0.072	0.070	0.073	0.028	0.065	0.061	300	250	450	350	.....	135
0.078	0.088	0.086	0.089	0.032	0.081	0.077	475	400	700	550	.....	225
0.094	0.105	0.103	0.106	0.038	0.097	0.093	700	550	1,000	800	375	300
0.109	0.120	0.118	0.121	0.038	0.112	0.108	950	750	1,400	1,125	525	425
0.125	0.138	0.136	0.139	0.044	0.129	0.124	1,250	1,000	2,100	1,700	675	550
0.166	0.171	0.168	0.172	0.048	0.160	0.155	1,925	1,550	3,000	2,400	1,100	875
0.188	0.205	0.202	0.207	0.055	0.192	0.185	2,800	2,250	4,400	3,500	1,500	1,200
0.219	0.238	0.235	0.240	0.065	0.224	0.217	3,800	3,000	5,700	4,600	2,100	1,700
0.250	0.271	0.268	0.273	0.065	0.256	0.247	5,000	4,000	7,700	6,200	2,700	2,200
0.312	0.337	0.334	0.339	0.090	0.319	0.308	7,700	6,200	11,500	9,200	4,440	3,500
0.375	0.403	0.400	0.405	0.095	0.383	0.370	11,200	9,000	17,600	14,000	6,000	5,000
0.438	0.469	0.466	0.471	0.095	0.446	0.431	15,200	13,000	22,500	18,000	8,400	6,700
0.500	0.535	0.532	0.537	0.110	0.510	0.493	20,000	16,000	30,000	24,000	11,000	8,800
0.625	0.661	0.658	.....	0.125	0.635	0.618	31,000	25,000	46,000	37,000	.....	.....
0.750	0.787	0.784	.....	0.150	0.760	0.743	45,000	36,000	66,000	53,000	.....	.....

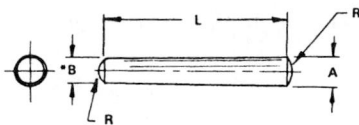

Pin Size Number and Basic Pin Dia		A Major Diameter (Large end)				R End Crown Radius Maximum
		Commercial Class		Precision Class		
		Max	Min	Max	Min	
7/0	0.0625	0.0638	0.0618	0.0635	0.0625	0.072
6/0	0.0780	0.0793	0.0773	0.0790	0.0780	0.088
5/0	0.0940	0.0953	0.0933	0.0950	0.0940	0.104
4/0	0.1090	0.1103	0.1083	0.1100	0.1090	0.119
3/0	0.1250	0.1263	0.1243	0.1260	0.1250	0.135
2/0	0.1410	0.1423	0.1403	0.1420	0.1410	0.151
0	0.1560	0.1573	0.1553	0.1570	0.1560	0.166
1	0.1720	0.1733	0.1713	0.1730	0.1720	0.182
2	0.1930	0.1943	0.1923	0.1940	0.1930	0.203
3	0.2190	0.2203	0.2183	0.2200	0.2190	0.229
4	0.2500	0.2513	0.2493	0.2510	0.2500	0.260
5	0.2890	0.2903	0.2883	0.2900	0.2890	0.299
6	0.3410	0.3423	0.3403	0.3420	0.3410	0.351
7	0.4090	0.4103	0.4083	0.4100	0.4090	0.419
8	0.4920	0.4933	0.4913	0.4930	0.4920	0.502
9	0.5910	0.5923	0.5903	0.5920	0.5910	0.601
10	0.7060	0.7073	0.7053	0.7070	0.7060	0.716
11	0.8600	0.8613	0.8593	.....	...	0.870
12	1.0320	1.0333	1.0313	.....	.....	1.042
13	1.2410	1.2423	1.2403	.....	.....	1.251
14	1.5210	1.5223	1.5203	.....	.....	1.531

* B dimension varies per length. Length L to be selected from manufacturer's catalogs.

Figure 14.28   Taper-pin dimensions.

or part is required. Most of these types of pins contain a push-button release action that allows the pin to be removed by a straight pulling action after the release button is pressed. The quick-release pin series is available in carbon steel, alloy steels, and stainless steels. Applications include tool engineering, aerospace vehicles, and many other applications where a strong, quick-release fastener is required. The quick-release pin is normally used for shear loading applications only.

(Text continued on page 903.)

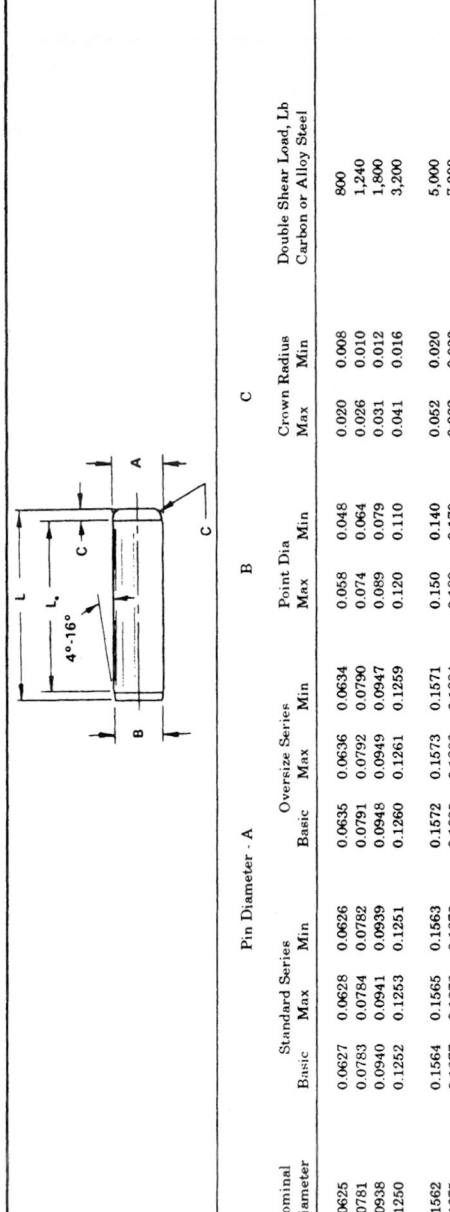

Nominal Diameter	Pin Diameter - A						B Point Dia		C Crown Radius		Double Shear Load, Lb Carbon or Alloy Steel
	Standard Series			Oversize Series							
	Basic	Max	Min	Basic	Max	Min	Max	Min	Max	Min	
0.0625	0.0627	0.0628	0.0626	0.0635	0.0636	0.0634	0.058	0.048	0.020	0.008	800
0.0781	0.0783	0.0784	0.0782	0.0791	0.0792	0.0790	0.074	0.064	0.026	0.010	1,240
0.0938	0.0940	0.0941	0.0939	0.0948	0.0949	0.0947	0.089	0.079	0.031	0.012	1,800
0.1250	0.1252	0.1253	0.1251	0.1260	0.1261	0.1259	0.120	0.110	0.041	0.016	3,200
0.1562	0.1564	0.1565	0.1563	0.1572	0.1573	0.1571	0.150	0.140	0.052	0.020	5,000
0.1875	0.1877	0.1878	0.1876	0.1885	0.1886	0.1884	0.180	0.170	0.062	0.023	7,200
0.2500	0.2502	0.2503	0.2501	0.2510	0.2511	0.2509	0.240	0.230	0.083	0.031	12,800
0.3125	0.3127	0.3128	0.3126	0.3135	0.3136	0.3134	0.302	0.290	0.104	0.039	20,000
0.3750	0.3752	0.3753	0.3751	0.3760	0.3761	0.3759	0.365	0.350	0.125	0.047	28,700
0.4375	0.4377	0.4378	0.4376	0.4385	0.4386	0.4384	0.424	0.409	0.146	0.055	39,100
0.5000	0.5002	0.5003	0.5001	0.5010	0.5011	0.5009	0.486	0.471	0.167	0.063	51,000
0.6250	0.6252	0.6253	0.6251	0.6260	0.6261	0.6259	0.611	0.595	0.208	0.078	79,800
0.7500	0.7502	0.7503	0.7501	0.7510	0.7511	0.7509	0.735	0.715	0.250	0.094	114,000
0.8750	0.8752	0.8753	0.8751	0.8760	0.8761	0.8759	0.860	0.840	0.293	0.109	156,000
1.0000	1.0002	1.0003	1.0001	1.0010	1.0011	1.0009	0.980	0.960	0.333	0.125	204,000

Note: Sizes 0.0781 and 0.1562 diameter not recommended for new design.

L = Total pin length; L$_e$ = Length of engagement.

Dowel pins listed are available in nominal lengths from 0.1875 to 6.000 inches.

Consult the manufacturer's catalogs for available lengths.

Figure 14.29 Dowel-pin dimensions.

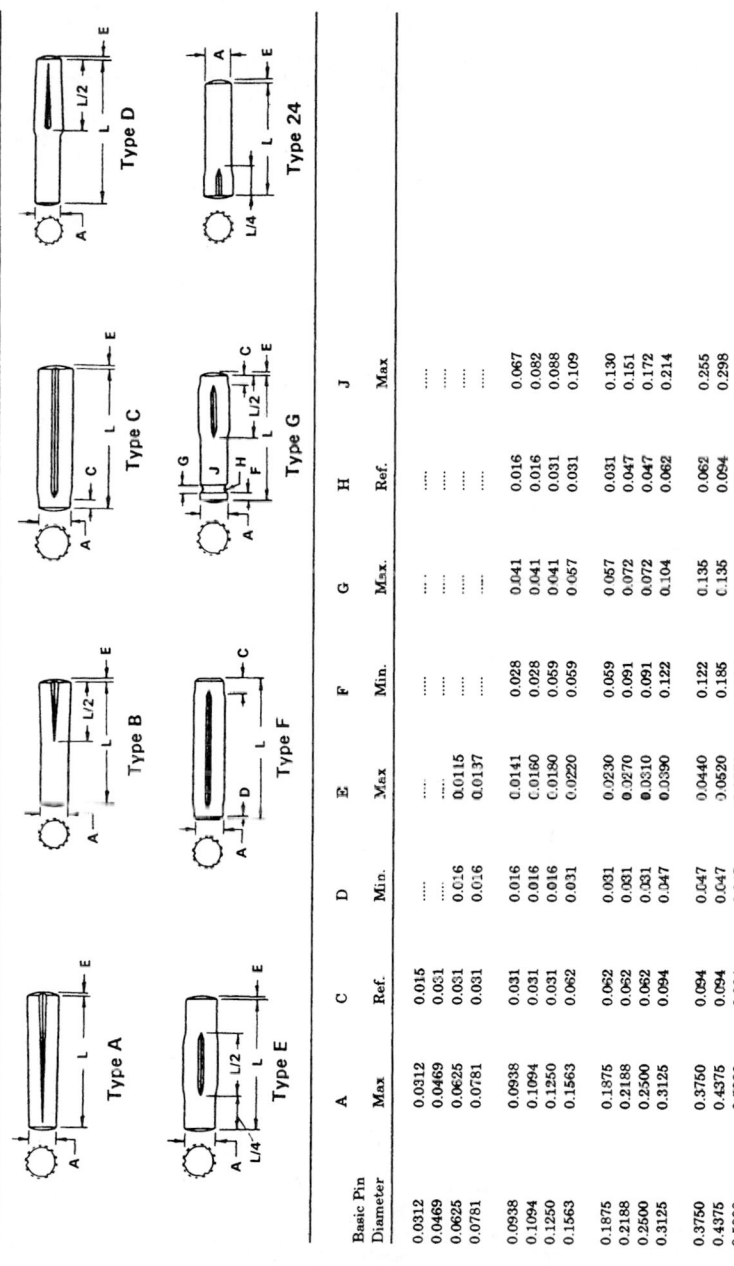

Type A · Type B · Type C · Type D · Type E · Type F · Type G · Type 24

Basic Pin Diameter	A Max	C Ref.	D Min.	E Max	F Min.	G Max.	H Ref.	J Max
0.0312	0.0312	0.015	....	....	....	....	....	....
0.0469	0.0469	0.031	....	....	....	....	....	....
0.0625	0.0625	0.031	0.016	0.0115	....	....	....	....
0.0781	0.0781	0.031	0.016	0.0137	....	....	....	....
0.0938	0.0938	0.031	0.016	0.0141	0.028	0.041	0.016	0.067
0.1094	0.1094	0.031	0.016	0.0160	0.028	0.041	0.016	0.082
0.1250	0.1250	0.031	0.016	0.0180	0.059	0.041	0.031	0.088
0.1563	0.1563	0.062	0.031	0.0220	0.059	0.057	0.031	0.109
0.1875	0.1875	0.062	0.031	0.0230	0.059	0.057	0.031	0.130
0.2188	0.2188	0.062	0.031	0.0270	0.091	0.072	0.047	0.151
0.2500	0.2500	0.062	0.031	0.0310	0.091	0.072	0.047	0.172
0.3125	0.3125	0.094	0.047	0.0390	0.122	0.104	0.062	0.214
0.3750	0.3750	0.094	0.047	0.0440	0.122	0.135	0.062	0.255
0.4375	0.4375	0.094	0.047	0.0520	0.185	0.135	0.094	0.298
0.5000	0.5000	0.094	0.047	0.0570	0.185	0.135	0.094	0.317

Figure 14.30 Grooved-pin dimensions.

901

Nominal Length	Nominal Size														
	1/32	3/64	1/16	5/64	3/32	7/64	1/8	5/32	3/16	7/32	1/4	5/16	3/8	7/16	1/2
1/8	Y	Y	Y												
1/4	Y	Y	Y	Y	Y	Y	Y								
3/8	Y	Y	Y	Y	X	X	X	X	X						
1/2	Y	Y	Y	Y	X	X	X	X	X	X	X				
5/8		Y	Y	Y	X	X	X	X	X	X	X	X			
3/4		Y	Y	X	X	X	X	X	X	X	X	X	X		
7/8			Y	Y	X	X	X	X	X	X	X	X	X	X	
1			Y	Y	X	X	X	X	X	X	X	X	X	X	X
1-1/4					X	X	X	X	X	X	X	X	X	X	X
1-1/2							X	X	X	X	X	X	X	X	X
1-3/4							X	X	X	X	X	X	X	X	X
2							X	X	X	X	X	X	X	X	X
2-1/4								X	X	X	X	X	X	X	X
2-1/2									X	X	X	X	X	X	X
2-3/4									X	X	X	X	X	X	X
3									X	X	X	X	X	X	X
3-1/4											X	X	X	X	X
3-1/2												X	X	X	X
3-3/4													X	X	X
4													X	X	X
4-1/4													X	X	X
4-1/2														X	X

Note: Carbon steel pins are normally available in the marked sizes by X and Y. X designates all types of pins; Y designates all types except type G. Other lengths may be available from different manufacturers.

Figure 14.31 Sizes and lengths of standard grooved pins.

Recommended Hole Sizes For Grooved Pins			
Nominal Pin Size	Drill Size	Hole Diameter	
		Max	Min
1/32	1/32	0.0324	0.0312
3/64	3/64	0.0482	0.0469
1/16	1/16	0.0640	0.0625
5/64	5/64	0.0798	0.0781
3/32	3/32	0.0956	0.0938
7/64	7/64	0.1113	0.1094
1/8	1/8	0.1271	0.1250
5/32	5/32	0.1587	0.1563
3/16	3/16	0.1903	0.1875
7/32	7/32	0.2219	0.2188
1/4	1/4	0.2534	0.2500
5/16	5/16	0.3166	0.3125
3/8	3/8	0.3797	0.3750
7/16	7/16	0.4428	0.4375
1/2	1/2	0.5060	0.5000

Figure 14.32 Hole sizes for grooved pins.

14.5   Retaining Rings

Retaining rings have many uses in the design and maintenance of modern equipment, including

- Shaft retention
- Bearing retention
- Retention of parts on shafts
- Spring retention
- Vertical and horizontal shaft support

The standard retaining ring is normally made of spring steel, although other materials such as beryllium-copper alloys and stainless steels are sometimes used. Figure 14.33 shows some of the main retaining types.

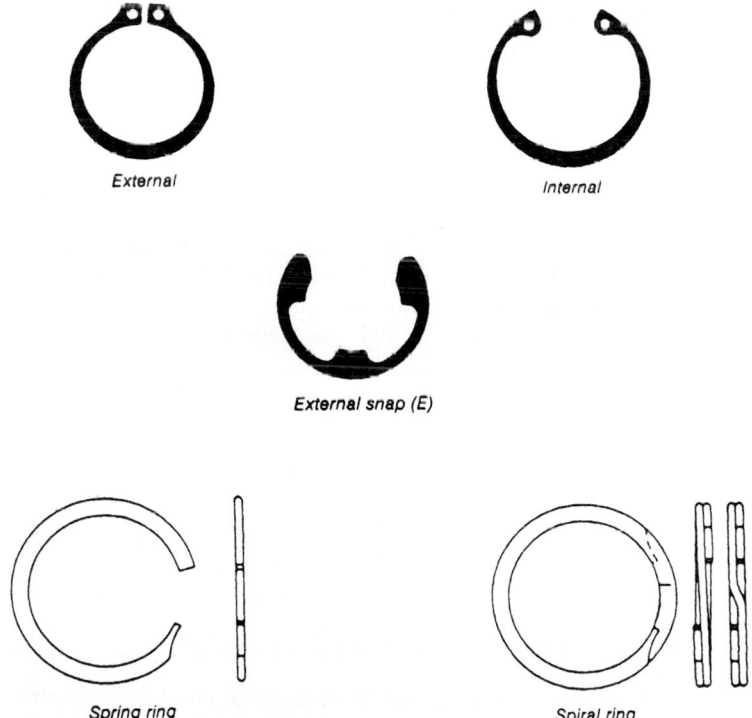

External

Internal

External snap (E)

Spring ring

Spiral ring

Figure 14.33 Retaining-ring styles.

The allowable thrust loads for each type of retaining ring must be obtained from the retaining ring manufacturers' handbooks. The calculation procedures for various applications are also shown in these handbooks. There are at least 25 to 30 different types of retaining rings. The critical dimensions for machining the grooves that hold the retaining ring in place are also obtained from the manufacturers' handbooks. Figure 14.34 shows a typical page from the Waldes retaining ring handbook. This sample page is for the 5100 external type ring, and the typical machining dimensions for the retaining groove may be seen under "Groove Dimensions" in the table. Dimensions and application data are also shown.

Note that electroplating thickness can interfere with the proper functioning of the ring in the groove. Overplating can cause the ring to come out of the retaining groove during operation of the mechanism on which the ring is used. Therefore, specify a maximum plating thickness of 0.0002 in (0.2 mil) or allow additional clearance for the ring in the groove when the part is to be electroplated.

Figure 14.35 shows a retaining-ring interchangeability chart for various manufacturers and also lists the appropriate military standard numbers.

14.5.1 X-washers (split washers)

An X-washer is a unique type of retaining ring. It is used in applications similar to those for the standard spring retaining ring except that it may be installed with a pair of common pliers, as shown in Fig. 14.36. The dimensions and sizes of the presently available X-washer series are also shown in this figure. The X-washer is sometimes called a *split washer* and may be listed as such in fastener catalogs, although that is an incorrect name.

An X-washer normally is used for a one-time-only application. Once the washer is installed on a shaft and removed with pliers, it should not be reused. Although applicable only for one-time installations, these devices are suitable and economical for many applications in product design and manufacturing. The fact that these washers may be installed and removed with common pliers, and not special tools, is an asset in their application.

14.6 Set-Collars, Clamp-Collars, and Split-Collars

These devices have many uses in product design and manufacturing. They are used to retain shafts, to withstand high-thrust loads, and to space parts, and they are often welded to plate cams, plate

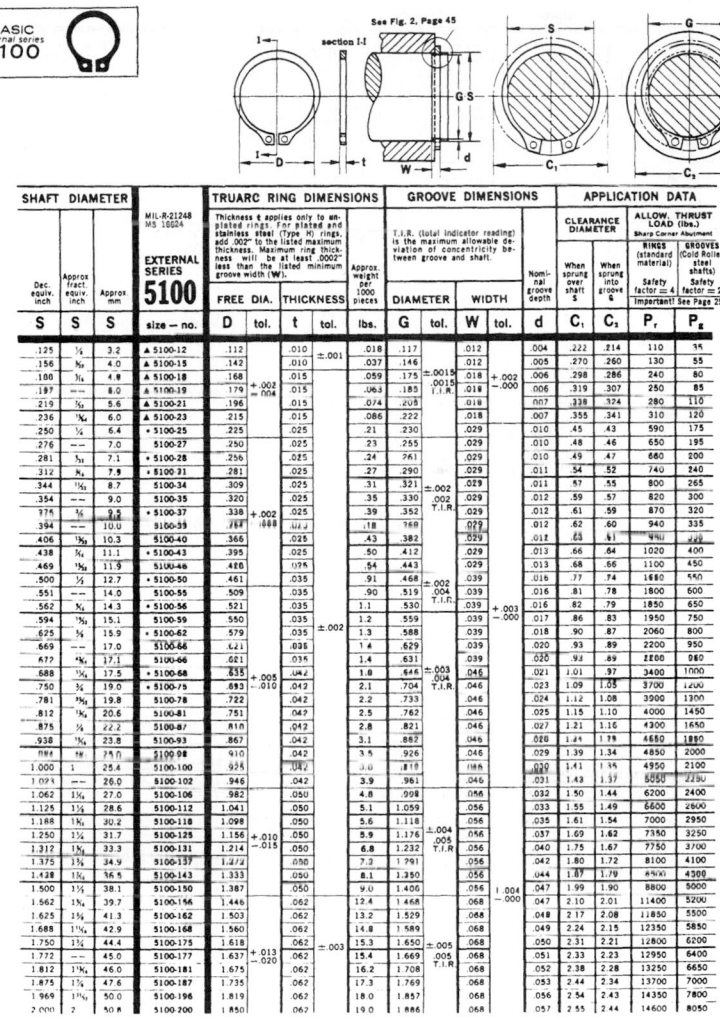

Figure 14.34 Sample Truarc ring data sheet.

sprockets, and indexing plates to form a hub by which the part is retained on a shaft. The standard set-collar is simply a machined ring with set screws that hold the ring on a shaft or cylindrical end of a part. Figure 14.37 shows the standard set-collar sizes and dimensions. These collars normally are made from low-carbon steel with zinc plating as a finish.

		Roto Clip	Waldes	I.R.R.	Anderton	Mil Standard
![E-ring symbol]	E	5133	1000	1500	16633
 ![BE-ring symbol]	BE	5131	1001	1501	16634
![RE-ring symbol]	RE	5144	1200	1540	3215
 ![C-ring symbol]	C	5103	2000	1800	16632
![HO-ring symbol]	HO	5000	3000	1300	16625
 ![BHO-ring symbol]	BHO	5001	3001	1301	16629
![VHO-ring symbol]	VHO	5002	—	1302	16631
 ![SH-ring symbol]	SH	5100	3100	1400	16624
![BSH-ring symbol]	BSH	5101	3101	1401	16628
![VSH-ring symbol]	VSH	5102	—	1402	16630
 ![HOI-ring symbol]	HOI	5008	4000	1308	16627
![SHI-ring symbol]	SHI	5108	4100	1408	16626
 ![SHF-ring symbol]	SHF	5555	7100	1440	90707
![SHR-ring symbol]	SHR	5160	—	1460	3217
 ![SHM-ring symbol]	SHM	5560	—	—	—

Figure 14.35   Retaining-ring interchangeability table.

Clamp-collars are similar to set-collars, except that they have a slit through one wall and are clamped on a shaft by tightening the clamp screw provided on the collar. These types of collars are also made with internal threads for adjustment and a more positive clamping force. Clamp-collars normally are made of carbon steels,

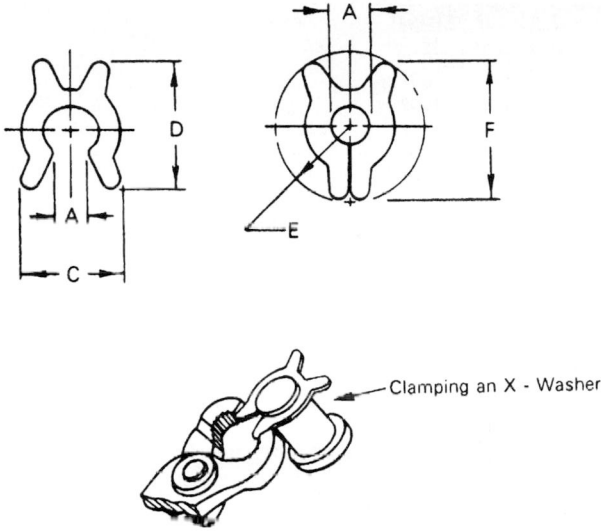

Clamping an X - Washer

X - Washer Dimensions

A	B*	C	D	E	F
.086	.025	.320	.406	.210	.406
.098	.055	.364	.490	.297	.475
.130	.055	.430	.575	.359	.556
.164	.065	.523	.687	.422	.665
.190	.065	.593	.745	.437	.730
.222	.075	.622	.776	.469	.775
.256	.075	.698	.905	.500	.890
.285	.075	.822	.986	.563	.984
.317	.089	.872	1.100	.609	1.078
.347	.089	.948	1.190	.688	1.188
.381	.089	1.060	1.297	.797	1.281

NOTE: B* = thickness, inches

Figure 14.36 X-washer dimensional sizes. X-washers are often called *split washers*.

DIMENSIONS			
BORE	**O. D.**	**WIDTH**	**SET SCREW**
3/16	7/16	1/4	8-32
1/4	1/2	9/32	10-32
5/16	5/8	11/32	10-32
3/8	3/4	3/8	1/4-20
7/16	7/8	7/16	1/4-20
1/2	1"	7/16	1/4-20
9/16	1"	7/16	1/4-20
5/8	1-1/8	1/2	5/16-18
11/16	1-1/4	9/16	5/16-18
3/4	1-1/4	9/16	5/16-18
13/16	1-5/16	9/16	5/16-18
7/8	1-1/2	9/16	5/16-18
15/16	1-5/8	9/16	5/16-18
1"	1-5/8	5/8	5/16-18
1-1/16	1-3/4	5/8	5/16-18
1-1/8	1-3/4	5/8	5/16-18
1-3/16	2"	11/16	3/8-16
1-1/4	2"	11/16	3/8-16
1-5/16	2-1/8	11/16	3/8-16
1-3/8	2-1/8	3/4	3/8-16
1-7/16	2-1/4	3/4	3/8-16
1-1/2	2-1/4	3/4	3/8-16
1-9/16	2-1/2	13/16	3/8-16
1-5/8	2-1/2	13/16	3/8-16
1-11/16	2-1/2	13/16	3/8-16
1-3/4	2-3/4	7/8	1/2-13
1-13/16	2-3/4	7/8	1/2-13
1-7/8	2-3/4	7/8	1/2-13
1-15/16	3"	7/8	1/2-13
2"	3"	7/8	1/2-13
2-1/16	3"	7/8	1/2-13
2-1/8	3"	7/8	1/2-13
2-3/16	3-1/4	15/16	1/2-13
2-1/4	3-1/4	15/16	1/2-13
2-5/16	3-1/4	15/16	1/2-13
2-3/8	3-1/4	15/16	1/2-13
2-7/16	3-1/2	1"	1/2-13
2-1/2	3-1/2	1"	1/2-13
2-9/16	3-3/4	1-1/8	1/2-13
2-5/8	3-3/4	1-1/8	1/2-13
2-11/16	4"	1-1/8	1/2-13
2-3/4	4"	1-1/8	1/2-13
2-13/16	4-1/4	1-1/8	1/2-13
2-7/8	4-1/4	1-1/8	1/2-13
2-15/16	4-1/4	1-1/8	1/2-13
3"	4-1/4	1-1/8	1/2-13

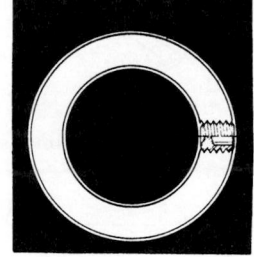

Figure 14.37   Set-collars. *(Ruland, Inc.)*

aluminum alloys, and stainless steels. Figure 14.38 shows the sizes and dimensions of standard clamp-collars, and Fig. 14.39 shows the data for the internally threaded clamp-collar series.

Split-collars and threaded split-collars are also available and used widely in machine design. The data for these types of clamping collars are shown in Figs. 14.40 and 14.41.

Note: Figures 14.37 through 14.41 were extracted from the Ruland catalog of collars and couplings (Ruland Manufacturing Company, Inc., Watertown, MA 02172).

14.7 Machinery Bushings, Shims, and Arbor Spacers

14.7.1 Machinery bushings

Machinery bushings are a special form of flat washer commonly made of low-carbon mild steel. They are used as spacers between gears, pulleys, and sprockets and as filler spacers for parts mounted on shafts. These bushings are manufactured in the following gauges and diameters:

- 18 gauge, 0.048 in
- 14 gauge, 0.075 in
- 10 gauge, 0.134 in
- ³⁄₁₆ in, 0.1875 in

Inside diameters range from 0.500 through 3.00 in.

14.7.2 Steel shims

Steel shims are thin steel rings with a plain center hole that are used for building up gears and bearings and to provide proper clearance between mating parts. Figure 14.42 lists the sizes and thicknesses normally available for steel shims.

14.7.3 Steel arbor spacers

Steel arbor spacers are thin steel rings with a keyway center hole that are used for accurately spacing milling cutters, slitter knives, and gang saws on keyway arbors. Steel shims and steel arbor spacers are made of AISI 1010, fully hardened, cold-rolled low-carbon steel. Figure 14.42 also lists the sizes and thicknesses of steel arbor spacers.

DIMENSIONS			
BORE	**O.D.**	**WIDTH**	**CLAMP SCREW**
1/8	1/2	.235	4-40
3/16	9/16	.235	4-40
1/4	5/8	.281	4-40
5/16	11/16	.281	4-40
3/8	7/8	.343	6-32
7/16	15/16	.343	6-32
1/2	1-1/8	.406	8-32
9/16	1-1/4	.437	10-32
5/8	1-5/16	.437	10-32
11/16	1-3/8	.437	10-32
3/4	1-1/2	1/2	1/4-28
13/16	1-5/8	1/2	1/4-28
7/8	1-5/8	1/2	1/4-28
15/16	1-3/4	1/2	1/4-28
1	1-3/4	1/2	1/4-28
1-1/16	1-7/8	1/2	1/4-28
1-1/8	1-7/8	1/2	1/4-28
1-3/16	2-1/16	1/2	1/4-28
1-1/4	2-1/16	1/2	1/4-28
1-5/16	2-1/8	9/16	1/4-28
1-3/8	2-1/4	9/16	1/4-28
1-7/16	2-1/4	9/16	1/4-28
1-1/2	2-3/8	9/16	1/4-28
1-9/16	2-3/8	9/16	1/4-28
1-5/8	2-5/8	11/16	5/16-24
1-11/16	2-3/4	11/16	5/16-24
1-3/4	2-3/4	11/16	5/16-24
1-13/16	2-7/8	11/16	5/16-24
1-7/8	2-7/8	11/16	5/16-24
1-15/16	3	11/16	5/16-24
2	3	11/16	5/16-24
2-1/16	3-1/8	3/4	5/16-24
2-1/8	3-1/4	3/4	5/16-24
2-3/16	3-1/4	3/4	5/16-24
2-1/4	3-1/4	3/4	5/16-24
2-5/16	3-3/8	3/4	5/16-24
2-3/8	3-1/2	3/4	5/16-24
2-7/16	3-1/2	3/4	5/16-24
2-1/2	3-3/4	7/8	3/8-24
2-9/16	3-7/8	7/8	3/8-24
2-5/8	3-7/8	7/8	3/8-24
2-11/16	4	7/8	3/8-24
2-3/4	4	7/8	3/8-24
2-13/16	4-1/4	7/8	3/8-24
2-7/8	4-1/4	7/8	3/8-24
2-15/16	4-1/4	7/8	3/8-24
3	4-1/4	7/8	3/8-24

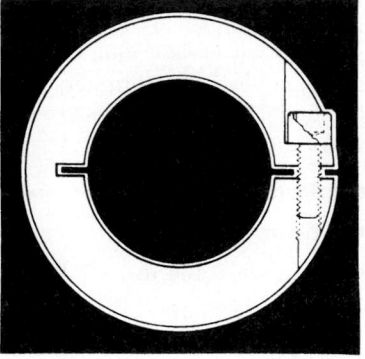

Figure 14.38 Clamp-collars. *(Ruland, Inc.)*

DIMENSIONS			
THREADED BORE	O.D.	WIDTH	CLAMP SCREW
8-32	1/2	.235	4-40
10-24	9/16	.235	4-40
10-32	9/16	.235	4-40
1/4-20	5/8	.281	4-40
1/4-28	5/8	.281	4-40
5/16-18	11/16	.281	4-40
5/16-24	11/16	.281	4-40
3/8-16	7/8	.343	6-32
3/8-24	7/8	.343	6-32
7/16-14	15/16	.343	6-32
7/16-20	15/16	.343	6-32
1/2-13	1-1/8	.406	8-32
1/2-20	1-1/8	.406	8-32
5/8-11	1-5/16	.437	10-32
5/8-18	1-5/16	.437	10-32
3/4-10	1-1/2	1/2	1/4-28
3/4-16	1-1/2	1/2	1/4-28
7/8-9	1-5/8	1/2	1/4-28
7/8 14	1-5/8	1/2	1/4-28
1-8	1-3/4	1/2	1/4-28
1-12	1-3/4	1/2	1/4-28
1-14	1-3/4	1/2	1/4-28
1-1/8-7	1-7/8	1/2	1/4-28
1-1/8 12	1-7/8	1/2	1/4-28
1-1/4-7	2-1/16	1/2	1/4-28
1-1/4-12	2-1/16	1/2	1/4-28
1-3/8-6	2-1/4	9/16	1/4-28
1-3/8-12	2-1/4	9/16	1/4-28
1-1/2-6	2-3/8	9/16	1/4-28
1-1/2-12	2-3/8	9/16	1/4-28
1-3/4-16	2-3/4	11/16	5/16-24
2"-12	3	11/16	5/16-24

Figure 14.39   Threaded clamp-collars. *(Ruland, Inc.)*

14.8   Specialty Fasteners

The specialty fastener component lines available today are great in numbers and types. This section will detail only those specialty fasteners which have become common and which are used widely in new product design and manufacturing.

A partial listing of some of the common specialty fasteners would include

- Acorn nuts
- Floating nuts

DIMENSIONS			
BORE	O.D.	WIDTH	CLAMP SCREW
1/8	1/2	.235	4-40
3/16	9/16	.235	4-40
1/4	5/8	.281	4-40
5/16	11/16	.281	4-40
3/8	7/8	.343	6-32
7/16	15/16	.343	6-32
1/2	1-1/8	.406	8-32
9/16	1-1/4	.437	10-32
5/8	1-5/16	.437	10-32
11/16	1-3/8	.437	10-32
3/4	1-1/2	1/2	1/4-28
13/16	1-5/8	1/2	1/4-28
7/8	1-5/8	1/2	1/4-28
15/16	1-3/4	1/2	1/4-28
1	1-3/4	1/2	1/4-28
1-1/16	1-7/8	1/2	1/4-28
1-1/8	1-7/8	1/2	1/4-28
1-3/16	2-1/16	1/2	1/4-28
1-1/4	2-1/16	1/2	1/4-28
1-5/16	2-1/8	9/16	1/4-28
1-3/8	2-1/4	9/16	1/4-28
1-7/16	2-1/4	9/16	1/4-28
1-1/2	2-3/8	9/16	1/4-28
1-9/16	2-3/8	9/16	1/4-28
1-5/8	2-5/8	11/16	5/16-24
1-11/16	2-3/4	11/16	5/16-24
1-3/4	2-3/4	11/16	5/16-24
1-13/16	2-7/8	11/16	5/16-24
1-7/8	2-7/8	11/16	5/16-24
1-15/16	3	11/16	5/16-24
2	3	11/16	5/16-24
2-1/16	3-1/8	3/4	5/16-24
2-1/8	3-1/4	3/4	5/16-24
2-3/16	3-1/4	3/4	5/16-24
2-1/4	3-1/4	3/4	5/16-24
2-5/16	3-3/8	3/4	5/16-24
2-3/8	3-1/2	3/4	5/16-24
2-7/16	3-1/2	3/4	5/16-24
2-1/2	3-3/4	7/8	3/8-24
2-9/16	3-7/8	7/8	3/8-24
2-5/8	3-7/8	7/8	3/8-24
2-11/16	4	7/8	3/8-24
2-3/4	4	7/8	3/8-24
2-13/16	4-1/4	7/8	3/8-24
2-7/8	4-1/4	7/8	3/8-24
2-15/16	4-1/4	7/8	3/8-24
3	4-1/4	7/8	3/8-24

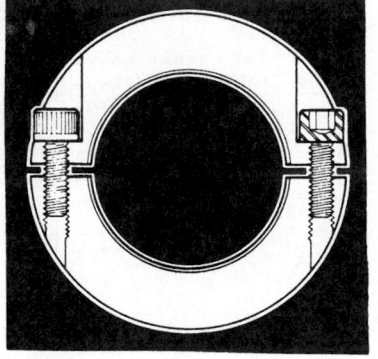

Figure 14.40 Split-collars. *(Ruland, Inc.)*

DIMENSIONS			
BORE	**O.D.**	**WIDTH**	**CLAMP SCREW**
8-32	1/2	.235	4-40
10-24	9/16	.235	4-40
10-32	9/16	.235	4-40
1/4-20	5/8	.281	4-40
1/4-28	5/8	.281	4-40
5/16-18	11/16	.281	4-40
5/16-24	11/16	.281	4-40
3/8-16	7/8	.343	6-32
3/8-24	7/8	.343	6-32
7/16-14	15/16	.343	6-32
7/16-20	15/16	.343	6-32
1/2-13	1-1/8	.406	8-32
1/2-20	1-1/8	.406	8-32
5/8-11	1-5/16	.437	10-32
5/8-18	1-5/16	.437	10-32
3/4-10	1-1/2	1/2	1/4-28
3/4-16	1-1/2	1/2	1/4-28
7/8-0	1-5/8	1/2	1/4-28
7/8-14	1-5/8	1/2	1/4-28
1-8	1-3/4	1/2	1/4-28
1-12	1-3/4	1/2	1/4-28
1-14	1-3/4	1/2	1/4-28
1-1/8-7	1-7/8	1/2	1/4-28
1-1/8-12	1-7/8	1/2	1/4-28
1-1/4-7	2-1/16	1/2	1/4-28
1-1/4-12	2-1/16	1/2	1/4-28
1-3/8-6	2-1/4	9/10	1/4-28
1-3/8-12	2-1/4	9/16	1/4-28
1-1/2-6	2-3/8	9/16	1/4-28
1-1/2-12	2-3/8	9/16	1/4-28
1-3/4-16	2-3/4	11/16	5/16-24
2"-12	3	11/16	5/16-24

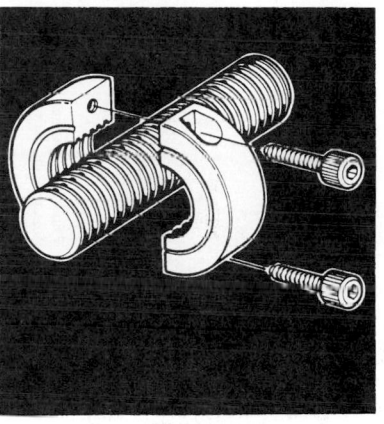

Figure 14.41   Split threaded clamp-collars. *(Ruland, Inc.)*

- Plastic bolts
- Split lock nuts
- SEMs
- Weld nuts
- Various plastic washers
- Sealing washers
- T-slot nuts and bolts
- Push nuts (pal nuts)

Sizes and Thicknesses of Steel Shims - (Inches)

		Thickness Ranges														
ID	OD	0.001	0.002	0.003	0.004	0.005	0.006	0.007	0.008	0.010	0.012	0.015	0.020	0.025	0.031	0.047
0.375	0.625	0.0062	0.093	0.125 (These thicknesses available in all ID and OD sizes)												
0.500	0.750															
0.625	1.000															
0.750	1.125															
0.875	1.375															
1.000	1.500															
1.125	1.625															
1.250	1.750															
1.375	1.875															
1.500	2.125															
1.750	2.750															
2.000	2.750															

Sizes and Thicknesses of Steel Arbor Spacers - (Inches)

		Thickness Ranges					
ID	OD	0.001	0.002	0.003	0.004	0.005	0.006 (These thicknesses available for diameters listed)
0.500	0.750						
0.625	1.000						
0.750	1.125						
0.875	1.375						
1.000	1.500						
1.250	1.750						
1.500	2.125						
2.000	2.250						
		0.007	0.008	0.010	0.012	0.015	0.020 (These thicknesses available for diameters listed)
0.750	1.125						
0.875	1.375						
1.000	1.500						
1.250	1.750						
1.500	2.125						
2.000	2.750						
		0.025	0.031	0.047	0.062	0.093	0.125 (These thicknesses available for diameters listed)
1.000	1.500						
1.250	1.750						
1.500	2.125						
2.000	2.750						

Figure 14.42 Steel shims and arbor spacers.

- Various types of weld studs
- Sheet metal nuts
- Nylok bolts
- Flanged "whiz" bolts
- Turnbuckles

14.8.1 Specialty fasteners in common use

Figure 14.43 shows the different types of SEMs (screw and captive washer assemblies) available today. Note that on the SEM, the screw is either thread-forming or thread-tapping. This makes this class of fastener useful and economical in rapid-assembly applications such as automotive equipment manufacturing. SEMs are specified in ANSI Standard ANSI/ASME B18.13.

Figure 14.44 shows some of the widely used Tinnerman types of speed nuts, which are made of high-carbon, spring-tempered steel. These types of speed nuts are produced in sizes from 6-32 through $\frac{5}{16}$-18 or larger in special cases. The Tinnerman type U and J nuts are used widely to fasten sheet metal screw covers onto sheet metal enclosures. The flat and round types are used on through-bolt sheet metal applications, such as automotive equipment and electronic chassis work. These are economical, efficient fasteners whose applications are limitless.

Another specialty type of fastener that is used widely is the swage nut. The swage nut is produced in several different styles, one of which is shown in Fig. 14.45a. The swage nut is extremely useful in applications where the thread cannot be produced efficiently or effectively in a parent metal that must be fastened to another part. Swage nuts are used in switch-gear equipment where copper bus bars are fastened together, and it is not practical to tap the soft copper bars for bolting. These nuts are also used on thin sheet metal parts where a strong joint is required, and not enough material thickness is available for tapping the sheet metal. The swage nut normally is made from carbon steel with zinc or cadmium plating, stainless steels, and aluminum alloys. Figure 14.45a shows a typical PEM-type nut.

The Rivnut and Plusnut, which are produced by B. F. Goodrich Company, are shown in Fig. 14.45b and c. These types of "blind" fasteners have countless applications in industry and are also produced with sealed ends for liquid-proofing applications. The Rivnut is used widely in the aerospace industry.

SEMS - by SHAKEPROOF
(Pre-assembled lock washer and screw units)

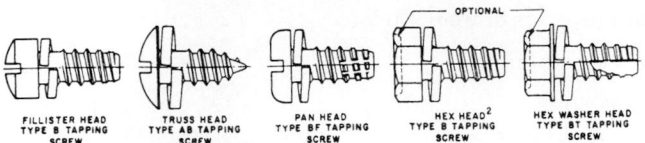

FILLISTER HEAD
TYPE B TAPPING
SCREW

TRUSS HEAD
TYPE AB TAPPING
SCREW

PAN HEAD
TYPE BF TAPPING
SCREW

HEX HEAD[2]
TYPE B TAPPING
SCREW

HEX WASHER HEAD
TYPE BT TAPPING
SCREW

REPRESENTATIVE EXAMPLES OF HELICAL SPRING LOCK WASHER SEMS

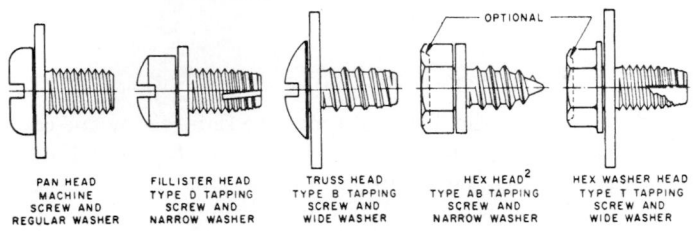

PAN HEAD
MACHINE
SCREW AND
REGULAR WASHER

FILLISTER HEAD
TYPE D TAPPING
SCREW AND
NARROW WASHER

TRUSS HEAD
TYPE B TAPPING
SCREW AND
WIDE WASHER

HEX HEAD[2]
TYPE AB TAPPING
SCREW AND
NARROW WASHER

HEX WASHER HEAD
TYPE T TAPPING
SCREW AND
WIDE WASHER

REPRESENTATIVE EXAMPLES OF PLAIN WASHER SEMS

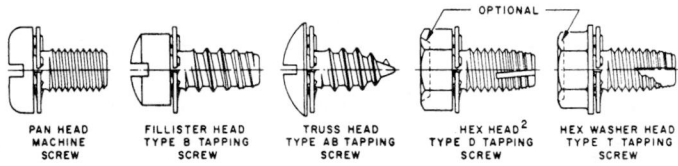

PAN HEAD
MACHINE
SCREW

FILLISTER HEAD
TYPE B TAPPING
SCREW

TRUSS HEAD
TYPE AB TAPPING
SCREW

HEX HEAD[2]
TYPE D TAPPING
SCREW

HEX WASHER HEAD
TYPE T TAPPING
SCREW

REPRESENTATIVE EXAMPLES OF INTERNAL TOOTH LOCK WASHER SEMS

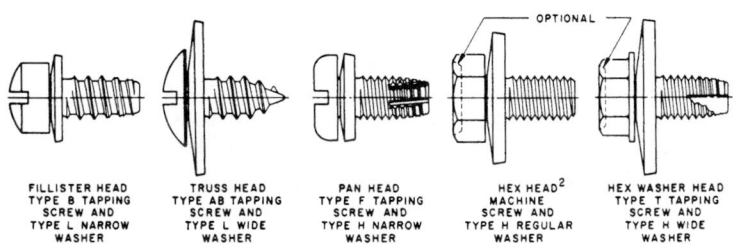

FILLISTER HEAD
TYPE B TAPPING
SCREW AND
TYPE L NARROW
WASHER

TRUSS HEAD
TYPE AB TAPPING
SCREW AND
TYPE L WIDE
WASHER

PAN HEAD
TYPE F TAPPING
SCREW AND
TYPE H NARROW
WASHER

HEX HEAD[2]
MACHINE
SCREW AND
TYPE H REGULAR
WASHER

HEX WASHER HEAD
TYPE T TAPPING
SCREW AND
TYPE H WIDE
WASHER

REPRESENTATIVE EXAMPLES OF CONICAL SPRING WASHER SEMS

Figure 14.43   Types of SEMs.

14.8.2   Electroplating fasteners

High-quality fasteners such as the Unbrako series of socket-head cap screws and shoulder bolts, which use the UNR thread profile, may be precision plated according to Table 14.2. Other types of fasteners also may use the plating specifications shown in Table 14.2.

(Text continued on page 922.)

"U" Type

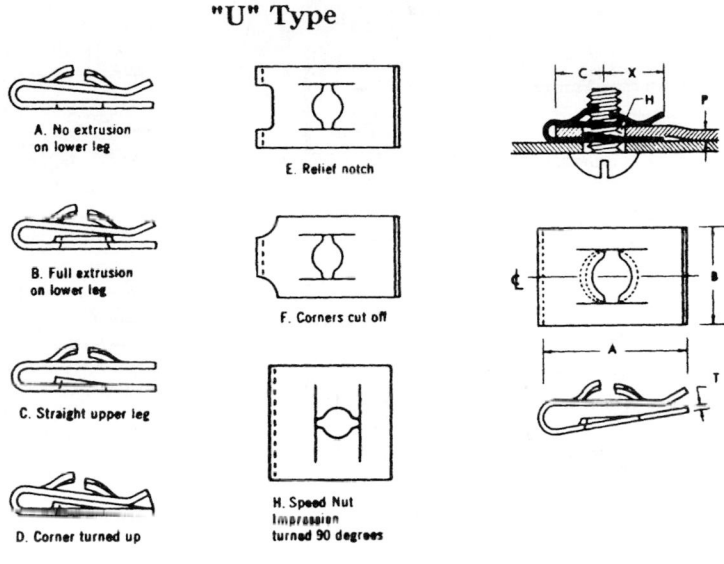

Flat Type

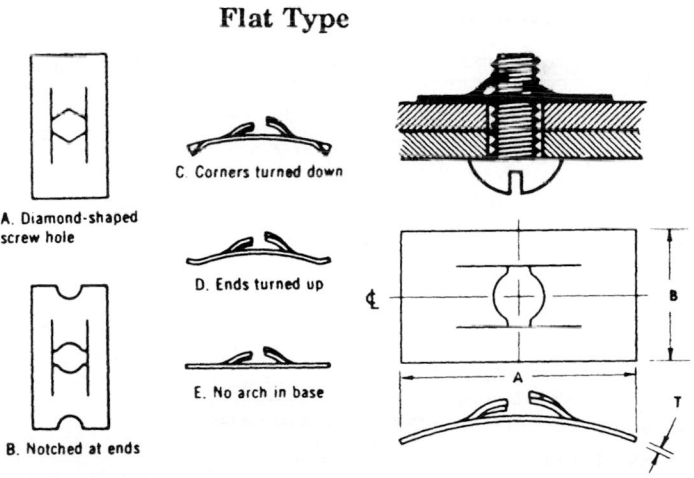

Figure 14.44   Tinnerman speed nuts.

"J" Type

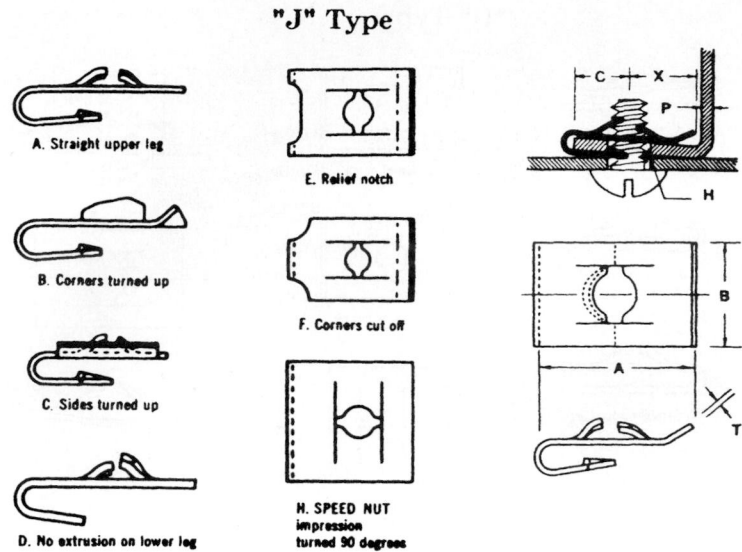

A. Straight upper leg

B. Corners turned up

C. Sides turned up

D. No extrusion on lower leg

E. Relief notch

F. Corners cut off

H. SPEED NUT impression turned 90 degrees

Round Type

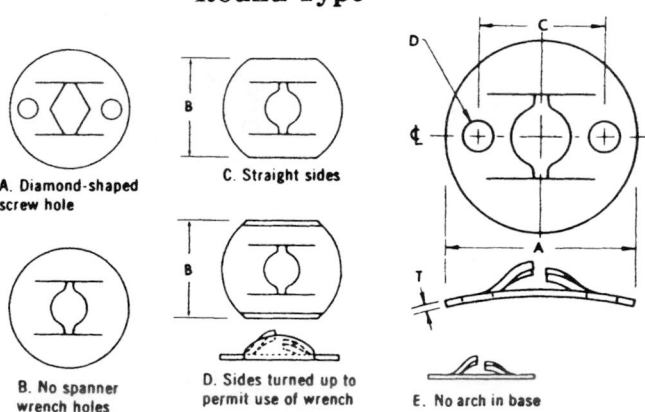

A. Diamond-shaped screw hole

B. No spanner wrench holes

C. Straight sides

D. Sides turned up to permit use of wrench

E. No arch in base

Figure 14.44 *(Continued)*

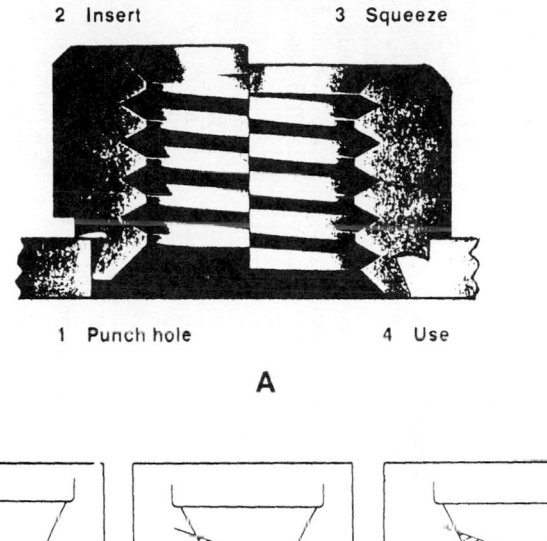

2 Insert   3 Squeeze

1 Punch hole   4 Use

A

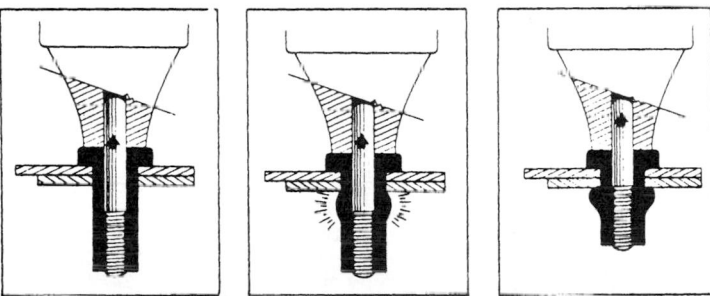

B

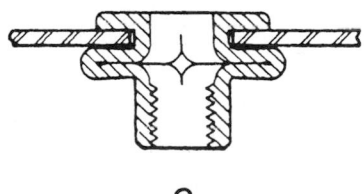

C

Figure 14.45   (*a*) Clinch or swage nut. (*b*) Rivnut installation. (*c*) Installed Plusnut.

TABLE 14.2 Precision Plating Specifications for Fasteners

Type	Thickness (minimum), in			Pre- or postplate treatments or instructions	Typical specifications
	A	B	C		
Cadmium	0.0002	0.0003	0.0005	Clear postplate dip	AMS 2400 and QQ-P-416, type I
Cadmium	0.0002	0.0003	0.0005	Olive drab chromate	QQ-P-416, type II
Cadmium	0.0002	0.0003	0.0005	Iridescent dichromate	QQ-P-416, type II
Zinc	0.0002	0.0003	0.0005	Clear bright	ASTM B633, type III
Zinc	0.0002	0.0003	0.0005	Olive drab chromate	ASTM B633, type II
Zinc	0.0002	0.0003	0.0005	Iridescent dichromate	ASTM B633, type II
Zinc	0.0002	0.0003	0.0005	Supplementary phosphate	ASTM B633, type IV
Silver	0.0002	0.0003	0.0005	Nickel strike	AMS 2410, AMS 2411
Black oxide	Alloy or carbon steel	18-8 stainless			AMS 2485, Mil-C-13924
Dull nickel	0.0002	0.0003	0.0005		AMS 2403
Copper	0.0002	0.0003	0.0005		AMS 2418
Tin	0.0002	0.0003	0.0005		AMS 2408, Mil-T-10727, type I
Phosphate (class A) (Parker-Lubrite)	Dry	Nondrying oil		Manganese phosphate	AMS 2481, DOD-P-16232
Phosphate (class B) (Parkerizing)	Dry	Drying oil		Zinc phosphate	AMS 2480, DOD-P-16232
Cadmium	0.0002	0.0003	0.0005	Black dye over olive drab chromate	QQ-P-416 type II except color

Coating	0.0002	0.0003	0.0005	Remarks	Specification
Cadmium	0.0002	0.0003	0.0005	Fluoborate bath, bake at 375° for 23 h, iridescent dichromateSilver	NAS 672
0.0005 Nickel Cadmium	Copper strike 0.0002 NI 0.0001 CD	AMS 2412	0.0005	Thermal treat 630°	0.0002 0.0003 AMS 2416
Vacuum cadmium	0.0002	0.0003	0.0005	Supplementary phosphate	Mil-C-8837, type I
Cadmium	0.0002	0.0003	0.0005		QQ-P-416, type III
Vacuum cadmium	0.0002	0.0003	0.0005	Iridescent dichromate "Cronak" or equivalent	Mil-C-8837, type II
Molydisulfide coating				Available with a variety of carriers, concentrations, and treatments to customer requirements	
Passivation				For austenitic series stainless steel	QQ-P-35
Passivation				For 400 series stainless steel	
Cadmium Sermetel	0.0002	0.0003	0.0005	Clear postplate dip	AMS 2401 AMS 2506
"Metric" blue dye IVD aluminum	0.0010	0.0005	0.0003	Supplementary chromate	Mil-C-83488, type II

For more complete information on electroplating, see Chap. 13, "Plating Practices and Finishes."

14.9 Welding, Brazing, and Soldering

Welding, brazing, and soldering are all important methods of joining and fastening metals and alloys. This section will detail the various methods or processes and materials used in these three types of joining techniques.

14.9.1 Welding

Welding is a fusion process for joining metals. The heat of application causes mixing of the joint metals or of the filler metal and the joint metal. The resulting joint is as strong as the parent metal, provided the weld is made correctly.

Numerous welding methods or techniques are in common use for a vast array of applications. Modern welding methods or techniques are categorized in Fig. 14.46, which shows the process and the American Welding Society (AWS) designation. Both welding and cutting processes for metals are shown in the figure.

Although the list of processes shown in Fig. 14.46 is extensive, the majority of welding is done by the following methods:

- Stick welding (fluxed rod)—SMAW
- MIG (metal inert gas) welding—GMAW
- TIG (tungsten inert gas) welding—GTAW
- Stud arc welding—SW

Whether the welding process is gas or arc, the welder may proceed to do the weld in the *forward* or *backward* direction, as shown in Fig. 14.47. The direction of welding is left to the judgment of the welder based on the configuration of the object being welded. When the welder is looking down at the welded joint, this is normally called *in-position welding*. When the welder is looking up at the joint, this is called *out-of-position welding*. Any orientation not looking down at the weld can be considered as out-of-position welding.

Each of the various welding processes produces physical characteristics that allow the process to be identified by direct eye inspection. Figure 14.48 shows an assembly welded using the MIG process (GMAW). Here, the weld is rather rough looking and was difficult

Welding Processes and Designations

Process	Designation
Arc welding (AW)	
Atomic-hydrogen welding	AHW
Bare-metal arc welding	BMAW
Carbon arc welding	CAW
-gas	CAW-G
-shielded	CAW-S
-twin	CAW-T
Flux-cored arc welding	FCAW
-electrogas	FCAW-EG
Gas-metal arc welding	GMAW
-electrogas	GMAW-EG
-pulsed arc	GMAW-P
-short circuiting arc	GMAW-S
Gas-tungsten arc welding	GTAW
-pulsed arc	GTAW-P
Plasma arc welding	PAW
Plasma gas-metal arc welding	PAW-GMAW
Shielded-metal arc welding	SMAW
Stud arc welding	SW
Submerged arc welding	SAW
-series	SAW-S
Solid-State Welding (SSW)	
Cold welding	CW
Diffusion welding	DFW
Explosive welding	EXW
Forge welding	FOW
Friction welding	FRW
Hot press welding	HPW
Roll welding	ROW
Ultrasonic welding	USW
Other Welding	
Electron-beam welding	EBW
Electroslag welding	ESW
Flow welding	FLOW
Induction welding	IW
Laser beam welding	LBW
Thermite welding	TW
Oxyfuel Gas Welding (OFW)	
Air acetylene welding	AAW
Oxyacetylene welding	OAW
Oxyhydrogen welding	OHW
Pressure gas welding	PGW
Resistance Welding (RW)	
Flash welding	FW
High frequency resistance welding	HFRW
Percussion welding	PEW
Projection welding	RPW
Resistance seam welding	RSEW
Resistance spot welding	RSW
Upset welding	UW

Metal Cutting Processes and Designations

Process	Designation
Thermal Oxygen Cutting (OC)	
Chemical flux cutting	FOC
Metal powder cutting	POC
Oxyfuel gas cutting	OFC
-oxyacetylene	OFC-A
-oxyhydrogen	OFC-H
-oxynatural gas	OFC-N
-oxypropane	OFC-P
Oxygen arc cutting	AOC
Oxygen lance cutting	LOC
Arc cutting (AC)	
Air-carbon arc cutting	AAC
Carbon arc cutting	CAC
Gas-metal arc cutting	GMAC
Gas-tungsten arc cutting	GTAC
Metal arc cutting	MAC
Plasma arc cutting	PAC
Shielded-metal arc cutting	SMAC
Other Cutting	
Electron-beam cutting	EBC
Laser beam cutting	LBC

Recognized by The American Welding Society (AWS)

Figure 14.46 Welding processes and designations.

to clean mechanically. The welding heat setting and the diameter of the weld wire play important roles in producing a neat, clean weld that is also mechanically sound. Figure 14.49 shows the same type of joint (using the same parts) that has been welded using the TIG process (GTAW). It is immediately apparent that the TIG process is more advantageous in this application, from the point of view of both strength and cosmetic appearance. This joint is comprised of parts manufactured from AISI 304 type stainless steel. Figure 14.49 shows the welded assembly sand-blasted after welding. The TIG weld in this application was made efficiently and correctly by a skilled welder. Welding is an art as well as a science, and a good welder needs a great deal of practice and experience.

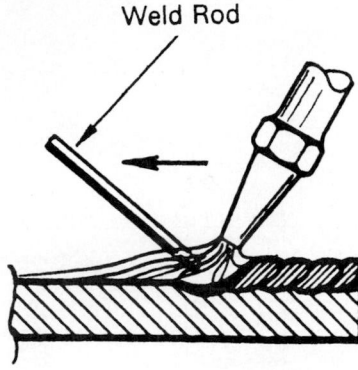

a Forward Welding

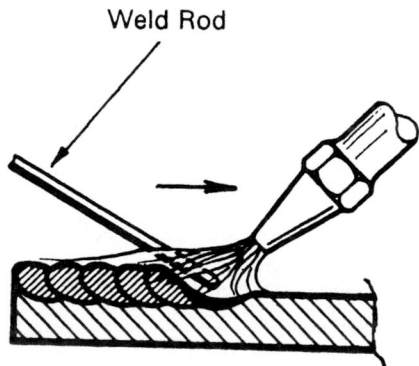

b Backward Welding

Figure 14.47 (*a*) Forward. (*b*) Backward.

In a welded assembly of numerous parts, the welder's skill and experience play an important role in producing an acceptable final weld assembly. Many welded assemblies require additional machining after the welding process because of the strains produced when the welded joints cool to room temperature. Allowance must be made in the design of such an assembly to accommodate these welding distortions, which are sometimes unavoidable.

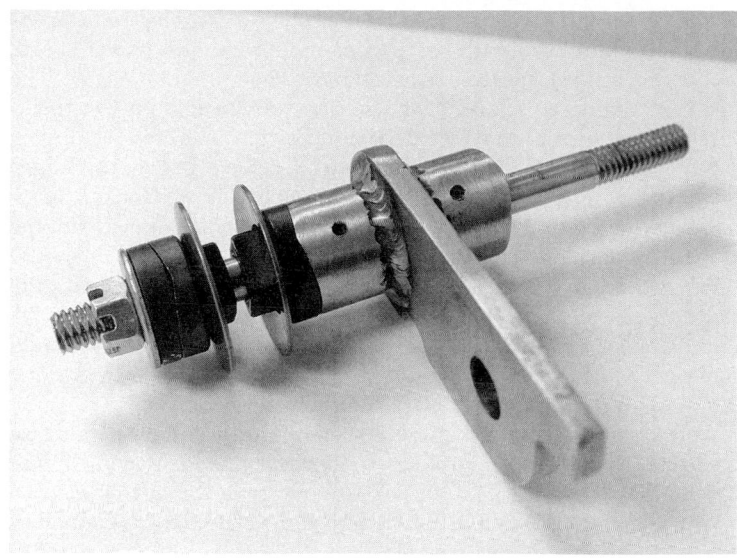

Figure 14.48 MIG-welded assembly.

Figure 14.49 TIG-welded assembly.

Welding procedures and electrode sizes.   The correct electrode size is one that, when used with the proper amperage and travel speed, produces a weld of the required size in the least amount of time. The electrode diameter selected for use depends largely on the thickness of the material to be welded, the position in which welding is to be performed, and the type of joint to be welded. In general, larger electrodes will be selected for applications involving thicker materials and for welding in the flat position in order to take advantage of their higher deposition rates.

For welding in the horizontal, vertical, and overhead positions, the molten weld metal tends to flow out of the joint owing to gravity. This can be controlled by using small electrodes to reduce the weld pool size. Electrode manipulation and increased travel speed along the weld joint also aid in controlling the weld pool size.

Weld groove design also must be considered when electrode size is selected. The experience of the welder often has a bearing on the size of the electrode. This is particularly true for out-of-position welding because the welder's skill determines the size of the molten pool that the welder can control.

Welds that are larger than necessary are more costly and, in certain instances, actually are harmful to the joint. Any sudden change in section size or contour of a weld, such as that caused by overwelding, creates stress concentrations. An improperly welded assembly of parts will distort on cooling, and an experienced welder can prevent this to a certain degree by applying the welds at the correct points and in the correct sequences. Tooling fixtures and weld jigs play an important role in controlling the distortion of welded assemblies when they are designed and applied properly.

Shielded-metal arc welding can be accomplished with either alternating current (ac) or direct current (dc) when the appropriate electrode is used. The melting rate of any given electrode is directly related to the electrical energy supplied to the arc via the welding controller apparatus.

Direct current (dc) always supplies a steadier arc and smoother metal transfer than alternating current (ac). Most covered electrodes operate better on reverse polarity (electrode positive), although some are suitable for and are even designed for straight-polarity welding (electrode negative). Reverse polarity produces deeper weld penetration, whereas straight polarity produces a higher electrode melting rate. Direct current is particularly suitable for thin-section welding and is preferred for vertical and overhead welding (out-of-position welding). If *arc blow* is a problem when welding with dc, change the current to ac.

For SMAW, ac offers advantages over dc; one is the absence of arc blow, and the other is the cost of the power source for producing the weld (the welding machine or apparatus). Without arc blow, larger electrodes and higher welding currents can be used.

Welding technique. In SMAW welding, you first select the proper equipment, materials, and tools for the job. The type of welding current and its polarity then must be selected, and the power source set accordingly. The power source must be set to give the proper voltampere (VA) characteristic for the size and type of electrode being used.

Strike the arc, for the weld to begin, by tapping the end of the electrode near the beginning of the weld joint; then quickly move it to produce an arc of proper length at the beginning of the weld joint. Then move the electrode uniformly along the joint, keeping a constant arc length as the electrode melts to produce the welded joint. A good deal of practice is required, especially in out-of-position type weld joints. To break the welding arc, when the weld joint is completed, stop the forward motion of the electrode and abruptly withdraw the electrode away from the joint or move the electrode into the weld pool quickly to kill the arc and then abruptly remove the electrode, thus breaking the arc.

Note: Complete welding procedures and data for all welding processes are prepared by the American Welding Society (AWS) and are available in handbook form in its *Welding Handbook*, eighth edition, in three volumes.

Weld strength: Related equations and tables. Here we will present the equations for determining the approximate strengths of welded joints. Any equation involving a process with many variables, such as welding, can only be an approximation. With this in mind, therefore, one should allow a factor of safety when designing and calculating the strengths of welded joints.

Fillet welds. Refer to Fig. 14.50a. The basic welding equations for the fillet weld are as follows:

$$h = 0.707l$$

$$hL = \frac{P}{S_i}$$

$$A = \frac{P}{S_i}$$

$$L = \frac{P}{S_i h}$$

where l = leg dimension of fillet weld, in
L = length of fillet weld, in
h = weld throat height, in
P = load, lb
S_i = induced stress, lb/in^2
A = throat area, in^2

Butt welds (primary).   Refer to Fig. 14.50b. The tensile stress in a butt weld induced by a tensile load P is

$$S_t = \frac{P}{td}$$

where S_t = tensile stress, lb/in^2
P = tensile load, lb

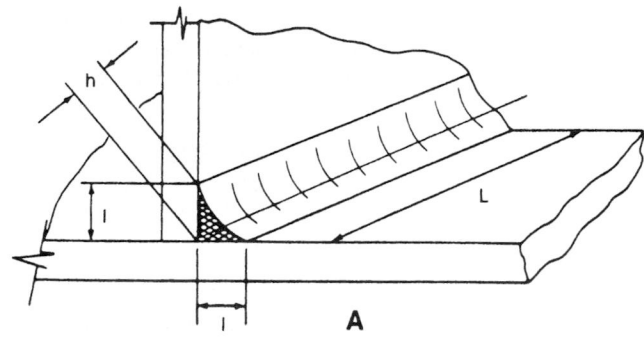

A
Fillet Weld

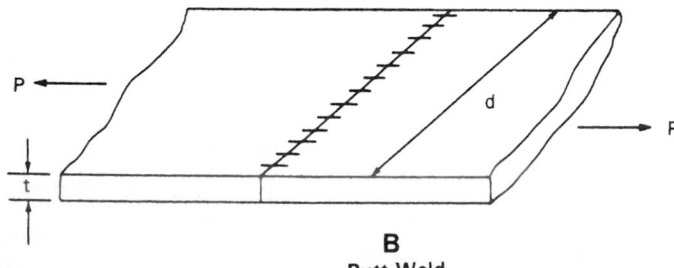

B
Butt Weld

Figure 14.50   (a) Fillet weld. (b) Butt weld.

t = material thickness, in
d = width of butt welded joint, in

For minimum leg sizes for fillet welds, see Figure 14.51b. For allowable design strengths and shear forces for fillet welds and partial penetration groove welds, see Fig. 14.51a.

Example: The allowable unit force for a fillet weld with a 0.25-in leg, using 80,000-lb/in² weld rod or wire, is 16,500 × 0.25 = 4125 lb/in. Thus, if the weld joint is 3 in long, the force allowable is 4125 × 3 = 12,375 lb.

Plug welds. Plug welds (Fig. 14.52) are useful in sheet metal and structural design applications. Plug welds are used primarily for shear loads, although they are not limited to this type of load. A plug weld may be subjected to a combination of shear and tensile loads. The typical sizes of plug welds are shown in Fig. 14.51c for various applications or combinations of material thicknesses. Figure 14.52 illustrates a typical plug weld.

Note: It should be noted that the weld-strength figures and tables, allowables, and examples are for *static* loads only. When the welded members are dynamically or cyclically loaded, a factor of safety should be applied. A safety factor of 3 should be applied for general dynamic conditions. In other words, if the weld joint was calculated to withstand a load of 3000 pounds force, for dynamic conditions this load should be reduced to a 1000 pounds force maximum. (Divide the calculated load by 3 to arrive at the allowable load with the factor of safety applied.)

Specifying welds. The type of welding, weld-rod strength and type, fillet or bead size, location, and length of welds all must be specified on the welding drawings of a part or assembly. Standard weld symbols recognized by the American Welding Society (AWS) should be used on the engineering drawings (see "Standard Weld Symbols" below).

Thin-section parts or any part or assembly that may pose a weld distortion problem should be reviewed in coordination with the welding department or welder prior to final design or beginning the work. Experienced welders usually know or can determine welding sequences to prevent distortion or keep it to a minimum. Welding sequence instructions may be required on the welding drawing.

Secondary machining operations usually are performed on a welded part or assembly after the welding operation to correct unavoidable distortion or dimensional changes that take place

Design-Allowable Strengths and Shear Forces

Strength of Weld* Rod or Wire Metal, psi	Allowable Shear Stress on Throat *(h)*, psi	Allowable Unit Force, lb/Linear inches
60,000	17,000	12,500 x *l*
80,000	23,000	16,500 x *l*
100,000	29,000	21,000 x *l*
120,000	35,000	25,000 x *l*

* For intermediate weld-rod strengths, interpolation may be used. In the above table, h = *weld throat*, l = *length of fillet leg*.

(a)

Minimum Leg Size, (Fillet Weld)

Thickness of Thicker Plate *(t)*, inches	Minimum* Leg Size *(l)*, inches
Up to 1/4	1/8
> 1/4 to 1/2	3/16
> 1/2 to 3/4	1/4
> 3/4 to 1-1/2	5/16
> 1-1/2 to 2-1/2	3/8

*Also minimum throat *(h)* of partial penetration groove weld. Leg of weld *(l)* should not exceed thickness of thinner plate.

(b)

Plug Weld Sizes, (Diameters)

Gauge or Thickness of Thinner Member, (Inches)	Plug-Weld Hole Diameter, (Inches)
1/16 or # 16 gauge	1/4 to 3/8
3/32 or # 13 gauge	1/2
1/8 or # 11 gauge	5/8
3/16 or # 7 gauge	5/8
1/4 plate	3/4
3/8 plate	1

(c)

Figure 14.51 (*a*) Allowable shear. (*b*) Weld leg sizes. (*c*) Plug weld sizes.

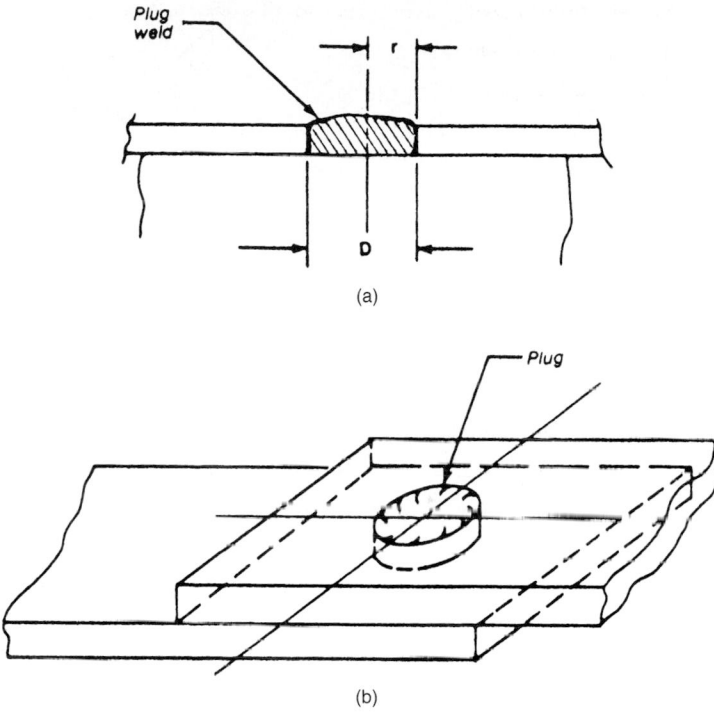

Figure 14.52 (*a*) Section of plug. (*b*) Plug-welded plates.

during welding. To reduce cost and save welding time, the amount of welding on a part or assembly should be kept to a minimum, in accordance with the strength requirements of the design or sealing requirements.

Standard weld symbols. The basic weld symbols shown in Fig. 14.53 should be used on all welding drawings, especially if the welded part is sent to an outside vendor or subcontractor. If in-house symbols are used, these should be noted on the welding drawings so that outside vendors or subcontractors know their exact meaning. The symbols shown in the figure are those recognized by the American Welding Society (AWS), the American Iron and Steel Institute (AISI), the American Society of Mechanical Engineers (ASME), the Society of Automotive Engineers (SAE), and other authorities and specification agencies.

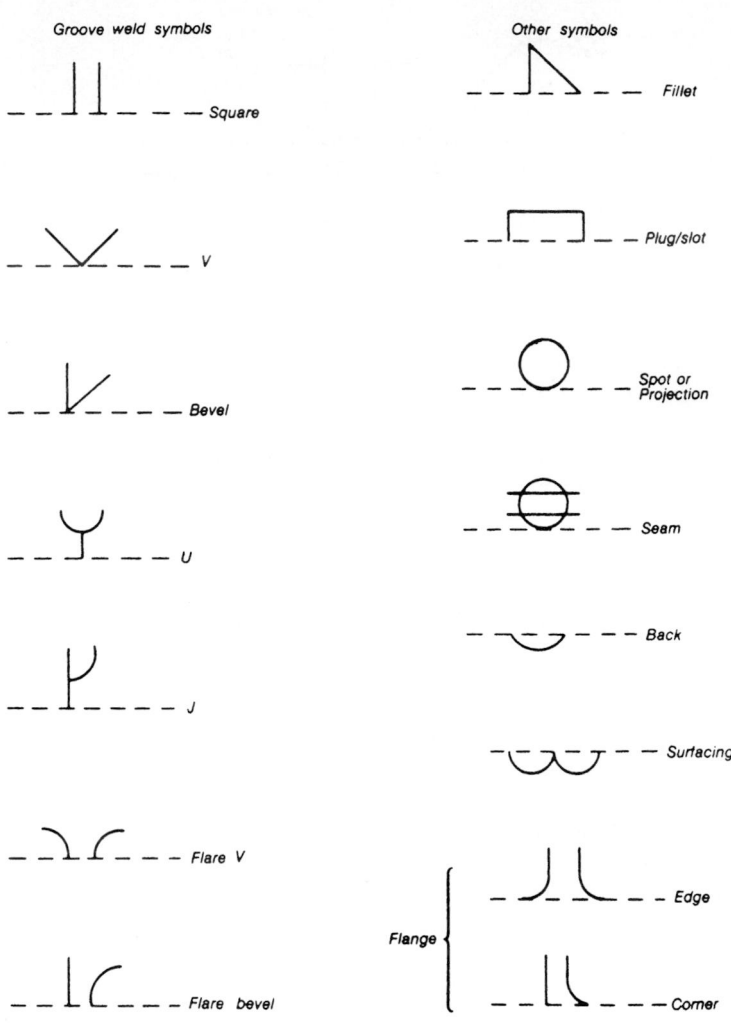

Figure 14.53 Weld symbols.

Elements of the welding symbol. When a weld is specified on an American standard engineering drawing, it should conform to certain characteristics, which are shown in Fig. 14.54. In this way, uniformity and complete understanding are maintained between the welder and the design engineer. Typical welding drawing call-outs

Standard locations of elements of a welding symbol

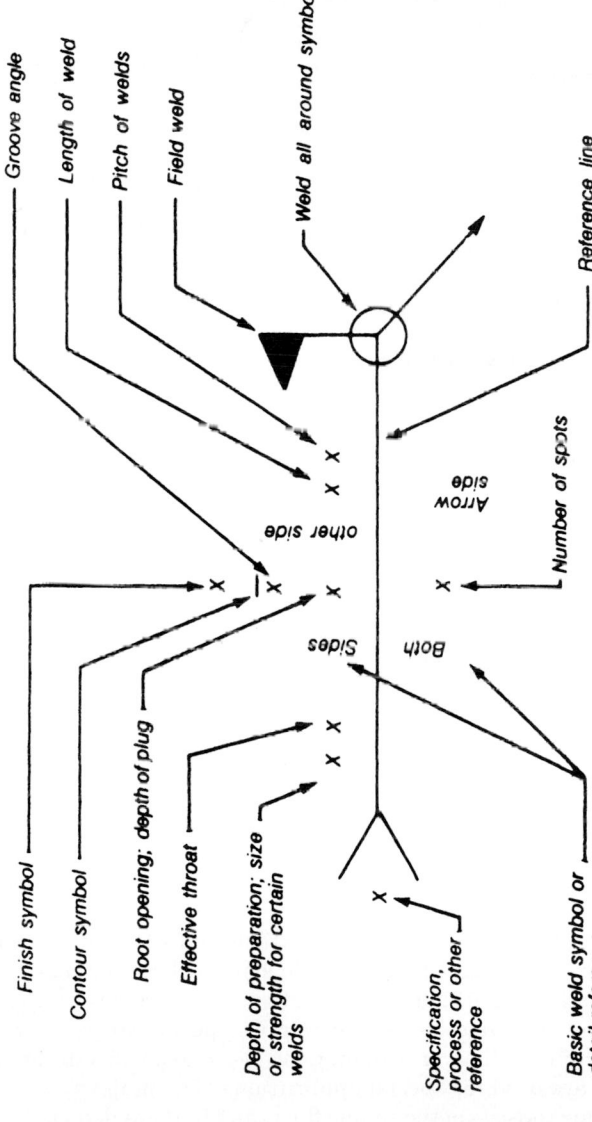

Groove angle

Length of weld

Pitch of welds

Field weld

Weld all around symbol

Reference line

other side

Arrow side

Number of spots

Sides

Both

Finish symbol

Contour symbol

Root opening; depth of plug

Effective throat

Depth of preparation; size or strength for certain welds

Specification, process or other reference

Basic weld symbol or detail reference

Figure 14.54 Elements of the welding symbol.

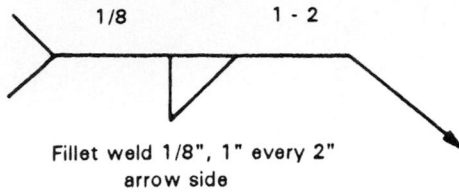

Fillet weld 1/8", 1" every 2"
arrow side

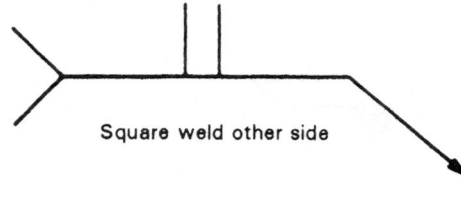

Square weld other side

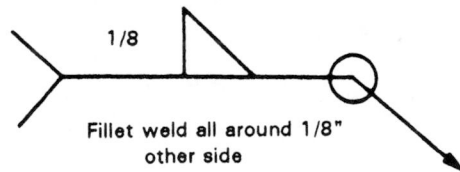

Fillet weld all around 1/8"
other side

Figure 14.55 Typical weld call-outs (welding drawings).

or symbols are shown in Fig. 14.55 with an explanation of their meaning.

Types of weld joints.   There are many types of weld joints or designs, and the basic ones are shown in Fig. 14.56. The various joints have been designed for different applications and strengths. Other characteristics are designed into the weld joint, such as minimal outgassing, dynamic strength, deep penetration, pressure-vessel applications, and others. Weld joints that require special preparation, such as machining, filing, or grinding, are more expensive to produce and thus are used for special applications. The majority of industrial welding consists of the simple fillet- and butt-welded joints, followed by the single-V and double-V joints.

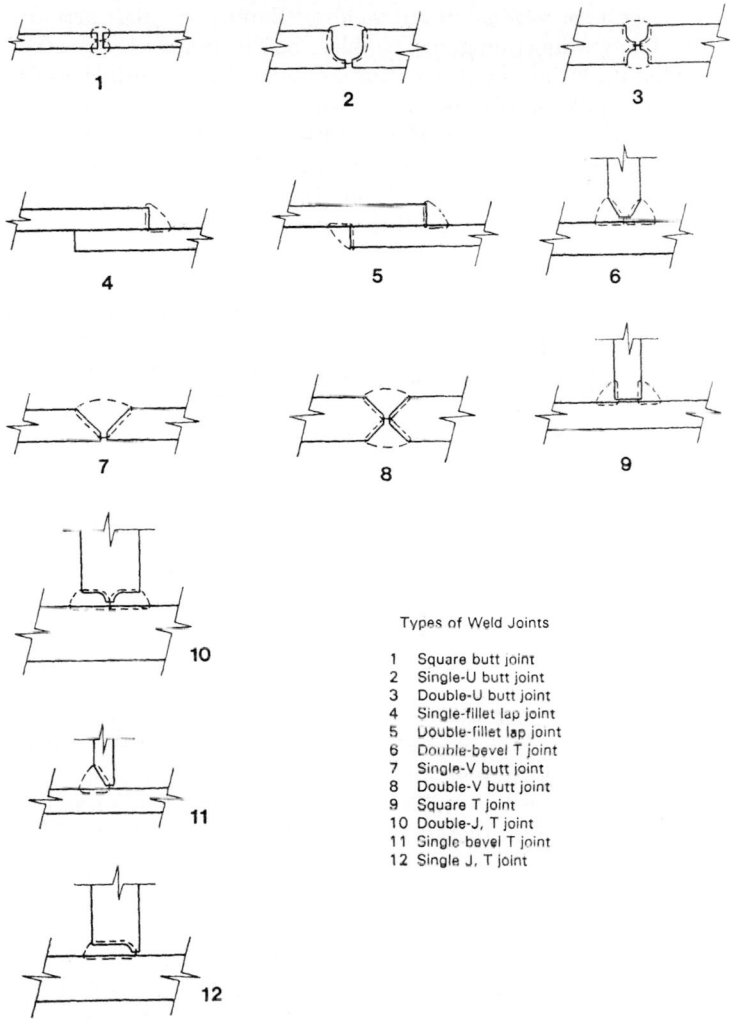

Figure 14.56 Basic weld-joint geometry.

Types of Weld Joints

1 Square butt joint
2 Single-U butt joint
3 Double-U butt joint
4 Single-fillet lap joint
5 Double-fillet lap joint
6 Double-bevel T joint
7 Single-V butt joint
8 Double-V butt joint
9 Square T joint
10 Double-J, T joint
11 Single bevel T joint
12 Single J, T joint

Welding applications data. Welding is one of the most common and important means of fastening. Its applications are limitless, and the technology is constantly changing. Given here is a listing of various welding process applications that are helpful in design as well as welding work.

Thin-gauge metal welding. Small welding flames or small arcs are required for welding thin-gauge metals. The TIG process is especially useful for producing small, accurate welds on thin materials, under 11 gauge (0.1196 in). To prevent buckling of large-area, thin materials, *heat sinks* are useful. Heat sinks may take the form of wet burlap bags or large blocks of metal clamped to the welded parts or sections. Applying *tack welds* in a specified sequence also may help to prevent buckling of large, thin sections prior to beginning the final seams.

Preheating. Large sections or masses of metal usually require a preheat stage, where the parts are heated a few hundred degrees Fahrenheit prior to beginning the welding process. This prevents thermal shock and minimizes distortion and possible cracking of the welds.

Air cooling. Welded parts or assemblies normally are allowed to cool to ambient temperature after the welding process is completed. Do not water quench welded parts immediately after welding. Because of the high temperatures generated in the welding process, cracking or distortion may occur. Changes in the grain structure of the metal may occur if the hot, welded part is cooled suddenly by water quenching.

Welding bases or platforms. A flat, level area is required for welding large assemblies. This is usually provided by structural beams embedded in the weld shop floor. The beams must be straight and leveled with a transit or leveling instrument. For smaller welded parts and assemblies, the standard welding table is used. Figure 14.57 shows a typical steel-grid welding table, which is used in many welding departments. This type of table is level and has square openings where different types of clamping and squaring tools may be attached to hold the welded assembly prior to the welding operation.

Notice the screening and plastic shielding located around the welding table area. This shielding prevents the intense ultraviolet radiation generated by the welding arc from reaching the eyes of other personnel. The intensity of the welding arc radiation is high enough to damage the membrane covering the human cornea and eyeball. It is not necessary to look directly at the welding arc for damage to occur to the eye. The arc rays can penetrate the side of the eye indirectly and cause damage. The usual effect of looking at a weld-

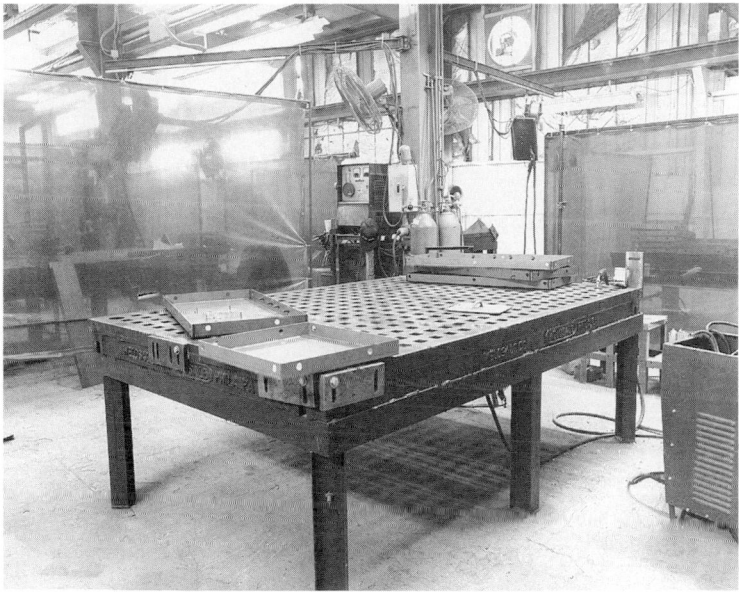

Figure 14.57   A cast iron grid welding table.

ing arc is the feeling that sand has entered the eye. This usually begins to show some hours after exposure to the arc, either directly or indirectly, and is the result of scar tissue formation in the damaged eye.

Figure 14.58 shows a welder performing the TIG welding process on an aluminum electrical reactor core. This is a precision process requiring great skill by the welder. Figure 14.59 shows a welder using a pneumatic grinder to smooth the edges of a welded sheet metal assembly that was completed using the MIG process. In many companies, smoothing of the completed welds by grinding is considered part of the welding process and is performed by the welders. Small subassemblies as well as large welded assemblies are treated in this manner.

Welding stainless steels.   The electrode used to weld stainless steels also should be stainless steel, matching the application. Types AISI 300 through 303 should not be welded because these are machining grades. Type AISI 304 is a preferred stainless steel for welding

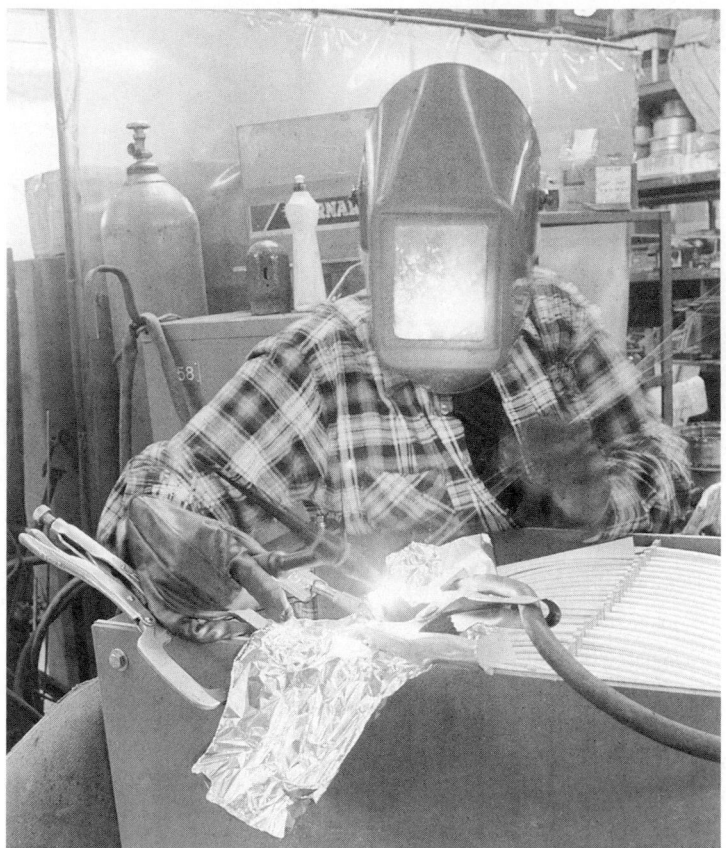

Figure 14.58 A welder welding an aluminum reactor core using the TIG process on aluminum alloy.

applications, with AISI type 304L, a special low-carbon grade for critical applications such as those used on aerospace vehicles. Too much carbon or sulfur in a stainless steel produces cracking during the welding process. AISI types 308, 309S, 310S, 316, and 316L are also suitable for welding applications because they are low in carbon content. AISI types 316 and 316L stainless steels are used widely in highly corrosive environments such as chemical and food-processing applications. The 300 series stainless steels are austenitic (nonmagnetic) and cannot be hardened by heat-treatment procedures. See Chap. 4 for more information on stainless steels.

Figure 14.59 A welder grinding and smoothing the edges of a welded sheet metal assembly for an electrical power distribution product.

Welding carbon and alloy steels. The low-carbon grades from AISI 1010 through 1020 are readily weldable because they are low in carbon content. Some of the medium-carbon grades are weldable with caution, whereas the high-carbon grades are not recommended for welding (AISI 1045–1095). The higher carbon content in the steel causes cracking during the welding process. Low-alloy- and alloy-grade steels are weldable according to type and welding process employed. See Chap. 4 for more data on the low-alloy and alloy steels, including tool steels.

Cutting metals. Oxyacetylene cutting (OFC-A) is the most common welding process for cutting ferrous metals. Most ferrous metals cut cleanly using this method, except the stainless steels, high-manganese steels, and special alloy steels. To cut these difficult materials, special techniques are required wherein the molten metal can

Figure 14.60   Laser-cut steel parts.

be blown away from the flame or arc during the cutting process. Laser beam cutting (LBC) is a modern method of cutting difficult materials and extremely thin materials. Figure 14.60 shows small parts made of carbon steel that were cut using the laser beam method. These parts are approximately 2 in long and 0.18 in thick. The quantity required did not justify building a stamping die for this part, and milling the part from stock was expensive; therefore, laser beam cutting was selected for the method of manufacture.

Many branches of industry rely on the welding processes for producing their equipment and machinery economically. Figure 14.61 shows a typical electrical power distribution industry lineup of switch gear that relies heavily on the welding processes to fasten and join the many sheet metal parts and structures required.

14.9.2   Brazing

Brazing employs a nonferrous filler metal, usually in wire or paste form, to join metal parts at a temperature above approximately 800°F but below the melting point of the base metals being joined. Various fluxes are used in the brazing process to remove oxides on

Figure 14.61 Switch-gear equipment fabricated from sheet steel. *(Powercon Corporation, Severn, MD.)*

the base-metal parts so that the filler metal can adhere to them strongly. Sal ammoniac is a good general-purpose flux for brazing copper, phosphor-bronze, and stainless steels.

The brazing of tungsten alloys to various base metals is accomplished by pretinning the brazing surfaces of the tungsten alloy part in a nitrogen-atmosphere furnace before joining the tungsten alloy part with the base-metal part. This is done because it is extremely difficult to remove oxide layers that form on the tungsten part unless it is pretinned and protected from the atmosphere. Many of the various designs of electrical contact points are brazed to their base-metal mountings, although riveting is sometimes employed.

Brazing heat. Heat for brazing parts may be supplied by the following methods:

- Torch brazing
- Furnace brazing
- Resistance brazing

- Dip brazing

- Vacuum-furnace brazing

- Induction brazing

When there is an unusual condition in the brazing method or process, a brazing specialist should be consulted. So many brazing materials and fluxes are available that only an expert in the field may be able to solve your particular brazing problem.

In the normal brazing process, a suitable brazing wire or paste is selected, and the parts to be brazed are set up or clamped into position (brazing fixtures sometimes are employed) after the joint has been fluxed with the proper flux for the application. Heat is applied to the parts to be joined, and when they are hot enough, the filler wire is fed into the brazed joint, where a wetting action takes place, and the joint is filled with the melted brazing filler. The joint is then allowed to air cool to room temperature. The method of fluxing and applying the filler metal is critical for a well-brazed joint. It requires considerable skill and practice to braze a difficult joint or complicated setup. Excess brazing filler should not be used on any brazed joint because it wastes material and produces an inefficient joint.

Brazing pastes and alloys.   Brazing and soldering alloys are available as wire, strip, preformed shapes, and pastes. Brazing pastes are applied to the joint prior to heating and therefore are applicable for manual as well as automated processing. Flux is also compounded in some of the brazing pastes, making their use easily applicable to production operations.

Lucas-Milhaupt (Handy and Harman Company) produces an extensive array of pastes for soldering and brazing in their HandyFlo series 110, 120, 210, 310, 320, and 330, which are listed in Fig. 14.62. Other standard Handy and Harman brazing materials are outlined in Fig. 14.63. With some of the new brazing products, it is now possible to join ceramics, diamonds, and other difficult to wet materials.

14.9.3   Soldering

The use of lead or tin base alloys having melting points below 800°F for joining metal parts is known as *soft soldering*. Such joints usually do not have great mechanical strength, although they may be required to carry electric currents.

	110	120	210	310	320	330
HEATING METHODS						
Torch	•	•	•			
Induction						
Resistance	•	•	•			
Vacuum Brazing				•	•	•
Atmospheric Furnace	•	•	•	•	•	•
Infrared, Hot plate						
Oven						
TEMPERATURE RANGE						
300 - 500° F 149-260° C						•
500 - 1000° F 260-524° C						
1000 - 1500° F 524-815° C	•	•		•	•	•
1500 - 2000° F 815-1093° C			•	•	•	•
Includes Flux	•	•	•			
Can Be Boron Modified	•	•	•			
MATERIALS JOINED						
Copper	•	•	•	•	•	•
Brass	•	•		•	•	
Cold rolled steels	•	•	•	•	•	•
Stainless Steel	•	•	•	•	•	•
Non-ferrous alloys	•	•	•	•	•	
Tungsten Carbide	•	•	•			

Figure 14.62   Brazing paste properties.

The three basic standard soldering alloys are

- 60% tin, 40% lead—melting point 375°F
- 50% tin, 50% lead—melting point 420°F
- 40% tin, 60% lead—melting point 460°F

A number of other alloys are available. The standard 60-40 solder is useful for most electrical/electronic soldering, and a small amount of antimony may be added for improving quality and strength. The electronic wire solders are usually made with a rosin flux core. The flux is noncorrosive and nonconducting and may be removed after the soldering operation with naphtha, Varsol, or enamel paint thinners. Rosin-cored solders can be stored for extensive periods without degradation of the rosin flux core.

Preforms are also produced in soldering alloys. A *preform* is a stamped shape, made of solder alloy, that will fit a particular device or application. Some of the delicate soldering operations would be difficult or impossible without preforms.

Figure 14.64 shows a typical printed circuit board that had been soldered automatically in a "wave" soldering operation. This board

H & H Alloy Designations	AWS A 5.8	AMS Spec.	Federal Spec. QQ-B-654A	Solidus °F	°C	Liquidus °F	°C	Nominal Compositions % Ag	Cu	Zn	Others	Comments
Easy-Flo 45	BAg-1	4769B	VII	1125	605	1145	620	45	15	16	24Cd	Versatile alloy; for Fe&Ni base alloys
Easy-Flo	BAg-1a	4770F	IV	1160	625	1175	635	50	15.5	16.5	18Cd	For ferrous, non-ferrous, dissimilar metals.
Easy-Flo 35	BAg-2	4768D	VIII	1125	605	1295	700	35	26	21	18Cd	Gen. purpose for larger gaps.
Easy-Flo 30	BAg-2a			1125	605	1310	710	30	27	23	20Cd	For ferrous, non-ferrous, dissimilar metals.
Easy-Flo 3	BAg-3	4771D	V	1170	630	1270	690	50	15.5	15.5	16Cd/3Ni	For tungsten carbide brazing. Al-Bronze to steel. Stainless Steels.
Sil-Fos 5	BCuP-3			1190	643	1495	812	5	89		6P	Same as Sil-Fos.
Sil-Fos	BCuP-5		BCuP-5	1190	643	1475	801	15	80		5P	For Cu or Cu alloys — Not for Fe Alloys.
Sil-Fos 6	BCuP-4			1190	643	1325	718	6.0	86.5		7.5P	
Sil-Fos 18				1191	643	1191	643	17.75	75		7.25P	
69-014				1215	657	1270	687		86.7		6.3P/7Sn	
70-873				1245	674	1285	696		87.3		7.1P/5.6Sn	
Fos-Flo 7	BCuP-2			1310	709	1460	793		92.75		7.25P	Low cost alloy for copper & copper alloys.
Fos-Flo 8				1350	732	1350	732		91.5		8.5P	
Braze 300	BAg-20		BAg-20	1250	675	1410	765	30	38	32		For steel & non-ferrous alloys melting above 1450°F; Ni-Ag knife handle
Braze 380	BAg-34			1200	650	1330	720	38	32	28	2 Sn	Cd free. For small gaps, Fe, Cu alloys.
Braze 403	BAg-4		BAg-4	1220	660	1435	780	40	30	28	2Ni	For tungsten carbides — stainless food handling equip. (no Cd).
Braze 404				1220	660	1580	860	40	30	25	5Ni	Brazing carbides & stainless steels.
Braze 450	BAg-5		BAg-5	1225	665	1370	745	45	30	25		For band instruments (Cu alloys), brass lamps.
Braze 495	BAg-22			1260	680	1290	700	49	16	23	7.5Mn/4.5Ni	For brazing of carbides and stainless steel.
Braze 505	BAg-24			1220	660	1305	705	50	20	28	2Ni	For Ni & Fe base alloys — retards interface corrosion (no Cd).
Braze 541	BAg-13	4772E		1340	725	1575	855	54	40	5	1Ni	For stainless steels used at elevated temps.
Braze 559	BAg-13a	4765A	BAg-13a	1420	770	1640	895	56	42		2Ni	Atmosphere brazing for high temp (NoZn).
Braze 560	BAg-7		BAg-7	1145	620	1205	650	56	22	17	5Sn	Food handling equipment (Cd free).
Braze 600				1245	675	1325	720	60	25	15		For Monel & other nickel alloys.
Braze 603	BAg-18	4773B	BAg-18	1115	600	1325	720	60	30		10Sn	Low temp. seals on vacuum parts.
Braze 630	BAg-21	4774B		1275	690	1475	800	63	28.5		6Sn/2.5Ni	Stainless steel food equip./400 stainless.
Braze 650	BAg-9		BAg-9	1240	670	1325	720	65	20	15		For silverware, Fe and Ni alloys.
Braze 720	BAg-8			1435	780	1435	780	72	28			Electronics where Cd & Zn not tolerated.
Hi-Temp 095	4764B			1615	880	1770	925		52.5		38Mn, 9.5Ni	High strength for carbides, steels.
Hi-Temp 870				1760	960	1885	1030		87		10Mn, 3 Co	High temperature strength for carbides, tool steel, stainless and nickel alloys.
Premabraze 616	BVAg-29			1155	625	1305	705	61.5	24		14.5 In	For ferrous, non-ferrous in vacuum.
Premabraze 130	BAu4	4787B		1742	950	1742	950				82 Au, 18 Ni	Oxidation resistant to 1500°F, service.

Figure 14.63 Standard brazing materials.

was taken from a piece of automotive equipment that failed in service. All the copper foil lines on the printed circuit board are covered with solder, with the termination points neatly made.

Soldering aluminum. Aluminum soldering employs a soldering alloy of 60 to 75 percent tin, the remainder being zinc. Special aluminum fluxes are available, and the process is carried out at a temperature of 550 to 775°F.

Commercial and high-purity aluminums are easiest to solder, with wrought alloys that contain no more than 1 percent magnesium or manganese the next easiest. The heat-treatable alloys are more difficult. Forged and cast aluminum alloys are not recommended for soldering, although it may be possible to a limited extent. The low-temperature brazing alloys may do the job more efficiently, with a higher-strength joint possible.

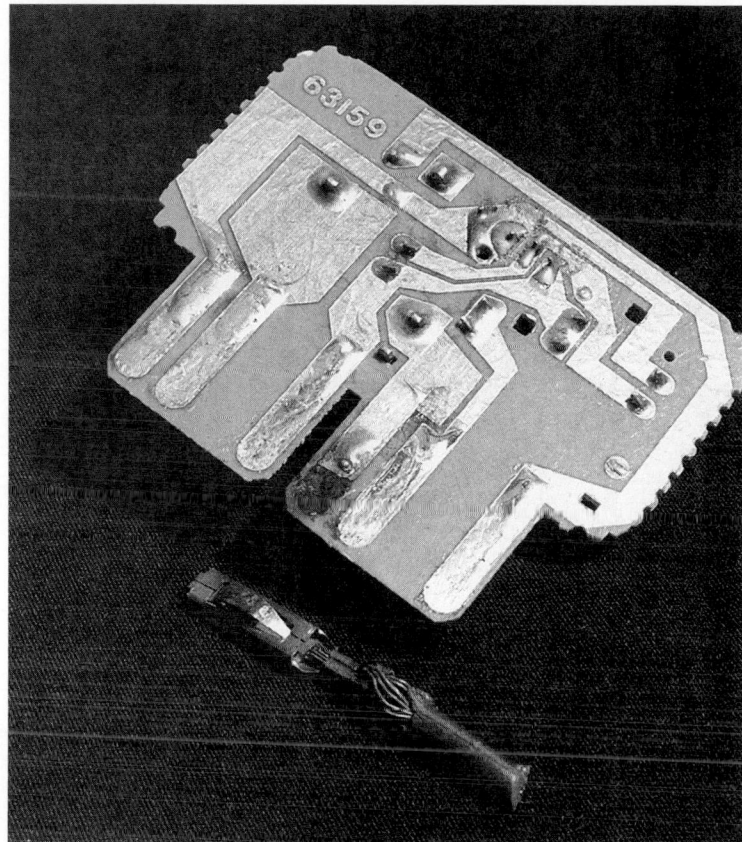

Figure 14.64 An electrically damaged PC board with solder-coated foil.

Ultrasonic soldering. The following metals and alloys may be soldered ultrasonically: aluminum, brass, copper, germanium, magnesium, silicon, and silver.

14.10 Adhesive Bonding

Adhesive bonding is used very frequently in modern manufacturing operations. The methods generally are fast, strong, and economical when used in the proper manner on appropriate articles. Entire fuselage sections are bonded or cemented on modern aircraft to

replace riveting. The method not only affords lighter weight but also is stronger and requires no maintenance. The bonding strength of cyanoacrylate adhesives is on the order of thousands of pounds of tension for 1 in^2 of bonded surface area.

Structural adhesives. Following is a list of the various types of adhesives used in industrial applications:

- Hot-melt adhesives, thermoplastic resins—100 percent solids

- Dispersion or solution adhesives, thermoplastic resins—20 to 50 percent solids

- Silicone adhesives, thermoset—100 percent solids (rubber-like with high impact and peel strengths)

- Anaerobic adhesives, thermoset—100 percent solids, liquid (thread-locking uses such as Loctite series adhesives)

- Phenolic or urea adhesives, thermoset resins—100 percent solids or solution

- Epoxy adhesives, thermoset—100 percent solids, liquid

- Polyurethane adhesives, thermoset—100 percent solids, liquid

- Cyanoacrylate adhesives, thermoset—100 percent solids, liquid (hazard of bonding skin on contact; almost instant set; 30-second to 5-minute cure; high strength)

- Modified acrylic adhesives, thermoset—100 percent solids, liquid, or paste (fast cure, 3 to 60 seconds)

Weldbonding. A combination of spot welding and adhesive bonding, this process is used on aerospace assemblies to cut costs, reduce weight, and increase strength. The method affords lighter weight than riveting, is stronger, yields air- and fuel-tight joints, and produces a smooth exterior surface that reduces wind resistance. The structural adhesives are applied prior to the spot-welding operation, and then the bonded and welded assembly is oven cured at an elevated temperature for a predetermined time interval.

15

Safety Practices
in Industry

The metalworking industries in the United States and throughout the world perform their functions and manufacturing practices under a variety of conditions, some of which are hazardous and may pose health problems to metalworkers. In order to protect the people involved in all classes of industrial manufacturing processes and situations, the Occupational Safety and Health Administration (OSHA) was created by the United States government.

OSHA is responsible for generating and administering the rules and regulations that were designed to protect American workers from industrial and manufacturing practices and materials that have proven to be safety and health hazards. For example, for many years, workers handled and processed asbestos without the slightest idea or knowledge that this material was dangerous to their health and in many instances caused fatal health problems. Likewise, processes for mercury and cadmium metals can pose serious health and environmental problems, as can the indiscriminate handling and disposal of toxic waste products.

In order to help protect the environment, the Environmental Protection Agency (EPA) was established. This agency generates and administers the rules and regulations designed to help protect the environment from poisonous and otherwise harmful chemicals and materials used throughout industry. It should be noted that many of the heavy metals and their compounds are toxic. Many

organic, carbon-based materials and chemicals also are toxic, some to an extreme extent. Many materials and compounds are also nonbiodegradable, which means that they will remain in the environment indefinitely as waste products.

Bearing the preceding in mind, we can readily see why these regulating agencies and administrations are necessary in the modern industrial world. Section 15.1.2 is a reference listing of the OSHA standards for use in any of the categories currently under regulation. The standard you are interested in checking may be obtained directly from OSHA or your company risk manager. All manufacturing installations are required to keep the OSHA and EPA regulating standards on their premises and make them available to any employee interested in obtaining information from these regulatory standards.

If you have doubts about the possible hazards of the materials, substances, and processes you use or come into contact with during your job duties, check the applicable OSHA or EPA standards. Modern American industrial employees are no longer required to forfeit their health or lives because of their job requirements, which may be in violation of the OSHA and EPA regulatory standards.

15.1 Typical Industrial Hazards

The following list of safety and health hazards is only a minute example of the conditions, materials, and substances that a metalworking employee could be subjected to:

- Welding over cadmium-plated parts (highly toxic fumes from the cadmium)

- Case-hardening steel parts with cyanide compounds without proper ventilation and equipment

- Fumes from electroplating processes

- Grinding and machining beryllium alloys (beryllium-copper, beryllium-nickel)

- Machining and grinding magnesium

- Machining plastics with fiberglass fillers without ventilation and evacuation equipment

- Skin contact with plating solutions

- Fumes from metal etching processes

- Foot and hand injuries when handling large sheet metal parts
- Injuries on machine tools not equipped with guards and shields
- Injuries on stamping dies owing to inadequate or missing shields
- Ultraviolet radiation burns and eye injuries from welding processes
- Paint fumes in undesignated painting areas

15.1.1 Safety practices when operating machine tools

Many grave injuries occur in the operation of machine tools, especially during manual operation. The following list of precautions will help to prevent injury to machine tool operators, machinists, and tool and die makers:

- Wear the proper eye protection.
- Know exactly what to expect your machine to do before you turn on the power switch.
- Do not wear neckties or any loose clothing while operating a machine tool; the sleeves of long-sleeve shirts also must be rolled up past the elbow.
- Do not wear rings, watches, neck chains, or wrist bracelets while operating machine tools.
- Do not wear gloves while operating machine tools.
- Know instinctively where the power switch is located on your machine tool so that it may be switched off quickly in an emergency.
- Keep your work area clear of obstructions.

15.1.2 OSHA standards by category

See Fig. 15.1.

15.2 Product Liability

Safety in product design should be of prime importance to all product designers. There have been over 500 awards of more than $1

(Text continued on page 955.)

PART 1910

OCCUPATIONAL SAFETY AND HEALTH STANDARDS

SUBPART A—GENERAL

1910.1 Purpose and scope.
1910.2 Definitions.
1910.3 Petitions for the issuance, amendment, or repeal of a standard.
1910.4 Amendments to this part.
1910.5 Applicability of standards.
1910.6 Incorporation by reference.
1910.7 Definition and requirements for a nationally recognized testing laboratory.

SUBPART B—ADOPTION AND EXTENSION OF ESTABLISHED FEDERAL STANDARDS

1910.11 Scope and purpose.
1910.12 Construction work.
1910.13 Ship repairing.
1910.14 Shipbuilding.
1910.15 Shipbreaking.
1910.16 Longshoring.
1910.17 Effective dates.
1910.18 Changes in established Federal standards.
1910.19 Special provisions for air contaminants.

SUBPART C—GENERAL SAFETY AND HEALTH PROVISIONS

1910.20 Access to employee exposure and medical records.

SUBPART D—WALKING—WORKING SURFACES

1910.21 Definitions.
1910.22 General requirements.
1910.23 Guarding floor and wall openings and holes.
1910.24 Fixed industrial stairs.
1910.25 Portable wood ladders.
1910.26 Portable metal ladders.
1910.27 Fixed ladders.
1910.28 Safety requirements for scaffolding.
1910.29 Manually propelled mobile ladder stands and scaffolds (towers).
1910.30 Other working surfaces.
1910.31 Sources of standards.
1910.32 Standards organizations.

SUBPART E—MEANS OF EGRESS

1910.35 Definitions.
1910.36 General requirements.
1910.37 Means of egress, general.
1910.38 Specific means of egress requirements by occupancy. [Reserved]

1

Figure 15.1 Table of OSHA standards.

1910.39 Sources of standards.
1910.40 Standards organizations.

SUBPART F—POWERED PLATFORMS, MANLIFTS AND VEHICLE-MOUNTED WORK PLATFORMS

1910.66 Power platforms for building maintenance.
1910.67 Vehicle-mounted elevating and rotating work platforms.
1910.68 Manlifts.
1910.69 Sources of standards.
1910.70 Standards organizations.

SUBPART G—OCCUPATIONAL HEALTH AND ENVIRONMENTAL CONTROL

1910.94 Ventilation.
1910.95 Occupational noise exposure.
1910.96 Ionizing radiation.
1910.97 Nonionizing radiation.
1910.98 Additional delay in effective date.
1910.99 Sources of standards.
1910.100 Standards organizations.

SUBPART H—HAZARDOUS MATERIALS

1910.101 Compressed gases (general requirements).
1910.102 Acetylene.
1910.103 Hydrogen.
1910.104 Oxygen.
1910.105 Nitrous oxide.
1910.106 Flammable and combustible liquids.
1910.107 Spray finishing using flammable and combustible materials.
1910.108 Dip tanks containing flammable or combustible liquids.
1910.109 Explosives and blasting agents.
1910.110 Storage and handling of liquified petroleum gases.
1910.111 Storage and handling of anhydrous ammonia.
1910.112 [Reserved]
1910.113 [Reserved]
1910.114 Effective dates.
1910.115 Sources of standards.
1910.116 Standards organizations.
1910.120 Hazardous Waste Operations and Emergency Response

SUBPART I—PERSONAL PROTECTIVE EQUIPMENT

1910.132 General requirements.
1910.133 Eye and face protection.
1910.134 Respiratory protection.
1910.135 Occupational head protection.
1910.136 Occupational foot protection.
1910.137 Electrical protective devices.
1910.138 Effective dates.
1910.139 Sources of standards.
1910.140 Standards organizations.

2

Figure 15.1 *(Continued)*

SUBPART J—GENERAL ENVIRONMENTAL CONTROLS

1910.141 Sanitation.
1910.142 Temporary labor camps.
1910.143 Nonwater carriage disposal systems.
1910.144 Safety color code for marking physical hazards.
1910.145 Specificatons for accident prevention signs and tags.
1910.146 [Reserved]
1910.147 Sources of standards.
1910.148 Standards organizations.
1910.149 Effective dates.

SUBPART K—MEDICAL AND FIRST AID

1910.151 Medical services and first aid.
1910.152 [Reserved]
1910.153 Sources of standards.

SUBPART L—FIRE PROTECTION

1910.155 Scope, application and definitions applicable to this part.
1910.156 Fire brigades.

PORTABLE FIRE SUPPRESSION EQUIPMENT

1910.157 Portable fire extinguishers.
1910.158 Standpipe and hose systems.

FIXED FIRE SUPPRESSION EQUIPMENT

1910.159 Automatic sprinkler systems
1910.160 Fixed extinguishing systems, general.
1910.161 Fixed extinguishing systems, dry chemical.
1910.162 Fixed extinguishing systems, gaseous agent.
1910.163 Fixed extinguishing systems, water, spray and foam.

OTHER FIRE PROTECTION SYSTEMS

1910.164 Fire detection systems
1910.165 Employee alarm systems
Appendices A–E

SUBPART M—COMPRESSED GAS AND COMPRESSED AIR EQUIPMENT

1910.166 Reserved
1910.167 Reserved
1910.168 Reserved

1910.169 Air receivers.
1910.170 Sources of standards.
1910.171 Standards organizations.

SUBPART N—MATERIALS HANDLING AND STORAGE

1910.176 Handling materials—general.
1910.177 Servicing multi-piece and single piece rim wheels.
1910.178 Powered industrial trucks.

3

Figure 15.1 *(Continued)*

4

Figure 15.1   (*Continued*)

Subpart T—Commercial Diving Operations

SUBPART Z—TOXIC AND HAZARDOUS SUBSTANCES

5

Figure 15.1 *(Continued)*

1910.1016	N-Nitrosodimethylamine.
1910.1017	Vinyl Chloride.
1910.1018	Inorganic arsenic.
1910.1025	Lead.
1910.1028	Benzene.
1910.1029	Coke Oven Emissions
1910.1043	Cotton dust.
1910.1044	1,2 - dibromo - 3 - chloropropane.
1910.1045	Acrylonitrile.
1910.1047	Ethylene Oxide
1910.1048	Formaldehyde
1910.1011	Asbestos
1910.1200	Hazard communication.
1910.1450	Occupational Exposure to Hazardous Chemicals in Laboratories.

AUTHORITY: The provisions of this Part 1910 issued under secs. 6(a), 8(g), 84 Stat. 1593, 1598; 29 U.S.C. 655, 657.

6

Figure 15.1 (*Continued*)

million granted in the United States since 1963 arising from product liability lawsuits brought against manufacturers. It is of prime importance for a product designer or design group never to sacrifice safety for profit. Some of the criteria to consider and make note of during the design stages of a product are

1. All possible hazards in using or misusing the product

2. The environment in which the product is used

3. The typical user of the product and the typical nonuser

4. All instructions and warnings that are to be presented with the product

Nearly all products can be dangerous to one degree or another if misapplied. This leaves the product designer with much to consider when designing a product. Some of the strongest safeguards available to designers and manufacturers against product liability lawsuits are

1. The use of prototypes and fully functional models to demonstrate a product's characteristics. Also important are quality-control tests of random-production-run samples of the product.

2. Accurate records of qualitative test results, including high-speed film records and laboratory test equipment readings.

3. The application of warning labels and the supplying of accurate instructions with the product in the form of instruction books and product brochures. Warning labels must attract the attention of the user, be clear and understandable, and finally, convey the nature and extent of the probable harm resulting from misuse or failure to follow the warnings and instructions issued with the product.

Following these safeguards, the product designer and manufacturer will be in a strong position if a product liability lawsuit is brought against the manufacturer in relation to one of its products.

Chapter

16

Societies, Associations, Institutes, and Specification Authorities

Following is a listing of recognized specification authorities, societies, and institutes from which the design engineer can obtain specifications and standards covering many areas of engineering design and manufacturing.

Many of the standards are revised periodically by these authorities to keep pace with changing technology. It is therefore suggested that copies of the American standards you require to perform your job function be obtained directly from the standards organizations that generate them. In this way, you are assured of using the most recent standard and its revisions. Updates to the standards listings are distributed periodically to those on the organization mailing lists, or annual standards listing manuals may be ordered directly from the standards organizations.

Many of the standards organizations publish handbooks and manuals related to their particular fields that are technically excellent. Indexes to the standards are also available for pinpointing your area of interest in the form of annual standards catalogs.

16.1 Standards Organizations and Their Acronyms

AAR—Association of American Railroads
Washington, DC 20001-1564
www.aar.org

ACA—American Chain Association
Naples, FL 34109
www.americanchainassn.org

ABMA—American Bearing Manufacturers Association
Washington, DC 20036
www.abma-dc.org

AGMA—American Gear Manufacturers Association
Alexandria, VA 22314-1581
www.agma.org

AIA—Aerospace Industries Association
Arlington, VA 22209-3901
www.aia-aerospace.org

AISC—American Institute of Steel Construction
Chicago, IL 60601-2001
www.aisc.org

AISI—American Iron and Steel Institute
Washington, DC 20036
www.steel.org

The Aluminum Association
Washington, DC 20006
www.aluminum.org

ANSI—American National Standards Institute, Inc.
Washington, DC 20036
www.ansi.org

API—American Petroleum Institute
Washington, DC 20005-4070
www.api.org

ASM International—American Society for
Materials International
Materials Park, OH 44073-0002
www.asminternational.org

ASME—American Society of Mechanical Engineers
New York, NY 10016-5990
www.asme.org

ASTLE—American Society of Tribologists and
Lubrication Engineers
Park Ridge, IL 60068
www.stle.org

ASTM International—American Society for Testing and Materials
West Conshohocken, PA 19428-2959
www.astm.org

AWS—American Welding Society
Miami, FL 33126
www.aws.org

BSA—Bearing Specialists Association
Glen Ellyn, IL 60137
www.bsahome.org

CDA—Copper Development Association, Inc.
New York, NY 10016
www.copper.org

EEI—Edison Electric Institute
Washington, DC 20004-2696
www.eei.org

FEMA—Farm Equipment Manufacturers Association
St. Louis, MO 63141-6369
www.farmequip.org

IEEE—Institute of Electrical and Electronics Engineers
New York, NY 10016-5997
www.ieee.org

IFI—Industrial Fasteners Institute
Cleveland, OH 44114
www.industrial-fasteners.org

IIE—Institute of Industrial Engineers
Atlanta, Norcross, GA 30092
www.iienet.org

ISA—Instrument Society of America
Research Triangle Park, NC 27709
www.isa.org

MPIF—Metal Powder Industries Federation
Princeton, NJ 08540
www.mpif.org

MPTA—Mechanical Power Transmission Association
Naples, FL 34109
www.mpta.org

NEMA—National Electrical Manufacturers Association
Rosslyn, VA 22209
www.nema.org

NFFS—Non-Ferrous Founder's Society
Park Ridge, IL 60068
www.nffs.org

NFPA—National Fire Protection Association
(National Electrical Code)
Quincy, MA 02169-7471
www.nfpa.org

NIBA—National Industrial Belting Association
Philadelphia, PA 19103
www.niba.org

NIST—National Institute of Standards and Technology
Gaithersburg, MD 20899
www.nist.gov

NLGI—National Lubricating Grease Institute
Kansas City, MO 64112
www.nlgi.com

NSC—National Safety Council
Itasca, IL 60143-3201
www.nsc.org

RMA—Rubber Manufacturers Association
Washington, DC 20005
www.rma.org

RSA—Robotics Society of America
www.robotics-society.org

SAE—Society of Automotive Engineers
Warrendale, PA 15096
www.sae.org

SME—Society of Manufacturing Engineers
Dearborn, MI 48128
www.sme.org

SMI—Spring Manufacturers Institute
Oak Brook, IL 60523-1335
www.smihq.org

UL—Underwriters Laboratories, Inc.
Northbrook, IL 60062-2069
www.ul.com

16.2 Quality-Control Systems

Various standards for controlling the quality of industrial and consumer products have been implemented both nationally and internationally. These standards are important to manufacturers, product design engineers, and engineering departments. In the United States, a system known as *total quality management* (TQM) has been in effect for some years. Other quality-control systems have been implemented nationally and on a company-level basis. These systems help the consumer as well as the manufacturer.

Internationally, the quality-control or quality-assurance system of most importance to American manufacturers is the International Standards Organization (ISO) 9000 system. On January 1, 1993, the European Community (EC), consisting of 12 countries, officially formed a unified market. This began the adoption of the ISO 9000 system. All U.S. companies wishing to sell products to the European market will use the ISO 9000 system as their quality standard.

ISO 9000 establishes a minimum quality standard that may be expanded on, based on future requirements. The U.S. Department of Defense may replace standards MIL-I-45208A and MIL-Q-9858A with the ISO 9000 system. After January 1, 1993, U.S. companies may be required to be in compliance with ISO 9000 in order to continue to do business with the EC nations. Note that the American TQM system will exceed the stated ISO 9000 guidelines. Sixty countries have adopted ISO 9000 as their national quality standard.

The EC Consumer Council Ministers have adopted a general product safety directive. This directive imposes general safety requirements on all consumer products sold in the EC market. U.S. exporters to the EC must familiarize themselves with the EC product liability regulations. Use of the ISO 9000 standard may be mandatory in the near future in order for U.S. companies to sell their products to the EC and other nations that have adopted the ISO 9000 standard.

16.3 ISO 9000 System

The ISO 9000 system is divided into a series of standards, including

- ISO 9001—model for quality assurance in design and development, production, installation, and servicing
- ISO 9002—model for quality assurance in production and installation

- ISO 9003—model for quality assurance in final inspection and test
- ISO 9004—guidelines to quality management elements

ISO 9001, 9002, and 9003 are audited quality-control standards, whereas ISO 9004 contains the guidelines to quality management elements and is not audited. The ISO standard your company may require depends on your manufacturing and product setup.

Additional information on the entire ISO 9000 system may be obtained from the following organizations:

American Society for Quality (ASQ)
Milwaukee, WI 53203
(800) 248-1946
www.asq.org

National Technical Information Service (NTIS)
Springfield, VA 22161
(703) 605-6000
www.ntis.gov

U.S. Government Printing Office (GPO)
Washington, DC 20402
(202) 783-3238
www.gpo.gov

17

American National Standards Applicable to Machinery, Machining, and Metalworking Practices

The American National Standards Institute (ANSI) issues standards in the form of published pamphlets that define the geometry, dimensions, inspection limits, test procedures and other control data, and specifications important to the design and manufacture of thousands of items. Components, materials, and specifications such as screw-thread systems, bolts, screws, washers, nuts, splines, pins, cutting tools, mechanical devices, various equipment and machinery, and a host of other items of importance to the machining, metalworking, and mechanical industries are all defined and specified in the various American national standards.

The purpose of these standards is to define the various physical, chemical, electrical, and mechanical characteristics of materials, components, systems, assemblies of equipment, and inspection and testing procedures. The American national standards provide a means for obtaining order and conformity among American manufactured products.

On August 24, 1969, the American Standards Association (ASA) was restructured as the United States of America Standards Institute, and standards that were approved as American standards were designated as USA standards. On October 6, 1969, the name

was then changed to American National Standards Institute (ANSI). The present standards designation is ANSI instead of ASA or USAS.

The American National Standards Institute (ANSI) works in collaboration with other national organizations such as the American Society of Testing and Materials (ASTM), the Society of Automotive Engineers (SAE), the American Welding Society (AWS), the American Society of Mechanical Engineers (ASME), the American Gear Manufacturers Association (AGMA), the American Iron and Steel Institute (AISI), the Institute of Electrical and Electronics Engineers (IEEE), and others in an effort to consolidate the American standards data and publications generated by these other national organizations. The metalworking industries and other industries nationwide depend on the combined national standards of all the various American societies, institutes, and associations in order to have guidelines, specifications, and design and test procedures for manufacturing their products.

Many products and materials are required by purchasing specifications to conform to the various American standards and will not be accepted by a purchaser unless they do conform to the specified standards. For example, when a design engineer specifies a material on his or her design drawing of a spring, the material as listed on the drawing may be 0.156-in-diameter music wire per ASTM-A228.

When the material is delivered to the spring manufacturer, it should conform to the specifications of this standard designation (ASTM-A228) both physically and chemically. If this specified material were to be tested and analyzed at a materials test laboratory, it would be required to conform to the ASTM-A228 specification. Failure of the material to conform to the specification could result in part failure in service.

The designations of the various ANSI standards that are of prime importance to the machining, metalworking, and mechanical industries are listed in this chapter by subject category. It was the author's and publisher's decision to not include extracts of the various ANSI standards in this *Handbook* for the following reasons:

- The basic standards applicable to the machining, metalworking, and mechanical industries are too extensive to republish.

- The standards are being revised constantly to keep pace with changing technologies.

- Whenever a standard is revised, in effect, a handbook containing these standards would be out of date and contain obsolete data.

Companies that rely on the data contained in the national standards published by ANSI and the other standards organizations should keep copies of the standards that apply to their work in the standards or engineering departments of their organizations. The ANSI standards listed in this chapter are considered by the author to be the main, basic standards required for the machining and metalworking industries. Standards from the ASTM and SAE for materials are extracted in Chap. 4, "Materials and Their Uses."

Most of the national associations, societies, and institutes that generate American standards applicable to the metalworking and electromechanical industries are listed in Chap. 16 of this *Handbook,* together with their current addresses. Standards may be purchased directly from these organizations, and most of the listed organizations have catalogs available that specify the various technical publications they produce.

17.1 Listing of ANSI Standards by Category

1. Thread systems

2. Fastening and joining devices

3. Machining practices

4. Tools and tooling

5. Mechanical components

6. Welding

7. Heat treatment

8. Tolerances and fits

9. Drawing symbols and formats

10. Gauging and inspection

Category 1 : Thread systems

ANSI B1.9-1973, *Buttress Inch Screw Threads*

ANSI B1.10-1958 (R1988), *Unified Miniature Screw Threads*

ANSI B1.1l-1958 (R1989), *Microscope Objective Thread*

ANSI B1.18M-1982 (R1987), *Metric Screw Threads for Commercial Mechanical Fasteners—Boundary Profile Defined*

ANSI B1.20.3-1976 (R1982), *Dry Seal Pipe Threads (Inch)*

ANSI B1.20.4-1976 (R1982), *Dry Seal Pipe Threads* (metric translation of B1.20.3-1976)

ANSI/ASME B1.1-1989, *Unified Inch Screw Threads (UN and UNR Thread Form)*

ANSI/ASME B1.5-1988, *Acme Screw Threads*

ANSI/ASME B1.7M-1984, *Screw Threads, Definitions and Letter Symbols*

ANSI/ASME B1.8-1988, *Stub Acme Screw Threads*

ANSI/ASME B1.12-1987, *Screw Threads—Class 5 Interference Fit Thread*

ANSI/ASME B1.13M-1983 (R1989), *Metric Screw Threads-M Profile*

ANSI/ASME B1.20.1-1983, *Pipe Threads (General Purpose), Inch*

ANSI/ASME B1.20.7-1966 (R1983), *Hose Coupling Screw Threads*

Category 2: Fastening and joining devices

ANSI B18.1.1-1972 (R1989), *Small Solid Rivets (0.4375 In Diameter and Under)*

ANSI B18.1.2-1972 (R1989), *Large Rivets (0.500 In Diameter and Over)*

ANSI B18.2.1-1981, *Square and Hex Bolts and Screws, Inch Series*

ANSI B18.2.3.1M-1979 (R1989), *Screws, Metric Hex Cap*

ANSI B18.2.3.2M-1979 (R1989), *Screws, Metric Formed Hex*

ANSI B18.2.3.3M-1979 (R1989), *Screws, Metric Heavy Hex*

ANSI B18.2.3.5M-1979 (R1989), *Bolts, Metric Hex*

ANSI B18.2.3.6M-1979 (R1989), *Bolts, Metric Heavy Hex*

ANSI B18.2.3.7M-1979 (R1989), *Bolts, Metric Heavy Hex Structural*

ANSI B18.2.3.8M-1981, *Screws, Metric Hex Lag*

ANSI B18.2.4.1M-1979 (R1989), *Hex Nuts, Style 1, Metric*

ANSI B18.2.4.2M-1979 (R1989), *Hex Nuts, Style 2, Metric*

ANSI B18.2.4.3M-1979 (R1989), *Hex Nuts, Slotted, Metric*

ANSI B18.2.4.4M-1982, *Nuts, Metric Hex Flange*

ANSI B18.2.4.5M-1979 (R1990), *Hex Jam Nuts, Metric*

ANSI B18.2.4.6M-1979 (R1990), *Hex Nuts, Heavy, Metric*

ANSI B18.5.2.1M-1981, *Bolts, Metric Round Head, Short Square Neck*

ANSI B18.6.1-1981, *Wood Screws, Inch Series*

ANSI B18.6.2-1972 (R1983), *Slotted Head Cap Screws, Square Head Set Screws, and Slotted Headless Set Screws*

ANSI B18.6.3-1972 (R1983), *Slotted and Recessed Head Machine Screws and Machine Screw Nuts*

ANSI B18.6.4-1981, *Screws, Tapping and Metallic Drive, Inch Series, Thread Forming and Cutting*

ANSI B18.7-1972 (R1980), *Semitubular Rivets, Full Tubular Rivets, Split Rivets, and Rivet Caps, General Purpose*

ANSI B18.8.1-1972 (R1983), *Clevis Pins and Cotter Pins*

ANSI B18.8.2-1978 (R1989), *Pins-Taper Pins, Dowel Pins, Straight Pins, Grooved Pins, and Spring Pins, Inch Series*

ANSI B18.9-1958 (R1989), *Plow Bolts*

ANSI B18.11-1961 (R1983), *Miniature Screws*

ANSI B18.17-1968 (R1983), *Wing Nuts, Thumb Screws, and Wing Screws*

ANSI B18.22M-1981, *Washers, Metric Plain*

ANSI B18.22.1-1965 (R198I), *Plain Washers*

ANSI/ASME B18.1.3M-1983 (RI989), *Metric Small Solid Rivets*

ANSI/ASME B18.2.2-1987, *Square and Hex Nuts, Inch Series*

ANSI/ASME B18.2.3.4M-1984, *Screws, Metric Hex Flange*

ANSI/ASME B18.2.3.9M-1984, *Metric Heavy Hex Flange Screws*

ANSI/ASME B18.3-1986, *Socket Cap, Shoulder and Set Screws, Inch Series*

ANSI/ASME B18.3.1M-1986, *Screws, Socket Head Cap, Metric Series*

ANSI/ASME B18.3.3M-1986, *Hexagon Socket Head Shoulder Screws, Metric Series*

ANSI/ASME B18.3.4M-1986, *Screws, Hexagon Socket Button Head Cap, Metric*

ANSI/ASME B18.3.5M-1986, *Hexagon Socket Flat Countersunk Head Cap Screws, Metric Series*

ANSI/ASME B18.3.6M-1986, *Screws, Hexagon Socket Set, Metric Series*

ANSI/ASME B18.5.2.2M-1982, *Bolts, Metric Round Head Square Neck*

ANSI/ASME B18.6.5M-1986, *Metric Thread Forming and Thread Cutting Tapping Screws*

ANSI/ASME B18.6.7M-1985, *Metric Machine Screws*

ANSI/ASME B18.13-1987, *Screw and Washer Assemblies, SEMs, Inch Series*

ANSI/ASME B18.15-1985, *Forged Eyebolts*

ANSI/ASME B18.21.1-1990, *Lock Washers*

ANSI/ASME B18.21.2M-1990, *Lock Washers, Metric Series*

Category 3: Machining

ANSI B5.8-1972 (R1988), *Chucks and Chuck Jaws*

ANSI B5.10-1981 (R1987), *Machine Tapers*

ANSI B5.16-1952 (R1986), *Accuracy of Engine and Tool Room Lathes*

ANSI B17.1-1967 (R1989), *Keys and Keyseats*

ANSI B17.2-1967 (R1978), Woodruff Keys and Keyseats

ANSI B74.2-1982, *Shapes and Sizes of Grinding Wheels and Shapes, Sizes, and Identification of Mounted Wheels, Specifications for*

ANSI B74.13-1990, *Markings for Identifying Grinding Wheels and Other Bonded Abrasives*

ANSI B94.2-1983 (R1988), *Reamers*

ANSI B94.3-1965 (R1984), *Straight Cutoff Blades for Lathes and Screw Machines*

ANSI B94.7-1980 (R1987), *Hobs*

ANSI B94.8-1967 (R1987), *Inserted Blade Milling Cutter Bodies*

ANSI B94.llM-1979 (R1987), *Twist Drills, Straight Shank and Taper Shank Combined Drills and Countersinks*

ANSI B94.21-1986 (R1987), *Gear Shaper Cutters*

ANSI B94.49-1975 (R1986), *Spade Drill Blades and Spade Drill Holders*

ANSI/ASME B5.1M-1985, *T-Slots—Their Bolts, Nuts, and Tongues*

ANSI/ASME B94.6-1984, *Knurling*

ANSI/ASME B94.9-1987, *Taps, Cut and Ground Threads*

ANSI/ASME B94.19-1985, *Milling Cutters and End Mills*

Category 4: Tools and tooling accessories

ANSI B5.25-1978 (R1986), *Punch and Die Sets, Inch*

ANSI B5.25M-1980 (R1986), *Punch and Die Sets, Metric*

ANSI B94.14-1968 (R1987), *Punches—Basic Head Type*

ANSI B94.14.1-1977 (R1984), *Punches—Basic Head Type, Metric*

ANSI B94.33-1974 (R1986), *Jig Bushings*

ANSI B107.0-1978 (R1987), *Box, Open End, Combination and Flare Nut Wrenches, Inch Series*

ANSI B107.9-1978 (R1987), *Box, Open End, Combination and Flare Nut Wrenches, Metric Series*

ANSI B107.10M-1982 (R1988), *Socket Wrenches, Handles and Attachments for Hand, Inch and Metric Series*

ANSI/ASME B107.5M-1987, *Socket Wrenches, Hand, Metric*

ANSI/ASME B107.8M-1984, *Adjustable Wrenches*

ANSI/ASME B107.19-1987, *Pliers, Retaining Ring*

Category 5: Mechanical components

ANSI B29.2M-1982 (R1987), *Inverted Tooth (Silent) Chains and Sprockets*

ANSI B29.6M-1983 (R1988), *Steel Detachable Link Chains and Sprockets*

ANSI B29.10M-1981 (R1987), *Heavy Duty Offset Sidebar Transmission Roller Chains and Sprocket Teeth*

ANSI B29.15-1973 (R1987), *Heavy Duty Roller Type Conveyor Chains and Sprocket Teeth*

ANSI B29.19-1976 (R1987), *A and CA550 and 620 Roller Chains, Attachments and Sprockets*

ANSI B92.1-1970 (R1982), *Involute Splines and Inspection, Inch Version*

ANSI B92.2M-1981 (R1989), *Involute Splines, Metric Module*

ANSI/ASME B29.1M-1986, *Precision Power Transmission Roller Chains, Attachments and Sprockets*

Category 6: Welding

ANSI/AWS D9.1-90, *Sheet Metal Welding Code*

ANSI/AWS D10.12-89, *Recommended Practices and Procedures for Welding Low-Carbon Steels*

ANSI/AWS D11.2-89, *Guide for Welding Iron Castings*

ANSI/AWS D14.2-86, *Machine Tool Weldments, Specification for Metal Cutting*

Category 7: Heat treatment

ANSI/SAE AMS 2728, *Heat Treatment of Wrought Copper Beryllium Alloy Parts*

ANSI/SAE AMS 2756, *Gas Nitriding of Steel Parts*

ANSI/SAE AMS 2757, *Gaseous Nitrocarburizing*

ANSI/SAE AMS 2759, *Heat Treatment of Steel Parts, General Requirements*

ANSI/SAE AMS 2759/3, *Heat Treatment of Precipitation Corrosion Resisting and Maraging Steel Parts*

ANSI/SAE AMS 2759/4, *Heat Treatment of Austenitic Corrosion Resistant Steel Parts*

ANSI/SAE AMS 2759/5, *Heat Treatment of Martensitic Corrosion Resistant Steel Parts*

ANSI/SAE AMS 2759/6, *Heat Treatment and Gas Nitriding of Low-Alloy Steel Parts*

ANSI/SAE AMS 2760A, *Heat Treatment-Carbon, Low-Alloy, and Specialty Steels*

ANSI/SAE AMS 2770D, *Heat Treatment of Aluminum Alloy Parts*

ANSI/SAE AMS 2775A, *Case Hardening of Titanium and Titanium Alloys*

Category 8: Tolerances and fits

ANSI B4.1-1967 (R1987), *Preferred Limits and Fits for Cylindrical Parts*

ANSI B4.2-1978 (R1984), *Preferred Metric Limits and Fits*

ANSI B4.3-1978 (R1984), *General Tolerances for Metric Dimensioned Products*

ANSI B89.3.1-1972 (R1988), *Out-of-Roundness, Measurement of*

ANSI B89.6.2-1973 (R1988), *Temperature and Humidity Environment for Dimensional Measurement*

ANSI Y14.5M-1982 (R1988), *Dimensioning and Tolerancing*

ANSI/ASME B1.22M-1985, *Gauges and Gauging Practice for M-J Series Metric Screw Threads*

ANSI/ASME B107.17M-1985, *Gauges, Wrench Openings, Reference*

Category 9: Drawing symbols and formats

ANSI Y10.20-1975 (R1988), *Mathematic Signs and Symbols for Use in Physical Sciences and Technology*

ANSI Y14.1-1980 (R1987), *Drawing Sheet Size and Format*

ANSI Y14.7.1-1971 (R1988), *Gear Drawing Standards, Part 1: For Spur, Helical, Double Helical, and Rack*

ANSI Y14.7.2-1978 (R1989), *Gear and Spline Drawing Standards, Part 2: Bevel and Hypoid Gears*

ANSI Y14.17-1966 (R1987), *Fluid Power Diagrams*

ANSI Y14.36-1978 (R1987), *Surface Texture Symbols*

ANSI Y32.10-1967 (R1987), *Graphic Symbols for Fluid Power Diagrams*

Category 10: Gaging and inspection

ANSI B4.4M-1981 (R1987), *Inspection of Work Pieces*

ANSI B89.3.1-1972 (R1988), *Out-of-Roundness, Measurement of*

ANSI/ASME B1.2-1983, *Gauges and Gauging for Unified Screw Threads*

ANSI/ASME B1.3M-1986, *Gauging Systems for Dimensional Acceptability, Inch and Metric Screw Threads (UN, UNR, UNJ, M, and MJ)*

ANSI/ASME B1.16M-1984, *Gauges and Gauging for Metric M Screw Threads*

ANSI/ASME B1.19M-1984, *Gauges for Metric Screw Threads for Commercial Mechanical Fasteners—Boundary Profile Defined*

ANSI/ASME B107.17M-1985, *Gauges, Wrench Openings, Reference*

17.2 Standards and Approval Agencies and Acronyms

AEC	Atomic Energy Commission
AGA	American Gas Association
AGMA	American Gear Manufacturers Association
AHAM	Association of Home Appliance Manufacturers
AMCA	Air Movement and Control Association
AN	Army-Navy Standard
ANSI	American National Standards Institute
ARA	American Refrigeration Association
ARL	Applied Research Laboratories
ASA	American Standards Association (Now ANSI)
ASHRAE	American Society of Heating, Refrigeration, and Air-Conditioning Engineers
ASME	American Society of Mechanical Engineers
ASSE	American Society of Sanitary Engineering
CEC	California Energy Commission
CGA	Canadian Gas Association

CSA	Canadian Standards Association
DIN	Deutschland Ingineering Normalization (German Engineering/Industrial Standard)
DOT	Department of Transportation
ETL	ETL Testing Laboratories
FM	Factory Mutual
FSEC	Florida Solar Energy Center
GAMA	Gas Appliance Manufacturers Association
HVI	Home Ventilating Institute
IAPMO	International Association of Plumbing and Mechanical Officials
IEC	International Electrotechnical Commission
ISO	International Standards Organization
JAN	Joint Army-Navy Standard
MS	Military Standard
NASA	National Aeronautics and Space Administration
NEC	National Electrical Code
NEMA	National Electrical Manufacturers Association
NFPA	National Fire Protection Association
NSF	National Sanitation Foundation
OPEI	Outdoor Power Equipment Institute
OSHA	Occupational Safety and Health Administration
SRCC	Solar Rating and Certification Corporation
UL	Underwriters' Laboratories
USAS	USA Standard (Now ANSI)
USDA	United States Department of Agriculture

17.3 Approval Associations and Their Trademarks

See Fig. 17.1.

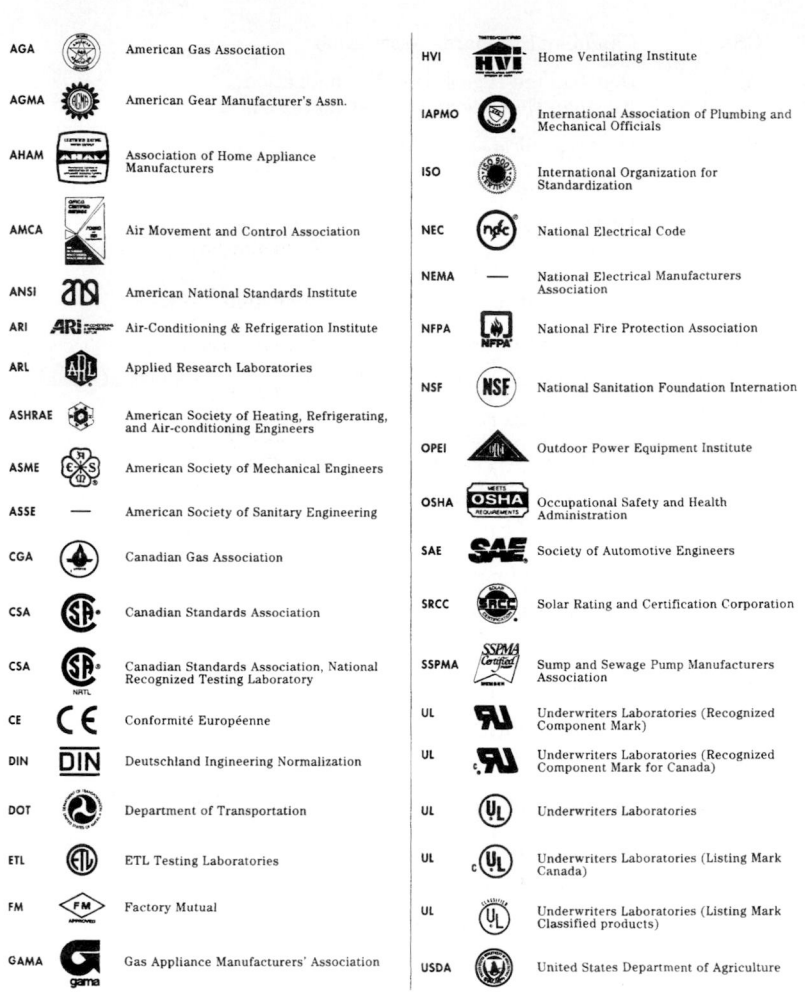

AGA		American Gas Association		HVI		Home Ventilating Institute
AGMA		American Gear Manufacturer's Assn.		IAPMO		International Association of Plumbing and Mechanical Officials
AHAM		Association of Home Appliance Manufacturers		ISO		International Organization for Standardization
AMCA		Air Movement and Control Association		NEC		National Electrical Code
ANSI		American National Standards Institute		NEMA	—	National Electrical Manufacturers Association
ARI		Air-Conditioning & Refrigeration Institute		NFPA		National Fire Protection Association
ARL		Applied Research Laboratories		NSF		National Sanitation Foundation Internation
ASHRAE		American Society of Heating, Refrigerating, and Air-conditioning Engineers		OPEI		Outdoor Power Equipment Institute
ASME		American Society of Mechanical Engineers		OSHA		Occupational Safety and Health Administration
ASSE	—	American Society of Sanitary Engineering		SAE		Society of Automotive Engineers
CGA		Canadian Gas Association		SRCC		Solar Rating and Certification Corporation
CSA		Canadian Standards Association		SSPMA		Sump and Sewage Pump Manufacturers Association
CSA		Canadian Standards Association, National Recognized Testing Laboratory		UL		Underwriters Laboratories (Recognized Component Mark)
CE		Conformité Européenne		UL		Underwriters Laboratories (Recognized Component Mark for Canada)
DIN		Deutschland Ingineering Normalization		UL		Underwriters Laboratories
DOT		Department of Transportation		UL		Underwriters Laboratories (Listing Mark Canada)
ETL		ETL Testing Laboratories		UL		Underwriters Laboratories (Listing Mark Classified products)
FM		Factory Mutual		USDA		United States Department of Agriculture
GAMA		Gas Appliance Manufacturers' Association				

Figure 17.1 Trademarks of approval associations.

Bibliography

The Aluminum Association. 1979. *Aluminum Data and Standards,* 6th ed., Washington, DC: The Aluminum Association.

Aluminum Company of America. 1960. *Alcoa Structural Handbook.* Pittsburgh, PA: Alcoa.

American Institute of Steel Construction. 1991. *Manual of Steel Construction,* 9th ed. Chicago, IL: AISC.

American Iron and Steel Institute. 1962. *Light Gage Cold-Formed Steel Design Manual.* New York: AISI.

American Society for Metals. 1961. *Metals Handbook,* Vol. 1: *Properties and Selection of Metals,* 8th ed. Metals Park, OH: ASM.

American Society for Metals. 1967. *Metals Handbook,* Vol. 3: *Machining,* 8th ed. Metals Park, OH: ASM.

American Society for Metals. 1969. *Metals Handbook,* Vol. 4: *Forming,* 8th ed. Metals Park, OH: ASM.

American Society for Testing and Materials. 1992. *Annual Book of ASTM Standards.* Philadelphia, PA: ASTM.

American Welding Society. 1991. *Welding Handbook,* 8th ed., Vol. 2. Miami, FL: AWS.

Bralla, J. 1999. *Design for Manufacturability Handbook,* 2d ed. New York: McGraw-Hill.

Carr Lane Manufacturing Company. 1995. *Jig and Fixture Design Handbook,* 2d ed. St. Louis: Carr Lane.

Copper Development Association. 1964. *Standard Handbook: Copper and Copper Alloys,* 5th ed. CDA publication number 101. New York: CDA.

Dallas, D. B. 1976. *Tool and Manufacturing Engineers Handbook,* 3d ed. New York: McGraw-Hill (Society of Manufacturing Engineers).

DataMyte. 1995. *DataMyte Hanbook,* 6th ed. Minnetonka, MN: DataMyte.

Durney, L. J. 1984. *Electroplating Engineering Handbook,* 4th ed. New York: Van Nostrand Reinhold.

Eves, H. 1969. *History of Mathematics,* 3d ed. New York: Holt, Reinhart & Winston.

Foster, L. 1994. *Geo-Metrics,* New York: Addison Wesley.

French, T. E., and Vierck, C. J. 1953. *Engineering Drawing,* 8th ed. New York: McGraw-Hill.

Gerolde, S. 1979. *Universal Conversion Factors.* Tulsa, OK: Petroleum Publishing Co.

Groover, M. 1996. *Fundamentals of Modern Manufacturing.* Upper Saddle River, NJ: Prentice-Hall.

Hicks, T. G. 1972. *Standard Handbook of Engineering Calculations.* New York: McGraw-Hill.

Industrial Fasteners Institute. 1983. *Metric Fastener Standards,* 2d ed. Cleveland, OH: IFI.

Industrial Fasteners Institute. 1988. *Fastener Standards,* 6th ed. Cleveland, OH: IFI.

Industrial Information Headquarters, Inc. 1991. *Bearing Manual Cyclopedia,* Vols. 1 and 2. Broadview, IL: IIH.

Jones, F. D., and Horton, H. L. 1988. *Machinery's Handbook,* 23d ed. New York: Industrial Press.

LeGrand, R. 1983. *The New American Machinist's Handbook.* New York: McGraw-Hill.

Middlemiss, R. R. 1952. *College Algebra.* New York: McGraw-Hill.

Parmley, R. O. 1977. *Standard Handbook of Fastening and Joining.* New York: McGraw-Hill.

Salmon, S. C. 1992. *Modern Grinding Process Technology.* New York: McGraw-Hill.

Selby, S. M., Weast, R. C., Shankland, R. S., and Hodgman, C. D. 1962. *Handbook of Mathematical Tables.* Cleveland, OH: Chemical Rubber Publishing Company.

Society of Automotive Engineers. 1992. *SAE Handbook,* Vol. 1: *Materials.* Warrendale, PA: SAE.

Wilson, F. W., and Harvey, P. D. 1965. *Die Design Handbook,* 2d ed. New York: McGraw-Hill (Society of Manufacturing Engineers).

Note: Boldface numbers indicate illustrations, italic *t* indicates a table.

ABOUT THE AUTHORS

Ronald A. Walsh (deceased) was an electromechanical design engineer for more than 40 years. He wrote several books, including *Machining and Metalworking Handbook*, Second Edition, *Electromechanical Design Handbook*, Third Edition, and *Engineering Mathematics Handbook*, Fourth Edition (co-author), all published by McGraw-Hill.

Denis R. Cormier has been a professor of Industrial Engineering at North Carolina State University since 1994. He is also an associate faculty member of the Integrated Manufacturing Systems Engineering Institute. He has published over 30 papers and book chapters pertaining to manufacturing processes and systems, and he was a 2003 recipient of the SME Outstanding Young Manufacturing Engineer award.

THE COMPLETE ESSAYS

'Dr Screech's principal achievement has been to render Montaigne
into contemporary English without quaintness, but also without
sacrifice of that flavour of the sixteenth century which is implicit
in Montaigne's thinking . . . We want the essence of the man in
a form accessible to modern readers, and that is what the
translator has so gracefully given us' – Robertson Davies

'An absolute treat . . . [Screech] is the master of Montaigne. He's
already written extremely eloquently about Michel Montaigne as
a melancholy man. There's a kind of liveliness, a vernacular about
the translation here that works very well' – Roy Porter on
Kaleidoscope, BBC Radio Four

'Of its [the translation's] limpidity and charm there can be no
question' – Simon Raven in the *Guardian*

'This thinking tome, edited by a fine scholar, is utterly readable as
fine scholars should be. It is more easily picked up than put
down, and should be on the bedside table of every *homme moyen
sensuel*, or lady for that matter' – Anthony Blond in the
Evening Standard

'Most of all, mention should be made of the other greatly original
feature of this translation, the commentary . . . [which]
constitutes a fascinating sixteenth-century *honnête homme*'s library.
For this reason the French reader will turn to the translation of
M. A. Screech, who takes his place among those who, crossing
cultural boundaries, enable each country to rediscover its
writers in a new light' – Jean-Robert Armogathe in
Bibliothèque d'Humanisme et Renaissance

'Anglophones of the next century will be deeply in [Dr Screech's]
debt' – Gore Vidal in *The Times Literary Supplement*

MICHEL EYQUEM, Seigneur de Montaigne, was born in 1533, the son and heir of Pierre, Seigneur de Montaigne (two previous children dying soon after birth). He was brought up to speak Latin as his mother tongue and always retained a Latin turn of mind; though he knew Greek, he preferred to use translations. After studying law he eventually became counsellor to the *Parlement* of Bordeaux. He married in 1565. In 1569 he published his French version of the *Natural Theology* of Raymond Sebond; his *Apology* is only partly a defence of Sebond and sets sceptical limits to human reasoning about God, man and nature. He retired in 1571 to his lands at Montaigne, devoting himself to reading and reflection and composing his *Essays* (first version, 1580). He loathed the fanaticism and cruelties of the religious wars of the period, but sided with Catholic orthodoxy and legitimate monarchy. He was twice elected Mayor of Bordeaux (1581 and 1583), a post held for four years. He died at Montaigne in 1592 while preparing the final, and richest, edition of his *Essays*.

M. A. SCREECH is an Honorary Fellow of Wolfson College and an Emeritus Fellow of All Souls College, Oxford (Fellow and Chaplain, 2001–3), a Fellow of the British Academy and of the Royal Society of Literature, a Fellow of University College London, and a corresponding member of the Institut de France. He long served on the committee of the Warburg Institute as the Fielden Professor of French Language and Literature in London, until his election to All Souls. He is a Renaissance scholar of international renown. He has edited and translated both the complete edition and a selection of the *Essays* for Penguin Classics and, in a separate volume, Montaigne's *Apology for Raymond Sebond*. His other books include *Erasmus: Ecstacy and the Praise of Folly* (Penguin, 1988), *Rabelais, Montaigne and Melancholy* (Penguin, 1991) and, most recently, *Laughter at the Foot of the Cross* (Allen Lane, 1998). All are acknowledged to be classics studies in their fields. He worked with Anne Screech on Erasmus' *Annotations on the New Testament*. Michael Screech was promoted Chevalier dans l'Ordre du Mérite in 1982 and Chevalier dans la Légion d'Honneur in 1992. He was ordained, in Oxford, a deacon in 1994 and a priest in 1995.

MICHEL DE MONTAIGNE

The Complete Essays

Translated and edited with an
Introduction and Notes by M. A. SCREECH

PENGUIN BOOKS

PENGUIN BOOKS

Published by the Penguin Group
Penguin Books Ltd, 80 Strand, London WC2R 0RL, England
Penguin Putnam Inc., 375 Hudson Street, New York, New York 10014, USA
Penguin Books Australia Ltd, 250 Camberwell Road, Camberwell, Victoria 3124, Australia
Penguin Books Canada Ltd, 10 Alcorn Avenue, Toronto, Ontario, Canada M4V 3B2
Penguin Books India (P) Ltd, 11 Community Centre, Panchsheel Park, New Delhi – 110 017, India
Penguin Books (NZ) Ltd, Cnr Rosedale and Airborne Roads, Albany, Auckland, New Zealand
Penguin Books (South Africa) (Pty) Ltd, 24 Sturdee Avenue, Rosebank 2196, South Africa

Penguin Books Ltd, Registered Offices: 80 Strand, London WC2R 0RL, England

www.penguin.com

Book II, Chapter 12 previously appeared as *An Apology for Raymond Sebond*,
published in Penguin Books 1987
The Complete Essays first published by Allen Lane The Penguin Press 1991
Reprinted with corrections and a new Chronology 2003

036

This translation and editorial material copyright © M. A. Screech 1987, 1991, 2003
All rights reserved

The moral right of the translator has been asserted

Printed in England by Clays Ltd, St Ives plc

ISBN-13: 978–0–140–44604–3

www.greenpenguin.co.uk

In Memory of

PHILIP EVELEIGH

Wit, poet, scholar
killed during the Allied
landings in Italy

Table of Contents

===

Introduction

Montaigne is one of the great sages of that modern world which in a sense began with the Renaissance. He is a bridge linking the thought of pagan antiquity and of Christian antiquity with our own. Colourful, practical and direct, and never intentionally obscure, he sets before us his modestly named *Essays*, his 'attempts' at sounding himself and the nature and duties of Man so as to discover a sane and humane manner of living. He enjoys a place apart among French Renaissance authors. Men and women of all sorts are fascinated by what they find in him. Many read him for his wisdom and humanity, for which he may be quoted in a newspaper as readily as in a history of philosophy. He writes about himself, but is no egocentric and is never a bore. He treats the deepest subjects in the least pompous of manners and in a style often marked by dry humour. His writings are vibrant with challenge; they are free from jargon and unnecessary technicalities. In the seventeenth century, Pascal, the great Jansenist author of the *Pensées* ('Thoughts' which owe much to Montaigne), was converted partly by reading him and was soon discussing the *Essays* at Port-Royal with his director, LeMaistre de Sacy (who had his reservations). Pascal gained, it is said, thirty years by reading Montaigne, thirty years of study and reflection.[1] Others, too, have felt the same. For Montaigne gives his readers the fruits of his own reading and of his own reflections upon it, all measured against his personal experience during a period of intellectual ferment and of religious and political disarray. Montaigne never let himself be limited by his office or station. As husband, father, counsellor, mayor, he kept a critical corner of himself to himself, from which he could judge in freedom and seek to be at peace with himself. He does not crush his reader under the authority of the great philosophers: he tries out their opinions and sees whether they work for him or for others. For he knew that opinions are not certainties, and that most human 'certainties' are in fact opinions.

Traces of Plato, Aristotle, Plutarch, Cicero, St Augustine or his own contemporaries can be found in every page he wrote, but they are skilfully

---

1. Blaise Pascal, *Pensées et Opuscules*, ed. L. Brunschvicg, 1909, p. 120 – an old study, but still useful.

interwoven into his own discourse, being renewed and humanized in the process. And he hardly ever names them when making such borrowings. That was because he was delighted to know that critics would be condemning an idea of Plato, Aristotle or Seneca, say, when they thought they were attacking merely an opinion of his own unimportant self.

After his beloved father died (18 June 1568), he succeeded to the title and the estates at Montaigne, in south-west France. (Provisions were made for his mother.) He was thirty-five, and three years married. Soon (1570) he was able to sell his charge as counsellor in the Parlement of Bordeaux (a legal office). His plan was, like cultured gentlemen in Ancient Roman times, to devote himself to learned leisure. He marked the event with a Latin inscription in his château – he had a taste for inscriptions, covering the beams and walls of his library with some sixty sayings in Greek and Latin, many of which figure in the *Essays*. His rejoicing at leaving *negotium* (business) for *otium* (leisure) was tempered by grief at the death of his friend, Etienne de la Boëtie (1563). (His children all died young, too, except a daughter, Léonor, who was deeply loved but could not, for a nobleman, replace a son and heir.)

Montaigne's project of calm study soon went wrong. He fell into an unbalanced melancholy; his spirit galloped off like a runaway horse; his mind, left fallow, produced weeds not grass. The terms he uses are clear: his complexion was unbalanced by an increase of melancholy 'humour'. His natural 'complexion' – the mix of his 'humours' – was a stable blend of the melancholic and the sanguine. So that sudden access of melancholy humour (brought on by grief and isolation) was a serious matter, for such an increase in that humour was indeed inimical to his complexion, tipping it towards chagrin, a depression touched by madness. Such chagrin induced *rêveries*, a term which then, and long afterwards, meant not amiable poetic musings but ravings. (The *Rêveries* of Jean-Jacques Rousseau, for example, are his 'ravings', not his 'day-dreams'.) So at the outset *otium* brought Montaigne not happy leisure and wisdom but instability. Writing the *Essays* was, at one period, a successful attempt to exorcize that demon. To shame himself, he tells us, he decided to write down his thoughts and his rhapsodies. That was the beginning of his *Essays*.[2] But he was not a professional scholar: he had no 'subject' to write about. He was not a statesman or a general. He soon decided to write about himself, the only subject he might know better than anyone else. This was a revolutionary decision, made easier, no doubt, by his bout of melancholy, for that

2. Cf. my study, *Montaigne and Melancholy*, Duckworth 1983; Penguin 1991.

humour encouraged an increased self-awareness. No one in Classical Antiquity had done anything like it. In the history of the known world only a handful of authors had ever broken the taboo against writing primarily about oneself, as an ordinary man. St Augustine had written about himself, but as a penitent in the *Confessions*; during the Renaissance, Girolamo Cardano wrote *On his Life* and *On his Books*, and Joachim Du Bellay lamented his Roman 'exile' in his poetic *Regrets*. But those works bear no resemblance to what the *Essays* were to become for Montaigne – 'tentative attempts' to 'assay' the value of himself, his nature, his habits and of his own opinions and those of others – a hunt for truth, personality and a knowledge of humanity through an exploration of his own reactions to his reading, his travels, his public and his private experience in peace and in Civil War, in health and in sickness. The *Essays* are not a diary but are of 'one substance' with their author: 'I am myself the matter of my book.' In the case of a questioning and questing mind like his this study became not a book on a 'subject' but Assays of Michel de Montaigne – 'assays' of himself by himself.

These essays were first divided into two books (a third followed later). Each book contains many chapters and each chapter contains many 'assays'. He himself never referred to his chapters as essays; his chapters were convenient groupings of several assays – primarily 'assays' of a man called Michel de Montaigne. He soon discovered that very short chapters did not allow him enough scope for all the assays he wanted to make. He let his chapters grow longer. In the process he discovered the joys of digression and freedom from an imposed order. And he found he could tackle deeper subjects more exhaustively.

Montaigne's method of writing makes it sometimes puzzling for the reader to follow the linkings of his thought. His chapters are not arranged in order of their composition. Within each chapter sentences and phrases written at widely different times were printed without any hint of dating. Moreover each chapter, no matter how long, was presented as one continuous slab of text. That was quite usual then, but for us it leads to a kind of intellectual indigestion. Modern editors introduce paragraphs as well as quotation marks, italics and a now more usual punctuation. That has been done here too. It makes it easier to pick up Montaigne and to put him down. That is a great advantage for what is one of Europe's great bedside books. But Montaigne warned us that we should be prepared to give him an hour or so at a stretch when necessary. Even that is easier when there are paragraphs, as well as some indication of what was written when.

As edition followed edition Montaigne changed a word here, a phrase

there, but above all he added more examples, more quotations and more arguments, as well as thoughts upon the thoughts he had formerly written. These all became more numerous in 1588 and even more so in the edition he was preparing for the press when he died (13 September 1592).

Until modern times there was no easy means of distinguishing the various layers of Montaigne's text. Pierre Villey pointed the way in his great edition. Now almost every editor uses [A] [B] [C] or similar signs to help the reader through the marquetry-cum-maze that the *Essays* eventually became. That has been done here. Knowing at least approximately what came when can make Montaigne not only more easy to follow but far more enjoyable.

Few noblemen knew Latin as Montaigne did. It was his native tongue. As soon as he was weaned his loving father had arranged for him to hear nothing but pure Classical Latin. As a child he at first spoke neither Gascon nor French. At an age when others delighted in tales of chivalry and rambling novels of love and adventure translated from the Spanish, he devoured Ovid's *Metamorphoses*. When he was eventually sent to school at the Collège de Guyenne in Bordeaux he chattered away in Latin so fluently that he scared the wits out of his schoolmasters, distinguished scholars though they were. One of them was so understanding, though, that he allowed his young pupil to read anything he liked, provided that he first did his prep.

Montaigne never acquired a similar fluency in Greek, so that even Plato and Aristotle (who influenced him deeply) he read in the Latin translations used throughout Europe. (Robert Burton, the author of *The Anatomy of Melancholy*, was to do the same.)

Montaigne revelled in the Latin poets. Quotations from them are strewn throughout the *Essays*, making wry points, opening windows on to beauty, providing authority or contrast or jests. Less obvious now – that is why footnotes are there to point them out – are Montaigne's numerous quiet, unheralded debts to the Classical moralists, philosophers, biographers, historians and statesmen. Since he read Latin with pleasure and such ease it was to Latin works above all that he turned for moral guidance and for insight into what human nature really is. But he did not turn to them exclusively: all historians delighted him, even naïve ones; not least he studied his near-contemporaries writing not only in Latin but in French, Italian or Spanish. It was in the light of such reading that he judged his own opinions and his own wide experience and sought to find out more about himself, about the 'human condition' (that is, about the characteristics which mankind was created with) and about the limits of human nature.

Montaigne was first, it seems as we read him, a Stoic, then a Sceptic, then an Epicurean. In fact he could hold all three philosophies in a kind of taut harmony. He realized that he was so open to influences from the sages of Antiquity that he took on the colour of whichever one he had just read. There is certainly a shift in his thought from a melancholic and stoic concern with dying to a full and joyful acceptance of life; a change of emphasis away from Seneca and towards the happier eclecticism of Cicero who, despite his verbosity, came close to guiding his maturer thought. But for Montaigne no author ever definitively banished or superseded any other; authors are not infallible; they can help us make 'assays' but they resolve nothing. Even the sage whom Montaigne most admired, Socrates, is eventually stripped of that saintly authority that Erasmus vested him with.

Gradually Montaigne realized that by studying and questioning the greater and lesser authors in the light of his own opinions and experience he was studying himself. Encouraged by the Classical sayings, which, in Erasmus' *Adages* for example, lie clustered around the commandment of the Delphic Oracle, 'Know Thyself', Montaigne was led to study his own self, as Socrates did his, coolly, probingly and without self-love. He was acutely aware that when doing so he was not gazing at a solid, stationary object, an evidently unified Ego, but at something ever-changing, ever-flowing. The self he discovered consisted in endless variations set in time, in series upon series of thoughts, feelings, desires, actions and reactions. Plato and Aristotle as then interpreted were excellent guides when he came to face up to that fact. Plato emphasized the primacy of the soul and yet, at least in some of his moods, did not despise the body. Aristotle taught Montaigne that individual persons belong to a genus and a species; so each man and woman individually possesses 'generic' and 'specific' qualities; and each of them has a specific human soul (or 'form'); it could vary in quality but not in nature. So any man or woman who remained human could at least partially understand any other, since all possessed a like soul. No virtue or no vice known to any individual human who remains sane should be totally incomprehensible to any other. Even the virtue of Socrates can be momentarily glimpsed, and indeed momentarily shared in, by a lesser member of his species. So too could the cruelties of a Tamberlane be understood by better men. All individual human beings (as the scholastic philosophers put it) bore in themselves the entire 'form' of the human race. To study one man is in a sense to study them all. Not that all are identical, but all are inter-related by species. And (more remarkably) Montaigne discovered that to think about women and their sexuality could also tell

you much about men and vice versa, since men and women are cast in the same mould: a quite revolutionary idea as Montaigne holds it.

What Montaigne discovered in himself – as others could do in their own cases too – was a self which was governed by a *forme maistresse*, a 'master-mould' which effectively resisted any attempt to change it by education or indoctrination. Without that mould Montaigne would have found in himself not personality but endless flux and change with no sense of identity.

It was this awareness of flux and change in all things human and sublunary which led him so staunchly to uphold the teaching authority of the Roman Church. Without it he could find nothing but uncertainty anywhere.

If he had been a don or a scholar Montaigne would doubtless have written in Latin. Encouraged though he was to write in French by the example of Bishop Jacques Amyot's lucid and elegant translation of Plutarch, he believed that by writing in the vulgar tongue which was continuously evolving he was in fact writing for a few readers and for a few years. His book would out-live him and keep him alive in the minds of those who knew him, but would soon become dated and hard to understand. In a sense he was right. His French did become harder to understand. But had he written in Latin few indeed would now take him down from the shelf.

Montaigne was a gentleman not a scholar. He was a man who knew the ways of diplomacy and the realities of the battlefield. He loved books but was no recluse. Among the qualities which he claimed to bring to his writing was a gentleman's loathing of the villein's vice of lying, as well as a soldier's love of bluntness and distaste for claptrap. He was not seeking for verbal subtleties but to portray himself in all truth, to find solid facts about what Man really is, and practical counsel about how he should live and die. That advice he properly and understandably sought not from theology but philosophy. For centuries Christendom had allowed philosophy to go largely its own way. Not that the Classical philosophers had ever been banished from Christian theology. From the very outset the theology of St Paul was indebted to Plato. And from the thirteenth century onwards Aristotle became *the* Philosopher. St Jerome in antiquity had rejoiced that the Stoics should hold so much in common with Christians. Seneca seemed indeed so close to Christian teachings that it was long believed that he had actually corresponded with St Paul; in Montaigne's own day Jacques Amyot, the Bishop of Auxerre, held that his much-admired Plutarch was so consonant with Christianity that his books could more profitably be

used to instruct Princes in their duties than Holy Writ itself, 'which seems peremptorily to command rather than graciously to persuade'. He says so quite straightforwardly in his dedication of Plutarch's *Oeuvres morales* 'to the most Christian King Charles the Ninth of that name'. Theologians such as Melanchthon strongly defended the claims of philosophy in its own domain. All agreed that philosophy's domain included large tracts of ethics. Much of the day-to-day ethics of Christendom derive directly from Aristotle and, directly or indirectly, from Plato. Christendom found it right and natural to draw for its ideas about virtue heavily on the School of Athens. Philosophy was a complement to theology. Even a Christian author such as Boethius, who wrote a tractate on the Trinity, also wrote what became a moral classic for medieval Christendom, his *Consolation of Philosophy*, which at no point betrays any awareness of theology's teachings. It was, after all, offering the reader the consolation of rational philosophy, not revealed theology. Again in Montaigne's own day the neo-Stoic Justus Lipsius, whom he much admired, became the darling of the Roman Catholic Church once he had returned from Reformation to the fold, yet his moral writings are a mosaic of Classical Stoicism, with no specific concessions whatsoever to theological verities. When necessary, philosophy had to yield to theology: it did not have constantly to compromise with it. The study of the Classical writers had made immense strides in the generation before Montaigne. The generation of Erasmus had seen Socrates for example as a kind of Christ-figure; Seneca's suicide was seen as close to Christian martyrdom. Montaigne, partly under the influence of the scholarship of Turnèbe (Adrian Turnebus) and of Denis Lambin, the editor of Lucretius, avoided such anachronisms. For Montaigne the attraction of Classical philosophy lay in its being philosophical. Lacking the authority of Christian revelation it was open to rational examination and discussion. Philosophy worked with its own tools: reason and experience; its domain was natural knowledge; such revelation as it enjoyed – if such revelation there be – was that kind which worked upon inspired poets, doctors, lawgivers, scientists and sages. But especially when philosophy ventured beyond physics into metaphysics it was not teaching but speculating: the 'essence' of being, truth and knowledge, is beyond reason and beyond experience. But we can enjoy hunting about for it.

The conventions of the time would have allowed Montaigne in the *Essays* to say nothing at all about his religion. He does indeed say nothing about Christian hopes and fears when writing of death. As a philosopher Montaigne was not concerned with being dead but with bearing with wisdom and fortitude the pain of dying as the soul is, often excruciatingly,

released from its body. Not that Montaigne disbelieved in the afterlife, but the splendour of the rewards awaiting redeemed Christian souls and, unimaginably, their bodies, is a matter of theology not of reasoned deduction or induction. The Christian heaven can only be imagined as unimaginable, thought of as unthinkable: to make that point authoritatively Montaigne based his case on the words of St Paul:

> Eye hath not seen, nor ear heard, neither have entered into the hearts of man, the things which God hath prepared for them that love him.[3]

There are areas where theology and philosophy overlap: so the *Essays* at times do touch upon religion, but always in the spirit of philosophy. The supreme example of this in the *Essays* is the longest chapter which Montaigne ever wrote (II, 12), 'An apology for Raymond Sebond'. Even judged by the length of the more developed chapters, 'An apology for Raymond Sebond' is in a class by itself. Its very length shows that it was a very special chapter indeed. Its topic could have afforded Montaigne, he felt, with matter to write upon for ever. It is an excellent chapter to study as a means of discovering how Montaigne reconciled throughout his *Essays* a questing, often sceptical, intelligence with a profound political conservatism, an unshakable respect for constitutional legality, a humane morality and an easy submission – in its proper sphere – to the teaching authority of the Roman Catholic Church. Those convictions helped Montaigne to remain tolerant, kind and loyal during the long, bitter, appallingly cruel Civil Wars of Religion which devastated the whole of France, not least the lands and villages of Gascony, including the domain of Montaigne itself. It is understandable that Montaigne should have written a considered defence of the *Natural Theology* of Raymond Sebond, since he himself had translated it into French. In the opening pages of the 'Apology' and in the dedication of the work to his father he tells us how he came to do so. Pierre Bunel, a Christian humanist from Toulouse (1499–1546), had once stayed at Montaigne and recommended Sebond's book as an antidote to the 'poison' of Lutheranism – a term often applied to protestantism generally. Bunel's visit may have occurred between 1538 and 1546; he was then living reasonably near Montaigne, first at Lavour and later in Toulouse. If so, Michel de Montaigne was still a child, perhaps not yet in his teens.

That Bunel should offer such a book to Montaigne's father makes good

---

3. The standard text of St Paul (I Corinthians 2:9) cited by theologians over the centuries. Montaigne quotes it to good effect when condemning the teachings about the afterlife found in Plato – teachings which he provocatively judged too corporeal. See 'An apology for Raymond Sebond', II, 12, note 212.

sense. Raymond Sebond was a local figure, possibly a Catalan. Montaigne refers to him as a Spaniard professing medicine in Toulouse. In fact he was a Master of Arts who professed both Medicine and Theology. His *Natural Theology* was written in Toulouse in the 1420s or early 1430s. It seems to have circulated fairly widely in manuscript. By Montaigne's time it had been printed more than once, as well as being adapted to dialogue form – still in Latin – by Petrus Dorlandus under the titles of *Violet of the Soul* or of *Dialogues concerning the Nature of Man: Exhibiting Knowledge of Christ and of Oneself.*

Apart from these Latin books Raymond Sebond had fallen into oblivion. When inquiries about him and his *Natural Theology* were addressed to Adrian Turnebus (Montaigne's scholarly friend 'who knew everything'), he could only say that the *Natural Theology* was a 'kind of quintessence drawn from Thomas Aquinas'. That may imply that Turnebus rightly considered it to have been influenced by another medieval Catalan theologian, Raymond Lull, the great *Doctor Illuminatus* who was himself held to be the Quintessence of Aquinas. Since Turnebus died in 1565, the *Natural Theology* of Sebond must have been in Montaigne's mind for several years before he published his translation.

In the 'Apology' Montaigne tells us that he translated Sebond at the request of his father in the 'last days' of his life. In the epistle in which he dedicated the translation to his father, Montaigne lets it be understood he had been working on the task at least some months before that. Since the *Theologia Naturalis* runs into nearly a thousand pages, a year or more is certainly likely. The finished translation was Montaigne's tribute to his beloved parent. The dedicatory epistle is addressed from Paris 'To My Lord, the Lord of Montaigne'; in it, he wishes his father long life: yet it is dated from the very day of his father's death – hardly a coincidence but rather a fitting tribute to a son's feelings of piety at the death of the 'best father that ever was'. It may well imply that he wished he had translated and published the work more speedily, to give his father joy in his lifetime.

The works of Sebond had been appreciated by high-born ladies in France long before Montaigne wrote his 'Apology' at the request of an unnamed patroness who may well have been Princess Margaret of France, the future wife of Henry of Navarre.[4] In 1551 Jean Martin had translated Dorlandus' version of Sebond's *Violet of the Soul* into highly latinate French for Queen Eleonora of Austria, the widow of King Francis I. In her absence from

4. Cf. 'An apology for Raymond Sebond', p. 628.

France the version was dedicated to the Cardinal de Lenincourt; there we read that the *Viola animae* is a book which could 'bring back atheists, if any there be, to the true light, while maintaining the faithful in the good way'. Clearly Pierre Bunel had every reason to give the original and full version of such a book to an intelligent but not formally educated nobleman such as Montaigne's father, who wanted to find an 'antidote' to Lutheranism.

The Catholic credentials of the *Natural Theology* of Raymond Sebond may appear to need no defence or apology. In the fifteenth century the scholarly and saintly Cardinal Nicolaus of Cusa had possessed a copy: it may have contributed to his doctrine of 'learned ignorance' – that Socratic, Evangelical *docta ignorantia* of the Christian who is content to own that all human knowledge is as nothing, compared to that infinity who is God; learned ignorance never claims to know, or to aspire to know, anything beyond the saving law of Christ. In the sixteenth century the French Platonizing humanist Charles de Bouelles also had a copy: he was a Christian apologist of real depth and power. But Montaigne was not mistaken in believing that the *Natural Theology* did need an apologist against criticisms arising within his own Church. In 1559 a work called the *Violetta del anima* appeared on a list of prohibited books drawn up by the Spanish Inquisitor Ferdinando de Valdés, Archbishop of Seville. It may refer to a Spanish version of the *Violet of the Soul*. More important, in 1558–9, the entry *Raymundus de Sabunde: Theologia Naturalis* appeared on the Index of Forbidden Books of Pope Paul IV.

So the Catholic Montaigne had translated a prohibited book! Or had he? His own translation was never condemned. On the contrary, it enjoyed a certain popularity well into the next century. After Montaigne's first and second editions in 1569 and 1581 (both in Paris) it was reprinted in Rouen in 1603, in Tournon in 1611, in Paris again, also in 1611 and finally in Rouen in 1641.

That fact can be easily explained. It was not to the *Natural Theology* that the censors took exception but to the short Prologue which accompanied it, as is shown by the definitive judgement of the Council of Trent; the Tridentine Index of Forbidden Books (1564) condemned the Prologue and nothing else. Shorn of the page and a half of Prologue, the Latin original of Sebond's *Natural Theology* circulated freely and was fully reprinted in Venice in 1581, in Frankfurt-on-the-Main in 1631 – with the Prologue – and finally in Lyons in 1648, by which time it was becoming dated. And even the Prologue was eventually removed from the Index in the nineteenth century.

This has not stopped Sebond's method of teaching the Catholic faith

from being thought of as somehow dangerous. Even the *New Catholic Encyclopedia* (which ought to know better) calls it heretical. It is not. But it was clearly a disturbing book – a good defence against heresy yet, for many, a work somehow not to be trusted. There were contemporaries of Montaigne who shared that opinion: hence his apology for it.

When Montaigne published his translation in 1569, he included with it a translation of the Prologue which proved quite acceptable to the Roman Catholic Church. No censor has ever said a word against it. He had clearly taken theological advice and had adapted the Prologue to meet the needs of the Faith. A comparison of his version and the original shows why the Latin Prologue appeared among the prohibited books, while the French version never did.

Sebond's original Prologue is dense and interesting. It is emphatic, trenchant and absolute. Its claims are such as were bound to appeal to intelligent Catholic ladies deprived of formal education and to laymen such as Montaigne's father. It claimed to 'illuminate' Christians with a knowledge of God and themselves. It required no previous knowledge of Grammar, Logic, nor any other deliberative art or science, nor of Physics nor of Metaphysics – no Aristotle, therefore. It offered a method applicable to both clergy and laity. It promised certain results, 'in less than a month, without toil and without learning anything off by heart. And once learned it is never forgotten.' The *Natural Theology* was said to lead not only to knowledge but to morality, making whoever studied it 'happy, humble, kind, obedient, loathing all vice and sin, loving all virtues, yet without puffing up with pride'.

Montaigne did not essentially lessen this appeal but introduced changes in the Prologue (and, indeed, in the work itself) which show a sensitivity to theological distinctions. Where the Prologue was concerned, his changes were few but vital enough to restore it to undoubted orthodoxy. For example, where Raymond Sebond had written of his art as '*necessary* to every man', Montaigne made it merely *useful*. When Sebond claimed that his method taught 'every duty' required for the student Montaigne changed that to 'nearly everything'. Sebond wrote:

> In addition this science teaches everyone really to know, without difficulty or toil, every truth necessary to Man concerning both Man and God; and all things which are necessary to Man for his salvation, for making him perfect and for bringing him through to life eternal. And by this science a man learns, without difficulty and in reality, whatever is contained in Holy Scripture.

*

Montaigne tones that down:

> In addition this science teaches everyone to see clearly, without difficulty or
> toil, *truth insofar as it is possible for natural reason*, concerning knowledge of
> God and of himself and of what he has need for his salvation and to reach life
> eternal; *it affords him access to understanding what is prescribed and commanded* in
> Holy Scripture.

The words in italics are vital. In Montaigne's hands the work of Sebond is
presented as a means of access to truths and duties prescribed in Scripture.
Sebond's original Prologue could be taken to mean that his method stood
alongside Scripture, independently. That of course would have been heretical if Sebond had been arguing from fallen natural reason. But he was not.

Today we are so used to commercialized religious charlatanism that the
claims of Sebond risk sounding like some slick, patent road to an illusory
salvation. That is far from the truth. The *Natural Theology* is a cogently
written work in scholastic Latin seeking to anchor the reader firmly within
the Roman Catholic Faith, free from all wavering and doubt. The Prologue
(in both the original and in Montaigne's translation) ends with an
uncompromising act of submission to the 'Most Holy Church of Rome,
the Mother of all faithful Christians, the Mistress of grace and faith, the
Rule of Truth'.

The method of Raymond Sebond is sufficiently complex to be misunderstood, not least by the many who were long deprived of his Prologue by
the folly of censorship. Obviously even quite a few moderns writing on
Montaigne have never been able to study it.[5]

Sebond firmly bases his method on 'illumination'. He does not claim
that human reason by itself can discover Christian truths. Quite the reverse.
Without 'illumination' reason can understand nothing fundamental about
the universe. But, duly illuminated, Man can come to know himself and
his Creator as well as his religious and moral duties, which he will then
love to fulfil. It is a method of freeing Man from doubts; it reveals the
errors of pagan antiquity and its unenlightened philosophers; it teaches
Catholic truth and shows up sects as errors and lies. It does all these things
by teaching the Christian the 'alphabet' which must be acquired if one is to
read Nature aright. The science 'teaches Man to know himself, to know
why he was created and by Whom; to know his good, his evil and his

5. A translation of Montaigne's version of the *Prologus* is given after this Introduction in an Appendix (p. lviii).

duty; by what and to Whom he is bound. What good are the other sciences to a man who is ignorant of such things?'

'The other sciences', when this basis is lacking, are but vanity. They lead to error, men not knowing 'whither they are going, whence they came' nor what Man is. Sebond shows Man how far he has fallen and how he can be reformed.

Raymond Sebond believes that God has given Man two Books, a metaphorical one and a real one. The first is the 'Book of all Creatures' or the 'Book of Nature'. The second is Holy Scripture. The first Book to be given Man – at the Creation – was the Book of Nature. In it all created things are like letters of the alphabet; they can be combined into words and sentences, teaching Man truths about God and himself. But with the Fall, Man was blinded to the sense of the Book of Nature. He could no longer read it aright. Nevertheless, that book remains common to all.

The Second Book, Holy Writ, is not common to all – 'to read the second book one must be a clerk'. Yet (unlike Scripture) the Book of Nature cannot be falsified; it cannot lead to heresy. Yet in fact both Books teach the same lesson (since the same God created all things in due order and revealed the Scriptures). They cannot contradict each other, even though the first is natural – of one nature with us Men – while the other is above all Nature, supernatural.

Now, Man was created in the beginning as a reasonable creature, capable of learning. But at his creation, Man – Sebond means Adam – knew nothing whatever. 'Since no doctrine can be acquired without books', it was most appropriate that Divine Wisdom should create this Book of Creatures in which Man, on his own, without a teacher, could study the doctrines requisite for him. It was the visible 'letters' of this Book – the 'creatures' placed in God's good order, not our own – that Man was intended to read, using the pre-lapsarian judgement which God had bestowed on him when he was newly created.

But since the Fall all that has changed. Man can no longer find God's truths in Nature, 'unless he is enlightened by God and cleansed of original sin.[6] And therefore not one of the pagan philosophers of Antiquity could read this science, because they were blinded concerning the sovereign good, even though they did read some sort of science in this Book and derived whatever they did have from it.' But the solid, true science which leads to life eternal – even though it was written there – they were unable to read.

---

6. This is the conviction of Pascal, *Pensées*, Brunschvicg no. 244.

In Montaigne's hands Raymond Sebond's method shows enlightened Christians that the revealed truths about God and man are consonant with the Book of Nature properly read. It reconciles observed nature with revealed truth and so can lead men to accept it without doubt or hesitation.

Montaigne's 'Apology' is a defence of this doctrine, and corresponds to the two assertions of Sebond: i: Man, when enlightened, can once again read the Book of Nature aright; ii: Man when not enlightened by God's grace can never be sure he has read it aright: Mankind has read 'some sort of science' in this Book of Nature but is 'unable to read' that 'true science which leads to life eternal'. This means that unenlightened Man, Man left to his own devices, can no longer 'read' God's creatures – and creatures covers not only plants and animals but the Universe and everything in it – the letters of that alphabet appear all jumbled up. No longer can Man be sure he has any knowledge of himself or of any created thing or being, from the highest heavens to the tiniest ant.

The two main sections of the 'Apology' are of widely different lengths. Montaigne dismisses fairly curtly, though courteously, the first of the two criticisms made of Sebond.

> The first charge is ... that Christians do themselves wrong in wishing to support their belief with human reason: belief is grasped only by faith and by private inspiration from God's grace.

Montaigne's reply is to accept 'that purely human means' are not enough; had they been so, 'many choice and excellent souls in ancient times' would have succeeded in reaching truth. But despite their integrity and their excellent natural faculties, the Ancients all failed in their ultimate quest: 'Only faith can embrace, with a lively certainty, the high mysteries of our religion' ('Apology', p. 492).

That is quite orthodox. At least from the time of Thomas Aquinas it was held that natural reason ought to bring Man to the preambles of the Faith – that there is one God, that he is good, that he can be known from revelation – but that specifically Christian mysteries are hidden until revealed.[7] Montaigne may seem to put even those preambles in doubt, only to vindicate them triumphantly at the end of the 'Apology' with the aid of Plutarch.

---

7. Thomas Aquinas, *Summa Theologica*, II*, II**, Q.I ad 5. Later this theme is briefly treated in ways relevant to an understanding of Montaigne in Daniel Huët, Bishop of Avranches, *De imbecillitate mentis humanae*, 1738, Bk. 2, chapter 1, or in the same chapter of the French original, *Traité philosophique de la foiblesse de l'esprit humain*, 1723. (It was already a standard doctrine long before Montaigne's time.)

But Montaigne contrasts the routine practising Christian, merely accepting the local religion of Germany or Périgord in casual devotion, with what illuminated Christians are really like when 'God's light touches us even slightly'. Such Christians emanate brightness ('Apology', p. 493). The apprentice Christian may not rise so high but, once his heart is governed by Faith, it is reasonable for Faith to draw on his other capacities to support him. Sebond's doctrine of illumination helps us to do so effectively and to draw religious strength from a knowledge of God's creation:

> [God] has left within these lofty works the impress of his Godhead: only our weakness stops us from discovering it. He tells us himself that he makes manifest his unseen workings through those things which are seen. ('Apology', p. 498)

Montaigne turns to a key text of Scripture which he suitably cites. Sebond could toil to show that, to the enlightened Christian, 'no piece within this world belies its Maker' precisely because Scripture gives Man that assurance:

> All things, Heaven, Earth, the elements, our bodies and our souls are in one accord: we simply have to find how to use them. If we have the capacity to understand, they will teach us. 'The invisible things of God,' says St Paul, 'are clearly seen from the creation of the world, his Eternal Wisdom and his Godhead being perceived from the things he has made.' ('Apology', p. 499)

That quotation, adapted from the Vulgate Latin text of Romans I:20, is the foundation of all natural theology in the Renaissance. That can be seen from author after author, since Montaigne had chosen his scriptural authority well. He had selected the obvious text. In 1606, for example, George Pacard published his own *Théologie Naturelle* and placed Romans I:20 firmly on his title page, lending its tone to his whole book. A generation later Edward Chaloner could defend the general thesis of Montaigne here, with precisely this verse, in a sermon preached at All Souls College in Oxford.[8]

8. Edward Chaloner, *The Gentile's Creede, or The Naturall Knowledge of God, in sixe sermons*, 1623, p. 223: 'The doctrine therefore which our *Apostles* in my Text doe insinuate unto us, when they say, that God left not himselfe to the Gentiles without witnesse, must needs be this. *That so much may be knowne of God by the Witness of Nature, as is sufficient to confirme unto us, though not his Persons, or workes of Redemption, yet his Godhead*, and also his *handie-worke in creating and governing the World*. God is himself invisible, and yet *The invisible things of him* (sayth the

To make this point clear, Montaigne uses an analogy taken from Aristotelian physics, in which any object is composed of inert matter and a form which gives it its being.

> Our human reasonings and concepts are like matter, heavy and barren: God's grace is their form, giving them shape and worth. ('Apology', p. 499)

Since men such as Socrates and Cato lacked God's grace, even their most virtuous actions are without shape or ultimate value; in the context of salvation they 'remain vain and useless'. So too with the themes of Sebond. By themselves they are heavy and barren. When Faith illuminates them, they become finger-posts setting man on the road which leads to his becoming 'capable' of God's grace.[9] In the light of the closing words of the 'Apology' that is a vital consideration.

The Renaissance thinker, like his forebears from the earliest Christian times, had to decide what to do about the great pious men of Ancient days. Were they saved by their loyalty to the Word (the *Logos*) before he was incarnate in Christ? One of the earliest theologians, Justin Martyr, thought they were. Or were they inevitably destined to eternal reprobation, since even their good actions were not directed to the right End? Were some, such as Socrates or Plato, vouchsafed special saving grace? Erasmus would like to think that God would make the same kind of understanding, graded concessions that he himself made to those Ancients who were pious, moral and sensitive to metaphysical realities.

Montaigne's admiration for the virtuous heroes of Antiquity was boundless: the moral system he was teasing out for Christian laymen like himself to supplement the Church's teaching owed nearly everything to them. He insisted nevertheless that they were great with human greatness only and in no wise proto-Christians. Yet the 'Apology' also shows by the careful use of theological language that Montaigne did not look on all the Ancients as an undifferentiate 'mass of damnation'. This is brought out by the way he cited Romans I:20, without the final clause, 'so that they [the pagans] are inexcusable'.

Many did attach this clause to St Paul's assertion that the invisible things

---

Apostle, Rom. I:20) *that is, his Eternall Power and Godhead, are seene by the creation of the World, being considered in their workes.* To resolve the members of which Verse, were to propose unto you a whole systeme of naturall Divinitie . . .' Cf. also Sir Walter Raleigh, *Historie of the World*, I, i, cited by E. M. W. Tillyard, *The Elizabethan World Picture*, London, 1963, p. 36; Tillyard gives a résumé of Raymond Sebond's *Viola animae*.

9. 'Apology', p. 500; cf. *Montaigne and Melancholy*, p. 49.

of God are accessible through the visible: George Pacard did precisely that in the title page of his *Théologie Naturelle*. But many did not; to cite only one example: Allessandro della Torre, Bishop of Sittià, cited this text of Romans three times in his Italian work, *The Triumph of Revealed Theology* (Venice, 1611): each time he omits that damning clause. By doing the same Montaigne and others could stress the human limitations of Socrates or Plato, while avoiding the Jansenist rigour which Pascal read back into the 'Apology':

> There is enough light [Pascal wrote] to lighten the Elect and enough darkness to make them humble. There is enough darkness to blind the Reprobate and enough light to damn them and render them inexcusable. *St Augustine, Montaigne, Sebond.*[10]

Montaigne follows Sebond in dwelling on the errors and the chaotic jumble of ideas expounded by those unenlightened wise men, vainly seeking certain truth with their human reason from the Book of Creatures: but he does not consider their opinions to be all equally 'inexcusable'. Nevertheless he asserted that 'human reason goes astray everywhere, but especially when she concerns herself with matters divine' ('Apology', p. 581). Christian mysteries they never grasped as Christians can. But what about God's 'Eternal Wisdom and his Godhead'?

A standard doctrine was, that a grasp of the elements of good morality was possible for all men, Christian or otherwise, though grace was always required for Salvation (even the Mosaic Law would not suffice by itself). That good morality was achieved by pagans is shown by Socrates or by other heroes of Montaigne, such as Epaminondas. (The great moral platitudes are never put in doubt anywhere in the Essays.)

Montaigne specifically finds pagan monotheism at its best not 'true' (in the sense of attaining with certainty to the Christian revelations) but nevertheless 'most excusable'. This is not a correction to St Paul's teaching in Romans I:20, but a gloss on it.[11]

10. *Pensées*, Brunschvicg no. 578.
11. 'Apology', p. 573; this can be conveniently seen from the gloss of a later scholar, Estius: in one sense even good pagans were 'not excusable' because of their ignorance; yet 'they can in some way be said to be excusable' by comparison with others who did less well. A hyper-orthodox preacher, Father Boucher, was to condemn Montaigne over this, but could only do so by distorting his thought. Cf. his *Triomphes de la Religion Chrestienne*, 1638, pp. 128–9; Boucher believed that Montaigne was advocating the pagan religion he was seeking to 'excuse'. That was because he was distressed to see Montaigne so influential over 'the *beaux esprits* of these times' that he attributed to him ideas he believed to be held by free-thinkers in his own day.

Montaigne touches so lightly on some crucial theological points that readers may miss their import. Yet they can be vital, not least in the 'Apology', which is centred on religious knowledge and doubt. In at least one respect, Montaigne's conception of God was that of St Augustine, of many medieval and Renaissance thinkers, and of Pascal: God is a Hidden God, a *Deus absconditus* who hides himself from Man and therefore can only be known from his self-revelation. Montaigne lightly but specifically attributes that concept to St Paul. When in Athens, Paul saw an altar dedicated to 'an unknown God' – Athenian philosophers could get that far. In the 'Apology' those words appear as 'a *hidden*, unknown God'. That enables Paul (in the 'Apology') to find the Athenian worshippers to be 'most excusable' ('Apology', p. 573). The same doctrine appears in the medieval theologian Nicolas of Lyra.[12]

Such deft and telling use of words should scotch the notion that Montaigne was theologically naïve. (No theologians who had studied his translation of Sebond could make such a gaffe.) And in this case it should help to undermine the curiously coarse interpretation of the 'Apology' as a work championing 'fideism', one, that is, which denies that there ever can be any rational basis for Christianity since all depends on unfettered faith – faith as trust and faith as credulity. For Montaigne there is a hierarchy of religious opinion among the pagans. (The 'Apology' ends with one of the most impressive of them all: Plutarch's.) Yet Montaigne held with Sebond that even the best of pagans failed to penetrate through to most of the vital truths contained in the Book of Creatures.[13]

The defence of Raymond Sebond against the second charge – that his arguments are weak – falls into several parts, all marked by varying degrees of scepticism. By turning his sceptical gaze on Man and his cogitations, Montaigne denies that it is possible to find better arguments than Sebond's anywhere whatsoever. This assertion is governed (as are all the long answers to the second objection) by a declaration of intent which applies to all the many pages which are to follow:

12. The expression *Hidden God* derives from Isaiah 45:15. Christians of many persuasions used the term to emphasize the need of grace and for revelation from God, who is his own interpreter. It was associated by Nicolas of Lyra with Romans I:20 in his gloss.
13. In this he remains orthodox. The notion of a Book of Nature (or of Creatures) in Sebond's and Montaigne's sense became quite common among theologians: cf. those mentioned in Reginald Pole's *Synopsis criticorum*, 1686, vol. 5, col. 21, line 45 f.; it was also pleasing to Francis Bacon, *Advancement of Learning* I, vi, 16.

Let us consider for a while Man in isolation – Man with no outside help, armed with no arms but his own and stripped of that grace and knowledge of God in which consist his dignity, his power and the very ground of his being. ('Apology', p. 502)

Today the very word scepticism implies for many a mocking or beady-eyed disbelief in the claims of the Church to intellectual validity. It did not do so then. You can be sceptical about the claims of the Church, or you can be sceptical about rational attempts to discredit them . . .

The unenlightened rivals to Sebond have both their hands tied firmly behind their back. Sebond has grace and illumination: they have not. In this second, longer part of the 'Apology', comments are occasionally addressed to this unilluminated ignorance on the basis of revealed wisdom, but the ignorance remains unilluminated and so can only fortuitously, randomly and hesitantly ever arrive at the goal gracefully reached by Sebond's natural theology. That is what makes the *Essays* as a whole so interesting. Instead of calmly orthodox certainty, we are exhilarated by following all the highways and byways and sidetracks travelled along by Man's questing spirit in his search for truth about God, Man and the Universe. Montaigne did his job thoroughly: that is why the *Essays* were pillaged for anti-Christian arguments by the *beaux esprits* of later centuries.

Montaigne is so lightly untechnical that it is easy to overlook that, in a fascinatingly personal and idiosyncratic way, he is saying what learned Latin treatises also taught about the opinions of fallen man. Since sixteenth-century Jesuits appreciated Montaigne, one could cite Cardinal Bellarmine, S. J., who (with the help of St Augustine's *City of God*) was struck by the 'monstrous opinions' of those unenlightened pagans who 'even went so far as to make gods of vines and garlick'.[14] But where Bellarmine finds bleak error Montaigne finds – also – fascinating and inevitable variety.

Montaigne answers the second lot of criticism of Sebond by first crushing human pride: no purely human reasons can show conclusively (as Sebond can) that Man – for all his 'reason' – is in any way higher than the other animals. They, too, like us, have reasoning powers. They have instincts, it is true, but so do we. For this crushing of Man's pride Montaigne first drew mainly on his favourite writer. It seems that Plutarch

14. Robert Bellarmine, S. J., *De Controversiis Christianae Fidei, adversus hujus temporis haereticos* (*Opera*, 1593, III; 'On the Loss of Grace and the State of Sin', col. 487 B; cf. p. 107).

so dominated the first outline of the 'Apology' that Montaigne could even assert that it owed everything to him, a remark he removed once he realized how far he had moved in indebtedness to Sextus, to Cicero, to Aristotle and to Plato ('Apology', p. 629, note 331).

Parts of this praise of the beasts to humble Man's pride have acquired a certain quaintness: zoology has been revolutionized since the Renaissance. Moreover, Montaigne, by long-established convention, cited the weeping war-horses of the poets or the tale of Androcles and the Lion as though they were zoological and historical fact. His loyal dogs commit suicide or haunt their masters' tombs. In his own day, however, his animal science was powerfully persuasive. (Well into the next century, his elephant lore is repeated by Salomon de Priezac in his *Histoire des Eléphants*, Paris, 1650.) As codified by his learned clerical disciple Pierre de Charron in his book *On wisdom*, Montaigne's attitude to the beasts became central to some of the great controversies among the most famous philosophers and theologians of the seventeenth century. In its own way it even had something of the appeal of Darwin. By a very different route it forced people to re-examine in anger or humility what place Man occupied in the Book of Nature among all the other creatures. And Montaigne emphasizes that the common examples of ants, bees and guide-dogs are just as persuasive as exotic rarities.

Pride is the sin of sins: intellectually it leads to Man's arrogantly taking mere opinion for knowledge. In terms which were common to many Renaissance writers, Montaigne emphasized that 'there is a plague (a *peste*) on Man: the opinion that he knows something.'[15]

This pride and this trust in opinion are all part of Man's vanity (of that vain emptiness evoked by Ecclesiastes, the Old Testament book from which were derived several sceptical inscriptions in Montaigne's library). The 'Apology' briefly contrasts such 'vanity' with the assurance supplied by 'Christian Folly' (which proclaims that God's true wisdom is to be found in the lowly, the simple, the humble and the meek).[16]

'Christian folly' was a major theme in Renaissance thought and had been

---

15. Cf. Melanchthon, *De Anima*: 'Hence arises other plagues (*pestes*): the soul loves itself and admires its own wisdom, fashions opinions about God and delights in this game and, in its distress, rails against God.' A century later Father Boucher is still using the same phrases: 'Presumptuousness of mind is the mother of error, the nurse of false opinions, the scourge of the soul, the plague (*peste*) of Man.'

16. Erasmus played a major role in spreading the doctrine of Christian Folly in the Renaissance. It was widely accepted by Christians of many persuasions.

long allied to scepticism. Montaigne was not writing the *Essays* in a void. More specifically, the general thrust of his defence of Sebond would have been evident to any reader of Henry Cornelius Agrippa's declamation *On the Weakness and Vanity of all Sciences and on the Excellence of the Word of God* (Cologne, 1530). It was reprinted in Montaigne's time; he drew on it heavily. It continues a tradition of Christian scepticism to be found in a fifteenth-century scholar such as Valla, who influenced Erasmus, but which is more fully developed in Gian-Francesco Pico della Mirandola's *Examination of the Vanity of the Doctrines of the Pagans and of the Truth of Christian Teachings* (Mirandola, 1520).[17]

These were major and successful books; Montaigne also drew heavily on a work of 1557, unsuccessful enough to be remaindered (freshened up with a new title page in 1587): the *Dialogues* of Guy de Brués. The magic of Montaigne's art in the *Essays* and the originality of his thought enabled him to take ideas and matter lying about in Latin tomes or even in unsaleable treatises and then metamorphose them into the very stuff of his most readable pages.[18]

That certainly applies to his scepticism.

Scepticism is a classical Greek philosophy. Its full force was rediscovered towards the end of the sixteenth century. As such it plays a vital role in Renaissance thought; but the essential doctrines of scepticism (including some of the basic arguments and examples which appear in the *Essays*) were known much earlier, from Cicero's *Academics* and from critical assessments of scepticism (sometimes associated with judgements on the proto-Sceptic Protagoras) in both Plato and Aristotle. Cicero's *Academics* is the easiest to read for lovers of Montaigne (who find that whole passages have been integrated into the *Essays*). So are major borrowings from other works of Cicero, including *On the Nature of the Gods* and the *Tusculan Disputations*. But the influence of Plato and Aristotle goes far deeper.

Up to a point Cicero was a good guide, but less exciting than Sextus

17. Fundamental scepticism, typified by the work of François Sanchez, a doctor in Toulouse, *Quod nihil scitur* ('That Nothing is Known', 1581), was also accessible to Montaigne. He may even have read this particular book in manuscript. (See *Francois Sanchez* [*Franciscus Sanchez*]: *That nothing is known* (*Quod nihil scitur*): Introduction, notes and bibliography by Elaine Limbrick; Latin text established, annotated and translated by D. F. S. Thomson; Cambridge University Press, 1988.)
18. The *Dialogues* of Guy de Brués were aimed against 'the new Academics' and sought to show 'that all does not consist in opinion'. The sceptics are allowed to state their case fairly.

Empiricus and the intellectual stimulus of Plato and Aristotle.[19] Clearly, Sextus' *Pyrrhonian Hypotyposes* dominates parts of the 'Apology', yet appears in no other chapter of the *Essays*. (This has helped support the contention that, when writing the 'Apology', Montaigne went through an acute crisis of scepticism, symbolized by his device of the poised scales with *Que sçay-je?*; What do I know?) By any standards the publication in 1562 by Henri Estienne of the first edition of the original Greek text of Sextus' account of Pyrrho's scepticism was a major event. (Montaigne probably relied chiefly on his Latin translation – also found in the second edition of 1567, but quotations from the original Greek enlivened his library.) Gentian Hervet in his introduction to Sextus' other work, *Against the Mathematicians* (or *Against the Professors*) (1569) helps us to read Montaigne in context. For Hervet, too, the works of Sextus are an excellent weapon against heretics: Pyrrho's scepticism, by reducing all Man's knowledge to opinion, deprives heretics of any criterion of truth. Montaigne did the same in the pages of the 'Apology' which follow upon his address to his patroness (p. 628).

However thorough-going the Pyrrhonism in these final pages, scepticism remained for Montaigne – as for many others – a weapon of last resort: a way of demolishing the arguments of would-be infallible adversaries. There was a price to pay, though. The Pyrrhonian method leaves you with no purely human certainties either! But only much later did that worry many Roman Catholics. Among writers variously attracted to Pyrrhonism were St Francis of Sales (who admired Montaigne's uprightness) and Maldonat (Montaigne's Jesuit friend).

Opinion is not knowledge. Pyrrhonist sceptics revelled in that fact. Sextus Empiricus systematized that contention into a powerful engine of doubt which helped a wise man to suspend his judgement and so to attain tranquillity of mind.

The rediscovery of the works of Sextus gave a fresh impetus to Renaissance scepticism, but it did not create it; Sextus fell on welcome ears: already in 1546 Rabelais has his wise old evangelical King delighted to find that all the best Philosophers are Pyrrhonists nowadays.

It is deliberately paradoxical that the poet who dominates the Pyrrhonist

---

19. Montaigne was irritably aware that Cicero was not an original thinker. More provocative for him were, say, Plato's hostility towards relativism, in the *Theaetetus* and similar passages in Aristotle, as well as his brief indirect account of scepticism and its arguments (*Metaphysics*, 1010 b), which resulted in scepticism being placed within the major philosophical contexts of the Renaissance, which was anchored in Aristotle.

pages of the 'Apology' should be Lucretius. That Latin poet of the first century BC was a follower of Epicurus and remains our principal source for Epicurean doctrine in the realm of physical nature. But Epicureanism is flatly opposed to Pyrrhonist scepticism. Far from asserting that all man's boasted knowledge is mere opinion, it holds that the senses give Man access to infallible certainty. The point is made clearly and sharply in Denis Lambin's edition of Lucretius, which Montaigne read with marked attention. (What seems to be Montaigne's own copy, annotated in his hand, was recently recognized as such by Paul Quarrie when he bought a Lambin *Lucretius* for Eton College library, where that book now is.) For Lucretius, truth about things must be accessible to our minds from sense-impressions: if they are not, all claims to know truth collapse. So even the Sun can be only a trifle larger than appears to our sight. If we cannot explain why, we must nevertheless make no concessions to those who deny this. Such a view flew in the face of traditional and solid scientific knowledge. Montaigne delights in citing Lucretius' own words to undermine Epicurean assertions.[20] But Lucretius also serves to undermine other ideas widely supposed to be true – and to warn against superstition.

Montaigne was perhaps first attracted to Lucretius by his arguments against that fear of dying which haunted his youth and young manhood. In the 'Apology', however, he chiefly cites him in order to reveal yet another source of darkness and error or, at best, of the kind of partial truths reached by unenlightened sages.

Particularly effective are his exploitations of precisely those verses in which Lucretius tried to refute those who hold that 'we can know nothing'. Denis Lambin in his edition praised Lucretius for his solid opposition to the doctrine that 'nothing can be known'. Montaigne eventually succeeds in exploiting the principal opponent of scepticism for sceptical ends![21]

On many matters, Montaigne and Lambin were in agreement. Especially interesting for the *Essays* is Lambin's dedication of Book III of Lucretius' poem to Germain Valence. It shows that the very failure of even Lucretius and the Epicureans to reach Christian certainties about the nature of the soul can be turned into yet another argument in favour of Christian revelation:

20. Cf. 'Apology', p. 634 ff., 664 ff.
21. This section begins with line 469 of Book IV of Lucretius: 'Moreover if anyone thinks nothing is to be known, he does not even know whether that can be known, as he says he knows nothing.' (Cf. ed. Lambin, 1563, p. 308 ff.)

Not unjustly we despise their unwise wisdom. We should congratulate ourselves that we have been taught by JESUS CHRIST ... (without being convinced or coerced by any human reasons or by any arguments, no matter how well demonstrated – not even by the Platonists) and so are persuaded that no opposing reasons, however sharp or compelling, however probable or verisimilitudinous, however firm or strong (let alone those of Lucretius, which are light and weak) could ever dislodge us from this judgement.[22]

The Renaissance was a period of new horizons: one was a vast increase in knowledge of the world and its inhabitants, as Europeans sailed the seas and discovered new lands, new peoples and moral and religious systems new to them; another was the rediscovery of Greek literature in its fullness. New horizons make local certainties seem wrong or parochial: they also open up whole treasure-houses of new facts and facets to the sceptic, who with their aid can increase the sense of the relativity of all Man's beliefs about himself and the universe in which he lives. Montaigne exploited Sextus Empiricus, but he also devoured the writing of the Spanish historians, including those who told of the horrors of the conquest of the New World. There were also compendia such as Johannes Boemus' *Manners of all Peoples* (Paris, 1538), as well as standard works such as Ravisius Textor's *Officina* ('Workshop') which contains chapters with such titles as 'Various opinions about God' and 'Divers morals and various rites of peoples' (Montaigne would have read in it a full account of Androcles and the Lion). New books gave him and the *Essays* a dimension and an actuality lacking to Agrippa and Pico. His universe was open to immense variety. He knew of Copernicus. If he wanted noble savages he could draw on the Indies as well as on the Golden Age; or he could try and talk to American Indians for himself (in Rouen) or question sailors.

But he did not stop there. If he had, he might indeed have been a fideist, claiming that only an arbitrary act of faith could make an irrational leap from a boundless sea of doubt to the rock of certain truth: the Church. Such a theology, never really convincing, was rarely less convincing than in the Renaissance and the nascent scientific world of the following century. If the leap is irrational, why leap to Catholicism and not to a sect or to any other of the teeming religions of the world? Truth must be the same everywhere.

This infinite variety of the world can be put to the service of Pyrrhonism and its universal doubt: it can also be put to the service of Catholic orthodoxy against sect and schism. If Catholic Christianity is true at all it

22. Lucretius, p. 190.

must be universally true, not merely true for Périgordians, Germans or successive English parliaments. Otherwise it is just one opinion among many. Ever since St Vincent of Lerins in the fifth century, Catholic truth was categorized as being *Quod semper, quod ubique, quod ab omnibus* ('What has been held always, everywhere and by all'). In the Renaissance the aspiration to make that a reality lay behind the vast, worldwide evangelism by Rome (which contrasted sharply with the local concerns of the rival Churches seeking to reform one City or one Kingdom). The Roman Catholic faith could indeed claim to be taught universally. Therein lay its strength for minds like Montaigne's.

For Montaigne, the strength of Raymond Sebond's *Natural Theology* also lay in universality. He believed that Sebond's illumination of the universal Book of Nature showed that all Nature everywhere was in strict conformity with Catholic truth.

At the end of Montaigne's Pyrrhonist pages we are brought to the very brink of uncertainty. Reason has been shaken. So have the senses. If sense-data are unsure, uncertain and often plainly misleading, that does not simply cut us off from any sure and solid knowledge of phenomena: it cuts us off from any sure and certain knowledge of 'being'. And so 'we have no communication with Being' – other than with our own transient one (perhaps).

> To conclude: there is no permanent existence either in our being or in that of objects. We ourselves, our faculty of judgement and all mortal things are flowing and rolling ceaselessly; nothing certain can be established about one from the other, since both judged and judging are ever shifting and changing.[23] ('Apology', p. 680)

But this – despite the words 'to conclude' (*finalement*) – is not the end of the 'Apology': it is the end of a chain of arguments which can leave man ignorant, or, on the contrary, show him a new way to proceed. If it had been Montaigne's conclusion, then Sextus Empiricus would literally have had the last word, for the Pyrrhonist basis is evident. But it is precisely here that Pyrrhonism joins Plato and Aristotle in joint hostility to a sophistical trust in individual subjectivity.

At the end of the long section which immediately precedes Montaigne's address to his Royal patroness, just as he was about to embark on his

23. Sextus, *Hypotyposes*, I, 217–19 (criticizing Protagoras for dogmatism and relativism). Some excellent reflections on this topic in Jean-Paul Dumont, *Le scepticisme et le phénomène* (especially Chapter 3); see also M. Burnyeat, *The Skeptical Tradition*.

Pyrrhonian arguments, Montaigne added an important comment in the margin of the Bordeaux copy of his works he was preparing for the press. It concerns Protagoras, the arch-Sophist who was trounced by Sextus, Plato and Aristotle in very similar terms and for identical reasons:

> And what can anyone understand who cannot understand himself? . . . Protagoras was really and truly having us on when he made 'Man the measure of all things' – Man, who has never known his own measurements.

Protagoras meant – that is what shocked Plato, Aristotle and Sextus Empiricus – that there is no universal standard of truth: each human being is severally and individually the sole criterion; all is opinion, and all opinions are equally true or false.

For Montaigne, Protagoras' 'measure of Man' is 'so favourable' to human vanity as to be 'merely laughable. It leads inevitably to the proposition that the measure and the measurer are nothing.'

Montaigne countered Protagoras, immediately, by citing Thales (the Greek sage to whom he himself had been likened): 'When Thales reckons that a knowledge of Man is very hard to acquire, he is telling him that knowledge of anything else is impossible.' ('Apology', p. 628) Hence the growing importance of the study of Man throughout the *Essays*, especially in Book III and in the hundreds of additions made to the chapters of the two previous Books when the new Book was written and the others were revised.

In the *Theaetetus*, Socrates treated Protagoras and his 'measure' as a clever man talking nonsense – otherwise how can the same wind be hot to one and cold to another? Nor would anyone maintain that, since a colour appears different to a dog, to other animals and to ourselves, that it differs in its essence.[24]

Montaigne made good use of such notions in the 'Apology': they can serve to show the fallibility of sense-data and also to place man where his unaided natural reason ought to place him: among the other creatures. But to go from there and make Truth itself the plaything of individual subjectivity, he never did.

Aristotle similarly mocked Protagoras and his Man-as-measure; his demonstration was adapted by Montaigne.[25]

---

24. Plato, *Theaetetus*, 152B; again, 152 CD. (Was this saying of Protagoras' only meant for the mob?) Arguments drawn from 153–4 are used by Montaigne.
25. Aristotle, *Metaphysics*, 1053b (misunderstanding Protagoras): 'Thus, seeming to say something unusual, he is really saying nothing.' More relevant to Montaigne (who uses some of the arguments) are *Metaphysics*, 1062b – 1063a.

Montaigne knew,[26] before he had read a word of Sextus – probably in his days at school in Bordeaux – that in the world of creation nothing ever is; it is only becoming. Plutarch reinforced this. But neither Plutarch nor Plato held that such doctrines cut Man off from a knowledge of God or obliged each person to plunge into pure subjectivism. There were, for Plato, divine revelations; and there was wisdom arising from knowing oneself as Man.

Within the flux of the created universe, Montaigne strove to follow the Delphic injunction, Know Thyself. He sought to discover the personal, individual, permanent strand in the transient, variegated flux of his experience and sensations, which alone gave continuity to his personality – to his 'being' as a Man.[27]

But this was not a merely subjective indulgence. By studying his own form (his soul within his body) he aspired to know Man – not just one odd individual example of humankind.[28]

The *Essays* as a whole do not end with the last words of the 'Apology'; much exploration of self and of Man remained to be done, but Montaigne had clearly seen that the characteristic property of the creature is impermanence. No creature ever *is*: a creature is always shifting, changing, becoming.

The Platonic background to such a conclusion – unlike the purely Pyrrhonian one – enabled Montaigne to pass from the impermanence of the everchanging creature to what he presents as a 'most pious' concept of the Godhead, accessible to purely human reason: the Creator must have those qualities which Man as creature lacks: he must have unity, not diversity; absolute Being, not mere 'becoming'. And since he created Time he must be outside it and beyond it.

It is strikingly right that this natural leap to the Eternal Being of God should be given not in Montaigne's own words – he is not a pagan – but in a long and unheralded transcription from Plutarch. Montaigne took it from the dense mystic treatise *On the E'i at Delphi*.

In this powerful work Plutarch grappled with the religious import of the word *E'i* inscribed on the temple of Apollo at Delphi. In Greek it can mean 'Five'; it can mean 'If': but above all it means 'Thou Art'. As such it declares that God has eternal Being. He is the eternal THOU to our transient I. Each individual human being is relative, contingent, impermanent. But

26. In his studies of Plato, e.g. *Timaeus*, 37D–38B; *Theaetetus*, 152DE, etc.
27. Cf. *Montaigne and Melancholy*, p. 125; 101 f.
28. Ibid., p. 104.

each 'I' can know itself; it can know Man through itself; and it can stretch out to Reality and say THOU ART.

In doing so, it recognizes God.[29]

The *Natural Theology* of Sebond taught each man to know himself and God. It is, in a sense, the key to that Delphic utterance: Know Thyself. Montaigne's translation of the *Natural Theology* is all of a piece with the self-exploration of the *Essays*. For both Plutarch and Raymond Sebond 'knowing oneself' is, properly understood, a complement to knowing God. Sebond says so on his title pages: Plutarch does so in the closing words of *On the E'i at Delphi*:

> Meanwhile it seems that the word *E'i* [THOU ART] is in one way an antithesis to that precept KNOW THYSELF, yet in another it is in agreement and accord with it. For one saying is a saying of awe and of adoration towards God as Eternal, ever in Being; the other is a reminder to mortal man of the weakness and debility of his nature.

Plutarch could reach that pious height: a Roman Stoic could also assert that if a man is to aspire towards God he must 'rise above himself'. So far so good.

We are doubtless stirred by such eloquent aspirations. But the final words of the chapter tip over the house of cards. If any human being is to rise up towards that Eternal Being glimpsed by Plutarch, it will not be through Greek philosophy or proud Stoic Virtue: it will be 'by grace' or, more widely, 'by purely heavenly means'. That will be an event 'extra-ordinary' – outside the natural *order* of the universe. In the process, the individual human being will not raise himself but *be* raised to a higher form. He will (in the last word of the chapter) be 'metamorphosed': transformed and transfigured.[30]

That leaves Montaigne free as always to continue to explore his 'master-mould'; to examine his relative 'being' – his body-and-soul conjoined.

Nowhere else in the *Essays* does Pyrrhonian scepticism make the running – it does not make all of it even in the 'Apology'. But to the solid

29. There is a striking parallel between Plutarch's conception of God and the Christian scholastic doctrines based upon God as revealed to Moses in the burning bush. (Where in English God says I AM THAT I AM, in the Greek and in scholastic theology he says I AM EXISTENCE or I AM THE EXISTING ONE.) Montaigne does not emphasize this: he lets it sink in.
30. 'Apology', p. 683. The full implications of this are not revealed until the last pages of the final chapter of the *Essays*: Book III, Chapter 13, 'On experience'.

bastion of his faith Montaigne added a shield of last resort, ever ready in reserve to use against those who sought to oppose his Church's infallibility by a rival one. As Edward Stillingfleet, Dean of St Paul's, perceived in the following century, Pyrrhonism comes into play only when men are not content to 'take in the assistance of Reason, which, though not Infallible, might give such Evidence, as afforded Certainty, where it fell short of Demonstration'. But as soon as 'Epicurus thought there could be no Certainty in Sense, unless it were made Infallible', he could only defend his hypothesis with absurdity: 'the Sun must be no bigger than a bonfire'.[31]

Of course Pyrrhonian scepticism shocked many. It always does. But when Montaigne's *Essays* were examined by a courteous censor in Rome, such little fuss there was at the time came from factions among the French. The Maestro del Sacro Palazzo, Sisto Fabri, told him to take no notice and do what he thought fit.[32]

In the following century Montaigne's respect for the beasts and his distrust of unaided human reason brought him many enemies among dogmatic philosophers and theologians; they brought him many friends as well, ranging from Francis Bacon to Daniel Huët, Bishop of Avranches. In his *Philosophical Treatise on the Weakness of the Human Spirit* (1723) Huët reminded his readers that when Pyrrhonism was rejected in Ancient times, it was nothing to do with Christians fearful for the Faith but of pagans fearful for their Science. What is dangerous to Christianity, he added, is not Pyrrhonism but Pride.[33]

But Montaigne had done his job well – well enough for many free thinkers including those of the Enlightenment to see him as a forerunner of their sceptical Deism or atheistic naturalism. This was in part inevitable: truth is one and unchanging while men are ever-changing. Truth cannot be set finally in words. It was a sound theologian, Bishop Wescott, who said, 'No formula which expresses clearly the thought of one generation can convey the same meaning to the generation which follows.' In a different climate of opinion, Montaigne's protestations of loyalty to his Church in

---

31. Edward Stillingfleet, *Nature and Grounds of the Certainty of Faith*, 1688, p. 35.
32. Montaigne, *Journal de voyage en Italie*, ed. Pierre Michel, 1974, pp. 287–8, 310.
33. Daniel Huët, *Traité philosophique*, 1723, III, 16. Cf. II, 6: 'What is the End Proposed by the Art of Doubting?' There are two ends. 'The proximate end is to avoid error, stubbornness and arrogance. The eventual end is to prepare one's spirit to receive Faith.' These are the professed aims of Montaigne.

several of his chapters were taken to be moonshine. Allusions to 'Christian folly' were interpreted as smirking and knowing ackowledgements that Christianity was silly or stupid, fit for fools. Read in this way, selectively, the *Essays* could, did and do provide weapons and delight to a variety of readers. This became more easily possible after Hellenistic philosophy lost its hold on many in the eighteenth century. Hellenistic Christianity (like Hellenistic Philosophy) accepts that the true nature of things lies behind their visible appearances, and beyond time and space. It holds with Plotinus that nothing that is can ever perish.[34] Such a conviction dominated the thought of Renaissance Christians including Ficino, Erasmus, Rabelais and Montaigne. Without such a conviction and a respect for its roots in Platonism, the end of the 'Apology' (and much else in the *Essays*) may seem purely arbitrary – arbitrary and ironic, or a meaningless tactical bow to authority.

But we know from Montaigne's *Journal* (discovered in 1772 and never intended for publication) that he was a practising Christian whose devotion was as superstitious as Newman's. He could attach great importance to the pious family ex-voto which he paid to be displayed in the Church of Santa Maria de Loreto (in the shrine of the Holy House of the Virgin, transported by angels to Loretto, was it not, on 2 December 1295) and to the miraculous cure there of Michel Marteau.[35]

This was a great shock to those *philosophes* and wits who had grown used to exploiting the *Essays* as an anti-Christian weapon-house. They had done so all the more mockingly after the *Essays* had been put on the Index in 1697. But that act of absurdity can be better attributed not to the *Essays* as such as to Jansenist zeal and to horror at the use made of them by the free-thinking libertins.[36]

In Montaigne's own day Rome knew better – and presumably does so now. The Vatican Manuscript No. 9693 records the granting of Roman citizenship to Montaigne. It states that it was granted to 'the French Socrates'.

And that Christian Socrates died in the bosom of his Church. (But, even then, André Gide persuaded himself that he only pretended to do so because of moral blackmail from his wife . . .)

*

34. W. R. Inge, Dean of St Paul's, in *Faith and Knowledge*, 1905, p. 245; *The Church and the World*, 1927, p. 191; *More Lay Thoughts of a Dean*, 1931, p. 160, aphorism no. 37: 'Know Thyself is really the sum of wisdom; for he who knows himself knows God.'

35. *Journal*, pp. 325–6, 330. Newman similarly puzzled and irritated many by this respect for the shrine of Our Lady at Walsingham and for Loretto.

36. Ibid. (note by Pierre Michel, p. 310 n. 184); and his Introduction.

Citing Plutarch at the end of the 'Apology' does more than vindicate Raymond Sebond: it vindicates St Paul. In retrospect it can be seen that Romans I:20 gave authority not only to Montaigne's defence of natural theology in his reply to the first charge against the book he had translated: it governs the long reply to the second charge. St Paul declared that what can be grasped by natural theology ('from the things that are made') are God's 'eternal power' and his 'divinity'. Plutarch shows that that is true: Plutarch did so. But Plutarch is nevertheless only one pagan voice among many, one ray of light in confused darkness.

Montaigne is exemplifying a tradition codified at least as early as Nicolas of Lyra, the thirteenth-century scriptural commentator who suggested that by the words 'from the creation of the world' (in the Vulgate Latin, *'a creatura mundi'*) Paul meant from Man (who is the 'creature of the world' *par excellence*). Montaigne does not say this explicitly, but his whole enterprise in the *Essays* is driven forward by a desire to know Man and his place in the Universe (not simply one example of Mankind, himself, though that is his means to the greater end). The seeking of God 'from the things that are made' is explained by Nicolas of Lyra to mean *'per creaturas'* ('through the creatures' – through all that God created). And what Man can discover concerns 'the divine Essence': 'from the creatures a man can learn that that eternal Essence is "One, Uncompounded and Infinite".'[37]

Of course, none of this 'natural theology' brings fallen Man effectively to the Triune God: that needs grace.[38]

An appreciation of the balance between religious certainty and rational doubt and inquiry which Montaigne struck in 'An apology for Raymond Sebond' is a great help in following his whole intellectual venture as he takes us through an astonishingly varied series of topics which lead us into the mind of a man who, though he lived four hundred years ago, yet remains fresh and stimulating and well able to speak for himself.

Montaigne's contemporaries were impressed by his reflections on his experience as a soldier and statesman. Both his Italian translator, G. Naselli, and his English one, John Florio, stress on their title pages that these *Essays* include moral, political and military discourses. For them they were

---

37. Cited from *Biblia Maxima*, Vol. XV. This exegesis was accepted by scholars of many schools and Churches: cf. for example Matthew Henry, *Exposition of the New Testament*, 1738, vol. 5, commentary on this verse.
38. Cf. Nicolas of Lyra on Romans 3:10 (*Biblia Maxima*, vol. XV: index, s.v. natura): a man can perform moral acts without grace: he cannot be justified.

primarily that. On matters of war and politics Montaigne was listened to throughout Europe as a gentleman who knew from experience, and not from book-learning alone, what he was talking about.[39] Up to our own time Montaigne has spoken directly to many who have had experience of war and of military life. Many who have been spared such experiences have at times undervalued the part played by the experience of battle, of parleys and of political negotiations in the formation of Montaigne's way of thinking.[40] That is a pity, for to undervalue this aspect of his life is to some extent to falsify the *Essays* which, where such matters are concerned, are not simply based upon hearsay. It is because Montaigne knew directly of barbaric cruelty that he could write so movingly of 'cannibals' and of the crimes of the Conquistadores. It is because he had seen and talked to 'savages' and, say, to women languishing in prisons under the accusation of witchcraft that he could write of them with such a sense of humanity. It is because he was privileged to experience a very special friendship with Estienne de La Boëtie that he could write on affectionate relationships more evocatively than even Cicero could. Only a man who loved poetry and had experienced the love of a wife and, especially in youth, of other women could have written so probingly on sexuality and its limitations, as well as on matrimony and the running of a household with their calmer joys and risks of daily irritations. And it is as a seasoned traveller that Montaigne wrote of his experience in Germany and Italy.

Nowadays a collection of essays can be read in any order, with each essay taken as a unit. Montaigne's *Essays* are not presented like that. His *Essays* form three Books, each Book divided not into self-contained units but into chapters. Book III, written unexpectedly after his first two, ends with a discourse 'On experience', which is not an 'essay' which happens to come last but the final chapter of the final book. It marks the end of Montaigne's quest. He was not, he tells us, a man over-given to bookish interests, but what he did seek from books and from experience he sought with passion and tenacity. 'On experience' (III, 13) gives us the distillation of his mature thought, showing us how to live our lives with gratitude.

39. In Italian, *Discorsi morali, politici et militari* (Ferrara, 1590). John Florio's title was: *The Essayes, or Morall, Politike and Millitarie Discourses . . . of Lo. Michaell de Montaigne . . .* (London, 1603).
40. The balance has been restored for more peaceful generations by an excellent book, happily in English, James Supple's learned and very readable study, *Arms versus Letters. The Military and Literary Ideals in the 'Essais' of Montaigne*, Oxford, Clarendon Press, 1984.

The *Essays* had begun with thoughts of ambiguity, sadness and of emotions which make men beside themselves: Montaigne, after a thousand or so pages of thought and after reflecting on a lifetime of experience, starts his final chapter with a ringing challenge. He alludes to the opening words of Aristotle's *Metaphysics*, words which every serious reader knew: 'All men naturally desire knowledge.' Why, even schoolboys knew them! Those words of Aristotle had for centuries evoked theological certainties, since they formed part of a standard chain of argument claiming to prove on Platonico-Christian grounds that the soul is immortal:

All men naturally desire knowledge;
But no man's desire for knowledge is satisfied in this life;
Yet Nature does nothing in vain;
Therefore there must be an afterlife in which that desire will be satisfied.

Aristotle wrote that first sentence of his *Metaphysics* to introduce the notion that experience, when collated, weighed and pondered over, can produce an 'art' (a *techné*) Such an art, he asserts, can help man towards knowledge in areas where pure reason proves inadequate. The two 'arts' most evidently based on such weighed experience are law and, above all, medicine, which was usually known in Montaigne's day as '*the* Art', or, by a corruption of the Greek, simply as '*Tegne*'.[41] But Montaigne, having throughout the *Essays* shown how fallible reason is, rejects any notion that certain knowledge can be based on fallible experience either; experience is finite: circumstances are infinitely varied. Hence the importance of judgement, of temperance and moderation, by the help of which the wise know how to think and to live in the midst of unresolvable uncertainties.

Montaigne was an excellent pupil of the School of Athens and of its Latin disciples. He realized that wisdom did not consist in simply studying Socrates, Plato, Aristotle and the schools of philosophy which the Ancient Greek and Latin world derived from them: wisdom consists in following precepts, not in knowing them off by heart. Aristotle may well point the way, but Montaigne was not content to know the words of his *Metaphysics* and his *Physics* (and even less merely to pick his way through a maze of commentaries upon them): more than any other object he studied his own 'self': that study was his metaphysics; that study was his physics; with their

41. For example, the expression 'Life is short but art is long' is the first of the *Aphorisms* of Hippocrates and means that life is all too brief for anyone who would study *the* Art – medicine.

aid he could judge whether or not Aristotle or anyone else was probably talking sense or nonsense.

His 'self', he found, was more than his soul. His 'being', like that of any person, consisted in a soul ('form') linked to a body ('matter'). (That was another scholastic axiom: 'Form gives being to matter.') But as his body aged it was racked with pain. (The colic paroxysms produced by his stone were recognized as being suicidal.[42]) Once he had grasped how 'wonderously corporeal' human beings are he saw that wisdom lies in keeping body and soul together in loving harmony, not in segregating the soul from the body to keep it pure and purely intellectual. Those Socratic injunctions to seek to 'practise dying' – to strive, that is, to enable the soul to leave the body for a while and 'go outside' – those ecstatic activities which send the soul soaring aloft from the body, are by most men to be rejected. As for those baser ecstasies which lead the body to wallow in the mire of lust and drunkenness, bereft of its soul, they are not 'bestial' (beasts do not act like that): they are sub-human.

Christian rapture is a great thing. Yet only a tiny handful of the Elect – only privileged ascetic contemplatives touched by grace – can safely neglect their bodies' transitory necessities while their souls feed by anticipation on lasting heavenly food. For ordinary folk to strive to ape them leads to madness: for madness, too, consists in the pulling apart of soul and body. Since Platonic times there was thought to be a hierarchy of souls within creation. Above the human souls were classed the souls of angels; below the human souls were classed the souls of beasts. But, concludes Montaigne, attempt, without a special gift of grace, to soar aloft and rank with the angels and you will end up a maniac: not an angel but below a beast; not supernally moral but subterrestrially immoral.

---

42. Montaigne's illness, 'the stone', came suddenly upon him in 1578. It, not unnaturally, changed his attitude to life and put his philosophy to the test. He frequently calls it simply 'the stone' or '_cholique_'. Both these terms (especially the second) risk misleading modern readers, who may fail to grasp their implications of dreadful internal pain and the retention of urine accompanied by paroxysms. I often render his (for us) at best neutral terms not by 'the stone' but by 'colic paroxysms' to drive home the ghastly pain from which he suffered and which (despite the promptings of Classical Stoicism which would have held suicide to be justified) he bore with resignation and fortitude. In his _Collection of Offices_, 1658, Jeremy Taylor included in his prayer for 'all them that roar and groane with intolerable paines and noisome diseases' those who are afflicted 'with the stone and with the gout, with violent colics and grievous ulcers'. Like Montaigne Jeremy Taylor saw such afflictions as 'the rod of God', a cause, indeed, for pity, but to be borne with patience by the sufferer.

Socrates and Plato, are, up to a point, good guides for that Elect: Aristotle is a safer guide for all the rest of us. And so (despite his own moral weakness and his inflated tongue) is Cicero. That Man should welcome his body and that his soul should love it, are ideas which Montaigne found in Cicero, in Erasmus, in Raymond Sebond and even, surprisingly, in Lucretius. From Raymond Sebond directly, no doubt, Montaigne derived the idea that the body and soul should live as in a loving marriage. Marriage he conceives of course as Christians did: as a mutually loving union of two unequals, each with duties to the other, each helping the other until death them do part. For either to neglect its duties, for either to regret or neglect its rightful pleasures or those of its partner, is to fall into the sin of ingratitude. During this life the soul needs the body, and the body needs the soul. As a Christian Montaigne knows that the body itself shares unimaginably in the afterlife. Except for a chosen few, the plain and explicit duty of each human being is to see that the body helps the soul; the soul (even more so), the body.

This civilized and humanizing concept of duty is supported by a long quotation from St Augustine's *City of God* (XIV, 5). That passage was well chosen, for it is drawn from a section in which Augustine censures the Manichees (who condemned matter, and hence the body, as evil). St Augustine also, as Montaigne does, draws support at this point from Cicero, whose treatises *On the ends of good and evil* and *On duties*, as well as the *Tusculan Disputations*, are alluded to here in Renaissance editions of the *City of God*. Those are specifically the treatises of Cicero on which Montaigne came to draw. Montaigne might not like Cicero's chatter, but he owed a great deal to his wisdom.

An elect group of Christian mystics are vouchsafed the gift of rapture. That gift of grace segregates them from all the rest of humanity, including philosophers and sages. Montaigne's conclusion is that all other human beings should acknowledge their humanity; acknowledge that even their greatest thoughts and discoveries are not all that important; acknowledge that there is ample time for the soul to enjoy its pabulum once the body has been fed and its few necessities wisely catered for. After all, even when a man is perched high on a lofty throne, what part of his body is he seated upon? Everything for mankind is '*selon*', an expression still current in popular French but strangely technical nowadays in English. Everything is *secundum quid*, 'according to something'. Montaigne wishes to be judged, he says, '*selon moy*', that is '*secundum me*', 'in accordance with myself', 'according to my standards'. If a man insists upon living in court he will have to dodge about and use his elbows, living 'according to this, according

to that and according to something else'. The wiser man will live (in harmony with creation, of which he knows he forms a part) *secundum naturam*, 'according to nature'. All schools of philosophy tell him to do so, but none now tells him how to do so, having obscured Nature's footsteps with their artifice. As always art or artifice is the antithesis of nature.

Classical philosophy, not least among the Latins, had taught men how to die. Yet the body and soul will know how to separate well enough when the time comes. Man needs to learn how to live! Meanwhile old age can be indulged and the Muses can bring joy and comfort. But the very last words of the *Essays* convey a warning: old men may go gaga. (Even the wisdom of Socrates, we were told, is at the mercy of the saliva of some slavering rabid cur.) At the end of his quest Montaigne gave, as a philosopher well might, the last word to Latin poetry, to Horace evoking the patron deity of health and the Muses. Montaigne had learned how to come to terms with ill-health and was grateful for pain-free interludes. He had schooled his soul to help its body over its bouts of anguish. He had gratefully discovered in old age that the Muses continued to make life worth living. The Muses, for a sick old man, meant mainly books and such social intercourse as still came his way, now that he had learned detachment and so prepared himself to part from those he loved. But Horace's words evoke the fear of fears for a man of Montaigne's turn of mind: senile dementia: and his last word of all encapsulates the dread of old folk throughout the ages: want − not in his case want of food or money or position but of what the Muses bring: *'nec cythera carentem'*.

ALL SOULS COLLEGE
OXFORD

ALL SOULS DAY, 1989

Note on the Text

There is no such thing as a definitive edition of the *Essays of Michel de Montaigne*. One has to choose. The *Essays* are a prime example of the expanding book.

The text translated here is an eclectic one, deriving mainly from the corpus of editions clustering round the impressive *Edition municipale* of Bordeaux (1906–20) edited by a team led by Fortunat Strowski. This was further edited and adapted by Pierre Villey (1924); V.-L. Saulnier of the Sorbonne again revised, re-edited and adapted the work for the Presses Universitaires de France (1965). Useful editions were also published by J. Plattard (Société 'Les Belles Lettres', 1947) as well as by A. Thibaudet and M. Rat for the Pléiade (1962). These editions largely supersede all previous ones and have collectively absorbed their scholarship.

I have also used the posthumous editions of 1595, 1598 and 1602 and, since it is good and readily available at All Souls, the *Edition nouvelle* procured in 1617 by Mademoiselle Marie de Gournay, the young admirer and bluestocking to whom Montaigne gave a quasi-legal status as a virtually adopted daughter, a *fille d'alliance*.

The Annotations

Marie de Gournay first contributed to the annotation of Montaigne by tracing the sources of his verse and other quotations, providing translations of them, and getting a friend to supply headings in the margins.

From that day to this, scholars have added to them. The major source has long been the fourth volume of the Strowski edition, the work of Pierre Villey. It is a masterpiece of patient scholarship and makes recourse to earlier editions largely unnecessary. Most notes of most subsequent editions derive from it rather than from even the fuller nineteenth-century editions subsumed into it. This translation is no exception, though I have made quite a few changes and added my own. Montaigne knew some of his authors very well indeed, but many of his *exempla* and philosophical sayings were widely known from compendia such as Erasmus' *Adages* and *Apophthegmata*. His judgements on women and marriage are sometimes paralleled in a widely read legal work on the subject, the *De legibus connubialibus* of Rabelais' friend Andreas Tiraquellus. Similarly some of his classical and scriptural quotations and philosophical arguments in religious contexts are to be found in such treatises as the *De Anima* of Melanchthon or in the theological books of clergymen of his own Church writing in his own day. I have taken care to point out some of these possible sources, since Montaigne's ideas are better understood when placed in such contexts.

References to Plato, Aristotle, Cicero and Seneca are given more fully than usual. Although Montaigne read Plato in Latin, references are given to the Greek text (except in 'An Apology for Raymond Sebond') since most readers will not have access to Ficino's Latin translation. References to Aristotle too are always given to the Greek: that will enable them to be more easily traced in such bilingual editions as the Loeb Classics. For Plutarch's *Moralia* detailed references are given to the first edition of Amyot's translation (*Les Oeuvres morales et meslées*, Paris, 1572); for Plutarch's *Lives* however only general references are given under their English titles (many may like to read them in North's *Plutarch*).

For historical writers of Montaigne's own time only brief references are given. All of them derive from Pierre Villey's studies in which the reader will find much relevant detail: *Les Livres d'histoire modernes utilisés par*

Montaigne, Paris, 1908, and *Les sources et l'évolution des Essais de Montaigne*, Paris, 1908 (second edition 1933). Those books are monuments of scholarship and have not been superseded.

The classical quotations (which from the outset vary slightly from edition to edition of the *Essays*) are normally given as they appear in the Villey/Saulnier edition: most readers discover that the quickest way to find a passage in another edition or translation is to hunt quickly through the chapter looking for the nearest quotation. Once found in the Villey/Saulnier edition a passage can be followed up in the Leake *Concordance* and traced to other standard editions.

My studies of Montaigne have been greatly helped by the kindness of the Librarian of University College London, the Reverend Frederick Friend, who has authorized several volumes to be made available to me on a very long loan. I am most grateful to him and to University College London.

I am most grateful to those readers who have suggested corrections or improvements, many of which have been included in this 1994 reprinting. A special word of thanks is due to Mr Jan Stolpe, the distinguished translator of Montaigne into Swedish, and to Donald Upton Esq., Dr Jon Haarberg, Dr Andrew Calder, M. Gilbert de Botton, Dr Bernard Curchod, Professor David Wiggins, Mrs Thalia Martin and Dr Jean Birrell.

*

Postscript:
Since my ordination by the Bishop of Oxford in 1993 I am often asked if I find Montaigne's arguments for his Church still convincing. Clearly not: I was not ordained in his Church, but I do think that Montaigne can still succeed in getting many to take Christianity – and religion in general – seriously.

M.A.S. All Souls College, Oxford.
June 2003.

Note on the Translation

I have tried to convey Montaigne's sense and something of his style, without archaisms but without forcing him into an unsuitable, demotic English. I have not found that his meaning is more loyally conveyed by clinging in English to the grammar and constructions of his French: French and English achieve their literary effect by different means. On the other hand I have tried to translate his puns: they clearly mattered to him, and it was fun doing so. Montaigne's sentences are often very long; where the sense does not suffer I have left many of them as they are. It helps to retain something of his savour.

It is seldom possible to translate one word in one language by one only in another. I have striven to do so in two cases vital for the understanding of Montaigne. The first is *essai, essayer* and the like: I have rendered these by *essay* or *assay* or the equivalent verbs even if that meant straining English a little. The second is *opinion*. In Montaigne's French, as often in English, *opinion* does not imply that the idea is true: rather the contrary, as in Plato.

Montaigne's numerous quotations are seldom integrated grammatically into his sentences. However long they may be we are meant to read them as asides – mentally holding our breath. I have respected that. To do otherwise would be to rewrite him.

When in doubt, I have given priority to what I take to be the meaning, though never, I hope, losing sight of readability.

Of versions of the Classics Jowett remarked that, 'the slight personification arising out of Greek genders is the greatest difficulty in translation.' In Montaigne's French this difficulty is even greater since his sense of gender enables him to flit in and out of various degrees of personification in ways not open to writers of English. Where the personification is certain or a vital though implied element of the meaning I have sometimes used a capital letter and personal pronouns, etc., to produce a similar effect.

Explanation of the Symbols

═══

[A] or '80: all that follows is (ignoring minor variations) what Montaigne published in 1580 (the first edition).

[A1]: all that follows was added subsequently, mainly in 1582 and in any case before [B].

[B] or '88: all that follows shows matter added or altered in 1588, the first major, indeed massive, revision of the *Essays*, which now includes a completely new Third Book.

[C]: all that follows represents an edited version of Montaigne's final edition being prepared for the press when he died. The new material derives mainly from Montaigne's own copy, smothered with additions and changes in his own hand and now in the Bibliothèque Municipale of Bordeaux.

'95: the first posthumous edition prepared for the press by Montaigne's widow and by Marie de Gournay. It gives substantially the text of [C] but with important variants. (The editions of 1598 and 1617 have also been consulted, especially the latter, which contains most useful marginal notes as well as French translations, also by Marie de Gournay, of most of Montaigne's quotations in Classical or foreign languages.)

Summary of the Symbols

[A] and '80: the text of 1580

[A1]: the text of 1582 (plus)

[B] and '88: the text of 1588

[C]: the text of the edition being prepared by Montaigne when he died, 1592

'95: text of the 1595 posthumous printed edition

In the notes there is given a selection of variant readings, including most abandoned in 1588 and many from the printed posthumous edition of 1595.

By far the most scholarly account of the text is that given in R. A. Sayce, *The Essays of Montaigne: A Critical Exploration*, 1972, Chapter 2, 'The Text of the *Essays*'.

Appendices

═══

I

Montaigne's dedication of his translation of Raymond Sebond's Natural Theology *to his father.*

TO MY LORD, MY LORD OF MONTAIGNE

My Lord, following the task you gave me last year at Montaigne, I have tailored and dressed with my hand a garment in the French style for Raymond Sebond, the great Spanish Theologian and Philosopher, divesting him (in so far as in me lay) of his uncouth bearing and of that barbarous stance that you were the first to perceive: so that, in my opinion, he now has sufficient style and polish to present himself in good company.

It may well be that delicate and discriminating people may notice here some Gascon usages or turns of phrase: that should make them all the more ashamed at having neglectfully allowed a march to be stolen on them by a man who is an apprentice and quite unsuited to the task.

It is, my Lord, right that it should appear and grow in credit beneath your name, since it is to you that it owes whatever amendment or reformation it now enjoys.

And yet I believe that if you would be pleased to reckon accounts with him, it will be you who will owe him more: in exchange for his excellent and most religious arguments, for his conceptions lofty and as though divine, you, for your part, have brought only words and language – a merchandise so base and vile that who has most is perhaps worth least.

My Lord, I beg God that he may grant you a most long and a most happy life.

From Paris: this 18th of June, 1568.
Your most humble and obedient son,
MICHEL DE MONTAIGNE.

II

Montaigne's translation and adaptation of the Prologus of Raymond Sebond.

Book of the Creatures of Raymond Sebond.
Translated from the Latin into French.
Preface of the Author.

To the praise and glory of the most high and glorious Trinity, of the Virgin Mary, and of all the heavenly Court: in the name of our Lord Jesus Christ, for the profit of all Christians, there follows the doctrine of the Book of the Creatures (or, Book of Nature): a doctrine of Man, proper to Man insofar as he is Man: a doctrine suitable, natural and useful to every man, by which he is enlightened into knowing himself, his Creator, and almost everything to which he is bound as Man. a doctrine containing the rule of Nature, by which also each Man is instructed in what he is naturally bound towards God and his neighbour: and not only instructed but moved and incited to do this, of himself, by love and a joyful will.

In addition this science teaches every one to see clearly, without difficulty or toil, truth insofar as it is possible for natural reason, concerning knowledge of God and of himself and of what he has need for his salvation and to reach life eternal; it affords him access to understanding what is prescribed and commanded in Holy Scripture, and delivers the human spirit from many doubts, making it consent firmly to what Scripture contains concerning knowledge of God and of oneself.

In this book the ancient errors of the pagans and the unbelieving philosophers are revealed and by its doctrine the Catholic Faith is defended and made known: every sect which opposes it is uncovered and condemned as false and lying.

That is why, in this decline and last days of the World it is necessary that Christians should stiffen themselves, arm themselves and assure themselves within that Faith so as to confront those who fight against it, to protect themselves from being seduced and, if needs be, joyfully to die for it.

Moreover this doctrine opens up to all a way of understanding the holy Doctors [of the Church]; indeed, it is incorporated in their books (even though it is not evident in them) as an Alphabet is incorporated in all writings. For it is the Alphabet of the Doctors: as such it should be learned

first. For which reason, to make your way towards the Holy Scriptures you will do well to acquire this science as the rudiments of all sciences; in order the better to reach conclusions, learn it before everything else, otherwise you will hardly manage to struggle through to the perfection of the higher sciences: for this is the root, the origin and the tiny foundations of the doctrine proper to Man and his salvation.

Whoever possesses salvation through hope must first have the root of salvation within him and, consequently, must furnish himself with this science, which is a fountain of saving Truth.

And there is no need that anyone should refrain from reading it or learning it from lack of other learning: it presupposes no knowledge of Grammar, Logic, nor any other deliberative art or science, nor of Physics nor of Metaphysics, seeing that it is this doctrine which comes first, this doctrine which ranges, accommodates and prepares the others for so holy an End – for the Truth which is both true and profitable to us, because it teaches Man to know himself, to know why he has been created and by Whom; to know his good, his evil and his duty; by what and to Whom he is bound.

What good are the other sciences to a man who is ignorant of such things? They are but vanity, seeing that men can only use them badly to their harm, since they know not where they are, whither they are going nor whence they came. That is why they are taught here to understand the corruption and defects of Man, his condemnation and whence it came upon him; to know the state in which he is now: the state in which he originally was: from what he has fallen and how far he is from his first perfection; how he can be reformed and those things which are necessary to bring this about.

And therefore this doctrine is common to the laity, the clergy and all manner of people: and yet it can be grasped in less than a month, without toil and without learning anything off by heart; no books are required, for once it has been perceived it cannot be forgotten. It makes a man happy, humble, gracious, obedient, the enemy of vice and sin, the lover of virtue – all without puffing him up or making him proud because of his accomplishments.

It uses no obscure arguments requiring deep or lengthy discourse: for it argues from things which are evident and known to all from experience – from the creatures and the nature of Man; by which, and from what he knows of himself, it proves what it seeks to prove, mainly from what each man has assayed of himself. And there is no need of any other witness but Man.

It may, meanwhile, at first appear contemptible, a thing of nothing, especially since its beginnings are common to all and very lowly: but that does not stop it from bearing great and worthwhile fruit, namely the knowledge of God and of Man. And the lower its starting-point, the higher it climbs, rising to matters high and celestial.

Wherefore, whosoever wishes to taste of its fruit, let him first familiarize himself with the minor principles of this science, without despising them: for otherwise he will never have that taste, no more than a child ever learns to read without a knowledge of the alphabet and of each individual letter. And, finally, let him not complain about this labour by which, in a few months, he becomes learned and familiar with many things, to know which it would be proper to spend long periods reading many books.

It alleges no authority — not even the Bible — for its end is to confirm what is written in Holy Scripture — and to lay the foundations on which we can build what is obscurely deduced from them. And so, in our case, it precedes the Old and New Testaments.

God has given us two books: the Book of the Universal Order of Things (or, of Nature) and the Book of the Bible. The former was given to us first, from the origin of the world: for each creature is like a letter traced by the hand of God: this Book had to be composed of a great multitude of creatures (which are as so many 'letters'); within them is found Man. He is the main, the capital letter.

Now, just as letters and words composed from letters constitute a science by amply marshalling different sentences and meanings, so too the creatures, joined and coupled together, form various clauses and sentences, containing the science that is, before all, requisite for us.

The second Book — Holy Scripture — was subsequently given in default of the first, in which, blinded as he was, he could make out nothing, notwithstanding that the first is common to all whereas the second is not: to read the second book one must be a clerk. Moreover, the Book of Nature cannot be corrupted nor effaced nor falsely interpreted. Therefore the heretics cannot interpret it falsely: from this Book no one becomes an heretic.

With the Bible, things go differently. Nevertheless both Books derive from the same Author: God created his creatures just as he established his Scriptures. That is why they accord so well together, with no tendency to contradict each other, despite the first one's symbolizing most closely with our nature and the second one's being so far above it.

Since Man, at his Birth, did not find himself furnished with any science (despite his rationality and capacity for knowledge) and since no science can be acquired without books in which it is written down, it was more

than reasonable (so that our capacity for learning should not have been given us in vain) that the Divine Intelligence should provide us with the means of instructing ourselves in the doctrine which alone is requisite, without a schoolmaster, naturally, by ourselves.

That is why that Intelligence made this visible world and gave it to us like a proper, familiar and infallible Book, written by his hand, in which the creatures are ranged like letters — not in accordance to our desires but according to the holy judgement of God, so as to teach us the wisdom and science of our salvation. Yet no one can [now] see and read that great Book by himself (even though it is ever open and present to our eyes) unless he is enlightened by God and cleansed of original sin. And therefore not one of the pagan philosophers of Antiquity could read this science, because they were all blinded concerning the sovereign good; even though they drew all their other sciences and all their knowledge from it, they could never perceive nor discover the wisdom which is enclosed within it nor that true and solid doctrine which guides us to eternal life.

Now, in anyone capable of discernment, there is engendered a true understanding from a combining together of the creatures like a well-ordered tissue of words. So the method of treating this subject in this treatise is to classify the creatures and to establish their relationships one with the other, taking into consideration their weightiness and what they signify and, after having drawn forth the divine wisdom which they contain, fixing it and impressing it deeply in our hearts and souls.

Now, since the Most Holy Church of Rome is the Mother of all faithful Christians, the Mother of Grace, the Rule of Faith and Truth, I submit to her correction all that is said and contained in this my work.

Chronology

1477 Ramond Eyquem, a rich merchant in Bordeaux trading in wines and salt fish, purchases the estates of Montaigne.

1497 Birth of Pierre Eyquem (Montaigne's father) at the family estates.

1519 Pierre Eyquem, as a result of deaths in the family, inherits the estates at Montaigne and leaves to fight in Italy, entailing an absence of several years.

1528 Pierre Eyquem marries Antoinette de Louppes, of a rich and politically influential family. The Louppes, a pious Christian family, were descended from Iberian Jews.

1530 Pierre Eyquem is *premier jurat* (first magistrate) and Provost of Bordeaux. Birth of Etienne de la Boëtie.

1533 *28 February,* birth of Michel Eyquem de Montaigne at the family estates.

1534 A brother, Thomas, is born.

1535 Montaigne's German tutor's aim is to make Latin his first language. This continues his father's scheme from the outset.
Another brother, Pierre, is born.

1536 A sister, Jeanne, is born.

1539/40 Montaigne enters the Collège de Guyenne at Bordeaux, where the tutors include Mathurin Cordier, Buchanan (the humanist playwright and future Scottish Reformer) and Elie Vinet. He stays there for six years. His understanding tutors encouraged his delight in Latin poetry. He acquired some Greek, but Latin was his literary language.

1546 Montaigne probably studies philosophy in the Faculty of Arts at Bordeaux.

1548 Civil disobedience and riots in Bordeaux, fiercely suppressed. Mayors now to be elected for periods of two years only. The Huguenots become established and numerous in the City and its environs.

1552 Birth of Montaigne's second sister, Lénor.

1554 Michel follows his father and becomes counsellor at the Cour des Aides at Périgueux. This Cour is suppressed three years later and the counsellors join the Parlement of Bordeaux. His father becomes Mayor of Bordeaux.
Birth of third sister, Marie.

1557/8 Montaigne meets Etienne de la Boëtie, also a member of the Parlement de Bordeaux; their deep and special friendship begins.

1559 Montaigne visits Paris, and follows King François II to Bar-le-Duc. Amyot's translation of Plutarch: it greatly influences Montaigne both in thought and style.

1560 Birth of Montaigne's brother, Bertrand.

1561 Second visit to Paris and the Royal Court, partly in connection with the serious religious strife in Guyenne.

1562 Proclamation of the *Edict of the Seventeenth of January 1562* granting limited rights of assembly to members of the 'so-called Reformed Church'. In June, Etienne de la Boëtie writes a *mémoire* on that Edict. Montaigne, still in Paris, makes a public profession of Roman Catholicism before the First President of the Parlement de Paris. In October he follows the Royal Army when Rouen is retaken from the Huguenots; he meets there Indians from Brasil. Massacre of Huguenots at Wassy.

1563 *February*: Montaigne returns to Bordeaux.
   18 August: death of La Boëtie at Germinant at the home of Montaigne's brother-in-law, Lestonnat. Montaigne writes of it to his father.
   Assassination of François de Guise.

1564 *16 October*: Montaigne finishes reading the *De Rerum Natura* of Lucretius and notes at the end the date and 31 (his age). The flyleaves are all covered with dense Latin notes. Several topics in the *Essays* go back to that initial reading. On a subsequent reading Montaigne made many notes on the pages of the text in French. This edition of Lucretius by Lambinus had been published either late in 1563 or early in 1564.

1565 *January*: visit of Charles IX to Bordeaux.
   Marriage of Montaigne to Françoise de la Chassaigne, the daughter of a colleague in the Parlement de Bordeaux.

1568 Death of Montaigne's father, Pierre. Montaigne becomes Seigneur de Montaigne and inherits the domain. (Difficulties with his mother over the inheritance.)

1569 Montaigne publishes his French translation of the *Theologia Naturalis* of Raimon Sebon (Raymundus de Sabunde), with the printer G. Chaudière of Paris.

1570 Montaigne sells his counsellorship of the Parlement de Bordeaux. Goes to Paris to publish works left by Etienne de la Boëtie (Latin, then French).
   Birth of his first daughter, Toinette, who dies three months later.

1571 Montaigne returns to his estates, to consecrate his life to the Muses: to

scholarship, philosphy and reflection. He receives the Ordre de Saint-Michel and is named Gentleman of the Chamber by Charles IX.

Birth of Léonor (the only one of his six daughters to live).

1572 *24 August*: massacre of Saint Bartholomew's Day. Uprisings at La Rochelle (a stronghold of the Reformed Church).

Publication of the French translation of the *Moral Works* of Plutarch by Bishop Amyot. It joins other authors studied by Montaigne in the tower of his château.

1572–4 During the civil wars Montaigne joins the royalist forces. Montaigne dispatched to Bordeaux to advise the Parlement to strengthen their defences.

1574 Anonymous publication (adapted to Reformed propaganda) of La Boëtie's short treatise *De la Servitude volontaire*.

1575 Reads Sextus Empiricus' *Hypotyposes*.

1576 Strikes a medal with the Greek motto *I abstain*. He is working on his *Apology of Raimon Sebon*.

1577 Henri de Navarre names Montaigne Gentleman of the Chamber.

About this time suffers his first attack of the stone.

1580–81 *1 March*: publication of the *Essays* (Simon Millanges, Bordeaux).

Montaigne leaves on his travels. At Paris he offers his book to Henri III, who is delighted with it. On his travels (partly to take the waters) Montaigne visits Plombières, Mülhauser, Basle, Baden, Augsburg, Munich, Innsbruck, the Tyrol, Padua, Venice, Ferrara and Rome (which was reached on 30 November). At Rome his books are impounded, but relations are good. The maestro di Palazzo offers suggestions for changes to be made by Montaigne in his *Essays*, without further interference. Montaigne has an audience of the Pope, Gregory XIII. On his way back he makes a pilgrimage to Loretto and has medals of the Virgin blessed for his wife and daughter as well as himself. Travels via Florence and Pisa and the baths at Lucca.

17 September: leaves on learning that royal approval requires him to become Mayor of Bordeaux.

30 September: arrives home.

1582 Second edition of the *Essays* published with the same publisher.

Gregory XIII reforms the calendar, a reform accepted in France, but not in England.

1583 Montaigne re-elected Mayor of Bordeaux for a further two years.

1582–5 During his Mayoralty Montaigne visits Paris and often stays on his estates. Henri de Navarre, now heir to the throne, visits Montaigne and stays in his château. Montaigne is concerned with high politics as well as

local affairs. In 1585 the plague ravages Bordeaux. Montaigne, absent, does not return to the town: he and his family are forced to leave their home, Montaigne, and wander about in search of a safe lodging.

1587 *24 October*: the King of Navarre dines at Montaigne.

1588 *16 February*: Montaigne, en route for Paris, is attacked and robbed by soldiers of *La Ligue*. His goods and freedom are restored to him. His third edition of the *Essays* is published in Paris by L'Angelier.

Mlle de Gournay sends him greetings from her lodgings in Paris. Montaigne visits her. She becomes eventually his *fille d'alliance*, virtually an adopted daughter.

June: publication of the greatly expanded edition of the *Essays*, which now includes a new third book (Paris, L'Angelier).

10 July: Montaigne is arrested in Paris and sent to the Bastille apparently to serve as a hostage. He is restored to freedom the same day by order of Catherine de' Medici.

1589 *2 August*: death of Henri III.

Montaigne begins working on a further expanded edition of the *Essays*.

1590 *18 June*: marriage of Montaigne's daughter Léonor to François de La Tour. Though ill, Montaigne writes to Henri de Navarre (now Henri IV), who replies to him (20 July) and invites him to come as (probably) his adviser.

1591 Birth of François de La Tour, Montaigne's grand-daughter.

1592 *13 September*: death of Montaigne during a Mass said in his bed-chamber.

1595 Montaigne's widow, Pierre de Brach and Marie de Gournay produce the first posthumous edition of the *Essays* incorporating Montaigne's last additions and changes.

1601 Death of Montaigne's mother.

1613 John Florio's translation of the *Essays*.

To the Reader

[A] You have here, Reader, a book whose faith can be trusted, a book which warns you from the start that I have set myself no other end but a private family one. I have not been concerned to serve you nor my reputation: my powers are inadequate for such a design. I have dedicated this book to the private benefit of my friends and kinsmen so that, having lost me (as they must do soon) they can find here again some traits of my character and of my humours. They will thus keep their knowledge of me more full, more alive. If my design had been to seek the favour of the world I would have decked myself out   [C]   better and presented myself in a studied gait.[1]   [A]   Here I want to be seen in my simple, natural, everyday fashion, without   [C]   striving[2]   [A]   or artifice: for it is my own self that I am painting. Here, drawn from life, you will read of my defects and my native form so far as respect for social convention allows: for had I found myself among those peoples who are said still to live under the sweet liberty of Nature's primal laws, I can assure you that I would most willingly have portrayed myself whole, and wholly naked.

And therefore, Reader, I myself am the subject of my book: it is not reasonable that you should employ your leisure on a topic so frivolous and so vain.

Therefore, Farewell:

From Montaigne;
this first of March, One thousand, five hundred and eighty.[3]

---

1. '80: myself out, *with borrowed beauties, or would have tensed and braced myself in my best posture.* Here I want . . .
2. '80: Without *study* or artifice . . .
3. Date as in [A] and [C]. In [B]: 12 June 1588.

BOOK I

1. We reach the same end by discrepant means

═══

[This first chapter treats of war and history, subjects appropriate to a nobleman. Montaigne introduces the irrational (astonishment, ecstasy and the fury of battle) and shows how unpredictable are the reactions of even great, brave and virtuous men. The verb to assay *is used three times; explanations of motives are mere conjecture – what 'could be said'; [A] cites the* exemplum *of Conrad III from the foreword to Bodin's* Method towards an Easy Understanding of History, *which Montaigne was reading about 1578. In [B] he adds his own reactions.]*

[A]   The most common way of softening the hearts of those we have offended once they have us at their mercy with vengeance at hand is to move them to commiseration and pity [C] by our submissiveness. [A]   Yet flat contrary means, bravery and steadfastness,[1] have sometimes served to produce the same effect.

Edward, Prince of Wales[2] – the one who long governed our Guyenne and whose qualities and fortune showed many noteworthy characteristics of greatness – having been offended by the inhabitants of Limoges, took their town by force. The lamentations of the townsfolk, the women and the children left behind to be butchered crying for mercy and throwing themselves at his feet, did not stop him until eventually, passing ever deeper into the town, he noticed three French noblemen who, alone, with unbelievable bravery, were resisting the thrust of his victorious army. Deference and respect for such remarkable valour first blunted the edge of his anger; then starting with those three he showed mercy on all the other inhabitants of the town.

1. '80: means, bravery, steadfastness *and resolution*, have . . .
2. The Black Prince (Limoges, 1370). Sources include Froissart, Paolo Giovio, *Vita di Scanderbeg*; Jean Bodin, *Methodus* (Preface); Plutarch (tr. Amyot), *Comment on peut se louer soy mesme*; *Dicts notables des Roys . . .*; *Instruction pour ceux qui manient les affaires d'estat*; Diodorus Siculus (tr. Amyot), *Histoires*; and Quintus Curtius, *Life of Alexander the Great*.

Scanderbeg, Prince of Epirus, was pursuing one of his soldiers in order
to kill him. The soldier, having assayed all kinds of submissiveness and
supplications to try and appease him, as a last resort resolved to await him,
sword in hand. Such resolution stopped his Master's fury short; having seen
him take so honourable a decision he granted him his pardon. (This
example will allow of a different interpretation only from those who have
not read of the prodigious strength and courage of that Prince.)

The Emperor Conrad III had besieged Guelph, Duke of Bavaria; no
matter how base and cowardly were the satisfactions offered him, the most
generous condition he would vouchsafe was to allow the noblewomen
who had been besieged with the Duke to come out honourably on foot,
together with whatever they could carry on their persons. They, with
greatness of heart, decided to carry out on their shoulders their husbands,
their children and the Duke himself. The Emperor took such great pleasure
at seeing the nobility of their minds that he wept for joy and quenched all
the bitterness of that mortal deadly hatred he had harboured against the
Duke; from then on he treated him and his family kindly.

[B]   Both of these means would have swayed me easily, for I have a
marvellous weakness towards mercy and clemency – so much so that I
would be more naturally moved by compassion than by respect. Yet for
the Stoics pity is a vicious emotion: they want us to succour the afflicted
but not to give way and commiserate with them.

[A]   Now these examples seem to me to be even more to the point in
that souls which have been assaulted and assayed by both those methods
can be seen to resist one without flinching only to bow to the other.

It could be said that for one's mind to yield to pity is an effect of
affability, gentleness – and softness (that is why weaker natures such as
those of women, children and the common-people are more subject to
them) – whereas, disdaining [C] tears [A] and supplications[3] and
then yielding only out of respect for the holy image of valour is the action
of a strong, unbending soul, reserving its good-will and honour for
stubborn, masculine vigour. Yet ecstatic admiration and amazement can
produce a similar effect in the less magnanimous. Witness the citizens of
Thebes: they had impeached their captains on capital charges for having
extended their mandates beyond the period they had prescribed and
preordained for them; they were scarcely able to pardon Pelopidas, who,
bending beneath the weight of such accusations, used only petitions and
supplications in his defence, whereas on the contrary when Epaminondas

3. '80: disdaining *prayers* and . . .

came and gloriously related the deeds he had done and reproached the people with them proudly [C] and arrogantly, [A] they had no heart for even taking the ballots into their hands: the meeting broke up, greatly praising the high-mindedness of that great figure.

[C] The elder Dionysius had captured, after long delays and extreme difficulties, the town of Rhegium together with its commander Phyton, an outstanding man who had stubbornly defended it. He resolved to make him into a terrible example of vengeance. Dionysius first told him how he had, the previous day, drowned his son and all his relations. Phyton merely replied that they were happier than he was, by one day. Next he had him stripped, seized by executioners and dragged through the town while he was flogged, cruelly and ignominiously, and plied with harsh and shameful insults. But Phyton's heart remained steadfast and he did not give way; on the contrary, with his face set firm he loudly recalled the honourable and glorious cause of his being condemned to death – his refusal to surrender his country into the hands of a tyrant – and he threatened Dionysius with swift punishment from the gods. Dionysius read in the eyes of the mass of his soldiers that, instead of being provoked by the taunts which this vanquished enemy made at the expense of their leader and of his triumph, they were thunder-struck by so rare a valour, beginning to soften, wondering whether to mutiny, and ready to snatch Phyton from the hands of his guards; so he brought Phyton's martyrdom to an end and secretly sent him to be drowned in the sea.

[A] Man is indeed an object miraculously vain, various and wavering. It is difficult to found a judgement on him which is steady and uniform. Here you have Pompey pardoning the entire city of the Mamertines, against whom he was deeply incensed, out of consideration for the valour and great-heartedness of Zeno,[4] a citizen who assumed full responsibility for the public wrong-doing and who begged no other favour than alone to bear the punishment for it. Yet that host of Sylla showed similar bravery in the city of Perugia and gained nothing thereby, neither for himself nor for the others.

[B] And, directly against my first examples, Alexander, the staunchest of men and the most generous towards the vanquished, stormed, after great hardship, the town of Gaza and came across Betis who commanded it; of his valour during the siege he had witnessed staggering proofs; now Betis was alone, deserted by his own men, his weapons shattered; all covered with blood and wounds, he was still fighting in the midst of several

4. Not Zeno, Stheno.

Macedonians who were slashing at him on every side. Alexander was irritated by so dearly won a victory (among other losses he had received two fresh wounds in his own body); he said to him: 'You shall not die as you want to, Betis! Take note that you will have to suffer every kind of torture which can be thought up against a prisoner!' To these menaces Betis (not only looking assured but contemptuous and proud) replied not a word. Then Alexander, seeing his haughty and stubborn silence said: 'Has he bent his knee? Has he let a word of entreaty slip out? Truly I will overcome that refusal of yours to utter a sound: if I cannot wrench a word from you I will at least wrench a groan.' And as his anger turned to fury he ordered his heels to be pierced[5] and, dragging him alive behind a cart, had him lacerated and dismembered.

Was it because   [C]   bravery was so usual for him[6] that   [B]   he was never struck with wonder by it and therefore respected it less?   [C]   Or was it because he thought bravery to be so properly his own that he could not bear to see it at such a height in anyone else without anger arising from an emotion of envy; or did the natural violence of his anger allow of no opposition? Truly if his anger could ever have suffered a bridle it is to be believed that it would have done so during the storming and sack of Thebes, at seeing so many valiant men put to the sword, men lost and with no further means of collective defence. For a good six thousand of them were killed, none of whom was seen to run away or to beg for mercy; on the contrary all were seeking, here and there about the streets, to confront the victorious enemy and to provoke them into giving them an honourable death. None was so overcome with wounds that he did not assay with his latest breath to wreak revenge and to find consolation for his own death in the death of an enemy. Yet their afflicted valour evoked no pity; a day was not long enough to slake the vengeance of Alexander: this carnage lasted until the very last drop of blood remained to be spilt; it spared only those who were disarmed – the old men, women and children – from whom were drawn thirty thousand slaves.

---

5. '80: pierced, *a rope threaded through them,* and, . . .
6. '80: because *strength of courage was so natural and usual to* him . . .

2. On sadness

====

[Chapters 2–18 (in their [A] version) seem to date from the earliest period, reflecting the influence of books which Montaigne was reading about 1572 – Guicciardini's History of Italy, Jean Bouchet's Mémoires d'Aquitaine and the Mémoires of the Du Bellay brothers. 'On sadness' shows Montaigne's concern with ecstasies produced by strong emotions and his impatience with merely fashionable tristesse (sadness) which sought to ape the abstracted, pensive depths of melancholy genius (as portrayed, for example, by Dürer).]

[B]   I am among those who are most free from this emotion;   [C]   I neither like it nor think well of it, even though the world, by common consent, has decided to honour it with special favour. Wisdom is decked out in it; so are Virtue and Conscience – a daft and monstrous adornment. More reasonably it is not sadness but wickedness that the Italians have baptised *tristezza*,[1] for it is a quality which is ever harmful, ever mad. The Stoics forbid this emotion to their sages as being base and cowardly.

[A]   But a story is told about Psammenitus, a King of Egypt. When he was defeated and captured by Cambises the King of Persia he showed no emotion as he saw his daughter walk across in front of him, dressed as a servant and sent to draw water. All his friends were about him, weeping and lamenting: he remained quiet, his eyes fixed on the ground. Soon afterwards he saw his son led away to execution; he kept the same countenance. But when he saw one of his household friends brought in among the captives, he began to beat his head and show grief.[2]

You can compare that with what we recently saw happen to one of our princes.[3] He was at Trent: first he heard the news of the death of his very special elder brother, the support and pride of his whole family; then came the death of his younger brother, their second hope. He bore both these blows with exemplary fortitude; yet, when a few days later one of his men happened to die, he let himself be carried away by this event; he abandoned

1. *Tristesse* in French means sadness.
2. Erasmus, *Apophthegmata; varie mixta: Diversum Graecorum*, IX (*Opera*, 1703–6, Vol. IV, col. 304EF).
3. Charles de Guise, Cardinal de Lorraine (at the Council of Trent).

his resolute calm and gave himself over to grief and sorrow – so much so
that some argued that only this last shock had touched him to the quick.
The truth is that he was already brimful of sadness, so the least extra
burden broke down the barriers of his endurance.

We could, I suggest, put the same interpretation on the story of
Psammenitus, except that the account goes on to tell us that Cambises
asked him why he had remained unmoved by the fate of his son and
daughter yet showed such emotion at the death of his friend. 'Only the last
of these misfortunes can be expressed by tears', he replied; 'the first two are
way beyond any means of expression.'

That may explain the solution adopted by a painter in antiquity.[4] He
had to portray the grief shown on the faces of the people who were present
when Iphigenia was sacrificed, giving each of them the degree of sorrow
appropriate to his feelings of involvement in the death of that fair and
innocent young woman. By the time he came to portray the father of
Iphigenia he had exhausted all the resources of his art, so he painted him
with his face veiled over, as though no countenance could display a grief so
intense.

That is why the poets feign that when Niobe lost seven sons and then
seven daughters she was overcome by such bereavements and was finally
turned into a rock:

> *Diriguisse malis.*

> [Petrified by such misfortunes.][5]

By this they expressed that sad, deaf, speechless stupor which seizes us
when we are overwhelmed by tragedies beyond endurance.

The force of extreme sadness inevitably stuns the whole of our soul,
impeding her freedom of action. It happens to us when we are suddenly
struck with alarm by some really bad news: we are enraptured, seized,
paralysed in all our movements in such a way that, afterwards, when the
soul lets herself go with tears and lamentations, she seems to have struggled
loose, disentangled herself and become free to range about as she wishes:

> [B]   *Et via vix tandem voci laxata dolore est.*

> [And then, at length, his grief can just force open a channel for his voice.][6]

---

4. Timanthes (Cicero, *De Oratore*, XXII; Quintillian, II, xiii, 12).
5. Ovid, *Metamorphoses*, VI, 304.
6. Virgil, *Aeneid*, XI, 151.

[C] In the war which King Ferdinand waged near Buda against the widow of King John of Hungary, there was a German officer called Raïsciac. As he saw men bringing back the body of a soldier slung across a horse, he joined in the general mourning for the man who had shown exceptional bravery in the clash of battle. Like the others he was curious to know who the man was. When they took off the armour he recognized his son. Amid all the public tears he alone stood dry-eyed, saying nothing, his gaze fixed on his son until the violent strain of that sadness froze his vital spirits and, just as he was, toppled him dead to the ground.[7]

[A] *Chi puo dir com' egli arde e in picciol fuoco –*

[He who can describe how his heart is ablaze is burning on a small pyre][8] –

that is what lovers say when they want to express an unbearable passion.

> *misero quod omnes*
> *Eripit sensus mihi. Nam simul te,*
> *Lesbia, aspexi, nihil est super mi*
> *Quod loquar amens.*
>
> *Lingua sed torpet, tenuis sub artus*
> *Flamma dimanat, sonitu suopte*
> *Tinniunt aures, gemina teguntur*
> *Lumina nocte.*

[How pitiable I am. Love snatches my senses from me. As soon as I see you, Lesbia, I can say nothing to you; I am out of my mind; my tongue sticks in my mouth; a fiery flame courses through my limbs; my ears are ringing and darkness covers both my eyes.][9]

[B] We cannot display our grief or our convictions during the living searing heat of the attack; the soul is then burdened by deep thought and

---

7. Paolo Giovio, *Historia sui temporis*, 1550, XXXIX.

  '95: John of Hungary, *a soldier was particularly remarked by everyone for showing outstanding personal bravery in a certain mêlée in which he fell, unidentified, but highly praised and pitied not least by a German lord called Raïsciac who was impressed by such great valour; when the body was brought back, that Lord, out of common curiosity, drew near to see who the man was. When the armour was stripped off the dead body he realized that it was his son. That increased the compassion of those present. He, without uttering a word or closing his eyes, remained standing, staring fixedly at his son until the vehement force of his sadness overwhelmed his vital spirits, and toppled him dead to the ground.*

8. Petrarch, Sonnet 137.

9. Catullus, LI, 5.

the body is cast down, languishing for love.   [A]   That is the source of
the occasional impotence which sometimes comes so unseasonably upon
men when making love, and of that chill produced, in the very lap of their
delight by excessive ardour. For pleasures to be tasted and then digested
they must remain moderate:

> *Curae leves loquuntur, ingentes stupent.*

[Light cares can talk: huge one are struck dumb.][10]

[B]   We can be equally stunned when surprised by joy unhoped for:

> *Ut me conspexit venientem, et Troïa circum*
> *Arma amens vidit, magnis exterrita monstris,*
> *Diriguit visu in medio, calor ossa reliquit,*
> *Labitur, et longo vix tandem tempore fatur.*

[As soon as she noticed me coming and saw the arms of Troy all about her, she
went out of her mind. As though terrified by some dreadful portents her gaze
became fixed upon them, the heat drained from her body; she fell to the ground
and for a long time uttered not a word.][11]

[A]   There was a Roman woman who was surprised by joy on seeing
her son return from the rout at Cannae and fell down dead; Sophocles and
Dionysius the Tyrant died of happiness; Talva died in Corsica upon
reading the news of the honours conferred on him by the Senate. Apart
from these it is claimed that in our own century Pope Leo X entered into
such an excess of joy upon being told of the capture of Milan (his desire for
which had been extreme) that he took fever and died.

And   there   is   an   even   more   noteworthy   witness   to   [C]
human   [A]   weakness:[12] the Ancients recorded that Diodorus the Dialecti-
cian 'fell in the field', overcome by an extreme sense of shame at being
unable to refute arguments put to him in public in the presence of his
followers.

[B]   Violent emotions like these have little hold on me. By nature my
sense of feeling has a hard skin, which I daily toughen and thicken by arguments.

10. '88: ardour − *an event with which I am not unacquainted.* For pleasures . . .
    Seneca (the dramatist), *Hippolitus*, II, iii, 607.
11. Virgil, *Aeneid*, III, 306.
12. '80: to *natural* weakness − (Pliny, *Hist. nat.*, 54, for both anecdotes.)

3. Our emotions get carried away beyond us

[Many of the exempla *in this chapter are rooted in war. They show, as do the more personal ones, how men fruitlessly worry about what happens to their bodies after death. Montaigne already states (as later he will insist) that a human being only is (only exists) when body and soul are conjoined.]*

[B] Those who reproach humanity with always gaping towards the future and who teach us to grasp present goods and to be satisfied with them since we have no hold over what is to come – less hold, even, than we have over the past – touch upon the most common of human aberrations (if we dare use the word 'aberration' for something towards which Nature herself brings us in the service of the perpetuation of her handiwork, [C] impressing this false thought upon us as she does many others, more ardently concerned as she is for us to do than to know). [B] We are never 'at home': we are always outside ourselves. Fear, desire, hope, impel us towards the future; they rob us of feelings and concern for what now is, in order to spend time over what will be – even when we ourselves shall be no more. [C] *'Calamitosus est animus futuri anxius'* [Wretched is a mind anxious about the future].[1]

'Do what thou hast to do, and know thyself' – that great precept is often cited by Plato;[2] each clause of it embraces our entire duty, generally, and similarly embraces its fellow. Whoever would do what he has to do would see that the first thing he must learn is to know what he is and what is properly his. And whoever does know himself never considers external things to be his; above all other things he loves and cultivates himself: he rejects excessive concerns as well as useless thoughts and resolutions. *'Ut stultitia etsi adepta est quod concupivit nunquam se tamen satis consecutam putat: sic sapientia semper eo contenta est quod adest, neque eam unquam sui poenitet.'* [Folly never thinks it has enough, even when it obtains what it desires, but

---

1. Seneca, *Epist. moral.*, XCVIII, 5–6.
2. Plato, *Timaeus*, 72a. Cf. Erasmus, *Adages, Nosce teipsum* (I.VII.XCV).

Wisdom is happy with what is to hand and is never vexed with itself.][3] Epicurus frees his Wise Man from anticipation and worry about the future.[4]

[B]   The most solid of our laws concerning the dead seems to me to be the one which requires the deeds of monarchs to be examined once their life is over. They are, if not the masters, then the companions of the laws: that which Justice could not visit upon their heads can rightly be visited upon their reputations and on the goods of their heirs – things we often prefer to life itself. It is a custom which brings many singular advantages to those peoples who observe it; it is something to be desired by good monarchs [C] who have cause to complain that the memory of the wicked is honoured just like their own. We owe subordination and obedience to all our kings equally, for that concerns their office; but we owe esteem and affection only to their virtue. Let us concede this much to the political order: to suffer kings patiently when unworthy, to hide their vices and to encourage their indifferent actions with our approbation – while their authority needs our support. But once our commerce with them is over, it is not reasonable to deprive Justice and our own freedom of the right to express our true feelings – and especially to deprive good subjects of the glory of having reverently and faithfully served a master whose imperfections were so well known to them, thus depriving posterity of so useful an example. Those who out of some private obligation wickedly espouse the memory of an unpraiseworthy monarch put private above public justice. Livy rightly says that the speech of men brought up under monarchies is always full of foolish pomposity and vain testimony, as each one of them elevates his king, regardless of merit, to the ultimate point of valour and sovereign greatness.[5]

One may reprove the greatness of soul of those two soldiers who answered Nero back to his face: one of them, when asked by Nero why he wished him ill, retorted: 'I loved you while you deserved it; but since you have become a parricide, a fire-raiser, a mountebank and a chariot-driver, I hate you as you deserve'; the other, asked why he wanted to kill him, replied, 'Because I can find no other remedy to your continual misdoings'; but how can anyone of sound judgement reprove those public and universal testimonies to his tyrannous and vile deeds which were rendered after his death – and always will be?[6]

3. Cicero, *Tusc. disput.*, V, xviii, 54 (replaced by a French translation in '95).
4. Cicero, *Tusc. disput.*, III, xv–xvi, 33–5.
5. A vague memory of Livy, not a direct allusion.
6. Tacitus, *Annals*, XV, lxvii.

It displeases me that such lying veneration should be found in so religious a regime as that of Sparta. On the death of their kings all their allies and neighbours and all the helots – men and women indiscriminately – slashed their foreheads in token of their grief, declaring in their cries and lamentations that the dead king was the best they had ever had, thus attributing to rank the praise which belongs to merit and attributing to the least and the lowest what belongs to the highest merit,[7]

Aristotle, who goes into everything, takes the saying of Solon that 'nobody can be termed happy before he is dead' and inquires whether even a man who has lived and died ordinately can be called happy if his reputation fares badly and if his descendants are wretched.[8] While we can move we can transport ourselves by anticipation wherever we may please: but once we have gone outside our being we have no commerce with that which *is*. It would be better to tell Solon that no man is ever happy, therefore, since he only is so when he *is* no more.[9]

[B] *Quisquam*
Vix radicitus e vita se tollit, et ejicit:
Sed facit esse sui quiddam super inscius ipse,
Nec removet satis a projecto corpore sese, et
Vindicat.

[A man does not tear himself from life by the roots and cast himself away: unawares, he dreams that some part of himself will still remain; he does not withdraw enough from that cast-off body nor free himself.][10]

[A]   Bertrand Du Guesclin died at the siege of Rancon castle near Le Puy in Auvergne. When the besieged later surrendered they were made to bring out the keys of the fortress borne on the body of that dead man. When Bartolomeo d'Alviano, the general of the Venetian army, had died in the service of their wars in Brescia, his corpse had to be brought back to Venice through the territory of their enemy, Verona; the majority in the army were in favour of asking the Veronese for a safe-conduct but Teodoro Trivulcio opposed it, choosing to pass through their lands by force of arms at the hazard of battle. 'It is not becoming,' he said, 'that he

7. Herodotus, VI, lxviii.
8. Aristotle, *Nicomachaean Ethics*, I, 10.
9. Cf. the last pages of 'On experience' (III, 13). Anyone whose soul is transported in ecstasy 'outside the body' ceases to exist as Man, since Man is body-plus-soul. His ecstatic soul may have commerce with Being; he, as Man, cannot.
10. Lucretius, III, 890–5 (adapted).

who never feared his enemy while alive should show fear of them when dead.'[11]

[B]   In a kindred matter, by the laws of the Greeks anyone who asked leave of the enemy to retrieve a corpse for burial had definitely given up any claim to victory; it was not licit to erect a trophy. For him to whom the request was made it was proof that he had won. That is how Nicias lost the advantage he so clearly had over the Corinthians, and how on the other hand Agesilaus rendered certain the doubtful advantage he had acquired over the Boeotians.[12]

[A]   These details might seem odd, were it not acceptable in all ages to project beyond this life the care we have for ourselves, and to believe moreover that divine favours often accompany us to the tomb and extend to our remains. There are so many ancient examples of this, let alone our own, that there is no need for me to dilate on them. Edward I, King of England, in the long wars with Robert, King of Scotland, had assayed how great an advantage his presence conferred on his affairs since he always won the victory when personally present at an engagement; when he was dying he bound his son by a solemn oath to boil his body as soon as he was dead, so as to separate his flesh from his bones and bury it; as for his bones, he was to keep them, carrying them with him in his army whenever he should happen to be fighting against the Scots – as though it were fated by Destiny that victory should reside in his joints.

[B]   John Vischa, who brought insurrection to Bohemia in defence of the errors of Wyclif, wished to be flayed after death and his skin to be made into a drum to bear in battle against his foes; he reckoned that that would prolong the superiority which he had known in the wars he had waged against them. Similarly certain Indians bore into battle against the Spaniards the bones of one of their leaders, out of respect for the good fortune which he had known in life. And other tribes in that same World bear in their war-train the bodies of their valiant men who had died in battle, to provide both good fortune and encouragement.[13]   [A]   My first examples were of men seeking to preserve in the tomb reputations acquired by previous acts: the later ones intended to convey that they still could act; but the deed of Captain Bayard constitutes a better compact:

---

11. Jean Bouchet, *Annales d'Acquitaine* (Poitiers, 1557) and Francesco Guicciardini, *L'Histoire d'Italie* (tr. Chomedey, Paris, 1568) XII.
12. Plutarch, *Lives* of Nicias and of Agesilaus.
13. Cf. Francisco Lopez de Gomara (tr. Fumée), *L'Histoire générale des Indes* (Paris, 1578), III, xxii. (The example of Vischa was a commonplace).

realizing he was fatally wounded in the body by a volley of harquebuses, he replied when advised to withdraw from the fray that he would not now at the end of his life start to turn his back on the enemy; having fought as long as he had strength, feeling himself faint and sliding from his saddle, he commanded his batman to lay him at the foot of a tree, but in such a fashion that he should die facing the foe. As he did.[14]

I must add this further example, which is as worthy of note in this connection as any of the foregoing. The Emperor Maximilian, the great-grandfather of the present King Philip, was a monarch fully endowed with great advantages; among others, he was singularly handsome. One of his humours was flat contrary to that of princes who, to get through important business, make a throne of their lavatory: he never allowed a valet such intimacy as to see him on his privy. He would even hide away to pass water, being as scrupulous as a [C] maiden [A] about uncovering, for a doctor or anyone else, those parts which are customarily kept hidden [15] [B] I myself, so shameless in speech, have nevertheless in my complexion a touch of such modesty: except when strongly moved by necessity or pleasure I rarely let anyone's eyes see those members or those actions which our customs ordain to be hidden. I find this all the more constraining in that I do not think it becoming in a man, above all in one of my calling.[16] But Maximilian became [A] so scrupulous that he expressly commanded in his will that linen drawers should be tied on him when he was dead. He should have added a codicil saying that the man who pulled them on ought to be blindfold!

[C] The order which Cyrus gave to his children (that neither they nor any others should see or touch his body once his soul had left it) I attribute to some personal vow; for among their other great qualities both he and his historian sowed broadcast through their lives a singular care and reverence for religion.[17]

[B] I was not pleased by a tale which a great prince told me about a member of a family allied to mine, a man well-known in peace and war: when very old, and dying within his court extremely tormented by the stone, he consumed all his last hours with vehement cares about the dignity and pomp of his funeral: he summoned all the nobility who visited him to promise to join his cortège. He urgently begged this very prince, who saw

---

14. Martin Du Bellay, *Mémoires*, Paris, 1569, II, p. 59.
15. '80: as a *girl* about . . . (Source of anecdote unknown.)
16. That is, unbecoming to an officer and a gentleman.
17. Xenophon, *Cyropaedia*, VIII, vii.

him in his last moments, to command that his family be present, employing many examples and arguments to prove that it was appropriate to a man of his station; and having extracted that promise and established to his liking the arrangements and order of his procession, he seemed to die happy.

I have rarely known vanity so persistent.

That opposite care (and my family does not lack an example of that either) seems to me to be cousin-german to the other: it consists in getting worried and worked up at this final stage about restricting the attendance (out of some private and unwonted frugality) to one servant and one lantern. I have seen this humour praised, as was the command of Marcus Aemilius Lepidus, who forbade his heirs to perform for him the ceremonies which were customary in such matters.[18] Is it still temperance and frugality if we avoid expenditures the use of which — and pleasures all knowledge of which — we are incapable of perceiving? An easy way to reform; and it costs little!

[C] If it were necessary to make arrangements for it, my decision would be that in this as in all other of life's actions each man should conform his principles to the size of his fortune; Lycon, the philosopher, wisely prescribed that his friends should lay his body where they thought best, and, as far as the funeral was concerned, should make it neither excessive nor niggardly.[19] [B] I shall leave it [C] purely to custom to order this ceremony; I shall entrust myself to the discretion of the first people this duty shall fall to.[20] *'Totus hic locus est contemnendus in nobis, non negligendus in nostris.'* [All this is a matter to be despised for ourselves but not neglected for our own.] And a saint put it a saintly way: *'Curatio funeris, conditio sepulturae, pompa exsequiarum magis sunt vivorum solatia quam subsidia mortuorum.'* [The arranging of funerals, the choosing of tombs and the pomp of obsequies are consolations for the living rather than supports for the dead.] That is why Socrates (when Crito asked him in his final moments how he wanted to be buried) replied, 'Just as you wish.'

[B] If I had to trouble myself further, I would find it more worthy to

---

18. Livy, *Hist.*, Epitome, XLVIII.
19. Diogenes Laertius, *Life of Lycon*.
20. '88: I shall leave it *rather* to custom to order this ceremony, *and, saving such things as are required in the service of my religion, if it be in a place where it be necessary to impose them,* I shall *willingly* entrust myself to the discretion of the first people this *burden* shall fall to . . . (The sense of the words struck out is supported by the three quotations added in [C], from Cicero, *Tusc. disput.*, I, XIV, 108; St Augustine, *City of God*, I, xii (Vives cites Socrates and other philosophers in his notes); and Cicero, ibid., I, xliii, 103.)

imitate those who set about enjoying the disposition and honour of their tombs while they are still alive and breathing, and who take pleasure in seeing their dead faces carved in marble. Happy are they who can please and delight their senses with things insensate – and who can live off their death.

[C] I can almost enter into an implacable hatred against all democratic rule (even though it seems to me to be the most natural and the most equitable) when I think of the inhuman injustice of the people of Athens, who sentenced to death, without remission, without even listening to their defence, those brave commanders who had just won that naval engagement against the Spartans off the Argunisae Islands; it was the most closely contested battle and the greatest that the Greek forces had ever fought at sea; but after that victory, rather than staying to gather up their dead and bury them, they had exploited the opportunities offered them by the laws of war. Diomedon's action made their execution even more odious: he was one of the condemned, a man of notable virtue in both war and politics; after hearing the judgement condemning them, he advanced to speak, only then obtaining a quiet hearing; instead of exploiting this for the good of his cause and for revealing the manifest injustice of so cruel a verdict, he showed only concern for the protection of his judges: he prayed the gods to turn this judgement to their advantage; and, lest failure to carry out the vows which he and his companions had made in gratitude for such glorious good fortune should draw down the wrath of the gods upon them, he then told them what those vows had been. Then, without another word and without bargaining, he strode courageously to his execution.

A few years later Fortune punished the Athenians by giving them sops from their own bread: for Chabrias the captain-general of their navy had got the upper hand over Pollis the Spartan admiral off the island of Naxos, but he lost the fruit of the victory clean outright – though it was of great consequence to them – out of fear induced by this exemplary punishment. Rather than lose a few dead bodies of his friends floating in the water, he allowed to sail away in safety a vast array of living enemies, who made them pay dearly later on for so grievous a superstition.[21]

> *Quaeris quo jaceas post obitum loco?*
> *Quo non nata jacent.*

> [You ask where you will lie after death?
> Why, where the unborn lie.][22]

21. Diodorus Siculus, XIII, xxxi–xxxii; XV, ix.
22. Seneca (the dramatist), *The Trojan Women*, II, 30.

On the other hand the following poet endows a body bereft of its soul with the ability to feel at rest:

> *Neque sepulchrum quo recipiat, habeat portum corporis,*
> *Ubi, remissa humana vita, corpus requiescat a malis.*

[May it have no tomb to welcome it, no harbour where, having surrendered human life, the body may find a rest from evils.][23]

23. Ennius, cited by Cicero, *Tusc. Disput.*, I, xliv, 107.
'95 has this addition: '*Nature thus shows us that several dead things still have some occult relationships with life: the wine in the cellar varies according to some of the changing seasons of the vine. And the meat of venison changes its character and flavour according to the laws governing the flesh of the living deer – so we are told*'. (Renaissance science attributed such changes to the forces of 'sympathy' or 'antipathy' inherent in all things.)

4. How the soul discharges its emotions against false objects when lacking real ones

═══

[As often in the Essays, *'soul' here includes all aspects of the human personality not strictly corporeal. Montaigne is especially concerned in this chapter with those irrational bursts of choler which are vented in wrath directed against inanimate or guiltless objects and which sweep over great generals every bit as much as over a girl distraught with grief for her brother or over a gouty old man. Our mind (our* esprit*) is ever like that: prone to be irrational as well as refractory to right rule.]*

|A| A local gentleman of ours who is marvellously subject to gout would answer his doctors quite amusingly when asked to give up salted meats entirely. He would say that he liked to have something to blame when tortured by the onslaughts of that illness: the more he yelled out curses against the saveloy or the tongue or the ham, the more relief he felt. Seriously though, when our arm is raised to strike it pains us if the blow lands nowhere and merely beats the air; similarly, if a prospect is to be made pleasing it must not be dissipated and scattered over an airy void but have some object at a reasonable distance to sustain it.

> [B] *Ventus ut amittit vires, nisi robore densæ*
> *Occurant silvæ spatio diffusus inani;*

[As winds, unless they come up against dense woods, lose their force and are distended into empty space;][1]

[A] it seems that the soul too, in the same way, loses itself in itself when shaken and disturbed unless it is given something to grasp on to; and so we must always provide it with an object to butt up against and to act upon. Plutarch says of those who dote over pet monkeys or little dogs that the faculty for loving which is in all of us, rather than remaining useless forges a false and frivolous object for want of a legitimate one.[2] And we

---

1. Lucan, *Pharsalia*, III, 362–3. (This poem is now frequently known by the better title of *The Civil War*.)
2. Plutarch, *Life of Pericles*.

can see that our souls deceive themselves in their emotions by erecting some false fantastical object rather than let there be nothing to act upon.   [B]   Animals are likewise carried away by anger: they attack the stone or piece of iron which has wounded them or else take vengeance on the pain they feel by biting themselves:

> Pannonis haud aliter post ictum saevior ursa
> Cum jaculum parva Lybis amentavit habena,
> Se rotat in vulnus, telumque irata receptum
> Impetit, et secum fugientem circuit hastam.

[Not otherwise does the bear in Pannonia: made more savage by the blow struck by the Libyan hunter with his dart tied to a leather thong, she rolls on her wound and attacks the weapon buried in her flesh and chases it round and round in circles as it flees from her.][3]

[A]   What causes do we not discover for the ills which befall us! What will we not attack, rightly or wrongly, rather than go without something to skirmish against? It is not those blond maiden tresses which you are tearing, nor the whiteness of that bosom which you are beating so cruelly in your distress, which killed your beloved brother with an unlucky musket-ball.   [C]   When the Roman army in Spain lost those two great commanders who were brothers, Pliny says *'flere omnes repente et offensare capita'*. [at once, they all start weeping and beating their heads.][4] A common practice. And was it not amusing of Bion the philosopher to ask of that king who was tearing out his hair in grief: 'Does he think that alopecia gives relief from sorrow?'[5]   [A]   And who has not seen a man sink his teeth into playing-cards and swallow the lot or else stuff a set of dice down his throat so as to have something to avenge himself on for the loss of his money! Xerxes flogged the waters   [C]   of the Hellespont, put them in shackles and heaped insults upon them   [A]   and wrote out a challenge defying Mount Athos; Cyrus kept an entire army occupied for several days in taking revenge on the river Gyndus for the fright it gave him when he was crossing it; and Caligula demolished a very beautiful house on account of the pleasure his mother had taken in it.[6]

[C]   In my younger days the country-folk used to tell how one of our neighbours' kings who had received a good cudgelling from God swore to

---

3. Lucan, *Pharsalia*, VI, 220–4.
4. Livy, XXV, xxxvii; of the brothers Publius and Cnaeus Scipio.
5. Cicero, *Tusc. disput.*, III, xxvi, 63 (not listed by Erasmus in the *Apophthegmata*).
6. Plutarch, *Comment il faut refrener la cholere*, 57 F–G; Herodotus, VII, xxxv.

get his revenge on him by ordering that, for ten years, nobody should pray to Him, mention Him nor (insofar as it lay in his power) even believe in Him. By this they meant to portray not so much the folly as the inborn arrogance of the nation about which this story was told.[7] Those vices always go together but, in truth, such actions are more beholden to overweening pride than to stupidity.

[A]   When Caesar Augustus had been battered by a storm, he began to defy Neptune, the god of the sea; to get his revenge during the ceremonies at the games in the Roman Circus he removed his statue from its place among the others. In that, he was less excusable than the generals mentioned above – and less than he himself was later on when, having lost in Germany a battle under Quintilius Varus, he kept beating his head against the wall in anger and despair, crying, 'Varus! Give me back my soldiers!' Those other cases surpass all folly since they add blasphemy to it when they address [C] themselves thus [A] to God – or even to Fortune as though she had ears subject to our assaults – [C] following the example of the Thracians who revenge themselves like a Titan during thunder and lightning by shooting darts into the sky, seeking to bring God to his senses by a shower of arrows.[8]

[A]   Yet as that old poet says in Plutarch:

> *Point ne se faut courroucer aux affaires:*
> *Il ne leur chaut de toutes nos choleres*

[There is no point in getting angry against events: they are indifferent to our wrath.][9]

[B]   But we shall never utter enough abuse against the unruliness of our minds.

7. Doubtless a King of Castille. Robert Burton cites this in the *Anatomy of Melancholy* after 'Montanus', but does not identify the king or country.
8. Suetonius, *Life of Caesar*; Herodotus, IV, xciv.
    '80: address *insults* thus . . .
9. Plutarch, *De la tranquillité de l'ame*, 69G.

5. Whether the governor of a besieged fortress should go out and parley

===

[This chapter, arising from Montaigne's reflections on his reading of Renaissance French and Italian historians in the light of his own experience of war, belongs to those chapters which he wrote near the beginning of his enterprise, when the Essays *appear to have been intended mainly as a gentleman's thoughts on matters military and political.]*

[A] In the war against Perseus, king of Macedonia, the Roman legate Lucius Marcius, wishing to gain the time he still needed to get his army ready, sowed hints of agreement by which the king was lulled into granting a truce for several days, thus furnishing his enemy with the opportunity and freedom to arm himself; because of this the king met his final downfall. Nevertheless [C] the old men in the Senate, mindful of the morals of their forefathers, condemned this action as being opposed to their practices [C] in ancient times which were, they said, to fight with valour not with trickery, surprise attacks or night encounters; nor did they use pretended flight or unexpected charges; they never made war before it was declared and seldom before announcing the time and place of the battle. From the same conscientious scruple they sent that treacherous doctor back to Pyrrhus and that wicked schoolmaster back to the Phalisci. Those were truly Roman ways of acting – not Grecian guile or Punic cunning, for which it is less glorious to win by might than by deceit. There may be a momentary advantage in deception, but only those men acknowledge that they are beaten who know that it was neither by ruse nor mischance but by valour, soldier against soldier in a legitimate and just war. It is clear from what those good men decided that they had not yet accepted [A] that fine saying:

> *dolus an virtus quis in hoste requirat?*

[Trickery or valour: what does it matter between enemies?][1]

---

1. '80: Nevertheless *the Roman* Senate, *for whom only superior virtue was deemed a just means of gaining victory, found this trick ugly and dishonourable, not yet having tingling in their ears that* fine saying ... (The 'fine saying' is from Virgil, *Aeneid*, II, 390, cited by Justus Lipsius, *Politici*, V, as are Polybius and Florus.)

[C] According to Polybius, the Achaeans detested all kinds of ruses in their wars, only holding it to be a victory when the hearts of their enemies were beaten low. Another writer said: *'Eam vir sanctus et sapiens sciet veram esse victoriam, quæ salva fide et integra aignitate parabitur.'* [A man who is pious and wise will know that a real victory is won only when integrity is safeguarded and greatness kept intact.]

> *Vos ne velit an me regnare hera quidve ferat fors*
> *Virtute experiamur.*

[Let us make trial by valour, to see whether my Lady Fortune wishes you to prevail or me.][2]

In the kingdom of Ternate, among those peoples whom we complacently dub barbarous, custom requires that they never start to fight without a declaration of war, to which is added a full statement of the means they have at their disposal: what they are, how many men they have, what munitions, what arms for both attack and defence. But once having done that, if their enemies do not give in or reach an agreement they permit themselves to do their worst, believing they cannot be reproached for treachery, for cunning or for any means leading to victory.[3]

The ancient Florentines were so far from wishing to get the better of their enemies by surprise attacks that a month before they sent their armies into the field they gave them warnings by continuously tolling the bell they called the *Martinella*.[4]

[A] We are less scrupulous: we hold that the honour of a war goes to him who wins by it, and following Lysander we say that when the lion's skin does not suffice we must sew on a patch from the fox's.[5] From such cunning derive the most usual opportunities for surprising the enemy: there is no hour when a commander ought to be more on his guard, we say, than during parleys and when treating for peace. That explains why it is a precept on the lips of all fighting-men of our time that no governor of a besieged fortress should ever personally go out to parley. In our fathers' days the Seigneurs de Montmord and de l'Assigny, when defending Mousson against the Count of Nassau, were blamed for doing so.

2. Ennius, in Cicero, *De officiis*, I, xii, 38.
3. Geronimo Osorio (da Fonseca) and others, *Histoire du Portugal*, tr. Simon Goulart (and sometimes attributed to him), Paris, 1587, XIV.
4. Cf. Giovanni Villani, *Croniche dell'origine di Firenze*, Venice, 1537, VI.
5. Erasmus, *Apophthegmata*, I; *Lysander*, XCI.

But by this reckoning a man would be justified if he went out in such a manner that the safety and advantage remained with his own side, as happened to Guy de Rangon when the Seigneur de l'Escut drew near to parley during the siege of Reggio (if we are to believe Du Bellay, that is, for Guicciardini said it happened to himself): Rangon clung so close to his fortress that when a disturbance broke out during the parley Monsieur de L'Escut and his troops who had advanced with him found themselves to be the weaker party, with the result that not only was Alessandro Trivulzio killed there but l'Escut himself was forced to take the Count at his word and, for greater safety, to dash after him into his citadel to shelter from the violence.[6]

[B]   Eumenes was pressed by Antigonus, who was besieging him in the town of Nora, to come out and parley; after several other considerations, Antigonus asserted that since he was the greater and the stronger it was only right that Eumenes should come out to him. Eumenes made this noble answer: 'I shall never reckon anyone to be greater than I am so long as I have the use of my sword.' He would not agree until Antigonus, as he had requested, had handed over his nephew Ptolomy as hostage.[7]

[A]   Yet some have done very well to trust in the word of their assailant and to come out. Witness Henry de Vaux, a knight from Champagne. He was under siege by the English in Commercy castle; Barthélemy de Bonnes, who was in charge of the operations, first sapped the greater part of the fortress so that all that was needed was a match and the besieged would be buried beneath the ruins; he then summoned Henry to come out to parley – for his own advantage. He was one of four who did so. When he was made to see with his own eyes that his destruction was inevitable, he felt singularly indebted to the enemy; once he had surrendered himself and his men into their power the fuse of the mine was lighted, the wooden props began to give way and the castle was blown up from roof to basement.[8]

[B]   I readily trust others: but I would only do so with difficulty if ever I were to give grounds for thinking that I was acting out of despair or from lack of courage rather than from frankness and trust in a man's word.

6. Martin Du Bellay, *Mémoires*, I, 22, on the siege of Mousson, and I, 29, on the siege of Reggio (cf. Guicciardini, *Histoire d'Italie*, 1568).

7. Plutarch, *Life of Eumenes*.

8. Jean Froissart, *Croniques*, I, ccix.

6. *The hour of parleying is dangerous*

===

[Montaigne wrote this when the Siege of Mussidan (April 1569) was fresh in his mind. Mussidan is less than twenty miles from Montaigne itself.]

[A] Nevertheless I recently saw during the siege of near-by Mussidan that those who had been forcibly dislodged by our army, as well as others of their faction, cried out as though it were treachery when, during the negotiations for an agreement, while the proceedings were still under way, they were taken unawares and hacked to pieces: an accusation which in another century might have seemed justified. But as I have just said, our ways are entirely removed from such rules: nowadays people must not trust each other before the very last binding seal has been affixed. And even that is not enough; [C] it is always a hazardous decision to trust that it will be the good pleasure of a victorious army to keep the promises made to a town which has just surrendered upon generous and favourable terms and to allow free entry to the heated soldiery. Lucius Aemilius Regillus, the Roman praetor, having made an assay of taking the town of Phocaea by force, but having wasted his time because of the outstanding prowess shown by the citizens in their defence, made a pact with them by which they would be accepted as Friends of the Roman People, while he would make an entry as into a confederate city; by which he removed all fear of a hostile action. Whereupon, in order to appear in greater pomp, he immediately brought his army in with him; but no matter what effort he employed it was not in his power to restrain his troops: before his very eyes they sacked a large section of the town, the rights of greed and vengeance overriding those due to his office and to army discipline.[1]

[A] Cleomenes maintained that, no matter what harm you inflicted on an enemy in war-time, that action was, before gods and men, always above the law and in no way subject to it. So having made a seven-day truce with the Argives, he fell upon them three nights later and killed them while they slept, maintaining that nothing had been said in his truce about night-time. But the gods took revenge on such crafty perfidy.[2]

1. Cf. Livy, XXXVII, xxxii.
2. Plutarch, *Les dicts notables des Lacedaemoniens*, 217 H–218 A.

[C] During a parley, while the citizens of Casilinum were dithering over their sureties their town was taken by surprise – yet that was during the age of Rome's justest commanders and of the perfection of the Roman art of war. For it is not said that we may not, at the right time and place, take advantage of the stupidity of our enemies just as we do of their cowardice. (War certainly has by its nature many privileges which are reasonable at reason's own expense. Here that rule does not apply, 'Neminem id agere ut ex alterius praedetur inscitia' [No one should prey on another's ignorance.]) But I am thunderstruck by the scope which Xenophon gives to those privileges in the plans and the deeds of his perfect general; Xenophon is a marvellously weighty authority on such matters, being a great commander and, as a philosopher, one of the foremost disciples of Socrates; but I do agree in all things everywhere with the measure he dispenses.[3]

[A] During the siege of Capua, after Monsieur d'Aubigny had given it a furious battering, Signor Fabrizio Colonna, the commander of the city, had begun to parley from the top of a bastion; as his men relaxed their guard, our men seized the town and tore it apart.[4] And, more fresh in our memory, Signor Giuliano Romero at Yvoy made the schoolboy howler of coming out to parley with my Lord the Constable, only to find when he got back that his fortress was taken![5] But we were not allowed to get off without due retribution: the Marquis of Pescara was besieging Genoa where Duke Octaviano Fregoso was in command under our protection; negotiations were so far advanced that it was regarded as if all was already over, when, just as they were about to be concluded, the Spaniards slipped into the city and treated it as though they were fully victorious. And since then, at Ligny-en-Barrois, where the Comte de Brienne was in command and where the siege was conducted by the Emperor in person, Bertheville, Brienne's lieutenant, came out to parley: it was during the bargaining that the town was taken.[6]

> Fu il vincer sempre mai laudabil cosa,
> Vincasi o per fortuna o per ingegno.

[Victory has ever been worthy of praise, even when due to Fortune or to trickery.][7]

---

3. Livy, XXIV, xix; Cicero, De officiis, III, xvii; Xenophōn, Cyropaedia.
4. Guicciardini, Histoire d'Italie, V, ii.
5. Montaigne confounds the sieges of Yvoy in the Ardennes with that of Dinan (1554).
6. Martin Du Bellay, Mémoires, II, and IX.
7. Ariosto, Orlando furioso, XV, 1.

They say that. Yet Chrysippus the philosopher would not have agreed: no more than I do. For, he said, those who contest a race must certainly make every effort to run fast, but it is in no ways allowable for them to lay their hand on a rival to stop him nor to thrust out a leg to trip him up.[8]   [B]   And nobler still was the answer made by Alexander the Great to Polypercon, who was urging him one night to take advantage of the darkness to launch an attack against Darius: 'Certainly not. I am not the man to thieve a victory and then follow it up!' – *'Malo me fortunae poeniteat, quam victoriae pudeat.'* [I would rather complain of Fortune than feel ashamed of victory.][9]

> *Atque idem fugientem haud est dignatus Orodem*
> *Sternere, nec jacta cæcum dare cuspide vulnus:*
> *Obvius, adversoque occurrit, seque viro vir*
> *Contulit, haud furto melior, sed fortibus armis.*

[Orodes did not deign to strike him in the back as he fled, nor to wound him with an unseen dart. He ran and confronted him, face to face; he fought with him man to man, proving himself superior not by trickery but by mighty arms.][10]

8. Cicero, *De officiis*, III, x, 42.
9. Quintus Curtius (Rufus), IV, xiii.
10. Virgil, *Aeneid*, X, 752.

7. *That our deeds are judged by the intention*

===

[The end of this chapter, written just before Montaigne died, turns fairly routine thoughts about motive into a personal declaration: Montaigne intends his death to be morally at one with his life.]

[A] 'Death,' they say, 'settles all obligations.' I know some who have taken that in a perverse sense. King Henry VII of England made an agreement with Don Felipe, the son of the Emperor Maximilian or (to situate him more nobly) the father of the Emperor Charles V, by which Don Felipe would hand over to him his enemy the Duke of Suffolk (of the White Rose, who had fled into hiding in the Low Countries) provided that he promise to make no attempt on the Duke's life. Yet as he lay a-dying Henry ordered his son in his testament to have the Duke killed as soon as his own death was over.[1]

More recently, in that tragedy put on for us by the Duke of Alba with the deaths of Count Horn and Count Egmont, there were many events worthy of note.[2] Among others was the fact that Count Egmont, on whose faith and assurances Count Horn had put himself into the hands of the Duke of Alba, insistently begged that he be executed first, so that his death should free him from the obligation he had incurred towards Count Horn.

It would seem that death had not freed King Henry from his sworn undertaking, but that Count Egmont had discharged his even before he died: we cannot be held to promises beyond our power or our means. That is why – since actions and performances are not wholly in our power and since nothing is really in our power but our will – it is on the will that all the rules and duties of Man are based and established. And so, since Count Egmont held his soul and his will to be debtors to his promise, he would without a doubt have been acquitted of his obligation even had he survived Count Horn, given that it was not in his power to carry it out. But the

---

1. Martin Du Bellay, etc., *Mémoires*, I, 7.
   '80: Henry *expressly* ordered . . .
2. Both were beheaded in 1568.

King of England, by breaking his word intentionally, cannot be absolved just because he put off the act of treachery until after his death – no more than that mason in Herodotus who loyally kept the secret of the treasures of the king of Egypt during his lifetime, only to reveal it to his children when he died.[3]

[C] I have seen many men in my time smitten in conscience for having withheld other men's goods who arrange in their testaments to put things right after they are dead. But it is valueless to fix a date for so urgent a matter or to wish to right wrongs without feeling or cost. They must pay with something which is truly theirs: the more burdensome and onerous their payment the more just and meritorious their atonement. Repentance begs for burdens.

Worse still are they who reserve for their last will and testament some hate-ridden provision affecting a near one, having concealed it during their lifetime. By stirring up against their memory the one they have offended they show scant regard for their reputations; and they show even less for their consciences since they cannot, even out of respect for death, make their animosities die, prolonging the life of them beyond their own. They are iniquitous judges, postponing judgement until they can no longer take cognizance of the case.

If I can, I will prevent my death from saying anything not first said by my life.

3. Herodotus, cited by Henri Estienne in his satire, *L'Apologie pour Hérodote*, XV.

8. On idleness

===

[The Essays *were started to tame melancholic delusions induced by Montaigne's withdrawal to his estates, when his thoughts galloped away with him much as Milton later describes in* Il Penseroso *as being typical of the melancholic in his lonely tower.]*

[A] Just as fallow lands, when rich and fertile, are seen to abound in hundreds and thousands of different kinds of useless weeds so that, if we would make them do their duty, we must subdue them and keep them busy with seeds specifically sown for our service; and just as women left alone may sometimes be seen to produce shapeless lumps of flesh but need to be kept busy by a semen other than her own in order to produce good natural offspring: so too with our minds.[1] If we do not keep them busy with some particular subject which can serve as a bridle to reign them in, they charge ungovernably about, ranging to and fro over the wastelands of our thoughts:

> [B] *Sicut aquae tremulum labris ubi lumen ahenis*
> *Sole repercussum, aut radiantis imagine Lunae*
> *Omnia pervolitat late loca jamque sub auras*
> *Erigitur, summique ferit laquearia tecti.*

[As when ruffled water in a bronze pot reflects the light of the sun and the shining face of the moon, sending shimmers flying high into the air and striking against the panelled ceilings.][2]

[A] Then, there is no madness, no raving lunacy, which such agitations do not bring forth:

> *velut ægri somnia, vanæ*
> *Finguntur species.*

[they fashion vain apparitions as in the dreams of sick men.][3]

---

1. The human egg not yet having been discovered, many believed with Galen that children were produced by an intermingling of a (weaker) female semen with the male's. By itself the female semen could at times produce *moles*, a misshapen lump. (Montaigne found the idea developed in Plutarch's *Matrimonial Precepts*, which La Boëtie translated, and which Montaigne published in 1571.)
2. Virgil, *Aeneid*, VIII, 22.
3. Horace, *Ars poetica*, 7.

When the soul is without a definite aim she gets lost; for, as they say, if you are everywhere you are nowhere.

[B]   *Quisquis ubique habitat, Maxime, nusquam habitat.*

[Whoever dwells everywhere, Maximus, dwells nowhere at all.][4]

Recently I retired to my estates, determined to devote myself as far as I could to spending what little life I have left quietly and privately; it seemed to me then that the greatest favour I could do for my mind was to leave it in total idleness, caring for itself, concerned only with itself, calmly thinking of itself. I hoped it could do that more easily from then on, since with the passage of time it had grown mature and put on weight.

But I find –

Variam semper dant otia mentis

[Idleness always produces fickle changes of mind][5]

– that on the contrary it bolted off like a runaway horse, taking far more trouble over itself than it ever did over anyone else; it gives birth to so many chimeras and fantastic monstrosities, one after another, without order or fitness, that, so as to contemplate at my ease their oddness and their strangeness, I began to keep a record of them, hoping in time to make my mind ashamed of itself.[6]

4. Martial, VII, lxxiii.
5. Lucan, *Pharsalia*, IV, 704.
6. Montaigne's terms are the technical ones of melancholy madness. Cf. for example Milton's *Ode to Melancholy*, where the English equivalents occur.

9. On liars

====

[Quintilian had said that a liar had better have a good memory: hence Montaigne's concern with memory before turning to lying – a vice particularly loathed by gentlemen and which Montaigne would discourage even in diplomatists.]

[A] There is nobody less suited than I am to start talking about memory. I can hardly find a trace of it in myself; I doubt if there is any other memory in the world as grotesquely faulty as mine is! All my other endowments are mean and ordinary: but I think that, where memory is concerned, I am most singular and rare, worthy of both name and reputation! [B] Apart from the natural inconvenience which I suffer because of this – [C] for memory is so necessary that Plato was right to call it a great and mighty goddess[1] – [B] in my part of the world they actually say a man 'has no memory' to mean that he is stupid. When I complain that my memory is defective they either correct me or disbelieve me, as though I were accusing myself of being daft. They see no difference between memory and intelligence. That makes my case worse than it is.

But they do me wrong. Experience shows us that it is almost the contrary: an outstanding memory is often associated with weak judgement. They also do me another wrong: I am better at friendship than at anything else, yet the very words used to acknowledge that I have this affliction are taken to signify ingratitude; they judge my affection by my memory and turn a natural defect into a deliberate one. 'We begged him to do this,' they say, 'and he has forgotten.' 'He has forgotten his promise.' 'He has forgotten his friends.' 'He never remembered – even for my sake – to say this, to do that or not to mention something else.' I certainly do forget things easily but I simply do not treat with indifference any charge laid on me by my friends. Let them be satisfied with my misfortune, without turning it into precisely the kind of malice which is the enemy of my natural humour.

---

1. [A] until [B]: reputation. *I could tell some remarkable tales about that, but, for the while, it is better to keep to my subject.* It is not for nothing . . .
   Then Plato, *Critias*, 108D.

I find ways of consoling myself. First, by arguing that [C] a poor memory is an evil which has enabled me to correct a worse one which might easily have arisen in me: ambition. A bad memory is an intolerable defect for anyone concerned with worldly affairs.

Moreover, Nature (as is shown by several similar examples of her ways of compensating) has strengthened other faculties of mine as this one has grown weaker. If, thanks to memory, other people's discoveries and opinions had been kept ever before me, I would readily have reached a settled mind and judgement by following other men's footsteps, failing as most people do to exercise my own powers.

Then again [B] I talk less; it is always easier to draw on the storehouse of memory than to find something original to say. [C] (If my memory had stood fast, I would have deafened my friends with my chatter, as the subjects themselves would have stimulated such gifts as I do have for arranging and exploiting them, and that would have encouraged and attracted my powers of argument.) [B] That it is pitiful I assay by the touchstone of some of my nearest and dearest: the more their memory furnishes them with full and ready matter the deeper they dig back when they tell us about it; they weigh it down with irrelevant circumstances, so that even if their story is interesting they smother the interest, and if it is boring you are cursing either their good memory or their bad judgement.

[C] Once you are off, it is hard to cut it short and stop talking. Nothing tells you more about a horse than a pronounced ability to pull up short. I have even known men who can speak pertinently, who want to stop their gallop but who do not know how to do so. While looking for a way of bringing their hoofs together they amble on like sick men, dragging out trivialities.

Old men are particularly vulnerable: they remember the past but forget that they have just told you! I have known several amusing tales become boring in one gentleman's mouth: his own people have had their fill of it a hundred times already.[2]

[B] A second advantage is that (as some Ancient writer put it) I remember less any insults received. [C] I would need an Official Reminder like Darius: in order not to forget an insult suffered at the hands of the Athenians he made a page intone three times in his ear as he sat at table: 'Remember the Athenians, Sire.'[3]

---

2. Montaigne probably means himself.
3. Cf. Cicero, *Pro Ligurio*; Herodotus, *Hist.*. V, cv.

[B]   Books and places which I look at again always welcome me with a fresh new smile.

[A]   It is not for nothing that it is said that he who does not feel his memory to be strong enough has no business lying.[4] I am well aware that grammarians make a distinction between 'to tell an untruth' and 'to lie'; they assert that 'to tell an untruth' is to say something false which one thinks to be true, and that the definition of *mentiri* [to lie] in Latin (the source of *mentir* in French) implies something like, 'to go against the testimony of one's knowledge', and so only applies to those who speak at variance with what they know. They are the people I am talking about.[5]

That kind of person either makes up the whole story or else disguises and pollutes some source of truth. When it is a case of disguising and changing something, you can normally hobble liars by making them tell the same tale several times over; since the real facts were lodged in their memory first, they make a deep imprint by means of awareness and knowledge; it is hard for those facts not to spring to mind and to dislodge the falsehoods (which cannot gain so settled and firm a footing there); hard too for the circumstances as they first learned them, by continually flowing into their minds, not to make them lose all memory of the false additions and distortions.

When the whole thing has been made up, liars might seem to have less reason to be afraid of getting things wrong, since there is no counter-impression to rival their falsehoods. Yet since such a lie is insubstantial and hard to get a grip on, it can easily slip out of a memory not extremely reliable.

[B]   Experience has often shown me that – amusingly so, at the expense of those kinds of men whose profession it is never to utter a word without trimming it to suit whatever business is being negotiated at the time, thereby pleasing the great ones with whom they are speaking. Such men are prepared to make their honour and conscience slaves to present circumstances: but circumstances are liable to frequent change, and their words must vary with them. They are obliged to call the very same thing first grey then yellow, saying one thing to this man, quite another to another. If the persons who receive such contrary advice happen to compare their haul, what becomes of their fine diplomacy?

4. Quintilian, *Institutio oratoria*, IV, ii, 91; the saying had become proverbial.
5. Pedro Mexia also treats this topic: cf. his *Diverses Leçons* (III, 'How we can tell lies'). His sources, like Montaigne's, are Aulus Gellius and Nonus.

Apart from that, they can be like a silly horse casting its own shoe; for what memory could ever suffice them, enabling them to remember all the various moulds they have invented for the same subject matter? In my time I have known several men who hankered after a reputation for this fine sort of prudence: they never can see that to have a reputation for it renders it ineffectual.

Lying is an accursed vice. It is only our words which bind us together and make us human. If we realized the horror and weight of lying we would see that it is more worthy of the stake than other crimes. I find that people normally waste time quite inappropriately punishing children for innocent misdemeanours, tormenting them for thoughtless actions which lead nowhere and leave no trace. It seems to me that the only faults which we should vigorously attack as soon as they arise and start to develop are lying and, a little below that, stubbornness. Those faults grow up with the children. Once let the tongue acquire the habit of lying and it is astonishing how impossible it is to make it give it up. That is why some otherwise decent men are abject slaves to it. One of my tailors is a good enough fellow, but I have never heard him once speak the truth, not even when it would help him if he did so.

If a lie, like truth, had only one face we could be on better terms, for certainty would be the reverse of what the liar said. But the reverse side of truth has a hundred thousand shapes and no defined limits. The Pythagoreans make good to be definite and finite; evil they make indefinite and infinite. Only one flight leads to the bull's-eye: a thousand can miss it.

I cannot guarantee that I could bring myself to tell a solemn, bare-faced lie, even to ward off some obvious and immense danger. One of the old Church Fathers says that even a dog we do know is better company than a man whose language we do not know. '*Ut externus alieno non sit hominis vice*'. [Just as any foreigner is not fully human.] How much less companionable than silence is the language of falsehood.[6]

[A] Francesco Sforza, the Duke of Milan, had an ambassador, Francesco Taverna, widely renowned for his knowledge of how to yap. King Francis I used to boast how he cornered him like a hare. Taverna's mission was to try and justify his lord to the King's Majesty over an action of great consequence, which was as follows.

The King wished to maintain some sources of confidential intelligence in

6. St Augustine, *City of God*, XIX, vii; Montaigne cites Pliny from J. L. Vives' note on this passage.

Italy (from which he had recently been driven) and especially in the Duchy of Milan; so he decided to keep a reliable man there, close to the Duke and virtually as ambassador but ostensibly as a private individual who pretended to be in Milan for his own affairs, since the Duke was far more dependent on the Emperor (especially since he was negotiating a marriage contract for his niece, the daughter of the King of Denmark, now Dowager Duchess of Lorraine). Without greatly harming his cause he could not let it openly be known that he was having any dealings or negotiations with the French.

A young Milanese nobleman, one of the King's equerries, was considered the right man for the mission. His name was Merveille. He was dispatched with secret credentials and documents as an ambassador, but for appearances' sake and as a disguise he also had letters of recommendation to the Duke concerning his 'private business'. He stayed so long in the Ducal court that some knowledge of this reached the Emperor who, we think, was the cause of what soon happened: namely, that the Duke arranged to have Merveille beheaded in the middle of the night; he had been charged with some murder or other and the trial had lasted only a couple of days.[7]

Signor Francesco arrived, duly prepared with a long distorted account of this event, for the King had written to all the rulers of Christendom, as well as to the Duke himself, demanding satisfaction.

He was granted an audience one morning; he proceeded to lay the foundations of his case and had drawn up several plausible reasons covering the affair, alleging that his lord had never considered our man Merveille to be anything but a private nobleman, one of his own subjects who had business in Milan and whose mode of life had suggested nothing else. He particularly denied any knowledge of his being on the establishment of the King's household, or even of his being known to the King, let alone taking him for an ambassador.

The King, for his part, plied the ambassador with a variety of questions and objections; he rounded upon him on every side and then cornered him, the point being, 'Why had the execution been carried out secretly, by night?' The wretched man, nonplussed but trying to be polite, replied that 'if such an execution had been carried out in daylight, why, the Duke – out of respect for his Majesty – would have been quite upset . . .' We can all imagine how he struggled to get up after coming such a cropper as that, under the very ample nose of Francis I.

Pope Julius II sent an ambassador to the King of England to rouse his

7. The murder of Captain Merveille became an international *cause célèbre*. It is narrated in the Du Bellay *Mémoires*.

animosity against King Francis. The ambassador having been heard out, the King of England in his reply dwelt on the difficulties he could see in making all the preparations which would be essential if war were to be waged against so mighty a monarch. He cited some of the reasons. The ambassador answered, most inappropriately, that he too had thought of them and had pointed them out to the Pope.

These words were so different from the case he had just put forward (which was to urge the English to go to war immediately) that the King of England began to suspect (what he later found to be actually true) that the private inclinations of the ambassador leant towards the French. The Pope, being informed of this, confiscated his property and the man all but lost his life.[8]

8. The original source of this account is the *De Lingua* of Erasmus. It is taken up by Henri Estienne in the *Apologie pour Hérodote*.

10. On a ready or hesitant delivery

=====

[Montaigne considers 'readiness' to speak in public, both in the sense of speaking easily and of being ready with a prepared text. These senses are contained in the Latin word promptus *which lies behind his French term for 'ready' speech:* prompt.*]*

[A]   *Onc ne furent à tous toutes grâces données.*

[It never was, that to every man was every gift vouchsafed.][1]

We can see that in the case of the gift of speaking well: some have such a prompt facility and (as we say) such ease in 'getting it out', that they are always ready anywhere: others, more hesitant, never speak without thinking and working it all out beforehand. Just as the rule given to ladies is to take up sports and exercises which show off their charms, so too, if I had to give similar advice where these two qualities are concerned, it seems to me that nowadays, when eloquence is mainly professed by preachers and barristers, the hesitant man had better be a preacher and the other man a barrister. Since the duties of a preacher allow him as much time as he wishes to make things ready, he runs an uninterrupted race from point to point, whereas the exigencies of a barrister require him to enter the fray at a moment's notice; the unforeseeable replies of the opposite party can throw him off his stride into a situation where a new decision has to be made in full course. Yet in that meeting between Pope Clement and King Francis at Marseilles the reverse applied:[2] Monsieur Poyet, a man whose whole life had been nurtured at the Bar and who was highly regarded, had the duty of making the oration before the Pope; he had given it long thought and (so it was said) had brought it from Paris already prepared; but on the very day that it was to be delivered the Pope (fearing that something might be addressed to him which could give offence to the other princes' ambassadors who were in attendance) conveyed to the King the topic which seemed most proper to that time and place – unfortunately a totally different one from

1. Etienne de La Boëtie, *Vers françois*, ed. Montaigne, Paris, 1572; sonnet XIV.
2. Pope Clement VII came to Marseilles to discuss heresy (and other matters) in 1533. Montaigne follows the account in the Du Bellay *Mémoires*.

what Monsieur Poyet had toiled over; his oration was now useless and he had to be quickly ready with another. But as he realized that he was incapable of doing that, My Lord the Cardinal Du Bellay had to take on the task.

[B] The role of a barrister is more demanding than that of a preacher, and yet in France at least we can find more tolerable barristers, in my opinion, than tolerable preachers.

[A] It seems that it is, rather, the property of Man's wit to act readily and quickly, while the property of the judgement is to be slow and poised. But there is the same measure of oddness in the man who is struck dumb if he has no time to prepare his speech and the man who cannot take advantage and speak better when he does have time. They say that Severus Cassius spoke better when he had not thought about it beforehand: that he owed more to Fortune than to hard work: that it was good for him to be interrupted, his opponents being afraid of provoking him, lest anger made him redouble his eloquence.[3]

I know from experience the kind of character which gets nowhere unless it is allowed to run happy and free and which by nature is unable to keep up vehemently and laboriously practising anything beforehand. We say that some books 'stink of lamp-oil', on account of the harshness and roughness which are stamped on writings in which toil has played a major part.[4]

In addition, a soul worrying about doing well, straining and tensely drawn towards its purpose, is held at bay – like water which cannot find its way through the narrow neck of an open gutter because of the violent pressure of its overflowing abundance. Moreover the particular character which I am speaking of does not want to be driven and spurred on by strong passions such as Cassius' anger (for such an activity would be too violent): it wants not to be shaken about but aroused; it wants to be warmed and awakened by events which are external, fortuitous and immediate. Leave it to act by itself and it will drag along and languish. Its life and its grace consist in activity.

[B] I cannot remain fixed within my disposition and endowments. Chance plays a greater part in all this than I do. The occasion, the company, the very act of using my voice, draw from my mind more than what I can find there when I exercise it and try it out all by

---

3. Marcus Annaeus Seneca (the rhetorician, not the philosopher), *Controversies*, III.
4. Horace, *Epistles*, I, xix, 6 (cf. Erasmus, *Adages*, IV, III, LVIII: '*Non est dithyrambus qui bibit aquam.*')

myself. [A]   And that is why the spoken word is worth more than the written – if a choice can be made between things of no value.[5]

[C]   This, too, happens in my case: where I seek myself I cannot find myself: I discover myself more by accident than by inquiring into my judgement. Suppose something subtle springs up as I write – I mean, of course, something which would be blunt in others but is acute in me. (Enough of these courtesies! When we say such things we all mean them to be taken in proportion to our abilities.) Later, I miss the point so completely that I do not know what I meant to say (some outsider has often rediscovered the meaning before I do). If every time that happened I were to start scraping out words with my eraser I would efface the whole of my *Essays*. Yet, subsequently, chance may make what I wrote clearer than the noon-day sun: it will be my former hesitations which then astonish me.

5. Cf. Montaigne's dedication of his translation of Raymond Sebond to his father (given in the Appendix to the Introduction, p. lvii).

11. On prognostications

[Christianity has banished most forms of prognostication. Those that remain are the sport of subtle credulous minds who could find hidden meanings anywhere. Socrates' daemon, which made him near-infallible, was in fact a natural impulse found to some extent in all of us. So the ecstasies of Socrates were at most 'natural' ones.]

[A]   Where oracles are concerned it is certain that they had begun to lose their credit well before the coming of Jesus Christ, since we can see Cicero striving to find the cause of their decline. [C]   These are his words: '*Cur isto modo jam oracula Delphis non eduntur non modo nostra ætate sed jamdiu, ut modo nihil possit esse contempsius?*' [Why are oracles no longer uttered thus at Delphi, so that not only in our own time but long before nothing could be held in greater contempt?]

[A]   But there were other prognostications, derived from the dissection of sacrificial animals – [C]   Plato held that the internal organs of those animals were partly created for that purpose – [A]   or from chickens scratching about, from the flight of birds – [C]   '*aves quasdam rerum augurandarum causa natas esse putamus*' [We think that some birds are born in order to provide auguries] – [A]   from lightning and from swirling currents in rivers – [C]   '*multa cernunt aruspices, multa augures provident, multa oraculis declarantur, multa vaticinationibus, multa somniis, multa portentis*' [the soothsayers divine many things; the augurs foresee many; many are revealed by oracles, many by predictions, many by dreams and many by portents]; [A]   and there were other similar ones on which the Ancient World grounded most of their undertakings, both public and private:[1] it was our religion which abolished them all.[2] There remain among us it is true some means of divination by the heavens, by spirits,

---

1. Cicero, *De divinatione*, II, lvii, 117; *De natura deorum*, II, lxiv, 160–1; lxv, 162 f. The theme is prominent in Plutarch's treatise on the cessation of oracles. There was a renewed interest in classical forms of prognostication during the Renaissance. (Cf. Rabelais, *Le Tiers Livre*, TLF, XXV.) Montaigne criticizes Plato's belief in divination from entrails in II, 12.
2. Cf. Robert Garnier, *Les Juifves*, final line: Christ 'Will come, to put an end to all prophecy'.

by bodily features, by dreams and so on: that is a remarkable example of the mad curiosity of our nature which wastes time trying to seize hold of the future as though it were not enough to have to deal with the present:

> [B]   *cur hanc tibi rector Olympi*
> *Sollicitis visum mortalibus addere curam,*
> *Noscant venturas ut dira per omina clades.*
> *Sit subitum quodcunque paras, sit cæca futuri*
> *Mens hominum fati, liceat sperare timenti!*

[O Ruler of Olympus, why did it please thee to add more care to worried mortals by letting them learn of future slaughters by means of cruel omens! Whatever thou hast in store, do it unexpectedly; let the minds of men be blind to their future fate: let him who fears, still cling to hope!][3]

[C] *'Ne utile quidem est scire quid futurum sit. Miserum est nihil proficientem angi;'* [It is not even useful to know what is to happen. It is wretched to suffer to no avail;][4]   [A]   nevertheless divination now has far less authority.

That is why the case of Francisco, Marquis of Saluzzo, struck me as so remarkable. He was the Lieutenant of King Francis' transalpine forces; he found endless favours at our French Court and was beholden to the King for his very marquisate, which had been forfeited by his own brother. There was no occasion for what he did: his own feelings ran counter to it; yet (as it was asserted) he let himself become terrified by the specious prognostications which were deliberately circulated everywhere in favour of the Emperor Charles V and to our own disadvantage – especially in Italy, where these insane prophecies gained such a footing that vast sums of money changed hands in the banks from the assumption of our overthrow. Having expressed grief to his friends over the ills which he saw inevitably in store for the Crown of France and for his French friends, he foresook all and changed sides. No matter what the stars portended, it proved greatly to his harm![5]

In this he acted like a man torn by conflicting emotions. For both the towns and the armies were under his control; the enemy forces led by

---

3. Lucan, *Pharsalia*, II, 4–6; 14–15.
4. Cicero, *De nat. deorum*, III, vi, 14.
5. The Du Bellay *Mémoires* relate this. For the context, cf. Rabelais, *Pantagruéline Prognostication*, TLF, Droz, 1974, pp. xviii–xxii.

Antonio de Leyva were only a few yards away; we suspected nothing: so he could have done us far more harm. Despite his treachery we lost not one single man nor any town except Fossano (and even that only after a long struggle).

> *Prudens futuri temporis exitum*
> *Caliginosa nocte premit Deus,*
> *Ridetque si mortalis ultra*
> *Fas trepidat. . .*

> *Ille potens sui*
> *Lætusque deget, cui licet in diem*
> *Dixisse, vixi, cras vel atra*
> *Nube polum pater occupato*
> *Vel sole puro . . .*

> *Lætus in præsens animus, quod ultra est,*
> *Oderit curare.*

[Wisely does God hide what is to come under the darkness of night, laughing if a mortal projects his anxiety further than is proper . . .

That man will be happy and master of himself who every day declares, 'I have lived. Tomorrow let Father Jove fill the heavens with dark clouds or with purest light' . . . Let your mind rejoice in the present: let it loathe to trouble about what lies in the future.][6]

[C] The following quotation contradicts that, but those who believe it are wrong: '*Ista sic reciprocantur, ut et, si divinatio sit, dii sint: et si dii sint, sit divinatio.*' [If there is divination there are gods, and conversely, if there are gods there is divination.][7]

Pacuvius was much more wise:

> *Nam istis qui linguam avium intelligunt,*
> *Plusque ex alieno jecore sapiunt, quam ex suo,*
> *Magis audiendum quam auscultandum censeo.*

[As for those who understand the language of the birds and who know the livers of animals better than their own, I believe it is better just to listen to them rather than pay attention to them.]

6. Horace, *Odes*, III, xxix, 29–32; 41–44; then II, xvi, 25.
7. Cicero, *De divinatione*, I, vi, 10; then I, lvii, 131, citing Pacuvius.

The birth of that famous Tuscan art of divination was on this wise: a ploughman ploughed his furrow deeply, from which arose Tages the demi-god; he had the face of a child but the wisdom of an old man. Everybody came running up; his words and wisdom were collected and kept for centuries; they contained the principles and practices of that art . . . A birth in conformity with its development . . .[8]

[B] I would rather order my affairs by casting dice, by lots, than by such fanciful nonsense.[9] [C] And truly all States have always attributed considerable authority to them. Plato, freely drawing up his constitution as he pleased, left many important decisions to lots, including the marriages of the good citizens; he attached such importance to these fortuitous matches that he decreed that the offspring of them be kept and brought up in the Republic, whilst those born to the wicked should be driven out; nevertheless if one of these banished children should happen to promise well as he grew up, he could be recalled; and if one of those who were kept turned out hopelessly in his youth, he was exiled.[10] ᐧ

[B] I know people who study their almanacs, annotate them and cite their authority as events take place. But almanacs say so much that they are bound to tell both truth and falsehood. [C] *'Quis est enim qui totum diem jaculans non aliquando conlineet?'* [For who can shoot all day without striking the target occasionally?][11] [B] I do not think any the better of them for seeing them happen to prove true on occasions; there would be more certainty in them if they had some right rule which made them always wrong. [C] Besides, nobody keeps a record of their erroneous prophecies since they are infinite and everyday; right predictions are prized precisely because they are rare, unbelievable and marvellous.

That explains the reply made by Diagoras, surnamed the Atheist, when he was in Samothrace: he was shown many vows and votive portraits from those who have survived shipwreck and was then asked, 'You, there, who think that the gods are indifferent to human affairs, what have you to say about so many men saved by their grace?' – 'It is like this,' he replied; 'there are no portraits here of those who stayed and drowned – and they are more numerous!' Cicero says that among the many philosophers who

---

8. Cicero, *De divinatione*, II, xxiii, 50–1.
9. A serious possibility, especially for students of Renaissance law; cf. Rabelais, *Tiers Livre*, TLF, XLIII–XLIIII.
10. Plato, *Republic*, V.
11. Cicero, *De divinatione*, II, lix, 121.

believed there were gods only Xenophanes of Colophon made an assay at uprooting all forms of divination.[12] It is less surprising, therefore, that we have occasionally [B] seen[13] some of our leading minds dwelling (often to their prejudice) on such empty nonsense.

[C] I would certainly like to have seen with my own eyes these two marvels: the book of the abbot Joachim of Calabria who predicted all the future popes with their names and styles; and that of the Emperor Leo who predicted all the Emperors and patriarchs of Greece[14] . . . But with my own eyes I *have* verified the following: that when men are stunned by their fate in our civil disturbances, they have resorted to almost any superstition, including seeking in the heavens for ancient portents and causes for their ills. In this they have been so strangely successful in my days that they have convinced me that (since this way of passing time is for acute yet idle minds) those who have been inducted into the subtle art of unwrapping portents and unknotting them would be able to find anything they wish in any piece of writing whatsoever: but their game is particularly favoured by the obscure, ambiguous, fantastical jargon of these prophecies, the authors of which never supply any clear meaning themselves so that posterity can give them any meaning it chooses.

[B] The daemon of Socrates was [C] perhaps [B] a certain thrust of the will which presented itself to him without waiting for rational argument.[15] It is likely that in a soul like his (well purified and prepared by the continual exercise of wisdom and virtue) such inclinations, albeit [C] bold and undigested, were nevertheless important and worthy to be followed. Everyone can sense in himself some ghost of such agitations, of a prompt, vehement, fortuitous opinion. It is open to me to allow them some authority, to me who allow little enough to human wisdom. And I have had some – equally weak in reason yet violent in persuasion or dissuasion but which were more common in the case of

12. Cicero, *De natura deorum*, III, xxxvii, 89; *De divinatione*, I, iii, 5.

13. '88: wrong. *I* have seen . . .

14. Joachim of Flora died about 1202. His *Prophecies* were in print (there is an edition, Venice 1589), but legends had attached to his name. The Emperor Leo's book is known only at second-hand.

15. '88: The daemon of Socrates was, *in my opinion*, a certain thrust . . .

(Socrates had a *daemon*, a good spirit, who enabled him to avoid error. Renaissance thinkers took this very seriously: Rabelais gives Pantagruel a similar *daemon*: *Quart Livre*, TLF, LXVI. Cf. Plato's *Apology for Socrates* and Plutarch's *Du Demon de Socrate*.)

Socrates[16] –   [B]   by which I have allowed myself to be carried away so usefully and so successfully, that they could have been judged to contain something of divine inspiration.[17]

16. '88: albeit *fortuitous* were always *good* and worthy to be followed. Everyone *has* in himself some ghost of such agitations. I have had some by which . . .
17. For some Renaissance thinkers Socrates' ecstasies made him into a forerunner of St Paul. Montaigne considered Socrates' ecstasies to be natural in origin and so quite unlike St Paul's privileged rapture. This became a standard opinion.

12. On constancy

=====

[Constancy is a Stoic virtue, but even Stoics have to confess that a Sage can be startled. Like Rabelais before him, Montaigne considers the limits of Stoic doctrine – basing himself partly on his own experience in the Wars of Religion.]

[A] Resolution and constancy do not lay down as a law that we may not protect ourselves, as far as it lies in our power to do so, from the ills and misfortunes which threaten us, nor consequently that we should not fear that they may surprise us. On the contrary, all honourable means of protecting oneself from evils are not only licit: they are laudable. The role played by constancy consists chiefly in patiently bearing[1] misfortunes for which there is no remedy. Likewise there are no evasive movements of the body and no defensive actions with any weapons in our hands which we judge wrong if they serve to protect us from the blows raining down on us.

[C] Many highly warlike nations included flight as one of their main tactical resources: when they turned their backs that was more risky to the enemy than when they showed their faces. The Turks still retain this to some extent.

In Plato Socrates mocked Laches for defining fortitude as 'standing firm in line in the face of the enemy'. 'What,' he said, 'would it be cowardice to defeat them by giving ground?' And he cited Homer who praised Aeneas for knowing when to flee. And once Laches had corrected himself and allowed that the Scythians did use that method as do cavalrymen in general, he then went on to cite the example of those foot-soldiers of Sparta, a nation trained above all to stand their ground: during the battle of Plataea they found that they could not penetrate the Persian phalanx and so decided to disengage and fall back in order that it should be thought that they were in full flight; that would lead to the breaking up of the Persians' dense formation which would fall apart as it pursued them. By which means they obtained the victory.[2]

1. [A] until [C]: patiently *and firm-footedly* bearing . . .
2. Plato, *Laches*, 190B–191D.

While on the subject of the Scythians, it is said that after Darius had set out to subjugate them, he sent many reproaches to their king when he saw him always withdrawing and avoiding battle. To this Indathyrsez (for that was the king's name) replied that he was not afraid of him nor of any man alive, but that this was the practice of his people, since they possessed no arable lands, no towns and no houses to defend for fear that an enemy might make use of them: but if Darius really was yearning to sink his teeth into a battle, then let him try to get near to their ancient burial grounds: he would find somebody to talk to there![3]

[A]   Nevertheless once a man's post is the target of cannon-fire (as the chances of war often require it to be) it is unbecoming for him to waver before the threatening cannon-balls, all the more so since we hold that they have such speed and such impetus that you cannot take evasive action. There are many cases of soldiers at least providing their comrades with a good laugh by shielding behind their arms or ducking their heads.

Yet in the expedition which the Emperor Charles V led against us in Provence, when the Marquis de Guast went to reconnoitre the city of Arles and suddenly appeared from behind a windmill under cover of which he had made his advance, he was spotted by the Seigneur de Bonneval and the Lord Seneschal d'Agenois who were strolling along the top of the amphitheatre. They pointed him out to the Seigneur de Villier, Master of the Ordnance, who aimed a culverin so accurately that if the Marquis had not seen the match applied to the fuse and jumped aside it was thought he would have been struck in the body.[4] Similarly a few years before, when Lorenzo de' Medici, the Duke of Urbino and the father of our Queen Mother, was laying siege to Mondolfo (a fortress in Italy in the territory they call the Vicariate) he saw the fire applied to a cannon which was pointing right at him and ducked; luckily for him, for otherwise the shot, which only grazed the top of his head, would have certainly struck him in the chest.[5]

To tell the truth I do not believe that such movements arise from reflexion: for in so sudden a matter how can you judge whether the aim is high or low? It is far easier to believe that fortune looked favourably on their fear but that another time they might have jumped into the path of the shot not out of it.

[B]   Personally I cannot stop myself from trembling if the shattering

3. Herodotus, *Hist.*, IV, cxxvii.
4. The invasion was in 1536. Cf. Du Bellay, *Mémoires*, VII, p. 129.
5. Francesco Guicciardini, *Historia d'Italia*, XIII, ii.

sound of a harquebus suddenly strikes my ear in a place where I could not have expected it; I have seen that happen to more valorous men than I am. [C] Not even the Stoics claim that their sage can resist visual stimuli or ideas when they first come upon him; they concede that it is, rather, part of man's natural condition that he should react to a loud noise in the heavens or to the collapse of a building by growing tense and even pale. So too for all other emotions, provided that his thoughts remain sound and secure, that the seat of his reason suffer no impediment or change of any sort, and that he in no wise give his assent to his fright or pain. As for anyone who is not a sage, the first part applies to him but not the second. For in his case the impress of the emotions does not remain on the surface but penetrates through to the seat of his reason, infecting and corrupting it: he judges by his emotions and acts in conformity with them.

The state of the Stoic sage is fully and elegantly seen in the following:

Mens immota manet, lachrimae volvuntur inanes.

[His mind remains unmoved: empty tears do flow.]

The Aristotelian sage is not exempt from the emotions: he moderates them.[6]

6. All [C] here from St Augustine, *City of God*, IX, iv (following Aulus Gellius and citing Virgil, *Aeneid*, IV, 449).

13. Ceremonial at the meeting of kings

[Here Montaigne still considers his Essays as a 'rhapsodie' (that is, a 'confused medley' of disparate pieces strung together). His term also suggests that there is an element of extravagant irrationality involved in his work.]

[A] No topic is so vain that it does not deserve a place in this confused medley of mine.

Our normal rules lay down that it would be a marked discourtesy towards an equal and even more so towards the great if we were to fail to be at home after he has warned that he must pay us a visit. Indeed Queen Margaret of Navarre asserts in this connection that it would be impolite for a nobleman to leave his house even (as is frequently done) to go and meet the person who is paying the visit, no matter how great he may be – since it is more civil and more respectful to wait to receive him when he does arrive – if for no other reason, for fear of mistaking the road he will come by: it suffices that we accompany him when he takes his leave.

[B] Personally I often neglect both these vain obligations: in my home I have cut out all formalities. Does anyone take offence? What of it? It is better that I offend him once than myself all the time – that would amount to servitude for life! What is the use in fleeing from the slavery of the Court if we then go and drag it back to our lairs?

[A] The normal rule governing all our interviews is that it behoves the lesser to arrive at the appointment first, since it is the privilege of the more prominent to keep others waiting. Yet at the meeting arranged between Pope Clement and King Francis at Marseilles, the King first made all necessary arrangements and then withdrew from the town, allowing the Pope two or three days to effect his entry and to rest before he then returned to find him.[1]

It was the same at the entry of Pope and Emperor into Bologna: the Emperor made arrangements for the Pope to be there first, himself arriving afterwards.[2]

1. Cf. I, 10, note 2.
2. In 1532. (Francesco Guicciardini makes similar remarks about their meeting in 1529.)

It is said to be the normal courtesy when princes such as these arrange a conference that the greatest should arrive at the appointed place before all the others, and especially before the one on whose territory the meeting takes place. We incline to explain this as a way of showing that it is the greater whom the lesser are coming to visit: they call on him, not he on them.

[C] Not only does every country have its own peculiar forms of politeness but so does every city and every profession. From childhood I was quite carefully trained in etiquette and I have always lived in sufficiently good company not to be ignorant of the rules of our French variety: I could even teach it. I like to keep to those rules, but not so abjectly as to constrict my daily life. Some forms of politeness are bothersome; provided they are omitted with discretion and not out of ignorance, there is no loss of elegance. I have often seen men rude from an excess of politeness, men boring you with courtesies.

Nevertheless to know how to be elegantly at ease with people is a useful accomplishment: like grace and beauty, it encourages the hesitant beginnings of fellowship and intimacy; as a result it opens the way to our learning from the examples of others and to ourselves providing and showing an example, if it is worth noting and passing on.

14. That the taste of good and evil things depends in large part on the opinion we have of them

=====

[The [A] text of this chapter (in which Montaigne reflects on standard philosophical arguments, especially Stoic paradoxes on pain and death) seems to date from about 1572. Later additions make it more personal and, after his own experience of pain and distress, weaken the force of the Classical commonplaces. Already in germ here are arguments developed in 'An apology for Raymond Sebond' and the final chapter, 'On experience'. The moral concerns are restricted to the domain of philosophy, a domain in which revealed religion properly has no part to play.]

[A] There is an old Greek saying that men are tormented not by things themselves but by what they think about them.[1] If that assertion could be proved to be always true everywhere it would be an important point gained for the comforting of our wretched human condition. For if ills can only enter us through our judgement it would seem to be in our power either to despise them or to deflect them towards the good: if the things actually do throw themselves on our mercy why do we not act as their masters and accommodate them to our advantage? If what we call evil or torment are only evil or torment insofar as our mental apprehension endows them with those qualities then it lies within our power to change those qualities. And if we did have such a choice and were free from constraint we would be curiously mad to pull in the direction which hurts us most, endowing sickness, poverty or insolence with a bad and bitter taste when we could give them a pleasant one, Fortune simply furnishing us with the matter and leaving it to us to supply the form. Let us see whether a case can be made for what we call evil not being an evil in itself or (since it amounts to the same) whether at least it is up to us to endow it with a different savour and aspect.

If the original essence of the thing which we fear could confidently

---

1. A saying of Epictetus, collected by Stobaeus in his *Apophthegms* and inscribed by Montaigne in his library.

lodge itself within us by its own authority it would be the same in all men. For all men are of the same species and, in varying degrees, are all furnished with the same conceptual tools and instruments of judgement. But the diversity of the opinion which we have of such things clearly shows that they enter us only by means of compromises: one man in a thousand may perhaps lodge them within himself in their true essence, but when the others do so they endow them with a new and contrary essence.

Our main enemies are held to be death, poverty and pain. Yet everyone knows that death, called the dreadest of all dreadful things, is by others called the only haven from life's torments, our natural sovereign good, the only guarantor of our freedom, the common and ready cure of all our ills;[2] some await it trembling and afraid: others   [C]   bear it more easily than life.[3]   [B]   One man complains that death is too available:[4]

> *Mors, utinam pavidos vita subducere nolles,*
> *Sed virtus te sola daret.*

[O Death! Would that thou didst scorn to steal the coward's life; would that only bravery could win thee.]

But leaving aside such boasting valour, Theodorus replied to Lysimachus who was threatening to kill him, 'Quite an achievement, that, matching the force of a poisonous fly!'[5] We find that most of the philosophers either deliberately went to meet death or else hastened and helped it along.   [A]   And how many of the common people[6] can we see, led forth not merely to die but to die a death mixed with disgrace and grievous torments, yet showing such assurance (some out of stubbornness, others from a natural simplicity) that we may perceive no change in their normal behaviour: they settle their family affairs and commend themselves to those they love, singing their hymns, preaching and addressing the crowd – indeed even including a few jests and drinking the health of their acquaintances every bit as well as Socrates did. When one man was being led to the gallows he asked not to be taken through such-and-such a street: there was a tradesman there who might arrest him for an old debt!

2. Aristotle considers death as something to be most dreaded: the Stoics believe that (since any man can take his own life) it is the ultimate means of escaping pain, disgrace, defeat or other evils. Montaigne's ideas here are influenced by Seneca.
3. '80: Others, *do they not welcome it with quite different countenance?* One . . .
4. [B] Until [C]: too *cheap and* available . . .
   The following verse: Lucan, *Pharsalia*, IV, 580.
5. Cicero, *Tusc. disput.*, V, xl, 117.
6. [A] until [C]: common *ordinary* people . . .

Another asked the executioner not to touch his throat: he was ticklish and did not want to burst out laughing! When the confessor promised another man that he would sup that day at table with Our Lord, he said, 'You go instead: I'm on a fast.' Yet another asked for a drink; when the executioner drank of it first, he declined to drink after him – 'for fear of the pox'! And everybody knows that tale of the man of Picardy who was on the scaffold when they showed him a young woman who was prepared to marry him to save his life (as our laws sometimes allow): he gazed at her, noticed that she had a limp, and said, 'Run up the noose: she's lame!' A similar story is told of a man in Denmark, who was condemned to be beheaded: they offered him similar terms, but he refused the young woman they brought because she had sagging jowls and a pointed nose.[7] In Toulouse when a man-servant was accused of heresy, the only justification he would give for his belief was to refer to that of his master, a young undergraduate who was in gaol with him: he preferred to die rather than accept that his master could be mistaken. When King Louis XI took Arras, many of the citizens let themselves be hanged rather than cry 'Long live the King.'[8]

[C] Even today in the Kingdom of Narsinga the wives of their priests are buried alive with their dead husbands. All other wives are burned alive at their husbands' funeral, not merely with constancy but with gaiety. And when they cremate the body of their dead king, all his wives and concubines, his favourites and a multitude of dignataries and servants of every kind, trip so lightly towards the pyre to cast themselves into it with their master that they apparently hold it an honour to be his comrades in death.[9]

[A] Among the lowly souls of Fools, some have been found who refused to give up clowning even in death. When the hangman sent one of them swinging from the rope he cried out his regular catch-phrase: 'Let her run with the wind.' Another jester lay dying on a palliasse in front of the fire; the doctor asked him where it hurt: 'Between that bench and this fire,' he replied. And when the priest was about to administer the last rites and was fumbling about to anoint his feet (which were all twisted up and retracted), 'You will find them,' he said, 'at the end of my legs!' When exhorted to charge someone to intercede with God, he inquired, 'Is anyone going to see Him?' When the other replied, 'You will soon, if God so wishes,' he exclaimed: 'Now, if I could only get there by tomorrow

---

7. Series of jests straight from Henri Estienne, *Apologie pour Herodote*, 1566, p. 175.
8. Jean Bouchet, *Annales d'Acquitaine*.
9. Simon Goulart, *Hist. du Portugal*, Paris, 1587, IV, ii.

evening, I . . .' – 'Just think about your intercessions,' continued the other; 'You will be there soon enough.' – 'In that case I had better wait,' he said, 'and deliver my intercessions in person.'[10]

I have heard my father tell how places were taken and retaken so many times in our recent wars in Milan that the people became weary of so many changes of fortune and firmly resolved to die: a tally of at least twenty five heads of family took their own lives in one single week. That incident was similar to what occurred in the city of the Xanthians who, besieged by Brutus, rushed out headlong, men, women and children, with so furious an appetite for death that to achieve it they omitted nothing that is usually done to avoid it; Brutus was able to save but a tiny number.[11]

[C] Any opinion is powerful enough for somebody to espouse it at the cost of his life. The first article in that fair oath that Greece swore – and kept – in the war against the Medes was that every man would rather exchange life for death than Persian laws for Greek ones. In the wars of the Turks and the Greeks how many men can be seen preferring to accept the cruellest of deaths rather than to renounce circumcision for baptism?

That is an example which all religions are capable of. When the Kings of Castile banished the Jews from their lands, King John of Portugal sold them sanctuary in his territories at eight crowns a head, on condition that they would have to leave by a particular day when he would provide vessels to transport them to Africa. The day duly arrived after which they were to remain as slaves if they had not obeyed: but too few ships were provided; those who did get aboard were treated harshly and villanously by the sailors who, apart from many other indignities, delayed them at sea, sailing this way and that until they had used up all their provisions and were forced to buy others from them at so high a price and over so long a period that they were set ashore with the shirts they stood up in. When the news of this inhuman treatment reached those who had remained behind, most resolved to accept slavery; a few pretended to change religion. When Emmanuel, ['95] John's successor, [C] came to the throne he first set them all free; then he changed his mind, giving them time to void his kingdom and assigning three ports for their embarkation. When the good-will he had shown in granting them their freedom had failed to convert them to Christianity, he hoped (said Bishop Osorius, the best Latin historian of our times) that they would be brought to it by the hardship of

10. Bonaventure Des Périers, *Nouvelles récréations et joyeux devis*; end of the first *nouvelle*.
11. Plutarch, *Brutus*.

having to expose themselves as their comrades had done to thievish seamen and of having to abandon a land to which they had grown accustomed and where they had acquired great wealth, in order to cast themselves into lands foreign and unknown. But finding his hopes deceived and the Jews determined to make the crossing, he withdrew two of the ports he had promised in order that the length and difficulty of the voyage would make some of them think again – or perhaps it was to pile them all together in one place so as the more easily to carry out his design, which was to tear all the children under fourteen from their parents and to transport them out of sight and out of contact, where they could be taught our religion. This deed is said to have produced a dreadful spectacle, as the natural love of parents and children together with their zeal for their ancient faith rebelled against this harsh decree: it was common to see fathers and mothers killing themselves or – an even harsher example – throwing their babes down wells out of love and compassion in order to evade that law. Meanwhile the allotted time ran out: they had no resources, so returned to slavery. Some became Christians: even today a century later few Portuguese trust in their sincerity or in that of their descendants, even though the constraints of custom and of long duration are as powerful counsellors as any other.[12] Cicero says: *'Quoties non modo ductores nostri sed universi etiam exercitus ad non dubiam mortem concurrerunt?'* [How often have not only our generals but entire armies charged to their death?][13]   [B]   I witnessed one of my friends energetically pursuing death with a real passion, rooted in his mind by many-faceted arguments which I could not make him renounce; quite irrationally, with a fierce, burning hunger, he seized upon the first death which presented itself with a radiant nimbus of honour.

[A]   In our own times there are many examples of even children killing themselves for fear of some slight setback. (In this connection one of the Ancients said, 'What shall we not go in fear of if we fear what cowardice itself has chosen for its refuge?')[14] If I were to thread together a long list of

12. Jeronimo Osorio (da Fonseca), *Historia de rebus Emmanuelis Lusitanae regis virtute gestis*, Cologne, 1574, pp.6r° and 13r°. (Montaigne was himself descended from Iberian Jews.)

From 1595 onwards, this chapter became I, 40.

['95]; as any other. *In the town of Castelnaudary fifty Albigensian heretics all suffered themselves to be burned together, with resolute hearts, in one fire, rather than to disown their opinions* . . . Cicero says . . .

13. Cicero, *Tusc. disput.*, I, xxxvii.

14. Perhaps Seneca, whose Epistle LXX is devoted to suicide and makes similar points.

people of all sorts, of both sexes and of all schools of thought, who even in happier times have awaited death with constancy or have willingly sought it – not merely to fly from the ills of this life but in some cases simply to fly from a sense of being glutted with life and in others from hope of a better mode of being elsewhere – I would never complete it: they are so infinite in number that, in truth, I would find it easier to list those who did fear death. One case only, the philosopher Pyrrho happened to be aboard ship during a mighty storm; to those about him whom he saw most terrified he pointed out an exemplary pig, quite unconcerned with the storm; he encouraged them to imitate it.[15] Dare we conclude that the benefit of reason (which we praise so highly and on account of which we esteem ourselves to be lords and masters of all creation) was placed in us for our torment? What use is knowledge if, for its sake, we lose the calm and repose which we would enjoy without it and if it makes our condition worse than that of Pyrrho's pig? Intelligence was given us for our greater good: shall we use it to bring about our downfall by fighting against the design of Nature and the order of the Universe, which require each creature to use its faculties and resources for its advantage?

Fair enough, you may say: your rule applies to death, but what about want? And what have you to say about pain which [C] Aristippus, Hieronymus and[16] [A] the majority of sages judge to be the [C] ultimate [A] evil?[17] Even those who denied this in words accepted it in practice: Possidonius was tormented in the extreme by an acutely painful illness; Pompey came to see him and apologized for having picked on so inappropriate a time for hearing him discourse on philosophy: 'God forbid,' said Possidonius, 'that pain should gain such a hold over me as to hinder me from expounding philosophy or talking about it.' And he threw himself into the theme of contempt for pain. Meanwhile pain played her part and pressed hard upon him. At which he cried, 'Pain, do your worst! I will never say you are an evil!'[18] A great fuss is made about this story, but what does it imply about his contempt for pain? He is arguing about words: if those stabbing pangs do not trouble him, why does he break off what he was saying? Why does he think it so important not to *call* pain an evil?

All is not in the mind in his case. We can hold opinions about other

15. A frequently cited example going back to Diogenes Laertius' *Life of Pyrrho*.
16. Cicero, *Tusc. disput.*, II, vi, 15.
17. '80: *sovereign evil* . . .
18. Cicero, *Tusc. disput.*, II, XXV, 61.

things: here the role is played by definite knowledge. Our very senses are judges of that:

> *Qui nisi sunt veri, ratio quoque falsa sit omnis.*

> [If they are not true, then reason itself is totally false.][19]

Are we to make our flesh believe that lashes from leather thongs merely tickle it, or make our palate believe that bitter aloes is *vin de Graves*? In this matter, Pyrrho's pig is one of us: it may not fear death, but beat it and it squeals and cries. Are we to force that natural universal and inherent characteristic which can be seen in every living creature under heaven: namely, that pain causes trembling? The very trees seem to shudder beneath the axe.

The act of dying is the matter of a moment: it is felt only by our powers of reason:

> *Aut fuit, aut veniet, nihil est praesentis in illa;*

> [Death either was or is to come: nothing of the present is in her;][20]

> *Morsque minus poenae quam mora mortis habet.*

> [There is less pain in Death than in waiting for Death.]

Thousands of beasts and men are dead before they are threatened. In truth, what   [C]   we say we   [A]   chiefly fear[21] in death is what usually precedes it: pain.

[C]   Yet if we are to believe a holy Father of the Church, *'Malam mortem non facit, nisi quod sequitur mortem.'* [Death is no evil, except on account of what follows it.][22] And I would maintain with greater likelihood that neither what precedes it nor what follows it appertains to death. Our self-justifications are false: I find from experience that it is our inability to suffer the thought of dying which makes us unable to suffer the pain of it, and that the pain we do suffer is twice as grievous since it threatens us with death. But as reason condemns our cowardice in fearing something so momentary, so unavoidable, so incapable of being felt as death is, we seize upon a more pardonable pretext.

---

19. Lucretius, IV, 485.
20. Etienne de La Boëtie, *Poèmes*, ed. Montaigne, p. 233; addressed to Montaigne.
21. Ovid, *Heroïdes*; 'Epistle of Ariana to Theseus', 82;
    [A]: what *the sages* chiefly fear . . .
22. St Augustine, *City of God*, I, ii (adapted).

We do not put on the danger list any painful ailment which comports no danger apart from the pain itself. Since toothache or gout, however painful, are not fatal nobody really counts them as illnesses. So let us concede that where dying is concerned we are chiefly concerned with the pain, [A] just as in poverty there is nothing to fear except the fact that it throws us into the embrace of pain by the thirst, hunger, cold and heat and the sleepless watches that we are made to suffer.

And so let us concern ourselves only with pain. I grant that pain is the worst disaster that can befall our being. I willingly do so, for, of all men in the world. I am the most ill-disposed toward pain and [C] flee[23] [A] it all the more for having had little acquaintance with it, thank God. But it lies within us not to destroy pain but at least to lessen it by patient suffering and, even if the body be disturbed by it, still to keep our reason and our soul well-tempered.

If this were not so, what could have brought us to respect manly courage, valour, fortitude, greatness of soul and determination? If there were no pain to defy, how could they play their part? *'Avida est periculi virtus'.* [Manly courage is avid for danger.][24] If we did not have to sleep rough, endure in full armour the midday sun, make a meal of horseflesh or donkey, watch as they cut us open to extract a bullet buried between our bones, allow ourselves to be stitched up again, cauterized and poked about, how could we ever acquire that superiority which we aspire to have over the common people? Fleeing pain and evil is not at all what the sages counsel – they say that among indifferent actions it is more desirable to perform the one which causes us most trouble. [C] *'Non enim hilaritate, nec lascivia, nec risu, aut joco comite levitatis, sed saepe etiam tristes firmitate et constantia sunt beati.'* [For happiness is to be found not in gaiety, pleasure, laughing, nor in levity the comrade of jesting: those are happy, often in sadness, who are constant and steadfast.][25] [A] That was why it was impossible to convince our forebears that conquests made by force of arms at the hazard of war were not superior to those safely won by intrigues and plotting:

> *Laetius est quoties magno sibi constat honestum.*

[Whenever virtue costs us dear, our joy is greater.][26]

23. '80: and *fear* it . . .
   [A] was written before the onset of Montaigne's colic paroxysms.)
24. Seneca, *De providentia*, VI.
25. Cicero, *De finibus*, II, xx, 65–6.
26. Lucan, *Pharsalia*, IX, 404.

In addition, it ought to console us that, by Nature, 'if pain is violent it is short; if long, light' — [C] 'si gravis brevis, si longus levis.'[27] [A] You will not feel it for long if you feel it grievously: either it will quench itself or quench you, which amounts to the same thing.[28] [C] If you find it unbearable, it will bear you away. 'Memineris maximos morte finiri; parvos multa habere intervalla requietis; mediocrum nos esse dominos: ut si tolerabiles sint feramus, sin minus, e vita, quum ea non placeat, tanquam e theatro exeamus.' [Remember that the greatest pains are ended by death, the smaller ones allow us periods of repose; and we are masters of the moderate ones, so that if they are bearable we shall be able to bear them; if they are not, when life fails to please us, we may make our exit as from the theatre.][29]

[A] What causes us to be so impatient of suffering is that we are not used to finding our [C] principal [A] happiness in the soul, [C] nor to concentrating enough on her, who alone is the sovereign Lady of our actions and of our mode of being. The body knows only differences of degree: otherwise it is of one uniform disposition: but the soul can be diversified into all manner of forms; she reduces all bodily sensations and all physical accidents to herself and to whatever her own state may be. That is why we must study her, inquire into her and arouse in her her almighty principles. No reasoning power, no commandment, no force can override her inclination or her choice. She is capable of inclining a thousand ways: let us endow her with an inclination which conduces to our rest and conversation: then we are not only protected from any shocks but, if it pleases her, we are delighted and flattered by those pains and shocks.

All things indifferently can be turned to profit by the soul: even errors and dreams can serve her as matter to be loyally used to protect us and to make us contented. It can easily be seen that what gives their edge to pain and pleasure is the hone of our mind. The beasts, since they leave them to the body while leading the mind by the nose, have feelings which are free, natural and therefore virtually the same in all species, as we can see from the similarity of their reactions. If we were to refrain from disturbing the jurisdiction which our members rightly have in such matters, it is to be believed that we would be better off and that Nature has endowed them with a just and moderate temperament towards pleasure and pain. Nature,

27. Cicero, De finibus, II, xxix, 95, translated in the text before quotation.
28. Seneca, Epist. moral., LXXVIII, 17.
29. Cicero, De finibus, I, xv, 49.

being equal and common to all, cannot fail to be just. But since we have unslaved ourselves from Nature's law and given ourselves over to the vagrant liberty of our mental perceptions, the least we can do is to help ourselves by making them incline towards the most agreeable direction. Plato is afraid of our bitter enslavement to pain and to pleasure, since they too firmly bind and shackle our souls to our bodies; I, on the contrary, because they release them and strike them free. [A] Just as[30] an enemy is made fiercer by our flight, so pain too swells with pride as we quake before her. If we withstand her she will make a compact on far better terms. We must brace ourselves against her. By backing away in retreat we beckon her on, drawing upon ourselves the very collapse which we are threatened by. [C] When tense, the body is firmer against attacks; so is the soul.

[A] But let us to come to those examples (which are proper hunting for weak-backed men like me) in which we find that it is with pain as with precious stones which take on brighter or duller hues depending on the foil in which they are set: pain only occupies as much space as we make for her. Saint Augustine says, '*Tantum doluerunt quantum doloribus se inseruerunt.*' [The more they dwelt on suffering, the more they felt it.][31] We feel the surgeon's scalpel ten times more than a cut from a sword in the heat of battle. The pangs of childbirth are reckoned to be great by doctors and by God himself;[32] many social conventions are there to help us to get through them: yet there are whole nations who take no heed of them. To say nothing of the women of Sparta, what difference does childbirth make to the wives of the Swiss guards in our infantry, except that today you can see them plodding after their husbands, bearing on their back the child they bore yesterday in their belly? And Gypsy women (not real Egyptians; but ones recruited from among ourselves) go and personally bathe their new-born infants in the nearest stream and then wash themselves. [C] Apart from the many sluts who daily conceive and deliver their children in secret, there was that good wife of the Roman patrician Sabinus who (for the sake of others) gave birth to twins and endured the labouring pains, alone, without help, without a word and without a groan.[33]

[A] Why, a little Spartan boy had stolen a fox: [C] (Spartans were

30. '80: happiness in the soul *and to have too much commerce with the body.* Just as . . .
    (Cf. Seneca, *Epist. moral.*, LXXVIII. In. [C], cf. Plato, *Phaedo*, 66B ff.
31. St Augustine, *City of God*, I, x (adapted).
32. *Esdras* 13:8; John 16:21.
33. Plutarch, tr. Amyot, *De l'amour*, XXXIV, p. 613C.

more afraid of being mocked for having botched a theft than we are of being punished for one):   [A]   he stuffed it[34] under his cloak and rather than betray himself let it gnaw into his belly. Another lad was carrying incense for the sacrifice when a live coal fell up his sleeve: he let it burn through to the bone so as not to disturb the ceremony. When, according to their educational practices, an assay was made of the bravery of boys at the age of seven, many let themselves be flogged to death rather than change their expressions.   [C]   Cicero saw crowds fighting each other with feet, fists and teeth, till they collapsed with exhaustion rather than admit defeat. 'Nunquam naturam mos vinceret: est enim ea semper invicta; sed nos umbris, deliciis, otio, languore, desidia animum infecimus; opinionibus maloque more delinitum mollivimus.' [Never could habit conquer Nature: Nature is unconquerable; yet we have corrupted our souls with unrealities, luxuries, leisure, idleness, listlessness and sloth; we have made them soft with opinions and evil habits.][35]

[A]   Everyone knows the story of Scaevola who had slipped into the enemy's camp to kill their leader: having failed in this attempt he thought of a strange way to complete his task and deliver his country: not only did he confess his purpose to Porsenna (the king he sought to kill) but added that he had a great number of Roman accomplices within the camp; to show what sort of man he was, a brazier was brought in: he suffered his arm to be grilled and roasted: he stood watching it until that enemy himself, in horror, ordered the brazier removed.[36]

What about that man who would not condescend to break off reading his book when they cut him open? And what of that other man who persisted in laughing at the ills done to him and in mocking them until his incensed and cruel torturers, having vainly invented new torments and increased them one after another, had to admit that he was the winner?[37]

'But then, he was a philosopher' − Yes, but when one of Caesar's gladiators suffered his wounds to be probed and cut open he kept on laughing.[38]   [C]   'Quis mediocris gladiator ingemuit; quis vultum mutavit unquam? Quis non modo stetit, verum etiam decubuit turpiter? Quis cum

34. '80: fox (for theft was a virtuous deed for them, but with the proviso that it was more disgraceful to be caught than it is with us): he stuffed . . .
35. Plutarch, tr. Amyot, Life of Lycurgus, xiv; Cicero, Tusc. disput., V, xxvii, 77. (Theme taken up again in the Essays, I, 23.)
36. Cf. Seneca, Ep. moral., XXIV, 5.
37. Seneca, Ep. moral., LXXVIII, 18–19.
38. Aulus Gellius, XII, xvii, 41.

decubuisset, ferrum recipere jussus, contraxit?' [What quite ordinary gladiator has ever made a groan? Which has ever changed his expression? Which has ever behaved shamefully whether still on his feet or beaten to the ground? And having fallen, which of them ever withdrew his neck when ordered to receive the sword?][39]

[A] Women can be brought in to this as well. Who has not heard of that woman of Paris who had herself flayed alive merely to acquire a fresh colour from a new skin? Women have been known to have good sound teeth extracted so as to rearrange them in a better order or in the hope of making their voices softer or fuller. How many examples of contempt for pain does that sex supply! Provided they can hope to enhance their beauty, what do they fear? What can they *not* do?

> [B] *Vellere queis cura est albos a stirpe capillos,*
>   *Et faciem dempta pelle referre novam.*

[Their labour consists in plucking out white hairs and scraping off skin to put on a new face.][40]

[A] I have known women who swallowed sand and ashes, specifically striving to ruin their digestions in order to acquire a pallid hue. And to appear slim in the Spanish fashion what tortures will they suffer, with corsets and braces cutting into the living flesh under their ribs – sometimes even dying from it.

[C] Many peoples in our own times commonly inflict deliberate wounds on themselves to give credit to their oaths – our own King tells of several memorable examples of this which he saw in Poland and in which he was involved.[41] I know some men who imitated that in France; apart from which I personally saw a girl who, to prove the earnestness of her promises as well as her constancy, took the pin she wore in her hair and jabbed herself four or five times in the arm, breaking the skin and bleeding herself in good earnest. The Turks sport great scars for their ladies; to make the marks permanent they immediately apply hot irons to their wounds to staunch the blood and form the scab, keeping them there an incredible time. Those who have seen it have written sworn depositions about it for

---

39. Cicero, *Tusc. disput.*, II, xvii, 41.
40. Tibullus, I, viii, 45–6 (adapted).
41. Written before the death of Henry II of France in 1589; he was King of Poland in 1573 and 1574.

me. Why, you can find Turks any day who will make a deep gash in their arms or their thighs for a mere ten aspers.[42]

[A] I am pleased to find martyrs nearer to hand where we need them more: Christendom provides us with plenty. Following the guidance of our Holy Ensample many, from devotion, have taken up the cross. From a most reliable witness we learn that Saint Louis, when king, wore a hair-shirt until his confessor gave him a dispensation in his old age; every Friday he made his priest flog his shoulders with five iron chains, which, for this purpose, he always bore about with him in a box.[43] Guillaume, our last Duke of Guyenne (father of that Aliénor who transmitted the Duchy to the houses of France and England) throughout the last ten or twelve years of his life, as a penance, continuously wore an armoured breast-plate under a monk's cloak.[44] Count Foulk of Anjou went all the way to Jerusalem to be scourged by two of his manservants as he stood with a rope round his neck before Our Lord's Sepulchre.[45] And in various places on Good Fridays do we not still find many men and women flagellating themselves, tearing into their flesh and cutting it to the bone? I have often seen it: there was no trickery. Since they wear masks some are said to witness to another's devotion in return for cash, showing an even greater contempt for pain in that the spur of devotion is greater than the spur of avarice.

[C] With calm faces, betraying no signs of grief, Quintus Maximus buried his son the Consul; Marcus Cato buried his son the Praetor elect; and Lucius Paulus, both of his sons within a few days of each other. Some time ago I said (as a quip) that one particular man had even cheated God of his justice: in one day three of his grown-up sons met violent deaths – which we may believe to have been sent to him as a bitter chastisement; yet he took it almost as a blessing.[46] I myself have lost two or three children, not without grief but without brooding over it; but they were still only infants. Yet hardly anything which can befall men cuts them

---

42. Guillaume Postel, *Des Histoires Orientales*, Paris, 1575, p. 228. The girl mentioned above is further situated in ['95] and could well be Mlle de Gournay.

['95]: apart from which, *when I came to those famous Estates meeting at Blois, I had seen* a girl *beforehand in Picardy* who, to prove . . .

43. This is confirmed by Joinville, *Histoire et cronique du Roy S. Loys*, XCIV.

44. Jean Bouchet, *Annales d'Acquitaine*, Poitiers, 1557, p. 75r°

45. Foulke III, who died in 1040.

46. Montaigne's diary suggests this was his friend the Comte de Foix, whose three sons were killed near Agen, 29 July 1587.

['95]: as a *special* blessing *from heaven. I do not follow such monstrous humours but* I myself . . .

more to the quick. I have observed several other misfortunes which commonly cause great affliction, but which I would hardly notice if they happened to me – and when they have done so I have been so contemptuous of some which other people consider to be hideous that I would prefer not to boast of it in public without managing a blush: *'Ex quo intelligitur non in natura, sed in opinione esse aegritudinem.'* [From which we may learn that grief lies not in nature but opinion][47]

[B] Opinion is a bold and immoderate advocate. Who ever sought security and repose as avidly as Alexander and Caesar sought insecurity and hardships? Teres, the father of Sitalces, used to say that when he was not waging war he felt that there was no difference between him and his stable-boy.[48]

[C] When Cato the Consul sought to secure a number of Spanish towns, many of their citizens killed themselves simply because he forbade them to bear arms: *'ferox gens nullam vitam rati sine armis esse'* [a fierce people who thought not to bear arms was not to live].[49]

[D] How many do we know of who have fled from the sweetness of a calm life at home among people they knew in order to undergo the horrors of uninhabitable deserts, throwing themselves into conditions abject, vile and despised by the world, delighting in them and going so far as to prefer them![50]

Cardinal Borromeo who recently died in Milan was surrounded by debauchery; everything incited him to it: his rank, his immense wealth, the atmosphere of Italy and his youth; yet he maintained a way of life so austere that the same garment served him winter and summer; he slept only on a palliasse; any time left over after the duties of his office he spent on his knees studying, with some bread and water set beside his book – which was all the food he ever took and the only time he did so.

I have even met men who have knowingly secured profit and preferment from letting themselves be cuckolded – yet that very word terrifies many.

Sight may not be the most necessary of our senses but it is the most pleasurable; the most useful and pleasurable of our limbs are those which serve to beget us, yet quite a few men have been seized with a mortal hatred for them simply because they do afford us such pleasure: they

47. Cicero, *Tusc. disput.*, III, xxviii, 71.
48. Plutarch, tr. Amyot, *Dicts notables des anciens Roys, princes et capitaines*, p. 189D.
49. Livy, XXXIV, xvii.
50. Montaigne is alluding to ascetic anchorites.

rejected them *because* of their value and worth; the man who plucked out his own eyes held the same opinion.[51]

[C]   An abundance of children is a blessing for the greater, saner, part of mankind: I and a few others find blessings in a lack of them. When Thales was asked why he did not get married, he replied that he did not want to leave any descendants.[52]

That it is our opinion which confers value can be seen from those many things which we do not even bother to look at when making our judgements, looking, rather, at ourselves: we consider neither their intrinsic qualities nor the uses they can be put to but only what it cost us to procure them – as if that were a part of their substance: in their case value consists not in what they give to us but in what we gave for them. While on this subject, I realize that where our expenditure is concerned we are good at keeping accounts: our outgoings cost us so much trouble, and we value them precisely because they do so; our opinion will never allow itself to be undervalued. What gives value to a diamond is its cost; to virtue, its difficulty; to penance, its suffering; to medicines their bitter taste.

[B]   To attain poverty one man cast his golden coins into that self-same sea which others ransack to net and catch riches.[53] Epicurus said that being rich does not alleviate our worries: it changes them.[54] And truly it is not want that produces avarice but plenty. I would like to tell you my experience of this. Since I grew up I have known three changes of circumstance: the first period (which lasted about twenty years) I spent with only a sporadic income, being at the orders of other people and dependent upon their help; I had no fixed allowance; nothing was laid down for me. I spent my money all the more easily and cheerfully precisely because it depended on the casualness of fortune. I have never lived better: never once did I find my friends' purses closed to me, since I had convinced myself that none of my other wants exceeded my wanting to pay back loans on the agreed date. Seeing the efforts I made to do so, the terms were extended hundreds of times; I acquired thereby a somewhat spurious reputation for punctilious husbandry. It is in my nature to like paying my debts, as though I were casting off my shoulders that very image of slavery, a weighty burden; in addition I experience a certain

51. Among those who had gelded themselves was Origen. Montaigne believed that Democritus had blinded himself (cf. I, 29; II, 12). Textor cites this after Lucretius in his *Officina* (s.v. *Caeci et Excaecati*).
52. Diogenes Laertius, *Life of Thales*, I, xxvi, 28.
53. Aristippus. Cf. Horace, *Satires*, II, iii, 99–110.
54. Seneca, *Ep. moral.*, XVII, 11.

pleasure in satisfying others and behaving justly. I make an exception however for the kind of repayments which involve haggling and bargaining: unless I can find somebody to do that job for me I wrongly and disgracefully put it off as long as possible, fearing the sort of quarrel which is totally incompatible with my humour and my mode of speaking. There is nothing I hate more than bargaining. It is a pure exchange of trickery and effrontery. after hours of arguing and haggling both sides go back on their pledged word to gain a few pence more. So I was always at a disadvantage when asking for a loan: I had no wish to make my request in person and relied on letters – a chancy business which is lacking in drive and actually encourages a refusal. Arrangements for my needs I used to leave light-heartedly to the stars – more freely than I later did to my own foresight and good sense.

Most thrifty people reckon that living in such uncertainty must be horrible. In the first place they fail to realize that most people have to do so. And how many honourable men have cast all their security overboard (and still do so) seeking favourable winds from Prince or Fortune? To become Caesar, Caesar borrowed a million in gold over and above what he possessed.[55] And how many merchants begin trading by selling up their farms and dispatching it all to the Indies,

> *Tot per impotentia freta!*
>
> [Over so many raging seas!][56]

And even now, in the present dearth of charity, countless thousands of religious houses live properly, expecting every day from the bounty of Heaven whatever they need for dinner.

In the second place, people fail to realize that they base themselves on a certainty which is hardly less uncertain and chancy than chance itself. I can see Want lurking beyond an income of two thousand crowns as readily as if she were right beside me. [C] Fate[57] [B] can make a thousand breaches for poverty to find a way into our riches; [C] often there is no intermediate state between the highest and the lowest fortunes:

> *Fortuna vitrea est: tunc cum splendet frangitur.*
>
> [Fortune is glass: it glitters, then it shatters.][58]

[B] Fate can send our dykes and ramparts a-toppling arse over tip;

55. Plutarch, *Life of Julius Caesar*.
56. Catullus, IV, 18.
57. '88: right beside me. *Fortune* can make . . . (*Sors* replacing *Fortune*)
58. Publius Syrus, cited by Justus Lipsius, *Politici*, V, xviii.

moreover I find that need, for a thousand diverse reasons, can make a home with those who have possessions as often as with those who have none; she is even perhaps less troublesome when she dwells with us alone than when we meet her accompanied by all our riches. [C] Riches are more a matter of ordinate living than of income: *'Faber est suae quisque fortunae.'* [By each man is his fortune wrought.]⁵⁹ [B] And it seems to me that a rich man who is worried, busy and under necessary obligations is more wretched than a man who is simply poor. [C] *'In divitiis inopes, quod genus egestatis gravissimum est.'* [Poverty amidst riches is the most grievous form of want.]⁶⁰ The greatest and richest of princes are regularly driven to extreme necessity by poverty and lack of cash: for what necessity is more extreme than that which turns them into tyrants, unjustly usurping the property of their subjects?

[B] In my second stage [C] I did have money. Becoming attached to it⁶¹ [B] I soon set aside savings which were considerable for a man of my station, never reckoning a man to have anything except what was over and above his regular outgoings and never believing that he should count on what he hopes to get, however clear that hope may be. 'What if such-and-such a mishap occurred,' I would say, 'and took me by surprise?' Then, as a result of these vain and vicious thoughts I would ingeniously strive to provide against all eventualities with what I had saved and put aside. I had a reply ready for anyone who maintained that the number of mishaps was infinitely great: 'I provided against some if not against all.' None of this happened without painful anxiety. [C] I made a secret of it: I, who dare talk so much about myself, never talked truthfully about my money – like those rich who act poor and those poor who act rich, their consciences dispensed from witnessing truly to what they own.

What ridiculous and shameful wisdom! [B] Was I setting out on a journey? I never thought I had made adequate provision. The heavier my money the heavier my worries, wondering as I did whether the roads were safe and then about the trustworthiness of the men in charge of my baggage; like others that I know, I was only happy about it when I had it before my eyes. When I left my strong-box at home, what thoughts and suspicions I had, sharp thorny ones and, what is worse, ones I could tell nobody about. My mind forever dwelt on it. [C] When you tot it all up, there is more trouble in keeping money than in acquiring it. [B] And

59. Sallust, *De republica*, I, 1; cited there as from Appuleius.
60. Seneca, *Epist. moral.*, LXXIV, 4 (adapted).
61. '88: stage, *I did have goods.* Becoming so *hotly* attached to *them*, I . . .

even if I did not actually do all I have just said, stopping myself from doing so cost me dear. I got little profit out of my savings: [C] I had more to spend, but the spending weighed [B] no less heavily on me, for as Bion said, when it comes to plucking out hairs, it hurts the balding no less than the hairy:[62] once you have grown used to having a pile of a certain size and have set your mind on doing so, you can no longer use it: [C] you do not even want to slice a bit off the top. [B] It is the kind of structure – so it seems to you – that would tumble down if you even touched it. For you to broach it, Necessity must have you by the throat. Formerly I would pawn my furniture and sell my horse far less unwillingly and with less regret than I would ever have made a breech in that beloved purse which I kept in reserve. The danger lies in its not being easy to place definite limits on such desires – [C] limits are hard to discover for things which seem good – [B] and so to know when to stop saving. You go on making your pile bigger, increasing it from one sum to another until, like a peasant, you deprive yourself of the enjoyment of your own goods: your enjoyment consists in hoarding and never actually using it [C] If this is 'using' money, then the richest in cash are the guards on the walls and gates of a goodly city! To my way of thinking, any man with money is a miser.

Plato ranks physical or human goods thus: health, beauty, strength, wealth. And wealth, he says, is not blind but extremely clear-sighted when enlightened by wisdom.[63]

[B] The Younger Dionysius acted elegantly in this regard.[64] They told him that one of the men in his city of Syracuse had buried a hidden treasure. He commanded him to bring it to him. Which he did, secretly keeping back a part which he went off to spend in another city, where he lost his taste for hoarding money and began to live more expensively. When Dionysius learned of this he sent him the remainder of his treasure, saying that he willingly returned it now he had learned how to use it.

I remained like this for [C] a few years; then some good *daemon* or other [B] cast me out[65] of it most usefully – like that man of Syracuse – and scattered all my parsimony to the winds, when the joyful undertaking of a certain very expensive journey sent all those silly notions tumbling down.

62. Seneca, *De Tranquillitate animi*, VIII.
63. Plato, *Laws*, I, 1, 631B–D.
64. Or rather, the Elder Dionysius: Plutarch (tr. Amyot), *Les dictz notables des anciens Roys, Princes, et grands capitaines*, p. 190E–F.
65. '80: for *four or five* years: some good *Fortune* or other cast me out . . .

That is how I dropped into a third way of life which – and I say what I really feel – is far more enjoyable, certainly, and also more orderly: I make my income and my expenditure run along in tandem: sometimes one pulls ahead, sometimes the other, but only drawing slightly apart. I live from day to day, pleased to be able to satisfy my present, ordinary needs: extraordinary ones could never be met by all the provision in the world.

[C] And it is madness to expect that Fortune will ever supply us with enough weapons to use against herself. We have to fight with our own weapons: fortuitous ones will let us down at the crucial moment.   [B] If I do save up now, it is only because I hope to use the money soon – not to purchase lands [C] that I have no use for [B] but to purchase pleasure.   [C] 'Non esse cupidum pecunia est, non esse emacem vectigal est.' [Not to want means money: not to spend means income.]⁶⁶   [B] I have no fear, really, that I shall lack anything: nor   [C] have I any wish for more. 'Divitiarum fructus est in copia, copiam declarat satietas.' [The fruit of riches consists in abundance: abundance is shown by having enough.]   [B] I particularly congratulate myself that this amendment of life should have come to me at an age which is naturally inclined to avarice, so ridding me of a vice⁶⁷ – the most ridiculous of all human madnesses – which is so common among the old.

[C] Pheraulas had experienced both kinds of fortune: he found that an increase in goods did not mean an increased appetite for eating, drinking, sleeping or lying with his wife; on the other hand he did find that the importunate cares of running his estates pressed heavy on his shoulders (as it does on mine); he decided to make one of his loyal friends happy – a poor young man always on the track of riches: he made him a present of all his own wealth, which was extremely great, as well as of everything which was daily accruing to him from the generosity of his good master Cyrus and also from the wars: the condition he made was that the young man should undertake to maintain him and feed him as an honoured guest and friend. Thus they lived thereafter in great happiness, both equally pleased with the change in their circumstances.⁶⁸

That is a course I would heartily love to imitate! And I greatly praise the lot of an old Bishop whom I know to have so purely and simply entrusted his purse, his revenue and his expenditures to a succession of chosen

---

66. Cicero, *Paradoxa*, VI, iii, 49.
67. Until [C]: a vice, *which I have always held to be the least excusable and* the most ridiculous . . .
68. Cf. Xenophon, *Cyropaedia*, VIII, iii, 40.

servants, that he has let many long years flow by, as ignorant as an outsider of the financial affairs of his own household.[69] Trust in another's goodness is no light testimony to one's own: that is why God looks favourably on it. And where that Bishop is concerned I know no household which is run more smoothly nor more worthily than his.

Blessed is the man who has ordered his needs to so just a measure that his riches suffice them without worrying him or taking up his time, and without the spending and the gathering breaking into his other pursuits which are quiet, better suited and more to his heart.

[B] So ease or indigence depend on each man's opinion: wealth, fame and health all have no more beauty and pleasure than he who has them lends to them. [C] For each man good or ill is as *he* finds. The man who is happy is not he who is believed to be so but he who believes he is so: in that way alone does belief endow itself with true reality.

Neither good nor ill is done to us by Fortune: she merely offers us the matter and the seeds: our soul, more powerful than she is, can mould it or sow them as she pleases, being the only cause and mistress of our happy state or our unhappiness. [B] Whatever comes to us from outside takes its savour and its colour from our internal attributes, just as our garments warm us not with their heat but ours, which they serve to preserve and sustain. Shelter a cold body under them and it will draw similar services from them for its coldness: that is how we conserve snow and ice.[70] [A] Study to the lazy, like abstinence from wine to the drunkard, is torture; frugal living to the seeker after pleasure, like exercise to the languid idle man, is torment: so too for everything else. Things are not all that painful nor harsh in themselves: it is our weakness, our slackness, which makes them so. To judge great and lofty things we need a mind which is like them: otherwise we attribute to them the viciousness which belongs to ourselves. A straight oar seems bent in water. It is not only seeing which counts: how we see counts too.[71]

Come on then. There are so many arguments persuading men in a variety of ways to despise death and to endure pain: why do we never find a single one which applies to ourselves? Thoughts of so many different kinds have persuaded others: why cannot we each find the one that suits

---

69. Perhaps Prevost de Sansac, Archbishop of Bordeaux, a contemporary of Montaigne's.
70. Cf. Plutarch, tr. Amyot, *Du vice et de la vertu*, I, 38B.
71. Themes developed in 'An apology for Raymond Sebond' (II, 12): cf. the bent oar. Here Montaigne is translating from Seneca, *Epist. moral.*, LXXI, 23–6.

our own disposition? If a man cannot stomach a strong purgative and root out his malady, why cannot he at least take a lenitive and relieve it? [C] *'Opinio est quædam effeminata ac levis, nec in dolore magis, quam eadem in voluptate: qua, cum liquescimus fluimusque mollitia, apis aculeum sine clamore ferre non possumus. Totum in eo est, ut tibi imperes.'* [As much in pain as in pleasure, our opinions are trivial and womanish: we have been melted and dissolved by wantonness; we cannot even endure the sting of a bee without making a fuss. Above all we must gain mastery over ourselves.][72]   [A]   We cannot evade Philosophy by immoderately pleading our human frailty and the sharpness of pain: Philosophy is merely constrained to [C] have recourse to her unanswerable counter-plea: [A] 'Living in necessity is bad: but at least there is no necessity that you should go on doing so.'[73]   [C]   No one suffers long, save by his own fault. If a man has no heart for either living or dying; if he has no will either to resist or to run away: what *are* we to do with him?

72. Cicero, *Tusc. disput.*, II, xxii, 52.
73. Seneca, *Epist. moral.*, XII, 10: the great Stoic commonplace making suicide the ultimate recourse of the wise man.
    Until [C]: constrained to *pay us with the following*: 'Living . . .

15. One is punished for stubbornly defending a fort without a good reason

[A brief consideration of the limits placed on stubborn bravery by the rules of contemporary warfare. Exceptionally, all Montaigne's examples are modern ones, with no reference to antiquity.]

[A] Like all other virtues valour has its limits: overstep them, and you tread the path of vice; consequently a man may go right through the dwelling-place of valour into rashness, stubbornness and madness if he does not know where those boundaries lie: yet at their margins they are not easy to pick out. From this consideration was born the custom observed in warfare of punishing even by death those who stubbornly persist in the defence of a fort when, by the very rules of war, it cannot be sustained. Otherwise if there were hope of escaping punishment whole armies would be held up by chicken-coops. At the siege of Pavia, My Lord the Constable de Montmorency was required to cross the Ticino and to take up position in the suburbs of San Antonio: he was delayed by a tower at the foot of a bridge which stubbornly held out until battered down: he hanged every man inside. Another time he accompanied My Lord the Dauphin on his Transalpine expedition. The castle at Villano was taken by force and all the defenders hacked to pieces by the soldiers in their frenzy, excepting only the Captain and his ensign; for the same reason he ordered both of them to be strangled to death by hanging. Captain Martin Du Bellay, when Governor of Turin in the same territory, similarly hanged Captain Saint-Bony after all his men had been massacred at the taking of his fort.[1] Yet judgements about the strength or weakness of a fort depend upon estimates of the relative strength of the attacking forces. A man could justly be obstinate when faced with a couple of culverins who would be out of his mind if he resisted thirty cannons. And where you have to take into account the greatness of the conquering prince, the reputation he has and the respect due to him, there is a risk of the balance being weighted in his

1. Several borrowings from the Du Bellay *Mémoires*, (II, 61 and VIII, 267).

favour; that is why some have so great an opinion of themselves and of their resources that they deem it unreasonable that anyone whatsoever be thought worthy of resisting them: when any are found doing so they put them all to the sword . . . while their fortune lasts. That can be seen in the form of defiance and the summonses to surrender used by Eastern potentates² and their successors today: they are proud, arrogant and full of barbaric assertiveness.   [C]   And in the regions where the Portuguese first penetrated into the Indies they discovered lands where it is universally held to be an inviolable law that an enemy defeated in the presence of the King or his Viceroy is excluded from any consideration of ransom or mercy.³

[B]   Above all, then, you must avoid (if you can!) falling into the hands of a judge who is your enemy, victorious and armed.

2. Until [C]: Eastern potentates, *the Tamberlanes, Mahomets* and their . . .
3. Simon Goulart, *Histoire du Portugal*, XIV, xv.

16. On punishing cowardice

[Renaissance Jurisconsults such as Tiraquellus were concerned to temper the severity of the Law by examining motives and human limitations. Montaigne does so here in a matter of great concern to gentlemen in time of war.]

[A] I once heard a prince, a very great general, maintain that a soldier should not be condemned to death for cowardice: he was at table, being told about the trial of the Seigneur de Vervins who was sentenced to death for surrendering Boulogne.

In truth it is reasonable that we should make a great difference between defects due to our weakness and those due to our wickedness. In the latter we deliberately brace ourselves against reason's rules, which are imprinted on us by Nature; in the former it seems we can call Nature herself as a defence-witness for having left us so weak and imperfect. That is why a great many[1] people believe that we can only be punished for deeds done against our conscience: on that rule is partly based the opinion of those who condemn the capital punishment of heretics and misbelievers as well as the opinion that a barrister or a judge cannot be arraigned if they fail in their duty merely from ignorance.

Where cowardice is concerned the usual way is, certainly, to punish it by disgrace and ignominy. It is said that this rule was first introduced by Charondas the lawgiver, and that before his time the laws of Greece condemned to death those who had fled from battle, whereas he ordered that they be made merely to sit for three days in the market-place dressed as women:[2] he hoped he could still make use of them once he had restored their courage by this disgrace − [C] '*Suffundere malis hominis sanguinem quam effundere.*' [Make the blood of a bad man blush not gush.][3]

[A] It seems too that in ancient times the laws of Rome condemned deserters to death: Ammianus Marcellinus tells how the Emperor Julian

---

1. Until [C] the misprint *peu de gens* (for *prou* de gens) made this read: *few* people (which inverts the sense).
2. Diodorus Siculus, *Histoires* (tr. Amyot), XII, ix.
3. Tertullian, cited by Justus Lipsius, *Adversus dialogistam*, III.

condemned ten of his soldiers to be stripped of their rank and then suffer death, 'following,' he said, 'our Ancient laws'. Elsewhere however Julian for a similar fault condemned others to remain among the prisoners under the ensign in charge of the baggage.[4] [C] Even the harsh sentences decreed against those who had fled at Cannae and those who in that same war had followed Gnaeus Fulvius in his defeat did not extend to death.

Yet it is to be feared that disgrace, by making men desperate, may make them not merely estranged but hostile.

[A] When our fathers were young the Seigneur de Franget, formerly a deputy-commander in the Company of My Lord Marshal de Châtillon, was sent by My Lord Marshal de Chabannes to replace the Seigneur Du Lude as Governor of Fuentarabia; he surrendered it to the Spaniards. He was sentenced to be stripped of his nobility, both he and his descendants being pronounced commoners, liable to taxation and unfit to bear arms. That severe sentence was executed at Lyons. Later all the noblemen who were at Guyse when the Count of Nassau entered it suffered a similar punishment; and subsequently others still.[5]

Anyway, wherever there is a case of ignorance so crass and of cowardice so flagrant as to surpass any norm, that should be an adequate reason for accepting them as proof of wickedness and malice, to be punished as such.

---

4. Ammianus Marcellinus, *Res gestae*, XXIV, iv, and XXV, i.
5. The Du Bellay *Mémoires*, II, 52; VII, 217.

17. The doings of certain ambassadors

[War and diplomacy, both noble subjects, dominate this chapter; topics are introduced, such as how to read history, which are later developed in 'On books' (II, 10) where the Du Bellays are further criticized. The folly of detailed laws and instructions is treated in 'On experience' (III, 13).]

[A] On my travels, in order to be ever learning something from my meetings with other people (which is one of the best of all schools), I observe the following practice: always to bring those with whom I am talking back to the subjects they know best.

> [A1]   *Basti al nocchiero ragionar de' venti,*
> *Al bifolco dei tori, e le sue piaghe*
> *Conti'l guerrier, conti'l pastor gli armenti.*

[Let the sailor talk but of the winds, the farmer of oxen, the soldier of his own wounds and the herdsman of his cattle.]¹

[A] For the reverse usually happens, everyone choosing to orate about another's job rather than his own, reckoning to increase his reputation by so doing; witness the reproof Archidamus gave to Periander: that he was abandoning an excellent reputation as a good doctor to acquire the reputation of a bad poet.² [C] Just observe how Caesar spreads himself when he tells us about his ingenuity in building bridges and siege-machines: in comparison he is quite cramped when he talks of his professional soldiering, his valour or the way he conducts his wars. His exploits are sufficient proof that he was an outstanding general: he wants to be known as something rather different: a good engineer.

The other day a professional jurist was taken to see a library furnished with every sort of book including many kinds of legal ones. He had nothing to say about them. Yet he stopped to make blunt comments, like

---

1. Verses derived from Propertius and translated in a recent Italian book of etiquette, Stefano Guazzo's *La civil conversatione*, which had at least five editions between 1574 and 1600. Cf. note 6, below.
2. Plutarch, tr. Amyot, *Les Dicts notables des Lacedaemoniens*, p. 215G.

an expert, on a defence-work fixed to the head of a spiral staircase in that library; yet hundreds of officers and soldiers came across it every day without comment or displeasure.

The elder Dionysius, as befitted his fortune, was a great leader in battle, but he strove to become mainly famed for his poetry – about which he knew nothing.

[A]   *Optat ephippia bos piger, optat arare caballus.*

[The lumbering ox yearns for the saddle: the nag yearns for the plough.][3]

[C] Follow that way and nobody achieves anything worthwhile. [A] So we should always lead[4] architects, painters, cobblers and so on to talk of their own business.

While on this subject, when reading history (which is anybody's business) I habitually turn my attention to the authors: if they are persons whose only profession is writing I chiefly learn points of style and language from them; if they are doctors I most readily believe them when they tell us about the climate, the health and humours of princes, of wounds and illnesses; when they are jurisconsults you should concentrate on legal controversies, laws, the bases of systems of government and the like; when Theologians, on Church affairs, ecclesiastical censures, dispensations and marriages; when courtiers, on manners and ceremonial; if warriors, on whatever concerns war and chiefly on detailed accounts of the exploits at which they were actually present; when ambassadors, on intrigues, understandings or negotiations, and how they were conducted – matters with which the Seigneur de Langey was fully conversant: that is why I noted and weighed in his *Mémoires* something I would have skipped over in another's:[5] he first gave an account of the remarkable formal statement made by the Emperor Charles V before the Roman Consistory Court in the presence of our ambassadors the Bishop of Mâcon and the Seigneur Du Velly; included in it were several outrageous remarks addressed to us French: among other things he declared that, if his own officers and soldiers had been no more loyal or skilled in warfare than our King's were, then he would have put a halter round his own neck and gone and begged

3. Diodorus Siculus, XV, ii, p. 179r°; Horace, *Epistles*, I, xiv, 43.
4. Until [C]: always *strive to* lead . . .
5. Guillaume Du Bellay was the Seigneur de Langey. The *Mémoires* (often attributed to Martin Du Bellay) were the work of Guillaume, Jean, René and Martin Du Bellay. (Cf. here, *Mémoires*, pp. 152–6.)

our King for mercy. (It seems he may have to some extent really meant this, for he uttered the same words two or three times in the course of his life.) He then challenged the King to single combat, with sword or poniard, in a boat, wearing only a doublet. Continuing his account the Seigneur de Langey added that when the two ambassadors sent their dispatch to the King, they reported the greater part of all this inaccurately and even hid the first two articles from him.

Now I found it very odd that an ambassador should have the power to choose what he should tell his sovereign, especially in a matter of such moment, coming from such a person and spoken before so large an assembly. It would seem to me that the duty of a servant is fully and faithfully to report events just as they occurred, so that his master can be free to arrange, judge and select for himself. To alter the truth and hide it from someone out of fear that he might take it otherwise than he should and be driven to make an unwise decision (meanwhile leaving him ignorant of his own affairs) would seem to belong to the monarch not the subject, to a responsible schoolmaster not to him who should consider himself not merely subordinate in authority but also in wisdom and counsel. Anyway, even in petty affairs such as mine I would not care to be served that way.

[C] Under some pretext or other we are always ready to withdraw our obedience and to usurp the mastery. Everyone so naturally aspires to freedom and authority that, to a superior, no quality should be dearer in those who serve him than simple, straightforward obedience.

The right to command is corrupted when we obey at our discretion not from subordination. Publius Crassus (the one the Romans considered to be 'five-times blessed') was Consul in Asia when he wrote to a Greek engineer ordering him to bring him the larger of two ship's masts which he had seen in Athens in order to use it in a siege-engine he wanted to make. The engineer, on the strength of his scientific knowledge, permitted himself to decide to bring the smaller one which, by the rules of his art, was the more suitable. Crassus listened to his arguments patiently, then had him soundly flogged, judging that the interests of discipline outweighed those of his machine.[6]

Nevertheless we should consider on the other hand that so strict an obedience is appropriate only to precise orders previously given. The charge of ambassadors leaves them with a freer hand, much depending

6. Aulus Gellius, I, xiii, 24. These facts, and a similar discussion based on Aulus Gellius, occur in another famous book of court etiquette, Castiglione's *Book of the Courtier*, which was written for King Francis I of France.

directly on their own judgement; they do not merely carry out their Master's will, they form that will and dress it by their counsel. In my time I have seen persons in authority criticized for having obeyed the King's dispatches to the letter rather than adapting them to changing local circumstances. Men of judgement still condemn the practice of the kings of Persia who used to break down their orders into such detail that their agents and representatives had to refer back for rulings on the most trivial matters; such delays, over so wide an empire, often proved strikingly prejudicial to their affairs.

As for Crassus, when he wrote to a specialist and actually told him what the mast was to be used for, did he not seem to be entering into a discussion about his intentions, inviting him to use his own discretion?

18. On fear

===

[Montaigne discusses fear, partly in the light of his own experience in war, partly from exempla. He sees it as often leading to mad, ecstatic behaviour: it was indeed to be classed as a case of rapture or of madness, the frightened man being 'beside himself'.]

[A]   *Obstupui, steteruntque comae, et vox faucibus haesit.*

[I stood dumb with fear; my hair stood on end and my voice stuck in my throat.][1]

I am not much of a 'natural philosopher' – that is the term they use; I have hardly any idea of the mechanisms by which fear operates in us; but it is a very odd emotion all the same; doctors say that there is no emotion which more readily ravishes our judgement from its proper seat. I myself have seen many men truly driven out of their minds by fear, and it is certain that, while the fit lasts, fear engenders even in the most staid of men a terrifying confusion.

I leave aside simple folk, for whom fear sometimes conjures up visions of their great-grandsires rising out of their graves still wrapped in their shrouds, or else of chimeras, werewolves or goblins; but even among [C] soldiers,[2] [A] where fear ought to be able to find very little room, how many times have I seen it change a flock of sheep into a squadron of knights in armour; reeds or bulrushes into men-at-arms and lancers; our friends, into enemies; a white cross into a red one.

When Monsieur de Bourbon captured Rome, a standard-bearer who was on guard at the Burgo San Pietro was [C] seized by [A] such terror[3] at the first alarm that he leapt through a gap in the ruins and rushed out of the town straight for the enemy still holding his banner; he thought he was running into the town, but at the very last minute he just managed to see the troops of Monsieur de Bourbon drawing up their ranks ready to

---

1. Virgil, *Aeneid*, II, 774.
2. '80: even among *warriors* where . . . (Many melancholics were prone to visions of chimeras and bugaboos).
3. '80: was *held* by such terror . . . (Du Bellay, *Mémoires*, III, 75.)

resist him (it was thought that the townsfolk were making a sortie); he realized what he was doing and headed back through the very same gap out of which he had just made a three-hundred-yards' dash into the battlefield.

But the standard-bearer of Captain Juille was not so lucky when Saint-Pol was taken from us by Count de Bures and the Seigneur de Reu; for fear had made him so distraught that he dashed out of the town, banner and all, through a gun-slit and was cut to pieces by the attacking soldiers. There was another memorable case during the same siege, when fear so strongly seized the heart of a certain nobleman, freezing it and strangling it, that he dropped down dead in the breach without even being wounded.[4]

[C] Such fear can sometimes take hold of a great crowd. [B] In one of the engagements between Germanicus and the Allemani two large troops of soldiers took fright and fled opposite ways, one fleeing to the place which the other had just fled from.[5]

[A] Sometimes fear as in the first two examples puts wings on our heels; at others it hobbles us and nails our feet to the ground, as happened to the Emperor Theophilus in the battle which he lost against the Agarenes; we read that he was so enraptured and so beside himself with fear, that he could not even make up his mind to run away: [B] *'adeo pavor etiam auxilia formidat'* [so much does fear dread even help].[6] [A] Eventually Manuel, one of the foremost commanders of his army, shook him and pulled him roughly about as though rousing him from a profound sleep, saying, 'If you will not follow me I will kill you; the loss of your life matters less than the loss of the Empire if you are taken prisoner.'

[C] Fear reveals her greatest power when she drives us to perform in her own service those very deeds of valour of which she robbed our duty and our honour. In the first pitched battle which the Romans lost to Hannibal during the consulship of Sempronius, an army of ten thousand foot-soldiers took fright, but seeing no other way to make their cowardly escape they fought their way through the thick of the enemy, driving right through them with incredible energy, slaughtering a large number of

---

4. Du Bellay, *Mémoires*, VIII, 255.
5. '88: Such *madness* can sometimes take hold of *entire armies*; – [B] (until [C]): Allemani, *fear being spread among their army*, two . . . (Tacitus, *Hist.*, I, lxiii.)
6. Quintus Curtius, III, ii. The general account is from Joannes Zonaras, *Historia*, III.

Carthaginians but paying the same price for a shameful flight as they should have done for a glorious victory.[7]

It is fear that I am most afraid of. In harshness it surpasses all other mischances. ['95] What emotion could ever be more powerful or more appropriate than that felt by the friends of Pompey who were aboard a ship with him and witnessed that horrible massacre of his forces? Yet even that emotion was stifled by their fear of the Egyptian sails as they began to draw nearer; it was noticed that his friends had no time for anything but urging the sailors to strive to save them by rowing harder; but after they touched land at Tyre their fear left them and they were free to turn their thoughts to the losses they had just suffered and to give rein to those tears and lamentations which that stronger emotion of fear had kept in abeyance.

> *Tum pavor sapientiam omnem mihi ex animo expectorat.*
>
> [Then fear banishes all wisdom from my heart.][8]

[C] Men who have suffered a good mauling in a military engagement, all wounded and bloody as they are, can be brought back to the attack the following day; but men who have tasted real fear cannot be brought even to look at the enemy again. People with a pressing fear of losing their property or of being driven into exile or enslaved also lose all desire to eat, drink or sleep, whereas those who are actually impoverished, banished or enslaved often enjoy life as much as anyone else. And many people, unable to withstand the stabbing pains of fear, have hanged themselves, drowned themselves or jumped to their deaths, showing us that fear is even more importunate and unbearable than death.

The Greeks acknowledged another species of fear over and above that fear caused when our reason is distraught; it comes, they say, from some celestial impulsion, without any apparent cause.[9] Whole peoples have been seized by it as well as whole armies. Just such a fear brought wondrous desolation to Carthage: nothing was heard but shouts and terrified voices; people were seen dashing out of their houses as if the alarm had been sounded; they began attacking, wounding and killing each other, as though they took each other for enemies come to occupy their city. All was

---

7. Livy, *Annal.*, XXI, lvi.
8. Cicero: *Tusc. disput.*, III, xxvii, 66. (The event figures in Shakespeare's *Antony and Cleopatra*.) Then ibid., IV, viii, 19, citing Ennius.
9. Diodorus Siculus, XV, vii.

disorder and tumult until they had calmed the anger of their gods with prayer and sacrifice.

Such outbursts are called 'Panic terrors'.[10]

---

10. Cf. Erasmus, *Adages*, III, VIII, III, *Panicus casus*; also *Apophthegmata*, V; *Epaminondas*, I.

19. *That we should not be deemed happy till after our death*

====

[A preoccupation with death was expected from melancholics: in Montaigne's case this was heightened by the deaths of La Boëtie and his own father, as well as by the murderous Wars of Religion. 'Death' is considered in the sense of the act of dying, not as the state of the soul in the after-life. As such it is the concern of philosophy not of religion. 'Happiness' in this context includes notions of blessedness and of good fortune. The influence of Stoic commonplaces is clear but not exclusively important; in [B] the aim is less a noble death than a quiet one.]

[A]   *Scilicet ultima semper*
Expectanda dies homini est, dicique beatus
Ante obitum nemo, supremaque funera debet.

[You must always await a man's last day: before his death and last funeral rites, no one should be called happy.][1]

There is a story about this which children know; it concerns King Croesus: having been taken by Cyrus and condemned to death, he cried out as he awaited execution, 'O Solon, Solon!' This was reported to Cyrus who inquired of him what it meant. Croesus explained to him that Solon had once given him a warning which he was now proving true to his own cost: that men, no matter how Fortune may smile on them,[2] can never be called happy until you have seen them pass through the last day of their life, on account of the uncertainty and mutability of human affairs which lightly shift from state to state, each one different from the other. That is why Agesilaus replied to someone who called the King of Persia happy because he had come so young to so great an estate, 'Yes: but Priam was not wretched when he was that age.'[3] Descendants of Alexander the Great,

1. Ovid, *Metamorphoses*, III, 135.
2. Cf. Erasmus, *Apophthegmata*, VII; *Solon Salaminius*, III (citing Herodotus).
    ([A] until [C]): smile on them, *how much treasure, how many Kingdoms and Empires might be seen in their hands,* can never . . .
3. Plutarch, tr. Amyot, *Dicts notables des Lacedaemoniens*, p. 211C.

themselves kings of Macedonia, became cabinet-makers and scriveners in Rome; tyrants of Sicily became schoolteachers in Corinth.[4] A conqueror of half the world, a general of numerous armies, became a wretched suppliant to the beggarly officials of the King of Egypt: that was the cost of five or six more months of life to Pompey the Great.[5] And during our fathers' lifetime Ludovico Sforza, the tenth Duke of Milan, who for so long had been the driving force in Italy, was seen to die prisoner at Loches – but (and that was the worst of it) only after living there ten years. [C] The fairest Queen, widow of the greatest King in Christendom, has she not just died by the hand of the executioner?[6] [A] There are hundreds of other such examples. For just as storms and tempests seem to rage against the haughty arrogant height of our buildings, so it could seem that there are spirits above us, envious of any greatness here below.

> *Usque adeo res humanas vis abdita quædam*
> *Obterit, et pulchros fasces sævasque secures*
> *Proculcare, ac ludibrio sibi habere videtur.*

[Some hidden force apparently topples the affairs of men, seeming to trample down the resplendent fasces and the lictor's unyielding axe, holding them in derision.][7]

Fortune sometimes seems precisely to lie in ambush for the last day of a man's life in order to display her power to topple in a moment what she had built up over the length of years, and to make us follow Laberius and exclaim: *'Nimirum hac die una plus vixi, mihi quam vivendum fuit.'* [I have lived this day one day longer than I ought to have lived.][8]

The good counsel of Solon could be taken that way. But he was a philosopher: for such, the favours and ill graces of Fortune do not rank as happiness or unhappiness and for them great honours and powers[9] are non-essential properties, counted virtually as things indifferent. So it seems likely to me that he was looking beyond that, intending to tell us that happiness in life (depending as it does on the tranquillity and contentment of a spirit well-born and on the resolution and assurance of an ordered soul) may never be attributed to any man until we have seen him act out

4. Dionysius the Tyrant became a pedagogue.
5. Cf. Cicero, *Tusc. disput.*, I, xxxv, 86.
6. Ludovico Sforza, ousted in 1500, spent *eight* years in the dungeon at Loches; Mary Stuart (widow of Francis II of France) was beheaded in 1587.
7. Lucretius, V, 1233. (The *fasces* and *axes* were Roman symbols of State.)
8. Macrobius, *Saturnalia*, II, vii.
9. [A] until [C]: honours, *riches* and powers . . .

the last scene in his play, which is indubitably the hardest.[10] In all the rest he can wear an actor's mask: those fine philosophical arguments may be only a pose, or whatever else befalls us may not assay us to the quick, allowing us to keep our countenance serene. But in that last scene played between death and ourself there is no more feigning; we must speak straightforward French; we must show whatever is good and clean in the bottom of the pot:

> *Nam veræ voces tum demum pectore ab imo*
> *Ejiciuntur, et eripitur persona, manet res*

[Only then are true words uttered from deep in our breast. The mask is ripped off: reality remains.][11]

That is why all the other actions in our life must be tried on the touchstone of this final deed. It is the Master-day, the day which judges all the others; it is (says one of the Ancients)[12] the day which must judge all my years now past. The assay of the fruits of my studies is postponed unto death. Then we shall see if my arguments come from my lips or my heart.

[B] I note that several men by their death have given a good or bad reputation to their entire life. Scipio, Pompey's father-in-law, redeemed by a good death the poor opinion people had had of him until then. And when asked which of three men he judged most worthy of honour, Chabrias, Iphicrates or himself, Epaminondas replied, 'Before deciding that you must see us die.'[13] (Indeed Epaminondas would be robbed of a great deal if anyone were to weigh his worth without the honour and greatness of his end.)

In my own times three of the most execrable and ill-famed men I have known, men plunged into every kind of abomination, died deaths which were well-ordered and in all respects perfectly reconciled: such was God's good pleasure.

[C] Some deaths are fine and fortunate. I knew a man[14] whose thread of life was progressing towards brilliant preferment when it was snapped;

---

10. Erasmus, *Apophthegmata*, V; *Epaminondas*, XXIII.
11. Lucretius, III, 57.
12. Seneca, *Epist. moral.*, XXIV and XXVI, parts of which are translated and paraphrased at length in this chapter.
13. The reference to Quintus Caecilius Metellus Pius Scipio (not Scipio Africanus) is from Seneca, *Epist. moral.*, XXIV, 9. Erasmus, *Apophthegmata*, V; *Epaminondas*, XXIII. For Montaigne, Epaminondas was the greatest of virtuous soldiers and a model to be followed.
14. Etienne de La Boëtie.

his end was so splendid that, in my opinion, his great-souled search after honour held nothing so sublime as that snapping asunder: the goal he aimed for he reached before he had even set out; that was more grand and more glorious than anything he had wished or hoped for. As he fell he surpassed the power and reputation towards which his course aspired.

[B] When judging another's life I always look to see how its end was borne: and one of my main concerns for my own is that it be borne well – that is, in a quiet and   [C]   muted   [B]   manner.[15]

15. '88: in a quiet and *assured* manner . . . (i.e., *seurement* corrected to *sourdement*)

20. *To philosophize is to learn how to die*

=====

[Montaigne comes to terms with his melancholy, now somewhat played down. He remains preoccupied with that fear of death — fear that is of the often excruciating act of dying — which in older times seems to have been widespread and acute. His treatment is rhetorical but not impersonal. The [C] text may be influenced by the advice of the Vatican censor. The philosophical presuppositions of this chapter are largely overturned at the end of the Essays (in III, 13, 'On experience'). Montaigne is on the way to discovering admirable qualities in common men and women. His starting-point here is Socratic: philosophy (by detaching the soul from the body) is a 'practising of death'; [C] introduces an Epicurean concern with pleasure.]

[A] Cicero says that philosophizing is nothing other than getting ready to die.[1] That is because study and contemplation draw our souls somewhat outside ourselves, keeping them occupied away from the body, a state which both resembles death and which forms a kind of apprenticeship for it; or perhaps it is because all the wisdom and argument in the world eventually come down to one conclusion; which is to teach us not to be afraid of dying.

In truth, either reason is joking or her target must be our happiness; all the labour of reason must be to make us live well, and at our ease, as Holy [C] Scripture [A] says.[2] All the opinions in the world reach the same point, [C] that pleasure is our target [A] even though they may get there by different means; otherwise we would throw them out

---

1. Cicero, *Tusc. disput.*, I, xxx, 74–xxxi, 75. In Plato (*Phaedo* 67D) for Socrates, whom Cicero is following, *to philosophize* is *to practise dying*. However, Cicero translates 'practice' not by *meditatio*, which means that, but by *commentatio*, which means a careful preparation. Montaigne is here echoing Cicero, not Socrates directly, and so lessens the element of ecstasy implied by Socrates.

2. '80: as the Holy *Word* says . . .

Montaigne is at best paraphrasing not citing Scripture: cf. Ecclesiastes 3:12; 5:17; 9:7; also Ecclesiasticus 14:14 (no New Testament text is relevant). Several inscriptions in Montaigne's library prove that he was citing either or both of Ecclesiastes and Ecclesiasticus from some untraced intermediary source.

immediately, for who would listen to anyone whose goal was to achieve for us   [C]   pain and suffering?[3]

In this case the disagreements between the schools of philosophy are a matter of words. *'Transcurramus solertissimas nugas.'* [Let us skip quickly through those most frivolous trivialities.][4] More stubbornness and prickliness are there than is appropriate for so dedicated a vocation, but then, no matter what role a man may assume, he always plays his own part within it.

Even in virtue our ultimate aim – no matter what they say – is pleasure. I enjoy bashing people's ears with that word which runs so strongly counter to their minds. When pleasure is taken to mean the most profound delight and an exceeding happiness it is a better companion to virtue than anything else; and rightly so. Such pleasure is no less seriously pleasurable for being more lively, taut, robust and virile. We ought to have given virtue the more favourable, noble and natural name of pleasure not (as we have done) a name derived from *vis* (vigour).[5]

There is that lower voluptuous pleasure which can only be said to have a disputed claim to the name not a privileged right to it. I find it less pure of lets and hindrances than virtue. Apart from having a savour which is - fleeting, fluid and perishable, it has its vigils, fasts and travails, its blood and its sweat; it also has its own peculiar sufferings, which are sharp in so many different ways and accompanied by a satiety of such weight that it amounts to repentance.[6]

Since we reckon that obstacles serve as a spur to that pleasure and as seasoning to its sweetness (on the grounds that in Nature contraries are enhanced by their contraries) we are quite wrong to say when we turn to virtue that identical obstacles and difficulties overwhelm her, making her austere and inaccessible, whereas (much more appropriately than for voluptuous pleasure) they ennoble, sharpen and enhance that holy, perfect pleasure which virtue procures for us. A man is quite unworthy of an acquaintance with virtue who weighs her fruit against the price she exacts; he knows neither her graces nor her ways. Those who proceed to teach us

---

3. '80: for us *our torment. Now there are no means of reaching this point, of fashioning a solid contentment, unless it frees us from the fear of death.*   [A]   That is why . . .
4. Seneca, *Epist. moral.*, CXVII, 30.
5. On Cicero's authority (*Tusc. disput.*, II, xviii, 43), *virtus*, the Latin word for virtue, was normally derived from *vir* (man) not from *vis* (strength). True virtue, in this sense, was 'manliness'. (Same etymology: *Essays*, II, 7.)
6. Philosophical pleasure (quite ascetic in Epicurus) is contrasted here with sexual pleasure.

that the questing after virtue is rugged and wearisome whereas it is delightful to possess her can only mean that she always lacks delight.[7] (For what human means have ever brought anyone to the joy of possessing her?) Even the most perfect of men have been satisfied with aspiring to her – not possessing her but drawing near to her. The contention is wrong, seeing that in every pleasure known to Man the very pursuit of it is pleasurable: the undertaking savours of the quality of the object it has in view; it effectively constitutes a large proportion of it and is consubstantial with it. There is a happiness and blessedness radiating from virtue; they fill all that appertains to her and every approach to her, from the first way in to the very last barrier.

Now one of virtue's main gifts is a contempt for death, which is the means of furnishing our life with easy tranquillity, of giving us a pure and friendly taste for it; without it every other pleasure is snuffed out. [A] That is why all rules meet and concur in this one clause.[8] [C] It is true that they all lead us by common accord to despise pain, poverty and the other misfortunes to which human lives are subject, but they do not do so with the same care. That is partly because such misfortunes are not inevitable. (Most of Mankind spend their lives without tasting poverty; some without even experiencing pain or sickness, like Xenophilus the musician, who lived in good health to a hundred and six.) It is also because, if the worse comes to worse, we can sheer off the bung of our misfortunes whenever we like: death can end them.[9] But, as for death itself, that is inevitable.

> [B] *Omnes eodem cogimur, omnium*
> *Versatur urna, serius ocius*
> *Sors exitura et nos in æter-*
> *Num exitium impositura cymbæ.*

[All of our lots are shaken about in the Urn, destined sooner or later to be cast forth, placing us in everlasting exile via Charon's boat.][10]

7. In the great myth of Hesiod, the father of Greek mythology (*Works and Days*, 289), the upward path to Virtue is steep and rugged: once attained, her dwelling-place is a delightful plateau. (Cf. Rabelais, *Quart Livre*, LVII, Joachim Du Bellay, *Regrets*, TLF, 3. 3.) Montaigne is rare in challenging the truth of the myth: most accepted it, often with a Christian sense.
8. '80: That is why all *Schools of Philosophy* meet and concur in this one clause, *teaching us to despise it* [i.e., death]. It is true . . .
9. The last resort of the Stoic: suicide. (Xenophilus' longevity was proverbial.)
10. Horace, *Odes*, II, iii, 25.

[A]   And so if death makes us afraid, that is a subject of continual torment which nothing can assuage.   [C]   There is no place where death cannot find us – even if we constantly twist our heads about in all directions as in a suspect land: *'Quae quasi saxum Tantalo semper impendet.'* [It is like the rock for ever hanging over the head of Tantalus.][11]   [A]   Our assizes often send prisoners to be executed at the scene of their crimes. On the way there, take them past fair mansions and ply them with good cheer as much as you like –

> [B]   . . . *non Siculæ dapes*
> *Dulcem elaborabunt saporem,*
> *Non avium cytharæque cantus*
> *Somnum reducent –*

[even Sicilian banquets produce no sweet savours; not even the music of birdsong nor of lyre can bring back sleep] –

[A]   do you think they can enjoy it or that having the final purpose of their journey ever before their eyes will not spoil their taste for such entertainment?

> [B]   *Audit iter, numeratque dies, spacioque viarum*
> *Metitur vitam, torquetur peste futura*

[He inquires about the way; he counts the days; the length of his life is the length of those roads. He is tortured by future anguish.][12]

[A]   The end of our course is death.[13] It is the objective necessarily within our sights. If death frightens us how can we go one step forward without anguish? For ordinary people the remedy is not to think about it; but what brutish insensitivity can produce so gross a blindness? They lead the donkey by the tail:

> *Qui capite ipse suo instituit vestigia retro.*

[They walk forward with their heads turned backwards.][14]

---

11. Cicero, *De finibus*, I, xviii, 60; Erasmus, *Adages*, II, IX, VII, *Tantali lapis* (a boulder was ever about to fall on Tantalus' head but never did, keeping him in suspense for all eternity).
12. [A] until [C]: past *all* fair mansions of *France*, and ply them . . . (Horace, *Odes*, I, xviii; Claudian, *In Ruffinum*, II, 137.)
13. Contrast III, 12, in which Montaigne denies that death is the end to which our life aims (its '*but*') but merely its ending ('*bout*').
14. Lucretius, IV, 472.

No wonder that they often get caught in a trap. You can frighten such people simply by mentioning death (most of them cross themselves as when the Devil is named); and since it is mentioned in wills, never expect them to draw one up before the doctor has pronounced the death-sentence. And then, in the midst of pain and terror, God only knows what shape their good judgement kneads it into!

[B] (That syllable 'death' struck Roman ears too roughly; the very word was thought to bring ill-luck, so they learned to soften and dilute it with periphrases. Instead of saying *He is dead* they said *He has ceased to live* or *He has lived*. They [C] found consolation in [B] living, even in a past tense! Whence our 'late' (*feu*) So-and-So: 'he was' So-and-So.)[15]

[A] Perhaps it is a case of, 'Repayment delayed means money in hand', as they say; I was born between eleven and noon on the last day of February, one thousand five hundred and thirty-three (as we date things nowadays, beginning the year in January);[16] it is exactly a fortnight since I became thirty-nine: 'I ought to live at least as long again; meanwhile it would be mad to think of something so far off'. – Yes, but all leave life in the same circumstances, young and old alike. [C] Everybody goes out as though he had just come in. [A] Moreover, however decrepit a man may be, he thinks he still has another [C] twenty years [A] to go[17] in the body, so long as he has Methuselah ahead of him. Silly fool, you! Where your life is concerned, who has decided the term? You are relying on doctors' tales; look at facts and experience instead. As things usually go, you have been living for some time now by favour extraordinary. You have already exceeded the usual term of life; to prove it, just count how many more of your acquaintances have died younger than you are compared with those who have reached your age. Just make a list of people who have ennobled their lives by fame: I wager that we shall find more who died before thirty-five than after. It is full of reason and piety to take as our example the manhood of Jesus Christ: his life ended at thirty-

15. Montaigne believed that *feu* ('the late') derived from *fut* ('he was'). That is a false etymology. But the Romans could indeed say *vixit* ('he has lived') to mean, 'he is dead' or 'he has died'.

[B]: They *were happy with* living . . .

16. Traditionally the year began at Easter (or thereabouts). Dating the year from the first of January, a Roman practice, was decreed in France in 1565 and generally applied in 1567.

17. '80: another *year more* to go . . .

three.[18] The same term applies to Alexander, the greatest man who was simply man.

Death can surprise us in so many ways:

> *Quid quisque vitet, nunquam homini satis*
> *Cautum est in horas.*

[No man knows what dangers he should avoid from one hour to another.][19]

Leaving aside fevers and pleurisies, who would ever have thought that a Duke of Brittany was to be crushed to death in a crowd, as one was during the state entry into Lyons of Pope Clement, who came from my part of the world! Have you not seen one of our kings killed at sport? And was not one of his ancestors killed by a bump from a pig? Aeschylus was warned against a falling house; he was always on the alert, but in vain: he was killed by the shell of a tortoise which slipped from the talons of an eagle in flight. Another choked to death on a pip from a grape; an Emperor died from a scratch when combing his hair; Aemilius Lepidus, from knocking his foot on his own doorstep; Aufidius from bumping into a door of his Council chamber. Those who died between a woman's thighs include Cornelius Gallus, a praetor; Tigillinus, a captain of the Roman Guard; Ludovico, the son of Guy di Gonzaga, the Marquis of Mantua; and – providing even worse examples – Speucippus the Platonic philosopher, and one of our Popes.[20]

Then there was that wretched judge Bebius; he was just granting a week's extra time to a litigant when he died of a seizure: his own time had run out. Caius Julius, a doctor, was putting ointment on the eyes of a patient when death closed his.[21] And if I may include a personal example, Captain Saint-Martin, my brother, died at the age of twenty-three while playing tennis; he was felled by a blow from a tennis-ball just above the right ear. There was no sign of bruising or of a wound. He did not even sit

18. Christ incarnate was God and Man, immortal as touching his Godhead, mortal as touching his Manhood. (Thirty-three is a traditional age of Christ at the Crucifixion.)
19. Horace, *Odes*, II, xiii, 13–14.
20. Lists like these were common in Renaissance compilations and handbooks. Montaigne is partly following here Ravisius Textor's *Officina* ('Workshop'). The lecherous Pope was Clement V (early fourteenth century); the French king killed in a tournament (1559) was Henry II; his ancestor killed by a pig was Philip, the crowned son, who never reigned, of Louis the Fat.
21. Two *exempla* from Pliny, VII, liii.

down or take a rest; yet five or six hours later he was dead from an apoplexy caused by that blow.

When there pass before our eyes examples such as these, so frequent and so ordinary, how can we ever rid ourselves of thoughts of death or stop imagining that death has us by the scruff of the neck at every moment?

You might say: 'But what does it matter how you do it, so long as you avoid pain?' I agree with that. If there were any way at all of sheltering from Death's blows – even by crawling under the skin of a calf – I am not the man to recoil from it. It is enough for me to spend my time contentedly. I deal myself the best hand I can, and then accept it. It can be as inglorious or as unexemplary as you please:

> *prætulerim delirus inersque videri,*
> *Dum mea delectent mala me, vel denique fallant,*
> *Quam sapere et ringi.*

[I would rather be delirious or a dullard if my faults pleased me, or at least deceived me, rather than to be wise and snarling.][22]

But it is madness to think you can succeed that way. They come and they go and they trot and they dance: and never a word about death. All well and good. Yet when death does come – to them, their wives, their children, their friends – catching them unawares and unprepared, then what storms of passion overwhelm them, what cries, what fury, what despair! Have you ever seen anything brought so low, anything so changed, so confused?

We must start providing for it earlier. Even if such brutish indifference could find lodgings in the head of an intelligent man (which seems quite impossible to me) it sells its wares too dearly. If death were an enemy which could be avoided I would counsel borrowing the arms of cowardice. But it cannot be done. [B] Death can catch you just as easily as a coward on the run or as an honourable man:

> [A] *Nempe et fugacem persequitur virum,*
> *Nec parcit imbellis juventæ*
> *Poplitibus, timidoque tergo;*

[It hounds the man who runs away, and it does not spare the legs or fearful backs of unwarlike youth;]

22. Horace, *Epistles*, II, ii, 126–8.

[B]   no tempered steel can protect your shoulders;

> *Ille licet ferro cautus se condat ære,*
> *Mors tamen inclusum protrahet inde caput;*

[No use a man hiding prudently behind iron or brass:
Death will know how to make him stick out his cowering head;][23]

[A]   we must learn to stand firm and to fight it.

To begin depriving death of its greatest advantage over us, let us adopt a way clean contrary to that common one; let us deprive death of its strangeness; let us frequent it, let us get used to it; let us have nothing more often in mind than death. At every instant let us evoke it in our imagination under all its aspects. Whenever a horse stumbles, a tile falls or a pin pricks however slightly, let us at once chew over this thought: 'Supposing that was death itself?' With that, let us brace ourselves and make an effort. In the midst of joy and feasting let our refrain be one which recalls our human condition. Let us never be carried away by pleasure so strongly that we fail to recall occasionally how many are the ways in which that joy of ours is subject to death or how many are the fashions in which death threatens to snatch it away. That is what the Egyptians did: in the midst of all their banquets and good cheer they would bring in a mummified corpse to serve as a warning to the guests:[24]

> *Omnem crede diem tibi diluxisse supremum.*
> *Grata superveniet, quæ non sperabitur hora.*

[Believe that each day was the last to shine on you. If it comes, any unexpected hour will be welcome indeed.][25]

We do not know where death awaits us: so let us wait for it everywhere. To practise death is to practise freedom. A man who has learned how to die has unlearned how to be a slave. Knowing how to die gives us freedom from subjection and constraint.   [C] Life has no evil for him who has thoroughly understood that loss of life is not an evil.   [A] Paulus Aemilius was sent a messenger by that wretched King of Macedonia who was his prisoner, begging not to be led in his triumphant procession. He replied: 'Let him beg that favour from himself.'

It is true that, in all things, if Nature does not lend a hand art and

---

23. Horace, *Epistles*, III, ii, 14–17; Propertius, IV, xviii, 25.
24. Plutarch (tr. Amyot), *Banquet des Sept Sages*, 1515A.
25. Horace, *Epistles*, I, iv, 13–14. Then echoes of Seneca's *Epist. moral.*, I, lxxxviii, 25, and of Plutarch's *Life of Paulus Aemilius*.

industry do not progress very far. I myself am not so much melancholic as an idle dreamer: from the outset there was no topic I ever concerned myself with more than with thoughts about death – even in the most licentious period of my life.

[B]　*Jucundum cum aetas florida ver ageret.*

[When my blossoming youth rejoiced in spring.][26]

[A]　Among the games and the courting many thought I was standing apart chewing over some jealousy or the uncertainty of my aspirations: meanwhile I was reflecting on someone or other who, on leaving festivities just like these, had been surprised by a burning fever and [C] his end, [A] with his head[27] full of idleness, love and merriment – just like me; and the same could be dogging me now:

[B]　*Jam fuerit, nec post unquam revocare licebit.*

[The present will soon be the past, never to be recalled.][28]

[A]　Thoughts such as these did not furrow my brow any more than others did. At first it does seem impossible not to feel the sting of such ideas, but if you keep handling them and running through them you eventually tame them. No doubt about that. Otherwise I would, for my part, be in continual terror and frenzy: for no man ever had less confidence than I did that he would go on living; and no man ever counted less on his life proving long. Up till now I have enjoyed robust good health almost uninterruptedly: yet that never extends my hopes for life any more than sickness shortens them. Every moment it seems to me that I am running away from myself. [C] And I ceaselessly chant the refrain, 'Anything you can do another day can be done now.'

[A]　In truth risks and dangers do little or nothing to bring us nearer to death. If we think of all the millions of threats which remain hanging over us, apart from the one which happens to appear most menacing just now, we shall realize that death is equally near when we are vigorous or feverish, at sea or at home, in battle or in repose. [C]　'Nemo altero fragilior est:

---

26. Catullus, LXVIII, 26. On Montaigne's melancholic humour, which was modified by the sanguine, cf. II, 17. (His comportment corresponds to the symptoms associated with melancholy.)
27. '80: fever and *death*, with his head . . .
28. Lucretius, III, 195.

nemo in crastinam sui certior.' [No man is frailer than another: no man more certain of the morrow.][29]

[A]   If I have only one hour's work to do before I die, I am never sure I have time enough to finish it. The other day someone was going through my notebooks and found a declaration about something I wanted done after my death. I told him straight that, though I was hale and healthy and but a league away from my house, I had hastened to jot it down because I had not been absolutely certain of getting back home.   [C]   Being a man who broods over his thoughts and stores them up inside him, I am always just about as ready as I can be: when death does suddenly appear, it will bear no new warning for me.   [A]   As far as we possibly can we must always have our boots on, ready to go; above all we should take care to have no outstanding business with anyone else.

> [B]   *Quid brevi fortes jaculamur ævo*
> *Multa?*

[Why, in so brief a span do we find strength to make so many projects?][30]

[A]   We shall have enough to do then without adding to it.

One man complains less of death itself than of its cutting short the course of a fine victory; another, that he has to depart before marrying off his daughter or arranging the education of his children; one laments the company of his wife; another, of his son; as though they were the principal attributes of his being.

[C]   I am now ready to leave, thank God, whenever He pleases, regretting nothing except life itself – if its loss should happen to weigh heavy on me. I am untying all the knots. I have already half-said my adieus to everyone but myself. No man has ever prepared to leave the world more simply nor more fully than I have. No one has more completely let go of everything than I try to do.

> [B]   *Miser o miser, aiunt, omnia ademit*
> *Una dies infesta mihi tot præmia vitæ.*

['I am wretched, so wretched,' they say: 'One dreadful day has stripped me of all life's rewards.']

[A]   And the builder says:

29. Seneca, *Epist. moral.*, XCI, 16.
30. Horace, *Odes*, II, xvi, 17.

> *Manent opera interrupta, minaeque*
> *Murorum ingentes.*

[My work remains unfinished; huge walls may fall down.][31]

We ought not to plan anything on so large a scale – at least, not if we are to get all worked up if we cannot see it through to the end.

We are born for action:[32]

> *Cum moriar, medium solvare inter opus.*

[When I die, may I be in the midst of my work.]

I want us to be doing things,   [C]   prolonging life's duties as much as we can;   [A]   I want Death to find me planting my cabbages, neither worrying about it nor the unfinished gardening. I once saw a man die who, right to the last, kept lamenting that destiny had cut the thread of the history he was writing when he had only got up to our fifteenth or sixteenth king!

> [B]   *Illud in his rebus non addunt, nec tibi earum*
> *Jam desiderium rerum super insidet una!*

[They never add, that desire for such things does not linger on in your remains!][33]

[A]   We must throw off such humours; they are harmful and vulgar.

Our graveyards have been planted next to churches, says Lycurgus, so that women, children and lesser folk should grow accustomed to seeing a dead man without feeling terror, and so that this continual spectacle of bones, tombs and funerals should remind us of our human condition:[34]

> [B]   *Quin etiam exhilarare viris convivia cæde*
> *Mos olim, et miscere epulis spectacula dira*
> *Certantum ferro, sæpe et super ipsa cadentum*
> *Pocula respersis non parco sanguine mensis;*

[It was once the custom, moreover, to enliven feasts with human slaughter and to entertain guests with the cruel sight of gladiators fighting: they often fell among the goblets, flooding the tables with their blood;]

---

31. Lucretius, III, 898–9 (Lambin); Virgil, *Aeneid*, IV, 88.
32. [A] until [C]: for action: *and I am of the opinion that not only an Emperor, as Vespasian said, but any gallant man should die on his feet:* Cum moriar . . . Then Ovid, *Amores*, II, x, 36.
33. Lucretius, III, 900.
34. By 'churches' here Montaigne means pagan temples. Then, Silius Italicus, *The Punic War*, XI, li.

[C] so too, after their festivities the Egyptians used to display before their guests a huge portrait of death, held up by a man crying, 'Drink and be merry: once dead you will look like this';[35] [A] similarly, I have adopted the practice of always having death not only in my mind but on my lips. There is nothing I inquire about more readily than how men have died: what did they say? How did they look? What expression did they have? There are no passages in the history books which I note more attentively. [C] That I have a particular liking for such matters is shown by the examples with which I stuff my book. If I were a scribbler I would produce a compendium with commentaries of the various ways men have died. (Anyone who taught men how to die would teach them how to live.) Dicearchus did write a book with some such title, but for another and less useful purpose.[36]

[A] People will tell me that the reality of death so far exceeds the thought that when we actually get there all our fine fencing amounts to nothing. Let them say so: there is no doubt whatsoever that meditating on it beforehand confers great advantages. Anyway, is it nothing to get even that far without faltering or feverish agitation?

But there is more to it than that: Nature[37] herself lends us a hand and gives us courage. If our death is violent and short we have no time to feel afraid: if it be otherwise, I have noticed that as an illness gets more and more hold on me I naturally slip into a kind of contempt for life. I find that a determination to die is harder to digest when I am in good health than when I am feverish, especially since I no longer hold so firmly to the pleasures of life once I begin to lose the use and enjoyment of them, and can look on death with a far less terrified gaze. That leads me to hope that the further I get from good health and the nearer I approach to death the more easily I will come to terms with exchanging one for the other. Just as I have in several other matters assayed the truth of Caesar's assertion that things often look bigger from afar than close to,[38] I have also found that I was much more terrified of illness when I was well than when I felt ill. Being in a happy state, all pleasure and vigour, leads me to get the other state quite out of proportion, so that I mentally increase all its discomforts by half and imagine them heavier than they prove to be when I have to bear them.

---

35. Herodotus, II, lxxviii; Erasmus, *Apophthegmata*, VI; *varie mixta*, LXXXIV.
36. Cicero, *De officiis*, II, V, 16. Dicearchus' book was called *The Perishing of Human Life*. It has not survived.
37. '80: than that. *I realize from experience that* Nature . . .
38. Caesar, *Gallic Wars*, VII, lxxxiv.

I hope that the same will apply to me when I die. [B] It is normal to experience change and decay: let us note how Nature robs us of our sense of loss and decline. What does an old man still retain of his youthful vigour and of his own past life?

> *Heu senibus vitae portio quanta manet.*

[Alas, what little of life's portion remains with the aged.][39]

[C] When a soldier of Caesar's guard, broken and worn out, came up to him in the street and begged leave to kill himself, Caesar looked at his decrepit bearing and said with a smile: 'So you think you are still alive, then?'[40]

[B] If any of us were to be plunged into old age all of a sudden I do not think that the change would be bearable. But, almost imperceptibly, Nature leads us by the hand down a gentle slope; little by little, step by step, she engulfs us in that pitiful state and breaks us in, so that we feel no jolt when youth dies in us, although in essence and in truth that is a harsher death than the total extinction of a languishing life as old age dies. For it is not so grievous a leap from a wretched existence to non-existence as it is from a sweet existence in full bloom to one full of travail and pain.

[A] When our bodies are bent and stooping low they have less strength for supporting burdens. So too for our souls: we must therefore educate and train them for their encounter with that adversary, death; for the soul can find no rest while she remains afraid of him. But once she does find assurance she can boast that it is impossible for anxiety, anguish, fear or even the slightest dissatisfaction to dwell within her. And that almost surpasses our human condition.

> [B] *Non vultus instantis tyranni*
> *Mente quatit solida, neque Auster*
> *Dux inquieti turbidus Adriæ,*
> *Nec fulminantis magna Jovis manus.*

[Nothing can shake such firmness: neither the threatening face of a tyrant, nor the South Wind (that tempestuous Master of the Stormy Adriatic) nor even the mighty hand of thundering Jove.][41]

---

39. Pseudo-Gallus, *Elegies*, I, 16. (Like his contemporaries Montaigne attributed to Cornelius Gallus poems later attributed to Maximianus.)
40. Seneca, *Epist. moral.*, LXXVII, 19. The Emperor was Gaius Caesar (Caligula), not Julius Caesar.
41. Horace, *Odes*, III, iii, 3–6.

[A]   She has made herself Mistress of her passions and her lusts, Mistress of destitution, shame, poverty and of all other injuries of Fortune. Let any of us who can gain such a superiority do so: for here is that true and sovereign freedom which enables us to cock a snook at force and injustice and to laugh at manacles and prisons:

> *in manicis, et*
> *Compedibus, sævo te sub custode tenebo.*
> *Ipse Deus simul atque volam, me solvet: opinor,*
> *Hoc sentit, moriar. Mors ultima linea rerum est.*

['I will shackle your hands and feet and keep you under a cruel gaoler.' – 'God himself will set me free as soon as I ask him to.' (He means, I think, 'I will die': for death is the last line of all.)][42]

Our religion has never had a surer human foundation than contempt for life; rational argument (though not it alone) summons us to such contempt: for why should we fear to lose something which, once lost, cannot be regretted? And since we are threatened by so many kinds of death is it not worse to fear them all than to bear one?[43]   [C]   Death is inevitable: does it matter when it comes? When Socrates was told that the Thirty Tyrants had condemned him to death, he retorted, 'And nature, them!'[44]

How absurd to anguish over our passing into freedom from all anguish. Just as our birth was the birth of all things for us, so our death will be the death of them all. That is why it is equally mad to weep because we shall not be alive a hundred years from now and to weep because we were not alive a hundred years ago. Death is the origin of another life. We wept like this and it cost us just as dear when we entered into this life, similarly stripping off our former veil as we did so. Nothing can be grievous which occurs but once; is it reasonable to fear for so long a time something which lasts so short a time? Living a long life or a short life are made all one by death: *long* and *short* do not apply to that which is no more. Aristotle says that there are tiny creatures on the river Hypanis whose life lasts one single day: those which die at eight in the morning die in youth; those which die at five in the evening die of senility.[45] Which of us would not laugh if so momentary a span counted as happiness or unhappiness? Yet if we compare our own span against eternity or even against the span of mountains, rivers, stars, trees or, indeed, of some animals, then saying *shorter* or *longer* becomes equally ridiculous.

42. Horace, *Epistles*, I, xvi, 76–9.
43. St Augustine, *City of God*, I, xi.
44. Erasmus, *Apophthegmata*, III; *Socratica*, LII.
45. Cicero, *Tusc. disput.*, I, xxxix, 94.

[A]   Nature drives us that way, too:[46] 'Leave this world,' she says, 'just as you entered it. That same journey from death to life, which you once made without suffering or fear, make it again from life to death. Your death is a part of the order of the universe; it is a part of the life of the world:

[B]   *inter se mortales mutua vivunt . . .*
Et quasi cursores vitai lampada tradunt.

[Mortal creatures live lives dependent on each other; like runners in a relay they pass on the torch of life.][47] –

[A]   Shall I change, just for you, this beautiful interwoven structure! Death is one of the attributes you were created with; death is a part of you; you are running away from yourself; this *being* which you enjoy is equally divided between death and life. From the day you were born your path leads to death as well as life:

Prima, quae vitam dedit, hora, carpsit.

[Our first hour gave us life and began to devour it.]

Nascentes morimur, finisque ab origine pendet.

[As we are born we die; the end of our life is attached to its beginning.][48]

[C]   All that you live, you have stolen from life; you live at her expense. Your life's continual task is to build your death. You are *in* death while you are *in* life; when you are no more *in* life you are after death. Or if you prefer it thus: after life you are dead, but during life you are dying: and death touches the dying more harshly than the dead, in more lively a fashion and more essentially.

[B]   'If you have profited from life, you have had your fill; go away satisfied:

Cur non ut plenus vitae conviva recedis?

[Why not withdraw from life like a guest replete?]

But if you have never learned how to use life, if life is useless to you, what does it matter if you have lost it? What do you still want it for?

---

46. The main source of what follows is Nature's soliloquy in Lucretius, III.
47. Lucretius, II, 76 and 79; cf. Erasmus, *Adages*, I, II, XXXVIII, *Cursu lampada tradunt.*
48. Seneca (the dramatist), *Hercules furens*, III, 874; Manilius, *Astronomica*, IV, xvi.

> Cur amplius addere quæris
> Rursum quod pereat male, et ingratum occidat omne?

[Why seek to add more, just to lose it again, wretchedly, without joy?][49]

[C] Life itself is neither a good nor an evil: life is where good or evil find a place, depending on how you make it for them.[50]

[A] 'If you have lived one day, you have seen everything. One day equals all days. There is no other light, no other night. The Sun, Moon and Stars, disposed just as they are now, were enjoyed by your grandsires and will entertain your great-grandchildren:

> [C]   Non alium videre patres: aliumve nepotes
> Aspicient.

[Your fathers saw none other: none other shall your progeny discern.][51]

[A] And at the worst estimate the division and variety of all the acts of my play are complete in one year. If you have observed the vicissitude of my four seasons you know they embrace the childhood, youth, manhood and old age of the World. Its [C] play [A] is done.[52] It knows no other trick but to start all over again. Always it will be the same.

> [B]   Versamur ibidem, atque insumus usque;

[We turn in the same circle, for ever;]

> Atque in se sua per vestigia volvitur annus.

[And the year rolls on again through its own traces.]

[A] I have not the slightest intention of creating new pastimes for you.

> Nam tibi præterea quod machiner, inveniamque
> Quod placeat, nihil est, eadem sunt omnia semper

[For there is nothing else I can make or discover to please you: all things are the same forever.][53]

Make way for others as others did for you.   [C]   The first part of equity is equality. Who can complain of being included when all are included?[54]

---

49. Lucretius, III, 938; 941–2.
50. Seneca, *Epist. moral.*, XCIX, 12.
51. Manilius cited by Vives (Commentary on St Augustine's *City of God*, XI, iv).
52. '80: Its *role* is done . . .
53. Lucretius, III, 1080; Virgil, *Georgics*, II, 402; Lucretius, III, 944–5.
54. Seneca, *Epist. moral.*, XXX, 11.

[A] 'It is no good going on living: it will in no wise shorten the time you will stay dead. It is all for nothing: you will be just as long in that state which you fear as though you had died at the breast;

> *licet, quod vis, vivendo vincere secla,*
> *Mors æterna tamen nihilominus illa manebit.*

[Triumph over time and live as long as you please: death eternal will still be waiting for you.]

[B] 'And yet I shall arrange that you have no unhappiness:

> *In vera nescis nullum fore morte alium te,*
> *Qui possit vivus tibi te lugere peremptum,*
> *Stansque jacentem.*

[Do you not know that in real death there will be no second You, living to lament your death and standing by your corpse.]

"You" will not desire the life which now you so much lament.

> *Nec sibi enim quisquam tum se vitamque requirit . . .*
> *Nec desiderium nostri nos afficit ullum.*

[Then no one worries about his life or his self; . . . we feel no yearning for our own being.]

Death is less to be feared than nothing – if there be anything less than nothing:

> *multo mortem minus ad nos esse putandum*
> *Si minus esse potest quam quod nihil esse videmus.*

[We should think death to be less – if anything is 'less' than what we can see to be nothing at all.][55]

[C] 'Death does not concern you, dead or alive; alive, because you are: dead, because you are no more.

[A] 'No one dies before his time; the time you leave behind you is no more yours than the time which passed before you were born;[56]   [B]   and does not concern you either:

> *Respice enim quam nil ad nos ante acta vetustas*
> *Temporis æterni fuerit.*

---

55. Lucretius, III, 1090 (within a wider Lucretian context); III, 885 (adapted); III, 919; 922; 926.
56. Seneca, *Epist. moral.*, LXIX, 6; then Lucretius, III, 972–3.

[Look back and see that the aeons of eternity before we were born have been nothing to us.]

[A] 'Wherever your life ends, there all of it ends.   [C]   The usefulness of living lies not in duration but in what you make of it. Some have lived long and lived little. See to it while you are still here. Whether you have lived enough depends not on a count of years but on your will.

[A] 'Do you think you will never arrive whither you are ceaselessly heading?   [C]   Yet every road has its end.   [A]   And, if it is a relief to have company, is not the whole world proceeding at the same pace as you are?

> [B]   *Omnia te vita perfuncta sequentur.*

[All things will follow you when their life is done.][57]

[A]   Does not everything move with the same motion as you do? Is there anything which is not growing old with you? At this same [C] instant [A] that you die[58] hundreds of men, of beasts and of other creatures are dying too.

> [B]   *Nam nox nulla diem, neque noctem aurora sequuta est,*
> *Quæ non audierit mistos vagitibus ægris*
> *Ploratus, mortis comites et funeris atri.*

[No night has ever followed day, no dawn has ever followed night, without hearing, interspersed among the wails of infants, the cries of pain attending death and sombre funerals.][59]

[C] 'Why do you pull back when retreat is impossible? You have seen cases enough where men were lucky to die, avoiding great misfortunes by doing so: but have you ever seen anyone for whom death turned out badly? And it is very simple-minded of you to condemn something which you have never experienced either yourself or through another. Why do you complain of me[60] or of Destiny? Do we do you wrong? Should you

---

57. Several echoes of Seneca: *Epist. moral.*, LXXVII, 20, 13 (etc.); XLIX; LXI, LXXVII. Then, Lucretius, III, 968.
58. '80: same *hour* that you die . . .
   Further borrowings, Seneca, *Epist. moral.*, LXXVII.
59. Lucretius, II, 578–80.
60. Nature is still speaking and the inspiration is still Senecan; cf. *Epist. moral.*, XCIII, 2 ff.

govern us or should we govern you? You may not have finished your stint but you have finished your life. A small man is no less whole than a tall one. Neither men nor their lives are measured by the yard. Chiron refused immortality when he was told of its characteristics by his father Saturn, the god of time and of duration.[61]

'Truly imagine how much less bearable for Man, and how much more painful, would be a life which lasted for ever rather than the life which I have given you. If you did not have death you would curse me, for ever, for depriving you of it.

'Seeing what advantages death holds I have deliberately mixed a little anguish into it to stop you from embracing it too avidly or too injudiciously. To lodge you in that moderation which I require of you, neither fleeing from life nor yet fleeing from death, I have tempered them both between the bitter and the sweet.

'I taught Thales, the foremost of your Sages, that living and dying are things indifferent. So, when asked "why he did not go and die then," he very wisely replied: "Because it *is* indifferent."[62]

'Water, Earth, Air and Fire and the other parts of this my edifice are no more instrumental to your life than to your death. Why are you afraid of your last day? It brings you no closer to your death than any other did. The last step does not make you tired: it shows that you are tired. All days lead to death: the last one gets there.'

[A] Those are the good counsels of Nature, our Mother.[63]

I have often wondered why the face of death, seen in ourselves or in other men, appears incomparably less terrifying to us in war than in our own homes — otherwise armies would consist of doctors and cry-babies — and why, since death is ever the same, there is always more steadfastness among village-folk and the lower orders than among all the rest. I truly believe that what frightens us more than death itself are those terrifying grimaces and preparations with which we surround it — a brand new way of life: mothers, wives and children weeping; visits from people stunned and beside themselves with grief; the presence of a crowd of servants, pale and tear-stained; a bedchamber without daylight; candles lighted; our bedside besieged by doctors and preachers; in short, all about us is horror and terror. We are under the ground, buried in our graves already!

61. Cf. Lucian, *Dialogues of the Dead*, XXVI; Ovid, *Metamorphoses*, II, 649 ff.
62. Diogenes Laertius, *Life of Thales*, XXX.
63. Seneca, *Epist. moral.*, CVII, and CXX. The entire speech of Nature, who adds her arguments to Reason's in support of 'our religion's contempt for life' is a patchwork of quotation, at first from Lucretius and subsequently from Seneca.

Children are frightened of their very friends when they see them masked. So are we. We must rip the masks off things as well as off people. Once we have done that we shall find underneath only that same death which a valet and a chambermaid got through recently, without being afraid.[64] Blessed [65] the death which leaves no time for preparing such gatherings of mourners.

---

64. Seneca, *Epist. moral.*, XXIV, 14.
65. [A] until [C]: Blessed, *and thrice blessed*, the death . . . (Doubtless an echo of Aeneas' evocation in Virgil, *Aeneid*, I, 94.)

21. On the power of the imagination

[Imagination, *the faculty of evoking mental images, traditionally included, as here, much of what we nowadays classify as 'thinking': thoughts, concepts, ideas, opinions as well as mental pictures. Religious authorities were divided about the power of the imagination to produce ecstasies as well as 'natural miracles'. Montaigne's ideas are controversial without ceasing to be orthodox. The additions in* [C] *are more personal, less dominated by* exempla, *and include a development on male sexuality and on that impotence during the marriage-night which was widely thought to be caused by sorcery. Montaigne, who had studied law, gives a mock-legal savour to his defence of the Penis.*]

[A] '*Fortis imaginatio generat casum,*' [A powerful imagination generates the event,] as the scholars say.[1] I am one of those by whom the powerful blows of the imagination are felt most strongly. Everyone is hit by it, but some are bowled over.[2] [C] It cuts a deep impression into me: my skill consists in avoiding it not resisting it. I would rather live among people who are healthy and cheerful: the sight of another man's suffering produces physical suffering in me, and my own sensitivity has often misappropriated the feelings of a third party. A persistent cougher tickles my lungs and my throat.

The sick whom I am duty-bound to visit I visit more unwillingly than those with whom I feel less concerned and less involved. When I contemplate an illness I seize upon it and lodge it within myself: I do not find it strange that imagination should bring fevers and death to those who let it act freely and who give it encouragement.

In his own time Simon Thomas was a great doctor. I remember that I happened to meet him one day at the home of a rich old consumptive; he told his patient when discussing ways to cure him that one means was to provide occasions for me to enjoy his company: he could then fix his eyes on the freshness of my countenance and his thoughts on the overflowing cheerfulness and vigour of my young manhood; by filling all his senses

1. Medieval philosophical axiom. Cf. the scholastic dictionary of Erasmus Sarcerius.
2. '80: Everyone is *struck* by it, but some are *transformed* . . .

with the flower of my youth his condition might improve. He forgot to add that mine might get worse.[3]

[A] Gallus Vibius so tensed his soul to understand the essence and impulsions of insanity that he toppled his own judgement from its seat and was never able to restore it again: he could boast that he was made a fool by his own wisdom.

Some there are who forestall the hand of their executioners; one man was on the scaffold, being un-blindfolded so that his pardon could be read to him, when he fell down dead, the blow being struck by his imagination alone. When imaginary thoughts trouble us we break into sweats, start trembling, grow pale or flush crimson; we lie struck supine on our feather-beds and feel our bodies agitated by such emotions; some even die from them. And boiling youth grows so hot in its armour-plate that it consummates its sexual desires while fast asleep in a dream –

> *Ut quasi transactis sæpe omnibus rebus profundant*
> *Fluminis ingentes fluctus, vestemque cruentent.*

[So that, as though they had actually completed the act, they pour forth floods of semen and pollute their garments.][4]

It is no new thing for a man to wake up with cuckold's horns which he never had when he went to bed, but it is worth remembering what happened to Cyppus, a king in Italy: he had been very excited by a bullfight one day and his dreams that night had filled his head full of bulls' horns: thereupon horns grew on his forehead by the sheer power of his imagination.[5]

Nature had denied the power of speech to the son of Croesus: passion gave it to him; [C] Antiochus [A] fell into a fever from the beauty of Stratonice, which was too vigorously imprinted on his soul; Pliny says that, on the very day of the wedding, he saw Lucius Cossitius change from woman to man; Pontanus and others tell of similar metamorphoses which have happened in Italy in recent centuries. And, since both Iphis' own desires and her mother's were so vehement,

---

3. ['95] adds that this event took place in Toulouse. The following *exemplum* concerns Gallus Vibius, an orator; his case is recorded by Marcus Annaeus Seneca (the rhetorician): *Controversiae*, 9, and was well-known from such compendia as Ravisius Textor's *Officina* (s.v. *maniaci et furiosi*) and Coelius Richerius Rhodiginus' *Antiquae Lectiones*, VI, 35.
4. Lucretius, IV, 1035–6.
5. Pliny, XI, xlv.

Vota puer solvit, quae foemina voverat Iphis.

[Iphis fulfilled as a boy vows made as a girl.][6]

- [B] I was travelling though Vitry-le-François[7] when I was able to see a man to whom the Bishop of Soissons had given the name of Germain at his confirmation: until the age of twenty-two he had been known by sight to all the townsfolk as a girl called Marie. He was then an old man with a full beard; he remained unmarried. He said that he had been straining to jump when his male organs suddenly appeared. (The girls there still have a song in which they warn each other not to take great strides lest they become boys, 'like Marie Germain'.) It is not surprising that this sort of occurrence happens frequently. For if the imagination does have any power in such matters, in girls it dwells so constantly and so forcefully on sex that it can (in order to avoid the necessity of so frequently recurring to the same thoughts and harsh yearnings) more easily make that male organ into a part of their bodies.

[A] The scars of King Dagobert and of Saint Francis are attributed by some to the power of their imagination:[8] and they say that by it bodies are sometimes transported from their places; Celsus gives an account of a priest whose soul was enraptured in such an ecstasy that for a considerable period his body remained breathless and senseless.[9] [C] Saint Augustine gives the name of another priest who only needed to hear lamentations and plaintive cries to fall into a swoon, being carried so vigorously outside himself that, until he came back to life again, in vain would you shake him about, shout at him, pinch him or sear his flesh: the priest said he heard their voices, but as though coming from afar; he was also aware of the bruising and branding. That this was no stubborn concealing of his sense-impressions is shown by his being, during this time, without pulse or breath.[10]

[A] It is likely that the credit given to miracles, visions, enchantments and such extraordinary events chiefly derives from the power of the

6. Current examples drawn from Lucian, the *Goddess of Syria*, I; Pliny, VII, iv; then, for Iphis, the Cretan girl who became a youth, Ovid, *Metamorphoses*, IX, 793 ff – For *Antiochus*, until [C], *Antigonus* (wrongly). Pontanus is Johannes Jovinianus Pontanus, a Renaissance scholar and philosopher.

7. Episode related in Montaigne's *Journal de Voyage* for September 1580.

8. Robert Burton later cites these examples, which Henry Cornelius Agrippa 'supposeth to have happened by force of imagination' (*Anatomy of Melancholy*, Part I, Sect. 2, Memb. 3, Subsection 2).

9. H. C. Agrippa, *De occulta philosophia*, I, 1xiv.

10. St Augustine, *City of God*, XIV, xxiv. The priest was called Restitutus. These *exempla* are in Coelius Richerius Rhodiginus, *Antiquae Lectiones*, XX, xvi.

imagination acting mainly on the more impressionable souls of the common people. Their capacity to believe has been so powerfully ravished that they think they see what they do not see. I am moreover of the opinion that those ridiculous attacks of magic impotence by which our society believes itself to be so beset that we talk of nothing else can readily be thought of as resulting from the impress of fear or apprehension.[11] I know this from the experience of a man whom I can vouch for as though he were myself: there is not the slightest suspicion of sexual inadequacy in his case nor of magic spells; but he heard one of his comrades tell how an extraordinary impotency fell upon him just when he could least afford it; then, on a similar occasion, the horror of this account struck his own imagination so brutally that he too incurred a similar fate;   [C] from then on he was subject to relapses, the ignoble memory of his misadventure taunting him and tyrannizing over him. He found that this madness he could cure by another kind of madness: he admitted beforehand that he was subject to this infirmity and spoke openly about it, so relieving the tensions within his soul; by bearing the malady as something to be expected, his sense of constriction grew less and weighed less heavily upon him; then (his thoughts being unencumbered and relaxed) when an occasion arose to its liking, his body, finding itself in good trim for first sounding itself out, seizing itself and taking itself by surprise with its partner in the know, clean cured itself of that condition. Except for genuine impotence, never again are you incapable if you are capable of doing it once.

[A]   This misfortune is to be feared only in adventures where our souls are immoderately tense with desire and respect; especially when the opportunity is pressing and unforeseen, there is no means of recovering from this confusion. I know one man who found it useful to bring to it a

---

11. Some scholars, as well as popular superstition, attributed such impotence to diabolical magic.

12. [A], replaced by [C]: A body *from* elsewhere. *For the man who has time to compose himself and to recover from this trouble, my advice is that he should divert his mind to other thoughts (if he can, for it is difficult) and that he should escape from such ardour and tension of imagination. I know of some who have found it useful to bring to the job a body which they had quietened and tamed elsewhere. And in the case of the man who is frightened of an attack of magic impotence, you should extricate him by persuading him that you can furnish him with counter-enchantments of miraculous and certain effect. But it is also requisite that the women whom one may legitimately approach should drop these ritual and affected manners of severity and refusal, and that they should constrain themselves a little to conform to the exigencies of our wretched century. For the heart of an attacker . . .*

body [C] on the point of being [A] satisfied [C] elsewhere,[12] in order to quieten the ardour of this frenzy and who, growing older, finds himself less impotent for being less potent.

Yet another found it helpful when a friend assured him that he was furnished with a counter-battery of enchantments certain to preserve him. I had better tell how that happened.

A highly placed Count with whom I was intimate was marrying a most beautiful lady who had long been courted by a guest present at the festivities; those who loved him were worried about him – especially one of his relations, an old lady who was presiding over the marriage (which was being held in her house): she feared there might be sorcery about and told me of it. I begged her to put her trust in me. I happened to have in my strong-boxes a certain little flat piece of gold on which were engraved celestial symbols, protecting against sunstroke and relieving headaches when correctly applied to the cranial suture; it was sewn on to a ribbon to be tied under the chin to keep it in place – a piece of lunacy akin to the one we are talking of. This peculiar present had been given me by Jacques Peletier: I decided to get some good out of it.[13] I told the Count that he might well incur the same misfortune as others and that there were those who would willingly see that he did so: but he should go to the marriage-bed confidently since I would do him a friendly turn, not failing in his moment of need to perform a miracle which lay within my power, provided that he promised me on his honour to keep it most faithfully secret, simply giving me a sign if things had gone badly when we rushed in with the festive supper. Both his soul and his ears had received such a battering that, because of his troubled imagination, he had indeed been incapable of an erection: so he gave me the sign. He was then to get up (I had told him) under pretence of chasing us out, playfully seize the night-shirt I was holding (we were much the same size) and wear it until he had followed my prescription – which was as follows: as soon as we had left the room he was to withdraw to pass water: he was then to say certain prayers three times and make certain gestures: each time he was to tie round himself the ribbon I had put into his hand and carefully lay the attached medallion over his kidneys, with the figure in a specified position. Having done so, he should draw the ribbon tight so that it could not come undone:

13. Such magico-medical medallions were favoured by Ficino and other Renaissance Platonists. Jacques Peletier, the mathematician, is mentioned again in 'An apology for Raymond Sebond'. Among his Latin treatises on mathematics is one *On the meeting of lines* (1579) and one on the mystical meanings of numbers (1560).

then he was to go back and confidently get on with the job, not forgetting to throw my night-shirt over his bed in such a way as to cover them both.

It is such monkeyings-about which mainly produce results: our thoughts cannot free themselves from the convictions that such strange actions must derive from some secret lore. Their weight and respect come from their inanity. In short the figures on my talisman proved to have more to do with Venus than with the Sun, more potent in action than as a prophylactic.

I was led to do this deed (which is so foreign to my nature) by a rash and troubled humour. I am opposed to all feigned and subtle actions; I hate sleight of hand not only in games but even when it serves a purpose. The way is vicious even if the deed is not.

Amasis, a King of Egypt, wed Laodice, a very beautiful Grecian maiden. He was a pleasant companion in every other way, but he was incapable of lying with her; he threatened to kill her, thinking there had been some witchcraft. Appropriately enough where mental apprehensions are concerned, she deflected his attention towards invocations: having made his vows and prayers to Venus, he found that very night, after his sacrificial oblations, that he had been divinely restored.[14]

Women are wrong to greet us with those affected provocative appearances of unwillingness which snuff out our ardour just as they kindle it. The daughter-in-law of Pythagoras used to say that a woman who lies with a man should doff her modesty with her kirtle and don it again with her shift.[15] [A] The heart of an attacker is easily dismayed when disturbed by calls to arms which are many and diverse; it is a bad start, once imagination makes a man suffer this shame (which she only does in those first encounters, since they are tempestuous and eager: it is in the first encounter that one most fears a defeat); this occurrence then puts him into a feverish moodiness which persists when subsequent opportunities arise.[16]

[C] Married folk have time at their disposal: if they are not ready they should not try to rush things. Rather than fall into perpetual wretchedness by being struck with despair at a first rejection, it is better to fail to make it properly on the marriage-couch, full as it is of feverish agitation, and to wait for an opportune moment, more private and less challenging. Before

---

14. Herodotus, II, clxxxi.
15. Cf. Plutarch (tr. Amyot), *Preceptes de mariage*, VIII.
16. [A] (instead of [C]): arise, *and this fearfulness increases and redoubles on all subsequent occasions: and without some counter-mine you cannot easily get the better of it.* One man, perhaps . . .

possessing his wife, a man who suffers a rejection should make gentle assays and overtures with various little sallies; he should not stubbornly persist in proving himself inadequate once and for all. Those who know that their member is naturally obedient should merely take care to out-trick their mental apprehensions.

We are right to note the licence and disobedience of this member which thrusts itself forward so inopportunely when we do not want it to, and which so inopportunely lets us down when we most need it; it imperiously contests for authority with our will: it stubbornly and proudly refuses all our incitements, both mental and manual. Yet if this member were arraigned for rebelliousness, found guilty because of it and then retained me to plead its cause, I would doubtless cast suspicion on our other members for having deliberately brought a trumped-up charge, plotting to arm everybody against it and maliciously accusing it alone of a defect common to them all. I ask you to reflect whether there is one single part of our body which does not often refuse to function when we want it to, yet does so when we want it not to. Our members have emotions proper to themselves which arouse them or quieten them down without leave from us. How often do compelling facial movements bear witness to thoughts which we were keeping secret, so betraying us to those who are with us? The same causes which animate that member animate – without our knowledge – the heart, the lungs and the pulse: the sight of some pleasant object can imperceptibly spread right through us the flame of a feverish desire. Is it only the veins and muscles of that particular member which rise or fall without the consent of our will or even of our very thoughts? We do not command our hair to stand on end with fear nor our flesh to quiver with desire. Our hands often go where we do not tell them; our tongues can fail, our voices congeal, when *they* want to. Even when we have nothing for the pot and would fain order our hunger and thirst not to do so, they never fail to stir up those members which are subject to them, just as that other appetite does: it also deserts us, inopportunely, whenever it wants to. That sphincter which serves to discharge our stomachs has dilations and contractions proper to itself, independent of our wishes or even opposed to them; so do those members which are destined to discharge the kidneys.

To show the limitless authority of our wills, Saint Augustine cites the example of a man who could make his behind produce farts whenever he would: Vives in his glosses goes one better with a contemporary example of a man who could arrange to fart in tune with verses recited to him; but that does not prove the pure obedience of that member, since it is normally

most indiscreet and disorderly.[17] In addition I know one Behind so stormy and churlish that it has obliged its master to fart forth wind constantly and unremittingly for forty years and is thus bringing him to his death.[18]

Yet against our very will (on behalf of whose rights we have drawn up this bill of accusation) can be brought a prima-facie charge of sedition and rebellion because of its own unruliness and disobedience. Does it always wish what we want it to? Does it not often wish what we forbid it to – and that to our evident prejudice? Is it any more subject to the determinations of our reason? Finally, on behalf of my noble client, may it please the Court to consider that, in this matter, my client's case is indissolubly conjoined to a consort from whom he cannot be separated. Yet the suit is addressed to my client alone, employing arguments and making charges which (granted the properties of the Parties) can in no wise be brought against the aforesaid consort.[19] By which it can be seen the manifest animosity and legal impropriety of the accusers. The contrary notwithstanding, Nature registers a protest against the barristers' accusations and the judges' sentences, and will meanwhile proceed as usual, as one who acted rightly when she endowed the aforesaid member with its own peculiar privilege to be the author of the only immortal achievement known to mortals. For which reason, generation is held by Socrates to be god-like, and Love, that desire for immortality, to be himself a *Daemon* and immortal.[20]

[A] One man, perhaps by the power of his imagination, leaves in France the very scrofula which his fellow then takes back into Spain.[21] That is why it is customary to insist in such matters that the soul lend her consent. Why do doctors first work on the confidence of their patient with so many fake promises of a cure if not to allow the action of the imagination to make up for the trickery of their potions? They know that

---

17. St Augustine, *City of God*, XIV, xxiv, incorporating the comments of Vives.
18. '95: death. *And would to God that I knew only from the history books how often our stomach, by the refusing of one single fart, may bring us to the very gates of a most excruciating death. And if only that Emperor who gave us liberty to fart in any place had also given us the power to do so!* Yet against . . .
The Emperor who intended to make this decree was Claudius.
19. Since the 'consort' (the female organ) has no erections.
'95: *For the action of the aforesaid is sometimes to invite inopportunely but never to refuse, inviting moreover wordlessly and quietly.* By which . . .
20. Love (Eros, Cupid) is a *daemon* in Plato's *Symposium*.
21. Until the eighteenth century the Kings of France (and of England) were credited with the power to cure scrofula (the 'King's evil').

one of the masters of their craft told them in writing that there are men for whom it is enough merely to look at a medicine for it to prove effective.[22]

That sudden whim of mine all came back to me because of a tale told me by one of my late father's servants who was an apothecary. He was a simple man – a Swiss (a people little given to vanity and lying). He had had a long acquaintance with a sickly merchant in Toulouse who suffered from the stone; he had frequent need of enemas and made his doctors prescribe him various kinds, depending on the symptoms of his illness. When the enemas were brought in, none of the usual formalities were omitted: he often used to finger them to see if they were too hot. There he was, lying down and turned on his side; all the usual preliminaries were gone through . . . except that no clyster was injected! After this ceremony the apothecary withdrew; the patient was treated as though he had taken the clyster and the result was the same as for those who had. If the doctor found that the treatment did not prove effective he gave him two or three other enemas – all of the same kind! Now my informant swears that the sick man's wife (in order to cut down expenses, since he paid for these clysters as though he had really had them) assayed simply injecting warm water; that proved to have no effect: the trickery was therefore discovered but he was obliged to return to the first kind.

There was a woman who believed she had swallowed a pin in her bread; she yelled and screamed as though she felt an insufferable pain in her throat where she thought she could feel it stuck; but since there was no swelling nor external symptoms, one clever fellow concluded that it was all imagination and opinion occasioned by a crust that had jabbed her on the way down; he made her vomit and secretly tossed a bent pin into what she had brought up. That woman believed that she had vomited it out and immediately felt relieved of the pain.

I know of a squire who had entertained a goodly company in his hall and then, four or five days later, boasted as a joke (for there was no truth in it) that he had made them eat cat pie; one of the young ladies in the party was struck with such horror at this that she collapsed with a serious stomach disorder and a fever: it was impossible to save her.

Even the very beasts are subject to the power of the imagination just as we are. Witness dogs, which grieve to death when they lose their masters. We can also see dogs yapping and twitching in their dreams, while horses whinny and struggle about.[23]

22. Apparently the doctor cited by Pedro Mexia in his *Silva de varia lecion*, II, vii.
23. This theme is taken up again in 'An apology for Raymond Sebond'.

But all this can be attributed to the close stitching of mind to body, each communicating its fortunes to the other. It is quite a different matter that the imagination should sometimes act not merely upon its own body but on someone else's. One body can inflict an illness on a neighbouring one (as can be seen in the case of the plague, the pox and conjunctivitis which are passed on from person to person):

> *Dum spectant oculi læsos, læduntur et ipsi:*
> *Multaque corporibus transitione nocent.*

[Looking at sore eyes can make your own eyes sore; and many ills are spread by bodily infection.][24]

Similarly when the imagination is vehemently shaken it sends forth darts which may strike an outside object. In antiquity it was held that when certain Scythian women were animated by anger against anybody they could kill him simply by looking at him. Tortoises and ostriches hatch out their eggs by sight alone – a sign that they emit certain occult influences.[25] And as for witches, they are said to have eyes which can strike and harm:

> *Nescio quis teneros oculus mihi fascinat agnos*

[An eye, I know not whose, has bewitched my tender lambs.][26]

For me magicians provide poor authority. All the same we know from experience that mothers can transmit to the bodies of children in their womb marks connected with their thoughts – witness that woman who gave birth to a blackamoor. And near Pisa there was presented to the Emperor Charles, King of Bohemia, a girl all bristly and hairy whom her mother claimed to have conceived like this because of a portrait of John the Baptist hanging above her bed. It is the same with animals: witness Jacob's sheep and those partridges and hares which are turned white by the snow in the mountains.[27] In my own place recently a cat was seen watching a bird perched high up a tree; they stared fixedly at each other for some little time when the bird tumbled dead between the paws of the cat: either its own imagination had poisoned it or else it had been drawn by the cat's

24. Ovid, *De remedio amoris*, 615–16.
25. Pliny, VII, ii, and IX, x.
26. Virgil, *Eclogue*, III, 103.
27. Standard *exempla* given by Coelius Richerius Rhodiginus, *Antiquae lectiones* (XX, xv) explaining the power over the body of the rational soul and of the faculty of imagination. For Jacob's ewes which produced variegated lambs, cf. Genesis, 30:36–9 and St Augustine, *City of God*, XII, xxv.

force of attraction. Those who are fond of hawking know the tale of the falconer who fixed his gaze purposefully on a kite as it flew and bet he could bring it down by the sheer power of his sight. And he did.

Or so they say: for when I borrow *exempla* I commit them to the consciences of those I took them from. [B] The discursive reflexions are my own and depend on rational proof not on experience: everyone can add his own examples; if anyone has none of his own he should not stop believing that such *exempla* exist, given the number and variety of occurrences.[28] [C] If my *exempla* do not fit, supply your own for me. In the study I am making of our manners and motives, fabulous testimonies – provided they remain possible – can do service as well as true ones. Whether it happened or not, to Peter or John, in Rome or in Paris, it still remains within the compass of what human beings are capable of; it tells me something useful about that. I can see this and profit by it equally in semblance as in reality. There are often different versions of a story: I make use of the one which is rarest and most memorable. There are some authors whose aim is to relate what happened: mine (if I could manage it) would be to relate what can happen. When details are lacking Schoolmen are rightly permitted to posit probabilities. I do not: where this is concerned I excel all historical fidelity in my devoted scrupulousness. Whenever my *exempla* concern what I have heard, what I have said or what I have done, I have not dared to allow myself to change even the most useless or trivial of circumstances. I do not know about my science, but not one jot has been consciously falsified.

While on this topic I often wonder how Theologians or philosophers and their like, with their exquisite consciences and their exacting wisdom, can properly write history. How can they pledge their own trustworthiness on the trustworthiness of ordinary people? How can they vouch for the thoughts of people they have never known and offer their own conjectures as sound coinage? They would refuse to bear sworn witness in Court about complex actions which actually occurred in their presence; there is no man so intimate with them that they would undertake to give a full account of all his thoughts.

I think it less risky to write about the past than the present, since the author has only to account for borrowed truth. Some have invited me to write about contemporary events, reckoning that I see them with eyes less vitiated by passion than others do and that I have a closer view than they, since Fortune has given me access to the various leaders of the contending

28. Until [C]: *human* occurrences . . .

parties. What they do not say is that I would not inflict such pain upon myself for all the fame of Sallust (being as I am the sworn enemy of binding obligations, continuous toil and perseverance), nor that nothing is so foreign to my mode of writing than extended narration. I have to break off so often from shortness of wind that neither the structure of my works nor their development is worth anything at all; and I have a more-than-childish ignorance of the words and phrases used in the most ordinary affairs. That is why I have undertaken to talk about only what I know how to talk about, fitting the subject-matter to my capacities. Were I to choose a subject where I had to be led, my capacities might prove inadequate to it. They do not say either that, since my freedom is so very free, I could have published judgements which even I would reasonably and readily hold to be unlawful and deserving of punishment. Of his own achievement Plutarch would be the first to admit that if his *exempla* are wholly and entirely true that is the work of his sources: his own work consisted in making them useful to posterity, presenting them with a splendour which lightens our path towards virtue.

An ancient account is not like a doctor's prescription, every item in it being tother or which.

22. *One man's profit is another man's loss*

=====

[Montaigne's principal source in this short chapter is Seneca's treatise De beneficiis.*]*

[A] Demades condemned a fellow Athenian whose trade was to sell funeral requisites on the grounds that he wanted too much profit from it and that this profit could only be made out of the deaths of a great many people.[1]

That judgement seems ill-founded since no profit is ever made except at somebody else's loss: by his reckoning you would have to condemn earnings of every sort. The merchant can only thrive by tempting youth to extravagance; the husbandman, by the high price of grain; the architect, by the collapse of buildings; legal officials, by lawsuits and quarrels between men; the very honorariums and the fees of the clergy are drawn from our deaths and our vices. 'No doctor derives any pleasure from the good health even of his friends' (as was said by an ancient author of Greek comedies);[2] 'neither does the soldier from peace in his city': and so on for all the others. And what is worse, if each of us were to sound our inner depths he would find that most of our desires are born and nurtured at other people's expense.

When I reflected on this the thought came to me that Nature here was not belying her general polity, for natural philosophers hold that the birth, nurture and increase of each thing is at the expense and corruption of another.

> *Nam quodcunque suis mutatum finibus exit,*
> *Continuo hoc mors est illius, quod fuit ante.*

[For when anything is changed and sallies forth from its confines, it is at once the death of something which previously existed.][3]

---

1. Seneca, *De beneficiis*, VI, xxxviii.
2. Philemon the Younger, cited in John Stobaeus, *Apophthegmata* (with Latin version by Varinus Favorinus). He wrote many comedies, all of which are lost.
3. Lucretius, II, 753; III, 519.

23. On habit: and on never easily changing a traditional law

===

[Montaigne called this chapter De la coustume . . . *'Custom' for him has essentially the same sense as* ἔθος *(ĕthos) for Aristotle (which is not our word* ēthos *but means custom, usage, manners, habit). No one English word now covers all these senses. Here any of the above may be used, especially habit and custom. Similarly* law *for Montaigne embraces not only legislation but religious and moral traditions.*

'On habit' (or custom) was much expanded in later editions; some of the themes are further developed in II, 12, 'An apology for Raymond Sebond'. The fresh discovery of new and old cultures overseas reinforced Montaigne's sceptical conservatism: habit and custom, however arbitrary, may be the cement of society.]

[A] The power of habit was very well understood, it seems to me, by the man who first forged that tale of a village woman who had grown used to cuddling a calf and carrying it about from the time it was born: she grew so accustomed to doing so that she was able to carry it when a fully grown bull.[1] For, in truth, Habit is a violent and treacherous schoolteacher. Gradually and stealthily she slides her authoritative foot into us; then, having by this gentle and humble beginning planted it firmly within us, helped by time she later discloses an angry tyrannous countenance, against which we are no longer allowed even to lift up our eyes.

At every turn we find habit infringing the rules of Nature: [C] *'Usus efficacissimus rerum omnium magister.'* [Custom, in all things, is a most effectual schoolmaster.][2] I trust here the Cave in Plato's Republic.[3] [A] I trust the doctors who often yield to the authority of habit the reasonings of their Art; I trust that king who, by means of habit, brought his stomach to

1. Erasmus, *Adages,* I, II, LI, *Taurum tollet, qui vitulum sustulerit* (stressing importance of childhood habits). Cf. also IV, IX, XXV, *Usus est altera natura.*
2. Pliny, XXVI, ii.
3. For Plato (*Republic,* VII) all mankind are like men born and bred in a cave, who are convinced that shadows on the wall projected by spiritual realities outside their cave are those realities themselves. Only the inspired philosopher can hope to enlighten them.

draw nourishment from poison,[4] and that maiden who, as Albertus relates, accustomed herself to live on spiders. [B] Why, in the new world of the Indies great nations were discovered in widely different climates who lived on spiders; they kept them and fed them, as they also did grasshoppers, ants, lizards and bats: when food was short a toad sold for six crowns. They cook them and serve them up in various sauces. Other peoples were found for whom our meats and viands were deadly poisonous.[5] [C] *'Consuetudinis magna vis est. Pernoctant venatores in nive; in montibus uri se patiuntur. Pugiles coestibus contusi ne ingemiscunt quidem.'* [Great is the power of habit: huntsmen spend nights in the snow and endure sunburn in the mountains; boxers, bruised by their studded gloves, do not even groan.][6]

These examples are from strange lands, but there is nothing strange about them, if only we consider what we assay by experience every day: how habit stuns our senses. There is no need to go in search of what is said about those who dwell near the cataracts of the Nile; nor what the philosophers deduce about the music of the spheres: that those solid material circles rub and lightly play against each other and so cannot fail to produce a wondrous harmony (by the modulations and mutations of which are conducted the revolutions and variations of the dance of the stars) yet none of the creatures in the whole Universe can hear it, loud though it is, since (as in the case of those Egyptians) our sense of hearing has been dulled by the continuity of the sound.[7] Blacksmiths, millers and armourers could not put up with the noise impinging upon their ears if they were stunned by it as we are. My scented waistcoat is at first scented for me: if I wear it for three days, only other people notice the scent.

What is stranger still is that habit can combine and stabilize the effects of its impressions on our senses despite long gaps and intervals: that has been assayed by those who live near belfries. At home I live in a tower where, at daybreak and sundown, a great bell tolls out the *Ave Maria* every day. My very tower is a-tremble at the din. At first I found it unbearable; a brief time was enough to break me in so that I can now hear it without annoyance and often without even being roused from sleep.

4. Mithridates. This, and the reference to Albertus Magnus, from Pedro Mexia, *Varia lecion*, I, xxvi.
5. From Francisco Lopez de Gomara (tr. Fumée), *Hist. générale des Indes* (Paris, 1578).
6. Cicero, *Tusc. disput.*, II, xvii, 40.
7. Those dwelling near the cataracts grow used to the noise and therefore cannot hear it: so too mankind cannot hear the music of the spheres. (Cicero, *Dream of Scipio*, XI, xix.)

Plato chided a boy for playing knuckle-bones; he replied, 'You are chiding me for something unimportant.' 'Habit,' said Plato, 'is not unimportant.'[8] I find that our greatest vices do acquire their bent during our most tender infancy, so that our formation is chiefly in the hands of our wet-nurses. Mothers think their boys are playing when they see them wring the neck of a chicken or find sport in wounding a dog or a cat. Some fathers are so stupid as to think that it augurs well for a martial spirit if they see their son outrageously striking a peasant or a lackey who cannot defend himself, or for cleverness when they see him cheat a playmate by some cunning deceit or a trick. Yet those are the true seeds by which cruelty, tyranny and treachery take root; they germinate there and then shoot up and flourish, thriving in the grip of habit. And it is a most dangerous start to education to make excuses for such low tendencies because of the weakness of childhood or the unimportance of the subject. In the first place, it is Nature speaking, whose voice is then all the more loud and clear for being yet unbroken. Secondly, the ugliness of cheating does not depend on the difference between money and counters: it depends on the cheating.

'If he cheats over counters, why should he not cheat over money?' I find it more just to argue that way than the way others do: 'They are only counters: he would not do that with money.'

We must carefully teach children to detest vices for what they consist in; we must teach them their natural ugliness, so that they flee them not only in their deeds but in their minds: the very thought of them should be hateful, whatever mask they hide behind.

I was trained from boyhood always to stride along the open highway and to find it repugnant to introduce cunning and deceit into my childish games. I am well aware that (since we should note that games are not games for children but are to be judged as the most serious things they do) there is no pastime so trivial that I do not bring to it (from an inner, natural and unstudied propensity) an extreme repugnance against cheating. If I am playing cards I treat pennies like double-doubloons, just as much when playing with my wife and daughter (when winning or losing hardly matters to me) as when I am gambling in earnest. Everywhere and in everything my own eyes suffice to keep me to my duty; no eyes watch me more closely: there are none I regard more highly.

[A]   I have recently seen in my home a native of Nantes, a man small and born without arms: he has so fashioned his feet to serve him as hands should do that they have, in truth, half forgotten their natural duties. He

---

8. Recorded by Diogenes Laertius, *Life of Plato*, III, xxxviii.

calls them his hands moreover; he cuts his food with them, loads a pistol and fires it with them; he threads a needle, sews, writes, doffs his hat, combs his hair, plays cards and plays dice (shaking them as skilfully as anyone else could). I gave him some money (for he earns his living by showing what he can do): he carried it away in his foot, just as we do in our hand. When I was a boy I saw another such who had no hands but wielded a two-handed sword and a halberd by hunching his neck: he tossed them up in the air, caught them again, threw a dagger – and cracked a whip as well as any carter in France.

But we can discover the effects of habit far better from the impressions which she imprints on our souls, in which she encounters less resistance. Where our judgements and beliefs are concerned, what can she *not* do? Is there any opinion so [C] bizarre [A] – (and[9] I am leaving aside that coarse deceit of religions which, as we can see, has intoxicated so many great nations and so many learned men: since that concern lies beyond human reason a man may be excused if he goes astray over that, whenever he is not by divine favour enlightened above the natural order)[10] – but in other opinions, are there any so strange that habit has not planted them and established them by laws, anywhere she likes, at her good pleasure? [C] And that ancient exclamation is totally right: '*Non pudet physicum, id est speculatorem venatoremque naturæ, ab animis consuetudine imbutis quærere testimonium veritatis!*' [Is it not a disgrace that the natural philosopher, that observer and tracker of Nature, should seek evidence of the truth from minds stupefied by habit!][11]

[B] I reckon that there is no notion, however mad, which can occur to the imagination of men of which we do not meet an example in some public practice or other and which, as a consequence, is not propped up on its foundations by our discursive reason. There are nations where you greet people by turning your backs on them and where you never look at anyone you wish to honour. There are nations where, when the king spits, the court favourite holds out her hand; in another nation the most eminent of those about him stoop down and gather up his faeces in a linen cloth.[12]

[C] Let us steal a little room here for a story. A certain French

9. '80: so *fantastical* – and . . .
10. Standard Christian doctrine: true belief requires prevenient grace, which cannot be merited.
11. Cicero, *De natura deorum*, I, xxx, 83.
12. All from Francisco Lopez de Gomara (tr. Fumée), *Hist. générale des Indes*, as is all of [B] after the following [C]

nobleman always used to blow his nose with his fingers, something quite opposed to our customs. Defending his action (and he was famous for his repartee) he asked me why that filthy mucus should be so privileged that we should prepare fine linen to receive it and then, going even further, should wrap it up and carry it carefully about on our persons; that practice ought to excite more loathing and nausea than seeing him simply excrete it (wherever it might be) as we do all our other droppings. I considered that what he said was not totally unreasonable, but habit had prevented me from noticing just that strangeness which we find so hideous in similar customs in another country.

Miraculous wonders depend on our ignorance of Nature not on the essence of Nature. Our judgement's power to see things is lulled to sleep once we grow accustomed to anything. The Barbarians are in no wise more of a wonder to us than we are to them, nor with better reason – as anyone would admit if, after running through examples from the New World, he concentrated on his own and then with good sense compared them. Human reason is a dye spread more or less equally through all the opinions and all the manners of us humans, which are infinite in matter and infinite in diversity.

I now return to the subject. There are peoples   [B]   where no one but his wife and his children can address the king except through an intermediary: in one and the same country virgins openly display their private parts whilst the married women carefully cover them and hide them; and there is another custom somewhat related to it: in this case chastity is only valued in the service of matrimony: girls can give themselves to whom they wish and, once pregnant, can openly abort themselves with special drugs. Elsewhere, if the bridegroom is, say, a merchant, all the merchants invited to the wedding lie with the bride before he does: the more numerous they are the greater the honour for her and the greater her reputation for staying-power and sexual capacity. The same applies if it is an official who is getting married; the same, too, if it is a nobleman, and so on; but if it is a peasant or anyone of low estate, the duty falls to his lord; and yet they continue to urge strict fidelity during the marriage. There are countries where there are public brothels of men and where men can marry each other; where women accompany their husbands to the wars and play a role not merely in the fighting but in the high command; where not only rings are worn through the nose, lips, cheeks and toes, but heavy golden rods are worn through bosoms and buttocks; where they wipe their fingers when eating on their thighs, on their balls or on the soles of their feet; where children do not inherit, but brothers and nephews do; (elsewhere only nephews do, except for the royal succession;) where, to oversee the

community of goods observed there, certain sovereign magistrates have entire control over the cultivation of the land and the distribution of crops, every man according to his need; where they bewail the death of children and rejoice at the death of the elderly; where ten or a dozen men and their wives share the same bed; where wives who lose their husbands by a violent death may remarry but not the others; where womanhood is rated so low that they kill the girls who are born there and buy women from their neighbours when they need them; where husbands can repudiate wives without stating the cause, but wives cannot do so whatever the cause; where husbands can sell their wives if they are barren; where they boil their dead and pound them into a sort of gruel which they then mix with wine and drink down; where the most desirable form of sepulture is to be eaten by dogs or birds; where they believe that the souls of the blessed dwell in pleasant pastures in complete freedom, furnished with all good things (they it is who make the echoes that we hear); where men fight in the water and aim their bows accurately while swimming; where they shrug their shoulders and bow their heads as a sign of subjection, taking off their shoes when they enter the king's apartments, where the eunuchs who guard the women who are vowed to religion lack noses and lips to make them unlovable and where the priests poke out their eyes to seek the acquaintance of their demons and to consult the oracles; where each makes a god of whatever he likes – the huntsman of a lion or a fox, the fisherman of a particular fish – and make idols of everything that humans do and have done to them (the Sun, Moon and the Earth are their principal gods: their form of oath is to touch the ground whilst looking at the Sun); where they eat fish and flesh raw; [C] where the greatest oath is to swear by the name of a dead man who had a great reputation in that country while touching his tomb;[13] where the New Year's gift sent from the king to his vassal princes is fire: when the ambassador arrives bearing it, the old fires are doused throughout the habitation and the people subject to that vassal must each come and bear away new fire for himself upon pain of *lèse-majesté*; where whenever a king lays down his charge for reasons of devotion (as he often does), his immediate heir is obliged to do the same, the right to rule in the kingdom passing to the third in line; where they change their form of polity as affairs require: they depose their king whenever it pleases them, appointing elders to govern the state and sometimes leaving it in the hands of the general public; where both men

13. The borrowings from Gomara end here: there follows a borrowing from Herodotus and a series of borrowings from Simon Goulart's *Histoire du Portugal*.

and women receive like baptism and are circumcised; where the warrior who manages to present his king with the heads of seven enemies killed in one or more battles is granted nobility; [B] where they live with that opinion, [C] so rare and so uncouth, [B] that souls[14] are mortal; where women give birth without groans and without fear; [C] where the women wear copper shin-guards on both their legs; where it is a sign of high breeding to bite the louse that has bitten them and where girls dare not marry until they have asked the king if he wants to take their maidenhead; [B] where greetings consist in putting one finger on the ground and then raising it heavenwards; where it is the men who carry things on their heads and the women who carry them on their shoulders; where the women stand up to piss and the men squat down; where they send each other their blood as an act of affection and burn incense to the men they would honour as they do to their gods; where the forbidden affinities in marriage extend not merely to the fourth degree but to the most distant relationships; where children are suckled for four years and often for twelve, yet where it is held to be fatal to give a child suck on its very first day; where fathers have the responsibility of punishing male children and the mothers the female, quite apart (the punishment consisting in hanging them by the feet over smoke); where they circumcise females; where they eat herbs of every kind, simply rejecting the ones which seem to them to smell badly; where everything is open and where the houses, however fair and opulent they may be, are without doors, windows or strong-boxes, and where thieves are punished twice as severely as elsewhere; where they kill lice as monkeys do, with their teeth and find it horrible to see them crushed between fingernails; where neither hair nor nails are cut for the whole of one's life; elsewhere they only cut the nails on their right hand, cultivating those on the left as a proof of gentility; [C] where they let all their hair grow on the left of their body and keep all the other side clean-shaven – and in one of the neighbouring provinces they let the hair grow in front and in another behind, shaving the other side; [B] where, in return for money, fathers let guests have sexual enjoyment of their children, and husbands of their wives; where it is honourable to have children by one's mother, for fathers to have sexual intercourse with their daughters and with their sons, [C] and where, when they gather for a festival, all can lie with each other's children.

14. '88: that opinion, *so unnatural*, that souls . . .
    (Belief in the immortality of the soul was thought to be virtually universal.)

[A]   Here they live on human flesh;[15] there, it is a pious duty to kill one's father at a particular age; elsewhere the fathers decide, when the children are still in the womb, which will be kept and brought up and which will be killed and abandoned; elsewhere aged husbands lend their wives to younger men to enjoy; elsewhere again there is no sin in having wives in common – indeed in one country the women, as a mark of honour, hem their skirts with a fringe of tassels to show how many men they have lain with.

And did not habit found a state composed only of women? Did it not place weapons in their hands, make them raise armies and fight battles?[16]

And does not habit teach the roughest of the rough something which the whole of philosophy fails to implant in the heads of the wisest of men? For we know of whole nations   [C]   where death is not merely despised but rejoiced in;   [A]   where seven-year-old boys[17] let themselves be flogged to death without changing their expression;[18] where riches are held in such contempt that the most wretched of their citizens would not deign to stoop and pick up a purse full of crowns. And we know of regions where every kind of food grows in abundance but where both the usual and the most savoury dish is bread and mustard-cress with water.[19]

[B]   And did not custom produce a miracle in Chios where, for seven hundred years, no one ever recalled a woman or girl who lost her honour?

[A]   To sum up then, the impression I have is that there is nothing that custom may not do and cannot do; and Pindar rightly calls her (so I have been told) the Queen and Empress of the World.[20]

[C]   The man found beating his father replied that such was traditional in his family; that his father had beaten his grandfather; his grandfather, his great-grandfather: 'And this boy will beat me once he has reached my age.' Yet the father whom the son was pulling about and dragging along the road commanded him to stop at a certain doorway, for he had not dragged his own father beyond that point, that being the boundary for the hereditary bashing of fathers customary among the sons of their line.[21]

15. Details follow, from Herodotus, Xenophon, Plutarch, etc.
16. The Amazons, described, for example, by the historian Justinus, II, iv.
17. '80: Where *not only the horror of* death is despised but *the hour of its coming even to the dearest persons one has is* rejoiced in *with great merriment; and as for pain, we know others* where seven-year-old boys . . .
18. In Sparta. A much cited and admired example of self-sacrifice.
19. In Persia (Xenophon, *Cyropaedia*, I, ii, 11).
20. Plutarch (tr. Amyot), *Des Vertueux faits des femmes*, and Herodotus, II, xii.
21. Aristotle, *Nicomachaean Ethics*, VII, vi, 1149 b; then follows a direct allusion to VII, v, 1148 b, on morbid desires arising from *ethos*.

It is by custom, says Aristotle, as often as from illness that women tear out their hair, gnaw their nails, eat earth and charcoal: just as it is as much by custom as by Nature that males lie with males.

The laws of conscience which we say are born of Nature are born of custom; since man inwardly venerates the opinions and the manners approved and received about him, he cannot without remorse free himself from them nor apply himself to them without self-approbation.

[B] In the past, when the Cretans wished to curse someone, they prayed the gods to make him catch a bad habit.

[A] But the principal activity of custom is so to seize us and to grip us in her claws that it is hardly in our power to struggle free and to come back into ourselves, where we can reason and argue about her ordinances. Since we suck them in with our mothers' milk and since the face of the world is presented thus to our infant gaze, it seems to us that we were really born with the property of continuing to act that way. And as for those ideas which we find to be held in common and in high esteem about us, the seeds of which were planted in our souls by our forefathers, they appear to belong to our genus, to be natural.   [C]   That is why we think that it is reason which is unhinged whenever custom is – and God knows how often we unreasonably do that! If (as those of us have been led to do who make a study of ourselves) each man, on hearing a wise maxim, immediately looked to see how it properly applied to him, he would find that it was not so much a pithy saying as a whiplash applied to the habitual stupidity of his faculty of judgement. But the counsels of Truth and her precepts are taken to apply to the generality of men, never to oneself: we store them up in our memory not in our manners, which is most stupid and unprofitable.

But let us get back to custom's imperial sway.

Peoples nurtured on freedom and self-government judge any other form of polity to be deformed and unnatural. Those who are used to monarchy do the same: as soon as they have rid themselves of the exactions of one master, no matter what opportunities for change Fortune may give them they rush to implant an equally difficult one in his place, incapable as they are of resolving to hate the over-mastery as such.[22]

[A] Darius asked some Greeks what it would take to persuade them to adopt the Indian custom of eating their dead fathers (for that was the ritual

22. ['95] over-mastery as such. *It is by means of custom that each is pleased with the place in which Nature has planted him: the savages of Scotland have no time for Touraine, nor the Scythians for Thessalia.* Darius . . .

among Indians who reckoned that the most auspicious burial they could give their fathers was within themselves): they replied that nothing on earth would make them do it. Then he made an assay at persuading those Indians to abandon their way and adopt that of the Greeks (which was to cremate their fathers' corpses): he horrified them even more.[23]

We all do likewise: usage hides the true aspect of things from us:

> *Nil adeo magnum, nec tam mirabile quicquam*
> *Principio, quod non minuant mirarier omnes*
> *Paulatim.*

[There is nothing which at first seems so great or so wondrous which we do not all gradually wonder at less and less.][24]

I once had the duty of justifying one of our practices which, far and wide around us, is accepted as having established authority; I did not wish to maintain it (as is usually done) exclusively by force of law and *exempla* so I traced it back to its origins: I found its basis to be so weak[25] that I all but loathed it – I who was supposed to encourage it in others.

[C] This was the remedy that Plato prescribed to banish the unnatural loves of his age – he considered it a basic, sovereign remedy: public opinion should condemn them; poets and everyone else should give dreadful accounts of them. By this remedy even the fairest daughters would not attract the lust of their fathers, nor would outstandingly handsome brothers that of their sisters, since the myths of Thyestes, of Oedipus and of Macareus would have planted moral beliefs in the tender minds of children by the charm of the poetry.[26] Indeed, chastity is a fair virtue; its usefulness is well recognized: yet it is as hard to treat it and to justify it from Nature as it is easy to do so from tradition, law and precept. Basic universal precepts of reason are difficult to investigate thoroughly: dons skim through them quickly or do not even dare to handle them, throwing themselves straightway into the sanctuary of tradition, where they can preen themselves on easy victories.

Those who refuse to be drawn away from the beginnings and sources fail even worse and find themselves bound to savage opinions – as Chrysippus was, who often strewed throughout his writings the little account he took of incestuous unions of any kind.[27]

23. After Herodotus, III, xii.
24. Lucretius, II, 1023–5 (Lambin).
25. [A], until [C]: so *wretched and* weak . . .
26. Plato, *Laws*, VIII, vi.
27. Diogenes Laertius, *Life of Chrysippus*.

[A]   A man who wished to loose himself[28] from the violent foregone conclusions of custom will find many things accepted as being indubitably settled which have nothing to support them but the hoary whiskers and wrinkles of attendant usage; let him tear off that mask, bring matters back to truth and reason, and he will feel his judgement turned upside-down, yet restored by this to a much surer state.

I will ask him, for example, what could be stranger than seeing a people obliged to obey laws which they have never understood;[29] in all their household concerns, marriages, gifts, wills, buying and selling, they are bound by laws which they cannot know, being neither written nor published in their own language: they needs must pay to have them interpreted and applied   [C]   – not following in this the ingenious notion of Isocrates[30] (who advised his king to make all trade and business free, unfettered and profitable but all quarrels and disputes onerous, loading them with heavy taxes); they prefer the monstrous notion of making a trade of reason itself and treating laws like merchandise.   [A]   I am pleased that it was (as our historians state) a Gascon gentleman from my part of the country whom Fortune led to be the first to object when Charlemagne wished to impose Imperial Roman Law on us.[31]

What is more uncouth than a nation[32] where, by legal custom, the office of judge is openly venal and where verdicts are simply bought for cash? where, quite legally, justice is denied to anyone who cannot pay for it, yet where this trade is held in such high esteem that there is formed a fourth estate in the commonwealth, composed of men who deal in lawsuits, thus joining the three ancient estates, the Church, the Nobility and the People? where this fourth estate, having charge of the laws and sovereign authority over lives and chattels, should be quite distinct from the nobility, with the result that there are two sets of laws, the law of honour and the law of justice which are strongly opposed in many matters (the first condemns an unavenged accusation of lying: the other condemns the revenge; a gentleman who puts up with an insult is, by the laws of arms, stripped of his rank

28. [A] until [C]: wished *similarly to assay himself and* to loose . . .
29. The French, many of whose laws were in Latin or medieval French.
30. Isocrates, *Ad Nicoclem*, VI, xviii (a treatise on government).
31. According to Paulus Aemilius this Gascony gentleman's name was, simply, 'Gascon'.
32. France. (Such criticisms were long current. In Molière's *Le Misanthrope* Alceste will appear laughable for objecting to such practices.)

and nobility: one who avenges it incurs capital punishment; if he goes to law to redress an offence against his honour, he is dishonoured; if he acts independently he is chastised and punished by the Law); where these two estates, so different from each other, both derive from a single Head, yet one is responsible for peace, the other for war; the first acquires profit, the second, honour; the first, learning, the second, virtue; the first, words, the second, deeds; the first, justice, the second valour; the first reason, the second, fortitude; the first the long gown, the second, the short?

Take things indifferent, such as clothing: if anyone cared to refer clothing back to its true purpose (which is its usefulness and convenience for the body – its original grace and comeliness depends on that), I would concede to him that the most monstrous clothes imaginable include, to my taste, our doctoral bonnets, that long tail of pleated velvet hanging down from the heads of our womenfolk with its motley fringes, and that silly codpiece uselessly modelling a member which we cannot even decently call by its name yet which we make a parade of, showing it off in public.

Nevertheless such considerations do not deter a man of intelligence from following the common fashion;[33] it seems to me on the contrary that all idiosyncratic and outlandish modes derive less from reason than from madness and ambitious affectation; it is his soul that a wise man should withdraw from the crowd, maintaining its power and freedom freely to make judgements, whilst externally accepting all received forms and fashions.

The government of a community has no right to our thoughts, but everything else such as our actions, efforts, wealth and life itself should be lent to it for its service or even given up when the community's opinions so require, [A1] just as that great and good man Socrates refused to save his life by disobeying [B] the magistrate, [A1] a most unjust and iniquitous magistrate. For the Rule of rules, the general Law of laws, is that each should observe those of the place wherein he lives.[34]

Νόμοις ἕπεσθαι τοῖσιν ἐγχώροις καλόν

[It is right to obey one's country's laws.]

33. The Stoic attitude: cf. Rabelais, *Tiers Livre*, TLF, VII and, in the following century, Molière in *Le Misanthrope*.
34. This was the golden political rule of Etienne de la Boëtie (cf. the end of I, 28, 'On affectionate relationships'). The following verse is from a fragment of Greek tragedy.

And here is one drawn from a different barrel: it is greatly to be doubted whether any obvious good can come from changing any traditional law, whatever it may be, compared with the evil of changing it; for a polity is like a building made of diverse pieces interlocked together, joined in such a way that it is impossible to move one without the whole structure feeling it. He who gave the Thurians their laws[35] ordained that if any man wished to abolish an ancient one or establish a new one he should appear before the people with a rope round his neck, so that he could be hanged at once if anyone failed to approve of his novelty. And the lawgiver to the Spartans spent his life in persuading the citizens to make a solemn promise not to break any of his ordinances. The Spartan Magistrate[36] who roughly cut the two extra strings which Phrynis added to his lyre was not worried about whether the music was improved or whether the chords were more ample: it sufficed him to condemn them because it was a departure from the traditional style. (That was the meaning of that rusty sword of justice hanging in Massilia.)[37]

[B] I abhor novelty, no matter what visage it presents, and am right to do so, for I have seen some of its disastrous effects. That novelty which has [C] for so many years [B] beset us is not solely responsible,[38] but one can say with every likelihood that it has incidentally caused and given birth to them all. Even for the evils and destruction which have subsequently happened without it and despite it it must accept responsibility.

Heu patior telis vulnera facta meis.

[Alas, I suffer wounds made by my own arrows.][39]

[C] Those [B] who shake[40] the State are easily the first to be engulfed in its destruction. [C] The fruits of dissension are not gathered by the one who began it: he stirs and troubles the waters for other men to fish in. [B] Once the great structure of the monarchy is shaken by novelty and its interwoven bonds torn asunder – especially in its old age – the gates are opened as wide as you wish to similar attacks. [C] It is

---

35. Zaleucus; known from Diodorus Siculus, XII, iv.
36. Lycurgus the lawgiver of Sparta. Cf. Plutarch's *Life of Lycurgus.*
37. [A] until [C]: that *old* rusty sword . . .
    Cf. *Valerius Maximus*, II, vi–7. Massilia (now Marseilles) was a Greek colony.
38. '88: which has beset us, *for the last twenty-five or thirty* years is not solely responsible . . .
    (The 'novelty' was the Reformation and the Wars of Religion.)
39. Ovid, *Heroides* (Epistle of Phyllis to Demophon, 48).
40. '88: *The first* who shake . . .

harder for Regal Majesty, said an ancient, to decline from the summit to a middling place than to plunge from there to the very bottom.[41]

But if innovators do most harm, those who copy them are more at fault for rushing to follow examples after they have experienced the horror of them and punished them. And if there are degrees of honour even in the doing of evil, then they must concede to the others the glory of innovation and the courage to make the first attempt.

[B] From this first and abundant source all kinds of new depravity go [C] happily [B] drawing ideas and models for disturbing our system of government. In the very laws which were made to remedy the original evil we can find introductions to all sorts of wicked actions and excuses for them: what is happening to us is what Thucydides said of the civil wars of his own time: that to flatter public vices and to justify them, gentler names were given to them, rejecting true indictments of them as spurious or else mitigating them.[42] Yet the intention is to reform our consciences and our beliefs! '*Honesta oratio est!*' [The plea is fine enough!][43] But even the best of [C] pretexts [B] for novelty are exceedingly dangerous: [C] '*adeo nihil motum ex antiquo probabile est*' [so true is it that no change from ancient ways can be approved].[44]

[B] To speak frankly, it seems to me that there is a great deal of self-love and arrogance in judging so highly of your opinions that you are obliged to disturb the public peace in order to establish them, thereby introducing those many unavoidable evils and that horrifying moral corruption which, in matters of great importance, civil wars and political upheavals bring in their wake – introducing them moreover into your own country. [C] Is it not bad husbandry to encourage so many definite and acknowledged vices in order to combat alleged and disputable error? Is there any kind of vice more wicked than those which trouble the naturally recognized sense of community?

When the Roman Senate had differences with the people about the service of their religion it dared to palm them off with this evasion: '*Ad deos id magis quam ad se pertinere, ipsos visuros ne sacra sua polluantur.*' [That this was less a matter for them than for the gods, who would see that their rites were not profaned.][45] That concurs with what the oracle replied to

---

41. Source not identified.
42. In Plutarch (tr. Amyot), *Comment on peut discerner le flatteur d'avec l'amy*, 44 E.
43. Terence, *Andria*, I, i, 114.
44. Livy, XXXIV, liv.
     '88: the best of *alleged reasons* for novelty . . . (*titre* replaced by *praetexte*)
45. Livy, X, vi.

the men of Delphi in their war against the Medes: fearing a Persian invasion they asked the god what they should do with the holy treasures in his temple, hide them or bear them away. He told them to move nothing; they should look after themselves: he was able to provide for his own.[46]

[B]   The Christian religion bears all the signs of the highest justice and utility, but none is more obvious than the specific injunction to obey the powers that be and to uphold the civil polity.[47] What a wondrous example was bequeathed to us by the Wisdom of God[48] when, in order to establish the salvation of the human race and forward his glorious victory over death and sin, he willed to do so only within the context of our political order. He submitted the course and conduct of so sublime an enterprise, so rich in salvation, to the blind injustice of our human usages and custom, permitting the innocent blood of so many of the Elect, his favoured ones, to flow; he suffered many long years to be lost in order to bring that inestimable fruit to ripeness.

There is a huge gulf between the man who follows the conventions and laws of his country and the man who sets out to regiment them and to change them.

In excuse the former can cite simplicity, obedience and example: whatever he may do, it cannot be from ill-will, only (at worse) from misfortune;   [C]   *'Quis est enim quem non moveat clarissimis monumentis testata consignataque antiquitas?'* [Who would not be moved by the most illustrious records witnessed and sealed by antiquity?] – apart from what Isocrates said, that imperfection has a greater interest in moderation than excess does.[49]

[B]   The other is in a much tougher position,[50]   [C]   since anyone who undertakes to chop and change usurps the right to judge and must pride himself on seeing the defect in what he would get rid of and the good in what he would bring in. The following principles are banal

---

46. Herodotus, VIII, xxxvi.

47. Cf. Titus 3:1; Romans 13:1–7.

48. Christ in his apparent 'foolishness' is the 'Wisdom of the Father' (I Corinthians 1:30); his trial and crucifixion took place according to State law. Christians are the 'elect' (those *chosen* by God for salvation) and often find salvation through martyrdom. Christianity is spread by accepting injustice not by rebellion against the State. These are standard Catholic arguments, accepted by many from their reading of Erasmus.

49. Cicero, *De divinatione*, I, xl, 87; Isocrates, *Ad Nicoclem*, IX, xxxiii.

50. '88: position: *one cannot change anything without judging whatever one abandons to be bad, and whatever one adopts to be good.* God does know . . .

enough but they did strengthen me in my position and even put a bridle on my rasher youth: never to load my shoulders with the heavy burden of claiming knowledge of so important a science;[51] Never to venture to do in such a matter what I would never dare to do in the easiest of those disciplines in which I had been instructed and where facile decisions do no harm, seeming to me as it does to be most iniquitous to wish to submit immovable public regulations and observances to the instability of private ideas (private reasoning having jurisdiction only in private matters) and to undertake against divine ordinances something that no State would tolerate against civil ones (even though human reason is far more involved in civil law, the Law is the sovereign judge of its judges; judicial discretion is limited to explaining and extending accepted usage: it cannot deflect it or make innovations). Though divine Providence has sometimes passed beyond the rules to which we are bound by necessity, it was not dispensing us from them. Such cases are blows from God's hand which we are not to imitate but to greet with amazement; they are inordinate examples, expressly marked by signs of special approval, the same in kind as the miracles which Providence gives us in witness of his omnipotence, miracles far above our order-of-being and our capacities which it is madness and impiety to assay to imitate. We are required not to follow them but to contemplate them in ecstasy. They are acts of his Person, not ours.[52]

Cotta's testimony is relevant here: '*Quum de religione agitur T. Coruncanium, P. Scipionem, P. Scævolam, pontifices maximos, non Zenonem aut Cleanthem aut Chrysippum sequor.*' [In matters of religion I do not follow Zeno, Cleanthes or Chrysippus but Titus Coruncanius, Publius Scipio, Publius Scaevola and the Supreme Pontiffs.][53]

[B] God does know, in our present quarrel, when a hundred articles of religion are to be removed or restored − great and profound ones − how many men are there who can proudly claim to have mastered in detail the reasons and fundamental positions of both sides. They amount to a number (if you could call them a number!) who would not have much power to disturb us. But all the rest of the crowd, where are they heading for? They rush to take sides, but under what banner? Their remedy is like other remedies when weak and badly prescribed: those humours in us which it

51. Theology.
52. Miracles are exceptions and make bad law. (For the consequence of such a conviction, cf. Montaigne on witchcraft, I, 21, 'On the power of the imagination'.)
53. Cicero, *De natura deorum*, III, ii, 5.

was meant to purge have been inflamed, irritated and aggravated by the conflict: yet the potion remains in the body. It could not purge us: it was too weak; but it has weakened us. So we cannot evacuate it, yet all we get out of its workings on our inwards is prolonged pain.

[A] Nevertheless Fortune ever reserves her authority far above our arguments; she sometimes presents us with a need so pressing that the laws simply must find room for it.   [B] If you are resisting the growth of an innovation which has recently been introduced by violence, it is a dangerous and unfair obligation to be restrained by rules everywhere and all the time in your struggle against those who run loose, for whom anything is licit which advances their cause, and for whom law and order means seeking their own advantage: [C]   *'Aditum nocendi perfido praestat fides.'* [To trust an untrustworthy man is to give him power to harm.]⁵⁴   [B] The normal restraints in a healthy State do not provide for such abnormal occurrences: they presuppose a body where the main links and functions cohere together, as well as a common consent to acknowledge and obey it.   [C]   The way of the law is weighty, cold and constrained; it is no good for resisting ways which are lawless and wild.

[A]   We all know that two great figures in two civil wars – Octavius against Sylla and Cato against Caesar – are still reproached for having preferred to let their country suffer any extremity rather than to come to its aid by changing anything whatever at the expense of the law. For truly in dire extremities, when it is impossible to hold on, it would perhaps be wiser to bow your head a little and to let the blow fall, rather than to cling impossibly on, refusing to concede anything whatsoever and so giving violence the occasion to trample everything underfoot; it would be better, since the Laws cannot do what they wish, to force them to wish what they can do.⁵⁵ That was the solution of the man who ordered them to go to sleep for four-and-twenty hours,⁵⁶ and of that other man who made a second May out of the month of June.⁵⁷ The Spartans religiously observed the ordinances of their country, but, when they were caught between a law forbidding them to elect the same admiral twice and a pressing emergency requiring Lysander to reassume that office, even they elected someone

---

54. Seneca (the tragedian), *Oedipus*, III, 686.
55. An echo of Terence applied politically. Cf. II, 19, 'On freedom of conscience', end.
56. Agesilaus: the first of a series of borrowings from the relevant *Lives* of Plutarch.
57. Alexander the Great, in Plutarch's *Life*. When he was told that Macedonian custom forbade their armies to take to the field during June, he commanded June to be renamed The Second May.

called Aracus as Admiral and Lysander as 'Superintendent of the Navy'![58] And similar acuteness was shown by a Spartan ambassador who was dispatched to the Athenians to negotiate a change in one of their laws, only to find Pericles testifying that it was forbidden to remove a tablet once a law had been inscribed on it: he counselled him – since *that* was not forbidden – simply to turn the tablet round.

Plutarch praises Philopoemen for being born to command, knowing how to issue commands by the laws and, when public necessity required it, to the laws.[59]

58. Plutarch's *Lysander*.
59. Plutarch's *Parallel Lives* of Flaminius and Philopoemen.

24. Same design: differing outcomes

====

[This chapter, while continuing to stress the uncertainty in all human affairs, shows Montaigne's concern with the role of Fortune in all human arts, not only in the art of Medicine but in the ecstatic creativity found in the fine arts and in the art of war. In conception it owes something to Plutarch's Parallel Lives. *]*

[A] I was given the following true account by Jacques Amyot, the Grand Almoner of France. (It is to the honour of one of our Princes, who is entitled to be counted as 'one of ours' even though he was born abroad.)[1] At the siege of Rouen during the first of our civil commotions this Prince had been warned by the Queen Mother[2] of a plot hatched against him; he was specifically informed by letter of the name of the nobleman who was to carry it out – a nobleman from Anjou or Maine who was frequenting the household of this Prince for this very purpose. Our Prince, telling nobody of this warning, was on Saint Catherine's Mount: our cannon-fire was being directed from there against Rouen, for it was during the siege of that town.[3] Walking with the Grand Almoner and another Bishop he espied this nobleman (who had been pointed out to him) and summoned him to appear before him. When the man was in his presence the Prince saw him turn suddenly pale and begin to tremble as his conscience alarmed him; 'My Lord So-and-so,' the Prince said to him, 'you know well enough what I want you for: your face shows it. There is nothing you can hide from me: I am so thoroughly informed of your enterprise that if you assay covering things up you will make a bad bargain worse. You know quite well such-and-such a thing – and this too (mentioning the salient points of the most secret elements of the conspiracy); for your very life you had better tell me the whole truth about this scheme.' When the wretched man realized he had been caught and found out (for one of his accomplices had revealed everything to the Queen) all he could do was to clasp his hands together and pray the Prince for pardon and

1. François de Guise, born outside France (in Lorraine).
2. Catherine de' Medici.
3. 1562.

mercy; he intended to throw himself at his feet but the Prince stopped him, going on to say: 'Come here! Have I ever done you wrong? Have I, out of private hatred, ever done any wrong to any of your family? I have known you for a mere three weeks: what reason can have induced you to plot my death?' The nobleman replied that he had no private cause only the general interest of his faction, since some had persuaded him that eradicating so mighty an enemy of their religion[4] would be a deed full of piety, no matter how it was done. 'Now,' continued the Prince, 'I would like to show you how much milder is the religion I hold than the one you profess. Your religion counsels you to kill me, unheard, even though I have done you no wrong: mine commands me to forgive you, guilty though you are of having wanted to murder me without cause. Get out. Clear off! Let me never set eyes on you again. And from now on, if you are wise you will seek men of better counsel to guide your actions.'

When the Emperor Augustus was in Gaul he received conclusive evidence of a conspiracy that Lucius Cinna was cooking up against him.[5] He decided to avenge himself and called a Council of his friends for the following morning. But he spent the night in great distress, reflecting that the young man to be put to death was of a good family, the nephew of Pompey the Great. He groaned out contradictory arguments like this: 'What! Shall men say that I live in fear, for ever on my guard, leaving the man who would murder me to go about at his ease? I have safely borne this head of mine through so many battles on land and sea: shall he go scot-free despite his attempt against me? And now that I have brought peace to the whole world, shall he be left free, after having determined not merely to murder me but to make a sacrificial victim of me?' – (The conspiracy was to kill him while performing a sacrifice.) – He remained silent for a while and then berated himself in a firmer voice: 'Why go on living if it matters to so many people that you should die? Will there never be an end of your cruel acts of vengeance? Is your life so valuable that such great harm must be done to preserve it?'

His wife, Livia, perceived his anguish and asked, 'Will you accept womanly advice? Do what the doctors do when the usual remedies fail to work; they then assay contrary ones. Up till now severity has profited you nothing: after Salvidienus there was Lepidus; after Lepidus, Murena; after

---

4. The Reformed Church. Guise was a Roman Catholic.
5. Seneca, *De clementia*, I, ix. Montaigne doubtless savoured the fact that John Calvin had edited this text (Paris, 1532). It later provides the subject of Corneille's tragedy, *Cinna, ou la clémence d'Auguste*.

Murena, Caepio; after Caepio, Egnatius. Begin again and find out how mildness and clemency succeed. Cinna is found guilty: grant him pardon. He can harm you no more, but he can contribute to your glory.'

Augustus was happy indeed to have found an advocate after his own heart; having thanked his wife he rescinded the order for his friends to come to Council and commanded Cinna to appear before him quite alone. Having made everyone else leave the Council chamber, he had a chair provided for Cinna and addressed him thus: 'In the first place I ask you, Cinna, to hear me out in silence. Do not interrupt what I have to say: I will allow you time and give you leave to reply. You are aware, Cinna, that I plucked you from the camp of my enemies, you who were not merely turned into my enemy but born an enemy; yet I saved you and handed all your goods back to you. In short I helped you and eventually made you so prosperous that the victors envied the condition of the vanquished. You asked to be appointed Pontifex: I granted that to you, although I had refused it to others whose fathers had always fought at my side. I have bound you so strongly to me: yet you have planned to kill me.'

At this, Cinna exclaimed that he was far from any such a wicked thought.

Augustus continued: 'You are not keeping your promise: you assured me that I would not be interrupted. Yes, you have planned to kill me, in such-and-such a place, in such-and-such a time, in such-and-such company and in such-and-such a manner.'

Cinna was paralysed by this news, remaining silent – not so as to keep his bargain but because his conscience was overwhelmed. Augustus saw this; he added: 'Why do such a thing? Is it to become Emperor? The State must truly be in a bad way if there is nothing but myself between you and the Imperial office. You cannot even look after your own household, having recently lost a lawsuit through the intervention of a mere freedman. Are you able to do nothing except take on Caesar? If I am the only one frustrating your hopes, then I give up. Do you believe you will be tolerated by Paulus? by Fabius? by the Cossii and the Servilii, or by that great band of noblemen who are not merely noble in name but who honour nobility itself by their deeds?'

He said a great deal more, speaking to him for a good two hours; then he added:

'Now go, Cinna. I once gave you your life as an enemy: I give it you now as a traitor and a parricide. From this day forth let there be loving-friendship between us: let us see who acts in better faith, I, in granting you your life, or you in accepting it.'

And with that he left him.

Some time later he granted the consulship to Cinna, reproaching him for not asking for it. Cinna subsequently became a firm friend and the sole heir to all his property.

After this incident (which occurred when Augustus was in the fortieth year of his age) there was no further plot or conspiracy against him and he received a just reward for his clemency.

But the same did not apply to our French Prince: for his mildness could not save him from falling into the snare of another similar act of treachery.[6] So vain and worthless is human wisdom: despite all our projects, counsels and precautions, the outcome remains in the possession of Fortune.

We talk of 'lucky' doctors who see a case successfully through, as though there was nothing in that Art of theirs which can stand firm, since its foundations are too fragile to hold it up by its own strength, and as though it alone needed a helping hand from Fortune to make it work. Say what you like about that Art of theirs – good or ill – and I will believe you: thank God we have nothing to do with each other. Contrary to other people, I always despise that Art when I am well but never make a truce with it when I am ill: I then begin to hate it and to fear it. I tell those who urge me to take medicine at least to wait until I am well and have got my strength back in order to have the means of resisting the hazardous effects of their potions. I let Nature run her course: I take it for granted that she is armed with teeth and claws to protect herself from attacks launched against her, so maintaining our fabric and avoiding its disintegration. Instead of going to her help when she is wrestling at close grips with the illness, I fear we help her adversary instead and load extra tasks upon her.

I maintain that not only in medicine but in many of the surer arts Fortune plays a major part. Take those creative ecstasies which transport a poet and carry him outside himself in rapture:[7] why do we not attribute them to good luck, since he himself confesses that they surpass his own strength and capacities, acknowledging that they come from without, being in no wise within his own power – no more than in the case of those adepts at oratory, who claim that in their art, too, there are stirrings and perturbations, outside the natural order, which impel them well beyond what they had planned. The same applies to painting, which sometimes escapes free from the brush-strokes of the painter's hand, surpassing his

6. François de Guise was assassinated in 1563.
7. The Platonic concept of poetic rapture widely accepted in Montaigne's time, especially by the *Pléiade*. The main source is Plato's dialogue, *Io*.

own conceptions and artistry and bringing him to an ecstasy of astonishment which leaves him thunderstruck. Why, Fortune herself reveals to us even more clearly the part she plays in all such works as these by the evidence of that grace and beauty which are found in them not only without the artist's intention but without his knowledge. A competent reader can often find in another man's writings perfections other than those which the author knows that he put there, and can endow them with richer senses and meanings.

As for military exploits, anyone can see that Fortune plays a major part in them: even in our very reflections and deliberations there certainly has to be an element of chance and good luck mingled in with them; all that our wisdom can do does not amount to much: the more acute and lively she is the more frailty she finds within herself and the more she distrusts herself. [A] I share Sylla's opinion:[8] [A] when I pay close attention to the most glorious exploits in war I see, I think that the leaders engage in deliberation and reflection merely as a pure formality, surrendering the best part of their undertaking to Fortune and, trusting in her aid, constantly going way beyond any bounds of rational decision. In the midst of their deliberations there comes upon them Fortune's joyful rapture and, from beyond them, inspired frenzies which as often as not push them towards the least likely of decisions and swell their hearts above the reach of reason. That explains why many great ancient Captains, in order to lend plausibility to their bolder decisions, claimed to their men that they had been bidden to reach them by some inspiration or other, by some sign or prognostic.

That is also why, in the state of indecision and perplexity brought upon us by our inability to see what is most advantageous and to choose it (on account of the difficulties which accompany the divers unforeseeable qualities and circumstances which events bring in their train) the surest way in my opinion, even if no other considerations brought us to do so, is to opt for the course in which is found the more honourable conduct and justice; and [A1] since we doubt which is the shorter road, we should keep going straight ahead. [A] That applies to the two examples which I have just narrated: there is no doubt that it was fairer and nobler in the one who was offended not to act otherwise but to forgive. If it turned out badly for the first of them that is no reason to condemn his good intention; and we do not know, even if he had taken the opposite decision, that he would have escaped the end to which his destiny called him; but he would have lost the glory of such a memorable good deed.

---

8. In Plutarch's *Life of Sylla*.

History tells of many people who, faced with such fears, have chosen the way of hastening to greet any conspiracies laid against them with vengeance and punishments; yet I can see hardly any who were well served by this remedy. Many Roman Emperors bear witness to that. Anyone who finds himself in this peril should not count much on his might or his vigilance; for how hard it is to protect oneself from an enemy who is hidden behind the face of the most dutiful friend we have, or to know the inner thoughts and wishes of those who surround us! In vain does a man employ foreign nations in his personal guard, always surrounded by a hedge of armed men: anyone who holds his own life cheap is always master of the life of another man. And then the continual suspicion which leads a Prince to distrust everyone must torment him strangely. [B] Which explains why, when Dion was told that Callipus was on the lookout for ways to kill him, he had no heart to find out more, saying that he would rather die than live in the misery of having to be on guard not only against his enemies but against his friends as well.[9]

Alexander acted this out even more clearly and rigorously: when he received a letter from Parmenion warning him that his beloved doctor Philip had been suborned by money from Darius to poison him, he handed the letter to Philip to read and, at the same time, swallowed down the medicine that he had just handed to him. Was he not showing his resolve to abet his friends if they wished to kill him? Alexander is the supreme model of daring deeds, but I doubt whether there is anything in his whole life which showed a firmer resolve than this nor a beauty shimmering with such lustre.[10]

Those who under pretext of their security teach princes so watchful a distrust teach them their downfall and their shame. Nothing noble is achieved without risk. I know one [C] whose mind is of a most martial and positive complexion [B] whose good fortune is daily corrupted by such arguments as urge him to remain surrounded by his own men; not to hear of any reconciliation with his former enemies; to keep aloof and never entrust himself to stronger hands, no matter what promises are made nor what advantage he might gain from doing so.[11] [C] (I know another[12] who has unexpectedly improved his fortune by having taken quite contrary advice. When need arises, that bravery which men seek so avidly may be

9. Erasmus, *Apophthegmata*, V; *Dion*, I.
10. Plutarch, *Life of Alexander*.
11. Perhaps Henry III of France, but it could be Henry IV.
12. Perhaps Henry IV, but it could be Duc Henri de Guise.

shown as magnificently in a doublet as in armour, in a closet as on the battlefield, when our hands are folded as when our fist is raised.) [B] So sensitive and circumspect a wisdom is the mortal enemy of great undertakings.   [C] (To gain the support of Syphax, Scipio[13] knew how to leave his army, quit Spain while still doubtful of that new conquest, cross over to Africa with only two small vessels and, in a hostile land, entrust his fate to a Barbarian King whose faith was untried; he was without bond or hostage, simply trusting surely in the greatness of his own heart, in his good fortune and in the promise of his high hopes: *'habita fides ipsam plerumque fidem obligat.'* [Our own trust frequently binds the trust of others.])[14]

[B] For a life ambitious for fame, a man must, on the contrary, yield little to suspicions and keep them on a tight rein: fear and distrust attract hostile actions: it invites them.

The most mistrustful of our kings made himself secure mainly by voluntarily surrendering his life and entrusting his liberty to the hands of his foes, showing complete trust in them so that they might learn trust from him.[15] Against legions, mutinous and under arms, Caesar simply opposed the authority of his countenance and his proud words; he trusted so much in himself and in his good fortune that he did not fear to yield and entrust them to a rebellious and seditious army.

> [C]   *Stetit aggere fulti*
> *Cespitis, intrepidus vultu, meruitque timeri*
> *Nil metuens.*

[With intrepid face he stood upon a mound of turf, deserving to be feared since he feared nothing.][16]

[B] It is however quite true that this strong confidence can only be manifested, natural and entire, by those who are not terrified by the thought of death or of the worst that can happen to them in the end: for to manifest it tremulously, still doubting and unsure, contributes nothing of value towards a great reconciliation. It is an excellent way to win the heart and mind of another man to go and trust him, putting yourself in his

---

13. Publius Cornelius Scipio Africanus (Livy, XXVIII, 17). Syphax was a King of Numidia during the Second Punic War.
14. Livy, XXII, 22.
15. Louis XI, who, according to Commines, entrusted his life to Charles the Bold at Conflans.
16. Lucan, *Pharsalia*, V, 316–18.

power – provided it be done freely, quite unconstrained by necessity, and on condition that the trust we bring is clear and pure, and that at least our brow is not weighed down by hesitations.

When a boy I saw the commander of a great city, a nobleman, who was in real difficulties from the violence of an enraged populace;[17] in order to snuff out this disturbance from the start he decided to leave the very safe place he was in and to put himself in the power of that mutinous mob; things went badly for him and he was ignominiously killed. But to me his error lay not in going out to them – the blame usually attached to his memory – but in adopting the way of submissiveness and weakness, wishing to appease that frenzy more by following than by giving a lead, by begging than by remonstrating; I believe that a military bearing full of assurance and confidence, a gracious severity becoming his rank and the dignity of his office, would have succeeded better, and at least more honourably and more fittingly. Nothing is less to be hoped from that monster[18] thus aroused than mildness or humanity; it will be more open to awe and to fear. I would also reproach him in that, having made a decision which to my taste was more brave than foolhardy, he cast himself into that stormy sea of furious men, weak and in his doublet; he ought to have drunk the cup to the dregs and never given up the part he was playing: whereas when he saw the danger at close quarters he did flinch, subsequently changing the modest ingratiating look he had assumed into one of terror, his voice and his eyes burdened with amazement and contrition. By trying to creep away and hide he set them ablaze and invited them to attack him.

It was decided once to hold a review of the various troops under arms[19] – such being just the place for secret plans of revenge: nowhere can you, in such security, carry them out. There were notorious public signs that it would be most unsafe for some of those on whom the obligation of reviewing the troops mainly fell. Several different pieces of advice were given, as was to be expected in a difficult matter of such weight and consequence. My own advice was not to give any sign of apprehension but to go there and walk between the ranks, faces frank and heads erect; rather than cut anything out (the direction towards which the majority opinion tended) we should on the contrary invite the captains to advise their men

17. Probably during the riots against the salt-tax in Bordeaux (1548), when the King's representative was murdered.
18. The mob.
19. At Bordeaux, in 1585, when Montaigne was mayor. Some of the soldiers were thought to be disloyal.

to make their welcoming volleys fair and hearty, not sparing their powder. This pleased the troops which we had had our doubts about, and it engendered from then on a most useful mutual confidence.

[A]   I find the most beautiful of all courses was that adopted by Julius Caesar. First he assayed making even his enemies love him by mildness and clemency: when conspiracies were uncovered he simply let it be known that he had been told about them; then, he made the very noble resolve to await the outcome without worry or fear, surrendering himself to the protection of the gods and entrusting himself to Fortune: such was the position when he was murdered.[20]

[B]   There was a foreigner who noised abroad that, in return for a good sum of money, he could teach Dionysius, the Tyrant of Syracuse, an infallible way of perceiving and uncovering any plots which his subjects should contrive against him. Dionysius was told of this and summoned him to come and enlighten him about an art so indispensable for his protection. The stranger told him that his art merely consisted in accepting half a hundredweight of silver from him and then boasting of having revealed such a very special secret to him. Dionysius approved of the idea and had six hundred crowns paid over to him: for it was not believable that he should have given so large a sum of money to an unknown man except as a recompense for being initiated into some very useful art; this consideration served to make his enemies fear him.[21]

Princes are wise to publish any information which they receive warning them of plots against their life so as to make people believe that they are indeed well-informed and that nothing can be undertaken without their having wind of it.   [C]   The Duke of Athens did many silly things when consolidating his recent Tyranny over Florence, but the most noteworthy was when he first received warning from Matteo Morozo, one of the accomplices, of the conspiracies that people were plotting against him: he put him to death, to suppress news of this warning and to prevent it being known that anyone in that city could be discontented with his upright rule![22]

[A]   I read an account once, I remember, of a Roman of high rank who was fleeing from the tyranny of the Triumvirate; he had already escaped his pursuers hundreds of times by subtle tricks he had invented. One day a troop of horsemen responsible for arresting him passed close by some

---

20. After Suetonius' *Life of Twelve Caesars.*
21. Plutarch, tr. Amyot: *Dicts des anciens Roys*, XXII.
22. Related by Giovanni Villani in his *Historia di suoi tempi.*

bushes behind which he was crouching; they failed to spot him. But he thought at this juncture of the toil and hardships he had so long undergone to save himself from the endless searches they were diligently making for him everywhere; of the little joy he could hope from such a life; of how much better it would be to die once than to remain forever in such dread: so he called them back and let them see where he was hiding. He voluntarily gave himself up to their cruelty to relieve both them and himself of further hardship.[23]

Issuing invitations to the hands of an enemy is a rather rash decision, yet I believe it would be better to take it than to remain in a continual sweat over an outcome which cannot be remedied. But since such provisions as we can make are full of uncertainty and anguish, it is better to be ready to face with fair assurance anything that *can* happen, while drawing some consolation from not being sure that it will.

23. Appianus of Alexandria, *De civilibus Romanorum bellis*, widely read during the Civil Wars in the French translation of Claude de Seyssel.

25. On schoolmasters' learning

[*This chapter – Du Pédantisme in French – is not limited to what we mean by pedantry today. Its main butt was originally dominees and dons who may impress the young but are parrots unfitted for real life; they know things off by rote but are not wise; they resemble the sophists mocked in Antiquity rather than true philosophers who, even then, were laughed at (but for reasons which did them honour). The later editions, especially [C], emphasize that true philosophers are an elite who know the limits of their knowledge.*

Montaigne, writing consciously as a gentleman, partly has in mind Baldassari Castiglione's Book of the Courtier, which was written in Italian for Francis I of France as a means of making his court more elegant.

The unnamed German who taught Montaigne to speak Latin as his first language was Albert Horstanus, some of whose letters have been preserved (cf. Hartmann, Amerbachkorrespondenz, IX – 2, p. 504).]

[A] When I was a schoolboy I was often upset when I saw schoolmasters treated as buffoons in Italian comedies – (and among us French the title of *Magister* can scarcely be said to imply much more respect).[1] Placed as I was under their control and tutelage, the least I could do was to be jealous of their reputation. I tried to make excuses for them in terms of the natural conflict between the common man and men of rare judgement and outstanding learning – an inevitable one since their courses run flat opposite to each other. But the effort was wasted: it was the most civilized of men who held them in the greatest contempt; witness our excellent Du Bellay:

> *Mais je hay par sur tout un sçavoir pédantesque.*

> [But most of all I loathe schoolmasterish erudition.][2]

[B] This attitude goes back to the Ancients: for Plutarch says that

---

1. A Sorbonne professor was addressed as *Magister Noster* ('Our Master'), a title already mocked by Erasmus and in the *Epistolae Obscurorum Virorum*. Schoolmasters were often addressed as *Magister* by their pupils.
2. Joachim Du Bellay, *Regrets* – the 'punchline' of Sonnet LXVIII.

scholar and *Greek* were terms of abuse among the Romans; they were insults.[3]

[A]   As I grew older I found that they were absolutely right and that '*magis magnos clericos non sunt magis magnos sapientes*' ['them most biggest clerks ain't the most wisest'].[4] Yet how it can happen that a soul enriched by so much knowledge should not be more alert and alive, or that a grosser, commonplace spirit can without moral improvement lodge within itself the reasonings and judgements of the most excellent minds which the world has ever produced: that still leaves me wondering.

[B]   A young woman, the foremost of our Princesses,[5] said to me of a particular man that, by welcoming in as he did the brains of others, so powerful and so numerous, his own brain was forced to squeeze up close, crouch down and contract in order to make room for them all!

[A]   I would like to suggest that our minds are swamped by too much study   [C]   and by too much matter   [A]   just as plants are swamped by too much water   [C]   or lamps by too much oil;   [A]   that our minds, held fast and encumbered by so many diverse preoccupations, may well lose the means of struggling free, remaining bowed and bent under the load; except that it is quite otherwise: the more our souls are filled, the more they expand; examples drawn from far-off times show, on the contrary, that great soldiers and statesmen were also   [C]   great   [A]   scholars.[6]

Those philosophers who did withdraw from all affairs of state were indeed mocked by the comic licence of their times[7]   [C]   since their opinions and manners made them look ridiculous: can you expect men like that to judge of rights in a law-suit or to judge a man's deeds? How fit they are to do that, I must say! They are still trying to find out whether there is such a thing as life or motion; whether Man differs from Ox; what is

3. The Roman mob applied those terms to Cicero, according to Plutarch in his *Life of Cicero*. The words used were *Graikos* and *scholastikos* (which Xylander (863B) rendered as *Graecus* and *otiosus*, since in Latin *otium* means both leisure and learned study).

4. Dreadful Latin: cited by the jolly, ignorant Benedictine monk, Frère Jean, in Rabelais (*Gargantua*, TLF, XXXVII, 95).

5. Perhaps Catherine de Bourbon, sister of Henry of Navarre.

6. A great many changes in [C].   [A] reads: I would like to suggest that our minds are swamped by too much study, just as plants are swamped by too much water: that our minds, seized and encumbered by so many diverse preoccupations . . .; also: *the best* scholars . . .

7. Socrates, for example, was mocked by Aristophanes. The rest of the paragraph paraphrases Plato's *Theaetetus*, XXIV, 173–5 in which the speaker is Socrates.

meant by active and passive; what sort of creatures law and justice are! When they talk of or to a man in authority they show an uncouth and disrespectful licence. Do they hear a king or their own ruler praised? To them he is but an idle shepherd who spends his time exploiting his sheep's wool and milk, only more harshly than a real shepherd does. Do you think a man may be more important because he possesses as his own a couple of thousand acres? They laugh at that, used as they are to treating the whole world as their own. Do you pride yourself on your nobility, since you reckon to have seven rich forebears? They do not think much of you: you have no conception of the universality of Nature – nor of the great many forebears each of us has – rich ones, poor ones, kings, lackeys, Greeks, Barbarians . . . Even if you were fiftieth in line from Hercules they would think you frivolous to value such a chance endowment. And so the common man despised *them*, as men who knew nothing about basic everyday matters or as men ignorant and presumptuous.

But that portrait drawn from Plato is far removed from what is lacking in the kind of people we are talking about.   [A]   The others were envied for being above the common concerns, as being contemptuous of public duties, and as men who had constructed a way of life which was private, inimitable, governed by definite, high and unusual principles; the men we are talking about are despised as inferior to the common model, as incapable of public duties, as men dragging their lives and their base vile morals way behind the common sort of men.

> [C]   *Odi homines ignava opera, philosopha sententia.*

> [I hate men whose words are philosophical but whose deeds are base.][8]

[A]   Those other philosophers, I say, were great in learning, greater still in activities of every kind. As in the tale of that geometrician of Syracuse[9] who was interrupted in his contemplations in order to put some of them to practical use in the defence of his country: he set about at once producing frightful inventions, surpassing human belief; yet he himself despised the work of his hands, thinking that he had compromised the dignity of his art, of which his inventions were but apprentice-toys: so too with them; when they were at times put to the test of action they were seen to fly aloft on so soaring a wing that it was clear that their understanding had indeed wondrously enriched their hearts and minds. But   [C]   some, observing

---

8. Aulus Gellius, *Attic Nights*, XIII, viii.
9. Archimedes (in Plutarch's *Life of Marcellus*, Xylander, 307B–D).

that the fortress of political power had been taken over by incompetents, withdrew: the man who asked Crates how long one had to go on philosophizing, was told, 'Until our armies are no longer led by mule-drivers.' Heraclitus made over his kingdom to his brother; and to the citizens of Ephesius who reproached him for spending his time playing with the children in front of the temple he retorted: 'Is doing that not more worthwhile than sharing the control of affairs with the likes of you?'[10] [A] Others, who had their thoughts set above the fortunes of this world, found the seats of Justice and the very thrones of kings to be base and vile: [C] Empedocles rejected the offer of kingship made by the men of Agrigentum. [A] When Thales condemned preoccupations with thrift and money-making he was accused of sour grapes like the fox. It pleased him, for fun, to make a revealing experiment; for this purpose he debased his knowledge in the service of profit and gain, setting up a business which in one year brought in as much wealth as the most experienced in the trade were hard put to match in a lifetime.[11]

[C] Aristotle tells of some people who called Thales, Anaxagoras and their like wise but not prudent, in that they did not concern themselves enough with the more useful matters;[12] I cannot easily swallow that verbal distinction, but apart from that it provides no excuse for the people I am talking about: judging from the base and needy lot they are satisfied with, they are both not wise and not prudent.

[A] But leaving aside this first explanation, I think it is better to say that the evil arises from their tackling the sciences in the wrong manner and that, from the way we have been taught, it is no wonder that neither master nor pupils become more able, even though they do know more. In truth the care and fees of our parents aim only at furnishing our heads with knowledge: nobody talks about judgement or virtue. When someone passes by, try exclaiming, 'Oh, what a *learned* man!' Then, when another does, 'Oh, what a *good* man!' Our people will not fail to turn their gaze respectfully towards the first. There ought to be a third man crying, 'Oh, what blockheads!'[13]

10. Erasmus, *Apophthegmata*, VII, *Crates Thebanus Cynicus*, XIII, and *Heraclitus Ephesus*, XV.
11. Diogenes Laertius, *Life of Empedocles*; Erasmus, *Apophthegmata*, VII, *Milesii Thaletis*, XIX, after Cicero, *De divinatione*, I, xlix.
12. Aristotle, *Nicomachaean Ethics*, VI, vii, 5.
13. Seneca, *Epist. moral.*, LXXXVIII, 39, where he values training in virtue well above the liberal and the useful arts.

We readily inquire, 'Does he know Greek or Latin?' 'Can he write poetry and prose?' But what matters most is what we put last: 'Has he become better and wiser?' We ought to find out not who understands most but who understands best. We work merely to fill the memory, leaving the understanding [C] and the sense of right and wrong [A] empty. Just as birds sometimes go in search of grain, carrying it in their beaks without tasting it to stuff it down the beaks of their young, so too do our schoolmasters go foraging for learning in their books and merely lodge it on the tip of their lips, only to spew it out and scatter it on the wind.

[C] Such foolishness fits my own case marvellously well. Am I for the most part not doing the same when assembling my material? Off I go, rummaging about in books for sayings which please me – not so as to store them up (for I have no storehouses) but so as to carry them back to this book, where they are no more mine than they were in their original place. We only know, I believe, what we know now: 'knowing' no more consists in what we once knew than in what we shall know in the future.

[A] But what is worse, their pupils and their little charges are not nourished and fed by what they learn: the learning is passed from hand to hand with only one end in view: to show it off, to put into our accounts to entertain others with it, as though it were merely counters, useful for totting up and producing statements, but having no other use or currency. [C] *'Apud alios loqui didicerunt, non ipsi secum'* [They have learned how to talk with others, not with themselves]: *'Non est loquendum, sed gubernandum.'* [We do not need talk but helmsmanship.][14]

Nature, to show that nothing beneath her sway is really savage, has brought forth among peoples whom art has least civilized things which rival the best that art can produce. There is a Gascon proverb, drawn from a country flute-song, which has just the right nuance for my purpose: *'Bouha prou bouha, mas a remuda lous ditz qu'em.'* ['Puff and blow as you will: what concerns us is the movement of the fingers.']

[A] We know how to say, 'This is what Cicero said'; 'This is morality for Plato'; 'These are the *ipsissima verba* of Aristotle.' But what have *we* got to say? What judgements do *we* make? What are *we* doing? A parrot could talk as well as we do.[15]

Such behaviour puts me in mind of a rich Roman who had, at great expense, taken care to obtain the services of experts in all branches of

14. Cicero, *Tusc. disput.*, V, xxxvi, 103 (adapted); Seneca, *Epist. moral.*, CVIII, 37.
15. Seneca, *Epist. moral.*, XXXIII, 7 (adapted).

learning;[16] he kept them always about him so that, when some topic or other should happen to come up when he was with friends, each would bring supplies to his market, ready to furnish him with a brace of arguments or a verse bagged from Homer, depending on what kind of game they traded in. He thought that that knowledge was his because it was in the heads of people who were in his pay – as is the case of those men whose learned abundance consists in owning sumptuous libraries.

[C] Whenever I ask a certain acquaintance of mine to tell me what he knows about anything, he wants to show me a book: he would not venture to tell me that he has scabs on his arse without studying his lexicon to find out the meanings of *scab* and of *arse*.

[A] All we do is to look after the opinions and learning of others: we ought to make them our own. We closely resemble a man who, needing a fire, goes next door to get a light, finds a great big blaze there and stays to warm himself, forgetting to take a brand back home.[17] What use is it to us to have a belly full of meat if we do not digest it, if we do not transmute it into ourselves, if it does not make us grow in size and strength? Do we imagine that Lucullus, whom reading, not experience, made [C] and fashioned [A] into so great a captain, treated reading as we do?[18] [B] We allow ourselves to lean so heavily on other men's arms that we destroy our own force. Do I wish to fortify myself against fear of death? Then I do it at Seneca's expense. Do I want to console myself or somebody else? Then I borrow from Cicero: I would have drawn it from my own resources if only I had been made to practise doing so. I have no love for such competence as is borne off and begged.

[A] Learned we may be with another man's learning: we can only be wise with wisdom of our own:

> Μίσω σοφιστὴν, ὅστις οὐχ αὑτῷ σοφος
>
> [I hate a sage who is not wise for himself.][19]

[C] *'Ex quo Ennius: Nequicquam sapere sapientem, qui ipse sibi prodesse non quiret'* [Hence what Ennius said: 'That Sage is in no way wise who seeks not self-improvement']. . .

16. The millionaire Calvisius Sabinus, in Seneca, *Epist. moral.*, XXVII, 5–6.
17. Plutarch (tr. Amyot), *Comment il faut ouïr*, p. 30H.
18. Cicero, *Academica*, II, i.2. Lucius Lucullus, a tiro in military matters, was dispatched against Mithridates, read up history on the way, and became an outstanding general.
19. Euripides (translated by Montaigne in [B] but not in [C]). From John Stobaeus, III, *De Prudentia: Sententiae monostichae.*

[B]   ... *Si cupidus, si*
Vanus et Euganea quantumvis vilior agna

[If he is avaricious and vain, or scraggier than a ewe in Euganea];

[C]   '*Non enim paranda nobis solum, sed fruenda sapientia est.*' ['We must not only obtain Wisdom: we must enjoy her.']²⁰ Dionysius used to laugh at professors of grammar who did research into the bad qualities of Ulysses yet knew nothing of their own; at musicians whose flutes were harmonious but not their morals; at orators whose studies led to talking about justice, not to being just.²¹

[A]   If our souls do not move with a better motion and if we do not have a healthier judgement, then I would just as soon that our pupil should spend his time playing tennis: at least his body would become more agile. But just look at him after he has spent some fifteen or sixteen years studying: nothing could be more unsuited for employment. The only improvement you can see is that his Latin and Greek have made him more conceited and more arrogant than when he left home.   [C]   He ought to have brought back a fuller soul: he brings back a swollen one; instead of making it weightier he has merely blown wind into it.

These *Magisters* (as Plato says of their cousins, the Sophists) are unique in promising to be the most useful of men while being the only ones who not only fail to improve what is entrusted to them (yet carpenters or masons do so) but actually make it worse. And then they charge you for it.²²

Were we to accept the terms put forward by Protagoras²³ – that either he should be paid his set fee or else his pupils should declare on oath in the temple what profit they reckoned they had gained from what he had taught them and remunerate him accordingly – these pedagogues of mine would be in for a disappointment if they had to rely on oaths based upon my experience.

[A]   In my local Périgord dialect these stripling *savants* are amusingly called *Lettreferits* ('word-struck'), as though their reading has given them, so to speak, a whack with a hammer. In truth, as often as not they appear to have been knocked below common-sense itself. Take a peasant or a cobbler: you can see them going simply and innocently about their business, talking only of what they know: whereas these fellows, who want

20. Cicero, *De officiis*, X, III, xiv, 62; Juvenal, VIII, xiv; Cicero, *De finibus*, I, i, 3.
21. Not Dionysius but Diogenes: Erasmus, *Apophthegmata*, III, *Diogenes Cynicus*, XVI.
22. Plato, *Meno*, XXVIII, 91.
23. Plato, *Protagoras*, XVI, 328.

to rise up [C] and fight [A] armed with knowledge which is merely floating about on the surface of their brains, are for ever getting snarled up and entangled. Fine words break loose from them: but let somebody else apply them! They know their Galen but not their patient. They stuff your head full of prescriptions before they even understand what the case is about. They have learned the theory of everything: try and find one who can put it into practice.

In my own house a friend of mine had to deal with one of these fellows; he amused himself by coining some nonsensical jargon composed of disconnected phrases and borrowed passages, but often interlarded with terms bearing on their discussion: he kept the fool arguing for one whole day, thinking all the time that he was answering objections put before him. Yet he had a reputation for learning – [B] and a fine gown, too.

> *Vos, o patritius sanguis, quos vivere par est*
> *Occipiti, cæco, posticæ occurrite sannæ.*

[O ye men of patrician blood! You have no eyes in the back of your heads: beware of the faces which are pulled behind your backs.][24]

[A] Whoever will look closely at persons of this sort – and they are spread about everywhere – will find as I do that for the most part they understand neither themselves nor anyone else and that while their memory is very full their judgement remains entirely hollow – unless their own nature has fashioned it for them otherwise, as I saw in the case of Adrian Turnebus who had no other profession but letters (in which he was, in my opinion, the greatest man for a millennium) yet who had nothing donnish about him except the way he wore his gown and some superficial mannerisms which might not be elegant *al Cortegiano* but which really amount to nothing.[25] [B] And I loathe people who find it harder to put up with a gown askew than with a soul askew and who judge a man by his bow, his bearing and his boots. [A] For, within, Turnebus was the most polished of men. I often intentionally tossed him into subjects remote from his experience: his insight was so lucid, his grasp so quick and his judgement so sound that it would seem that he had never had any other business but war or statecraft.

Natures like that are fair and strong:

---

24. Persius, *Satires*, I, lxi.
25. Adrian Turnebus, for Montaigne, was the scholar 'who knew everything', even though he might not be elegant after the style of Castiglione's famous *Book of the Courtier*. Cf. II, 12, 'An apology for Raymond Sebond'.

[B]   *queis arte benigna*
Et meliore luto finxit præcordia Titan;

[Whose minds are made by Titan with gracious art and from a better clay;][26]

[A]   they keep their integrity even through a bad education. Yet it is not enough that our education should not deprave us: it must change us for the better.

When our Courts of Parliament have to admit magistrates, some examine only their learning: others also make a practical assay of their ability by giving them a case to judge. The latter seem to me to have the better procedure, and even though both those are necessary and both needed together, nevertheless the talent for knowledge is less to be prized than that for judging. Judgement can do without knowledge: but not knowledge without judgement. It is what that Greek verse says:

Ὡς οὐδέν ἡ μάθησις, ἤν μὴ νοῦς παρῇ

– 'what use is knowledge if there is no understanding?'[27] Would to God for the good of French justice that those Societies should prove to be as well furnished with understanding and integrity as they still are with knowledge!   [C]   *'Non vitae sed scholae discimus.'* [We are taught for the schoolroom not for life.][28]

[A]   Now we are not merely to stick knowledge on to the soul: we must incorporate it into her; the soul should not be sprinkled with knowledge but steeped in it.[29] And if knowledge does not change her and make her imperfect state better then it is preferable just to leave it alone. Knowledge is a dangerous sword; in a weak hand which does not know how to wield it it gets in its master's way and wounds him,   [C]   *'ut fuerit melius non didicisse'* [so that it would have been better not to have studied at all].[30]

[A]   Perhaps that is why we French do not require much learning in our wives (nor does Theology) and why, when Francis Duke of Brittany, the son of John V, was exploring the possibility of a marriage to Isabella, a princess of Scotland, and was told that she had been brought up simply and

26. Juvenal, *Satires*, XIV, 35. (Here *Titan* means Prometheus, the grandson of Titan. He fashioned men from clay and gave them souls made from fire stolen from the heavens.)
27. John Stobaeus, *Sententiae*, III, *De Prudentia: Sententiae monostichae*. Translated in the text.
28. Seneca, *Epist. moral.*, CVI, 12.
29. Seneca, *Epist. moral.*, XXXVI, 3–4.
30. Cicero, *Tusc. disput.*, II, iv, 12.

never taught to read, he replied that he liked her all the better for it and that a wife is learned enough when she can tell the difference between her husband's undershirt and his doublet.[31] And it is not as great a wonder as they proclaim it to be that our forebears thought little of book-learning and that even now it is only found by chance in the chief councils of our monarchs; for without the unique goal which is actually set before us (that is, to get rich by means of jurisprudence, medicine, paedagogy, and Theology too, a goal which does keep such disciplines respected) you would see them still as wretched as they ever were. If they teach us neither to think well nor to act well, what have we lost? [C] *'Postquam docti prodierunt, boni desunt.'* [Now that so many are learned, it is good men that we lack.][32] All other knowledge is harmful in a man who has no knowledge of what is good.

But the reason that I was looking for just now, could it not also arise from the fact that since studies in France have virtually no other end than the making of money, few of those whom nature has begotten for duties noble rather than lucrative devote themselves to learning; or else they do so quite briefly, withdrawing (before having acquired a taste for learning) to a profession which has nothing in common with books; normally there are few left to devote themselves entirely to study except people with no money, who do strive to make their living from it. And the souls of people like that – souls of the basest alloy by nature, by their home upbringing and by example – bear but the false fruits of knowledge. For learning sheds no light on a soul which lacks it; it cannot make a blind man see: her task is not to furnish him with sight but to train his own and to put it through its paces – if, that is, it has legs and hoofs which are sound and capable.

Learning is a good medicine: but no medicine is powerful enough to preserve itself from taint and corruption independently of defects in the jar that it is kept in. One man sees clearly but does not see straight: consequently he sees what is good but fails to follow it; he sees knowledge and does not use it. The main statute of Plato in his *Republic* is to allocate duties to his citizens according to their natures.[33] Nature can do all, does do all: the lame are not suited to physical exercises, nor are lame souls suited to spiritual ones: misbegotten and vulgar souls are unworthy of philosophy. When we see a man ill-shod, we are not surprised when he

31. Known from Gilles Corrozet, *Propos memorables de nobles hommes de la chrestienté.*
32. Seneca, *Epist. moral.*, XCV, 13.
33. Plato, *Republic*, III, 415A.

turns out to be a cobbler! In the same way it would seem that experience often shows us that doctors are the worst doctored, theologians the most unreformed and the learned the least able.

In Ancient times Ariston of Chios was right to say that philosophers do harm to their hearers, since most souls are incapable of profiting from such teaching, which when it cannot do good turns to bad: *'asotos ex Aristippi: acerbos ex Zenonis schola exire.'* [debauchees come from the school of Aristippus; little savages from Zeno's.][34]

[A] In that excellent education that Xenophon ascribed to the Persians,[35] we find that they taught their children to be virtuous, just as other peoples teach theirs to read. [C] Plato says that the eldest son in their royal succession was brought up as follows: at birth he was entrusted not to women but to eunuchs holding highest authority in the king's entourage on account of their virtue. They accepted responsibility for making his body fair and healthy; when he was seven they instructed him in riding and hunting. When he reached fourteen they placed him into the hands of four men: the wisest man, the most just man, the most temperate man and the most valiant man in all that nation. The first taught him religion; the second, to be ever true; the third to be master of his desires; the fourth to fear nothing.[36]

[A] It is a matter worthy of the highest attention that in that excellent constitution which was drawn up by Lycurgus and was truly prodigious in its perfection, the education of the children was the principal duty, yet little mention was made of instruction even in the domain of the Muses; it was as though those great-hearted youths despised any yoke save that of virtue, so that they had to be provided not with Masters of Arts but Masters of Valour, of Wisdom and of Justice – [C] an example followed by Plato in his *Laws*. [A] Their mode of teaching consisted in posing questions about the judgements and deeds of men: if the pupils condemned or praised this or that person or action, they had to justify their statement: by this means they both sharpened their understanding and learned what is right.

In Xenophon, Astiages asked Cyrus for an account of his last lesson.[37] 'In our school,' he said, 'a big boy had a tight coat; he took a coat away from a classmate of slighter build, because it was on the big side, and gave him his. Teacher made me judge of their quarrel and I judged that things

---

34. Cicero, *De natura deorum*, III, xxxi, 77.
35. In his *Cyropaedia*. Cf. John Stobaeus, *Sermo* LXXXIV, 30 f.
36. Plato, *Alcibiades*, I; John Stobaeus, *Sermo* LXXXIV, 10–20.
37. Xenophon, *Cyropaedia*, I, iii, 15.

were best left as they were, since both of them were better off by what had been done. He then showed me that I had judged badly, since I had confined myself to considering what seemed better, whereas I should first have dealt with justice, which requires that no one should be subjected to force over things which belonged to him.' He then said he was beaten, just as we are in our village schools for forgetting the first aorist of *tuptō* ['I thrash']. (A dominie would have to treat me to a fine harangue in the demonstrative mode before he would convince me that his school was worth that one!) Those Persians wanted to shorten the journey, and since it is true that study, even when done properly, can only teach us what wisdom, right conduct and determination consist in, they wanted to put their children directly in touch with actual cases, teaching them not by hearsay but by actively assaying them, vigorously moulding and forming them not merely by word and precept but chiefly by deeds and examples, so that wisdom should not be something which the soul knows but the soul's very essence and temperament, not something acquired but a natural property.

While on this subject, when Agesilaus was asked what he thought should be taught to children he replied, 'What they should do when they are grown up.'[38] No wonder that education such as that should have produced such astonishing results. They used to go to other Grecian cities in search of rhetoricians, painters and musicians: the others came to Sparta for law-givers, statesmen and generals. In Athens they learned to talk well: here, to act well; there, to unravel sophistries and set at nought the hypocrisy of words craftily intertwined; here, to free themselves from the snares of pleasure and to set at nought great-heartedly the menaces of fortune and of death; the Athenians were occupied with words: the Spartans with things; there, it was the tongue which was kept in continuous training; here, there was a continuous training of the soul. That is why it was not odd that when Antipater demanded fifty of their sons as hostages they replied (quite the opposite to what we would) that they preferred to give twice as many grown-up men, so high a value did they place on depriving the boys of their national education.[39] When Agesilaus urged Xenophon to send his sons to be brought up in Sparta, it was not to learn rhetoric there nor dialectic but, he said, to learn the finest subject of all: namely how to obey and how to command.[40]

---

38. Plutarch (tr. Amyot), *Dicts notables des Lacedaemoniens*, 212F.
39. Ibid., 225A.
40. Erasmus, *Apophthegmata*, I, *Agesilaus*, XLIX.

[C] It is most pleasing to see Socrates in his own way poking fun at Hippias, who was telling him how he had earned a great sum of money as a schoolmaster in some little towns in Sicily whereas he could not earn a penny in Sparta since Spartans are stupid people who cannot measure or count, who do not esteem grammar or prosody, merely spending their time learning by heart the list of their kings, stories about the founding and decline of states and similar nonsense. When he had finished Socrates, by bringing him to admit in detail the excellence of the Spartans' political constitution and the happiness and virtue of their lives, let him anticipate his conclusion: that it was his own arts which were quite useless.[41]

Both in that martial government and in all others like it examples show that studying the arts and sciences makes hearts soft and womanish rather than teaching them to be firm and ready for war. The strongest State to make an appearance in our time is that of the Turks; and the Turkish peoples are equally taught to respect arms and to despise learning. I find that Rome was more valiant in the days before she became learned. In our time the most warlike nations are the most rude and ignorant: the Scythians, the Parthians and Tamburlane serve to prove that. When the Goths sacked Greece, what saved their libraries from being burned was the idea spread by one of the marauders that such goods should be left intact for their enemies: they had the property of deflecting them from military exercises while making them spend time on occupations which were sedentary and idle.

When our own King Charles V found himself master of the kingdom of Naples and of a large part of Tuscany without even drawing his sword, he attributed such unhoped for ease of conquest to the fact that the Italian princes and nobility spent more time becoming clever and learned than vigorous and soldierly.[42]

---

41. Plato, *Hippias Major*, 285; John Stobaeus, *De justitia*, sermo IX.
42. Several of the above examples are given in the anonymous *Tesoro politico* and appear to have been well-known at least afterwards.

26. On educating children

====

[The previous chapter, 'On schoolmasters' learning', was read in manuscript by visitors. Montaigne was encouraged to write at greater length on how to bring up boys. He had no son of his own but wrote partly for his friend and admirer Diane de Foix (who married in 1579 and was pregnant, hoping for a son and heir). Montaigne tells of his own upbringing by the best of fathers. Emphasizing the importance of things over words brings him to write of his own 'brain-child': the Essays, a matter of words. Thinking of his father's gentle methods based on exciting a child's love and enthusiasm for learning and good morals, he makes a diversion on kings and magistrates, who as 'fathers of the people', ought to use similar methods. The additions marked [C] show his new and growing respect for Plato (for whom books were indeed the preferred 'children' of superior minds). Montaigne launches a frontal attack (without naming him) on Hesiod, who made the path to virtue sweaty, painful and rough. For Montaigne, even children can find the paths to virtue lovely and delightful.]

For Madame Diane de Foix, Countess of Gurson

[A] I have never known a father fail to acknowledge his son as his own, no matter how [C] scurvy or crook-backed [A] he may be.[1] It is not that he fails to see his infirmities (unless he is quite besotted by his affection): but the thing is his, for all that! The same applies to me: I can see – better than anyone else – that these writings of mine are no more than the ravings of a man who has never done more than taste the outer crust of knowledge – even that was during his childhood – and who has retained only an ill-formed generic notion of it: a little about everything and nothing about anything, in the French style. For, in brief, I do know that there is such a thing as medicine and jurisprudence; that there are four parts to mathematics: and I know more or less what they cover. [C] (Perhaps I do also know how the sciences in general claim to serve us in our lives.) [A] But what I have definitely not done is to delve deeply into them, biting my nails over the study of Aristotle,[2] [C] that monarch of

---

1. '80: how *crooked-back or lame* he may be . . .
2. [A] until [C]: study of *Plato or* Aristotle . . . (A significant deletion).

the doctrine of the Modernists,[3] [A] or stubbornly persevering in any field[4] of learning. [C] I could not sketch even the mere outlines of any art whatsoever; there is no boy even in the junior forms who cannot say he is more learned than I am: I could not even test him on his first lesson, at least not in detail. When forced to do so, I am constrained to extract (rather ineptly) something concerning universals, against which I test his inborn judgement – a subject as unknown to the boys as theirs is to me.[5]

I have fashioned no sustained intercourse with any solid book except Plutarch and Seneca; like the Danaïdes I am constantly dipping into them and then pouring out: I spill some of it on to this paper but next to nothing on to me.[6]

[A] My game-bag is made for history [C] rather, [A] or poetry, which I love, being particularly inclined towards it;[7] for (as Cleanthes said) just as the voice of the trumpet rings out clearer and stronger for being forced through a narrow tube so too a saying leaps forth much more vigorously when compressed into the rhythms of poetry, striking me then with a livelier shock. As for my own natural faculties which are being assayed here, I can feel them bending beneath their burden. My concepts and judgement can only fumble their way forward, swaying, stumbling, tripping over; even when I have advanced as far as I can, I never feel satisfied, for I have a troubled cloudy vision of lands beyond, which I cannot make out. I undertake to write without preconceptions on any subject which comes to mind, employing nothing but my own natural resources: then if (as happens often) I chance to come across in excellent authors the very same topics I have undertaken to treat (as I have just done recently in Plutarch about the power of the imagination) I acknowledge myself to be so weak, so paltry, so lumbering and so dull compared with such men, that I feel scorn and pity for myself. I do congratulate myself,

---

3. *Modernists* (often a pejorative term, as in Spiegel's *Lexicon Juris Civilis*) was applied to Nominalists – neo-Aristotelians who refused to seek philosophical truths in revelation, restricting revealed truth to Christian theology.
4. [A] until [C]: any *solid* field. . .
5. Celio Calcagnini stressed that the young can be knowledgeable about 'universals' but not about 'particulars', which depend on experience. Cf. also Aristotle, *Nicomachaean Ethics*, VI, 8, 5–8.
6. Those daughters of Danaus who killed their husbands and were condemned to fill a leaky jar with water in Hades.
7. [A] until [C]: made, *where books are concerned*, for history, or poetry . . .
     Then, Cleanthes in Seneca, *Epist. moral.*, CVIII, 10.

however, that my opinions frequently coincide with theirs [C] and on the fact that I do at least trail far behind them murmuring 'Hear, hear'. [A] And again, I do know (what many do not) the vast difference there is between them and me. What I myself have thought up and produced is poor feeble stuff, but I let it go on, without plastering over the cracks or stitching up the rents which have been revealed by such comparisons.[8] [C] You need a strong backbone if you undertake to march shoulder to shoulder with fellows like that.

[A] Those rash authors of our own century who scatter whole passages from ancient writers throughout their own worthless works, seeking to acquire credit [C] thereby,[9] [A] achieve the reverse; between them and the Ancients there is an infinite difference of lustre, which gives such a pale sallow ugly face to their own contributions that they lose far more than they gain.

– [C] There were two opposing concepts. Chrysippus the philosopher intermingled not merely passages from other authors into his writings but entire books: in one he cited the whole of the *Medea* of Euripides! Apollodorus said that if you cut out his borrowings his paper would remain blank. Epicurus on the other hand left three hundred tomes behind him: not one quotation from anyone else was planted in any of them.[10] –

[A] The other day I chanced upon such a borrowing. I had languished along behind some French words, words so bloodless, so fleshless and so empty of matter that indeed they were nothing but French and nothing but words. At the end of a long and boring road I came upon a paragraph which was high, rich, soaring to the clouds. If I had found a long gentle slope leading up to it, that would have been pardonable: what I came across was a cliff surging up so straight and so steep that I knew I was winging my way to another world after the first half-a-dozen words. That was how I realized what a slough I had been floundering through beforehand, so base and so deep that I did not have the heart to sink back into it.

If I [C] were to stuff one of my chapters with such rich spoils, that chapter [A] would reveal[11] all too clearly the silliness of the others.

---

8. '80: comparisons; *for otherwise I would have given birth to monstrosities, as do* those rash authors . . .
9. '80: credit *by their theft* achieve . . .
10. Diogenes Laertius, *Life of Chrysippus* and *Life of Epicurus.*
11. '80: If I *used such rich paintings as make-up for a chapter of mine that* would reveal . . .

[C] Reproaching other people for my own faults does not seem to me to be any more odd than reproaching myself for other people's (as I often do). We must condemn faults anywhere and everywhere, allowing them no sanctuary whatsoever. Yet I myself know how valiantly I strive to measure up to my stolen wares and to match myself to them equal to equal, not without some rash hope of throwing dust in the eyes of critics who would pick them out (though more thanks to the skill with which I apply them than to my skill in discovering them or to any strengths of my own). Moreover I do not take on those old champions all at once, wrestling with them body to body: it is a matter of slight, repeated, tiny encounters. I do not cling on: I merely try them out, going less far than I intended when haggling with myself over them. If I should prove merely up to sparring with them it would be a worthy match, for I only take them on when they are toughest.

But what about the things I have caught others doing? They bedeck themselves in other men's armour, with not even their fingertips showing. As it is easy for the learned to do on some commonplace subject, they carry through their projected work with bits of what was written in ancient times, patched together higgledy-piggledy. In the case of those who wish to hide their borrowings and pass them off as their own, their action is, first and foremost, unjust and mean: they have nothing worthwhile of their own to show off so they try to recommend themselves with someone else's goods; secondly it is stupid to be satisfied with winning, by cheating, the ignorant approbation of the crowd while losing all credit among men of understanding: their praise alone has any weight, but they look down their noses at our borrowed plasterwork. For my part there is nothing that I would want to do less: I only quote others the better to quote myself.

None of this applies to *centos* published as such; I have seen some very ingenious ones in my time, including one under the name of Capilupi, not to mention those of the ancients. Their authors show their wits in both this and other ways, as did Justus Lipsius in his *Politics*, with its industriously interwoven erudition.[12]

12. A *cento* was a literary poem, entirely, and often ingeniously, composed of lines from other authors and made to apply to a different subject. Lelio Capilupi's *cento* was a work on monks, entirely composed of lines of Virgil. Justus Lipsius, an author much admired by Montaigne, who knew his *Politics* well and borrowed much from it, did not write *centos* but did at times make his works into a patchwork of borrowings from ancient writers, especially the Stoics.

[A]   Be that as it may; I mean that whatever these futilities of mine may be, I have no intention of hiding them, any more than I would a bald and grizzled portrait of myself just because the artist had painted not a perfect face but my own. Anyway these are *my* humours, *my* opinions: I give them as things which *I* believe, not as things to be believed. My aim is to reveal my own self, which may well be different tomorrow if I am initiated into some new business which changes me. I have not, nor do I desire, enough authority to be believed. I feel too badly taught to teach others.

Now in my home the other day somebody read the previous chapter and told me that I ought to spread myself a bit more on the subject of children's education. If, My Lady, I did have some competence in this matter I could not put it to better use than to make a present of it to that little man who is giving signs that he is soon to make a gallant sortie out of you. (You are too great-souled to begin other than with a boy.) Having played so large a part in arranging your marriage I have a rightful concern for the greatness and prosperity of all that springs from it, quite apart from that long enjoyment you have had of my service to you, by which I am indeed bound to desire honour, wealth and success to anything that touches on you. But in truth I know nothing about education except this: that the greatest and the most important difficulty known to human learning seems to lie in that area which treats how to bring up children and how to educate them.

[C]   It is just as in farming: the ploughing which precedes the planting is easy and sure; so is the planting itself: but as soon as what is planted springs to life, the raising of it is marked by a great variety of methods and by difficulty. So too with human beings; it is not much trouble to plant them, but as soon as they are born we take on in order to form them and bring them up a diversity of cares, full of bustle and worry.   [A]   When they are young they give such slight and obscure signs of their inclinations, while their promises are so false and unreliable, that it is hard to base any solid judgement upon them.   [B]   Look at Cimon, Themistocles and hundreds of others; think how unlike themselves they used to be! Bear–cubs and puppies manifest their natural inclinations but humans immediately acquire habits, laws and opinions; they easily change or adopt disguises.[13]

[A]   Yet it is so hard to force a child's natural bent. That explains why, having chosen the wrong route, we toil to no avail and often waste years training children for occupations in which they never achieve anything. All

---

13. Cimon and Mnesiphilus Themistocles were, as young men, 'debauched and dissolute', then they reformed: Plutarch (tr. Amyot), *Si l'homme sage doit entremettre et mesler des affaires publiques*, 186v°.

the same my opinion is that, faced by this difficulty, we should always guide them towards the best and most rewarding goals, and that we should attach little importance to those trivial prognostications and foretellings we base on their childish actions.   [C]   Even Plato seems to me to give too much weight to them in his *Republic.*

[A]   Learning, My Lady, is a great ornament and a useful instrument of wondrous service, especially in those who are fortunate to live in so high an estate as yours. And in truth she does not find her true employment in hands base and vile. She is far more proud to deploy her resources for the conducting of a war, the commanding of a nation and the winning of the affection of a prince or of a foreign people than for drawing up dialectical arguments, pleading in a court of appeal or prescribing a mass of pills. And, therefore, My Lady, since I believe you will not overlook this aspect of the education of your children, you who have yourself tasted its sweetness and who belong to a family of authors – for we still possess the writings of those early de Foix, Counts from whom both the present Count, your husband, is descended and you yourself, while your uncle François, the Sieur de Candale, gives birth every day to new ones, which will spread an awareness of this family trait to many later centuries[14] – I want to tell you of one thought of mine which runs contrary to normal practice. That is all I am able to contribute to your service in this matter.

The responsibilities of the tutor you give your son (and the results of the education he provides depend on your choice of him) comprise many other elements which I do not touch upon since I have nothing worthwhile to contribute; as for the one subject on which I do undertake to give him my advice, he will only accept what I say insofar as it seems convincing to him. The son of the house is seeking book-learning[15] not to make money (for so abject an end is unworthy of the grace and favour of the Muses and anyway has other aims and depends on others) nor for external advantages, but rather for those which are truly his own, those which inwardly enrich and adorn him. Since I would prefer that he turned out to be an able man not an erudite one, I would wish you to be careful to select as guide for him a tutor with a well-formed rather than a well-filled brain. Let both be looked for, but place character and intelligence before knowledge; and let him carry out his responsibilities in a new way.

---

14. Gaston III, (*'Phœbus'*) Comte de Foix (†1391) wrote a famous book on hunting. It was published in Paris c. 1510. François de Candale, Bishop of Aire, translated Hermes Trismegistus and Euclid in 1578–9.
15. [A] until [C]: book-learning *and instruction*, not . . .

Teachers are for ever bawling into our ears as though pouring knowledge down through a funnel: our task is merely to repeat what we have been told. I would want our tutor to put that right: as soon as the mind in his charge allows it, he should make it show its fettle by appreciating and selecting *things* – and by distinguishing between them; the tutor should sometimes prepare the way for the boy, sometimes let him do it all on his own. I do not want the tutor to be the only one to choose topics or to do all the talking: when the boy's turn comes let the tutor listen to his pupil talking. [C] Socrates and then Archesilaus used to make their pupils speak first; they spoke afterwards. *'Obest plerumque iis qui discere volunt authoritas eorum qui docent.'* [For those who want to learn, the obstacle can often be the authority of those who teach.][16]

It is good to make him trot in front of his tutor in order to judge his paces and to judge how far down the tutor needs to go to adapt himself to his ability. If we get that proportion wrong we spoil everything; knowing how to find it and to remain well-balanced within it is one of the most arduous tasks there is. It is the action of a powerful elevated mind to know how to come down to the level of the child and to guide his footsteps. Personally I go uphill more firmly and surely than down.

Those who follow our French practice and undertake to act as schoolmaster for several minds diverse in kind and capacity, using the same teaching and the same degree of guidance for them all, not surprisingly can scarcely find in a whole tribe of children more than one or two who bear fruit from their education.

[A] Let the tutor not merely require a verbal account of what the boy has been taught but the meaning and the substance of it: let him judge how the child has profited from it not from the evidence of his memory but from that of his life. Let him take what the boy has just learned and make him show him dozens of different aspects of it and then apply it to just as many different subjects, in order to find out whether he has really grasped it and made it part of himself, [C] judging the boy's progress by what Plato taught about education. [A] Spewing up food exactly as you have swallowed it is evidence of a failure to digest and assimilate it; the stomach has not done its job if, during concoction, it fails to change the substance and the form of what it is given.[17]

---

16. For the importance of *things* not *words*, cf. Erasmus, *Apophthegmata*, III, *Socratica*, LXXXIII; then, Cicero, *De natura deorum*, I, v, 10.
17. '80: given. *They are only seeking a reputation for erudition. When they can say 'He's a learned man', they think they have said it all. Their souls . . .*

Our [B] souls are moved only at second-hand, being shackled and constrained to what is desired by someone else's ideas; they are captives, enslaved to the authority of what they have been taught. We have been so subjected to leading-reins that we take no free steps on our own. Our drive to be free has been quenched.   [C] *'Nunquam tutelae suae fiunt.'* [They are never free from tutelage.][18]   [B] In Pisa I met, in private, a decent man who is such an Aristotelian that the most basic of his doctrines is that the touchstone and the measuring-scale of all sound ideas and of each and every truth lie in their conformity with the teachings of Aristotle, outside of which all is inane and chimerical: Aristotle has seen everything, done everything. When that proposition was taken too widely and unfairly interpreted, for a long time he had a great deal of trouble from the Roman Inquisition.[19]

[A] Let the tutor pass everything through a filter and never lodge anything in the boy's head simply by authority, at second-hand. Let the principles of Aristotle not be principles for him any more than those of the Stoics or Epicureans. Let this diversity of judgements be set before him; if he can, he will make a choice: if he cannot then he will remain in doubt.   [C] Only fools have made up their minds and are certain:

[A1]   *Che non men che saper dubbiar m'aggrada.*

[For doubting pleases me as much as knowing.][20]

For if it is by his own reasoning that he adopts the opinions of Xenophon and Plato, they are no longer theirs: they are his.   [C] To follow another is to follow nothing: *'Non sumus sub rege: sibi quisque se vindicet.'* [We are under no king: let each man act freely.][21] Let him at least know what he does know.   [A] He should not be learning their precepts but drinking in their humours. If he wants to, let him not be afraid to forget where he got them from, but let him be sure that he knows how to appropriate them. Truth and reason are common to all: they no more belong to the man who first put them into words than to him who last did so.   [C] It

18. Seneca, *Epist. moral.*, XXXIII, 10.
19. Known from Montaigne's *Journal de Voyage* to be Dr Girolamo Borro, released from the prisons of the Inquisition on Papal authority. He wrote important books on motion and on the tides.
20. Dante, *Inferno*, XI, 93.
21. Seneca, *Epist. moral.*, XXXIII, 4 (adapted). Romans hated kings: here Seneca virtually means, 'We are under no despot.'

is no more *secundum Platonem* than *secundum me*: Plato and I see and understand it the same way. [A] Bees ransack flowers here and flowers there: but then they make their own honey, which is entirely theirs and no longer thyme or marjoram. Similarly the boy will transform his borrowings; he will confound their forms so that the end-product is entirely his: namely, his judgement, the forming of which is the only aim of his toil, his study and his education.

[C] Let him hide the help he received and put only his achievements on display. Pillagers and borrowers make a parade of what they have bought and built not of what they have filched from others! You never see the 'presents' given to a Parliamentary lawyer: what you see are the honours which he obtains for his children, and the families they marry into. Nobody puts his income on show, only his possessions. The profit we possess after study is to have become better and wiser.

[A] As Epicharmus said, that which sees and hears is our understanding; it is our understanding which benefits all, which arranges everything, which acts, which is Master and which reigns.²² We indeed make it into a slave and a coward by not leaving it free to do anything of itself. Which tutor ever asks his pupil what he thinks about [B] rhetoric or grammar or [A] this or that statement of Cicero? They build them into our memory, panelling and all, as though they were oracles, in which letters and syllables constitute the actual substance. [C] 'Knowing' something does not mean knowing it by heart; that simply means putting it in the larder of our memory. That which we rightly 'know' can be deployed without looking back at the model, without turning our eyes back towards the book. What a wretched ability it is which is purely and simply bookish! Book-learning should serve as an ornament not as a foundation – following the conclusion of Plato that true philosophy consists in resoluteness, faithfulness and purity, whereas the other sciences, which have other aims, are merely cosmetic.

[A] Take Palvel and Pompeo, those excellent dancing-masters when I was young: I would like to have seen them teaching us our steps just by watching them without budging from our seats, like those teachers who seek to give instruction to our understanding without making it dance – [C] or to have seen others teach us how to manage a horse, a pike or a lute, or to sing without practice, as these fellows do who want to teach us

22. Plutarch (tr. Amyot), *De la Fortune ou vertu d'Alexandre*, 313E (cf. *Quels animaux*, 508H).

to judge well and to speak well but who never give us exercises in judging or speaking.   [A]   Yet for such an apprenticeship everything we see can serve as an excellent book: some cheating by a page, some stupidity on the part of a lackey, something said at table, all supply new material.

For this purpose mixing with people is wonderfully appropriate. So are visits to foreign lands: but not the way the French nobles do it (merely bringing back knowledge of how many yards long the Pantheon is, or of the rich embroidery on Signora Livia's knickers); nor the way others do so (knowing how much longer and fatter Nero's face is on some old ruin over there compared with his face on some comparable medallion) but mainly learning of the humours of those peoples and of their manners, and knocking off our corners by rubbing our brains against other people's. I would like pupils to be taken abroad from their tenderest years, mainly (so as to kill two birds with one stone) to any neighbouring peoples whose languages, being least like our own, are ones which our tongue cannot get round unless you start bending it young. And, besides, it is a universally received opinion that it is not sensible to bring up a boy in the lap of his parents. Natural affection makes parents too soft, too indulgent – even the wisest of them. They are incapable of either punishing his faults or of bringing him up as roughly and as dangerously as he ought to be. They could not bear to see him riding back from his training all dirty and sweaty,   [C]   drinking this hot, drinking that cold,   [A]   nor to see him on a fractious horse, or up against a tough opponent foil in hand, nor with his first arquebus. But there is no other prescription: anyone who wants to be absolutely certain of making a real man of him must not spare his youth and must frequently flout the laws of medicine.

[B]   *vitamque sub dio et trepidis agat*
 In rebus.

[Let him camp in the open, amidst war's alarms.][23]

[C]   Nor is it enough to toughen up his soul: you must also toughen up his muscles.[24] The Soul is too hard-pressed if she is not seconded. She has too much to do herself to think of taking on the duties of both. I know how my own soul groans in her fellowship with a body so soft and sensitive to pain and which relies too heavily on her; and I have noticed in my reading that in their writings my moral guides pass off as examples of greatness of soul or strength of mind things which really belong to a tough

23. Horace, *Odes*, III, ii, 5.
24. Cf. Erasmus, *Apophthegmata*, VII, *Plato*, XVII.

skin or to strong bones. I have known men, women and children who are so constituted that a good beating means less to them than a pinch does to me, and who stir neither tongue nor eyebrow under the blows. When athletes play the philosopher in endurance it is strength of muscle rather than strength of mind. Now learning to endure toil is learning to endure pain: *'labor callum obducit dolori'* [toil puts callouses on our minds, against pain].[25] Pain and discomfort in training are needed to break him in for the pain and discomfort of dislocated joints, of the stone and of cauterizings – and of dungeons and tortures as well for, seeing the times we live in, those two may concern the good man as much as the bad. We are experiencing that now: whoever bears arms against Law threatens the best of men with the cat-o'-nine-tails and the rope.

[A] And then the authority of the tutor, which must be sovereign for the boy, is hampered and interrupted by the presence of his parents. Add to which the respect paid to the boy by his household and his awareness of the resources and dignity of his family are not in my opinion trivial disadvantages at that particular age.

Yet in the school of conversation among men I have often noticed a perversion: instead of learning about others we labour only to teach them about ourselves and are more concerned to sell our own wares than to purchase new ones. In our commerce with others, silence and modesty are most useful qualities. Train the lad to be sparing and reticent about his accomplishments (when he eventually has any) and he will not take umbrage when unlikely tales and daft things are related in his presence – for it is unmannerly and impolite to criticize everything which is not to our liking. [C] Let him be satisfied with correcting himself without being seen to reproach others for doing things he would not do himself and without flouting public morality: *'Licet sapere sine pompa, sine invidia.'* [One should be wise without ostentation or ill-will.][26] Let him shun any semblance of impolitely laying down the law, as well as that puerile ambition to wish to appear clever by being different or to earn a name for criticizing or flaunting novelties. It is only appropriate for great poets freely to break the rules of poetry: similarly it is intolerable for any but great and illustrious souls to give themselves unaccustomed prerogatives: – *'Si quid Socrates et Aristippus contra morem et consuetudinem fecerint, idem sibi ne arbitretur licere: magnis enim illi et divinis bonis hanc licentiam assequebantur.'* [Although Socrates and Aristippus sometimes

25. Cicero, *Tusc. disput.*, II, xv, 36.
26. Seneca, *Epist. moral.*, CIII, 5.

flouted normal rules and customs, one should not feel free to do the same: they obtained that privilege by qualities great and sublime.][27]

[A] The boy will be taught not to get into a discussion or a quarrel except when he finds a sparring-partner worth wrestling with – and even then not to employ all the holds which might help him but merely those which help him most. Teach him a certain refinement in sorting out and selecting his arguments, with an affection for relevance and so for brevity.

Above all let him be taught to throw down his arms and surrender to truth as soon as he perceives it, whether that truth is born at his rival's doing or within himself from some change in his ideas. He will never be up in a pulpit reading out some prescribed text: he only has to defend a case when it has his approbation. He is not going to take up the kind of profession in which freedom to think again, or to admit mistakes, has been traded for ready cash. [C] *'Neque, ut omnia que præscripta et imperata sint defendat, necessitate ulla cogitur.'* [He is under no obligation to support all precepts and assertions.][28]

If the tutor's complexion is like mine he will so form the will of the boy that he will become a loyal subject of his monarch as well as a devoted and brave one, but he will throw cold water on any desire to be attached to him except through public service. Apart from several other disadvantages which cripple our freedom, when a man's judgement is pledged and purchased by private obligations, either it is partial and less free or else he can be taxed with unwisdom and ingratitude. A courtier can have neither the right to speak nor the desire to think other than favourably of a Master who from among so many thousands of his subjects has chosen to favour him with his own hand and to elevate him. Not unreasonably such favour and preferment will corrupt his freedom and dazzle him. That is why what that lot have to say on the topic is habitually at variance with all others in the State and little to be trusted.

[A] As for our pupil's talk, let his virtue and his sense of right and wrong shine through it [C] and have no guide but reason. [A] Make him understand that confessing an error which he discovers in his own argument even when he alone has noticed it is an act of justice and integrity, which are the main qualities he pursues; [C] stubbornness and rancour are vulgar qualities, visible in common souls whereas to think again, to change one's mind and to give up a bad case in the heat of the argument are rare qualities showing strength and wisdom.

27. Cicero, *De officiis*, I, xli, 148.
28. Cicero, *Academica*, II (*Lucullus*), iii, 8 (adapted).

[A]   When in society the boy will be told to keep his eyes open: I find that the front seats are normally taken as a right by the less able men and that great inherited wealth is hardly ever associated with ability; while at the top end of the table the talk was about the beauty of a tapestry or the bouquet of the malmsey, I have known many witty remarks at the other end pass unnoticed. He will sound out the capacity of each person: of a herdsman, a mason, a wayfarer: he must use what he can get, take what a man has to sell and see that nothing goes wasted: even other people's stupidity and weakness serve to instruct him. By noting each man's endowments and habits, there will be engendered in him a desire for the good ones and a contempt for the bad.

Put into his mind a decent, careful spirit of inquiry about everything: he will go and see anything nearby which is of singular quality: a building, a fountain, a man, the site of an old battle, a place which Caesar or Charlemagne passed through:

[C]   *Quæ tellus sit lenta gelu, quæ putris ab æstu,*
      Ventus in Italiam quis bene vela ferat.

[what land is benumbed with the cold, which dusty with heat, which favourable winds blow sails towards Italian coasts.][29]

[A]   He will inquire into the habits, means and alliances of various monarchs, things most pleasant to study and most useful to know. In his commerce with men I mean him to include – and that principally – those who live only in the memory of books. By means of history he will frequent those great souls of former years. If you want it to be so, history can be a waste of time: it can also be, if you want it to be so, a study bearing fruit beyond price –   [C]   the only study, Plato said, which the Spartans kept as their share.[30]   [A]   Under this heading what profit will he not get out of reading the *Lives* of our favourite Plutarch! But let our tutor remember the object of his trust, which is less to stamp   [C]   the date of the fall of Carthage on the boy as the behaviour of Hannibal and Scipio; less to stamp   [A]   the name of the place where Marcellus died as how his death there showed him unworthy of his task. Let him not so much learn what happened as judge what happened.   [C]   That, if you ask me, is the subject to which our wits are applied in the most diverse of manners. I have read hundreds of things in Livy which another has not found there. Plutarch found in him hundreds of things which I did not see

29. Propertius, IV, iii, 39–40.
30. Plato, *Greater Hippias* (beginning).

(and which perhaps the author never put there). For some Livy is purely a grammatical study; for others he is philosophy dissected, penetrating into the most abstruse parts of our nature. [A] There are in Plutarch developed treatises very worth knowing, for he is to my mind the master-craftsman at that job; but there are also hundreds of points which he simply touches on: he merely flicks his fingers towards the way we should go if we want to, or at times he contents himself with a quick shot at the liveliest part of the subject: those passages we must rip out and put out on display. [B] For example that one saying of his, 'that the inhabitants of Asia were slaves of one tyrant because they were incapable of pronouncing one syllable: NO,' may have furnished La Boëtie with the matter and moment of his book *De la Servitude volontaire*.[31] [A] Seeing Plutarch select a minor action in the life of a man, or an apparently unimportant saying, is worth a treatise in itself. It is a pity that intelligent men are so fond of brevity: by it their reputation is certainly worth all the more, but we are worth all the less. Plutarch would rather we vaunted his judgement than his knowledge, and he would rather leave us craving for more than bloated. He realized that you could say too much even on a good subject, and that Alexandridas rightly criticized the orator whose address to the ephors was good but too long, saying, 'Oh, Stranger, you say what you should, but not the way that you should!'[32] [C] People whose bodies are too thin pad them out: those whose matter is too slender pad it out too, with words.

[A] Frequent commerce with the world can be an astonishing source of light for a man's judgement. We are all cramped and confined inside ourselves: we can see no further than the end of our noses. When they asked Socrates where he came from he did not say 'From Athens', but 'From the world'.[33] He, whose thoughts were fuller and wider, embraced the universal world as his City, scattered his acquaintances, his fellowship and his affections throughout the whole human race, not as we do who only look at what lies right in front of us. When frost attacks the vines in my village my parish priest talks of God being angry against the human race: in his judgement the Cannibals are already dying of the croup! At the sight of our civil wars, who fails to exclaim that the world is turned upside

31. Plutarch (tr. Amyot), *De la mauvaise honte*, 79B. La Boëtie's book circulated under the title of *Contr'un* (*Against* [the rule of] One) after his death and was used as Protestant propaganda against the French King.
32. Plutarch (tr. Amyot), *Dicts notables des Lacedaemoniens*, 214F.
33. Plutarch (tr. Amyot), *Du bannissement, ou de l'exil*, 125D–E.

down and that the Day of Judgement has got us by the throat, forgetting that many worse events have been known in the past and that, in thousands of parts of the world, they are still having a fine old time! [B] Personally I am surprised that our wars turn out to be so mild and gentle, given their unpunished licentiousness. [A] When the hail beats down on your head the entire hemisphere seems stormy and tempestuous. Like that peasant of Savoy who said that if only that silly King of France had known how to use his luck properly he could have become the Duke's chief steward eventually! His mind could not conceive of any degree of grandeur above that of his Duke. [C] We are all caught in that same error without realizing it: a harmful error of great consequence. [A] Only a man who can picture in his mind the mighty idea of Mother Nature in her total majesty; who can read in her countenance a variety so general and so unchanging and then pick out therein not merely himself but an entire kingdom as a tiny, feint point: only he can reckon things at their real size. This great world of ours (which for some is only one species within a generic group) is the looking-glass in which we must gaze to come to know ourselves from the right slant. To sum up then, I want it to be the book which our pupil studies. Such a variety of humours, schools of thought, opinions, laws and customs teach us to judge sanely of our own and teach our judgement to acknowledge its shortcomings and natural weakness. And that is no light apprenticeship. So many revolutions, so many changes in the fortune of a state, teach us to realize that our own fortune is no great miracle. So many names, so many victories and conquests lying buried in oblivion, make it ridiculous to hope that we shall immortalize our names by rounding up ten armed brigands or by storming some hen-house or other known only by its capture. The proud arrogance of so many other nations' pomp and the high-flown majesty of the grandeur of so many courts strengthen our gaze to look firmly and assuredly, without blinking, at the brilliance of our own. So many millions upon millions of men dead and buried before us encourage us not to be afraid of going to join such a goodly company in the world to come.

And so on.

[C] Our life, said Pythagoras,[34] is like the vast throng assembled for the Olympic Games: some use their bodies there to win fame from the contests; others come to trade, to make a profit; still others – and they are by no means the worst – seek no other gain than to be spectators, seeing

---

34. Erasmus, *Apophthegmata*, VII, *Pythagoras* VII (from Cicero, *Tusc. disput.*, V, iii, 9).

how everything is done and why; they watch how other men live so that they can judge and regulate their own lives.   [A]   All the most profitable treatises of philosophy (which ought to be the touchstone and measure of men's actions) can be properly reduced to examples. Teach the boy this:

> [B]   *quid fas optare, quid asper*
> *Utile nummus habet; patriæ charisque propinquis*
> *Quantum elargiri deceat: quem te Deus esse*
> *Jussit, et humana qua parte locatus es in re;*
> *Quid sumus, aut quidnam victuri gignimur;*

[what he may justly wish for; that money is hard to earn and should be used properly; the extent of our duty to our country and to our dear ones; what God orders you to be, and what place He has assigned to you in the scheme of things; what we are and what we shall win when we have overcome;][35]

teach him   [A]   what knowing and not knowing means (which ought to be the aim of study); what valour is, and justice and temperance; what difference there is between ordinate and inordinate aspirations; slavery and due subordination; licence and liberty; what are the signs of true and solid happiness; how far we should fear death, pain and shame:

> [B]   *Et quo quemque modo fugiatque feratque laborem;*
>
> [How we can flee from hardships and how we can endure them;][36]

[A]   what principles govern our emotions and the physiology of so many and diverse stirrings within us. For it seems to me that the first lessons with which we should irrigate his mind should be those which teach him to know himself, and to know how to die ... and to live.   [C]   Among the liberal arts, start with the art which produces liberal men. All of them are of some service in the regulation and practice of our lives, just as everything else is; but let us select the one which leads there directly and professes to do so.

If we knew how to restrict our life's appurtenances to their right and natural limits, we would discover that the greater part of the arts and sciences as now practised are of no practical use to us, and that, even in those which are useful, there are useless wastes and chasms which we would do better to leave where they are; following what Socrates taught, we should set limits to our study of subjects which lack utility.

35. Persius, *Satires*, III, 69–73.
36. Virgil, *Aeneid*, III, 459.

[A]   *sapere aude,*
Incipe: vivendi qui recte prorogat horam,
Rusticus expectat dum defluat amnis; at ille
Labitur et labetur in omne volubilis ævum.

[Dare to be wise. Start now. To put off the moment when you will start to live justly is to act like the bumpkin who would cross but who waits for the stream to dry up; time flows and will flow for ever, as an ever-rolling stream.][37]

There is great folly in teaching our children

[B]   *quid moveant Pisces, animosaque signa Leonis,*
Lotus et Hesperia quid Capricornus aqua,

[what influences stem from Pisces and the lively constellation of Leo or from Capricorn which plunges into the Hesperian Sea,][38]

about the heavenly bodies   [A]   and the motions of the Eighth Sphere before they know about their own properties.

Τί πλειώδευσι χάμυί,
Τί δ 'ἀστράσι βοώτεω!

[What do the Pleiades or the Herdsman matter to me!][39]

[C]   Writing to Anaximenes, Pythagoras asked: 'What mind am I supposed to bring to the secrets of the heavens, having death and slavery ever present before my eyes?' (At that time the kings of Persia were preparing for war against his country.)[40] We could all ask the same: 'Assaulted as I am by ambition, covetousness, rashness and superstition, and having such enemies to life as that within me, should I start wondering about the motions of the Universe?'

[A]   Only after showing the boy what will make him a wiser and a better man will you explain to him the elements of Logic, Physics, Geometry and Rhetoric. Since his judgement has already been formed he will soon get to the bottom of any science he chooses. His lessons will sometimes be discussion, sometimes reading from books; at times the tutor will provide him with extracts from authors suited to his purposes: at others the tutor will pick out the marrow and chew it over for him. If the

37. Horace, *Epistles*, I, ii, 40–3.
38. Propertius, IV, i, 85–6.
39. In Ptolomaic astronomy, the Eighth Sphere contained the fixed stars (Anacreon, *Odes*, XVII, x).
40. Diogenes Laertius, *Life of Anaximenes.*

tutor is not sufficiently familiar with those books to find the discourses in them which serve his purposes you could associate with him a scholar who could furnish him, as the need arises, with material for him to arrange and dispense to the growing boy.

Who can doubt that such lessons will be more natural and easy than those in Theodore Gaza,[41] whose precepts are prickly and nasty, and whose words are hollow and fleshless, with nothing to get hold of or to quicken the mind. Here then is nourishment for the soul to bite on. The fruit is incomparably more plentiful and will ripen sooner.

Oddly, things have now reached such a state that even among men of intelligence philosophy means something fantastical and vain, without value or usefulness, [C] both in opinion and practice. [A] The cause lies in chop-logic which has captured all the approaches. It is a great mistake to portray Philosophy with a haughty, frowning, terrifying face, or as inaccessible to the young. Whoever clapped that wan and frightening mask on her face! There is nothing more lovely, more happy and gay – I almost said more amorously playful. What she preaches is all feast and fun. A sad and gloomy mien shows you have mistaken her address.

Some philosophers were sitting together in the temple at Delphi one day. 'Either I am mistaken,' said Demetrius the grammarian, 'or your calm happy faces show that you are not having an important discussion.' One of them, Herakleon of Megara, retorted: 'Furrowed brows are for grammarians telling us whether *ballō* takes two *l*s in the future, researching into the derivation of the comparatives *keiron* and *beltion* and of the superlatives *keiriston* and *beltiston*: philosophical discussions habitually make men happy and joyful not frowning and sad.'[42]

> [B] *Deprendas animi tormenta latentis in ægro*
> *Corpore, deprendas et gaudia: sumit utrumque*
> *Inde habitum facies.*

[You can detect in a sickly body the hidden torments of the mind; you can detect her joys as well: the face reflects them both.][43]

[A] The soul which houses philosophy must by her own sanity make for a sound body. Her tranquillity and ease must glow from her; she must fashion her outward bearing to her mould, arming it therefore with gracious pride, a spritely active demeanour and a happy welcoming

---

41. Author of a fifteenth-century Greek grammar.
42. Plutarch (tr. Amyot), *Des oracles qui ont cessé*, 338A.
43. Juvenal, *Satires*, IX, 1879.

face. [C] The most express sign of wisdom is unruffled joy: like all in the realms above the Moon, her state is ever serene. [A] *Baroco* and *Baralipton* have devotees reeking of filth and smoke.[44] She does not. They know her merely by hearsay. Why, her task is to make the tempests of the soul serene and to teach hunger and fever how to laugh – not by imaginary epicycles but by reasons, [C] natural and palpable.[45] Her aim is virtue, which is not (as they teach in schools) perched on the summit of a steep mountain, rough and inaccessible. Those who have drawn nigh her hold that on the contrary she dwells on a beautiful plateau, fertile and strewn with flowers; from there she clearly sees all things beneath her; but if you know the road you can happily make your way there by shaded grassy paths, flower-scented, smooth and gently rising, like tracks in the vaults of heaven.[46]

This highest virtue is fair, triumphant, loving, as delightful as she is courageous, a professed and implacable foe to bitterness, unhappiness, fear and constraint, having Nature for guide, Fortune and Pleasure for her companions: those who frequent her not have, after their own weakness, fashioned an absurd portrait of her, sad, shrill, sullen, threatening and glowering, perching her on a rocky peak, all on her own among the brambles – a spectre to terrify people.

This tutor of mine, who knows that his duty is to fill the will of his pupil with at least as much love as reverence for virtue, will know how to tell him that our poets are following commonplace humours: he will make him realize that the gods place sweat on the paths to the chambers of Venus rather than of Pallas.[47] And when he comes to know his own mind and is faced with a choice between Bradamante to court and enjoy or Angelica[48] – one with her natural beauty, active, noble, virile though not mannish, contrasting with the other's beauty, soft, dainty, delicate and all artifice; the one disguised as a youth with a shining helmet on her head, the other robed as a maiden with pearls in her headdress: then his very passion will be deemed manly if he chooses flat contrary to that effeminate Phrygian

44. Mnemonics representing by their vowels: i) the fourth mood of the Second figure of syllogisms; ii) the first indirect mood of the Second figure of syllogisms. (Here used to mock dry scholastic logic.)
45. '80: reasons *gross, manageable* and palpable. Since Philosophy . . .
46. In the myth of Hesiod, Virtue dwells on a fair plateau reached by a rugged and toilsome path. Cf. I, 20, 'To philosophize is to learn how to die'; note 7. (Seneca denied it, *De ira*, III, xiii.)
47. Venus, the goddess of love; Pallas, of wisdom.
48. Heroines in Ariosto's *Orlando furioso*.

shepherd.[49] The tutor will then be teaching him a new lesson: what makes true virtue highly valued is the ease, usefulness and pleasure we find in being virtuous: so far from it being difficult, children can be virtuous as well as adults; the simple, as well as the clever. The means virtue uses is control not effort. Socrates, the foremost of her darlings, deliberately renounced effort so as to glide along with her easy natural progress. She is a Mother who nurtures human pleasures: by making them just she makes them sure and pure; by making them moderate they never pant for breath or lose their savour; by cutting away those which she denies us she sharpens our appreciation of those she leaves us – an abundance of all those which Nature wills for us; Mother-like, she provides them not until we are satiated but until we are satisfied (unless, that is, we claim that her rule is the enemy of pleasure because she ordains drinking without drunkenness, eating without indigestion, and sex without the pox). If Virtue should lack the ordinary share of good fortune, she evades or does without it, or else she forges a private happiness of her own, neither floating nor changeable. She knows how to be rich, powerful and learned and how to lie on a perfumed couch; she does love life; she does love beauty, renown and health. But her own peculiar office is to know how to enjoy those good things with proper moderation and how to lose them with constancy: an office much more noble than grievous; without it the whole course of our life becomes unnatural, troubled, deformed; then you can indeed tie it to those rocky paths, those brambles and those spectres.

Were our pupil's disposition so bizarre that he would rather hear a tall story than the account of a great voyage or a wise discussion; that at the sound of the drum calling the youthful ardour of his comrades to arms he would turn aside for the drum of a troop of jugglers; that he would actually find it no more delightful and pleasant to return victorious covered with the dust of battle than after winning a prize for tennis or dancing: then I know no remedy except that his tutor should quickly strangle him when nobody is looking or apprentice him to make fairy-cakes in some goodly town – even if he were the heir of a Duke – following Plato's precept that functions should be allocated not according to the endowments of men's fathers but the endowments of their souls.[50]

---

49. Virgil, *Aeneid*, VII, 363: Paris who, in his famous judgement between Aphrodite, Hera and Athene, chose the more dainty and artificial Aphrodite (Venus). Montaigne opts for Bradamante.
50. This passage is toned down in ['95]. Cf. Plato, *Republic*, 415 BC; then, Persius, *Satires*, III, 23–4.

[A]   Since philosophy is the art which teaches us how to live, and since children need to learn it as much as we do at other ages, why do we not instruct them in it?

> [B]   *Udum et molle lutum est; nunc nunc properandus et acri*
> *Fingendus sine fine rota.*

[The clay is soft and malleable. Quick! hurry to fashion it on that potter's wheel which is for ever spinning.]

[A]   They teach us to live when our life is over. Dozens of students have caught the pox before they reach the lesson on temperance in their Aristotles.   [C]   Cicero said that even were he to live two men's lives he would never find enough time to study the lyric poets.[51] I find these chop-logic merchants even more gloomily useless. Our boy is too busy for that: to school-learning he owes but the first fifteen or sixteen years of his life: the rest is owed to action. Let us employ a time so short on things which it is necessary to know.   [A]   Get rid of those thorny problems of dialectics – they are trivial: our lives are never amended by them; take the simple arguments of philosophy: learn how to select the right ones and to apply them. They are easier to grasp than a tale in Boccaccio: a boy can do it as soon as he leaves his nanny; it is much easier than learning to read and write. Philosophy has arguments for Man at birth as well as in senility.

I share Plutarch's conviction[52] that Aristotle never spent much of the time of his great pupil Alexander on the art of syllogisms nor on the principles of geometry: he taught him, rather, sound precepts concerning valour, prowess, greatness of soul and temperance, as well as that self-assurance which fears nothing. With such an armoury he sent him still a child to conquer the empire of the world with merely thirty thousand foot-soldiers, four thousand horsemen and forty-two thousand crowns. As for the other arts and sciences, Plutarch says that he held them in esteem, praising their excellence and their nobility; but whatever pleasure he found in them he did not allow himself to be surprised by a desire to practise them himself.

> [B]   *Petite hinc, juvenesque senesque,*
> *Finem animo certum, miserisque viatica canis.*

[Seek here, young men and old, a lasting purpose for your mind and a provision for white-haired wretchedness.][53]

---

51. According to Seneca, *Epist. moral.*, XLIX, 5.
52. Plutarch (tr. Amyot), *De la fortune ou vertu d'Alexandre*, 308GH.
53. Persius, *Satires*, V, 64–5.

[C]   That is what Epicurus says at the beginning of his letter to Meniceus: 'Let the youngest not reject philosophy nor the oldest tire of it. Whoever does otherwise seems to be saying that the season for living happily has not yet come or is already past.'[54]

[A]   Despite all this I do not want to imprison the boy.[55] I do not want him to be left to the melancholy humour of a furious schoolmaster. I do not want to corrupt his mind as others do by making his work a torture, slaving away for fourteen or fifteen hours a day like a porter.   [C]   When you see him over-devoted to studying his books because of a solitary or melancholy complexion, it would not be good I find to encourage him in it: it unfits boys for mixing in polite society and distracts them from better things to do. And how many men have I known in my time made as stupid as beasts by an indiscreet hunger for knowledge! Carneades was turned so mad by it that he could not find time to tend to his hair or his nails.[56]

[A]   Nor do I wish to have his noble manners ruined by the uncouthness or barbarity of others. In antiquity French wisdom was proverbially good at the outset, but lacking in staying power.[57] And truly, still now, nothing is more gentlemanly than little French children; but they normally deceive the hopes placed in them, being in no ways outstanding once they are grown up. I have heard men of wisdom maintain that it is those colleges which parents send children to – and we have them in abundance – which make them so stupid.

For our boy any place and any time can be used to study: his room, a garden; his table, his bed; when alone or in company; morning and evening. His chief study will be philosophy, that Former of good judgement and character who is privileged to be concerned with everything.

It was an orator Isocrates who, being begged to talk about his art at a feast, replied (rightly we all think): 'What I can do, this is no time for: what this is time for, I cannot do!'[58] To present harangues and rhetorical debates to a company gathered for laughter and good cheer would be to mix together things too discordant.

You can say the same of all the other disciplines, but not of that part of

54. Diogenes Laertius, *Life of Epicurus*.
55. [A] until [C]: boy *in a college*. I do not . . .
56. Diogenes Laertius, *Life of Carneades*.
57. The Romans said this of the French (the Gauls') fighting-power (Erasmus, *Apophthegmata*, VI, *varie mixta*, CIII).
58. Plutarch (tr. Amyot), *Premier livre des Propos de Table*, 359F.

Philosophy which treats of Man, his tasks and his duties: by the common consent of all the wise, she should not be barred from sports nor feastings seeing that commerce with her is sweet. And Plato having invited her to his *Banquet*, we can see how she entertained the guests in a relaxed manner appropriate to time and place even when treating one of her most sublime and most salutary themes:[59]

> *Æque pauperibus prodest, locupletibus æque;*
> *Et, neglecta, æque pueris senibúsque nocebit.*

[She is equally helpful to the poor and the rich: neglect her, and she equally harms the young and the old.][60]

In this way he will certainly lie fallow much less than do others. Now we can take three times as many steps strolling about a long-gallery and still feel less tired than on a walk to a definite goal: so too our lessons will slip by unnoticed if we apparently happen upon them, as, restricted to neither time or place, they intermingle with all our activities. The games and sports themselves will form a good part of his studies: racing, wrestling, [C] music-making, [A] dancing, hunting and the handling of arms and horses. I want his outward graces, his social ease [C] and his physical dexterity [A] to be moulded step by step with his soul. We are not bringing up a soul; we are not bringing up a body: we are bringing up a man. We must not split him into two. We must not bring up one without the other but, as Plato said, lead them abreast like a pair of horses harnessed together to the same shaft. [C] And does not Plato when you listen to him appear to devote more time and care to exercising the body, convinced that the mind may be exercised with the body but not vice versa?

[A] This education is to be conducted, moreover, with a severe gentleness, not as it usually is.[61] Instead of children being invited to letters as guests, all they are shown in truth are cruelty and horror. Get rid of violence and force: as I see it, nothing so fundamentally stultifies and bastardizes a well-born nature.

If you want the boy to loathe disgrace and punishment do not harden him to them. Harden him to sweltering heat and to cold, to wind and sun

59. Plato's *Symposium* in French, and sometimes in English, is known as the *Banquet*: its theme is the nature of love. Then Plutarch, ibid., 360E.

60. Horace, *Epistles*, I, i, 25–6.

61. [A] until [C]: *usually is in colleges, where* instead ... (Cf. Plutarch (tr. Amyot), *Regles et Preceptes de Santé*, 302B.)

and to such dangers as he must learn to treat with contempt. Rid him of all softness and delicacy about dress and about sleeping, eating and drinking. Get him used to anything. Do not turn him into a pretty boy or a ladies' boy but into a boy who is fresh and vigorous. [C] Boy, man and now old man, I have always thought this. But I have always disliked, among other things, the way our colleges are governed. Their failure would have been less harmful, perhaps, if they had leant towards indulgence. They are a veritable gaol for captive youth. By punishing boys for depravity before they are depraved, you make them so.

Go there during lesson time: you will hear nothing but the screaming of tortured children and of masters drunk with rage. What a way to awaken a taste for learning in those tender timorous souls, driving them to it with terrifying scowls and fists armed with canes! An iniquitous and pernicious system. And besides (as Quintilian justly remarked)[62] such imperious authority can lead to dreadful consequences – especially given our form of flogging.

How much more appropriate to strew their classrooms with leaf and flower than with blood-stained birch-rods. I would have portraits of Happiness there and Joy, with Flora and the Graces, as Speucippus the philosopher did in his school.[63]

When they have something to gain, make it enjoyable. Health-giving foods should be sweetened for a child: harmful ones made to taste nasty.

It is amazing how concerned Plato is in his *Laws* with the amusements and pastimes of the youths of his City and how he dwells on their races, sports, singing, capering and dancing, the control and patronage of which has been entrusted, he said, in antiquity to the gods, to Apollo, the Muses and Minerva. His care extends to over a hundred precepts for his gymnasia, yet he spends little time over book-learning; the only thing he seems specifically to recommend poetry for is the music.[64]

[A] In our manners and behaviour any strangeness and oddness are to be avoided as enemies of easy mixing in society – [C] and as monstrosities. Who would not have been deeply disturbed by Alexander's steward Demophon whose complexion made him sweat in the shade and shiver in the sun? [A] I have known men who fly from the smell of apples rather than from gunfire; others who are terrified of a mouse, who vomit at the

62. Quintilian, *Institutio*, I, iii, 13–16. Dismissing Chrysippus' belief in the value of flogging, Quintilian held that it can produce mental depression.
63. Diogenes Laertius, *Life of Speucippus*.
64. Plato, *Laws*, VII; then, for Demophon, Sextus Empiricus, *Hypotyposes*, I, xiv, and for Germanicus, Plutarch (tr. Amyot), *De l'envie et la haine*, 108A.

sight of cream or when a feather mattress is shaken up (like Germanicus who could not abide cocks or their crowing). Some occult property may be involved in this, but, if you ask me, if you set about it young enough you could stamp it out.

One victory my education has achieved over me (though not without some trouble, it is true) is that my appetite can be brought to accept without distinction any of the things people eat and drink except beer. While the body is still supple it should, for that very reason, be made pliant to all manners and customs. Provided that he can restrain his appetites and his will, you should not hesitate to make the young man suited to all peoples and companies, even, should the need arise, to immoderation and excess.

[C] His practice should conform to custom. [A] He should be able to do anything but want to do only what is good. (The very philosophers do not approve of Calesthenes for falling from grace with his master Alexander the Great, by declining to match drink for drink with him. He will laugh, fool about and be unruly with his Prince.) I would want him to outstrip his fellows in vigour and firmness even during the carousing and that he should refrain from wrongdoing not because he lacks strength or knowledge but because he does not want to do it. [C] *'Multum interest utrum peccare aliquis nolit aut nesciat.'* [There is a great difference between not wanting to do evil and not knowing how to.][65]

[A] My intention was to honour a nobleman who is as far removed from such excesses as any man in France when I asked him, in the presence of guests, how many times in his life he had had to get drunk while serving the King in Germany. He took it in the right spirit and said he had done it three times; and he told me about them. (I have known people who have run into real difficulties when frequenting that nation because they lacked this ability.)

I have often noted with great astonishment the extraordinary character of Alcibiades who, without impairing his health, could so readily adapt to diverse manners: at times he could outdo Persians in pomp and luxury; at others, Spartans in austerity and frugal living.[66] He was a reformed man in Sparta, yet equally pleasure-seeking in Ionia:

> *Omnis Aristippum decuit color, et status, et res.*

> [On Aristippus any colour, rank or condition was becoming.]

65. Seneca, *Epist. moral.*, XC, 46.
66. Plutarch, *Life of Alcibiades.*

Thus would I fashion my pupil:

> *quem duplici panno patientia velat*
> *Mirabor, vitæ via si conversa decebit,*
> *Personámque feret non inconcinnus utramqué.*

[One who is patiently clad in rags yet could also adapt to the opposite extreme, playing both roles becomingly: him I will admire.][67]

Such are my lessons.[68] [C] For him who draws most profit from them, they are acts, not facts. To see his deeds is to hear his word: to hear his word is to see his deeds. 'God forbid,' says someone in Plato,[69] 'that philosophy should mean learning a lot of things and then talking about the arts: *'Hanc amplissimam omnium artium bene vivendi disciplinam vita magis quam literis persequuti sunt'.* [The fullest art of all – that of living good lives – they acquired more from life than from books.][70]

Prince Leon of the Phliasians inquired of Heraclides of Pontus which art or science he professed. 'I know none of them', he replied; 'I am a philosopher.' Diogenes was reproached for being ignorant yet concerned with philosophy. 'My concern is all the more appropriate,' he replied. When Hegesias begged him to read a certain book he replied, 'How amusing of you. You prefer real figs to painted ones, so why not true and natural deeds to written ones?'[71]

My pupil will not say his lesson: he will do it. He will rehearse his lessons in his actions. You will then see whether he is wise in what he takes on, good and just in what he does, gracious and sound in what he says, resilient in illnesses, modest in his sports, temperate in his pleasures, [A] indifferent to the taste of his food, be it fish or flesh, wine or water; [C] orderly in domestic matters: *'Qui disciplinam suam, non ostentationem scientiæ, sed legem vitæ putet, quique obtemperet ipse sibi, et decretis pareat'* [as a man who knows how to make his education into a rule of life not a means of showing off; who can control himself and obey

---

67. Horace, *Epistles*, I, xvii, 23, 25, 26, 29.
68. '80: lessons, *in which doing goes with saying. For what is the use of preaching at his mind if deeds do not go along with it? You will see from what he undertakes whether there is any wisdom there: if there is any goodness in his actions, if he is* indifferent . . .
69. Plato, *The Lovers* (*Erastai*) 137 A B (which shows Socrates discussing the nature of philosophy with schoolboys).
70. Cicero, *Tusc. disput.*, IV, iii, 5.
71. Cicero, ibid., IV, iii, 8; then, Diogenes Laertius, *Life of Diogenes*.

his own principles].[72] The true mirror of our discourse is the course of our lives.

[A]   To a man who asked him why the Spartans never drew up written rules of bravery and gave them to their children, Zeuxidamus replied that they wished to accustom them to deeds not   [C]   words.   [A]   After[73] fifteen or sixteen years compare with that one of our college latinizers who has spent precisely as long simply learning to talk!

The world is nothing but chatter: I have never met a man who does not say more than he should rather than less. Yet half of our life is spent on that; they keep us four or five years learning the meanings of words and stringing them into sentences; four or five more in learning how to arrange them into a long composition, divided into four parts or five; then as many again in plaiting and weaving them into verbal subtleties.

Let us leave all that to those who make it their express profession.

When I was travelling to Orleans one day, on the plain this side of Cléry I met two college tutors who were coming from Bordeaux; there was about fifty yards between them; I could also make out some troops further away still, led by their officer (who was the Count de la Rochefoucault). One of my men asked the first of these tutors who was 'that gentleman coming behind him'? The tutor had not noticed the party following behind them and thought they were talking about his companion: 'He is not a gentleman,' he amusingly replied, 'but a grammarian. And I am a logician.' Now we who, on the contrary, are trying to form a gentleman not a grammarian or a logician should let them waste their own time: we have business to do elsewhere. Provided that our student be well furnished with *things*, words will follow only too easily: if they do not come easily, then he can drag them out slowly.

I sometimes hear people who apologize for not being able to say what they mean, maintaining that their heads are so full of fine things that they cannot deliver them for want of eloquence. That is moonshine. Do you know what I think? It is a matter of shadowy notions coming to them from some unformed concepts which they are unable to untangle and to clarify in their minds: consequently they cannot deliver them externally. They themselves do not yet know what they mean. Just watch them giving a little stammer as they are about to deliver their brain-child: you can tell that they have labouring-pains not at childbirth   [C]   but during

72. Cicero, *Tusc. disput.*, II, iv, 11.
73. Plutarch (tr. Amyot), *Dicts notables des Lacedaemoniens*, 217A.
   '80: deeds not *writings*. After . . .

conception! [A] They are merely licking an imperfect lump into shape.[74] For my part I maintain – [C] and Socrates is decisive – [A] that whoever has one clear living thought in his mind will deliver it even in Bergamask.[75] Or if he is dumb he will do so by signs.

Verbaque prævisam rem non invita sequentur.

[Once you have mastered the things the words will come freely.][76]

And as another said just as poetically though in prose: '*Cum res animum occupavere, verba ambiunt.*' [When things have taken hold of the mind, the words come crowding forth.][77] [C] And another one: '*Ipsae res verba rapiunt.*' [The things themselves ravish the words.][78]

[A] 'But he does not know what an ablative is, a conjunctive, a substantive: he knows no grammar!' Neither does his footman or a Petit-Pont fishwife[79] yet they will talk you to death if you let them and will probably no more stumble over the rules of their own dialect than the finest Master of Arts in France.

'But he knows no rhetoric nor how to compose an opening *captatio benevolentiae* for his gentle reader!'[80] He does not need to know that. All those fine 'colours of rhetoric' are in fact easily eclipsed by the light of pure and naïve truth. Those elegant techniques (as Afer shows in Tacitus) merely serve to entertain the masses who are unable to [C] take [A] heavier solid meat.[81]

Ambassadors from Samos came to King Cleomenes of Sparta with a long prepared speech to persuade him to go to war against Polycrates the Tyrant. He let them have their say and then replied: 'As for your preamble

---

74. Like bears, the offspring of which were thought to be born without form but 'licked into shape' by their parents.
75. Bergamask – the dialect of Bergamo (in Venice). The inhabitants and their language were considered rustic and uncouth. (In the context of imagery drawn from childbirth there is possibly also a play on *boucler à la bergamasque*, to shut up one's wife in a chastity-belt.)
76. Horace, *Ars poetica*, 311.
77. Marcus Annaeus Seneca, *Controversiae*, III.
78. Cicero, *De finibus*, III, v. 19.
79. *Petit-Pont* – the 'Billingsgate' of Paris.
80. A *captatio benevolentiae* is a literary device designed to catch the reader's sympathetic attention. It was taught as part of rhetoric and dialectic.
81. [A]: unable to *appreciate* heavier . . .
Tacitus, *Dialogus an sui saeculi oratores antiquioribus concedant*, XIX.

and preface, I no longer remember it; nor of course your middle bit. As for your conclusion, I will do none of it.'[82] An excellent answer, it seems to me, with a blow on the nose of those speechifiers.

[B] And what about this other case. The Athenians had to choose between two architects to take charge of a large building project. The first one was the more fly and presented himself with a fine prepared speech about the job to be done; he won the favour of the common people. The other architect merely spoke two or three words: 'Gentlemen of Athens: what he said, I will do.'[83]

[A] At the height of his eloquence Cicero moved many into ecstasies of astonishment. But Cato merely laughed:'Quite an amusing consul we have,' he said.[84]

Now a useful saying or a pithy remark is always welcome wherever it is put. [C] If it is not good in the context of what comes before it or after it, it is good in itself. [A] I am not one of those who hold that good scansion makes a good poem: let the poet lengthen a short syllable if he wants to. That simply does not count. If the invention of his subject matter is happy and if his wit and judgement have done their jobs, then I shall say: 'Good poet: but bad versifier.'

> *Emunctæ naris, durus componere versus.*
>
> [He has the flair, though his verses are harsh.]

Take his work (says Horace) and pull its measured verse apart at the seams —

> [B] *Tempora certa modosque, et quod prius ordine verbum est,*
> *Posterius facias, præponens ultima primis,*
> *Invenias etiam disjecti membra poetæ*

[Take away rhythm and measure; change the order of the words putting the first last and the last first: you will still find the poet in those scattered remains] –

[A] he will still not belie himself for all that: even bits of it will be beautiful.[85]

That is what Menander replied when the day came for his promised comedy and people chided him for not yet putting it in hand: 'It is already

---

82. Plutarch (tr. Amyot), *Dicts notables des Lacedaemoniens*, 218B.
83. Plutarch (tr. Amyot), *Instructions pour ceux qui manient les affaires d'Estat*, 163F.
84. Erasmus, *Apophthegmata*, V, *Cato Uticensis*, III.
85. Horace, *Satires*, I, iv, 8; 58–60.

composed,' he said. 'All I have to do is to put it into verse.'[86] Having thought the things through and arranged them in his mind, he attached little importance to the remainder. Since Ronsard and Du Bellay have brought renown to our French poetry, every little apprentice I know is doing more or less as they do, using noble words and copying their cadences. [C] *'Plus sonat quam valet.'* ['More din than sense.'][87] [A] Ordinary people think there never have been so many poets. But easy as it has proved to copy their rhymes they all fall short when it comes to imitating the rich descriptions of the one and the delicate invention of the other.

'Yes. But what will he do when they harass him with some sophistical syllogistic subtlety:

Bacon makes you drink;
Drinking quenches your thirst:
Therefore bacon quenches your thirst?

[C] Let him simply laugh at it: it is cleverer to laugh at it than to answer it. Or let him borrow Aristippus's amusing rejoinder: 'Why should I unravel that? It is bad enough all knotted up!' Someone challenged Cleanthes with dialectical trickeries; Chrysippus said to him: 'Go and play those conjuring tricks on children: do not interrupt the serious thoughts of a grown-up.'[88] [A] If this verbal jugglery – [C] *'contorta et aculeata sophismata'* [these contorted prickly sophisms][89] – [A] should persuade him to accept an untruth, that is dangerous: but if they do not result in action and simply move him to laughter, I really do not see why he should pay attention to them.

Some people are so daft that they will go a mile or so out of their way to hunt for a good word: [C] *'aut qui non verba rebus aptant, sed res extrinsecus arcessunt, quibus verba conveniant!'* [they do not fit words to things but look for irrelevant things to fit to their words!] Or again: *'Sunt qui alicujus verbi decore placentis vocentur ad id quod non proposuerant scribere.'* [There are authors who are led by the beauty of some attractive word to write what they never intended.][90] I myself am more ready to distort

---

86. Plutarch (tr. Amyot), *Si les Atheniens ont esté plus excellents en armes qu'en lettres*, 525DE.
87. Seneca, *Epistles*, XL, 5.
88. Both from Diogenes Laertius, *Life of Aristippus*. (Cf. also Erasmus, *Apophthegmata*, III, *Aristippus*, XIII.)
89. Cicero, *Academica*, II (*Lucullus*), xxiv, 75.
90. Quintilian, *Institutiones oratoriae*, VIII, iii, 30 (adapted); then, Seneca, *Epist. moral.*, LIX, 5.

a fine saying in order to patch it on to me than to distort the thread of my argument to go in search of one.

[A] It is, on the contrary, for words to serve and to follow: if French cannot get there, let Gascon do so. I want *things* to dominate, so filling the thoughts of the hearer that he does not even remember the words. I like the kind of speech which is simple and natural, the same on paper as on the lip; speech which is rich in matter, sinewy, brief and short; [C] not so much titivated and refined as forceful and brusque –

Haec demum sapiet dictio, quae feriet

[The good style of speaking is the kind which strikes home][91] –

[A] gnomic rather than diffuse, far from affectation, uneven, disjointed and bold – let each bit form a unity – not schoolmasterly, not monkish, not legalistic, but soldierly, rather as Sallust described Julius Caesar's [C] (though I do not quite see why he did so).[92] [B] I like to imitate the unruly negligence shown by French youth in the way they are seen to wear their clothes,[93] [C] with their mantles bundled round their neck, their capes tossed over one shoulder or [B] with a stocking pulled awry: it manifests a pride contemptuous of the mere externals of dress and indifferent to artifice. But I find it even better applied to speech. [C] All affectation is unbecoming in a courtier, especially given the hearty freedom of the French: and under a monarchy every gentleman is inevitably schooled in court manners. So we do well to lean towards the careless and natural. [A] I have no love for textures where the joins and seams all show (just as you ought not to be able to count the ribs or the veins in a beautiful body). [C] *'Quae veritati operam dat oratio, incomposita sit et simplex.'* [Speech devoted to truth should be straightforward and plain.][94] *'Quis accurate loquitur, nisi qui vult putide loqui?'* [Who can speak carefully unless he wants to sound affected?][95]

91. From the Epitaph of Lucan (the poet of the *Pharsalia*).
92. A very perspicacious judgement. Suetonius' alleged remark arises from a poor manuscript reading of a passage in his *Life of Caesar* which is corrected in modern editions but was accepted during the Renaissance.
  [B] until [C]: Julius Caesar's. *Let us boldly hold against him what was held against Seneca: his style was quick-lime, but without the sand.* I like to . . .
93. '88: clothes, *letting themselves be taken for German mercenaries, wearing a cape and with a stocking* . . .
94. Seneca, *Epist. moral.*, XL, 4 (on the style fit for a philosopher).
95. Ibid., LXXV, 1.

When eloquence draws attention to itself it does wrong by the substance of *things*.

Just as in dress it is the sign of a petty mind to seek to draw attention by some personal or unusual fashion, so too in speech; the search for new expressions and little-known words derives from an adolescent schoolmaster-ish ambition. If only I could limit myself to words used in Les Halles in Paris! That grammarian Aristophanes did not know the first thing about it when he criticized Epicurus for his simple words and for having perspicuity of language as his sole rhetorical aim.[96]

To imitate speech is easy: an entire nation can do it: to imitate judgement and the research for your material takes rather more time! Most readers think similar styles, when they find them, clothe similar bodies. But you cannot borrow strength or sinews: you can borrow mantles and finery. Most of the people who haunt my company talk like these *Essays*:[97] I cannot tell whether they think like them . . .

[A]   Athenians (says Plato) take copious and elegant speech as their share; Spartans, brevity; Cretans, fecundity of thought not speech – and they are the best.[98] Zeno said that he had two sorts of followers: those he termed *philologous*, who cared for real learning (they were his favourites) and those he termed *logophilous*, whose concern was with words.[99]

That does not mean that speaking well is not a fine thing or a good thing, but that it is not as good as we make it out to be. It irritates me that our life is taken up by it. I would prefer first to know my own language well and then that of the neighbours with whom I have regular dealings.

There is no doubt that Greek and Latin are fine and great accomplish-ments; but they are bought too dear. I will tell you a cheaper way of buying them: it was assayed on me. Anyone is welcome to use it. My late father, after having made all possible inquiries among the learned and the wise about the choicest form of education, was warned about the disadvantages of the current system: they told him that the length of time we spend learning languages, [C] which cost the Ancients nothing, [A] is the sole reason why we cannot attain to the greatness of mind and knowledge of those old Greeks and Romans. I do not believe that to be the sole reason. Nevertheless the expedient found by my father

96. Diogenes Laertius, *Life of Epicurus*.
97. That is, they talk in French (not Gascon).
98. Plato, *Laws*, I, 641E.
99. A pun known only from John Stobaeus' compendium of Greek sayings (xxxvi). There is a play on the two senses of *logos* in Greek: *reason* and *word*.

was to place me, while still at the breast and before my tongue was untied, in the care of a German (who subsequently died in France as a famous doctor); he was totally ignorant of our language but very well versed in Latin. He had been brought over expressly and engaged at a very high fee: he had me continuously on his hands. There were two others with him, less learned: their task was to follow me about and provide him with some relief. They never addressed me in any other language but Latin. As for the rest of the household, it was an inviolable rule that neither he nor my mother nor a manservant nor a housemaid ever spoke in my presence anything except such words of Latin as they had learned in order to chatter a bit with me. It is wonderful how much they all got from it. My father and my mother learned in this way sufficient Latin to understand it and acquired enough to be able to talk it when they had to, as did those other members of the household who were most closely devoted to my service. In short we became so latinized that it spilled over into the neighbouring villages, where, resulting from this usage, you can still find several Latin names for tools and for artisans. As for me, I was six years old before I knew French any more than I know the *patois* of Périgord or Arabic. And so, without art, without books, without grammar, without rules, without whips and without [C] tears,[100] [A] I had learned Latin as pure as that which my schoolteacher knew – for I had no means of corrupting it or contaminating it. So if they wanted me to assay writing a prose (as other boys do in the colleges by translating from French) they had to give me some bad Latin to turn into good. And Nicholas Grouchy (who wrote *De comitiis Romanorum*), Guillaume Guerente (who wrote a commentary on Aristotle), George Buchanan, that great Scottish poet, [A1] Marc-Antoine Muret [C] whom France and Italy acknowledge to be the best prose-writer in his day, [A] who were my private tutors, have often told me that as in my infancy I had that language so fluent and so ready that they were afraid to approach me.[101]

Buchanan, whom I subsequently met in the retinue of the late Lord Marshal de Brissac, told me that he was writing a book on educating children and was taking my education as his model, for he was then the tutor of Count de Brissac whom we have since seen so valiant and brave.

As for Greek (which I scarcely understand at all) my father planned to have it taught to me as methodically, but in a new way, as a sort of game

100. '80: without *constraint*, I had learned . . .
101. All these great Latinists were masters at the Collège de Guyenne in Bordeaux, to which Montaigne was sent after studying at home.

or sport. We would bounce declensions about, rather like those who use certain board-games as a means of learning arithmetic and geometry. For among other things he had been counselled to bring me to love knowledge and duty by my own choice, without forcing my will, and to educate my soul entirely through gentleness and freedom. He was so meticulous about this that since some maintain that it disturbs the tender brains of children to wake them up with a start and to snatch them suddenly and violently out of their sleep (in which they are far more deeply plunged than we are) he would have me woken up by the sound of a musical instrument; [A1] and I was never without someone to do this for me.[102]

This example will suffice to judge all the rest by, and also to emphasize the wisdom and love of so good a father who is by no means to be criticized because he harvested no fruits worthy of so choice a husbandry. There were two causes for that: first, my soil was ill-suited and barren, for, though I enjoyed solid good health together with a quiet and amenable nature, I was, despite that, so heavy, passive and dreamy that nobody could drag me out of my idleness, not even to make me play. Whatever I could perceive I saw well and, beneath my heavy complexion, I nursed bold ideas as well as opinions old for my age. My mind was lazy and would only budge so long as it was led; I was slow to understand and my inventiveness was [C] slack.[103] [A] To top it all, my memory was incredibly unreliable.

Considering all that it is no wonder that my father could make nothing of me.

The second reason was as follows: just as people who are frantic about finding a cure go and consult anybody, that good man was extremely frightened of failure in a matter which meant so much to him: he finally let himself be carried away by the common opinion (which always merely follows the leader as cranes do); he fell in with standard practice (no longer having about him the men who had given him his original educational ideas, which he had brought back from Italy) and sent me, at the age of six, to the Collège de Guyenne, then in full flourish as the best school in France. There too it is impossible to exaggerate the trouble he took over choosing good personal tutors for me and over all the other details of my education, preserving several idiosyncrasies opposed to the usual practices

102. [A]: for me: *and there was always at hand someone who played the spinet for this purpose*. This example . . .
103. '80: was *dull*. To top . . .

of the College. But for all that, it was still school. My Latin was at once corrupted and, since then, I have lost all use of it from lack of practice. And all my novel education merely served to enable me to stride right into the upper forms: I left College at thirteen, having 'completed the course' (as they put it); and in truth I now have nothing to show for it.

My first taste for books arose from enjoying Ovid's *Metamorphoses*; when I was about seven or eight I used to sneak away from all other joys to read it, especially since Latin was my mother-tongue and the *Metamorphoses* was the easiest book I knew and the one most suitable by its subject to my tender age. (As for *Lancelot du lac*, [B] *Amadis*, [A] *Huon de Bordeaux* and so on, trashy books which children spend time on, I did not even know their titles – and still do not know their insides – so exact was the way I was taught.)

This rendered me a bit slacker about studying my set-books. I was particularly lucky at this stage to have to deal with an understanding tutor who adroitly connived at this and other similar passions: for I read in succession Virgil (the *Aeneid*), Terence, Plautus and the Italian comedies, ever seduced by the attractiveness of their subjects. Had he been mad enough to break this succession I reckon that I would have left school hating books, as most French aristocrats do. He acted most ingeniously. Pretending not to notice anything, he sharpened my appetite by making me devour such books in secret, while gently requiring me to do my duty by the other, prescribed, books. For the chief qualities which my father looked for in those who had charge of me were affability and an easy-going complexion: and my own complexion had no vices other than sluggishness and laziness. The risk was not that I should do wrong but do nothing. Nobody forecast that I would turn out bad, only useless. What they foretold was idleness not wickedness.

[C] I am aware that that is the way things have turned out. The complaints which ring in my ears confirm it: 'Lazy! No warmth in his duties as friend, relation or public official! Too much on his own!' Even the most insulting accusers never say, 'Why did he go and take that?' or 'Why has he never paid up?' What they say is, 'Why will he not write it off?' or 'Why will he not *give* it away?'

I would consider it flattering if people found me wanting only in such works of supererogation. Where they are unjust is in requiring that I exceed my obligations – more rigorously, indeed, than they require of themselves their mere fulfilment. By so requiring they destroy the gratuitous nature of the deed and therefore the gratitude which would be my due; whereas any active generosity on my part should be more highly

appreciated, seeing that I have never been the recipient of any. I am all the more free to dispose of my fortune in that it is thoroughly mine. Yet if I were a great burnisher-up of my actions I might well beat off such reproaches. I would teach some of these people that they are not annoyed because I do not do a lot for them but because I could do a lot more.

[A]   For all that, my soul was not wanting in powerful emotions of her own,   [C]   nor in sure and open judgements about subjects which she knew,   [A]   digesting them alone, without telling anyone else. Among other things, I truly believe that she was incapable of surrendering to force and violence.

[B]   Should I include another of my characteristics as a child? I had an assured countenance and a suppleness of voice and gesture when I undertook to act in plays; for, in advance of my age,

> *Alter ab undecimo tum me vix ceperat annus,*

> [The following year had scarcely plucked me from my eleventh,][104]

I played the chief characters in the Latin tragedies of Buchanan, Guerente and Muret, which were put on in our Collège de Guyenne with dignity.[105] In such matters Andreas Gouveanus, our principal, was incomparably the best principal in France, as he was in all other aspects of his duties; and I was held to be a past master. Acting is an activity which is not unpraiseworthy in the children of good families; I have subsequently seen our Princes actively involved in it (following the example of the ancients) and winning honour and praise.   [C]   In Greece it was open even to gentlemen to make acting their profession: *'Aristoni tragico actori rem aperit: huic et genus et fortuna honesta erant; nec ars, quia nihil tale apud Græcos pudori est, ea deformabat.'* [He disclosed his project to Ariston, the tragic actor, a gentleman respected for his birth and his fortune; his profession in no ways impaired this respect, since nothing like that is a source of shame among the Greeks.][106]

[B]   Those who condemn such entertainments I have even accused of lack of perspicacity; and of injustice, those who deny entry into our goodly towns to worthwhile troops of actors, begrudging the people such public festivities. Good governments take the trouble to bring their citizens

---

104. Virgil, *Eclogue*, VIII, 39 (adapted).
105. These plays were all in Latin; they included no doubt Muret's *Julius Caesar* and Buchanan's *Jephthes*. Guerante's plays are not known.
106. Livy, XXIV, xxiv.

together and to assemble them for sports and games just as they do for serious acts of worship: a sense of community and good-will is increased by this. And you could not allow citizens any amusements better regulated than those which take place in the presence of all and in full view of the magistrate. I would find it reasonable that the magistrate or the monarch should occasionally offer such amusements to the people for nothing, with a kind of fatherly goodness and affection,   [C]   and that in the bigger towns there should be places set aside and duly appointed for such spectacles, which would be a diversion from worse and secret goings-on.

[A]   Now, to get back to my subject, there is nothing like tempting the boy to want to study and to love it: otherwise you simply produce donkeys laden with books. They are flogged into retaining a pannierful of learning; but if it is to do any good, Learning must not only lodge with us: we must marry her.

27. *That it is madness to judge the true and the false from our own capacities*

====

[Curiosity when applied to strange or miraculous events is both vain and arrogant. Since men are lulled by habit, they cease to wonder at the glory of the heavens yet they claim to know the limits of the whole order of Nature – and so to judge from their own parochial experience what is miraculous and what is not. Only the authority of the Church and of God's saints can recognize miracles for what they are and vouch for them. Once the Church has decided any issue of fact or doctrine, Roman Catholics must never deviate from her teachings. A man may reject her authority altogether, but he is not free to pick and choose among doctrines, especially during discussions with heretics.]

[A]   It is not perhaps without good reason that we attribute to simple-mindedness a readiness to believe anything and to ignorance the readiness to be convinced, for I think I was once taught that a belief is like an impression stamped on our soul: the softer and less resisting the soul, the easier it is to print anything on it: [C]   *'Ut necesse est lancem in libra ponderibus impositis deprimi, sic animum perspicuis cedere.'* ['For just as a weight placed on a balance must weigh it down, so the mind must yield to clear evidence.']¹ The more empty a soul is and the less furnished with counterweights, the more easily its balance will be swayed under the force of its first convictions.   [A]   That is why children, the common people, women and the sick are more readily led by the nose.

On the other hand there is a silly arrogance in continuing to disdain something and to condemn it as false just because it seems unlikely to us. That is a common vice among those who think their capacities are above the ordinary.

I used to do that once: if I heard tell of ghosts walking or of prophecies, enchantments, sorcery, or some other tale which I could not get my teeth into –

> *Somnia, terrores magicos, miracula, sagas,*
> *Nocturnos lemures portentaque Thessala*

1. Cicero, *Academica*, II, ii, 127.

[Dreams, magic terrors, miracles, witches, nocturnal visits from the dead or spells from Thessaly][2]

— I used to feel sorry for the wretched folk who were taken in by such madness. Now I find that I was at least as much to be pitied as they were. It is not that experience has subsequently shown me anything going beyond my original beliefs (nor is it from any lack of curiosity on my part), but reason has taught me that, if you condemn in this way anything whatever as definitely false and quite impossible, you are claiming to know the frontiers and bounds of the will of God and the power of Nature our Mother; it taught me also that there is nothing in the whole world madder than bringing matters down to the measure of our own capacities and potentialities.

How many of the things which constantly come into our purview must be deemed monstrous or miraculous if we apply such terms to anything which outstrips our reason! If we consider that we have to grope through a fog even to understand the very things we hold in our hands, then we will certainly find that it is not knowledge but habit which takes away their strangeness;

> [B] *jam nemo, fessus satiate videndi,*
> *Suspicere in cœli dignatur lucida templa;*

[Already now, tired and satiated with seeing, nobody bothers to gaze up at the shining temples of the heavens:]

[A] such things, if they were newly presented to us, would seem as unbelievable as any others;

> *si nunc primum mortalibus adsint*
> *Ex improviso, ceu sint objecta repente,*
> *Nil magis his rebus poterat mirabile dici,*
> *Aut minus ante quod auderent fore credere gentes.*

[supposing that now, for the first time, they were suddenly shown to mortal men: nothing could be called more miraculous; such things the nations would not have dared to believe.][3]

He who had never actually seen a river, the first time he did so took it for the ocean, since we think that the biggest things that we know represent the limits of what Nature can produce in that species.

2. Horace, *Epistles*, II, ii, 208–9.
3. Lucretius, II, 1037–8; 1032–5.

[B]   *Scilicet et fluvius, qui non est maximus, eii est*
Qui non ante aliquem majorem vidit, et ingens
Arbor homoque videtur;   [A]   *et omnia de genere omni*
Maxima quæ vidit quisque, hæc ingentia fingit.

[B]   Just as a river may not be all that big, but seems huge to a man who has never seen a bigger one, so, too, for the biggest tree or biggest man;   [A]   and the biggest thing of any kind which we know is considered huge by us.]

[C]   *'Consuetudine oculorum assuescunt animi, neque admirantur, neque requirunt rationes earum rerum quas semper vident.'* [When we grow used to seeing anything it accustoms our minds to it and we cease to be astonished by it; we never seek the causes of things like that.][4] What makes us seek the cause of anything is not size but novelty.

[A]   We ought to judge the infinite power of [C] Nature [A] with more reverence[5] and a greater recognition of our own ignorance and weakness. How many improbable things there are which have been testified to by people worthy of our trust: if we cannot be convinced we should at least remain in suspense. To condemn them as impossible is to be rashly presumptuous, boasting that we know the limits of the possible.   [C]   If we understood the difference between what is impossible and what is unusual, or between what is against the order of the course of Nature[6] and what is against the common opinion of mankind, then the way to observe that rule laid down by Chilo, *Nothing to excess*, would be, Not to believe too rashly: not to disbelieve too easily.

[A]   When we read in Froissart that the Comte de Foix knew the following morning in Béarn of the defeat of King John of Castille at Juberoth, and when we read of the means he is alleged to have used, we can laugh at that;[7] we can laugh too when our annals tell how Pope Honorius, on the very same day that King Philip-Augustus died at Mante, celebrated a public requiem for him and ordered the same to be done

4. Lucretius, VI, 674–7; Cicero, *De natura deorum*, II, XXXVIII, 96.
5. '80: power of *God* with more reverence . . .
6. In Christian theology it is only an event which occurs against the *whole order* of Nature which constitutes a miracle.
7. In 1385 the Comte de Foix took to his rooms and then was able to announce that there had just occurred in Portugal a huge slaughter of soldiers from Béarn. It was believed that he had a familiar spirit, either one called Orthon or another like him, who, in an earlier period, had deserted the local *curé* to serve the Seigneur de Corasse (Froissart, III, 17).

throughout Italy;[8] for the authority of such witnesses is not high enough to rein us back.

But wait. When Plutarch (leaving aside the many examples which he alleges from Antiquity) says that he himself knows quite definitely that, at the time of Domitian, news of the battle lost by Antony several days' journey away in Germany was publicly announced in Rome and spread through all the world on the very day that it was lost; and when Caesar maintains that it was often the case that news of an event actually anticipated the event itself: are we supposed to say that they were simple people who merely followed the mob and who let themselves be deceived because they saw things less clearly than we do![9]

Can there be anything more delicate, clear-cut and lively than the judgement of Pliny when he pleases to exercise it? Is there anything further from triviality? (I am not discussing his outstanding erudition; I put less store by that: but in which of those two qualities are we supposed to surpass him?) And yet every little schoolboy convicts him of lying and lectures him about the march of Nature's handiwork.[10]

When we read in Bouchet about miracles associated with the relics of Saint Hilary we can shrug it off:[11] his right to be believed is not great enough to take away our freedom to challenge him. But to go on from there and condemn all similar accounts seems to me to be impudent in the extreme. Such a great saint as Augustine swears that he saw:[12] a blind child restored to sight by the relics of Saint Gervaise and Saint Protasius at Milan; a woman in Carthage cured of a cancer by the sign of the cross made by a woman who had just been baptised; his close friend Hesperius driving off devils (who were infesting his house) by using a little soil taken from the sepulchre of our Lord, and that same soil, borne into the Church, suddenly curing a paralytic; a woman who, having touched the reliquary of Saint Stephen with a posy of flowers during a procession, rubbed her eyes with them afterwards and recovered her sight which she had recently lost – as well as several other miracles which occurred in his presence. What are we to accuse him of – him and the two holy bishops, Aurelius and Maximinus, whom he calls on as witnesses? Is it of ignorance, simple-

8. Nicole Gilles, *Annales des moderateurs des belliqueuses Gaulles*; the event 'happened' in 1233.

9. Plutarch, *Life of Paulus Aemilius*. The reference to Caesar is puzzling.

10. Such works as the *De Plini erroribus* of Nicolaus Leonicenus had helped spread criticisms of Pliny.

11. Jean Bouchet, *Annales d'Acquitaine*, Poitiers, 1567 etc., pp. 21–30.

12. St Augustine, *City of God*, XII, viii.

mindedness, credulity, deliberate deception or imposture? Is there any man in our century so impudent as to think he can be compared with them for virtue, piety, scholarship, judgement and ability?   [C]   *'Qui, ut rationem nullam afferent, ipsa authoritate me frangerent.'* [Why, even if they gave no reasons, they would convince me by their very authority.][13]

[A]   Apart from the absurd rashness which it entails, there is a dangerous boldness of great consequence in despising whatever we cannot understand. For as soon as you have established the frontiers of truth and error with that fine brain of yours and then discover that you must of necessity believe some things even stranger than the ones which you reject, you are already forced to abandon these frontiers.

Now it seems to me that what brings as much disorder as anything into our consciences during our current religious strife is the way Catholics are prepared to treat some of their beliefs as expendable. They believe they are being moderate and well-informed when they surrender to their enemies some of the articles of faith which are in dispute. But, apart from the fact that they cannot see what an advantage you give to an adversary when you begin to yield ground and beat a retreat, or how much that excites him to follow up his attack, the very articles which they select as being less weighty are sometimes extremely important ones.

We must either totally submit to the authority of our ecclesiastical polity or else totally release ourselves from it. It is not for us to decide what degree of obedience we owe to it.

Moreover I can say that for having assayed it; in the past I made use of that freedom of personal choice and private selection in order to neglect certain details in the observances of our Church because they seemed to be rather odd or rather empty; then, when I came to tell some learned men about it, I discovered that those very practices were based on massive and absolutely solid foundations, and that it is only our ignorance and animal-stupidity which make us treat them with less reverence than all the rest.

Why cannot we remember all the contradictions which we feel within our own judgement, and how many things which were articles of belief for us yesterday are fables for us today?

Vainglory and curiosity are the twin scourges of our souls. The former makes us stick our noses into everything: the latter forbids us to leave anything unresolved or undecided.

---

13. Cicero, *Tusc. disput.*, I, xxi, 49, adapted: Cicero wrote, 'For even though Plato gave no reasons – note what tribute I pay to him – he would convince me by his very authority.'

28. On affectionate relationships

[*This chapter, 'De l'amitié', is traditionally called 'On friendship'. But in Renaissance French* amitié *includes many affectionate relationships, ranging from a father's love for his child (or for his brain-child) to the friendly services of a doctor or lawyer, to that conjugal love felt by Montaigne for his wife, and to that rarest of lasting friendships which David shared with Jonathan, Roland with Oliver or Montaigne with La Boëtie. Several terms are needed in English to render these different senses; they include friendship, loving-friendship, benevolence, affection, affectionate relationships and love. The basic meaning of* amitié *is rooted in* aimer *(to love); but it often excluded* amour, *love between the sexes, and always* folle amour *('mad love') which was sexual and extra-marital. The first syllable of* amitié *was fully nasalized in Renaissance French: it therefore sounds like* âme *(soul). Since ancient times philosophy had classified love between the sexes as at least primarily an affair of the other, lower, part of Man: the body; some Renaissance Platonists were concerned to modify this stark dichotomy between soul-love and body-love. Much was written on* parfaite amitié, *a 'perfect' loving relationship which could arise between a man and a woman in which physical love was relegated to a vital but second place. Montaigne does not underplay the role of sexual love (cf. III, 5, 'On some lines of Virgil', and III, 3, 'On three kinds of social intercourse'); but despite Classical precedent he does wonder whether a fully sexual love plus a fully soul-centred* amitié *could not bind an exceptional man to an exceptional woman. If it could, then it would engage the whole individual person, body and soul. That would indeed be 'perfect love',* parfaite amitié. *Male homosexual love, which did from Socratic times claim to do just that, does not disturb nor preoccupy Montaigne: he dismisses it as 'justly abhorrent to our manners' and as a parody of heterosexual love. But philosophical homosexuality shows,* mutatis mutandis, *what the love of man and woman could ideally be: a marriage of bodies and souls.*

Montaigne's main concern is with the very special loving-friendship which he shared with Etienne de La Boëtie. La Boëtie's youthful treatise De la Servitude volontaire *(On Willing Slavery) which praised the polity of the Republic of Venice at the expense of monarchy was used seditiously after his death by those who had taken up arms against their King in the Wars of Religion. Montaigne is at pains to show that a rare and exemplary friendship has ever been consonant with loyalty to the State and that both he and La Boëtie were loyal to each other and, therefore, loyal to their country.*]

[A]   I was watching an artist on my staff working on a painting when I felt a desire to emulate him. The finest place in the middle of a wall he selects for a picture to be executed to the best of his ability; then he fills up

the empty spaces all round it with *grotesques*, which are fantastical paintings whose attractiveness consists merely in variety and novelty. And in truth what are these *Essays* if not monstrosities and *grotesques* botched together from a variety of limbs having no defined shape, with an order sequence and proportion which are purely fortuitous?

> *Desinit in piscem mulier formosa superne.*
>
> [A fair woman terminating in the tail of a fish.][1]

I can manage to reach the second stage of that painter but I fall short of the first and better one: my abilities cannot stretch so far as to venture to undertake a richly ornate picture, polished and fashioned according to the rules of art. So I decided to borrow a 'painting' from Etienne de La Boëtie, which will bring honour to the rest of the job: I mean the treatise to which he gave the title *On Willing Slavery* but which others, not knowing this, very appropriately baptised afresh as *Against One*.[2] He wrote it, while still very young,[3] as a kind of essay against tyrants in honour of freedom. It has long circulated among men of discretion – not without great and well-merited esteem, for it is a noble work, as solid as may be. Yet it is far from being the best he was capable of. If, at the age when I knew him when he was more mature, he had conceived a design such as mine and written down his thoughts, we would now see many choice works bringing us close to the glory of the Ancients; for, particularly where natural endowments are concerned, I know nobody who can compare with him. Yet nothing of his survives apart from this treatise – and even that is due to accident: I do not think he ever saw it again once he let go of it – and some *Considerations* on that *Edict of January* which our civil wars have made notorious: I may [C] perhaps [A] still find a place for it elsewhere.[4] That is all I have been able to recover of his literary remains, [C] I the heir to whom, with death on his lips, he so lovingly willed his books and

---

1. Horace, *Ars poetica*, 4. (Poets can create monsters at will; say a fair maid with the tail of a fish, that is, a mermaid.)
2. Edited and translated by Malcolm Smith as *Slaves by Choice*, Runnymede Books, RHBNC, Egham, 1988.
3. '80: young, *not having reached the age of eighteen years*, as . . .
4. Cf. E. de La Boëtie: *Mémoire sur la pacification des troubles*, ed. Malcolm Smith, TLF, Droz, Geneva, 1983. This work antedates the Royal Edict of 17 January 1562 (which afforded limited toleration to Protestants and recognized the 'Allegedly Reformed Church'. Montaigne published neither of these in his collection of the works of La Boëtie, F. Morel, Paris, 1571, since (as the Preface says) the time was

his papers – [A] apart from the slim volume of his works which I have had published already.

Yet I am particularly indebted to that treatise, because it first brought us together: it was shown to me long before I met him and first made me acquainted with his name; thus preparing for that loving-friendship between us which as long as it pleased God we fostered so perfect and so entire that it is certain that few such can even be read about, and no trace at all of it can be found among men of today. So many fortuitous circumstances are needed to make it, that it is already something if Fortune can achieve it once in three centuries. There seems to be nothing for which Nature has better prepared us than for fellowship – [C] and Aristotle says that good lawgivers have shown more concern for friendship than for justice.[5] [A] Within a fellowship the peak of perfection consists in friendship; for [C] all forms of it which are forged or fostered by pleasure or profit or by public or private necessity are so much the less beautiful and noble – and therefore so much the less 'friendship' – in that they bring in some purpose, end or fruition other than the friendship itself. Nor do those four ancient species of love conform to it: the natural, the social, the hospitable and the erotic.[6]

[A] From children to fathers it is more a matter of respect; friendship, being fostered by mutual confidences, cannot exist between them because of their excessive inequality; it might also interfere with their natural obligations: for all the secret thoughts of fathers cannot be shared with their children for fear of begetting an unbecoming intimacy; neither can those counsels and admonitions which constitute one of the principal obligations of friendship be offered by children to their fathers. There have been peoples where it was the custom for children to kill their fathers and others for fathers to kill their children to avoid the impediment which each can constitute for the other: one depends naturally on the downfall of the other.[7]

'too unpleasant'. This chapter is an apology for La Boétie, a defence of his ideas and a rejection of the smear that such loyal friendships can entail disloyalty to the State (a question already raised in antiquity).

5. For Aristotle, *Nicomachaean Ethics*, VIII, 1, a good fellowship (or society) is one which fosters 'friendship' in all of its senses.

6. Cf. C. S. Lewis, *The Four Loves*, Collins, Fount Paperbacks, 1960.

7. Montaigne mentions this in I, 23: 'On habit: and on never easily changing a traditional law'.

[A]: the other. *Friendship never gets to such a point*. There . . .

There have been philosophers who held such natural bonds in contempt – witness [C] Aristippus: when he was being pressed about the affection which he owed to his children since they had sprung from him, he began to spit, saying that that sprang from him too, and that we also engender lice and worms.⁸ [A] And there was that other one whom Plutarch sought to reconcile with his brother but who retorted: 'He matters no more to me for coming out of the same hole.'⁹

The name of brother is truly a fair one and full of love: that is why La Boëtie and I made a brotherhood of our alliance. But sharing out property or dividing it up, with the wealth of one becoming the poverty of the other, can wondrously melt and weaken the solder binding brothers together. Brothers have to progress and advance by driving along the same path in the same convoy: they needs must frequently bump and jostle against each other. Moreover, why should there be found between them that congruity and affinity which engender true and perfect friendship? Father and son can be of totally different complexions: so can brothers. 'He is my son, he is my kinsman, but he is wild, wicked or daft!' And to the extent that they are loving relationships commanded by the law and the bonds of nature, there is less of our own choice, less 'willing freedom'.¹⁰ Our 'willing freedom' produces nothing more properly its own than affection and loving-friendship. It is not that I have failed to assay all that the other kind can afford, having had the best father who ever was, and the most indulgent even into extreme old age, and coming as I do from a family renowned and exemplary from generation to generation in the matter of brotherly harmony:

[B]   *et ipse*
Notus in fratres animi paterni.

[And myself known for my fatherly concern for my brothers.]¹¹

[A]   You cannot compare with friendship the passion men feel for women, even though it is born of our own choice, nor can you put them in the same category. I must admit that the flames of passion –

---

8. [A]: much the same as [C], but Aristippus not named. (Cf. Erasmus, *Apophthegmata*, III, *Aristippus*, LV, the probable source of   [C]).
9. Plutarch (tr. Amyot), *De l'amitié fraternelle*, 82E. Montaigne coarsens the terms of the bad brother (a philosopher) who in Plutarch simply refers to 'the same natural organ'.
10. The antithesis of 'willing slavery', the subject of La Boëtie's book.
11. Horace, *Odes*, II, ii, 6–7 (adapted to apply to Montaigne).

neque enim est dea nescia nostri
Que dulcem curis miscet amaritiem

[for I am not unacquainted with that goddess who mingles sweet bitterness with love's cares][12] –

are more active, sharp and keen. But that fire is a rash one, fickle, fluctuating and variable; it is a feverish fire, subject to attacks and relapses, which only gets hold of a corner of us. The love of friends is a general universal warmth, temperate moreover and smooth, a warmth which is constant and at rest, all gentleness and evenness, having nothing sharp nor keen. What is more, sexual love is but a mad craving for something which escapes us:

> *Come segue la lepre il cacciatore*
> *Al freddo, al caldo, alla montagna, al lito;*
> *Ne piu l'estima poi che presa vede,*
> *Et sol dietro a chi fugge affretta il piede.*

[Like the hunter who chases the hare through heat and cold, o'er hill and dale, yet, once he has bagged it, he thinks nothing of it; only while it flees away does he pound after it.][13]

As soon as it enters the territory of friendship (where wills work together, that is) it languishes and grows faint. To enjoy it is to lose it: its end is in the body and therefore subject to satiety. Friendship on the contrary is enjoyed in proportion to our desire: since it is a matter of the mind, with our souls being purified by practising it, it can spring forth, be nourished and grow only when enjoyed. Far below such perfect friendship those fickle passions also once found a place in me – not to mention in La Boëtie, who confesses to all too many in his verses. And so those two emotions came into me, each one aware of the other but never to be compared, the first maintaining its course in a proud and lofty flight, scornfully watching the other racing along way down below.

As for marriage, apart from being a bargain where only the entrance is free (its duration being fettered and constrained, depending on things outside our will), it is a bargain struck for other purposes; within it you soon have to unsnarl hundreds of extraneous tangled ends, which are enough to break the thread of a living passion and to trouble its course, whereas in friendship there is no traffic or commerce but with itself. In

12. Catullus, *Epigrams*, LXVI, 17–18.
13. Ariosto, *Orlando furioso*, X, vii.

addition, women are in truth not normally capable of responding to such familiarity and mutual confidence as sustain that holy bond of friendship, nor do their souls seem firm enough to withstand the clasp of a knot so lasting and so tightly drawn. And indeed if it were not for that, if it were possible to fashion such a relationship, willing and free, in which not only the souls had this full enjoyment but in which the bodies too shared in the union − [C] where the whole human being was involved − it is certain[14] [A] that the loving-friendship would be more full and more abundant. But there is no example yet of woman attaining to it [C] and by the common agreement of the Ancient schools of philosophy she is excluded from it.

[A] And that alternative licence of the Greeks is rightly abhorrent to our manners; [C] moreover since as they practised it it required a great disparity of age and divergence of favours between the lovers, it did not correspond either to that perfect union and congruity which we are seeking here. *'Quis est enim iste amor amicitiae? Cur neque deformem adolescentem quisquam amat, neque formosum senem?'* [What is this 'friendship-love'? Why does nobody ever fall in love with a youth who is ugly or with a beautiful *old* man?][15] For even the portrayal of it by the Academy will not I think belie me when I say this about it: that the original frenzy inspired by Venus' son[16] in the heart of the Lover towards the bloom of a tender youth (in which they allow all the excessive and passionate assaults which an immoderate ardour can produce) was simply based on physical beauty, a false image of generation in the body (for it could not have been based on the mind, which had yet to show itself, which was even then being born, too young to sprout); that if so mad a passion took hold of a base mind the means of pursuing it were riches, presents, favouritism in advancement to high office and such other base traffickings which the Academy condemned; if it lighted on a more noble mind its inducements were likewise more noble: instruction in philosophy; lessons teaching reverence for religion, obedience to the law and dying for the good of one's country; examples of valour, wisdom, justice, with the Lover striving to make himself worthy of acceptance by the graciousness and beauty of his soul (that of his body having long since faded) and hoping by this mental alliance to strike a

14. '80: union, *it is likely* that . . .
15. Cicero, *Tusc. disput.*, IV, xxxiii, 70. (In Greek philosophical homosexuality the older man was the Lover; the younger, the Beloved, showed admiration, or gratitude for instruction.)
16. Cupid. (The 'Academy' was the School of Plato.)

more firm and durable match. When this suit produced its results – in due season (for while they did not require the Lover to devote time and discretion to this undertaking they strictly required it of the Beloved, since he had to reach a judgement about a kind of beauty which is internal, difficult to recognize and concealed from discovery) – there was then born in that Beloved the desire mentally to conceive through the medium of the beauty of the mind. For him this beauty was pre-eminent: that of the body, secondary and contingent – quite the opposite from the Lover. For this reason they held the Beloved in higher esteem and proved that the gods do so too; they severely rebuked the poet Aeschylus for having given, in the love of Achilles and Patroclus, the role of the Lover to Achilles, who was the fairest of all the Greeks, in the first verdure of unbearded youth.[17]

Once this general communion had been established, with the more worthy aspect of it fulfilling its duties and predominating, they said that it produced fruits useful for private and public life; that it was the strength of those countries where it was the accepted custom and the main defence of right conduct and freedom – witness the loves of Hermodius and Aristogiton. That is why they call it sacred and divine. By their reckoning only the violence of tyrants and the baseness of the people are opposed to it.[18] Yet when all is said and done the only point we can concede to the Academy is that it was a love-affair which ended in friendship – which conforms well enough to the Stoic definition of love: '*Amorem conatum esse amicitiae faciendae ex pulchritudinis specie.*' [Love is the striving to establish friendship on the external signs of beauty.][19]

I now return to a kind of love more equable and more equitable: '*Omnino amicitiae, corroboratis jam confirmatisque ingeniis et aetatibus, judicandae sunt.*' [Such only are to be considered friendships in which characters have been confirmed and strengthened with age.][20]

[A] Moreover what we normally call friends and friendships are no more than acquaintances and familiar relationships bound by some chance or some suitability, by means of which our souls support each other. In the friendship which I am talking about, souls are mingled and confounded in

17. In Plato's *Symposium* (or *Banquet*), the main general source of all of [C] here.
18. Ibid.: tyrannies do not favour (homosexual) friendship-love; Hipparchus, tyrant of Athens, was therefore assassinated by the friends Harmodius and Aristogiton. (Cf. Aulus Gellius, *Attic Nights*, XVII, 21: 7; Cicero, *Tusc. disput.*, I, xlix, 116.)
19. Cicero, *Tusc. disput.*, IV, xxiv, 71.
20. Cicero, *De amicitia*, XX, 74.

so universal a blending that they efface the seam which joins them together so that it cannot be found. If you press me to say why I loved him, I feel that it cannot be expressed   [C]   except by replying: 'Because it was him: because it was me.'   [A]   Mediating this union there was, beyond all my reasoning, beyond all that I can say   [C]   specifically   [A]   about it, some [C]   inexplicable   [A]   force of destiny.[21]

[C]   We were seeking each other before we set eyes on each other – both because of the reports we each had heard (which made a more violent assault on our emotions than was reasonable from what they had said, and, I believe, because of some decree of Heaven: we embraced each other by repute, and, at our first meeting, which chanced to be at a great crowded town-festival, we discovered ourselves to be so seized by each other, so known to each other and so bound together that from then on none was so close as each was to the other. He wrote an excellent Latin *Satire*, which has been published,[22] by which he defends and explains the suddenness of our relationship which so quickly reached perfection. Having so short a period to last, having begun so late (for we were both grown men – he more than a few years older than I) – it had no time to waste on following the pattern of those slacker ordinary friendships which require so much prudent foresight in long preliminary acquaintance. This friendship has had no ideal to follow other than itself; no comparison but with itself.   [A]   There is no one particular consideration – nor two nor three nor four nor a thousand of them – but rather some inexplicable quintessence of them all mixed up together which, having captured my will, brought it to plunge into his and lose itself   [C]   and which, having captured his will, brought it to plunge and lose itself in mine with an equal hunger and emulation.   [A]   I say 'lose itself' in very truth; we kept nothing back for ourselves: nothing was his or mine.

In the presence of the Roman Consuls (who, after the condemnation of Tiberius Gracchus were prosecuting those who had been in his confidence) Laelius eventually asked Caius Blosius, the closest friend of Gracchus, how much he would have done for him. He replied: 'Anything.' – 'What, anything?' Laelius continued: 'And what if he had ordered you to set fire to our temples?' – 'He would never have asked me to,' retorted Blosius. 'But supposing he had,' Laelius added. 'Then I would have obeyed,' he replied.[23] Now if he really were so perfect a friend of Gracchus as history

21. '80: some *divine* force of destiny . . .
22. Published in the 1571 edition of La Boëtie's works by Montaigne.
23. Cicero, *De amicitia*, XI, 33–9.

asserts, he had no business provoking the Consuls with that last rash assertion and ought never to have abandoned the certainty he had of the wishes of Gracchus.[24] But those who condemn his reply as seditious do not fully understand the mystery of friendship and fail to accept the premiss that he had Gracchus' intentions in the pocket of his sleeve, both by his influence and by his knowledge. [C] They were more friends than citizens; friends, more than friends or foes of their country or friends of ambition and civil strife. Having completely committed themselves to each other, they each completely held the reins of each other's desires; grant that this pair were guided by virtue and led by reason (without which it is impossible to harness them together) Blosius' reply is what it should have been. If their actions broke the traces, then they were, by my measure, neither friends of each other nor friends of themselves. Moreover [A] that reply sounds no different than mine would be, if I were interrogated thus: 'If your will commanded you to kill your daughter would you kill her?' and I said that I would. For that is no witness that I would consent to do so, because I do not doubt what my will is, any more than I doubt the will of such a friend. All the arguments in the world have no power to dislodge me from the certainty which I have of the intentions and decisions of my friend. Not one of his actions could be set before me – no matter what it looked like – without my immediately discovering its motive. Our souls were yoked together in such unity, and contemplated each other with so ardent an affection, and with the same affection revealed each to each other right down to the very entrails, that not only did I know his mind as well as I knew my own but I would have entrusted myself to him with greater assurance than to myself.

Let nobody place those other common friendships in the same rank as this. I know about them – the most perfect of their kind – as well as anyone else, [B] but I would advise you not to confound their rules: you would deceive yourself. In those other friendships you must proceed with wisdom and caution, keeping the reins in your hand: the bond is not so well tied that there is no reason to doubt it. 'Love a friend,' said Chilo, 'as though some day you must hate him: hate him, as though you must love him.'[25] That precept which is so detestable in that sovereign master-friendship is salutary in the practice [C] of friendships which are

24. '80: the wishes of Gracchus, *for which he could answer as for his own.* But . . .
25. Chilo's chilling judgement was well known (cf. Du Bellay, *Regrets*, 140). It was normally attributed to Bias, one of the Seven Sages of Greece. Cf. Aulus Gellius, *Attic Nights*, I.3.30; Cicero, *De amicitia*, XVI, 49; Aristotle, *Rhetoric*, II, 14.

common and customary,²⁶ in relation to which you must employ that saying which Aristotle often repeated: 'O my friends, there is no friend!'²⁷

[A]   In this noble relationship, the services and good turns which foster those other friendships do not even merit being taken into account: that is because of the total interfusion of our wills. For just as the friendly love I feel for myself is not increased – no matter what the Stoics may say – by any help I give myself in my need, and just as I feel no gratitude for any good turn I do to myself: so too the union of such friends, being truly perfect, leads them to lose any awareness of such services, to hate and to drive out from between them all terms of division and difference, such as good turn, duty, gratitude, request, thanks and the like. Everything is genuinely common to them both: their wills, goods, wives, children, honour and lives;   [C]   their correspondence is that of one soul in bodies twain, according to that most apt definition of Aristotle's,²⁸   [A]   so they can neither lend nor give anything to each other. That is why those who make laws forbid gifts between husband and wife, so as to honour marriage with some imagined resemblance to that holy bond, wishing to infer by it that everything must belong to them both, so that there is nothing to divide or to split up between them. In the kind of friendship I am talking about, if it were possible for one to give to the other it is the one who received the benefaction who would lay an obligation on his companion. For each of them, more than anything else, is seeking the good of the other, so that the one who furnishes the means and the occasion is in fact the more generous, since he gives his friend the joy of performing for him what he most desires.   [C]   When Diogenes the philosopher was short of money he did not say that he would ask his friends to give him some but to give him some back!²⁹   [A]   And to show how this happens in practice I will cite an example – a unique one – from Antiquity.

Eudamidas, a Corinthian, had two friends: Charixenus, a Sicyonian, and Aretheus, also a Corinthian. As he happened to die in poverty, his two friends being rich, he made the following testament: 'To Aretheus I bequeath that he look after my mother and maintain her in her old age; to Charixenus, that he see that my daughter be married, providing her with the largest dowry he can; and if one of them should chance to die I appoint the survivor to substitute for him.' Those who first saw his will laughed at

26. '88: in *common practice*, in relation . . .
27. Erasmus, *Apophthegmata*, VII, *Aristoteles Stagirites*, XXVIII.
28. Erasmus, ibid., VII, *Aristoteles Stagirites*, XIX.
29. Erasmus, ibid., III, *Diogenes Cynicus*, LXXXII.

it; but, when those heirs learned of it, they accepted it with a unique joy. One of them, Charixenus, did die five days later; the possibility of substitution was thus opened in favour of Aretheus, and he looked after the mother with much care; then, of five hundred weight of silver in his possession, he gave two and a half for the marriage of his only daughter and two and a half for the daughter of Eudamidas, celebrating their weddings on the same day.[30]

This example is a most full one, save for one circumstance: there was more than one friend. For the perfect friendship which I am talking about is indivisible: each gives himself so entirely to his friend that he has nothing left to share with another: on the contrary, he grieves that he is not two-fold, three-fold or four-fold and that he does not have several souls, several wills, so that he could give them all to the one he loves.

Common friendships can be shared. In one friend one can love beauty; in another, affability; in another, generosity; in another, a fatherly affection; in another, a brotherly one; and so on. But in this friendship love takes possession of the soul and reigns there with full sovereign sway. that cannot possibly be duplicated.   [C]   If two friends asked you to help them at the same time, which of them would you dash to? If they asked for conflicting favours, who would have the priority? If one entrusted to your silence something which it was useful for the other to know, how would you get out of that? The unique, highest friendship loosens all other bonds. That secret which I have sworn to reveal to no other, I can reveal without perjury to him who is not another: he *is* me. It is a great enough miracle for oneself to be redoubled: they do not realize how high a one it is when they talk of its being tripled. The uttermost cannot be matched. If anyone suggests that I can love each of two friends as much as the other, and that they can love each other and love me as much as I love them, he is turning into a plural, into a confraternity, that which is the most 'one', the most bound into one. One single example of it is moreover the rarest thing to find in the world.

[A]   The rest of that story conforms well what I was saying: for Eudamidas bestows a grace and favour on his friends when he makes use of them in his necessity. He left them heirs to his own generosity, which consists in putting into their hands the means of doing him good. And there is no doubt that the force of loving-friendship is more richly displayed in what he did than in what Aretheus did. To sum up, these are deeds which surpass the imagination of anyone who has not tasted

30. From Lucian of Samosata, *Toxaris, or, On friendship*, XXII.

them; [C] they make me wondrously honour the reply of that young soldier when Cyrus inquired of him how much he would take for a horse which had enabled him to win the prize in the races: 'Would he sell it for a kingdom?' – 'No, indeed, Sire; but I would willingly give it away to gain a friend, if I could find a man worthy of such an alliance.'[31] Not badly put, that, 'If I could find'; for you can easily find men fit for a superficial acquaintanceship. But for our kind, in which we are dealing with the innermost recesses of our minds with no reservations, it is certain that all of our motives must be pure and sure to perfection.

In those alliances which only get hold of us by one end, we need simply to provide against such flaws as specifically affect that end. It cannot matter to me what the religion of my doctor or my lawyer is: that consideration has nothing in common with the friendly services which they owe to me. And in such commerce as arises at home with my servants I act the same way: I make few inquiries about the chastity of my footman: I want to know if he is hard-working; I am less concerned by a mule-driver who gambles than by one who is an idiot, or by a cook who swears than by one who is incompetent. It is not my concern to tell the world how to behave (plenty of others do that) but how I behave in it:

> *Mihi sic usus est; tibi, ut opus est facto, face.*
>
> [This is what I do: do what serves you.][32]

For the intimate companionship of my table I choose the agreeable not the wise; in my bed, beauty comes before virtue; in social conversation, ability – even without integrity. And so on.

[A] Just as that philosopher[33] playing with his children and riding astride a hobby-horse told the man who surprised him at it not to make comments before he had children of his own, judging that the emotions which would then arise in his soul would make him a good judge of such behaviour: so too I could wish that I were speaking to people who had assayed what I am talking about; but realizing how far removed from common practice is such a friendship – and how rare it is – I do not expect to find one good judge of it. For the very writings which Antiquity have left us on this subject seem weak to me compared to what I feel. In this case the very precepts of philosophy are surpassed by the results:

---

31. Xenophon, *Cyropaedia*, VIII, iii, 270.
32. Terence, *Heautontimorumenos*, I, i, 28.
33. Agesilaus (Cf. Erasmus, *Apophthegmata*, I, *Agesilaus*, LXVIII).

Nil ego contulerim jucundo sanus amico.

[Whilst I am in my right mind, there is nothing I will compare with a delightful friend.][34]

In Antiquity Menander pronounced a man to be happy if he had merely encountered the shadow of a friend.[35] He was certainly right to say so, especially if he had actually tasted friendship. For in truth if I compare all the rest of my life – although by the grace of God I have lived it sweetly and easily, exempt (save for the death of such a friend) from grievous affliction in full tranquillity[36] of mind, contenting myself with the natural endowments which I was born with and not going about looking for others – if I compare it, I say, to those four years which it was vouchsafed to me to enjoy in the sweet companionship and fellowship of a man like that, it is but smoke and ashes, a night dark and dreary.

Since that day when I lost him,

> *quem semper acerbum,*
> *Semper honoratum (sic, Dii, voluistis) habebo,*

[which I shall ever hold bitter to me, though always honour (since the gods ordained it so),][37]

I merely drag wearily on. The very pleasures which are proffered me do not console me: they redouble my sorrow at his loss. In everything we were halves: I feel I am stealing his share from him:

> *Nec fas esse ulla me voluptate hic frui*
> *Decrevi, tantisper dum ille abest meus particeps.*

[Nor is it right for me to enjoy pleasures, I decided, while he who shared things with me is absent from me.][38]

I was already so used and accustomed to being, in everything, one of two, that I now feel I am no more than a half:

> [B]   *Illam meæ si partem animæ tulit*
> *Maturior vis, quid moror altera,*
> *Nec charus æque, nec superstes*
> *   Integer? Ille dies utramque*
> *Duxit ruinam.*

---

34. Horace, *Satires*, I, v, 44.
35. Plutarch (tr. Amyot), *De l'amitié fraternelle*, 82C–D.
36. '80: full *happiness and* tranquillity.
37. Virgil, *Aeneid*, V, 49–50.
38. Terence, *Heautontimorumenos*, I, 1, 97–8.

[Since an untimely blow has borne away a part of my soul, why do I still linger on less dear, only partly surviving? That day was the downfall of us both.][39]

[A]   There is no deed nor thought in which I do not miss him – as he would have missed me; for just as he infinitely surpassed me in ability and virtue so did he do so in the offices of friendship:

> *Quis desiderio sit pudor aut modus*
> *Tam chari capitis? . . .*
>   *O misero frater adempte mihi!*
> *Omnia tecum una perierunt gaudia nostra,*
>   *Quæ tuus in vita dulcis alebat amor.*
> *Tu mea, tu moriens fregisti commoda, frater;*
>   *Tecum una tota est nostra sepulta anima,*
> *Cujus ego interitu tota de mente fugavi*
>   *Hæc studia atque omnes delicias animi.*
> *Alloquar? audiero nunquam tua verba loquentem?*
>   *Nunquam ego te, vita frater amabilior,*
> *Aspiciam posthac? At certe semper amabo.*

[What shame or limit should there be to grief for one so dear? . . . How wretched I am, having lost such a brother! With you died all our joys, which your sweet love fostered when you were alive. You, brother, have destroyed my happiness by your death: all my soul is buried with you. Because of your loss I have chased all thoughts from my mind and all pleasures from my soul . . . Shall I never speak to you, never hear you talking of what you have done? Shall I never see you again, my brother, dearer than life itself? But certainly I shall love you always.][40]

Let us hear a while this   [C]   sixteen-year-old   [A]   boy.[41]

Having discovered that this work of his has since been published to an evil end by those who seek to disturb and change the state of our national polity without worrying whether they will make it better, and that they have set it among works of their own kidney, I have gone back on my decision to place it here. And so that the author's reputation should not be harmed among those who cannot know his opinions or his actions, I tell them that this subject was treated by him in his childhood purely as an

---

39. Horace, *Odes*, II, xvii, 5–9.
40. Catullus, LXVIII, 20 f.; LXV, 9 f. (adapted).
41. '80: this *eighteen-year-old* boy. (Montaigne was planning to publish here, as the central 'painting' enhanced by his fringe of 'grotesques', La Boëtie's essay 'On Willing Slavery'. It had been exploited by Protestants as an anti-monarchist pamphlet, so he reluctantly omits it.)

exercise; it is a commonplace theme, pawed over in hundreds and hundreds of books. I have no doubt that he believed what he wrote, for he was too conscientious to tell untruths even in a light-hearted work. And I know, moreover, that if he had had the choice he would rather have been born in Venice than in Sarlat. Rightly so. But he had another maxim supremely imprinted upon his soul: to obey, and most scrupulously submit to, the laws under which he was born. There never was a better citizen, one more devoted to his country's peace or more opposed to the disturbances and novelties of his time. He would have used his abilities to snuff them out, not to provide materials to stir them up. The mould of his mind was cast on the model of centuries different from ours.

So instead of that serious work I will substitute another one, more gallant and more playful, which he wrote in the same season of his life.[42]

42. '80: life. *It consists of twenty-nine sonnets which the Sieur de Poiferré, a man both practical and understanding who knew him long before me, has found by chance at home among his other papers and has just sent to me: for which I am much beholden to him; and I would wish that others who possess other fragments of his writings scattered here and there would do the same.*

29. Nine-and-twenty sonnets of Estienne de La Boëtie

===

[This chapter was designed to introduce sonnets by Montaigne's especial friend La Boëtie, the subject of the previous chapter, and did indeed do so in all editions published during Montaigne's lifetime. All the previous allusions which lead us to expect them here are kept, not least the promise to print them in compensation for his decision to omit the text of De la Servitude volontaire. *In the Bordeaux copy Montaigne simply struck them all out – leaving his own text as 'grotesques' surrounding an absent masterpiece. No attempt is made to conceal the omission: the gaps are like blank columns in a censored newspaper. Montaigne had just defended his friend, and himself, from suspicion of seditious republicanism. His respect for the magistrature would have led him to consent to the excision (if pressure was in fact put on him) but not to change his loyalty or his judgement. His action can be compared to his refusing (III, 10) to concede 'to the magistrature itself the right to condemn a book (his own* Essais) *for having classed a heretic (Beza) among the best poets of this century'.*

Montaigne had published some Sonnets *of La Boëtie in 1572 (Fédéric Morel, Paris) and dedicated them to the Count de Foix. The sonnets which were printed here do not figure in them. This chapter is dedicated to Diane, wife of the Count of Grammont and Guiche (a good friend of his) and subsequently mistress of the protestant Henry of Navarre (the future King Henry IV). She was surnamed Corisande d'Andoins from a character in* Amadis de Gaule, *a vast many-volumed novel in which she delighted.]*

[A]   To Madame de Grammont, Countess of Guiche

Madame, I am offering you nothing of mine, either because it is yours already or else because I deem none of it worthy of you. But I have wanted these verses, wherever they may be read, to be headed by your name because it would honour them to have the great Corisande d'Andoins to guide them on their way. This gift seemed appropriate to you, inasmuch as there are few ladies in France who are better judges of poetry or who can more rightly take advantage of it. And since not one of them can sing poetry more vividly or more animatedly than you can with that tuneful voice so full and fair with which Nature has endowed you among a million other graces, these verses deserve that you, madame, should encourage them: for you will share my opinion that none have come out of

Gascony which are better contrived or more refined, or which bear witness of deriving from a richer hand. You must not feel jealous because you have merely received the remainder of what I have already had printed, dedicated to your good kinsman, Monsieur de Foix, for these have something more indescribably lively and overflowing, having been written in his verdant youth in the heat of a fair and noble passion which I will one day, Madame, whisper in your ear. The others were written later for his wife when he was courting her; somehow they already have the cooler savour of marriage. Personally I am one who holds that poetry is never more gay than when treating a subject unruly and wanton.

[C]   These verses can be found elsewhere.[1]

1. The printed versions are less abrupt: '95: wanton. *These nine-and-twenty sonnets of Estienne de La Boëtie which were placed here have since been printed with his works.*
    This edition of La Boëtie's *Oeuvres* remains untraced, but the sonnets themselves were printed in all the editions.

30. On moderation

=====

[Moderation is a Classical virtue. This chapter is a vital step in Montaigne's thought, especially in the light of his later comments marked by [C]. It continues his reflections on love and marriage, moving from banter to seriousness. It examines why it is that the clergy and the doctors (whose duty is to cure souls and bodies) seek their remedies through pain and bitter medicines. Both remedies are often immoderate, in some ways akin to the ghastly sacrifices of the Incas, which also show how little most men value the gifts of Nature, including the gift of life and the natural pleasures.]

[A] It is as though our very touch bore infection: things which in themselves are good and beautiful are corrupted by our handling of them. We can seize hold even of Virtue in such a way that our action makes her vicious if we clasp her in too harsh and too violent an embrace. Those who say that Virtue knows no excess (since she is no longer Virtue if there is excess within her) are merely playing with words.

> *Insani sapiens nomen ferat, æquus iniqui,*
> *Ultra quam satis est virtutem si petat ipsam*

[The name of 'insane' is borne by the Sage and the name of 'unjust' is borne by the Just, if in their strivings after Virtue herself they go beyond what is sufficient.]¹

That is a subtle observation on the part of philosophy: you can both love virtue too much and [B] behave with excess [A] in an action which itself is just. The [B] Voice [A] of God adapts itself fittingly to that bias: 'Be not more wise than it behoveth, but be ye soberly wise.'²

[C] I have seen one of our great noblemen harm the reputation of his

---

1. Horace, *Epistles*, I, vi, 15–16.
2. Romans 12:3, following the Vulgate Latin version in which Montaigne read his Bible. (The Greek original talks not of 'moderation' but of a sober estimate of one's unimportance.) The text was inscribed in Montaigne's library.
   '88: playing with *the subtlety of words*; behave *immoderately* in; just *and virtuous*; The word of God . . . (By both 'word' and 'voice' of God Montaigne means Holy Scripture.)

religion by showing himself religious beyond any example of men of his rank.[3]

I like natures which are temperate and moderate. Even when an immoderate zeal for the good does not offend me it still stuns me and makes it difficult for me to give it a Christian name. Neither Pausanias' mother (who made the first accusation against her son and who brought the first stone to wall him up for his death) nor Posthumius (the Dictator who had his own son put to death because he had been carried away by youthful ardour and had fought – successfully – slightly ahead of his unit) seem 'just' to me: they seem odd.[4] I neither like to advise nor to imitate a virtue so savage and so costly: the archer who shoots beyond his target misses it just as much as the one who falls short; my eyes trouble me as much when I suddenly come up into a strong light as when I plunge into darkness.

Callicles says in Plato[5] that, at its extremes, philosophy is harmful; he advises us not to go more deeply into it than the limits of what is profitable: taken in moderation philosophy is pleasant and useful, but it can eventually lead to a man's becoming vicious and savage, contemptuous of religion and of the accepted laws, an enemy of social intercourse, an enemy of our human pleasures, useless at governing cities, at helping others or even at helping himself – a man whose ears you could box with impunity. What he says is true, for in its excesses philosophy enslaves our native freedom and with untimely subtleties makes us stray from that beautiful and easy path that Nature has traced for us.

[A] The affection which we bear towards our wives is entirely legitimate: yet Theology nevertheless puts reins on it and restrains it. Among the reasons which Saint Thomas Aquinas[6] cites in condemnation of marriages between relatives who are within the forbidden affinities I think I once read the following: There is a risk that the love felt for such a wife might be immoderate; for if the marital affection between them is full and entire (as it ought to be) and then you add on to it the further affection proper among kinsfolk, there is no doubt that such an over-measure would ravish such a husband beyond the limits of reason.[7]

---

3. Perhaps King Henry III.
4. Diodorus Siculus, XI, x; XII, xix.
5. Plato, *Gorgias*, 484C–D.
6. Thomas Aquinas, *Summa theologica*, II^a, II^{ae}, 154, art. 9: the standard reference; cf. A. Tiraquellus, *De legibus connubialibus*, VII, 46.
7. '80: reason, *either in loving-affection or in the practices of pleasure.* Those . . .

Those sciences which govern the morals of mankind, such as   [C]   Theology and   [A]   philosophy, make everything their concern: no activity is so private or so secret as to escape their attention or their jurisdiction.   [C]   Only mere beginners criticize their freedom to do so: they are like the kind of women whose organs are as accessible as you wish for copulation but who are too bashful to show them to the doctor.   [A]   On behalf of these sciences I therefore want to teach husbands the following[8] –   [C]   if, that is, there are any who are still too eager: [A]   even those very pleasures which they enjoy when lying with their wives are reproved if not kept within moderation; you can fall into licence and excess in this as in matters unlawful.[9]   [C]   All those shameless caresses which our first ardour suggests to us in our sex-play are not only unbecoming to our wives but harmful to them when practised on them. At least let them learn shamelessness from some other hand! They are always wide enough awake when we need them. Where this is concerned what I have taught has been natural and uncomplicated.

[A]   Marriage is a bond both religious and devout: that is why the pleasure we derive from it must be serious, restrained and intermingled with some gravity; its sensuousness should be somewhat wise and dutiful. Its chief end is procreation, so there are those who doubt whether it is right to seek intercourse when we have no hope of conception, as when the woman is pregnant or too old.[10]   [C]   For Plato that constitutes a kind of homicide.   [B]   There are whole peoples,   [C]   including the Mahometans,   [B]   who abominate intercourse with women who are pregnant, and others still during monthly periods. Zenobia admitted her husband for a single discharge; once that was over she let him run wild throughout her pregnancy, giving him permission to begin again only once it was over. There was a fine and noble-hearted marriage for you![11]

[C]   It was from some yearning sex-starved poet that Plato borrowed his story about Jupiter's making such heated advances to his wife one day that he could not wait for her to lie on the bed but tumbled her on the floor, forgetting the great and important decisions which he had just

8. '80: following (*since there is a great danger that they may lose themselves in these excesses*): even those . . .

9. '80: matters *strange and* unlawful . . .

10. '80: old; *and I hold it to be certain that it is much holier to abstain.* There is a people who abominate . . .

11. Plato, *Laws*, VIII, 838A ff.; Guillaume Postel, *Histoire des Turcs*; for Zenobia, Tiraquellus, *De legibus connubialibus*, IX, 88.

reached with the other gods in his celestial Court and boasting that he had enjoyed it as much as when, hidden from her parents, he had first taken her maidenhead.[12]

[A] The kings of Persia did invite their wives as guests to their festivities, but once the wine had seriously inflamed them so that they had to let their lust gallop free, they packed them off to their quarters so as not to make them accomplices of their immoderate appetites, sending instead for other women whom they were not bound to respect.[13]

[B] It is not every pleasure or favour that is well lodged in people of every sort. Epaminondas had a dissolute boy put in prison: Pelopidas, for his own purposes, begged for his freedom; Epaminondas refused but granted it to one of his whores who also begged for it, saying that it was a favour due to a mistress but not to a captain. [C] Sophocles, when a Praetor with Pericles, happened to see a handsome youth go by: 'What a handsome boy,' said he to Pericles. 'That', said Pericles, 'would be all right coming from anyone but a Praetor, who must not only have pure hands but pure eyes.'[14]

[A] When the wife of the Emperor Aelius Verus complained of his permitting himself [C] affairs with [A] other women,[15] he replied that he acted thus for reasons of conscience, marriage being a term of honour and dignity not of wanton and lascivious lust. [C] And our old Church authors make honourable mention of a wife who rejected her husband since she had no wish to be a partner to his lascivious and immoderate embraces.[16]

[A] In short there is no pleasure, however proper, which does not become a matter of reproach when excessive and intemperate.

But, seriously though, is not Man a wretched creature? Because of his natural attributes he is hardly able to taste one single pleasure pure and entire: yet he has to go and curtail even that by arguments; he is not wretched enough until he has increased his wretchedness by art and assiduity.

---

12. Plato, *Laws*, III, 390 BC, after Homer, *Iiad*, XIV, 294–341.
13. Plutarch (tr. Amyot), *Preceptes de mariage*, 146E.
   [A]: their *unruly and* immoderate appetites . . .
14. Plutarch (tr. Amyot), *Instruction pour ceux qui manient les affaires d'Estat*, 167 H; Cicero, *De officiis*, I, xl, 144, distinguishing between moderation (*modestia*) and orderly conduct (*eutaxia*).
15. '80: permitting himself *loving-friendships* with other women . . . (i.e. '80: *amitié*; [C]: *amour*.)
16. E.g., Eusebius (*Pamphilus*), *Ecclesiastical History*, IV.

[B]   *Fortunae miseras auximus arte vias.*

[The wretched paths of Fortune we make worse by art.]¹⁷

[C]   Human wisdom is stupidly clever when used to diminish the number and sweetness of such pleasures as do belong to us, just as she employs her arts with diligence and fitness when she brings comb and cosmetics to our ills and makes us feel them less. If I had founded a school of philosophy I would have taken another route – a more natural one, that is to say a true, convenient and inviolate one; and I might have made myself strong enough to know when to stop.

[A]   Consider the fact that those physicians of our souls and bodies, as though plotting together, can find no other way to cure us and no other remedy for our illnesses of soul and body than by torment, pain and tribulation. Vigils, fasting, hair-shirts and banishments to distant solitary places, endless imprisonments, scourges and other sufferings have been brought in to that end: but only on condition that the suffering is real and should cause bitter pain,   [B]   and that there should not befall what happened to a man called Gallio who was banished to the island of Lesbos: Rome was told that he was enjoying himself there and that what had been inflicted as a punishment was turning into a pleasure, at which he was ordered back to wife and home and commanded to stay put, so as to adapt the punishment to his real feelings.¹⁸   [A]   For if a man's health and happiness were made keener by fasting, or if he found fish more tasty than meat, it would cease to be a salutary prescription: just as drugs prescribed by the other kind of doctor have no affect on anyone who swallowed them with pleasure and enjoyment. The bitter taste and the hardship are attributes which make them work. A constitution which could regularly stand rhubarb would spoil its efficacity: to cure our stomachs it must be something which hurts it: and here the usual axiom that 'contraries cure contraries' breaks down;¹⁹ for in this case illness cures illness.

[B]   This notion is somewhat like that other very ancient one which was universally embraced by all religions and which leads us to think that we can please Heaven and Nature by our murders and our massacres.

[C]   Even in our fathers' time Amurath, when he conquered the

---

17. Propertius, III, vii, 32.
18. The Senator Junius Gallo; cf. Tacitus, *Annals*, VI, iii.
19. A Renaissance medical axiom. It led doctors to recommend, for example, that the cold of Montaigne's favourite fruit, melons, be 'cured' by the heat of ham, pepper or ginger; but it applied to most illnesses too.

Isthmus, sacrificed six hundred Greek youths for the soul of his father, so that their blood might serve as a propitiation, expiating the sins of that dead man.[20] [B] And in those new lands discovered in our own time, lands pure and virgin compared with ours, the practice is accepted virtually everywhere: all their idols are slaked with human blood, not without various examples of dreadful cruelty. Men are burned alive; when half-roasted they are withdrawn from the fire so that their hearts and entrails can be plucked out; others, even women, are flayed alive: their skin, all bloody, serves as a cloak to mask others; and there are no less examples of constancy and determination. For those wretches who are to be immolated, old men, women and children, beg for alms a few days beforehand as offertories at their sacrifice, and present themselves to the slaughter singing and dancing with the congregation. The ambassadors from the King of Mexico, to make Fernando Cortez realize the greatness of their master, first told him that he had thirty vassal-lords, each one of whom could muster a hundred thousand fighting men, and that he dwelt in the strongest fairest city under Heaven; they then added that he had fifty thousand men sacrificed to the gods every year. It is truly said that he cultivated war with some great neighbouring peoples not merely to train the youth of his country but chiefly to furnish prisoners of war for his sacrifices. In another place there was a town where they welcomed Cortez by sacrificing fifty men at the same time. And I will relate one more account: when Cortez had conquered some of these peoples they sent messengers to find out about him and to seek his friendship. They offered him three sorts of gifts in this wise: 'Lord, here are five slaves; if thou art a fierce god who feedest on flesh and blood, eat them and we shall bring thee more. If thou art a kindly god, here are feathers and incense; if thou art human, accept these birds and these fruits.'[21]

20. Related by Laonicus Chalcocondylas (tr. Blaise de Vigenère), *Histoire de la décadence de l'empire grec*, VII, iv.
21. All from Francisco Lopez de Gomara, *Historia de Mexico*, Antwerp, 1554 (tr. A. de Cravaliz as *Historia del Capitano Don Fernando Cortes*, Rome, 1556).

31. On the Cannibals

[The cannibals mentioned in this chapter lived on the coasts of Brazil. Montaigne had read many accounts of the conquest of the New World, including Girolamo Benzoni's Historia del mondo novo (Venice, 1565) in the French translation by Urbain Chauveton, the very title of which emphasizes the dreadful treatment of the natives by the Conquistadores: A New History of the New World containing all that Spaniards have done up to the present in the West Indies, and the harsh treatment which they have meted out to those peoples yonder . . . Together with a short History of a Massacre committed by the Spaniards on some Frenchmen in Florida (two editions in 1579).

Montaigne's 'primitivism' (his respect for barbarous peoples and his admiration for much of their conduct, once their motives are understood) has little in common with the 'noble savages' of later centuries. These peoples are indeed cruel: but so are we. Their simple ways have much to teach us: they can serve as a standard by which we can judge Plato's Republic, the myth of the Golden Age, the cruelty, the corruption and the culture of Europe, and show up that European insularity which condemns peoples as barbarous merely because their manners and their dress are different.]

[A]   When King Pyrrhus crossed into Italy, after noting the excellent formation of the army which the Romans had sent ahead towards him he said, 'I do not know what kind of Barbarians these are' (for the Greeks called all foreigners Barbarians) 'but there is nothing barbarous about the ordering of the army which I can see!' The Greeks said the same about the army which Flaminius brought over to their country,   [C]   as did Philip when he saw from a hill-top in his kingdom the order and plan of the Roman encampment under Publius Sulpicius Galba.[1]   [A]   We should be similarly wary of accepting common opinions; we should judge them by the ways of reason not by popular vote.

I have long had a man with me who stayed some ten or twelve years in that other world which was discovered in our century when Villegaignon made his landfall and named it *La France Antartique*.[2] This discovery of a

1. Plutarch, *Life of Pyrrhus* and *Life of Flaminius*.
2. Durand de Villegagnon struck land, in Brazil, in 1557. Cf. *Lettres sur la navigation du chevalier de Villegaignon es terres de l'Amérique*, Paris, 1557, by an author who calls himself simply N.B.

boundless territory seems to me worthy of reflection. I am by no means sure that some other land may not be discovered in the future, since so many persons, [C] greater than we are, [A] were wrong about this one! I fear that our eyes are bigger than our bellies, our curiosity more³ than we can stomach. We grasp at everything but clasp nothing but wind.

Plato brings in Solon to relate that he had learned from the priests of the town of Saïs in Egypt how, long ago before the Flood, there was a vast island called Atlantis right at the mouth of the Straits of Gibraltar, occupying an area greater than Asia and Africa combined; the kings of that country, who not only possessed that island but had spread on to the mainland across the breadth of Africa as far as Egypt and the length of Europe as far as Tuscany, planned to stride over into Asia and subdue all the peoples bordering on the Mediterranean as far as the Black Sea. To this end they had traversed Spain, Gaul and Italy and had reached as far as Greece when the Athenians withstood them; but soon afterwards those Athenians, as well as the people of Atlantis and their island, were engulfed in that Flood.⁴

It is most likely that that vast inundation should have produced strange changes to the inhabitable areas of the world; it is maintained that it was then that the sea cut off Sicily from Italy –

> [B] *Hæc loca, vi quondam et vasta convulsa ruina,*
> *Dissiluisse ferunt, cum protinus utraque tellus*
> *Una foret.*

[Those places, they say, were once wrenched apart by a violent convulsion, whereas they had formerly been one single land.]⁵

– [A] as well as Cyprus from Syria, and the island of Negropontus from the Boeotian mainland, while elsewhere lands once separated were joined together by filling in the trenches between them with mud and sand:

> *sterilisque diu palus aptaque remis*
> *Vicinas urbes alit, et grave sentit aratrum.*

---

3. '80: our bellies, *as they say, applying it to those whose appetite and hunger make them desire more meat than they can manage: I fear that we too have curiosity far more . . .*
4. Plato, *Timaeus*, 24E etc., and Girolamo Benzoni, *Historia del mondo novo*, Venice 1565. Cf. also Plato, *Critias*, 113 A ff.
5. Virgil, *Aeneid*, III, 414–17.

[Barren swamps which you could row a boat through now feed neighbouring cities and bear the heavy plough.][6]

Yet there is little likelihood of that island's being the New World which we have recently discovered, for it was virtually touching Spain; it would be unbelievable for a flood to force it back more than twelve hundred leagues to where it is now; besides our modern seamen have already all but discovered that it is not an island at all but a mainland, contiguous on one side with the East Indies and on others with lands lying beneath both the Poles – or that if it is separated from them, it is by straits so narrow that it does not deserve the name of 'island' on that account.

[B] It seems that large bodies such as these are subject, as are our own, to changes, [C] some natural, some [B] feverish.[7] When I consider how my local river the Dordogne has, during my own lifetime, been encroaching on the right-hand bank going downstream and has taken over so much land that it has robbed many buildings of their foundation, I realize that it has been suffering from some unusual upset: for if it had always gone on like this or were to do so in the future, the whole face of the world would be distorted. But their moods change: sometimes they incline one way, then another: and sometimes they restrain themselves. I am not discussing those sudden floodings whose causes we know. By the coast-line in Médoc, my brother the Sieur d'Arsac can see lands of his lying buried under sand spewed up by the sea: the tops of some of the buildings are still visible: his rents and arable fields have been changed into very sparse grazing. The locals say that the sea has been thrusting so hard against them for some time now that they have lost four leagues of land. These sands are the sea's pioneer-corps: [C] and we can see those huge shifting sand-dunes marching a half-league ahead in the vanguard, capturing territory.

[A] The other testimony from Antiquity which some would make relevant to this discovery is in Aristotle – if that little book about unheard wonders is really his.[8] He tells how some Carthaginians struck out across the Atlantic beyond the Straits of Gibraltar, sailed for a long time and finally discovered a large fertile island entirely clothed in woodlands and watered by great deep rivers but very far from any mainland; they and others after them, attracted by the richness and fertility of the soil,

6. Horace, *Ars poetica*, 65–6.
7. '88: changes *sickly* and feverish. When . . .
8. The *Secreta secretorum* is supposititious. Montaigne is following Girolamo Benzoni.

emigrated with their wives and children and started living there. The Carthaginian lords, seeing that their country was being gradually depopulated, expressly forbade any more to go there on pain of death and drove out those new settlers, fearing it is said that they would in time increase so greatly that they would supplant them and bring down their State.

But that account in Aristotle cannot apply to these new lands either.

That man of mine was a simple, rough fellow – qualities which make for a good witness: those clever chaps notice more things more carefully but are always adding glosses; they cannot help changing their story a little in order to make their views triumph and be more persuasive; they never show you anything purely as it is: they bend it and disguise it to fit in with their own views. To make their judgement more credible and to win you over they emphasize their own side, amplify it and extend it. So you need either a very trustworthy man or else a man so simple that he has nothing in him on which to build such false discoveries or make them plausible; and he must be wedded to no cause. Such was my man; moreover on various occasions he showed me several seamen and merchants whom he knew on that voyage. So I am content with what he told me, without inquiring what the cosmographers have to say about it.

What we need is topographers who would make detailed accounts of the places which they had actually been to. But because they have the advantage of visiting Palestine, they want to enjoy the right of telling us tales about all the rest of the world! I wish everyone would write only about what he knows – not in this matter only but in all others. A man may well have detailed knowledge or experience of the nature of one particular river or stream, yet about all the others he knows only what everyone else does; but in order to trot out his little scrap of knowledge he will write a book on the whole of physics! From this vice many great inconveniences arise.

Now to get back to the subject, I find (from what has been told me) that there is nothing savage or barbarous about those peoples, but that every man calls barbarous anything he is not accustomed to; it is indeed the case that we have no other criterion of truth or right-reason than the example and form of the opinions and customs of our own country. There we always find the perfect religion, the perfect polity, the most developed and perfect way of doing anything! Those 'savages' are only wild in the sense that we call fruits wild when they are produced by Nature in her ordinary course: whereas it is fruit which we have artificially perverted and misled from the common order which we ought to call savage. It is in the first

kind that we find their true, vigorous, living, most natural and most useful properties and virtues, which we have bastardized in the other kind by merely adapting them to our corrupt tastes. [C] Moreover, there is a delicious savour which even our taste finds excellent in a variety of fruits produced in those countries without cultivation: they rival our own. [A] It is not sensible that artifice should be reverenced more than Nature, our great and powerful Mother. We have so overloaded the richness and beauty of her products by our own ingenuity that we have smothered her entirely. Yet wherever her pure light does shine, she wondrously shames our vain and frivolous enterprises:

> [B] *Et veniunt ederæ sponte sua melius,*
> *Surgit et in solis formosior arbutus antris,*
> *Et volucres nulla dulcius arte canunt.*

[Ivy grows best when left untended; the strawberry tree flourishes more beautifully in lonely grottoes, and birds sing the sweeter for their artlessness.][9]

[A] All our strivings cannot even manage to reproduce the nest of the smallest little bird, with its beauty and appropriateness to its purpose; we cannot even reproduce the web of the wretched spider. [C] Plato says that all things are produced by nature, fortune or art, the greatest and fairest by the first two, the lesser and least perfect by the last.[10]

[A] Those peoples, then, seem to me to be barbarous only in that they have been hardly fashioned by the mind of man, still remaining close neighbours to their original state of nature. They are still governed by the laws of Nature and are only very slightly bastardized by ours; but their purity is such that I am sometimes seized with irritation at their not having been discovered earlier, in times when there were men who could have appreciated them better than we do. It irritates me that neither Lycurgus nor Plato had any knowledge of them, for it seems to me that what experience has taught us about those peoples surpasses not only all the descriptions with which poetry has beautifully painted the Age of Gold[11] and all its ingenious fictions about Man's blessed early state, but also the very conceptions and yearnings of philosophy. They could not even imagine a state of nature so simple and so pure as the one we have learned about from experience; they could not even believe that societies of men could be maintained with so little artifice, so little in the way of human

9. Propertius, I, ii, 10–12.
10. Plato, *Laws*, X, 889 A–C.
11. Cf. Elizabeth Armstrong, *Ronsard and the Age of Gold*, Cambridge, 1968.

solder. I would tell Plato that those people have no trade of any kind, no acquaintance with writing, no knowledge of numbers, no terms for governor or political superior, no practice of subordination or of riches or poverty, no contracts, no inheritances, no divided estates, no occupation but leisure, no concern for kinship – except such as is common to them all – no clothing, no agriculture, no metals, no use of wine or corn. Among them you hear no words for treachery, lying, cheating, avarice, envy, backbiting or forgiveness. How remote from such perfection would Plato find that Republic which he thought up – [C] *'viri a diis recentes'* [men fresh from the gods].[12]

[B] *Hos natura modos primum dedit.*

[These are the ways which Nature first ordained.][13]

[A] In addition they inhabit a land with a most delightful countryside and a temperate climate, so that, from what I have been told by my sources, it is rare to find anyone ill there;[14] I have been assured that they never saw a single man bent with age, toothless, blear-eyed or tottering. They dwell along the sea-shore, shut in to landwards by great lofty mountains, on a stretch of land some hundred leagues in width. They have fish and flesh in abundance which bear no resemblance to ours; these they eat simply cooked. They were so horror-struck by the first man who brought a horse there and rode it that they killed him with their arrows before they could recognize him, even though he had had dealings with them on several previous voyages. Their dwellings are immensely long, big enough to hold two or three hundred souls; they are covered with the bark of tall trees which are fixed into the earth, leaning against each other in support at the top, like some of our barns where the cladding reaches down to the ground and acts as a side. They have a kind of wood so hard that they use it to cut with, making their swords from it as well as grills to cook their meat. Their beds are woven from cotton and slung from the roof like hammocks on our ships; each has his own, since wives sleep apart from their husbands. They get up at sunrise and have their meal for the day as soon as they do so; they have no other meal but that one. They drink

12. Seneca, *Epist. moral.*, XC, 44. (This epistle is a major defence of the innocence of natural man before he was corrupted by philosophy and progress.)
13. Virgil, *Georgics*, II, 208.
14. One of Montaigne's sources was Simon Goulart's *Histoire du Portugal*, Paris, 1587, based on a work by Bishop Jeronimo Osorio (da Fonseca) and others.

nothing with it, [B] like those Eastern peoples who, according to Suidas,[15] only drink apart from meals. [A] They drink together several times a day, and plenty of it. This drink is made from a certain root and has the colour of our claret. They always drink it lukewarm; it only keeps for two or three days; it tastes a bit sharp, is in no ways heady and is good for the stomach; for those who are not used to it it is laxative but for those who are, it is a very pleasant drink. Instead of bread they use a certain white product resembling coriander-cakes. I have tried some: it tastes sweet and somewhat insipid.

They spend the whole day dancing; the younger men go off hunting with bow and arrow. Meanwhile some of the women-folk are occupied in warming up their drink: that is their main task. In the morning, before their meal, one of their elders walks from one end of the building to the other, addressing the whole barnful of them by repeating one single phrase over and over again until he has made the rounds, their building being a good hundred yards long. He preaches two things only: bravery before their enemies and love for their wives. They never fail to stress this second duty, repeating that it is their wives who season their drink and keep it warm. In my own house, as in many other places, you can see the style of their beds and rope-work as well as their wooden swords and the wooden bracelets with which they arm their wrists in battle, and the big open-ended canes to the sound of which they maintain the rhythm of their dances. They shave off all their hair, cutting it more cleanly than we do, yet with razors made of only wood or stone. They believe in the immortality of the soul: souls which deserve well of the gods dwell in the sky where the sun rises; souls which are accursed dwell where it sets. They have some priests and prophets or other, but they rarely appear among the people since they live in the mountains. When they do appear they hold a great festival and a solemn meeting of several villages – each of the barns which I have described constituting a village situated about one French league distant from the next. The prophet then addresses them in public, exhorting them to be virtuous and dutiful, but their entire system of ethics contains only the same two articles: resoluteness in battle and love for their wives. He foretells what is to happen and the results they must expect from what they undertake; he either incites them to war or deflects them from it, but only on condition that if he fails to divine correctly and if things turn out other than he foretold, then – if they can catch him – he is condemned as a false prophet and hacked to pieces. So the prophet who gets it wrong once is seen no more.

15. Suidas, *Historica, caeteraque omnia quae ad cognitionem rerum spectant*, Basle, 1564.

[C] Prophecy is a gift of God.[16] That is why abusing it should be treated as a punishable deceit. Among the Scythians, whenever their soothsayers got it wrong they were shackled hand and foot and laid in ox-carts full of bracken where they were burned.[17] Those who treat subjects under the guidance of human limitations can be excused if they have done their best; but those who come and cheat us with assurances of powers beyond the natural order and then fail to do what they promise, should they not be punished for it and for the foolhardiness of their deceit?

[A] These peoples have their wars against others further inland beyond their mountains; they go forth naked, with no other arms but their bows and their wooden swords sharpened to a point like the blades of our pig-stickers. Their steadfastness in battle is astonishing and always ends in killing and bloodshed: they do not even know the meaning of fear or flight. Each man brings back the head of the enemy he has slain and sets it as a trophy over the door of his dwelling. For a long period they treat captives well and provide them with all the comforts which they can devise; afterwards the master of each captive summons a great assembly of his acquaintances; he ties a rope to one of the arms of his prisoner [C] and holds him by it, standing a few feet away for fear of being caught in the blows, [A] and allows his dearest friend to hold the prisoner the same way by the other arm: then, before the whole assembly, they both hack at him with their swords and kill him. This done, they roast him and make a common meal of him, sending chunks of his flesh to absent friends. This is not as some think done for food — as the Scythians used to do in antiquity — but to symbolize ultimate revenge. As a proof of this, when they noted that the Portuguese who were allied to their enemies practised a different kind of execution on them when taken prisoner — which was to bury them up to the waist, to shoot showers of arrows at their exposed parts and then to hang them — they thought that these men from the Other World, who had scattered a knowledge of many a vice throughout their neighbourhood and who were greater masters than they were of every kind of revenge, which must be more severe than their own; so they began to abandon their ancient method and adopted that one. It does not sadden me that we should note the horrible barbarity in a practice such as theirs: what does sadden me is that, while judging correctly of their wrong-doings we should be so blind to our own. I think there is more

16. Cf. Cicero, *De divinatione*, I, i.1; I Peter 1:2; I Corinthians 12:20; 13:2.
17. Herodotus, *History*, IV, lxix.

barbarity in eating a man alive than in eating him dead; more barbarity in
lacerating by rack and torture a body still fully able to feel things, in
roasting him little by little and having him bruised and bitten by pigs and
dogs (as we have not only read about but seen in recent memory, not
among enemies in antiquity but among our fellow-citizens and neighbours –
and, what is worse, in the name of duty and religion) than in roasting him
and eating him after his death.

Chrysippus and Zeno, the leaders of the Stoic school, certainly thought
that there was nothing wrong in using our carcasses for whatever purpose
we needed, even for food – as our own forebears did when, beleaguered by
Caesar in the town of Alesia, they decided to relieve the hunger of the
besieged with the flesh of old men, women and others who were no use in
battle:

> [B]   *Vascones, fama est, alimentis talibus usi*
> *Produxere animas.*

[By the eating of such food it is notorious that the Gascons prolonged their
lives.][18]

[A]   And our medical men do not flinch from using corpses in many
ways, both internally and externally, to cure us.[19] Yet no opinion has ever
been so unruly as to justify treachery, disloyalty, tyranny and cruelty,
which are everyday vices in us. So we can indeed call those folk barbarians
by the rules of reason but not in comparison with ourselves, who surpass
them in every kind of barbarism. Their warfare is entirely noble and
magnanimous; it has as much justification and beauty as that human
malady allows: among them it has no other foundation than a zealous
concern for courage. They are not striving to conquer new lands, since
without toil or travail they still enjoy that bounteous Nature who furnishes
them abundantly with all they need, so that they have no concern to push
back their frontiers. They are still in that blessed state of desiring nothing
beyond what is ordained by their natural necessities: for them anything
further is merely superfluous. The generic term which they use for men of
the same age is 'brother'; younger men they call 'sons'. As for the old men,
they are the 'fathers' of everyone else; they bequeath all their goods,
indivisibly, to all these heirs in common, there being no other entitlement

---

18. Sextus Empiricus, *Hypotyposes*, III, xxiv; Caesar, *Gallic Wars*, VII, lvii–lviii;
Juvenal, *Satires*, XV, 93–4.
19. Mummies were imported for use in medicines. (Othello's handkerchief was
steeped in 'juice of mummy'.)

than that with which Nature purely and simply endows all her creatures by bringing them into this world. If the neighbouring peoples come over the mountains to attack them and happen to defeat them, the victors' booty consists in fame and in the privilege of mastery in virtue and valour: they have no other interest in the goods of the vanquished and so return home to their own land, which lacks no necessity; nor do they lack that great accomplishment of knowing how to enjoy their mode of-being in happiness and to be content with it. These people do the same in their turn: they require no other ransom from their prisoners-of-war than that they should admit and acknowledge their defeat – yet there is not one prisoner in a hundred years who does not prefer to die rather than to derogate from the greatness of an invincible mind by look or by word; you cannot find one who does not prefer to be killed and eaten than merely to ask to be spared. In order to make their prisoners love life more they treat them generously in every way,[20] but occupy their thoughts with the menaces of the death awaiting all of them, of the tortures they will have to undergo and of the preparations being made for it, of limbs to be lopped off and of the feast they will provide. All that has only one purpose: to wrench some weak or unworthy word from their lips or to make them wish to escape, so as to enjoy the privilege of having frightened them and forced their constancy.[21]

Indeed, if you take it the right way, true victory[22] consists in that alone:

[C] *victoria nulla est*
Quam quæ confessos animo quoque subjugat hostes.

[There is no victory unless you subjugate the minds of the enemy and make them admit defeat.][23]

In former times those warlike fighters the Hungarians never pressed their advantage beyond making their enemy throw himself on their mercy. Once having wrenched this admission from him, they let him go without injury or ransom, except at most for an undertaking never again to bear arms against them.[24]

[A] Quite enough of the advantages we do gain over our enemies are mainly borrowed ones not truly our own. To have stronger arms and legs

20. '80: generously in every way, *and furnish them with all the comforts they can devise* but . . .
21. '80: their *virtue and their* constancy . . .
22. '80: true *and solid* victory . . .
23. Claudian, *De sexto consulatu Honorii*, 248–9.
24. Nicolas Chalcocondylas (tr. Blaise de Vigenère), *De la décadence de l'empire grec*, V, ix.

is the property of a porter not of Valour; agility is a dead and physical quality, for it is chance which causes your opponent to stumble and which makes the sun dazzle him; to be good at fencing is a matter of skill and knowledge which may light on a coward or a worthless individual. A man's worth and reputation lie in the mind and in the will: his true honour is found there. Bravery does not consist in firm arms and legs but in firm minds and souls: it is not a matter of what our horse or our weapons are worth but of what we are. The man who is struck down but whose mind remains steadfast,   [C] *'si succiderit, de genu pugnat'* [if his legs give way, then on his knees doth he fight];[25]   [B]   the man who relaxes none of his mental assurance when threatened with imminent death and who faces his enemy with inflexible scorn as he gives up the ghost is beaten by Fortune not by us: [C] he is slain but not vanquished.[26]   [B] Sometimes it is the bravest who may prove most unlucky.   [C] So there are triumphant defeats rivalling victories; Salamis, Plataea, Mycale and Sicily are the fairest sister-victories which the Sun has ever seen, yet they would never dare to compare their combined glory with the glorious defeat of King Leonidas and his men at the defile of Thermopylae.[27] Who has ever run into battle with a greater desire and ambition for victory than did Captain Ischolas when he was defeated? Has any man ever assured his safety more cleverly or carefully than he assured his destruction?[28] His task was to defend against the Arcadians a certain pass in the Peleponnesus. He realized that he could not achieve this because of the nature of the site and of the odds against him, concluding that every man who faced the enemy must of necessity die in the battlefield; on the other hand he judged it unworthy of his own courage, of his greatness of soul and of the name of Sparta to fail in his duty; so he chose the middle path between these two extremes and acted thus: he saved the youngest and fittest soldiers of his unit to serve for the defence of their country and sent them back there. He then determined to defend that pass with men whose loss would matter less and who would, by their death, make the enemy purchase their breakthrough as dearly as possible. And so it turned out. After butchering the Arcadians who beset them on every side, they were all put to the sword. Was ever a trophy raised to a victor which was not better due to those who were vanquished?

25. Seneca, *De constantia*, II.
26. '80: by us: *he is vanquished in practice but not by reason; it is his bad luck which we may indict not his cowardice.* Sometimes . . .
27. Cf. Cicero, *Tusc. disput.*, I, xli, 100 for the glory of Leonidas' death in the defile of Thermopylae.
28. Diodorus Siculus, XV, xii.

True victory lies in your role in the conflict, not in coming through safely: it consists in the honour of battling bravely not battling through.

[A] To return to my tale, those prisoners, far from yielding despite all that was done to them during the two or three months of their captivity, maintain on the contrary a joyful countenance: they urge their captors to hurry up and put them to the test; they defy them, insult them and reproach them for cowardice and for all the battles they have lost against their country. I have a song made by one such prisoner which contains the following: Let them all dare to come and gather to feast on him, for with him they will feast on their own fathers and ancestors who have served as food and sustenance for his body. 'These sinews,' he said, 'this flesh and these veins – poor fools that you are – are your very own; you do not realize that they still contain the very substance of the limbs of your forebears: savour them well, for you will find that they taste of your very own flesh!' There is nothing 'barbarous' in the contriving of that topic. Those who tell how they die and who describe the act of execution show the prisoners spitting at their killers and pulling faces at them. Indeed, until their latest breath, they never stop braving them and defying them with word and look. It is no lie to say that these men are indeed savages – by our standards; for either they must be or we must be: there is an amazing gulf between their [C] souls [A] and ours.[29]

The husbands have several wives: the higher their reputation for valour the more of them they have. One beautiful characteristic of their marriages is worth noting: just as our wives are zealous in thwarting our love and tenderness for other women, theirs are equally zealous in obtaining them for them. Being more concerned for their husband's reputation than for anything else, they take care and trouble to have as many fellow-wives as possible, since that is a testimony to their husband's valour.

– [C] Our wives will scream that that is a marvel, but it is not: it is a virtue proper to matrimony, but at an earlier stage. In the Bible Leah, Rachel, Sarah and the wives of Jacob all made their fair handmaidens available to their husbands; Livia, to her own detriment, connived at the lusts of Augustus, and Stratonice the consort of King Deiotarus not only provided her husband with a very beautiful chambermaid who served her but carefully brought up their children and lent a hand in enabling them to succeed to her husband's rank.[30]

29. '80: their *constancy* and ours . . .
30. Standard examples: cf. Tiraquellus, *De legibus connubialibus*, XIII, 35, for all these un-jealous wives. (But Leah and Sarah were in fact Jacob's wives.)

− [A] Lest anyone should think that they do all this out of a simple slavish subjection to convention or because of the impact of the authority of their ancient customs without any reasoning or judgement on their part, having minds so dulled that they could never decide to do anything else, I should cite a few examples of what they are capable of.

Apart from that war-song which I have just given an account of, I have another of their songs, a love-song, which begins like this:

> O Adder, stay: stay O Adder! From your colours
>   let my sister take the pattern for a girdle
>   she will make for me to offer to my love;
> So may your beauty and your speckled hues be for
>   ever honoured above all other snakes.

This opening couplet serves as the song's refrain. Now I know enough about poetry to make the following judgement: not only is there nothing 'barbarous' in this conceit but it is thoroughly anacreontic.[31] Their language incidentally is [C] a pleasant one with an agreeable sound [A] and has terminations[32] rather like Greek.

Three such natives, unaware of what price in peace and happiness they would have to pay to buy a knowledge of our corruptions, and unaware that such commerce would lead to their downfall − which I suspect to be already far advanced − pitifully allowing themselves to be cheated by their desire for novelty and leaving the gentleness of their regions to come and see ours, were at Rouen at the same time as King Charles IX.[33] The King had a long interview with them: they were shown our manners, our ceremonial and the layout of a fair city. Then someone asked them what they thought of all this and wanted to know what they had been most amazed by. They made three points; I am very annoyed with myself for forgetting the third, but I still remember two of them. In the first place they said (probably referring to the Swiss Guard) that they found it very odd that all those full-grown bearded men, strong and bearing arms in the King's entourage, should consent to obey a boy rather than choosing one of themselves as a Commander; secondly − since they have an idiom in their language which calls all men 'halves' of one another − that they had noticed that there were among us men fully bloated with all sorts of

---

31. Anacreon was the great love-poet of Teos (*fl.* 540 B C).
32. '80: their language *is the pleasantest language in the world; its* sound *is agreeable to the ear* and has terminations . . .
33. In 1562, when Rouen was retaken by Royalist forces.

comforts while their halves were begging at their doors, emaciated with poverty and hunger: they found it odd that those destitute halves should put up with such injustice and did not take the others by the throat or set fire to their houses.

I had a very long talk with one of them (but I used a stupid interpreter who was so bad at grasping my meaning and at understanding my ideas that I got little joy from it). When I asked the man (who was a commander among them, our sailors calling him a king) what advantage he got from his high rank, he told me that it was to lead his troops into battle; asked how many men followed him, he pointed to an open space to signify as many as it would hold – about four or five thousand men; questioned whether his authority lapsed when the war was over, he replied that he retained the privilege of having paths cut for him through the thickets in their forests, so that he could easily walk through them when he visited villages under his sway.

Not at all bad, that. – Ah! But they wear no breeches . . .

32. Judgements on God's ordinances must be embarked upon with prudence

===

[The theme that God's counsel is a secret which Man should not try to scan was a common one in the Renaissance. Montaigne applies that dogma to the ups and downs of the Wars of Religion: we cannot say that God is on the side of the victors in battle. Montaigne asserts that even the pagan Indians of the New World know that better than warring Christians do.]

[A]   The real field and subject of deception are things unknown: firstly because their very strangeness lends them credence; second, because they cannot be exposed to our usual order of argument, so stripping us of the means of fighting them.   [C]   Plato says that this explains why it is easier to satisfy people when talking of the nature of the gods than of the nature of men: the ignorance of the hearers provides such hidden matters with a firm broad course for them to canter along in freedom.[1]   [A]   And so it turns out that nothing is so firmly believed as whatever we know least about, and that no persons are more sure of themselves than those who tell us tall stories, such as alchemists and those who make prognostications: judicial astrologers, chiromancers, doctors and *'id genus omne'* [all that tribe].[2] To which I would add if I dared that crowd of everyday chroniclers and interpreters of God's purposes who claim to discover the causes of everything that occurs and to read the unknowable purposes of God by scanning the secrets of His will; the continual changes and clash of events drive them from corner to corner and from East to West, but they still go on chasing the tennis-ball and sketching black and white with the same crayon.   [B]   In one Indian tribe they have a laudable custom: when they are worsted in a skirmish or battle they publicly beseech the Sun their god for pardon for having done wrong, attributing their success or failure to the divine mind, to which they submit their own judgement and discourse.

[A]   For a Christian it suffices to believe that all things come from God,

---

1. Plato, *Critias*, 107B.
2. Horace, *Satires*, I, ii, 2.

to accept them with an acknowledgement of His holy unsearchable wisdom and so to take them in good part, under whatever guise they are sent to him. What I consider wrong is our usual practice of trying to support and confirm our religion by the success or happy outcome of our undertakings. Our belief has enough other foundations without seeking sanction from events: people who have grown accustomed to such plausible arguments well-suited to their taste are in danger of having their faith shaken when the turn comes for events to prove hostile and unfavourable. As in the religious wars which we are now fighting, after those who had prevailed at the battle of La Rochelabeille had had a great feast-day over the outcome, exploiting their good fortune as a sure sign of God's approval for their faction, they then had to justify their misfortunes at Moncontour and Jarnac as being Fatherly scourges and chastisements:[3] they would soon have made the people realize (if they did not have them under their thumb) that that is getting two kinds of meal from the same bag and blowing hot and cold with the same breath. It would be better to explain to the people the real foundations of truth.

That was a fine naval engagement which we won against the Turk a few months ago, led by Don John of Austria: yet at other times it has pleased God to make us witness other such battles which cost us dear.[4]

In short it is hard to bring matters divine down to human scales without their being trivialized. Supposing someone sought to explain why Arius and Leo his Pope (who were the main proponents of the Arian heresy) both died at different times of deaths so strange and similar, for they both had to leave the debates because of pains in their stomach and go to the lavatory, where both promptly died; and supposing they emphasized God's vengeance by insisting on the nature of the place where this happened: well, they could add the example of the death of Heliogabalus who was also killed in a privy. Why, Irenaeus himself met the same fate.[5]

[C] God wishes us to learn that the good have other things to hope for and the wicked other things to fear than the chances and mischances of this

---

3. The Reformers won at La Rochelabeille (1562) and lost at Jarnac and Moncontour (1569). Both sides attributed their defeats to God's 'fatherly' chastisement, on the authority of II Samuel 7:14 and Hebrews 12:5–6.
4. Don John of Austria's Catholic Spanish navy won at Lepanto (1571); but the Spanish Invincible Armada was scattered and defeated in 1588, a defeat attributed throughout Protestant Europe to God's intervention on the side of true religion.
5. Ravisius Textor in his *Officina* lists under the heading *Dead or killed in latrines* Heliogabalus and also the martyrs Irenaeus and Albundius who were tossed alive into the latrines by Valerianus, where they died.

world, which his hands control according to his hidden purposes: and so he takes from us the means of foolishly exploiting them. Those who desire to draw advantage from them by human reason delude themselves. For every hit which they make, they suffer two in return. St Augustine amply proved that against his opponents: the arms which decide that wrangle are not those of reason but of memory.[6]

[A]   We must be content with the light which the Sun vouchsafes to shed on us by its rays: were a man to lift up his eyes to seek a greater light in the Sun itself, let him not find it strange if he is blinded as a penalty for his presumption.   [C]   *'Quis hominum potest scire consilium dei? aut quis poterit cogitare quid velit dominus?'* [For what man can know the counsel of God: or who shall conceive what the Lord willeth?][7]

6. Cf. St Augustine, *City of God*, I, viii.
7. Wisdom of Solomon 9:13.

33. On fleeing from pleasures at the cost of one's life

[This chapter shows how Christianity in early times was closer to Stoicism than many realized. The anecdote of St Hilary derives from the Annales d'Aquitaine *of Jean Bouchet (a friend of Rabelais and a fluent moral poet and historian of the generation before Montaigne).]*

[A]   I had already noted that the majority of ancient opinions agree on the following: that it is time to die when living entails more ill than good, and that preserving our life to our anguish or prejudice is to infringe the very laws of Nature − as these old precepts put it:

> Ἡ ζῆν ἀλύπως, ἢ θανεῖν εὐδαιμόνως.
> Καλὸν θνήσκειν οἷς ὕβριν τὸ ζῆν φέρει.
> Κρεῖσσον τὸ μὴ ζῆν ἐστὶν ἢ ζῆν ἀθλίως.
>
> [Either a quiet life or a happy death.
> It is good to die for those who find life a burden.
> Better not to live than to live in wretchedness.][1]

But to carry contempt for death to the point of using it to dissuade us from honours, riches, great offices and the rest of what we call Fortune's goods and favours, as if reason did not already have cause enough to persuade us to abandon them without adding this fresh attack, is something I had never seen enjoined or practised until I happened upon a passage in Seneca; in it he advises Lucilius, a man of power and of great authority in the Emperor's court, to renounce his life of pomp and pleasure and to withdraw from such worldly ambition into a solitary life, tranquil and philosophical. When Lucilius cited some of the difficulties, Seneca said: 'My counsel is that either you should quit that life or life itself; I do indeed advise you to adopt the easier course of slipping the bonds which you have wrongly tied

1. Three Greek poetic proverbs; taken it seems from the Greek anthology of *Georgica, Bucolica* and *Gnomica* published by Jean Crespin of Geneva about 1570 (copy in Cambridge University Library).

rather than breaking them, providing that you do break them if you cannot do otherwise. No man is such a coward that he would not rather make one fall than to be forever on the brink.'[2]

I would have found that counsel in keeping with the asperity of the Stoics, but it is rather odd that Seneca could borrow it from Epicurus, who on this subject writes to Idomeneus in the very same way.[3] Nevertheless I think I have noted a similar precept among our own people, but with Christian moderation. St Hilary, the Bishop of Poitiers and a famous enemy of the Arian heresy, was in Syria when he was told that his only daughter Abra, whom he had left overseas with her mother, was being courted by some of the most notable lords of the land since she was very well brought up, a maiden fair, rich and blooming. He wrote to her (as we know) that she should get rid of her love of the pleasures and favours that were being offered her, saying that he had found for her during his journey a Suitor who was far greater and more worthy, a Bridegroom of very different power and glory, who would vouchsafe her a present of robes and jewels of countless price. His aim was to make her lose the habit and taste of worldly pleasures and to wed her to God; but since the most sure and shortest way seemed to him that his daughter should die, he never ceased to beseech God in his prayers, vows and supplications that he should take her from this world and call her to Himself. And so it happened: soon after his return she did die, at which he showed uncommon joy. He seems to have outstripped those others in that he had immediate recourse to a means which they keep in reserve; and besides it concerned his only daughter.

But I would not omit the end of this story: when St Hilary's wife heard from him how the death of their daughter had been brought about by his wish and design, and how much happier she was to have quitted this world than to have remained in it, she too took so lively a grasp on that eternal life in Heaven that she besought her husband, with the utmost urgency, to do the same for her. Soon after, when God took her to Himself in answer to both their prayers, the death was welcomed with open arms and with an uncommon joy which both of them shared.[4]

---

2. Seneca, *Epist. moral.*, XXII, 3.
3. *Idem*, 14. (That Epicurus was writing to Idomeneus we know from XXI, 7.)
4. From Jean Bouchet, *Annales d'Acquitaine*, Poitiers, 1557, pp. 16–21.

34. *Fortune is often found in Reason's train*

===

[The Roman censor was not too happy about Montaigne's writing about Fortune (as distinct from Providence) – strangely so, since fickle Fortune and Fortune's Wheel were centuries-old commonplaces. (The word Fortune *itself occurs some 350 times in the* Essais. *Montaigne explains why he finds it right to use words such as Fortune and Destiny in I, 56, 'On prayer'.)]*

[A]   The changeableness of Fortune's varied dance means that she must inevitably show us every kind of face. Has any of her actions ever been more expressly just than the following? The Duke of Valentinois decided to poison Adrian the Cardinal of Corneto, to whose home in the Vatican he and his father Pope Alexander VI were coming to dine; so he sent ahead a bottle of poisoned wine with instructions to the butler to look after it carefully. The Pope, chancing to arrive before his son, asked for a drink; that butler, who thought that the wine had been entrusted to him merely because of its quality, served it to him; then the Duke himself, arriving just in time for dinner and trusting that nobody would have touched his bottle, drank some too, so that the father died suddenly while the son, after being tormented by a long illness, was reserved for a worse and different fortune.[1]

Sometimes it seems that Fortune is literally playing with us. The Seigneur d'Estrées (who was then ensign to Monseigneur de Vendôme) and the Seigneur de Licques (a lieutenant in the forces of the Duke of Aerschot) were both suitors of the sister of the Sieur de Fouquerolle – despite their being on opposite sides, as often happens with neighbours on the frontier. The Seigneur de Licques was successful. However, on his very wedding-day and, what is worse, before going to bed, the bridegroom desired to break a lance as a tribute to his new bride and went out skirmishing near St Omer; there, he was taken prisoner by the Seigneur d'Estrées who had proved the stronger. To exploit this advantage to the full, d'Estrées compelled the lady –

---

1. Francesco Guicciardini, *L'Historia d'Italia*, IV.

> *Conjugis ante coacta novi dimittere collum,*
> *Quam veniens una atque altera rursus hyems*
> *Noctibus in longis avidum saturasset amorem.*

[Forced to release her embrace of her young husband before the long nights of a couple of winters had sated her eager love][2] –

personally to beg him, of his courtesy, to surrender his prisoner to her. Which he did, the French nobility never refusing anything to the ladies . . .

[C] Was the following not Fate apparently playing the artist? The Empire of Constantinople was founded by Constantine son of Helena: many centuries later it was ended by another Constantine son of Helena!

[A] Sometimes it pleases Fortune to rival our Christian miracles. We hold that when King Clovis was besieging Angoulême, by God's favour the walls collapsed of themselves; Bouchet borrows from some other author an account of what happened when King Robert was laying siege to a certain city: he slipped off to Orleans to celebrate the festival of St Aignan; while he was saying his prayers, at a certain point in the Mass the walls of the besieged city collapsed without being attacked.[3] But Fortune produced quite opposite results during our Milanese wars: for after Captain Renzo had mined a great stretch of the wall while besieging the town of Arona for us French it was blown right up in the air, only to fall straight back on to its foundations all in one piece so that the besieged were no worse off.[4]

Sometimes Fortune dabbles in medicine. Jason Phereus was given up by his doctors because of a tumour on the breast; wishing to rid himself of it even by death, he threw himself recklessly into battle where the enemy was thickest; he was struck through the body at precisely the right spot, lancing his tumour and curing him.

Did Fortune not surpass Protogenes the painter in mastery of his art? He had completed a portrait of a tired and exhausted dog; he was pleased with everything else but could not paint its foaming slaver to his own satisfaction; irritated against his work, he grabbed a sponge and threw it at it, intending to blot everything out since the sponge was impregnated with a variety of paints: Fortune guided his throw right to the mouth of the dog and produced the effect which his art had been unable to attain.[5]

---

2. Catullus, LXVIII, 81–3. (The event is narrated by the Du Bellays in their *Mémoires*, II.)

3. Jean Bouchet, *Annales d'Acquitaine*.

4. The Du Bellays' *Mémoires*, II. Cf. Simon Goulart, *Histoires admirables*, 1610–14, IV, p. 686.

5. Both anecdotes derive from Pliny (*Hist. nat.*, VII and XXXV).

Does she not sometimes direct our counsels and correct them? Queen Isabella of England had to cross over to her kingdom from Zealand with her army to come to the aid of her son against her husband; she would have been undone if she had landed at the port she had intended, for her enemies were awaiting her there; but Fortune drove her unwillingly to another place, where she landed in complete safety.[6] And that Ancient who chucked a stone at a dog only to hit his stepmother and kill her could he not have rightly recited this verse:[7]

Ταυτόματον ἡμῶν καλλίω βουλεύεται.

'Fortune has better counsel than we do.'[8]   [C]   Icetes had bribed two soldiers to murder Tomoleon during his stay in Adrana in Sicily. They chose a time when he was about to make some sacrifice or other; they mingled with the crowd; just as they were signalling to each other that the time was right for their deed, along comes a third soldier who landed a mighty sword-blow on the head of one of them and then ran away. His companion, believing he was discovered and undone, ran to the altar begging for sanctuary and promising to reveal all the truth. Just as he was giving an account of the conspiracy the third man was caught and was being dragged and manhandled through the crowd towards Timoleon and the more notable members of the congregation: he begged for mercy, saying that he had rightly killed his father's murderer, immediately proving by witnesses which good luck had conveniently provided that his father had indeed been murdered in the town of the Leontines by the very man against whom he had taken his revenge. He was granted ten Attic silver-pounds as a reward for his good luck in saving the life of the Father of the Sicilian People while avenging the death of his own father. Such fortune surpasses in rightness the right-rules of human wisdom.[9]

[B]   To conclude. Does not the following reveal a most explicit granting of her favour as well as her goodness and singular piety? The two Ignatii, father and son, having been proscribed by the Roman Triumvirate, nobly decided that their duty was to take each other's life and so frustrate the cruelty of those tyrants. Sword in hand they fell on each other. Fortune guided their sword-points, made both blows equally mortal and honoured

---

6. Cf. Froissart, *Histoire et cronique*, Lyons, 1559, I, x.
7. Plutarch (tr. Amyot), *De la tranquillité de l'esprit*, p. 70A.
8. Menander (translated in the text).
9. Cf. Plutarch, *Life of Timoleon*.

the beauty of such a loving affection by giving them just enough strength to withdraw their forearms from the wounds, blood-stained and still grasping their weapons, and to clasp each other, there as they lay, in such an embrace that the executioners cut off both their heads at once, allowing their bodies to remain nobly entwined together, wound against wound, lovingly soaking up each other's life-blood.[10]

10. Appian (of Alexandria), *Des guerres civiles des Romains* (translated from the Greek by Claude de Seyssel), Lyons, 1544.

35. Something lacking in our civil administrations

===

[Montaigne's father had sound ideas on practical charity and on running a household. The link between the two parts of this short chapter is the word polices *(polity: civil administration) which applies both to the running of a country and the running of an estate. Montaigne shows the breadth of his charity, Sebastian Castalio being in bad odour because of his translation of the Bible and for his use of Lutheran works to preach religious tolerance.]*

[A] My late father, a man of a decidedly clear judgement, based though it was only on his natural gifts and his own experience, said to me once that[1] he had wished to set a plan in motion leading to the designation of a place in our cities where those who were in need of anything could go and have their requirements registered by a duly appointed official; for example: [C] 'I want to sell some pearls'; or 'I want to buy some pearls.' [A] 'So-and-so wants to make up a group to travel to Paris'; 'So-and-so wants a servant with the following qualifications'; 'So-and-so seeks an employer'; 'So-and-so wants a workman'; each stating his wishes according to his needs.

It does seem that this means of mutual advertising would bring no slight advantage to our public dealings; for at every turn there are bargains seeking each other but, because they cannot find each other, men are left in extreme want.

I have just learnt something deeply shameful to our times; under our very eyes two outstanding scholars have died for want of food, Lilius Gregorius Giraldus in Italy and Sebastian Castalio in Germany; and I believe that there are hundreds of people who would have invited them to their houses on very favourable terms [C] or sent help to them where they were, [A] if only they had known.[2]

---

1. [A] until [C]: that, *among the instructions which fell into his hands*, he had wished . . .
2. Giraldi, famous for his erudite works on the gods of Antiquity and on their burial customs, died in poverty (Ferrara, 1552); Châteillon (Castalio) died in Basle, 1563.

The world is not so completely corrupt that we cannot find even one man who would not gladly wish to see his inherited wealth able to be used (as long as Fortune lets him enjoy it) to provide shelter for great men who are renowned for some particular achievement but who have been reduced to extreme poverty by their misfortunes; he could at least give them enough assistance that it would be unreasonable for them not to be satisfied.

[C] In his administration of his household affairs my father had a rule which I can admire but in no ways follow. In addition to keeping a record of household accounts entrusted to the hands of a domestic bursar (making entries for small bills and payments or transactions which did not need the signature of a lawyer) he told the man who acted as his secretary to keep a diary covering any noteworthy event and the day-to-day history of his household. It is very pleasant to consult, once time begins to efface memories; it is also useful for clearing up difficulties. When was such-and-such a job begun? When was it finished? Who called at Montaigne with their retinues? How many came to stay? It notes our journeys, absences, marriages and deaths, the receipt of good or bad news; changes among our chief servants – things like that: an ancient custom which I would like to be revived by each denizen in his own den. I think I am a fool to have neglected it.

36. On the custom of wearing clothing

===

[*In this chapter Montaigne makes a pun on the French taste for* bigarures, *which means, as Cotgrave explains it in his* Dictionarie of the French and English Tongues *(1632) both a medley of 'sundry colours mingled together' and a discourse 'running odly and fantastically, from one matter to another'. This chapter is an example of such a colourful medley, hopping from thoughts on Man's natural nakedness to examples of extraordinary cold.*]

[A] Whichever way I want to go I find myself obliged to break through some barrier of custom, so thoroughly has she blocked all our approaches. During this chilly season I was chatting about whether the habit of those newly discovered peoples of going about stark naked was forced on them by the hot climate, as we say of the Indians and the Moors, or whether it is the original state of mankind. Since the word of God says that 'everything under the sun' is subject to the same law,[1] in considerations such as these, where a distinction has to be made between natural laws and contrived ones, men of understanding regularly turn for advice to the general polity of the world: nothing can be counterfeit there. Now, since everything therein is exactly furnished with stitch and needle for maintaining its being, it is truly unbelievable that we men alone should have been brought forth in a deficient and necessitous state, a state which can only be sustained by borrowings from other creatures. I therefore hold that just as plants, trees, animals and all living things are naturally equipped with adequate protection from the rigour of the weather –

1. The phrase 'under the sun' occurs as a refrain in Ecclesiastes (and nowhere else in the Bible). On the beams of his library Montaigne inscribed, *'Omnium quae sub sole sunt fortuna et lex par est, Eccl.ix.'* [Of everything which is under the sun the fortune and law are equal, Ecclesiastes 9.] The word *fortuna* occurs but once in the Latin Bible (Isaiah 65). Montaigne's 'quotation' is apparently a loose paraphrase of Ecclesiastes 9:2 (Vulgate), 9:3 (AV), *'Hoc est pessimum inter omnia quae sub sole fiunt, quia eadem cunctis eveniunt.'* ['Among all things done under the sun, this is the worst: that the same outcome awaits all men.'] Ecclesiastes stresses that you cannot tell from their earthly fate the good from the bad.

Proptereaque fere res omnes aut corio sunt,
Aut seta, aut conchis, aut callo, aut cortice tectæ

[Wherefore virtually everything is protected by hides, silks, shells, tough skin or bark][2]

– so too were we; but like those who drown the light of day with artificial light, we have drowned our natural means with borrowed ones. It can easily be seen that custom makes possible things impossible for us: for some of the peoples who have no knowledge of clothing live under much the same climate as ourselves – and even we leave uncovered the most delicate parts of our bodies: [C] our eyes, mouth, nose, ears and, in the case of our peasants and forebears, the chest and the belly. If we had been endowed at birth with undergarments and trousers there can be no doubt that Nature would have armed those parts of us which remained exposed to the violence of the seasons with a thicker skin, as she has done for our fingertips and the soles of our feet.

[C] Why should this seem so hard to believe? The gulf between the way I dress and the way my local peasant does is wider than that between him and a man dressed only in his skin. In Turkey especially many go about naked for the sake of their religion.[3]

[A] In midwinter somebody or other asked one of our local tramps who was wearing nothing but a shirt yet remained as merry as a man swaddled up to his ears in furs how he could stand it. 'You, Sir,' he replied, 'have your face quite uncovered: myself am all face!'

The Italians tell a tale about (I think it was) the Duke of Florence's jester. He was poorly clad; his master asked him how he managed to stand the cold, which he himself found very troublesome. 'Do as I do,' he said, 'and you won't feel the cold either. Pile on every stitch you've got!'

Even when very old, King Massinissa could not be persuaded to wear anything on his head, come cold, wind or rain.[4] [C] And the same is told about the Emperor Severus.

Herodotus says that both he and others noted that, of those who were left dead in the battles between the Egyptians and the Persians, the Egyptians had by far the harder cranium: that was because the Persians always kept their heads covered first with boys' caps and then with turbans,

---

2. Lucretius, IV, 936–7. (The theme of the weakness of man was commonplace. Cf. for a comic use of it, Rabelais, *Tiers Livre*, VIII.)
3. After Guillaume Postel, the Renaissance authority on the Turks.
4. Cicero, *De Senectute*, X, 34.

whereas the Egyptians went close-cropped and bareheaded from child-hood.[5]

[A]   And King Agesilaus wore the same clothes, summer and winter, until he was decrepit. According to Suetonius, Caesar always led his armies, normally bare-headed and on foot, in sunshine as in rain. The same is said of Hannibal:

> *tum vertice nudo*
> *Excipere insanos imbres cœlique ruinam.*

[Bare-headed he withstood the furious rainstorms and the cloudbursts.][6]

[C]   A Venetian just back from the Kingdom of Pegu where he had spent a long time writes that the men and women there cover all the rest of their body, but always go barefoot even on their horses. And Plato enthusiastically advises that, for the health of our entire body, we should give no other covering to head or foot than what Nature has put there.[7]

[A1]   The man whom the Poles elected King after our own monarch[8] (and he is truly one of the greatest of princes) never wears gloves and never fails to wear the same hat indoors, no matter what the winter weather.

[B]   Whereas I cannot bear to go about with my buttons undone or my jacket unlaced, the farm-labourers in my neighbourhood would feel shackled if they did not do so. Varro maintains that when mankind was bidden to remain uncovered in the presence of gods and governors it was for our health's sake and to help us to endure the fury of the seasons rather than out of reverence.[9]

[A]   While on the subject of cold, since the French are used to a medley of colours – not me though: I usually wear black and white like my father – let me switch subject and add that Captain Martin Du Bellay relates how he saw it freeze so hard during the Luxembourg expedition that the wine-ration had to be hacked at with axes, weighed out to the soldiers and carried away in baskets.[10] Ovid is but a finger's breadth from that:

> *Nudaque consistunt formam servantia testæ*
> *Vina, nec hausta meri, sed data frusta bibunt.*

5. Herodotus, III, xii.
6. Plutarch (tr. Amyot), *Dicts notables des Lacedaemoniens*, 21OF; Pedro Mexia (tr. Gruget), I, xvi; Silius Italicus, *De bello punico*, I, 250–1.
7. Plato, *Laws*, XII, 942D.
8. Stephen Bathory.
9. Pliny, *Hist. nat.*, XXVII, 6.
10. Du Bellay, *Mémoires*, X.

[The naked wine stands straight upright, retaining the shape of the jar: they do not swallow draughts of wine but chunks of it.][11]

[B]   It freezes so hard in the swampy distributaries of Lake Maeotis that in the very same spot where Mithridates' lieutenant fought dry-shod against his enemies and defeated them, he defeated them again, when summer came, in a naval engagement.

[C]   In their battle against the Carthaginians near Placentia, the Romans were at a great disadvantage since they had to charge while their blood was nipped and their limbs stiff with the cold, whereas Hannibal had caused fires to be lit throughout his camp to warm his soldiers and had also distributed an embrocation oil to his troops to rub in, thaw out their muscles and limber up, while clogging their pores against the penetrating blasts of the prevailing bitter wind.

The Greeks' homeward retreat from Babylon is famous for the hardships and sufferings they had to overcome. One was their encountering a dreadful snowstorm in the Armenian mountains; they lost all their bearings in that country and its roads; they were so suddenly beset that, with most of their mule-train dead, they went one whole day and night without food or drink; many of them met their deaths or were blinded by the hailstones and the glare of the snow; many had frostbitten limbs and many others remained conscious but were frozen stiff and unable to move.[12]

Alexander came across a people where they bury their fruit trees in winter to protect them from the frost.[13]

[B]   While on the subject of clothing, the King of Mexico changed four times a day and never wore the same clothes twice; his cast-off garments were constantly used for gifts and rewards; similarly no pot, plate, kitchen-ware or table-ware was ever served him twice.[14]

11. Ovid, *Tristia*, III, X, 23–4. There follow anecdotes from Livy, Xenophon, Diodorus Siculus and Lopez de Gomara.
12. Xenophon, *Anabasis*, IV.
13. Diodorus Siculus, *Alexander*, XVII, xviii.
14. Lopez de Gomara (tr. Fumée). *Histoire générale des Indes*, II, xxxiii.

37. On Cato the Younger

========

[Cato the Younger was a philosophical and moral hero for many Renaissance Christians despite his having preferred suicide to ignominy. (In Dante he is Beatrice's guide to the Heavenly regions.) Montaigne protests against those who reduce the 'forms' (the souls) of great men to their own mean level: the condign reaction to greatness of soul is not a niggling desire to diminish but that kind of ecstasy produced by wonder and amazement – admiratio. Poetry, conceived much as Plato conceived it in his dialogue Io (a source of Ronsard's theories too), is playing its rightful role when, by its beauty, it stuns the reader, performer or listener into just such a condign ecstasy of amazement. At least at this stage in the Essays, Montaigne sees the ascetic Christian Feuillants and Capuchins – heroes of Christian virtue – as remaining within the general form of Man.]

[A] I do not suffer from that common failing of judging another man[1] [C] by me: I can easily believe that others have qualities quite distinct from my own. Just because I feel that I am pledged to my individual form, I do not bind all others to it as everyone else does: I can conceive and believe that there are thousands of different ways of living and, contrary to most men, I more readily acknowledge our differences than our similarities. I am as ready as you may wish to relieve another human being of my attributes and basic qualities and to contemplate him simply as he is, free from comparisons and sculpting him after his own model. I am not sexually continent, but that does not stop me from sincerely acknowledging the continence of the Feuillants and Capuchins nor from thinking well of their way of life: in thought, I can readily put myself in their place. Indeed I love and respect them all the more for being different from me.[2] My one desire is that each of us should each be judged apart and that conclusions about me should not be drawn from routine *exempla*.

1. '80: another man by me *and of reducing characteristics of others to my own. I easily believe of others many things which my own powers cannot attain.* My own weakness . . .
2. Montaigne was buried with the religious order of the Feuillants of Bordeaux, to whom his widow entrusted his working copy of the *Essays* with its [C] additions and changes and which is the basis of the Edition municipale.

[A]   My own weakness in no way affects the opinion which I should have of the strength and vigour of those who merit it.   [C]   '*Sunt qui nihil laudent, nisi quod se imitari posse confidunt.*' [There are those who praise nothing except what they are sure they can match.][3]   [A]   I crawl in earthy slime but I do not fail to note, way up in the clouds, the matchless height of certain heroic souls. It means a great deal to me to have my judgement rightly controlled even if my actions cannot be so, and to maintain at least that master-part of me free from corruption.[4] Even when my legs let me down it is something that my will is sound. At least in our latitudes, the century we live in is so leaden that   [C]   it lacks not only the practice of virtue but the very idea of it:[5]   [A]   *virtue* seems to be no more than scholastic jargon:

> [A1]   *virtutem verba putant, ut*
> *Lucum ligna:*

[they think that virtue is but a word and that sacred groves are mere matchwood.][6]

[C]   '*Quam vereri deberent, etiam si percipere non possent.*' [Even if they cannot understand it, they should revere it.][7] It is a gewgaw to hang up in a display-case, or to have dangling from your tongue just as an earring dangles from your ear.

[A]   Virtuous actions are no longer there to be recognized: those which have the face of virtue do not have her essence, since we are led to do them from profit, reputation, fear, custom and other similar motives. Such justice, valour and graciousness as we practise then can be termed so in the view of others from the face they put on in public, but they are by no means virtuous to the doer: a different end was aimed at;   [C]   there was a different motivation.   [A]   Virtue acknowledges nothing which is not done by her and for her alone.

[C]   When, following their custom, the victors in that great battle of Potidaea (which the Greeks under Pausanias won against Mardonius and the Persians)[8] had to divide the glory of that exploit among themselves,

3. Cicero, *Tusc. disput.*, II, i, 3 (adapted).
4. '80: corruption *and licentiousness.* Even . . .
5. '80: so leaden that it lacks *the very taste of virtue*: virtue . . .
6. Horace, *Epistles*, I, vi, 31–2.
7. Cicero, *Tusc. disput.*, V, ii, 6 (adapted).
8. Not Potidaea but Plataeae, a city of Boeotia, famous for the Greek victory over the Persians (Herodotus, IX, lxx).

they awarded pre-eminence in valour on the field to the Spartan people. Then, when those excellent judges of virtue, the Spartans, had to decide which of their men should individually hold the honour of having done best that day, they decided that Aristodemus had the most courageously exposed himself to risk: yet they never awarded him the prize because his valour had been spurred on by his wish to purge himself of the reproach he had incurred in the battle of Thermopylae and by a desire to die bravely to atone for his past disgrace.

[A] Our judgements follow the depravity of our morals and remain sick. I note that the majority of ingenious men in my time are clever at besmirching the glory of the fair and great-souled actions of ancient times, foisting some base interpretation on them and devising frivolous causes and occasions for them. [B] What great subtlety! Why, show me the most excellent and purest deed there is and I can go and furnish fifty vicious but plausible motives for it! What a variety of concepts, God knows, can be foisted on to our inner wills if anyone wishes to work on them in detail! [C] Such men are clever in their denigration, yet not so much maliciously as heavily and clumsily. The same pains that they take to detract from those great reputations I would readily take to lend a shoulder to enhance them. Those rare persons who have been hand-picked by the wise to be exemplary to us all I will not hesitate, on my part, to load with honour, insofar as my material allows, by interpreting their characteristics favourably. But we must believe that, for all our striving, our thoughts fall well below what the great deserve. It is the duty of good men to depict virtue as beautiful as possible; and it would not be inappropriate if our emotions should make us ecstatic under the influence of souls so august. What these people do, on the contrary, [A] they do, as I have just said, either out of malice or from that defect which reduces what they believe to what they can grasp, or else (as I am inclined to think) because their perception is not strong and clear enough to comprehend the splendour of virtue in her native purity, since they have not trained it to do so. Plutarch states that some men in his time attributed the death of Cato the Younger to his fear of Caesar; this rightly incensed him – by which one can judge how more indignant he would have been at those who attributed it to ambition.[9] [C] Idiots! Cato would rather have done a fair and noble deed which brought him shame than to do it for glory. [A] That great

9. Plutarch (tr. Amyot), *De la malignité d'Herodote*, 649H–650A.

man was truly a model which Nature chose to show how far human virtue and fortitude can reach.

But I am not up to treating so rich a subject here. I simply wish to make verses from five Latin poets rival each other in their praise of Cato,   [C] both in the interest of Cato and secondarily in their own.

Now a well-educated boy ought to find the first two feeble compared to the third; the third, more young and vigorous but ruined by its own excessive power; he ought to reckon that there is room for two or three degrees of ingenuity before we reach the fourth, at which point he will clasp his hands in wonder. When he comes to the final one, which far outdistances the others, by a distance that he will swear no human wit can cover, he will be thunderstruck and moved to ecstasy.

Here is something of a marvel: we now have far more poets than judges and connoisseurs of poetry. It is far easier to write poetry than to appreciate it. At a rather low level you can judge it by the rules of art: but good, enrapturing, divine poetry is above reason and rules. Whoever can distinguish its beauties with a firm and settled gaze does not in fact see it all, no more than we can see the brilliance of a flash of lightning. It does not exercise our judgement, it ravishes it and enraptures it; the frenzy which sets its goads in him who knows how to discern it also strikes a third person who hears him relate and recite it, just as a magnet not only attracts a needle but also pours into it the faculty of attracting others. It can more easily be seen in the theatre that the sacred inspiration of the Muses, having first seized the poet with anger, grief or hatred and driven him outside himself whither they will, then affects the actor through the poet and then, in succession, the entire audience – needle hanging from needle, each attracting the next one in the chain.[10]

From my earliest childhood poetry has had the power to transpierce and transport me. But this living feeling, which is innate to me, has been variously affected by the variety of poetic forms – it is not a matter of higher or lower (for each was the highest of its kind) but of a difference of lustre: first came a gay and genial flowing; then a keen and sublime subtlety; and finally a ripe and constant power. Examples will convey this better: Ovid, Lucan, Virgil.[11]

---

10. The famous image of poetry's magnetic power in Plato's *Io*, widely known from Ronsard's ode, *A Michel de L'Hospital*.
11. This order is not kept. The first poet is Martial, VI, xxxii.

But here are our poets waiting to compete:

> [A]   *Sit Cato, dum vivit, sane vel Cæsare major,*
>
> [Let Cato while he lives be greater even than Cæsar,]

says one of them.

> *Et invictum, devicta morte, Catonem,*
>
> [Then undefeated, death-defeating Cato,]

says another.[12] And the next, telling of the civil wars between Caesar and Pompey:

> *Victrix causa diis placuit, sed victa Catoni.*
>
> [The cause of the victors pleased the gods: that of the vanquished, Cato.][13]

And the fourth, when praising Caesar:

> *Et cuncta terrarum subacta*
> *Praeter atrocem animum Catonis.*
>
> [The whole world conquered, save for the unyielding soul of Cato.][14]

And then the master of the choir,[15] having listed and displayed the names of all the greatest of the Romans, ends in this wise:

> *his dantem jura Catonem.*
>
> [and then – a law to them all – Cato.]

---

12. Manilius, IV, 87.
13. Lucan, *Pharsalia*, I, 128.
14. Horace, *Odes*, II, i, 23.
15. Virgil, *Aeneid*, VIII, 670.

38. How we weep and laugh at the same thing

====

[An understanding of the complexity of conflicting emotions helps us to avoid trivial interpretations of great men and their grief.]

[A] When we read in our history books that Antigonus was severely displeased with his son for having brought him the head of his enemy King Pyrrhus who had just been killed fighting against him and that he burst into copious tears when he saw it;[1] and that Duke René of Lorraine also lamented the death of Duke Charles of Burgundy whom he had just defeated, and wore mourning at his funeral; and that at the battle of Auroy which the Count de Montfort won against Charles de Blois, his rival for the Duchy of Brittany, the victor showed great grief when he happened upon his enemy's corpse: we should not at once exclaim,

> *Et cosi aven che l'animo ciascuna*
> *Sua passion sotto et contrario manto*
> *Ricopre, con la vista hor' chiara hor bruna.*

[Thus does the mind cloak every passion with its opposite, our faces showing now joy, now sadness.][2]

When they presented Caesar with the head of Pompey our histories say[3] that he turned his gaze away as from a spectacle both ugly and displeasing. There had been such a long understanding and fellowship between them in the management of affairs of State, they had shared the same fortunes and rendered each other so many mutual services as allies, that we should not believe that his behaviour was quite false and counterfeit – as this other poet thinks it was:

> *tutumque putavit*
> *Jam bonus esse socer; lachrimas non sponte cadentes*
> *Effudit, gemitusque expressit pectore læto.*

1. Plutarch, *Life of Pyrrhus*; then an allusion to the defeat of Charles the Bold by René II, 1477. The battle of Auroy is narrated by Froissart.
2. Petrarch, Sonnet 81.
3. Plutarch, *Life of Caesar*.

[And now he thought it was safe to play the good father-in-law; he poured out tears, but not spontaneous ones, and he forced out groans from his happy breast.][4]

For while it is true that most of our actions are but mask and cosmetic, and that it is sometimes true that

> *Hæredis fletus sub persona risus est;*
>
> [Behind the mask, the tears of an heir are laughter;][5]

nevertheless we ought to consider when judging such events how our souls are often shaken by conflicting emotions. Even as there is said to be a variety of humours assembled in our bodies, the dominant one being that which normally prevails according to our complexion, so too in our souls: although diverse emotions may shake them, there is one which must remain in possession of the field; nevertheless its victory is not so complete but that the weaker ones do not sometimes regain lost ground because of the pliancy and mutability of our soul and make a brief sally in their turn. That is why we can see that not only children, who artlessly follow Nature, often weep and laugh at the same thing, but that not one of us either can boast that, no matter how much he may want to set out on a journey, he still does not feel his heart a-tremble when he says goodbye to family and friends: even if he does not actually burst into tears at least he puts foot to stirrup with a sad and gloomy face. And however noble the passion which enflames the heart of a well-born bride, she still has to have her arms prised from her mother's neck before being given to her husband, no matter what that merry fellow may say:

> *Est ne novis nuptis odio venus, anne parentum*
> *Frustrantur falsis gaudia lachrimulis,*
> *Ubertim thalami quas intra limina fundunt?*
> *Non, ita me divi, vera gemunt, juverint.*

[Is Venus really hated by our brides, or do they mock their parents' joy with those false tears which they pour forth in abundance at their chamber-door? No. So help me, gods, their sobs are false ones.][6]

And so it is not odd to lament the death of a man whom we would by no means wish to be still alive.

When I rail at my manservant I do so sincerely with all my mind: my

---

4. Lucan, *Pharsalia*, 1037–9.
5. Publius Syrus *apud* Aulus Gellius, XVII, 14.
6. Catullus, *De coma Berenices*, LXVI, 15.

curses are real not feigned. But once I cease to fume, if he needs help from me I am glad to help him: I turn over the page.   [C]   When I call him a dolt or a calf I have no intention of stitching such labels on to him for ever: nor do I believe I am contradicting myself when I later call him an honest fellow. No one characteristic clasps us purely and universally in its embrace. If only talking to oneself did not look mad, no day would go by without my being heard growling to myself, against myself, 'You silly shit!' Yet I do not intend that to be a definition of me.

[B]   If anyone should think when he sees me sometimes look bleakly at my wife and sometimes lovingly that either emotion is put on, then he is daft. When Nero took leave of his mother whom he was sending to be drowned, he nevertheless felt some emotion at his mother's departure and felt horror and pity.[7]

[A]   The sun, they say, does not shed its light in one continuous flow but ceaselessly darts fresh rays so thickly at us, one after another, that we cannot perceive any gap between them:

> [B]   *Largus enim liquidi fons luminis, ætherius sol*
> *Inrigat assidue cœlum candore recenti,*
> *Suppeditatque novo confestim lumine lumen.*

[That generous source of liquid light, the aethereal sun, assiduously floods the heavens with new rays and ceaselessly sheds light upon new light.][8]

So, too, our soul darts its arrows separately but imperceptibly.

[C]   Artabanus happened to take his nephew Xerxes by surprise. He teased him about the sudden change which he saw come over his face. But Xerxes was in fact thinking about the huge size of his army as it was crossing the Hellespont for the expedition against Greece; he first felt a quiver of joy at seeing so many thousands of men devoted to his service and showed this by a happy and festive look on his face; then, all of a sudden his thoughts turned to all those lives which would wither in a hundred years at most: he knit his brow and was saddened to tears.[9]

[A]   We have pursued revenge for an injury with a resolute will; we have felt a singular joy at our victory . . . and we weep: yet it is not for

---

7. Agrippina, Nero's mother, shouted as she was killed: 'Stab the belly which brought forth such a monster.' Boethius (*De consolatione philosophiae*, II, vi, metre), says that 'Nero shed no tears.' Tacitus' account in the *Annals*, xiv, 9, states that 'some say, but others deny' that he looked at her dead body and praised its beauty.
8. Lucretius, V, 282–4.
9. After Herodotus, VII, xlv, and Valerius Maximus, IX, xiii.

.that that we weep. Nothing has changed; but our mind contemplates the matter in a different light and sees it from another aspect: for everything has many angles and many different sheens. Thoughts of kinship, old acquaintanceships and affections suddenly seize our minds and stir them each according to their worth: but the change is so sudden that it escapes us:

> [B]   *Nil adeo fieri celeri ratione videtur*
> *Quam si mens fieri proponit et inchoat ipsa.*
> *Ocius ergo animus quam res se perciet ulla,*
> *Ante oculos quarum in promptu natura videtur.*

[Nothing can be seen to match the rapidity of the thoughts which the mind produces and initiates. The mind is swifter than anything which the nature of our eyes allows them to see.][10]

[A]   That is why we deceive ourselves if we want to make this never-ending succession into one continuous whole. When Timoleon weeps for the murder which, with noble determination, he committed, he does not weep for the liberty he has restored to his country; he does not weep for the Tyrant: he weeps for his brother.[11] He has done one part of his duty: let us allow him to do the other.

10. Lucretius, III, 183–6.
11. Plutarch, *Life of Timoleon*.

39. On solitude

====

[Montaigne himself had withdrawn in solitude to his estates, as many an ancient philosopher and statesman had done, with leisure to seek after wisdom, goodness and tranquillity of mind. His advice that we should set aside for ourselves a 'room at the back of the shop' is a reminder that true solitude is a spiritual withdrawal from the world. Living in solitude did not mean living as a hermit but living with detachment – if possible away from courts and the bustle of the world. Living as though always in the presence of a great and admired figure was a Renaissance practice (Sir Thomas More lived as though always in the company of the elder Pico). Montaigne draws a sharp distinction between the solitude of rare saintly ecstatics and that of ordinary men.]

[A] Let us leave aside those long comparisons between the solitary life and the active one;[1] and as for that fine adage used as a cloak by greed and ambition, 'That we are not born for ourselves alone but for the common weal,'[2] let us venture to refer to those who have joined in the dance: let them bare their consciences and confess whether rank, office and all the bustling business of the world are not sought on the contrary to gain private profit from the common weal. The evil methods which men use to get ahead in our century clearly show that their aims cannot be worth much.

Let us retort to ambition that she herself gives us a taste for solitude, for does she shun anything more than fellowship? Does she seek anything more than room to use her elbows?

The means of doing good or evil can be found anywhere, but if that quip of Bias is true, that 'the evil form the larger part', or what Ecclesiasticus says, 'One good man in a thousand have I not found'[3] –

> [B] *Rari quippe boni: numero vix sunt totidem, quot*
> *Thebarum portæ, vel divitis ostia Nili.*

[Good men are rare: just about as many as gates in the walls of Thebes or mouths to the fertile Nile.] –

1. From early Christian times such comparisons were legion.
2. The great Platonic adage spread by Cicero in its Latin form and stating that 'No man is born for himself alone, but partly for his country and partly for those whom he loves.' (Erasmus, *Adages*, IV, VI, VIII, *Nemo sibi nascitur.*)
3. Ecclesiasticus 7:28; then, Juvenal, *Satires*, XIII, 26–7.

[A]   then contagion is particularly dangerous in crowds. Either you must loathe the wicked or imitate them. It is dangerous both to grow like them because they are many, or to loathe many of them because they are different.

[C]   Sea-going merchants are right to ensure that dissolute, blasphemous or wicked men do not sail in the same ship with them, believing such company to be unlucky. That is why Bias jested with those who were going through the perils of a great storm with him and calling on the gods for help: 'Shut up,' he said, 'so that they do not realize that you are here with me.'⁴ And (a more pressing example) when Albuquerque, the Viceroy of India for Emmanuel, King of Portugal, was in peril from a raging tempest, he took a boy on his shoulders for one reason only: so that by linking their fates together the innocence of that boy might serve him as a warrant and intercession for God's favour and so bring him to safety.

[A]   It is not that a wise man cannot live happily anywhere nor be alone in a crowd of courtiers, but Bias says that, if he has the choice, the wise man will avoid the very sight of them. If he has to, he will put up with the former, but if he can he will choose the other. He thinks that he is not totally free of vice if he has to contend with the vices of others. [B]   Those who haunted evil-doers were chastised [C]   as evil [A]   by Charondas.⁵

[C]   There is nothing more unsociable than Man, and nothing more sociable: unsociable by his vice, sociable by his nature. And Antisthenes does not seem to me to have given an adequate reply to the person who reproached him for associating with the wicked, when he retorted that doctors live among the sick: for even if doctors do help the sick to return to health they impair their own by constantly seeing and touching diseases as they treat them.⁶

[A]   Now the end I think is always the same: how to live in leisure at our ease. But people do not always seek the way properly. Often they think they have left their occupations behind when they have merely changed them. There is hardly less torment in running a family than in running a whole country. Whenever our soul finds something to do she is

---

4. Diogenes Laertius, *Life of Bias*. (The subsequent references to Bias are also from this work.) His remark became proverbial; cf. Erasmus, *Apophthegmata*, VII, *Bias Prienaeus*, II. Then, Simon Goulart, *Histoire du Portugal*, VIII, ix.
5. Charondas the lawgiver of Sicily and follower of Pythagoras (Seneca, *Epist. moral.*, XC, 6).
  '80: chastised *with great punishments* by . . .
6. Erasmus, *Apophthegmata*, VII, *Antisthenes Atheniensis*, XXII.

there in her entirety: domestic tasks may be less important but they are no less importunate. Anyway, by ridding ourselves of Court and market-place we do not rid ourselves of the principal torments of our life:

> *ratio et prudentia curas,*
> *Non locus effusi late maris arbiter, aufert.*

[it is reason and wisdom which take away cares, not places affording wide views over the sea.]⁷

Ambition, covetousness, irresolution, fear and desires do not abandon us just because we have changed our landscape.

> *Et post equitem sedet atra cura.*

[Behind the parting horseman squats black care.]⁸

They often follow us into the very cloister and the schools of philosophy. Neither deserts nor holes in cliffs nor hair-shirts nor fastings can disentangle us from them:

> *haerit lateri letalis arundo.*

[in her side still clings that deadly shaft.]⁹

Socrates was told that some man had not been improved by travel. 'I am sure he was not,' he said. 'He went with himself!'¹⁰

> *Quid terras alio calentes*
> *Sole mutamus? patria quis exul*
> *Se quoque fugit?*

[Why do we leave for lands warmed by a foreign sun? What fugitive from his own land can flee from himself?]¹¹

If you do not first lighten yourself and your soul of the weight of your burdens, moving about will only increase their pressure on you, as a ship's cargo is less troublesome when lashed in place. You do more harm than good to a patient by moving him about: you shake his illness down into the sack,   [A1]   just as you drive stakes in by pulling and waggling them about.   [A]   That is why it is not enough to withdraw from the mob,

---

7. Horace, *Epistles*, I, xi, 25–6.
8. *Odes*, III, i, 40.
9. Virgil, *Aeneid*, IV, 73.
10. Seneca, *Epist. moral.*, CIV, 7, Erasmus, *Apophthegmata*, III, *Socrates*, XLIV.
11. Horace, *Odes*, II, xvi, 18–20. (The ideas in general are indebted here to Seneca.)

not enough to go to another place: we have to withdraw from such attributes of the mob as are within us. It is our own self we have to isolate and take back into possession.

> [B]   *Rupi jam vincula dicas:*
> *Nam luctata canis nodum arripit; attamen illi,*
> *Cum fugit, a collo trahitur pars longa catenæ.*

['I have broken my chains,' you say. But a struggling cur may snap its chain, only to escape with a great length of it fixed to its collar.][12]

We take our fetters with us; our freedom is not total: we still turn our gaze towards the things we have left behind; our imagination is full of them.

> *Nisi purgatum est pectus, quæ prælia nobis*
> *Atque pericula tunc ingratis insinuandum?*
> *Quantæ conscindunt hominem cuppedinis acres*
> *Sollicitum curæ, quantique perinde timores?*
> *Quidve superbia, spurcitia, ac petulantia, quantas*
> *Efficiunt clades? quid luxus desidiesque?*

[But if our breast remains unpurged, what unprofitable battles and tempests we must face, what bitter cares must tear a man apart, and then what fears, what pride, what sordid thoughts, what tempers and what clashes; what gross gratifications; what sloth!][13]

[A]   It is in our soul that evil grips us: and she cannot escape from herself:

> *In culpa est animus qui se non effugit unquam.*

[That mind is at fault which never escapes from itself.][14]

So we must bring her back, haul her back, into our self. That is true solitude. It can be enjoyed in towns and in kings' courts, but more conveniently apart.

Now since we are undertaking to live, without companions, by ourselves, let us make our happiness depend on ourselves; let us loose ourselves from the bonds which tie us to others; let us gain power over ourselves to live really and truly alone – and of doing so in contentment.

Stilpo had escaped from the great conflagration of his city in which he had lost wife, children and goods; when Demetrius Poliorcetes saw him in the midst of so great a destruction of his homeland, yet with his face

12. Persius, *Satires*, V, 158–60.
13. Lucretius, V, 43–8.
14. Horace, *Epistles*, I, xiv, 13.

undismayed, he asked him if he had suffered no harm. He said, No. Thank God he had lost nothing of his.[15]   [C]   The philosopher Antisthenes put the same thing amusingly when he said that a man ought to provide himself with unsinkable goods, which could float out of a shipwreck with him.[16]

[A]   Certainly, if he still has himself, a man of understanding has lost nothing.

When the city of Nola was sacked by the Barbarians, the local Bishop Paulinus lost everything and was thrown into prison; yet this was his prayer: 'Keep me O Lord from feeling this loss. Thou knowest that the Barbarians have so far touched nothing of mine.' Those riches which did enrich him and those good things which made him good were still intact.[17]

There you see what it means to choose treasures which no harm can corrupt and to hide them in a place which no one can enter, no one betray, save we ourselves. We should have wives, children, property and, above all, good health . . . if we can: but we should not become so attached to them that our happiness depends on them. We should set aside a room, just for ourselves, at the back of the shop, keeping it entirely free and establishing there our true liberty, our principal solitude and asylum. Within it our normal conversation should be of ourselves, with ourselves, so privy that no commerce or communication with the outside world should find a place there; there we should talk and laugh as though we had no wife, no children, no possessions, no followers, no menservants, so that when the occasion arises that we must lose them it should not be a new experience to do without them. We have a soul able to turn in on herself; she can keep herself company; she has the wherewithal to attack, to defend, to receive and to give. Let us not fear that in such a solitude as that we shall be crouching in painful idleness:

[B]   *in solis sis tibi turba locis.*

[in lonely places, be a crowd unto yourself.][18]

[C]   'Virtue,' says Antisthenes, 'contents herself, without regulations, words or actions.'   [A]   Not even one in a thousand of our usual activities has anything to do with our self.

---

15. Seneca, *Epist. moral.*, IX, 18.
16. Diogenes Laertius, *Life of Antisthenes* (with later references also to this work).
17. St Augustine, *City of God*, I, x.
18. Tibullus, IV, xiii, 12 (adapted).

That man you can see over there, furiously beside himself, scrambling high up on the ruins of that battlement, the target of so many volleys from harquebuses; and that other man, all covered with scars, wan, pale with hunger, determined to burst rather than open the gate to him: do you think they are in it for themselves? It could well be for someone they have never seen, someone plunged meanwhile in idleness and delights, who takes no interest in what they are doing. And this man over here, rheumy, filthy and blear-eyed, whom you can see coming out of his work-room at midnight! Do you think he is looking in his books for ways to be better, happier, wiser? Not a bit. He will teach posterity how to scan a verse of Plautus and how to spell a Latin word, or else die in the attempt.

Is there anyone not willing to barter health, leisure and life itself against reputation and glory, the most useless, vain and counterfeit coinage in circulation? Our own deaths have never frightened us enough, so let us burden ourselves with fears for the deaths of our wives, children and servants. Our own affairs have never caused us worry enough, so let us start cudgelling and tormenting our brains over those of our neighbours and of those whom we love.

> *Vah! quemquamne hominem in animum instituere, aut*
> *Parare, quod sit charius quam ipse est sibi?*

[Eh? Should a man prepare a settled place in his soul for something dearer than himself!]¹⁹

[C] It seems to me that solitude is more reasonable and right for those who, following the example of Thales, have devoted to the world their more active, vigorous years.

[A] We have lived quite enough for others: let us live at least this tail-end of life for ourselves. Let us bring our thoughts and reflections back to ourselves and to our own well-being. Preparing securely for our own withdrawal is no light matter: it gives us enough trouble without introducing other concerns. Since God grants us leave to make things ready for our departure, let us prepare for it; let us pack up our bags and take leave of our company in good time; let us disentangle ourselves from those violent traps which pledge us to other things and which distance us from ourselves. We must unknot those bonds and, from this day forth, love this or that but marry nothing but ourselves. That is to say, let the rest be ours, but not so

19. Terence, *Adelphi*, I, i, 13–14.

glued and joined to us that it cannot be pulled off without tearing away a piece of ourselves, skin and all. The greatest thing in the world is to know how to live to yourself.

[C] It is time to slip our knots with society now that we can contribute nothing to it. A man with nothing to lend should refrain from borrowing. Our powers are failing: let us draw them in and keep them within ourselves. Whoever can turn round the duties of love and fellowship and pour them into himself should do so. In that decline which makes a man a useless encumbrance importunate to others, let him avoid becoming an encumbrance, importunate and useless to himself. Let him pamper himself, cherish himself, but above all control himself, so respecting his reason and so fearing his conscience that he cannot stumble in their presence without shame: *'Rarum est enim ut satis se quisque vereatur.'* [It is rare for anybody to respect himself enough.][20] Socrates says that youth must get educated; grown men employ themselves in good actions; old men withdraw from affairs, both civil and military, living as they please without being bound to any definite duties.[21]

[A] There are complexions more suited than others to these maxims [C] about retirement. [A] Those who hold on to things slackly and weakly, and whose will and emotions are choosy, accepting neither slavery nor employment easily – and I am one of them, both by nature and by conviction – will bend to this counsel better than those busy active minds which welcome everything with open arms, which take on everything, get carried away about everything and which are always giving themselves, offering themselves, putting themselves forward. When any good things happen to come to us from outside we should make use of them, so long as they remain pleasurable; we must not let them become our principal base, for they are no such thing: neither reason nor Nature will have them so. Why do we go against Nature's laws and make our happiness a slave in the power of others?

Yet to go and anticipate the injuries of Fortune, depriving ourselves of such good things as are still in our grasp, as several have done out of devotion and a few philosophers out of rational conviction, making slaves of themselves, sleeping rough, poking out their own eyes, chucking their wealth into rivers, going about looking for pain – the first to acquire blessedness in the next life because of torment in this one, the others to ensure against tumbling afresh by settling for the bottom rung – are actions

20. Quintilian, X, 7.
21. The source of this saying is unknown to me.

of virtue taken to excess. Let tougher sterner natures make even their hiding-places glorious and exemplary.

> *tuta et parvula laudo,*
> *Cum res deficiunt, satis inter vilia fortis:*
> *Verum ubi quid melius contingit et unctius, idem*
> *Hos sapere, et solos aio bene vivere, quorum*
> *Conspicitur nitidis fundata pecunia villis.*

[When I lack money, I laud the possession of a few things which are sure; I show fortitude enough among paltry goods: but – still the same person – when anything better, more sumptuous, comes my way, then I say that only they are wise and live right well whose income is grounded in handsome acres.][22]

I have enough to do without going that far. When Fortune favours me, it is enough to prepare for her disfavour, picturing future ills in comfort, to the extent that my imagination can reach that far, just as we train ourselves in jousts and tournaments, counterfeiting war in the midst of peace.   [C]   I do not reckon that Arcesilaus the philosopher had reformed his mind any the less because I know he used such gold and silver vessels as the state of his fortune allowed: for using them frankly and in moderation I hold him in greater esteem than if he had got rid of them.

[A]   I know how far our natural necessities can extend; and when I reflect that the indigent beggar at my door is often more merry and healthy than I am, I put myself firmly in his place and make an assay at giving my soul a slant like his. Then by running similarly through other examples, though I may think that death, poverty, contempt and sickness are dogging my heels, I can readily resolve not to be terrified by what a man of lesser estate than mine can accept with such patience. I cannot believe that a base intelligence can do more than a vigorous one or that reason cannot produce the same effects as habit. And since I realize how insecure these adventitious comforts are, my sovereign supplication, which I never fail to make to God, is that, even while I enjoy them fully, He may make me content with myself and with such goods as are born within me. I know healthy young men who travel with a mass of pills in their baggage to swallow during an attack of rheum, fearing it less since they know they have a remedy to hand. That is the way to do it, only more so: if you know yourself subject to some grave affliction, equip yourself with medicines to benumb and deaden the part concerned.

The occupation we must choose for a life like this one should be neither

22. Horace, *Epistles*, I, xv, 42–6.

toilsome nor painful (otherwise we should have vainly proposed seeking such leisure). It depends on each man's individual taste. My taste is quite unsuited to managing my estates: those who do like it, should do it in moderation:

> *Conentur sibi res, non se submittere rebus.*

[They should try to subordinate things to themselves, not themselves to things.][23]

Otherwise management, as Sallust puts it, is a servile task.[24] (Some aspects of it are more acceptable, such as an interest in gardening – which Xenophon attributes to Cyrus.)[25] A mean can be found between that base unworthy anxiety, full of tension and worry, seen in those who immerse themselves in it, and that profound extreme neglect one sees in others, who let everything go to rack and ruin:

> *Democriti pecus edit agellos*
> *Cultaque, dum peregre est animus sine corpore velox.*

[Democritus left his herds to ravage fields and crops, while his speeding soul was wandering outside his body.][26]

But let us just listen to the advice about solitude which Pliny the Younger gave to his friend Cornelius Rufus: 'I counsel you in that ample and thriving retreat of yours, to hand the degrading and abject care of your estates over to those in your employ, and to devote yourself to the study of letters so as to derive from it something totally your own.'[27] By that, he means a good reputation, his humour being similar to Cicero's who said he wanted to use his withdrawal and his repose from the affairs of State to gain life everlasting through his writings!

> [B]   *Usque adeo ne*
> *Scire tuum nihil est, nisi te scire hoc sciat alter?*

[Does *knowing* mean nothing to you, unless somebody else knows that you know it?][28]

[C]   It seems logical that when you talk about withdrawing from the

---

23. Horace, *Epistles*, I, i, 19.
24. Sallust, *Catilenae conjuratio*, IV.
25. Cicero, *De Senectute*, XVI, 59.
    [A]: more *noble* and acceptable . . .
26. Horace, *Epistles*, I, xii, 12–13.
27. Pliny the Younger, *Epistles*, I, i. no. 3.
28. Persius, *Satires*, I, xxiii.

world you should be contemplating things outside it; they only half do that: they do indeed arrange their affairs for when they will no longer be in the world, yet the fruits of their project they claim to draw from the world they have left: a ridiculous contradiction.

The thought of those who seek solitude for devotion's sake, filling their minds with the certainty of God's promises for the life to come, is much more sane and appropriate. Their objective is God, infinite in goodness and power: the soul can find there matters to slake her desires in perfect freedom. Pains and afflictions are profitable to them, being used to acquire eternal healing and joy; death is welcome as a passing over to that perfect state. The harshness of their Rule is smoothed by habit; their carnal appetites are rejected and lulled asleep by their denial – nothing maintains them but practising them and using them. Only this end, another life, blessedly immortal, genuinely merits our renunciation of the comforts and sweetnesses of this life of ours. Whoever can, in reality and constancy, set his soul ablaze with the fire of this lively faith and hope, builds in his solitude a life of choicest pleasures, beyond any other mode of life.

[A] Neither the end, then, nor the means of Pliny's counsel satisfy me: we are always jumping from feverish fits into burning agues. Spending time with books has its painful side like everything else and is equally inimical to health, which must be our main concern; we must not let our edge be blunted by the pleasure we take in books: it is the same pleasure as destroys the manager of estates, the miser, the voluptuary and the man of ambition.

The wise men teach us well to save ourselves from our treacherous appetites and to distinguish true wholesome pleasures from pleasures diluted and crisscrossed by pain. Most pleasures, they say, tickle and embrace us only to throttle us, like those thieves whom the Egyptians called *Philistae*.[29] If a hangover came before we got drunk we would see that we never drank to excess: but pleasure, to deceive us, walks in front and hides her train. Books give pleasure: but if frequenting them eventually leads to loss of our finest accomplishments, joy and health, then give up your books. I am one who believes that their fruits cannot outweigh a loss such as that.

As men who have long felt weakened by illness in the end put themselves at the mercy of medicine and get that art to prescribe a definite diet never to be transgressed: so too a man who withdraws pained and disappointed with the common life must rule his life by a diet of reason, ordering it and arranging it with argument and forethought. He should have taken leave of toil and travail, no matter what face they present, and should flee from

---

29. Seneca, *Epist. moral.*, LI, 13; the *Philistae* (or *Philetai*) were assassins.

all kinds of passion which impede the tranquillity of his body and soul, [B]   and choose the way best suited to his humour.

> *Unusquisque sua noverit ire via.*
>
> [Let each man choose the road he should take.]³⁰

[A]   Whether we are running our home or studying or hunting or following any other sport, we should go to the very boundaries of pleasure but take good care not to be involved beyond the point where it begins to be mingled with pain. We should retain just enough occupations and pursuits to keep ourselves fit and to protect ourselves from the unpleasantness which comes in the train of that other extreme: slack and inert idleness.

There are branches of learning both sterile and prickly, most of them made for the throng: they may be left to those who serve society. Personally I only like pleasurable easy books which tickle my interest, or those which console me and counsel me how to control my life and death.

> *Tacitum sylvas inter reptare salubres,*
> *Curantem quidquid dignum sapiente bonoque est.*
>
> [Walking in silence through the healthy woods, pondering questions worthy of the wise and good.]³¹

Wiser men with a strong and vigorous soul can forge for themselves a tranquillity which is wholly spiritual. Since my soul is commonplace, I must help sustain myself with the pleasures of the body – and since age has lately robbed me of those more pleasing to my fancy I am training and sharpening my appetite for those which are left, more suited to my later season. We must cling tooth and claw to the use of the pleasures of this life which the advancing years, one after another, rip from our grasp.

> [B]   *Carpamus dulcia; nostrum est*
> *Quod vivis: cinis et manes et fabula fies.*
>
> [Let us pluck life's pleasures: it is up to us to live; you will soon be ashes, a ghost, something to tell tales about.]³²

30. Propertius, II, xxv, 38.
31. Horace, *Epistles*, I, iv, 4–5.
32. Persius, *Satires*, V, 151–2.
    '80 (instead of this quotation): grasp, *and prolong them with all our power: Quamcunque Deus tibi fortunaverit horam, Grata sume manu, nec dulcia differ in annum* [Whatever happy hour God has allotted you, accept with a grateful hand and do not put off delights for a year] . . . (Did Montaigne strike out this because he had confused, in his quotation from Horace, *Epistles* I, xi, 22, *God* with *Fortuna*? All editions of Horace read *Fortuna* not *Deus*.)

[A]   As for glory – the end proposed by Pliny and Cicero – that is right outside my calculations. Ambition is the humour most contrary to seclusion. Glory and tranquillity cannot dwell in the same lodgings. As far as I can see, those authors have withdrawn only their arms and legs from the throng: their souls, their thoughts, remain even more bound up with it.

[B]   *Tun', vetule, auriculis alienis colligis escas?*

[Now then, old chap, are you collecting bait to catch the ears of others?][33]

[A]   They step back only to make a better jump, and, with greater force, to make a lively charge through the troops of men.

Would you like to see how they fall just a tiny bit short of the target? Let us weigh against them the counsels of two philosophers – and from two different schools at that – one of them writing to his friend Idomeneus and the other to his friend Lucilius, to persuade them to give up the management of affairs of state and their great offices and to withdraw into solitude:[34]

'You have (they said) lived up to the present floating and tossing about; come away into the harbour and die. You have devoted your life to the light: devote what remains to obscurity. It is impossible to give up your pursuits if you do not give up their fruits. Renounce all concern for name and glory. There is the risk that the radiance of your former deeds may still cast too much light upon you and pursue you right into your lair. Among other gratifications give up the one which comes from other people's approval. As for your learned intelligence, do not worry about that: it will not lose its effect if you yourself are improved by it. Remember the man who was asked why he toiled so hard at an art which few could ever know about: "For me a few are enough; one is enough; having none is enough." He spoke the truth. You and one companion are audience enough for each other; so are you for yourself. For you, let the crowd be one, and one be a crowd. It is a vile ambition in one's retreat to want to extract glory from one's idleness. We must do like the beasts and scuff out our tracks at the entrance to our lairs. You should no longer be concerned with what the world says of you but with what you say to yourself. Withdraw into yourself, but first prepare yourself to welcome yourself there. It would be madness to entrust yourself to yourself, if you did not know how to govern yourself. There are ways of failing in solitude as in society. Make

---

33. Persius, *Satires*, I, 19–20.
34. The first is Epicurus. The second is Seneca. The following epistle is largely composed of borrowings from various epistles of Seneca.

yourself into a man in whose sight you would not care to walk awry; feel shame for yourself and respect for yourself, − [C] *"observentur species honestae animo"* [let your mind dwell on examples of honour];[35] until you do, always imagine that you are with Cato, Phocion and Aristides, in whose sight the very madmen would hide their faults; make them recorders of your inmost thoughts, which, going astray, will be set right again out of reverence for them.[36]

'The path they will keep you on is that of being contented with yourself, of borrowing all from yourself, of arresting and fixing your soul on thoughts contained within definite limits where she can find pleasure; then, having recognized those true benefits which we enjoy the more the more we know them, content yourself with them, without any desire to extend your life or fame.'

That is the advice of a philosophy which is natural and true, not like that of those other two,[37] all verbiage and show.

---

35. Cicero, *Tusc. disput.*, II, xxii, 52.
36. Modelled on Seneca, *Epist. moral.*, XXV, 6. The 'companions' proposed there are Cato, Scipio and Laelius. Montaigne prefers Phocion, the great Athenian general, and Aristides, a statesman renowned for his integrity.
37. Pliny the Younger and Cicero, condemned above for seeking glory from their withdrawal from the world.

40. Reflections upon Cicero

[*This chapter continues the reflections of the previous one, by further comparisons between 'those two couples': that is, between Cicero and Pliny the Younger on the one hand, and Epicurus and Seneca on the other. Montaigne sides with Epicurus and Seneca because like him they give priority to matter over style. Montaigne has a gentleman's contempt for mere style and formalities. His preference for the 'comic style' – that of Terence and of himself – is explained in the later chapter 'On books'.*]

[A] One more point of contrast between those two couples: from the writings of Cicero and of the younger Pliny (who in my judgement is [C] not much[1] [A] like his uncle in character) there can be drawn a great many details which witness to an overweeningly ambitious nature: among others that they publicly urged contemporary historians not to forget them in their chronicles; and Fortune – as though moved by pique – has made the vanity behind those requests last to our own day, while the chronicles themselves have long since been lost.[2] But what surpasses all vulgarity of mind in people of such rank is to have sought to extract some major glory from chatter and verbiage, using to that end even private letters written to their friends; when some of their letters could not be sent as the occasion for them had lapsed they published them all the same, with the worthy excuse that they did not want to waste their long nights of toil! How becoming in two Roman consuls, sovereign governors of the commonwealth which was mistress of the world, to use their leisure to construct and nicely clap together some fair missive or other, in order to gain from it the reputation of having thoroughly mastered the language of their nanny! What more could some wretched schoolteacher do, who earned his money by it!

If the eloquent language of Xenophon and Caesar had not been far surpassed by their deeds I do not believe they would ever have written about them. They sought to commend their actions not their style. And if

1. '80: is *in no ways* like his uncle . . .
2. Cicero wrote to Luxeius, and Pliny the Younger to Tacitus, asking for a place in their histories.

a perfect mastery of language could contribute anything worthy of a great public figure, Scipio and Laelius would certainly not have allowed the credit for their comedies, with all their grace and delightful language, to be attributed to an African slave – for the beauty and excellence of those works are adequate proof that they are really theirs, and Terence himself admits it.³   [B]   I would be deeply displeased to have that belief of mine shaken.

[A]   It is a kind of mockery and insult to value a man for qualities unbecoming to his rank, even if they are otherwise commendable, or for qualities which should not be his chief ones – as though we were to praise a monarch for being a good painter or a good architect, or even for being good with the harquebus or at tilting in the jousting-ring; such praises bring no honour unless they are put forward, among many others, after those which are proper to his rank: after justice, that is, and knowing how to lead his people in peace and war. In this way Cyrus can be honoured by his farming and Charlemagne for his eloquent style and his knowledge of literature.   [C]   More strangely I have known in my time great men whose professional reputation is indeed based on writing but who disown their indentured skills, corrupt their style and affect an ignorance of so menial a quality, which Frenchmen think is hardly ever found in clever men; they prefer to commend themselves by better qualities.

[B]   The comrades of Demosthenes during their embassy to Philip praised that monarch for his beauty, eloquence and his taste in wine: Demosthenes commented that praises such as those belonged rather to a woman, a barrister and a sponge than to a king.⁴

> *Imperet bellante prior, jacentem*
> *Lenis in hostem.*

[Let him be first triumphant over the enemy, then generous to the defeated.]⁵

It is not his profession to know how to hunt well or to dance well.

---

3. Not really. Terence *may* have been a Carthaginian slave freed by Terentius Lucanus. In the *Prologue* to the *Adelphi* (15–21) he says he is flattered by the imputation that great men helped him write his comedies, which may or may not mean what Montaigne thinks it does.
4. Plutarch, *Life of Demosthenes*.
5. Horace, *Carmen Saeculare*, 51–2.

Orabunt causas alii, cælique meatus
Describent radio, et fulgentia sidera dicent;
Hic regere imperio populos sciat.

[Others shall make better legal pleas, trace the paths of the heavens with their measuring-rods and tell which stars are rising: you must remember to rule the nations with authority.][6]

[A] Plutarch goes further, asserting that to appear to excel in such unnecessary accomplishments is to bear witness against yourself of time ill-spent on leisure and study which ought to be better spent on things more necessary and more useful. So Philip, King of Macedonia, when he heard his son Alexander the Great singing at a feast and rivalling the best musicians, remarked: 'Are you not ashamed of singing so well?' And to this same Philip, a musician with whom he was arguing about his art, said, 'God forbid, Sire, that you should ever have the ill fortune to understand such things better than I do.' [B] A king should be able to reply like Iphicrates did to the ambassador who was haranguing him with invectives and saying,'What have you got to boast about? Are you an infantryman? Are you an archer? Are you a pikesman?' – 'None of these,' he replied. 'But I am the one who can lead them all.' [A] Antisthenes took it as evidence of Ismenias' lack of valour when he was praised for being an outstanding flautist.[7]

[C] What I do know is that when I hear anyone lingering over the language of these *Essays* I would rather he held his peace: it is not a case of words being extolled but of meaning being devalued; it is all the more irritating for being oblique. I may be wrong but there are not many writers who put more matter in your grasp than I do and who, with such concern for this matter, scatter at least the seeds of it so thickly over their paper. To make room for more, I merely pile up the heads of argument: if I were to develop them as well I would increase the size of this tome several times over. And how many tacit *exempla* have I scattered over my pages which could all give rise to essays without number if anyone were to pluck them apart with a bit of intelligence. Neither they nor my quotations serve always as mere examples, authorities or decorations: I do not only have regard for their usefulness to me: they often bear the seeds of a richer, bolder subject-matter; they often sound a more subtle note on the side,

6. Virgil, *Aeneid*, VI, 849–51.
7. Plutarch: *Life of Pericles* (twice) and (tr. Amyot) *Dicts notables des anciens Roys, Princes et grands Capitaines*, 192C.

both for me, who do not wish to press more out of them, and also for those who get my gist.

To return to verbosity: I cannot find much difference between always handling words badly and knowing nothing save how to handle them well: 'Non est ornamentum virile concinnitas.' [An elegant garb is no manly adornment.][8]

[A]   Wise men say that the only qualities which are proper to every rank and class in general are, where knowledge is concerned, a love of wisdom and, where deeds are concerned, virtue.

There is something to the same effect in the other two, Epicurus and Seneca (since they do promise their friends lasting fame from the letters they pen to them). But it is in a different guise, making compromises – for a good end – with the other people's vanity; for they do urge that if an anxious concern with renown and for making themselves known to future centuries keeps their correspondents still managing the affairs of state and makes them afraid of the solitude and withdrawal to which they would summon them, such things should trouble them no more, since they have sufficient influence over posterity to guarantee that their correspondents' names will be as well-known and as famous from these very letters as they could ever be from their public duties.[9]

But even allowing for that difference, their letters are not skinny empty ones, propped up merely by a nice choice of words amassed and arranged with an elegant rhythm, but are fully fleshed out with arguments both beautiful and wise, by which we acquire not eloquence but wisdom, instructing us not how to talk well but how to act well.

Shame on all eloquence which leaves us with a taste for itself not for its substance – unless you could say in Cicero's case that the ultimate perfection of his style gives it a substance of its own.

I will add a tale which we can read on this topic about Cicero which lets us put our finger on the kind of man he was. He had to deliver a public oration and was a bit short of time to get himself conveniently ready. One of his slaves called Eros came and told him that the case had been put off till the following day: he was so delighted at this good news that he gave him his freedom.[10]

[B]   On the subject of letters, I would like to note that it is a genre in

8. Seneca, *Epist. moral.*, XCV, 2–3.

9. Ibid., XXI, 4–5.

10. Plutarch (tr. Amyot), *Dicts notables des anciens Roys, Princes et grands Capitaines*, 208A.

which my friends say I show some ability. [C] If I had somebody to write to I would readily have chosen it as the means of publishing my chatter. But I would need some definite correspondent, as I used to have,[11] who would draw me out, sustain me and keep me going. For to correspond with thin air as others do is something I could only manage in my dreams; nor, being the sworn enemy of all deception, could I treat serious matters under made-up names. I would have been more observant and confident if I were addressing one strong and beloved friend than I am now when I need to have regard for a many-sided public. Unless I deceive myself my achievement then would have been greater.

[B] My natural style is that of comedy, but one whose form is personal to me, a private style unsuited to public business – as is my language in all its aspects, being too compact, ill-disciplined, disjointed and individual;[12] and I know nothing about formal letter-writing where the substance consists in merely stringing courtly words together. I have neither the gift nor the taste for all those long drawn-out offers of affection and service. I do not believe in them much and dislike going much beyond what I do believe. That is far removed from present-day practice: there never was so servile and abject a prostitution of formal courtesies: my life, my soul, devotion, adoration, serf, slave – all such words are so current and common that when anyone wishes to convey a more explicit intention, one showing more respect, he has no means left to express it.

I hate unto death to sound like a flatterer; which means that I naturally adopt a dry, blunt, raw kind of language which to anyone who does not otherwise know me may seem somewhat haughty.[13] [C] I pay most honour to those to whom I show it least: when my soul is happily cantering along I forget all the conventional paces. [B] I present myself meagrely and proudly to those to whom I am really devoted; [C] and I commend myself least to those to whom I have given myself most; [B] I feel that they ought to be able to read all that in my mind, as well as the fact that my verbal expressions do wrong to my thoughts.

[C] When welcoming people, taking my leave, thanking them, greeting them, expressing my devotion, as well as in all those verbose compliments required by the rules of courtesy in our etiquette, I know no one

11. Etienne de La Boëtie.
12. '80: disjointed and *difficult*; and I know . . . (Montaigne sees his style as marked by the dry, everyday language of Latin comedy. Cf. Seneca, *Epist. moral.*, C, 10.)
13. '80: haughty. *Those whom I love cause me pain if I have to tell them I do so.* I present myself . . .

who is so stupid and bereft of words as I am. I have never been asked to write letters of support or recommendation without those for whom I was doing it finding them lukewarm and desiccated.

[B] The Italians are great printers of their letters. I believe I have a hundred separate volumes of them; the best seem to me to be those of Annibal Caro. If all the paper were still in existence which I had once scribbled upon for the ladies when my pen was really carried away by my passion, you might have found a page or two there which deserved to be read by idle youth befuddled with such madness.

I always write my letters at the gallop, with so headlong a dash that I prefer to write them by hand than to dictate them (despite my appalling writing) since I can never find anyone who can keep up with me; I never have them copied out neatly. I have accustomed the great men who know me to put up with my scratchings-out and erasings as well as with paper which is not folded double or which has no margins. The most useless of my letters are those which cost me most trouble: as soon as I flag, that is a sign that my heart is not in it. I prefer to begin without a plan, the first phrase leading on to the next. Letters nowadays are more full of lace borders and prefaces than of matter. Just as I would rather write two letters than fold and seal up one and always leave that job to somebody else, so too, when I have said what I have to say, I would like to be able to make someone else responsible for those long formulas, those offers of service and I-beg-you-Sirs which we place at the end; I wish some new custom would liberate us from them, as well as from having to address our letters with a list of qualities and titles. For fear of tripping up over them I have often not written at all, especially to men of law and finance; there are so many changes of function, such difficulty of arrangement and in giving everyone his various honorific titles; and they have cost them so dear that you cannot mix them up or forget them without causing offence.

I find it particularly bad grace to load them on to the title page and frontispieces of any books we send to be printed.

41. On not sharing one's fame

[*A series of exempla showing rare examples of selflessness over fame and amusing examples of casuistry.*]

[A]   Of all the lunacies in this world the most accepted and the most universal is concern for reputation and glory, which we espouse even to the extent of abandoning wealth, rest, life and repose (which are goods of substance and consequence) in order to follow after that image of vanity and that mere word which had no body, nothing, to hold on to.

> La fama, ch' invaghisce a un dolce suono
> Gli superbi mortali, & par si bella,
> E un echo, un sogno, anzi d'un sogno un ombra
> Ch' ad ogni vento si dilegua et sgombra.

[That fame, which enchants proud mortals with its fair words and which seems so beautiful, is but an echo, a dream, nay, the shadows of a dream, dissolved and scattered by each breath of wind.]

And among all the irrational humours of men, it seems that even philosophers free themselves from this one later and more reluctantly than from all others.   [B]   It is the most tetchy and stubborn lunacy of them all: [C]   '*Quia etiam bene proficientes animos tentare non cessat*' [since it never ceases to tempt even those souls who are advancing in virtue].[1]   [B]   None of the others is more clearly accused of vanity by reason, but its roots are so active within us that I doubt if anyone has managed to cast it clean off. When you have said everything to disavow it, and believed all of it, it still marshals such an inner persuasion against your arguments that you have scant means of holding out against it.

[A]   For, as Cicero says, even those who fight it still want their books against it to bear their name in the title and hope to become famous for despising fame.[2] Everything else is subject to barter: we will let our friends

---

1. Torquato Tasso, *Gierusalemme liberata*, XIV, 63; St Augustine, *City of God*, V, xiv.
2. Cicero, *Tusc. disput.*, I, xv, 34–5.

have our goods and our lives if needs be: but a case of sharing our fame and making someone else the gift of our reputation is hardly to be found.

In the war against the Cimbrians, Catulus Luctatius made every effort to stop his soldiers who were fleeing before their enemies: he then joined the rout and pretended to be a coward himself so that they might appear to be following their commander rather than fleeing from the enemy.

When the Emperor Charles V invaded Provence in 1537, it is believed that Antonio de Leyva, seeing that his monarch was quite determined on this expedition and believing that it would wonderfully add to his fame, spoke against it and counselled him not to do it; his sole aim was that all the fame and honour of the decision should be attributed to his monarch, with everyone saying that his judgement and his foresight had been such as to carry through so fair an enterprise against everybody's opinion. That was to honour his master to his own detriment.

When the Thracian ambassadors were consoling Argelionidis over the death of her son Brasidas and praising him so highly as to lament that there was no one like him left, she rejected such private praise of one individual and rendered it general: 'Do not say that to me,' she replied. 'I know that the city of Sparta has many a citizen greater and more valiant than my son was.'

In the battle of Crécy the Prince of Wales, a youngster still, was leading the vanguard; the main thrust of the battle was concentrated against it. The lords who accompanied him, finding the fighting tough, sent a dispatch asking King Edward to come to their aid: he inquired how his son was doing: when he was told that he was alive and in the saddle, he said, 'I would do him wrong to come and rob him now of the honour of the victory in this battle where he has held out so well; whatever the risk, that honour will be his alone.' And he would not go himself nor would he send help, well aware that if he did so men would say that without his succour all had been lost, and that the credit for this exploit would have been attributed to himself: [C] *'semper enim quod postremum adjectum est, id rem totam videtur traxisse'* [the last forces to be thrown in always seem to have done it all themselves].[3]

[B] Several people in Rome thought, as was commonly said, that the chief of Scipio's fine achievements were [C] partly [B] due to Laelius who nevertheless was ever moving and seconding the honour and greatness of Scipio, taking no care of his own.

---

3. The Du Bellay *Mémoires*, VI; Plutarch (tr. Amyot), *Dicts notables des Lacedaemoniens*, 216B; Froissart, *Chroniques*; Livy, XXVII, xlv.

To the man who told Theopompus King of Sparta that the citizens were at his feet because he was so good at giving orders he replied, 'It is rather because they are so good at obeying them.'[4]

[C] Just as, despite their sex, women who succeeded to peerages had the right to attend and give their opinion in cases falling within the jurisdiction of the peers of the realm, so too the lords spiritual, despite their calling, were required to assist our kings in their wars not only with their allies and retainers but also in person. The Bishop of Beauvais was with Philip Augustus at the battle of Bouvines and fought very bravely in that encounter; but it did not seem right to him to win gain or glory from such a violent and bloody action. He personally took several of his enemies that day, but gave them to the first gentleman he came across, who was allowed to do what he liked with them, either cut their throats or keep them prisoner; in this way he handed Count William of Salisbury over to Messire Jean de Nesle. By a refinement of conscience similar to the above he was prepared to knock a man senseless but not to slash at him: that is why he fought only with a club.[5] Somebody in my own time was criticized by the King for 'laying hands on a clergyman'; he strongly and firmly denied it: all he had done was to thrash him and to trample on him.

---

4. Plutarch (tr. Amyot), *Instruction pour ceux qui manient affaires d'Estat*, 166BC; 172H–173A.
5. Bishop Jean Du Tillet, *La chronique des Roys de France*.

42. On the inequality there is between us

===

[Wisdom not rank constitutes the only inequality that matters. The changes and additions made to this chapter show Montaigne's growing sympathy for the common peasant.]

[A] Plutarch says somewhere that he finds less distance between beast and beast than between man and man. He was talking of mental powers and inner qualities.[1] Truly, I find Epaminondas, as I conceive him to be, so far above some men I know – I mean men in their right mind[2] – that I would go farther and say that there is a greater distance between this man and that one than between this man and that beast:

[C]   *Hem vir viro quid praestat.*

[Hmm! How far one man excels another.][3]

There are as many degrees of intelligence as there are fathoms 'twixt heaven and earth.

[A]   While on the subject of men it is astonishing that everything except ourselves is judged by its own properties: we praise a horse for its vigour and dexterity –

[B]   *volucrem*
Sic laudamus equum, facili cui plurima palma
Fervet, et exultat rauco victoria circo,

[it is the swift horse that we praise, the one which, to the noisy shouts of the spectators, easily wins the prize;][4]

---

1. Plutarch (tr. Amyot), *Que les bestes usent de la Raison*, 274AB.
   '80: qualities. *For as concerns bodily shape it is evident that the species of beasts are distinguished by a more evident difference than we are from each other.* Truly . . .
2. '80: mind – *for fools and those made witless by accident are not complete men*, that I would go . . .
3. Terence, *Eunuch*, II, iii, 1, adapted.
   and that beast, *meaning that the most excellent of the animals is nearer to a man of lowest degree than that man is to another man, great and excellent.* He . . .
4. Juvenal, *Satires*, VIII, lvii.

– we do not praise it for its harness. We praise a greyhound for its speed not for its neck-band; a hawk, for its wing not for its bells and its leg-straps. So why do we not similarly value a man for qualities which are really his? He may have a great suite of attendants, a beautiful palace, great influence and a large income: all that may surround him but it is not *in* him. You would never buy a cat in a bag. If you are haggling over a horse, you strip off its trappings and examine it naked and bare – or if it does wear an ornamental cover as used to be the case for horses offered for sale to royalty, it was only spread over the inessentials, so that you should not waste time over its handsome coat or its broad crupper but mainly concentrate on its legs, eyes and hooves – the parts which really matter:

> *Regibus hic mos est: ubi equos mercantur, opertos*
> *Inspiciunt, ne, si facies, ut sæpe, decora*
> *Molli fulta pede est, emptorem inducat hiantem,*
> *Quod pulchræ clunes, breve quod caput, ardua cervix.*

[This is how kings do it: when they buy horses they inspect them in their caparisons lest they as buyers may be tempted (as often happens with lame horses with a fine mane) to gape at their broad cruppers, their neat heads or their proud necks.][5]

Why do you judge a man when he is all wrapped up like a parcel? He is letting us see only such attributes as do not belong to him while hiding the only ones which enable us to judge his real worth. You are trying to find out the quality of the sword not of the scabbard: strip it of its sheath and perhaps you would not give twopence for it. You must judge him not by his finery but by his own self. As one of the old writers amusingly put it: 'Do you know why you think he is so tall? You are including his high-heels!' The plinth is no part of the statue.[6] Measure his height with his stilts off: let him lay aside his wealth and his decorations and show us himself in his shimmy. Is his body functioning properly? Is it quick and healthy? What sort of soul does he have? Is his soul a beautiful one, able, happily endowed with all her functions? Are her riches her own or are they borrowed? Has luck had nothing to do with it? Does she face drawn

5. Horace, *Satires*, I, ii, 86.
6. Seneca, *Epist. moral.*, LXXVI, 31. (There are a great many echoes of this and other *Epistles* of Seneca in this section.)

swords with steady gaze? Does it not bother her whether she expires with a
sigh or a slit throat? Is she calm, unruffled and contented? That is what we
need to know; that is what the immense distances between us men should
be judged by.

Is he,

> *sapiens, sibique imperiosus,*
> *Quem neque pauperies, neque mors, neque vincula terrent.*
> *Responsare cupidinibus, contemnere honores*
> *Fortis, et in seipso totus teres atque rotundus,*
> *Externi ne quid valeat per læve morari,*
> *In quem manca ruit semper fortuna?*

[wise, lord of himself, not terrified of death, poverty or shackles? Is he a man who
stoutly defies his passions, who scorns ambition? Is he entirely self-sufficient? Is he
like a smooth round sphere which no foreign object can adhere to and which
maims Fortune herself if she attacks him?]

That kind of man is miles above kingdoms and dukedoms. He is an empire
unto himself.[7]

> [C]   *Sapiens pol ipse fingit fortunam sibi.*

[Why, the wise man shapes his own destiny.]

What more can he desire?

> [A]   *Non ne videmus*
> *Nil aliud sibi naturam latrare, nisi ut quoi*
> *Corpore sejunctus dolor absit, mente fruatur,*
> *Jucundo sensu cura semotus metuque?*

[Can we not see that Nature demands nothing for herself except a body free from
pain and a mind rejoicing in a happy disposition, remote from fear and worry?][8]

Compare with him the mass of men nowadays, senseless, base, servile,
unstable, continually bobbing about in a storm of conflicting passions
which drive them hither and thither, men totally dependent upon others:
they are farther apart than earth and sky. But so blind are our habitual

---

7. Horace, *Satires*, II, vii, 83–8.
   '80: an empire *and riches* unto himself; *he lives satisfied, content and happy.* And
whoever has that, what more *is there?* 'non ne videmus . . .
8. Plautus, *Trinummus*, II, ii, 84; Lucretius, II, 16.

ways that we take little or no account of such things; when we come to consider a peasant or a monarch, [C] a nobleman or a commoner, a statesman or a private citizen, a rich man or a poor man, [A] we find therefore an immense disparity between men who, it could be said, differ only by their breeches.

[C] (In Thrace, the king was distinguished from his people in a most amusing and extravagant manner: he had his own separate religion, a god all to himself whom his subjects had no right to adore — Mercury it was; Mars, Bacchus and Diana were the people's gods, whom he despised.)[9]

Such things are only so much paint: they do not make for differences of essence. [A] For as you see actors in plays imitating on the trestles dukes or emperors, only to return suddenly to their original natural position of wretched valets and drudges: so too with that Emperor whose pomp in public dazzles you —

> [B] *Scilicet et grandes viridi cum luce smaragdi*
> *Auro includuntur, teriturque Thalassina vestis*
> *Assidue, et Veneris sudorem exercita potat;*

[Because his huge green emeralds are set in gold, and he assiduously dresses in sea-green garments drenched in the sweat of Venus' games;][10] —

[A] draw back the bed-curtains and look at him: he is but a commonplace man, baser perhaps than the least of his subjects. [C] '*Ille beatus introrsum est. Istius bracteata felicitas est*' [That man is inwardly blessed; the other's happiness is merely gold-plated]:[11] [A] he is wracked like another man by cowardice, wavering, ambition, anger and envy;

> *Non enim gazæ neque consularis*
> *Summovet lictor miseros tumultus*
> *Mentis et curas laqueata circum*
> *Tecta volantes.*

[For it is not treasures nor even the consul's lictor that can banish wretched storms of passion from our minds nor banish those anguished cares which flutter about beneath fretted ceilings.]

[B] Even when surrounded by his armies, anxiety and fear can have him by the throat.

---

9. Herodotus says the same, without the irony (V, vii).
10. Lucretius, IV, 1123–5.
11. A combination of two phrases in Seneca: *Epist. moral.*, CXIX, 12 and CXV, 9.

> *Re veraque metus hominum, curæque sequaces,*
> *Nec metuunt sonitus armorum, nec fera tela;*
> *Audacterque inter reges, rerumque potentes*
> *Versantur, neque fulgorem reverentur ab auro.*

[The fears and dogging cares of men are not themselves afraid of fierce swords nor the sounds of war: they boldly come to kings and powerful men and have no reverence for the gleam of gold.][12]

[A]   Do fever, headache or gout spare him any more than us? When old age is on his back, will the archers of his guard carry it for him? When he is paralysed by dread of dying, will he be calmed by the presence of the gentleman-in-waiting of his bedchamber? When he is jealous and jumpy, will our doffed hats cure him? The roof of his four-poster may be stuffed with gold and pearls but it has no virtue to assuage the anguished paroxysms of a lively attack of the stone.

> *Nec calidæ citius decedunt corpore febres,*
> *Textilibus si in picturis ostroque rubenti*
> *Jacteris, quam si plebeia in veste cubandum est.*

[Nor do burning fevers quit your body sooner if you lie under embroidered bedclothes in your purple than if you are covered by plebeian sheets.]

Flatterers were bringing Alexander the Great to believe that he was the Son of Jove; but when he was wounded one day and saw the blood pour out of the gash he said, 'What do you say about this, then? Is this blood not red and thoroughly human? It is not the same colour as the blood which Homer has flowing from the wounds of gods!'[13]

Hermodorus the poet wrote verses in honour of Antigonus in which he called him Offspring of the Sun; he retorted, 'The man who slops out my chamber-pot knows nothing about that!'[14] After all what we have is a man; and if he himself is born awry then ruling the world will not put him right.

---

12. Horace, *Odes*, II, xvi, 9–12; Lucretius, II, 47–50.
13. Lucretius, II, 34–6; Erasmus, *Apophthegmata*, IV, *Alexander Macedo*, XVI. (Alexander was echoing Homer's account of Venus wounded by Diomedes.)
14. Plutarch (tr. Amyot), *De Isis et Osiris*, 323F; Erasmus, *Apophthegmata*, IV, *Antigonus Rex Macedonum*, VII. (The poet's name was Hermodotus not Hermodorus. The error is Montaigne's. In the *Quart Livre* of Rabelais, as in Erasmus, he is correctly named.)

[B] *Puellæ*
Hunc rapiant; quicquid calcaverit hic, rosa fiat;

[Let girls fight over him; let roses grow where'er his feet have trod;]

but what does that amount to if his soul is coarse and doltish? Even joy and sensual pleasure are not perceptible without vigour and wit:

> *hæc perinde sunt, ut illius animus qui ea possidet,*
> *Qui uti scit, ei bona; illi qui non utitur recte, mala.*

[Such things are like the mind which possesses them; good for the mind which knows how to use them rightly, but for the mind which knows not, bad.]

[A] The goods of Fortune (all of them, such as they are) cannot be savoured without tasting them: what makes us happy is not possessing them but enjoying them:

> *Non domus et fundus, non æris acervus et auri*
> *Ægroto domini deduxit corpore febres,*
> *Non animo curas: valeat possessor oportet,*
> *Qui comportatis rebus bene cogitat uti.*
> *Qui cupit aut metuit, juvat illum sic domus aut res,*
> *Ut lippum pictæ tabulæ, fomenta podagram.*

[It is not house and lands nor piles of bronze and gold which banish fevers from their owner's sickly body nor anxieties from his sickly mind. Their owner must be well if he wants to enjoy his acquisitions. When a man is full of fears or cravings, house and goods are as enjoyable as paintings are to blear eyes or hot fomentations to the gout.][15]

He is a fool: then his taste is flat and dull; he no more enjoys the sweet savour of Greek wine than a man with the snuffles, or than a horse enjoys the rich harness with which men bedeck it; [C] exactly as Plato says that health, beauty, strength, riches and all other things termed 'good' are bad to the unjust but, equally, are good to the just; and vice versa for 'bad' things.[16]

[A] And then, when your body and mind are in a bad state, what is the use of those external advantages, seeing that the merest pinprick or a passion of the soul are enough to take away the pleasure of being ruler of the world? At the first anguished pain of the gout [B] it is no help to be called Sire and Majesty,

15. Persius, *Satires*, II, 38–9; Terence, *Heautontimorumenos*, I, iii, 21–6; Horace, *Epistles*, I, ii, 47–52.
16. Plato, *Laws*, II, 661C–D.

Totus et argento conflatus, totus et auro;

[With everything cast in gold and silver;]

[A]   does he not lose all memory of his grandeur and his palaces? And if he is in a temper, does his kingdom stop him from turning red, then livid, and grinding his teeth like a madman?

Now if he is a clever man and well endowed, his royal state will add [C] little [A] to his happiness:

Si ventri bene, si lateri est pedibusque tuis, nil
Divitiæ poterunt regales addere majus;

[If your stomach, lungs and feet are all right, then a king's treasure can offer you no more;][17]

he knows it to be deception and vanity. Yes, and he may perhaps agree with the opinion of King Seleucus, that if a man knew the weight of a sceptre he would not bother to pick it up if he found it lying on the ground – he said that because of the great and painful responsibilities weighing on a *good* king.[18] Indeed it is no little thing to have to rule others, since there are so many difficulties in ruling ourselves. As for being in command – which appears so pleasant – I am strongly of the opinion (given the weakness of man's judgement and the difficulty of making choices in new and doubtful matters) that it is far more easy and agreeable to be led than to lead, and that there is great peace of mind to be found in merely having to follow the road you are told to and in being responsible for no one but yourself:

[B]   *Ut satius multo jam sit parere quietum*
Quam regere imperio res velle.

[So that it is far better quietly to obey than to seek to rule in state.]

Added to which Cyrus said that no man has any right to give orders if his worth is not greater than those who receive them.[19]   [A]   But King Hieron, in Xenophon, goes farther and maintains that in the very enjoyment of pleasures kings are in a worse condition than private citizens, since ease and accessibility robs them of that bittersweet pain we find in them:

17. Tibullus, I, i, 71; Horace, *Epistles*, I, xii, 5.   [A]   until   [C]:   add *nothing* to.
18. Plutarch (tr. Amyot), *Si l'homme d'aage doit encore mesler des affaires*, 183D.
19. Lucretius, V, 1126–7; then Erasmus, *Apophthegmata*, V, *Cyrus Major*, I.

[B]   *Pinguis amor nimiumque potens, in tædia nobis*
   Vertitur, et stomacho dulcis ut esca nocet.

[Too strong and rich a love-affair soon turns loathsome, just as sweet food sickens the stomach.]

[A]   Do we believe that choirboys greatly enjoy the music or rather that, being glutted with it, they find it boring? Feasting and dancing, masquerades and tournaments give delight to those who do not often see them and who were yearning to see them; but for a man who attends them regularly they become tasteless and disagreeable. Nor do women excite a man who has enjoyed them until his *mind* is sated; if a man does not give himself time to get thirsty he will never enjoy drinking.[20] We enjoy farces: they are drudgery to the travelling players. As proof of this, it is a treat and feast for princes to put on disguises occasionally and to drop into the way of living of the ordinary common people.

   Plerumque gratæ principibus vices,
   Mundæque parvo sub lare pauperum
   Cænæ, sine aulæis et ostro,
      Solicitam explicuere frontem.

[Often a change is pleasant to princes; a clean and frugal meal beneath a poor man's modest roof, without tapestries and purple, has smoothed the worried brow.][21]

[C]   Nothing cloys and impedes like abundance. What appetite would not be put off by the sight of three hundred accessible women such as the Grand Seigneur has in his harem? And what appetite for what kind of hunting did one of his ancestors keep up, who never took to the field with fewer than seven thousand falconers?
   [A]   Moreover I believe that the splendour of greatness brings quite a few impediments to the enjoyment of even the sweetest pleasures; they are too brightly illuminated, too much on show.
   [B]   And I do not know why, but we expect kings to cover up their faults more and to hide them better. What is a misdemeanor in us is, in them, considered an act of tyranny by the people, as disdain and contempt for the law; any tendency to vice apart, they look as if they are taking additional pleasure in scornfully trampling public decency underfoot.

20. Borrowings here and later from Xenophon's *Hieron* (*On Kingship*), also Ovid, *Amores*, II, xix, 25–6.
21. Horace, *Odes*, III, xxix, 12–15.

[C]   Indeed Plato in his dialogue *Gorgias* defines a tyrant as a man who, in his city, is free to do anything he wants.[22]   [B]   So, often, the flaunting of their vice in public hurts more than the vice itself. Every man loathes being spied on and having his actions recorded: but kings *are* spied on, down to their facial expressions and their thoughts, the entire people reckoning that they have the right and privilege of making judgements upon them. The higher and brighter the spot, the bigger the stain: a mole or wart on your forehead shows up more than a scar does elsewhere.

[A]   That is why poets feign that Jupiter conducted his love-affairs disguised as something else: among all the amorous adventures which they credit him with, there is not one, I think, where he appears in might and majesty.

But let us get back to Hieron. He tells of all the inconveniences he experiences in his royal state arising from the impossibility of going freely about on his travels (which makes him a prisoner within the frontiers of his own country) and from always being hemmed in by a troublesome crowd. Indeed when seeing our own monarchs sitting alone at their tables, besieged by so many unknown talkers and gazers, I have often felt more pity for them than envy.   [B]   King Alfonso said that donkeys were better off than kings: their drivers let them at least feed in peace, whereas kings cannot get even their servants to let them do so.   [A]   And the idea has never occurred to me that it was a special privilege for a man of intelligence to have a score of witnesses standing round his lavatory-seat, nor that it was more pleasant and agreeable to be waited on by a man worth ten thousand a year or by a soldier who had taken Casale or defended Siena than by a good and experienced manservant.

[B]   Most royal prerogatives are virtually imaginary: each degree of wealth has some image of royalty in it. The term used by Caesar for all the lords who held sway in the France of his time was 'little kings';[23] and in truth, apart from the title *Sire*, you can all but live like a king. Just consider for example those provinces lying far from the Court – Brittany, say. Take a lord who lives at home on his estates there and who has been brought up among his men-servants: note his retinue, his subjects, his officers-of-state, his pastimes, the way he is served, his ceremonial; then see how high his thoughts can soar. Nothing could be more royal. His

22. Plato, *Gorgias*, 468C–469C. Ensuing anecdote: Erasmus, *Apophthegmata*, VIII, *Alphonsus Rex Aragonum*, XVII.
23. A slip of memory: Livy says somewhat similar things of the Spanish (XXXVII, 25).

own feudal master is mentioned, like the King of Persia, about once in a twelve-month; he acknowledges him merely because of some ancient cousinship recorded in his secretary's archives. In very truth our laws are in no wise repressive: the weight of the sovereign power is felt by your average French nobleman about twice in a lifetime. Real effective subordination only concerns those who welcome it and who love to gain honour and wealth by such servitude: the man who is content to squat by his hearth and who knows how to govern his household without squabbles or law-suits is as free as the Duke of Venice. [C] '*Paucos servitus, plures servitutem tenent.*' [Slavery holds on to few: many hold on to it.]²⁴

[A] But Hieron regrets above all that he finds himself deprived of mutual friendship and companionship, in which consists the most perfect and the sweetest fruit of human life: 'For what proof of love or affection can I draw from a man who, whether he wants to or not, owes me everything in his power? Can I attach importance to his humble address and his courteous respect, seeing that he cannot refuse them to me? The honour we receive from those who fear us is not honour at all: their respect is due to my royal state not to myself:'

> [B]   *maximum hoc regni bonum est,*
> *Quod facta domini cogitur populus sui*
> *Quam ferre tam laudare.*

[The greatest advantage of being a king is that his people are not only forced to put up with whatever their Master does: they must praise it.]²⁵

[A] 'Can I not see the same respect shown to the good king and to the bad, to the king they hate and to the one they love? The same outward show and the same ceremonial were offered to my predecessor and will be offered to my successor. If my subjects do nothing to displease me that is no proof of their good-will towards me: why should I take it to be so, since they could not do anything else even if they should wish to? No one follows me for any love that is between us, since loving-friendship cannot be plaited together when there is so little contact and correlation. My high rank has put me outside all human relationships: there is too great a disparity and disproportion. It is by convention and custom that they follow me –   [C]   not so much me as my fortune, so as to amass fortunes

---

24. Seneca, *Epist. moral.*, XXII, 11. (The *Duke* of Venice is the *Doge*.)
25. Seneca (the dramatist), *Thyestes*, II, i, 30.

of their own.[26]   [A]   All they say and do for me is merely cosmetic. Since their freedom is everywhere bridled by the mighty power I wield over them, I can see nothing around me but hypocrisy and disguise.'

Courtiers were praising the Emperor Julian one day for administering such good justice: 'I would be prepared to be proud of such praises,' he said, 'if they came from persons who could dare to condemn and censure any actions of mine when they were contrary to justice.'[27]

[B]   All the real prerogatives of monarchs are held in common by all men of moderate wealth. (It is for gods to mount wingèd horses and to sup on ambrosia!) They have no other sleep, no other appetites but ours; their steel is not better tempered than that of our swords; their crown does not protect them from sun or from rain. Diocletian, who wore a crown of such honour and good omen, gave it up to withdraw to the pleasures of private life; some time later when a crisis of state solicited him to return and take up his burden, he replied to those who were begging him to do so: 'If only you could see the ordered beauty of the trees I have planted in my garden and the fine melons I have sown there you would not try and persuade me.'

The conviction of Anacharsis was that the happiest establishment for a State would be one in which all else being equal, degrees of honour went according to virtue and those of reprobation to vice.[28]

[A]   When King Pyrrhus was planning to cross over into Italy his wise counsellor Cyneas, wishing to make him realize the inanity of his ambition, asked him, 'Well now, Sire, what end do you propose in planning this great project?' – 'To make myself master of Italy,' came his swift reply. 'And when that is done?' – 'I will cross into Gaul and Spain.' – 'And then?' – 'I will go and subjugate Africa.' – 'And in the end?' – 'When I have brought the whole world under my subjection, I shall seek my repose, living happily at my ease.' Cyneas then returned to the attack: 'Then by God tell me, Sire, if that is what you want, what is keeping you from doing it at once? Why do you not place yourself now where you say you aspire to be, and so spare yourself all the toil and risk that you are putting between you and it?'

26. '80: follow me, *or to draw from it their own individual aggrandisement and advantages.* All they say . . .
27. Ammianus Marcellinus, XXII, 10; of Julian the Apostate.
28. Diocletian's reluctance to rule was proverbial (cf. Erasmus, *Apophthegmata*, VI, *Diocletianus*, I). For Anacharsis, cf. Plutarch (tr. Amyot), *Banquet des sept sages*, 155B. (The ensuing anecdote, from Plutarch, *Life of Pyrrhus*.)

> *Nimirum quia non bene norat quæ esset habendi*
> *Finis, et omnino quoad crescat vera voluptas.*

[It is because he does not seem to know the bounds one should set to desire nor how far true pleasure can extend.][29]

I am going to close this chapter with an ancient phrase which I find particularly beautiful and apt: '*Mores cuique sui fingunt fortunam.*' [Each man's morals shape his destiny.][30]

29. Erasmus, *Apophthegmata*, V, *Pyrrhus*, XXIV; Lucretius, V, 1431–2.
30. Erasmus, *Adages*, II, IV, XXX, *Sui cuique mores fingunt fortunam*, citing Cornelius Nepos' *Life of Pomponius Atticus*, together with similar sayings of Menander (in Plutarch) and many others.

43. On sumptuary laws

[*A whole series of sumptuary laws sought to restrain rash expenditure on clothing and jewels and to limit extravagances in eating and dressing to certain classes of society. These laws were reiterated under Francis I, Henry II and Charles IX. Montaigne was inspired to write on the subject by Amyot's translation of Diodorus Siculus. His conservatism is deep-rooted and based on moral commitment. Dress and so on are 'matters indifferent', but constant giddy change undermines the very foundations of a culture.*]

[A] The way our laws make an assay at limiting insane and inane expenditure on table and clothing seems to run contrary to their end. The right way would be to engender in men a contempt for gold and silk as things vain and useless: we increase their honour and esteem, which is a most inappropriate way of putting people off them. For to declare that only princes may   [C] eat turbot and   [A] wear velvet and gold braid, forbidding them to the people, what is that but enhancing such things and making everyone want to have them? Let kings stoutly renounce such symbols of greatness: they have others enough;   [B] such excess is more pardonable in anyone else but a king.   [A] We can learn from the example of many a nation plenty of better ways of indicating our different ranks and distinctions – something which I do indeed think to be requisite in a state – without encouraging such manifest corruption and evil. It is wonderful how quickly and easily custom plants her authoritative foothold in matters so indifferent. In mourning for Henry II we have been wearing plain-cloth at Court for barely a year now, yet it is already certain that silk has become so unaristocratic that if you do see anyone wearing it you   [C] immediately take him for one of the townsfolk.[1]   [A] It has become the lot of doctors and barber-surgeons. Even if everyone dressed more or less identically there would still be enough other ways of showing differences of rank.

[B] (How quickly do muddy doublets of chamois-leather or coarse-cloth come to be honoured by our soldiers in the field, and rich elegant clothes bring reproach and contempt.)

---

1. '80: now yet *you at once infer that he is a man of little importance*. It . . .

[A]   Let our kings start giving up spending money on such things and it would be all over in a month, without edict or ordinance: we will all follow suit. The Law ought to state, on the contrary, that purple and goldsmithery are forbidden to all ranks of society except whores and travelling-players.

It was with astuteness like that that Zeleucus reformed the debauched customs of the Locrians. He ordained as follows: 'That no free born woman be attended by more than one chambermaid, except when she be drunk; That no woman leave the city by night or wear any golden jewellery about her person nor any richly embroidered dress, unless she be a public prostitute; That except for such as live on immoral earnings, no man shall wear gold rings on his fingers nor any elegant robes such as those tailored from cloth woven in Miletus.' Thus, with those shaming exceptions, he cleverly diverted the inhabitants of his city away from pernicious superfluities and luxuries.   [B]   That was a most useful way to bring men to obedience by honour and ambition.[2]

French kings are all-powerful over the reformation of such externals: their fancy is law.   [C]   *'Quidquid principes faciunt, praecipere videntur.'* [Any actions of princes seem like commands.][3] The rest of the country[4] adopts as canon the canons of the Court.   [B]   Let the Court stop liking those vulgar codpieces which make a parade of our [C] hidden [A] parts, those heavily padded doublets which make our shape look different and our armour so hard to put on; those long effeminate tresses; the custom of kissing any gift offered to our companions, and our hands, too, when we greet them (an honour formerly due only to princes); allowing a nobleman to appear in respectable company with no sword at his side, untidy and unbuttoned, as though he had just come straight from the privy; the custom (something contrary to the practice of our forefathers and the express privilege of the nobility of this Kingdom) of remaining hat-in-hand even when at some distance from our monarchs, wherever they happen to be (as well as in the presence of dozens of others, so many tercelets and quartlets of kings do we have);[5] and so on for similar recent

2. Diodorus Siculus, *Historia*, XII, cited Tiraquellus, *De legibus connubialibus*, III, §13.
3. Quintilian, *Declamationes*, III.
4. '80: The rest of the country *adopts as its model whatever is done,* in court: *those vicious fashions are born close to it.* Let . . .
   '80: our *shameful* parts, those *monstrously* padded. . .
5. A tercelet is a male falcon (one-third smaller than the female). Montaigne invents the word 'quartlet' for even smaller kinglets.

and depraved innovations: then they would soon all vanish in disapproval. Such defects may be all on the surface, but they augur badly: when we see cracks in the plaster and the cladding of our walls it warns us that there are fissures in the actual masonry.

[C] In his *Laws* Plato concludes that no plague in this world can do more damage to his city than allowing liberty to the young to change from fashion to fashion in their dress, comportment, dances, sports and songs, constantly changing the basis of their ideas this way and that, running after novelties and honouring those who invent them; by such things are morals corrupted and all ancient principles brought into disdain and contempt. In all things – except quite simply for those which are evil – change is to be feared, including changes of seasons, winds, diets and humours; and no laws are truly respected except those to which God has vouchsafed so long a continuance that no one knows how they were born or that they had ever been different.[6]

6. Plato, *Laws*, VII. In Ficino's Latin translation Plato talks not of customs 'to which God has vouchsafed' continuance but of those to which 'some divine Fortune' has done so.

44. On sleep

[Classical philosophy tends to see the Sage as a man untouched by emotion. Montaigne, preoccupied as often by war, treats of sleep in the context of exempla relating to great men in wartime.]

[A] Reason ordains that we should keep to the same road but not to the same rate; and although the wise man must never allow his human passions to make him stray from the right path, he may without prejudice to his duty certainly quicken or lessen his speed, though never plant himself down like some fixed and impassive Colossus. If Virtue herself were incarnate I believe that even her pulse would beat faster when attacking the foe than when attacking a dinner — indeed it is necessary that she should be moved and inflamed. That is why I have noted as something quite rare the sight of great persons who remain so utterly unmoved when engaged in high enterprises and in affairs of some moment that they do not even cut short their sleep.[1]

On the day appointed for his desperate battle against Darius, Alexander the Great slept so soundly and so late that when the hour of battle was pressing close, Parmenion was obliged to enter his chamber and call out his name two or three times to wake him up.[2]

The very same night that the Emperor Otho had resolved to end his life, he put his private affairs in order, distributed his money between his followers, sharpened the edge of the sword he intended to use for his blow and then, waiting only to know that each one of his friends had withdrawn to safety, fell so soundly asleep that his servants of the bedchamber heard him snoring.

The death of that Emperor has much in common with the death of the great Cato, and especially that feature; for when Cato was ready to take his

---

1. An echo of Seneca, *Epist. moral.*, XX, 2–3: the wise man acts consistently, his deeds always in harmony with his words.
2. Cf. Erasmus, *Apophthegmata*, VI, *Alexander Magnus*, LXIV. (Borrowings follow from i) Plutarch's *Lives* of Alexander, Otho, Sylla and Paulus Aemilius; and ii) Suetonius, *Life of Augustus*.)

own life, he was waiting for news to be brought that the Senators he had sent away had sailed out of the port of Utica when he fell into so deep a sleep that his breathing could be heard in the neighbouring room; and when the man he had sent to the port woke him to tell him of the storm which had prevented the Senators from sailing away in safety, he dispatched another and, settling down in his bed, he went off to sleep again until the man came back and told him that they had left.

And we can again compare him to Alexander, when during the Cataline Conspiracy there was such a storm over the treachery of Metellus the tribune who was determined to publish the decree summoning Pompey and his army back to Rome. Cato alone opposed that decree and he and Metellus had exchanged gross insults and great threats in the Senate; but the decision had to be carried out the following morning in the public Forum. Metellus was to come there, favoured by the plebs as well as by Caesar who was then allied to Pompey's interests: he was to be accompanied by a crowd of foreign mercenaries and gladiators who would fight to the last; Cato was to come supported by nothing but his own constancy. His family and friends and many others were deeply anxious about this: some of them spent the night together with no desire to sleep, drink or eat because of the danger they saw awaiting him; his wife and his sisters especially did nothing but fill his home with weeping and wailing; he on the contrary reassured everyone there; having dined as usual, he went to lie down and slept a deep sleep until morning, when one of his fellow tribunes came to wake him up to enter the affray. What we know of the courage of [C] this man from the rest of his life[3] [A] enables us to judge with absolute certainty that what he did proceeded from a soul high above such events, which he did not deign to take to heart more than any ordinary occurrence.

In that sea-fight which Augustus won against Sextus Pompeius off Sicily, he was just about to go into battle when he was overcome by so sound a sleep that his friends had to come and wake him up to get him to give the signal for the engagement. That provided Mark Antony later on with an excuse for accusing him of not having the will even to look straight in the eye of the troops he had drawn up for battle and of not daring to face his soldiers before Agrippa came to tell him the news of his own victory over his enemies.

But to turn to young Marius, he did worse: for on the day of his last encounter with Sylla, he drew up his army, gave the signal for battle, then

3. [A] until [C]: courage of *those three men* enables . . .

went to lie down for a rest in the shade of a tree where he fell so fast asleep that he could scarcely be awakened by the rout of his fleeing soldiers, having seen nothing of the combat; they say it was from being so exhausted by fatigue and want of sleep that nature could stand no more.

While on this topic it is for the doctors to decide whether sleep is such a necessity that our very life depends on it: for we are certainly told that King Perseus of Macedonia, when a prisoner in Rome, was done to death by being prevented from sleeping.

[C] Herodotus mentions nations where men sleep and wake a half-year at a time. And the biographer of Epimenides the Wise says that he slept for fifty-seven years in a row.[4]

4. Herodotus, *History*, IV, xxv; Diogenes Laertius, *Life of Epimenides*.

45. On the Battle of Dreux

[The Battle of Dreux, 19 December 1562, between the victorious Duc de Guise (for the Roman Catholics) and the Constable Montmorency (for the Reformed Church) evokes a comparison with analogous exempla, *in Plutarch's* Life of Philopoemen *and* Life of Agesilaus.]*

[A]   There was a full bag of remarkable incidents in our battle at Dreux; but those who are not strongly inclined towards the reputation of Monsieur de Guise like to allege that he cannot be forgiven for having called a halt and marking time with the forces under his command while the Constable, who was leading his army, was being battered by our artillery: it would have been better to have exposed himself to risk and to have attacked the enemy's flank than to have waited to see his rear, so incurring a heavy loss. But apart from what is proved by the outcome, anyone who will debate the matter dispassionately will, I think, readily concede that the target in the sights of any soldier, let alone a commander, must be overall victory and that no events, no matter what their importance to individuals, should divert him from that aim.

Philopoemen, in his encounter with Machanidas, advanced a good troop of archers and spearmen to open the affray; his enemy knocked them about and, after this success, spent time galloping after them slipping right along the flank of the company commanded by Philopoemen who, despite his soldiers' excitement, decided not to budge from his positions and not to offer the enemy battle even to save those men; but after allowing them to be hunted down and cut to pieces before his eyes, he opened an attack against the enemy foot-soldiers once he saw that they had been quite abandoned by their cavalry. And even though they were Spartans he quickly achieved his end, especially because he surprised them at a time when they thought they had already won and were beginning to break ranks. Only when that was done did he set about pursuing Machanidas.

That case is germane to that of Monsieur de Guise.

[B]   In that harsh battle of Agesilaus against the Boeotians (which Xenophon who was there said was the cruellest he had ever seen) Agesilaus refused the opportunity which Fortune gave him – even though he foresaw

certain victory from it – of letting the Boeotian battalion slip through and then charging their rear; he considered there was more art in that than valour. And so, to display his prowess, he preferred by an extraordinary act of ardent courage to make a frontal attack. But he was thoroughly beaten and wounded; he was obliged to disengage and accept the opportunity he had first rejected: he split his ranks and let the Boeotians pour through. Once they had all done so, he noted that they were marching in some disorder like men who thought they were out of danger: he commanded them to be pursued and attacked on their flanks. Even then he was unable to make them retreat in a headlong rout: they withdrew foot by foot, still showing their teeth until they had reached safety.

46. On names

[Platonic philosophy attached a real power to names; Montaigne is sceptical. Even a great 'name' (a well-deserved reputation after death) is an empty thing for the dead heroes themselves. This chapter is in many ways a diptych to I:37, 'On Cato the Younger'.]

[A]   No matter how varied the greenstuffs we put in, we include them all under the name of salad. So too here: while surveying names I am going to make up a mixed dish from a variety of items.

I do not know why but every nation has some names which are taken in a bad sense: we do so with *Jean*, *Guillaume* and *Benoît*.

Item: in the genealogy of kings there seem to be some names beloved by Fate, as Ptolomey was for kings in Egypt, Henry in England, Charles in France, Baldwin in Flanders and, in our own Aquitania in olden times, Guillaume, from which they say is derived the name of Guyenne – a poor enough pun were there not equally crude ones in Plato himself.[1]

Item: a trifling matter, but nevertheless worthy of remembrance because of its oddness and its being vouched for by an eye-witness, is the fact that when Henry Duke of Normandy, son of King Henry II of England, held a great feast in France, such a huge crowd of the nobility had gathered together that it was decided for amusement to divide people up into groups bearing similar names: the first troop consisted of the Guillaumes, comprising one hundred and six knights of that name seated at table, without counting the ordinary gentlemen and servants.   [B]   Just as amusing as seating guests at table according to their names was the idea of the Emperor Geta who arranged his bill of fare according to the first letter of the name of each dish, serving up together those which begin with M, such as mutton, marcassin, merle, marsouin and so on.

[A]   Item: they say that it is a good thing to have a good name (meaning renown and reputation); but it is also a real advantage to have a fine one which is easy to pronounce and to remember, since kings and the great can then recognize us more easily and less wilfully forget us. Even where our servants are concerned we usually summon for a job those

1. In the *Cratylus*, where several etymologies do indeed appear fanciful nowadays.

whose names come most readily to our tongue. I noticed that King Henry II was never able to call a gentleman from our part of the world by his right name; and he even decided to call one of the Queen's maidservants by her family name because the Christian name given her by her father seemed too awkward. [C] Socrates himself thought it was worth a father's while to take trouble to give his children beautiful-sounding names.

[A] Item: it is said that the origin of the founding of Notre-Dame-La-Grand at Poitiers was the discovery by a local dissolute youth that the girl he had just picked up and whose name he had asked was called Mary; he felt such vivid awe and respect sweep over him on hearing the name of the Most Blessed Mother of our Saviour, that not only did he send the girl packing but brought amendment to the rest of his life; in consideration of this miracle a chapel was built to Our Lady on the square where the youth's house stood, and subsequently there was built the church we can see there today.[2]

[C] That conversion by word and hearing, being pious, struck straight at the soul; the following is similar but was subtly introduced through the physical senses: Pythagoras was in the company of some young men: he heard them plotting, when they were inflamed with wine, to go and rape some chaste women in their own home: he ordered the minstrel-girl to change her musical mode, so that, by a weighty and grave tune meant for solemn drinking, he gently charmed away their hot lust and calmed it down.

[A] And will not posterity say that our present-day Reformation has been scrupulous and dainty indeed! It not only fought against error and vice, filling our whole world with piety, humility and obedience, with peace and with virtues of every kind, but it also went so far as to fight against our ancient Christian names of Charles, Louis and François, so as to people the earth with Methuselahs, Ezekiels and Malachis, names so much more redolent of our Faith! . . .

One of my neighbouring gentlemen, when listing the superiorities of former times over our own, did not forget to mention the proud and magnificent names of the noblemen in those days; by simply hearing names such as Don Grumedan, Quedragan or Agesilan, he felt they had been men of a different kind than our Pierres, Guillots and Michels.

Item: I am deeply grateful to Jacques Amyot for leaving Latin names as

2. Jean Bouchet, *Annales d'Acquitaine*.

they were in the course of his French prose, without altering and changing
their colour by giving them French endings. It seemed a bit harsh at first,
but usage, because of the authority of his *Plutarch*, has removed their
strangeness for us. And I have often wished that those who write our own
history in Latin would leave French names alone, for when they make
Vaudemont into *Vallemontanus* and change the shape of our names so as to
robe them in the Greek or Latin style we no longer know where we are
and cannot understand them any more.

To end my account, it is a custom worthy of villeins – and of great
consequence for this France of ours – that we call people by the name of
their lands and lordships: nothing in the world is so responsible for
confusing and confounding our family trees. The younger son of a good
family, having received as his portion lands by whose name he is honoured
and known, cannot honourably go and dispose of them; but ten years
after he is dead they do pass to a stranger, who then acts the same way.
You can guess how far we get when we try to identify those men. We
need to look no further for examples of this than to our own royal
house: so many portions, so many surnames. Meanwhile we have lost
the original stem.

[B]   These mutations are allowed such licence that I know nobody in
my own time who has had the good fortune to be elevated to some
extraordinarily high rank who has not been immediately endowed with
new genealogical styles of which his father knew nothing, or failed to be
grafted on to some illustrious stock. Luckily it is the obscurer families
which best lend themselves to such falsifications. How many mere gentlemen are there in France who are of royal stock ... by their own
reckoning! More I think than of any other rank.

Was this habit not put to shame with good grace by one of my friends?
Several gentlemen had gathered together on account of a dispute between a
lord and a certain gentleman who had in truth some precedence by title
and alliance which did raise him above the ordinary men of his rank. On
the subject of this precedence, all the other gentlemen strove to make
themselves his equal, each alleging this origin or that, or some similarity of
name or arms or some old family document: even the least among them
proved to be remotely descended from some king of Outremer![3] When
dinner was served that Lord, instead of taking his seat, walked backwards
bowing deeply, begging the assembled company to pardon his temerity for

3. *Outremer* ('Overseas') was the collective name of French Crusader kingdoms in
the Middle East.

having heretofore lived with them as an equal: but now, having been informed of their ancient lineages he would start honouring each according to his degree: it was not for him to take a seat in the presence of so many princes. When this farce was over he addressed a great many rebukes to them: 'In God's name let us be content with [C] what contented our ancestors and with [B] whatever we are; if we can sustain that, we are good enough. Let us not disown the fortune and circumstances of our forefathers; let us get rid of such stupid fancies: they will never run out for such as are impudent enough to allege them.'

Coats-of-arms are no more reliable than our family names. Say I sport *Azure, semee of trefoils, or; lions rampant, also or; armed, fesse gules*. What privilege is accorded to this design to remain specific to my house? A son-in-law will transport it to some other family; some wretched man will buy it for his first coat-of-arms: such changes and confusion can be found nowhere else.

[A] This consideration drags me into another subject. Let us make our soundings go a little deeper and for God's sake look at the foundations on which we build all that honour and glory for which the world is thrown into chaos. To what do we attach the reputation which we seek after with such labour? Why, it is a man called Pierre or Guillaume who enjoys it, guards it and who is touched by it. [C] (Oh what a sagacious faculty is hope, which for a moment arrogates infinity, immensity and eternity to a mortal creature! What a nice little toy Nature has given us there.)

[A] In the first place: this *Pierre* or this *Guillaume*, what is it, if you come to think of it, but a spoken noun or three or four pen-strokes so easily corrupted that you may well wonder who actually did get the honour of all those victories: was it *Guesquin*, was it *Glesquin*, was it *Gueaquin*?[4] (There are better grounds for a law-suit here between Letter S and Letter T than there are in Lucian,[5] for,

> *non levia aut ludicra petuntur*
>> *Præmia;*
>
> [the prize they seek is no light or trivial one;][6]

this is serious.) The question is, which letters of the alphabet in those names

---

4. There are as many spellings of the name of the great medieval constable, Bertrand Du Guesclin, as that of Shakespeare.
5. Cf. Lucian of Samosata, *Lawsuit between the Vowels*.
6. Virgil, *Aeneid*, XII, 764.

are to be credited with all those sieges, battles, wounds, imprisonments and duties undertaken on behalf of the Crown of France by her famous Constable?

Nicolas Denisot was only concerned with the letters of his name: he strung them together in a different arrangement as the 'Conte d'Alsinois', to which name he gave all the glory of his poetry and painting.[7] But the historian Suetonius was attached only to the meaning of his; his father's name was *Lenis* ('Calm'): he disowned it and bequeathed his own reputation as a writer to *Tranquillus*.[8] Who would ever have believed that the fame of Capitaine Bayard is all borrowed from the deeds of a man called Pierre Terrail, or that the name Antoine Escalin should allow itself to be robbed, with its eyes wide open, of all its voyages over land and sea undertaken by a Capitaine Poulin and a Baron de la Garde?[9]

In the second place: those pen-strokes are shared by hundreds of men. How many people are there of the same kindred who all bear the same name and surname? [C] And how many are there of different kindred, periods and countries? History has known three men called Socrates, five Platos, eight Aristotles, seven Xenophons, twenty Demetriuses and twenty Theodores; just guess how many she has never known!

[A] What can stop my ostler calling himself Pompey the Great?[10] When all is said and done, what means or links are there which can securely attach that glorious spoken name or pen-strokes either to my ostler, once he is dead, or to that other man whose head was severed in Egypt, in such a way that they can profit by them?

> [A1]   *Id cinerem et manes credis curare sepultos?*
>
> [Do you think that bothers spirits and ashes in their tombs?][11]

[C] What can they feel now, the following two heroes who share in

---

7. Nicolas Denisot, poet, novelist and portrait-painter as well as intelligence-agent and diplomatist, was known by the anagram of his name and regularly addressed as *Comte* (Count) (Cf. Margaret Harris, *A Study of Théodose Valentinian's 'Amant Resuscité' (by Nicolas Denisot?)*, Geneva, 1966).

8. Suetonius' *cognomen* was probably *Tranquillus*.

9. The 'good chevalier Bayard' (on whom Jacques de Mailles wrote a popular book) was indeed really called Pierre Du Terrail. Escalin, Baron de la Garde, was nicknamed Captain Poulin.

10. The Renaissance cult of Classical names adds force to Montaigne's point. (When Erasmus first heard of Julius Caesar Scaliger he thought the name was fictional.)

11. Virgil, *Aeneid*, IV, 34.

fellowship the highest bravery known to me? Can Epaminondas hear that glorious verse about him so frequently on our lips:

> *Consiliis nostris laus est attonsa Laconum*

[My counsels clipped the praise of Sparta]?[12]

Can Scipio Africanus hear these:

> *A sole exoriente supra Mœotis paludes*
> *Nemo est qui factis me æquiparare queat*

[There is no man from where the eastern sun rises above the marshes of the Scythian Lake who can match my deeds]?[13]

It is those who survive who are moved by the sweetness of those sounds; stirred by a desire to rival those dead men, without reflection they mentally attribute their own emotions to them and deceive themselves into thinking that they too will be able to feel them in their turn. God knows that is true.

[A] Nevertheless:

> ad hæc se
> *Romanus, Graiusque, et Barbarus Induperator*
> *Erexit, causas discriminis atque laboris*
> *Inde habuit, tanto major famæ sitis est quam*
> *Virtutis.*

[For this were Roman, Greek and Barbarian chiefs aroused; this was the motive of their risks and labours, so much more did they thirst for fame than virtue.][14]

12. Cicero, *Tusc. disput.*, V, xvii, 49; translated from the Greek epitaph of Epaminondas.
'95: on our lips *for so many centuries* . . .
13. From Ennius' epitaph on Scipio Africanus.
14. Juvenal, *Satires*, X, 137–41.

47. On the uncertainty of our judgement

===

[Renaissance education in both rhetoric and dialectic gave a large place to arguments pro et contra. Montaigne's own arguments suggest that this is no mere schoolboy practice but of vital interest in war. As well as Classical exempla *of diametrically opposed decisions leading to similar results, Montaigne gives towards the end, in a long, rambling sentence, reflections attributed to King Francis I in the* Mémoires *of the brothers Du Bellay.]*

[A]   As this verse rightly says,

> ’Επέων δὲ πολὺς νόμος ἔνθα καὶ ἔνθα

'there is every possibility of speaking for and against anything'.[1] For example:

> *Vinse Hannibal, et non seppe usar' poi*
> *Ben la vittoriosa sua ventura.*

[Hannibal won battles, but he never knew how to profit from his victories.][2]

If anyone wants to defend that position and to persuade our side that it was wrong not to have followed up our recent victory at Montcontour, or if he should want to criticize the King of Spain for not knowing how to exploit the advantage he won over us at Saint-Quentin, he may say as follows: that these mistakes proceeded from a soul drunk with good fortune and from a mind which, having gorged itself full on such a happy beginning, had lost all appetite for more, finding it hard to digest what it already had; the fellow has his arms full: he cannot take anything else; he is not worthy that Fortune should have placed such a favour in his hands: what has he gained if he then goes and provides his enemy with the means of recovery? What hope can a man have of daring to attack his enemies later, after they have rallied and re-mustered and are newly armed with vengeance and anger, when he did not dare to hunt them down when terrified and routed?

---

1. Homer, *Iliad*, XX, 249 (translated in text).
2. Petrarch, Sonnet 82 (83).

Dum fortuna calet, dum conficit omnia terror.

[When Fortune is aroused and Terror in control.][3]

And after all, what better opportunity can he expect than the one he has just lost? War is not like a fencing-match where you can win on points: so long as your enemy is on his feet you must begin to attack him again harder; while the war is not ended there is no victory. In that skirmish in which Caesar was worsted near the town of Oricum he shamed Pompey's soldiers by saying that he would have lost everything if their general had only known how to win; and he made Pompey clap on his spurs to quite other effect when his own turn came!

But why do they not also state the opposite, as follows: that it is the action of a headlong and insatiable mind not to know when to set a limit to what it covets; that it is to abuse God's favours to wish to strip them of that moderation which he has prescribed for them; that to plunge into danger after a victory is to put it again at the mercy of Fortune; that one of the wisest rules of the art of war is never to drive your foes to despair. In the war between the allies, when Sylla and Marius had defeated the Marsi and spotted a group of survivors about to make a desperate attack like beasts driven mad, they did not think it wise to await them. If Monsieur de Foix had not been led by his ardour to pursue the remnant so relentlessly after the victory of Ravenna he would not have sullied it by his death. (Nevertheless the memory of his example was still fresh enough to preserve Monsieur d'Enghien from a similar mistake at Cérisoles.) It is hazardous to go and attack a man when you have deprived him of all means of escape save his weapons, for Necessity is a ferocious teacher: [C] *'Gravissimi sunt morsus irritatæ necessitatis.'* [When Necessity is aroused her bites are most grievous.]

[B]   *Vincitur haud gratis jugulo qui provocat hostem.*

[It will not cost you nothing to defeat a man if you are threatening to slit his throat.][4]

[C]   That is why Pharax prevented the King of Sparta, who had just won the day against the men of Mantinea, from confronting several hundred Argives who had escaped unscathed from their defeat, persuading him to let them flee without hindrance so as not to have to assay what virtue is like when goaded and outraged by misfortune.

---

3. Lucan, *Pharsalia*, VII, 734.
4. Portius Latro *apud* Justus Lipsius, *Politici*, V, xviii; Lucan, *Pharsalia*, IV, 275.

[A]   After his victory King Clodomire of Aquitania was pursuing the fleeing King Gondemar of Burgundy whom he had defeated, when he forced him to turn about and face him: his stubborn determination robbed him of the fruits of his victory, for he was slain.

Similarly, supposing you had to choose between keeping your soldiers armed richly and sumptuously or armed only with the bare necessities. In favour of the first side – that of Sertorius, Philopoemen, Brutus, Caesar and others – the following argument can be found: it is always a spur to honour and glory for a soldier to be splendidly armed as well as an encouragement to him to fight more stubbornly, seeing that he has to safeguard his weapons which constitute his inherited wealth.   [C]   That, says Xenophon, was the reason why the Asians used to bring their wives, their concubines and their richest jewels and treasures with them when they fought.[5]

[A]   But in favour of the other side there is found the following: rather than encouraging it, we should remove from the soldier any thought of preserving his life; such a practice would redouble his fears of exposing himself to risks; with such rich spoils you increase the enemy's lust for victory; it was noticed on other occasions that it was wonderful how hopes of spoil put heart into the Romans in their encounters with the Samnites.   [B]   When Antiochus was showing off to Hannibal the army which he was making ready to fight the Romans, with all its splendid and magnificent equipment of every kind, he asked: 'Will this army be enough for the Romans?' – 'Will it be enough for them! I should say it would,' he replied, 'no matter how greedy they may be!'   [A]   Lycurgus not only forbade his own men to have luxurious equipment but even to despoil their vanquished enemies: he wished, he said, that it was their poverty and frugality that should outshine all else on the battlefield.[6]

During sieges and the like when circumstances bring us close to our enemies we readily allow our soldiers full freedom to defy them and to taunt and insult them with all manner of abuse: and that can seem to be reasonable, for it is no little achievement to deprive our own men of any hope of mercy or of reconciliation by showing them that they no longer have any cause to expect such things from enemies they have so strongly provoked; there is only one remedy: victory.

5. Anecdotes from Diodorus Siculus, Jean Bouchet (*Annales d'Acquitaine*), Plutarch's *Lives*, Suetonius and Xenophon (*Cyropaedia*).
6. Anecdotes from Livy, XI, xl, Aulus Gellius (*Attic Nights*, V, v) and Plutarch (tr. Amyot), *Dicts notables des Lacaedaemoniens*, 221 C.

But in the case of Vitellius that all went awry. He was confronting Otho, who was weaker than he was because his soldiers were no longer used to actual fighting, being debilitated by the pleasures of Rome; but he maddened them so with his stabbing insults, mocking them for their weakness and their regrets at leaving the feastings and women of Rome, that – something which no exhortations had managed to do – he put new heart into them so that while no one could drive them to engage him he led them to do so. And in truth when such insults touch a man to the quick they can soon make someone who was going slackly about his duty on behalf of his king start doing it with a far different emotion on behalf of himself.[7]

Considering how vital it is to safeguard an army's leader whose head his enemies have constantly in their sights since all his men cling to him and depend on him, it would seem impossible to cast doubt on the decision which we have seen taken by many great military leaders to disguise their apparel at the moment of battle; yet the disadvantages of this practice are no less than those we think we avoid: for when their general cannot be recognized by his own men the courage they derive from his example and his presence begins at once to fail them; they miss the sight of his usual symbols and insignia and think he is killed or has despaired of victory and fled.

As for experience, it can be seen to favour now one party in the dispute now the other. What happened to Pyrrhus in the battle waged against the consul Levinus in Italy can be cited by us on either side: by deciding to disguise himself under the armour of Demogacles and to give him his he undoubtedly saved his life; but he all but fell into the other disadvantage: that of losing the day.[8]   [C]   Alexander, Caesar and Lucullus liked to stand out on the battlefield in their rich equipment and armour, with their own particular colour gleaming: Agis, Agesilaus and the mighty Gylippus on the other hand went to war in dark colours, not dressed like the man in command.

[A]   Among other things which were held against Pompey at the battle of Pharsalia was his bringing his army to a firm halt and awaiting the enemy. 'That is because' (and here I will steal the very words of Plutarch, which are worth more than my own) 'such tactics reduce the ferocious power which the act of charging gives to the opening blows and also removes that shock of combatant against combatant which, more than

7. Plutarch, *Life of Otho*.
8. Plutarch, *Life of Pyrrhus*.

anything else, regularly fills soldiers with a furious madness as they stoutly dash at each other, their shouts as they run giving them more heart; the other tactics can be said to dampen their ardour and to chill it.' That is what Plutarch says on the subject.[9]

But supposing Caesar had lost; could not anyone just as easily have asserted the contrary: that the most effective and most firm posture is to stand stock still; that whoever comes to a halt, sparing his energy for when needed and storing it up in himself, has a great advantage over the man in motion who has already used up half his breath in the charge. An army, moreover, being made up of many different individuals, it is not possible for it to manoeuvre so accurately during the frenzy of battle that its ranks be not weakened or broken, the more agile soldier already at grips with the enemy before his comrade can come to his support.

[C] In that ignoble battle between two Persian brothers,[10] Clearchus of Sparta who commanded the Greeks on the side of Cyrus led them to make a controlled and unhurried advance; then, when fifty yards away, he ordered his men to advance at the run, hoping that such a short distance would spare their breath and maintain their ranks while giving them the advantage of impulse both for their bodies and their javelins.

[A] In their own armies others have resolved that dilemma this way: if the enemy charges, stand firm; if he stands firm, charge.

During the invasion of Provence by the Emperor Charles V King Francis was able to choose between going to confront him in Italy or waiting for him in his own territory.[11] And although he took into consideration what an advantage it is to keep the homeland clean, unsullied by the tumult of war, so that with its resources intact it can go on furnishing treasure and succour when needed; although he considered that the exigencies of war constantly oblige armies to lay waste, something which cannot be easily done in one's own lands, the peasants moreover not putting up with such devastation so patiently when done by their own side rather than the enemy, so that it is easy to kindle seditious disturbances among us; that permission to rob and to pillage cannot be allowed in one's own country, yet is a great compensation for the hardship of fighting; that it is difficult to keep a man to his duty when he has nothing to hope for but his pay and is only a few yards away from wife and hearth; that he

9. Plutarch, *Life of Pompey*.
10. The battle of Cunaxa between Artaxerxes and Cyrus the Great, 401 BC. Cf. Xenophon, *Anabasis*.
11. In 1536. The discussion is influenced by the Du Bellay *Mémoires*.

who orders the dinner pays the bill; that there is more joy in attack than in defence; that the shock of losing a battle within the guts of our land is so violent that it is hard to stop it shaking the whole body, seeing that there is no passion so contagious as fear nor caught so easily by hearsay nor spread so quickly; and that, when towns have heard the crashing of such storms at their very gates and let in their own officers and soldiers still quivering and breathless, there is a risk that the townsfolk may in the heat of the moment leap to some evil decision: nevertheless King Francis chose to recall his transalpine forces and to watch the enemy approach.

For he may have thought, on the contrary, that being at home among people who loved him, he could not fail to have plenty of supplies (the rivers and passes being devoted to him would bring him provisions and treasure in complete safety without need of escort); that his subjects would be all the more loyal to him for having the danger nearer at hand; that, having so many towns and fortified places to rely on, it was for him to order the fighting when it was opportune and advantageous to himself; that if he decided to play for time, he could stay comfortably at ease and watch his enemy hanging about and defeating himself in his battle against hardships, caught in a hostile land where everything was at war with him in front, at his rear and on his flanks, with no means of resting his army nor of segregating his soldiers when illness came among them nor of sheltering his wounded; no money, no provisions save at lance-point; no time for repose and to get back his breath; no knowledge of the terrain or of the countryside which could save him from ambush and surprise attacks; and if it did come to a defeat, no means of saving the survivors.

And there was no lack of examples on either side.

Scipio found it wiser to go and assault the lands of his enemy in Africa than to defend his own and fight in Italy where he was; things turned out well in his case. But Hannibal, on the contrary, in that very war, ruined his chances by giving up his conquest of a foreign land to go and defend his own.

The Athenians left the enemy in their own lands and crossed over to Sicily: Fortune went against them; yet Agathocles, King of Syracuse, found Fortune favourable when he crossed into Africa leaving war at home.

And so as we often say, rightly, events and their outcomes depend, especially in war, mainly on Fortune, who will not submit to our reasoning nor be subject to our foresight – as these lines put it:

> *Et male consultis pretium est: prudentia fallax,*
> *Nec fortuna probat causas sequiturque merentes;*

> *Sed vaga per cunctos nullo discrimine fertur;*
> *Scilicet est aliud quod nos cogatque regatque*
> *Majus, et in proprias ducat mortalia leges.*

[Badly conceived projects are rewarded; foresight fails, for Fortune does not examine causes nor follow merit but meanders through everything without distinction. Clearly there is Something greater which drives and controls us and subjects the concerns of men to laws of its own.][12]

But it seems, if you take it aright, that our counsels and decisions too depend just as much on Fortune and that she [C] involves in her turbulence and uncertainty even our reasoning. 'We argue rashly and unadvisedly,' says Timaeus in Plato, 'because in our reasoning as in ourselves, a great part is played by chance.'[13]

---

12. Manilius, *Astronomica*, IV, 95–9.
13. '80: Fortune, and that she *is as* uncertain *and random as* our reasoning.
   Plato, *Timaeus*, 34C.

48. On war-horses

===

[Montaigne, as a gentleman who loved riding and enjoyed soldiering, lets himself go in this formless chapter, collecting anecdotes about a subject which interested him: horses, especially horses in war. The seeds of later chapters are found here, including III, 6, 'On coaches' and doubtless the chapter comparing the armaments of the ancients and moderns which is mentioned below but was stolen by a manservant and never rewritten.]

[A]   I have never learned any language except by using it and I still do not know what an adjective is nor a subjunctive nor an ablative: yet here I am, turning into a grammarian. I believe I have heard it said that the Romans had horses called FUNALES or DEXTRARII (which were trace-horses either accompanying them on their right-hand side or stationed at relays, so that they were quite fresh when needed) and that that explains why we call our war-horses *destriers*[1] (Our French romances also regularly use *adestrer* to mean *to accompany*.) The Romans also used the term DESULTARII EQUI [leaping horses] for horses which had been so trained that when they were galloping at full force coupled together but without bridle or saddle the nobles riding them would leap from one to the other in full career, clad in their armour.   [C]   The Numidian cavalry, so as to change horses in mid battle, kept a second one handy: *'quibus, desultorum in modum, binos trahentibus equos, inter acerrimam saepe pugnam in recentem equum ex fesso armatis transsultare mos erat: tanta velocitas ipsis, tamque docile equorum genus.'* [Their custom was to have two horses on traces just as our acrobats do and to leap from the one they were riding on to their fresh one, often in the bitterest moment of the battle, such was their own dexterity and the fitness for training of that breed of horses.][2]

Many horses are taught to come to their master's assistance, to run down anyone who threatens them with a naked sword and to kick and to bite all those who make or come straight for them: but they succeed in doing more harm to friends than foes. Moreover you cannot pull them off when

---

1. *Destrier* does indeed derive from the Latin for right-hand (*dexter*).
2. Livy, XXIII, xxix.

you want to once they have become engaged: you have to wait and see what happens in the fight. It turned out disastrously for Artibius, the general of the Persian army, to have been mounted on a horse which had been trained in such a school when he was fighting hand to hand against Onesilus the King of Salamis: that horse was the cause of his death, his knife-bearing equerry slashing it between its shoulderblades just as it was rearing up over its master.[3]

The Italians tell how our King was saved[4] at the battle of Fornova when his horse trampled down several enemies who were pressing him hard; without that, he would have been killed. That was a great stroke of luck, if it is true.

['95]   The Mamelukes boast of having the most skilful horses of any knights in the world: they say that their nature and training are such that they can be brought to identify and recognize the enemy against whom they are to charge using teeth and hoofs, following the word of command or the signal given to them. They similarly pick up in their mouths lances and darts lying on the ground and give them to their masters when he tells them to.[5]

[A]   Among other outstanding qualities both Caesar and Pompey the Great were said to be fine horsemen; and Caesar is said in his youth to have ridden, bareback and without bridle, at full gallop with his hands behind his back. You could say that, just as Nature intended both that great person and Alexander as well to be miracles within the art of war, she also took pains to see that they should be equipped in ways which surpass the natural order: as everyone knows, Alexander's horse Bucephalus had a head somewhat like a bull's; would allow nobody but its master to mount it; would allow only him to train it; was granted honours at its death; and that a city was built in its name. Caesar also had a horse which had forefeet like a man's, the horn of its hoofs being divided into toes; it too could be ridden by no one but Caesar who, after its death, dedicated a statue of it to Venus.[6]

Once I am in the saddle I never willingly dismount, for, whether well or ill, I feel better in that position.   [C]   Plato recommends it as good for

3. Herodotus, *History*, VIII.
4. '95: King *Charles* was saved . . .
   (According to Bishop Paolo Giovio, *Historiae sui temporis*.)
5. The [C] text of Bordeaux is damaged here. It is slightly different, where readable, from the '95 posthumous text given here.
6. Aulus Gellius, *Attic Nights*, V, 11; Suetonius, *Life of Caesar*.

your health, [A] and Pliny says it is good for your stomach and your joints.[7] But since we have got this far, let us press on.

In Xenophon we can read the enactment[8] forbidding anyone with a horse to go on foot. Trogus and Justinus say that the Parthians customarily rode their horses not only to war but to all their public and private engagements: trading, discussing, conversing and simply going out for pleasure, they add that the most striking difference between the freemen and serfs among them is that one lot rides and the other lot walks: [C] a practice conceived and enacted by King Cyrus.[9]

[A] There are several examples in Roman history of captains (Suetonius mentions it particularly of Caesar) who would order their horsemen to dismount when they were hard pressed, to remove from the soldiers any hope of flight – [C] and also for the advantage they expected from fighting on foot – *'quo haud dubie superat Romanus'* [in which the Roman undoubtedly excels], as Livy says.[10]

All the same, the first precautionary measure they used to take to curb any rebellion among conquered peoples was to confiscate their arms and their horses: that is why we so often find in Caesar: *'arma proferri, jumenta produci, obsides dari jubet'* [he orders them to surrender their weapons, hand over their horses and deliver their hostages]. To this day the Grand Seigneur allows no Christian or Jew under his rule to have his own horse.

[A] Our forebears, especially during the English wars, mostly fought on foot in all formal battles and in fixed encounters so as not to have to rely, where things as dear as life and honour were concerned, on anything but their own might, their stout hearts and their own limbs. You link – |C| no matter what Chrysanthus says in Xenophon[11] – [A] your own valour and fortune to that of your horse: its wounds and its death involve your own; its fear or its impetuosity make you too either cowardly or foolhardy; if it does not respond to bit or spur it is your honour which has to answer for it. That is why I do not find it strange that battles fought on foot should have been more bitter and more ferocious than those fought on horses:

7. Plato, *Laws*, VII, 789A ff.; Pliny, *Hist. nat.*, XXVIII, xiv.
8. '80: the enactment *of Cyrus*, forbidding . . .
9. Xenophon, *Cyropaedia*; Justinus, *Historia* (an extract of Trogus Pompeius).
10. Livy, IX, xxii.
11. In the *Cyropaedia* he praises the role of cavalry.

[B]   *Cedebant pariter, pariterque ruebant*
Victores victique, neque his fuga nota neque illis.

[Locked together they yielded ground; locked together, they advanced, both victor and vanquished; neither side knew the meaning of flight].[12]

[C]   Their battles were far better contested than ours are: nowadays we only have routs: *'primus clamor atque impetus rem decernit'* [the first yell and the first onslaught decide the battle].[13]

[A]   Anything which we invite to share our great hazards with us must, as far as is feasible, remain under our control: so I would always advise anyone to choose the shortest weapons and those which we can be most answerable for. It is far more likely that we can rely on the sword we hold in our hand than on a bullet which is discharged from a pistol, since that pistol comprises several elements, the powder, the flint and the striker; if the least of them fails then so does your fortune.

[B]   When your blow has to travel through the air you are less sure of your aim.

> *Et quo ferre velint permittere vulnera ventis:*
> *Ensis habet vires, et gens quæcunque virorum est,*
> *Bella gerit gladiis.*

[They let the winds decide where their wounds are made. The soldier's weapon is the blade: the custom of all manly peoples is to make war with the sword.][14]

[A]   But as for the pistol, I will speak of it more fully when I compare the arms of former times with our own.[15] Except for the deafening noise – and we have all been broken in to that – it is an ineffectual weapon and I hope we shall   [C]   one day   [A]   give up using it.

[C]   The Italians of old employed a fiery projectile which was indeed most formidable: what they called a PHALARICA was a kind of javelin with a three-foot iron tip, enough to go right through a man in armour; it was hurled either by hand on the field of battle or, when defending places under siege, by catapult-machines. Its shaft was draped in wadded flax soaked in oil and pitch which caught fire as it flew through the air; it stuck to the body or shield and made it impossible for you to use your limbs or

12. *Aeneid*, X, 756–7.
13. Livy, XXV, xli.
14. Lucan, *Pharsalia*, VIII, 384–6.
15. The chapter in which this was treated was stolen by a manservant. (Cf. II, 9 and 37.)

your weapons. But it seems to me that, once the armies were joined in battle, these weapons would cause just as much trouble to the attackers and that a battlefield strewn with those blazing shafts must have been an equal hazard to all in the melee –

> *magnum stridens contorta phalarica venit*
> *Fulminis acta modo.*

[the whirling phalarica hissed through the air and struck like a flash of lighting.][16]

They had other weapons in which they were skilled through practice. Unbelievable though they may seem to us because we have no experience of them, they compensated for their lack of our bullets and gunpowder. Their javelins were hurled with such force that they went through two men at a time, stitching them together, shields and all, as with a needle. The shot from their slings were no less accurate than our bullets and carried just as far: '*Saxis globosis funda mare apertum incessentes: coronas modici circuli, magno ex intervallo loci, assueti trajicere: non capita modo hostium vulnerabant, sed quem locum destinassent.*' [Being practised in hurling their round stones over the open sea and hitting tiny circles a long way off, they could not only wound their enemies in the head but in any part of the head that they chose.]

Their battering-pieces made as much din as our own weapons: '*Ad ictus mænium cum terribili sonitu editos pavor et trepidatio cepit.*' [Fear and panic seized the inhabitants at the terrifying noise of their walls being battered].

Our Gaulish cousins in Asia hated those treacherous flying weapons, trained as they were to fight most courageously hand to hand: '*Non tam patentibus plagis moventur: ubi latior quam altior plaga est, etiam gloriosius se pugnare putant: idem, cum aculeus sagittæ aut glandis abditæ introrsus tenui vulnere in speciem urit, tum, in rabiem et pudorem tam parvæ perimentis pestis versi, prosternunt corpora humi.*' [They are not so much moved by the size of their wounds: they think they have fought with all the more glory when their wounds are wide and deep: consequently when an arrow-head or a bolt from a sling buries itself in their flesh and leaves only a small hole in their skin, the very idea of dying from so trivial a wound drives them mad with shame and they roll about on the ground.][17] A description close indeed to a shot from a harquebus.

Those two thousand Greeks in their famous prolonged retreat came

16. Virgil, *Aeneid*, IX, 704–5.
17. Livy, XXXVIII, xxix; v; xxi. (Both the Ancient Galatians and the Turks were believed to be cousins of the French.)

across a nation which did them wondrous harm with great powerful bows
shooting arrows so long that you could grab them up and hurl them back
like javelins: they could go right through a shield or a man in armour. The
war-machines which were invented by Dionysius, the Tyrant of Syracuse,
for launching huge heavy spears and rocks of a horrifying size with great
force over a huge distance were much like our own inventions.[18]

[A] I must not overlook the amusing way of sitting on [C] his
mule [A] adopted by a certain Master Pierre Pol, Doctor of Theology.
Monstrelet tells that he used to go about the city of Paris riding sidesaddle
like the ladies. Elsewhere he says that the Gascons used to have terrifying
horses trained to turn about at the gallop; the French, the Picards, the
Flemings and the men of Brabant, not having seen anything like it,
thought this was quite miraculous.

Those are his own words.[19]

Caesar, writing of the Swedes, says that during cavalry engagements
they often leap to the ground and fight on foot, having taught their horses
not to budge from a spot which they quickly run back to whenever there
is need. He adds that, according to their customs, nothing is so cowardly or
so base as using saddles or padded cushions; since they despise those who do
so, even a tiny group of them are never afraid of attacking a largish
number of men so mounted.[20]

[B] I used to marvel at seeing horses trained to do all sorts of
manoeuvres at the touch of a wand while their reins drooped loose below
their ears: that was current practice among the Massilians who rode their
horses without saddle or bridle:

> Et gens quæ nudo residens Massilia dorso
> Ora levi flectit, frænorum nescia, virga.
> [C]   Et Numidæ infræni cingunt;

[The people of Massilia, seated on their horses' naked backs, know nothing of
bridles: they guide them with a light rod. They were surrounded by Numidians
who use no reins;][21]

---

18. Xenophon, *Anabasis* (the Greek retreat from Asia); Diodorus Siculus for the
huge catapults.

19. '80: on his *horse* adopted . . .

'80: words. *I do not know what manoeuvre this might be, unless it were one of our
'passades'.* Caesar . . . (Both anecdotes from the *Chroniques d'Enguerran de Monstrelet*,
which continue those of Froissart.)

20. Caesar, *Gallic Wars*, IV (of the Suevi of north-east Germany).

21. Virgil, *Aeneid*, IV, 41–3.

'*equi sine freni, deformis ipse cursus, rigida cervice et extento capite currentium*' [their horses, which are not bridled, lope along; their necks are held stiff and their heads are stretched forwards as though they were running].[22]

[A]   King Alfonso – the one who first founded the Chivalric Order of the Band (or Scarf) – forbade the Knights, among other rules, ever to ride he-mule or she-mule on pain of a fine of one silver mark; I have just learned that fact from Guevara's so-called *Golden Letters* – those who gave them that name judged them very differently than I do.   [C]   And *Il Corteggiano* states that it was formerly a disgrace for a gentleman to ride one.[23] Yet, on the contrary, the nobler the Abyssinians are and the more closely related they are to Prester John their ruler, the more they esteem it an honour to ride on a mule.

According to Xenophon, the Assyrians always hobbled their horses in their stables, so ferocious were they and unpredictable; it took so long to unhobble and harness them that the Assyrians never stayed in an encampment unless surrounded by ditches and ramparts lest, under war conditions, the delay should act against them if they were surprised by their enemies and taken unprepared. Xenophon's Cyrus, such a past-master in such matters, treated his horses as comrades and never ordered them to be fed before they had earned it by sweating through some exercises.[24]

[B]   When pressed by necessity in war, the Scythians drew blood from their horses and drank it for nourishment:

> *Venit et epoto Sarmata pastus equo.*
>
> [Then comes the Sarmatian, fed on draughts from his horses.][25]

When the men of Crete were besieged by Metellus they were so short of anything to drink that they were forced to use their horses' urine.[26]

[C]   To demonstrate how much more economically the Turkish armies are managed and maintained than ours are, they say that not only do their soldiers drink nothing but water and eat nothing but rice and salt-meat ground to powder (each soldier easily carrying his ration for a month), but, like Tartars and Muscovites, they also know how to live on the blood of their horses, which they salt.

22. Livy, XXXV, xi.
23. Both the letters and the *Book of Marcus Aurelius* by Bishop Antonio de Guevara were termed 'Golden'. Castiglione's *The Book of the Courtier* was written in Italian for the Court of Francis I.
24. Xenophon, *Cyropaedia*, III, iii.
25. Martial, *Epigrams*: *Spectacula* III, 4.
26. Valerius Maximus, VII.

[B]   When the Spaniards made their landfall, those new people of the
Indies thought that both the men and the horses were either gods or
animate creatures of a nobler or higher nature than theirs. When those
Indians were defeated some, coming to seek peace and pardon from the
men, brought offerings of gold and food which they did not omit to
offer to the horses as well, addressing speeches to them exactly as to the
humans, interpreting their whinnying as the language of compromise and
truce.

In the East Indies in ancient times the first degree of honour, the King's,
was to ride an elephant; the second, to ride in a coach drawn by four
horses; the third, to ride a camel, and the last and basest degree was to ride
or be drawn by a single horse.

[C]   One of our contemporary writers says that he saw in that region
lands where they ride on oxen equipped with small packsaddles, stirrups
and bridles; he found them comfortable to sit on.[27]

When Quintus Fabius Maximus Rutilianus was fighting the Samnites,
he saw that his horsemen, after two or three charges, had failed to
penetrate the enemy battalion; he decided that they should unbridle their
horses and dig in their spurs as hard as they could; since nothing could stop
them, they opened a gap for his foot-soldiers right through the scattered
men and weapons, so achieving a most bloody defeat.

Quintus Fulvius Flaccus gave similar orders when fighting against the
Celtiberians: '*Id cum majore vi equorum facietis, si effrenatos in hostes equos
immittitis; quod sæpe romanos equites cum laude fecisse sua, memoriæ proditum est.
Detractisque frenis, bis ultro citroque cum magna strage hostium, infractis omnibus
hastis, transcurrerunt.*' ['The shock of your horses will be greater if before
throwing them against the enemy you take off their bridles. We recall that
Roman horsemen have often done that with great honour.'] They removed
the bridles and charged through the enemy and back again, slaughtering
many and shattering all their lances.[28]

[B]   In former days the Duke of Muscovy owed the following mark of
respect to the Tartars when they dispatched ambassadors to him: he went
to meet them on foot and presented them with a bowl of mare's milk
(which is a delicacy for them); and if a drop of it fell on one of their horse's
manes as they drank it he was obliged to lick it off with his tongue.

The army which the Emperor Bajazet sent to Russia was overwhelmed

27. Paolo Giovio, *Disciplinae Turcae militis*; Lopez de Gomara, *Historia de Capitano
Don Ferdinando Cortes*; Flavius Arrianus, *De rebus gestis Alexandri Magni*.
28. Livy, VIII, xxx; XL, xl.

by such a dreadful snowstorm that many sought to shelter themselves from the cold and to save their lives by slaughtering their horses, slitting open their bellies and crawling quickly inside to enjoy their vital heat.[29]

[C] When Bajazet was broken in battle by Tamburlane he would have saved himself as he was speeding away on his arab mare if he had not been forced to let her drink her fill when fording a stream; that made her so limp and so shivery that his pursuers easily caught up with him. It is certainly said that horses are weakened by letting them piss, but I would have thought that such drinking would have refreshed her and given her more strength.

When Croesus was skirting the city of Sardis he found some pastures full of snakes which his pack-horses gobbled up; that, says Herodotus, was a bad omen for his enterprise.[30]

[B] An 'entire horse' is a stallion with ears and mane; no other will pass muster. After defeating the Athenians in Sicily, the Spartans were returning in pomp from their victory to the city of Syracuse when, among other insults, they cut the manes off the defeated enemy's horses and led them like that in their triumph.[31]

Alexander fought a people called the Dahae; they went armed into battle riding two to a horse; in the melee one of them jumped down; they then took it in turns to fight mounted or on foot.[32]

[C] No other nation surpasses us, I think, in skill and grace when riding. In the idiom of our language, 'a good horseman' seems to refer not so much to skill as to courage.

The man known to me who was most expert, reliable and elegant at training a horse was, to my taste, the Sieur de Carnavalet, whose skills were at the service of King Henry II.

I have seen a man ride at speed with both feet in the saddle, throw the saddle to the ground, return to pick it up, strap it on again and sit in it, with the reins hanging loose as he galloped. He rode over a hat, shot backwards at it with his bow, hitting it repeatedly; he picked up anything he liked, with one foot on the ground and the other still in the stirrup — and many other tricks by which he earned his living.

29. Jan Herbut (tr. F. Baudouin), *Histoire des Roys de Pologne*, Paris, 1573 (Latin edn, Basle, 1571).
30. Nicolas Chalcocondylas, *Histoire de la décadence de l'Empire grec* (tr. V. de Vigenère); Herodotus, *History*, I, lxxxviii.
31. Plutarch, *Life of Nicias*.
32. Quintus Curtius, VII, viii.

[B]   There have been seen in my own time, in Constantinople, two men on one horse, galloping at full speed and taking it in turns to jump down to the ground then up to the saddle; another put bridle and harness on his horse using nothing but his teeth; another, riding astride two horses, one foot on each saddle, carried a second man on his shoulders while going full tilt; that second man, standing erect, shot accurately from his bow as they raced along; several riders stood on their heads in the saddle with their feet in the air between the points of scimitars fixed to their harness.[33]

When I was a boy the Prince of Sulmona in Naples, while putting an untrained horse through all sorts of manege, used to hold pieces of eight under his knees as if they had been nailed there   [C]   to show the firmness of his seat.

---

33. Henry Porsius and George Lebelski, *Histoire de la guerre de Perse* [of 1578], *Avecques la description des jeux . . . à Constantinople* [of 1582] (Paris, 1583).

49. On ancient customs

===

[This is doubtless one of the chapters written early: its lists of ancient customs became the raw material for deeper reflections on the relativity of much that passes as natural in various societies. As it stands it has something in common with earlier works, such as the Officina *('Workshop') of Ravisius Textor or the* Ancient Readings *of Richerius Rhodiginus, which, with their successors, were appreciated especially by readers who had access to few books and enjoyed compendia.]*

[A]   I am prepared to forgive our own people for having no other model or rule of perfection but their own manners and behaviour, for it is a common failing not only of the mob but of virtually all men to set their sights within the limitations of the customs into which they were born. I can accept that a man, if he met them, should find the appearance of Fabricius or   [C]   Laelius   [A]   barbaric,[1] seeing that they did not wear clothing tailored to our fashion, but I do complain of his singular lack of judgement if he lets himself be so thoroughly taken in and blinded by the authority of contemporary modes that he is capable of changing his mind and his opinions every four weeks if fashion demands it, and of making mutually exclusive judgements about himself. Even if the waistband of his doublet were worn high up the chest he would heatedly maintain that that is where it ought to be; then, when it was lowered a few years later down below his thighs, he would ridicule the other fashion and find it absurd and intolerable. Today's fashion leads him immediately to condemn the old one with so great and universal a certainty that you could say there was some species of mania making his mind do somersaults.

Because changes of cut are so quick and sudden and the inventiveness of all the tailors in the world insufficient to provide enough novelty, it is inevitable that styles once despised should come back into fashion, only to fall out of fashion again a little while later.

I also complain that one and the same mind should, for a period of some fifteen or twenty years, hold with such unbelievable and frivolous inconstancy two or three opinions which are not merely divergent but

---

1. '80: or *Scipio* barbaric . . .

incompatible. [C] None of us is so clever as not to be made a mockery of by such contradictions, allowing our sight and our insight to be dazzled without realizing it.

[A] I intend here to make a pile of a few of the customs of the Ancients which I have stored in my memory, some of them resembling ours, some of them different, so that by keeping in mind the continual changes in human affairs, our judgements on them may be more firm and more enlightened.

Fighting with cloak and sword (as we call it) was already the custom among the Romans; Caesar says: '*Sinistris sagos involvunt, gladiosque distringunt.*' [They wrap their garments over their left arms and draw their swords.] And he already comments on one of our national failings (which is still with us): that we stop any travellers we meet on the road, require them to tell us who they are and take it as an insult and as a pretext for a quarrel if they decline to answer.[2]

The Ancients used to bath every day before dinner; it was as usual as our washing our hands; at first they only bathed their arms and legs but subsequently (by a custom which lasted several centuries among most of the peoples in the Roman world) they stripped naked and washed in a mixture of water and perfume (so that to mean that someone lived very simply they would say that he washed in water). The more elegant and refined among them perfumed their bodies three or four times a day. They removed all their hair with tweezers (just as French women, some time ago now, started to pluck out the hairs on their forehead) –

> *Quod pectus, quod crura tibi, quod brachia vellis;*

[You tweeze out your hairs from chest, thighs and arms;] –

despite having had special ointments for that purpose:

> *Psilotro nitet, aut arida latet oblita creta.*

[She glistens with a depilatory and hides behind a mask of dry plaster.][3]

They loved to lie softly and took sleeping on a palliasse as a sign of toughness. They reclined on beds for their meals, in more or less the same posture as the Turks of today:

---

2. Julius Caesar, *Gallic Wars*, IV, v.
3. Martial, II, lxii, 1; VI, xciii, 9.

Inde thoro pater Æneas sic orsus ab alto.

[Then, from his lofty bed, Father Aeneas spoke.][4]

It is said of Cato the Younger that, after the battle of Pharsalia, grieving for the lamentable state of public affairs, he adopted a more austere way of life and always took his meals sitting down. They kissed the hands of the great to honour them and to show their devotion; friends greeted each other with a kiss as Venetians do now:

Gratatusque darem cum dulcibus oscula verbis.

[I would wish you well with kisses and sweet words.][5]

[C] When greeting a great man or begging his favour, they would tap him on the knee. The brother of Crates, Pasicles the philosopher, instead of placing his hand on the knee placed it on the genitals. The great man thus addressed pushed him rudely aside. 'Come now,' Pasicles replied, 'are they not as much yours as your knees are?'[6]

[A] Like us they ate fruit at their last course. They wiped their arses with a sponge (we can leave silly superstitions about words to the women). That is why SPONGIA is an uncouth word in Latin. The sponge was fixed to the end of a stick: that is shown by the account of the man who was being led off to be thrown to the beasts in the sight of the plebs and who asked permission to answer a call of nature; having no other means of killing himself he thrust stick and sponge down his throat and suffocated.[7]

They wiped their cocks with perfumed wool after they had had a go with them:

At tibi nil faciam, sed lota mentula lana.

[I'll do nothing to you till you've washed your tool with wool.][8]

At the street-corners in Rome they kept jars and demijohns for passers-by to piss in:

Pusi sæpe lacum propter, se ac dolia curta
Somno devincti credunt extollere vestem.

[And sleepy children often dream that they are lifting their robes and pissing in the public urinals.][9]

---

4. Virgil, *Aeneid*, II, 2.
5. Ovid, *Ex Ponto*, IV, ix, 13.
6. Plutarch, *Life of Crates*.
7. Seneca, *Epist. moral.*, LXX, 20.
8. Martial, XI, lxviii, 11.
9. Lucretius, IV, 1020–1.

They used to eat between meals. In summer there were men who sold snow for cooling the wine. Some even used snow in winter too, not finding their wine cool enough even then. Great men had their servers to carve and to bring them their wine, as well as fools to amuse them. In winter dishes were brought to the table on food-warmers; they also had portable kitchens –   [C]   I have seen some myself –   [A]   in which they carried about everything needed for preparing a meal:

> *Has vobis epulas habete lauti;*
> *Nos offendimur ambulante cæna.*

[Keep your old feasts you gluttons; we dislike your portable food.][10]

They had fresh clear water flowing along small aqueducts through the basements beneath them in which they kept live fish which guests chose and caught in their hands and then had them prepared to their taste. Fish always did enjoy their present privilege of having the great take trouble to find out how to prepare them; their flesh has a more refined taste than meat, at least to my liking.

We certainly do our utmost to equal the Ancients in every sort of ostentation, in debauchery and in the devising of gratifications, in comforts and in luxuries, for our wills are as vitiated as theirs were but our ingenuity cannot bring it off. Our powers are no more capable of competing with them in vice than in virtue, both of which derive from a vigour of mind which was incomparably greater in them than in us: the weaker the souls, the less able they are to do anything really good or really bad.

For them the middle place at table was the most honoured. No order of precedence was entailed in putting names first or second, either in speech or writing; that can be seen clearly from their books: they will as readily say 'Oppius and Caesar' as 'Caesar and Oppius', or indifferently say, 'me and you' or 'you and me'. That is why I was once struck by a passage of the *Life of Flaminius* in our French Plutarch in which, treating of the rivalry between the Aetolians and the Romans over who deserved the glory for the victory which they had jointly won, Plutarch attaches some weight to the fact that in the songs of the Greeks the Aetolians were named before the Romans – unless, that is, there is some ambiguity in the French.

Ladies in the public baths would receive men and even got their men-servants to rub oil into them:

10. Martial, VII, xlviii, 4.

Inguina succinctus nigra tibi servus aluta
Stat, quoties calidis nuda foveris aquis.

[Whenever you take your naked bath a slave stands by, girt with a black leathern apron above his groin.][11]

They sprinkled a kind of powder over themselves to stop the sweat.

Sidonius Apollinaris says that the ancient Gauls wore their hair long on the front of their heads and shaved close at the neck; that is precisely the hairstyle which has been brought back into fashion by the slack and womanish mode of our own century.

The Romans used to pay their boatmen for their ferries as soon as they came on board: we do it only when we have reached harbour:

dum as exigitur, dum mula ligatur,
Tota abit hora.

[what with collecting the fares and tying up the mules, a whole hour is wasted.][12]

Women used to sleep on the side of the bed nearer the wall; that is why Caesar was called '*spondam Regis Nicomedis*' [King Nicomedes' wall-side bed-frame].[13]

[B] They took a gulp of breath when they drank. They watered their wine:

quis puer ocius
Restinguet ardentis falerni
Pocula prætereunte lympha?

[what slave-boy will swiftly temper the bowls of fiery Falernian wine with water from the flowing stream?][14]

Those insolent looks we see on our lackeys' faces were known to them too:

O Jane, a tergo quem nulla ciconia pinsit,
Nec manus auriculas imitata est mobilis albas,
Nec linguæ quantum sitiet canis Apula tantum!

[O Janus, behind whose back no mocking gestures are made and no quick hands form signs of asses' ears and no tongue is poked out, long as a thirsty dog's from Apulia!][15]

11. Martial, VII, xxxv, 1–2.
12. Horace, *Satires*, I, v, 13–14.
13. Suetonius, *Life of Caesar*, xlix.
14. Horace, *Odes*, II, xi, 18–20.
15. Persius, *Satires*, I, 58–60.

The women of Argos and of Rome used to wear white for mourning, as was once the custom of our women – and would be still if I were listened to. But there are whole books on that question.[16]

16. Cf. Rabelais, *Gargantua*, TLF, IX, 'On what is signified by the colours white and blue'.

50. On Democritus and Heraclitus

[Contrasts between Heraclitus who wept over the tragic situation of Man and Democritus who laughed at Man's comic predicament go back to Antiquity and were renewed in the early sixteenth century by Antonio Fregoso in The Laughter of Democritus and the Tears of Heraclitus. *Montaigne like Rabelais sees laughter as the 'property' of Man, his specific characteristic; like Rabelais too (in the Prologue to the* Third Book of Pantagruel*) he sees Diogenes the Cynic as a guide to amused satirical comment and condemnation. This chapter is concerned to show Montaigne's developing conception of what his Essays really are (cf. I, 8, 'On idleness' and II, 37, 'On the resemblance of children to their fathers'). Some of the ideas found here are developed in 'An apology for Raymond Sebond' (II, 12).]*

[A] Our power of judgement is a tool to be used on all subjects; it can be applied anywhere. That is why I seize on any sort of occasion for employing it in the assays I am making of it here. If it concerns a subject which I do not understand at all, that is the very reason why I assay my judgement on it; I sound out the ford from a safe distance: if I find I would be out of my depth, then I stick to the bank: the realization that I cannot get further across is one effect of its action; indeed, it is the effect that judgement is especially proud of. Sometimes, when the subject is trivial and vain, I assay whether my judgement can find anything substantial in it, anything to shore it up and support it. Sometimes I employ it on some elevated, well-trodden subject where it can discover nothing new, since the path is so well beaten that our judgement can only follow in another's tracks. In that case it plays its role by selecting what appears the best route: out of hundreds of paths it says this one or that one is the best to choose.[1]

[C] I take the first subject Fortune offers: all are equally good for me. I never plan to expound them in full for I do not see the whole of anything:

---

1. '80: choose. *Meanwhile I leave* Fortune *to furnish me with subjects. Since* all are equally good for me, *and I do not undertake to treat them fully or to scrape the barrel; of the hundreds of features which each of them* has; I take the one *which pleases me: I grasp them preferably by some extraordinary aspect: I could well select richer, fuller ones if I had some other objective. Every act is appropriate for making ourselves known:* that same soul of Caesar's . . .

neither do those who promise to help us to do so! Everything has a hundred parts and a hundred faces: I take one of them and sometimes just touch it with the tip of my tongue or with my fingertips, and sometimes I pinch it to the bone. I jab into it, not as wide but as deep as I can; and I often prefer to catch it from some unusual angle. I might even have ventured to make a fundamental study if I did not know myself better. Scattering broadcast a word here, a word there, examples ripped from their contexts, unusual ones, with no plan and no promises, I am under no obligation to make a good job of it nor even to stick to the subject myself without varying it should it so please me; I can surrender to doubt and uncertainty and to my master-form, which is ignorance.

Anything we do reveals us. [A] The same soul of Caesar's which displayed herself in ordering and arranging the battle of Pharsalia is also displayed when arranging his idle and amorous affrays. You judge a horse not only by seeing its paces on a race-track but by seeing it walk – indeed, by seeing it in its stable.

[C] The soul has her lower functions: anyone who does not know her in those does not know her thoroughly. And you may perhaps get to know her better when she is ambling along. It is in her loftier sites that the winds of passion batter her about. Besides she throws herself wholly into every matter, and never treats more than one at a time: moreover she treats it not its way but her way.

Things external to her may have their own weight and dimension: but within inside us she gives them such measures as she wills: death is terrifying to Cicero, desirable to Cato, indifferent to Socrates. Health, consciousness, authority, knowledge, beauty and their opposites doff their garments as they enter the soul and receive new vestments, coloured with qualities of her own choosing: brown or green; light or dark; bitter or sweet, deep or shallow, as it pleases each of the individual souls, who have not agreed together on the truth of their practices, rules or ideas. Each soul is Queen in her own state. So let us no longer seek excuses from the external qualities of anything: the responsibility lies within ourselves. Our good or our bad depends on us alone. So let us make our offertories and our vows to ourselves not to Fortune: she has no power over our behaviour; on the contrary our souls drag Fortune in their train and mould her to their own idea.

Why shall I not judge Alexander chatting and drinking his fill at his table? Or if he were playing chess, what mental chord is not touched and employed in that silly childish game, which I hate and avoid because there is not enough *play* in it, feeling ashamed to give it such attention as would

suffice to achieve something good: Alexander was not more preoccupied with planning his magnificent expedition into India, nor is this other man with explaining a knotty passage on which depends the salvation of the human race. Notice how our soul gives weight and depth to that silly pastime: how all her sinews are strained; how amply she provides each of us with the means of knowing himself or of judging himself aright. There is no other activity in which I can see and explore myself so thoroughly. What passion does not try us in that game: anger, vexation, hatred, impatience and an ambition to win such as carries the mind away – and that, in something where a more pardonable ambition would be to lose! For to show a rare and extraordinary excellence in frivolous pursuits is unworthy of a man of honour. And what I say in this example can also be said of all the others: every constituent of a man, each occupation, tells us about him and reveals him as well as any other.

[A]   Democritus and Heraclitus were both philosophers; the former, finding our human circumstances so vain and ridiculous, never went out without a laughing and mocking look on his face: Heraclitus, feeling pity and compassion for these same circumstances of ours, wore an expression which was always sad, his eyes full of tears.

[B]   *Alter*
Ridebat, quoties a limine moverat unum
Protuleratque pedem; flebat contrarius alter.

[One, whenever he put a foot over his doorstep, was laughing: the other, on the contrary, wept.][2]

[A]   I prefer the former temperament, not because it is more agreeable to laugh than to weep but because it is more disdainful and condemns us men more than the other – and it seems to me that, according to our deserts, we can never be despised enough. Lamentation and compassion are mingled with some respect for the things we are lamenting: the things which we mock at are judged to be worthless. I do not think that there is so much wretchedness in us as vanity; we are not so much wicked as daft; we are not so much full of evil as of inanity; we are not so much pitiful as despicable. Thus Diogenes who frittered about all on his own trundling his barrel and cocking a snook at Alexander,[3] accounting us as no more than flies or bags of wind, was a sharper and harsher judge (and consequently,

2. Juvenal, *Satires*, X, xxviii.
3. He asked Alexander the Great to get out of his light. On his trundling of his barrel, cf. the *Prologue* to the *Tiers Livre* of Rabelais.

for my temperament, a juster one) than Timon who was surnamed the misanthropist. For what we hate we take to heart. Timon wished us harm; passionately desired our downfall; fled our company as dangerous, as that of evil men whose nature was depraved. Diogenes thought us worth so little that contact with us could neither trouble him nor corrupt him: he avoided our company not from fear of associating with us but from contempt. He thought us incapable of doing good or evil.

Statilius' reply was of a similar character when Brutus spoke to him about joining in their plot against Caesar: he thought the enterprise to be just but did not find that men were worth taking any trouble over; [C] which is in conformity with the teaching of Hegesias (who said the wise man should do nothing except for himself, since he alone is worth doing anything for) and the teaching of Theodorus, that it is unjust that the wise man should hazard his life for the good of his country, so risking his wisdom for fools.[4]

Our own specific property is to be equally laughable and able to laugh.[5]

---

4. Plutarch, *Life of Brutus*; Diogenes Laertius, *Life of Aristippus*.
5. Laughter is the 'property' – the specific quality – of Man. Cf. Rabelais, *Gargantua*, preliminary poem.

51. On the vanity of words

===

[Montaigne, despite his own mastery of language, despised words and admired deeds or 'matter'. He showed this before he embarked on the Essays *in the dedicatory letter of his translation of Raymond Sebond's* Natural Theology, *addressed to his father. What Montaigne admired in ancient Sparta – and what he found lacking in his own day – was a genuine respect for action over rhetoric.]*

[A] In former times there was a rhetorician who said his job was to make trivial things seem big and to be accepted as such. [A1] He is a cobbler who can make big shoes fit little feet. [A] In Sparta they would have had him flogged for practising the art of lying and deception. [B] And I am sure that Archidamus their king did not hear without amazement the answer given by Thucydides when he asked him whether he was better at wrestling than Pericles: 'That,' Thucydides replied, 'would be hard to prove: for after I have thrown him to the ground in the match he persuades the spectators that he did not have a fall and is declared the winner.'[1] [A] Those who hide women behind a mask of make-up do less harm, since it is not much of a loss not to see them as they are by nature, whereas rhetoricians pride themselves on deceiving not our eyes but our judgement, bastardizing and corrupting things in their very essence. Countries such as Crete and Sparta which maintained themselves in a sound and regulated polity did not rate orators very highly.

[C] Ariston wisely defined rhetoric as the art of persuading the people; Socrates and Plato, as the art of deceiving and flattering; and those who reject this generic description show it to be true by what they teach. The Mahometans will not allow their children to be taught it because of its uselessness. And the Athenians, despite the fact that the practice of it was esteemed in their city, realizing how pernicious it was, ordained that the main part of it which is to work on the emotions should be abolished, together with formal introductions and perorations.[2]

---

1. Plutarch (tr. Amyot), *Dicts notables des Lacedaemoniens*, 209F; also listed among Erasmus' Spartan apophthegmata. Then, Plutarch, *Life of Pericles*.
2. Guillaume Postel, *Histoire des Turcs*; Quintilian, *Institutiones*, II.

[A] It is a means invented for manipulating and stirring up the mob and a community fallen into lawlessness; it is a means which, like medicine, is used only when states are sick; in states such as Athens, Rhodes and Rome where the populace, or the ignorant, or all men, held all power and where everything was in perpetual turmoil, the orators flooded in. And in truth few great men in those countries managed to thrust themselves into positions of trust without the help of eloquent speech: Pompey, Caesar, Crassus, Lucullus, Lentulus and Metellus all made it their mainstay for scrambling up towards that grandiose authority which they finally achieved, helped more by rhetoric than by arms, [C] contrary to what was thought right in better times. For Lucius Volumnius, making a public address in favour of the candidates Quintus Fabius and Publius Decius during the consular elections, declared, 'These are great men of action, born for war; they have Consular minds, uncouth in verbal conflict. Subtle, eloquent, learned minds are good but for Praetors, administering justice in the City.'[3]

[A] Rhetoric flourished in Rome when their affairs were in their worst state and when they were shattered by the storms of civil war, just as a field left untamed bears the most flourishing weeds.

It would seem that polities which rely on a monarch have less use for it than the others: for that animal-stupidity and levity which are found in the masses, making them apt to be manipulated and swayed through the ears by those sweet harmonious sounds without succeeding in weighing the truth of anything by force of reason – such levity, I repeat, is not so readily found in one individual man; and it is easier to protect him by a good education and counsel from being impressed by that poison. No famous orator has ever been seen to come from Macedonia or from Persia.

What I have just said was prompted by my having talked with an Italian who served as chief steward to the late Cardinal Caraffa until his death. I got him to tell me about his job. He harangued me on the art of feeding with a professional gravity and demeanour as though he were explaining some important point of Theology. He listed differences of appetite: the appetite you have when you are hungry, the one you have after the second and third courses; what means there are of simply satisfying it or of sometimes exciting it and stimulating it; how to govern the commonwealth of sauces, first in general then in particular, listing the qualities of every ingredient and its effects; the different green-stuffs in their season, the ones

3. Livy, X, xxii.

which must be served hot, the ones which must be served cold as well as the ways of decorating them and embellishing them to make them look even more appetizing. After all that he embarked upon how the service should be ordered, full of fine and weighty considerations:

> [B]   *nec minimo sane discrimine refert*
> *Quo gestu lepores, et quo gallina secetur!*

[For it is of no small importance to know how to carve a hare or a chicken!][4]

[A]   And all this was inflated with rich and magnificent words, the very ones we use to discuss the government of an empire. I was reminded of that man in the poem:

> *Hoc salsum est, hoc adustum est, hoc lautum est parum,*
> *Illud recte; iterum sic memento; sedulo*
> *Moneo quæ possum pro mea sapientia.*
> *Postremo, tanquam in speculum, in patinas, Demea,*
> *Inspicere jubeo, et moneo quid facto usus sit.*

['This is too salty; this has been burned; this needs to be properly washed; this is excellent – remember that next time.' I advise them carefully as far as my wisdom allows; finally I tell them, Demea, to polish the dishes until they can see their faces in them as in a mirror. I tell them the lot.][5]

Even the Greeks after all highly praised the order and arrangement which were observed in the banquet which Paulus Aemilius threw for them on his return from Macedonia; but I am not talking here of deeds but of words.[6]

I cannot tell if others feel as I do, but when I hear our architects inflating their importance with big words such as pilasters, architraves, cornices, Corinthian style or Doric style, I cannot stop my thoughts from suddenly dwelling on the magic palaces of Apollidon:[7] yet their deeds concern the wretched parts of my kitchen-door!

[B]   When you hear grammatical terms such as metonymy, metaphor and allegory do they not seem to refer to some rare, exotic tongue? Yet they are categories which apply to the chatter of your chambermaid.

[A]   It is a similar act of deception to use for our offices of state the same grandiloquent titles as the Romans did, even though they have no

---

4. Juvenal, V, 123–4.
5. Terence, *Adelphi*, III, iii, 71–5.
6. Plutarch, *Life of Paulus Aemilius*.
7. In *Amadis de Gaule*, II.

similarity of function and even less authority and power.[8] Similar too –
and a practice which will, in my judgement, bear witness one day to the
singular ineptitude of our century – is our unworthily employing for
anybody we like those glorious cognomens with which Antiquity honoured
one or two great men every few hundred years. By universal acclaim Plato
bore the name *divine*, and nobody thought to dispute it with him:[9] now
the Italians, who rightly boast of having in general more lively minds and
saner discourse than other peoples of their time, have made a gift of it to
Aretino, in whom (apart from a style of writing stuffed and simmering
over with pointed sayings, ingenious it is true but fantastical and far-
fetched, and apart from his eloquence – such as it is) I can see nothing
beyond the common run of authors of his century, so far is he from even
approaching that 'divinity' of the Ancients.

And the title Great we now attach to kings who have nothing beyond
routine greatness.

8. Cf. the language of Montaigne's title to Roman Citizenship cited in III, 9, 'On
vanity'.
9. But cf. the closing pages of III, 13, 'On experience'.

52. On the frugality of the Ancients

[*Frugality in public and private matters was admired by the sterner Ancients (cf. Seneca, Epistulae morales, I, 5, etc.). This is an example of one of the earlier compilations of Montaigne which failed to grow into a larger chapter.*]

[A]   Attilius Regulus, the commander-in-chief of the Roman Army in Africa, at the height of his reputation for his victories over the Carthaginians wrote to the Roman State saying that one of his ploughmen whom he had left in sole charge of his estates (which consisted of some seven acres of land all told) had run off with his farm equipment: he asked for leave to go home and see to things, lest his wife and children should suffer want. (The Senate decided to appoint another man to manage the property and to make good what had been stolen, and decreed that his wife and children should be cared for at public expense.)

The elder Cato, when returning as Consul from Spain, sold his working horse to spare the expense of shipping it back to Italy; and when he was Governor of Sardinia he made his inspections on foot; his retinue consisted of one officer-of-state bearing his robes and a sacrificial vessel; and most of the time he carried his baggage himself. He was proud of never having any clothing which cost more than ten crowns and of never having spent more than tenpence a day in the market; and as for his houses in the country, not one was pointed and plastered on the outside.

Scipio Aemilianus, after having had two Triumphs and two Consulships, went on an embassy with just ten servants. They say that Homer only had one; Plato, three; Zeno the head of the Stoic sect, not even one.

[B]   When Tiberius Gracchus went on an official government mission he was voted fivepence-halfpenny a day: he was then the highest man in Rome.[1]

---

1. The anecdotes are taken from Valerius Maximus IV, Plutarch's *Lives* of Cato the Censor and of Tiberius Gracchus, and Seneca's *De consolatione ad Albinam*.

53. On one of Caesar's sayings

[*This short chapter, concerned as it is with that* contextura corporis, *that 'bodily structure', which interested Lucretius, is one of many which contributed thoughts and ideas to 'An apology for Raymond Sebond' (II, 12).*]

[A]   If we were occasionally to linger over an examination of ourselves and were to save the time which we spend on finding out about others and in learning about externals so as to use it to make soundings of ourselves, we would soon realize how this structure of ours is made up of weak and deficient elements. Is it not a peculiar sign of our imperfections that we cannot settle our happiness on any single thing, and that even in our wishes and our thoughts we are incapable of choosing the things which we need? Corroboration of this fact is provided by that great dispute which has ever divided philosophers over Man's sovereign good: it still goes on, and will go on for ever, with no conclusion and no agreement:

> [B]   *dum abest quod avemus, id exuperare videtur*
> *Cætera; post aliud cum contigit illud avemus,*
> *Et sitis æqua tenet,*

[as long as we do not have it, the object of our desire seems greater than anything else: as soon as we enjoy it, we long for something different with an equal craving.][1]

[A]   No matter what falls within our knowledge, no matter what we enjoy, it fails to make us content and we go gaping after things outside our knowledge, future things, since present goods never leave us satisfied – not in my judgement because they are inadequate to satisfy us but because we clasp them in a sick and immoderate grip:

> [B]   *Nam, cum vidit hic, ad usum quæ flagitat usus,*
> *Omnia jam ferme mortalibus esse parata,*
> *Divitiis homines et honore et laude potentes*
> *Affluere, atque bona natorum excellere fama,*
> *Nec minus esse domi cuiquam tamen anxia corda,*

1. Lucretius, III, 1095–7.

Atque animum infestis cogi servire querelis:
Intellexit ibi vitium vas efficere ipsum,
Omniaque illius vitio corrumpier intus,
Quæ collata foris et commoda quæque venirent.

[For when Epicurus saw that almost everything necessary for Man's life is at his disposal; when he saw men who were replete with honour and wealth and reputation and who were proud of their sons' good fame, not one of whom was not full of inner anxiety or whose mind was not racked by grievous lamentations: then he realized that the fault was in the vessel itself, corrupting internally any good which came in to it from the outside.][2]

[A] Our appetite lacks decision and is uncertain: it can neither have anything nor enjoy anything in the proper way. Man, reckoning that the defect lies in those things themselves, feeds to the full on other things which he neither knows nor understands, and honours and reveres them; as Caesar says: '*Communi fit vitio naturæ ut invisis, latitantibus atque incognitis rebus magis confidamus, vehementiusque exterreamur.*' [By a defect of nature common to all men, we place our trust, rather, in things unseen, hidden and unknown, and are terrified to distraction by them.][3]

2. Lucretius, VI, 9–17.
3. *Gallic Wars*, II, iv. (Until [C] this was accompanied by a French translation.)

54. On vain cunning devices

===

[Montaigne is brought to wonder what his Essays are worth and to draw an important distinction between good naïf Christians or good Christian mystics (who both make excellent believers) and the middling mediocre minds which do so much harm but might appreciate his Essays. The background of his distinction between good Christians and mediocre ones derives from the vital commonplaces of Christian 'folly'.]

[A] There are those kinds of cunning devices, frivolous and vain, through which a reputation is sought by some men, such as those poets who compose entire works from lines all beginning with the same letter; and we can see that by increasing or shortening the length of their lines the ancient Greeks would form poems of various shapes such as eggs, balls, wings and axe-heads.[1] Of such a kind was the art of the man who spent his time counting the number of ways in which he could arrange the letters of the alphabet and found that they came to that incredible number we can find in Plutarch.[2]

I agree with the opinion of the man to whom was presented another man who was an expert at throwing grains of millet so cleverly that they infallibly went through the eye of a needle; he was asked afterwards to bestow a reward for such a rare ability: whereupon he commanded – very amusingly and correctly, if you ask me – that the man who did it should be given two or three baskets of millet so that so fine a skill should not remain unpractised![3] It is a [C] wonderful [A] testimony of the weakness of Man's judgement that things which are neither good nor useful it values on account of their rarity, novelty and, even more, their difficulty.

At home we have just been playing at who can find most things which meet at extremes – such as *Sire*, which is the title given to the highest

1. Lines all beginning with the same letter were affected by some early Renaissance poets (the 'Grands Rhétoriqueurs'). Poems with lines of varying lengths arranged to make shapes were known in Late Antiquity. Herbert's 'Easter-Wings' is an example in English.
2. One hundred million two hundred thousand ways, according to Xenophon (Plutarch, tr. Amyot, *Propos de table*, 430C).
3. An example in Quintilian vulgarized by Castiglione, *Book of the Courtier* II, 31.

person in our State, the king, and also to common folk such as tradesmen but is never used for anyone in between. Women of the nobility are called *Dames*; middle-ranking women are called *Damoiselles*; and we use *Dames* again for the lowest women of all. [B] Canopies are hung over tables only in princely houses and in taverns.

[A] Democritus said that gods and beasts have senses more acute than men, who are at the stage in between. The Romans wore the same clothes for days of mourning and for festival-days. It is certain that extreme cowardice and extreme bravery disturb the stomach and are laxative. [C] The nickname of *Trembler* given to King Sancho XII of Navarre⁴ serves as a reminder that boldness can make your limbs shake just as much as fear. And the man whom his squires assayed to reassure by minimizing the dangers as they helped him into his armour and saw his flesh a-quiver said to them: 'You know me badly: if my skin realized where my heart was soon to take it, it would fall flat on the ground in a faint.'

[A] That incapacity which comes over us in the sports of Venus from lack of ardour or attraction can also do so from too ecstatic an ardour or too unruly a passion. Food can be roasted and cooked by extreme cold as well as extreme heat: Aristotle says that lead ingots will melt and turn liquid with the cold in a rigorous winter as readily as in an intensely hot summer. [C] The stages above pleasure and below pleasure can be filled with pain by both desire and satiety. [A] Animal-stupidity and wisdom converge in the way they feel and resist the misfortunes men must endure: wise men bully misfortune and master it: the others ignore it; the latter are on this side of misfortune so to speak: the former are beyond it; they first weigh and consider what misfortunes are and then judge them for what they are; they leap above them by the force of a vigorous mind; they despise them and trample them underfoot; they have souls so strong and so solid that when the arrows of Fortune strike against them they can only bounce back and be blunted, having met an obstacle which they cannot dent. Men of ordinary middling capacities are lodged between these two extremes, which is where men perceive adversities, feel them and find them unbearable. Babyhood and extreme old age meet in mental imbecility; so do avarice and profligacy, in their like desire to grab and acquire.

[B] It may be plausibly asserted that [C] there is an infant-school ignorance which precedes knowledge and another doctoral ignorance which comes after it, an ignorance made and engendered by knowledge just as it

---

4. Or rather to Garcia V, son of Sancho Garcia.

unmade and slaughtered the first kind.   [B]   Good Christians are made
from simple minds, incurious and unlearned, which out of reverence and
obedience have simple faith and remain within prescribed doctrine. It is in
minds of middling vigour and middling capacity that are born erroneous
opinions, for they follow the apparent truth of their first impressions and
do have a case for interpreting as simplicity and animal-stupidity the sight
of people like us who stick to the old ways, fixing on us who are not
instructed in such matters by study. Great minds are more settled and see
things more clearly: they form another category of good believers; by long
and reverent research they penetrate through to a deeper, darker light of
Scripture and know the sacred and mysterious secret of our ecclesiastical
polity. That is why we can see some of them arrive at the highest level via
the second, with wondrous fruit and comfort, reaching as it were the
ultimate bounds of Christian understanding and rejoicing in their victory
with alleviation of sorrow, acts of thanksgiving, reformed behaviour and
great modesty. I do not intend to place in that rank those other men who,
to rid themselves of the suspicion of their past errors and to reassure us
about themselves, become extremists, men lacking all discretion and unjust
in the way they uphold our cause, besmirching it with innumerable
reprehensible acts of violence.

   [C]   The simple peasants are honest people; honest, too, are
philosophers, insofar as we have any nowadays with natures strong and
clear, enriched by wide learning in the useful sciences. Half-breeds who
have turned with contempt from the first state (illiterate ignorance) and
who are incapable of reaching the other (their arses between two stools, like
me and lots of others) are dangerous, absurd and troublesome: such men
bring disturbances to the world. That is why, for my part, I draw back as
far as I can into that first and natural state, which I had vainly made an
assay at leaving behind.

   Popular and purely natural poetry has its naïf charms and graces by
which it can stand comparison with that chief of beauties we find in
artistically perfect poetry. That can be seen from our Gascony *villanelles*
and from those songs which have been reported from nations which have
no knowledge of any science nor even of writing. But that middling
poetry which remains between the two is despised and is without honour
or price.

   [A]   But, because I have discovered that once our mind has found an
opening we have, as usual, mistaken for a difficult task and a rare topic
something which is nothing of the sort, and that once our capacity for
research has been aroused we can find an infinite number of like examples,

I will merely add the following: that if these *Essays* were worthy of being judged, it could turn out in my opinion that they will hardly please common vulgar minds nor unique and outstanding ones: the former would never get enough of their meaning; the latter would understand them only too easily. These *Essays* might eke out an existence in the middle region.

55. On smells

==

[An early compilation which progressively becomes more personal: the topic itself may have been suggested by a commonplace of the Querelle des femmes (the centuries-long series of works for and against women and marriage).].

[A]   Of some such as Alexander the Great it is said that their sweat smelt nice (because of some rare complexion outside the natural Order, the cause of which was sought by Plutarch and others).[1] But the normal fashioning of our bodies works contrary to that: the best characteristic we can hope for is to smell of nothing. The sweetness of the purest breath consists in nothing more excellent than to be without any offensive smell, as the breath of healthy children. That is why Plautus says, '*Mulier tum bene olet, ubi nihil olet*', 'A woman smells nice when she smells of nothing,' [B]   just as we say that the best perfume for her actions is for her to be quiet and discreet.[2]   [A]   And when people give off nice odours which are not their own we may rightly suspect them, and conclude that they use them to smother some natural stench. That is what gives rise to those adages of the ancient poets which claim that the man who smells nice in fact stinks:

> *Rides nos Coracine, nil olentes.*
> *Malo quam bene olere, nil olere.*

[You laugh at us, Coracinus, because we emit no smell: I would rather smell of nothing than smell sweetly.]

And again,

> *Posthume, non bene olet, qui bene semper olet.*

[A man who always smells nice, Posthumus, actually stinks.][3]

---

1. Plutarch (tr. Amyot), *Propos de table*, 366C.
2. Plautus, *Mustellaria*, I, iii, 117; cf. Tiraquellus, *De legibus connubialibus*, III, §§9–10.
3. Martial, VI, lv, 4–5; II, xii, 4 (both in Tiraquellus, loc. cit.).

[B]   However I am myself very fond of living amongst good smells and I immeasurably loathe bad ones, which I sense at a greater distance than anyone else:

> *Namque sagacius unus odoror,*
> *Polypus, an gravis hirsutis cubet hircus in alis,*
> *Quam canis acer ubi lateat sus.*

[I have a nose with more flair, Polypus, for sensing the goaty smell of hairy armpits than any hound on the track of a stinking boar.][4]

[C]   The simpler, more natural smells seem to me to be the most agreeable. A concern for smells is chiefly a matter for the ladies. In deepest Barbary the Scythian women powder themselves after washing and smother their whole face and body with a certain sweet-smelling unguent, native to their soil; when they take off this cosmetic they find themselves smooth and nice-smelling for an approach to their menfolk.

[B]   Whatever the smell, it is wonderful how it clings to me and how my skin is simply made to drink it in. The person who complained that Nature left Man with no means of bringing smells to his nose was in error: smells do it by themselves. But, in my particular case the job is done for me by my thick moustache: if I bring my glove or my handkerchief anywhere near it, the smell will linger there all day. It gives away where I have just come from. Those close smacking kisses of my youth, [C] gluey and greedy,   [B]   would stick to it and remain there for hours afterwards. Yet I find myself little subject to those mass illnesses which are caught by social intercourse and spring from infected air; and I have been spared those of my own time, of which there have been several kinds in our towns and among our troops.   [C]   We read that although Socrates never left Athens during several recurrences of the plague which so often racked that city, he alone suffered no harm.[5]

[B]   It seems to me that doctors could make better use of smells than they do, for I have frequently noticed that, depending on which they are, they variously affect me and work upon my animal spirits;[6] which convinces me of the truth of what is said about the invention of odours and incense in our Churches (a practice so ancient and so widespread among all nations

4. Horace, *Epodes*, XII, 4–7.
5. Diogenes Laertius, *Life of Socrates*.
6. 'Animal spirits' are the elements in man, separable from the body, which it animates.

and religions): that it was aimed at making us rejoice, exciting us and purifying us so as to render us more capable of contemplation.

[C]   In order to judge it I wish I had been invited to experience the culinary art of those chefs who know how to season wafting odours with the savour of various foods, as was particularly remarked in our time in the case of the King of Tunis who landed at Naples for face to face talks with the Emperor Charles. His meats were stuffed with sweet-smelling ingredients, so luxuriously that a peacock and two pheasants cost a hundred ducats to prepare in their manner. And when those birds were cut up they filled not merely the hall but all the rooms of his palace and even the neighbouring houses with a delicious mist which was slow to evaporate.

[B]   When choosing where to stay, my principal concern is to avoid air which is oppressive and stinking. My liking for those fair cities Venice and Paris is affected by the pungent smell of the marshes of one and the mud of the other.[7]

---

7. In Venice the stench of the canals produced 'bad air' (*malaria*). As for Paris, Joachim Du Bellay emphasizes how its mud struck him on his return from Rome (*Regrets*, 138).

56. On prayer

[We are given here a deeper insight into the austerer, rigorist side of Montaigne's Catholicism. The additions, which are numerous, beginning with those of 1582, marked [A1], are partly designed to meet the criticisms raised by the Maestro di Palazzo at the Vatican about Montaigne's assertion 'that a man when he prays must be free of sinful inclinations during that time'. Such a doctrine savours of that 'puritanism' of which the Roman Catholic Church was ever suspicious. Together with III, 2, 'On repenting' we can see here how demanding Montaigne's Catholicism was beneath its urbane exterior. We can also understand his work better: he is writing philosophy not theology; and philosophy has its own rules and its own language. As usual Montaigne is suspicious of words, even liturgical words, without deeds. To many in his Church his theological position appeared rigorous to the point of heresy where sin-free prayer was concerned. But he himself presents his thoughts as a kind of disputabilis opinio, *that is, as analogous to an unresolved topic or paradox, subject to open debate in the universities.]*

[A1] The notions which I am propounding have no form and reach no conclusion. (Like those who advertise questions for debate in our Universities I am seeking the truth not laying it down.) I submit them to the judgement of those whose concern it is to govern not only my actions and my writings but my very thoughts. Both condemnation and approbation will be equally welcome, equally useful, [C] since I would loathe to be found saying anything ignorantly or inadvertently against the holy teachings of the Church Catholic, Apostolic and Roman, in which I die and in which I was born.[1] [A1] And so, while ever submitting myself to the authority of their censure, whose power over me is limitless, I am emboldened to treat all sorts of subjects – as I do here.

[A] I may be mistaken but, seeing that we have been granted by special grace and favour a set form of prayer prescribed and dictated to us, word by word, by God's own mouth, it has always seemed to me that we

---

1. All churches claim to be catholic. Roman Catholics in Montaigne's time often stressed the 'Roman' so as to avoid any ambiguity.

should use it more commonly.[2] If it depended on me I would like to
see Christians saying the Lord's Prayer as a grace before and after meals,
when we get up and go to bed and on all those special occasions where
we normally include prayers,   [C]   saying it always if not exclusively.
[A]   The Church may lengthen or vary prayers according to her need
to instruct us; for I am well aware that the matter is identical and always
substantially the same. But this prayer ought to have the prerogative
of being on people's lips at all times, since it is certain that it says everything necessary and that it is always most appropriate on all occasions.
[C]   It is the only prayer that I say everywhere; instead of varying it
I repeat it. That explains why it is the only prayer I can ever remember.

[A]   I was wondering recently how the error arose which leads us to
have recourse to God in all our doings and designs,   [B]   calling upon
him in every kind of need and in any place whatsoever where our
weakness needs support, without once considering whether the occasion is
just or unjust. No matter how we are or what we are doing – however
sinful it may be – we invoke God's name and power.   [A]   He is of
course our only and unique Protector,   [C]   able to do anything whatever
to help us;   [A]   but even though he does vouchsafe to grant us that
sweet honour of being our Father by adoption,[3] he is as just as he is
good   [C]   and powerful; but he uses his justice more often than his
power;   [A]   and he grants us his favours according to   [C]   its criteria
not our petitions.[4]

In his *Laws*, Plato lists three kinds of belief which are insulting to the
gods: that there are none; that they do not concern themselves with our
affairs; that they never refuse to answer our prayers, oblations and sacrifices.
He believes that the first error never remains stable in anyone from
childhood to old age but that the other two do allow of constancy.[5]

[A]   God's power and his justice are inseparable. If we implore him to
use his power in a wicked cause it is of no avail. Our soul must be pure, at
least for that   [C]   instant   [A]   when we make our prayer, free from

2. The Lord's Prayer (Matthew 6:9–13; Luke 11:2–4). Montaigne always presents
the Bible as divinely inspired by the Holy Ghost. Here, by special grace, the
incarnate Son ensures the absolute verbal accuracy of the central prayer of
Christendom.
3. Romans 8:14–17, etc.
4. '80: According to *the* criteria *of his justice, not according to our inclinations and
wishes.* God's . . .
5. Plato, *Laws*, X, 885 B–C.

the weight of vicious passions; otherwise we offer him rods for our own chastisement.[6] Instead of amending our faults we redouble them by offering God (from whom we ought to be begging forgiveness) emotions full of irreverence and hatred. That is why I do not approve of those whom I see praying to God frequently and regularly if deeds consonant with their prayers do not bear me witness of some reformation and amendment[7] –

[B]   *si, nocturnus adulter,*
Tempora Sanctonico velas adoperta cucullo.

[if, for your nightly adultery, you hide beneath an Aquitanian cowl.][8]

[C]   The position of a man who mingles devotion with a detestable life seems somehow to deserve condemnation more than that of a man who is self-consistent, dissolute in everything. That is why our Church daily excludes all stubborn notorious evildoers from entry into our fellowship.

[A]   We say our prayers out of habit and custom, or to put it better, we merely read and utter the words of our prayers. It amounts, in the end, to [C]   outward show.   [B]   And it displeases me to see a man making three signs of the cross at the *Benedicite* and three more at grace – displeasing me all the more since [C]   it is a sign which I revere and continually employ, not least when I yawn – [B]   only[9] to see him devoting every other hour of the day to [C]   hatred, covetousness and injustice.[10]   [B]   Give vices their hours, then one hour to God – a sort of barter or arrangement! What a miracle it is to see actions so incompatible proceeding at so even a course that at the very point where they pass from one to the other you can notice no break or hesitation.

[C]   What monstrous a conscience it is that can find rest while nurturing together in so peaceful and harmonious a fellowship both the crime and the judge in the same abode. If a man has his head full of the demands of lechery, judging it to be something most odious in the sight of God, what

---

6. '80: at least for that *time* when . . .   (This passage was raised by the Maestro del Palazzo. Consult Malcolm Smith, *Montaigne and the Roman Censors*, Geneva, 1981. Montaigne's assertion is rigorist and neo-Augustinian. Some still judge it hyperorthodox.)

7. Cf. Matthew 3:8.

8. Juvenal, *Satires*, VIII, 144–5.

9. The *Benedicite* precedes dinner; grace follows it.
   '80: It amounts in the end to *pretence*. And it . . .
   '88: since *they are practices which I honour and often imitate*, only . . .

10. '80: to *usury, venality and lechery*. Give . . . [Montaigne strengthens his case, replacing sinful practices by the infinitely more serious inward sins of the mind.]

does he say to God when he tells him of it? He repents, only to fall again –
at once. If as he claims the concept of divine Justice really did strike home,
scourging and chastising his soul, then however short his repentance fear
itself would force him to cast his mind back to it, making him thenceforth
master of those bloated vices which were habitually his.

And what about those men whose whole life reposes on the fruits and
profits of what they know to be a mortal sin? How many trades and vocations are there which gain acceptance, yet whose very essence is vicious?

And then there is the man who confided to me how, all his life, he had
professed and practised a religion which he believed to be damnable, quite
opposite to the one dear to him, so as not to lose favour or the honour of
his appointments. How did he defend such reasoning in his mind? When
men address God's Justice on such matters, what do they say? Since their
repentance requires a visible and tangible reparation, they forfeit all means
of pleading it before God or men. Do they go so far as to dare to beg
forgiveness without making satisfaction, without repentance? I hold that
the first ones I mentioned are in the same state as these; but their obstinacy
is far less easy to overcome.

Those sudden violent changes and veerings of opinion that they feign for
us are a source of wonder to me. They reveal a state of unresolved conflict.
And how fantastical seem to me the conceptions of those who, in recent
years, have habitually accused anyone who showed a glimmer of intelligence yet professed the Catholic faith of only feigning to do so – even
maintaining, to do him honour, that whatever he might actually say for
show, deep down inside he could not fail to hold the religion as 'reformed'
by their standards! What a loathsome malady it is to believe that you are so
right that you convince yourself that nobody can think the opposite. And
most loathsome still, to convince yourself that such a mind may prefer
some chance but present advantage to the hopes and fears of eternal life.
They can take my word for it: if anything could have tempted me in
youth, a large part would have been played by an ambition to share in the
hazards and hardships intendant upon that fresh young enterprise.

[A]   It is not without good reason, it seems to me, that the Church has
forbidden the indiscriminate, thoughtless and indiscreet use of those venerable sacred songs which the Holy Ghost dictated through David.[11] We

11. '80: the *Catholic* Church has forbidden . . . (Psalm-singing, often in the
translation of the French poet Clément Marot, had been a practice in the Court of
Margaret of Navarre but had become for many the sign of the Reformed
Church.)

must only bring God into our activities with reverence and attentiveness full of honour and respect. That Word is too holy to serve merely to exercise our lungs and to please our ears; it must be rendered by our hearts not by our tongues. It is unreasonable to permit some shop-boy to amuse himself playing about with it while his mind is on silly frivolous matters. [B] Nor, certainly, is it right to see the Sacred Book of the holy mysteries of our faith dragged about through hall or kitchen[12] – [C] they used to *be* mysteries: now they serve as amusements and pastimes.

[B] A study so serious, a subject so revered, should not be handled incidentally or hurriedly. It should always be a considered calm activity, prefaced as in our liturgy by the *Sursum corda;*[13] we should bring to it even our bodies disposed in such attitudes as bear witness to a special attentiveness and reverence. [C] It is not a study for just anybody: it is a study for those who are dedicated to it, for people whom God calls to it. It makes the wicked and the ignorant grow worse. It is not a story to be told but a story to be reverenced, feared, adored.

How silly they are who think they have made it accessible to the vulgar simply by translating it into the vulgar tongues. When people fail to understand everything they read is it only the fault of the words! I would go further. By bringing Scripture that little bit nearer they actually push it further away. Pure ignorance, leaving men totally dependent on others, was much more salutary and more learned than such vain verbal knowledge, that nursery of rashness and presumption.

[B] I also believe that the liberty everyone takes of[14] broadcasting so religious and so vital a text into all sorts of languages is less useful than dangerous. Jews, Mahometans and virtually all the others have reverently espoused the tongue in which their mysteries were first conceived; any changes or alterations are forbidden; not, it seems, without reason. Can we be sure that in the Basque country or in Brittany there are enough good judges, men adequate enough to establish the right translation in their languages? Why, the Catholic Church has nothing more difficult to do than to decide such matters – and nothing more solemn. When it is a case of preaching or speaking our translations can be vague, free, variable and partial: that is not at all the same thing.

---

12. '80: kitchen, *in the hands of everybody*. A study . . .
13. 'Lift up your hearts' – the liturgical summons to prayer.
14. '88: of *translating and* broadcasting . . .

[C] One of our Greek historians[15] justly accused his own time of having so scattered the secrets of the Christian religion about the market-place and into the hands of the meanest artisans that everybody could argue and talk about them according to his own understanding: 'It is deeply shameful,' he added, 'that we who by God's grace enjoy the pure mysteries of our pious faith should allow them to be profaned in the mouths of persons ignorant and base, seeing that the Gentiles forbade even Socrates, Plato and the wisest men to talk or to inquire about matters entrusted to the priests at Delphi.' He also said that, where Theology is concerned, the factions of princes are armed with anger not with zeal; that zeal itself does partake of the divine Reason and Justice when it behaves ordinately and moderately but that it changes into hatred and envy whenever it serves human passions, producing then not wheat and the fruit of the vine but tares and nettles. And there was another man who rightly advised the Emperor Theodosius that debates never settled schisms in the Church but rather awakened heresies and put life into them; therefore he should flee all contentiousness and all dialectical disputations, committing himself to the bare prescriptions and formulas of the Faith established of old. And when the Emperor Andronicus came across two great men verbally skirmishing in his palace against Lopadius over one of the more important points of our religion, he reprimanded them, even threatening to have them thrown into the river if they still went on.

Nowadays women and children read lectures about ecclesiastical law to the oldest and most experienced of men whereas the first of Plato's laws forbids them to inquire even into the reason for merely civil ones, which must be regarded as divine ordinances; he allowed only the older men to discuss laws among themselves and with the Magistrate – adding, 'provided that it is not done in the presence of the young and the uninitiated'.[16]

A bishop has testified in writing[17] that there is, at the other end of the world, an island which the Ancients called Dioscorides, fertile and favoured with all sorts of fruits and trees and a healthy air; the inhabitants are Christian, having Churches and altars which are adorned with no other images but crosses; they scrupulously observe feast-days and fasts, pay their

---

15. All this paragraph of Nicetas comes directly from Justus Lipsius' *De una religione*.

16. Plato forbade youths, not women, to discuss the laws (Plato, *Laws*, I, 634 D– E). Here, as often in the Renaissance, *Law* includes religion. (Christianity was termed 'the law of Christians' from medieval times.)

17. Bishop Jeronimo Osorio (da Fonseca), *De rebus Emanuëlis Lusitaniae Regis gestis*, Cologne, 1581 (1586).

tithes meticulously and are so chaste that no man ever lies with more than one woman for the whole of his life; meanwhile, so happy with their lot that, in the middle of the ocean, they know nothing about ships, and so simple that they do not understand a single word of the religion which they so meticulously observe – something only unbelievable to those who do not know that pagans, devout worshippers of idols, know nothing about their gods apart from their statues and their names. The original beginning of Euripides' tragedy *Menalippus* went like this:

> *O Juppiter, car de toy rien sinon*
> *Je ne connois seulement que le nom . . .*

[O Juppiter – for I know nothing of thee but thy Name . . .][18]

[B] I have also seen in my time criticisms laid against some books for dealing exclusively with the humanities or philosophy without any admixture of Theology. The opposite case would not be totally indefensible, namely: that Christian Doctrine holds her rank better when set apart, as Queen and Governor; that she should be first throughout, never ancillary nor subsidiary; that Grammar, Rhetoric and Logic should [C] perhaps [B] choose their examples from elsewhere not from such sacred materials, as also should the subjects of plays for the theatre, farces and public spectacles; that Divinity is regarded with more veneration and reverence when expounded on its own style rather than when linked to human reasoning; that the more frequent fault is to see Theologians writing like humanists rather than humanists like Theologians (Philosophy, says St Chrysostom, has long been banished from the School of Divinity as a useless servant judged unworthy of glimpsing, even from the doorway when simply passing by, the sanctuary of the holy treasures of sacred doctrine); that the language of men has its own less elevated forms and must not make use of the dignity, majesty and authority of the language of God. I myself let it say – [C] *verbis indisciplinatis* [using undisciplined words] – [B] fortune, destiny, accident, good luck, bad luck, the gods and similar phrases, following its own fashion.[19]

[C] I am offering my own human thoughts as human thoughts to be considered on their own, not as things established by God's ordinance,

---

18. Euripides *apud* Plutarch (tr. Amyot), *De l'amour*, 604B.
19. St Augustine, *City of God*, X, xxix. This was current Renaissance practice. For some reason the Maestro di Palazzo raised the question of the use of 'fortune' in the *Essays*. Montaigne changed a few passages but held his ground and explains why. (The passage of Chrysostom remains untraced.)

incapable of being doubted or challenged; they are matters of opinion not matters of faith: what I reason out *secundum me*, not what I believe *secundum Deum*[20] – like schoolboys reading out their essays, not teaching but teachable, in a lay not a clerical manner but always deeply devout.

[B] And might it not be said, apparently reasonably, that a decree forbidding anyone to write about religion (except very reservedly) unless expressly professing to do so would not lack some image of usefulness and justice – as perhaps would one requiring me too to hold my peace on the subject?

[A] I have been told that for reasons of reverence even those who are not of our Church forbid the use among themselves of the name of God in their everyday speech.[21] They do not want it to be used as a kind of interjection or exclamation, nor to support testimony nor when making contracts; in that I consider they are right. Whenever we bring God's name into our affairs or our society let it be done seriously and devoutly.

I believe there is a treatise in Xenophon somewhere in which he shows that we ought to pray to God less often, since it is not easy for us to bring our souls so frequently into that controlled, reformed and supplicatory state needed to do so; without that, our prayers are not only vain and useless: they are depraved. 'Forgive us', we say, 'as we forgive them that trespass against us.' What do those words mean if not that we are offering God our souls free from vengeance and resentment? Yet we call on God and his help to connive at wrongdoings   [C]   and to invite him to be unjust:

[B]   *Quæ, nisi seductis, nequeas committere divis.*

[Things which you would not care to entrust to the gods, except when drawing aside.][22]

[A]   The miser prays God for the vain and superfluous preservation of his hoard; the ambitious man, for success and the achievement of his desires; the thief uses God to help him overcome the dangers and difficulties which obstruct his nefarious designs or else thanks God when he finds it easy to slit the gizzard of some passer-by.   [C]   At the foot of the mansion which

20. Montaigne's terms are technical. He is giving his *opinions* (i.e. his unproven notions) 'according to himself', '*selon moy*' (*secundum me*). Anything which is said *secundum quid* ('according to anything') is not stated *simpliciter* (absolutely, simply) but in some partial respect only. Anything stated '*selon Dieu*', 'according to God' (*secundum Deum*) would be infallible and a matter of absolute faith.
21. That was the practice of the Reformed Church. (Cf. Joachim Du Bellay, *Regrets*, 136, on the Genevan Calvinists.)
22. Persius, *Satires*, II, 4; glossing a petition from the Lord's Prayer.

they are about to climb into and blow up, men say their prayers, while their purposes and hopes are full of cruelty, lust and greed.

> [B] *Hoc ipsum quo tu Jovis aurem impellere tentas,*
> *Dic agedum, Staio, pro Juppiter, ô bone clamet,*
> *Juppiter, at se se non clamet Juppiter ipse?*

[Try telling Statius what you are up to, what you have just whispered to Jove: 'By Jove!' he'll say: 'How dreadful!' – 'Well, cannot Jove say *By Jove!* to Himself?']²³

[A] Queen Margaret of Navarre relates the tale of a young 'prince' – and, even though she does not name him his exalted rank is quite enough to make him recognizable; whenever he was out on an assignation (lying with the wife of a Parisian barrister) he would take a short-cut through a church and never failed to make his prayers and supplications in that holy place, both on the way there and on the way back. I will leave you to judge what he was asking God's favour for when his soul was full of such fair cogitations! Yet she cites that as evidence of outstanding devotion. But that is not the only proof we have of the truth that it hardly befits women to treat Theological matters.

A devout reconciliation with God, a true prayer, cannot befall a soul which is impure and, at that very time, submissive to the domination of Satan. A man who calls God to his aid while he is actually engaged in vice is like a cutpurse calling on justice to help him or like those who produce the name of God to vouch for their lies:

> [B] *tacito mala vota susurro*
> *Concipimus.*

[we softly murmur evil prayers.]²⁴

[A] Not many men would care to submit to view the secret prayers they make to God:

> *Haud cuivis promptum est murmurque humilesque susurros*
> *Tollere de templis, et aperto vivere voto.*

[It is hardly everyone who could take his murmured prayers whispered within the temples and say them aloud outside.]

That is why the Pythagorians believed that prayer should be public and

---

23. Persius, *Satires*, II, 21–31. The young monarch (or '*prince*') in the next paragraph is Francis I (cf. Margaret of Navarre, *Heptaméron*, III, 25). *Prince* regularly means *King* in the Renaissance, as a current Latinism.
24. Lucan, *Pharsalia*, V, 104–5.

heard by all, so that God should not be begged for things unseemly or unjust – like the man in this poem:

> *clare cum dixit: Apollo!*
> *Labra movet, metuens audiri: pulchra Laverna,*
> *Da mihi fallere, da justum sanctumque videri.*
> *Noctem peccatis et fraudibus objice nubem.*

[he first exclaims, 'Apollo!' loud and clear; then he moves his lips, addressing the goddess of Theft and fearing to be overheard: 'O fair Laverna: do not let me get found out; let me appear to be just and upright; cloak my sins with night and my lies with a cloud.']²⁵

[C] The gods heavily punished the unrighteous prayers of Oedipus by granting them: he prayed that his children should fight among themselves to decide who should succeed to his inheritance, he was wretched enough to be taken literally.

We should not ask that all things should comply with our will but that they should comply with wisdom.

[A] It really does seem that we use prayer [C] as a sort of jingle and [A] like those who exploit God's holy words in sorcery and practical magic.²⁶ As for their effect, we apparently count on their structure, their sound and the succession of words, [A1] or on our outward appearance. [A] For, with our souls still full of concupiscence, untouched by repentance or by any fresh reconciliation with God, we offer him such words as memory lends to our tongue, hoping in that way to obtain the expiation of our sins.

Nothing is so gentle, so sweet, so gracious as our Holy Law:²⁷ she calls us to her, all sinful and abominable as we are; she stretches forth her arms and clasps us to her bosom, however base, vile and besmirched we may be now and shall be once again. But we on our part must look favourably upon her. We must also receive her absolution with thanksgiving and – at least for that instant when we address ourselves to her – have a soul loathing its own shortcomings and hostile to those [C] passions [A] which²⁸ drove us to offend her.

---

25. Persius, *Satires*, II, 6–7; Horace, *Epistles*, III, i, 16–19.
26. For Oedipus, cf. Plato, *Second Alcibiades*, 138 B–C. Then, for prayer, cf. the 'magic' prayers of Panurge during the Storm in the *Quart Livre* of Father Rabelais. Montaigne's point is theologically sound and, at the time, not difficult to grasp.
27. That is, Christianity.
28. '80: those *concupiscences* which . . .

[C] Neither the gods nor good men, Plato says, accept gifts from a wicked man:[29]

> [B] *Immunis aram si tetigit manus,*
> *Non sumptuosa blandior hostia*
> *Mollivit aversos Penates,*
> *Farre pio et saliente mica.*

[If the hands which have touched the altar are undefiled, then, even when they are not commended by some costly sacrifice, they can appease the hostile household gods with a simple cake of meal sprinkled with salt.][30]

29. Plato, *Laws*, IV, 717E.
30. Horace, *Odes*, III, xxxiii, 13–16.

57. On the length of life

═══

[*Montaigne, who published the first two books of his* Essays *when he was forty-seven, looks back at youth and sees thirty as the watershed dividing vigour from decline. The last word of this chapter, and so of Book I, is 'apprenticeship'. At thirty a wise man's 'apprenticeship' should doubtless be over, but, for those who make good use of their time, can knowledge and experience grow with the years?*]

[A]   I cannot accept the way we determine the span of our lives.[1] I note that wise men shorten it considerably compared to the common opinion. 'What!' said Cato the Younger to those who wanted to stop him killing himself: 'Am I still at the age when you can accuse me of leaving life too soon?'[2] Yet he was only forty-eight. He reckoned, considering how few men reach it, that his age was fully mature and well advanced. And those who keep themselves going with the thought that some span of life or other which they call 'natural' promises them a few years more could only do so provided that there was some ordinance exempting them personally from those innumerable accidents (which each one of us comes up against and is subject to by nature) which can rupture the course of life which they promise themselves.

What madness it is to expect to die of that failing of our powers brought on by extreme old age and to make that the target for our life to reach when it is the least usual, the rarest kind of death. We call that death, alone, a natural death, as if it were unnatural to find a man breaking his neck in a fall, engulfed in a shipwreck, surprised by plague or pleurisy, and as though our normal condition did not expose us to all of those harms. Let us not beguile ourselves with such fine words: perhaps we ought, rather, to call natural anything which is generic, common to all and universal. Dying of old age is a rare death, unique and out of the normal order and therefore less natural than the others. It is the last, the uttermost way of dying; the farther it is from us, the less

1. Presumably the biblical 'three-score years and ten', held to be the norm.
2. Plutarch, *Life of Cato of Utica.*

we can hope to reach it; it is indeed the limit beyond which we shall not go and which has been prescribed by Nature's law as never to be crossed: but it is a very rare individual law of hers which makes us last out till then. It is an exemption which she grants as an individual favour to one man in the space of two or three centuries, freeing him from the burden of those obstacles and difficulties which she strews along the course of that long progress.

Therefore my opinion is that we should consider whatever age we have reached as an age reached by few. Since in the normal course of events men never reach that far, it is a sign that we are getting on. And since we have crossed the accustomed limits – and that constitutes the real measure of our days – we ought not to hope to get much farther beyond them; having escaped those many occasions of death which have tripped up all the others, we ought to admit that an abnormal fortune such as that which has brought us so far is indeed beyond the usual procedure and cannot last much longer.

It is a defect in our very laws to hold that false idea, for they do not admit that a man be capable of managing his affairs before the age of twenty-five, yet he can scarcely manage to make his life last that long! Augustus lopped five years off the old Roman ordinances and decreed that it sufficed to be thirty for a man to assume the office of judge. Servius Tullius exempted knights who had passed the age of forty-seven from obligatory war-service; Augustus remitted it at forty-five.[3] Sending men into inactivity before fifty-five or sixty does not seem very right to me. I would counsel extending our vocations and employments as far as we could in the public interest; the error is on the other side, I find: that of not putting us to work soon enough. The man who had power to decide everything in the whole world at nineteen[4] wanted a man to be thirty before he could decide where to place a gutter!

Personally I reckon that our souls are free from their bonds at the age of twenty, as they ought to be, and that by then they show promise of all they are capable of. No soul having failed by then to give a quite evident pledge of her power ever gave proof of it afterwards. By then – or never at all – natural qualities and capacities reveal whatever beauty or vigour they possess.

3. Suetonius, *Life of Augustus*.
4. The Emperor Augustus.

[B]　*Si l'espine nou pique quand nai,*
　　A peine que pique jamai

[If a thorn pricks not at its birth,
It will hardly prick at all]

as they say in Dauphiné.

[A]　Of all the fair deeds of men in ancient times and in our own which have come to my knowledge, of whatever kind they may be, I think it would take me longer to enumerate those which were made manifest before the age of thirty than after. [C] Yes, and often in the lives of the very same men: may I not say that with total certainty in the case of Hannibal and his great adversary Scipio? They lived a good half of their lives on the glory achieved in their youth: they were great men later compared with others, but not great compared with themselves. [A] As for me, I am convinced that, since that age, my mind and my body have not grown but diminished, and have retreated not advanced.

It may well be that (for those who make good use of their time) knowledge and experience grow with the years but vitality, quickness, firmness and other qualities which are more truly our own, and more important, more ours by their essence, droop and fade.

[B]　*Ubi jam validis quassatum est viribus ævi*
　　Corpus, et obtusis ceciderunt viribus artus,
　　Claudicat ingenium, delirat linguaque mensque.

[When the body is shattered by the mighty blows of age and our limbs shed their blunted powers, our wits too become lame and our tongues and our minds start to wander.][5]

Sometimes it is the body which is the first to surrender to old age, sometimes too the soul; and I have known plenty of men whose brains grew weak before their stomachs or their legs; and it is all the more dangerous an infirmity in that the sufferer is hardly aware of it and its symptoms are not clear ones.

But now [A] I am complaining not that the laws allow us to work so late but that they are so late in putting us to work.

It seems to me that, considering the frailty of our life and the number of

5. Lucretius, III, 452–4.

natural hazards to which it is exposed, we should not allow so large a place in it to being born, to leisure and to our apprenticeship.[6]

6. Montaigne's next-to-last noun, *oisiveté* probably renders the classical Latin word *otium*; in which case he is not thinking of 'idleness' but of that 'leisure' time, when learning, study and culture took precedence over 'business' (*negotium*), which included all duties and employments.

BOOK II

1. On the inconstancy of our actions

[In Montaigne's French inconstance is a term which includes fickleness and variability as well as inconsistency of conduct. In Latin, constantia (inner consistency and steadfast constancy) were the ideals of Stoic philosophy. Montaigne, having finished Book I with the notion of apprenticeship, now moves more boldly into new areas of exploration of himself and the nature of Man, both of which he finds subject to fickleness and marked by inconsistent qualities.]

[A] Those who strive to account for a man's deeds are never more bewildered than when they try to knit them into one whole and to show them under one light, since they commonly contradict each other in so odd a fashion that it seems impossible that they should all come out of the same shop. Young Marius now acts like a son of Mars, now as a son of Venus. They say that Pope Boniface VIII took up his duties like a fox, bore them like a lion and died like a dog. And who would ever believe that it was Nero, the very image of cruelty, who when they presented him with the death-sentence of a convicted criminal to be duly signed replied, 'Would to God that I had never learned to write!' so much it oppressed his heart to condemn a man to death?[1]

Everything is so full of such examples (indeed each man can furnish so many from himself) that I find it strange to find men of understanding sometimes taking such trouble to match up the pieces, seeing that vacillation seems to me to be the most common and blatant defect of our nature: witness the famous line of Publius the author of farces:

> Malum consilium est, quod mutari non potest!

> [It's a bad resolution which can never be changed!][2]

[B] It seems reasonable enough to base our judgement of a man on the more usual features of his life: but given the natural inconstancy of our behaviour and our opinions it has often occurred to me that even sound authors are wrong in stubbornly trying to weave us into one invariable and solid fabric.

1. Plutarch, *Life of Marius*; Bouchet, *Annales d'Acquitaine*; Seneca, *De Clementia*.
2. Publius Syrus cited by Aulus Gellius, *Attic Nights*, XVII, 14.

They select one universal character, then, following that model, they classify and interpret all the actions of a great man; if they cannot twist them the way they want they accuse the man of insincerity. Augustus did get away from them: for there is in that man throughout his life a diversity of actions so clear, so sudden and so uninterrupted that they had to let him go in one piece, with no verdict made on him by even the boldest judges. Of Man I can believe nothing less easily than invariability: nothing more easily than variability. Whoever would judge a man in his detail, [C] piece by piece, separately, [B] would hit on the truth more often.

[A] It is difficult to pick out more than a dozen men in the whole of Antiquity who groomed their lives to follow an assured and definite course, though that is the principal aim of wisdom. To sum it all up and to embrace all the rules of Man's life in one word, 'Wisdom,' said an Ancient, 'is always to want the same thing, always *not* to want the same thing.' I would not condescend to add, he said, 'provided that your willing be right. For if it is not right, it is impossible for it to remain ever one and the same.'³

I was once taught indeed that vice is no more than a defect and irregularity of moderation, and that consequently it is impossible to tie it to constancy. There is a saying attributed to Demosthenes: the beginning of all virtue is reflection and deliberation: its end and perfection, constancy. If by reasoning we were to adopt one definite way, the way we chose would be most beautiful of all; but nobody has thought of doing that.

> *Quod petiit, spernit, repetit quod nuper omisit;*
> *Æstuat, et vitae disconvenit ordine toto.*

[Judgement scorns what it yearned for, yearns again for what it recently spurned; it shifts like the tide and the whole of life is disordered.]⁴

Our normal fashion is to follow the inclinations of our appetite, left and right, up and down, as the winds of occasion bear us along. What we want is only in our thought for the instant that we want it: we are like that creature which takes on the colour of wherever you put it. What we decided just now we will change very soon; and soon afterwards we come back to where we were: it is all motion and inconstancy:

---

3. Seneca, *Epistles*, XX, 5.
4. Demosthenes (?), *On the Fallen at Chaeronea*; then, Horace, *Epistles*, I, i, 98–9.

> *Ducimur ut nervis alienis mobile lignum.*

[We are led like a wooden puppet by wires pulled by others.]

We do not *go*: we are borne along like things afloat, now bobbing now lashing about as the waters are angry or serene.

> [B]   *Nonne videmus*
> *Quid sibi quisque velit nescire, et quærere semper,*
> *Commutare locum, quasi onus deponere possit?*

[Surely we see that nobody knows what he wants, that he is always looking for something, always changing his place, as though he could cast off his burden?]

[A]   Every day a new idea: and our humours change with the changes of weather:

> *Tales sunt hominum mentes, quali pater ipse*
> *Juppiter auctifero lustravit lumine terras.*

[The minds of men are such as Father Juppiter changes them to, as he purifies the world with his fruitful rays.]⁵

[C]   We float about among diverse counsels: our willing of anything is never free, final or constant.

[A]   If a man were to prescribe settled laws for a settled government established over his own brain, then we would see, shining throughout his whole life, a calm uniformity of conduct and a faultless interrelationship between his principles and his actions.

–   [C]   (The defect in the Agrigentines noted by Empedocles was their abandoning themselves to pleasure as though they were to die the next day, while they built as though they would never die at all.)⁶ –

[A]   It would be easy enough to explain the character of such a man; that can be seen from the younger Cato: strike one of his keys and you have struck them all; there is in him a harmony of sounds in perfect concord such as no one can deny. In our cases on the contrary every one of our actions requires to be judged on its own: the surest way in my opinion would be to refer each of them to its context, without looking farther and without drawing any firm inference from it.

During the present debauchery of our wretched commonwealth I was told about a young woman near where I then was who had thrown herself from a high window to avoid being forced by some beggarly soldier

---

5. Horace, *Satires*, II, vii, 82; Lucretius, III, 1070–3; Homer, cited in Latin by St Augustine, *City of God*, V, xxxviii.
6. Diogenes Laertius, *Life of Empedocles*. Also cited in Erasmus' *Apophthegmata*.

billeted on her. She was not killed by her fall and repeated her attempt by trying to slit her own throat with a knife; she was stopped from doing so, but only after she had given herself a nasty wound. She herself admitted that the soldier had not yet gone beyond importuning her with requests, solicitations and presents, but she was afraid that he would eventually use force. And above all this, there were the words she used, the look on her face and that blood testifying to her chastity, truly like some second Lucretia. Now I learned as a fact that both before and after this event she was quite wanton and not all that hard to get. It is like the moral in that tale: 'However handsome and noble you may be, when you fail to get your end in do not immediately conclude that your lady is inviolably chaste: it does not mean that the mule-driver is not having better luck with her.'

Antigonus had grown to love one of his soldiers for his virtue and valour and ordered his doctors to treat him for a malignant internal complaint which had long tormented him; he noticed that, once the soldier was cured, he set about his work with much less ardour and asked him who had changed him into such a coward. 'You yourself, Sire,' he replied, 'by freeing me from the weight of those pains which made me think life was worth nothing.'[7]

Then there was the soldier of Lucullus who had been robbed of everything by the enemy and who, to get his own back, made a fine attack against them. After he had plucked enough enemy feathers to make up for his loss Lucullus, who had formed a high opinion of him, began urging some hazardous exploit upon him with all the fairest expostulations he could think of:

> *Verbis quae timido quoque possent addere mentem.*

> [With words enough to give heart to a coward.]

'You should try urging that,' he replied, 'on some wretched soldier who has lost everything' –

> *quantumvis rusticus ibit,*
> *Ibit eo, quo vis, qui zonam perdidit, inquit*

[yokel though he was, he replied: 'The man who will go anywhere you like is the one who has just lost his money-belt']–

and he absolutely refused to go.[8]

---

7. Erasmus, *Apophthegmata*, IV, *Antigonus Rex Macedonum*, XXXIII.
8. Horace, *Epistles*, II, ii, 36; 26–40 (where the soldier's tale is told).

[C] When we read that after Mechmet[9] had insulted and berated Chasan the chief of his Janissaries for allowing his line of battle to be broken by the Hungarians and for fighting faint-heartedly, Chasan's only reply was, alone and just as he was, weapon in hand, to charge madly against the first group of enemy soldiers to come along, who promptly overwhelmed him: that may well have been not so much an act of justification as a change of heart; not so much natural bravery as a new feeling of distress.

[A] That man you saw yesterday so ready to take risks: do not think it odd if you find him craven tomorrow. What had put heart into his belly was anger, or need, or his fellows, or wine, or the sound of a trumpet. His heart had not been fashioned by reasoned argument: it was those factors which stiffened it; no wonder then if he has been made quite different by other and contrary factors.

[C] The changes and contradictions seen in us are so flexible that some have imagined that we have two souls, others two angels who bear us company and trouble us each in his own way, one turning us towards good the other towards evil, since such sudden changes cannot be accommodated to one single entity.[10]

[B] Not only does the wind of chance events shake me about as it lists, but I also shake and disturb myself by the instability of my stance: anyone who turns his prime attention on to himself will hardly ever find himself in the same state twice. I give my soul this face or that, depending upon which side I lay it down on. I speak about myself in diverse ways: that is because I look at myself in diverse ways. Every sort of contradiction can be found in me, depending upon some twist or attribute: timid, insolent; [C] chaste, lecherous; [B] talkative, taciturn; tough, sickly; clever, dull; brooding, affable; lying, truthful; [C] learned, ignorant; generous, miserly and then prodigal – [B] I can see something of all that in myself, depending on how I gyrate; and anyone who studies himself attentively finds in himself and in his very judgement this whirring about and this discordancy. There is nothing I can say about myself as a whole simply and completely, without intermingling and admixture. The most universal article of my own Logic is DISTINGUO.[11]

9. That is, Mechmet II. Cf. Nicolas Chalcocondylas (tr. Blaise de Vigenère), *De la décadence de l'empire grec*, 1584.

10. That each individual is swayed by a good guardian angel and a bad angel derives from platonizing interpretations of Matthew 18:10; Rabelais accepts it (*Tiers Livre*, TLF, VII). (Cf. Erasmus, *Adages*, I, I, LXXII, *Genius malus*.)

11. 'I make a distinction', a term used in formal debates to reject or modify an opponent's assertion.

[A] I always mean to speak well of what is good, and to interpret favourably anything that can possibly be taken that way; nevertheless, so strange is our human condition that it leads to our being brought by vice itself to 'do good', except that 'doing good' is to be judged solely by our intentions. That is why one courageous action must not be taken as proof that a man really is brave; a man who is truly brave will always be brave on all occasions. If a man's valour were habitual and not a sudden outburst it would make him equally resolute in all eventualities: as much alone as with his comrades, as much in a tilt-yard as on the battlefield; for, despite what they say, there is not one valour for the town and another for the country. He would bear with equal courage an illness in his bed and a wound in battle, and would no more fear dying at home than in an attack. We would never see one and the same man charging into the breach with brave assurance and then raging like a woman over the loss of a lawsuit or a son. [C] If he cannot bear slander but is resolute in poverty; if he cannot bear a barber-surgeon's lancet but is unyielding against the swords of his adversaries, then it is not the man who deserves praise but the deed. Cicero says that many Greeks cannot even look at an enemy yet in sickness show constancy: the Cimbrians and the Celtiberians on the contrary; *'nihil enim potest esse æquabile, quod non a certa ratione proficiscatur.'* [For nothing can be called constant which does not arise out of a fixed principle.][12]

[B] There is no valour greater in its kind than Alexander's; yet it is but one kind of valour; it is not in all cases sufficiently whole or all-pervasive. [C] Absolutely incomparable it may be, but it has its blemishes, [B] with the result that we see him worried to distraction over the slightest suspicion he may have that his men are plotting against his life, and see him conducting his investigations with an injustice so chaotic and ecstatic and with a fear which overturned his natural reason. Then there is the superstition from which he so markedly suffered: it bears some image of faint-heartedness. [C] And the excessive repentance he showed for murdering Clytus is another testimony to the inconstancy of his mind.[13]

[A] We are fashioned out of oddments put together – [C] *'voluptatem contemnunt, in dolore sunt molliores; gloriam negligunt, franguntur infamia'* [they despise pleasure but are rather weak in pain; they

---

12. Cicero, *Tusc. disput.*, II, xxvii, 65.
13. Ibid., IV, xxxvii, 79. Alexander murdered Clitus when drunk.

are indifferent to glory, but are broken by disgrace][14] – [A] and we wish to win honour under false flags. Virtue wants to be pursued for her own sake: if we borrow her mask for some other purpose then she quickly rips it off our faces. Virtue, once the soul is steeped in her, is a strong and living dye which never runs without taking the material with her.

That is why to judge a man we must follow his tracks long and carefully. If his constancy does not rest firmly upon its own foundations; [C] *'cui vivendi via considerata atque provisa est'*; [the path which his life follows having been thought about and prepared for beforehand;] [A] if various changes make him change his pace – I mean his *path*, for his pace may be hastened by them or made heavy and slow – then let him go free,[15] for that man will always 'run with the wind', *A vau le vent*, as the crest of our Lord Talbot puts it.

No wonder, said an Ancient, that chance has so much power over us, since it is by chance that we live. Anyone who has not groomed his life in general towards some definite end cannot possibly arrange his individual actions properly. It is impossible to put the pieces together if you do not have in your head the idea of the whole. What is the use of providing yourself with paints if you do not know what to paint? No man sketches out a definite plan for his life; we only determine bits of it. The bowman must first know what he is aiming at: then he has to prepare hand, bow, bowstring, arrow and his drill to that end. Our projects go astray because they are not addressed to a target.[16] No wind is right for a seaman who has no predetermined harbour. I do not agree with the verdict given in favour of Sophocles in the action brought against him by his son, which argued, on the strength of seeing a performance of one of his tragedies, that he was fully capable of managing his domestic affairs.[17] [C] Neither do I agree that the inferences drawn by the Parians sent to reform the Milesian government justified the conclusion they reached: visiting the island they looked out for the best-tended lands and the best-run country estates and, having noted down their owners' names, summoned all the citizens of the town to assemble and appointed those owners as the new governors and magistrates, judging that those who took care of their private affairs would do the same for the affairs of state.[18]

14. Cicero, *De officiis*, I, xxi, 71.
15. Cicero, *Paradoxa*, V, i; Seneca, *Epist. moral.*, XX, 2–3.
16. Several echoes of Seneca, *Epist. moral.*, LXXI and XCII and of other Epistles throughout this chapter.
17. Cicero, *De senectute*, VII.
18. Herodotus, *Historia*, V, xxix.

[A]   We are entirely made up of bits and pieces, woven together so diversely and so shapelessly that each one of them pulls its own way at every moment. And there is as much difference between us and ourselves as there is between us and other people.   [C]   *'Magnam rem puta unum hominem agere'* [Let me convince you that it is a hard task to be always the same man.]¹⁹   [A]   Since ambition can teach men valour, temperance and generosity – and, indeed, justice; since covetousness can plant in the mind of a shop-boy, brought up in obscurity and idleness, enough confidence to cast himself on the mercy of the waves and angry Neptune in a frail boat, far from his hearth and home, and also teach him discernment and prudence; and since Venus herself furnishes resolution and hardiness to young men still subject to correction and the cane, and puts a soldier's heart into girls still on their mothers' knees:

> [B]   *Hac duce, custodes furtim transgressa jacentes,*
> *Ad Juvenem tenebris sola puella venit:*

[With Venus as her guide, the maiden, quite alone, comes to the young man, sneaking carefully through her sleeping guardians:]²⁰

it is not the act of a settled judgement to judge us simply by our outward deeds: we must probe right down inside and find out what principles make things move; but since this is a deep and chancy undertaking, I would that fewer people would concern themselves with it.

19. Seneca, *Epist. moral.*, CXX, 22. In the following sentence 'ambition', as often, means inordinate ambition; so too covetousness ('*avarice*' in the French original) means an inordinate desire to obtain, and retain, not only wealth but honour: its sense is close to that of inordinate ambition. Montaigne holds that bad motives can produce admirable qualities.
20. Tibullus, II, i, 75–6.

2. On drunkenness

[Drunkenness was considered a form of ecstasy, in which body and soul became separated or loosely joined. From Ancient times it was associated with the higher ecstasies (those of mystics, poets, prophets and lovers) as well as with the ecstasy of wonder, of bravery and of fear. (In his Paraphrases on the New Testament *Erasmus has a long section explaining the rapture of the disciples at Pentecost by analogy with the effects of drunkenness, of which the disciples were accused.) Montaigne is wary of ecstasy and despises excessive drinking, which is for him a rapture not of the mind but the body.]*

[A]   The world is all variation and dissimilarity. Vices are all the same in that they are vices – and doubtless the Stoics understand matters after that fashion: but even though they are equally vices they are not equal vices. That a man who has overstepped by a hundred yards those limits

> *quos ultra citraque nequit consistere rectum,*

> [beyond which, and short of which, there is no right way,]

should not be in a worse condition than a man who has only overstepped them by ten yards is not believable; nor that sacrilege should be no worse than stealing a cabbage from our garden:

> *Nec vincet ratio, tantumdem ut peccet idemque*
> *Qui teneros caules alieni fregerit horti,*
> *Et qui nocturnus divum sacra legerit.*

[Reason cannot convince me that there is equal sinfulness in trampling down someone's spring cabbages and in robbing the temple-treasures in the night.][1]

There is as much diversity in vice as in anything else.

[B]   It is dangerous to confound the rank and importance of sins: murderers, traitors and tyrants gain too much by it. It is not reasonable that they should be able to salve their consciences because somebody else is lazy, lascivious or not assiduous in his prayers. Each man comes down heavily on

---

1. For Stoics all vices are equally evil; all virtues equally good. Horace (as cited) denies that: *Satires*, I, i, 107; I, iii, 115–17.

his neighbours' sins and lessens the weight of his own. Even the doctors of the Church often rank sins badly to my taste.

[C]   Just as Socrates said that the prime duty of wisdom is to distinguish good from evil,[2] we, whose best always partakes of vice, should say the same about knowing how to distinguish between the vices: if that is not done exactingly, the virtuous man and the vicious man will be jumbled unrecognizedly together.

[A]   Now drunkenness, considered among other vices, has always seemed to me gross and brutish. In others our minds play a larger part; and there are some vices which have something or other magnanimous about them, if that is the right word. There are some which are intermingled with learning, diligence, valour, prudence, skill and *finesse*: drunkenness is all body and earthy. Moreover the grossest nation of our day is alone in honouring it.[3] Other vices harm our intellect: this one overthrows it;   [B]   and it stuns the body:

> *cum vini vis penetravit,*
> *Consequitur gravitas membrorum, præpediuntur*
> *Crura vacillanti, tardescit lingua, madet mens,*
> *Nant oculi; clamor, singultus, jurgia gliscunt.*

[when the strength of the wine has sunk in, our limbs become heavy, we stagger and trip over our legs; our speech becomes slow; our mind, sodden; our eyes are a-swim. Then comes the din, the hiccoughs and the fights.][4]

[C]   The worst state for a man is when he loses all consciousness and control of himself.

[A]   And among other things they say that, just as the must fermenting in the wine-jar stirs up all the lees at the bottom, so too does wine unbung the most intimate secrets of those who have drunk beyond measure:

> [B]   *tu sapientium*
>   *Curas et arcanum jocoso*
>   *Consilium retegis Lyæo.*

[in those jolly Bacchic revels you, my wine-jar, uncover worries and the secret counsels of the wise.][5]

---

2. Erasmus, *Apophthegmata*, III, *Socratica*, XXXIII.
3. The Germanic peoples.
4. Lucretius, III, 475–8.
5. Seneca, *Epist. moral.*, LXXXIII, 16; Horace, *Odes*, III, xxi, 14–17.

[A] Josephus[6] tells how he wheedled secrets out of an ambassador sent to him by his enemies by making him drink a lot. Nevertheless Augustus confided his most private secrets to Lucius Piso, the conqueror of Thrace, and was never let down; nor was Tiberius let down by Cossus on whom he unburdened all of his plans: yet we know that those two men were so given to drinking that they had often to be carried out of the Senate, both drunk,[7]

> *Externo inflatum venas de more Lyæo.*

> [With veins swollen with others' wine, as usual.][8]

[C] And the plan to kill Caesar was well kept when confided to Cassius, who drank water, but also when confided to Cimber, who often got drunk; which explains his joking reply: 'Should I bear the weight of a tyrant, when I cannot bear the weight of my wine!'[9] [A] Even our German mercenaries when drowned in their wine remember where they are quartered, the password and their rank:

> [B]   *nec facilis victoria de madidis, et*
> *Blæsis, atque mero titubantibus.*

[it is not easy to beat them, even when they are sodden-drunk, incoherent and staggering about.][10]

[C] I would never have thought anybody could be buried so insensibly in drunkenness if I had not read the following in the history books. With the purpose of inflicting on him some notable indignity, Attalus invited to supper that Pausanias who, on this very subject, later killed Philip King of Macedonia (a king whose fine qualities nevertheless bore witness to the education he had received in the household and company of Epaminondas). He got him to drink so much that he could bring him, quite unaware of what he was doing, to abandon his fair body to mule-drivers and to many of the most abject scullions in his establishment, as if it were the body of some whore in a hedgerow.[11]

And then there is the case told me by a lady whom I honour and hold in

6. Flavius Josephus (the Jewish historian): *De vita sua.*
7. Seneca, *Epist. moral.*, LXXXIII, 14–15 (for both Piso and Cossa).
8. Virgil, *Bucolica*, VI, 15 (adapted).
9. Seneca, *Epist. moral.*, LXXXIII, 12–13.
10. Juvenal, *Satires*, XV, 47–8.
11. Diodorus Siculus, XV, xxvi.

the greatest esteem: towards Castres, near Bordeaux, where her house is, there was a village woman, a widow of chaste reputation, who, becoming aware of the first hints that she might be pregnant, told the women of the neighbourhood that if only she had a husband she would think she was expecting. But as the reason for her suspicions grew bigger every day and finally became evident, she was reduced to having a declaration made from the pulpit in her parish church, stating that if any man would admit what he had done she promised to forgive him and, if he so wished, to marry him. One of her young farm-labourers took courage at this proclamation and stated that he had found her one feast-day by her fireside after she had drunk her wine freely; she was so deeply and provocatively asleep that he had been able to have her without waking her up. They married each other and are still alive.

[A]   Antiquity, certainly, did not greatly condemn this vice. The very writings of several philosophers speak of it indulgently; even among the Stoics there are those who advise you to let yourself drink as much as you like occasionally and to get drunk so as to relax your soul:

> [B]   *Hoc quoque virtutum quondam certamine, magnum*
> *Socratem palmam promeruisse ferunt.*

[They say that Socrates often carried off the prize in this trial of strength too.][12]

[C]   That Censor and corrector of others,[13]   [A]   Cato was reproached for his heavy drinking:

> [B]   *Narratur et prisci Catonis*
> *Sæpe mero caluisse virtus.*

[It is told how the virtue of old Cato was often warmed with wine].[14]

[A]   Such a famous King as Cyrus cited among the praiseworthy qualities which made him preferable to his brother Artaxerxes the fact that he knew how to drink better. Even among the best regulated and best governed peoples it was very common to assay men by making them drunk. I have heard one of the best doctors in Paris, Silvius, state that it is a good thing once a month to arouse our stomachs by this excess so as to stop their powers from getting sluggish and to stimulate them in order to prevent

---

12. Pseudo-Gallus, I, 47–8.
13. [A]: *That true portrait of Stoic virtue*, Cato. . . (Montaigne had first confused Cato of Utica with Cato the Censor).
14. Horace, *Odes*, III, xxi, 11–12.

their growing dull. [B] And we can read that the Persians discussed their most important affairs after drinking wine.[15]

[A] My taste and my complexion are more hostile than my reason to this vice. For, leaving aside the fact that I readily allow my beliefs to be captive to the Ancients, I find this vice base and stultifying but less wicked and a cause of less harm than the others, which virtually all do more direct public damage to our society. And if, as they maintain, we can never enjoy ourselves without it costing us something, I find that this vice costs our conscience less than the others: besides it is not a negligible consideration that it is easy to provide for and easy to find.

[C] A man advanced in years and rank told me that he counted drink among the three main pleasures left to him in this life.[16] But he set about it in the wrong way; for fine palates and an anxious selecting of wine are to be absolutely avoided. If you base your pleasure on drinking good wine you are bound to suffer from sometimes drinking bad. Your taste ought to be more lowly and more free. To be a good drinker you must not have too tender a palate. The Germans enjoy drinking virtually any wine. Their aim is to gulp it rather than to taste it. They get a better bargain. Their pleasure is more abundant and closer at hand.

Secondly, to drink in the French style at both meals, but moderately for fear of your health, is too great a restraint on the indulgence of god Bacchus: more time and constancy are required. The Ancients spent entire nights in this occupation and often went on into the next day. So we should train our habit in wider firmer ways. I have seen in my time a great lord, a person famous for his successes in several expeditions of high importance, who effortlessly and in the course of his ordinary meals never drank less than two gallons of wine and who, after that, never showed himself other than most sage and well-advised in the conduct of our affairs.

We should allow more time to that pleasure which we wish to count on over the whole of our lives. Like shop-apprentices and workmen we ought to refuse no opportunity for a drink; we ought always to have the desire for one in our heads: it seems that we are cutting down this particular one all the time and that, as I saw as a boy, dinner parties, suppers, and late-

15. '88: dull. *Plato attributes to it the same effect on the mind.* [B] And we can . . . (Cf. Erasmus, *Adages*, IV, III, LVIII, *Non est dithyrambus qui bibit aquam*; Rabelais, *Tiers Livre*, TLF, *Prologue*, 175ff.; Plutarch (tr. Amyot), *Propos de Table*, 364B; 420A.) Joannes Sylvius (Dubois) was a doctor and pharmacologist of note. He died in 1576.

16. '95: life: *and where do you hope more rightly to find them among the natural pleasures?* But . . .

night feasts used to be much more frequent and common in our houses than they are now. Could we really be moving towards an improvement in something at least! Certainly not. It is because we throw ourselves into lechery much more than our fathers did. Those two occupations impede each other's strength. On the one hand lechery has weakened our stomachs: on the other, sober drinking has rendered us vigorous and lively in our love-making.

It is wonderful what accounts I heard my father give of the chastity of his times. He had the right to say so, as he was both by art and nature most graceful in the company of ladies. He talked little and well; he intermingled his speech with elegant references to books in the vernacular, especially Spanish, and among the Spanish he frequently cited the so-called *Marco Aurelio*.[17] His face bore an expression of gentle seriousness, humble and very modest; he took particular care to be respectable and decent in his person and his dress both on horse and on foot. He was enormously faithful to his word and, in all things, conscientious and meticulous, tending rather towards over-scrupulousness. For a small man he was very strong, straight and well-proportioned; his face was pleasing and rather brown; he was skilled and punctilious in all gentlemanly sports. I have also seen some canes filled with lead with which he is said to have exercised his arms for throwing the bar and the stone or for fencing, as well as shoes shod with lead to improve his running and jumping. Folk recall little miracles of his at the long-jump. When he was over sixty I remember him laughing at our own agility by vaulting into the saddle in his furry gown, by putting his weight on his thumb and leaping over a table and by never going up to his room without jumping three or four steps at a time. But more to my subject, he said that there was hardly one woman of quality in the whole province who was ill-spoken of, and he would tell of men – especially himself – who were on remarkably intimate terms with decent women without a breath of suspicion. In his own case he solemnly swore that he came virgin to his marriage-bed; and yet he had long done his bit in the transalpine wars, leaving a detailed diary of events there, both public and personal. And he married on his return from Italy in 1528 at the mature age of thirty-three.

Let us get back to our bottles.

[A] The disadvantages of old age (which has need of support and renewal) could reasonably give birth to a desire for drink, since a capacity for wine is virtually the last pleasure which the passing years steal from us.

17. The *Libro aureo del emperador Marco Aurelio* of Bishop Antonio de Guevara.

According to our drinking fraternity natural heat first gets a hold on our feet; that concerns our childhood; from there it rises to our loins where it long settles in, producing there if you ask me the only true bodily pleasures of this life: [C] in comparison, the other pleasures are half asleep. [A] Finally, like a mist rising and evaporating, it lands in the gullet and makes there its last abode.

[B] For all that, I do not understand how anyone can prolong the pleasure of drinking beyond his thirst, forging in his mind an artificial appetite which is contrary to nature. My stomach would never get that far: it has enough bother dealing with what it takes in for its needs. [C] I am so constituted that I care little for drink except at dessert; that is why my last draught is usually my biggest. Anacharsis was amazed that the Greeks should drink out of bigger glasses at the end of their meals;[18] it was I think for the same reason that the Germans do: that is when they start their drinking contests.

Plato forbids young people to drink before the age of eighteen and to get drunk before forty. But men over forty he tells to enjoy it and to bring copiously into their banquets the influence of Dionysius, that kind god who restores gaiety to grown men and youth to the old ones, who calms and softens the passions of the soul just as iron is softened by the fire. And in his *Laws* he considers convivial drinking to be useful (provided that the group has a leader to ensure that order is maintained), since getting drunk is a good and certain trial of each man's character and, at the same time, has the property of giving older men the idea of enjoying themselves in music and dancing, useful pastimes which they would not dare to engage in when of settled mind. Wine also has the capacity of tempering the soul and giving health to the body. Nevertheless he liked the following restrictions, partly borrowed from the Carthaginians: that it should be done without on military expeditions; that all statesmen and judges should abstain when about to perform their duties and to deliberate on matters of public concern; that the daytime should be avoided – that is owed to other activities – as well as any night when we intend to beget children.[19]

They say that the philosopher Stilpo, weighed down by old age, deliberately hastened his death by drinking his wine without water. A similar cause suffocated the failing powers of the aged philosopher Arcesilaus, but that was unintentional.[20]

---

18. Diogenes Laertius, *Life of Anacharsis*.
19. Cf. Tiraquellus, *De legibus connubialibus*, XIII, §147, citing Plato's *Laws*.
20. Diogenes Laertius, *Lives of Stilpo* and of *Arcesilaus*.

[A]   Whether the soul of a wise man should be such as to surrender to the power of wine is an old and entertaining question:

> *Si 'munitae adhibet vim sapientiae'.*

[Whether 'wine should be able to make an assault on secure wisdom'.][21]

To what inanities are we driven by that good opinion we men have of ourselves! The best governed Soul in the world has quite enough to do to stay on her feet and to keep herself from falling to the ground from her own weakness. Not one in a thousand can stand up calm and straight for one instant in her life; it can even be doubted, given her natural condition, whether she ever can. But if you add constancy as well, then that is her highest perfection: I mean if nothing should shake it, something which hundreds of events can do. It was no good that great poet Lucretius philosophizing and bracing himself: a love-potion drove him insane. Do they think that an apoplexy will not make Socrates lose his wits as much as a porter? Some have forgotten their own names by the force of an illness, and a light wound has struck down the judgement of others. A man can be as wise as he likes: he is still a man; and what is there more frail, more wretched, more a thing of nothing, than man? Wisdom cannot force our natural properties:

> [B]   *Sudores itaque et pallorem existere toto*
> *Corpore, et infringi linguam, vocemque aboriri,*
> *Caligare oculos, sonere aures, succidere artus,*
> *Denique concidere ex animi terrore videmus.*

[Then we see sweat and pallor take over his whole body, his tongue grows incoherent, his voice fails, his eyes are troubled, his ears begin to ring, his legs give way and he falls to the ground, as panic seizes his mind.][22]

[A]   When he is threatened with a blow nothing can stop a man closing his eyes, or trembling if you set him on the edge of a precipice,   [C]   just like a child, Nature reserving to herself these signs of her authority, signs slight but unattackable by reason or Stoic virtue, in order to teach Man that he is mortal and silly.   [A]   He becomes livid with fear; he reddens with shame; he bewails an attack of colic paroxysms if not with a loud cry of despair at least with a cry which is broken and wheezing.

> *Humani a se nihil alienum putet!*

[Let him realize that nothing human is a stranger to him!][23]

21. Horace, *Odes*, III, xxviii, 4.
22. Lucretius, III, 155–8.
23. Terence, *Heautontimorumenos*, I, i, 25.

Poets [C] who can make up anything they like [A] dare not relieve their heroes even of the burden of weeping:

> *Sic fatur lachrymans, classique immittit habenas.*

[Thus spoke Aeneas through his tears and his fleet sailed unbridled away.]²⁴

It suffices that a man should rein in his affections and moderate them, for it is not in his power to suppress them. And my very own Plutarch – so perfect, so outstanding a judge of human actions – when confronted by Brutus and Torquatus killing their children was led to doubt whether virtue could really get that far, and whether those great men had not in fact been shaken by some passion or other.²⁵ All actions which exceed the usual limits are open to sinister interpretations, since higher things are no more to our taste than inferior ones.

[C] Let us leave aside that other School which makes an express profession of pride.²⁶ Yet even in that third School which is reckoned to be the most indulgent of them all we hear similar boastings from Metrodorus:²⁷ '*Occupavi te, Fortuna, atque cepi; omnesque aditus tuos interclusi, ut ad me aspirare non posses.*' [I have forestalled you, O Fortune and I have caught you; I have blocked off all your approaches, you cannot get near me.]

When Anaxarchus, on the orders of Nicocreon, Tyrant of Cyprus, was put into a stone mortar and beaten to death with blows from an iron pestle, he never ceased to cry, 'Go on! Strike, bash on, you are not pounding Anaxarchus but his casing';²⁸ [A] when we hear our Christian martyrs shouting out to the tyrant from the midst of the flames, 'It is well roasted on this side; chop it off and eat it; it is cooked just right: now start on the other side'; when we hear in Josephus²⁹ of the boy who was torn to pieces with clawed pincers and bored through by the bradawls of Antiochus, yet who still defied him, crying out in a firm assured voice: 'Tyrant! You are wasting your time! I am still here, quite comfortable! Where is this pain, where are those tortures you were threatening me with? Is this all you can do? My constancy hurts you more than your cruelty hurts me!

24. Virgil, *Aeneid*, VI, 1.
25. Plutarch (tr. Amyot), *Publicola*, III.
26. The Stoics.
27. The Epicureans; Cicero, *Tusc. disput.*, V, ix, 27, citing Metrodorus the pupil of Epicurus.
28. Diogenes Laertius, *Lives of Philosophers*, I, civ.
29. Flavius Josephus, *De Macabaeorum martyrio.*

You cowardly beggar! It is you who are surrendering: I am growing stronger! Make me lament, make me give way, make me surrender, if you can! Goad on your henchmen and your hangmen: they have lost heart and can do nothing more! Give them weapons! Egg them on!' – then we have to admit that there is some change for the worse in their souls, some frenzy, no matter how holy.

When we hear such Stoic paradoxes as, 'I would rather be raging mad than a voluptuary' [C] – that is the saying of Antisthenes,[30] [A] *Μανείειν μᾶλλον ἢ ἡθείειν* – when Sextius tells us that he would rather be transfixed by pain than by pleasure; when Epicurus decides to treat gout as though it were tickling him, refuses rest and good health, light-heartedly defies ills and, despising less biting pains, will not condescend to struggle in combat against them but summons and even wishes for pains which are strong and anguishing and worthy of him:

> *Spumantemque dari pecora inter inertia votis*
> *Optat aprum, aut fulvum descendere monte leonem;*

[Amidst his placid flock he prays to be vouchsafed some slavering boar, or that some wild lion will come down from the mountain;][31]

who does not conclude that those are the cries of a mind which is leaping out of its lodgings? Our Soul cannot reach so high while remaining in her own place. She has to leave it and rise upwards and, taking the bit between her teeth, bear her man off, enrapture him away so far that afterwards he is amazed by what he has done; just as in war, the heat of the combat often makes the valiant soldiers take such hazardous steps that they are the first to be struck with astonishment once they have come back to themselves; so too the poets are often seized by amazement by their own works and no longer recognize the defiles through which they had passed at so fine a gallop. In their case too it is called frenzy and mania. And just as Plato says that a sedate man knocks in vain at poetry's door, so too Aristotle says that no outstanding soul is free from a mixture of folly.[32] He is right to call *folly* any leap – however praiseworthy it might be – which goes beyond our reason and our discourse. All the more so in that wisdom is a controlled handling of our soul, carried out, on our Soul's responsibility, with measure and proportion.

---

30. Erasmus, *Apophthegmata*, VII, *Antithenes Atheniensis*, III; other examples from Aulus Gellius, IX, v, and Sextus Empiricus, *Hypotyposes*, III, xx.
31. Virgil, *Aeneid*, IV, 158–9.
32. Seneca, *De tranquillitate*, XV (a major borrowing).

[C]   Plato contends that the faculty of prophesying is 'above ourselves'; that we must be 'outside ourselves' when we accomplish it; our prudence must be darkened by some sleep or illness, or else snatched out of its place by a heavenly rapture.[33]

33. Plato, *Timaeus*, 71D–72A.

3. A custom of the Isle of Cea

===

[Montaigne shows with examples and pro et contra *arguments that philosophy has its own way of favouring self-destruction and of opposing it with equally strong reasons. Traditionally, theology classed suicide as a crime (we 'commit' suicide). That was because it is defined as the prime example of despair, whereas hope is one of the three theological virtues: Montaigne (after due submission to the will of God) shows that suicide does not always arise from despair: it can be provoked by many motives including hope. He is often said to be bold or even anti-Christian in his attitudes. That judgement cannot stand a comparison between what Montaigne writes and what was written on the subject by Jesuit casuists and theological students of morals, some of whom he had evidently read and who use the same arguments and* exempla *as he does.]*

[A] If, as they say, to philosophize is to doubt, then, *a fortiori*, to fool about and to weave fantasies as I do must also be to doubt. For it is the role of apprentices to ask questions and to debate: the professor provides the solutions from his chair. My professor is the authority of God's Will, which undeniably governs us and which ranks way above vain human controversies.

When Philip had entered the Peleponnesus with his army, somebody told Damidas that the Spartans would have sufferings in plenty if they did not get back into his favour. 'Coward,' he replied; 'what can men suffer who do not fear death?' Agis was similarly asked how a man could live in freedom: 'By holding death in contempt,' he replied. These and a thousand similar assertions which agree on this matter evidently mean something more than merely patiently waiting for death to come. For in life there are many events harder to suffer than death itself. Witness that Spartan boy who was captured by Antigonus then sold as a slave: when his master pressed him to perform some abject task he said: 'I will show you what you have bought; it would be shameful for me to be a slave when freedom is at hand.' And so saying, he jumped to his death from the top of the house. When Antipater was uttering bitter threats against the Spartans to force them to acquiesce in one of his demands, their answer was: 'If you are threatening us with something worse than death, we will be all the more

willing to die.'¹ [C] And when Philip wrote to them that he would thwart all their undertakings, 'What,' they said, 'will you stop us from dying?'

[A] The saying goes that a wise man lives not as long as he can but as long as he should, and that the greatest favour that Nature has bestowed on us, and the one which removes all grounds for lamenting over our human condition, is the one which gives us the key to the garden gate; Nature has ordained only one entrance to life but a hundred thousand exits.²

[B] We may not have enough land to live off but (as Boiocalus said to the Romans) we shall never lack land to die on. [A] Why raise plaints about this world? It has no hold on you; if you live in anguish the cause lies with your cowardice: to end your life you need only the will to do so:

> *Ubique mors est: optime hoc cavit Deus,*
> *Eripere vitam nemo non homini potest;*
> *At nemo mortem: mille ad hanc aditus patent.*

[Death can be found everywhere. It is a great favour from God that no man can wrest death from you, though he can take your life; a thousand open roads lead to it.]³

And it is not the prescription for one single illness: death is the prescription for all our ills. Death is an assured haven, never to be feared, often to be sought. It comes to the same thing if a man makes an end to himself or passively accepts it; whether he runs to meet his last day or simply awaits it; wherever death comes from, it is always his death; no matter where the thread may break, the whole thread is broken: there is no more life on the spindle.

The fairest death is one that is most willed. Our lives depend on the will of others: our death depends on our own. In nothing whatever should we bow to our humour more than in this. Reputation has nothing to do with such an undertaking: to take it into account is madness. Living is slavery if the freedom to die is wanting.

Cures are normally effected at the expense of life: we are cut about and cauterized; they lop off our limbs, they deprive us of food or of blood: one more step and they have cured you once and for all! Why is the vein in our

---

1. Several examples, all from Plutarch (tr. Amyot), *Dicts notables des Lacedae-moniens.*
2. Cicero, *Tusc. disput.,* V, XIV, 42; then many borrowings from Seneca, *Epist. moral.,* LXIX–LXXVIII, especially LXX.
3. Tacitus, *Annales,* XIII, lvi; Seneca, *Phoenissae,* 151–3.

gullet not as much at our command as the vein used for bleeding? Strong diseases need strong remedies. When Servius the grammarian suffered from gout, the best thing he could do, he decided, was to rub in poison and kill off his legs. [C] Let them be as gouty as they liked, as long as he could not feel them. [A] God gives us ample leave to go when he reduces us to the state where living is worse than dying. [C] It is weakness to give in to evils, but madness to tend them.

According to the Stoics, 'living in conformity with Nature' means that the wise man can even depart from this life while still enjoying good fortune, provided that he does so opportunely; but it also means that the fool can remain alive even when he is wretched, provided that he still has the benefit of most of the things which they define as being 'in accord with Nature'.[4]

Just as I break no laws against theft when I make off with my own property or cut my own purse, nor the laws against arson if I burn my own woods, so too I am not bound to the laws against murder if I take my own life. Hegesias said that both the circumstances of our life and the circumstances of our death should depend on our choice. And when Diogenes met Speucippus the philosopher, long afflicted with dropsy and being borne on a litter, he was greeted thus: 'I wish you good health, Diogenes'; but he retorted, 'No good health to you, who allow yourself to live in such a condition.' And, truly, soon afterwards Speucippus did have himself put to death, distraught by the painful circumstances of his life.[5]

[A] That does not go by without opposition. For [A1] many hold [A] that[6] we may not leave our guard-duty in this world without the express commandment of Him who has posted us here; that it is for God (who has sent us here not for ourselves alone but for his glory and for the service of others) to grant us leave-of-absence when he wishes; it is not ours for the taking; [C] that we were not born for ourselves alone but for our country also; that the law can sue us for damages and bring an action for homicide against us; [A] otherwise, as deserters from our duty we are punished in this world and the next:[7]

4. Cicero, *De finibus*, III, xviii, 60.
5. Diogenes Laertius, *Lives of Aristippus* and *of Speucippus* (up to this point, Montaigne's position is that of Seneca).
6. [A] originally read: For, *apart from that authority which, when forbidding murder, included self-murder in it*, many *philosophers* hold . . .
7. The great commonplace from Plato's *Phaedo*: see St Augustine, *City of God*, I, xxii; Erasmus, *Adages*, IV, VI, LXXXI, *Nemo sibi nascitur*; Tiraquellus' discussion for and against suicide in *De nobilitate*, XXXI (where Plato is cited, §561).

> *Proxima deinde tenent mæsti loca, qui sibi lætum*
> *Insontes peperere manu, lucemque perosi*
> *Projecere animas.*

[Then, nearby, was the region where, overwhelmed with sadness, stand the just who had killed themselves by their own hand and, loathing the light of day, had thrown away their souls.][8]

There is more constancy in wearing out our chains than in breaking them and a greater test of firmness in Regulus than in Cato.[9] It is rashness and impatience which hasten our steps. No mishap can make living Virtue turn her back: she goes looking for ills and pains and feeds on them. The threats of tyrants, torture and executioners are life and soul to her:

> *Duris ut ilex tonsa bipennibus*
> *Nigræ feraci frondis in Algido,*
> *Per damna, per cædes, ab ipso*
> *Ducit opes animumque ferro.*

[Like an oak-tree lopped of its leafy boughs by harsh axes on dark-leaved Mount Algidus: its wounds, its losses, the very iron which strikes it, give it fresh vigour.][10]

Or, as they say:

> *Non est, ut putas, virtus, pater,*
> *Timere vitam, sed malis ingentibus*
> *Obstare, nec se vertere ac retro dare . . .*

[Virtue is not as you think, Father, fearing life; it is confronting huge evils without turning one's back or retreating   ][11]

> *Rebus in adversis facile est contemnere mortem:*
> *Fortius ille facit qui miser esse potest.*

[In adversity it is easy to despise death: stronger is the man who can live in misery.][12]

It is the role of Cowardice not Virtue to avoid the blows of Fortune by crouching in a hollow grave beneath a massive tombstone. Virtue never breaks off her journey or slackens her pace, no matter what the storm.

8. Virgil, *Aeneid*, IV, 434–7:
9. St Augustine, *City of God*, I, xxii and xxiv.
10. Horace, *Odes*, IV, iv, 57–60.
11. Seneca (the dramatist), *Phoenissae*, 190–93.
12. Martial, XI, lvi, 15–16.

> *Si fractus illabatur orbis,*
> *Inpavidam ferient ruinæ.*

[If the world were to shatter and fall on him, its ruins would strike him but fear would not.][13]

As often as not, flying from other ills brings us to this one; indeed, flying from death often means running towards it:

> [C]   *Hic, rogo, non furor est, ne moriare, mori!*

[I ask you! Is it not madness to perish in order to avoid death!][14]

[A]   It is like those who are afraid of heights and then jump off the edge:

> *multos in summa pericula misit*
> *Venturi timor ipse mali; fortissimus ille est,*
> *Qui promptus metuenda pati, si cominus instent,*
> *Et differe potest. . .*

[The very fear of future ills have driven many into great dangers; strongest of all is the man who can brave dangers when they come but who knows how to avoid them when possible. . .][15]

> *Usque adeo, mortis formidine, vitæ*
> *Percipit humanos odium, lucisque videndæ,*
> *Ut sibi consciscant mærenti pectore lethum,*
> *Obliti fontem curarum hunc esse timorem.*

[Fear of dying can even bring men to hate life and the very sight of the light so that, with heavy heart, they arrange their own deaths, forgetting that the source of all their distress was their fear of dying.][16]

[C]   In his *Laws* Plato ordains an ignominious funeral for any man who has deprived his nearest and dearest (namely himself) of his life and of his destined course when not compelled by the sentence of the public court, by some sad circumstance of Fortune which cannot be avoided or by some unbearable shame, but only by the cowardice and weakness of a timorous soul.[17]

---

13. Horace, *Odes*, III, iii, 7–8.
14. Martial, II, lxxv, 2.
15. Lucan, *Pharsalia*, VII, 104–7.
16. Lucretius, III, 79–82.
17. Plato, *Laws*, 9. (See Tiraquellus, *De nobilitate*, XXXI, §561.)

[A]   Moreover the opinion which holds our life in contempt is a
ridiculous one. For, in the end, life is our being and our all. Creatures who
enjoy a being richer and nobler than we do may well criticize ours, but it is
unnatural that we should despise ourselves or care little for ourselves; it is a
sickness peculiar to Man to hate and despise himself; it is found in no other
animate creature.

It is a similar vain desire which makes us want to be something other
than what we are. The fruits of such desires can never be of concern to us
since that desire is self-contradictory; it works against itself. Anyone who
wishes to be changed from man to angel does nothing at all for himself: *he*
would gain nothing by it. Who is supposed to be feeling that amendment
for him and rejoicing at it? *He* is no more:

> [B]   *Debet enim, misere cui forte ægreque futurum est,*
> *Ipse quoque esse in eo tum tempore, cum male possit*
> *Accidere.*

[If anyone must perhaps be wretched and suffer pain in the future, then he himself
must exist in that future when such evil occurs.][18]

[A]   Freedom from care, from pain and from emotion, together with
freedom from the evils of this life, if purchased by our deaths can bring no
advantage to us. Avoiding war means nothing if *you* cannot enjoy the
peace: fleeing pain means nothing to a man who has no means of savouring
the respite.

Among those who maintained the first alternative there was considerable
uncertainty over what occasions could fully justify anyone deciding to take
his own life. (They called that an εὔλογον ἐξαγωγὴν [a *reasonable exodus*]).[19]
They say in fact that one ought to end one's life for quite minor causes,
since the causes which keep us alive are not very strong either; but there
has to be a degree of moderation.

There have been fantastical and baseless humours which have driven not
only individual men but whole peoples to do away with themselves. I have
already cited some examples; we can read in addition of those maidens of
Miletus who conspired in their frenzy to hang themselves one after another
until the magistrates considered the matter and commanded that any found
hanging should be dragged by the same rope naked through the city.[20]

18. Lucretius, III, 862–4.
19. A concept attributed to Zeno the philosopher.
20. Tiraquellus, *De legibus connubialibus*, IV, §32 (after Plutarch's *Famous Women*).

Threicion urged Cleomenes[21] to kill himself because of the sorry state of his affairs: as he had fled from a most honourable death in the battle he had just lost, he should accept this other one which abounds in honour for him, and give the victors no opportunity of making him suffer a shameful death or a shameful life. Cleomenes, with a Stoic Spartan courage, rejected this counsel as weak and effeminate: 'That is a remedy,' he said, 'that I will never be without but which no one should use while there remains a finger's breadth of hope,' adding that to go on living sometimes requires valour and constancy and that he wished his very death to be of service to his country; he intended to make it an honourable and virtuous deed. Threicion took his own advice and killed himself. Cleomenes did the same later on, but only after assaying the very worse that Fortune can do.

All ills are not worth our avoiding them by death. Moreover, there are so many sudden reversals in the affairs of men that it is not easy to judge at what point it is right to abandon hope:

> [B]   *Sperat et in sæva victus gladiator arena,*
> *Sit licet infesto pollice turba minax.*

[Even when lying vanquished on the cruel sand, while the menacing crowd in the arena turn their thumbs round, the gladiator still hopes on.][22]

[A]   There is an ancient saying that anything can be hoped for while a man is still alive. But Seneca replies, 'Ah yes; but why should I recall that Fortune can do all things for one who remains alive rather than that other saying, that Fortune can impose nothing on one who knows how to die?'[23]

We can read how Josephus[24] was involved in a danger so clear and so imminent (with an entire nation in revolt against him) that he could not reasonably hope for relief; yet, as he tells us, he was advised at this juncture to do away with himself but was right as it turned out to cling stubbornly to hope, for Fortune so changed the entire situation beyond any human foresight that he found himself delivered from danger quite unharmed. Brutus and Cassius on the other hand, by the precipitous haste with which they killed themselves before the time or circumstances were right, brought about the final loss of the remnants of that Roman freedom which it was

21. Plutarch, *Life of Cleomenes*. (The man's name was *Therycion*.)
22. In the *Saturnalia* of Justus Lipsius, attributed to Pentadius.
23. Seneca, *Epist. moral.*, LXX, 7.
24. Flavius Josephus, *De vita sua*.

their duty to protect.[25] – [C] I have seen hundreds of hares escape from the very jaws of the greyhounds: *'Aliquis carnifici suo superstes fuit.'* [A man has been known to outlive his executioner.][26]

> [B] *Multa dies variusque labor mutabilis ævi*
> *Rettulit in melius; multos alterna revisens*
> *Lusit, et in solido rursus fortuna locavit.*

[Time in her wavering course has often produced great changes for the better; and Fortune, altering her course, has sported with men and restored them again to solid prosperity.][27]

[A] Pliny lists three kinds of illness which man can justly avoid by killing himself: the harshest of them all is a stone in the bladder with retention of urine;[28] [C] Seneca only allows those illnesses which chronically affect the faculties of the soul. [A] Others maintain that death is always permitted at man's discretion, to avoid a worse one.[29]

[C] Damocritus, the leader of the Aetolians, was led prisoner to Rome; one night he succeeded in escaping but being pursued by his guards he fell on his sword before they could recapture him.[30] When the city of Epirus was reduced to the last extremity by the Romans, Antinous and Theodotus advised mass suicide; but once the counsel to surrender prevailed they went and sought death, rushing upon the enemy, intent on striking blows not on protecting themselves.[31]

A few years ago when the island of Gosso was stormed by the Turks, a Sicilian with two beautiful nubile daughters killed them both and then killed their mother who came running up at their death. Once he had done that, he went out into the street with a crossbow and a harquebus; with two shots he killed the first two Turks who came near his door; he then grabbed a sword and threw himself furiously into a skirmish where he was

---

25. An addition by Montaigne has gone astray from the Bordeaux copy. In '95 we read: protect. *In the battle of Serisolles Monsieur d'Enghien made two assays at slashing his throat with his sword, despairing of the fortune of a battle, which, where he was, was going badly, and in his haste nearly deprived himself of the pleasure of so fair a victory.* I have. . .
26. Seneca, *Epist. moral.*, XIII, 11.
27. Virgil, *Aeneid*, I, 425–7.
28. Pliny, *Hist. nat.*, XXV. The stone was Montaigne's complaint.
29. Seneca, *Epist. moral.*, LVIII, 36.
30. Livy, XXXVII, xlvi.
31. Ibid., XLV, xxvi.

quickly surrounded and cut to pieces, saving himself from slavery after having first delivered his family from it.[32]

[A] Fleeing from the cruelty of Antiochus Jewish women, after circumcizing their infants, jumped to their deaths with them.[33]

I was told this tale about a prisoner from a good family in one of our French gaols: his parents, upon hearing that he would certainly be condemned to death, avoided such an ignominious end by procuring a priest to tell him that he had a sovereign way of escape: he should commend himself to a particular saint, making such and such a vow, then go a whole week without food, no matter how weak or faint he felt. He trusted him and, without realizing what he was doing, rid himself of life and subjection.

Scribonia advised her nephew Libo to kill himself rather than await the hand of Justice, telling him he was doing other people's work for them if he preserved his life merely to surrender it three or four days later into the hands of those who would come looking for it: he would be serving his enemies if he kept his own life-blood to be thrown to their dogs.[34]

We read in the Bible of Nicanor, a persecutor of God's law, who sent his guards to seize the good old man named Raxias, who in honour of his virtue 'was called the Father of the Jews'. When that good man saw no other way, once his gate was in flames and the enemy about to seize him, 'he struck himself with his sword, choosing to die nobly rather than to fall into the hands of the wicked and to be treated like a dog, in a manner unworthy of his noble birth: but whereas through haste he missed giving himself a sure wound, he ran to the wall through the throng and threw himself down into the crowd; but, as they made room for his fall, he fell straight on his head. Nevertheless, feeling there was still some life in him, he inflamed his heart, staggered to his feet all bloody under the weight of the blows, ran through the crowd and charged towards a certain rock, steep and precipitous, where with no strength left he thrust both hands through a wound, grasped his bowels, tore them out and squashed them together and cast them at his pursuers,' calling God's vengeance down upon them and bearing witness to it.[35]

Of all the violences done to the conscience the one most to be avoided, it

---

32. Narrated by Guillaume Paradin, *Histoire de son temps*.
33. Flavius Josephus, *Jewish Antiquities*, XII, v.
34. Seneca, *Epist. moral.*, LXX, 10.
35. II Maccabees 14:37–46 – virtually word for word from the Latin Vulgate. (The English Geneva Bible warns the reader that there are occasions when Biblical *exempla* are not to be followed: this suicide is one of them.)

seems to me, is violence against the chastity of women, since an element of bodily pleasure is naturally in it for them. For this reason their resistance cannot be abolutely complete and it would seem that the rape may be mingled with a kind of willingness. Pelagia and Sophronia have both been canonized:[36] the first cast herself and her mother and sisters into the river to avoid rape by a group of soldiers, while the other killed herself to avoid being raped by the Emperor Maxentius. [C] Ecclesiastical history reveres several examples of devout persons who sought death as a protection from outrages against their conscience prepared by tyrants.

[A] Future centuries may honour us for having a learned author in our days (a Parisian be it noted) who has gone to some pains to persuade the ladies today to take any other way out rather than to accept such a horrifying counsel of despair.[37] I am only sorry he did not know the story I heard in Toulouse so that he could include it in his tales; it concerns a woman who had passed through the hands of a group of soldiers: 'God be praised,' she said, 'that at least once in my life I have been satisfied without sin.'

But such cruelties are truly unworthy of French courtesy; thank God our climate has been thoroughly purged of them since that sound piece of advice – the rule of good old Marot: it is enough for women to say 'No, no!' while doing it.[38]

History is full of people who have, in thousands of ways, exchanged a pain-filled life for death. [B] Lucius Aruntius killed himself, 'to escape', he said, 'from the future *and* the past'.[39] [C] Granius Silvanus and Statius Proximus killed themselves after being pardoned by Nero so as not to live by the grace of so wicked a man, or else so as not to have to beg for a second pardon seeing the ease with which he suspected and accused all men of honour.[40]

Spargapises, the son of Queen Tomyris, was taken prisoner by Cyrus; released of his bonds, he exploited this very first favour that Cyrus had granted him to kill himself, never having intended any other profit from his freedom than to atone with his life for the shame of his capture.[41]

---

36. Cf. St Augustine, *City of God*, I, xxv–xxvi; he feared that some virgins might, despite themselves, enjoy rape. Nevertheless, except when individually counselled to do so by God, desire to avoid such pleasure does not justify suicide. Vives in his notes cites Montaigne's examples of Pelagia and Sophronia, after Eusebius' *Ecclesiastical History*.

37. Allusion to some *conteur*, not a theologian.

38. Clément Marot, *De nenny* (ed. Guiffrey), IV, 241.

39. Tacitus, *Annals*, V.

40. Ibid., XV.

41. Herodotus, I, ccxiii.

Bogez, who was Governor of Eon on behalf of King Xerxes, was besieged by the Athenian army under the leadership of Cimon but refused the suggested terms of a safe-conduct to Persia for him and his goods since he could not bear to survive the loss of what his Master had placed under his guard; and after having defended his city to the very end when there was nothing more left to eat, he first threw into the river Strymon all the gold and everything else which he thought the enemy might take as booty; then, having ordered a huge pyre to be lighted and the throats of his wife, children, concubines and servants to be slit, he cast them, and then himself, into the flames.[42]

Ninachetuen, an Indian Lord, when he first got wind of the Portuguese Viceroy's intention to strip him, for no apparent reason, of the office he filled in Malacca so as to bestow it on the King of Campar, privately resolved to act as follows: he had a scaffold erected, longer than it was wide, supported on columns, royally carpeted with flowers and decorated with an abundance of sweet-smelling woods. Then, having donned a robe of cloth-of-gold laden with precious stones of great price, he issued forth into the street and mounted the steps of the scaffold in the corner of which burned a pyre of aromatic wood. Everyone ran out to see what these unusual preparations might portend. With a countenance both brave and angry, he recalled what the Portuguese people owed to him; how faithfully he had carried out his duties; how, for the sake of others, he had often borne witness, weapon in hand, that honour was much dearer to him than life; he was not going to give up caring for honour in his own case; but, although Fortune denied him any way of resisting the insult they intended to do him, his mind told him to remove his power of feeling it or of serving as a fable to the people and as a triumph for persons less worthy than himself. So saying he threw himself into the fire.[43]

[B] Sextilia the wife of Scaurus, and Paxea the wife of Labeo, to encourage their husbands to avoid the dangers which beset them and in which they personally were not concerned except as loving wives, voluntarily took their own lives so as to serve them as examples in their dire necessity and to keep them company.

What they did for their husbands Coceius Nerva did for his country, less usefully but with just as much love. That great jurisconsult, in the full

42. Herodotus, VII, cvii.
43. Simon Goulart, *Histoire du Portugal*. Examples follow from Tacitus, *Annals*, Livy, Quintus Curtius and Plutarch (tr. Amyot), *Du trop parler*, 93D–E.

bloom of health, wealth, reputation and respect from the Emperor, killed himself for no other reason than compassion for the wretched condition of the Roman Republic.

Nothing could surpass the delicacy shown by the death of the wife of Fulvius, the close friend of Augustus. One morning Augustus, having learned that Fulvius had let out a vital secret entrusted to him, gave him a meagre welcome when he came to see him. Fulvius returned home in despair and told his wife piteously that he had resolved to kill himself for having fallen into this misfortune. She frankly replied: 'That is only right, seeing that you have had enough experience of the indiscipline of my tongue, yet it did not put you on your guard. But wait; let me kill myself first.' Then without more ado she thrust the sword through her body.

[C] Vibius Virius, despairing of saving his city, Capua, which was besieged by the Romans or of obtaining mercy for it, spoke up in the last debate held in their Senate, made several exhortations suggesting his conclusion and ended by declaring that the finest thing to do was to escape their fortune by their own hands: their enemies would hold them in honour and Hannibal would realize what faithful allies he had deserted. He invited those who approved of his counsel to come and partake of a good supper already prepared in his home and then, after making good cheer, they would all drink together from the cup he would offer them: 'It is a drink that will deliver our bodies from torment, our souls from insults and our eyes and our ears from knowledge of the base evils which the vanquished have to suffer from enemies, cruel and incensed. I have,' he said, 'arranged for there to be men able to throw us on to a funeral pyre before my door once we have breathed our last.'

Many gave their approval to this high resolution but few imitated him. Twenty-seven senators did follow him and, after assaying to stifle their dreadful thoughts in wine, they finished their meal with that deadly drink; they lamented together their country's misfortunes and embraced each other; then some withdrew to their homes while the others remained behind to be laid with Vibius on his flaming pyre. All of them were so long a-dying, since the fumes of the wine had blocked their arteries and retarded the effects of the poison, that some came within an hour of seeing their enemies in Capua (which was taken the next day) and of incurring the very miseries they had fled from at such a cost.

When the Consul Fulvius was returning from the disgraceful butchering of two hundred and twenty-five senators, Taurea Jubellius, another citizen from those parts, called him back by name fiercely, made him stop, then

said: 'Command them to add me to so great a massacre so that you can at least boast of killing a man more valiant than you are.' Fulvius treated him with disdain, as a madman (since his hands were bound by a letter just arrived from Rome condemning the inhumanity of his action); Jubellius went on: 'My country is occupied; my friends are dead; although I have killed my wife and children by my own hand to save them from the desolation of this defeat, it is forbidden me to die the same death as my fellow-citizens; so let Virtue lend me the means to take vengeance on this odious life.' And drawing a sword which he had concealed he ran it through his bosom and fell dying at the consul's feet.

[B] Alexander was besieging a town in India: those within the town, finding themselves hard-pressed, vigorously resolved to deprive him of the joy of victory, and – despite his humanity – they all set fire to their town and burned themselves to death. A new kind of war: the enemy fought to save them: they, to destroy themselves; to ensure that they died they did all that men normally do to protect their lives.

[C] Astapa, a town in Spain, had walls and defence-works too weak to withstand the Romans so the inhabitants made a pile of their valuables and movable goods in the market-place and placed their wives and children on top of the heap, surrounding it with wood and other materials which catch fire easily; then, leaving behind fifty younger men to carry out their plan, they made a sortie during which, as they had sworn to do, they all sought death, not being able to win the battle. The fifty young men, having first massacred every living soul scattered about their town, set fire to the heap and then threw themselves upon it, so bringing their great-hearted freedom to an end in insentience rather than in shame and sorrow; they showed their enemies that if it had pleased Fortune they would have been as brave in wresting victory from them as they had been in frustrating them of a victory which was horrifying and indeed mortal to those who had fallen for the bait of the glittering gold melting in those flames and who had crowded round it, only to be suffocated and burned to death, unable to draw back because of the crowd behind them.

The citizens of Abydoss, invested by Philip, made the same resolution. But they had too little time. King Philip, horrified by the desperate haste of their preparations (and having already seized the treasures and the portable property they had each condemned to destruction by fire or water) withdrew his soldiers and allowed the townsfolk three days' grace to kill themselves, days which they filled with blood and murder exceeding any enemy's cruelty; not one person was saved who had power over himself.

There are countless similar examples of mass resolution: they seem all the more horrible for applying to everyone; but they are less horrible in fact than when done individually. What reason cannot do for each man separately it can do for them all together, their enthusiasm as a group ravishing each individual judgement.

[B] In the time of Tiberius the condemned men who waited to be executed forfeited their property and were denied funeral rites: those who anticipated it by killing themselves were buried and allowed to make a will.

[A] But sometimes we can desire death out of hope for a greater good: 'I want', said St Paul, 'to be loosened asunder so as to be with Jesus Christ,' and, 'Who shall deliver me from these bonds?' Cleombrotus Ambraciota, having read the *Phaedo* of Plato, entered into so great a yearning for the life to come that, without further cause, he cast himself into the sea. [C] That clearly shows how incorrect we are to call this deliberate 'loosening asunder' despair: we are often brought to it by a burning hope – often, also, by a calm and certain propensity of our judgement.[44]

[A] During the journey to Outremer made by Saint Louis, Jacques du Chastel the Bishop of Soissons saw that the King and the whole army were preparing to return to France leaving their religious affairs unfinished; he resolved, rather, to leave for Paradise: having said God speed to his friends, he charged single-handed into the enemy in full view of everyone and was cut to pieces.[45]

[C] In one particular kingdom in a recently discovered country[46] there is a day of solemn procession during which the idol that is worshipped there is carried in public on a festival-car of astonishing dimensions; many can be seen cutting off chunks of their living flesh to offer to the idol and, in addition, a number of others prostrate themselves in the main square to be crushed and smashed to pieces by the wheels in order to win such veneration as saints after their death as is indeed rendered to them.

---

44. St Paul, Philippians 1:23; Romans 7:24; Cicero, *Tusc. disput.*, I, xxxiv, 84. Contemporary theologians, philosophers and jurisconsults used these texts to show that suicide is often both reasonable and natural, but forbidden by God's ordinance which supersedes both reason and nature. (Cf. Bartholomew of Medina, *Expositio in Secundam Secundae* (of Thomas Aquinas), Salamanca, 1588; Tiraquellus, *De nobilitate*, XXXI, §§ 512–13.
45. Jean de Joinville, *Histoire et cronique de Saint Louis*, LI. (*Outremer*: the Crusader Kingdoms, and the Near East generally.)
46. Orissa. This is an account of the Juggernaut (Krishna's idol dragged in a huge carriage, beneath whose wheels pilgrims were said to immolate themselves).

The death of that Bishop, arms in hand, has more nobility but implies less pain, since the zeal of battle would have partly deadened his sense of feeling.

[A]   Some forms of government have been concerned to decide when suicide may be legal and opportune. In our own city of Marseilles in former times they used to keep a supply of a poison based on hemlock always available at public expense to all those who wished to hasten their days; they first had to get their reasons approved by their Senate (called the Six Hundred); it was not permissible to lay hands on oneself, save by leave of the magistrate and for lawful reasons.[47]

This same law was also found elsewhere. When sailing to Asia, Sextus Pompeius went via the island of Cea in the Aegean. As one of his company tells us, it chanced when Pompeius was there that a woman of great authority, who had just explained to the citizens why she had decided to die, begged him to honour her death with his presence; which he did; and having long vainly assayed to deflect her from her purpose with his eloquence (at which he was wonderfully proficient) and with his powers of dissuasion, he finally allowed her to do what pleased her. She had lived to be ninety, blessed in mind and body; now she was lying on her bed (made more ornate than usual) and was propped up on her elbow. 'Sextus Pompeius,' she said, 'may the gods be kind to you (especially the gods I leave behind rather than those I am about to discover) for you did not despise being my counsellor in life and my witness in death. For my part I have assayed only the kindlier face of Fortune; fearing that the desire to go on living might make me see an adverse one, I am with this happy death giving leave of absence to the remnant of my soul and leaving behind me two daughters and a legion of grandchildren.' She then addressed her relations, urging them to agree in peace and unity; divided her property and commended her household gods to her elder daughter; then with a steady hand she took the cup containing the poison; and having addressed her vows to Mercury, praying to be taken to some seat of happiness in the next world, she abruptly swallowed that mortal potion. She then kept the company informed of the progress of the poison as it worked through her body, telling how her limbs grew cold, one after another, until she was finally able to say it had reached her inward parts and her heart; whereupon she called on her daughters to do one last duty: to close her eyes.

Pliny gives an account of a certain Hyperborean people whose climate is

47. This and the following episode from Valerius Maximus, *Memorabilia*, II, vi, 7 and 8. (Cea or Ceos is an island of the Cyclades.)

so temperate that the inhabitants do not usually die before they actually want to; when they become weary, having had their fill of life and reached an advanced age, they hold a joyful celebration and then leap into the sea from a high cliff set aside for this purpose.[48]

[B]   Of all incitements   [C]   unbearable   [B]   pain and a worse death seem to me the most pardonable.

48. Pliny, *Hist. nat.*, IV, xii.

4. 'Work can wait till tomorrow'

[Montaigne discovered Amyot's French translation of Plutarch's Lives and his Moral Works after he had embarked on the Essays. His respect for Plutarch's wisdom and style does not stop him from drawing different moral conclusions from his examples. On the contrary: he is moved to imitate and rival Plutarch's admired wisdom and judgement. Amyot's translation was, even at the time, criticized for inaccuracy: Montaigne, without comparing it in scholarly detail with the Greek, is sure that Amyot had grasped the essence of Plutarch's mind. Both French and English readers have indeed found in Amyot's Plutarch a lasting source of pleasure and wisdom – North's famous English Plutarch is translated from Amyot's French, not from the original Greek. Scholars can, did and do find errors in them both, but it is they who are read, not Xylander or Cruserius or even more recent translators from among the scholars.]

[A] It seems to me that I am justified in awarding the palm, above all our writers in French, to Jacques Amyot, not merely for the simplicity and purity of his language in which he excels all others, nor for his constancy during such a long piece of work, nor for the profundity of his knowledge in being able to disentangle an author so complex and thorny (for you can say what you like: I cannot understand the Greek, but everywhere in his translation I see a meaning so beautiful, so coherent and so consistent with itself that either he has definitely understood the true meaning of his author or else, from a long frequentation with him, he has planted in his own soul a vigorous generic Idea of Plutarch's, and has at least foisted upon him nothing which belies him or contradicts him); but above all I am grateful to him for having chosen and selected so worthy and so appropriate a book to present to his country. Ignorant people like us would have been lost if that book had not brought us up out of the mire: thanks to it, we now dare to speak and to write – and the ladies teach the dominies; it is our breviary.

If that good man is still alive I would assign him Xenophon to do just as well with: that is an easier task – one therefore all the more suited to his advanced years. And it somehow seems to me that, even though Amyot can slip very briskly and neatly round some tight corners, his style is nevertheless more at home when it is untrammelled and can roll easily along.

I recently came upon the passage where Plutarch tells us that when he

himself was delivering a declamation in Rome Rusticus, who was in the audience, received a packet of letters from the Emperor and put off opening it until the end. For which, he says, all the audience most highly praised the dignity of that man. Indeed, since Plutarch was on the subject of curiosity (that avid passion, greedy for news, which leads us to drop everything indiscriminately and impatiently so as to entertain every newcomer and, losing all sense of respect and politeness, to tear open the letters they bring us no matter where we may be) he was right to praise Rusticus' dignity; and he could also have gone on to praise his decency and his courtesy in not wanting to interrupt his declamation.[1]

But I doubt whether Rusticus could be praised for his wisdom: for since he was receiving those letters unexpectedly, and from an Emperor at that, to put off reading them might have had grave consequences.

The opposite vice to curiosity is lack of concern, [B] which my complexion manifestly inclines me to, and [A] which is so extreme in many men I have known that you can find them with unopened letters in their pockets brought three or four days earlier.

[B] I not only never open any letter entrusted to me but not even any which Fortune may pass through my hands; I feel guilty if, when standing beside some great man, my eyes inadvertently thieve some knowledge from the important letter he is reading. Never was anyone less inquisitive, less given to poking about in another man's affairs.

[A] In our fathers' time Monsieur de Boutières nearly lost Turin because he was enjoying good company at dinner and put off reading a warning sent to him of some treachery being plotted against that town, which was under his command. And I also learned from Plutarch that Julius Caesar would have saved himself if he had read a note which was handed to him that day on his way to the Senate where the conspirators killed him.[2]

Plutarch relates too how Archias, the Theban Tyrant, on the evening before Pelopidas executed his plan to kill him and so restore freedom to his country, was written to by another Archias, an Athenian, to inform him point by point of what was being prepared against him. This missive was delivered to him during dinner; he put off opening it, saying words which later became a Greek proverb: 'Work can wait till tomorrow.'[3]

---

1. Plutarch (tr. Amyot), *De la curiosité*, 67 G–H.
2. Plutarch, *Life of Caesar*.
3. Erasmus, *Adagia*, IV, VII, LX, *In crastinum seria* (after Plutarch's *Life of Pelopidas*; cf. also Plutarch's *Du démon de Socrates*, 647G–648C.

In my opinion a wise man can (out of concern for others, such as not impolitely interrupting a social event, as was Rusticus' case, or so as not to break into some other affair of importance) put off reading any news brought to him; but, particularly if he holds some public office, to do so for his own interest or pleasure – not interrupting his dinner or even his sleep – is unpardonable. And in ancient Rome the 'consular place' as they called it was the most honoured seat at table, since it was the one most readily accessible to those who might come [C] and consult the man seated there.[4] [A] Which shows that even at table the Romans did not cut themselves off from dealing with other matters and with unexpected occurrences.[5]

But when all has been said, it is not easy in any human activity to lay down a rule so well grounded on reasoned argument that Fortune fails to maintain her rights over it.

---

4. '80: Come *either to bring news* to the man seated there *or to whisper some warning in his ear.* Which shows . . .
5. Plutarch (tr. Amyot), *Propos de table*, 363 E–H (again citing Achias' saying).

5. On conscience

===

[Conscience originally meant connivance. Conscience in the sense of our individual consciousness of right and wrong or of our own guilt or rectitude fascinated Montaigne. It became a vital concern of his during the Wars of Religion with their cruelties, their false accusations and their use of torture on prisoners. Such moral basis as there was for the 'question' (judicial torture) seems, curiously enough, to have been a respect for the power of conscience – of a man's inner sense of his guilt or innocence which would strengthen or weaken his power to withstand pain. A major source of Montaigne's ideas here is St Augustine and a passionate note by Juan Luis Vives in his edition of the City of God designed to undermine confidence in torture.]

[A]   During our civil wars I was travelling one day with my brother the Sieur de la Brousse when we met a gentleman[1] of good appearance who was on the other side from us; I did not know anything about that since he feigned otherwise. The worst of these wars is that the cards are so mixed up, with your enemy indistinguishable from you by any clear indication of language or deportment, being brought up under the same laws, manners and climate, that it is not easy to avoid confusion and disorder. That made me fear that I myself would come upon our own troops in a place where I was not known, be obliged to state my name and wait for the worst. [B]   That did happen to me on another occasion: for, from just such a mishap, I lost men and horses. Among others, they killed one of my pages, pitifully: an Italian of good family whom I was carefully training; in him was extinguished a young life, beautiful and full of great promise.

[A]   But that man of mine was so madly afraid! I noticed that he nearly died every time we met any horsemen or passed through towns loyal to the King; I finally guessed that his alarm arose from his conscience. It seemed to that wretched man that you could read right into the very secret thoughts of his mind through his mask and the crosses on his greatcoat.[2] So

---

1. '80: *an honourable* gentleman.
2. Reformers often considered the cross, when used as a symbol, to be idolatrous and blasphemous. Here it is used as a disguise.

wondrous is the power of conscience! It makes us betray, accuse and fight against ourselves. In default of an outside testimony it leads us to witness against ourselves:

> *Occultum quatiens animo tortore flagellum.*

> [Lashing us with invisible whips, our soul torments us.][3]

The following story is on the lips of children: a Paeonian called Bessus was rebuked for having deliberately destroyed a nest of swallows, killing them all. He said he was right to do so: those little birds kept falsely accusing him of having murdered his father! Until then this act of parricide had been hidden and unknown; but the avenging Furies of his conscience made him who was to pay the penalty reveal the crime.[4]

Hesiod corrects that saying of Plato's, that the punishment follows hard upon the sin. He says it is born at the same instant, with the sin itself; to expect punishment is to suffer it: to merit it is to expect it. Wickedness forges torments for itself,

> *Malum consilium consultori pessimum,*

> [Who counsels evil, suffers evil most,][5]

just as the wasp harms others when it stings but especially itself, for it loses sting and strength for ever:

> *Vitasque in vulnere ponunt.*

> [In that wound they lay down their lives.][6]

The Spanish blister-fly secretes an antidote to its poison, by some mutual antipathy within nature. So too, just when we take pleasure in vice, there is born in our conscience an opposite displeasure, which tortures us, sleeping and waking, with many painful thoughts.[7]

---

3. Juvenal, *Satires*, XIII, 195 (adapted).
4. Plutarch (tr. Amyot), *Pourquoy la justice divine differe la punition des malefices*, 261 E–G (a major borrowing).
5. Erasmus, *Adages*, I, II, XIV, *Malum consilium*.
6. Virgil, *Georgics*, IV, 238. Montaigne wrote *Mousches guespes* (wasps), but clearly means 'bees'.
7. This Spanish fly was particularly poisonous. Cf. Cicero, *Tusc. disput.*, V, xl, 117; Pliny, XXIX, iv, 30; XI, xxv, 41.

[B]   *Quippe ubi se multi, per somnia sæpe loquentes,*
Aut morbo delirantes, procraxe ferantur,
Et celata diu in medium peccata dedisse.

[Many indeed, often talking in their sleep or delirious in illness, have proclaimed, it is said, and betrayed long-hidden sins.][8]

[A]   Apollodorus dreamed that he saw himself being flayed by the Scythians then boiled in a pot while his heart kept muttering, 'I am the cause of all these ills.' No hiding-place awaits the wicked, said Epicurus, for they can never be certain of hiding there while their conscience gives them away.[9]

                    Prima est hæc ultio, quod se
        Judice nemo nocens absoluitur.

[This is the principal vengeance: no guilty man is absolved: he is his own judge.][10]

Conscience can fill us with fear, but she can also fill us with assurance and confidence.   [B]   And I can say that I have walked more firmly through some dangers by reflecting on the secret knowledge I had of my own will and the innocence of my designs.

[A]   *Conscia mens ut cuique sua est, ita concipit intra*
Pectora pro facto spemque metumque suo.

[A mind conscious of what we have done conceives within our breast either hope or fear, according to our deeds.][11]

There are hundreds of examples: it will suffice to cite three of them about the same great man.

When Scipio was arraigned one day before the Roman people on a grave indictment, instead of defending himself and flattering his judges he said: 'Your wishing to judge, on a capital charge, a man through whom you have authority to judge the Roman world, becomes you well!'

Another time his only reply to the accusations made against him by a Tribune of the People was not to plead his cause but to say: 'Come, fellow citizens! Let us go and give thanks to the gods for the victory they gave me over the Carthaginians on just such a day as this!' Then as he started to walk towards the temple all the assembled people could be seen following after him – even his prosecutor.

8. Lucretius, V, 1157–9.
9. Plutarch (tr. Amyot), *Pourquoi la justice divine diffère*, 262 D–E; Seneca, *Epist. moral.*, XCVII, 13.
10. Juvenal, *Satires*, XIII, 2–3.
11. Ovid, *Fasti*, I, 485–6. Cf. also Cognatus, *Adages, Conscientia crimen prodit.*

Again when Petilius, under the instigation of Cato, demanded that Scipio account for the monies that had passed through his hands in the province of Antioch, Scipio came to the Senate for this purpose, took his account-book from under his toga and declared that it contained the truth about his receipts and expenditure; but when he was told to produce it as evidence he refused to do so, saying that he had no wish to act so shamefully towards himself; in the presence of the Senate he tore it up with his own hands. I do not believe that a soul with seared scars could have counterfeited such assurance. [C] He had, says Livy, a mind too great by nature, a mind too elevated by Fortune, even to know how to be a criminal or to condescend to the baseness of defending his innocence.[12]

[A] Torture is a dangerous innovation; it would appear that it is an assay not of the truth but of a man's endurance. [C] The man who can endure it hides the truth: so does he who cannot. [A] For why should pain make me confess what is true rather than force me to say what is not true? And on the contrary if a man who has not done what he is accused of is able to support such torment, why should a man who has done it be unable to support it, when so beautiful a reward as life itself is offered him?

I think that this innovation is founded on the importance of the power of conscience. It would seem that in the case of the guilty man it would weaken him and assist the torture in making him confess his fault, whereas it strengthens the innocent man against the torture. But to speak the truth, it is a method full of danger and uncertainty. What would you *not* say, what would you *not* do, to avoid such grievous pain?

> [C]   *Etiam innocentes cogit mentiri dolor.*
>
> [Pain compels even the innocent to lie.]

This results in a man whom the judge has put to the torture lest he die innocent being condemned to die both innocent and tortured.[13] [B] Thousands upon thousands have falsely confessed to capital charges. Among them, after considering the details of the trial

12. Plutarch (tr. Amyot), *Comment on se peut louer soy-mesme*, 139 F; Aulus Gellius, *Attic Nights,* IV, xviii; Livy, *Annales*, XXXVIII. Erasmus gives these anecdotes s.v. *Scipio Africanus Major* in his *Apophthegmata*.
13. St Augustine, *City of God*, XIX, vi (against torture) with Vives' comments (in which Vives cites *Etiam innocentes* [from Publius Syrus] and apologizes for turning a commentary into a plea against torture). Montaigne is deeply indebted to him for what follows.

which Alexander made him face and the way he was tortured, I place Philotas.[14]

[A] All the same it is [C], so they say, [B] the least bad[15] [A] method that human frailty has been able to discover. [C] Very inhumanely, however, and very ineffectually in my opinion. Many peoples less barbarous in this respect than the Greeks and the Romans who call them the Barbarians reckon it horrifying and cruel to torture and smash a man of whose crime you are still in doubt.[16] That ignorant doubt is yours: what has it to do with him? You are the unjust one, are you not? who do worse than kill a man so as not to kill him without due cause! You can prove that by seeing how frequently a man prefers to die for no reason at all rather than to pass through such a questioning which is more painful than the death-penalty itself and which by its harshness often anticipates that penalty by carrying it out.

I do not know where I heard this from, but it exactly represents the conscience of our own Justice: a village woman accused a soldier before his commanding general – a great man for justice – of having wrenched from her little children such sops as she had left to feed them with, the army having laid waste all the surrounding villages. As for proof, there was none. That general first summoned the woman to think carefully what she was saying, especially since she would be guilty of perjury if she were lying; she persisted, so he had the soldier's belly slit open in order to throw the light of truth on to the fact. The woman was found to be right.[17] An investigatory condemnation!

14. Quintus Curtius, VI ff.
15. '80: it is the *best* method that . . .
16. Vives (cf. note 13 above).
17. Anecdote from Froissart in H. Estienne's *Apologie pour Hérodote*.

6. On practice

═══

[This chapter discusses a key event in Montaigne's life: the brave but stupid act of one of his labourers who knocked him senseless from his horse in a minor encounter during the Wars of Religion. Reflecting on it led him to lose that philosophic fear of the act of dying which had obsessed him (and so many others before him). The major addition at the end shows that evil self-love (philautia as it was called) is the essence of pride; 'knowing oneself', on the contrary, is the essence of wisdom. Philosophy was conceived by Socrates as 'practising dying' (that is, by training, to practise the separation of the soul from the body, which will be achieved in death). Montaigne, while still claiming to follow Socrates, shifts the ground towards 'practising living'.]

[A]　Even when our trust is readily placed in them, reasoning and education cannot easily prove powerful enough to bring us actually to do anything, unless in addition we train and form our Soul by experience for the course on which we would set her; if we do not, when the time comes for action she will undoubtedly find herself impeded.[1] That explains why those among the philosophers who wished to attain to some greater excellence were never content to await the rigours of Fortune in shelter and repose for fear that Fortune might take them unawares, inexperienced and untried in battle; they preferred to go forth to meet her and deliberately threw themselves into the trial of hardships. Some renounced wealth to practise voluntary poverty; some sought toil and the austerity of a laborious life so as to harden themselves against ills and travail; some stripped themselves of those parts of their bodies which were most dear – their eyes, say, or their organs of generation – fearing that their use, being too pleasurable and too enervating, might weaken and relax the firmness of their souls. But practice is no help in the greatest task we have to perform: dying. We can by habit and practice strengthen ourselves against pain, shame, dire poverty and other occurrences: but as for dying, we can only assay that once; we are all apprentices when it comes to that.

Men were found in ancient days so excellent at using their time that they even assayed tasting and savouring their own death: they bent their minds

---

1. '80: impeded, *no matter how good a will she may have.* That . . .

on discovering what that crossing-over really was: but they have not come back to tell us about it.

> *Nemo expergitus extat*
> *Frigida quem semel est vitai pausa sequuta.*

[Who once has felt the icy end of life awakes not again.]

Canius Julius, a noble Roman of particular virtue and steadfastness, was condemned to death by that [C] blackguard[2] [A] Caligula; apart from the many wondrous proofs he gave of his determination, there was the moment when he was about to feel the hand of his executioner: one of his philosopher friends asked him, 'Well, Canius, how goes it with your Soul at present? What is she doing? What are your thoughts dwelling on?' – 'What I am thinking about is preparing and bracing myself with all my might to see whether, in that short brief moment of death, I can perceive anything of the Soul's departure and whether she herself has any sensation of issuing forth, so that if I do find out anything I may come back if I can to inform my friends.' He was philosophizing not merely unto death but into death. What assurance was that, what a proud mind, to wish that even his death could teach him something and to feel free to think of anything else but that great event!

> [B] *Jus hoc animi morientis habebat.*

[Even when dying he had such sway over his mind.][3]

[A] Yet it does seem that we have some means of breaking ourselves in for death and to some extent of making an assay of it. We can have experience of it, not whole and complete but at least such as not to be useless and to make us more strong and steadfast. If we cannot join battle with death we can advance towards it; we can make reconnaissances and if we cannot drive right up to its stronghold we can at least glimpse it and explore the approaches to it. It is not without good cause that we are brought to look to sleep itself for similarities with death.

[C] How easily we pass from waking to falling asleep! And how little we lose when we become unconscious of the light and of ourselves! It could perhaps even seem that our ability to fall asleep, which deprives us of all action and sensation, is useless and unnatural were it not that Nature by

---

2. Lucretius, III, 942–3.
    '80: that *monster* Caligula. . .
3. Seneca, *De tranquillitate*, XIV; Lucan, *Pharsalia*, VIII, 636.

sleep teaches us that she has made us as much for dying as for living and, already in this life, shows us that everlasting state which she is keeping for us when life is over, to get us accustomed to it and to take away our terror.

[A]   But those who have fallen into a swoon after some violent accident and have lost all sensation, have been in my opinion very close to seeing Death's true and natural face, for it is not to be feared that the fleeting moment at which we pass away comports any hardship or distress, since we cannot have sensation without duration. For us, suffering needs time; and time is so short and precipitate when we die that death must be indiscernible. What we have to fear is Death's approaches: they can indeed fall within our experience.

Many things appear greater in thought than in fact. I have spent a large part of my life in perfect good health: it was not only perfect but vivacious and boiling over. That state, so full of sap and festivity, made thinking of illness so horrifying that when I came to experience it I found its stabbing pains to be mild and weak compared with my fears.

[B]   Here is an everyday experience of mine: if I am sheltered and warm in a pleasant room during a night of storm and tempest, I am dumb-struck with affliction for those then caught out in the open; yet when I am out there myself I never even want to be anywhere else.

[A]   The mere thought of being always shut up indoors used to seem quite unbearable to me. Suddenly I was directed to remain there for a week or a month, all restless, distempered and feeble; but I have found that I used to pity the sick much more than I find myself deserving of pity now I am sick myself, and that the power of my imagination made the true essence of actual sickness bigger by half. I hope the same thing will happen with death and that it will not be worth all the trouble I am taking to prepare for it, nor all the aids I am gathering together and invoking to sustain my struggle. But whatever may happen, we can never give ourselves too many vantages!

During the third of our disturbances (or was it the second, I do not remember which) I was out riding one day about one league from my home, which is situated at the very hub of the disturbances in our French Civil Wars; I reckoned I was quite safe and so near my dwelling that I had no need of better protection and had taken an undemanding but not very reliable horse. On my way back there suddenly arose an occasion to use that horse for a task to which it was not much accustomed; one of my men, a big strong fellow, was on a powerful farm-horse with a hopeless mouth but also fresh and vigorous. He wanted to show off and to get

ahead of the others, but he happened to ride it full pelt right in my tracks and came down like a colossus upon me, a little man on a little horse, striking us like a thunderbolt with all his roughness and weight, knocking us over with our legs in the air. So there was my horse thrown down and lying stunned, and me, ten or twelve yards beyond, stretched out dead on my back, my face all bruised and cut about, the sword I had been holding lying more than ten yards beyond that, my belt torn to shreds; and me with no more movement or sensation than a log.

To this day that is the only time I have ever lost consciousness. Those who were with me, having assayed every means in their power to bring me round, thought I was dead; they took me in their arms and struggled back with me to the house, which was about half a French league away.

After having been taken for dead for two good hours, on the way I began to make movements and to inhale because such a great quantity of blood had been discharged into my stomach that my natural powers had to be restored for me to void it. They got me on my feet, when I threw up a bucketful of pure clotted blood; and I had to do the same several times on the way. With that I began to get a bit of life back into me, but only little by little and over so long a stretch of time that at first my sensations were closer to death than to life:

> [B]   *Perche, dubbiosa anchor del suo ritorno,*
> *Non s'assecura attonita la mente.*

[Because the mind, struck with astonishment, still doubts it will return and remains unsure.]

[A]   The memory of this, being deeply planted in my soul, paints for me the face of Death and her portrait so close to nature that it somewhat reconciles me to her.

When I did begin to see anything, my sight was so dead and so weak that I could make out nothing but light:

> *come quel ch'or apre or chiude*
> *Gli occhi, mezzo tra'l sonno è l'esser desto.*

[as one who now opens his eyes, now shuts them, half sleeping, half awake.][4]

As for the faculties of my soul, they progressively came back to life with those of my body. I could see myself covered with blood since my doublet was spattered with the blood I had brought up. The first thought that

---

4. Torquato Tasso, *Gierusalemme liberata*; XII; lines from stanzas 74 and 26.

occurred to me was that I had been shot in the head by a volley of harquebuses; and indeed several were being fired around us. To me it seemed as though my life was merely clinging to my lips. It seemed, as I shut my eyes, as though I was helping to push it out, and I found it pleasant to languish and to let myself go. It was a thought which only floated on the surface of my soul, as feeble and delicate as everything else, but it was, truly, not merely free from unpleasantness but tinged with that gentle feeling which is felt by those who let themselves glide into sleep.

I believe that those whom we see failing from weakness in the throes of death find themselves in that same state, and I maintain that we pity them without cause, thinking that they are troubled by grievous pains or have their souls full of distressing thoughts. It has always been my belief (despite the opinion of others including Etienne de La Boëtie) that those whom we see lying prostrate in a coma at the approach of death, or overwhelmed by the length of their illness or by an apoplectic fit or by the falling sickness –

> [B] . . . *vi mòrbi sæpe coactus*
> *Ante oculos aliquis nostros, ut fulminis ictu,*
> *Concidit, et spumas agit; ingemit, et fremit artus;*
> *Desipit, extentat nervos, torquetur, anhelat,*
> *Inconstanter et in jactando membra fatigat;*

[often, before our very eyes, a man is struck down by illness as if by lightning; he foams at the mouth; he groans and he twitches; he is delirious; he stretches out his legs, he twists and turns; he pants for breath and tires his limbs as he throws himself about;][5]

– [A] or by a wound in the head, and whom we can hear groaning and sometimes uttering penetrating sighs which we take for signs indicating that they seem to retain some remnant of consciousness, have, I repeat – no matter what bodily movements they make – both their body and soul buried in stupor.

> [B] *Vivit, et est vitæ nescius ipse suæ.*

[He lives, unconscious of his own life.][6]

[A] And I could never believe, after so great a stunning of the limbs and so great a weakening of the senses, that their souls could sustain any inward

5. Lucretius, III, 485–9.
6. Ovid, *Tristia*, I, iii, 12.

powers of self-cognition; and consequently that those men had any thoughts to torment them and to make them feel, or be aware of, their miserable condition; and that in consequence they were not much to be pitied.

[B] I can think of no state more horrifying or more intolerable for me than to have my Soul alive and afflicted but with no means of expressing herself; I would say the same of those who are sent to be executed with their tongues cut out, were it not that the most becoming death of that sort is one that is mute, provided that it is accompanied by a firm and grave countenance; the same applies to those wretched prisoners-of-war who fall into the clutches of those vile hangmen–soldiers of these times, by whom they are tortured with every kind of cruel mistreatment to compel them to pay some huge impossible ransom, being held meanwhile under such conditions and in such a place that they have no means of expressing their thoughts or of giving sign of their misery.

[A] The poets had imaginary gods favourable to the deliverance of such who thus dragged out a lingering death:

> *hunc ego Diti*
> *Sacrum jussa fero, teque isto corpore solvo.*

[to Dis I bear, as he decreed, this lock of hair, and free thee from thy body.][7]

Even such brief words and incoherent replies as we extort from the dying by yelling in their ears and storming about, even the gestures which seem to bear some relation to the questions we put to them, are, for all that, no testimony to their being alive, at least not fully alive. The same thing happens to us when we are hesitantly drifting off to sleep, before sleep has taken us over completely: we are aware of what is going on about us as in a dream, and we follow any words spoken with a cloudy uncertain sense of hearing which seems to touch only the edges of our soul; and, to the last words spoken to us which we could follow, we make replies more marked by chance than by sense.

Now that I have actually experienced it I have no doubt whatsoever that I have been right all the time. First of all, when I was unconscious I strove to rip my doublet half-open with my nails – I was not wearing armour – and I know that in my mind I felt none of the wounds: for many of our movements do not arise from any command of ours:

7. Virgil, *Aeneid*, IV, 702.

[B]   *Semianimesque micant digiti ferrumque retractant.*

[Half-dead fingers twitch and grasp the sword again.][8]

[A]   By some natural impulse, when we trip over we throw out our arms *before* we fall, which shows that our limbs spontaneously come to each other's aid   [B]   and have movements independent of our reasoning:

> *Falciferos memorant currus abscindere membra,*
> *Ut tremere in terra videatur ab artubus id quod*
> *Decidit abscissum, cum mens tamen atque hominis vis*
> *Mobilitate mali non quit sentire dolorem.*

[They tell how chariots with scythes on their wheels can cut so quickly that severed limbs are writhing on the ground before the mind has the power to feel the pain.][9]

[A]   My stomach was swollen with clotted blood; my hands rushed to it of their own accord, as they often do against the counsel of our will to a part which is itching. There are many animals, and men too, who are seen to contract their muscles and move after they are dead. Every man knows from his own experience that he has a part of his body which often stirs, erects and lies down again without his leave. Now such passive movements, which only touch our outsides, cannot be called ours. For them to be ours the whole man must be involved: any pain which our foot or our hand feels while we are asleep does not belong to us.

As I was nearing my home, to which news of my fall had already run quickly, and after members of my family had greeted me with the cries usual in such circumstances, not only did I answer a word or two to their questions but they say that I was determined to order a horse to be provided for my wife whom I saw struggling and stumbling along the road, which is difficult and steep. It might appear that such thoughts must have arisen from a soul which is awake: nevertheless I played no part in them: they were empty acts of apparent thinking provoked by sensations in my eyes and ears: they did not arise from within me. I had no idea where I was coming from nor where I was going to; nor could I weigh attentively what I was asked. My reactions were trivial ones, produced by my senses themselves, doubtless from habit. Any contribution from my soul, which was only very lightly involved and as though licked by the dew of some light impression of the senses, came only in a dream. Meanwhile my

8. Virgil, *Aeneid*, X, 396.
9. Lucretius, III, 642–5.

condition was truly most agreeable and peaceful: I felt no affliction either for myself or for others; it was a kind of lassitude and utter weakness, without any pain. I saw my house but I did not recognize it. When they got me into bed, I experienced a feeling of infinite rest and comfort, for I had been dreadfully pulled about by those poor fellows who had taken the trouble to carry me in their arms over a long and very poor road and who, one after another, had tired themselves out two or three times.

I was offered several medicines: I would not take any of them, being convinced that I was fatally wounded in the head. It would have been – no lying – a very happy way to die, for the feebleness of my reasoning powers kept me from judging anything, and that of my body from feeling anything. I felt myself oozing away so gently, in so gentle and pleasing a fashion, that I can think of hardly any action   [C]   less grievous than   [A] that was.[10]

When I began to come back to life and regained my strength,

[B]   *Ut tandem sensus convaluere mei,*
[As my senses at last regained their health,][11]

[A]   which was two or three hours later, only then did I feel myself all at once linked with pain again, having all my limbs bruised and battered by my fall; and I felt so ill two or three nights later that I nearly died a second time, but of a livelier death! And I can still feel the effects of that battering.

I must not overlook the following: the last thing I could recover was my memory of the accident itself; before I could grasp it, I got them to repeat several times where I was going to, where I was coming from, what time it happened. As for the manner of my fall, they hid it from me for the sake of the man who had caused it and made up other explanations. But some time later the following day when my memory happened to open up and recall to me the circumstances which I found myself in on that instant when I was aware of that horse coming at me (for I had seen it at my heels and already thought I was dead, but that perception had been so sudden that fear had no time to be engendered by it), it appeared to me that lightning had struck my soul with a jolt and that I was coming back from the other world.

This account of so unimportant an event is pointless enough but for the

10. '80: action *as pleasant as* that was . . .
11. Ovid, *Tristia*, I, iii, 14.

instruction I drew from it for my own purposes: for in truth, to inure yourself to death all you have to do is to draw nigh to it. Now, as Pliny says, each man is an excellent instruction unto himself provided he has the capacity to spy on himself from close quarters.[12]

Here you have not my teaching but my study: the lesson is not for others; it is for me.   [C]   Yet, for all that, you should not be ungrateful to me for publishing it. What helps me can perhaps help somebody else. Meanwhile I am not spoiling anything: I am only using what is mine. And if I play the fool it is at my own expense and does no harm to anybody. Such foolishness as I am engaged in dies with me: there are no consequences. We have reports of only two or three Ancients who trod this road and we cannot even say if their manner of doing so bore any resemblance to mine since we know only their names.[13] Since then nobody has leapt to follow in their traces. It is a thorny undertaking – more than it looks – to follow so roaming a course as that of our mind's, to penetrate its dark depths and its inner recesses, to pick out and pin down the innumerable characteristics of its emotions. It is a new pastime, outside the common order; it withdraws us from the usual occupations of people – yes, even from the most commendable ones. For many years now the target of my thoughts has been myself alone; I examine nothing, I study nothing, but me; and if I do study anything else, it is so as to apply it at once to myself, or more correctly, within myself. And it does not seem to me to be wrong if (as is done in other branches of learning, incomparably less useful) I share what I have learned in this one, even though I am hardly satisfied with the progress I have made. No description is more difficult than the describing of oneself; and none, certainly, is more useful. To be ready to appear in public you have to brush your hair; you have to arrange things and put them in order. I am therefore ceaselessly making myself ready since I am ceaselessly describing myself.

Custom has made it a vice to talk about oneself and obstinately prohibits it, hating the boasting which always seems to be attached to any testimony about oneself. Instead of wiping the child's nose you cut it off!

In vitium ducit culpæ fuga.

[Flying from a fault, we fall into a vice.][14]

12. Cf. Pliny, cited Erasmus, *Adages*, I, VII, XCIV, *In tuum ipsius sinum inspue.*
13. It is not certain who these 'two or three Ancients' were. They may have included Lucillius, the 'father of satire'.
14. Horace, *Ars poetica*, 31.

I find more evil than good in that remedy. But even if it should be true that engaging people in talk about oneself is inevitably presumption, still, if I am to carry out my plan I must not put an interdict on an activity which makes that sickly quality public, since it *is* in me and I must not hide that defect; I do not merely practise it: I make a profession of it. Anyway, my belief is that it is wrong to condemn wine because many get drunk on it. You can abuse things only if they are good. I believe that that prohibition applies only to the popular abuse. It is a bridle made to curb calves: it is not used as a bridle by the Saints, who can be heard talking loudly about themselves, nor by philosophers nor by theologians;[15] nor by me though I am neither one nor the other. If they do not literally write about themselves, when the occasion requires it they do not hesitate to trot right in front to show off their paces. What does Socrates treat more amply than himself? And what does he most often lead his pupils to do, if not to talk about themselves – not about what they have read in their books but about the being and the movement of their souls? We scrupulously talk of ourselves to God and to our confessors, just as our neighbours do before the whole congregation.[16] 'But,' somebody will reply, 'we talk then only of our offences.' In that case we say it all: for our very virtue is faulty and needs repentance.

My business, my art, is to live my life. If anyone forbids me to talk about it according to my own sense, experience and practice, let him also command an architect to talk about buildings not according to his own standard but his next-door neighbour's, according to somebody else's knowledge not his own. If publishing one's own worth is pride, why does not Cicero puff the eloquence of Hortensius, and Hortensius that of Cicero?[17]

Perhaps they mean that I should witness to myself by works and deeds not by the naked word alone. But I am chiefly portraying my ways of thinking, a shapeless subject which simply does not become manifest in deeds. I have to struggle to couch it in the flimsy medium of words. Some

---

15. Montaigne may be thinking, among other works, of St Augustine's *Confessions*, but there are signs that he never read that particular work, though one would have expected him to have done so.

16. The Reformed Church rejected private confession to priests but encouraged a sinner to confess his sins to the assembled Church.

17. Montaigne's gibe is unfair. Quintus Hortensius was a famous orator of Cicero's time; Cicero named his treatise on oratory after him. Quintilian (XI, iii, 8) held his oratory to be inferior to Cicero's.

of the wisest of men and the most devout have lived their lives avoiding any sign of activity. My activities would tell you more about Fortune than about me. They bear witness to their own role not to mine, unless it be by uncertain conjecture: they are samples and reveal only particulars. I am *all* on display, like a mummy on which at a glance you can see the veins, the muscles and the tendons, each piece in its place. Part of me is revealed – but only ambiguously – by the act of coughing; another by my turning pale or by my palpitations. It is not what I do that I write of, but of me, of what I *am*. I hold that we must show wisdom in judging ourselves, and, equally, good faith in witnessing to ourselves, high and low indifferently. If I seemed to myself to be good or wise – or nearly so – I would sing it out at the top of my voice. To say you are worse than you are is not modest but foolish. According to Aristotle, to prize yourself at less than you are worth is weak and faint-hearted. No virtue is helped by falsehood; and the truth can never go wrong. To say we are better than we are is not always presumption: it is even more often stupidity. In my judgement, the substance of that vice is to be immoderately pleased with yourself and so to fall into an injudicious self-love.

The sovereign remedy to cure self-love is to do the opposite to what those people say who, by forbidding you to talk about yourself, as a consequence even more strongly forbid you to think about yourself. Pride lies in our thoughts: the tongue can only have a very unimportant share in it. They think that to linger over yourself is to be pleased with yourself, to haunt and frequent yourself is to hold yourself too dear. That can happen. But that excess arises only in those who merely finger the surface of themselves; who see themselves only when business is over; who call it madness and idleness to be concerned with yourself; for whom enriching and constructing your character is to build castles in the air; who treat themselves as a third person, a stranger to themselves.

If anyone looks down on others and is drunk on self-knowledge let him turn his gaze upwards to ages past: he will pull his horns in then, discovering many thousands of minds which will trample him underfoot. If he embarks upon some flattering presumption of his own valour let him recall the lives of the two Scipios and all those armies and peoples who leave him so far behind. No one individual quality will make any man swell with pride who will, at the same time, take account of all those other weak and imperfect qualities which are in him and, finally, of the nullity of the human condition.

Because Socrates alone had taken a serious bite at his god's precept to

'know himself' and by such a study had reached the point of despising himself, he alone was judged worthy of being called The Sage.[18]

If any man knows himself to be thus, let him boldly reveal himself by his own mouth.

18. Socrates maintained that men should be concerned not with cosmology but with self-knowledge and morals. He followed Apollo's revealed commandment, 'Know Thyself'. (Cf. Erasmus, *Apophthegmata*, III, *Socratica*, XII and XXXVI; *Adages*, I, VI, XCV, *Nosce teipsum*.)

7. On rewards for honour

[The historic Order of St Michael, of which Montaigne was a knight, had become debased, partly as the result of an inflation of awards during the Wars of Religion. The new Order of the Holy Ghost was instituted by Henry III of France in 1578, ceremonies creating the new knights taking place in December 1578 and January 1579. Montaigne's reflections lead him to thoughts on the origin of inequality among men.]

[A] The biographers of Augustus Caesar picked out this point to emphasize in his military discipline: he was wonderfully free with his gifts to those who deserved it; but where rewarding honour itself was concerned he was equally sparing. Yet before he had ever gone to war himself he had had bestowed on him by his uncle all the military awards.

It was a fine innovation practised by most of the systems of government in the world to establish certain vain and, in themselves, valueless decorations, in order to honour and reward virtue, such as crowns of laurel oak or myrtle leaves, certain forms of dress, the privilege of riding through the city in a coach or with torch-bearers by night, a special seat at public meetings, the prerogative to certain special names and titles, to certain symbols on their coats-of-arms and such-like things; this system was operated differently according to each nation's set of values, and still is.

For our part, like many of our neighbours, we have Orders of Chivalry which were instituted for this express purpose. It is, in truth, a very good and beneficial custom to have found a way of recognizing the worth of rare outstanding men and to please and to satisfy them with rewards which are no charge on the people and which cost the monarch nothing. It was always recognized by the experience of the Ancients – and was formerly seen to be so among us French – that men of distinction were more zealous for such rewards than for those which brought gain and profit: that was not unreasonable nor without evident justification. If you introduce other advantages and riches into a prize which should be for honour alone, instead of increasing the prestige you prune it back and degrade it.

The Order of Saint Michael, which was so long held in high esteem among us, had no greater advantage than its being in no ways associated

with any other advantage. As a result there used to be no office or estate whatsoever to which the nobility aspired with so much longing and yearning as they did to that Order, nor was there any rank which comported more respect and dignity, since Virtue more readily aspires to embrace such recompense as is truly her own, more glorious than useful. The other rewards which are bestowed do not have the same dignity; they are[1] employed on all sorts of occasions: money rewards the services of a manservant, the diligence of a messenger, dancing, vaulting, talking and the meanest services done for us; yes, and we use it to reward vice, flattery, pimping and treachery. No marvel[2] then if Virtue desires and accepts that sort of common currency less willingly than the one which is proper and peculiar to herself. Augustus was right to be much more niggardly and sparing over this one than the other, especially since honour is a privilege, the main essence of which is its rarity. So, too, for Virtue.

Cui malus est nemo, quis bonus esse potest?

[For him who thinks no man is bad, can any man be good?][3]

We do not pick out for praise a man who takes trouble over the education of his children, since however right that is it is not unusual, [C] no more than we pick out a tall tree in a forest where all the trees are tall. [A] I do not think that any citizen of Sparta boasted of his valour, for it was the virtue of all the people of their nation; nor did he boast of his reliability or of his contempt for riches. No matter how great it may be, no recompense is allotted to any virtue which has passed into custom: I doubt if we would ever call it great once it was usual.

Since such distinctions have no other value or prestige than the fact that few men enjoy them, to make them worthless you simply have to be generous with them. Even if there were more men nowadays who merited our Order it still ought not to have its prestige debased.

And it could easily happen that more deserve it since not one of the virtues can spread so easily as military valour. There is valour of another kind, true, perfect, philosophical (I am not speaking of it here: I use the word *valour* in accordance with our own usage); it is greater than our kind, it is more ample: it consists in fortitude and assurance of soul, despising all hostile accidents equally; it is calm, uniform and constant; our own kind is

---

1. '80: dignity, *being coin which buys any sort of traded goods*; they are . . .
2. '80: treachery *and such-like which we exploit for our own ends by the intermediary of others.* No marvel . . .
3. Martial, *Epigrams*, XII, lxxxii.

but a glimmer of it. Habit, education, example and custom are all-powerful in establishing the valour I am talking about, and can easily make it common, as can be readily seen from our experience in our Civil Wars. [B] If anyone could unite us now and arouse our whole people for some common emprise we would make our ancient [C] military [B] reputation flower again.

[A] It is certain that in former times this Order was not concerned with valour by itself: it looked much further. It was never earned by a brave soldier but by a famous⁴ leader: knowing how to obey orders never deserved so honourable a reward then. In former times they were looking out for a more general expert knowledge of warfare, embracing the greater part of the greatest parts of the fighting man – [C] *'Neque enim eædem militares et imperatoriæ artes sunt'* [For the skills of a soldier and those of a commander are not the same]⁵ – [A] they sought a man whose circumstances also were worthy of such an honour. But, as I was saying, even if more men were judged worthy than were found in former times, we still must not be more liberal with it, and it would have been better to fail to bestow it on everyone to whom it was due than for ever to lose in practice so useful an innovation. No great-minded man deigns to see any advantage in what he holds in common with many others; and today those who merit it least are the first to affect to despise it in order to range themselves with those who were wronged when a decoration which was peculiarly theirs was unworthily extended and debased.

Now to wipe out this Order, to abolish it with the expectation of giving a new and sudden prestige to some similar decoration, is an undertaking inappropriate to so licentious and diseased a period as our own present one; what will happen is that the latest Order will, from its inception, incur the same disadvantages which have just ruined the other. To give it authority, the rules governing the awarding of this new Order would need to be extremely tight and restrictive, whereas our troubled times are not susceptible to a short governing-rein. Apart from that, before it could be given any prestige we should need to have lost all memory of the former Order and of the contempt into which it has fallen.

My topic could lend itself to a discussion of Valiance and of its differences from other virtues. But since Plutarch has often touched on that theme⁶ I

4. '80: famous *and noble* leader . . .
5. Livy, XXV, xix.
6. '80: theme, *and since it is so familiar to us from the French appearance which has been given to it, so accomplished and so pleasing,* I would . . .

would be wasting my time, repeating here what he has already said about it.[7] But it is worth considering that our own nation gives the first place among the virtues to valiance as its name shows, *vaillance* deriving from *valeur*, worth. By our usage, in the language of the Court and of the nobility, when we say that a man *vaut beaucoup* ('has great worth') or is an *homme de bien* ('a good man') we mean he is a valiant one. The custom of the Romans was similar: they derived their general term for virtue (*virtus*) from the word for strength (*vis*).[8]

The only, essential, proper form of nobility in France is the profession of arms. It is probable that the first of the virtues to appear among men, giving some of them superiority over others, was the one by which the stronger and the more courageous made themselves masters of the weaker and so acquired individual rank and reputation, from which derive our terms of honour and dignity; or else those nations, being most warlike, gave the prize and the title highest in dignity to the virtue which they were most familiar with. So too our passion, our feverish concern, for the chastity of women results in *une bonne femme* ('a good woman'), and *une femme d'honneur, ou de vertu* ('a woman of honour' or 'of virtue') in reality meaning for us a chaste woman – as though, in order to bind them to that duty, we neglected all the rest and gave them free rein for any other fault, striking a bargain to get them to give up that one.

---

7. Plutarch (tr. Amyot), *Les dicts notables des Anciens Roys*, 199 C; valiance is the proper virtue of beasts not men (*Que bestes brutes usent de Raison*, 271 A–H).
8. In fact Cicero derives *virtus* (virtue) from *vir* (man), not from *vis* (strength) (*Tusc. disput.*, II, XVIII, 43), adding that 'Man's proper virtue is fortitude.'

8. On the affection of fathers for their children

[*This is one of the most moving and revealing of the chapters: it starts with the bout of melancholy which upset Montaigne's complexion and led him to write his* Essays; *it ends with thoughts of the mad frenzy which can lead fathers to fall in love with their own children or brain-children. Some of the examples he cites of strange behaviour concern* chagrin *(manic-depression) and melancholy itself. The shift from real children to brain-children (a vital platonic commonplace) is given greater urgency by the fact that Montaigne's children all died in infancy, one daughter excepted. His final examples emphasize that great deeds and books can be not only a man's 'sons' but his 'daughters' too.*

The irritability which transpires through his discussion of wills and inheritances reminds us of tensions between him and his mother over the dispositions in Pierre de Montaigne's last will and testament. An earlier will (1560–61) had left great financial authority to the mother; the last will simply followed the relevant practices of the customary law of Bordeaux, which treated widows generously, though less so than some other legal systems within France. The widow (who died in 1601) harboured resentment until the end, in her own will bitterly accusing Michel de Montaigne's only daughter, her own granddaughter, of enjoying wealth which ought to have been hers. (See R. M. Calder, 'Montaigne and Customary Law', in Bibliothèque d'Humanisme et Renaissance, XLVII, 1985, pp. 79–85). *In this chapter we are far from that balanced serenity which Montaigne often achieves. There are deletions in the Bordeaux manuscript probably not made by the author himself. Two have been reinserted here.*

Incidentally, Michel de Montaigne's own marriage-settlement, doubtless principally drawn up by his father, Pierre, did not follow the customary law of Bordeaux and was less generous in its provisions for his widow than customary law allowed.]

For Madame d'Estissac

[A]   Madame: unless I am saved by oddness or novelty (qualities which usually give value to anything) I shall never extricate myself with honour from this daft undertaking; but it is so fantastical and presents an aspect so totally unlike normal practice that it may just get by.

It was a melancholy humour (and therefore a humour most inimical to my natural complexion) brought on by the chagrin caused by the solitary retreat I plunged myself into a few years ago, which first put into my head this raving concern with writing.[1] Finding myself quite empty, with nothing to write about, I offered myself to myself as theme and subject matter. It is [C] the only book of its kind in the world, [A] in its conception wild and [C] fantastically eccentric.[2] [A] Nothing in this work of mine is worthy of notice except that bizarre quality, for the best craftsman in the world would not know how to fashion anything remarkable out of material so vacuous and base.

Now, Madame, having decided to draw a portrait of myself from life, I would have overlooked an important feature if I had failed to portray the honour which I have always shown you for your great merits.[3] I particularly wanted to do so at the head of this chapter, since of all your fine qualities one of the first in rank is the love you show your children.

Anyone who knows how young you were when your husband Monsieur d'Estissac left you a widow; the proposals which have been made to you by such great and honourable men (as many as to any lady of your condition in France); the constancy and firmness of purpose with which you have, for so many years and through so many difficulties, carried the weight of responsibility for your children's affairs (which have kept you busy in so many corners of France and still besiege you); and the happy prosperity which your wisdom or good fortune have brought to those affairs: he will readily agree with me that we have not one single example of maternal love today more striking than your own.

I praise God, Madame, that your love has been so well employed. For the great hopes of himself raised by your boy, Monsieur d'Estissac, amply assure us that when he comes of age you will be rewarded by the duty and gratitude of an excellent son.[4] But he is still a child, unable to appreciate the innumerable acts of devotion he has received from you: so I should like him, if this book should fall into his hands one day, to be able to learn something from me at a time when I shall not even have a mouth to tell it

---

1. Montaigne's *complexion* (balance of humours) was melancholy modified by sanguine elements. An access of melancholy *humour* would unbalance his complexion, plunging him into a depression (*chagrin*).
2. '80: wild and *monstrous*. Nothing . . .
3. '80: until [C]: the honour *and particular reverence* which [. . .] merits *and virtues*. I . . .
4. Montaigne took him as a youth to Italy.

to him – something I can vouch for quite truthfully and which will be made even more vigorously evident, God willing, by the good effects he will be aware of in himself: namely, that there is no nobleman in France who owes more to his mother than he does, and that in the future he will be able to give no more certain proof of his goodness and virtue than by acknowledging your qualities.

If there truly is a Law of Nature – that is to say, an instinct which can be seen to be universally and permanently stamped on the beasts and on ourselves (which is not beyond dispute) – I would say that, in my opinion, following hard on the concern for self-preservation and the avoidance of whatever is harmful, there would come second the love which the begetter feels for the begotten. And since Nature seems to have committed this love to us out of a concern for the effective propagation of the successive parts of the world which she has contrived, it is not surprising if love is not so great when we go backwards, from children to fathers.   [C]   To which we may add a consideration taken from Aristotle,[5] that anyone who does a kindness to another loves him more than he is loved in return; that anyone to whom a debt is owed feels greater love than the one by whom the debt is owed; and that every creator loves what he has made more than it would love him if it were capable of emotions. This is especially true because each holds his *being* dear: and *being* consists in motion and activity; in a sense, therefore, everyone is, to some degree, within anything he does: the benefactor has performed an action both fair and noble: the recipient, on the other hand, has only performed a useful one, and mere usefulness is less lovable than nobility. Nobility is stable and lasting, furnishing the one who has practised it with a constant satisfaction. Usefulness, however, can easily disappear or diminish, and the memory of it is neither so refreshing nor so sweet. The things which have cost us most are dearest to us – and it costs us more to give than to receive.

[A]   Since it has pleased God to bestow some slight capacity for discursive reasoning on us so that we should not be slavishly subject to the laws of Nature as the beasts are but should conform to them by our free-will and judgement, we should indeed make some concessions to the simple authority of the common laws of Nature but not allow ourselves to be swept tyrannously away by her: Reason alone must govern our inclinations.

For my part, those propensities which are produced in us without the command and mediation of our judgement taste strangely flat. In the case

5. Aristotle, *Nicomachaean Ethics*, IX, vii, 4–6.

of the subject under discussion, I am incapable of finding a place for that emotion which leads people to cuddle new-born infants while they are still without movements of soul or recognizable features of body to make themselves lovable. [C] And I have never willingly allowed them to be nursed in my presence. [A] A true and well-regulated affection should be born, and then increase, as children enable us to get to know them; if they show they deserve it, we should cherish them with a truly fatherly love, since our natural propensity is then progressing side by side with reason; if they turn out differently, the same applies, *mutatis mutandis*: we should, despite the force of Nature, always yield to reason.

In fact, the very reverse often applies; we feel ourselves more moved by the skippings and jumpings and babyish tricks of our children than by their activities when they are fully formed, as though we had loved them not as human beings but only as playthings [C] or as pet monkeys. [A] Some fathers will give them plenty of toys when they are children but will resent the slightest expenditure on their needs once they have come of age. It even looks, in fact, as if we are jealous of seeing them cut a figure in the world, able to enjoy it just when we are on the point of leaving it, and that this makes us miserly and close-fisted towards them: it irritates us that they should come treading on our heels, [C] as if to summon us to take our leave. [A] Since in sober truth things are so ordered that children can only have their being and live their lives at the expense of our being and of our lives, we ought not to undertake to be fathers if that frightens us.

For my part, I find it cruel and unjust not to welcome them to a share and fellow-interest in our property – giving them full knowledge of our domestic affairs as co-partners when they are capable of it – and not to cut back on our own interests, economizing on them so as to provide for theirs, since we gave them birth for just such a purpose. It is unjust to see an aged father, [B] broken [A] and only half alive,[6] stuck in his chimney-corner with the absolute possession of enough wealth to help and maintain several children, allowing them all this time to waste their best years without means of advancement in the public service and of making themselves better known. They are driven by despair to find some way, however unjust, of providing for their needs: I have seen in my time several young men of good family so addicted to larceny that no punishment could turn them from it. I know one young man, very well connected, with whom I had a word about just such a matter at the earnest

6. '80: father, *in his dotage* and only half alive . . .

request of his brother, a brave and most honourable nobleman. In reply the young man admitted quite openly that he had been brought to such vile conduct by the unbending meanness of his father, adding that he had now grown so used to it that he could not stop himself. He had just been caught stealing rings from a lady whose morning reception he was attending with several others. It reminded me of a story I had heard about another nobleman who had so adapted himself to the exigencies of that fine profession that when he did become master of his inheritance and decided to give up this practice he nevertheless could not stop himself from stealing anything he needed when he passed by a stall, despite the bother of having to send somebody to pay for it later. I have known several people so trained and adapted to thieving that they regularly steal from their close companions things which they intend to return.

[B] I may be a Gascon but there is no vice I can understand less. My complexion makes me loathe it rather more than my reason condemns it: I have never even wanted to steal anything from anyone. [A] It is true that my part of the world is rather more infamous for theft than the rest of our French nation: yet we have all seen in our time, on several occasions, men of good family from other provinces convicted of many dreadful robberies. I am afraid that we must partly attribute such depravity to the fault of their fathers.

If anyone then tells me, as a very intelligent nobleman once did, that the only practical advantage he wanted to get from saving up all his money was to be honoured and courted by his children (since now that age had deprived him of strength that was the only remedy he had left against being treated with neglect and contempt by everybody, and so maintaining his authority over his family – [C] and truly, not only old age but all forms of weakness are, according to Aristotle, great encouragements to miserliness)[7] – [A] then there is something in that. But it is medicine to cure an illness the birth of which ought to have been prevented. A father is wretched indeed if he can only hold the love of his children – if you can call it love – by making them depend on his help.

We should make ourselves respected for our virtues and our abilities and loved for our goodness and gentlemanliness. The very ashes of a rare timber have their value, and we are accustomed to hold in respect and reverence the very bones and remains of honourable people. In the case of someone who has lived his life honourably, no old age can be so decrepit and smelly that it ceases to be venerable – especially to the children, whose

7. Aristotle, *Nicomachaean Ethics*, IV, i, 37.

souls should have been instructed in their duty not by need and want, nor by harshness nor force, but by reason:

> *et errat longe, mea quidem sententia,*
> *Qui imperium credat esse gravius aut stabilius*
> *Vi quod fit, quam illud quod amicitia adjungitur.*

[if you ask my opinion, it is quite untrue that authority is firmer or more stable when it relies on force than when it is associated with affection.][8]

[B] I condemn all violence in the education of tender minds which are being trained for honour and freedom. In rigour and constraint there is always something servile, and I hold that you will never achieve by force what you cannot achieve by reason, intelligence and skill.[9]

That was the way I was brought up. They tell me that I tasted the [C] rod [B] only twice[10] during all my childhood, and that was but lightly. I owed the same treatment to the children born to me; they all die, though, before they are weaned. But [C] Leonor, [B] an only daughter who has escaped that calamity, has reached the age of six or more (her mother's gentleness readily predisposing her that way) without our having used in her upbringing and in the punishment of her childish faults anything but words – gentle ones at that. And even if my hopes for her turn out to be frustrated, there are other causes in plenty to blame for that without finding fault with my method of upbringing, which I know to be just and natural.

I would have been even more punctilious with boys, who are less born to serve and whose mode-of-being is freer: I would have loved to make their hearts overflow with openness and frankness. I have never seen caning achieve anything except making souls more cowardly or more maliciously stubborn.

[A] Do we want to be loved by our children? Do we want to remove any occasion for their wishing us dead? – though no occasion for so horrible a wish could ever be right or pardonable: [C] *'nullum scelus rationem habet'* [no crime has rational justification][11] – then let us within reason enrich their lives with whatever we have at our disposal. To achieve

---

8. Terence, *Adelphi*, I, i, 40–3.
9. The gentlemanly idea of education, as in Rabelais, who also loathed corporal punishment.
10. '80: tasted the *whip* only twice . . .
11. Cf. *Adagia*, Frankfurt, 1656, *Appendix Erasmi*, p. 313, *Scelera non habent consilia*, cited after Livy, XXVIII, xxviii.

that we ought not to get married so young that our adult years almost become confounded with theirs. Such unseemliness can plunge us into many great difficulties – I mean especially in the case of the nobility, whose way of life is one of leisure and who can live, as we say, on their income. In other cases, where life is a struggle for money, the fellowship of a great many children is a help to the whole family; they are so many new ways and means of helping to enrich it.

[B] I was thirty-three when I married; and I approve of thirty-five – the opinion attributed to Aristotle.   [C] Plato does not want any man to marry before thirty; he is also right to laugh at spouses who lie together after fifty-five, judging their offspring unworthy to live and eat.[12]

It was Thales who gave the right ages; his mother pressed him to get married when he was young: 'Too soon,' he said. When he was older: 'Too late!' Accept no time as opportune for any inopportune activity!

[A]   The Ancient Gauls reckoned it to be extremely reprehensible for a man to lie with a woman before he was twenty, particularly advising those who wanted to train for war to remain chaste well into adulthood, [A1]   because sexual intercourse makes minds soft and deflects them.

> Ma hor congiunto a giovinetta sposa,
> Lieto homai de' figli, era invilito
> Ne gli affetti di padre e di marito.

[But now, married to a young wife, happy to have children, he was weakened by his love as father and husband.][13]

[C]   The history of Greece notes how Iccus of Tarentum, Chryso, Astylus, Diopompus and others deprived themselves of any sort of sexual activity during all the time they were getting their bodies in trim for the races, wrestling and other contests at the Olympic Games.[14]

Muley Hassan, the Dey of Tunis (the one whom the Emperor Charles V restored to his throne) was critical of his father's memory because he was always with his wives, calling him a weak effeminate spawner of children.

[B]   In a certain province in the Spanish Indies men were allowed to marry only after forty, yet girls could marry at ten.[15]

12. Aristotle, *Politics*, VII, xvi (age of thirty-seven not thirty-five); Plato, *Republic*, V, 460A ff.; cf. Tiraquellus, *De legibus connubialibus*, VI, §§ 44–7; 52.
13. Plutarch, *Life of Thales*; Caesar, *Gallic Wars*, VI (cf. Tiraquellus, ibid., VI, § 47); Torquato Tasso, *Gierusalemme liberata*, X, 39–41.
14. Tiraquellus, ibid., XV, § 26, citing Plato, *Laws*, VIII, 839E–840A.
15. Paolo Giovio, *Historia sui temporis*, on 'Muleasses' (Muley Hassan); Lopez de Gomara, *Histoire générale des Indes*.

[A] If a nobleman is only thirty-five it is too soon for him to make way for a twenty-year-old son: he has still got to achieve a reputation in military expeditions or at the Court of his monarch: he needs his cash; he should allow his son a share but not forget himself. Such a man can rightly give the answer which fathers often have on their lips: 'I have no wish to be stripped bare before I go and lie down'. But a father who is brought low by age and illness, whose weakness and ill-health deprive him of ordinary human fellowship, does wrong to himself and to his family if he broods over a great pile of riches. If he is wise, he has reached the period when he really ought to want to get stripped and lie down – not stripped to his shirt but down to a nice warm dressing-gown. He has no more use for all the remaining pomp: he should give it all away as a present to those whom it ought to belong to by Nature's ordinance.

It is right that he should let them use what Nature deprives him of: otherwise there is certainly an element of malice and envy. The finest gesture the Emperor Charles V ever made was when, [C] in imitation of some ancient holders of his rank, [A] he was able to recognize that reason clearly commands us to strip off our garments when they weigh us down and get in our way, and to go and lie down when our legs fail us.[16] Once he began to feel deficient in the strength and energy needed to continue to conduct his affairs with the glory he had earned, he handed over his wealth, his rank and his power to his son:

> *Solve senescentem mature sanus equum, ne*
> *Peccet ad extremum ridendus, et ilia ducat.*

[Be wise enough to unharness that tired old nag lest it ends up short-winded, stumbling while men jeer at it.][17]

This defect of not realizing in time what one is, of not being aware of the extreme decline into weakness which old age naturally brings to our bodies and our souls – to them equally in my opinion unless the soul actually has the larger share – has ruined the reputation of most of the world's great men. I have seen in my lifetime and intimately known great men in authority who had clearly declined amazingly from their former

16. Charles V resigned his crown and entered a monastery in 1557 (cf. J. Du Bellay, *Regrets*, 111).
17. Horace, *Epistles*, I, i, 8. (The 'old nag' is his Muse: hence the following development.)

capacities, which I knew of from the reputation they had acquired in their better years. For their honour's sake I would deeply have wished that they had withdrawn to their estates, dropping the load of public or military affairs which were no longer meant for their shoulders.

There was a nobleman whose house I used to frequent who was a widower, very old but still with some sap in him. He had several daughters to marry and a son already old enough to enter society, so that his house was burdened with considerable expenditure and quite a lot of outside visitors; he took little pleasure in this, not only out of concern for economy but even more because, at his age, he had adopted a mode of life far different from ours. In that rather bold way I have[18] I told him one day that it would be more becoming if he made room for us youth, leaving his principal residence to his son (for it was the only one properly equipped and furnished) and withdrew to a neighbouring estate of his where nobody would trouble his rest, since, given his children's circumstances, there was no other way he could avoid our unsuitable company. He later took my advice and liked it.

That is not to say we should make a binding gift of our property and not be able to go back on it. I am old enough to have to play that role now, and would leave the young the use of my house and property but be free to withdraw my consent if they gave me cause. I would let them have use of them because they no longer gave me pleasure, but I would retain as much general authority over affairs as I wanted to, for I have always thought that it must be a great happiness for an old father to train his own children in the management of his affairs; he could then, during his lifetime, observe how they do it, offering advice and instruction based on his own experience in such things, and personally arranging for the ancient honour and order of his house to come into the hands of his successors, confirming in this way the hopes he could place in their future management of them.

To do this I would not avoid their company; I would like to be near so as to watch them and to enjoy their fun and festivities as much as my age permitted. Even if I did not live among them (as I could not do without embarrassing the company by the gloominess of my age and by my being subject to illnesses – and also without being forced to restrict my own rules and habits), I would at least like to live near them in some corner of my house – not the fanciest but the most comfortable. Not (as I saw a few

18. [A] until [C]: I have *of bringing forth whatever comes to my lips* I told him . . .

years ago) a dean of St-Hilaire-de-Poitiers[19] brought to such a pitch of solitude by the troublesome effects of his melancholy that, when I went into his room, he had not set foot outside it for twenty-two years; yet he could still move about freely and easily, apart from a rheumatic flux discharging into his stomach. He would let scarcely anyone in to see him even once a week; he always stayed shut up in that room all by himself except for a valet who brought him his food once a day and who merely went in and out. His only occupation was to walk about reading a book (for he had some acquaintance with literature), obstinately determined as he was to die in those conditions – as soon afterwards he did.

I would try to have gentle relations with my children and so encourage in them an active love and unfeigned affection for me, something easily achieved in children of a well-born nature; of course if they turn out to be wild beasts   [C]   (which our century produces in abundance)   [A]   then you must hate them and avoid them as such.

I am against the custom   [C]   of forbidding children to say 'Father' and requiring them to use some other, more respectful title, as though Nature had not sufficiently provided for our authority. We address God Almighty as Father and scorn to have our own children call us by that name.

It is also unjust, and mad, to   [A]   deprive our grown-up children of   [C]   easy relations   [A]   with their[20] fathers by striving to maintain   [C]   an austere and contemptuous   [A]   frown, hoping[21] by that to keep them in fear and obedience. That is a quite useless farce which makes fathers loathsome to children and, what is worse, makes them ridiculous. Since youth and vigour are in their children's hands they enjoy the current favour of the world; they treat with mockery the fierce tyrannical countenance of a man with no blood left in his veins or his heart – scarecrows in a field of flax! Even if I were able to make myself feared I would rather make myself loved.[22]

[B]   There are so many drawbacks in old age, so much powerlessness; it so merits contempt that the best endowment it can acquire is the fond love of one's family: its arms are no longer fear and commands.

I know one man who had a most imperious youth. Now that old age is

---

19. Jean d'Estissac, who died in 1576. Such symptoms of melancholy as Montaigne describes are not rare in Renaissance medical treatises.
20. '80: children of *private intercourse and easy understanding* with . . .
21. '80: maintain *a severe and distant frown, full of rancour and contempt,* hoping . . .
22. Cf. Erasmus' similar reaction in his *Adages,* II, IX, LXII, *Oderint dum metuant.*

coming upon him, despite trying to accept it as well as he can, he slaps and bites and swears – [C] the stormiest master in France; [B] he frets himself with cares and watchfulness: but it is all a farce which the family conspire in; the others have access to the best part of his granary, his cellar and even his purse: meanwhile he keeps the keys in his pouch, dearer to him than sight itself. While he is happy to keep so spare and thrifty a table, in various secret places in his house all is dissipation, gambling, prodigality and tales about his fits of temper and his precautions. Everybody is on the lookout against [C] him.²³ [B] If some wretched servant happens to become devoted to him, suspicion is immediately thrown on to him – a quality which old age is only too ready to ruminate upon. How many times has that man boasted to me of keeping his family on a tight rein, of the meticulous obedience and reverence he received because of it, and of the lucid watch he kept over his affairs:

> *Ille solus nescit omnia!*
>
> [He alone is unaware of the lot!]²⁴

No man of my acquaintance can claim more qualities, natural and acquired, proper for maintaining his mastery; yet he had failed completely, like a child. That is why I have picked him out as an example from several other cases that I know.

[C] It would make a good scholastic debate: whether or not he is better off as he is. In his presence, all things defer to him; his authority runs its empty course: nobody ever resists him; they believe what he says, they fear and respect him . . . as much as he could wish! Should he dismiss a servant he packs his bag and is off at once – but only out of his presence. Old people's steps are so slow and their senses so confused that the valet can live a full year in the house doing his duty without their even noticing it. At the appropriate time a letter arrives from distant parts, a pitiful one, a submissive one, full of promises to do better in the future; the valet then finds himself back in favour.

Does my Lord strike a bargain and send a missive which the family do not like? They suppress it, sometimes inventing afterwards reasons to explain the lack of action or reply. Since no letters from outside are ever brought to him first, he only sees the ones which it seems convenient for him to know. If he happens to get hold of any, he always has to rely on

23. '88: against *that poor man*. If . . .
24. Terence, *Adelphi*, IV, ii, 9.

somebody else to read them for him, so they invent things on the spot: they are always pretending that someone is begging his pardon in the very letter that contains abuse. In short he sees his affairs only through some counterfeit image designed to be as pleasing to him as they can make it so as not to awake his spleen or his anger.

Under various guises, but all to the same effect, I have seen plenty of households run long and steadily in this way.

[B] Wives are always disposed to disagree with their husbands.[25] [C] With both hands they grasp at any pretence for contradicting them; any excuse serves as full justification. I know one who used to rob her husband wholesale – in order, she told her confessor, to 'fatten up her almsgiving'. (There's a religious spendthrift for you to trust!) Whatever their husbands agree to never provides them with enough dignity. To give it grace and authority they must have usurped it by ruse or by force, but always unjustly. When, as in the case I am thinking of, they are acting against some poor old man on behalf of the children, they seize on this pretext and are honoured for serving their own passions; and, as though they were all slaves together, readily plot against his sovereignty and government. [B] If the children are male and grown-up, in the bloom of youth, then their mothers gang up with them and corrupt the steward, the bursar and everyone else by force or favour.

Old men without wives and children fall into this evil less easily but more cruelly and with less dignity. [C] Cato the Elder already said in his time, 'So many valets: so many enemies.' Given the gulf separating the purity of his century from ours, just think whether he was not really warning us that wife, sons and valet are all 'so many enemies' in our case.[26]

[B] It is a good thing that decrepitude furnishes us with the sweet gifts of inadvertency, ignorance and a readiness to be cheated. If we were to resist, what would happen to us, especially nowadays when the judges who settle our quarrels are usually on the side of the children – and venal?

[C] The cheating may escape my sight, but it does not escape my sight that I am very cheatable. ★★ Thrice and four times blessed is he who can entrust his pitiful old age into the hands of a friend. ★★[27] And

---

25. '88: husbands, *especially if they are old and irascible: but when it is a matter of favouring their children they grasp that pretext and glory in it.* If the children . . .

26. Cf. Seneca, *Epist. moral.*, XLVII, 5; but it was not Cato who said it. (The proverb applied to slaves, not valets or servants.)

27. This and the following passage between stars have been restored. In the Bordeaux manuscript they are deleted, but not certainly by Montaigne himself.

shall we have ever said enough about the value of a friend and how totally different it is from bonds based on contracts! Even that counterpart to a friend which I see between beasts, how devoutly I honour it!   ** Am I better or worse off for having savoured a friend? Better off, certainly. My regret for him consoles me and honours me. Is it not a most pious and pleasant task in life to be ever performing his obsequies? Can any pleasure possessed equal that pleasure lost? I would readily let myself be rapt insensible lingering over so caressing a notion.   **

[C] Others may deceive me, but at least I do not deceive myself into thinking that I can protect myself against it; nor do I cudgel my brains for ways of making myself able to do so. Only in my own bosom can I find salvation from treachery like this – not in disquieting and tumultuous inquisitiveness but in diversion and constancy. Whenever I hear of the state that some other man is in, I waste no time over that but immediately turn my eyes on to myself to see how I am doing. Everything which touches him touches me too. What has happened to him is a warning and an alert coming from the same quarter. Every day, every hour, we say things about others which ought more properly to be addressed to ourselves if only we had learned to turn our thoughts inward as well as widely outward. Similarly many authors inflict wounds on the cause they defend by dashing out against the attackers, hurling shafts at their enemies which can properly be hurled back at them.

[A] The late Monsieur de Monluc, the Marshal, when talking to me of the loss of his son (a truly brave gentleman of great promise who died on the island of Madeira), among other regrets emphasized the grief and heartbreak he felt at never having revealed himself to his son and at having lost the pleasure of knowing and savouring him, all because of his fancy to appear with the gravity of a stern father; he had never told him of the immense love he felt for him and how worthy he rated him for his virtue. 'And all that poor boy saw of me,' he said, 'was a frowning face full of scorn; he is gone, believing I was unable to love him or to judge him as he deserved. Whom was I keeping it for, that knowledge of the special love I harboured for him in my soul! Should not he have felt all the pleasure of it, and all the bonds of gratitude? I forced myself, I tortured myself, to keep up that silly mask, thereby losing the joy of his company – and his good-will as well, which must have been cold towards me: he had never received from me anything but brusqueness or known anything but a tyrannous façade.'

I find that lament to be reasonable and rightly held: for as I know only too well from experience when we lose those we love there is no consolation

sweeter than the knowledge of having remembered to tell them everything and to have enjoyed the most perfect and absolute communication with them.

[B] As much as I can I open myself to my own folk, and am most ready to tell them or anyone else what I intend towards them and what is the judgement I make on them. I hasten to reveal myself, to make myself known, for I do not want them to be misled about me in any way whatsoever.

[A] According to Caesar, among the customs peculiar to our ancient Gauls there was the following: sons were not presented to their fathers and never dared to appear in public with them until they had begun to bear arms, as if to signify that the time had now come for the fathers to admit them to their intimate acquaintance.[28]

Yet another abuse of paternal discretion which I have seen in my time is when fathers are not content with having deprived their children of their natural share of the property during their long lifetime, but then go and leave authority over all of it after their death to their widows, free to dispose of it at their pleasure. One lord I have known (among the highest officers of the Realm) could rightfully have expected to come into property worth fifty thousand crowns a year: yet he died in need, overwhelmed with debts at the age of fifty, while his mother, despite advanced senility, still enjoyed rights over the entire property under the will of his father, who himself had lived to be eighty.

To me that seems in no way reasonable.

[B] For all that, I cannot see it helps much when a man whose affairs are prospering goes and seeks a wife who burdens him with a large dowry: no debt contracted outside the family is more ruinous to a household. My ancestors have all followed this precept, most fittingly; so have I.

[C] Yet those who warn us against marrying rich wives out of fear that they might be less beholden to us and more difficult wrongly lose a real advantage for a frivolous conjecture.[29] If a woman is unreasonable it costs her no more to jump over one reason than another. Such women are most pleased with themselves when they are most in the wrong: it is the injustice which allures them; whereas for good women it lies in their virtuous deeds: the richer they are the more gracious they are, just as beautiful women are more willingly and more triumphantly chaste.

---

28. Caesar, *Gallic Wars*, VI, xviii.
29. Cf. Tiraquellus, *De legibus connubialibus*, V, § 1 ff., repeating Aristotle's warning against wives who dominate because of their wealth.

[A] It is reasonable to let mothers run affairs until the sons are legally old enough to assume the responsibility; but the father has brought them up wrongly if (considering the normal weakness of the female) he could not expect them to be wiser and more competent than his wife once they have reached that age. But it would be even more unnatural to make mothers depend on the discretion of their sons. They should be given a provision generous enough to maintain their state according to the condition of their family and their age, especially since want or indigence are far more difficult for them to bear with decorum than for males: that burden ought to be put on the sons rather than on the mother.

[C] On the whole, the soundest way of sharing out our property when we die is (I believe) to follow local customary law. The Law has thought it out better than we have, so it is better to let the Law make the wrong choice than rashly hazard doing so ourselves. The property does not really belong to us personally, since without our leave it is entailed by civil law to designated heirs. And even though we have some discretion as well, I hold that it would take a great and very clear reason to justify our depriving anyone of what he was entitled to by the fortune of his birth and of what common law leads him to expect; it would be an unreasonable abuse of that freedom to make it serve whims both frivolous and private.

Fate has been kind, sparing me opportunities which might have tempted me to change my predilection for the dictates of common law. I know people whom it would be a waste of time to serve long and dutifully: one word taken the wrong way can wipe out ten years of merit. Anyone able to butter them up when they are just about to go is lucky indeed! The latest action scoops the lot: it is not the best and most frequent services which prove efficacious but recent ones, present ones.

There are people who exploit their wills as sticks and carrots to punish or reward every action of those who may claim an interest in the inheritance. But this is a matter of long-lasting consequence; it is too weighty to be changed from moment to moment: wise men settle it once and for all – and have regard for the reasonable customs of the community.

We are a little too fond of male entail; we foresee a ridiculous eternity for our family name and attach too much weight to silly conjectures about the future based on the minds of little boys. Somebody might easily have been unjust to me, ousting me from my place because I was more lumpish, more leaden, more slow and more unwilling to learn than any of my brothers (indeed, than all the children in my province), whether I was being taught to exercise mind or body. It is madness to make such selections, interrupting the succession on the faith of such fortune-telling

which has so often deceived us. If we can ever infringe that rule and correct the choice of heirs made by destiny, it would probably be out of consideration for some huge noticeable physical deformity, of a permanent and incurable kind, one which those of us who are great admirers of beauty believe to be highly deleterious.

There is an agreeable dialogue between Plato's Lawgiver and his citizens which may honour my pages here: 'What!' they say, feeling their end draw near: 'Can we not bequeath our own property to anyone we please? What cruelty, O gods, that it be not lawful to give more or to give less just as we like, depending on how our heirs have helped us in our affairs, our illnesses or our old age!' The Legislator made this reply: 'My dear friends; you are certain to die soon: so it is difficult for any of you to "Know Thyself" (according to that Delphic inscription) and to know what is yours. I make these laws and maintain that you do not belong to yourselves, nor do the things of which you enjoy the use actually belong to you. You and your goods belong to your family, both past and future. And still more do your family and your goods belong to the commonwealth. Therefore if on your sickbed or in your old age some flatterer tries to persuade you to make an unjust will (or if a fit of temper does) I will protect you. Out of respect for the general concerns of our City and of your family, I will establish laws which make it known that private interests must reasonably yield to those of the community. Go, gently and willingly, whither human necessity bids you. It is for me, who favour all things equally and who take care of the people in general, to take care also of what you leave behind you.'[30]

To return to my subject, [A] it seems to me right, somehow, that women should have no mastery over men save only the natural one of motherhood – unless it be for the chastisement of those who have wilfully submitted to them out of some feverish humour; but that does not apply to old women, the subject of our present discussion. It is the manifest truth of this consideration which has made us so ready to invent and entrench that Salic Law – which nobody has ever seen – which debars women from succeeding to our throne;[31] and though Fortune has lent it more credence in some places than others, there is scarcely one jurisdiction in the world where that law is not cited as here, because of the genuine appearance of reason which gives it authority.

30. Plato, *Laws*, XI, 922 D–924 A.
31. The English claim to the French crown was based on the irrelevance of the mythical Salic Law. (Guillaume Postel maintained that it specifically applied to France, its real name being the 'Gallic' Law: *La Loi Salique*, Paris, 1552.)

It is dangerous to leave the superintendence of our succession to the judgement of our wives and to their choice between our sons, which over and over again is iniquitous and fantastic. For those unruly tastes and physical cravings which they experience during pregnancy are ever-present in their souls. They regularly devote themselves to the weakest and to the feeblest, or to those (if they have any) who are still hanging about their-necks. Since women do not have sufficient reasoning-power to select and embrace things according to their merits they allow themselves to be led to where natural impressions act most alone – like animals, which only know their young while they are still on the teat.[32]

Incidentally, experience clearly shows us that the natural love to which we attach such importance has very shallow roots. For a very small sum of money we daily tear their own children out of women's arms and get them to take charge of our own; we make them entrust their babes to some wretched wet-nurse to whom we have no wish to commit our own or else to a nanny-goat; then we forbid them not only to give suck to theirs no matter what harm it might do them but even to look after them; they must devote themselves entirely to the service of our children. And then we see that in most cases custom begets a kind of bastard love more distracted than the natural kind; they are far more worried about the preservation of those foster-children than of the children who really belong to them.

I mentioned nanny-goats because the village-women where I live call in the help of goats when they cannot suckle their children themselves; I have now two menservants who never tasted mothers' milk for more than a week. These nanny-goats are trained from the outset to suckle human children; they recognize their voices when they start crying and come running up. They reject any other child you give them except the one they are feeding; the child does the same to another nanny-goat. The other day I saw an infant who had lost its own nanny-goat as the father had only borrowed it from a neighbour: the child rejected a different one which was provided and died, certainly of hunger.

The beasts debase and bastardize maternal affection as easily as we do.

[C] Herodotus tells of a certain district of Libya where men lie with women indiscriminately, but where, once a child can toddle, it recognizes its own father out of the crowd, natural instinct guiding its first footsteps.[33] There are frequent mistakes, I believe . . .

---

32. '80: their young, *or savour their kinship* while . . .
33. Tiraquellus, *De legibus connubialibus*, VII, § 51; Herodotus, *History*, IV.

[A]   Now once we consider the fact that we love our children simply because we begot them, calling them our second selves, we can see that we also produce something else from ourselves, no less worthy of commendation: for the things we engender in our soul, the offspring of our mind, of our wisdom and talents, are the products of a part more noble than the body and are more purely our own. In this act of generation we are both mother and father; these 'children' cost us dearer and, if they are any good, bring us more honour. In the case of our other children their good qualities belong much more to them than to us: we have only a very slight share in them; but in the case of these, all their grace, worth and beauty belong to us. For this reason they have a more lively resemblance and correspondence to us.   [C]   Plato adds that such children are immortal and immortalize their fathers – even deifying them, as in the case of Lycurgus, Solon and Minos.[34]

[A]   Since our history books are full of exemplary cases of the common kind of paternal love, it seemed to me not inappropriate to cite a few examples of this other kind too.

[C]   Heliodorus, that good bishop of Tricca, preferred to forego the honour of so venerable a bishopric with its income and its dignity rather than to destroy his 'daughter', who still lives on – a handsome girl but attired perhaps with a little more care and indulgence than suits the daughter of a priest, of a clerk in holy orders – and fashioned in too erotic a style.[35]

[A]   In Rome there was a figure of great bravery and dignity called Labienus;[36] among other qualities he excelled in every kind of literature; he was, I think, the son of that great Labienus who was the foremost among captains who served under Caesar in the Gallic Wars, subsequently threw in his lot with Pompey the Great and fought for him most bravely until Caesar defeated him in Spain. There were several people who were jealous of the Labienus I am referring to; he also probably had enemies among the courtiers and favourites of the contemporary Emperors for his frankness and for inheriting his father's innate hostility towards tyranny, which we

34. Plato, *Phaedrus*, 258 C, dealing with a man's writings, his 'brain-children'; but Montaigne has transcribed *Minos* for *Darius*.
35. His Greek novel, *An Ethiopian History*, tells of the loves of Theagenes and Chariclea. It was translated into French by Amyot (Paris, 1547) and often reprinted.
36. Labienus was, for the ferocious nature of his controversial style, nicknamed *Rabienus* (the Fierce One). (Cf. Marcus Annaeus Seneca, *Controversiae*, 10, Preface; Suetonius, *Caligula*, 16.)

may believe coloured his books and writing. His enemies prosecuted him before the Roman magistrates and obtained a conviction, requiring several of the books he had published to be burnt. This was the very first case of the death-penalty being inflicted on books and erudition; it was subsequently applied at Rome in several other cases. We did not have means nor matter enough for our cruelty unless we also let it concern itself with things which Nature has exempted from any sense of pain, such as our renown and the products of our minds, and unless we inflicted physical suffering on the teachings and the documents of the Muses.

Labienus could not bear such a loss nor survive such beloved offspring; he had himself borne to the family vault on a litter and shut up alive; there he provided his own death and burial. It is difficult to find any example of fatherly love more vehement than that one. When his very eloquent friend Cassius Severus saw those books being burnt, he shouted that he too ought to be burnt alive with them since he actively preserved their contents in his memory.

[B] A similar misfortune happened to Greuntius Cordus who was accused of having praised Brutus and Cassius in his books.[37] That slavish base and corrupt Senate (worthy of a worse master than Tiberius) condemned his writings to the pyre: it pleased him to keep his books company as they perished in the flames by starving himself to death.

[A] Lucan was a good man, condemned by that blackguard Nero; in the last moments of his life, when most of his blood had already gushed from his veins (he had ordered his doctors to kill him by slashing them) and when cold had already seized his hands and feet and was starting to draw near to his vital organs, the very last thing he remembered were some verses from his *Pharsalian War*; he recited them, and died with them as the last words on his lips. Was that not saying farewell to his children tenderly and paternally, the equivalent of those adieus and tender embraces which we keep for our children when we die, as well as being an effect of that natural instinct to recall at our end those things which we held dearest to us while we lived?

When Epicurus lay dying, tormented they say by the most extreme colic paroxysms, he found consolation only in the beauty of the philosophy he had taught to the world;[38] are we to believe that he would have found happiness in any number of well-born, well-educated children (if he had

---

37. Or rather, *Cremutius* Cordus, an historian honoured for his frankness: Tacitus, *Annals*, IV, xxxiv; Marcus Annaeus Seneca, *Suasiora*, VII; Quintilian, X, i, 104.
38. Cicero, *De finibus*, II, xxx, 96.

had any) to equal what he found in the abundant writing which he had brought forth? And if he had had the choice of leaving either an ill-conceived and deformed child behind him or a stupid and inept book, would – not he alone but any man of similar ability – have preferred to incur the first tragedy rather than the other?

It would probably have been impious of Saint Augustine (for example) if someone had obliged him to destroy either his children (supposing he had had any) or else his writings (from which our religion receives such abundant profit) and he had not preferred to destroy his children.[39]

[B] I am not at all sure whether I would not much rather have given birth to one perfectly formed son by commerce with the Muses than by commerce with my wife. [C] As for this present child of my brain, what I give it I give unconditionally and irrevocably, just as one does to the children of one's body; such little good as I have already done it is no longer mine to dispose of; it may know plenty of things which I know no longer, and remember things about me that I have forgotten; if the need arose to turn to it for help, it would be like borrowing from a stranger. It is richer than I am, yet I am wiser than it.

[A] Few devotees of poetry would not have been more gratified at fathering the *Aeneid* than the fairest boy in Rome, nor fail to find the loss of one more bearable than the other. [C] For according to Aristotle, of all artists the one who is most in love with his handiwork is the poet.[40]

[A] It is hard to believe that Epaminondas (who boasted that his posterity consisted in two 'daughters' who would bring honour to their father one day – he meant his two noble victories over the Spartans) would have agreed to exchange them for daughters who were the most gorgeous in the whole of Greece; or that Alexander and Caesar had ever wished they could give up the greatness of their glorious feats in war in return for the pleasure of having sons and heirs however perfect, however accomplished; indeed I very much doubt whether Phidias or any other outstanding sculptor would have found as much delight in the survival and longevity of his physical children as in some excellent piece of sculpture brought to completion by his long-sustained labour and his skill according to the rules of his art.

And as for those raging vicious passions which have sometimes inflamed fathers with love for their daughters, or mothers for their sons, similar ones

---

39. St Augustine did have an illegitimate son. If Montaigne had read the *Confessions* he would have known of him.
40. Aristotle, *Nicomachaean Ethics*, IX, vii, 3.

can be found in this other kind of parenthood: witness the tale of Pygmalion who, having carved the statue of a uniquely beautiful woman, was so hopelessly ravished by an insane love for his own work that, for the sake of his frenzy, the gods had to bring her to life:

> *Tentatum mollescit ebur, positoque rigore*
> *Subsedit digitis.*

[He touches the ivory statue; it starts to soften; its hardness gone, it yields to his fingers.][41]

---

41. Ovid, *Metamorphoses*, X, 243 ff., citing 283–4.

9. On the armour of the Parthians

====

[French knights, thinking more of protecting their bodies than going over to the attack, wore increasingly heavy plate-armour. Montaigne had little trust in armour, not least when worn by men out of training.]

[A] The vile and thoroughly enervating practice of our noblemen today is never to don their armour until the very last second when absolutely necessary, and to throw it off as soon as there is the slightest sign of the danger being past. This results in chaos. What with everyone rushing about calling for his armour at the very moment of the attack, some are still lacing up their breast-plates after their companions have already been routed. Our forebears used to have their helmets, lances and gauntlets carried for them, but kept on the rest of their armour until they had finished their stint. Our cavalry units are in confusion and disorder, all mixed up together with the baggage-train and the batmen who cannot go far from their officers because they are carrying their armour for them [C]. Livy was talking of our soldiers when he said: '*Intolerantissima laboris corpora vix arma humeris gerebant*' [Their bodies being utterly incapable of toil, their shoulders can hardly bear the weight of their armour.][1]

[A]   Several peoples used to go to war unprotected or wearing things which afforded no protection; they still do:

[B]   *Tegmina queis capitum raptus de subere cortex.*

[With helmets made of cork stripped from the tree.][2]

Alexander, the most daring captain ever, rarely wore armour.   [A]   And those among us who despise it do not much weaken their bargaining-power! Although we do see a man killed occasionally for want of armour, we hardly find fewer who were killed because they were encumbered by it, slowed down by its weight, rubbed sore or worn out by it, struck by a blow glancing off it, or in some other way. It would seem indeed, given

1. Livy, XXVII, xlviii.
2. Virgil, *Aeneid*, VII, 742.

the weight and thickness of our armour, that we have no thought of anything but defending ourselves, [C] and that we are not so much covered as laden with it. [A] Impeded and constrained by it, we have enough to do to support its weight,³ as if fighting merely consisted in receiving blows on our armour [A1] and as if we were not equally beholden to defend it as it is to defend us.

[B] Tacitus amusingly describes the warriors of our Ancient Gaul, armed so as not to yield ground but no more, having no means of striking a blow nor of being struck by one nor of getting up once they were down. When Lucullus saw certain Median men-at-arms, drawn up facing the army of Tigranes clad in heavy awkward armour as though in an iron prison, he formed the opinion that he could easily defeat them and began his victory by charging against them.⁴ [A] And now that our musketeers are so highly prized, I think that we will discover some new invention to wall us up against them, making us drag ourselves off to war enclosed in little forts such as those which the Ancients made their elephants carry.

Such a humour is far removed from that of Scipio the Younger,⁵ who harshly rebuked his soldiers for having sown spiked cavalry-traps under water at the spot where the inhabitants of the town under siege could make sorties against him through the moat: he said assailants should have thoughts not of dread but of plans of attack, [C] rightly fearing that their precautions might deaden their vigilance during their guard-duty. [B] And he also said to a young man who was showing off his shield: 'It is a very fine one, my lad: but a Roman soldier must have more trust in his right arm than his left.'⁶

[A] Now what makes our armour an intolerable burden to us is want of habit:

> L'husbergo in dosso haveano, e l'elmo in testa,
> Dui di quelli guerrier, de i quali io canto.
> Ne notte o di, doppo ch'entraro in questa
> Stanza, gli haveanò mai mesi da canto,
> Che facile a portar comme la vesta
> Era lor, perche in uso l'avean tanto.

3. [A] until [C]: weight, *without taking on anything else, hampered and constrained without movement or manoeuvring*, as if . . .
4. Tacitus, *Annals*, III; Plutarch, *Life of Lucullus*.
5. [A] until [C]: the Younger, *surnamed Aemilianus*, who . . .
6. Plutarch (tr. Amyot), *Dicts notables des Anciens Roy, Princes et grands Capitaines*, 204 F; 205 E.

[The two warriors of whom I sing both were clad in hauberks with helmets on their heads; since entering their redoubt they had never taken them off, night or day, wearing them as easily as their clothing, so accustomed had they grown to them.][7]

[C] And the Emperor Caracalla marched in full armour through the countryside at the head of his troops. [A] The Roman foot-soldiers not only bore iron helmets, swords and shields (for, says Cicero, they were so used, where their equipment was concerned, to have it ever on their backs, that it bothered them no more than their limbs did – [C] 'arma enim membra militis esse dicunt' [for a soldier's weapons are called his very limbs] [A] but they also had to carry their rations for a fortnight and a fixed quantity of stakes to make defence-works, [B] weighing up to sixty pounds. And the soldiers of Marius, thus laden,[8] were trained to march five leagues in five hours – or six leagues when it was necessary to hurry. [A] Their military training was much tougher than ours and produced very different results. It is wonderfully instructive in this connection that a Spartan soldier was criticized for having been seen sheltering in a house while on a military expedition: they were so trained to hardship that it appeared shameful to be seen sheltering beneath any roof but the sky, no matter what the weather. [C] Scipio the Younger, when he was reforming his army in Spain, commanded his soldiers to eat only on their feet and to eat nothing cooked. [A] We would not get our men to go very far at that rate![9]

Moreover Marcellinus, a man brought up in the Roman wars, carefully noted the Parthian way of bearing arms, all the more so as it was very different from that of the Romans.[10] 'They have,' he said, 'armour plaited together like fine plumage which did not impede the movements of their bodies; yet it was so strong that our darts bounced off it when they

7. Ariosto, Orlando furioso, XII, 30–5.
8. [A] until [C]: laden, marching into battle, were trained . . .
   Cicero, Tusc. disput., II, xvi, 31.
9. Plutarch, Dicts notables des Princes . . ., 205 D. (Cf. Erasmus, Apophthegmata, II, on the Spartan discipline, and Plutarch, Dicts notables des Lacedaemoniens.)
10. '80: Romans. Now as it seems to me that their way was very close to our own, I once copied out the following passage from its author, having formerly taken the trouble to state much more fully what I knew about the comparison between our armour and that of the Romans; but that bit of my scribblings having been stolen with some others by one of my serving-men, I will not deprive him of the profit which he hopes to get out of it; besides it would be hard for me to chew over the same stuff twice. 'They have . . .
   The passages cited are from Ammianus Marcellinus, XIV and XV.

happened to strike it.' (The armoured scales which our ancestors made much use of were just like that.) Elsewhere he says: 'They had strong tough horses, clad in thick leather; they themselves were armed from head to foot in thick iron-plating so skilfully arranged that it lent itself to movement at their joints. You could have taken them for iron men; so closely fitted was the armour of their head, reproducing the shape of their facial features, that there was no means of landing a blow except through the tiny round holes which corresponded to their eyes and let in a little light and through the slits where their nostrils were, through which with some difficulty they could breathe':

> [B]   *Flexilis inductis animatur lamina membris,*
> *Horribilis visu; credas simulachra moveri*
> *Ferrea, cognatoque viros spirare metallo.*
> *Par vestitus equis: ferrata fronte minantur,*
> *Ferratosque movent, securi vulneris, armos.*

[The flexile iron-plating is brought to life by the limbs it encloses. A sight to strike terror: you could believe that iron statues were moving, the metal incorporate and breathing. Their horses are similarly armed: their iron-clad foreheads threaten, and they move their flanks, safely protected from wounds by iron.][11]

[A]   There you have a description which strongly resembles the way a French knight is equipped with all his bits of armour. Plutarch[12] says that Demetrius had made for himself and for Alcinus, the foremost soldier about him, two complete suits of armour each weighing one hundred and twenty-five pounds, whereas the normal armour weighed only sixty.[13]

11. Claudius Claudianus, *In Rufinum*, II, 358–62.
12. '80: armour. *I want to say the following words in conclusion.* Plutarch . . .
13.Plutarch, *Life of Demetrius*.

10. On books

===

[Montaigne gives himself to us in this chapter; especially the [C] additions show how he had moved from studying himself as a particular man to studying also Man in general, and how he, as a man, should live. The framework of his judgement on books (which is clearly implied but was then so well-known that it did not need to be spelled out) was Horace's division of good authors in his Ars Poetica into the author who 'simply delights' us and the very great one 'qui miscuit utile dulci' – who 'mixes the useful with the sweet'. This notion was so current in the Renaissance that great authors such as Rabelais and Ronsard were often called utiles-doux ('useful–delightful'). In this context, useful always meant 'useful for learning moral lessons'. For Montaigne all good historians are both delightful and useful, in this sense, but there are very few of them.]

[A]   I do not doubt that I often happen to talk of things which are treated better in the writings of master-craftsmen, and with more authenticity. What you have here is purely an assay of my natural, not at all of my acquired, abilities. Anyone who catches me out in ignorance does me no harm: I cannot vouch to other people for my reasonings: I can scarcely vouch for them to myself and am by no means satisfied with them. If anyone is looking for knowledge let him go where such fish are to be caught: there is nothing I lay claim to less. These are my own thoughts, by which I am striving to make known not matter but me. Perhaps I shall master that matter one day; or perhaps I did do so once when Fortune managed to bring me to places where light is thrown on it. But   [C]   I no longer remember anything about that. I may be a man of fairly wide reading, but I retain nothing.[1]

[A]   So I guarantee you nothing for certain, except my making known[2]   [A1]   what point I have so far reached in my knowledge[3]

---

1. '80: But *I have a memory which is unable to store for three days at a stretch any provisions which I have given into its keeping. So . . .*
2. '88: known *what I think,* Excutienda damus praecordia, *and* what point . . .
   Citing Persius, *Satires,* V, 22: later cited in III, 9, 'On vanity'.
3. '80: knowledge *of what I am treating. Do not linger over the things I talk about but over the fashioning I give to them when talking about them. What I steal from others I do not wish to make mine: I claim no part in them except for my reasoning and judgement:*

[C] of it. Do not linger over the matter but over my fashioning of it. Where my borrowings are concerned, see whether I have been able to select something which improves my theme: I get others to say what I cannot put so well myself, sometimes because of the weakness of my language and sometimes because of the weakness of my intellect. I do not count my borrowings: I weigh them; if I had wanted them valued for their number I would have burdened myself with twice as many. They are all, except for very, very few, taken from names so famous and ancient that they seem to name themselves without help from me. In the case of those reasonings and original ideas which I transplant into my own soil and confound with my own, I sometimes deliberately omit to give the author's name so as to rein in the temerity of those hasty criticisms which leap to attack writings of every kind, especially recent writings by men still alive and in our vulgar tongue which allow anyone to talk about them and which seem to convict both their conception and design of being just as vulgar. I want them to flick Plutarch's nose in mistake for mine and to scald themselves by insulting the Seneca in me. I have to hide my weakness beneath those great reputations. I will love the man who can pluck out my feathers – I mean by the perspicacity of his judgement and by his sheer ability to distinguish the force and beauty of the topics. Myself, who am constantly unable to sort out my borrowings by my knowledge of where they came from, am quite able to measure my reach and to know that my own soil is in no wise capable of bringing forth some of the richer flowers that I find rooted there and which all the produce of my own growing could never match.

[A] What I am obliged to answer for is for getting myself tangled up, or if there is any inanity or defect in my reasoning which I do not see or which I am incapable of seeing once it is pointed out to me. Faults can often escape our vigilance: sickness of judgement consists in not perceiving them when they are revealed to us. Knowledge and truth can lodge within us without judgement; judgement can do so without them: indeed, recogniz-

---

the rest is not my role. I ask for nothing except that you should see whether I have been capable of selecting what can be rightly linked to my topic. The fact that I sometimes deliberately hide the name of the author in the things which I cite is intended to rein in the *frivolousness of those who are concerned to make judgements upon whatever is offered them but who, having no flair for savouring the things in themselves, stop at the name of the workman or his reputation. I wish them to scald themselves by condemning the Cicero and Aristotle in me. What I am obliged . . .*

ing our ignorance is one of the surest and most beautiful witnesses to our judgement that I can find.

I have no sergeant-major to marshal my arguments other than Fortune. As my ravings present themselves, I pile them up; sometimes they all come crowding together: sometimes they drag along in single file. I want people to see my natural ordinary stride, however much it wanders off the path. I let myself go along as I find myself to be, anyway the matters treated here are not such that ignorance of them cannot be permitted nor talking of them casually or rashly. I would very much love to grasp things with a complete understanding but I cannot bring myself to pay the high cost of doing so. My design is to spend whatever life I have left gently and unlaboriously. I am not prepared to bash my brains for anything, not even for learning's sake however precious it may be. From books all I seek is to give myself pleasure by an honourable pastime: or if I do study, I seek only that branch of learning which deals with knowing myself and which teaches me how to live and die well:

[B] *Has meus ad metas sudet oportet equus.*

[This is the winning-post towards which my sweating horse must run.][4]

[A] If I come across difficult passages in my reading I never bite my nails over them: after making a charge or two I let them be. [B] If I settled down to them I would waste myself and my time, for my mind is made for the first jump. What I fail to see during my original charge I see even less when I stubborn it out.

I can do nothing without gaiety: persistence [C] and too much intensity [B] dazzle my judgement, making it sad and weary. [C] My vision becomes confused and dissipated: [B] I must tell it to withdraw and then make fresh glancing attacks, just as we are told to judge the sheen of scarlet-cloth by running our eyes over it several times, catching various glimpses of it, sudden, repeated and renewed.

[A] If one book wearies me I take up another, applying myself to it only during those hours when I begin to be gripped by boredom at doing nothing. I do not have much to do with books by modern authors, since the Ancients seem to me to be more taut and ample; nor with books in Greek, since my judgement [C] cannot do its job properly on the basis of a schoolboy, apprenticed [A] understanding.[5]

4. Propertius, IV, i, 70.
5. '80: my judgement *is not satisfied with a mediocre* understanding . . .

[A]   Among books affording plain delight, I judge that the *Decameron*
of Boccaccio, Rabelais and the *Basia* of Johannes Secundus (if they are to
be placed in this category)[6] are worth spending time on. As for the *Amadis*
and such like, they did not have enough authority to captivate me even in
childhood.[7] I will also add, boldly or rashly, that this aged heavy soul of
mine can no longer be tickled by good old Ovid (let alone Ariosto): his
flowing style and his invention, which once enraptured me, now hardly
have the power of holding my attention.

I freely say what I think about all things – even about those which
doubtless exceed my competence and which I in no wise claim to be
within my jurisdiction. When I express my opinions it is so as to reveal the
measure of my sight not the measure of the thing. When I find that I have
no taste for the *Axiochus* of Plato – a weak book, considering its author[8] –
my judgement does not trust itself: it is not so daft as to oppose the
authority of so many   [C]   other judgements, famous and ancient, which
it holds as its professors and masters: rather is it happy to err with
them.[9]   [A]   It blames itself, condemning itself either for stopping at
the outer rind and for being unable to get right down to the bottom of
things, or else for looking at the matter in some false light. My judgement
is quite content merely to protect itself from confusion and unruliness:
as for its weakness, it willingly acknowledges it and avows it. What it
thinks it should do is to give a just interpretation of such phenomena as its
power of conception presents it with: but they are feeble ones and imperfect. Most of Aesop's fables have several senses and several ways of being
understood. Those who treat them as myths select some aspect which squares
well with the fable; yet   [B]   in most cases   [A]   that is only the first surface facet of them: there are other facets, more vivid, more of their
essence, more inward, to which they never manage to penetrate; that is
what I do.

But to get on: it has always seemed to me that in poetry Virgil,

6. '80: category, *and from the centuries rather earlier than our own, the Ethiopian
History*, are worth . . .
    (For this work, cf. II, 8, note 13). Johannes Secundus' *Kisses* were much
appreciated and imitated.
7. *Amadis de Gaule*, a Spanish novel translated into French in twenty-one volumes,
had the success of a high-class soap-opera.
8. The *Axiochus* was already considered supposititious in the Renaissance.
9. '80: many *better* judgements, *nor does it rashly give itself the right to arraign them*. It
blames . . .

Lucretius, Catullus and Horace rank highest by far – especially Virgil in his *Georgics*, which I reckon to be the most perfect achievement in poetry; by a comparison with it one can easily see that there are passages in the *Aeneid* to which Virgil, if he had been able, would have given a touch of the comb. [B] And in the *Aeneid* the fifth book seems to me the most perfect. [A] I also love Lucan and like to be often in his company, not so much for his style as for his own worth and for the truth of his opinions and judgements. As for that good poet Terence – the grace and delight of the Latin tongue – I find him wonderful at vividly depicting the emotions of the soul and the modes of our behaviour; [C] our own actions today constantly bring me back to him. [A] However often I read him I always find some new grace and beauty in him.

Those who lived soon after Virgil's time complained that some put Lucretius on a par with him. My opinion is that such a comparison is indeed between unequals; yet I have quite a job confirming myself in that belief when I find myself enthralled by one of Lucretius' finer passages. If they were irritated by that comparison what would they say of the animal stupidity and barbarous insensitivity of those who now compare Ariosto with him? And what would Ariosto himself say?

[A1]   *O seclum insipiens et infacetum!*

[O what a silly, tasteless age!]¹⁰

[A] I reckon that the Ancients had even more reason to complain of those who put Plautus on a par with Terence (who savours much more of the nobleman) than of those who did so for Lucretius and Virgil. [C] It does much for Terence's reputation and superiority that the Father of Roman Eloquence has him – alone in his class – often on his lips, and so too the verdict which the best judge among Roman poets gave of his fellow-poet.¹¹

[A] It has often occurred to me that those of our contemporaries who undertake to write comedies (such as the Italians, who are quite good at it) use three or four plots from Terence or Plautus to make one of their own. In one single comedy they pile up five or six tales from Boccaccio. What makes them so burden themselves with matter is their lack of confidence in their ability to sustain themselves with their own graces: they need

---

10. Catullus, XLIII, 8.
11. Cicero, the Father of Eloquence; Horace (the 'best' judge) prefers Terence to Plautus, *Epistles*, II, i, 55 ff.

something solid to lean on; not having enough in themselves to captivate us they want the story to detain us. In the case of my author, Terence, it is quite the reverse: the perfections and beauties of the fashioning of his language make us lose our craving for his subject: everywhere it is his elegance and his graciousness which hold us; everywhere he is so delightful –

liquidus puroque simillimus amni

[flowing exactly like a pure fountain][12]

– and he so fills our souls to the brim with his graces that we forget those of his plot.

Considerations like these encourage me to go further: I note that the good poets of Antiquity avoided any striving to display not only such fantastic hyperboles as the Spaniards and the Petrarchists do but even those sweeter and more restrained acute phrases which adorn all works of poetry in the following centuries. Yet not one sound judge regrets that the Ancients lacked them nor fails to admire the incomparable even smoothness and the sustained sweetness and flourishing beauty of the epigrams of Catullus above the sharp goads with which Martial enlivens the tails of his. The reason for this is the same as I stated just now, and as Martial said of himself: '*Minus illi ingenio laborandum fuit, in cujus locum materia successerat.*' [He had less need to strive after originality, its place had been taken by his matter.][13] Those earlier poets achieve their effects without getting excited and goading themselves on; they find laughter everywhere: they do not have to go and tickle themselves! The later ones need extraneous help: the less spirit they have, the more body they need.   [B]   They get up on their horses because they cannot stand on their own legs.   [A]   It is the same with our dancing: those men of low estate who teach it are unable to copy the deportment and propriety of our nobility[14] and so try to gain favour by their daring footwork and other strange acrobatics.   [B]   And it is far easier for ladies to cut a figure in dances which require a variety of intricate bodily movements than in certain other stately dances in which they merely have to walk with a natural step and display their native bearing and their usual graces.   [A]   Just as some excellent clowns whom I have seen are able to give us all the delight which can be drawn from their art

12. Horace, ibid., II, ii, 120.
13. Martial, *Epigrams*, VIII, dedication to the Emperor Domitian.
14. '80: nobility, *to compensate for that grace which they are unable to imitate*, try to . . .

while wearing their everyday clothes, whereas to put us in a laughing mood their apprentices and those who are less deeply learned in that art have to put flour on their faces, dress up in funny clothes and hide behind silly movements and grimaces.

Better than any other way this idea as I conceive it can be understood from a comparison between the *Aeneid* and the *Orlando furioso*. We can see the *Aeneid* winging aloft with a firm and soaring flight, always pursuing its goal: the *Orlando furioso* we see hopping and fluttering from tale to tale as from branch to branch, never trusting its wings except to cross a short distance, seeking to alight on every hedge lest its wind or strength should give out,

> *Excursusque breves tentat.*
>
> [Trying out its wings on little sorties.][15]

So much, then, for the authors who delight me most on that kind of subject.

As for my other category of books (that which mixes a little more usefulness with the delight[16] and from which I learn how to control my humours and my qualities), the authors whom I find most useful for that are Plutarch (since he has become a Frenchman)[17] and Seneca. They both are strikingly suited to my humour in that the knowledge that I seek from them is treated in pieces not sewn together (and so do not require me to bind myself to some lengthy labour, of which I am quite incapable). Such are the *Moral Works* of Plutarch, as well as the *Epistles* of Seneca which are the most beautiful part of his writings and the most profitable. I do not need a great deal of preparation to get down to them and I can drop them whenever I like, for one part of them does not really lead to another. Those two authors are in agreement over most useful and true opinions; they were both fated to be born about the same period; both to be the tutors of Roman Emperors; both came from foreign lands and both were rich and powerful.[18] Their teachings are some of the cream of philosophy and are presented in a simple and appropriate manner. Plutarch is more

15. Virgil, *Georgics*, V, 194.
16. Horace, *Ars poetica*, 343: the author who 'wins every vote', *'miscuit utile dulci'* (mixes moral usefulness with delight).
17. That is, since Bishop Amyot translated him into French.
18. Seneca was born 4 BC, Plutarch half a century later; Seneca was Nero's tutor; Plutarch (it was thought) Trajan's. (Here Montaigne writes a brief parallel life, in the style of Plutarch.)

uniform and constant: Seneca is more diverse and comes in waves. Seneca
stiffens and tenses himself, toiling to arm virtue against weakness, fear and
vicious appetites; Plutarch seems to judge those vices to be less powerful
and to refuse to condescend to hasten his step or to rely on a shield.
Plutarch holds to Plato's opinions, which are gentle and well-suited to
public life: Seneca's opinions are Stoic and Epicurean, farther from common practice but in my judgement more suited   [C]   to the individual
[A]   and firmer. It seems that Seneca bowed somewhat to the tyranny of
the Emperors of his day, for I hold it for certain that his judgement was
under duress when he condemned the cause of those great-souled murderers
of Caesar; Plutarch is a free man from end to end. Seneca is full of pithy
phrases and sallies; Plutarch is full of matter. Seneca enflames you and stirs
you: Plutarch is more satisfying and repays you more.   [B]   Plutarch
leads us: Seneca drives us.

[A]   As for Cicero, the works of his which are most suitable to my
projects are those which above all deal with moral philosophy. But to tell
the truth boldly (for once we have crossed the boundaries of insolence
there is no reining us in) his style of writing seems boring to me, and so
do all similar styles. For his introductory passages, his definitions, his
sub-divisions and his etymologies eat up most of his work; what living
marrow there is in him is smothered by the tedium of his preparations. If I
spend an hour reading him (which is a lot for me) and then recall what
pith and substance I have got out of him, most of the time I find nothing
but wind, for he has not yet got to the material which serves my purposes
and to the reasoning which actually touches on the core of what I am
interested in. For me, who am only seeking to become more wise not
more learned   [C]   or more eloquent,   [A]   all those marshallings of
Aristotelian logic are irrelevant; I want authors   [C]   to begin with their
conclusion: [A]   I know[19] well enough what is meant by death or
voluptuousness: let them not waste time dissecting them; from the outset I
am looking for good solid reasons which teach me how to sustain their
attacks. Neither grammatical subtleties nor ingenuity in weaving words or
arguments help me in that. I want arguments which drive home their first
attack right into the strongest point of doubt: Cicero's hover about the pot
and languish. They are all right for the classroom, the pulpit or the Bar
where we are free to doze off and find ourselves a quarter of an hour later
still with time to pick up the thread of the argument. You have to talk like

19. '80: authors *to come quickly to the point.* I know . . .

that to judges whom you want to win over whether you are right or wrong, or to schoolboys and the common people [C] to whom you have to say the lot and see what strikes home. [A] I do not want authors to strive to gain my attention by crying *Oyez* fifty times like our heralds. The Romans in their religion used to cry *Hoc age!* [This do!], [C] just as in our own we cry *Sursum corda* [Lift up your hearts];[20] [A] for me they are so many wasted words. I leave home fully prepared: I need no sauce or appetizers: I can eat my meat quite raw; and instead of whetting my appetite with those preliminaries and preparations they deaden it for me and dull it.

[C] Will the licence of our times excuse my audacious sacrilege in thinking that even Plato's *Dialogues* drag slowly along stifling his matter, and in lamenting the time spent on those long useless preparatory discussions by a man who had so many better things to say? My ignorance may be a better excuse, since I can see none of the beauty of his language.

In general I ask for books which use learning not those which trim it up.

[A] My first two, as well as Pliny and their like, have no *Hoc age*: they want to deal with people who are already on the alert – or if they do have one it is an *Hoc age* of substance with its own separate body.

I also like reading Cicero's *Letters to Atticus*,[21] not only because they contain much to teach us about the history and affairs of his time but, even more, so as to find out from them his private humours. For as I have said elsewhere I am uniquely curious about my authors' soul and native judgement. By what their writings display when they are paraded in the theatre of the world we can indeed judge their talents, but we cannot judge them as men nor their morals.

I have regretted hundreds of times that we have lost the book which Brutus wrote about virtue: it is a beautiful thing to learn the theory from those who thoroughly know the practice; yet seeing that the preacher and the preaching are two different things, I am just as happy to see Brutus in Plutarch as in a book of his own. I would rather have a true account of his chat with his private friends in his tent on the eve of a battle than the oration which he delivered next morning to his army, and what he did in his work-room and bedroom than what he did in the Forum or Senate.

As for Cicero, I share the common opinion that, erudition apart, there

---

20. In pagan Rome, *Hoc age* was the order to commence the sacrificial slaughter; *Sursum corda* figures in the Christian liturgy of the Eucharist.
21. '80: Letters, *and especially those to* Atticus . . .

was little excellence in his soul. He was a good citizen, affable by nature as fat jolly men like him frequently are; but it is no lie to say that his share of weakness and ambitious vanity was very great. I cannot excuse him for reckoning his poetry worth publishing; it is no great crime to write bad verses but it was an error of judgement on his part not to have known how unworthy they were of the glory of his name.

As for his eloquence it is beyond compare; I believe no one will ever equal it.[22] The younger Cicero, who resembled his father only in name, when in command of Asia found there were several men whom he did not know seated at his table: among others there was Caestius at the foot of it, where people often sneak in to enjoy the open hospitality of the great. Cicero asked one of his men who he was and was told his name; but, as a man whose thoughts were elsewhere and who kept forgetting the replies to his questions, he asked it him again two or three times. The servant, to avoid the bother of having to go on repeating the same thing and so as to enable Cicero to identify the man by something about him, replied, 'It is that man called Caestius who is said not to think much of your father's eloquence compared to his own.' Cicero, suddenly provoked by that, ordered his men to grab hold of that wretched Caestius and, in his presence, to give him a good flogging. A most discourteous host![23]

Even among those who reckoned that his eloquence was, all things considered, beyond compare, there were some who did not omit to draw attention to some defects in it; such as his friend the great Brutus who said it was an eloquence *'fractam et elumbem'* – 'broken and dislocated'.[24] Orators living near his own time criticized him for the persistent trouble he took to end his periods with lengthy cadences, and noted that he often used in them the words *'esse videatur'* [it would seem to be].

Personally I prefer cadences which conclude more abruptly, cut into iambics. He too can, very occasionally, mix his rhythms quite roughly: my own ears pointed this sentence out to me: *'Ego vero me minus diu senem esse mallem, quam esse senem, antequam essem.'* [I indeed hold being old less long better than being old before I am.][25]

---

22. '80: equal it. *Yet he was not able to exploit his superiority as clearly as Virgil did in his poetry: for soon after him many thought they could equal him or surpass him, though under fake colours: but no poet since Virgil had dared to compare himself to him; and I would like to add another story on this topic.* The younger Cicero . . .

23. Marcus Annaeus Seneca, *Suasiorae*, VIII.

24. Tacitus, *Oratores*, XVIII.

25. Indeed a rough bit of Latin! (Cicero, *De Senectute*, X, 32.)

The historians play right into my court. They are pleasant and delightful; and at the same time[26] [C] Man in general whom I seek to know appears in them more alive and more entire than in any other sort of writing, showing the true diversity of his inward qualities, both wholesale and retail, the variety of ways in which he is put together and the events which menace him.

[A] Now the most appropriate historians for me are those who write men's lives, since they linger more over motives than events, over what comes from inside more than what happens outside. That is why, of historians of every kind, Plutarch is the man for me.

I am deeply sorry that we do not have Diogenes Laertiuses by the dozen, or that he himself did not spread himself more widely [C] or more wisely, for I consider the lives and fortunes of the great teachers of mankind no less carefully than their ideas and doctrines.[27]

[A] In this genre – the study of history – we must without distinction leaf our way through all kinds of authors, ancient and modern, in pidgin and in French, so as to learn about the *matter* which they treat in their divergent ways. But Caesar seems to me to deserve special study, not only to learn historical facts but on his own account, since his perfection excels that of all others, even including Sallust.

I certainly read Caesar with rather more reverence and awe than is usual for the works of men, at times considering the man himself through his deeds and the miracle of his greatness, at others the purity and the inimitable polish of his language which not only surpassed that of all other historians, as Cicero said, but [C] perhaps [A] that of Cicero himself.[28] There is such a lack of bias in his judgement when he talks of his enemies[29] that the only thing you can reproach him with, apart from the deceptive colours under which he seeks to hide his bad cause and the filth

26. '80: The historians *are the true game that my study would bag*; they are pleasant and delightful, and at the same time, *reflections on the natures and circumstances of various men and on the customs of different nations are the real subject of ethics.* Now . . .
27. '80: for me. *I most carefully seek out not only the various opinions and arguments on my endeavour of ancient philosophers of all the schools, but also their morals, fates and lives.* I am deeply sorry that we do not have Diogenes Laertiuses by the dozen, or that he himself did not spread himself more widely. In this genre . . .
Diogenes Laertius' compendium of the lives and doctrines of philosophers is indeed incomplete and unoriginal.
28. '80: Cicero himself *and all the yap there ever was*. There . . .
Montaigne echoes Cicero's frequently cited praise of Caesar in *Brutus or the Orator*.
29. '80: enemies, *and so much truth*, that . . .

of his pernicious ambition, is that he talks of himself too sparingly. For so many great things cannot have been done by him without he himself contributing more to them than he includes in his books.

I like either very simple historians or else outstanding ones. The simple ones, who have nothing of their own to contribute, merely bringing to their task care and diligence in collecting everything which comes to their attention and chronicling everything in good faith without choice or selection, leave our judgement intact for the discerning of the truth. Among others there is for example that good man Froissart who strides with such frank sincerity through his enterprise that when he has made an error he is never afraid to admit it and to correct it at whatever point he has reached when told about it; and he relates all the various rumours which were current and the differing reports that were made to him. Here is the very stuff of history, naked and unshaped: each man can draw such profit from it as his understanding allows.

The truly outstanding historians are capable of choosing what is worth knowing; they can select which of two reports is the more likely; from the endowments and humours of princes they can draw conclusions about their intentions and attribute appropriate words to them. Such historians are right to assume the authority of controlling what we accept by what they do: but that certainly belongs to very few.

Those who lie in between (as most historians do) spoil everything for us: they want to chew things over for us; they give themselves the right to make judgements and consequently bend history to their own ideas: for once our judgement leans to one side we cannot stop ourselves twisting and distorting the narration to that bias. They take on the task of choosing what is worth knowing, often hiding from us some speech or private action which would have taught us much more; they leave out things they find incredible because they do not understand them, and doubtless leave out others because they do not know how to put them into good Latin or French. Let them make a display of their rhetoric and their arguments if they dare to; let them judge as they like: but let them leave us the means of making our own judgements after them; let them not deprave by their abridgements nor arrange by their selection anything of material substance, but rather let them pass it all on to us purely and wholly, in all its dimensions.[30]

30. '80: dimensions. *Those historians are also very commendable who have knowledge of the events they write about either because they played a part in doing them or because they were privy to those who were in charge. For* as often as not . . .

As often as not, and especially in our own times, historiographers are appointed from among quite commonplace people, simply on account of their knowing how to write well, as though we wanted to learn grammar! They are right, having been paid to do that and having nothing but chatter to sell, to worry mainly about that aspect. And so with many a fine phrase they spin a web of rumours gathered at the crossroads of our cities.

The only good histories are those written by men who were actually in charge of affairs or who played some part in that charge, [C] or who at least were fortunate enough to have been in charge of others of a similar kind. [A] Such were virtually all the Greek and Roman historians. For, with several eye-witnesses having written on the same subject (as happened in those days when greatness and learning were [C] commonly [A] found together), if an error were made it must have been wonderfully slight, or concern some incident itself open to great doubt.[31]

What can we hope from a doctor who writes about war, or a schoolboy writing about the designs of kings?

To realize how scrupulous the Romans were over this, we need only one example: Asinius Pollio found even in Caesar's histories some mistakes into which he had fallen because he had not been able to look with his own eyes at every part of his army and had believed individual men who had reported to him things which were often inadequately verified, or else because he had not been carefully enough informed by his commanders-delegate of their conduct of affairs during his absence.[32]

We can see from that example what a delicate thing our quest for truth is when we cannot even rely on the commander's knowledge of a battle he has fought nor on the soldiers' accounts of what went on round them unless, as in a judicial inquiry, we confront witnesses and accept objections to alleged proofs of the finer points of every occurrence. Truly, the knowledge we have of our own affairs is much slacker. But that has been adequately treated by Bodin, and in conformity with my own ideas.[33]

To help my defective and treacherous memory a little – and it is so extremely bad that I have more than once happened to pick up again, thinking it new and unknown to me, a book which I had carefully read

31. '80: were *always* found . . .
'80: doubt. *Though they did not write about what they had seen, they at least had experienced the managing of similar affairs which rendered their judgement more sound. For* what can we . . .
32. Suetonius, *Life of Caesar*, LVI.
33. Jean Bodin, *Methodus ad facilem historiarum cognitionem*, Paris, 1566.

several years earlier and scribbled all over with my notes – I have for some time now adopted the practice of adding at the end of each book (I mean of each book which I intend to read only once) the date when I finished reading it and the general judgement I drew from it, in order to show me again at least the general idea and impression I had conceived of its author when reading it. I would like to transcribe here some of those annotations.

Here is what I put about ten years ago on my Guicciardini (for no matter what language is spoken by my books, I speak to them in my own):[34] 'He is an industrious writer of history, from whom in my judgement we can learn the truth about the affairs of his time more accurately than from any other; moreover he played a part in most of them, holding an honoured position. There is no sign that he ever disguised anything through hatred, favour or vanity; that is vouched for by the unfettered judgements he makes of the great, especially of those by whom he had been promoted to serve in responsible positions, such as Pope Clement VII. As for the quality in which he seems to want most to excel, namely his digressions and reflections, some are excellent and enriched by beautiful sketches; but he enjoyed them too much: he did not want to leave anything out, yet his subject was a full and ample one – infinite almost – and so he can become sloppy and somewhat redolent of academic chatter. I have also been struck by the following: that among all his judgements on minds and actions, among so many motives and intentions, he attributes not one of them to virtue, religious scruple or conscience, as if those qualities had been entirely snuffed out in our world; and among all those deeds, no matter how beautiful they might seem in themselves, he attributes their cause to some evil opportunity or gain. It is impossible to conceive that among the innumerable actions on which he makes a judgement there were not at least some produced by means of reason. No corruption can have infected everyone so totally that there was not some man or other who escaped the contagion. That leads me to fear that his own taste was somewhat corrupted: perhaps he happened to base his estimates of others on himself.'

This is what I have on my Philippe de Commines: 'You will find the language here smooth and delightful, of a natural simplicity; the narration pure, with the good faith of the author manifestly shining through it;

---

34. Guicciardini wrote his *History of Italy* in Italian: Montaigne's notes on it were in French.

himself free from vanity when talking of himself, and of favour and of envy when talking of others, together with a fine zeal for truth rather than any unusual acuteness; and from end to end, authority and weight showing him to be a man of good extraction and brought up to great affairs.'

And on the *Memoirs* of Monsieur Du Bellay,[35] the following:

'It is always a pleasure to see things written about by those who had assayed how to manage them, but there is no denying that in these two noblemen[36] there is clearly revealed a great decline from that shining frankness and freedom in writing found in older authors of their rank such as the Seigneur de Joinville (the close friend of Saint Louis), Eginhard (the Chancellor of Charlemagne) and more recently Philippe de Commines. This is not history so much as pleading the case of King Francis against the Emperor Charles V. I am unwilling to believe that they altered any of the major facts, but they make it their job to distort the judgement of events to our advantage, often quite unreasonably, and to pass over anything touchy in the life of their master: witness the fall from grace of the Seigneur de Montmorency and the Seigneur de Brion, which is simply omitted: indeed the very name of Madame d'Estampes is not to be found in them![37] Secret deeds can be hushed up, but to keep silent about things which everyone knows about, especially things which led to public actions of such consequence, is a defect which cannot be pardoned. In short if you take my advice you should look elsewhere for a full account of King Francis and the events of his time; what can be profitable are the particulars given of the battles and military engagements when these noblemen were present; a few private words and deeds of a few princes of their time; and the transactions and negotiations conducted by the Seigneur de Langey which are chock full of things worth knowing and of uncommon reflections.'[38]

35. *Monsieur* ('My Lord') Du Bellay is Martin Du Bellay, under whose name the *Memoirs* were published. They include matter from other Du Bellays: Guillaume (the Seigneur de Langey), Bishop Jean and René.
36. Guillaume and Martin Du Bellay.
37. Philip Chabot (Brion) was disgraced in 1540; Montmorency, who was in part responsible, was disgraced in his turn. The King's acknowledged mistress, the Duchesse d'Estampes, was influential in their downfall.
38. Langey (Guillaume Du Bellay) sought to reconcile German Lutherans by reforming Roman Catholicism; he pacified the Piedmont and was Rabelais' heroic statesman-scholar.

11. On cruelty

[In the previous chapter Montaigne had praised Froissart for admitting that he had changed his mind at whatever point of his book that the change actually occurred. In this chapter Montaigne follows suit, letting us see how he suddenly realized that virtue as conceived by Hesiod or by Cato is inadequate to explain the virtue of Socrates, which Montaigne had come to prefer to the sterner kind. Cruelty, which Montaigne loathed, is not one of the seven deadly sins and was not widely considered wrong in itself. Montaigne sees cruelty as arising from ecstasies of anger or from ecstatic sexual encounters. Even worse are cruelty and torture done for the fun of it.

The extension of Montaigne's sensitivity to the rest of creation, especially to Man's fellow-creatures the animals, is a skilful preparation for one of the major themes of the following chapter, 'An apology for Raymond Sebond'.]

[A] It seems to me that virtue is something other, something nobler, than those tendencies towards the Good which are born in us. Such souls as are well-endowed and in control of themselves adopt the same gait as virtuous ones and, in their actions, present the same face: but the word virtue has a ring about it which implies something greater and more active than allowing ourselves to be gently and quietly led in reason's train by some fortunate complexion.

A man who, from a naturally easy-going gentleness, would despise injuries done to him would do something very beautiful and praiseworthy; but a man who, stung to the quick and ravished by an injury, could arm himself with the arms of reason against a frenzied yearning for vengeance, finally mastering, it after a great struggle, would undoubtedly be doing very much more. The former would have acted well: the latter, virtuously; goodness is the word for one of these actions; virtue, for the other; for it seems that virtue presupposes difficulty and opposition, and cannot be exercised without a struggle. That is doubtless why we can call God good, mighty, bountiful and just, but we cannot call him virtuous: his works are his properties and cost him no struggle.

Among the philosophers take the Stoics, and even more so the Epicureans – and I borrow that 'even more so' from the common opinion, which is wrong, [C] despite the clever retort which Arcesilaus made to the

philosopher who reproached him with the fact that many people crossed over from his school to the Epicurean one, but never the other way round: 'I am sure that is so,' he said; 'you can make plenty of cocks into capons, but never capons into cocks!'[1] – [A] for in truth the Epicurean School in no wise yields to the Stoic in firmness of opinion and rigour of doctrine. A Stoic (who showed better faith than those disputants who, to oppose Epicurus and to make the game easy for themselves, put into his mouth things he never even thought of, sinisterly twisting his words and by the rules of grammar claiming to find other senses in his way of speaking, and beliefs different from those which he showed in mind and manners) declared that he ceased being an Epicurean for this reason among others, that he found their path too steep and unapproachable; [C] '*et ii qui φιλήδονοι vocantur, sunt φιλόκαλοι et φιλοδίκαιοι, omnesque virtutes et colunt et retinent.*' [and those who were called 'lovers of pleasure' are in fact 'lovers of honour' and 'lovers of justice', cultivating and practising all the virtues].[2]

[A] Among the Stoic and Epicurean philosophers there were, I say, many who judged that it was not sufficient to have our soul in a good state, well under control and ready for virtue; that it was not sufficient to have our powers of reason and our thoughts above all the strivings of Fortune, but that we must do more, seeking occasions to put them to the test. They wish to go looking for pain, hardship and contempt, in order to combat them and to keep our souls in fighting trim: [C] '*multum sibi adjicit virtus lacessita.*' [virtue gains much by being put to the proof.][3]

[A] That is one of the reasons why Epaminondas, who belonged to a third School, rejected the wealth which Fortune put in his hands in the most legitimate of ways, in order, he said, to have to fence against poverty; and he remained extremely poor unto the end. Socrates, it seems to me, assayed himself even more roughly: to exercise his virtue he put up with the malevolence of his wife, which is to assay yourself in good earnest.[4]

Metellus alone among all the Roman Senators undertook to withstand by the force of his virtue the violence of Saturninus (the Tribune of the People of Rome, intent on forcing through an unjust law in favour of the

1. Erasmus, *Apophthegmata*, VII, *Arcesilaus*, II.
2. Cicero, *Epistulae ad familiares*, XV, 19.
3. Seneca, *Epist. moral.*, XIII, 3.
4. Epaminondas was a Pythagorean; Socrates' wife Xanthippe was, for Plato, the archetypal shrew.

plebs); having incurred the death penalty which Saturninus had decreed for those who rejected it, he conversed with those who were escorting him to the Forum in this extremity, saying that to act badly was too easy and too cowardly; to act well when there was no danger, too commonplace; but to act well when danger threatened, was the proper duty of a virtuous man.[5]

Those words of Metellus show us clearly what I wanted to prove: that virtue rejects ease as a companion, and that the gentle easy slope up which are guided the measured steps of a good natural disposition is not the path of real virtue. Virtue demands a rough and thorny road:[6] she wants either external difficulties to struggle against (which was the way of Metellus) by means of which Fortune is pleased to break up the directness of her course for her, or else inward difficulties furnished by the disordered passions [C] and imperfections [A] of our condition.

I have got this far quite easily. But by the end of the above argument the thought occurs to me that the soul of Socrates, which is the most perfect to have come to my knowledge, would be by my reckoning a soul with little to commend it, for I cannot conceive in that great man any onslaught from vicious desires. I cannot imagine any difficulty or any constraint in the progress of his virtue; I know that his reason was so powerful and sovereign within him that it would never have even let a vicious desire be born in him. I cannot put anything face to face with so sublime a Virtue as his: it seems I can see her striding victoriously and triumphantly along, stately and at her ease, without let or hindrance. If Virtue can only be resplendent when fighting opposing desires are we therefore to say that she cannot manage without help from vice, to whom she at least owes the fact that she is held in esteem and honour? And what would become of that bold and noble-minded Pleasure of the Epicureans, who prides herself on nursing Virtue gently in her lap and making her sport there, giving her shame and fevers and poverty and death and tortures to play with? If I postulate that perfect Virtue makes herself known by fighting pain and bearing it patiently, by sustaining attacks from the gout without being shaken in her seat; if I make her necessarily subject to hardship and difficulty, what becomes of that Virtue who has reached such a pinnacle that she not only despises pain but delights in it, taking the stabbings of a strong colic paroxysm as tickling pleasures? Such was the virtue established by the Epicureans, of which several of them have left us by their actions proofs which are absolutely certain.

5. Plutarch, *Life of Marius*.
6. As in the myth of Hesiod, *Works and Days*, 289 f.

So have many others whom, in their actions, I find surpassing the very rules of their doctrines. Witness the Younger Cato. When I see him dying and ripping out his entrails I cannot be satisfied with the belief that he then simply had his soul totally free from trouble and dismay; I cannot believe that he merely remained in that state which the rules of the Stoic School ordained for him: calm, without emotion, impassible. There was, it seems to me, in the virtue of that man too much panache and green sap for it to stop there. I am convinced that he felt voluptuous pleasure in so noble a deed and that he delighted in it more than in anything else he did in his life: [C] '*Sic abiit e vita ut causam moriendi nactum se esse gauderet.*' [He quitted this life, rejoicing that he had obtained a pretext for dying.][7] [A] I am so convinced by this that I begin to doubt whether he would have wished the opportunity for so fine an exploit to be taken from him. And, if that goodness which led him to embrace public interests rather than his own did not rein me back, I would readily concur with the opinion that he was grateful to Fortune for having put his virtue to so fine a proof and for having favoured that brigand[8] who was to trample the ancient freedom of his country underfoot. I seem to read in his action some unutterable joy in his Soul, an access of delight beyond the usual order and a manly pleasure, when she considered the sublime nobility of his deed:

[B]   *Deliberata morte ferocior;*

[She was all the more ferocious for having chosen death;][9]

[A] she was not pricked on by any hope of glory (as the base and womanish judgements of some men have opined) for such a consideration is too low to touch so generous a mind, so high and unbending; he did it for the beauty of the thing itself, which he, who could handle such motives better than we can, saw much more clearly in all its perfection.

[C] It pleases me that philosophy decreed that so beautiful a deed would become no other life than Cato's and that it was for his life alone to end that way. That is why he rightly ordered his son and the senators who bore him company to provide some other way in their own case: '*Catoni cum incredibilem natura tribuisset gravitatem, eamque ipse perpetua constantia roboravisset, semperque in proposito consilio permansisset, moriendum potius quam tyranni vultus aspiciendus erat.*' [To Cato Nature had attributed an

7. Cicero, *Tusc. disput.*, I, xxx, 74.
8. Julius Caesar; defeated by him at Pharsalia, Cato killed himself later at Utica.
9. Horace, *Odes*, I, xxxvii, 29.

unbelievable dignity; he himself had strengthened it by his unfailing constancy; he had remained ever loyal to the principles which he had adopted, so he had to die rather than look on the face of a tyrant.][10]

Every man's death should be one with his life. Just because we are dying we do not become somebody else. I always interpret a man's death by his life. And if I am given an account of an apparently strong death linked to a weakling life, I maintain that it was produced by some weakling cause in keeping with that life.

[A]   So are we to say that the ease with which Cato died, and that power which he acquired by the strength of his soul to do so without difficulty, should somehow dim the splendour of his virtue? And who among those whose brain is even slightly tinged with true philosophy can be satisfied with imagining a Socrates merely free from fear and anguish when his lot was prison, shackles and a verdict of guilty? And who fails to recognize in him not merely firmness and constancy (that was his ordinary state) but some new joy and a playful rapture in his last words and ways?   [C]   When he scratched his leg after the shackles were off he trembled with pleasure: does that not suggest a similar sweet joy in his Soul at being unshackled from her past hardships and capable of entering into a knowledge of things to come?   [A]   Cato must please forgive me: his death is more taut and more tragic but Socrates' is somehow even more beautiful.   [C]   To those who deplored it, Aristippus replied: 'May the gods send me one like it!'[11]

In the souls of those two great men and in those who imitated them (for I very much doubt if any were actually like them) you can see such a perfect acquisition of the habit of virtue that it became a matter of their complexion. It was no longer a painful virtue nor a virtue ordained by reason, virtues which they had to stiffen their souls to maintain: it was the very being of their souls, their natural ordinate manner. They rendered them thus by a long practice of the precepts of philosophy encountering beautiful and richly endowed natures. Those vicious passions which are born in us can find no entry into them; the force and rectitude of their souls stifle and snuff out concupiscence as soon as it begins to stir.

That it is more beautiful to prevent the birth of temptations by a sublime and god-like resolve and to be so fashioned to virtue that even the seeds of vices have been uprooted rather than to prevent their growing by active force and, once having been surprised by the first stirrings of the

10. Cicero, *De officiis*, I, xxxi, 112.
11. Erasmus, *Apophthegmata*, III, *Aristippus*, XXXV.

passions, to arm and tense oneself to halt their progress and to vanquish them; or that this second action is nevertheless more beautiful than to be simply furnished with an easy affable nature which of itself finds indulgence and vice distasteful: cannot I think be doubted. For this third and last manner may seem to produce an innocent man but not a virtuous one; a man exempt from doing evil but not one apt for doing good. Added to which such a mode of being is so close to imperfection and weakness that I cannot easily unravel and distinguish what separates them. That is why the very terms 'goodness' and 'innocence' are somewhat pejorative. I note that several virtues – chastity, sobriety and temperance – can come to us as our bodies grow weaker. Staunchness in the face of danger (if that is the right name for it), together with contempt for death and patience in affliction, can come to men (and are often found in them) by a defect in their assessment of such misfortunes, by a failure to conceive them as they are. A lack of intelligence or even animal-stupidity can counterfeit virtuous deeds: I have often seen men praised for deeds which deserved blame.

An Italian nobleman once spoke as follows in my presence at the expense of his nation: the subtlety of the Italians and the vividness of their minds are so great that they can foresee far ahead the dangers and mishaps that may befall them, so that in war we should not consider it strange if we often find them providing for their safety even before reconnoitring the danger; whereas we French and the Spaniards, who were nothing like as subtle, would press on; we had to be made to see danger with our own eyes and to handle it before we took fright; then there was no holding us; whereas the Germans and the Swiss, grosser and stolider men, scarcely had enough sense to change their ideas even when they were being struck down. Perhaps it was said for a laugh. It is nevertheless true that apprentices to the craft of war often leap into dangers more thoughtlessly than they do once they have been mauled:

> [B] *haud ignarus quantum nova gloria in armis,*
> *Et prædulce decus primo certamine possit.*

[I was not unaware of what can be achieved by a man coming fresh to battle seeking glory and by the sweet honour of a first engagement.][12]

[A] That then is why, when we make a judgement of any individual action, we must consider a great many circumstances as well as the man as a whole who performed it before we give it a name.

Now a word about myself. [B] I have sometimes seen my friends

12. Virgil, *Aeneid*, XI, 154–5.

speak of wisdom in me when it was really luck, or attribute something to my courage and endurance when it was really due to my judgement or to my opinion, attributing one quality to me instead of another, sometimes to my advantage, sometimes to my detriment. Meanwhile, [A] so far am I from having reached that first degree of excellence where virtue becomes an acquired habit that I have hardly given any proof of the second. I have not made much of a struggle to bridle any of my pressing desires. My virtue is a virtue – or rather a state of innocence – which is incidental and fortuitous. If I had been born with a more unruly complexion I am afraid my case would have been deserving of pity. Assays of myself have not revealed the presence in my soul of any firmness in resisting the passions whenever they have been even to the slightest degree ecstatic. I do not know how to sustain inner conflicts and debates. So I cannot congratulate myself much if I do find that I am exempt from many of the vices:

> *si vitiis mediocribus et mea paucis*
> *Mendosa est natura, alioqui recta, velut si*
> *Egregio inspersos reprehendas corpore nævos;*

[if, in my nature, which is otherwise straight, there are a few trivial vices, just as you might criticize an otherwise beautiful body for having a few moles;][13]

I owe that more to my Fortune than to my reason.

Fortune caused me to be born from a stock famous for its honourable conduct and from an excellent father. Did some of his humours flow into me? Was it the examples in the home and the good education I received as a boy which contributed to it without my knowledge? Was it due to some other accident of birth? I cannot tell:

> [B]   *Seu libra, seu me scorpius aspicit*
> *Formidolosus, pars violentior*
>   *Natalis horæ, seu tyrannus*
>   *Hesperiæ Capricornus undæ?*

[Was I born under the constellation of the Balance? Or was it dread Scorpio with violent power over the hour of birth? Or was it Capricorn, who rules as tyrant over the Hesperian waves?][14]

[A] it is at all events true that, of my own self, I am horrified by most of the vices. [C] ('To unlearn evil', the reply which Antisthenes made to

13. Horace, *Satires*, I, vi, 65–7.
14. Horace, *Odes*, II, xvii, 13–16. (To be born under the equable Balance, Libra, was to be learned and judicious: cf. Manilius, *Astronomica*, IV, 202 ff.)

the man who asked him what was the best way to be initiated, seems to centre on that idea.)[15] I am, I repeat, horrified by them, [A] out of a native conviction so thoroughly my own that I have retained the impulses and character which I bore away with me when I was weaned; no other factors have made me worsen them – not even my own arguments which, since they have in some things broken ranks and left the common road, would readily license actions in me which my natural inclinations make me loathe. [B] I shall be saying something monstrous but I will say it all the same: I find, [C] because of this, in many cases [B] more rule and order in my morals than in my opinions, and my appetites less debauched than my reason.

[C] Aristippus laid down opinions about pleasure and riches which were so bold that the whole of philosophy rose and stormed against him. Yet where his morals were concerned, when Dionysius the Tyrant presented him with three beautiful young women to choose from, he said he would choose all three, since things had gone badly for Paris when he preferred one woman to her two companions: but having escorted them to his home he sent them away without laying a finger on them. And when his man-servant found the load of coins he was carrying too heavy to manage, he told him to pour out as many of them as he found too heavy.[16]

Epicurus too, whose doctrines are free from religious scruple and favour luxury, in fact behaved in real life most devoutly and most industriously. He wrote to a friend of his that he lived on nothing but coarse bread and water, asking him to send him a bit of cheese for when he wanted to give himself an extra special treat.[17]

Can it possibly be true that to be good in practice we must needs be so from some inborn, all-pervading property hidden within us, without law, without reason and without example?

[A] By God's mercy any excesses in which I have found myself implicated have not been of the worst. In my own case, I have condemned them at their true value, since my judgement was never infected by them. I have made the case for the prosecution against myself more rigorously than against anyone else. But that is not the whole story: despite all this, I bring too little resistance to bear on them, letting myself readily come down on the opposite side of the scales, except that I do control my vices, preventing them from being contaminated by others: for unless you are on your guard

15. Erasmus, *Apophthegmata*, VII; *Antisthenes Atheniensis*, XXVII.
16. Erasmus, *Apophthegmata*, III; *Aristippus*, III and XXXVII.
17. Diogenes Laertius, *Life of Epicurus*.

one vice leads to another; and most support each other. I prune my own vices and train them to be as isolated and as uncomplicated as possible.

[B]   *Nec ultra*
Errorem foveo.

[Beyond that point I do not indulge my faults.][18]

[A]   As for the opinion of the Stoics, who say that when the Wise Man acts he acts through all his virtues together even though there is one virtue which is more in evidence depending on the nature of the action (and a comparison with the human body is of some service to them in that, since choler cannot be exercised in action unless all our humours come to our aid, even though the choler predominates): if they then proceed to draw the parallel consequence that when the bad man does wrong he does so through all his vices together, then I do not believe it to be simple – or else I fail to understand what they mean, for I know the contrary to be true by experience.[19]

[C]   Such are the insubstantial pin-point subtleties which philosophy occasionally lingers over.

Some vices I follow: others I flee as much as any saint could do. And the Peripatetics reject the idea of any such indissoluble interconnection and bonding: Aristotle maintains that a man may be wise and just yet intemperate and lacking in restraint.[20]   [A]   Socrates confessed to those who recognized in his physiognomy some inclination towards vice that such was indeed his natural propensity but he had corrected it by discipline. [C]   And the intimate friends of Stilpo the philosopher said that he was born subject to wine and women but had trained himself to be most abstemious in both by study.[21]   [A]   Any good that I may have in me I owe on the contrary to the luck of my birth. I do not owe it to law, precept or apprenticeship.   [B]   Such innocence as there is in me is an unfledged innocence: little vigour, no art.

[A]   Among the vices, both by nature and judgement I have a cruel

18. Juvenal, *Satires*, VIII, 164–5.
19. For Stoics the virtues are individually impossible without all the others. Cf. Cicero, *De finibus*, IV, xxviii, 77 ff. Augustine, *Catalogus hereseon* considers that this doctrine favours the Jovinian heresy.
20. Diogenes Laertius, *Life of Aristotle*.
21. Zopyrus the Physiognomist judged from Socrates' features that he was lecherous and a dullard. Socrates agreed: he was born such, but had 'reformed' his soul: see Erasmus, *Apophthegmata*, III; *Socratica*, LXXX; and Cicero, *De fato*, V, 10 for both Socrates and Stilpo.

hatred of cruelty, as the ultimate vice of them all. But I am so soft that I cannot even see anyone lop the head off a chicken without displeasure, and cannot bear to hear a hare squealing when my hounds get their teeth into it, even though I enjoy the hunt enormously.

Those who have to write against sensual pleasure like to use the following argument to show that it is entirely vicious and irrational: when its force is at its climax it overmasters us to such an extent that reason has no way to come into it; they go on to cite what we know of that from our experience of lying with women –

> *cum jam præsagit gaudia corpus,*
> *Atque in eo est venus ut muliebria conserat arva*

[as when the body already anticipates its joy, and Venus is about to scatter seeds broadcast in the woman's furrows][22] –

in which it seems to them that the delight so transports us outside ourselves that our reason could not possibly perform its duty then, being entirely transfixed and enraptured by the pleasure.

But I know that it can go otherwise and that, if we have the will, we can sometimes manage, at that very instant, to bring our soul back to other thoughts. But we must vigilantly ensure that our soul is taut and erect. I know it is possible to master the force of that pleasure; and [C] I am quite knowledgeable about the subject; I have never found Venus to be as imperious a goddess as several people, chaster than I am, attest her to be. [A] I do not regard it as a miracle, as the Queen of Navarre does in one of the tales in her *Heptaméron* (which is a noble book for its cloth), nor even as a matter of extreme difficulty, to spend nights at a time with a mistress long yearned for, in complete freedom and with every opportunity, while keeping my promised word to her to content myself with simple kisses and caresses.[23]

[C] I think a more appropriate example would be that of hunting (in which there is less pleasure but more ecstasy and rapture by which our reason is stunned, so losing the ability of preparing and bracing itself for the encounter), [A] when,[24] after a long chase, our quarry suddenly

22. Lucretius, IV, 1099–10.
23. Margaret of Navarre, *Heptaméron*, IIIe *Journée, conte* 30; she states that St Ambrose had to forbid such tests of virtue.
24. '80: I think a more appropriate *comparison would be with* hunting, *in which there seems to be more rapture: not in my opinion that the pleasure in itself is greater but because it affords us no leisure to brace and prepare ourselves against it, and that it surprises us when . . .*

pops up and reveals itself where we were perhaps least expecting it. The shock of this   [C]   and the heat of the view-halloo strike us so, that it would be difficult for those who love this sort of hunt to bring their thoughts at this point back to anything else. That is why the poets [A]   make Diana[25] to triumph over Cupid's flames and arrows:

> *Quis non malarum, quas amor curas habet,*
> *Haec inter obliviscitur?*

[Is there anyone who, in the joys of the hunt, does not forget the ills which love's cares bring?][26]

To return to my subject, I feel a most tender compassion for the afflictions of others and would readily weep from fellow-feeling − if, that is, I knew how to weep at anything at all.   [C]   Nothing tempts my tears like tears − not only real ones but tears of any kind, in feint or paint.

[A]   I scarcely ever lament for the dead: I would be more inclined to envy them; but I do make great lamentations for the dying. Savages do not upset me so much by roasting and eating the bodies of the dead as those persecutors do who torture the bodies of the living. However reasonable lawful public executions may be, I cannot even look fixedly at them. Someone, having to bear witness to the clemency of Caesar, wrote the following:[27] 'He was so mild in his vengeance that, having forced surrender on the pirates who had formerly taken him prisoner and held him to ransom, he did indeed condemn them to be crucified since he had threatened them with that fate, but he first had them strangled. His secretary Philemon, who had tried to poison him, he punished with nothing severer than simple death.' Without my even naming the author who ventures to allege as evidence of clemency the mere killing of those who have injured us, it is easy enough to guess that he was shocked by the base and horrifying examples of cruelty which the Roman tyrants introduced. As for me, even in the case of Justice itself, anything beyond the straightforward death-penalty seems pure cruelty, and especially in us Christians who ought to be concerned to dispatch men's souls in a good state, which cannot be so when we have driven them to distraction and despair by unbearable tortures.

25. '80: The shock of this *pleasure* strikes us so *furiously* that it would be difficult for those who love the hunt to bring their *soul* at this point back *from its rapture. Love gives way to the pleasure of the chase, say the poets: that is why they* make Diana . . .
26. Horace, *Epodes*, II, 37–8.
27. The author is Suetonius (*Life of Julius Caesar*). Related by Erasmus, *Apophthegmata*, IV; *Julius Caesar*, I.

['95] A few days ago[28] a soldier in prison noticed from the tower in which he was held that a crowd was gathering in the square and that the carpenters were at work constructing something; he concluded that this was for him; he determined to kill himself, but found nothing which could help him to do so save a rusty old cart-nail which Fortune offered to him. He first of all gave himself two big jabs about the throat, but finding that this was not effective he soon afterwards gave himself a third one in his stomach, leaving the nail protruding. The first of his gaolers to come in found him in this state, still alive but lying on the ground weakened by the blows. So as not to waste time before he swooned away, they hastened to pronounce sentence on him. When he had heard it and learned that he was to be decapitated, he seemed to take new heart; he accepted the wine which he had previously refused and thanked his judges for the unhoped for mildness of their sentence, saying that he had made up his mind to appeal personally to death because he had feared a death more cruel and intolerable, having formed the opinion that the preparations which he had seen being made in that square meant that they wanted to torture him with some horrifying torment. This change in the way he was to die seemed to him like a deliverance from death.

[A] My advice would be that exemplary severity intended to keep the populace to their duty should be practised not on criminals but on their corpses: for to see their corpses deprived of burial, boiled or quartered would strike the common people virtually as much as pains inflicted on the living, though in reality they amount to little or nothing – [C] as God says, *'Qui corpus occidunt, et postea non habent quod faciant.'* [Who kill the body and after that have nothing that they can do.][29] And the poets particularly emphasize the descriptions of such horrors as something deeper than death.

> *Heu! relliquias semiassi regis, denudatis ossibus,*
> *Per terram sanie delibutas fæde divexarier.*

[O grief! that the remains of a half-burnt king, his flesh torn to the bone, and spattered with mud and blood, should be dragged along in shame.][30]

---

28. The text of the Bordeaux manuscript addition is partly damaged, but clearly tells of the same event in much the same words. Here ['95] replaces [C] as being more reliable.
29. Luke 12:4. (Christ's own words, but cited inexactly from memory).
30. Cicero, *Tusc. disput.*, I, xliv, 106 (citing Ennius)

[A1]   I found myself in Rome at the very moment when they were dispatching a notorious thief called Catena. The crowd showed no emotion when he was strangled, but when they proceeded to quarter him the executioner never struck a blow without the people accompanying it with a plaintive cry and exclamation, as if each person had transferred his own feelings to that carcass.[31]

[B]   Such inhuman excesses should be directed against the dead bark not the living tree. In somewhat similar circumstances Artaxerxes tempered the harshness of the ancient laws of Persia: he ordained that noblemen who had failed in their tasks should not be whipped as they used to be but stripped naked and their clothes whipped instead, and that whereas they used to have their hair torn out by the root they should merely be deprived of their tall headdresses.[32]

[C]   The scrupulously devout Egyptians reckoned that they adequately satisfied divine justice by sacrificing swine in figure and effigy; it was a bold innovation to wish to pay in shadow and effigy God whose substance is very essence.[33]

[A]   I live in a season when unbelievable examples of this vice of cruelty flourish because of the licence of our civil wars; you can find nothing in ancient history more extreme than what we witness every day. But that has by no means broken me in. If I had not seen it I could hardly have made myself believe that you could find souls so monstrous that they would commit murder for the sheer fun of it; would hack at another man's limbs and lop them off and would cudgel their brains to invent unusual tortures and new forms of murder, not from hatred or for gain but for the one sole purpose of enjoying the pleasant spectacle of the pitiful gestures and twitchings of a man dying in agony, while hearing his screams and groans. For there you have the farthest point that cruelty can reach: [C] '*Ut homo hominem, non iratus, non timens, tantum spectaturus occidat.*' [That man should kill man not in anger or in fear but merely for the spectacle.][34]

[A]   As for me, I have not even been able to witness without displeasure an innocent defenceless beast which has done us no harm being hunted to the kill. And when as commonly happens the stag, realizing that it has

---

31. Described in Montaigne's *Journal de Voyage.*
32. Erasmus, *Apophthegmata,* V, *Artoxerxes,* XVIII. (Similarly cited in Amyot's Plutarch, but as Artaxerxes).
33. Herodotus, *History,* II, xlvii.
34. Seneca, *Epist. moral.,* XC, 45.

exhausted its breath and its strength, can find no other remedy but to surrender to us who are hunting it, throwing itself on our mercy which it implores with its tears:

> [B]　*quæstuque, cruentus*
> *Atque imploranti similis;*

[all covered with blood, groaning, and seeming to beg for grace]³⁵

[A]　that has always seemed to me the most disagreeable of sights.

[B]　I hardly ever catch a beast alive without restoring it to its fields. Pythagoras used to do much the same, buying their catches from anglers and fowlers:

> [A]　*primoque a cæde ferarum*
> *Incaluisse puto maculatum sanguine ferrum.*

[it was, I think, by the slaughter of beasts in the wild that our iron swords were first spattered with warm blood.]³⁶

Natures given to bloodshed where beasts are concerned bear witness to an inborn propensity to cruelty.

[B]　In Rome, once they had broken themselves in by murdering animals they went on to men and to gladiators. I fear that Nature herself has attached to Man something which goads him on towards inhumanity. Watching animals playing together and cuddling each other is nobody's sport: everyone's sport is to watch them tearing each other apart and wrenching off their limbs.

[A]　And lest anyone should laugh at this sympathy which I feel for animals,³⁷ Theology herself ordains that we should show some favour towards them; and when we consider that the same Master has lodged us in this palatial world for his service, and that they like us are members of his family, Theology is right to enjoin upon us some respect and affection for them.

Pythagoras borrowed his metempsychosis from the Egyptians, but it was subsequently accepted by many peoples including our Druids:³⁸

35. Virgil, *Aeneid,* VII, 501.
36. Erasmus, *Adages,* I, I, II, *Amicitia aequalis*; section *Pythagorae Symbolae: A pisces abstineto*; then, Ovid, *Metamorphoses,* XV, 106–7.
37. '80: sympathy *and love* [amitié] which I confess that I feel for *them . . .*
　　An echo of the Pythagorean adage of Erasmus, *Amicitia aequalis* (see note 36).
38. The Druids were the priests and philosophers of the Ancient Gauls: Caesar, *Gallic Wars,* V, xiii ff.

> *Morte carent animæ; semperque, priore relicta*
> *Sede, novis domibus vivunt, habitantque receptæ.*

[Souls have no death: they live for ever welcome in new abodes, having left their former ones.][39]

The religion of our Ancient Gauls included the belief that souls, being eternal, never cease changing and shifting from one body to another. In addition the Gauls attached to this idea some concern with divine justice: they said that for the Soul which had made her home in, say, Alexander there was ordained by God, depending on how she had behaved, a different body, more [C] painful [A] or less so,[40] according to her behaviour:

> [B]   *muta ferarum*
> *Cogit vincla pati, truculentos ingerit ursis,*
> *Prædonesque lupis, fallaces vulpibus addit;*
> *Atque ubi per varios annos, per mille figuras*
> *Egit, lethæo purgatos flumine, tandem*
> *Rursus ad humanæ revocat primordia formæ.*

[He compels those souls to accept the mute fetters of the beasts: the merciless are imprisoned in bears; thieves, in wolves; cheats in foxes; then, having driven them over many a year through thousands of shapes, He at last purges them in the waters of Lethe and summons them back to their original human shape].[41]

[A]   If the Soul had been valiant, she was lodged in the body of a lion; if a voluptuary, in a pig's; if a coward, in a stag's or a hare's; if cunning, in a fox's; and so on until, purified by such chastisement, she took on the shape of some other man.

> *Ipse ego, nam memini, Trojani tempore belli*
> *Panthoides Euphorbus eram.*

[For I, Pythagoras, as I remember well, was Euphorbus, son of Pantheus during the Trojan war.][42]

---

39. Ovid, *Metamorphoses*, XV, 106–7. The Egyptian origin of metempsychosis is mentioned by Ovid's commentators (e.g., among many, the Venice edition, 1586, p. 295).
40.   [A]: body more *vile*, or less so . . .
41. Claudius Claudianus, *In Ruffinum*, II, 482–7.
42. Ovid, *Metamorphoses*, XV, 160–1 – from the verses which sympathetically expound Pythagoras' ideas.

I do not attach much importance to such cousinship between us and the beasts;[43] nor to the fact that many nations, particularly some of the oldest and noblest, not only welcomed animals to companionship and fellowship with themselves but even ranked them far above themselves, sometimes reckoning that they were the familiar friends and favourites of their gods, respecting them and reverencing them as above mankind, sometimes acknowledging no other god nor godhead but them: [C] *'belluæ a barbaris propter beneficium consecratæ.'* [beasts were sacred to the Barbarians because of the blessings they bestowed.][44]

> [B]   *Crocodilon adorat*
> *Pars hæc, illa pavet saturam serpentibus Ibin;*
> *Effigies hic nitet aurea cercopitheci;*
> *hic piscem fluminis, illic*
> *Oppida tota canem venerantur.*

[This region worships the crocodile; another trembles before the ibis, gorged with snakes; here on the altar stands a golden image of a long-tailed monkey; in this town they venerate a river-fish; in another, a dog.][45]

[A]   And the actual interpretation which Plutarch makes of this error (which is a very sound one) is to their honour. For he states that it was not the cat or (for example) the bull which the Egyptians worshipped: what they worshipped in those beasts was an image of the divine attributes: in the bull, patience and utility; in the cat, quickness,[46] [C] or, like our neighbours the Burgundians as well as all the Germans, its refusal to let itself be shut in: by the cat they represented that freedom which they loved and adored above any other of God's attributes. And so on.

[A]   But when among other more moderate opinions I come across arguments which assay to demonstrate the close resemblance we bear to the animals, and how much they share in our greatest privileges and how convincingly they can be compared to us, I am led to abase our presumption considerably and am ready to lay aside that imaginary kingship over other creatures which is attributed to us.

---

43. Such 'cousinship' is briefly mentioned by Brassicanus in his remarks on Pythagoras' adage *Ab animalibus abstine*, with an allusion to Ovid's 'truly golden' verses in the *Metamorphoses*, XV, which, throughout the Renaissance, is the source always cited or followed.
44. Cicero, *De nat. deorum*, I, xxxvi, 101.
45. Juvenal, *Satires*, XV, 2–6.
46. Plutarch (tr. Amyot), *De Isis et Osiris*, 333F–334H.

Even if all of that remained unsaid, there is a kind of respect and a duty in man as a genus which link us not merely to the beasts, which have life and feelings, but even to trees and plants. We owe justice to men: and to the other creatures who are able to receive them we owe gentleness and kindness. Between them and us there is some sort of intercourse and a degree of mutual obligation. [C] I am not afraid to admit that my nature is so childishly affectionate that I cannot easily refuse an untimely gambol to my dog wherever it begs one.

[A1] The Turks have charities and hospitals for their beasts. [A] The Romans had a public duty to care for geese, by the vigilance of which their Capitol had been saved;[47] the Athenians commanded that the he-mules and she-mules which had been used in building the temple named the Hecatompedon should be set free and allowed to graze anywhere without hindrance.[48]

[C] It was the usual practice of the citizens of Agrigentum to give solemn burial to the beasts they loved, such as to horses of some rare merit, to working birds and dogs or even to those which their children had played with. And their customary magnificence in all things was particularly paraded in the many splendid tombs which they erected for that purpose; they remained on display many centuries afterwards. The Egyptians buried wolves, bears, crocodiles, dogs and cats in hallowed places; they embalmed their corpses and wore mourning at their deaths. [A] Cimon gave honourable burial to the mules which had thrice won him the prize for racing in the Olympic Games. In Antiquity Xantippus had his dog buried on a coastal headland which has borne its name ever since.[49] And Plutarch says that it offended his conscience to make a little money by sending to the slaughter-house an ox which had long been in his service.[50]

---

47. Plutarch (tr. Amyot), *Les demandes des choses Romaines*, 475E. The geese heard the Barbarians scaling the walls while the guard-dogs slept (*Quels animaux sont les plus advisez*, 514 D–E).
48. The Hecatompedon ('the Hundred-feet long') was the regular name for the Parthenon (the temple of Athena Parthenos in the citadel of Athens). It was rebuilt by Pericles on the site of a previous temple of that name.
49. Same examples in Ravisius Textor, *Officina* (*Bruta animalia honorata sepulchris aut statuis*).
50. Plutarch, *Life of Cato*.

12. An apology for Raymond Sebond

[The chapter which follows is by far the longest one which Montaigne ever wrote. It is discussed in the Introduction (pp. xx–xliv). In the Appendices to that Introduction are given a translation of Montaigne's dedication to his father of his French version of Raymond Sebond's Theologia naturalis, *as well as a translation of Montaigne's French version of the Prologue of Raymond Sebond himself].*

[A]   Truly, learning is a most useful accomplishment and a great one. Those who despise it give ample proof of their animal-stupidity. Yet I do not prize its worth at that extreme value given to it by some, such as the philosopher Erillus who lodged Supreme Good in it, holding that it was within the power of learning to make us wise and contented.[1] That, I do not believe – nor what others have said: that learning is the Mother of virtue and that all vice is born of Ignorance. If that is true, it needs a lengthy gloss.[2]

My house has long been open to erudite men and is well known to them, since my father, who had the ordering of it for fifty years and more, all ablaze with that new ardour with which King Francis I embraced letters and raised them in esteem, spent a great deal of trouble and money seeking the acquaintance of the learned, welcoming them into his house as holy persons who had been granted private inspiration by Divine Wisdom; he collected their sayings and their reasonings as though they were oracles – with all the more awe and devotion in that he had less right to judge: he had no acquaintance with literature, [A1]   any more than his forebears did.   [A]   I like learned men myself, but I do not worship them.

---

1. Commonplace; cf. Cicero, *De fin.*, II. xiii. 43; Erillus, though a pupil of Zeno the Stoic, was close to Plato (Cicero, *Acad.*, II. xlii. 129).
2. The Platonic contention. Cf. Socrates in Aristotle, *Nicomachaean Ethics*, VII, i. 6–ii, 7 (a commonplace: cf. Cognatus' Adages, *Indocto nihil iniquius* and *Nil scientia potentius*); vulgarized by Erasmus' *Apophthegmata* (*Socrates*, XXXIII): 'He said knowledge is the only good, ignorance the only evil.' The intemperate, say, believe inordinate reactions to be ordinate. 'The *summum bonum* is therefore knowledge of what is to be sought or avoided.'

Among others there was Pierre Bunel, a man who, in his own time, enjoyed a great reputation for learning.[3] He and other men of his kind stayed several days at Montaigne in my father's company; when leaving, Bunel gave him a book called *Theologia Naturalis sive Liber creaturarum magistri Raymondi de Sabonde* – *Natural Theology, or, The Book of Creatures by Master Raymond Sebond.* My father was familiar with Italian and Spanish and so, since the book is composed in a kind of pidgin – Spanish with Latin endings – Bunel hoped that my father could profit by it with only very little help. He recommended it to him as a book which was very useful for the period in which he gave it to him: that was when the novelties of Luther were beginning to be esteemed, in many places shaking our old religion. He was well advised, clearly deducing that this new disease would soon degenerate into loathsome atheism. The mass of ordinary people[4] lack the faculty of judging things as they are, letting themselves be carried away by chance appearances. Once you have put into their hands the foolhardiness of despising and criticizing opinions which they used to hold in the highest awe (such as those which concern their salvation), and once you have thrown into the balance of doubt and uncertainty [C] any [A] articles of their religion, they soon cast all the rest of their beliefs into similar uncertainty. They had no more authority for them, no more foundation, than for those you have just undermined; and so, as though it were the yoke of a tyrant, they shake off all those other concepts which had been impressed upon them by the authority of Law and the awesomeness of ancient custom.

[B] *Nam cupide conculcatur nimis ante metutum.*

[That which once was feared too greatly is now avidly trampled underfoot.][5]

[A] They then take it upon themselves to accept nothing on which they have not pronounced their own approval, subjecting it to their individual assent.

Now, my father, a few days before he died, happened to light upon this book beneath a pile of old papers; he ordered me to put it into French for him. It is good to translate authors like these, where there is little to express apart from the matter. Authors much devoted to grace and elegance of

---

3. A distinguished scholar and tutor from Toulouse (1499–1546). Similar praise in Lambin's dedication to him of Lucretius, *De nat. rerum*, V.
4. '88: ordinary people (*and virtually everybody is in that category*) lack . . .
5. Lucretius, V, 1140 (alluding to regicide).

language are a dangerous[6] undertaking, [C] especially when you are turning them into a weaker language. [A] It was a strange and novel occupation for me, but, happening to be at leisure and never being able to refuse any command from the best father that ever was, I did what I could and finished it. He took particular delight in it and gave instructions to have it printed. They were carried out after his death.[7]

I found the concepts of Sebond to be beautiful, the structure of his book well executed and his project full of piety. Many people spend time reading it – especially ladies, to whom we owe greater courtesy. I have often been able to help them by relieving this book of the weight of the two main objections made against it. (Sebond's aim is a bold and courageous one, since he undertakes to establish against the atheists and to show by human, natural reasons the truth of all the articles of the Christian religion.)

Frankly, I find him so firm and so successful in this, that I do not think it is possible to do better on this topic and I do not believe that anyone has done so well.

It seemed too rich and too fine a book for an author whose name is so obscure – all we know of him is that he was a Spaniard professing medicine in Toulouse some two hundred years ago; so I once asked Adrian Turnebus – who knew everything – what he made of it.[8] He replied that he thought it was a quintessence distilled from St Thomas Aquinas, only a wit like Thomas's, full of infinite learning and staggering subtlety, being capable of such concepts. Anyway, whoever it was who conceived and wrote this book (and it is not reasonable to deprive Sebond of his title without greater cause), he was a most talented man, having many fine accomplishments.

The first charge made against the book is that Christians do themselves wrong by wishing to support their belief with human reasons: belief is grasped only by faith and by private inspiration from God's grace.

A pious zeal may be seen behind this objection; so any assay at satisfying those who put it forward must be made with gentleness and respect. It is really a task for a man versed in Theology rather than for me, who know nothing about it. Nevertheless, this is my verdict: in a matter so holy, so

---

6. '88: a *difficult* undertaking.

7. '88: death, *with the carelessness which you can see from the infinite number of misprints left in by the printer, who alone was responsible for its execution* . . . Montaigne struck out his first printer's liminary material for the second edition.

8. He is also highly praised by Montaigne in I, 25, 'On schoolmasters' learning', and II, 17, 'On presumption'.

sublime, so far surpassing Man's intellect as is that Truth by which it has pleased God in his goodness[9] to enlighten us, we can only grasp that Truth and lodge it within us if God favours us with the privilege of further help, beyond the natural order.

I do not believe, then, that purely human means have the capacity to do this; if they had, many choice and excellent souls in ancient times – souls abundantly furnished with natural faculties – would not have failed to reach such knowledge by discursive reasoning. Only faith can embrace, with a lively certainty, the high mysteries of our religion.[10]

But that is not to imply that it is other than a most fair and praiseworthy undertaking to devote to the service of our faith those natural, human tools which God has granted us. It is not to be doubted that it is the most honourable use that we could ever put them to and that there is no task, no design, more worthy of a Christian than to aim, by assiduous reflection, at beautifying, developing and clarifying the truth of his beliefs. We are not content merely to serve God with our spirits and our souls: we owe him more than that, doing him reverence with our bodies; we honour him with our very members, our actions and with things external. In the same way we must accompany our faith with all the reason that lies within us – but always with the reservation that we never reckon that faith depends upon ourselves or that our efforts and our conjectures can ever themselves attain to a knowledge so supernatural, so divine.

If faith does not come and dwell within us as something infused, beyond the natural order; if she comes in, not just by reasoning but by any human means, then she is not there in her dignity and splendour. And yet I fear that we do only enjoy her presence in that way. If we held fast to God by means of a lively faith; if we held fast to God by God, not by ourselves; if our footing and our foundation were divine: then human events would not have the power to shake us which they do have; our fortress would not be for surrendering to so feeble a battery; the love of novelty, the constraint of Princes, the good luck of one party or rash and fortuitous changes in our own opinions, would have no power to shake our beliefs or modify them. We would not let our faith be troubled at the mercy of some new argument or by persuasion – not by all the rhetoric there ever was. We would withstand such billows with a firmness, unbending and unmoved:

---

9. '88: his *sacrosanct* goodness . . .
10. A 'lively' faith shows itself in good works; Christian 'mysteries' are not accessible to unaided human reason: that is standard orthodox doctrine.

> *Illisos fluctus rupes ut vasta refundit,*
> *Et varias circum latrantes dissipat undas*
> *Mole sua . . .*

[As a mighty rock, by its very mass, withstands the lashing waves, pouring them back and breaking up the waters raging round about it . . .][11]

If a ray of God's light touched us even slightly, it would be everywhere apparent: not only our words but our deeds would bear its lustre and its brightness. Everything emanating from us would be seen shining with that noble light. We ought to be ashamed: among the schools of human philosophy there never was an initiate who did not make his conduct and his life conform, at least in some respect, to their teachings, however difficult or strange: and yet so holy and heavenly an ordinance as ours only marks Christians on their tongues.

[B] Do you want to see that for yourself? Then compare our behaviour with a Moslem's or a pagan's: you always remain lower than they are. Yet, given the advantage of our own religion, our superiority ought to outshine them, far beyond any comparison. Men ought to say: 'Are they really so just, so loving, so good? Then these people must be Christians.'[12] [C] All other manifestations are common to all religions: hope, trust, deliverances, ceremonies, penances and martyrdoms. The distinctive mark of the Truth we hold ought to be virtue, which is the most exacting mark of Truth, the closest one to heaven and the most worthy thing that Truth produces.

[B] That is why our good Saint Louis was right, when the Tartar king who was converted to Christianity planned to come to Lyons to kiss the Pope's feet and to study the holiness he hoped to find in our behaviour, to turn him away from it at once, fearing that our disordered way of life would sour his taste for so sacred a belief.[13]

The actual outcome, on the other hand, was different for that later convert who went to Rome for the same purpose: seeing the dissolute life of the prelates and people there at that time, he became even more firmly attached to our religion: he considered how much strength and holiness it

11. Anon. The poem (based on *Aeneid*, VII, 587 ff.) praises the staunchly Catholic Ronsard and accompanies his reply to Protestant critics, *Response aux injures et calomnies*, 1563.
12. Guillaume Postel, the French orientalist, highly praised the fervour of Moslem believers. He believed that, once converted, they would be the most exemplary of Christians.
13. Cf. Joinville, *Histoire*, XIX.

must have to be able to maintain its dignity and splendour in the midst of corruption so great, in hands so vicious![14]

[A] The Word of God says that if we had one single drop of faith we would 'move mountains':[15] our actions, guided and accompanied by God, would not be simply human: they would partake of the miraculous, just as our belief does. [C] *'Brevis est institutio vitae honestae beataeque, si credas'* [Laying the principles for an honourable and blessed life is soon done . . . if you believe].[16]

Some people make the world believe that they hold beliefs they do not hold. A greater number make themselves believe it, having no idea what 'believing' really means, once you go deeply into the matter. [A] We find it strange when, in the wars now besetting our country, we see the outcome of events drifting and changing in a manner marked by nothing unusual or beyond the natural order. That is because we bring to it nothing beyond ourselves. There is Justice on one of the sides, but only as a decoration and a cloak — often cited but never received, welcomed and truly wedded. Justice is lodged as in the mouth of a lawyer, not as in the heart and emotions of the man whose suit it is. God owes help — beyond the natural order — to our faith, to our religion: he does not owe it to our passions.[17] Men take the lead in them, making use of religion: things ought to be clean contrary.[18]

[C] Think whether we do not take religion into our own hands and twist it like wax into shapes quite opposed to a rule so unbending and direct. Has that ever been seen more clearly than in France today? Some approach it from this side, some from the other; some make it black, others make it white: all are alike in using religion for their violent and ambitious schemes, so like each other in managing their affairs with excess and injustice, that they make you doubt whether they really do hold different opinions over a matter on which depends the way we conduct and regulate our lives. Could you find behaviour more like, more closely identical even, coming from the same teaching in the same school? Just see the horrifying impudence with which we toss theological arguments to and fro and how irreligiously we cast them off or take them up again, whenever we happen to switch places in these civil tumults. Take that most formal proposition:

14. Boccaccio, *Decameron*, day I, tale 2.
15. Matthew 17:20.
16. Quintilian, XII, 11, 12 – enjoining men to will to achieve natural virtue.
17. '88: to *men*. Men take . . .
18. J.-A. de Thou in his *Historia sui temporis* relates how Montaigne made similar remarks to him directly.

Whether it be permitted for a Subject to rebel and to take up arms against his Ruler, in defence of his religion? First, remember which side, only last year, was mouthing the affirmative, making it the buttress of their faction, and what side was mouthing the negative, making their buttress out of that. Then listen from what quarter come voices defending which side now, and judge whether they are rattling their swords less for this side than they did for the other![19] We burn people at the stake for saying that Truth must bow to our necessities: and, in France, how much worse is what we do than what we say!

[A] Let us confess the truth: pick out, even from the lawful, moderate army,[20] those who are fighting simply out of zeal for their religious convictions; then add those who are concerned only to uphold the laws of their country and to serve their King: you would not have enough to form one full company of fighting men. How does it happen that so few can be found who maintain a consistent will and action in our civil disturbances? How does it happen that you can see them sometimes merely ambling along, sometimes charging headlong – the very same men sometimes ruining our affairs by their violence and harshness and at other times by their lukewarmness, their softness and their sloth? It must be that they have been motivated by private concerns, [C] by ones due to chance; [A] as these change, so do they.

[C] It is evident to me that we only willingly carry out those religious duties which flatter our passions. Christians excel at hating enemies. Our zeal works wonders when it strengthens our tendency towards hatred, enmity, ambition, avarice, evil-speaking . . . and rebellion. On the other hand, zeal never makes anyone go flying towards goodness, kindness or temperance, unless he is miraculously pre-disposed to them by some rare complexion. Our religion was made to root out vices: now it cloaks them, nurses them, stimulates them.

[A] There is a saying: 'Do not try to palm off sheaves of straw on God.' If we believed in God – I do not mean by faith but merely with bare credence, indeed (and I say it to our great shame) if we believed him and knew him just as we believe historical events or one of our companions, then we would love him above all other things, on account of the infinite goodness and beauty shining within him: at the very least he would march equal in the ranks of our affections with riches, pleasure, glory and

19. Many Roman Catholics and Protestants switched positions as their rival candidates drew near to the throne. The Catholic Henry III, assassinated 2 August 1589, was succeeded by the Protestant Henry IV, who became a Roman Catholic in 1593.
20. '88: from *our armies* those . . .

friends.[21]   [C]   The best among us does not fear to offend him as much as offending neighbour, kinsman, master. On this side there is the object of one of our vicious pleasures: on the other, the glorious state of immortality, equally known and equally convincing – is there anyone so simple-minded as to barter one for the other? And yet we often give it up altogether, out of pure contempt; for what attracts us to blasphemy except, perhaps, the taste of the offence itself?

Antisthenes, the philosopher, was being initiated into the Orphic mysteries; the priest said that those who make their religious profession would receive after death joys, perfect and everlasting. He replied: 'Why do you not die yourself then?' Diogenes' retort was more brusque (that was his fashion) and rather off our subject: when the priest was preaching at him to join his order so as to obtain the blessings of the world to come, he replied: 'Are you asking me to believe that great men like Agesilaus and Epaminondas will be wretched, whilst a calf like you will be happy, just because you are a priest?'[22]

[A]   If we were to accept the great promises of everlasting blessedness as having the same authority as a philosophical argument, no more, we would not hold death in such horror as we do:

> [B]   *Non jam se moriens dissolvi conquereretur;*
> *Sed magis ire foras, vestemque relinquere, ut anguis,*
> *Gauderet, praelonga senex aut cornua cervus.*

[The dying man would not then complain that he is being 'loosened asunder', but would, rather, rejoice to be 'going outside', like a snake casting off its skin, or an old stag casting off his over-long antlers.][23]

[A]   'I wish to be loosened asunder', he would say, 'and to be with Jesus Christ.' The force of Plato's dialogue on the immortality of the soul led some of his disciples to kill themselves, the sooner to enjoy the hopes which he gave them.[24]

---

21. Historical faith (by which one believes historical facts) is a low form of faith, quite insufficient for salvation; Montaigne's contemporaries fail (he suggests) even to have that.
22. Diogenes Laertius, *Lives* (VI, 4 and 39), a major source of Montaigne's knowledge of scepticism. (Both anecdotes in Erasmus' *Apophthegmata*.)
  '95: like you *who does nothing worthwhile?* . . .
23. Lucretius, III, 612 f. (Lambin, 1563, p. 230), alludes to the *De divino praemio*, VII, of the Christian writer Lactantius for an answer to these words. Montaigne provides an answer in his own way.
24. Paul (Philippians 1:23) becomes an answer to Lucretius. For the highly orthodox association of Paul with Platonizing suicides, see my study, *Montaigne and Melancholy*, chapter 5, § 1.

All this is a clear sign that we accept our religion only as we would fashion it, only from our own hands – no differently from the way other religions gain acceptance. We happen to be born in a country where it is practised, or else we have regard for its age or for the authority of the men who have upheld it; perhaps we fear the threats which it attaches to the wicked or go along with its promises. Such considerations as these must be deployed in defence of our beliefs, but only as support-troops. Their bonds are human. Another region, other witnesses, similar promises or similar menaces, would, in the same way, stamp a contrary belief on us. [B] We are Christians by the same title that we are Périgordians or Germans.

[A] Plato said few men are so firm in their atheism that a pressing danger does not bring them to acknowledge divine power;[25] such behaviour has nothing to do with a true Christian; only mortal, human religions become accepted by human procedures. What sort of faith must it be that is planted by cowardice and established in us by feebleness of heart! [C] What an agreeable faith, which believes what it believes, because it is not brave enough to disbelieve it! [A] How can vicious passions, such as inconstancy and sudden dismay, produce in our souls anything right?

[C] Plato says that people first decide, by reasoned judgement, that what is told about hell and future punishment is just fiction. But when they have the opportunity really to find out, by experience, when old age or illness brings them close to death, then the terror of it fills them with belief again, out of horror for what awaits them.

To impress such ideas upon people is to make them timorous of heart: that is why Plato in his *Laws* forbids any teaching of threats such as these or of any conviction that ill can come to Man from the gods. (When it does happen, it is for man's greater good or like a medical purgation.)[26]

They tell that Bion, infected by the atheistic teachings of Theodorus, used to mock religious men; but eventually, when death approached, he gave himself over to the most extreme superstitions, as though the gods took themselves off and brought themselves back according to the needs of Bion.[27]

Plato – and these examples – lead to the conclusion that either love or

25. '88: pressing danger, *extreme pain or closeness of death* do not . . . Idea taken possibly from Plato, *Laws*, X (cf. Montaigne in I, 56, 'On prayer') and Plato, tr, Ficino, *Republic*, I, 330, 532.

26. Plato, *Republic* (Ficino, III, 391; cf. II, 379).

27. Diogenes Laertius, *Lives*, Bion.

force can bring us back to a belief in God. Atheism, as a proposition, is a monstrous thing, stripped, as it were, of natural qualities. It is awkward and difficult to fix it firmly in the human spirit, however impudent or however unruly. We have seen plenty of people who are egged on by vanity and pride to conceive lofty opinions for setting the world to rights; to put themselves in countenance they affect to profess atheism: but even if they are mad enough to try and plant it in their consciousness, they are not strong enough to do so. Give them a good thrust through the breast with your sword and they never fail to raise clasped hands to heaven. And when fear or sickness has cooled down the licentious fever-heat of that transient humour, they never fail to come back to themselves again, letting themselves be reconciled to recognized standards and beliefs. Seriously digested doctrine is one thing: these surface impressions are quite another. They are born of a mind unhinged, in the spirit of debauchery; they drift rashly and erratically about in the fancies of men. What wretched, brainless men they are, trying to be worse than they can be!

[A] That great soul [C] of Plato [A] – great, however, with merely human greatness – was led into a neighbouring mistake by the error of paganism and his ignorance of our holy Truth: he held that it is children and old men who are most susceptible to religion, as if religion were born of human weakness and drew her credibility from it.[28] [A] The knot which ought to attach our judgement and our will and to clasp our souls firmly to our Creator should not be one tied together with human considerations and strengthened by emotions: it should be drawn tight in a clasp both divine and supernatural, and have only one form, one face, one lustre; namely, the authority of God and his grace.

But, once our hearts and souls are governed by Faith, it is reasonable that she should further her purposes by drawing upon all of our other parts, according to their several capacities. Moreover, it is simply not believable that there should be no prints whatsoever impressed upon the fabric of this world by the hand of the great Architect, or that there should not be at least some image within created things relating to the Workman who made them and fashioned them. He has left within these lofty works the impress of his Godhead: only our weakness stops us from discovering it. He tells us himself that he makes manifest his unseen workings through those things which are seen. Sebond toiled at this honourable endeavour, showing

28. Cf. Erasmus, *In Praise of Folly*, LXVI.

us that there is no piece within this world which belies its Maker. God's goodness would be put in the wrong if the universe were not compatible with our beliefs. All things, Heaven, Earth, the elements, our bodies and our souls are in one accord: we simply have to find how to use them. If we have the capacity to understand, they will teach us. [B] For this world is a most holy Temple into which Man has been brought in order to contemplate the Sun, the heavenly bodies, the waters and the dry land – objects not sculpted by mortal hands but made manifest to our senses by the Divine Mind in order to represent intelligibles. [A] 'The invisible things of God', says St Paul, 'are clearly seen from the creation of the world, his Eternal Wisdom and his Godhead being perceived from the things he has made.'[29]

> *Atque adeo faciem coeli non invidet orbi*
> *Ipse Deus, vultusque suos corpusque recludit*
> *Semper volvendo; seque ipsum inculcat et offert,*
> *Ut bene cognosci possit, doceatque videndo*
> *Qualis eat, doceatque suas attendere leges.*

[God himself does not begrudge to the world the sight of the face of heaven, which, ever-rolling, unveils his countenance, his incorporate being inculcating and offering himself to us, so that he may be known full well; he teaches the man who contemplates to recognize his state, teaches him, also, to wait upon his laws.][30]

Our human reasonings and concepts are like matter, heavy and barren: God's grace is their form, giving them shape and worth. The virtuous actions of Socrates and of Cato remain vain and useless, since they did not have, as their end or their aim, love of the true Creator of all things nor obedience to him: they did not know God; the same applies to our concepts and thoughts: they have a body of sorts, but it is a formless mass, unenlightened and without shape, unless accompanied by faith in God and by grace. When Faith tinges the themes of Sebond and throws her light

29. Plutarch, tr. Amyot, *De la tranquillité de l'âme*, I, 76; Romans I:20; cf. Introduction, p. xxvii.
30. Manilius, IV, 907;

'88 (after quotation, referring to his translation of Sebond): *If my printer were so enamoured of those studied, borrowed prefatory-pieces with which (according to the humour of this age) there is no book from a good publishing-house but has its forehead garnished, he should make use of verses such as these, which are of a better and more ancient stock than the ones he has planted there.*

upon them, she makes them firm and solid. They then have the capacity of serving as a finger-post, as an elementary guide setting an apprentice on the road leading to knowledge such as this; they fashion him somewhat into shape and make him capable of God's grace, which then furnishes out our belief and perfects it.

I know a man of authority, a cultured, educated man, who admitted to me that he had been led back from the errors of disbelief by means of the arguments of Sebond. Even if you were to strip them of their ornaments and of the help and approbation of Faith – even if you were to take them for purely human notions – you would find, when it comes to fighting those who have plunged down into the dreadful, horrible darkness of irreligion, that they still remain more solid and more firm than any others of the same kind which you can set up against them. We rightly can say to our opponents, '*Si melius quid habes, accerse, vel imperium fer*' [If you have anything better, produce it, or submit]:[31] let them allow the force of our proofs or else show us others, elsewhere, on another subject, as closely woven or of better stuff.

Without thinking I have already half-slipped into the second of the charges which I set out to counter on behalf of Sebond.

Some say that his arguments are weak and unsuited to what he wants to demonstrate; they set out to batter them down with ease. People like those need to be shaken rather more roughly, since they are more dangerous than the first and more malicious. [C] We are only too willing to couch other men's writings in senses which favour our settled opinions: an atheist prides himself on bringing all authors into accord with atheism, poisoning harmless matter with his own venom.[32] [A] Such people have some mental prepossession which makes Sebond's reasons seem insipid. Moreover it seems to them that they have been allowed an easy game, with freedom to fight against our religion with purely human weapons: they would never dare to attack her in the full majesty of her imperious authority. The means I use and which seem more fitted to abating such a frenzy is to trample down human pride and arrogance, crushing them under our feet; I make men feel the emptiness, the vanity, the nothingness of Man, wrench-

31. Horace, *Epistles*, V, 6.
32. '88: malicious. *Anyone who is already imbued with a belief more readily accepts arguments which support it than does a man who has drunk draughts from a contrary opinion, as do these people here. Some mental predisposition makes Sebond's reasons . . .*

'95: opinions. *For an Atheist all writings lean towards atheism. He infects* harmless matter . . .

ing from their grasp the sickly arms of human reason, making them bow their heads and bite the dust before the authority and awe of the Divine Majesty, to whom alone belong knowledge and wisdom; who alone can esteem himself in any way, and from whom we steal whatever worth or value we pride ourselves on: Οὐ γὰρ ἐᾶ φρονεῖν ὅ θεὸς μέγα ἄλλον ἢ ἑωτόν [God permits no one to esteem himself higher].[33]

[C] Let us smash down such presumption. It is the very foundation of the tyrannous rule of the Evil Spirit: '*Deus superbis resistit; humilibus autem dat gratiam*' [God resisteth the proud, and giveth grace to the humble]. 'There is intelligence in all the gods,' says Plato, 'but very little of it in men.'[34]

[A] Yet it is a great source of consolation to a Christian man to see our transitory mortal tools so properly matched to our holy and divine faith that when we use them on subjects which, like them, are transitory and mortal, it is precisely then that they are most closely and most powerfully matched. Let us try and see, then, whether a man has in his power any reasons stronger than those of Sebond – whether, indeed, it is in man to arrive at any certainty by argument and reflection.

[C] St Augustine, pleading his case against presumptuous people, has cause to criticize their injustice when they consider those parts of our faith to be false which human reason is unable to establish. In order to show that many things can exist or have had existence, even though their nature and causes have no foundation which can be fixed by rational discourse, he advances various indubitable, recognized experiences, for which Man admits he can see no explanation. Augustine does this, as he does all things, after careful and intelligent search.[35] We must do even more, teaching such people the lesson that the weakness of their reason can be proved without our having to marshal rare examples; that reason is so inadequate, so blind, that there is no example so clear and easy as to be clear enough for her; that the easy and the hard are all one to her; that all subjects and Nature in general equally deny her any sway or jurisdiction.

[A] What is Truth teaching us, when she preaches that we must fly from the wisdom of this world; when she so frequently urges that what seems wise to Man is but foolishness to God; that of all vain things, Man is

33. Herodotus, VII, 10, *apud* John Stobaeus, *Apophthegmata*, 22. This was inscribed by Montaigne on a beam in his library.
34. I Peter: 5. Cf. Augustine, *City of God*, XVII, 4; Plato, tr. Ficino, *Timaeus*, 1546, p. 715.
35. *City of God*, XXI, 5.

the most vain; that a man who dares to presume that he knows anything, does not even know what knowledge is; that Man, who is nothing yet thinks he is something, misleads and deceives himself? These are verdicts of the Holy Ghost;[36] they express so clearly and so vividly what I myself wish to uphold that I would need no other proof to use against people who, with due submission and obedience, would surrender to his authority. But these people simply ask to be whipped, and will not let us fight their reason, save by reason alone.

So let us consider for a while Man in isolation – Man with no outside help, armed with no arms but his own and stripped of that grace and knowledge of God in which consist his dignity, his power and the very ground of his being.[37] Let us see how much constancy there is in all his fine panoply. Let Man make me understand, by the force of discursive reason, what are the grounds on which he has founded and erected all those advantages which he thinks he has over other creatures and who has convinced him that it is for his convenience, his service, that, for so many centuries, there has been established and maintained the awesome motion of the vault of heaven, the everlasting light of those tapers coursing so proudly overhead or the dread surging of the boundless sea? Is it possible to imagine anything more laughable than that this pitiful, wretched creature – who is not even master of himself, but exposed to shocks on every side – should call himself Master and Emperor of a universe, the smallest particle of which he has no means of knowing, let alone swaying! Man claims the privilege of being unique in that, within this created frame, he alone is able to recognize its structure and its beauty; he alone is able to render thanks to its Architect or to tot up the profit or loss of the world . . . But who impressed his seal on such a privilege? If Man has been given so great and fair a commission, let him produce documents saying so.   [C]   Were they drawn up in favour of wise men only? (They apply to few enough!)[38] Are fools and knaves worthy of a favour so far exceeding the normal order – the worst thing in the world exalted above all others? Are we supposed to believe that fellow who wrote: '*Quorum igitur causa quis dixerit effectum esse mundum? Eorum scilicet animantium quae ratione utuntur. Hi sunt dii et homines,*

---

36. Colossians 2:8; I Corinthians 3:19; I Corinthians 8:2; Galatians 6:3 (the last two inscribed in Montaigne's library). For Montaigne, the Bible is the Holy Ghost speaking through men.
37. From here to the last page, revealed wisdom is left aside. See Introduction, p. xxv ff.
38. Cicero, *De nat. deorum*, I, ix, 23.

quibus profecto nihil est melius' [Who will tell for whose sake this world has been brought about? Why, for the sake of beings having souls able to use reason, those most perfect of beings, gods and men].[39] Coupling gods and men together! We can never do enough to batter down such impudence.

[A] Poor little wretch! What is there in man worthy of such a privilege?

Consider the sun, moon and stars, with their lives free from corruption, their beauty, their grandeur, their motions ever proceeding by laws so just:

> *cum suspicimus magni caelestia mundi*
> *Templa super, stellisque micantibus Aethera fixum,*
> *Et venit in mentem Lunae Solisque viarum;*

[When we gaze upwards to the celestial temples of this great Universe, to the Aether with its fixed and twinkling stars, and when there comes to mind the courses of the Moon and of the Sun . . .][40]

then consider the dominion and power which those bodies have, not only over our lives and the settled detail of our fortunes –

> *Facta etenim et vitas hominum suspendit ab astris*

[For he made the deeds and lives of man to depend upon the Sun, the Moon and the Stars][41] –

but over our very inclinations, our discursive reasoning and our wills, which are all governed, driven and shaken at the mercy of their influences. Our reason tells us that and finds it to be so;

> *speculataque longe*
> *Deprendit tacitis dominantia legibus astra,*
> *Et totum alterna mundum ratione moveri,*
> *Fatorumque vices certis discernere signis.*

[it gazes in the distance, grasping that the heavenly bodies govern us by silent laws, that all the world is moved by periodic causes; and it discerns changing Fate in fixed and certain signs.]

Then see how not merely one man or one king is sent reeling by the slightest motion of the heavenly bodies, but whole monarchies, empires and all this lower world:

---

39. Ibid., II, liii, 133 (where the idea is attributed to Balbus the Stoic).
40. Lucretius, V, 1203 f.
41. Manilius, III, 58 (Montaigne mistranscribed *Fata* (fate) as *facta* (deeds). *Fata* makes better sense); then, I, 60–63; I, 55 and IV, 93; IV, 79 and 118.

> *Quantaque quam parvi faciant discrimina motus:*
> *Tantum est hoc regnum, quod regibus imperat ipsis!*

[When such small motions produce such changes, how great must be the kingdom which rules over kings themselves!]

Then allow that our reason judges that our virtues and our vices, our competencies, our knowledge, and this very discourse we are making here and now about the power of the heavenly bodies, comes to us by their means and by their favour:

> *furit alter amore,*
> *Et pontum tranare potest et vertere Trojam;*
> *Alterius sors est scribendis legibus apta;*
> *Ecce patrem nati perimunt, natosque parentes;*
> *Mutuaque armati coeunt in vulnera fratres:*
> *Non nostrum hoc bellum est; coguntur tanta movere,*
> *Inque suas ferri poenas, lacerandaque membra;*
> *Hoc quoque fatale est, sic ipsum expendere fatum.*

[One man, mad with love, can cross the sea and topple Troy: another's lot is to be apt at prescribing laws. Look: children kill parents: parents, children; brothers bear arms and clash to wound each other. Such wars do not belong to men alone. Men are compelled to do such things, compelled to punish themselves, to tear their limbs apart. And when we ponder thus on Fate, that too is fated . . .]

If we are dependent upon the disposition of the heavens for such little rationality as we have, how can our reason make us equal to the Heavens? How can their essence, or the principles on which they are founded, be subjects of human knowledge? Everything that we can see in those bodies produces in us ecstatic wonder. [C] *'Quae molitio, quae ferramenta, qui vectes, quae machinae, qui ministri tanti operis fuerunt?'* [What engineering, what tools, what levers, what contrivances, what agents were used in such an enterprise?][42]

[A] Why do we deprive the heavenly bodies of souls, life or rationality? Have we, who have no dealings with them beyond pure obedience, been able to recognize in them some kind of stupor, motionless and insensible? [C] Shall we say that we have seen no other creature but Man possessed of a rational soul? What do we mean? Have we ever seen anything like the Sun? And just because we have seen nothing like it, does it cease to be; or, since we have seen nothing like its movements, shall they, too, cease to be? If things we have not actually seen do not exist, then our

---

42. Cicero, *De nat. deorum*, I, viii, 19.

knowledge is wondrously diminished! *'Quae sunt tantae animi angustiae'* [What narrow defiles has our mind].[43]

[A] What vain human dreams, to make the Moon into some celestial Earth, [C] dreaming up, like Anaxagoras, mountains and valleys for it, [A] planting human dwellings and habitations on it and, like Plato and Plutarch, settling colonies there for our convenience: and then to make our own Earth into a brightly shining star: [C] *'Inter caetera mortalitatis incommoda et hoc est, calligo mentium, nec tantum necessitas errandi sed errorum amor'* [Among the other disorders of our mortal condition there is that mental darkness which not only compels us to go wrong but makes us love to do so]. *'Corruptibile corpus aggravat animam, et deprimit terrena inhabitatio sensum multa cogitantem'* [For the corruptible body is a load upon the soul, and the earthly habitation presseth down the mind that museth on many things].[44]

[A] The natural, original distemper of Man is presumption. Man is the most blighted and frail of all creatures and, moreover, the most given to pride.[45] This creature knows and sees that he is lodged down here, among the mire and shit of the world, bound and nailed to the deadest, most stagnant part of the universe, in the lowest storey of the building, the farthest from the vault of heaven; his characteristics place him in the third and lowest category of animate creatures, yet, in thought, he sets himself above the circle of the Moon, bringing the very heavens under his feet. The vanity of this same thought makes him equal himself to God; attribute to himself God's mode of being; pick himself out and set himself apart from the mass of other creatures; and (although they are his fellows and his brothers) carve out for them such helpings of force or faculties as he thinks fit. How can he, from the power of his own understanding, know the hidden, inward motivations of animate creatures? What comparison between us and them leads him to conclude that they have the attributes of senseless brutes?

[C] When I play with my cat, how do I know that she is not passing time with me rather than I with her?[46]

---

43. Ibid., I, xxxi, 87 and 88 (refuting Epicurus).
44. Plutarch, tr. Amyot, *De la face qui apparoist dedans le rond de la Lune*; Diogenes Laertius, II, viii, 100; Seneca, *De ira*, II, ix; *Wisdom of Solomon* 9:15, *apud* Augustine, *City of God*, XII, 15.
45. '88: moreover, *says Pliny*, the most given ... (This quotation is used by Montaigne to conclude II, 14, 'How our mind tangled itself up'; it was cited in Montaigne's library.)
46. '95: her? *We entertain ourselves with mutual monkey-tricks. If I have times when I want to begin or to say no, so does she.*

In his description of the Golden Age under Saturn, Plato counted among one of the principal advantages which Man then had his ability to communicate with the beasts; inquiring and learning from them, Man knew what they were really like and how they differed from each other. By this means Man used to acquire a full understanding and discretion, leading his life far more happily than we ever can now. After that, do we need a better proof of the impudence of Man towards beast? Well, that great author then opined that Nature mainly gave the beasts their bodily forms to enable the men in his time to foretell the future![47]

[A] Why should it be a defect in the beasts not in us which stops all communication between us? We can only guess whose fault it is that we cannot understand each other: for we do not understand them any more than they understand us. They may reckon us to be brute beasts for the same reason that we reckon them to be so. It is no great miracle if we cannot understand them: we cannot understand Basques or Troglodytes! –

[A1] Some have boasted, though, that they could understand the beasts: Apollonius of Thyana, [B] Melampus, Tiresias, Thales [A1] and others. [B] And since there are nations (so the cosmographers tell us) who acknowledge a dog as their king, they must interpret its bark and its movements as having some definite meaning.[48] [A] We ought to note the parity there is between us. We have some modest understanding of what they mean: they have the same of us, in about equal measure. They fawn on us, threaten us and entreat us – as we do them. Meanwhile we discover that they manifestly have converse between themselves, both whole and entire: they understand each other, not only within one species but across different species.

> [B]   *Et mutae pecudes et denique secla ferarum*
> *Dissimiles suerunt voces variasque cluere,*
> *Cum metus aut dolor est, aut cum jam gaudia gliscunt.*

[And dumb cattle and, finally, the generations of wild beasts customarily make sounds having various meanings, when they feel fear or pain or when joy overflows.][49]

[A] A horse knows there to be anger in a given bark of a dog; but that

47. Plato, tr. Ficino, *Politics*, p. 206; *Timaeus*, p. 274 (cf. Montaigne in I, 11, 'On prognostications').
48. Benedetto Varchi, *L'Hercolano. Dialogo nel qual si ragiona. . . delle lingue*; Richerius Rhodiginus, *Antiquae Lectiones* XVII, xiii (disapprovingly); Pliny, *Hist. nat.*, VI, xxxv, etc.
49. Lucretius, V, 1058.

horse does not take fright when the same dog makes some other meaningful cry. Even in beasts who cannot utter meaningful sounds we can readily conclude that there is some other means of communication between them, from the way they work purposefully together; [C] their very movements serve as arguments and ideas.

> [B] *Non alia longe ratione atque ipsa videtur*
> *Protrahere ad gestum pueros infantia linguae.*

[In a not dissimilar way, the very inability to speak leads infants to make gestures.]⁵⁰

[A] And why not? Our deaf-mutes have discussions and arguments, telling each other stories by means of signs.⁵¹ I have seen some who are so nimble and so practised at this that they truly lack nothing necessary for making themselves perfectly understood. After all, lovers quarrel, make it up again, beg favours, give thanks, arrange secret meetings and say everything, with their eyes.

> [A1] *E'l silentio ancor suole*
> *Haver prieghi e parole.*

[Silence itself can talk and beg requests.]⁵²

[C] And what about our hands? With them we request, promise, summon, dismiss, menace, pray, supplicate, refuse, question, show astonishment, count, confess, repent, fear, show shame, doubt, teach, command, incite, encourage, make oaths, bear witness, make accusations, condemn, give absolution, insult, despise, defy, provoke, flatter, applaud, bless, humiliate, mock, reconcile, advise, exalt, welcome, rejoice, lament; show sadness, grieve, despair; astonish, cry out, keep silent and what not else, with a variety and multiplicity rivalling the tongue.

What of the head? We summon, dismiss, admit, reject, deny, welcome, honour, venerate, disdain, request, refuse, rejoice, lament, fondle, tease, submit, brave, exhort, menace, affirm and inquire.

And what of our eyebrows or our shoulders? None of their movements fails to talk a meaningful language which does not have to be learned, a language common to us all. This suggests (given the variety and different

---

50. Ibid., V, 1029.
51. '88: by means of *gestures*. I have . . . (Cf. Rabelais, *Tiers Livre*, TLF, XIX–XX and notes.)
52. Torquato Tasso, *Aminta*, II, 34.

usage among spoken languages) that it is, rather, sign-language that should be judged the 'property' of Man.[53]

I shall leave aside what Necessity can suddenly teach men in individual cases of particular need, as well as finger-alphabets, grammars of gesture and those branches of learning conducted and expressed through them and, finally, those peoples who, according to Pliny, have no other tongue.[54]   [B]   An ambassador from the city of Abdera, after delivering a long address to King Agis of Sparta, asked him: 'Sire, what reply do you want me to bear back to our citizens?' – 'That I allowed you to say all you wanted, for as long as you wanted, without uttering a word.' Was that not an eloquent and most intelligible silence?[55]

[A]   After all, what aspects of our human competence cannot be found in the activities of animals? Is there any form of body politic more ordered, more varied in its allocation of tasks and duties or maintained with greater constancy than that of the bees? Can we conceive that an allocation of tasks and activities, so striking for its orderliness, should be conducted without reasoned discourse and foresight?

> *His quidam signis atque haec exempla sequuti,*
> *Esse apibus partem divinae mentis et haustus*
> *Aethereos dixere.*

[From such signs and examples men conclude that bees have been given some part of the divine Mind and have drunk Aethereal draughts.][56]

Take the swallows, when spring returns; we can see them ferreting through all the corners of our houses; from a thousand places they select one, finding it the most suitable place to make their nests: is that done without judgement or discernment? And then when they are making their nests (so beautifully and so wondrously woven together) can birds use a square rather than a circle, an obtuse angle rather than a right angle, without knowing their properties or their effects? Do they bring water and then clay without realizing that hardness can be softened by dampening? They cover the floors of their palaces with moss or down; do they do so

---

53. Quintilian, XI, iii, 66, 85–7; 68, 71–2; 78–86. Laughter and/or speech were normally considered the 'specific characteristic' (the 'property') of Man.

54. Pliny, VI, 30; cf. Rabelais, *Pantagruel*, TLF, XIII; *Tiers Livre*, XXX; J.-B. della Porta, *De furtivis litterarum notis*, 1563; etc.

55. Plutarch, tr. Amyot, *Les Dicts notables des Lacedaemoniens*, I, 214 A.

56. Virgil, *Georgics*, IV, 219 f. For what follows, cf. Plutarch, tr. Amyot, *Quels sont les animaux les plus advisez ceulx de la terre ou ceulx des eaux?* 512 CD.

without foreseeing that the tender limbs of their little ones will lie more softly there and be more comfortable? Do they protect themselves from the stormy winds and plant their dwellings to the eastward, without recognizing the varying qualities of those winds and considering that one is more healthy for them than another? Why does the spider make her web denser in one place and slacker in another, using this knot here and that knot there, if she cannot reflect, think or reach conclusions?

We are perfectly able to realize how superior they are to us in most of their works and how weak our artistic skills are when it comes to imitating them. Our works are coarser, and yet we are aware of the faculties we use to construct them: our souls use all their powers when doing so. Why do we not consider that the same applies to animals? Why do we attribute to some sort of slavish natural inclination works that surpass all that we can do by nature or by art?

In this, we thoughtlessly give them a very great superiority over us: we make Nature take them by the hand and guide them with a mother's gentle care in all the actions and advantages of their lives; we, on the other hand, are abandoned by Nature to chance and to Fortune, obliged to seek, by art, all things necessary for our conservation; meanwhile, Nature refuses us the very means which would enable us to reach, by education or intelligent application, the level reached by the natural industry of other creatures. In this way we make their brutish stupor have every advantage over our divine intelligence![57]

In truth, on this account, we would be right to treat Nature as a very unjust stepmother. But it is not so. We do not live under so misshapen or so lawless a constitution:[58] Nature clasps all her creatures in a universal embrace; there is not one of them which she has not plainly furnished with all means necessary to the conservation of its being.

There are commonplace lamentations which I hear men make (as the unruly liberty of their opinions raises them above the clouds and then tumbles them down lower than the Antipodes): We are, they say, the only animal abandoned naked on the naked earth; we are in bonds and fetters, having nothing to arm or cover ourselves with but the pelts of other creatures; Nature has clad all others with shells, pods, husks, hair, wool, spikes, hide, down, feathers, scales, fleece or silk, according to the several necessities of their being; she has armed them with claws, teeth and horns for assault and defence; and, as is proper to them, has herself taught them

57. '88: over our *invention and our arts . . .*
58. '88: so *monstrous* a constitution . . .

to swim, to run, to fly or to sing. Man, on the other hand, without an apprenticeship, does not know how to walk, talk, eat or to do anything at all but wail:[59]

> [B]   *Tum porro puer, ut saevis projectus ab undis*
> *Navita, nudus humi jacet, infans, indigus omni*
> *Vitali auxilio, cum primum in luminis oras*
> *Nixibus ex alvo matris natura profudit;*
> *Vagituque locum lugubri complet, ut aequum est*
> *Cui tantum in vita restet transire malorum.*
> *At variae crescunt pecudes, armenta, feraeque,*
> *Nec crepitacula eis opus est, nec cuiquam adhibenda est*
> *Almae nutricis blanda atque infracta loquella;*
> *Nec varias quaerunt vestes pro tempore coeli;*
> *Denique non armis opus est, non moenibus altis,*
> *Queis sua tutentur, quando omnibus omnia large*
> *Tellus ipsa parit, naturaque deadala rerum.*

[Then the child, like a sailor cast up by raging seas, lies naked on the earth, unable to talk, bereft of everything that would help him to live, when Nature first tears him struggling from his mother's womb and casts him on the shore of light. He fills the place with his mournful cries – rightly, for one who still has to pass through so many evils. Yet all sorts of cattle, farm animals as well as wild beasts, thrive; they need no rattles nor the winsome baby language of the gentle nurse; they do not need clothing varying with the weather; and finally they need no weapons nor lofty walls to make them safe, since Earth herself and skilful Nature give all of them, amply, everything they need.][60]

[A]   Such plaints are false. There are more uniform relationships and greater fairness in the constitution of this world.[61] Our skin, like theirs, is adequately provided with means to resist intemperate weather with firmness: witness those many peoples who have yet to acquire a taste for clothing.   [B]   Our ancient Gauls wore hardly any clothes: nor do the Irish, our neighbours, under a sky so cold.

[A]   But we can judge that from ourselves; all parts of the body which we are pleased to leave uncovered to air and wind prove able to endure it: face, feet, hands, legs, shoulders, head, as custom suggests. If there be a part

---

59. Commonplace deriving from Pliny, VII. Erasmus exploited it (Adage, *Dulce bellum inexpertis*); Rabelais satirized it (*Tiers Livre*, TLF, VIII).
60. Lucretius, V. 223; cf. Lambin, 389.
61. '88: this world: *our feebleness at birth is found, more or less, at the birth of the other creatures. Our skin . . .*

of us so weak that it does seem that it has to fear the cold it is our belly, in which digestion takes place: yet our forefathers left it uncovered – and in our society ladies (however soft and delicate they are) occasionally go about with it bare down to the navel. Binding and swaddling up children is not necessary. The mothers of Sparta used to bring up their children with complete freedom of movement for their limbs, without binders or fastenings.⁶² Infant cries are common to most other animals, nearly all can be seen wailing and whining long after they are born; such behaviour is quite appropriate to the helplessness that they feel. As for eating, it is natural to us and to them; it does not have to be learned.

[B] *Sentit enim vim quisque suam quam possit abuti.*

[For every creature feels the powers at its disposal.]⁶³

[A] Does anyone doubt that a child, once able to feed himself, would know how to go in search of food? And Earth, with no farming and with none of our arts, produces quite enough for his needs and offers it to him – perhaps not at all seasons, but neither does she do that for the beasts: witness the stores we can see ants or others provide for the barren season of the year. Those peoples we have recently discovered, so abundantly furnished with food and natural drinks needing no care or toil, have taught us that there is other food beside bread and that Mother Nature can provide us plenteously, without ploughing, with all we need – indeed (as is likely) more straightforwardly and more richly than she does nowadays, when we have brought in our artificial skills.

> *Et tellus nitidas fruges vinetaque laeta*
> *Sponte sua primum mortalibus ipsa creavit;*
> *Ipsa dedit dulces foetus et pabula laeta,*
> *Quae nunc vix nostro grandescunt aucta labore,*
> *Conterimusque boves et vires agricolarum.*

[And Earth herself first willingly provided grain and cheerful vines; she gave sweet produce and good pastures, such as, with all our increased toil, we can but scarcely make to grow; we wear out oxen and the strength of farmers . . .]⁶⁴

The lawless flood of our greed outstrips everything we invent to try and slake it.

62. Plutarch, tr. Amyot, *Lives*, Lycurgus, XIII.
63. Lucretius, V, 1032.
64. Ibid., II, 1157.

As for armaments, we have more natural ones than most other animals do, as well as a greater variety in our movements; we draw greater service from them, too – naturally, without being taught. Men trained to fight naked throw themselves into danger just as our men do. Although some beasts are better armed than we are, we are better armed than others. And we are given to covering the body with acquired means of protection because Nature teaches us to do so instinctively.

To see that this is true, note how the elephant sharpens to a point the teeth which it uses to fight with (for it has special teeth reserved for fighting, and never used for other tasks); when bulls come out to fight they throw up dust and scatter it round about; wild boars whet their tusks; and the ichneumon, before coming to grips with the crocodile, takes mud, kneaded and compressed, and smears it over itself as a crust to serve as body-armour. Why do we not say, therefore, that arming ourselves with sticks and iron bars is equally natural?[65]

As for the power of speech, it is certain that, if it is not natural, then it cannot be necessary. And yet I believe (though it would be difficult to assay it) that if a child, before learning to talk, were brought up in total solitude, then he would have some sort of speech to express his concepts; it is simply not believable that Nature has refused to us men a faculty granted to most other animals; we can see they have means of complaining, rejoicing, calling on each other for help or inviting each other to love; they do so by meaningful utterances: if that is not talking, what is it?   [B]   How could they fail to talk among themselves, since they talk to us and we to them? How many ways we have of speaking to our dogs and they of replying to us! We use different languages again, and make different cries, to call birds, pigs, bulls and horses; we change idiom according to each species.

> [AI]   *Così per entro loro schiera bruna*
> *S'ammusa l'una con l'altra formica*
> *Forse à spiar lor via, e lor fortuna.*

[As one ant from their dark battalion stops to talk to another, perhaps asking the way or how things are faring.]

And does not Lactantius appear to attribute not only speech to animals, but laughter too?

[A]   The different varieties of speech found among men of different countries can be paralleled in animals of the same species. On this subject

---

65. Plutarch, tr. Amyot, *Quels animaux?*, 512 CD.

Aristotle cites the ways in which the call of the partridge varies from place to place.

> [B] *variaeque volucres*
> *Longe alias alio jaciunt in tempore voces,*
> *Et partim mutant cum tempestatibus una*
> *Raucisonos cantus.*

[At different times some birds utter highly different sounds, some even making their songs more raucous with changes in the weather.]

[A] But we do not know what language an isolated child would actually speak and the guesses made about it all seem improbable.[66]

If anyone challenges my opinion, citing the fact that people who are born deaf never learn to talk at all, I have an answer to that: it is not simply because they are unable to receive instruction in speech through the ear but rather because of the intimate relationship which exists between the faculty of hearing (the power they are deprived of) and the faculty of speech, which are by their nature closely sutured together. Whenever we talk, we must first talk as it were to ourselves: our speech first sounds in our own ears, then we utter it into the ears of other people.

I have gone into all this to emphasize similarities with things human, so bringing Man into conformity with the majority of creatures. We are neither above them nor below them. 'Everything under the Sky', said the Wise Man, 'runs according to like laws and fortune.'[67]

> [B] *Indupedita suis fatalibus omnia vinclis.*

[All things are enchained in the fetters of their destiny.]

[A] Some difference there is: there are orders and degrees: but always beneath the countenance of Nature who is one and the same.

> [B] *res quaeque suo ritu procedit, et omnes*
> *Foedere naturae certo discrimina servant.*

66. Commonplace; for Herodotus, II, 2, Phrygian is Man's natural language. Principal sources: Aristotle, *Hist. animal.*, IV, ix; Varchi, *L'Hercolano* (citing Dante, *Purgatorio*, XXXVI, 34) and L. Joubert, *Erreurs populaires au faict de la médecine*, 1578, *ad fin.*, (Lucretius, V. 1077, cited directly, and according to Lactantius, *Div. institut.*, III). Same scepticism, Rabelais, *Tiers Livre*, XIX. If some animals can laugh, then laughter is not the 'property' of Man.
67. Already cited by Montaigne in I, 36 ('On the custom of wearing clothing'); inscribed in Montaigne's library and attributed to 'Eccl. IX'.

[Each thing proceeds after its own manner, and all things maintain their distinctive qualities by the fixed compact of Nature.][68]

[A]   Man must be restrained, with his own rank, within the boundary walls of this polity: the wretch has no stomach for effectively clambering over them; he is trussed up and bound, subject to the same restraints as the other creatures of his natural order. His condition is a very modest one. As for his essential being, he has no true privilege or pre-eminence: what he thinks or fancies he has, has no savour, no body to it. Granted that, of all the animals, Man alone has freedom to think and such unruly ways of doing so that he can imagine things which are and things which are not, imagine his wishes, or the false and the true! but he has to pay a high price for this advantage – and he has little cause to boast about it, since it is the chief source of the woes which beset him: sin, sickness, irresolution, confusion and despair.

To get back to the subject, there is, I say, no rational likelihood that beasts are forced to do by natural inclination the selfsame things which we do by choice and ingenuity. From similar effects we should conclude that there are similar faculties. Consequently, we should admit that animals employ the same method and the same reasoning as ourselves when we do anything.[69] Why should we think that they have inner natural instincts different from anything we experience in ourselves? Added to which, it is more honourable that we be guided towards regular, obligatory behaviour by the natural and ineluctable properties of our being: that is more God-like than rash and fortuitous freedom; it is safer to leave the driver's reins in Nature's hands, not ours. Our empty arrogance makes us prefer to owe our adequacies to our selves rather than to the bounty of Nature; we prefer to lavish the natural goods on other animals, giving them up so as to flatter and honour ourselves with acquired properties. We do that, it seems to me, out of some simple-minded humour. Personally I value graces which are mine since I was born with them more than those which I have had to beg and borrow as an apprentice. It is not within our power to acquire a higher recommendation than to be favoured by God and Nature.

68. Lucretius, V, 874; 921 (Lambin, pp. 430–4).
69. '95: similar faculties, *and from richer effects, richer faculties*. Consequently we should admit that the animals employ the same method *or some better one* and the same reasoning . . . (*Imagination* in Montaigne can include *thought*. Sebond, LXIII, champions a contention rejected here by Montaigne: it is not convincing to unaided human reason.)

Consider the fox which Thracians employ when they want to cross the ice of a frozen river; with this end in view they let it loose. Were we to see it stopping at the river's edge, bringing its ear close to the ice to judge from the noise how near to the surface the current is running; darting forward or pulling back according to its estimate of the thickness or thinness of the ice, would it not be right to conclude that the same reasoning passes through its head as would pass through ours and that it ratiocinates and draws consequences by its natural intelligence like this: 'That which makes a noise is moving; that which moves is not frozen; that which is not frozen is liquid; that which is liquid bends under weight'? Attributing all that exclusively to its keen sense of hearing, without any reasoning or drawing of consequences on the part of the fox, is unthinkable, a chimera. The same judgement should apply to all the ingenious ruses by which beasts protect themselves from our schemes against them.

Should we pride ourself on our ability to capture them and make them work for us? But that is no more than the advantage we have over each other: our slaves are in the same condition. [B] Were not the *Climacides* Syrian slave-women who went down on all fours to serve as steps or ladders for the ladies to climb up into their coaches? [A] Even the majority of free men and women, for very slight advantages, place themselves in the power of others. [C] Thracian wives and concubines beg to be selected for slaughter over the dead husband's tomb. [A] Have tyrants ever failed to find men sworn and devoted to them – even though some require them to follow them in death as in life? [B] Whole armies have been bound to their captains that way.

The form of oath used in that rough school which trained gladiators to fight to the finish included the vows: 'We swear to let ourselves be fettered, burned, beaten or killed by the sword, suffering all that true gladiators suffer at the hands of their Master'; they most scrupulously bound themselves, body and soul, to his service:

> *Ure meum, si vis, flamma caput, et pete ferro*
> *Corpus, et intorto verbere terga seca.*

[Burn my head, if you will, with fire, plunge your iron sword through my body or lash my back with your twisted thongs.]

It was a real, binding undertaking. And yet, one year, ten thousand men were found to enter that school and perish there.

[C] When the Scythians buried their king, over his body they strangled his favourite concubine, his cup-bearer, his ostler, his chamberlain, the guard to his bedchamber and his cook. And on the anniversary of his death

they would take fifty pages mounted on fifty horses and kill them, impaling them from behind, from spine to throat, and leaving them dead on parade about his tomb.[70]

[A] The men who serve us do so more cheaply than our falcons, our horses or our hounds; and they are less carefully looked after – [C] what menial tasks will we not bow to for the convenience of those animals! The most abject slaves, it seems to me, will not willingly do for their masters what princes are proud to do for such creatures. When Diogenes saw his parents striving to purchase his freedom he exclaimed: 'They must be fools: my Master looks after me and feeds me; he is my servant!'[71] So too those who keep animals can be said to serve them, not be served by them.

[A] There is as well a nobility in animals such that, from want of courage, no lion has ever been enslaved to another lion; no horse to another horse. We go out to hunt animals: lions and tigers go out to hunt men; each beast practises a similar sport against another: hounds against hares; pike against tenches; swallows against grasshoppers; sparhawks against blackbirds and skylarks:

> [B]　*serpente ciconia pullos*
> *Nutrit, et inventa per devia rura lacerta,*
> *Et leporem aut capream famulae Jovis, et generosae*
> *In saltu venantur aves.*

[The stork feeds her young on snakes and on lizards found in trackless country places; eagles, those noble birds, servants of Jupiter, hunt hares and roes in the forests.][72]

We share the fruits of the chase with our hounds and our hawks, as well as its skill and hardships. In Thrace, above Amphipolis, huntsmen and wild falcons each share a half of the booty, very exactly, just as the fisherman by the marshes of the Sea of Azov sets aside, in good faith, half of his catch for the wolves: if not, they go and tear his nets.

[A] We have a kind of hunting conducted more with cunning than with force, as when we use gin-traps, hooks and lines. Similar things are found among beasts. Aristotle relates that the cuttle-fish casts a line of gut

---

70. Plutarch, *Quels animaux?*, 513 G; *Comment on pourra discerner le flatteur d'avec l'ami*, 41A; Herodotus, IV, 71–2; Petronius, *Satyricon*; and Tibullus, I, ix, 21, cited by Justus Lipsius, *Saturnalia*, II, 5.
71. Diogenes Laertius, *Lives*, Diogenes. The following pages are largely based on Plutarch, *Quels animaux?* and *Que les brutes usent de la raison*, with additions from Pliny, X, 43, and Plutarch's *Life of Sylla*, etc. Cf. n. 94, below.
72. Juvenal, *Satires*, XIV, 74; 81.

from its neck, pays it out and lets it float. When it wants to, it draws it in. It spots some little fish approaching, remains hiding in the sand or mud and allows it to nibble at the end of the gut and gradually draws it in until that little fish is so close it can pounce on it.

As for force, no animal in the world is liable to so many shocks as Man. No need for a whale, an elephant, a crocodile or animals like that, any one of which can destroy a great number of men. Lice were enough to make Sylla's dictatorship vacant; and the heart and life-blood of a great and victorious Emperor serve as breakfast for some tiny worm.

Why do we say, in the case of Man, that distinguishing plants which are useful for life or for medicines from those which are not (recognizing, say, the virtues of rhubarb or polypody) is a sign that he has scientific knowledge based on skill and reason? Yet the goats of Candia can be seen picking out dittany from a million other plants when they are wounded by spears; if a tortoise swallows a viper it at once goes in search of origanum as a purge; the dragon wipes its eyes clear and bright with fennel; storks give themselves salt water enemas; elephants can remove darts and javelins thrown in battle from their own bodies, from those of their fellows and even from those of their masters (witness the elephant of that King Porus who was killed by Alexander); they do so with more skill than we ever could while causing so little pain. Why do we not call it knowledge and discretion in their case? To lower them in esteem we allege that Nature alone is their Schoolmaster; but that is not to deprive them of knowledge or wisdom: it is to attribute them to them more surely than to ourselves, out of respect for so certain a Teacher.

In all other cases Chrysippus was as scornful a judge of the properties of animals as any philosopher there ever was, yet he watched the actions of a dog which came upon three crossroads – it was either looking for its master or chasing some game fleeing before it; it tried first one road then a second; then, having made sure that neither of them bore any trace of what it was looking for, it charged down the third road without hesitation. Chrysippus was forced to admit that that dog at least reasoned this way: 'I have tracked my master as far as these crossroads; he must have gone down one of these three paths; not this one; not that one; so, inevitably, he must have gone down this other one.' Convinced by this reasoned conclusion, it did not sniff at the third path; it made no further investigations but let itself be swayed by the power of reason. Here was pure dialectic: the dog made use of disjunctive and copulative propositions and adequately enumerated the parts. Does it matter whether he learned all this from himself or from the *Dialectica* of George of Trebizond?

Yet beasts, like us, are not incapable of instruction. Blackbirds, ravens, magpies and parrots can be taught to speak:[73] we recognize in them a capacity for making their voice and their breath subtle and pliant enough for us to mould and restrict them to a definite number of letters and syllables. That capacity witnesses to an inward power of reasoning which makes them teachable – and willing to learn. We have all had our fill I expect of the sort of monkey-tricks which minstrels teach their dogs to do: those dances in which they never miss a note they hear or those varied jumps and movements which they perform on command. But I am much more moved to wonder by the action of the guide-dogs used by the blind in town and country, common enough as they are. I have watched those dogs stop at certain doors where people regularly give alms, and seen how, even when there is room enough to squeeze through themselves, they still avoid encounters with carts and coaches; I have seen one, following the town trench but abandoning a level, even path for a worse one, in order to keep its master away from the ditch. How was that dog brought to realize that it was its duty to neglect its own interests and to serve its master? How does it know that a path might be wide enough for itself but not wide enough for a blind man? Could all that be grasped without thought and reasoning?

I should not overlook what Plutarch tells us about a dog he saw with the elder Vespasian, the Emperor, in the theatre of Marcellus in Rome. This dog served a juggler who was putting on a play with several scenes and several parts. The dog had its own part: it had to pretend, among other things, to swallow some poison and to lie dead for a while. First it swallowed the supposedly poisoned bread; then it began to shake and tremble as though it were dizzy; finally, it lay down and stiffened as though it were dead. It let itself be pulled about and dragged from one place to another, as the plot required. Then, when it knew the time was right, it began to stir very gently, as though awakening from a deep sleep and raised its head, looking from side to side in a way which made the audience thunderstruck.

Oxen were used to water the Royal Gardens of Susa: they had to draw up the water by turning large wheels with buckets attached – you can see plenty of them in Languedoc. Each one had been ordered to do one hundred turns of the wheel a day. They grew so used to this number that nothing would force them to do one more; when their alloted task had been done they stopped dead. Yet we have reached adolescence before we

---

73. Persius, *Choliambics,* which often appear as a preface or postscript to the *Satires.*

can count up to a hundred; and we have just discovered peoples with no knowledge of numbers at all.

You need still greater powers of reason to teach others than to be taught yourself. Democritus thought, and proved, that we had been taught most of our arts by animals: the spider taught us to weave and to sew and the swallow to build; the swan and nightingale taught us music and many other animals taught us by imitation the practice of medicine. Moreover, Aristotle maintains that nightingales teach their young to sing, spending time and trouble doing so: that explains why the song of nightingales brought up in cages, with no freedom to be schooled by their parents, loses much of its charm. [B] From that we may conclude that any improvement is due to learning and study.

Even nightingales born free do not all sing one and the same song: each one sings according to its capacity to learn. They make jealous classmates, squabbling and vying with each other so heartily that the vanquished sometimes drops down dead, not from lack of song but lack of breath. The youngest birds ruminate thoughtfully and then begin to imitate snatches of song; the pupils listen to the lessons of their tutors and then give an account of themselves, taking it in turns to stop their singing. You can hear their faults being corrected; some of the criticisms of their tutors are perceptible even to us.

Arrius[74] said that he once saw an elephant with cymbals hanging from each thigh and a third on its trunk; the other elephants danced round in a ring, rising and falling to the cadences of this musical instrument, which was harmonious and pleasant to listen to. [A] In the great spectacles of Rome it was quite usual to see elephants trained to execute dance steps to the sound of the human voice; such performances comported several intricate movements, interlacings, changes of step and cadenzas, all very hard to learn. Some were seen revising their lessons in private, practising and studying so as to avoid being beaten or scolded by their masters.

But strange indeed is the account of a female magpie vouched for by Plutarch, no less. It lived in a barber's shop in Rome and was wonderfully clever at imitating any sounds it heard. It happened one day that some musicians stopped quite a while in front of the shop, blasting away on their trumpets. Immediately the magpie fell pensive, mute and melancholic, remaining so all the following day. Everyone marvelled, thinking that the blare of the trumpets had frightened and confused it, making it lose both hearing and song at the same time. But they eventually found that it had

74. Or rather, Flavius Arrianus, tr. Vuitart, *Les faicts d'Alexandre*, 1581, XIV.

been deeply meditating and had withdrawn into itself; it had been inwardly practising, preparing its voice to imitate the noise of those trumpeters. The first sound it did make was a perfect imitation of their changes, repetitions and stops; after this new apprenticeship it quit with disdain all that it was able to do before.

I do not want to leave out another example of a dog, also seen by Plutarch. (I realize I am digressing, showing no sense of order, but I can no more observe order when arranging these examples than I can in the rest of my work.) Plutarch was on board ship when he saw a dog which wanted to lap up some oil in the bottom of a jar; it could not get its tongue right down into the vessel because the neck was too narrow, so it went in search of pebbles which it dropped into the jar until the oil rose near to the top where it could get at it. What is that if not the actions of a very subtle intelligence? It is said that Barbary ravens do the same when the water they want to drink is too low to get at.

The above action is somewhat akin to what is related by Juba (a king in elephant country): hunters cunningly prepare deep pits hidden beneath a cover of undergrowth; when an elephant is trapped in one, its fellows promptly bring a great many sticks and stones to help it clamber out.

But so many of their actions bring elephants close to human capacities that if I wanted to relate in detail everything that experience has shown us about them, I would easily win one of my regular arguments: that there is a greater difference between one man and another than between some men and some beasts.

An elephant-driver in a private household in Syria used to steal half the allotted rations at every feed. One day the master himself wanted to attend to things; he tipped into the elephant's manger the right measure of barley, as prescribed. The elephant glared at its driver and, with its trunk, set half the ration aside, to reveal the wrong done to it. Another elephant, whose driver used to adulterate its feed with stones, went up to the pot where he was stewing meat for his own dinner and filled it with ashes. Those are special cases, but we all know from eye-witnesses that the strongest elements in the armies based in the Levant were elephants; their effectiveness surpassed what we can obtain nowadays from our artillery, which more or less replaces elephants in line of battle (as can be easily judged by those who know their ancient history).

> [B]   *siquidem Tirio servire solebant*
> *Annibali, et nostris ducibus, regique Molosso,*
> *Horum majores, et dorso ferre cohortes,*
> *Partem aliquam belli et euntem in praelia turmam.*

[Their sires served Hannibal of Carthage, as well as our generals and the Molossian King, bearing on their backs into the fray cohorts and squadrons, and taking part in the battle themselves.][75]

[A] To make over to them like this the vanguard of their army soldiers must have seriously relied on the trustworthiness of these beasts and on their powers of reason; because of their size and bulk the slightest stoppage on their part or else the slightest panic making them head back towards their own side would be enough to undo everything. There are fewer examples of their turning and charging their own troops than of us men charging back on each other in rout. They were entrusted not with one simple manoeuvre but with several different roles in combat.

[D] The Spaniards, likewise, employed dogs in their recent conquest of the American Indies; they paid them like soldiers and gave them a share in the booty. Those animals displayed eagerness and fierceness but no less skill and judgement, whether in pursuing victory or in knowing when to stop, in charging or withdrawing as appropriate, and in telling friend from foe.[76]

[A] Much more than everyday things, far-off things move us to wonder; they impress us more; otherwise I would not have spent so much time over this long catalogue; for, in my opinion, anyone who took careful note of the everyday animals we see living among us would find them doing things just as astonishing as the examples we gather from far-off times and places.[77] [C] Nature is One and constant in her course. Anybody who could adequately understand her present state could draw reliable conclusions about all the future and all the past.

[A] I once saw men brought to us from distant lands overseas. We could understand nothing of their language; their manners and even their features and clothing were far different from ours. Which of us did not take them for brutes and savages? Which of us did not attribute their silence to dullness and brutish ignorance? After all, they knew no French, were unaware of our hand-kissings and our low and complex bows, our bearing and our behaviour – such things must, of course, serve as a pattern for the whole human race . . .

75. Juvenal, XII, 107.
76. Lopez de Gomara, tr. Fumée, *Hist. générale des Indes*, 1584, II, 9. Cf. G. Bouchet, *Sérées*, I, 7.
77. '88: places. *We live, both they and ourselves, under the same roof and breathe the same air. There is, save for more or less, a perpetual similarity between us.* I once saw . . .

Everything which seems strange we condemn, as well as everything we do not understand; that applies to our judgements on animals. Many of their characteristics are related to ours; that enables us to draw conjectures from comparisons. But they also have qualities peculiar to themselves: what can we know about that? Horses, dogs, cattle, sheep, birds and most other animals living among men recognize our voices and are prepared to obey them. Why, Crassus even had a lamprey which came to him when he called it, and there are eels in the fountain of Arethusa which do the same. [B] I have seen stews in plenty where the fish, on hearing a particular cry from those who tend them, all rush to be fed.

> [A]   *nomen habent, et ad magistri*
> *Vocem quisque sui venit citatus.*

[They have a name and each comes to its master when he calls them.][78]

Such evidence we can judge.

We can also go on to say that elephants have some notion of religion since, after ablutions and purifications, they can be seen waving their trunks like arms upraised, while gazing intently at the rising sun; for long periods at fixed times in the day (by instinct, not from teaching or precept) they stand rooted in meditation and contemplation; there may be no obvious similarities in other animals, but that does not allow us to make judgements about their total lack of religion. When matters are hidden from us, we cannot in any way conceive them.

We can partly do so in the case of an activity noticed by Cleanthes the philosopher, because it resembles our own. He saw, he said, ants leave their own ant-hill for another one, bearing the body of a dead ant. Several others came out to meet them, as if to parley. They remained together for some time; then the second group of ants went back to consult, it was thought, their fellow-citizens. They made two or three such journeys, because of hard bargaining. In the end, the newcomers brought a worm out from their heap, apparently as a ransom for the dead ant. The first lot loaded it on their shoulders and carried it back, leaving the body of the dead ant with the others.

That is the interpretation given by Cleanthes; it witnesses to the fact that voiceless creatures are not deprived of mutual contact and communication; if we cannot share in it, that is because of a defect in us; we would be very stupid indeed to have any meddlesome opinions on the matter.

---

78. Cf. I, 31, 'On the Cannibals' (*ad fin.*); Martial, *Epigrams*, IV, xxix, 6.

Animals do many actions which surpass our understanding; far from being able to imitate them we cannot even conceive them in our thoughts. Many hold that in that last great sea-fight which Antony lost against Augustus, the flag-galley was stopped dead in its course by the fish which is called Remora ('Hindrance') since it has the property of hindering any ship it clings to. When the Emperor Caligula was sailing along the coast of Romania with a large fleet, his galley alone was pulled up short by this very fish. Attached as it was to the bottom of his vessel, he caused it to be seized, angry that so small a creature – it is a shellfish – could just cling by its mouth to his galley and outdo the combined might of the sea, the winds and all his oarsmen. Understandably, he was even more amazed to learn that, once it was brought aboard ship, it no longer had the power it had had in the water.

A citizen of Cyzicum once acquired a reputation as a good mathematical astrologer from noticing the practice of the hedgehog: its den is open in various places to various winds; it can foretell from which direction the wind will blow and plugs up the hole on the windward side. Observing that, he supplied the town with reliable forecasts about the direction of the winds.

The chameleon takes on the colour of its surroundings, but the octopus assumes whatever colour it likes to suit the occasion, hiding, say, from something fearful or lurking for its prey. The chameleon changes passively, the octopus actively. We change hue as well, from fear, anger, shame and other emotions which affect the colour of our faces. That happens to us, as to the chameleon, passively. Jaundice, not our will, has the power to turn us yellow.

Such characteristics in other animals which we realize to surpass our own show that they have, to an outstanding degree, a faculty which we classify as 'occult'. Similarly, animals probably have many other characteristics and powers [C] which are in no way apparent to us.

[A] Of all the omens of former times, the most ancient and the most certain were those drawn from the flight of birds. We have nothing corresponding to that, nothing as wonderful. The beatings of the birds' wings, from which consequences were drawn about the future, show rule and order: only some very special means could produce so noble an activity: to attribute so great an effect entirely to some ordinance of Nature, without any understanding, agreement and thought on the part of the creatures which perform it, is to be taken in by words; such an opinion is evidently false. Here is proof of that: the torpedo is a fish with the property of benumbing the limbs of anyone who directly touches it; in

addition it can even send a numbing torpor into the hands of anyone touching it or handling it indirectly through a net or something similar. They even say that, if you pour water on to it, you can feel this effect working upwards, numbing your sense of touch through the water. This force is worth marvelling at, but is not without its usefulness to the torpedo; that fish knows it has it and uses it to trap its prey when hunting; it snuggles down into the mud: other fish gliding overhead, struck by its cold torpor, are benumbed and fall into its power.

Cranes, swallows and other birds of passage which change dwellings with the seasons, clearly show that they are aware of their ability to foretell and put it to good use.

Hunters assure us that the way to choose from a litter the puppy which will turn out best is simply to make the bitch choose it herself: take the puppies out of their kennel and the first one she brings back will always prove the best; or else make a show of putting a ring of fire around their kennel; then take the first puppy she dashes in to rescue. From that it is obvious that either bitches have powers of foresight which we lack or else that they have a capacity for judging their young which is more lively than our own.[79]

Beasts are born, reproduce, feed, move, live and die in ways so closely related to our own that, if we seek to lower their motivations or to raise our own status above theirs, that cannot arise from any reasoned argument on our part. Doctors recommend us to live and behave as animals do – and ordinary people have ever said:

> Tenez chauts les pieds et la teste;
> Au demeurant, vivez en beste.

> [Keep feet and head warm:
> Then live like the beasts.]

Sexual generation is the principal natural action. Our human members are rather more conveniently arranged for that purpose; and yet we are told that if we want to be really effective we should adopt the position and posture of the animals:

> more ferarum
> Quadrupedumque magis ritu, plerumque putantur
> Concipere uxores; quia sic loca sumere possunt,
> Pectoribus positis, sublatis semina lumbis.

79. '88: own, for in our own children it is certain that until they are nearly grown up, we can find nothing to go on but their physical form.

[Most think that wives conceive more readily in the posture of wild animals and four-footed beasts; that is because the semen can find its way better when the breasts are low down and the loins up-raised.]

[A1] All those immodest and shameless movements that women have invented out of their own heads are condemned as positively harmful; women are advised to return to the more modest and poised comportment of animals of their sex.

> *Nam mulier prohibet se concipere atque repugnat,*
> *Clunibus ipsa viri Venerem si laeta retractet,*
> *Atque exossato ciet omni pectore fluctus.*
> *Ejicit enim sulci recta regione viaque*
> *Vomerem, atque locis avertit seminis ictum.*

[For the woman hinders or averts conception when passion leads her to withdraw Venus and her buttocks from the man, diverting the flow entirely over her yielding belly; she makes the plough-share leap out of its furrow and broadcasts the seed where it does not belong.][80]

[A] If justice consists in rendering everyone his due, then animals which serve, love and protect those that treat them well and which attack strangers and those that do them harm show some resemblance to aspects of our own justice; as they also do by maintaining strict fair-shares for their young.

As for loving affection, theirs is incomparably more lively and consistent than men's. King Lisimachus had a dog called Hircanus. When its master died it remained stubbornly by his bed, refusing to eat or drink; when the day came to cremate the body, it ran dashing into the fire and was burned to death. The dog of a man called Pyrrhus did the same: from the moment he died it would not budge off its master's bed, and when they bore the body away, it let itself be carried off too, finally throwing itself into the pyre as they were burning its master's corpse.

There are also inclinations where our affection arises not from reasoned counsel but by that random chance sometimes called *sympathy*. Animals are capable of it too. We can see horses grown so attracted to each other that we can hardly get them to live or travel apart. We can see them attracted to a particular kind of coat among their fellow horses, as we are to

80. Lucretius, IV, 1261 f.; 1266 f. (cited with approval by Tiraquellus, *De Legibus Connubialibus* who is similarly disapproving of women's provocatory movements: see his Law XV, *in toto*).

particular faces; whenever they come across it they straightway approach it with pleasure and display their affection, whereas they dislike or hate a different kind of coat.

Animals, like us, have a choice of partners and select their females. Nor are they free from our jealousies and great irreconcilable hatreds.

Desires are either natural and necessary, like eating and drinking; natural and not necessary, such as mating with a female; or else neither natural nor necessary, like virtually all human ones, which are entirely superfluous and artificial. Nature needs wonderfully little to be satisfied and leaves little indeed for us to desire. The activities of our kitchens are not Nature's ordinance. Stoics say that a man could feed himself on one olive a day. The choiceness of our wines owes nothing to Nature's teachings, any more than do the refinements we load on to our sexual appetites:

> neque illa
> Magno prognatum deposcit consule cunnum.

[That does not demand a cunt descended from some great consul.][81]

False opinions and ignorance of the good have poured so many strange desires into us that they have chased away almost all the natural ones, no more nor less than if a multitude of strangers in a city drove out all the citizens who were born there, snuffed out their ancient power and authority, seized the town and entirely usurped it.

Animals obey the rules of Nature better than we do and remain more moderately within her prescribed limits – though not so punctiliously as to be without something akin to our debaucheries. Just as there have been mad desires driving humans to fall in love with beasts, so beasts have fallen in love with us, admitting monstrous passions across species: witness the elephant which was the rival of Aristophanes the Grammarian for the affection of a young Alexandrian flower-girl and which was every bit as dutiful in its passion as he was: when walking through the fruit market it took fruit in its trunk and brought it to her. It never took its eyes off her except when it had to and sometimes slipped its trunk into her bosom through her neckband and stroked her breasts. We are also told of a dragon which fell in love with a maiden; of a goose enamoured of a boy in the town of Asopus, and of a ram which sighed for Glaucia the minstrel-girl – and baboons falling madly in love with women are an everyday occurrence. You can also see some male animals falling for males of their own kind.

81. Horace, *Satire* I, 2, 69. In the final pages of the *Essays* sex *is* considered a 'necessity' for the vast majority of humankind.

Oppianus[82] and others relate some examples to show that beasts in their couplings respect the laws of kinship, but experience frequently shows us the contrary:

> *nec habetur turpe juvencae*
> *Ferre patrem tergo; fit equo sua filia conjux;*
> *Quasque creavit init pecudes caper; ipsaque cujus*
> *Semine concepta est, ex illo concipit ales.*

[The heifer feels no shame if covered by the sire nor does the mare; the billy-goat goes on to the nanny-goats he has fathered, and birds conceive from the semen that begot them.][83]

Has there ever been a more express case of subtle malice than that of the mule of Thales the philosopher? Laden with salt, it chanced to stumble when fording a river, so wetting the sacks; noticing that the salt dissolved and lightened its load, it never failed, whenever it could, to plunge fully loaded into a stream. Eventually its master discovered its trick and ordered it to be laden with wool. Finding its expectations deceived, it gave up that trick.

Some animals so naturally mirror the face of human avarice that you can see them stealing anything they can and hiding it carefully, even though they never have any use for it.

As for household management beasts surpass us in the foresight necessary to gather and store for the future, and also possess many of the kinds of knowledge required to do so. When ants notice their grain or seeds going mouldy and smelling badly, they stop them from spoiling or going rotten by spreading them on the ground outside their storehouses, airing, drying and freshening them up. But the measures and precautions they take to gnaw out their grains of corn surpass any imaginable human foresight. Corn does not always stay dry and wholesome but gets soft, flabby and milky, as a step towards germinating and sprouting anew; to stop it turning to seed-corn and losing its nature and properties as grain in store for future use, ants gnaw off the end which does the sprouting.

As for war – the most grandiose and glorious of human activities – I would like to know whether we want to use it to prove our superiority or, on the contrary, to prove our weakness and imperfection. We know how to defeat and kill each other, to undermine and destroy our own species: not much there, it seems, to make them want to learn from us.

---

82. Oppianus was translated into Latin by both Adrian Turnebus and Jean Bodin, scholars admired by Montaigne.
83. Ovid, *Metam.*, X, 325.

[B]   *quando leoni*
Fortior eripuit vitam leo? quo nemore unquam
Expiravit aper majoris dentibus apri?

[When has a stronger lion ever torn life from a weaker lion? In what woodlands has a wild boar ever died at the teeth of a stronger?][84]

[A]   They are not universally free from this, though – witness the furious encounters of bees and the enterprises of their monarchs in the opposing armies:

saepe duobus
Regibus incessit magno discordia motu,
Continuoque animos vulgi et trepidantia bello
Corda licet longe praesciscere.

[Often there arises great strife between two King bees; great movements are afoot; you may imagine the passion and the warlike frenzy which animates the populace.][85]

I can never read that inspired account without thinking that I am reading a description of human vanity and ineptitude.

The deeds of those warriors which ravish us with their horror and their terror; those tempestuous sounds and cries:

[B]   *Fulgur ibi ad coelum se tollit, totaque circum*
Aere renidescit tellus, subterque virum vi
Excitur pedibus sonitus, clamoreque montes
Icti rejectant voces ad sidera mundi;

[There, armour glitters up to heaven and all the surrounding fields shimmer with bronze; the earth shakes beneath the soldiers' tread; the mountains re-echo to the stars above, the clamour striking against them;][86]

[A]   that dread array of thousands upon thousands of soldiers bearing arms; such bravery, ardour, courage: be pleased to consider the pretexts, many and vain, which set them in motion and the pretexts, many and frivolous, which make them cease.

Paridis propter narratur amorem
Graecia Barbariae diro collisa duello.

84. Juvenal, *Satires*, XV, 160.
85. Virgil, *Georgics*, IV, 67. For the Ancients, Queen bees were Kings.
86. Lucretius, II, 325 (Lambin, p. 127).

[They narrate how Greece, for the love of Paris, made fatal war against the Barbarians.][87]

It was because of the lechery of Paris that all Asia was ruined and destroyed: one man's desires, the annoyance and pleasure of one man, one single family quarrel – causes which ought not to suffice to set two fishwives clawing at each other's throats – were the soul, the motive-force, of that great discord.

Do we want to trust the word of those who were the main authors and prime movers of wars like these? Then let us listen to Augustus, the greatest, most victorious and most powerful Emperor there ever has been, sporting and jesting (most amusingly and wittily) about several battles risked on land and sea, the life and limb of the five hundred thousand men who followed his star, and the might and treasure of both parts of the Roman world, exhausted in the service of his adventures:

> *Quod futuit Glaphyran Antonius, hanc mihi poenam*
> *Fulvia constituit, se quoque uti futuam.*
> *Fulviam ego ut futuam? Quid, si me Manius oret*
> *Paedicem, faciam? Non puto, si sapiam.*
> *Aut futue, aut pugnemus, ait. Quid, si mihi vita*
> *Charior est ipsa mentula? Signa canant!*

[Because Antony fucked Glaphyra, Fulvia decided I had to fuck her – as revenge. Me, fuck Fulvia! Supposing Manius begged me to bugger him? Not if I can help it! 'Fuck or we fight,' she said. What if my cock is dearer than life to me? . . . Sound the war trumpets!]

(I quote my Latin with freedom of conscience! You, my Patroness, have given me leave.)[88]

Now this mighty Body, War, with so many facets and movements, which seems to threaten both earth and heaven –

> [B] *Quam multi Lybico volvuntur marmore fluctus,*
> *Saevus ubi Orion hybernis conditur undis,*
> *Vel cum sole novo densae torrentur aristae,*
> *Aut Hermi campo, aut Lyciae flaventibus arvis,*
> *Scuta sonant, pulsuque pedum tremit excita tellus.*

---

87. Horace, *Epistle* I, 2, 6.
88. Verses attributed to Augustus, in Martial, *Epigrams*, XI, 20. The patroness may be Margaret of France, the future wife of Henry of Navarre.

[As the waves innumerable which roll in the Libyan sea, when fierce Orion plunges into the billows as winter returns; or, as when the summer sun bakes the thick shooting corn on the plains of Hermus or the golden fields of Lycia: so clash the shields, and the stricken land trembles beneath their feet] –

[A] this mad Monster with all its many arms and legs, is only Man: weak, miserable, wretched Man. An ant-hill disturbed and hot with rage!

> *It nigrum campis agmen.*

[The black battalion advances in the plain.][89]

A contrary wind, the croak of a flight of ravens, a stumbling horse, an eagle chancing by, a dream, a word, a sign, a morning mist, all suffice to cast him down and bring him to the ground. Let a ray of sunlight dazzle him in the face, and there he lies, limp and faint. Let a speck of dust blow into his eyes (as our poet Virgil writes of the bees), and all our ensigns, all our legions, even with Pompey the Great himself at the head of them, are broken and shattered . . . (I believe it was Pompey who was defeated by Sertorius in Spain with such fine arms as these, [B] which also served a turn for others – for Eumenes against Antigonus, and for Surena against Crassus:

> [A] *Hi motus animorum atque haec certamina tanta*
> *Pulveris exigui jactu compressa quiescent.*

[These passionate commotions and these great battles are calmed down with a handful of dust.][90]

[C] Send out a detachment made up of a couple of bees: they will be strong and brave enough to topple the Monster of war. We still recall how the Portuguese were investing the town of Tamly in their territory of Xiatime when the inhabitants, who had hives in plenty, carried a great many of them to their walls and smoked the bees out so vigorously that their enemies were unable to sustain their stinging attacks and were all put to rout. They owed the freedom of their town and their victory to such novel reinforcements – and with so happy an outcome that not one bee was reported missing.[91]

---

89. Virgil, *Aeneid*, VII, 718 f., IV, 404; here cited with Seneca in mind (Preface to *Quaestiones Naturales*).
90. Plutarch, *Lives*, Sertorius (but it was not Pompey); Virgil, *Georgics*, IV, 86.
91. S. Goulard, *Histoire du Portugal*, 1581 (1587), VIII, 19, 244v°.

[A]   The souls of Emperors and of cobblers are cast in the same mould. We consider the importance of the actions of Princes and their weight and then persuade ourselves that they are produced by causes equally weighty, equally important. In that we deceive ourselves. They are tossed to and fro by the same principles as we are. The reasons that make us take issue with a neighbour lead Princes to start a war; the same reason which makes us flog a lackey makes kings lay waste a province.   [B]   They can do more but can wish as lightly.   [A1]   The same desires trouble a fleshworm and an elephant.

[A]   As for faithfulness, there is no animal in the world whose treachery can compete with Man's. Our history books tell of certain dogs which vigorously reacted to the murders of their masters. King Pyrrhus once came across a dog guarding the body of its dead master; when he was told the dog had done this duty for three days, he ordered the corpse to be buried and took the dog away with him. Later, when he was making a general review of his troops the dog recognized the murderers of its master and ran at them barking loudly and angrily. This was the first piece of evidence leading to its master's murder being avenged; justice was soon done in the courts. The dog of Hesiod the Wise did the same, leading to the sons of Ganistor (a man from Naupactus) being convicted of the murder of its master.

Another dog was guarding a temple in Athens when it spotted a thief sacrilegiously making off with the finest jewels. It began barking at him as loud as it could, but the temple sextons never woke up; so the dog started to trail the thief and, when day broke, hung behind a little without losing him from sight. When the thief offered it food, it refused to take anything from him, whilst accepting it from others who passed by, treating them all to a good wagging of its tail. When the thief stopped to sleep, so did the dog, in the same place. News of this dog reached the sextons of that church; they set out to find it; by making enquiries about the colour of its coat, they eventually caught up with it at Cromyon. The thief was there too; they brought him back to Athens, where he was punished. In recognition of its good sense of duty, the judges awarded the dog a fixed measure of wheat out of public funds to pay for its keep and ordered the priests to look after it. This happened in Plutarch's own time and he himself asserts that the account was very thoroughly vouched for.

As for gratitude – and it seems to me that we could well bring this word back into repute – one example will suffice. Apion relates it as something he had seen himself. He tells how, one day, the people of Rome were given the pleasure of watching several strange animals fight – mainly, in

fact, unusually big lions; one of these drew the eyes of the entire audience by its wild bearing, the strength and size of its limbs and its proud and terrifying roar. Amongst the slaves presented to the populace to fight with these beasts was Androdus, a slave from Dacia, belonging to a Roman lord of consular rank. This lion, seeing him from afar, first pulled up short, as though struck with wonder; it then came gently towards him; its manner was soft and peaceful, as if it expected to recognize an acquaintance. Then, having made certain of what it was looking for, it began to wag its tail as dogs do when fondly greeting their masters; it kissed and licked the hands and thighs of that poor wretch, who was beside himself, ecstatic with fear. The gracious behaviour of the lion brought Androdus back to himself so that he fixed his gaze on it, staring at it and then recognizing it. It was a rare pleasure to see the happy greetings and blandishments they lavished on each other. The populace raised shouts of joy; the Emperor sent for the slave to learn how this strange event had come about. He gave him an account, novel and wonderful: 'My Master', he said, 'was a proconsul in Africa; he treated me so cruelly and so harshly, flogging me every day, that I was forced to steal myself from him and run away. I found the quickest way to hide myself safely from a person having such great Provincial authority was to make for that country's uninhabited sandy deserts, fully resolved, if there was no means of keeping myself in food, to kill myself. The midday sun was so fierce and the heat so intolerable that when I stumbled on a hidden cave, difficult of access, I plunged into it. Soon afterwards this lion came in, its paw all wounded and bloody; it was groaning and whining with pain. I was very frightened when it arrived but, when it saw me hiding in a corner of its lair, it came gently up to me and showed me its wounded paw, as though asking for help. I removed a great splinter of wood; when I had made it a little more used to me, I squeezed out the filthy pus that had collected in the wound, wiped it and made it as clean as I could. The lion, aware that things were better and that the pain had been relieved, began to rest, falling asleep with its paw in my hands. After that we lived together in that cave for three whole years; we ate the same food since the lion brought me choice morsels of the animals it had killed in the hunt; I had no fire but I fed myself by cooking the meat in the heat of the sun. In the end I grew disgusted with this savage, brutish life and so, when the lion had gone out one day on its usual quest for food, I slipped away. Three days later I was surprised by soldiers who brought me from Africa to Rome and handed me over to my master. He promptly condemned me to die by being exposed to the beasts in the arena. I realize now that the lion was also captured soon afterwards and that it wanted to repay me for my kindness in curing its wound.'

That is the account which Androdus told to the Emperor and which he also spread from mouth to mouth. Androdus was given his freedom by general acclaim and relieved of his sentence; by order of the people he was made a gift of the lion.

Ever since, says Apion, we can see Androdus leading the lion about on a short leash, going from tavern to tavern in Rome collecting money, while the lion lets itself be strewn with flowers. All who meet them say: 'There goes the Lion, host to the Man: there goes the Man, doctor to the Lion.'[92]

[B] We often shed tears at the loss of animals which we love: they do the same when they lose us:

> *Post, bellator equus, positis insignibus, Aethon*
> *It lachrymans, guttisque humectat grandibus ora.*

[Then comes Aethon, the war-horse, stripped of its insignia, weeping and drenching its face in mighty tears.][93]

Some peoples hold their wives in common while in others each man has a wife of his own; can we not see the same among the beasts? Do they not have marriages better kept than our own?

[A] As touching the confederations and alliances which animals make to league themselves together for mutual succour, oxen, pigs and other animals can be seen rushing in to help when one of their number is being attacked and rallying round in its defence. If a scar-fish swallows a fisherman's hook, its fellows swarm around and bite through the line; if one of them happens to get caught in a wicker trap, the others dangle their tails down into it from outside while it holds on grimly with its teeth. In this way they drag it right out. When a barbel-fish is hooked, the others stiffen the spine which projects from their backs; it is notched like a saw; they rub it against the line and saw it through.

As for the special duties we render to each other in the service of life, there are several similar examples amongst the animals. The whale, it is said, never travels without a tiny fish like a sea-gudgeon swimming ahead of it (for this reason it is called a 'guide-fish'). The whale follows it everywhere, allowing itself to be directed and steered as easily as a rudder turns a boat. Everything else – beast or ship – which falls into the swirling chaos of that creature's mouth is straightway lost and swallowed up: yet

92. Aulus Gellius, *Attic Nights*, V, 15, etc. This tale of 'Androdus' and the lion is related in Ravisius Textor's *Officina*, which is a probable source of some of Montaigne's animal lore throughout the 'Apology'.
93. Virgil, *Aeneid*, XI, 89.

that little fish can retire there and sleep in its mouth in complete safety. While it is asleep, the whale never budges, but as soon as it swims out, the whale constantly follows it; if it should chance to lose its guide-fish it flounders about all over the place, often dashing itself to pieces against the rocks like a rudderless ship. Plutarch testifies to having seen this happen on the island of Anticyra.

There is a similar companionship between the tiny wren and the crocodile: the wren stands guard over that big creature; when the crocodile's enemy, the ichneumon, closes in for a fight, this little bird is afraid that its companion may be caught napping, so it pecks it awake and sings to warn it of danger. The wren lives on the leftovers of that monstrous crocodile, which welcomes it into its jaws and lets it pick at the meat stuck between its teeth. If it wants to shut its mouth it warns the wren to fly out by gradually closing its jaws a little, without squashing it or harming it in any way.

The shellfish called a nacre lives in similar company with the pinnothere, a kind of small crab which serves it as tout and doorkeeper; squatting by the orifice which the nacre always keeps half-open, it waits until some little fish worth catching swims into it. The crab then slips into the nacre, pinching its living flesh to make it close its shell. Having imprisoned the fish they both set about eating it.

Three parts of Mathematics are particularly well known to tunny-fish: the way they live shows that.

First, Astrology; it is they who teach it to men: wherever they may be when surprised by the winter solstice, there they remain until the following equinox (which explains why even Aristotle readily allows them a knowledge of that science).

Next Geometry and Arithmetic: tunny-fish always form up in the shape of a cube, equally square on all sides. Drawing themselves up into a solid battalion, a corps enclosed and protected all round by six faces of equal size, they swim about in this order, square before, square behind – so that if you count one line of them you have the count of the whole school, since the same figure applies to their depth, breadth and length.

As for greatness of spirit, it would be hard to express it more clearly than that great dog did which was sent to King Alexander from India. It was first presented with a stag, next with a boar, then with a bear: it did not deign to come out and fight them, but as soon as it saw a lion it leaped to its feet, clearly showing that it thought such an animal was indeed worthy of the privilege of fighting against it.

[B]   Touching repentance and the acknowledging of error, they tell of

an elephant which killed its master in a fit of anger; its grief was so intense that it refused to eat and starved itself to death.

[A]   As for clemency, they tell of a tiger – the most inhuman of all beasts – which was given a goat to eat. It fasted for two days before being even tempted to harm it; by the third day, it considered the goat as a familiar guest, so, rather than attack it, it broke out of its cage and sought food elsewhere.

As for rights bred of familiarity and friendly converse, it is quite normal to train cats, dogs and hares to live tamely together.

But surpassing all human imagination is what experience has taught travellers by sea – especially those in the sea of Sicily – about the halcyons. Has Nature ever honoured any creature as she has honoured these king-fishers in their procreation, lying-in and birth? The poets feign that one single island, Delos, was a floating land before being anchored so that Latona might give birth upon it. But God himself has wished the entire sea to be settled, smooth and calm, free from wave and wind and rain, on those halcyon days when these creatures produce their young. (This befalls, precisely, about the shortest day of the year, the solstice. This privilege of theirs gives us seven days and nights at the very heart of the winter, when, without danger, we can sail the seas.) Each female knows no male but its own; it helps it all its life and never forsakes it. If the male is weak or crippled the female carries it everywhere on her back, serving it till death.

But no ingenuity has ever fathomed the miraculous artifice by which the halcyons build their nest for their young nor divined its fabric. Plutarch saw several of them and handled them. He thinks they may be composed of the bones of certain fish, joined, bound and interwoven together, some lengthwise, some crosswise; bent and rounded struts are then added, eventually forming a coracle ready to float upon the water. The female halcyon then brings them where they can be lapped around by the waves of the sea. The salt water gently beats upon them, showing her where ill-fitting joints need daubing and where she needs to strengthen the sections where her construction is coming loose or pulling apart at the beating of the sea. On the other hand this battering by the waves binds all the good joints up tight and knits them so close that they can only with difficulty be smashed, broken or even damaged by blows with stone or iron. Most wonderful of all are the shape and proportions of the concave hold, for it is shaped and proportioned to admit only one creature snugly: the one who made it. To everything else it is closed, barred and impenetrable. Nothing can get in, not even sea water.

That is a fine description of this construction, taken from a fine book. Yet even that, it seems to me, fails to enlighten us adequately about the

difficulty of such architecture. What silly vanity leads us to take products we can neither imitate nor understand, range them beneath us and treat them with disdain.[94]

Let us go further into such equalities and correspondences between us and the beasts. The human soul takes pride in its privilege of bringing all its conceptions into harmony with its own condition: everything it conceives is stripped of its mortal and physical qualities; it compels everything which it judges worthy of notice to divest itself completely of such of its own conditions as are corruptible – of all physical accidents such as depth, length, breadth, weight, colour, smell, roughness, smoothness, hardness, softness; it casts them aside like old garments; it clothes everything in its own condition, spiritual and immortal: the Rome or the Paris which exists in my soul – the Paris imagined in thought – is conceived in my imagination without size, without place, without stone, without plaster, without wood. Well, that self-same privilege seems evidently shared with the beasts; for, asleep on its litter, a war-horse accustomed to trumpet, harquebus and combat can be seen twitching and trembling as though in the thick of battle: clearly its mind is conceiving a drum without drum-beats, an army without arms, without physical body.

> *Quippe videbis equos fortes, cum membra jacebunt*
> *In somnis, sudare tamen, spirareque saepe,*
> *Et quasi de palma summas contendere vires.*

[You can, indeed, see vigorous racehorses, resting their limbs in sleep, yet often sweating and panting as though disputing the prize with all their might.]

The greyhound imagines a hare in a dream: we can see it panting after it in its sleep as it stretches out its tail, twitches its thighs and exactly imitates its movements in the chase: that hare has no coat and no bones.

> *Venantumque canes in molli saepe quiete*
> *Jactant crura tamen subito, vocesque repente*
> *Mittunt, et crebras reducunt naribus auras,*
> *Ut vestigia si teneant inventa ferarum.*
> *Expergefactique sequuntur inania saepe*
> *Cervorum simulachra, fugae quasi dedita cernant:*
> *Donec discussis redeant erroribus ad se.*

---

94. The long series of borrowings from Plutarch on animals ends here (cf. n. 71, above). The paragraphs which follow are indebted to Sebond, chapters 217 and 293.

[Often hunting dogs lying quietly asleep, suddenly paw about, bark out loud and sharply draw their breath as if they were on the track of their prey. Even after they have started out of their sleep they still pursue that empty ghost of a stag as though they could see it fleeing before them, until the error fades and they come back to themselves.]

Guard dogs can be found growling in their sleep, then yapping and finally waking with a start as though they saw some stranger coming: that stranger which their souls can see is a spiritual man, not perceptible to the senses, without dimensions, without colour and without being.

> *consueta domi catulorum blanda propago*
> *Degere, saepe levem ex oculis volucremque soporem*
> *Discutere, et corpus de terra corripere instant,*
> *Proinde quasi ignotas facies atque ora tueantur.*

[The dog, that fawning creature at home in our houses, often quivers its eyelids in winged sleep and starts to its feet as if it saw the faces and features of strangers.]⁹⁵

As for physical beauty, before I can go any further I need to know if we can agree over its description. It seems we have little knowledge of natural beauty or of beauty in general, since we humans give so many diverse forms to our own beauty; [C] if it had been prescribed by Nature, we would all hold common views about it, just as we all agree that fire is hot. We give human beauty any form we fancy:

> [B] *Turpis Romano Belgicus ore color.*

[On the face of a Roman a Belgian's colour is ugly.]⁹⁶

[A] For a painter in the Indies beauty is black and sunburnt, with thick swollen lips and broad flat noses; [B] there, they load the cartilage between the nostrils with great rings of gold, so that it hangs right down to the lips; the lower lip is similarly weighed down to the chin with great hoops studded with precious jewels; for them it is elegant to lay their teeth bare [C] exposing the gum below their roots. [B] In Peru, big ears are beautiful: they stretch them as far as they can, artificially. [C] A man still alive today says that he saw in the East a country where this custom of stretching ears and loading them with jewels is held in such esteem that he was often able to thrust his arm, clothes and all, through the holes women pierced in their lobes. [B] Elsewhere there are whole

95. Lucretius, IV, 988 f., 992 f., 999 f. (Lambin, p. 345).
96. Propertius, II, 18, 26.

nations who carefully blacken their teeth and loathe seeing white ones. Elsewhere they dye them red. [C] Not only in the Basque country do they prefer beautiful women to have shaven heads; the same applies elsewhere – even, according to Pliny, in certain icy lands. [B] The women of Mexico count low foreheads as a sign of beauty: so, while they pluck hair from the rest of their body, there they encourage it to grow thick and propagate it artificially. They hold large breasts in such high esteem that they affect giving suck to their children over their shoulders.[97]

[A]   We would fashion ugliness that way.

Italians make beauty fat and heavy; Spaniards gaunt and skinny; some of us French make it fair, others dark; some soft and delicate; others strong and robust; some desire grace and delicacy; others proud bearing and majesty. [C] Similarly, while Plato considered the sphere to be the perfection of beauty[98] the Epicureans preferred the pyramid or the square, finding it hard to swallow a god who was shaped like a ball!

[A]   Anyway, Nature has no more given man privileges in beauty than in any other of her common laws. If we judge ourselves fairly we will find some animals less favoured than we are, others (more numerous) which are more so: [C] 'a multis animalibus decore vincimur' [we are surpassed in beauty by many of the beasts][99] – especially among our fellow-citizens, the denizens of dry land. As for the creatures of the sea, we can leave their beauty of form aside, since it has no point of comparison with ours; we are thoroughly beaten by them in colour, brightness, sheen and the general disposition of our members; beaten by the birds of the air, too, in all qualities. And   [A]   then there is that privilege the poets stress – the fact that we hold ourselves erect, gazing up to heaven, from whence we came:

> Pronaque cum spectent animalia caetera terram,
> Os homini sublime dedit, coelumque videre
> Jussit, et erectos ad sydera tollere vultus.

[The other animals look downwards to the ground; God gave Man a face held high and ordered him to look towards heaven and raise his eyes towards the sun, moon and stars.][100]

97. Lopez de Gomara, II: XX, 73; LXXXIV, 170 f.; IV: III, 276; Pliny, *Hist. Nat.* VI, xiii; Gasparo Balbi, *Viaggio dell'Indie Orientali*, 1590, 76; Pliny, *Hist. Nat.* VI, xiii.
98. Cf. Cicero, *De nat. deorum*, I. x. 24.
99. Seneca, *Ep. moral.* 124, 22 (reading *multis for mutis*).
100. Ovid, *Metam.*, I, 84; it was often (as by Sebond) taken very seriously (cf. J. Du Bellay, *Regrets*, TLF, Sonnet 53, notes), but it does not commend itself to unaided, or unilluminated, human reason.

That privilege is well and truly poetic! Some quite small animals gaze up to heaven all the time; camels and ostriches seem to me to have necks straighter than ours and more erect. [C] And which are these animals which are supposed not to have faces in front and on top, not to look straight ahead as we do nor, in their normal posture, to see as much of heaven and earth as we do? What characteristics of man's body as described by Plato and Cicero do not equally apply to a thousand other animals![101] [A] The animals most like us are the worst and the ugliest of the bunch: the one with an outward appearance and face closest to ours is the baboon;

> [C] *Simia quam similis, turpissima bestia, nobis!*
>
> [That vilest of beast, the monkey – how like us!][102]

[A] the one with inwards and vital organs closest to ours is the pig.[103]

When I think of the human animal, stark naked, with all its blemishes, natural weaknesses and flaws, I find that we have more cause to cover ourselves up than any other animal. (That even applies to the female sex which seems to have a greater share of beauty.) We could be excused for having borrowed from those which Nature has favoured more than us, decking ourselves in their beauty,[104] hiding ourselves in their coats: wool, feathers, hide or silk.

We may note *en passant* that we are the only animals whose physical defects are offensive to our fellows; we are also the only ones to hide from others of our species when answering the calls of Nature. Also worth considering is the fact that those who know prescribe for lovesickness a good look at the totally naked body which is so much desired. To cool amorous passion, all you need to do is to be free to look at the one you love!

> *Ille quod obscoenas in aperto corpore partes*
> *Viderat, in cursu qui fuit, haesit amor.*

[It has been known for a man to see his mistress's private parts and to find his ardour pulled up short.][105]

101. Cicero, *De nat. deorum*, II, liv, 133 ff. (A long praise of the Immortals' care in shaping Man. It is indebted to Plato's *Timaeus*.)
102. Ennius, *apud* Cicero, ibid., I, xxxv, 97.
103. '88: vital, *noble organs* closest to ours is, *according to the doctors*, the pig . . .
104. '88: beauty. *And since Man did not have the wherewithal to present himself naked to the sight of the world, he was right to hide himself behind the coats of others*: wool, feathers, hide or silk, *and other borrowed commodities* . . .
105. Ovid, *Remedia amoris*, 429.

It is true that this prescription may result from a cool and delicate humour in Man; nevertheless it is a striking sign of our weakness that it is enough for us to frequent and know each other for us to feel disgust. [B] Ladies are circumspect and keep us out of their dressing-rooms before they have put on their paint and decked themselves out for public show: that is not so much modesty as skill and foresight.

> [A1]   *Nec veneres nostras hoc fallit: quo magis ipsae*
> *Omnia summopere hos vitae post scenia celant,*
> *Quos retinere volunt adstrictoque esse in amore.*

[Fair women know this: they are all the more careful to hide the changing-rooms of their lives from those lovers they wish to hold and bind to them.][106]

Yet we like all the parts of some animals, finding them so pleasing to our tastes that from their very droppings, discharges and excreta we make dainty things to eat as well as ornaments and perfumes.

Such arguments apply only to the common order of men; they are not sacrilegious enough to want to include those beauties, supernatural and beyond the common order, which can sometimes be seen shining among us like stars beneath a bodily and earthly veil.

Now even that share in Nature's favour which we do concede to the animals is much to their advantage. To ourselves we attribute goods which are purely imaginary and fantastical; future, absent goods, which it exceeds our human capacity, of itself, to vouch for; or else they are goods which our unruly opinions attribute to ourselves quite wrongly, such as knowledge, rationality or pre-eminence. We abandon to animals a share in solid, palpable goods which really do exist: peace, repose, security, innocence, health . . . Health! the fairest and finest gift that Nature can bestow. That is why even Stoic Philosophy dares to assert that Heraclitus (who had dropsy) and Pherecydes (who had been infected by lice) would have been right, if they could, to barter their wisdom against a cure. By weighing and comparing wisdom against health they make it even more splendid than in another of their assertions. Supposing Circe (they say) had presented Ulysses with two different potions, one to make a madman wise, the other a wise man mad: rather than allow her to transform him from

106. Lucretius, IV, 1182 (Lambin, p. 359 f.).

human to beast, he ought to have accepted the one that would make him mad. Wisdom herself, they say, would have argued like this: 'Leave me, forsake me, rather than lodge me in the bodily shape of an ass.' What? Will philosophers forsake Wisdom, great and divine, to cleave to the veil of this earthy body?[107] So we do not, after all, excel over beasts by wit and our power of reason but merely by our physical beauty, our beautiful colour, the beautiful way our members are arranged! For things like that we must forsake our intellect, our moral wisdom and what not!

Well, that is a frank and artless admission and I accept it. At least philosophers have admitted that all those qualities they make such a fuss about are fantastic and vain: even if beasts had all the virtue, knowledge, wisdom and contentment of the Stoic [C] they would still be beasts, [A] in no way to be compared to any man, however wretched, wicked or daft! [C] In fine, nothing is worth anything if it does not look like us. Even God has to become like us, to be appreciated – I shall go into that later.[108] It is clear from this that [A] we do not place ourselves above other animals and reject their condition and companionship by right reason but out of stubbornness and insane arrogance.

To get back to the subject: we have been allotted inconstancy, hesitation, doubt, pain, superstition, worries about what will happen (even after we are dead), ambition, greed, jealousy, envy, unruly, insane and untameable appetites, war, lies, disloyalty, backbiting and curiosity. We take pride in our fair, discursive reason and our capacity to judge and to know, but we have bought them at a price which is strangely excessive if it includes those passions without number which prey upon us. [B] Unless, that is, we choose, like Socrates, to pride ourselves on the one noteworthy prerogative we do have over the beasts: Nature lays down limits and seasons to their lusts, but gives us a full rein – anytime, any place.

[C] *Ut vinum aegrotis, quia prodest raro, nocet saepissime, melius est non adhibere omnino, quam, spe dubiae salutis, in apertam perniciem incurrere: sic haud scio an melius fuerit humano generi motum istum celerem cogitationis, acumen, solertiam, quam rationem vocamus, quoniam pestifera sint multis, admodum paucis salutaria, non dari omnino, quam tam munifice et tam large dari.*

---

107. Plutarch, tr. Amyot, *Des conceptions communes contre les Stoiques*, 577 AB; cf. Erasmus, *In Praise of Folly*, XXXV and XI.

'88: to the *mask* of . . .

108. Cf. p. 573.

'88: daft. *All our perfection, then, consists in being men.* We do not . . .

[Wine is often bad and rarely good for the sick, so it is better to let them have none at all than to run known risks for a doubtful remedy. So too with that mental agility, shrewdness and ingenuity which we call *reason*: it is baleful to many and good for only a few. It would have been better for Man not to have been given it at all than to have been given it with such great munificence.][109]

[A]   What good did their great erudition do for Varro and Aristotle? Did it free them from human ills? Did it relieve them of misfortunes such as befall a common porter? Could logic console them for the gout – and did they feel it any the less because they knew how that humour lodged in their joints? Did it help them to come to terms with death, knowing that whole tribes take delight in it? Did they not mind being cuckolded, since they knew that in some place or other men have wives in common? Not at all. Varro among the Romans and Aristotle among the Greeks were ranked first for knowledge at a time when learning was flourishing and at its best. Yet nobody says that their lives were particularly outstanding. There are, in fact, notorious stains on the life of the Greek one, which he cannot easily escape.[110]   [B]   Have we discovered that health and pleasure taste better if you know astrology or grammar –

> *Illiterati num minus nervi rigent?*

[Men who cannot read do not find it harder to get an erection, do they?]

– or that shame and poverty become more bearable?

> *Scilicet et morbis et debilitate carebis,*
> *Et luctum et curam effugies, et tempora vitae*
> *Longa tibi post haec fato meliore dabuntur.*

[You will doubtless be free from ills and weakness and be free from grief and care, and a long life will be granted you, one with a better destiny.][111]

I have seen in my time hundreds of craftsmen and ploughmen wiser and happier than University Rectors – and whom I would rather be like. Among the necessities of life learning seems to me to rank with fame,

---

109. Socrates: Xenophon, *Memorabilia*, I, iv. 12; Cicero, *De nat. deorum*, III, xxvii, 69.
110. Epicureans, especially, accused Aristotle of disloyalty and of a misspent youth.
111. Horace, *Epodes*, VIII, 17; Juvenal, XIV, 156.

noble blood and dignity[112] [C] or, at most, with beauty, riches [A] and such other qualities which do indeed contribute a great deal to life, but from a distance and somewhat more in the mind than in nature.

[C] We hardly need more duties, laws and rules of conduct in human society than cranes or ants do in theirs: they have no learning, yet live their lives quite ordinately. If Man were wise he would gauge the true worth of anything by its usefulness and appropriateness to his life.

[A] If anyone were to tot up our deeds and our actions he would find more outstanding men among the ignorant than among the wise – outstanding in virtues of every kind. Old Rome seems to me to have borne many men of greater worth, both in peace and war, than the later, cultured Rome which brought about its own downfall. Even if everything else were identical, at very least valour and uprightness would still tilt the balance towards Old Rome, for they make uniquely good bedfellows with simplicity.

But I will let this subject drop; it would draw me further on than I want to go. I will merely add this: only humility and submissiveness[113] can produce a good man. We must not let everyone work out for himself what his duties are. Duty must be laid down for him, not chosen by him from his own reasoning; otherwise, out of the weakness and infinite variety of our reasons and opinions, we will – as Epicurus said – end up forging duties for ourselves which will have us eating each other. The first commandment which God ever gave to Man was the law of pure obedience. It was a bare and simple order, leaving Man no room for knowing or arguing [C] – since the principal duty of a reasonable soul which acknowledges a Superior and a Benefactor in heaven is to obey him. All other virtues are born of submission and obedience, just as all other sins are born of pride. [B] The first temptation came to humankind from the opposite extreme: the Devil first poured his poison into our ears with promises about knowledge and understanding: *'Eritis sicut dii, scientes bonum et malum'* [Ye shall be as Gods, knowing good and evil]. [C] In Homer, when the Sirens wished to deceive Ulysses, draw him into their dangerous snares and so destroy him, they offered him the gift of knowledge.[114]

[A] There is a plague on Man: his opinion that he knows something.

112. '88: for: '*Among . . . dignity*', this sentence reads: *Learning is even less necessary in the service of life than glory and such other qualities.*

113. '88: only *obedience* can . . .

114. References to the Fall, Genesis, III, and to Homer *apud* Cicero, *De fin.*, V, xviii, 49. Cf. Plutarch (tr. Amyot), *Contre Colotes*, 597 FG, for the remark attributed to Epicurus.

That is why ignorance is so strongly advocated by our religion as a quality appropriate to belief and obedience. [C] '*Cavete ne quis vos decipiat per philosophiam et inanes seductiones secundum elementa mundi*' [Beware lest any man cheat you through philosophy and vain deceptions, according to the rudiments of the world].[115]

[A] All the philosophers of all the sects are in general accord over one thing: that the sovereign good consists in peace of mind and body. [B] But where are we to find it?

> [A]   *Ad summum sapiens uno minor est Jove: dives,*
> *Liber, honoratus, pulcher, rex denique regum:*
> *Praecipue sanus, nisi cum pituita molesta est.*

[To sum up then: the wise man has only one superior – Jupiter – and is rich, free, honourable, beautiful, the king of kings in fact ... especially when well and not troubled by snot!]

It does seem true that Nature allotted us one thing only to console us for our pitiful, wretched condition: arrogance. Epictetus agrees, saying that Man has nothing properly his own except his opinions. For our portion we have been allotted wind and smoke.[116]

[B] Philosophy asserts that gods enjoy health as it really is, though they can understand illness; Man, on the contrary, enjoys his goods only in fantasy, but knows ills as they really are.[117]   [A] We have done right to emphasize our imaginative powers: all our goods exist only in a dream.

Man is a wretched creature, subject to calamities;[118] but just listen to him bragging: 'There is no occupation', says Cicero, 'so sweet as scholarship; scholarship is the means of making known to us, while still in this world, the infinity of matter, the immense grandeur of Nature, the heavens, the lands and the seas. Scholarship has taught us piety, moderation, greatness of heart; it snatches our souls from darkness and shows them all things, the high and the low, the first, the last and everything between; scholarship furnishes us with the means of living well and happily; it teaches us how to

---

115. '88: he knows something. That is why *simplicity and ignorance are* so strongly advocated by our religion as *elements properly conducive to subjection*, belief and obedience. All the philosophers ... (Colossians 2:8. Cf. Augustine, *City of God*, VIII, ix, a key text for Christian folly, the praise of which is soon to be taken up by Montaigne.)
116. Horace, *Epistles*, I, i. 106; John Stobaeus, *Apophthegmata*, Sermo 21.
117. Plutarch, *Contre les Stoïques,* 578 G.
118. Plutarch, *Que les bestes brutes usent de raison*, 270 F.

spend our lives without discontent and without vexation' . . .[119] Is this fellow describing the properties of almighty and everlasting God! In practice, thousands of little women in their villages have lived lives more gentle, more equable, more constant than his.

> [AI]   *Deus ille fuit, Deus, inclute Memmi,*
> *Qui princeps vitae rationem invenit eam, quae*
> *Nunc appellatur sapientia, quique per artem*
> *Fluctibus e tantis vitam tantisque tenebris*
> *In tam tranquillo et tam clara luce locavit.*

[It was a god, noble Memmius, yes, a god who first discovered that rule of life which we now call Wisdom and who, through his skill, brought our lives out from storm and darkness and fixed them in such tranquillity and light.]

Beautiful, magnificent words, those! Yet, despite the god who taught him such divine wisdom, a minor accident reduced the wits of the fellow who wrote them to a state worse than that of the meanest shepherd![120]

[A]   Of similar impudence are   [C]   that promise of Democritus in his preface: 'I am going to write about Everything'; the stupid title Aristotle bestows on us men: 'Mortal Gods';[121] and   [A]   Chrysippus' judgement that Dion was as virtuous as God. And even Seneca, my favourite, asserts that, by God's gift he is living: but living well he owes to himself   [C]   – which conforms to what that other fellow said: '*In virtute vere gloriamur; quod non contingeret, si id donum a deo, non a nobis haberemus*' [We rightly glory in our virtue; that would not arise if it were a gift of God and not of ourselves]. This is in Seneca, too: 'The wise man has fortitude similar to God's, but since he has it within human weakness, he surpasses God.'[122]

[A]   There is nothing more common than rash quips like these. We are so much more jealous of our own interests than of those of our Creator that not one of us is more shocked when he sees himself made equal to God than reduced to the ranks of the other animals. We must trample down this

119. Cicero: *Tusc. disput.*, V, xxxvi.
120. Lucretius, V, 8; Montaigne discusses his madness in II, 2, 'On drunkenness': 'That great poet Lucretius vainly philosophizes and braces himself: there he was, driven out of his senses by a love potion.'
121. Cicero, *Acad.: Lucullus*, II, xxiii, 73; *De finibus*, II, xiii, 40: 'As Aristotle says: Man is born for thought and action: he is, as it were, a mortal god.'
122. Plutarch, *Contre les Stoïques*, 583 E; cited with Seneca in La Primaudaye, *Académie françoyse*, 1581, p. 5; Cicero, *De nat. deorum*, III, xxxvi, 87; Seneca, *Epist. moral.*, LIII, 11–12.

stupid vanity, violently and boldly shaking the absurd foundations on which we base such false opinions. So long as Man thinks he has means and powers deriving from himself he will never acknowledge what he owes to his Master. All his geese will be swans, as the saying goes. So we must strip him down to his shirt-tails. Let us look at some notable examples of what his philosophy actually produces.

Possidonius was beset with an illness so painful that it made him twist his arms and grind his teeth; he thought he could cock a snook at Pain by crying out at her: 'It's no good; whatever you do I will never admit that you are evil.' He boasts that he will at least contain his speech within the rules of his sect, yet he feels exactly the same pain as my footman.[123] [C] *'Re succumbere non oportebat verbis gloriantem'* [If you boast in words you should not surrender in fact].

Arcesilaus was suffering from gout. Carneades came to see him and was just going sadly away when he called him back; he pointed from his feet to his heart and said, 'Nothing has passed from here to there.' There is a little more elegance in that: he admits to pain and would gladly be rid of it; it is an evil, all right, but his heart is neither cast down nor weakened by it. That other fellow clings to his position, which is, I fear, more a matter of words than of reality. When Dionysius of Heraclea was nearly driven out of his mind by stabbing pains in his eyes, he was forced to give up such Stoical assertions.[124]

[A] But supposing knowledge actually could produce the effects claimed for it, actually could blunt and reduce the pangs of the misfortunes which beset us: even then, what does it really achieve over and beyond what ignorance does — more purely and more evidently? When Pyrrho, the philosopher, was exposed to the hazards of a mighty tempest, he could set no better example before his companions than the indifference of a pig on board ship with them: it gazed at the storm quite free from fear. When Philosophy has run out of precepts she sends us back to athletes and mule-drivers. Such men are usually less apprehensive of death, pain and other misfortunes. They also show more steadfastness than scholarship affords to any man not already predisposed to it by birth and by a duly cultivated natural talent.[125] What is it if not ignorance which allows our surgeons to make incisions in the tender limbs of children more easily than in our

---

123. '88: footman. *It is all wind and words.* But supposing . . .
124. Cicero, *Tusc. disput.*, II, xiii; *De fin.*, V, xxxi, 94.
125. '88: natural talent. *Knowledge sharpens our feelings for ills rather than lightening them.* What . . .

own? [C] (The same applies to horses.) [A] How many men have been made ill by the sheer force of imagination? Is it not normal to see men bled, purged and swallowing medicines to cure ills which they feel only in their minds? When we run out of genuine ills, Learning will lend us some of her own: this or that colour are symptoms of a catarrh you *will* have; this heat-wave threatens you with some turbulent fever; this break in the line of life on your left hand warns you of some grave and imminent illness . . . Finally Learning openly makes assaults against health itself: that youthful vigour and liveliness of yours cannot remain stable for long! Better bleed away some of their force in case it turns against you . . .

Compare the life of a man enslaved by such fantasies with the life of a ploughman who, free from learning and prognostics, merely follows his natural appetites and judges things as they feel at present. He only feels ill when he really is ill; the other fellow often has stone in the mind before stone in the kidney. As though it were not time enough to suffer pain when it really comes along, our thoughts must run ahead and meet it.

What I say about medicine applies to erudition in general – hence that ancient philosophical opinion that sovereign good lies in recognizing the weakness of our powers of judgement. My ignorance can supply as good a cause to hope as to fear; for me, the only rule of health lies in the example of other people and how I see them fare in similar circumstances; but since I can find all sorts of examples, I dwell on the comparisons which are most favourable to me! Health, full, free and entire, I welcome with open arms. I whet my appetites so that I can truly enjoy it, all the more so since health is not usual to me any more, but quite rare. Far be it from me to trouble the sweet repose of health with bitterness arising from a new regime based on restraint. The very beasts can show us that illness can be brought on by mental agitations.

[C] The natives of Brazil are said to die only of old age; they attribute that to the serenity and tranquillity of the air: I would attribute it to the serenity and tranquillity of their souls; they are not burdened with intense emotions and unpleasant tasks and thoughts: they pass their lives in striking simplicity and ignorance. They have no literature, no laws, no kings and no religion of any kind.[126]

[A] Experience shows that gross, uncouth men make more desirable and vigorous sexual partners; lying with a mule-driver is often more welcome than lying with a gentleman. How can we explain that except by assuming that emotions within the gentleman's soul undermine the strength

of his body, break it down and exhaust it,   [A1]   just as they exhaust and harm the soul itself? Is it not true that the soul can be most readily thrown into mania and driven mad by its own quickness, sharpness and nimbleness – in short by the qualities which constitute its strength?   [B]   Does not the most subtle wisdom produce the most subtle madness? As great enmities are born of great friendships and fatal illnesses are born of radiant health, so too the most exquisite and delirious of manias are produced by the choicest and the most lively of the emotions which disturb the soul. It needs only a half turn of the peg to pass from one to the other. [A1]   When men are demented their very actions show how appropriate madness is to the workings of our souls at their most vigorous. Is there anyone who does not know how imperceptible are the divisions separating madness from the spiritual alacrity of a soul set free or from actions arising from supreme and extraordinary virtue? Plato says that melancholics are the most teachable and the most sublime; yet none has a greater propensity towards madness. Spirits without number are undermined by their own force and subtlety. There is an Italian poet, fashioned in the atmosphere of the pure poetry of Antiquity, who showed more judgement and genius than any other Italian for many a long year; yet his agile and lively mind has overthrown him; the light has made him blind; his reason's grasp was so precise and so intense that it has left him quite irrational; his quest for knowledge, eager and exacting, has led to his becoming like a dumb beast; his rare aptitude for the activities of the soul has left him with no activity . . . and with no soul. Ought he to be grateful to so murderous a mental agility? It was not so much compassion that I felt as anger when I saw him in so wretched a state, surviving himself, neglecting himself (and his works, which were published, unlicked and uncorrected; he had sight of this but no understanding).[127]

Do you want a man who is sane, moderate, firmly based and reliable? Then array him in darkness, sluggishness and heaviness.   [C]   To teach us to be wise, make us stupid like beasts; to guide us you must blind us.

[A]   If you say that the convenience of having our senses chilled and blunted when tasting evil pains must entail the consequential inconvenience of rendering us less keenly appreciative of the joys of good pleasures, I agree. But the wretchedness of our human condition means we have less to relish than to banish: the most extreme pleasures touch us less than the lightest of pains: [C]   '*Segnius homines bona quam mala sentiunt*' [Men feel

127. Aristotle, *Problems*, 30–I. (For Tasso's madness, see *Montaigne and Melancholy*, p. 371ff.)

pleasure more dully than pain].   [A]   We are far less aware of perfect health than of the slightest illness:

> *pungit*
> *In cute vix summa violatum plagula corpus,*
> *Quando valere nihil quemquam movet. Hoc juvat unum,*
> *Quod me non torquet latus aut pes: caetera quisquam*
> *Vix queat aut sanum sese, aut sentire valentem.*

[A man feels the slightest prick which scarcely breaks his skin; yet he remains unmoved by excellent health. Personally I feel delight in simply being free from pain in foot or side, while another scarcely realizes he is well and remains unaware of his good health.]

For us, being well means not being ill. So that philosophical school which sets the highest value on pleasure reduces it to the mere absence of pain. To be free from ill is the greatest good that Man can hope for.   [C]   As Ennius puts it,

> *Nimium boni est, cui nihil est mali*

[Ample good consists in being free from ill].[128]

[A]   For even that tickling excitement which accompanies certain pleasures and which seems to exalt us above mere good health and freedom from pain, that shifting delight, active, inexplicably biting and sharp, aims in the end at freedom from pain. The appetite which enraptures us when we lie with women merely aims at banishing the pain brought on by the frenzy of our inflamed desires; all it seeks is rest and repose, free from the fever of passion.

The same applies to all other appetites. I maintain, therefore, that if ignorant simplicity can bring us to an absence of pain, then it brings us to a state which, given the human condition, is very blessedness.

[C]   Yet we should not think of a simplicity so leaden as to be unable to taste anything. Crantor was right to attack 'freedom from pain' as conceived by Epicurus, insofar as it was built upon foundations so deep that pain could not even draw near to it or arise within it. I have no words of praise for a 'freedom from pain' which is neither possible nor desirable. I am pleased enough not to be ill but, if I am ill, I want to know; if you cut me open or cauterize me, I want to feel it. Truly, anyone who could uproot all knowledge of pain would equally eradicate all knowledge of pleasure and

128. Livy, XXX, xxi; La Boëtie (ed. Bonnefon, 1892, p. 234); Ennius, *apud* Cicero, *De fin.*, II, xiii, 41.

finally destroy Man: *'Istud nihil dolere, non sine magna mercede contingit immanitatis in animo, stuporis in corpore'* [That 'freedom from pain' has a high price: cruelty in the soul, insensate dullness in the body]. For Man, ill can be good at times; it is not always right to flee pain, not always right to chase after pleasure.

[A]   It greatly advances the honour of Ignorance that Learning has to throw us into her arms when powerless to stiffen our backs against the weight of our ills; she has to make terms, slipping the reins and giving us leave to seek refuge in the lap of Ignorance, finding under her protection a shelter from the blows and outrages of Fortune.

Learning instructs us to [C]   withdraw our thoughts from the ills which beset us now and to occupy them by recalling the good times we have known;   [A]   to make use of the memory of past joys in order to console ourselves for present sorrows, or to call in the help of vanished happiness to set against the things which oppress us now –   [C]   *'Levationes aegritudinum in avocatione a cogitanda molestia et revocatione ad contemplandas voluptates ponit'* [He found a way to lessen sorrows by summoning thoughts away from troubles and calling them back to gaze on pleasure] – [A]   when Learning runs out of force, she turns to cunning; when strength of arm and body fails, she resorts to conjuring tricks and nimble footwork; if that is not what is meant, what does it mean? When any reasonable man, let alone a philosopher, feels in reality a blazing thirst brought on by a burning fever, can you buy him off with memories of the delights of Greek wine?   [B]   That would only make a bad bargain worse.

> *Che ricordarsi il ben doppia la noia.*
>
> [Recalling pleasure doubles pain.]

[A]   Of a similar nature is that other counsel which Philosophy gives us: to keep only past pleasures in mind and to wipe off the sorrows we have known – as if we had the art of forgetfulness in our power.   [C]   Anyway, such advice makes us worse:

> *Suavis est laborum praeteritorum memoria.*
>
> [Sweet is the memory of toils now past.]

[A]   Philosophy ought to arm me with weapons to fight against Fortune; she should stiffen my resolve to trample human adversities underfoot; how has she grown so weak as to have me bolting into burrows with such

cowardly and stupid evasions? Memory reproduces what she wants, not what we choose. Indeed there is nothing which stamps anything so vividly on our memory as the desire not to remember it: the best way to impress anything on our souls and to make them stand guard over it, is to beg them to forget it.

– [C] The following is false: *'Est situm in nobis, ut et adversa quasi perpetua oblivione obruamus, et secunda jucunde et suaviter meminerimus'* [There is within us a capacity for consigning misfortunes to total oblivion, while remembering favourable things with joy and delight].

The following is true: *'Memini etiam quae nolo, oblivisci non possum quae volo'* [I remember things I do not want to remember and I cannot forget things I want to forget].[129] –

[A] Whose advice have I just cited? Why, that of the man [C] *'qui se unus sapientem profiteri sit ausus'* [who, alone, dared to say he was wise];

> [A] *Qui genus humanum ingenio superavit, et omnes*
> *Praestrinxit stellas, exortus uti aetherius sol.*

[who soared above human kind by his genius and who, like the Sun rising in heaven, obscured all the stars.]

Emptying and stripping memory is, surely, the true and proper road to ignorance. [C] *'Iners malorum remedium ignorantia est'* [Ignorance is an artless remedy for our ills].[130]

[A] We find several similar precepts permitting us, when strong and lively Reason cannot suffice, to borrow the trivial pretences of the vulgar, provided that they make us happy or provide consolation. Those who cannot cure a wound are pleased with palliatives which deaden it. If philosophers could only find a way of adding order and constancy to a life which was maintained in joy and tranquillity by weakness and sickness of judgement, they would be prepared to accept it. I do not think they will deny me that.

> *Potare et spargere flores*
> *Incipiam, patiarque vel inconsultus haberi!*

[I may appear silly, but I am going to start drinking and strewing flowers about!][131]

---

129. Cicero, *Tusc. disput.*, III, vi, 12; III, xv, 33; *De fin.*, II, xxxii, 105 (citing, in translation, Euripides' *Andromeda*); I. xvii, 57; II, xxxii, 104 – and contexts. The Italian verse is otherwise unknown.
130. Epicurus (in Cicero, *De fin.*, II, iii, 7 and in Lucretius, III, 1043–4); Seneca (the dramatist) *Oedipus*, III, 17.
131. Horace, *Epist.* I, v, 14.

You would find several philosophers agreeing with Lycas: he was a man of very orderly habits, living quietly and peaceably at home; he failed in none of the duties he owed to family and strangers; he guarded himself effectively from harm; however, some defect in his senses led him to imprint a mad fantasy on his brain: he always thought he was in the theatre watching games, plays and the finest comedies in the world. Being cured of this corrupt humour, he nearly took his doctors to court to make them restore those sweet fantasies[132] to him:

> pol! me occidistis, amici,
> Non servastis, ait, cui sic extorta voluptas,
> Et demptus per vim mentis gratissimus error.

['You have killed me, my friends, not cured me,' he said. 'You have wrenched my pleasure from me and taken away by force that most delightful wandering of my mind.']

Thrasilaus, son of Pythodorus, had a similar mad fantasy; he came to believe that all the ships sailing out of the port of Piraeus or coming in to dock there were working for him alone. When good fortune attended their voyages he rejoiced in it and welcomed them with delight. His brother Crito brought him to his senses, but he sorely missed his former condition, which had been full of happiness, not burdened by troubles.

A line of Ancient Greek poetry says 'There is great convenience in not being too wise': Ἐν τῷ φρονεῖν γὰρ μηδὲν ἥδιστος βίος. So does Ecclesiastes: 'In much wisdom there is much sadness, and he that acquireth knowledge acquireth worry and travail.'

Philosophy in general agrees[133] that there is an ultimate remedy to be prescribed for every kind of trouble: namely, ending our life if we find it intolerable. [C] 'Placet? Pare. Non placet? Quacunque vis, exi.' [All right? Then put up with it. Not all right? Then out you go, any way you like.] – 'Pungit dolor? Vel fodiat sane. Si nudus es, da jugulum; sin tectus armis Vulcaniis, id est fortitudine, resiste.' [Does it hurt? Is it excruciating? If you are defenceless, get your throat cut; if you are armed with the arms of

132. '88: vain fantasies. (What follows is virtually all from Erasmus' adage, In nihil sapiendo jucundissima vita (including references to Horace, Epist., II, ii, 138; Sophocles, Ajax, 554; Ecclesiastes I:18). Also, Erasmus, In Praise of Folly, XXXVII.)
133. '88 onwards: All Philosophy agrees . . . (The remedies of Philosophy are not of course those of revealed religion (which supersedes them when there is a clash). But Christianity welcomes Philosophy. For the usual view, see Melanchthon, On the First Book of the Ethics of Aristotle, 'On the distinction between Philosophy and the Christian Religion', Opera, 1541, IV, 127.)

Vulcan (that is, fortitude) then fight it!] As the Greeks said at their banquets: 'Let him drink or be off!' (*'Aut bibat, aut abeat!'*) – That is particularly apt if you pronounce Cicero's language like a Gascon, changing your 'B's to 'V's: *Aut vivat* – 'Let him *live . . .*'

> [A] *Vivere si recte nescis, decede peritis;*
> *Lusisti satis, edisti satis atque bibisti;*
> *Tempus abire tibi est, ne potum largius aequo*
> *Rideat et pulset lasciva decentius aetas.*

[If you do not know how to live as you should, give way to those who do. You have played enough in bed; you have eaten enough, drunk enough: it is time to be off, lest you start to drink too much and find that pretty girls rightly laugh at you and push you away.]

But what does this consensus amount to, if not to a confession of power-lessness on the part of Philosophy? She sends us for protection not merely to ignorance but to insensibility, to a total lack of sensation, to non-being.

> *Democritum postquam matura vetustas*
> *Admonuit memorem motus languescere mentis,*
> *Sponte sua leto caput obvius obtulit ipse.*

[When mature old age warned Democritus that he was losing his memory and his mental faculties, he spontaneously offered his head to Destiny.]

As Antisthenes said: We need a store of intelligence, to understand; failing that, a hangman's rope. In this connection Chrysippus used to quote from Tyrtaeus the poet: *'Draw near to virtue . . . or to death.'* [C] Crates used to say that love was cured by time or hunger; those who like neither can use the rope. [B] Sextius – the one whom Seneca and Plutarch talk so highly of – gave up everything and threw himself into the study of philosophy; he found his progress too long and too slow, so he decided to drown himself in the sea. In default of learning, he ran to death.

Philosophy lays down the law on this subject in these words: If some great evil should chance upon you – one you cannot remedy – then a haven is always near: swim out of your body as from a leaky boat; only a fool is bound to his body, not by love of life but by fear of death.[134]

134. Seneca, *Epist. moral.* LXX, 15–16 (adapted); Cicero, *Tusc. disput.,* V, xli; Horace, *Epist.,* II, ii, 213; Lucretius, III, 1039 (Lambin, pp. 266–7); Plutarch, *Contre les Stoïques,* 564 CD; Diogenes Laertius, *Lives,* Crates; Plutarch, tr. Amyot, *Comment l'on pourra apparcevoir si l'on amende et profite en l'exercice de la vertu,* 114 EF; for Seneca's praise of Quintus Sextius the Elder, cf. Seneca, *Epist. moral.,* XCVIII, 13.

[A] Just as life is made more pleasant by simplicity, it is also made better and more innocent (as I was about to say earlier on). According to St Paul, it is the simple and the ignorant who rise up and take hold of heaven, whereas we, with all our learning, plunge down into the bottomless pit of hell.[135] I will not linger here over two Roman Emperors, Valentian – a sworn enemy of knowledge and scholarship – and Licinius, who called them a poison and a plague within the body politic;[136] nor over Mahomet who, [C] I am told, [A] forbade his followers to study. What we must do is to attach great weight to the authoritative example of a great man, Lycurgus, as well as to the respect we owe to Sparta, a venerable, great and awe-inspiring form of government, where letters were not taught or practised but where virtue and happiness long flourished. Those who come back from the New World discovered by the Spaniards in the time of our fathers can testify how those peoples, without magistrates or laws, live lives more ordinate and more just than any we find in our own countries, where there are more laws and legal officials than there are deeds or inhabitants.

> Di cittatorie piene e di libelli,
> D'esamine e di carte, di procure,
> Hanno le mani e il seno, e gran fastelli
> Di chiose, di consigli e di letture:
> Per cui le faculta de poverelli
> Non sono mai ne le citta sicure;
> Hanno dietro e dinanzi, e d'ambi ilati,
> Notai procuratori e advocati.

[Their hands and their law-bags are full of summonses, libels, inquests, documents and powers-of-attorney; they have great folders full of glosses, counsels' opinions and statements. For all that, the poor are never safe in their cities but are surrounded, in front, behind and on both sides, by procurators and lawyers.][137]

A later Roman senator meant much the same when he said that the breath of their forebears stank of garlic but inwardly they smelt of the musk of a good conscience; men of his time, on the contrary, were doused in perfume yet inwardly stank of every sort of vice.[138] In other words he agrees with

---

135. H.C. Agrippa, *De Vanitate omnium scientiarum et de excellentia verbi Dei*, 1537, I. St Paul is loosely paraphrased here, not quoted, on Christian Folly.
136. *Idem* (where *Valentian* also appears for *Valentinian*).
137. Ariosto, *Orlando furioso*, XIV, § 84.
138. Varro *apud* Nonius Marcellus, *Opera*, 201, 6.

me: they had ample learning and ability but were very short of integrity. Lack of refinement, ignorance, simplicity and roughness go easily with innocence, whereas curiosity, subtlety and knowledge have falsehood in their train; the main qualities which conserve human society are humility, fear and goodness: they require a soul which is empty, teachable and not thinking much of itself.[139]

In Man curiosity is an innate evil, dating from his origins: Christians know that particularly well. The original Fall occurred when Man was anxious to increase his wisdom and knowledge: that path led headlong to eternal damnation. Pride undoes man; it corrupts him; pride makes him leave the trodden paths, welcome novelty and prefer to be the leader of a lost band wandering along the road to perdition; prefer to be a master of error and lies than a pupil in the school of Truth, guided by others and led by the hand along the straight and beaten path. That is perhaps what was meant by that old Greek saying, that Superstition follows Pride and obeys it as a father: *ἡ δεισιδαιμονία κατάπερ πατρὶ τῷ τυφῷ πείτεται*.[140]

[C] 'Oh Pride! How thou dost trammel us!' When Socrates was told that the god of Wisdom had called him wise, he was thunderstruck; he ransacked his mind and shook himself out but could find nothing to base this divine judgement upon. He knew other men who were as just, temperate, valiant and wise as he was: others he knew to be more eloquent, more handsome, more useful to their country. He finally concluded that, if he was different from others and wiser, it was only because he did not think he was; that his God thought any human who believed himself to be knowledgeable and wise was a singularly stupid animal; that his best teaching taught ignorance and his best wisdom was simplicity.[141]

[A] The Word of God proclaims that those of us who think well of ourselves are to be pitied: *Dust and ashes* (it says to them) *what have ye to boast about?* And elsewhere: *God maketh man like unto a shadow: who will judge him when the light departeth and the shadow vanisheth?*[142]

In truth we are but nothing.

It is so far beyond our power to comprehend the majesty of God that

---

139. '88: not thinking *anything* of itself.
140. Genesis; then Socrates *apud* John Stobaeus, *Apophthegmata*, Sermo XXII (a saying inscribed in Montaigne's library).
141. Plato, *Apology for Socrates*, 6.
142. Sayings inscribed on Montaigne's library; the first from Ecclesiasticus 10:9; the second, ascribed to 'Eccl. 7', may perhaps be a paraphrase of Ecclesiastes 7:1 (Vulgate) or a loose rendering of the Septuagint.

the very works of our Creator which best carry his mark are the ones we least understand. To come across something unbelievable is, for Christians, an opportunity to exercise belief; it is all the more reasonable precisely because it runs counter to human reason.   [B] If it were reasonable, it would not be a miracle: if it followed a pattern, it would not be unique.   [C] *'Melius scitur deus nesciendo'* [God is best known by not knowing], said St Augustine. And Tacitus says, *'Sanctius est ac reverentius de actis deorum credere quam scire'* [It is more holy and pious to believe what the gods have done than to understand them].[143] Plato reckons that there is an element of vicious impiety in inquiring too curiously about God and the world or about first causes. As for Cicero, he says: *'Atque illum quidem parentem hujus universitatis invenire difficile; et, quum jam inveneris, indicare in vulgus, nefas'* [It is hard to discover the Begetter of this universe; and when you do discover him, it is impious to disclose him to the populace].[144]

[A] We confidently use words like might, truth, justice. They are words signifying something great. But what that 'something' is we cannot see or conceive.   [B] We say that God 'fears', that God 'is angry', that God 'loves':

> *Immortalia mortali sermone notantes.*
>
> [Denoting immortal things in mortal speech.][145]

But they are disturbances and emotions which in any form known to us find no place in God. Nor can we imagine them in forms known to him.   [A] God alone can know himself; God alone can interpret his works.   [C] And he uses improper, human, words to do so, stooping down to the earth where we lie sprawling.

Take Prudence; that consists in a choice between good and evil; how can that apply to God? No evil can touch him. Or take Reason and Intelligence, by which we seek to attain clarity amidst obscurity; there is nothing obscure to God. Or Justice, which distributes to each his due and which was begotten for the good of society and communities of men; how can that exist in God? And what about Temperance? It moderates bodily pleasures which have no place in the Godhead. Nor is Fortitude in the face of pain, toil or danger one of God's qualities: those three things are

---

143. Augustine, *De ordine,* II, xvi, and Tacitus, *De Moribus Germanorum,* XXXIV, both cited in Justus Lipsius, *Politicorum sive Civilis Doctrinae,* 1584, I, ii.
144. Plato, *Laws,* VII (Ficino, 1546, p. 837); tr. Cicero, *Timaeus,* II (*in Fragmentis*).
145. Lucretius, V, 121 (Lambin, pp. 383–4).

unknown to him. That explains why Aristotle held that God is equally as free from virtue as from vice. *'Neque gratia neque ira teneri potest, quod quae talia essent, imbecilla essent omnia'* [He can experience neither gratitude nor anger; such things are found only in the weak].[146]

[A] Whatever share in the knowledge of Truth we may have obtained, it has not been acquired by our own powers. God has clearly shown us that: it was out of the common people that he chose simple and ignorant apostles to bear witness of his wondrous secrets; the Christian faith is not something obtained by us: it is, purely and simply, a gift depending on the generosity of Another. Our religion did not come to us through reasoned arguments or from our own intelligence: it came to us from outside authority, by commandments. That being so, weakness of judgement helps us more than strength; blindness, more than clarity of vision. We become learned in God's wisdom more by ignorance than by knowledge. It is not surprising that our earth-based, natural means cannot conceive knowledge which is heaven-based and supernatural; let us merely bring our submissiveness and obedience: 'For it is written: I will destroy the wisdom of the wise and bring to nothing the prudence of the prudent. Where is the wise? Where is the scribe? Where is the disputer of this world? Hath God not made the wisdom of this world like unto the foolishness as of beasts? For seeing that the world, through wisdom, knew not God, it pleased God through the vanity of preaching to save them that believe.'[147]

But is it within the capacity of Man to find what he is looking for? Has that quest for truth which has kept Man busy for so many centuries actually enriched him with some new power or solid truth? Now, at last, it is time to look into that question.

I think Man will confess, if he speaks honestly, that all he has gained from so long a chase is knowledge of his own weakness.[148] By long study we have confirmed and verified that ignorance does lie naturally within us. The truly wise are like ears of corn: they shoot up and up holding their heads proudly erect – so long as they are empty; but when, in their maturity, they are full of swelling grain, their foreheads droop down and they show humility. So, too, with men who have assayed everything, sounded everything; within those piles of knowledge and the profusion of so many diverse things, they have

---

146. Cicero, *De nat. deorum*, III, xv, 38, and quoting from I, xvii, 45. (Aristotle, *Nicomachaean Ethics*, VII, i. 1–2 may be in mind also.)
147. I Corinthians, 1:19–21, a key text for Christian Folly (cf. Erasmus, *In Praise of Folly*, LXV).
148. '88: his own *vileness and his* weakness . . .

found nothing solid, nothing firm, only vanity. They then renounce arrogance and recognize their natural condition.[149]

[C]   For that is what Velleius reproached Cotta and Cicero with: they had learned from Philo that they had learned nothing.[150]

When one of the Seven Sages of Greece, Pherecides, lay dying, he wrote to Thales saying, 'I have commanded my family, once they have buried me, to send you all my papers; if you and the other Sages are satisfied with them, publish them; if not, suppress them: they contain no certainties which satisfy me. I make no claim to know what truth is nor to have attained truth. Rather than lay subjects bare, I lay them open.'[151]

[A]   The wisest man that ever was, when asked what he knew, replied that the one thing he did know was that he knew nothing.[152] They say that the largest bit of what we do know is smaller than the tiniest bit of what we do not know; he showed that to be true. In other words, the very things we think we know form part of our ignorance, and a small part at that.   [C]   We know things in a dream, says Plato; we do not know them as they truly are.[153]

'Omnes pene veteres nihil cognosci, nihil percipi, nihil sciri posse dixerunt; angustos sensus, imbecillos animos, brevia curricula vitae' [Virtually all the Ancients say that nothing can be understood, nothing can be perceived, nothing can be known; our senses are too restricted, our minds are too weak, the course of our life is too short].[154]

[A]   Cicero himself, who owed such worth as he had to his learning, was said by Valerius to have begun to think less of literary culture in his old age.[155]   [C]   And even while he was still writing he felt bound to no sect; he followed the teachings of this school or that as seemed to him most probable, remaining always within that Doubt taught by the Academy: *'Dicendum est, sed ita ut nihil affirmem: quaeram omnia, dubitans plerumque et mihi diffidens'* [I have to write, but in such a way as to vouch for nothing; I shall always be seeking, mostly doubting, rarely trusting myself].[156]

---

149. Plutarch, *Comment l'on peut apparcevoir si l'on amende et profite en l'exercice de la vertu*: 116 EF.
150. Cicero, *De nat. deorum*, I, vii, 17.
151. Diogenes Laertius, *Lives*, Pherecides, I, 122.
152. Socrates; cf. Plato, *Apology for Socrates*, Lucretius, ed. Lambin, 309, etc.
    '88: ever was *(and who had no other just cause to be called wise apart from this saying)*, when . . .
153. Plato, *Politicus*, 19, 277.
154. Cicero, *Academica*, I, xii, 44.
155. According to H. C. Agrippa, *De Vanitate omnium scientiarum*, I.
156. Cicero, *De divinatione*, II, iii, 8.

[A] It would be too easy a game if I limited myself to the ordinary run of men considered *en masse*; I would be justified in doing so by Man's curious convention that votes are not to be weighed but counted. But let us leave aside the ordinary people,

> *Qui vigilans stertit,*
> *Mortua cui vita est prope jam vivo atque videnti;*

[Who snore whilst they are awake and whose lives are dead even while they live and keep their eyes open;][157]

they have no self-awareness; they never judge themselves and let most of their natural faculties stand idle. I want to take Man in his highest state. Let us consider only that tiny number of outstanding, handpicked men who are born with a fine natural endowment peculiar to themselves and who then take care to strengthen and sharpen it by skill and study; by such means they raise it to the highest point [C] of wisdom [A] that it can attain to. They mould their souls in ways which keep them open on every side to every tendency; they assist their souls with the help of every appropriate outside support; they adorn them and enrich them with every advantage which they can discover both within and beyond this world. The highest possible form of human nature finds its home in such men. These are men who have given laws and constitutions to the world; it is their arts and sciences which have taught the world; so, too, the example of their astounding moral integrity. I will take account of the testimony and experience only of men such as these. Let us see how far they got and what they concluded. They form a fellowship such that any ills and defects found in them can confidently be accepted by the world as inherent ones.

Whoever sets out to find something eventually reaches the point where he can say that he has found it, or that it cannot be found, or that he is still looking for it. The whole of Philosophy can be divided into these three categories; her aim is to seek true, certain knowledge.

Peripatetics, Epicureans, Stoics[158] and others think they have discovered it. They founded the accepted disciplines and expounded their knowledge as certainties.

Clitomachus, Carneades and the Academics despaired of their quest; they conclude that Truth cannot be grasped by human means. Their conclusion

---

157. Lucretius, III, 1048; 1046 (Lambin pp. 266–8).
158. '88: knowledge. *Aristotle, Epicurus*, Stoics . . .

is one of weakness, of human ignorance. This school has had the greatest number of adherents and some of the noblest.[159]

As for Pyrrho and the other Sceptics or Ephectics,   [C]   (whose teachings many of the Ancients derived from Homer, the Seven Sages, Archilochus and Euripides, and associated with Zeno, Democritus and Xenophanes),   [A]   they say they are still looking for Truth. They hold that the philosophers who think they have found it are infinitely wrong. They go on to add that the second category – those who are quite sure that human strength is incapable of reaching truth – are overbold and vain. To determine the limits of our powers and to know and judge the difficulty of anything whatsoever constitutes great, even the highest, knowledge. They doubt whether Man is capable of it.

> *Nil sciri quisquis putat, id quoque nescit*
> *An scire possit quo se nil scire fatetur.*

[Any man who thinks that 'nothing can be known', does not know whether he can know even that thing by which he asserts that he knows nothing.][160]

Ignorance which is aware of itself, judges itself, condemns itself, is not complete ignorance: complete ignorance does not even know itself. Consequently the professed aim of Pyrrhonians is to shake all convictions, to hold nothing as certain, to vouch for nothing. Of the three functions attributed to the soul (cogitation, appetite and assent) the Sceptics admit the first two but keep their assent in a state of ambiguity, inclining neither way, giving not even the slightest approbation to one side or the other.

[C]   It was by gesture that Zeno illustrated his conception of the three functions of the soul: a hand stretched out open meant probability; half-closed, with the fingers bent over, meant assent; clenched, it meant understanding; with the other hand pressing it tighter still, it meant knowledge.[161]

[A]   Now the Pyrrhonians make their faculty of judgement so unbending and upright that it registers everything but bestows its assent on nothing. This leads to their well-known *ataraxia*: that is a calm, stable rule of life, free from all the disturbances (caused by the impress of opinions, or

159. Sextus Empiricus, *Hypotyposes*, I, i, I; xix, xxii, xxiii. With the opening words of this book Montaigne begins his first major borrowing from one of the main sources of scepticism.
160. Lucretius, IV, 469–70 (Lambin, p. 308). With these words begin Lucretius' dense criticism of scepticism. Montaigne borrows much from him and the commentary of Lambin.
161. Cicero, *Acad.: Lucullus*, II, xlvii, 144–45.

of such knowledge of reality as we think we have) which give birth to fear, acquisitiveness, envy, immoderate desires, ambition, pride, superstition, love of novelty, rebellion, disobedience, obstinacy and the greater part of our bodily ills. In this way, they even free themselves from passionate sectarianism, for their disputes are mild affairs and they are never afraid of the other side having its say. When they assert that heavy things tend to fall downwards, they would be most upset if you believed them. They want you to contradict them in order to achieve their end: doubt and suspense of judgement. They only put forward propositions of their own in order to oppose the ones they think we believe in. Accept theirs, and they will gladly maintain the opposite. It is all the same to them: they take no sides. If you maintain that snow is black, they will argue that it is, on the contrary, white. If you say that it is neither, their task is to say that it is both. If you conclude that you definitely know nothing, they will maintain that you do know something. Yes, and if you present your doubt as axiomatic, they will challenge you on that too, arguing that you are not in doubt, or that you cannot decide for certain and prove that you are in doubt. This is doubt taken to its limits; it shakes its own foundations; such extremes of doubt separate them completely from many other theories including those which in many ways do indeed teach doubt and ignorance.[162]

[B] If some Dogmatists call green what others call yellow, why, they ask, cannot they doubt both of them? Can there be any proposition capable of acceptance or rejection which it is not right to consider ambiguous?

Other people are prejudiced by the customs of their country, by the education given them by their parents or by chance encounter: normally, before the age of discretion, they are taken by storm and, without judgement or choice, accept this or that opinion of the Stoic or Epicurean sects. There they stay, mortgaged, enslaved, caught on a hook which they cannot get off – [C] *'ad quamcumque disciplinam velut tempestate delati, ad eam tanquam ad saxum adhaerescunt'* [they cling to any old teaching, like sailors washed up on a rock]. [B] But why should people like these not also be allowed their freedom, making up their own minds without bonds and slavery? [C] *'Hoc liberiores et solutiores quod integra illis est judicandi potestas'* [They are all the more independent and free in that they enjoy the full power of judgement].[163] There is some advantage, surely, in being detached

162. Sextus Empiricus, *Hypotyposes*, I, XII, 30; XIII, 33; cf. Rabelais, *Tiers Livre*, TLF, XXXVI.
163. Cicero, *Acad.: Lucullus*, II, iii, 8–9, the source of both quotations.

from the reins of the Necessity which curb others.   [B]   Is it not better to remain in doubt, than to get entangled in the many errors produced by human fantasy? Is it not better to postpone one's adherence indefinitely than to intervene in factions, both quarrelling and seditious?

[C]   'What ought I to choose?' – 'Anything you wish, so long as you choose something.' A daft enough reply! Yet it seems to be the one reached by every kind of dogmatism which refuses us the right not to know what we do not know.

[B]   Try siding with the school enjoying majority support: but it will never be safe enough: to defend it you will have to attack opponents by the hundreds. Is it not better to keep out of the fray altogether? You allow yourself to espouse, like honour and dear life, Aristotle's beliefs about the eternity of the soul; to do that you must reject and contradict Plato. In that case, why should others be forbidden simply to go on doubting?[164]

[C]   Panaetius was legally permitted to suspend judgement about dreams, oracles, prophecies and divination by entrails; yet his school, the Stoics, never doubted them. Why cannot a wise man dare to doubt anything and everything, if Panaetius could dare to doubt doctrines which were taught by his own masters and founded on the common consent of the school he adhered to and whose doctrines he claimed to profess?

[B]   If it is a child who makes the judgement, he does not know enough about the subject: if it is a learned man, then he has made up his mind already! – Pyrrhonians have given themselves a wonderful strategic advantage by shrugging off the burden of self-defence. It does not matter who attacks them, as long as somebody does. Anything serves their purpose: if they win, your argument is defective; if you do, theirs is. If they lose, they show the truth of Ignorance; if you lose, you do. If they can prove that nothing is known: fine. If they do not succeed in proving it, that is fine too.   [C]   '*Ut quum in eadem re paria contrariis in partibus momenta inveniuntur, facilius ab utraque parte assertio sustineatur*' [So that by finding equally good cases, for and against, on the very same subject, it is easier to suspend one's judgement about either side].[165]

They make it their pride to be far more ready to find everything false than anything true and to show that things are not, rather than that they are. They prefer to proclaim what they do not believe, rather than what they do.   [A]   Their typical phrases include: 'I have settled nothing'; 'It is no more this than that'; 'Not one rather than the other'; 'I do not

164. For Plato, *Forms* are created: for Aristotle, they exist from all eternity.
165. Cicero, *Acad.: Lucullus*, II, xxxiii, 107, and I, xii, 45–6.

understand'; 'Both sides seem equally likely'; 'It is equally right to speak for and against either side'. [C] To them, nothing seems true which cannot also seem false. [A] They have sworn loyalty to the word ἐπέχω: 'I am in suspense'; I will not budge.[166]

These sayings, and others like them, form refrains which lead to a pure, whole, complete suspension of their judgement, which is kept permanently in abeyance. They use their reason for inquiry and debate but never to make choices or decisions. If you can picture an endless confession of ignorance, or a power of judgement which never, never inclines to one side or the other, then you can conceive what Pyrrhonism is.

I have tried to explain this notion as clearly as I can, because many find it hard to grasp, and its very authors present it somewhat diversely and rather obscurely.

Where morals are concerned, they conform to the common mould. They find it appropriate to yield to natural inclinations, to the thrust and constraints of their emotions, to established laws and customs and to the traditional arts.[167] [C] *'Non enim nos Deus ista scire, sed tantummodo uti voluit'* [For God did not want us to know such things: merely to make use of them]. [A] They let their everyday activities be guided by such considerations, neither assenting nor adhering to anything. That is why I cannot square with these conceptions what is told about Pyrrho himself. They[168] describe him as emotionless and virtually senseless, adopting a wild way of life, cut off from society, allowing himself to be bumped into by wagons, standing on the edge of precipices and refusing to conform to the law. That goes well beyond his teaching. He[169] was not fashioning a log or a stone but a living, arguing, thinking man, enjoying natural pleasures and comforts of every sort and making full use of all his parts, bodily as well as spiritual − [C] in, of course, a right and proper way. [A] Those false, imaginary and fantastic privileges usurped by Man, by which he claims to profess, arrange and establish the truth, were renounced and abandoned by Pyrrho, in good faith.

− [C] Yet there is not one single school of philosophy which is not

166. These and similar aphorisms from Sextus Empiricus were inscribed in Montaigne's library; *Hypotyposes*, I, 6, 21, 23, 26 and 27.
167. Sextus Empiricus, *Hypotyposes*, I, xi, 23–24, followed by quotation from Cicero, *De divinat.*, I, xviii, 35.
168. '88: himself. *Laertius in the Life of Pyrrho says (and both Lucianus and Aulus Gellius incline the same way)* describe him ... (Laertius' *Life* was printed in Montaigne's copy of Sextus.)
169. Major borrowings follow from Cicero, *Acad.: Lucullus*, II, xxxi, 99–101.

forced to allow its Sage (if he wishes to live) to accept a great many things which he cannot understand, perceive or give his assent to. Say he boards a ship. He carries out his design, not knowing whether it will serve his purpose; he assumes the vessel to be seaworthy, the pilot to be experienced and the weather to be favourable. Such attendant details are, of course, merely probable: he is obliged to let himself be guided by appearances, unless they are expressly contradicted. He has a body. He has a soul. He feels the impulsions of his senses and the promptings of his spirit. He cannot find within himself any sign specifically suggesting that it be appropriate for him to make an act of judgement: he realizes he must not bind his consent to anything, since something false may have every appearance of particular truth. Despite all this, he never fails to do his duty in this life, fully and fittingly.

How many disciplines are there which actually profess to be based on conjecture rather than on knowledge, and which, being unable to distinguish truth from falsehood, merely follow what seems likely? Pyrrhonians say that truth and falsehood exist: within us we have means of looking for them, but not of making any lasting judgement: we have no touchstone.

We would be better off if we dropped our inquiries and let ourselves be moulded by the natural order of the world. A soul safe from prejudice has made a wondrous advance towards peace of mind. People who judge their judges and keep accounts of what they do fail to show due submissiveness. Among people who are amenable to the legitimate teachings of religion and politics, there are more simple and uninquisitive minds than minds which keep a schoolmasterly eye on causes human and divine. –

[A] No system discovered by Man has greater usefulness nor a greater appearance of truth [than Pyrrhonism] which shows us Man naked, empty, aware of his natural weakness, fit to accept outside help from on high: Man, stripped of all human learning and so all the more able to lodge the divine within him, annihilating[170] his intellect to make room for faith; [C] he is no scoffer, [A] he holds no doctrine contrary to established custom; he is humble, obedient, teachable, keen to learn – and as a sworn enemy of heresy he is freed from the vain and irreligious opinions introduced by erroneous sects. [B] He is a blank writing-tablet, made ready for the finger of God to carve such letters on him as he pleases. The more we refer ourselves to God, commit ourselves to him and reject

170. '88: greater *probability* nor a greater appearance . . .
    '88: the divine *instruction and belief*, annihilating . . .

ourselves, the greater we are worth. Ecclesiastes says: 'Accept all things in good part, just as they seem, just as they taste, day by day. The rest is beyond thy knowledge':[171] [C] *'Dominus novit cogitationes hominum, quoniam vanae sunt'* [The Lord knoweth the thoughts of men, that they are vanity].

[A] And so two out of the three generic schools of Philosophy make an express profession of doubt and ignorance; it is easy to discover that most who belonged to the third school, the Dogmatists, put on an assured face merely because it looks better. They did not really think that they had established any certainties, but wanted to show us how far they had advanced in their hunt for Truth, [C] *'quam docti fingunt, magis quam norunt'* [which the learned feign rather than know]. When Timaeus had to reveal to Socrates what he knew about the Gods, the world and mankind, he determined to speak of such things as one man to another: it would be enough if the reasons he gave had as much probability as anyone else's, since precise reasons were neither in his grasp nor in the grasp of any mortal man.[172]

One of the followers of his school imitated him in these words: *'Ut potero, explicabo: nec tamen, ut Pythius Apollo, certa ut sint et fixa, quae dixero; sed, ut homunculus, probabilia conjectura sequens'* [I will unravel things as best I may. What I shall say is neither fixed nor certain: I am no Pythian Apollo; I am a little man seeking the probable through conjecture]. Yet he was merely treating a common, not supernatural theme: contempt for death! In another place he translates Timaeus directly from Plato: *'Si forte, de deorum natura ortuque mundi disserentes, minus id quod habemus animo consequimur, haud erit mirum. Aequum est enim meminisse et me qui disseram, hominem esse, et vos qui judicetis; ut, si probabilia dicentur, nihil ultra requiratis'* [If we are unable to achieve what we have in mind to do when we set out to treat the nature of the Gods and the origin of the world, that will not be surprising. It is right to remember that both I who am speaking and you who are judging are men. If what I say is probable, you can demand nothing more].[173]

[A] Aristotle regularly piles up many different opinions and beliefs, so as to evaluate his own against them. He shows how much farther he has

171. This 'quotation' from Ecclesiastes figures in Latin in Montaigne's library as *'Fruere jucunde praesentibus, caetera extra te'*. (Its actual source is unknown.) Then, [C], Psalm 94 (93): II.
172. Plato, *Timaeus*, 29 (Ficino, p. 705). The Latin quotation is from Livy, *Hist.*, xxvi, 22, 14. A marginal note authorized by Marie de Gournay reads, 'Perhaps Seneca in *Epistles*' – a wrong guess.
173. Cicero: *Tusc. disput.*, I, ix.; *Timaeus*, III (*in Fragmentis*).

gone and how much nearer he has approached to probability – Truth not being something we should accept on authority or from the testimony of others. [C] (That is why Epicurus scrupulously avoided citing such evidence in his writings.) [A] Aristotle is the Prince of the Dogmatists; and yet it is from him we learn that greater knowledge leads to further doubt. You can often find him hiding behind a deliberate obscurity,[174] so deep and impenetrable that you cannot make out what he meant. In practice it is Pyrrhonism cloaked in affirmation.

[C] Just listen to this assertion of Cicero, explaining to us another's notion by his own: '*Qui requirunt quid de quaque re ipsi sentiamus, curiosius id faciunt quam necesse est. Haec in philosophia ratio contra omnia disserendi nullamque rem aperte judicandi, profecta a Socrate, repetita ab Arcesila, confirmata a Carneade, usque ad nostram viget aetatem. Hi sumus qui omnibus veris falsa quaedam adjuncta esse dicamus, tanta similitudine ut in iis nulla insit certe judicandi et assentiendi nota.*' [Those who want to know what my personal opinions are on each of these subjects are more inquisitive than they ought to be. Up to now it has been a principle of philosophy to argue against anything but to decide nothing. This principle was established by Socrates; Arcesilaus repeated it; Carneades strengthened it further. I am one of those who hold that there is, in all truths, an admixture of falsehood so like Truth that there is no way of deciding or determining anything whatever with complete certainty.][175]

[B] Not only Aristotle but most philosophers aim at being hard to understand; why? – if not to emphasize the vanity of their subject-matter and to give our minds something to do! Philosophy is a hollow bone with no flesh on it: are they providing us with a place to feed in, where we can chew on it?[176]

[C] Clitomachus maintained that he could not tell from Carneades' books what his opinions were.[177] [B] That is why Epicurus avoided perspicuity in his writings and why Heraclitus was surnamed Σκοτεινὸς, 'Dark'. Difficulty is a coin [C] which the learned conjure with, so as not to reveal the vanity of their studies and [B] which human stupidity is keen to accept in payment.

---

174. '88: obscurity, (*as for example on the subject of the immortality of the soul*) so deep . . .
175. Cicero, *De nat. deorum*, I, v, 10 (adapted).
176. Plato called a dog 'philosophical' since it strives to get at the marrow of a bare bone (*Republic*, III, 375E; cf. Rabelais, *Gargantua*, TLF, p. 13).
177. Cicero, *Acad.: Lucullus*, II, xlv, 139.

> *Clarus, ob obscuram linguam, magis inter inanes,*
> *Omnia enim stolidi magis admirantur amantque*
> *Inversis quae sub verbis latitantia cernunt.*

[Clear was his fame, especially among the empty-headed, simply because his language lacked clarity: for stupid people are filled with awe and wonder when they find ideas wrapped up in words turned inside out.][178]

[C] Cicero reproached some of his friends with being accustomed to give more time than they were worth to such subjects as astrology, law, dialectic and geometry: it kept them away from the more useful and honourable of life's duties. The Cyrenaic philosophers held physics and dialectic in equal contempt. At the very beginning of his books on the *Republic* Zeno pronounced all liberal disciplines to be useless. [A] Chrysippus said that what Plato and Aristotle wrote about logic must have been written for sport or as an exercise; he could not believe that they had anything serious to say on so empty a subject. [C] Plutarch makes a similar remark about metaphysics.[179] [A] Epicurus would have spoken similarly about rhetoric, grammar, [C] poetry, mathematics and all subjects of study other than physics – [A] and Socrates, about every one of them, with the sole exception of the study of how we should behave in this life. [C] Whatever question Socrates was asked, he first made the speaker give a detailed account of his way of life, both present and past; he made that the basis of his inquiries and judgements, believing as he did that any other approach was secondary to that and superfluous.

'*Parum mihi placeant eae literae quae ad virtutem doctoribus nihil profuerunt*' [I take no pleasure in the kind of writings which do not increase the virtue of those who teach them].[180]

[A] Learning[181] itself has despised most disciplines, but men have thought it not inappropriate to train and entertain their minds even by studying subjects where nothing solid is to be gained. Moreover, some have classified Plato as a Dogmatist; some, as a Doubter; others as both, depending on the subject.

[C] Socrates, who takes the lead in the *Dialogues*, always asks questions

---

178. Lucretius, I, 639–42, incorporating matter in Lambin, p. 63 (Vitruvius, Cicero, etc.).
179. Cicero, *De officiis*, I, vi, 19: Diogenes Laertius, *Lives*: Aristippus, II, 91; Zeno, VII, 32; Plutarch, tr. Amyot, *Life of Alexander*.
180. Sallust *apud* Justus Lipsius, *Politicorum*, 1584, I, 10.
181. '88: Learning *and philosophy have* despised . . . (Sextus Empiricus, *Hypotyposes*, I, XXXI, 221.)

designed to provoke discussion: he is never satisfied and never reaches any conclusion. He says that the only thing he knew how to do was to make objections.

All schools of philosophy derive their foundations from Homer, but it was a matter of indifference to him what direction we then took; to show that, he gave equally good foundations to all of them. They say that ten distinct schools sprang from Plato. And indeed, as I see it, if his teachings are not faltering and unaffirmative, then I do not know whose are![182]

Socrates said that midwives were *Sage-women* who stop producing children of their own once they help others to do so; when, therefore, the gods conferred the title *Sage* on him, he too gave up his capacity for producing brain-children of his own by acts of manly love, in order to encourage and help other men to deliver theirs: he opened the genitals of their minds, lubricated the passages and made it easier for their child to issue forth; he then made an appreciation of that child, washed it, nursed it, strengthened it, swaddled it up and circumcised it. He used and exercised his own ingenuity: the others faced the perils and the risks.[183]

[A] What I said just now is true of most other philosophers in the third category, [B] as the Ancients already noted in the writings of Anaxagoras, Democritus, Parmenides, Zenophanes and others: [A] their substance induces doubt; their purpose is inquiry rather than instruction, even though, in their works, they do at times interlard[184] [C] their style with Dogmatic cadences. Is that not equally true of both Seneca and Plutarch? Go into it closely and you see they are constantly talking from different points of view. As for those jurisconsults whose task it is to harmonize the various legal authorities, they first ought to harmonize each authority with himself.

Plato seems to me to have quite knowingly chosen to treat philosophy in the form of dialogues: he was better able to expound the diversity and variety of his concepts by putting them appropriately into the mouths of divers speakers. Variety of treatment is as good as consistency. Better in fact: it means being more copious and more useful.

---

182. Cf. Seneca, *Epist*, LXXXVIII; Diogenes Laertius, *Lives*, Socrates (*ad fin.*).
183. Plato, *Theaetetus*, 150–1.
  '95: *circumscribed* for *circumcised*.
184. '88 (in place of [C]) they do interlard them *often with traits, dogmatist in form. In whom can one see that more clearly than in our Plutarch? How differently he treats the same subjects! How many times does he present us with two or three incompatible causes and divers reasons for the same subject, without selecting the one we ought to follow?* What else can that refrain mean . . .

Let us take one example from our own society. The highest degree of dogmatic and conclusive speaking is reached in parliamentary rescripts. Of the judicial decrees which French Parliaments hand down to the people, the ones which are most exemplary (and the most proper to encourage the respect which is rightly due to such high office, mainly on account of the ability of those who exercise it) do not draw their beauty from their decisions as such. Decisions are everyday affairs, common to all judges. Their beauty lies in the disquisitions and that pursuit of varied and opposing arguments which legal matters can so well accommodate.

When philosophers find fault with each other, their widest field of action lies in the internal contradictions and inconsistencies which entangle them all – either deliberately (so as to show the vacillations of the human mind over any subject whatever) or else quite unintentionally, because all matters are shifting and elusive.

[A] What else can that refrain mean: 'In slippery, shifting places, let us suspend our judgement'? For, as Euripides said, 'The works of God, in divers ways, perplex us,'[185] [B] which is similar to the words which Empedocles strewed throughout his books when he was shaken as by divine mania and compelling truth: 'No, no! We feel nothing: we know nothing! All things are hidden from us: we can determine the nature of nothing whatsoever,' [C] words which conform to that holy saying: *'Cogitationes mortalium timidae et incertae adinventiones nostrae et providentiae'* [For the thoughts of mortal men are timorous, and our devices and foresight prone to fail].[186]

[A] We ought not to find it strange that people who despair of the kill should not renounce the pleasure of the hunt: study is, in itself, a delightful occupation, so delightful that, among the forbidden pleasures which need to be held on a tight rein, the Stoics include pleasure arising from exercising the mind.[187] [C] They find intemperance in knowing too much.

[A] Democritus ate some figs which tasted of honey. He at once began to rack his brains to try and explain this unusual sweetness. He was about to abandon his dinner and set out to trace and examine the place where the figs had been picked, when his servant-girl heard the cause of the

185. Cited from Plutarch, tr. Amyot, *Des oracles qui ont cessé*, 348B: *'Les oeuvres de Dieu en diverses/Façons nous donnent des traverses.'*
186. Cicero, *Acad.: Lucullus*, II, V, 14; Wisdom of Solomon 9:14. Cf. Augustine, *City of God*, XII, 16.
187. '88: mind; *and desire moderation*. Democritus . . . (Seneca, *Epist.*, LXXXVIII, 36.)

commotion and began to laugh; she told him to stop worrying about all that, since she had put the figs in a jar which had previously held honey. He flew into a rage with her because she had deprived him of the chance of finding things out for himself and had robbed his curiosity of something to work on: 'Go away,' he said, 'you have offended me. I shall continue to look for the cause as though it were to be found in Nature.' [C] And he did manage to find some sort of 'true' explanation for a false and imaginary fact!

[A] This story about a great and famous philosopher clearly illustrates that passion for study which keeps us occupied, hunting after things we can never hope to catch. Plutarch relates a similar anecdote about a man who did not want anyone to enlighten him on a subject of doubt, so as not to lose the pleasure of the search – like that other man, who would not allow his doctor to cure a thirst brought on by fever, so as not to lose the pleasure of quenching it! [C] 'Satius est supervacua discere quam nihil' [Better to learn something useless than nothing at all].[188] It is the same with food of all kinds. Sometimes we eat just for pleasure: there are things we eat which are neither nutritious nor sustaining. So too for the pabulum which our spirits draw from erudition: it may be neither nutritious nor sustaining, but it gives great pleasure.

[B] This is how they put it: contemplating Nature supplies good food to the spirit: it replenishes it, helps it to soar aloft, makes it despise low and earthly things by comparing them with heavenly things. It is delightful merely to study great and abstruse subjects: that remains true even of the man who acquires nothing from study except a sense of awe and a fear of making judgements on such matters.[189]

That, in a few words, is what they profess.

An express image of the vanity of such sickly curiosity can be better seen from another example, which philosophers are always honouring themselves by quoting. Eudoxus prayed to the gods, hoping to be allowed to have just one sight of the sun from close at hand, so as to apprehend its shape, grandeur and beauty. Even if it meant being burned alive, he would pay the price. He wanted to learn, at the cost of his life, something he would lose as soon as he had acquired it. For such a brief and fleeting glimpse of knowledge he was prepared to surrender all the knowledge he already had or could later have acquired.[190]

188. Plutarch, tr. Amyot; *Propos de Table*, 368 G-H.; King Philopappus (Plutarch, *loc. cit.*); Seneca, *Epist.*, LXXXVIII, 45.
189. Cicero, *Acad.: Lucullus*, II, xli, 127.
190. Plutarch, tr. Amyot: *Que l'on ne sçauroit vivre selon la doctrine d'Epicurus*, 282H-283A.

[A] I cannot really convince myself that Epicurus, Plato and Pythagoras genuinely wanted us to accept their Atoms, Ideas and Numbers as valid currency. They were too wise to base the articles of their belief on foundations so shaky and so challengeable. Each of these great figures strove to bring some image of light into the dark ignorance of this world; they applied their minds to concepts which had at least some subtle and pleasing appearance of truth, [C] their only proviso being that they could stand up to hostile objections: *'unicuique ista pro ingenio finguntur, non ex scientiae vi'* [such theories are fictions, produced not from solid knowledge but from their individual wits].[191]

[A] One of the Ancients was reproved for not judging philosophy to be of much account yet continuing to profess it; 'That is what being a philosopher means,' he replied.[192] Such men wanted to weigh everything in their mental balances; there is curiosity in all of us: this, they found, was a proper way to keep it occupied. Part of what they wrote was simply designed to meet the social needs of the general public – their accounts of their religion, for example.[193] With that end in view it was reasonable not to strip popularly held opinions of their living feathers. They had no wish to spawn ideas which would disturb the people's obedience to the laws and customs of their land.

[C] When treating religion Plato plays a very open game. Writing in his own name he lays down nothing as certain, but, whenever he acts as Lawgiver, he adopts an assertive professorial style. Even then, he is bold enough to work in a few of his most fantastic notions (which were as useful for convincing the people as they were ridiculous for convincing himself), well aware how receptive our minds are to any impressions, especially to the wildest and most extraordinary ones. That explains why, in the *Laws*, Plato is careful to allow no poetry to be recited in public unless its fables and fictions serve some moral end: it is so easy to impress fancies on the human mind that it is not right to feed minds on useless, harmful lies, when you can feed them on profitable ones. In the *Republic* he says quite bluntly that you must often deceive the people for their own good.[194]

You soon discover that some schools of philosophy were chiefly

191. Marcus Annaeus Seneca, *Suasoriae*, IV.
192. Diogenes, cf. Diogenes Laertius *apud* Guy de Brués, *Dialogues, contre les nouveaux Academiciens, que tout ne consiste point en opinion*, 1557, p. 46.
193. '88: public, their *account of religions*, for example: *for it is not forbidden for us to draw advantage even from a lie, if needs be*. With that . . .
194. Diogenes Laertius, *Lives*, Plato, II, lxxx; Plato, *Republic*, II (end), III (beginning); ibid., V, p. 459, tr. Ficino, p. 591.

concerned to pursue truth, and others – gaining credit thereby – moral usefulness. Our human condition is pitiable: often, the things which strike our imagination as the most true are ones which appear least useful for the purposes of life. Even the most audacious of the schools, the Epicureans, the Pyrrhonians and the New Academy, are constrained in the end to bow to the laws of society.

[A] There are other subjects which philosophers toss to and fro in their sieves, trying to dredge them (whether they deserve it or not) into some appearance of likelihood. Having discovered nothing so profound as really to be worth talking about, they are obliged to forge some weak and insane conjectures of their own, treating them not as bases for truth but for studious exercises.   [C] *'Non tam id sensisse quod dicerent, quam exercere ingenia materiae difficultate videntur voluisse'* [They do not seem to believe what they say, but, rather, to exercise their wits on difficult material].[195]

[A] If you will not take it that way, how else can we explain the obvious inconstancy, diversity and vanity of the opinions produced by such excellent and, indeed, awesome, minds? What can be more vain, for example, than trying to make guesses about God from human analogies and conjectures which reduce him and the universe to our own scale and our own laws, taking that tiny corner of intellect with which it pleases God to endow the natural Man and then employing it at the expense of his Godhead? And since we cannot stretch our gaze as far as the seat of his Glory, are we to drag him down to our corruption and our wretchedness?

Of all the ancient opinions of men touching religion, it seems to me that the most excusable and verisimilitudinous was the one which recognized God as some incomprehensible Power, the Origin and Preserver of all things, of all goodness and of all perfection, who took and accepted in good part, the honour and reverence which human beings rendered him, under any guise, under any name and in any way whatsoever.

> [C]   *Jupiter omnipotens rerum, regumque deumque*
> *Progenitor genitrixque.*

[Almighty Jupiter, Father and Mother of the world, of rulers and of gods.][196]

---

195. Quintilian, II, 17, 4.
196. Valerius Soranus *apud* Augustine, *City of God*, VII, 11.
     '88 (in place of   [C]   ): *For the deities to which men have wished to give a form of their own invention are harmful, full of errors and impiety.* That *is why of* all the religions . . .

Such devotion has always been regarded by Heaven with favour.

All forms of government have profited from their allegiance to it; under it, men and impious deeds have met their just deserts; even pagan histories acknowledge the dignity, order and justice of the portents and oracles manifested in their fabulous religions for the benefit and instruction of men. With such temporal benefits as these God in his mercy may perhaps have deigned to protect those tender principles of rough-and-ready knowledge of Himself which Natural Reason affords us, amid the false imaginings of our dreams. But there are religions Man has forged entirely on his own: they are not only false but impious and harmful.

[A] Of all the religions which St Paul found honoured in Athens, the most excusable, he thought, was the one dedicated to a hidden, 'unknown God'.[197]

[C] Pythagoras closely adumbrated truth when he concluded that any conception we have of that First Cause, of that Being of beings, must be free of limits, restrictions or definitions; it was in fact the utmost striving of our intellect towards perfection, each of us enlarging the concept according to his capacity.

But if Numa really did attempt to make his people's worship conform to this model, tying them to an entirely cerebral religion with no object set up before their eyes and no material elements mixed in with it, then his undertaking could serve no purpose.[198] The human mind cannot stand such wanderings through an infinity of shapeless thoughts: they must be brought together into some definite concept modelled on man. The very majesty of God allows itself to be, in some sense, circumscribed for us within physical limits: God's sacraments are supernatural and celestial, yet they bear signs of our own condition, which is earthly; and we express our adoration in words and duties perceptible to the senses. After all, it is Man who does the believing and the praying.

I shall leave aside other arguments marshalled on this topic; consider the sight of our crucifixes and the piteous chastisement which they portray; the ornaments and moving ceremonial in our churches; the voices so aptly fitted to the reverent awe of our thoughts, and all the stirring of our emotions: you will have a hard time making me believe that such things do

197. Paul's sermon in Acts 17:23: 'I found also an altar with this inscription "TO AN UNKNOWN GOD".' By adding *hidden* Montaigne links this text to God as *Deus absconditus* (Introduction, p. xxx). Even good natural religion requires grace if it is to take root and grow.
198. Plutarch, tr. Amyot, *Life of Numa*.

not set whole nations' souls ablaze with a passion for religion, with very useful results.

[A]   Of all the deities to which bodies have been ascribed (as necessity required during that universal blindness), I think[199] I would have most willingly gone along with those who worshipped the Sun:

> *la lumiere commune,*
> *L'œil du monde; et si Dieu au chef porte des yeux,*
> *Les rayons du Soleil sont ses yeux radieux,*
> *Qui donnent vie à tous, nous maintiennent et gardent,*
> *Et les faicts des humains en ce monde regardent:*
> *Ce beau, ce grand soleil qui nous faict les saisons,*
> *Selon qu'il entre ou sort de ses douze maisons;*
> *Qui remplit l'univers de ses vertus connues;*
> *Qui, d'un traict de ses yeux, nous dissipe les nues:*
> *L'esprit, l'ame du monde, ardant et flamboyant,*
> *En la course d'un jour tout le Ciel tournoyant;*
> *Plein d'immense grandeur, rond, vagabond et ferme;*
> *Lequel tient dessoubs luy tout le monde pour terme;*
> *En repos sans repos; oysif, et sans sejour;*
> *Fils aisné de nature et le pere du jour.*

[. . . the Common Light, the Eye of the World; if God himself has eyes they are radiant ones made of the Sun's rays which give life to all, protect and guard us men, gazing down upon our actions in this world; this fair, this mighty Sun who makes the seasons change according to his journey through his dozen Mansions; who floods the earth with his acknowledged power; who, with a flicker of his eye disperses clouds; the Spirit and Soul of the World, ardent and aflame, encompassing the world in the course of one single day; full of immense grandeur, round, wandering and firm; who holds beneath him the boundaries of the world; resting, unresting; idle, never staying; the eldest Son of Nature and the Father of Light.][200]

Even leaving its grandeur and beauty aside, the Sun is the most distant part of the universe which Man can descry, and hence so little known that those who fell into reverent ecstasies before it were excusable.

[C]   Thales[201] was the first to inquire into such matters: he thought God was a Spirit who made all things out of water; Anaximander said that the gods are born and die with the seasons and that there are worlds infinite

199. '88: required, *because of the people's conception*), I think . . .
200. Ronsard, *Remonstrance au peuple de France*, 64 f.
201. There follows a massive borrowing, condensed, from Cicero, *De nat. deorum*, I, X, 25–xii, 30 (with some errors), with additions from *ibid.*, I, viii, 18 f.; xxiv, 63, and *De divinat.*, II, XVII, 40.

in number; Anaximenes said God was Air, immense, extensive, ever moving; Anaxagoras was the first to hold that the delineation and fashioning of all things was directed by the might and reason of an infinite Spirit; Alcmaeon attributed Godhead to the Sun, the Moon, the stars and to the soul; Pythagoras made God into a Spirit diffused throughout all nature and from whom our souls are detached; for Parmenides God was a circle of light surrounding the heavens and sustaining the world with its heat; Empedocles made gods from the four natural elements of which all things are compounded; Protagoras would not say whether the gods existed or not or what they are if they do; Democritus sometimes asserted that the constellations and their circular paths were gods, sometimes that God was that Nature whose impulse first made them move; then he said our knowledge and our intellect were God; Plato's beliefs are diffuse and many-sided: in the *Timaeus* he says that the Father of the world cannot be named; in the *Laws* he forbids all inquiry into the proper being of God: elsewhere, in these very same books, he makes the world, the sky, the heavenly bodies, the earth and our souls into gods, recognizing as well all the gods accepted by ancient custom in every country. Xenophon records a similar confusion in the teachings of Socrates: sometimes he has Socrates maintaining that no inquiry should be made into the properties of God; at other times he has him deciding that the Sun is God, that the soul is God, that there is only one God and then that there are many. The nephew of Plato, Speusippus, holds God to be a certain animate Power governing all things; Aristotle sometimes says that God is Mind and sometimes the World; at times he gives the world a different Master and sometimes makes a god from the heat of the sky. Zenocrates has eight gods: five are named after the planets; the sixth has all the fixed stars as his members, the seventh and eighth being the Sun and Moon. Heraclides of Pontus meanders along beneath these various notions and ends up with a God deprived of all sensation; he has him changing from one form to another and finally asserts that he is heaven and earth. Theophrastus is similarly undecided, wandering about between his many concepts, attributing the government of the world sometimes to Intelligence, sometimes to the sky and sometimes to the stars; Strato says God is Nature, giving birth, making things wax and wane, but itself formless and insensate; Zeno makes a god of Natural Law: it commands good, forbids evil and is animate; he dismisses the gods accepted by custom – Jupiter, Juno and Vesta; Diogenes of Apollonia says God is Time. Xenophanes makes God round, able to see and hear but not to breathe and having nothing in common with human nature; Ariston thinks that the form of God cannot be grasped: he deprives him of senses and cannot tell

whether he is animate or something quite different. For Cleanthes God is sometimes Reason, sometimes the World, sometimes the Soul of Nature, sometimes absolute Heat surrounding and enveloping all things. Perseus, a pupil of Zeno's, maintained that the name *god* was bestowed on people who had contributed some outstandingly useful improvements to the life of Man – or even on the improvements themselves. Chrysippus made a chaotic mass of all these assertions and included among his thousand forms of gods men who had been immortalized. Diagoras and Theodorus bluntly denied that gods exist. Epicurus has shiny gods, permeable to wind and light, who are lodged between two worlds which serve as fortresses protecting them from being battered; they are clothed in human shape, with limbs like ours which are quite useless.

> *Ego deûm genus esse semper duxi, et dicam coelitum;*
> *Sed eos non curare opinor, quid agat humanum genus.*

[Personally I have always thought, and will always say, that a race of gods exists in heaven. But I do not think that they care about the actions of the human race.][202]

So much din from so many philosophical brainboxes! Trust in your philosophy now! Boast that you are the one who has found the lucky bean in your festive pudding!

I have drawn some profit from the confusion of forms in the customs of the world: manners and concepts different from mine do not so much annoy me as instruct me; comparing them does not puff me up with pride but humbles me. There is for me no such thing as a privileged choice, except one coming expressly from the hand of God.

I shall not go into monstrous and unnatural vice but on that subject the legislatures of this world are no less contradictory than the rival schools of philosophy. From that we can learn that Fortune herself is not more varied, fickle, blind and ill-advised than human reason.

[A]   Things we know least about are the ones we find most proper to deify.[203]   [C]   Making gods of men   [A]   as Antiquity did surpasses even the most extreme imbecility of reason. I would rather have followed those who worship the serpent, the dog and the bull; since the natures of such animals are less known to us, we are free to imagine them as we like and to endow them with extraordinary qualities. But the Ancients

---

202. Ennius *apud* Cicero, *De divinat.* II, 1, 104.
203. '88: deify: *for adoring things of our own kind, sickly, corruptible and mortal,* as all Antiquity did, *of men whom they had seen living and dying and disturbed by our passions* surpasses . . .

attributed to their gods our own condition – the imperfections of which we ought to know; they gave them desire, wrath, acts of vengeance, marriages, powers of generation and family trees, love, jealousy, bones and limbs like ours, our own feverish passions and pleasures, [C] our deaths and funerals. [A] The human intellect must have been astonishingly drunk to produce all that!

> [B] *Quae procul usque adeo divino ab numine distant,*
> *Inque Deum numero quae sint indigna videri.*

[Things far removed from numinous deity, unworthy to appear among the Gods.]

[C] '*Formae, aetates, vestitus, ornatus noti sunt; genera, conjugia, cognationes omniaque traducta ad similitudinem imbecillitatis humanae: nam et perturbatis animis inducuntur; accipimus enim deorum cupiditates, aegritudines, iracundias*' [We know their faces, their ages, their vestments and their adornments. Their families, their marriages and their kinships are all reduced to the model of human weakness. They are even given troubled minds. We hear of the desires of the gods, of their sicknesses and of their fits of anger].

[A] Similarly they made gods [C] not only of Faith, Virtue, Honour, Concord, Freedom, Victory, Piety, but even of Pleasure, Fraud, Death, Envy, Old Age and Misery, [A] of Fear, Fever, Ill-Fortune and the other evils which beset our fragile and decaying lives:

> [B] *Quid juvat hoc, templis nostros inducere mores?*
> *O curvae in terris animae et coelestium inanes!*

[What pleasure can be found from introducing our manners into our temples? O souls bowed earthwards, entirely void of things celestial!][204]

[C] With what unwise wisdom did the Egyptians forbid, under pain of hanging, that anyone should let it be known that their Gods Serapis and Isis had once been human: everybody knew then that they had been so! According to Varro effigies of these Gods were carved with their fingers on their lips to signify this mysterious command to their priests to hush up their mortal origins (otherwise all worship of them would inevitably be brought to naught).[205]

---

204. Lucretius, V, 123 (Lambin, pp. 383–4); Cicero, *De nat. deor.*, II, xxviii, 70 (cited with Augustine, *City of God*, IV, xxx, in mind); Cicero, *De nat. deorum*, II, xxiii, 59 ff.; I, xi. 28; xvi, 42; Persius, *Satires*, II, 62 and 61.
205. St Augustine, *City of God*, XVIII, v.

[A]   Since man was so desirous of making himself the equal of God, it would have been better, said Cicero, to bring the properties of God down to earth and to turn them into human attributes rather than to send our wretchedness and corruption up to heaven.[206] But if you look at it aright, equally vain opinions have led Man, in various ways, to do both.

When philosophers go into the hierarchy of their gods and rush to distinguish the alliances, attributes and powers of each of them, I cannot believe they are serious.

When Plato deciphered for us the myth of the 'Orchard of Dis'[207] telling us of the physical pleasures and pains awaiting us (after our bodies have decayed into nothing!); when he associated them with sensations experienced in this present life –

> *Secreti celant calles, et myrtea circum*
> *Sylva tegit; curae non ipsa in morte relinquunt.*

[They hide away in secret glades, screened by myrtle groves on every side; even when dead their troubles do not leave them] –

and when Mahomet promised his followers a paradise decked out with tapestries and carpets, with ornaments of gold and precious stones, furnished with voluptuous nymphs of outstanding beauty, with wines and choice foods to eat: I realized that they were both laughing at us, stooping low to tempt our brutish stupidity with sweet allurements, enticing us with notions and hopes appropriate to our mortal appetites.

[C]   Even some of our Christians have fallen into a similar error, promising themselves an earthly life after our resurrection, a life within time, accompanied by all kinds of worldly pleasures and comforts.   [A] Plato's thoughts were all of heaven; his familiarity with things divine was so great that the surname *Divine* has clung to him ever since;[208] are we to believe that even he thought there was, in a wretched creature like Man, something able to approach such incomprehensible Power? Did he believe that we, with our feeble grasp, could actually have a share in eternal

---

206. Cicero, *De nat. deorum*, I, xxxii, 90; also *Tusc. disput.*, I, xxvi, 65, *apud* Augustine, *City of God*, IV, xxvi.
207. Plato, *Gorgias*, 524a; *Repub.*, 614E; Plutarch, *De la face qui apparoist dedans le ronde de la lune*, 626 CD. For the implications of this passage for Montaigne's conception of the after-life, see *Montaigne and Melancholy*, pp. 131–2, and note 1.
208. '88: has *justly* clung to him . . . (From Antiquity onwards we find the term *Divinus Plato*: in the Renaissance it acknowledges Plato's inspiration and sometimes his preoccupation with the world of the soul.)

blessedness or reprobation, or that our senses were robust enough to do so?[209]

This is what we ought to say to him, on behalf of human reason:

If the pleasures you offer me in the next life are related to ones I have experienced here on earth, that can have nothing to do with the Infinite. Even if my five natural senses were overwhelmed with joy; even if this soul of mine were seized of all the happiness she could ever hope for or desire, we know her limitations:[210] that would amount to nothing. Where there remains anything of mine, there is nothing divine. If your promises merely relate to what can exist in our present condition, they cannot enter into the reckoning. [C] All the pleasures of mortals are mortal. [A] Take recognizing parents, children and friends in the next world: if that can touch us and titillate us, if we grasp at such pleasures as that, then we still remain within earthbound,[211] finite pleasures. We cannot condignly conceive those high, divine promises if we are able to conceive them at all. To imagine them condignly, we must imagine them unimaginable, unutterable, incomprehensible [C] and entirely different from our own wretched experiences. [A] 'Eye cannot see', says St Paul, 'nor can there rise up in the heart of man, what God has prepared for his own.'[212] And if (as you assert, Plato, with your 'purifications') we have to modify our being in order to render ourselves capable of celestial joy, that would mean a change so extreme and so total that (as we know from Physics) we would cease to be ourselves:

> [B] *Hector erat tunc cum bello certabat; at ille,*
> *Tractus ab Aemonio, non erat Hector, equo.*

[Hector was killed in battle: but it was not Hector who was dragged along by Achilles' horse.]

[A] Something else would receive our rewards.

> [B] *quod mutatur, dissolvitur; interit ergo:*
> *Trajiciuntur enim partes atque ordine migrant.*

209. '88: a *vile* creature like man . . . our *languishing* grasp . . . or that *our taste was firm* enough to do so?
210. '88: hope for or *can do, we know the weakness and inadequacy of her forces:* that . . .
211. '88: within *mortal,* finite . . .
212. I Corinthians II:9, adapted. (*The* text for Pauline ecstasy; see *Erasmus: Ecstasy and the Praise of Folly,* pp. 174–9; *Montaigne and Melancholy,* p. 131.)

[When what is changed is loosened asunder, that is death. The elements are displaced and change their ordered places.][213]

[A] Pythagoras thought up his metempsychosis in which souls change their dwelling-places: are we to think that the lion which is now housing the soul of Caesar has espoused the passions which moved Caesar, [C] or that it really is Caesar? And if it really were Caesar, then victory would lie with those who opposed Plato over this opinion, pointing out, among other absurdities, that a son might well find himself astride his mother, now clothed in the body of a mule.[214]

Do we doubt [A] that, in such transmigrations as may take place within the same species, the newcomers are different from their forebears? They say that from the ashes of the Phoenix there is born first a worm and then another Phoenix; can anyone think that the second Phoenix is no different from the first? The worm which produces silk for us can be seen dying and shrivelling up: then, from that same body a butterfly appears; that produces another worm: it would be absurd to think it was still the first one. That which once ceases to be no longer exists.

> *Nec si materiam nostram collegerit aetas*
> *Post obitum, rursumque redegerit, ut sita nunc est,*
> *Atque iterum nobis fuerint data lumina vitae,*
> *Pertineat quidquam tamen ad nos id quoque factum,*
> *Interrupta semel cum sit repetentia nostra.*

[If Time, after we are dead, should gather our matter together and make it as it now is; if the light of life were again granted to us – even that would not concern us, once the thread of our memory has been snapped asunder.]

You assert somewhere or other, Plato, that rewards in the life to come concern the spiritual part of man, but that remains just as unlikely.

> [B]   *Scilicet, avolsis radicibus, ut nequit ullam*
> *Dispicere ipse oculus rem, seorsum corpore toto.*

[For an eye torn from its socket and removed from its body can see nothing whatsoever.]

[A]   By your reckoning, it would no longer be Man who is touched by such Joy – no longer us: for we are built of two principal parts,

213. Ovid, *Tristia*, III, 11, 27; Lucretius, III, 756–7 (Lambin, p. 241).
214. Porphyry in St Augustine, *City of God*, X, xxx.

which together form our being; to separate them is death and the collapse of our being.

[B] *Inter enim jecta est vitai pausa, vageque*
Deerrarunt passim motus ab sensibus omnes.

[For life has been interrupted. No motions can affect our senses now; they are quite lost.]

When the limbs a man had in life are eaten by worms and turning to dust we never say that the man is feeling pain:

Et nihil hoc ad nos, qui coitu conjugioque
Corporis atque animae consistimus uniter apti.

[That is nothing to us. We are a union formed from the marriage and embrace of body and soul.][215]

Moreover, what just grounds do the gods have for noting and rewarding, after death, a man's good, virtuous deeds? Within that man it is the gods themselves who nurtured and produced them. And why are the gods offended by his vicious deeds? Why do they punish them? They themselves brought him forth in this faulty state; with a mere nod of their will they can prevent his failure.[216]

Surely Epicurus, with every appearance of human rationality, could have raised such objections to Plato, [C] had he not already covered himself by his oft-repeated conclusion: 'From mortal nature nothing certain can be inferred about the Immortal.'

[A] Human reason goes astray everywhere, but especially when she concerns herself with matters divine. Who knows that better than we do? For we have supplied Reason with principles which are certain and infallible; we light her steps with the holy lamp of that Truth which God has been pleased to impart to us; yet we can see, every day, that as soon as she is allowed to deviate, however slightly, from the normal path, turning and straying from the beaten track traced for us by the Church, she immediately stumbles and becomes inextricably lost; she whirls aimlessly about, bobbing unchecked on the huge, troubled, surging sea of human opinion. As soon as she misses that great public highway she disintegrates and scatters in hundreds of different directions.

Man cannot be other than he is; he cannot have thoughts beyond his

215. Lucretius, III, 846 (Lambin, pp. 247–51); III, 563–4 (Lambin, pp. 227–8); III, 860 (Lambin, pp. 251–4); III, 845 (Lambin, pp. 247–50).
216. '88: *our* vicious deeds . . . brought *us* forth . . . prevent *our* failure?

reach. [B] Plutarch says that it is greater arrogance for mere men to start talking and arguing about gods and demi-gods than for a man who knows nothing whatever about music to start criticizing singers, or for a man who has never been on a battlefield to try and argue about arms and war, from some trivial conjecture presuming to understand an art which far exceeds his knowledge.[217]

[A] I believe that, in the Ancient World, men thought they were actually enhancing the greatness of God when they made him equal to Man, clothed him with Man's faculties and made him a present of Man's fair humours [C] and even of his most shameful necessities; [A] they offered him our food to eat, [C] our dances, mummeries and farces to amuse him; [A] our vestments to clothe him and our houses to dwell in, courting him with odours of incense, sounds of music and garlands of flowers; [C] they made him conform to our own vicious passions, subverting his justice in the name of inhuman vengeance, causing him to rejoice in the smashing and wasting of the very things he had created and protected (as Tiberius Sempronius did when he made a burned sacrifice to Vulcan of the arms and treasures seized as booty from his enemies in Sardinia; as Paul Aemilius did, when he sacrificed the spoils of Macedonia to Mars and Minerva, and as Alexander did, when he reached the shores of the Indian Ocean and sought the favour of Thetis by casting many huge golden jars into the sea).[218] More. They loaded his altars with butchered carcasses – not only of innocent beasts but of men, [A] following the established custom of many peoples – including our own; no nation, I believe, is exempt: all have assayed it.

> [B]   *Sulmone creatos*
> *Quattuor hic juvenes, totidem quos educat Ufens,*
> *Viventes rapit, inferias quos immolet umbris.*

[He took alive four young men, begot by Sulmo, and another four bred by Ufens, to immolate them as sacrifices to the Shades.][219]

[C] The Getae think they are immortal; for them, dying is but a

217. Plutarch, tr. Amyot, *Pourquoy la justice divine differe quelquefois ses malefices*, 259 C.
218. Livy, XLI, 16; XLV, 33; Arrian, *Alexander*, VI, 19.
   '88 (in place of [C]): flowers: *once with the pleasure of a blood-drenched vengeance – witness that widely received notion of sacrifices: and that God took pleasure in murder, and in the torture of things made, preserved and created by him, and that he can rejoice in the blood of innocent souls, not only of animals, which are powerless,* but of men . . .
219. Julius Caesar, *De bello gallico*, VI, xvi; Virgil, *Aeneid*, X, 517.

journey to their God Zamolxis. Every five years they dispatch one of their number to him to ask for what they need. The ambassador is chosen by lot. The actual dispatching takes this form: the man is told of his charge by word of mouth; three of those present hold three javelins upright, the others toss the man on to them. If some vital organ is impaled and he dies at once, that is a clear indication of divine approval. If he escapes death, he is thought to be evil and accursed, so another ambassador is similarly dispatched.

On one occasion, when Amestris the mother of Xerxes had grown old and wished to appease some god of the Underworld, she caused fourteen young men from the best families in Persia to be buried alive, in accordance with the religious rites of that country. And even today the cement used to make the idols of Themistitan is mixed with the blood of little children, since the only sacrifices they relish are the pure souls of little boys: Justice hungry for innocent blood!

> *Tantum relligio potuit suadere malorum.*

> [So great are the evils Religion has encouraged.][220]

[B] The Carthaginians sacrificed their own children to Saturn; those who had none of their own, bought some; their fathers and mothers had to attend the service, looking happy and contented. [A] It is a strange notion to seek to requite divine Goodness with our human affliction; the Spartans did: they courted that Diana of theirs with the suffering of boys who were flogged for her sake – sometimes flogged to death. It was a savage humour which sought to please the Architect by ruining what he had built; to ward off the punishment due to the guilty by punishing the innocent; or to believe that that poor wretched Iphigenia, by her sacrificial death in the port of Aulis, could free the Greek army of the weight of offences they had committed against God.

> [B] *Et casta inceste, nubendi tempore in ipso,*
> *Hostia concideret mactatu moesta parentis.*

[At the very time of her wedding, the pure was impurely slaughtered, a victim sadly murdered by her father.]

[C] And there were the fair and noble souls of the two Decii, father and son, who threw themselves wildly into the thick of the enemy, as a propitiation to make the gods favour the affairs of Rome: '*Quae fuit tanta*

---

220. Herodotus, IV, 94; VII, 114; Plutarch, tr. Amyot, *De la superstition*, 124 A; Lucretius, I, 102 (Lambin, pp. 13–15). The reference to Themistitan is untraced.

deorum iniquitas, ut placari populo Romano non possent, nisi tales occidissent'
[What great wickedness on the part of the gods to refuse to favour the
Roman People unless such men were killed!].[221]

[A]   We might add that it is not for the criminal to decide how and
when he will be whipped: it is for the judge, who can only take account of
such chastisements as he himself has ordered and who cannot treat as
punishment anything that is pleasing to the sufferer. Both for the sake of its
own justice and of our punishment, God's vengeance must presuppose our
complete resistance to it.

[B]   It was an absurd caprice on the part of Polycrates, the tyrant of
Samos, to cast his most precious jewel into the sea to atone for his
continuous run of good fortune by interrupting its course; he thought to
placate the turning Wheel of Fortune with a carefully arranged disaster.
[C]   Fortune, to mock such absurdity, caused the jewel to be returned into
his hands through the belly of a fish.

[A]   And then   [C]   what is the use of all those lacerations and lop-
pings off of limbs practised by the Corybantes and Maenads, or, in our
own day, by the Mahometans who slash their faces, their bellies and their
limbs, to please their prophet, seeing that   [A]   the offence lies in the will
not   [C]   in the breast, the eyes, the genitals, a well-rounded belly or
in   [A]   the shoulders or the throat.   [C]   *'Tantus est perturbatae mentis et
sedibus suis pulsae furor, ut sic Dii placentur, quemadmodum ne homines quidem
saeviunt'* [Such is their frenzy, arising from minds disturbed and forcibly
unhinged, that it is thought the gods can be placated by surpassing even
our human cruelty].

How we treat the natural fabric of our bodies concerns not only
ourselves but the service of God and of other men. It is not right to harm it
deliberately, just as it is wrong to kill ourselves on any pretext whatsoever.
There is, it seems, both great treachery and great cowardice in whipping
and mutilating the servile, senseless functions of our bodies in order to
spare our souls the trouble of governing them reasonably: *'Ubi iratos deos
timent, qui sic propitios habere merentur? In regiae libidinis voluptatem castrati
sunt quidam; sed nemo sibi, ne vir esset, jubente domino, manus intulit'* [What do
they think the gods are angry about, when they believe they can propitiate
them thus? Some have been castrated to serve the lust of kings, but no one
has ever emasculated himself, even at the command of his master].   [A]   In
this way they filled their religion with many bad deeds,

---

221. Plutarch, *De la superstition*, 123 G–124 A; *Les Dicts notables des Lacedaemoniens*,
227 EF; Lucretius, I, 98; Cicero, *De nat. deorum*, III, vi, 15.
      '88: to requite divine *justice* with our *torment and our suffering*; the Spartans . . .

saepius olim
Relligio peperit scelerosa atque impia facta.

[Too often in the past, religion has given birth to impious and wicked actions.][222]

Nothing of ours can be compared or associated with the Nature of God, in any way whatsoever, without smudging and staining it with a degree of imperfection. How can infinite Beauty, Power and Goodness ever suffer any juxtaposition or comparison with a thing as abject as we are, without experiencing extreme harm and derogating from divine Greatness? [C] *'Infirmum dei fortius est hominibus, et stultum Dei sapientius est hominibus'* [The weakness of God is stronger than men and the foolishness of God is wiser than men].[223]

Stilpon the philosopher was asked whether the gods took pleasure in our homage and sacrifices: 'You are most indiscreet,' he replied; 'if you want to talk about that, let us draw aside.'[224] [A] And yet we prescribe limits in the Infinite and besiege his mighty power with those reasons of ours (I call our ravings and our dreamings 'reasons', under the general dispensation of Philosophy who maintains that even the fool and the knave act madly 'from reason' – albeit from one special form of reason).[225]

We wish to make God subordinate to our human understanding with its vain and feeble probabilities; yet it is he who has made both us and all we know. 'Since nothing can be made from nothing: God could not construct the world without matter.' What! Has God placed in our hands the keys to the ultimate principles of his power? Did he bind himself not to venture beyond the limits of human knowledge? Even if we admit, O Man, that you have managed to observe some traces of his acts here in this world, do you think that he has used up all his power by filling that work with every conceivable Form and Idea? You only see – if you see that much – the order and government of this little cave in which you dwell; beyond, his Godhead has an infinite jurisdiction. The tiny bit that we know is nothing compared with ALL:

omnia cum coelo terraque marique
Nil sunt ad summam summai totius omnem.

222. Much from St Augustine, *City of God*, VI, 10 (citing a lost book of Seneca's *Against Superstition*). Also, Lucretius, I, 82 (Lambin, pp. 12–15).
223. I Corinthians I:25, a central text for Christian Folly since Augustine, not least for Erasmus.
224. Diogenes Laertius, *Lives*, Stilpon, II, 117.
225. For Platonizing thinkers the fool's soul (being divine in origin) remains rational; the knave reasons incorrectly about what is good but is not irrational (cf. n. 2). With what follows, cf. Ronsard, *Remonstrance*, 119 f.

[The entire heavens, sea and land are nothing compared with the greatest ALL of all.][226]

The laws you cite are by-laws: you have no conception of the Law of the Universe. You are subject to limits: restrict yourself to them, not God. He is not one of your equals; he is not a fellow-citizen or a companion. He has revealed a little of himself to you, but not so as to sink down to your petty level or to make himself accountable for his power to you. The human body cannot fly up to the clouds – that applies to you! The Sun runs his ordered course and never stops still; the boundaries of sea and land can never be confounded; water is yielding and not solid; a material body cannot pass through a solid wall; a man cannot stay alive in a furnace; his body cannot be present in heaven, on earth and in a thousand places at once. It is for you that he made these laws; it is you who are restricted by them. God, if he pleases, can be free from all of them: he has made Christians witnesses to that fact. And in truth, since he is omnipotent, why should he restrict the measure of his power to definite limits? In whose interest ought he to give up being a Law unto himself?

That Reason of yours never attains more likelihood or better foundations than when it succeeds in persuading you that there are many worlds:

> [B]   *Terramque, et solem, lunam, mare, caetera quae sunt*
> *Non esse unica, sed numero magis innumerali.*

[The earth, the sun, the moon and all that exists are not unique, but numerous beyond numbering.]

[A]   That belief was held by the most famous minds of former ages (and still is by some today), on grounds which, to purely human reason, seem compelling, because nothing else within the fabric of the universe stands unique and alone:

226. Lucretius, VI, 678 (Lambin, pp. 508–10, reading *sint* not *sunt*). A lesson against homocentricity, inscribed in Montaigne's library. Platonic-Christian arguments are marshalled against Aristotle's denial of a creation *ex nihilo*. Allusions follow to biblical miracles: Elijah's rapture to heaven (II Kings 2:11) and/or to Christ's bodily Ascension; the halting of the Sun in Joshua 10:12; the Flood in Genesis 6–9 (cf. Genesis 1:9, 7:4); Psalm 104 (103):6–9; Christ's walking on the water (Matthew 14:25); Christ's appearing in an enclosed room (John 20:19 ff.); Shadrach, Meshach and Abednego in the fiery furnace (Daniel 3:22–7). The final miracle is the Real Presence of Christ's risen body in Heaven and in each Eucharist. In the background is the Platonic doctrine of the great chain of being (God created all possible forms). With the *cave* Montaigne exploits the central Platonic myth: man, living as it were in a cave, mistakes shadows on the wall for the reality outside his cave which casts those shadows.

> [B]   *cum in summa res nulla sit una,*
> *Unica quae gignatur, et unica solaque crescat.*

[Since nothing born of Nature is unique, nor when it grows is anything unique or all alone.]

[A] There is some element of multiplicity within every species; it seems unlikely, therefore, that God made only this one universe and no other like it, or that all the matter available for this Form should have been exhausted on this one Particular:

> [B]   *Quare etiam atque etiam tales fateare necesse est*
> *Esse alios alibi congressus materiai,*
> *Qualis hic est avido complexu quem tenet aether.*

[Such things must be said again and again: there are, elsewhere, other material aggregates than the one which the air enfolds in her keen embrace.]

[A] That is especially the case if the universe has a soul – something which its movements make credible, [C] so credible that Plato was sure of it; many Christians, too, either allow it or dare not disallow it, any more than the ancient opinion that the heavens, all heavenly bodies and other constituent parts of the universe are creatures composed of body and soul, subject to mortality, being composite, but immortal by the decree of their Maker.[227]

[A] Now, if there are several worlds, as [C] Democritus, [A] Epicurus and almost the whole of philosophy have opined, how do we know whether the principles and laws which apply to this world apply equally to the others? Other worlds may present different features and be differently governed. [C] Epicurus thought of them as being both similar and dissimilar.[228] [A] Even within our own world we can see how mere distance produces infinite differences and variety. Neither

---

227. '80: the most famous *and noble* minds ... movements make *more* credible. Now, if there are several worlds, as *Plato*, Epicurus ... Lucretius, II, 1085 f., 1077 f., 1064 f. (Lambin pp. 180–2). Montaigne echoes the commentary ('There is no verisimilitude in this world's having been created alone' etc.) and the commentary on pp. 178–79 (allusions to Democritus after Cicero, *De fin.*, and *Acad.: Lucullus*). In the *Timaeus* (31 AB; 55 DE) Plato *defends* (against the atomists) the essential unity of the Universe but believes in a world-soul, as did the Christian Origen (St Augustine, *City of God*, XI, 23). Augustine (XIII, 16 and 17) did not reject Plato's contention (*Timaeus* 41D–42A) that the stars had souls and could be rendered immortal. Echoes in Montaigne of Plutarch, tr. Amyot, *Des Opinions des Philosophes*, 446A-F.

228. Diogenes Laertius, *Lives*, Democritus, IX, 45; cf. Epicurus, IX, 85.

wheat nor wine was found in those New Lands discovered by our fathers, nor any of our animals: everything there is different. [C] And only think of those parts of the world which, in times gone by, had no knowledge of Bacchus' grapes or Ceres' corn.

[A] Should anyone care to believe Pliny [C] and Herodotus,[229] [A] there are species of men, in some places, which have very little resemblance to our own; [B] there are some ambiguous, mongrel forms, between the human and the beast; there are lands where men are born without heads, having eyes and mouths in their chests; there are androgynous creatures and creatures who walk on all fours, have only one eye in the middle of their forehead, or have a head more like a dog's than our own; some are fishes below the waist and live in water; some have wives who give birth at five and die at eight; other men have skin on their forehead and on the rest of their cranium so hard that iron spears cannot dent it but simply blunt themselves; there are men without beards, [C] peoples without the use or knowledge of fire and others who ejaculate black semen.

[B] What about those people who, by natural means, can change into wolves [C] and mares [B] and back again? And [A1] even if you were to accept as true [A] what Plutarch says (that somewhere in the Indies there are men without mouths who sustain themselves by inhaling certain smells) how many of our own descriptions today are certainly wrong![230] If laughter were no longer the property of Man and if Man were no longer a political animal able to reason, our conception of what our inner disposition and causations are would be largely irrelevant . . .[231]

To go further, we have imposed our own commandments on Nature and carved them in stone: yet how many things do we know which defy those fine rules of ours! And yet we try to bind God by them!

How many things are there which we call miraculous or contrary to Nature? [C] All men and nations do that according to the measure of their ignorance. [A] How many quintessences, how many occult properties have we discovered! For us, following Nature means following our

229. What follows derives from Pliny, *Hist. Nat.*, VI, 2; VIII, 22; Herodotus, III, 101; IV, 191. Pliny's 'errors' and Herodotus' 'lies' were often evoked in the Renaissance.

230. Plutarch, *De la face qui apparoist dedans le rond de la Lune*, 623 F (producing amused laughter from the hearers).

231. The standard definitions of Man, as a thinking, laughing or 'political' animal, could not apply to men without brains in their heads or mouths to laugh with or cities to live in (as political animals).

own intelligence as far as it is able to go and as far as we are able to see.[232]
Everything else is a monster, outside the order of Nature! By that reasoning
the cleverest and wisest men would find everything monstrous, since they
are convinced that reason has no foundation to stand on, not even to
determine [C] whether snow is white (Anaxagoras said it was black), or
whether there are such things as knowledge and ignorance (Metrodorus of
Chios denied that Man could ever know), [A] or even whether we are
alive: Euripides hesitates, 'Is *life* this life that we live now? Or is *life* really
what we call death?' That is:

> Τὶς δ'οἶδεν εἰ ζῆν τοῦθ' ὸ κέκληται θανεῖν,
> τὸ ζῆν δὲ θνείσκειν ἔστι[233]

[B] There is a degree of probability in that alternative: for why do we
give the name *existence* to that instant which amounts to no more than a
flash of lightning against the infinite course of eternal light, or to that tiny
break which interrupts the condition which is naturally ours for all
eternity, [C] since death fills everything before that moment and
everything which comes afterwards as well as a large part of the moment
itself?

[B] Some swear that nothing moves and that there is no such thing at
all as motion – [C] as was believed by the followers of Melissus (since,
as Plato proves, there is no place for spherical motion within strict Unity,
nor even for movement from one place to another) – [B] or that there
is, in Nature, no generation and no corruption. [C] Protagoras says that
in Nature nothing exists but doubt: that everything is equally open to
discussion, including the assertion that everything is equally open to
discussion; Nausiphanes holds that among phenomena there is nothing
which *is* rather than *is not*: that nothing is certain but uncertainty. For
Parmenides, within the world of phenomena there is no such thing as
genus: there is only Unity. For Zeno, there is not even Unity, only
Nothing: for if Unity exists it must exist either within another or within
itself; if it exists in another, that makes two; if it exists within itself, that
still makes two – the container and the thing contained.

---

232. A miracle is, for Christians, an event 'against the whole order of Nature'. To
recognize such an event by natural reason requires, therefore, a true knowledge of
the limits of Nature.
233. Cicero, *Acad.: Lucullus*, II, xxxi, 100, cf. xxxiii, 105–8; the verses from
Euripides were inscribed in Montaigne's library; they are cited by Sextus Empiricus,
Hypotyposes, III, 229, but in a different form; Montaigne's version derives from
Stobaeus, Sermo 119, but there are minor variations in many editions of this text.

According to these tenets, Nature is but a shadow, false or vain.[234]

[A] It has always seemed to me that certain expressions are too imprudent and irreverent for a Christian: 'God cannot die'; 'God cannot change his mind'; 'God cannot do this or cannot do that'. I find it unacceptable that the power of God should be limited in this way by the rules of human language; these propositions offer an appearance of truth, but it ought to be expressed more reverently and more devoutly. Our speech, like everything else, has its defects and weaknesses. Most of the world's squabbles are occasioned by grammar! Law-suits are born from disputes over the interpretation of laws; most wars arise from our inability to express clearly the conventions and treaties agreed on by monarchs. How many quarrels, momentous quarrels, have arisen in this world because of doubts about the meaning of that single syllable *Hoc*.[235]

[B] Take the proposition which Logic asserts to be the clearest of all. If you say 'The weather is fine' and you say it truly, then the weather is fine. That seems to be clear enough; and yet such a formula can lead us astray. You can see that from the following example: if you say, 'I lie', and you say it truly, then you lie! In both cases, the art, reason and force of the conclusion are the same: yet the second leaves you stogged in the mud![236]

[A] Pyrrhonist philosophers, I see, cannot express their general concepts in any known kind of speech; they would need a new language: ours is made up of affirmative propositions totally inimical to them – so much so that when they say 'I doubt', you can jump down their throats and make them admit that they at least know one thing for certain, namely that they doubt. To save themselves they are constrained to draw an analogy from medicine: without it their sceptical humour would never get purged! When they say *I know not* or *I doubt* that affirmation purges itself (they maintain) along with all the others, exactly like a dose of rhubarb, which evacuates all our evil humours, itself included.[237]

---

234. Plato, *Theaetetus*, 180E-183E; Seneca, *Epist.*, LXXXVIII, 43–6; Plato, *Parmenides*, 138.

'88 (In place of [C]): *I do not know whether Ecclesiastical teaching judges otherwise – and I submit myself, in all things everywhere to its ordinance, but* it has always seemed to me . . .

235. Matthew 26:26. Disputes over the eucharistic formula 'This (*Hoc*) is my body' are central to Christian controversy. Cf. H. C. Agrippa, *On the Vanity of all Learning*, III.

236. Cicero, *Acad.: Lucullus*, xxix, 95.

237. Diogenes Laertius, *Lives*, Pyrrho, IX, 76 (for 'rhubarb' the text gives *medicamenta*).

[B] (Scepticism can best be conceived through the form of a question: 'What do I know?' – *Que sçay-je*, words inscribed on my emblem of a Balance.)[238]

[A] See how people avail themselves of such forms of speech as are full of irreverence. In our present religious strife, if you press your adversaries too hard they will bluntly reply that it exceeds God's power to make his body be in paradise and in several places on earth all at the same time. How that scoffer[239] among the Ancients exploited similar assertions! 'At least', he said, 'it is no light comfort for Man to know that God cannot do everything! God cannot kill himself when he wants to (which is the greatest prerogative attached to the human condition); he cannot bring the dead back to life; he cannot make someone not to have lived who has lived, or not to have received honour who has received honour; he has no jurisdiction over the past other than to make it merge into oblivion; finally (so that this equality of status in God and Man can be further strengthened with amusing examples), God cannot even stop ten and ten from making twenty!' That is what he says – and what should never pass a Christian's lips. Whereas, on the con⸱rary, men seem to me to go looking for such insane and arrogant terms in order to cut God down to their own size:

> *cras vel atra*
> *Nube polum pater occupato,*
> *Vel sole puro; non tamen irritum*
> *Quodcumque retro est, efficiet, neque*
> *Diffinget infectumque reddet*
> *Quod fugiens semel hora vexit.*

[Tomorrow the Father can cover the pole with black clouds or with pure sunlight, but he cannot change the past, he cannot undo or annul anything that fleeting time has borne away.]

When we say that countless ages – ages past and ages yet to come – are but a moment to God and that God's essence consists in goodness, wisdom, power, we utter words, but our intelligence cannot grasp the sense. Despite that, we, in our arrogance, want to force God through human filters. All the raving errors that this world possesses are bred from trying to squeeze

238. In 1576 (doubtless under the influence of Pyrrho), Montaigne struck a medal with a Balance, poised, bearing the device *Que sçay-ie?*
239. '88: that *scoffer* Pliny exploited . . . (Pliny, *Hist. Nat.*, II, 7; the two following quotations are from Horace, *Odes*, III, 29, 43; Pliny, ibid., II, 23.)

on to human scales weights far beyond their capacity: [C] *'Mirum quo procedat improbitas cordis humani, parvulo aliquo invitata successu'* [It is astonishing how far the impudence of the human heart can go, once encouraged by the least success].

How insolently the Stoics taunt Epicurus for holding that essential goodness and happiness belong to God alone, so that the Sage can only possess some shadowy likeness of them. [A] How rashly they subject God to Destiny (would that some who bear the name of Christians did not do so still);[240] Thales, Plato and Pythagoras even make God the slave of Necessity. This fierce desire to scan the Divine through human eyes even brought one of our own great Christian figures to endow God with a corporeal shape;[241] [B] it also explains why we daily assign to God a peculiar responsibility for any event, the outcome of which seems important to us. We attach particular weight to such events, so God must do so too, paying more attention to them than to others which seem unimportant to us or simply part of the regular order: [C] *'magna dii curant, parva negligunt!'* [The gods take care of great matters and neglect the small!] Listen to the example given and you will see more clearly what is meant: *'nec in regnis quidem reges omnia minima curant'* [Even kings in their kingdoms do not concern themselves with every tiny detail]. As though it were more difficult for God to shake an empire than to shake a leaf, or as though his Providence were exercised differently when influencing the outcome of a battle and the jump of a flea.

The hand of God's governance supports all things with an equal and unchanging sway, with the same order, the same power. Our concerns contribute nothing to this; our human activities and standards are quite irrelevant: *'Deus ita artifex magnus in magnis, ut minor non sit in parvis'* [In great things God is a great artificer, but in such a way that he is no less great in little things].

Our arrogance constantly finds fresh ways of blasphemously equating man with God: our jobs are a burden to us men, so Strato endows the gods – and their priests – with complete immunity from work! For Strato it is Nature who produces and maintains all things, Nature who constructs every part of the universe with her weights and her forces. In this way he frees mankind of a burden: the fear of divine judgement: *'Quod beatum*

240. Seneca, *Epist.*, XCII, 275. The Stoics 'subject God to destiny': the Christians who are alleged to do so are doubtless, for Montaigne, Calvinists – cf. Cicero, *Acad.: Lucullus*, II, 29.
241. Tertullian, apparently, while still a Catholic; he became a Montanist.

aeternumque sit, id nec habere negotii quicquam, nec exhibere alteri' [A blessed and eternal Being has no duties and imposes none on others].[242]

'Nature's will is that like things should have like correlatives; for example: the fact that mortals are innumerable leads to the conclusion that the immortals are too; the vast number of things which kill or do harm leads to the conclusion that an equivalent number preserve and do good'; so, just as the souls of the gods have no tongue, eyes or ears yet can understand each other and also judge what we are thinking: so too the souls of men, when free from the bonds of the body in sleep or any kind of ecstasy, have powers of divination, can foretell the future and see such things as they could never see when joined to their bodies . . . [A] 'Men', says St Paul, 'have become fools, professing to be wise, and have changed the glory of the incorruptible God into the image of corruptible man'.[243]

[B] Only consider what jugglers' farces those 'deifications' were among the Ancients. After the stately pride and pomp of the funeral procession, just as the fire was taking hold of the apex of the pyre and about to engulf the litter with the dead man on it, they would release an eagle which flew upwards, representing the soul making its way to Paradise. We still possess a thousand medallions (above all, the one of that – oh, so honourable – woman Faustina) where the eagle is portrayed bearing off the deified soul, which is slung over its shoulder just like a dead goat! It is pitiful the way we deceive ourselves with the monkey-tricks that we invent;

Quod finxere timent

[They are terrified of their own creations]

– like children who are scared by the very face of the friend they have just daubed with black. [C] *'Quasi quicquam infelicius sit homine cui sua figmenta dominantur'* [As though anything were more pitiful than a man overmastered by his own figments].[244]

We are far from honouring him who made us when we honour a creature we ourselves have made.

242. Cicero, *De nat. deorum*, II, lxvi, 167; III, xxxv, 86; St Augustine, *City of God*, XI, 22; Cicero, *Acad.*, II, xxviii, 121.
243. Epicurus' principle of *isonomia* (Cicero, *De nat. deorum*, I, xix, 50) and the contentions of Cicero's brother in *De divinat.* I, lvii, 129, are here countered by Romans I:22–23.
244. Lucan, *Pharsalia*, I, 486. (For ancient deifications and medals, cf. G. du Choul, *De la religion des anciens Romains*, 1556, p. 75, etc.; also Joachim Du Bellay, *Regrets*, TLF, *Songe* XI and illustration.) Seneca, *Epist.*, XXIV, 13; St Augustine, *City of God*, VIII, 23–4.

[B]   Augustus had more temples than did Jupiter, in which he was
served with just as much devotion and just as much belief in his miracles.

The Thasians, wishing to repay the benefits they had received from
Agesilaus, came to tell him that they had put him on the canonical list of
their gods. 'Are you a people', he asked, 'who have the power to make a
god of anyone you please? Just to see, first make a god of one of yourselves; and then,   [C]   when I have learned how he has prospered,   [B]
I will come and thank you heartily for your offer.'

[C]   Man is indeed out of his mind. He cannot even create a fleshworm, yet creates gods by the dozen. Just listen to Hermes Trismegistus
praising our sufficiency: 'Among all the things which can astonish us, one
thing has surpassed astonishment itself: Man's capacity to discover what the
Divine nature is and then proceed to create it.'

[B]   Here are some arguments from the very school in which Philosophy
learned her lessons:

> Nosse cui Divos et coeli numina soli,
> Aut soli nescire, datum

[Philosophy, she to whom alone it is given to know the gods and the numinous
powers of heaven, or, alone, to know that they cannot be known!][245]

– If God exists, (she says) he is animate; if he is animate he has senses; if he
has senses, he is subject to corruption! If he is incorporeal, he has no soul
and consequently is without activity; if he is corporeal, then he is mortal!
What a triumph! –   [C]   We could never make this world; therefore a
Nature *even more excellent than ours* must have taken the task in hand! – It
would be stupid arrogance to esteem ourselves the most perfect object in
the universe: there must therefore be *one* thing better: God! – When you
see a rich and stately dwelling you may not know who the master of it is,
but at least you could say that it was not built for rats: take the divine
architecture of the palaces of heaven, which we ourselves can see; does it
not oblige us to believe that it is the dwelling-place of a Master greater *even
than we are*? – Is not the higher always more worthy – and we are at the
bottom. – Nothing without reason and soul can beget an animate creature
capable of reason: the world has begotten us: therefore it has both reason
and soul! – Each part of us is less than the whole: we are part of the world:
the world is, therefore, provided with wisdom and reason more abundantly

245. Plutarch, *Les Dicts notables des Lacedaemoniens*, 210 GH; Hermes Trismegistus,
Asclepius, 37, *apud* St Augustine, *City of God*, VIII, 24; Lucan, *Pharsalia*, I, 452
(adapted).

than we are. – It is a fair thing to hold great powers of government: the government of the world must, therefore, belong to some happy Nature. – The heavenly bodies do us no harm: they are, therefore, full of goodness. – [B] We need food: so do the gods, who feed on vapours rising up from here below! – [C] Worldly goods are not goods to God: therefore they are not goods to us. – To do harm and to experience harm are equal proofs of weakness: it is therefore mad to be afraid of God! – God is good by nature, man by industry: which makes man superior! – There is no difference between divine wisdom and human wisdom, except that the divine is eternal: but time adds nothing to the quality of wisdom: therefore we and God are on equal footing! [B] We enjoy life, reason, freedom and we esteem goodness, love and justice: therefore these qualities must be in God!

In short, both constructively and destructively, we forge for ourselves the attributes of God, taking ourselves as the correlative. What a model, what a pattern! Take human qualities and stretch them, raise them, magnify them as much as you please! Wretched little Man, puff yourself up as much as you like! More. More. More still: *'Non si te ruperis, inquit'* . . . ['Not even', he said, 'if you burst.']. [C] *'Profecto non Deum, quem cogitare non possunt, sed semet ipsos pro illo cogitantes, non illum sed se ipsos non illi sed sibi comparant'* [Indeed, Men cannot conceive of God, so they base their conceptions on themselves instead; they do not compare themselves to him, but him to themselves].[246]

[B] Even within Nature, effects barely suggest half their causes. But what of this Cause? God is a Cause completely above the order of Nature. His mode of being is too high, too distant, too magisterial to allow our logical conclusions to judge or to bind him. We shall never get that far by our own efforts: our path is too lowly. We are no nearer the heavens on the top of Mount Cenis than we are at the bottom of the sea. Your astrolabe will tell you that.

Yet men even reduce God to having sexual intercourse with women, noting how often he did it and for how many births.

Paulina, the wife of Saturninus, was a Roman matron of great reputation; she thought she was lying with a god, Serapis, but through the pimping of

---

246. Several Stoic commonplaces and major borrowings from Cicero (*De nat. deorum*, II, vi, 16–VIII, 22) and others (cf. Pontus de Tyard, *Second Curieux in Discours philosophiques*, 1587, 310); Horace's fable of the puffed-up frog (*Satires*, II, iii, 319); finally St Augustine, *City of God*, XII, 18.

the temple-priests she found herself in the arms of a lover.[247] [C] In his treatises on theology, Varro, the most subtle and learned of Latin authors, wrote of a sexton in the temple of Hercules who cast dice with both hands, one for himself, the other for Hercules. The stakes were a supper and a woman: if he won, he paid for them out of the collection; if he lost, he paid for them himself. He lost; so the cost of the woman and dinner fell to himself. Now the woman was called Laurentina; lying that night with this 'god' in her arms, she heard him volunteer the remark that the first man she met when she left in the morning would see that she received from heaven the money she had just earned. She did in fact meet a rich young man called Taruntius who took her back home and eventually left her all his money. She in her turn, hoping to do an action pleasing to this god, left her inheritance to the Roman People, who then bestowed divine honours upon her.[248]

As though it were simply not enough that Plato should be descended, on both sides, from the gods, with Neptune as the common ancestor, it was believed as a fact in Athens that, when Ariston had wished to consummate his love for the fair Perictione, he could not bring it off; he was warned in a dream by the god Apollo not to deflower her but to leave her a virgin until she had given birth . . . And they were Plato's father and mother![249]

How many other accounts are there of similar cuckoldries procured by the gods against wretched human beings, or of husbands unjustly defamed to honour their children! In the religion of Mahomet the people believe that there are 'Merlins' in plenty – children, that is, begot without fathers, spiritual children divinely conceived in virgins' wombs. (They are given a special name which, in their language, means just that.)[250]

[B] We should note that no creature holds anything dearer than the kind of being that it is [C] (lions, eagles, dolphins value nothing above their own species) [B] and that every species reduces the qualities of everything else to analogies with its own. We can extend our characteristics or reduce them, but that is all we can do, since our intellect can do nothing and guess nothing except on the principle of such analogies; it is impossible for it to go beyond that point. [C] That explains Ancient philosophical

247. Commonplace deriving from Josephus, *Jewish Antiquities*, XVIII, 4 (but in the temple of Anubis not Serapis).
248. Varro *apud* St Augustine, *City of God*, VI, 7; tale current since Antiquity.
249. Diogenes Laertius, *Lives*, Plato, III, ii, 185.
250. Guillaume Postel, *Des Histoires Orientales* (*De la République des Turcs*), 1575, 919 r.°

conclusions such as these: Man is the most beautiful of all forms, so God must also have that form! – No one can be happy without virtue; virtue cannot be without reason: no reason can dwell elsewhere but in the human shape: therefore God is clad in a human shape! *'Ita est informatum, anticipatum mentibus nostris ut homini, cum de Deo cogitet, forma occurrat humana'* [The mould and prejudice of our minds are such that when we think of God it is the human form which occurs to them].[251]

[B]   That is why Xenophanes said with a smile that if the beasts invent gods for themselves, as they probably do, they certainly make them like themselves, glorifying themselves – as we do.[252] For why should a gosling not argue thus: 'All the parts of the universe are there for me: the earth serves me to waddle upon, the sun to give me light; the heavenly bodies exist to breathe their influences upon me; the winds help me this way, the waters, that way: there is nothing which the vault of Heaven treats with greater favour than me. I am Nature's darling: does not Man care for me, house me, serve me? It is for me that Man sows and grinds his corn; it is true that he eats me, but he also eats his fellow-men, and I eat the worms which kill him and eat him.'

A crane could say the same – even more majestically on account of the freedom of its flight and its secure enjoyment of those fair and higher regions: [C]   *'Tam blanda conciliatrix et tam sui est lena ipsa natura'* [So flattering a procuress is Nature, such a seductress of herself].[253]

[B]   Well, if that is how it goes, the Universe and the Fates are all for us! The lightning flashes for us; the thunder crashes for us; the Creator and all his creatures exist just for us. We are the end which the entire Universe is aiming towards. Just examine the records of celestial affairs which Philosophy has kept for two thousand years and more: the gods have acted and spoken only for Man. Philosophy attributes no other concern to them, no other employment: they go to war against us,

> *domitosque Herculea manu*
> *Telluris juvenes, unde periculum*
> *Fulgens contremuit domus*
> *Saturni veteris.*

---

251. Cicero, *De nat. deorum*, I, xxvii, 76–78.
252. Eusebius Pamphilus, *Preparatio evangelica*, XIII, 13, perhaps via Ph. Duplessis-Mornay, *De la Verité de la religion chrestienne*, chapters I (end), 4 (beginning).
253. Developments inspired by Cicero, *De nat. deorum*, I, xxvii, 78: 'Suppose animals possessed reason: would they not attribute superiority to their own kind?' Latin quotation: ibid., 77.

[The Sons of Earth, those Titans at whose assault the shining house of ancient Saturn shook with fear, are defeated by the hand of Hercules.]

The gods side with us in our civil disturbances,   [C]   to return our services, since we have so often taken sides in theirs:

> [B]   *Neptunus muros magnoque emota tridenti*
> *Fundamenta quatit, totamque a sedibus urbem*
> *Eruit. Hic Juno Scaeas saevissima portas*
> *Prima tenet.*

[With his mighty trident Neptune shakes the walls of Troy to their foundations and dashes the whole city to the ground; here, implacable Juno holds the Scaean gates.]

[C]   On their feast-days, the Caunians, jealous for the hegemony of their own gods, load weapons on their shoulders and charge around the outskirts of their city stabbing their swords into the air, fighting the foreign gods to the finish and driving them out of their lands.[254]

[B]   The powers of the gods are tailored to meet our human needs: this one cures horses, another, men;   [C]   this one, the plague,   [B]   that one, the ring-worm, that one, the cough;   [C]   this one cures one sort of mange; that one, another: *'adeo minimis etiam rebus prava religio inserit deos'* [Thus does religion, when depraved, bring the gods even into the most trivial affairs];   [B]   this god makes grapes to grow, another, garlic; this god is responsible for lechery, that one, for trade,   [C]   (each tribe of craftsmen has its god!);   [B]   this god's sway and reputation lie in the East; that god's lie in the West.

> *hic illius arma,*
> *Hic currus fuit;*

[Here were her arms, here stood her chariot;]

> [C]   *O Sancte Apollo, qui umbilicum certum terrarum obtines;*

[O holy Apollo, thou that holdest sway in the Navel of the world;]

> *Pallada Cecropidae, Minoia Creta Dianam,*
> *Vulcanum tellus Hipsipilea colit,*

---

254. Horace, *Odes*, II, 12, 6; Virgil, *Aeneid*, II, 610; Herodotus, I, 172. (For the gods of grapes and garlic, cf. Cardinal Robert Bellarmine, *On the Loss of Grace and the State of Sin*, book X, chapter ix, 'An enumeration of the maladies and wounds of the human mind', § 6, in *Opera*, 1593, 487B.)

Junonem Sparte Pelopeiadesque Mycenae;
Pinigerum Fauni Maenalis ora caput;
Mars Latio venerandus.

[The descendants of Cecrops worship Pallas in Athens; Minoan Crete worships Diana; Lemnos, Vulcan; Sparta and Peloponnesian Mycenae, Juno. Pan, crowned with pine leaves, is venerated in Maenalus; and Mars in Latium.]

[B] This god has only a single town or family under his sway,   [C]   that one lives alone, but the other one, willingly or from necessity, lives with his peers:

Junctaque sunt magno templa nepotis avo.

[The grandson's temple is amalgamated with the temple of his grandsire.]²⁵⁵

[B] Some of these gods are so mean and so lowly (for their number amounts to thirty-six thousand) that you need a pile of five or six of them to make a grain of corn – their various names are taken from this – [C] you need three for a door (one for the wood, one for the hinge, one for the doorstep); then you need four for an infant (protecting its cradle, its drink, its food and its sucking). The functions of some are uncertain and doubtful; others are not allowed into Paradise yet:

Quos quoniam coeli nondum dignamur honore,
Quas dedimus certe terras habitare sinamus.

[Since some are not yet worthy to be honoured with paradise, we at least allow them to dwell in the lands we have given them.]

There are nature-gods, poetic gods, civic gods; there are intermediary beings, half-way between the divine nature and the human, who are mediators, doing business between us and God and worshipped with an inferior, second-grade worship; they have innumerable titles and duties. Some are good: some are bad.   [B]   There are gods who are old and decrepit; there are gods who are mortal; for Chrysippus considered that all gods died in the last great conflagration of the world, except Jupiter.

[C] Man invents a thousand amusing links of fellowship between

255. Livy, XXVII, xxiii; Virgil, *Aeneid*, I, 16; Anon., cited Cicero, *De divinatione*, II, Ivi, 115; Ovid, *Fasti*, III, 81 and I, 294.

himself and God. Is God not a fellow-countryman! '*Jovis incunabula Creten*' [Crete, cradle of Jupiter].²⁵⁶

Here is the justification given after reflection by Scaevola, a great Pontifex, and by Varro, a great theologian (both 'great' in their time): it is necessary (they said) that people should not know many things which are true and should believe many things which are false, '*cum veritatem qua liberetur, inquirat, credatur ei expedire, quod fallitur*' [since man only wants to find such truth as sets him free, it can be thought expedient for him to be deceived].²⁵⁷

[B]   Human eyes can only perceive things in accordance with such Forms as they know.   [C]   We forget what a tumble the wretched Phaëton took when, with a mortal hand, he tried to manage the reins of his father's horses: our rashness causes our minds to take a similar plunge and to be bruised and broken as he was.²⁵⁸   [B]   Ask Philosophy what the Sky and the Sun are composed of; what will she answer, if not iron, or,   [C]   with Anaxagoras,   [B]   stone, or some such everyday material? If you ask Zeno what Nature is, he replies Fire – an artificer having as its properties generative powers and regularity; if you ask Archimedes (the master of geometry, that science which grants itself precedence over all others in matters of truth and certainty) he replies that the Sun is a god of burning iron. What a fine idea to come out of geometrical demonstrations, with their beauty and compelling necessities! Not so compelling   [C]   and useful, though,   [B]   but that [C]   Socrates thought you only need to know enough geometry to survey any land given or acquired;   [B]   the illustrious Polyaenus (formerly a famous teacher of geometry) came to despise its demonstrations as false and manifestly vain; that was after tasting the sweet fruits of the idle gardens of Epicurus.²⁵⁹

[C]   In Antiquity Anaxagoras was believed to have excelled all others in treating matters celestial and divine; but in Xenophon, Socrates, talking of

---

256. Echoes of St Augustine, *City of God*, IV, 8; VI, 5 and 7; III, 12 etc.; quotation from Ovid, *Metam.*, I, 194 in Vivès's commentary (ibid., III, 12); Plutarch, *Contre les Stoïques*, 583A (cf. Rabelais, *Quart Livre*, TLF, XXVII, p. 135); Ovid, ibid., VIII, 99.

257. St Augustine, *City of God*, IV, xxxi and xxxvii.

258. Phaëton was the son of Helios and Clymene. Seeking to reach the heavens he was drowned: the symbol of hubris. (The 'forms', or 'Ideas', exist in the heavenly regions; Man only knows those which God makes accessible to him: to try and discover more is to court disaster.)

259. Xenophon, *Memorabilia*, IV, vii, 2; Cicero, *De nat. deorum*, II, xxii, 57–58; for Archimedes and the compelling power of geometry, Cicero, *Acad.: Lucullus*, II, xxxvii, 116–17 (influenced by a reading of S. Bodin, *De la démonomanie des sorciers*); Guy de Brués, *Dialogues*, p. 90.

his teaching, said that the brain of Anaxagoras finally became disturbed: that often happens to those who immoderately pore over matters which do not appertain to them.[260]

As for Anaxagoras' making the Sun a burning stone, he failed to realize that stone does not glow in the fire, or, what is worse, that it is consumed by fire; as for his making the Sun and Fire one, he further failed to realize that fire does not blacken those who simply look at it, that we can gaze fixedly at fire, or that fire *kills* plants and grasses. Socrates' verdict — and mine as well — is that the best judgement you can make about the heavens is not to make any at all.[261]

When Plato in the *Timaeus* was about to talk about *daemons* he declared: This is an undertaking which is beyond our range; we are obliged to have faith in men of old who said they were born of *daemons*: it is not reasonable to refuse to believe these children of the gods — even though what they say is not supported by compelling reasons or by verisimilitude — since they swear they are talking about matters known within their homes and families . . .'[262]

[A] Now let us see whether we have a little more light than that concerning our knowledge of Man and Nature.

When treating objects which, by our own admission, exceed our knowledge, is it not stupid to go forging bodies for them and imposing on them false Forms of our own invention? — as in the case of the movement of the planets: since our minds cannot manage to conceive what makes them move naturally, we impose on them our own heavy corporeal, material principles:

> *temo aureus, aurea summae*
> *Curvatura rotae, radiorum argenteus ordo.*

[The shaft was of gold; so too the rim of the wheels and the spokes were made of silver.][263]

It is almost as though we had sent coach-smiths, carpenters [C] and painters [A] up there, preparing mechanical contrivances with diverse movements [C] and then, in accordance with Plato's instructions, arranging, round about the spindle of Necessity, sets of wheels and interlaced courses for the heavenly bodies, variously painted.[264]

260. Xenophon, *Memorabilia*, IV, vii, 7.
261. Ibid., IV, vii, 7; Socrates' verdict was proverbial (Erasmus, *Adages, Quae supra nos, nihil ad nos*).
262. Plato, *Timaeus*, 40 DE (not evidently ironical in Ficino's Latin rendering, p. 710).
263. Ovid, *Metam.*, II, 107.
264. Plato, *Republic*, X, xii, 616.

[B]   *Mundus domus est maxima rerum,*
Quam quinque altitonae fragmine zonae
Cingunt, per quam limbus pictus bis sex signis
Stellimicantibus, altus in obliquo aethere, lunae
Bigas acceptat.

[The Universe is an edifice, immense, encircled by five thundering belts and crossed obliquely by an aethereal sash, decorated with twice half-dozen constellations and the paired horses of the Moon.][265]

These are dreams [C] and frantic folly. [A] If only Nature would deign to open her breast one day and show us the means[266] and the workings of her movements as they really are [C] (first preparing our eyes to see them). [A] O God, what fallacies and miscalculations we would find in our wretched science! [C] Either I am quite mistaken or our science has not put one single thing squarely in its rightful place, and I will leave this world knowing nothing better than my own ignorance. It was in Plato (was it not?) that I came across the inspired adage, 'Nature is but enigmatic poetry,' as if to say that Nature is intended to exercise our ingenuity, like a painting veiled in mists and obscured by an infinite variety of wrong lights.[267] *'Latent ista omnia crassis occultata et circumfusa tenebris, ut nulla acies humani ingenii tanta sit, quae penetrare in coelum, terram intrare possit'* [All things lie hidden, wrapped in a darkness so thick that no human mind is sharp enough to pierce the heavens or to sound the earth]. Certainly, philosophy is poetry adulterated by Sophists. Where do all those Ancient authors get their authority from, if not from the poets? The original authorities were themselves poets; they treated philosophy in terms of poetic art. Plato is but a disjointed poet. As an insult, Timon called him a great contriver of miracles.[268]

[A] When their natural teeth are missing, women use false ones made of ivory; they replace their real complexion by one contrived from

265. Varro: known only from Probus' commentary on Virgil, *Eclogue*, VI.
266. '88: principles (*ressorts*) for *moiens* (means).
267. Plato, *Alcibiades*, II, 147: 'For poetry as a whole is inclined to be enigmatic'; Ficino's Latin rendering (p. 47) is ambiguous, giving rise to Montaigne's rendering, also found (for example) in Cognatus' adage, *'Multa novit, sed male novit omnia'* (cf. *Adagia, id est proverbiorum . . . omnium*, Wechel, 1643, *index rerum s.v. natura*).
268. Cicero, *Acad.: Lucullus*, II, xxxix, 122.
    '95: disjointed poet. *All superhuman sciences bedeck themselves in the style of poetry.* When their natural . . . (Timon of Athens' insult, repeated by Montaigne in II, 16, 'On glory'; Diogenes Laertius, *Lives*, Plato, III, xxvi, 119.)

borrowed materials; they pad out their thighs with cloth or felt, round out their bellies with cotton-wool and, as everyone knows and sees, enhance themselves with a false and borrowed beauty.

Learning does the same;   [B]   even our system of Law, they say, bases the truth of its justice upon legal fictions. Learning pays us in the coin of suppositions which she confesses she has invented herself. Those eccentric and concentric epicycles by which Astrology tries to make sense out of the motions of the heavenly bodies are presented to us merely as the best she can produce; all Philosophy does the same, presenting us not with what really is, nor even with what she believes to be true, but with the best probabilities and elegancy she has wrought.[269]   [C]   Take Plato, explaining the attributes of the bodies of men and beasts, 'We would be certain that what we say is true, if we could have it confirmed by an oracle; as it is, we can only be certain that I have spoken with the greatest appearance of truth that I can find.'[270]

[A]   Philosophy does not only impose her ropes, wheels and contrivances on to the high heavens. Just think for a while what she says about the way we humans are constructed. For our tiny bodies she has forged as many retrogradations, trepidations, conjunctions, recessions and revolutions as she has for the stars and the planets. They are right to call our bodies *Microcosms* ('little worlds') seeing all the various pieces and angles they need to build them up and cement them together. To house all the activities which they find in Man and all the various functions and faculties which we are aware of within us, think of all the sections into which they have subdivided our souls and how many organs they have ascribed to them; think of all the storeys and levels and all the duties and activities they have assigned to us, over and above the natural ones which our poor humanity can actually perceive! They have invented an entire Republic! Man is an object to be seized and handled. Each philosopher, according to his fancy, has been left entirely free to unstitch him, rearrange him, put him together again and furnish him out afresh.

Yet even now they have not overmastered him. They cannot even dream up an ordinance for Man – let alone find out a true one – without there being some sound or cadence which they cannot quite fit in, however

269. Astronomy, for example, was concerned to 'save the appearances' – that is, to account for observed phenomena; it did not claim to be describing fact but 'appearances' (*phenomena*), which may or may not really be true.
270. Plato, *Timaeus*, 72D (Ficino, p. 724).

abnormal[271] they make their contrivance and however much they try and botch it up with a thousand false and fantastical patches.   [C]   It is wrong to find excuses for them. We do indeed condone artists who represent the sky and far-off lands, seas, mountains or islands with a few slight brush-strokes; we do not know what they are like so are happy with the shadowy imitations that they feign; but when they paint from nature on a known subject – one which we are familiar with – we require of them a perfect, detailed representation of the lines and colours. If they fail, we despise them.[272]

[A]   I have always felt grateful to that girl from Miletus who, seeing the local philosopher Thales with his eyes staring upwards, constantly occupied in contemplating the vault of heaven, made him trip over, to warn him that it was time enough to occupy his thoughts with things above the clouds when he had accounted for everything lying before his feet. It was certainly good advice she gave him, to study himself rather than the sky;   [C]   for, as Democritus says through the mouth of Cicero, *'Quod est ante pedes, nemo spectat: coeli scrutantur plagas'* [Nobody examines what is before his feet: they scrutinize the tracts of the heavens].

[A]   But in fact, the human condition is such that, where our understanding is concerned, the things we hold in our hands are as far above the clouds as the heavenly bodies are!   [C]   As Socrates says in Plato, you can make against anyone concerned with Philosophy exactly the same reproach as that woman made against Thales: he fails to see what lies before his feet. No philosopher understands his neighbour's actions nor even his own; he does not even know what either of them is in himself, beast or Man.[273]

[A]   These people, now, who find Sebond's arguments to be too feeble, these know-alls who are ignorant of nothing and make rules for the whole Universe –

> *Quae mare compescant causae; quid temperet annum;*
> *Stellae sponte sua jussaeve vagentur et errent;*
> *Quid premat obscurum lunae, quid proferat orbem;*
> *Quid velit et possit rerum concordia discors;*

[What limits the seas to their confines, what regulates the years: whether the heavenly bodies travel and wander freely or by constraint; what makes the dark

---

271. '88: monstrous (*monstrueuse*) for abnormal (*enormale*).
272. Plato, *Critias*, 107, CD (adapted) (Ficino, p. 107).
273. Erasmus, *Adages: Ad pedes* (but the servant-girl did not trip him up: he fell); Cicero, *De div.*, II, xiii, 30 (a verse from the *Iphigeneia* of Ennius); Plato, *Theaetetus*, 174 B (Ficino, p. 149).

orb of the Moon to wax or wane, or what the discordant concord of all things can mean or bring about][274] –

have they never, among all their books, plumbed the difficulties which confront them in understanding their own being? Some things can be seen easily enough: our finger and foot are capable of motion; some of our members move on their own while others move only when we make them do so; certain impressions produce a blush, others pallor; some thoughts act on the spleen, others on the brain; some make us laugh, others weep; some stun our minds into ecstasies and arrest the movements of our limbs; [C] there are objects which make our gorges rise, others which raise up something lower down. [A] But no man has yet discovered how purely mental impressions like these can effect such deep incursions into objects as massively solid as our bodies nor the nature of the linking sutures by which these astonishing stimuli are transmitted: [C] *'Omnia incerta ratione et in naturae majestate abdita'* [All things remain unknown to reason and are hidden in the majesty of Nature], says Pliny; and St Augustine: *'Modus quo corporibus adhaerent spiritus, omnino mirus est, nec comprehendi ab homine potest: et hoc ipse homo est'* [How the spirit adheres to the body is entirely a matter of wonder and cannot be understood by Man; nevertheless this union of body and spirit is Man].[275]

[A] And yet everybody knows the answer! Merely human opinions become accepted when derived from ancient beliefs, and are taken on authority and trust like religion or law! We parrot whatever opinions are commonly held, accepting them, as truths, with all the paraphernalia of supporting arguments and proofs, as though they were something firm and solid; nobody tries to shake them; nobody tries to refute them. On the contrary, everybody vies with each other to plaster over the cracks and prop up received beliefs with all his powers of reason – a supple instrument which can be turned on the lathe into any shape at all. Thus the world is pickled in stupidity and brimming over with lies.

We do not doubt much, because commonly received notions are assayed by nobody. We never try to find out whether the roots are sound. We argue about the branches. We do not ask whether any statement is true, but what it has been taken to mean. We ask whether Galen said this or said that: we never ask whether he said anything valid.

It is understandable that this curb on our freedom of judgement and this tyranny over our beliefs should spread to include the universities and the

274. Horace, *Epistles*, I, xii, 16.
275. Pliny, *Hist. nat.*, II, xxxvii; St Augustine, *City of God*, XXI, 10.

sciences: Aristotle is the god of scholastic science: it is heresy to discuss his commandments (as it once was to discuss those of Lycurgus in Sparta). What Aristotle taught is professed as law – yet like any other doctrine it may be false. Where the first principles of Nature are concerned I cannot see why I should not accept, as soon as the opinions of Aristotle, the 'Ideas' of Plato, the atoms of Epicurus, the plenum and vacuum of Leucippus and Democritus, the water of Thales, the infinity of Nature of Anaximander, or the aether of Diogenes, the numbers and symmetry of Pythagoras, the infinity of Parmenides, the Unity of Musaeus, the fire and water of Apollodorus, the homogeneous particles of Anaxagoras, the discord and concord of Empedocles, the fire of Heraclitus, or any other opinion drawn from the boundless confusion of judgement and doctrines produced by our fine human reason, with all its certainty and perspicuity, when it turns its attention to anything whatever.

Aristotle based the principles of Nature on three elements: matter, form and privation. Yet what is more silly than actually to make a vacuum into one of the causes of the production of material objects? Privation is a negative: what fanciful humour led Aristotle to make it the original cause of objects which actually exist? Yet, except as an exercise in logic, nobody dares to shake that belief. Nobody debates anything to increase doubt but only to defend the founding author of their school against outside objections; his authority marks the goal; beyond it, no further inquiry is permitted.[276]

Base yourself on admitted postulates and you can build up any case you like; from the rules which order the original principles the remainder of your construction will follow on easily without self-contradiction.

This method allows us to bowl our arguments with the jack in view (and so be satisfied that our foundations are rational ones); before they even begin, our professors (like geometricians with their postulated axioms) establish such a hold over our beliefs that they can subsequently reach any conclusion they want. We give them our agreement and consent: they can then pull us this way and that way, spinning us about at will. Once we accept anyone's postulates he becomes our professor and our god: for his foundations he will grab territory so ample and so easy that, if he so wishes, he will drag us up to the clouds. In the practice and business of scholarship we have accepted Pythagoras' contention as legal tender: every

276. Criticism of Aristotle's doctrine of the creative force of privation was current: e.g. in Ramus and in Guy de Brués, *Dialogues*, 161. Cf. also Cicero, *Acad.*: *Lucullus*, II, xxxvii (118–19); *De nat. deorum*, I, X, xi.

expert, he says, must be believed in his own speciality. So, for the meaning of words the logician turns to the grammarian; for the matter of his arguments the rhetorician borrows from the logician; the poet takes his rhythms from the musician; the geometer takes his propositions from the arithmetician; the metaphysicians make their foundations out of the conjectures of physics. For their principles, all branches of learning take admitted postulates, which restrain human judgement on every side. If you come up against the barrier behind which their error of principle is sheltering, they have an axiom ready on their lips: Never argue with those who deny first principles.[277] But there can be no first principles unless God has revealed them; all the rest – beginning, middle and end – is dream and vapour.

Whenever a case is fought from preliminary assumptions, to oppose it take the very axiom which is in dispute, reverse it and make that into your preliminary assumption. For any human assumption, any rhetorical proposition, has just as much authority as any other, unless a difference can be established by reason. So they must all be weighed in the balance – starting with general principles and any tyrannous ones. [C] To be convinced of certainty is certain evidence of madness and of extreme unsureness: no people are more insane or less philosophical than the 'lovers of opinion' whom Plato dubbed *philodoxoi*.[278] [A] We want to find out by reason whether fire is hot, whether snow is white, whether anything within our knowledge is hard or soft. There are ancient stories of the replies made to the man who doubted whether heat exists – they told him to jump into the fire – or to the one who doubted whether ice is cold – they told him to slip some into his bosom: but a reply like that is quite unworthy of the professed aims of philosophy. Philosophers could have spoken in this way only if they had left us in a state of nature, simply accepting external appearances as they offer themselves to our senses, or if they had left us to follow our basic appetites, governed only by such modes of being as we are born with. But they themselves have taught us to make judgements about the universe; they themselves have fed us with the notion that human reason is the Comptroller-General of everything within and without the

277. H. C. Agrippa, *De Vanitate*, III (*ad fin.*). The axiom cited above was not Pythagorean: cf. Cognatus' adage, '*Peritis in sua arte credendum*'.
278. Plato, *Republic*, V. 480 A. (For what follows, cf. Erasmus, *Apophthegmata*, III, *Diogenes*, L: when Zeno was proving 'by most acute arguments that there is no such thing as motion', Diogenes got up and walked away. 'What are you doing, Diogenes?' asked Zeno in surprise. 'I am confuting your arguments,' he replied.)

vault of heaven; they themselves say that it can embrace everything, do everything and is the means by which anything is known or understood. Such replies would be good among the Cannibals who live long and happy lives, in peace and tranquility, without the benefits of Aristotle's precepts and without even knowing what the word 'physics' means. Perhaps such a reply could even be better and more firmly based than all the ones which philosophers owe to reason or discovery. Such arguments would be within the capacity of ourselves, of all the animals and of all for whom the pure and simple law of Nature still holds sway. But they themselves have renounced such arguments. They must not tell me: 'This is true; you can see it is; you can feel it is.' What they must tell me is whether I really and truly feel what I think I feel; and if I do feel it, they must go on and tell me why and how and what: let them tell me the name, origin, connections and frontiers of heat or of cold and what qualities are found in the agents and patients of heat and of cold. Otherwise, let them abandon their professional intention, which is to accept nothing and approve nothing except by following the ways of reason. When they have to assay anything, reason is their touchstone. But it is, most surely, a touchstone full of falsehood, error, defects and feebleness. How better to test that than by reason itself. If we cannot trust reason when talking about itself, it can hardly be a judge of anything outside itself.

If human reason knows anything at all, it must be its own essence and its own domicile. It is domiciled within the soul, being either a part of it or one of its activities — as for the permanent home of that true and essential Reason, whose name we steal under false colours, it is in the bosom of God: that is the habitation where it dwells; that is where it comes from when it pleases God to allow us to have a glimmer of Reason, like Pallas leaping from the head of her Father to make herself known unto the world.

Now let us see what human reason can tell us about itself and about the soul! [C] I am not talking now of that generic soul, in which virtually all philosophy makes the heavenly bodies and the elements to share; nor of that soul which Thales, prompted by his study of the magnet, attributes to objects normally considered inanimate; I am concerned with the soul which belongs to us, the one we should know best:

> [B] *Ignoratur enim quae sit natura animai,*
> *Nata sit, an contra nascentibus insinuetur,*
> *Et simul intereat nobiscum morte dirempta,*
> *An tenebras orci visat vastasque lacunas,*
> *An pecudes alias divinitus insinuet se.*

[The nature of the soul is not known; whether it is innate or, on the contrary, slipped into creatures at the moment of their birth; does it die when we die, does it visit the darkness and the vast depths of Orcus, or else does it, under divine guidance, slip into animals different from ouselves?][279]

[A] Reason taught Crates and Dicaearchus that there is no soul (bodies being endowed with natural power of movement); it taught Plato that the soul is a self-moving substance; Thales, that soul is a natural substance, never in repose; Asclepiades, an exercising of the senses; Hesiod and Anaximander, a substance composed of fire and water; Parmenides, of earth and fire; Empedocles, of blood:

> *– Sanguineam vomit ille animam*

[He vomits up his soul of blood]–

Possidonius, Cleanthes and Galen, that the soul is heat or a hot complexion –

> *Igneus est ollis vigor, et coelestis origo*

[Souls have a fiery vigour and a heavenly origin] –

Hippocrates, a spirit spread throughout the body; Varro, air, infused through the mouth, warmed in the lungs, refreshed in the heart and spread throughout the body; Zeno, the quintessence of four elements; Heraclides of Pontus, light; Xenocrates and the Egyptians, number in motion; the Chaldeans, a power of indeterminate form:

> [B] *habitum quemdam vitalem corporis esse,*
> *Harmoniam Graeci quam dicunt.*

[There is a certain life-giving quality in the body which the Greeks call Harmony.][280]

[A] And let us not overlook Aristotle, who said the soul was that power which naturally moved the body and which he called *entelechia* (as dull an

---

279. Lucretius, I, 112 (Lambin, p. 16). The following list of opinions combines commonplaces from Sextus Empiricus, Cicero and, especially, H. C. Agrippa, *De Vanitate*, II. But one of the most influential studies of the soul in the Renaissance was Melanchthon's *De anima*. Some of the matter of the following pages can be found there or may derive from there.
280. Virgil, *Aeneid*, IX, 349 and VI, 730. Both cited in Melanchthon, *De anima* (*Opera*, 1541, III, 9); Lucretius, III, 99 (Lambin, pp. 198–9).

idea as anyone else's, for he does not mention the essence, origin or nature of the soul but merely notes what it does). Lactantius, Seneca and the better part of the Dogmatists all confessed that they did not know what it was. [C] And after running through all these opinions, Cicero comments: '*Harum sententiarum quae vera sit, deus aliquis viderit*' [It is up to some god or other to say which of these is true]. [A] 'I know from myself', said St Bernard, 'how incomprehensible God is: I cannot even comprehend the constituents of my own being.' [C] Heraclitus held that everything is full of souls and daemons; he nevertheless maintained that whatever advances we may make in our knowledge of the soul, we would never get to the end, since its essence is too profound.[281]

[A] There is just as much disagreement and argument about the seat of the soul: Hippocrates and Hierophilus lodge it in the ventricle of the brain; Democritus and Aristotle, throughout the body –

> [B]   *Ut bona saepe valetudo cum dicitur esse*
> *Corporis, et non est tamen haec pars ulla valentis.*

[As we often say that a man has a healthy body, without implying that health is part of a healthy man.] –

[A]   Epicurus lodges it in the stomach –

> [B]   *Hic exultat enim pavor ac metus, haec loca circum*
> *Laetitiae mulcent;*

[For terror and fear make the stomach tremble, while joys soothe its pains;][282]

[A]   the Stoics lodge it within and around the heart; Erasistratus, adjoining the membrane of the epicranium; Empedocles, in the blood – like Moses, who for this reason forbade men to 'eat the blood' of beasts (whose soul is within the blood);[283] Galen thought that each part of the body had its own soul; Strato lodged it between the eyebrows: [C] '*Qua facie quidem sit animus, aut ubi habitet, ne quaerendum quidem est*' [As for the aspect of the

---

281. For entelechy (*actuality* or *activity*) as principle of soul, see Aristotle, *De anima*, 2, I and *Metaph.*, 8. 3; discussed, similarly, in Melanchthon, *De anima*, II ff. (cf. Tertullian, *De anima*, 32); Cicero, *Tusc. disput.*, I, xi; St Bernard, *De anima seu meditationes devotissimae*, I, *in princ.*; Diogenes Laertius, *Lives*, Heraclitus, IX, vii. What follows may be influenced by H. C. Agrippa, *De Vanitate*, LII; for Renaissance scholarship, see Melanchthon, *De anima*, 17 ff. (*Quid est organum?*).
282. Lucretius, III, 102; 142 (Lambin, pp. 198–99, 201–204) where stomach = breast.
283. A basic interdict of the Law of Moses, e.g. Leviticus 7:26–27; but it is the *anima* (life) not *animus* (mind) which is 'in the blood': ibid., 17:11. Cf. Melanchthon, *De anima*, 16.

soul and the place wherein it dwells, we should not even try to inquire]. – I gladly let that fellow Cicero use his own words (should I dare to contaminate the utterances of Eloquence!) and there is little to gain from stealing the substance of his own ideas, which are neither frequent, sound nor unknown.[284]

[A] But the reason which led Chrysippus and others of his sect to make a case out for the heart is not to be forgotten: it is (he says) because, when we want to swear an oath, we place our hand upon our bosom, and when we want to pronounce the word ἔγω (which means 'I') we lower our jaw towards our chest. This passage should not be allowed to slip by without a remark about such silliness in so great a person. Even if you leave aside the total lack of weight in the argument as such, his last proof could only convince Greeks that their soul is where he said it is. No man's judgement is so alert as never to nod off to sleep![285] [C] Why are we afraid to say so? Here are the Stoics, the fathers of human wisdom, finding that, when a man is buried under the weight of a fallen building, his soul cannot extricate itself but makes lengthy struggles to get free – like a mouse in a trap!

Some[286] maintain that the world was made specifically to give bodies to souls, as a punishment for having wilfully fallen from their original purity; at first they were simply incorporeal; they are given light or heavy bodies, depending upon how far they have withdrawn from their original spiritual state (which explains the great variety of created matter). The spirit who, as a punishment, was invested with the body of the Sun must have fallen off in some very rare and special way!

The frontiers of our research are lost in dazzling light. Plutarch, writing of the fountain-heads of history, says that when we push our investigations to extremes, they all fall into vagueness, rather like maps where the margins of known lands are filled in with marshes, deep forests, deserts and uninhabitable places.[287] That explains why the most gross and puerile of rhapsodies are to be found among thinkers who penetrate most deeply into the highest matters: they are engulfed by their curiosity and their arrogance.

284. Cicero, *Tusc. disput.*, I, xxvii, 67. Montaigne used Cicero as a source, but he was impatient with his wordiness and credited him with no originality as a thinker.
285. Galen, *De placitis Hippocratis et Platonis*, II, ii; Stoics, rejected by Seneca, *Epist.*, LVII, 7–8. In the original French, Montaigne confusingly uses *estomach* in both its Latin sense (stomach) and its Greek sense (breast).
286. Platonists, including Origen (criticized by St Augustine, *City of God*, XI, 23).
287. Plutarch, *Life of Theseus*, I, 1.

The beginnings and the ends of our knowledge are equally marked by an animal-like stupor: witness Plato's soarings aloft in clouds of poetry and the babble of the gods to be found in his works. Whatever was he thinking about when he [A] defined Man as an animate creature with two legs and no feathers? He furnished those who wanted to laugh at him with an amusing opportunity for doing so. For, having plucked a live capon, they went about calling it 'Plato's Man'.[288]

And the Epicureans too. With what simple-mindedness they first imagined that the universe had been formed by their atoms (which, they said, were bodies having some weight and a natural downward movement) until their opponents reminded them that, by their own description, it was impossible for these atoms to link up together: their fall, being straight and perpendicular, could only be effected along parallel lines. This obliged them to add a quite fortuitous sideways motion, and to furnish their atoms with curved hooks on their tails by which they could link themselves firmly to each other. [C] Even then, they were in trouble from others, who hounded them with another consideration: if atoms do, by chance, happen to combine themselves into so many shapes, why have they never combined together to form a house or a slipper? By the same token, why do we not believe that if innumerable letters of the Greek alphabet were poured all over the market-place they would eventually happen to form the text of the *Iliad*?

That which is capable of reasoning, argued Zeno, is superior to that which is not: nothing is superior to the Universe, therefore the Universe is capable of reasoning. Cotta used the same argument to make the Universe into a mathematician and another argument of Zeno's to make it into a musician – an organist. The whole is greater than the part: we, who are parts of the Universe, are capable of wisdom: therefore the Universe is wise.[289]

[A]   One can find innumerable examples[290] of similar arguments which are not only false but inept and unable to hold together, emphasizing that their inventors were not so much ignorant as silly; you can find them in the criticisms which philosophers make of each other in their clashes of opinion and in the disagreements between Schools.

288. Diogenes Laertius, *Lives*, Diogenes, VI, 40.
     '80: sleep. And then Plato defined man . . .
289. Cicero, *De fin.*, I, v, 13–vi, 21; *De nat. deorum*, II, xxxvi, 93–4 (adapted); III, ix, 20–3. Cotta is mocking Zeno.
290. '88: find *many similar* examples. . .; (in place of [C],   below): schools, *as you can see in the infinite examples in Plutarch, against the Epicureans and Stoics: and in Seneca against the Peripatetics.* We . . .

[C] Anyone who made an intelligent collection of the asinine stupidities of human Wisdom would have a wondrous tale to tell. I like collecting such things as evidence which, from some angles, can be studied as usefully as sane and moderate opinions. [A] We can judge what we should think of Man, of his sense and of his reason, when we find such obvious and gross errors even in these important characters who have raised human intelligence to great heights. Personally I prefer to believe that they treated knowledge haphazardly, sporting with it, in any fashion, like a toy and that they played with reason as if it were some vain and frivolous instrument, putting forward all kinds of thoughts and fantasies, some forceful, others, weak. The selfsame Plato who defined Man as a capon, elsewhere follows Socrates and says that, in truth, he does not know what Man is, and that Man is one of the hardest things in the world to understand.²⁹¹ With such varied and unstable opinions they lead us tacitly by the hand to inconclusive conclusions. They profess that they do not present the face of their thought openly and unveiled; they hide it beneath obscurities of poetic fable or behind some other mask. Our imperfection is such that raw meat is not always proper food for our stomachs: it first has to be dried, treated or hung. They do the same: they sometimes take their straightforward opinions and judgements and hide them behind obscurity [C] and season them with falsehood, [A] so as to prepare them for public consumption. They do not want to make an express avowal of the ignorance and weakness of human reason — [C] they want to avoid frightening the children — [A] but they give us a good glimpse of it beneath the appearance of confused and unstable erudition.

[B] When I was in Italy, I advised a man who was at pains to learn Italian that if it were merely to be understood, without excelling in any other way, he should simply use the first words which came to his lips, Latin, French, Spanish or Gascon, and stick an Italian ending on them; he would never fail to hit on some local dialect, Tuscan, Roman, Venetian, Piedmontese or Neapolitan: there are so many forms that he was bound to coincide with one of them. I say the same about Philosophy. She has so many faces, so much variety and has been so garrulous, that all our ravings and our dreams may be found within her. Human fancy can conceive nothing, good or evil, which is not there already. [C] 'Nihil tam absurde dici potest quod non dicatur ab aliquo philosophorum' [Nothing can be so absurd that it has not already been said by one of the philosophers].²⁹² [B] So I

---

291. Plato, *Alcibiades*, I, 129 A.
292. Cicero, *De divinat.*, II, lviii, 119.

am all the more ready to give a free run to my own whims in public: I know they were born to me, not modelled on others, but you can always find some Ancient or other whose fantasies are akin to them. There will always be somebody to say, 'Look, he got it from there.'

[C] My ways of life are natural to me: in forming them I have never called in the help of any erudite discipline; but when I was seized with the desire to give a public account of them, weak as they are, I made it my duty to help them along with precepts and examples, so that I could publish them more decorously. I was then astonished myself to find that, by sheer chance, they were in conformity with so many philosophical examples and precepts. Only after my life was settled in its activity did I learn which philosophy was governing it! A new character: a chance philosopher, not a premeditated one!

[A] To get back to our souls,[293] Plato placed reason in the brain, anger in the heart, desire in the liver; but that probably resulted from an interpretation of the emotions of the soul rather than from any desire to divide the soul up into separate parts; it was more like one body with several members. The most likely of all these opinions states that the human soul is one single entity with the faculties for ratiocinating, remembering, comprehending, judging and desiring; it exercises its other functions through the instrumentality of the various parts of the body (just as the seaman sails his vessel according to his experience of it, at times tightening or slackening a sheet, at others hoisting the yard or pulling the oar, one single power organizing all these actions); the seat of this power is the brain, as is clearly shown by the fact that wounds and accidents affecting the head immediately harm the faculties of the soul; it is not inappropriate, therefore, that this power should extend from the brain to the rest of the body[294] −

> [C]   *medium non deserit unquam*
> *Coeli Phoebus iter; radiis tamen omnia lustrat.*

[Phoebus never deserts his path through the sky, yet bathes all things with light from his rays] −

just as the Sun in the sky pours out its light and its powers and fills the whole universe:

293. '88: souls, (*for I have chosen this one example as being the most convenient for witnessing to our feebleness and vanity*) Plato. . .(Cf. Melanchthon, *De anima*, 29 f.)
294. Diogenes Laertius, *Lives*, Plato, III, lxvii, 224 *apud* Guy de Brués, p. 79 f.

> *Caetera pars animae per totum dissita corpus*
> *Paret, et ad numen mentis nomenque movetur.*

[The remainder of the soul, scattered throughout the body, obeys, and is activated by the majesty and authority of the mind.][295]

Some said that there is a general Soul, like some huge body, from which individual souls were extracted, later returning there to be re-absorbed in that universal matter:

> *Deum namque ire per omnes*
> *Terrasque tractusque maris coelumque profundum:*
> *Hinc pecudes, armenta, viros, genus omne ferarum,*
> *Quemque sibi tenues nascentem arcessere vitas;*
> *Scilicet huc reddi deinde, ac resoluta referri*
> *Omnia: nec morti esse locum;*

[For God is said to spread through all lands, all tracts of sea and highest heaven; from him all flocks and herds and men and every race of beast all take, at birth, their tenuous lives, and to him all things eventually return, when they are loosened asunder: and so there is no place at all for death;][296]

others said that the individual souls merely rejoined this general Soul – attached to it, but as individuals; others said that souls were produced from the divine substance itself; others, from fire and water, by angels; some said they existed from the earliest times; others, that they were created only when actually required. Some said they came down from the circle of the Moon and later returned there. Most of the Ancients held that, exactly like all other natural things, they were engendered from father to son, adducing as an argument the resemblance of sons to their fathers:[297]

> *Instillata patris virtus tibi:*
> *Fortes creantur fortibus et bonis.*

[Your father's virtue is transmitted to you; strong men are born from strong men and good.][298]

---

295. Claudian, cited in the *Politici* of Justus Lipsius, IV, ix; then Lucretius, III, 143 (Lambin, pp. 201–2). Montaigne misreads *momen* (impulse) as *nomen* (name, authority) despite Lambin's explanation.
296. Aristotelian opinions, backed by Virgil, *Georgics*, IV, 221.
297. This doctrine (traducianism) is discussed by Melanchthon, *De anima*, along with other notions mentioned by Montaigne.
298. First line, anon., second, Horace, *Odes*, IV, iv, 29.

Not only physical characteristics were held to flow like this from father to son but similar humours, complexions and inclinations of the soul:

> Denique cur acris violentia triste leonum
> Seminium sequitur; dolus vulpibus, et fuga cervis
> A patribus datur, et patrius pavor incitat artus;
> Si non certa suo quia semine seminioque
>   Vis animi pariter crescit cum corpore toto?

[Finally, why is impetuous ferocity the hereditary mark of the dire lion family, trickery of the fox and swiftness of the deer (which inherits the paternal instinct towards timorous flight) if not because the soul is born from semen and grows with the rest of the body?]

This was held to be the basis of divine Justice which readily visits upon the children the sins of the fathers, because the pollution of the fathers' vices is to some extent imprinted upon the souls of their children, who are influenced by their fathers' unruly desires.[299]

Moreover if souls do not come from natural succession but by some other way — if, say, they existed beforehand as entities independent of their bodies — they would have had some memory of their former state, given that reflection, reason and memory are the natural properties of the soul:

> [B]   si in corpus nascentibus insinuatur,
>   Cur superante actam aetatem meminisse nequimus,
>   Nec vestigia gestarum rerum ulla tenemus?

[If souls are only introduced into the bodies at birth, why cannot we fully remember what happened before nor retain any trace of the things which we did?][300]

[A]   If we are to give the value we wish to the attributes of our souls, we are obliged to assume that, even in their natural simplicity and purity, they

---

299. Lucretius, III, 741 (Lambin, 241–2). Cf. Andreas Tiraquellus, De legibus connubialibus, VII, I–4. It was accepted that sensitive and vegetative souls could be transmitted in semen: the human rational soul was individually created (Melanchthon, De anima, 15).

300. Lucretius, III, 671 (Lambin, p. 235: criticism of Pythagoreans, citing Aristotle). There follows criticism of the Platonic doctrine that all learning is recollection of knowledge pre-dating the imprisonment of the soul in the body (Phaedo, XVIII, 73E). Similar refutations are found elsewhere (e.g. in L. Joubert's Erreurs populaires, 1578 (Preface), exploited above, note 66, on natural language). Christianity avoids the problem of rewards and punishments in the afterlife by making them depend on the presence or absence of imputed merits (Christ's not Man's).

are full of knowledge; free from the prison of the body, our souls, therefore, must be such, before they entered their bodies, as we hope they will be once they have gone forth from them; so, while they are in the body, they must continue to remember that knowledge: hence Plato's assertion that whatever we learn is really the recollection of what we once knew. But we all know that to be false from our own experience. First: we remember nothing save what we have been taught; if memory did its duty 'purely', it would at least hint at something beyond our apprenticed knowledge. Second: what the soul knew in her pure state was true knowledge: since her intelligence was divine, she knew things as they really are; here below you can make the soul accept lies and errors, if you teach them to her. She cannot be using her powers of recollection in that case, since she had never accommodated such Forms and concepts!

But to say that her imprisonment in the body smothers her native faculties so completely as to snuff them right out, runs, first of all, contrary to that other belief: that we can recognize her powers to be so great, and those of her workings which we are conscious of in this life to be so wonderful, that they allow us to conclude that she is divine, has existed from all eternity and will enjoy immortality.

> [B] *Nam, si tantopere est animi mutata potestas*
> *Omnis ut actarum exciderit retinentia rerum,*
> *Non, ut opinor, ea ab leto jam longior errat.*

[For if all the faculties of the soul are so completely changed that no memory of the past remains, that seems to me to be no different from extinction.][301]

[A] Moreover, the powers and actions of our souls must be examined not elsewhere but here, at home in our bodies. Any other perfections they may have are useless and irrelevant; it is for their present state that their whole immortality will receive its acknowledged rewards: each is entirely accountable for the life of a human being. But it would be an act of gross injustice to lop off the soul's powers and resources, to strip her of all her weapons and then to take the very time when she lies weak and ill in prison – a time of repression and constraint – and to make that the basis for a judgement leading to endless, everlasting punishment; it would be unjust to limit consideration to so short a span, to a life that may have lasted a mere two hours or, at the very worst a hundred years – an instant in proportion to infinity – and then, from that momentary interlude, to order and establish, once and for all, the whole state of her future existence. To reward or

301. Lucretius, III, 674 (Lambin, pp. 265–7, reading *longior* for *longiter*).

punish on the basis of so short a life would be disproportionate and iniquitous.

[C] To get out of this difficulty, Plato wants future rewards and punishments never to exceed a hundred years and always to be proportionate to the actual length of a man's life. Quite a few Christians too have imposed temporal limits on to them.[302]

[A] As a result of all this men followed Epicurus and Democritus (whose opinions were most widely received); they concluded that the generation and life of the soul shared all the usual characteristics of things human. Many striking features make this seem probable: they could see that the soul was born precisely when the body was capable of receiving her; that her strength increased as the body's did: it was observed that the soul was weak in infancy and then, eventually, experienced a vigorous maturity, a decline into old age and, finally, decrepitude:

> *gigni pariter cum corpore, et una*
> *Crescere sentimus, pariterque senescere mentem.*

[We can feel that the soul is born with the body, grows up with it and then grows old.][303]

Man perceived that the soul can experience various passions and be disturbed by several emotions which subject her to pain and lassitude; she is capable of change, including change for the worse; she is capable of joy, tranquillity, languor; like the stomach or the foot, she is subject to wounds and illness:

> [B] *mentem sanari, corpus ut aegrum*
> *Cernimus, et flecti medicina posse videmus.*

[We see that the mind can be cured like the body and be modified by drugs.]

[A] She can be confused and dazed by the powers of wine, be upset by the vapours of a burning fever; be lulled to sleep by certain drugs and aroused by others:

> [B] *corpoream naturam animi esse necesse est,*
> *Corporeis quoniam telis ictuque laborat.*

---

302. Plato, *Republic*, X, 615. Origen and the Universalists held that, eventually, Hell would be empty and all would be saved. Montaigne may also be alluding to misconceptions of Purgatory (as a modification of Hell, rather than of Heaven).
303. A series of sustained borrowings from Lucretius, III, 445 f.; 510 f.; 175 f.; 499–501; 492 f.; 463 f.; 800 f.; 458; 110 f. Throughout, the comments of Lambin are relevant (pp. 190–272). For a Christian answer in the dedication of Book III of Lucretius, see the Introduction, p. xxxvii.

[The nature of the mind is necessarily corporeal, for it can be hurt by physical cuts and blows.]

[A]   Men saw that all the soul's faculties can be stunned and overthrown by the mere bite of a sick dog; that the soul has no way of avoiding any of these accidents, even by showing the utmost firmness of mind or any moral quality or virtue, by philosophical determination or by any straining of her forces. Let the saliva of some wretched dog slaver over the hand of Socrates and they knew that it would put a sudden end to all his wisdom and to all his mighty, disciplined thought, reducing them to nothing, so that no trace whatever would remain of his original awareness:

[B]   *vis animai*
Conturbatur, . . . et divisa seorsum
Disjectatur, eodem illo distracta veneno.

[The power of the soul is disturbed and its parts are broken up and dispersed by that same poison.]

[A]   They knew that the poison would find no greater powers of resistance in his soul than in a four-year-old's: if Philosophy herself became incarnate, such a poison would make her lose her senses and drive her insane. Cato could wring the neck of Death and Destiny, but if ever he had been bitten by a mad dog and contracted that illness which doctors call *hydroforbia*,[304] even he would have been overcome with fear and terror, quite unable to bear the sight of water or a looking glass.

[B]   *vis morbi distracta per artus*
Turbat agens animam, spumantes aequore salso
Ventorum ut validis fervescunt viribus undae.

[The power of the disease spreading through one's limbs drives the soul to distraction, like stormy winds lashing the waves of the troubled sea.]

[A]   While we are on this subject, Philosophy has armed Man well against all the other ills which may befall him, teaching him either to bear them or else, if the cost of that is too high, to inflict certain defeat on them by escaping from all sensation. But such methods can only be of service to a vigorous soul in control of herself, a soul capable of reason and decision: they are no use in a disaster such as this, where the soul of a philosopher becomes the soul of a madman, confused, lost and deranged. This can happen from several causes: by some excessive emotion which snatches the

304. Ignorant medical deformation of *hydrophobia*.

mind away; by some strong passion engendered by the soul herself; by a wound in certain parts of the body; by a gastric vapour subjecting the soul to giddiness and confusion:

> [B]   *morbis in corporis, avius errat*
> *Saepe animus: dementit enim, deliraque fatur;*
> *Interdumque gravi lethargo fertur in altum*
> *Aeternumque soporem, oculis nutuque cadenti.*

[During physical illness, the soul often goes astray, becoming mad and talking deliriously; sometimes it plunges into a deep lethargy, into a perpetual sleep, as the eyes close and the head droops down.]

[A] Philosophers, it seems to me, have hardly begun to pluck that particular chord; [C] no more than another one of similar importance. To console us in our mortal state they constantly present us with the following dilemma: the soul is either mortal or immortal; if mortal, she will be without pain; if immortal she will go on improving. But they never touch on the other alternative. What if she goes on getting worse! They simply hand threats of further punishment over to the poets. But that game is far too easy.

I am often struck by these two omissions in their argument: I now go back to the first. [A] The deranged soul loses all taste for the Sovereign Good of the Stoics, so constant and so resolute. On this point our wisdom, fair though she is, really must surrender and lay down her arms.

Meanwhile the vanity of human reason led philosophers to conclude that a composite being, linking in fellowship two elements as diverse as mortal body and immortal soul, is quite inconceivable.

> *Quippe etenim mortale aeterno jungere, et una*
> *Consentire putare, et fungi mutua posse,*
> *Desipere est. Quid enim diversius esse putandum est,*
> *Aut magis inter se disjunctum discrepitansque,*
> *Quam mortale quod est, immortali atque perenni*
> *Junctum, in concilio saevas tolerare procellas?*

[It is mad to think that the mortal is able to be joined to the eternal, to agree together and each to help the other. What can we possibly conceive more different, or, rather, more contrary and incompatible, than these two elements, one mortal, the other immortal and eternal, which you would join together to ride out the wildest storm?]

Moreover the soul, like the body, was thought to be involved in death,

[B]   *Simul aevo fessa fatiscit.*

[She droops down, tired out with age.]

[C]   According to Zeno this is shown to us clearly by the image of sleep (which he thought was both the soul and the body dropping down in a faint): *'Contrahi animum et quasi labi putat atque concidere'* [He conceived that the soul contracts, as it were, collapses and falls down in a swoon].[305]

[A]   It was recognized that the soul may sometimes retain her force and vigour to the end; that was explained by the different varieties of illness, just as some men retain one or other of their senses intact to the end – their hearing, say, or their sense of touch – nobody being so enfeebled as to have absolutely no part vigorous and whole.

[B]   *Non alio pacto quam si, pes cum dolet aegri,*
In nullo caput interea sit forte dolore.

[In the same way, a sick man's feet may feel sharp pains, without his head feeling anything.][306]

Our mental insight is to Truth what an owl's eyes are to the splendour of the sun. Aristotle says that. Is there any better way of convicting ourselves than by noting such total blindness in so clear a light?

[A]   Now for the contrary opinion: that the soul is in fact immortal.   [C]   Cicero says that, at least as far as books are concerned, it was first introduced by Pherecides of Scyros in the time of King Tullus. Some others attribute it to Thales, and there are other candidates.[307]   [A]   This branch of human learning is treated with the greatest reservation and doubt. On this matter, even the most confirmed Dogmatists are mainly constrained to shelter behind the shadowy teachings of Plato's Academy. On this subject, nobody knows what Aristotle's conclusions were,   [C]   no more than those of the Ancients in general, who handle the matter with a kind of vacillating belief: *'rem gratissimam promittentium magis quam probantium'* [a thing most pleasing, but more in promise than in proof].[308]   [A]   Aristotle hid behind a cloud of difficult and incomprehensible words and meanings, leaving his followers arguing as much about what he meant as about the matter itself.

---

305. Cicero, *De divinat.*, II, lviii, 119. Montaigne takes some of these arguments up again in III, 13, 'On experience'.
306. The last of this series of borrowings from Sextus Empiricus; then Aristotle, *Metaphysics*, II, I, 993 b (a *bat* not an *owl*).
307. Cicero, *Tusc. disput.*, I, xvi.
308. Seneca, *Epist.*, CII, 2 (a major treatment of the theme of immortality, influencing the following argument).

Two considerations made this opinion plausible to them: first, that without the immortality of the soul, fame would have no secure basis and so be hoped for in vain. (By the standards of the world that is a consideration of wonderful importance.) The second is one of utility: it is useful that people should be convinced, [C] as Plato says, [A] that even when vices escape the dark and uncertain vigilance of human justice, they still remain exposed to that of divine Justice which will pursue them even after the death of the guilty.[309] [C] Man takes extreme care to prolong his being, providing for it by all possible means: he has tombs to preserve his body and fame to preserve his soul.

Dissatisfied with his lot, Man has given free run to his opinions, building himself up into something else and propping himself up with his own ingenuity. The soul can never find a sure footing; she is too confused and weak for that. She roams about seeking bases for her hopes and consolations in conditions which are foreign to her nature. She clings to them and puts down roots. These notions which she ingeniously forges for herself may be ever so frivolous and fantastic, but she can find repose in them more surely than in herself, and much more willingly. [A] But it is a source of wonder that even those who are most obstinately attached to so just and clear a persuasion as spiritual immortality fall short, being powerless to establish it by their human ability. [C] One Ancient writer said, '*Somnia sunt non docentis, sed optantis*' [They are not the dreams of one who demonstrates but of one who desires].[310] [A] From this evidence Man realizes that such truth as he does find out for himself is due to Fortune and to chance. Even when truth drops into his hands, Man has no means of seizing hold of it; his reason does not have power enough to establish any rights over it. Every single idea which results from our own reflections and our own faculties – whether it is true or false – is subject to dispute and uncertainty. In bygone days God produced the confusion and disorder of the Tower of Babel as a chastisement of our pride, to teach us our wretchedness and our inadequacy. Everything we undertake without God's help, everything we try and see without the lamp of his grace, is vanity and madness. The essence of Truth is to be constant and uniform: when Fortune arranges for a little of it to come into our possession, out of weakness we corrupt it and debase it. Any course a man may adopt on his own is allowed by God to lead to this same confusion, the idea of which is so vividly portrayed in the just punishment which God visited upon the

309. Plato, *Laws*, X, 907.
310. Cicero, *Acad.: Lucullus*, II, xxxviii, 121 (citing Democritus).

arrogance of Nembroth, bringing to nought his vain attempts to build that pyramidal Tower: [C] *'Perdam sapientiam sapientium et prudentiam prudentium reprobabo'* [I will destroy the wisdom of the wise, and the understanding of the prudent I will reject]. [A] That diversifying of tongues and language by which God threw confusion over the enterprise of Babel, what else does it signify if not the infinite, endless altercation over discordant opinions and arguments which accompanies the vain structures of human knowledge, enmeshing them in confusion. [C] Usefully enmeshing them! If we actually possessed one grain of knowledge, there would be no holding us back. I like what that Saint said: *'Ipsa utilitatis occultatio, aut humilitatis exercitatio est, aut elationis attritio'* [Even that which is useful has been rendered obscure: that provides an occasion for exercising our humility and restraining our pride]. To what degree of arrogance and insolence do we not carry our blindness and our brutish stupidity.[311]

[A] But to get back to our subject: it is truly reasonable that we should be beholden to God alone, to the benefit of his grace, for the truth of so excellent a belief: it is from God's bountiful liberality that we receive the fruition of everlasting life, which is the enjoyment of eternal blessedness.

[C] We should freely admit that God alone tells us this, and faith.[312] It is not a lesson we have been taught by Nature or Reason. Anyone who makes repeated examinations of himself, internally and externally, as a human being, with human powers but bereft of the divine privilege of grace; anyone who sees Man as he is, without flattery, will find no quality or faculty in Man which is not redolent of death and dust. The more we attribute, grant and refer to God, the more Christianly we act. Would the Stoic philosopher not be better advised to owe to God what he said he owed to the chance agreement of the Voice of the People? *'Cum de animarum aeternitate disserimus, non leve momentum apud nos habet consensus hominum aut timentium inferos, aut colentium. Utor hac publica persuasione'* [When treating the immortality of the soul we attach no little weight to the general agreement among those who fear or worship the gods of the Underworld. I make good use of this general conviction].[313]

311. Nembroth (Nimrod) was King of Babel; the Tower of Babel, sometimes portrayed as pyramidal, sought to 'reach unto heaven'; God overthrew it and confounded men's language, 'that they may not understand another's speech': Genesis 10:9–11:9; then I Corinthians 1:19; St Augustine, *City of God*, XI, 22.
312. Points made in Lambin's dedication of Book III of Lucretius to 'Germano Valenti Pimpuntio': no human arguments assure us of immortality, not even Plato's: only Christ does. Cf. Introduction, p. 25 xxiv ff.
313. Seneca, *Epist.*, CXVII, 6.

[A]   The feebleness of human reasoning on this subject is particularly
noticeable from the fabulous details which men have added to it in their
efforts to discover the characteristics of our future immortality.   [C]   We
may leave aside the Stoics, who grant that souls do have a future life, but
only a finite one: *'usuram nobis largiuntur tanquam cornicibus: diu mansuros
aiunt animos; semper, negant'* [They allow us to live as long as crows: our
souls last a long time, they say, but not for ever].³¹⁴

[A]   The most universally received opinion (which still subsists today in
some places) was the one attributed to Pythagoras − (not that he was the
first to hold it, but because his approval and authority gave great weight
and credence to it); it was that our souls, when they depart from us, go the
rounds from one body to another, from a lion, say, to a horse; from a
horse, to a king, ceaselessly driven from one abode to another.³¹⁵   [C]
Pythagoras said he distinctly remembered having previously been
Aethalides, then Euphorbus, then Hermotimus and finally Pyrrhus, before
his soul eventually passed into himself, with recollections covering two
hundred and six years.

Some added that these souls sometimes go back to heaven, and then
come down again:

> *O pater, anne aliquas ad coelum hinc ire putandum est*
> *Sublimes animas iterumque ad tarda reverti*
> *Corpora? Quae lucis miseris tam dira cupido?*

[O Father, must we believe that some exalted souls go from here to heaven and
then come back again to sluggish bodies? Why do those wretches still yearn for the
light of day?]³¹⁶

Origen has souls everlastingly shuttling back and forth between wretchedness and bliss. Varro's opinion relates how souls rejoin their original bodies
after four hundred and forty years have rolled; for Chrysippus that happens
after an undefined period.³¹⁷ Plato says that it was from Pindar and the old

---

314. Cicero, *Tusc. disp.*, I, xxxi; cf. Rabelais, *Quart Livre*, XXVII, *ad fin.*
315. Diogenes Laertius, *Lives*, Diogenes, VIII, 526.
'88 (in place of [C]): another. *Socrates, Plato and virtually all those who
wished to believe in the immortality of souls, allowed themselves to be convinced by that
discovery, as well as whole nations, our own among them.* But . . . (Cf. Caesar, *De bello
gallico*, VI, 18.)
316. Virgil, *Aeneid*, VI, 719 (cf. St Augustine, *City of God*, XIV, 5). Platonic
teachings: cf. Plutarch, *De la face qui apparoist dedans le rond de la Lune*, 626 C-H
(the 'orchard of Dis').
317. St Augustine, *City of God*, XXI, 16–17; XXII, 28 (including note by Vivès).

poets that he acquired his belief in the endless succession of changes by which the soul is purified (in the World to Come her rewards and punishments are temporary, since her life on earth is lived within time); he drew the conclusion that the soul must possess a detailed knowledge of the affairs of heaven, hell and earth (having sojourned in them during her many journeys to and fro): for her, it is a matter of recollection.[318]

Elsewhere, the soul's progression is like this: if a man has lived well he joins the star to which he is assigned; if badly, he becomes a woman; if even then he does not amend, he changes once more, this time into a beast with attributes appropriate to his vices; he will know no end to his punishments until he returns to his native condition, having rid himself, by force of reason, from all the gross, dull and material qualities within him.[319]

[A] But I must not forget the objection raised by the Epicureans against this transmigration of souls from body to body. It is quite entertaining. They pose the question: What order could be maintained if the crowds of the dying proved greater than the number being born? The souls turned out of house and home would all be jostling each other, trying to be the first to get into their new containers! They also ask how souls would spend their time while waiting for their new lodgings to be got ready. The Epicureans maintain that if, at the other extreme, more animate creatures were born than died, their bodies would be in a parlous state, having to wait for souls to be poured into them: some would die before they had started to live:

> *Denique connubia ad veneris partusque ferarum*
> *Esse animas praesto deridiculum esse videtur,*
> *Et spectare immortales mortalia membra*
> *Innumero numero, certareque praeproperanter*
> *Inter se, quae prima potissimaque insinuetur.*

[It seems absurd that souls should have to wait for the connubial embraces and parturitions of beasts – innumerable immortal beings looking out for mortal limbs and struggling among themselves to see who is strong enough to slip in first.][320]

Others make our souls remain in the body after death, so as to animate the snakes, worms and other creatures which are said to be produced by spontaneous generation in our rotting flesh or even from our ashes. Others

---

318. Plato, *Meno*, 82 (Ficino, p. 19).
319. Plato, *Timaeus*, 42 E D (Ficino, p. 710).
320. Lucretius, III, 776 f. (Lambin, pp. 243–5). The following passage draws on III, 712–40 (Lambin, pp. 237–41).

split the soul into two parts, mortal and immortal. Others make it corporeal yet immortal. Others make it immortal, but without knowledge or awareness. There have been those who thought that the souls of the damned become devils [C] (and some of us Christians have thought that, too). [A] Similarly, Plutarch thinks that those who are saved become gods. There are few things that Plutarch asserts with more conviction (everywhere else his manner is one of sustained doubt and indecision). 'We must think', he says, 'and firmly believe that the souls of men who have been virtuous by the standards of Nature and divine Justice, change from men into saints and from saints into demi-gods; finally these demi-gods become gods, once they are perfectly cleansed and purified (as in the sacrifices of purgation), and delivered in this way from death and passability. They do not become gods by some decree of the Senate but are gods in very truth, such as one could rationally expect them to be, full and perfect gods, to whom is granted a most blessed and most glorious apotheosis.'[321] Plutarch is the most reticent and most moderate of the whole bunch, but if you would like to see him indulging in some bolder skirmishing and spinning some miraculous yarns about all this, I refer you to his treatises *On the Moon* and *On the Daemon of Socrates*; there, more clearly than anywhere, you can confirm that the mysteries of philosophy have plenty of oddities in common with poetry; human understanding in its strivings to plumb the depths of everything and to give an account of it, destroys itself, just as we ourselves, tired and exhausted by life's long race, fall back into childishness.[322]

With that we come to the end of all the fine doctrines which we can distill from human science about our souls.

There is no less rashness in what science tells us about our bodily parts. We had better choose one or two examples, otherwise we shall drown in the vast and troubled sea of medical error. We can at least find out whether there is any agreement over the material from which Man reproduces himself.[323]

321. Plutarch, *Life of Romulus*, XIV, *ad. fin.*
322. In Amyot's Plutarch, *De la face qui apparoist dedans le rond de la Lune*, 614–27, and *Du Demon ou esprit familier de Socrates* (636–49). (This is a reminder of a revolution in thought; the generation of Rabelais still sought mystical religious truths in these treatises.)
323. Discussion of the body, and of the various theories of human reproduction form a major element in Melanchthon's *De anima* (cf. 39 ff.). Since the human egg had yet to be discovered, all theories of generation turned on the nature of semen and of the womb. Rival schools, especially those of Hippocrates and Galen,

[C] As for the way Man was originally produced, that is a very deep and ancient problem: small wonder, then, that it leaves the human mind troubled and distraught. Archelaus the natural philosopher of whom Socrates was the disciple (and, according to Aristoxenus, the paramour) taught that men and animals were made of milky sludge, exuded from the earth under the influence of heat.

[A] Pythagoras said that our semen is the foam of our purest blood; Plato, a liquid draining from the marrow of the spinal column (supporting this with the argument that our backs are the first of our members to feel tired when we are on the job); Alcmeon says it is a part of the substance of the brain (proving this by the fact that men's vision becomes troubled when they work immoderately at that particular exercise); for Democritus it is a substance extracted from the whole mass of the body; for Epicurus, a substance extracted from the soul as well as from the body; for Aristotle, the final excretion drawn from the nutriment of blood which spreads through all our limbs; for others it is concocted blood, digested by heat in the testicles – because extreme exertions can make us ejaculate drops of blood: there may be a little more probability here if, that is, any probability at all can be drawn from confusion so infinite.

How does this semen achieve its purpose? Opinions are as numerous and as contradictory. Aristotle and Democritus hold that women have no semen, but only a kind of sweat which they exude when they bounce about in the heat of their enjoyment: it plays no role in generation. Galen, on the contrary, and those who follow him assert that generation can only occur when semen from male and female come into contact.

And then, see how the doctors, philosophers and lawyers are all disputing and quarrelling with our women about how long a pregnancy can last! Personally I support, from my own case, those who assert that a pregnancy can last eleven months: the whole world is full of such experiences; any simple, uneducated woman could give advice on these disputed questions. And still we cannot reach agreement![324]

---

clashed from Antiquity (cf. Rabelais, *Tiers Livre*, TLF, VIII; XXXIII; and notes). Montaigne draws on H.C. Agrippa, *De Vanitate*, LXXXII, and Plutarch, tr. Amyot, *Des opinions des philosophes*, 456 G–459 D. Cf. also Tiraquellus, *De legibus connubialibus*, XV, 10–11.

324. The duration of pregnancies was a question of great actuality: in general doctors accepted as legitimate children born after eleven (or even thirteen) months; some lawyers denied the possibility. Cf. Rabelais, *Gargantua*, TLF, III and notes. Also discussed in Melanchthon. Montaigne was born after a prolonged pregnancy of eleven months.

That suffices to demonstrate that Man has no more knowledge of his own body than of his own soul. We have shown Man to himself – and his reason to his reason, to see what it has to tell us. I have succeeded in showing, I think, how far reason is from understanding even itself.

[C] And what can anyone understand who cannot understand himself? *'Quasi vero mensuram ullius rei possit agere, qui sui nesciat'* [As though one could measure anything and not know how to measure oneself].[325] Protagoras was really and truly having us on when he made Man the measure of all things – Man, who has never known even his own measurements. If Man cannot have it, then his dignity will not let any other creature have it: yet Man is so full of contradictions and his ideas are so constantly undermining each other that so favourable a proposition is simply laughable: it leads to the inevitable conclusion that both measure and the measurer are nothing.[326]

When Thales reckons that a knowledge of Man is very hard to acquire, he is telling him that knowledge of anything else is impossible.[327]

For your sake, Patroness,[328] I have abandoned my usual practice and have taken some pains to make this into a very long chapter. Sebond is your author: you will, of course, continue to defend him with the usual forms of argument in which you are instructed every day; that will exercise your mind and your scholarship. The ultimate rapier-stroke which I am using here must only be employed as a remedy of last resort. It is a desperate act of dexterity, in which you must surrender your own arms to force your opponent to lose his. It is a covert blow which you should only use rarely and with discretion. It is rashness indeed to undo another by undoing yourself. [B] We must not seek to die as an act of revenge, as Gobrias did when locked in close combat with a Persian nobleman: Darius arrived on the scene, sword in hand, but was afraid to strike for fear of killing him; Gobrias shouted to him to strike boldly, even if he had to run both of them through.[329]

325. Pliny, II, I.
326. For Protagoras, the arch-Sceptic and agnostic who introduced total relativism by making each individual man the measure of all things, see Plato, *Theaetetus*, 152 A–C: 166D; 174 A–B; Aristotle, *Metaph.*, XV, v, 6, (1062 b). Later, Montaigne draws on these pages as well as on Sextus, *Hypotyposes*, I, XXXII, 216 ff.
327. Thales (Diogenes Laertius, *Lives*, Thales, I, XXXV, 36), as cited by Erasmus in his Socratic adage *Nosce teipsum*. (For Justus Lipsius, Montaigne was 'our Thales'.)
328. See above, p. 529. Montaigne undermines the case of deriving knowledge from sense-data – a central contention of Pyrrhonism.
329. Herodotus, III, 73, cited by Plutarch, tr. Amyot, *Comment on pourra discerner le flatteur d'avec l'amy*, 41 B–C.

[C] I have seen the proffered terms of a duel condemned in cases where the weapons or the circumstances left no room for hope that either of the combatants could survive.

The Portuguese took fourteen Turkish prisoners in the Indian Ocean, who, impatient of their captivity, decided to reduce themselves, their masters and the vessel to ashes; they succeeded in doing so by rubbing some of the ship's nails together until a spark fell among the barrels of gunpowder which were there.[330]

[A] Here we have now reached the limits and very boundaries of knowledge, where (as in the case of Virtue) extremes become vices. [A1] Keep to the beaten track: it can hardly be good to be so subtle and so clever. Remember the Tuscan proverb, '*Chi troppo s'assottiglia si scavezza*' [He who becomes too clever is lost]. [A] My advice to you is to cling to moderation and temperance, as much in your opinions and arguments as in your conduct, fleeing what is merely new or odd. All roads which wander from the norm displease me. You, by the authority of your high rank as well as by virtue of qualities which are more strictly your own, can, with a glance, command anyone you please; you ought to have entrusted this task to a professional scholar, who would have been able to make a very different defence of these ideas and to have enriched them more effectively.[331] Nevertheless there is ample material here for what you have to do.

When talking of Law, Epicurus said that even the harshest laws were necessary: without them men would start eating each other. [C] Plato is a mere finger's breadth away from that; he says that, without laws, we would live like wild animals: and he makes a good assay at proving that true.[332] [A] Our minds are dangerous tools, rash and prone to go astray: it is hard to reconcile them with order and moderation. We have seen during my lifetime virtually all outstanding men, all men of abnormally lively perception, breaking out into licentiousness of opinion or behaviour. It is a miracle if you find one who is settled and civilized. We

330. Cf. S . Goulart, *Hist. du Portugal*, XII, xxiii, 366r°; similar but not identical account.
'95: gunpowder, which were *in the place where they were kept*. Here we have now . . .
331. Petrarch, *Canzoniere*, XXII, 48.
'88: effectively *and who would have used, in piling up his case, other authors besides our Plutarch. When* . . . (Cf. Erasmus' adages *Ne quid nimis* and *Medium sequere*.)
332. Epicurus, cf. p. 543; Plato, *Laws*, 874 (tr. Ficino, p. 862).

are right to erect the strictest possible fences around the human mind. In the march of scholarship or anything else the mind must needs have its footsteps counted and regulated; you must supply artificial hedges and make it hunt only within them. [A1] We rein it in, neck and throat, with religions, laws, customs, precepts, rewards and punishments (both mortal and immortal), and we still find it escaping from all these bonds, with its garrulousness and laxity. It is an empty vessel: we can neither grasp it nor aim it; it is bizarre and misshapen and suffers no knot and no grapple.

[B] Certainly few souls are so powerful, so law-abiding and so well endowed that we can trust them to act on their own, allowing them liberty of judgement to sail responsibly and moderately beyond accepted opinion. It is more expedient to keep them under tutelage. What an outrageous sword [C] the mind is, even for its owner, [A] unless he knows how to arm himself ordinately and with discretion. [C] No beast more rightly needs blinkers to compel it to restrict its gaze to what lies before its feet, and to stop it from wandering about, this way and that, outside the ruts which custom and law have trodden out for it. [A] That is why it would be better for you to keep closely to your usual ways, whatever they may be, rather than to fly off like this with such frantic licence. Nevertheless, if one of those newfangled 'doctors' comes into your presence and starts acting clever, putting your spiritual health at risk as well as his own, you can, in the last resort, call on this remedy as a prophylactic against the deadly plague which is daily spreading through your courts: it will stop that poisonous contagion from infecting you and those about you.[333]

The freedom and vigour of minds in Antiquity created many Schools holding different opinions in philosophy and the humanities; before taking sides, each individual was responsible for judging and choosing for himself. But nowadays [C] men are all in step, *'qui certis quibusdam destinatisque sententiis addicti et consecrati sunt, ut etiam quae non probant, cogantur defendere'* [bound by vows to certain definite opinions, so that they are forced to defend even those which have not won their assent];[334] [A] our studies are accepted according to the decrees of civil authority, [C] with the result that our Schools have only one model, all having the same

---

333. R. Sebond is a prophylactic against the 'poison' of Lutheranism (see p. 490 ff). The rest of the *Apology* uses scepticism as the ultimate defence of catholicism.

'88: a *dangerous* sword . . .

334. Cicero, *Tusc. disput.* II, ii.

circumscribed form of basic instruction and teaching; [A] we now no longer try and find out what weight and value such coins have: each of us in his turn accepts them at the going rate with the generally approved value. Nobody defends the alloy, only its currency. Every discipline becomes equally acceptable. Medicine is accepted as though it were as valid as geometry; jiggery-pokery, enchantments, magic spells producing impotence, communication with the spirits of the dead, prognostications, casting horoscopes and even that absurd hunt for the philosopher's stone, all pass without contradiction. You merely have to know that the seat of Mars lies at the centre of the triangle of the palm, Venus in the thumbs and Mercury in your little finger; or know that, if the line of Fortune cuts across the protuberance of the forefinger, that is a sign of cruelty, but when it stops short at a point below the middle finger and the median line forming an angle with the line-of-life just below it, that is the sign of a pitiful death; in the case of a woman, if the line-of-nature is 'open' (not forming an angle with the line-of-life) that portends unchastity. Witness for yourself whether a mastery of this particular science does not win a man favour and respect in any company.

Theophrastus said that the human intellect, guided by the senses, could go only so far towards understanding natural causes; but when it reaches the original first causes it proves blunt and has to stop, either because of its own weakness or else because of the difficulty of the subject.[335]

That is a moderate and modest opinion which holds that our intellect is adequate enough to bring us to the knowledge of some things but that there are definite limits to its power, beyond which it is rash to use it.

It is a plausible opinion, set forth by conciliatory men (but it is difficult to fix boundaries for the human mind: it is avidly curious and sees no more reason for stopping after a mile than after fifty yards); it says: 'The assays of experience have taught me that where one man fails another succeeds; that what is unknown to one century is clarified by the next; that the sciences and the arts are not just cast in a mould all at once, but have to be gradually shaped by repeated handling and polishing, just as the mother-bear takes time to lick her cub into shape; I may not be strong enough to uncover anything but I can still take soundings and make assays; by kneading and working the dough of this new subject-matter, by blending it and warming it through, I make it easier for my successor to enjoy it at leisure; I render it more pliable for him, more manageable.

335. Cf. H. C. Agrippa, *De Vanitate*, I.

ut hymettia sole
Cera remollescit, tractataque pollice, multas
Vertitur in facies, ipsoque fit utilis usu.

[As wax from Mt Hymettos can be softened in the sun and kneaded with the thumb to form various shapes, becoming more useful with usage.][336]

A second man will do the same for the third: that is why no difficulty should drive me to despair – nor should my own powerlessness, for it is merely my own; Man is capable of understanding everything as well as something.'

Yes; but if Man admits, like Theophrastus, that he has no knowledge of first causes and principles, then let him boldly give up all the rest of his knowledge; without foundations, his argument collapses; discussion and inquiry have only one aim: to establish first principles; if Man's course is not stopped by his reaching that goal, he is thrown into boundless uncertainty.   [C]   *'Non potest aliud alio magis minusve comprehendi, quoniam omnium rerum una est definitio comprehendedi'* [One thing cannot be better understood, or less understood, than another: 'understanding' anything always means the same].[337]

[A]   It is probable that if the soul knew anything, she would first know herself; then, if she knew anything outside herself, she would first of all know her bodily sheath. Yet we can see the gods of the medical schools still quarrelling over human anatomy:

> *Mulciber in Trojam, pro Troja stabat Apollo.*

> [Vulcan against Troy: Apollo for Troy.][338]

Can we ever expect them to agree! We are closer to ourselves than to the whiteness of snow or the weight of a stone: if Man does not know himself, how can he know what his properties and powers are? Some true knowledge may perhaps find lodgings in us; if so, that is by chance, since error is received into the soul in the same way and in the same fashion; souls have no means of telling one from the other, no means of separating truth from falsehood.

The Academic philosophers accepted that our balance of judgement may be swayed one way or the other; they found it too crude to say that it is no

336. Ovid, *Metam.*, X, 284.
337. Cicero, *Acad.: Lucullus*, II, xli, 128 (adapted).
338. Ovid, *Tristia*, I, ii, 5.

more likely that snow be white than black, or that we no more understand the movement of a stone thrown by our own hand than the movement of the Eighth Sphere. These are bizarre difficulties and our intellect can hardly find room for them (even though they had established that we are incapable of knowing anything and that Truth is swallowed up in deep abysses where Man's vision cannot penetrate); to avoid them they admitted that some things are more likely than others and concede to judgement the power to incline towards one probability rather than another. They grant it this propensity, but they deny it conclusions.

The Pyrrhonists' idea is bolder, yet, at the same time, more true-seeming.[339] For what is this Platonic *inclination*, this *propensity* towards one proposition rather than another, than the recognition of there being more apparent truth in this than in that? But if our minds could grasp the form, lineaments, stance and face of Truth, then they would see whole truths as easily as partial truths, nascent and imperfect. Take that apparent verisimilitude which makes the scales incline to the left rather than to the right – then increase it; take that ounce of verisimilitude which turns the scales: multiply it a hundredfold or a thousandfold; in the end the balance will come down definitely on one side, deciding on one choice, on one whole truth.

But how can they bring themselves to yield to verisimilitude if they cannot recognize verity? How can they know there to be a resemblance to something the essence of which they do not know? We judge entirely, or entirely not. If our intellectual faculties and our senses have no foundation to stand on but only float about in the wind, then it is pointless to allow our judgement to be influenced by their operation, no matter what 'probabilities' it seems to present us with;[340] and so the surest position for our intellect to adopt, and the happiest, would be the one where it could remain still, straight, inflexible, without motion or disturbance. [C] *'Inter visa vera aut falsa ad animi assensum nihil interest'* [Where the assent of the mind is concerned, there is no difference between true impressions and false ones].[341]

[A] Things do not lodge in us with their form and their essence; they do not come in by the force of their own authority: we can see that clearly; if they did, we would all react to them in the same way: wine would taste

339. '88: more *true and more firm*. For . . .
340. St Augustine advanced such arguments against Academic theories of probability (*Contra academicos*, e.g., II, 7); they had long been current.
341. Cicero, *Acad.: Lucullus*, II, xxviii, 90.

the same in the mouth of a sick man and a healthy one; a man whose hands were calloused or benumbed would find the same hardness in the timber or iron he was handling as anyone else. External objects therefore throw themselves on our mercy; we decide how we accept them.[342]

Now, if we, for our part, could receive anything without changing it, if our human grasp were firm and capable of seizing hold of truth by our own means, then truth could be passed on from hand to hand, from person to person, since those means are common to all men. Among so many concepts we could find at least one which all would believe with universal assent. But the fact that there is no single proposition which is not subject to debate or controversy among us, or which cannot be so, proves that our natural judgement does not grasp very clearly even what it does grasp, since my judgement cannot bring a fellow-man's judgement to accept it, which is a sure sign that I did not myself reach it by means of a natural power common to myself and to all men.

Let us leave aside that infinite confusion of opinions which we can see among the philosophers themselves and that endless, world-wide debate about knowledge. It really is the truest of presuppositions that men – I mean the most learned, the best-endowed and the cleverest of men – never agree about anything, not even that the sky is above our heads. Those who doubt everything doubt that too. Those who deny that we can ever know anything say we cannot know whether the sky is above our heads or not. Those two opinions are by far the strongest, numerically.

Apart from this infinite diversity and disagreement, we can easily see that the foundations of our powers of judgement are insecure from the worry it personally causes us and from the lack of certainty each man feels within himself. How our judgements vary! How frequently we change our ideas! What I hold and believe today, I hold and believe with the totality of my belief. All my faculties, all my resources hold tight to that opinion and vouch for it with all their might. It would be impossible for me to embrace and maintain any truth more strongly. I am wholly for it, truly for it. But – not once, not a hundred times, not a thousand times, but every day – have I not embraced something else with the same resources and under the same circumstances, only to be convinced later that it was wrong? At least we should acquire wisdom at our own expense! If this appearance has once deceived me, if my touchstone regularly proves unreliable and my scales

---

342. From here Montaigne takes on Lucretius, the defender of the senses as true guides. Cf. Introduction, p. xxxv ff. He relies mainly on his own experience, in sickness and in health, against which he judges the established Classical authorities.

wrong and out of true, why should I trust them this time, rather than all the others? Is it not stupid to let oneself be deceived so often by the same guide? Fortune may shift us five hundred times, may treat our powers of belief like a pot to be endlessly emptied and filled with ever-differing opinions: nevertheless, the present one, the last one, is always sure and infallible! For this last one we must abandon goods, honour, life, health, everything.

> *posterior res illa reperta,*
> *Perdit, et immutat sensus ad pristina quaeque.*

[When we find something new, the recent destroys the older and makes us change our taste for it.][343]

[B] Whatever people preach to us and whatever we may learn from them, never forget that the giver is a man and so is the taker; a mortal hand presents it to us: a mortal hand takes it from him. Only such things as come to us from Heaven have the right and the authority to carry conviction; they alone bear the mark of Truth; but even they cannot be seen with our human eyes, nor do we obtain them by our own means: so great and so holy an Image could never dwell in so wretched a dwelling, unless God first makes it ready for that purpose, unless he forms it anew and fortifies it by his special grace and supernatural favour.

[A] Our condition is subject to error: that ought, at very least, to lead us to be more moderate and restrained in making changes. We ought to admit that, no matter what we allow into our understanding, it often includes falsehoods which enter by means of the same tools which have often proved contradictory and misleading.

It is not surprising that they should prove contradictory, since they are so easily biased and twisted by the lightest of occurrences. It is certain that our conceptions, our judgement and our mental faculties in general are all affected by the changes and alterations of the body. Those alterations are ceaseless. Are our minds not more alert, our memory more ready, our reasoning powers more lively when we are well rather than ill? Does not everything present a different aspect to our minds under the influence of joy and gaiety or of chagrin and melancholy? Do you think that the poems of Catullus or Sappho delight a miserable old miser as they do a vigorous and ardent youth? [B] Cleomenes the son of Anaxandridas being ill, his friends reproached him with having new and unaccustomed humours and ideas. 'I am not surprised,' he replied; 'I am not the same

343. Lucretius, V, 1414 (Lambin, pp. 462–3 – explained with Montaigne's sense).

person when I am well: being different, my opinions and ideas are different too.'³⁴⁴

[A] There is a saying current in the legal chicanery of our law-courts applied to a criminal who comes before judges who happen to be in a good, gentle, generous mood: GAUDEAT DE BONA FORTUNA 'Let him enjoy this good luck': for it is certain that we sometimes come across minds whose judgement is prickly, sharp and poised to condemn and which, at other times, are less difficult, more affable, more given to finding excuses. A judge may leave home suffering from the gout, jealous, or incensed by a thieving valet: his entire soul is coloured and drunk with anger: we cannot doubt that his judgement is biased towards wrath. [B] The august Senate of the Areopagus held their sessions at night, lest the sight of the plaintiff should influence their justice. [A] The very air and calm weather have power to change us – as that Greek poem says which Cicero cited:

> *Tales sunt hominum mentes, quali pater ipse*
> *Juppiter auctifera lustravit lampade terras.*

[The minds of men are such as Father Jupiter wills them to be, as he bathes the earth in fruitful light.]³⁴⁵

It is not only fevers, potions and great events which upset our judgement: the lightest thing can send it spinning. If a continual fever lays our minds prostrate, you can be sure that a three-day fever will have a proportionately bad effect on them, even though we are not aware of it. If apoplexy can dim and totally snuff out our mental vision, you can be sure that even a cold will confuse it. Consequently, there can hardly be found a single hour in an entire lifetime when our powers of judgement are settled in their proper place; our bodies are subject to so many sustained changes and are composed of so many kinds of principles that there is always one pulling the wrong way – I trust the doctors over that!

This malady, moreover, is not so easy to detect unless it is extreme and past all cure; Reason always hobbles, limps and walks askew, in falsehood as in truth, so that it is hard to detect when she is mistaken or unhinged.

By *reason* I always mean that appearance of rationality which each of us

---

344. Plutarch, *Les Dicts notables des Lacedaemoniens*, 218 C; also a general influence of Pyrrhonism (*Hypotyposes*, I, xxxii, 217–19 etc.). *Chagrin* was a technical word for melancholic depression.
345. Homer, *Odyssey*, XVIII, 135, translated by Cicero, *apud* St Augustine, *City of God*, V, 8. (Montaigne has already cited this in II, I, 'On the inconstancy of our actions'.)

constructs for himself – the kind of reason which can characteristically have a thousand contrary reactions to the same subject and is like a tool of malleable lead or wax: it can be stretched, bent or adapted to any size or to any bias; if you are clever, you can learn to mould it.

Take a judge; however well-intentioned he may be, he must watch himself carefully (and not many people spend much time doing that), otherwise some inclination toward friend, relation, beauty or revenge (or even something far less weighty, such as that chance impulse which leads us to favour one thing rather than another, or which enables us to choose, without any sanction of reason, between two identical objects – or even some more shadowy cause, equally vain) will encourage some sneaking sympathy or hostility toward one of the parties to slip, unnoticed, into his judgement and tip the balance.

I spy closely on myself and keep my eyes constantly directed on myself alone – I do not have much else to do:

> *quis sub Arcto*
> *Rex gelidae metuatur orae,*
> *Quid Tyridatem terreat, unice*
> *Securus*

[Quite indifferent to what ruler of the frozen North inspires great fear, or what dangers frighten Tiridates][346] –

yet even I hardly dare to tell of the vanity and the weakness which I find in myself. I have such wobbly legs, I am so unsteady on my feet, I totter about so and cannot even trust my eyesight, with the result that I feel quite a different person before and after a meal; when good health and a fine sunny day smile at me, I am quite debonair; give me an ingrowing toe-nail, and I am touchy, bad-tempered and unapproachable. [B] My horse's gait seems sometimes rough, sometimes gentle; the very same road, now short, now much longer, and the same form of action more agreeable or less so. [A] Now, I am ready to do anything; later, ready to do nothing; what is nice now can be nasty later on. [A1] A thousand chance emotions, unbidden, are in turmoil within me; sometimes a melancholic humour gets hold of me; at others, a choleric one; sometimes grief or joy dominate me, for reasons of their own. [A] I pick up some books: I may have discovered outstanding beauties in a particular passage which really struck home: another time I happen upon the same passage and it remains an unknown, shapeless lump for me, however much I twist it, and pat it and bend it or turn it. [B] Even in the case of my own

346. Horace, *Odes*, I, xxvi, 3.

writings I cannot always recover the flavour of my original meaning; I do not know what I wanted to say and burn my fingers making corrections and giving it some new meaning for want of recovering the original one – which was better. I go backwards and forwards: my judgement does not always march straight ahead, but floats and bobs about,

> *velut minuta magno*
> *Deprensa navis in mari vesaniente vento.*

[Like a tiny boat buffeted on the ocean by a raging tempest.][347]

Many's the time I have taken an opinion contrary to my own and (as I am fond of doing) tried defending it for the fun of the exercise: then, once my mind has really applied itself to that other side, I get so firmly attached to it that I forget why I held the first opinion and give it up. Almost any inclination, no matter which, takes me with it and carries me along by my own weight. Almost anybody could say much the same of himself if he watched himself [C] as I do. [B] Preachers know that the emotion which comes upon them as they speak moves them towards belief; and we know that when we are in a temper we devote ourselves to defending an assertion, impressing it upon ourselves and embracing it with furious approbation, far more than we ever do in cold-blooded calm.

You give your lawyer a simple statement of your case; he replies, hesitantly, doubtfully: you feel that he is quite indifferent which side he is to defend. But if you offer him a good fee to get stuck into it and all worked up about it, does he not begin to take a real interest and, once his will is inflamed, do not his arguments and forensic skills become inflamed as well? A clear and indubitable truth comes and presents itself to his understanding. He finds that your case sheds quite a new light: he really believes in it and convinces himself accordingly. I even wonder whether ardour, born of despite and of obstinacy, when confronted by pressure from a magistrate or by violent threats – [C] (or even simply a concern for reputation) – [B] has not brought some men to be burned in defence of an opinion for which, when at liberty among friends, they would never even have burned their finger-tips.

[A] The jolts and shocks which our soul receives from the passions of the body greatly affect her, but her own proper passions do so even more. They have such a hold on her that it could perhaps be maintained that her motions and propulsion come from her own tempests: without those agitations she would be becalmed like a ship on the open sea, abandoned by the helpful winds.[348] Anyone who did maintain that, [C] following

347. '88: does not always *get better, but floats and rolls about* . . . (Catullus, XXV, 12.)
348. Cf. Plutarch, tr. Amyot, *De la Vertu Morale*, 37 F–G.

the Peripatetics, [A] would do us little wrong, since it is recognized that most of the finer actions of the soul require – and can only arise from – such passionate impulses. It is said that valour cannot be achieved without the help of anger –

[C] *Semper Ajax fortis, fortissimus tamen in furore*

[Ajax was always brave, but bravest when mad with fury][349] –

that we do not attack the wicked or our foes vigorously enough, unless we are angry; and that, to get justice out of judges, counsel must move them to anger. Strong desires motivated Themistocles; they motivated Demosthenes and forced philosophers to travel far and work late: and they lead us too towards useful ends: honour, learning, health.

In addition, our soul's weakness when faced with pain and suffering serves to nurture repentance and remorse within our conscience and to feel the chastisements with which God scourges us as well as the chastisements of political punishment. [A] Compassion acts as a stimulus to [B] clemency; prudent self-preservation[350] [C] and self-control [B] are awakened by our fear; and how many fair actions are awakened by ambition? And how many by arrogance? [A] In short, not one eminent or dashing virtue can exist without some strong, unruly emotion. Was this one of the considerations which moved the Epicureans to relieve God of all care and concern for the affairs of men, since even the very actions of his goodness could not be directed towards us without disturbing his repose with passions – which are the goads and the incitements which drive the soul towards virtuous actions? [C] Or else did they think differently, taking the passions to be like storms, shamefully deflowering the soul of her tranquillity? '*Ut maris tranquillitas intelligitur, nulla ne minima quidem aura fluctus commovente: sic animi quietus et placatus status cernitur, quum perturbatio nulla est qua moveri queat*' [We know the sea is tranquil when not even the slightest breath of wind ruffles the surface; so too the soul is calm and at peace when there is no emotion seeking to disturb it].[351]

---

349. Cicero, *Tusc., disput.*, IV, xxiii; the rest of [C] follows closely ibid., xix. For the role of passion and anger in bravery, cf. Aristotle, *Eudemian Ethics*, III, 15–19, 1229a.

350. '80 (in place of [B]): stimulus to *liberality and justice* . . .

351. Cicero, *Tusc. disput.*, V, vi.

'88 (in place of [C]): actions? *At least we know only too well that the passions produce innumerable and ceaseless changes in our soul and tyrannize over it wondrously; is the judgement of an angry man or a fearful one the same judgement as he will have later when he has calmed down?* What varied . . .

[A]   What varied thoughts and reasons, what conflicting notions, are presented to us by our varied passions! What certainty can we find in something so changeable and unstable as the soul, subject by her condition to the dominance of perturbations,   [C]   and who never moves except under external constraint.   [A]   If our judgement is in the hands of illness itself and of turbulence; if it is obliged to receive its impressions from foolhardiness and madness: what certainty can we expect from it?

[C]   Is it not somewhat bold of Philosophy to think that men perform their greatest deeds, those nearest to the divine, when they are beside themselves, frenzied and out of their senses? Our amendment comes when our reason slumbers or when we are deprived of it; the two natural ways of entering into the council chamber of the gods and to have foreknowledge of Destiny are sleep and frenzy.[352]

Here is a pleasant thought: when the passions bring dislocation to our reason, we become virtuous; when reason is driven out by frenzy or by sleep, that image of death, we become prophets and seers. I have never been more inclined to believe Philosophy! It was a pure enthusiasm – breathed into the spirit of Philosophy by holy Truth herself – which wrenched from her, against her normal teaching, that the tranquil state of our soul, the quiet state, the sanest state that Philosophy can obtain for her, is not her best state. Our waking sleeps more than our sleeping; our wisdom is less wise than our folly; our dreams are worth more than our discourse; and to remain inside ourselves is to adopt the worst place of all.

But does Philosophy not realize that we are clever enough to notice that that maxim which makes the spirit so great, so perfect, and so clear-sighted when detached from Man, and yet so dark, so ignorant and so earthy when it remains in Man, is produced by the very spirit which itself forms part of dark, ignorant and earthy Man. And so, for that very cause, is neither to be trusted nor believed?[353]

[A]   Being of a soft and heavy complexion, I do not have much experience of those disturbances which bear the mind away and which mostly take our souls by surprise without giving them time to know

---

352. The ideal of *tranquillity of mind* is indeed, for Platonizing philosophers, subordinated to visions, dreams and philosophical ecstasy; cf. Rabelais, *Tiers Livre*, TLF, XIIII and XXXVII.
353. That is, philosophical ecstasy cannot claim to reveal infallible truth. Montaigne proceeds to emphasize the 'asinine' aspect of his own melancholy complexion (an antidote to all melancholic ecstasies).

themselves. But there is a passion in the heart of the young (induced, they say, by idleness); those who have assayed resisting its power, even when it takes an untrammelled, moderate course, find that it gives a good idea of the abrupt changes and deteriorations which our judgement can suffer. There was a time when I tensed myself to resist and parry its assaults (for I am so far from being one who welcomes vices, that I never give in to them unless they compel me to); despite my resistance, I would feel it within me as it was born, and as it grew and developed; I was lively: my eyes were open. Yet it would seize me, possess me. It was like a kind of drunkenness; everything took on an unaccustomed appearance; I would see the woman I yearned for becoming manifestly more attractive, her qualities swelling and growing as the wind of my imagination blew upon them; the difficulties facing my courtship would seem to become easy and smooth; my reason and conscience would withdraw into the background. Then, with lightning speed, at the very instant when my fire had burned itself out, my soul would recover another state, another judgement, another way of looking at things; it was now the difficulties of getting out of it which seemed immense and insurmountable; the very same things took on very different tastes and appearances from the ones offered me by inflamed desire.

Which was right? Pyrrho knows nothing about that!

We are never free from illness: fevers blow hot and cold; we drop straight from symptoms of a burning passion into symptoms of a shivery one. [B] The more I jumped forward, the more I now leap back:

> *Qualis ubi alterno procurrens gurgite pontus*
> *Nunc ruit ad terras, scopulisque superjacit undam,*
> *Spumeus, extremamque sinu perfundit arenam;*
> *Nunc rapidus retro atque aestu revoluta resorbens*
> *Saxa fugit, littusque vado labente relinquit.*

[Thus does the sea with alternate tides now dash up the beach, covering the rocks with its foaming billows, and seeking out the deep recesses of the sand; and then it quickly turns, sucking back the shingle and fleeing the rocks, as its sinking waters relinquish the beach.][354]

[A] This very awareness of my mutability has had the secondary effect of engendering a certain constancy in my opinions. I have hardly changed any of my first and natural ones, since whatever likelihood novelty may

354. Virgil, *Aeneid*, XI, 624.

appear to have, I do not change easily, for fear of losing in the exchange. As I do not have the capacity for making a choice myself, I accept Another's choice and remain where God put me. Otherwise I would not know how to save myself from endlessly rolling.

[AI]   And thus, by God's grace, without worry or a troubled conscience, I have kept myself whole, within the ancient beliefs of our religion, through all the sects and schisms that our century has produced.   [A]   The writings of the Ancients – I mean the good, ample, solid ones – tempt me and stir me almost at will; the one I am reading always seems the most firm. All appear right in their turn, even though they do contradict each other. The ease with which good minds can make anything they wish seem likely, so that there is nothing so strange but that they will set about lending it enough colour to take in a simple man like me, shows how weak their proofs really are. For three thousand years the skies and the stars were all in motion; everyone believed it; then   [C]   Cleanthes of Samos or, according to Theophrastus, Nicetas of Syracuse   [A]   decided to maintain that it was the Earth which did the moving,[355]   [C]   revolving on its axis through the oblique circle of the Zodiac;   [A]   and in our own time Copernicus has given such a good basis to this doctrine that he can legitimately draw all the right astronomical inferences from it. What lesson are we to learn from that, except not to worry about which of the two opinions may be true? For all we know, in a thousand years' time another opinion will overthrow them both.[356]

> Sic volvenda aetas conmmutat tempora rerum:
> Quod fuit in pretio, fit nullo denique honore;
> Porro aliud succedit, et e contemptibus exit,
> Inque dies magis appetitur, floretque repertum
> Laudibus, et miro est mortales inter honore.

[Thus the rolling years give various things their time; what used to be highly esteemed is now worthless; something else comes out from discredit and succeeds the old; it is daily sought for; everyone praises it and it is wondrously honoured among mortal men.][357]

---

355. Plutarch, *De la face qui apparoist dedans le rond de la Lune*, 615 E; Cicero, *Acad.: Lucullus*, II, xxxix, 123 (reading *Nicetas* for *Hicetas*). Montaigne's *three thousand years* means from the Creation (dated about 4000 BC) to the time of Cleanthes and Nicetas.

356. The theory of Copernicus 'saved the appearances' as did that of Ptolemy: but Galileo later claimed to describe reality.

357. Lucretius, V, 1276 (Lambin, pp. 454–5).

Thus, whenever some new doctrine is offered to us we have good cause for distrusting it and for reflecting that the contrary was in fashion before that was produced; it was overturned by this later one, but some third discovery may overturn that too, one day. Before the principles which Aristotle introduced came into repute, other principles satisfied human reason just as his satisfy us now. What letters-patent do Aristotle's principles have, what exclusive privilege, that the course of our inquiries should stop with them and that they have the right to our assent for all time? They are not exempt: they can be kicked out as their predecessors were. When some new argument presses me hard, it is up to me to decide whether someone else may find a satisfactory reply even if I cannot; for to believe everything that may look true just because we ourselves cannot refute it, is very simple-minded. From that it would follow that the belief of the common people – [C] and all of us are common people – [A] would blow about like a weathercock: for their minds, soft and non-resistant, would constantly be forced to accept different impressions, each one effacing the trace of the other. Anyone who feels too weak to resist should follow legal practice and reply that he will consult counsel – or refer to the wiser heads who trained him.

How long has medicine been in the world? They say that some newcomer called Paracelsus is changing or reversing the entire order of the old rules, maintaining that, up to the present, medicine has merely served to kill people. He will be able to prove that easily enough, I believe, but it would not be very wise for me, I think, to test his new empiricism at the risk of my life. [A1] 'Believe nobody,' as the saying goes. 'Anyone can say anything.'[358]

One of those men who champion novelties and reformations in natural science told me recently that all the Ancients had evidently been wrong about the nature and movements of the winds; if I would only listen he would make me clearly see the palpable truth. After showing some patience in hearing his arguments (which looked extremely probable) I said, 'What! Those who were navigating according to the rules of Theophrastus, were they really going West when steering East? Were they sailing sideways or astern?' – 'That is as may be,' he replied, 'but they certainly got it wrong.' I then retorted that I would rather be guided by results than by reason – for they are always clashing! I have even been told that in geometry (which claims to have reached the highest degree of

358. Paracelsus (1493–1541). His works appeared posthumously (1575–88). He scorned traditional medicine absolutely.

certainty among the sciences) there are irrefutable demonstrations which overturn truth based on experience. Jacques Peletier, for example, in my own home, told me how he had discovered two lines drawing ever closer together but which, as he could prove, would meet only in infinity.[359] And the sole use Pyrrhonists have for their arguments and their reason is to undermine whatever experience shows to be probable; it is wonderful how far our supple reason will go along with their project of denying factual evidence: they can prove that we do not move, that we do not speak and that there is no such thing as weight or heat, with the same force of argument as we have when we prove the most likely things to be true.

Ptolemy was a great figure; he established the boundaries of the known world; all the ancient philosophers thought they had the measure of it, save for a few remote islands which might have escaped their knowledge. A thousand years ago, if you had questioned the data of cosmography, you would have been accused of Pyrrhonizing – of doubting opinions accepted by everybody; [B] it used to be heresy to allow the existence of the Antipodes![360] [A] But now that in our century new discoveries have revealed, not the odd island or the odd individual country, but an infinite land-mass, almost equal in size to the part we already knew, geographers today proceed to assure us that everything has really been seen and discovered this time.

> *Nam quod adest praesto, placet, et pollere videtur.*
>
> [For we are pleased with what is to hand; it works its spell.][361]

Since Ptolemy was once mistaken over his basic tenets, would it not be foolish to trust what moderns are saying now?[362] [C] Is it not more likely that this huge body which we call the Universe is very different

359. Peletier, a poet and mathematician, doubtless explained the conic hyperbola and asymptotes (lines which draw ever nearer to a given curve but do not meet it within a finite distance). He was actively opposed to the renewal of Pyrrhonism.

360. Cicero suspended judgement over the Antipodes (*Acad.: Lucullus*, II, xxxix, 123); St Augustine rejected the idea (*City of God*, XVI, 9); but it never was heretical to believe in them.

361. Lucretius, V, 1412, (Lambin, pp. 462–3).

362. '88 (in place of [C]): saying now? *Aristotle says that all human opinions have existed in the past and will do so in the future an infinite number of other times: Plato, that they are to be renewed and come back into being after thirty-six thousand years.* Epicurus . . . (Taken from Varchi, *L'Hercolano.* Montaigne replaced this with authorities taken from St Augustine or thought of because of him.)

from what we think? Plato holds that its entire aspect changes – that there comes a point when the heavens, the stars and the sun reverse the motions which we can see there and actually rotate from East to West.[363] The Egyptian priests told Herodotus that since the time of their first king, some eleven thousand years ago – (and they showed him the statues of all these kings, portrayed from life) – the Sun had changed its course four times, and the sea and land had changed places. They also said that no date within time can be ascribed to the origin of the world;[364] Aristotle and Cicero agree with that; and one of our own people maintained that the world exists from all eternity but has a cycle of deaths and rebirths; he cited Solomon and Isaiah as witnesses, his aim being to counter objections to God's having been a Creator who had once never created anything, an idle God who only cast aside his idleness when he set his hand to this enterprise and therefore a God subject to change.[365]

In the most famous of the Greek Schools of Philosophy the Universe is considered to be a god made by a greater one; it is composed of a body, with a soul situated in the centre but extending to the circumference by means of musical Numbers; it is divine, most blessed, most great, most wise and eternal. Within this 'god' there are other gods (the earth, the sea, the heavenly bodies) all maintained by the harmonious and perpetual movement of a sacred dance as they draw together then draw apart, hide then reveal themselves, or move to and fro and change their rows.[366]

Heraclitus laid down that the Universe was composed of fire and was destined one day to burst into flames and burn itself out: it would be born again some other time. Apuleius said that Men were *'sigillatim mortales, cunctim perpetui'* [individually mortal, collectively eternal]. Alexander gave his mother the written record which one of the Egyptian priests had taken from their monuments; it bore witness to the boundless antiquity of that people and included a true account of the birth and growth of other countries. Cicero and Diodorus say that, in their own days, the Chaldaeans kept records going back some four hundred thousand years; Aristotle, Pliny and others date Zoroaster six thousand years before the time of Plato.

363. Plato, *Politicus*, XIII, 270 AC; cf. St Augustine, *City of God*, XII, 14.
364. Herodotus, II, 142–3 (cf. St Augustine, *City of God*, XII, 13; J. Bodin, *Methodus ad Hist. cognit.*, 1595, p. 293).
365. Origen, *De Princ*, 3, 5, 3; cf. St Augustine, *City of God*, XII, 14 (citing Solomon and Ecclesiastes; Isaiah is in the notes of Vivès), and XI, 23. The doctrine of a Creator who had not yet created was rejected by Neo-Platonists such as Proclus.
366. Plato, in the *Timaeus*, 33D–41E.

Plato says that the citizens of Sais possess written records covering eight thousand years, adding that the city of Athens was built a thousand years before the foundation of that city.[367]

[B]   Epicurus taught that there exist in several other worlds objects very like the ones we can see here, fashioned the same way.[368] He would have said that with even greater assurance if he could have seen those strange examples of past and present similarities and resemblances to be found between our world and that New World of the West Indies.

[C]   In truth, when I consider what we know about the course of social life on this earth, I have often been struck with wonder at the resemblances there are – separated by immense spaces of place and time – between many savage beliefs or fantastic popular opinions which, whatever way you look at them, do not seem to arise from our natural reasoning. The human mind is a great forger of miracles, we know that: but this relationship has something abnormal about it which I cannot define; you can even see it in names, events and thousands of other ways.   [B]   For we have newly discovered peoples who, as far as we know, have never heard of us, yet where they believe in circumcision; where countries or great states are entirely governed by women, without men; where you can find something like our Lenten fasts, with the addition of sexual abstinence. We have found peoples where our crosses are honoured in various ways (in one place they even displayed them prominently on their graves); in another crosses were used (especially the cross of St Andrew) to ward off nocturnal visions; they also put them on their children's beds against enchantments. Elsewhere was discovered a wooden cross, immensely tall, which was worshipped as the god of rain – and that was very far from the coast. Also found there were the express image of our penitents, the use of mitres, the practice of priestly celibacy, the art of divination from the entrails of sacrificed animals,   [C]   a total abstention from all kinds of fish and flesh,   [B]   the custom for priests to make liturgical use of a special tongue not the common one; the idea that the first god was driven away by a second, his younger brother; the belief that they were created with all kinds of advantages which were subsequently cut off because of their sin; their land changed and their natural condition made harsher; they were submerged by a heaven-sent flood, only a few families being saved who had taken refuge in high mountain caves, which they blocked up to stop

367. Texts cited after St Augustine, *City of God,* VIII, 5; XII, 10, 11, including the notes of Vivès.
368. Plutarch, *Des oracles qui ont cessé,* 342 D.

the waters getting in; various species of animals were shut in there too; when they thought the rain had ceased, dogs were sent out: they came back dripping wet and clean, so it was judged that the waters had only begun to subside; later other dogs were sent out. When they returned all covered in mud the humans emerged to re-people the world, which they found to be full of nothing but snakes.

In one case the inhabitants were convinced of a Day of Judgement. When the Spaniards scattered the bones of their dead about as they plundered their graves in search of treasure, they were beside themselves with anger, declaring that such scattered bones could not easily be put together again. They have trade by barter (but no other) with fairs and markets for this purpose; they have dwarves and deformed people to enliven the banquets of their princes; falconry they have, but with their own native birds; they have tyrannous taxation, elegant gardens, acrobats, dancing, musical instruments, coats-of-arms, tennis-courts, games of dice and chance – at which they get so carried away that they stake themselves and their freedom; they have medicine based entirely on magic charms, pictorial writing, a belief in one first man who was father of all peoples; they have the worship of a god who once lived as a Man in perfect celibacy, abstinence and penitence, preaching the law of Nature and liturgical ceremonies and who disappeared from the world without a natural death; a belief in giants, the custom of getting drunk on their local drink and seeing who can down the most, religious ornaments painted with bones and death's heads, surplices, holy water and aspergilla, women and servants who gaily volunteer to be burnt or buried alive with their husband or masters, laws of inheritance which leave everything to the eldest son and set nothing but obedience aside for the younger one, the custom that a man promoted to high rank adopts a new name and abandons his old one, the custom of sprinkling chalk on the knee of a new-born babe, saying to him: 'Dust thou art, and to dust thou shalt return'; and they have the art of augury.

Such vain shadows of our religion as may be seen in some of these examples witness to its dignity and holiness: it has penetrated into infidel nations on our side of the world by a kind of imitation, but to those natives of far-off lands it came by a shared supernatural inspiration. For we found a belief in Purgatory but of a different style: they attribute to cold what we attribute to heat, thinking that the souls of the dead are punished and purged by the rigours of extreme cold.

That reminds me of another pleasing example of diversity: some peoples like to uncover the end of the penis, circumcizing the foreskin like Jews or

Moslems, whereas others have such conscientious objections to ever uncovering it that, lest the top of it should ever see the light of day, they scrupulously stretch the foreskin right over it and tie it together with little cords.

And here is another one: just as we honour kings and festive days by putting on our best clothes, there are regions where they emphasize the disparity between themselves and their king and mark their total submission to him by appearing in their shabbiest clothing; as they go into the palace they put a tattered robe over their good one, so that all pomp and glory should belong to the king alone.[369]

But to get on.

[AI] If Nature includes among her normal activities – along with everything else – the beliefs, judgements and opinions of men; and if such things have their cycles, seasons, births and deaths, every bit as much as cabbages do, the heavens changing them and influencing them at will: what permanent, magisterial authority should we go on attributing to them?[370]

[B] Now if experience makes it clear that the very form of our being – not only our colour, build, complexion and behaviour but our mental faculties as well – depends upon our native air, climate and soil ([C] as Vegetius said: *'et plaga coeli non solum ad robur corporum, sed etiam animorum facit'* [the heavenly regions contribute not only to the strength of men's bodies but of their souls as well]);[371] and if the goddess who founded Athens chose for her city a country of temperate climate which made men wise – that is what the priests of Egypt told Solon: *'Athenis tenue coelum, ex quo etiam acutiores putantur Attici; crassum Thebis, itaque pingues Thebani et valentes'* [the air of Athens is not oppressive, which is why the Athenians are considered most intelligent; that of Thebes is oppressive, therefore the Thebans are considered heavy and tough][372] – [B] then men must vary as flora and fauna do: whether they are more warlike, just, equable, clever or dull, depends on where they were born. Here they are addicted to wine: there, to robbery and lechery; here they are inclined towards superstition: there to disbelief; [C] here, to freedom: there, to slavery; [B] they may be more suited to learn one particular art of science than another; they may be slow or intelligent, obedient or rebellious, good or bad, all depending

369. All the above compiled from Lopez de Gomara, *L'Histoire générale des Indes*.
370. A regular theme for reflection. Cf. J. Bodin, *Methodus*, V.
371. Vegetius, I, ii, *apud* Justus Lipsius, *Politicorum*, V, 10.
372. Cicero, *De fato*, IV, 7.

on inclinations arising from their physical environment. Change their location, and, like trees, they take on a new character. That was why Cyrus refused to allow the Persians to give up their squat and rugged land and emigrate to softer plains; [C] he said that rich soft lands make for soft men, that fertile lands make for barren minds.[373] [B] Now, if we can see that the influence of the stars makes an art or an opinion to flourish; and if a particular age produces a particular kind of nature and inclines the human race towards some particular trait of character (their spirits producing good crops then lean crops, as fields do): what happens to all those special privileges which we pride ourselves upon? A wise man can be mistaken; a hundred men can; indeed, according to us, the whole human race has gone wrong for centuries at a time over this or that: so how can we be sure that human nature ever stops getting things wrong, [C] and that she is not wrong now, in our own period?

[A] Among other considerations witnessing to Man's weakness, it seems to me that we should not overlook that even his desires cannot lead him to discover what he needs; I am not talking about fruition, but about thinking and wishing: we cannot even agree on what we need to make us contented. Even if we let our thoughts tailor everything to their wishes, they cannot even desire what is proper to them [C] and so be satisfied:

> [B]   *quid enim ratione timemus*
> *Aut cupimus? quid tam dextro pede concipis, ut te*
> *Conatus non poeniteat votique peracti?*

[Is it reason that governs our fears and our desires? What have you ever conceived, even auspiciously, without being sorry about the outcome – even of its success?][374]

[A] That is why [C] Socrates prayed the gods to give him only what they knew to be good for him. The Spartans, in public as in private, simply prayed that good and beauteous gifts be vouchsafed to them; they left the choice and selection to the gods:[375]

> [B]   *Conjugium petimus partumque uxoris; at illi*
> *Notum qui pueri qualisque futura sit uxor.*

373. Plutarch, tr. Amyot, *Les Dicts notables des Anciens Roys* . . . 188E; Herodotus, IX, 121. Erasmus, *Apophthegmata, Cyrus Major*, II.
374. Juvenal, *Sat.*, X. 4.
375. Xenophon, *Memorabilia*, II, iii, 2: Plato, *Alcibiades*, II, 148 B–C.
   '88: That is why *the Christian, wiser and more humble and more aware of what he is, refers himself to his Creator to choose and command what he needs.* Conjugium . . .

[We pray to have a wife and children, yet only Jupiter knows what the children and that wife will be like.][376]

[A]   In his supplications the Christian says, 'Thy will be done', in order not to suffer that unseemly state which poets feign for King Midas: he prayed to God that all he touched should turn to gold. His prayer was granted: his wine was gold, his bread was gold, so were the very feathers in his bed, his undershirt and all his garments. In this way he found that the enjoyment of his desires crushed him and that he had been granted a boon no man could bear. He had to unpray his prayers:

> *Attonitus novitate mali, divesque miserque,*
> *Effugere optat opes, et quae modo voverat, odit.*

[Thunderstruck by so new an evil, rich and wretched both at once, he hates what once he prayed for.][377]

[B]   I can cite my own case. When I was young I begged Fortune, as much as anything, for the Order of St Michael: it was then the highest mark of honour for the French nobility, and very rare. Fortune granted it to me, but with a smirk: instead of elevating me, instead of lifting me up so that I could reach it, she used greater condescension: she debased the Order, and brought it right down to my neck – lower still in fact.

[C]   Cleobis and Bito asked their god, Trophonius and Agamedes their goddess, for rewards worthy of their piety; the gift they were given was death: so different from ours, where our needs are concerned, are the opinions of heaven.[378]

[A]   It is sometimes to our detriment that God vouchsafes us riches, honour, life and health itself: the things which please us are not always good for us. If, instead of a cure, he sends us death or a worsening of our ills – '*Virga tua et baculus tuus ipsa me consolata sunt*' [Even thy rod and thy staff do comfort me] – God acts thus by reason of his Providence, which knows our deserts far more accurately than we can ever do; whatever comes from a hand most loving and omniscient we must accept as good:

> *si consilium vis*
> *Permittes ipsis expendere numinibus, quid*
> *Conveniat nobis, rebusque sit utile nostris:*
> *Charior est illis homo quam sibi.*

376. Juvenal, *Sat.*, X, 352. Then [B]: *he* says . . . done' *and may chance* not to . . .
377. The Lord's Prayer ('Thy will be done') glossed with Ovid, *Metam.*, XI, 128.
378. Cicero, *Tusc. disput.*, I, xlvii.

[If you want my advice, allow the gods to judge what is best for us and most advantageous for our affairs; a man is dearer to them than he is to himself.][379]

For to ask the gods for honour and high office is like begging them to send you into battle, into a game of dice or into some other situation where the outcome is unknown and the gain dubious.[380]

[A]   No quarrel among philosophers is more violent or so bitter as the one which looms over the question of Man's sovereign good;   [C]   according to Varro's calculation, 288 sects were produced by it:[381] *'Qui autem de summo bono dissentit, de tota philosophiae ratione dissentit'* [Whoever disagrees over the sovereign good disagrees about the whole of philosophy].[382]

> [A]   *Tres mihi convivae prope dissentire videntur,*
> *Poscentes vario multum diversa palato:*
> *Quid dem? quid non dem? Renuis tu quod jubet alter;*
> *Quod petis, id sane est invisum acidumque duobus.*

[For me it resembles three men disagreeing at a feast, each liking very different dishes and asking for them. What am I to give them? What am I not to give them? You reject what delights another: what you like is tart and unpleasant to the other two.][383]

That is the way Nature ought to answer their disputes and their quarrels.

There are those who say that our good is to be found in virtue; some who say in pleasure; some, in conforming to Nature; one says in knowledge [C] or freedom from pain;[384]   [A]   another, in not letting oneself be deceived by appearances, a notion rather like that other one [B] taught by Pythagoras of old:

> [A]   *Nil admirari prope res est una, Numaci,*
> *Solaque quae possit facere et servare beatum;*

[Be astonished by nothing; it is almost the one and only way, Numacius, which leads to lasting happiness;][385]

---

379. Psalm 23 (22): 4; Juvenal, *Sat.*, X, 346.
380. Xenophon, *Memorabilia*, I, iii, 2.
381. St Augustine, *City of God*, XIX, I – also exploited in the following paragraphs (*ibid.*, 1–4).
382. Cicero, *De fin.*, V, v, 14.
383. Horace, *Ep.*, II, ii, 61.
384. Cicero, *De fin.*, V, v, 14, citing Hieronymus, the pupil of Aristotle.
385. Horace, *Ep.*, I, vi, I.

that is the aim of the Pyrrhonists. [C] (To be astonished by nothing is, for Aristotle, the attribute of greatness of soul.)[386] [A] Archesilaus said that suspending the judgement and keeping it upright and inflexible are good actions, whereas acts of consent and commitment are vicious and bad. It is true that he left his Pyrrhonism behind when he erected that axiom into a certainty![387] Pyrrhonists say that the sovereign good is *Ataraxia*, which consists in a total immobility of judgement; they consider that not to be a positive affirmation but simply an inner persuasion such as makes them avoid precipices and protect themselves from the chill of the evening; it presents them with this notion and makes them reject any other.

[B] How I wish that, during my lifetime, someone like Justus Lipsius (the most learned man left, a polished and judicious mind, a veritable brother to my dear Turnebus), had the health, the will and sufficient leisure to compile an honest and careful account which listed by class and by category everything we can find out about the opinions of Ancient philosophy on the subject of our being and our morals; it would include their controversies and their reputations, it would tell us who belonged to which school, and how far the founders and their followers actually applied their precepts on memorable occasions which could serve as examples. What a beautiful and useful book that would be![388]

[A] Moreover, if we draw our moral rules from ourselves, what confusion we cast ourselves into! For the most convincing advice we get from reason is that each and every man should obey the laws of his own country;[389] [C] that is Socrates' precept, inspired (he said) by divine counsel.[390] [A] But what does that mean, except that our rules of conduct are based on chance? Truth must present the same face everywhere. If Man could know solid Rectitude and Justice in their true Essences, he would never restrict them to the customary circumstances of this place or of that; Virtue would not be fashioned from whatever notions happen to be current in Persia or in India.

---

386. Greatness of Soul is the subject of *Nicomachaean Ethics*, IV, iii, and of *Eudemian Ethics*, III, v (1232a f.).

387. Sextus Empiricus, *Hypotyposes*, I, xxxiii, 223–34 (for Archesilas), I, iv, 8; vi, 12; xii, 25–30 (for *Ataraxia*).

388. Justus Lipsius, the neo-Stoic moralist (1547–1606) was read by Montaigne and admired by him. After a period of conforming to Protestantism he became a Roman Catholic fundamentalist. For Turnebus, see pp. 157 and 491.

389. '88: country, *as Socrates' oracle had taught him, that to do punctiliously one's duty of piety according to the uses of one's nation is equivalent to serving God. But . . .*

390. Aristotle, *Nicomachaean Ethics*, V, vii, 1–3. Cf. La Boëtie on p. 219.

Nothing keeps changing so continuously as the Law. Since I was born I have seen our neighbours, the English, chopping and changing theirs three or four times, not only on political matters (where we may wish to do without constancy) but on the most important subject there ever can be: religion.[391] It makes me feel sad and ashamed, since the English are a people with whom we used to be so familiarly acquainted in my part of the world that traces of their former kinship can still be seen in my own house.

[C] And closer to home, I have seen capital offences made lawful; such are the uncertainties and the fortunes of war that any one of us may eventually be found guilty of *lèse-majesté* against God and the King, simply for holding fast to different ideas of legitimacy, once our Justice were to fall to the mercy of Injustice (which, after a few years of possession, would change its essence).[392] Could that ancient god have more clearly emphasized the place of ignorance within our human knowledge of the divine Being, or taught us that religion is really no more than a human invention, useful for binding societies together, than by telling those who came before his Tripod to beg for instruction that the true way of worship is the one hallowed by custom in each locality?[393]

Oh God, how bound we are to the loving-kindness of our sovereign Creator for making our belief grow up out of the stupidities of such arbitrary and wandering devotions, establishing it on the changeless foundation of his holy Word![394]

[A] But what has Philosophy to teach us in this plight? Why, that we should follow the laws of our country! – laws which are but an uncertain sea of opinions deriving from peoples or princes, who will paint it in as many different colours and present it, reformed, under as many different faces as they have changes of heart. I cannot make my judgement as flexible as that. What kind of Good can it be, which was honoured yesterday but not today [C] and which becomes a crime when you cross a river! What kind of truth can be limited by a range of mountains, becoming a lie for the world on the other side![395]

[A] Philosophers can hardly be serious when they try to introduce

391. Allusion to religious settlements by Parliaments under Henry VIII, Edward VI, Bloody Mary and Elizabeth I.
392. Allusions to changing alliances and legitimacies in the French Wars of Religion.
393. Apollo (n. 389 above); Xenophon, *Memorabilia*, I, iii, I.
394. The conviction of Lambin also; cf. Introduction, p. xxxv ff.
395. Cf. Erasmus, *The Complaint of Peace* (*Opera*, 1703–1706, IV, 628 DE).

certainty into Law by asserting that there are so-called Natural Laws, perpetual and immutable, whose essential characteristic consists in their being imprinted upon the human race. There are said to be three such laws; or four; some say less, some say more: a sign that the mark they bear is as dubious as all the rest. How unlucky they are – (for what else should I call it but bad luck, seeing that out of laws so infinite in number, they cannot find even one which luck   [C]   or accidental chance   [A]   has allowed to be universally accepted by the agreement of all peoples). They are so pitiful that there is not one of these three – or four – selected laws, which has not been denied and disowned by several nations, not just one. Yet universal approval is the only convincing indication they can cite in favour of there being any Natural Laws at all. For whatever Nature truly ordained, we would, without any doubt, all perform, by common consent: not only all nations but all human beings individually would be deeply aware of force or compulsion when anyone tried to make them violate it. Let them show me just one law with such characteristics: I would like to see it.[396]

Protagoras and Ariston said that the essential justice of any law consists in the will of the lawgiver: without it, *good* and *honourable* lose their qualities, simply lingering on as empty words for things indifferent.

In Plato, Thrasymachus thinks that there is no right other than the advantage of the superior.[397]

Nothing in all the world has greater variety than law and custom. What is abominable in one place is laudable somewhere else – as clever theft was in Sparta. Marriages between close relations are capital offences with us: elsewhere they are much honoured:

> *gentes esse feruntur*
> *In quibus et nato genitrix, et nata parenti*
> *Fungitur, et pietas geminato crescit amore.*

[They say there are peoples where the son lies with his mother, the daughter with her father, where family piety is enhanced by a double affection.][398]

Murdering children, murdering fathers, holding wives in common, making

---

396. Aristotle's doctrine of Natural Law came in for increased criticism as new peoples were discovered, but also because of inner inconsistencies; cf. Jeremy Taylor, *Ductor Dubitantium*, 1660, p. 221.

397. Protagoras was allegedly banished for atheistic impiety: Cicero, *De nat. deorum*, I, xxiii, 63; Ariston of Chios was a Stoic inclined to cynicism; Thrasimacus, in Plato, *Republic*, 338 (Ficino, p. 535).

398. Ovid. *Metam.*, X, 331, in Tiraquellus, *De legibus connubialibus*, VII, 38. For context cf. Sextus Empiricus, *Hypotyposes*, III, xxiv, 203–17.

a business out of robbery, giving free rein to lusts of all sorts – in short there is nothing so extreme that it has not been admitted by the custom of some nation or other.

[B] It is quite believable that natural laws exist: we can see that in other creatures. But we have lost them; that fine human reason of ours is always interfering, seeking dominance and mastery, distorting and confounding the face of everything according to its own vanity and inconsistency. [C] '*Nihil itaque amplius nostrum est: quod nostrum dico, artis est*' [Nothing of ours is left: what I call *ours* is really artificial].[399]

[A] Any object can be seen in various lights and from various points of view: it is chiefly that which gives birth to variety of opinion: one nation sees one facet, and stops there; another sees another.

Nothing can be imagined more horrible than eating one's father: yet the peoples who followed this custom in the Ancient world looked on it as a mark of piety and love, seeking to provide their ancestors with the most worthy and honourable of obsequies, finding a home for their father's remains in their own person, in the very marrow of their bones; they were giving them a kind of new life; they were born again, as it were, by being transmuted into their living flesh as their children ate and digested them. It is easy to think what abominable cruelty it would be for men deeply imbued in such a superstition to leave their parents' remains to rot in the earth, food for beasts and worms.[400]

The aspects of theft which struck Lycurgus were the quickness, the industry, the boldness and the skill necessary to steal something from a neighbour, as well as of the public good which came from each man carefully guarding his own property. He believed that this gave a grounding in the twin subjects of assault and defence, both of which are useful for training soldiers (the principal virtue and science which he wished to instil into that nation). That outweighed the disorder and injustice of carrying off other people's property.

The tyrant Dionysius offered Plato a long, perfumed, damask robe, fashionable in Persia. Plato refused it saying that, since he was born a man, he would not willingly wear women's clothing. Aristippus, however, accepted it, replying that no apparel could corrupt a chaste heart;[401] and [C] when his friends taunted him with cowardice for taking so

---

399. Cicero: *De fin.*, V, xxi, 60 (now parsed differently).
400. Cf. 'On habit: and on never easily changing a traditional law', I, 23, after Herodotus, III, xii, etc.
401. Sextus Empiricus, *Hypotyposes*, III, xxiv, 204.

little offence when Dionysius spat in his face, he replied: 'Merely to catch a gudgeon fishermen suffer the waves to bespatter them from head to foot.' Diogenes was washing some cabbage leaves when he saw Aristippus go by: 'If you knew how to live on cabbage,' Diogenes said, 'you would not be courting a tyrant.' Aristippus retorted: 'You would not be here washing cabbages, if you knew how to live among men.'[402]

[A]   That is how Reason can make different actions seem right.   [B] Reason is a two-handled pot: you can grab it from the right or the left.

> *bellum, o terra hospita, portas;*
> *Bello armantur equi, bellum haec armenta minantur.*
> *Sed tamen iidem olim curru succedere sueti*
> *Quadrupedes, et frena jugo concordia ferre;*
> *Spes est pacis.*

[You are threatening war; what a hospitable land! Horses are armed for war: war is what these beasts portend! – Yet those same animals are often yoked to carts, plodding tranquilly in harness; there is hope for peace.][403]

[C]   When they lectured Solon for shedding vain and useless tears at the death of his son, he replied, 'It is precisely because they are vain and useless that I am right to shed them.' Socrates' wife exclaimed, increasing her grief: 'Those wretched judges have condemned him to death unjustly!' But Socrates replied, 'Would you really prefer that I were justly condemned?'[404]

[A]   We pierce our ears: the Greeks held that to be a mark of slavery. When we lie with our wives we hide away: the Indians lie with them in public. The Scythians used their temples to execute foreigners: elsewhere temples serve as sanctuaries:[405]

> [B]   *Inde furor vulgi, quod numina vicinorum*
> *Odit quisque locus, cum solos credat habendos*
> *Esse Deos quos ipse colit.*

[The fury of the mob is aroused since everyone hates his neighbours' gods, convinced that the gods he adores are the only true ones.][406]

402. Erasmus, *Apophthegmata, Aristippus* V and I.
403. Virgil, *Aeneid*, III, 539.
404. Diogenes Laertius, *Lives*, Solon, I, lxiii, 53; Erasmus, *Apophthegmata: Socratica* LIII.
    '88 (in place of [C]): *From this diversity of aspects there arises the fact that judgements are variously applied to the choice of objects.*
405. Sextus Empiricus, *Hypotyposes*, III, xxiv, 200–203.
406. Juvenal, *Sat.*, XV, 36.

[A] I have heard tell of a judge who, whenever he came across in his lawbooks a thorny disagreement between Bartolus and Baldus or a subject marked by conflicting interpretations, wrote in the margin, *Question for friend*, meaning by that that the truth was so entangled in controversy that in a similar case he could favour whichever party he wanted to. It was only lack of wit and intellect which stopped him from writing *Question for friend* all over the place! Counsel and judges today find enough bias in their lawsuits to bowl them any way they please. A field of study so limitless, dependent on the authority of so many opinions and subject to such arbitrariness, is bound to give rise to an extreme confusion of judgements. There is no case so clear that it does not provoke controversy. One court judges this way: another reverses the verdict and then, on a later occasion, reverses its own judgement. Familiar examples of this can be seen in an astonishing abuse which stains the splendour and ceremonial authority of our judicial system: the verdict of the parties is *not* to settle for the verdict of the Court: they dash from one judge to another for a decision on the same case.

As for the licence of philosophical opinion about vice and virtue, there is no need to go lengthily into that; it is better to pass over some of the notions in silence than to trumpet them abroad [C] before weaker intellects. [B] Arcesilaus said that in lechery proclivities [C] and occasions [B] were irrelevant.[407] '*Et obscoenas voluptates . . . si natura requirit, non genere, aut loco, aut ordine, sed forma, aetate, figura metiendas Epicurus putat*' [Epicurus thinks that when Nature demands to be satisfied by lascivious pleasures, we need not consider family origin, position or rank but only beauty, youth and figure]. '*Ne amores quidem sanctos a sapiente alienos esse arbitrantur*' [They even think that forbidden *affaires* are not incompatible with being a Sage]. '*Quaeramus ad quam usque aetatem juvenes amandi sint*' [Let us investigate up to what age it is proper to love young men]. The last two quotations are Stoic; together with the reproach which Dicaearchus made to Plato himself on this subject, they show how far even the sanest philosophy will go in tolerating quite excessive licence, far from common practice.[408]

407. Plutarch, tr. Amyot, *Les Regles et preceptes de Santé*, 295 D E (condemning all vicious sexuality).
408. Cicero, *Tusc. disput.*, V, xxxiii, 94; *De fin.*, III, xx, 68; Seneca, *Epist.* CXXIII, 15 (condemning Stoic indiscretions); Cicero, *Tusc. disput.*, IV, xxxiv, 71. Dicaearchus reproached Plato for his *Symposium* and *Phaedrus*; Montaigne takes all these quotations as allusions to irregular *affaires*; Marie de Gournay translates *amores sanctos* by *amours illicites* ('illicit love-*affaires*') which is, I think, the sense.
'95 '98, etc.: for *Dicaearchus*, 'Diogarchus'.

[A] Laws gain their authority from actual possession and custom: it is perilous to go back to their origins; laws, like our rivers, get greater and nobler as they roll along: follow them back upstream to their sources and all you find is a tiny spring, hardly recognizable; as time goes by it swells with pride and grows in strength. But just look at those Ancient concerns which gave the original impulse to that mighty stream, famed, full of dignity, awesome and venerable: you then see them to be so light and so delicate that it is not surprising that these people here – philosophers who weigh everything and reduce everything to reason, never accepting anything on authority and trust – reach verdicts far removed from those of the generality. These people, who model themselves on their concept of Nature as she originally was, not surprisingly stray from the common path in most of their opinions. Few of them for example would have approved of the constraints we impose on marriage;   [C] most of them wanted a community of wives without binding obligations. [A] Courteous conventions like ours they rejected.[409] Chrysippus said that, for a dozen olives, a philosopher will turn a dozen somersaults in public, even with his breeches off.[410] [C] He could hardly have advised Clisthenes against giving his fair daughter Agarista to Hippoclides, just because he saw him stand on his head on a table with his legs wide apart in the air.[411]

In the midst of a discussion, and in the presence of his followers, Metrocles rather injudiciously let off a fart. To hide his embarrassment he stayed at home until, eventually, Crates came to pay him a visit; to his consolations and arguments Crates added the example of his own licence: he began a farting match with him, thereby removing his scruples and, into the bargain, converting him to the freer Stoic school from the more socially oriented Peripatetics whom he had formerly followed.[412] What we call 'honourable'

409. '88: rejected. *Everyone had heard tell of the shameless way of life of the Cynic philosophers.* Chrysippus . . .
410. Cf. Plutarch, tr. Amyot, *Contredicts des philosophes Stoïques*, 569 B ('In the VIIth Book of his *Offices* he goes further, saying he will do a somersault three times, provided he be given a talent.').
'88 (in place of [C]): breeches off. *And that 'honesty' and 'reverence', as we call them, which make us hasten to hide some of our natural and rightful actions, not to dare to call things by their name or to fear to mention things we are allowed to do, could they not be said to be a guileful wantonness, invented in Venus' own chambers so as to give more value and stimulus to her games? Is it not an allurement, a bait and a stimulus to voluptuousness? For usage makes us evidently feel that ceremony, modesty and difficulties are means of sharpening and inflaming such fevers as those.* That is why some say . . .
411. Herodotus, VI, cxxix; Aelian, *Var. hist.*, XII, 24.
412. Erasmus, *Apophthegmata*, VII, *Crates Thebananus Cynicus*, XVII.

behaviour – not to dare to perform openly actions which are 'honourable' when done in private – they termed silliness. As for ingeniously concealing or disowning those of our actions which Nature, custom and our very desires publish and proclaim abroad, they reckoned that to be a vice. They thought it a desacralizing of Venus' mysteries to take them out from the discreet sanctuary of her temple and exhibit them to the public gaze: draw back the curtains, and her sports are debased. (Shame has a kind of weight: concealment, dissimulation and constraint form part of our esteem.) They thought that it was most ingenious that Lust, out of regret for the dignity and convenience of her traditional bedchambers, should don the mask of Virtue, seeking to avoid being prostituted at the crossroads and trampled underfoot before the eyes of the mob. That is why [A] some say that abolishing the public brothels would not merely take the fornication at present restricted to such places and spread it everywhere, but would also stimulate that vice in men by making it more difficult:

> *Moechus es Aufidiae, qui vir, Corvine, fuisti;*
> *Rivalis fuerat qui tuus, ille vir est.*
> *Cur aliena placet tibi, quae tua non placet uxor?*
> *Nunquid securus non potes arrigere?*

[Corvinus! You used to be the husband of Aufidia; she has married your rival and you are her lover. Now she has become the wife of another, she pleases you (she never did when she was your own). Why? Are you unable to get it up without risking a beating?][413]

You can find a thousand variations on that experience.

> *Nullus in urbe fuit tota qui tangere vellet*
> *Uxorem gratis, Caeciliane, tuam,*
> *Dum licuit; sed nunc, positis custodibus, ingens*
> *Turba fututorum est. Ingeniosus homo es.*

[Caecilianus: when you left your wife free, nobody in the whole of Rome wanted to touch her: now you have put guards round her, she is besieged by a huge crowd of fucking admirers. Clever chap!][414]

---

413. Martial, III, lxx.
414. Martial, I, lxxiv, cited Tiraquellus, *De legibus connubialibus*, XVI, II.

Once a philosopher was surprised in the very act; asked what he was doing, he coldly replied: 'I am planting a man'; he no more blushed than if he had been caught planting garlic.[415]

[C] It is, I think, too tender and respectful an opinion when one of our great religious authors holds that Necessity actually compels this act to be carried out in modest seclusion: he could not convince himself that the Cynics actually consummated it in their licentious embraces, but were content with imitating lascivious motions in order to display that absence of shame which formed part of their teachings. He thought that they had to find a secluded place later on, so as to be able to ejaculate what shame had constrained them to hold back. But he had insufficiently plumbed the depths of the Cynics' debauchery: for when Diogenes was masturbating in the presence of crowds of bystanders, he specifically said he wanted to give his belly *complete* satisfaction by rubbing it up like this. To those who asked why his 'hunger' had to be satisfied in the street, not in some more suitable place, he replied, 'I was in the street when I felt hungry.' Women philosophers who joined this school joined in with their bodies – everywhere and indiscriminately: Hipparchia was only admitted into the group of disciples around Crates on condition that she followed the customary practices and rules in every particular.[416]

These philosophers attached the highest value to virtue; they rejected all other disciplines except morals; nevertheless, they attributed ultimate authority, above any law, to the decisions of their Sage: they decreed no restraints on pleasure [A] except moderation and the respect for the freedom of others.

Heraclitus and Protagoras noted that wine tastes bitter when you are sick, delightful when you are well, and that an oar looks crooked in the water but straight out of it; from these and similar contradictory appearances they argued that every object contains within itself the causes of such appearances: that there was a bitterness in wine which was related to the

415. Source unknown.

'88: planting *cabbages. Solon is said to have been the first to give women freedom in his Laws to profit publicly from their bodies. And the philosophical school which most honoured Virtue did not in short impose any bridle on the practising of lust of all sorts except moderation . . .* (Transferred by Montaigne to III, 5, 'On some lines of Virgil'.)

416. St Augustine, *City of God*, XIV, 20 (defending the notion that shame is natural); Diogenes Laertius, *Lives*, Diogenes, VI, lxix and lviii (cf. Erasmus, *Apophthegmata*, III, *Diogenes Cynicus*, XLVII) and *Lives*, Hipparchia, VI, cxvi. The same associations, with additional material, are found in Tiraquellus, *De legibus connubialibus*, XV, 159.

taste of the sick man; a quality of bentness in the oar which was related to whoever was looking at it in the water; and so on, for all the rest. That is equivalent to saying that everything is in everything; from which it follows that nothing is in anything: for where everything is, nothing is.[417]

It was this opinion which reminded me of an experience which we have all had, that once you start digging down into a piece of writing there is simply no slant or meaning – straight, bitter, sweet or bent – which the human mind cannot find there.

Take that clearest, purest and most perfect Word there can ever be: how much falsehood and error have men made it give birth to! Is there any heresy which has not discovered ample evidence there for its foundation and continuance? That is why there is one proof which the founders of such erroneous doctrines will never give up: evidence based upon exegesis of words.

A man of some rank, deeply immersed in the quest for the philosopher's stone, wanted to justify it to me recently on authority: he cited five or six Biblical texts which he said were the ones he chiefly relied on to salve his conscience (for he is in holy orders). The choice of texts he produced was not only amusing but most applicable to the defence of that egregious science.

That is how divinatory nonsense comes to be believed in. Provided that a writer of almanacs has already gained enough authority for people to bother to read his books, examining his words for implications and shades of meaning, he can be made to say anything whatever – like Sybils. There are so many ways of taking anything, that it is hard for a clever mind *not* to find in almost any subject something or other which appears to serve his point, directly or indirectly. [C] That explains why an opaque, ambiguous style has been so long in vogue. All an author needs to do is to attract the concern and attention of posterity. (He may achieve that not so much by merit as by some chance interest in his subject-matter.) Then, whether out of subtlety or stupidity, he can contradict himself or express himself obscurely: no matter! Numerous minds will get out their sieves, sifting and forcing any number of ideas through them, some of them relevant, some off the point, some flat contradictory to his intentions, but all of them doing him honour. He will grow rich out of his students' resources – like dons being paid their midsummer fees at the *Lendit* fair.

417. Cf. Sextus Empiricus, *Hypotyposes*, I, xxix, 210–11; xxxii, 218; Cicero, *Acad.*: *Lucullus*, II, vi, 79. (The refraction of a 'bent oar' was a major argument for sceptics.) Cf. I, 14, note 71.

[A] This has lent value to many a worthless piece, making several books seem valuable by loading on to them anything at all; one and the same work is susceptible to thousands upon thousands of diverse senses and nuances – as many as we like. [C] Is it possible that Homer really wanted to say all that people have made him say,[418] and that he really did provide us with so many and so varied figurative meanings that theologians, military leaders, philosophers and all sorts of learned authors (no matter how different or contradictory their treaties) can refer to him and cite his authority as the Master General of all duties, works and craftsmen, the Counsellor General of all enterprises?

[A] Anyone on the lookout for oracles and predictions has found plenty of material there! I have a learned friend who is astonishingly good at producing wonderfully apt passages from Homer in favour of our religion: he cannot be easily prised from the opinion that Homer actually intended them (yet he knows Homer as well as any man alive). [C] And the very things he finds favouring our religion were thought in ancient times to favour theirs.

See how Plato is tossed and turned about. All are honoured to have his support, so they couch him on their own side. They trot him out and slip him into any new opinion which fashion will accept. When matters take a different turn, then they make him disagree with himself. They force him to condemn forms of behaviour which were quite licit in his own century, just because they are illicit in ours. The more powerful and vigorous the mind of his interpreters, the more vigorously and powerfully they do it.

[A] Democritus took the very foundations of Heraclitus – his assertion that things bear within themselves all the features we find in them – and drew the contrary conclusion, namely, that objects have none of the qualities we find in them: from the fact that honey is sweet to some and bitter to others, he concluded that it was neither sweet nor bitter. The Pyrrhonists said that they did not know whether it is sweet or bitter or neither or both, for they always reach the highest summit of doubt.[419] [C] The Cyrenaics held that nothing is perceptible which

---

418. Cf. Rabelais, *Gargantua*, TLF, Prologue, 87 f.

'88 (in place of [C]): like. *Homer is as great as you wish, but it is not possible that he intended to represent as many ideas as people attribute to him. Law-givers have divined in him instructions without number for their own concerns; so have military men; so have those who treat of the arts.* Anyone on the . . .

419. Sextus Empiricus, *Hypotyposes*, I, xxx, 213–14. What follows is from Cicero, *Acad.: Lucullus*, II, xxiv, 76; xlvi, 142.

comes from without: the only things perceptible are those which affect us inwardly, such as pain and pleasure. They did not even recognize the existence of tones or colours, but only certain emotional impulses produced by them; on these alone Man must base his judgement: Protagoras thought that whatever appears true is true for the man concerned; the Epicureans place judgement – in the case of both knowledge and pleasure – in the senses. Plato wanted judgements about Truth, and Truth herself, to be independent of opinion and the senses, belonging only to the mind and thought.[420]

[A] Such discussion has brought me to the point where I must consider the senses: they are the proof as well as the main foundation of our ignorance.

Without a doubt, anything that is known is known by the faculty of the knower; for, since judgement proceeds from the activity of a judge, it is reasonable that he perform that activity by his own means and by his will, not by outside constraint (as would be the case if the essence of an object were such that it forced us to know it). Now knowledge is conveyed through the senses: they are our Masters:

> [B] *via qua munita fidei*
> *Proxima fert humanum in pectus templaque mentis.*

[the highway by which conviction penetrates straight to men's hearts and to the temple of their minds.][421]

[A] Knowledge begins with them and can be reduced to them. After all, we would have no more knowledge than a stone if we did not know that there exist sound, smell, light, taste, measure, weight, softness, hardness, roughness, colour, sheen, breadth, depth. They form the foundations and principles on which our knowledge is built. [C] Indeed, for some thinkers, knowledge is sensation. [A] Anyone who can force me to contradict the evidence of the senses has got me by the throat: he cannot make me retreat any further. The senses are the beginning and the end of human knowledge.

---

420. Plato, cited Cicero (note 419), and *Theaetetus*, 186: knowledge is not in sensation but in reasoning upon sensation. Truth is 'perceived', not apprehended; it is not attainable from 'opinion'.
421. Lucretius, V, 102 (Lambin, p. 382).

> *Invenies primis ab sensibus esse creatam*
> *Notitiam veri, neque sensus posse refelli.*
> *Quid majore fide porro quam sensus haberi*
> *Debet?*

[You realize that the conception of truth is produced by the basic senses; the senses cannot be refuted. What should we trust more than our senses, then?][422]

Attribute as little to them as you can, but you will have to grant them this: that all the instruction we receive is conveyed by them and through them. Cicero says that when Chrysippus assayed denying their force and power, so many contrary arguments and overwhelming objections occurred to him that he could not answer them. Whereupon Carneades, who maintained the opposite side, boasted of fighting him with his own words and weapons, exclaiming, 'Wretch! You have been defeated by your own strength!'[423]

For us there is absolutely nothing more absurd than to say that fire is not hot; that light does not illuminate; that iron has no weight or resistance. Those are notions conveyed to us by our senses. There is no belief or knowledge in man of comparable certainty.

Now, on the subject of the senses, my first point is that I doubt that Man is provided with all the natural senses.[424] I note that several creatures live full, complete lives without sight; others, without hearing. Who can tell whether we, also, lack one, two, three or more senses? If we do lack any, our reason cannot even discover that we do so. Our senses are privileged to be the ultimate frontiers of our perception: beyond them there is nothing which could serve to reveal the existence of the senses we lack. One sense cannot reveal another:

> [B]   *An poterunt oculos aures reprehendere, an aures*
> *Tactus, an hunc porro tactum sapor arguet oris,*
> *An confutabunt nares, oculive revincent?*

---

422. Lucretius, IV, 478, 482 (Lambin, pp. 308–11). This section of Lucretius is aimed at anyone who dares to think that 'nothing is known' (*nil sciri*); Lucretius, 469 ff. This fact lends piquancy to what follows: Montaigne, like Carneades, is about to use his opponent's weapons against him.

423. Cicero, *Acad.: Lucullus*, II, xxvii, 87 and Plutarch, *Contredicts des philosophes Stoïques*, 562H–563A.

424. Sextus Empiricus, *Hypotyposes*, I, xiv, 96–7. The whole of this section (36–163) forms the background to these pages.

[Can the ears correct the eyes; the ears the touch? Can the tastes in our mouths correct the touch? Or will our nostrils and our eyes prove touch to be wrong?]

[A] They all form, each one of them, the ultimate boundary of our faculty of knowledge:

> *seorsum cuique potestas*
> *Divisa est, sua vis cuique est.*

[For each has received its share and power, quite separate from the others.][425]

A man born blind cannot be made to *understand* what it is not to see; he cannot be made to wish he had sight and to regret what he is lacking. (Therefore we ought not to take comfort from our souls' being happy and satisfied with the senses we do have; if we are deprived and imperfect, our souls have no way of sensing it.) It is impossible to say anything to that blind man by reason, argument or comparison, which will fix in his understanding what light, colour and sight really are. There is nothing beyond the senses which can supply evidence of them. We do find people who are born blind expressing a wish to see: that does not mean that they know what they are asking for. They have learned from us that they lack something which we have, and they wish that they had it; [C] they name it all right, as well as its effects and its consequences; [A] but they do not know what it is, for all that; they cannot even get near to grasping what it is.

I have met a nobleman of good family who was born blind, or, at least, blind enough not to know what sight is. He has so little knowledge of what he is lacking that he is always using words appropriate to seeing, just as we do; he applies them in his own peculiar way. When he was presented with one of his own godchildren, he took him in his arms and said: 'My God, what a handsome child. How nice to see him! What a happy face he has.' He will say (like one of us): 'What a lovely view there is from this room! What a clear day. How bright the sun is.' And that is not all. Hearing how much we enjoy the sports of hunting, tennis and shooting, he likes them, too; he tries to join in and believes that he can take part like us. He gets carried away, has a great deal of fun and yet has no knowledge at all of these sports, except through the ears. On open ground, where he can use his spurs, somebody shouts, 'There goes a hare.' Then somebody says, 'Look, the hare has been caught.' You will see him as proud of the kill as other men he has heard.

At tennis he takes the ball in his left hand and hits it with his racket. As

425. Lucretius, IV, 486, 490.

for the harquebus, he shoots at random, and is delighted when his men tell him he has shot too high or too wide.

How do we know that the whole human race is not doing something just as silly? We may all lack some sense or other; because of that defect, most of the features of objects may be concealed from us. How can we know that the difficulties we have in understanding many of the works of Nature do not derive from this, or that several of the actions of animals which exceed our powers of understanding are produced by a sense-faculty which we do not possess? Perhaps some of them, by such means, enjoy a fuller life, a more complete life than we do.

We need virtually all our senses merely to recognize an apple: we recognize redness in it, sheen, smell and sweetness. An apple may well have other qualities than that: for example powers of desiccation or astringency, for which we have no corresponding senses.[426]

Take what we call the occult properties of many objects (such as the magnet attracting iron).[427] Is it not likely that there are certain senses known to Nature which furnish the faculties necessary for perceiving them and understanding them, and that the lack of such faculties entails our ignorance of their true essence? There may be some peculiar sense which tells cocks when it is midday or midnight and makes them crow, [C] or which teaches hens (before any practical experience) to fear the sparrow-hawk but not larger animals like geese or peacocks; which warns chickens of the innate hostility of cats but tells them not to fear dogs; which puts them on their guard against a *miaou* (quite a pleasing sound, really) but not against a bark (a harsh and aggressive sound);[428] which tells hornets, ants and rats how to select the best cheese and the best pear, before they even taste them; [A] which leads stags [C] elephants and snakes [A] to recognize herbs necessary to cure them.

There is no sense which is not dominant and which does not have the means of contributing vast amounts of knowledge. If we had no comprehension of sounds, harmony and the spoken word, that would throw all the rest of our knowledge into inconceivable confusion. For, quite apart from all that arises from the properties of each individual sense, think of the

426. Sextus Empiricus, *Hypotyposes*, I, xiv, 95–6.
427. These qualities were classified as 'sympathies' and 'antipathies' within nature and were fundamental to Renaissance science; cf. G. Fracastoro, *De sympathia et antipathia rerum,* 1554. For the magnet, cf. Rabelais, *Quart Livre*, TLF, LXII; for animals recognizing medical simples, *ibid.*, LXII (drawing on Plutarch and Celio Calcagnini).
428. Seneca, *Epist.*, CXXI, 19.

arguments, consequences and conclusions which we infer by comparing one sense with another. Let an intelligent man imagine human nature created, from the beginning, without sight; let him reflect how much ignorance and confusion such a defect would entail, how much darkness and blindness there would be in our minds. We can see from that how vital it would be for our knowledge of truth if we lacked another sense, or two or three senses. We have fashioned a truth by questioning our five senses working together; but perhaps we need to harmonize the contributions of eight or ten senses if we are ever to know, with certainty, what Truth is in essence.

Those schools which attack Man's claim to possess knowledge base themselves mainly on the fallibility and weakness of our senses: for, since all knowledge comes to us through them and by them, we have nothing left to hold on to if they fail in their reports to us, if they change and corrupt what they convey to us from outside, or if the light which filters through to our mind from them is darkened in the process.

This ultimate difficulty has given rise to many strange notions: that a given object does have all the qualities we find in it; that it has none of the qualities which we think we find in it;[429] or, as the Epicureans contend, that the Sun is no bigger than our sight judges it to be –

> [B]   *Quicquid id est, nihilo fertur majore figura*
> *Quam nostris oculis quam cernimus, esse videtur*

[Be that as it may, its size is no bigger than it seems when we behold it][430]

[A]   – or, that those appearances which make an object look big when you are close to it and smaller when you are farther from it, are both true –

> [B]   *Nec tamen hic oculis falli concedimus hilum*
> *Proinde animi vitium hoc oculis adfingere noli*

[We do not at all concede that the eyes can be deceived. Do not attribute to the eyes the errors of the mind][431]

429. Sextus Empiricus, *Hypotyposes*, I, xxix, 210–11.
430. Lucretius, V, 577 (of the Moon, not the Sun; but the section starts (564) '*Nec nimio solis major rota*' [The wheel of the Sun cannot be much larger than as perceived by our senses]). Lambin (p. 410) classes as 'the most stolid and silly of the opinions of Epicurus that the Sun, Moon and Stars have the size they appear to have.' He cites Cicero, *Acad.: Lucullus*, II, xxxix, 124 (cf. Introduction, p. xli and Cicero, ibid., xxvi, 82).
431. Lucretius, IV, 379;386. (Lambin, pp. 300–2, explains: 'Lucretius says that, if we are deceived in our seeing things, that is a defect of our minds, not of our eyes ... For Epicurus wished the senses to be certain and true; see Cicero [*Acad.*] *Lucullus*, II [142 f.]; later we add material from Lucretius himself.')

[A] – or, conclusively, that there is no deception whatsoever in our senses, so that we must throw ourselves on their mercy and seek elsewhere the justification for any differences and contradictions which we find in them: that, indeed, we should invent some lie or raving lunacy (yes, they get as far as that!) rather than condemn our senses.

[C] Timagoras said that he did not really see the candle-flame double when he squeezed his eye-ball sideways, but that this appearance arose from a defect of opinion not of vision.[432] [A] The absurdest of all absurdities [C] for Epicureans [A] is to deny [C] the effective power of [A] the senses:

> *Proinde quod in quoque est his visum tempore, verum est.*
> *Et, si non potuit ratio dissolvere causam,*
> *Cur ea quae fuerint juxtim quadrata, procul sint*
> *Visa rotunda, tamen praestat rationis egentem*
> *Reddere mendose causas utriusque figurae,*
> *Quam manibus manifesta suis emittere quoquam,*
> *Et violare fidem primam, et convellere tota*
> *Fundamenta quibus nixatur vita salusque.*
> *Non modo enim ratio ruat omnis, vita quoque ipsa*
> *Concidat extemplo, nisi credere sensibus ausis,*
> *Praecipitesque locos vitare, et caetera quae sint*
> *In genere hoc fugienda.*

[Consequently, whatever, at any time, has seemed to the senses to be true, is in fact true. If reason cannot unravel the causes which explain why things that are square when you are close to them appear round at a distance, it is better to find some untrue explanation of these two different impressions than to let the evidence of our senses slip through our fingers, violate first principles and shake the foundations on which our lives and their preservation are built. For, if we could no longer trust our senses and so avoid the giddy heights and other dangers Man must shun, not only would our Reason collapse in ruins but our lives as well.][433]

[C] That is a counsel of despair. It is quite unphilosophical. It reveals that human knowledge can only be supported by an unreasonable Reason, by mad lunatic ravings; that, if Man is to make himself worth anything, it is better to exploit 'Reason' such as this or any other remedy, no matter how

---

432. Cicero, *Acad., Lucullus*, II, xxv, 79–80; for the importance of the contention, cf. Aristotle, *Metaph.*, XI, vi, 7 (1063a), a criticism of 'Man as measure' which, if accepted, would imply the truth of the notions for which Lucretius is to be cited – with disapproval.

433. Lucretius, IV, 499 (Lambin, pp. 300–2).

fantastic it may be, rather than to admit so unflattering a truth that he is, of necessity, as stupid as a beast. Man cannot avoid the fact that his senses are both the sovereign regents of his knowledge, and yet, in all circumstances, uncertain and fallible. So here they must fight to a finish; if legitimate weapons fail us – and they do – they must use stubbornness, foolhardiness or cheek!

[B] Should what the Epicureans say be true (namely, that if the senses play us false we have no knowledge at all);[434] and should what the Stoics say be equally true (that sensible appearances are so deceptive that they can give rise in us to no knowledge whatever); then we are forced to conclude, at the expense of the two great schools of Dogmatists, that there is no such thing as knowledge.

[A] Anybody can provide as many examples as he pleases of the ways our senses deceive or cheat us, since so many of their faults or deceptions are quite banal: a trumpet sounds a league behind us, but an echo in a valley may make it seem to come from in front:

> [B] *Extantesque procul medio de gurgite montes*
> *Iidem apparent longe diversi licet*
> *Et fugere ad puppim colles campique videntur*
> *Quos agimus propter navim*
> *ubi in medio nobis equus acer obhaesit*
> *Flumine, equi corpus transversum ferre videtur*
> *Vis, et in adversum flumen contrudere raptim.*

[Distant mountains beetling over the sea may appear as one, yet are in fact many; as we sail along, hills and plains appear to be rushing towards our prow; if we look down when our horse stops in mid-stream, the river seems to be forcing it to go up-stream against the current.][435]

[A] Hold a musket-ball beneath your second finger, with your middle finger entwined over it: you will have to force yourself to admit that there is only one ball, so decidedly do you sense it to be two. We can see every day that our senses have mastery over our reason, forcing it to receive impressions which it knows to be false and judges to be false.

I will not go into the sense of touch. Its effects are immediate, lively and concrete; many a time, as a result of the pain which it causes the body, it overthrows all those fine Stoic axioms. It takes a man who has resolutely

434. Cicero, *Acad.: Lucullus*, II, xxxii, 101.
435. Lucretius, IV, 397; 389; 421 (Lambin, pp. 300–2; but in 390 reading *praeter* as *propter*); 'defects of the mind are not defects of the senses'.

made up his mind that colic paroxysms are a thing indifferent (like any other pain or disease) and that they have no power to affect the blessed state of supreme felicity in which the Sage has been lodged by his Stoic Virtue – and makes him yell about his belly.

No heart is so flabby that the sounds of our drums and trumpets do not set it ablaze, nor so hard that sweet music does not tickle it and enliven it; no soul is so sour that it does not feel touched by some feeling of reverence[436] when it contemplates the sombre vastness of our Churches, the great variety of their decorations and our ordered liturgy, or when it hears the enchantment of the organ and the poised religious harmony of men's voices. Even those who come to scoff are brought to distrust their opinion by a shiver in their heart and a sense of dread.

[B] As for me, I do not think I would be strong enough to remain unmoved even by verses of Horace or Catullus, if well sung by a good voice coming from a fair young mouth! [C] Zeno was right to claim that the voice is Beauty's flower.[437] Some people have even tried to make me believe that a famous man known to all Frenchmen had impressed me unduly with a recital of some of his verses, which seem very different seen on paper than heard in the air, and that my eyes would contradict my ears, so great is the power of eloquent delivery to endow any work which accepts its sway with value and style.

While on the topic, Philoxenus' reaction was not without charm: he heard a piece he had composed being sung badly, so he jumped on some of the singer's tiles and smashed them. 'I spoil your things,' (he said) 'you despoil mine!'[438]

[A] Why did even those who had firmly decided to die avert their gaze from the very blow which they ordered to be struck? Why do those who have freely agreed to cauterizations and incisions for the sake of their health find that they cannot stand the sight of all the preparations, of the surgical instruments or of the actual operation? Sight does not share in the pain.

Are not these appropriate examples for demonstrating the authority of our senses over our powers of reason? – Even though we know that a lady's tresses are borrowed from a page or a lackey; that her rosy colour comes from Spain and her smooth whiteness from the ocean, we still find

436. '88: of *religious* reverence . . .
437. Erasmus, *Apophthegmata*, VII, *Zeno*, XXIV.
438. Attributed by Diogenes Laertius to Arcesilas (*Lives*, IV, xxxvi, 270).

her person more attractive and agreeable – quite unreasonably, though, for in all that nothing is her own:

> *Auferimur cultu; gemmis auroque teguntur*
> *Crimina: pars minima est ipsa puella sui.*
> *Saepe ubi sit quod ames inter tam multa requiras:*
> *Decipit hac oculos Aegide, dives amor.*

[We are carried away by clothing; ugliness is hidden behind gems and gold; the smallest part of herself is the actual girl! You can often look in vain for the girl you love under all these gewgaws. This is the shield with which the rich deceive a lover's eyes.][439]

What great power our poets attribute to the senses, when they make Narcissus enamoured of his own reflection:

> *Cunctaque miratur, quibus est mirabilis ipse;*
> *Se cupit imprudens; et qui probat, ipse probatur;*
> *Dumque petit, petitur; pariterque accendit et ardet.*

[He is enchanted by his own enchantments; unawares, he loves himself; he both praises and is praised; he yearns and is yearned for; the passion he kindles enflames himself.]

Similarly, Pygmalion's mind was disturbed by the visual impact of his ivory statue: he fell in love with it and sighed for it:

> *Oscula dat reddique putat, sequiturque tenetque,*
> *Et credit tactis digitos insidere membris;*
> *Et metuit pressos veniat ne livor in artus.*

[He kisses her, and believes his kisses are returned; he waits on her, embraces her; he believes her limbs respond to the touch of his fingers; he fears that in his ardour he may bruise her.][440]

Take a philosopher, put him in a cage made from thin wires set wide apart; hang him from one of the towers of Notre Dame de Paris. It is evident to his reason that he cannot fall; yet (unless he were trained as a steeplejack) when he looks down from that height he is bound to be terrified and beside himself. It is hard enough to feel safe at the top of a church tower, even behind open-work ramparts of stone: some people cannot even bear thinking about it.

---

439. Ovid, *Remedia amoris*, 343. ('From the ocean': that is, from pulverized sea-shells, used as face 'powder'.)
440. Ovid, *Metam,,* III, 424; X, 256.

Take a beam wide enough to walk along: suspend it between two towers: there is no philosophical wisdom, however firm, which could make us walk along it just as we would if we were on the ground.

I am not particularly afraid of heights, but when I was on the French side of the Italian Alps I made an assay and found that I could not suffer the sight of those boundless depths without a shiver of horror; I was at least my own height away from the edge and could not have fallen over unless I deliberately exposed myself to danger: yet my knees and thighs were trembling. I also noticed that, whatever the height, it was comforting and reassuring if there happened to be some tree or rock jutting out on the slope which could hold our gaze and interrupt our vision: it was as though they could have helped us if we fell. But when the precipices were sheer and smooth we could not even look at them without feeling giddy, [C] *'ut despici sine vertigine simul oculorum animique non possit'* [such that no one could look down without vertigo in eyes and mind].[441]

Which shows how sight can deceive us.

One fine philosopher even poked out his eyes so as to free his mind from visual debauchery; he could then go on philosophizing in freedom. But by the same standard he ought to have blocked up his ears[442] – [B] which Theophrastus says are the most dangerous of all our organs when it comes to receiving violent impressions capable of changing and disturbing us.[443] [A] Eventually he would have to deprive himself of every other sense (tantamount to life and being), for all the senses can have this dominant power over our reason and our soul: [C] *'Fit etiam saepe specie quadam, saepe vocum gravitate et cantibus, ut pellantur animi vehementius; saepe etiam cura et timore'* [Some visual feature, some grave voice or incantations may often strike the mind most vehemently: worry and care may often do that too].[444]

[A] Doctors maintain that people with some complexions can be driven mad by certain sounds or instruments. I have known people who could not even hear a bone being gnawed under their table without losing control; and there is hardly a person who is not upset by the sharp rasping

441. Livy, XLIV, 6.
442. Democritus (whom Montaigne already mentions in I, 14: 'That the taste of good and evil things depends in large part on the opinion we have of them', and I, 39: 'On solitude'). Cf. Aulus Gellius, *Attic Nights*, X, xvii; Cicero, *De fin.*, V, xxix, 87 (hesitating to believe it).
443. Plutarch, tr. Amyot, *Comment il fault oïr*, 24H–25A.
444. Cicero, *De divinat*, XXXVI, 80.

sound of a file against iron. Some people are moved to anger or even hatred by hearing somebody chewing nearby or talking with some obstruction of their throat or nose.

Gracchus had a prompter who was a flautist; he conducted the voice of his master, softening it or making it firm:[445] what use was he if the rhythm and quality of the sounds did not have the power of moving and swaying the judgement of the listeners? We have good enough reason to make a fuss about this judgement of ours: it lets itself be affected and managed by the modulations and properties of so light a breath of wind!

The senses deceive our intellect; it deceives them in their turn. Our soul sometimes gets her own back: [C] they both vie with each other in lying and deceiving. [A] When we are moved to anger, we do not hear things as they are:

> *Et solem geminum, et duplices se ostendere Thebas.*

[We see twin suns; two Thebes.][446]

Love someone and she appears more beautiful than she is:

> [B] *Multimodis igitur pravas turpesque videmus*
> *Esse in delitiis, summoque in honore vigere.*

[Many ugly and deformed women are deeply loved, enjoying, as we see, the highest favour.][447]

[A] And anyone we dislike appears more ugly. When a man is in pain and affliction, the very light of day seems sombre and dark. Our senses are not only changed for the worse, they are knocked quite stupid by the passions of the soul. How many things do we see which we do not even notice when our minds are preoccupied with other matters?

> *In rebus quoque apertis noscere possis,*
> *Si non advertas animum, proinde esse, quasi omni*
> *Tempore semotae fuerint, longeque remotae.*

[Even in the case of things which are clearly visible, you know that if you do not turn your mind to them, it is as though they had never been there or were far away.][448]

---

445. Plutarch, tr. Amyot, *Comment il fault refrener la colere*, 57H–58A.
446. Virgil, *Aeneid*, IV, 470.
447. Lucretius, IV, 1155 (Lambin, pp. 358–9).
448. Lucretius, IV, 811 (Lambin, pp. 331–3, citing Cicero, *Tusc. disput.*, in support).

It seems, then, that the soul draws the powers of the senses right into herself and makes them waste their time.

And so, both within and without, man is full of weakness and of lies.

[B]   Those who have compared our lives to a dream are right – perhaps more right than they realized. When we are dreaming our soul lives, acts and exercises all her faculties neither more nor less than when she is awake, but she does it much more slackly and darkly; the difference is definitely not so great as between night and the living day: more like that between night and twilight. In one case the soul is sleeping, in the other more or less slumbering; but there is always darkness, perpetual Cimmerian darkness.

[C]   We wake asleep: we sleep awake. When I am asleep I do see things less clearly but I never find my waking pure enough or cloudless. Deep sleep can sometimes even put dreams to sleep; but our waking is never so wide awake that it can cure and purge those raving lunacies, those waking dreams that are worse than the real ones.

Our rational souls accept notions and opinions produced during sleep, conferring on activities in our dreams the same approbation and authority as on our waking dreams: why should we therefore not doubt whether our thinking and acting are but another dream; our waking, some other species of sleep?

[A]   If the senses are our basic judges, we should not merely call upon our own for counsel: where this faculty is concerned, the animals have as much right as we do, or even more. Some certainly have better hearing, sight, smell, touch or taste. Democritus said that the gods and the beasts have faculties of sense far more perfect than Man does.

Now there are extreme differences between the action of their senses and ours: our saliva cleanses and dries up our wounds: it kills snakes.[449]

> *Tantaque in his rebus distantia differitasque est,*
> *Ut quod aliis cibus est, aliis fuat acre venenum.*
> *Saepe etenim serpens, hominis contacta saliva,*
> *Disperit, ac sese mandendo conficit ipsa.*

[There are so many differences and variations: one man's food is another man's bitter poison. Indeed if a snake comes into contact with human saliva, it begins to bite its own tail and dies.][450]

So what quality are we to give to saliva? Do we follow our own senses or the snake's? We are trying to discover the truth about its true essence:

---

449. Cf. Rabelais, *Quart Livre*, TLF, LXIV, derived from Celio Calcagnini.
450. Lucretius, IV, 636 (Lambin, p. 619).

which of the two will tell us? Pliny says that there are certain 'sea-hares' in the Indies which are poison to us and we to them: a touch kills them.[451] Which is truly poisonous, the fish or the man? Which should we believe: the effect of the fish on the man or the man on the fish? [B] The quality of one kind of air is infectious to Man but not to cattle; another has the quality of being infectious to cattle but harmless to men. Which of the two has truly and naturally the quality of being infectious? [A] Sufferers from jaundice see everything paler and yellower than we do:

> [B] *Lurida praeterea fiunt quaecunque tuentur Arquati.*

[Those ill from 'rainbow-yellow' see everything in sallow colours.][452]

[A] There is a suffusion of blood under the skin around the eye which doctors call *Hyposphragma* – those who suffer from it see everything blood-red.[453] How do we know that these humours, which can affect the workings of Man's eyesight, are not the dominant norm among beasts? Some animals, as we know, have yellow eyes exactly like sufferers from jaundice and others have eyes which are blood-red. It is probable that the colours of objects appear different to them and to us. Who judges them right? Nobody claims that the essence of anything relates only to its effect on Man. Hardness, whiteness, depth, bitterness – such qualities are of service to animals and are known to them as to ourselves: Nature has granted that they be useful to animals as well as to us men.

If we squeeze one of our eyes, the objects we look at appear thinner and elongated: many beasts have eyes which are always squeezed up like that. For all we know, that elongated form is the true one, not what our eyes see in their normal state. [B] If we press up our eyes from the bottom, we see double:

> *Bina lucernarum florentia lumina flammis,*
> *Et duplices hominum facies, et corpora bina.*

[The lamp has twin flowerings of light, men have twin faces and twin bodies.][454]

[A] If our ears are blocked up or if the auricular passage is constricted we hear sounds differently from normal: animals have hairy ears or, in some

451. Pliny, *Hist. Nat..,* XXXII, I.
452. Lucretius, IV, 333 (Lambin, pp. 296–7).
453. Medical deformation of *hyposphagma*; cited after Sextus Empiricus, *Hypotyposes*, I, xiv, 45. The following is from ibid., 45–7.
454. Lucretius, IV, 450 (Lambin, pp. 305–7, who alludes to Aristotle, *Problemata*, 3, for the explanation); Sextus Empiricus, *Hypotyposes*, I, 47; Plato, *Theatetus*, 153b–154a.

cases, merely a little hole instead of an ear: consequently, they do not hear what we hear and the sound is perceived differently.[455]

At banquets or in the theatre, when various shades of coloured glass are placed in front of the torches, we know that they can make everything appear green, yellow or violet:

> [B]   *Et vulgo faciunt id lutea russaque vela*
> *Et ferruginea, cum magnis intenta theatris*
> *Per malos volgata trabesque trementia pendent:*
> *Namque ibi consessum caveai subter, et omnem*
> *Scenai speciem, patrum, matrumque, deorumque*
> *Inficiunt, coguntque suo volitare colore.*

[When yellow, red or rust-brown awnings are stretched over our vast theatres, flapping about in the wind on their poles and their frames, it is quite usual for them to impart their colours to the stage and to the whole assembly seated in their seats, to senators and matrons and to the statues of the gods, as their colours dance about.][456]

[A]   It seems likely that the different coloured eyes which we can notice in some animals may impart corresponding colours to what the animals see.

If we want to judge the activities of the senses we should agree with the animals and then among ourselves. We are far from doing that. Quarrels are constantly arising because one person hears, sees or tastes something differently from another. As much as anything, we quarrel over the diversity of the images conveyed to us by our senses.[457]

A child, a man of thirty, a sexagenarian, each hears and sees things differently: that is a normal law of Nature. Similarly for taste. Some people's senses are dullish and dimmer: others are more open and acute. We perceive objects to be like this or that in accordance with our own state and how they seem to us.[458] But *seeming*, for human beings, is so uncertain and so controvertible that it is no miracle if we are told that we may acknowledge that snow seems white to us but cannot guarantee to establish that it is truly so in essence. And once you shake that first principle, all the knowledge in the world is inevitably swept away.

---

455. Sextus Empiricus, *Hypotyposes*, I, xiv, 50–1.
456. Lucretius, IV, 74 (Lambin, pp. 278–81) reading *volitare* for *fluitare*.
457. Cf. Sextus Empiricus, *Hypotyposes*, I, xiv, 78–9; 106.
458. Ibid., I, xiii, 33–34.
   '88: acute. *Sick people lend a bitter taste to sweet things; from which it transpires that we do not receive things as they are but*, like this or that ... (From Aristotle, *Metaph.*, IV, v, 27 – dropped as a repetition.)

What about our very senses hampering each other? A painting may seem to have depth, but feels flat. Musk is pleasant to the smell but offensive to the taste: should we call it pleasant or not? There are herbs and ointments suitable to one part of the body but injurious to another; honey is pleasant to taste, unpleasant to look at.[459] Take those rings wrought in the shape of plumes which are called in heraldry *Feathers without Ends*. Can any eye ever be sure how wide they are and avoid being taken in by the optical illusion? For they seem to get wider on one side, narrower and more pointed on the other, especially if you turn them round your finger; yet to your touch they all appear to have the same width all the way round.

– [C] (In the ancient world some men increased their lust by the use of distorting mirrors which enlarged whatever was put before them, so that the organs used on the job pleased them more, because they looked as though they had grown bigger. But which sense did they allow to win? Was it their sight, which showed them their members as thick and big as they liked, or was it their touch, which showed the same members to be tiny and despicable?) – [460]

[A] Is it our senses which endow the object with these diverse attributes, whereas, in reality, objects only have one? Rather like bread when we eat it; it is one thing, bread, but we turn it into several: bones, blood, flesh, hair and nails.

[B]   *Ut cibus, in membra atque artus cum diditur omnes,*
Disperit, atque aliam naturam sufficit ex se.

[Like food, which spreads to all our limbs and joints, destroys itself and produces another substance.][461]

[A] Moisture is sucked up by the roots of a tree: it becomes trunk, leaf and fruit; air is one, but when applied to a trumpet it is diversified into a thousand kinds of sound: is it our senses (I say) which similarly fashion such objects with diverse qualities or do they really have such qualities? Then, given that doubt, what conclusion can we reach about their true essence?

And then, to go further still: the attributes of illness, madness or sleep make things appear different from what they do to the healthy, the sane

459. Sextus Empiricus, *Hypotyposes*, I, xiii, 91–2.
460. Ibid., I, xiv, 48–9; Seneca, *Quaest. Nat.*, I, xvi.
461. Sextus Empiricus, *Hypotyposes*, I, xiv, 33; Lucretius, III, 703 (Lambin, pp. 237–8).

and the waking man:[462] is it not likely therefore that our rightful state and our natural humours also have attributes which can endow an object with a mode of being corresponding to their own characteristics, making it conform to themselves, just as our disordered humours do? [C] Why should a temperate complexion not endow objects with a form corresponding to itself just as our distempers can, stamping its own imprint upon them?[463] On to his wine the queasy man loads tastelessness; the healthy man, a bouquet; the thirsty man, sheer delight.

[A] Now, since our state makes things correspond to itself and transforms them in conformity with itself, we can no longer claim to know what anything truly is: nothing reaches us except as altered and falsified by our senses. When the compasses, the set-square and the ruler are askew, all the calculations made with them and all the structures raised according to their measurements, are necessarily out of true and ready to collapse.

The unreliability of our senses renders unreliable everything which they put forward:

> *Denique ut in fabrica, si prava est regula prima,*
> *Normaque si fallax rectis regionibus exit,*
> *Et libella aliqua si ex parte claudicat hilum,*
> *Omnia mendose fieri atque obstipa necessum est,*
> *Prava, cubantia, prona, supina, atque absona tecta,*
> *Jam ruere ut quaedam videantur velle, ruantque*
> *Prodita judiciis fallacibus omnia primis.*
> *Hic igitur ratio tibi rerum prava necesse est*
> *Falsaque sit, falsis quaecumque a sensibus orta est.*

[It is as when a building is erected: if the ruler is false from the outset, or the set-square deceptive and out of true, if the level limps a bit to one side, then the building is necessarily wrong and crooked; it is deformed, pot-bellied, toppling forwards or backwards and quite disjointed; some parts seem about to fall down now: all will fall down soon, betrayed by the original mistakes of calculation; similarly every argument that you base on facts will prove wrong and false, if the facts themselves are based on senses which prove false.][464]

---

462. Sextus Empiricus, *Hypotyposes*, I, xiv, 100–4.

   '88: Waking man: *Since that particular state, by endowing objects with a being different from the one they have, and since a jaundiced humour changes everything to yellow,* is it not likely . . . (Then, for *rightful* state, *ordinary* state.)

463. Ibid., xiv, 102.

464. Lucretius, IV, 513 (Lambin, pp. 309–11).

And meanwhile who will be a proper judge of such differences? It is like saying that we could do with a judge who is not bound to either party in our religious strife, who is dispassionate and without prejudice. Among Christians that cannot be.[465] The same applies here: if the judge is old, he cannot judge the sense-impressions of old age, since he is a party to the dispute; so too if he is young; so too if he is well; so too if he is unwell, asleep or awake.[466] We would need a man exempt from all these qualities, so that, without preconception, he could judge those propositions as matters indifferent to him.

On this reckoning we would need such a judge as never was.

We register the appearance of objects; to judge them we need an instrument of judgement; to test the veracity of that instrument we need practical proof; to test that proof we need an instrument. We are going round in circles.[467]

The senses themselves being full of uncertainty cannot decide the issue of our dispute. It will have to be Reason, then. But no Reason can be established except by another Reason. We retreat into infinity.[468] Our mental faculty of perception is never directly in touch with outside objects – which are perceived via the senses, and the senses do not embrace an outside object but only their own impressions of it; therefore the thought and the appearance are not properties of the object but only the impressions and feelings of the senses. Those impressions and that object are different things. So whoever judges from appearances judges from something quite different from the object itself.

If you say that these sense-impressions convey the quality of outside objects to our souls by means of resemblances, how can our rational soul make sure that they are resemblances, since it has no direct contact of its own with the outside objects? It is like a man who does not know Socrates; if he sees a portrait of him he cannot say whether it resembles him or not.[469]

But supposing, nevertheless, that anyone did wish to judge from appearances, he cannot do so from all of them, since (as we know from experience) they all mutually impede each other because of contradictions and discrepancies. Will he select only some appearances to control the

465. Both sides in the religious wars claim to be the one true Church, so no Christian anywhere can remain impartial.

466. Sextus Empiricus, *Hypotyposes*, I, xiv, 104–6.

467. Ibid., 115–17.

468. Ibid., II, vii, 89.

469. Ibid., II, vii, 72–5. A similar argument appealed to St Augustine (*Contra academicos*, II, 7); cf. also Sextus Empiricus, *Against the Mathematicians*, II, 58–9.

others? But the first one selected will have to be tested for truth against another one selected, and that one against a third: the end will therefore never be reached.⁴⁷⁰

To conclude: there is no permanent existence either in our being or in that of objects. We ourselves, our faculty of judgement and all mortal things are flowing and rolling ceaselessly: nothing certain can be established about one from the other, since both judged and judging are ever shifting and changing.⁴⁷¹

'We have no communication with Being;⁴⁷² as human nature is wholly 'situated, for ever, between birth and death, it shows itself only as a dark 'shadowy appearance, an unstable weak opinion. And if you should 'determine to try and grasp what Man's *being* is, it would be exactly like 'trying to hold a fistful of water: the more tightly you squeeze anything the 'nature of which is always to flow, the more you will lose what you try to 'retain in your grasp. So, because all things are subject to pass from change 'to change, Reason is baffled if it looks for a substantial existence in them, 'since it cannot apprehend a single thing which subsists permanently, 'because everything is either coming into existence (and so not fully 'existing yet) or beginning to die before it is born.' Plato said that bodies never have existence, though they certainly have birth, [C] believing that Homer made Oceanus Father of the Gods and Thetis their Mother, to show that all things are in a state of never-ending inconstancy, change and flux (an opinion, as he says, common to all the philosophers before his time, with the sole exception of Parmenides, who denied that anything has motion – attaching great importance to the force of that idea).⁴⁷³

[A] Pythagoras taught that all matter is labile and flowing;⁴⁷⁴ the

---

470. Ibid., II, ix, 88–9: the climax to Sextus' denial that appearances can be judged as probable, let alone true. It rules out dialectic as a means of telling truth from error (ibid., 94) and continues suspension of judgement (95).
471. This Platonic assertion forces man to go beyond the transient flux of things and to seek the unchanging Reality lying behind it. From now to the last paragraph Montaigne transcribes, with minor adaptations, a very large borrowing from Amyot's translation of Plutarch: *Que signifioit ce mot E'i* (456H–357E); this is indicated here by continuous quotation marks: in the original no indication of any kind shows that this is a borrowing. (Even Marie de Gournay did not recognize it as such.) Departures from the original version by Amyot are indicated below. (Amyot's French version differs markedly from modern interpretations of the original Greek of Plutarch.)
472. Plutarch, 356H: with *true* Being . . .
473. Plato, *Theaetetus*, 180E.
474. Not *Pythagoras* but *Protagoras*: cf. Sextus Empiricus, *Hypotyposes*, I, xxxii, 217.

Stoics, that there is no such thing as the present (which is but the joining and the coupling together of the future and past);[475] 'Heraclitus, that no man ever stepped twice into the same river' –([B] Epicharmus, that a man who borrowed money in the past does not owe it now, and that a man invited to breakfast yesterday evening turns up this morning uninvited, both having become different people).[476] – [A] Heraclitus 'that no 'mortal substance can ever be found twice in an identical state because the 'rapidity and ease of its changes make it constantly disperse and reassemble; 'it is coming and going, so that whatever begins to be born never achieves 'perfect existence, since its delivery is never complete and never stops as 'though it had come to the end; but, ever since the seeds of it were sown, it 'is continually modifying and changing from one thing to another; just as 'from the human seed there first springs a shapeless embryo in the mother's 'womb, then a human shape, then, once out of the womb, a suckling child, 'then a boy, then, in due course, a youth, a mature man, an elderly and 'then a decrepit, aged man, so that each subsequent age to which birth is 'given is for ever undoing and destroying the previous one.'

> [B] *Mutat enim mundi naturam totius aetas,*
> *Ex alioque alius status excipere omnia debet,*
> *Nec manet ulla sui similis res: omnia migrant,*
> *Omnia commutat natura et vertere cogit.*

[For Time changes the nature of all things in the world; each stage must be succeeded by another, nothing remains as it was; all things depart and Nature modifies all things and compels them to change.][477]

[A] 'And after that we men stupidly fear one species of death, when we 'have already passed through so many other deaths and do so still; yet, as 'Heraclitus said, not only is the death of fire the birth of air, and the death 'of air the birth of water, but we may see it even more clearly in ourselves: 'the flower of our life withers and dies into old age; but youth ended in that 'adult flower, as childhood in youth and as that embryonic stage died into 'childhood; yesterday dies into today, and this day will die into tomorrow. 'Nothing lasts; nothing remains forever one.'[478]

475. Plutarch, *Des communes conceptions contre les Stoïques*, 586B–C. For Heraclitus, see Aristotle, *Metaph.*, IV, v, 1010a.
476. Plutarch, tr. Amyot, *Pourquoi la justice divine diffère quelquefois la punition des malefices*, 264. (Some small changes to Amyot's French here, to accommodate the interpolations; grammar and clarity suffer.)
477. Lucretius, V, 828 (Lambin, p. 426).
478. Five words of Amyot omitted and a phrase adapted (357B).

To prove that this is so: 'if we remained forever one and the same, how 'is it that we can delight in one thing now and later in another? How can 'we each be *one* if we love or hate contradictory things, first praising them, 'then condemning them?[479] How can we have different emotions, no 'longer retaining the same sentiment within the same thought? For it is not 'likely that we can experience different reactions unless we ourselves have 'changed; but whoever suffers change is no longer the same *one*: he no 'longer is. For his *being*, as such, changes when his *being one* changes, as each 'personality ever succeeds another. And, consequently, it is of the nature of 'our senses to be misled and deceived. Because they do not know what *being* 'is, they take *appears to be* for *is*.

'What is it then which truly is? That which is eternal – meaning that 'which has never been born; which will never have an end; to which Time 'can never bring any change. For Time is a thing of movement, appearing 'like a shadow in the eternal flow and flux of matter, never remaining stable 'or permanent;[480] to Time belong the words *before* and *after; has been* and '*shall be*, words that show at a glance that Time is evidently not a thing 'which is. For it would be great silliness and manifest falsehood to say that 'something is which has not yet come into being or has already ceased to 'be.

'With the words "*Present*", "*This instant*", "*Now*", we above all appear 'to support and stabilize our understanding of Time: but Reason strips it 'bare and at once destroys it: for Reason straightway cleaves *Now* into two 'distinct parts, the future and the past, as needing of necessity to see it thus 'divided into two parts.

'The same applies to Nature (which is measured) as to Time (which 'measures her): for there is nothing in Nature, either, which lasts or subsists; 'in her, all things are either born, being born, or dying.[481]

'It would therefore be a sin to say *He was* or *He will be* of God, who is 'the only ONE who IS. For those terms are transitions, declensions and 'vicissitudes in things which cannot endure nor remain in Being.

'From which we must conclude that God alone is: not according to any 'measure known to Time, but according to an unchanging and immortal 'eternity, not measured by Time, not subject to any declension; before

---

479. Small omission from Amyot (357B).
480. Omission: Amyot, 357C ('like a sinking ship in which are contained generation and corruption').
481. Montaigne adds the words 'or born' (*ou nées*) and omits, 'intermingled with Time' (357D).

'Whom nothing *is*, neither will there be anything after Him, nor anything 'newer or more recent; but ONE, existing in reality, He fills Eternity with a 'single Now; nothing really IS but He alone; of Him you cannot say *He* '*was* or *He will be*: He has no beginning and no end.'[482]

To that very religious conclusion of a pagan I would merely add one more word from a witness of the same condition, in order to bring to a close this long and tedious discourse which could furnish me with matter for ever. 'Oh, what a vile and abject thing is Man,' he said, 'if he does not rise above humanity.'[483]

[C] A pithy saying; a most useful aspiration, but absurd withal. For [A] to make a fistful bigger than the fist, an armful larger than the arm, or to try and make your stride wider than your legs can stretch, are things monstrous and impossible. Nor may a man mount above himself or above humanity: for he can see only with his own eyes, grip only with his own grasp. He will rise if God proffers him – [C] extraordinarily – [A] His hand; he will rise by abandoning and disavowing his own means, letting himself be raised and pulled up by purely heavenly ones.[484]

[C] It is for our Christian faith, not that Stoic virtue of his, to aspire to that holy and miraculous metamorphosis.[485]

---

482. The long borrowing from Plutarch ends here. The concluding words of the treatise *On the E'i at Delphi* emphasize its connection with Montaigne's themes of self-knowledge and the abasement of Man: 'And meanwhile it seems that this word *E'i* is somewhat opposed to the precept *Know Thyself* and also in some ways accordant and agreeable to it: the one is a kind of verbal astonishment and adoration before God, as being Eternal and Ever in Being, while the other is a warning and reminder to mortal man of the weakness and debility of his nature' (358C).

483. Seneca, *Quaest. nat.*, I (Preface), cited by Sebond, tr. Montaigne, 186r°.

'88: humanity.' *There is in all his Stoic school no saying truer than that one: but* to make . . .

484. '88: pulled up by *divine grace: but not otherwise*. (The closing words of the *Apology* until [C].)

485. *Metamorphose* may imply 'transfiguration': it certainly implies 'transformation' – the theme of the final pages of the last chapter (III, 13, 'On experience').

13. On judging someone else's death

[After the Christian climax of 'An apology for Raymond Sebond' which stresses that Stoic virtue cannot lead to grace and salvation, we are shown how splendid Cato's glorious suicide was in human, philosophical terms. But, as the closing words quietly recall, Cato's self-destruction was actually an act of murder.]

[A]   When we judge the assurance shown by a person as he is dying – and dying is without doubt the most noteworthy action in a man's life – there is one thing we must always take into account: it is hard for anyone to believe that he himself has reached that point. Few die convinced that their last hour has come; nowhere else does deceiving Hope take up more of our time. She never stops making our ears ring with thoughts such as, 'Others have been much more ill without dying,' or, 'My condition is not as hopeless as they think'; and, if the worst comes to the worst, 'God has performed plenty of other miracles.'

This happens because we set too much store by ourselves. It appears to us that the whole universe in some way suffers when we are obliterated and that it feels compassion for our predicament, especially since our perception has been affected and sees things accordingly: as our vision fails we think that it is they which are failing: just as for those travelling by sea the mountains, fields, cities, sky and land all go by at the same speed as they do:[1]

[B]   *Provehimur portu, terræque urbesque recedunt.*

[We sail out of harbour and the land and its cities withdraw.][2]

Who has ever seen an old man who did not praise former times and condemn the present, loading on to the world the weight of his own wretchedness and on to the manners of men his own melancholy!

---

1. An idea of Lucretius, already exploited in II, 12, 'An apology for Raymond Sebond'.
2. Virgil, *Aeneid*, III, 72.

Jamque caput quassans grandis suspirat arator,
Et cum tempora temporibus præsentia confert
Præteritis, laudat fortunas sæpe parentis,
Et crepat antiquum genus ut pietate repletum.

[The grand old ploughman shakes his head, contrasting the past with the present; he constantly praises his father's good fortune and croaks on about folk in former days being overflowing with piety.][3]

We drag everything along with us.

[A]   And so it follows that we reckon our death to be a great event, something which does not happen lightly nor without solemn consultations among the heavenly bodies: [C]   *'tot circa unum caput tumultuantes deos!'* [all those gods in a tumult over one capital punishment!][4]   [A]   And the higher we rate our worth the more we think that way.   [C]   What! Should so much learning be lost, should so much harm be done, without the especial concern of the Fates! Can so rare, so model a soul as mine be killed as cheaply as a useless common one! Is such a life as mine, which is the mainstay of so many others, upon which so many others depend, which has activities giving employment to so many people and which occupies so many offices, to be displaced like a life which has no attachments save one single knot! None of us gives enough thought to his being only *one*.

[A]   Hence those words addressed by Caesar to the captain of his ship, words running prouder than the sea which threatened him:

Italiam si, cœlo authore, recusas,
Me pete: sola tibi causa hæc est justa timoris,
Vectorem non nosse tuum; perrumpe procellas,
Tutela secure mei.

[If by Heaven's command you refuse to sail for Italy, then turn to me: this fear of yours is only justified if you do not know who your passenger is! Battle through those waves. Trust in my protection.][5]

And there is this as well:

credit jam digna pericula Cæsar
Fatis esse suis: Tantusque evertere, dixit,
Me superis labor est, parva quem puppe sedentem
Tam magno petiere mari

---

3. Lucretius, *De nat. rerum*, II, 1165–8.
4. Marcus Annaeus Seneca, *Suasoriae*, I, iv.
5. Lucan, *Pharsalia*, V, 579–82; cf. Erasmus, *Apophthegmata*, IV, *Julius Caesar*, IX.

[Caesar now believed the perils to be worthy of his destiny: 'Is it so great a labour for the gods to topple me, seeking me out where I sit on a huge sea in a tiny boat!']⁶

[B] And there was that mad official belief that, for one whole year, the Sun's face was in mourning out of grief for Caesar's death:

> *Ille etiam, extincto miseratus Cæsare Romam,*
> *Cum caput obscura nitidum ferrugine texit;*

[And likewise the Sun itself pitied Rome with Caesar's light put out, veiling its radiant forehead in purple darkness;]⁷

and there are hundreds of others by which this world of ours deceives itself, reckoning that our troubles can bring changes to the face of Heaven [C] and that the heavens' infinity is passionately concerned with our piddling distinctions. *'Non tanta coelo societas nobiscum est, ut nostro fato mortalis sit ille quoque siderum fulgor!'* [There is not such a fellowship between the heavens and ourselves that when we are fated to perish the splendour of the stars should perish also!]⁸

[A] Now to judge the resolution and constancy of a man who does not believe with certainty that the peril is upon him, even though it is, is not reasonable; it is not enough that he did die with such resolute constancy unless he rightly adopted it to perform that action. It happens that most men stiffen their countenance and their words to acquire a reputation which they still hope to live to enjoy. [C] In all the deaths that I have witnessed, it was Fortune which arranged that countenance, not the man's designs.

[A] And even among those who killed themselves in ancient times there is a great distinction to be made between a quick death and one which took time. That cruel Roman Emperor who would say of his prisoners that he wanted them to *feel* death, would comment, if one of them killed himself while in prison, 'That one got away!'⁹ He wished to prolong their dying and to make them feel what it is through torture:

> [B] *Vidimus et toto quamvis in corpore cæso*
> *Nil animæ letale datum, moremque nefandæ*
> *Durum sævitiæ pereuntis parcere morti.*

---

6. Lucan, ibid., V, 653–6.
7. Virgil, *Georgics*, I, 466–7.
8. Pliny, *Hist. nat.*, II, viii.
9. Caligula's cruelty was legendary, but the saying is that of Tiberius: Erasmus, *Apophthegmata*, VI, *Tiberius Caesar*, X: 'Carvillius has got away.'

[We saw his body all covered with wounds, but no lethal one was allowed it, by a custom of atrocious cruelty which kept death from the dying.][10]

[A] It is not at all difficult to say when you are quite well and quite calm that you have decided to kill yourself: it is easy to act the formidable fighter before you come to grips; so Heliogabalus,[11] the most unmanly man in the world, in the midst of his vile debaucheries planned to end his life [C] daintily [A] whenever circumstances should force him to; and so that his death should not belie the rest of his life, he had caused to be built a gorgeous tower, the base and façade of which were enriched with gold and jewels, expressly to throw himself down from it. He made ropes of gold, and of crimson silk as well, to strangle himself with, and a sword of beaten gold to run himself through with; and he kept potions in vessels of emerald and topaz to poison himself with, so that he could choose one or other of these ways of dying as his fancy moved him:

[B] *Impiger et fortis virtute coacta*

[Ready to die and strong — by an enforced valour.][12]

[A] However in his case the delicacy of his preparations renders it likely that when it came to the crunch he would have started snivelling blood!

Yet even in those more vigorous men who had made up their minds to carry it out, we must (I insist) look to see if it was to be by a blow which removed any possibility of their feeling its effect; for if they were to see their life dripping away drop by drop, with their body's awareness mingling with that of their soul and offering them the means for a change of heart, it is a matter of conjecture whether we would find them stubborn and constant in so perilous an intent.

During Caesar's civil wars, Lucius Domitius was captured in Abruzzi, poisoned himself and then changed his mind.[13] In our own days there was the case of a man who had decided to die but with his first assay at it he did not go deep enough since his quivering flesh made his arm flinch; he did give himself two or three wounds afterwards, but could never bring himself to thrust his blows right home.

10. Lucan, *Pharsalia*, II, 178–80.
11. Ever since Lampridius' *Life* of him, the Emperor Heliogabalus, the son of Antonius Caracalla, was infamous for his effeminacy and luxurious ways.
12. Lucan, *Pharsalia*, IV, 798.
13. Plutarch, *Life of Caesar*.

[C]   When Plantius Sylvanus was on trial his grandmother Urgulania sent him a dagger; he could not manage to kill himself with it but got his servants to slash his veins.[14]   [B]   In the time of Tiberius, Albucilla tried to kill herself but the blow was too light; she thus gave her enemies the means of taking her prisoner and killing her their own way.[15] Much the same happened to Demothenes (the captain) after his defeat in Sicily.[16]   [C] Caius Fimbria also struck himself too weak a blow and got his manservant to finish him off.[17]   On the other hand Ostorius, who was unable to use his own arm, disdained to use that of his servant except for holding the dagger straight and firm: he ran on to it, offering his throat and stabbing it through.[18]

[A]   Meat such as this must, in truth, be swallowed unchewed, unless you have a gizzard paved with frost-nails! The Emperor Hadrian got his doctor to mark with a circle the exact spot round his tit where a blow would prove fatal; the man he made responsible for killing him had to aim at that target.[19] Which explains why Caesar, when asked what kind of death he found most desirable, replied, 'The least anticipated and the quickest.'[20]   [B]   If Caesar dared to say it I can no longer be a coward for thinking the same.

[A]   'A quick death,' says Pliny, 'is the sovereign blessing of human life.'[21] People hate reconnoitring death. No man can be said to be resolute in death who refuses to haggle with it and who cannot look at it with his eyes open. Those men at the gallows whom we see running to their end, hastening and hurrying towards it, are not doing so because they are resolute: they want to deprive themselves of time to think about it:

> *Emori nolo, sed me esse mortuum nihili aestimo.*

[I think nothing of being dead: it is the dying that I dislike.][22]

I know from experience that I could attain to that degree of steadfastness, like men who dive into dangers as into the sea – with their eyes closed.

---

14. Tacitus, *Annals*, IV, xxii.
15. Tacitus, ibid., VI, xlviii.
16. Ravisius Textor, *Officina*; chapter headed 'Mortem qui sibi consciverunt'.
17. Cf. Ravisius Textor, ibid.
18. Tacitus, *Annals*, XVI, xv.
19. Anecdote from Xiphilinus' *Life of Hadrian*.
20. Plutarch (tr. Amyot), *Dicts des anciens Roys*, 209 F.
21. Pliny, *Hist. nat.*, VII, liii.
22. Cited by Cicero, *Tusc. disput.*, I, viii, 15.

[C] According to my standards there is nothing more glorious in the life of Socrates than his having had thirty whole days to chew over his death and his having digested it, all that time, with a most certain hope, without fuss, without alteration and with a line of conduct and conversation subdued and relaxed by the weight of that thought rather than heightened and tensed.

[A] When he was ill, Pomponius Atticus (to whom Cicero addressed his epistles)[23] summoned his son-in-law Agrippa and two or three other friends and told them that he had essayed it and knew that he had nothing to gain from wanting to be cured: everything he was doing to prolong his life was both prolonging and increasing his suffering; so he had decided to end them both. He begged them to approve of his decision, or at least not to waste their efforts on trying to dissuade him. Whereupon, having chosen to die by starvation, by accident his illness was cured! The remedy he had chosen to end his life restored him to health. His doctors and his friends feasted such a happy outcome and were rejoicing in his presence but they were much mistaken: for all that, they did not find it possible to make him go back on his decision: he said that he had to go through with it some time or other and that, having got thus far, he wanted to rid himself of the trouble of starting all over again on another occasion. That man, having had leisure to make a reconnaissance of death, not only was not disheartened at joining battle with it, he was keen to do so; once he had been satisfied by his reasons for entering the fight, he spurred himself on bravely to see the end of it.[24]

It is to go far beyond having no fear of death actually to want to taste it, to savour it.

[C] The account of what happened to Cleanthes the philosopher is a close parallel. His gums were swollen and rotting; the doctors advised extreme abstinence. After two days of fasting he made such a good recovery that they pronounced him cured and allowed him to return to his usual way of life. He on the other hand already savouring a kind of sweetness in his failing powers, determined not to retreat and crossed that boundary towards which he had so firmly advanced.[25]

[A] Tullius Marcellinus, a Roman youth, wishing to forestall his fatal hour so as to rid himself of an illness which was battering him more than he was prepared to put up with even though his doctors promised him a

23. His *Epistulae ad Atticum.*
24. Cornelius Nepos, *Life of Atticus.*
25. Diogenes Laertius, *Life of Cleanthes.*

certain, but not a quick, cure, called his friends together to consider the matter. 'Some,' says Seneca, 'gave him the advice which they would have cowardly chosen for themselves; others, out of flattery, the advice which they thought would be most pleasing to him; but a Stoic said the following: "Do not toil over it, Marcellinus, as if you were considering anything important: it is no great thing to be alive: your servants and the animals are; the great thing is to die honourably, wisely and with constancy. Think how long you have been doing the same things – eating, drinking and sleeping: drinking, sleeping and eating. We are for ever going round in that circle; not only bad and intolerable mishaps but merely being sated with living gives us a desire for death." '

Marcellinus – he went on – did not need anyone to advise him: he wanted someone who could help him. His servants were frightened of getting mixed up with it; but that Stoic philosopher made them understand that a man's domestic servants fall under suspicion only when there is reason to doubt that their master's death was deliberate; therefore they would set as bad an example by hindering him as by murdering him, since

Invitum qui servat idem facit occidenti.

[To save a man against his will is the same as murdering him.][26]

He then suggested to Marcellinus that, just as when we have finished our dinners we leave what is left on the tables for those who have waited on us, so too, having finished his life, it would not be inappropriate to distribute something among those who were to help him. Now Marcellinus was of a frank and generous mind; he caused a certain sum to be shared among his servants and comforted them. For the rest, he needed neither blade nor bloodshed: he undertook not to run away from this life but to take leave of it; not to escape from this life but to assay death. And to give himself leisure to haggle with it, he gave up all food; three days later he had himself sprinkled with warm water; he failed away gradually, not, judging from what he said, without a feeling of pleasure. Indeed those who have experienced such failings away of the mind brought on by weakness say that they felt no pain but rather indeed a certain kind of pleasure, like dropping off to sleep and resting.[27]

---

26. Horace, *Ars poetica*, 467.
27. Seneca's *Epist. moral.*, XXIX, is devoted to the illness of Marcus Tullius Marcellinus, a friend of his, whose suicide is related in LXXVII, 5 ff.

There you have deaths which have been carefully prepared for and digested. But so that Cato alone should furnish a complete model of virtue it seems that his good Destiny gave him some trouble in the arm with which he dealt himself the blow, in order to afford him leisure to confront Death and to fall about its neck, strengthening his courage in that peril not weakening it. And if it had been up to me to portray him in his most exalted posture, it would have shown him all covered with blood and tearing out his entrails, rather than sword in hand as did the sculptors of his time. For that second murder was more ecstatic than the first.[28]

28. That is, Cato of Utica (the defender of the Republic against Julius Caesar) 'murdered himself' in a manner more exalted than that of Marcellinus and it strikes us with more ecstatic amazement (Plutarch, *Life of Cato of Utica*).

14. How our mind tangles itself up

==

[Stoic philosophers were in a quandary about adiaphora, (that is) things which are 'indifferent' – neither good nor bad in themselves. How can the wise man possibly choose between them? Montaigne is led to conclude this short chapter with a lesson about human pride and the weakness of reason.]

[A]   It is a pleasant thought to imagine a mind exactly poised between two parallel desires, for it would indubitably never reach a decision, since making a choice implies that there is an inequality of value; if anyone were to place us between a bottle and a ham when we had an equal appetite for drink and for food there would certainly be no remedy but to die of thirst and of hunger![1]

In order to provide against this difficulty the Stoics, when you ask them how our souls manage to choose between two things which are indifferent and how we come to take one coin rather than another from a large number of crowns when they are all alike and there is no reason which can sway our preference, reply that this motion in our souls is extraordinary and not subject to rules, coming into us from some outside impulse, incidental and fortuitous.

It seems to me that we could say that nothing ever presents itself to us in which there is not some difference, however slight: either to sight or to touch there is always an additional something which attracts us even though we may not perceive it.

Similarly if anyone would postulate a cord, equally strong throughout its length, it is impossible, quite impossible, that it should break. For where would you want it to start to fray? And it is not in nature for it all to break at once.

Then if anyone were to follow that up with those geometrical propositions which demonstrate by convincing demonstrations that the container is greater than the thing contained and that the centre is as great as the

---

1. The dilemma of Buridan's ass: it starved to death when equidistant from identical food.

circumference, and which can find two lines which ever approach each other but can never meet,[2] and then with the philosopher's stone and the squaring of the circle, where reason and practice are so opposed, he would perhaps draw from them arguments to support the bold saying of Pliny: *'Solum certum nihil esse certi, et homine nihil miserius aut superbius.'* [There is nothing certain except that nothing is certain, and nothing more wretched than Man nor more arrogant.][3]

2. The mathematician Jacques Peletier du Mans had puzzled Montaigne with conic asymptotes which, towards the end of II, 12 ('An apology for Raymond Sebond') Montaigne assimilated to Pyrrhonist arguments which undermine reason and experience. (Such *asymptotes* are lines which ever approach a given curve but never touch it within infinity.)

3. A saying of Pliny's (*Hist. nat.*, II, vii) which Montaigne inscribed in his library; until [C] he translated it in his text.

15. That difficulty increases desire

[The opening words of this chapter are a Pyrrhonist saying inscribed in Montaigne's library. Montaigne sees the principle of contrariness working in all things, in virtue as in vice, in politics as in God's Church. We are shown also that, in a matter of the greatest importance, Montaigne lived in accordance with his principles. The area around his estates at Montaigne was fiercely fought over and often controlled by his opponents, but he never fortified his manor-house nor hid his spoons.]

[A] 'No reason but has its contrary,' says the wisest of the Schools of Philosophy.[1]

I have just been chewing over that other fine saying which one of the Ancient philosophers cites as a reason for holding life in contempt: 'No good can bring us pleasure except one which we have prepared ourselves to lose';[2] [C] *'In aequo est dolor amissae rei et timor amittendae;'* [Sorrow for something lost is equal to the fear of losing it;][3] he wanted to show by that that the fruition of life can never be truly pleasing if we go in fear of losing it.[4]

But we could, on the contrary, say that we clasp that good in an embrace which is all the fonder and all the tighter in that we see it as less surely ours, and fear that it may be taken from us. For we know from evidence that the presence of cold helps fire burn brighter and that our wills are sharpened by flat opposition:

> [B] *Si nunquam Danaen habuisset ahenea turris,*
> *Non esset Danae de Jove facta parens.*

[Danae would never have had a child by Juppiter had she never been shut up in a tower of bronze.][5]

---

1. The sceptics. Cf. Sextus Empiricus, *Hypotyposes*, I, vi, 12.
2. '80: *most beautiful and very fine* saying . . .
   Seneca, *Epist. moral.*, IV, 6.
3. Ibid., XCVIII, 6.
4. Ibid., IV, 5–6.
5. Ovid, *Amores*, II, xix, 27–8.

[A]   We see also that by nature there is nothing so contrary to our tastes than that satiety which comes from ease of access; and nothing which sharpens them more than rareness and difficulty: '*Omnium rerum voluptas ipso quo debet fugare periculo crescit.*' [In all things pleasure is increased by the very danger which ought to make us flee from them.][6]

> *Galla, nega: satiatur amor, nisi gaudia torquent.*

[Say 'No' to him, Galla: Love is soon sated unless joys meet torments.][7]

To keep love in trim Lycurgus ordained that married couples in Sparta should only have intercourse with each other by stealth, and that it should be as much a disgrace for them to be discovered lying together as lying with others.[8] The difficulty of arranging trysts, the danger of being surprised, the embarrassment on the morning after,

> *et languor, et silentium,*
> *Et latere petitus imo spiritus*

[and listlessness and no word spoken and the sigh coming from the depth of our bosom][9]

– that is what gives smack to the sauce.   [C]   How many pleasant and very stimulating verbal frolics arise from the chaste and modest vocabulary we use when talking of sexual intercourse.   [A]   Pleasure itself seeks stimulation from pain.   [A1]   It tastes far more sweet when it hurts and takes your skin off.   [A]   Flora, the courtesan, said that she had never lain with Pompey without making him bear the marks of her teeth:[10]

> *Quod petiere premunt arcte, faciuntque dolorem*
> *Corporis, et dentes inlidunt sæpe labellis:*
> *Et stimuli subsunt, qui instigant lædere idipsum,*
> *Quodcunque est, rabies unde illæ germina surgunt.*

[The object of their desire they tightly hug, hurting each other's body; they keep sinking their teeth into each other's lips; some hidden goads prick them on to give pain to the very thing, whatever it is, from which spring the seeds of their ecstasy.][11]

---

6. Seneca, *De beneficiis*, VII, ix.
7. Martial, *Epigrams*, IV, xxxvii.
8. Plutarch, *Life of Lycurgus*.
9. Horace, *Epodes*, XI, 9–10 (adapted).
10. Plutarch, *Life of Pompey the Great*.
11. Lucretius, *De nat. rerum*, IV, 1076–9.

So it is with everything: it is difficulty which makes us prize things.

[B] The people of the Marches of Ancona more readily go to Saint James of Compostela to make their vows: those of Galicia, to Our Lady of Loreto. At Liège they sing the praises of the baths at Lucca: in Tuscany, of those of Spa-by-Liège. You hardly ever see a Roman in the fencing school of Rome: it is full of Frenchmen! Great Cato tired of his wife – just like the rest of us – while she was his: when she belonged to another he yearned for her.[12]   [C]   I had an old stallion which I put out to stud: there was no holding it back when it scented the mares. The ease of it all soon sated it where its own mares were concerned; but with other mares, as soon as one passes by its paddock it returns to its incessant neighings and its frenzied passions just as before.

[A]   Our appetite scorns and passes over what it holds in its hand, so as to run after what it does not have:

> *Transvolat in medio posita, et fugientia captat.*

[He leaps over what lies fixed in his path, to chase after whatever runs away.][13]

To forbid us something is to make us want it:

> [B]   *Nisi tu servare puellam*
> *Incipis, incipiet desinere esse mea!*

[Unless you start looking after that girl of yours better, I shall soon stop wanting her!][14]

[A]   To hand it over to us completely is to breed contempt for it in us. To Want and Plenty befall identical misfortunes.[15]

> *Tibi quod superest, mihi quod defit, dolet.*

[You have too much of it, and that pains you: what pains me is that I do not have enough.][16]

---

12. Cato of Utica lent his second wife, Marcia, to Hortensius. This was much commented on by Christian writers. Cf. Tiraquellus, *De legibus connubialibus*, VII, 28.

13. Horace, *Satires*, I, ii, 108 (a huntsman comparing his course of love to his pursuit of a hare).

14. Ovid, *Amores*, II, xix, 47–8.

15. For Plato it is Want (*Poros*) and Plenty who together give birth to love: neither does by itself. Cf. Plutarch (tr. Amyot), *De Isis et Osiris*, 330H–331B.

16. Terence, *Phormio*, I, iii, 10.

We are equally troubled by desiring something and by possessing it.
[A1] Coldness in mistresses is most painful, but in very truth compliance
and availability are even more so; that is because the yearning which is
born in us from the high opinion in which we hold the object of our love
sharpens our love, and the choler similarly makes it hot: but satiety
engenders a feeling of insipidness; our passion then is blunted, hesitant,
weary and half-asleep:

[B]   *Si qua volet regnare diu, contemnat amantem.*

[If any mistress wants to go on reigning over her lover, then let her scorn him.][17]

Contemnite, amantes,
Sic hodie veniet si qua negavit heri.

[Scorn your mistress, young lovers: then she will come back today for what she
denied you yesterday.][18]

[C]   Why did Poppaea hit on the idea of hiding the beauties of her face
behind a mask if not to make them more precious to her lovers?[19]
[A1]   Why do women now cover up those beauties – right down
below their heels – which every woman wants to display and every man
wants to see? Why do they clothe with so many obstacles, layer upon
layer, those parts which are the principal seat of our desires – and of theirs?
And what use are those defence-works with which our women have
started to arm their thighs, if not to entrap our desires and to attract us by
keeping us at a distance?

Et fugit ad salices, et se cupit ante videri;

[She flees into the willow trees – but wants you to see her first;]

[B]   *Interdum tunica duxit operta moram.*

[Sometimes she delays me by letting her dress get in the way.][20]

[A1]   What is the purpose of that artful maidenly modesty, that poised
coldness, that severe countenance, that professed ignorance of things which
they know better than we do who are teaching them to them, if not to
increase our desire to vanquish, overcome and bend to our passion all those
conventional obstacles? For there is not only pleasure in making that sweet

---

17. Ovid, *Amores*, II, xix, 33.
18. Propertius, II, xiv, 19–20.
19. The mistress, then wife, of Nero. Tacitus, *Annals*, XIII, xlv.
20. Virgil, *Eclogues*, III, 65, then, Propertius, II, xv, 6.

gentleness and that girlish modesty go mad with sensual desire but glory as well in reducing a proud and imperious gravity to the mercy of our ardour.

There is glory, they say, in triumphing over coldness, modesty, chastity and moderation, and those who counsel ladies against such qualities betray both the ladies and themselves. We need to believe that their minds are quivering with fear; that the sound of our words offends the purity of their ears; that they hate us for it and yield to our insolence with an enforced fortitude.

Beauty, however powerful it may be, has no way of making itself savoured without such preliminaries. See how in Italy – where there are more beautiful women on sale, and finer ones too[21] – Beauty still has to seek extraneous means and other artifices to make herself attractive: and yet, in truth, being public and buyable she remains weak and languishing: [A] just as in virtue, even out of two similar actions, we hold the one to be more beautiful and more highly prized in which there are more difficulties and hazards to be faced.

It is an act of God's Providence to allow his Holy Church to be, as we can see she now is, shaken by so many disturbances and tempests, in order by this opposition to awaken the souls of the pious and to bring them back from the idleness and torpor in which so long a period of calm had immersed them. If we weigh the loss we have suffered by the numbers of those who have been led into error against the gain which accrues to us from our having been brought back into fighting trim, with our zeal and our strength restored to new life for the battle, I am not sure whether the benefit does not outweigh the loss.

We thought we were tying our marriage-knots more tightly by removing all means of undoing them;[22] but the tighter we pulled the knot of constraint the looser and slacker became the knot of our will and affection. In Rome, on the contrary, what made marriages honoured and secure for so long a period was freedom to break them at will. Men loved their wives more because they could lose them; and during a period when anyone was quite free to divorce, more than five hundred years went by before a single one did so:

---

21. [A1] until [C]: and *more perfect than in any other nation.* Beauty . . .
22. Throughout the Roman Empire divorce was permitted by law. The Roman Catholic Church forbade it utterly, though it did allow *divortium* (legal separation) and annulment.

> *Quod licet, ingratum est; quod non licet, acrius urit.*

[What is allowed has no charm: what is not allowed, we burn to do.][23]

There is an opinion of an Ancient philosopher which we could add on this subject: punishments sharpen our vices rather than blunt them; [24] [B] they do not engender a concern to do good (which is the result of reason and self-discipline) but only a concern not to be found out doing wrong:

> *Latius excisæ pestis contagia serpunt.*

[The contagious sore is cut out; the infection spreads imperceptibly wider.][25]

[A] I do not know whether that opinion is true, but this I do know from experience: no polity has ever been reformed by such means. To bring order and rule to our morals we must depend on some other method.

[C] The Greek histories mention some neighbours of the Scythians, the Argippaei, who do not even have sticks or clubs for weapons; not only does no one ever set out to attack them but because of their virtuous holy lives, any man who seeks refuge with them is quite safe: no one would dare to come and lay hands on him. Recourse is had to them to settle any disputes which arise among men elsewhere.[26]

[B] And there is a nation where the gardens and fields which they want to protect are bounded by cotton-thread: it proves more secure and reliable than our hedges and ditches.[27] [C] '*Furem signata sollicitant . . . Aperta effractarius præterit.*' [Locked houses invite the thief: the burglar passes them by when they are wide open.][28]

Perhaps it is ease of access, among other things, which serves to protect my dwelling from the violence of our civil wars. Defences attract offensives; defiance, attacks. I have weakened any designs which the soldiers may have on it by removing from such an exploit all the dangers and occasions for military glory which usually provide them with a pretext and an excuse. At times when justice is dead, anything done courageously is always done honourably: I make the taking of my house something cowardly and

23. Ovid, *Amores*, II, xix, 3.
24. Seneca, *De clementia*, I, xxiii.
25. Claudius Rutilius (of Numantia; fl. AD 410), *De reditu suo*, 397.
26. See Charles Estienne, *Dictionarium historicum*, s.v. 'Argippei', when the same details are given. (The eventual source is Herodotus.)
27. Lopez de Gomara, *Histoire des Indes*, III, xxx.
28. Seneca, *Epist. moral.*, LXVIII, 4.

treacherous. It is closed to no one who knocks. My entire protection consists of an old-fashioned courteous porter who serves not so much to protect my door as to welcome anyone to it with becoming grace. I have no guard, no watch, save that which the heavenly bodies provide for me. A gentleman is wrong to give the appearance of being defended unless his defences are complete. Whoever is exposed on the flank is exposed overall. Our fathers had no thought of building defensive manor-houses. The means of storming and surprising our houses – I mean even without cannons and armies – increase every day, exceeding our means of safeguarding them. Good minds are working that way all the time; invading a house touches all men: protecting it, only the rich.

My own house was a stronghold for the time it was built. In that respect I have added nothing to it, fearing that its strength could be turned against me. Peaceful times moreover will require us to unfortify our houses again. There is also the risk that we would never be able to retake them; yet it is hard to render them safe, for, where civil wars are concerned, your man-servant may be on the side you go in fear of. And once religion serves as pretext, you cannot even trust such kinsmen as may veil themselves behind a pretence of justice.

Our home-garrisons are not paid for out of the public exchequer, which would be exhausted by doing so. We ourselves have not the means of paying for them without ruining ourselves or (more inappropriately and unjustly) without ruining our people. And my position will be no worse if I do lose my house; for if you lose it when defended, even those who love you will spend less time on sympathy than on criticisms of your lack of vigilance and foresight, of your ignorance and neglect of the duties of a soldier.

The fact that so many protected houses have been lost while this house of mine goes on makes me suspect that they were lost precisely because they were protected: protective-works provide an attacker with both the desire and the excuse. All kinds of protection look belligerent. If God so wills it, let any man burst into my home: all the same, I shall never invite him to do so. It is my place of retreat, to rest from the wars. I assay to steal this corner from the public storms, as I do for another corner in my soul. Our war can change its patterns, multiply and diversify into new factions; but to no avail: as for me, I do not budge.

In the midst of so many fortified houses, I, alone of my rank in the whole of France as far as I know, have entrusted mine entirely to the protection of Heaven. I have never removed from it either silver spoon or title-deed. I will never fear for myself, nor save myself, by halves. If God's

favour is acquired by a complete confidence in it, it will endure unto the end for me; if not I have myself already endured long enough to render that length of time remarkable and worth recording. What! It has been thirty years or more![29]

29. The first form of this chapter dates from about 1576. But Montaigne's long reflection here was written on the Bordeaux copy just before he died in 1592.

16. On glory

[*This chapter shows how Montaigne's moral interests were based more on experience than on books. A Classical concern with 'honour' – a good reputation after death – was widely adopted in the Renaissance. By his own experience in the civil wars and by his own reflections on virtue in both men and women, Montaigne is led to a Christian insistence on the primacy of conscience over reputation, as well as in [C] to a jaundiced view of even Socrates and Plato who evoked special revelation when at a loss for argument.*

The opening lines, with their sharp distinction between words and the reality which they signify is a current Renaissance distinction (not accepted by most Platonists) which derives from Aristotle. We are reminded that the Civil Wars of Religion had as great an effect on the minds of men in Montaigne's day as two world wars have had in our own time on those who were caught up in them.

Some of the ideas in this chapter are derived from the Théologie naturelle *of Raymond Sebond.*]

[A] There are names and there are things. A name is a spoken sound which designates a thing and acts as a sign for it. The name is not part of that thing nor part of its substance: it is a foreign body attached to that thing; it is quite outside it.[1] God, who is the plenitude and ultimate of all perfection, cannot himself either increase or grow: but his name can increase and grow through the praises and thanksgivings which we bestow on His works, which are external to him.[2] Now those praises cannot be incorporated into the substance of God, in whom there can be no increase of good, so we attribute them to his Name, which is the external quality which is nearest to him. That is why it is to God alone that belong all honour and glory[3] and why there is nothing so remotely unreasonable as to go seeking them for ourselves; for since we are wanting and necessitous within (our essence being imperfect and having a continual need of improvement) we should be attending to that. We are all hollow and empty:

---

1. Opinion deriving from Aristotle's treatise *On Interpretation*.
2. Praising and exalting God's 'name' is a *leitmotiv* of the psalms.
3. Cf. I Timothy 1:17; I Chronicles 29:11–13.
   [A] until [C]:   nothing *so vain and* so remotely. . .

it is not with wind and spoken sounds that we have to fill ourselves: to restore ourselves we need a substance more solid. A starving man would be a simpleton if he went in search of fine clothes rather than a good meal: we must run to our most pressing needs. As our common prayers put it: '*Gloria in excelsis Deo, et in terra pax hominibus.*' [Glory to God in the highest: and in earth, peace to men.][4] We are wanting in beauty, health, wisdom, virtue and other qualities of our essence: external ornaments we shall seek for only after we have provided for our necessities.

Theology treats this subject fully and more pertinently than I do, but I am not well versed in it.

Chrysippus and Diogenes were the first and most decisive authorities to hold that glory is to be disdained;[5] they said that of all the pleasures none was more dangerous nor more to be fled than the pleasure which comes to us from other men's approval. And, truly, experience shows us that its deceptions can often be very harmful.

Nothing poisons monarchs more than flattery: nothing, either, by which bad men can more easily gain credit in their courts; nor is there any pimping more common nor more apt for corrupting the chastity of women than feeding them and entertaining them with their praises. [B] The first enchantment which the Sirens used to deceive Ulysses was of such a nature:

> *Deça vers nous, deça, ô treslouable Ulisse,*
> *Et le plus grand honneur dont la Grece fleurisse.*

[Come hither to us, come hither, O Ulysses, most worthy of praise and the greatest in that honour which flourishes in Greece.][6]

[A] Those philosophers I mentioned said that all the glory in the world was not worth that a man of discretion should merely stretch out a finger to acquire it – [7]

> [B] *Gloria quantalibet quid erit, si gloria tantum est?*

[Make glory as great as you will, yet what is it but glory?][8] –

[A] I mean, to acquire it for its own sake; for it does bring in its train

---

4. The paeon of the angelic host at the Nativity (Luke 1:14).
5. Cicero, *De finibus*, III, xvii, 57.
6. Translated from Homer, *Odyssey*, XII, 184.
7. From the same section of Cicero's *De finibus* as in note 5.
8. Juvenal, *Satires*, VII, 81.

several advantages which can make it desirable. Glory brings us good-will; it makes us less exposed to insult and injury than other men; and so on.

That was also one of the principal doctrines of Epicurus: for that precept of his School, *Conceal thy life* (which enjoins men not to lumber themselves with business and affairs) also necessarily presupposes a contempt for glory, which is the world's approbation of such of our actions as we make public.[9] That philosopher who orders us to conceal ourselves and to care for no one but ourselves and who wishes us to remain unknown to others, wants us even less to be held in honour and glory by them. He also advised Idomeneus in no wise to govern his actions by reputation or by common opinion, except to avoid such incidental disadvantages as the contempt of men might bring him.[10] Those words are infinitely true, in my opinion, and are reasonable. Yet within ourselves we are somehow double creatures, with the result that what we believe we do not believe, what we condemn we cannot rid ourselves of. Look at the last words of Epicurus, said when he was dying: they are great words, worthy of such a philosopher: nevertheless they bear some sign of a concern for his reputation and of the very humour which he had denounced in his precepts.

Here is a letter which he dictated just before he breathed his last:

EPICURUS TO HERMACHUS: Greetings:

'I wrote this while I was spending the happiest day of my life, which is also my last, accompanied however by such pain in the bladder and the intestines that nothing additional could make it greater. But it is outweighed by the pleasure brought to my soul by the remembrance of my solutions and arguments. You now should welcome the task of looking after the children of Metrodorus, as required by the love you have from your childhood felt for me and for philosophy.'[11]

That is his letter.

What leads me to conclude that the pleasure which he says that he feels in his soul from his solutions is in some way connected with the reputation he hoped to acquire after death is a clause in his will requiring his heirs, Aminomachus and Timocrates, to furnish every January on his birthday such monies as Hermachus should require to celebrate it, and also such

---

9. Plutarch (tr. Amyot), *Si ce nom commun est bien dict, Cache ta vie*, 291 A ff.
10. Seneca, *Epist. moral.*, XXI, 3 ff.
11. Quoted from Cicero, *De finibus*, II, xxx, 96–7, to prove how far apart were Epicurus' words and his practices.

expenses which were incurred in entertaining his philosopher-friends who would assemble on the twentieth day of each moon to honour the memory of Metrodorus and himself.[12]

Carneades was the leader of the opposing School[13] and maintained that glory was desirable for itself, in the same way that we are attached for their own sake to those who come after us even though we enjoy no knowledge of them. That opinion has not failed to be widely accepted, as opinions which are most adapted to our inclinations readily are. [C] Aristotle gives glory the first rank among external goods: 'Avoid, as two vicious extremes, immoderately seeking glory or fleeing it.'[14] [A] I believe that if we had the books which Cicero wrote on the subject he would have spun us some good ones! For that fellow was so raging mad with a passion for glory that, if he had dared, he would readily have fallen into the extreme which others fell into: that even Virtue herself is only desirable for the honour which ever attends her.

> *Paulum sepultæ distat inertiæ*
> *Celata virtus:*

[Little does concealed virtue differ from slumbering idleness:][15]

which is an opinion so false that it irks me that it could ever have entered the mind of a man who bore the honoured name of philosopher. If that were true, we ought to be virtuous only in public; and as for those workings of our soul (which is the true seat of virtue) we would never need to keep them in due order under control except when they would come to the notice of others.

[C] Is it only a matter, then, of being sly and subtle about our failings? 'If,' says Carneades, 'you know that a snake is hidden in a place where a man who is unaware of it and by whose death you hope to profit is about to sit down, and you do not warn him of it, you act wickedly.'[16] All the more so if your deed could be known only to yourself. Unless we draw the rules of right-conduct from within ourselves and if to us justice means not

---

12. Same conclusion in Cicero, ibid., 101 (where the heir is normally called Amynochus).
13. He was leader of the New Academy and a declared opponent of the Stoics. His ideas are expounded by Cicero in *De finibus*, II, 35–59.
14. Aristotle, *Nicomachaean Ethics*, II, vii (1107b), during a general discussion of the Mean.
15. Horace, *Odes*, IV, ix, 29–30. (In context Horace means that heroes need poets to sing of their glories.)
16. Cicero, *De finibus*, II, xviii, 59.

being punished, how many kinds of wicked deeds must we daily abandon ourselves to! What Sextus Peducaeus did when he faithfully returned what Gaius Plotius had entrusted to him, he alone knowing it – something I often do – I do not so much find laudable as I should find any failure to do so execrable.[17] And I consider it good and useful to recall the case of Publius Sextilius Rufus, whom Cicero condemns for having accepted an inheritance despite what he knew to be right, although he acted not merely without illegality but through the law.[18] Then there were Marcus Crassus and Quintus Hortensius who had been invited by a foreigner to accept certain inheritances from the provisions of a false will, so that by means of their power and authority he could be sure of his own share; they were quite happy to play no part in the forgery yet did not refuse to profit by it; they felt safe enough if they could be protected from prosecutors, witnesses and law-suits.[19] *'Meminerint Deum se habere testem, id est [ut ego arbitror] mentem suam.'* [Let them remember that there is a witness, God: that is (as I understand it), their own minds.][20]

[A] Virtue is a vain and frivolous thing if she draws her commendation from glory: then, for nothing should we undertake to make her hold her rank apart and detach her from Fortune: for what is there more fortuitous than reputation? [C] *'Profecto fortuna in omni re dominatur: ea res cunctas ex libidine magis quam ex vero celebrat obscuratque.'* [Indeed Fortune dominates over all things: she makes all things celebrated or obscure by her own whim not by truth.][21] [A] To make deeds seen and known is purely the work of Fortune.

[C] Chance it is which bestows glory on us according to her fickle will: I have often seen it marching ahead of merit, and often outstripping merit by a long chalk. The man who first recognized the resemblance between shadow and glory did better than he intended.[22] Both are things exceedingly vain. Sometimes the shadow is thrown ahead of its body; and sometimes it greatly exceeds it in length.

[A] Those who teach noblemen to seek only honour from valour, [C] *'quasi non sit honestum quod nobilitatum non sit'* [as if no deed

17. Cicero, ibid., II, xviii, 58: Cicero adds that 'you yourself would undoubtedly have done the same'.
18. Ibid., II, xvii, 55.
19. Those men, praised by Cicero (*De finibus*, II, xviii, 57), are condemned for the same reason as Montaigne in *De officiis*, III, 73.
20. Cicero, *De officiis*, III, x, 44 (adapted).
21. St Augustine, *City of God*, VII, iii; citing Sallust.
22. Cicero held that glory 'follows virtue like a shadow': *Tusc. disput.*, I, xlv, 110.

is distinguished unless it receive some distinction],[23] [A] what do they achieve by it except teaching them never to hazard themselves if nobody is looking, and to take care to see that there are witnesses who can bring back news of their valour, whereas there are hundreds of occasions for acting well without anyone ever noticing us for it? How many beautiful individual deeds are buried in the throng of a battle? Whoever spends time noting down what another is doing in such an engagement cannot have much to do himself, and so the testimony he renders to the achievements of his comrades is produced against himself. [C] *'Vera et sapiens animi magnitudo honestum illud quod maxime naturam sequitur, in factis positum, non in gloria, judicat.'* [A truly great and wise mind judges that honour – which is its nature's greatest aim – is found not in glory but in deeds.][24]

All the glory I claim for my life is to have lived a tranquil one – not tranquil according to the standards of Metrodorus or Arcesilas or Aristippus but my own. Since Philosophy has been able to discover no good method leading to tranquillity which is common to all men, let each man seek his own one as an individual.[25]

[A] To whom do Caesar and Alexander owe the measureless greatness of their renown if not to Fortune? How many men has Fortune snuffed out at the very start of their careers and of whom we have no knowledge at all, yet who would have brought to those careers a mind as good as theirs if the mischance of their lot had not stopped them short at the birth of their expeditions! I cannot recall reading that Caesar, in the course of so many and so extreme dangers, was ever wounded. Hundreds have died from lesser perils than [C] the least of [A] those which he passed through safely. Fair deeds without number must be wasted, unwitnessed, before one of them proves profitable. We are not always at the spearhead of a breakthrough nor at the forefront of our army in full view of our general as on a stage. We are taken by surprise between the hedge and the ditch; we must tempt Fortune by attacking a chicken-coop; we have to flush four wretched men armed with harquebuses out of a barn; we must draw away from our unit and go it alone, as necessity requires. And if you take notice you will find from experience that the less spectacular opportunities are the

---

23. [A] until [C]: who teach *our fighting-men to have honour as their target and to seek nothing* from valour *but reputation*, what do they achieve . . .
   Cicero, *De officiis*, I, iv, 14 (adapted).
24. Ibid., I, xix, 65.
25. Achieving tranquillity of mind was the aim of many classical philosophers.

most dangerous ones and that in the wars which have happened in our own times more good men have been killed during trivial unimportant actions – fighting over some shack or other – than in places of honour and dignity.

[C] Anyone who holds that his death is wasted except on some conspicuous occasion, instead of making his death illustrious is deliberately casting a shadow over his life, meanwhile letting many just occasions for taking risks slip by. And all just occasions are illustrious enough: each man's conscience can trumpet them – to himself: *'Gloria nostra est testimonium conscientiae nostrae.'* [Our glory is the testimony of our conscience.][26]

[A] Whoever acts worthily only when others can know of it (and think better of him when they do), whoever never wishes to act well in circumstances where his virtue cannot come to the knowledge of men, is not a man who will be of much use to you:

> *Credo che'l resto di quel verno cose*
> *Facesse degne di tenerne conto;*
> *Ma fur sin' a quel tempo si nascose,*
> *Che non è colpa mia s'hor' non le conto:*
> *Perche Orlando a far opre virtuose,*
> *Piu ch'a narrarle poi, sempre era pronto,*
> *Ne mai fu alcun' de li suoi fatti espresso,*
> *Senon quando hebbe i testimonii apresso.'*

[I believe that during the rest of that winter Roland did deeds worth the telling. It is not my fault if I do not tell them for they have so far remained secret because Roland was ever more ready to do valiant deeds than to relate them afterwards: none of his exploits ever came to light except when there happened to be witnesses present.][27]

We must go to war as a duty: the reward we should expect is one which cannot fail any noble action, however obscure it may be: we should not even think of virtue but of the satisfaction which a well-governed conscience derives from acting well. We must be valiant for our own sakes, and for the advantages of having our minds lodged in a place which is firm and secure against the assaults of Fortune.

> [B]   *Virtus, repulsæ nescia sordidæ,*
> *Intaminatis fulget honoribus,*
> *Nec sumit aut ponit secures*
> *Arbitrio popularis auræ.*

26. II Corinthians 1:12.
27. Ariosto, *Orlando furioso*, XI, lxxxi.

[Virtue ignores all squalid slights: it gleams with unstained honour; it neither accepts the insignia of Consul nor lays them down at the whim of the plebs.][28]

[A]   Our soul must act her part not when on parade but at home within us where no eyes but our own can penetrate. There she shields us from fear of death, of pain, of shame even; she gives us assurance to face the loss of our children, of those whom we love and of our chattels; and when the opportunity arises, she also leads us into the hazards of war: [C]   *'non emolumento aliquo, sed ipsius honestatis decore.'*   [not for any sort of gain, but for the seemliness of honour itself.][29]   [A]   Such profit is much grander and more worthy to be wished for and hoped for than honour and glory, which are no more than the favourable judgement men make of us.

[B]   To adjudicate an acre of land we have to select a dozen men out of an entire nation; yet when it comes to adjudicating our propensities and our actions – the most difficult and most important matter of all – we have recourse to the votes of the common people and of the mob, that mother of ignorance, of injustice and of inconstancy.   [C]   Is it reasonable to make the life of a man depend on the judgement of idiots? *'An quidquam stultius quam quos singulos contemnas, eos aliquid putare esse universos?'* [Can anything be more stupid than to value collectively those whom we despise as individuals?][30]   [B]   Whoever aims to please that lot will never finish: such a target is shapeless and cannot be reached.   [C]   *'Nihil tam inaestimabile est quam animi multitudinis.'* [Nothing is less worth esteeming than the mind of the many.][31] Demetrius put it amusingly: he set no more store by the voice of the people when it came out of their tops than out of their bottoms.[32] And this one goes further: *'Ego hoc judico, siquando turpe non sit, tamen non esse non turpe, quum id a multitudine laudetur.'* [My judgement is that, even when a deed is not actually base, it cannot be entirely free from baseness when it is praised by the mob.][33]

[B]   No skill, no mental agility, could direct our footsteps if we were to follow so unruly a guide, one so far off course. Amidst that windy babble of popular rumour, report and opinions blowing down upon us, no valid course can be fixed on. Let us not look before us towards a goal so floating

28. Horace, *Odes*, III, ii, 17–20.
29. Cicero, *De finibus*, I, x, 36.
30. Cicero, *Tusc. disput.*, V, xxxvi, 104.
31. Livy, XXXI, xxxiv.
32. Seneca, *Epist. moral.*, XCI, 19 (adapted).
33. Cicero, *De finibus*, II, xv, 49. (A different reading is now current.)

and wavering: let us follow after reason with constancy: let public approval – if it can – follow us thither; but since it depends entirely on Fortune we are no more entitled to expect it than if we adopt a different route. Even if I did not follow the right road for its rightness, I would still follow it because I have found from experience that, at the end of the day, it is usually the happiest one and the most useful. [C] *'Dedit hoc providentia hominibus munus, ut honesta magis juvarent.'* [Honourable conduct is the most profitable: that is Providence's gift to men.]³⁴

[B] There was of old a seaman who addressed Neptune thus during a violent storm: 'O God, if it pleaseth thee thou wilt save me; if it pleaseth thee thou wilst destroy me: but I will ever hold straight to my helm.'

I have known in my time hundreds of men more devious, more supple more equivocal – and doubtless more worldly-wise – than I am, who destroyed themselves while I was saved:

> *Risi successu posse carere dolos.*

> [I laughed when I saw how trickery could fail.]³⁵

[C] When Paulus Aemilius set out on his glorious Macedonian expedition he told all the people in Rome not to talk freely about his actions while he was away.³⁶ Licence in judging such things is a great distraction in affairs of public concern, since not every man has the same determination as Fabius did in the face of opposing and harmful popular counsels: he preferred his high reputation to be torn to shreds by the frivolous notions of men rather than to carry out his responsibilities less well – thereby earning approval and popular support.³⁷

[B] There is an indescribable pleasure in being praised, but we value it far too much.

> *Laudari haud metuam, neque enim mihi cornea fibra est;*
> *Sed recti finemque extremumque esse recuso*
> *Euge tuum et belle!*

[I am not afraid of being praised; my sensitivities are not horny-hard; but I refuse

---

34. Quintilian, I, xii, 19.
35. Ovid, *Heroïdes*, I, 18.
36. Livy, LIV, xxii.
37. Fabius' delaying tactics in the war against Carthage earned him the hostile nickname *Cunctator* (the Delayer). It later became a title of praise (Livy, XXX, xxvi).

to accept that the final goal of right-conduct should be, 'Hooray! How fine!']³⁸

[A] I am not so much worried about how I am in the minds of other men as how I am to myself. I want to be enriched by me not by borrowings from others. Those outside us only see events and external appearances: anyone can put on a good outward show while inside he is full of fever and fright. They do not see my mind: they only see the looks on my face.

We are right to denounce play-acting in war: what is easier for a cunning man than to dodge danger, acting the fierce fighter while his heart is full of weakness. There are so many ways of avoiding occasions for exposing ourselves to personal risk that we shall have deceived everybody a thousand times before getting into dangerous straits; and even then, once we are caught in them, we can manage to put on a good face for the occasion and speak confident words while our soul is a-tremble within us. [C] And quite a few people, if they had the use of that ring of Plato's which made the man who wore it on his finger invisible if he gave it a twist towards the flat of his hand,³⁹ would go into hiding just when they ought to be most in evidence and would regret being exposed in a place of such honour, where it is Necessity which makes them valiant:

> [A] *Falsus honor juvat, et mendax infamia terret*
> *Quem, nisi mendosum et mendacem?*

[Who rejoices in unmerited honours or goes in fear of lying infamy save the deceiver and the liar?]⁴⁰

That is why all those judgements which are based on external appearances are unbelievably unreliable and dubious, and why there is no more reliable witness than each man is to himself.

[A1] Where external appearances are concerned, how many batmen are our companions in glory! The man who stands firm in a trench once it is dug, what is he doing which was not done before him by fifty wretched men of the pioneer-corps who open the way for him and protect him at the risk of their bodies for twopence halfpenny a day?

> [B] *Non, quicquid turbida Roma*
> *Elevet, accedas, examenque improbum in illa*
> *Castiges trutina: nec te quæsiveris extra.*

38. Persius, *Satires*, I, 47–9.
39. King Gyges' ring (Cicero, *De officiis*, III, xix, 78).
40. Horace, *Epistles*, I, xvi, 39–40.

[Do not accept whatever turbulent Rome decides: do not attempt to rectify her faulty scales: do not seek to base yourself on such externals.][41]

[A]   When we spread our name by scattering it into many mouths we call that 'increasing our renown'; we wish our name to be favourably received there and that it may gain from such an increase. That is what is most pardonable in such a design. But carried to excess this malady makes many seek to be on others' lips, no matter how. Trogus Pompeius says of Herostratus, and Livy says of Manlius, that they were more desirous of a wide reputation than a good one.[42] That is a common vice. We are more concerned that men should talk of us than of how they talk of us; and we are far more concerned that our name should run from mouth to mouth than under what circumstances it should do so.

It seems that to be known is in some way to have our life and our enduring fame under the protection of others. As for me, I only exist 'at home' (in myself); and as for that other life of mine which lies in what those who love me know of me,   [C]   considered naked and simply in itself,   [A]   I am well aware that I feel no fruit or joy from it, other than from the vanity of an imagined opinion.

And when I am dead, I shall feel it far, far less   [C]   and I shall lose completely those true advantages which sometimes happen to attend it; [A]   I shall then have no hands to grip hold of reputation or to hang on to it by, no means by which it can touch me or get through to me. As for expecting my name to receive it, well, first of all I have no name which is sufficiently my own. Of the two that I do have, one is common to the whole tribe of us; indeed, to others as well. There is a family in Paris and in Montpellier with the surname Montaigne. There is another in Brittany and another in Saintonge called De la Montaigne: change but one syllable and it will so tangle the threads of our destinies that I shall share in their glory and they, perhaps, in my disgrace; then again, my folks were formerly surnamed Eyquem, a surname which is still of concern to a well-known family in England.[43]

As for my Christian name, it is there for anyone who wishes to adopt it. So instead of myself I may bring honour to a porter.

41. Persius, Satires, I, 5–7.
42. To make himself famous Herostratus set fire to the temple of Diana at Ephesus; Lucius Manlius the dictator sought renown from his imperious bullying (Livy, VII, iii). (Often cited together.)
43. There are no famous Eyquems in England, though links between families in the Bordeaux region and England were strong ever since both formed part of the Norman domains.

And then, if I did have a label which was particular to me, what could it label when I am no more? Can it designate and commend nothingness?

> [B] *Nunc levior cyppus non imprimit ossa?*
> *Laudat posteritas: nunc non e manibus illis,*
> *Nunc non e tumulo fortunataque favilla*
> *Nascuntur violæ?*

[Does my tombstone press less on my bones now? There is the praise of posterity: for all that, no violets grow now from my remains in this tomb nor from my fortunate ashes.]

[A] But I have talked of that elsewhere.[44]

Moreover, of the ten thousand men who are maimed or killed in a battle, there are not fifteen whom we ever talk about. There must needs be some towering greatness, or some consequence of importance that Fortune has attached to it, to make any personal deed appreciated – not merely an infantryman's but even a general's. For to kill a man or two, or even ten, to expose oneself courageously to death, means something to each of us as individuals since our all is at risk: but for everyone else they are such everyday things, so ordinary, and we need so many of them to produce any noticeable results, that we can expect no individual commendation for them.

> [B] *casus multis hic cognitus ac jam*
> *Tritus, et e medio fortunæ ductus acervo.*

[. .a fate known to many, already well-worn, picked from the middle of Fortune's heap.][45]

[A] Of the thousands upon thousands of valiant men who have died in France, arms at the ready, over the last fifteen years, not a hundred have come to our knowledge. The memory not only of the leaders but of the battles, of the victories, lies buried. [C] The destinies of half the world stay where they are and, for want of record, do not last but vanish. If I had in my possession all the events which are unknown, I think I could easily supplant the ones we do know in examples of every kind. [A] Why, amid so many writers, so many witnesses and so many rare and noble exploits, few have come down to us even from the Romans and the Greeks.

---

44. Persius, *Satires*, I, 37–40. (Cf. I, 46, 'On names'.)
45. Juvenal, *Satires*, XIII, 9–10.

[B]　　*Ad nos vix tenuis famae perlabitur aura.*

[There scarcely wafts to us a thin breath of their fame.]⁴⁶

[A]　It will already be something if, a hundred years from now, people roughly remember that in our time there were civil wars in France.

[B]　Before going into battle the Spartans would make sacrifices to the Muses, praying that their deeds be well and worthily written about, reckoning it to be a divine and no common favour that beautiful deeds should find witnesses who knew how to make them live on in memory.⁴⁷

[A]　Do we think that at every volley from harquebuses which concerns us, at every risk that we run, there suddenly appears a clerk to keep a record of it? And, besides, a hundred clerks can jot it down whose accounts will not last three days and will come to nobody's attention. We do not possess a thousandth part of the writings of the Ancients: it is Fortune's favour which grants them a short life or a long one　[C]　(and we may well have cause to wonder if we have the worst part, since we have not even seen the rest).　[A]　Nobody writes histories about such trivial events: you have to be the head man in conquering an Empire or a Kingdom; you have to have won fifty-two set-piece battles, always with inferior forces, as Caesar did. Ten thousand fine comrades and many great Captains died following him valiantly and courageously whose names lasted only as long as their wives and children lived:

[B]　　*quos fama obscura recondit.*

[whom a darkened fame has hid.]⁴⁸

[A]　Even those we see acting well are no more talked of three months, or three years, after they left their bodies on the field than if they had never been. Whoever will reflect, with due measure and proportion, on what kind of people and what kind of glory are kept in remembrance through books, will find that very few of the deeds and very few of the men of our century may claim a place in them.

How many valiant men have we seen outliving their reputations, men who, while they are still alive, have seen and suffered the eclipse of the honour and glory which they so justly acquired in their younger days? And shall we go and lose that true life which is our essence and plunge ourselves into everlasting death for three years of that fancied imaginary life? Wise

46. Virgil, *Aeneid*, VII, 646.
47. Plutarch (tr. Amyot), *Dicts notables des Lacedaemoniens*, 216H–217A.
48. Virgil, *Aeneid*, V, 302.

men set up a more beautiful, a juster end for so important an undertaking: [C] '*Recte facti, fecisse merces est;*' [The reward for acting properly is to have done so;] '*officii fructus ipsum officium est.*' [the recompense of duty is duty done.][49]

[A] It might perhaps be pardonable for a painter or a craftsman, or even a rhetorician or a grammarian, to labour to acquire a name through his works; but virtuous deeds are too noble in themselves to seek any other reward than their own intrinsic worth, and especially to seek it from the vanity of human judgements. And yet if that false opinion serves the public good by keeping men to their duty; [B] if the people are incited to virtue by it; if rulers are influenced by the sight of men blessing Trajan's memory and abominating Nero's; if it affects them to see the name of that great criminal, once so fearsome and so formidable, so freely cursed and slighted by the first schoolboy who takes him on: [A] then let it boldly flourish and may it be fostered among us as much as is in our power.

[C] Even Plato, employing every means to make his citizens virtuous, also counsels them not to disdain a good repute in the judgement of the nations and says that, through some divine inspiration, it turns out that even the wicked can often, in speech and thought, justly distinguish the good people from the bad. That person and his pedagogue are marvellous and bold workmen at introducing divine operations and revelations, anywhere and everywhere, when human strength gives out:[50] '*ut tragici poetae confugiunt ad deum, cum explicare argumenti exitum non possunt.*' [just as the writers of tragedies resort to a *deus ex machina* when they cannot disentangle their threads at the end of their plays.] Perhaps that explains why Timon attacked Plato as a great maker of miracles.[51]

[A] Since men are not intelligent enough to be adequately paid in good coin let counterfeit coin be used as well. That method has been employed by all the lawgivers. And there is no polity which has not brought in some vain ceremonial honours, or some untruths, to serve as a bridle to keep the people to their duties; that is why most of them have fables about their origins and have beginnings embroidered with supernatural mysteries. That is what has lent credence to bastard religions and led them to find favour among men of understanding; and it explains

49. Seneca, *Epist. moral.*, LXXXI, 20; Cicero, *De finibus*, II, xxii, 73.
50. Plato, *Laws*, XII, 950B–C. Plato's 'paedagogue' is Socrates.
51. Cicero, *De nat. deorum*, I, xx, 53; Diogenes Laertius, *Life of Plato*, II, xxvi, 199 (tr. Timon).

why Numa and Sertorius fed men on the following idiotic tales to make
them put more trust in them: the former, that the nymph Egeria, the latter,
that a white hind of his, brought them counsels from the gods, which they
then followed.

[C] And the same authority which Numa gave to his laws by citing
the patronage of the goddess Egeria was given to him by Zoroaster, the
lawgiver of the Bactrians and the Persians, in the name of his god
Oromasis; by Trismegistus, the lawgiver of the Egyptians, in the name
of Mercury; by Zamolxis, the lawgiver of the Scythians, in the name
of Vesta; by Charondas, the lawgiver of the Chalcidians, in the name of
Saturn; by Minos, the lawgiver of the people of Candy, in the name of
Juppiter; by Lycurgus, the lawgiver of Sparta, in the name of Apollo; and
by Draco and Solon, lawgivers of the Athenians, in the name of Minerva.
And all polities have a god at their head, truly so in the case of the one
drawn up by Moses for the people of Judaea on leaving Egypt;[52] the rest,
falsely so.

[A] The religion of the Bedouins, as the Sire de Joinville relates, had as
one of its beliefs that each one of them who died for his monarch went
straight into a more blessed body, stronger and more beautiful; because of
this they were much more ready to hazard their lives:

> [B]   *In ferrum mens prona viris, animæque capaces*
> *Mortis, et ignavum est redituræ parcere vitæ.*

[The minds of these warriors defy the iron blade; their hearts embrace their deaths;
it is for them cowardice to save lives which are to be given back to them.][53]

[A] There you have a belief which, however vain it may be, results in
much good. Every nation can provide its own similar examples; but that
subject would merit separate treatment.

To add just one word more on my original topic: I do not advise ladies
to call their duty honour: [C] *'ut enim consuetudo loquitur, id solum dicitur
honestum quod est populari fama gloriosum;'* [just as in everyday speech, the
term 'honourable' is used only for what brings glory in the opinion of the
people;][54] their duty is the core: their honour, only the skin. [A1] Nor
do I advise them to pay us for their refusals by citing honour as an
excuse: [A] for I suppose that their intentions, their desire and their will
(which are qualities which their reputation has nothing to do with since

52. Exodus 20:1 f.
53. Jean de Joinville, *Cronique de Saint Loys*, LVI; Lucan, *Pharsalia*, I, 461–2.
54. Cicero, *De finibus*, II, xv, 48.

they are in no wise apparent on the surface) are even better moderated than their acts:

> *Quæ, quia non liceat, non facit, illa facit.*

[She who does not do it 'because it's not allowed' does it really.]

The offence against God and their conscience would be just as great if they wanted to do it as if they had carried it out.[55] And then we are dealing with an activity which is in itself hidden and secret; it would be quite easy for ladies to hide one such case from the knowledge of those other people on which their 'honour' depends, if they did not also have regard for their duty and a love leading to chastity for its own sake.

[C] Any honourable person prefers to sully his honour than to sully his conscience.[56]

55. Ovid, *Amores*, III, iv, 4. (Cf. Christ's warning in Matthew 5:28.)
56. Montaigne's discussion of honour echoes in general Aristotle's conception of the great-souled man (*Nicomachaean Ethics*, IV, iii, 1124a–b).

17. On presumption

═══

[Montaigne moves straight from glory to vainglory. Presumption is a mark of vainglory and of vicious self-love (philautia, as it was called): so Montaigne describes himself honestly, without that blindness to his faults or distortion of home-truths associated with self-love. His self-portrait, with all its honesty, is associated (as was Du Bellay's in the Regrets), with the Latin satirists, the father of whom was Lucilius. Through knowledge of himself Montaigne sought also a wider knowledge of Man.]

[A]   There is another kind of 'glory': the over-high opinion we conceive of our own worth. It is an imprudent affection by which we hold our own self dear, presenting ourself to ourself other than we are, just as passionate love lends grace and beauty to the person it embraces and leads to those who are enraptured by it being disturbed and confused in their judgement, so finding their Beloved other than she is, and more perfect.

Now I have no wish that a man should underestimate himself for fear of erring in this direction, nor that he should think he is worth less than he is. In all matters our judgement must maintain its rights. It is reasonable that, in this as in any other matter, it should perceive whatever truth presents it with. If he is Caesar, then let him frankly acknowledge that he is the greatest Captain in all the world. We are nothing but etiquette. We are carried away by it and neglect the substance; we cling to branches and let go of trunk and body. We have taught ladies to blush at the mere mention of something which they do not have the slightest fear of doing. We dare not call our private parts by their proper names yet are not afraid to use them for all sorts of debauchery. Etiquette forbids us from expressing in words things which are licit and natural: and we believe it. Reason forbids us to do things which are bad and illicit: and nobody believes it. Here I find myself bogged down in the laws of etiquette, which do not allow a man to speak well of himself nor ill of himself. I shall put all that aside for a while.

People whom Fortune (good or bad, whichever you want to call it) has caused to live their lives in some exalted position or other bear witness to themselves by their public deeds; but those whom Fortune has set to work merely among the crowd   [C]   and whom no one would ever talk about

if they did not talk about themselves, [A] can be excused if they do indeed dare to talk about themselves for the sake of those who have an interest in getting to know them, following the example of Lucilius:

> *Ille velut fidis arcana sodalibus olim*
> *Credebat libris, neque, si male cesserat, usquam*
> *Decurrens alio, neque si bene: quo fit ut omnis*
> *Votiva pateat veluti descripta tabella*
> *Vita senis.*

[He used to confide his secrets to his books as to trusted companions; he never turned anywhere else, whether things went well or ill; so that when he was old his entire life lay revealed as though written down on votive tablets.]

Lucilius committed to paper his deeds and his thoughts and portrayed himself as he knew himself to be. [C] '*Nec id Rutilio et Scauro citra fidem aut obtrectationi fuit.*' [Neither were Rutilius and Scaurus disbelieved nor vilified for doing so.][1]

[A] I can remember, then, that from my tenderest childhood people noticed in me some indefinable way of holding myself and some gestures which bore witness to a sort of vain silly pride. But first of all I would like to say this: it is not inappropriate that we should have some characteristics and propensities so proper to us and so physically part of us that we ourselves have no means of being aware of them nor of recognizing them; and such innate dispositions produce, without our knowledge or consent, a kind of bodily quirk. It was a certain mannerism appropriate to their beauty that made the head of Alexander lean a little to one side and Alcibiades to speak with a slight lisp; Julius Caesar used to scratch his head with his finger – which is the comportment of a man overflowing with troublesome thoughts; and Cicero, I seem to recall, used to wrinkle his nose – which signifies an innate tendency to mockery. Such gestures can root themselves in us imperceptibly.

There are also other gestures which are cultivated – and I am certainly not talking about them – such as bowing to people and ways of greeting them, by which we acquire, as often as not wrongly, the honour of being thought humble and courteous: [C] you can be humble out of pride! [B] I am fairly lavish with raising my hat, especially in summer, and I never receive such a greeting without returning it whatever the social status of the man may be, unless I pay his wages. I could wish that some princes whom I know were more sparing and discriminating over this, for

1. Horace, *Satires*, II, 1, 30–4; Tacitus, *Agricola*, 1.

such gestures lose all meaning when they are spread about without distinction. If they are made with no regard for status they are without effect.

Among odder affectations   [A]   let us not overlook the haughty mien of Constantius (the Emperor) who always held his head quite straight in public, neither turning it to right or left nor inclining it even to acknowledge those who were bowing to him from the side, keeping his body fixed and unmoving, without even swaying with the motion of his coach, without daring to spit or to wipe his nose or mop his brow in front of other people.[2]

I do not know whether those gestures which were noticed in me were characteristics of that first kind nor whether I really did have some hidden propensity to that vice of pride, as may well be the case; I cannot answer for the activities of my body; as for those of my soul, I want to confess now what I know about them.

In this kind of 'glory' there are two parts: namely, to rate oneself too high and to rate others too low. As for the former   [C]   I think we should take account of the following consideration: I am aware that I am troubled by an aberration of my soul which displeases me as iniquitous and even more as inappropriate; I make assays at correcting it, but as for eradicating it, I cannot: it consists in diminishing the real value of the things which I possess, simply because it is I who possess them, and in overvaluing whatever things are foreign to me, lacking in me or are not mine. This is a very widespread humour. Thus the man's prerogative of authority leads husbands to regard their own wives with a vicious disdain and leads many fathers to do the same to their own children; so too with me: out of two equal achievements I always come down against my own. It is not so much that a jealous concern to do better or to amend my ways disturbs my judgement and stops me from being satisfied with myself as that our mastery over anything engenders a contempt for what we hold under our sway. I am impressed by remote systems of government and of manners; so too for languages: I am aware that Latin by its dignity seduces me to favour it beyond what is appropriate to it, as it does in the case of children and the common people. My neighbour's domestic arrangements, his house and his horse, though equal to my own are better than my own because they are not mine.

Besides, I am most ignorant about myself. I marvel at the assurance and

2. Ammianus Marcellinus, XXI, xvi.

confidence everyone has about himself, whereas there is virtually nothing that I *know* that I know and which I would dare to guarantee to be able to perform. I do not have my capacities listed and classified; I only find out about them after the event, being full of doubt about myself as about everything else. The result is that if I happen to do a job in a praiseworthy fashion, I attribute that more to my good fortune than to my ability, especially since all my plans for it were made haphazardly and tentatively.

So too, [A] in a general way, the following applies to me as well: of all the opinions which [C] *grosso modo,* [A] Antiquity held about Man, the ones which I most readily embrace and to which I am most firmly attached are those which most despise us men, bring us low and treat us as nought. Philosophy never seems to me to have a better hand to play than when she battles against our presumption and our vanity; when in good faith she acknowledges her weakness, her ignorance and her inability to reach conclusions. It seems to me that the false opinion which is the mother suckling all the others, both in public and private, is the over-high opinion which Man has of himself. Those people who perch astride the epicycle of Mercury, [C] and who see so far into the heavens, [A] are an excruciating pain in the neck: for in the study that I am undertaking, the subject of which is Man, I find such extreme variation of judgement, such a deep labyrinth of difficulties one on top of another, so much disagreement and uncertainty in the very School of Wisdom, that you will understand that, since those fellows have not been able to reach any knowledgeable conclusions about themselves and their own mode of being (which is continuously before their eyes and which is within them) and since they do not understand the motions which they themselves set in action, nor how to describe and decipher the principles which they themselves hold in their hands: I cannot believe them, can I, about the cause[3] of the ebb and flow of the Nile!

An eager desire to know things was given to man as a scourge, says [C] Holy [A] Writ.[4]

---

3. [A] *until* [C]: the cause *of the movement of the Eighth Sphere and* of the . . .

Epicycles form part of the system of Ptolomaic astronomy. Rabelais makes a similar point about Empedocles: *Pantagruel,* TLF, X, 24.

4. Inscribed, in Latin, in Montaigne's library and attributed there to Eccl. I. This is, at best, but a paraphrase of Ecclesiastes I. (There is nothing relevant in Ecclesiasticus I.)

[A]: says *the sacrosanct* Writ . . .

But to come to myself as an individual, it seems to me that it would be hard for anyone to esteem himself less than I do. [C] I think that I am an ordinary sort of man, except in considering myself to be one; I am guilty of the failings of the lowest ranks of the common people but I neither disown my failings nor make excuses for them. I pride myself only on knowing what I am worth. If I have an element of vainglory it is superficial, treacherously diffused in me by my complexion but having nothing substantial enough for it to be summoned to appear before my judgement. I am sprinkled all over with it but not dyed in it. [A] For in truth, whatever form they may take, where the products of my mind are concerned nothing has come forth which has fully satisfied me – and other people's approbation is no reward.

My taste is discriminating and hard to please, especially where I myself am concerned: I [C] am constantly making disclaimers and [A] feel myself to be [C] everywhere [A] floating and bending from feebleness.[5] Nothing of mine that I possess satisfies my judgement. My insight is clear and balanced but when I put it to work it becomes confused: I have most clearly assayed that in the case of poetry. I have a boundless love for it; I know my way well through other men's works; but when I set my own hand to it I am truly like a child: I find myself unbearable. You may play the fool anywhere else but not in poetry:

> Mediocribus esse poetis
> Non dii, non homines, non concessere columnæ.

[Poets are never allowed to be mediocre by the gods, by men or by publishers.][6]

Would to God that the following saying was written up above our printers' workshops to forbid so many versifiers from getting in:

> verum
> Nil securius est malo Poeta.

[truly nothing is more self-assured than a bad poet.]

[C] Why are not our people like these Greeks? Dionysius (the elder) thought more highly of his poetry than of anything else of his; at the season of the Olympic Games, as well as sending chariots surpassing all others in magnificence he also sent golden awnings and royally tapestried

---

5. [A] until [C]: feebleness. *I know myself so well that if anything came from me which pleased me, I would owe it certainly to Fortune.* Nothing of mine . . .
6. Horace, *Ars poetica*, 272–3, then Martial, *Epigrams*, XII, lxiii.

marquees for the musicians and poets who were to recite his verses. When they were performed, the charm and excellence of the way they were recited at first attracted the attention of the people; but when a little later they came to weigh the incompetence of the work itself, they began to show contempt for it; as their judgement grew more harsh they threw themselves into a frenzy and angrily rushed to knock over all his marquees and to tear them to shreds. And the fact that his chariots achieved nothing worthwhile in the races either, and that the ship which was bringing his men home missed Sicily and was driven by the storm against the coast of Tarentum and smashed to pieces, was taken by the people as certain proof that this was the wrath of the gods, as angry as they were over that bad piece of poetry. And the very sailors who escaped the shipwreck accepted the opinion of the people, to which the oracle which had predicted his death gave some support: it declared that Dionysius' end would be near when he had 'vanquished those who were worth more than he was'. Dionysius took that to refer to the Carthaginians who surpassed him in strength. So whenever he had to encounter them he often avoided victory or played things down so as not to meet the fate mentioned by the oracle. But he got it wrong: for that god was referring to the time when by favour and corruption he would be preferred in Athens to tragic poets better than he was. He entered the competition with a tragedy of his called *The Lenæans*; he won but immediately died partly because of the excessive joy he derived from this.[7]

[A] That I find my own work pardonable is not so much for itself or its true worth as from a comparison with other writings which are worse – things which I can see people taking seriously. I envy the happiness of those who can find joy and satisfaction in their own works, for it is an easy way to give oneself pleasure, [B] deriving as it does from oneself, [C] especially if they show a little confidence in their self-deception. I know one poet to whom, in both the crowd and in his drawing-room, the mighty, the humble and the very earth and heavens all cry out that he knows nothing about poetry. For all that he will not in any way lower the status which he has carved out for himself: he is for ever beginning again, having second thoughts, never giving up, all the more strong in his opinion, all the more inflexible, since he has to maintain it all alone.

---

7. Putarch, *Dionysius*. (The *Lenæa* were festivals of Bacchus in Athens with contests between dramatists.)

[A]   My own works, far from smiling on me, irritate me every single time I go over them again:

> [B]   *Cum relego, scripsisse pudet, quia plurima cerno,*
> *Me quoque qui feci judice, digna lini.*

[When I read it over, I am ashamed to have written it, because even I who wrote it judge it worth erasing.][8]

[A]   I always have in my soul an Ideal form, [C]   some vague pattern, [A]   which presents me, [C]   as in a dream, [A]   with a better form than the one I have employed; but I can never grasp it nor make use of it. And   [C]   even that Ideal is only of medium rank. I argue from this that the products of   [A]   those great fertile minds[9] of former times greatly surpass the farthest stretch of my imagination and my desires. Their writings do not merely satisfy me and leave me replete: they leave me thunderstruck and throw me into an ecstasy of wonder. I can judge their beauty and can see it, if not through and through at least penetrating so deep that I know it is impossible for me to aspire that far. No matter what I undertake, I owe a sacrifice to the Graces to gain their favour (as Plutarch says of someone or other).

> *Si quid enim placet,*
> *Si quid dulce hominum sensibus influit,*
> *Debentur lepidis omnia gratiis.*

[If anything pleases, if it infuses any delight into the minds of men, all is owed to the elegant Graces.][10]

But the Graces are always deserting me. In my case everything is coarse: there is a lack of charm and beauty. I cannot manage to give things their full worth; and my style adds nothing to my matter. That is why I need my matter to be solid, with plenty to get hold of, matter shining in its own right.

[C]   When I seize upon more popular or more cheerful matter it is to follow my own bent: I have no love as the world has for gloomy formal wisdom; I do it to cheer up myself not to cheer up my style, which prefers grave and serious matters – that is if I ought to use the term style for my formless way of speaking, free from rules and in the popular idiom,

8. Ovid, *Ex ponto*, I, v, 15–16; written in exile on the Black Sea.
9. '88: And even *in my imagination I do not conceive things in their greatest perfection. From which I know that what I see produced by* those great fertile minds . . .
10. Plutarch (tr. Amyot), *Preceptes de mariage*, 147F (Plato tells the severe Xenocrates to 'sacrifice to the Graces': a goodwife should do the same). The author and source of the verse are, however, untraced.

proceeding without definitions, subdivisions and conclusions, confused like that of Amafanius and Rabirius.[11] [A] I have no idea how to please, delight or titillate; the best tale in the world withers in my hand and loses its sparkle. I can talk only when I am in earnest; I am quite devoid of that fluent discourse which I notice in many of my companions who are able to entertain every newcomer, to keep an entire crowd in suspense or to gain the ear of a monarch on all sorts of topics without boring him and without ever running out of things to say, because of their gift of exploiting the first matter which comes along, by adapting it to the humour and intelligence of those with whom they are dealing. [B] Princes are not very fond of solid arguments: and I am not very fond of spinning yarns. [A] Take the easiest and the most basic arguments (which are also usually the most readily grasped): I have no ideas how to use them — [C] I am bad at preaching to the common man. On any topic I like starting with my conclusions. Cicero reckons that the hardest part of a philosophical treatise is the beginning.[12] Since that is so I tackle the end. [A] Yet we have to [C] tune [A] our string to all kinds of modes: and the most acute mode is the one which is most infrequently played. There is at least as much achievement in enhancing an empty subject as in bearing up under a weighty one. Sometimes we must treat only the surface arguments; at other times we must go deeper. I am well aware that most men keep to that lower level because they are unable to conceive anything beyond the outer skin; but I am also aware that the greatest masters such as [C] Xenophon and [A] Plato can often be found slackening their string for that baser, more popular style of speaking and of treating their subjects, sustaining their style with their never-failing graces.

Meanwhile there is nothing fluent or polished about my language; it is rough [C] and disdainful, [A] with rhetorical arrangements which are free and undisciplined. And I like it that way, [C] by inclination if not by judgement. [A] But I fully realize that I sometimes let myself go too far in that direction, striving to avoid artificiality and affectation only to fall into them at the other extreme:

> *Brevis esse laboro:*
> *obscurus fio.*

11. Popular Epicurean writers, all of whose works are lost. Montaigne uses Cicero's description in the preceding lines (*Academica*, I, ii, 5). The first writer was Amafinius not Amafanius.
12. In his Latin translation of Plato's *Timaeus*, II. Then, '80: to *slacken* our string . . .

[I try to be brief and become obscure.]¹³

[C]   Plato says that neither length nor concision are properties which add anything to one's language or detract from it.

[A]   Even if I were to try to follow that other smooth-flowing well-ordered style I could never get there; and though the abrupt cadences of Sallust best correspond to my humour, I nevertheless find Caesar a greater writer and one less easy to reproduce stylistically. Although my own bent leads me to imitate rather the spoken style of Seneca, I nevertheless esteem Plutarch's more highly. In doing as in writing, I simply follow my natural form: which perhaps explains why I am better at speaking than I am at writing. Gestures and movements animate words, especially in the case of those who gesticulate brusquely as I do and who get excited. Our bearing, our facial expressions, our voice, our dress and the way we stand can lend value to things which in themselves are hardly worth more than chatter. In Tacitus Messala complains of the restrictive accoutrements of his time and the construction of the benches which orators had to speak from which weakened their eloquence.¹⁴

[A1]   In pronunciation, among other things, my French is corrupted by home-grown barbarisms; I have never known a man from our part of the world who did not obviously reek of dialect and who did not offend pure French ears. Yet that is not because I have a wide knowledge of my local Périgordian speech, for I am no more fluent in that than in German; it does not concern me much.   [C]   It is a dialect like the others here and there around me – those of Poitou, Saintonge, Angoûmois, Limoges and the Auvergne – soft, drawling and squittering.   [A]   Towards the mountains way above where we live there is indeed a form of Gascon which I find singularly beautiful, dry, concise and expressive, a language more truly manly and soldierly than any other I know,   [C]   as sinewy, forceful and direct as French is graceful, refined and ample.

[A]   As for Latin, which was vouchsafed me as my mother-tongue, I have through lack of practice lost the readiness I had for talking it –   [C]   yes, and for writing it too, for which I was once called a clever Johnny. [A]   Which shows what little I am worth from that angle.

In commerce between men beauty is a quality of great price; it is the first means of reconciling men to each other; there is no man so barbarous or uncouth as not to feel himself at least a little struck by its sweetness. The

13. Horace, *Ars poetica*, 25–6. Then Plato, *Laws*, X, 887 B.
14. Tacitus, *De Oratoribus*, XXXIX.

body is a major part of our being; it ranks greatly within it; that is why the way it is built up and composed is most justly worth attention. Those who wish to take our two principal pieces apart and to sequester one from the other are wrong. We must on the contrary couple and join them closely together. We must command the soul not to withdraw to its quarters, not to entertain itself apart, not to despise and abandon the body (something which it cannot do anyway except by some monkey-like counterfeit) but to rally to it, take it in its arms and cherish it, help it, look after it, counsel it, and when it strays set it to rights and bring it back home again. It should in short marry the body and serve as its husband, so that what they do should not appear opposed and divergent but harmonious and uniform. Christians have their own special teaching about this bonding, for they know that God's justice embraces this joint fellowship of body and soul (going so far as to make the body able to enjoy everlasting rewards) and that God sees the deeds of the whole man, willing that the whole man should receive rewards or punishments according to his merits.[15]

[C]   The Peripatetic School, the school most concerned with civility attributes to wisdom only one task: to obtain and procure the common good of these two parts in fellowship; it demonstrates that the other schools, by not being adequately devoted to the concerns of this liaison, have taken sides, one for the body the other for the soul, equally erroneous in having pulled apart their object (which is Man) and their Guide (which, for the genus Man, they swear to be Nature).

[A]   The first sign of distinction among men and the first consideration which gave some of them pre-eminence over others was in all likelihood superior beauty.

> [B]   *Agros divisere atque dedere*
> *Pro facie cujusque et viribus ingeniόque:*
> *Nam facies multum valuit viresque vigebant.*

[They divided up their lands and granted them to each according to his beauty, his strength and his intelligence; for beauty had great power, and strength was respected.][16]

15. Unlike the pagan Greeks, Christians believe in the resurrection of the dead not in the immortality of the soul permanently freed from the body. The major source of Montaigne's important concept of the 'marriage' of body and soul is Raymond Sebond. A secondary influence is doubtless Lucretius. In general, cf. Cicero, *De finibus*, IV, vii, 16–17.
16. Lucretius, V, 1109–11.

[A]   Now my build is a little below the average. This defect is not only ugly but unbecoming, especially in those who hold commands and commissions since they lack the authority given by a handsome presence and a majestic body.   [C]   Gaius Marius never willingly accepted soldiers who were under six foot. For the gentleman whom he is grooming *Il Cortegiano* is quite right to desire a medium height rather than any other, and to reject for him any oddity which made him conspicuous. But when that medium is lacking, to go and choose that he should fall short of it rather than exceed it is something I would not do in the case of a fighting-man. Aristotle says a small man may well be pretty but not beautiful;[17] as a great soul is manifested in its greatness, so beauty is known from a body great and tall.   [A]   'The Ethiopians and Indians,' he says, 'when they select their kings and magistrates take account of the beauty and height of the individuals.'[18] And they were right, for a man's followers feel respect and the enemy feels dismay upon seeing a leader with a splendid beautiful stature marching at the head of his troops:

> [B]   *Ipse inter primos præstanti corpore Turnus*
> *Vertitur, arma tenens, et toto vertice supra est.*

[Turnus himself, outstanding in body, is in the foremost rank, weapon in hand, head and shoulders above the others.]

Our great and holy heavenly King, every circumstance of whom should be noted with care, devotion and reverence, did not spurn the advantage of bodily beauty: *'speciosus forma præ filiis hominum'*. [fairer than the children of men.]   [C]   And as well as temperance and fortitude, Plato desired beauty in the guardians of his Republic.[19]

[A]   It is highly irritating if you are asked in the midst of your own servants, 'Where is your Master?' and if, when hats are doffed, you get only the tail-end of it, after your barber or your secretary. As happened to the wretched   [A1]   Philopoemen.[20]   [A]   When he was the first of his

---

17. Gaius Marius, the conqueror of Jugurtha (Vegetius, *De re militari*, I, v); Baldassare Castiglione, *Courtier*; Aristotle, *Nicomachaean Ethics*, IV, iii, 1123b.
18. Aristotle, *Politics*, IV, xliv; then Virgil, *Aeneid*, VII, 783–4, replacing '80: *Colloque tenus supereminet omnes* [He stood head and neck above them]. Ovid, *Metamorphoses*, II, 275.
19. Psalm 44 (45):3. (The application of this psalm to Christ is traditional.) Plato, *Republic*, VII, 535.
20. Montaigne had first written *Phocion*. (Anecdote from Plutarch's *Life of Philopoemen*.)

troops to arrive where he was to lodge and where he was expected, his
hostess, who did not recognize him and saw him looking rather shabby,
made him go and help her women-folk to draw water and to poke the fire,
'to prepare things for [A1] Philopoemen'! [A] When the gentlemen
of his entourage arrived, came upon him labouring at this handsome task
(for he had not failed to obey the orders given him) and asked him what he
was up to, he replied, 'I am paying the price of my ugliness.' Other
beauties are for the women: the only masculine beauty is beauty of stature.
When a man is merely short, neither the breadth and smoothness of a
forehead nor the soft white of an eye nor a medium nose nor the smallness of an ear or mouth nor the regularity or whiteness of teeth nor the
smooth thickness of a beard, brown as the husk of a chestnut, nor curly
hair nor the correct contour of a head nor freshness of hue nor a pleasing
face nor a body without smell nor limbs justly proportioned can make him
beautiful.

Meanwhile my build is tough and thick-set, my face is not fat but
full; my complexion is [B] between the jovial and the melancholic,
moderately [A] sanguine and hot;

> *Unde rigent setis mihi crura, et pectora villis;*

> [Whence my hairy legs and my hirsute chest;]

my health is sound and vigorous and until now, when I am well on in
years, [B] rarely troubled by illness – [A] I used to be like that, for I
am not considering myself as I am now that I have entered the approaches
to old age, having [A1] long since [A] passed forty.

> [B] *Minutatim vires et robur adultum*
> *Frangit, et in partem pejorem liquitur ætas.*

> [Bit by bit age smashes their vigour and their adult strength, and they drift into a
> diminished existence.]

[A] From now on, what I shall be is but half a being; it will no longer be
me,

> *Singula de nobis anni prædantur euntes.*

> [One by one things are stolen by the passing years.][21]

---

21. Martial, *Epigrams*, II, xxxvi, 5; then Lucretius, II, 1131–2, and Horace, *Epistles*,
II, ii, 55. (In the next sentence: [A]: son of *the most* agile father *to be seen in his
time*, with an energy . . .)

Skill and agility I have never had; yet I am the son of   [C]   a very agile father,   [A]   with an energy which lasted into his extreme old age. There was hardly anyone of his rank to equal him in all the physical exercises, just as I have found hardly anyone who could not do better than me except at running (at which I was among the average). As for music, either vocal (for which my voice is quite unsuited) or instrumental, nobody could ever teach me anything. At dancing, tennis and wrestling I have never been able to acquire more than a slight, vulgar skill; and at swimming, fencing, vaulting and jumping, no skill at all. My hand is so clumsy that I cannot even read my own writing, so that I prefer to write things over again rather than to give myself the trouble of disentangling my scribbles.   [C]   And my reading aloud is hardly better: I can feel myself boring my audience. That apart, I am quite a good scholar!   [A]   I can never fold up a letter neatly, never sharpen a pen, never carve passably at table,   [C]   nor put harness on horse, nor bear a hawk properly nor release it, nor address hounds, birds or horses.

[A]   My bodily endowments are, in brief, in close harmony with my soul's. There is no agility, merely a full firm vigour; but I can stick things out, provided that I set myself to it and as long as I am guided by my own desires:

> *Molliter austerum studio fallente laborem.*
>
> [The pleasure hides the austerity of the toil.][22]

Otherwise if I am not attracted by some pleasure and if I have any guide but my own will, pure and free, then I am no good at all. For as I am now, except for life and health there is nothing   [C]   over which I am willing to chew my nails or   [A]   which I am willing to purchase at the cost of a tortured spirit or constraint —

> [B]   *tanti mihi non sit opaci*
> *Omnis arena Tagi, quodque in mare volvitur aurum;*

[at such a price I would not buy all the sand of the muddy Tagus nor the gold which it carries down to the sea;]

[C]   being extremely idle and extremely free both by nature and by art. I would as soon give the blood of my veins as to take any pains.

[A]   My Soul is herself's alone and used to acting after her own fashion. Since up till now I have never had anyone giving me orders or any forced

22. Horace, *Satires*, II, ii, 12; then Juvenal, *Satires*, III, 54–5.

master I have gone my way just as far and just as fast as I liked. That has made me slack and useless to serve others; it has made me good for nothing but myself. And for myself there was no need to force my heavy, lazy, dilatory nature. Finding myself since birth with such a degree of fortune that I had cause to remain as I was, and with such a degree of intelligence as to make me appreciate that fact, I have sought nothing – and have taken nothing either:

> *Non agimur tumidis velis Aquilone secundo;*
> *Non tamen adversis ætatem ducimus austris:*
> *Viribus, ingenio, specie, virtute, loco, re,*
> *Extremi primorum, extremis usque priores.*

[I do not scud with bellying sails before the good North Wind, nor does an adverse gale from the south stay my course: in strength, in wit, in beauty, virtue, birth and goods I am the last of the first and the first of the last.][23]

The only talent I needed was to be content with myself – [C] which is nevertheless an ordering of the soul (if you understand it aright) equally hard in any sort of circumstances and which in practice we can find more readily in want than in plenty, perhaps because (as is the way with our other passions too) the hunger for riches is more sharpened by having them than by lacking them, while the virtue of moderation is rarer than that of endurance. All I needed was [A] gently to enjoy such good things as God in his bounty has placed in my hands. I have never tasted [C] excruciating [A] toil of any kind. [C] I have had to manage little apart from my own affairs; or if I have had to do anything else, it was in circumstances which let me manage things in my own time and in my own way, delegated to me by such as trusted me, never bothered me and knew me. For experienced men can even get some service out of a skittish wheezing horse. My very boyhood was spent [A] in a manner slack and free,[24] exempt from rigorous subjection. All of which formed a fastidious complexion for me, one incapable of supporting worry – to the extent that I prefer people to hide my losses and my troubles from

23. Horace, *Epistles*, II, ii, 201–4.
  Until [C]: priores, *having been born such that I did not have to go in quest of other advantages.* The only talent . . .
24. '80: of any kind: *I am very badly schooled in self-constraint, unskilled at any sort of business or painful negotiations, having never had to manage anything but myself and being brought up from boyhood* in a manner slack and free . . .
  Following verse from Horace, *Epistles*, I, vi, 45–6.

me: under the heading *Expenditure* I include whatever my indifference costs
me for its board and lodging:

> *hæc nempe supersunt,*
> *Quæ dominum fallant, quæ prosint furibus.*

[superfluities which the Master never knows about and which profit the thieves.]

I prefer not to know about my estate-accounts so as to feel my losses less
exactly. [B] Whenever those who live with me lack affection and its
duties I beg them to deceive me, paying me by putting a good face on
things. [A] I do not have firmness enough to put up with the
importunate demands of those adverse accidents which we are subject to
and I cannot brace myself to control and manage my affairs; so, by
abandoning myself to Fortune, I nurture in me as much as I can the
opinion which always sees the worst of everything; and then I resolve to
bear that worst gently and patiently. That is the only thing I do work at: it
is the goal towards which I direct all my arguments.

[B] Faced with danger I do not reflect on how to escape but on how
little it matters that I do so. If I remained in danger what would it matter?
Not being able to control events I control myself: if they will not adapt to
me then I adapt to them. I have hardly any of the art of knowing how to
cheat Fortune, of escaping her or compelling her, nor of dressing and
guiding affairs to my purpose by wisdom. I have even less powers of
endurance for sustaining the bitter painful care which is needed to do so.
And the most anguishing position for me is to remain in suspense among
pressing troubles, torn between fear and hope. It bothers me to make up
my mind even about the most trivial things, and I feel my spirits more
hard-pressed in suffering the swings of doubt and the diverse shocks of
decision-making than in remaining fixed, resigned to any outcome whatsoever once the dice have been thrown. Few emotions have ever disturbed
my sleep, yet even the slightest need to decide anything can disturb it for
me. For my journey I avoid steep slippery downward slopes and leap into
the most muddy and mirey of beaten tracks from which I can slip no
lower, and find assurance there: so too I prefer misfortunes to be unalloyed,
ones which do not try me, nor trouble me further about whether they can
be put right, but which immediately drive me straight into suffering.

> [C]   *Dubia plus torquent mala.*

[Uncertain evils most torment us.][25]

25. Seneca, the dramatist; *Agamemnon*, III, i, 29.

[B]   In events I act like a man: in the conduct of events, like a boy. The dread of a tumble gives me more anguish than the fall. That game is not worth the candle: the miser with his passion fares worse than the poor man, and the jealous husband worse than the cuckold; and there is often less harm in losing your vineyard than in pleading for it in court. The bottom step is the surest: it is the seat of Constancy; there, you need only yourself, and she can make it her base and rely on herself alone.

The following example of a gentleman known to many has something philosophical about it, has it not? He married when well on in years, having spent his youth as a good drinking-companion; he was a great raconteur and a great lover of jests. Recalling how the subject of cuckoldry had provided him with matter for stories and gibes against others, he sought protection by marrying a wife whom he chose in a place where anyone can find a woman for money and he established terms of acquaintance with her: 'Good morning, Mistress Whore!' – 'Good morning, Master Cuckold!' In his home there was no subject which he more frequently and openly entertained his visitors with than this plan of his, by which he bridled the secret gossip of the mockers and blunted the sharp points of his disgrace.

[A]   As for Ambition (which is neighbour to Presumption, or rather her daughter), to find me advancement Fortune would have needed to seek me out by the hand; as for striving for an uncertain hope and submitting myself to all the difficulties which accompany those who seek to thrust themselves forward at the start of their careers, I never could have done it.

[B]   *Spem pretio non emo.*

[I will not pay cash for some hope in the future.][26]

I bind myself to what I can see and to what I can hold on to; and I scarcely venture far from the harbour:

Alter remus aquas, alter tibi radat arenas.

[Let one oar sweep the water and the other sweep the strand.]

And then, few land such advancements without first hazarding their goods; moreover I am convinced that, as soon as a man has enough to keep the estate which he was born to and brought up for, it is madness to give up

26. Terence, *Adelphi*, II, iii, 11; then Propertius, III, iii, 23 and Seneca, the dramatist, *Agamemnon*, II, i, 47.

what he holds in his hand for the uncertain chance of increasing it. A man to whom Fortune refuses the means of not being footloose and of establishing a calm and tranquil life can be excused if he stakes all that he has on chance, since either way necessity sends him out on a quest:

> [C]   *Capienda rebus in malis præceps via est.*
>
> [In misfortune dangerous paths must be taken.]

[B]   And I can excuse a younger son for chancing his inheritance more than I can a man who is responsible for the honour of a household which may fall into want only through his fault.

[A]   On the advice of good friends of mine in former times I have indeed found the shorter easier road to ridding myself of such desires and to remaining quiet,

> *cui sit conditio dulcis sine pulvere palmæ;*

[a man whose pleasant lot is to gain the palms without struggling in the dusty arena;][27]

making also a healthy judgement that my powers are incapable of great achievements, and also remembering that quip of the late Chancellor Olivier, that the French are like monkeys which go scrambling up a tree from branch to branch, never ceasing until they reach the top; then, once they are there, they show you their arses.

> [B]   *Turpe est, quod nequeas, capiti committere pondus,*
> *Et pressum inflexo mox dare terga genu.*

[It is shameful to heap loads on your head which you cannot bear, only to bend your knees and show your back.]

[A]   The very qualities that I do have which do not deserve reproach are useless, I find, in this century. The affability of my manners would be called slackness and weakness; my faith and my conscience would be thought over-scrupulous and superstitious; my frankness and freedom, inconsiderate and audacious. Evil fortune does have some use: it is a good thing to be born in a century which is deeply depraved, for by comparison with others you are reckoned virtuous on the cheap. Nowadays if you have merely murdered your father and committed sacrilege you are an honest honourable man:

---

27. Horace, *Epistles*, I, i, 51: then, Propertius III, ix, 5–6 and Juvenal, XIII, 60–3.

[B]   *Nunc, si depositum non inficiatur amicus,*
Si reddat veterem cum tota ærugine follem,
Prodigiosa fides et Tuscis digna libellis,
Quæque coronata lustrari debeat agna.

[If a friend nowadays does not deny that you entrusted money to him and returns your old purse full of rusty coins, he is a prodigy of trustworthiness, meriting a place on the Etruscan Kalendar and the sprinkled blood of a sacrificial lamb.]

And there never was a time and place in which princes could find greater or surer reward given to their generosity and justice. Unless I am mistaken, the first prince to make himself favoured and trusted in that way will, at little cost, outstrip his companions. Might and violence can achieve something, but not always and not everything.

[C]   Where valour and the art of war are concerned we can see tradesmen, village wise men and artisans matching the nobility: they fight honourably in open combat and in duels; they do battle and defend cities in these wars of ours. A prince's reputation is smothered in such a throng. Let him shine forth by his humanity, truthfulness, loyalty, temperance and above all by justice – which are rare tokens now, unknown and driven abroad. It is only by the good-will of the people that he can carry on his business: no other qualities can gratify that good-will more than these, since they are more useful to them than the others are.

Nihil est tam populare quam bonitas.

[Nothing is more pleasing to the people than affability.][28]

[A]   By such comparisons I would have found myself [C]   a giant and unusual, just as I find myself a pygmy and quite commonplace in comparison with some former times in which it was indeed considered commonplace (if other stronger qualities did not accompany it) to find a man [A]   moderate in revenge, slow to take offence, punctilious in keeping his word, neither treacherous nor pliant, nor accommodating his trust to the will of others and to circumstances. I would rather let affairs go hang than to warp my trustworthiness in their service.

As for that novel virtue of deceit and dissimulation that is now much honoured I hate it unto death, and among all the vices I can find none which bears better testimony to cowardice and to baseness of mind. It is an abject and a slave-like humour to go disguising and hiding yourself behind

28. Cicero, *Pro Ligario*, X.

a mask and not to dare to let yourself be seen as you are. That way, men of our time are trained for perfidy: [B] being used to utter words of falsehood, to break their word they do not scruple. [A] A noble mind must not belie its thoughts: it wants its inward parts to be seen: [C] everything there is good — or at least humane. Aristotle reckons that magnanimity has the duty to hate and to love openly, to speak with total frankness and to think nothing of other men's approval or disapproval compared with the truth. [A1] Apollonius said that it was for slaves to lie and for free-men to speak the truth.[29] [C] Truth is the first and basic part of virtue. It must be loved for its own sake. A man who tells the truth because he is otherwise bound to do so or because it serves him to do so, yet who is not afraid to tell lies when it does not matter to anyone, is not truthful enough. My soul's complexion is such that it flees from lying and hates even to think of it. I have an inward sense of shame and a stabbing remorse if a lie escapes me — as it does sometimes, when occasions take me by surprise and disturb me unawares.

[A] We should not always say everything: that would be stupid; but what we do say must be what we think: to do otherwise is wicked. I do not know what princes expect to get out of constantly pretending and lying, except not to be believed even when they do tell the truth. It may deceive people once or twice; but to profess your dissimulation and to boast as some of our princes have done that they would toss their very shirt on to the fire if it knew of their real intentions (which is a saying of an Ancient, Metellus of Macedon); to declare that a man who knows not how to feign knows not how to reign is to forewarn those who have to deal with them that what they say is all cheating and lies.[30] [C] *'Quo quis versutior et callidior est, hoc invisior et suspectior, detracta opinione probitatis.'* [The more crafty and artful a man is, the more he is loathed and mistrusted, once he has lost his reputation for probity.] [A] A man would be very simple to let himself be deceived by the looks or words of a man who, as Tiberius did, thought it important to appear outside always different from what he was inside; and I do not know what commerce such people can have with men when they proffer nothing which you can accept at its face-value. [B] A man who is disloyal to truth is disloyal to lies as well.

---

29. Aristotle, *Nicomachaean Ethics*, IV, iii, 1124b; Apollonius' remark has not been traced, but cf. Plutarch, *Comment il fault nourrir les enfants*, 6H.
30. Charles VIII. Cf. Plutarch (tr. Amyot), *Du trop parler*, 92E–F; Then, Cicero, *De Officiis*, II, ix, 34.

[C] Those writers nowadays who, when drawing up the duties of a prince, have considered only what is good for the affairs of State, placing that before his fidelity and conscience, might have something to say to a prince whose fortune had so arranged his affairs that he could for ever secure them by one single act of deception, one failure to keep his word. But things do not happen that way: princes stumble again into similar bargains: they make more than one peace, more than one treaty in their lifetime. The profit tempts them when they first prove untrustworthy – and virtually always some profit is on offer, as in every act of wickedness (sacrilege, murder, rebellion and treachery are done for some kind of gain); but that first profit entails infinite subsequent losses, putting that prince, by his first breach of trust, beyond all negotiations, beyond any mode of agreement.

When I was a boy Suleiman, of the family of the Ottomans, a family not scrupulous in keeping promises and agreements, landed his army at Otranto and he learned that Mercurino de' Gratinare and the citizens of Castro were, despite the stipulations in the treaty, still kept prisoner after surrendering their fortress. He ordered them to be released, saying that he was engaged in other great expeditions in that region and that even though such a breach of faith might have some appearance of present advantage it would bring upon him discredit and distrust which would be infinitely damaging in the future.[31]

[A] Now, as for me, I prefer to be awkward and indiscreet rather than to flatter and dissemble. [B] I confess that there is an element of pride and stubbornness in remaining open and all of a piece, with no consideration for others, and it seems to me that I become a little too free when I least ought to be so and that I react to the duty of respect by growing more heated. It may also be that for lack of art I just follow my nature. When I use then that same liberty of tongue and expression that I bring to my household, I feel how much it sinks towards a lack of discretion and rudeness. But apart from the fact that I am made that way, my wit is not supple enough to dodge a sudden question and to escape down some side-road, nor to pretend that something is true. My memory is not good enough to remember that pretence nor reliable enough to maintain it: so I act the brave out of weakness. I therefore entrust myself to simplicity, always saying what I think; by temperament and by conviction I leave the

---

31. Montaigne has Machiavelli's *Prince* in mind throughout this chapter. This anecdote is from the anonymous *Thesoro Politico cioè relationi, instruttioni . . . di multo importanza per li disegni di principe* (II, v), a major source in several chapters.

outcome to Fortune. [C] Aristippus said that to speak freely and openly to all men was the chief fruit he derived from philosophy.³²

[A] Memory is an instrument of wondrous service, without which judgement is hard put to it to do its duty. In me it is entirely lacking. If you want to propound anything to me you must do it bit by bit. It is beyond my ability to answer propositions in which there are several heads of argument. I could not take on any commission without my jotter. And when I myself have anything of importance to propound, if it is at all long-winded I am reduced to the abject and pitiful necessity of learning off by heart, [C] word by word, [A] what I have to say: otherwise I would have neither shape nor assurance, being ever fearful that my memory would play a dirty trick on me. [C] But for me that method is no less difficult. It takes me three hours to learn three lines of verse; and then, in a composition of my own, an author's freedom to switch the order and to change a word, forever varying the matter, makes the work harder to learn. [A] Now the more I mistrust my memory, the more confused it gets; it serves me best when I take it by surprise; I have to address requests to it somewhat indifferently, for it becomes paralysed if I try to force it, and once it has started to wobble the more I dig into it the more it gets tied up and perplexed; it serves me in its own time not in mine.

[A1] What I feel in the case of my memory I feel in many other aspects of myself. I flee from all orders, obligations and constraints. Even things I do easily and naturally I cannot do once I order myself to do them with an express and prescribed command. The very parts of my body which have a degree of freedom and autonomy sometimes refuse to obey me if I plan to bind them to obligatory service at a certain time and place. Such tyrannical and preordained constraint disgusts them: they cower from fear and irritation and swoon away.

[B] I was once in a place where it is barbarously rude not to drink when you are invited to do so: I was left completely free, but I tried to be a good fellow to please the ladies who by local custom were in the party. We had a fine old time: for this anticipated threat of having to make myself go beyond my nature and custom so blocked my gullet that I could not gulp down one single drop and I was even deprived of the wine I wanted for my dinner. All the drink that I had already taken in imagination had quenched my thirst and I had had enough!

[A1] This effect is more evident in those whose imagination gets strongly carried away: it is nevertheless quite natural; there is nobody who

32. Erasmus, *Apophthegmata*, III, *Aristippus*, VI.

does not feel it to some extent. An excellent bowman was condemned to death, but offered a chance to live if he would agree to demonstrate some noteworthy proof of his skill. He refused to make an assay, fearing that the excessive strain on his will would make his hand go wrong and that instead of saving his life he would also lose the reputation that he had acquired as an archer.

When a man is walking up and down anywhere, if his thoughts are on something else he will never fail – give an inch or so – to make the same number of equal strides; but if he goes to that place with the intention of counting and measuring his strides, he will find that he will never achieve so exactly by design what he had done naturally and by chance.

[A] My library, which is a fine one as village libraries go, is sited at one of the corners of my house. If an idea occurs to me which I want to go and look up or write down, I have to tell somebody else about it in case it slips out of my mind as I merely cross my courtyard. If I am rash enough to interrupt the thread of what I am saying, I never fail to lose it: which means that in talking I become constrained, dry and brief. Even my serving-men I have to call by the name of their office or the place which they come from, for it is hard for me to remember their names. [B] (I can tell you well enough that it has three syllables, is hard on the ear or begins with such and such a letter.) [A] And if I lasted for long I do not doubt that I would forget my own name, as others have done. [B] Messala Corvinus lived two years without any trace of memory; [C] and the same is said of George of Trebizond;[33] [B] so in my own interest I often chew over what sort of life they had and whether, without that faculty, there would be enough of me left to maintain my identity at all easily; and if I look at it closely I am afraid that, if this defect were complete, all the activities of my soul would be lost. [C] *'Memoria certe non modo philosophiam, sed omnis vitæ usum omnesque artes una maxime continet.'* [It is certain that memory alone is what retains not only our philosophy but also the whole of life's practices and all the arts and sciences.][34]

[A] *Plenus rimarum sum, hac atque illac effluo.*

[I am full of cracks and leaking everywhere.]

[A1] More than once I have forgotten the password [C] for the watch [A1] which [C] but three hours previously [A1] another

---

33. Same examples in Ravisius Textor, *Officina*, s.v. *Obliviosi.*
34. Cicero, *Academica*, II (Lucullus), vii, 22; then, Terence, *Eunuch*, I, ii, 25.

man had told me or had learnt from me – [C] and, no matter what Cicero says, I have even forgotten where I had hidden my purse.[35] Anything I hide away privately I am helping myself to mislay. [A] Now memory is the coffer and store-box of knowledge: mine is so defective that I cannot really complain if I know hardly anything. I do know the generic names of the sciences and what they mean, but nothing beyond that. I do not study books, I dip into them: as for anything I do retain from them, I am no longer aware that it belongs to somebody else: it is quite simply the material from which my judgement has profited and the arguments and ideas in which it has been steeped: I straightway forget the author, the source, the wording and the other particulars.

[B] I am so outstanding a forgetter that, along with all the rest, I forget even my own works and writings. People are constantly quoting me to me without my realizing it. If anyone wanted to know the sources of the verse and *exempla* that I have accumulated here, I would be at a loss to tell him, and yet I have only gone begging them at the doors of well-known and famous authors, not being satisfied with splendid material if it did not come from splendid honoured hands. In them, authority and reason coincide.   [C]   No wonder that my own book incurs the same fate as the others and that my memory lets go of what I write as of what I read; of what I give as of what I receive.

[A]   I have other defects apart from memory which greatly contribute to my ignorance. My wits are sluggish and blunt: the slightest fog will arrest their thrust, so that (for example) they can never unravel the easiest of puzzles which I set them. The vainest of subtleties can embarrass me. I have only the roughest idea of games such as chess, cards, draughts and so on in which the wits play a part. My power of understanding is slow and confused, but once it has grasped anything, as long as it continues to do so it holds on to it well, hugging it tightly, deeply and in its entirety. My eyesight is sound, whole and good at distances, but when I work it easily tires and grows lazy. That explains why I cannot have any lengthy commerce with books except through the assistance of somebody else. Those who have not made an assay of this can learn from their Younger Pliny how much such a slowing down matters to those who are given to this occupation.[36]

35. Cicero (*De senectute*, vii, 22), 'had never known an old man forget where he had hidden his treasure!'
36. From the *Letters* of Gaius Plinius Caecilius Secundus, the adopted son of Pliny the Elder.

Nobody's soul is so brutish and wretched that, within it, some peculiar faculty cannot be seen to shine; no soul is buried so deep that some corner of it cannot break out. How it happens that a soul which is blind and dull to everything else is found to be lively, clear and outstanding in some definite activity peculiar to itself is something you will have to inquire from the experts. But the most beautiful of souls are those universal ones which are open and ready for anything, [C] untaught perhaps but not unteachable. [A] And I say that to indict my own: for whether by weakness or indifference – and it is far from being part of my beliefs that we should be indifferent to what lies at our feet, is ready to hand or closely regards the conduct of our lives – no soul is so unfit or ignorant as mine concerning many commonplace matters of which you cannot be ignorant without shame.

I must relate a few examples.

I was born and brought up in the country, surrounded by agriculture; farming and its concerns have been in my hands ever since those who previously owned the lands which I enjoy moved over for me; yet I cannot do sums with either abacus or pen. Most of our coins I do not recognize; unless it is all too obvious I do not know the difference between one grain and another, neither in the ground nor in the barn; and in my vegetable garden I can scarcely tell my cabbages from my lettuces. I do not even know the names of the most elementary farming implements nor even the most basic principles of agriculture such as children know; [B] still less do I know anything about the manual arts, about the nature of merchandise and its trade, about the natural qualities and varieties of fruit, wine and foodstuffs, about training a hawk or curing a horse or a hound. [A] And since I must reveal the whole of my shame, only a month ago I was caught not knowing that yeast is used to make bread [C] and what was meant by 'fermenting' wine. [A] It was inferred once in Athens that a man had an aptitude for mathematics when it was seen how he arranged a pile of brushwood into faggots. They would certainly draw the opposite conclusion from me: give me a complete set of kitchen equipment and I would still go famished![37]

From these details of my confession you can imagine others to my disadvantage. But no matter how I may appear when I make myself known, provided that I do make myself known such as I am, I have done

---

37. Aulus Gellius, V, iii (Democritus judging Protagoras' ability).

[A] until [C]: famished! *And, were I to be given a horse with its gear, I very much doubt whether I would know how to harness it for my service. From these...*

what I set out to do. Yet I make no apology for daring to commit to writing such ignoble and frivolous matters: the ignobility of my subject [C] restricts me to them. If you will you may condemn my project: but the way I do it, you may not. [A] I can see well enough, without other people telling me, how little all this weighs and is worth and the [C] madness [A] of my design.[38] It is already something if my judgement, of which these are the assays, does not cast a shoe in the process.

> *Nasutus sis usque licet, sis denique nasus,*
>   *Quantum noluerit ferre rogatus Athlas,*
> *Et possis ipsum tu deridere Latinum,*
>   *Non potes in nugas dicere plura meas,*
> *Ipse ego quam dixi: quid dentem dente juvabit*
>   *Rodere? carne opus est, si satur esse velis.*
> *Ne perdas operam: qui se mirantur, in illos*
>   *Virus habe; nos hæc novimus esse nihil.*

[Go on: wrinkle your nose – a nose so huge that Atlas would not carry it if you asked him – mock the famous mocker Latinus if you can, yet you will never succeed in saying more against my trifles than I have said myself. What use is there in champing your teeth? To be satisfied you need to sink them into meat. Save your energy. Keep your venom for those who admire themselves: I know my work is worthless.]

I am under no obligation not to say daft things, provided that I do not deceive myself in recognizing them as such. It is so usual for me to know I am going wrong that I hardly ever go wrong any other way: I never go wrong by chance. It is a slight thing for me to attribute my silly actions to the foolhardiness of my humour, since I know no way of avoiding regularly attributing my vicious ones to it.

I saw one day in Bar-Le-Duc King Francis II being presented with a self-portrait by King René of Sicily as a souvenir of him. Why is it not equally permissible to portray yourself with your pen as he did with his brush?[39]

---

38. '80: ignobility of my subject, *which is myself, cannot tolerate any fuller or more solid ones: and in addition it is a new and fantastical humour which impels me and we must let it run.* [. . .] worth and the *boldness and rashness* of my design . . .

Following lines from Martial, *Epigrams*, XIII, 2, attacking censorious critics and know-alls.

39. '80: brush? *And may I not portray what I find out about myself, whatever it may be?* . . .

Following line from Petrarch, Sonnet 135.

So, unfit though it is to be brought out in public, I have no wish to overlook another of my scabs: my inability to reach decisions, a most inconvenient defect when transacting the world's business. When there are doubts about an enterprise I do not know what decision to make:

[A1]   *Ne si, ne no, nel cor mi suona intero.*

[My mind neither says firmly Yea nor firmly Nay.]

[B]   I can defend an opinion all right, but I cannot select it.

[A]   No matter what side we incline to in the affairs of men, many likely arguments come to confirm our choice –   [C]   that is why Chrysippus the philosopher said that he merely wanted to learn the dogmas of his Masters Zeno and Cleanthes: as for proofs and reasons he would supply them himself[40] –   [A]   Therefore no matter what side I turn to I can furnish myself with cause and true-sounding reasons for remaining there. So I maintain within me my doubt and my freedom to choose until the occasion becomes urgent, when, to tell the truth, I 'toss a feather to the wind' (as the saying goes) and put myself at the mercy of Fortune: the slightest of inclinations or circumstances then carries me away.

Dum in dubio est animus, paulo momento huc atque illuc impellitur.

[When the mind is in doubt, it wavers hither and thither at the merest impulse.][41]

The indecision of my judgement is so equally balanced in most encounters that I would willingly have recourse to deciding by lots and by dice. And I note as having great relevance to our human weakness the examples which Holy Writ itself has left us of this practice of referring decisions of choice in matter of doubt to the hazard of Fortune: *'Sors cecidit super Mathiam.'* [The lot fell upon Matthias.][42]

–   [C]   Human reason is a dangerous two-edged sword. Just see how, even in the hands of Socrates, its most familar and intimate friend, it is a stick with a great many ends to it.

[A]   And so I am fitted only for following, and easily allow myself to be persuaded by the crowd. I do not trust enough in my powers to undertake to guide or to command. I am very happy to find my path trodden out for me by others. If I must run the hazard of an uncertain

40. Diogenes Laertius, *Life of Chrysippus.*
41. Terence, *Andrea*, I, vi, 32.
42. Acts 1:26. To choose between Joseph Barsabbas and Matthias as a successor to the apostolate of Judas Iscariot, the Church drew lots.

choice, I prefer that that choice should be guided by someone who trusts to his opinions and is more wedded to them than I am to my own, [B] the very foundations and basis of which I find inclined to slip. Yet I do not easily change my opinions, since I find the opposite ones equally weak. [C] *'Ipsa consuetudo assentiendi periculosa esse videtur et lubrica.'* [The very habit of giving one's assent seems to be slippery and dangerous.][43] [A] Notably in affairs of state there is a wide field for vacillation and controversy:

> *Justa pari premitur veluti cum pondere libra*
> *Prona, nec hac plus parte sedet, nec surgit ab illa.*

[As when a just balance weighs equal weights, neither comes down on this side nor rises on the other.]

The discourses of Machiavelli, for example, were solid enough, given their subject, yet it was extremely easy to attack them; and those who have done so left it just as easy to attack theirs too.[44] On such a subject there would always be matters for counter-arguments, counter-pleas, replications, triplications, fourth surrejoinders and that endless web of argument which our chicanery has stretched out as far as may be in favour of legal actions —

> *Cædimur, et totidem plagis consumimus hostem,*

[The foe hits at us and we return blow for blow,]

— because such reasoning has no other basis than experience while the diversity of events offers us an infinity of examples of every kind of type.[45] A great learned person of our own time says that, when our almanacs predict cold or wet, if anyone were to put warm or dry, always saying the opposite to their predictions, he would not care which side he was on if he had to bet on the outcome except for forecasts which permit no doubt (such as predicting extreme heat at Christmas or the rigours of winter on Midsummer Day). I think the same about debates on politics: whatever role you are given your game is as easy as your opponent's, provided that you do no violence to the most obvious and evident of principles. That is

---

43. Cicero, *Academica*, II (Lucullus), xxi, 68; then Tibullus, IV, i, 40–1.
44. The leader of the opposition to the ruthless *raisons d'état* of *Il Principe* was Innocent Gentillet, whose treatise on government (Geneva?, 1576) was often called *l'Anti-Machiavel*.
45. Horace, *Epistles*, II, ii, 97. For the argument cf. III, 13, 'On experience'.

why, for my humour, there is no system so bad (provided it be old and durable) as not to be better than change and innovation. Our manners are corrupt in the extreme and wondrously inclined to get worse; many of our French laws and customs are monstrous and barbaric: yet, because of the difficulty of putting ourselves into a better state, and because such is the danger of collapse into ruin, if I could jam the brake on our wheel and stop it dead at this point I would happily do so.

[B]   *nunquam adeo fœdis adeoque pudendis*
Utimur exemplis ut non pejora supersint.

[none of the examples which we cite is so infamous and shameful that there be not worse to come.][46]

[A]   I find that the worst aspect of the state we are in is our lack of stability and that our laws cannot adopt one fixed form any more than our fashions can. It is easy enough to condemn a polity as imperfect since all things mortal are full of imperfection; it is easy enough to generate in a nation contempt for its ancient customs: no man has ever tried to do so without reaching his goal; but as for replacing the conditions you have ruined by better ones, many who have tried to do that have come to grief.

[C]   In my own activities I allow but a small part to my intelligence: I readily let myself be led by the public order of this world. Blessed are they who, without tormenting themselves about causes, do what they are told rather than tell others what to do; who, as the Heavens roll, gently roll with them. When a man reasons and pleads causes, his obedience is neither tranquil nor pure.

[A]   To get back to myself, the only quality, in short, for which I reckon I am worth anything is the one which no man ever believed he lacked; what I commend in myself is plebeian, commonplace and ordinary, for whoever thought he lacked [C] sense? [A] That would be an assertion which implied its own contradiction; [C] lack of sense is a malady which never exists if you can see it: it is tenacious and strong, yet the first ray darting from the sufferer's eye pierces it and dispels it as the face of the sun dispels a dense mist; in this case to bring a charge is to grant a discharge, [A] and to condemn yourself would be to acquit yourself. Never was there hodman nor womanling but thought they had sense enough for their needs. In others we readily acknowledge superior courage, [C] physical [A] strength, experience, agility and beauty: but superior judgement we concede to none. And such arguments in

46. Juvenal, *Satires*, VIV, 183–4.

another as derive from pure inborn wit we think that   [C]   we would
have discovered too if only we had looked at things from the same
angle.   [A]   The erudition, the style and such-like that we see in the
works of others we can easily acknowledge when they surpass our own;
but when it comes to the pure products of the intelligence, each man thinks
that he has it in him to hit upon exactly the same things; only with
difficulty can he perceive the weight and labour of it all   [C]   unless, that
is, it is incomparably and extremely beyond him – and even then, only
just.   [A]   So this is an exercise of judgement from which I must expect
very little praise and honour;   [C]   it is a kind of writing of little
renown.[47]

Moreover, whom are you writing for? The scholars whose concern it is
to pass judgement on books recognize no worth but that of learning and
allow no intellectual activity other than that scholarship and erudition.
Mistake one Scipio for the other, and you have nothing left worth saying,
have you! According to them, fail to know your Aristotle and you fail to
know yourself. But as for souls which are commonplace and ordinary, they
cannot perceive the grace and the weight of sustained elegant discourse.
And those two species occupy the whole world! Men of the third species,
the one which falls to your lot, composed of minds which are strong and
well-adjusted, arc so rare that, precisely, they have no name or rank among
us; to aspire and strive to please them is time half wasted.

[A]   It is commonly held that good sense is the gift which Nature has
most fairly shared among us, for there is nobody who is not satisfied with
what Nature has allotted him.   [C]   And is that not reasonable? Anyone
who would peer beyond it would be peering beyond what his sight can
reach.   [A]   I believe that my opinions are sound and good (who does
not?). One of the best proofs that I have of that's being true is that I do not
think highly of myself; for if these opinions had not been firm and assured,
they would easily have been led astray by the singular affection I have for
myself, referring as I do virtually everything back to myself and not
squandering much affection on others. All that affection which other men
scatter over a boundless multitude of friends and acquaintances (and over
their glory and greatness) I devote entirely to my peace of mind and to *me*.

---

47. Many changes from [A]: thought he lacked *judgement*? [. . .] agility and
beauty *and nobility*: but superior judgement [. . .] we think that *they are ours*. The
erudition [. . .] it is a *nature* of writing of little *credit*. *The stupidest man in the world
thinks he has as much understanding as the cleverest. That is why* [A] It is commonly
held. . .

Whatever does escape from me elsewhere is strictly not according to the dispensation of my reasoning:

> *mihi nempe valere et vivere doctus.*

> [I am learned in living and flourishing for myself.][48]

Now as for my opinions I find them infinitely blunt and tenacious in condemning me for inadequacy. Truly, more than any other, that is also a subject in which I exercise my judgement. All men gaze ahead at what is confronting them: I turn my gaze inwards, planting it there and keeping it there. Everybody looks before himself: I look inside myself; I am concerned with no one but me; without ceasing I reflect on myself, I watch myself, savour myself. Other men (if they really think about it) always forge straight ahead:

> *nemo tentat in sese descendere.*

> [no one attempts to go down into himself.]

I turn round and round in myself. I owe chiefly to myself the capacity – [A1] such as it is in me [A] – for sifting the truth and my freeman's humour for not easily enslaving my beliefs: for the firmest universal reasons that I have were, so to say, born in me. They are natural ones and entirely mine. I brought them forth crude and uncomplicated – products which are bold and strong but somewhat confused and imperfect. I subsequently confirmed and strengthened them by other men's authority and by the sound reasonings of those Ancients with whom I found myself in agreement in judgements; they made my hold on them secure and gave me the full enjoyment of their possession.

[B] Everyone seeks a reputation for a lively ready mind: I claim a reputation for steadiness; they seek a reputation for some conspicuous and signal activity or for individual talent: I claim one for the ordinate quality, the harmony and the tranquillity of my opinions and morals: [C] *'Omnino, si quidquam est decorum, nihil est profecto magis quam æquabilitas universæ vitæ, tum singularum actionum: quam conservare non possis, si, aliorum*

---

48. Lucretius, V, 959; Montaigne is about to follow the proverbial wisdom of Socrates; cf. Erasmus, *Adages*, I, VII, LXXVI, *In se descendere*, which contains Montaigne's quotation from Persius, *Satires*, IV, 23 – a major moral commonplace; also *Nosce teipsum* (I, VII, XCV) and *In tuum ipsius sinum inspue* (I, VII, XCIV). These adages form the cream of Socratic wisdom for many Renaissance moralists.

naturam imitans, omittas tuam.' [If anything at all is becoming, then nothing is more so than the even consistency of your entire life and of every one of its activities: and you cannot maintain that if you imitate other men's natures and neglect your own.][49]

[A] There then you have the extent to which I feel guilty of that first characteristic which I attributed to the vice of presumption.[50] As for the second, which consists in not thinking highly enough of others, I do not know that I can plead so innocent to that – for, cost me what it will, I am determined to tell things as they are.

Perhaps because of the constant commerce I have with the humours of the Ancients and of the ideal I have formed of the richly endowed souls of the men of former times, I feel a distaste for others and for myself. Perhaps we really do live in a time which begets nothing but the mediocre. However that may be, I know nothing worthy of any great ecstasy of admiration nowadays; moreover I know hardly any men with the intimacy needed to judge them; the ones whom my circumstances commonly bring me among are not on the whole concerned with cultivating their higher faculties but are men to whom has been proposed no beatitude but honour and no perfection but valour.

Whatever of beauty I do find in others I am most ready to praise and to value: indeed I often go farther than I really think, and to that extent permit myself to lie, not being able, though, to invent falsehoods entirely. I readily bear witness to those I love of what I find praiseworthy in them: if they are worth a foot I make it a foot and a half; but what I cannot do is to attribute qualities to them which they do not have; nor can I frankly defend their imperfections.

[B] Whatever witness I owe to the honour of my very enemies I bear unambiguously. [C] My sympathies change but not my judgement; [B] I do not confound my quarrel with other circumstances which have nothing to do with it and I am so jealous for my freedom of judgement that I find it hard to give it up for any passion whatsoever. [C] By telling lies I harm myself more than the one I lie about. Attention is drawn to the laudable and noble custom of the people of Persia who, both in speaking of their mortal enemies and in waging

49. Cicero, *De officiis*, I, xxxi, lll.
50. That is, blind self-love, *philautia*, which leads a man to flatter himself and to condemn others. Erasmus' adages cited in note 48 support Montaigne's contention that the wise man, by 'descending into himself', far from being selfish can avoid the vices of self-love.

total war against them, do so with such honour and equity as their virtue deserves.[51]

[A]   I know plenty of men with specific endowments of great beauty: some have intelligence; others courage, skill, conscience, eloquence, learning and so on. But as for the kind of man who is great over all and who has so many beautiful qualities all at once   [A1],   or one of them to such a degree of excellence,   [A]   that we should be thunderstruck by him and compare him with those from times past whom we hold in honour, I have not had the good fortune to meet even one. And the greatest man I ever knew during his lifetime – I mean great for the inborn qualities of his soul and his natural endowments – was Etienne de La Boëtie; his was indeed an ample soul, beautiful from every point of view, a soul of the Ancient mould which would have brought forth great achievements if his fate had so allowed, having greatly added to its natural richness by learning and assiduous study. I do not know how it happens   [C]   (though it certainly does)   [A]   but there is more triviality and weakness of understanding in those who profess to have most ability, who engage in the literary professions and whose responsibilities are concerned with books than in any other kind of person; it is because we demand more from them and expect more, so that we cannot pardon everyday defects in them; or is it because their reputation for knowledge makes them bolder in displaying and revealing themselves so intimately that they give themselves away and condemn themselves? A craftsman gives surer proof of his stupidity when he has some rich substance in his hands and prepares it and mixes it contrary to the rules of his art than when he is working on some cheap stuff; and we are more offended by defects in a statue made of gold than in one made of plaster; so too with the learned: when they exploit materials which in themselves and in their right place would be good they use them without discernment, honouring their power of memory rather than their understanding. It is Cicero, Galen, Ulpian and St Jerome that they honour: themselves they make ridiculous.[52]

I gladly come back to the theme of the absurdity of our education: its end has not been to make us good and wise but learned. And it has succeeded. It has not taught us to seek virtue and to embrace wisdom: it

51. According to Plutarch, the Persians held the two greatest vices to be borrowing and lying (*Qu'il ne faut point emprunter à usure*, 131C). Perhaps the origin of this assertion.
52. Philosophers study Cicero; doctors, Galen; lawyers, Ulpian; theologians, Jerome. Montaigne is criticizing all the university disciplines.

has impressed upon us their derivation and their etymology. We know how to decline the Latin word for virtue: we do not know how to love virtue. Though we do not know what wisdom is in practice or from experience we do know the jargon off by heart. Yet we are not content merely to know the stock, kindred and intermarriages of our neighbours: we want to love them and to establish commerce and communication with them: our education has taught us the definitions, divisions and subdivisions of virtue as though they were the surnames and the branches of a family-tree, without any concern for establishing between us and it any practice of familiarity or personal intimacy. For our apprenticeship it has not prescribed the books which contain the soundest and truest opinions but those which are written in the best Greek and Latin, and in the midst of words of beauty it has poured into our minds the most worthless humours of Antiquity. A good education changes a boy's judgement and morals, as happened in the case of a dissipated young Greek called Polemon who happened to attend a lecture [C] by Xenocrates [A] and who did not only take note of the eloquence and expertise of the lectures nor merely go back home bearing some knowledge of a beautiful subject but bearing more evident and solid fruit, namely the sudden change and amendment of his former life. Who has ever experienced a similar effect from the way we are taught?

> *faciasne quod olim*
> *Mutatus Polemon? ponas insignia morbi,*
> *Fasciolas, cubital, focalia, potus ut ille*
> *Dicitur ex collo furtim carpsisse coronas,*
> *Postquam est impransi correptus voce magistri?*

[would you do what was formerly done by Polemon on his conversion? Would you cast aside the marks of your distemper, that is your padded legs, your cushioned elbows and your cissy scarves, as he quietly ripped the garland from his drunken neck when the words of his fasting teacher chided him?][53]

[C] It seems to me that the sorts of men who are simple enough to occupy the lowest rank are the least worthy of contempt and that they show us relationships which are better ordered. The morals and the speech of the peasants I find to be more in conformity with the principles of true philosophy than those of the philosophers: '*Plus sapit vulgus, quia tantum*

53. Horace, *Satires*, II, ii, 254–8.

quantum opus est sapit.' [The common people know best: they know as much as they need to.]⁵⁴

[A]   The men whom I have judged most notable from their outward appearances (for to judge them my own way would entail seeing them more closely in a better light) are: in war and military ability, the Duc de Guise who died at Orleans and the late Marshal Strozzi; for men of ability and uncommon virtue, François Olivier and Michel de l'Hôpital, both Chancellors of France. Poetry too, it seems to me, has had its successes in our times. We have had plenty of good craftsmen in that mystery: Dorat, Beza, Buchanan, Michel de l'Hôpital, Montaureus and Turnebus. As for poets writing in French, I think that they have raised poetry as high as it ever will be and that in those qualities in which Ronsard and Du Bellay excel I find them close to the perfection of the Ancients. Adrian Turnebus knew more, and knew it better, than any man of his own day and for many a long year.   [B]   The lives of the Duke of Alva who died recently and of our own Constable Montmorency were noble ones and in some ways unusually similar in their fortunes: but the beauty and glory of the death of the Constable, in full view of Paris and of his King while serving them against his nearest kinsmen at the head of an army which he had led to victory – a victory coming suddenly as it did in his extreme old age – seems worthy to me of a place among the most notable events of our time.   [C]   So too the constant goodness, gentleness of manners and scrupulous courtesy of Monsieur de La Noue, who had been brought up amidst the injustice one finds among armed factions (a real school of treachery, inhumanity and brigandage) yet was a great and experienced warrior.⁵⁵

*

['95]   I have been delighted to declare in several places the hopes I put in my adopted daughter Marie de Gournay, who is loved by me with a more than fatherly love and included in my solitary retirement as one of the better parts of my being. She is the only person in the world I have regard

---

54. Lactantius, *Divinarum Institutiones*, III, v. Also cited by Justus Lipsius, *Politici*, V, x.

55. In his list Montaigne includes men opposed to him in war or doctrine, e.g. the Spanish Duke of Alba who fought against France, and Theodore Beza, the erotic poet who became Calvin's great successor as leader of the Reformed Church. General Anne de Montmorency died, aged seventy-four, at the Battle of St-Denis, 1567. The following passage between asterisks is suspect and may have been added to the posthumous printed editions by Marie de Gournay, the subject of its praise, who edited the *Essais*.

for. If youth is any omen her soul will be capable of great things one day –
among other things of that most perfect hallowed loving-friendship to
which (so we read) her sex has yet been unable to aspire: the purity and
solidity of her morals already suffice for this and her love for me is more
than overflowing, such, in short, as to leave nothing to desire, if only the
dread of my death (seeing that I was fifty-five when I first met her) were to
torment her less cruelly. The judgement she made on my original *Essays*,
she, a woman, in this century, so young and the only one to do so in her
part of the country, as well as the known enthusiasms of her long love for
me and her yearning to meet me simply on the strength of the esteem she
had for me before she even knew me, are particulars worthy of special consideration.

*

[A]   The other virtues are little valued nowadays, or not valued at all, but
bravery has become commonplace through our Civil Wars: where that
quality is concerned there are among us men whose souls are perfectly
unshakeable, so numerous indeed that no selection is possible. That is all
that I have come across till now of uncommon and exceptional greatness.

18. On giving the lie

[The first version of this chapter (which is indebted to the Roman satirists) insists that the self-portrait of Montaigne is destined for friends and descendants. It has been printed not for the public but because printing is more easy than copying manuscripts. The additions in [C] take a different line, as will the chapter 'On repenting' (III, 2): Montaigne insists on the moral value of his work and of telling the truth.]

[A] Yes. But somebody will tell me that my project of using myself as a subject to write about would be pardonable in exceptional, famous men who by their reputations had given us the desire to know them. That is certainly true: I admit it; I am aware that a mere craftsman will scarcely glance up from his work to look at a man of the common mould, whereas shops and work-places are emptied to look at a great and famous personality arriving in town. It is unseemly for anyone to make himself known except he who can provide some example and whose life and opinions can serve as a model. Caesar and Xenophon could firmly base their narrations on the greatness of their achievements which formed a just and solid foundation. So we can regret the loss of the diaries of Alexander the Great and of the commentaries on their own actions which Augustus, [C] Cato, [A] Sylla, Brutus and others left behind them. We love to study the faces of such men even in bronze and stone.

That rebuke is very true: but it hardly touches on me:

> *Non recito cuiquam, nisi amicis, idque rogatus,*
> *Non ubivis, coramve quibuslibet. In medio qui*
> *Scripta foro recitent, sunt multi, quique lavantes.*

[I do not read this to anyone except my friends; even then they have to ask me; I do not do so anywhere or to anyone. Some men read their works to the public in the Forum or in the baths!][1]

I am not preparing a statue to erect at a city crossroads nor in a Church or some other public place:

1. Horace, *Satires*, I, iv, 73–5; then Persius, *Satires*, V, 19–21.

[B]   *Non equidem hoc studeo, bullatis ut mihi nugis*
Pagina turgescat.
Secreti loquimur.

[I do not intend to puff up my pages with inflated trifles: we are talking in private.]

[A]   It is   [C]   for some corner of a library and as a pastime   [A]   for a neighbour,[2] a relative or a friend who will find pleasure in meeting me and frequenting me again through this portrait. Those others took heart to speak of themselves because they found their subject rich and worthwhile: I on the contrary because I find it so sterile and meagre that no suspicion of ostentation can fall upon me.   [C]   I readily make judgements on other men's actions: I give little grounds for judging mine because of their nothingness.   [B]   I do not find so much good in me that I may not tell of it without blushing.

[A]   What happiness it would afford me to hear someone giving me such an account of the manners,   [B]   the look and the expressions, the ordinary talk   [A]   and the fortunes of my forebears! How attentive I would be. It would indeed come from an evil nature if we were to despise the actual portraits of our beloved ancestors,   [C]   the style of their clothes and their armour. I preserve the escritoire, the seal, the prayer-book and a special sword which they used, and I have never banished from my own room the long canes that my father used to hold in his hands. '*Paterna vestis et annulus tanto charior est posteris, quanto erga parentes major affectus.*' [A father's clothes or ring are dearer to his descendants the more they loved him.][3]

[A]   However, if my own descendants have different tastes, I shall have the means of giving as good as I get, since when that time comes they cannot possibly have less concern for me than I will for them!   The only commerce I have with the public at large is my borrowing their printing-tools, which are more ready and convenient.[4] In exchange   [C]   I may

---

2. Until   [C]: It is *to be hidden in* some corner of a library and as a pastime for *anyone who has a private interest in knowing me*; for a neighbour . . .

3.   [A]   until   [C]: beloved ancestors *and to disdain them. A dagger, a suit of armour, a sword which served them, I preserve, out of love for them, as well as I can, from the injuries of time.* However . . .

Quotation from St Augustine, *City of God*, I, xiii.

4.   [A]   until   [C]: convenient. *I had to cast this portrait in print to free myself from the bother of making several manuscript copies. In return for this convenience which I have borrowed from the public I hope to do it the service of providing* wrapping-paper . . .

Then, Martial, *Epigrams*, XIII, i; Catullus, XCIV, 8.

provide wrapping-paper to stop some slab of butter from melting in the market:

[A]   *Ne toga cordyllis, ne penula desit olivis;*

[Lest they are short of wrappings for their tunny-fish or their olives;]

[B]   *Et laxas scombris saepe dabo tunicas.*

[And I shall often provide a loose garment to wrap up their mackerel.]

[C]   Even if nobody reads me, have I wasted my time when I have entertained myself during so many idle hours with thoughts so useful and agreeable?

Since I was modelling this portrait on myself, it was so often necessary to prepare myself and to pose so as to draw out the detail that the original has acquired more definition and has to some extent shaped itself. By portraying myself for others I have portrayed my own self within me in clearer colours than I possessed at first. I have not made my book any more than it has made me − a book of one substance with its author, proper to me and a limb of my life. Have I wasted my time by so continuously and carefully telling myself of myself? Those who merely think and talk about themselves occasionally do not examine the basics and do not go as deep as one who makes it his study, his work and his business, who with all good faith and with all his might binds himself to keeping a long-term account. The most delightful of pleasures are inwardly digested: they refuse to leave their spoor behind and refuse to be seen not only by the many but even by one other. How frequently has this task diverted me from painful thoughts! And all trivial thoughts should be counted as painful. Nature has vouchsafed us a great talent for keeping ourselves occupied when alone and often summons us to do so in order to teach us that we do owe a part of ourselves to society but that the best part we owe to ourselves. With the aims of teaching my mental faculty even to rave with some order and direction and so as to stop it losing its way and wandering in the wind, I need simply to give it body and to keep detailed accounts of my petty thoughts as they occur to me. How often when I have been irritated by some action which politeness and prudence forbid me from openly censuring have I unburdened myself here − not without the design of giving a public reproof.[5] And, indeed, those scourgings by the poet −

5. Cf. Joachim Du Bellay's reasons for writing personal poetry (*Regrets*, 4, 14, etc.). Then, Clément Marot, *Epistre de Fripelipes* against Sagon, punning on his name Sagon (*sagouin*, lout).

Zon dessus l'euil, zon sur le groin
Zon sur le dos du Sagoin.

[Bong in the eye, bong on the snout,
Bong on the back of Sagon the Lout.]

are even better when imprinted on paper than on the living flesh.

And what if I now lend a more attentive ear to the books I read, being on the lookout to see whether I can thieve something with which to decorate and support my own? I have never studied so as to write a book, but I have done some study because I have written one, if studying a little means lightly touching this author or that and tweaking his head or his foot – not so as to shape my opinions but, long after they have taken shape, to help them, to back them up and to serve them.

[A]   But during a time so debased, what man are we to trust when he speaks of himself, seeing there are few, perhaps none, whom we can trust when they speak of others, where they have less to gain from lying? The first sign of corrupt morals is the banishing of truth: for as Pindar says, being truthful is the beginning of any great virtue,   [C]   and it is the first item that Plato required in the governor of his Republic.[6]   [A]   Truth for us nowadays is not what is, but what others can be brought to accept: just as we call money not only legal tender but any counterfeit coins in circulation. Our nation has long been accused of this vice: Salvianus of Massilia, who lived in the time of the Emperor Valentinian, says that lying and perjury are not a vice for the French but a figure of speech![7] If you wanted to outbid that testimony you could say that at the present time it is for them a virtue. People train themselves for it and practise for it as for some honoured pursuit: dissimulation is one of the most striking characteristics of our age. So I have often reflected on what could have given birth to our scrupulously observed custom of taking bitter offence when we are accused of that vice which is more commonplace among us than any of the others, and why for us it should be the ultimate verbal insult to accuse us of lying. Whereupon I find it natural for us to protect ourselves from those failings with which we are most sullied. It seems that by resenting the accusation and growing angry about it we unload some of the guilt; we are guilty, in fact, but at least we condemn it for show.

6. Pindar, in Plutarch, *Life of Marius*; Plato, *Republic*, VI, 489e ff.
7. Presbyter Salvianus of Massilia, *De gubernatione Dei*, I, i, xiv (a work printed in Paris in 1580).

[B] Could it not also be because this accusation seems to imply cowardice and faintness of heart? Is anything more expressly cowardly than to deny one's word – nay, to deny what we ourselves know to be so?

[A] Lying is a villein's vice, a vice which an Ancient paints full shamefully when he says that it gives testimony to contempt for God together with fear of men.[8] It is not possible to show more richly the horror of it, its vileness and its disorderliness. For what can one imagine more serf-like than to be cowardly before men and defiant towards God? Our understanding is conducted solely by means of the word: anyone who falsifies it betrays public society. It is the only tool by which we communicate our wishes and our thoughts; it is our soul's interpreter: if we lack that, we can no longer hold together; we can no longer know each other. When words deceive us, it breaks all intercourse and loosens the bonds of our polity.

[B] Certain peoples of the new-found Indies (and there is no point in emphasizing their names which are no more, since – an amazing example, the like of which has never been heard – the utter devastation of that Conquest extended even to the total destruction of names and of all ancient knowledge of places) used to offer to their gods human blood, drawn exclusively from their ears and tongue, in expiation of the sin of both hearing and of telling lies.[9]

[A] That jolly fellow from Greece declared that boys play with knuckle-bones and men play with words.[10]

As for our various conventions for giving the lie, the laws of honour governing them and the changes they have undergone, I will put off saying what I have to say about that to some other time; meanwhile I will find out if I can from what period dates our custom of exactly weighing and measuring words and making that a question of honour. For it is easy to see that it was not like that in Ancient times among the Greeks and Romans. It often seemed strange and new to me to watch them giving each other the lie and insulting each other without it starting a brawl. Their laws of duty took some other road than ours. Caesar was variously called a thief and a drunkard to his very beard.[11] We can see the freedom

8. Plutarch, *Life of Lysander*.

9. Lopez de Gomara (tr. Fumée), *Histoire generale des Indes*, II, xxviii. (These new-found 'Indies' are the Americas.)

10. Androclidas criticizing Lysander, in Plutarch's *Life of Lysander*.

11. Cf. II, 33, 'The tale of Spurina', and also the insulting by Christians of the Emperor Julian in II, 19, 'On freedom of conscience'.

of invective which they used against each other (and I mean by *they* the greatest war-leaders in both those nations) where words were avenged by words alone, with no further consequence.

19. On freedom of conscience

[Freedom of conscience – freedom of worship and association granted to a rival sect of Christians claiming to be the one true Church – was a new idea, only reluctantly accepted by the Kings of France (or, indeed, of England). Montaigne regards it as a pis-aller, forced on the government by the condition of France, exhausted by the Wars of Religion. Montaigne's concern to present fairly the anti-Christian Emperor Julian the Apostate (which raised some eyebrows in the Vatican) shows how we can be just even to enemies of our religion. In fact Montaigne's judgement is that of the Christian poet Prudentius, whose childhood was spent under Julian. This chapter continues the reflections of the previous one on the great Ancients' indifference to invective. It ends with a quip borrowed from Montaigne's favourite writer of Latin comedies, Terence. It had long been going the rounds in a Pasquinade; here it applies to the stalemate which led to the proclamation of Henry III and of Catherine de' Medici in 1576, tolerating the Huguenots, except in Paris – since they could not be crushed.]

[A] It is quite normal to see good intentions, when not carried out with moderation, urging men to actions which are truly vicious. In the present quarrel which is driving France to distraction with its civil wars, the better and more wholesome party is certainly the one upholding the religion and constitution of our country. Now among the men of honour who support it (for I am not talking about people who use it as a pretext for settling private scores, satisfying their greed or courting the favour of princes but about those who support it out of true zeal for their religion and a sacred desire to defend the peace and good estate of their homeland) even among such men as these you can find many who, once passion drives them beyond the bounds of reason, take decisions which are unjust, violent and rash.

It is certain that, in those early days when our religion began to be backed by the authority of law, zeal provided many with weapons to use against all sorts of pagan books, causing the learned public to suffer staggering losses. I reckon that this inordinate zeal caused more harm to literature than all the fires started by the Barbarians.

Cornelius Tacitus can bear witness to this. His kinsman the Emperor Tacitus expressly commanded all the libraries of the world to be furnished with copies of his *Histories*, yet not a single one of them wholly escaped the

meticulous search of those who sought to destroy them simply because they contain five or six wretched sentences hostile to our religion.[1]

They went further, heaping false praise upon all the Emperors who favoured us and completely condemning all the actions of our adversaries. That can readily be seen from the case of the Emperor Julian, dubbed the Apostate. He was a truly great and outstanding person, appropriate enough for a man whose mind was steeped in philosophical argument by which he claimed to order all his activities. And indeed he left behind examples of model behaviour in every single field of virtue.

As for his chastity, his whole life affords clear testimony of it. A similar characteristic is ascribed to him as to Alexander and Scipio: he did not even want to look at any of the many beautiful women he captured. And that was in the flower of his manhood, as when the Parthians killed him he was only thirty-one.[2]

As for justice, he took care to hear the contending parties himself. He was curious about what religion was professed by those who appeared before him and asked them about it; yet the hatred he bore against our own never turned the scales of his justice. He personally enacted several good laws and severely pruned the taxes and imposts raised by his predecessors.

We have two good historians who were eye-witnesses of his actions. One of them, Ammianus Marcellinus, bitterly reproaches him several times in his *History* for barring Christian rhetoricians and grammarians from the institutes of learning and forbidding them to teach. Marcellinus said that he could wish that deed were buried in silence. It is probable that if Julian had done anything harsher against us Marcellinus would not have overlooked it, since he was well disposed towards our side.

Julian was an enemy harsh towards us, it is true, but not cruel. Even our own side tell the following tale about him: when he was walking one day near the town of Chalcedon the local bishop, Maris, dared to rail at him as a traitor to Christ. He simply replied, 'Go away, you wretched man, and lament the loss of your eyesight!' The bishop retorted: 'I thank Jesus Christ for having taken away my sight; it stops me seeing your insolent face!'

---

1. Only two manuscripts of part of Tacitus' *Annals* survived. Montaigne read Tacitus through at one go (III, 8, 'On the art of conversation') and had studied the *Commentaria* on Tacitus of Justus Lipsius. For some, Tacitus was very much to be condemned since in his account of Nero's persecution (*Annals*, XV, 44) he refers to Christianity as 'a pernicious superstition'.
2. Ammianus Marcellinus, XXXV, iv.

Julian, so they say, was simply acting the patient philosopher.

In any case what he did then cannot be squared with the cruelties he is said to have used against us. According to Eutropius, my other witness, he was an enemy of Christianity but without shedding blood.

To return to his justice: the only reproach to be made against it is the severe treatment he meted out at the beginning of his reign to those who supported the party of Constantius, his predecessor.

As for sobriety, he always lived a soldierly life. Even in times of total peace he dined as though he were in training and accustoming himself to the austerities of war. He was so watchful that he divided the night into three or four parts, giving only the smallest of them over to sleep. The remainder he devoted either to checking up in person on his army and his Imperial guard, or else to study. Among his other rare qualities he greatly excelled in all branches of literature.

It is said of Alexander the Great that when he lay down for a rest he kept sleep from debauching his thinking and studies by having a basin placed beside him; he then held one of his hands outside the couch clasping a little copper ball. If he fell asleep his fingers let go of the ball which clanged into the basin and woke him up.[3] Julian's mind was so intent on what he was about and (thanks to his exceptional abstemiousness) so unclouded, that he could do without such tricks.

As for his competence in military matters, he was astonishingly endowed with all the requisites of a great general. He spent most of his time engaged in fighting, mostly together with us here in France against the Germans and the Franks.

There is hardly a man on record who experienced more danger or who risked his own life more often. His death was something like that of Epaminondas, since he was struck by a dart and tried to pull it out.[4] He would have done so, only the edge was sharp, cutting his hand and weakening his grasp. He kept insisting that he be carried as he was into the thick of the fray to encourage his soldiers. Even without him they fought that battle most courageously until nightfall parted the armies.

To philosophy he owed his remarkable contempt for his own life and for all things human. He firmly believed in the immortality of the soul.

In matters of religion he was altogether vicious.[5] He was named the Apostate for having abandoned ours, but the most likely opinion seems

---

3. Ibid., XVI, v.
4. Ibid., XXV, iii. (Epaminondas was a model hero for Montaigne.)
5. Cf. Prudentius, *Apotheosis*, 448–53.

to me to be that he never took it to his heart, merely pretending to do so and obeying the law until he had the Empire under his thumb. In his own religion he was so superstitious that even his contemporaries laughed at him: they said that if he had managed to gain victory over the Parthians his sacrifices would have exhausted the world's entire stock of bulls!

He was besotted with the art of divination, lending his authority to every sort of augury. As he lay dying he said, among other things, that he was grateful to the gods for not wanting death to take him by surprise (having long since warned him of the place and time of his end) and for not giving him a soft relaxed death more suitable for idle delicate people, nor yet a death which was long, languishing and painful; he thanked them for having found him worthy of dying in that noble fashion, in the flush of his victories and the flower of his glory. He had a vision such as that of Marcus Brutus: it first came to threaten him in Gaul and appeared to him again in Persia when he was on the point of dying.

[C] These words have been attributed to him as he was struck down: 'Thou hast conquered, Nazarean!' or sometimes, 'Be satisfied, Nazarean!'[6] But if my authorities had believed that, they would not have overlooked them: they were present in his army and noted the slightest of his final words and gestures. Nor would they have overlooked certain miracles now associated with his death.

[A]   To get back to the theme of my subject: Marcellinus says that Julian had long nursed paganism in his heart but dared not disclose this fact, since his army was made up of Christians. When at last he found himself strong enough to dare to proclaim his intentions, he ordered the temples of the gods to be reopened and he assayed every means of restoring the worship of idols.

Finding the laity of Constantinople torn apart and the bishops of the Christian Church divided, to achieve his purposes he made them appear before him in his palace, warned them to damp down the civil strife at once and commanded that every person, without let or fear, should follow his own religion.[7]

He urged his case strongly, hoping that the licence he gave them would increase their divisions and schismatic plottings, so preventing the people

6. Theodoret, Bishop of Cyprus, for the first version, Zonaras for the second. Montaigne's authorities who witnessed Julian's death are Ammianus Marcellinus and Eutropius.
7. Ammianus Marcellinus, XXI, v.

from uniting together and strengthening their resistance to him by their harmony and unanimity. He had assayed from his experience with some of the Christians that no beast in the world is more to be feared by Man than Man.[8]

Those are approximately his very words.

It is worth considering that, in order to stir up the flames of civil strife, the Emperor Julian exploited the self-same remedy of freedom of conscience which our kings now employ to stifle them.

On the one side you could say that to slacken the reins and allow the parties to hold on to their opinions is the way to sow dissension broadcast: it is all but equivalent to lending a hand to increase it, since there is no obstacle to bar its course and no legal constraint to rein it back.

For the other side you could say that to slacken the reins and allow the parties to hold on to their opinions is to soften and weaken them by ease and laxity; it blunts the goad, whereas rareness, novelty and difficulty sharpen it.

Yet for the honour and piety of our kings I prefer to believe that, since they could not do what they wished, they pretended to wish to do what they could.[9]

8. Cf. Erasmus, *Adages*, I, I, LXX, *Homo homini lupus*.
9. A line from Terence (*Andria*, II.i.6–7), satirically applied long before Montaigne to the King of France acting under compulsion, e.g. 'Pasquillus on the King of France compelled to make peace: *Quoniam id fieri quod vis non potest, velis id quod possis*' ['Since you cannot do what you wish, wish what you can'], in *Pasquillus novus Terentianus*, 1546 (no place of printing).

20. We can savour nothing pure

[A chapter particularly interesting for the light it throws on melancholy. Some of Montaigne's quotations derive directly from Justus Lipsius.]

[A] The feebleness of our condition means that we can make habitual use of nothing in its natural unsophisticated purity. The very elements which we enjoy are corrupt: so too are the metals – even gold must be alloyed with some other substance to make it serviceable to us. [C] Nor could the simple virtue which Ariston and Pyrrho, and the Stoics too, taught as the aim of our life serve that end without some admixture, any more than the hedonism of Aristippus and the Cyrenaics. [A] Of the pleasures and goods which we enjoy, not one is exempt from being compounded with some evil and injury.

> [B]   *medio de fonte leporum*
> *Surgit amari aliquid, quod in ipsis floribus angat.*

[from the very fount of our delights there surges something bitter which gives us distress even among the flowers.][1]

The greatest of our pleasures has an air of groaning and lamentation. Could you not say that it was languishing from affliction? Indeed when we forge images of it at its highest reach we paint its face with sickly epithets and dolorous qualities: languor, faintness, weakness, debility, *morbidezza*,[2] which greatly witnesses to their common blood and consubstantiality. [C] Deep joy has more gravity than gaiety; the highest and fullest happiness, more calm than playfulness. *'Ipsa fœlicitas, se nisi temperat, premit.'* [Even joy overwhelms us, unless it be tempered.][3] Ease crushes us.   [A] That is what is meant by that line of ancient Greek poetry: 'The gods sell us all the

---

1. Lucretius, IV, 1130–1. (The ensuing 'greatest of our pleasures' is sexual intercourse.)
2. Italian word meaning delicate flesh-tints. Montaigne sees in it the Latin word *morbidus* (disease, unwholesome).
3. Seneca, *Epist. moral.*, LXXIV, 18. (The following translated Greek verse appears in John Stobaeus' *Sententiae*.)

pleasures which they give us'; that is to say, none that they give us is pure and perfect: we can only buy them at the price of some suffering. [C] Pleasure and travail, so unlike in their natures, are yet fellows by some inexplicable natural relationship. Socrates said that some god or other made an assay at fusing pain and pleasure into one mass: when he could not achieve this he decided at least to couple them by their tails.[4]

[B] Metrodorus said that sadness was not unalloyed with a certain pleasure. I do not know whether he meant something else: personally I can readily think that there is an element of purpose, consent and complacency in feeding oneself on melancholy – I mean, quite apart from ambition, which can also be mixed up with it. There is some hint as of delicate sweetmeats which smiles at us and flatters us in the very bosom of melancholy. Are there not some complexions which make it their only food?

> *Est quædam flere voluptas.*

> [There is a certain pleasure in our tears.]

[C] And Attalus says in Seneca that the memory of those loved ones we have lost tastes pleasant, like the bitterness of very old wine:

> *Minister vetuli, puer, falerni,*
> *Ingere mi calices amariores!*

> [Butler serving Falernian wine! Pour me out your bitterest cups!]

– like those apples which are both sharp and sweet.[5]

[B] Nature reveals this alloy to us; painters hold that the same wrinkling movements of our faces which serve to show weeping also show laughter. Indeed. Watch the picture in progress before either emotion has been finally delineated: you are in doubt towards which it is tending. And the extremes of laughter are mixed with tears.

[C] '*Nullum sine auctoramento malum est.*' [There is no evil without its compensations.][6] When I picture a man besieged by all the enjoyments which he could desire – say that all his members were forever seized of a

---

4. Plato, *Phaedo*, 60B; then Seneca, *Epist. moral.*, XCIX, 25 (Metrodorus, cited with disapproval by Seneca.)

5. Ovid, *Tristia*, IV, iii, 27; Catullus, XXVII, 1–2; Seneca, *Epist. moral.*, LXIII, 5, citing his Stoic teacher Attalus.

6. Seneca, *Epist. moral.*, LXIX, 4.

pleasure equal to that of sexual intercourse at its climax – I see him collapsing under the weight of his joy; and I can perceive him quite incapable of bearing pleasure so pure, so constant and so total: truly, once there, he runs away and naturally hastens to escape from it as from some narrow passage where he cannot find solid ground and fears to be engulfed.

[B] When I scrupulously make my confession to myself I find that the best of the goodness in me has some vicious stain. And I am afraid that Plato, even in his most flourishing virtue – (and I say this who am the most genuine and loyal admirer of it, as of all virtues of similar stamp) if he had put his ear close to it   [C]   (and he did put his ear close to it)   [B] – he would have heard in it some sinister sound of a human alloy, even though it were a muffled sound which only he could detect. Man, totally and throughout, is but patches and many-coloured oddments.

[A] The very laws of justice cannot subsist without some admixture of injustice; and Plato says that those who claim to remove all the improprieties and inconsistencies from the laws are undertaking to cut off the Hydra's head.[7] Tacitus says: '*Omne magnum exemplum habet aliquid ex iniquo, quod contra singulos utilitate publica rependitur.*' [Every case of exemplary punishment is unfair to individuals: that is counterbalanced by the public good.]

[B] It is likewise true that for the usages of the life and service of the common weal there can be an excess of purity and discernment in our wits; such penetrating clarity has too much subtleness and inquisitiveness. We must weigh down our wits and blunt their edge to render them more obedient to precedent and practice; we must coarsen them and darken them to give them the proportions of this earthy darksome life. That is why the more commonplace and less tense of wits are more appropriate to the conduct of affairs and more successful. The high inquisitive opinions of philosophy prove unsuited in practice. Such sharp vigour of soul and such supple restless whirring motions trouble our negotiations. We must manage the affairs of men more rough-and-readily, more superficially, leaving a good and better share to the rights of Fortune. There is no need to cast light so deeply and keenly on to our affairs. You lose yourself in them by contemplating so much varied brilliance and such diverse forms: [C] '*Voluntantibus res inter se pugnantes obtorpuerant animi.*' [Minds wallowing in mutual contradictions are benumbed.][8]

---

7. Erasmus, *Adages*, I, X, IX, *Hydram secus* (citing Plato, *Republic*, IV, 426E–427A); then Tacitus, *Annals*, XIV, xliv.
8. Livy, XXXII, xx.

That is what the Ancients said of Simonides. When King Hiero posed him a question to answer which he had several days to meditate upon, his powers of thought presented him with so many keen and subtle considerations that, doubting which was the most likely, he totally despaired of the truth.[9]

[B] He who seeks out all the circumstances and grasps their consequences impedes his choice. A modest talent suffices and can equally well carry into execution matters of great and little weight. Note how those who best manage their estates are the least able to explain how they do so, while the most skilful talkers are as often as not useless at it. I know one man who is excellent at talking about all kinds of estate-management and at describing it but who has let a hundred thousand pounds of income slip through his fingers. I know another who speaks and deliberates better than any man in his council-chamber; never in the world was there a more beautiful display of intelligence and of competence: yet when it comes to practice his servants find he is quite other than that – I mean, even leaving aside bad luck.

9. Condensed from Cicero, *De nat. deorum*, I, xxii, 60, on 'What is the Being and Nature of God?'

21. Against indolence

[*This chapter reveals Montaigne's resentment as a soldier towards princes who seek honour from battles in which they had not fought and conquests they had not led. Montaigne finds his heroes in Ancient times and in monarchs other than French.*]

[A]   When the Emperor Vespasian was ill of the illness which killed him, he never ceased to want to learn of the condition of the Empire; from his very bed he ceaselessly dealt with many matters of consequence. His doctor chid him for it as something harmful to his health. 'An Emperor,' he replied, 'should die on his feet.'[1] There you have a fine epigram, in my opinion one worthy of a great ruler. The Emperor Hadrian later used the same expression. And we ought often to remind kings of it to make them realize that the great charge entrusted to them is no idle one and that there is nothing which can make a subject more rightly lose his taste for exposing himself to trouble and danger in the service of his prince than to see him meanwhile indolently engaged in occupations base and frivolous, nor lose his concern for his protection than to see him indifferent to ours.

[C]   Should anyone wish to maintain that it be better for a prince to conduct his wars through someone other than himself, Fortune will furnish him with plenty of examples of princes whose lieutenants have successfully concluded great campaigns and also of others whose presence would have been more harmful than useful. But no virtuous and courageous prince could tolerate being given such shameful counsels. Under colour of saving his head for the well-being of the state, as though he were some plaster saint, they specifically demote him from his *imperium*, which consists entirely in military activity, and declare him incapable of it. I know one prince who would rather be worsted in battle than allow others to fight for him while he slept, and who never saw without some envy even his own followers achieve anything great in his absence.[2] Selim I was very right, it seems to me, to say that no victories won in the leader's absence are ever

1. Erasmus, *Apophthegmata*, VI, *Vespasianus Pater*, XVII.
2. Doubtless Henry IV.

unqualified. And he would have all the more readily maintained that the leader ought to blush with shame to claim a part in them for his own renown when he had contributed nothing to the task but his voice and his thinking – not even that, seeing that in tasks such as these the counsel and commands which bring men their glory are exclusively those which are given on the spot in the midst of the action. No pilot can perform his duty on dry land.

The Princes of the Ottoman nation (the first nation in the world in the fortunes of war) have enthusiastically embraced this opinion. Bajazet II and his son who departed from it and spent their time on erudition and other indoor occupations dealt severe blows to their Empire. And the one who reigns at present, Amurath III, by following their examples has made a good start at proving the same. Was it not Edward III, King of England, who made this quip about our Charles V: 'Never was there king who less donned his armour. Yet never was there king who gave me more trouble!' And he was right to find that strange, being the result of luck rather than of reason.[3] And let those seek supporters other than me who want to number the Kings of Castile and Portugal among the great-souled conquerors in war because, at twelve hundred leagues from their idle dwellings, they made themselves masters (risking the skin of their factors) of both the Indies; we still do not know whether they would simply have had courage enough to go and take possession of them in person.

[A] The Emperor Julian went further: a philosopher and a gallant man should not pause for breath – meaning that they ought to concede nothing to their bodily necessities except what could not be denied them, since they are ever keeping both body and soul occupied in great and beautiful deeds of virtue. He was ashamed to be seen even spitting or sweating in public (as has also been said of the youth of Sparta, and by Xenophon of the youth of Persia) because he reckoned that exercise, continuous toil and sober living ought to have burnt dry all such excess fluids. What Seneca said would not fit badly here: the Ancient Romans kept their youth on their toes, teaching their boys nothing which had to be learned sitting down.[4]

[C] It is a noble desire that even one's death should be manly and useful: but that action lies not in our sound resolve but in our good

3. Details from the anonymous *Tesoro politico per li disegni de Principe* and from Froissart.
4. Johannes Zonaras; Xenophon, *Cyropaedia*, I, ii; Seneca, *Epist. moral.*, LXXXVIII, 19.

fortune. Hundreds who have intended to win, or to die fighting, have failed at both, wounds or prisons blocking this design by granting them a compulsory existence. There are maladies which strike to the ground our wishes and our consciousness.

['95] Fortune did not feel obliged to favour the vanity of the Roman legions who bound themselves by oath to vanquish or to die: '*Victor, Marce Fabi, revertar ex acie: si fallo, Jovem Patrem Gradivumque Martem aliosque iratos invoco Deos.*' [I shall, O Marcus Fabius, return victorious from the line of battle. If I fail, I invoke the anger of Father Jove, of Marching Mars and the other gods.] The Portuguese tell how in one place during their conquest of the Indies they confronted warriors who had doomed themselves with horrifying oaths to accept no terms other than death or victory; as a sign of that vow they had cropped their heads and shaved off their beards.[5]

It is no use stubborning it out and taking risks: it seems that blows avoid those who gaily expose themselves to them and never willingly land on those who too willingly face them and thus spoil their intention.

Many a man, unable to manage to get killed by the might of the enemy, despite assaying every way to keep his vow to return with victory or not at all, has been constrained to kill himself in the very heat of battle. There are other examples, but here is one: Philistus, the commander of the navy of the younger Dionysius against the men of Syracuse, engaged the enemy, the battle being bitterly contested since their forces were equal. He got the upper hand at first because of his daring; but the Syracusans took up position round his galley and besieged it: he personally did many gallant deeds-of-arms in an attempt to break through, then, despairing of any other solution, took his own life which he had so freely – and so unsuccessfully – exposed to the hands of his enemies.[6]

[C] Muley Moloch, the King of Fez who has just beaten Sebastian, King of Portugal, on that famous day which saw the death of three kings and the transfer of that mighty crown to the crown of Castile, was already gravely ill when the Portuguese invaded his territory. Thereafter he daily declined nearer and nearer to a death which he clearly foresaw. Never did a man exert himself more energetically nor with greater glory. He realized he was too weak to stand the ceremonial entry into his camp – an entry which is traditionally full of magnificence and chock-full of action – so he

---

5. Livy, II, xlv; Simon Goulart, *Histoire du Portugal*, V, vii.
6. Diodorus Siculus, *Philistus*; the historian–admiral killed himself after being defeated by Dion (356 BC).

surrendered that honour to his brother. But it was also the only duty of a Commander that he did give up: all the other duties, the necessary and the useful ones, he carried out most rigorously and most punctiliously; he allowed his body to lie down but he kept his mind on its toes and his heart firm until he drew his last breath – indeed a little beyond. He would have been able to sap the strength of his enemies who had imprudently thrust deep into his lands, so it grieved him terribly that, for lack of a little more life and also for the lack of anyone who could take his place in waging that war and guiding his troubled kingdom, he had to go in search of a hazardous bloody victory when a clear and certain one was within his grasp. However he made a wonderful use of his remaining time: he led the enemy to exhaust himself by drawing them far from their navy and the maritime fortresses which they had established on the African coast; that he did until his last day of life which he had kept as a reserve force to cast into that battle. Deploying his troops in a ring he invested the Portuguese army on all sides, closing the circle and squeezing them tight; the fighting was very bitter because of the valour of the young king of the invaders, but the enemy were hampered by having to face attacks from all directions and were unable to flee after they were routed; they were therefore constrained to charge into their own ranks (*'coarcervanturque non solum cæde, sed etiam fuga'*) [the dead lay in heaps not only from the slaughter but from the retreat], men, pile upon pile, furnishing the victors with a total and murderous victory. As he was dying he was carried about from place to place wherever need called him; passing through the ranks he exhorted his officers and men one after another. But when the enemy broke through his troops in one sector he could not be dissuaded from mounting his saddle, sword in hand. He strove to join the affray; his men stopped him by clinging to his bridle, to his clothes and his stirrups; but the effort finally overwhelmed what little life he had left. He was laid down again. When all his other faculties were failing he started out of his swoon to warn that his death must be kept quiet – which was indeed the most necessary order which he still had to give – so that news of it should not arouse despair among his troops; he then died, holding his finger to his sealed lips (the common gesture meaning, Keep quiet).[7]

What man has ever lived so far and so deep into his own death? What man ever died more on his feet!

The ultimate degree of treating death courageously, and the most natural

7. Jeronimo Conestaggio, *Dell'unione del regno di Portogallo alla . . . Castiglia* (Genoa, 1585); interpolated quotation from Livy, II, iv.

one, is to face it not only without amazement but without worry, extending the ordinary course of your life right into death. As Cato did, who spent his time in sleep and study while keeping present in his head and heart that violent bloody death and holding it in his palm.[8]

8. Cf. I, 37, 'On Cato the Younger' (of Utica). Plutarch's *Life* is the main source.

22. On riding 'in post'

=====

[*From the early sixteenth century, generals and statesmen 'laid posts' (at first temporary ones but later permanent ones) along 'post-routes'. At each post-stage horses were kept and it was the duty of the post-master or courier to ride at all speed to the next post with the dispatches (or 'post'). 'To ride in post' meant to ride literally as such a postman or else at express speed as a sport, normally in relay-races. It was this sport which Montaigne used to be good at. The military, political and financial advantages of rapid communication also led to the reintroduction of carrier-pigeons into Europe: Rabelais used this fact in his* Sciomachie *of 1549 to explain otherwise miraculously rapid spreadings of news, especially between bankers. By Montaigne's time they were more usual but still a source of curiosity.*]

[B] I have not been one of the weakest at this sport, which is suited to men of my stocky short build; but I am giving up such business: it makes too great an assay of our strength to keep it up for long.

[A] I was reading just now that King Cyrus, in order to facilitate the reception of news from all parts of his very wide Empire, found out how far horses could get in a day at one stretch, and then at such distances stationed men with responsibility for holding horses in readiness to furnish to those who were travelling to see him.[1]   [C] Some maintain that the speed of such journeys is that of cranes in flight.

[A] Caesar said that when Lucius Vibulus Rufus was hurrying to bring a warning to Pompey, he remained night and day on the road, changing horses so as to travel more swiftly. And according to Suetonius, Caesar himself covered a hundred miles a day in a hired chariot. But he was a mad courier! For whenever rivers cut across his road he swam across them,   [C] turning off neither left nor right to search for a bridge or a ford.   [A] When Tiberius Nero went to see his brother Drusus who was ill in Germany, he covered two hundred miles in twenty-four hours in three chariots.

[C] Livy says that during the war which the Romans fought against King Antiochus, Titus Sempronius Gracchus '*per dispositos equos prope*

---

1. Xenophon, *Cyropaedia*, VIII, vi, 17–18.

incredibili celeritate ab Amphissa tertio die Pellam pervenit' [using relays of horses travelled on the third day with almost unbelievable speed from Amphissa to Pella]. And if you look at the context it is clear that it refers to permanent posts, not ones newly established for that ride.[2]

[B] Even faster was Caecinus' new way of sending news to those at home: he took swallows with him which he released to fly back to their nests whenever he wished to send his news home, staining them with the coloured mark appropriate to his message according to a code which he had agreed on with his family.

In the Roman theatres the paterfamilias kept pigeons in the breast of his toga and attached messages to them whenever he wished to ask those at home to do something for him; they were moreover trained to bring back the answers. Decimus Brutus made use of them when under siege at Mutina; others have done so elsewhere.[3]

In Peru the couriers rode on men who bore them in litters on their shoulders with such agility that the first porters relayed their burden to the next team at the run without missing a step.[4]

[C] I have been given to understand that the Wallachians, the couriers of the Grand Seigneur, make the fastest speeds of all, since they have the right to force anyone whom they meet travelling on their road to dismount and to exchange his horse for their exhausted one, and also because they wear a tight broad band round their waists to stop them from tiring,[5] [95] as quite a few others do. I have found no relief in this method.

2. Julius Caesar, *De Bello Gallico*, III, iii; Suetonius, *Caesar*, LVII; Pliny, *Hist. nat.*, VII, xx; Livy, XXXVII, vii.
3. Pliny, *Hist. nat.*, X, XIV, xxxviii.
4. Lopes de Gomara, *Hist. des Indes*, V, vii.
5. Nicolas Chalcocondylas (tr. Vigenère), *Histoire de la décadence de l'Empire Grec et établissement de celuy des Turcs*, Paris, 1585.

23. On bad means to a good end

[Reflections on the health and sickness of States, mainly arising from reading Jean Bodin.]

[A]   Throughout the whole system governing the works of Nature there can be found an amazing analogy and correspondence which shows that it is neither fortuitous nor controlled by a variety of Masters. The maladies and the characteristics of our bodies can also be found in States and polities; like us, kingdoms and republics are born, flourish and fade into decrepitude. We are subject to a surfeit of humours which serves no purpose and is harmful. The humours themselves may be good (and the doctors fear them particularly: they say that since nothing within us remains stable, health when perfect can be too positive and vigorous and should be tamed and diminished by the Art of medicine for fear that our nature, being unable to remain fixed in any one place yet having no possibility of further improvement, should suddenly collapse in disorder: that is why they prescribe for athletes purgations and bleedings so as to draw off that superabundance of health); the humours may be also bad, which is the usual cause of illness.

Ailing political systems may often show a similar surfeit, and various sorts of purges are normally used for it. Sometimes, to take the load off the country, a great multitude of families are given leave to seek better conditions elsewhere, to some other nation's detriment. It was in this way that our ancient Franks left the depths of Germany and came and took over Gaul, driving out the original inhabitants. Thus too were formed those huge waves of humanity which poured into Italy under Brennus and others; so too the Goths and the Vandals, like the peoples who at present hold Greece, abandoned their native lands to settle more spaciously elsewhere. There are scarcely two or three corners in the world which have not experienced such migrations.

That was the way the Romans built their colonies: when they thought that their City was becoming excessively big they relieved it of the people they needed least, sending them off to inhabit and farm the lands which they had conquered. And sometimes they deliberately kept up wars with

some of their enemies, not only to keep their men in training, fearing that idleness the Mother of decadence might bring some worse trouble upon them –

[B]   *Et patimur longæ pacis mala; sævior armis,*
Luxuria incumbit

[We are suffering the ills of a prolonged peace: luxury, more savage than war, is crushing us][1]

– [A]   but also to serve as a good phlebotomy for the Republic and to ventilate a little of the excessively mind-stirring heat of their young men, pruning and pollarding the branches of a stock growing rampant from too much energy. They sometimes used their war against the Carthaginians for this purpose.

King Edward III of England would not include in the general peace established with our French King at the Treaty of Bretigny their quarrel over the Duchy of Brittany: he wanted somewhere to unload his fighting-men and to dissuade the multitude of Englishmen who had served him across the Channel from pouring back into England.

One of the reasons why King Philip agreed to dispatch his son Jean to the wars in Outremer was that he could take with him a large number of the hot-blooded young men to be found in his army.

There are many today who use similar arguments, wishing that the heat of the civil commotions among us could be diverted into some war against our neighbours, fearing that those aberrant humours which now dominate the body politic would, if not decanted elsewhere, continue to maintain our troubles at fever-pitch, finally entailing our complete collapse. And indeed a foreign war is a distemper much less harsh than a civil war: but I do not believe that God would look favourably on so wicked an enterprise as our attacking and quarrelling with a neighbour simply for our own convenience.

[B]   *Nil mihi tam valde placeat, Rhamnusia virgo,*
Quod temere invitis suscipiatur heris.

[O Nemesis, ye Rhamnusian Virgin, grant that I may desire nothing so much that I should wrench it from its rightful owner.][2]

[A]   Yet so wretched is our condition that we are often driven to the

---

1. Juvenal, *Satires*, VI, 291–2.
2. Catullus, LXVIII, 77–8.

necessity of using evil means to a good end. Lycurgus, the most virtuous and perfect lawgiver there ever was, introduced a most iniquitous way of training his Spartan citizens in temperance: he compelled their slaves the Helots to get drunk so that the Spartans should see them lost and wallowing in their wine and so hold the excesses of that vice in horror.[3]

Even more wrong were those who in Ancient times permitted that criminals who had been condemned to any kind of death might be cut up alive by doctors so as to reveal our inner organs in their natural state and so establish greater certainty in their Art; for if we really must indulge in depravity, we are more to be excused if we do so for the good of the soul than for the good of the body: as did the Romans who trained their citizens in valour and in contempt for death and danger by those frenzied spectacles of gladiators and swordsmen who fought to the death, hacking at each other and killing each other while they looked on:

> [B] *Quid vesani aliud sibi vult ars impia ludi,*
> *Quid mortes juvenum, quid sanguine pasta voluptas?*

[For what else can be meant by those mindless impious shows, by those slaughterings of young men and that pleasure fed on blood?]

Such slaughter lasted until the time of the Emperor Theodosius:

> *Arripe dilatam tua, dux, in tempora famam,*
> *Quodque patris superest, successor laudis habeto.*
> *Nullus in urbe cadat cujus sit pæna voluptas.*
> *Jam solis contenta feris, infamis arena*
> *Nulla cruentatis homicidia ludat in armis.*

[O thou, our Leader, succeed to your father's glory and grasp such honour as he set aside for our times . . . Let no man be any longer killed in Rome to provide entertainment . . . Let the infamous arena be content with wild beasts alone and no more make a sport of murder wrought with blood-stained weapons.]

[A] It was indeed a wonderful and very fruitful example for training the people that they should have every day before their eyes a hundred, two hundred or even a thousand pairs of men bearing arms, hacking each other to pieces with such extreme strength of courage that never was heard a single word of weakness or of pity, never a back was turned, never was an opponent's blow cowardly dodged even but rather were necks offered to swords and presented to blows. Several of them who were mortally

---

3. Erasmus, *Apophthegmata*, II, *Prisca Lacedaemorum instituta*, XXV.

covered with many a wound, before lying down to die in the arena sent messages to the spectators to inquire whether they were pleased with their service. It was not enough that they should fight and die with constancy: they had to do it cheerfully: with the result that if they were seen to be reluctant to die there was booing and cursing.

[B]   The very maidens egged them on.

> *Consurgit ad ictus;*
> *Et, quoties victor ferrum jugulo inserit, illa*
> *Delitias ait esse suas, pectusque jacentis*
> *Virgo modesta jubet converso pollice rumpi.*

[The vestal virgin jumps to her feet with each blow and every time the victor lunges his sword through his opponent's throat she cries, 'Oh, what fun!' And when one of the men is struck to the ground, she twists her thumb round to have him dispatched.][4]

[A]   To provide such examples the earlier Romans used criminals only; afterwards they used innocent slaves and even freemen who sold themselves for this purpose –   [B]   they included Senators and Roman knights; and women too.

> *Nunc caput in mortem vendunt, et funus arenæ,*
> *Atque hostem sibi quisque parat, cum bella quiescunt.*

[Now they each sell their own persons to die in the arena: when all is at peace they find a foe to attack.]

> *Hos inter fremitus novosque lusus,*
> *Stat sexus rudis insciusque ferri,*
> *Et pugnas capit improbus viriles.*

[Among these tumultuous new sports you see women, clumsy and unused to arms, fighting frenetically with the men.][5]

[A]   That is something that I would have found most strange and unbelievable were it not that in our Civil Wars we have become daily accustomed to seeing thousands of foreigners pledging for money their very life-blood in quarrels which are no concern of theirs at all.

---

4. Prudentius, *Contra Orationem Symmachi*, I, 382–3; II, 1122–6; II, 1096–9.
5. Manlius, *Astronomica*, IV, 225–6; Statius, *Sylvae*, I, vi, 51–3.

24. On the greatness of Rome

[A series of exempla partly arising from reading an edition of Julius Caesar, and starting with a major borrowing from Cicero's Epistulae familiares, *'Familiar letters', which many, including Montaigne, thought to be better called* Epistulae ad familiares, *'Letters to his friends'.]*

[A] I only want to say one word on this inexhaustible subject in order to show the silliness of those who compare the wretched greatness of our times to that of Rome. In the seventh book of Cicero's *Epistulae familiares* (and our grammarians if they wish can indeed remove the epithet *familiar,* which is not really appropriate, while those who wish to replace *familiares* by *ad familiares* [to his friends] can find some support from Suetonius, who in his *Life of Caesar* says that he had a volume of his *Epistulae ad familiares),*[1] there is a letter from Cicero to Caesar, then in Gaul, in which he repeats words from another letter which Caesar had written to him: 'As for Marcus Furius whom you have recommended to me, I will make him King of Gaul; and if you want me to advance some other friend of yours, send him to me.'[2] It was no new thing for a simple Roman citizen, as Caesar then was, to dispose of kingdoms, since he relieved King Dejotarus of his to bestow it on a nobleman of the town of Pergamo who was called Mithridates. And his biographers mention several other kingdoms which he sold; Suetonius says that he extorted from King Ptolemy three million six hundred thousand crowns at one go — which was tantamount to selling it to him!

[B]   *Tot Galatæ, tot Pontus eat, tot Lydia nummis.*

[For Galatia, so much, Pontus, so much, Lydia, so much.]

Mark Antony said that the greatness of the Roman people was not so much revealed by what they took away as by what they gave

1. Suetonius, *Caesar*, XVI.
2. Cicero, *Epist. fam.*, VII, v.

away.[3] [C] Yet among other things, a good century before Antony they took away something with such a wonderful show of authority that I do not know any single event in all of their history which raises higher the credit of the name of Rome: Antiochus had subdued the whole of Egypt and was preparing to conquer Cyprus and other outposts of its Empire; in the flood of his victories Gaius Popilius journeyed to him on behalf of the Senate and, from the outset, refused to clasp his hand until he had read the letter he had brought. King Antiochus read it and said he would think about it; whereupon Popilius drew a circle round him with his baton and said: 'Before you step out of that circle give me an answer to take back to the Senate.' Antiochus was thunderstruck by the roughness of so pressing an order; he reflected for a while and then said: 'I shall do whatever the Senate commands me.' Thereupon Popilius greeted him as a friend of the Roman People.[4] When his fortunes were prospering thus he gave up so great a monarchy under the impact of three lines of writing! He was indeed right, as he later did, to inform the Senate by his ambassadors that he had received their command with the same respect as if it had come from the immortal gods.

[B] All the kingdoms which Augustus conquered by right of war he either restored to those who had lost them or bestowed on foreigners.

[A] In this connection Tacitus, talking of the English King Cogidunus, has a marvellous remark which makes us feel Rome's infinite power. 'The Romans,' he says, 'from the earliest times have been accustomed to leave kings whom they have vanquished in the possession of their kingdoms but under their authority, so that they might have even kings as tools of servitude – 'ut haberet instrumenta servitutis et reges'.[5]

[C] It is likely that Solyman, whom we have seen generously giving away the Kingdom of Hungary and other states,[6] was moved more by that consideration than by the one he usually cited: namely that he was sated by so many monarchies ['95] and overburdened by such dominion acquired by his own virtue or that of his forebears.

3. Cicero, De divinatione, I, xv, 26–7; for Mithridates, the anonymous De Bello Alexandrino, XXVI; Suetonius, Caesar, LIV; Claudianus, In Eutropium, I, 203; Plutarch, Mark Antony, VIII.
4. Livy, LXV, xii ff.
5. Tacitus, Agricola, XIV.
6. Solyman entrusted the Kingdom to Elizabeth of Hungary as Regent.

25. On not pretending to be ill

[Molière may have been thinking of this chapter when his Imaginary Invalid wondered: 'Is there not some danger in pretending to be ill?']

[A]   There is an epigram of Martial's – a good one, for there are all kinds in his book – in which he amusingly tells of Coelius who pretended to suffer from gout in order to avoid having to pay court to some of the Roman grandees, be present at their levees, wait upon them and join their followers. To make his excuse more plausible he would cover his legs with ointment, wrap them in bandages and in every way counterfeit the gait and appearance of sufferers from the gout. In the end Fortune favoured him by giving it to him:

> Tantum cura potest et ars doloris,
> Desiit fingere Cœlius podagram.

[So much can care and the art of pain! Coelius has no longer to feign to be gouty.][1]

I have read somewhere in Appian   [C]   I think   [A]   a similar tale of a man who sought to escape from a declaration of outlawry by the Roman Triumvirate and to hide from his pursuers; he remained in hiding, took on a disguise, deciding in addition to pretend to be blind in one eye. When he was able to recover a little liberty and wanted to rid himself of the plaster which he had worn so long over his eye, he found that he had actually lost the sight of that eye while under the mask. It is possible that his power of sight had been weakened by not having being exercised for such a long time and that his visual powers had all transferred to the other eye: for we can plainly feel that when we cover one eye it transfers to its fellow some part of its activity so that the remaining eye grows and becomes swollen; similarly for that gouty man in Martial: lack of use, together with the heat of his ointments and bandages, may well have concentrated upon his leg some gouty humour.

Since I read in Froissart[2] of the vow taken by a troop of some young

1. Martial, Epigrams, VII, xxxix; then Appianus, IV, vi.
2. Froissart, I, xxix.

English noblemen to keep their left eyes covered until they had crossed into France and achieved some great deed of arms against us, I have often been obsessed by the thought that it may have befallen them as it did to those others and that when they came back to greet the ladies for whose sake they had done such deeds they would all have become blind in one eye.

Mothers are right to scold their children when they play at being one-eyed, limping or squinting or having other such deformities; for, leaving aside the fact that their tender bodies may indeed acquire some bad habit from this, it seems to me that Fortune (though I do not know how) delights in taking us at our word: I have heard of many examples of people falling ill after pretending to be so.

[C] Whether riding or walking I have always been used to burdening my hand with a cane or a stick, even affecting an air of elegance by leaning on it with a distinguished look on my face. Several people have warned me that one day Fortune may change this affectation into a necessity. I comfort myself with the thought that, if so, I would be the first of my tribe to get the gout!

[A] But let us stretch out this chapter and stick on to it a different coloured patch concerned with blindness. Pliny tells of a man who, never having been ill before, dreamt he was blind and woke up next morning to find that he was.[3] The force of imagination could well have contributed to that, as I have said elsewhere, and Pliny seems to share that opinion: but it is more likely that the dream was produced by the same internal disturbances as his body experienced and which deprived him of his sight; if they want to, the doctors will find their cause . . .

Now let us add another closely similar account which Seneca gives in one of his letters.[4] 'You know Harpasté, my wife's female idiot,' he wrote to Lucilius. 'She is staying in my house as I have inherited the burden of looking after her. I loathe such freaks; if I ever want to laugh at a fool I do not have to look far: I can laugh at myself. She has suddenly become blind. It may seem incredible but it is true that she does not realize she is blind: she keeps begging her keeper to take her away; she thinks that my house is too dark. What in her we laugh at I urge you to believe to apply to each one of us. No one realizes he is miserly; no one realizes he is covetous. At least the blind do ask for a guide: we wander off alone. "I am not an ambitious man," we say, "but you can live in Rome no other way. I am no spendthrift, but it costs a lot merely to live in Rome." "It is not my fault if I

3. Pliny, *Hist. nat.*, VII, i; cf. I, 21, 'On the power of the imagination'.
4. Seneca, *Epist. moral.*, L, 2 ff., 9.

get angry or if I have not yet definitely settled down: it is the fault of my youth." Let us not go looking elsewhere for our evils: they are at home in us, rooted in our inward parts. We make the cure harder precisely because we do not realize we are ill. If we do not soon start to dress our wounds, when shall we ever cure them and their evils? Yet Philosophy provides the sweetest of cures: other cures are enjoyed only after they have worked: this one cures and gives joy all at once.'

That is what Seneca says; he carried me off my subject, but there is profit in the change.

26. On thumbs

=====

[Renaissance etymologies are often very fanciful, but in the case of the French and Latin words for thumb (pouce, pollex) philologists today continue to accept the derivations advanced by Montaigne and his contemporaries. Our own word 'thumb' derives also, it seems, from a Sanskrit word meaning 'the strong one'.]

[A]   Tacitus relates that it was the custom among certain Barbarian kings to make a treaty binding by pressing their right hands together and interlocking their thumbs until they had squeezed the blood to their tips, whereupon they lightly pricked them with a needle and sucked each other's blood.[1]

Doctors say that our thumb is our master-finger and that our French word for it, *pouce*, derives from the Latin verb *pollere* [to excel in strength].[2] The Greeks called it *anticheir*, 'another hand', so to speak. And the Latins seem occasionally to use it to mean the whole of the hand:

> *Sed nec vocibus excitata blandis,*
> *Molli pollice nec rogata, surgit.*

[Neither sweet words of persuasion nor the help of her thumb can get it erect.]

In Rome it was a sign of approval to turn your thumbs and twist them downwards –

> *Fautor utroque tuum laudabit pollice ludum*

[Your fans admire your play by turning down both their thumbs]

– and of disapproval to raise them and extend them outwards:

> *converso pollice vulgi*
> *Quemlibet occidunt populariter.*

---

1. Tacitus, *Annals*, XII, xlvii.
2. *Pollex*, the Latin for thumb, 'the strong one', was indeed derived from the verb 'to be strong'. Cf. Macrobius, *Saturnalia*, VII, xiii. The Greek etymology is fanciful.

[when the mob twist their thumbs round, anyone at all is slaughtered to their acclaim.][3]

The Romans exempted from war-service those who had injured thumbs since they could no longer firmly grasp their weapons. Augustus confiscated the estates of a Roman knight who had craftily cut off the thumb of two of his sons to stop them being mobilized into the army. Before that, during the Italian Wars, the Senate had sentenced Caius Vatienus to life imprisonment and confiscated all his estates for having deliberately cut off his left thumb to get out of an expedition. Some general or other (I cannot remember his name) cut off the thumb of his defeated enemies after winning a naval engagement so as to deprive them of the means of fighting and of pulling on the oar.[4] [C] The Athenians did the same to the men of Aegina to deprive them of their naval superiority.[5] [B] In Sparta the schoolmaster punished his pupils by biting their thumbs.

3. Martial, *Epigrams*, XII, xcviii, 8; Horace, *Epist.*, I, xviii, 66; Juvenal, III, 36. (Our 'thumbs up' was 'thumbs down' for the Romans.)
4. Suetonius, *Augustus*, XXIV; Valerius Maximus, V; Plutarch, *Life of Lysander.* Philoctetes left them *able* to row (in the galleys).
5. Cicero, *De officiis*, III, xi, 46; then, Plutarch, *Life of Lysander.*

27. On cowardice, the mother of cruelty

[Montaigne returns to the theme of cruelty (cf. 'On conscience', II, 5; 'On cruelty', II, 11; and 'On coaches', III, 6.) He loathed torture, then widely practised as a justifiable means of interrogation, being accepted as such by Roman Law, and like many, including Michel de l'Hôpital and French kings at least from Charles IX, disliked duelling. Montaigne's opinion that torture, or indeed anything beyond straightforward execution, amounted to cruelty caused some disquiet in the Vatican, but Montaigne held his ground.]

[A] I have often heard it said that cowardice is the mother of cruelty. [B] And I have learnèd from experience that that harsh rage of wicked inhuman minds is usually accompanied by womanish weakness. I have known the cruellest of men to cry easily for the most frivolous of causes. The Tyrant Alexander of Pheres could not bear to hear tragedies performed in the amphitheatre for fear that the citizens might see him, who had without pity put many to death every day, blubbering over the misfortunes of Hecuba and Andromache.[1] Can it be a weakness in their soul which makes such men susceptible to every extreme? [A] Valour (which acts only to overcome resistance) –

Nec nisi bellantis gaudet cervice juvenci

[And which takes no delight in killing even a bull unless it resists][2] –

stops short when it sees the enemy at its mercy. But pusillanimity, so as to join in the festivities even though it could not have any role in the first act, chooses its role in the second: that of blood and slaughter. Murders after victory are normally done by the common people and the men in charge of the baggage-train; and what makes us witness so many unheard of cruelties in these people's wars of ours is that the common riff-raff become used to war and swagger about, up to their arms in blood, hacking at a body lying at their feet since they can conceive of no other valour:

---

1. Plutarch, *Life of Pelopidas*.
2. Claudianus, *Ad Hadriam*, 30.

[B]   *Et lupus et turpes instant morientibus ursi,*
Et quæcunque minor nobilitate fera est,

[The wolves and base bears fall on the dying, and so do all the more ignoble beasts,][3]

[A]   like the cowardly curs which, in our homes, snap and tear at the skins of wild beasts which they would not dare to attack in the field.

What is it that makes all our quarrels end in death nowadays? Whereas our fathers knew degrees of vengeance we now begin at the end and straightway talk of nothing but killing. What causes that, if not cowardice? Everyone knows that there is more bravery in beating an enemy than in finishing him off; more contempt in making him bow his head than in making him die; that, moreover, the thirst for vengeance is better slaked and satisfied by doing so, since the only intention is to make it felt. That is why we do not attack a stone or an animal if it hurts us, since they are incapable of feeling our revenge. To kill a man is to shield him from our attack.

[B]   And just as Bias cried out to a wicked man, 'I know you will be punished sooner or later, but I am afraid I shall never live to see it'; and just as he sympathized with the Orchomenians because the chastisement of Lyciscus' treachery against them came at a time when there was nobody left who had suffered by it whom such chastisement would have gratified the most: vengeance is at its most wretched when it is wreaked upon someone who has lost the means of feeling it; for, as the one who seeks revenge wishes to see it if he is to enjoy it, the one who receives it must see it too if he is to suffer the pain and be taught a lesson.[4]

[A]   'He'll be sorry for it,' we say. Do we really think he is sorry for it once we have shot him through the head? Quite the contrary: if we look closely we will find him cocking a snook as he falls: he does not even hold it against us. That is a long way from feeling sorry!   [C]   And we do him one of the kindest offices of this life, which is to let him die quickly and painlessly.   [A]   He is at rest while we have to scuttle off like rabbits, running away from the officers of the watch who are on our trail. Killing is all right for preventing some future offence but not for avenging one already done.   [C]   It is a deed more of fear than of bravery; it is an act of caution rather than of courage; of defence rather than of attack.   [A]   It

3. Ovid, *Tristia*, III, v, 35–6.
4. Plutarch (tr. Amyot), *Comment il faut uoir* (30 G) and *Pour quoy la justice divine differe . . . la punition* (258 E–G).

is clear that by acting thus we give up both the true end of vengeance and all care for our reputation: we show we are afraid that if we let the man live he will do it again.   [C]   By getting rid of him you act not against him but against yourself.

In the Kingdom of Narsinga their way of doing things would be no use to us. There, not only soldiers but even artisans settle their quarrels with their swords; their king never denies the field to any who would fight a duel, and in the case of men of quality he honours it with his presence and bestows a golden chain on the victor. But the very first man who wants that chain can dispute it with the wearer who, by having rid himself of one duel, finds himself with several more on his hands.[5]

[A]   If we had thought that we had for ever overcome our enemy by valour and could dominate him as we pleased, we would be sorry indeed if he were to escape: he does that when he dies. We do want to beat him, but with more security than honour,   [C]   and we seek not so much glory through our quarrel but the end of that quarrel.

For a man of honour Asinius Pollio also made a similar mistake: he wrote invectives against Plancus but waited until he died before he published them. That was like poking out your tongue to a blind man, shouting insults at a deaf one or hitting a man who cannot feel it, rather than risking his resentment. And they said of him that only the shades should shadow-box with the dead. Anyone who waits to see an author dead before attacking his writings, what does he reveal except that he is both weak and quarrelsome?[6] Aristotle was told that someone had spoken ill of him: 'Let him do worse,' he replied, 'let him scourge me – as long as I am not there!'

[A]   Our fathers were content to avenge an insult by a denial; avenge a denial by a slap in the face; and so on in due order. They were valiant enough not to be afraid of an enemy who was outraged but living. We tremble with fear while we see him still on his feet. As proof of that, is it not one of our beautiful practices today to hound to death not only the man who has offended us but also the man we have offended?

[B]   It is also a reflection of our cowardice which has brought into our single combats the practice of our being accompanied by seconds – and thirds and fourths. Once upon a time there were duels: nowadays there are clashes and pitched battles. The first men who introduced such practices

5. S. Goulart, *Histoire du Portugal*, IV, xii.
6. From a note of Vives on Augustine, *City of God*, V, xxvii; then, Diogenes Laertius, *Aristotle*, V, xviii.

were afraid of acting on their own, [C] *'cum in se cuique minimum fiduticæ esset'* [since neither had the slightest confidence in himself].[7] [B] For it is natural that company of any sort brings comfort and solace in danger. Once upon a time third parties were brought in to guard against rule-breaking and foul play [C] and also to bear witness to the result of the duel; [B] but now that it has come to such a pass that anyone who is invited along involves himself in the quarrel, he can no longer remain a spectator for fear that it was from lack of engagement or of courage. Apart from the injustice and baseness of such an action which engages in the defence of your honour some other might or valour than your own, I find it derogatory to anyone who does fully trust in himself to go and confound his fortune with that of another. Each of us runs risks enough for himself without doing so for another: each has enough to do to defend his life on behalf of his own valour without entrusting so dear a possession into the hands of third parties. For unless it be not expressly agreed to the contrary, the four of them form one party under bond. If your second is downed you are faced, by the rules, with two to contend with; you may say that that is unfair. And indeed it is – like charging well-armed against a man who has only the stump of his sword, or when you are still sound against a man who is already grievously wounded. However, when you have won such advantages in battle you can exploit them without dishonour. Inequality and disproportion weigh in our consideration only at the outset, when battle is joined: thereafter you can rail against Fortune! And even if you find yourself one against three after your two companions have been killed, they do you no more wrong than I would do if, in the wars with a similar advantage, I were to strike a blow with my sword at one of the enemy whom I found attacking one of our men. The nature of our alliances entails that when we have group against group (as when our Duke of Orleans challenged Henry, King of England, one hundred against one hundred; [C] or three hundred against three hundred like the Argives against the Spartans; or three against three like the Horatii against the Curatii),[8] [B] whatever crowd there may be on either side they are regarded as one man. And whenever you have companions the chance of the outcome is confused and uncertain.

I have a private interest to declare in this discussion: for my brother the Seigneur de Matecoulom was summoned to Rome to act as second for a gentleman he hardly knew, who was the defender, having been challenged

7. Livy, xxxiv, 28.
8. Enguerrand de Monstrelet, *Chroniques*, I, ix; Herodotus, I, lxxxii; Livy, I, xxiv.

by another. By chance he found himself face to face with a man who was closer and better known to him (I would like to see somebody justify these 'laws of honour' which are so often opposed in hostility to the laws of reason). Having dispatched his opponent and seeing the two principals in the quarrel still unharmed on their feet, he went to the relief of his companion. What less could he do? Ought he to have remained quiet and watched the man defeated, if such was his lot, for whose defence he had come to Rome? All he had achieved so far was of no avail: the quarrel had still to be decided. The courtesy which you yourself can and must show to your enemy when you have reduced him to a sorry state and have him at a great disadvantage, I cannot see how you can show it when it concerns somebody else, when you are but the second, when the quarrel is not yours. He could neither be just nor courteous at the expense of the one to whom he had lent his support. So he was released from prison in Italy by the swift formal request of our King.

What a stupid nation we are. We are not content with letting the world know of our vices and follies by repute, we go to foreign nations in order to show them to them by our presence! Put three Frenchmen in the Libyan deserts and they will not be together for a month without provoking and clawing each other: you would say that one of the aims of these journeys is expressly to make spectacles of ourselves before foreigners – especially those who take delight in our misfortunes and laugh at them.

We go to Italy to learn fencing,   [C]   and then put it into practice at the expense of our lives before we have learnt how.   [B]   Yet, by the rules of instruction, theory should come before practice: we betray that we are mere apprentices:

> *Primitiæ juvenum miseræ, bellique futuri*
> *Dura rudimenta.*

[Wretched first fruits of mere youth: harsh training for the future wars.][9]

I know that fencing is an art   [C]   which achieves what it sets out to do: in the duel in Spain between two Princes who were cousins german, the elder, says Livy, easily overcame the reckless force of the younger by strategy and skill with his weapons.[10] And as I myself know from experience it is an art   [B]   which has raised the hearts of some above their natural measure; yet that is not really valour since it draws its support from

9. Virgil, *Aeneid*, XI, 156–7.
10. Livy, *Annals*, XXVIII.

skill and has some other foundation than itself. The honour of combat consists in rivalry of heart not of expertise; that is why I have seen some of my friends who are past masters in that exercise choosing for their duels weapons which deprived them of the means of exploiting their advantage and which depend entirely on fortune and steadfastness, so that nobody could attribute their victory to their fencing rather than to their valour. When I was a boy noblemen rejected a reputation for fencing as being an insult; they learned to fence in secret as some cunning craft which derogated from true inborn virtue:

> *Non schivar, non parar, non ritirarsi*
> *Voglion costor, ne qui destrezza ha parte.*
> *Non danno i colpi finti, hor pieni, hor scarsi:*
> *Toglie l'ira e il furor l'uso de l'arte.*
> *Odi le spade horribilmente urtarsi*
> *A mezzo il ferro; il pie d'orma non parte:*
> *Sempre è il pie fermo, è la man sempre in moto;*
> *Ne scende taglio in van, ne punta à voto.*

[They have no wish to dodge, to parry nor to make tactical retreats: skill has no part to play in their encounter; they make no feints, nor blows oblique, nor shamming lunges; anger and fury strips them of their art. Just listen to the terrifying clash of striking swords, iron against iron; no foot gives way but stays ever planted firm: it is their arms which move; every thrust strikes home and no blows fall in vain.][11]

Our forebears' training was a true image of martial combat: target-practice, tournaments and the tilting-yard; that other skill is all the more ignoble in that it has nothing but a private end, teaching us to destroy each other against all law and justice and, whatever else happens, always producing harmful effects. It is much more meet and right to practise such arts as defend our polity not those which undermine it, such as have regard for national security and the glory of the common weal.

Publius Rutilius when consul was the first to train soldiers in handling their weapons with skill and technique and to couple art and valour:[12] but that was for the wars and contentions of the Roman People – [C] official fencing for citizens in common. And, leaving aside Caesar's example when he ordered his men to aim principally at the faces of Pompey's men during the battle of Pharsalia, hundreds of other leaders of men in war have

11. Tasso, *Gierusalemme liberata*, XII, 55–62.
12. Valerius Maximus, II, iii.

decided to employ new kinds of weapons and new ways of attack and defence according to the exigencies of the moment.[13] [B] But just as Philopoemen condemned wrestling (in which he excelled) because the basic skills learned in that sport were quite different from those which appertain to military training on which alone he reckoned that men of honour should spend their time, it seems to me also that those feints and tricks and that agility which young men acquire for their limbs in this new-fangled school are not merely useless for fighting wars but are hostile and harmful to it. [C] Moreover people today normally use special weapons, specifically destined for fencing; I have noticed that it is hardly considered proper that a gentleman challenged to sword and dagger should turn up armed like a soldier.

It is worth considering that in Plato, Laches, talking about a kind of apprenticeship in weapon-training just like ours today, says that he has never seen any great soldier come out of such a school – and especially not from among the instructors![14] (As for that lot, our own experience teaches us the same!) We can also certainly at least assert that we are dealing with accomplishments which are quite unrelated and distinct. And in this system of education for the boys of his Republic Plato forbids fisticuffs (which was introduced by Amycus and Epeius) as well as wrestling (introduced by Antaeus and Cercyo) since they have some other aim than rendering youth more apt for service in war and contribute nothing to it. [B] But I am wandering away from my theme.

[A] The Emperor Maurice,[15] having been warned by his dreams and several omens that he was to be killed by a certain Phocas, a soldier then unknown to him, inquired of his son-in-law Philip who this Phocas was, what he was like and how he behaved; when Philip told him that Phocas was among other things cowardly and fearful, the Emperor straightway concluded from this that he was therefore murderous and cruel. What is it that makes tyrants so lust for blood? It is their worries about their own safety and the fact that when they fear a scratch their cowardly minds can furnish them with no other means of security save exterminating all those who simply have the means of hurting them, women included.

> [B]   *Cuncta ferit, dum cuncta timet.*

> [Fearing all, he strikes at all.][16]

13. Plutarch, *Life of Caesar*; then, *Life of Philopoemen*.
14. Plato, *Laches*, 183 B–C; then *Laws*, 796.
15. Of the Eastern Empire. Zonaras, III.
16. Claudianus, *In Eutropium*, I, 182.

[C]   The first acts of cruelty are done for their own sake; from them there is born fear of a just revenge; that produces a succession of fresh cruelties, each intended to smother each other. Philip, King of Macedon, who had many a crossed thread to untangle with the Roman People, was shaken with terror by the murders committed on his orders; since he could not find a means of delivering himself from so many families harmed at various times, he decided to seize all the children of those he had put to death so as to kill them off, one by one, day after day . . . and so find rest.

Beautiful topics can always hold their own, no matter where you strew them. I who am more concerned with the weight and usefulness of my writings than with their order and logical succession must not be afraid to place here, a little off the track, an account of great beauty.[17] Among the others condemned by Philip there was a certain Herodicus, Prince of the Thessalians. After him it was the turn of his two sons-in-law to be killed, each leaving a baby son. Their widows were called Theoxena and Archo. Theoxena was much courted but could not be brought to remarry. Archo married the leading man among the Aenians called Poris and had a number of sons by him who were all young when she died. Theoxena, feeling the urge to mother her nephews, wedded Poris. Then the King's edict was proclaimed. That courageous mother, fearing both the cruelty of Philip and the abusive lust of his underlings, boldly stated that she would kill them with her own hands rather than hand them over. Poris was terrified by this declaration of hers and promised to steal secretly away with them to Athens and place them under the protection of some of his faithful vassals. Taking advantage of a yearly feast celebrated in Aenia in honour of Aeneas, they set about it. After being present during the daytime at the ceremonies and the public banquet, they slipped away by night to a ship which was waiting to put some space between them. But there was a contrary wind; the following morning they were still in sight of the land where they had left their moorings and were pursued by the harbour-guards. When they were overhauled, while Poris was busy urging the sailors to flee faster Theoxena, raving mad with love for the children and for vengeance, returned to her original plan; she got weapons and poison ready; she then showed them to them saying, 'Come now, my children; from henceforth death is your sole means of defence and of remaining free;

17. '95: beauty. *When such accounts are richly beautiful in themselves and can sustain themselves in isolation, I am content to link them to my argument with a scrap of hair.* Among . . .
    Then Livy, XL, iii.

and it will provide the gods with something to work their hallowed justice upon. These drawn swords and these goblets open the way to it for you. Be brave. And you, my oldest son, grasp this blade and die the bolder death.' The children, with this staunch counsellor on one side and the enemy at their throats on the other, frantically ran to whatever goblet was nearest to hand and were thrown still half-dead into the sea. Theoxena, proud of having so gloriously saved all her children, threw her arms passionately round her husband and said, 'Let us follow these boys, my love, and let us enjoy the same grave with them.' Clasped thus in each other's embrace, they plunged headlong into the sea. And so that boat was brought back to land empty of its masters.

[A] Tyrants, to do two things at once (killing, and making their anger felt), have exhausted their ingenuity in inventing means of prolonging the death. They want to do in their enemies all right, but not so quickly that they have no time to spare for savouring their vengeance. In this they are greatly perplexed; for if the tortures are intense they are short: if they are long they are not painful enough to their liking; so they have to tread carefully with machinery of torture.

We can see hundreds of examples of this in Antiquity – and I wonder whether we do not still retain traces of such barbarity without our realizing it. Everything which goes beyond mere death seems to me to be cruelty. Our justice cannot hope that a man who will not be kept from wrongdoing by fear of death on the block or the gallows may yet be deterred by the thought of pincers or a slow fire or the racking-wheel. And for all I know, during this time we drive them to despair: for in what state can a man's soul be as he lies waiting for death for twenty-four hours, broken on the wheel, or in the Ancient fashion nailed to a cross? Josephus relates how, during the Roman wars in Judaea, he was passing by the place where some Jews had been crucified three days before, when he recognized three of his friends and was allowed to take them away. 'Two of them died,' he says, 'and the other is still alive.'[18]

[C] Chalcocondylas, a reliable man, left memoirs of events which happened in his own time and near where he was; in them he relates as the ultimate in punishments the practice of the Emperor Mahomet who, with one blow from a scimitar, often had men sliced in two through their middle just above the abdomen so that they died as it were two deaths at once; 'And,' he adds, 'you could see both parts, still alive long afterwards, twitching and writhing in torment.' I am not convinced that those twitch-

18. Josephus, *De vita sua.* (Torture was a legacy of Roman Law.)

ings imply much pain. Tortures which are most ghastly to see are not always the harshest to suffer. More atrocious I find are the accounts in other historians of what he did to some of the noblemen of Epirus: he had them flayed alive, bit by bit, following a procedure so evilly devised that, for a whole fortnight, they lived to endure such anguish.[19]

And there are those two others as well: Croesus, having seized one of his brother's intimate supporters called Pantalcon, dragged him off to a wool-carder's shop where he had him so excoriated with the carder's combs and teasles that he died from it;[20] George Sechel (the leader of those Polish peasants who wrought such havoc under the pretext of a Crusade) was defeated and captured in battle by the Voivode of Transylvania; he was strapped for three days, naked, to a wooden rack and subjected to every kind of torture which anyone at all could devise for him. During this time the other prisoners were given neither food nor drink. In the end, while he was still alive and able to see it, they compelled his dear brother Lucat to quench his thirst in his blood (but he went on praying for Lucat's safety, taking upon himself all the hatred aroused by their crimes); then they made twenty of his most intimate captains eat him, tearing at his flesh with their bare teeth and swallowing it down. Once he was dead they boiled his remaining flesh and entrails and gave it to others of his followers to eat.

19. Nicolas Chalcocondylas, *Hist. de la decadence de l'Empire grec et l'etablissement de celui des Turcs*, X, ii; Jacques Lavardin, *Scanderbeg, Roi d'Albanie* (1576), 446.
20. Plutarch (tr. Amyot), *Malignité de Herodote*, 651 C; then Bishop Paolo Giovio, *Historia sui temporis*, XIII.

28. *There is a season for everything*

===

[Marcus Porcius Cato, the elder, surnamed Censorius (the Censor) was, since Classical times, associated with his descendant, also called Marcus Porcius Cato (who fought against Julius Caesar and killed himself after his defeat at Pharsalia). Both were cited as twin examples of great patriotism, sound judgement and stern morality. (Cf. Erasmus' adage, Tertius Cato.) Montaigne shows considerable originality here in his criticism of the Elder Cato (the Censor): in the Renaissance, that Cato's learning of Greek in his old age was normally held up as an example to be followed. For Montaigne, the Younger Cato's suicide was one of the highest peaks that philosophical (as distinct from theological) morality could reach.]

[A]   Those who compare Cato the Censor to the Younger Cato, the self-murderer,[1]   [C]   are indeed comparing two beautiful natures with closely similar souls. Cato the Censor displayed his nature in many more of its aspects and outstrips the younger in military exploits and in the usefulness of his service to the public. But as for the virtue of the Younger Cato, apart from the fact that it is sacrilege to compare its living fortitude to that of anyone else's whatsoever, his was far more pure. For could anyone absolve the Censor's virtue from its load of envy and ambition, seeing that he dared to attack the honour of Scipio, who in goodness and in all excellent endowments far excelled him and all other men of his time?   [A]   What they tell of Cato the Censor, that among other things, when he was well advanced in years, he set about learning Greek with a burning craving as though he were satisfying some long-felt thirst, does not seem to me to be greatly to his honour. That is exactly what we mean by tumbling into second childhood. There is a season for all things – all, including the good: even my Lord's Prayer may be said at an inappropriate time,   [C]   as was the case of Titus Quintius Flaminius who was arraigned because, as general of the army, he had been seen

---

1. Livy, XXXVIII.
  [A]: self-murderer, *do great honour to the former, in my opinion, for I find them very wide apart.* What they tell . . .

when the fighting began, to draw apart to pray to God in a battle (which he won).[2]

[B]   *Imponit finem sapiens et rebus honestis.*

[The wise man sets limits even to things which are good.]

[A]   When Eudomidas saw Xenocrates working hard at his school lessons when he was very old he remarked: 'When will this man know anything if he is still learning!'   [B]   As Philopoemen said to those who were singing the praises of King Ptolomy for daily strengthening his body by the practice of arms: 'It is not very praiseworthy in a king of his age to be practising arms: he should be really using them now!'

[A]   'Youth should make provisions: Old Age should enjoy them,' say the wise.[3] And the greatest flaw which they find in our nature is that our desires are for ever renewing their youth. We are constantly beginning our lives all over again. Our zeal and our desire should sometimes smell of old age. We already have one foot in the grave yet our tastes and our pursuits are always just being born.

[B]   *Tu secanda marmora*
Locas sub ipsum funus, et sepulchri
Immemor, struis domos.

[You go cutting marble and are about to die: yet you forget your own tomb and start building houses.][4]

[C]   The longest of my projects are for less than a year; I think only of bringing things to a close; I free myself from all fresh hopes and achievements; I say my last farewell to all the places I am leaving and daily rid myself of my belongings. '*Olim jam nec perit quicquam mihi nec acquiritur . . . Plus superest viatici quam viœ.*' [I have long since ceased to lose or gain: I have more rations than road left.]

Vixi, et quem dederat cursum fortuna peregi.

[I am dead: I have run the course which Fortune gave.]

In short all the comfort I find in my old age is that it deadens within me

---

2. Plutarch: *Parallel lives of Flaminius and Philopoemen*; then, Juvenal, VI, 444; Plutarch (tr. Amyot) *Dicts notables des Lacedaemoniens*, 216 F; *Life of Philopoemen*, VIII.
3. Seneca, *Epist. moral.*, XXXVI, 4.
4. Horace, *Odes*, II, xviii, 17–19; then, Seneca, *Epist. moral.*, LXXVII, 3; Virgil, *Aeneid*, IV, 653.

many of the desires and worries which trouble our lives: worry about the way the world is going; worry about money, honours, erudition, health . . . and me. [A] Cato the Censor was learning to talk just when he ought to be learning to shut up forever. [C] We can always continue our studies but not our *school*-work: what a stupid thing is an old man learning his alphabet!

> [B]   *Diversos diversa juvant, non omnibus annis*
> *Omnia conveniunt.*

[Divers men, divers tastes: nor are all things fit for all ages.][5]

[A]   If study we must, let us study something suitable to our circumstances, so that we can make the same reply as that man who was asked what use were his studies in decrepit old age: 'That I may better and more happily leave it behind,' he said.[6]

Such when he felt his end was near was the study of the Younger Cato, which brought him to Plato's discussion of the immortality of the soul. Not (as we must believe) that he was not long since furnished with every sort of provision for his soul's departure: of assurance, resolute will and preparedness he had more than did Plato in his writings: his knowledge and his heart were in this respect above philosophy. He occupied himself thus, not so as to help himself die but as one who would not even trouble his sleep by dwelling on the importance of such reflections; he continued his studies, as he did all the customary activities of his life, neither chopping nor changing.

[C]   The night the Praetorship was refused him he spent in play: the night he was destined to die, he spent in reading. It was all one to him whether he lost life or office.

5. Pseudo-Gallus, I, 104–5.
6. Seneca, *Epist. moral.*, LXVIII, 14.

29. On virtue

====

[More considerations on virtue and its relationship to ecstasy and constancy, as well as a fresh consideration of how Fate and Classical determinism may be reconciled with God's omnipotence and human freedom. There are suggestions that this chapter was written in its first form before 'An apology for Raymond Sebond' (II, 12), since it contains an elementary exposition of Pyrrhonism.]

[A]   I find from experience that there is a difference between the leaps and sallies of the soul and a settled constant habit: and I am well aware that there is nothing we cannot do (indeed, even surpassing the Divinity, as somebody once said, since it is a greater thing to make oneself impassible than to be so as a property of one's being) even combining the frailty of Man with the resolution and assurance of God.[1] But only spasmodically. Sometimes there is in the lives of those heroes in Ancient times miraculous flashes which appear far to exceed our natural powers: but, truly, flashes they are; it is hard to believe that we can so steep and dye our soul in such elevated attributes that they become ordinary and natural to her. It happens even to us who are mere abortions of men that we can occasionally enrapture our Soul far beyond her ordinary state when she is awakened by the words or examples of another man: but it is a kind of passion which impels her, disturbs her and ravishes her somewhat outside ourselves; for once that whirlwind is over, we can see that she spontaneously relaxes and comes down, not perhaps down to the lowest stage of all but at least to less than she was, so that we can be moved to anger more or less like any ordinary man by the loss of a hawk or by a broken glass.

[C]   Ordinate conduct, moderation, constancy apart, I believe that anything at all can be done, even by a man who, taken overall, is lacking and deficient.   [A]   That is why the wise men say that to judge a man properly we must principally look at his routine activities and surprise him in his everyday dress.

---

1. Seneca, *Epist. moral.*, LIII, 11–12 (mocked in II, 12, 'An apology for Raymond Sebond', as are the following anecdotes about Pyrrho).

Like all other true philosophers, Pyrrho, the man who built up ignorance into so pleasing a science, made an assay at conforming his life to his doctrine. And because he maintained that the feebleness of human judgement was so extreme as to be unable to incline towards any decision or persuasion and wanted to keep it forever hanging in the balance, regarding and welcoming all things as adiaphora, stories are told how he always maintained the same manner and expression: when he had started to say anything he never failed to go on to the end, even if the man he was speaking to had walked off; he never swerved from his path for any obstacle whatsoever, protected only by his friends from precipices or from being bumped into by carts, and similar accidents; for to fear or to avoid anything would have shocked his own principles, which remove all choice and election even from the senses. On occasions he allowed himself to be cut open or cauterized with such steadfastness that he never batted an eyelid.

Now it is one thing to bring your soul to accept such ideas: it is quite another to combine theory and practice. Yet it is not impossible. But what is virtually incredible is that you should combine them with such perseverance and constancy as to make it your regular routine in actions so far from common custom. That is why, when he was once surprised in his home quarrelling bitterly with his sister and reproached for having thereby forgotten his adiaphorism, he retorted: 'What! Must even this silly woman serve to prove my rules?' On another occasion he was seen defending himself against a dog; 'It is,' he said, 'very difficult to cast off the Man entirely, and we must make it our duty to strive to fight against things first by deeds or, as second best, by reason and argument.'[2]

About seven or eight years ago, some two leagues from here, there was a villager, who is still alive; his brain had long been battered by his wife's jealousy; one day he came home from work to be welcomed by her usual nagging; it made him so mad that, taking the sickle he still had in his hand he suddenly lopped off the members which put her into such a fever and chucked them in her face.

It is also told how one of our young local gentry who was desperately in love at last succeeded, by sheer perseverance, in softening the heart of his beautiful lady; he was thrown into despair when about to make his sally to find that it was he who was the soft and yielding one, and that

2. Diogenes Laertius, *Pyrrho*, IX.

non viriliter
Iners senile penis extulerat caput;

[without virility his sluggish penis raised its senile head;][3]

he went straight back home and cut it off, sending it as a cruel and bloody victim to atone for his offence. If that had been done rationally for religion, like the priests of Cybele, what would we not have said of so sublime an action!

A few days ago at Bergerac, about five leagues up the Dordogne from my house, there was a wife who had been battered and beaten the previous night by her husband, a man melancholic and irritable by complexion; she resolved to escape from his brutality at the cost of her life. She got up and gossiped with her neighbours as usual, slipping in a word or two entrusting her affairs to them; she took a sister of hers by the hand and led her to the bridge; after saying goodbye as though it were a game, with no other shift or change of expression, she threw herself off it down into the river, where she perished. What is more remarkable in her case is that this project had matured in her brain all night long.

It is quite another matter with women in India. It is the custom there for husbands to have several wives and for the one he loves most to kill herself after him: during the whole of their lives they each scheme to gain this vantage point over the others; the kindnesses which they do to their husband aim at no other reward than to be selected to accompany him in death:

> [B]   *Ubi mortifero jacta est fax ultima lecto,*
> *Uxorum fusis stat pia turba comis;*
> *Et certamen habent lethi, quæ viva sequatur*
> *Conjugium; pudor est non licuisse mori.*
> *Ardent victrices, et flamme pectora præbent,*
> *Imponuntque suis ora perusta viris.*

[When the last torch is cast on the funeral pyre, the wives remain there with their hair in disarray and begin their mortal combat over which of them, alive, may join their husband in death; for it is a disgrace not to be allowed to die. Those who emerge victorious offer their bosoms to the flames and press their scorched lips on their husband.][4]

3. Tibullus, *De inertia inguinis*. (Story from H. Estienne, *Apologie pour Hérodote*, XV, xxix.)
4. Propertius, II, xiii, 17–22.

[C]   One writer says that even in our own times he has seen the custom honoured by those Eastern peoples among whom not only the wives are buried with their husbands but also the slave-girls he had enjoyed. This is the way it is done. When her husband is dead the widow can if she wishes – but few do – ask for two or three months to arrange her affairs. When the day arrives she is arrayed as for a wedding, a looking-glass in her left hand and a wand in the other; she mounts a horse and with a happy face as though, she says, she were going to lie asleep beside her husband. Having paraded in pomp accompanied by her friends and relations and by a crowd in festive mood, she eventually comes to a public place devoted to such spectacles – a great square in the midst of which is a ditch filled with wood, having next to it a mound four or five steps high on to which she is escorted where she is served a sumptuous repast. After which she begins to dance and to sing; when the moment seems right to her she commands that the fire be lit. That done, she comes down and, taking her husband's nearest kinsman by the hand, they go together to the neighbouring river where she strips herself naked, distributes her clothes and jewels among her friends and plunges into the water as though to lave away her sins. She comes out and wraps herself in a yellow cloth fourteen yards long; then, again offering her hand to her husband's kinsman, they return to the mound from which she addresses the people and, if she has any, entrusts her children to their care. Between the ditch and the mound they are willing to draw a curtain to hide that burning furnace from her view; many widows forbid them to do so, to show greater courage. When she has finished what she has to say, a woman presents her with a cruse of oil to anoint her head and her whole body; after having done so, she casts it into the fire and then immediately leaps in herself, whereupon the people throw a great many faggots on top of her to save her from a lingering death; then their joy turns to mourning and sadness. If they are people of meaner stuff the dead body of the husband is taken to the spot where it is to be buried; there it is placed in a sitting position; the widow kneels before it and embraces it tightly. She remains in that posture while they build a wall around them; when it reaches just up to the widow's shoulders, one of her family grabs her head from behind and twists her neck; once her spirit has departed the wall is quickly raised and covered over and there they remain entombed.[5]

[A]   In that same country there was a similar practice among their

---

5. From Classical times suttee was known from Cicero, *Tusc. disput.*, V, xxvii, 77 and its commentators; Montaigne clearly used another source as well.

Gymnosophists for, not by the constraint of others nor by a sudden caprice but by the express terms of their profession, their custom was, as they were approaching a certain age or realized that they were threatened by some disease, to have a pyre built for them, on top of which was, placed a bed richly adorned; then, after having joyfully feasted their friends and acquaint-ances, they settled themselves firmly on that bed, resolved that when the fire was put to it no man should see them stir hand or foot. Thus did one of them, Calanus, die before the entire army of Alexander the Great.[6] [B] And no man among them was reckoned holy or blessed unless he killed himself that way, having dispatched his soul, purged and purified by fire, after all that was mortal and earthy in him had been consumed.

[A] What makes it a miracle is that stable, lifelong premeditation. For intermingled with it is, among our other debates, the question of fate, FATUM. For if we bind things to come and our very wills to a definite and inevitable necessity, we are still on that age-old argument: since God foresees, as he undoubtedly does, that all must happen thus, happen thus they must. To which *Magistri Nostri*[7] reply that to see something happen as we do — and God, too (since all is present to him; he sees rather than foresees) — is not to force it to happen. Indeed, we see the things because they happen: they do not happen because we see them. The event produces the knowledge, not the knowledge the event. What we see happen *is* happening: but it could have happened otherwise. And God, in the book of the causes of events which he has in his foreknowledge, also includes such causes as we term fortuitous and voluntary, those which depend on the liberty which he has given to our free-will: he knows that we will go astray because we shall have willed to do so. Now I have seen plenty of nations encouraging their troops with this necessity of Fate. For if our hour is bound to come at a particular point, neither volleys from enemy harquebuses nor our own rashness nor our running away nor our cowardice can advance it or retard it. That is a beautiful saying: but find a man who will act on it! And if it is the case that a strong and lively belief brings in its train analogous actions, then that faith which our mouths are so full of is

6. Plutarch, *Life of Alexander*.
7. 'Our Masters': the title of Professors of Theology in the Sorbonne. Their explanation of God's foreknowledge is the standard Platonico-Christian one: God, the Creator of time, alone has an absolute existence outside time. For God, all things past, present and future are seen in an eternal present. But to see an event is not to cause it; neither, for God therefore, is 'foreseeing' necessarily causative. (Cf. the end of II, 12, 'An apology for Raymond Sebond'.)

wondrously light in our own age, unless it be that her contempt for works
makes her despise their company![8] All the same, while on this subject, the
Sire de Joinville, as good a witness as any other, tells us that the Bedouin, a
people living among the Saracens with whom our King Saint Louis had
some trouble in the Holy Land, believe so firmly by their religion that the
days of each man have been numbered in advance by a preordained
inevitability that they go bare to the wars except for a sword of Turkish
fashion and a white linen garment. And the worst malediction they always
had on their lips when they were angry with one of their own men was,
'Cursed be thou like he who wears armour for fear of death!' That is a very
different proof of belief and faith than ours is.[9]

We can rank with it the proof given by two monks of Florence in the
time of our fathers. Opposed over some point of doctrine they agreed that
both of them should be burned in the public square before all the people,
each one wishing to prove he was right. All the preparations had been
made and the deed on the very point of being done when it was interrupted
by some unforeseen incident.[10]

[C] A young Turkish lord had personally performed some remarkable
feat of arms before the armies of both Amurath and Hunyadi who were
ready to join battle; when Amurath asked him how a youth so young and
inexperienced (for it was the first war he had seen) had been filled with
such noble and valliant courage, he replied that his sovereign tutor in
valour had been a hare. 'I was out hunting one day,' he said, 'when I came
across a hare lying in its form; although I had two excellent greyhounds at
my side it seemed to me better, so as not to lose it, to use my bow, for it
made a very good target. I started shooting off my arrows – I had some
forty odd in my quiver – but did not hit it, let alone disturb it. In the end I
let loose my hounds: they could do nothing either. This made me realize

8. A theological quip. A 'lively' faith, a faith informed with charity, manifests itself
through the works of charity. Otherwise it is dead. Theological controversy led
Reformers and Evangelicals to segregate faith and works into discrete compart-
ments: a man may have the 'true' faith yet do no corresponding good works
which prove it to be a true and lively one. Both sides in the Civil Wars could be
misled into contempt for good works, prizing orthodoxy above all else. Hence
(for Montaigne) the decadence and the atrocities of his age in which rival credal
orthodoxies took precedence over works of charity.

9. Joinville, *Vie de Saint Louis*, XXX. (Guillaume Postel, the Renaissance expert on
Turkish affairs, was struck by the religion and piety of the Turks and by their
valour.)

10. Cf. Innocent Gentillet, *Discours . . . de bien gouverner*, II, xii. Then, Nicolas
Chalcocondylas, *De la décadence . . .*, VII, viii.

that that hare was protected by destiny and that swords or arrows only strike home by leave of our fate, which it is not in our power to retard or advance.'

That tale should teach us *en passant* how bendable our reason is to all sorts of conceptions. A man of importance, great in years, in glory, in dignity and in doctrine, boasted to me that he had been led to make a most important change of faith by some monition coming to him, one so bizarre and incidentally so inconclusive that I found that it tended, rather, the opposite way. He called it a miracle. So did I – in a different sense.

The historians of the Turks say that this conviction that their days are numbered by the unbending decision of Fate is so widespread among the people that it manifestly is seen to give them assurance in danger. And I know a great Prince who may nobly draw some profit from it, if Fortune continues to give him a shove.[11]

[B] In living memory there has been seen no more strikingly resolute act than that of the two men who plotted the death of the Prince of Orange.[12] It is a marvel how anyone could have so enflamed the second of them, who brought it off, for an undertaking in which his comrade had fared so badly despite doing all that he could: to go and follow in his tracks, and with the same weapons to take on a nobleman, freshly armed with a lesson in mistrust and strong in bodily strength and in his retinue of friends, in his own hall, surrounded by his guards, and in a town devoted to him! He indeed had brought to bear a most determined hand and a mind moved by a stalwart passion. A dagger is surer to land a blow; but since it requires a bigger movement and more strength of arm than does a pistol, its blow is more susceptible to being warded off or intercepted. I have no great doubt that that man knew he was running to a certain death: for any hopes which he could have been brought to entertain could find no lodging in a settled intelligence – and the way he executed his deed shows that he had no lack of that, no more than of courage.

The motives for such a powerful conviction may well be various, since our faculty of perception does with us, and with itself, whatever it likes.

That other assassination carried out near Orleans was nothing like it:[13]

11. Doubtless Henry of Navarre (Henry IV).

'95: profit from it, *should he either believe it or else use it as justification to take extraordinary risks, provided that Fortune does not tire too soon of giving him a leg up.* [B] In living . . .

12. The would-be assassins were Jeaureguy (1582) and Balthasar Gérard (1584).

13. The murder of the Duc de Guise (1563) by Poltrot de Méré.

there was far more chance in it than vigour; the blow itself would not have been fatal if chance had not rendered it so; and the design of shooting from the saddle, and at a distance, and at a man who was jogging up and down with his horse, was the plan of a man who would rather lose his chance than lose his life. What happened afterwards proves this. For the thought of having killed someone so exalted made him dazed and transfixed so that he completely lost his wits and was too disturbed to guard his escape, or his tongue under questioning. What did he have to do, beyond galloping back to his friends across a river! It is a method I have leapt to in lesser dangers and which I do not consider very hazardous, however wide the crossing, provided that your horse readily finds its footing and you can see ahead to an easy place to land on the other side, depending on the current. As for that other man,[14] when they pronounced his dreadful sentence, he replied, 'I was ready for that: I will amaze you by my endurance.'

[C] The Assassins, who are a people dependent on Phoenicia, are considered by the Mahometans to be sovereignly devout and pure in morals. They hold that the surest way to merit paradise is to kill someone of an opposing religion. They therefore show contempt for all personal danger and are often to be found singly or in pairs, carrying out such profitable executions at the cost of their certain death, appearing before an enemy in the midst of his troops to 'assassinate' him – (it is from them that we have borrowed that word). Our own Count Raymond of Tripoli was killed this way in his own city.[15]

---

14. Balthasar Gérard.
15. '95: city, *during our expeditions in the Crusades. So too Conrad, Marquess of Montfarat, whose murderers were all brought to the scaffold full of elation and proud of so beautiful a masterpiece* . . . Cf. Bernard de Girard, *Hist. des Roys de France.*

30. On a monster-child

['Monsters' were widely held, even by professional men of all kinds, to be 'demonstrations' – portents of God's will. Montaigne personally examined two such cases: some Siamese twins and a malformed shepherd. His original chapter left all discussion to the doctors, many of whom, even the great Dr Ambrose Paré, believed that at least some 'monsters' are monstra, omens showing divine anger or approval. In his final text, [C], Montaigne explains 'monsters' in platonic terms as rare examples of the infinite forms existing in God's created Nature, vast numbers of which are unknown to Man.]

[A] This tale will go its simple way, for I shall leave all the discussion to the doctors.

I saw the day before yesterday an infant child that two men and a wet-nurse (who said they were its father, uncle and aunt) were travelling about with and exhibiting for its strangeness, so as to make a penny or two out of it.

In every other way that boy was of the normal form and could stand up on his own legs, walking and warbling more or less like other children of his age: he had not yet been willing to accept any food other than from his nurse's breasts: what they assayed putting into his mouth, in my presence, he chewed for a while then spat out without swallowing anything. There certainly seemed something peculiar about the way he cried; he was then just fourteen months old. Just below his breast he was firmly attached to another child with no head and with the spinal canal blocked, though the rest of the body was entire: one arm was in fact shorter than the other, but that was accidentally broken at birth. They were joined facing each other, looking as though a slightly smaller child were trying to put his arm round the neck of a slightly bigger one. The area joining them together was merely about four fingers wide, so that if you raised up that imperfect child you could see the other one's navel underneath: the join was therefore found between his nipple and his navel. There was no sign of a navel in the imperfect child, though all the rest of the belly was there: the parts of that imperfect child which were not attached, such as the arms, buttocks, thighs and legs, dangled down loosely over the other one, and in length could reach down to his knees. The wet-nurse said the monster urinated through

both places: indeed the limbs of the imperfect child were as much alive, as well fed and in the same condition as the other's, except that they were smaller and thinner.

This double body and these sundry limbs all depending on one single head could well provide us with a favourable omen that our king will maintain the sundry parties and factions of our State in unity under his laws; but for fear lest the outcome should belie it we should let that happen first, for there is no divining like divining about the past! [C] *'Ut quum facta sunt, tum ad conjecturam aliqua interpretatione revocantur.'* [Once things have happened we can find some interpretation of them which turns them into prophecies.] [B] As was said of Epimenides: he always prophesied backwards.[1]

I have just seen a shepherd in Médoc: he is about thirty years old and has no sign of any genitals, having three holes through which he ceaselessly makes water. He wears a beard and enjoys the touch of women.

[C] What we call monsters are not so for God who sees the infinite number of forms which he has included in the immensity of his creation: it is to be believed that the figure which astonishes us relates to, and derives from, some other figure of the same genus unknown to Man. God is all-wise; nothing comes from him which is not good, general and regular: but we cannot see the disposition and relationship: *'Quod crebro videt, non miratur, etiam si cur fiat nescit. Quod ante non vidit, id, si evenerit, ostentum esse censet.'* [What a man frequently sees never produces wonder in him, even though he does not know how it happens. But if something occurs which he has never seen before, he takes it as a portent.][2]

Whatever happens against custom we say is against Nature, yet there is nothing whatsoever which is not in harmony with her. May Nature's universal reason chase away that deluded ecstatic amazement which novelty brings to us.

---

1. Cicero, *De divinatione*, II, xxxi, 56; Aristotle, *Rhetorica*, III, viii. On Epimenides the Greek philosopher and thaumaturge, cf. Cicero, *De legibus*, II, 11, 28; *De divinatione*, I, xviii; Pliny, VII, 48–53.
2. Cicero, *De divinatione*, II, xxii, 49. (The Platonic notion of the 'great chain of being' held that God in his infinite power created all possible forms. Man, being finite, can know only a few of them.)

31. On anger

===

[Montaigne first read the Moral Works of Plutarch (as distinct from his Parallel Lives) in Amyot's great French translation during 1573. This chapter shows how philosophy is not merely a matter of argument and abstractions but of basic practical morals affecting wives and children as much as generals and statesmen. That a true philosopher should not give way to anger was a commonplace, emphasized by the Stoics and taken over by many Christians – in Le Tartuffe Molière will make the servant-girl laugh at Orgon with the taunt: 'Ah! You are devout: and you are angry!' Anger was believed to be caused by choler, one of the four humours, which made a man bilious and irascible. Montaigne also associated it with chagrin, that grievous vexation brought on by melancholy.]

[A]   Plutarch is amazing in every respect but especially where he makes judgements on men's actions. In his parallel lives of Lycurgus and Numa we can see the beauty of what he says when treating of our great stupidity in abandoning children to the responsibility and control of their fathers. [C]   The majority of our polities, as Aristotle says, are like the Cyclops, abandoning the guidance of the women and children to each individual man according to his mad and injudicious ideas: hardly any, except the polities of Sparta and of Crete, have entrusted the education of children to their laws.[1]   [A]   Anyone can see that all things within a State depend upon the way it educates and brings up its children. Yet quite injudiciously that is left to the mercy of the parents, no matter how mad or wicked they may be.

How many times have I been tempted, among others things, to make a dramatic intervention so as to avenge some little boys whom I saw being bruised, knocked about and flayed alive by some frenzied father or mother beside themselves with anger. You can see fire and rage flashing from their eyes –

1. Aristotle, *Nicomachaean Ethics*, X, ix, 1180a (with, for Crete, I, xiii, 1102a). The educational ideas of Sparta so impressed Erasmus that he devoted a whole section of the *Apophthegmata* to them, remarking as how Christians can learn from them.

[B]   *rabie jecur incendente, feruntur*
Præcipites, ut saxa jugis abrupta, quibus mons
Subtrahitur, clivoque latus pendente recedit

[they are carried away by burning wrath, like boulders wrenched free from the cliff crashing down the precipitous slope]

(according to Hippocrates the most dangerous of distempers are those which contort the face)² – [A] as with shrill wounding voices, they scream at children who are often barely weaned. Children are crippled and knocked stupid by such batterings: yet our judicial system takes no note of it, as though it were not the very limbs of our State which are thus being put out of joint and maimed.

[B]   *Gratum est quod patriæ civem populoque dedisti,*
Si facis ut patriæ sit idoneus, utilis agris,
Utilis et bellorum et pacis rebus agendis.

[It is good to have given a citizen to the people and the State – provided that you make him fit for his country, good at farming, good in war and peace.]³

[A]   No passion disturbs the soundness of our judgement as anger does. No one would hesitate to punish with death a judge who was led to condemn his man as a criminal out of anger: then why is it any more permissible for fathers and schoolmasters to punish and flog children in anger? That is no longer correction, it is vengeance. For a child punishment is a medicine: would we tolerate a doctor who was animated by wrath against his patient? We ourselves, if we would act properly, should never lay a hand on our servants as long as our anger lasts. While our pulse is beating and we can feel the emotion, let us put off the encounter: things will really and truly look different to us once we have cooled off a bit and quietened down. Until then passion is in command, passion does all the talking, not us. [B] Faults seen through anger are like objects seen through a mist: they appear larger. If a man is hungry, then let him eat food: but he should never hunger and thirst for anger if he intends to chastise.

[A]   And then punishments applied after being judiciously weighed are more acceptable and more useful to the sufferer. Otherwise he does not think that he has been justly condemned by a man shaking with anger and

2. Juvenal, VI, 647–9; Hippocrates, in Plutarch (tr. Amyot), *Comment il fault refrener la colere* 579–H and, later, 60 E.
3. Juvenal, XIV, 70–3.

fury; he cites in his own justification the inflamed face of the schoolmaster, his unaccustomed swearing, his mental disturbance and his precipitate haste.

> [B]   *Ora tument ira, nigrescunt sanguine venæ,*
> *Lumina Gorgoneo sævius igne micant.*

[His face swells with anger, the blood darkens in his veins and his eyes flash with fire more savage than a Gorgon's.][4]

[A]   Suetonius relates that Lucius Saturninus, after being condemned by Caesar, was helped in winning his case before the People (to whom he had appealed) above all by the bitter *animus* that Caesar brought to his verdict.

Saying is one thing: doing is another; we must consider the preaching apart and the preacher apart. Those who in our own time have made an assay at shaking the truth taught in our Church by citing the vices of her ministers have given themselves an easy game. Her testimonies are drawn from elsewhere. That way of arguing is stupid: it would throw everything into confusion. A man of good morals may hold false opinions: a wicked man can preach the truth – yes, even truths he does not believe. It is most certainly harmonious and beautiful when saying and doing go together; I have no wish to deny that saying has more authority and efficacity when followed by doing – as Eudamidas remarked on hearing a philosopher discoursing about war: 'Beautiful words: but the man who spoke them cannot be believed since his ears are not used to the sound of the trumpet.' And when Cleomenes heard a professor of rhetoric declaiming about valour he burst out laughing; the professor was scandalized but Cleomenes replied: 'I would do the same if it were a swallow speaking: now if it were an eagle, I would willingly listen.'[5]

It seems to me that I can perceive from the writings of the Ancients that the man who says what he really thinks drives it home in a livelier way than he who only pretends. Listen to Cicero talking about the love of liberty: then listen to Brutus! The very writings declare that Brutus was the man to purchase liberty at the cost of his life. Let Cicero, the father of eloquence, treat the theme of contempt for death; let Seneca treat it too: Cicero drags it out lifelessly and you can feel that he wants to make you resolute about something for which he himself has no resolution at all. He cannot put heart into you: he has none to give. But Seneca rouses you and inflames you.

4. Ovid, *De arte amandi*, III, 503–4. (Echoes of Seneca's *De ira*, III, xxxii, and of Plutarch's (tr. Amyot) *Comment il fault refrener la colere*, and of Suetonius' *Caesar*.)
5. Plutarch (tr. Amyot), *Dicts notables des Lacedaemoniens*, 216–18.

I never read an author, especially one treating of virtue and duty, without curiously inquiring what sort of man he was. [B] The Ephors of Sparta, on seeing a dissolute man giving useful advice in a speech before the people, ordered him to stop and requested a man of honour to sponsor the new idea and to speak for it.[6]

[A] The writings of Plutarch if you savour them well adequately reveal him – and I believe that I know Plutarch, penetrating even into his soul. Yet I could wish that we had some personal memoirs. If I have let myself go in this digression it is because of the gratitude I feel towards Aulus Gellius for having bequeathed to us in his writings the following account of his manners which touches again on my subject: anger.

One of Plutarch's slaves, a bad, wicked man whose ears had however drunk in a few lectures in philosophy, had been stripped for some crime by order of Plutarch; at first, while he was being flogged, he snarled about its not being right and that he had not done anything wrong; but in the end he started to shout abuse at his master in good earnest, accusing him of not really being a philosopher as he boasted, since he had often heard him say that it was ugly to get angry and had even written a book on the subject; the fact that he was now immersed in anger and having him cruelly flogged completely gave the lie to his writings. To which Plutarch, quite without heat and completely calm, replied: 'What makes you think, you ruffian, that I am angry at this time? Does my face, my voice, my colouring or my speech bear any witness to my being excited? I do not think my eyes are wild, my face distorted nor my voice terrifying. Is my face inflamed? Am I foaming at the mouth? Do words escape me which I will later regret? Am I all a-tremble? Am I shaking with wrath? Those, I can tell you, are the true symptoms of anger.' Then turning towards the man who was doing the flogging he said, 'Carry on with your job, while this man and I are having a discussion.'

That is the account in Aulus Gellius.

On returning from a war in which he had served as Captain-General, Archytas of Tarentum found in his house every sign of mismanagement and his lands lying fallow through the neglect of his steward. He summoned him before him and said, 'Go. If I were not so angry I would give you a good going over.' So too Plato: when he was inflamed against one of his slaves he handed him over to Speucippus for punishment, apologizing for not laying hands on him himself since he was angry. Charillus, a Spartan, said to a helot who was behaving most insolently and audaciously toward

6. Plutarch (tr. Amyot), *Comment il fault ouir*, 26G. Then, Aulus Gellius, I, xxvi.

him: 'By the gods! If I were not so angry I would have you put to death at once.'[7]

Anger is a passion which delights in itself and fawns on itself. How often, if we are all worked up for some wrong reason and then offered some good defence or excuse, we are vexed against truth and innocence itself! I can recall a marvellous example of this from Antiquity. Piso, a great man in every other way, noted for his virtue, was moved to anger against one of his soldiers. Because that soldier had returned alone after foraging and could give no account of where he had left his comrade, Piso was convinced that he had murdered him and at once condemned him to death. When he was already on the gallows, along comes the lost comrade! At this the whole army was overjoyed and, after many a hug and embrace between the two men, the executioner brought both of them into the presence of Piso; all those who were there were expecting that Piso himself would be delighted. Quite the contrary: for, through embarrassment and vexation, his fury, which was still very powerful, suddenly redoubled and, by a quibble which his passion promptly furnished him with, he found three men guilty because one had just been found innocent, and had all three of them executed: the first soldier because he was already sentenced to death; the second, the one who had gone missing, because he had caused the death of his comrade; the hangman for failing to obey orders.[8]

[B] Those who have had to deal with obstinate women may have made an assay of the raging madness that they are thrown into when you confront their agitated minds with silence and coldness and do not condescend to feed their bad temper. Celius the orator was of a marvellously choleric nature. There was, dining in his company, a man of mild and gentle manners who, so as not to provoke him, decided to approve of everything he said and always to agree with him; but Celius could not tolerate that his evil temper should thus pass unfed and exclaimed: 'For the gods' sake challenge something that I say, so that there can be two of us!'[9] Similarly women get angry only to make us angry in turn, imitating the laws of love. Phocion, when a man kept interrupting what he was saying

7. Plutarch (tr. Amyot), *Comment il fault nourrir les enfans*, 6D–E; *Dicts notables des anciens Roys*, 198 F–G. Both anecdotes are well-known from Erasmus' *Apophthegmata*, VII, *Plato*, VII; I, *Charillus seu Charilaus*, XLV; cf. also VIII, *Architas*, XXXII.

8. Seneca, *De ira*, I, xvi.

9. Seneca, *De ira*, III, viii; then, Plutarch, *Instruction pour ceulx qui manient affaires d'Estat*, 169 B.

with bitter insults, simply stopped talking, giving him enough time to exhaust his choler; when that was over, without mentioning the disturbance, he took up his speech just where he had left off. No retort goads a man more sharply than disdain such as that.

Of the most choleric man in France (and it is always a defect, though pardonable in a fighting man since in the exercise of that profession there are certainly situations where it cannot be dispensed with) I often say that he is in fact the most long-suffering man I know in restraining his choler. It shakes him with such violence and frenzy –

> *magno veluti cum flamma sonore*
> *Virgea suggeritur costis undantis aheni,*
> *Exultantque æstu latices; furit intus aquaï*
> *Fumidus atque alte spumis exuberat amnis;*
> *Nec jam se capit unda; volat vapor ater ad auras*

[as when, beneath a brazen cauldron, the fire roars noisily into flame and licks its sides, the water boils with the heat and, madly foaming in its prison, breaks over the edge and can contain itself no longer, sending black fumes off into the air]

– that, to moderate it, he has to keep himself under cruel restraint.[10] Personally I know of no passion of mine for which I could ever make so great an effort to hide and withstand. I would not care to rate wisdom at so high a price as that. I do not so much look at what that man does, as what it costs him not to do worse.

Another great man boasted to me of the gentle correctness of his manners, which was truly unique. I replied that, especially in one of so eminent a rank and on whom all eyes were turned, it was indeed something to present oneself always moderate to the world, but that the main thing was to provide inwardly for oneself: to my taste a man was not managing his business well if he was eating his insides out. I am afraid that he was doing just that, so as to maintain the mask of that outward appearance of correctness.

By hiding our choler we drive it into our bodies: as Diogenes said to Demosthenes, who kept drawing back further inside so as not to be spotted in a tavern: 'The more you draw back, the further in you go!'[11] I would advise you to give your valet a rather unseasonable slap on the cheek rather than to torture your mind so as to put on an appearance of wisdom; I would rather make an exhibition of my passions than brood over them to

10. Virgil, *Aeneid*, VII, 462–6. (The man is unidentified.)
11. Erasmus, *Apophthegmata*, III, *Diogenes Cynicus*, XXXIII.

my cost: express them, vent them, and they grow weaker; it is better to let them jab outside us than be turned against us: [C] *'Omnia vitia in aperto leviora sunt; . . . et tunc perniciosissima cum simulata sanitate subsidunt.'* [All defects are lighter in the open: . . . they are most pernicious when concealed beneath a pretence of soundness.][12]

[B] I advise those of my family who have the right to show their anger, firstly to be sparing of their choler and not to scatter it abroad no matter what the cost, since that thwarts its action and its weight; even the anger you vent on a servant for a theft makes no impression then: it is the same anger he has seen you use against him a hundred times already, for a glass badly rinsed or a stool left out of place. Secondly, let them not get angry in the void; let them see that their reprimand falls to the one they are complaining about, for as a rule they are yelling before he has answered their summons; and they go on doing so for ages after he has gone:

> *et secum petulans amentia certat.*

> [petulant madness turns against itself.][13]

They go at their own shadows and bluster about in places when nobody is punished or affected by it, except such as cannot stand their din.

I similarly blame those who boast and bluster about in quarrels where there is no adversary: let them keep such rodomontades for when they can have a target:

> *Mugitus veluti cum prima in prælia taurus*
> *Terrificos ciet atque irasci in cornua tentat,*
> *Arboris obnixus trunco, ventósque lacessit*
> *Ictibus, et sparsa ad pugnam proludit arena.*

[Thus roars the bull fresh to the combat. With terrifying bellows it tries out its anger by dashing its horns against a tree-trunk; it lashes out at the air and paws the sand in the arena as a prelude to battle.]

When I get angry it is as lively, but also as short and as secret, as I can make it. I lose control quickly and violently, but not with such turmoil that I go gaily hurling about all sorts of insults at random and fail to lodge my goads pertinently where I think they can do the most damage: for I normally use only my tongue. My servants get off more cheaply in serious

---

12. Seneca, *Epist. moral.*, LVI, 10 – reading *leniora* (more gentle) not *leviora* (more light).
13. Claudianus, *In Eutropium*, I, 237; then Virgil, *Aeneid*, XII, 103–6.

cases than in little ones. The little ones take me by surprise: unfortunately, once you are over the edge, no matter what gave you the shove, you go right down to the bottom; the very fall, of itself, presses on in haste and confusion. In the serious cases I am satisfied with their being so obvious that everybody expects me to give birth to justified anger: I glory in disappointing their expectations. I prepare and brace myself against those serious cases: they dig into my brain and threaten to carry me too far if I follow where they lead. No matter how violent the cause it is easy to prevent myself from giving way to the impulsion of that passion, and I am strong enough to resist it, provided I am expecting it. But if it takes me unawares and once gets a hold on me I am carried away, no matter how trivial the cause.

This is the bargain I strike with those who may have to contend with me: when you see I am the first to get worked up, just let me go on, right or wrong: I will do the same in return. The storm is engendered only by the confluence of cholers, both prone to beget the other: and they are not both born at the same instant. Let us allow each one to run its course: then we always have peace.

A useful prescription but difficult in practice.

It sometimes happens that, without any real emotion, I put on an act of being angry in order to govern my household. But as my age renders my humours more and more acrid I strive to oppose them; if I can, I will see that, from this time forth, the more justification and inclination I have, the less I shall be chagrined and difficult – although I have been among the least so up till now.

[A]   One more word to close this chapter. Aristotle says that choler sometimes serves virtue and valour as a weapon.[14] That is most likely; nevertheless those who deny it have an amusing reply: it must be some new-fangled weapon; for we wield the other weapons: that one wields us; it is not our hand that guides it: it guides our hand; it gets a hold on us: not we on it.

14. Aristotle, *Nicomachaean Ethics*, III, viii, 1167b, commented on by Seneca, *De ira*, III, viii, in Montaigne's sense.

32. In defence of Seneca and Plutarch

[In this chapter Montaigne reveals not only how he reads his books but dares to give the great Bodin, the author of the famous Method *for studying history, a lesson in historical interpretation. That makes this one of his more personally revealing chapters, as well as once again emphasizing Montaigne's lasting preoccupation with philosophical and moral ecstasy.]*

[A]   My intimacy with those two great men and the help they give to me in my old age,   [C]   as well as to my book which is built entirely out of their spoils,   [A]   bind me to espouse their honour.

As for Seneca, among the thousands of little pamphlets that those of the Religion Allegedly Reformed[1] circulate in defence of their cause (which come sometimes from the hands of good writers which it is a pity not to find occupied on a better subject) I saw one, long ago, which extended and filled out the similitude it intended to establish between the rule of our poor late King Charles IX and that of Nero, by comparing the late Cardinal of Lorraine to Seneca – including their destinies (which made them both first men in the governments of their monarchs) their morals, endowments and conduct.[2] In this, in my opinion, he does too much honour to my Lord the Cardinal; for while I am one of those who rate highly his intelligence, his eloquence, his zeal for religion and for the King's service as well as his good fortune in being born in an age when it was so new, so rare and so necessary for the public good to have a great Churchman of such nobility, so worthy and capable of his office: nevertheless, to tell the truth, I do not think that his ability was anywhere near Seneca's nor that his virtue was as pure and as inflexible as his.

---

1. The official French Roman Catholic name for the religion of the Reformed Church of the 'Calvinists' was *la Religion Prétendue Réformée*, often abbreviated to RPR.
2. When Nero became Emperor in AD 54, Seneca, who had been his tutor, became his counsellor and minister; the Cardinal of Lorraine was counsellor to Charles IX.

Now that pamphlet which I am talking about,[3] so as to attain its purpose, has a description (which is deeply insulting) borrowed from the strictures on Seneca by Dion the historian, whose testimony I simply do not believe; for Dion, apart from being inconsistent in first calling Seneca very wise and also a mortal enemy of Nero's vices, nevertheless makes him mean, given to usury, ambitious, cowardly, pleasure-seeking and a counterfeit philosopher under false colours. Seneca's virtue is so evidently alive and vigorous in his writings, which themselves provide such a manifest defence against such insinuations as his being excessively rich and spendthrift, that I could never accept any witness to the contrary. Moreover in matters such as these it is more reasonable to trust the Roman historians than foreign Greek ones.[4] Now Tacitus speaks most honourably of his life and of his death, portraying him in all things as a great man, most excellent and most virtuous. And it will be enough for me to make no criticism but this of Dion's power of judgement – an unavoidable one: his judgement of matters Roman was so diseased that he ventured to champion the causes of Julius Caesar against Pompey, and of Antony against Cicero.

Now for Plutarch.

Jean Bodin is a good contemporary author, endowed with far better judgement than the mob of scribblers of his time: he merits our own considered judgement. I find him a bit rash in that passage of his *Method of History* where he accuses Plutarch not only of ignorance (on that he can say what he likes: I do not hunt that game) but also of frequently writing 'things which are incredible and entirely fabulous' (those are his very words).[5] If he had simply said 'things otherwise than they are', that would have been no great censure, since we have to take on trust from the hands of others things we have not ourselves witnessed, and I can see that he occasionally relates the same event differently, well aware that he is doing so: for example, the judgement of the three best Captains that there ever were, which Hannibal made, appears differently in his life of Flaminius and in his life of Pyrrhus. But to charge him with having accepted as valid currency things unbelievable and impossible is to accuse the most judicious author in the world of lack of judgement.

---

3. Perhaps the *Memoires de l'Estat de France, sous Charles Neufiesme* of Simon Goulart.
4. Dion Cassius' censures in his Greek *Roman History* (which was widely read in Xylander's Latin translation) are normally accepted now as justified. (But cf. Tacitus, *Annals*, XIII, 1, XIV, liii, etc.)
5. Jean Bodin, *Methodus ad facilem historiarum cognitionem*, 1566, IV, p. 58.

Here is Bodin's example. 'As,' he says, 'when he relates that a Spartan boy allowed his entire stomach to be torn out by a fox-cub which he had stolen and kept hidden under his tunic until he died rather than reveal his theft.' In the first place I find that a badly chosen example, since it is hard indeed to prescribe limits to the powers of the faculties of our souls, whereas in the case of bodily strength we have more means of knowing them and of setting bounds to them. For that reason if I had to choose an example I would rather have taken one from the second category, where some facts are harder to believe – among others what he narrates about Pyrrhus: that, gravely wounded as he was, he gave so great a blow with his sword to an enemy clad in full armour that, from the top of his head downwards, he clove him in two.

In Bodin's own example I find no great miracle; nor do I accept the excuse that he makes for Plutarch, that he added the words 'So they say', to warn us to keep a bridle on our credulity. For apart from such things as are accepted on the authority of Antiquity or out of respect for religion, Plutarch would not himself have accepted to believe things intrinsically incredible nor would he have proposed that we should. And as for the phrase, 'So they say', he does not employ it in this context with that sense: that is easy to see, since he relates elsewhere other examples touching the powers of endurance of the boys of Sparta which happened in his own time and which are even harder to accept, such as the one to which Cicero bore witness before him, having, he says, been there himself: some boys were undergoing that test of endurance by which the Spartans assayed them before the altar of Diana; they allowed themselves to be flogged until they were all over blood, without uttering cry or groan, some having sufficient strength of will to lose their lives there.[6]

Then there is the one which Plutarch relates with a hundred other witnesses; during the sacrifice a hot coal slipped up the sleeve of a Spartan boy while he was swinging the incense; he let the whole of his arm be burnt until the smell of cooked flesh reached the congregation.

By Spartan custom nothing more directly affected your reputation nor made you suffer more shame and disgrace than being caught out stealing.[7] I am so imbued with the greatness of those men of Sparta that not only

6. Cicero, *Tusc. disput.*, II, xiv, 34; cf. Erasmus, *Apophthegmata*, II, *Prisca Lacedaemoniorum Instituta*, XXXIV.
7. Spartan boys were underfed and taught to steal food: i) to increase their hardihood and skill at foraging in war; ii) to make Spartans defend their property. Any boy *caught* stealing was flogged. (Erasmus, *Apophthegmata*, XII.)

does it not seem incredible to me as it does to Bodin: it does not even seem rare or unusual. [C] The history of Sparta is full of hundreds of harsher and rarer examples: by Bodin's standards it is all miracle. [A] On this subject of theft, Marcellinus reports that nobody in his time had yet found any kind of torture which could force any Egyptians surprised in this crime to tell you even their own name.[8]

[B] A Spanish peasant who was put to the rack to make him reveal his accomplices in the murder of the praetor Lucius Piso yelled out in the midst of his tortures that his friends should not go away but stay and watch in full confidence, since it was not in the power of pain to force a single word of confession from him. And on the first day that was all they did get out of him. The next day, as they were escorting him back to start torturing him again, he struggled violently in the hands of his guards and killed himself by bashing his head against a wall.

[C] Epicharis, having glutted and exhausted the cruelty of Nero's attendants and withstood for one full day their burning brands, their beatings and their instruments of torture without revealing a word of her conspiracy, was brought back to the rack the next day with her limbs all shattered: she slipped the cord from her dress through the arm of a chair, made a running knot, thrust her head through it and hanged herself by the weight of her body. Having as she did the courage to die thus after having endured those first tortures, does she not appear to have deliberately lent herself to that trial of her endurance in order to mock that tyrant and to encourage others to make a plot against him similar to her own?

[A] And if anyone would go and ask our mounted riff-raff about the experiences which they have had in these civil wars of ours, he will hear of acts of endurance, of obstinate resistance and of stubbornness even among that rabble – effeminate though it is with a more than Egyptian sensuality[9] – worthy of being compared which we have just rehearsed of Spartan valour.

I know that there are cases of simple peasants who were prepared to allow the soles of their feet to be burnt, their fingertips to be smashed with the butt of a pistol, their eyes to be forced all bloody from their sockets by having a thick cord twisted tight around their foreheads, before they would even think about putting themselves to ransom. I myself saw one who was left for dead, naked in a ditch, with his neck all swollen and bruised by a halter which still dangled down from it and by which he had

8. Ammanius Marcellinus, XXII, xvi; then Tacitus, *Annals*, IV, xlv and XV, lvii.
9. Cf. Cognatus' Adage, *Miles Romane, Aegyptum cave*.

been dragged all night behind a horse, his body stabbed through by daggers in a hundred places – not to kill him but to make him feel pain and fear – and who had suffered all that until he had lost all power of speech, all consciousness, determined (as he told me) to die a thousand deaths (as indeed, so far as suffering is concerned, he had died one whole death already) rather than promise them anything; yet he was one of the richest husbandmen of the entire region. How many have we seen patiently suffering to be roasted or burnt for opinions which, without understanding or knowledge, they have taken from others!

[B] I have known hundreds and hundreds of women (for they say that Gascon heads have some special gift for this) whom you would have more easily made to bite a red-hot iron than made to let go of an opinion conceived in a fit of choler once they have got their teeth into it. Women are rendered intractable by blows and constraint. That man who forged the tale of the goodwife who would not stop calling her husband lice-ridden however much she suffered correction by threats and cudgelings, who was thrown into a pond and, even while she was drowning, thrust her hands out of the water high above her head and made the sign for squashing lice, forged indeed a tale the express image of which we can see every day in the stubbornness of women.[10] And stubbornness is the sister of constancy, in vigour and inflexibility at least.

[A] We must not judge what is possible and impossible according to what seems credible or incredible to our own minds (as I have said elsewhere). It is nevertheless a major fault into which most people fall – [C] and I do not say that of Bodin – [A] to make difficulties about believing of another anything which they could not [C] or would not [A] do themselves.[11] It seems to each man that the master Form of Nature is in himself, as a touchstone by which he may compare all the

---

10. A well-known tale in Poggio's *Facetiae*.

11. A reworked passage revealing Montaigne's conception of philosophical ecstasy:
i) '80: do themselves. I consider *some of those souls of the Ancients to be raised up to Heaven when valued against mine;* and even though I realize that I am powerless to follow them, I do not give up judging the principles which raise *and lift* them thus aloft. I admire . . .

ii) ['95] . . . that the master Form of *human* nature is in himself *and that all the others must be regulated in accordance with it. Attitudes which do not correspond to his own are* feigned and *false. Do you set before him some details of the deeds or capacities of another man? The first thing which he calls upon to guide his judgement is himself as a standard: as things go with him, thus must they go with the Order of the world. O dangerous and intolerable asininity!* I consider . . .

other forms. Activities which do not take his form as their model are feigned and artificial. What brute-like stupidity! I consider some men, particularly among the Ancients, to be way above me and even though I clearly realize that I am powerless to follow them on my feet I do not give up following them with my eyes and judging the principles which raise them thus aloft, principles the seeds of which I can just perceive in myself, as I also can that ultimate baseness in minds which no longer amazes me and which I do not refuse to believe in either. I can clearly see the spiral by which those great souls wind themselves higher. [A] I admire the greatness of those souls; those ecstasies which I find most beautiful I clasp unto me; though my powers do not reach as far, at least my judgement is most willingly applied to them.

The other example which Bodin cites of 'things which are incredible and entirely fabulous' in Plutarch is the statement that Agesilaus was condemned to pay a fine by the Ephors for having attracted to himself the hearts and minds of his fellow-citizens. I do not know what mark of falsehood he discovers in that, but at any rate Plutarch on this occasion was writing of things which must have been far better known to him than to us, and it was no novelty in Greece for men to be punished and exiled merely because they were too well-liked by their citizens: witness their ostracism and their petalism.[12]

In the same passage there is another accusation which irritates me on Plutarch's behalf: it is where Bodin says that Plutarch showed good faith in his parallels between Roman and Roman or Greek and Greek but not between Roman and Greek. Witness, he says, Demosthenes and Cicero; Cato and Aristides; Sylla and Lysander; Marcellus and Pelopidas; Pompey and Agesilaus, reckoning as he does that he favoured the Greeks by matching them so unfairly. That is precisely to attack what is most excellent and commendable in Plutarch: for in those parallel lives (which are the most admirable part of his works and to my mind the one he took most pleasure in) the faithfulness and purity of his judgements equals their weight and profundity. He is a philosopher who teaches us what virtue is. Let us see whether we can save him from this accusation of falsehood and prevarication.

---

12. Bodin, *Methodus*, IV, 58 (here and also later in the chapter). Over-popular leaders were indeed banished for five or ten years: i) by *ostracism* in Athens, signified by writing the leader's name on a potsherd; ii) by *petalism* in Syracuse, signified by writing the name on an olive leaf.

The only thing I can think of which can have given occasion for Bodin's judgement is that great and dazzling lustre of the Roman names which we have in our heads. It does not seem possible to us that Demosthenes could ever equal the glory of a man who was Consul, Proconsul and Quaestor of that great Republic. But whoever would consider the truth of the matter and the men themselves (which was Plutarch's chief aim, namely to weigh against each other their morals, their natures, their competencies rather than their destinies) will find, I think, contrary to Bodin, that Cicero and the Elder Cato weigh lighter than their parallels.

For Bodin's purpose I would have chosen the parallel between Cato the Younger and Phocion; for in that pair there could with verisimilitude be found an inequality – to the advantage in the Roman.

As for Marcellus, Sylla and Pompey, I quite see that their exploits in war are more expansive, more glorious and more splendid than those of the Greeks whom Plutarch puts in parallel to them; but, no less in war as elsewhere, the most beautiful and most virtuous deeds are not always the most celebrated ones. I often find the names of Captains overshadowed by the splendour of other names of lesser merit: witness Labienus, Ventidius, Telesinus and many others. And if I had to look at things in such a way as to complain on behalf of the Greeks, might I not say that Camillus is far less to be compared to Themistocles; the Gracchi to Agis and Cleomenes; Numa to Lycurgus?

But it is lunacy to wish to judge from one aspect things which present so many facets. When Plutarch compares men he does not thereby make them equal. Who could ever bring out their differences more clearly and conscientiously! When he comes to match the victories, the martial exploits and the might of the armies led by Pompey, and his triumphs, against those of Agesilaus, this is what he says: 'I do not believe that even Xenophon, had he been alive, would have dared to judge them comparable to those of Agesilaus, even if he had been allowed to write all he wished in his favour.' Does he talk of matching Lysander to Sylla? 'There is,' he says, 'no comparison, neither in the number of victories nor the hazards run in battle: for Lysander only won two naval victories . . .' and so on.

In that, he is not cheating the Romans out of anything: he cannot have wronged them merely by placing them beside the Greeks, no matter what disparity there was between them. Plutarch does not weigh them in the lump; he does not prefer one to the other over all: one after the other he matches piece against piece, circumstance against circumstance. So if you wanted to convict him of partiality you would have to take one particular judgement of his and tease it out or else make a general criticism: that he

was wrong to match this Roman against that Greek since there were others which more closely resembled each other and were better fitted for comparison.

33. The tale of Spurina

[*Reflections on Julius Caesar, similar to those we find in Shakespeare, lead Montaigne to compare the powers of bodily vices and those of the mind. Moderation is preferred even to most acts of virtue when they are marked by rapture or ecstasy.*]

[A] Philosophy believes she has not made a bad use of her resources when she has bestowed on Reason sovereign mastery over our soul and authority to bridle our appetites. Those who judge that there are no appetites more violent than the ones engendered by love have on their side the facts that they partake of both body and soul and that every man is swayed by them in such a way that his very health depends on them, so that even medicine is sometimes constrained to serve them as a pimp.

But on the opposite side we could also say that this bodily element somewhat lessens them and weakens them, for bodily appetites are subject to satiety and are susceptible to material remedies. Several men who wished to deliver their souls from the continual alarms caused by bodily appetite have resorted to gelding or castrating those parts which were stirred or depraved.[1] Others have entirely subdued the powers and ardour of those members by frequent compresses of cold things such as snow or vinegar. For this purpose our ancestors used their *haire*, a stuff made of woven horsehair which some made into undershirts and others into girdles to torment their loins.[2] Not long ago a prince told me that in his youth, on a solemn feast-day in the court of King Francis I when everyone was wearing their finery, he felt the desire to wear one of these *haires* which he had at home and which had belonged to my lord his father; but however devout he was he could not endure waiting for night to come to take it off, and it made him ill for a long period. He added that he did not think that there was any youthful lust so sharp as not to be mortified by the use of such a remedy. But perhaps he had not made an assay of the most stinging

---

1. For example, the theologian Origen in Christian antiquity.
2. From *haire* (hair-cloth, a kind of sack-cloth) were made 'hair-shirts'.

lusts, for experience shows us that such an emotion can often subsist beneath rough and filthy garments and that horsehair does not always make sages of the men who wear it.

Xenocrates set about it more vigorously: his disciples, to make an assay of his continence, smuggled into his bed Lais, that beautiful and famous courtesan, quite naked apart from her 'love-filtres', that is, her beauty and her wanton charms. Xenocrates felt that, despite his doctrine and his rules of conduct, his intractable body was beginning to mutiny; so he seared those members of his which had lent an ear at that rebellion.[3] Yet when the passions are all in the soul, as in ambition, covetousness and the rest, they are much more troublesome to reason, for reason cannot be succoured save by her own means: and those passions are not susceptible to satiety – indeed they are sharpened and increased by our enjoyment of them.

The example of Julius Caesar, all by itself, can show us the inequality of these two sets of appetites, since never was there a man more addicted to sexual pleasure. The peculiar care he took over his person is one testimony to that: he even went so far as to make use of the most lascivious methods then current, such as plucking the hairs from his entire body and plastering it with the choicest perfumes. And he was himself quite handsome, white-skinned, with a beautiful slim waist, a full face with lively brown eyes – if we can believe Suetonius, for the statues of him to be seen in Rome do not correspond much to that description. Besides his wives, whom he changed four times (and not counting his youthful affaire with the King of Bithynia, Nicomedes) he had the maidenhead of Cleopatra, that so famous Queen of Egypt – witness little Caesarion who was born thereby. He also made love to Eunoe, Queen of Mauritania; in Rome to Posthumia, the wife of Servius Sulpitius; Lollia, the wife of Gabinus; to Tertulla, the wife of Crassus, and even to Mutia the wife of Pompey the Great (which was the cause, say the Roman historians, of her husband's repudiating her – something which Plutarch admits he did not know). And the two Curios, father and son, later reproved Pompey when he married Caesar's daughter for becoming the son-in-law of the man who had cuckolded him and whom he himself had regularly nicknamed *Aegisthus*.[4]

In addition to all these he kept Servilia, the sister of Cato and mother of

3. Diogenes Laertius, *Xenocrates*, IV, ii.
4. Virtually all the anecdotes and judgements about Caesar in this chapter derive from Suetonius' *Caesar*. (Aegisthus lived adulterously with Clytemnestra, whose husband he had murdered.)

Marcus Brutus, which explains (everyone says) his deep affection for Brutus, who was born at a time when it was probable that he was the father.

I am right therefore, it seems to me, to take him for a man given to the extremes of sexuality, a man of an exceedingly amorous complexion. But he was infinitely infected by another passion: ambition. When that clashed with the former, ambition forced it to give way at once.

[C] In this connection I can think of no better case of these two passions being evenly balanced than that of Mechmet – the one who brought Constantinople under his yoke and finally extinguished the renown of Greece:[5] he was equally indefatigable as both womanizer and soldier. But when these two passions occurred together in his life, his hot lust for fighting dominated his hot lust for women, which did not regain its full authority until it was out of its natural season and he was very old indeed, no longer capable of supporting the burden of warfare.

A contrary example often cited is that of Ladislaus, King of Naples, and it is indeed worthy of note. He was a good general; he was courageous and he was ambitious. The main target of his ambition was to put his lust to work by enjoying some woman or other of rare beauty. His death was in keeping with this end. He reduced the town of Florence to such straits by a well executed siege that the citizens were ready to make terms to concede victory. But he conceded victory to them only on condition that they would deliver up to him a maiden of surpassing beauty of whom he had heard tell. They had no option but to grant her to him, averting public catastrophe by private outrage. She was the daughter of a doctor, famous in his day, who, finding himself trapped by so cruel a necessity, resolved on a momentous design. When everyone was occupied in dressing his daughter and arraying her in jewels and adornments which could make her pleasing to that new lover of hers, he too gave her something: an exquisitely perfumed lace-work handkerchief for her to use when the couple first lay together. In those parts ladies rarely forget to furnish themselves with one. That handkerchief was as poisoned as his art of medicine knew how. When the couple happened to wipe the open pores of their passionate flesh with it, it so suddenly filled them with its noxious fluid that at once their hot sweat turned cold and they died there in each other's arms.

Now to get back to Caesar.

[A] His pursuit of pleasure never made him steal one single minute,

---

5. Mahomet II. This anecdote, and the following one about Ladislaus, from Nicolas Chalcocondylas' *De la décadence de l'Empire Grec*, V, xi.

never deflected him one inch, from any opportunity which was offered him to aggrandize himself. His passionate ambition ruled so sovereignly over all other passions and possessed his soul with such total authority, that, wherever it wanted to go, it carried him there. That vexes me when I reflect on the grandeur of that great man in all other respects and on his marvellous gifts; there was in him so great a competence in every sort of learning that there is virtually no field which he did not write about. He was an orator such that many rated his eloquence above Cicero's. And he himself, in my judgement, did not think himself much inferior to him as far as that endowment goes: his two works against Cato were mainly written as a counterweight to Cicero's fine words in his *Cato*. And, as for the rest, was there ever a man's soul so vigilant, so active and so long-suffering in toil as his was? For without doubt it was rendered beautiful by many a rare seed of virtues – living, natural ones I mean, not counterfeit. He was uniquely lacking in self-indulgence and so undemanding about food that Oppius tells how, one day, when he was served with some oil-of-physic in mistake for salad oil, he used it copiously so as not to embarrass his host. On another occasion he had his baker whipped for supplying him with other than coarse bread. Even Cato used to say that he was the first abstemious man to set out on a road which was, where his country was concerned, the road to ruin.

Cato did call him a drunkard once:[6] but that happened in this way: they were both in the Senate debating the conspiracy of Cataline when a sealed letter was brought to Caesar (who was suspected of being implicated in the conspiracy). Cato, concluding that it was some warning from the conspirators, summoned him to hand it over – which Caesar was forced to do to avoid further suspicion. It chanced to be a love letter which Cato's sister Servilia had written to him. Cato read it and tossed it back to him saying, 'Here, drunkard!' That was a term of anger and contempt rather than an express accusation of drunkenness: we too often insult those who have irritated us with the first insults which come to our tongue, even though they may be in no wise deserved by those we apply them to. Added to which that vice which Cato accused him of is a wondrously close neighbour to the vice he had surprised in Caesar; as the proverb says, 'Venus and Bacchus are readily found together.'[7] [B] Though in my own case Venus is more lively when accompanied by abstinence.

6. Erasmus, *Apophthegmata*, V, *Cato Uticensis*, IV.
7. Erasmus, *Adages*, II, III, XCVII, *Sine Cerere et Baccho friget Venus*; Tiraquellus, *De legibus connubialibus*, IX, 208.

[A]   Examples of his kindness and clemency towards those who had harmed him are numberless – I mean not counting those he provided during the period when the Civil Wars were still in progress: he himself makes us realize clearly in his writings that he exploited those cases to woo his opponents and make them less fearful of his victory and of his future dominance. Yet even if we must say that those particular examples do not suffice to prove to us his native clemency, they do at least show us in that great man a marvellous self-assurance and grandeur. It frequently happened that, once he had defeated them, he sent entire armies back to the enemy without even condescending to make them swear binding oaths that, even if they would not support him, they would at least refrain from making war on him. He captured certain of Pompey's Captains three or four times, and as many times set them free again. Pompey declared that all those who were not his companions-in-arms were his enemies: Caesar had it proclaimed that all those who stayed put and did not actually take up arms against him were his friends. If his Captains sneaked away to seek other employment, Caesar sent them their arms, their horses and their equipment. The towns which he had captured by force of arms he left free to take whatever decision they liked, leaving no garrison behind save the memory of his kindness and clemency. On his great day of battle at Pharsalia he forbade anyone, except as an ultimate extremity, to lay hands on a Roman citizen.

In my judgement you have there some very chancy strokes: no wonder that, in the Civil Wars which we know, those who attack the ancient constitution of their country have not imitated his example. They are abnormal methods which it behoved only Caesar's good fortune and Caesar's foresight to manage auspiciously. When I reflect on the incomparable greatness of his soul I can pardon Victory for not distancing herself from him even in a cause so unjust and so iniquitous.

To come back to his clemency, we have many simple examples of it during the time of his ascendancy when, having everything under his thumb, he no longer needed to dissemble. Caius Memmius had written against him some very sharp criticisms to which he himself replied very sharply: that did not prevent him soon afterwards from helping to make him Consul. When Caius Calvus had composed several insulting epigrams about him and then enjoined his friends to bring about a reconciliation, Caesar decided to write to him first. Our excellent Catullus had given him some rough treatment, coupling his name with Mamurra's: when he came to make apologies Caesar invited him to dinner that very day. When he had been informed that some were talking ill of him, all he did was to

announce in one of his public speeches that he had been so informed. He feared his enemies even less than he hated them. When certain conspiracies and cabals were revealed to him, he was satisfied with an edict stating they were known to him, without further prosecuting those responsible. As for his concern for his friends, once, when Caius Oppius was taken ill while travelling with him, he gave up the only available bed and slept out hard in the open. As for his justice, although no one had lodged any complaint, he had a slave whom he particularly liked put to death for lying with the matron of a Roman knight.

Never was there man who showed more moderation in victory nor more resolution in adversity. Yet all these beautiful dispositions were stifled and corrupted by that frenzied passion of ambition by which he permitted himself to be so totally carried away that it is easy to show that it was the rudder which steered all his actions. It changed a generous man into a plunderer of the State to furnish the wherewithal for his profuse scattering of gifts; it brought him to make that base and iniquitous assertion that if the most wicked and abandoned men in the world had done him faithful service in his advancement to greatness he would cherish them and use his power to promote their interests just as he would in the case of the best of men; it made him so drunk with a vanity so extreme that he dared to boast in the presence of his fellow-citizens that he had stripped the great name of the Roman Republic of body and soul, and to declare that from thenceforth his replies must serve as laws; when the corps of the Senators came to greet him he dared to remain seated; he allowed himself to be worshipped as a god and that divine offices should be celebrated to him in his presence. To sum up, that one vice alone, in my judgement, undid the most beautiful and the most richly endowed nature there ever was, making his name abominable to all good men for having willed to seek his own glory from the destruction and overthrow of his country, the most powerful and flourishing commonwealth that the world will ever see.

On the opposite side many examples of great public figures such as Mark Antony and so on could be found whose lust made them forget the conduct of affairs of state; but whenever sexual love and ambition were to be evenly balanced and come to blows with similar forces, I am in no doubt whatsoever that the former would win the advantage and dominate.

Now to pick up my track again, it is a great thing to rein in our own appetites by reasoned argument or violently to compel our own members to keep to their duty: but to flog ourselves because of our concern for others, not merely ridding our own selves of that sweet passion which excites us and of the pleasure we feel when we find ourselves attractive to

others and loved and courted by everyone, but to loathe and abhor our very qualities which provoke such things, damning our own beauty because it arouses somebody else: well, I do not find many examples of that. But here is one of them.

In Tuscany there was a youth called Spurina,

> [B] *Qualis gemma micat, fulvum quæ dividit aurum,*
> *Aut collo decus aut capiti, vel quale, per artem*
> *Inclusum buxo aut Oricia terebintho,*
> *Lucet ebur.*

[like a jewel set in yellow gold fit for a necklace or a diadem: shining ivory, inlaid in box-wood or wood from the terebinth trees of Illyria.][8]

[A] He was gifted with singular beauty, so extreme that the chastest of eyes could not chastely suffer its brilliance. He never found it sufficient merely not to encourage the flames of feverish passion which he everywhere set ablaze: he conceived a raging loathing for himself and for those rich gifts with which Nature had endowed him, as though the faults in others should be blamed on those gifts: so he slashed his face, deliberately disfiguring with the scars of his wounds that perfect disposition and proportion which Nature had so carefully followed in making it.[9]

[C] To tell you how I judge that: such actions stun me rather than make me honour them; those extremes are inimical to my rules. His intentions were beautiful and loyal to his conscience, but in my judgement somewhat lacking in wisdom. What? Supposing his ugliness later served to provoke others to the sins of scorn, or of hatred and envy of his great repute, or of calumny – interpreting this humour of his as insane ambition? Is there any concept in which, if it wishes to, vice cannot find occasions for displaying itself one way or another? It would have been more just, and also more to his glory, if he had made those gifts of God into a means of exemplary virtue and orderly living.

To my taste, those men who steal away from common obligations and from that infinity of thorny, many-sided conventions which a punctiliously decent man treats as binding when living in society spare themselves a great deal, no matter what singular penance they inflict upon themselves. That is to die a little so as to flee the pain of a life well lived. They may win some other prize, but never, it seems to me, the prize for difficulty; for where

---

8. Virgil, *Aeneid*, X, 134–7.
9. Tiraquellus, *De legibus connubialibus*, II, 12; after Valerius Maximus, and stating that St Ambrose cited Spurina as an example for Christians.

hardship is concerned there is nothing worse than standing upright amid the floods of this pressing world, loyally answering and fulfilling all the duties of one's charge.

It is perhaps easier to do without women altogether than duly and scrupulously to restrict yourself to the company of your wife: a man has more means of living an unworried life in poverty than in duly controlled abundance; behaviour duly governed by reason is more thorny than abstinence. Moderation is a virtue which makes more demands on you than suffering does. The Younger Scipio's way of living aright has a thousand forms: Diogenes' has but one. Diogenes' life is as superior in its innocence to ordinary lives, as choice lives which have achieved much are superior to his in usefulness and fortitude.

34. Observations on Julius Caesar's methods of waging war

===

[An interesting example of how Montaigne read his Classics. He often wrote his opinion of the books he had just read on their flyleaves. Several extracts from what he wrote on his Caesar are transcribed in this chapter. As always the Roman wars, especially the Civil Wars, evoke comparisons with the Wars of Religion in France.]

[A] We read that many leaders in war held particular books in special esteem: Alexander the Great esteemed Homer; [C] Scipio Africanus, Xenophon; [A] Marcus Brutus, Polybius; Charles V, Philippe de Commines. And it is said that in these our days there are others who still think highly of Machiavelli, though the late Marshal Strozzi who took Caesar for his book had without any doubt made the much better choice; truly Caesar ought to be the breviary of every fighting-man: he was the true and sovereign model for the art of war. And in addition God knows with what grace and beauty Caesar painted up such rich material in a written style so pure, so refined and so perfect that there are, to my taste, no writings in all the world which can compare in style with his.

Here I intend to list certain special and particular details about Caesar on the subject of war which have remained in my memory.[1]

His army was in some dismay because of the rumour then current about the great forces which King Juba was leading against him. Instead of playing down the opinions which his soldiers had formed or minimizing the resources of his enemy, he had the troops assemble to reassure them and to put heart into them; but he adopted a course quite opposite to what we are used to: he told them not to bother any more about attempts to find out what forces his enemy was leading, as he had received a definite report. He then told them that the enemy's numbers were far greater than they really were and than what they were rumoured to be among his troops, doing so more readily since (according to the judgement of Cyrus and

---

1. As in Chapter 33, most details derive from Suetonius' *Caesar*, incorporating Renaissance footnotes, commentaries and further details from Caesar's own writings, mainly from the *Gallic Wars*.

Xenophon), such deception is less reprehensible when we eventually find the enemy to be weaker than we had expected than when we find that in reality he is stronger.

Caesar trained all his soldiers simply to obey orders, without being concerned to criticize or discuss his plans as their Captain; he never informed them of his plans before the moment came to carry them out; and if any leaks did occur he delighted in changing his decision on the spot in order to keep his men guessing; for this purpose, after having determined to camp at a certain place, he would march right past it and, especially in bad rainy weather, lengthen that day's march.

At the beginning of his wars in Gaul the Swiss sent envoys to him asking for leave to cross through Roman territory; he had already decided to stop them by force from doing so, but he put on an affable expression and delayed a few days in giving his reply so as to give him time to concentrate his troops. Those simple folk had no idea how good he was at using time: he himself repeated on several occasions that the most sovereign qualities in a commander are knowing how to seize opportunities at the right moment, which in the case of his campaigns was truly unparalleled, and acting with speed, which in his case was truly incredible.

He did not show much of a conscience in seizing the advantage over his enemies under pretext of a treaty of concord; but then neither did he do so in never requiring any virtue from his soldiers but valour, punishing hardly any vice except mutiny and the failure to obey orders. After a victory he would often allow them unbridled licence; he would even free them for a while from the rules of military discipline, adding that he had soldiers who were so well formed that even when smelling of musk and scent they would still go and fight like mad. He genuinely preferred them to have splendid weapons, getting them to wear armour which was engraved in gold and silver, so that they would fight more bitterly out of a concern not to lose them. When he addressed them he called them his companions – a term we still use; but his successor, Augustus, changed all that, reckoning that Caesar had merely done so because his affairs required him to flatter the minds of his followers, who were all volunteers:

> [B]   *Rheni mihi Cæsar in undis*
> *Dux erat, hic socius: facinus quos inquinat, æquat.*

[When Caesar crossed the waters of the Rhine he was my Leader; here in Rome he is my companion, since aiders and abettors are equal in crime.][2]

2. Lucan, *Pharsalia*, V, 289–90.

[A]   As such a usage was too lowly for the dignity of an Emperor, of a Commander of armies, he restored the practice of simply calling them soldiers.

Caesar however intermingled such courtesies with a great severity in keeping men down. When his Ninth Legion revolted near Placentia he smashed it into ignominy; even though Pompey was still on his feet he restored it to favour only after many an entreaty. He appeased his men more by an audacious use of authority than by being conciliatory.

When he talks of his crossing of the Rhine into Germany he states that he considered that it was unworthy of the dignity of the Roman people to do so in boats, so he caused a bridge to be made to enable his men to march across it dry-shod. It was then that he built that astonishing bridge, over the construction of which he goes into such detail; for he never more willingly lingers over his achievements than in describing for us the skill of his inventions in similar sorts of engineering.

I have also noted how he attached great importance to his exhortations to his soldiers before battle, so that whenever he wants to indicate that he was taken by surprise or obliged to hurry he always mentions that he did not even have time to address his men. He says that, before the great battle against the Turones, 'Caesar, having seen to everything else, ran at once wherever Fortune led him and encouraged his men; when he came across the Tenth Legion he merely had time to tell them to remember their usual valour, not to be thrown into confusion and boldly to withstand the enemy's charge. Then, as the enemy were already within bow-shot, he gave the signal to engage; he at once crossed the field to encourage the others, but found that they had already joined battle.'

That is what he says at this point.

It is undeniable that his tongue served him notably well on several occasions; even at the time his eloquence on the field was so highly esteemed that many in his armies took down his speeches; by which means there were compiled several volumes which long outlived him. His style of speaking had a grace of its own – so much so that, whenever readings were made from that compilation, those who knew him well, including Augustus, recognized even words and phrases which were not his own.

The first time that he left Rome with an official command he reached the banks of the Rhone in just over a week; in front of him in his coach he kept a secretary or two ceaselessly taking down what he said: behind him was the man who had charge of his sword. And certainly, even if you had nothing to do but make the journey, you could hardly equal the speed

with which, ever victorious, he left Gaul, followed Pompey to Brundisium, subjugated Italy in eighteen days, and then went on from Brundisium to Rome; from Rome he went off to the remotest parts of Spain where he surmounted the greatest of difficulties in the war against Affranius and Petreius, and then on to besiege Massilia. From there he proceeded to Macedonia, defeated the Roman army at Pharsalia; crossed over to Egypt in pursuit of Pompey; subjugated it; went on from there to Syria and to the country round Pontus where he fought Pharnaces; then he went on to Africa where he defeated Scipio and Juba; and finally he returned through Italy into Spain where he defeated the sons of Pompey:

[B]   *Ocior et cæli flammis et tigride foeta.*

[Swifter than lightning and a tigress defending her young.]

> *Ac veluti montis saxum de vertice præceps*
> *Cum ruit avulsum vento, seu turbidus imber*
> *Proluit, aut annis solvit sublapsa vetustas,*
> *Fertur in abruptum magno mons improbus actu,*
> *Exultatque solo, silvas, armenta virosque*
> *Involvens secum.*

[It was like a landslide rushing down the mountain slopes when land is uprooted by the wind or loosened by the lashing rain or undermined by the force of passing years: as the huge mass crashes down into the void, it makes the earth tremble and bears away forests with their herds and herdsmen.][3]

[A]   Talking of the siege of Avaricum he tells how it was his practice to remain night and day with the men whom he kept toiling at the siege-works. In all his important campaigns he did his own reconnoitring and never sent his army anywhere without first seeing the place for himself. And, on Suetonius' authority, when he had led his campaign across the Channel into Britain he was the first to leap down to test the depth of the water. He would say that he preferred his victories to be won by thought than by might; when Fortune presented him with a clear chance of gaining the advantage during the war against Petreius and Affranius he rejected it, hoping, he says, to finish off his enemies by taking more time but less risk.

[B]   And it was a marvellous stroke when he ordered his entire army to swim across the river when nothing compelled him to do so:

3. Lucan, V, 405; Virgil, *Aeneid*, XII, 684–9.

> *rapuitque ruens in prælia miles,*
> *Quod fugiens timuisset, iter; mox uda receptis*
> *Membra fovent armis, gelidosque a gurgite, cursu*
> *Restituunt artus.*

[the soldier, hastening to the fray, takes a route which he would have dreaded to take in flight: his drenched limbs glow as he puts his armour back on and he runs to warm up his blood, frozen by the swirling current.][4]

[A]   In his campaigns I find more restraint and reflection than in those of Alexander, who seems to go looking for dangers and charging at them like a rushing torrent which indiscriminately batters and unselectively attacks anything it meets.

> [B]   *Sic tauri-formis volvitur Aufidus,*
> *Qui Regna Dauni perfluit Appuli,*
> *Dum sævit, horrendamque cultis*
> *Diluviem meditatur agris.*

[Thus does the river Aufidus charge like a bull as it flows through the realm of Daunus of Apulia: it rages along, threatening the ploughed fields with a dreadful flood.]

[A]   Alexander was active in the flower and first ardour of manhood, whereas Caesar began when already ripe and fully mature. Moreover Alexander was of a more sanguine complexion, choleric and ardent, and he further stimulated that humour with wine, whereas Caesar was very abstemious. But whenever the present occasion necessitated it, when the action itself required it, never was there a man who put less value on his own person.

It seems to me that in many of his exploits Caesar showed a definite resolve to get killed so as to flee from the disgrace of being beaten. At that great battle which he fought against the Turones, when he saw the advanced thrust of his army giving way he rushed out just as he was, offering himself to the oncoming enemy without his shield. That happened several times. After he had learned that some of his men were surrounded, he passed in disguise through the enemy ranks so as to fortify them by his presence. Once, after he had crossed over to Dyrracchium with a very few troops and had realized that the rest of his army which he had entrusted to Antony to lead was slow in following after, he personally undertook to sail back during a very great storm; he had to steal away to resume command

4. Lucan, IV, 151–4; then Horace, *Odes*, IV, xiv, 25–8.

of the rest of his forces because the harbours and the entire seaway were in the hands of Pompey.

As for exploits carried out with a handful of troops, there are several in which the dangers he took exceed any reasoned military argument: for with what puny resources did he undertake to subjugate the Kingdom of Egypt and thereafter to go and attack the forces of Scipio and Juba, which were ten times greater.

Those two men had some inexplicable, more-than-human confidence in their fortunes. [B] And Caesar said that great campaigns are not to be deliberated upon but waged.[5]

[A] After the battle of Pharsalia, when he had sent his army ahead into Asia, he was crossing the straits of Hellespont with a single ship when he met Lucius Cassius sailing with ten heavy warships; he had the courage not simply to wait for them but to head straight for them and to summon him to surrender.

He was already engaged in that frenzied siege of Alexia, where the defenders numbered eighty thousand, when the whole of Gaul rose to attack him and raise the siege, gathering an army of a hundred and nine horses and two hundred and forty thousand infantry. What boldness, what insane confidence, he showed by deciding not to give up the siege he had undertaken and by determining to take on two such problems at the same time! And he did withstand them: after he had won that great battle against the forces outside he soon reduced to submission those he held under siege. (The same happened to Lucullus at the siege of Tigranocerta against King Tigranes, but under different circumstances, given the weakness of the enemy with whom Lucullus had to deal.)[6]

I would like to emphasize here two rare and extraordinary events concerning that siege of Alexia: one was that the Gauls, who had assembled to confront Caesar, first counted all their troops and then decided in council to cut out a goodly part of that huge crowd, fearing that they might produce chaos.

This fear that your forces are too numerous provided a new example: yet if you look at it the right way it is indeed likely that a body of soldiers should be only moderately big and limited to some definite size, both for the difficulty in feeding them and also for the difficulty of leading them or

---

5. '80: More-than-human confidence, *beyond the natural order*, in their fortunes . . .

    A significant excision in the light of the end of III, 13, 'On experience'. (Plutarch (tr. Amyot), *Dicts des anciens Roys*, 208 D.)

6. Plutarch, *Life of Lucullus*.

keeping them in formation. At least it is easy to prove that those monstrously large armies have rarely achieved anything worthwhile, [C] which agrees with the saying of Cyrus in Xenophon, that the advantage lies not in numbers of men but in numbers of good men, the rest being less a help than a hindrance.[7] And Bajazet based his decision to dispute the field with Tamberlane against the advice of his Captains mainly on the fact that the uncountable numbers of the enemy gave him certain hope of their falling into confusion. Scanderbeg, a good and very experienced judge, often said that ten or twelve thousand faithful combat troops should suffice a competent war-leader to guarantee his reputation in every sort of military need.

[A] The other point which seems to be contrary to both the reason and usages of war is that Vercingetorix (who had been named General commanding all the areas of Gaul which were under revolt) should have taken the decision to go and shut himself up in Alexia. For a man in command of an entire country must never, except in a case of extreme necessity, so hem himself in that the fight becomes his last stand, his only hope lying in defence; in other cases he must ensure his liberty, so as to have the means of providing for things in general over all the regions he controls.

To get back to Caesar. As time passed he became a little more slow and deliberate, as Oppius who was intimate with him shows; Caesar later judged that he should not really hazard the honour acquired from so many victories, which one single disaster could lose for him. That concords with what is said by the Italians when they wish to reprove that rash bravery found in younger men by calling them *bisognosi d'honore*, 'needy of honour': they say that since they are still hungry for reputation, which is hard to come by, they are right to go and look for it at any price – something which ought not to be done by those who have already acquired a store of it. In this appetite as in any other there can indeed be found a just moderation between desire for glory and satiety.

Caesar was far removed from the scruples of the Ancient Romans who wished to exploit only their simple, straightforward valour, yet he brought more conscience to bear than we would nowadays; he would not have approved of acquiring the victory by any sort of means. In the war against Ariovistus he was parleying with him when a disturbance broke out between the two armies, started by the horsemen of the enemy; during the

---

7. Xenophon, *Cyropaedia*, II, ii; then Nicolas Chalcocondylas, *De la décadence de l'Empire Grec*, III, xi (for Bajazet), and Jacques Lavardin, *Histoire de Scanderbeg* (1576), 444 r°.

confusion Caesar found himself in a position of real advantage over Ariovistus: but he had no wish to exploit it, fearing that he could have been accused of having acted throughout in bad faith.

He customarily wore in battle his most splendid equipment, brilliantly coloured so as to make himself stand out. When approaching the enemy he kept his soldiers on a shorter, tighter rein.

When the Ancient Greeks wished to accuse anyone of extreme inadequacy they used the common proverb: 'He can neither read nor swim.'[8] He shared their opinion that to know how to swim was most useful in war, and he derived many advantages from it: when he needed to hurry he normally swam across any rivers he encountered for, like Alexander the Great, he preferred to travel on foot. When he was in Egypt he was forced to escape in a small boat; so many jumped in with him that it was in danger of going under; he therefore preferred to jump into the sea and swim out to his fleet which was about two hundred yards away, holding his writing-tablets above the water in his left hand and dragging his armour along with his teeth to prevent the enemy getting hold of it. He was then well on in years.

Never did a leader in war inspire greater trust in his soldiers: at the beginning of the Civil Wars, each of his centurions offered to pay out of his own purse for one soldier and his equipment, while his foot-soldiers offered to serve him at their own expense, those who were better off undertaking to defray the expenses of the poorer ones.

The late Admiral de Chastillon recently provided a case similar to that in our own Civil Wars, for the French in his army furnished the pay of the mercenaries in their units out of their own purses. You could not find many examples of such burning and spontaneous devotion among those who march under our old regime, under our ancient religious polity.[9] [C] Emotion dominates us more vigorously than reason. Yet in the war against Hannibal, following the free-born example of the People of Rome in their City, the soldiers and captains refused their pay; in the camp of Marcellus those who did accept it were dubbed mercenaries.

[A] When Caesar's soldiers were worsted near Dyracchium, they so spontaneously offered themselves to be chastised and punished that he needed to console them rather than to berate them. One single cohort of his men withstood the legions of Pompey for over four hours until

---

8. Cited as a proverbial saying by Aristotle, *Laws*, III, 689 D.
9. An example of fairness to enemies, Gaspard de Coligny (Chastillon) being a Protestant leader; those 'under the old regime' are the French Roman Catholics.

virtually all were put out of action by arrow-wounds; over one hundred and thirty thousand arrows were found in their trenches; one soldier called Scaeva, who was in charge of an entry-slit, remained undefeated at his post despite having one eye transfixed, one shoulder and thigh shot through and a shield pierced in two hundred and thirty places. It often happened that any soldiers of Caesar's who were made prisoner accepted to be killed rather than to change sides. When Granius Petronius was taken prisoner in Africa, Scipio had his companions put to death and then sent to inform him that he was granting him his life because of his rank of quaestor. Petronius retorted that the soldiers of Caesar were used to granting life to others not to receiving it themselves; he at once killed himself with his own hands.[10]

There are innumerable examples of their loyalty; but we must not overlook the bold stroke of the men who were besieged at Salona (a town which had declared for Caesar against Pompey); for the event which took place there was a rare one: Marcus Octavius was investing Salona; the besieged were in every way reduced to the ultimate extremity; to supply what they lacked in men (since most of them were killed or wounded) they had freed all their slaves; to be able to use their catapults, they were reduced to cutting off their wives' tresses to make into ropes; apart from that there was a staggering shortage of food. They were determined, nevertheless, not to surrender. When they had dragged this siege out to such a length that Octavius had grown careless and was paying less attention to his campaign, they picked one day just before noon, stationed their women and children on the walls so that things should look normal, then made such a frenzied sortie against the besiegers that they broke through the first rank of their guards, then the second, then the third, then the fourth, then the rest, forcing them entirely to abandon their entrenchments and driving them right back to their ships. Octavius himself fled to Dyrrachium, where Pompey was staying.[11]

I cannot for the while recall having come across any other examples of the besieged routing the mass of the besiegers and winning mastery of the field, nor of a sortie leading to a pure and total victory in battle.

10. Plutarch, *Caesar*.
11. Caesar relates this himself in his *Civil Wars*, III, ix.

35. On three good wives

[A chapter which in some ways is a pendant to II, 10, 'On books'. History, true history, can be a source of both aesthetic delight and of moral profit. It is potentially a valuable alternative to moral fiction, to tales (such as those of Boccaccio). Montaigne's preoccupation with great-souled suicides in the Stoic mould is rarely more visible than in this chapter; it is given prominence by coming near the end of Book II and so having (until Book III was published) the air of leading up to the conclusion.]

[A] As every man knows, they are not counted in dozens, especially in performing their matrimonial obligations: for marriage is a business full of so many thorny conditions that a woman cannot keep her intentions in it for long. Even the men (who are there under slightly better terms) find it hard to do so.

[B] The touchstone of a good marriage, the real test, concerns the time that the association lasts, and whether it has been constant – sweet, loyal and pleasant. In our century wives usually reserve their displays of duty and vehement love for when they have lost their husbands; [C] then at least they bear witness to their good intentions – a laggardly, unseasonable witness, by which they prove that they love their husbands only once they are dead. [B] Life is full of inflammatory material: death, love and social duties. Just as fathers hide their love for their sons so as to keep themselves honoured and respected, so do wives readily hide theirs for their husbands. That particular mystery-play is not to my taste! It is no good widows tearing their hair and clawing their faces: I go and whisper straight in the ear of their chambermaid or private secretary, 'How did they get on? What were they like when living together?' I always remember that proverbial saying: *'Jactantius moerent, quae minus dolent.'* [Women who weep most ostentatiously grieve least.][1] Their lamentations are loathed by living husbands and useless to the dead ones. We husbands will willingly let them laugh afterwards if they will only laugh with us while we are alive.

1. Tacitus, *Annals*, II, lxxvii.

[C]   Is it not enough to raise a man from the dead out of vexation, if a wife who had spat in my face while I was still there were to come and massage my feet once I am beginning to go!   [B]   If some honour resides in weeping for husbands it belongs to widows who laughed with them in life; let those widows who wept when they were alive laugh outwardly and inwardly once they are dead. Moreover, take no notice of those moist eyes and that pitiful voice: but do note the way they carry themselves and the colour of those plump cheeks beneath their veils! That way they speak to us in the kind of French we can understand! There are few widows who do not go on improving in health: and health is a quality which cannot lie. All that dutiful behaviour does not regard the past as much as the future: it is all profit not loss. When I was a boy an honest and most beautiful lady, a prince's widow who is still alive, began to wear some little extras not allowed by our convention of widowhood. To those who reproached her with this she replied, 'It is because I meet no new suitors now: I have left behind the desire to remarry.'

So as not to be totally out of keeping with our customs, I have selected three wives who, on the death of their husbands, did show the force of their goodness and their love. They are however rather diverse examples of pressing cases which resulted in a bold sacrifice of life.

[A]   In Italy Pliny the Younger had a man living near one of his houses who was appallingly tormented by ulcers which appeared on his private parts. His wife watched him languishing in pain; she begged him to allow her enough time to examine the symptoms of his disease: she would then tell him more frankly than anyone else what hope he could have. She obtained this of him and carefully examined him; she found that it was impossible for him to be cured and that all he could expect was, over a long period, to drag out a painful and languishing life. And so she advised him, as the surest, sovereign remedy, to kill himself. Finding him a little hesitant about so stark a deed she said: 'You must never think, my Beloved, that the pains which I see you suffer do not affect me as much as you, or that to deliver myself from them I am unwilling to use the same remedy that I am prescribing for you. I wish to be your companion in your cure as I am in your illness: lay aside your fears and think only that we shall have the pleasure of that journey into death which must free us from such torments. We shall go happily away together.' Having finished speaking and bringing new warmth to her husband's heart, she resolved that they should cast themselves into the sea from a window in their house which gave on to it. And so as to maintain unto the end that loyal and vehement love by which she had clung to him in life, she wanted him also

to die in her arms. But fearing that those arms might fail her and that the clasp of her embrace might be loosened by the terror of the fall, she had herself tied to him, tightly bound by their waists. And thus she gave up her life for the repose of her husband.[2]

That woman was from a lowly class; among people of that condition it is not all that new to find signs of rare goodness.

> *Extrema per illos*
> *Justitia excedens terris vestigia fecit.*

[When Justice finally left this earth, she left her last vestiges with them.][3]

The other two are rich and noble; examples of virtue rarely make their home among people like that.

Arria was the wife of Caecinna Paetus, a great man of consular rank; she was the mother of another Arria, the wife of Thrasea Paetus who was so renowned for his virtue during the time of Nero; through this son-in-law she was the grandmother of Fannia. The similarity of name and fortune of these men and women has often led to confusion. This first Arria (when her husband Caecinna Paetus had been taken prisoner by the supporters of the Emperor Claudius after the defeat of Scribonianus whose faction he had supported) begged the men who were transferring their prisoner to Rome to take her aboard their ship, where she would be much less expense and trouble than the many people they would need to look after her husband since she alone would take care of his room, his cooking and all other chores. They refused this to her; so she leapt into a fisherman's boat which she had immediately hired and in this manner followed her husband from Sclavonia.

One day in Rome in the presence of the Emperor she was familiarly approached by Junia, the widow of Scribonianus, because of their shared misfortunes; but she roughly thrust her away with these words: 'Should I even talk to you or listen to you when Scribonianus, the husband of your bosom, is dead. Yet you are still alive!' Such words and several other indications brought her relations to realize that, unable to endure her husband's misfortune, she intended to do away with herself.

On hearing those words her son-in-law Thrasea begged her not to desire to kill herself, saying: 'What? If I incurred a similar misfortune to Caecinna's, would you want my wife, your daughter, to do likewise?' – 'What do you mean, would I!' she replied. 'Yes. Yes of course I would, if she had lived as long and as peacefully together with you as I did with my

---

2. Pliny the Younger, *Epist.*, VI, xxiv.
3. Virgil, *Georgics*, II, 473–4 (of happy rustics).

husband.' Such answers increased their worries about her and led to their watching her behaviour closely.

One day she said to those who were set to guard her: 'It is no good, you know. You can force me to make the death I die much harsher: you cannot stop me from dying.' She madly darted out of the chair she was sitting in and, with all her might, bashed her head against the nearby wall. The blow felled her to the ground, severely wounded and unconscious. They just managed to bring her round with great difficulty. 'I told you plainly,' she said, 'that if you refuse me the means to kill myself easily, then I shall choose some other way, no matter how hard it might be.'

The end of so amazing a virtue came like this: by himself Paetus her husband did not have courage enough to kill himself, as the Emperor's cruelty would force him to do some day or other; so having first used the appropriate arguments and exhortations for the counsel which she was giving him to bring him to do so, she seized the dagger which her husband was wearing, drew it, held it in her hand and concluded her exhortation thus: 'This is the way to do it, Paetus.' And that same instant, having struck herself a mortal blow in the bosom, she wrenched the dagger from her wound and offered it to him, ending her life as she did so with these noble, great-souled, immortal words: *'Paete, non dolet.'* Those three words so full of beautiful meaning were all she had time to utter: 'You see, Paetus: it doesn't hurt.'[4]

> *Casta suo gladium cum traderet Arria Pæto,*
> *Quem de visceribus traxerat ipsa suis:*
> *Si qua fides, vulnus quod feci, non dolet, inquit;*
> *Sed quod tu facies, id mihi, Pæte, dolet.*

[When chaste Arria proffered the blade to Paetus which she had torn from her very entrails, she said: 'Believe me, that wound I have given myself does not hurt me. What hurts me, Paetus, is the wound you will give to yourself.']

But it has much more living force in the original and a much richer meaning. Far from being depressed by the thought of her husband's wound and death, or of her own, she was the one who advised and encouraged them; so, having performed that high courageous deed solely in the interest of her husband, even with the final words of her life her only thought was of removing from him his fear of following her by taking his life. Paetus at once struck himself through with that same blade, feeling shame, in my judgement, at having needed so costly and so precious a lesson.

4. Retold after Pliny the Younger. *Epistles*, III, xvi; then, Martial, *Epigrams*, I, xiv.

There was a young and very high-born Roman matron called Pompeia Paulina. She had wedded Seneca in his extreme old age. Nero, that fine pupil of his, sent one of his courtiers to him to announce that he was sentenced to death.[5] (Such sentences used to be executed in this way: when the Emperors of Rome had condemned any man of quality, they dispatched their officials to tell him to choose which death he would prefer and to see that he carried it out within such time as they caused to be prescribed, shorter or longer depending on how finely tempered their choleric humour was: it was a concession designed to allow him to put his affairs in order, though too short on occasions to permit him to do so. If the condemned person resisted their command they brought in suitable men to carry it out, either by slashing the veins in his arms and legs or forcing him to swallow poison. Men of honour did not wait for such compulsion but used their own doctors and surgeons to do the deed.) With a peaceful resolute expression Seneca listened to the order brought by Nero's henchmen, then asked for paper to write his will. That was refused by the Captain, so Seneca turned to those who loved him and said: 'Since I can bequeath you nothing else, out of gratitude for what I owe you I shall at least bequeath you the most beautiful thing I possess: the portrait of my morals, of my life, which I pray you to conserve in your memory; by doing so you will acquire the reputation of ones who loved me purely and truly.' At the same time with gentle words he quietened the bitter anguish which he saw that they were suffering, though sometimes speaking more firmly to rebuke them: 'Where are all those beautiful precepts of philosophy?' he asked. 'What has happened to that store which we have set aside over so many years against the accidents of Fortune? Did we not know of Nero's cruelty? What could we expect from a man who had killed his mother and his brother, except that he would also kill his tutor who had looked after him and brought him up?'

Having addressed them all in general, he turned aside to his wife; and since her heart and strength were yielding under the weight of her grief he held her tight in his arms; he prayed her that, for love of him, she should bear this misfortune a little more patiently, since the hour had come when he had to show the fruit of his studies not by speeches and arguments but by deeds, and since he, without the slightest doubt, was welcoming death not merely without grief but with joy. 'Wherefore my Beloved do not dishonour it by your tears,' he said, 'lest it should seem that you love yourself more than my reputation. Quieten your grief and console yourself

5. Retold from Tacitus, *Annals*, XV, lvii–lxiv.

with the knowledge that you have of me and of my actions, consecrating the rest of your life to those honourable occupations to which you are so devoted.'

Paulina replied, having somewhat recovered her composure and brought warmth again to her magnanimous heart by her noblest love: 'No, Seneca. I am not one to leave you companionless in such great need. I do not want you to think that the virtuous examples of your life have not yet taught me to know how to make a fine death. When could I ever die better, or more honourably, or more as I would wish to, than together with you? Rest assured that it is with you that I shall go.' Whereupon Seneca, welcoming such a beautiful and glorious resolve in his wife, and also to rid himself of his fear of leaving her to the tender mercies of his enemies after his death, replied: 'I once taught you, Paulina, such things as served you to live your life contentedly. Now you prefer the honour of death: truly I will never begrudge you that. The constancy and the resolve of our common end may be equal: but allow that on your side the beauty and the glory are greater.'

That done they both together slashed the veins in their arms; but since Seneca's veins had become constricted by old age[6] and abstemious diet they merely allowed the blood to trickle out slowly, so he gave orders to slash the veins in his thighs as well. Then, fearing that the torment he was suffering might sadden the heart of his wife, and also to deliver himself from the grief he bore at seeing her in so pitiful a state, after taking leave of her most lovingly he begged her to permit them to carry him away to another room; which they did. But as all those incisions were still insufficient to cause his death, he commanded Statius Annaeus his doctor to administer the poisoned drink; that too had little effect, since it could not reach his heart because his limbs were weak and chill. So they further prepared a very hot bath for him; as he felt his end approaching he continued, as long as he had breath, to deliver most excellent discourses on the subject of his present state, which his secretaries took down as long as they could hear his voice; and for years afterwards his final words remained honoured and respected, circulating in the hands of men. (It is a most regrettable loss that they have not come down to us.) As he felt the last pangs of death he took the blood-drenched waters of the bath and asperged his head with them saying, 'This water I consecrate to Jove the Liberator.'

Nero, warned of all this, fearing that he might be criticized for the death of Paulina (who was one of the most nobly-connected of Roman matrons)

6. [A] Until [A]: old age *(for he was then about one hundred and forty years old)* and . . .

and having no particular reason to hate her, sent back orders with all speed that her wounds were to be bound. Her people did so – without her knowledge, since she was half-dead already and quite without sensation. And so against her own design she lived her remaining span most honourably, as behoved her virtue, showing by the pallor of her face how much life-blood she had shed through her wounds.

There you have my three very true tales, which I find as pleasing in their tragedy as those fictions which we forge at will to give pleasure to the many. I am amazed that those who engage in that activity do not decide to choose some of the ten thousand beautiful historical accounts to be found in our books. In that they would have less toil and would afford more pleasure and profit. If any author should wish to construct them into a single interconnected unity he would only need to supply the links – like soldering metals together with another metal. He could by such means make a compilation of many true incidents of every sort, varying his arrangement as the beauty of his work required, more or less as Ovid in his *Metamorphoses* made a patchwork of a great number of varied fables.

In the case of my last couple it is also worth pondering on the fact that Paulina willingly gave up her life for love of her husband, and that formerly, for love of her, he had once given up dying. There is little equivalence in that for the likes of us: but to his Stoic humour I believe that he thought he had done as much for her by prolonging his life to please her as if he had died for her. In one of his letters to Lucilius,[7] after telling him how he had caught a fever in Rome and promptly climbed into his coach to go off to one of his houses in the country against the wish of his wife who wanted to prevent him, he tells how he replied that his fever was not physical but geographical. He went on: 'She then let me go, telling me to look after my health. So I, who know that her life is lodged in mine, now begin to take care of myself so as to take care of her. The privilege which old age had bestowed on me, making me more firmly resolved on many things, I am losing now that I remember that, in this old man, there is a young woman to whom I am of some use. Since I cannot bring her to love me more courageously, she brings me to love myself more carefully. For we must allow some place to honourable affections; so sometimes when opportunities pressingly invite us the other way, we must summon our life back – yes, even in torment. We must cling by our teeth to our souls since, for moral men, the law of life is not 'as long as they please' but 'as long as

7. Seneca, *Epist. moral.*, CIV, 2–6.

they should'. The man who does not think enough of his wife or of his friend to prolong his life for them and who is determined to die is too fastidious, too self-indulgent. Our souls must order themselves to die when the interests of our dear ones require it. Sometimes we must make a loan of ourselves to those we love: even when we should wish to die for ourselves we should break off our plans on their account. It is a sign of greatness of mind to lay hold of life again for the sake of others, as several great and outstanding men have done. And it is a mark of particular goodness to prolong one's old age (the greatest advantage of which is to be indifferent to its duration and to be able to use life more courageously and contemptuously) if one knows that such a duty is sweet, delightful and useful to someone who loves us dearly. And we ourselves receive a most delightful recompense: for what can be more delightful than to be so dear to your wife that you become dearer to yourself for her sake? Thus my Paulina has laid upon me not only her fears for me but my fears for myself as well. It is not enough for me to consider with what resolution I could die: I also have to consider how irresolutely she would bear it. So I have compelled myself to go on living. Sometimes there is magnanimity in doing so.'

Those are his words, [C] as excellent as are his deeds.

36. On the most excellent of men

[Homer, Alexander the Great and Epaminondas form this most outstanding trio (with Alcibiades as a more attainable model for the average decent man). The praise of Alexander is dominated by one long and grammatically chaotic sentence which betrays both enthusiasm and confusion on Montaigne's own part. This praise of Alexander towards the end of Book II leads to the criticism of him in the final pages of Book III and helps to emphasize the force of the concluding pages of the Essays.]

[A] If I were asked my pick of all the men who have come to my notice, I would find three I think who excel all others.

One of them is Homer: it is not that Aristotle or Varro for example might not perhaps have been as learned as he was; not that Virgil may not possibly be compared to him even as an artist: I leave that to be judged by those who know both those poets; I who know only one of them[1] may say that as far as I am able to tell the very Muses could not reach further than he did:

> [B] *Tale facit carmen docta testudine, quale*
> *Cynthius impositis temperat articulis.*

[On his learned lyre he sings verses such as Cynthian Apollo chants when he attunes his strings to his plucking fingers.][2]

[A] However in making such a judgement I should never overlook that it is from Homer, his guide and his teacher, that Virgil derives his skill, nor that one single incident in the *Iliad* supplied the bulk of the material for that great and divine *Aeneid*. But that is not the way I do my sums. I marshal other qualities, ones which make that great Homer amazing to me, as though he were above our human condition. And in truth I am often struck with wonder that he, who by his authority created so many gods and made them honoured in this world, has not himself been deified. Poor

---

1. Virgil. Montaigne realized that his Greek was not up to appreciating the real merits of Homer.
2. Propertius, II, xxxiv, 79–80.

and blind, living as he did before learning had been codified into rules and definite precepts, he had mastered it all so well that those who have subsequently undertaken to establish forms of government, to conduct wars or to write on religion and on philosophy – [C] no matter what School they belong to – [A] or about the arts and crafts, have accepted him as their master, most perfect in all things, and taken his books as a seed-bed for every kind of knowledge.

> *Qui quid sit pulchrum, quid turpe, quid utile, quid non,*
> *Plenius ac melius Chrysippo ac Crantore dicit.*

[Better and more fully than Chrysippus and Crantor he teaches us what beauty is, what ugliness, what is profitable, what is not.]

And another poet says:

> *A quo, ceu fonte perenni,*
> *Vatum Pyeriis labra rigantur aquis.*

[From his unfailing stream the poets come and wet their lips in the Pierian waters.]

Yet another:

> *Adde Heliconiadum comites, quorum unus Homerus*
> *Astra potitus.*

[To these add those companions of the Muses, among whom Homer alone was made into a star.]

And one more:

> *Cujusque ex ore profuso*
> *Omnis posteritas latices in carmina duxit,*
> *Amnemque in tenues ausa est deducere rivos,*
> *Unius fœcunda bonis.*

[From whose abundant source all posterity have drawn their songs, dividing his one river into their many rivulets, each poet rich in the wealth of one single man.][3]

It was against the order of Nature for Homer to have brought forth the most [C] excellent [A] work there can ever be. In Nature's order things are imperfect at birth: they grow up, and become stronger as they grow. He made the childhood of poetry and of several other arts to be

---

3. Horace, *Epistles*, I, ii, 3–4; then, Ovid, *Amores*, III, ix, 25–6; Lucretius, III, 1050–51; Manilius, *Astronomica*, II, 8–11.
    Then, '80: the most *noble* work [. . .] are *feeble and* imperfect . . .

adult, complete and mature. That is why, following that beautiful testimony to him which Antiquity has bequeathed to us, he can be called 'the first poet and the last': since before him there was none whom he could imitate: after him, none who could imitate him. According to Aristotle, his words alone have properties of movement and of action: they are the only words which are endowed with substance.[4] When Alexander the Great came across a costly jewel-box among the spoils of Darius, he commanded that it be set aside for him to keep his copy of Homer in, saying that it was his best and most faithful counsellor on the subject of armies. For the same reason Cleomenes son of Anaxandridas said that Homer was the poet of the Spartans, since he was an excellent instructor in the art of warfare. There has also come down to us a unique and individual tribute: in Plutarch's judgement he is the only author in the world who has never sated his readers nor grown insipid to them, since he ever seems different to them, ever blossoming into new graces. That whimsical Alcibiades asked a man with pretentions to culture to show him his Homer: when he could not produce one he boxed his ears – it would be like finding one of our priests with no breviary! Xenophanes once complained to Hiero, the Tyrant of Syracuse, that he was too poor to provide for two servants: 'How is that?' he replied: 'Homer was far poorer than you yet, dead though he is, he provides grist for over ten thousand!' [C] And what more could be said when Panaetius called Plato 'the Homer of the Philosophers?'[5]

[A] Besides, what renown can be compared with his? Nothing lives like his fame and his works on the lips of men: nothing is so known or accepted as Troy, Helen and Homer's wars – though they may never have existed. Our children are still given names which he invented over three thousand years ago. Who has not heard of Hector and Achilles? Not only individual families but most of the nations seek their origins in what Homer created. When Mahomet II, the Turkish Emperor, wrote to our Pope Pius II, he said, 'I am amazed that the Italians should band against me, since we both have a common origin in the Trojans and, like the Italians, I have an interest in avenging the blood of Hector on the Greeks whom they

---

4. Aristotle, after Plutarch (tr. Amyot), *Des oracles de la prophetisse Pythie*, VIII, 629 E; then, Plutarch, *Life of Alexander*; *Dicts notables des Lacedaemoniens*, 217 H; *De trop parler* D–E; *Dicts des anciens Roys*, 196 H. (For these well-known and authoritative sayings, cf. Erasmus, *Apophthegmata*, I, *Cleomenes*, I; IV, *Alexander Magnus*, LIV; V, *Alcibiades*, III. For Homer's 'winged words,' cf. Rabelais, *Quart Livre*, TLF, LV, 63 ff.)

5. Cicero, *Tusc. disput.*, I, xxxii, 79 (condemning the Stoic Panaetius for not believing in the immortality of the soul).

are supporting against me.'[6] Homer provides a noble farce in which over the centuries Kings, Republics and Emperors all play their parts and for which this great universe serves as the theatre.

Seven towns of Greece squabbled over his birthplace, so much honour was brought him by his obscure origins:

> *Smyrna, Rhodos, Colophon, Salamis, Chios, Argos,*
> *Athenae.*

[Smyrna, Rhodes, Colophon, Salamis, Chios,
Argos and Athens.]

My second example is Alexander the Great.

For let a man consider Alexander's age when he set out on his expeditions; the meagre resources with which he achieved so glorious a design; the authority he won as a mere boy over so many of the greatest and most experienced Captains in the world who followed him; the extraordinary favour with which Fortune embraced him and favoured his hazardous – I almost said rash – exploits.

> [B]   *impellens quicquid sibi summa petenti*
> *Obstaret, gaudensque viam fecisse ruina;*

[toppling everything in the way of his ambition, glorying in marking his route with destruction;][7]

[A]   his greatness in having passed victorious through all this inhabitable earth by the age of thirty-three;   [B]   his having attained, in but half a lifespan, the ultimate limit of human nature, so that you cannot imagine him living the normal span or continuing to grow in virtue or good fortune to the natural term of a man's life without imagining something surpassing our humanity;   [A]   his making so many royal branches sprout from among his soldiers; the world divided at his death among his four successors – simple Captains in his armies whose descendants subsequently long endured, maintaining such great dominions – let a man consider so many excellent virtues in him,   [B]   justice, temperance, liberality, faithfulness to his word; love for his people, humanity towards the

---

6. Innocent Gentillet, *Discours sur le moyen de bien gouverner*, III; then a line of Greek poetry as translated by Aulus Gellius into Latin, III, xi.

7. Lucan, *Pharsalia*, I, 149–50. The main original sources of Montaigne's long, grammatically confusing eulogy of Alexander are Plutarch's *Alexander* and the *Fortunes of Alexander*. Later borrowings, from Flavius Arrianus' *Deeds of Alexander* and Quintus Curtius' work with the same title.

vanquished [A] for his character seems to have justly been beyond reproach [B], though not some of his rarer, untypical, isolated actions: but it is not possible to head such great movements and always act according to the rules of justice: men such as he need to be judged overall, by the dominant aim of their activities: his destruction of Thebes and the murders of Menander, of the doctor Ephestion, of so many Persian prisoners at one stroke, of a troop of Indian soldiers (not without impugning his pledged word), of the Cosseians, including their children, are ecstasies a little hard to excuse; but in the case of Clytus he made amends far beyond the gravity of the offence – and that action as much as any other bears witness to a generous complexion, which was a complexion excellently formed for goodness: [C] it was cleverly said of him that he owed his virtues to Nature, his vices to Fortune; [B] as for the fact that he was a bit of a boaster, a bit too impatient of hearing ill said of himself, and that he scattered his mangers, arms and bridle-bits all over India, well, that kind of thing seems pardonable to me in a man of his age and of his [C] strangely [A] prosperous Fortune); whoever will also at the same time reflect on his many military virtues – speed, foresight, endurance, self-discipline, subtlety, magnanimity, resolve, good luck – in which he was the first among men (even if Hannibal had not pointed it out to us); [A] the rare beauty and endowments of his person which touched the miraculous; [B] his carriage and that venerable bearing of his beneath a face so young,[8] so flushed with radiance –

> *Qualis, ubi Oceani perfusus lucifer unda,*
> *Quem Venus ante alios astrorum diligit ignes,*
> *Extulit os sacrum cælo, tenebrasque resolvit;*

[Shining like that morning star which Venus loves above all others when, bathed in Ocean's waves, it raises up its sacred face in the heavens and drives away the darkness;]

[A] the excellence of his knowledge and his capacities; the duration and grandeur of his glory, pure as it was, free from spot or envy; [B] the fact that long after his death it was a pious belief to hold that his medallions brought good luck to those who wore them; that more kings and princes have written of his deeds than other historians have written of any king or prince that has ever been; [C] that even today the Mahometans who despise all other biographies accept and honour his alone by a special dispensation: [A] let a man consider all that and he will admit, if he

8. Livy, XXXV, xiv; then, Virgil, *Aeneid*, VIII, 589–91.

lumps it all together, that I was right to prefer Alexander even to Caesar, who alone was able to make me hesitate over my choice. [B] It cannot be denied that there is more of Caesar in Caesar's exploits: more of Fortune in Alexander's. [A] In many things they were equal; Caesar may even have been greater in a few.

[B] They were two conflagrations, two torrents, flooding through the world in divers places:

> *Et velut immissi diversis partibus ignes*
> *Arentem in silvam et virgulta sonantia lauro;*
> *Aut ubi decursu rapido de montibus altis*
> *Dant sonitum spumosi amnes et in æquora currunt,*
> *Quisque suum populatus iter.*

[Like two forest-fires raging in different parts of a dry forest of laurel trees full of crackling twigs; or like two foaming torrents rushing down the mountain-sides with a roar, charging across the plains, having swept away everything before them.][9]

But even if Caesar's ambition were more moderate, it was still disastrous: it had as its vile objective the collapse of his country and the debasement of the entire world, so that, [A] when all is put together and weighed in the balance, I cannot do other than to come down on the side of Alexander.

My third example, and to my mind the most distinguished, is Epaminondas.[10]

As for glory, he is far from having such renown as the others (nor is glory a quality of the substance of anything); as for resolution and valour – not the kind which is sharpened by ambition but the kind which wisdom and reason can implant in a well-ordered soul – he had all that can ever be imagined. As for proof of his valour, he provided as much of it in my judgement as even Alexander and Caesar; for though his exploits in war were neither so numerous nor so grandiose, if you consider them thoroughly and in all their circumstances, they do not cease, for all that, to be any less weighty or impressive; they provide equal proof of [C] bravery and [A] military skill. The Greeks did him the honour of unanimously naming him the first man among them; but to be first among

9. Virgil, *Aeneid*, XII, 21–5.
10. *The* hero for Montaigne. His main sources here are again Plutarch, *Life of Pelopidas*; also Cornelius Nepos' *Epaminondas* or Erasmus' *Apophthegmata*, V, *Epaminondas*.

the Greeks is to be easily the best in the world. As for his knowledge and skill, an ancient verdict has come down to us, that never did man know more nor talk less. [C] For he belonged to the Pythagorean School. What he did say, no one put better: an excellent and very convincing orator.

[A] But as for his morals and his sense of right and wrong, he far surpassed in that all those who have ever engaged in the affairs of state. For in that quality (which we must consider to be the principal one, [C] which alone truly reveals what we are and which, all by itself, outweighs for me all the other qualities put together) [A] he yields to no philosopher whatsoever, not even to Socrates himself. [B] In Epaminondas integrity is the dominant quality proper to him, constant, equable, incorruptible. Against his standard, the sense of right and wrong in Alexander seems subordinate, hesitant, spasmodic, weak and subject to chance.

[C] In Antiquity's judgement anyone who studied all the other Captains in the smallest detail would find in each of them one special quality which rendered him illustrious; but in Epaminondas alone there is virtue ever abundant, an unchanging competence which never leaves anything to be desired in any of a man's duties in this life, whether in political or private occupations, in peace or in war, or, when it comes to dying, in greatness and glory. I know of no man's form or fortune that I can regard with such honour and love. It is true that I find his stubbornly persisting in poverty as portrayed by his best friends somewhat over-scrupulous. That action alone, sublime and most admirable though it be, is rather too grim, I feel, for me even to wish that I could desire to imitate it.

Only one man could make me hesitate in his case: Scipio Aemilianus – if we could provide him with as proud and illustrious an end and with as deep and universal a knowledge of the arts and the sciences.[11] What a grievous loss to me it is that Time has robbed our eyes of precisely the foremost parallel lives, the lives of the noblest pair in Plutarch, those of these two great men who were, respectively, by common consent, the greatest of the Greeks and the greatest of the Romans. What a subject! And what a craftsman!

For a man who was no saint but (as they say) a gentlemanly kind of fellow, with the manners of a good public figure and citizen who was

11. Scipio Africanus Minor, the son of Paulus Aemilius, often called Scipio Aemilianus; he achieved great renown in the Third Punic War. Cicero idealized him in his *De Republica*, *De Senectute* and *De Amicitia*.

moderately distinguished, the most richly varied life that I know to have been lived (as we say) among the living, a life crammed with splendid and desirable qualities is (all things considered and to my liking) that of Alcibiades.

But to return to Epaminondas, [A] I would like to cite a few of his opinions so as to provide an example of his excellent goodness.

[B] He swore that the greatest satisfaction he ever had in his life was to have given pleasure to his father and mother by his victory at Leuctra. It is greatly to his favour that he should prefer their pleasure in such a glorious battle to his own full and rightful pleasure in it.

[A] He did not think it was permissible to kill any man without understanding why, not even to restore freedom to one's country. That explains why he was so cold towards the campaign to deliver Thebes led by his companion Pelopidas. He also held that in battle a man should spare anyone he loves on the opposing side and fly from encounters with him. [C] His humane treatment of his enemies as well made him suspect to the Boeotians: he had by some miracle forced the Spartans to open to him at Morea (near Corinth) a pass which they had taken up arms to defend; he was content to strike straight through their middle without hounding them to death. For that he was relieved of his post as Captain-General – very honourably so, seeing that it was for such a cause, and also because the Boeotians were soon shamed by the necessity of having to reinstate him and to admit how much their glory and their safety were due to him, since whenever he led his forces victory followed him like a shadow.[12]

The prosperity of his country died as it had been born: with him.

12. Montaigne's principal sources are Plutarch's *Pelopidas*, Cornelius Nepos' *Epaminondas*, Erasmus' *Apophthegmata* and Plutarch (tr. Amyot), *Esprit familier de Socrate*.

37. On the resemblance of children to their fathers

═══════

[This is the final chapter of Book II and so, until 1588, the final chapter of the whole work, which ended therefore with two dominant notions: that the Essays are a portrait of Montaigne's character, opinions and bearing destined for his immediate descendants and friends; that the most marked characteristic of Nature is diversity and discordance.

Montaigne was convinced that he had inherited from his forefathers not only an antipathy to medicine but also the stone (that is, to the suicide pains of colic paroxysms). He explains how he fortified his inherited antipathy to the art of medicine with often contrived arguments, so giving us insights into his mind and incidentally providing a lively picture of life in watering-places. Spa-waters, being natural, might cure the stone and can probably do no harm. But how experimental medicine is ever supposed to be led to a cure for melancholy is another matter . . .

A major source of Montaigne's scepticism here about professional arts and sciences is Henry Cornelius Agrippa's book On the Vanity of all Sciences and on the Excellence of the Word of God.]

[A]   All the various pieces of this faggot are being bundled together on the understanding that I am only to set my hand to it in my own home and when I am oppressed by too lax an idleness. So it was assembled at intervals and at different periods, since I sometimes have occasion to be away from home for months on end. Moreover I never correct my first thoughts by second ones –   [C]   well, except perhaps for the odd word, but to vary it, not to remove it.   [A]   I want to show my humours as they develop, revealing each element as it is born. I could wish that I had begun earlier, especially tracing the progress of changes in me.

One of the valets I used for dictation stole several pages of mine which were to his liking and thought he had acquired great plunder. It consoles me that he will no more gain anything by it than I shall lose.

Since I began I have aged by some seven or eight years – not without some fresh gain, for those years have generously introduced me to colic paroxysms. Long commerce and acquaintance with the years rarely proceed without some such benefit! I could wish that, of all those gifts which the years store up for those who haunt them, they could have chosen a present

more acceptable to me, for they could not have given me anything that since childhood I have held in greater horror. Of all the misfortunes of old age, that was precisely the very one I most dreaded. I often thought to myself that I was travelling too far and that on such a long road I was eventually bound to be embroiled in some nasty encounter; I realized, and much proclaimed, that it was time for me to go. Following the surgeon's rule when he cuts off a limb, [C] I declared that life should be amputated at the point where it is alive and healthy; he who repays not his debt to Nature in good time usually finds she exacts interest with a vengeance.

[A] But my declarations were in vain. I was so far from being ready to go then that even now, after about eighteen months in this distasteful state, I have already learnt how to get used to it. I have made a compact with this colical style of life; I can find sources of hope and consolation in it. So many men have grown so besotted with their wretched existence that no circumstances are too harsh, provided that they can cling on. [C] Just listen to Maecenas:

> *Debilem facito manu,*
> *Debilem pede, coxa,*
> *Lubricos quate dentes:*
> *Vita dum superest bene est.*

[Lop off a hand; lop off a foot and a thigh; pull out all my teeth: I am all right though: I am still alive.][1]

And it was with the philanthropy of a lunatic that Tamberlane cloaked his arbitrary cruelty against lepers when he put to death all those that came to his knowledge – 'In order,' he said, 'to free them from so painful a life.' Any of them would rather have been thrice a leper than to cease to be.[2] When Antisthenes the Stoic was extremely ill he cried out, 'Who will make me free from these ills?' Diogenes, who had come to see him, gave him a knife: 'If you so desire, this soon will,' he said. There came the reply: 'I never said *from this life*: I said, *from these ills.*'

[A] Sufferings which touch the soul alone afflict me much less than they do most men; that is partly from judgement (for the majority think many things to be dreadful and to be avoided even at the cost of their life

1. Seneca, *Epist. moral.*, CI, ll, citing 'the most vile prayer of Maecenas'.
2. Nicolas Chalcocondylas, *De la décadence de L'Empire Grec*, III, x; then, Erasmus, *Apophthegmata*, III, *Diogenes Cynicus*, CCXVII.

which are almost indifferent to me); it is also partly because of my stolid complexion which is insensitive to anything which does not come straight at me; I believe that complexion to be one of the best of my natural characteristics. But bodily sufferings – which are very real – I feel most acutely. And yet, formerly, when I used to foresee them through eyes made weak, fastidious and flabby by the enjoyment of that long and blessed health and ease which God had lent me for the greater part of my life, I thought of them as so unbearable that in truth my fear of them exceeded the suffering they now cause me: that fact further increases my belief that most of the faculties of our soul,   [C] as we employ them, [A]   disturb our life's repose rather than serve it.

I am wrestling with the worst of all illnesses, the most unpredictable, the most painful, the most fatal and the most incurable. I have already assayed five or six very long and painful attacks. Yet either I am flattering myself or else, even in this state, a man can still find things bearable if his soul has cast off the weight of the fear of dying and the weight of all the warning threats, inferences and complications which Medicine stuffs into our heads. Even real pain is not so shrill, harsh and stabbing that a man of settled temperament must go mad with despair. I draw at least one advantage from my colic paroxysms: whatever I had failed to do to make myself familiar with death and reconciled to it that illness will do for me: for the more closely it presses upon me and importunes me the less reason I shall have to be afraid to die. I had already succeeded in holding on to life only for what life has to offer: my illness will abrogate even that compact; and may God grant that at the end, if the harsh pain finally overcomes my strength, it may not drive me to the other extreme (no less wrong) of loving and yearning to die.

Summum nec metuas diem, nec optes.

[Neither be afraid of your last day nor desire it.][3]

Both emotions are to be feared, though one has its remedy nearer at hand.

Moreover I have always considered that precept to be sheer affectation which so rigorously and punctiliously ordains that, when we are enduring pain, we must put on a good countenance and remain proud and calm. Why should Philosophy (whose concerns are with deeds and with inner

3. Martial, *Epigrams*, X, xlvii, 3. (Having inscribed this maxim in his library, Montaigne later painted another over it.)

motions) waste her time over external appearances?[4]   [C]   Let her leave such worries to actors in farces and to masters of rhetoric, who make such a fuss about our gesticulations. Let Philosophy have enough courage to concede that pain may act cowardly so long as the cowardice remains a matter of words, being neither heartfelt nor visceral. Let her classify such plaints (even if they do come from our will) with those sighs, sobs, tremblings and drainings of colour which Nature has placed beyond our control. So long as our minds know no terror and our words no despair, let Philosophy be contented. What does it matter if our arms flay about as long as our thoughts do not? Philosophy put us through our training not for others but for ourselves, so that we may *be* thus, not seem thus.

[A]   Let her limit herself to controlling our intellect, which she has undertaken to instruct. Against the onslaught of colic paroxysms let her enable us to have souls capable of knowing themselves and following their accustomed courses, souls fighting pain and sustaining it, not shamelessly grovelling at her feet, souls stirred and aroused for battle, not cast down and subdued,   [C]   able to communicate and to some extent able to converse.

[A]   In such extreme misfortunes it is cruelty to require of us too studied a comportment. If we play our role well, it matters little if we put a bad face on things! If the body finds relief in lamentations, let it; if it wants to toss about, let it writhe and contort as much as it likes; if the body believes that some of the pain can be driven off as vapour by forcing out our cries – or if doing so distracts us from the anguish, as some doctors say it helps pregnant women in their deliveries – just let it shout out.   [C]   Do not order the sound to come but allow it to do so. Epicurus does not merely allow his wise man to yell out in torment, he counsels him to: '*Pugiles etiam, quum feriunt in jactandis coestibus, ingemiscunt, quia profundenda*

---

4. '80 . . . Philosophy (whose concerns are with *life and substance*) waste her time over external appearances *as though she was rehearsing men for the actions of a play, or as though it was of her jurisdiction to restrain movements and changes which we are required by Nature to accept? Let her restrain Socrates, then, from blushing with emotion or shame, from blinking when threatened with a blow, or trembling and sweating under the shakings of a fever: the descriptions of Poetry (who is free and freely willed) dare not deprive of tears even those persons whom she would present as perfect and complete: 'Et se n'aflige tanto/Che se mordi le man, morde le labbia,/Sperge le guancie di continuo pianto': Philosophy should leave such a duty to those whose profession it is to rule our deportment and outward show.* Let her limit . . . (The Italian verse means: 'Her pain is such that she wrings her hands, bites her lips, while her cheeks are bathed in a flood of tears.)'

voce omne corpus intenditur, venitque plaga vehementior.' [Even the wrestlers grunt when lashing out with their boxing-gloves, because uttering such sounds makes the whole body tense, driving the blow home with greater vehemence.]⁵ [A] We have pangs enough from the pain without the pangs caused by clinging to superfluous rules.

It is usual to see men thrown into turmoil by the [C]   attacks and   [A]   assaults of this illness; it is for them that I have said all this; for in my own case I have up till now put on a slightly better countenance: not that I take any trouble to maintain a decent appearance, for I do not think much of such an achievement and, in this respect, concede whatever my illness demands: but either the pain in my case is not so excessive or else I can show more steadfastness than most. I moan and groan when the stabbing pains hurt most acutely but I do not   [C]   lose control like this fellow:

> *Ejulatu, questu, gemitu, fremitibus*
> *Resonando multum flebiles voces refert.*

[Re-echoing with his tearful voice, wailing, groaning, lamenting, sighing.]⁶

At the darkest moment of the paroxysm I explore myself and have always found that I am still capable of talking, thinking and replying as sensibly as at any other time but not as imperturbably, since the pain disturbs me and distracts me. When those around me start to spare me, thinking that I am at my lowest ebb, I often assay my strength and broach a subject as completely removed as possible from my condition. I can bring off anything with a sudden effort. But do not ask it to last . . .

If only I were like that dreamer in Cicero who dreamed he had a woman in his arms and had the faculty of ejaculating his gallstone in the bedclothes!⁷ My own gallstones monstrously unlecher me!

[A] In the intervals between these extremes of anguish, [C] when my urinary ducts are sick but without the stabbing pains, [A] I return at once to my accustomed form, since my soul knows no call to arms without bodily feeling – I definitely owe that to the care I once took to prepare myself by reason for such misfortunes:

> [B]   *laborum*
> *Nulla mihi nova nunc facies inopinaque surgit;*
> *Omnia præcepi atque animo mecum ante peregi.*

5. Diogenes Laertius, *Epicurus*, X, cxviii; then, Cicero, *Tusc. disput.*, II, xxiii, 50.
6. Cicero, *De finibus*, II, xxix, 94; equally condemned by Cicero.
   '80: show despair and rage . . .
7. Cicero, *De divinatione*, II, lxix, 143.

[No toils present themselves new or unforeseen: I have seen them coming and been through them already in my mind.][8]

[A]   But I have been assayed rather too roughly for an apprentice; the blow was indeed sudden and rough, for I fell all at once from a most gentle and happy mode of life into the most painful and distressing one imaginable: for, leaving aside the fact that the stone is an illness itself to be dreaded, its onset was in my case unusually difficult and harsh. Attacks recur so frequently that nowadays I hardly ever feel perfectly well. Yet if only I can add duration to the state in which I now maintain my spirits, I shall be in very much better circumstances than hundreds of others who have no fever nor illness except the ones which they inflict on themselves by defect of reason.

There is a certain kind of wily humility which is born of presumption. This for instance: we admit there are many things we do not understand; we confess frankly enough that within the works of Nature there are some qualities and attributes which we find incomprehensible, the means or the causes of which cannot be discovered by capacities such as ours. With so frank and scrupulous an admission we hope to make people believe what we say about the ones we do claim to understand. Yet there is no need to go picking over strange problems or miracles; it seems to me that among the things which we see quite regularly there are ones so strange and incomprehensible that they surpass all that is problematic in miracles.

What a prodigious thing it is that within the drop of semen which brings us forth there are stamped the characteristics not only of the bodily form of our forefathers but of their ways of thinking and their slant of mind. Where can that drop of fluid lodge such an infinite number of Forms? [B]   How does it come to transmit these resemblances in so casual and random a manner that the great-grandson is like his great-grandfather, the nephew like his uncle? In the family of Lepidus in Rome there were born three children (not all at once; there were gaps between them) with cartilage over the very same eye.[9] There was a whole family in Thebes whose members all bore birthmarks shaped like a lance-head; any child who did not do so was held to be illegitimate. According to Aristotle there is a certain nation where they have wives in common and where children were assigned to fathers by resemblances.

[A]   We can assume that it is to my father that I owe my propensity to the stone, for he died dreadfully afflicted by a large stone in the bladder.

---

8. Virgil, *Aeneid*, VI, 103–5.
9. Pliny, *Hist. nat.*, VII, xii; then, Plutarch (tr. Amyot) *Pourquoy la justice divine differe* . . . 267 B; Aristotle, *Politics*, II, i, 1262 a (this nation was in Upper Libya).

He was not aware of it until he was sixty-seven; he had experienced no sign or symptom of it beforehand, in his loins or his sides or anywhere else. Until then he had not been subject to much illness and had in fact enjoyed excellent health; he lasted another seven years with that affliction, lingering towards a very painful end.

Now I was born twenty-five years and more before he fell ill, during his most vigorous period: I was his third child. During all that time where did that propensity for this affliction lie a-brooding? When his own illness was still so far off, how did that little piece of his own substance which went to make me manage to transmit so marked a characteristic to me? And how was it so hidden that I only began to be aware of it forty-five years later – so far the only one to do so out of so many brothers and sisters, all from the same mother? If anyone can tell me how this comes about I will trust his explanations of as many other miracles as he likes – providing that he does not fob me off (as they usually do) with a theory which is more difficult and more fanciful than the thing itself.

Doctors will have to pardon my liberty a while, but from that same ejaculation and penetration I was destined to receive my loathing and contempt for their dogmas: my antipathy to their Art is hereditary; my father lived to seventy-four, my grandfather to sixty-nine, my great-grand-father to nearly eighty, none having swallowed any kind of drug. 'Medicine' for them meant anything they did not use regularly.

The Art of Medicine is built from examples and experience. So are my opinions. Have I not just cited an experience both relevant and convincing? I doubt if the annals of medicine can provide an example of three generations born, bred and dying in the same home under the same roof who have lived under doctor's orders as long as they did. Doctors will have to concede that on my side there is either reason or luck. And with them luck is a more valuable commodity than reason . . .

But they must not take advantage of me now, and certainly not threaten me after I have been struck down: that would not be fair. I have truly won a solid victory over them with that example of the rest of my family, even if it stops with them. Human affairs allow of no greater constancy: we have assayed our beliefs now for two centuries minus eighteen years: my great-grandfather was born in the year one thousand four hundred and two. It is only right that this experiment of ours should begin to run out on us. Let them not quote against me the illness which has got a stranglehold on me now. Is it not enough that even I stayed healthy for forty-seven years? Even if it should prove to be the end of our course, it has been longer than most.

My forebears disapproved of medicine because of some unexplained

natural inclination. The very sight of medicine horrified my father. The Seigneur de Gaviac was one of my uncles on my father's side; he was in holy orders, a weakling from birth, who nevertheless struggled on to sixty-seven; once he did fall victim of a grave and delirious attack of Continual Fever; the doctors ordered that he be informed that he would definitely die if he did not call in aid − (what they call 'aid' is more often than not an impediment). Terrified though he was by this dreadful sentence of death, that good man replied: 'I am dead then.' But soon afterwards God showed the vanity of their prognosis.

[B] I had four brothers; the youngest, born a long time after the others, was the Sieur de Bussaguet; he was the only one to submit to the Art of medicine, doing so I think because of his contacts with practitioners of other arts, since he was counsellor in the Court of Parliament. It turned out so badly for him that, despite apparently having the strongest of complexions, he died way before all the others with the sole exception of the Sieur de Saint-Michel.

[A] Though it is possible that I inherited this natural aversion from my ancestors I would have assayed ways of countering it if that had been the only factor, since all non-rational inborn tendencies are a kind of disease which ought to be fought against. It may well be that I inherited the disposition, but I have supported it, fortified it, and corroborated my opinions, by reasoned argument: I loathe such motives as refusing medicine just because it tastes bitter. My temperament is not at all like that: I believe health to be so precious that I would buy it at the cost of the most agonizing of incisions and cauterizations. [C] Following Epicurus I believe pleasures are to be avoided if they result in greater pain, and pain is to be welcomed if it results in greater pleasure.[10]

[A] Health is precious. It is the only thing to the pursuit of which it is truly worth devoting not only our time but our sweat, toil, goods and life itself. Without health all pleasure, scholarship and virtue lose their lustre and fade away. The most firmly supported arguments against this that Philosophy seeks to impress on us can be answered by this hypothesis: imagine Plato struck down by epilepsy or apoplexy; then challenge him to get any help from all those noble and splendid faculties of his soul.

No road leading to health can be called rough or expensive for me. But there are other likely reasons too which make me suspicious of all such trafficking. I do not deny that there may be an element of art in medicine. It is quite certain that among all the works of Nature things may be found with properties which can preserve our health. [B] I mean that there are

10. Cicero, *Tusc. disput.*, V, xxxiii, 95.

simples which moisten and desiccate; I know from experience that horse-radish produces flatulence and that senna-pods act as an aperient. Experience has taught me other things too; so that I know that mutton nourishes me and wine warms me (Solon used to say that eating was like other remedies: it was a cure for a disease called hunger).[11] I do not reject practices drawn from the natural world; I do not doubt the power and fecundity of Nature nor her devotion to our needs. I can see that the pike and the swallows do well under her. What I am suspicious of are the things discovered by our own minds, our sciences and by that Art of theirs in favour of which we have abandoned Nature and her rules and on to which we do not know how to impose the limits of moderation.

[C] What we call justice is a farrago of any old laws which fall into our hands, dispensed and applied often quite ineptly and iniquitously; those who mock at this and complain of it are not reviling that noble virtue itself but only condemning the abuse and the profanation of that venerable name of justice. So too with medicine: I honour its glorious name, its aim and its promises, so useful to the human race; but what that name actually designates among us I neither honour nor esteem.

[A] In the first place experience makes me afraid of it, for as far as I can see no tribe of people are more quickly ill nor more slowly well than those who are under the jurisdiction of medicine. The constraints of their diets impair and corrupt their health. Doctors are not content with treating illness; they make good health ill too so as to stop us ever escaping from their jurisdiction. Do they not assert that long and continuous good health argues future illness?

I have been ill quite frequently; without help from doctors I have found my illnesses – and I have assayed virtually all of them – quite easy to bear and as short-lasting as anyone else's; and I have done this without bringing in the bitter taste of their prescriptions. My health is complete and untrammelled, with no rule but my habits, no discipline but my good pleasure. Any place is good enough for me to stay in: I need no more comforts when I am ill than when I am well. I do not get worked up because there is no doctor or no apothecary nearby to come to my aid (something which I can see to be a greater affliction for some people than the illness itself). Yet are the lives of doctors themselves so long and so happy that they can witness to the manifest effectiveness of their discipline?

Every nation existed without medicine for centuries (that was the first age of Man, the best and the happiest centuries); even now less than a tenth of the world makes use of it. Nations without number have no knowledge

11. Diogenes Laertius, *Epicurus*, CXXIX.

of medicine and live longer and more healthily than we do here. And among us the common folk manage happily without it. The Roman People were six hundred years old before they adopted it; then, having assayed it, they drove it out of their city at the instance of Cato the Censor who showed how easily he could do without it, having lived to be eighty-five himself and helping his wife to live to an extreme old age – not without medicine but without medical practitioners. (Anything at all which promotes good health can be called medicine.)

Plutarch says that Cato kept his family in good health by making use, [A1] it appears, [A] of the hare, just as the Arcadians, according to Pliny, cured all illnesses with cow's milk.[12] [C] Herodotus asserts that the Libyan people all enjoy a rare degree of good health owing to their custom of searing the veins in the head and temples of their children with cauteries at the age of four, thus blocking the way for the rest of their lives to all morbid defluxions of mucous. [A] And the villagers round here when they are ill never use anything but the strongest wine they can get, mixed with plenty of saffron and spice. And they all work equally well.

Truly, among all that confusing diversity of prescriptions is there any practical result except the evacuation of the bowels? Hundreds of homely simples can produce that. [B] And I am not convinced that the action of the bowels is as beneficial as they claim; perhaps our nature needs, up to a point, the residue of its excreta just as wine must be kept on its lees if you want to preserve it. You can often see healthy men succumbing, from some external cause, to attacks of vomiting or diarrhoea: they have a big turn-out of excrement without any prior need or subsequent benefit: indeed it does harm; they get worse. [C] It is from the great Plato himself that I recently learned that of the three motions which apply to men, the last and the worst is the motion of purgations; no man, unless he is a fool, should undergo one except of extreme necessity.[13] We set about disturbing and activating our illnesses by fighting them with contraries: yet it ought to be our way of life which gently reduces them and brings them to an end. Those violent clashes between the illness and the medicine always cost us dear since the quarrel is fought out in our inwards, while drugs give us unreliable support, being by their nature the enemy of our health and gaining access into our estates only through disturbances.

Let us leave things alone for a while: that Order which provides for the flea and the mole also provides for all men who suffer themselves to be

---

12. Henry Cornelius Agrippa's *De Vanitate omnium scientiarum et de excellentia verbi Dei* LXXXIII, is a major source here.
13. Plato, *Timaeus*, 8 B.

governed by it as the flea and the mole are. We shout *Gee up* in vain: it
will make our throats sore but not make that Order go faster, for it is
proud and knows no pity. Our fear and despair repel it and delay its help
for us rather than summoning it. It owes it to disease as to health that each
should run its course. It will not be bribed to favour one at the expense of
the rights of the other: for then it would become Disorder. For God's sake
let us follow. I repeat, follow. That Order leads those who follow: those
who will not follow will be dragged along,[14] medicine, terror and all. Get
them to prescribe an aperient for your brain; it will be better employed
there than in your stomach.

[A] When a Spartan was asked what made him live so long, 'Ignorance of medicine,' he replied. And the Emperor Hadrian kept repeating
as he lay dying that 'all those doctors' had killed him.[15] [B] When a
bad wrestler became a doctor Diogenes said, 'That's the spirit. You are
right. Now you can pin to the ground all those who used to do it to
you.' [A] But doctors are lucky [B] according to Nicocles: [A] the
sun shines on their successes and the earth hides their failures; on top
of that they have a way of turning anything which happens to their
own advantage: medicine claims the right to take credit for every improvement or cure brought about by Fortune, Nature or any other external
cause (and the number of those is infinite). When a patient is under
doctors' orders anything lucky which happens to him is always due to
them. Take those opportune circumstances which have cured me and
hundreds of others who never call in medical help: in the case of their
patients doctors simply usurp them. And when anything untoward happens
they either disclaim responsibility altogether or else blame it on the
patient, finding reasons so vacuous that they need never fear they will
ever run out of them: 'he bared his arm'; [B] 'he heard the noise of a
coach'—

> *rhedarum transitus arcto*
> *Vicorum inflexu;*
> [wagons passing at the bends in narrow streets;][16]

---

14. The great Renaissance commonplace, deriving from Seneca, *Epist. moral.*,
CVIII, 11: Greek verses of Cleanthes, translated by Seneca and ending, 'The Fates
lead the willing but drag the unwilling.' St Augustine cites them (*City of God*, V,
8) in the context of the will of God.
15. Quips from H. C. Agrippa's *De Vanitate*, LXXXIII (from which work several
subsequent borrowings are made); also, Diogenes Laertius, *Diogenes*, VI, lxii.
16. Juvenal, *Satires*, III, 236–7.

– [A] 'somebody opened a window'; 'he has been lying on his left side'; 'he has let painful thoughts run through his head'. In short a word, a dream, a glance, all appear to be sufficient excuses for shrugging off the burden of responsibility.

Or when we get worse they take advantage of that too if they want to, profiting from another ploy which can never fail: when their poultices merely help to inflame the illness they palm us off with assertions that without their remedies things would have been even worse. They take a man with a bad cold, turn it into a recurrent fever, then claim that without them it would have been a continual fever. No need to worry that business should be bad: when an illness grows worse it means greater profits for them. They are certainly right to require their patients to favour them with their trust. It truly has to be trust – and a pliant trust too – to cling to notions so hard to believe.

[B] Plato put it well when he allowed freedom to lie to no one but doctors, since their promises are empty and vain but our health depends on them.[17]

[A] Aesop is an author of the choicest excellence, though few people discover all his beauties; he agreeably portrays the tyrannous authority which doctors usurp over wretched souls weakened by sickness and prostrated by fear when he tells how a patient was asked by his doctor what effects he felt from a medicine he had given him: 'I sweated a lot,' said the patient. 'Good,' said the doctor. Another time he asked him how he had fared since then: 'I felt extremely cold and shivery,' he said. – 'Good,' replied the doctor. On a third occasion he again asked him how he felt: 'All puffy and swollen up,' he said, 'as though I had dropsy.' – 'Excellent!' said the doctor. Then one of the patient's close friends came to ask how things were with him. 'I am dying of good health, my friend,' he replied.[18]

They used to have a more equitable contract in Egypt: for the first three days the doctor took on the patient at the patient's risk and peril: when the three days were up, the risks and perils were the doctor's. Is it right that Aesculapius, the patron of medicine, should have been struck down by a thunderbolt for having brought the dead ['95] Hippolytus[19] [A] back to life –

17. Plato, *Republic*, 389 BC (cf. 382 D).
18. Aesop, *Fables*.
19. Until ['95]: not *Hippolytus* but, erroneously, *Helen*. (Then, Virgil, *Aeneid*, VII, 770–3.)

[B]   *Nam pater omnipotens, aliquem indignatus ab umbris*
Mortalem infernis ad lumina surgere vitæ,
Ipse repertorem medicinæ talis et artis
Fulmine Phæbigenam stygias detrusit ad undas

[For the Father Almighty, angry that a mortal should rise from the Shades of the Underworld to the light of the living, struck down the discoverer of the Art of Medicine, the son of Apollo, and with his thunderbolt cast him into the waters of Styx]

– [A] while his followers who send so many souls from life to death find absolution! [B] A doctor was boasting to Nicocles that his Art had great prestige. Nicocles retorted: 'It must indeed be so, if you can kill so many people with impunity.'

[A] Meanwhile if they had asked my advice I would have rendered their teachings even more mysterious and awesome. They began well but did not keep it up to the end. A good start that, making gods and *daemons* the authors of their doctrines and then adopting a specialized language and style of writing – [C] even though Philosophy may think that it is madness to give a man good counsel which is unintelligible: '*Ut si quis medicus imperat ut sumat: "Terrigenam, herbigradam, domiportam, sanguine cassam"* . . .' [As though a doctor's prescription for a diet should say: 'Take terrigenous herbigressive autodomiciled desanguinated gasteropods . . .']²⁰

[A] It has proved a rule good for the Art (found in all vain fantastical supernatural arts) that the patient must first trust in the remedy with firm hope and assurance before it can work effectively. They cling to that rule so far as to hold that a bad doctor whom a patient trusts is better than the most experienced one whom he does not know.

The very constituents selected for their remedies recall mystery and sorcery: the left foot of a tortoise, the urine of a lizard, the droppings of an elephant, the liver of a mole, blood drawn from under the right wing of a white pigeon; and for those of us with colic paroxysms (so contemptuously do they abuse our wretchedness) triturated rat-shit and similar apish trickery which look more like magic spells than solid knowledge. I will not even mention pills to be taken in odd numbers; the designation of particular days and festivals as ominous; the prescribing of specific times for gathering the herbs for their ingredients; and the severe, solemn expression on doctors' faces which even Pliny laughs at.

Where doctors went wrong (I mean after such a good start) is that they did not also make their assemblies more religious and their deliberations

20. Cicero, *De divinatione*, II, lxiv, 133. The doctor is prescribing a diet of snails!

more secret: no profane layman ought to have access to them, no more than to the secret ceremonies of Aesculapius. Because of this error their uncertainties and the feebleness of their arguments, of their guesswork and of their premises, as well as the bitterness of their disagreements (full of hatred, of envy and personal considerations), have all been revealed to everybody, so that a man must be wondrously blind if he does not feel at risk in their hands.

Did you ever find a doctor taking over a colleague's prescription without putting in something extra or cutting something out? That gives their Art away and reveals that they are more concerned with their own reputation (and therefore with their fee) than with the well-being of their patients. The wisest of them all was he who decreed that each patient should be treated by only one doctor:[21] for if he does no good the failure of one single man would be no great reproach to the whole Art of medicine, but if on the contrary he does strike lucky, then great is the glory; whereas when many are involved, they discredit their trade at every turn, especially since they normally manage to do more harm than good. They ought to have remained satisfied with the constant disagreements to be found among the opinions of the great masters and ancient authorities of their Art – only the bookish know about those – without letting everybody know of their controversies and the intellectual inconsistencies which they still foster and prolong among themselves.

Do we want to see an example of medical disagreement among the Ancients? Hierophilus locates the original cause of illness in the humours; Erasistratus, in arterial blood; Asclepiades, in invisible atoms flowing through the pores; Alcmaeon, in the exuberancy or deficiency of bodily strength; Diocles, in the imbalance of our corporeal elements and the balance of the air that we breathe; Strato, in the quantity, crudity and decomposition of the food we eat; and Hippocrates locates it in our spirits.[22]

A friend of the doctors, whom they know better than I do, exclaims in this connection that it is a great misfortune that the most important of all the sciences we use, the one with responsibility for our health and preservation, should be the most uncertain, the most unstable and the one shaken by the most changes.

There is no great harm done if we miscalculate the height of the sun or

21. Rasis (Al Razi); after H. C. Agrippa.
22. Again, from H. C. Agrippa, as is the following list of medical variations; also Pliny (*Hist. nat.*, XXIX, i). The following 'friend of the doctors' is Pliny.

the fractions in some astronomical computation: but here, when it is a matter of the whole of our being, there is no wisdom in abandoning ourselves to the mercy of so many contrary gales.

Nothing much was heard of this science before the Peloponnesian War. It was brought into repute by Hippocrates. Everything he established was overturned by Chrysippus; everything Chrysippus wrote was then overturned by Erasistratus, the grandson of Aristotle. After that lot there came the Empirics who in their Art adopted a method quite different from the Ancients; when their reputation began to grow shaky, Herophilus succeeded in getting a new kind of medicine accepted, which Asclepiades came and attacked, destroying it in his turn. Then successively the opinions of Themison gained authority, then Musa's, then later still those of Vexius Valens (the doctor famous for his intimacy with Messalina). At the time of Nero, the empire (of medicine) fell to Thessalus, who condemned and destroyed everything taught before him. His teachings were subsequently struck down by Crinas of Massilia, whose new contribution was to regulate all the workings of medicines by ephemerides and astral movements, making men eat, sleep and drink at the times which suited the Moon or Mercury. His authority was soon supplanted by that of Charinus, also a doctor in Massilia; he not only fought against Ancient medicine but also against the centuries-old public institution of hot baths. He made his patients take cold baths even in winter, immersing the sick in streams of fresh-water.

Before Pliny's time no Roman had ever condescended to practise medicine; that was done by Greeks and foreigners – as among us French it is practised by spouters of Latin. As a very great doctor has said, we do not easily accept treatments we can understand, any more than we [C] trust the simples we ourselves gather.[23] [A] If those nations where we find our guaiacum, sarsaparilla and china-root have doctors of their own, just think how exoticism and costliness must make them esteem our cabbages and our parsley: who would dare to despise plants sought in such distant lands at the risk of long and perilous journeys!

Since those medical upheavals among the Ancients there have been innumerable others up to our own times, mostly total fundamental revolu-

23. Latin was the usual language of doctors throughout the sixteenth century. For the first doctor in Rome see Polydore Vergil, *De Inventoribus rerum*, I, xx, citing Livy, XVI.

'80: we *could value drugs which are known to us: if a drug does not come from overseas and has not been brought to us from far-off regions it has no efficacy. If* ... (I cannot identify the 'very great doctor'.)

tions like those recently produced by Paracelsus, Fioravanti and Argenterius; I am told that they do not only change the odd prescription but the woof and web and the government of the medical corpus, accusing those who professed it before them of being ignorant charlatans.[24]

I leave you to imagine where that leaves the wretched patient.

If we could only be sure that their mistakes did us no harm even if they did no good it would be a reasonable bet to chance gaining something without risk of losing everything.   [B]   But Aesop tells how a man bought a Moorish slave and thought that his colour was incidental, brought on by ill-treatment from his former master; so he had him carefully physicked with baths and medical concoctions. As a result the Moor was not cured of swarthiness but he did lose his good health. [A]   How often have we found doctors blaming each other for the deaths of their patients! I remember the local epidemic a few years ago: it was fatally dangerous. When the storm was over (having swept away innumerable people) one of the most celebrated doctors in the land published a booklet on the subject in which he regretted having prescribed bloodletting, admitting that it was one of the principal sources of the harm that was done.[25]

Moreover their authors maintain that there is no medicine without harmful side-effects: if those which do us some good do us some harm as well, what must the other ones do when applied to us quite abusively?

As for those who loathe the taste of medicine, I personally feel that, even if for no other reason, it would be dangerous and harmful for them to make themselves force it down at so inappropriate a time: just when they need rest it constitutes, I think, an unacceptable assay of their strength. Besides when I consider the factors which are said to occasion our illnesses I find them so slight and so specific that I am forced to conclude that even a tiny error in the prescribed dosage could do us great harm.

Now things go very bad indeed for us if our doctor's mistake is a dangerous one, for it is difficult for him not to go on falling into yet more errors. To aim at the right target his treatment must embrace very many

24. The most famous (and infamous) of these was Bombast von Hohenheim (Paracelsus) whose chemical and mystical therapeutics reject medical tradition. One of Leonardo Fioravanti's books appeared in English in 1582 as *A compendium of Rational Secrets . . .*; Johannes Argenterius wrote critical commentaries on Galen and Hippocrates.
25. Aesop's fable gave rise to the expression, 'to wash an Ethiop white'. Ambroise Paré questioned the validity of phlebotomy in his treatise on the plague.

factors, circumstances and elements: he must know his patient's complexion, his temperament, his humours, his inclinations, his actions, even his thoughts and his ideas; he must take into account external circumstances such as the nature of the locality, the condition of the air and the weather, the position of the planets and their influences; then he must know what causes the illness, its symptoms and their effects and the day when the crisis is reached. Where the drugs themselves are concerned he must know their dosage and their strength, their country of origin, appearance and maturity as well as the right prescription. And then he must know how to combine those elements together in the right proportions so as to produce a perfect balance. If he gets any one of them slightly wrong, or if one of his principles is slightly awry, that is enough to undo us. Only God knows how difficult it is to understand most of these elements; for example, how can a doctor discover the proper symptoms of your illness when each illness can comport an infinite number of them? How many hesitations and disputes do they have over the analysis of urines? Otherwise, how could we explain their ceaseless wranglings over their diagnoses? How else could we excuse their 'mistaking sables for foxes' – the fault they fall into so often? In such illnesses that I have had, as soon as there was the slightest complication I never found three doctors to agree.

I am most impressed by the examples which could affect me.

Recently there was a nobleman in Paris who was cut on doctor's orders: the surgeon found he no more had a stone in his bladder than in the palm of his hand.

Then there was a close friend of mine, a bishop; most of the doctors he consulted urgently pressed him to be cut; trusting in the others, I too joined in the persuasion; once he was dead they opened him up and found he only had vague kidney trouble. They have less excuse in the case of the stone, which, to some extent, can be felt by probing. That is why surgery always seems to me to be more exact: it sees and feels its way along; there is less conjecture and guesswork: medicine has no vaginal prod which can open up the passages of our brains, our lungs or our livers.

The very promises of medicine pass belief; for doctors have to treat several maladies which afflict us at the same time and which, almost of necessity, are interconnected – for example, a heated liver and a chilled stomach; they try to persuade us that some of their ingredients warm the stomach while others refresh the liver: one is said to go straight to the liver or even to the very bladder without displaying its powers on the way and while conserving its efficacity and virtues throughout that long journey with all its pitfalls, until it arrives where its occult properties are destined to

apply. This one desiccates the brain while the other humidifies the lungs. Is it not a kind of raving madness to think you can mix a draught out of all those remedies and then hope that all the virtues of the drugs contained in that chaotic mixture will split themselves up, sort themselves out and rush each to its divergent task? I would have endless fears that the instructions on their labels might get lost or switched round so that they confuse their destinations! And how can anyone think that the various properties in that liquid jumble would not corrupt, counteract and subvert each other? Then again the prescription has to be made up by another expert and our lives placed at the mercy of his good faith too.

[C] When it comes to clothing ourselves we have tailors specializing in doublets or breeches; they serve us better because each sticks to his trade and his restricted area of competence; when it comes to food, great households employ cooks who are specialists in soups or in roasts: no cook with an overall responsibility can make them so exquisitely. The same applies to cures: the Egyptians were right to reject general practitioners and to split the profession up, one man working on each illness and each part of the body, which were then treated more appropriately and less indiscriminately since each doctor was only concerned with his speciality.[26] Our doctors never realize that he who provides for all provides for none and that the overall organization of our human microcosm is too much for them to digest. While they were frightened to stop a dysentery lest it brought on a fever, they killed a friend of mine who was worth more than the lot of them, however many there may be. They attach more weight to their guesses about the future than to present illnesses; so as not to cure the brain at the expense of the stomach, they harm the stomach and aggravate the brain with their tumultuous and dissident drugs.

[A] The rational bases of this particular Art are more feeble, clearly, and contradictory than those of any other: *aperient substances are good for a patient suffering from colic paroxysm*, because they dilate and distend the tubes and so facilitate the passage of the glutinous matter which can build up into gravel or stone, so evacuating whatever is beginning to gather and to harden in the kidneys; *aperient substances are dangerous for a patient suffering from colic paroxysm*, because by dilating and distending the tubes they facilitate the passage towards the kidneys of substances whose property is to build up the gravel, for which the kidneys have a propensity, so that it is

---

26. Cf. Polydore Vergil, *De inventoribus rerum*, I, xx: a probable source of part of Montaigne's erudition and arguments about medicine. (The following case is the death of Etienne de La Boëtie.)

difficult to stop them retaining much of what passes through them; moreover if there should happen to be some solid body a trifle too large to pass through the narrows which have to be navigated if the gravel is to be expelled, this body may be set in motion by the aperient and forced into these narrow channels, bunging them up and causing inevitable and painful death.

They show the same certainty in the advice they give us about healthy living: *it is good to pass water frequently*, for experience shows us that by allowing it to stand we let its lees and impurities settle; they then serve to build up stones in the bladder; *it is good not to pass water frequently*: since the heavier impurities borne along in the urine will be discharged only if evacuated violently (we know from experience that a rushing torrent scours the bed it passes through more thoroughly than a sluggish, debilitated stream). Similarly: *it is good to lie frequently with our wives*, because it dilates the tubes and carries away the sand and gravel; *but it is bad*, since it overheats the kidneys, tires them and weakens them.[27]

27. [A] until [A1] (instead of the next seven paragraphs): weakens them. *To sum up, they have no reasons which do not allow of such counter-arguments. As for the judgement about the effectiveness of the drugs, it is as much or more uncertain. I have twice been to drink the hot waters of our local mountains, accepting to do so because it is a natural drink, simple, with no additives, which is at least not dangerous even if useless and which fortunately turned out to be not inimical to my taste (it is true that I take it according to my own rules, not the doctors'); moreover the pleasure of visiting several relatives and friends on the way and of the company which gathers there, as well as the beauty of the countryside, attract me there. Those waters, without a doubt, work no miracle. And I do not believe all the wonderful effects told about them: for while I was there several rumours were spread which I discovered to be false when I informed myself rather carefully about them. But people deceive themselves easily about things they desire. You should not, though, deny that the waters stimulate appetite, aid digestion and give you a new gaiety, provided that you do not go there weak and exhausted. But I never went there, and was determined never to go there, other than hale and happy. Now, as for what I was saying about the difficulty which presents itself in judging their effectiveness, here is an example: I first went to Aigues-Caudes; from the waters I felt no effect, no evident purgation while I was there: but, for a year after my return I was without any pain from the colic on account of which I went there. Since then I went to Bagnères: those waters made me void a deal of gravel and left me long afterwards with a very loose stomach. Yet they did not protect my health more than two months, after which I was most maltreated by my malady. I would ask my doctor which of the two waters he considers, on this evidence, I should put my faith in, having as we do opposing arguments and circumstances for each of them. People should stop yelling against those who, in such uncertainty, let themselves be guided by their inclination and by the simple advice of Nature. Thus, when they themselves advise one water rather than another, prescribing aperients such as those hot waters or forbidding them, they act with the same uncertainty; without a doubt they entrust to the mercy of Fortune the*

[A1]   *It is good to take hot baths at the spas*, because they relax and soften the places where the sand or stone is lurking: *it is bad*, because such an application of external heat encourages the kidneys to concoct, harden and then petrify the matter which is deposited therein. Once you are at the spa, *it is healthier to eat little during the evenings*, so that when you take the waters in the morning they can work more effectively because they encounter a stomach empty and unclogged; but, on the contrary, *it is better not to eat much at midday*, so as not to confuse the workings of the spa-water which are not yet completed and so as not to overload the stomach too soon after such a labour: it is wiser to allow food to digest overnight, which is better than the daytime, during which the mind and body are in ceaseless movement and agitation.

That is how they juggle and trifle with reason – to our detriment. [B]   They cannot give me a single proposition against which I could not construct an opposing one equally valid.   [A1]   Stop railing then at those who, amid such confusion, allow themselves to be gently led by their feelings and by the counsels of Nature, entrusting themselves to common fate.

My travels have provided occasions for seeing virtually all the famous baths of Christendom; I have been using them for some years now, for I reckon that bathing in general is salubrious and I believe that our health has suffered several quite serious inconveniences since we lost the habit (which was formerly observed by virtually all peoples and still is by many) of washing our bodies every day; I can only think that we are all the worse for having our limbs encrusted and our pores blocked up with filth.

As for drinking the waters, fortunately that is in the first place not inimical to my taste; secondly it is both natural and simple and so, at the very least, not dangerous even if it does no good. To support that I can refer to the huge crowds which assemble there, people of every condition and every complexion. Although I have never seen any miraculous or extraordinary cures there – on the contrary whenever I have bothered to investigate a little more carefully than is usual I have found all the rumours of cures which are scattered about such places to be ill-founded and false

---

outcome of their advice, since it is not within their power nor their Art to answer for the quantity of gravelly substances which are being nurtured in our loins, whereas a very slight difference in their size can produce contrasting results affecting our health. You can judge their form of argument from this example. But to press them more vigorously, you would need a man who was not so ignorant of their Art as I am. Poets . . .

(despite their being believed, since people easily deceive themselves when they want to) – nevertheless I have also hardly met anyone who was made worse by taking the waters; and you cannot honestly deny that they stimulate the appetite, help the digestion and liven us up a bit (unless you are already too weak when you go there, something I would advise you against). They cannot rebuild massy ruins but they can shore up a tottering wall or forestall the threat of something worse.

If you cannot come with enough spriteliness to enjoy the company gathered there or the walks and relaxations to which we are tempted by the beauty of the countryside in which most of these spas are situated, you certainly lose the better and surer part of their effect. For this reason I have so far chosen to stay and take the waters at the more beautifully situated spas where you find more pleasant lodgings, food and company, such as the baths at Bagnères in France, Plombières on the border between Germany and Lorraine, Baden in Switzerland and Lucca in Tuscany (especially the Spa at Della Villa, which I have used most often and at various seasons).

Each country has its own peculiar opinions about how to make use of the waters as well as their own rules and methods. In my experience the effects are virtually identical. In Germany it is not done to drink them; for all illnesses people stay in the waters from sunrise to sunset like frogs. In Italy, for every nine days they drink they bathe at least thirty; they usually drink the waters mixed with additional medicinal substances to help them to work. Here in France people are told to go for a walk to help digest the water; elsewhere they make you stay in the bed where you drank it until you have voided it, while keeping your stomach and feet warm. The peculiarity of the Germans is to use cupping-horns or cupping-glasses in the bath, accompanied by scarification; the Italians have their *doccie*, which are showers of hot water conveyed through pipes; for a whole month they douse their head or their stomach, or whatever part is to be treated, for an hour in the morning and the same in the afternoon. There are innumerable differences in the customs of each country or, more correctly, virtually no agreement whatsoever between them.

So much then for the only branch of medicine which I have frequented; it is the least artificial but has its fair share of the confusion and uncertainty you see everywhere else in that Art.[28]

28. The long re-writing in 1582 from seven paragraphs back ends here. The following epigrams are from Ausonius, LXXIV, and Martial, VI, liii.

[A]   Poets can choose how to say it with more eloquence and grace; witness these two epigrams:

> *Alcon hesterno signum Jovis attigit. Ille,*
> *Quamvis marmoreus, vim patitur medici.*
> *Ecce hodie, jussus transferri ex æde vetusta,*
> *Effertur, quamvis sit Deus atque lapis.*

[Alcon touched Jove's statue yesterday. It was of marble, but it felt that doctor's power! You see, god of stone though it was, they bore it out of its hallowed temple today and buried it.]

And the second one:

> *Lotus nobiscum est hilaris, cænavit et idem,*
> *Inventus mane est mortuus Andragoras.*
> *Tam subitæ mortis causam, Faustine, requiris?*
> *In somnis medicum viderat Hermocratem.*

[Andragoras was laughing and bathing with us yesterday; he dined with us too. This morning he was found dead. Do you want to know, Faustinus, why he died so suddenly? He had a dream about Dr Hermocrates.]

I can add two stories to that.

The Baron de Caupène, in Chalosse, is joint patron with me of a benefice called Lahontan at the foot of our mountains; it covers a wide area. There befell to the inhabitants of this region what they also tell of the inhabitants of the valley of Angrougne. Once upon a time they lived cut off, with their own peculiar ways, dress and manners, ruled by their own peculiar institutions and governed by their own customs which were handed down from father to son and to which they were bound by no constraint other than respect for tradition. This tiny state had lasted from ancient times in such happy circumstances that no neighbouring judge had ever been troubled to inquire into the affairs of its inhabitants, no lawyer had ever earned fees by giving them advice, no outsider had ever been called in to settle a quarrel, and nobody in the whole region was ever known to beg. They avoided all leagues and dealings with the outside world so as not to soil the purity of their institutions, until eventually, so they tell, one of their number, as their fathers could remember, was spurred on by an ambition for nobility and decided to increase the honour and reputation of his name by educating one of his sons to become a lawyer, Maître Jean or Maître Pierre. He had him taught to write in a neighbouring town and turned him into a fine village notary-public. As he

rose higher he began to despise their ancient customs and to stuff the locals' heads with thoughts of the glorious world beyond. The first of his companions to be tricked out of a goat he counselled to seek satisfaction from the King's judges nearby; then, from one thing on to another he bastardized everything.

This corruption, so they tell, was soon followed by another with graver results when a doctor conceived a desire to marry one of their maidens and to dwell among them. First he taught them the names of their fevers, rheums and swellings; he told them where their hearts were, their livers and their intestines (something quite unknown to them before). Instead of the garlic which they had formerly used to cure any ills, no matter how harsh or serious they were, he taught them to take strange mixtures for their coughs and colds and began to do good business not only out of their health but out of their deaths.

They swear that only after he came did they realize that the air at nightfall gives them heavy heads,[29] that drinking when overheated does them harm, that the winds of autumn are more unhealthy than those of spring. Only since they started following this medicine of his have they found themselves overwhelmed by a legion of unaccustomed maladies and noticed a general decline in their former vigour. Their lives have been shortened by half.

That is the first of my tales.

The other is that, before I fell victim to the stone, I heard a fuss being made about billy-goats' blood, which many considered to be like manna from heaven vouchsafed to Man in recent centuries to protect and preserve our human lives; many intelligent people talked of it as an infallible wonder-cure. Being a man convinced that I may well fall prey to any misfortune which strikes another, it was my pleasure to produce this wonder while I was yet in full health; I ordered a billy-goat on my farm to be fed as prescribed; it has to be segregated during the hottest months of summer and given only aperient herbs to eat and white wine to drink. I happened to return home the very day it was slaughtered; they came and told me that my cook discovered in its paunch, among all the edible parts, two or three large balls which rattled together. I was careful to have all the entrails brought to me and got them to slit open the great heavy paunch; three large objects fell out; light as sponges, they looked hollow yet were hard and tough on the outside and mottled in several dullish colours. One

29. Contemporary French gentlefolk feared the *serein*, the cool dewy air of a summer evening. The mass of peasants ignored it!

was perfectly round, as big as a bowling-wood: the other two were smaller, not so perfectly round but apparently still growing. When I asked those who regularly slaughter such beasts I was told that such things are unusual and rarely found. It is probable that they were stones related to our own kind, and that it is vain for a sufferer from the gravel to hope to be cured by the blood of a goat about to die of a similar illness. As for asserting that the blood itself is not affected by such contact and that its usual virtues are not impaired, it is more likely that nothing is engendered within a body but by the conspiring of all the parts working together; the whole mass is involved, although one member may contribute more than another depending on the various ways they work. It seems very probable that in all the members of that billy-goat there was some quality of petrification.

I was curious about this experiment, not so much   [C]   for myself or from fears of the future   [A]   but because,[30] in my home as in many others, the womenfolk make a store of such remedies to help the local people, prescribing the same remedy for some fifty illnesses; they never take it themselves yet exult when it works well.

In the meanwhile I honour doctors, less as the precept goes 'for the need' (since against that may be set the example of the prophet reproving King Asa for having 'sought to the physicians') but because I like the men themselves, having known several honourable and likeable ones.[31] I have nothing against doctors, only against their Art; I do not blame them much for taking advantage of our follies: most people do; many vocations, both less honourable and more so, have no other base or stay than the abuse of a trusting public. When I am ill I call them in if they happen to be around at the right time: I ask them for treatment and pay up like anyone else. I grant them leave to order me[32]   [C]   to wrap up warm . . . if I prefer it that way;   [A]   they can prescribe either leeks or lettuce to make me my broth or can limit me to white wine or claret – and so on, for anything

30. '80: petrification. *And if that beast is subject to that malady, I find that it was badly chosen to serve us as a medicine for it.* I was curious about this experiment not so much *for my own use* but . . .
(This account and its argument incensed some doctors.)
31. The famous praise of medicine in Ecclesiasticus 38:1, 'Honour a physician according to the need' (*propter necessitatem* – that is, 'according to thy need of him'). For Asa's equally famous counter-example, see II Chronicles 16:12; Asa 'was diseased in his feet, until his disease was exceeding great: yet in his disease he sought not to the Lord but to the physicians.'
32. '80: order me to *sleep on my right side, if I like that as well as sleeping on my left*; they can prescribe . . .

which my appetite or habits judge indifferent. [A1] I know that that does not really help them, since bitterness and rareness are essential properties of all medicines. Why did Lycurgus order sick Spartans to drink wine? Because when in health they hated it, just as a gentleman in my neighbourhood uses wine as a successful cure for fever precisely because by nature he hates the very taste of it.

[A] How many of the doctors we know share this humour of mine, never condescending to use medicine themselves but living untrammelled lives, flat contrary to what they prescribe for others? If that is not openly mocking our simple-mindedness what is? For their life and health are as dear to them as ours to us and they would practise what they preach if they did not know it to be false. What blinds us is our fear of pain and death, our inability to put up with illness and an insane indiscriminate thirst for cures; what makes our credulity so pliant and impressionable is pure funk. [C] Even then, most people do not so much believe in it as tolerate it; I hear them talking and complaining about medicine as I do; but they end up saying, 'What else can I do?' As though lack of endurance were superior to endurance.

[A] Is there a single case of anyone who subjects himself to such wretchedness who does not also give way to all sorts of imposters, putting himself at the mercy of anyone shameless enough to promise him a cure?

[C] The Babylonians used to carry their sick into the market-place: the people were their doctors, each passer-by asking how they felt and giving them advice on getting better based on their own experience.[33] We do much the same. [A] There is hardly one silly little woman whose spells and amulets we fail to use; and my own humour would lead me to accept such remedies (if I had to accept any): at least there are no ill-effects to fear from them. [C] What Homer and Plato said of Egyptians, that they were all doctors, applies to all nations; there is nobody who does not boast of his nostrum and and risk it on a neighbour who trusts him.

[A] The other day I was in company when a fellow-sufferer brought news of a new kind of pill, compounded from literally over a hundred ingredients. He made quite a to-do about it and felt singularly alleviated: for what boulder could withstand the blows of such a numerous battering! Yet from those who assayed those pills I understand not even the tiniest grain of gravel deigned to be dislodged.

I cannot give up this piece of paper without saying just one more word about the way doctors guarantee the reliability of their cures by citing

33. Herodotus, I, lxxxviii. Then, Diogenes Laertius, *Plato*, III, vii; Homer, *Odyssey*, IV, 231. (Cf. Polydore Vergil, *De inventoribus rerum*, I, xx.)

personal experience. The greater part (perhaps over two-thirds) of the virtues of medicines consists in the latent properties or the quintessences of simples; only practical usage can tell us about that, for quintessence is, precisely, a quality the cause of which our reason cannot explain.[34]

Those of their proofs which doctors say they owe to revelations from some *daemon* or other,[35] I am content just to accept (I never touch miracles); the same goes for proofs based on things we use every day for other purposes – if for example they stumble on to some latent powers of desiccation in the wool we use to clothe us and then cure the blisters on our heels with it; or they may discover some aperient action in the horseradish we eat every day for [C] food. Galen gives an account [A] of a leper[36] cured by drinking wine from a jar into which a viper had chanced to slip. That example shows how our experimental knowledge is likely to increase, as do those cures which doctors claim to have been put on to by the example of certain animals. But for most of the rest of the experimental knowledge to which they claim to have been guided by fortune or luck, I find it impossible to believe that they actually advanced their knowledge that way. I think of a doctor looking round at the infinite variety of matter: plants, animals, metals. What should he assay, to start with? I cannot tell. Supposing his thought first lights, say, on an elk's horn – and one's credulity must be soft and compliant to suppose that![37] His next task is equally difficult. He has to confront so many illnesses, so many attendant circumstances, that before he can advance to the point where his experiment reaches certainty the human intellect runs out of words. Before he can discover, among the infinite number of objects, what that horn actually is; then, among the infinite number of illnesses, what epilepsy actually is; then,

34. The 'fifth essence' (quintessence) is the substance of which the heavenly bodies were thought to be composed; it was held to be latent in all things and extractable by distillation.

35. Such claims were made, for example, for Hippocrates, who, it was claimed, learned by inspiration the way that semen was produced by the brain. This led his Renaissance followers to claim that, while not omniscient, he was incapable of error or of misleading others.

36. '80: for *its taste: just as* Galen gives an account (*so I have been told*) of a leper . . .
I do not know the source of Galen's alleged account, but Polydore Vergil (*De Inventoribus rerum*, I, xxi) gives accounts of how the medical qualities of herbs were discovered which support Montaigne's contentions.

37. This doctor's researches are concerned with epilepsy (a 'holy sickness' in ancient Rome), melancholy and the conjunction of Venus with Saturn (which aggravates melancholy). Montaigne starts with the horn of an *ellend* (elk), an animal described by Cotgrave in his *Dictionarie* as a 'fearful melancholike beast, much troubled by the falling sicknesse', that is, by epilesy.

from among all the complexions, identify melancholy, then, from all the seasons, winter; then, from so many peoples, the French; then, from so many stages of life, old age; then, from so many motions in the heavens, the conjunction of Venus and Saturn; then, from so many parts of the body, the finger: being guided in all that not by argument, not by conjecture, not by example, not by divine inspiration but only as moved by Fortuna: well, before all that can happen, he would need a Fortuna who was a perfect practitioner of the Art, with her rules and method.

And then, even if a cure is achieved, how can the doctor be certain that the malady had not simply run its course or that it was not a chance effect or produced by something else which the patient had eaten, drunk or touched that day – or by the merits of his grandmother's prayers?

Furthermore, even if that proof were absolutely convincing, how many times was it repeated and how often was the doctor able to string such chance encounters together again, so as to establish a rule?[38]

[B]   And if that rule is to be established, who does it? Only three men out of so many millions are concerned to keep records of their experiments. Did Chance happen to come across precisely one of those? Supposing another man – or a hundred men – make opposing experiments. Perhaps we could see a little light if all the judgements and reasonings of all men were known to us. But that three witnesses – three doctors – should make rules for the whole human race is not reasonable: for that to happen our human Nature would have to select them, depute them and then have them declared our Syndics,   [C]   by express letters of procuration.[39]

[A] TO MADAME DE DURAS.[40]

My Lady:
    You caught me just at that point when you called to see me recently. Since these ineptitudes may fall into your hands one day I would also like them to testify that their author feels most honoured by the favour you will be doing them. In them you will find the same mannerisms and attitudes which you have known in your commerce with him. Even if I could have adopted some style other than my usual one or some form better or more honourable I would not have done so; I want nothing from these writings except that they should recall me to your memory as

38. Aristotle insists at the outset in his *Metaphysics* that an 'art' (such as medicine) is not based on experience (or experiment) as such, but on reflection on experience, by which general rules are established.
39. Renaissance medicine was, in the orthodox schools, still dominated by Hippocrates, Galen and Avicenna.
40. Marguerite d'Aure de Gramont, an intimate of Marguerite de Valois.

sketched from nature. I want to take those very same characteristics and attributes which you, My Lady, have known, welcoming them with more honour and courtesy than they deserve, and lodge them (without change or alteration) within some solid body which is able to outlive me by a few years – or a few days – in which you will be able to find them again, refreshing your memory of them whenever you want to, without having the burden of otherwise keeping them in mind (they would not be worth that). I desire that you will go on favouring me with your affection for the same qualities which first aroused it. I have no wish to be better loved or better valued when dead than when alive.   [B]   That humour of Tiberius which made him[41] more concerned to be widely honoured in the future than to make himself esteemed or liked in his own day is ridiculous, though common enough.   [C]   If I were one of those to whom the world may owe a debt of praise, I would rather be paid in advance, please, and wipe off the debt. Let praise rush to pile up round me, thickly not thinly spread, plentiful rather than long-lasting. Then, when its sweet voice can strike my ears no more, it can be bold enough to disappear with my own consciousness.   [A]   Now that I am ready to give up all commerce with men it would be an insane humour to parade myself before them decked in some new subject of esteem. I will not acknowledge any receipt for goods not delivered for use during my lifetime. Whatever I may be like, I have no desire to exist only on paper! My art and industry have been employed to make this self of mine worth something; my studies, to teach me not to write but to act. All my effort has gone into the forming of my life: that is my trade and vocation. Any other job is more for me than the scribbling of books. I have wanted merely to be clever enough for my present and essential comforts, not to store up a reserve for my heirs.

[C]   If anyone is worth anything, let it appear in his behaviour, in his ordinary talk when loving or quarrelling, in his pastimes, in bed, at table, in the way he conducts his business and runs his house. Those men whom I see writing good books while wearing torn breeches would first mend their breeches if they took my advice. Ask a Spartan if he would rather be a good orator or a good soldier; why, I would rather be a good cook, if I did not have a fine one serving me already.

[A]   Good God, My Lady, how I would hate the reputation of being clever at writing but stupid and useless at everything else! I would rather be stupid at both than to choose to employ my good qualities as badly as that. Far from expecting to acquire some new honour by this silly nonsense, I

---

41. '80: which made him, *said Tacitus*, more concerned . . .
   Tacitus, *Annals*, VI, xlvi.

shall have achieved a lot if it does not make me lose the little I have. Leaving aside the fact that this dumb *nature morte* will be an impoverished portrait of my natural being, it is not even drawn from my state at its best but only after it has declined from its original joy and vigour, now seeming withered and rancid. I have reached the bottom of the barrel which readily stinks of lees and sediment.

Moreover I would never have dared, My Lady, to be bold at disturbing the mysteries of Medicine (seeing the trust that you and so many others place in her), if I had not been helped along by the medical authorities themselves. There are only two among the Latins: Pliny and Celsus. If you take a look at them sometime you will find that they treat Medicine more rudely than I do. I give her a pinch: they slit her gizzard. Among other things Pliny mocks doctors who, when they have come to the end of their tether, have found a fine way of ridding themselves of their patients after they have racked them with their potions and tormented them with their diets, all to no avail: they pack some of the sick off to be succoured by vows and miracles, and the rest of them to hot-spring resorts. (Do not be offended, My Lady: he did not mean the ones on our own mountain-slopes which are all under the protection of your family, all devoted to the Gramonts.)

Doctors have a third way[42] of getting rid of us, driving us away and freeing themselves of the weight of our reproaches for the lack of improvement in our illnesses which they have been treating for so long that they can devise nothing new to spin out more time: they send us away to some other region to discover how good the air is there!

Enough of this, My Lady. You will allow me now to pick up the thread of my subject which I had digressed from in order to converse with you.

It was I think Pericles who was asked how he was getting on and replied, 'You can tell from all this,' pointing to the amulets tied to his neck and his arms. He wanted to imply that, since he had been reduced to having recourse to such silly things and to allow them to be used as protection, he was ill indeed.[43]

I do not mean that I may not one day be swept away by the ridiculous idea of entrusting my life to the mercy of the doctors and my health to their ordinances; I might well fall into such raving madness (I cannot vouch for my future constancy); but if I do, like Pericles I shall say to anyone who

42. Pliny, *Hist. nat.*, XXIX, xlvi.
     '80: Gramonts.) *Our own doctors are bolder still: for they* have a third way . . .
43. Plutarch, *Life of Pericles*.

asks how I am, 'You can tell from all this,' showing him my palm burdened with six drams of opiate. That will be a manifest symptom of violent illness.[44] My judgement will have miraculously flown off the handle. If fear and intolerance of pain ever make me do that, you may diagnose a very harsh fever in my soul.

I have bothered to plead this case (which I do not well understand) in order to lend a little support and reinforcement to that natural aversion for our medical drugs and practices which has been handed on to me by my ancestors, so that it should be more than some thoughtless, senseless tendency but an aversion with a little more form. I also want those who see me firmly set against the persuasions and menaces addressed to me when my afflictions oppress me not to take it for pure stubbornness nor be so hasty as to conclude that I am pricked on by vainglory. What a well-placed blow that would be, to wish to squeeze honour from a practice common to me, my gardener and my mule-driver![45] I certainly do not have a mind so distended and flatulent that I would go and swap a solid flesh-and-marrow joy like health for some fancied joy all wind and vapour. For a man of my humour even glory such as that of the *Four Sons of Aymon* is purchased too dear if the price is three good attacks of the stone. Health! For God's sake!

Those who like our medicine may also have their own good, great, powerful arguments: I do not loathe ideas which go against my own. I am so far from shying away when others' judgements clash with mine, so far from making myself unsympathetic to the companionship of men because they hold to other notions or parties, that, on the contrary, just as the most general style followed by Nature is variety – [C] even more in minds than in bodies, since minds are of a more malleable substance capable of accepting more forms – [A] I find it much rarer to see our humours and [C] purposes [A] coincide. In the whole world there has never been two identical opinions, any more than two identical [C] hairs or seeds. [A] Their most universal characteristic is diversity.[46]

44. '80: violent illness, *which will have disturbed the seat of my understanding and my reason.* My judgement . . .
45. Neither of whom could afford doctors' fees and so went without doctors.
46. '80: humours and *thoughts* coincide. *And perhaps* there has never been two identical opinions any more than two identical *faces.* Their most universal characteristic is diversity *and discordance* . . .
In [C], *hairs or seeds* replace *faces* under the influence of Cicero, *Academica*, II (Lucullus), xxvi, 85, where a case is marshalled against the assertion made here, which is presented as a Stoic one. With this phrase Montaigne discreetly emphasizes the Stoic savour of his argument.

BOOK III

1. On the useful and the honourable

===

[Montaigne's conception of the 'useful' is, as it often was in his day, one at home in moral philosophy: here it embraces notions which include both what is profitable to a man or to his country and every sort of public and private interest. The moral dilemma caused by the clash between private morality, piety, benevolence and social ethics on the one hand, and raisons d'état on the other is always a problem in war, and never more so than in civil wars. Cicero considers such problems in De officiis *(On Duties) in which he weighs the duties of goodness, expediency and their conflicting claims. Montaigne strongly defends the claims of kinship, loving friendship, personal integrity and humanity, even during the horrors of the Wars of Religion which were marked by almost unparalleled acts of cruelty and treachery. His hero Epaminondas is presented as a model who can serve to counteract the examples of impiety which, in his more barbarous times, risk becoming the norm.*

In the Renaissance the word honneste *had many interlocking meanings including 'honourable' and 'decent'.]*

[B]   No one is free from uttering stupidities. The harm lies in doing it meticulously:

> *Nae iste magno conatu magnas nugas dixerit.*

> [Of course that chap will make enormous efforts to say enormous trifles!][1]

That does not apply to me. My trifles escape me with as little gravity as they deserve. Good luck to them for that. I would part with them at once, however low their price. I do not buy and sell them for more than they weigh. I speak to my writing-paper exactly as I do to the first man I meet. Here is proof that what I say is true.

Is there anyone for whom treachery should not be loathsome when even Tiberius rejected it at some cost to himself! Tiberius received word from Germany that, if he approved, Ariminius could be got rid of by poison. (Ariminius was the most powerful of the enemies facing the Romans: he had humiliated them under Varus and alone was preventing Tiberius from

---

1. Terence, *Heautontimorumenos*, III, v, 8 (adapted).

extending his dominion over that territory.) He replied that the Roman People were in the habit of avenging themselves on their enemies sword in hand, by overt means not by trickery and covert ones.[2]

He renounced what was useful for what was honourable.

You may reply that he was a hypocrite. I believe he was – hardly a miracle in a man of his line of business. But virtue carries no less far for being professed on the lips of a man who loathes it: indeed truth tears it from him by force, so that even if he does not welcome it inwardly he hides behind it as an adornment.

Both in public and in private we are built full of imperfection. But there is nothing useless in Nature – not even uselessness. Nothing has got into this universe of ours which does not occupy its appropriate place. Our being is cemented together by qualities which are diseased. Ambition, jealousy, envy, vengeance, superstition and despair lodge in us with such a natural right of possession that we recognize the likeness of them even in the animals too – not excluding so unnatural a vice as cruelty; for in the midst of compassion we feel deep down some bitter-sweet pricking of malicious pleasure at seeing others suffer. Even children feel it:

> *Suave, mari magno, turbantibus æquora ventis,*
> *E terra magnum alterius spectare laborem.*

[Sweet it is during a tempest when the gales lash the waves to watch from the shore another man's great striving.][3]

If anyone were to remove the seeds of such qualities in Man he would destroy the basic properties of our lives. So, too, in all polities there are duties which are necessary, yet not merely abject but vicious as well: the vices hold their rank there and are used in order to stitch and bind us together, just as poisons are used to preserve our health. If vicious deeds should become excusable insofar as we have need of them, necessity effacing their true qualities, we must leave that role to be played by citizens who are more vigorous and less timorous, those prepared to sacrifice their honour and their consciences, as men of yore once sacrificed their lives: for the well-being of their country. Men like me are too weak for that: we accept roles which are easier and less dangerous. The public interest requires men to betray, to tell lies [C] and to massacre;[4] [B] let us assign that commission to such as are more obedient and more pliant.

---

2. Tacitus, *Annals*, II, lxxxiii.
3. Lucretius, II, 1–2.
4. Perhaps an allusion to the Massacre of St Bartholomew's Day.

I have certainly been moved to anger at seeing judges use fraud and false hopes of favour or of pardon to tempt criminals to reveal what they have done, even using barefaced lies. It would be helpful to justice (and to Plato, too, who is in favour of that practice)[5] to furnish me with other methods, more in keeping with myself. Such justice is crafty: I reckon that it is no less wounded by others than by itself. Not long ago I replied that I would hardly be one to betray my Prince for a private citizen when I would be deeply grieved to betray any private citizen for my Prince; I not only loathe to deceive, I also loathe others to be deceived about me: I am unwilling even to provide matter or occasion for it. In the little I have had to do with negotiations between our Princes during these disputes and sub-disputes which tear us apart nowadays, I have scrupulously stopped anyone from 'running himself through with my visor' – from being deceived by my position. Those in the business hide as much as they can: they present themselves as being as moderate as possible and pretend that their views are very close. For my part I recommend myself by my liveliest opinions and by the manner which is most truly mine. I am a tender novice at negotiating: I would rather let down my negotiations than let down myself. I have been very lucky though so far – and luck certainly plays the major part in this; few men have gone from one armed band to another with less suspicion or more favour and courtesy.

I have an open manner, readily striking up acquaintance and being trusted from the first encounter. Simpleness and unsullied truth are always opportune and acceptable in any period whatsoever. And then frank speech is less suspect or offensive in men who are not working for some private gain and who can with truth make the reply that Hyperides made to the Athenians who complained of his blunt way of speaking: 'Gentlemen, do not consider only my frankness but that I am frank without having anything to gain, without restoring my own fortunes.'[6] My own frankness, by its vigour, has quickly freed me too from suspicion of deceitfulness (since I do not spare men anything, however hurtful or oppressive, which could be put worse behind their backs); and also by showing my frankness to be simple and unbiased. All I want to gain from doing anything is the fact of having done it: I do not attach distant corollaries and pleadings to it; each thing I do does its job separately: let it succeed if it can.

I feel, by the way, no driving passion about the great of the land, neither

5. Plato permitted the magistrates or governor to lie 'as a medicine' in the interests of the State and morality. Cf. *Republic*, 389 b; 459c.
6. Plutarch (tr. Amyot), *Comment on pourra discerner le flatteur d'avec l'amy*, 51 A.

love nor hatred: nor has my will in this matter been throttled by private injury or obligation. [C] I think of our Kings with the simple loyal affection of a subject, neither encouraged nor discouraged by personal interest. I feel pleased with myself over that. [B] I am only moderately devoted to public affairs, and only dispassionately to just ones. I am not enslaved by deep-seated pledges and intimate engagements. Anger and hatred go beyond the duty of justice; they are passions which merely serve those who are not held to their duty uniquely by reason. All loyal and equitable purposes are loyal and equitable in themselves; if they are not so they are soon corrupted into sedition and disloyalty.

That is what makes me stride forward, head erect, open-faced and open-hearted. I tell you truly that I am not afraid to admit that, if only I could, I would readily follow that old crone's plan and offer a candle to St Michael and another to his dragon.[7] I shall support the good side as far as (but, if possible, excluding) the stake: let Montaigne, my seat, be engulfed in the collapse of the commonwealth if needs be; but, if needs not be, I shall be grateful to Fortune for preserving it. Was it not Atticus who held to the just side, to the losing side, yet saved himself by his moderation in that universal shipwreck of the world among so many schisms and upheavals?

It is easier for private citizens like he was: in such sorts of turmoil I find that you can, with justice, not be ambitious to get involved unless you are invited to. But I find that to remain vacillating and mongrel, or to keep one's affections in check, unmoved by civil strife in one's country and having no preference when the State is divided, is neither beautiful nor honourable: [C] *'Ea non media, sed nulla via est, velut eventum expectantium quo fortunae consilia sua applicent.'* [That is not the way of moderation: it is no way at all. It is simply awaiting the outcome so as to support those who happen to win.][8] That can be permissible towards the affairs of neighbouring countries: Gelon, the Tyrant of Syracuse, refrained from supporting either side in the war of the Barbarians against the Greeks, keeping an envoy in readiness at Delphi, bearing gifts but waiting to see which side Fortune would favour before seizing the occasion when it was ripe for an alliance with the victor. But it would be a species of treachery to act thus in civil strife at home, in which of necessity [B] we must decide to join one side or other. But (even though I do not exploit it myself) I do find it

7. A tale told by John Calvin in his *Traité des Reliques*; then, Cornelius Nepos, *Life of Atticus*.
8. Livy, XXXI, xxi; then, Herodotus, VII, clxiii (for Gelon).

to be more excusable in a man who has received no express command or office if he does not actually get embroiled in the strife, except in the case of foreign wars (in which however, by our own laws, no man is involved save by choice). Nevertheless even those who become totally committed can still do so with such order and moderation that the storm may pass over their heads without battering them. Were we not right to think that way about the late Bishop of Orleans, the Sieur de Morvilliers?[9] And some others that I know, who are now struggling valiantly, have manners which are so equable and gentle that they are the kind who will remain upright no matter what destructive upheavals and collapses Heaven may have in store for us.

I hold that it is the property of kings alone to feel animosity towards other kings, and I laugh at the types of mind which gaily volunteer for quarrels which are so disproportionate: for a man has no private quarrel with a prince when he marches openly and courageously against him, honourably doing his duty. He may not love that great person but he does something better: he esteems him. And there is always this in favour of the cause of legitimacy, of the defence of the traditional institution: the very ones who disturb it for their personal ends can excuse those who defend it, even though they do not honour them. But we must not (as we do every day) give the name of duty to an inward bitter harshness born of self-interested passion, nor that of courage to malicious and treacherous dealings. What they call zeal is their propensity to wickedness and violence: it is not the cause which sets them ablaze but self-interest: they stoke up war not because it is just but because it is war.

Nothing stops us from behaving properly even when among mutual enemies – nor loyally either. Comport yourself among them not with an equal good will (for good-will can allow of varying degrees) but at least with a temperate one, so that you do not become so involved with one of those mutual enemies that he can demand of you your all. Be satisfied too with a modest degree of their favour: do not fish in troubled waters, glide through them!

The other way, that of offering one's services to both sides, savours even less of wisdom than it does of morality. The man to whom you betray another's secrets although you are equally favoured by both realizes, does he not, that you will do the same by him when his turn comes? He listens to you, gets what he can out of you, turns your treachery to his advantage,

9. As Chancellor of France he showed, despite his bishopric, an understanding for the Protestants.

but regards you as a bad man: men of duplicity are useful for what they bring, but mind you see that they take as little away as possible!

I never say anything to one side which I cannot say to the other when the time comes, merely changing the emphasis a little. I bring only such information as is already available, or indifferent or useful to all in common. There is no advantage whatsoever for which I would permit myself to lie to them.

I scrupulously conceal whatever has been entrusted to my silence, but I take care to have as little as possible to conceal. Guarding the secrets of princes when it is not your job to do so is far too much bother. The bargain I am prepared to offer is that, as long as they make few confidences to me, they can certainly place full confidence in whatever I bring with me.

I have always known more about such things than I wanted to. [C] Open talk opens the way to further talk, as wine does or love. [B] Phillipides replied wisely to King Lysimachus who asked him, 'Which of my possessions shall I share with you?' – 'Whatever you like, provided it be none of your secrets.'[10]

I know that everyone rebels if the deeper implications of the negotiations he is employed on are concealed from him and if some ulterior motive is secreted away. Personally I am glad if princes tell me no more than they want me to get on with; I have no desire that what I know should impede or constrain what I have to say. If I have to serve as a means of deception let at least my own conscience be safeguarded. I do not want to be judged so loyal and loving a servant that I would be good for betraying any man. If a man does not keep faith with himself he can pardonably not do so to his master. But these princes will not accept half a man and despise services limited by conditions. There is no other remedy than frankly to state where your boundaries lie: only to Reason should I be a slave – and I can barely do that properly.   [C] They arc also wrong to require a free man to be as abjectly bound to their service as a man they have bought and made, or whose fate is expressly and individually tied to theirs. [B]   Our laws have freed me from great anguish: they have chosen my party for me and have given me a master: all other superior authority is related to the authority of that law; all other obligations are restrained by it. That does not mean that if my affections inclined to the other side that I would immediately lend it my support: our wills and desires are laws unto themselves but our actions must accept law as ordained by the State.

10. Plutarch (tr. Amyot), *De la curiosité*, 64 F.

This way of mine of proceeding jars a bit with our customs; it is not made to achieve great effects nor to endure very long. Innocence herself could not have commerce among us without deception, nor do her business without lying. So public employments are not for my game-bag. Whatever my profession requires of me in such matters I provide in the most private way I can. As a boy I was immersed in politics right up to my ears: it succeeded all right, but I quickly struggled free. Subsequently I have often got out of such engagements, rarely accepted them and never begged for them; I keep my back turned towards ambition – not perhaps like oarsmen who actually proceed backwards but in such a way that my not having embarked upon such a career is less due to my resolve than to my good fortune. For there are paths which are less inimical to my taste and more in conformity with my capacities: if Fortune had ever summoned me to follow those paths towards political service and advancement in worldly renown I know that I would have skipped over my reasoned opinions and followed her.

Those who counter what I profess by calling my frankness, my simplicity and my naturalness of manner mere artifice and cunning – prudence rather than goodness, purposive rather than natural, good sense rather than good hap – give me more honour than they take from me. They certainly make my cunning too cunning. If any one of those men would follow me closely about and spy on me, I would declare him the winner if he does not admit that there is no teaching in his sect which could counterfeit my natural way of proceeding and keep up an appearance of such equable liberty along such tortuous paths, nor of maintaining so uncompromising a freedom of action along paths so diverse, and concede that all their striving and cleverness could never bring them to act the same. The way of truth is one and artless: the way of private gain and success in such affairs as we are entrusted with is double, uneven and fortuitous.

I have often seen that counterfeit artful frankness in practice: it is most often unsuccessful. It readily recalls that donkey in Aesop which, to rival the dog, went and gaily threw both its forefeet round its master's neck: but for such a welcome the wretched donkey received twice as many blows as the dog did caresses.[11] [C] *'Id maxime quemque decet quod est cujusque suum maxime.'* [What best becomes a man is whatever is most peculiarly his own.]

[B] I do not want to deprive wiliness of its rank: that would be to misunderstand the world. I know that it has often proved profitable and

11. Aesop, *Fables*, 293; then Cicero, *De officiis*, I, xxi, 113.

that it feeds and maintains most of the avocations of men. Some vices are legal, just as some deeds are good or pardonable yet illegal. That justice which of itself is natural and universal is ordered differently and more nobly than that other sort of justice, which is [C] particular to one nation and [B] confined by our political necessities. [C] *'Veri juris germanœque justitiœ solidam et expressam effigiem nullam tenemus; umbra et imaginibus utimur.'* [We possess no expressly sculptured portrait of true Law and absolute Justice: we enjoy mere sketches and shadows];[12] [B] so that when Dandamys the Wise heard accounts of the lives of Socrates, Pythagoras and Diogenes, he said that they were in every way great personalities, except for their being too subject to venerating the Law: for, to support Law with its authority, true virtue must doff much of its original vigour; and many vicious deeds are done not merely with the Law's permission but at its instigation:[13] [C] *'Ex senatusconsultis plebisque scitis scelera exercentur.'* [There are crimes authorized by decrees of the Senate and by plebiscites.] [B] I adopt the ordinary usage which differentiates between things useful and things decent and which leads to certain natural functions, which are not merely useful but necessary, being termed indecent or foul.

But let us get on with exemplifying treachery.

Two pretenders to the kingdom of Thrace had fallen into a quarrel over their claims. The Emperor stopped their coming to blows; but one of them, under the pretext of a meeting to establish loving harmony between them, arranged for his rival to feast in his house; he then had him imprisoned and killed. Justice required that the Romans should avenge this crime, but difficulties lay in doing so the normal way: what the Romans could not legally achieve without the hazard of war they therefore undertook to do by treachery. They could not do so 'honourably', but they did so 'usefully'. A certain Pomponius Flaccus was deemed the very man for the job; he ensnared that other pretender with feigned words and assurances and, instead of the honour and favour which he promised him, he dispatched him to Rome bound hand and foot. Here we have one traitor betraying another, which goes against the usual pattern, for traitors are full of mistrust and it is hard to catch them out by cunning like their own – witness the painful experience we have just had.[14]

Let whoever will be a Pomponius Flaccus – and there are plenty who

12. Cicero, *De officiis*, III, xvii, 769.
13. Plutarch, *Life of Alexander*, then Seneca, *Epist moral.*, XCV, 30.
14. Tacitus, *Annals*, II, lxv–lxvii.

would. In my case my word and my bond, like all the rest, form limbs of our commonwealth: they are best employed in serving the State. I take that as granted. But if I were commanded to assume responsibility for the Palace of Justice and its pleas I would reply: 'I know nothing at all about such things'; if commanded to oversee a corps of pioneers I would say: 'I am called to play a more honourable role.' Similarly if anyone should wish to employ me to tell lies, to be treacherous or to perjure myself in some important cause (not to mention assassinations or poisonings), I would say, 'If I have robbed anyone or stolen anything, send me rather to the galleys.' It is licit for a man of honour to speak as the Spartans did when, defeated by Antipater, they were agreeing terms with him: 'You may command us to accept conditions which are as grievous and as damaging as you please: but you will waste your time if you command us to accept shameful and dishonourable ones.'[15]

Each of us ought to have sworn to himself the oath which the kings of Egypt made their judges solemnly swear: that as judges they would never stray from their conscience for any command which even they their kings might give.

There are such evident signs of disapprobation and ignominy in those other commissions; the one who gives them to you is condemning you and, if you grasp it aright, is giving it to you as an accusation and a punishment. The more the affairs of State are mended by your exploit, the worse it goes for your own affairs: the better you do, the worse it is. And it would not be for the first time if the very man who set you the task chastised you for doing it – not without some appearance of justice.

[C] In some particular case betrayal of trust may be excusable, but only when used to betray and punish another betrayal of trust. [B] There are plenty of treacherous deeds which have been not only disowned but punished by the very ones on whose behalf they were perpetrated. Who does not know the judgement which Fabricius pronounced against the physician of Pyrrhus?[16] But further still, there are cases when the very one who ordered the deed has exacted rigorous revenge on the man whom he employed to do it, disclaiming to have had such authority and power and disowning so abandoned a servility and so cowardly an obedience.

---

15. Plutarch (tr. Amyot), *Comment on pourra discerner le flatteur d'avec l'amy*, 49 D–E; then, *Les dicts notables des anciens Roys*, 189 C and Erasmus, *Apophthegmata*, V, *Aegyptii*, XXXIII.
16. The doctor wrote to Fabricius offering to poison Pyrrhus, to whom Fabricius forwarded the letter, telling him to choose his friends better.

A Russian duke called Jaropelc bribed an Hungarian nobleman to betray the King of Poland, Boleslaus, either by killing him or by providing the Russians with the means of doing him some resounding harm. That Hungarian acted the honest man and devoted himself to the service of that king; he succeeded in becoming one of his advisers – among the most trusted. Taking advantage of this and choosing an opportune moment when his master was absent, he betrayed Wielickzka to the Russians; that great and flourishing city was entirely burnt and sacked by them; not only did they slaughter the entire population of whatever age or sex but also a large number of noblemen from the surrounding area whom he had assembled there with that end in view.

Jaropelc, his vengeance and his anger assuaged – and they were not unjustified, since Boleslaus had greatly injured him in a similar manner – was satiated by the fruits of that treacherous deed. He came to reflect on its naked, simple ugliness, seeing it with a saner vision no longer obscured by passion; he was seized with so great revulsion and remorse that he put out the eyes of the perpetrator, cut off his tongue and gelded him.[17]

Antigonus persuaded the Argyraspidian guards of his adversary Eumenes (who was their Captain-General) to betray him to him. No sooner did he have him killed after being delivered into his power than he himself desired to become the agent of divine Justice in punishing so loathsome a crime; he gave written orders for those guards to be handed over to the Provincial Governor, expressly commanding him to wipe them out, making their end as horrible as he could. Out of that great multitude not one ever breathed again the air of Macedonia. The greater the service they had done him, the more wicked he judged it to be and the more punishable.

[C] The slave who betrayed the hiding-place of his master Publius Sulpicius was set free, in accordance with the promise of Sylla's proclamation; but in accordance with State policy, freeman as he was, he was cast down from the Tarpeian Rock. They have such men hanged with their gains slung in a purse round their necks: they first fulfil their secondary, special sort of promise, and then they fulfil the general one, the primary one.[18]

Mahomet II, jealous of his power in accordance with his family tradition, wanted to rid himself of his brother. He employed one of his officers to do so, who choked him by forcing him to imbibe a great quantity of water all at once. After this was done Mahomet II, to expiate the crime, handed the

---

17. Jean Hubert-Fulstin, *Hist. des Roys, et Princes de Pologne,* 1573. Then, Plutarch, *Life of Eumenes.*
18. From the *Epitome* of Florus, often printed before Livy, XXVII.

murderer over to the dead man's mother – they were brothers only by their father. She, in his presence, slit open the murderer's bosom and, hot with passion, fumbled inside, tore out his heart and threw it to the dogs.[19] And our own King Clovis had the three servants of Cannacre hanged after they had betrayed their master: yet he had bribed them to do so. [B] Even worthless men find it pleasant, once they have profited from a vicious deed, to go and quite safely fasten upon it some mark of goodness and of justice, as though their conscience wished to make up for it and put things right. [C] To which may be added the fact that the agents of such horrific wickedness are a reproach to them: so they seek through their deaths to smother any knowledge witnessing to such conspiracies.

[B] Now even if it should happen that you do get a reward so as not to deprive *raison d'état* of so extreme and desperate a remedy, the one who rewards you will not fail to regard you as a man accursed and abominable – unless that is he is one himself. And he will think you a bigger traitor than does the man you betrayed; for he proves the malevolence of your heart at the touchstone of your hands, with no possible denial or objection. He exploits you just as we do those degraded men, Public Executioners of High Justice – an office as useful as it is shameful.

Apart from the baseness of such commissions, there is the prostitution of your conscience. Since the daughter of Sejanus was a virgin and therefore not punishable by death according to a specific judicial formula in Rome, in order to have scope to apply the law she was raped by the hangman before he strangled her: not merely his hand but his mind was the slave of the interests of State.[20]

[C] When Amurath I, to increase the severity of his punishment of those subjects who had supported the sacrilegious revolt of his son against him, commanded that their closest kinsmen should take part in their execution, I find it most honourable in some of them to have preferred to be held, with gross injustice, guilty of another's sacrilege rather than to serve another's justice by sacrilege of their own. And in my own time when I have seen some rabble, after we have stormed their wretched hovels, saving their own skins in return for the hanging of their friends and relatives, I have always thought that they were worse off than the ones we hanged. It is said that in former days Prince Vitold of Lithuania proclaimed

19. Jacques Laverdin, *Scanderbeg*. Then, for Clovis, Du Haillant, *Histoire des Roys de France*.
20. Tacitus, *Annals*, V, ix; then, Nicolas Chalcocondylas, *De la decadence de l'Empire Grec*.

as law that condemned criminals must execute their sentences by their own hand, finding it monstrous that a third party, who was innocent of their crime, should be burdened with the task of homicide.

[B]   As for a prince, whenever some urgent necessity or some violent unforeseeable event affecting the needs of his State obliges him to go back on his pledged word, or otherwise forces him from the ordinate path of duty, he must consider it as a scourging by the rod of God; vice it is not, for he has abandoned his own right-reason for a more powerful universal one: but it is indeed a calamity. So when I was asked, 'What remedy is there?' I replied, 'None: if the prince was really torn between those two extremes, then he had to do it' – [C] *'Sed videat ne quaeratur latebra perjurio'* [But let him be sure not to seek any pretexts for such perjury][21] – [B] 'But if he had no regrets about doing it, if it did not weigh upon him, then that is a sign that his conscience has gone astray.'

[C]   (Even if a Prince could be found with so tender a conscience that no cure seemed worth such a grievous remedy, I would not think any the less of him. He could never lose out more excusably or more decorously. We cannot do everything. Come what may we are often obliged to commit our ship to the sole guidance of Heaven as our ultimate refuge. For what more just necessity is that Prince keeping himself in store? Is there anything more impossible for him to do than what he can only do at the expense of his faith and his honour, attributes which ought perhaps to be dearer to him than his own preservation – and indeed the preservation of his people? If he should simply fold his arms and call on God for help, would he not have grounds to hope that God in his goodness is not such as to refuse the favour of his hand, beyond the normal Order, to a hand so pure and just?)

[B]   Those are dangerous examples, rare and sick exceptions to our rules of nature. Yield to them we must, but with great moderation and circumspection. No private good is worth our doing such violence to our consciences; the common good: well, all right, when it is most apparent and when it really matters.

[C]   Timoleon with the tears he shed rightly saved himself from the monstrous quality of his deed, remembering that he had killed the tyrant with the hand of a brother; and it rightly pricked his conscience that he had been obliged to purchase the common good at the price of his moral honour. The very Senate which he served to free from slavery dared not plainly make up its mind about so deep a deed which presented two such

21. Cicero, *De officiis*, III, xxix, 106.

grievous and contrary faces: when the citizens of Syracuse opportunely, at that very moment, send to beg protection from the Corinthians, asking for a leader worthy of restoring their city to its former splendour and of cleansing Sicily of the many petty tyrants which oppressed it, the Senators deputed Timoleon, declaring, with a new ruse, that their decision would be in favour of the liberator of his country or against the murderer of his brother depending on whether he acquitted himself of his charge well or badly.[22]

That fanciful verdict did however have the excuse of the dangerous nature of his example and the implications of so self-contradictory a deed; they did well to free their judgement of such a burden and to base it on some other independent considerations. Now Timoleon's behaviour during that expedition soon threw light on his case, so worthily and so virtuously did he act in every way; and the good fortune which accompanied him during the hardships he had to overcome in that noble task seemed to them to have been sent by the gods, united in favour of vindicating him.

If ever an aim was worthy of pardon, that aim was Timoleon's. But the convenience of increasing the State revenue, which served as a pretext to the Roman Senate in the filthy decree that I am about to relate, is not strong enough to warrant such an injustice. Certain cities had ransomed themselves for cash and regained their freedom from Lucius Scylla by the permission and decree of the Senate. Their case, it so happened, had to be considered afresh: the Senate condemned them to be taxable as before, declaring that the money used for their ransom should be forfeited.[23]

Civil wars often produce base examples of our punishing private citizens for trusting in us when we once thought differently, and the very same magistrate makes someone who had nothing to do with it bear the penalty of his own change of mind. The master flogs the pupil because he was *willing* to learn, and the guide flogs the blind man. A horrifying image of our justice. There are rules in philosophy which are false and weak. The example which it propounds to us to enable private advantage to prevail over our plighted troth is not sufficiently justified by the weight of the attendant circumstances: 'Thieves have captured you; they have set you free after exacting from you an oath to pay a certain sum.' It would be quite wrong to say that a good man, once out of their hands, would be free of his oath without paying up! He is nothing of the sort! Whatever fear has

22. Cf. I, 38, 'How we weep and laugh at the same thing'.
23. Cicero, *De officiis*, III, xxii, 87. (In the next sentence I follow the reading of '95, etc.: *changement* (change of mind), not *jugement*.)

made me want to do once, I am obliged to want to do when freed from that fear. And if fear had merely forced my tongue without my will, I am still bound by my word down to the last farthing. In my own case when my tongue has, without reflection, gone beyond my intentions, it has been a point of conscience not to disavow it for that reason. Otherwise, step by step, we will reach the point where it will overthrow any right that a third party acquires by our promises and our oaths: *'Quasi vero forti viro vis possit adhiberi.'* [As though force could be used against a man of fortitude.]²⁴ In one thing alone does private interest excuse our failure to keep a promise: if we have promised something which is wicked and iniquitous in itself; for the right of virtue must take precedence over the rights of our obligation.

I have already placed Epaminondas among the foremost ranks of outstanding men and I have no wish to unsay what I said.²⁵ How far would he go, out of consideration for his private duty? He never killed a man he had vanquished; he scrupled to kill, without due form of law, a tyrant   [C]   or his accomplices,   [B]   even for the inestimable good of restoring freedom to his country; he thought it wicked of a man, no matter how good a citizen he might be, if he did not spare his friend and host among his enemies even in battle. There you have a soul compounded of noble elements! To the harshest and most violent of human activities he married goodness and humanity – indeed the most exquisite to be found in the school of philosophy. That mind so great, so rigid and so obstinate in the face of pain, death and poverty: was it nature or art which had made it tender to the point of extreme gentleness and of affability of humour? Terrifying with blood and sword, he goes smashing and shattering a nation unbeatable save by him alone, only to turn aside in the midst of the melee when he comes upon his friend and host. Truly that man was genuinely in command of War when he compelled her mouth to answer to the bit of his kindness at the highest point of her most blazing ardour, all enflamed as she was and foaming with frenzy and slaughter. It is a miracle to bring even the image of Justice into actions such as that: to the righteous Epaminondas alone it belonged to bring in mildness, most gentle-mannered benevolence   [C]   and pure innocence.²⁶

[B]   Whereas one leader said to the Mammertines that statute-law did

---

24. Cicero, *De officiis*, III, xxx, 110.
25. Cf. II, 36, 'On the most excellent of men'.
26. Montaigne's veneration of Epaminondas is shared by Plutarch (his principal source of the details given) throughout his *Oeuvres Morales*: cf. Amyot's index s.v. *Epaminondas*.

not apply to men under arms; whereas another said to the Tribune of the People that the times of war and of justice were two different things, while a third declared that the din of arms prevented his hearing the voice of the laws:[27] Epaminondas was never prevented from hearing the laws of kindness and of unsullied courtesy. Had he not borrowed from his enemies the practice of sacrificing to the Muses as he went to war in order to temper by their gentleness and gaiety the harshness and frenzy of Mars?[28] After so great a preceptor let us not fear to think   [C]   that some things are unlawful even when done to enemies or   [B]   that the common interest cannot require all men to sacrifice all private interest always,   [C]   *'manente memoria etiam in dissidio publicorum foederum privati juris'* [the memory of individual rights subsisting even in the strife of public abominations];[29]

[B]   *et nulla potentia vires*
Præstandi, ne quid peccet amicus, habet;

[no might has the power to authorize a friend to act wickedly;]

and that not all things are legitimate to a man of honour at the service   [C]   of his king or   [B]   of the cause of the commonwealth and its laws.   [C]   *'Non etiam patria praestat omnibus officiis, et ipsi conducit pios habere cives in parentes.'* [The claims of our country are not paramount over all other duties: it is good for it to have citizens who are dutiful to their kindred.]

[B]   There you have a lesson proper to our own times. It is enough that the ironplate of our armour should give us calloused shoulders: there is no need to allow it to make our minds callous as well; it is enough to plunge our pens in ink without plunging them in blood. If it is greatness of mind and a deed of rare and special virtue to hold in contempt the bonds of love, our private obligations, our word and our kinsfolk in the interests of the common good and of obedience to officers of State, then for us to decline such greatness it suffices that it cannot find lodging within the greatness of mind of Epaminondas.

---

27. Plutarch, lives of *Caesar* and of *Marius*.
28. Before battle the Spartans (the enemy of Epaminondas) tamed their wrath by listening to flute music: Plutarch (tr. Amyot), *Comment il faut refrener la colere*, 59F; cf. 51 G–H.
29. Livy, XXV, xviii; then Ovid, *Ex ponto*, I, vii, 37–8, and Cicero, *De officiis*, III, xxiii, 90.

I hold in abomination the frenetic exhortations of that other man with his disordered mind:

> *dum tela micant, non vos pietatis imago*
> *Ulla, nec adversa conspecti fronte parentes*
> *Commoveant; vultus gladio turbate verendos.*

[while your weapons flash, let no thought of duty to your parents move you, nor the sight of your fathers on the other side: slash with your swords at the faces which you should venerate.][30]

Let us deprive wicked treacherous natures, athirst for blood, of such a pretext of justification. Let us cast aside such abnormal and insane justice and cling to models which are more humane. Think what examples can do over time! In an engagement against Cinna during the Civil War, one of Pompey's soldiers unintentionally killed his brother on the other side; from shame and sorrow he killed himself then and there on the field; yet a few years later, in another Civil War between the same nations, a soldier killed his brother and then asked his officers for a reward for doing so.[31] We wrongly adduce the honour and beauty of an activity from its usefulness, and our conclusion is wrong if we reckon that all are bound to perform it,   [C]   and that it is honourable for each to do so,   [B]   provided it be useful:

> [C]   *Omnia non pariter rerum sunt omnibus apta.*
>
> [Not all things are equally fitted to all men.]

[B]   Select the most necessary, the most useful activity of human society: that will be marriage. Yet the counsel of the Saints finds the opposing party to be more worthy of honour and excludes from marriage the vocation which is most to be revered among men, just as we assign to our studs the beasts we value less.[32]

---

30. Lucan, *Pharsalia*, VII, 321–3 (a poet much read because of his subject during the French Civil Wars of Religion).
31. Tacitus, *Hist.*, III, l; and III, li; then, Propertius, III, ix, 7.
32. That is, individual priests and monks are required to be celibate, despite the acknowledged prime usefulness of marriage. Both Plato and Aristotle ranked marriage among the most useful institutions; Stobaeus (*Sermo* LXV) has a long eulogy on the subject from Hierocles' book *On Marriage*.

2. On repenting

[Montaigne does not deal here primarily with the sacrament of repentance but with the act of repenting in domains religious, moral and practical. In this sense repenting consists not in regret but in denying the rightness of what one had formerly willed. Like Rabelais's good giant Gargantua, Montaigne knows that a man may live as a Christian gentleman: 'without reproach though not of course without sin'. And Montaigne's sense of sin is not a matter of wishing in old age that he had not committed the sins (especially the sensual sins) of his youth nor the worse sins of old men; neither is it a matter of wishing that he had been vouchsafed a higher Form than Man (that of an angel) or a better human Form than his own botched one (a Form like Cato's). In practical affairs, however they turn out, Montaigne sees no cause for repenting of decisions honourably made within Man's limitations; in his dealings with others in peace and civil war he knows he has acted as an honourable gentleman, far better than most. Where sins against the Christian God are concerned, Montaigne never hid from himself the ugly face which lurks behind their stormy beauty; that is where repentance comes in; real repentance – of the demanding, ultimate kind which alone moves Montaigne – is an agonizing matter: we must see ourselves throughly, as with the eyes of God who searches the reins and the bowels and from whom no secrets are hidden. To do that 'God must touch our hearts' – an act of grace which became the superscription of a religious emblem.]

[B]   Others form Man; I give an account of Man and sketch a picture of a particular one of them who is very badly formed and whom I would truly make very different from what he is if I had to fashion him afresh. But it is done now. The brush-strokes of my portrait do not go awry even though they do change and vary. The world is but a perennial see-saw. Everything in it – the land, the mountains of the Caucasus, the pyramids of Egypt – all waver with a common motion and their own.[1] Constancy itself is nothing but a more languid rocking to and fro. I am unable to stabilize my subject: it staggers confusedly along with a natural drunkenness. I grasp it as it is now, at this moment when I am lingering over it. I am not portraying being but becoming: not the passage from one age to another

1. Propertius, II, i, 69. (For the theme, cf. Erasmus, *Opera*, 1703–6, V, 488F–461E. Montaigne's theme of the perennial flux of all things is Heracleitan.)

(or, as the folk put it, from one seven-year period to the next) but from day to day, from minute to minute. I must adapt this account of myself to the passing hour. I shall perhaps change soon, not accidentally but intentionally. This is a register of varied and changing occurrences, of ideas which are unresolved and, when needs be, contradictory, either because I myself have become different or because I grasp hold of different attributes or aspects of my subjects. So I may happen to contradict myself but, as Demades said, I never contradict truth.[2] If my soul could only find a footing I would not be assaying myself but resolving myself. But my soul is ever in its apprenticeship and being tested. I am expounding a lowly, lacklustre existence. You can attach the whole of moral philosophy to a commonplace private life just as well as to one of richer stuff. Every man bears the whole Form of the human condition.[3] [C] Authors communicate themselves to the public by some peculiar mark foreign to themselves; I – the first ever to do so – by my universal being, not as a grammarian, poet or jurisconsult but as Michel de Montaigne. If all complain that I talk too much about myself, I complain that they never even think about their own selves.

[B] But is it reasonable that I who am so private in my habits should claim to make public this knowledge of myself? And is it also reasonable that I should expose to a world in which grooming has such credit and artifice such authority the crude and simple effects of Nature – and of such a weakling nature too? Is writing a book without knowledge or art not like building a wall without stones and so on? The fancies of the Muses are governed by art: mine, by chance. But I have one thing which does accord with sound teaching: never did man treat a subject which he knew or understood better than I know and understand the subject which I have undertaken: in that subject I am the most learned man alive! Secondly, no man ever [C] went more deeply into his matter, ever stripped barer its own peculiar members and consequences, or ever [B] reached more precisely or more fully the goal he had proposed for his endeavour. To finish the job I only need to contribute fidelity: and fidelity is there, as clean and as pure as can be found. I tell the truth, not enough to make me replete but as much as I dare – and as I grow older I dare a little more, for

2. Plutarch, *Life of Demosthenes*.
3. Aristotle's opinion was normative: all human beings have the same form (soul), the form of Man. What distinguishes each individual person is the union of one particular example of that form with one particular body.

custom apparently concedes to old age a greater licence to chatter more indiscreetly about oneself. What cannot happen here is what I often find elsewhere: that the craftsman and his artefact thwart each other: 'How can a man whose conversation is so decent come to write such a scurrilous book?' or 'How can such learned writings spring from a man whose conversation is so weak?'

[C] When a man is commonplace in discussion yet valued for what he writes that shows that his talents lie in his borrowed sources not in himself. A learned man is not learned in all fields: but a talented man *is* talented in all fields, even in ignorance. [B] Here, my book and I go harmoniously forward at the same pace. Elsewhere you can commend or condemn a work independently of its author; but not here: touch one and you touch the other. Anyone who criticizes it without knowing that will harm himself more than me; anyone who does know it has satisfied me completely. I shall be blessed beyond my merit if public approval will allow me this much: that I have made intelligent people realize that I would have been capable of profiting from learning if I had had any and that I deserved more help from my memory.

Let me justify here what I often say: that I rarely repent [C] and that my conscience is happy with itself – not as the conscience of an angel is nor of a horse, but as behoves the conscience of a man[4] – [B] ever adding this refrain (not a ritual one but one of simple and fundamental submission): that I speak as an ignorant questioning man: for solutions I purely and simply abide by the common lawful beliefs.[5] I am not teaching, I am relating.

There is no vice that is truly a vice which is not odious and which a wholesome judgement does not condemn; for there is so much evident ugliness and impropriety in it that perhaps those philosophers are right who maintain that it is principally the product of stupidity and ignorance, so hard it is to imagine that anyone could recognize it without loathing it.[6] [C] Evil swallows most of its own venom and poisons itself. [B] Vice leaves repentance in the soul like an ulcer in the flesh which is forever scratching itself and bleeding.[7] For reason can efface other

---

4. In the 'chain of being', Man comes between the beasts and the angels.
5. 'Lawful' by the law of the Church.
6. Socrates and his fellows. Cf. II, 12, 'An apology for Raymond Sebond' (beginning); then [C], from Seneca, *Epist. moral.*, LXXXI, 22.
7. Image and development from Plutarch (tr. Amyot), *De la tranquillité de l'ame*, 75 G.

griefs and sorrows, but it engenders those of repentance which are all the more grievous for being born within us, just as the chill and the burn of our fevers are more stinging than such as come to us from outside. I hold to be vices (though each according to its measure) not only those vices which are condemned by reason and nature but even those which have been forged by the opinions of men, even when false or erroneous, provided that law and custom lend them their authority.

Likewise there is no goodness which does not rejoice a well-born nature. There is an unutterable delight in acting well which makes us inwardly rejoice; a noble feeling of pride accompanies a good conscience. A soul courageous in its vice can perhaps furnish itself with composure but it can never provide such satisfaction and happiness with oneself. It is no light pleasure to know oneself to be saved from the contagion of a corrupt age and to be able to say of oneself: 'Anyone who could see right into my soul would even then not find me guilty of any man's ruin or affliction, nor of envy nor of vengeance, nor of any public attack on our laws, nor of novelty or disturbance, nor of breaking my word. And even though this licentious age not only allows it but teaches it to each of us, I have nevertheless not put my hand on another Frenchman's goods or purse but have lived by my own means, in war as in peace; nor have I exploited any man's labour without due reward.' Such witnesses to our conscience are pleasant; and such natural rejoicing is a great gift: it is the only satisfaction which never fails us.

Basing the recompense of virtuous deeds on another's approbation is to accept too uncertain and confused a foundation − [C]   especially since in a corrupt and ignorant period like our own to be in good esteem with the masses is an insult: whom would you trust to recognize what was worthy of praise! May God save me from being a decent man according to the self-descriptions which I daily see everyone give to honour themselves: '*Quae fuerant vitia, mores sunt.*' [What used to be vices have become morality.][8]

Some of my friends have occasionally undertaken to lay bare my heart, to charge me and put me through the assizes, either on their own initiative or else summoned by me; of all the offices of friendship that is not only the most useful for a well-turned mind but also the sweetest. I have always welcomed it with the most courteous and grateful of embraces. But speaking of it now in all conscience I have often found such false measure in their praise and blame that, judging from their standards, I would not have been wrong to do wrong rather than right.

8. Seneca, *Epist. moral.*, XXXIX, 6.

[B]   Especially in the case of people like us who live private lives which only go on parade before ourselves, we must establish an inner model to serve as touchstone of our actions, by which we at times favour ourselves or flog ourselves. I have my own laws and law-court to pass judgement on me and I appeal to them rather than elsewhere. I restrain my actions according to the standards of others, but I enlarge them according to my own. No one but you knows whether you are base and cruel, or loyal and dedicated. Others never see you: they surmise about you from uncertain conjectures; they do not see your nature so much as your artifice. So do not cling to their sentence: cling to your own.   [C]   *'Tuo tibique judicio est utendum.'* [You must use your own judgement of yourself.][9] *'Virtutis et vitiorum grave ipsius conscientiae pondus est: qua sublata, jacent omnia.'* [Your own conscience gives weighty judgement on your virtues and vices: remove that, and all lies sprawling.]

[B]   Yet the saying that 'repentance follows hard upon the sin' does not seem to me to concern sin in full apparel, when lodged in us as in its own home. We can disown such vices as take us by surprise and towards which we are carried away by our passions; but such vices as are rooted and anchored in a will which is strong and vigorous brook no denial. To repent is but to gainsay our will and to contradict our ideas; it can lead us in any direction. It makes that man over there disown his past virtue and his continence!

> *Quæ mens est hodie, cur eadem non puero fuit?*
> *Vel cur his animis incolumes non redeunt genæ*

[Alas! Why did I not want to do as a young man what I want to do now? Or why, thinking as I do now, cannot my radiant cheeks return?][10]

Rare is the life which remains ordinate even in privacy. Anyone can take part in a farce and act the honest man on the trestles: but to be right-ruled within, in your bosom, where anything is licit, where everything is hidden – that's what matters. The nearest to that is to be so in your home, in your everyday actions for which you are accountable to nobody; there is no striving there, no artifice.[11] That is why Bias when portraying an excellent family state said it was one where the head of the family was of his own

---

9. Cicero, *Tusc. disput.*, II, lxiii; then, *De nat. deorum*, III, xxxv.
10. Horace, *Odes*, IV, x, 7–8.
11. Cf. the adage attributed variously to Socrates and to Diogenes: *'Aedibus in nostris quae prava aut recta geruntur'* (It is in our own home that good or evil are done): Erasmus, *Adages*, I, VI, LXXXV.

volition, the same indoors as he was outdoors for fear of the law and the comments of men: and it was a worthy retort of Julius Drusus to the builders who offered for three thousand crowns to re-plan his house so that his neighbours could no longer see in as they did: 'I will give you six thousand, and you can arrange for them to see in everywhere!' We comment with honour on Agesilas' practice of taking up lodgings in the temples when on a journey, so that the people and the very gods could see what he did in private.[12] A man may appear to the world as a marvel: yet his wife and his manservant see nothing remarkable about him. Few men have been wonders to their families.

[C] 'No man has been a prophet not only in his own home but in his own country,' says the experience of history.[13] The same applies to trivialities. You can see an image of greater things in the following lowly example: in my own climate of Gascony they find it funny to see me in print; I am valued the more the farther from home knowledge of me has spread. In Guienne I pay my printers: elsewhere, they pay me. That consideration is the motive of those who hide away when alive and present, so as to enjoy a reputation when they are dead and gone. I would rather have a lesser one: I throw myself upon the world for the one that I can enjoy now. Once I am gone I acquit the world of its debt.

[B]   That man over there is escorted to his door ecstatically by a public procession: he doffs that role when he doffs his robes; the higher he has climbed the lower he falls. Once at home he is all tumult and baseness within. And even if right-rule is to be found in him, you need a quick and highly selected judgement to perceive it in his humble private actions. Besides, to be ordinate is a glum and sombre virtue. Storming a breach, conducting an embassy, ruling a nation are glittering deeds. Rebuking, laughing, buying, selling, loving, hating and living together gently and justly with your household – and with yourself – not getting slack nor belying yourself, is something more remarkable, more rare and more difficult. Whatever people may say, such secluded lives sustain in that way duties which are at least as hard and as tense as those of other lives.   [C]   And Aristotle says that private citizens serve virtue as highly and with as much difficulty as those who hold office.[14]   [B]   We prepare

---

12. Plutarch (tr. Amyot), *Le banquet des sept sages*, 155 E (Bias), and *Instruction pour ceulx qui manient affaires d'Estat*, 162 G (Julius Caesar); then, *Life of Agesilas*.
13. Matthew 13:57; Mark 6:4; Luke 4:24; John 4:44.
14. Aristotle, *Nicomachaean Ethics*, X, vii, 10 (1179 a).

ourselves for great occasions more for the glory than for good conscience. [C] The quickest road to glory would be to do for conscience what we do for glory. [B] And the virtue of Alexander seems to me to act out less virtue on its stage than that of Socrates in his humble obscure role. I can easily conceive of Socrates in Alexander's place: but Alexander in Socrates' place, I cannot. Ask Alexander what he can do and he will reply: 'Subdue the whole world.' Ask Socrates, and he will answer, 'Live the life of man in conformity with his natural condition': knowledge which is much more general, onerous and right.

The soul's value consists not in going high but in going ordinately. [C] Its greatness is not displayed in great things but in the Mean.

Just as those who judge us by the touchstone of our motives do not rate highly the sparkle of our public deeds and see that it is no more than thin fine jets of water spurting up from the depths (which are moreover heavy and slimy), so too those who judge us from our brave outward show conclude that our inward disposition corresponds to it: they cannot couple ordinary talents just like their own with those other talents, so far beyond their ken, which amaze them. That is why we give savage shapes to demons. And who does not give Tamberlane arching eyebrows, gaping nostrils, a ghastly face and an immense size proportionate to the idea we have conceived of him from the spreading of his name? Once, if anyone had brought me to meet Erasmus it would have been hard for me not to take for adages and apophthegms everything he said to his manservant or to his innkeeper's wife. We can with more seemliness imagine an artisan on his jakes or on his wife than a great lord chancellor venerated for his dignity and wisdom. It seems to us that they never come down from their lofty thrones, even to live.

[B] As vicious souls are often incited to do good by some outside instigation, so are virtuous souls to do evil. We must therefore judge souls in their settled state, when they are at home with themselves – if they ever are – or at least when they are nearest to repose in their native place. Natural tendencies are helped and reinforced by education, but they can hardly be said to be altered or overmastered. In my lifetime hundreds of natures have escaped towards virtue, or vice, despite teaching to the contrary:

> *Sic ubi desuetæ silvis in carcere clausæ*
> *Mansuevere feræ, et vultus posuere minaces,*
> *Atque hominem didicere pati, si torrida parvus*

Venit in ora cruor, redeunt rabiesque furorque,
Admonitæque tument gustato sanguine fauces;
Fervet, et a trepido vix abstinet ira magistro.

[As when wild beasts, shut up in a cage, forget their forests and are tamed, losing their menacing looks and learning to be ruled by men, yet if a tiny drop of blood falls on their avid lips, back come their snarls and their ragings; they have tasted blood; their jaws yawn wide; they are in turmoil and can hardly be stopped from venting their wrath on their trembling tamer.][15]

You cannot extirpate the qualities we are originally born with: you can cover them over and you can hide them. Latin is a native tongue for me: I understand it better than French; yet it is forty years now since I used it for speaking or writing. Nevertheless on those two or three occasions in my life when I have suffered some extreme and sudden emotion – one was when my perfectly healthy father collapsed back on to me in a dead faint – the first words which I have dredged up from my entrails have always been in Latin – [C] nature, against long nurture, breaking forcibly out and finding expression. [B] And this example applies to many others.

Those who have sought in my time to improve the morals of the world with their new opinions reform the vices which show: the essential vices they leave us as they are – if they do not make them grow bigger. And such growth is to be feared: we are ready to take a holiday from all other good deeds on the strength of those uncertain surface reformations which cost us less and which gain us more esteem; and we thereby cheaply give satisfaction to our other vices: those which are inborn, of one substance with us and visceral.

Just take a little look at what our own experience shows. Provided that he listen to himself there is no one who does not discover in himself a form entirely his own, a master-form which struggles against his education as well as against the storm of emotions which would gainsay it. In my case I find that I am rarely shaken by shocks or agitations; I am virtually always settled in place, as heavy ponderous bodies are. If I should not be 'at home' I am always nearby. My indulgences do not catch me away very far: there is nothing odd or extreme about them, though I do have some sane and vigorous changes of heart.

The real condemnation which applies to the common type of men nowadays is that their very retreat is full of filth and corruption, that their amendment of life is vague, and their repentance nearly as sickly and guilt-

15. Lucan, *Pharsalia*, 237–42.

ridden as their sinning. Some of them are so stuck to their vices by long habit or some natural bonding that they no longer find them ugly. There are others – and I am one of that regiment – for whom vice does have some weight but who counterbalance it by the pleasure it gives or by some other factor; they put up with it and give themselves over to it, but at a definite price – viciously though and basely. Yet a vastly disproportionate measure could be imagined between the vice and the price, one where the pleasure could with justice compensate for the sin (as expediency is said to do) – not when the pleasure is incidental, forming no part of the sin, as in theft, but as in lying with women where the pleasure resides in performing the sin and where the drive is violent and, so it is said, irresistible.

The other day when I was in Armagnac on the estates of one of my relations I met a peasant whom everybody called Pincher. He gave me this account of his life: being born to beggary and finding that he would never succeed in earning his bread and warding off indigence by the labour of his hands, he took the decision to become a thief and had spent his entire youth safely in that trade because he was so physically strong; for he used to harvest the corn and grapes on other men's lands, but so far off and in such huge quantities that it was unthinkable that one man could have loaded so much on his back in one single night. He also took care to spread the damage equally about, so that each of his victims found the loss less hard to bear. Now, in his old age, he is rich for a man of his station – thanks to that trafficking, which he openly admits. To come to terms with God for his gains he declares that, by making free-gifts, he is always keen to compensate the heirs of all the men he robbed, and that if he does not finish this (for he simply cannot provide for all at once) he will charge his heirs to do so, based on the knowledge which he alone has of the evil he had done to each individual. From this account, be it true or false, that man regards theft as a dishonest deed; and he hates it . . . less than he hates poverty. He indeed repents of the theft as such, but he does not feel any repentance for its being counterbalanced and counterweighed. We do not find in this case that habitual practice which makes us fellows-incorporate with vice and brings our mind itself to conform to it; nor is it that violent gale which batters and blinds our soul and sweeps us for a while into the power of vice, judgement and all.

My custom is to be entirely given to what I do, marching forward all of a piece. There is hardly an emotion in me which sneaks away and hides from my reason or which is not governed by the consent of almost all my parts, without schism or inner strife. The entire blame or praise for that belongs to my judgement; and once it accepts that blame it has it for ever,

because virtually since birth it has always been one: the same bent, the same route, the same strength. And as for all my general opinions, I have since childhood lodged me where I was to remain.

There are sins which are violent, quick and sudden. Let us leave them aside. But as for those other sins, so often repeated, deliberated and meditated upon, those sins which are rooted in our complexions [C] and, indeed in our professions or vocations, [B] I cannot conceive that they could be rooted so long in one identical heart without the reason and conscience of him who is seized of them being constant in his willing and wanting them to be so; and the repentance which he boasts to come to him at a particular appointed instant is hard for me to imagine or conceive. [C] I cannot follow the Pythagorean dogma that men take on a new soul when they draw near to the statues of the gods to gather up their oracles, unless Pythagoras meant that their soul must actually be a new one, foreign to them and lent for the occasion, since their own soul showed so little sign of being cleansed by purification and condign for that duty.[16] [B] What they do is flat contrary to the Stoics' precepts, which do indeed command us to correct any vices or imperfections which we acknowledge to be in us but forbid us to be sorry or upset about them. But these men would have us believe that they do feel deep remorse and regret within; yet no amendment or improvement, [C] no break, [B] ever becomes apparent. But if you do not unburden yourself of the evil there has been no cure. If repentance weighed down the scales of the balance it would do away with the sin. I can find no quality so easy to counterfeit as devotion unless our morals and our lives are made to conform to it; its essence is hidden and secret: its external appearances are easy and ostentatious.

As for me, I can desire to be entirely different, I can condemn my universal form and grieve at it and beg God to form me again entirely and to pardon my natural frailty. But it seems to me that that should not be called repenting any more than my grieving at not being an angel or Cato.[17] My doings are ruled by what I am and are in harmony with how I was made. I cannot do better: and the act of repenting does not properly touch such things as are not within our power – *that* is touched by regretting. I can imagine countless natures more sublime and better ruled than my own: by doing that I do not emend my own capacities, any more

16. Seneca, *Epist. moral.*, XCIV, 42.
17. Angels have souls higher than men's in the chain-of-being: Cato had a soul higher than Montaigne's within the human scale.

than my arm or my intelligence become more strong because I can imagine others which are. If imagining and desiring actions nobler than ours made us repent of our own we would have to go repenting of our most innocent doings, since we can rightly judge that they would have been brought to greater perfection and grandeur in a nature far excelling our own. When I reflect on my behaviour as a young man and as an old one I find that I have mainly behaved ordinately *secundum me*.[18] My power of resistance can do no more. I do not flatter myself: in like circumstances I would still be thus. It is no spot but a universal stain which soils me. I do not know any surface repentance, mediocre and a matter of ceremony. Before I call it repentance it must touch me everywhere, grip my bowels and make them yearn – as deeply and as universally as God does see me.[19]

In my business dealings several good opportunities have escaped me for want of the happy knack of conducting them: yet my decisions were well chosen *secundum quid* (that is, according to the events which they ran up against); my decisions are so fashioned as always to take the easiest and the surest side. I find that I proceeded wisely, according to my rule, in my previous deliberations given the state of the subject as set before me: and in the same circumstances I would do the same a thousand years from hence. I pay no regard to what it looks like now but to how it was when I was examining it.

[C] The force of any advice depends upon the time: circumstances endlessly alter and matters endlessly change. I have made some grievous mistakes in my life – important ones – for want of good luck not for want of good thought. In the subjects which we handle, and especially in the natures of men, there are hidden parts which cannot be divined, silent characteristics which are never revealed and which are sometimes unknown even to the one who has them but which are awakened and brought out by subsequent events. If my wisdom was unable to penetrate through to them and foresee them I bear it no grudge: there are limits to its obligations. What defeats me is the outcome, and   [B]   if it favours the side I rejected, that cannot be helped. I do not find fault with myself: I blame not what I did but my fortune. And that is not to be called repenting.

18. That is, even ordinate actions and reactions are relative insofar as they must be judged 'according to' one's capacities and judgements.
19. Each man is, in God's sight, sinful (Romans 3:23; 5:12), and God is the *scrutator cordium*, 'He who searches all hearts' (I Chronicles 28:9); 'He who searcheth the heart and knoweth the mind' (Romans 8:27); 'He that searcheth the reins and the heart' (Revelations 2:23).

Phocion gave a certain piece of advice to the Athenians which was not acted upon. When the affair turned out successfully against his advice somebody asked him, 'Well now, Phocion, are you pleased that things are going so well?' 'Of course,' he said, 'I am happy that it has turned out this way, but I do not repent of the advice that I gave.'[20] When my friends come to me for advice I give it freely and clearly, without (as nearly everyone does) dwelling on the fact that, since the matter is chancy, things can turn out contrary to what I think, so that they may well have cause to reproach me for my advice. That never bothers me, for they will be in the wrong: I *ought* not to have refused them such service.

[C] I have hardly any cause to blame anyone but myself for my failures or misfortunes, for in practice I rarely ask anyone for advice save to honour them formally; the exception is when I need learned instruction or knowledge of the facts. But in matters where only my judgement is involved, the arguments of others rarely serve to deflect me though they may well support me; I listen to them graciously and courteously – to all of them. But as far as I can recall I have never yet trusted any but my own. According to my standards they are but flies and midges buzzing over my will. I set little store by my own opinions but just as little by other people's. And Fortune has treated me worthily. I receive little counsel: I give even less. I am very rarely asked for it: I am even less believed, and I know of no public or private undertaking which has been set right or halted on my advice. Even such persons as chance to be somewhat dependent on my advice have readily allowed themselves to be swayed by some completely different mind. Since I am just as jealous of my right to peace and quiet as of my right to authority, I prefer it that way. By leaving me out they are acting on my own principles, which consist in being settled and contained entirely within myself: it is a joy for me to be detached from others' affairs and relieved of protecting them.

I have few regrets for affairs of any sort, no matter how they have turned out, once they are past. I am always comforted by the thought that they had to happen that way: there they are in the vast march of the universe and in the concatenation of Stoic causes; no idea of yours, by wish or by thought, can change one jot without overturning the whole order of Nature, both past and future.[21]

Meanwhile I loathe that consequential repenting which old age brings.

20. Plutarch (tr. Amyot), *Dicts notables des anciens Roys . . .*, 197 E.
21. For the Stoics, causation was absolute: everything is fated and unalterable.

That Ancient who said that he was obliged to the passing years for freeing him from sensual pleasures held quite a different opinion from mine: I could never be grateful to infirmity for any good it might do me. [C] *'Nec tam aversa unquam videbitur ab opere suo providentia, ut debilitas inter optima inventa sit.'* [And Providence will never be found so hostile to her work as to rank debility among the best of things.]²² [B] Our appetites are few when we are old: and once they are over we are seized by a profound disgust. I can see nothing of conscience in that: chagrin and feebleness imprint on us a lax and snotty virtue. We must not allow ourselves to be so borne away by natural degeneration that it bastardizes our judgement. In former days youth and pleasure never made me fail to recognize the face of vice within the sensuality: nor does the distaste which the years have brought me make me fail to recognize now the face of pleasure within the vice.

I have nothing to do with it now, but I judge it as though I did. [C] Personally, when I give my reason a lively and attentive shake, I find that [B] it is just the same as in my more licentious years, except that it has perhaps grown more feeble and much worse with age; [C] and I find that, although it declines to stoke up such pleasures out of consideration for the interests of my physical health, it would not do that, even now, any more than it once did, for the sake of my spiritual health. [B] I do not think it any braver for seeing it drop out of the battle. My temptations are so crippled and enfeebled that they are not worth opposing. I can conjure them away by merely stretching out my hands. Confront my reason with my former longings and I fear that it will show less power of resistance than once it did. I cannot see that, of itself, it judges in any way differently now than it did before, nor that it is freshly enlightened. So if it has recovered it is a botched recovery. [C] A wretched sort of cure, to owe one's health to sickliness.

It is not for our wretchedness to do us that service: it is for the happy outcome of our judgement. As for whacks and afflictions, you can make me do nothing but curse them: they are meant for men whose desires are aroused only by a good whipping. Indeed my reason runs freer when things go well: it is far more distracted and occupied when digesting misfortunes than pleasures. I can see much more clearly when the weather is serene. Health counsels me both more actively and usefully than illness does. I had progressed as far as I could towards right-rule and reformation

---

22. Sophocles, criticized by Epicurus: Plutarch (tr. Amyot), *Que l'on ne sçauroit vivre heureusement selon la doctrine d'Epicurus*, 283 DE; Quintilian, V, xii.

when I had health to enjoy. I would be ashamed and jealous if the wretched lot of my decrepitude were to be preferred above the years when I was healthy, aroused and vigorous, and if men had to esteem me not for what I was but for ceasing to be like that. It is my conviction that what makes for human happiness is not, as Antisthenes said, dying happily but living happily.[23] I have never striven to make a monster by sticking a philosopher's tail on to the head and trunk of a forlorn man, nor to make my wretched end disavow and disclaim the more beautiful, more whole-some and longer part of my life. I want to show myself to have been uniform and to be seen as such. If I had to live again, I would live as I have done; I neither regret the past nor fear the future. And unless I deceive myself, things within have gone much the same as those without. One of my greatest obligations to my lot is that the course of my physical state has brought each thing in due season. I have known the blade, the blossom and the fruit; and I now know their withering. Happily so, since naturally so. I can bear more patiently the ills that I have since they come in due season, and since they also make me recall with more gratitude the long-lasting happiness of my former life.

My wisdom may well have had the same stature in both my seasons, but it was far more brilliant and graceful then, green-sprouting, gay and naïve; now it is bent double, querulous and wearisome.

I disclaim those incidental reformations based on pain.   [B]   God must touch our hearts.[24] Our conscience must emend itself by itself, by the strengthening of our reason not by the enfeebling of our appetites. Sensual pleasure, of itself, is neither so pale nor so wan as to be perceived by bleared and troubled eyes. We must love temperance for its own sake and out of respect for God who has commanded it to us; and chastity too: what we are presented with by rheum, and what I owe to the grace of my colic paroxysms, are neither chastity nor temperance.[25]

You cannot boast of despising and of fighting pleasure if you cannot see her and if you do not know her grace and power, or her beauty at its most

23. Plutarch, *Life of Antisthenes*; Erasmus, *Apophthegmata*, VII, *Antisthenes*, XIV.

24. I Samuel 10: 26, 'whose hearts God hath touched'. Adapted for the motto of her emblematic picture FRUSTRA ('in Vain') by the Protestant author Georgette de Montenay in her *Emblemes ou devises chrestiennes* (Lyons, 1571).

25. Temperance is Aristotle's *sōphrosyne* (the Mean between two vices, one of excess and one of defect) (*Nicomachaean Ethics*, II, vi, 3). This Classical virtue, as well as the four Cardinal virtues, were held to apply to Christians, though all needed completing by the three theological virtues (Faith, Hope and Charity). St Paul in Philippians 4:5 counselled, 'Let your moderation be known to all men.'

attractive. I know them both: and I am the one to say so. But it seems to me that our souls are subject in old age to ills and imperfections more insolent than those of youth. I said so when I was young, and they cast my beardless chin in my teeth. And I still say so now that my [C] grey [B] hair lends me credit. What we call wisdom is the moroseness of our humours and our distaste for things as they are now. But in truth we do not so much give up our vices as change them – for the worse, if you ask me. Apart from silly tottering pride, boring babble, prickly unsociable humours, superstition and a ridiculous concern for wealth when we have lost the use of it, I find that there are more envy and unfairness and malice; age sets more wrinkles on our minds than on our faces. You can find no souls – or very few – which as they grow old do not stink of rankness and of rot. It is the man as a whole that marches towards his flower and his fading.

[C] When I see the wisdom of Socrates and several of the circumstances surrounding his condemnation, I would venture to conclude that to some degree he connived at it and deliberately put up a sham defence, since at seventy years of age he soon had to suffer the benumbing of his splendid endowments and the clouding over of his habitual clarity.

[B] What transformations do I daily see wrought by old age in those I know. It is a powerful illness which flows on naturally and imperceptibly. You must have a great store of study and foresight to avoid the imperfections which it loads upon us – or at least to weaken their progress. I know that, despite all my entrenchments, it is gaining on me foot by foot. I put up such resistance as I can. But I do not know where it will take me in the end. Yet come what may, I should like people to know from what I shall have declined.

3. On three kinds of social intercourse

[One of the most personal of the chapters so far. The 'trois commerces' examined by Montaigne are the three forms of social intercourse which enrich his private life and make it worth living: 1) loving-friendship – even though ordinary friendships become rather insipid when judged against his perfect friendship with La Boëtie; 2) loving relationships with 'ladies', beautiful and, if possible, intelligent; 3) reading books. The one adjective common to the friends, women and books discussed here and in 'On books' is honnête (honourable and decent). Montaigne's ideal social intercourse would engage the whole man, body and soul. By themselves none of these three fully does so, and the first two engage the body and the soul in widely differing proportions, while books hardly engage the body at all.

Montaigne speaks of women in a gruffly humorous way, but it will be noted that the reading he would concede to them corresponds closely to what he says of his own reading in the chapter 'On books'.

There is an important insistence that sexual intercourse is more than a physical 'necessity' and so not merely a hunger to be satisfied physically without the involvement of the higher faculties.

The disease Montaigne caught from prostitutes was syphilis.]

[B] We should not nail ourselves so strongly to our humours and complexions. Our main talent lies in knowing how to adapt ourselves to a variety of customs. To keep ourselves bound by the bonds of necessity to one single way of life is to be, but not to live. Souls are most beautiful when they show most variety and flexibility. [C] Here is a testimony which honours Cato the Elder: *'Huic versatile ingenium sic pariter ad omnia fuit, ut natum ad id unum diceres, quodcumque ageret.'* [His mind was so versatile, and so ready for anything, that whatever he did you could say he was born for that alone].[1]

[B] If it was for me to train myself my way, there would be no mould in which I would wish to be set without being able to throw it off. Life is a rough, irregular progress with a multitude of forms. It is to be no friend of yourself – and even less master of yourself – to be a slave endlessly following yourself, so beholden to your predispositions that you cannot

1. Livy, XXXIX, xl.

stray from them nor bend them. I am saying this now because I cannot easily escape from the state of my own Soul, which is distressing in so far as she does not usually know how to spend her time without getting bogged down nor how to apply herself to anything except fully and intensely. No matter how trivial the subject you give her she likes to magnify it and to amplify it until she has to work at it with all her might. For this reason her idleness is an activity which is painful to me and which damages my health. Most minds have need of extraneous matter to make them limber up and do their exercises: mine needs rather to sojourn and to settle down: *'Vitia otii negotio discutienda sunt'* [We must dispel the vices of leisure by our work];[2] my own mind's principal and most difficult study is the study of itself. [C] For it, books are the sort of occupation which seduces it from such study. [B] With the first thoughts which occur to it it becomes agitated and makes a trial of its strength in all directions, practising its control, sometimes in the direction of force, sometimes in the direction of order and gracefulness, [C] controlling, moderating and fortifying itself. [B] It has the wherewithal to awaken its faculties by itself: Nature has given it (as she has given them all) enough matter of its own for its use and enough subjects for it to discover and pass judgement upon.

[C] For anyone who knows how to probe himself and to do so vigorously, reflection is a mighty endeavour and a full one: I would rather forge my soul than stock it up. No occupation is more powerful, or more feeble, than entertaining one's own thoughts — depending on what kind of soul it is. The greatest of souls make it their vocation, *'quibus vivere est cogitare'* [for them, to think is to live];[3] there is nothing we can do longer than think, no activity to which we can devote ourselves more regularly nor more easily: Nature has granted the soul that prerogative. It is the work of the gods, says Aristotle, from which springs their beatitude and our own.[4] Reading, by its various subjects, particularly serves to arouse my discursive reason: it sets not my memory to work but my judgement. [B] So, for me, few conversations are arresting unless they are vigorous and powerful. It is true that grace and beauty occupy me and fulfil me as much or more as weight and profundity. And since I doze off during any sort of converse and lend it only the outer bark of my attention, it often happens that during polite conversation (with its flat,

2. Seneca, *Epist. moral.*, LVI, 9 (adapted).
3. Cicero, *Tusc. disput.*, V, XXXVIII, 113 (of the learned and erudite).
4. Aristotle, *Nicomachaean Ethics*, X, viii, 1178 b (referring to *theōrētikē*, contemplation, intellectual activity).

well-trodden sort of topics) I say stupid things unworthy of a child, or make silly, ridiculous answers, or else I remain stubbornly silent which is even more inept and rude. I have a mad way of withdrawing into myself as well as a heavy, puerile ignorance of everyday matters. To those two qualities I owe the fact that five or six true anecdotes can be told about me as absurd as about any man whatsoever.

Now to get on with what I was saying: this awkward complexion of mine renders me fastidious about mixing with people: I need to handpick my companions; and it also renders me awkward for ordinary activities. We live and deal with the common people; if their commerce wearies us, if we disdain to apply ourselves to their humble, common souls – and the humble, common ones are often as well-governed as the most refined [C] (all wisdom being insipid which does not adapt to the common silliness) –   [B]   then we must stop dealing with our own affairs and anyone else's: both public and personal business involves us with such people. The most beautiful motions of our soul are those which are least tense and most natural: and the best of its occupations are the least forced. O God! What good offices does Wisdom do for those whose desires she ranges within their powers! No knowledge is more useful. 'According as you can' was the refrain and favourite saying of Socrates, a saying of great substance.[5] We must direct our desires and settle them on the things which are easiest and nearest. Is it not an absurd humour for me to be out of harmony with the hundreds of men to whom my destiny joins me and whom I cannot manage without, in order to restrict myself to one or two people who are beyond my ken? Or is it not rather a mad desire for something I cannot get?

My mild manners, which are the enemies of all sharpness and contentiousness, may easily have freed me from the burden of envy and unfriendliness: never did man give more occasion – I do not say to be loved but certainly not to be hated? But the lack of warmth in my converse has rightly robbed me of the good-will of many, who can be excused for interpreting it differently, in a worse sense.

Most of all I am able to make and keep exceptional and considered friendships, especially since I seize hungrily upon any acquaintanceship which corresponds to my tastes. I put myself forward and throw myself into them so eagerly that I can hardly fail to make attachments and to leave

5. Cited by Xenophon (*Memorabilia*, I, iii, 3). In the Latin form *'secundum quod potes'* it lends force to Montaigne's conviction that all is not simple but *secundum quid*.

my mark wherever I go. I have often had a happy experience of this. In commonplace friendships I am rather barren and cold, for it is not natural to me to proceed except under full sail. Besides, the fact that as a young man I was brought to appreciate the delicious savour of one single perfect friendship has genuinely made the others insipid to me and impressed on my faculty of perception that (as one ancient writer said) friendship is a companiable, not a gregarious, beast.[6] I also, by nature, find it hard to impart myself by halves, with limitations and with that suspicious vassal-like prudence prescribed to us for our commerce with those multiple and imperfect friendships[7] – prescribed in our time above all, when you cannot talk to the world in general except dangerously or falsely.

Yet I can clearly see that anyone like me whose aim is the good things of life (I mean those things which are of its essence) must flee like the plague from such moroseness and niceness of humour. What I would praise would be a soul with many storeys, one of which knew how to strain and relax; a soul at ease wherever fortune led it; which could chat with a neighbour about whatever he is building, his hunting or his legal action, and take pleasure in conversing with a carpenter or a gardener. I envy those who can come down to the level of the meanest on their staff and make conversation with their own servants.   [C] I have never liked Plato's advice to talk always like a master to our domestics, without jests or intimacy, whether addressing menservants or maidservants.[8] For, apart from what my own reason tells me, it is ill-bred and unjust to give such value to a trivial privilege of Fortune: the most equitable politics seem to me to be those which allow the least inequality between servants and masters.

[B]   Other men study themselves in order to wind their minds high and send them forth: I do so in order to bring mine lower and lay it down. It is vitiated only when it reaches out:

> *Narras, et genus Æaci,*
> *Et pugnata sacro bella sub Ilio :*
> *Quo Chium pretio cadum*
> *Mercemur, quis aquam temperet ignibus,*

6. Plutarch (tr. Amyot), *De la pluralité des amis*, 103 B–C, stressing that great friendships come in pairs, not in groups.
7. The most famous prudential maxim was, 'So have a friend that he may be your enemy.' Aristotle attributes it to Bias, one of the Seven Sages of Greece. In I, 28, Montaigne attributes it to Chilon.
8. Plato, *Laws*, VI, 778 A (of slaves, not servants).

> *Quo præbente domum, et quota,*
> > *Pelignis caream frigoribus, taces.*

[You sing of Aeacus' line and the wars beneath the sacred walls of Ilium: but you do not say how much I must pay for a jar of Chian wine, who will heat my water on his fire, where I shall find shelter and when I shall escape from the cold of the Pelignian mountains.][9]

Thus, just as Spartan valour needed moderating by the gentle gracious playing of flutes to calm it down in war lest it cast itself into rashness and frenzy[10] (whereas all other peoples normally employ shrill sounds and powerful voices to stir and inflame the hearts of their warriors), so it seems to me that, in exercising our minds, we for the most part – contrary to normal practice – have greater need of lead-weights than of wings, of cold repose rather than hot agitation.

Above all, to my mind, it is to act like a fool to claim to be in the know amidst those who are not, and to be ever speaking guardedly – *'favellar in punta di forchetta'* [speaking daintily, 'with the prongs of your fork']. You must come down to the level of those you are with, sometimes even affecting ignorance. Thrust forceful words and subtleties aside: when dealing with ordinary folk it is enough if you maintain due order. Meanwhile, if they want you to, creep along at ground-level. That is the stone which scholars frequently trip up over. They are always parading their mastery of their subject and scattering broadcast whatever they have read. Nowadays they have funnelled so much of it into the ears of the ladies in their drawing-rooms that, even though those ladies of ours have retained none of the substance, they look as though they have: on all sorts of topics and subjects, no matter how menial or commonplace, they employ a style of speaking and writing which is newfangled and erudite:

> *Hoc sermone pavent, hoc iram, gaudia, curas,*
> *Hoc cuncta effundunt animi secreta; quid ultra?*
> *Concumbunt docte.*

[This is the style in which they express their fears, their anger, their joy and their cares. This is the style in which they pour forth all their secrets; why, they even lie with you eruditely.][11]

9. Horace, *Odes*, III, xix, 3–8.
10. Cf. III, 1, 'On the useful and the honourable', note 28.
11. Juvenal, *Satires*, VI, 189–91 (adapted).

They cite Plato and St Thomas Aquinas for things which the first passer-by could serve to support. That doctrine which they have learned could not reach their minds so it has stayed on their tongues. However well-endowed they are, they will, if they trust me, be content to make us value the natural riches proper to them. They hide and drape their own beauties under borrowed ones. There is great simpleness in such smothering of their own light so as to shine with borrowed rays; they are dead and buried under artifice: [C] *'De capsula totae.'* [All out of the clothes-press.][12]

[B] That is because they do not know enough about themselves: there is nothing in the whole world as beautiful; they it is who should be lending honour to art and beauty to cosmetics. What more do they want than to live loved and honoured? They have enough, and know enough, to do that. All that is needed is a little arousing and enhancing of the qualities which are in them. When I see them saddled with rhetoric, judicial astrology, logic and such-like vain and useless trash, I begin to fear that the men who counsel them to do so see it as a way of having a pretext for manipulating them. For what other excuse can I find for them?

It suffices that ladies (without our having to tell them how) can attune the grace of their eyes to gaiety, severity and gentleness; season a 'No! No!' with rigour, doubt or favour; and seek no hidden meanings in the speeches with which we court them. With knowledge like that it is they who wield the big stick and dominate the dominies and their schools.

Should it nevertheless irk them to lag behind us in anything whatsoever; should they want a share in our books out of curiosity: then poetry is a pastime rightly suited to their needs: it is a frivolous, subtle art, all disguise and chatter and pleasure and show, like they are. They will also draw a variety of benefits from history; and in philosophy – the part which helps us to live well – they will find such arguments as train them to judge of our humours and our attributes, to shield them from our deceptions, to control the rashness of their own desires, to cultivate their freedom and prolong the pleasure of this life, and to bear with human dignity the inconstancy of a suitor, the moroseness of a husband and the distress of wrinkles and the passing years. That sort of thing.[13]

That – at most – is the share of learning that I would assign to them.

12. Seneca, *Epist. moral.*, CXV, 2, condemning the affected style of clothes and speech of dandies.
13. Much the same reading as Montaigne likes himself (II, 10, 'On books'), though doubtless presupposing that the books are in French not Latin.

Some natures are withdrawn, enclosed and private. The proper essence of my own form lies in imparting things and in putting them forth: I am all in evidence; all of me is exposed; I was born for company and loving relationships. The solitude which I advocate is, above all, nothing but the bringing of my emotions and thoughts back to myself, restricting and restraining not my wandering footsteps but my anxiety and my desires, abandoning disquiet about external things and fleeing like death from all slavery and obligation, [C] and running away not so much from the throng of people as from the throng of affairs.

[B] To tell the truth, localized solitude makes me reach out and extend myself more: I throw myself into matters of State and into the whole universe more willingly when I am alone. In a crowd at the Louvre I hold back and withdraw into my skin; crowds drive me back into myself and my thoughts are never more full of folly, more licentious and private than in places dedicated to circumspection and formal prudence. It is not our folly which makes me laugh: it is our wisdom.

I am not by complexion hostile to the jostlings of the court: I have spent part of my life there and am so made that I can be happy in large groups provided that it be at intervals and at my own choosing. But that lax judgement I am speaking of forces me to bind myself to solitude even in my own home, in the midst of a crowded household which is among the most visited. I meet plenty of people there, but rarely those whom I love to converse with; and I reserve an unusual degree of liberty there for myself and for others. There we have called a truce with all etiquette, welcomings and escortings and other such painful practices decreed by formal courtesy. (Oh what servile and distressing customs!) Everybody goes his own way; anyone who wants to can think his own thoughts: I remain dumb, abstracted and inward-looking – no offence to my guests.

I am seeking the companionship and society of such men as we call honourable and talented: my ideal of those men makes me lose all taste for the others. It is, when you reflect on it, the rarest of all our forms; and it is a form which is mainly owed to nature. The ends of intercourse with such men are simply intimacy, the frequenting of each other and discussion – exercising our souls with no other gain. In our conversation any topic will do: I do not worry if they lack depth or weight: there is always the grace and the appropriateness: everything in it is coloured by ripe and sustained judgement mingled with frankness, goodwill, gaiety and affection. Our minds do not merely show their force and beauty on the subject of entailed property or our kings' business: they show it just as well in our private discussions together. I recognize my kind of men by their very silences or

their smiles; and I perhaps discover them better at table than in their work-rooms. Hippomachus said that he could tell a good wrestler simply by seeing him walk down the street.[14]

If Erudition wants to mingle in our discussions, then she will not be rejected, though she must not be, as she usually is, professorial, imperious and unmannerly, but courting approval, herself ready to learn. We are merely seeking a pastime: when the time comes to be lectured to and preached at we will go and seek her on her throne. Let her be kind enough to come down to us on this occasion, please! For, useful and desirable as she is, I presume that if we had to we could get on quite well in her absence and achieve our effect without her. A well-endowed Soul, used to dealing with men, spontaneously makes herself totally agreeable. Art is but the register and accounts of the products of such souls.

There is for me another delightful kind of converse: that with [C] beautiful and [B] honourable women: [C] *'Nam nos quoque oculos eruditos habemus.'* [For we too have well-taught eyes.][15] [B] Though there is less here for our souls to enjoy than in the first kind, our physical senses, which play a greater part in this one, restore things to a proportion very near to the other – though for me not an equal one. But it is a commerce where we should remain a bit on our guard, especially men like me over whom the body has a lot of power. I was scalded once or twice in my youth and suffered all the ragings which the poets say befall men who inordinately and without judgement let go of themselves in such matters. It is true that I got a beating which taught me a lesson:

> *Quicunque Argolica de classe Capharea fugit,*
> *Semper ab Euboicis vela retorquet aquis.*

[Anyone in the Grecian fleet who escaped from that shipwreck on the promontory of Caphareus ever thereafter turns his sails away from the waters of Euboea.][16]

It is madness to fix all our thoughts on it and to engage in it with a frenzied singleminded passion. On the other hand to get involved in it without love or willing to be bound, like actors, so as to play the usual part expected from youth, contributing nothing of your own but your words, is indeed to provide for your safety; but it is very cowardly, like a man who would

---

14. Plutarch, *Life of Dion*; Hippomachas was a teacher of athletics.
15. Cicero, *Paradoxa Stoicorum*, V, 38 (the wise can appreciate objects of artistic beauty but should not be enslaved by them).
16. Ovid, *Tristia*, I, i, 83–4.

jettison his honour, goods and pleasure from fear of danger. For one thing is certain: those who set such a snare can expect to gain nothing by it which can affect or satisfy a soul of any beauty. We must truly have desired any woman we wish truly to enjoy possessing; I mean that, even though fortune should unjustly favour play-acting – as often happens, since there is not one woman, no matter how ugly she may be, who does not think herself worth loving   [C]   and who does not think herself attractive for her laugh, her gestures or for being the right age, since none of them is universally ugly any more than universally beautiful. (When the daughters of the Brahmans have nothing else to commend them, the town-crier calls the people together in the market-place expressly for them to show off their organs of matrimony to see whether they at least can be worth a husband to them.)   [B]   It follows that there is not one who fails to let herself be convinced by the first oath of devotion sworn by her suitor. Now from the regular routine treachery of men nowadays there necessarily results what experience already shows us: to escape us, women turn in on themselves and have recourse to themselves or to other women; or else they, on their side, follow the example we give them, play their part in the farce and join in the business without passion concern or love.   [C]   *'Neque affectui suo aut alieno obnoxiae'* [Beholden to no love, their own or anyone else's];[17] following the conviction of Lysias in Plato and reckoning that the less we love them the more usefully and agreeably they can devote themselves to it.   [B]   It will go as in comedies: the audience will have as much pleasure as the comedians, or more.

As for me, I no more know Venus without Cupid than motherhood without children: they are things whose essences are interdependent and necessary to each other. So such cheating splashes back on the man who does it. The *affaire* costs him hardly anything, but he gets nothing worthwhile out of it either.

Those who turned Venus into a goddess considered that her principal beauty was not a matter of the body but of the spirit: yet the 'beauty' such men are after is not simply not human, it is not even bestial. The very beasts do not desire it so gross and so earth-bound: we can see that imagination and desire often set beasts on heat and arouse them before their body does; we can see that beasts of both sexes choose and select the object of their desires from among the herd and that they maintain long affectionate relationships. Even beasts which are denied physical powers by old age still quiver, whinny and tremble with love. We can see them full of hope

17. Tacitus, *Annals*, XIII, xlv; then, Plato, *Phaedrus*, 227 B–228 C.

and fire before copulation, and, once the body has played its part, still tickling themselves with the sweet memory of it; some we see which swell with pride as they make their departure and which produce songs of joy and triumph, being tired but satisfied. A beast which merely wished to discharge some natural necessity from its body would have no need to bother another beast with such careful preparations: we are not talking about feeding some gross and lumpish appetite.[18]

Being a man who does not ask to be thought better than I am, I will say this about the errors of my youth: I rarely lent myself to venal commerce with prostitutes, not only because of the danger [C] to my health (though even then I did not manage to escape a couple of light anticipatory doses) [B] but also because I despised it. I wanted to sharpen the pleasure by difficulties, by yearning and by a kind of glory; I liked the style of the Emperor Tiberius (who in his love-affairs was attracted more by modesty and rank than by any other quality)[19] and the humour of Flora the courtesan (who was also attracted by a dictator, a consul or a censor, delighting in the official rank of her lovers). Pearls and brocade certainly add to the pleasure; so do titles and retainers. Moreover I set a high value on wit, provided however that the body was not wanting; for if one of those two qualities had to be lacking, I must admit in all conscience that I would have chosen to make do without the wit; it has use in better things. But where love is concerned – a subject which is mainly connected with sight and touch – you can achieve something without the witty graces but nothing without the bodily ones.

Beauty is the true privilege of noblewomen. [C] It is so much more proper to them than ours is to us men, that even though ours requires slightly different traits, at its highest point it is boyish and beardless, and therefore confounded with theirs. They say that in the place of the Grand Seigneur males chosen to serve him for their beauty – and they are countless in number – are sent away at twenty-two at the latest.[20] [B] Reasoning powers, wisdom and the offices of loving-friendship are rather to be found in men: that is why they are in charge of world affairs.

---

18. Philosophy classified sexual intercourse among the physical necessities. Montaigne does not deny that it is so, but insists that sexual fulfilment is more than the physical slaking of an appetite.
19. Tacitus, *Annals*, VI, i; then a tale of Flora recounted among others by Brantôme in *Les Dames Galantes* (Deuxième Discours).
20. Guillaume Postel, *Histoire des Turcs*.

Those two forms of converse[21] depend on chance and on other people. The first is distressingly rare, the second withers with age, so they could not have adequately provided for the needs of my life. Converse with books (which is my third form) is more reliable and more properly our own. Other superior endowments it concedes to the first two: its own share consists in being constantly and easily available with its services. This converse is ever at my side throughout my life's course and is everywhere present. It consoles me in my old age and in my retreat; it relieves me of the weight of distressing idleness and, at any time, can rid me of boring company. It blunts the stabs of pain whenever the pain is not too masterful and extreme. To distract me from morose thought I simply need to have recourse to books; they can easily divert me to them and rob me of those thoughts. And yet there is no mutiny when they see that I only seek them for want of other benefits which are more real, more alive, more natural: they always welcome me with the same expression.

It is all very well, we say, for a man to go on foot when he leads a ready horse by the bridle! And our James, King of Naples, manifested a kind of austerity which was still delicate and vacillating, when, young, handsome and healthy, he had himself wheeled about the land on a bier, lying on a cheap feather-pillow, clad in a robe of grey cloth with a bonnet to match, followed meanwhile by great regal pomp with all sorts of litters and horses to hand, and by officers and noblemen.[22] 'No need to pity an invalid who has a remedy up his coat-sleeve!' All the profit which I draw from books consists in experiencing and applying that proverb (which is a very true one). In practice I hardly use them more than those who are quite unacquainted with them. I enjoy them as misers do riches: because I know I can always enjoy them whenever I please. My soul is satisfied and contented by this right of possession. In war as in peace I never travel without books. Yet days and even months on end may pass without my using them. 'I will read them soon,' I say, 'or tomorrow; or when I feel like it.' Thus the time speeds by and is gone, but does me no harm; for it is impossible to describe what comfort and peace I derive from the thought that they are there beside me, to give me pleasure whenever I want it, or from recognizing how much succour they bring to my life. It is the best protection which I have found for our human journey and I deeply pity men of intelligence who lack it. I on the other hand can accept any sort of pastime, no matter how trifling, because I have this one which will never fail me.

21. Intercourse with friends and with ladies.
22. Olivier de La March, *Mémoires*, 1561.

At home I slip off to my library a little more often; it is easy for me to oversee my household from there. I am above my gateway and have a view of my garden, my chicken-run, my backyard and most parts of my house. There I can turn over the leaves of this book or that, a bit at a time without order or design. Sometimes my mind wanders off, at others I walk to and fro, noting down and dictating these whims of mine.

[C] It is on the third storey of a tower. The first constitutes my chapel; the second, a bed-chamber with a dressing-room, where I often sleep when I want to be alone. Above that there is a large drawing-room. It was formerly the most useless place in my house: I spend most days of my life there, and most hours of each day, but I am never there at night. It leads on to quite an elegant little chamber which can take a fire in winter and agreeably lets in the light. If I feared the bother as little as the expense — and the bother drives me away from any task — I could erect a level gallery on either side, a hundred yards long and twelve yards wide, having found all the walls built (for some other purpose) at the required height. Every place of retreat needs an ambulatory. My thoughts doze off if I squat them down. My wit will not budge if my legs are not moving — which applies to all who study without books.

My library is round in shape, squared off only for the needs of my table and chair; as it curves round it offers me at a glance every one of my books ranged on five shelves all the way along. It has three splendid and unhampered views and a circle of free space sixteen yards in diameter. I am less continuously there in winter since my house is perched on a hill (hence its name) and no part of it is more exposed to the wind than that one. By being rather hard to get at and a bit out of the way it pleases me, partly for the sake of the exercise and partly because it keeps the crowd from me. There I have my seat. I assay making my dominion over it absolutely pure, withdrawing this one corner from all intercourse, filial, conjugal and civic. Everywhere else I have but a verbal authority, one essentially impure. Wretched the man (to my taste) who has nowhere in his house where he can be by himself, pay court to himself in private and hide away! Ambition well rewards its courtiers by keeping them always on display like a statue in the market-place: *'Magna servitus est magna fortuna.'* [A great destiny is great slavery.][23] They cannot even find privacy on their privy! I have never considered any of the austerities of life which our monks delight in to be harsher than the rule that I have noted in some of their foundations:

23. Seneca, *Consolatio ad Polybium*, XXVI.

to be perpetually with somebody else and to be surrounded by a crowd of people no matter what they are doing. And I find that it is somewhat more tolerable to be always alone than never able to be so.

[B] If anyone says to me that to use the Muses as mere playthings and pastimes is to debase them, then he does not know as I do the value of pleasure, [C] plaything or pastime. [B] I could almost say that any other end is laughable. I live from day to day; and, saving your reverence, I live only for myself. My plans stop there. In youth I studied in order to show off; later, a little, to make myself wiser; now I do it for amusement, never for profit. A silly spendthrift humour that once I had for furnishing myself with books, [C] not to provide for my needs but three paces beyond that, [B] so as to paper my walls with them as decorations, I gave up long ago.

Books have plenty of pleasant qualities for those who know how to select them. But there is no good without ill. The pleasure we take in them is no purer or untarnished than any other. Reading has its disadvantages — and they are weighty ones: it exercises the soul, but during that time the body (my care for which I have not forgotten) remains inactive and grows earth-bound and sad. I know of no excess more harmful to me in my declining years, nor more to be avoided.

There you have my three favourite private occupations. I make no mention of the ones I owe to the world through my obligations to the state.

4. On diversion

===

[From personal experience Montaigne learnt that grief and pain cannot always be cured but can often be diverted into less anguished channels. In this the body plays a major part. The soul has to be watched: human beings are so made that they can be moved to ecstasies of anger by insubstantial dreams and raving lunacies. Quintilian's teaching that an orator first rouses an emotion in himself and then transfers it to his audience is accepted as proof of the power of wilful self-deception – a useful quality for a man who would divert his thoughts from pain, but also proof of the nothingness of Man.]

[B] Once I was charged with consoling a lady who was feeling distress – genuinely (mostly their mourning is affected and ritualistic):

> *Uberibus semper lachrimis, semperque paratis*
> *In statione sua, atque expectantibus illam,*
> *Quo jubeat manare modo.*

[*A woman has a reserve of abundant tears ever ready to flow, ever awaiting her decision to make them do so.*][1]

To oppose such suffering is the wrong way to proceed, for opposition goads the women on and involves them more deeply in their sadness; zeal for argument makes a bad condition worse. (We can see that from commonplace discussions: if anyone challenges some casual statement of mine I become all formal and wedded to it; more so if it is a matter of concern to me.) And then, by acting that way you set about your cure in a rough manner, whereas the first greetings which a doctor makes to his patient must be cheerful, pleasing and full of grace: nothing was ever achieved by an ugly uncouth doctor. So from the outset you must, on the contrary, encourage women's lamentations and show that they are justified and have your approval. This understanding between you will earn you the trust needed to proceed further; then you can glide down an easy and imperceptible slope to the more steadfast arguments appropriate for curing

1. Juvenal, *Satires*, VI, 272–4.

them. Personally, since my main desire was to escape from the bystanders who all kept their eyes on me, I decided in this difficult case to plaster over the cracks. And so I found out by experience that when it came to persuasion I was unsuccessful and heavy-handed: I either offer my arguments too pointedly and drily or else too brusquely, showing too little concern. After I had sympathized with her anguish for a while, I made no assay at curing it by powerful vigorous arguments (because I never had any, or perhaps because I thought I could achieve my effect better by another way); [C] and I did not start choosing any of the various methods which philosophy prescribes for consoling grief,[2] saying like Cleanthes for example that what we are lamenting is not an evil; nor did I say like the Peripatetics that it is but a light one; nor like Chrysippus that such plaints are neither just nor laudable; nor did I follow Epicurus' remedy (which is close neighbour to my own), that of shifting her mind away from painful thoughts to pleasant ones; nor did I attack her grief with the weight of all those arguments put together, dispensing them as required like Cicero: [B] but by gently deflecting our conversation and gradually leading it on to the nearest subject, and then on to slightly more remote ones depending on how she answered me, I imperceptibly stole her from her painful thoughts; and as long as I remained with her I kept her composed and totally calm.

I made use of a diversion. But those who came to help her after me found no improvement in her, since I had not set my axe to the root of the trouble.

[C] I have doubtless touched elsewhere on the kind of diversion used in politics.[3] And the practice of military diversions (such as those used by Pericles in the Peloponnesian Wars and by hundreds of others in order to tempt the enemy forces from their lands) is very common in the history books.

[B] It was an ingenious diversion by which the Sieur de Himbercourt saved himself and others in the town of Liège, which the Duke of Burgundy, who was besieging it, had obliged him to enter so as to draw up agreed terms of surrender. The citizens assembled for this purpose by night but began to rebel against what had previously been agreed; several decided to fall upon the negotiators whom they had in their power. He heard the rumble of the first wave of citizens who were coming to break

---

2. The following taken from Cicero, *Tusc. disput.*, III, xxxi, 77 (where Cicero alludes also to his own (now lost) *Consolatio* on the death of his daughter).
3. II, 23, 'On bad means to a good end'.

into his apartments, so he at once dispatched two of the inhabitants – there were several with him – bearing new and milder conditions to put before their town council; he had made them up for the occasion, then and there. These two men calmed the original storm and led that excited mob to the Hôtel de Ville to hear the terms they were charged with and to deliberate upon them. The deliberation was brief; whereupon a second storm was unleashed, as animated as the first; so he dispatched four new mediators similar to the first two, protesting that he now wanted to announce much more tempting conditions which would entirely please and satisfy them; by this means he drove the citizens back to their conclave. In short, by managing to waste their time that way he diverted their frenzy, dissipated it in vain deliberations and eventually lulled it to sleep until daybreak – which had been his main concern.[4]

My next story is in the same category. Atalanta was a maiden of outstanding beauty and wonderfully fleet of foot; to rid herself of a crowd of a thousand suitors all seeking to wed her, she decreed that she would accept the one who could run a race as fast as she could, provided that all those who failed should lose their lives. There were found plenty who reckoned the prize worth the hazard and who incurred the penalty of that cruel bargain. Hippomenes' turn to make an assay came after the others; he besought the goddess who protects all amorous passion to come to his aid. She answered his prayer by furnishing him with three golden apples and instructing him in their use. As the race was being run, when Hippomenes felt his lady pressing hard on his heels he dropped one of the apples as though inadvertently. The maiden was arrested by its beauty and did not fail to turn aside to pick it up.

> *Obstupuit virgo, nitidique cupidine pomi*
> *Declinat cursus, aurumque volubile tollit.*

[The maiden was seized by ecstasy and desire for the smooth apple: she turns from the race and picks up the golden ball as it rolls along.]

At the right moment he did the same with the second and the third apples, finally winning the race because of those distractions and diversions.[5]

When our doctors cannot purge a catarrh they divert it towards another part of us where it can do less harm. I have noticed that to be also the most usual prescription for illnesses of our soul: [C] '*Abducendus etiam non-nunquam animus est ad alia studia, solicitudines, curas, negotia; loci denique*

---

4. Related by Philippe de Commines, *Mémoires*, II, iii.
5. Ovid, *Metamorphoses*, X, 666–7 and context.

mutatione, tanquam ægroti non convalescentes, sæpe curandus est.' [The mind is often to be deflected towards other anxieties, worries, cares and occupations; and finally it is often cured (like the sick when slow to recover) by a change of place.]⁶ [B] Doctors can rarely get the soul to mount a direct attack on her illness: they make her neither withstand the attack nor beat it off, parrying it rather and diverting it.

The next example is too grand and too difficult; only the highest category of men can stop to take a pure look at the phenomenon itself, reflecting on it and judging it. It behoves none but Socrates to greet death with a normal countenance, training himself for it and sporting with it. He seeks no consolation not inherent to the deed: dying seems to him a natural and neutral event; he justly fixes his gaze upon it and, without looking elsewhere, is resolved to accept it. Whereas the disciples of Hegesias (who were excited by his beautiful discourses during his lectures and who starved themselves to death   [C]   in such quantities that King Ptolemy forbade him to defend such murderous doctrines in his School)   [B]   were not considering the dying as such and were definitely not making a judgement about it; it was not on dying that they fixed their thoughts: they had a new existence in view and were dashing to it.⁷ Those poor wretches to be seen on our scaffolds, filled with a burning zeal to which they devote, as far as they are able, all their senses – their ears drinking in the exhortations they receive, while their arms and their eyes are lifted up to Heaven and their voices raised in loud prayer full of fierce and sustained emotion – are certainly performing a deed worthy of praise and proper to such an hour of need. We must praise them for their faith but not strictly for their constancy. They flee the struggle; they divert their thoughts from it (just as we occupy our children's attention when we want to use a lancet on them). Some I have seen occasionally lowering their gaze on to the horrifying preparations for their death which are all about them: then they fall into a trance and cast their frenzied thoughts elsewhere.

Those who have to cross over some terrifyingly deep abyss are told to close their eyes or to avert them.

[C]   On Nero's orders Subrius Flavius was condemned to be put to

6. Cicero, *Tusc. disput.*, IV, xxxv, 74–5.
7. Ibid., I, xxxiv, 83–4. Hegesias the Cyrenaic's pupils who committed suicide are linked by Cicero to Cleombrotus Ambraciotes, who did so after reading Plato; his example is mentioned in II, iii, 'A custom of the Isle of Cea', and linked to St Paul's yearning to die so as to be with Christ.

death at the hands of Niger. Both were military commanders. When he was escorted to the field of execution he saw that the grave which Niger had ordered to be dug for him was uneven and shoddily made; turning to the soldiers about him he snapped, 'You could not do even this according to your military training!' And when Niger urged him to keep his head straight, he retorted, 'I hope you can strike as straight!' And he guessed right: Niger's arms were all a-tremble and he needed several blows to chop his head off. Now there was a man who did fix his attention directly on the object.[8]

[B] A soldier who dies in the melee, his weapons in his hand, is not contemplating death: he neither thinks of it nor dwells on it; he is carried away by the heat of battle. An honourable man that I know was struck to the ground after entering the lists to do battle; while he was down he felt his enemy stab him nine or ten times with a dagger. Everybody present yelled at him to make peace with his conscience, but he told me later that although their words touched his ears they did not get through to him; he had no thought but of struggling loose and avenging himself; and he did kill his man in that very fight.

[C] The soldier who brought news of his sentence to Lucius Silanus did him a great service; having heard Silanus reply that he was prepared to die but not at such wicked hands, the man rushed at him with his soldiers to take him by force, while he, all unarmed as he was, stoutly resisted with fists and feet. They killed him in the struggle. By his quick and stormy anger he destroyed the pain he would have felt from the long-drawn-out death awaiting him to which he had been destined.[9]

[B] Our thoughts are always elsewhere. The hope of a better life arrests us and comforts us; or else it is the valour of our sons or the future glory of our family-name, or an escape from the evils of this life or from the vengeance menacing those who are causing our death:

> *Spero equidem mediis, si quid pia numina possunt,*
> *Supplicia hausurum scopulis, et nomine Dido*
> *Sæpe vocaturum . . .*
> *Audiam, et hæc manes veniet mihi fama sub imos.*

[I hope that if the righteous deities can prevail you will drink the cup of my vengeance, driven on the rocks in the midst of the sea, constantly crying out the

---

8. Tacitus, *Annals*, XV, lxvii.
9. Ibid., XVI, ix.

name of Dido ... I shall hear it, and its fame will reach me in the deepest Underworld.][10]

[C]   Crowned in the victor's garland Xenophon was performing his sacrificial rites when he was told of the death of Gryllus his son at the battle of Mantinea. His first reaction to this news was to throw down his garland; but then, when he heard of the very valorous style of his son's death, he picked it up from the ground and placed it back on his head.

[B]   When he was dying, even Epicurus found consolation in the eternity and moral usefulness of his writings:[11]   [C]   *'Omnes clari et nobilitati labores fiunt tolerabiles'* [All labours are bearable which bring fame and glory]; and (says Xenophon) the identical wound and travail do not grieve a General as much as an Other Rank. Epaminondas accepted death much more cheerfully for being told that his side was victorious. *'Haec sunt solatia, haec fomenta summorum dolorum.'* [Such things bring solace and comfort to the greatest of sufferings.]

[B]   Other similar circumstances can divert and distract us from considering the thing in itself.   [C]   In fact the arguments of philosophy are constantly skirting the matter and dodging it, scarcely grazing the outer surface with its fingertips. The great Zeno, the leading figure in the leading school of philosophy which dominates all the others,[12] says this concerning death: 'No evil is to be honoured; death is honoured: therefore death is no evil'; and he says of drunkenness, 'No one confides his secrets to a drunkard; each man trusts the wise man: therefore the wise man will not be a drunkard.' Do you call that hitting the bull's-eye! I delight in seeing those first-rate minds unable to free themselves from fellowship with the likes of us! Perfect men though they may be, they always remain grossly human.

[B]   Vengeance is a sweet passion deeply ingrained in us by our nature; I can see that clearly, even though I have never experienced it. Recently, having to draw a young prince away from it, I did not start by saying that when anyone strikes you on one cheek you must, as a work of charity, turn the other,[13] nor did I draw a picture of the tragic results which poets attribute to that passion. I left vengeance aside and spent my time making

10. Virgil, *Aeneid*, IV, 382–4; 387; then, Diogenes Laertius, *Life of Xenophon*.
11. Cicero, *De finibus*, II, xxx, 96; then   [C]: *Tusc. disput.*, II, xxvi, 62 (twice); II, xxiv, 59.
12. Zeno was a Stoic; the following criticism of his arguments, from Seneca, *Epist. moral.*, LXXXII, 9, and LXXXIII, 9. Seneca considers them 'Greek absurdities'.
13. The ideal Christian reaction (Matthew 5:39), but not to be pressed at the wrong psychological moment.

him savour the beauty of the opposite picture: the honour, acclaim and goodwill he would acquire from clemency and bounty.

I diverted him towards ambition. That is how we get things done.

If when in love your passion is too powerful, dissipate it, they say. And they say truly: I have often usefully made the assay. Break it down into a variety of desires, one of which may rule as master if you like, but enfeeble it and delay it by subdividing it and diverting it, lest it dominate you and tyrannize over you:

> *Cum morosa vago singultiet inguine vena,*
> *Conjicito humorem collectum in corpora quæque.*

[When the peevish vein gurgles in your vagrant groin, ejaculate the gathered fluid into any bodies whatever.][14]

And see to it quickly, lest you find yourself in trouble once it has seized hold of you,

> *Si non prima novis conturbes vulnera plagis,*
> *Volgivagaque vagus venere ante recentia cures*

[unless you befuddle those first wounds by new ones, effacing the first by roaming as a rover through vagrant Venus.]

Once upon a time I was touched by a grief, powerful on account of my complexion and as justified as it was powerful. I might well have died from it if I had merely trusted to my own strength. I needed a mind-departing distraction to divert it; so by art and effort I made myself fall in love, helped in that by my youth. Love comforted me and took me away from the illness brought on by that loving-friendship. The same applies everywhere: some painful idea gets hold of me; I find it quicker to change it than to subdue it. If I cannot substitute an opposite one for it, I can at least find a different one. Change always solaces it, dissolves it and dispels it. If I cannot fight it, I flee it; and by my flight I made a diversion and use craft; by changing place, occupation and company I escape from it into the crowd of other pastimes and cogitations, in which it loses all track of me and cannot find me.

That is Nature's way when it grants us inconstancy; for Time, which she has given us as the sovereign doctor of our griefs,[15] above all achieves its ends by furnishing our power of thought with ever more different concerns,

14. Persius, *Satires*, VI, 73, linked to Lucretius, IV, 1062; then, Lucretius, 1063–4.
15. Cf. Erasmus, *Adages*, II, V, V, *Dies adimit aegritudinem*, citing Iphiclus, 'Time cures all our ills,' and Euripides on time as 'doctor' of men's problems.

so dissolving and breaking up the original concept however strong it may be. A wise man can see his dying friend scarcely less clearly after five-and-twenty years than after the first year, [C] and according to Epicurus not a jot less, for he attributed no lessening of our sufferings either to our anticipating them or to their growing old.[16] [B] But so many other thoughts cut across the first one that in the end it grows tired and weary.

To change the direction of current gossip Alcibiades lopped off the ears and tail of his beautiful dog and then chased it out into the square, so that by giving the populace something else to chatter about they would leave his other activities in peace.[17] I have known women too who have hidden their true affections under pretended ones, in order to divert people's opinions and conjectures and to mislead the gossips. But one I knew got well and truly caught: by feigning a passion, she quitted her original one for the feigned one. From her I learned that lovers who are well received ought not to consent to such mummery: since overt greetings and meetings are reserved for that decoy of a suitor, believe you me he will not be very clever if he does not eventually take your place and give you his. [C] That really is cobbling and stitching a shoe for another to wear.

[B] We can be distracted and diverted by small things, since small things are capable of holding us. We hardly ever look at great objects in isolation: it is the trivial circumstances, the surface images, which strike us – the useless skins which objects slough off,

> *Folliculos ut nunc teretes æstate cicadæ*
> *Linquunt.*

[such as those smooth eggshells which the cicadas cast off in summer.][18]

Even Plutarch laments his daughter by recalling her babyish tricks as a child.[19] We can be afflicted by the memory of a farewell, of a gesture of some special charm or a last request. Caesar's toga threw all Rome into turmoil – something which his death did not achieve. Take the forms of address which stay ringing in our ears – 'My poor Master'; or 'My dear friend'; or 'Dear papa' or 'My darling daughter': if I examine them closely when their repetition grips me, I discover that the grief lies in grammar and phonetics! What affects me are the words and the intonation (just as it

---

16. Cicero, *Tusc. disput.*, III, xv, 32.
17. Plutarch, *Life of Alcibiades*.
18. Lucretius, V, 801–2.
19. Plutarch (tr. Amyot), *consolation envoyée à sa femme*, 256 A; then, his *Life of Antony*.

is not the preacher's arguments which most often move a congregation but his interjections – like the pitiful cry of a beast being slaughtered for our use); during that time I cannot weigh the mass of my subject or penetrate to its real essence:

> *His se stimulis dolor ipse lacessit;*

> [With goads such as these grief wounds its own self;][20]

yet they are the foundations of our grief.

[C] The stubborn nature of my stones, especially when in my prick, has sometimes forced me into prolonged suppressions of urine during three or four days; they bring me so far into death that, given the cruelty of the strain which that condition entails, it would have been madness to hope to avoid dying or even to want to do so. (Oh what a past master of the art of torment was that fair Emperor who used to bind his criminals' pricks and make them die for want of pissing!)[21] Having got that far I would consider how light were the stimuli and the objects of my thought which could nurse a regret for life in me, and what minutiae served to construct in my soul the weight and difficulty of her departure; I would consider how frivolous are the images we find room for in so great a matter – a hound, a horse, a book, a wine-glass and what-not had their role in my loss. Others have their ambitious hopes, their money-bags or their erudition, which to my taste are no less silly. When I looked upon death as the end of my life, universally, then I looked upon it with indifference. Wholesale, I could master it: retail, it savaged me; the tears of a manservant, the distributing of my wardrobe, the known touch of a hand, a routine word of comfort discomforted me and made me weep.

[B] In the same way we disturb our souls with fictional laments; the plaints of Dido and Ariadne in Virgil and Catullus arouse the feelings of the very people who do *not* believe in them. [C] To experience no emotion from them is to be like Polemon (of whom that is told as a miracle) and to serve as an example of a hard and inflexible heart – but Polemon of course did not even blench when a mad dog chewed off his calf![22]

[B] By inquiry no wisdom can draw so close towards understanding the condition of a living, total grief but that it will be drawn closer still by physical presence, when ears and eyes (organs which can be stirred by inessentials only) can play their part.

20. Lucan, *The Civil War*, II, 42.
21. Tiberius, in Plutarch's *Life*.
22. Diogenes Laertius, *Life of Polemon*, IV, xxvii.

Is it right for the arts to serve our natural weakness and to let them profit from our inborn animal-stupidity? The orator (says Rhetoric) when acting out his case will be moved by the sound of his own voice and by his own feigned indignation; he will allow himself to be taken in by the emotion he is portraying. By acting out his part as in a play he will stamp on himself the essence of true grief and then transmit it to the judges (who are even less involved in the case than he is); it is like those mourners who are rented for funerals and who sell their tears and grief by weight and measure: for even though they only borrow their signs of grief, it is nevertheless certain that by habitually adopting the right countenance they often get carried away and find room inside themselves for real melancholy.

With several other of his friends I once had to escort the body of the Sieur de Gramont from La Fère, where he was killed in the siege, to Soissons.[23] I reflected that wherever we passed it was by the sheer display of the pomp of our procession that we filled the populace with tears and lamentations, since they had never even heard of his name!

[C] Quintilian says that he had known actors to be so involved in playing the part of a mourner that they were still shedding tears after they had returned home; and of himself he says that, having accepted to arouse grief in somebody else, he had so wedded himself to that emotion that he found himself surprised not only by tears but by pallor of face and by the stoop of a man truly weighed down by grief.[24]

[B] In a country place hard by our mountains the women play both priest and clerk, like Father Martin. They magnify their grief for their lost husbands by recalling their good and agreeable qualities but at the same time (to counterbalance this, it seems, and to divert their pitiful feelings towards contempt) they also list and proclaim all their failings – [C] with far better grace than we have when we lose a mere acquaintance and pride ourselves on bestowing on him novel and fictitious praises, turning him, once he is lost to sight, into something quite different from what he appeared to be when we used to see him – as though regret taught us something new and tears could lave our minds and bring enlightenment to them. Here and now I renounce any flattering eulogies you may wish to make of me, not because I shall not have deserved them but because I shall then be dead!

---

23. In 1580.
24. Quintilian, VI, ii – the standard view eventually challenged by Diderot in his *Paradoxe sur le comédien.*

[B]   If you ask that man over there, 'How does this siege concern you?' he will reply: 'I am concerned to give an example of routine obedience to my Prince; I do not expect to gain any benefit from it. And as for glory, I know what a small share of it can concern a private individual like me. I feel no passion; I make no claims.' Yet look at him the following morning; there he is, ready for the assault in his place in the ranks; he is entirely changed, boiling, flushed with yellow bile. What has sent this new determination and hatred coursing through his veins is the glint of so much steel, the flashes of our cannon and the din of our kettle-drums.

'A frivolous cause,' you will say. What do you mean, cause? To excite our souls we need no causes: they can be controlled and excited by some raving disembodied fancy based on nothing. When I throw myself into building castles in the air my imagination forges me pleasures and comforts which give *real* delight and joy to my soul. How often do we encumber our spirits with yellow bile or sadness by means of such shadows? And we put ourselves into fantastical rages, deleterious to our souls and bodies! [C]   What confused, ecstatic, madly laughing grimaces can be brought to our faces by such ravings! What jerkings of our limbs and trembling of our voices! That man over there is on his own, but does he not seem to be deceived by visions of a crowd of other men whom he has to deal with, or else to be persecuted by some devil within him?[25]

[B]   Ask yourself where is the object which produced such an alteration: apart from us men, is there anything in nature which is sustained by inanities or over which they have such power? Cambyses dreamt in his sleep that his brother was to become King of Persia; so he killed him — a beloved brother whom he had always relied on! Aristodemus, King of the Messenians, on account of an idea put into his head of some ill omen read into the howling of his dogs, killed himself. King Midas did the same, disturbed and worried by some unpleasant dream he had had.[26]

Abandoning your life for a dream is to value it for exactly what it is worth.[27] Listen [C] though [B] to our soul triumphing over her wretched body and its frailty, as the butt of all indispositions and degradations. A fat lot of reason she has to talk!

> *O prima infœlix fingenti terra Prometheo!*
> *Ille parum cauti pectoris egit opus.*

25. Montaigne is likening the ecstasy of battle to that of melancholy madness.
26. Plutarch (tr. Amyot), *De l'amitié fraternelle*, 88 E–F; *De la superstition*, 122 C–D; Ravisius Textor, *Officina, Fratrum et Sororum interfectores*.
27. Cf. Erasmus, *Adages*, II, III, XLVIII, *Homo bulla*.

Corpora disponens, mentem non vidit in arte;
Recta animi primum debuit esse via.

[O wretched clay which Prometheus first moulded! How unwisely he wrought!
By his art he arranged the body but saw not the mind. The right way would have
been to start off with the soul.][28]

28. Propertius, III, J, 7–10.

5. On some lines of Virgil

———

[Montaigne now breaks totally new ground. A concern for marriage and human sexuality was widespread in the Renaissance, partly because of the Reformation with its respect for marriage and the demands made on it, partly because of ferment within the Roman Catholic Church, the universities, legal and medical circles and among moralists. (A good example of such ferment in a comic setting is The Third Book of Pantagruel *by Rabelais.) Montaigne's achievement can be compared and contrasted with that of a friend of Rabelais, the great jurisconsult Andreas Tiraquellus in his ever-expanding Latin* Laws of Marriage. *But Montaigne is partly making a general confession; partly (for the first time ever) giving a self-portrait in which the sexual drive is openly portrayed; partly showing how old age may come to terms with dwindling physical potency yet powerful erotic dreams and memories. The development of sexuality in his own time from (in Montaigne's view) the courteous chastity of his father's days to his own youth with its tolerance of the courtly service of love in extramarital love-affairs (especially between young unmarried gentlemen and married ladies) to the brutality which he believed to mark French sexuality in his declining years was doubtless (if true) one of the results of the moral collapse brought about by the Wars of Religion. Montaigne, as usual, sees men and women as body-plus-'soul' (or 'spirit' or 'mind'). Love-affairs, primarily but by no means exclusively, concern the body. The love,* amour, *which Montaigne discusses here is not* amitié, *that loving-friendship proper to marriage at its best; after his own wedding he himself was much more loyal to his marriage-vows than he had ever dreamt possible. Virgil and Lucretius lead him to stress the poetry of erotic love and to contrast and compare it with the outspoken quasi-pornographic verses of the classical Priapics and their Renaissance imitators, who included religious leaders such as Beza. The chapter is marked by statements of anti-feminism and of jaundiced views of marriage: these are in fact often humorous in ways not always clear to modern readers. Medieval and Renaissance convention often made such attitudes comic or ironical: there is much of that here; but Horace is cited: 'What can stop us from telling the truth with a laugh!' Montaigne was warned before publication that his ironies might be taken seriously. That did not worry him: this is a self-portrait and he was indeed given to irony. But while Montaigne presents men and women as a case of 'us' and 'them', he frequently gives examples of men to support a statement of allegedly female vice or virtue, and of women to exemplify allegedly masculine ones. In Rabelais or Tiraquellus, men and women are almost different creatures, their sexual drives deriving from different causes and producing different effects (men being able to control their sexuality without risk to life and health, women not). Montaigne goes back to the very passage of Plato's* Timaeus *where doctors had for a millennium and a half found justification for that conviction and quietly shows that Plato made men and women equally subject to analogous*

sexual drives. The conclusion of Montaigne is an arresting one: women should be allowed more freedom: men and women share a common 'mould' – both have the common form of human kind. And that is nowhere more obvious than in our sexuality.

The element of confession in this chapter is emphasized by Montaigne's reminder that God sees through society's conventions and what are nowadays called taboos, seeing us not clad in evasive words but in the cankered nakedness of soul and body, 'with our tattered rags ripped off our pudenda.]

[B] The more our moral thoughts are abundant and solid the more engrossing they are and oppressive. Vice, death, poverty, illness are weighty subjects and they do indeed weigh on us. We need our Soul to be instructed in the means of sustaining evils and of fighting them off, instructed too in the rules of right-living and right-believing; and we need to awaken her to practise so fine an endeavour. But in the case of a soul of the common sort this must be done with moderation and some laxity: keep her continually tensed and you drive her mad. In my youth I needed to arouse myself and counsel myself if I were to remain dutiful: liveliness and good-health do not agree all that well, [C] they say, [B] with serious and sagacious discourse. Nowadays I am in a different state: the properties of old age give me too many counsels, making me wise and preaching at me. I have fallen from excessive gaiety into excessive seriousness which is more bothersome. That is why I deliberately go in for a bit of debauchery at times by employing my Soul on youngish wanton thoughts over which she can linger a while. From now on I am all too stale, heavy and ripe. Every day the years read me lectures on lack of ardour and on temperance. My body flees from excess: it is afraid of it. It is its turn now to guide my mind towards amendment of life. It is its turn now to act the professor, and it does so more harshly and imperiously. For one single hour, sleeping or waking, it never allows me to take time off from learning about death, suffering and penitence. I now defend myself against temperance as I used to do against voluptuousness. Now it is my body which pulls me back, to the point of numbness. Yet I want to be in every way master of myself. Wisdom has its excesses and has no less need of moderation than folly. So, fearing that in the intervals which my ills allow me, I may be desiccated, dried up and weighed down by wisdom –

mens intenta suis ne siet usque malis

[Lest my mind should dwell intensely on its ills][1]

1. Ovid, *Tristia*, IV, i, 4 (adapted); then, Petronius, *Satyricon*, 128.

I turn very gently aside and make my eyes steal away from such stormy, cloud-wracked skies as lie before me: which, thanks be to God, I can contemplate without terror but not without strain and effort; and I find myself spending my time recalling periods of my past youth:

> *animus quod perdidit optat,*
> *Atque in præterita se totus imagine versat*

[My mind prefers what it has lost and gives itself entirely over to by-gone memories.]

Let babes look ahead, old age behind: is that not what was meant by the double face of Janus?[2] The years can drag me along if they will, but they will have to drag me along facing backwards. While my eyes can still make reconnaissances into that beautiful season now expired, I will occasionally look back upon it. Although it has gone from my blood and veins at least I have no wish to tear the thought of it from my memory by the roots:

> *hoc est*
> *Vivere bis, vita posse priore frui.*

[To be able to enjoy your former life again is to live twice.][3]

[C] Plato tells old men to go and watch the exercises, dancing and sports of the young, to enjoy in others that beauty and suppleness of body which they have no longer and to recall to their memory the grace and privileges of those years of bloom; and he desires that they should award the victory in those sports to the young man who has given most joy and gladness to the greatest number of the old.

[B] Once upon a time I used to mark as exceptional the dark, depressing days: those days are now my routine ones; it is the ones which are beautiful and serene which are extraordinary now. I am close to the point when I shall jump for joy and accept anything which does not actually hurt as some new favour. Tickle myself I may, but I cannot force a laugh out of this vile body. I make myself delight in dream and fantasy so as to divert by ruse the chagrin of old age. But it would take a different remedy to cure it. What a feeble struggle of art against nature!

There is great silliness in extending by anticipation our human ills; I do

2. Janus, the god of the beginning of the year, had two faces, one looking back, the other forward (Ovid, *Fasti*, I, 345 etc.).
3. Martial, *Epigrams*, X, xxiii, 7; then, Plato, *Laws*, II, 657 D–E.

not want to be old before my time; I prefer to be old for a shorter one. I grab hold of even the slightest occasions of pleasure that I come across. I know from hearsay that there are several species of pleasure which are wise, strong and laudable; but rumour has not enough power over me to arouse an appetite for them in me. [C] I do not so much want noble, magnificent and proud pleasures as sweetish ones, easy and ready to hand: '*A natura discedimus; populo nos damus, nullius rei bono auctori*' [We are departing from what is natural, surrendering ourselves to the plebs who are never a good guide in anything.][4]

[B] My philosophy lies in action, in natural [C] and present [B] practice, and but little in ratiocination. Would that I could enjoy tossing hazelnuts and whipping tops!

Non ponebat enim rumores ante salutem

[Not for him did common report take precedence over his welfare.][5]

As a quality, pleasure-seeking is not very ambitious; of itself it reckons it is rich enough without bringing in the prize of reputation; it likes itself more in the shadows. If a man spends time savouring the tastes of wine and sauces when he is young, we ought to give him a good hiding. There is nothing I knew or valued less. I am learning about them now, I am ashamed to say: but what else can I do? I am even more ashamed and angry at the causes which drive me to it. It is for us to act the madman over trifles: young men ought to stand to their reputation and in the best places; youth is making its way forward in the world and seeking a name: we are on our way back. [C] '*Sibi arma, sibi equos, sibi hastas, sibi clavam, sibi pilam, sibi natationes et cursus habeant; nobis senibus, ex lusionibus multis, talos relinquant et tesseras.*' [Let them have their arms, their horses, their spears and their fencing-foils; let them toss balls and swim and race: and from the many pastimes let old men choose dice and knuckle-bones.][6] [B] The very laws send us back to our homes. The least I can do on behalf of this wretched state into which my age has thrust me is to furnish it, as we do childhood, with toys and playthings: for that is what we are declining into. Wisdom and folly both will have plenty to do if they are to support and succour me alternately in disastrous old age:

---

4. Seneca, *Epist. moral.*, CXIX, 17.
5. Cicero, *De officiis*, I, xxiv, 84: from lines of Ennius, the ancient Latin poet. In context the word *salutem* means not 'his welfare' but 'the safety' of the State.
6. Cicero, *De senectute*, XVI, 58.

Misce stultitiam consiliis brevem.

[Mix a little brief folly in your counsels.][7]

I similarly flee from the slightest pin-pricks: those which once would have scarcely scratched me now run right through me. My mode of being is beginning to like dwelling on the pain. [C] *'In fragili corpore odiosa omnis offensio est.'* [To a frail body every shock is vexatious.][8]

[B] *Mensque pati durum sustinet ægra nihil.*

[A mind that is ill can tolerate no hardships whatsoever.][9]

I have always been delicately sensitive to attacks of pain; I am more tender still now and in every way defenceless.

Et minimæ vires frangere quassa valent.

[The least shock will shatter a cracked vessel.]

My judgement prevents me from kicking and muttering against the indignities which Nature orders me to tolerate, but it does not stop me from feeling them. I would run from one end of the world to the other to seek a single twelve-month of gay and pleasant tranquillity: I have no other end but to live and enjoy myself. There is enough sombre and dull tranquillity for me now, but it sends me to sleep and dulls my brain: I can never be satisfied by it. If there is any man or any good fellowship of men in town or country, in France or abroad, sedentary or gadabout, whom my humours please and whose humours please me, they have but to whistle through their fingers and I'll come to them, furnishing them with 'essays' in flesh and blood.

Since it is the privilege of the mind to escape from old age I counsel it to do so with all my might: let it meanwhile sprout green and flourish, if it can, like mistletoe on a dead tree. But it is a traitor, I fear: it is so closely bound in brotherhood to the body that it is constantly deserting me to follow my body in its necessity. In vain do I try to divert it from this attachment; I set before it Seneca and Catullus and the ladies and their *dances royales*: but if its comrade has colic paroxysms it thinks it has them too! The very activities which are proper and peculiar to it cannot then

7. Horace, *Odes*, IV, xii, 27.
8. Cicero, *De senectute*, XVIII, 65.
9. Ovid, *Ex Ponto*, I, v, 18, and *Tristia*, IV, xi, 22.

raise it up: they too manifestly reek of snot. There is no alacrity about what the mind brings forth when there is none in its body at the very same time.

[C] *Magistri Nostri*[10] are wrong when they seek to explain the extra-ordinary transports of our spirit. Leaving aside the attribution of some of them to divine rapture, to love, to the harshness of war, to poetry, to wine, they do not allow the part played in them by good health, by boiling vigorous health, whole and idle, such as from time to time in former days my verdant years, so free from care, provided for me. That joyful fire gives rise to flashes in our spirit; they are lively and bright beyond our natural reach; they are some of our most lively enthusiasms, even though they are not the most frenzied. No wonder then if the opposite state overburdens my spirit, hammers it down and produces opposite results:

[B]   *Ad nullum consurgit opus, cum corpore languet.*

[No task can make it struggle to its feet: it languishes with the body.][11]

Furthermore my spirit wants me to be beholden to it for its allegedly showing much less complicity in all this than is usually the practice among men. Let us at least drive away ills and hardships from our human intercourse while we are enjoying a truce:

Dum licet, obducta solvatur fronte senectus.

[So while it can, let old age smooth away the wrinkles on its brow.]

'*Tetrica sunt amoenenda jocularibus.*' [Gloomy thoughts should be made pleasant by jests.] I like the kind of wisdom which is gay and companion-able; I fly from grating manners and from sourness; I am suspicious of grim faces.

[C]   *Tristemque vultus tetrici arrogantiam;*

[The sad arrogance of a gloomy face;]

[B]   *Et habet tristis quoque turba cynaedos.*

[And buggers too are found in groups of sombre men.]

---

10. The title of university professors, especially theologians. Here they are explain-ing various forms of ecstasy and rapture.
11. Pseudo-Gallus, I, 125; then, Horace, *Epodes*, XIII, 7; Bishop Caius Sollius Apollinaris (Sidonius), *Epist.*, I, ix; George Buchanan, *Joannes Baptista* (prologue); Martial, *Epigrams*, VII, lvii, 8.

[C] I wholeheartedly believe Plato when he says that great portents of the goodness or evil of a soul are easy or difficult humours. Socrates had a set expression but a serene and laughing one: it was not set as was that of the aged Crassus who was never known to laugh.[12] [B] As a quality virtue is pleasing and gay. I know that few of those who will glower at the unrestrained freedom of my writings do not have greater cause to glower at the unrestrained freedom of their thoughts. I am certainly in harmony with their sentiments: it is their eyes I offend! What a well-ordered mind that is which can gloss over the writings of Plato burying all knowledge of his alleged affairs with Phaedo, Dion, Stella and Archeanassa! *'Non pudeat dicere quod non pudeat sentire.'* [Let us be not ashamed to say whatever we are not ashamed to think.][13]

[B] I loathe a morose and gloomy mind which glides over life's pleasures but holds on to its misfortunes and feeds on them – like flies which cannot get a hold on to anything highly polished and smooth and so cling to rough and rugged places and stay there; or like leeches which crave to suck only bad blood.[14] I have moreover bidden myself to dare to write whatever I dare to do: I am loath even to have thoughts which I cannot publish. The worst of my deeds or qualities does not seem to me as ugly as the ugly cowardice of not daring to avow it. Everybody is circumspect about confessing, whereas they ought to be circumspect about doing: daring to do wrong is to some extent counterweighted and bridled by the courage needed to confess it. [C] Any man who would bind himself to tell all would bind himself to do nothing which we are forced to keep quiet about. God grant that my excessive licence may draw men nowadays to be free, rising above those cowardly counterfeit virtues which are born of our imperfections, and also grant that I may draw them to the pinnacle of reason at the expense of my own lack of moderation! If you are to tell of a vice of yours you must first see it and study it. Those who conceal it from others usually do so from themselves as well: they hold that it is not sufficiently hidden if they can see it, so they disguise it and steal it from their own moral awareness. *'Quare vitia sua nemo confitetur? Quia etiam nunc in illis est; somnium narrare vigilantis est.'* [Why does nobody confess his

---

12. Cicero, *Tusc. disput.,* III, xv, 31; Ravisius Textor, *Officina* (for both Socrates and Crassus): *Severissimi et maxime tetrici.*

13. Source not identified.

14. Plutarch (tr. Amyot), *De la tranquillité de l'âme,* 73 H (for the flies); *Du banissement, ou de l'exil,* 125 A B (for the leeches).

faults? Because even now he remains within them: only after men have awakened can they relate their dreams.][15]

The body's ills become clearer as they grow bigger: we discover that what we called a sprain or a touch of rheumatism is the gout. But as the soul's ills grow in strength they are wrapped in greater obscurity: the more ill a man is, the less he realizes it. That is why the maladies of the soul need to be often probed in daylight, cut and torn from our hollow breasts by a pitiless hand. What applies to the benefactions we receive applies to the evils that we do: sometimes the only way to requite them is to acknowledge them. Is there some ugliness in our wrong-doing which dispenses us from the duty of acknowledging it?

[B] I suffer such pains whenever I dissemble that I avoid being entrusted with another man's secret, having no mind to deny what I know. I can keep quiet about it but I cannot deny it without strain and unease. To be really able to keep a secret you need to be made that way by nature, not doing so because you are under bond. When serving princes it is not enough to keep a secret: you need to be a liar as well. To the man who inquired of Thales of Milesia whether he should deny on oath that he had been a lecher I would have replied that he should not do so, for lying has always seemed worse to me than lechery. Thales gave quite different advice, telling him to swear the oath so as to cloak a bad vice by a lesser one. Yet this counsel means not so much choosing between vices as increasing their number.[16]

Be it said *en passant* that if you present a man of conscience with the need to weigh an awkward situation against a vice he can easily strike the right bargain, but if you imprison him between two vices you oblige him to make a harsh choice – as happened to Origen who had either to commit idolatry or submit to being carnally assaulted by an ugly great Ethiopian paraded before him. He suffered the first alternative. Wrongly, it is said. So those women nowadays who protest to us that they would rather have ten lovers on their conscience than a single Mass would – by their false standards – not be making a bad choice.[17]

---

15. Seneca, *Epist. moral.*, LIII, 8; he continues: 'Similarly a confession of one's evils is proof of a healthy mind'; Montaigne then develops LIII, 6.
16. Erasmus, *Apophthegmata*, VII, *Milesii Thaletis*, VII. (Erasmus is also puzzled by this counsel.)
17. Nicephoros Callistos Xanthopoullos, *Ecclesiastical History*, V, who asserts that Origen uselessly damned his soul by this act. Montaigne compares Origen's choice to that of those women of the Reformed Church (the 'Calvinists'), who would rather consent to commit fornication than consent to the 'idolatry' of the

There may be a lack of discretion in publishing one's defects this way but there is no great danger of it becoming customary by example, for Ariston said that the winds which men most fear are those which uncover them.[18] We must truss up those silly rags which cover over our morals. Men dispatch their consciences to the brothels and regulate their appearances. Even traitors and murderers are wedded to the laws of etiquette and dutifully stick to them. Yet it is not for injustice to complain of discourtesy [C] nor for wickedness to complain of indiscretion. It is a pity that a wicked man should not also be a boor and that his vice should be palliated by politeness. Such stucco belongs rightly to good healthy walls which are worth whitening or preserving.

[B] As a courtesy to the Huguenots who damn our private auricular confession I make my confession here in public, sincerely and scrupulously. St Augustine, Origen and Hippocrates publicly admitted the error of their opinions; I do more; I include my morals.[19] I hunger to make myself known. Provided I do no truly I do not care how many know it. Or, to put it better, I hunger for nothing, but I go in mortal fear of being mistaken for another by those who happen to know my name. If a man does all for honour and glory what does he think he gains by appearing before the world in a mask, concealing his true being from the people's knowledge? If you praise a hunchback for his fine build he ought to take it as an insult. Are people talking about *you* if they honour you for valour when you are really a coward? They mistake you for somebody else. It would amuse me as much if such a person were to be gratified when men raised their caps to him, thinking that he was the master of the band when he was merely one of the retainers. When King Archelaus of Macedonia was going along the street somebody threw water over him. His entourage wanted to punish the man. 'Ah yes,' he replied, 'but he never threw it at me but at the man he mistook me for.'[20] [C] When somebody told Socrates that people were gossiping about him he said, 'Not at all. There is nothing of me in what they are saying.'[21] [B] In my case, if a man were

---

Roman Catholic mass, which was indeed often assimilated by their ministers (in Old Testament terms) to 'whoremongering after strange gods'. All theologians of all Churches agreed that physical sins are far, far less serious than spiritual ones.

18. Plutarch (tr. Amyot), *De la curiosité*, 64 C–D.
19. This may well imply that Montaigne had never read the *Confessions* of St Augustine, though he knew the *City of God* in detail.
20. Erasmus, *Apophthegmata*, V, *Archelaus*, V.
21. Diogenes Laertius, *Life of Socrates*.

to praise me for being a good navigator, for being very proper or very chaste I would not owe him a thank you. Similarly, if anyone should call me a traitor, a thief or a drunkard I would not think that it was me he attacked. Men who misjudge what they are like may well feed on false approval: I cannot. I see myself and explore myself right into my inwards; I know what pertains to me. I am content with less praise provided that I am more known.   [C]   People might think that I am wise with the kind of wisdom which I hold to be daft.

[B]   It pains me that my *Essays* merely serve ladies as a routine piece of furniture – something to put into their *salon*. This chapter will get me into their private drawing-rooms; and I prefer my dealings with women to be somewhat private: the public ones lack intimacy and savour.

When saying our goodbyes we feel warmer affection than usual for whatever we are giving up. I am taking a last farewell of this world's sports: these are our final embraces. But now let us get round to my subject.

The genital activities of mankind are so natural, so necessary and so right: what have they done to make us never dare to mention them without embarrassment and to exclude them from serious orderly conversation? We are not afraid to utter the words *kill*, *thieve* or *betray*; but those others we only dare to mutter through our teeth. Does that mean that the less we breathe a word about sex the more right we have to allow it to fill our thoughts?

[C]   It is interesting that the words which are least used, least written and the least spoken are the very ones which are best known and most widely recognized. No one of any age or morals fails to know them as well as he knows the word for bread. They are printed on each one of us without being published; they have no voice, no spelling. It is interesting too that they mean an act which we have placed under the protection of silence, from which it is a crime to tear it even to arraign it and to judge it. We dare not even flog it except by periphrasis and similitude. A criminal is greatly favoured if he is so abominable that even the laws think it illicit to touch him or to see him: he is freed by the beneficence of his condemnation and saved by its severity. Is it not the same concerning books, which become more saleable and publicized once they are suppressed? Personally I intend to take Aristotle's advice literally: he says that coyness serves as an ornament in youth and a defect in old age.[22]

22. Aristotle, *Nicomachaean Ethics*, IV, ix, 1128 b. (His term, aidōs, covers modesty, bashfulness and shamefacedness. It keeps young men in check: old men should not need it, since they should do nothing shameful.)

[B] In the school of the Ancients – the school I cling to far more than to the modern, [C] its virtues seeming greater to me and its vices less – [B] they preach these words at you:

> [B] *Ceux qui par trop fuyant Venus estrivent*
> *Faillent autant que ceux qui trop la suivent.*

[Those who excessively strive to flee from Venus fail just like those who follow her excessively.][23]

> *Tu, Dea, tu rerum naturam sola gubernas,*
> *Nec sine te quicquam dias in luminis oras*
> *Exoritur, neque fit lætum nec amabile quicquam.*

[Thou alone, O goddess, rulest over the totality of nature; without thee nothing comes to the heavenly shores of light, nothing is joyful, nothing lovable.]

I do not know who managed to make Pallas and the Muses fall out with Venus and chill their ardour for Cupid;[24] yet I can find no deities who become each other more or who owe more to each other. Anyone who removed their amorous thoughts from the Muses would rob them of the most beauteous entertainment they provide and of the noblest subject-matter of their works; and anyone who made Cupid lose contact with poetry and its services would weaken him by depriving him of his weapons. In that way we charge both the god of sexual relationships and of tenderness, and the tutelary goddesses of elegance and justice, with the vices of ingratitude and churlishness.

I have been struck off the roll of Cupid's attendants but not for so long that my memory is not still imbued with his powers and his values:

> *agnosco veteris vestigia flammæ.*

[I can recognize the tracks of my former passions.][25]

There are still some traces of heat and emotion after the fever,

> *Nec mihi deficiat calor hic, hiemantibus annis*

[And let me not lack that warmth in my winter years.]

23. Plutarch (tr. Amyot), *Qu'il fault qu'un Philosophe converse avec les Princes*, 134 C; then, Lucretius, I, 6 and 23–4.
24. Among others Joachim Du Bellay regretted that Ronsard devoted so much time and genius to love-poetry; cf. *Regrets*, XXIII.
25. Virgil, *Aeneid*, IV, 23; then, Johannes Secundus, *Elegies*, III, 29; Tasso, *Gierusalemme liberata*, XII, 63–6; Juvenal, *Satires*, VI, 196.

All gross and dried up as I am, I can still feel some lukewarm remnants from that bygone ardour:

> *Qual l'alto Ægeo, per che Aquilone o Noto*
> *Cessi, che tutto prima il vuolse et scosse,*
> *Non s'accheta ei pero: ma'l sono e'l moto,*
> *Ritien de l'onde anco agitate è grosse.*

[As the Aegean sea when the North Wind and the South have dropped, which first had whipped and churned it up, does not at once grow calm but retains the roar and surge of the waves, huge still and thrashing.]

To the best of my knowledge the powers and values of that god are found more alive and animated in poetry than in their proper essence:

> *Et versus digitos habet.*

[Poetry has playful fingers too.]

Poetry can show us love with an air more loving than Love itself. Venus is never as beautiful stark naked, quick and panting, as she is here in Virgil:

> *Dixerat, et niveis hinc atque hinc diva lacertis*
> *Cunctantem amplexu molli fovet. Ille repente*
> *Accepit solitam flammam, notusque medullas*
> *Intravit calor, et labefacta per ossa cucurrit.*
> *Non secus atque olim tonitru cum rupta corusco*
> *Ignea rima micans percurrit lumine nimbos.*
> *. . . Ea verba loquutus,*
> *Optatos dedit amplexus, placidumque petivit*
> *Conjugis infusus gremio per membra soporem.*

[Venus fell silent; and as he hesitates she encircles him in her snow-white arms and warms him in her soft embrace. Soon he was welcoming the accustomed flame; its well-known heat struck him to the marrow and coursed through the bones of his trembling limbs. It was like unto the brilliant lightning which, with a thunder-clap, flashes through the clouds ... He spoke to her, gave her the embraces that she yearned for, and then his limbs sought quiet repose as he lay flowing around his wife's bosom.][26]

What I find worth stressing is that Virgil in these lines portrays her as a little too passionate for a married Venus. Within that wise contract our

---

26. These are the lines of Virgil alluded to in the chapter heading (*Aeneid*, VIII, 387–92; 404–6). Cf. below, note 99.

sexual desires are not so madcap; they are darkened and have lost their edge. Cupid hates that couples should be held together except by himself, and only slackly comes into partnerships such as marriage which are drawn up and sustained by different title-deeds. In marriage, alliances and money rightly weigh at least as much as attractiveness and beauty. No matter what people say, a man does not get married for his own sake: he does so at least as much (or more) for his descendants, for his family. The customary benefits of marriage go way beyond ourselves and concern our lineage. That is why I like the practice of having marriages arranged at the hands of a third party rather than our own, not by our own judgement but by someone else's. How contrary all that is to amorous compacts. Moreover there is a kind of lewdness (as I think I have said already) in deploying the rapturous strivings of Love's licentiousness within such a relationship, which is sacred and to be revered.[27] Aristotle says that we should approach our wives wisely and gravely for fear lest we unhinge their reason by arousing them too lasciviously. What he says for our moral sense the doctors say for our health's sake, namely that too hot, voluptuous and unremitting a pleasure is deleterious to the sperm and impedes conception.[28] They go on to say that in the case of the kind of intercourse which is feeble by nature (as the married kind is) we should undertake it rarely, at stated intervals, so as to fill it with a just and fruitful heat,

> *quo rapiat sitiens venerem interiusque recondat.*

[by which the mare avidly seizes on Venus' seed and buries it deep inside her.][29]

I know no marriages which fail and come to grief more quickly than those which are set on foot by beauty and amorous desire. Marriage requires foundations which are solid and durable; and we must keep on the alert. That boiling rapture is no good at all.

Those who think to honour marriage by associating passion with it are like those (it seems to me) who to promote virtue hold rank to be none other than a virtue: there is some cousinship between rank and virtue but great differences as well; there is no gain in confusing their names and title-deeds: we wrong them both by confounding them that way. Noble rank is a beautiful quality and was rightly instituted; but, since it is a quality

27. Then a commonplace of traditional Christian morality.
28. Cf. Andreas Tiraquellus, *De legibus connubialibus*, XV, 23 ff.; but the reference to Aristotle is puzzling.
29. Virgil, *Georgics*, III, 137.

dependent on others and can fall to a vicious man of naught, it is well
below virtue in esteem. It is a 'virtue' – if indeed it be one – which is
artificial and visible, dependent on time and fortune, differing in style in
various countries; it lives, yet is mortal, having no more origin than the
river Nile. Genealogical and not individual, it depends on succession; it is
drawn from sequency – and a feeble sequency at that! Knowledge, fortitude,
goodness, beauty, riches, indeed all other qualities, are subject to communication and sharing; rank is self-devouring and of no utility in the service of
others. It was explained to one of our kings that a choice had to be made
between two candidates for the same office: one of them was a nobleman,
the other certainly not. He commanded that they should choose, irrespective
of rank, the man with the greater merit; but should they prove to be of
exactly equal worth, they should in that case take rank into account. That
was to assign to it its just importance. When a young unproven man asked
Antigonus for the position held by his father (a valiant man who had just
died), he replied: 'My friend, in such promotions I do not so much have
regard for the rank of my soldiers as for their prowess.'[30]

[C] It really should not be done as it was for the office-holders of the
kings of Sparta – trumpeters, minstrels and cooks – who were succeeded in
their charges by their sons, no matter how ignorant they might be, taking
precedence over men best skilled at the craft.[31] The people of Calicut make
their nobility into a species higher than Man. Marriage is forbidden them,
as is any profession but war. They can have their fill of concubines, and
their women may have as many studs; jealousy is unknown between them;
but it is an unforgivable crime punishable by death to lie with anyone of a
different rank; they feel defiled if they are even touched by them as they go
by; and since their noble state is marvellously polluted and tainted by it,
they slaughter those who draw even a little too close to them; the
untouchables are therefore forced to cry out at street corners as they walk
along, like gondoliers in Venice, to avoid colliding. And persons of rank
can order them to get out of their way whenever they want to. By such
means the nobility avoid a disgrace which they consider indelible; the
others avoid certain death. No stretch of time, no princely favour, no
office, valour or wealth can entitle a commoner to become a nobleman.
This is reinforced by their custom of forbidding marriages across trades: a
woman descended from cobblers cannot marry a woodworker; and parents

30. Erasmus, *Apophthegmata*, V, *Antigonus Secundus*, IV.
31. Herodotus, VI, lx.

are under the obligation of training their sons for their father's calling – exactly that one: no other will do. By such means they maintain permanent distinctions in their lot.[32]

[B] A good marriage (if there be such a thing) rejects the company and conditions of Cupid: it strives to reproduce those of loving-friendship. It is a pleasant fellowship for life, full of constancy, trust and an infinity of solid useful services and mutual duties. No wife who has ever savoured its taste –

optato quam junxit lumine tæda

[whom the marriage-torch has joined with its long-desired light][33]

– would ever wish to be the beloved mistress of her husband. If she is lodged in his affection as a wife then her lodging is far more honourable and secure. Even when he is swept off his feet with passion for another, just ask him whether he would prefer some disgrace to befall his wife or his mistress; whose misfortune would grieve him more? for which of them he desires the greater respect? In a healthy marriage such questions admit of no doubt. The fact that one sees so few good ones is a token of its value and price. Shape it and accept it rightly and there is no more beautiful element in our society. We cannot do without it yet we go and besmirch it, with the result that it is like birds and cages: the ones outside despair of getting in: the ones inside only care to get out. [C] When Socrates was asked whether it was more appropriate to take or not to take a wife, he replied, 'Whichever you do you will be sorry.'[34] [B] It is a contractual engagement to which can be exactly applied the proverb: Man is god or wolf to Man. Many elements have to coincide to construct it. In our times it is considered to be more rewarding for those with uncomplicated everyday souls which are not so troubled by frivolity, curiosity and sloth. Roving humours such as mine which loathe all forms of tie or bond are not so proper for it:

Et mihi dulce magis resoluto vivere collo.

[For me too it is sweeter far to live with no chain about my neck.][35]

32. Montaigne's account of the Hindu caste-system is based on Simon Goulart's *Histoire du Portugal*, II, iii.
33. Catullus, LXIV, 79.
34. Erasmus, *Apophthegmata*, III, *Socratica*, XL; then, *Adages*, I, I, LXIX, *Homo homini lupus*, and I, I, LXX, *Homo homini Deus*.
35. Pseudo-Gallus, I, 61.

By my own design I would have fled from marrying Wisdom herself if she would have had me. But no matter what we may say, the customs and practices of life in society sweep us along. Most of my doings are governed by example not choice. Nevertheless I did not, strictly speaking, invite myself to the feast: I was led there, brought to it by external considerations.

There is nothing so awkward – in fact nothing at all, no matter how ugly, vitiated or repugnant – but can become bearable under certain conditions and in certain circumstances, so vain is our human situation. When I was borne into marriage I was less broken in and more recalcitrant than I am now that I have made an assay at it. And, womanizer though I am held to be, I have, in truth, more rigidly observed the laws of matrimony than I ever vowed or hoped. It is no longer the time for kicking over the traces once they have tied your legs together! We should tend our freedom wisely; but once we have submitted to the marriage-bond we must stay there under the laws of our common duty (or at least strive to). The actions of those husbands who accept the bargain and then show hatred and contempt are harsh and unjust. Equally unfair and intolerable is that fine counsel which I see passed from hand to hand among our women:

> *Sers ton mary comme ton maistre,*
> *Et t'en guarde comme d'un traistre.*

[Serve him like a master: watch him like a traitor.]

That is a challenge and a call to battle, meaning, 'Act towards him with a constrained respect, hostile and suspicious.'

I am too easy-going for such prickly designs. To tell the truth I have yet to attain to that perfect intellectual elegance and cunning which confound right and injustice and which ridicule any rule and order which may not accord with my desires. Just because I loathe superstition I do not go straightway mocking religion. Though we may not always do our duty we must always at least love and acknowledge it. [C] To take a wife without espousing her is treachery.

[B] Let us get on.

Our poet Virgil portrays a marriage full of concord and harmony, in which however there is not much fidelity. Did he mean to say that it is not impossible to surrender to the attacks of Cupid and yet nevertheless to keep a sense of duty towards one's marriage; that one may injure marriage without tearing it totally apart? [C] A valet can diddle his master without hating him!

[B]   A wife may be attracted to an unknown man by beauty, opportuneness and destiny – for destiny plays its role in it:

> *fatum est in partibus illis*
> *Quas sinus abscondit: nam, si tibi sidera cessent,*
> *Nil faciet longi mensura incognita nervi*

[The privy parts hidden in your toga are fated: if the stars forsake you, it will do you no good to have a tool of unprecedented size][36]

yet she may not be so totally attracted that there remain no bonds still holding her to her husband. We are dealing with two projects which each go their own distinct separate ways. A wife may give herself to another man whom – not because of the state of his finances but because of his very personality – she would never wish to marry. Few men have married their mistresses without repenting of it.   [C]   That even applies to the other world! What a wretched household, that of Jupiter and a wife whom he had seduced and had enjoyed having little affairs with! That, as the saying goes, is shitting in the basket and then plonking it on your head.

[B]   I have in my time seen a highly placed love-affair shamefully and dishonourably cured by a marriage. The motives of both are quite distinct. We can, without difficulty, love two very different and incompatible things.

Isocrates said that the City of Athens was pleasing in the same way as a mistress served for love: all men took pleasure in spending their time and walking with her, but no man loved her well enough to wed her (that is, to make his home and habitation there).[37]

It has angered me to see husbands hating their wives precisely because they are doing them wrong: at very least we should not love them less when the fault is ours; at very least they ought to be made dearer to us by our regrets and our sympathy.

Isocrates meant that, while the ends were different, they were in certain circumstances not incompatible. For its part marriage has usefulness, justice, honour and constancy: a level but more universal pleasure. A love-affair

---

36. Juvenal, *Satires,* IX, 32–4.
37. Isocrates, the pupil of Gorgias and the friend of Plato. I do not know what Montaigne drew upon for his saying, unless it be a confused memory of Zeno and Cleanthes' reasons for not becoming citizens of Athens (Plutarch tr. Amyot), *Contredicts des Stoïques,* 561 F).

is based on pleasure alone: and in truth its pleasure is more exciting, lively and keen: a pleasure set ablaze by difficulties. It must have stabs of pain and anguish. Without darts and flames of desire Cupid is Cupid no longer. In marriage the ladies are so lavish with their presents that they dull the edge of our passion and desire.   [C]   You merely need to see the trouble that Lycurgus and Plato give themselves in order to avoid this incongruity.

[B]   Women are not entirely wrong when they reject the moral rules proclaimed in society, since it is we men alone who have made them. There is by nature always some quarrelling and brawling between women and men: the closest union between us remains turbulent and tempestuous. In the opinion of our poet we treat women without due consideration. That is seen by what follows.

We realize that women have an incomparably greater capacity for the act of love than we do and desire it more ardently – and we know that this fact was attested in Antiquity by that priest who had been first a man and then a woman:

> *Venus huic erat utraque nota.*
>
> [He knew Venus from both angles.][38]

Moreover we have learned from their own lips such proof as in former ages was provided by an Emperor and Empress of Rome, both infamous past masters on the job: he managed to deflower ten captive Sarmatian virgins in one night, but she in one night furnished the means of five-and-twenty engagements, changing her partners according to her needs and preferences:

> *adhuc ardens rigide tentigine vulve,*
> *Et lassata viris, nondum satiata, recessit.*

[at last she retired, inflamed by a cunt stiffened by tense erections, exhausted by men but not yet satisfied.]

Then there was that plea lodged in Catalonia by a wife as plaintiff against her husband's excessively assiduous love-making: not I think because she was actually troubled by it (except within the Faith I believe in

---

38. Tiresias, who changed sex; a frequently cited example: cf. Ovid, *Metamorphoses*, III, 323; then, Juvenal, *Satires*, VI, 128–9; the Emperor was Proculus; the Empress, Messalina, the consort of Claudius. (Cf. Tiraquellus, *De legibus connubialibus*, IX, 94 for Messalina, and XV, 92 for Proculus.)

no miracles) but rather to have a pretext for pruning back and curbing the authority of husbands over their wives even in the very deed which forms the basic act of marriage, and also to show that the nagging and spitefulness of wives extend over the marriage-bed and trample under heel the sweet delights of Venus. Her husband, a really depraved brute of a fellow, made the rejoinder that even on days of abstinence he could not manage with less than ten times. Whereupon intervened that notable judgement of the Queen of Aragon: after mature deliberation in her counsel that good Queen (wishing to provide for all time an example of the moderation required in a proper marriage and a measuring-rod for temperance) ordained that it is necessary to limit and restrict intercourse to six times a day – sacrificing much of women's needs and surrendering many of their desires in order to establish a scale which would be unexacting and therefore durable and unchanging.[39] At which the doctors exclaim: 'If that is the rate assessed by a reasoned moral reformation, what must be the lusts and the appetites of women?' [C] Just think of the disparity of judgements on our appetites: Solon, the head of the school of lawgivers, with the aim of avoiding failure, sets the rate for such conjugal intimacy at three times a month.[40]

We believe all that and teach all that. And then we go and assign sexual restraint to women as something peculiarly theirs, under pain of punishments of the utmost severity. No passion is more urgent than this one, yet our will is that they alone should resist it – not simply as a vice with its true dimensions but as an abomination and a curse,[41] worse than impiety and parricide. Meanwhile we men can give way to it without blame or reproach.

Those men who have made an assay at overcoming it, employing purely material remedies to cool down the body, to weaken it and to subdue it, have adequately vouched for the difficulty, or rather the impossibility, of achieving it. Yet we men on the other hand want our wives to be in good health, energetic, radiant, buxom . . . and chaste at the same time, both hot and cold at once.

As for marriage (which has the duty, we say, of stopping them from burning)[42] it brings them but little respite given our manners: if they do

---

39. Tiraquellus, *De legibus connubialibus*, XV, 1.
40. Plutarch (tr. Amyot), *De l'amour*, 612 C; Tiraquellus, XV, 83.
41. Perhaps an echo of Leviticus 26.
42. St Paul, I Corinthians 7:9; then, Martial, *Epigrams*, XII and Diogenes Laertius, *Life of Polemaon*.

take a husband in whom the vigour of youth is still a-boil he will boast of scattering it elsewhere:

> *Sit tandem pudor, aut eamus in jus:*
> *Multis mentula millibus redempta,*
> *Non est hæc tua, Basse; vendidisti.*

[A little more propriety, please, or I'll take you to law. I paid a few thousand for your cock. It is not yours now, Bassus: you sold it to me.]

[C] And Polemon the philosopher rightly received a legal summons from his wife because he scattered on a barren field the fruitful seed he owed to her fertile one.

[B] If they take one of those broken-down husbands, there they are, fully wed yet worse off than virgins and widows. (We assume that they are furnished with all they need because they have a man about the place, just as the Romans assumed that a Vestal Virgin called Clodia Laeta had been raped simply because Caligula had made an approach to her, even though it was proved that he had done no more than that.)[43] Their needs are then not satisfied but increased, since their ardour, which would have remained calm in their single state, is awoken by contact with any male company whatsoever. That explains why those monarchs of Poland, Boleslaus and Kinge his consort, agreed together to take the vow of chastity on their very wedding-day as they lay side by side, maintaining it in the teeth of the pleasure which marriage offers: such considerations and circumstances made their chastity more meritorious.[44]

We train women from childhood for the practices of love: their graces, their clothes, their education, their way of speaking regard only that one end. Those in charge of them impress nothing on them but the face of love, if only to put them off it by continually portraying it to them. My daughter − I have no other children − is of an age when the more passionate girls are legally allowed to marry. She is slender and gentle; by complexion she is young for her age, having been quietly brought up on her own by her mother; she is only just learning to throw off her childish innocence. She was reading from a French book in my presence when she came across the name of that well-known tree *fouteau* [a beech].[45] The

---

43. Clodia Laeta was buried alive. The emperor was in fact Caracalla.
44. Jan Herburt, *Histoire des Roys de Pologne*, 1573.
45. *Fouteau* evoked *foutre*, then the usual vulgar word for 'to have sexual intercourse', a meaning almost submerged by other usages in modern French.

woman she has for governess pulled her up short rather rudely and made her jump over that awkward ditch. I let her be, so as not to interfere with women and their rules, for I play no part at all in that sort of education: feminine polity goes its own mysterious way: we must leave it entirely to them. But unless I am mistaken the company of twenty lackeys would not in half a year have imprinted on her mind an understanding of what those naughty syllables mean, how they are used and what they imply, as did that good old crone by her one reprimand and prohibition.

> *Motus doceri gaudet Ionicos*
> *Matura virgo, et frangitur artubus*
> *Jam nunc, et incestos amores*
> *De tenero meditatur ungui.*

[The marriageable maiden loves to learn the steps of the Ionic dance; she twists her limbs and from a tender age trains herself for unchaste loves.][46]

Just let them dispense with a little ceremony and become free to develop their thoughts: in knowledge of such things we are babes compared with them. Just listen to them describing our pursuit of them and our rendezvous with them. They will soon show you that we contribute nothing but what they have known and already assimilated independently of us. [C] Could Plato be right when he said that in a former existence girls had been lascivious boys![47]

[B] I happened to be one day in a place where my ear could unsuspectedly catch part of what they were saying to each other. I wish I could tell you! 'By our Lady,' I said, 'let us go, after this, and study the language of Amadis and tales in Boccaccio and Aretino so as to appear sophisticated.' What a good use of our time! There is no word, no exemplary tale and no stratagem which women do not know better than our books do. The doctrines which nature, youth and good health (those excellent schoolmasters) ceaselessly inspire in their souls are born in their veins:

> *Et mentem Venus ipsa dedit.*

[Venus herself inspired their frenzy.][48]

They do not need to learn them: they give birth to them.

46. Horace, *Odes*, III, vi, 21–4.
47. Cf. Plato, *Timaeus*, 42 B–C.
48. Virgil, *Georgics*, 42 B–C.

> *Nec tantum niveo gavisa est ulla columbo*
> *Compar, vel si quid dicitur improbius,*
> *Oscula mordenti semper decerpere rostro,*
> *Quantum præcipue multivola est mulier.*

[Never did white dove nor any more lascivious bird which you could name invite love's kisses with its pecking beak as much as a woman yearning for a host of men.][49]

If the ferocity of their desires were not somewhat reined in by that fear for their honour with which all women are endowed, we would all be laughing-stocks. The whole movement of the world tends and leads towards copulation. It is a substance infused through everything; it is the centre towards which all things turn. We can still read some of the ordinances made by that wise Rome of old to regulate love-affairs, as well as Socrates' precepts for the education of courtesans.[50]

> *Nec non libelli Stoici inter sericos*
> *Jacere pulvillos amant.*

[And there are little books which love to lie strewn about in silken cushions: some of them are Stoic ones.]

There are enactments among Zeno's Laws covering penetration and opening up for deflowering.[51]   [C] I wonder what was the drift of that book by Strato the philosopher entitled *On carnal knowledge*;[52] what did Theophrastus treat of in those books of his which bore the titles *The Lover* and *On Love-affairs*; and what did Aristippus treat in his work *On Antique Delights*? What was Plato's intention in his long and vivid descriptions of the most controversial love-affairs of the day? Then there are *The Book of the Love-maker* by Demetrius Phalereus; *Cliniasor, or the Lover Raped*, by Heraclides of Pontus; *On Marriage: or How to make Children*, and another, *On Master and Lover*, by Antisthenes; *On Amorous Exploits* by Ariston; two by Cleanthes, *The Art of Loving* and *On Love*; *Lovers' Dialogues* by Sphaerus;

---

49. Catullus, LXVI, 125–8.
50. For the laws of Rome, cf. Tiraquellus, *De legibus connubialibus*, XIII, 12 ff. But I do not know what Socrates' precepts were. Then, Horace, *Epodes*, VIII, 15–16 (adapted).
51. Cf. Plutarch (tr. Amyot), *Propos de table*, III, question 6, p. 384 C (blaming Zeno).
52. All these books are lost. (Cf. Tiraquellus, *De legibus connubialibus*, XV, 91.)

The Fable of Jupiter and Juno, intolerably pornographic, by Chrysippus, with his *Fifty Lecherous Letters*. And I am not counting the writings of philosophers who followed the Epicurean School. [B] In bygone days fifty gods were tied to this job; and a nation was discovered who kept male and female prostitutes in their temples all ready to be enjoyed, so as to lull to sleep the lusts of those who came to worship there. [C] '*Nimirum propter continentiam incontinentia necessaria est; incendium ignibus extinguitur.*' [Sexual excesses are doubtless needed for sexual restraint, as fire is doused by fire.][53]

[B] In most parts of the world that member of our male bodies was turned into a god.[54] In a single province some peeled off the skin and consecrated part of it as an oblation while others offered up their sperm and consecrated it. In another province the youths bored holes through it in public, prised gaps between the flesh and the skin and then threaded through them the longest thickest skewers which they could stand. They afterward made a bonfire of those skewers as an offering to their gods, and if they were stunned by the violence of the ferocious pain they were reckoned unchaste and lacking in vigour. Elsewhere the revered symbol of the most hallowed magistrate was the sexual organ; and in many processions an effigy of it was borne in pomp, in honour of a variety of gods. During the feast of Bacchus the ladies of Egypt wore such an effigy about their necks; it was of wood, exquisitely fashioned and as big and heavy as each could manage. In addition the statue of their god had a carved member which was bigger than the rest of his body. The married women near my place twist their headscarves into the shape of one to revel in the enjoyment they derive from it; then on becoming widows they push it back and bury it under their hair. The wisest of the Roman matrons were granted the honour of offering crowns of flowers to the god Priapus; when their maidens came to marry, they were required to squat over its less decent parts.[55]

I even wonder whether I have not seen in my own lifetime practices recalling similar devotions: what was the sense of that silly flap on our fathers' flies which you can still see worn by our Swiss guards?[56] Why do

53. Tertullian (known to Montaigne only at second-hand?); cf. Villey, *Sources et évolution des 'Essais' de Montaigne*, p. 256.
54. Cf. Coelius Richerius Rhodiginus, *Antiquae lectiones*, VII, xvi, *Dionysiorum ritus. Qui sunt phalle. Phallogogia Sacra.*
55. St Augustine, *City of God*, VI, ix; note of J. L. Vives on this passage.
56. The codpiece.

we parade our genitals even now behind our loose-breeches, and, what is worse, cheat and deceive by exaggerating their natural size? [C] I would like to believe that such styles of clothing were invented in better and more moral times so that people should in fact not be deceived, each man gallantly rendering in public an account of his endowments; the more primitive peoples do still display it somewhere near its real size. In those days they supplied details of man's working member just as we give the measurements of our arm or foot.

[B] That fine fellow who when I was young castrated so many beautiful ancient statues in his City so as not to corrupt our gaze,[57] [C] following the counsel of that other fellow in Antiquity:

> – *Flagitii principium est nudare inter cives corpora*

[Baring the body among our citizens is the beginning of shameful deeds] –

[B] ought to have recalled that (as in the mysteries of the *Bona Dea* in which all signs of the male were banned) nothing is achieved unless you also geld horses, donkeys and finally everything in nature:

> *Omne adeo genus in terris hominumque ferarumque,*
> *Et genus æquoreum, pecudes, pictæque volucres,*
> *In furias ignemque ruunt.*

[All species on earth, both man and brute, and dwellers in the sea, and flocks and painted birds, all dash madly into the flames of desire.][58]

[C] The gods, says Plato, have furnished men with a rebellious and tyrannical member which tries to force everything to submit to its appetite like an animal on the rampage. So too the women have an animal, avid and greedy: if you deny it in due season, it becomes frenzied and can brook no delay; its own raging madness is inhaled into their bodies; it stops all respiration by blocking up the tubes, so causing hundreds of kinds of illness which last until after it has drawn inwards with its breath the product of our common desire and scattered it broadcast, planting it in the ground of the womb.[59]

---

57. Just possibly Pope Paul IV; then, Cicero, *Tusc. disput.*, IV, xxxiii, 70, citing Ennius.
58. Virgil, *Georgics*, III, 242–4. *Bona Dea* (the Good Goddess) was worshipped by Roman women as the patron of fertility and chastity. No man might enter her temple.
59. Plato made both men and women subject to sexual organs which were deaf to reason. Ancient medical writers isolated the women in this context, with the result

[B]   Now that that lawgiver of mine[60] ought also to have recalled that it is perhaps a more chaste and fruitful practice to bring women to learn early what the living reality is rather than to allow them to make conjectures according to the licence of a heated imagination: instead of our organs as they are their hopes and desires lead them to substitute extravagant ones three times as big.   [C]   And one man I know lost out by exposing his somewhere while they were still unready to perform their most serious task.

[B]   What great harm is done by those graffiti of enormous genitals which boys scatter over the corridors and staircases of our royal palaces! From them arise a cruel misunderstanding of our natural capacities. [C]   Who knows whether that explains why Plato decreed (following the practice of other states with sound institutions) that both men and women, old and young, should appear naked before each other during exercises in the gymnasia?[61]   [B]   Those Indian women who see their men in the nude have at least cooled off their visual senses.

[C]   The women of that great Kingdom of Pegu wear below the belt nothing but a kirtle slit in the front and so tight that, no matter what formal decency they may seek to preserve, they reveal everything they have got with every step they take. They maintain that this fashion was created in order to attract the men to them and to distract them from that taste for males to which that nation has entirely surrendered. Yet it could be said that they lose more than they gain and that a complete hunger is more cruel than one where at least the eyes are satisfied.[62]   [B]   Livia said, moreover, that to a moral woman a naked man means no more than a statue.   [C]   And the women of Sparta, who as wives were more virginal than our daughters, saw every day the young men of their city take everything off for their exercises; they themselves were not very particular

---

that women – but not men – were, on the highest medical authority, for centuries thought to be subject to an irrational 'animal' (the womb), the frustrations of which could cause a form of hysteria ('womb-disease') all but indistinguishable from death. Rabelais makes this medical belief central to his doctor's judgement on women in the *Tiers Livre du Pantagruel*, XXXII. Montaigne, unlike Rabelais, shows great independence of mind by going back to Plato himself (*Timaeus*, 91 B–C), so putting men and women essentially on a par, sexually speaking, both being subject to the irrational demands of their genitalia (which were defined in both sexes as 'animals' in accordance with criteria long accepted by doctors).
60. The 'fine fellow' who put fig leaves on the Roman statues.
61. Plato, *Republic*, V, 452.
62. G. Balbi, *Viaggio del' Indie*, then, for Livia, Dion Cassius, *Life of Tiberius*.

about keeping their thighs covered as they went about, believing, says Plato, that they were sufficiently veiled with virtue without needing a 'virtue-guard'.[63] Yet Saint Augustine is our witness for there once having been men who attributed such wonderful powers of temptation to nudity that they doubted whether, at the General Resurrection, women would rise again as women rather than in our sex so as not to go on tempting us in that blessed state![64]

[B] In short we bait and lure women by every means. We are constantly stimulating and overheating their imagination. And then we gripe about it.

Let us admit it: there is hardly one of us who is not more afraid of the disgrace which comes to him from his wife's immorality than from his own; hardly one who is not so amazingly charitable that he worries more about his dear wife's conscience than he does about his; hardly one who would not rather commit theft and sacrilege – or that his wife were a murderer or a heretic – than to have her be no chaster than he is.

And our women would much rather volunteer to go and earn their fees in the law-courts or their reputations on the battlefield than to have to mount so difficult a guard in the midst of idle pleasures. Our women can see, can they not, that there is no merchant, no barrister, no soldier who does not drop what he is doing so as to hurry and get on with 'the job' – no porter or cobbler either, however weary with toil or faint with hunger.

> *Num tu, que tenuit dives Achœmenes,*
> *Aut pinguis Phrygiæ Mygdonias opes,*
> *Permutare velis crine Licinniæ,*
> *   Plenas aut Arabum domos,*

> *Dum fragrantia detorquet ad oscula*
> *Cervicem, aut facili sævitia negat,*
> *Quæ poscente magis gaudeat eripi,*
> *   Interdum rapere occupet?*

[Would you really exchange – even for all the wealth of Achaemenes or all the riches of Mygdon, King of fertile Phrygia, or the treasure-boxes of Araby – a

---

63. The *vertugade* (farthingale) was a structure worn beneath the skirts. Obviously, it 'got in the way'. Montaigne therefore derives *vertugade* from '*virtue-guard*'.
64. St Augustine, *City of God*, XXII, xvii; St Paul (Romans 8:29) teaches that God will raise Christians from the dead to be 'conformed to the image of His Son'. Augustine denies that this means that all Christians, male and female, will arise again as males.

single one of Licinnia's tresses when she bends her neck towards you for a fragrant kiss or when, with sweet severity, she denies what she in fact desires far more than you do, and will soon be snatching from you?]⁶⁵

[C] We do not weigh the vices fairly in our estimation. Both men and women are capable of hundreds of kinds of corrupt activities more damaging than lasciviousness and more disnatured. But we make things into vices and weigh them not according to their nature but our self-interest: that is why they take on so many unfair forms. The ferocity of men's decrees about lasciviousness makes the devotion of women to it more vicious and ferocious than its characteristics warrant, and engages it in consequences which are worse than their cause.

[B] I am not even sure that the campaigns of Caesar and Alexander surpass the stern resolve of a beautiful young woman, brought up our way, in the light of society's social norms and battered by numerous examples to the contrary, who, in the midst of hundreds of unending and forceful suitors yet remains pure. No attaining so bristles with difficulties as her abstaining; nor is any more active. I think it easier to keep on a suit of armour all your life than to keep a maidenhead. And so the vow of virginity is the noblest of all the vows and also the harshest.   [C]   As Saint Jerome says, *'Diaboli virtus in lumbis est.'* [The Devil's power is in the loins.]⁶⁶

[B] We have certainly assigned to the ladies the most exacting and arduous of human duties and we let them have all the glory. It ought to serve them as a singular goad to help them stubborn it out that this is a subject in which they can challenge that vain pre-eminence in virtue and valour which men claim over them and can trample it underfoot. If they take care over it, they will find that not only are they most highly thought of but also better loved. No gentleman abandons his suit because he is refused, provided that the refusal is based on chastity not on preference for another. In vain do we swear oaths and make menaces and lamentations. We lie. We love them all the better for it. There is no lure like wise conduct when not brusque and glowering. There is cowardice and a lack of feeling in stubbornly continuing despite loathing and contempt: but when up against a constant and virtuous resolve mingled with an appreciative good-will it is an exercise fit for a noble and magnanimous soul.

65. Horace, *Odes*, II, xii, 21–8 (the text is corrected from the posthumous printed editions of Montaigne).
66. St Jerome, *Contra Jovinianum*, II – a work so rhetorically hostile to marriage that Erasmus prefaced it with an 'Antidote'.

They can, up to a point, show their appreciation of our courtship and make us realize that in all honour they do not disdain us. [C] For that rule which ordains that they must detest us because we worship them and hate us because we love them is indeed cruel, if only for the hardship it causes. Why should ladies not lend an ear to our requests and offers of service provided we do not go beyond the bounds of propriety, and why do we go on assuming that their doing so suggests some inner licentiousness of thought? A Queen in our own days wittily said that to exclude such advances was a sign of frailty and an indication of one's own levity, adding that no lady who had not been tempted could boast of her chastity.

[B] The boundaries of honour are by no means so narrowly drawn. There are means of being relaxed and showing some initiative without infringing them. Along its frontiers there is a stretch of neutral territory where a woman is free to show some discretion. If a man has been able to pursue her honour and to bring it to bay in its own corner of its fortress, then he is a silly fellow if he is not satisfied with his fortune. The prize of victory is valued for its difficulty. Do you want to know what impact your courtship and your merits have had on her heart? Measure it by her morals. Some women grant much who grant little: it is entirely in relation to the will of the one who grants it that we judge her gratitude for a kindness. The other attributes which apply to love's favours are fortuitous and are deaf and dumb. That little which one lady grants you costs her more than it costs her companions to grant you her all. If rarity is worth esteeming in anything it must be so in this case: do not consider the smallness of the favour but the small number of those who receive it. Money is valued according to its stamp and hallmark. Whatever some men may be brought to say by frustration and bad judgement at the height of their distress, truth and virtue always regain the advantage.

I have known ladies whose reputation was unjustly compromised over a long period but who, without careful planning, were later restored to the unanimous esteem of mankind by their constancy alone. Everybody is sorry and denies what he once believed. After being young women who were just a little suspect they now hold the foremost rank among good and honoured noblewomen. When someone said to Plato, 'They are all gossiping about you,' he said, 'Let them. I will so live that I will compel them to change their style.'[67] But apart from the fear of God and the winning of

67. Source unknown. Cf. (not very close!) Plutarch (tr. Amyot), *Comment on pourra recevoir utilité de ses ennemis*, 110 EF.

the prize of so rare a glory (which must incite women to protect themselves) the corrupt state of our century drives them to do so; and if I were in their place there is nothing I would not do rather than commit my reputation to such dangerous hands. In my day the pleasure of telling of an affair (a pleasure scarcely less delightful than having one) was conceded only to such as had one single faithful friend; nowadays the most usual talk at table and when men get together turns to boasting about favours received and the secret bounties of the ladies, who really do show abject baseness of mind to allow such tender gifts to be thus cruelly hunted, grabbed and plundered by men so ungrateful, so indiscreet and so inconstant.

It is our exaggerated and improper harshness towards this vice which gives birth to jealousy, the most vain and turbulent distemper which afflicts our human souls:

> *Quis vetat apposito lumen de lumine sumi?*

[Whatever stops us lighting one torch from another's light?][68]

> *Dent licet, assidue, nil tamen inde perit.*

[They can go on giving, on and on: they lose nothing in the process.]

Jealousy and Envy her sister seem to me to be the most absurd of the bunch. About envy I can say virtually nothing: that passion which is portrayed as so powerful and violent has no hold on me (and I thank her for it). As for Jealousy, I know her – by sight at least. Beasts can feel it too. When the shepherd Crastis fell in love with a nanny-goat her billy charged him while he lay asleep, butting his head and smashing it.[69]

We have raised the temperature of jealousy's fevered climax, following in that some of the Barbarian nations. The better educated nations have been touched by jealousy – that is reasonable – but not caught away by it:

> *Ense maritali nemo confossus adulter*
> *Purpureo stygias sanguine tinxit aquas.*

[There, never did adulterer stain with his blood the waters of Styx while he lay pierced by a husband's sword.][70]

Lucullus, Caesar, Pompey, Antony, Cato and other fine men were all cuckolds and knew it: they never made a commotion about it. In those

68. Ovid, *Ars amandi*, III, 93; then a verse from the *Priapeia*.
69. A tale related, after Aelianus, by Coelius Richerius Rhodiginus, *Antiquae Lectiones*, XXV, xxxii.
70. Johannes Secundus, *Elegiae*, I, vii, 71–2.

days there was only one man who died of distress over it: Lepidus; and he was a fool:[71]

> *Ah! tum te miserum malique fati,*
> *Quem attractis pedibus, patente porta,*
> *Percurrent mugilesque raphanique.*

[Ah! You wretched man caught out on the job! They will bind your legs together and stuff mullet and Greek radishes up your back passage.]

But when that god in our poet[72] surprised one of his comrades lying with his wife he was satisfied with exposing them both to shame;

> *atque aliquis de Diis non tristibus optat*
> *Sic fieri turpis!*

[but one of the other gods, not the most severe, wished he was shamed as well!]

And that did not stop him from being inflamed by the sweet kisses she gave him as she lamented that, for so little a thing, she had begun to doubt his love for her:

> *Quid causas petis ex alto, fiducia cessit*
> *Quo tibi, diva, mei?*

[Why, my goddess, do you seek such far-fetched arguments? Have you lost your faith in your husband?][73]

More. She begs him a favour for one of her bastards –

> *Arma rogo genitrix nato*

[I, a mother for her son, am begging you for his armour]

– and it is generously granted to her, Vulcan speaking honourably of Aeneas:

> *Arma acri facienda viro.*

[Arms must be forged for such a man.]

---

71. Plutarch, *Life of Pompey* (Lepidus intercepted a love-letter and died of grief); then, Catullus, XV, 17–19. (For this use of mullet to punish adulterers, cf. Juvenal, *Satires*, X, 317.)
72. Vulcan, in the verse of Virgil cited, p. 958; then, Ovid, *Metamorphoses*, IV, 187–8.
73. Virgil, *Aeneid*, VII, 395–6; then, VIII, 383; VIII, 441.

Humane kindness surpassing humankind! And I do agree that we can leave such excessive bounty to the gods.

Nec divis homines componier æquum est

[Nor is it right to compare men to deities.][74]

As for the confounding of children, [C] apart from the fact that the gravest of lawgivers want it and legislate for it in their republics,[75] [B] it does not affect the women, yet it is precisely in them that jealous passion is somehow more at home.

Sæpe etiam Juno, maxima cœlicolum,
Conjugis in culpa flagravit quotidiana.

[Even Juno, the greatest goddess among the dwellers in heaven, feels the scourge of jealousy over her consort's daily wrongs.][76]

When jealousy seizes hold of the feeble, defenceless souls of such women it is pitiful to see how it bowls them over and cruelly tyrannizes them. It slips into them, under the title of loving affection: but as soon as it gets possession of them, those same causes which served as a basis for benevolence now serve as a basis for deadly hatred. [C] Of all the spiritual illnesses, jealousy is the one which has more things which feed it and fewer things which cure it. [B] The manly virtue, the health, the merit and the reputation of their husbands then kindle the flames of their wives' maleficent frenzy:

Nullæ sunt inimicitiæ, nisi amoris, acerbæ.

[No hatreds so bitter than those of love.][77]

It is a feverish passion which turns all that is beautiful in them ugly and corrupts what is good; in a jealous woman, no matter how chaste and thrifty she may be as a wife, there is nothing which does not reek of bitterness and savagery. It is an insane perturbation which drives them to the other extreme, to the contrary of what causes it.

An interesting example of this was a man called Octavius in Rome. After lying with Pontia Posthumia, his delight in it so increased his love that he persistently begged her to marry him. When he could not win her

74. Catullus, LXVIII, 141.
75. Above all, Plato and, presumably, those who follow him.
76. Catullus, LXVIII, 138–9.
77. Propertius, II, viii, 3.

over, his extreme love hurled him headlong into deeds of most cruel and mortal hatred; and he killed her.[78]

Similarly the regular symptoms of this kind of love-sickness are domestic discord, plottings and conspiracies –

> *notumque furens quid fœmina possit*
>
> [we all know what a woman's rage can do][79]

– and a fury which is all the more gnawing for being compelled to justify itself by loving affection.

Now the duty of chastity is wide-ranging. What is it that we want women to bridle? Their wills? But the will is a seductive and active quality: it is too quick to let itself be restrained. Supposing their dreams sometimes so hold them in pawn that they cannot redeem them? It is not in their power to protect themselves from sexual desire and lust – not even perhaps in the power of Chastity herself: she is a woman. So if our sole concern is with their will, where do we stand? Just think of the press of assignations if a man were to have the privilege of being borne on wings (with no eyes to see him and no tongue to gossip) to the lap of every woman who would have him!

[C] The Scythian women used to poke out the eyes of all their slaves and prisoners of war in order to avail themselves of them more freely and secretly.[80]

Oh, what a mad advantage lies in the opportune moment! If anyone were to ask me what is the first quality needed in love I would reply: knowing how to seize an opportunity. It is the second and the third as well. It is the factor which can achieve anything. I have often lacked good fortune but also occasionally lacked initiative. God help those who can mock me for it! In our days you need to be more inconsiderate – which our young men justify under the pretence of ardour; but if women looked into it closely they would find that it arises rather from lack of respect. I myself devoutly feared to give offence and am always inclined to respect whomever I love. Besides in this sort of business if you remove the respect

---

78. Tacitus, *History*, IV, xliv.

79. Virgil, *Aeneid*, V, 6.

80. A misunderstanding of Herodotus, IV; cf. Plutarch (tr. Amyot), *Que la vertu se peult enseigner et apprendre*, 399: the Scythian women blinded slaves to stop them from stealing milk. (Montaigne had certainly read this passage, the following sentence of which concerning Iphicrates he used in I, 40 'Reflections on Cicero'.)

you dowse the lustre. I like a lover to play the timid youth serving his lady. Not in this situation precisely but in other ones, I do have something of that awkward shyness which Plutarch speaks of;[81] the course of my life has been in varying ways bespattered and harmed by it. It is a quality which ill becomes my overall character: but then, what are we but dissension and discord?

I am as sensitive about giving a refusal as receiving one, and my eyes show it. It so weighs on me to weigh on others that when duty forces me to assay the intentions of a man in a matter of doubt which could cost him some bother I hold back and skimp it. But if it concerns my own interests – [C] though Homer says truly that in a beggar shyness is a stupid virtue[82] – [B] I usually charge a third person to blush in my stead. I find it equally difficult to deny those who ask a service of me: I have occasionally had the will to refuse but not the capacity.

It is therefore madness to assay restraining [C] so blazing [B] a desire, so natural to women. And when I hear them boasting that their very wills are coldly chaste and virginal I laugh at them: that really is backing away too far. It may still not be credible, but there is at least some appearance of plausibility in the case of a toothless old hag or a young girl wasted by consumption. But women who are still alive and breathing worsen the terms of the bargain by saying so, since ill-advised excuses serve as accusations. Like one of the gentlemen in my neighbourhood who was suspected of impotence:

> *Languidior tenera cui pendens sicula beta*
> *Nunquam se mediam sustulit ad tunicam.*

[whose tiny dagger, drooping like a flabby parsnip, never stuck halfway up his underwear.][83]

Two or three days after his wedding, to prove his masculinity he went about boasting that he had ridden his wife twenty times the previous night. That was cited later to convict him of absolute ignorance and to annul the marriage.

Besides, those women are saying nothing worthwhile: for where there is no struggle there is neither continence nor virtue. 'That is true,' they should say, 'but I have no intention of giving way.' The very saints put it thus.

81. Plutarch sees it as the sign of a good marriage (*Les preceptes de mariage*, 146 A).
82. Homer, *Odyssey*, XVII, 347, cited by Plato (*Charmides*, 161 A).
83. Catullus, LXVII, 21–2.

I am of course talking of women who seriously boast of their cold chastity and indifference, who keep a straight countenance and want us to believe what they say. For when they put on a studied countenance (with eyes which belie their looks) and make their profession with cant phrases which imply the contrary to what they say, I like that. I am the obedient servant of naïve frankness: nevertheless I cannot refrain from saying that, unless it is absolutely innocent and childlike, it does not become a lady and is inappropriate to courtship: it at once slips into provocativeness. Women's affectations and grimaces deceive only idiots. Lying is then in the seat of honour: it is a diversion which brings us to the right truth through the wrong door.

Now if we cannot bridle their thoughts, what is it we want from women? Action? But plenty of their actions which corrupt chastity escape the knowledge of others:

> *Illud sæpe facit quod sine teste facit.*

> [She often does it without testes to testify.][84]

Such actions as we fear the least are perhaps the most to be feared: silent sins are the worst:

> *Offendor mæcha simpliciore minus.*

> [A straightforward whore offends me less.]

[C] And then there are actions by which women can lose their maidenheads without their maidenhood – and, what is more, without their knowing it: '*Obstetrix, virginis cujusdam integritatem manu velut explorans, sive malevolentia, sive inscitia, sive casu, dum inspicit, perdidit.*' [Sometimes the obstetrician while examining with her fingers whether the hymen is intact, has ruptured it – by ignorance or malice or bad luck.][85] Some maidens have lost their maidenhead while feeling for it: others have ruptured it while out riding.

[B] We could never delimit precisely what are the actions we forbid to them. We must frame our law in vague general terms.

The very ideal which men forge of their chastity is ridiculous: among the most extreme models of it that I know are Fatua the wife of Faunus,

---

84. Martial, *Epigrams*, VII, lxi, 6; then, VI, vii, 6.
85. St Augustine, *City of God*, I, xviii (stressing that modesty is a matter of the mind not the body).

who after her wedding never let herself be seen by any man whatever, and the wife of Hiero, who never realized that her husband's breath stank, thinking that it was a quality common to all men.[86]

To satisfy us they have to be invisible and insensate.

So now let us admit that the crucial element in judging this duty in women lies mainly in the intention. There have been husbands who have suffered adultery not only without feeling reproach or hostility for their wives but specifically bound to acknowledge their virtue. Many a woman who loved her honour more than her life has nevertheless prostituted herself to the insane lusts of a deadly enemy in order to save her husband's life, doing for him what she would never have done for herself. This is not the place to dwell on such *exempla*. They are too splendid and sublime to be rehearsed in the light of this chapter: let us keep them for a nobler place.

[C] But to give some examples here which do shine with a more vulgar light, are there not wives who daily lend their bodies to others solely to help on their husbands – and with their express command and pandering? In ancient times, for ambition's sake, Phaulius of Argos offered his wife to King Philip;[87] so too when Galba was entertaining Maecenas to dinner he noticed that his wife and his guest were beginning to ogle and to make signs and advances to each other, so he slipped down on his cushions and acted like a man heavy with sleep in order, for hospitality's sake, to lend a hand to their arrangements. And he let this be known, not without some elegance: for when the wine-steward ventured to reach out for the wine-jars on the table he shouted: 'Can you not see, you dolt, that I have only fallen asleep for Maecenas?'

[B] A woman may behave loosely yet have a will which she has reformed more than another whose conduct is hidden by a more orderly appearance: just as we know of women who complain that they were dedicated to chastity before the age of discretion, I know of some who sincerely complain that, before the age of discretion, they were dedicated to debauchery. Vicious parents may be the cause, or the force of necessity which is a cruel counsellor.

In the East Indies, although chastity is singularly valued there, custom

---

86. Fatua's case was a commonplace; Plutarch tells of Hiero's wife (*Comment on pourra recevoir utilité de ses ennemis*, 111 D–E). So does Tiraquellus, *De legibus connubialibus*, IV, 1.

87. Plutarch (tr. Amyot), *De l'amour*, 606 E–F; then, for Galba, 606 D–E and Erasmus, *Apophthegmata*, VI, *Varie mixta*, LVIII.

suffers a married woman to give herself to any man who presents her with an elephant – and not without glory for being so highly prized.[88]

[C] A man of good family, Phaedo the philosopher, when his country of Elis was captured, professionally prostituted his youthful beauty (as long as it lasted) to anyone who would pay for it, so as to earn his living.[89] And Solon, they say, was the first legislator in Greece to give women the right to provide for the necessities of life at the expense of their modesty, a practice which Herodotus however says was accepted earlier by several polities.

[B] Then what do we hope to gain from such painful disquiet: for however justified the jealousy we still have to see whether that passion enraptures us to any purpose! Is there one man who believes that he is clever enough to buckle up his women?

> *Pone seram, cohibe; sed quis custodiet ipsos*
> *Custodes? Cauta est, et ab illis incipit uxor!*

[Lock her up; shut her in. But who will guard your guardians? Your wife is clever: she will start with them!]

In so ingenious a century any occasion will suffice.

Curiosity is always a fault; here it is baleful. It is madness to want to find out about an ill for which there is no treatment except one which makes it worse and exacerbates it; one the shame of which is spread abroad and augmented chiefly by our jealousy; one which to avenge means hurting our children rather than curing ourselves. You wither and die while hunting for such hidden truth. How wretched are those husbands in my days who manage to find out!

If the man who warns you of it does not also at once supply a remedy and his help, his warning is noxious, deserving your dagger more than if he called you a liar. We mock the husband who cannot put things right no less than the one who knows nothing about it. Cuckoldry has an indelible stamp: once a man is branded with it he has it for ever; chastising cuckoldry emphasizes it more than the defect. A fine thing to tear our private misfortunes from the shadow of doubt and trumpet them abroad like tragedians on the trestles – especially misfortunes which hurt only when they are related. Marriages and wives are called good not because they *are* good but because they are not talked about.

---

88. Flavius Arrian, *Alexander the Great*, VII.
89. The usual accounts say Phaedo was compelled to do so (cf. Aulus Gellius, *Attic Nights*, II, xviii, 1). Then, cf. for Solon, Coelius Richerius Rhodiginus, *Antiquae lectiones*, XIV, iv, and Juvenal, *Satires*, VI, 347–8.

We should use our ingenuity to avoid making such useless discoveries which torture us. It was the custom of the Romans when returning home from a journey to send a messenger ahead to announce their arrival to their womenfolk so as not to take them unawares. That is why there is a certain people where the priest welcomes the bride and opens the proceedings on the wedding-night to remove from the groom any doubts and worries about whether she came to him virgin or already blighted by an *affaire*.[90]

'Yes. But people talk!' I know a hundred men who are cuckolds yet honoured and not unrespected. A decent man is sympathized with for it, not discredited by it. See to it that your misfortune is smothered by your virtue, so that good folk curse the cause of it and the man who wrongs you trembles to think of it.

And then who is never gossiped about for this, from the least to the greatest?

> *Tot qui legionibus imperitavit, . . .*
> *Et melior quam tu multis fuit, improbe, rebus!*

[Even the general who commanded all those legions . . . and was a far better man than you, you reprobate!][91]

When so many honourable men have been included in this opprobrium in your presence, do you think you are spared elsewhere?

'But even the ladies will laugh at me!' Well, what do they laugh at nowadays more readily than a peaceful, orderly marriage? [C] Each one of you has cuckolded somebody: and Nature is ever like, alternating and balancing accounts. [B] The frequency of this misfortune ought by now to have limited its bitter taste: why, it will soon be customary.

In addition that wretched misery is one you cannot even tell anyone about:

> *Fors etiam nostris invidit questibus aures.*

[Even Fortune refuses to listen to our woes.][92]

For what friend can you dare to confide your worries to? Even if he does not laugh at you, will he not be put on the track and shown how to join in the kill?

90. Plutarch (tr. Amyot), *Demandes des choses Romaines*, IX, 462 B–L; S. Goulart, *Hist. générale des Indes*, in which the priests are called *Piates*.
91. Lucretius, III, 1041 (adapted) and III, 1039.
92. Catullus, LXIV, 170.

[C]   Wise men keep secret both the sweets of marriage and its bitternesses. For a talkative man like me, of all the distressing disadvantages of marriage one of the principal is the fact that custom has made it indecorous and obnoxious to discuss with anyone whatever all that we know and feel about it.

[B]   It would be a waste of time to give women the same advice in order to make jealousy distasteful to them. Their essence is so pickled in suspicion, vanity and curiosity that you must not hope to do so by legitimate means. They often cure this infirmity by a species of well-being which is more to be feared than the malady. Just as there are magic spells which can only remove an evil by loading it on to someone else, so too wives readily pass this fever of jealousy on to their husbands, once they themselves have lost it.

All the same, to tell the truth, I do not know whether one can ever suffer anything worse than their jealousy: it is the most dangerous of their characteristics, as the head is of the anatomy. Pittacus said that every man has his curse: his was his wife's bad temper; if it were not for that he would think himself entirely happy. Seeing that so just, so wise, so valiant, so great a man should feel the whole state of his life corrupted by it, it must indeed be a grievous clog.[93] So what are we to do about it, little men like us!

[C]   The Senate of Marseilles[94] was right to accede to the request of a husband for permission to kill himself so as to escape his wife's petulance, for it is an evil which can never be removed except by removing the whole limb: you can make no worthwhile arrangement with it except by fleeing from it or putting up with it: both are fraught with difficulties.   [B]   That man knew what he was talking about, it seems to me, who said that a good marriage needs a blind wife and a deaf husband.[95]

We also need to ensure that the great and intense harshness of the obligations which we lay on women should not produce two results hostile to our ends: namely, that it does not whet the appetites of their suitors nor make the wives more ready to surrender. As for the first point, by raising the value of a redoubt we raise the value of conquering it and the desire to do so. May not Venus herself cunningly have raised the cost of her merchandise by making the laws pimp for her, realizing that it is a silly

93. Plutarch (tr. Amyot), *De la tranquillité de l'ame et de l'esprit*, 72 C.
94. Cf. II, 3, 'A custom of the Isle of Cea', p. 406.
95. Erasmus, *Apophthegmata*, VIII, *Alphonsus Aragonum Rex*, IV, commented upon in Montaigne's sense.

pleasure for anyone who does not enhance it by imagination and by buying it dear?

In short, as Flaminius' host said, 'it is all pork with different sauces.'[96] Cupid is a mischievous god: his sport is to wrestle with loyalty and justice; glory for him means clashing his strength against all others' strength, all rules yielding to his.

> *Materiam culpæ prosequiturque suæ.*

> [He is always hunting for occasion to do wrong.][97]

And as for my second point, would we be cuckolded less often if we were less afraid of being so, thus conforming to the complexion of women? For interdicts provoke and incite them.

> *Ubi velis, nolunt; ubi nolis, volunt ultro.*

> [What you want they don't: what you don't, they do.]

> *Concessa pudet ire via.*

> [They feel disgraced if they go the way we permit them.]

What better interpretation can we find for the case of Messalina? At the start she cuckolded her husband in secret, as one does; but as she carried on her affairs too easily because of her husband's dull unawareness, she suddenly felt contempt for that practice. So there she was being openly courted, acknowledging her lovers, welcoming them and granting her favours in sight of everyone. She was determined that he should know of it. When that dull brute could not even be aroused by all that (so rendering her pleasures weak and insipid by his excessive complaisance, which seemed to permit them and to legitimize them) what else could she do? Well, one day when her husband was out of the City, she – the consort of an Emperor alive and in good health, at noon, in Rome the theatre of the world, with public pomp and festivity – married Silius, the man she had long since enjoyed.

Does it not appear that either she had set herself on the road to becoming chaste because of the indifference of her husband, or else that she had sought another husband who would stimulate her desire by his jealousy [C] and excite her by standing up to her?

[B] However, the first trouble she had to face was also her last. That

96. Plutarch (tr. Amyot), *Dicts notables des anciens Roys . . .*, 203 B.
97. Ovid, *Tristia*, IV, i, 34; then, Terence, *Eunuch*, IV, viii, 43 and Lucan, *Pharsalia*, II, 446.

brute of hers did wake up with a start. You often get the worst treatment from such dozing dullards. Experience has shown me that such excessive tolerance once it bursts apart produces the harshest of vengeances, for then wrath and frenzy fuse into one and fire their whole battery during the first assault;

> *irarumque omnes effundit habenas.*

> [it looses anger's every rein.][98]

He put her to death, together with a large number of those who were in complicity with her, even including some who had had no option, having been driven to her marriage-bed with leathern scourges.

What Virgil sings of Venus and Vulcan, Lucretius sings more fittingly of stolen joys between her and Mars:

> *belli fera mœnera Mavors*
> *Armipotens regit, in gremium qui sæpe tuum se*
> *Rejicit, æterno devinctus vulnere amoris:*
> *Pascit amore avidos inhians in te, Dea, visus,*
> *Eque tuo pendet resupini spiritus ore:*
> *Hunc tu, diva, tuo recubantem corpore sancto*
> *Circunfusa super, suaveis ex ore loquelas*
> *Funde.*

[Mars, mighty in arms, ruler of the savage works of war, now wounded by an everlasting wound of love, flees to thy bosom. He feeds his eyes on thee with gaping lips, O goddess, his breath now hanging on thy mouth. While he rests upon thy sacred body as it flows around him, pour from thine own lips, O goddess, thy sweet complaints.][99]

When I chew over those words, *rejicit, pascit, inhians,* and then *molli fovet, medullas, labefacta, pendet, percurrit,* and Lucretius' noble *circunfusa* mother to Virgil's elegant *infusus,* I feel contempt for those little sallies and verbal sports which have been born since then. Those fine poets had no need for

98. Virgil, *Aeneid,* XII, 499. The standard source about Messalina is Tacitus, XI, xvi–xvii. She is given as an example of 'prodigious lust' by Tiraquellus and, indeed, by almost everyone.

99. Lucretius, I, 33–40. The first three of the following Latin words are from Lucretius, and so is *pendet.* The rest are from the lines of Virgil which are alluded to in the title of this chapter and cited above (cf. p. 958). Montaigne believed that Lucretius' use of the word *circunfusa* (literally 'poured like water around' the body of Mars in a close embrace) was imitated by Virgil when he used *infusus* in a similar sense.

smart and cunning word-play; their style is full, pregnant with a sustained and natural power. With them not the tail only but everything is epigram: head, breast and feet. Nothing is strained. Nothing drags. Everything progresses steadily on its course: [C] '*Contextus totus virilis est; non sunt circa flosculos occupati.*' [The whole texture of their work is virile: they were not concerned with little purple passages.][100] [B] Here is not merely gentle eloquence where nothing offends: it is solid and has sinews; it does not so much please you as invade you and enrapture you. And the stronger the mind the more it enraptures it. When I look upon such powerful means of expression, so dense and full of life, I do not conclude that it is said well but thought well. It is the audacity of the conception which fills the words and makes them soar: [C] '*Pectus est quod dissertum facit.*' [It is the mind which makes for good style.][101] [B] Nowadays when men say judgement they mean style, and rich concepts are but beautiful words.

Descriptions such as these are not produced by skilful hands but by having the subject vividly stamped upon the soul. Gallus writes straightforwardly because his concepts are straightforward. Horace is not satisfied with some superficial vividness; that would betray his sense; he sees further and more clearly into his subject: to describe itself his mind goes fishing and ferreting through the whole treasure-house of words and figures of speech; as his concepts surpass the ordinary, it is not ordinary words that he needs. Plutarch said that he could see what Latin words meant from the things which they signified.[102] The same applies here: the sense discovers and begets the words, which cease to be breath but flesh and blood. [C] They signify more than they say. [B] Even the weaker brethren have some notion of this: when I was in Italy I could express whatever I wanted to say in everyday conversation, but for serious purposes I would not have dared to entrust myself to a language which I could neither mould nor turn on my lathe beyond the common idiom. I want to add something of my own.

What enriches a language is its being handled and exploited by beautiful minds – not so much by making innovations as by expanding it through more vigorous and varied applications, by extending it and deploying it. It is not words that they contribute: what they do is enrich their words, deepen their meanings and tie down their usage; they teach it unaccustomed rhythms, prudently though and with ingenuity.

100. Seneca, *Epist. moral.*, XXXIII, 1.
101. Quintilian, X, vii, 15.
102. Plutarch, *Life of Demosthenes*.

That such a gift is not vouchsafed to everybody can be seen from many of the French authors of our time. They are bold enough and proud enough not to follow the common road; but their want of invention and power of selection destroys them. All we can see is some wretched affectation of novelty, cold and absurd fictions which instead of elevating their subject batter it down. Provided they are clad in new-fangled apparel they care nothing about being effective. To seize on some new word they quit the usual one which often has more sinew and more force.

In our own language there is plenty of cloth but a little want of tailoring. There is no limit to what could be done with the help of our hunting and military idioms, which form a fruitful field for borrowing; locutions are like seedlings: transplanting makes them better and stronger. I find French sufficiently abundant but not sufficiently [C] tractable and [B] vigorous. It usually collapses before a powerful concept. If you are taut as you proceed, you can often feel it weakening and giving way under you; in default your Latin comes to your aid – and Greek to the aid of others.

It is hard for us to perceive the power of some of the words I have just selected because use has somewhat cheapened their grace, and familiarity has made it commonplace. So too in our vulgar tongue there are some excellent expressions whose beauty is fading with age and metaphors whose colour is tarnished by too frequent handling. But by that they lose nothing of their savour for a man who has a good nose for them; nor does it detract from the glory of those ancient authors who were (as seems likely) the first to shed such lustre on those words.

Erudite works treat their subjects too discreetly, in too artificial a style far removed from the common natural one. My page-boy can court his lady and understands how to do so. Read him Leone Ebreo and Ficino: they are talking about him, about what he is thinking and doing. And they mean nothing to him![103] I cannot recognize most of my ordinary emotions in Aristotle: they have been covered over and clad in a different gown for use by the schoolmen. Please God they know what they are doing! If I were in that trade, [C] just as they make nature artificial, I would make art natural.[104]

103. Authors of treatises on Renaissance Platonic love: Ficino, *Commentary on Plato's Symposium*; Leone Ebreo (Judah Abravanel), *Dialogues of Love*.
104. '88: trade, I would *treat art as naturally as I could*. Let us . . .
    Allusions follow to Pietro Bembo, *Gli Asolani* and Mario Equicola, *On the Nature of Love*: two more Renaissance Platonists.

[B]   Let us skip over Bembo and Equicola.

When I am writing I can well do without the company and memory of my books lest they interfere with my style. Also (to tell the truth) because great authors are too good at beating down my pretensions: they dishearten me. I am tempted to adopt the ruse of that painter who, having wretchedly painted a portrait of some cocks, forbade his apprentices to let any natural cock enter his workshop.[105]   [C]   And to lend me some lustre I would need to adopt the device of Antinonides the musician[106] who, whenever he had to perform, arranged that, either before him or after him, his audience should have their fill of some bad singers.   [B]   But I cannot free myself from Plutarch so easily. He is so all-embracing, so rich that for all occasions, no matter how extravagant a subject you have chosen, he insinuates himself into your work, lending you a hand generous with riches, an unfailing source of adornments. It irritates me that those who pillage him may also be pillaging me: [C]   I cannot spend the slightest time in his company without walking off with a slice of breast or a wing.

[B]   For this project of mine it is also appropriate that I do my writing at home, deep in the country, where nobody can help or correct me and where I normally never frequent anybody who knows even the Latin of the Lord's Prayer let alone proper French. I might have done it better somewhere else, but this work would then have been less mine: and its main aim and perfection consists in being mine, exactly. I may correct an accidental slip (I am full of them, since I run on regardless) but it would be an act of treachery to remove such imperfections as are commonly and always in me. When it is said to me, or I say to myself: 'Your figures of speech are sown too densely'; 'This word here is pure Gascon'; 'This is a hazardous expression' – I reject no expressions which are used in the streets of France: those who want to fight usage with grammar are silly – 'Here is an ignorant development'; 'Here your argument is paradoxical'; 'This one is too insane';   [C]   'You are often playing about; people will think that you are serious when you are only pretending':   [B]   'Yes,' I reply, 'but I correct only careless errors not customary ones. Do I not always talk like that? Am I not portraying myself to the life? If so, that suffices! I have achieved what I wanted to: everyone recognizes me in my book and my book in me.'

105. Plutarch (tr. Amyot), *Comment on peult discerner le flatteur d'avec l'amy*, 49 H.
106. Or rather, *Antigenides*. Cf. Coelius Richerius Rhodiginus, *Antiquae Lectiones*, XV, x.

Now I have a tendency to ape and to imitate: when I took up writing verse – I wrote it exclusively in Latin – it always manifestly betrayed who was the last poet I had been reading; and some of my earliest essays are somewhat redolent of others' work. [C] When in Paris I talk rather differently than at Montaigne. [B] Anyone I look at with attention easily stamps something of his on me. Whatever I contemplate I make my own – a silly expression, a nasty grimace, a ridiculous turn of speech. Faults, even more so: as soon as they strike me they cling to me and will not leave me unless shaken off; I have more often been heard using swear-words from conformity than by complexion.

[C] Such imitation kills, like that of those monkeys terrifying in strength and size which King Alexander had to confront in a certain country in India.[107] He would have found it hard to get the better of them, but they showed him the way to do so by their tendency to imitate everything they saw being done. This inspired those who were hunting them to put on their boots, tying many knots in the laces, while the monkeys looked on; then to deck themselves in headgear with dangling nooses and to pretend to daub their eyes with bird-lime. And so those poor creatures were led to their doom by their apish complexions: they too daubed themselves with bird-lime, tied themselves in knots and garotted themselves. Yet the talent for cleverly imitating intentionally the words and gestures of another is no more in me than in a tree-stump. When I swear my own way it is always 'By God' – which is the most direct of all the oaths. They say that Socrates used to swear 'By dog'; Zeno 'By goats' (the same exclamation used today by the Italians, *Cappari*); Pythagoras, 'By air and by water.'

[B] I am marked so easily by surface impressions that, having *Sire* or *Your Majesty* [C] thoughtlessly [B] on my lips for three days in a row, those terms slip out a full week later instead of *Your Excellency* or *My Lord*. And any expression which I have fallen into saying in jest or for fun I will say the following day seriously. That is why I am loath to write on well-trodden topics: I am afraid I might treat them with another man's substance. All topics are equally productive to me. I could write about a fly! (God grant that the topic I now have in hand be not chosen at the behest of a will which is as light as a fly's.) I may begin with any subject I please, since all subjects are linked to each other.

But what displeases me about my soul is that she usually gives birth

107. Diodorus Siculus, XVII, xxv.

quite unexpectedly, when I am least on the lookout for them, to her profoundest, her maddest ravings which please me most. Then they quickly vanish away because, then and there, I have nothing to jot them down on; it happens when I am on my horse or at table or in bed – especially on my horse, the seat of my widest musings.

When speaking I have a fastidious zeal for attention and silence if I am in earnest; should anyone interrupt me he stops me dead. On journeys the very exigencies of the roads cut down my conversation; moreover I most often journey without the proper company for sustained conversation, which enables me to be free to think my own thoughts. What happens is like what happens to my dreams: during them I commend them to my memory (for I often dream I am dreaming); next morning I can recall their colouring as it was – whether they were playful or sad or weird – but as for all the rest, the more I struggle to find it the more I bury it in forgetfulness. It is the same with those chance reflections which happen to drop into my mind: all that remains of them in my memory is a vague idea, just enough to make me gnaw irritably away, uselessly seeking for them.

Well now, leaving books aside and talking more simply and plainly, I find that sexual love is nothing but the thirst for the enjoyment of that pleasure [C] within the object of our desire, and that Venus is nothing but the pleasure of unloading our balls;[108] it becomes vitiated by a lack either of moderation or discretion:[109] for Socrates love is the desire to beget by the medium of Beauty.[110]

[B] Reflecting as I often do on the ridiculous excoriations of that pleasure, the absurd, mindless, stupefying emotions with which it disturbs a Zeno or a Cratippus,[111] that indiscriminate raging, that face inflamed with frenzy and cruelty at the sweetest point of love, that grave, severe, ecstatic face in so mad an activity, [C] the fact that our delights and our waste-matters are lodged higgledy-piggledy together; [B] and that its highest

---

108. '95: balls, *analogous to the pleasure which Nature vouchsafes to us when we are unloading other organs of ours;* it becomes. . .

Montaigne's word for balls, *vases,* represents the Latin word *vas (tool)* used in this sense in the Priapics and, for example, by Plautus, *Poenulus* IV, ii.

109. Aristotle, *Nicomachaean Ethics,* II, ii, 1104 a ff.

110. Plato, *Symposium,* 203 ff.

111. Zeno, the founder of the Stoic School; Cratippus, the Peripatic who taught the son of Cicero; both admitted the effects of terrifying emotion: cf. St Augustine, *City of God,* IX, iv.

pleasure has something of the groanings and distraction of pain, I believe   [C]   that what Plato says is true: [B]   Man is the plaything of the gods[112] –

> *quænam ista jocandi*
> *Sævitia!*

[what a ferocious way of jesting!]

– and that it was in mockery that Nature bequeathed us this, the most disturbing of activities, the one most common to all creatures, so as to make us all equal, bringing the mad and the wise, men and beasts, to the same level.

When I picture to myself the most reflective and the most wise of men in such postures, I hold it as an effrontery that he should claim to be reflective and wise; like the legs on a peacock, they humble pride;

> *ridentem dicere verum*
> *Quid vetat?*

[what can stop us telling the truth with a laugh?][113]

[C]   Those who reject serious opinions in the midst of fun are, it is said, like the man who refuses to venerate the statue of a saint because it wears no drapery.

[B]   We eat and drink as the beasts do, but those activities do not hamper the workings of our souls. So in them we keep our superiority over the beasts. But that other activity makes every other thought crawl defeated under the yoke; by its imperious authority it makes a brute of all the theology of Plato and a beast of all his philosophy. Everywhere else you can preserve some decency; all other activities accept the rules of propriety: this other one can only be thought of as flawed or ridiculous. Just try and find a wise and discreet way of doing it! Alexander said that he acknowledged he was a mortal because of sleep and this activity: sleep stifles and suppresses the faculties of our souls; the 'job' similarly devours and disperses them.[114] It is indeed a sign of our original Fall, but also of our inanity and ugliness. On the one hand Nature incites us to it, having attached to this desire the most noble, useful and agreeable of her labours: on the other hand she lets us condemn it as immoderate and flee it as indecorous, lets us blush at it and recommend abstaining from it.

112. Plato, *Laws*, VII, 803 E and I, 644 D; then, Claudius Claudianus, *In Eutropium*, I, 24.
113. Horace, *Satires*, I, i, 24.
114. Tiraquellus, *De legibus connubialibus*, XV, 63–4.

[C] Are we then not beasts to call the labour which makes us bestial?

[B] In their religions all peoples have several similarities which coincide, such as sacrifices, lights, incense, fastings, offertories and, among others, the condemnation of this act. All their opinions come to it, not to mention the widespread practice of cutting off the foreskin [C] which is a punishment for it. [B] Perhaps we are right to condemn ourselves for giving birth to such an absurd thing as a man; right to call it an act of shame and the organs which serve to do it shameful. [C] (It is certain that mine may now properly be called shameful and wretched.)

The Essenes whom Pliny mentions were maintained for several centuries without wet-nurses or swaddling-clothes by the arrival of outsiders who, attracted by the beauty of their doctrines, constantly joined them. An entire people risked self-extermination rather than engage in woman's embraces, risked having no successors rather than create one.[115] It is said that Zeno lay with a woman only once in his entire life; and that that was out of politeness, so as not to seem to have too stubborn a contempt for that sex.[116]

[B] No man likes to be in on a birth: all men rush to be in on a death. [C] To unmake a human being we choose an open field in broad daylight: to make one, we hide away in a dark little hollow. When making one we must hide and blush: but glory lies in unmaking one, and it produces other virtues. One act is unwholesome: the other, an act of grace, for Aristotle says that in his country there is a saying 'To do a man a favour', which means to kill him.[117] The Athenians showed those two activities to be equally blemished when they were required ritually to purge the island of Delos and to seek reconciliation with Apollo: within its coasts they forbade both childbirth and burial:[118]

[B] *Nostri nosmet pœnitet.*

[We are embarrassed by our very selves.]

115. The Essenes forbade procreation, depending on proselytes to continue their community (Pliny, V, xvii).

116. Diogenes Laertius, *Life of Zeno.*

117. Plutarch (tr. Amyot), *Demandes des choses Romaines,* 469 A: not a general statement, but Aristotle's gloss on a term in a peace-treaty between the Arcadians and the Spartans.

118. Diodorus Siculus, XII, xvii; then, Terence, *Phormio,* I, iii, 20.

'88: poenitet. *We condemn in hundreds of ways the circumstances of our being.* There are . . .

[C]   We regard our very being as vitiated.

[B]   There are some nations where they hide to eat. I know one lady (among the greatest) who shares the opinion that chewing distorts the face, derogating greatly from women's grace and beauty; when hungry she avoids appearing in public. And I know a man who cannot tolerate watching people eat nor others watching him do so: he shuns all company even more when he fills his belly than when he empties it.   [C]   In the Empire of the Grand Turk you can find many men who, to rise above their fellows, never allow themselves to be seen eating a meal; they eat but once a week; they slash and disfigure their faces and limbs and never talk to anyone – ['95] fanatics   [C]   all – folk who believe they are honouring their nature by defacing it; who pride themselves on their contempt; who seek to make themselves better by making themselves worse.

[B]   What a monstrosity of an animal,[119] who strikes terror in himself,   [C]   whose pleasures are a burden to him and who thinks himself a curse.   [B]   Those there are who hide their existence –

> *Exilioque domos et dulcia limina mutant*

> [They give up their homes and domestic delights to go into exile][120]

– stealing away from the sight of other men; they shun health and happiness as harmful and inimical qualities. There are not merely several sects but whole peoples for whom birth is a curse, death a blessing.   [C]   And some there are who loathe the sunlight and worship the darkness.

[B]   We show our ingenuity only by ill-treating ourselves: that is the real game hunted by the power of our mind –   [C]   an instrument dangerous in its unruliness.

> [B]   *O miseri! quorum gaudia crimen habent.*

> [O pitiful men, who hold their joys a crime.]

Alas, wretched Man, you have enough   [C]   necessary   [B] misfortunes[121] without increasing them by inventing others. Your condition is wretched enough already without making it artificially so. You

---

119. '88: What a *disnatured* animal . . .
   (Cf. the similar change in note 121.)
120. Virgil, *Georgics*, II, 511.
121. Pseudo-Gallus, I, 180.
   '88: enough *natural* misfortunes . . . ('Necessary' misfortunes are those entailed by the human condition and its *necessitates*.)

have uglinesses enough which are real and of your essence without fabricating others in your mind. [C] Do you really think that you are too happy unless your happiness is turned to grief? [B] Do you believe that you have already fulfilled all the necessary duties in which Nature involves you and that, unless you bind yourself to new ones, Nature is [C] defective and [B] idle within you? You are not afraid to infringe her universal and undoubted laws yet preen yourself on your own sectarian and imaginary ones: the more particular, [C] uncertain and [B] controverted they are, the more you devote your efforts to them. [C] The arbitrary laws of your own invention – your own parochial laws – engross you and bind you: you are not even touched by the laws of God and this world. [B] Just run through a few *exempla* of that assertion: why, all your life is there.

Those lines of our two poets,[122] treating sexual pleasure as they do with reserve and discretion, seem to me to reveal it and throw a closer light upon it. Ladies cover their bosoms with lace-work; priests similarly cover many sacred objects; painters paint shadows in the pictures to emphasize the light; and it is said that the sun and wind beat down more heavily on us when deflected than when they come direct. When that Egyptian was asked, 'What are you carrying there, hidden under your cloak?' he gave a wise reply: 'It is hidden under my cloak so that you should not know what it is.'[123] Nevertheless some things are hidden in order to reveal them more.

Just listen to this man writing more openly:

> *Et nudam pressi corpus adusque meum.*
>
> [Nude against my body did I press her.][124]

I can feel him gelding me!

Let Martial, as he does, pull up Venus' skirts: he does not succeed in revealing her all that completely. The poet who tells all, gluts us and puts us off: the one who is timid about expressing his thoughts leads us in our thoughts to discover more than is there. There are revelations in that sort of modesty; especially when, as they do, they half-open such a beautiful highway for our imagination. Both that act and its portrayal should savour of theft.[125]

122. Virgil and Lucretius, cited earlier.
123. Plutarch (tr. Amyot), *De la curiosité*, 64 C.
124. Ovid, *Amores*, I, v, 24.
125. As, for example, in 'stolen' kisses.

For the Spaniard and the Italian sex-love is more timid and respectful, more coy and less open: I like that. (In ancient times someone or other wished that his throat was as long as the neck of a crane so as to have more time to taste what he was swallowing.[126] Such a wish is more appropriate to this hasty and headlong pleasure, especially for natures such as mine whose fault is to be too quick.) For them, so as to stop its flight and to let it expand itself on preliminaries, everything serves as a grace and reward: a loving glance, a bow of the head, a word, a gesture.

Would anyone who could actually dine on the smell of roast beef not be making a fine saving?[127] Well, this is a passion which mingles very little essential solids with plenty of vanity and feverish madness: we should reward it and treat it accordingly. Let us instruct our ladies how to make themselves valued and esteemed, to keep us waiting and to be sweet deceivers. We French always make our last attack the first: there is always that impetuosity of ours.[128] If only our ladies were to string out love's favours, offering them retail, then each one of us, according to his worth and merit, would get a scrap even in our pitiful old age. A man who only enjoys enjoying a woman, a man who only wins if he takes the lot and who, in hunting, only likes the kill, is not made for joining our sect. The more the steps the greater the height, and the more the rungs the greater the honour, of that ultimate bastion. We should take delight in being conducted there as through splendid palaces, by varied portals and corridors, long and pleasant galleries and many a winding way. Such stewardship would turn to our advantage; there we would linger and love longer: without hope and desire we no longer achieve anything worthwhile. Women should infinitely fear our overmastery and entire possession. Their position is pretty perilous once they have totally thrown themselves on the mercy of our faith and constancy; those virtues are rare and exacting; as for the women, as soon as we have them, they no longer have us:

> *postquam cupidæ mentis satiata libido est,*
> *Verba nihil metuere, nihil perjuria curant.*

126. Aristotle, *Nicomachaean Ethics*, III, x, 1118a; Aristophanes, *The Frogs*, 934.
127. Allusion to a famous legal tale related by Rabelais (*Tiers Livre*, TLF, xxxvii, after Tiraquellus, *De legibus connubialibus*, XI, 5): a chef complained that a poor man was savouring the smell of his roast beef: a fool, called in to judge, ordered the smell to be paid by the jangle of coins.
128. An ancient Roman gibe against the Gauls (referring to military not amorous ventures): Erasmus, *Apophthegmata*, VI, *Varie mixta*, CIII.

[as soon as eager longing is satisfied, our minds fear not for their pledged word nor care about perjury.][129]

[C]   A young Greek called Thrasonides was so in love with love that, having won his lady's heart, he refused to enjoy her so as not to weaken, glut and deaden by the joy of lying with her that unquiet ardour in which he gloried and on which he fed.

   [B]   Foods taste better when they are dear. Think how far kisses, the form of greeting peculiar to our nation, have had their grace cheapened by availability: Socrates thought they were most powerful and dangerous at stealing our hearts.[130] Ours is an unpleasant custom which wrongs the ladies who have to lend their lips to any man, however ugly, who comes with three footmen in his train.

> *Cujus livida naribus caninis*
> *Dependet glacies rigetque barba:*
> *Centum occurrere malo culilingis.*

[Cold leaden snot drips from his dog-like conk and bedews his beard. Why, I would a hundred times rather go and lick his arse.][131]

And we men gain little from it: for as the world is made we have to kiss fifty ugly women for every three beauties. And for the delicate gullets of men of my age, a bad kiss outweighs a good one.

   In Italy they play the swooning suitor even with women who sell their favours. They defend themselves thus: there are degrees in enjoying a woman; by such courtship they want to obtain for themselves the fullest enjoyment of all. Such women sell only their bodies; their wills cannot be up for sale: they are too free, too autonomous. It is her will that the Italians are after, they say. And they are right. What must be courted and ensnared is the will. I am horrified by the thought of a body given to me but lacking love. To me such raging madness is analogous to that of the boy who sullied with his love that beautiful statue of Venus sculpted by Praxiteles, or to that of the Egyptian madman who was inflamed with love for the corpse of a dead woman he was embalming while wrapping it in its shroud, and who gave rise to the law subsequently proclaimed in Egypt that the corpses of beautiful young women and of women of noble families

---

129. Catullus, LXIV, 147–8; then, Diogenes Laertius, *Life of Zeno*.
130. Platonic theories of mutual love held that by kissing one another lovers exchange souls and so literally 'live in' each other. Ficino had made such a belief current during the Renaissance.
131. Martial, *Epigrams*, VII, cxv, 10–12.

should be kept for three days before being handed over to those whose task it was to bury them. Periander acted more horrifyingly still when he prolonged his conjugal love (itself most proper and legitimate) by enjoying his departed wife Melissa.[132]

[C]   And was Luna's humour not clearly lunatic when, being unable to enjoy in any other way her beloved Endymion, she went and put him to sleep for several months, feasting herself on the enjoyment of a boy who never stirred but in her dreams?[133]

[B]   I claim that we are similarly loving a body deprived of soul and sensation when we make love to one without its agreement and desire. All enjoyings of women are not the same. Some are thin and languid: hundreds of causes other than tenderness can obtain that privilege from women. It is not in itself a sufficient proof of affection: deceiving can be found in that as in anything else; sometimes they only set about it with one cheek of their arse:

> *tanquam thura merumque parent:*
> *Absentem marmoreamve putes.*

[as cool as though preparing an offertory of incense and wine; you would think she was somewhere else, or made of marble.][134]

Some ladies I know would rather lend you 'that' than their carriage: it is the only way they know how to converse. You need to see whether your company pleases them for some other end also (or, as does some hulking great stable-boy, only for 'that'), and in what rank, and at what price, you are accepted:

> *tibi si datur uni*
> *Quo lapide illa diem candidiore notet.*

[whether she gives herself to you alone, and marks that day with her whitest milestone.]

What if she is eating your bread with a sauce derived from more pleasing thoughts!

> *Te tenet, absentes alios suspirat amores.*

132. Ravisius Textor, *Officina: Animalium et aliarum rerum amatores* (for the statue); *amor conjugalis* (for Periander); Herodotus, II, lxxxix (for the Egyptian law).
133. Erasmus, *Adages*, I, IX, LXIII, *Endymionis somnium dormis*, alluding to the tale of the shepherd Endymion in Cicero, *Tusc. disput.*, I, xxxviii, 92; Plato, *Phaedo*, 72 C; Aristotle, *Nicomachaean Ethics*, VI, viii.
134. Martial, X, ciii; XI, lix; then, Catullus, LXVIII, 147–8 and Tibullus, I, vi, 35.

[She holds you close while sighing for the loves of an absent lover.]

What! Do we not know of a man who in our own day used this activity as a means of horrifying vengeance, so as to inject poison into a decent woman and kill her?[135]

Those who know Italy will never find it odd if, while on this subject, I do not go anywhere else for *exempla*, since that nation can claim to be the world's professor in such matters. They have more routinely beautiful women than we do and fewer ugly ones, though for rare and outstanding beauties we are on a par. And I think the same applies to wit: of routinely fine ones they have more and it is obvious that brutish stupidity is incomparably more rare. But in matchless minds, those of the highest rank, we owe them [C] nothing.[136] [B] Were I to have to extend that comparison it could probably be said, on the contrary, that, by their standards, valour is commonplace and natural with us: yet sometimes you can see it so full and vigorous as they handle it that it surpasses all the stern examples which we have. Italian marriages are crippled: by their customs, so harsh and slavish a rule is imposed on their wives that the slightest acquaintance with another man is as capital an offence as the most intimate. The result of this rule is that any approach to their wives becomes, of necessity, basic; and since whatever they do amounts to the same, the choice is made for them already.   [C]   And once they have broken out of their pens, believe you me, they are all ablaze: *'luxuria ipsis vinculis, sicut fera bestia, irritata, deinde emissa.'* [sexual desire then breaks loose, like a wild beast first provoked and then set free.][137]   [B]   They really ought to give them a little more rein.

> *Vidi ego nuper equum, contra sua frena tenacem,*
> *Ore reluctanti fulminis ire modo.*

[Of late I saw a horse, straining at the bit, pulling with its mouth and careering along like lightning.]

We can weaken the desire for such companionship by allowing them a mite of freedom.[138]

---

135. Brantôme relates a case of a French nobleman who poisoned his wife through her genitals in the hope of marrying another woman (*Dames galantes*, ed. M. Rat, Paris, 1947, pp. 14–15).
136. '88: We owe them *hardly anything . . .*
137. Cato, in Livy, XXIV, iv; then, Ovid, *Amores*, III, iv, 13–14.
138. '88: freedom. *They are, in their social life, ladies of many parts. We put them on the way to using the ultimate one, since we rate them all the same.* Both run . . .

Both run more or less equal risks. They are excessive in restraint: we, in freedom. One of the fine customs of our nation is that the boys of good families are taken in as pages to be educated and brought up, schooled for nobility. It is said to be rude and discourteous to refuse a young gentleman. I have noted (but there are as many fashions as there are different homes) that ladies who have sought to impose the most austere of rules on the girls in their entourage have not produced any better results. What we need is moderation. We should leave a good bit of the behaviour of girls to their own discretion; whatever you do, there is no training that can bridle them in all the time; but what is true is that a girl who has bolted, bag and baggage, from a dressage in freedom inspires much more confidence than one who emerges with propriety from an austere prison of a school.

Our forefathers trained their daughters' countenances to be bashful and timorous; their minds and desires were alike: we, knowing nothing about the matter, train them to be bold. [C] That is for Sauromatians who are forbidden to lie with a man until they have killed one with their own hands in war.[139] [B] It suffices me (who have no rights in the matter except to be heard) that they retain me as a counsellor, according to the privilege of my age. So I would counsel them – [C] and us too – [B] to refrain; but if this age is too inimical to that, at least to show discretion and moderation. [C] As in the story told of Aristippus: some young men blushed at seeing him go in to the house of a courtesan: he said to them: 'The error lies not in going in but in never coming out.'[140] [B] If a woman cannot save her conscience let her at least save her reputation: even if the base is not worth it let appearances hold out. I advocate gradualness and stringing things out when dispensing of love's favours. [C] Plato demonstrates that surrendering easily or quickly is forbidden to the defenders in loves of all kinds.[141] [B] To yield all, so inadvisedly and so hastily, is a sign of voracity,[142] which they must hide with all their art. By acting ordinately and with measure when distributing their gifts they succeed far better in tempting our desires and hiding their own. Let them ever flee before us – I mean even those who intend to be

---

139. Also called *Sarmatae*; cf. Herodotus, IV, cxvii; Coelius Richerius Rhodiginus, *Antiquae lectiones*, IX, xii, who assimilates them to the Amazons.
140. Erasmus (*Apophthegmata*, III, *Aristippus*, XIII), who adds a caution, restricting the saying to legitimate relationships.
141. Implied in Plato's *Symposium*.
142. '88: voracity *and hunger*, which . . .

caught: like the Scythians they beat us best when retreating. By the law which Nature gives them, it is truly not for them to wish and to desire: their role is to accept, to obey, to consent. That is why Nature has made them able to do it at any time: we men are only able to do it occasionally and unreliably. The time is always right for them, so that they will be always ready when our time comes along: [C] *'pati natae'* [they are born to be passive].[143] [B] And whereas Nature has so arranged it that men's desires should declare themselves by a visible projection, theirs are hidden and internal and she has furnished them with organs [C] unsuited to making a display and [B] strictly defensive.

[C] We should leave to the licence of the Amazons events like the following: when Alexander was marching through Hircania, Queen Thalestris of the Amazons came to meet him with three hundred warriors of her sex, well mounted and well armed, having left beyond the nearby mountains the rest of a big army which followed her leadership; she told him, aloud and in public, that the rumour of his victories and of his valour had brought her there to see him and to offer him her might and her support to forward his campaigns; she added that as she found him to be so beautiful, young and full of vigour she, who was perfection itself in all her qualities, advised him that they should lie together, so that there should be born from the most valiant woman in the world and the most valiant man then alive some great and rare offspring for the future. For the rest Alexander merely thanked her kindly, but he remained for thirteen days to allow time to fulfil her last request, days which he celebrated with all possible eagerness to please so courageous a princess.[144]

[B] In virtually everything we men are as unjust judges of women's actions as they are of ours – I confess the truth when it goes against me just as when it serves me. It is a base disorder which drives them to change so frequently and which impedes them from settling their affections firmly on any person whatsoever; as we can see in that goddess Venus to whom is attributed so many changes of lovers. Yet it is true that it is against the nature of sex-love not to be impetuous, and it is against the nature of what is impetuous to remain constant: so those men who are amazed by this and who denounce and seek the causes of this in women as unbelievable and unnatural, ought to ask themselves why that distemper finds acceptance in

143. Seneca, *Epist. moral.*, CXV, 21.
144. Diodorus Siculus, XVII, xvi.

themselves, without their being stunned as by a miracle. It would perhaps be more odd to find any fixity in it. It is not a passion of the body alone. Just as there is no end to covetousness and ambition, so there is no end to lust. It still lives on after satiety: you can prescribe to it no end, no lasting satisfaction: it always proceeds beyond possession. And fickleness is perhaps somewhat more excusable in them than in us. Like us they can cite in their defence the penchant we both have for variety and novelty; secondly they can cite, what we cannot, that they buy a pig in a poke [C] (Queen Joanna of Naples caused her first husband Andreosso to be hanged from the grill of her window by a gold and silver cord, plaited by her own hands, once she discovered that neither his organs nor his potency corresponded to the hopes she had conceived of his matrimonial duties from his stature, his beauty, his youth and his disposition, by which he had won her and deceived her);[145] [B] they can also cite the fact that since the active partner is required to make more effort than the passive one, they at least can always provide for this necessity while we cannot. [C] That is why Plato wisely established in his laws that those making a judgement on the suitability of a marriage should see the youths who were ambitious to marry stark naked but the maidens naked only down to the girdle.[146] [B] By assaying us that way the women might perhaps find us not worth the choosing:

> *Experta latus, madidoque simillima loro*
> *Inguina, nec lassa stare coacta manu,*
> *Deserit imbelles thalamos.*

[She deserts his impotent bed after exploring his thighs and his prick which, like a damp leather thong, refuses an erection to her exhausted hand.][147]

It is not enough to have the will to drive straight up: in law impotence and an inability to consummate annul a marriage –

> *Et querendum aliunde foret nervosius illud,*
> *Quod posset zonam solvere virgineam*

[You had to look elsewhere for a more sinewy one, capable of unsealing her maidenly girdle][148]

145. Jacques de Lavardin, *Scanderbeg*; cf. Tiraquellus, *De legibus connubialibus*, IX, 99.
146. Plato, *Laws*, XI, 925 A.
147. Martial, *Epigrams*, VII, lvii, 3–5.
148. Catullus, LXVII, 27–8.

– so why should a proportionately more wanton and active sexual skill not do so,

> *si blando nequeat supresse labori?*
>
> [if it proves unequal to its pleasant task?][149]

But it is most unwise (is it not?) to bring our inadequacy and our weaknesses to a place where what we would leave behind is a good reputation and a good impression. For the little that I need nowadays –

> *ad unum*
> *Mollis opus*
>
> [limp, even for one go]

– I would not embarrass any lady whom I should hold in reverence and awe:

> *Fuge suspicari,*
> *Cujus heu denum trepidavit aetas,*
> *Claudere lustrum.*
>
> [Suspect not a man whose life has staggered to its fiftieth year.][150]

Nature ought to be satisfied with making that age pitiful without making it ridiculous as well. I hate to see old age with an inch of paltry vigour which arouses it three times a week dashing about and bragging with the same vehemence as if it had a good day's legitimate work in its belly. Straw on fire![151] Truly. [C] And I am always shocked when its lively and quivering fire is promptly quenched and frozen cold. That appetite was meant for the flower of beauteous youth. [B] Just to see, try relying on old age to further that tireless, full constant and great-souled ardour that is in you! It will leave you stranded halfway there! Venture to cede it to some gawky gentle dazzled youth, still quaking before his wand and blushing at it,

> *Indum sanguineo veluti violaverit ostro*
> *Si quis ebur, vel mista rubent ubi lilia multa*
> *Alba rosa*

---

149. Virgil, *Georgics*, III, 127 (adapted).
150. Horace: *Epodes*, XII, 15; then, *Odes*, II, iv, 22–4. The text of '95 reads *undenum*, not *heu denum*, that is, 'fifty-five', not 'alas fifty'. Horace wrote *octavum* (forty). Horace is counting by five-year units (*lustra*).
151. Allusion to the proverb (listed by Cotgrave): A whore's love is but straw on fire (*Amour de putain, feu d'estoupe*).

[like Indian ivory stained blood-red, or even as white lilies arranged among red roses reflect their hue.][152]

Any man who can without dying of shame await the morning which brings disdain from a pair of lovely eyes, conscious of his flaccidity and irrelevance,

> *Et taciti fecere tamen convitia vultus,*
>
> [her silent features eloquent with loud reproach,]

has never known the happy pride of turning them glazed and dim by the vigorous exercises of a fulfilled and active night. When I have found a woman discontented with me I have not immediately gone and railed at her fickleness: I have asked myself, rather, whether I would be right to rail against Nature.

> *Si non longa satis, si non bene mentula crassa,*
>
> [Should my cock be not long enough nor good and thick,][153]

then Nature has indeed treated me unlawfully and unjustly –

> *Nimirium sapiunt, videntque parvam*
> *Matronae quoque mentulam illibenter*
>
> [Even good matrons know all too well and do not gladly see a tiny cock]

– [C] and inflicted the most enormous injury. Every one of my members, each as much as another, makes me myself: and none makes me more properly a man than that one. I owe to the public my portrait complete.

The wisdom to be found in my account lies in truth, in frankness and in essentials – entirely; it disdains to count among its real duties those little made-up rules based on provincial custom; it is natural, unvarying, universal; its daughters are indeed courtesy and respect, but they are bastard ones. Apparent defects we shall get the better of all right once we have got the better of those which are of the essence. After we have finished with the latter here, we will fall upon the others – if we find we still need to do so. For there is a danger that we will think up imaginary new duties so as to excuse our neglect of our natural ones and to jumble them up together.

152. Virgil, *Aeneid*, XII, 67–9; then, Ovid, *Amores*, I, vii, 21.
153. *Priapeia*, LXXX, 1; then VIII, 4–5.

That can be shown: you can see that wherever peccadillos are treated as crimes, crimes are treated as peccadillos; that among the peoples whose laws of politeness are fewest and slackest, the more basic laws, those common to all, are best observed since the countless multitude of those other obligations smother our concern, weaken it and disperse it. Applying ourselves to petty things diverts us from the pressing ones. Oh what an easy, favoured route such superficial men follow compared with ours! Such things are but shadowy pretences with which we bedaub each other and repay our mutual debts; but we cannot repay with them, but increase rather, the debt owed to that Great Judge who rips our tattered rags from off our pudenda and really sees us through and through, right down to our innermost and most secret filth. Our maidenly bashfulness would be useful and fitting if it could order that Judge not to uncover us!

To sum up: whoever could make Man grow out of an over-nice dread of words would do no great harm to this world. Our life consists partly in madness, partly in wisdom: whoever writes about it merely respectfully and by rule leaves more than half of it behind. I address no apologies to myself; were I to do so I would apologize for those apologies more than anything else. My apology is addressed to those of certain kinds of temperament (who are I believe numerically greater than those siding with me). I would like to please everyone, even though it is a difficult thing '*esse unum hominem accommodatum ad tantam morum ac sermonum et voluntatum varietatem*' [for one single man to conform to so great a variation in manners, speech and intentions];[154] so out of consideration for them I will add this, that they cannot justifiably complain that I am putting words into the mouths of authors accepted with approval for many centuries, nor can they deny me, because I lack verse, the freedom enjoyed by some of the greatest clerical cocks-of-the-walk of our own days. Here are two examples:

> *Rimula, dispeream, ni monogramma tua est;*
>
> [Strike me dead if your slit is more than one sketchy line;][155]

---

154. Cicero, *De petitione consulatus*, xiv. Montaigne's manuscript jottings at this point, eventually crossed out, show that he was aware of going beyond the limits of decency which he had set himself in his Preface: that was because he had been emboldened by the welcome given to his book.

155. A line from the erotic *Juvenilia* of Theodore Beza, the great Reformer and successor to Calvin. The next is by Mellin de Saint-Gelais, a Roman Catholic cleric and court poet.

and:

> *Un vit d'amy la contente et bien traicte.*
>
> [A lover's cock services and delights her.]

And what about all the others?

I like modesty. It is not my judgement which makes me choose this shocking sort of talk: Nature chose it for me. I am no more praising it than I am praising any behaviour contrary to the accepted norms; but I *am* defending it, lessening the indictment by citing individual and general considerations.

Let us get on.

Similarly,   [B]   from what do you derive that sovereign authority you assume over any ladies who, to their own cost, grant you their favours –

> *Si furtiva dedit nigra munuscula nocte*
>
> [If she gives you some little stolen present in the black of night][156]

– so that you immediately invest yourselves with rights, cold disapproval and husbandly authority? It is a covenant freely entered into: why do you not stick to it if you want to hold them to it?   [C]   Voluntary agreements grant no prescriptive rights.

[B]   It was not good form, but nevertheless true, that in my day I kept this bargain (as far as its nature allows) as conscientiously as any other one, and with a sort of justice, since I never showed more affection to the woman than I felt, portraying to them in all simplicity its decline, its flourishing period and its birth, its accesses of fever and its relapses. We do not go about such things with an even stride. I was so mean with my promises that I think I kept more than I ever vowed or owed. They found faithfulness there, even to the extent of my serving their inconstancy – and I mean inconstancy admitted and at times repeated. I never broke with one of them as long as I was held there even by the tail-end of a thread. And no matter what occasions they gave me, I never broke it off even for hatred or disdain: for such intimacies still oblige me to show some kindness even when acquired by the most discreditable of covenants. I did sometimes show my choler and a somewhat undiscerning impatience at the high point of their trickery, their evasions and our quarrels, but then I am by complexion subject to sudden distempers which, despite being short and

156. Catullus, LXVIII, 145.

light, are often prejudicial to my affairs. If they wanted to make an assay of my freedom of judgement, I never baulked at giving them bitingly paternal advice and lancing them where it hurts. If I left them any room to complain of me, it is rather for having found me to be, by modern standards, a ridiculously scrupulous lover. I kept my word in cases where anyone at all would have readily released me from it: women yielded in those days while saving their reputations by terms of surrender which they would readily have allowed their conqueror to infringe. In the interests of their honour I have more than once made my pleasure strike its sails at the point of a climax, and, when reason urged me, I have even armed them against me, so well indeed that they acted more safely and soberly by my rules, once they had frankly accepted them, than they would have done by their own.

[C] As far as in me lay I personally assumed all the risks of our assignations so as to take the load off them; and I managed our intrigues in the most difficult and unforeseeable of ways, for they are the least open to suspicion and, in my opinion, the most practical. Assignations are most overt when they seem the most covert. What is least feared is least protected, least observed; it is easy to dare what nobody thinks you will: the difficulty makes it easy.

[B] No man's advances were ever more saucily genital. The way of courting I have described is more in harmony with the rules: but does anyone know better than I do how ridiculous it appears to folk nowadays and how unsuccessful it is! Yet I shall never be brought to gainsay it: I have nothing more to lose by it now,

> *me tabula sacer*
> *Votiva paries indicat uvida*
> *Suspendisse potenti*
> *Vestimenta maris Deo.*

[As is shown by my votive tablet, I have hung up my dripping garments on the temple wall and dedicated them to the god of the sea.][157]

It is time, now, to talk of this openly. But as I might say to someone now, 'You are raving mad, my friend: love in these days of yours has nothing to do with fidelity and loyalty' —

> *hæc si tu postules*
> *Ratione certa facere, nihilo plus agas,*
> *Quam si des operam, ut cum ratione insanias*

157. Horace, *Odes*, I, v, 13–16; then, Terence, *The Eunuch*, I, i, 16–18.

[if you try to reduce all this to rational rules you will simply give yourself the task of going rationally insane]

– so, on the other hand, if I had to start again, I would certainly adopt the same course and the same method, however fruitless that might prove for me. [C] Inexpertise and silliness are praiseworthy in an activity which deserves no praise. [B] The further I go from others' humours in this, the nearer I draw to my own.

Incidentally, I never allowed all of myself to be totally devoted to this business. I took delight in it but I never forgot *me*: both in the ladies' service and in mine I conserved, in its entirety, such little sense and discretion as Nature had allotted me: some passion but no raging madness. My conscience was compromised by it so far as to include lasciviousness and licentiousness, though never ingratitude, treachery, wickedness or cruelty. These are prices which I would not pay for the pleasures of this vice: I was happy to pay its proper honest price: [C] *'Nullum inter se vitium est.'*[158] [No vice is self-enclosed.] I have a virtually equal loathing of all cowering torpid idleness and all prickly painful bustle. One cuts into me, the other knocks me senseless: and I am no more fond of cuts than of bruises, of slashing blows than of blunt ones. In these affairs, when I was more fit for them, I found a just moderation between those two extremes. Love is a lively emotion, light-hearted and alert: I was neither confused nor afflicted by it but I was thrown into a heat by it and troubled. There you must stop: it is harmful only for fools.

When a youth asked Panaetius the philosopher whether it became a wise man to be in love, 'Let us leave aside the wise,' he replied, 'neither you nor I are that; but let us not pledge ourselves to an activity so violent and disturbing, one which makes us the slave of another and despicable to ourselves.'[159] He was telling the truth when he said that something so intrinsically impulsive should not be entrusted to a man's soul if it has no means of withstanding its assaults and of disproving by its deeds the assertion of Agesilaus, that wisdom and love cannot live together.[160]

It is a vain pastime, it is true, indecorous, shaming and wrong; but I reckon that, treated in this fashion, it is health-bringing and appropriate for loosening up a sluggish mind and body; as a doctor I would order it for a

---

158. Seneca, *Epist. moral.*, XCV, 33 (that is, one vice leads to another).
159. Ibid., CXVI, 5. (Panaetius was a Stoic.)
160. Erasmus, *Apophthegmata*, I, *Agesilaus*, XIX; Plutarch (tr. Amyot), *Dicts notables des Lacedaemoniens*, 210 EF (when duty required Agesilaus to leave a sick friend).

man of my mould and disposition as readily as any other prescription so as to liven him up and keep him in trim until he is well on in years and to postpone the onset of old age. While we are still only in its outskirts, while there is still life in our pulse,

> *Dum nova canities, dum prima et recta senectus,*
> *Dum superest Lachesi quod torqueat, et pedibus me*
> *Porto meis, nullo dextram subeunte bacillo,*

[while the hair is but newly grey, while old age is still fresh and erect, while there is still some yarn for Lachesis to spin, while I can stand on my own feet without leaning on a stick,][161]

we have need of being stirred and thrilled by some such perturbation as that: just think how it restored youth, vigour and merriness to wise Anacreon. And Socrates, when older than I am, said, in talking of someone he loved, 'When we touched shoulders and brought our heads together while looking at the same book I felt, I can assure you, a sudden jab in my shoulder like an insect's sting: it went on irritating for five whole days and poured into my mind a ceaseless longing.'[162] – A mere touch, by chance, on the shoulder, was enough to warm and disturb a soul chilled and enervated by age, a soul which was foremost among all human souls in its re-formation.[163] [C] And why not? Socrates was a man: he never wanted to be, or to seem to be, anything else.

[B] Philosophy does not do battle against such pleasures as are natural, provided that temperance accompanies them:[164] [C] she teaches moderation in such things not avoidance; [B] her powers of resistance are used against bastard unnatural pleasures. She says that the body's desires must not be augmented by the mind and cleverly warns us [C] not to seek to stimulate our hunger by sating it, not to seek to stuff our bellies instead of filling them, as well as to avoid any enjoyment which brings us to penury,

161. Juvenal, *Satires*, I, 26–8.
162. Xenophon, *Symposium*, IV, 27–8. Socrates was consulting a book-scroll with Cleinias, bare shoulder to bare shoulder. Sage though he was, he was disturbed for five days as though he had been bitten by a wild beast. In his innocence he did not realize why, until Charmides twitted him about it.
163. '88: human souls, *in rule and* in its re-formation . . .
Socrates, as he told Zopyrus the physiognomist, had been born with a vicious, lecherous inferior 'form' (soul), but had re-formed it.
164. The Classic Aristotelian teaching (e.g. *Nicomachaean Ethics*, II, vii, 3; VIII, 2 ff.; III, x–xii, etc.).

all meats which increase hunger and all drinks that increase thirst, [B] just as[165] in the service of love she orders us to take a person who simply satisfies the needs of the body and who does not disturb the soul; the soul must not make love its concern, but follow nakedly along, accompanying the body.[166]

But am I not right to think that these precepts – which are by my standard nevertheless a trifle rigorous[167] – concern a body which is functioning properly, and that for a broken-down body (as for a prostrate stomach) we are allowed to use the art of medicine to prop it up and put a little heat into it by means of our imagination so as to restore its appetite and joy, since, left to itself, it has lost them for good? May we not say that there is nothing in us during this earthly prison either purely corporeal or purely spiritual and that it is injurious to tear a living man apart; and that it seems reasonable that we should adopt towards the enjoyment of pleasure at least as favourable an attitude as we do towards pain? Pain for example was vehement to the point of perfection in the Soul of the saints doing penance; the body naturally took part in it by right of the links binding it to her; yet it could have had little part in the cause.[168] But the saints were by no means content that the body should 'follow nakedly along, accompanying' the afflicted soul: they afflicted such horrifying punishment on it as was proper to it, in order that both body and soul should emulate each other, plunging the whole man into pain, most salutary when most atrocious.

[C] So, in the parallel case of bodily pleasures, is it not unjust to chill the Soul towards them and to maintain that she should be dragged towards them as to some compelling obligation or some slavish need? It is for the Soul, rather, to keep them warm like a broody hen and, since she has the responsibility of governing them, to come forward and welcome them;

165. '88: warns us *to avoid* all meats which increase hunger, *that is, which make us desire to be hungry afresh,* just as . . .
   Cf. Plutarch (tr. Amyot), *De la curiosité,* 67 p.
166. Cf. the advice of the giant heroes in Rabelais, *Tiers Livre,* TLF, XXXV, 46 ff. and notes. Cf. Tiraquellus, *De legibus connubialibus,* XV, 56 ff, with references to Thomas Aquinas, etc.
167. '88: rigorous *and inhumane* – concern . . .
   The ensuing notion that the soul is 'imprisoned' in the body is a Platonic commonplace. (The usual corollary was that the soul should strive, in ecstasy and rapture, to escape from the body. Montaigne does not accept it for most men.)
168. The temptations of saints are not so much grossly corporeal as spiritual and mental.

just as in my opinion it is also her duty in the case of such pleasures as are proper to her to inject and pour into the body every sense-impression which their attributes allow and to see that they are made sweet to it and salutary. For it is, as they say, right that the body should never follow its appetites to the prejudice of the Soul. Why is it not right, then, that the Soul should not follow hers to the prejudice of the body?

[B] I have absolutely no other passion but love to keep me going. What covetousness, ambition, quarrels and lawsuits do for men who, like me, have no other allotted task, love would do more suitably: it would restore me to vigilance, sober behaviour, graceful manners and care about my person; love would give new strength to my features so that the distortions of old age, pitiful and misshapen, should not come and disfigure them; [C] it would bring me back to wise and healthy endeavours by which I could make myself better esteemed and better loved, banishing from my mind all sense of hopelessness about itself and about its application, while bringing it to know itself again: [B] it would divert me away from a thousand painful thoughts, [C] from a thousand melancholy sorrows [B] which idleness burdens us with in old age, [C] as does the poor state of our health; [B] it would, at least in dream, restore some heat to my blood – this blood of mine which Nature is foresaking; it would lift up my chin and unbuckle my sinews [C] as well as the vigour and exhilaration of the soul [B] for this poor fellow who is on his way out, rushing towards disintegration.

But I am well aware that love is a good thing very hard to recover. Our tastes have, through weakness, become more delicate and, through experience, more discriminating. We demand more when we have less to offer: we want the maximum of choice just when we least deserve to find favour. Realizing we are thus, we are less bold and more suspicious; knowing our own circumstances – and theirs – nothing can assure us we are loved.

I feel shame for myself to be found among fresh-green, boiling youth

> *Cujus in indomito constantior inguine nervus,*
> *Quam nova collibus arbor inhæret.*

[in whose indomitable groin there is a tendon firmer far than a young tree planted on the hillside.][169]

Why should we go and show our wretchedness among such eager joy,

169. Horace, *Epodes*, XII, 19–20; then, *Odes*, IV, xiii, 26–8.

Possint ut juvenes visere fervidi,
Multo non sine risu,
Dilapsam in cineres facem?

[so that burning youth, not without many a laugh, may see our nuptial torch decayed into ashes?]

They have strength and reason on their side; let us make room for them; we can hold out no longer.

[C] That sprig of budding beauty will not suffer itself to be handled by hands benumbed, nor seduced by purely material means. For, as that ancient philosopher replied to one who was laughing at him for being unable to win the favour of some tendril he was pursuing: 'My friend, the hook will not bite when the curd is so fresh.'[170]

[B] Now love is a commerce which requires inter-relationship and reciprocity. We can show our appreciation of the other pleasures we receive by recompenses of a different nature: this one can only be repaid in the same coin. [C] Truly in this one the pleasure that I give stimulates my imagination more sweetly than the pleasure I receive. [B] A man who can receive pleasure when he gives none at all is in no wise generous: it is a base soul which will owe the lot and is pleased to nurse contacts with women who do all the paying. There is no beauty nor grace nor intimacy so exquisite that a gentleman should want them at that price. If they can only do us a good turn out of pity, then I would dearly prefer not to live at all than to live on charity. Would that I had the right to ask it of them in the style which I have seen beggars use in Italy: *'Fate ben per voi'* [Do a good turn for yourself]; [C] or in the manner which Cyrus adopted to exhort his soldiers: 'He who loves himself, let him follow me.'[171]

[B] Someone will say to me: 'Go back again to women who are now in the same state as you are: fellowship in the same misfortune will make them easier to get.' What absurd and dull terms for a truce!

Nolo
Barbam vellere mortuo leoni!

170. Erasmus, *Apophthegmata*, VII, *Bion Borysthenites*, II. Cheese, curd, *Caseus*, was a Latin term of amorous endearment. (Erasmus chastely holds this expression to mean that Philosophy cannot 'hook' tender minds; Montaigne, more literally, that ageing philosophers cannot 'hook' tender lovers.)
171. A famous saying, parodied by Rabelais (*Gargantua*, XXXI, end) to mock Picrochole, his foolish, choleric monarch.

[I have no desire to pluck hairs from a dead lion's beard!][172]

[C]   One of the reproaches and accusations that Xenophon makes about
Meno is that in his love-affairs he only got on the job with partners past
their bloom.[173] The sight of a young couple appropriately united in a
tender embrace – or even the contemplation of it in imagination – contains
I believe more sensual pleasure than being the second partner in a sad
misshapen union.   [B]   I leave that fanciful appetite to the Emperor
Galba who devoted himself only to tough and ancient flesh – or to that
other pitiful wretched man:

> *O ego di' faciant talem te cernere possim,*
> *Charaque mutatis oscula ferre comis,*
> *Amplectique meis corpus non pingue lacertis!*

[O would the gods let me see you as you are, tenderly kiss your fading hair and
clasp your withered body in my embrace!][174]

[C]   And I count among the principal forms of ugliness all beauties due
to artifice and constraint. A young lad of Chio called Hemon, hoping that
fine clothes would procure him that handsomeness which Nature had
denied him, came to the philosopher Arcesilaus and asked him if a
philosopher could ever find himself in love. 'Oh yes,' he replied, 'provided
it be not with a dishonest dressed-up beauty such as yours.'[175] An ugly old
age when openly avowed is in my opinion less old and less ugly than one
smoothed out and painted over.

[B]   Shall I say it, on condition that you do not jump down my throat?
Love never seems to me to be properly and naturally seasonable except in
the age nearest boyhood:

> *Quem si puellarum insereres choro,*
> *Mille sagaces falleret hospites*
> *Discrimen obscurum, solutis*
> *Crinibus ambiguoque vultu.*

[A youth such that, if you put him among a band of maidens, those who knew

172. Martial, *Epigrams*, X, xc, 10–11.
173. Xenophon, *Anabasis*, II, vi; then, Suetonius, *Life of Galba*, XXII.
174. Ovid, wretched in unending exile on the orders of Augustus Caesar (*Ex Ponto*, I, iv, 49–51).
175. Diogenes Laertius, *Life of Arcesilaus*. (Cf. Erasmus, *Apophthegmata*, VII, *Arcesilaus*, VI.)

him not, for all their perspicacity, would fail to pick him out with his flowing hair and his hermaphrodite's face.][176]

[C]   Nor handsomeness, either. For Plato himself noted that Homer prolongs it until there is a shadow of a beard on the chin, but remarks that such a flower is rare. (We all know why Dion the Sophist jokingly called the mossy beards of adolescence Aristogitons and Harmodians!)[177]

[B]   I find love already out of place in adult manhood let alone in old age.

> *Importunus enim transvolat aridas*
> *Quercus.*

[For Cupid disdainfully flies past the withered oak.][178]

[C]   Queen Margaret of Navarre (just like a woman) greatly extends the privileges of women when she ordains that it is time for them to change the title *beautiful* for *good* after they have reached thirty.[179]

[B]   The shorter the tenancy we grant to Cupid in our lives the better off we are. Look at his deportment! And his chin is as smooth as a boy's! Who is unaware that in Cupid's school you do everything contrary to good order? There the novices are the professors: study, practice and experience lead to failure.   [C]   *'Amor ordinem nescit.'* [Cupid knows no order.][180]   [B]   The way Cupid conducts things is most in fashion when mingled with ingenuousness and awkwardness; mistakes and failures lend it charm and grace; provided it is sorrowful and yearning, it little matters whether it shows prudence. See how Cupid stumbles along, tripping over merrily; to guide him by art and wisdom is to clamp him in the stocks: you constrain his divine freedom when you lay hairy calloused hands upon him.

Moreover I often hear women portraying a relationship as being entirely of the mind, disdaining to take into consideration the interests which our senses have in it.[181] Everything helps in this case, but I should add that

---

176. Horace, *Odes*, II, v, 21–4; then, Plato, *Protagoras*, 309 AB, alluding to Homer, *Iliad*, XXIV, 348.

177. Conspirators who freed Athens from the tyranny of the Pisistratids. Similarly, a sprouting beard freed youths from the 'tyranny' of homosexual advances: Plutarch (tr. Amyot), *De l'amour*, 613 AB. Saying of Bion (not Dion).

178. Horace, *Odes*, IV, xiii, 9–10.

179. Margaret of Navarre, *Heptaméron*, Journée 4, nouvelle 35 (an unfair remark: Margaret does not 'ordain' it, but notes that it is usual).

180. St Jerome, *Letters, Ad Chromatium* (identified by Marie de Gournay).

181. This was the general drift of Renaissance 'platonic' love.

though I have often found that we men have overlooked weaknesses in their minds on account of the beauty of their bodies, I have yet to see one woman willing, on account of the beauty of a man's mind, however mature and wise, to lend a helping hand to his body once it has even begun to decline. Why is not one of them ever moved by desire for that noble [C] Socratic [B] bargain of body for mind, [C] purchasing at the price of her thighs a philosophical relationship and procreation through the soul – the highest price she could ever get for them![182]

Plato decrees in his laws that a man who has achieved some signal and useful exploit in a war may not, for the duration of that conflict, irrespective of his age or ugliness, be refused a kiss or any other of love's favours from anyone he pleases [183] Can what he finds so just in commendation of a warrior's worth not also be used to commend worth of another kind? And why is no woman ever moved [B] to win, before her fellow-women do, the glory of a love so chaste? Yes, I do indeed say chaste:

> *nam si quando ad prælia ventum est,*
> *Ut quondam in stipulis magnus sine viribus ignis*
> *Incassum furit.*

[for when it comes to the clinch, its frenzied love serves no purpose; like burning stubble: lots of flame but no force.][184]

We do not rank among our worst vices those whose fire is smothered in our minds.

To bring to an end these infamous jottings which I have loosed in a diarrhoea of babble – a violent and at times morbid diarrhoea –

> *Ut missum sponsi furtivo munere malum*
> *Procurrit casto virginis e gremio,*
> *Quod miseræ oblitæ molli sub veste locatum,*
> *Dum adventu matris prosilit, excutitur,*
> *Atque illud prono præceps agitur decursu;*
> *Huic manat tristi conscius ore rubor*

[as when an apple, secretly given by her admirer breaks loose from the chaste bosom of a maiden as she starts to her feet on hearing her mother's footstep,

---

182. Cf. I, 28, 'On affectionate relationships'; Socratic philosophers paid with their teachings of virtue and wisdom for the homage of youthful disciples. Philosophers beget ideas (brain-children) rather than real children.
183. In Plato's *Republic* (not his *Laws*), V, 468.
184. Virgil, *Georgics*, III, 98–100 (of an aged stallion).

forgetting she had concealed it beneath her flowing robes; it lies there on the ground while a blush suffuses her troubled face and betrays her fault][185]

– I say that male and female are cast in the same mould: save for education and custom the difference between them is not great. [C] In *The Republic* Plato summons both men and women indifferently to a community of all studies, administrations, offices and vocations both in peace and war;[186] and Antisthenes the philosopher removed any distinction between their virtue and our own.[187]

[B]   It is far more easy to charge one sex than to discharge the other. As the saying goes: it is the pot calling the kettle smutty.

185. Catullus, LXV, 19–24.
186. Plato, *Republic*, V, where no sex distinctions are allowed to affect eligibility for the offices of State.
187. Erasmus, *Apophthegmata*, VII, *Antisthenes*, LVII. Erasmus comments: 'So too did Socrates think women to be no less apt for instruction in all the duties of wisdom than men, provided they receive the same education. Yet the mob condemn women as though they cannot be taught virtue.'

6. On coaches

[A favourite chapter, linking the ideas of fantastic luxury, generosity and princely magnificence with fantastic cruelty, vulgarity and ostentation. Coaches (which for Montaigne means all sorts of wheeled vehicles including Roman chariots) were the symbols of luxury. They are contrasted with the simplicity of those American Indian cultures which had never invented the wheel, had no horses and used gold for its beauty alone. Their simplicity emphasized the horrors of the Spanish conquest of Peru, with its naked cruelty and avarice.

Montaigne's three main sources are a work of Pietro Crinito, De honesta disciplina; another, by Justus Lipsius, De amphitheatro; a third by Francisco Lopez de Gomara, one of the Conquistadores, whom he read in the French translation by J. Fumée: Histoire générale des Indes.]

[B] It is very easy to prove that, when great authors write about causes, they not only marshal those which they reckon to be true but also those which they do not believe, provided that they have some [C] originality and [B] beauty.[1] If what they say is ingenious they think that their words are sufficiently useful and true. We cannot be sure of the master-cause, so we pile cause upon cause, hoping that it may happen to be among them:

> *namque unam dicere causam*
> *Non satis est, verum plures, unde una tamen sit.*

[since it suffices not to give one single cause, many must be given, one of which only may be true.][2]

You ask me: 'What is the origin of our custom of saying *Bless you* when people sneeze?' Well, we break three sorts of wind: the one which issues lower down is very dirty; the one which issues from the mouth comports an element of reproach for gluttony; and the third is sneezing, to which, since it issues from the head and is blameless, we give that honourable greeting.

1. '88: some *appositeness* and beauty . . .
2. Lucretius, VI, 704–5.

Do not mock such subtle reasoning: it is (so they say) from Aristotle . . .³

I came across, in Plutarch I think (and he is of all the authors I know the one who has best blended art with nature and judgement with erudition), the explanation that the vomiting from the stomach which befalls men on sea-voyages is to be attributed to fear.⁴ (He had already found some reason or other to prove that fear can produce such an effect.) Now I am very subject to seasickness and I know that that cause does not apply to me; and I know it not by argument but compelling experience. I shall not cite what I have been told, that animals, especially pigs, which have no conception of danger, get seasick; nor what one of my acquaintances has told me about himself: he is much subject to it yet on two or three occasions when he was obsessed by fear during a great storm the desire to vomit disappeared – [C]   as it did to that man in Antiquity: *'Pejus vexabar quam ut periculum mihi succurreret.'* [I was too shaken for the danger to occur to me.]⁵ [B]   Though many occasions for being afraid have arisen (if you count death as one) I have never felt, on water nor anywhere else, such fear as to confuse or to daze me. Fear can arise from lack of judgement as well as from lack of courage. All such dangers as I have encountered have been with my eyes open, with my sight free, sound and whole: besides, to feel fear you also need to have courage. Once when I did have to flee, I was able to manage my flight well and, compared with others, to maintain some order because I did so   [C]   if not without fear nevertheless [B]   without ecstatic terror; fear was aroused, but not the kind which is thunderstruck or insane. The souls of great men can go far beyond that, showing us retreats which were not merely tranquil and sane but marked by pride.

Here let me quote the flight which Alcibiades relates: it concerns Socrates, his companion in arms:⁶ 'I came across him (he said) after the rout of our army; he and Laches were the last to retreat. I could watch him at leisure and in safety, since I was on a good horse while he was on foot; that is the way we had fought. I noted first his presence of mind and the resolve which he showed in contrast with Laches; next it was his confident walk, in no ways different from his usual one; then the controlled and steady eyes with which he weighed and evaluated what was going on about him, staring now at some who were friends, now at others who

3. In the *Problemata*, XXXIII, 9, attributed to Aristotle.
4. Plutarch (tr. Amyot), *Causes naturelles*, 536H–537A.
5. Seneca, *Epist. moral.*, LIII, 3 (of his own experience).
6. Plato, *Symposium*, 221A–B.

were foes, encouraging the friends and showing the others that he was a man to sell his life-blood very dear should any assay to take it from him. That saved them, for you do not willingly attack men like that: you hunt the fearful.'

There you have the testimony of a great Captain, teaching us (what we can assay every day) that nothing casts us into dangers so much as a rash hunger to get out of them: [C] *'Quo timoris minus est, eo minus ferme periculi est.'* [As a rule, where you feel less fear you experience less danger.][7]

[B] People today are wrong to say 'That man is frightened of dying,' when they really mean that he dwells on it and anticipates it. Anticipation equally concerns whatever affects us, for good or evil. In some ways, weighing and evaluating a danger is the opposite of being thrown into amazement by it. I do not think I am strong enough to sustain the violent onslaught of fear nor of any other passion which disturbs the mind. If ever I were once to be vanquished and thrown to the ground by it I would never wholly get up again; should anything make my soul lose her footing I could never set her back straight in place again. She is ever probing and feeling herself too vigorously and examining herself too deeply; consequently she would never allow the wound which had pierced her to grow together and become strong. It is a good thing for me that no malady has so far overthrown her. Each onslaught against me I confront and oppose equipped in full armour, so the first to get the better of me would leave me without resources. There is no question of doing anything twice: let the storm once breach my dyke anywhere and all of me is open, irremediably drowned. [C] Epicurus asserts that no wise man can become the opposite of wise.[8] I know something about that judgement the other way round: no man who has been a real fool once will ever be really wise again!

[B] God sends us cold according to our garment; he sends me emotions according to my means of sustaining them. Nature, having exposed me on one flank has covered me on the other: having stripped me of fortitude she has equipped me with an inability to feel and with blunted balanced powers of anticipation.

Now I cannot put up for long with coach, litter or boat (and could do so less still in my youth). I loathe all means of conveyance but the horse, both for town and country. But litters I can tolerate less than coaches; and for

7. Livy, XXII, v.
8. Diogenes Laertius, *Life of Epicurus.*

the same reason I can better tolerate being thrown about on a rough sea – which produces fear – than I can the motion experienced during calm weather. Just as I cannot suffer a rickety chair under me, similarly I cannot suffer that slight jerk made by the oars as they pull the boat from under us without it somehow disturbing my brain and my stomach. Now, when sail or current bears us smoothly along or when we are towed, the unified motion in no wise bothers me; what upsets me is that series of broken movements, the more so when it is slow. I cannot describe its characteristics any other way. Doctors have prescribed binding a towel as a compress round the lower part of my belly; I have never assayed it, being used to fighting against my defects and vanquishing them by myself.

[C] If my memory were adequately furnished with them I would not regret time spent listing here the infinite variety of historical examples of the applications of coaches to the service of war, varying as they do from nation to nation and century to century; they are, it seems to me, most effective and very necessary. It seems a marvel to me that we have forgotten all about it. I will merely say this: quite recently in our fathers' time the Hungarians put them to excellent use against the Turks; in each coach a soldier with a round buckler was stationed beside a musketeer, together with a number of harquebuses in racks, already loaded. They clad the sides of each coach with rows of shields rather like a frigate. They drew up a line in front of their troops consisting of three thousand such coaches; after the cannon had played their part they either sent them ahead towards the enemy who had to swallow that salvo as a foretaste of what was to come (no slight advantage), or else they threw them against the enemy squadrons to break them up and open a way through. In addition there was the help they could give in covering the flanks of their troops when marching through ticklish country or in speedily defending an encampment by turning it into a fort.[9]

In my own day there was a gentleman living on one of our frontiers; he was an invalid and could find no horse able to bear his weight; he was involved in a feud and campaigned in a coach such as I have described and managed very well. But let us finish with those war-coaches.

The kings of our first Gaulish dynasty used to travel the land in a cart drawn by four oxen.[10] [B] Mark Antony was the first to be drawn

9. Nicolas Chalcocondylas, *Décadence de l'empire grec*, VII, vii (tr. Blaise de Vigenère).
10. Du Haillant, *Hist. des Roys de France*, II; then a series of examples from Pietro Crinito, *De honesta disciplina*, XVI, v.

through Rome – with a minstrel-girl beside him – by lions harnessed to a coach. Heliogabalus did the same somewhat later, claiming to be Cybele the Mother of the gods; then, drawn by tigers, he pretended to be the god Bacchus. On other occasions he harnessed two stags to his coach; once it was four dogs; then he stripped naked and was drawn in solemn procession by four naked girls. The Emperor Firmus had his coach drawn by ostriches of such extraordinary size that he seemed to fly rather than to roll along. The oddness of such novelties leads me on to the idea that it is a sort of lack of confidence in monarchs, a sign of not being sure of their position, to strive to make themselves respected and glorious through excessive expenditure. It would be pardonable abroad but among his subjects, where he is the sovereign power, the highest degree of honour to which he can attain is derived from the position he holds. Similarly it seems to me that it is superfluous for a gentleman to take a lot of trouble over how he dresses when at home: his house, his servants, his cuisine are enough to vouch for him there.

[C] Isocrates' advice to his king does not seem to lack good sense: let his furniture and his tableware be magnificent, for such expenditure is of lasting value and is passed on to his successors: let him avoid all magnificence which drains away immediately from use or memory.[11]

[B] When I was a young man, in default of other glories I gloried in fine clothes. In my case they were quite becoming; but there are folk on whom fine clothes sit down and cry.

There are tales of the extraordinary meanness of some of our kings over both personal expenditure and donations – and they were kings great in reputation, wealth and fortune. Demosthenes fought unsparingly against one of his city's laws which authorized monies to be spent on parades of athletes and festivals (he wanted his city's greatness to be displayed in the number of its well-armed fighting-ships and in good, well-equipped forces). [C] And Theophrastus is rightly condemned for asserting the opposite doctrine in his book *On Riches*, in which he maintained that expenditure on festivals was the true fruit of opulence. Such pleasures, says Aristotle, have an effect only on the lowest of the low; they immediately vanish from their memory as soon as they have had enough of them; no serious man of judgement can hold them in esteem.[12]

11. Isocrates, *Nicocles*, VI, xix.
12. [C] all from Cicero, *De officiis*, II, xvi, 56–7. Aristotle's judgement otherwise unknown.

Such funds would seem to me to be more regal, useful, sensible and durable if spent on ports, harbours, fortifications and walls, on splendid buildings, on churches, hospitals and colleges, and on repairing roads and highways. In my time Pope Gregory XIII left a favourable reputation behind him by so doing; and, by so doing, our own Queen Catherine would for many a long year to come leave witnesses to her natural generosity and munificence, if only her means were sufficient for her desires.[13] Fortune deeply distressed me by interrupting the construction in our capital city of the Pont neuf, a beautiful bridge, so cheating me of the hope of seeing it in regular use before I die.

[B] Moreover to their subjects who form the spectators of these festivities, it seems that it is their own wealth that is being flaunted and that they are being feasted at their own expense. Their peoples are always ready to assume about kings what we assume about our servants: that their job is to provide abundantly for everything that we want but never to spend anything on themselves. That is why the Emperor Galba, when he was delighted by a musician during dinner, called for his chest, plunged in his hand and gave him a fistful of crowns saying, 'This is my own money not the government's.'[14] Be that as it may, the people are usually right: money earned to feed their bellies is used instead to feed their eyes.

Even munificence is not truly resplendent from a sovereign's hands: it more rightly belongs to private citizens; for strictly speaking a king has nothing which is properly his own: even his person belongs to others. [C] Sentences are not passed in the interests of the judge but of the plaintiffs. We never appoint our superiors for their own advantage but for that of their inferiors; we appoint a doctor for his patients not for himself. All public offices, like all professional skills, aim at something beyond themselves: *'nulla ars in se versatur'* [no art is concerned with itself].[15]

[B] That is why those tutors of youthful princes who pride themselves on impressing upon them that there is virtue in lavishness, who exhort them not to know what it means to reject anything and to hold that money is never better spent than when given away (teaching, greatly honoured, I know, in my own lifetime), are either thinking more of their own good than that of their own master or else they do not know what they are talking about. It is all too easy to stamp ideas of generosity on a

13. The Queen Mother, Catherine de' Medici.
14. Plutarch, *Life of Galba*.
15. Cicero, *De finibus*, V, vi, 16.

man who has the means of fulfilling them with other people's money.
[C] And since generosity is measured not against the gift but the means
of the giver, in such powerful hands it always proves useless. To be
generous, they discover, they have to be prodigal. [B] So it is not
highly honoured compared to the other kingly virtues: it is, said Dionysius
the Tyrant, the only virtue to be fully compatible with tyranny itself.[16] I
would rather teach a king this line from one ancient ploughman:

Τῇ χειρὶ δεῖ σπείρειν, ἀλλὰ μὴ ὅλῳ τῷ θυλάκῳ

that is, 'If you want a good crop, you must broadcast your seed not pour it
from your sack.'[17] [C] Seed must be drilled not spilled. [B] So
when a king has to make gifts or, to put it better, has to make payments to
so many persons for services rendered, he should distribute royally but
advisedly. If a prince's generosity is indiscriminate and immoderate I would
like him better as a miser.

It is in justice that kingly virtue seems mainly to consist. And what most
distinguishes a king is that kind of justice which is the companion of
generosity; kings readily dispense all other kinds of justice through
intermediaries: that one they reserve to themselves.

Liberality without moderation is a feeble means of acquiring good-will,
since it offends more people than it seduces. [C] *'Quo in plures usus sis,
minus in multos uti possis. Quid autem est stultius quam quod libenter facias,
curare ut id diutius facere non possis?'* [The more people you have helped by
it, the fewer you can help in the future . . . Is there a greater folly than
doing something you like in such a way that you can do it no
longer?][18] [B] And if it is exercised without due regard for merit, it
embarrasses the recipient, who receives it without gratitude. There have
been tyrants who have been sacrificed to the people's hatred by the very
men they have unjustly advanced, since [C] men of that sort [B]
reckon that[19] they can insure their possession of ill-gotten gains by showing
hatred and contempt for the one they got them from; in that way they
seek to placate the judgement and opinions of the people.

The subjects of a prince who is lavish in giving become lavish in their
demands. They base their assessments not on reason but example. We

16. Plutarch (tr. Amyot), *Dicts notables des anciens Roys*, 190 D–E.
17. In Amyot's Plutarch (525 F) this verse of Corinna's is cited in French, not
Greek. The original appears in Justus Lipsius, *De amphitheatro*.
18. Cicero, *De officiis*, II, xv, 52–3; 54.
19. '88: since *clowns, pimps, fiddlers and other such riff-raff* reckon that . . .

certainly ought often to blush at our shamelessness. We are already overpaid by just standards once the reward is equal to our services. Do we owe nothing to our princes by natural obligation? If our prince meets our expenses he has already done a great deal. Should he contribute to them, that is enough: anything above that is called a bounty: as such it cannot be demanded. (The very word liberality has the sound of liberty.) By our fashion there is no end to it: goods already received do not figure in our accounts: we only love future liberality. So the more a prince exhausts his wealth in giving, the poorer he is in friends.   [C]   How could he possibly slake desires which grow bigger the more he pours wealth into them? The man whose thoughts are set on getting thinks no longer of what he has got. The property of covetousness is, above all, ingratitude.[20]

The example of Cyrus would not fit in badly here to serve our kings today as a touchstone for discovering whether their gifts are well or ill bestowed (and to show them that that Emperor distributed his gifts better than they do; by their extravagance they are reduced to raising loans from subjects unknown to them or from those whom they have harmed rather than from those whom they have helped, receiving 'gratuities' from them which have nothing gratuitous about them but the name). Croesus reproached Cyrus for his bounty, calculating what his treasure would have amounted to if he had restrained his hands a little more. Cyrus sought to justify his liberality: so he dispatched messengers all over the place to those magnates of his empire whose interests he had individually advanced, begging each of them to help him with as much money as they could for some urgent need and to write to him disclosing the amount. When all the letters of credit were brought to him, none of his friends had reckoned that it was enough to offer merely as much as they had received from his munificence but included much of their own wealth. He found that the sum amounted to far more than Croesus' economies. Whereupon Cyrus said to him, 'I love riches no less than other princes do; if anything I am more sparing. You can see by what little outlay I have acquired the countless riches of so many friends, and how much better Chancellors of the Exchequer they are than hired men would be with no bonds of affection, and how my wealth is better lodged with them than in my own treasure-chests, calling down upon me the hatred, envy and contempt of other princes.'[21]

   [B]   The Roman Emperors justified the lavishness of their public games

20. Seneca, *Epist. moral.*, LXXIII, 2–3.
21. Xenophon, *Cyropaedia*, VIII, ii.

and parades by the fact that their authority in some ways depended (in appearance at least) on the will of the people, who had ever been accustomed to be courted by such extravagant spectacles. Yet it was private citizens who had encouraged this custom of pleasing their fellow-citizens and their equals with such a profusion of magnificence drawn mainly from their own purses. It took on a quite different savour when their masters came to imitate them. [C] *'Pecuniarum translatio a justis dominis ad alienos non debet liberalis videri.'* [Taking money from rightful owners and giving it to others ought not to be regarded as liberality.][22] When his son assayed winning the support of the Macedonians by sending them gifts, Philip reprimanded him in a letter with these words: 'What? Do you desire that your subjects should consider you not their King but their bursar? If you want to seduce, seduce them by deeds of virtue not by deeds of your purse-strings.'

[B] Yet there was beauty in providing a great quantity of mature trees, with thick green branches, and in planting them beautifully and symmetrically in the arena to make a great shady forest, and then, on the first day, in releasing within it a thousand ostriches, a thousand stags, a thousand wild boars and a thousand deer and in handing it over to the populace to pillage; then, on the following day, in killing off before them a hundred full-grown lions, a hundred leopards and three hundred bears; then, on the third day, in having three hundred pairs of gladiators fight to the finish, as did the Emperor Probus.[23] Beautiful too to see those great amphitheatres incrusted on the outside with marble and decorated with works of art and statuary, the inside gleaming with rare and precious stones –

> *Baltheus en gemmis, en illita porticus auro*

[Here is the circular partition clad in gems; here, the portico, daubed with gold]

– with all the sides surrounding that vast space completely encircled from top to bottom with sixty to eighty tiers of seats, also of marble, covered with cushions –

> *exeat, inquit,*
> *Si pudor est, et de pulvino surgat equestri,*
> *Cujus res legi non sufficit;*

22. Cicero, *De officiis*, I, xiv, 43 (on the liberality of Sylla and Gaius Caesar); then, II, xv, 53–4 (on Philip of Macedonia).
23. Related after Pietro Crinito, *De honesta disciplina*, XII, vii, with interpolated verses from Calpurnius' *Bucolica*, VII, 47; Juvenal, *Satires*, III, 153–5 and Calpurnius, *Bucolica*, VII, 64–75, taken (with much else) from Justus Lipsius' *De amphitheatro*.

['Shame him out', they say: 'he has only paid for the cheapest seats, not for the cushioned ones of the knights]

— where you could seat a hundred thousand men in comfort; beauty, too, to have the base of the arena where the games took place dug up and divided into caverns representing lairs which spewed forth the animals destined for the spectacle; subsequently to flood it with a deep sea of water, sweeping along many a sea-monster and bearing armed warships to enact a naval engagement; then, thirdly, to flatten it and drain it out afresh for the gladiatorial combats; and then, for the fourth act, to strew it, not with sand but with vermilion and aromatic resin in order to prepare upon it a formal banquet for that infinite crowd of people — the final scene on one single day!

> *Quoties nos descendentis arenæ*
> *Vidimus in partes, ruptaque voragine terræ*
> *Emersisse feras, et iisdem sæpe latebris*
> *Aurea cum croceo creverunt arbuta libro.*
> *Nec solum nobis silvestria cernere monstra*
> *Contigit, æquoreos ego cum certantibus ursis*
> *Spectavi vitulos, et equorum nomine dignum,*
> *Sed deforme pecus.*

[How often have we beheld a section of the arena drop down, forming a gaping chasm from which emerged wild beasts and whole forests of golden trees with barks of saffron! Not only have we seen the denizens of the forests in our amphitheatres but sea-beasts set in the midst of fighting bears and those monstrous hippopotamuses honoured by the name of 'river-horses'.]

Sometimes they produced in the arena a great mountain covered with green trees, many bearing fruit, and a river running from its summit as from the source of a flowing stream. Sometimes they had a great ship sail into the arena; it opened up and fell apart automatically, spewed forth from its belly four or five hundred beasts of combat, reassembled itself unaided and vanished from sight. Sometimes down there in the arena they produced fountains and water-jets which spouted immensely high, sprinkling perfume over that vast multitude. To protect themselves from the hot weather they caused that immense area to be covered either with awnings of purple needlework or with variously coloured silks, which they drew or withdrew at will:

> *Quamvis non modico caleant spectacula sole,*
> *Vela reducuntur, cum venit Hermogenes.*

[Although the fierce sun beats down on the amphitheatre they draw back the awnings whenever Hermogenes appears.][24]

Even the netting erected in front of the crowd to protect them from the ferocity of the wild beasts once they were loosed was plaited with gold:

> *auro quoque torta refulgent*
> Retia.

[The very nets glisten with woven gold.]

If anything can justify such excesses, it is the cases where the amazement was caused not by the expense but by the originality and ingenuity.

Even in vanities such as these we can discover how those times abounded in more fertile minds than ours. The same applies to that sort of fertility as to any other which Nature produces. Which is not to say that she then employed her utmost forces.[25] [C] We cannot be said to progress but rather to wander about this way and that. We follow our own footsteps. [B] I am afraid that our knowledge is in every sense weak; we cannot see very far ahead nor very far behind; it grasps little, lives little, skimped in terms of both time and matter.

> *Vixere fortes ante Agamemnona*
> *Multi, sed omnes illachrimabiles*
> *Urgentur ignotique longa*
> Nocte.

[Great heroes lived before Agamemnon; many they were, yet none is lamented, being swept away unknown into the long night.]

> *Et supera bellum Trojanum et funera Trojæ,*
> *Multi alias alii quoque res cecinere poetæ.*

[Before the Trojan War and the death of Troy many other poets have sung of other wars.][26]

[C] And while on this subject I think we should not reject the testimony

24. Martial, *Epigrams*, XII, xxix, 15–16; then, Calpurnius, *Bucolica*, VII, 53–4, with other matter from Justus Lipsius.
25. '88: forces. *There is verisimilitude in saying that we neither go forward nor backwards, rolling, rather, spinning and changing.* I am afraid . . . Then, Horace, *Odes*, IV, ix, 25–8.
26. Lucretius, V, 327–8.

of Solon's account of how he had learned from the priests of Egypt the long history of their State and their way of teaching and preserving the history of other peoples: '*Si interminatam in omnes partes magnitudinem regionum videremus et temporum, in quam se injiciens animus et intendens ita late longeque peregrinatur, ut nullam oram ultimi videat in qua possit insistere: in hac immensitate infinita vis innumerabilium appareret formarum.*' [If we were vouchsafed a sight of the infinite extent of time and space stretching away in every direction, and if our minds were allowed to wander over it far and wide, ranging about and hastening along without ever glimpsing a boundary where it could halt: from such an immensity we would grasp what almighty power lies behind those innumerable forms.][27]

[B]   Even if everything that has come down to us about the past by report were true and known to someone, that would be nothing compared with what we do not know. And against the idea of a universe which flows on while we are in it, how puny and stunted is the knowledge of the most inquisitive men. A hundred times more is lost for us than what comes to our knowledge, not only of individual events (which sometimes are turned by Fortune into weighty *exempla*) but of the circumstances of great polities and nations. When our artillery and printing were invented we clamoured about miracles: yet at the other end of the world in China men had been enjoying them over a thousand years earlier.[28] If what we saw of the world were as great as the amount we now cannot see, it is to be believed that we would perceive an endless [C] multiplication and [B] succession of forms. Where Nature is concerned, nothing is unique or rare: but where our knowledge is concerned much certainly is, which constitutes a most pitiful foundation for our scientific laws, offering us a very false idea of everything.

Just as we vainly conclude today that the world is declining into decrepitude using arguments drawn from our own decline and decadence –

Jamque adeo affecta est ætas, affectaque tellus

[Our age lacks vigour now: even the soil is less abundant][29]

27. Cicero, *De natura deorum*, I, xx, 54 (changing Cicero's *atomorum* to *formarum*, thus linking the concept less to Lucretius than to Plato's Great Chain of Being).
28. Many, including Rabelais, believed that printing was invented under the inspiration of the Holy Ghost, so as to counteract the Devil's invention of gunpowder and artillery (cf. *Pantagruel*, TLF, VIII, 92–5). Knowledge of China was being spread especially by the Jesuits.
29. Lucretius, II, 1136; then, V, 331–5.

– so that same poet concluded that the world was yet newly born and young, from the vigour of the minds of his day, fertile in new inventions and the creation of various arts:

> *Verum, ut opinor, habet novitatem summa, recensque*
> *Natura est mundi, neque pridem exordia cœpit:*
> *Quare etiam quœdam nunc artes expoliuntur,*
> *Nunc etiam augescunt, nunc addita navigiis sunt*
> *Multa.*

[In my opinion our universe is new; the origin of the world is recent: it is but newly born. That is why some arts are still developing nowadays and growing still; the art of navigation is even now progressing.]

Our world has just discovered another one: and who will answer for its being the last of its brothers, since up till now its existence was unknown to the daemons, to the Sybils, and to ourselves? It is no less big and full and solid than our own; its limbs are as well developed: yet it is so new, such a child, that we are still teaching it its ABC; a mere fifty years ago it knew nothing of writing, weights and measures, clothing, any sort of corn or vine. It was still naked at the breast, living only by what its nursing Mother provided. If we are right to conclude that our end is nigh, and that poet is right that his world is young, then that other world will only be emerging into light when ours is leaving it. The world will be struck with the palsy: one of its limbs will be paralysed while the other is fully vigorous, yet I fear we shall have considerably hastened the decline and collapse of that young world by our contagion and that we shall have sold it dear our opinions and our skills.

That world was an infant: we whipped it and subjected it to our teaching, but not from any superior worth of ours or our natural energy; we neither seduced it by our justice and goodness nor subjugated it by our greatness of soul. Most of the responses of its peoples, and most of our negotiations with them, witness that they are in no ways beholden to us where aptitude and natural clarity of mind are concerned. The awe-inspiring magnificence of the cities of Cuzco and Mexico and, among similar things, the gardens of that king where all the trees and fruits and all the plants were, in size and arrangement, as in a normal garden, but all excellently wrought in gold, as were in his museum all the creatures which are born in his estates or in his seas;[30] the beauty of their

30. The Inca garden and museum described on hearsay by Lopez de Gomara (tr. Fumée), *Histoire générale des Indes*, V, xiii. Much of what follows is from that work.

works of art in precious stones, feathers and cotton as well as in painting shows that they were not behind us in craftsmanship either.

And as for their piety, observance of the laws, goodness, liberality, loyalty and frankness: well, it served us well that we had less of that than they did; their superiority in that ruined them, sold them and betrayed them.

As for bravery and courage; as for resolution, constancy and resistance to pain, hunger and death, I would not hesitate to compare the examples provided by them with the most celebrated ones of the Ancients written in the annals of our own world on this side of the seas.

As regards those men who subjugated them, were you to take from them the trickery and sleight-of-hand which they used to deceive them, the justified ecstasy of amazement which struck those peoples at the sight of the totally unexpected landing of bearded men, differing from them in language, religion, build and facial features, coming from a world so remote and from regions in which they had never even dreamed that there were any humans dwelling whatsoever; men mounted on big unknown monsters confronting men who had never seen not merely horses but any animal whatsoever trained to be ridden by man or to bear any other burden; men whose skin was shining and hard, men armed with a glittering cutting-instrument confronting men who would barter a vast wealth of gold or pearls for a looking-glass or a knife, the sheen of which to them appeared miraculous; men who, even if they had had the time, had neither the knowledge nor the materials to discover ways of piercing our steel; to which add the lightning flashes of our cannons, the thundering of our harquebuses (able to confuse the mind of Caesar himself in his day if they had surprised him when he was as ignorant of them as they were) opposed to people who were naked except in those areas which had been reached by the invention of a kind of woven cotton-cloth; peoples with no arms except (at the most) bows, stones, staves [C] or wooden shields; [B] peoples who, under pretence of friendship and good faith, were caught off their guard by their curiosity to see things strange and unknown: remove (I say) from the Conquistadores such advantages and you strip them of what made so many victories possible.

When I reflect on the indomitable ardour with which so many thousands of men, women and children came so many times and threw themselves into certain danger in defence of their gods and their freedom, and when I reflect on that great-souled stubborn determination to suffer any extremity, any hardship including death, rather than to submit to the domination of those who had so disgracefully deceived them – some of them preferring

once they were captured to die slowly of hunger than to accept food from the hands of enemies so vilely victorious: I maintain that, if they had been attacked equal to equal in arms and experience and numbers, then the conflict would have been as hazardous (or more so) as any other that we know of.

Oh why did it not fall to Alexander and those ancient Greeks and Romans to make of it a most noble conquest; why did such a huge transfer of so many empires, and such revolutions in the circumstances of so many peoples, not fall into hands which would have gently polished those peoples, clearing away any wild weeds while encouraging and strengthening the good crops that Nature had brought forth among them, not only bringing to them their world's arts of farming the land and adorning their cities (in so far as they were lacking to them) but also bringing to the natives of those countries the virtues of the Romans and the Greeks? What a renewal that would have been, what a restoration of the fabric of this world, if the first examples of our behaviour which were set before that new world had summoned those peoples to be amazed by our virtue and to imitate it, and had created between them and us a brotherly fellowship and understanding. How easy it would have been to have worked profitably with folk whose souls were so unspoiled and so hungry to learn, having for the most part been given such a beautiful start by Nature. We, on the contrary, took advantage of their ignorance and lack of experience to pervert them more easily towards treachery, debauchery and cupidity, toward every kind of cruelty and inhumanity, by the example and model of our own manners. Whoever else has ever rated trade and commerce at such a price? So many cities razed to the ground, so many nations wiped out, so many millions of individuals put to the sword, and the most beautiful and the richest part of the world shattered, on behalf of the pearls-and-pepper business! Tradesmen's victories! At least ambition and political strife never led men against men to such acts of horrifying enmity and to such pitiable disasters.

While sailing along the coast on the lookout for the natives' mines there were some Spaniards who went ashore in a fertile, pleasant and densely populated countryside; they gave the inhabitants their usual warning, declaring that they were men of peace, coming to them after sailing far across the seas, sent on behalf of the King of Castile, the greatest monarch in the inhabited world, to whom the Pope, as Vicar of God on earth, had granted dominion over all the Indies; that, if they would pay that King tribute they would be most kindly treated; then they asked for victuals to eat, and for gold . . . which they needed as a medicine; they incidentally

insisted that there is only one God, that our religion is the true one which they advised them to adopt – adding a menace or two.

In reply they were told that, as for their being men of peace, if they were they did not look it; as for their King, he must be poor and needy since he came a-begging; as for that man who had apportioned that tribute, he was a man who loved dissension since he gave to a third party something which was not his to give, seeking to pick a quarrel with those who had long possessed it; as for victuals, they would supply some; as for gold, they had very little; it was something they did not highly value since it was of small practical use in life, whereas their aim was to live their lives in happiness and contentment; so the Spaniards could readily have whatever gold they could find, except the gold which was used in the service of their own gods. As for there being only one God, they were pleased by the argument but did not intend to change their religion, having so profitably followed their own for such a long time and being unaccustomed to taking advice from anyone but their friends and acquaintances. As for their menaces, it was a sign of lack of judgement in them to go about threatening people the nature of whose resources was unknown to them; so let them get out of their country, quickly, for they were not accustomed to take in good part such courtesies from armed men and warnings from foreigners. They would do to them what they had done to others – and they indicated the heads of men condemned to death and displayed about their city.

There is an example of their baby-talk for you!

So the Spaniards neither remained nor campaigned in that place nor in many others where they found none of the merchandise they were after, no matter what other delights could be found there. Witness my cannibals.[31]

The last two kings whom the Spaniards hounded were kings over many kings, the most powerful kings in that new world and perhaps also in our own.

The first was the King of Peru. He was captured in battle and put to so huge a ransom that it defies all belief; he paid it faithfully and showed by his dealings that he was of a frank, noble and steadfast heart, a man of honest and tranquil mind. The Conquistadores, having already extracted gold weighing one million three hundred and twenty-five thousand five hundred ounces (not counting silver and other booty amounting to no less,

---

31. I, 31, 'On the Cannibals', above, pp. 79–92.

so that afterwards they even used solid gold to shoe their horses) were seized with the desire to discover what remained of the treasures of that king, no matter what it cost them in bad faith, [C] and to make free with whatever he had kept back. [B] They fabricated false evidence, accusing him of planning to get his territories to rise up in revolt and to set him free. Whereupon – a beautiful sentence, delivered by those who had got up this act of treachery! – he was condemned to be publicly hanged until he was dead, having first been compelled to buy off the agony of being burned alive at the stake by accepting baptism – which was administered to him while he was being tortured.

A horrifying, unheard-of action, which he nevertheless bore without demeaning, by look or word, his truly regal gravity and comportment. And then to placate the people who were stunned into an ecstasy of amazement by so outlandish a deed, they counterfeited great grief at his death and arranged a costly funeral.

The second was the King of Mexico: he had long held out during the siege of his city, showing (if ever a people did so) what can be achieved by endurance and constancy, yet he had the misfortune to fall alive into the hands of his enemies, but on terms of being treated like a king. (And during his captivity he showed nothing unworthy of that title.) But the Spaniards, not finding after that victory as much gold as they had anticipated, pillaged and ransacked everything and then proceeded to seek information by inflicting on the prisoners they had taken the most painful tortures that they could devise. But since nothing of value could be extorted from them, their hearts being stronger than the tortures, the Spaniards finally fell into such a fit of madness that, contrary to their word and to the law of nations, they sentenced the King and one of the chief lords of his court to be tortured in each other's sight. That lord, overcome with pain, surrounded by blazing braziers, finally turned his gaze piteously towards his sovereign, as if to beg [C] forgiveness because he could stand it no longer. [B] That King[32] proudly and severely fixed his eyes on him to reproach him for his cowardice and faint-heartedness and simply said these words in a firm hoarse voice: 'What about me? Am I having a bat'? Am I any more at ease than you are?' Straightway afterwards that lord succumbed to the pain and died where he was. The King was borne away, half-roasted, not so much out of pity (for what pity could ever touch the souls of men who, for dubious information about some golden

32. '88: to beg *leave to tell what he knew to redeem himself from the unbearable pain.* That King . . .

vessel or other that they would pillage, would grill a man before their very eyes, not to mention a King of so great a destiny and merit) but because his constancy rendered their cruelty more and more humiliating.

When he afterwards made a courageous attempt to effect an armed escape from so long a captivity and slavery, they hanged him; he made an end worthy of a prince so great of soul.

On another occasion they set about burning, at one time and in the same pyre, four hundred and sixty men – every one of them alive – four hundred from the common people, sixty from the chief lords of the land, all straightforward prisoners of war.

These accounts we have from the Spaniards themselves.[33] They do not merely confess to them, they [C] boast of them and proclaim them. [B] Could it be[34] in order to witness to their justice or to their religious zeal? Such ways are certainly too contrary, too hostile, to so holy a purpose. If their intention had simply been to spread our faith, they would have thought upon the fact that it grows not by taking possession of lands but of men, and that they would have had killings enough through the necessities of war without introducing indiscriminate slaughter, as total as their swords and pyres could make it, as though they were butchering wild animals, merely preserving the lives of as many as they intended to make pitiful slaves to work and service their mines: so that several of the leaders of the Conquistadores were punished by death in the very lands they had conquered by order of the Kings of Castile, justly indignant at their dreadful conduct, while virtually all the others were loathed and hated.[35] To punish them God allowed that their vast plunder should be either engulfed by the sea as they were shipping it or else in that internecine strife in which they all devoured each other, most being buried on the scene, in no wise profiting from their conquest.

The gold actually received, even into the hands of a wise and thrifty Prince, corresponds so little to the expectations aroused in his predecessors

33. Montaigne's main source throughout is Francisco Lopez de Gomara (tr. Fumée), *L'Histoire générale des Indes* (1578 and 1587). It is not known whether he had also read the blistering attacks on the Conquistadores or on Spanish policy by Bishop Bartolome de las Casas, e.g. his *Brevissima relación de la destruyción de Las Indias* (Seville, 1552) or the account of his dispute entitled *Aqui se contiene una disputa entre B. de las Casas y G. de Sepulveda* (Seville, 1552), with which he would have been in agreement.

34. '88: they *preach and proclaim them*. Could it be . . .

35. These included Pizarro, condemned to death in 1548.

and to the abundant riches discovered when men first came to these new lands (for while they draw great profit from them we can see that it is nothing compared with what they could have expected); that is because the Indians knew nothing about the use of coinage. Consequently all their gold was gathered in one place, used only for display and parade; their gold was moveable-goods handed on from father to son by several puissant kings who always worked their mines merely to make great quantities of vessels and statues to decorate their palaces and their temples. All our gold circulates in trade. We break it down, change it in a thousand ways, spread it about and so disperse it. Just imagine what it would be like if our kings, over several centuries, had likewise piled up all the gold they could find and kept it idle.

The peoples of the Kingdom of Mexico were somewhat more urban and more cultured than the other peoples over there.[36] In addition, like us, they judged that the world was nearing its end, taking as a portent of this the desolation that we visited upon them. They believed that the world's existence was divided into five periods, each as long as the life of five successive suns. Four suns had already done their time, the one shining on them now being the fifth. The first sun perished with all other creatures in a universal Flood; the second, by the sky falling on mankind and choking every living thing (to which age they ascribed giant men, showing the Spaniards bones of men of such proportion that they must have stood twenty spans high); the third, by a fire which engulfed and burnt everything; the fourth, by a rush of air and wind which flattened everything including several mountains; human beings were not killed by it but changed into baboons (what impressions cannot be stamped on the receptive credulity of men!). After the death of that fourth sun the world was in perpetual darkness for twenty-five years, during the fifteenth of which was created a man and a woman who remade the human race. Ten years later, on a particular day which they observe, the sun appeared, newly created; they count their years from that day. On the third day after it was created their old gods died; new gods were subsequently born from time to time. My authority[37] could learn nothing about how they believed this fifth sun

36. Montaigne's term *plus civilisez* probably means not 'more civilized', but 'more urban and hence more given to civic virtues' than the pastoral Indians; similarly his term *plus artistes* probably means 'more cultured' rather than 'more artistic': they had more developed arts and sciences.

37. Francisco Lopez de Gomara, *Histoire générale des Indes*, II, lxxv and (for the Royal road described later) V, lxxxvii.

will die. But their dating of that fourth change tallies with that great conjunction of the planets which (eight hundred years ago, according to the reckoning of our astrologers) produced many great changes and innovations in the world.[38]

As for that ostentatious magnificence which led me to embark on this subject, neither Greece nor Rome nor Egypt can compare any of their constructions, for difficulty or utility or nobility, with the highway to be seen in Peru, built by their kings from the city of Quito to the city of Cuzco — three hundred leagues, that is — dead-straight, level, twenty-five yards wide, paved, furnished on either side with a revetment of high, beautiful walls along which there flow on the inside two streams which never run dry, bordered by those beautiful trees which they call *molly*. Whenever they came across mountains and cliffs they cut through them and flattened them, filling in whole valleys with chalk and stone. At the end of each day's march there are beauteous palaces furnished with victuals and clothing and weapons, both for troops and travellers who have to pass that way.

My judgement on this construction takes account of the difficulty, which in that place is particularly relevant since they build using blocks never less than ten-foot square; they have no means of transporting them except to drag them along by the force of their arms: they do not even have the art of scaffolding, knowing no other method than to pile up earth against a building as it rises and then to remove it afterwards.

But let us drop back to those coaches of ours.

Instead of using coaches or vehicles of any kind they have themselves carried on the shoulders of men. The day he was captured, that last King of Peru[39] was in the midst of his army, borne seated on a golden chair suspended from shafts of gold. The Spaniards in their attempts to topple him (as they wanted to take him alive) killed many of his bearers, but

38. According to the teaching of Alkindi, Albumasar and other Islamic astrologers widely accepted in medieval and Renaissance Europe, when a 'great conjunction' (that of the planets Saturn and Jupiter) occurs in the first degree of the zodiacal sign of the Ram, it produces one single outstanding prophet, teacher or lawgiver. Such a great conjunction was calculated to occur every 960 years. Both Islamic and Christian astrologers often held that a great conjunction heralded the birth of Moses, Jesus and Mahomet. Cf., for example, Petrus de Abano, *Conciliator* (*Diff.* XVIII). The great conjunction mentioned by Montaigne was the one preceding the birth of the Prophet of Islam. The theory of the influence of conjunctions was, of course, challenged by many.
39. Attabalipa.

many more vied to take the places of the dead, so that, no matter how many they slaughtered, they could not bring him down until a mounted soldier dashed in, grabbed hold of him and yanked him to the ground.

7. *On high rank as a disadvantage*

===

[The kind of outspoken judgement on monarchs which seems to have brought Montaigne the respect of the future Henry IV.]

[B] Since we cannot attain it, let us get our own back by disparaging it! Not that you are disparaging anything in its entirety when you find defects in it: there are defects in all things, no matter how beautiful or desirable they may be.

In general high rank has one obvious advantage: it can lay itself aside whenever it wants to; it is virtually free to choose either condition. All forms of greatness are not brought low uniquely by a fall: some there are which allow you to stoop low without falling.

It does seem to me that we set too high a value on it, as we also do on the determination of those whom we have seen or heard refusing it or resigning it at their own volition. In its essence the advantage of it is not so self-evident that it takes a miracle to reject it.

What I find hard is striving to bear misfortune. There does not seem to be much involved in being content with a modest measure of wealth and avoiding greatness; that is a virtue which I think even I could reach without a great deal of exertion, and I am only a fledgling. So what must become possible for men who would put to their account as well the glory which accompanies such a rejection (in which there may be more ambition than in the actual possession of the desired greatness, since ambition is never acting more in accord with its nature than when it adopts some unusual road, somewhat off the beaten track).

I whet my mind to face endurance:[1] I enfeeble it towards desire. I can wish as well as the next man and I allow great freedom and indiscretion to my wishes; yet I have never found myself wishing for imperial or royal rank nor for the prominence of those high destinies where men command.

---

1. Montaigne's term *j'esguise mon courage* echoes Cicero's *acuant mentem* (*Tusc. disput.*, I, xxxiii, 80), where Cicero stresses the influence of body on mind and congratulates himself (as Montaigne often does) on being slow-witted rather than a volatile, melancholy genius.

My aims do not tend that way: I love myself too much for that. When I think of growing in constancy or wisdom or health or beauty, or even wealth, it is in a modest way, with a timid constricted growth appropriate to myself; but my imagination is oppressed by great renown or mighty authority. Contrary to what was said by that other chap,[2] I would rather be one whose lot was to be second or third in Périgueux than first in Paris – or at least, to tell no lie, third in Paris, rather than the one in charge. I want neither to be a wretched nobody arguing with doorkeepers nor one who causes crowds to part with awe as I pass through. By [C] lot [B] and also by taste[3] I am accustomed to a middling rank. [C] In the conduct of my life and of anything I have undertaken, I have shown that I have fled rather than sought means of stepping above the degree of fortune in which God has placed me at birth. Anything established by Nature is as just as it is pleasant.

[B] I have a soul so lazy that I do not measure my fortune by its height: I measure it by its pleasantness. [C] But though I do not have all that great a mind, I do have one which is correspondingly open, one which orders me to dare to publish its weaknesses. You might ask me to compare two lives. The first is that of Lucius Thorius Balbus, a gentleman who was handsome, learned, healthy, intelligent and abounding in all sorts of talents and pleasures, leading a quiet existence which was entirely his own, with a soul fully armed against death, superstition, pain and the other burdens of our human distress, who finally died in battle, weapon in hand, in the defence of his country. The second is the life of Marcus Regulus, so great and sublime that we all know of it, with his death so worthy of admiration. One of those men was without rank or reputation; the other, amazingly glorious and exemplary. I would say of them the same as Cicero (if I could talk as well as he could).[4] If I had to lay those lives against my own, I would say that the former is as much in harmony with my abilities (and with my desires, which I make to conform to my abilities) as the latter far outstrips them; I can only approach the latter with veneration; I could readily approach the other in actual practice.

2. Julius Caesar; cf. Erasmus, *Apophthegmata*, IV *C. Julius Caesar*, V. (Caesar would rather be the first man in an alpine hamlet than second in Rome.)
3. '88: by *fortune* and also by taste . . .
4. Cicero (*De finibus*, II, XX, 63–4) compares, as does Montaigne, Balbus (who despite a certain greatness, 'knew no limit but satiety') with Regulus and judged him a less happy example. Cicero also prefers Lucretia, who took her own life, and Lucius Verginius, a poor man who killed his virgin daughter rather than have her defiled by Appius Claudius.

Now let us get back to my starting point, temporal greatness.

[B] I dislike all domination, by me or over me.   [C]   Otanes, one of the Seven who had rightful claims to the throne of Persia, took a decision which I could well have taken myself. To his rivals he abandoned his rights to be elected or chosen by lot, on condition that he and his family could live in that empire free from all domination, and from all subordination except to those of the ancient laws, and should enjoy every freedom not prejudicial to those laws, since he found it intolerable both to give or to accept commands.[5]

[B]   The harshest and most difficult job in the world, in my judgement, is worthily to act the king. I can excuse more shortcomings in kings than men commonly do, out of consideration for the horrifying weight of their office, which stuns me. It is difficult for such disproportionate power to act with a sense of proportion. Yet even for men of less outstanding character it is a singular incitement to virtue for them to be placed where you can do no good deed which is not noted and chronicled; where the slightest good action affects so many people and where your talents (like those of preachers) are mainly addressed to the populace – not an exacting judge, one easily duped and easily contented.

There are few matters on which we can give an unbiased judgement because there are few in which we do not have a private interest some way or other. Superior or inferior rank, the role of ruler or subject, are bound to each other by natural rivalry and competition: they need to be always pillaging each other. I never believe either's case against its yoke-mate: let reason judge of it (when we can prevail upon her): she cannot be swayed and is exempt from passion. Less than a month ago I was turning over the pages of a couple of Scottish books on this subject; the people's man makes the king's position worse than a carter's: the monarchist places him in sovereignty and power a few yards higher than God.[6]

Now the disadvantage of great rank (which I have taken as the subject of my remarks here since some event called it to my attention) is the following: nothing perhaps in the whole of our dealings with others is more pleasant than those assays which we make of each other as rivals for

---

5. Herodotus, III, lxxxiii
6. George Buchanan ('the people's man') the future Scottish reformer, had taught Montaigne at the Collège de Guienne in Bordeaux. His *De jure regni apud Scotos* appeared in 1579. This was answered by Adam Blackwood's *Apologia* for Mary Stuart against Buchanan. Both works were translated into French. Jean Dorat wrote a prefatory poem for Blackwood's book.

honour in physical sports and for esteem in those of the mind – and in which a sovereign can take no real part. It has often seemed true to me that the force of respect leads to our actually treating princes disdainfully and insultingly.

Something which infinitely annoyed me as a boy was when those who played sports against me dispensed themselves from making any serious attempt at beating me, finding me an opponent not worth the effort; princes see that happen every day, each partner finding himself unworthy of striving to beat him. Whenever anyone perceives that princes have the slightest desire to win, there is no partner who does not labour to see that they do so, preferring to betray his own glory rather than to attack theirs: we merely make just enough effort to enhance their reputation. What part can they play in a friendly skirmish if everyone in it is on their side? It recalls those paladins in days of yore who entered jousts and combats with enchanted bodies and weapons. When Brisson was racing against Alexander he merely pretended to run swiftly. Alexander did rebuke him for it, but he ought to have had him flogged.[7]

That is why Carneades said that the only thing which the sons of princes really learned properly was horsemanship, since in all other sports men yield to them and allow them to win whereas a horse is neither a flatterer nor a courtier: it will throw a king's son as soon as a porter's.[8] Homer was compelled to allow Venus, so gentle and inviolable a deity, to be ever so lightly wounded at the siege of Troy so as to attribute boldness and courage to her, qualities which do not fall to the lot of those who are exempt from risk of harm.[9] Gods are made to get angry, feel fear and flee, [C] to be jealous, [B] to lament and to feel passion, in order to honour them with virtues which among us humans are constructed from our imperfections.

Anyone who has no part in the danger and difficulty can make no claim to a share in the honour and delight which ensue upon the dangerous deed. It is pitiful to have such power that it results in everything giving way to you. Then your destiny removes you too far from the fellowship and companionship of men; you are stuck there, too remote. The unchallenging and facile ease with which you can make everything bow down before you is the enemy of every sort of pleasure. That is not walking but gliding; not

7. Plutarch (tr. Amyot), *De la tranquillité de l'ame*, 72 G.
8. Erasmus, *Apophthegmata*, VII, *Carneades*, XXXII; Plutarch (tr. Amyot), *Comment on peult disarner le flatteur de l'amy*, 46 A–B).
9. Venus (or rather Aphrodite) in the *Iliad* (V).

living, but sleeping. (Just imagine Man to be endowed with omnipotence: you throw him into an abyss; his being and his well-being are in dire necessity: he has to beg you of your charity for obstacles and opposition.)

Even such men's good qualities are dead and gone, for qualities are known only by comparison, and such men are beyond compare; they have little knowledge of true praise, being battered by continual and uniform acclaim. Even if they are up against the most stupid of their subjects they have no way of showing they are better than he is; he only has to say, 'I did that because he is my King, you see,' and he then believes he has said enough to imply that he contributed to his own defeat.

This kingly quality stifles and annihilates their other qualities, their real ones which are of their essence: they lie buried under their royal state. That leaves them with no means of showing their worth except actions which directly touch upon their royal state or which contribute to it, namely the duties of their rank. Which means that such a one is so entirely a king that he has no other existence. That radiance which surrounds him is not him, but it hides and conceals him from us: the rays from our eyes strike against it and are scattered, being overwhelmed and arrested by the strong light.[10] The Senate voted to award the prize for eloquence to Tiberius: he declined it, believing that, even if it were justified, he could take no pleasure in a verdict so unfreely reached.[11]

As we concede every advantage of honour to princes we confirm them in their defects and, not merely by our approval but by our imitation, we give warrant to their defects and their vices. All of Alexander's courtiers used to twist their heads to one side as he did; those who flattered Dionysius used to bump into each other when he was present, stumbling against whatever was under their feet and knocking it over, to suggest that they were as short-sighted as he was. Even having a rupture has at times helped a man to advancement and favour! I have known men pretend to be deaf; and Plutarch knew courtiers who repudiated wives – wives whom they loved – because their lord hated his. Further still, lechery has been in fashion and every kind of licentiousness, as also have disloyalty, blasphemy, cruelty, as well as heresy and superstition, irreligion and decadence, and even worse things if worse there be, so providing thereby an example even more dangerous than that of Mithridates' flatterers: their lord yearned to be

10. Renaissance science believed that we see objects by means of rays leaving our eyes, not by rays striking the retina.
11. Perhaps a confused memory of an event related in Erasmus, *Apophthegmata*, VI, *Varie mixta*, XXVIII, when Tiberius rebuked a flattering senator.

honoured as a good doctor so they offered him their limbs to be cut open and cauterized; but that other lot allowed a nobler and more tender part to be cauterized: their soul.[12]

But to end where I began: when the Emperor Hadrian was discussing the meaning of a word with Favorinus the philosopher, Favorinus quickly let him win the argument. When his friends criticized him for it he replied, 'You are joking! Would you want him to be less learned than I am? He is in command of thirty legions!' After Augustus had written some verses against Asinius Pollio, Pollio said: 'I am keeping my mouth shut. It is not wise to skirmish with him who can banish.'[13] And he was right. For, as Dionysius could not equal Philoxenus in poetry or Plato in prose, he condemned one to the quarries and sent the other to the island of Aegina to be sold as a slave.[14]

---

12. Above *exempla* from Plutarch, *Comment on pourra discerner le flatteur d'avec l'amy*, 42 G, 43 A, 43 B, 45 E.
13. Both *exempla* from Pietro Crinito, *De honesta disciplina*, XII.
14. Plutarch (tr. Amyot), *De la tranquillité de l'ame*, 72 E; cf. *De la fortune d'Alexandre*, 312 E.

8. On the art of conversation

====

[French children know that Pascal referred to Montaigne as 'the incomparable author of "The art of conversation"'. That has given this chapter a special place in French culture. It is further valued for the light it throws on to Montaigne's character. The conversation in this chapter turns to Tacitus and shows us how Montaigne had conversations with himself about the books he was reading.]

[B]   It is a custom of our justice to punish some as a warning to others. [C]   For to punish them for *having done* wrong would, as Plato says, be stupid: what is done cannot be undone. The intention is to stop them from repeating the same mistake or to make others avoid their error.[1]   [B]   We do not improve the man we hang: we improve others by him. I do the same. My defects are becoming natural and incorrigible, but as fine gentlemen serve the public as models to follow I may serve a turn as a model to avoid:

> *Nonne vides Albi ut male vivat filius, utque*
> *Barrus inops? magnum documentum, ne patriam rem*
> *Perdere quis velit*

[You can see, can't you, how wretchedly Albus' son is living and how poor Barrus is? An excellent lesson in not squandering your inheritance.][2]

The act of publishing and indicting my imperfections may teach someone how to fear them. (The talents which I most esteem in myself derive more [C] honour [B] from[3] indicting me than praising me.) That is why I so often return to it and linger over it. Yet, when all has been said, you never talk about yourself without loss: condemn yourself and you are always believed: praise yourself and you never are.

There may be others of my complexion who learn better by counter-example than by example, by eschewing not pursuing. That was the sort of

---

1. Plato, *Laws*, XI, 934 A–B.
2. Horace, *Satires*, I, iv, 109–11.
3. '88: more *advantage* from . . .

instruction which the Elder Cato was thinking of when he said that the wise have more to learn from the fools than the fools from the wise;[4] as also that lyre-player in antiquity who, Pausanias says, used to require his students to go and listen to some performer who lived across the street so that they would learn to loathe discords and faulty rhythms.[5] My horror of cruelty thrusts me deeper into clemency than any example of clemency ever could draw me. A good equerry does not make me sit up straight in the saddle as much as the sight of a lawyer or a Venetian out riding, and a bad use of language corrects my own better than a good one. Every day I am warned and counselled by the stupid deportment of someone. What hits you affects you and wakes you up more than what pleases you. We can only improve ourselves in times such as these by walking backwards, by discord not by harmony, by being different not by being like. Having myself learned little from good examples I use the bad ones, the text of which is routine.   [C]   I strove to be as agreeable as others were seen to be boring; as firm as others were flabby; as gentle as others were sharp. But I was setting myself unattainable standards.[6]

[B]   To my taste the most fruitful and most natural exercise of our minds is conversation. I find the practice of it the most delightful activity in our lives. That is why, if I were now obliged to make the choice, I think I would rather lose my sight than my powers of speech or hearing. In their academies the Athenians, and even more the Romans, maintained this exercise in great honour. In our own times the Italians retain some vestiges of it – greatly to their benefit, as can be seen from a comparison of their intelligence and ours. Studying books has a languid feeble motion, whereas conversation provides teaching and exercise all at once. If I am sparring with a strong and solid opponent he will attack me on the flanks, stick his lance in me right and left; his ideas send mine soaring. Rivalry, competitiveness and glory will drive me and raise me above my own level. In conversation the most painful quality is perfect harmony.

Just as our mind is strengthened by contact with vigorous and well-ordered minds, so too it is impossible to overstate how much it loses and deteriorates by the continuous commerce and contact we have with mean and ailing ones. No infection is as contagious as that is. I know by

---

4. Erasmus, *Apophthegmata*, V, *Cato Senior*, XXXIX.
5. Anecdote not traced. Perhaps a confusion with the practice of the ancient musician Timotheus of Miletus. Cf. Quintilian, II, iii, 3.
6. '88: routine: *the routine sight of thieving and perfidiousness has guided and restrained my morals.* To my taste . . .

experience what that costs by the ell. I love arguing and discussing, but with only a few men and for my own sake: for to serve as a spectacle to the great and indulge in a parade of your wits and your verbiage is, I consider, an unbecoming trade for an honourable gentleman.

Stupidity is a bad quality: but to be unable to put up with it, to be vexed and ground down by it (as happens to me) is another, hardly worse in its unmannerliness than stupidity. And that is what at present I wish to condemn in myself.

I embark upon discussion and argument with great ease and liberty. Since opinions do not find in me a ready soil to thrust and spread their roots into, no premise shocks me, no belief hurts me, no matter how opposite to my own they may be. There is no idea so frivolous or odd which does not appear to me to be fittingly produced by the mind of man. Those of us who deprive our judgement of the right to pass sentence look gently on strange opinions; we may not lend them our approbation but we do readily lend them our ears. When one scale in the balance is quite empty I will let the other be swayed by an old woman's dreams: so it seems pardonable if I choose the odd number rather than the even, or Thursday rather than Friday; if I prefer to be twelfth or fourteenth at table rather than thirteenth; if I prefer on my travels to see a hare skirting my path rather than crossing it, and offer my left foot to be booted before the right. All such lunacies (which are believed among us) at least deserve to be heard. For me they only outweigh an empty scale, but outweigh it they do. Similarly the weight of popular and unfounded opinions has a natural existence which is more than nothing. A man who will not go that far perhaps avoids the vice of superstition by falling into the vice of stubbornness.

So contradictory judgements neither offend me nor irritate me: they merely wake me up and provide me with exercise. We avoid being corrected: we ought to come forward and accept it, especially when it comes from conversation not a lecture. Whenever we meet opposition, we do not look to see if it is just but how we can get out of it, rightly or wrongly. Instead of welcoming arms we stretch out our claws. I can put up with being roughly handled by my friends: 'You are an idiot! You are raving!' Among gentlemen I like people to express themselves heartily, their words following wherever their thoughts lead. We ought to toughen and fortify our ears against being seduced by the sound of polite words. I like a strong, intimate, manly fellowship, the kind of friendship which rejoices in sharp vigorous exchanges just as love rejoices in bites and scratches which draw blood. [C] It is not strong enough nor

magnanimous enough if it is not argumentative, if all is politeness and art; if it is afraid of clashes and walks hobbled. *'Neque enim disputari sine reprehensione potest.'* [It is impossible to debate without refuting.][7]

[B] When I am contradicted it arouses my attention not my wrath. I move towards the man who contradicts me: he is instructing me. The cause of truth ought to be common to us both. – What will his answer be? The passion of anger has already wounded his judgement. Turbulence has seized it before reason can. – It would be a useful idea if we had to wager on the deciding of our quarrels, useful if there were a material sign of our defeats so that we could keep tally on them and my manservant say: 'Last year your ignorance and stubbornness cost you one hundred crowns on twenty occasions.'

I welcome truth, I fondle it, in whosesoever hand I find it; I surrender to it cheerfully, welcoming it with my vanquished arms as soon as I see it approaching from afar. [C] And provided that they do not set about it with too imperious and schoolmasterish a frown I will put my shoulder to the wheel to help along the criticisms that people make of my writings: I have often made changes more for reasons of politeness than to effect reasonable corrections, preferring to please and encourage people's freedom to criticize me by my readiness to give way – yes, even when it cost me something. Yet it is difficult to attract men to do that in our days They have no stomach for correcting because they have no stomach for suffering correction, always dissembling when talking in each other's presence.

I take such great pleasure in being judged and known that it is virtually indifferent to me which of the two forms it takes. My thought so often contradicts and condemns itself that it is all one to me if someone else does so, seeing that I give to his refutation only such authority as I please. But I fall out with anyone who is too high-handed, like one man I know who laments the fact that he gave you advice if you do not accept it and takes it as an insult if you shy at following it.

Socrates always laughingly welcomed contradictions made to his arguments. It could be said that since his arguments were the stronger the advantage would always fall to him and that he welcomed them as matter for fresh triumphs: but we, on the contrary, find that there is nothing which makes us more susceptible than convictions about our own surpassing excellence, our contempt for our adversary, and about its being reasonable

---

7. Cicero, *De finibus*, I, viii, 28 (Torquatus defending Epicurus' style of conversation).

for the weaker to be willing to accept refutations which set him back on his feet and redress him.

[B] I do truly seek to frequent those who manhandle me rather than those who are afraid of me. It is a bland and harmful pleasure to have to deal with people who admire us and defer to us. Antisthenes commanded his sons never to give thanks or show gratitude to anyone who praised them.[8] I feel far prouder of the victory I win over myself when I make myself give way beneath my adversary's powers of reason in the heat of battle than I ever feel gratified by the victory I win over him through his weakness. In short I admit and acknowledge any attacks, no matter how feeble, if they are made directly, but I am all too impatient of attacks which are not made in due form. I care little about what we are discussing; all opinions are the same to me and it is all but indifferent to me which proposition emerges victorious. I can go on peacefully arguing all day if the debate is conducted with due order.   [C] It is not so much forceful and subtle argument that I want as order – the kind of order which can be found every day in disputes among shepherds and shop-assistants yet never among us. If they go astray it is in lack of courtesy. So do we. But their stormy intolerance does not make them stray far from their theme: their arguments keep on course. They interrupt each other. They jostle, but at least get the gist. To answer the point is, in my judgement, to answer very well.   [B] But when the discussion becomes turbulent and lacks order, I quit the subject-matter and cling irritably and injudiciously to the form, dashing into a style of debate which is stubborn, ill-willed and imperious, one which I have to blush for later.

[C] It is impossible to argue in good faith with a fool. Not only my judgement is corrupted at the hands of so violent a master, so is my sense of right and wrong. Our quarrels ought to be outlawed and punished as are other verbal crimes. Since they are always ruled and governed by anger, what vices do they not awaken and pile up on each other? First we feel enmity for the arguments and then for the men. In debating we are taught merely how to refute arguments; the result of each side's refuting the other is that the fruit of our debates is the destruction and annihilation of the truth.[9] That is why Plato in his *Republic* prohibits that exercise to ill-endowed minds not suited to it.[10]

8. Plutarch (tr. Amyot), *De la mauvaise honte*, 81 B.
9. Renaissance rhetoric and dialectic in school and university did indeed often encourage *pro et contra* debates rather than a search for truth.
10. Plato, *Republic*, 539 A–C.

[B] You are in quest of [C] what *is*.[11] [B] Why on earth do you set out to walk that road with a man who has neither pace nor style? We do no wrong to the subject-matter if we depart from it in order to examine the way to treat it — I do not mean a scholastic donnish way, I mean a natural way, based on a healthy intellect. But what happens in the end? One goes east and the other west; they lose the fundamental point in the confusion of a mass of incidentals. After a tempestuous hour they no longer know what they are looking for. One man is beside the bull's eye, the other too high, the other too low. One fastens on a word or a comparison; another no longer sees his opponent's arguments, being too caught up in his own train of thought: he is thinking of pursuing his own argument not yours. Another, realizing he is too weak in the loins, is afraid of everything, denies everything and, from the outset, muddles and [C] confuses the argument, or else, at the climax of the debate he falls into a rebellious total silence, affecting, out of morose ignorance, a haughty disdain or an absurdly modest desire to avoid contention, [B] Yet[12] another does not care how much he drops his own guard provided that he can hit you. Another counts every word and believes they are as weighty as reasons. This man merely exploits the superior power of his voice and lungs. And then there is the man who sums up against himself; and the other who deafens you with useless introductions and digressions. [C] Another is armed with pure insults and picks a groundless 'German quarrel' so as to free himself from the company and conversation of a mind which presses hard on his own.

[B] Lastly, there is the man who cannot see reason but holds you under siege within a hedge of dialectical conclusions and logical formulae. Who can avoid beginning to distrust our professional skills and doubt whether we can extract from them any solid profit of practical use in life when he reflects on the use we put them to? *'Nihil sanantibus litteris.'* [such erudition as has no power to heal.][13] [B] Has anyone ever acquired intelligence through logic? Where are her beautiful promises? [C] *'Nec ad melius vivendum nec ad commodius disserendum.'* [She teaches neither how to live a better life nor how to argue properly.] [B] Is there more of a hotchpotch in the cackle of fishwives than in the public disputations of men who profess logic? I would prefer

11. '88: of *the truth*: why . . .
12. '88: muddles *and ruffles the debate*. Yet another . . .
13. Seneca, *Epist. moral.*, LIX, 15; then, Cicero, *De finibus*, I, xix, 63, criticizing Epicurean logic.

a son of mine to learn to talk in the tavern rather than in our university yap-shops.

Take an arts don; converse with him. Why is he incapable of making us feel the excellence of his 'arts' and of throwing the women, and us ignoramuses, into ecstasies of admiration at the solidity of his arguments and the beauty of his ordered rhetoric! Why cannot he overmaster us and sway us at his will? Why does a man with his superior mastery of matter and style intermingle his sharp thrusts with insults, indiscriminate arguments and rage? Let him remove his academic hood, his gown and his Latin; let him stop battering our ears with raw chunks of pure Aristotle; why, you would take him for one of us – or worse. The involved linguistic convolutions with which they confound us remind me of conjuring tricks: their sleight-of-hand has compelling force over our senses but it in no wise shakes our convictions. Apart from such jugglery they achieve nothing but what is base and ordinary. They may be more learned but they are no less absurd.

I like and honour erudition as much as those who have it. When used properly it is the most noble and powerful acquisition of Man. But in the kind of men (and their number is infinite) who make it the base and foundation of their worth and achievement, who quit their understanding for their memory, [C] *'sub aliena umbra latentes'* [hiding behind other men's shadows],[14] [B] and can do nothing except by book, I loathe (dare I say it?) a little more than I loathe stupidity.

In my part of the country and during my own lifetime school-learning has brought amendment of purse but rarely amendment of soul. If the souls it meets are already obtuse, as a raw and undigested mass it clogs and suffocates them; if they are unfettered, it tends to purge them, strip them of impurities and volatilize them into vacuity. Erudition is a thing the quality of which is neither good nor bad, almost: it is a most useful adjunct to a well-endowed soul: to any other it is baleful and harmful; or rather, it is a thing which, in use, has great value,[15] but it will not allow itself to be acquired at a base price: in one hand it is a royal sceptre, in another, a fool's bauble.

But to get on: what greater victory do you want than to teach your enemy that he cannot stand up to you? Get the better of him by your argument and the winner is the truth; do so by your order and style, then you are the winner!

[C]   I am persuaded that, in both Plato and Xenophon, Socrates debates

---

14. Seneca, *Epist. moral.*, XXXIII, 7.
15. '88: great *nobility and* value . . .

more for the debater's than for debating's sake; more to teach Euthydemus and Protagoras their own absurdity than the absurdity of their sophists' art. He seizes hold of the first subject which comes to hand, as a man who has a more useful aim than to throw light on his subject as such: namely, to enlighten the minds which he accepts to train and to exercise. [B] The game which we hunt is the fun of the chase: we are inexcusable if we pursue it badly or foolishly: it is quite another thing if we fail to make a kill. For we are born to go in quest of truth: to take possession of it is the property of a greater Power.[16] Truth is not (as Democritus said) hidden in the bottom of an abyss: it is, rather, raised infinitely high within the knowledge of God.[17]

[C] This world is but a school of inquiry. [B] The question is not who will spear the ring but who will make the best charges at it. The man who says what is true can act as foolishly as the one who says what is untrue: we are talking about the way you say it not what you say. My humour is to consider the form as much as the substance, and the barrister as much as his case, as Alcibiades told us to.[18] [C] Every day I spend time reading my authors, not caring about their learning, looking not for their subject-matter but how they handle it; just as I go in pursuit of discussions with a celebrated mind not to be taught by it but to get to know it.

[B] Any man may speak truly: few men can speak ordinately, wisely, adequately. And so errors which proceed from ignorance do not offend me: absurdity does. I have often broken off discussing a bargain, even one advantageous to me, because of the silly claims of those I was bargaining with. For their mistakes I do not lose my temper above once a year with any of those who are subject to my authority, but when the point is the stupidity of their assertions or the obstinacy of their asinine excuses and their daft defences, then we are daily at each other's throats. They understand neither why nor what they are told: they answer accordingly. It is enough to make you despair. It is only when my head bangs against another head that I feel a big bump: I can come to terms with the failings

16. The theme of III, 13, 'On experience'.
17. For Democritus, cf. Cicero, *Academica*, I, xii, 44: a celebrated saying of Democritus, cited similarly to Montaigne by the Christian theologian Lactantius, *Institutiones divinarum* III, 28, a reference given in the adage *Veritas in profundo* (*Appendix Erasmi*, in *Adagia id est Proverbiorum collectio absolutissima*, Frankfurt, 1656, p. 453).
18. Perhaps an echo of the similar remark attributed to him in Henry Estienne's *Apophthegmata*, 1588, pp. 110–11.

of my servants better than with their thoughtlessness, insolence and downright silliness. Let them do less, provided that they can do something! You live in hope of making their wills warm to their work: but there is nothing to get from a blockhead, nothing to hope for.

Yes, but what if I myself am taking things for other than they are? That may well be: that explains first of all why I condemn my inability to put up with it, holding it to be equally a defect in those who are right and those who are wrong, since there is always an element of tyrannical bad temper in being unable to tolerate characters different from your own. Secondly, there is in truth no greater silliness, none more enduring, than to be provoked and enraged by the silliness of this world – and there is none more bizarre. For it makes you principally irritated with yourself: that philosopher of old would never have lacked occasion for his tears if he had concentrated on himself.[19]   [C]   One of the Seven Sages, Myson, was of the same humour as Timon and Democritus: when asked what he was laughing at all by himself, he replied, 'At the fact that I am laughing all by myself.'

[B]   How many statements and replies do I make every day which are silly by my norms – so even more frequently, to be sure, by the standards of others![20]   [C]   If I bite my lips for them, what must the others be doing! To sum up, we have to live among the living and let the stream flow under the bridge without worrying about it or, at very least, without making ourselves ill over it.   [B]   Indeed, why can we encounter a man with a twisted deformed body without getting irritated, yet are unable to tolerate a deranged mind without flying into a rage?[21] Such harshness is vitiated and derives from the critic rather than the fault. Let us always have Plato's saying on our lips: [C]   'If I find ill in something may it not be because I myself am ill?   [B]   Am I not the one at fault? May my own criticism not be turned against me?' A wise and inspired refrain which chastises the most common and universal error of mankind.   [C]   It is not merely the reproaches which we make to each other which can be regularly turned against us but also our reasons and our arguments in matters of controversy: we run ourselves through with our own

19. Heraclitus, the Sage who wept at the folly of the world; normally coupled with Democritus, who laughed at it. Followed by the most famous saying of Myson (Erasmus, *Apophthegmata*, VII, *Myson*, I).
20. Literally silly '*selon moy*' (that is, by my own terms of reference), even sillier 'according to others' (by their terms of reference).
21. Plutarch (tr. Amyot), *Comment on pourra recevoir utilité de ses ennemis*, 110 E–F (and for Plato's saying about to be quoted).

swords. [B] As it was ingeniously and aptly put by the man who first said it: *'Stercus cuique suum bene olet.'* [Everyone's shit smells good to himself.][22]

[C] Our eyes see nothing behind us.[23] A hundred times a day when we go mocking our neighbour we are really mocking ourselves; we abominate in others those faults which are most manifestly our own, and, with a miraculous lack of shame and perspicacity, are astonished by them. Only yesterday I was able to watch an intelligent nobleman making jokes, as good as they were pertinent, about the silly way in which another nobleman went bashing everyone's ear about his family-tree and his family alliances, more than half of which were false, that kind of man being most inclined to launch out on such stupid subjects when his escutcheon is more dubious and least certain: yet he too, if he had stood back and looked at himself, would have discovered that he was hardly less extravagant in broadcasting and less boring in stressing the claims to precedence of his wife's family. What a dangerous arrogance with which a wife is seen to be armed at the hands of her very husband! If they understood Latin we ought to say to such people:

Age! si hæc non insanit satis sua sponte, instiga!

[That's the way! If she is not mad enough herself, egg her on!][24]

I do not mean that nobody should make indictments unless he is spotless; if that were so no one would make them. What I mean is that when our judgement brings a charge against another man over a matter then in question, it must not exempt us from an internal judicial inquiry. It is a work of charity for a man who is unable to weed out a defect in himself to try, nevertheless, to weed it out in another in whom the seedling may be

22. Erasmus, *Adages*, III, IV, II. Erasmus links the saying to Aristotle's *Nicomachaean Ethics*, and to the complementary adage, *Suum cuique pulchrum* (one's own is beautiful to oneself) (I, II, XV), further linked with Plato, Aristotle and Horace as a condemnation of *philautia* (self-love).

23. Another authoritative condemnation of self-love, in Aesop's *Beggar's Wallet*: we put our neighbours' faults in the front pocket where we can see them, our own in the back one where we cannot. (Cf. Rabelais, TLF, *Tiers Livre*, TLF, XV, note 108, citing Erasmus' *Adages* and Raymond Sebond.)

'88: olet. *To sum up, we must live among the living and let each man follow his fashion without our worrying or without making ourselves ill about it.* (In [C] changed and placed earlier.)

24. Terence, *Andria*, IV, ii, 9.

less malignant and stubborn. And it never seems to me to be an appropriate answer to anyone who warns me of a fault in me to say that he has it too. What difference does that make? The warning remains true and useful. If we had sound nostrils our shit ought to stink all the more for its being our own. Socrates was convinced that if there was a man who, together with his son and a stranger, was found guilty of violence or injury, that man should begin with himself, first presenting himself to be sentenced by the judge and to beg for expiation at the hands of the executioner; next, he should present his son; then the stranger.[25] If that precept pitches it rather too high, at least he should be the first to be presented before his own conscience for punishment.

[B] Our first judges are properly our senses, which perceive things only by their external accidents. No wonder then that in all the elements which contribute to our society there is such a constant and universal addition of surface appearances and ritual; with the result that the best and most effective part of our polities consists in that. We are always dealing with Man, whose nature is wondrously corporeal. Those who in recent years have wished to build up for us so contemplative and non-material an exercise of worship should not be astonished if there are those who think that it would have slipped and melted through their fingers if it did not keep a hold among us as a mark, sign and means of division and of faction rather than for itself.[26]

It is the same in discussion: the gravity, academic robes and rank of the man who is speaking often lend credence to arguments which are vain and silly. Who could believe that so redoubtable a lord with so great a retinue does not have within him some more-than-ordinary talent, or that a man who is entrusted with so many missions and offices of state, a man so disdainful and so arrogant, is not cleverer than another man who bows to him from afar and whom nobody ever employs! Not only the words of such people but their very grimaces are watched and put to their account, each man striving to give them some fine solid significance. If they condescend to join in ordinary discussions and you show them anything but approval and reverence, they clobber you with the authority of their experience: they have heard this; they have seen that; they have done this: you are overwhelmed with cases. I would like to tell such men that the

25. Plato, *Gorgias*, 480 B–C.
26. Perhaps a reference to the members of the Reformed Church; it is often taken to be so. But is it not rather an allusion to ascetic movements within the Roman Catholic Church tending to devalue the body and elevate asceticism?

fruit of a surgeon's experience lies not in a recital of his operations nor in his reminding us that he has cured four patients of the plague and three of the gout, unless he knows how to extract from them material for forming his judgement and unless he knows how to convince us that he has been made wiser by the practice of his medical art.[27] [C] So, in a consort of instruments, we do not hear the lute, the spinet and the flute but a global harmony, the fruit resulting from the combination of the entire group.

[B] If they have been improved by their missions and their travels that should appear in the products of their understanding. It is not enough to relate our experiences: we must weigh them and group them; we must also have digested them and distilled them so as to draw out the reasons and conclusions they comport. There never were so many writing history! It is always good and profitable to listen to them, for they furnish us with ample instruction, fine and praiseworthy, from the storehouse of their memory; that is certainly of great value in helping us to live. But we are not looking for that at the moment: we are trying to find out whether the chroniclers and compilers are themselves worthy of praise.

I loathe all tyranny, both in speech and action. I like to brace myself against those trivial incidentals which cheat our judgement via our senses; and by keeping a watchful eye on men of extraordinary rank I have discovered that they are, for the most part, just like the rest of us:

> *Rarus enim ferme sensus communis in illa*
> *Fortuna.*

[Common sense is rare enough in that high station.][28]

Perhaps we esteem them and perceive them for less than they are, because they undertake to do more and so reveal themselves more. The porter must be stronger and tougher than his load. The man who has not had to use all his strength leaves you to guess whether he has any more in reserve, whether he has been assayed to the ultimate point: the man who succumbs under the weight betrays his limitations and the weakness of his shoulders. That is why, more than other people, so many of the learned can be seen to have inadequate souls. They could have been good farmers, good merchants, good craftsmen: their natural forces were tailored to such proportions. Knowledge is a very weighty thing: they sink beneath it.

27. Aristotle's contention in *Metaphysics*, I, 1, 980b–981a. Experience and experiments as such do not constitute the *art* of medicine: the *art* consists in a general inference drawn from it by a man's judgement.
28. Juvenal, *Satires*, VIII, 73–4.

Their mental apparatus has not enough energy nor skill to display that noble material and to apportion its strength, to exploit it and to make it help them. Knowledge can lodge only in a powerful nature: and that is very rare. [C] Feeble minds, said Socrates, corrupt the dignity of philosophy when they handle it; she appears to be useless and defective when sheathed in a bad covering.[29]

[B]   That is how they grow rotten and besotted,

> *Humani qualis simulator simius oris,*
> *Quem puer arridens pretioso stamine serum*
> *Velavit, nudasque nates ac terga reliquit,*
> *Ludibrium mensis.*

[like an ape, that imitator of the human face, which a boy dresses up, for a laugh, in precious silken robes, leaving the cheeks of its backside bare to amuse the guests at table.]

It is the same for those who rule over us and give orders, who hold the world in their hands: it is not enough for them to have an ordinary intelligence, to be able to achieve what we can. They are far beneath us if they are not way above us. Since they promise more, they owe more too; that is why keeping silent is not, in their case, merely a courteous and grave demeanour; it is also more often a profitable and gainful one. For when Megabysus went to see Appelles in his studio, he long remained silent. But when he began to discourse on the works of art, he received this rude reprimand: 'While you kept silent you appeared to be a great Somebody because of your chains-of-office and your retinue, but now we have heard you talk the very apprentices in my workshop despise you.'[30] Those magnificent decorations, that grand estate would not tolerate ordinary plebeian ignorance in him, nor inappropriate comments on paintings: he should have maintained that outward presumed connoisseurship. For how many men in my time has a cold, taciturn mien served their silly souls as signs of wisdom and ability!

Of necessity dignities and offices are bestowed more by fortune than by merit: you often do wrong to blame kings for that. On the contrary, it is a wonder that they have such good luck, enjoying as they do so few ways of finding out.

[C]   *Principis est virtus maxima nosse suos,*

---

29. Perhaps a reference to Plato, *Republic*, VI, 495 C–D.
30. Erasmus, *Apophthegmata*, VI, *Diversorum Graecorum*, XXXII.

[For a prince, the chief merit is to know his subjects,][31]

[B] for Nature has not given them eyes which can extend over so many peoples, distinguishing pre-eminence and seeing into our bosoms, where is lodged the knowledge of our will and of our better qualities. They have to select us by fumbling guesses: by our family, our wealth, our learning and the voice of the people – the feeblest of arguments. Anyone who could discover the means by which men could be justly judged and reasonably chosen would, at a stroke, establish a perfect form of commonwealth.

'Yes. But he brought this great matter to a successful conclusion.' – That means something, but not enough; for we rightly accept the maxim which says that plans must not be judged by results. [C] The Carthaginians punished bad counsels in their captains even when they were put right by a happy outcome. And the Roman people often refused to mark great and beneficial victories because the qualities of leadership of the commander were inferior to his good luck. [B] In this world's activities we often notice that Fortune rivals Virtue: she shows us what power she has over everything and delights in striking down our presumption by making the incompetent lucky since she cannot make them wise. She loves to interfere, favouring those performances whose course has been entirely her own. That is why we can see, every day, the simplest among us bringing the greatest public and private tasks to successful conclusions.

Siramnes the Persian replied to those who were amazed that his enterprises turned out so badly, seeing that his projects were so wise, by saying that he alone was master of his projects while Fortune was mistress of the outcome of his enterprises: they too could make the same reply to explain the opposite tendency.[32]

Most of this world's events happen by themselves:

Fata viam inveniunt.

[The Fates find a way.][33]

The outcome often lends authority to the most inept leadership. Our intervention is virtually no more than a habit, the result of tradition and example rather than of reason. I was once astounded by the greatness of a venture; I then learnt from those who had brought it to a successful

31. Martial, *Epigrams*, VIII, 15.
32. Cited by Amyot in his Prologue to *Les Vies de Plutarque*.
33. Virgil, *Aeneid*, III, 395; then, Horace, *Odes*, I, ix, 9.

conclusion what their motives were and what methods they used: I found nothing but ordinary notions.

Indeed the most ordinary usual ones are also perhaps the most reliable and the most suitable in practice if not for show. What if the most lowly reasons are the most solidly based? What if the [C] most humble, most lax and [B] best-trodden ones are the most suited to our concerns? If we are to safeguard the authority of the Privy Council we do not need laymen participating in it nor seeing further than the first obstacle. If we want to maintain its reputation it must be taken on trust, as a whole.

My thought sketches out the matter for a while and dwells lightly on the first aspects of it: then I usually leave the principal thrust of the task to heaven.

> *Permitte divis cætera.*
>
> [Entrust the rest to the gods.]

To my mind Good Luck and Bad Luck are two sovereign powers. There is no wisdom in thinking that the role of Fortune can be played by human wisdom. What he undertakes is vain if a man should presume to embrace both causes and consequences and to lead the progress of his action by the hand; and it is especially vain in counsels of war. Never were there [C] more military circumspection and prudence than I sometimes see practised among us [B]:[34] perhaps we fear that we shall get lost en route, and therefore keep ourselves in reserve for the climax in the final act!

I will go on to say that our very wisdom and mature reflections are for the most part led by chance. My will and my reasoning are stirred this way and that. And many of their movements govern themselves without me. My reason is daily subject to incitements and agitations [C] which are due to chance:

> [B]   *Vertuntur species animorum, et pectora motus*
>   *Nunc alios, alios dum nubila ventus agebat,*
>   *Concipiunt.*

[Their minds' ideas are ever turning round; the emotions in their breasts are driven hither and thither like clouds before the wind.][35]

---

34. '88: never were there *such* military circumspection and prudence, *especially in our nation as* I see practised: perhaps . . .
35. Virgil, *Georgics*, I, 420–2.

Look and see who wield most power in our cities; who do their jobs best. You will find that they are usually the least clever. There have been cases when women, children and lunatics have ruled their states equally as well as the most talented princes. [C] Coarse men more usually succeed in such things, says Thucydides, better than the subtle ones do.[36] [B] We ascribe the deeds of their good fortune to their wisdom.

[C]   *Ut quisque fortuna utitur*
Ita præcellet, atque exinde sapere illum omnes dicimus.

[Each outstanding man is raised by his good fortune; we then say that he is clever.]

[B] That is why I insist that, in all our activities, their outcomes provide meagre testimony of our worth and ability.

Now I was just about to say that it merely suffices for us to see a man raised to great dignity; even though we knew him three days before to be a negligible man, there seeps into our opinions, unawares, a notion of greatness, of talents, and we convince ourselves that by growing in style and reputation he has grown in merit. Our judgements of him are not based on his worth but (as is the case with the counters of an abacus) on the tokens of rank. Let his luck turn again, let him have a fall and be lost in the crowd again, then we all ask in wonder what had made him soar so high! 'Is this the same man?' we ask. 'Did he not know more about it when he was up there? Are princes satisfied with so little? We were in good hands, indeed we were!'

That is something I have seen many times in my own days.

Why, even the mask of greatness which is staged in our plays affects us somewhat and deceives us. What I worship in kings is the crowd of their worshippers. Everything should bow and submit to our kings – except our intelligence. My reason was not made for bending and bowing, my knees were.

When Melanthius was asked how Dionysius' tragedy appeared to him, 'I never saw it,' he replied. 'It was obscured by the words!' So, too, most of those who judge what the great have to say ought to answer: 'I never heard his words: they were too much obscured by his dignity, grandeur and majesty.'[37]

One day, when Antisthenes urged the Athenians to command that donkeys be used, as their horses were, to plough their fields, he was told

---

36. Thucydides, cited (with others of the above) from Justus Lipsius' *Politici*, as is the following, from Plautus' *Pseudolus*.
37. Plutarch (tr. Amyot), *Comment il faut ouïr*, 64 H.

that donkeys were not born for such a service. 'That does not matter,' he retorted. 'It all depends on your issuing the order: for the most ignorant and incompetent men whom you put in command of your wars never fail to become suddenly most worthy of command, because it is you who employ them!'[38]

Related to this is the practice of so many people to sanctify the kings whom they have chosen from among themselves. They are not contented with honouring them: they need to worship them. The people of Mexico dare not look at the face of their king once they have completed the rites of his enthronement, but as though they had deified him by his royal state they make him swear not merely to maintain their religion, laws and liberties and to be valiant, just and debonair, he must also swear to cause the sun to run shining with its accustomed light, the clouds to break in due season, the rivers to flow in their courses and the earth to bring forth all things needful for his people.[39]

I am opposed to that widespread fashion and I most doubt a man's ability when I see it accompanied by great rank and public acclaim. We should remember what it means to a man to be able to speak when he wants to, to choose the right moment, to break off the discussion or switch the subject with the authority of a master, to defend himself against objections with a shake of the head, a little smile or with silence, in front of courtiers who tremble with reverence and respect.

A monstrously rich man, when some trivial matter was being aired casually over dinner, joined in the discussion and began with these very words: 'Anyone who says otherwise is either ignorant or a liar,' and so on. You had better follow up that philosophical thrust with a dagger in your hand!

Here is another warning, which I find most useful: in debates and discussions we should not immediately be impressed by what we take to be a man's own *bons mots*. Most men are rich with other men's abilities. It may well be that such-and-such a man makes a fine remark, a good reply or a pithy saying, advancing it without realizing its power. [C] (That we do not grasp everything we borrow can doubtless be proved from my own case.) [B] We should not always give way, no matter what beauty or truth it may have. We should either seriously attack it or else, under pretence of not understanding it, retreat a little so as to probe it thoroughly and to discover how it is lodged in its author. We may be helping his

38. Erasmus, *Apophthegmata*, VII, *Antisthenes*, XXX.
39. Lopez de Gomara (tr. Fumée), *Histoire générale des Indes*, II, lxxvii.

sword-thrust to carry beyond his reach, running on to it ourselves. There have been times when, pressed by necessity in the duel of words, I have made counter-attacks which struck home more than I ever hoped or expected. I was counting their number: they were accepted for their weight.

When I am disputing with a man of strong arguments I enjoy anticipating his conclusions; I save him the bother of explaining himself; I make an assay at forestalling his ideas while they are still unfinished and being formed (the order and stretch of his intelligence warn me and threaten me from afar). Similarly, with those others I mentioned I do quite the opposite: we should suppose nothing, understand nothing but what they explain. If their judgements are apposite but expressed in universals — 'This is good: that is bad' — find out whether it is luck which makes them apposite. [C] Make them circumscribe and restrict their verdict a little: 'Why is it good? How is it good?' Those universal judgements (which I find so common) say nothing. They are like those who greet people as a mass or a crowd: those who have genuine knowledge of them greet them by name and distinguish them as individuals.[40] But it is a chancy business. Which explains why, on average more than once a day, I have seen men with ill-founded minds trying to act clever by showing me some beautiful detail in the book they are reading, but choosing so badly the point on which they fix their admiration that instead of revealing the excellence of their author they reveal their own ignorance.

When you have just listened to a whole page of Virgil you can safely exclaim, 'Now that is beautiful!' The cunning ones escape that way. But to undertake to go back over the detail of a good author, to try to indicate with precise and selected examples where he surpasses himself and where he flies high by weighing his words and his locutions and his choice of materials one after another: not many try that. *'Videndum est non modo quid quisque loquatur, sed etiam quid quisque sentiat, atque etiam qua de causa quisque sentiat.'* [We should not only examine what each one says, but what are his opinions and what grounds he has for holding them.][41] Day after day I hear stupid people uttering words which are not stupid. [B] They say something good; let us discover how deeply they understand it and where they got hold of it. They do not own that fine saying or that fine reasoning, but we help them to use it. They are only looking after it. Perhaps they only produced it fortuitously, hesitantly: it is we who

40. Plutarch (tr. Amyot), *De l'esprit familier de Socrates*, 636 BC.
41. Cicero, *De officiis*, I, xli, 147.

give it credit and value. You are lending them a hand. But why? They feel no gratitude towards you for it and become all the more silly. Do not support them; let them go their own way: they will handle that material like a man who fears getting scalded: they dare not show it in a different light or context nor to deepen it. Give it the tiniest shaking and it slips away from them: then, strong and beautiful though it be, they surrender it to you. They have beautiful weapons, but the handles are loose! How often have I learnt that from experience!

Now, if you come and clarify and reinforce it for them, they immediately take advantage of your interpretation and rob you of it: 'That is what I was about to say,' or, 'That is how I understand it, exactly,' or, 'If I did not put it that way it was because I could not find the right words.' – Bluster on! We should use even cunning to punish such arrogant stupidity.

[C] Hegesias' principle that we should neither hate nor blame but instruct is right elsewhere but not here.[42]   [B]   There is neither justice nor kindness in helping a man to get up who does not know how to use your help and who is all the worse for it. I like to let them sink deeper in the mire and to get even more entangled – so deeply that, if possible, even they finally realize it!

You cannot cure silliness and unreasonableness by one act of warning. [C] Of that sort of cure we can properly say what Cyrus replied to the man who urged him to give an exhortation to his troops at the moment of battle: that men are not made courageous warriors on the battlefield by a good harangue any more than you can become a good musician by hearing a good song.[43] Apprenticeships must be served, before you set hand to anything, by long and sustained study.

[B]   It is to our own folk that we owe this obligation to be assiduous in correcting and instructing; but to go preaching at the first passer-by or to read lectures on ignorance and silliness to the first man we come across is a practice which I loathe. I rarely do it during discussions in which I am involved; I prefer to let it all go by rather than to resort to such remote and donnish lecturing.   [C]   My humour is unsuited, both in speaking and writing, to those who are learning first principles.   [B]   But however false or absurd I judge things to be which are said in company or before a third party, I never leap in to interrupt them by word or gesture.

Meanwhile nothing in stupidity irritates me more than its being much more pleased with itself than any reasonableness could reasonably be. It is a

42. Diogenes Laertius, *Life of Aristippus*.
43. Xenophon, *Cyropaedia*, III, iii, 49–50.

disaster that wisdom forbids you to be satisfied with yourself and always sends you away dissatisfied and fearful, whereas stubbornness and foolhardiness fill their hosts with joy and assurance. It is the least clever of men who look down at others over their shoulders, always returning from the fray full of glory and joyfulness. And as often as not their haughty language and their happy faces win them victory in the eyes of the bystanders who are generally feeble in judging and incapable of discerning real superiority. [C] The surest proof of animal-stupidity is ardent obstinacy of opinion. Is there anything more certain, decided, disdainful, contemplative, grave and serious, than a donkey?

[B] Perhaps we may include in the category of conversation and discussion those short pointed exchanges which happiness and intimacy introduce among friends when pleasantly joking together and sharply mocking each other. That is a sport for which my natural gaiety makes me rather well-suited; and if it is not as tensely serious as the other sport I have just described, it is no less keen and clever. [C] nor, as it seemed to Lycurgus, any less useful.[44] Where I am concerned I contribute more licence than wit, being more happy in that than in finding my material; but I am a perfect target, for I can put up with retaliation without getting angry not merely when sharp but even when rude. When I am suddenly attacked, if I cannot at once find a good repartee I do not waste time following up that thrust with vague boring contestations akin to stubbornness but I let it go by, cheerfully flapping down my ears and waiting for a better moment to get my own back. No huckster wins every haggle.

Most people, when their arguments fail, change voice and expression, and instead of retrieving themselves betray their weaknesses and susceptibility by an unmannerly anger. In the excitement of jesting we can sometimes nip those secret chords of one another's imperfections which we cannot even pluck without offence when we are calm; we warn each other profitably of each other's faults. There are other sports, physical ones, rash and harsh in the French manner, which I hate unto death. I am touchy and sensitive about such things: in my lifetime I have seen two princes of the blood [C] royal [B] laid in their graves because of them. [C] It is an ugly thing to fight for fun.[45]

44. Perhaps a vague recollection of Plutarch (tr. Amyot), *Du trop parler*, 95 BC, or of Lycurgus' forbidding of hand-to-hand sports among citizens (Henry Estienne, *Apophthegmata*, 1568, pp. 416–17).
45. Henry II was killed while jousting; Henry, Marquess of Beaupréau died of wounds received in a tournament. There were other cases as well.

[B] In addition when I want to judge another man I ask him to what extent he is himself satisfied; how far he is happy with what he has said or written. I want him to avoid those fine excuses: 'I was only playing at it' –

Ablatum mediis opus est incudibus istud

[It was taken off the anvil only half finished][46]

– 'I only spent an hour on it'; 'I have not seen it since'. – 'All right,' I say: 'let us leave those examples. Show me something which does represent you entirely, something by which you are happy to be measured.' And then I say, 'What do you consider the most beautiful aspect of your work? Is it this quality or that quality? Is it its gracious style, its subject-matter, your discovery of the material, your judgement, your erudition?'

For I normally find that men are as wrong in judging their own work as other people's, not simply because their emotions are involved but because they lack the ability to understand it and to analyse it. The work itself, by its own momentum and fortune, can favour the author beyond his own understanding and research; it can run ahead of him. There is no work that I can judge with less certainty than my own: the *Essays* I place – very hesitantly and with little assurance – sometimes low, sometimes high.

Many books are useful for their subject-matter: their authors derive little glory from them. And there are good books which as far as good workmanship is concerned are a disgrace to their authors. I could write about our style of feasting, about our clothing – and I could write it gracelessly; I could publish contemporary edicts and the letters of princes which come into the public domain; I could make an abridgement of a good book (and every abridgement of a good book is a daft one) and then the book itself could chance to get lost. Things like that. From such compilations posterity would derive unique assistance: but what honour would I derive from them except for being lucky? A good proportion of famous books fall in that category.

When I was reading a few years ago Philippe de Commines – a very good author, certainly – I noted the following saying as being above average: 'We should be wary of doing such great services to our master that we render him unable to reward them justly.' I should have praised not him but his discovery of a topic. Not long ago I came upon this

46. Ovid, *Tristia*, I, vii, 9.

sentence in Tacitus: *'Beneficia eo usque læta sunt dum videntur exolvi posse; ubi multum antevenere, pro gratia odium redditur.'* [Good turns are pleasing only in so far as they seem repayable. Much beyond that we repay with hatred not gratitude.] [C] Seneca puts it forcefully: *'Nam qui putat esse turpe non reddere, non vult esse cui reddat.'* [He for whom not to repay is a disgrace wants his benefactor dead.] Quintus Cicero, with a laxer turn of phrase, writes: *'Qui se non putat satisfacere, amicus esse nullo modo potest.'* [He who cannot repay his debt to you can in no wise love you.][47]

[B] An author's subject can, when appropriate, show him to be erudite or retentive, but if you are to judge what qualities in him most truly belong to him and are the most honourable (I mean the force and beauty of his soul) you must know what is really his and what definitely is not; and in that which is not, how much we are indebted to him for his selection, disposition, ornamentation and the literary quality of what he had contributed. Supposing he has taken somebody else's matter and then ruined the style, as often happens! People like us who have little experience of books are in difficulties when we come across some fine example of ingenuity in a modern poet or some strong argument in a preacher. We dare not praise them for it before we have learned from a scholar whether that item is original to them or taken from another. Until I have done that I remain suspicious.

I have just read through at one go Tacitus' *History* (something which rarely happens to me: it is twenty years since I spent one full hour at a time on a book. I did it on the recommendation of a nobleman highly esteemed in France both for his own virtue and for that sustained quality of ability and goodness which he is seen to share with his many brothers). I know of no author who combines a chronicle of public events with so much reflection on individual morals and biases.[48] [C] And it appears to me (contrary to what appears to him) that, as he has the particular task of following the careers of the contemporary Emperors (men so odd and so extreme in their various characters) as well as the noteworthy deeds which they provoked in their subjects above all by their cruelty, he has a more striking and interesting topic to relate and discourse upon than if he had to tell of battles and world revolutions. Consequently I find

47. Montaigne is contrasting *inventio* (the discovery of arguments or topics) with original powers of judgement. Philippe de Commines, III, xii; Tacitus, *Annals*, IV, xviii; Seneca, *Epist. moral.*, LXXXVI, 32; Cicero, *De petitione consultatus*, ix.
48. '88: biases. *In that he is no less careful and diligent than Plutarch, who made an express claim to do so.* This manner . . .

him unprofitable when he dashes through those fair, noble deaths as
though he were afraid of tiring us by accounts both too long and too
numerous.

[B] This manner of history is by far the most useful. The unrolling of
public events depends more on the guiding hand of Fortune: that of private
ones, on our own.[49] Tacitus' work is more a judgement on historical
events than a narration of them. There are more precepts than accounts. It
is not a book to be read but one to be studied and learnt. It is so full of
aphorisms that, apposite or not, they are everywhere. It is a seed-bed of
ethical and political arguments to supply and adorn those who hold high
rank in the governing of this world. He pleads his case with solid and
vigorous reasons, in an epigrammatic and exquisite style following the
affected manner of his century. (They were so fond of a high style that
when they found no wit or subtlety in their subject-matter they resorted to
witty subtle words.) He is not all that different from Seneca, but while he
seems to have more flesh on him Seneca is more acute. Tacitus can more
properly serve a sickly troubled nation like our own is at present: you
could often believe that we were the subject of his narrating and berating.
Those who doubt his good faith clearly betray that they resent him from
prejudice. He has sound opinions and inclines to the right side in the affairs
of Rome. I do regret though that, by making Pompey no better than
Marius and Scylla only more secretive, he judged him more harshly than is
suggested by the verdict of men who lived and dealt with him.[50] True,
Pompey's striving to govern affairs has not been cleared of ambition nor a
wish for vengeance: even his friends feared that victory might make him
go out of his mind, though not to the extremes of insanity of those other
two. Nothing in his life suggests to us the menace of such express tyranny
and cruelty. Besides we ought never to let suspicions outweigh evidence: so
on this point I do not trust Tacitus.

That the accounts which he gives are indeed simple and straight can
perhaps be argued from the very fact that they do not exactly fit his
concluding judgements, to which he is led by the slant he had adopted;
they often go beyond the evidence which he provides – which he had not
deigned to bias in the slightest degree. He needs no defence for having
assented to the religion of his day, in accordance with the laws which bade

49. '88: our own. *Yet he did not overlook what he owed to the other aspect.* Tacitus'
work . . .
50. Tacitus, *Histories*, II, xxxviii.

him to do so, and for being ignorant of the true religion. That is his misfortune not his fault.[51]

What I have chiefly been considering is his judgement: I am not entirely clear about it. For example, take these words from the letter sent to the Senate by the aged ailing Tiberius: 'What, Sirs, should I write to you, what indeed should I not write to you at this time? I know that I am daily nearing death; may the gods and goddesses make my end worse if I know what to write.' I cannot see why he applies them with such certainty to a poignant remorse tormenting Tiberius' conscience. Leastways when I came across them I saw no such thing.[52]

It also seemed to me a bit weak of him when he was obliged to mention that he had once held an honourable magistracy in Rome to go on and explain that he was not referring to it in order to boast about it. That line seemed rather shoddy to me for a soul such as his: not to dare to talk roundly of yourself betrays a defect of thought. A man of straight and elevated mind who judges surely and soundly employs in all circumstances examples taken from himself as well as from others, and frankly cites himself as witness as well as third parties. We should jump over those plebeian rules of etiquette in favour of truth and freedom. [C] I not only dare to talk about myself but to talk of nothing but myself. I am wandering off the point when I write of anything else, cheating my subject of *me*. I do not love myself with such lack of discretion, nor am I so bound and involved in myself, that I am unable to see myself apart and to consider myself separately as I would a neighbour or a tree. The error is the same if you fail to see the limits of your worth or if you report more than you can see. We owe more love to God than to ourselves.[53] We know him less, yet talk about him till we are glutted.

[B] If Tacitus' writings tell us anything at all about his character, he was a very great man, upright and courageous, whose virtue was not of the superstitious kind but philosophical and magnanimous. You could find

---

51. Once more a judgement *secundum quid* (in this case according to the standard of the laws of Tacitus' day). It was not Tacitus' fault, since a knowledge of Christian truth requires prevenient grace, which by definition cannot be in any way earned or deserved.

52. Tacitus, *Annals*, VI, vi.

53. Montaigne apparently accepts the contention of Duns Scotus (and others) that when a man loves himself or any other creature properly he loves God even more. Luther and many others denied this (*Weimarer Ausgabe*, XL, p. 461). Montaigne's contention is more traditionally Catholic than Humanist.

some of his testimony rather rash; for example he maintains that when a soldier's hands grew stiff with the cold while carrying a pile of wood they adhered to his load, broke away from his arms and stuck there dead.[54] In similar cases my custom is to bow to the authority of such great witnesses. When he says that, by favour of Serapis the god, Vespasian cured a blind woman in Alexandria by anointing her eyes with his saliva and also performed some additional miracle or other, he was following the dutiful example of all good historians who keep a chronicle of important happenings: included among public events are popular rumours and opinions. Their role is to give an account of popular beliefs, not to account for them: which part is played by Theologians and philosophers as directors of consciences. That is why his fellow-historian, great man as he was, most wisely said: *'Equidem plura transcribo quam credo: nam nec affirmare sustineo, de quibus dubito, nec subducere quae accepi.'* [I do indeed pass on more than I believe. I cannot vouch for the things which I doubt, nor can I omit what I have been told by tradition.] And another says: *'Haec neque affirmare, neque refellere operae pretium est: famae rerum standum est.'* [These things are neither to be vouched for nor denied: we must cling to tradition.][55] Tacitus, writing during a period in which belief in portents was on the wane, says that he nevertheless does not wish to fail to provide a foothold for them, and so includes in his *Annals* matters accepted by so many decent people with so great a reverence for antiquity.

[B] That is very well said. Let them pass on their histories to us according to what they find received, not according to their own estimate. I, who am monarch of the subject which I treat and not accountable for it to anyone, do not for all that believe everything I say. Sometimes my mind launches out with paradoxes which I mistrust [C] and with verbal subtleties which make me shake my head; [B] but I let them take their chance. [C] I know that some men gain a reputation from such things. It is not for me alone to judge them. I describe myself standing up and lying down, from front and back, from right and left and with all my inborn complexities. [B] Even [C] minds[56] [B] of sustained power are not always sustained in their application and discernment.

---

54. Tacitus, *Annals*, XIII, xxxv; then, IV, lxxi (seen by some as a parody of Christ's curing the blind man in Mark 8:23).
55. Quintus Curtius, IX, i; Livy, VIII, vi.
56. '88: even *judgements* which are . . .

That is, *grosso modo*, the Tacitus which is presented to me, vaguely enough, by my memory.   [C]   All *grosso-modo* judgements are lax and defective.[57]

57. 88: All *universal* judgements are lax and *dangerous* . . .

9. On vanity

===

[Montaigne justifies his digressions and expresses his admiration for the 'motley' style of Plato's dialogue Phaedrus, *with its varied themes, its 'party-coloured' subject-matter; there is the suggestion that such a style is particularly appropriate to Man who is (in the final words) 'the jester of the farce'. 'On vanity' is just such a motley, with abrupt changes of subject from vanity (in general and particular) to travel and to political morality. Although philosophy aims at 're-forming' a man (that is, at improving and remoulding his soul as Socrates did) it hardly ever succeeds in its aim. Montaigne's own soul is unreformed (in that sense) and so never content. Wisdom consists in learning how to accept that fact and to welcome such palliatives as inquiry and travel.]*

[B] Perhaps there is no more manifest vanity than writing so vainly about it. That which the Godhead has made so godly manifest should be meditated upon by men of intelligence anxiously and continuously.[1] Anyone can see that I have set out on a road along which I shall travel without toil and without ceasing as long as the world has ink and paper. I cannot give an account of my life by my actions: Fortune has placed them too low for that; so I do so by my thoughts. Thus did a nobleman I once knew reveal his life only by the workings of his bowels: at home he paraded before you a series of seven or eight days' chamber-pots. He thought about them, talked about them: for him any other topic stank. Here (a little more decorously) you have the droppings of an old mind, sometimes hard, sometimes squittery, but always ill-digested. And when shall I ever have done describing some commotion and revolution of my thoughts, no matter what subject they happen upon, when Diomedes wrote six thousand books on the sole subject of philology?[2] What can babble produce when the stammering of an untied tongue smothered the world under such a dreadful weight of volumes? So many words about

1. Allusion to the 'Vanity of vanities' of Ecclesiastes 1:2 and 14; 3:19; 11:8; 12:8 and the leitmotiv 'vanity' and 'vain' throughout this, the most sceptical book of the Bible.
2. This awesome philologist was in fact Didymus, who wrote four thousand books. His nickname was 'Brazen-bowels', which doubtless explains his being cited here. Montaigne's error was already in Bodin's *Methodus* (dedicatory epistle).

nothing but words! O Pythagoras! Why couldest thou not conjure such turbulence![3]

A certain Galba in days gone by was criticized for living in idleness. He replied that everyone should have to account for his actions but not for his free time. He was deceiving himself: for justice also takes note and cognizance of those who are not employed. The Law ought to impose restraints on silly useless writers as it does on vagabonds and loafers. Then my own book and a hundred others would be banished from the hands of our people. I am not joking. Scribbling seems to be one of the symptoms of an age of excess. When did we ever write so much as since the beginning of our Civil Wars? And whenever did the Romans do so as just before their collapse? Apart from the fact that to make minds more refined does not mean that a polity is made more wise, such busy idleness arises from everyone slacking over the duties of his vocation and being enticed away. Each individual one of us contributes to the corrupting of our time: some contribute treachery, other (since they are powerful) injustice, irreligion, tyranny, cupidity, cruelty: the weaker ones like me contribute silliness, vanity and idleness. When harmful things are compelling then, it seems, is the season for vain ones; in an age when so many behave wickedly it is almost praiseworthy merely to be useless. I console myself with the thought that I shall be one of the last they will have to lay hands on. While they are dealing with the more urgent cases I shall have time to improve, for to me it seems contrary to reason to punish minor offences while we are ravished by great ones. Philotimus, a doctor, recognized the symptoms of an ulcerated lung from the features and breath of a patient who brought him his finger to be dressed. 'My friend,' he said, 'this is no time to be worrying about fingernails!'[4]

While on this subject, a few years ago a great man, whom I recall with particular esteem, in the midst of our great ills, when there was no justice, law or magistrate functioning properly any more than today, went and published edicts covering some wretched reform or other of our clothing, eating and legal chicanery.[5] Such things are tidbits on which we feed an ill-

3. Cf. Erasmus, *Adagia*, IV, III, LXXII, *Taciturnior Pythagoreis*. (Pythagoras imposed five years of silence on his disciples.) Then, for Galba, Erasmus, *Apophthegmata*, VIII, *Thrasea*, XLVII.

4. Plutarch (tr. Amyot), *Comment il fault ouïr*, 54 G.

5. Perhaps a reference to Charles IX's law on the shortening of legal actions (13 December 1563), and on his sumptuary laws controlling superfluous clothing (17 January – 10 February 1563/4) and hotels and restaurants (20 January 1563). All were printed by Robert Estienne in Paris. Some think it is an allusion to Michel de l'Hospital.

governed people to show that we have not entirely forgotten them. Others do the same when they issue detailed prohibitions of swear-words, dances and sports for a people sunk in detestable vices of every kind.[6] It is not the time to wash and to get the dirt off once you have caught a good fever. [C] It is right only for Spartans about to rush into some extreme mortal danger to start combing and dressing their hair.[7]

[B] I have a worse habit myself: if one of my shoes is askew then I let my shirt and my cloak lie askew as well: I am too proud to amend my ways by halves. When my condition is bad I cling violently to my illness: I abandon myself to despair and let myself go towards catastrophe, [C] casting as they say the haft after the axe-head; [B] stubbornly, I want to get worse and think myself no longer worth curing. Either totally well or totally ill.

It is a boon for me that the forlorn State of France should correspond to the forlorn age I have reached. It is easier for me to accept that my ills should be augmented by it than that such good things as I have should be troubled by it. The words I utter when wretched are words of defiance: instead of lying low my mind bristles up. Contrary to others I find I am more prayerful in good fortune than in bad. Following Xenophon's precept, though not his reasoning, I am more ready to make sheep's eyes at Heaven in thanksgiving than in supplication.[8] I am more anxious to improve my health when it beams upon me than to restore it when I have lost it; prosperous times serve to discipline me and instruct me, as rods and adversities do to others. [C] As though good fortune were incompatible with a good conscience, men never become moral except when fortune is bad. For me good luck[9] [B] is a unique spur to measure and moderation. Entreaties win me over: menaces I despise; [C] good-will makes me bow: fear makes me unbending.

[B] Among men's characteristics this one is common enough: to delight

---

6. Cf. 'The ordinance of the King [Charles IX] and of Monsieur de Losse forbidding blasphemy and playing or singing dissolute songs' (promulgated 5 December 1564).

7. Plutarch (tr. Amyot), *Dicts notables des Lacedaemoniens*, 221B. Cf. p. 905, note 28.

8. Erasmus, *Apophthegmata*, VII, *Xenophon*, XXVI. (A man should above all worship the gods when things go well, so that he can confidently appeal to them as friends when in sore straits. Erasmus approves of Xenophon's saying, stating that most men act to the contrary.)

9. [B] instead of [C]: others. For me the good is a unique spur to measure and moderation . . .

more in what belongs to others than to ourselves and to love variation and change:

> *Ipsa dies ideo nos grato perluit haustu*
> *Quod permutatis hora recurrit equis.*

[Even the daylight only pleases us because the hours run by on changing steeds.][10]

I have my share of that.

Those who go to the other extreme, who are happy with themselves, who esteem above all else whatever they possess and who recognize no form more beautiful than the one they behold, may not be wise as we are but they are truly happier. I do not envy them their wisdom but I do envy them their good fortune.

My avid humour for things new and unknown helps to foster in me my yearning to travel, though plenty of other circumstances contribute to it as well. I am most willing to turn aside from ruling my house. There is some pleasure in being in charge, if only of a barn, and in being obeyed in one's household, but it is too uniform and listless a pleasure; it also necessarily involves you in many troublesome thoughts. You are distressed when your tenants suffer from famine, when your neighbours quarrel among themselves or encroach on you.

> *Aut verberatæ grandine vineæ,*
> *Fundusque mendax, arbore nunc aquas*
> *Culpante, nunc torrentia agros*
> *Sidera, nunc hyemes iniquas.*

[Either the hail has ravaged your vineyards, or the soil deceives your hopes, or your fruit trees are lashed by the rain, or the sun scorches your fields. And there are the rigours of winter.][11]

Then there is the fact that barely once in six months will God send you weather which totally satisfies your steward: if it is good for the vines it is bad for the pastures:

> *Aut nimiis torret fervoribus ætherius sol,*
> *Aut subiti perimunt imbres, gelidæque pruinæ,*
> *Flabraque ventorum violento turbine vexant.*

[Either the blazing sun shrivels your harvests or else they are ruined by sudden rainstorms or frosts, or ravaged by violent whirlwinds.]

10. Petronius (fragment).
11. Horace, *Odes*, I, 29–32; then, Lucretius, V, 216–18.

Then there is that shoe of the man of yore: new and shapely but pinching the foot;[12] no outsider ever understands how much it costs you, and how much it takes out of you, to keep up that appearance of order to be seen in your household and which perhaps is bought too dearly.

I came late to managing my estates. Those whom Nature had given birth to before me long relieved me of that burden. I had already acquired a different bent, one more in keeping with my complexion. Nevertheless, from what I have seen of it, it is an occupation more time-consuming than difficult: if a man has the ability to do other things, then he can do it easily. If I were seeking to get rich, that way would have seemed too long; I would have served kings – a business which produces better crops than any other. Since I [C] aim only to acquire the reputation for having acquired nothing, and squandered nothing either (in conformity with the rest of my life, which is as ill-suited to doing evil as good) and [B] seek only to get by, I can do it without paying much attention.

'If the worst comes to worst, forestall poverty by cutting down expenses.' That is what I try to do, changing my ways before poverty compels me to. Meanwhile I have established enough gradations in my soul to allow me to do with less than I have – and I mean contentedly. [C] *'Non æstimatione census, verum victu atque cultu, terminatur pecuniæ modus.'* [Your degree of wealth is not measured against your income but against your expenditure on food and luxuries.][13] [B] My real need does not so exactly take up all my income as to leave nothing for Fortune to get her teeth into without biting me to the quick. My presence, ignorant and disdainful though it be, does give a strong shove to the business of my home-estates. I do work at it, albeit grudgingly. And you can say this for me at home: while I do burn my end of the candle on my own, the other end does not have to cut down on anything.

[C] My travels only hurt me by their expense, which is considerable and exceeds my resources. Used as I am to travel not merely with an adequate retinue but an honourable one, I have to make my journeys shorter and less frequent, spending only the froth of my savings, putting things off and spinning them out as the money comes in. I have no wish that the pleasure of roaming should mar the pleasure of repose; on the contrary, I intend that each should nourish and encourage the other. Fortune has helped me in that; my chief aim in life being to live it lazily

---

12. Erasmus, *Apophthegmata*, V, *Paulus Aemilius*, XVI (explaining why he divorced a beautiful wife).
13. Cicero, *Paradoxa*, VI, iii.

and leisurely rather than busily, she has taken from me the need to proliferate in wealth to provide for a proliferation of heirs. For a single heir, if what has been plenty enough for me is not enough for him, that is just too bad. His foolishness would not justify my wishing him more.[14] Following the example of Phocion, every man provides enough for his children insofar as he provides for characters not dissimilar to his.[15] I would in no wise favour what Crates did: he left his money with a banker to give to his children if they turned out to be fools, but to share between the simpletons among the people if they turned out to be clever. As if fools are better able to use money because they are less able to do without it!

[B] Anyhow such harm as may be done by my absence does not seem to me to merit my refusing to accept, while I can afford it, such occasions as come along to withdraw my irksome presence. Something is always going awry there. You are always tugged at by business concerning this house or that. You survey everything at too close quarters: there your sharp-sightedness is harmful to you, as often enough elsewhere. I shun all occasions for annoyance and keep myself from learning about things going wrong, yet not so successfully as to avoid stumbling at home upon things which displease me. [C] And the mean tricks they hide from me are the ones I know best: you have to help to conceal some of them yourself so that they hurt you the less! [B] Vain little jabs – [C] well, vain sometimes – [B] but jabs all the same.[16] It is the smallest, finest cuts which are the most piercing; just as the smallest print tires and hurts your eyes so do the smallest concerns stab you most. [C] A multitude of petty ills beset you more than the violence of a single one, no matter how big. [B] The finer and more frequent those domestic thorns the more sharply and unexpectedly they bite into us, easily taking us by surprise.[17]

14. Montaigne had only one child, his daughter Léonor, who could not inherit as could a son and heir. He talks here of a male heir, either thinking of the entailed property of his estates or perhaps of a son-in-law.
15. Cornelius Nepos, *Life of Phocion*, I; then, Diogenes Laertius, *Life of Crates*, VI, lxxxviii.
16. '88: same, *and shaming ones*. It is . . .
17. '88: surprise. *Now Homer shows us plainly enough what advantage is given by surprise, when he portrays Ulysses weeping over the death of his dog and not weeping over the tears of his mother: the first event, slight though it was, overwhelmed him since he was unexpectedly assailed by it; he withstood the second more violent one because he was prepared for it. The reasons may be trivial, yet they disturb our lives: our life is a delicate thing, easy to wound.* Once my face . . .
Cf. Plutarch (tr. Amyot), *De la tranquillité de l'ame*, 74FG. Perhaps omitted because close to the exemplum of Psammenitus in I, 2, 'On sadness'.

[C] I am no philosopher: ills crush me in proportion to their weight, and they weigh as much by their manner as their matter, often more. I know them better than ordinary people do and so bear them better; but in the end, though they do not wound me they do strike me. Life is a delicate thing, easy to disturb. *'Nemo enim resistit sibi cum coeperit impelli'* [No one can stop himself once he yields to the first impulse]:[18]   [B]   once my face is turned towards chagrin, no matter how silly the cause which brought me to be so, I goad my humour in that direction. Thereafter it nourishes itself, provoking itself under its own impetus, drawing to itself and piling up matter upon matter on which to feed:

> *Stillicidi casus lapidem cavat.*

> [Water dripping drop by drop makes fissures in a stone.]

Those everyday fissures eat into me.   [C]   Everyday irritations are never slight. They are constant and irremediable, particularly when they arise from the cares of your estates, which are constant and unavoidable.

[B] When I consider my affairs overall and from a distance I find (perhaps because my memory of them is hardly a detailed one) that they have, up to the present, gone on prospering beyond my projections or calculations; I seem to be getting more out than is there: their happy state misleads me. But once I am involved in the job and watching the progress of all the details –

> *Tum vero in curas animum diducimur omnes*

> [Our souls torn asunder by all our cares]

– thousands of things cause me to hope or to fear. It is exceptionally easy for me to abandon them completely: dealing with them without anguish is exceptionally hard. It is wretched to be in a place where everything you see makes work for you and concerns you. I believe I am more happy when enjoying the pleasures of someone's else's house, and that I bring a more innocent taste to them.   [C]   When asked what kind of wine he thought best, Diogenes replied, 'Someone else's'.[19] I agree with that.

[B] My father loved building at Montaigne, where he was born. In all my government of my domestic affairs I like to follow his precept and

---

18. Seneca, *Epist. moral.*, XIII, 13; then Lucretius, I, 314, and Virgil, *Aeneid*, V, 720.
19. Diogenes Laertius, *Life of Diogenes*, VI, liv. Listed also s.v. *Diogenes* in Erasmus' *Apophthegmata*.

example, and as far as I can I will impose that duty on my successors. If I could do better for him I would. I glory in the fact that his wishes are still effective and implemented by me. God forbid that I should allow to fail in my hands any ghost of life which I could give to so good a father. The fact that I have bothered to complete some old section of wall and repair some botched bit of building has certainly been more out of regard for his intention than my contentment. [C] And I reproach my own laziness for not having gone on to complete the fine things he started in this house of his, the more so since I am most likely to be the last of my stock to own it and to give it a final touch. [B] As for my own inclinations, neither the pleasures of building (which are supposed to be so attractive) nor of hunting nor of laying out gardens, nor the other pleasure of life in the country, can keep me much occupied. I think ill of myself for this, as I do for all opinions which are disadvantageous to me. I do not so much care about having vigorous and informed opinions as having easy ones, convenient to live with; [C] they are true and sound enough if they are useful and pleasant

[B] Those who, when they hear me tell of my inadequacies for the tasks of managing my estates, proceed to yell in my ears that it is due to disdain and that I cannot be bothered to learn the names of the tools used in husbandry, nor about its seasons and succession of tasks, nor how my wines are made, how grafting is done, the names of plants and fruits and the ways of preparing them for the table, [C] nor the names and quality of the cloth I wear, [B] because my mind is full of some higher knowledge, do me mortal wrong. That would be silly, more stupid than glorious. I would[20] rather be a good equerry than a good logician:

> *Quin tu aliquid saltem potius Quorum indiget usus,*
> *Viminibus mollique paras detexere junco?*

[Why do you not do something useful, like making baskets of wickerwork or pliant reeds?][21]

[C] We confuse our thoughts with generalities, universal causes and processes which proceed quite well without us, and leave behind our own concerns for *Michel*,[22] which touch us even more intimately than Man.

---

20. '88: would *not be a contempt, it* would be silly . . .
21. Virgil, *Eclogues*, II, 71–2.
22. That is, for ourselves under our Christian names as individual persons.

[B]   Now usually I do remain at home; but I could wish that I were happier there than elsewhere.

> *Sit meæ sedes utinam senectæ,*
> *Sit modus lasso maris, et viarum,*
> *Militiæque.*

[May it be my final haven when I am weary of the sea, of roaming and of war.][23]

I do not know whether I shall manage to struggle through. I wish that, in lieu of some other part of his inheritance, my father had bequeathed me that passionate love for the running of his estates which he had in his old age. He was most successful in limiting his desires to his means and in knowing how to be content with what he had. If only I can acquire the taste for it as he did, then political philosophy can, if it will, condemn me for the lowliness and barrenness of my occupation. I do believe that the most [C] honourable [B] vocation[24] is to serve the commonwealth and to be useful to many.   [C]   *'Fructus enim ingenii et virtutis omnisque præstantiæ tum maximus accipitur, cum in proximum quemque confertur.'* [The fruits of intellect and virtue and of all outstanding talents are best employed when shared with one's neighbour.][25]   [B]   But where I am concerned I renounce my share, partly from self-awareness (which enables me to see both the weight attached to such vocations and the scant means I have of providing for them) –   [C]   even that master-theoretician of all political government Plato did not fail to abstain from it himself –   [B]   partly from laziness. I am content to enjoy the world without being over-occupied with it and to lead a life which is no more than excusable, neither a burden to myself nor to others.

No man ever entrusted his affairs more fully and passively into the care and control of another than I would do if only I had someone available. One of my wishes now would be to find me a son-in-law who would fill my beak, comfort my final years and lull them to sleep, into whose hands I could resign the control and use of my goods, with complete sovercignty to do with them as I do, getting out of them what I do now – provided that he brought to it a truly grateful and loving affection. Yes: but we live in a world where the loyalty of one's own children is unheard of.

When I am on my travels, whoever has my purse has full charge of it without supervision. He could cheat me just as well if I kept accounts,

23. Horace, *Odes*, II, vi, 6–8.
24. '88: most *noble and just* vocation . . .
25. Cicero, *De amicitia*, XIX, 70.

and, unless he is a devil, by such reckless trust I oblige him to be honest. [C] *'Multi fallere docuerunt, dum timent falli, et aliis jus peccandi suspicando fecerunt.'* [Many by their fear of being cheated have taught others to cheat; others have found justification for wrong-doing in suspicion thrown upon them.][26] [B] The surety I most usually have for my servants is my own ignorance. (I never assume defects until I have seen them, and I trust the young more, reckoning that they are less corrupted by bad example.) I prefer hearing after two months that I have spent four hundred crowns than having my ears battered every morning with three, five or seven. Yet [C] by larcenies of that kind [B] I have been as little robbed as anyone. True, I lend my ignorance a helping hand. I consciously encourage my knowledge of my money to be somewhat vague and uncertain; up to a point I am pleased to be unsure about it. You should leave a little room for the improvidence or dishonesty of your manservant. On condition that there should remain, by and large, enough for us to do what we want, let us allow the surplus of Fortune's liberality to flow on a little farther at her behest — [C] the gleaner's portion.[27] After all I do not prize the faithfulness of my men more than I disprize their wronging me. [B] Oh, what a servile and silly care is care for your money, loving to handle it, weigh it, count it over. That is the way miserliness makes its advances.

I have been in charge of property for the last eighteen years but have never yet got myself to look into my title-deeds nor into my principal affairs which must needs be transacted with my knowledge and attention. This is no philosophical contempt for the transitory things of this world: my taste has not been so purified as that. At the very least I value such things at their worth. It is a case, most certainly, of inexcusable and puerile[28] [C] laziness and negligence. What would I not do to avoid reading through a contract and shaking the dust off piles of papers, a slave to my affairs and, worse still, a slave to other people's, like so many folk who do it for the money! For me nothing is expensive save toil and worry: all I want is to be indifferent and bovine.

[B] I was made, I think, more for living off somebody else, if that could be done without servitude and obligation. And when I look at things closely I am not sure whether, for a man of my temperament and station,

26. Seneca, *Epist. moral.*, III, 3 (adapted).
27. Leviticus 19:10, which commands reapers to leave the gleanings for the poor and the stranger.
28. '88: of inexcusable and puerile *sloth and flabbiness*. I was made . . .

what I have to put up with from business and agents and servants does not entail more degradation, bother and bitterness than there would be in following a man born greater than I who would give me a bit of guidance and comfort. [C] *'Servitus obedientia est fracti animi et abjecti.'* [Slavery is the obedience of a weak and despondent mind lacking in will.][29]

[B] When Crates, to rid himself of the cares and indignities of his home, jumped into the freedom of poverty, he made things worse.[30] That I would never do. I loathe poverty on a par with pain. But I would indeed exchange that first sort of existence for another less grand and less busy.

Once I am away I slough off all such preoccupations: I would feel it less then if a tower collapsed than I feel the fall of a tile when I am there. Once I am away my soul can easily find detachment: when I am there she frets like a wine-grower's. [C] A twisted rein on my horse or a stirrup-strap knocking against my leg can put me out of humour for an entire day. [B] In face of difficulties I can lift up my thoughts but not my eyes.

> *Sensus, O superi, sensus.*
>
> [Feelings, ye gods! Feelings!][31]

It is I who am responsible when anything goes wrong at home. There are few masters – I mean of my middle station (and if there are any at all the luckier they are) who are able to rely on anyone else without retaining most of the load. That [C] somewhat detracts from the way I treat visitors (though I may have made the odd one stay on, as bores do, more for my cuisine than for my charm); and it [B] considerably detracts from the pleasure I ought to take in visits and gatherings of friends in my house.

A gentleman in his own home never looks so [C] silly [B] as when[32] he is seen to be preoccupied with the arrangements, having a word in a manservant's ear or casting threatening glances at another: such arrangements should flow unnoticed and suggest a normal pattern. And I

---

29. Cicero, *Paradoxa*, V, i.
30. Cf. Plutarch (tr. Amyot), *De la tranquillité de l'esprit*, 69 F.
31. Source unknown. Montaigne is contrasting his *courage* (his mind, that is, or his thoughts or his faculty of thought) with the power of his feelings (*sensus*). His meaning is perhaps parallel to Democritus' assertion 'that there is more sensation [or, sense] in the brute beasts – and in the wise'.
32. '88: so *inept and cheap* as when . . .

find it ugly to discuss with your guests the way you are treating them, either to apologize or to boast.

Order and cleanliness I love –

> *et cantharus et lanx*
> *Ostendunt mihi me*
>
> [I can see my reflection in tankard and plate][33]

– on a par with abundance; in my own home I am punctilious about necessities but have little regard for ostentation. When you are in somebody else's house and a servant brawls or a dish is spilled you simply laugh; and while My Lord settles tomorrow's arrangements for you with his butler you can doze off.

[C] I am speaking for myself: I do not fail to realize how great a pastime it generally is for certain natures to run their households quietly and prosperously, all done with regularity and order. I do not wish to attribute my own mistakes and shortcomings to the thing itself nor to contradict Plato's contention that the happiest occupation for any man is to manage his private concerns without injustice.[34]

[B] When on my travels I have to think only of me – and how to spend my money (one injunction can see to that). To amass a fortune you need too many talents: I know nothing about that. I know a bit about spending it and making a good show of my expenditure – which is indeed its principal use – but I strive a bit too ambitiously over it, which makes my spending uneven and misshapen, given to excess at both extremes. If it makes a parade, if it serves a purpose, I let myself be carried away injudiciously; and just as injudiciously I close up tight if it has no gleam and does not beam on me.

Whether it is art or nature which stamps on us that characteristic of living by what others say, it does us much more harm than good. We cheat ourselves of what is rightly useful to us in order to conform our appearances to the common opinion. We are not so much concerned with what the actual nature of our being is within us, as with how it is perceived by the public. Even wisdom and the good things of the mind seem fruitless to us if we enjoy them by ourselves, if they are not paraded before the approving eyes of others. Men there are whose gold flows unnoticed, swishing through great caverns underground: others spread theirs widely – all sheets of gold-leaf – so that the pennies of some are worth the guineas of

33. Horace, *Epistles*, I, v, 23–4.
34. Plato, Epistle IX, to Archytas.

others and vice versa, the world judging worth and expenditure by their show.

All attentive care for riches reeks of covetousness, as do spending when too ordinate and generosity when too contrived. They are not worth anxious attention and worry. Anyone who wants to make his expenditure just right makes it constricted and confined. Keeping and spending are in themselves indifferent: they take on the colour of good or evil depending upon how we apply our wills to them.[35]

The other cause which invites me to travel is my incompatibility with our present political morality. So far as the public interest is concerned I could reconcile myself easily enough to that corrupt condition:

> *pejoraque sæcula ferri*
> *Temporibus, quorum sceleri non invenit ipsa*
> *Nomen, et a nullo posuit natura metallo;*

[worse than that Age of Iron in which crimes lacked a name, an age which Nature could find no metal to describe;][36]

where my own interests are concerned, I cannot. It presses too hard upon me individually. For in my neighbourhood the prolonged licence of our Civil Wars has already hardened us to a form of government so overflowing with evil –

> *Quippe ubi fas versum atque nefas*

[Where right and wrong are all confounded][37]

– that it is a miracle that it can endure.

> *Armati terram exercent, semperque recentes*
> *Convectare juvat prædas et vivere rapto.*

[Men bear arms while ploughing the fields, thinking only of grabbing fresh plunder and living by rapine.]

---

35. A classic Stoic contention: *adiaphora* (things indifferent) become good or bad according to our attitude towards them. (Cf. Rabelais, *Tiers Livre*, TLF, VIII, 45–53.)

36. Juvenal, *Satires*, XIII, 28–30. The Age of Iron was the cruellest known to the Ancients, marking the decline from the happy innocence of mankind during the Golden Age, through the Silver and Bronze Ages, to the Age of Iron, when men's weapons and hearts were hard.

37. Virgil, *Georgics*, I, 505; then, *Aeneid*, VII, 748–9.

In short I learn from our example that, whatever the cost, human society remains cobbled and held together. No matter what position you place them in, men will jostle into heaps and arrange themselves in piles, just as odd objects thrust any-old-how into a sack find their own way of fitting together better than art could ever arrange them. King Philip made just such a pile from the most wicked and depraved men he could find. He built them a city which bore their name and sent them there.[38] I reckon that out of their very vices they wove for themselves a political fabric and an advantageous lawful society.

It is not one deed that I see, not three, not a hundred, but morals, now commonly accepted, so monstrous in their inhumanity and above all in their disloyalty (which are for me the worst species of vice), that my mind cannot conceive of them without horror. Almost as much as with loathing they strike me with amazement. The practice of such remarkable wickedness is as much a sign of vigour and power in the soul as of error and unruliness. Necessity associates men and brings them together; afterwards that fortuitous bond is codified into laws; for there have been societies as ferocious as any that human opinion can spawn which have nevertheless kept their structures as sound and as durable as any which Plato or Aristotle could ever have founded. And indeed such descriptions of fictional and artificial polities are ridiculous and silly when it comes to putting them into practice.[39] All those solemn long debates about the best form of society and the laws most suitable for bonding us together are appropriate only for exercising our minds. Among our arts disciplines there are several subjects, the essence of which consists in disputing and arguing and which, apart from that, have no existence.

Such political theories might be applied in some new-made world, but we have to take men already fashioned and bound to particular customs: we are not begetting them anew like Pyrrha and Cadmus.[40] We may have the right to use any means to arrange them and to set them up afresh, but we can hardly ever wrench them out of their acquired bent without destroying everything. Solon was asked whether he had drawn up the very

38. The infamous *Poneropolis*, the town of the Wicked: Plutarch (tr. Amyot), *De la Curiosité*, 66 D.

39. The most famous in modern times is More's *Utopia*, but there were several others.

40. Pyrrha, the wife of Deucalion; after the classical Flood this couple repeopled the world by casting over their shoulders stones which turned into men and women. Cadmus sowed the dragon's teeth, which produced a crop of soldiers who all slaughtered each other.

best laws which he could for the Athenians: 'Yes, indeed'; he replied, 'the best that they would accept.'[41]

[C] Varro pleaded a similar excuse: if he had to write on religion as something new he would tell us what he believed, but since it is already fashioned and accepted, he will talk about it following custom rather than its nature.

[B] Not as a matter of opinion but of truth, the best and most excellent polity for each nation is the one under which it has been sustained. Its form and its essential advantages depend upon custom. It is easy for us to be displeased with its present condition; I nevertheless hold that to yearn for an oligarchy in a democracy or for another form of government in a monarchy is wrong and insane.

> *Ayme l'estat tel que tu le vois estre:*
> *S'il est royal, ayme la royauté;*
> *S'il est de peu, ou bien communauté,*
> *Ayme l'aussi, car Dieu t'y a faict naistre.*

[Love the constitution of your State as you find it; if a kingdom, love kingship; if the rule of the few or of the many, love them too: for God caused you to be born under it.]

Those verses are by that good man Monsieur de Pibrac[42] whom we have just lost, a man of so noble a mind, so sound opinions, so gentle in his ways. His loss, and that of Monsieur de Foix which we suffered at the same time, are losses which matter to our Crown. I do not know whether there remains in France another pair of gentlemen who, for integrity and ability, could take the place of those two Gascons as counsellors to our Kings. Their souls were beautiful in different ways, each, in a time like ours, not only beautiful but rare in its own form. But whoever lodged in this age souls so unsuited to our corruption and so disproportionate to our tempestuous times?

Nothing crushes a State save novelty. Change alone provides the mould for injustice and tyranny. When some part works loose we can prop it up; we can resist being swept away from our original principles by the

41. Plutarch, *Life of Solon*, IX; then St Augustine, *City of God*, VI, iii–iv; in vi, Varro is praised as one of the wisest of men.
42. Verses of Guy du Faur de Pibrac (†1584), cited in Louis Le Caron's *De la tranquillité de l'esprit*, Paris, 1588, a source of several ideas in this chapter. Paul de Foix (†1584) was the oecumenical Privy Counsellor to whom Montaigne dedicated his edition of *Les Vers françois d'Estienne de La Boëtie*.

corruption and degradation natural to all things. But to undertake to recast such a huge [C] lump, [B] to shift[43] the foundations of so great an edifice, is a task for those [C] for whom cleaning means effacing, [B] who seek to emend individual defects by universal disorder and to cure illnesses by death, [C] *'non tam commutandarum quam evertendarum rerum cupidi'* [yearning not so much to change as to overthrow the constitution].[44]

[B] The world is not good at curing itself: it is so impatient of pressure that it can think of nothing but breaking loose from it without counting the cost. We know from hundreds of examples that it normally cures itself at the expense of itself. To throw off the burden of a present evil is no cure unless the general condition is improved. [C] The surgeon's aim is not to cause the death of foetid flesh: that is merely the means which lead to the cure. He looks beyond that, to making natural flesh grow back again and to restoring the limb to its proper state. Anyone who proposes merely to remove what is irking him falls short, for good does not necessarily succeed evil. Another evil can succeed it – as befell Caesar's killers who threw the Republic into such a crisis that they had cause to regret their intervention. The same has happened to many others down to our own times. My own contemporaries here in France could tell you a thing or two about that! All great revolutions convulse the State and cause disorder. Anyone who was aiming straight for a cure, and would reflect about it before anything was done, would soon cool his ardour for setting his hand to it.

Pacuvius Calavius corrected that defective procedure, so providing a memorable example.[45] His fellow-citizens had revolted against their magistrates. He was an important man with great authority in his city of Capua. One day he found the means of locking the Senate in their palace; calling the citizens together in the marketplace he told them that the time had come when they were fully at liberty to take their revenge on the tyrants who had so long oppressed them. He had those tyrants in his power, disarmed and isolated. His advice was that they should summon them out one at a time by lots, decide what should be done to each of them and immediately carry out the sentence, provided that they should at the same time decide to put some honourable man in the place of the man they had condemned, so that the office should not remain unfilled. No sooner had they heard the name of the first Senator than there arose shouts

43. '88: such a huge *contrivance and* to shift . . .
44. Cicero, *De officiis*, II, i, B.
45. Taken from Livy, XXIII, iii.

of universal disapproval. 'Yes, I can see,' said Pacuvius, 'that we shall have to get rid of that one. he is a wicked man. Let us put a good man in his place.' An immediate silence fell, everyone being embarrassed over whom to choose. When the first man was rash enough to name his choice there was an even greater consensus of voices yelling out a hundred defects, and just causes for rejecting him. As those opposing humours became inflamed, the second and third senators fared even worse, with as much discord over the elections as agreement over the rejections. Having uselessly exhausted themselves in this quarrel they gradually began to slip this way and that out of the meeting, each going off convinced in his mind that an older, better-known evil is more bearable than a new and untried one.

[B]   I see we are in pitiful disarray – for what have we *not* done?

> *Eheu cicatricum et sceleris pudet,*
> *Fratrumque: quid nos dura refugimus*
> *Ætas? quid intactum nefasti*
> *Liquimus? unde manus juventus*
> *Metu Deorum continuit? quibus*
> *Pepercit aris?*

[We are alas disgraced by scars and crimes and fratricide. In this cruel age what atrocities have we not committed? Have our young men ever stayed their hand for fear of the gods? What altar have they spared?][46]

Yet I do not immediately conclude that

> *ipsa si velit salus,*
> *Servare prorsus non potest hanc familiam.*

[even the goddess Deliverance could not save this family if she tried.]

For all that, we may perhaps not yet have reached our own final period. The preservation of states is probably something which surpasses our understanding.   [C]   As Plato says, civic polities are strong, and difficult to break asunder.[47] They can endure mortal illnesses in their guts and survive the injury of unjust laws, despite tyranny and despite the immorality and ignorance of their governors and the seditious licence of their peoples.   [B]   In all our misfortunes we compare ourselves with whatever is above us, looking towards those who are better off. Let us take our

---

46. Horace, *Odes*, I, xxxv, 33–8; then, Terence, *Adelphi*, IV, vii, 43–4.
47. Plato, *Republic*, VIII, 545E–546A (but in Plato a less generalized statement than in Montaigne).

measure from what is below: there is no one so ill-fated as not to find hundreds of examples to console him. [C] Our crime is to be ever less willing to see people get ahead of us than trailing behind us. [B] Yet Solon[48] said that if you were to gather all ills into a pile, there is nobody who would not rather bear away from that pile the ills he now has than to arrange to divide them equally between all other men, each taking his fair share.

Our polity is sick: yet some have been sicker still without dying. The gods use us for games of tennis, knocking us about in numerous ways.

> *Enimvero Dii nos homines quasi pilas habent.*

> [To the gods we are indeed like balls to play with.][49]

The stars fatally decreed that the Roman State should be the example of what they can achieve in this category, comprising every sort of fortune which can befall a State, all that order can do to it and chaos, every chance and mischance. Seeing the shocks and revolutions which shook it and which it survived, what State should despair of its condition? If the well-being of a State depends upon the extent of its dominions – which I in no wise accept, [C] liking as I do what Isocrates taught Nicocles, not to envy rulers who held sway over wide dominions but those who know how to look after those which they have inherited[50] – [B] then Rome was never more flourishing than when its malady was greatest. You can scarcely recognize the ghost of a polity under the first few emperors: it was the densest and most dreadful confusion that man can conceive. Yet Rome endured it and survived it, preserving, not one single kingdom driven back to its frontiers, but such a great number of peoples, so diverse, so far scattered, so disaffected, so chaotically governed and so unjustly conquered.

> *Nec gentibus ullis*
> *Commodat in populum terræ pelagique potentem,*
> *Invidiam fortuna suam.*

[Fortune allows no nation to pay off some private score against a people who rule both land and sea.][51]

---

48. Anecdote attributed by Erasmus to Socrates (*Apophthegmata*, III, *Socrates*, XCI).
49. Plautus, cited by Justus Lipsius, *Saturnalia*, I, i.
50. Socrates, *Ad Nicoclem*, IX, xxvi.
51. Lucan, *Pharsalia*, I, 82–4; then, I, 138–9.

All that totters does not collapse. More than one nail holds up the framework of so mighty a structure. Its very antiquity can hold it up, like old buildings which, without cement or cladding, are propped up by their own mass:

> *nec jam validis radicibus hærens,*
> Pondere tuta suo est.

[No longer does it cling to the earth with its mighty roots: it is saved by its own weight.]

Besides it is not good practice to reconnoitre only your flank and trench: to judge the security of a fort you must note where the enemy can break through and what is the condition of your attackers. Few ships founder by their own weight without outside violence.

Now let us gaze all round us: all about us is collapsing; take all the great States which we know, in Christendom and elsewhere, and look at them: you will find a manifest threat of change and collapse:

> *Et sua sunt illis incommoda, parque per omnes*
> *Tempestas.*

[They too have their misfortunes and a similar tempest threatening them all.][52]

The astrologers have an easy time warning us as they do of great changes and mutations soon to come; what they foretell is present and palpable: no need to turn to the heavens for that! We should not only derive consolation from this universal fellowship in evil and menace: we should derive some hope that our State will endure, since in nature, when everything falls in unison, nothing falls. Universal illness means individual health. Uniformity is a quality hostile to disintegration. Personally I am not reduced to despair and it seems to me that there are ways of saving us:

> *Deus hæc fortasse benigna*
> *Reducet in sedem vice.*

[Perhaps God of his kindness will restore things to their former state.][53]

Who knows whether God's will may not be that the same should happen to us as to bodies which are purged and restored to a better state by those long and grievous maladies which bring to them a fuller purer health than what they took away?

What depresses me most is that when I run through the symptoms of

---

52. Virgil, *Aeneid*, XI, 422–3 (adapted).
53. Horace, *Epodes*, XIII, 7–8.

our malady I find as many natural ones and as many sent by the heavens and proper to that malady as ones attributable to our disorder and unwisdom. [C] It seems that the very stars ordain that we have lasted beyond the normal limits. And what also depresses me is that the most immediate evil which threatens us is not change *within* the whole solid lump, but our ultimate dread: disintegration and tearing asunder.

[B] In these ravings of mine, what I fear is that my treacherous memory should make me inadvertently record the same thing twice. I hate going over my writings and only unwillingly probe a topic again once it has got away. I have no freshly learned doctrines; these are my normal ideas. Having doubtless conceived them a hundred times I am afraid that I may have mentioned them already. Repetition is always a bore, even if it were in Homer, but it is disastrous in works which only make a superficial and passing impression. I hate persistent admonition even when it serves a purpose as in Seneca, [C] and I dislike the practice of the Stoic School of repeating copiously and at length, for each individual subject, the principles and postulates which apply over all, ever citing afresh their general arguments and universal reasons.

[B] My memory is growing cruelly worse every day:

> *Pocula Lethæos ut si ducentia somnos*
> *Arente fauce traxerim.*

[As though my parched throat had drunk long draughts of the forgetful waters of Lethe.][54]

Now — for thank God nothing has gone wrong up till now — whereas others seek time and occasion to think over what they have to say, I avoid preparation for fear of assuming an obligation from which I then have to extricate myself. I get lost when I am under an obligation, as I do when I depend on an instrument as feeble as my memory.

I never read the following account without being struck by a proper and natural resentment. Lyncestes was accused of conspiring against Alexander. On the day that he was brought to appear before the army, as was customary, to be heard in his defence, he, having learned off by heart a prepared speech, stammered out a few hesitant words. As he became more and more confused, fumbling and struggling with his memory, he was suddenly struck dead by blows from the pikes of the nearest soldiers who believed he had convicted himself. His dazed silence served them as a confession. Since he had time in prison to prepare himself, it was not his

54. Horace, *Epodes*, XIV, 3–4.

memory that was defective, they thought, but a case of guilt bridling his tongue and making him so feeble.[55] What a good argument! Even when you merely aim to speak well you can be dazed by the place, the audience and their expectations. What can happen when you have to make an harangue on which your life depends!

For me the very fact of being tied down to what I have to say is enough to make me forget it. Once I have wholly committed and entrusted myself to my memory, I lean on it so heavily that I overwhelm it and it becomes afraid of its burden. As long as I rely upon it I lose control of myself, so much so that my very coherence is assayed. There was one day when I was hard put it to hide the servitude in which I was entangled, whereas my intention is always to suggest a deep indifference when speaking, making apparently fortuitous and unprepared gestures arising from the actual circumstances, preferring to say nothing at all of consequence rather than to show that I have come prepared to make a fine speech – something especially unbecoming in a man like me, a professed soldier, [C] and too much of an obligation for one who cannot retain much: preparation arouses greater hopes than it can satisfy. You often stupidly don your doublet, only to leap no better than in your smock. *'Nihil est his qui placere volunt tam adversarium quam expectatio.'* [Nothing is more adverse to those who would please than aroused expectation.][56]

[B] It is written of Curio the orator that after he had announced that he would divide his speech into three parts or four, or had stated the number of his arguments and reasons, he would often forget one of them or add one or two more.[57] I have always taken care not to fall into that trap, loathing all such promises and outlines, not simply out of distrust for my memory but also because that style is too donnish: [C] *'Simpliciora militares decent'* [In soldiers more bluntness is appropriate.]

[B] It is enough that from this day forth I have promised myself never again to accept the task of speaking in formal situations.

As for reading from a prepared script, that is not only a monstrosity but greatly to the disadvantage of those who by nature are capable of achieving anything directly. And as for throwing myself on the mercy of improvisation, that is even less acceptable: my powers of improvisation are stolid and confused and could never respond to sudden emergencies of any consequence.

55. Quintus Curtius, VII, i.
56. Cicero, *Academica*, II (*Lucullus*), IV, 10 (adapted).
57. Caius Scribonius Curio, a friend and correspondent of Cicero's (cf. Cicero, *Brutus*, LX); then, Quintilian, XI, i, 32–3.

Reader: just let this tentative essay, this third prolongation of my self-portrait, run its course. I make additions but not corrections: firstly, that is because when a man has mortgaged his book to the world I find it reasonable that he should no longer have any rights over it. Let him put it better elsewhere if he can, not corrupt the work he has already sold. From such folk you should buy nothing until they are dead. Let them do their thinking properly before they publish. Who is making them hurry? [C] My book is ever one: except that, to avoid the purchaser's going away quite empty-handed when a new edition is brought out, I allow myself, since it is merely a piece of badly joined marquetry, to tack on some additional ornaments. That is no more than a little extra thrown in, which does not damn the original version but does lend some particular value to each subsequent one through some ambitious bit of precision. From this there can easily arise however some transposition of the chronological order, my tales finding their place not always by age but by opportuneness.

[B] My second reason is this: I fear that I will personally lose by the change. My mind does not always move straight ahead but backwards too. I distrust my present thoughts hardly less than my past ones and my second or third thoughts hardly less than my first. We are often as stupid when correcting ourselves as others.[58]   [C] My first edition dates from fifteen hundred and eighty: I have long since grown old but not one inch wiser. 'I' now and 'I' then are certainly twain, but which 'I' was better? I know nothing about that. If we were always progressing towards improvement, to be old would be a beautiful thing. But it is a drunkard's progress, formless, staggering, like reeds which the wind shakes as it fancies, haphazardly.

Antiochus had written vigorously in support of the Academy. In old age he took a different line. Would I not be following Antiochus whichever I followed? After having established doubt he wished to establish the validity of human opinions: that amounted, did it not, to establishing doubt not validity, suggesting that if longer life were granted him he would have been ready for some new upset, not so much better as different.

[B] The approval of the public has made me a little more adventurous than I expected; but what I most fear is to surfeit. Like a certain scholar of my time, I would rather provoke than bore. Praise is always pleasant, no

---

58. '88: ourselves as others. *I have aged by eight years since my first publication but I doubt whether I have amended myself* by one inch. The approval . . .

matter why it comes or from whom it comes; but genuinely to delight in it you need to discover its cause: even defects have ways of finding favour. The approval of ordinary common folk rarely hits the point, and I am mistaken if, in my own time, it is not the worst books which come top in popular approbation. I am indeed grateful to those gentlemen who deign to take my feeble efforts in good part. Nowhere are defects of style more obvious than when the subject-matter itself has little to commend it.

I do not, Reader, accept responsibility for misprints which slip in through the carelessness or fantasy of the various craftsmen; each hand introduces his own. I do not concern myself with the spelling (merely telling them to follow the traditional one) nor with punctuation: I am expert in neither. Even where they completely destroy my meaning, that does not worry me over-much: they at least take some weight off me; but when (as they often do) they substitute a false meaning and deflect me towards their own conception, they destroy *me*. So whenever the thought does not measure up to my own standard a gentleman should decline to accept it as mine. Anyone who knows how little industrious I am, and how far I am cast in a mould of my own, will not find it hard to believe that I would more readily compose as many essays again than subject myself to going through them once more to make schoolboy corrections.

I said just now that, being set in the deepest mine of that new metal,[59] not only am I deprived of close contact with people whose manners and opinions hold them together by a bond which allows no other and which differs from mine, but I also run some risk by living among people who think that all deeds are equally lawful, most of whom have debts to pay to our justice which could not be made worse – whence arises the ultimate degree of licence. When I tot up all the details which concern me as an individual, I find that there is no man hereabouts to whom the defence of our laws costs more than it does to me, 'either' (as the law-clerks say) 'in gains forgone or damages incurred'.   [C]   Some there are who boast of their zeal and toughness who, if you weigh things properly, do far less than I do.

[B]   My house, being always open, easily approached and ever ready to welcome all men (since I have never let myself be persuaded to turn it into a tool for a war in which I play my part most willingly when it is farthest from my neighbourhood) has earned quite a lot of popular affection, so that it would be difficult to challenge me on my own dunghill. It is, I judge, a miraculous and exemplary achievement that it should remain

---

59. That is, in an age worse than the Age of Iron. Cf. note 36.

unspotted by blood or sack during so long a tempest and so many upheavals and changes hereabouts. For to tell the truth it would have been possible for a man of my complexion to escape the effects of pressure of any kind, provided that it was constant and continuous, but these alternating invasions and incursions, these reversals and vicissitudes of Fortune round about me have, to date, hardened the temper of the local people rather than softened it, loading upon me insurmountable dangers and hardships. I escape, but it displeases me that I do so by Fortune and, indeed, by my cleverness rather than by justice; it displeases me to be outside the safeguard of our laws and under any other protection but theirs. As things stand I live more than half by somebody else's favour, which is a harsh obligation. I do not want to owe my safety to the bounty and good-will of great men who respect my loyalty and independence, nor to the affable manners of my forebears or of myself. Supposing I had been different! And if my conduct and the frankness of my dealings do impose obligations on my neighbours and kinsmen, there is cruelty in their being able to pay off their debt by letting me stay alive and in their being able to say: 'We allow him[60] [C] to continue freely to have divine service in the chapel of his house now that we have pillaged and smashed all the neighbouring churches;[61] and we allow him to keep his property and his life, [B] since, when the need arises, he protects our wives and our cattle.' (We are old hands in my home at sharing in the praise given to Lycurgus of Athens, that he was the guardian and general depository of the purses of his fellow-citizens.)[62]

I maintain that we ought to live by the authority of the law, not by [C] recompense and [B] favour. How many gallant men have preferred to lose their life rather than to owe it to anyone. I avoid any sort of obligation, but above all the kind which binds me by a debt of honour. For me nothing costs dearer than what is vouchsafed to me and for which my will remains mortgaged under the title of gratitude: I prefer to receive services which are up for sale. And I should think so too! For the latter I give mere money: for the others I give myself. Such knots as bind me by the laws of honour seem tighter to me and heavier than the knots of civil constraint. A lawyer ties me in his knots more loosely than I do myself.

60. [B] instead of [C]: we allow him *his life and his house*, since, when the need arises . . .
61. Montaigne, a Roman Catholic, lived in an area dominated by members of the Reformed Church, to which several members of his family adhered.
62. Plutarch, *Life of Lycurgus*.

And is it not reasonable that my conscience should be under a far greater obligation when anyone has put simple trust in it. In other cases my trustworthiness owes them nothing: they never lent it anything. Let them seek help from the trust and reliance which they placed in others than me. I would much rather break the restrictions of walls or of laws than of my word. [C] Being nice to the point of superstition over keeping my promises, I prefer on all subjects to make them conditional and provisional. To unimportant promises I attach weight because I keep jealously to my rule, which racks me and burdens me out of concern for itself. Why, even in such undertakings as are freely and entirely my own, once I have declared my intention I feel that I have ordered myself to carry it out, and that, by letting others into the know, I have prescribed it to myself. It seems to me that to state it is to promise it. That is why I do not give much wind of my projects.

[B] Any sentence which I pass on myself is far stiffer and more rigorous than any given by judges who can seize me only by aspects of common obligation, whereas my conscience is stricter and more severe. But in the case of duties towards which they would drag me if I would not go willingly, I pursue them but slackly: [C] *'Hoc ipsum ita justum est quod recte fit, si est voluntarium.'* [The essence of a just deed lies in being voluntary.][63] [B] If the deed has none of the splendour of freedom it has neither grace nor honour:

> *Quod me jus cogit, vix voluntate impetrent.*

[You will not easily get me to do what the law says I must.]

When necessity compels me, I like to slacken my will, *'quia quicquid imperio cogitur exigenti magis quam praestanti acceptum refertur'* [because when anything is commanded, gratitude is given to the one who issues the order not the one who obeys it].

I know some who adopt that position to the point of unfairness: they would rather give away than return, and lend out rather than repay, doing good most meanly to those to whom they are most beholden. I do not go that far, but I get close to it.

I am so fond of ridding myself of the weight of obligations that I have occasionally counted as gains such attacks or insults or acts of ingratitude as came from those to whom, by nature or accident, I owed some duty of

63. Cicero, *De officiis*, I, ix, 28; then, Terence, *Adelphi*, III, v, 44, and Valerius Maximus, II, ii, 6.

affection, taking their offence, as it occurred, as so much towards the settling or discharge of my debt. Even when I continue to pay them the visible courtesies which society requires, I still find it a great saving [C] to do for justice what I used to do for affection and [B] to alleviate a little the inward stress and anxiety of my will [C] *'Est prudentis sustinere ut cursum, sic impetum benevolentiae.'* [Wise men should stop a rush of benevolence as they would a runaway chariot.][64] [B] I have a will which, when I yield to it, is rather too impulsive and pressing, at least for a man who wishes never to be under any pressure. My restraint can reconcile me to the imperfections of those who are in contact with me: I am sorry that they are worth the less for it but I can nevertheless economize a little over my attachment and engagement towards them. I approve of the man who loves his son[65] less if he is scabby or a hunchback, not merely when he is wicked but also when he is unfortunate or ill-endowed (for God has himself, to that extent, reduced his natural worth and value), provided that he behave, in his absence of warmth, with moderation and scrupulous fairness. In my own case a close relationship does not lighten defects: it tends to aggravate them.

After all that, insofar as I understand the subject of beneficence and gratitude (which is a delicate and most useful science) I know no one more free and under less obligation than I am so far. I owe whatever I do owe to common natural obligations: no one is more purely unindebted:[66]

> *nec sunt mihi nota potentum*
> *Munera.*

[and as for presents from powerful men, I know them not.]

Princes [C] give me plenty if they take nothing from me and [B] do me enough good if they do me no harm. That is all I ask of them. Oh how beholden I am to God that it should have pleased Him that I should receive all I have directly from His grace and for His reserving all my debt to Him alone! [C] How urgently I beg God of His mercy that I may never owe a fundamental 'Thank you' to any man. Blessed freedom, which has guided me thus far! May it last to the end.

[B] I try to have no express need of anyone: [C] *'in me omnis spes est*

---

64. Cicero, *De amicitia*, XVII, 63.
65. '88: his son *or his cousin* less . . .
66. '88: more purely unindebted *towards obligations and benefits from others*: nec . . .
    Then Virgil, *Aeneid*, XII, 519–10 (adapted).

mihi' [all my hope is in myself].⁶⁷   [B]   That is something all can do, but it is easier for those whom God has protected from pressing natural needs. To depend upon another is pitiful and hazardous. Even our own self (which is the most secure and right place to turn to) does not provide adequate security. I own nothing but myself, yet even my possession of that is partly imperfect and defective. I husband myself⁶⁸ [C] and put heart into myself (which is more important) while still fortunate, [B] so as to find there the wherewithal to satisfy me when all else should abandon me.

[C]   Eleus Hippias did not equip himself solely with learning so as to be able, if needs be, to withdraw happily from all other company into the lap of the Muses, nor solely with philosophy so as to teach his soul to be content with itself, manfully doing without all external goods when Fate demands it: he took care to learn to be his own cook and barber, to make his own clothes, shoes and rings so as to be able to rely as far as possible entirely on himself and to relieve himself of the need of others' help.⁶⁹   [B]   You can enjoy more freely and contentedly the use of good things which do not derive from yourself when your enjoyment of them is not bound and constrained by necessity and when your will has the power, and your financial resources the means, of doing without them.

[C]   I know myself well, but it is hard for me to conceive of any act of kindness from anyone or any hospitality so frank and free but that, if I were to become involved in it out of necessity, it would be to me painful, tyrannical and stained with reproach. Just as giving is a pretentious quality, a prerogative, receiving is an act of subordination − witness Bajazet's insulting and bellicose rejection of the gifts sent to him by Tamberlane;⁷⁰ and the gifts sent on the part of the Emperor Soleiman put the Emperor of Calicut in such a rage that he not only bluntly rejected them, saying that neither he nor his predecessors were accustomed to take, it being their place to bestow, but he also had the envoys who had been sent with them cast into a dungeon.

Aristotle says that when Thetis flatters Jupiter, and the Spartans the Athenians, they do not start reminding them again of all that they

---

67. Terence, *Phormio*, 139.
68. '88: husband myself *and augment myself with all my care*, so as . . .
69. Cicero (*De oratore*, III, xxxii, 127), who thinks that Hippias 'went too far'. Hippias referred not to 'rings' but to 'the ring he was wearing'.
70. Nicolas Chalcocondylas, *De la décadence de l'empire grec*, II, xii, then, Simon Goulart, *Hist. du Portugal*, XIX, vi.

themselves have done for them – that is always odious – but of all they have received from them.[71] Those whom I see readily using the good offices of each and everyone and pawning themselves to them would not do so if they attached the weight which wise men should to the bond of an obligation: it can sometimes be repaid but never untied – a cruel trussing-up for anyone who likes to give his freedom elbow-room everywhere. Those who are acquainted with me (both those above and below me) know whether they have ever met anyone who puts fewer burdens upon others. If I am excessive about this by today's standards, that is no great marvel, since so many elements in my character contribute to it: a little innate pride, the inability to bear a refusal, my restricted needs and my lack of flair for any kind of business – and my most cherished characteristics: idleness and frankness. For all of which reasons I have a mortal hatred of being beholden to anyone or through anyone but myself. Under any circumstances whatever, before I will make use of another's kindly services, no matter how trivial or unimportant, I make vigorous use of every means of doing without them. Those whom I hold in affection distress me hugely when they beg me to beg a favour for them from a third party. If I make use of anyone, it seems to cost me no less to redeem what he has in pawn to me than, if he owes me nothing, to pawn myself to him on behalf of others. But apart from that condition and the next (that they do not want anything from me which requires anxious bargaining, for I have declared a war unto death against bother of any sort), I am easily accessible to the needs of everyone.[72]

[B] But even more than seeking to bestow I have fled from all receiving – [C] which Aristotle says is an easier thing to do.[73] [B] My Fortune has not allowed me to give much to others, and the little she has allowed me has been lodged with the very poor.

If Fortune had brought me into this world to hold high rank among men I would have been ambitious to be loved, not feared or held in awe. Shall I express it more cheekily? I would have been more concerned to please than to bring moral improvement. [C] Cyrus said most wisely (through the mouth of an excellent captain and better philosopher)[74] that

71. Aristotle, *Nicomachaean Ethics*, IV, iii, 25–6.
72. [B]: instead of [C]: of doing without them. *I have most readily sought the opportunity to do good and to bind others to me; and it seems to me that there is no sweeter use of our resources.* But even more than . . .
73. Aristotle, *Nicomachaean Ethics*, IX, vii, 6–7.
74. Xenophon, *Cyropaedia*, VIII, iv, 8; then, Livy, XXXVII, vi (for Scipio).

he reckoned that his generosity and benefactions far excelled his valour and his conquests in war. And whenever Scipio the Elder wants to make himself esteemed he rates his affability and humanity above his bravery and victories, and always has this proud saying on his lips: he had given his foes as much reason to love him as his friends.

[B]   What I mean, then, is that if I must owe anyone anything it should be for some other more legitimate pretext than the one I mentioned just now, in which I am implicated by the laws of this wretched war, one where the debt does not amount to my entire preservation. Such a debt overwhelms me. I have gone to bed in my own home hundreds of times thinking that I would be betrayed and killed that night, bargaining with Fortune that the event should not be terrifying and long drawn-out. And after reciting my Lord's Prayer I have exclaimed,

> *Impius hæc tam culta novalia miles habebit!*

[Some impious soldier, then, will get these well-farmed lands!][75]

What remedy is there? I was born in this place and so were most of my ancestors. They have entrusted their love and reputation unto it. We get hardened to anything to which we are accustomed. And in wretched circumstances such as ours now it is a most kindly gift of Nature that we do grow accustomed to it, so that it deadens our sense of suffering many evils.

What makes civil wars worse than other wars is that each man is on sentry-guard over his own home.

> *Quam miserum porta vitam muroque tueri,*
> *Vixque suæ tutum viribus esse domus.*

[How pitiful it is to need gates and walls to protect your life and scarcely to be able to trust in the strength of your own home.]

It is to be in great extremity to be hard-pressed even within your very house, in the quiet of your home.[76] The place where I dwell is always the first and the last to be pounded by our strife: peace never shows her full face there:

> *Tum quoque cum pax est, trepidant formidine belli.*

[Even when there is peace we tremble for fear of war.][77]

---

75. Virgil, *Eclogues*, I, 71; then Ovid, *Tristia*, IV, i, 69–70.
76. '88: your home. *This misfortune affects me more than it does anyone else, because of the characteristics of its site.* The place . . .
77. Ovid, *Tristia*, III, x, 67; then, Lucan, *Pharsalia*, I, 256–7; 251–2.

> *Quoties pacem fortuna lacessit,*
> *Hac iter est bellis. Melius, fortuna, dedisses*
> *Orbe sub Eoo sedem, gelidaque sub Arcto,*
> *Errantesque domos.*

[Every time that Fortune strikes at peace, that is the road to war. O Fortune, you would have been better advised to settle me in lands beneath the Morning Star or the wandering planets of the frozen North.]

Sometimes I find in indifference and languor the means of firming myself against such reflections – for they too can make us somewhat resolute. It often happens that I think with some pleasure of those mortal dangers and wait for them: I lower my head and plunge, devoid of sensation, into death, neither contemplating it nor exploring it, as into some voiceless, darkling deep, which swallows me up at one jump and in an instant overwhelms me with a powerful sleep entirely lacking any sensation or suffering. And what I foresee to follow upon those short and violent deaths consoles me more than their reality disturbs me. [C] (Life, they say, is no better for being long: death is better for not being so.) [B] I do not recoil from being dead but, rather, I become reassured about dying. I wrap up and crouch down during the storm, which, with one quick attack, one unfelt blow, must blind me and ravish me in its frenzy.

Just as some gardeners say that roses and violets spring up more sweet-scented near garlic and onions which attract and draw to themselves all that is foul-smelling in the soil,[78] suppose those depraved characters similarly suck up all the venom in the air of our climate, rendering me better and purer by their proximity, so that all is not loss. But things are not so. Yet there may be something in the following: goodness is more beautiful and attractive when it is rare, while the determination to act well is stiffened by contradiction and concentrated in us by opposition, being enflamed by glory and a jealous desire to resist.

[C] Robbers, of their courtesy, do not have it in for me personally. Do I not return the compliment? I would need to have it in for too many people! ['95] Under various kinds of dress [C] are lodged[79] similar consciences, similar cruelty, treachery and robbery, and they are all the worse when they are more cowardly and safe for being better hidden behind the shadow of the law. Avowed injuries I hate less than treacherous

78. Plutarch (tr. Amyot), *Comment on pourra recevoir utilité de ses ennemis*, 112 F.
79. [C]: under *diverse* kinds of *fortune* are lodged . . .
  Then, Virgil, *Georgics*, I, 506.

ones, and those of war less than those of peace – ['95] judicial ones. [C] This fever of ours has occurred in a body which it has hardly made worse: the fire was there already: the flames had already taken. The din is much greater, the evil but little more.

[B] When people ask why I go on my travels I usually reply that I know what I am escaping from but not what I am looking for. If they tell me that there may be [C] just as little soundness [B] among foreigners[80] and that their morals may be no better than ours, I reply: first, that that would not be easy:

> *Tam multae scelerum facies.*

> [Our wickedness has assumed so many faces.]

Secondly, that there is always gain in changing a bad condition for an uncertain one, and that the ills of others do not need to sting us as our own do.

And I do not want to omit that I am never such an enemy of France that I fail to look kindly on Paris: Paris has had my heart since boyhood. And as happens with all incomparable things, the more beautiful the other towns I have seen the more the beauty of Paris gains power over my affections. I love her for herself, more when left alone than overloaded with extra ornaments. I love her tenderly, warts, stains and all. That great city alone makes me a Frenchman,[81] a city great in citizens, great in its happy choice of site, but great above all and incomparable in the variety and diversity of its attractions; it is the glory of France and one of the world's great splendours. God drive our divisions far from her! When entire and united she is safe from other violence. The worst of all decisions, by my counsel, would be one which brought discord to her. I fear nothing for her but herself. And I certainly fear for her more than for any part of our State. While she endures I shall not lack a lair in which I can die at bay, one enough to make me lose all regret for any other.

Not because Socrates said it but because it truly corresponds to my humour (and is perhaps not free from excess): I reckon all men my fellow-citizens,[82] embracing a Pole as I do a Frenchman, placing a national bond after the common universal one. I do not particularly hanker after the sweetness of my native soil. Acquaintances which are entirely new and

---

80. '88: may be *similar maladies* among foreigners . . .
81. As distinct from a Gascon.
82. Cicero, *Tusc. disput.*, V, xxxvii, 108.

entirely mine seem to me to be worth just as much as the other common kind, casually based on neighbourhood. Those loving relationships which are purely our own achievement normally outweigh those to which we are bound by ties of place or blood. Nature brought us forth free and unbound: we imprison ourselves in particular confines, like those kings of Persia who bind themselves to drink no water but that of the river Choaspes, foolishly renouncing their right to use all other waters, making, so far as they are concerned, all the rest of the world a desert.[83]

[C] When Socrates was near his end he judged that a sentence of exile was for him worse than a sentence of death. As far as I can tell I could never be so broken in, nor so narrowly accustomed to my part of the world, as to say that. Those heaven marked lives have many traits which I embrace more with esteem than emotion. They also have other traits so soaring and inordinate that I cannot even do so with esteem, since I am quite unable to conceive them. That was a very delicate humour in a man who considered the whole world his city! It is true that he despised travel and had hardly set foot outside Attic territory. And what about his sparing his friends' money with which they would have saved his life, and his refusal to escape from prison through the intercession of anyone at all, so as not to disobey the laws at a time when they were highly corrupt? Those examples fall into my first category: there are others to be found in that great man which fall into my second one. Many such examples surpass my power of action, but some surpass even my power of judgement.

[B] In addition to such reasons, travel seems to me to be an enriching experience. It keeps our souls constantly exercised by confronting them with things new and unknown; and (as I have often said) I know of no better school for forming our life than ceaselessly to set before it the variety found in so many other lives, [C] concepts and customs, [B] and to give it a taste of the perpetual diversity of the forms of human nature. The body is neither idle nor exhausted by it: the moderate exercise keeps it in good trim. Even suffering from the stone as I do, I can stay in the saddle, without dismounting, for eight or ten hours at a stretch:

> *Vires ultra sortemque senectae.*

> [strength beyond the lot of old age.][84]

---

83. Plutarch (tr. Amyot), *Du bannissement ou l'exil*, 125 G–H: then, also for Socrates; together with echoes of Plato's *Apology for Socrates*, etc.
84. Virgil, *Aeneid*, VI, 114.

No weather is inimical to me except the harsh heat of a blazing sun (for those parasols which Italy has used since the Ancient Romans put more weight on your arm than they take off your head). [C] I would love to know how hard it was for the Persians, so long ago at the very birth of luxury, to produce at will cool winds, as Xenophon says they did, and patches of shade.[85]

[B] I take to rain and mud like a duck. A change of air and weather does not disturb me: to me all climates are the same. The only things which do batter me are such internal disturbances as I produce within me – and they occur less during my travels.

It is hard to get me moving, but once I have started I will go on as far as you like. I resist little expeditions [C] as much as [B] big ones,[86] and equipping myself for a day-trip or a visit to a cousin [C] as much as [B] for a real journey. I have learned to do each day's journey in the Spanish style, all at one go, a long but reasonable day. When it is extremely hot I travel by night, from sunset to sunrise. (The other way – stopping to eat *en route*, in chaos and haste over your post-house dinner – is disagreeable, especially when the days are drawing in.) My horses are all the better for it. No horse which can get through the first day's journey with me has ever let me down. I water them everywhere, merely taking the precaution of having enough road left for them to work it off. My own reluctance to get up allows my retinue to breakfast at leisure before we set off. I myself never dine very late. Appetite comes to me only with eating;[87] except at table I never feel hungry.

Some complain at my delight in continuing this practice as a man married and old.[88] They are wrong. The best time to leave our family is after we have set it on course to proceed without us, after we can leave behind such order as does not belie its former character. It is far more imprudent to go off if you leave your home in charge of a protectress who is less reliable and who may take less trouble to provide for your needs. The most useful science and the most honourable occupation for a wife is home-management. I am aware of more than one wife who is mean but of

---

85. *Cyropaedia*, VIII, viii is now considered an addition to the work and not to be by Xenophon. It treats of the birth of luxury and decadence among the subjects of Cyrus' Persian Empire but does not describe the amenities mentioned by Montaigne.
86. '88: expeditions *more than* big ones [. . .] cousin *more than* for a real journey . . .
87. A proverb best known from Rabelais' Good Drinkers, who attribute it in jest to the Bishop of Le Mans (*Gargantua*, TLF, IV, 85 var.).
88. '88: married and *soon* old . . .

few who are good managers. Yet to be one is a wife's chief virtue, the one that we should look for first as the only dowry which may either save our households or ruin them. [C] There is no need to lecture me on the subject: experience has taught me to seek one virtue above all others in a married woman: the virtue of sound housekeeping. [B] I enable my wife to do this properly when, by my absence, I leave the government of my house in her hands. It irritates me to see in many a household my lord coming home about noon, all grimy and tetchy from business worries, while my lady is still in her dressing-room, dolling herself up and doing her hair. That is for queens – and I am not sure even then. It is unjust and absurd that our wives should be maintained in idleness[89] by our sweat and toil. [C] As far as it lies with me, nobody shall have a more serene enjoyment of my goods than I do, one more quit and more quiet. [B] Though the husbands provide the matter, Nature herself wills that the wives provide the form.[90]

As for the duties of conjugal love which are thought to be infringed by such absences, I do not believe that they are. On the contrary: such intercourse can easily be cooled by too continuous a presence and impaired by assiduity: every other woman seems charming then! Everyone knows that seeing each other all the time cannot provide the same pleasure as is given by alternately going away and coming together. [C] Such intervals fill me with fresh love for my family and restore me to a more agreeable use of my home. Alternation sharpens my appetite for both home and travel. [B] Loving affection, as I know, has arms long enough to stretch from one end of the world to the other and meet – especially conjugal love, for it comports a continuous exchange of duties which reawaken our memory of the tie. The Stoics say that there are such great bonds of interdependence and interconnection between the wise, that he who dines in France nourishes his fellow in Egypt and that, wherever he may be, if he merely raises a finger to help, all the wise men on this habitable earth feel the benefit.[91]

Enjoyment – possession – belongs mainly to the mind. [C] It more ardently embraces whatever it goes a-seeking than anything we actually

89. '88: in *pomp and* idleness . . .
90. Cf. the maxim 'woman desires man as matter desires form'; it was taken from Aristotle but traditionally misunderstood. (Cf. Tiraquellus, *De legibus connubialibus*, IX, 92.)
91. Plutarch (tr. Amyot), *Des communes conceptions contre les Stoïques*, 579 F. (an effect attributed to *amitié*, 'loving affection').

hold, and it does so more continuously. Note how you spend your time every day: you will find that you are most absent from the one you love when he is present: your attentiveness is released by the fact that he is there; that gives your thoughts freedom to go absent at any time, on any pretext.

[B]   Outwards from Rome I control and govern my household and the good things I have left there. Just as when I am there, I know within an inch or two how my walls, my trees or my rents are growing or declining:

> *Ante oculos errat domus, errat forma locorum.*

[Before my eyes there floats a vision of my home and the places I have left.][92]

If we only enjoy things when we touch them, then goodbye to our golden sovereigns when they are in our money-chests – and to our sons when they are out hunting. We want them nearer. They are in our grounds: is that 'far'? Is half a day's journey 'far'? How about ten leagues? Is that 'far' or 'near'? If near, how about eleven leagues, twelve, thirteen and so on, pace by pace? Truly, if any wife can lay down for her husband how many paces make 'far' and how many paces make 'near', my counsel is to make him stop half-way –

> *excludat jurgia finis.*
> *Utor permisso, caudæque pilos ut equinæ*
> *Paulatim vello, et demo unum, demo etiam unum,*
> *Dum cadat elusus ratione ruentis acervi*

['Let us set limits and end this domestic strife!' . . . Yes, but I take whatever you allow and (like plucking hair after hair, one by one, from my horse's tail) I take yard after yard until you are cheated by my accumulated sophisms][93]

– and let those wives dare to call Philosophy to their aid. But someone will object that Philosophy can only judge very vaguely where the middle point lies: she can descry neither of the limits linking too much and too little, long and short, light and heavy, since she can recognize neither their end nor beginning: [C] *'Rerum natura nullam nobis dedit cognitionem finium.'* [Nature has given us no faculty which can know the boundaries of anything.]

[B]   Are they not still the wives and beloveds of the dead who are not

---

92. Ovid, *Tristia*, III, iv, 57.
93. Horace, *Epistles*, II, i, 38; 45–7; then, Cicero, *Academica* (*Lucullus*), II, xxix, 92.

at the end of this world but in the next? Our arms enfold not only our absent ones but also those who have died or are yet to be. When we married each other we did not contract to be ever attached to each other's tails like some little creatures or other we know of,[94] [C] or doggy-fashion, like those bewitched couples of Karenty.[95] Moreover a wife should not have her eyes so hungrily fixed on her husband's foreparts that when the need arises she cannot bear to see his backside.

[B] Perhaps this jest from a most excellent portrayer of wives' humours would not be out of place here to describe the cause of their complaints:

> *Uxor, si cesses, aut te amare cogitat,*
> *Aut tete amari, aut potare, aut animo obsequi,*
> *Et tibi bene esse soli, cum sibi sit male.*

[You are late coming home. Your wife assumes that you are in love with somebody, or somebody with you, that you are getting drunk and having a good time without her while she feels miserable.][96]

Or would it not be, perhaps, because they like opposing and thrive on contradiction, happy enough if they can make you unhappy?

In a truly loving relationship – which I have experienced – rather than drawing the one I love to me I give myself to him.[97] Not merely do I prefer to do him good than to have him do good to me, I would even prefer that he did good to himself rather than to me: it is when he does good to himself that he does most good to me. If his absence is either pleasant or useful to him, then it delights me far more than his presence. And it is not strictly absence when there are means of keeping in touch. In former times I found advantages and pleasure in our being far apart. By going our separate ways we possessed life more fully and widely. It was for me that he lived, and saw and enjoyed things: and I for him – more fully than if he had been there. When we were together part of us remained idle: we were merged into one. Geographical separation rendered more rich the union of our wills. That insatiable hunger for physical presence reveals a certain weakness in the enjoyment of our souls.

---

94. Perhaps creatures such as the *mustellae* (a kind of weasel) which Ravisius Textor describes as being bound to the female by their testicles (*Officina, Animalia diversa,* s.v.).
95. Known from Saxo Grammaticus; when the couples lay together this way they could not be separated and became a laughing-stock.
96. Terence, *Adelphi,* I. i, 7–9.
97. Montaigne is generalizing from his love for La Boëtie.

As for my old age, which they cite against me, it is on the contrary for youth to be enslaved by common opinions and to restrain itself for someone else. Youth has plenty enough to provide for itself and others: we have too much to do to provide for ourselves. As natural pleasures fail us, let us support ourselves by artificial ones. It is unfair to forgive youth for pursuing its pleasures, while forbidding old age even to look for any. [C] When I was young I veiled my playful passions behind wisdom: now I am old, I disperse my gloomy ones by excess. Though Plato's laws forbid foreign travel before forty or fifty so as to make it more useful and instructive, I would more readily subscribe to the second article in those same laws, which prohibits it after sixty.[98]

[B] 'But at your age you will never return from so long a road' – What does that matter to me? I did not set out either to return or to complete. I set out merely to keep on the move while moving pleases me. [C] I travel for travelling's sake. They do not run for sport who course after hares or benefices: they run for sport who gallop in tournaments for the joy of the coursing.

[B] My itinerary can be interrupted at any point; it is not based on great expectations: each day's journey is complete in itself. My life's journey is conducted the same way. Yet I have seen enough far-off places where I would have liked to have been retained. And why ever not, when so many [C] wise [B] men of the most glowering sect[99] – Chrysippus, Cleanthes, Diogenes, Zeno and Antipater – abandoned their homeland, having no cause to complain of it but merely to enjoy a different clime. Indeed what most displeases me in my peregrinations is that I cannot bring with me the right to make my home wherever I please and that (adapting myself to the common prejudice) I must always intend to come back. If I were afraid of dying anywhere else but where I was born, and if I thought that I would die less at my ease when far from my family, not only would I hardly ever go out of France without terror, I would hardly go out of my parish: I feel death all the time, jabbing at my throat and loins. But I am made otherwise: death is the same for me anywhere. If I were allowed to choose I would, I think, prefer to die in the saddle rather than in my bed, away from home and far from my own folk. There is more heartbreak than comfort in taking leave of those we love. I am inclined to neglect that social duty: for of all the obligations of loving affection it alone is

98. Plato, *Laws*, XI, 950 D; 951 D (treating of commissioners sent out officially to report on foreign lands).
99. The Stoics.
   '88: so many *decent* men . . .

displeasing. I would willingly therefore neglect to bid that great and everlasting farewell. Although some advantage may be drawn from the presence of others, there are hundreds of disadvantages. I have seen several men die most wretchedly, besieged by all that activity; they are suffocated by the crowd. It is undutiful, and a sign of slight care and affection, to let you die in peace! Someone is messing about with your eyes, another with your ears and another with your tongue: you have no limb nor sense which they are not badgering. Your heart is racked with pity at hearing the lamentations of those who love you – and perhaps with anger at hearing other lamentations, feigned and hypocritical. Anyone with a taste for gentleness has it more when he is weak. In such great straits he needs a soft hand to scratch him precisely where it itches. Otherwise, leave him alone. If we need a 'wise-woman' to midwife us into this world we need an even wiser man to get us out of it. We ought to pay a high price to have such a man, a friend, for such an event.

I have not attained to that vigorous contempt which fortifies itself and which nothing can help, nothing disturb. I am one peg below that. Not from fear but from cunning, I want to go to earth like a rabbit and steal off as I pass away. It is not my intention to test or to display my constancy during that action. For whom would it be? Then all my right to reputation and all my concern for it will be at an end. I am satisfied with a death which will withdraw into itself, a calm and lonely one, entirely my own, one in keeping with my life – retiring and private. Contrary to Roman superstition (according to which a man was held wretched if he died without speaking and without his nearest kinsfolk to close his eyes)[100] I have enough to do to console myself without having to console others; enough thoughts in my mind without fresh ones evoked by my surroundings; enough to think about without drawing on others. This event is not one of our social engagements: it is a scene with one character. Let us live and laugh among our own folk, but let us die, grinding our teeth, among strangers. Provided you can pay, you can always find someone to turn your head and massage your feet, and who will leave you alone as much as you like, showing you an unconcerned face and letting you think and moan in your own way.

Every day I argue myself out of that childish and unkindly humour which makes us desire that our own ills should arouse compassion and mournful thoughts in those we love. So as to bring on their tears we exaggerate our misfortunes beyond all measure. And that steadfastness in

100. Plutarch (tr. Amyot), *Contredicts des Philosophes Stoïques*, 561 D.

supporting ill-fortune which we eulogize in everyone else, we arraign and condemn in close relatives when the ill-fortune is ours. We are not content that they should sympathize with our ills unless they are also afflicted by them. Joy we should spread: sadness, prune back as much as we can. [C] Whoever evokes pity without cause is not to be pitied when cause there is. To be always lamenting is to have none to lament you; so often to look pitiful arouses pity in nobody. Act dead when you are living, and you are likely to be treated as alive when you are dying. I have known it get the goat of some invalids if you said they had a healthy colour or a regular pulse; they would hold back their laughter since it would betray that they were cured; they hated good health because it aroused no compassion. And what is more, they were not women either.

[B] I present my maladies, at most, for what they are and I avoid studied groans and words of foreboding. If not merriness at least composure is appropriate for those attending a sick wise man. Just because he knows he is in the opposite condition himself he picks no quarrel with health: he delights in contemplating in others health, strong and whole, at least enjoying it through their company. Just because he knows that he is sinking, he does not reject all thoughts of life or avoid ordinary conversation. I want to study illness when I am well: when it is present it makes a real enough impact without my imagination helping it. We prepare ourselves beforehand for such journeys as we are resolved to undertake, but the hour when we should be climbing into the saddle we devote to those about us and prolong it in their favour.

I realize that there is an unexpected benefit from this publication of my manners: in some ways it serves me as a rule. Occasionally the thought comes over me that I should not prove disloyal to [C] this account of my life.[101] [B] This public disclosure obliges me to stick to my path and not to belie the portrayal of my qualities, which are, on the whole, less deformed and objectionable than is commonly thought by the malice and distemper of present-day judgements. The consistency and straightforwardness of my ways produce an outward appearance which is easy to interpret, but because my style is rather novel and unusual it gives slander too easy a time. Yet it seems to me that anyone who wanted to criticize me honestly would find in my avowed and admitted imperfections quite enough to get his teeth into and to satisfy him without fencing with the wind. If it seems to such a man that, by forestalling his criticisms and revelations, I have made his bite toothless, it is reasonable that he should arrogate to himself

101. '88: disloyal to *my portrait*. This public . . .

the right to amplify and extend them (since offensives do have the right to go beyond justice) and he can take those defects whose roots in me I have revealed and magnify them into trees, using to that end not only such defects as have got a hold on me but also those which threaten me. Both in quality and quantity they are iniquitous: let him batter me with them. [C] I could frankly welcome the example of Dion the philosopher:[102] Antigonus was trying to provoke him on the subject of his origins. He cut him short and retorted: 'I am the son of a butcher – a branded slave – and of a prostitute whom my father married because of the baseness of his fortune. Both were punished for such-and-such a crime. When I was a youth an orator took a fancy to me and bought me. When he died he left me all his possessions. I transferred them here to Athens and devoted myself to philosophy. Biographers do not need to bother to seek news about me, for I will tell them how things stand.'

Free and open avowal robs rebuke of its sinews and strips insult of its weapons. [B] Nevertheless when all is said and done it appears that I am as often praised as disparaged beyond reason. It appears to me that I have, since my boyhood, been afforded a degree of rank and honour above what is mine rather than below. [C] I would feel more at ease in a land where such rankings were either regulated or held in contempt. Among men, as soon as a legal altercation about the order of precedence in processions or seating exceeds a triple rejoinder, it is discourteous. To avoid such churlish disputes I am never afraid to take or yield precedence unjustly: no man has ever challenged my precedence without my letting him take it.

[B] Apart from that profit which I derive from writing about myself, there is another which I hope for: if it chances before I die that my humours should please and suit some decent man, he might try to bring us together. I am meeting him more than half-way, since all that he could have gained from a long acquaintance and intimacy with me, he could get more reliably and minutely in three days from my account. [C] A pleasing fancy: many things that I would not care to tell to any individual man I tell to the public, and for knowledge of my most secret thoughts I refer my most loyal friends to a bookseller's stall:

Excutienda damus præcordia.

[We give them our inner hearts to ransack.][103]

102. Not *Dion* but *Bion*. Cf. Erasmus, *Apophthegmata*, VII, *Bion Borysthenites*, I (after Diogenes Laertius, *Life of Bion*).
103. Persius, *Satires*, V, 32.

[B] If on equally good evidence I knew a man who was right for me I would certainly go far to find him, for in my judgement the sweetness of well-matched and compatible fellowship can never cost too dear. O! a friend! How true is that ancient judgement, that the frequenting of one is more sweet than the element water, more necessary than the element fire.[104]

To get back to my narration: there is no great evil in dying alone and afar. [C] Indeed we reckon that it is a duty to seek seclusion for natural functions less ugly than that one and less repulsive. [B] And, farther, those who are reduced by their sufferings to drag out a long existence should perhaps not wish to burden a large family with their misery. [C] (That is why the Indians in one particular territory thought it right to kill anyone who had fallen into such distress, while in another they would abandon him alone to save himself as best he could.) [B] Is there anyone for whom those long a-dying are not in the end an intolerable burden? Our duties to each other do not extend that far. Inevitably you teach cruelty to those who love you best, making your wife and children, by long accustoming, grow callous, no longer feeling or pitying your afflictions. (The groans of my colic paroxysms no longer bring distress to anybody.) And even if we were to derive some pleasure from their company (which is not always the case, because of the dissimilarity of our circumstances which readily produces contempt and hatred towards anyone whomsoever) is it not an abuse to make it last an entire age? The more I were to see them generously constraining themselves for my sake the more I should regret the trouble they were taking. We have a right to lean on others, but not to lie that heavily on top of them, supporting ourselves by their collapse – like that man who had little boys' throats slit so as to use their blood to cure an illness of his, or like that other man who was supplied with little mites to warm his old limbs at night and to mingle the sweetness of their breath with the heavy sourness of his own.

As an asylum for such a condition and so feeble an existence I would be inclined to prescribe myself Venice. [C] Decrepitude is a solitary quality: I am sociable to the point of excess, yet from this day forward it seems reasonable that I should withdraw my importunity from the sight of the world and brood over it myself, retreating and shrinking into my shell as tortoises do. I am learning to see people without clinging to them – that would be an outrage on so steep a decline. It is time to turn my back on company.

104. Plutarch (tr. Amyot), *Comment on pourra discerner le flatteur d'avec l'amy*, 41.

[B] 'But on so long a journey you will end up in some thieves' kitchen where you will lack everything.' – I carry most of my necessities with me. At all events we have no way of avoiding Fortune if she undertakes to fall upon us. When I am ill I want nothing beyond the natural order: what Nature cannot work in me I do not want some quack's pill to do. While I am still in one piece and next-door to health, at the very onset of any fever or sickness which strikes me down, I reconcile myself to God by the last rites of Christianity; I find myself liberated and relieved by them, seeming to have got so much the better of my illness. Of lawyer and counsel I have even less need than of the doctor: do not expect me when I am ill to settle any affairs not already settled when I was in good health. What I intend to do to prepare for my death is done already; I would not dare to put it off for one single day. So if something remains undone, that means either that doubt has made me defer a decision (for sometimes the best decision is not to make one) or that quite simply I have wanted to do nothing about it.

My book I write for a few men and for a few years. If it had been on a lasting subject I would have entrusted it to a more durable language. Judging from the constant changes undergone by our own tongue up to the present, who can hope that its contemporary form will be current fifty years from now? [C] (It goes flowing through our fingers every day, and during my lifetime half of it has changed. We say that it is perfect now: each age says that of its own. I do not think it has reached perfection while it is still running away and changing form. It is up to good and useful writings to buckle French on to themselves, and its reputation will follow the fortunes of our State.)

[B] That is why I am not afraid to put in several personal details the currency of which will be exhausted during the lifetime of those who are alive today and which touch upon the private knowledge of some folk, who will see further into them than the general public can. When all is said and done I have no wish (as I know often happens whenever the dead are recalled to memory) that people should start arguing, claiming 'This is how he thought; this is how he lived'; 'If only he had uttered a few last words he would have said this or given away that'; 'I knew him better than anyone else.' Here I make known, as far as propriety allows, my feelings and inclinations. I do so more freely and readily by word of mouth for any who want to know; nevertheless if you look into these memoirs of mine you will find that I have said everything or intimated everything. What I have been unable to express in words I point towards with my finger:

Verum animo satis hæc vestigia parva sagaci
Sunt, per quæ possis cognoscere cætera tute.

[Those slight traces are enough for a keen-scented mind and will safely lead you to discover the rest.][105]

About myself nothing is wanting and there is nothing to guess. If you must discuss such things, I want it to be done truly and fairly. I would willingly come back from the next world to refute anyone who, even to do me honour, would fashion me other than I was. I know that people make even the living different when they talk of them. Had I not with all my might come to the defence of a friend whom I had lost, they would have ripped him into hundreds of incompatible little features.[106]

To finish talking of my foibles, I admit that I hardly ever arrive at my lodgings during my travels without the question passing through my mind whether I could be ill and die there comfortably, lodged as I like in a place entirely to my taste – no noise, not filthy, smoky or stuffy. By such trivial amenities I seek to cajole death or (to put it better) to relieve myself of all other impediments to enable me to concentrate on death alone: it will probably weigh heavily enough on me without adding to the burden. I want it to have a share in the comforts and conveniences of my life. Death forms a big chunk of it, an important one: from this day forth I hope it will not belie my past.

Some forms of death are easier than others: death takes on qualities which differ according to each man's way of thinking. Among natural deaths, pleasant and easy it seems to me is the one which comes from our growing torpid and weak. Among violent ones, I find it far harder to think of a precipice than the collapse of a wall, a slash from a sword than a volley from harquebuses; and I would rather have drunk Socrates' poison than to have run myself through like Cato. And although it all comes to the same, my imagination can feel a difference as great as life from death between jumping into a fiery furnace and into the stream of a smooth-flowing river – [C] so absurdly does our fear look more at the means than the result. [B] It only takes a moment, but I would give several days of my

105. Lucretius, I, 403–4.
106. The lost friend is La Boëtie.
   '88 features. *I well know that I shall leave behind me no guarantor even approximately as devoted to my case, and as knowledgeable, as I was to his. There is nobody with whom I would exchange vows to portray me: he alone had the privilege of my true portrait, which he took with him. That is why I explain my secrets so punctiliously.* to finish . . .

life to spend that moment in my own fashion. Since each man's fancy can find greater or less harshness in it, since each has some preference between ways of dying, let us assay going a little further and finding one quite free from unpleasantness. Might we not even make death luxurious like Antony and Cleopatra, those fellows in death? I leave aside as harsh the efforts devised by philosophy, and as ideal those devised by religion; but among lesser men we find a certain Petronius and a certain Tigillinus in Rome, who were required to kill themselves, lulling death to sleep,[107] so to speak, by their voluptuous preparations. They made death flow gently along, slipping it in among their usual wanton pastimes, between their girls and their drinking-companions: no mention of consolation, no mention of wills, no ambitious show of constancy, no talk of their condition in the life to come, but amidst games and festivities, jokes and common everyday conversation, music and love-poetry. Could we not imitate their resolve, with a more honourable restraint? Since there are deaths good for fools and others for sages, let us find some which are good for people in between. [C] My imagination can present me with a kind of death which is easy and (since we have to die) desirable. The Roman tyrants virtually spared a criminal's life when they allowed him to choose how he would die.

Yet was not a philosopher as subtle, modest and wise as Theophrastus forced by reason to recite the verse which Cicero put into Latin as:

> *Vitam regit fortuna, non sapientia.*

> [Our life is governed by Fortune not philosophy.][108]

How Fortune helps me now to rate my life at bargain-price, having reduced it to the point where nobody needs it and nobody is inconvenienced by it! That is a situation which I would have accepted at any period of my existence, but at this time, when I must fold up my garments and pack my bags, I find a special pleasure in causing no one when I die either pleasure or displeasure. By skilfully balancing the accounts, Fortune has made those who have a claim to some material gain from my death also conjoint heirs to some material loss. Death often oppresses us in as much as it weighs on others: we are virtually as concerned for their concerns as for our own — sometimes more or entirely so.

---

107. Both are mentioned by Tacitus: *Annals*, XVI, xix; *History*, I, lxxii.

'88: required *by the Emperors* to kill themselves, *according to the laws of that time* lulling death to sleep . . .

108. Cicero, *Tusc. disput.*, V, ix, 25; cf. Plutarch (tr. Amyot), *De la Fortune*, 106 B.

[B] Among the qualities I look for in my lodgings I do not include grandiose spaciousness – I hate it rather – but a simple individual charm more often met in places where there is less artifice and which Nature honours with some loveliness all her own: *'Non ampliter sed munditer convivium.'* *'Plus salis quam sumptus.'* [An elegant not a copious feast. More wit than waste.][109]

Moreover it is for those whose business drags them up over the Grisons in midwinter to be surprised on the highway by their own life's end. I, who most often travel for my own pleasure, am not all that bad a guide. If it looks nasty to the left I turn off to the right; if I find myself unfit to mount the saddle, I stop where I am. By acting thus I really do see nothing which is not as pleasant and agreeable to me as my home. It is true that I always do find superfluity superfluous and that I am embarrassed by delicacy, even, and by profusion. Have I overlooked anything which I ought to have seen back there? Then I go back to it: it is still on my road. I follow no predetermined route, neither straight nor crooked. Supposing when I do go to some place that I do not find there what I was told to expect: since others' judgements do not agree with mine (I have more often proved them wrong) I do not regret my exertion; I have learned that something which they told me about is not there!

My physical predisposition is as flexible, and my tastes as catholic, as any man's in the world. The diversity of custom between one nation and another touches me only by the pleasure of variety; each has its reason. Let the dishes be of pewter, wood or earthenware, consist of boiled meats or roasts, with butter, chestnut oil or olive oil, be hot or cold: it is all the same to me – so much so that, now I am getting old, I condemn such magnanimous facility and shall need discernment and selection to put a stop to my appetite's lack of discrimination and to look after my stomach occasionally.

[C] When I have been elsewhere than in France and people have courteously inquired whether I want to eat French cooking, I have always laughed at the idea and hastened straight for the sideboards most crowded with foreigners. [B] I am ashamed at the sight of our Frenchmen befuddled by that stupid humour which shies away from fashions which conflict with their own. Once out of their villages they feel like fish out of water. Wherever they go they cling to their ways and curse foreign ones. If they come across a fellow-countryman in Hungary, they celebrate the

109. Cited after i) Justus Lipsius, *Saturnalia*, I, vi; ii) – the second sentence – after Cornelius Nepos, *Life of Atticus*.

event: there they are, hobnobbing and sticking together and condemning every custom in sight as barbarous. And why not barbarous since they are not French! And those are the cleverer ones: as they speak ill of those customs, they have at least noticed them. Most go abroad merely to return. With a morose and taciturn prudence they travel about wrapped up in their cloaks and protecting themselves from the contagion of an unknown clime.

What I have said about them recalls something similar which I have noticed at times among some of our young courtiers. They mix only with their own kind, staring at us with disdain and pity as men from some other world. Strip them of their talk about the mysteries of the court and they are outside their hunting-grounds, as raw and awkward to us as we are to them. It is true what men say: a proper gentleman is a man of parts.

I on the contrary, as one who has had his fill of our customs, do not go looking for Gascons in Sicily – I have left enough of them at home. I look for Greeks, rather, or Persians. I make their acquaintance and study them. That is what I devote myself to and work on. And, what is more, I seem hardly ever to have come across any customs which are not worth quite as much as our own. I am not risking much by that assertion: I have hardly been out of sight of my own weathercocks.[110]

Meanwhile, most of the companions you chance to meet on the road are more an encumbrance than a pleasure: I never latch on to them – even less so nowadays when old age singles me out and sets me somewhat apart from the usual pattern. Either you are putting up with them or they with you. Both awkwardnesses weigh heavy, but the latter seems harsher to me. It is a rare stroke of fortune, but an inestimable pleasure, to have a gentleman who likes to accompany you, a man with manners which conform to your own. I have greatly missed one on all my travels. But such a companion must be selected and secured from the outset. No pleasure has any taste for me when not shared with another: no happy thought occurs to me without my being irritated at bringing it forth alone with no one to offer it to. [C] '*Si cum hac exceptione detur sapientia ut illam inclusam teneam nec enuntiem, rejiciam.*' [If even wisdom were granted me on condition that I shut it away unspoken, I would reject it.][111] This next author raised that a tone higher: '*Si contigerit ea vita*

---

110. That is, he has remained in his 'parish' – Western Europe – never having travelled to such exotic places as Greece or Persia, let alone China or the Americas.
111. Seneca, *Epist. moral.*, VI, 4; then, Cicero, *De officiis*, I, xliii, 153.

sapienti ut, omnium rerum affluentibus copiis, quamvis omnia quæ cognitione digna sunt summo otio secum ipse consideret et contempletur, tamen si solitudo tanta sit ut hominem videre non possit, excedat e vita.' [Supposing it were granted to a sage to live in every abundance, his time entirely free to study and reflect upon everything worth knowing: yet if his solitude were such that he could never meet another man he would quit this life.]

[B] I agree with the opinion of Archytas that there would be no pleasure in travelling through the heavens among those great immortal celestial bodies without the presence of a companion.[112] Yet it remains better to be alone than in silly boring company. Aristippus preferred to live as an alien everywhere.

> *Me si fata meis paterentur ducere vitam*
> *Auspiciis,*

[As for me, if the fates were to allow me to spend my life as I pleased,][113]

I would choose to spend it with my arse in the saddle,

> *visere gestiens,*
> *Qua parte debacchentur ignes,*
> *Qua nebulæ pluviique rores.*

[happy to see where the heat rages, or the clouds or the dripping rain.]

'Do you not have easier ways of spending your time? What do you lack? Is your house not set in a fine healthy climate; is it not adequately furnished and more than adequately spacious? [C] The King's Majesty in his splendour more than once put up with it![114] [B] Has your family not left behind many more families whose standards are below it than it has families above it in eminence? Is there something about the place so inordinate and [C] indigestible [B] that it gives you an ulcer,[115]

> *quæ te nunc coquat et vexet sub pectore fixa?*

[and which, rooted in your stomach, burns you and distresses you?]

Where do you think you can ever be without fuss and bother? *"Nunquam*

---

112. Cicero, *De amicitia*, XXIII, 88.
113. Virgil, *Aeneid*, IV, 340–1; then, Horace, *Odes*, III, iii, 34–6.
114. Henry of Navarre twice visited and stayed at Montaigne.
115. '88: so inordinate and so *uncurable* that . . .
   Then, Cicero, *De senectute*, I, 1 (the opening verse, from Ennius), and Quintus Curtius, IV, xxiv.

simpliciter fortuna indulget." [Fortune never sends unmixed blessings.] You really should realize that nobody is in your way but yourself, and that you will be following yourself about everywhere, and moaning to yourself everywhere,[116] for there is no contentment here below except for souls like those of beasts or gods.[117] Where can a man expect to find contentment if he is not content when he has such good cause? How many thousands of men are there whose aspirations do not exceed such circumstances as yours? Simply remould your Form: in such a matter you can do anything, whereas in face of Fortune you have no right but to endure. [C] *"Nulla placida quies est, nisi quam ratio composuit."* [There is no tranquil calm unless soothed by reason.]'[118]

[B] I can see the reasonableness of such counsel, see it very well. But it would have been quicker and more apposite simply to say to me one thing: 'Be wise!' Such a solution as yours lies the other side of wisdom: wisdom makes it and produces it. It is as though a doctor kept yelling at a wretched, languishing patient to feel merry: he would be prescribing a little less stupidly if he said, 'Get well!' As for me, I am merely a man [C] with a base Form.[119]

[B] 'Be content with what is yours' (that is, 'with reason'). That is a sound precept, definite and easy to understand; but sages can no more put it into effect than I can. There is a saying – popular, but appalling in its extent (what is not included in it?) – 'All things are subject to qualification and [C] limitation.'[120]

[B] I am well aware that, taken literally, this delight in travelling bears witness to restlessness and inconstancy. But those are indeed our dominant master-qualities. Yes. I admit it. Even in my wishes and dreams I can find nothing to which I can hold fast. The only things I find rewarding (if anything is) are variety and the enjoyment of diversity. When on my travels the very fact that I can stop without hindrance and conveniently make a diversion bolsters me up.

I love living a private life because I do so by my own choice, not because

116. Cf. Socrates' quip to the man who had not been improved by travel: 'Not surprising. You took yourself with you.' (Erasmus, *Apophthegmata*, III, *Socrates*, XLIV.)
117. The souls (or forms) of beasts are too low in the chain of being, those of divinities too high, to experience discontent. Man is in between.
118. Seneca, *Epist. moral.*, LVI, 6.
119. '88: man *of the common sort*. 'Be content . . .
120. '88: qualification and *moderation* [*modification*, 'limitation,' replacing *mesure*, 'moderation] . . .

I am unsuited to a public one (which doubtless equally accords with my complexion). I serve my Prince all the more happily because that is the free choice of my judgement and reason, [C] without any private obligation, [B] and because I am not constrained or forced back to it by being unacceptable to all the other parties or disliked by them. And so on. I detest such helpings as necessity carves for me. Any advantage would have me by the throat if I had to rely on it alone.

> *Alter remus aquas, alter mihi radat arenas.*

> [Let one oar sweep the water and the other sweep the strand.]¹²¹

One cord is never enough to hold me in place.

'There is vanity,' you say, 'in such a pastime.' – Yes. Where is there not? Those fine precepts are all vanity, and all wisdom is vanity: [C] *'Dominus novit cogitationes sapientium, quoniam vanae sunt.'* [The Lord knoweth the thoughts of the wise, that they are vain.]¹²² [B] Those exquisite subtleties are only good for sermons: they are themes which seek to drive us into the next world like donkeys. But life is material motion in the body, an activity, by its very essence, imperfect and unruly: I work to serve it on its own terms.

> *Quisque suos patimur manes.*

> [Each suffers his own torments.]¹²³

[C] *'Sic est faciendum ut contra naturam universam nihil contendamus; ea tamen conservata, propriam sequamur.'* [We must so live as not to struggle against Nature in general; having safeguarded such things, we should follow our own nature.]

[B] What is the use of those high philosophical peaks on which no human being can settle and those rules which exceed our practice and our power? I am well aware that people often expound to us ideas about life which neither the speaker nor the hearers have any hope of following or (what is more) any desire. The judge filches a bit of the very same paper on which he has just written the sentence on an adulterer in order to send a

---

121. Propertius, III, iii, 23.
122. I Corinthians 3:20, citing Psalm 94 (93):11.
123. Virgil, *Aeneid*, VI, 743. (Virgil's sense is by no means clear: the ancient commentator Servius explained *Manes*, 'spirits' here as 'punishments' or 'torments', an interpretation I have followed.) Then, Cicero, *De officiis*, I, xxxi, 110.

billet-doux to the wife of a colleague. [C] The woman you have just been having an illicit tumble with will soon, in your very presence, be screaming harsher condemnations of a similar fault in a friend of hers than Portia would. [B] Some condemn people to death for crimes which they do not actually believe to be even mistakes. When I was a youth I saw a fine gentleman offering to the public, with one hand, poetry excelling in beauty and eroticism both, and with the other, at the same instant, the most cantankerous reformation of theology that the world has had for breakfast for many a long year.[124]

That is the way humans proceed. We let the laws and precepts go their own way: we take another – not only because of unruly morals but often because of contrary opinions and judgement. Listen to the recital of a philosophical discourse: its invention, eloquence and appositeness at once strike your attention and move your emotions. But there is nothing there which stings or pricks your conscience: it was not addressed to it, was it? Yet Ariston said that neither a bath nor a lecture bears any fruit unless they cleanse you and get the filth off.[125] You can linger over the lido, but only after extracting the marrow, just as it is only after we have drunk the wine that we examine the engravings and workmanship of a beautiful goblet.

In all the chambers of the ancient philosophers you will find that the same author, at the same time, publishes rules for temperance and works of love and debauchery. [C] Xenophon wrote against Aristippus' concept of pleasure while lying in the lap of Clinias. [B] Those were not miraculous conversions sweeping over them in waves. First it is Solon presenting himself in the guise of a lawgiver, and then as himself: at one time he is speaking for the many, at another for himself alone and (certain as he is that he is firmly and totally well) he takes for himself the free and natural rules:

> *Curentur dubii medicis majoribus ægri!*
>
> [Let the dangerously ill call in great doctors!][126]

[C] Antisthenes allows his sage to like anything he finds appropriate, and to do it in his own fashion without heeding the laws, since he has a better judgement than they do and a greater knowledge of virtue. His disciple Diogenes said that we should counter perturbations by reason; fortune, by

---

124. Probably Theodore Beza, the erotic poet and successor to Calvin.
125. Plutarch (tr. Amyot), *Comment il fault ouïr*, 27 CD.
126. Juvenal, *Satires*, XIII, 124.

courage; laws, by nature.[127] [B] It is for tender stomachs that we have restricted, artificial diets: [C] sound ones simply follow the prescriptions of their natural appetite. [B] Thus do our doctors eat melons and drink cool wine while keeping their patients on syrups and pap.

'I know nothing of their books,' said Laïs the courtesan, 'nor of their wisdom and philosophy, but those fellows come knocking at my door as often as anyone.'[128] Since our licence always takes us beyond what is lawful and permissible, we have often made the precepts and laws for our lives stricter than universal reason requires.

> *Nemo satis credit tantum delinquere quantum*
> *Permittas.*

[Nobody thinks that his own transgressions exceed what is allowable.][129]

It would be preferable if there were more proportion between commands and obedience. A target we cannot reach appears unfair. No man is so moral but that, if he submitted his deeds and thoughts to cross-examination by the laws, he would be found worthy of hanging on ten occasions in his lifetime – yes, even the kind of man whom it would be a great scandal to punish and a great injustice to execute.

> *Olle, quid ad te*
> *De cute quid faciat ille, vel illa sua?*

[What concern is it of yours, Ollus, what he does with his own skin and she with hers?]

And one who deserves no praise as a man of virtue [C] and whom philosophy could most justly cause to be flogged [B] may well break no laws, so confused and unfair is the correspondence between law and virtue. We do not care to be decent folk by the standards of God: we could never be so by our own. Human wisdom has never managed to live up to the duties which it has prescribed for itself; and if it had done so, it would have prescribed itself more, further beyond them still, towards which it could continue to strive and aspire, so hostile is our condition to immobility. [C] Man commands himself to be necessarily at fault. It is not

---

127. Erasmus, *Apophthegmata*, VII, *Antisthenes*, XLVIII; then, III, *Diogenes Cynicus*, XLV.
128. Cf. Brantôme, *Dames Galantes*, IV (Garnier edition p. 219 and note). The tale is told by Bishop Antonio de Guevara in his work translated as *Les Epistres dorées, moralles et familieres*.
129. Juvenal, *Satires*, XIV, 233–4; then, Martial, *Epigrams*, VII, ix, 1–2.

very clever of him to tailor his obligations to the standards of a different kind of being. He expects no one to do it, so whom is he prescribing it for? Is it wrong of Man not to do what is impossible for him to do? The very laws which condemn us to be unable blame us for being so.

[B] If the worst comes to the worst, that deformed licence to present themselves in two ways, their actions in one fashion and their rhetoric in another, may be conceded to those who tell of *things*: it cannot apply to those who tell of themselves as I do; my pen must go the same way as my feet. A life lived in society must bear some relationship to other lives. Cato's virtue was excessively rigorous by the standards of his age; and in a man occupied in governing others and destined to serve the commonwealth, we could say that his justice, if not unjust, was at least vain and unseasonable. [C] My own manners deviate from current morality by hardly more than an inch, yet even that makes me untractable for this age and unsociable. I do not know whether I am unreasonable in losing my taste for the society I frequent, but I do know that it would be unreasonable if I complained that it had lost its taste for me more than I for it.

[B] The virtue allotted to this world's affairs is a virtue with many angles, crinkles and corners so that it can be applied and joined to our human frailty; it is complex and artificial, not straight, clear-cut, constant, nor purely innocent. To this very day our annals criticize one of our kings for allowing himself to be too naïvely influenced by the persuasions which his confessor addressed to his conscience.[130] Affairs of state have their own bolder precepts:

exeat aula

Qui vult esse pius.

[he who would be pious should quit the court.][131]

Once I made an assay at using in the service of some political manoeuvrings, such opinions and rules of life as were born in me or instilled into me by education – rough, fresh, unpolished and unpolluted ones, the virtues of a schoolboy or a novice, which I practise, [C] if not [B] conveniently [C] at least surely, [B] in my private life. I found that they were inapplicable and dangerous. Anyone who goes into the throng must be prepared to side-step, to squeeze in his elbows, to dodge to and fro and, indeed, to abandon the straight path according to

130. Perhaps Henry II, whose confessor the Cardinal de Lorraine persuaded him to persecute the members of the Reformed Church.
131. Lucan, *Pharsalia*, VIII, 493–4.

what he encounters; he must live not so much by his norms but by those of others; not so much according to what he prescribes to himself but to what others prescribe to him, and according to the time, according to the men, according to the negotiations . . .[132]

[C] Plato says that anyone who escapes with unsmirched linen from the management of the world's affairs does so by a miracle. He also says that when he laid it down that his philosopher should rule the state he was not speaking of corrupt polities such as that of Athens (and even less of ones like our own, faced with which even Wisdom would forget her Latin), since a seedling transplanted into a soil very different in character from itself conforms itself to it rather than reforming it.[133]

[B] If I had thoroughly to prepare myself for such occupations, I know that I would need many changes and adjustments. Even if I could manage it (and why should I not do so, given time and trouble?) I would not want to. The little I have assayed of such a vocation was quite enough to put me off. Sometimes I do feel some temptations towards ambition smouldering in my soul, but I tense myself and obstinately resist.

> *At tu, Catulle, obstinatus obdura!*
>
> [Come on Catullus! Be obstinately obdurate!][134]

I am rarely summoned: and I just as seldom volunteer.   [C] My master qualities, liberty and laziness, are qualities which are diametrically opposed to such a trade.   [B] We do not know how to distinguish the faculties of men: they have fine divisions and their boundaries are hard to select. To infer a capacity for the affairs of State from a capacity for private affairs is to make a bad inference. A man may control himself but not others,   [C] being able to produce Essays but nothing effective; another   [B]   may organize a good siege but not a battle; he may speak well in private but badly in public or before his prince. Indeed, evidence that he can do one perhaps suggests that he cannot do the other.

[C] I find that higher intellects are hardly less suited to lowlier matters than lowly intellects are to the higher. Who would ever have expected Socrates to have furnished the Athenians with a good laugh at his expense because he was never able to add up the votes of his tribe and report them

---

132. Everything in public life is *secundum quid dependens*, ever (in Montaigne's repeated word) *selon*; 'it all depends' on something else or on someone else.
133. Plato, *Republic*, VI, 492 E and 497 A–C.
134. Catullus, VII, 19.

to the Council?[135] The veneration that I feel for the perfections of that great man certainly deserves that it should be his fortune to supply such a magnificent example to excuse my chief imperfections!

[B] Our ability is chopped up into little bits. My own has no breadth, and is also numerically weak. Saturninus said to those who had conferred on him the supreme command: 'You have lost a fine captain, Comrades, to make a poor general.'

Anyone who, in an ailing time like ours, boasts that he can bring a naïve and pure virtue to this world's service either has no idea what virtue is, since our opinions are corrupted along with our morals – indeed, just listen to them describing it; listen to most of them vaunting of their deeds and formulating their rules: instead of describing virtue they are describing pure injustice and vice, and they present it, thus falsified, in the education of princes – or else, if he does have some notion of it, he boasts wrongfully and, say what he will, does hundreds of things for which his conscience condemns him. In similar circumstances Seneca's account of his experience I would readily believe, provided that he would talk to me about it unreservedly. In such straits the most honourable mark of goodness consists in freely acknowledging your defects and those of others, while using your powers to resist and retard the slide towards evil, having to be dragged down that slope, while hoping for improvement and desiring improvement.

During the divisions into which we are fallen, tearing France limb from limb, each man, I notice, strives to defend his cause, but even the best of them with deception and lies. Anyone who wrote bluntly about it would do so inadequately and ill-advisedly, since even the juster party is itself a limb of that rotten, worm-eaten body. Yet in such a body the least affected limb is termed healthy – rightly so: since our qualities are valid only by comparison, civil integrity is measured according to time and place. I would like to see, I must say, Agesilaus praised as follows in Xenophon!

Having been asked by a neighbouring Prince with whom he had formerly been at war for permission to pass through his domains, he granted it to him, affording him passage through the Peleponnesus: and not only did he not take him prisoner nor poison him, despite having him thus at his mercy, but he welcomed him courteously without doing him injury.[136]

135. Plato, *Gorgias*, XXIX, 474A (following the sense of Ficino's Latin rendering).
136. Montaigne parodies the kind of praise heaped on Francis I for allowing Charles V to pass in safety through his domains in 1539–40, despite the French humiliation at Pavia in 1524.

Given the characters of people then, such things were taken for granted: elsewhere, and at other times, men will tell of the noble frankness and magnanimity of that deed! Why, our be-caped baboons of the Collège de Montaigue would laugh at him for it, so little does our French integrity resemble that of the Spartans. We still have men of virtue . . . but by our norms.[137] Whoever has morals fixed to rules above the standards of his time must either distort and blunt his rules or (as I would advise him, rather) draw apart and having nothing to do with us. What would he gain from us?

> *Egregium sanctumque virum si cerno, bimembri*
> *Hoc monstrum puero, et miranti jam sub aratro*
> *Piscibus inventis, et fœtæ comparo mulæ.*

[When I come across an outstandingly moral man, he seems to me like a kind of freak, like a two-headed child, like fish turning up under an astonished farmer's ploughshare, or like a pregnant mule.][138]

We can regret better times but we cannot escape from the present; we can wish for better men to govern us but we must nevertheless obey those we have. There is perhaps more merit in obeying the bad than the good. While the ghost of the traditional ancient laws of this our monarchy glows in a corner somewhere, you will see me planted there. If those laws should, to our misfortune, become mutually exclusive or contradictory, producing a hard and dubious choice between two factions, my preference would be for hiding and escaping from that tempest. In the meanwhile, Nature may lend me a hand; so may the hazards of war.

Between Caesar and Pompey I would have declared myself frankly. But if the choice lay between those three crooks who came after them,[139] then I would either have fled into hiding or gone the way the wind blew (which I judge to be legitimate, once reason no longer guides us).

> *Quo diversus abis?*

[Where are you heading, so far off course?][140]

This padding is rather off my subject. I get lost, but more from licence than carelessness. My ideas do follow on from each other, though sometimes at a distance, and have regard for each other, though somewhat obliquely. [C] I have just looked through one of Plato's dialogues.[141] It is parti-

---

137. That is, *secundum nos, selon nous.*
138. Juvenal, *Satires*, XIII, 64–6.
139. Mark Antony, Octavius and Lepidus, the Triumvirate.
140. Virgil, *Aeneid*, V, 166.
141. The *Phaedrus.*

coloured, a motley of ideas: the top deals with love and all the bottom with rhetoric. They were not afraid of such changes, and have a marvellous charm when letting themselves be blown along by the wind, or appearing to be so. [B] The names of my chapters do not always encompass my subject-matter: often they merely indicate it by some token, like those other [C] titles, *Andria* or *The Eunuch*, or like those other [B] names Sylla, Cicero and Torquatus.[142]

I love the gait of poetry, all jumps and tumblings. [C] Poetry, says Plato, is an art which is light, winged and inspired by daemons.[143] There are works of Plutarch in which he forgets his theme, or in which the subject is treated only incidentally, since they are entirely padded out with extraneous matter: witness how he proceeds in *The Daemon of Socrates*. My God! what beauty there is in such flights of fancy and in such variation, especially when they appear fortuitous and casual. It is the undiligent reader who loses my subject not I. In a corner somewhere you can always find a word or two on my topic, adequate despite being squeezed in tight. [B] I change subject violently and chaotically. [C] My pen and my mind both go a-roaming. [B] If you do not want more dullness you must accept a touch of madness, [C] so say the precepts of our past masters and, even more so, their example. [B] There are hundreds of poets who drag and droop prosaically, but the best of ancient prose – [C] and I scatter prose here no differently from verse – [B] sparkles throughout with poetic power and daring, and presents the characteristics of its frenzy. We must certainly cede to poetry the mastery and pre-eminence in prattle. [C] The poet, says Plato, seated on the tripod of the Muses, pours out in rapture, like the gargoyle of a fountain, all that comes to his lips, without weighing it or chewing it; from him there escape things of diverse hue, contrasting substance and jolting motion.[144] Plato

142. Two comedies of Plautus, the titles of which merely hint at their subject. Then, in [C], the surnames given to Lucius Cornelius, the dictator: *Sylla* ('Freckles'); to Mark Tully: *Cicero* ('Chickpea'); and Titus Manlius: *Torquatus* (a nickname drawn from *torque*, a Gaulish necklace which he once wore as booty).

143. In Plato's dialogue *Ion* – a major source of the French Pléiade's conception of poetic inspiration, and especially of Ronsard's (who did not perceive Plato's irony) in his famous *Ode à Michel de l'Hospital*.

144. Plato, *Laws*, IV, 719 CD, contrasting inspired poets, who in their daemon-inspired *mimesis* (imitation of nature) pour forth verbal inconsistencies, with lawgivers, whose writings must be consistent and coherent. Montaigne is defending the ecstatic, enraptured, enthusiastic element in high poetry. Like Sir Philip Sidney in his *Defence of poesy* Montaigne includes both prose and verse in the category of poetry.

himself is entirely poetic; and the scholars say that the ancient theology was poetry, as also the first philosophy.[145] Poetry is the original language of the gods.

[B] I intend my subject-matter to stand out on its own: it can show well enough where changes occur, where the beginnings are and the ends, and where it picks up again, without an intricate criss-cross of words, linking things and stitching them together for the benefit of weak and inattentive ears, and without my glossing myself. Where is the author who would rather not be read at all than to be dozed through or dashed through? [C] *'Nihil est tam utile, quod in transitu prosit.'* [Nothing really useful can be casually treated.][146] If taking up books were to mean taking them in; if glancing at them were to mean seeing into them; and skipping through them to mean grasping them: then I would be wrong to make myself out to be quite so totally ignorant as I am.

[B] Since I cannot hold my reader's attention by my weight, *manco male* [it is no bad thing] if I manage to do so by my muddle. 'Yes, but [C] afterwards [B] he will be sorry he spent time over it.' I suppose so: but still he would have done it! And there are humours so made that they despise anything which they can understand and which will rate me more highly when they do not know what I mean. They will infer the depth of my meaning from its obscurity – a quality which (to speak seriously now) I hate [C] most strongly; [B] I would avoid it if there were a way of [C] avoiding [B] myself.[147]   [B] Aristotle somewhere congratulates himself on affecting it: a depraved [C] affectation![148]

Because the very frequent division into chapters which I first adopted seemed to me to break the reader's attention before it was aroused and to loosen its hold so that it did not bother for so slight a cause to apply itself and to concentrate, I started making longer chapters which require a decision to read them and time set aside for them. In this kind of occupation, whoever is not prepared to give a man one hour is prepared to give him nothing; and you do nothing for a man if you only do it while doing something else. Besides I may perhaps have some personal quality

145. Homer and Hesiod were treated as both poets and philosophers from the earliest times: Plato's title 'Divine' emphasized the role of poetic inspiration in his philosophy.
146. Seneca, *Epist. moral.*, II, 3.
147. '88: were a way of *counteracting* myself . . .
148. '88: a depraved *conception*. It remains to add . . .
   For Aristotle's wilful opacity, cf. Aulus Gellius, *Attic Nights*, XX, iv.

which obliges me to half-state matters and to speak confusedly and incompatibly.

[B] It remains for me to add that I wish no good to that chattering buffoon of a reason, and that, while those fantastic speculations and those oh-so-subtle notions may contain some truth, I find it too dear and too troublesome.[149] I, on the contrary, strive to give worth to vanity itself – [C] to doltishness – if it affords me pleasure, [B] and I follow my natural inclinations without accounting for them thus closely.

– I have 'already seen elsewhere ruined palaces and sculptures of things in heaven and on earth: and it is ever the work of Man'. That is quite true. Yet, however often I were to revisit the tomb of that great and mighty City, I would feel wonder and awe. We are enjoined to care for the dead: and since infancy I was brought up with those dead. I knew about the affairs of Rome long before those of my family; I knew of the Capitol and its site long before I knew of the Louvre, and of the Tiber before the Seine. My head was full of the characters and fortunes of Lucullus, Metellus and Scipio rather than of any of our own men. – 'They are *dead!*' So is my father, every bit as dead as they: in eighteen years he has gone as far from life and me as they have done in sixteen hundred, yet I do not cease to cherish his memory nor experience his love and fellowship in a perfect union, fully alive. Indeed, of my own humour, it is to the dead that I am most dutiful: since they can no longer help themselves I consider that they need my help the more. It is precisely then that gratitude shines forth resplendent. A favour is less richly bestowed when it can be returned or reflected back.

When Arcesilaus was visiting the [C] ailing Ctesibius,[150] [B] he realized that he was badly off, so he gave him money, slipping it under his pillow. By concealing it from him he was also giving him a quittance from

149. [B] : to vanity itself – to *dullness* – if it affords me *contentment* and I allow . . .

The argument here continues that of the [B]-text, ignoring the interpolated [C]-text. It was a little clearer before Montaigne replaced 'depraved conception (*imagination*)' by 'depraved *affectation*'. Montaigne calls Aristotle's search for obscurity a *raison trouble-feste*, a 'trouble-feast' then meaning an importunate buffoon whose idle chatter spoils a merry feast. It was a word which implied a silly incessant talker, not a 'wet-blanket'.

150. '88: visiting the *sick Appelles* he realized . . .

Plutarch says it was Appelles: *Comment on peult discerner le flatteur d'aveques l'amy* 48 GH; Montaigne corrected him after reading Diogenes Laertius.

a debt of gratitude. Those who have deserved my love and thanks have never lost anything for being no longer with me: I have repaid them better and more punctiliously when they were absent and unaware. I speak all the more affectionately of those I love when they no longer have any way of knowing it. So I have begun dozens of quarrels in defence of Pompey or the cause of Brutus. Acquaintanceship still endures between us; why, even things present are grasped only by a faculty of the mind.

Finding myself useless for this present age I fall back on that one. I am such a silly baboon about it that the state of Ancient Rome, free and just and flourishing (for I like neither its birth nor its decline), is of passionate concern to me. That is why I could never so often revisit the site of their streets and their palaces, and their ruins stretching down to the Antipodes, without lingering over them.   [C]   Is it by nature or an aberrant imagination that the sight of places which we know to have been frequented or inhabited by those whose memory we hold dear moves us somewhat more than hearing a recital of their deeds or reading their writings?[151] *'Tanta vis admonitionis inest in locis. Et id quidem in hac urbe infinitum: quacunque enim ingredimur in aliquam historiam vestigium ponimus.'* [Such powers of evocation are inherent in those places [. . .] And in this City there is no end to them: wherever we go we walk over history.]

[B]   I like thinking about their faces, their bearing and their clothing. I mutter their great names between my teeth and make them resound in my ears.   [C]   *'Ego illos veneror et tantis nominibus semper assurgo.'* [I venerate them, and on hearing such names I leap always to my feet.][152] [B]   Whenever there are qualities in things which are great and awesome, I feel awe for their ordinary ones as well. I would love to see those men talking, walking and eating. It would be ungrateful to neglect the remains and ghosts of so many honoured and valiant men whom I have watched live and die and who, by their example, provide us with instructions in what is good if we know how to follow them.

And then this very Rome, the one that we see now, deserves our love as having been so long and by so many titles an ally of our Crown and the only city common to all men and universal. The sovereign magistrate who rules there is similarly acknowledged everywhere; it is the mother city of all Christian peoples: both Frenchman and Spaniard are at home there. To become princes of that state you merely need to belong to Christendom,

---

151. A major debt to Cicero, *De finibus*, V, i, 2 (translated tacitly by Montaigne in the text and then cited; also V, ii, 5, actually alluding to Athens, not Rome.
152. Seneca, *Epist. moral.*, LXIV, 10.

no matter where. There is nowhere here below upon which the heavens have poured influences so constantly favourable. Even in ruins it is glorious and stately:

> [C]   *Laudandis preciosior ruinis.*

> [More precious for her ruins which deserve our praise.][153]

[B]   Even in her tomb she still retains the signs and ghost of empire: [C]   *'ut palam sit uno in loco gaudentis opus esse naturae'* [so that it should be obvious that in this one place Nature delights in her work].

[B]   A man might condemn himself and inwardly rebel for feeling stirred by so vain a pleasure. Yet our humours, if they do afford pleasure, are not too vain; whatever they may be, if they afford constant delight to a man capable of common feelings, I would be of no mind to feel sorry for him.

I am deeply indebted to Fortune in that, up to present, she has done me no outrage,   [C]   at least, none above what I can bear.[154]   [B]   (Might it not be her style to leave in peace those who do not pester her?)

> *Quanto quisque sibi plura negaverit,*
> *A Diis, plura feret. Nil cupientium*
> *Nudus castra peto . . .*
>   *. . . Multa petentibus*
> *Desunt multa.*

[The more a man denies himself, the more he will receive from the gods. I am naked but put myself in the camp of those who want nothing . . . Those who want much, lack much.][155]

If she continues she will dispatch me content and well satisfied:

> *nihil supra*
> *Deos lacesso.*

> [for nothing more do I harass the gods.]

But watch out for the snag! Hundreds founder within the harbour.

---

153. Bishop Caius Sollius Apollinaris Sidonius: *Carmina*, XXIII, 62; then, Pliny, III, v.
154. '88: outrage, *beyond my strength*. (Might it not . . .
155. Horace, *Odes*, III, xvi, 21–3; 42–3; then, II, xviii, 11–12; Ovid, *Metamorphoses*, II, 140.

I can easily find consolation over what will happen here below once I am gone: present concerns keep me busy enough:

fortunæ cætera mando.

[the rest I entrust to Fortune.]

Besides I do not have that strong link which is said to bind a man to the future by sons who bear his name and rank – and if that is what makes sons desirable I should perhaps desire them all the less: of myself I am only too bound to this world and this life.[156] I am content to be at grips with Fortune through attributes which are strictly necessary to my being without extending her jurisdiction over me in other ways; and I have never thought that not having sons made life less perfect and less satisfying. There are advantages too in the vocation of childlessness. Sons are to be counted among things which do not have much to make them desired, especially at this moment when it would be hard to make them good – [C] *'Bona jam nec nasci licet, ita corrupta sunt semina'* [Good things are not born now: the seed is so corrupt][157] – but which, once acquired, are rightly to be regretted by those who lose them.

He who left me responsible for my household forecast that I would ruin it, seeing how little stay-at-home my humour is. He was wrong. Here I am, just as I inherited it, or perhaps a little better off, yet without appointment or benefice.

Howbeit, though Fortune has done me no unusually violent outrage, neither has she done me any favour. Whatever gifts of hers are to be found in our home have been there for a hundred years before my time. Not one solid essential good thing do I personally owe to her generosity. To me she has vouchsafed some honorary titular favours, all wind and no substance; and (God knows!) she did not so much vouchsafe them to me as offer them to me – to me who am wholly material, who seek satisfaction in realities (solid ones at that) and who (if I dared to admit it) would scarcely find covetousness any less pardonable than ambition; pain, any less to be avoided than disgrace; health, any less desirable than learning; and wealth than noble rank. Among her vain favours I have none more pleasing to that silly humour in me which feeds on it than an authentic Bull of Roman Citizenship which was granted to me recently when I was there, resplendent

156. In this passage (as often elsewhere) *enfants* means sons, not infants, children of either sex.
157. Tertullian, *De pudicitia*.

with seals and gilded letters, granted moreover with all gracious generosity.[158]

Since they are given in a variety of styles, with more favour or less, and since before I had seen one myself I would very much like to have been shown one drawn up in due form, I want to transcribe it here *in extenso*, to satisfy anyone suffering from the same curiosity as I had.[159]

HORATIUS MAXIMUS, MARTUS CECIUS, ALEXANDER MUTUS, CONSERVATORS OF OUR KINDLY CITY, HAVING REPORTED UNTO THE SENATE CONCERNING THE GRANTING OF ROMAN CITIZENSHIP TO THE MOST ILLUSTRIOUS MICHAEL MONTANUS, KNIGHT OF SAINT MICHAEL AND GENTLEMAN-IN-WAITING TO THE MOST CHRISTIAN KING: THE ROMAN SENATE AND PEOPLE HEREBY DECREE:

Whereas by Ancient custom and law, men have ever been received among us with eagerness and ardour when, outstanding for their virtue and nobility, they have either done great service to our Republic and enhanced it or may so do in the future: We, aroused by the authority and example of our Forefathers, decree that we should imitate and maintain so noble a custom: Wherefore: whereas the illustrious Michael Montanus, Knight of Saint Michael and Gentleman-in-Waiting to the Most Christian King, is most devoted to the name of Rome and is found most worthy, by the reputation and the splendour of his family and by the merit of his own virtue, to be admitted to Roman Citizenship by the highest judgement of the Roman People and Senate: it has pleased the SPQR that the most illustrious Michael Montanus, in all things most honoured, and most dear to this renowned People, be inscribed, him and his descendants, as Roman Citizens, and be further honoured by all those rewards and distinctions which such enjoy who are Roman Citizens and Patricians by birth or by legal processes duly thereanent. Which doing, the SPQR do not esteem that they are granting him these Rights of Citizenship of their bounty so much as repaying a debt, granting him no greater benefit than he has conferred upon them by accepting this

158. Montaigne's *Journal de Voyage* tells that he received it from the Pope by the intercession of Philippo Musotti; he was proud that it was couched in the same terms as that of one of the Pope's sons.

159. Montaigne gives the text in the original Latin, which is not given here but translated. The style is that of Ancient Rome – SPQR standing for *Senatus Populusque Romanus*, the formula used by the Roman Republic for a *senatusconsultum* (decree).

their Citizenship, by which this their City is particularly honoured and enhanced.

Which *Senatusconsultum* the aforesaid Conservators, by their authority, hereby cause to be immatriculated by the scribes of the Roman Senate and People and deposited in the Roman Curia; and have caused this Document to be duly drawn up, sealed with the accustomed seal of the City. In the Year from the Foundation of the City Two Thousand Three Hundred and Thirty-one, and in the Year of Our Lord One Thousand Five Hundred and Eighty-one: The Third of the Ides of March.

HORATIUS FUSCUS: Scribe to the Holy Senate and People of Rome;
VINCENTIUS MARTHOLUS: Scribe to the Holy Senate and People of Rome.

Not being the citizen of any city, I am delighted to have been made one of the noblest City there ever was or ever shall be.

If others were to look attentively into themselves as I do, they would find themselves, as I do, full of emptiness and tomfoolery. I cannot rid myself of them without getting rid of myself. We are all steeped in them, each as much as the other; but those who realize this get off, as I know, a little more cheaply.

That commonly approved practice of looking elsewhere than at our own self has served our affairs well! Our self is an object full of dissatisfaction: we can see nothing there but wretchedness and vanity. So as not to dishearten us, Nature has very conveniently cast the action of our sight outwards. We are swept on downstream, but to struggle back towards our self against the current is a painful movement; thus does the sea, when driven against itself, swirl back in confusion. Everyone says: 'Look at the motions of the heavens, look at society, at this man's quarrel, that man's pulse, this other man's will and testament' – in other words always look upwards or downwards or sideways, or before or behind you. That commandment given us in ancient times by that god at Delphi was contrary to all expectation: 'Look back into your self; get to know your self; hold on to your self.' Bring back to your self your mind and your will which are being squandered elsewhere; you are draining and frittering your self away. Consolidate your self; rein your self back. They are cheating you, distracting you, robbing you of your self.[160]

160. Montaigne is drawing upon the heights of Greek philosophical wisdom, evoking the most famous of all precepts inscribed on the portal of the temple of

Can you not see that this world of ours keeps its gaze bent ever inwards and its eyes ever open to contemplate itself? It is always vanity in your case, within and without, but a vanity which is less, the less it extends.

Except you alone, O Man, said that god, each creature first studies its own self, and, according to its needs, has limits to its labours and desires. Not one is as empty and needy as you, who embrace the universe: you are the seeker with no knowledge, the judge with no jurisdiction and, when all is done, the jester of the farce.

Apollo at Delphi: *Gnōthi seauton* (*Nosce teipsum*, Know Thyself). Cf. Plato, *Charmides* 164 E ff.; *Alcibiades I*, 129 E ff.) It was precisely because Socrates was not yet able to satisfy the Delphic inscription to know himself that he judged it ridiculous to investigate anything irrelevant to self-knowledge (*Phaedrus*, 229 D – 230 A). Erasmus' explanation of Know Thyself in his *Adages* (I, VI, XCV) was standard; he associates it (as does Montaigne) with other precepts: *In se descendere* (Go down into your self: ibid., LXXXVI); *Tecum habita* (Dwell with your self, LXXXVII), *Aedibus in nostris quae prava aut recta geruntur* (Things are done right or wrong in our *own* dwellings, LXXXV); *In tuum ipsius sinum inspue* (It is your own bosom you should spit upon – that is, criticize, XCIV); *Quae supra nos, nihil ad nos* (What is above us [i.e. astronomy and so on] does not concern us, I, VI, LXIX); and several others. Erasmus' explanations (like Plato's) make it plain that we are not being encouraged to cultivate self-love but self-knowledge. Montaigne gives to these precepts his own startlingly original twist.

10. On restraining your will

[Montaigne justifies his two unremarkable periods as Mayor of Bordeaux. His ideal is that of a tranquil mind based on a sense of having fulfilled his moral obligations to Church, State, family and city, without excessive emotional involvement. Both Socratic ideals and Stoicism blend with his Christianity. The idea that wisdom begins 'at home', that is, with our own self, is Socratic. Montaigne supports his case with echoes of the previous chapter as well as a saying of the Holy Ghost's and a petition from the Lord's Prayer.]

[B]   Compared with the common run of men, few things touch me or, to speak more correctly, get a hold on me (it being reasonable for things to touch us provided that they do not take us over). I exercise great care to extend by reason and reflection this privileged lack of emotion, which is by nature well advanced in me. I am wedded to few things and so am passionate about few. My sight is clear but I fix it on only a few objectives; my perception is scrupulous and receptive, but I find things hard to grasp and my concentration is vague. I do not easily get involved. As far as possible I work entirely on my self, but even on that subject I prefer to rein back my emotion so as to stop it from plunging right in, since it is a subject which I possess at the mercy of Another – Fortune having more rights over it than I do.

I value health most highly: but it follows that I ought not to seek or desire even that so frenetically that I find illness unbearable.   [C]   We should follow the Mean between hatred of pain and love of pleasure: Plato prescribes a way of life midway between the two.[1]   [B]   But there are emotions which drag me from myself and tie me up elsewhere: those I oppose with all my might. In my opinion we must lend ourselves to others but give ourselves to ourselves alone. Even if my will did find it easy to

---

1. It was axiomatic to Plato that 'Know Thyself' was an injunction to be temperate and to follow moderation (*Charmides*, 146 C ff.). The explicit injunction alluded to here is in Plato, *Laws*, VII, 792 E – 793 A.

pawn and bind itself to others, I could not persevere: by nature and habit I am too fastidious for that:

fugax rerum, securaque in otia natus.

[fleeing from obligations and born for untroubled leisure.][2]

Stubborn earnest arguments which ended in victory for my opponent, as well as results which made me ashamed of my hot pursuit, might indeed most cruelly gnaw at me. If I were then to bite back as others do my soul would never find the strength to support the alarms and commotions which attend those who embrace so much: it would straightway be put out of joint by such internal strife. If I am occasionally pressed into taking in hand some business foreign to me, then it is in hand that I promise to take it, not in lung nor in liver! I accept the burdens but I refuse to make them parts of my body. Take trouble over them: yes; get worked up about them: never. I look after them, but not like a broody hen. I have enough to do to order and arrange those pressing affairs of my own which lie within my veins and vitals without having a jostling crowd of other folk's affairs lodged there and trampling all over me; I have enough to do to attend to matters which by nature belong to my own being without inviting in outsiders. Those who realize what they owe to themselves, and the great duties which bind themselves to themselves, discover that Nature has made that an ample enough charge and by no means a sinecure. Do not go far away: you have plenty to do 'at home!'[3] Men put themselves up for hire. Their talents are not for themselves but for those to whom they have enslaved themselves. They are never 'at home': their tenants are there! That widespread attitude does not please me. We should husband our soul's freedom, never pawning it, save on occasions when it is proper to do so – which, if we judge soundly, are very few.

Just watch people who have been conditioned to let themselves be enraptured and carried away: they do it all the time, in small matters as in great, over things which touch them and those which touch them not at all. They become involved, indiscriminately, wherever there is a task [C] and obligations; [B] they are not alive without bustle and bother. [C] *'In negotiis sunt negotii causa'.* [They are busy so as to *be* busy.][4] The only reason why they seek occupations is to be occupied. It is

2. Ovid, *Tristia*, III, ii, 9.
3. The Socratic injunction, *Aedibus in nostris*: cf. III, 9, note 160.
4. Seneca, *Epist. moral.*, XXII, 8 (adapted).

not a case of wanting to move but of being unable to hold still, just as a rock shaken loose cannot arrest its fall until it lies on the bottom. For a certain type of man, being busy is a mark of competence and dignity. [B] Their minds seek repose in motion, like babes in a cradle. They can say that they are as useful to their friends as they are bothersome to themselves. Nobody gives his money away to others: everyone gives his time. We are never more profligate than with the very things over which avarice would be useful and laudable.

The complexion which I adopt is flat contrary to that. I keep within myself; such things as I do want I usually want mildly. And I want very few. I rarely become involved in anything; if I am busy I am calmly so. What others want or do, they want with all their will, frantically. There are so many awkward passages that the surest way is to glide rather lightly over the surface of this world.   [C]   We should slide over it, not get bogged down in it.   [B]   Pleasure itself is painful in its deeper reaches:

> *incedis per ignes*
> *Suppositos cineri doloso*

[You are walking through fires hidden beneath treacherous ashes.][5]

The Jurors of Bordeaux elected me mayor of their city when I was far from France, and even farther from such a thought. I declined; but I was brought to see that I was wrong, since the King had also interposed his command.[6]

It is an office which should seem all the more splendid for having no salary or reward other than the honour of doing it. It lasts two years, but can be extended by a second election. That very rarely happens. It did in my case; and to two others previously: some years ago to Monsieur de Lanssac and more recently to Monsieur de Biron, Marshal of France, to whose place I succeeded. My own place I yielded to another Marshal of France, Monsieur de Matignon, taking pride in such noble company:

> [C]   *uterque bonus pacis bellique minister.*
>
> [both good officers in peace and war.][7]

[B]   By those particular circumstances which she contributed herself,

---

5. Horace, *Odes*, II, i, 7–8.
6. Henry III enjoined him to return to France (from Della Villa Spa) and to take up the office of Mayor (or Governor) of Bordeaux.
7. Virgil, *Aeneid*, XI, 658.

Fortune decided to play a part in my preferment. Nor was it entirely vain, since Alexander [C] showed contempt for [B] the ambassadors[8] of Corinth who offered him the citizenship of their city, but when they happened to explain that Bacchus and Hercules were also on the roll of honour he accepted it graciously.[9]

As soon as I arrived I spelled out my character faithfully and truly, just as I know myself to be – no memory, no concentration, no experience, no drive; no hatred either, no ambition, no covetousness, no ferocity – so that they should be told, and therefore know, what to expect from my service. And since the only thing which had spurred them to elect me was what they knew of my father and his honoured memory, I very clearly added that I would be most distressed if anything whatsoever were to make such inroads upon my will as the affairs of their city had made on my father's while he was governing it in the very same situation to which I had been summoned. I can remember seeing him when I was a boy: an old man, cruelly troubled by the worries of office, forgetting the gentle atmosphere of his home (to which he had long been confined by the weakness of advancing years) as well as his estates and his health, thinking little of his own life (which he nearly lost, having been involved for them in long and arduous journeys). That was the kind of man he was: and his character arose from great natural goodness. Never was there a soul of man more charitable, more devoted to the people.

Such ways I praise in others: but do not like to follow them myself: not without some justification. He had heard it said that one should forget oneself on behalf of one's neighbour and that, compared to the general, the individual is of no importance.

Most of the world's rules and precepts do adopt such an attitude, driving us outside ourselves and hounding us into the forum in the interests of the public weal. They thought they were doing some fair deed by diverting us and withdrawing us from ourselves, taking it for granted that we were clinging too much to ourselves by a bond which was all too natural. And they left nothing to that purpose unsaid. It is no novelty that clever men should preach not things as they are but things such as might serve them. [C] Truth has its difficulties, its awkwardnesses and its incompatibilities with us. It is often necessary to deceive us so as to stop us from deceiving ourselves, hooding our eyes and dazzling our minds so as to train

8. '88: Alexander *wrinkled his nose at* the ambassadors . . .
9. Erasmus, *Apophthegmata*, IV, *Alexander Magnus*, LXV, mentioning Hercules but not Bacchus.

them and cure them. '*Imperiti enim judicant, et qui frequenter in hoc ipsum fallendi sunt, ne errent.*' [Those who judge are inexperienced: they must needs be deceived precisely to stop them from going wrong.][10]
[B]   When they tell us to prefer to ourselves three, four or fifty categories of objects, they are imitating the art of the bowman, who, so as to hit his target, raises his sights way above it. To straighten a piece of bent wood we bend it right over backwards.

I reckon that in the temple of Pallas (as can be seen to be the case in all other religions) there were open secrets, to be revealed to the people, and other hidden   [C]   higher   [B]   ones,[11] to be revealed only to initiates. It is likely that the true degree of love which each man owes to himself is found among the latter: not a   [C]   false   [B]   love   [C]   which makes us embrace glory, knowledge, riches and such-like with an immoderate primary passion, as though they were members of our being, nor a love   [B]   which is easy-going and random, acting like ivy which cracks and destroys the wall which it clings to, but a healthy, measured love, as useful as it is pleasant. Whoever knows its duties and practises them is truly in the treasure-house of the Muses: he has reached the pinnacle of human happiness and of man's joy. Such a man, knowing precisely what is due to himself, finds that his role includes frequenting men and the world; to do this he must contribute to society the offices and duties which concern him.   [C]   He who does not live a little for others hardly lives at all for himself: '*Qui sibi amicus est, scito hunc amicum omnibus esse.*' [Know that a man who feels loving-friendship for himself does so for all men.][12]   [B]   The chief charge laid upon each one of us is his own conduct: [C]   that is why we are here.   [B]   For example, any man who forgot to live a good and holy life himself, but who thought that he had fulfilled his duties by guiding and training others to do so, would be stupid: in exactly the same way, any man who gives up a sane and happy life in order to provide one for others makes (in my opinion) a bad and unnatural decision.

I have no wish that anyone should refuse to his tasks, when the need arises, his attention, his deeds, his words, or his sweat and blood:

> *non ipse pro charis amicis*
> *Aut patria timidus perire.*

10. Quintilian, II, xvii, 28.
11. '88: hidden, *more noble* ones . . .
12. Seneca, *Epist. moral.*, VI, 7 (adapted).
   Then, '88: The chief *and most legitimate* charge . . .

[personally I am not afraid of dying for those whom I love dearly or for my country.][13]

But it will be in the form of an incidental loan, his mind meanwhile remaining quiet and sane – not without activity but without distress, without passion. Straightforward action costs him so little that he can do it in his sleep. But it must be set in motion with discernment, for whereas the body accepts whatsoever is loaded upon it according to its real weight the mind expands it and makes it heavier, often to its own cost, giving it whatever dimensions it thinks fit. With different efforts and different straining of our wills we achieve similar things. One thing does not imply the other: for how many soldiers put themselves at risk every day in wars which they care little about, rushing into danger in battles the loss of which will not make them lose a night's sleep: meanwhile another man in his own home and far from that danger (which he would never have dared to face) is more passionate about the outcome of the war, and has his soul in greater travail over it, than the soldier who is shedding his life-blood there. I have been able to engage in public duties without going even a nail's breadth from myself, [C] and to give myself to others without taking myself away from *me*.

[B] Such a rough and violent desire is more of a hindrance than a help in carrying out our projects; it fills us with exasperation in the face of results which are slow to come or which turn against us, and with bitterness and suspicion towards those with whom we are negotiating. We can never control well any business which obsesses and controls us:

> [C] *male cuncta ministrat*
> *Impetus.*

[violent impulses serve everything badly.][14]

[B] Anyone who brings only his judgement and talents to the task sets about it more joyfully; totally at his ease, he feints, parries or plays for time as need arises; he can fail to strike home without torment or affliction, ready and intact for a fresh encounter; when he walks he always retains the bridle in his hands. In a man who is bemused by violent and tyrannical strain there can, of necessity, be seen a great deal of unwisdom and injudiciousness. The impetus of his desire carries him along: such a motion is rash and (unless Fortune contributes much) is hardly fruitful. When we

13. Horace, *Odes*, IV, ix, 51–2.
14. Statius, *Thebaid*, X, 704 (read in Justus Lipsius).

punish any injuries which we have received, philosophy wants us to avoid choler, not so as to diminish our revenge but (on the contrary) so that its blows may be weightier and better aimed: philosophy considers violent emotion to be an impediment to that.[15] [C] Choler does not simply confuse: of itself it tires the arms of those who inflict chastisement; its flames confound and exhaust their strength. [B] When you are in a dashing hurry, *'festinatio tarda est'* [haste causes delay].[16] Haste trips over its own feet, tangles itself up and comes to a halt. [C] *'Ipsa se velocitas implicat.'* [The very haste ties you in knots.] [B] For example, from what I can see to be usually the case, covetousness knows no greater hindrance than itself: the more tense and vigorous it is, the less productive it is. It commonly snaps up riches more quickly when masked by some semblance of generosity.

A gentleman, an excellent fellow and one of my friends, nearly drove himself out of his mind by too much strain and passionate concern for the affairs of a prince, his master: yet that self-same master[17] described himself to me as one who can see the weight of a setback as well as anyone else but who resolves to put up with it whenever there is no remedy; in other cases he orders all the necessary measures to be taken (which he can do promptly because of his quick intelligence) then quietly waits for the outcome. And indeed I have seen him doing it, remaining very cool in his actions and relaxed in his expression throughout some important and ticklish engagements. I find him greater and more able in ill fortune than in good; [C] his defeats are more glorious to him than his victories: his mortifications more glorious than his triumphs.

[B] Consider how even in vain and trivial pursuits such as chess or tennis matches, the keen and burning involvement of a rash desire at once throws your mind into a lack of discernment and your limbs into confusion: you daze yourself and tangle yourself up. A man who reacts with greater moderation towards winning or losing is always 'at home': the less he goads himself on, and the less passionate he is about the game, the more surely and successfully he plays it.

Moreover we impede our soul's grip and her grasp by giving her too much to embrace. Some things should be merely shown to her; some

---

15. The standard *exemplum* is that of Plato, who said to Xenophon, 'Beat this boy, for I myself am angry.' (Erasmus, *Apophthegmata*, VII, *Plato Atheniensis*, VII.) It was also a Stoic commonplace that the Sage avoids anger.
16. Quintus Curtius, IX, ix, 12; then, Seneca, *Epist. moral.*, XLIV, 7.
17. Doubtless Henry of Navarre (Henry IV).

affixed to her and others incorporated into her. The soul can see and know all things, but she should feed only on herself; she should be taught what properly concerns her, what goods and substances are properly hers. The Laws of Nature teach us what our just needs are. The wise first tell us that no man is poor by Nature's standards and that, by opinion's standards, every man is; they then finely distinguish between desires coming from Nature and those coming from the unruliness of our thoughts: those whose limits we can see are hers; those which flee before us and whose end we can never reach are our own. To cure poverty of possessions is easy: poverty of soul, impossible.[18]

> [C] *Nam si, quod satis est homini, id satis esse potesset,*
> *Hoc sat erat: nunc, cum hoc non est, qui credimus porro*
> *Divitias ullas animum mi explere potesse?*

[This would be enough, if enough could really be enough for any man. Since it never is, why should we believe that any wealth can glut my mind?]

When Socrates saw a great quantity of wealth (valuable jewels and ornaments) being borne in procession through the city, he exclaimed: 'How many things there which I do not want!' [B] Metrodorus lived on twelve ounces a day; Epicurus on less; Metrocles slept among his sheep in the winter and, in summer, in the temple porticos; [C] *'Sufficit ad id natura, quod poscit.'* [What nature demands, she supplies.] Cleanthes lived by his hands and boasted that 'Cleanthes, if he so wished, could support another Cleanthes.'[19]

[B] If what Nature precisely and basically requires for the preservation of our being is too little (and how little it is and how cheaply life can be sustained cannot be better expressed than by the following consideration: that it is so little that it escapes the grasp and blows of Fortune) then let us allow ourselves to take a little more: let us still call 'nature' the habits and endowments of each one of us; let us appraise ourselves and treat ourselves by that measure:[20] let us stretch our appurtenances and our calculations as

---

18. Borrowings from Seneca, *Epist. moral.*, XVI; then, Lucilius as cited by Nonius Marcellus, *De proprietate sermonis*, V. (This work was published in Paris in 1583.)

19. Erasmus, *Apophthegmata*, III, *Socratica*, XXVIII and VII; *Cleanthes Assius*, II. For Epicurus and his follower Metrodorus cf. Seneca, *Epist. moral.*, XVIII, 9. The quotation is from Seneca, XC, 19 (praising the simple life before the advent of luxury and civilization, the general theme of these pages).

20. Cf. Erasmus, *Adages*, VII, LXXXVIII, *Tuo te pede metire* (Measure yourself by your own yardstick), associated by Erasmus with *Nosce teipsum*, etc., as the conduct of the wise man.

far as that. For as far as that, it does seem that we have a good excuse: custom is a second nature and no less powerful;[21]   [C]   if I lack anything which I have become used to, I hold that I truly lack it.   [B]   I would just as soon (almost) that you took my life than have you restrict it or lop it much below the state in which I have lived it for so long. I am not suited any more to great changes nor to throwing myself into some new and unaccustomed way of life – not even a richer one. It is no longer the time to become different. And if some great stroke of luck should fall into my hands now, how sorry I would be that it did not come when I could have enjoyed it.

> *Quo mihi fortuna, si non conceditur uti?*
>
> [What is a fortune to me if I am not able to use it?][22]

[C]   I would similarly regret any new inward attainment. It is almost better never to become a good man at all than to do so tardily, understanding how to live when you have no life ahead. I am on the way out: I would readily leave to one who comes later whatever wisdom I am learning about dealing with the world. I do not want even a good thing when it is too late to use it. Mustard after dinner! What use is knowledge to a man with no brain left? It is an insult and disfavour of Fortune to offer us presents which fill us with just indignation because they were lacking to us in due season. Take me no farther; I can go on no more. Of all the qualities which sufficiency possesses, endurance alone suffices. Try giving the capabilities of an outstanding treble to a chorister whose lungs are diseased, or   [B]   eloquence to a hermit banished to the deserts of Arabia! No art is required to decline.   [C]   At the finish of every task the ending makes itself known. My world is over: my mould has been emptied; I belong entirely to the past; I am bound to acknowledge that and to conform my exit to it. This I will say ['95] to explain what I

---

21. Erasmus, *Adages*, IV, IX, XXV, *Usus est altera natura*.
   '88: less powerful: *and to my humour* I would just as soon . . .
22. Horace, *Epistles*, I, v, 12.
   '88: uti. *Similarly I do not reform myself in wisdom by frequenting and dealing with the world without regretting that the amendment came to me so late that I no longer have time to enjoy it: from henceforth I need no other talent than that of endurance before death and old age. What is the use of a new art of living in such a decline and of a new assiduity to guide me along that road along which I have only a few steps to take? Go and teach* eloquence to a *man banished to the deserts of Arabia. No art is required to decline. Here I am in short . . .*

mean: [C] the recent suppression of ten days by the Pope has brought me so low that I really cannot wear it:[23] I belong to those years when we computed otherwise: so ancient and long-established a custom claims me and summons me back to it. Since I cannot stand novelty even when corrective, I am constrained to be a bit of a heretic in this case. I grit my teeth, but my mind is always ten days ahead or ten days behind; it keeps muttering in my ears: 'That adjustment concerns those not yet born.'

Although health – oh so sweet! – comes and finds me spasmodically, it is so as to bring me nostalgia, not right of possession. I no longer have anywhere to put it. Time is quitting me: without time there is no right of possession. What little value would I attribute to those great elective offices-of-state which are bestowed only on those who are on the way out! No one is concerned there with whether you will perform them properly but how short a time you have to fill them. From the moment of your entry they are thinking of your exit.

[B] Here, I am in short putting the finishing touches to a particular man, not making another one instead. By long accustoming this form of mine has passed into substance and my fortune into nature. So I maintain that each wretched one of us may be pardoned for reckoning as his whatever is comprised within the measure of custom, and also that, beyond those limits, there is nought but confusion. It is the widest extent that we can allow to our rights: the more we increase our needs and possessions the more we expose ourselves to adversities and to the blows of Fortune. The course run by our desires must be circumscribed and restricted to the narrow limits of the most accessible and contiguous pleasures. Moreover their course should be set not in a straight line terminating somewhere else but in a circle both the start and finish of which remain and terminate within ourselves after a short gallop round: any action carried through without such a return on itself – and I mean a quick and genuine one – is [C] wayward[24] [B] and diseased: such are those of covetous and ambitious men and of so many others who dash towards a goal, careering ever on and on.

Most of our occupations are farcical: *'Mundus universus exercet histrionem.'* [Everybody in the entire world is acting a part.][25] We should play our role

23. The reformed Gregorian calendar (jumping in fact eleven days) introduced in France in 1582.
24. '88: is *vain* and . . .
25. Petronius (fragment) cited after Justus Lipsius' *De constantia*.

properly, but as the role of a character which we have adopted. We must not turn masks and semblances into essential realities, nor adopted qualities into attributes of our self. We cannot tell our skin from our shimmy! [C] It is enough to plaster flour on our faces without doing it to our minds. [B] I know some who transubstantiate and metamorphose themselves into as many new beings and forms as the dignities which they assume: they are prelates down to their guts and livers and uphold their offices on their lavatory-seat. I cannot make them see the difference between hats doffed to them and those doffed to their commissions, their retinue or their mule.[26] *'Tantum se fortunae permittunt, etiam ut naturam dediscant.'* [They allow so much to their Fortune that they unlearn their own natures.][27] They puff up their souls and inflate their natural speech to the height of the magistrate's bench.

The Mayor and Montaigne have always been twain, very clearly distinguished. Just because you are a lawyer or a financier you must not ignore the trickery there is in such vocations: a man of honour is not accountable for the crimes or stupidities of his profession, nor should they make him refuse to practise it; such is the custom of his country: and he gets something from it. We must make our living from the world and use it as it is. Yet even an Emperor's judgement should be above his imperial sway, seeing it and thinking of it as an extraneous accessory. He should know how to enjoy himself independently of it, talking (at least to himself) as Tom, Dick or Harry.

I cannot get so deeply and totally involved. When my convictions make me devoted to one faction, it is not with so violent a bond that my understanding becomes infected by it. During the present confusion in this State of ours my own interest has not made me fail to recognize laudable qualities in our adversaries nor reprehensible ones among those whom I follow. [C] People worship everything on their own side: for most of what I see on mine I do not even make excuses. A good book does not lose its beauty because it argues against my cause. [B] Apart from the kernel of the controversy, I have remained balanced and utterly indifferent: [C] *'Neque extra necessitates belli praecipuum odium gero.'* [And I act with no special hatred beyond what war requires.][28] [B] I congratulate

---

26. The mule was the animal usually ridden on formal occasions by the higher clergy.
27. Quintus Curtius, IV, xxv.
28. Pliny, XXVII, 22 (adapted). Then, Cicero, *Tusc. disput.*, IV, xxv, 55, speaking of the irrational man.

myself for that: it is usual to fall into the opposite extreme: [C] *'Utatur motu animi qui uti ratione non potest'* [If he cannot be reasonable, let him indulge his emotions!]

[B] Those who extend their anger and hatred beyond their concerns (as most men do) betray that their emotion arises from something else, from some private cause, just as when a man is cured of his ulcer but still has a fever that shows that it arises from some other and more secret origin. [C] The fact is that they feel no anger at all for the general cause in so far as it inflicts wounds on the interests of all men and on the State: they resent it simply because it bruises their private interest. That is why they goad themselves into a private passion which goes beyond public justice and reason: *'Nam tam omnia universi quam ea quae ad quomque pertinent singuli carpebant.'* [They did not carp about the terms as a whole but about how they affected them as individuals.][29]

[B] I want us to win, but I am not driven mad if we do not. [C] I am firmly attached to the sanest of the parties, but I do not desire to be particularly known as an enemy of the others beyond what is generally reasonable. I absolutely condemn such defective arguments as, 'He belongs to the League because he admires the grace of Monsieur de Guise'; 'He is a Huguenot: the activity of the King of Navarre sends him into ecstasies'; 'He finds such-and-such lacking in the manners of the King: at heart he is a traitor.' I did not concede to the magistrate himself that he was right to condemn a book for having named a heretic among the best poets of the age.[30] Should we be afraid to say that a thief has nice shins! ['95] Must a whore smell horrid? [C] In wiser ages did they revoke Marcus Manlius' proud title *Capitolinus*, awarded him earlier as saviour of the liberty and religion of the State? Did they smother the memory of his generosity, of his feats of arms and the military honours awarded for his valour, because he subsequently hankered after kingship, to the prejudice of the laws of his land?[31]

Some start hating a barrister: by next morning they are saying that he is a poor speaker! (I have touched elsewhere on how zeal has driven decent

---

29. Livy, XXXIV, xxxvi.
30. Montaigne was criticized by Sisto Fabri in the Vatican for placing Theodore Beza, the successor to Calvin, among the best contemporary Latin poets. He stood by his opinion (cf. II, 17, 'On presumption').
31. In the Gallic War he saved the Capitol but, suspected of monarchical ambitions, was thrown from the Tarpeian Rock. Livy, V, xlvii; VI, xi and Cicero, *De Republica*, II, xxvii, 49.

men to similar errors. For my part I can easily say, 'He does this wickedly, that virtuously.') Similarly, when the outlook or the outcome of an event is unfavourable, they want each man to be blind and insensible towards his own party, and that our judgement and conviction should serve not the truth but to project our desires. I would rather err to the other extreme, for I fear that my desires may seduce me. Added to which I have a rather delicate mistrust of anything I desire. I have seen in my time amazing examples of the indiscriminate and prodigious facility which peoples have for letting their beliefs be led and their hopes be manipulated towards what has pleased and served their leaders, despite dozens of mistakes piled one upon another and despite illusions and deceptions. I am no longer struck with wonder at those who were led by the nose by the apish miracles of Apollonius and Mahomet:[32] their thoughts and their minds had been stifled by their emotions. Their power of discernment could no longer admit anything save that which smiled upon them and favoured their cause.

I thought this had attained its highest degree in the first of our feverish factions: that other one, born subsequently, imitated it and surpasses it.[33] From which I conclude that it is a quality inseparable from mass aberrations: all opinions tumble out after the first one, whipped along like waves in the wind. You do not belong if you can change your mind, if you do not bob along with all the rest. Yet we certainly do wrong to just parties when we would support them by trickery. I have always opposed that. It only works for sick minds: for sane ones there are surer ways (not merely more honourable ones) of sustaining courage and explaining setbacks.

[B] The heavens have never seen strife as grievous as that between Caesar and Pompey, and never will again. Yet I believe I can detect in both their fair, noble souls a great moderation towards each other. Their rivalry over honour and command did not sweep them into frenzied and indiscriminate hatred. Even in their harshest deeds I can discover some remnants of respect and good-will, which leads me to conclude that, had it been possible, each of them would have wished to achieve his ends without the downfall of his fellow rather than with it.

32. Apollonius of Tyana claimed to have risen from the dead and to have performed miracles. He, and Mahomet, were by many thought of as would-be rivals and imitators of Christ. Montaigne's term *singeries* (monkey-tricks) implies miracles worked by the Devil, the Ape of God.
33. First, the war-party of the Reformed Church; then their confederate Roman Catholic opponents in *La Ligue*.

Between Marius and Sylla how different things were! Take warning.[34] We should not dash so madly after our emotions and selfish interests. When I was young I resisted the advances of love as soon as I realized that it was getting too much hold over me; I took care that it was not so delightful to me that it finally took me by storm and held me captive entirely at its mercy: on all the other occasions upon which my will seizes too avidly I do the same: I lean in the opposite direction when I see it leaping in and wallowing in its own wine; I avoid so far fuelling the advance of its pleasure that I cannot retake it without loss and bloodshed.

There are souls which, through insensitivity, see only half of anything; they enjoy the good fortune of being less bruised by harmful events. That is a leprosy of the mind which has some appearance of sanity and of such a sanity as philosophy does not entirely despise; for all that, it is not reasonable to call it wisdom, as we often do. There was a man in antiquity who for just such an affectation mocked Diogenes who, to assay his powers of endurance, went out stark naked and threw his arms round a snowman. He came across him in that attitude. 'Feeling very cold just now?' he asked. 'Absolutely not,' replied Diogenes. 'In that case,' continued the other, 'what is there hard and exemplary, do you think, about hanging on out there?'[35]

To measure steadfastness we must know what is suffered. But let those souls which have to experience the adversities and injuries of Fortune in all their depth and harshness and which have to weigh them at their natural weight and taste them according to their natural bitterness employ their arts to avoid being involved in what causes them and to deflect their approaches. What was it that King Cotys did? He paid handsomely when some beautiful and ornate tableware was offered to him, but since it was unusually fragile he immediately smashed the lot, ridding himself in time of an easy occasion for anger against his servants.[36]

[C] I have likewise deliberately avoided confusion of interests; I have not sought properties adjoining those of close relatives or belonging to folk to whom I should be linked by close affection; from thence arise estrangements and dissension.

---

34. The murderous and atrocious civil wars between Marius and Sylla are recapitulated with horror and burning indignation by Lucan in the *Pharsalia*, II, 42–233. Montaigne saw close parallels with the French Civil Wars of Religion.

35. Plutarch (tr. Amyot), *Dicts notables des Lacedaemoniens*, 223 F (but Plutarch says it was a bronze statue).

36. Plutarch (tr. Amyot), *Dicts notables des anciens Roys*, 189 D – E. (Cotys realized he was prone to fits of anger.)

[B]   I used to like games of chance with cards and dice. I rid myself of them long ago – for one reason only: whenever I lost, no matter what a good face I put on, I still felt a stab of pain. A man of honour, who must take it deeply to heart if he is insulted or given the lie   [C]   and not be one to accept some nonsense to pay and console him for his loss,   [B]   should avoid letting controversies grow as well as stubborn quarrels. I avoid like the plague morose men of gloomy complexions, and I do not engage in any discussions which I cannot treat without self-interest or emotion, unless compelled to do so by duty: [C]   *'Melius non incipient, quam desinent.'* [Better that they should never begin than to leave off.][37]   [B]   The safest way is to be prepared before the event. I am well aware that there have been sages who have adopted a different course: they were not afraid to sink their hooks deep, engaging themselves in several objectives. Those fellows are sure of their fortitude, beneath which they can shelter against all kinds of hostile events, wrestling against evils by the power of their endurance:

> *velut rupes vastum quæ prodit in æquor,*
> *Obvia ventorum furiis, expostaque ponto,*
> *Vim cunctam atque minas perfert cælique marisque,*
> *Ipsa immota manens.*

[as a cliff, jutting out into the vast expanse of ocean, exposed to furious winds and confronting the waves, braves the menaces of sea and sky and itself remains unmoved.][38]

Let us not attempt to follow such examples: we shall never manage it. Such men have made up their minds to watch resolutely and unmoved the destruction of their country, which once held and governed all their affection.[39] For common souls like ours there is too much strain, too much savagery in that. Cato gave up for his country the most noble life there ever was; little men like us should flee farther from the storm; we should see that there are no pains to feel, no pains to endure, dodging blows not parrying them.   [C]   When Zeno saw Chremonides, a young man whom he loved, coming to sit beside him, he jumped up. Cleanthes asked why. 'I understand,' he replied, 'that when any part of the body starts to swell the

37. Seneca, *Epist. moral.*, LXXII, 11.
38. Virgil, *Aeneid*, X, 693–6.
39. For example, when Rome fell St Augustine remarked that all is transitory and vanity succeeds to vanity.

doctors chiefly prescribe rest and forbid emotion.'[40]   [B]   Socrates never says, 'Do not surrender to the attraction of beauty; resist it; struggle against it.' He says, 'Flee it; run from its sight and from any encounter with it, as from a potent poison which can dart and strike you from afar.'   [C]   And that good disciple of his, describing either fictionally or historically (though in my opinion more historically than fictionally) the rare perfections of Cyrus the Great, shows him distrusting his ability to resist the attractions of the heavenly beauty of his captive the illustrious Panthea: it was to a man who was less at liberty than he was that he gave the tasks of visiting her and guarding her.[41]   [B]   And the Holy Ghost likewise says, '*Ne nos inducas in tentationem* ' [Lead us not into temptation.][42] We pray, not that our reason may not be assailed and overcome by worldly desires, but that it may not even be assayed by them, that we be not led into a position where we have even merely to withstand the approaches, blandishments and temptations of sin, and we beseech our Lord to keep our consciences quiet, wholly and completely delivered from commerce with evil.[43]

[C]   Those who say that they have got the better of their vindictive feelings or of some other species of blameworthy passion often speak truly of things as they are but not as they were. They are talking to us now that the causes behind their error have been advanced and promoted by themselves. But push farther back; summon those causes back to their first principles: there you will catch them napping. Do they expect their faults to be trivial just because they are older, and that the outcome of an unjust beginning should be just?

[B]   Whoever would wish his country well (as I do) without getting ulcers about it or wasting away will, when he sees it threatening either to collapse in ruin or to continue in a no-less-ruinous state, be unhappy about it but not knocked senseless. O wretched ship of State, 'hauled in different direction by the waves, the winds and the man at the wheel':

---

40. Diogenes Laertius, *Life of Zeno*, with a priapic Latin pun on *tumor* (swelling, erection); then, Socrates in Xenophon, *Memorabilia*, I, iii, 13. The 'poison' of beauty is that of a scorpion; but it can reach one not only through a kiss but when beauty is seen from afar.
41. Erasmus, *Apophthegmata*, V, *Cyrus Major*, III, citing Socrates' disciple Xenophon (*Cyropaedia*, V, i, 17; VI, i, 31. Panthea, the wife of Abradatas, was the most beautiful woman of Asia, IV, vi, 11).
42. Once again Montaigne cites the Bible as verbally inspired by the Holy Ghost (here, particularly, in the Lord's Prayer given in Matthew 6:13).
43. Echoing the final clause of the Lord's Prayer, *Libera nos a malo* (Deliver us from evil).

in tam diversa magister,
Ventus et unda trahunt.[44]

Whoever does not gape after the favours of princes as something he cannot live without is not greatly stung by the coldness of their reception nor the fickleness of their wills. A man who does not brood over his children or his honours with a [C] slavish [B] propensity[45] does not cease to live comfortably after he has lost them. Whoever acts well mainly for his own satisfaction is not much put out when he sees men judging his deeds contrary to his merit. A quarter of an ounce of endurance can provide for such discomforts. I find that the remedy which works for me is, from the outset, to purchase my freedom at the cheapest price I can get; I know that I have by this means escaped much travail and hardship. With very little effort I stop the first movement of my emotions, giving up whatever begins to weigh on me before it bears me off. [C] If you do not stop the start, you will never stop the race. If you cannot slam the door against your emotions you will never chase them out once they have got in. If you cannot struggle through the beginning, you will never get through the end; nor will you withstand the building's fall, if you cannot stand its being shaken. *'Etenim ipsæ se impellunt ubi semel a ratione discessum est; ipsaque sibi imbecillitas indulget, in altumque provehitur imprudens, nec reperit locum consistendi.'* [Once they have departed from reason the emotions drive themselves on; their very weakness indulges itself, venturing imprudently on to the deep and finding no place in which it can heave to.][46]

[B] I can feel in time the tiny breezes which come fondling me and rustling within me, as forerunners of gales: [C] *'Animus, multo antequam opprimatur, quatitur.'* [The mind is lashed well before it is engulfed.]

[B]   *Ceu flamina prima*
Cum deprensa fremunt sylvis, et cæca volutant
Murmura, venturos nautis prodentia ventos.

[Thus when the light breeze is pent up in the woodlands, it swirls about and makes a sullen roar, warning seamen that a storm is nigh.][47]

---

44. Translated in the text by Montaigne. Apparently, verses from Buchanan's *Franciscanus,* incorrectly cited from memory.
45. '88: a *tyrannical* propensity . . .
46. Cicero, *Tusc. disput.,* IV, xviii, 42. (Montaigne's general context owes much here to Seneca.) The next quotation is attributed to Seneca by Marie de Gournay and is indeed from *Epist. moral.,* LXXIV, 33.
47. Virgil, *Aeneid,* X, 97–9.

How frequently have I done myself an evident injustice so as to avoid the risk of receiving a worse one from the judges after years of agony and of vile and base machinations which are more hostile to my nature than the rack or pyre. [C] *'Convenit a litibus quantum licet, et nescio an paulo plus etiam quam licet, abhorrentem esse. Est enim non modo liberale, paululum nonnunquam de suo jure decedere, sed interdum etiam fructuosum.'* [It is seemly to avoid lawsuits as far as you should, and even a little bit further. It is not only gentlemanly to waive one's rights a little: it is sometimes also profitable.][48] If we were truly wise we should delight in it and boast about it, like the innocent son of a great house whom I heard happily welcoming each guest with, 'Mother has just lost her case!' as though her case were a cough or a fever or some other thing which it is grievous to have. Even such advantages as Fortune has favoured me with – namely kinships and ties with men who have supreme authority over matters of that kind – I have consciously striven hard to avoid exploiting to the detriment of anyone else or to inflate my rights beyond their rightful worth. In short [B] I am happy to say that I have spent all my days virgin of lawsuits (even though they have not failed frequently to offer themselves to my service on many a just pretext if only I would listen) and virgin of actions against me. So I shall soon have spent a long life without serious harm given or received, and without being called anything worse than my name: a rare gift of Heaven.

Our greatest commotions arise from laughable principles and causes. What ruin befell our last Duke of Burgundy because of an action against him for a cartload of sheep-skins. And was not the engraving on a seal the original and main cause of the most horrifying disaster that the fabric of this world has ever suffered?[49] (For Pompey and Caesar are only side-shoots, consequent upon the first two rivals.) And in my own day I have seen the wisest heads in this Kingdom assembled with great ceremony and at great public expense to make treaties and agreements, while the details of them depended on sovereign chatter in the ladies' drawing-room and on the inclination of some slip of a woman. [C] The poets understood that

48. Cicero, *De officiis*, II, xviii, 64.
49. Cf. the cause of the Picrocholine War in the *Gargantua* of Rabelais: a brawl over buns. Commines gives the cause of the Duke's war; in his *Life of Marius* Plutarch gives as the first and enduring cause of the great Roman civil strife Marius' resentment over the triumphant engraving on the ring which Sylla had made to celebrate the capture of Jugurtha.

rightly enough when they put all Greece and Asia to fire and bloody strife
for the sake of an apple.[50]

[B]   Think why that man over there takes his sword and dagger and
risks his life and honour; let him tell you the source of the quarrel: the
occasion was so trivial that he cannot tell you of it without blushing. When
it is starting to ferment, all you need is a little wisdom. Once you have
embarked, all the hawsers pull tight: then, great precautions are needed,
much more difficult and important ones.

[C]   How much easier it is never to get in than to get yourself
out!   [B]   We should act contrary to the reed which, when it first
appears, throws up a long straight stem but afterwards, as though it were
exhausted and had lost its wind, makes several dense nodules, as so many
respites which indicate that it no longer has its original vigour and drive.[51]
We must rather begin gently and coolly, saving our breath for the
encounter and our vigorous thrusts for finishing the job off. In their
beginnings it is we who guide affairs and hold them in our power; but
once they are set in motion, it is they which guide us and sweep us along
and we who have to follow.

[C]   Yet that does not mean that this stratagem of mine has relieved me
of all difficulties or that I have not often found it very hard to master or
bridle my emotions. They cannot always be restrained to the measure of
their causes, and even their beginnings can be harsh and aggressive.
Nevertheless there are fair savings to be derived from it, and some fruits
too except by those whom no fruit can satisfy when no honour is to be
had. For in truth such an action can only be valued by each man himself.
You yourself are happier but you are not more esteemed, since you
reformed yourself before you took to the floor, before the matter could be
seen. However there is this as well: not merely in this case but in all other
of life's duties, the way of those who aim at honour is different indeed
from that followed by those whose objective is the ordinate and reasonable.

[B]   I find that some dash thoughtlessly and furiously into the lists only
to slow down during the charge. Plutarch says that those who suffer from
excessive diffidence readily and easily agree to anything but also readily
break their word and go back on what they have said; so, similarly, anyone
who enters lightly upon a quarrel is liable to be equally light in getting out

---

50. Allusion to the judgement of Paris, who awarded the golden apple for her
beauty to Venus (who promised him Helen), thus arousing the wrath of Juno and
Minerva. By carrying off Helen to Troy he brought about the Trojan War.
51. Comparison inspired by Plutarch, (tr. Amyot), *Comment on peult appercevoir si
l'on profite en l'exercise de la vertu*, 114 B.

of it.[52] The same difficulty which stops me from broaching anything would spur me on once I was heated and excited. What a bad way to do it: once you are in, you must go on or burst! [C] 'Undertake relaxedly,' said Bias, 'but pursue hotly.'[53]

[B] But what is even less tolerable, for want of wisdom we decline into want of bravery.

Today most settlements of our disputes are shameful and lying: we merely seek to save appearances, while betraying and disowning our true thoughts. We plaster over facts; we know how we said it and what we meant by it; the bystanders know it; so do our friends to whom we wished to prove our superiority. We disavow our thoughts at the expense of our frankness and our reputation for courage, seeking bolt-holes in falsehoods so as to reach a conciliation. We give the lie to ourselves in order to get out the fact that we gave the lie to somebody else. You ought not to be considering whether your gesture or words may be given a different meaning: from now on it is your true and honest meaning that you should be seeking to defend, no matter what the cost. At stake are your morality and your honour: those are not qualities for you to protect behind a mask. Let us leave such servile shifts and expediences to the chicanery of the law-courts. Every day I see excuses and reparations made to purge an indiscretion which seem uglier to me than the indiscretion itself. It would be better to offend your adversary afresh than to commit an offence against yourself by making him such a reparation as that. You were moved to anger when you defied him: now that you are cooler and more sensible, you are going to appease him and fawn on him! That way, you retreat further than you ever advanced. I reckon that nothing which a gentleman says can seem worse than the shame of his unsaying it under duress from authority: stubbornness in a gentleman is more pardonable than pusillanimity.

For me passions are as easy to avoid as hard to moderate: [C] *'Abscinduntur facilius animo quam temperantur.'* [They are more easily cut out from the mind than tempered.][54]

[B] If a man cannot attain to that noble Stoic impassibility, let him hide in the lap of this peasant insensitivity of mine. What Stoics did from virtue I teach myself to do from temperament. Storms lodge in the middle

---

52. Plutarch (tr. Amyot), *De la mauvaise honte*, 79 A–C (after warning that a passion for honour frequently leads to deeds of dishonour).

53. Diogenes Laertius, *Life of Bias*, I, lxxxvii (not listed by Erasmus (*Apophthegmata*, VII, *Bias Prienaeus*). Erasmus asserts there that most of the sayings of the seven sages are fabulous and that many are too trite to be attributed to sages.

54. Attributed by Marie de Gournay to Seneca, but not traced.

regions; philosophers and country bumpkins – the two extremes – meet in peace of mind and happiness.

> *Fœlix qui potuit rerum cognoscere causas,*
> *Atque metus omnes et inexorabile fatum*
> *Subjecit pedibus, strepitumque Acherontis avari.*
> *Fortunatus et ille Deos qui novit agrestes,*
> *Panaque, sylvanumque senem, nymphasque sorores.*

[Blessed the man who can find out causes, who can trample down all fears of inexorable Fate and the howls of the close-fisted Underworld: blessed, too, he who knows the rustic gods, Pan, old Sylvanus and the sister nymphs.][55]

The infancies of all things are feeble and weak. We must keep our eyes open at their beginnings; you cannot find the danger then because it is so small: once it has grown, you cannot find the cure. While chasing ambition I would have had to face, every day, thousands of irritations harder to digest than the difficulty I had in putting a stop to my natural inclination towards it.

> *jure perhorrui*
> *Late conspicuum tollere verticem.*

[I was right to abhor raising my head and attracting attention.][56]

All public deeds are liable to ambiguous and diverse interpretations since so many heads are judging them. Now about this municipal office of mine (and I am delighted to say a word about it, not that it is worth it but to show how I behave in such matters): some say that I bore myself as a man who shows too little passion and whose zeal was too slack. As far as appearances go, they were not all that wrong: I assay keeping my soul and my thoughts in repose: [C] *'Cum semper natura, tum etiam aetate jam quietus'* [Always tranquil by nature, I now am also so by my age];[57] [B] if they turn riotous from some deep and disturbing impression that, in truth, is against my intention. Yet from this natural languor of mine one should not draw evidence of incapacity (since lack of worry and lack of wit are two different things) and even less of ingratitude or of lack of appreciation towards those citizens who went to every available extreme to please me, both before and after they knew me – for they did far more

55. Virgil, *Georgics*, II, 490–4.
56. Horace, *Odes*, III, xvi, 80–1.
57. Cicero, *De petitione consolatus*, II.

for me in re-electing me to office than in electing me in the first place. I wish them all possible good: and indeed, if the occasion had arisen, there is nothing that I would have spared in their service. I bestirred myself as much for them as I do for myself. They are a fine people, good brave fighting-men, able therefore to accept discipline and obedience and to serve a good cause when well led.

People also say that my period of office passed without trace or mark. Good. They accuse me of being dilatory at a time when nearly everyone else was convicted of doing too much. I [C] paw the ground when my will bolts away with me: [B] but that trait[58] is the enemy of persever-ance. Should anyone wish to use me as I am, let him give me tasks which require vigour and frankness, as well as straightforward, brief and hazardous execution. I could do something then. But if it needs to be subtle, toilsome, clever and tortuous, better ask somebody else.

Not all important commissions are difficult. I would have been prepared to work a little harder had that been very necessary: I am capable of doing somewhat more than I do or like to do. To the best of my knowledge I never left undone any action that duty seriously required of me; but I readily overlooked those where ambition mingles with duty and uses it as a pretext: it is those which, more often than not, fill men's eyes and ears and please them; they are satisfied not with realities but appearances. If they do not hear a sound they think you are asleep! My own humours are opposed to noisy ones: I could certainly remain undisturbed while quelling a disturbance, and could punish a riot without losing my temper. Should I need a little choler and fire, then I borrow some to mask me. My manners are unabrasive, more insipid than sharp; I do not bring actions against an official who dozes, provided that those whom he administers can doze quietly with him. That is the way the laws doze.

Personally I favour an obscure mute life which slips by: [C] *'neque submissam et abjectam, neque se efferentem'* [neither submissive and mean nor puffed up].[59] [B] That is how my Fortune wills it: I was born into a family which has flowed on without brilliance or turbulence, one long remembered as being particularly ambitious for probity. Nowadays men are so conditioned to bustle and ostentation that we have lost the feel of goodness, moderation, even-temper, steadfastness and other

---

58. '88: I *have an excitable way of reacting towards that to which my will is drawing me,* but that trait . . .
59. Cicero, *De officiis,* I, xxxiv, 124 – on the right conduct for the good private citizen.

such [C] quiet [B] and unpretentious[60] qualities; rough objects make themselves felt: smooth ones can be handled without sensation. Illness is felt: good health, little or not at all; neither do we feel things which flatter us, compared with those which batter us.

If we postpone something which could be done in the council-chamber until it is done in the market-square, keeping back till noon something which could have been finished the night before, or if we are anxious to do personally something which a colleague could have done just as well, then we are acting for the sake of our own reputation and for private advantage, not for the Good. (That is what some barber-surgeons used to do in ancient Greece, performing their operations on a daïs in view of passers-by so as to enlarge their practices and the number of patients.)[61] They think that good regulations can only be heard when announced with a fanfare.

Ambition is not a vice fit for little fellows or for enterprises such as ours. Alexander was told: 'Your father will leave you wide dominions, peaceful and secure.' But that lad wanted to rival his father's victorious and righteous government.[62] He had no wish to enjoy ruling the entire world undemandingly and peacefully. [C] (Alcibiades in Plato says he prefers to die young as a beautiful, rich, noble and exceedingly learned youth than to stay fixed in those qualities.)[63]

[B] Ambition is doubtless a pardonable malady in a strong and full soul such as Alexander's. But when petty, dwarfish souls start aping them, believing that they can scatter their renown abroad by having judged one matter rightly or for having arranged the changing of the guard at the town gate, then the higher they hope to raise their heads the more they bare their arses. Such petty achievements have no body, no life; they start evaporating on the first man's lips and never get from one street-corner to another. Have the effrontery to talk about them to your son or your man-servant, like that old fellow who had nobody else to listen to his praises or to acknowledge his worth and so boasted to his chambermaid: 'Oh, what a gallant and clever man you have for a master, Perrette!'[64] If the worse

---

60. '88: other such *lack-lustre* and unpretentious . . .

61. Plutarch (tr. Amyot), *Comment on pourra discerner le flatteur d'avec l'ami*, 53 D.

62. Erasmus, *Apophthegmata*, IV, *Alexander Magnus*, I.

63. Actually, it is Socrates who says this of Alcibiades (Plato, *Alcibiades* I, 105 A), where Alexander is mentioned also in this context.

64. Plutarch (tr. Amyot), *Comment on pourra apparcevoir, si l'on profite en l'exercice de la vertu*, 116 D: 'See how unboastful and unarrogant I am, Dionysia!' By changing the servant-girl's name to Perrette Montaigne gives his allusion the tone of a French farce.

comes to worst, talk about it to yourself, like a King's Counsel I know who, having (with extreme exertion and extreme absurdity) disgorged a boatload of legal references, withdrew from the council-chamber to the court piss-house, where he was heard devoutly muttering through his teeth: *'Non nobis, Domine, non nobis, sed nomine tuo da gloriam.'* [Not unto us, O Lord, not unto us, but unto Thy Name be the glory.][65] If you cannot get it from somebody else's purse, get it from your own!

Fame does not play the whore for so base a price. Those rare and exemplary deeds to which fame is due would not tolerate the company of such a countless mob of petty everyday actions. Marble can boast your titles as much as you like for having repaired a stretch of wall or cleaned up some public gutter, but men of sense will not. Renown does not ensue upon anything done well unless difficulty and unusualness are involved. Indeed, according to the Stoics, simple esteem is not due to every action born of virtue: they would not even faintly praise a man for having abstained from some sore-eyed old whore for temperance' sake![66]   [C] Those who already knew of the astonishing qualities of Scipio Africanus rejected the 'glory' which Panaetius gave him for refusing bribes: that glory was not his alone but belonged to his entire age.[67]

[B]   We have pleasures appropriate to our station: let us not usurp those of greatness: ours are more natural and are the more solid and certain for being more humble. Let us reject ambition out of ambition, since we do not do so out of a sense of right and wrong; let us despise that base beggarly hunger for renown and honour which makes us solicit them from all kinds of people by abject means, no matter how vile the price: [C]   *'Quae est ista laus quae possit e macello peti?'* [What kind of praise is it that you can order from the butcher's?][68]   [B]   To be honoured thus is a dishonour.

Let us learn to be no more avid for glory than we deserve. Boasting of every useful or blameless action is for men in whom such things are rare and unusual: they want them to be valued at what it cost them! The more glittering the deed the more I subtract from its moral worth, because of the suspicion aroused in me that it was exposed more for glitter than for

---

65. Psalm 115 (113) 1.
66. Plutarch (tr. Amyot), *De communes conceptions, contre les Stoïques*, 575 D, citing a book about Zeus (now lost) by Chysippus.
67. Cicero, *De officiis*, II, xxii, 76, repeating Cicero's own judgement. (Panaetius of Rhodes was a Stoic philosopher).
68. Cicero, *De finibus*, II, xv, 50.

goodness: goods displayed are already half-way to being sold. The most elegant deeds are those which slip from the doer's hand nonchalantly and without fuss, and which some man of honour later picks out and saves from obscurity, bringing them to light for their own sake. [C] '*Mihi quidem laudabiliora videntur omnia, quae sine venditatione et sine populo teste fiunt*' [Personally I always find more praiseworthy whatever is done without ostentation and without public witnesses] – says the vainest man in the world![69]

[B] I had nothing to do except to preserve things and to keep them going; those are dull and unnoticeable tasks. There is a great deal of splendour in innovation, but that is under a ban nowadays when it is by novelties alone that we are oppressed, against novelties alone that we must defend ourselves. [C] Although it is less in the daylight, refraining from action is often more noble than action: what little I am worth is virtually all on that side. [B] In short, my opportunities while in office accorded with my temperament. I am most grateful to them for it. Is there any man who wants to be ill so as to provide work for his doctor? Ought we not to whip a doctor who hoped for the plague so as to practise his Art? Although that wicked humour is common enough, I have never hoped that trouble and distemper in this city might increase the glory and honour of my mayoralty. I put my shoulder loyally to the wheel to make things smooth and easy.

Even he who would not show me gratitude for the gentle and muted calm which accompanied my administration cannot at least deprive me of that share which does belong to me by title of my good fortune. And I am so made that I would as soon be fortunate as wise, owing my success simply to God's grace rather than to the intervention of my labours. I had proclaimed most eloquently to the whole world my inadequacy for handling such public affairs. And I have something worse than that inadequacy: the fact that I hardly find it displeasing and, given the kind of life that I have sketched out for myself, that I hardly even attempt to cure it.

Now I was not satisfied, either, with my conduct of affairs: but I did achieve – more or less – what I promised myself I would, and I far exceeded what I promised to those whom I was dealing with, since I prefer to promise rather less than I can do and hope to do. I am sure I left no injury or hatred behind me: as for leaving any regret or desire for me, I do at least know that I never much [C] cared [B] for that.[70]

---

69. Cicero's vanity was indeed great. (Montaigne cites his *Tusc. disput.*, II, xxvi, 64.)
70. '88: much *hoped* for that . . .

> *Mene huic confidere monstro,*
> *Mene salis placidi vultum fluctusque quietos*
> *Ignorare?*

[Me! put faith in such a monster! Me! not realize that the sea simply happens to be calm and to look peaceful!][71]

71. Virgil, *Aeneid*, V, 849, 848 (two lines of the *Aeneid* with the words rearranged and adapted).

11. On the lame

[*The human mind is capable of great self-deception. It can find reasons for anything —
even for non-existent phenomena and unreal 'facts'. Experience is no guard against
error: it can be conditioned by prior expectations. That is one of the considerations
which led Montaigne never to discuss alleged miracles and to remain unimpressed by
judicial certainties.*

*For us today Montaigne's scepticism about the reality of the powers of male and
female witches is arresting. (He was not alone in holding such views, though he
remained in the minority.) But he is determined to subordinate his own opinions to the
teachings of the Roman Catholic Church. For him the value of his opinions is that
they are opinions and that they are his: they tell us of his* forma mentis. *But since
men's opinions are never certainties, should we ever burn people on account of them,
unless God directly intervenes to order us to do so?*]

[B]   In France, some two or three years ago now, they shortened the year
by ten days.[1] What changes were supposed to result from that reform! It
was, quite literally, to move both the heavens and the earth at the same
time. Yet nothing has been shoved out of place. My neighbours find that
seed-time and harvest, auspicious times for business, as well as ill-omened
or propitious days, come at precisely the same second to which they have
ever been assigned. The error of our practices was never felt beforehand:
no amendment is felt there now, so much uncertainty is there everywhere,
so gross is our faculty of perception,   [C]   so darkened and so blunt.

[B]   They say that this adjustment could have been made less awkwardly
by following the example of Augustus and omitting the extra day over a
period of several leap-years — it is a source of trouble and confusion
anyway — until we had paid back the missing time (something which we
have not even achieved by this correction: we are still a day or two in
arrears). By this means we could also have provided for the future,
declaring that after a specified number of years had rolled by that extra day
would be banished for ever, with the result that our miscalculation from
then on could not exceed twenty-four hours.

1. A further allusion to the Gregorian reform of the calendar (1582). Cf. III, 10,
note 23. The previous reform was that of the Emperor Augustus.

Years are the only measure we have for time. The world has been using 'years' for many centuries, yet it is a unit which we have never succeeded in standardizing, so that we live in daily uncertainty about the incompatible forms given to it by other nations, and about how they apply them.

And what if (as some say) the heavens as they grow old are contracting downwards towards us, thereby casting our very hours and days into confusion? And what of our months too, since Plutarch says that even in his period the science of the heavens had yet to fix the motions of the moon?[2]

A fine position we are in to keep chronicles of past events!

I was recently letting my mind range wildly (as I often do) over our human reason and what a rambling and roving instrument it is. I realize that if you ask people to account for 'facts', they usually spend more time finding reasons for them than finding out whether they are true. They ignore the *whats* and expatiate on the *whys*. [C] Wiseacres!

To know causes belongs only to Him who governs things, not to us who are patients of such things and who, without penetrating their origin or essences, have complete enjoyment of them in terms of our own nature. Wine is no more delightful to the man who knows its primary qualities. Quite the reverse: by bringing in pretensions to knowledge the body infringes, and the soul encroaches upon, the rights which both of them have to enjoy the things of this world. To define, to know and to allow belong to professors and schoolmasters: to enjoy and to accept belong to inferiors, subordinates and apprentices.

Let us get back to that custom of ours.

[B] They skip over the facts but carefully deduce inferences. They normally begin thus: 'How does this come about?' But does it do so? That is what they ought to be asking. Our reason has capacity enough to provide the stuff for a hundred other worlds, and then to discover their principles and construction! It needs neither matter nor foundation; let it run free: it can build as well upon the void as upon the plenum, upon space as upon matter:

> *dare pondus idonea fumo.*

> [meet to give heaviness even to smoke.][3]

---

2. Plutarch (tr. Amyot), *Demandes des choses Romaines*, 464 B, drawing the same conclusions as Montaigne.
3. Persius, *Satires*, I, 20.

I find that we should be saying virtually all the time, 'It is not at all like that!' I would frequently make that reply but I dare not, since folk bellow that it is a dodge produced by ignorance and by weakness of intellect; so I am usually obliged to be a mountebank for the sake of good company, and to discuss trivial subjects and tales which I totally disbelieve. Moreover it is rather rude and aggressive flatly to deny a statement of fact; and (especially in matters where it is difficult to convince others) few people fail to assert that 'they have seen it themselves' or to cite witnesses whose authority puts a stop to our contradictions. By following this practice we know the bases and causes of hundreds of things which never were; the world is involved in duels about hundreds of questions where both the for and the against are false: [C] *'Ita finitima sunt falsa veris, ut in præcipitem locum non debeat se sapiens committere.'* [The false and the true are in such close proximity that the wise man should not trust himself to so steep a slope.][4]   [B]   Truth and falsehood are both alike in form of face and have identical stances, tastes and demeanours. We look on them with the same eye. I find that we are not merely slack about guarding ourselves from dupery, but we actually want to fall on its sword. We love to be entangled with vanity, since it corresponds in form to our own being.

I have seen in my time the birth of several miracles. Even if they are smothered at birth, that does not stop us from predicting the course they would have taken if they had grown up! We only need to get hold of the end of the thread: we then reel off whatever we want. Yet the distance is greater from nothing to the minutest thing in the world than it is from the minutest thing to the biggest. Now when the first people who drank their fill of the original oddity come to spread their tale abroad, they can tell by the opposition which they arouse what it is that others find difficult to accept; they then stop up the chinks with some false piece of oakum.   [C]   Moreover, *'insita hominibus libidine alendi de industria rumores'* [by man's inborn tendency to work hard at feeding rumours][5] we naturally feel embarrassed if what was lent to us we pass on to others without some exorbitant interest of our own. At first the individual error creates the public one: then, in its turn, the public error creates the individual one.   [B]   And so, as it passes from hand to hand, the whole fabric is padded out and reshaped, so that the most far-off witness is better informed about it than the closest one, and the last to be told more convinced than the first. It is a natural progression. For whoever believes anything reckons

4. Cicero, *Academica*, II (Lucullus), XXI, 68.
5. Livy, XXVIII, xxiv.

that it is a work of charity to convince someone else of it; and to do this he is not at all afraid to add, out of his own invention, whatever his story needs to overcome the resistance and the defects which he thinks there are in the other man's ability to grasp it. I myself am particularly scrupulous about lying and can scarcely be bothered to quote authority for what I say in order to make it believable: yet even I notice that when I get heated about a matter I have in hand, [C] either because of another's resistance to it or else because of the excitement of the actual telling, [B] I increase the importance of my subject and puff it up by tone of voice, gestures, powerful and vigorous words – and also by stretching it a bit and exaggerating it, not without some damage to native truth. But I do so with the proviso that I immediately give up the attempt for the first man who summons me back and demands the truth, bare and bold, which I then give to him without exaggeration, without bombast and without embroidery. [C] A loud and lively gab, such as mine habitually is, soon flies off into hyperbole.

[B] There is nothing over which men usually strain harder than when giving free run to their opinions: should the regular means be lacking, we support them by commands, force, fire and sword. It is wretched to be reduced to the point where the best touchstone of truth has become the multitude of believers, at a time when the fools in the crowd are so much more numerous than the wise: [C] '*quasi vero quidquam sit tam valde quam nil sapere vulgare*' [as though anything whatsoever were more common than lack of wisdom].[6] '*Sanitatis patrocinium est, insanientium turba.*' [A mob of lunatics now form the authority for sane truth.]

[B] It is hard to stiffen your judgement against widely held opinions. At first simple folk are convinced by the event itself: it sweeps over them. From them it spreads to the more intelligent folk by the authority of the number and the antiquity of the testimonies. Personally, what I would not believe when one person says it, I would not believe if a hundred times one said it. And I do not judge opinions by their age.

Not long ago one of our princes, whose excellent natural endowments and lively constitution had been undermined by the gout, allowed himself to be so strongly convinced by the reports which were circulating about the wonderful treatments of a priest who, by means of words and gestures, cured all illnesses, that he made a long journey to go and consult him. By the force of his imagination he convinced his legs for a few hours to feel no

---

6. Cicero, *De divinatione*, II, xxxix, 81; then, St Augustine, *City of God*, VI, x. (Montaigne's context echoes Seneca, *Epist. moral.*, LXXXI, etc.)

pain, so that he made them serve him as they had long since forgotten how to do. If Fortune had allowed some five or six such events to happen one on top of the other, they would have sufficed to give birth to a miracle. Afterwards, there was found such simplemindedness and such little artifice in the inventor of this treatment that he was not judged worthy of any punishment. We would do the same for most such things if we examined them back in their burrows. [C] *'Miramur ex intervallo fallentia.'* [We are astounded by things which deceive us by their remoteness.][7] [B] Thus does our sight often produce strange visions in the distance which vanish as we draw near. *'Nunquam ad liquidum fama perducitur.'* [Rumour never stops at what is crystal-clear.]

It is wonderful how such celebrated opinions are born of such vain beginnings and trivial causes. It is precisely that which makes it hard to inquire into them: for while we are looking for powerful causes and weighty ends worthy of such great fame we lose the real ones: they are so tiny that they escape our view. And indeed for such investigations we need a very wise, diligent and subtle investigator, who is neither partial nor prejudiced.

To this hour all such miracles and strange happenings hide away when I am about. I have not seen anywhere in the world a prodigy more expressly miraculous than I am. Time and custom condition us to anything strange: nevertheless, the more I haunt myself and know myself the more my misshapenness amazes me and the less I understand myself.

The right to promulgate and to publish such phenomena is mainly reserved to Fortune. The day before yesterday I was on my way through a village two leagues from home when I found the market-place still hot and excited about a miracle which had just come to grief there. All the neighbourhood had been preoccupied with it for months; the excitement had spread to the neighbouring provinces and great troops of people of all classes came pouring in. One night a local youth had larked about at home, imitating the voice of a ghost; he intended no trickery beyond enjoying the immediate play-acting. He succeeded somewhat beyond his hopes, so, to heighten the farce and thicken the plot, he brought in a village maiden who was absolutely stupid and simple. Eventually there were three of them, all of the same age, all equally stupid. From sermons in people's homes they progressed to sermons in public, hiding under the altar in church, delivering them only at night and forbidding any lights to be brought in. They started with talk directed towards the conversion of the

7. Seneca, *Epist. moral.*, CVIII, 7; then, Quintus Curtius, IX, ii.

world and the imminence of the Day of Judgement (for imposture can more readily crouch behind our reverence for the authority of such subjects) and then progressed on to several visions and actions so silly and laughable there is hardly anything more crude in the games of little children. Yet if Fortune had chosen to lend them a little of her favour, who knows what that play-acting might have grown into? Those poor devils are even now in gaol and may easily have to pay the penalty for the public's gullibility. And who knows whether some judge or other may not revenge his own upon them?

This incident has been uncovered: we can see clearly into it this time; but in many similar kinds of case which surpass our knowledge I consider that we should suspend our judgement, neither believing nor rejecting. Many of this world's abuses are engendered – [C] or to put it more rashly, all of this world's abuses are engendered – [B] by our being schooled to fear to admit our ignorance [C] and because we are required to accept anything which we cannot refute. [B] Everything is proclaimed by injunction and assertion. In Rome, the legal style required that even the testimony of an eye-witness or the sentence of a judge based on his most certain knowledge had to be couched in the formula, 'It seems to me that . . .'[8]

You make me hate things probable when you thrust them on me as things infallible. I love terms which soften and tone down the rashness of what we put forward, terms such as 'perhaps', 'somewhat', 'some', 'they say', 'I think' and so on. And if I had had sons to bring up I would have trained their lips to answer with [C] inquiring and undecided [B] expressions such as, 'What does this mean?' 'I do not understand that', 'It might be so', 'Is that true?' so that they would have been more likely to retain the manners of an apprentice at sixty than, as boys do, to act like learned doctors at ten. Anyone who wishes to be cured of ignorance must first admit to it: [C] Iris is the daughter of Thaumantis: amazement is the foundation of all philosophy; inquiry, its way of advancing; and ignorance is its end.[9]

[B] Yes indeed: there is a kind of ignorance, strong and magnanimous, which in honour and courage is in no wise inferior to knowledge;

---

8. Cicero, *Academica*, II (Lucullus), xlvii, 146.
9. The Scholastic axiom, *Admiratio parit scientiam*. (Consult Signoriello, *Lexicon peripateticum philosophico-theologicum*, s.v. *Admiratio*, citing Thomas Aquinas.) The saying derives from Plato, *Theaetetus*, 155 D. (Plato derived the name of Thaumas, Iris' father, from *thauma*, wonder, prodigy. Montaigne's name for him, Thaumantis, is in fact the name of Isis herself.)

[C] you need no less knowledge to beget such ignorance than to beget knowledge itself.

[B] When I was a boy I saw the account of a trial of a strange event printed by Coras, a learned counsel in Toulouse, concerning two men who each passed himself off for the other. What I remember of it (and I remember nothing else) is that it seemed to me at the time that Coras had made the impersonation on the part of the one he deemed guilty to be so miraculous and so far exceeding our own experience and his own as judge, that I found a great deal of boldness in the verdict which condemned the man to be hanged.[10] Let us (more frankly and more simply than the judges of the Areopagus, who when they found themselves hemmed in by a case which they could not unravel decreed that the parties should appear before them again a hundred years later) accept for a verdict a formula which declares, 'The Court does not understand anything whatever about this case.'[11]

My local witches go in risk of their lives, depending on the testimony of each new authority who comes and gives substance to their delusions. The Word of God offers us absolutely certain and irrefragable examples of such phenomena,[12] but to adapt and apply them to things happening in our own times because we cannot understand what caused them or how they were done needs a greater intelligence than we possess. It may perhaps be the property of that almighty Witness[13] alone to say to us: 'This is an example of it; so is that; this is not.' We must believe God – that really is right – but not, for all that, one of ourselves who is amazed by his own

10. The case of Martin Guerre (now well-known from a film thanks to the scholarship of Professor Nathalie Zemon Davies). Cf. the *Arrest memorable du Parlement de Tholose contenant une histoire prodigieuse d'un supposé mary, advenüe de nostre temps . . . par M. Iean de Coras,* Paris, 1582. Coras (p. 129) justifies the sentence of strangulation by hanging followed by the public burning of the body but (pp. 130–3) makes a passionate plea against burning anyone alive and against cruel torturings as unworthy of Christians, since they are partly based on a desire to purge one's own guilt.

11. The Areopagus in Athens had to judge a wife who murdered her second husband who, with his own son, had murdered her child by her dead husband. (This became the classical example of a *casus perplexus,* a case with the maximum degree of moral difficulty.) The Areopagus decreed that the parties concerned were to return to the Court, in person, one hundred years later! Tiraquellus evokes this well-known *exemplum* in his treatise *De poenis temperandis* (*Opera,* 1597, VII, 14). Cf. Rabelais (*Tiers Livre,* TLF, XLIIII, 6–44).

12. Cf. II Chronicles 33; II Kings 9; I Samuel 28 (the Witch of Endor consulted by Saul).

13. The Holy Ghost who, for Montaigne, was the author of Scripture.

narration − necessarily amazed if he is not out of his senses − whether testifying about others or against himself. I am a lumpish fellow and hold somewhat to solid probable things, avoiding those ancient reproaches: '*Majorem fidem homines adhibent iis quae non intelligunt*' [Men place more trust in whatever they do not understand] and, '*Cupidine humani ingenii libentius obscura creduntur.*' [There is a desire in the mind of Man which makes it more ready to believe whatever is obscure.][14] I am well aware that folk get angry and and forbid me to have any doubts about witches on pain of fearsome retribution. A new form of persuasion! Thanks be to God my credo is not to be managed by thumps from anyone's fists. Let them bring out the cane for those who maintain that their opinions are wrong; I merely maintain that their opinions are bold and hard to believe, and I condemn a denial as much as they do, though less imperiously. [C] '*Videantur sane: ne affirmentur modo.*' [Let us grant that things so appear, provided they be not affirmed.][15]

[B] Any man who supports his opinion with challenges and commands demonstrates that his reasons for it are weak. When it is a question of words, of scholastic disputations, let us grant that they apparently have as good a case as that of their objectors: but in the practical consequences that they draw from it the advantages are all with the latter. To kill people, there must be sharp and brilliant clarity; this life of ours is too real, too fundamental, to be used to guarantee these supernatural and imagined events.

As for the use of compounds and potions, I leave it out of account: that is murder of the worst sort.[16] Yet even there it is said that we should not always be content with the confessions of such folk, for they have been known to accuse themselves of killing people who have later been found alive and well. As for those other accusations which exceed the bounds of

14. The second from Tacitus, *Hist.*, I, xxii; the first is attributed by Marie de Gournay to Pliny, but remains untraced.
15. Cicero, *Academica*, II (Lucullus), xxvii, 87.
16. In law a *maleficus* (a witch, an 'evil-doer') was taken in general as one who harmed another and was not necessarily restricted to *incantatores* (workers of spells). (Cf. Spiegel, *Lexicon Juris*, s.v.) Montaigne here excludes those not allegedly working their evil through magic; thus strengthening and limiting his argument. The crucial biblical authority is Exodus 22:18, 'Thou shalt not suffer a witch to live.' But what does it mean? The Greek Septuagint uses the word *pharmakous* here, the Clementine Vulgate uses *maleficos*. Both words apply to both sexes. But Hebraists, since at least Nicolas de Lyra, insisted that the original term *kashaph* is used in the feminine. Liberal theologians clung to the Greek term and insisted that it means sorcerers who use potions to produce their wicked effects.

reason I would like to say that it is quite enough for any man – no matter how highly esteemed he is – to be believed about matters human: in the case of whatever is beyond his comprehension and produces supernatural results he should be believed only when supernatural authority confirms it.

That privilege which God has granted to some of our testimonies must not be debased or lightly made common.[17] They have battered my ears with hundreds of stories like this: three men saw him in the east on a particular day; the following morning, in such-and-such a time and place and dress, he was seen in the west. I would certainly never trust my own testimony over such a matter: how much more natural and probable it seems to me that two men should lie, rather than that, in twelve hours, one man should go like the wind from east to west; how much more natural that our mind should be enraptured from its setting by the whirlwind of our own deranged spirit than that, by a spirit from beyond, one of us humans, in flesh and blood, should be sent flying on a broomstick up the flue of his chimney. We, who are never-endingly confused by our own internal delusions, should not go looking for unknown external ones. It seems to me that it is excusable to disbelieve any wonder, at least in so far as we can weaken its 'proof' by diverting it along some non-miraculous way. I am of Saint Augustine's opinion, that in matters difficult to verify and perilous to believe, it is better to incline towards doubt than certainty.[18]

A few years ago I was passing through the domains of a sovereign prince who, as a courtesy to me and to overcome my disbelief, graciously allowed me to see, in a private place when he was present, ten or a dozen of this kind of prisoner, including one old woman, truly a witch as far as ugliness and misshapenness was concerned, and who had long been most famous for professing witchcraft. I was shown evidence and voluntary confessions as well as some insensitive spot or other on that wretched old woman;[19] I talked and questioned till I had had enough, bringing to bear the most sane attention that I could – and I am hardly the man to allow my judgement to

17. As Montaigne is about to talk of physical rapture from one place to another he is doubtless thinking of the rapture of Philip (Acts 8:39) when the 'Spirit of the Lord caught away' Philip from the road to Gaza so that he was found at Azotus.
18. St Augustine, *City of God*, XIX, xviii, contrasting scriptural truth with human testimony. Vives comments that no human knowledge, since it is known through the senses, can have the certainty of Scripture.
19. The so-called *witches' spot*; when pricked the true witch felt no sensation there. Inquisitors made painful searches for such a spot on the body of anyone charged with witchcraft.

be muzzled by preconceptions – but in the end, and in all honesty, I would have prescribed not hemlock for them but hellebore:[20]   [C]   *'Captisque res magis mentibus, quam consceleratis similis visa.'* [Their case seemed to be more a matter of insane minds rather than of delinquents.][21]   [B]   Justice has its own remedies for such maladies.[22]

As for the objections and arguments put to me there, and often elsewhere, by decent men, none ever seemed to tie me fast: all seemed to have a solution more convincing than their conclusions. It is true, though, that I never attempt to unknot 'proofs' or 'reasons' based on   [C]   experience nor on   [B]   a fact: they have no ends that you can get hold of; so, like Alexander cutting his knot, I often slice through them.[23] After all, it is to put a very high value on your surmises to roast a man alive for them.

[C]   Praestantius – and we have various examples of similar accounts – tells how his father fell into a profound sleep, deeper far than normal sleep at its best: he thought that he was a mare, serving soldiers as a beast of burden. And he actually became what he thought he was.[24] Now even if wizards dream concrete dreams like that; even if dreams can at times take on real bodies: still I do not believe that our wills should be held responsible to justice for them.   [B]   I say that, as one who am neither a king's judge nor counsellor, and who consider myself far from worthy of being so; I am an ordinary man, born and bred to obey State policy in both word and deed. Anyone who took account of my ravings, to the prejudice of the most wretched law, opinion or custom of his village, would do great wrong to himself and also to me.   [C]   I warrant you no certainty for whatever I say, except that it was indeed my thought at the time . . . my vacillating and disorderly thought. I will talk about anything by way of

20. Hemlock (*cicuta*) was used by the Greeks to poison criminals – hence Socrates' death by it; hellebore was used to purge madness.
21. Livy, VIII, xviii.
22. From the earliest times, Roman law placed the insane in the primary care of their blood relations.
23. It was said that whoever undid the untieable knot in the temple of Gordius would conquer the East: Alexander sliced it through with his sword. Cf. Erasmus, *Adages*, I, I, VI, *Nodum solvere* and, I, IX, XLVIII, *Heraculanus nodus*. (Throughout this passage Montaigne plays on the double meaning of *solutio* in Latin: 'unloosening' and 'resolving'.)
24. St Augustine, *City of God*, XVIII, xviii, suggesting that the cause was diabolical deception working through a Platonizing philosopher. Vives has a long theological note on the subject, rejecting as fictional Apuleius' metamorphosis into a donkey in his *Golden Ass*.

conversation, about nothing by way of counsel. *'Nec me pudent, ut istos, fateri nescire quod nesciam.'* [Nor, like those other fellows, am I ashamed to admit that I do not know what I do not know.][25]

[B]   I would not be so rash of speech if it were my privilege to be believed on this matter. And I replied thus to a great nobleman who complained of the sharpness and tension of my exhortations: 'Knowing that you are braced and prepared on one side, I set out the other side for you as thoroughly as I can, not to bind your judgement but to give it some light. God holds sway over your mind: he will allow you a choice. I am not so presumptuous as to desire that my opinions should weigh even slightly in a matter of such importance: it is not my lot to groom them to influence such mighty and exalted decisions.'

It is certain that I have not only a great many humours but also quite a few opinions which I would willingly train a son of mine to find distasteful, if I had one that is. Why! What if even the truest of them should not always be the most appropriate for Man, given that his make-up is so barbarous?

On the point or off the point, no matter; it is said as a common proverb in Italy that he who has not lain with a lame woman does not know Venus in her sweet perfection. Chance, or some particular incident, long ago put that saying on the lips of the common people. It is applied to both male and female, for the Queen of the Amazons retorted to the Scythian who solicited her: Ἄριστα χολός οἰφεῖ: 'The lame man does it best.'[26]

In that Republic of women, in order to avoid the dominance of the male, they crippled their boys in childhood – arms, legs and other parts which give men the advantage over women – and exploited men only for such uses as we put women to in our part of the world.

Now I would have said that it was the erratic movements of the lame woman which brought some new sensation to the job and some stab of pleasure to those who assayed it: but I have just learned that ancient philosophy itself has decided the matter: it says that the legs and the thighs of lame women cannot receive (being imperfect) the nourishment which is their due, with the result that the genital organs which are sited above them become more developed, better fed and more vigorous. Alternatively, since this defect discourages exercise, those who are marked by it dissipate

25. Cicero, *Tusc. disput.*, I, xxv, 60.
26. Erasmus, *Adages*, II, IX, XLIX, *Claudus optime virum agit.* Cf. also Septalius' note in his edition of Aristotle's (or Pseudo-Aristotle's) *Problemata* X, 25 (26); Coelius Richerius Rhodiginus, *Antiquae Lectiones*, XIV, v, *Cur claudi salaciores.* Cf. also Erasmus, *Apophthegmata*, VIII, *Thrasea*, second hundred, XXI.

their strength less and so come more whole to Venus' sports which is also why the Greeks disparaged women who worked at the loom, saying they were lustier than others because of their sedentary occupation which is without much physical exertion.

At this rate, what can we *not* reason about! Of those women weavers I could just as well say that the shuttling to and fro which their work imposes on them while they are squatting down stimulates and arouses them just as the jerking and shaking of their coaches do for our ladies.

Do not these examples serve to prove what I said at the outset: that our reasons often run ahead of the facts and enjoy such an infinitely wide jurisdiction that they are used to make judgements about the very void and nonentity. Apart from the pliancy of our inventive powers when forging reasons for all sorts of idle fancies, our imagination finds it just as easy to receive the stamp of false impressions derived from frivolous appearances: for on the sole authority of the ancient and widespread currency of that saying, I once got myself to believe that I had derived greater pleasure from a woman because she was deformed, even counting her deformity among her charms.

In his comparison between France and Italy Torquato Tasso says that he had noticed that we have skinnier legs than the gentlemen of Italy and attributes the cause of it to our being continually on our horses. Now that is the very same 'cause' which leads Suetonius to the opposite conclusion: for he says, on the contrary, that Germanicus had fattened his legs by the constant practice of that same exercise![27]

There is nothing so supple and eccentric as our understanding. It is like Theramenes' shoe: good for either foot.[28] It is ambiguous and faces both ways; matters, too, are ambiguous and facing both ways: 'Give me a silver penny,' said a Cynic philosopher to Antigonus. 'That is no present from a king,' he replied. 'Give me half a hundredweight of gold then' – 'That is no present for a Cynic!'[29]

> *Seu plures calor ille vias et cæca relaxat*
> *Spiramenta, novas veniat qua succus in herbas;*
> *Seu durat magis et venas astringit hiantes,*
> *Ne tenues pluviæ, rapidive potentia solis*
> *Acrior, aut Boreæ penetrabile frigus adurat.*

27. Torquato Tasso, *Paragon dell'Italia alla Francia*; Suetonius, *Life of Caligula*, III.
28. Cf. Erasmus, *Adages*, I, I, XCIV, *Cothurno versatilior*. Theramenes was an Athenian rhetorician who could find arguments for either party.
29. Erasmus, *Apophthegmata*, IV, *Antigonus Rex Macedonum*, XV.

[It is either because the heat opens up new ways through the secret pores in the soil, along which the sap rises to the tender plants, or else because it hardens that soil and constricts its gaping veins, thus protecting it from the drizzling rain, the heat of the burning sun and the penetrating cold of the north wind.][30]

'*Ogni medaglia ha suo riverso.*' [Every medal has its obverse.] That is why Clitomachus said in ancient times that Carneades had surpassed the labours of Hercules by having wrenched assent away from Man (that is, conjecturing and rashness in judging).[31]

That idea of Carneades – such a vigorous one – was born, I suggest, in antiquity because of the shamelessness of those whose profession was knowledge and their overweening arrogance.

Aesop was put on sale with two other slaves. The purchaser asked the first what he could do: he, to enhance his value, answered mountains and miracles: he could do this and he could do that. The second said as much or more of himself. When it was Aesop's turn to be asked what he could do he said, 'Nothing! These two have got in first and taken the lot: they know everything!'[32]

That is what happened in the school of philosophy. The arrogance of those who attributed to Man's mind a capacity for everything produced in others (through irritation and emulation) the opinion that it has a capacity for nothing. Some went to the same extreme about ignorance as the others did about knowledge, so that no one may deny that Man is immoderate in all things and that he has no stopping-point save necessity, when too feeble to get any farther.

---

30. Virgil, *Georgics*, I, 89–93 (two of several reasons why burning stubble is good for crops).
31. Translated from Cicero, *Academica*, II (*Lucullus*), xxxiv, 108: '*adsensionem, id est, opinationem et temeritatem.*'
32. From Maximus Planudes' *Life of Aesop*, frequently printed with the *Fables*.

12. On physiognomy

===

[Renaissance books on physiognomy all gave pride of place to Zopyrus the Physiognomist, who judged by his art that Socrates was a bad man and a born womanizer. (Socrates admitted this, adding that he had 're-formed' his soul.) Montaigne compares and contrasts himself to Socrates and shows how his own frank expression served him well. This chapter corrects much of what had been said in I, 20 ('To philosophize is to learn how to die') and takes even farther Montaigne's respect for Nature and the wisdom of the beasts expounded in 'An apology for Raymond Sebond'. In this most personally anecdotal of chapters, Montaigne has discovered the moral greatness of simple folk faced with certain death. And he hints at his hopes that Henry of Navarre will bring peace to France.]

[B]   Virtually all the opinions which we have are held on authority and trust. That is no bad thing: in so ailing a time as this we could do nothing worse than to make our own choices. That portrait of the conversations of Socrates which his friends have bequeathed to us receives our approbation only because we are overawed by the general approval of them. It is not from our own knowledge, since they do not follow our[1] practices: if something like them were to be produced nowadays there are few who would rate them highly. We can appreciate no graces which are not pointed, inflated and magnified by artifice. Such graces as flow on under the name of naïvety and simplicity readily go unseen by so coarse an insight as ours: they have a delicate, secret beauty: to uncover their hidden light requires sight which is purged and pure. For us, is not naïvety close kin to simplemindedness and a quality worthy of reproach?[2] Socrates makes his soul move with the natural motion of the common people: thus speaks a peasant; thus speaks a woman.   [C]   He has nothing on his lips but draymen, joiners, cobblers and masons.   [B]   His inductions and comparisons are drawn from the most ordinary and best-known of men's activities; anyone can understand him. Under so common a form we today would never have discerned the nobility and splendour of his astonishing concepts; we   [C]   who judge any which are not swollen up by erudition

1. '88: our *tastes and* practices . . .
2. '88: reproach *and insult*? Socrates . . .

to be base and commonplace and   [B]   who are never aware of riches
except when pompously paraded. Our society has been prepared to appreciate nothing but ostentation: nowadays you can fill men up with
nothing but wind and then bounce them about like balloons. But
this man, Socrates, did not deal with vain notions: his aim was to provide
us with matter and precepts which genuinely and intimately serve our
lives:

> *servare modum, finemque tenere,*
> *Naturamque sequi.*

[to keep the mean; to hold fast to the limit; and to follow nature.][3]

He was ever one, ever the same: he raised himself up to the highest level of
vigour not by sallies but by complexion. Or (to put it better) he raised
nothing, but rather brought it down and back to its natural and original
level, by which he moderated vigour, hardships and difficulties.

In the case of Cato we can clearly see that his manner is strained far
above the normal: in the brave actions of his life and death we know that
he is riding high as his tallest horses. Socrates however keeps his feet on the
ground, dealing with the most useful subjects at a quiet and everyday pace,
advancing at the rate of human life towards both death and the harshest
ordeals that can ever occur. Fortunately it turned out that the man most
worthy of being known and of being set before the world as an example
was precisely the one we have the surest knowledge about.[4] He was
observed by the most observant men there ever have been: the testimonies
that we have of him are astonishing by their fidelity and their skill.
Happily for us he could so order the purest and most child-like thoughts
that, without stretching them or perverting them, he could produce by
them the most beautiful actions of our souls. He portrays the soul as neither
high-soaring nor abundantly endowed: he portrays it simply as sane,
though with a pure and lively sanity. From such commonplace natural
principles, from such ordinary everyday ideas, without being carried away
and without goading himself on, he formed beliefs, actions and morals
which were not simply the best regulated but also the most sublime and
most forceful that ever have been.   [C]   He it was who brought human
wisdom back from the heavens where she was wasting her time and
returned her to mankind, in whom lies her most proper and most demand-

---

3. Lucan, *Pharsalia,* II, 381–2, praising Cato.
4. '88: about *either for judging or comparing.* He was . . .

ing task as well as her most useful one.[5] [B] See him pleading his case before his judges; see with what arguments he awakens his mind for the hazards of war; see what reasons strengthen his endurance when confronted by lies, tyranny and death, as well as by his wife's pig-headedness. Nothing there is lifted from the arts or sciences: the simplest folk can recognize in him their own means and strengths. It is not possible to be less pretentious or more lowly. He did a great favour to human nature by showing how much she can do by herself. We are richer than we think, each one of us. Yet we are schooled for borrowing and begging! We are trained to make more use of other men's goods than of our own.

In nothing does Man know how to halt at the point of his need; be it pleasure, wealth or power, he clasps at more than he can hold: his greed is not susceptible to moderation. It is the same, I find, with his curiosity for knowledge: he hacks out for himself much greater tasks than he needs or can achieve, [C] making the extent of knowledge and the usefulness of knowledge co-equal: '*Ut omnium rerum, sic littorarum quoque intemperantia laboramus.*' [In learning as in everything else, we suffer from lack of temperance.][6] And Tacitus is right to praise the mother of Agricola for having restrained in her son too seething an appetite for knowledge:[7] like the rest of men's goods, knowledge is one which, if we look at it steadily, has much inherent vanity and natural feebleness. And it costs us dear. To acquire such pabulum is more hazardous than the acquiring of other food or drink;[8] for in other cases whatever food we have bought we can carry home in containers – which gives us time to decide on its worth, and on how much of it we shall take and when. But from the outset all kinds of learning can be put into no container but our soul: as we buy them we ingest them, leaving the market-place either already contaminated or else improved. Some of them, instead of nourishing us, burden us and hamper us; others still, under pretence of curing us, poison us.

[B] I have taken pleasure in hearing of men somewhere or other who, from piety, make vows of ignorance similar to vows of chastity, poverty and penance. To take the edge off that cupidity which goads us towards

5. In the Renaissance this was summed up in the Socratic adage, *Quae supra nos, nihil ad nos* (What is above us is nothing to do with us). As Erasmus points out (I; VI; LXIX) the early Christian writer Lucius Lactantius considered it to be 'famous and approved by all'.

6. Seneca, *Epist. moral.*, CVI, 12.

7. Tacitus, *Agricola*, I, x.

8. The notion that the soul, like the body, needs *pabulum* (food, nourishment) is Platonic.

the study of books, and to deprive our souls of that pleasurable self-satisfaction which thrills us with the opinion that we know something is farther to castrate our disordered desires.   [C]   And it is to fulfil the vow of poverty abundantly to be also poor in spirit.[9]

[B]   We need but little doctrine to live at our ease. And Socrates teaches us that it lies within us, as well as how to find it there and how to make it help us.[10] All that capacity of ours for exceeding what is   [C]   natural is more or less   [B]   vain and superfluous:[11] it is much if it does not burden and bother us more than it serves us: [C]   *'Paucis opus est litteris ad mentem bonam.'* [To produce a good mind you need only a few books.]   [B]   They are the feverish excesses of our mind, a confused and disquieted tool.

Contemplate yourself. You will find within you Nature's arguments concerning death – true arguments, most fit to serve you in your need: they it is which make a farm-labourer, as well as entire nations, die with as much constancy as a philosopher.[12]   [C]   Would I have died any the less happily before reading the *Tusculan Disputations*? I judge that I would not. And now that I find that I must really face death, I realize that my tongue has been enriched by them but not at all my mind, which is as Nature forged it for me: its buckler in that combat is to approach it as do the common people. Books have been useful to me less for instruction than as training. What if   [B]   erudition, while making an assay at arming us with new defences against natural ills, should have imprinted on our thoughts the weight of those ills and their size rather than her subtle arguments for protecting us against them!   [C]   For subtle arguments they are, by which erudition most vainly alerts us. Just see how writers – even the most wise and succinct of them – strew additional trivial arguments round about one good one, arguments which, if you look at them closely, have no body in them. They are nothing but verbal contortions by which we are deceived. Yet, in so far as they may serve a purpose, I have no wish to pluck them any barer. Here and there within these covers there are enough arguments of that sort, either borrowed or imitated. Nevertheless

---

9. Christian 'fools' often combined real or pretended ignorance and madness with their other ascetic ideals and practices. An echo of Matthew 5:3, 'Blessed are the poor in spirit' (i.e., the foolish).

10. Cf. for example the adages of Erasmus mentioned in III, 9, 'On vanity', note 160.

11. '88: what is *common and* natural is vain and superfluous . . .
    Then, Seneca, *Epist. moral.*, CVI, 11 (adapted).

12. [B] instead of [C]: philosopher. Erudition, while making an assay . . .

we must be careful not to give the name of fortitude to what is but the conduct of a gentleman, nor call solid what is but clever, nor good what is but beautiful – *'quae magis gustata quam potata delectant'* [things which are more pleasant to sip than to quaff].[13] And, *'ubi non ingenii sed animi negotium agitur'* [whenever we are concerned with the soul not the mind], not everything that we fancy feeds us.

[B] To see the exertions that Seneca imposed upon himself in order to steel himself against death, to see him sweat and grunt in order to stiffen and reassure himself during his long struggles on his pedestal, would have shaken his reputation for me if he had not sustained it with such valour as he was dying. His burning emotion, [C] so oft repeated, shows that he himself was ardent and impetuous. (We must convict him out of his own mouth: *'Magnus animus remissius loquitur et securius'* [A great mind speaks with more calm and assurance]; *'Non est alius ingenio, alius animo color.'* [There is not one colour for the wit another for the mind].) And it also [D] shows that[14] he was to some extent hard pressed by his adversary. The style of Plutarch, being more detached and relaxed, is for me more manly and persuasive: I would find it easier to believe that his soul's emotions were more assured and steady. Seneca, more [C] lively [B], puts in the goad[15] and wakes us up with a start; he stimulates, rather, our wit: Plutarch, more [C] settled, [B] constantly[16] reassures and strengthens us; he stimulates, rather, our understanding. [C] Seneca enraptures our judgement: Plutarch wins it.

I have likewise seen even more hallowed writings which, in their portrayal of the conflict sustained against the prickings of the flesh, show them to be so sharp, so strong and invincible, that the likes of us, who are but the off-scourings of the commonality, are as struck with wonder by the strangeness and unknown power of the temptations as by the resistance put up to them.[17]

[B] Why do we go on stiffening our morale by such learned maxims?[18]

13. Cicero, *Tusc. disput.*, V, v, B (criticizing Stoic arguments); Seneca, *Epist. moral.*, LXXV, 5 (adapted).
14. '88: His burning emotion, *so animated*, show that . . .
   In [C], Seneca, *Epist. moral.*, CXV, 2, and CXIV, 3; in both cases contrasting mere words and style with solid moral action.
15. '88: more *pointed*, puts in the goad . . .
16. '88: more *solid*, constantly . . .
17. Such temptations as St Jerome in the desert or the fearsome hallucinations of St Antony.
18. '88: such *subtleties and* learned maxims . . .

Let us look to the land and to the wretched people we can see scattered over it, bending low over their toil, ignorant of Aristotle, Cato, example and precept: from them Nature draws every day deeds of constancy and steadfastness which are purer and more unbending than those which we so carefully study in our schools. How many country-folk do I see ignoring poverty; how many yearning for death or meeting it without panic or distress? That man over there who is trenching my garden has, this morning, buried his father or his son. The very names by which they call our afflictions soften them and sweeten their bitter taste: for them consumption is 'the cough'; dysentery, a 'runny stomach'; pleurisy, 'a chill'. And as they give them mild names they endure them better too. Ills have to be grievous indeed to interrupt their habitual toil. They take to their beds only to die: [C] *'Simplex illa et aperta virtus in obscuram et solertam scientiam versa est.'* [Virtue, simple and open, has been converted into obscure and subtle erudition.][19]

[B] I wrote this round about the time when the huge burdens of our civil disturbances were for several months pressing right down on me with all their weight. I had the enemy at my gates on one side and on the other side a worse enemy, marauders: [C] *'non armis sed vitiis certatur'* [not with arms is the fight but with crimes].[20] [B] I was being assayed by every kind of military outrage all at once.

> *Hostis adest dextra levaque a parte timendus,*
> *Vicinoque malo terret utrumque latus.*

[A redoubtable enemy I have to left and right: on either side immediate danger threatens.]

What a monstrosity this war is! Other wars are external: this one also gnaws at itself and destroys itself with its own poison. Its nature is so malign and so destructive that it destroys itself along with everything else, tearing itself limb from limb in its frenzy. As often as not we can see it falling apart, more by itself than from lack of any necessary commodity or by enemy action. All military lore flees it: it came to cure sedition, yet it is full of it; it seeks to punish disobedience, and is an example of it; it is used to defend our laws, but it takes part in a rebellion against its own. Where has it got us? Our medicines are infected:

19. Seneca, *Epist. moral.*, XCV, B.
20. Attributed by Marie de Gournay to 'Seneca's Epistles', but untraced. Then, Ovid, *Ex Ponto*, I, iii, 57–8.

Nostre mal s'empoisonne
Du secours qu'on luy donne.

[Our illness draws Venom from the succour we bring it.][21]

Exuperat magis aegrescitque medendo;

[The illness grows greater and more sickly with the cure;]

Omnia fanda, nefanda, malo permista furore,
Justificam nobis mentem avertere Deorum.

[Now that right and wrong have been confounded by our wicked frenzy, it has brought the gods to turn away their righteous will from us.]

When such distempers attack a whole people, at first you can tell the sound from the sickly; but when they come to drag on as ours do, the whole body politic feels it from head to heel: no organ is free from corruption. For there is no air which is so greedily breathed as the air of licence, nor which spreads and permeates as far. Our armies are bound and held together now only by the cement of foreign mercenaries: you can no longer make up one reliable and and disciplined body of soldiers. How ignominious! There is only as much discipline as our hired soldiers care to show us: every man Jack of us follows his own discretion not that of his commander, who has more trouble with his own troops than with the enemy. The General it is who must follow, woo and bow: he alone has to obey; all the rest are free and untrammelled.[22]

I take pleasure in seeing how much baseness and faint-heartedness there are in ambition, and how much abject servitude it requires to achieve its end. But what displeases me is daily to see decent characters who are capable of justice corrupted by their management and command of this disorder. Long sufferance begets habit: habit, acceptance and imitation. We have enough men with ill-endowed souls without spoiling the good and generous ones. If we go on this way there will scarcely be one man left to whom we could entrust the welfare of our State, should Fortune ever restore its health to us.

Hunc saltem everso Juvenem succurrere seclo
Ne prohibite.

21. Source unknown; then, Virgil, *Aeneid*, XII, 46; Catullus, *Epithalamia Thetis et Pelei*, 406–7.
22. Doubtless an echo of the saying of Alcibiades that an army should be organized under a Head, as is the human body.

[At least do not prevent this Youth from bringing succour to this prostrate world.][23]

[C] What has become of that old axiom that soldiers should go more in fear of their captain than of their enemy? And what of that wonderful example of an apple tree which happened to be enclosed within the limits of a Roman army-camp, yet, when the camp was struck next day, was still there, leaving its owner with his full complement of ripe and delicious apples?[24] I wish that our young men, instead of the time they spend on less useful tours and less honourable apprenticeships, would devote half of it to watching the war at sea under some good Captain-Commander of Rhodes, and the other half to studying the discipline of the Turkish armies, for it has many superiorities and advantages over ours. One is, that whereas our soldiers become more disorderly during our campaigns, there they become more self-controlled and circumspect, since such offences and larcenies against the common people as are punishable by cudgelling in times of peace become capital in time of war. There is a pre-established tariff; for one egg taken without payment: fifty strokes of the cane; for anything else, no matter how trivial, which cannot be used as food: immediate impaling or beheading. It amazed me to read in the history of Selim, the cruellest conqueror there has ever been, that, when he had subdued Egypt, those wonderful gardens which surround the City of Damascus and abound in delicacies remained unsullied by the hands of his soldiers. Yet those gardens were all open and unfenced.

[B] But is there any affliction in a polity which it is worth tackling with so fatal a cure? According to Favonius, not even the usurping of a tyrannous hold upon the State.[25]   [C] Similarly Plato does not allow that violence be done against the peace of one's country even to cure it, and will accept no correction which costs the blood and ruination of the citizens, laying down that it is a good man's duty in such a case to leave things as they are, simply praying God to exceed the normal Order and to bring His hand to bear. He appears to have criticized his close friend Dion for having acted just a little out of line.[26] I was a Platonist in that way

---

23. Virgil, *Georgics*, I, 500, applied almost certainly by Montaigne to the then Protestant Henry of Navarre, who, as the Roman Catholic Henri Quatre, did indeed bring comparative peace and moral government to France.
24. Both cited by Justus Lipsius, *Politici*, V, xiii.
25. Plutarch, *Life of Brutus*.
26. Plato, *Letters*, VII, 331.

before I ever knew there had been a Plato in this world. And although that great man must be simply excluded from our communion – he who by the purity of his conscience deserved so well of God's favour as to penetrate through the widespread darkness of his time deeply into the light of Christianity – I do not think it well becomes us to let ourselves be taught by a pagan how impious it is to expect from God no succour whatsoever which is His alone, requiring no cooperation on our part.[27]

I often doubt whether, among all those who engage in such disorders, there has ever been found one man who was so feeble of understanding that he actually let himself be convinced that he was advancing towards reformation by way of the ultimate in deformation; that he was ensuring his salvation by means of the most explicit causes we know of most certain damnation;[28] or that, by overthrowing the constitution, the authorities and the laws under the tutelage of which God has placed him, and by the dismembering of his motherland (tossing parts of her to be gnawed by her ancient foes, filling brotherly hearts with parricidal hatreds and summoning up devils and the Furies to help him) he can somehow bring succour to the most holy loveliness and justice of God's word.[29] [B] Ambition, greed, cruelty, revenge do not have enough natural violence of their own, so let us light the match and stir the fire under the glorious pretext of justice and devotion! No worse state of affairs can be imagined than one in which wickedness becomes lawful, donning, by leave of the magistrate, the mantle of authority. [C] '*Nihil in speciem fallacius quam prava relligio, ubi deorum numen praetenditur sceleribus.*' [Nothing is more deceitful than a depraved piety by which the will of the gods serves as a pretext for crimes.][30] According to Plato, the ultimate species of injustice is when what is unjust is held to be just.

[B] The common people suffered then not merely present depravations –

27. How far man should 'work together' with God is a major theological problem, much quarrelled over during the Renaissance. Montaigne gives the prime and indispensable role to divine grace but expects man to work together with God. His theology is orthodox, as is his treating Plato as a pagan – for Erasmus he was a proto-Christian.

28. Sedition is a sin for Christians. St Paul classifies 'seditions' with 'heresies' as works of the flesh (Galatians 5:20).

29. In Roman Law parricide was not limited to killing fathers but used of all foul murders to mark the height of their impiety. (Cf. Spiegel's *Lexicon juris civilis*.)

30. Livy, XXXIX, xvi; then, Plato, *Republic*, II, 361 A. (Cf. Cicero, *De officiis*, I, xiii, 41.)

undique totis
Usque adeo turbatur agris,

[all fields devasted everywhere,][31]

– but future ones as well: the living had to suffer; so too those who were yet unborn. They were robbed of everything, even of hope (and so, in consequence, was I): they ravished from them all the means they possessed, which for long years would have provided their livelihood:

Quæ nequeunt secum ferre aut abducere perdunt,
Et cremat insontes turba scelesta casas.

[They smash whatever they cannot carry or cart away; the mob of ruffians burn down innocent cottages.]

Muris nulla fides, squallent populatibus agri.

[There is no safety within city walls: outside, the fields are ravaged.]

Apart from that attack, I suffered others as well. I incurred the penalties which moderation entails during such disorders. I was trounced on every hand: I was Guelph to the Ghibelline, Ghibelline to the Guelph. (One of our poets says just that, but I do not remember where.) The fact that my home is where it is, coupled with my affability towards the men of my neighbourhood, made me appear one thing, my life and actions another.[32] No formal indictments were made: folk had nothing to get their teeth into. I never break the laws: if a proper investigation had been made, any remaining doubt would have been owed to me. But there were unspoken suspicions [C] circulating underhand [B] for[33] which there is never any lack of pretext in so confused a chaos, no more than there is any lack of envious minds or silly ones. [C] I usually have a way of aggravating any harmful inferences which Fortune strews against me by refusing to justify, defend or explain myself, reckoning that to plead for my good conscience is to compromise it:[34] '*Perspicuitas enim argumentatione elevatur.*'

31. Virgil, *Eclogues*, I, 11–12; then, Ovid, *Tristia*, III, x, 65–6, and Claudianus, *In Eutropium*, I, 244.
32. Montaigne lived in a region dominated by the Reformed Church; he was an active Roman Catholic who never hid his allegiance.
33. '88: unspoken *and hidden* suspicions, for which . . .
34. Montaigne's term *conscience*, like *conscientia* often in Latin, means not conscience here but a good conscience, the consciousness of having done right. In Montaigne as in the Renaissance generally it rarely means what it now does in English.

[Argument merely removes the perspicuity.]³⁵ And so, as though each man can see into me as clearly as I can, instead of distancing myself from an accusation I advance towards it, improving upon it by an ironic and mocking admission of guilt – if, that is, I do not flatly keep silent about it, as being unworthy of a reply.

But those who take that for some excessive arrogance on my part wish me scarcely less harm than those who take it for the weakness of an indefensible case – especially those great lords for whom the ultimate crime is lack of submissiveness and who are insolent in face of any justice which knows what is what, and which fails to be humble, submissive and begging. I have often bumped up against that pillar. Be that as it may, over what befell me then   [B]   an ambitious man would have hanged himself; so would a covetous man. I am in no wise acquisitive –

> *Sit mihi quod nunc est, etiam minus, ut mihi vivam*
> *Quod superest ævi, si quid superesse volent dii.*

[Let me keep what I have now – or less even – so that I may live the rest of my life for myself (if the gods grant me any more life to live).]³⁶

– yet such losses as do befall me through another's wrong-doing, be it larceny or violence, pain me just about as much as they do a man sick and tortured by covetousness. The affront is immeasurably more bitter than the loss.

Hundreds of different kinds of misfortune rushed upon me one after another: if they had come together I could have borne them more cheerfully. I had already thought about entrusting my impoverished and straitened old age to one of my loved ones, but after letting my eyes rove over all my affairs, I realized that I was reduced to my shirt. To plummet down from such a height you need to be caught by firmly loving arms, solid and favoured by fortune. Such arms are rare – if there be any at all. In the end I realized that the surest way was to entrust my needs and my person to myself and that, if I should chance to be coldly treated by Fortune's favour, then I should commend myself even more strongly to my own, clinging to myself and becoming more intimately beholden to myself.   [C]   In all their concerns men dash to seek props from others so as to spare their own, which alone, for anyone who knows how to arm himself with them, are certain and strong. Each man rushes elsewhere and towards the future, since no man has reached his own self.

35. Cicero, *De nat. deorum*, III, iv, 9 (applied when a case is self-evident).
36. Horace, *Epistles*, I, xviii, 107–8.

[B]   I concluded that my afflictions were useful ones. Firstly, because bad pupils, when reason proves inadequate, have to be taught by a good hiding, [C] just as we straighten back wood, when it has become warped, by driving in wedges over a fire.   [B]   For such a long time now I have been lecturing myself about holding to myself and keeping apart from matters external, yet I still go on turning my gaze sideways: I am tempted by a nod, by a gracious word from some great man or by an encouraging face. (God knows there is a dearth of those nowadays, and how little they imply!) I can still hear without a frown the seductions of those who would try to put me up for auction, resisting them so feebly that it appears that I would really rather be convinced by them. A mind so unwilling to learn requires flogging: I am a cask which is splitting apart, leaking and failing in its duty: it needs knocking together and tightening up with good whacks from the mallet.

Secondly, because my misfortune served me as practice, preparing me for the worst (should I, who by the bounty of Fortune and the properties of my character hope to be among the last, happen to be among the first to be caught up in this tempest), teaching me in good time to limit my way of life, and to order it for a new estate. True freedom is to have power over oneself to do *anything* with oneself.   [C]   '*Potentissimus est qui se habet in potestate.*' [Most powerful is he who has himself in his power.][37]   [B]   In ordinary tranquil times we prepare ourselves for moderate and common ills. But during the disorders in which we have lived these last thirty years, every man in France sees himself, both individually and collectively and hour by hour, on the point of having his entire fate reversed: all the more reason, then, to keep one's mind supplied with stronger and more manly provender. Let us be grateful to our fate for having made us live in an age which is neither soft and idle nor lazy: nowadays a man who would never otherwise have become famous may do so because of his misfortunes.

[C]   I rarely read in my history books about the disorders in other States without regretting that I could not have been there to study them more closely: so, too, my desire for knowledge leads me to find at least some satisfaction in being able to see with my own eyes this remarkable spectacle of the death of our institutions, the manner of it and its symptoms. Since I cannot retard it, I am happy to be destined to be present and to learn from it. After all, we make great efforts so that we can eagerly witness performances of fictional portrayals of the tragedies of human fortune; it is not that we lack sympathy for what we hear there but that we

37. Seneca, *Epist. moral.*, XC, 34.

delight in awakening our grief by the exceptional nature of those pitiable events. Nothing thrills without hurting. Good historians avoid telling of calm events – still waters and dead seas – in order to sail again into wars and seditions, to which (as they know) we summon them.

I doubt whether I can properly admit how little it has cost me in terms of my life's repose and tranquillity to have passed more than a half of my days during the collapse of my country. Faced with misfortunes which do not concern me directly, I buy my resignation a little too cheaply; as for lamenting on my own behalf, I have regard not so much for what has been taken from me but for what still remains to me, both within and without. There is some consolation in dodging, one after another, the successive evils which have us in their sights, only to strike elsewhere around us. Moreover, where public misfortunes are concerned, the more my compassion is spread overall the weaker it becomes. To which add that it is certainly more or less true that *'tantum ex publicis malis sentimus, quantum ad privatas res pertinet'* [from public ills we feel only as much as touches us directly],[38] and that our original health was such as to diminish any sorrow we ought to have felt for its loss. It was indeed 'health', but only by comparison with the malady which followed it. We did not have far to fall; least tolerable of all, it seems to me, are honoured corruption and institutionalized brigandry: there is less wrong in stealing from us in a forest than in a place of safety. The 'health' of our State concerned a body entirely composed of organs each rivalling one another in corruption, and (for the most part) of aged sores, no longer being cured nor wanting to be cured.

[B]   This shaking of the foundations stimulated me rather than flattened me, thanks to my sense of right and wrong which acted not merely peaceably but proudly, and I found nothing to reproach myself with. And since God never sends us pure evils any more than pure blessings, my own health held out better than usual throughout this period: and just as without health I can achieve nothing, with health there are few things which I cannot achieve. It provided me with the means of quickening my store of wisdom and of stretching forth my hand to parry blows which would readily have wounded more deeply. And in bearing my afflictions I found some means of withstanding Fortune and found that it would take some great shock to throw me from the saddle. (I do not say that to provoke her into making a more vigorous attack on me! I am her 'most obedient servant': my hands are raised in supplication: let her be satisfied, for God's sake!)

38. Livy, XXX, xliv.

Do I feel her assaults? Of course I do. As[39] those who are overwhelmed and obsessed by grief yet allow some pleasure to fondle them from time to time and to release a smile, so too I have enough hold over myself to make my usual state a peaceful one, free from the burden of painful reflections; yet I can allow myself occasionally to be surprised by those biting and unpleasant thoughts which, while I am arming myself to drive them off or struggle against them, come along and batter me.

Following hard upon the others a worse calamity befell me: the plague, of unique virulence, raged both inside my home and around it; for, just as healthy bodies fall prey only to the most serious of illnesses, which alone can get a hold on them, similarly the air around my estates (which in human memory had never given a foothold to contagion, even when it came very close) once it was corrupted produced strange effects[40] indeed:

> *Mista senum et juvenum densantur funera, nullum*
> *Sæva caput Proserpina fugit.*

[Young and old come in crowds to be buried: cruel Proserpine spares no one's head.]

I had to put up with a fine state of affairs: the very sight of my house was terrifying. Everything inside lay unprotected, left to anyone who wanted it. I, who am so hospitable myself, had to go in painful quest of a refuge for my family – a family of castaways, a source of fear to those who loved us and to itself, and of terror wherever it sought to settle, having to change quarters as soon as one of us got a slightly sore finger. All illnesses are then taken to be the plague: no time is allowed to probe them. And (best of all!) according to the rules of the Art, every time you are exposed to risk, you spend your quarantine in an ecstatic dread of that illness; your imagination meanwhile has its own way of agitating you, making your very health sweat with fever.

All of which would have touched me far less if I did not have to worry about others, spending six wretched months acting as guide for that caravan: for I myself bear within me my own prophylactics, namely determination and long-suffering. I am not much bothered by dread (which is particularly to be feared in this illness): and so, if I alone had sought to make an escape, it would have been a merrier and more distant one. It is not, I think, the worst of deaths: it is normally short, marked by numbness and lack of pain, comforted by being shared by many, without ritual and without a crowd of mourners.

---

39. '88: Of course I do. *But* as . . .
40. '88: strange *and unheard of* effects . . .   Then, Horace, *Odes*, I, xxviii, 11–12.

As for those who dwelt around us, not one in a hundred escaped:

> *videas desertaque regna*
> *Pastorum, et longe saltus lateque vacantes.*

[you may see the abandoned realms of the shepherds and, far and wide, the deserted pastures.][41]

Down here my income is mainly from farm-labour; now, the land which once had a hundred men on it working for me has long lain fallow. At that time what exemplary resignation did we see among all those simple folk. In general each one gave up worrying about his life. The grapes, the principal produce of the region, remained hanging on the vines, since everybody without exception was ready, awaiting death that night or next morning with voices and faces so little terrified that it seemed they had all made a pact with that unavoidable evil, and that the sentence upon them was universal and inevitable. That sentence always is! Yet our resolution in death hangs on so little: its being delayed by a few hours, or the mere factor of our having companions, make us [C] conceive of death [B] differently.[42] But just look at these folk: they are no longer amazed that, babes, children and old men, they are all to die the same month: they no longer weep for themselves. I saw some who were afraid that they would be left behind as in some ghastly wilderness; the only worry that I know they had concerned their burial: it disturbed them to see corpses scattered over the fields at the mercy of the beasts, which at once started to thrive there. [C] (How incompatible human notions are! The Neorites, a people subjugated by Alexander, abandon the bodies of their dead deep in their forests, there to be eaten – for them it is the only blessed form of sepulture.)[43] [B] One man, in good health, was already digging his grave: others would lie down in theirs while there was still life in them. And one of my day-labourers pulled the earth over himself as he lay dying, using his hands and feet. Was he not donning his own shroud so as to lay himself more comfortably at rest – [C] a deed in some ways as sublime as that of those Roman soldiers who, after the Battle of Cannae, were discovered to have dug holes in the ground, thrust in their heads, drawn in the soil and suffocated themselves?[44]

---

41. Virgil, *Georgics*, III, 476–7.
42. '88: make us *taste* death *quite* differently . . .
43. Diodorus Siculus, XVII, xxiii.
44. Livy, XXII, li.

[B] In short, an entire people, at a stroke and pragmatically, were brought to a state which yielded nothing in firmness of purpose to any studied philosophical steadfastness. Most of the teachings which schooling supplies us with to give us courage have more ostentation than fortitude, and are cultivated more for decoration than for profit. We have abandoned Nature and want to teach her own lessons to her who used to guide us so happily and surely. And yet such traces of her teachings and whatever little of her image remain by favour of ignorance stamped on the life of that crowd of uncultured country-folk, Erudition is compelled to go and beg from them, day in, day out, in order to supply patterns of constancy, simplicity and tranquillity for its own pupils. Fine it is to see the latter, full as they are of fair learning, having to imitate that untutored simplicity – imitating it moreover in the most basic acts of virtue; fine too that our wisdom must learn from the very beasts the lessons most useful for the greatest and most necessary aspects of our life: how we should live and die, manage our goods, love and educate our offspring and maintain justice. That is a singular witness that humanity is sick and that our reason (which we mould as we will, ever finding some novelty or some different approach) leaves behind in us no manifest trace of Nature. Men have done to Nature what makers of perfume have done to their essential oil: they have adulterated her with so many arguments and extraneous reasonings that she has become varied, different for each man,[45] having lost her own unchanging universal visage and so making us seek her testimony from the beasts, which are not subject to bias, corruption or diversity of opinion. For while it is indeed true that even they do not always exactly follow the path of Nature, yet they stray so little from it that you can always see Nature's rut. It is as with horses: when you lead them along they jump about, making little rebellions which extend no further than their leading-reins, meanwhile always following the steps of the man who is guiding them; and like the hawk which takes to flight, but always under the control of its string.

[C] *'Exilia, tormenta, bella, morbos, naufragia meditare, ut nullo sis mala tiro.'* [Practise banishments, torments, wars, diseases and shipwrecks, so that you may not be a tyro in any misfortune.][46] –   [B]   What is the use of

---

45. Echo of Plutarch (tr. Amyot), *De l'amour envers les enfans*, 100 F. (Also the general influence of *Que les bestes brutes usent de raison*, 271 A–273 G).
46. A *tiro* (a recruit or beginner) practises (*meditat*) the difficulties he must overcome. Cf. note 56, below. Two clauses of Seneca conflated: *Epist. moral.*, XCI, 8; CVII, 4: (Here begins a major rejection of aspects of Seneca's stoicism.)

that curious desire to anticipate all the ills that can befall human nature and to prepare ourselves even against those which may perhaps never touch us? [C] '*Parem passis tristiam facit, pati posse*' [The possibility of suffering makes one as sad as actual suffering]:[47] we are hit not only by the bullet but by its bang and its wind! [B] Or why, like the most fevered minds (for fever it is) do we ask to be whipped right now, just because it may be that Fortune will, perhaps, make you suffer a whipping some day? [C] Or why do you not don your fur coat on Midsummer's Day, because you will need it at Christmas!

[B] 'Cast yourself into experiencing such ills as *may* befall you, [C] especially [B] the more extreme ones:[48] test yourself against them,' men say, 'make absolutely certain.'

On the contrary; it would be more easy and more natural to free your very thoughts of such a burden. They will not come quick enough! Their true essence does not last long enough for us! And so, as though they did not weigh sufficiently upon our senses, our minds must go and extend them and prolong them, incorporating them within us beforehand. [C] 'They will weigh on us enough once they are there,' said one of the leaders, not of the tenderest school but the toughest. 'Meanwhile decide in your own favour: believe what suits you best. What use is it to you to go welcoming and anticipating your ill fortune, losing the present because of fear of the future, and being miserable now because you must be so eventually?'[49] Those are his very words.

[B] 'Learning certainly does us a good service by instructing us very precisely about the dimensions of all evils':

> *Curis acuens mortalia corda.*

> [Sharpening with cares the minds of men.][50]

What a pity if a little of their size should escape our sensations and our knowledge! It is certain that most preparations for death have caused more torment than undergoing it. [C] It was said in former times, most freely, by a most judicious author, '*minus afficit sensus fatigatio quam cogitatio.*' [Our senses are less affected by hardships than by hard thinking.][51]

47. Seneca, *Epist. moral.*, LXXIV, 4: Seneca used against himself.
48. '88: especially *all* such ills as may befall you, at *least* the more extreme ones.
49. Seneca, *Epist. moral.*, XIII, 12–13; 10; XXIV, 2 (conflated).
50. Virgil, *Georgics*, I, 123.
51. Quintilian, I, xii, 11.

The feeling that death is present is, of itself, sometimes enough to stir us to a quick resolve no longer to seek to avoid the inevitable. Several gladiators in former times were seen, after putting up a cowardly fight, to accept death most courageously, offering their throats to their opponents' swords and welcoming them; but contemplating a future death requires a more leisurely steadfastness, one more difficult therefore to supply.[52]

[B] If you do not know how to die, never mind. Nature will tell you how to do it on the spot, plainly and adequately. She will do this job for you most punctiliously: do not worry about it:

> *Incertam frustra, mortales, funeris horam*
> *Quæritis, et qua sit mors aditura via.*

[In vain, O mortals, do you strive to know the uncertain hour of your death and by which road it will come.][53]

> *Pæna minor certam subito perferre ruinam,*
> *Quod timeas gravius sustinuisse diu.*

[It is less painful to have to undergo sudden and sure destruction than long to anticipate what you fear the most.]

We confuse life with worries about death, and death with worries about life. [C] One torments us: the other terrifies us. [B] We are not preparing ourselves to die: that is too momentary a matter. [C] A quarter of an hour of pain, without after-effects, without annoyance, has no need of precepts of its own. [B] To speak truly, we prepare ourselves against our preparations for death! Philosophy first commands us to have death ever before our eyes, to anticipate it and to consider it beforehand, and then she gives us rules and caveats in order to forestall our being hurt by our reflections and our foresight! Thus do doctors tip us into illnesses in order that they may have the means of employing their drugs and their Art.

[C] If we have not known how to live, it is not right to teach us how to die, making the form of the end incongruous with the whole. If we have known how to live steadfastly and calmly we shall know how to die the same way. They may bluster as much as they like, saying that '*tota philosophorum vita commentatio mortis est*' [the entire life of philosophers is a

52. Seneca, *Epist. moral.*, XXX, 7.
53. Propertius, II, xxvii, 1–2 (adapted); then Pseudo-Gallus, *Elegeia*, I, 277–8.

preparation for death];[54] but my opinion is that death is indeed the ending of life, but not therefore its End: it puts an end to it; it is its ultimate point; but it is not its objective. Life must be its own objective, its own purpose. Its right concern is to rule itself, govern itself, put up with itself. Numbered among its other duties included under the general and principal heading, *How to live*, there is the sub-section, *How to die*. If our fears did not lend it weight, dying would be one of our lighter duties.

[B] Judging from their usefulness and naïve truth, the teachings of Simple-mindedness are not much inferior to those contrary ones which are lectured upon by Erudition. Men differ in tastes and fortitude: they must each be brought, by differing routes, to what is good for them, each according to his nature:

[C]   *Quo me cunque rapit tempestas, deferor hospes.*

[Wherever the storm may drive me, there I land and find a welcome.][55]

[B] I have never known even one of my neighbouring peasants embark upon thoughts about what countenance and steadfastness he will show in his final hour. Nature teaches him never to reflect on death except when he lies a-dying. Then he does it with better grace than Aristotle, who is doubly oppressed by death: by death itself and by his long [C] anticipation. [B] That is why Caesar opined that the happiest and least burdensome of deaths was the one least [C] thought about.[56] '*Plus dolet quam necesse est, qui ante dolet quam necesse est.*' [He who suffers before he needs to, suffers more than he needs to.]

The painfulness of such thoughts is born of our excessive interest. We

---

54. Cicero, *Tusc. disput.*, I, xxx, 74. Cicero is citing Socrates in Plato's *Phaedrus*, 67 D, but changes *meletema*, 'practising' dying, into *commentatio*, a 'diligent meditation' upon dying. Montaigne is correcting in the light of experience what he wrote in I, 20, 'To philosophize is to learn how to die'. He now believes that most of mankind should neither 'practise' dying nor 'meditate' upon dying.

55. Horace, *Epistles*, I, i, 15.

56. Suetonius, *Life of Caesar*, lxxxvii. In 1588 Montaigne wrote '. . . by death itself and his long premeditation. That is why Caesar opined that the happiest and least burdensome death is the least premeditated.' In [C] Montaigne twice replaces the notion of 'premeditation' by words which cannot evoke the philosophical meaning of *praemeditatio*, that is, an advance 'practising' of death in rapture or ecstasy. (Cf. the final pages of the last chapter, III, 13, 'On experience'.) The quotation is from Seneca, *Epist. moral.*, XCVIII, 8.

are always getting in our own way, wishing to forestall and overmaster Nature's prescriptions. Only dons ought to die more badly when they are well, glowering at the thought of death. Common folk need no remedy nor consolation save when the blow falls; and then they reflect on it all the more justly since they are feeling it. [B] We assert (do we not) that what gives the common folk their power to endure [C] present ills, [B] as well as their profound indifference towards inauspicious future events, is their insensitivity and [C] lack of [B] understanding[57] [C] and the fact that their souls, being crass and obtuse, are less open to penetration and disturbance. [B] If that is so, then for God's sake let us adhere, from now on, to that School of animal stupidity! It leads its pupils to the ultimate profit promised by the sciences; and does it gently. We shall not lack good professors to interpret that natural simplicity. Socrates for one. For, as far as I can recall, he says more or less the following to the judges who were deliberating about his life:

Gentlemen: I am afraid that if I were to beseech you not to put me to death I should impale myself on the denunciation of my accusers: namely that I claim to know more than everyone else, because I have some more [C] secret [B] knowledge[58] of things above us and of things below. I know that I have neither frequented death nor reconnoitred it; nor do I know anyone who, having assayed what it is like, can teach me about it. Those who fear death presuppose that they know it. As for me, I know neither what death is nor what the world to come is like. Death may be something indifferent or something desirable. [C] (We may believe, however, that it is a migration, a crossing from one place to another, and that there is some improvement in going to live among so many great men who have crossed that divide – and to be free from having to deal with wicked and corrupt judges! If death be a reduction of our being to nothingness, it is still an improvement to enter upon a long and peaceful night. We know of nothing in life sweeter than quiet rest and deep dreamless sleep.) [B] That which I know to be wicked, such as harming one's neighbour and disobeying a superior, be it God

---

57. '88: endure ills, *which is greater than ours* . . . insensitivity and *beast-like* understanding . . .
58. '88: some more *inner* knowledge . . .

or man, I scrupulously avoid. I cannot go in fear of things when I do not know whether they be good or evil.[59]

[C] If I go off to my death and leave you here alive, the gods alone know whether you or I will have the better of it. So, as far as it concerns me, you will please give such a sentence as suits yourselves. But following my way of giving just and useful counsel, I do say that, unless you can see more deeply into my case than I can, you would do better for your consciences' sake to set me free; and also that, having made your judgement in keeping with my past deeds (both public and private), and also in keeping with my intentions and in keeping with the profit which so many of our citizens, both young and old, daily derive from my conversation and the advantages I bring to you, to all of you: you cannot properly release yourselves of your debt towards my merit except by issuing an order that I be maintained in the Prytaneum – at public expense, given my poverty – something which I have often seen you grant, with less reason, to others.

Do not take it as stubbornness or contempt if I do not follow precedent and become a suppliant moving you to pity.

Being no more than anyone else 'engendered by sticks and stones', as Homer puts it, I have friends and relations well able to appear before you in tears and grief; and I have three weeping children who can move you to pity. But I would bring shame on our city if, at my age, and having that reputation for wisdom (with which I am now charged) I were to sink to such cowardly behaviour. What would people say about the other Athenians! I have always counselled those who listened to me never to ransom their life by a dishonourable deed. And in my country's wars, at Amphipolis, at Potidaea, at Delium, as well in others in which I played a part, I showed in practice how far I was from ensuring my safety by my shame.

Moreover I would be compromising your sense of duty and soliciting you to do something ugly: it is not for any supplications of mine to persuade you, but for pure and solid reasons of justice to do so. You have sworn to the gods to bear yourselves thus: it would seem that I were wishing to bring a counter-indictment, suspecting

---

59. '88: good or evil. *You will therefore issue such orders as you like.* As a plea . . .

In [C] Montaigne's Socrates stresses that he should be treated *secundum se*, 'in keeping' with his deeds and character.

you of not believing that there are any gods! And I too would bear witness against myself, showing that I did not believe in them as I ought to, either, since I distrusted their governance and did not entrust my case entirely to their hands. I have complete trust in them, convinced as I am that they will act in this matter as will be best for me and for you.

Good men, whether living or dead, have nothing to fear from the gods.[60]

[B] As a plea is that not [C] crisp and sensible, yet naïve and lowly,[61] [B] unimaginably sublime, [C] true, frank and incomparably right [B] – and made in such an hour of need! [C] It was reasonable indeed of Socrates to prefer it to the one which the great orator Lysias had written out for him, excellently couched in lawyers' language but unworthy of so noble an accused.[62] Should one ever hear a word of supplication from the lips of Socrates! Should such proud virtue strike sail precisely when it was being most vigorously displayed! Should his nature, noble and puissant, have entrusted his defence to art, and when it was being most highly assayed have renounced truth and simplicity, which were the ornaments of his speech, in order to bedizen itself with the cosmetic figures and fictions of a prepared address?

He acted most wisely and in keeping with himself by not corrupting the tenor of an incorruptible life, or so august a concept of the form of humankind, in order to prolong his old age by a year and so betray the immortal memory of that glorious end.

His duty in life was not to himself but to be an example to the world: would it not have been a public catastrophe if he had ended his life in some idle obscure manner?

[B] Indeed such a detached and quiet way of rating his death deserved that posterity should rate it more highly for him. And it did so. In the whole of justice nothing is more just than what Fortune ordained for its glory. The Athenians held those who were responsible for it in such loathing that they shunned them as persons accursed: anything which they

---

60. Based on Plato's *Apology for Socrates*. Cicero resumes it, with eulogies, in *De Oratore* I, liv, 232. (Cf. Erasmus, *Apophthegmata*, III, *Socratica*, LXVI.) Honoured guests were lodged, wined and dined in the *Pyrtaneum*. The quotation from Homer is from the *Odyssey*, XIX, 163, cited by Socrates, *Apology*, 23 B.
61. '88: not *child-like*, unimaginably sublime . . .
62. Cicero, *De oratore*, cited in note 60.

touched was held to be polluted; no one would bathe in the public baths with them; no one greeted them; no one approached them; so that, finally, no longer able to bear such public opprobrium, they went and hanged themselves.[63]

If anyone reckons that I chose a bad example from among so many of Socrates' speeches which could have served my purpose, and if he judges that Socrates' reasoning here is far above the opinion of common men, well, I chose it on purpose. For I judge otherwise and maintain that his reasoning here holds a more modest rank than even common opinions and that its naïve simplicity is less elevated; [C] within an unspoiled boldness quite without artifice, and with a childlike assurance, [B] it exhibits Nature's pure and primary [C] stamp and simplicity.[64] [B] While it is credible that we should have a natural fear of pain, it is not credible that we should fear dying as such, which is a part of the essence of our being, no less than living is. For what purpose would Nature have engendered within us a loathing and horror of dying, seeing that dying rates as something extremely useful, in that it ensures succession and substitution within Nature's works and also, within the [C] commonwealth [B] of this world,[65] serves birth and increase more than loss and destruction.

Sic rerum summa novatur

[Thus is totality renewed]

[C] *Mille animas una necata dedit.*

[One death gives rise to a thousand lives.]

[B] The failing of one life is the gate to a thousand other lives.[66] [C] Nature has stamped on the beasts a concern for themselves and their own conservation. They can get as far as being afraid of harm from knocking against things and so hurting themselves and of our tying them up and beating them – things which are within their sensations and experience. What they cannot fear is that we may kill them: they do not have the faculty of imagining death or thinking about it. In addition it is said that [B] one can see them not merely suffering death gladly (most

---

63. Plutarch (tr. Amyot), *De l'envie et de la haine*, 108 EF.
64. '88: exhibits Nature's primary *concept*. While . . .
65. '88: the *task* of this world . . .
   Then, Lucretius, II, 74; Ovid, *Fasti*, I, 330.
66. '88: lives. *Let us look at the beasts*: one can see . . .

horses whinny when dying, while swans [C] sing at [B] their deaths)[67] but even seeking it when necessary, as is shown by several examples of elephants.[68]

Moreover is not the style of argument which Socrates uses here one which stuns us equally by its simplicity and its ecstatic force? In truth it is far easier to talk like Aristotle and to live like Caesar than both to talk like Socrates and live like Socrates. In him is lodged the highest degree of perfection and of difficulty. Art cannot reach it. Moreover our own faculties are not trained that way. We neither assay them nor understand them: we clothe ourselves in those of others and allow our own to lie unused – and some may say that about me, asserting that I have merely gathered here a big bunch of other men's flowers, having furnished nothing of my own but the string to hold them together.

I have indeed made a concession to the taste of the public with these borrowed ornaments which accompany me. But I do not intend them to cover me up or to hide me: that is the very reverse of my design: I want to display nothing but my own – what is mine by nature. If I had had confidence to do what I really wanted, I would have spoken utterly alone, come what may.   [C]   Yet despite my projected design and my original concept (but following the whim of the age and the exhortation of others) I burden myself with more and more of them every day. That may not become me well: I think it does not, but never mind: it might be useful to somebody else.

[B]   There are men who quote Plato and Homer without ever setting eyes on them. (I too have often taken my quotations not from the originals but from elsewhere.) Since in the place where I write I am surrounded by one thousand volumes, I could easily, if I wanted to, now borrow without trouble or scholarship, from a dozen of the kind of botchers whose pages I hardly ever turn, quite enough to   [C]   put an enamel gloss on   [B]   this treatise[69] about physiognomy. To cram myself full of quotations all I would need would be the preliminary epistle of some German! And that is the way we go seeking tidbits of glory with which to diddle this foolish world!

67. '88: while swans *celebrate it* but even . . . (The swan-song, sung at death, has become proverbial. Erasmus, *Adages*, I, II, LV, *Cygnea Cantio*.)
68. Antiquity generally denied that the beasts have reason as Man has. The Roman populace believed however that elephants in the gladiatorial arena sometimes asked to die (cf. Chanet, *De l'instinct et la connoissance des animaux*, La Rochelle, 1640, p. 178).
69. '88: enough to *enrich* this treatise . . .

[C] Those meat-pies stuffed with commonplaces by which so many eke out their studies on the cheap are useless except for commonplace topics; they can be used to show off, but not for right conduct – just such a laughable fruit of learning as served as knock-about amusement for Socrates against Euthydemus.[70] I have known books made out of materials which have never been studied or understood, the author having entrusted the research for this and that needed to construct it to divers learned friends, being content for his part with having thought up the project and then having made an industrious compilation out of that bundle of unknown materials. At least the ink and paper are his. In all conscience that is not writing a book but purchasing one, borrowing one. It shows men – something of which they might have remained in doubt – that you are unable to write one. [B] A presiding judge boasted in my presence that he had amassed two hundred or so borrowed commonplaces and worked them into one of his presidential rescripts.[71] [C] By declaring that fact to all and sundry he seemed to me to be nullifying the glory he was being given for it. [B] A petty and ridiculous vanity for my taste in such a subject and in such a person.[72]

[C] Among my many borrowings I take delight in being able to conceal the occasional one, masking it and distorting it to serve a new purpose. At the risk of letting people say that it is because I failed to understand any of the meanings in context, I give that one some peculiar slant with my own hand, so that they may all be less purely and simply someone else's. [B] But those others put their larcenies on parade and into their accounts, thereby acquiring a better claim in law than I do! [C] Followers of Nature like me reckon that, in honour, invention takes incomparably higher precedence over quotation.

[B] If I had wanted to speak from erudition I would have done so sooner: I would have written at a time closer to my studies when I had more memory and Nous. And if I had wanted to make a trade out of writing I would have had more confidence in myself at that age than I do now. [C] Moreover one particular favour which Fortune may have

70. In Plato's dialogue bearing that name.
71. Probably not an exaggeration. Such legal works cited legal authorities and maxims by the hundreds.
72. [B] in place of [C] to the end of the paragraph: *I conceal my larcenies and disguise them: others put their larcenies on parade and into their accounts: thereby acquiring a better claim in law than I do; like those who disguise horses I stain their mane and their tail, and sometimes I poke out an eye: if their first master used them as amblers I make them trot; if used for the saddle, I use them for packs.* If I had wanted . . .

granted me by means of this book would then have occurred at a more propitious season. [B] Two of my acquaintances, men of great scholarship, have in my opinion lost half their value by declining to publish at forty and waiting until they were [C] sixty.[73] [B] Like youth, maturity has its defects: worse ones. And old age is as unsuited to work of that nature as to any other. Whoever submits his senile mind to the presses is mad if he hopes to extract anything which does not stink of a man who is ugly, raving and half-asleep. Our mind as it ages becomes constipated and squat. I reveal my ignorance with copious pomp: I reveal my learning meagrely and pitifully – [C] the latter as an accessory, a by-product: the former, as explicit and primary. Strictly, I treat nothing except nothing, and I treat not science but nescience. [B] I have selected the time when my life (which I have to portray) is laid out before me: whatever remains over has more to do with dying. The only news which I would willingly still give to the public as I pack my bags would concern my dying, if I found it, as others do, to be loquacious.

It vexes me that Socrates, who was the perfect exemplar of all the great qualities, should have chanced to have so ugly a face and body (as they say he did), one so unbecoming to the beauty of his soul, [C] he who was so much in love, so madly in love, with beauty. Nature did him an injustice there. [B] There is nothing more probable than the conformity and correspondence of the body and the mind.[74] [C] *'Ipsi animi magni refert quali in corpore locati sint: multa enim e corpore existunt quæ acuant mentem, multa quæ obtundant.'* [It matters much to souls in what sort of body they are lodged; for many of the body's qualities serve to sharpen the mind, and many others make it obtuse.][75] The author here is talking about unnatural ugliness and physical deformity. But we also use ugliness to mean an immediately recognizable uncomeliness, which is lodged primarily in the face and which we often find distasteful for quite trivial causes: for its colouring, a spot, a coarse expression or for some inexplicable reason even when the limbs are well-proportioned and whole. In that category was the ugliness which clothed the most beautiful soul of La Boëtie. Such surface ugliness, imperious though it may be, is less harmful in its effects on a man's mind and is not, in people's opinion, by any means a certain prognostic. The other kind, which is strictly speaking deformity, is more

---

73. '88: *seventy . . .*
74. [B], instead of [C]: mind. *It is not credible that such dissonance should occur without some accident which ruptured the normal development.* As Socrates said . . .
75. Cicero, *Tusc. disput.*, I, xxxiii, 80.

substantial and more inclined to turn its effects inwards. The shape of the foot is revealed not only by a shoe of fine polished leather but by any close-fitting one. [B] As Socrates said of his own ugliness: it would have revealed quite justly the ugliness of his soul, had he not corrected his soul by education.[76] [C] But in saying it I hold that he was jesting as usual: never did so excellent a soul make itself.

[B] I cannot repeat often enough how highly I rate beauty, which is a powerful and most beneficial quality. (Socrates called it a 'brief tyranny' [C] and Plato 'a privilege of Nature'.)[77] [B] We have no other qualities which surpass it in repute. It holds the highest rank in human intercourse: it runs ahead of the others, carries off our judgement and biases it with its great authority and its wondrous impact. [C] Phryne would have lost her case even in the hands of an excellent advocate if she had not corrupted her judges by the brilliance of her beauty as she parted her garment.[78] And I find that Cyrus, Alexander and Caesar, those three lords of the world, did not neglect it in order to execute their great endeavours. Nor did the elder Scipio. In Greek one and the same word embraces the beautiful and the good. And the Holy Ghost often calls things good when He means beautiful.[79] I would readily defend the hierarchy of goods taken

76. Socrates' famous reply to Zopyrus the physiognomist: he was indeed born with lecherous tendencies but had re-formed his soul. (Cicero, *De fato*, V, 10; *Tusc. disput.*, IV, 80; Erasmus, *Apophthegmata*, III, *Socratica*, LXXX. Cf. above, III, 5, 'On some lines of Virgil', note 163.)

77. Erasmus, *Apophthegmata*, VII, *Aristoteles Stagirites*, XV: '"Beauty," he said, "is more efficacious than any written testimonial." Some attribute that [not to Aristotle but] to Diogenes. Aristotle used to call beauty "a gift", because it approached the nature of grace. Socrates called it "a brief tyranny", because the grace of beauty soon wilted; Plato, a "privilege of nature" because it came to few. Theophrastus called it "a silent deception", since it persuaded without words; Theocritus an "ivory harm" since, though it was fair to view, it was the cause of many inconveniences; Carneades "a kingdom without protection", since the beautiful obtain whatever they will, no force impeding them. [Diogenes] Laertius relates this.' (In the *City of God*, XV, xxii, St Augustine makes beauty a gift of God, given to both good and evil persons.)

78. Cited probably after Erasmus by Tiraquellus (*De legibus connubialibus*, II, 61), with reference to Quintilian, Athenaeus, Pausanias, Diodorus Siculus, Propertius, etc., etc. That Phryne's parting her garment to reveal her bosom was more effective than the best rhetoric became proverbial.

79. The term *kalokagathos* in Greek combined *kalos* (beautiful) and *agathos* (good). 'Beautiful' is used, rather, for 'good' in the Greek Bible: e.g. the 'good' fish in Matthew 13:48 which were gathered into vessels while the bad were cast away are termed *ta kala* (the 'beautiful' ones); similarly the 'good' seed in the parable of

from an ancient poem and song which Plato says was popular: health, beauty, wealth.[80].

Aristotle says that the right to command belongs to the beautiful and that, whenever there are persons whose beauty approaches that of the portraits of the gods, like veneration is due to them. When someone asked him why men haunt the company of the beautiful both longer and more often, he replied: 'Only a blind man should ask that.'[81] Most of the philosophers, and the greatest, paid for their tuition and acquired their wisdom by the favour and agency of their beauty.

[B] Not only in the men but in the animals serving me I consider beauty to be only two fingers away from goodness. Yet to me it seems that those facial traits and features and those distinctive characteristics from which inner complexions are inferred as well as our future destinies are things which cannot be lodged simply and directly under the headings of beauty or ugliness: no more than in times of plague pleasant smells and a clear atmosphere can promise salubrity nor all kinds of oppressiveness and stench threaten infection.

Those who accuse ladies of contradicting their beauty by their morals do not always strike home: a face may not be very well-shaped yet have an air of probity and dependability; just as, on the contrary, I have read behind a pair of beautiful eyes warnings of a malicious and dangerous character. Some physiognomies augur well: in the thick of victorious enemies you would immediately, from among men unknown, pick out one rather than another to surrender to and to entrust with your life: and you would not have been influenced strictly speaking by beauty. Looks are a weak guarantee, yet they have some influence.

If my task were to administer floggings, I would do so more severely to criminals who belie and betray the promises which Nature had planted on their features: I would inflict harsher punishment on malice in a man who looked debonair. It appears that some faces are blessed, others unblessed, and there is I think an art which can distinguish between the debonair face and the simple one, the severe and the harsh the sullen and the chagrined,

---

the sower is *kalon sperma* ('beautiful' seed). There are many other examples, especially in the Septuagint (the ancient Greek translation of the Hebrew Scriptures, of which Montaigne possessed a copy).

80. Plato, *Gorgias*, VII, 452.

81. Diogenes Laertius, *Life of Aristotle*, V, 20. Cf. H. Estienne's *Apophthegmata*, s.v. *Aristoteles*.

the arrogant and the melancholic, and such other pairs of qualities.[82] Some forms of beauty are not merely proud but haughty; others are gentle, and others still are lifeless. As for forecasting the future from them, such [C] matters [B] I leave undecided.[83]

As I have said already, as regards myself I have simply adopted raw that ancient precept which says that we cannot go wrong by following Nature, and that the sovereign precept is to conform to her.[84] Unlike Socrates I have not corrected my natural complexions by the power of reason,[85] and I have in no wise let my inclinations become confused by artifice. I let myself go as I came in: I combat nothing; my two principal parts live graciously together in peace and harmony. But, thank God, my nurse's milk was moderately healthful and temperate.

[C] May I say *en passant* that I know there is a certain scholastic concept of morality – virtually the only one current – which is held in higher esteem among us than it is worth; it is a slave to precepts and bound by hopes and fears. I like the morality which laws and religions do not make up but make perfect and authoritative, one which knows that it has the means of sustaining itself without help, one which, rooted on its own stock, is born in us from the seed of that universal reason which is stamped upon every man who is not disnatured. That Reason which straightened out Socrates' vicious kink made him obedient to the men and the gods who commanded in his city and courageous in death not because his soul was immortal but because he himself was mortal. Any instruction which convinces people that religious belief alone, without morality, suffices to satisfy God's justice is destructive of all government and is far more harmful than it is ingenious and subtle. Men's practices reveal an extraordinary distinction between devotion and the sense of right and wrong.

[B] I have a [C] bearing [B] which,[86] both in beauty and as it is interpreted, is of good augury –

> *Quid dixi habere? Imo habui, Chreme!*
>
> [What am I saying, *have*, Chremes? I mean I *had*!]

---

82. This was the aim of the art or science of physiognomy, highly developed during the Renaissance.
83. '88: such *questions* I leave undecided . . .
84. The great precept of Classical philosophy. Cf. Cicero, *Laelius*, V, 19; XII, 42.
85. '88: my natural complexions by *education* and the power of reason . . .
86. '88: I have *a face* which is . . .

Heu tantum attriti corporis ossa vides!

[Alas! You now see only the bones of this worn-down body!]⁸⁷

– and has an appearance contrary to that of Socrates. It has often happened that people who have had no previous acquaintance with me, people going merely by my fine air and [C]   presence, [B]   have put⁸⁸ great trust in me either for their own affairs or my own. And in foreign countries I have received singular and rare favour because of it. The following two experiences are perhaps both worth narrating in detail.

There was a man who had determined to take me and my house by surprise. His trick was to come alone to my gate and to press to be admitted fairly urgently. I knew him by name and had occasion to put my trust in him as a neighbour who was to some degree related to me by marriage. I opened the gate for him [C]   as I do for everyone. [B]   There he was, looking quite terrified, with his horse winded and quite exhausted. He told me the following story: one of his enemies had just come across him some half a league away. (I knew that man too and had heard of their quarrel.) He said that this enemy had followed remark-ably close on his heels. He, having been taken by surprise [C]   in disarray [B]   and being weaker [C]   in numbers, [B]   had rushed to my gate for safety; he was very worried about his men whom he said he supposed were dead [C]   or taken.⁸⁹

[B]   Very naively I assayed to strengthen, reassure and reinvigorate him. Soon after, lo and behold! four or five of his soldiers appeared, looking equally frightened and wanting to be let in. More came; then still more, until there was some twenty-five or thirty of them, all armed and well-equipped and claiming to have their enemy at their heels.

[C]   This mystery-play began to awaken my suspicions: [B]   I had not forgotten what a time we were living in, nor how much my house⁹⁰ might be coveted; and I knew of several cases of acquaintances of mine who had had similar bad experiences. Nevertheless, I considered that there was nothing to be gained by having started out to be welcoming if I did not go through with it; so, not being able to defeat them without smashing up everything, I allowed myself to take the simplest and most natural course (as I always do) and ordered them to come in.

---

87. Terence, *Heautontimorumenos*, I, i, 42; Pseudo-Gallus, I, 238.
88. '88: air and *bearing* have put . . .
89. '88: dead *and defeated, having been come across when in disorder and widely separated from each other.* Very naively . . .
90. '88: my house, *notwithstanding the vain truce in which we then were*, might be . . .

Besides, by my nature I am neither very suspicious nor distrustful; and that is the truth. I have a strong tendency to find justifications and the kindest interpretation. I judge men according to the common order of Nature; I do not believe in perverted and disnatured tendencies, any more than in portents and miracles, unless I am forced to do so by some major piece of evidence. I am moreover a man inclined to trust myself to Fortune and to allow myself to dash into her arms. Up to the present I have had more reason to congratulate myself on that than to pity myself, and I have found Fortune [C] both better informed and better disposed towards my affairs than I am. [B] There[91] have been a few deeds in my life the handling of which could rightly be called difficult or, if you wish, wise. Allow even a third of those to be due to me: but two-thirds, certainly, were abundantly due to her.

– [C] Where we go wrong, if you ask me, is in not entrusting ourselves enough to Heaven and in expecting more from our own conduct of affairs than rightly belongs to us. That explains why our schemes so often go awry. Heaven is jealous of the scope which we allow to the rights of human wisdom to the prejudice of its own: the more we extend them the more Heaven cuts them back. –

[B] Those armed men remained mounted in my courtyard, while their leader was with me in my hall; he had not wished his horse to be stabled, saying that he would withdraw as soon as he had news of his men. He saw he was master of the situation and that the moment had come to execute his plan. Subsequently he often told me – for he was not afraid to tell his tale – that what wrenched his treachery from his grasp were my countenance and my frank behaviour. He got back into the saddle; his men, keeping their eyes constantly fixed on him to catch what signal he would give them, were amazed to see him ride out, surrendering his advantage.

On another occasion, trusting to some truce or other which had just been proclaimed between our forces, I was on the road travelling through some particularly ticklish terrain. As soon as wind of me got about, three or four groups of horsemen set out from different places to trap me. After three days one of them made contact with me and I was charged by fifteen or twenty masked gentlemen[92] followed by a wave of mounted bowmen. There I was, captured; having surrendered I was dragged off into the thick of some neighbouring forest, deprived of my horse and luggage, my

91. '88: Fortune *wiser than me.* There . . .
92. '88: masked gentlemen, *well mounted and well armed,* followed . . .

coffers ransacked, my strong-box seized. Horses and equipment were divided between their new owners. We haggled for some time in that thicket over my ransom, which they had pitched so high that it was obvious that they knew little about me. A great quarrel started between them over whether they would let me live. There were indeed several threatening circumstances which showed what a dangerous situation I was in:

[C]   *Tunc animis opus, Aenea, tunc pectore firmo.*

[Now, Aeneas, you need all your courage and a firm mind.][93]

[B]   I continued to hold out for the terms of my surrender: that I should give up to them only what they had won by despoiling me (which was not to be despised), with no promise of further ransom. We were there for two or three hours when they set me on a nag unlikely to want to bolt away and committed me, individually, to be brought along under the guard of some fifteen or twenty men armed with harquebuses, while my men were dispersed among other such soldiers, each with orders to escort us as prisoners along different routes. I had already covered the distance of some two or three harquebus shots,

Jam prece Pollucis, jam Castoris implorata,

[Having by then prayed to Pollux and implored Castor,][94]

when, all of a sudden, a most unexpected change came upon them. I saw their leader ride over to me, [C] using most gentle words and [B] putting[95] himself to the trouble of searching among his troops for my scattered belongings, which, insofar as he could find them, he returned to me, not excluding my strong-box.[96] In the end they gave me my best present, my freedom: the rest hardly affected me [C] at the time.

[B]   The true cause of so novel a volte-face, of such second thoughts which derived from no apparent impulsion, of so miraculous a reversal of intent, at such a time and in the course of such an operation which was fully thought through and deliberated upon and which custom had made

93. Virgil, *Aeneid*, VI, 261.
94. Catullus, LXVI, 65.
95. '88: ride over to me, *no longer with his threats but with* words *full of courtesy,* putting . . .
96. '88: find them, *the principal ones of which* he returned to me, not excluding my *purse and* my strong-box . . .

lawful (for from the outset I openly admitted which side I was on and the road I was taking), I certainly do not really know even now. The most prominent man among them took off his mask and informed me of his name;[97] he then told me several times that I owed my liberation to my countenance as well as to my freedom and firmness of speech which made me unworthy of such a misfortune; and he asked me to promise if necessary to return him the compliment.

It is possible that God in his goodness wished to make use of such trivial means to preserve me. (He protected me again the following day from other and worse [C] ambushes[98] [B] which these very men had warned me about.)

The man in the second of these incidents is still alive to tell the tale: the man in the first was killed a little while ago.

If my countenance did not vouch for me, if people did not read in my eyes the innocence of my intentions, I would never have endured so long without feud or offence, given my indiscriminate frankness in saying, rightly or wrongly, whatever comes into my mind and in making casual judgements. Such a style may rightly appear discourteous and ill-suited to our manners, but I have never found anyone who considered it abusive or malevolent or who, provided he had it from my own mouth, was stung by my frankness. (Reported words have both a different resonance and a different sense.) Besides I do not hate anybody; and I am such a coward about hurting people that I cannot do it even to serve a rational end: when circumstances have required me to pass sentences on criminals I have preferred not to enforce justice; [C] '*Ut magis peccari nolim quam satis animi ad vindicanda peccata habeam.*' [I wish the only crimes committed were those which I really had the heart to punish.][99] Aristotle was reproached with being too merciful to a wicked man. 'True,' he said. 'But I was merciful to the man not to the wickedness.'[100]

Judgements normally inflame themselves towards revenge out of horror for the crime. That is precisely what tempers mine: my horror for the first murder makes me frightened of committing a second, and my loathing for the original act of cruelty makes me loathe to imitate it. [B] I am only

97. '88: his name. *(I would love, in my turn, to assay what expression he would show in a similar event.)* He then . . .

98. '88: other and worse *dangers* which . . .

99. Not identified by Marie de Gournay or others. (The general theme is that of Tiraquellus in *De poenis legum temporandis.*)

100. Erasmus, *Apophthegmata*, VII, *Aristoteles Stagirites*, V.

a [C] Jack [B] of Clubs,[101] but you can apply to me what was said of Charillus King of Sparta: 'He cannot be good: he is not bad to the wicked.' Or (since Plutarch presents it, as he does hundreds of other things, in two opposite and contrasting manners) you can put it thus: 'He must be good: he is good even to the wicked.'[102]

When the deeds are not illegal, and those who do them dislike them, I am loath to act against them: so too if the deeds are illegal, and those who do them delight in them, then (to tell the truth) I am not over-scrupulous when acting against them.

101. '88: I am only a *Knave* of Clubs . . .
The gentleman in Montaigne had second thoughts about the term *varlet* (knave) even when used in a metaphor, so he replaced *varlet* by *escuyer*, squire (a knight's attendant). Both terms were used more or less indifferently, just as we use both *Jack* and *Knave* for the playing-cards.
102. Plutarch puts the contrasting point of view in his *Life of Lycurgus*, iv. On other occasions he condemns Charillus: cf. *De l'envie et de la haine*, 108 C; *Les dicts notables des Lacedaemoniens*, 215 D; *Comment on pourra discerner le flatteur d'avec l'amy*, 44 B. Erasmus (*Apophthegmata*, I, *Archidamas*, XXXVIII) also blames it, adding, 'That outstanding man Archidamas perceived that mercy needs to be associated with justice. Otherwise what is a prince's leniency towards offenders but cruelty toward the good?'

13. On experience

====

[The end of Montaigne's quest. See the Introduction, pp. xliv ff.]

[B]   No desire is more natural than the desire for knowledge.[1] We assay all the means that can lead us to it. When reason fails us we make use of experience –

> [C]   *Per varios usus artem experientia fecit:*
> *Exemplo monstrante viam.*

[By repeated practice, and with example showing the way, experience constructs an art.][2]

Experience is a weaker and   [C]   less dignified means: [B]   but truth[3] is so great a matter that we must not disdain any method which leads us to it. Reason has so many forms that we do not know which to resort to: experience has no fewer. The induction which we wish to draw from the   [C]   likeness   [B]   between events is unsure since they all show unlikenesses.[4] When collating objects no quality is so universal as diversity and variety.[5] As the most explicit example of likeness the Greeks, Latins and we ourselves allude to that of eggs, yet there was a man of Delphi among others who recognized the signs of difference between eggs and never mistook one for another;[6]   [C]   when there were several hens he could tell which egg came from which.   [B]   Of itself, unlikeness obtrudes into

---

1. The opening sentence of Aristotle's *Metaphysics*.
2. Manilius, *Astronomica*, I, 62–3.
3. '88: weaker and *baser* means: but truth . . .
4. '88: from the *comparison* between events . . .
   Montaigne is contesting Aristotle's assertion that arts and sciences derive from judgements upon experiences.)
5. 'Collating objects': Montaigne's term *image des choses* is technical and based on Latin usage: *imago* in this sense is the comparison of form with form by some likeness between them.
6. Erasmus, *Adages*, I, V, X, *Non tam ovum ovo simile* (as we say, 'As alike as two eggs'), citing Montaigne's example of the 'man at Delphi' (or, rather, the *men* at *Delos*) who had this skill, from Cicero, *Academica*, II, (Lucullus) xviii, 58–9.

anything we make. No art can achieve likeness. Neither Perrozet nor anyone else can so carefully blanch and polish the backs of his playing-cards without at least some players being able to tell them apart simply by watching them pass through another player's hands. Likeness does not make things 'one' as much as unlikeness makes them 'other':   [C]   Nature has bound herself to make nothing 'other' which is not unlike.

[B]   That is why I am not pleased by the opinion of that fellow who sought to rein in the authority of the judges with his great many laws, 'cutting their slices for them'.[7] He was quite unaware that there is as much scope and freedom in interpreting laws as in making them. (And those who believe that they can assuage our quarrels and put a stop to them by referring us to the express words of the Bible cannot be serious: our minds do not find the field any less vast when examining the meanings of others than when formulating our own – as though there were less animus and virulence in glossing than inventing!)

We can see how wrong that fellow was: in France we have more laws than all the rest of the world put together – more than would be required to make rules for all those worlds of Epicurus;   [C]   *'ut olim flagitiis, sic nunc legibus laboramus'* [we were once distressed by crimes: now, by laws].[8]   [B]   And, even then, we have left so much to the discretion and opinion of our judges that never was there liberty so licentious and powerful. What have our legislators gained by isolating a hundred thousand categories and specific circumstances, and then making a hundred thousand laws apply to them? That number bears no relationship to the infinite variations in the things which humans do. The multiplicity of our human inventions will never attain to the diversity of our cases. Add a hundred times more: but never will it happen that even one of all the many thousands of cases which you have already isolated and codified will ever meet one future case to which it can be matched and compared so exactly that some detail or some other specific item does not require a specific judgement. There is hardly any relation between our actions (which are perpetually changing) and fixed unchanging laws.

The most desirable laws are those which are fewest, simplest and most general. I think moreover that it would be better to have none at all than to have them in such profusion as we do now. Nature always gives us

---

7. Tribonian, the 'architect of the Pandects' of Justinian. He 'cut their slices' by carving up the Roman laws into gobbets. For an attack on him in the same terms, cf. Rabelais, *Tiers Livre*, TLF, XLIIII, 82–94.

8. Tacitus, *Annals*, III, xxv.

happier laws than those we give ourselves. Witness that Golden Age portrayed by the poets[9] and the circumstances in which we see those peoples live who have no other laws. There is a nation who take as the judge of their disputes the first traveller who comes journeying across their mountains; another which chooses one of their number on market-days and he judges their cases there and then.[10] Where would be the danger if the wisest men among us were to decide our cases for us according to the details which they have seen with their own eyes, without being bound by case-law or by established precedent? For every foot its proper shoe.

When King Ferdinand sent colonies of immigrants to the Indies he made the wise stipulation that no one should be included who had studied jurisprudence, lest lawsuits should pullulate in the New World – law being of its nature a branch of learning subject to faction and altercation: he judged with Plato that to furnish a country with lawyers and doctors is a bad action.[11]

Why is it that our tongue, so simple for other purposes, becomes obscure and unintelligible in wills and contracts? Why is it that a man who expresses himself with clarity in anything else that he says or writes cannot find any means of making declarations in such matters which do not sink into contradictions and obscurity? Is it not that the 'princes' of that art,[12] striving with a peculiar application to select traditional terms and to use technical language, have so weighed every syllable and perused so minutely every species of conjunction that they end up entangled and bogged down in an infinitude of grammatical functions and tiny sub-clauses which defy all rule and order and any definite interpretation? [C] *'Confusum est quidquid usque in pulverem sectum est.'* [Cut anything into tiny pieces and it all becomes a mass of confusion.][13]

[B] Have you ever seen children making assays at arranging a pile of quicksilver into a set number of segments? The more they press it and knead it and try to make it do what they want the more they exasperate the taste for liberty in that noble metal: it resists their art and proceeds to

9. The poets stressed that in the Golden Age, 'there was no mine and thine'; and Ovid, in the *Metamorphoses*, I, 89 ff., stresses that no law was needed since each was guided by his innocent natural sense of right and wrong.
10. Given Montaigne's assimilation of Indians to happy primitive tribes in the Golden Age, those nations are doubtless to be sought in the Americas.
11. Guillaume Bouchet, *Serées*, IX; Plato, *Republic*, III, 405 A.
12. Experts in the 'art' of law were often, even on the title-pages of their own books, referred to as 'princes'.
13. Seneca, *Epist. moral.*, LXXXIX, 3.

scatter and break down into innumerable tiny parts. It is just the same here: for by subdividing those subtle statements lawyers teach people to increase matters of doubt; they start us off extending and varying our difficulties, stretching them out and spreading them about. By sowing doubts and then pruning them back they make the world produce abundant crops of uncertainties and quarrels,   [C]   just as the soil is made more fertile when it is broken up and deeply dug: *'difficultatem facit doctrina'* [it is learning which creates the difficulty].[14]

[B]   We have doubts on reading Ulpian: our doubts are increased by Bartolo and Baldus.[15] The traces of that countless diversity of opinion should have been obliterated, not used as ornaments or stuffed into the heads of posterity. All I can say is that you can feel from experience that so many interpretations dissipate the truth and break it up. Aristotle wrote to be understood: if he could not manage it, still less will a less able man (or a third party) manage to do better than Aristotle, who was treating his own concepts. By steeping our material we macerate it and stretch it. Out of one subject we make a thousand and sink into Epicurus' infinitude of atoms by proliferation and subdivision. Never did two men ever judge identically about anything, and it is impossible to find two opinions which are exactly alike, not only in different men but in the same men at different times. I normally find matter for doubt in what the gloss has not condescended to touch upon. Like certain horses I know which miss their footing on a level path, I stumble more easily on the flat.

Can anyone deny that glosses increase doubts and ignorance, when there can be found no book which men toil over in either divinity or the humanities whose difficulties have been exhausted by exegesis? The hundredth commentator dispatches it to his successor prickling with more difficulties than the first commentator of all had ever found in it. Do we ever agree among ourselves that 'this book already has enough glosses: from now on there is no more to be said on it?' That can be best seen from legal quibbling. We give force of law to an infinite number of legal authorities, an infinite number of decisions and just as many interpretations. Yet do we ever find an end to our need to interpret? Can we see any

---

14. Quintilian, X, iii, 16 (explaining why peasants and uneducated folk speak more directly and less hesitantly).

15. Ulpian, the great second-century jurisconsult; the other two are Italian medieval glossators. Criticisms of such glossators was common in France among partisans of certain schools of legal methodology who included Guillaume Budé and Rabelais (cf. *Pantagruel*, TLF, IX *bis*, 76–100, etc.).

progress or advance towards serenity? Do we need fewer lawyers and judges than when that lump of legality was in its babyhood?

On the contrary we obscure and bury the meaning: we can no longer discern it except by courtesy of those many closures and palisades. Men fail to recognize the natural sickness of their mind which does nothing but range and ferret about, ceaselessly twisting and contriving and, like our silkworms, becoming entangled in its own works: *'Mus in pice.'* [A mouse stuck in pitch.][16] It thinks it can make out in the distance some appearance of light, of conceptual truth: but, while it is charging towards it, so many difficulties, so many obstacles and fresh diversions strew its path that they make it dizzy and it loses its way. The mind is not all that different from those dogs in Aesop which, descrying what appeared to be a corpse floating on the sea yet being unable to get at it, set about lapping up the water so as to dry out a path to it, [C] and suffocated themselves.[17] And that coincides with what was said about the writings of Heraclitus by Crates: they required a reader to be a good swimmer, so that the weight of his doctrine should not pull him under nor its depth drown him.[18]

[B] It is only our individual weakness which makes us satisfied with what has been discovered by others or by ourselves in this hunt for knowledge: an abler man will not be satisfied with it. There is always room for a successor – [C] yes, even for ourselves – [B] and a different way to proceed. There is no end to our inquiries: our end is in the next world.[19]

[C] When the mind is satisfied, that is a sign of diminished faculties or weariness. No powerful mind stops within itself: it is always stretching out and exceeding its capacities. It makes sorties which go beyond what it can achieve: it is only half-alive if it is not advancing, pressing forward, getting driven into a corner and coming to blows; [B] its inquiries are shapeless and without limits; its nourishment consists in [C] amazement, the hunt and [B] uncertainty,[20] as Apollo made clear enough to us by his speaking (as always) ambiguously, obscurely and obliquely, not glutting us but

16. Erasmus, *Adages*, II, III, LXVIII.
17. [B] instead of [C]: path to it, and *killed* themselves. It is . . .
18. Not Crates but Socrates, not the proverbially obscure Heraclitus, but a certain Delius; cf. Erasmus, *Adages*, I, III, XXXVI, *Davus sum non Oedipus*, linking the saying to Heraclitus and to Diogenes Laertius, *Life of Socrates*, II, xxii.
19. A step in the argument from the opening quotation from *Metaphysics*, I, i: see the Introduction, p. xlv.
20. '88: consists in *doubt* and uncertainty . . .

keeping us wondering and occupied.²¹ It is an irregular activity, never-ending and without pattern or target. Its discoveries excite each other, follow after each other and between them produce more.

> *Ainsi voit l'on, en un ruisseau coulant,*
> *Sans fin l'une eau apres l'autre roulant,*
> *Et tout de rang, d'un eternel conduict,*
> *L'une suit l'autre, et l'une l'autre fuyt.*
> *Par cette-cy celle-là est poussée,*
> *Et cette-cy par l'autre est devancée:*
> *Tousjours l'eau va dans l'eau, et tousjours est-ce*
> *Mesme ruisseau, et toujours eau diverse.*

[Thus do we see in a flowing stream water rolling endlessly on water, ripple upon ripple, as in its unchanging bed water flees and water pursues, the first water driven by what follows and drawn on by what went before, water eternally driving into water – even the same stream with its waters ever-changing.]²²

It is more of a business to interpret the interpretations than to interpret the texts, and there are more books on books than on any other subject: all we do is gloss each other. [C] All is a-swarm with commentaries: of authors there is a dearth. Is not learning to understand the learned the chief and most celebrated thing that we learn nowadays! Is that not the common goal, the ultimate goal, of all our studies?

Our opinions graft themselves on to each other. The first serves as stock for the second, the second for a third. And so we climb up, step by step. It thus transpires that the one who has climbed highest often has more honour than he deserves, since he has only climbed one speck higher on the shoulders of his predecessor.

[B] How often and perhaps stupidly have I extended my book to make it talk about itself: [C] stupidly, if only because I ought to have remembered what I say about other men who do the same: namely that those all-too-pleasant tender glances at their books witness that their hearts are a-tremble with love for them, and that even those contemptuous drubbings with which they belabour them are in fact only the pretty little rebukes of motherly love (following Aristotle for whom praise and dispraise of oneself often spring from the same type of pride).²³ For I am not sure that everyone will understand what entitles me to do so: that I must have

---

21. Cf. III, 11, 'On the lame', note 9. Apollo was surnamed *loxias*, 'obscure'.
22. Etienne de La Boëtie, *A Marguerite de Carle*.
23. Aristotle, *Nicomachaean Ethics*, II, vii, 12, 1108a.

more freedom in this than others do since I am specifically writing about myself and (as in the case of my other activities) about my writings.

[B]   I note that Luther has left behind in Germany as many – indeed more – discords and disagreements because of doubts about his opinions than he himself ever raised about Holy Scripture.[24] Our controversies are verbal ones. I ask what is nature, pleasure, circle or substitution. The question is about words: it is paid in the same coin. – 'A stone is a body.' – But if you argue more closely: 'And what is a body?' – 'Substance.' – 'And what is substance?' And so on; you will eventually corner your opponent on the last page of his lexicon. We change one word for another, often for one less known. I know what 'Man' is better than I know what is animal, mortal or reasonable.[25] In order to satisfy one doubt they give me three; it is a Hydra's head.[26] Socrates asked Meno what virtue is. 'There is,' said Meno, 'the virtue of a man, a woman, a statesman, a private citizen, a boy and an old man ' That's a good start,' said Socrates. 'We were looking for a single virtue and here is a swarm of them.'[27] We give men one question and they hand us back a hive-full.

Just as no event and no form completely resembles another, neither does any completely differ.   [C]   What an ingenious medley is Nature's: if our faces were not alike we could not tell man from beast: if they were not unalike we could not tell man from man.[28]   [B]   All things are connected by some similarity; yet every example limps and any correspondence which we draw from experience is always feeble and imperfect;[29] we can nevertheless find some corner or other by which to link our comparisons. And that is how laws serve us: they can be adapted to each one of our concerns by means of some   [C]   twisted,   [B]   forced[30] or oblique interpretation.

Since the moral laws which apply to the private duties of all individuals

24. A true statement. Geneseolutherans, Philippists (Melanchthonians), etc. formed hostile schools.
25. *The* example of a perfect definition, which can be used both ways: you can start from the definition and arrive at Man: start from Man and arrive at this definition: Priscian, *Opera*, 1527, XVII, 1180.
26. Cf. Erasmus, *Adages*, I, X, IX, *Hydram Secas*; you cut off one head of the serpent Hydra and several others grow in its place. (Well-known from Plato in the *Republic*, I, 427 A, where it is applied to the multiplicity of laws in an ill-governed state.)
27. Plutarch (tr. Amyot), *De la vertu*, 31, CD.
28. St Augustine, *City of God*, XXI, viii.
29. Cf. Cicero, *Academica*, II (*Lucullus*), 56.
30. '88: some *fine drawn-out*, forced . . .

are so difficult to establish (as we see that they are), not surprisingly those laws which govern collections of all those individuals are even more so. Consider the form of justice which has ruled over us: it is a true witness to the imbecility of Man, so full it is of contradiction and error. Wherever we find favouritism or undue severity in our justice – and we can find so much that I doubt whether the Mean between them is to be found as frequently – they constitute diseased organs and corrupt members of the very body and essence of Justice. Some peasants have just rushed in to tell me that they have, at this very moment, left behind in a wood of mine a man with dozens of stab-wounds; he was still breathing and begged them of their mercy for some water and for help to lift him up. They say that they ran away fearing that they might be caught by an officer of the law and (as does happen to those who are found near a man who has been killed) required to explain this incident; that would have ruined them, since they had neither the skill nor the money to prove their innocence. What ought I to have said to them? It is certain that such an act of humanity would have got them into difficulties.

How many innocent parties have been discovered to have been punished – I mean with no blame attached to their judges? And how many have never been discovered? Here is something which has happened in my time: some men had been condemned to death for murder; the sentence, if not pronounced, was at least settled and determined. At this juncture the judges were advised by the officials of a nearby lower court that they were holding some prisoners who had made a clean confession to that murder and thrown an undeniable light on to the facts. The Court deliberated whether it ought to intervene to postpone the execution of the sentence already given against the first group. The judges considered the novelty of the situation; the precedent it would constitute for granting stays of execution, and the fact that once the sentence had been duly passed according to law they had no powers to change their minds. In short those poor devils were sacrificed to judicial procedures. Philip or somebody provided for a similar absurdity in the following manner.[31] He had condemned a man to pay heavy damages to another and the sentence had been pronounced. Some time afterwards the truth was discovered and he realized that he had made an unjust judgement. On one side there were the interests of the case as now proven: on the other, the interests of judicial procedure. To some extent he satisfied both, allowing the sentence to stand while reimbursing from his own resources the expenses of the condemned

31. Plutarch (tr. Amyot), *Dicts notables des anciens Roys*, 192 B.

man. But he was dealing with a reparable situation: those men of mine were irreparably hanged. [C] How many sentences have I seen more criminal than the crime . . .

[B] All this recalls to my mind certain opinions of the Ancients: that a man is obliged to do retail wrong if he wants to achieve wholesale right, committing injustices in little things if he wants to achieve justice in great things; that human justice is formed on the analogy of medicine, by which anything which is effective is just and honourable;[32] that, as the Stoics held, Nature herself acts against justice in most of her works;[33] [C] or, what the Cyrenaics hold, that nothing is just *per se*, justice being a creation of custom and law; and what the Theodorians hold: that the wise man, if it is useful to him, may justifiably commit larceny, sacrilege and any sort of lechery.[34]

[B] It cannot be helped. My stand is that of Alcibiades: never, if I can help it, will I submit to be judged by any man on a capital charge, during which my life or honour depend more on the skill and care of my barrister than on my innocence.[35]

I would risk the kind of justice which would take cognizance of good actions as well as bad and give me as much to hope for as to fear: not to be fined is an inadequate reward to bestow on a man who [C] has achieved better than simply doing no wrong. [B] Our justice[36] offers us only one of her hands, and her left one at that. No matter who the man may be, the damages are against him. [C] In China (a kingdom whose polity and sciences surpass our own exemplars in many kinds of excellence without having had any contact with them or knowledge of them and whose history teaches me that the world is more abundant and diverse than

32. Plutarch (tr. Amyot): of. Jason, Tyrant of Thessalia; *Instruction pour ceulx qui manient affaires d'estat*, 173 F; *Pourquoy la justice divine differe quelquefois la punition des malefices*, 265 C (analogy with medicine).

33. A surprising statement. The Stoics took Nature as their standard of value. But their conception of Nature was paradoxical and, as such, attacked by Plutarch (tr. Amyot), *Que les Stoïques disent des choses plus estranges que les poëtes* (560C – 561A); *Les contredicts des Philosophes Stoïques* (561A – 574 C); *Des communes conceptions contre les Stoïques* (574 C – 588 F). Montaigne's assertion may possibly be read into such objections, but one would expect him to have some definite authority behind him.

34. From Diogenes Laertius, *Life of Aristippus*, II, xciii and xcix; Coelius Richerius Rhodiginus, *Antiquae lectiones*, XIV, vi.

35. Cf. Henry Estienne, *Apophthegmata*, s. v. Alcibiades.

36. '88: on a man who *is not merely free from evil-doing but who acts better than others.* Our justice . . .

either the ancients or we ever realized), the officials dispatched by the prince to inspect the condition of his provinces do punish those who act corruptly in their posts but also make *ex gratia* rewards to those who have behaved above the common norm and beyond the obligations of duty. You appear before them not simply to defend yourself but to gain something, and not simply to receive your pay but to be granted bounties.[37]

[B] Thank God no judge has ever addressed me *qua* judge in any case whatsoever, my own or a third party's, criminal or civil. No prison has had me inside it, not even to stroll through it. Thinking about it makes the very sight of a prison even from the outside distressing to me. I so hunger after freedom that if anyone were to forbid me access to some corner of the Indies I would to some extent live less at ease. And as long as I can find earth or sky open to me elsewhere I will never remain anywhere cowering in hiding. My God, how badly would I endure the conditions of those many people I know of who, for having had an altercation with our laws, are pinned to one region of this Kingdom, banned from entering our main cities or our courts or from using the public highways. If the laws that I obey were to threaten me only by their finger-tips I would be off like a shot looking for other laws, no matter where they might be. All my petty little wisdom during these civil wars of ours is applied to stop laws from interfering with my freedom to come and go.

Now laws remain respected not because they are just but because they are laws. That is the mystical basis of their authority. They have no other. [C] It serves them well, too. Laws are often made by fools, and even more often by men who fail in equity because they hate equality:[38] but always by men, vain authorities who can resolve nothing.

No person commits crimes more grossly, widely or regularly than do our laws. If anyone obeys them only when they are just, then he fails to obey them for just the reason he must![39] [B] Our French laws, by their chaotic deformity, contribute not a little to the confused way they are

37. China, increasingly known, especially from Jesuit sources, vastly widened the horizons of Renaissance moralists. Montaigne's account doubtless derives from Juan Gonzalez, whose *Historia de las cosas mas notables de la China* (Rome, 1585) was rapidly translated into French by L. de la Porte (Paris, 1588).
38. Cicero contrasts justice with equity (*De oratore*, I, lvi, 240). It was a legal contention that, in law, equity is above all to be observed (Spiegel, *Lexicon Juris Civilis*, s.v. *Aequitas*).
39. [B] instead of [C]: no other. If anyone *obeys the law because it is just, obeys it not.* Our French laws . . .

applied and the corrupt way in which they are executed. The fact that their authority is so vague and inconsistent to some extent justifies our disobeying them and our faulty interpretation, application and enforcement of them.

Whatever we may in fact get from experience, such benefit as we derive from other people's examples will hardly provide us with an elementary education if we make so poor a use of such experience as we have presumably enjoyed ourselves; that is more familiar to us and certainly enough to instruct us in what we need.

I study myself more than any other subject. That is my metaphysics; that is my physics.[40]

> *Qua Deus hanc mundi temperet arte domum,*
> *Qua venit exoriens, qua deficit, unde coactis*
> *Cornibus in plenum menstrua luna redit;*
> *Unde salo superunt venti, quid flamine captet*
> *Eurus, et in nubes unde perennis aqua?*

[By what artifice God governs this world, our home; where the moon comes from, where she does go and how she does bring her horns together month after month and so grow full; whence the gales spring which rule the salty sea, and what dominion does the South Wind enjoy; whence come those waters which are ever in the clouds?][41]

> [C]   *Sit ventura dies mundi quæ subruat arces?*

[And will there come a day when our hills shall be made low?]

> [B]   *Quaerite quos agitat mundi labor.*

[It is for those who are worried by problems about how the world works to inquire into that.]

I, unconcerned and ignorant within this universe, allow myself to be governed by this world's general law, which I shall know sufficiently when I feel it. No knowledge of mine will bring it to change its course: it will not take a different road for my sake. It is madness to wish it to; greater madness to be upset by that fact, since such law is, of necessity, unvarying, generic and applied to all. The goodness and sway of the Ruler should

---

40. That is, a study of his own self replaces a study of Aristotle's *Metaphysics* and *Physics*.
41. Propertius, III, v, 26–30, 31; then a line interpolated from Lucan, *Pharsalia*, I, 417.

purely and utterly free us from any weight of anxiety about His rule. Scientific investigations and inquiries serve merely to feed our curiosity. They have nothing to do with knowledge so sublime: the philosophers are very right to refer us to the laws of Nature, but they pervert them and present Nature's face too sophistically, painted in colours which are far too exalted, from which arise so many diverse portraits of so uniform a subject. As Nature has furnished us with feet to walk with, so has she furnished us with wisdom to guide us in our lives. That wisdom is not as clever, strong and formal as the one which they have invented, but it is becomingly easy and beneficial; in the case of the man who is lucky enough to know how to use it simply and ordinately (that is, naturally) it does – very well – what the other *says* it will. The more simply we entrust ourself to Nature the more wisely we do so. Oh what a soft and delightful pillow, and what a sane one on which to rest a well-schooled head, are ignorance and unconcern. [B] I would rather be an expert on me than on [C] Cicero.[42]

[B]   Were I a good pupil there is enough, I find, in my own experience to make me wise. Whoever recalls to mind his last bout of choler and the excesses to which that fevered passion brought him sees the ugliness of that distemper better than in Aristotle and conceives even more just a loathing for it. Anyone who recalls the ills he has undergone, those which have threatened him and the trivial incidents which have moved him from one condition to another, makes himself thereby ready for future mutations and the exploring of his condition. (Even the life of Caesar is less exemplary for us than our own; a life whether imperial or plebeian is always a life affected by everything that can happen to a man.) We tell ourselves all that we chiefly need: let us listen to it. Is a man not stupid if he remembers having been so often wrong in his judgement yet does not become deeply distrustful of it thereafter?

When I find that I have been convicted of an erroneous opinion by another's argument, it is not so much a case of my learning something new he has told me nor how ignorant I was of some particular matter – there is not much profit in that – but of learning of my infirmity in general and of the treacherous ways of my intellect. From that I can reform the whole lump.

With all my other mistakes I do the same, and I think this rule is of great use to me in my life. I regard neither a class of error nor an example of it as one stone which has made me stumble: I learn to distrust my trot in

42. '88: than on *Plato*. Were I . . .

general and set about improving it.   [C]   To learn that we have said or done a stupid thing is nothing: we must learn a more ample and important lesson: that we are but blockheads.

[B]   The slips by which my memory so often trips me up precisely when I am most sure of it are not vainly lost: it is no use after that its swearing me oaths and telling me to trust it: I shake my head. The first opposition given to its testimony makes me suspend judgement and I would not dare then to trust it over any weighty matter nor to stand warrant for it when another is involved. Were it not that[43] others do even more frequently from lack of integrity what I do from lack of memory, I would on matters of fact as readily accept that truth is to be found on another's lips not mine.

If each man closely spied upon the effects and attributes of the passions which have rule over him as I do upon those which hold sway over me, he would see them coming and slow down a little the violence of their assault. They do not always make straight for our throat: there are warnings and degrees:

> *Fluctus uti primo cœpit cum albescere ponto,*
> *Paulatim sese tollit mare, et altius undas*
> *Erigit, inde imo consurgit ad æthera fundo.*

[At first the gale whips up the foam-topped wavelets, then little by little the sea begins to heave, the billows roll and the sea surges from the deep to the very heavens.][44]

Within me judgement holds the rector's chair, or at least it anxiously strives to do so. It permits my inclinations to go their own way, including hatred and love (even self-love) without itself being worsened or corrupted. Though it cannot reform those other qualities so as to bring them into harmony with itself, at least it does not let itself be deformed by them: it plays its role apart.

It must be important to put into effect the counsel that each man should know himself, since that god of light and learning had it placed on the tympanum of his temple as comprising the totality of the advice which he had to give us.[45]   [C]   Plato too says that wisdom is but the executing of that command, and Socrates in Xenophon proves in detail that it is

---

43. '88: Were it not that *I see nothing but lying and that* others do . . .
44. Virgil, *Aeneid*, VII, 528–30.
45. The *Know Thyself* of the Temple of Apollo at Delphi. (Cf. III, 9, 'On vanity', note 160).

true.[46]　[B]　The difficulties and obscurities of any branch of learning can be perceived only by those who have been able to go into it; for we always need some degree of intelligence to become aware that we do not know: if we are to learn that a door is shut against us we must first give it a shove.　[C]　From which springs that Platonic paradox: those who know do not have to inquire since they know already: neither do those who do not know, since to find out you need to know what you are inquiring into.[47]　[B]　And so it is with this knowing about oneself: the fact that each man sees himself as satisfactorily analysed and as sufficiently expert on the subject are signs that nobody understands anything whatever about it – [C]　as Socrates demonstrates to Euthydemus in Xenophon.[48]　[B]　I who make no other profession but getting to know myself find in me such boundless depths and variety that my apprenticeship bears no other fruit than to make me know how much there remains to learn.

It is to my inadequacy (so often avowed) that I owe my tendency to moderation, to obeying such beliefs as are laid down for me and a constant cooling and tempering of my opinions as well as a loathing for that distressing and combative arrogance which has complete faith and trust in itself: it is a mortal enemy of finding out the truth. Just listen to them acting the professor: the very first idiocies which they put forward are couched in the style by which religion and laws are founded:　[C]　*'Nil hoc est turpius quam cognitioni et perceptioni assertionem approbationemque praecurrere.'* [There is nothing more shocking than to see assertion and approval dashing ahead of cognition and perception.][49]

[B]　Aristarchus said that in olden days there were scarcely seven wise men to be found in the whole world whereas in his own days there were scarcely seven ignoramuses.[50] Have we not more reason to say that than he did? Assertion and stubbornness are express signs of animal-stupidity. This man over here has bitten the ground a hundred times a day: but there he is strutting about crowing *ergo*, as decided and as sound as before: you would say that some new soul, some new mental vigour has been infused into him, and that he was like that Son of Earth of old who, when thrown down, found fresh resolve and strength:

46. Cf. Erasmus, *Adages*, I, VII, XCV, *Nosce teipsum* (citing Plato, *Charmides*, 164 D); Xenophon, *Memorabilia*, IV, ii, 24 ff, and his portrait of Socrates in general.
47. Plato, *Meno*, XIV, 80.
48. Xenophon, *Memorabilia*, IV, ii, 29–40.
49. Cicero, *Academica*, I, xii, 45. (The standard reading today is *adsensionem*, assent, not *assertionem*, assertion.)
50. Plutarch (tr. Amyot), *De l'amitié fraternelle*, 81 F.

cui, cum tetigere parentem,
Jam defecta vigent renovato robore membra.

[whose failing limbs, when they touched the earth, his Mother, took on new strength and vigour.][51]

The unteachable, stubborn fool! Does he believe that he assumes a new mind with each new dispute? It is from my own experience that I emphasize human ignorance which is, in my judgement, the most certain faction in the school of the world. Those who will not be convinced of their ignorance by so vain an example as me – or themselves – let them acknowledge it through Socrates. [C] He is the Master of masters; the philosopher Antisthenes said to his pupils, 'Let us all go to hear Socrates: you and I will all be pupils there.' And when he was asserting the doctrine of his Stoic school that, to make a life fully happy, virtue sufficed without need of anything else, he added, 'except the strength of Socrates'.[52]

[B] This application which I have long devoted to studying myself also trains me to judge passably well of others: there are few topics on which I speak more aptly or acceptably. I often manage to see and to analyse the attributes of my friends more precisely than they can themselves. There is one man to whom I told things about himself which were so apposite that he was struck with amazement. By having trained myself since boyhood to see my life reflected in other people's I have acquired a studious tendency to do so; when I give my mind to it, few things around me which help me to achieve it escape my attention: looks, temperaments, speech, I study the lot for what I should avoid or what I should imitate.

I similarly reveal to my friends their innermost dispositions by what they outwardly disclose. I do not however classify such an infinite number of diverse and distinct activities within genera and species, sharply distributing my sections and divisions into established classes or departments,

sed neque quam multæ species, et nomina quæ sint,
Est numerus.

[for there is no numbering of their many categories nor of the names given to them.][53]

---

51. Lucan, *Pharsalia*, IV, 599–60; of Anthaeus, one of the giants called Sons of Earth; cf. Du Bellay, *Antiquités de Rome*, TLF, 12 and 11.
52. Erasmus, *Apophthegmata*, VII, *Antisthenes* II and XLIV.
[B] instead of [C]: through Socrates, *the wisest man there ever was by the testimony of the gods and men.* This application . . .
53. Virgil, *Georgics*, II, 103–4.

[C]   The learned do arrange their ideas into species and name them in detail. I, who can see no further than practice informs me, have no such rule, presenting my ideas in no categories and feeling my way – as I am doing here now;   [B]   I pronounce my sentences in disconnected clauses, as something which cannot be said at once all in one piece. Harmony and consistency are not to be found in ordinary   [C]   base[54]   [B]   souls such as ours. Wisdom is an edifice solid and entire, each piece of which has its place and bears its hallmark: [C]   *'Sola sapientia in se tota conversa est.'* [Wisdom alone is entirely self-contained.][55]

[B]   I leave it to the graduates – and I do not know if even they will manage to bring it off in a matter so confused, intricate and fortuitous – to arrange this infinite variety of features into groups, pin down our inconsistencies and impose some order. I find it hard to link our actions one to another, but I also find it hard to give each one of them, separately, its proper designation from some dominant quality; they are so ambiguous, with colours interpenetrating each other in various lights.

[C]   What is commented on as rare in the case of Perses, King of Macedonia (that his mind, settling on no particular mode of being, wandered about among every kind of existence, manifesting such vagrant and free-flying manners that neither he nor anyone else knew what kind of man he really was), seems to me to apply to virtually everybody.[56] And above all I have seen one man of the same rank as he was to whom that conclusion would, I believe, even more properly apply: never in a middle position, always flying to one extreme or the other for causes impossible to divine; no kind of progress without astonishing side-tracking and back-tracking; none of his aptitudes straightforward, such that the most true-to-life portrait you will be able to sketch of him one day will show that he strove and studied to make himself known as unknowable.[57]   [B]   You need good strong ears to hear yourself frankly judged; and since there are few who can undergo it without being hurt, those who risk undertaking it do us a singular act of love, for it is to love soundly to wound and vex a

---

54. '88: ordinary *vile* souls . . .
55. Cicero, *De finibus*, III, vii, 24.
56. King Perses (or Perseus, as Livy calls him) was the last king of Macedonia and was conquered by Paulus Aemilius. For his character cf. Livy, XLI, xx.
57. This bold judgement is made on the character of a king, doubtless Henry of Navarre (Henri Quatre). A rejected manuscript reading in the Bordeaux copy is: '*I have since seen one other king to whom* . . .' Henry (King of Navarre, 1572–1610) became King of France in 1589. He is sure of himself enough, it is suggested, to accept frank criticism.

man in the interests of his improvement. I find it harsh to have to judge anyone in whom the bad qualities exceed the good. [C] Plato requires three attributes in anyone who wishes to examine the soul of another: knowledge, benevolence, daring.[58]

[B] Once I was asked what I thought I would have been good at if anyone had decided to employ me while I was at the right age:

> *Dum melior vires sanguis dabat, æmula necdum*
> *Temporibus geminis canebat sparsa senectus.*

[When I drew strength from better blood and when envious years had yet to sprinkle snow upon my temples.]

'Nothing,' I replied; 'and I am prepared to apologize for not knowing how to do anything which enslaves me to another. But I would have told my master some blunt truths and would, if he wanted me to, have commented on his behaviour – not wholesale by reading the Schoolmen at him (I know nothing about them and have observed no improvement among those who do), but whenever it was opportune by pointing things out as he went along, judging by running my eyes along each incident one at a time, simply and naturally, bringing him to see what the public opinion of him is and counteracting his flatterers.' (There is not one of us who would not be worse than our kings if he were constantly [C] corrupted by that riff-raff as they are.) [B] How else[59] could it be, since even the great king and philosopher Alexander could not protect himself from them?[60] I would have had more than enough loyalty, judgement and frankness to do that. It would be an office without a name, otherwise it would lose its efficacity and grace. And it is a role which cannot be held by all men indifferently, for truth itself is not privileged to be used all the time and in all circumstances: noble though its employment is, it has its limits and boundaries. The world being what it is, it often happens that you release truth into a Prince's ear not merely unprofitably but detrimentally and (even more) unjustly. No one will ever convince me that an upright rebuke may not be offered offensively nor that considerations of matter should not often give way to those of manner.

For such a job I would want a man happy with his fortune –

---

58. '88: without being hurt *and resentful*, those who risk . . .
   Then, Plato, *Gorgias*, 487 A; Virgil, *Aeneid*, V, 415–16.
59. '88: constantly *cheated and diddled* as they are. How else . . .
60. Cf. Erasmus, *Apophthegmata*, IV, *Alexander Magnus*, XV; LXIII, etc.

Quod si esse velit, nihilque malit

[Who would be what he is, desiring nothing extra][61]

– and born to a modest competence. And that, for two reasons: he would not be afraid to strike deep, lively blows into his master's mind for fear of losing his way to advancement; he would on the other hand have easy dealings with all sorts of people, being himself of middling rank.   [C] And only one man should be appointed; for to scatter the privilege of such frankness and familiarity over many would engender a damaging lack of respect. Indeed what I would require above all from that one man is that he could be trusted to keep quiet.[62]

A king   [B]   is not to be believed if he boasts of his steadfastness as he waits to encounter the enemy in the service of his glory if, for his profit and improvement, he cannot tolerate the freedom of a man who loves him to use words which have no other power than to make his ears smart, any remaining effects of them being in his own hands. Now there is no category of man who has greater need of such true and frank counsels than kings do. They sustain a life lived in public and have to remain acceptable to the opinions of a great many on-lookers: yet, since it is customary not to tell them anything which makes them change their ways, they discover that they have, quite unawares, begun to be hated and loathed by their subjects for reasons which they could often have avoided (with no loss to their pleasures moreover) if only they had been warned in time and corrected. As a rule favourites are more concerned for themselves than for their master: and that serves them well, for in truth it is tough and perilous to assay showing the offices of real affection towards your sovereign: the result is that not only a great deal of good-will and frankness are needed but also considerable courage.

In short all this jumble that I am jotting down here is but an account of the assays of my life: it is, where the mind's health is concerned, exemplary enough – if you work against its grain. But where the body's health is concerned no one can supply more useful experience than I, who present it pure, in no wise spoiled or adulterated by science or theory. In the case of medicine, experience is on its own proper dung-heap, where reason voids

---

61. Martial, *Epigrams*, X, xlvii, 12.
62. Henry IV did indeed ask Montaigne to become such a counsellor, but too late, for Montaigne was dying.
   '88: middling rank. A *prince* is not . . .

the field.[63] Tiberius said that anyone who had lived for twenty years ought to be able to tell himself which things are harmful to his health and which are beneficial and to know how to proceed without medicine.[64] [C] Perhaps he learned that from Socrates who when advising his followers to devote themselves assiduously, with a most particular devotion, to their health added that if a man of intelligence was careful about his eating, drinking and exercise, it would be difficult for him not to discern what was good or bad for him better than his doctor could.[65]

[B] Certainly medicine professes always to have experience as the touchstone of its performance. Plato was therefore right to say that to be a true doctor would require that anyone who would practise as such should have recovered from all the illnesses which he claimed to cure and have gone through all the symptoms and conditions on which he would seek to give an opinion.[66] If doctors want to know how to cure syphilis it is right that they should first catch it themselves! I would truly trust the one who did; for the others pilot us like a man who remains seated at his table, painting seas, reefs and harbours and, in absolute safety, pushing a model boat over them. Pitch him into doing the real thing and he does not know where to start. They give the kind of description of our maladies as the town-crier announcing a lost horse or hound: this colour coat, so many span high, this kind of ears: but confront him with it, and for all that he cannot identify it. By God let medicine provide me with some good and perceptible help some day and I will proclaim in good earnest,

Tandem efficaci do manus scientiæ!
[At last I yield to thy effective Art!][67]

Those disciplines which promise to maintain our bodies in health and our souls in health promise a great deal:[68] yet none keeps their promises less than they do; and those who profess those Arts in our own time show the effects of them less than any other men. The most you can say of them is that they trade in the *materia medica* of those healing Arts: that they are

63. That is Aristotle's position on all arts at the outset of his *Metaphysics*. Renaissance scholars applied it particularly but not exclusively to medicine, the Art *par excellence*.
64. Cf. Erasmus, *Apophthegmata* VI, *Tiberius*, XIII (but referring not to 'twenty years' but to the age of sixty).
65. Xenophon, *Memorabilia*, IV, vii, 9.
66. Plato, *Republic*, III, 408 D–E.
67. Horace, *Epodes*, XVIII, 1.
68. Medicine and philosophy.

healers you cannot say. I have lived long enough now to give an account of the regimen which has got me thus far. Should anyone want to try it, I have assayed it first as his taster. Here are a few items as memory supplies them. [C] (There is no practice of mine which has not been varied according to circumstances, but I note here those which, so far, I have most often seen at work and which are rooted in me.)

[B] My regimen is the same in sickness as in health: I use the same bed, same timetable, same food and same drink. I add absolutely nothing except for increasing and decreasing the measure depending on my strength and appetite. Health means for me the maintaining of my usual route without let or hindrance. I can see that my illness has blocked one direction for me: if I put trust in doctors they will turn me away from the other, so there I am off my route either by destiny or their Art; there is nothing that I believe so certainly as this: that carrying on with anything to which I have so long been accustomed cannot do me harm. It is for custom to give shape to our lives, such shape as it will – in such matters it can do anything. It is the cup of Circe which changes our nature as it pleases. How many peoples are there, not three yards from us, who think that our fear of the cool evening air – which 'so evidently' harms us – is ludicrous; and our boatsmen and our peasants laugh at us too.

You make a German ill if you force him to lie in bed on a straw mattress, as you do an Italian on a feather one, or a Frenchman without bed-curtains or a fire. The stomach of a Spaniard cannot tolerate the way we eat: nor can ours the way the Swiss drink. I was amused by a German in Augsburg who attacked our open hearths, emphasizing their drawbacks with the same arguments which we normally use against their stoves! And it is true that those stoves give out an oppressive heat and that the materials of which they are built produce when hot a smell which causes headaches in those who are not used to them: not however in me. On the other hand since the heat they give out is even, constant and spread over-all, without the visible flame, the smoke and the draught produced for us by our chimneys, it has plenty of grounds for standing comparison with ours. (Why do we not imitate the building methods of the Romans, for it is said that in antiquity their house-fires were lit outside, at basement level; from there hot air was blown to all the house through pipes set within the thickness of the walls which surrounded the areas to be heated. I have seen that clearly suggested somewhere in Seneca, though I forget where.)[69] That man in Augsburg, on hearing me praise the advantages and beauties

69. Seneca, *Epist. moral.*, XC, 25 (regretting the luxury of civilized man).

of his city (which indeed deserved it) started to pity me because I had to leave it; among the chief inconveniences he cited to me was the heavy head I would get 'from those open hearths yonder'. He had heard somebody make this complaint and linked it with us, custom preventing him from noticing the same thing at home.

Any heat coming from a fire makes me weak and drowsy. Yet Evenus maintained that fire was life's condiment.[70] I adopt in preference any other way of escaping the cold.

We avoid wine from the bottom of the barrel; in Portugal they adore its savour: it is the drink of princes. In short each nation has several customs and practices which are not only unknown to another nation but barbarous and a cause of wonder.

What shall we do with those people who will receive only printed testimony, who will not believe anyone who is not in a book, nor truth unless it be properly aged? [C] We set our stupidities in dignity when we set them in print. [B] For these people there is far more weight in saying, 'I have read that . . .' than if you say, 'I have heard tell that . . .' But I (who have the same distrust of a man's pen as his tongue; who know that folk write with as little discretion as they talk and who esteem this age as much as any other former one) as willingly cite a friend of mine as Aulus Gellius or Macrobius, and what I have seen as what they have written. [C] And just as it is held that duration does not heighten virtue,[71] I similarly reckon that truth is no wiser for being more ancient.

[B] I often say that it is pure silliness which sets us chasing after foreign and textbook exemplars. They are produced no less abundantly nowadays than in the times of Homer and Plato. But are we not trying to impress people by our quotations rather than by the truth of what they say? – as though it were a [C] greater thing [B] to borrow our proofs from the bookshops of Vascosan and Plantin than from our village?[72] Or is it that we do not have wit enough to select and exploit whatever happens in front of us or to judge it so acutely as to draw examples from it? For if we say that we lack the requisite authority to produce faith in our testimony we are off the point: in my opinion the most ordinary things, the most commonplace and best-known can constitute, if we know how to present

70. Plutarch (tr. Amyot), *Propos de table*, 410 B, etc. (cited several more times in the *Oeuvres morales*).
71. A Stoic contention.
72. '88: were a *more noble* thing to borrow . . .
   Vascosan and Plantin were two great printing-houses.

them in the right light, the greatest of Nature's miracles and the most amazing of examples, notably on the subject of human actions.[73]

Now on this topic of mine (leaving aside any examples I know from books   [C]   and what Aristotle said of Andros the Argive who traversed the arid sands of Lybia without once drinking),[74]   [B]   a nobleman who has acquitted himself with honour of several charges stated in my presence that he had journeyed without drinking from Madrid to Lisbon in the height of summer. He is vigorous for his age and there is nothing in his way of life which goes beyond the normal Order except that he can, so he told me, do without drinking for two or three months or even a year. He feels a little thirsty but lets it pass: he maintains that it is a craving which can easily weaken by itself. He drinks more on impulse than from necessity, or for enjoyment.

Here is another. Not long ago I came across one of the most learned men in France – a man of more than moderate wealth; he was studying in a corner of his hall which had been partitioned off with tapestries; around him were his menservants making the most disorderly racket. He told me – [C]   and Seneca said much the same of himself[75] –   [B]   that he found their hubbub useful: it was as though, when he was being battered by that din, he could withdraw and close in on himself so as to meditate, and that those turbulent voices hammered his thoughts right in. When he was a student at Padua his work-room was for so long subject to the clatter of wagons and the tumultuous uproar of the market-place that he had trained himself not merely to ignore the noise but to exploit it in the service of his studies.   [C]   When Alcibiades asked in amazement how Socrates could put up with the sound of his wife's perpetual nagging, he replied: 'Just like those who get used to the constant grating of wheels drawing water from the well.'[76]   [B]   I am quite the opposite: I have a mind which is delicate and easy to distract: when it withdraws aside to concentrate, the least buzzing of a fly is enough to murder it!

[C]   When Seneca was a young man, having been keenly bitten by the example of Sextius, he ate nothing that had been slaughtered. For a whole year he did without meat – with great pleasure as he relates. He did give

---

73. 'Miracles of Nature' were unusual and most rare events but not in any theological sense miraculous: they were sources of wonder.

74. Diogenes Laertius, *Life of Pyrrho*, IX, lxxxi. (The contemporary nobleman next mentioned is Marquis Jean de Vivonne.)

75. Seneca, *Epist. moral.*, LVI.

76. Erasmus, *Apophthegmata*, III, *Socratica*, LX.

up that diet, but only to avoid the suspicion of being influenced by certain new religions which were disseminating it. He had adopted at the same time one of the precepts of Attalus: never to lie on soft mattresses; until his death he continued to use the kinds which do not yield to the body.[77] That which the customs of his day led him to count as an austerity our own make us think of as an indulgence.

[B] Consider the diversity between the way of life of my farm-labourers and my own. Scythia and the Indies have nothing more foreign to my force or my form. And this I know: I took some boys off begging into my service: soon afterwards they left me, my cuisine and their livery merely to return to their old life. I came across one of them gathering snails from the roadside for his dinner: neither prayer nor menace could drag him away from the sweet savour he found in poverty. Beggars have their distinctions and their pleasures as do rich men, and, so it is said, their own political offices and orders.

Such are the effects of Habituation: she can not only mould us to the form which pleases her (that is why, say the wise, we must cling to the best form, which she will straightway make easy for us)[78] but also mould us for change and variation (which are the noblest and most useful of her crafts). Of my own physical endowments the best is that I am flexible and not stubborn: some of my inclinations are more proper to me than others, more usual and more agreeable, but with very little effort I can turn away from them and glide easily into an opposite style. A young man ought to shake up his regular habits in order to awaken his powers and stop them from getting lazy and stale. And there is no way of life which is more feeble and stupid than one which is guided by prescriptions and instilled habit:[79]

> *Ad primum lapidem vectari cum placet, hora*
> *Sumitur ex libro; si prurit frictus ocelli*
> *Angulus, inspecta genesi collyria quærit.*

[Does he want to be borne as far as the first milestone? Then he consults his almanack to find out the best time. Has he got a sore in the corner of an eye? Then he consults his horoscope before buying some ointment.][80]

---

77. Seneca, *Epist. moral.*, CVIII, 17 f.
78. Erasmus, *Adages*, IV, IX, XXV, *Usus est altera natura.*
79. By using *discipline* for instilled habit, Montaigne may be echoing the usage of the Roman comedies, where *disciplina* has this sense.
80. Juvenal, *Satires*, VI, 576–8.

If he trusts me a young man will often jump to the other extreme: if he does not, the least excess will undermine him: he makes himself disagreeable and clumsy in society. The most incompatible quality in a gentleman is to be over-nicely bound to one fixed idiosyncratic manner: and idiosyncratic it is, if it is not pliable and supple. There is disgrace in being incapable or afraid to do what your companions are up to. Such men should stay in their kitchens! Unbecoming it is, in everyone else: in a warrior it is vile and not to be endured; he, as Philopoemen said, must get accustomed to all kinds of this life's changes and hardships.[81]

Although I was brought up, as much as is humanly possible, for freedom and flexibility, nevertheless as I grow older I am becoming through indifference more fixed in certain forms (I am past the age for elementary schooling; now old age has no other concern than to look after itself); without my noticing it, custom has imprinted its stamp on me so well where some things are concerned that any departure from it I call excess; and I cannot, without turning it into an assay of myself, sleep by day, eat snacks between meals, nor eat breakfast, nor go to bed after supper without having a considerable gap, [C] say three hours or more, [B] nor have sexual intercourse except before going to sleep, nor do it standing up, nor remain soaking with sweat, nor drink either water or wine unmixed, nor remain for long with my head uncovered, nor have my haircut after dinner. I would feel just as ill at ease without gloves or shirt, or without a wash on leaving the table and when getting up in the morning, or lying in a bed without canopy and curtains, as I would if forced to do without things which really matter.

I could dine easily enough without a tablecloth, but I feel very uncomfortable dining without a clean napkin as the Germans do. I dirty my napkins more than they or the Italians and rarely seek the aid of spoon or fork. I regret that we have not continued along the lines of the fashion started by our kings, changing napkins like plates with each course.

We are told that as Marius grew older, tough old soldier though he was, he became choosy about his wine and would only drink it out of his own special goblet. [C] I too incline[82] towards glasses of a particular shape and I no more like drinking out of a common cup than I would like eating

---

81. Plutarch, *Life of Philopoemen*, I.
82. [B] instead of [C]: special goblet: *earthenware and silver displease me compared with glass, as does being served by hands which I am unused to or which are not in my employ, or from* a common cup, *and* I incline *to choose* glasses of a particular shape. Several such foibles . . .

out of common fingers; and I dislike all metals compared with clear transparent materials. Let my eyes too taste it to the full.

[B] Several such foibles I owe to habit: on the other hand Nature has contributed her own, such as my not being able to stand more than two proper meals a day without overloading my stomach, nor to go without a meal altogether without filling myself with wind, parching my mouth and upsetting my appetite; nor can I stand a long exposure to the evening dew. During these last few years when a whole night has to be spent (as often happens) on some military task, my stomach begins to bother me after five or six hours; I have splitting headaches and can never get through to morning without vomiting. Then, while the others go to breakfast, I have a sleep; after which I am quite happy again.

I had always been taught that evening dew formed only after night-fall, but upon frequenting a nobleman who was imbued with the belief that such dew is more dangerous and severe two or three hours before sunset (when he scrupulously avoids going out) he made such an impression on me that I almost not so much believed it as felt it. Well now, that very doubt and concern for our health can hammer our thought-process and change us. Those who slide precipitously down slopes such as that bring disaster upon themselves. There are several gentlemen for whom I feel pity: through the stupidity of their doctors they shut themselves up indoors while still young and healthy; it would be better to put up with a chill rather than forever to forgo joining in common everyday life outdoors. [C] What a grievous skill medicine is, disparaging for us the more delightful hours of the day. [B] Let us extend our hold on things by every means we possess. Usually if you stubborn things out you toughen yourself up, correcting your complexion by despising it and seducing it, as Caesar did his epilepsy. We should give ourselves, but not enslave ourselves, to the best precepts, except in such cases (if there be any) in which constraint and slavery serve a purpose.

Kings and philosophers shit: and so do ladies.[83] The lives of public figures are devoted to etiquette: my life, an obscure and private one, can enjoy all the natural functions: moreover to be a soldier and to come from Gascony are both qualities given to forthrightness. And so of that activity I shall say that it needs to be consigned to a set hour – not daytime – to which we should subject ourselves by force of habit, as I have done, but not (as applies to me now that I am growing old) subject to the pleasures of a

83. '88: and so do ladies; *others have tact and competence as their qualities: I, frankness and freedom.* The lives . . .

particular place and seat for this function, nor to making it uncomfortable by prolonging it or by being fastidious. All the same, is it not to some extent pardonable to require more care and cleanliness for our dirtiest functions? [C] *'Natura homo mundum et elegans animal est.'* [By Nature Man is a clean and neat creature.][84] Of all the natural operations, that is the one during which I least willingly tolerate being interrupted. [B] I have known many a soldier put out by the irregularity of his bowels. My bowels and I never fail to keep our rendezvous, which is (unless some urgent business or illness disturbs us) when I jump out of bed.

So, as I was saying, I can give no judgement about how the sick can be better looked after except that they should quietly hold to the pattern of life in which they have been schooled and brought up. Change of any kind produces bewilderment and trauma. Convince yourself if you can that chestnuts are harmful to the men of Périgord or Lucca, or milk and cheese to folk in the highlands! Yet the sick are constantly prescribed not merely a new way of life but an opposite one – such a revolution as could not be endured by a healthy man. Prescribe water for a seventy-year-old Breton; shut a sailorman up in vapour-bath; forbid a Basque manservant to go for walks! They are deprived of motion and finally of breath and the light of day:

> *An vivere tanti est?*
>
> [Is life worth that much?][85]
>
> *Cogimur a suetis animum suspendere rebus,*
> *Atque, ut vivamus, vivere desinimus.*
>
> *Hos superesse rear, quibus et spirabilis aer*
> *Et lux qua regimur redditur ipsa gravis?*

[We are compelled to deprive our souls of what they are used to; to stay alive we must cease to live! Should I count among the survivors those men for whom the very air they breathe and the light which lightens them have become a burden?]

If doctors do nothing else, they do at least in plenty of time prepare their patients to die, sapping and retrenching their contacts with life.

Sound or sick I willingly let myself follow such appetites as become pressing. I grant considerable authority to my desires and predispositions. I do not like curing one ill by another; I loathe remedies which are more importunate than the sickness: being subjected to colic paroxysms and then

---

84. Seneca, *Epist. moral.*, XCII, 12.
85. Untraced. Then verses from Pseudo-Gallus, *Elegeia*, I.

made to abstain from the pleasure of eating oysters are two ills for the price of one. On this side we have the illness hurting us, on the other the diet. Since we must risk being wrong, let us risk what gives us pleasure, rather. The world does the reverse, thinking that nothing does you good unless it hurts: pleasantness is suspect. In many things my appetite, of its own volition, has most successfully accommodated and adapted itself to the well-being of my stomach. When I was young I liked the tartness and sharp savour of sauces: my stomach being subsequently troubled by them, my taste for them at once followed its lead.

[C] Wine is bad for the sick: it is the first thing I lose my taste for, my tongue finding it unpleasant, invincibly unpleasant. [B] Anything the taste of which I find unpleasant does me harm: nothing does me harm if I swallow it hungrily and joyfully. I have never been bothered by anything I have done in which I found great pleasure. And that is why I have, by and large, made all medical prescriptions give way to what pleases me.

When I was young –

> *Quem circumcursans huc atque huc sæpe Cupido*
> *Fulgebat, crocina splendidus in tunica,*

[when shining Cupid flew here and there about me, resplendent in his saffron tunic,][86]

– I yielded as freely and as thoughtlessly as anyone to the pleasure which then seized hold of me:

> *Et militavi non sine gloria,*

[and I fought not without glory,]

making it last and prolonging it, however, rather than making sudden thrusts.

> *Sex me vix memini sustinuisse vices.*

[I cannot recall managing it more than six times in a row.][87]

There is indeed some worry and wonder in confessing at what tender an age I happened to fall first into Cupid's power – 'happened' is indeed right, for it was long before the age of discretion and awareness – so long ago that I cannot remember anything about myself then. You can wed my

---

86. Catullus, LXVI, 133–4; then, Horace, *Odes*, III, xxvi, 2.
87. Ovid, *Amores*, III, vii, 26 (who says nine, not six, times).

fortune to that of Quartilla, who could not remember ever having been a virgin.[88]

> *Inde tragus celeresque pili, mirandaque matri*
> *Barba meæ.*

[My armpits had precocious hairs and stank like a goat: Mother was astonished by my early beard.]

The doctors usually bend their rules – usefully – before the violence of the intense cravings which surprise the sick: such a great desire cannot be thought of as so strange or vicious that Nature is not at work in it. And then, what a great thing it is to satisfy our imagination. In my opinion that faculty concerns everything, at least more than any other does: the most grievous and frequent of ills are those which imagination loads upon us. From several points of view I like that Spanish saying: '*Defienda me Dios de my.*' [God save me from myself.] When I am ill what I lament is that I have no desire then which gives me the satisfaction of assuaging it: Medicine would never stop me doing so! It is the same when I am well: I have scarcely anything left to hope or to wish for now. It is pitiful to be faint and feeble even in your desires.

The art of medicine has not reached such certainty that, no matter what we do, we cannot find some authority for doing it. Medicine changes according to the climate, according to the phases of the moon, according to Fernel and according to Scaliger. If your own doctor does not find it good for you to sleep, to use wine or any particular food, do not worry: I will find you another who does not agree with his advice. The range of differing medical arguments and opinions embraces every sort of variety. I knew one wretched patient, weak and fainting with thirst as part of his cure, who was later laughed at by another doctor who condemned that treatment as harmful. Had his suffering been to some purpose? Well there is a practitioner of that mystery who recently died of the stone and who had used extreme abstinence in fighting that illness: his fellow-doctors say that, on the contrary, such deprivation had desiccated him, maturating the sand in his kidneys.

I have noted that when I am sick or wounded talking excites me and does me as much harm as any of my excesses. Speaking takes it out of me and tires me, since my voice is so strong and booming that when I have needed to have a word in the ear of the great on a matter of some

---

88. Known from Petronius. Cf. Tiraquellus, *De legibus connubialibus*, IX, 98; then, Martial, *Epigrams*, XI, xxii, 7–8.

gravity I have often put them to the embarrassment of asking me to lower it.

The following tale is worth a digression: there was in one of the schools of the Greeks a man who used to talk loudly as I do. The Master of debate sent to tell him to speak lower: 'Let him send and tell me what volume he wants me to adopt,' he said. The Master replied that he should pitch his voice to the ears of the man he was addressing.[89] Now that was well said, provided that he meant, 'Speak according to the nature of your business with your hearer.' For if he meant, 'It is enough if people can catch what you say,' or, 'Let yourself be governed by your hearer,' then I do not believe that he was right. Volume and intonation contribute to the expression of meaning: it is for me to control them so that I can make myself understood. There is a voice for instructing, a voice for pleasing or for reproving. I may want my voice not simply to reach the man but to hit him or go right through him. When I am barking at my footman with a rough and harsh voice, a fine thing it would be if he came and said to me, 'Speak more softly, Master. I can hear you quite well.' [C] *'Est quaedam vox ad auditum accommodata, non magnitudine sed proprietate.'* [There is a kind of voice which impresses the hearer not by its volume but its own peculiar quality.][90] [B] Words belong half to the speaker, half to the hearer. The latter must prepare himself to receive them according to such motion as they acquire, just as among those who play royal-tennis the one who receives the ball steps backwards or prepares himself, depending on the movements of the server or the form of his stroke.

Experience has also taught me that we are ruined by impatience. Illnesses have their life and their limits,[91] [C] their maladies and their good health. The constitution of illnesses is formed on the pattern of that of animals: from birth their lot is assigned limits, and so are their days. Anyone who makes an assay at imperiously shortening them by interrupting their course prolongs them and makes them breed, irritating them instead of quietening them down. I am of Crantor's opinion that we should neither resist illnesses stubbornly and rashly nor succumb to them out of weakness but yield to them naturally, according to our own mode of being and to theirs.[92] [B] We must afford them right-of-passage, and I find that they stay less long with me, who let them go their way; and through

89. Erasmus, *Apophthegmata*, VII, *Carneades*, XXXI.
90. Quintilian, XI, iii, 40.
91. '88: limits. We *should* afford them right-of-passage . . .
92. Paraphrased from Cicero, *Tusc. disput.*, III, v, 12.

their own decline I have rid myself of some which are held to be the most tenacious and stubborn, with no help from that Art and against its prescriptions. Let us allow Nature to do something! She understands her business better than we do. – 'But so-and-so died of it!' – So will you, of that illness or some other. And how many have still died of it with three doctors by their arses? Precedent is   [C]   an uncertain looking-glass,   [B] all-embracing,   [C]   turning all ways.[93]   [B]   If the medicine tastes nice, take it: that is so much immediate gain at least.   [C]   I will not jib at its name or colour if it is delicious and whets my appetite for it. One of the principal species of profit is pleasure.   [B]   Among the illnesses which I have allowed to grow old and die of a natural death within me are rheums, fluxions of gout, diarrhoeas, coronary palpitations and migraines, which I lost just when I was half-resigned to having them batten on me. You can conjure them away better by courtesy than by bravado. We must quietly suffer the laws of Man's condition. Despite all medicine, we are made for growing old, growing weaker and falling ill. That is the first lesson which the Mexicans teach to their children when, on leaving their mother's womb, they greet them thus: 'Child: thou hast come into this world to suffer: suffer, endure and hold thy peace.'

It is unfair to moan because what can happen to any has happened to one: [C]   *'indignare si quid in te inique proprie constitutum est'* [if anything is unjustly decreed against you alone, that is the time to complain].[94]

[B]   Here you see an old man praying God to keep him entirely healthy and strong – that is to say, to make him young again:

> *Stulte, quid hæc frustra votis puerilibus optas?*

[You fool. What do you hope to gain by such useless, childish prayers?][95]

Is it not madness? His mode of being does not allow it.   [C]   Gout, gravel and bad digestion go with long years just as heat, wind and rain go with long journeys. Plato does not believe that Aesculapius should trouble to provide remedies to prolong life in a weak and wasted body, useless to its country, useless to its vocation and useless for producing healthy robust sons: nor does he find such a preoccupation becoming to the justice and wisdom of God who must govern all things to a useful purpose.[96]   [B]   It

---

93. '88: Precedent is *a free and* all-embracing *pattern. If the medicine . . .*
94. Seneca, *Epist. moral.*, XCI, 15, after listing the normality of war, illness and death, and stressing that if we do not obey the laws of the world we should quit it.
95. Ovid, *Tristia*, III, viii, 11.
96. Plato, *Republic*, III, 407 C.

is all over, old chap: nobody can put you back on your feet; they will [C] at most [B] bandage and prop you up for a bit, [C] prolonging your misery an hour or so:

> [B] *Non secus instantem cupiens fulcire ruinam,*
> *Diversis contra nititur obicibus,*
> *Donec certa dies, omni compage soluta,*
> *Ipsum cum rebus subruat auxilium.*

[As a man, desiring to keep a building from collapsing, shores it up with various props until there comes the day when all the scaffolding shatters and the props collapse together with the building.][97]

We must learn to suffer whatever we cannot avoid. Our life is composed, like the harmony of the world, of discords as well as of different tones, sweet and harsh, sharp and flat, soft and loud. If a musician liked only some of them, what could he sing? He has got to know how to use all of them and blend them together. So too must we with good and ill, which are of one substance with our life. Without such blending our being cannot be: one category is no less necessary than the other. To assay kicking against natural necessity is to reproduce the mad deed of Ctesiphon who, to a kicking-match, challenged his mule.[98]

I do not go in much for consultations over such deterioration as I feel: once those medical fellows have you at their mercy they boss you about: they batter your ears with their prognostics. Once, taking advantage of me when I was weak and ill, they abused me with their dogmas and their masterly [C] frowns,[99] [B] threatening me with great suffering and then with imminent death. They did not succeed in knocking me down or dislodging me from my fortress, but I was jolted and jostled: my judgement was neither changed nor troubled by them but it was at least preoccupied, and that means so much agitation and strife. I treat my imagination as gently as I am able, freeing it if I can from the load of any pain and conflict. Anyone who can should help it, stroke it, mislead it. My wit is well suited to such service: it never runs out of specious arguments about anything. If it could convince as well as it preaches its help would be most welcome.

Would you like an example? It tells me: that it is for my own good that

---

97. Pseudo-Gallus, *Eclogues*, I, 171–4; then, a development inspired by Plutarch (tr. Amyot), *De la tranquillité de l'ame*, 74 A–D.
98. Erasmus, *Adages*, I, III, XLVI, *Contra stimulum calces*, explaining the Classical and biblical maxim, *To kick against the pricks*, by Plutarch's example of a choleric athlete named Ctesiphon, unknown except for this incident.
99. '88: masterly *countenances*, threatening me . . .

I have the gravel; that structures as old as I am are naturally subject to seepage (it is time they began to totter apart and decay; that is a common necessity, otherwise would not some new miracle have been performed just for me? I am paying the debt due to old age and could not get off more lightly); that I should be consoled by the fact that I have company, since I have fallen into the most routine illness for men of my age (on all sides I can see men afflicted by a malady of the same nature as mine and their companionship honours me since that malady willingly strikes the aristocracy: its essence is noble and dignified); and that, of the men who are stricken with it few get off more lightly – and even then it is at the cost of having the bother of following a nasty diet and of taking troublesome daily doses of medicine, whereas I owe everything to my good fortune. (As for the few routine concoctions of eryngo or burstwort[100] which I have swallowed twice or thrice thanks to those ladies who gave me half of their own to drink (their courtesy exceeding in degree the pain of my complaint) they seemed to me to be as easy to take as they were ineffectual in practice. For that easy and abundant discharge of gravel which I have often been vouchsafed by the bounty of Nature, those men had to pay a thousand vows to Aesculapius and as many crowns to their doctor.   [C]   (In normal company my comportment remains decorous, even, and is untroubled by my illness; and I can hold my urine for ten hours at a time – as long as the next man.)

[B] 'The fear of this illness,' (to go on), 'used to terrify you: that was when it was unknown to you; the screams and distress of those who make the pain more acute by their unwillingness to bear it engendered a horror of it in you. This illness afflicts those members of yours by which you have most erred. You are a man with some sense of right and wrong:

> *Quæ venit indigne pæna, dolenda venit.*

> [Only punishment undeserved comes with cause for anger.][101]

Reflect on this chastisement: it is mild indeed compared with others and shows a Fatherly kindness.[102] Reflect on how late it appeared: having first

---

100. Herbal laxatives and astringents.
101. Ovid, *Heroidum Epistolae*, V, 8.
102. Not least during the French Civil Wars of Religion, setbacks and afflictions were often seen as divinely sent punishments, proof of the Fatherly love of God correcting and purging his children with salutary chastisements. All could thus find strength and comfort in tribulation.

made a compact by which it gave free-play to the excesses and pleasures of your youth, it occupies with its vexations only that season of your life which, willy-nilly, is sterile and forlorn. The fear and pity felt by people for this illness gives you something to glory about (you may have purged your judgement and cured your reason of such glorying, but those who love you still recognize some stain of it within your complexion). There is pleasure in hearing them say about you: "There's fortitude for you! There's long-suffering!" They see you sweating under the strain, turning pale, flushing, trembling, sicking up everything including blood, suffering curious spasms and convulsions, sometimes shedding huge tears from your eyes, excreting frightening kinds of urine, thick and black, or finding that they are retained by some sharp stone, bristling with spikes which cruelly jab into the neck of your prick and skin it bare: you, meanwhile, chat with those about you, keeping your usual expression, occasionally clowning about with [C] your servants,[103] [B] defending your corner in a tense argument, apologizing for any sign of pain and understating your suffering.

'Do you remember those men of yore who greatly hungered after ills so as to keep their virtue in trim and practise it? Supposing Nature is pushing and shoving you into that [C] proud [B] Sect[104] into which you would never have entered on your own! If you tell me that yours is a dangerous, killing affliction, which of the others is not? For it is medical hocus-pocus to pick out some and say that they do not follow a direct line towards death: what does it matter if they only lead there incidentally, floundering along by-ways in the same direction as the road which leads us thither? [C] You are not dying because you are ill: you are dying because you are alive;[105] Death can kill you well enough without illness to help her. In some cases illnesses have postponed death, the sick living longer precisely because they thought they were a-dying; besides, just as there are some wounds which cure you or make you better, so too there are some illnesses. [B] Your colic is often no less tenacious of life than you are: we know of men in whom it has lasted from childhood to extreme old age: and it would have gone along with them further if they themselves had not deserted its company. Men kill the stone more than it

103. '88: With *the ladies*, defending . . .
104. '88: that *noble* sect . . .
    (Certain Stoics.)
105. Seneca, *Epist. moral.*, LXXVIII, 6 (with a wider influence on Montaigne's general context).

kills men. And if it did present you with the idea of imminent death, would it not be doing a good turn to a man of your age to bring him to meditate upon his end?

[C] 'And the worst of it is you have nobody left to be cured for. As soon as she likes, whatever you do, our common Fate is summoning you. [B] Reflect on how skilfully and gently your colic makes you lose your taste for life and detaches you from the world – not compelling you by some tyrannous subjection as do so many other afflictions found in old men which keep them continually fettered to weakness and unremittingly in pain but with intermittent warnings and counsels interspersed with long periods of respite, as if to give you the means to meditate on its lesson and to go over it again at leisure. And so as to give you the means to make a sound judgement and to be resolved like a sensible man, it shows you the state of the whole human condition, both good and bad, shows you, during one single day, a life at times full of great joy, at times unbearable. Although you may not throw your arms about Death's neck, you do, once a month, shake her by the hand. [C] That gives you more reason to hope that Death will snatch you one day without warning and that, having so often brought you as far as the jetty, one morning, unexpectedly, when you are trusting that you are still on the usual terms, you and your trust will have crossed the Styx. [B] You have no need to complain of ill-nesses which share their time fairly with health.

'I am obliged to Fortune for the fact that she so often uses the same sort of weapons to assail me: she forms me and schools me for them by habit, hardens me and makes me used to them: I more or less know now what it will cost me to be released from what I owe them. [C] (Lacking a natural memory I forge one from paper: whenever some new feature occurs in my affliction, I jot it down. And so by now, when I have gone through virtually every category of examples of such symptoms, whenever some appalling crisis threatens me I can without fail, by flipping through my notes (which are as loose as the leaves of the Sybils), find grounds for consolation in some favourable prognosis based on past experience.) [B] Such habituation helps me to hope for better things in the future: this way of voiding the stone has continued for such a long time now that it is probable that Nature will not change the way of it and that nothing worse will happen than what I already know.

'Moreover the properties of this Affliction of mine are not ill-suited to my complexion, which is quick and sudden. It is when she makes mild assaults on me that she frightens me, for that means a long spell: yet she is by nature a thing of violent and audacious bouts, giving me a thorough

shaking up for a day or two. My kidneys held out for [C] an age [B] without deterioration: it will soon be [C] another age, now, [B] since[106] they changed their condition. Ills as well as blessings run their courses. Perhaps this misfortune is near its end. Old age reduces the heat of my stomach, which therefore digests things less perfectly and dispatches waste matter to my kidneys: so why should the heat of my kidneys, after a stated period has rolled by, not similarly be reduced, rendering them unable to continue to petrify my phlegm and obliging Nature to find some other means of purging it? It is clear that the passing years have exhausted some of my discharges: why not then those excretions which furnish the raw material for my gravel?

'But is there anything so delightful as that sudden revolution when I pass from the extreme pain of voiding my stone and recover, in a flash, the beauteous light of health, full and free, as happens when our colic paroxysms are at their sharpest and most sudden? Is there anything in that suffered pain which can outweigh the joy of so prompt a recovery? Oh how much more beautiful health looks to me after illness, when they are such close neighbours that I can study both, each in her full armour, each in each other's presence, defying each other as though intending to stubborn it out and hold their ground. The Stoics say that the vices were introduced for a purpose – to second virtue and make her prized: we can say, with better justification and less bold conjecture, that Nature has lent us suffering in order that it may honour and serve the purposes of pleasure and of mere absence of pain. When Socrates was freed from the load of his fetters he enjoyed the delicate tingling in his legs that their pressure had produced and he delighted in thinking about the close confederacy that there is between pain and pleasure, so bound together in fellowship as they are by bonds of necessity that they succeed each other and mutually produce each other; and he exclaimed that that excellent man Aesop ought to have drawn from such factors the substance of a beautiful fable.[107]

'For the worst feature of other maladies is that they are less grievous in what they do at the time than in what comes later: you spend a whole year convalescing, all the time full of fear and debility. There is so much hazard in recovery, so many levels involved, that there is no end to it all: before they let you strip off your scarves and then your nightcaps, before they have allowed you to avail yourself again of fresh air, wine, your wife – and of melons – it is quite something if you have not had a relapse into some

106. '88: for *forty years* [. . .] soon be *fourteen years* since . . .
107. Plato, *Phaedo*, 60 B–E.

new wretchedness. My illness is privileged to make a clean break: the others lend each other a hand: they always leave some dent and weakness in you which render your body susceptible to some fresh woe. We can condone such illnesses as are content with their own rights-of-possession over us without introducing their brood: but those whose journey through us produces some useful result are courteous and gracious. Since my stone I find that I have been freed from the load of other ailments and that I seem to feel better than I did before. I have not had a temperature since! I reason that the frequent and extreme vomiting which I suffer purges me and that, from another aspect, the losses of appetite and bizarre fastings which I go through disperse my offending humours, Nature voiding with those stones all her noxious superfluities. And do not tell me that such medicine is bought at too high a price. What about those stinking possets, those cauterizations, incisions, sweat-baths, drainings of pus, diets and those many forms of treatment which often bring death upon us when we cannot withstand their untimely onslaught! So when I suffer an attack I consider it to be a cure: when freed from it, I consider that to be a durable and complete deliverance.

'Another specific blessing of my illness is that it all but gets on with its own business and (unless I lose heart) lets me get on with mine. I have withstood it, at the height of an attack, for ten hours at a time in the saddle. "Just put up with it, that's all! You need no other prescription: enjoy your sports, dine, ride, do anything at all if you can: your indulgences will do you more good than harm." Try saying that to a man with syphilis, the gout or a rupture! The constraints of other illnesses are more all-embracing: they are far more restricting on our activities, upsetting our normal ways of doing anything and requiring us to take account of them throughout the entire state of our lives. Mine does no more than pinch the epidermis: it leaves you free to dispose of your wit and your will as well as of your tongue, your hands and your feet. Rather than battering you numb, it stimulates you. It is your soul which is attacked by a burning fever, cast to the ground by epilepsy, dislodged by an intense migraine and, in short, struck senseless by those illnesses which attack all the humours and the nobler organs. Such are not attacked in my case: if things go ill for my soul, too bad for her! She is betraying, surrendering and disarming herself. Only fools let themselves be persuaded that a solid, massy substance concocted within our kidneys can be dissolved by draughts of medicine. So, once it starts to move, all you can do is to grant it right of passage: it will take it anyway.

'There is another specific advantage that I have noticed: it is an illness

which does not leave us guessing. It dispenses us from the turmoil into which other ills cast us because of uncertainties about their causes, properties and development – an infinitely distressing turmoil. We need have nothing to do with consulting specialists and hearing their opinions: our senses can show us what it is and where it is.'

With such arguments, both strong and feeble, I try, as Cicero did with that affliction which was his old age, to benumb and delude my power of thought and to put ointment on its wounds. And tomorrow, if they grow worse, we will provide other escape-routes for them.

[C] To show that that is true, since I wrote that, the slightest movements which I make have begun to squeeze pure blood from my kidneys again. Yet because of that I do not stop moving about exactly as I did before and spurring after my hounds with a youthful and immoderate zeal. And I find that I have got much the better of so important a development, which costs me no more than a dull ache and heaviness in the region of those organs. Some great stone is compressing the substance of my kidneys and eating into it: what I am voiding drop by drop – and not without some natural pleasure – is my life blood, which has become from now on some noxious and superfluous discharge.

[B] Can I feel something disintegrating? Do not expect me to waste time having my pulse and urine checked so that anxious prognostics can be drawn from them: I will be in plenty of time to feel the anguish without prolonging things by an anguished fear. [C] Anyone who is afraid of suffering suffers already of being afraid. And then the hesitation and ignorance of those who undertake to explain the principles by which Nature operates and her inner progression (as well as the false prognoses of their Art) oblige us to recognize that she keeps her processes absolutely unknown. In her promises and threats there is great uncertainty, variability and obscurity. With the exception of old age (which is an undoubted prognostic of the approach of death), in all our other maladies I can find few prognostics of the future on which we should base our predictions. [B] Judgements about myself I make from true sensation not from argument: what else, since all I intend to bring to bear are patience and endurance. 'What do I gain from that,' do you ask? Look at those who act otherwise and who rely on all that contradictory counsel and advice. How often does their imagination assail them, independently of the body! When safely delivered from a dangerous bout, I have often found pleasure in consulting doctors about it as though it were just starting. Fully at ease I would put up with the formulation of their terrifying diagnoses, and would remain that much more indebted to God for his mercy and better instructed in the vanity of that Art.

There is nothing which ought to be commended to youth more than being active and energetic. Our life is but motion: I am hard to budge and sluggish about everything, including getting up, going to bed and eating. For me, seven o'clock is early morning! And where I head the household I never lunch before eleven nor have supper after six. The causes of those feverish ailments which I formerly used to fall into I once ascribed to the heaviness and sluggishness brought on by prolonged sleep; and I have always regretted falling back to sleep again of a morning. [C] Plato is harder against excessive sleep than excessive drink.[108]

[B] I like a hard bed all to myself, indeed (as kings do) without my wife, with rather too many blankets. I never use a warming-pan, but, since I have grown old, whenever I need them they give me coverlets to warm my feet and stomach. The great Scipio was criticized for being a slug-a-bed, for no other reason, if you ask me, than that it irritated people that in him alone there was nothing to criticize.[109] If I am fastidious about an item in my regimen it is more about bed than anything else: but on the whole I yield to necessity as well as anyone [C] and adjust to it. [B] Sleeping has taken up a large slice of my life and even at my age I can sleep eight or nine hours at a stretch. I am finding it useful to rid myself of this propensity towards laziness and am clearly the better for it. I am feeling the shock of such a revolution, but only for two or three days. And I know hardly anyone who can do with less sleep when the need arises, who can keep on working more continuously or feel less than I do the weight of the drudgery of war. My body is capable of sustained exertions but not of sudden, violent ones. I avoid nowadays all violent activities including those which bring on sweat: before my limbs get hot they feel exhausted. I can be on my feet all day, and I never tire when walking. Over paved roads however, [C] since my earliest childhood [B] I have always preferred to go by horse:[110] when on foot I splatter mud right up to my backside; and in our streets little men are liable to being jostled [C] and elbowed aside, [B] for want of an imposing appearance. And I have always liked to rest, lying or seated, with my legs at least as high as the bench.

No occupation is as enjoyable as soldiering – an occupation both noble in its practice (since valour is the mightiest, most magnanimous and proudest of the virtues) and noble in its purpose: there is no service you can

108. On legislation against excessive sleep, cf. Plato, *Laws*, VII, 807 E–808 D; on milder condemnation of excessive drinking, cf. ibid. II, 673 E–674 D.
109. Plutarch (tr. Amyot), *Qu'il est requis qu'un Prince soit sçavant*, 137 A.
110. '88: however, I *can only* go by horse . . .

render more just nor more complete than protecting the peace and greatness of your country. You enjoy the comradeship of so many men who are noble, young and active, the daily sight of so many sublime dramas, the freedom of straightforward fellowship as well as a manly, informal mode of life, the diversions of hundreds of different activities, the heart-stirring sound of martial music which fills your ears and enflames your soul, as well as the honour of this activity,[111] its very pains and hardships, [C] which Plato rates so low in his *Republic* that he allocates a share in it to women and children. [B] You urge yourself to accept specific tasks or hazards, depending upon your judgement of their splendour or importance; [C] you are a volunteer [B] and can see when your life itself may justifiably be sacrificed to them:

> *pulchrumque mori succurrit in armis.*

> [it is indeed beautiful, I think, to die in battle.][112]

It is for a mind [C] weak [B] and base beyond all measure to be afraid of risks shared in common with a crowd of others, or not to dare to do what men of so many kinds of soul may dare. The comradeship gives confidence to the very boys. Others may surpass you in knowledge, grace, force or fortune: in that case you can put the responsibility for it on to a third party; but if you yield to them in fortitude of soul you alone are responsible. Death is more abject, lingering and painful in bed than in combat; fevers and catarrhs are as painful and as mortal as volleys from harquebuses. Any man who could bear with valour the mischances of ordinary life would have no need to be more courageous on becoming a soldier. [C] *'Vivere, mi Lucili, militare est.'* [To live, my dear Lucilius, is to do battle.][113]

I cannot recall ever having had scabies, but scratching is one of the most delightful of Nature's bounties: and it is always ready to hand! But its neighbour, inconveniently close, is regret for having done it. I mainly practise it on my ears, which from time to time itch inside.

[B] I was born with all my senses[114] intact and virtually perfect. My

---

111. '88: honour and *nobility* of this activity . . .
  Then Plato, *Republic*, V, etc.
112. Virgil, *Aeneid*, II, 317.
  Then '88: for a mind *vile* and base . . .
113. Seneca, *Epist. moral.*, XCVI, 5 (in Seneca a metaphor, not a statement about war).
114. '88: all my *bodily* senses . . .

stomach is as sound as you could wish; my head is, too: both usually remain so during my bouts of fever. The same applies to my respiration. I have exceeded [C] recently, by six years, that fiftieth birthday [B] which[115] some peoples have not unreasonably laid down as termination of life, one so just that nobody was permitted to go beyond it: yet I still have periods of reprieve which, despite being short and variable, are so flawless that they lack nothing of that pain-free health of my youth. I am not referring to liveliness and vigour: it is not reasonable that they should accompany me beyond their limits:

> *Non hæc amplius est liminis, aut aquæ*
> *Cælestis, patiens latus.*

[No longer can I endure waiting on my mistress's doorstep in the pouring rain.][116]

It is my face which gives the game away first; [C] so do my eyes: [B] all changes in me begin there, appearing rather more grim than they are in practice. I often find my friends pitying me before I am aware of any cause. My looking-glass never strikes me with terror, because even in my youth I would often take on a turbid complexion and a look which boded ill without much happening, with the result that the doctors, who could find no cause in my body which produced that outward deterioration, attributed it to my mind and to some secret passion gnawing away within me. They were in error. My body and I would have got on rather better if it had behaved *secundum me*, as did my Soul which was then not only free from turbidity but, better still, full of joy and satisfaction – as she usually is, half because of her complexion and half by design.

> *Nec vitiant artus ægræ contagia mentis.*

[The illnesses of my mind do not affect my joints.][117]

I maintain that this disposition of my Soul has repeatedly helped up my body after its falls: my body is often knocked low whereas she, even when not merry, is at least calm and tranquil. I once had a quartan fever for four or five months which put me right out of countenance, yet my mind still went not merely peacefully but happily on her way. Once the pain has gone I am not much depressed by weakness or lassitude. I know of several bodily afflictions which are horrifying even to name but which I fear less

---

115. '88: exceeded *the age at* which . . .
116. Horace, *Odes*, III, ix, 19–20.
117. Ovid, *Tristia*, III, viii, 25.

than hundreds of current disturbances and distresses of the mind. I have decided never again to run: it is enough for me if I can drag myself along. Nor do I lament the natural decline which has me in its grip:

> *Quis tumidum guttur miratur in Alpibus?*

> [In the Alps is anyone surprised to find goitres?][118]

– no more do I lament that my lifespan is not as long and massive as an oak's. I have no cause to complain of my thought-processes: few thoughts in my life have ever disturbed even my sleep, except when concerned with desire (which woke me up without distressing me). I do not dream much: when I do it is of grotesque things and of chimeras usually produced by pleasant thoughts, more laughable than sad. And although I maintain that dreams are loyal interpreters of our inclinations, there is skill in classifying them and understanding them.

> [C] *Res quæ in vita usurpant homines, cogitant, curant, vident*
> *Quæque agunt vigilantes, agitantque, ea sicut in somno accidunt*
> *Minus mirandum est.*

[It is no miracle that men should find again in their dreams things which occupy them in their lives, things which they think about, worry about, gaze upon and do when they are awake.][119]

Plato further adds that it is wisdom's task to extract from them information telling of future events. I know nothing about that except the wondrous experiences related by Socrates, Xenophon and Aristotle – great men of irreproachable authority.[120] The history books tell us that the Atlantes never dream;[121] they add that they never eat anything which has been slaughtered, a fact which I mention because it may explain why they do not dream, since Pythagoras prescribed a certain preparatory diet designed to encourage dreams.[122] My dreams are weak things: they occasion no twitching of the body, no talking in my sleep. I have known in my time some who have been astonishingly troubled by them. Theon the

118. Juvenal, XIII, 162. (Lack of iodine produced goitres among the Swiss.)
119. Cited by Cicero, *De divinatione*, I, 45, xxii from a lost work of Accius.
120. Cited together by Cicero in the same work, I, xxv, 52–3. (The work of Aristotle referred to by Cicero is lost.)
121. The example of the Atlantes was standard (cf. Rabelais, *Tiers Livre*, TLF, XIII, 56; Coelius Richerius Rhodiginus, XXVII, 16).
122. Cicero, *De divinatione*, II, lviii, 119.

philosopher walked while he dreamed (as did the manservant of Perides, on the tiles of the very roof-ridge of his house).[123]

[B]   At table I rarely exercise a choice, tackling the first and nearest dish; I do not like shifting about from one taste to another. I dislike a multitude of dishes and courses as much as any other multitude. I can be easily satisfied with a few items and loathe the opinion of Favorinus[124] that during a feast any dish you are enjoying should be whipped away from you and a new one always brought in instead, and also that it is a wretched supper at which the guests are not stuffed with rumpsteaks exclusively taken from a variety of birds – only the fig-pecker bird being worth eating whole. I frequently eat salted meats but prefer my bread unsalted: the baker in my own kitchen (contrary to local custom) serves no other at my table. When I was a boy I often had to be punished for refusing precisely those things which are usually best liked at that age: sweets, jams and pastries. My tutor opposed this hatred of fancy foods as being itself a kind of fancy. And indeed, no matter what it applies to, it is nothing but finicking over your food: rid a boy of a fixed private love of coarse-bread, bacon or garlic and you rid him of self-indulgence. There are men who groan and suffer for want of beef or ham in the midst of partridge! Good for them: that is to be a gourmet among gourmets: it is a weak ill-favoured taste which finds insipid those ordinary everyday foods, [C] *'per quae luxuria divitiarum taedio ludit'* [by the which luxury escapes from the boredom of riches].[125]   [B]   The essence of that vice consists in failing to enjoy what others do and in taking anxious care over your diet,

> *Si modica cœnare times olus omne patella.*
>
> [If you jib at an herb salad on a modest platter].

There is certainly a difference, in that it is better to shackle your appetite to whatever is easier to obtain: but such shackling is still a vice. I once called a relation of mine self-indulgent because he had forgotten, during a period in our galleys, how to undress at night and sleep in our beds.

If I had any sons I would readily wish them a fate like mine: God gave me a good father (who got nothing from me apart from my acknowledgement of his goodness – one cheerfully given); from the cradle he sent me to be suckled in some poor village of his, keeping me there until I was

---

123. Both cited together by Diogenes Laertius in his *Life of Pyrrho*.
124. Actually Favorinus criticized this view, which he reported (Aulus Gellius, *Attic Nights*, XV, viii).
125. Seneca, *Epist. moral.*, XVIII, 7. Then, Horace, *Epistles*, I, 52.

weaned – longer in fact, training me for the lowliest of lives among the people: [C]   *'Magna pars libertatis est bene moratus venter.'* [Freedom consists, for a large part, in having a good-humoured belly.][126]

[B]   Never assume responsibility for such upbringing yourself and even less allow your wives to do so: let boys be fashioned by fortune to the natural laws of the common people; let them become accustomed to frugal and severely simple fare, so that they have to clamber down from austerity rather than scrambling up to it. My father's humour had yet another goal: to bring me closer to the common-folk and to the sort of men who need our help; he reckoned that I should be brought to look kindly on the man who holds out his hand to me rather than on one who turns his back on me and snubs me. And the reason why he gave me godparents at baptism drawn from people of the most abject poverty was to bind and join me to them. His plan has not turned out too badly. I like doing things for lowly people, either because there is more glory in it or else from innate sympathy (which can work wonders with me).   [C]   The party I condemn in these wars of ours I would condemn more severely when it is flourishing and successful: it can almost reconcile me to it when I see it   [B]   wretched and overwhelmed.[127] How I love to reflect on that beautiful humour of Chelonis who was both daughter and wife of Kings of Sparta: while her husband Cleombrotus had the edge over her father Leonidas she was a good daughter, rallying to her father in his wretched exile and defying the victor. Then fortune veered about, did it not? Whereupon, as fortune changed she changed her mind, ranging herself courageously beside her husband, whom she followed no matter where his downfall drove him, having, it seems, no preference between them but leaping to the support of whichever party needed her more and to whom she could better show pity.[128] My nature is to follow the example of Flaminius (who lent his support to those who needed him, not to those who could help him) rather than that of Pyrrhus (who had the characteristic of being humble before the great and arrogant before the common-folk).

Long sittings at table   [C]   irritate me and   [B]   disagree with me, since, lacking restraint (doubtless because I formed the habit as a boy), I go

126. Seneca, *Epist. moral.*, CXXIII, 3.
127. '88: me.) *I condemn* in these *disturbances* of ours *the cause of one of the parties, but more so* when it is flourishing and successful: it [i.e. the cause] *has* almost *reconciled* me to it when I see it wretched and overwhelmed . . . [Pity, or sympathy, for the cause of the Reformers changes to pity for their faction.]
128. Condensed from Plutarch's *Life of Agis* and *Life of Cleomenes*; then, *Life of Flaminius* and *Life of Pyrrhus*.

on eating as long as I am there. That is why at home   [C]   (even though
our meals are among the shorter ones)   [B]   I like to come in   [C]   a
little   [B]   after the others, following the fashion of Augustus, although I
do not imitate him in leaving before the others. On the contrary: I like to
stay on a long time afterwards listening to the conversation, provided that I
do not join it since I find it as tiring and painful to talk on a full stomach as
I find it a healthy and pleasant exercise to argue and bellow before a
meal.   [C]   The ancient Greeks and Romans were more reasonable than
we are: unless some other quite unusual task intervened they assigned to
eating (which is one of the chief activities of our lives) several hours a day
and the best part of the night, eating and drinking less hurriedly than we
do who gallop through everything; they extended both the leisureliness of
this natural pleasure and its conviviality by interspersing it with various
social duties both useful and pleasant.

[B]   Those who   [C]   ought to take care of me, could,   [B]   at little
cost[129] to themselves, cheat me of whatever they think harmful to me, for
in such matters I neither want what is not there nor notice its absence: but
they also waste their breath if they lecture me on abstaining from whatever
is served. The result is that when I resolve to diet you have to put me apart
from the other diners, serving me precisely what is sufficient for a moderate
snack; for if I sit down at table I forget my resolution. When I order my
servants to change the way they are serving up a dish they know that that
means my appetite is gone and that I will not touch any. I prefer to eat rare
any flesh that lends itself to it. I like it to be well-hung, even in many
cases until it starts to smell high. Generally speaking toughness is the only
quality which irritates me (towards all others I am as indifferent and long-
suffering as anyone), so much so that, contrary to the usual whim, I find
even some fish too fresh and firm. That is nothing to do with my teeth
which have always been exceedingly good and which only now are
starting to be threatened by old age. Since boyhood I learned to rub them
on my napkin, both on rising and before and after meals.

God shows mercy to those from whom he takes away life a little at a
time: that is the sole advantage of growing old; the last death which you
die will be all the less total and painful: it will only be killing off half a
man, or a quarter. Look: here is a tooth which has just fallen out with no
effort or anguish: it had come to the natural terminus of its time. That part
of my being, as well as several other parts, are already dead: others are half-
dead, including those which were, during the vigour of my youth, the

129. '88: Those who take care of me *can* at little cost . . .

most energetic and uppermost. That is how I drip and drain away from myself. What animal-stupidity it would be if my intellect took for the whole of that collapse the last topple of an already advanced decline. I hope that mine will not.

[C] To tell the truth the principal consolation I draw from thoughts of my death is that it will be right and natural: from this day forth I could not beg or hope from Destiny any but a wrongful favour. People convince themselves that in former times man's lifespan, like his height, was bigger. Yet Solon, who belongs to those times, cuts off our extreme limit at three score years and ten.[130] I, who have in all things so greatly honoured that ἄριστον μέτρον [excellent Mean] of former ages and who have taken moderation as the most perfect measure, should I aspire to an immoderate and enormously protracted old age? Anything which goes against the current of Nature is capable of being harmful, but everything which accords with her cannot but be pleasant: '*Omnia quae secundum naturam fiunt, sunt habenda in bonis,*' [Everything that happens in accordance with Nature must be counted among the things which are good.][131] That is why Plato says that deaths caused by wounds and illnesses may be termed violent, but the death which, as Nature leads us toward her, takes us by surprise is of all deaths the lightest to bear and to some extent enjoyable. '*Vitam adolescentibus vis aufert, senibus maturitas.*' [Life is wrenched from young men: from old men it comes from ripeness.]

[B] Everywhere death intermingles and merges with our life: our decline anticipates its hour and even forces itself upon our very progress. I have portraits of myself aged twenty-five and thirty-five. I compare them with my portrait now: in how many ways is it no longer me! How far, far more different from them is my present likeness than from what I shall be like in death. It is too much an abuse of Nature to [C] flog[132] [B] her along so far that she is, for us, compelled to give up and abandon our guidance, our eyes, teeth, legs and so on to the mercy of remedies not our own but such as we can beg, relinquishing us, since she is weary of following us, into the hands of that 'Art'.

I am not over-fond of salads nor of any fruit except melons. My father loathed all kinds of sauces: I love them all. Overeating distresses me, but I am not aware that any food as such definitely disagrees with me, any more

---

130. Herodotus, I, xxxii.
131. Cicero, *De senectute*, xix, 71; then, Plato, *Timaeus*, 81E and Cicero, *De senectute*, ix, 71 (again).
132. '88: to *drag* her along . . .

than I take note of full or crescent moons or of spring or autumn. There are fickle inexplicable changes which occur in us: for example I first of all found that radishes agreed with me; then they did not; now they do again. I have found my stomach and my tastes varying like this over several foods: I have replaced white wine by red, then red by white. I delight in fish, so that my days of abstinence are days of plenty and my fast-days are feast-days. I believe what some say: that fish is more easily digestible than flesh. It goes against my conscience to eat flesh on fish-days and against my preference to mix fish and flesh: there seems to be too wide a difference between them.

Since I was a young man I have occasionally gone without my dinner, either to whet my appetite for the next day (for, while Epicurus went without food or ate little in order to accustom his sense of enjoyment to do without abundance, I on the contrary do so in order to train it to profit from abundance and to make merry with it); or so as to husband my strength in the service of some physical or mental activity (since both grow cruelly sluggish within me through repletion: and I loathe above all that silly yoking together of so sane and merry a goddess as Venus with that little belching dyspeptic Bacchus, all blown up by the fumes of his wine);[133] or else to cure a sick stomach, or for want of appropriate company (since with that same Epicurus I say that we should be less concerned with what we eat than with whom we eat,[134] and I approve of Chilo's refusal to promise to come to a banquet at Periander's before finding out who the other guests were). No recipe is so pleasing to me, no sauce so appetizing, as those which derive from the company.

I believe it is healthier to eat more leisurely, less, and at shorter intervals. But I would give precedence to appetite and hunger: I would find no pleasure in dragging through three or four skimped meals a day on doctor's orders: [C]   who could assure me that at suppertime I would find again that frank appetite I have this morning? Especially we old men should seize the first opportune moment which comes along. Let us leave the prognostics of propitious times to the scribblers of almanacks and to the doctors.[135]   [B]   The ultimate benefit of my feeling well is pleasure: let

133. Montaigne is rejecting proverbial Classical wisdom, which made food and wine the precursors of love-making. Cf. Erasmus, *Adages*, II, III, XCVII, *Sine Cerere et Baccho friget Venus*.

134. Seneca, *Epist. moral.*, XIX, 10; then for Chilo, Plutarch (tr. Amyot), *Banquet des sept Sages*, 150H–151C.

135. Medical astrological almanacks (a legal monopoly of the medical profession) marked particular dates as propitious for certain foods, treatments and so on.

us cling to the first pleasure which is present and known. I refuse to stick for long to any prescriptions limiting my diet. A man who wants a regimen which serves him must not allow it to go on and on; for we become conditioned to it; our strength is benumbed by it; after six months you will have so degraded your stomach that it will have profited you nothing: you will merely have lost your freedom to do otherwise without harm.

My legs and thighs I cover no more in winter than in summer, wearing simple silken hose. I did let myself go, keeping my head warmer to help my rheum and my stomach warmer to help my stone, but within a day or two my ailments grew used to this and showed contempt for such routine provisions: so I moved on from a cap to a head scarf and then from a bonnet to a fur hat. The padding of my doublet now only serves as decoration: it is pointless unless I add a layer of rabbit-fur or vulture-skin[136] and wear a skull-cap under my hat. Follow that gradation and you will go a long way! I will not do so and would willingly countermand what I have already done if only I dared. 'Are you feeling some fresh discomfort? Well, then, that reform of yours did you no good: you have grown used to it. Find another.' Thus are men undermined when they allow themselves to become encumbered with restricted diets and to cling to them superstitiously. They need to go farther and farther on, and then farther still. There is no end to it.

For both work and pleasure's sake it is far more convenient to do as the ancients did: go without lunch and, so as not to break up the day, put off the feast until the time comes to return home and rest. I used to do that once, but I have subsequently found from experience that, on the contrary, it is better for my health's sake to eat at lunchtime, since digestion is better when you are awake.

I rarely feel thirsty when I am in good health – nor when ill, though I do get a dry mouth then, yet without a thirst. Normally I drink only for the thirst which comes as I eat, well on into the meal. For a man of the ordinary sort I drink quite enough: even in summer and during an appetizing meal I not only exceed the limits set by Augustus (who drank exactly three glasses, no more), but so as not to infringe the rule of Democritus (who forbade you to stop at four as being an unlucky number) I down up to five if the occasion arises (that is about a pint and

136. Cotgrave's *Dictionarie of the French and English Tongues* confirms that vulture-skin was used in garments for warmth.

a quarter: for I favour smaller glasses and like draining them dry, something which others avoid as unseemly).[137] I water my wine, sometimes half and half, sometimes one-third water. When I am home I follow an ancient custom which my father's doctor prescribed for him (and for himself): I have what I need mixed for me in the buttery two or three hours before serving. [C] It is said that this custom of mixing wine and water was invented by Cranaus, King of Athens – I have heard arguments both for and against its usefulness. I think it more proper and more healthy that boys should not drink any wine until they are sixteen or eighteen. [B] The finest custom is the one most current and common: in my view all eccentricity is to be avoided; I would hate a German who put water in his wine as much as a Frenchman who drank it neat. The law in such things is common usage.

I am afraid of stagnant air and go in mortal fear of smells (the first repairs I hastened to make in my place were to the chimneys and lavatories – the usual flaws in old buildings and quite intolerable) and among the hardships of war I count those thick clouds of dust under which we are buried in summer during a long day's ride. My breath comes easily and freely and my colds usually clear away without affecting my lungs or giving me a cough.

The rigours of summer are more inimical to me than those of winter, for (apart from the inconvenience of the heat, less easy to remedy than the cold, and apart from sunstroke from the sun beating down on your head) my eyes are affected by any dazzling light: I could not lunch now facing a bright and flaming fire. At the time when I was more in the habit of reading I used to place a piece of glass over my book to soften the glare of the paper and found it quite a relief. Up till now[138] I have no acquaintance with spectacles and can see as well at a distance as ever I did or as anyone can. It is true that towards nightfall I begin to be aware that when reading my vision is weak and hazy: reading has strained my eyes at all times, but especially in the evening. [C] Though barely noticeable, that constitutes one step backwards. I shall take another step back, the second followed by a third, the third by a fourth, so gently that, before I am aware that my ageing sight is failing, I shall have become quite blind – so skilfully do the Fates spin the thread of our lives.

---

137. Erasmus, *Adages*, II, III, I, *Aut quinque bibis aut treis, aut ne quatuor*. Montaigne drinks three *démi-sétiés*. A *septier* (or *sétier*) was a variable measure, but for wine contained two Parisian *chopines*, each a little less than an English pint. Montaigne may have drunk as much as a pint and a half.
138. '88: now, *at the age of fifty-four*, I have ...

I am similarly unwilling to admit that I am on the point of becoming hard of hearing, and you will find that when I am half-deaf I shall still be blaming it on the voices of those who are speaking to me. If we want our Soul to be aware of how she is draining away we must keep her on the stretch.

[B] My walk is quick and steady and I do not know whether I have found it harder to fix my mind in one place or my body. Any preacher who can hold my attention throughout an entire sermon must be a good friend of mine! In the midst of ceremonial, where everyone else maintains a fixed expression and where I have seen ladies keep their very eyes still, I have never succeeded in stopping at least one of my limbs from jigging about: seated I may be, but sedate, never. [C] Just as the chambermaid said of her master the philosopher Chrysippus that only his legs were drunk[139] (for he had this same habit of fidgeting them about, no matter what position he sat in, and she said it of him when the wine was exciting the others while he alone felt none the worse for it), so too people have been able to say of me since boyhood that I have 'mad' or 'quicksilver' feet: no matter where I put them, they are restless and never still.

[B] To eat ravenously as I do is not only unseemly: it is bad for your health, and indeed for your pleasure. In my haste I often bite my tongue and occasionally bite my fingers. When Diogenes came across a boy who was eating like that he slapped his tutor.[140] – [C] There were instructors in Rome who taught how to masticate and perambulate graciously. – [B] By eating thus I lose an occasion for talking, which is such a fine [C] seasoning[141] [B] at table – provided that both the meal and the topics are pleasant and brief. There is jealousy and rivalry among our pleasures: they clash and get in each other's way. Alcibiades was a man who well understood good living: he specifically banished music from his table so that it should not interfere with the conversation, [C] justifying this with the reason which Plato ascribes to him, that it is the practice of common-place men to invite musicians and singers to their feasts since they lack that good talk and those pleasant discussions with which intelligent men understand how to delight each other.[142]

139. Erasmus, *Apophthegmata*, VII, *Chrysippus Solensis*, VI.
   '88: sedate, never: *and for gesticulation I am rarely to be found, on horse or on foot, without a stick in my hand.* To eat ravenously . . .
140. Erasmus, *Apophthegmata*, III, *Diogenes*, final hundred, XXIII.
141. '88: fine *condiment* at table . . .
142. Plato, *Protagoras*, 347.

[B]   The following are Varro's prescription for a banquet: an assembly of people of handsome presence who are agreeable to frequent and neither dumb nor talkative; clean and delightful food in a clean and delightful place; serene weather.[143]   [C]   An enjoyable dinner is a feast requiring no little skill and affording no little pleasure: neither great war-leaders nor great philosophers have declined to learn how to arrange one. My mind has entrusted to my memory three such feasts: they occurred at different times during the flower of my youth and chanced to give me sovereign pleasure (guests contributing to such sovereign delight according to the degree of good temper of body and soul in which each man chances to be). My present circumstances exclude me from such things.

[B]   I who am always down-to-earth in my handling of anything loathe that inhuman wisdom which seeks to render us   [C]   disdainful and   [B]   hostile towards the care of our bodies.[144] I reckon that it is as injudicious to set our minds against natural pleasures as to allow them to dwell on them.   [C]   Xerxes was an idiot to offer a reward to anyone who could invent some new pleasure for him when he was already surrounded by every pleasure known to Man:[145] but hardly less idiotic is the man who lops back such pleasures as Nature has found for him.   [B]   We should neither hunt them nor run from them: we should accept them. I do so with a little more zest and gratitude than that, and more readily follow the slope of Nature's own inclining.   [C]   There is no need for us to exaggerate their emptiness: that makes itself sufficiently known and sufficiently manifest, thanks to our morbid spoilsport of a mind which causes them all to taste as unpleasant to us as it does itself, treating both itself and everything it absorbs, no matter how minor, according to its own insatiable, roaming and fickle condition:

> *Sincerum est nisi vas, quodcunque infundis, acessit.*

> [If the jug is not clean, all you pour into it turns sour.][146]

I who boast that I so sedulously and so individually welcome the pleasures of this life find virtually nothing but wind in them when I examine them in detail. But then we too are nothing but wind. And the wind (more wise than we are) delights in its rustling and blowing, and is content with its

143. Aulus Gellius, *Attic Nights*, XIII, 11.
144. '88: care *and pleasure* of our bodies . . .
145. Cicero, *Tusc. disput.*, V, vii, 20.
146. Horace, *Epistles*, I, ii, 54.

own role without yearning for qualities which are nothing to do with it such as immovability or density.

Some say that the greatest pleasures and pains are those which, as was shown by the Balance of Critolaus, belong exclusively to the mind.¹⁴⁷ No wonder: the mind fashions them as it wills and tailors them for itself from the whole cloth. Everyday I see noteworthy and doubtless desirable examples of it. But I, whose constitution is composite and coarse, cannot so totally get a bite on such an indivisible single object that I do not tend heavily towards the immediate pleasures of that law of humans and their genus: things are sensed through the understanding, understood through the senses.¹⁴⁸

The Cyrenaic philosophers held that the most intense pleasures and pains are those of the body, virtually double and more right.¹⁴⁹ [B] There are [C] those who, from an uncouth insensibility hold (as Aristotle says) bodily pleasures in disgust.¹⁵⁰ I know some who do it from ambition. [B] Why do they not also give up breathing, so as to live on what is theirs alone,¹⁵¹ [C] rejecting the light of day because it is free and costs them neither ingenuity nor effort? [B] Just to see, let Mars, Pallas or Mercury sustain them instead of Venus, Ceres and Bacchus.¹⁵² [C] I suppose they think about squaring the circle while lying with their wives! [B] I hate being told to have our minds above the clouds while our

147. The balance of Critolaus, the peripatetic philosopher, always gave greater weight to the goods of the soul. (Cicero, *Tusc. disput.*, V, xvii, 51.)

148. For this much reworked sentence, I have followed the punctuation of ['95] etc. The general meaning is: Being a man (that is, body-plus-soul) and being weighted towards the body, Montaigne is unable fully to enjoy pure and simple intellectual pleasures. The law of Nature which applies to our genus (*animal*) makes the senses the gateway of cognition and cognition the means by which the ideas are appreciated. (The ideas are consonant with Epicureanism: cf. Cicero, *Tusc. disput.*, V, xxxiii, 95–8.)

149. Cicero, *De officiis*, III, xxxi, 116; *Academica*, II (Lucullus), xlii, 131 and xxiv, 76.

150. Probably an allusion to Aristotle, *Nicomachaean Ethics*, III, xi, 7 (1119a): men insensible to pleasure are very few and such insensibility is not human.

'88: There are *in our youth those who ambitiously claim to trample them underfoot*: why do they . . .

151. '88: theirs alone, *without help from their normal pattern.* Just to see, let Mars . . .

152. That is, let them live on war (Mars), wisdom (Pallas) or eloquence (Mercury) instead of sexual intercourse (Venus), corn (Ceres) and wine (Bacchus), the second three representing bodily 'necessities'.

'88: Bacchus. Such *vaunting humours can forge themselves some contentment (for what power can our minds not have over us!) but of wisdom they have no tincture.* I hate . . .

bodies are at the dinner-table. It is not that I want the mind to be nailed to it or wallowing in it but I do want it to apply itself to it,　[C]　to sit at table, not to lie on it. Aristippus championed only the body, as though we had no soul: Zeno embraced only the soul, as though we had no body. Both were flawed.[153] They say that Pythagoras practised a philosophy which was pure contemplation: Socrates one which was all deeds and morals; between them both Plato found the Mean. But they are pulling our legs. The true Mean is to be found in Socrates; Plato is far more Socratic than Pythagorean, and it better becomes him.[154]

[B]　When I dance, I dance. When I sleep, I sleep; and when I am strolling alone through a beautiful orchard, although part of the time my thoughts are occupied by other things, for part of the time too I bring them back to the walk, to the orchard, to the delight in being alone there, and to me. Mother-like, Nature has provided that such actions as she has imposed on us as necessities should also be pleasurable, urging us towards them not only by reason but by desire. To corrupt her laws is wrong.

When, in the thick of their great endeavours, I see Caesar and Alexander so fully enjoying pleasures which are[155]　[C]　natural and consequently necessary and right,　[B]　I do not say that their souls are relaxing but giving themselves new strength, by force of mind compelling their violent pursuits and burdensome thoughts to take second place to the usages of everyday life,　[C]　wise if they were to believe that to be their normal occupation and the other one abnormal.

What great fools we are! 'He has spent his life in idleness,' we say. 'I haven't done a thing today.' – 'Why! Have you not lived? That is not only the most basic of your employments, it is the most glorious.' – 'I would have shown them what I can do, if they had set me to manage some great affair.' – If you have been able to examine and manage your own life you have achieved the greatest task of all. Nature, to display and show her powers, needs no great destiny: she reveals herself equally at any level of life, both behind curtains or without them. Our duty is to bring order to our morals not to the materials for a book: not to win provinces in battle but order and tranquillity for the conduct of our life. Our most great and

153. Cicero's contention, *Academica*, II (Lucullus), xlv, 139.
154. Probably an echo of St Augustine, *City of God*, VIII, iv; but while Augustine makes Plato combine Socrates' virtues with those of Pythagoras, he does not write of his being the mean between them. Montaigne's term for the Mean, *tempérament*, represents Aristotle's term *sophrosyne*.
155. '88: pleasures which are *human and bodily*, I do not say . . .

glorious achievement is to live our life fittingly. Everything else – reigning, building, laying up treasure – are at most tiny props and small accessories. [B] I delight in coming across a general in the field, at the foot of a breach which he means soon to attack, giving himself whole-heartedly to his dinner while chatting freely with his friends, [C] or across Brutus, with heaven and earth conspiring against him and the liberty of Rome, stealing an evening hour from his rounds of duty to jot down notes on his Polybius as he read him with complete composure.[156]   [B] It is for petty souls overwhelmed by the weight of affairs to be unable to disentangle themselves for them completely, not knowing how to drop them and then take them up again:

> *O fortes pejoraque passi*
> *Mecum sæpe viri, nunc vino pellite curas;*
> *Cras ingens iterabimus æquor.*

[O ye strong men who have often undergone worse trials with me, banish care now with wine. Tomorrow we will sail again over the vast seas.][157]

Whether as a joke or in earnest, 'theological wine' and [C] 'Sorbonne [B] wine' have become proverbial, as have their gaudies;[158] but I find that the fellows are right to dine all the more indulgently and enjoyably in that they have seriously and usefully used their mornings for the concerns of their college: the shared awareness of having used those other hours well is a proper and piquant condiment for their table. That is how the sages lived. Thus too did the inimitable eager striving towards virtue which amazes us in both the Catos, as well as their severity of humour to the point of rudeness, mildly and happily submit to the laws of our human condition, to Venus and to Bacchus,[159]   [C] following the precepts of their School which required the perfect sage to be as experienced and knowledgeable about the use of the natural pleasures as about all the rest of life's duties: *'Cui cor sapiat, ei et sapiat palatus'* [To a discriminating mind let him ally a discriminating palate.][160]

[B] In a strong and great-souled man it is, it seems to me, wondrously honourable to be relaxed and approachable, and it is most befitting.

156. From Plutarch's *Life of Brutus*.
157. Horace, *Odes*, I, vii, 30–2.
158. '88: and *professorial* wine . . .
    The quality and quantity of the drinking in the Sorbonne (the Faculty of Theology) was indeed proverbial. Cf. Sainéan, *Langue de Rabelais*, I, 368.
159. Cf. Rabelais, *Tiers Livre*, TLF, *Prologue*, 182; Horace, *Odes*, III, xxi, 9–12.
160. Cicero, *De finibus*, II, viii, 24 (truncated and differently applied).

Epaminondas never thought that to join in the dance with the young men of his city, [C] to sing and strum with them, [B] and to bother to do it properly, in any way detracted from the honour of his glorious victories nor from the [C] perfect [B] reformation of morals [C] which was within him.[161] [B] And among all the remarkable actions of Scipio [C] – the grandfather, that great man worthy of having been thought to descend from the gods[162] – [B] none is more gracious than his having been seen idling along, unperturbed, choosing and collecting shells like a schoolboy, playing *Quick! Quick! Pick up sticks* with Laelius along the seashore and, when the weather was bad, passing his time enjoyably by writing comedies about the most plebeian and realistic activities of men;[163] [C] and, while his mind was full of that marvellous African campaign of his against Hannibal, visiting the philosophy schools in Sicily and attending the lectures, so providing his enemies at Rome with something to snap at in their blind envy.[164]

[B] Nor is there anything more striking about Socrates than his finding the time when he was old to learn how to dance and to play instruments, maintaining that it was time well spent. He was seen standing in an ecstatic trance for a day and a night in view of all the Grecian army, surprised and caught up by some deep thought. He was seen [C] to be the first of many brave men in that army to dash to the help of Alcibiades when he was overwhelmed by the enemy, shielding him with his body and pulling him out from under the weight of their numbers by the sheer force of his arms; and of all the people of Athens (outraged like him by such a shameful sight) he was the first to stand forth to rescue Theramenes from the Thirty Tyrants, whose henchmen were escorting him to his death and, although he was seconded by only two men, he did not give up that valiant attempt until Theramenes himself urged him to do so. When wooed by a person whose beauty had enthralled him, he was

---

161. Cornelius Nepos, *Life of Epaminondas*.
    '88: morals *there ever was in man.* And among . . .
162. '88: of Scipio *the Younger (when all is done the first man among the Romans)* none is . . .
163. Erasmus, *Adages*, V, II, XX, *Conchas legere*, citing, apropos of Scipio and Laelius, Valerius Maximus, VIII, viii, and Cicero, *De oratore*, II, vi. (Montaigne introduces a confusion in [C]: he means, as he first wrote, the Younger, not the Elder, Scipio. The error remains in the posthumous editions, with the result that anecdotes about Scipio Africanus Major and Scipio Aemilianus Africanus Minor are fused into one, as are these two Scipios themselves.)
164. Livy, XIX, xix, of Scipio Africanus Major.

seen to maintain, as was necessary, the strictest continence; he was seen helping up Xenophon at the battle of Delium, saving him when his horse had given him a tumble. He was seen [B] striding undeviatingly to war [C] trampling over the ice [B] in his bare feet; wearing the same gown in winter as in summer; surpassing all his comrades in his endurance of hardships and, at feasts, eating no differently from usual. [C] He was seen, unmoved in countenance, putting up for twenty-seven years with hunger and poverty, with loutish sons, with a cantankerous wife and finally with calumny, tyranny, imprisonment, leg-irons and poison. [B] Yet that very man, when the dictates of courtesy made him a guest at a drinking-match, was, from the entire army, the man who best acquitted himself. Nor did he refuse to play five-stones with the boys nor to run about with them astride a hobby-horse. And he did it with good grace: for Philosophy says that all activities are equally becoming in a wise man, all equally honour him. We have the wherewithal, so we should never tire of comparing the ideal of that great man against all patterns and forms of perfection.[165] [C] There are very few pure and complete exemplars of how to live; those who instruct us do wrong to set before us weak and faulty ones with scarcely a single good habitual quality, ones which are more likely to pull us backwards, corrupting us rather than correcting us.

[B] The many get it wrong: you can indeed, using artifice rather than nature, make your journey more easily along the margins, where the edges serve as a limit and a guide, rather than take the wide and unhedged Middle Way; but it is also less noble, less commendable. [C] Greatness of soul consists not so much in striving upwards and forwards as in knowing how to find one's place and to draw the line. Whatever is adequate it regards as ample; it shows its sublime quality by preferring the moderate to the outstanding. [B] Nothing is so beautiful, so right, as acting as a man should: nor is any learning so arduous as knowing how to live this life [C] naturally and [B] well. And the most uncouth of our afflictions is to [C] despise [B] our being.[166] If anyone desires to set his soul apart so as to free it from contagion, let him have the boldness to do so (if he can) while his body is unwell: otherwise, on the contrary, his soul should assist and applaud the body, not refuse to participate in its

---

165. A composite picture of Socrates from the standard sources: especially Plato's *Symposium*, 213A − 220D, with a borrowing from Diogenes Laertius' *Life of Socrates*.
166. '88: is to *hate and disdain* our being . . .

natural pleasures but delight in it as if it were its husband, contributing, if it is wise enough, moderation, lest those pleasures become confounded with pain through want of discernment.   [C]   Lack of temperance is pleasure's bane: temperance is not its chastisement but its relish. It was by means of temperance, which in them was outstanding and exemplary, that Eudoxus (who made pleasure his sovereign good) and his companions (who rated it at so high a price) savoured it in its most gracious gentleness.[167]

[B]   I so order my soul that it can contemplate both pain and pleasure with eyes equally   [C]   restrained – *'eodem enim vitio est effusio animi in laetitia quo in dolore contractio'* [for it is as wrong for the soul to dilate with joy as to contract with pain][168] – doing so with eyes equally   [B]   steady, yet looking merrily at one and soberly at the other and, in so far as it can contribute anything itself, being as keen to snuff out the one as to stretch out the other.   [C]   Look sanely upon the good and it follows that you look sanely upon evils: pain, in its tender beginnings, has some qualities which we cannot avoid: so too pleasure in its final excesses has qualities which we can avoid. Plato couples pain and pleasure together and wants it to be the duty of fortitude to fight the same fight against pain and against the seductive fascinations of immoderate pleasure. They form two springs of water: blessed are they, city, man or beast, that draw what they should, when they should and from the one they should. From the first we should drink more sparingly, as a medicine, as a necessity: from the second to slake our thirst, though not to the point of drunkenness. Pain and pleasure, love and hatred, are the first things a child is aware of: if, after Reason develops, they are guided by her, then that is virtue.[169]

[B]   I have a lexicon all to myself: I 'pass' the time when tide and time are sticky and unpleasant: when good, I do not want to 'pass' time, I   [C]   savour it and hold on to it.[170]   [B]   We must run the gauntlet through the bad and recline on the good. 'Pastimes' and 'to pass the time' are everyday expressions which correspond to the practice of those clever folk who think that they can use their life most profitably by letting it leak and slip away, by-passing it or avoiding it and (as far as they can manage to

---

167. Eudoxus maintained that pleasure is the Supreme Good, arguing that all creatures, rational and irrational, seek it and avoid pain. (Aristotle, *Nicomachaean Ethics*, X, ii, 1172 b.) Aristotle adds that Eudoxus had a reputation for exceptional temperance. (Cf. also ibid., I, xii.) His 'companions' are doubtless the Platonists, of whom he was an unorthodox associate.

168. Cicero, *Tusc. disput.*, IV, xxxi, 66.

169. Plato, *Laws*, I, 632C–634B; 6360; 653A–C.

170. '88: I *taste* it and *linger over* it. We must . . .

do so) ignoring it and fleeing from it as painful and contemptible. But I know life to be something different: I find it to be both of great account and delightful – even as I grasp it now [C] in its final waning; [B] Nature has given it into our hands garnished with such attributes, such agreeable ones, that if it weighs on us, if it slips uselessly from us, we have but ourselves to blame. [C] *'Stulti vita ingrata est, trepida est, tota in futurum fertur.'* [It is the life of the fool which is graceless, fearful and entirely sacrificed to the future.][171]

[B] That is why I so order my ways that I can lose my life without regret, not however because it is troublesome or importunate but because one of its attributes is that it must be lost. [C] Besides, finding it not unpleasant to die can only rightly become those who find life pleasant. [B] To enjoy life requires some husbandry. I enjoy it twice as much as others, since the measure of our joy depends on the greater or lesser degree of our attachment to it. Above all now, when I see my span so short, I want to give it more ballast; I want to arrest the swiftness of its passing by the swiftness of my capture, compensating for the speed with which it drains away by the intensity of my enjoyment. The shorter my lease of it, the deeper and fuller I must make it.

Others know the delight of happiness and well-being: I know it as they do, but not *en passant*, as it slips by. We must also study it, savour it, muse upon it, so as to render condign thanksgivings to Him who vouchsafes it to us. Other folk enjoy all pleasures as they enjoy the pleasure of sleep: with no awareness of them. Why, with the purpose of not allowing even sleep to slip insensibly away, there was a time when I found it worthwhile to have my sleep broken into so that I could catch a glimpse of it. I deliberate with my self upon any pleasure. I do not skim it off: I plumb it, and now that my reason has grown chagrin and squeamish I force it to accept it. Do I find myself in a state of calm? Is there some pleasure which thrills me? I do not allow it to be purloined by my senses: I associate my Soul with it, not so that she will [C] bind herself to it[172] [B] but take joy in it: not losing herself but finding herself in it; her role is to observe herself as mirrored in that happy state, to weigh that happiness, gauge it and increase it. She measures how much she owes to God for having her conscience and

171. '88: I grasp it now, in its *decadence*; Nature . . .
Seneca, *Epist. moral.*, XV, 9. (Seneca presents this saying as an 'excellent Greek proverb' uttered by Epicurus, warning that it applies not to the lives of obviously foolish men but to our own, with its unsatisfiable desires.)
172. '88: she will *get drunk on* it but take . . .

her warring passions at peace, with her body in its natural [C] state, [B] enjoying ordinately and [C] appropriately [B] those sweet and pleasant functions by which it pleases Him, through His grace, to counterbalance the pains with which His justice in its turn chastises us;[173] she gauges how precious it is to her to have reached such a point that, no matter where she casts her gaze, all around her the heavens are serene – no desire, no fear or doubt bring disturbing gales; nor is there any hardship, [C] past, present or future [B] on which her thoughts may not light without anxiety. This meditation gains a great splendour by a comparison of my condition with that of others. And so I [C] pass in review,[174] [B] from hundreds of aspects, those whom fortune or their own mistakes sweep off into tempestuous seas, as well as those, closer to my own case, who accept their good fortune with such languid unconcern. Those folk really do 'pass' their time: they pass beyond the present and the things they have in order to put themselves in bondage to hope and to those shadows and vain ghosts which their imagination holds out to them –

> *Morte obita quales fama est volitare figuras,*
> *Aut quæ sopitos deludunt somnia sensus*

[Like those phantoms which, so it is said, flit about after death or those dreams which delude our slumbering senses]

– the more you chase them, the faster and farther they run away. Just as Alexander said that he worked for work's sake –

> *Nil actum credens cum quid superesset agendum:*

[Believing he had not done anything, while anything remained to be done:]

– so too your only purpose in chasing after them, your only gain, lies in the chase.[175]

As for me, then, I love life and cultivate it as it has pleased God to vouchsafe it to us. I do not go yearning that it should be without the need to eat and drink: [C] indeed to wish that need redoubled would not seem to me a less pardonable error: '*Sapiens divitiarum naturalium quaesitor acerrimus*' [The wise man is the keenest of seekers after the riches of Nature];[176] nor [B] that we could keep up our strength by merely popping into our mouths a little of that drug by means of which Epimenides

---

173. '88: natural *health*, enjoying ordinately and *fully* those sweet . . .
174. '88: I *picture to myself*, from hundreds of aspects . . .
175. Virgil, *Aeneid*, X, 641–2; Lucan, *Pharsalia*, II, 657.
176. Seneca, *Epist. moral.*, CXIX, 5.

assuaged his appetite and kept alive;[177] nor that we could, without sensation, produce children by our fingers and our heels [C] but rather,
speaking with reverence, that we could also do it voluptuously with our
fingers and our heels as well; [B] nor that our body should be without
desire or thrills. Such plaints are [C] ungrateful and iniquitous. [B] I
accept wholeheartedly [C] and thankfully [B] what Nature has
done for me: I delight in that fact and am proud of it. You do wrong to that
great and almighty Giver to [C] refuse [B] His gift, to [C] nul-
lify [B] it or disfigure it. [C] Himself entirely Good, he has made
all things good: '*Omnia quae secundum naturam est, aestimatione digna sunt.*'
[All things which are in accordance with Nature are worthy of esteem.][178]

[B] I embrace most willingly those of Philosophy's opinions which are
most solid, that is to say, most human, most ours: my arguments, like my
manners, are lowly and modest. [C] To my taste she is acting like a
child when she starts crowing out *ergo*, preaching to us that it is a barbarous
match to wed the divine to the earthy, the rational to the irrational, the
strict to the permissive, the decent to the indecent; that pleasure is a bestial
quality, unworthy that a wise man should savour it; that the only
enjoyment he gets from lying with his beautiful young wife is the pleasure
of being aware that he is performing an ordinate action – like pulling on
his boots for a useful ride! May Philosophy's followers, faced with breaking
their wife's hymen, be no more erect, muscular nor succulent than her
arguments are![179]

That is not what Socrates says – Philosophy's preceptor as well as
ours. He values as he should the body's pleasure but he prefers that of
the mind as having more force, constancy, suppleness, variety and dignity.
And, according to him, even that pleasure by no means goes alone (he is
not given to such fantasies): it merely has primacy. For him temperance
is not the enemy of our pleasures: it moderates them.[180]

[B] Nature is a gentle guide but no more gentle than wise and
just: [C] '*Intrandum est in rerum naturam et penitus quid ea postulet*

177. Plutarch, (tr. Amyot), *Banquet des Sept Sages*, 156 G.
178. '88: plaints are *those of ingratitude*. I accept wholeheartedly and *thank her for it*,
what Nature ... Giver to *despise* His gift, to *debase* it or disfigure it – Echoes of
James 1:17, and of Genesis 1:25; then a conflation of phrases from Cicero, *De
finibus*, III, vi, 20.
179. Montaigne is, textually, condemning Seneca here (*Epist. moral.*, XCII, 7–8).
Cf. also Aristotle, *Nicomachaean Ethics*, III, x, 8–9; Cicero, *Paradoxes*, 1.
180. Erasmus, *Apophthegmata*, III, *Socrates*, LXXVI (among others); Plato, *Laws*,
728E; 892 AB; 896 C ff.

pervidendum.' [We must go deeply into the nature of things and find out precisely what Nature wants.] [B] I seek her traces everywhere: we have jumbled them together with the tracks of artifice; [C] and thereby that sovereign good of the Academics and Peripatetics, which is to live according to Nature, becomes for that very reason hard to delimit and portray; so too that of the Stoics which is a neighbour to it, namely, to conform to Nature.[181] [B] Is it not an error to reckon some functions to be less worthy because they are necessities? They will never beat it out of my head anyway that the marriage of Pleasure to Necessity [C] (with whom, according to an ancient, the gods ever conspire) [B] is a most suitable match.[182] What are we trying to achieve by taking limbs wrought together into so interlocked and kindly a compact and tearing them asunder in divorce? On the contrary let us tie them together by mutual duties. Let the mind awaken and quicken the heaviness of the body: let the body arrest the lightness of the mind and fix it fast: [C] *'Qui velut summum bonum laudat animae naturam, et tanquam malum naturam carnis accusat, profecto et animam carnaliter appetit et carnem carnaliter fugit, quoniam id vanitate sentit humana, non veritate divina.'* [He who eulogizes the nature of the soul as the sovereign good and who indicts the nature of the flesh as an evil desires the soul with a fleshly desire and flees from the flesh in a fleshly way, since his thought is based on human vanity not on divine truth.][183]

[B]   There is no part unworthy of our concern in this gift which God has given to us; we must account for it down to each hair. It is not a merely [C] formal [B] commission to Man to guide himself according to Man's [C] fashioning: it is expressly stated, [B] inborn, [C] most fundamental, [B] and the Creator gave it to us seriously [C] and strictly. Commonplace intellects can be persuaded by authority alone, and it has greater weight in a foreign tongue; so, at this point, let us make another charge at it: *'Stultitiae proprium quis non dixerit, ignave et*

181. Cicero, *De senectute*, iii, 5 *De finibus*, V, xxiv, 69; III, vi, 44.
   '88: with *bastard* tracks of artifice. Is it not . . .
182. Cf. Erasmus, *Adages*, II, III, XLI, *Adversum necessitatem ne dii quidem resistunt*, citing Simonides' saying and, above all, Plato. Montaigne is strongly influenced by Cicero (*De finibus*, II, xi, 34; IV, x, 25 – IV, xi, 27–9). In I, ii, 7 Cicero notes that the three schools mentioned by Montaigne, the Academics (the Platonists), the Peripatetics (the Aristotelians) and the Stoics have the virtual monopoly of ethics. Current distortions of their principles therefore pervert virtually the whole of moral philosophy. (Cf. also, *De finibus*, III, vi, 20–3; ix, 25–6; *Laelius*, V, 19; etc.) The debt to Cicero is fundamental.
183. St Augustine, *City of God*, XIV, v; stressing that even Plato devalued the body in the life of Man, who is body plus soul.

contumaciter facere quae facienda sunt, et alio corpus impellere, alio animum, distrahique inter diversissimos motus?' [Who would not say that it was really foolish to do in a slothful, contumacious spirit something which has to be done anyway, thrusting the body in one direction and the soul in another where it is torn between totally conflicting emotions?][184]

[B] Go on then, just to see: get that fellow over there to tell you one of these days what notions and musings he stuffs into his head, for the sake of which he diverts his thoughts from a good meal and regrets the time spent eating it. You will find that no dish on your table tastes as insipid as that beautiful pabulum of his soul (as often as not it would be better if we fell fast asleep rather than stayed awake for what we do it for) and you will find that his arguments and concepts are not worth your rehashed leftovers. Even if they were the raptures of Archimedes, what does it matter?[185]

Here, I am not alluding to – nor am I confounding with the [C] scrapings of the pot [B] that we are, and with the vain longings and ratiocinations which keep us musing – those revered souls which, through ardour of devotion and piety, are raised on high to a constant and scrupulous anticipation of things divine; [C] souls which (enjoying by the power of a quick and rapturous hope a foretaste of that everlasting food which is the ultimate goal, the final destination, that Christians long for) scorn to linger over our insubstantial and ambiguous pleasurable 'necessities' and easily assign to the body the bother and use of the temporal food of the senses. [B] That endeavour is a privilege.[186] [C] Among the likes of us there are two things which have ever appeared to me to chime particularly well together – supercelestial opinions: subterranean morals.

That great man [B] Aesop saw his master pissing as he walked along. 'How now,' he said. 'When we run shall we have to shit?'[187] Let us husband our time; but there still remains a great deal fallow and underused. Our mind does not willingly concede that it has plenty of other hours to

184. '88: merely a *farcical* commission . . . man's *natural* fashioning [. . .] it is *simple* and inborn [. . .] seriously *and expressly* . . .
   Seneca, *Epist. moral.*, LXXIV, 32 (adapted).
185. Archimedes was ecstatic when he discovered his famous principle. In the next sentence, for 'rabble', *voirie*, Montaigne substituted *marmaille*, a pejorative term recalling to the ear both monkey (*marmot*) and stew-pot (*marmite*).
186. '88: privilege. *Our endeavours are all worldly and among the worldly ones the most natural are the most right.* Aesop . . .
187. From Planudes' *Life of Aesop*, often printed with the *Fables*.

perform its functions without breaking fellowship during the short time the body needs for its necessities. They want to be beside themselves, want to escape from their humanity. That *is* madness: instead of changing their Form into an angel's they change it into a beast's; they crash down instead of winding high.   [C]   Those humours soaring to transcendency terrify me as do great unapproachable heights; and for me nothing in the life of Socrates is so awkward to digest as his ecstasies and his daemonizings, and nothing about Plato so human as what is alleged for calling him divine. [B] And of [C] our [B] disciplines it is those which ascend the highest which, it seems to me, are the most [C] base and [B] earth-bound. I can find nothing so [C] abject [B] and so mortal in the life of Alexander as his fantasies about [C] his immortalization. [B] Philotas, in a retort he made in a letter, showed his mordant wit when congratulating Alexander on his being placed among the gods by the oracle of Jupiter Ammon: 'As far as you are concerned I'm delighted,' he said, 'but there is reason to pity those men who will have to live with a man, and obey a man, who [C] trespasses beyond, and cannot be content with, [B] the measure of a man':[188]

[C]   *Diis te minorem quod geris, imperas.*

[Because you hold yourself lower than the gods, you hold imperial sway.][189]

[B]   The noble inscription by which the Athenians honoured Pompey's visit to their city corresponds to what I think:

> *D'autant es tu Dieu comme*
> *Tu te recognois homme.*

[Thou art a god in so far as thou recognizest that thou art a man.]

It is an accomplishment, absolute and as it were God-like, to know how to enjoy our being as we ought. We seek other attributes because we do not understand the use of our own; and, having no knowledge of what is

188. '88: of *human* disciplines [. . .] I can find nothing *so base* and so mortal . . . about his *deification*. Philotas . . . who *exceeds* the measure of a man. The noble inscription . . .
   ('Deification' was used by Christian mystics for the highest rapture. Montaigne replaced it, no doubt, as potentially misleading, Alexander's 'deification' not being an ecstasy but an act of flattery.) For Philotas, cf. Quintus Curtius, VI, 9.
189. Horace, *Odes*, III, vi, 5; then the inscription greeting Pompey as he left Athens, according to Plutarch. (Cited from Amyot's translation of his *Life of Pompey the Great*.)

within, we sally forth outside ourselves. [C] A fine thing to get up on stilts: for even on stilts we must ever walk with our legs! And upon the highest throne in the world, we are seated, still, upon our arses.

[B] The most beautiful of lives to my liking are those which conform to the common measure, [C] human and ordinate, without miracles though and [B] without rapture.

Old age, however, has some slight need of being treated more tenderly. Let us commend it to that tutelary god of health – and, yes, of wisdom merry and companionable:

> *Frui paratis et valido mihi,*
> *Latoe, dones, et, precor, integra*
> *Cum mente, nec turpem senectam*
> *Degere, nec cythara carentem.*

[Vouchsafe, O Son of Latona, that I may enjoy those things I have prepared; and, with my mind intact I pray, may I not degenerate into a squalid senility, in which the lyre is wanting.][190]

190. '88: common measure, without *marvel*, without rapture . . . more tenderly *and more delicately*. Let us commend . . .

Horace, *Odes*, I, xxxi, 17–20. Apollo, son of Jupiter and Latona, was the god of healing and presided over the Muses.

Index

Summary of the Symbols

[A] and '80: the text of 1580
[A1]: the text of 1582 (plus)
[B] and '88: the text of 1588
[C]: the text of the edition being prepared by Montaigne when he died, 1592
'95: text of the 1595 posthumous printed edition

In the notes there is given a selection of variant readings, including most abandoned in 1588 and many from the printed posthumous edition of 1595.

By far the most scholarly account of the text is that given in R. A. Sayce, *The Essays of Montaigne: A Critical Exploration*, 1972, Chapter 2, 'The Text of the *Essays*'.

PENGUIN ⊙ CLASSICS

The Classics Publisher

'Penguin Classics, one of the world's greatest series' JOHN KEEGAN

'I have never been disappointed with the Penguin Classics. All I have read is a model of academic seriousness and provides the essential information to fully enjoy the master works that appear in its catalogue' MARIO VARGAS LLOSA

'Penguin and Classics are words that go together like horse and carriage or Mercedes and Benz. When I was a university teacher I always prescribed Penguin editions of classic novels for my courses: they have the best introductions, the most reliable notes, and the most carefully edited texts' DAVID LODGE

'Growing up in Bombay, expensive hardback books were beyond my means, but I could indulge my passion for reading at the roadside bookstalls that were well stocked with all the Penguin paperbacks ... Sometimes I would choose a book just because I was attracted by the cover, but so reliable was the Penguin imprimatur that I was never once disappointed by the contents.

Such access certainly broadened the scope of my reading, and perhaps it's no coincidence that so many Merchant Ivory films have been adapted from great novels, or that those novels are published by Penguin' ISMAIL MERCHANT

'You can't write, read, or live fully in the present without knowing the literature of the past. Penguin Classics opens the door to a treasure house of pure pleasure, books that have never been bettered, which are read again and again with increased delight' JOHN MORTIMER

CLICK ON A CLASSIC
www.penguinclassics.com
The world's greatest literature at your fingertips

Constantly updated information on over 1600 titles, from Icelandic sagas to ancient Indian epics, Russian drama to Italian romance, American greats to African masterpieces

•

The latest news on recent additions to the list, updated editions and specially commissioned translations

•

Original scholarly essays by leading writers: Elaine Showalter on Zola, Laurie R. King on Arthur Conan Doyle, Frank Kermode on Shakespeare, Lisa Appignanesi on Tolstoy

•

A wealth of background material, including biographies of every classic author from Aristotle to Zamyatin, plot synopses, readers' and teachers' guides, useful web links

•

Online desk and examination copy assistance for academics

•

Trivia quizzes, competitions, giveaways, news on forthcoming screen adaptations

•

eBooks available to download

READ MORE IN PENGUIN

In every corner of the world, on every subject under the sun, Penguin represents quality and variety – the very best in publishing today.

For complete information about books available from Penguin – including Puffins and Penguin Classics – and how to order them, write to us at the appropriate address below. Please note that for copyright reasons the selection of books varies from country to country.

In the United Kingdom: *Please write to* Dept EP, Penguin Books Ltd, Bath Road, Harmondsworth, West Drayton, Middlesex UB7 0DA

In the United States: *Please write to* Consumer Services, Penguin Putnam Inc., 405 Murray Hill Parkway, East Rutherford, New Jersey 07073-2136. *VISA and MasterCard holders call 1-800-631-8571 to order Penguin titles*

In Canada: *Please write to* Penguin Books Canada Ltd, 10 Alcorn Avenue, Suite 300, Toronto, Ontario M4V 3B2

In Australia: *Please write to* Penguin Books Australia Ltd, 487 Maroondah Highway, Ringwood, Victoria 3134

In New Zealand: *Please write to* Penguin Books (NZ) Ltd, Private Bag 102902, North Shore Mail Centre, Auckland 10

In India: *Please write to* Penguin Books India Pvt Ltd, 11, Community Centre, Panchsheel Park, New Delhi 110017

In the Netherlands: *Please write to* Penguin Books Netherlands bv, Postbus 3507, NL-1001 AH Amsterdam

In Germany: *Please write to* Penguin Books Deutschland GmbH, Metzlerstrasse 26, 60594 Frankfurt am Main

In Spain: *Please write to* Penguin Books S. A., Bravo Murillo 19, 1°B, 28015 Madrid

In Italy: *Please write to* Penguin Italia s.r.l., Via Vittoria Emanuele 451a, 20094 Corsico, Milano

In France: *Please write to* Penguin France, 12, Rue Prosper Ferradou, 31700 Blagnac

In Japan: *Please write to* Penguin Books Japan Ltd, Iidabashi KM-Bldg, 2-23-9 Koraku, Bunkyo-Ku, Tokyo 112-0004

In South Africa: *Please write to* Penguin Books South Africa (Pty) Ltd, P.O. Box 751093, Gardenview, 2047 Johannesburg

MACHIAVELLI
The Prince

*'One must be a fox in order to recognize traps,
and a lion to frighten off wolves'*

The Prince shocked Europe on publication with its advocacy of
ruthless tactics for gaining absolute power and its abandonment
of conventional morality. Niccolò Machiavelli (1469–1527)
came to be regarded by some as an agent of the Devil and his
name taken for the intriguer 'Machevill' of Jacobean tragedy.
For his treatise on statecraft Machiavelli drew upon his own
experience of office under the turbulent Florentine republic,
rejecting traditional values of political theory and recognizing
the complicated, transient nature of political life. Concerned
not with lofty ideals but with a regime that would last, *The
Prince* has become the bible of realpolitik, and still retains its
power to alarm and to instruct.

In this edition Machiavelli's tough-minded and pragmatic
Italian is preserved in George Bull's clear, unambiguous
translation, while Anthony Grafton's introduction depicts
Machiavelli's world of power struggles and intrigue, and
discusses his role as political teacher of Europe.

Translated with notes by GEORGE BULL
With an introduction by ANTHONY GRAFTON

PLATO
The Republic

*'We are concerned with the most important of
issues, the choice between a good and an evil life'*

Plato's *Republic* is widely acknowledged as the cornerstone of
Western philosophy. Presented in the form of a dialogue
between Socrates and three different interlocutors, it is an
inquiry into the notion of a perfect community and the ideal
individual within it. During the conversation other questions
are raised: what is goodness; what is reality; what is knowledge?
The Republic also addresses the purpose of education and
the roles of both women and men as 'guardians' of the people.
With remarkable lucidity and deft use of allegory, Plato arrives
at a depiction of a state bound by harmony and ruled by
'philosopher kings'.

Desmond Lee's translation of *The Republic* has come to be
regarded as a classic in its own right. His introduction discusses contextual themes such as Plato's disillusionment with
Athenian politics and the trial of Socrates. This new edition also
features a revised bibliography.

Translated with an introduction by DESMOND LEE

TOCQUEVILLE

Democracy in America
and Two Essays on America

'A new political science is needed for a totally new world'

In 1831 Alexis de Tocqueville made a nine-month journey through eastern America. The result was *Democracy in America*, a monumental study of the strengths and weaknesses of the nation's evolving politics and institutions. Tocqueville looked to the flourishing democratic system in America as a possible model for post-revolutionary France, believing that the egalitarian ideals it enshrined reflected the spirit of the age – even that they were the will of God. His insightful work has become one of the most influential political texts ever written on America and an indispensable authority for anyone interested in the future of democracy. This volume includes the rarely translated 'Two Weeks in the Wilderness', an evocative account of Tocqueville's travels in Michigan among the Iroquois and Chippeway, and 'Excursion to Lake Oneida'.

This is the only edition that contains all Tocqueville's writings on America, and it includes a chronology, further reading and explanatory notes. Isaac Kramnick's introduction discusses Tocqueville's life and times, and the enduring significance of *Democracy in America*.

Translated by GERALD BEVAN
With an introduction and notes by ISAAC KRAMNICK

MADISON, HAMILTON AND JAY

The Federalist Papers

'The establishment of a Constitution, in a time of profound peace, by the voluntary consent of a whole people, is a PRODIGY'

Written at a time when furious arguments were raging about the best way to govern America, *The Federalist Papers* had the immediate practical aim of persuading New Yorkers to accept the newly drafted constitution in 1787. In this they were supremely successful, but their influence also transcended contemporary debate to win them a lasting place in discussions of American political theory. Acclaimed by Thomas Jefferson as 'the best commentary on the principles of government which ever was written', *The Federalist Papers* make a powerful case for power-sharing between state and federal authorities and for a constitution that has endured largely unchanged for more than two hundred years.

In his brilliantly detailed introduction, Isaac Kramnick sets the *Papers* in their historical and political context. This edition also contains the American constitution as an appendix.

'The introduction is an outstanding piece of work ... I am strongly recommending its reading' WARREN BURGER, former Chief Justice, Supreme Court of the United States

Edited with an introduction by ISAAC KRAMNICK

OSCAR WILDE
Complete Short Fiction

*'He saw a most wonderful sight. Through a
little hole in the wall the children had crept in ...
And the trees were so glad to have the children
back again that they had covered themselves
with blossoms'*

Fairy tales, ghost stories, detective fiction and comedies of manners – the stories collected in this volume made Oscar Wilde's name as a writer of fiction, showing breathtaking dexterity in a wide range of literary styles. Victorian moral justice is comically inverted in 'Lord Arthur Savile's Crime' and 'The Canterville Ghost', and society's materialism comes under sharp, humorous criticism in 'The Model Millionaire', while 'The Happy Prince' and 'The Nightingale and the Rose' are hauntingly melancholic in their magical evocations of selfless love. These small masterpieces convey the brilliance of Wilde's vision, exploring complex moral issues through an elegant juxtaposition of wit and sentiment.

This edition includes the complete texts of Wilde's three volumes of short fiction, together with 'The Portrait of Mr W. H.' Ian Small's introduction discusses Wilde's life, the cultural and literary background to his fiction and the complex ways in which it can be read.

Edited with an introduction by IAN SMALL

OSCAR WILDE
The Picture of Dorian Gray

*'The horror, whatever it was, had not yet entirely
spoiled that marvellous beauty'*

Enthralled by his own exquisite portrait, Dorian Gray
exchanges his soul for eternal youth and beauty. Influenced by
his friend Lord Henry Wotton, he is drawn into a corrupt
double life, indulging his desires in secret while remaining a
gentleman in the eyes of polite society. Only his portrait bears
the traces of his decadence. *The Picture of Dorian Gray* was a
succès de scandale. Early readers were shocked by its hints at
unspeakable sins, and the book was later used as evidence
against Wilde at his trial at the Old Bailey in 1895.

This definitive edition includes a selection of contemporary
reviews condemning the novel's immorality, and the introduction to the first Penguin Classics edition by Peter Ackroyd.

Edited with an introduction and notes by ROBERT MIGHALL

HUGO
Les Misérables

'He was no longer Jean Valjean, but No. 24601'

Victor Hugo's tale of injustice, heroism and love follows the fortunes of Jean Valjean, an escaped convict determined to put his criminal past behind him. But his attempts to become a respected member of the community are constantly put under threat: by his own conscience, when, owing to a case of mistaken identity, another man is arrested in his place; and by the relentless investigations of the dogged policeman Javert. It is not simply for himself that Valjean must stay free, however, for he has sworn to protect the baby daughter of Fantine, driven to prostitution by poverty. A compelling and compassionate view of the victims of early nineteenth-century French society, *Les Misérables* is a novel on an epic scale, moving inexorably from the eve of the battle of Waterloo to the July Revolution of 1830.

Norman Denny's introduction to his lively English translation discusses Hugo's political and artistic aims in writing *Les Misérables*.

'A great writer – inventive, witty, sly, innovatory' A. S. BYATT

Translated and with an introduction by NORMAN DENNY

DUMAS

The Count of Monte Cristo

*'On what slender threads do life and
fortune hang'*

Thrown in prison for a crime he has not committed, Edmond
Dantès is confined to the grim fortress of If. There he learns of a
great hoard of treasure hidden on the Isle of Monte Cristo and
he becomes determined not only to escape, but also to unearth
the treasure and use it to plot the destruction of the three men
responsible for his incarceration. Dumas's epic tale of suffering
and retribution, inspired by a real-life case of wrongful imprisonment, was a huge popular success when it was first serial-
ized in the 1840s

Robin Buss's lively English translation remains faithful to the
style of Dumas's original. This edition includes an introduction,
explanatory notes and suggestions for further reading.

'Robin Buss broke new ground with a fresh version of *Monte
Cristo* for Penguin' *Oxford Guide to Literature in English
Translation*

Translated with an introduction by ROBIN BUSS